ELEMENTS

Element	Symbol	Atomic Number	Element	Symbol	Atomic Number
Actinium	Ac	89	Mercury	Hg	80
Aluminum	Al	13	Molybdenum	Mo	42
Americium	Am	95	Neodymium	Nd	60
Antimony	Sb	51	Neon	Ne	10
Argon	Ar	18	Neptunium	Np	93
Arsenic	As	33	Nickel	Ni	28
Astatine	At	85	Niobium	Nb	41
Barium	Ba	56	(Columbrium)		
Berkelium	Bk	97	Nitrogen	N	7
Beryllium	Be	4	Nobelium	No	102
Bismuth	Bi	83	Osmium	Os	76
Boron	B	5	Oxygen	O	8
Bromine	Br	35	Palladium	Pd	46
Cadmium	Cd	48	Phosphorus	P	15
Calcium	Ca	20	Platinum	Pt	78
Californium	Cf	98	Plutonium	Pu	94
Carbon	C	6	Polonium	Po	84
Cerium	Ce	58	Potassium	K	19
Cesium	Cs	55	Praseodymium	Pr	59
Chlorine	Cl	17	Promethium	Pm	61
Chromium	Cr	24	Protactinium	Pa	91
Cobalt	Co	27	Radium	Ra	88
Copper	Cu	29	Radon	Rn	86
Curium	Cm	96	Rhenium	Re	75
Dysprosium	Dy	66	Rhodium	Rh	45
Einsteinium	Es	99	Rubidium	Rb	37
Erbium	Er	68	Ruthenium	Ru	44
Europium	Eu	63	Samarium	Sm	62
Fermium	Fm	100	Scandium	Sc	21
Fluorine	F	9	Selenium	Se	34
Francium	Fr	87	Silicon	Si	14
Gadolinium	Gd	64	Silver	Ag	47
Gallium	Ga	31	Sodium	Na	11
Germanium	Ge	32	Strontium	Sr	38
Gold	Au	79	Sulfur	S	16
Halfnium	Hf	72	Tantalum	Ta	73
Helium	He	2	Technetium	Tc	43
Holmium	Ho	67	Tellurium	Te	52
Hydrogen	H	1	Terbium	Tb	65
Indium	In	49	Thallium	Tl	81
Iodine	I	53	Thorium	Th	90
Iridium	Ir	77	Thulium	Tm	69
Iron	Fe	26	Tin	Sn	50
Krypton	Kr	36	Titanium	Ti	22
Lanthanum	La	57	Tungsten (Wolfram)	W	74
Lawrencium	Lr	103	Uranium	U	92
Lead	Pb	82	Vanadium	V	23
Lithium	Li	3	Xenon	Xe	54
Lutetium	Lu	71	Ytterbium	Yb	70
Magnesium	Mg	12	Yttrium	Y	39
Manganese	Mn	25	Zinc	Zn	30
Mendelevium	Md	101	Zirconium	Zr	40

HAZARDOUS CHEMICALS DESK REFERENCE

N. IRVING SAX
RICHARD J. LEWIS, SR.

VNR VAN NOSTRAND REINHOLD
_____ New York

Printed in the United States of America

Van Nostrand Reinhold
115 Fifth Avenue
New York, New York 10003

Van Nostrand Reinhold International Company Limited
11 New Fetter Lane
London EC4P 4EE, England

Van Nostrand Reinhold
480 La Trobe Street
Melbourne, Victoria 3000, Australia

Macmillan of Canada
Division of Canada Publishing Corporation
164 Commander Boulevard
Agincourt, Ontario MIS 3C7, Canada

16 15 14 13 12 11 10 9 8 7 6 5 4

Library of Congress Cataloging-in-Publication Data

Hazardous chemicals desk reference.

1. Hazardous substances—Handbooks, manuals, etc.
I. Sax, N. Irving (Newton Irving)   II. Lewis,
Richard J., Sr.
T55.3.H3H398   1987   604.7   87-2084
ISBN 0-442-28208-7

To Pauline and Grace
at our sides, as always

To Carol D. Wickell and William Mahn for their
professional assistance with this book.

Our thanks to Susan Munger and Alberta W. Gordon of
Van Nostrand Reinhold for their constant encouragement and
material assistance.

PREFACE

Reference works on hazardous materials seem to fall into two categories, the limited and the very detailed and comprehensive. The editors noted a need for a reference of moderate size which would serve the needs of many who must work with and evaluate the hazards of chemicals. This book was designed to fill that need.

To make the book useful, approximately 5000 materials were selected based upon their importance in industry, their toxicity or fire and explosion hazard, or upon widespread interest in the material. The actual entries were extracted from the 6th Edition of *Dangerous Properties of Industrial Materials*. The entries were shortened by removing citations to toxicity data and other less relevant information. The Toxic and Hazard Reviews, however, were mostly expanded and made more readable. The German Research Society's MAK values were added to assist in the design of safer workplaces.

Not only is it important to have specific data on hazardous materials but it is also important to convey information to employees handling the materials. For this reason, we have included five chapters of information on protective clothing, use of respirators, fire protection, storage and handling, and first aid. The primary purpose is to provide guidelines for managers and others responsible for maintaining a safety program. For example, major types of respirators and protective clothing are described and some explanation is given regarding recommendations for their use. The first aid chapter offers suggestions on what staff should be assigned responsibility in this area. The reader will need to consult other sources including government regulations, voluntary standards, and manufacturer literature for information regarding a specific chemical.

Two cross reference indices are provided as Appendices to permit rapid location of a material if either a Chemical Abstract Service (CAS) number or a synonym for the material is the point of entry.

There is an average of three synonyms for each entry.

N. Irving Sax
Richard J. Lewis, Sr.

INTRODUCTION

Entries in this book include basic chemicals, pesticides, dyes, detergents, lubricants, plastics, drugs, food additives, preservatives, ores, soaps, extracts from plant and animal sources, and industrial intermediates and waste products from production processes. Some of the information refers to materials whose composition is not precisely known. The chemical materials included are assumed to exhibit the reported toxic effect in their pure state unless otherwise noted. However, even in the case of a supposedly "pure" material, there is usually some degree of uncertainty as to its exact composition and the impurities which may be present. This possibility must be considered in attempting to interpret the data presented since the toxic effects observed could in some cases have been caused by a contaminant.

Excluded from our list are tradename products representing compounded or formulated proprietary mixtures available as commercial products. These exclusions are necessary because of difficulties in assessing the contribution of each component of a mixture to that material's total toxicity and because a product's formulation is often changed by varying the components, their concentration, or their purity. Commercial product tradenames as synonyms are included, particularly when they represent a single active chemical entity or a well-defined mixture of relatively constant composition. Radioactive materials are included but the effects reported are the chemically produced effects rather than the radiation effect.

For each material described the following data are provided when available: the material name, Hazard Rating (HR:), CAS number, RTECS number, molecular formula, molecular weight, selected properties including a description of the material (where necessary), synonyms, the U.S. Occupational Safety and Health Administration's (OSHA) air standards, the American Conference of Governmental Industrial Hygienists' (ACGIH) Threshold Limit Values, the German Research Society's (MAK) values, U.S. Department of Transportation (DOT) classifications, and the Toxic and Hazard Review (THR). Each data type is described below.

1. *Name* The name of each material is selected to facilitate recognition of the material. In many cases no single name will be recognized by a majority of readers. Extensive cross-indexing by synonyms is provided to aid in locating an entry.

2. *HR:* is the hazard rating assigned to the material on a scale of 1 to 3 that briefly identifies the level of the toxicity as follows:

The number "3" indicates an LD_{50} below 400 mg/kg.
The number "2" indicates an LD_{50} of 400-4,000 mg/kg.
The number "1" indicates an LD_{50} of 4,000-40,000 mg/kg.

3. *CAS:* is the American Chemical Society's Chemical Abstracts Service Registry Number. It is a numeric designation assigned by the Chemical Abstracts Service and uniquely identifies a

specific chemical compound. This entry allows one to conclusively identify a material regardless of the name or naming system used.

4. *DOT:* indicates a four digit hazard code assigned by the U.S. Department of Transportation. This code is recognized internationally and is in agreement with the United Nations coding system. The code is used on transport documents, labels, and placards. It is also used to determine the regulations for shipping the material.

5. *RTECS:* is the accession number used by the Registry of Toxic Effects of Chemical Substances produced by the National Institute for Occupational Safety and Health, U.S. Department of Health and Human Services, 4676 Columbia Pkwy., Cincinnati, Ohio 45226. The RTECS data base contains toxicity data and related information for over 85,000 substances and is useful for locating published toxicity data.

6. *mf:* (molecular formula) or *af:* (atomic formula) designates the elemental composition of the material and is structured according to the Hill System (see *Journal of the American Chemical Society*, 22(8): 478-494, 1900) in which carbon and hydrogen (if present) are listed first, followed by the other elemental symbols in alphabetical order. The formula for compounds that do not contain carbon are ordered strictly alphabetically by element symbol. Compounds such as salts or those containing waters of hydration have molecular formulas incorporating the CAS dot-disconnect convention, in which the components are listed individually and separated by a period. The individual components of the formula are generally given in order of decreasing carbon atom count, and the component ratios given. A lower case "x" indicates that the ratio is unknown. A lower case "n" indicates a repeating, polymer-like structure.

7. *mw:* (molecular weight) is calculated from the molecular formula using standard elemental molecular weights (carbon = 12.01).

8. *PROP:* (Properties) are selected to be useful for evaluating the hazard of a material and designing proper storage and use procedures. A definition of the material is included where necessary. The physical description of the material may include the form, color and odor to aid in positive identification. When available, the boiling point, melting point, density, vapor pressure, vapor density, and refractive index are given. The flash point, autoignition temperature, and lower and upper explosive limits are included to aid in fire protection and control. An indication is given of the solubility or miscibility of the material in water and common solvents.

9. *SYN(S)* (synonyms) for the material are listed alphabetically. Synonyms include other chemical names, common or generic names, foreign names (with the language in parentheses), or codes. Some synonyms consist in whole or in part of registered trademarks. These trademarks are not identified as such.

The reader is cautioned that some synonyms, particularly common names, may be ambiguous and refer to more than one material.

10. *OSHA PEL:* (Permissible Exposure Limits) are the air concentrations to which workers can be exposed for a normal 8-hour day, 40-hour work week without ill effects as defined by the U.S. Occupational Safety and Health Administration (OSHA), Department of Labor. These standards may also include the notation "CL" indicating a ceiling limit which must not be exceeded, or "Pk" indicating the maximum short time peak allowed above the ceiling value. These limits are found in 29 CFR (Code of Federal Regulations) 1910.1000. The CFR regulations also contain detailed requirements for control of some substances and special regulations for carcinogenic substances. Additional information is available from OSHA, Technical Data Center, U.S. Department of Labor, Washington, D.C. 20210.

11. *ACGIH TLV:* are the Threshold Limit Values of the American Conference of Governmental Industrial Hygienists (ACGIH). The TLV represents a time weighted average (TWA) air concentration to which workers can be exposed for a normal 8-hour day, 40-hour work week without ill effects. The notation "CL" indicating a ceiling limit which must not be exceeded. The notation "skin" indicates that the material penetrates intact skin, and skin contact should be avoided even though the TLV concentration is not exceeded. STEL indicates a short-term exposure limit which is a 15-minute time-weighted average which should not be exceeded. Biological Exposure Indices (BEI) are, according to the ACGIH, set to provide a warning level "...of biological response to the chemical, or warning levels of that chemical or its metabolic product(s) in tissues, fluids, or exhaled air of exposed workers..."

The latest annual TLV list is contained in the publication *Threshold Limit Values for Chemical Substances and Physical Agents in the Work Environment and Biological Exposure Indices with Intended Changes.* The values in this document should be consulted for future trends in recommendations. The ACGIH TLV's are adopted in whole or in part by many countries and local administrative agencies throughout the world. As a result, these recommendations have a major impact on the control of workplace contaminant concentrations. The ACGIH may be contacted for additional information at 6500 Glenway Ave., Cincinnati, Ohio 45211, USA.

12. *DFG MAK:* are the German Research Society's MAK value. Those materials which are classified as to workplace hazard potential by the German Research Society are noted on this line. The MAK values are also revised annually and discussions of materials under consideration for MAK assignment are included in the annual publication together with the current values. *BAT:* indicates Biological Tolerance Value for a Working Material which is defined as, "...the maximum permissible quantity of a chemical compound, its metabolites, or any deviation from the norm of biological parameters induced by these substances in exposed humans." *TRK:* values are Technical Guiding Concentrations for workplace control of carcinogens. For additional information, write to Deutsche Forschungsgemeinschaft (German Research Society), Kennedyallee 40, D-5300 Bonn 2, Federal Republic of Germany. The publication *Maximum Concentrations at the Workplace and Biological Tolerance Values for Working Materials* can be obtained from Verlag Chemie GmbH, Buchauslieferung, P.O. Box 1260 /1280, D-6940 Weinheim, Federal Republic of Germany, or Verlag Chemie, Deerfield Beach, Florida.

13. *DOT Classification:* is the hazard classification according to the U.S. Department of Transportation (DOT) or the International Maritime Organization (IMO.) This classification gives an indication of the hazards expected in transportation, and serves as a guide to the development of proper labels, placards, and shipping instructions. The basic hazard classes include compressed gases, flammables, oxidizers, corrosives, explosives, radioactive materials, and poisons. Although a material may be designated by only one hazard class, additional hazards may be indicated by adding labels or by using other means directed by DOT. Many materials are regulated under general headings such as "pesticides" or "combustible liquids" as defined in the regulations. These are not noted here as their specific concentration or properties must be known for proper classification. Special regulations may govern shipment by air. This information should serve *only as a guide* since the regulation of transported materials is carefully controlled in most countries by federal and local agencies. Because of frequent changes to regulations, it is recommended that the reader contact the applicable agency for information about the current standards for a particular material. United States transportation regulations are found in 40 CFR, Parts 100 to 189. Contact the U.S. Department of Transportation, Materials Transportation Bureau, Washington, D.C. 20590.

14. *THR* Under this heading the reader will find both a brief summary of the toxicity and a discussion of the symptoms caused by exposure. Materials incompatible with an entry are listed

here. Fire and explosion hazards are briefly summarized in terms of flash points and upper and lower explosive limits. Where feasible, fire-fighting materials and methods are discussed. A material with a flash point of 100°F or less is flammable and dangerous; if the flash point is from 100° to 200°F, it is combustible and of moderate hazard; if it is above 200°F, the material is combustible and of low fire hazard.

CONTENTS

SECTION I

SECTION I

1

SAFE STORAGE AND HANDLING OF CHEMICALS

Donald D. Hedberg
President
Lab Safety Supply Co.
Janesville, Wisconsin

Chemicals stored properly present a minimum of hazards. A flammable chemical retains its hazardous properties when stored in a safety can, but its likelihood of becoming involved in a fire is minimal simply because of the "designed in" safety features of a safety can. Likewise, a toxic chemical is harmless as long as its storage prevents inhalation, ingestion, or contact. The recent Hazard Communication Standard has placed greater importance on good storage than ever before because good storage is one of the first measures that any employer can take to protect workers from exposure to hazards.

The first step in organizing the storage of chemicals is to take an inventory of the chemicals. All too frequently, chemical storerooms resemble kitchen pantries. In a chemical storeroom, a forgotten container of a chemical may deteriorate on the shelf to something with a completely different chemical identity. Not only can this change in identity ruin a chemical experiment or procedure, it can also present an immediate hazard.

A tape recorder can be used to take your own chemical inventory. The identity of each chemical and the condition of the container, as well as its location, can be recited into the tape recorder. This eliminates the necessity of moving or picking up potentially dangerous containers to read their labels. Afterwards, with reference materials available, the hazard of each chemical can be accessed and proper safe handling and storage procedures established. A chemical storage area should be identified with a sign along with other important signs such as "No Smoking," "Fire Extinguisher," "Safety Shower," etc.

Typically, storage of chemicals involves the basic container, usually the original shipping container, and transfer containers. The basic container ranges from large tanks or drums to cans, bottles, and paper sacks. Transfer containers can be cans, pails, bottles, etc. Handling and storage guidelines for these containers are frequently found on the original shipping container and by referring to the Material Safety Data Sheet (MSDS) that should accompany each shipment. Before handling and storage procedures for a chemical can be established, its hazards must be known. If in doubt, put the container in a safe, secure location until a proper determination of its hazards can be made.

STORAGE BY HAZARD CLASS

Chemicals should not be stored alphabetically. If all sodium compounds are stored together, then strong oxidizers like sodium chromate might be stored next to strong reducers like sodium dithionite, and an accidental mixture could be devastating. Instead, chemicals should be stored by their respective hazard classes. That is, flammables should be stored with flammables, acids with acids, etc. The following table gives a list of hazard classes and storage guidelines.

Hazard Class	Storage Guidelines	Storage Color	Examples
Flammable	Store in flammable liquid storage area	red	ethanol, mineral spirits, gasoline
Flammable	Store in flammable liquid storage area but separate from other flammables	red with white stripes	benzoyl peroxide, sodium metal, lithium aluminum hydride
Reactive	Store separately and away from flammable and combustible materials	yellow	ammonium nitrate, sodium chlorate, hydrogen peroxide
Reactive	Store separately and away from flammable and combustible materials but separate from other reactive chemicals	yellow with white stripes	acrylamide, sodium dithionite, sodium hypophosphite, hydrazine
Contact	Store in corrosion-proof area	white	hydrochloric acid, iodine, titanium tetrachloride
Contact	Store in corrosion-proof area but separate from other contact hazards	white with black stripes	chlorosulfonic acid, sodium hydroxide
Health	Store in secured area	blue	sodium cyanide, mercury compounds, carbon tetrachloride
Moderate	Substances which are suitable for general storage area	orange	sodium chloride, dextrose, monoethanol amine

The above color-coded storage system was developed by J. T. Baker Chemical Co. Other chemical companies such as Fisher Scientific Co. and Mallinkrodt Chemical Co. have developed similar storage systems.

Some chemicals with multiple hazards fit into different storage classes. Phenol, for example, is flammable, toxic, and corrosive. The storage class it is put in depends upon the most likely hazard found in your workplace. If sources of ignition are present, perhaps it should be stored as a flammable. On the other hand, if contact is possible, it should be stored as a contact hazard, but separate from oxidizers like nitric acid. If the possibility of human contact is minimal, its storage as a toxic substance becomes less important.

LABELING

For safe handling and storage, all containers of chemicals should be properly labeled and the information on the label should also be found on the Material Safety Data Sheet. Take care to preserve the label on the original container, and if the contents are

ACETONE

DANGER!

EXTREMELY FLAMMABLE. HARMFUL IF SWALLOWED OR INHALED. CAUSES IRRITATION.

Keep away from heat, sparks, flame. Avoid contact with eyes, skin, clothing. Avoid breathing vapor. Keep in tightly closed container. Use with adequate ventilation. Wash thoroughly after handling.

EFFECTS OF OVEREXPOSURE: Contact with skin has a defatting effect, causing drying and irritation. Overexposure to vapors may cause irritation of mucous membranes, dryness of mouth and throat, headache, nausea and dizziness.

FIRST AID PROCEDURES: If inhaled, remove to fresh air. If not breathing, give artificial respiration. If breathing is difficult, give oxygen. If contacted, immediately flush eyes with plenty of water for at least 15 minutes. Flush skin with water. If swallowed, if conscious, immediately induce vomiting.

Consult MSDS for further hazardous information and instructions.   CAS NO. [67-64-1]

A label designed in an ANSI format.

"In plant" container labels.

transferred to a secondary container, it also must be labeled, provided it is used during different work shifts or stored on a shelf. Various labeling systems are available. Some are based upon the NFPA rating system of 0 (least) to 4 (extreme) and others are based on the ANSI format which contains written statements of warnings, precautions, storage guidelines, first aid, etc. Sax uses a rating of 0 (least) to 3 (extreme).

The recent Hazard Communication Standard has very specific requirements with respect to labeling containers. Make certain that the label on the original shipping container contains the same information supplied on the MSDS. This should include the name of the manufacturer, the hazards of the chemical, and the organs affected. The statement "Harmful if Inhaled" is not sufficient according to the Standard. Instead, the label should say why it is harmful, such as, "causes lung damage." The labeling requirements for secondary or "in plant" containers is less strict. However, containers can no longer be left unlabeled or identified with a simple name. These new labeling requirements should eliminate the questions "What's in that can?" or "Where should it be put?" Instead, the questions will be "What is it?" "Will it harm me?", and "How and where should it be stored?" If you still have casually placed containers in your facilities that are poorly identified, you have a problem that must be corrected immediately.

SHELVING

A chemical storeroom is equipped with shelving to hold various sized bottles and containers of chemicals. Check to make certain that the weight of the chemicals does not exceed the manufacturer's weight limitations and that the shelves are securely attached to prevent them from falling over. To stop containers from "creeping" off the shelf, a lip is frequently installed along the front edge. This is of particular importance in areas of

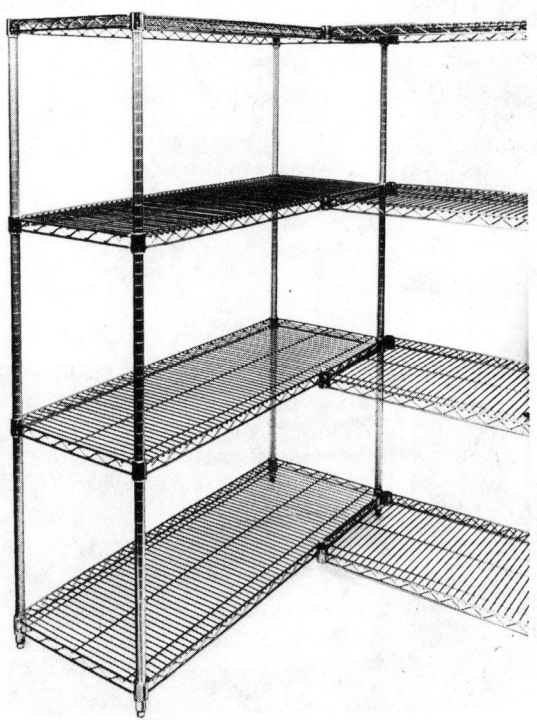

Open-type shelving permits free circulation of air.

heavy traffic or earthquake prone areas. Housekeeping measures should include preventing containers from protruding over shelf edges, removing empty containers or bottles from the shelves and keeping the shelving units clean and free of dust and chemical contamination. Large containers should always be on lower shelves and hygroscopic and water reactive chemicals stored in a polyethylene bag or covered by a polyethylene sheet.

Corrosion of metal is a problem where containers of corrosive chemicals are stored. Shelves should be lined with sheets of polypropylene or made of particle board laminated with chemically resistant Chemsurf®. Wire shelving permits free circulation of air, preventing the accumulation of vapors. Some wire shelving is available in stainless steel or with special coatings. Of course, improperly labeled containers should be removed, as well as bottles with frozen stoppers and lids. Stoppers in containers should be easy to remove from bottles, but must also provide an airtight seal.

Inspect your shelf storage. Store chemicals in their original shipping containers, which were designed to prevent breakage. If chemicals are stored alphabetically, they should all be in the same hazard class. Become familiar with incompatibilities of chemicals. Prepare an incompatibility chart for the chemicals in your storeroom. If you are unsure about the chemistry involved, get advice. The preparation of this chart could be a special project for your local college or university.

ACIDS

Acids create a unique set of hazards for storage and handling due to their corrosiveness. The corrosiveness of the compounds may affect containers, cabinets, and equipment, not to mention personnel.

There are two basic groups of acids: inorganic (mineral acids) and organic. Inorganic acids are nonflammable and include sulfuric acid, nitric acid, and hydrochloric acid to name a few. The organic acids contain carbon atoms and can be flammable. Examples of organic acids are acetic acid, formic acid, and oxalic acid.

Most accidents with acids involve contact with the skin. Depending on the acid, it may visibly destroy or alter human skin tissue at the site of contact. When handling acids in transport, transfer, or work processes, it is therefore critical that appropriate

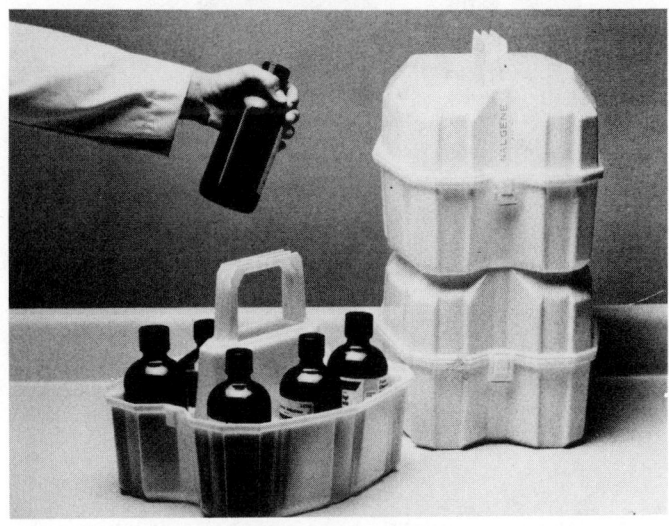

Safely transport acid solutions in carriers to prevent breakage.

personal protective equipment be used. This equipment may include gloves, safety glasses, goggles, face shields, aprons, and even full body protection as indicated.

For safe transport of small quantities of acid, bottle carriers should be used. The bottle carrier will not only protect the acid bottle from breakage during transport, it also will serve as a containment vessel for leaking bottles and will prevent contact with incompatible materials as well.

When transferring acids, either from a drum or smaller container, personal protective equipment must be selected as mentioned earlier, and the chemical compatibility of transfer pumps and receiving vessels must be checked. Original labels on containers should be preserved and maintained.

A spill control plan for an acid spill should be established and cleanup materials should be selected according to the type and volume of acid spill and its potential for damage to life and property. Special neutralizers are available for acid spills. They contain a color indicator to indicate when neutralization is complete. Other spill control supplies such as spill control pillows and dykes are available to prevent the spread of spills.

Acids may be segregated and stored in chemical storerooms on shelving or in specially constructed acid storage cabinets. Regardless of the type of storage equipment, there are general precautionary measures to follow with acids. They must be segregated from active metals such as sodium, potassium, or magnesium. Oxidizing acids should be separated from organic acids and from flammable and combustible materials. Bases and acids are not compatible. Segregate acids from chemicals which could generate

The corrosive nature of acids requires cabinets made of corrosion-resistant materials.

toxic or flammable gases upon contact, such as sodium cyanide, iron sulfide, or calcium carbide. Store large bottles of acids on low shelves or in acid cabinets on trays large enough to contain spillage or leakage. Concentrated acids should only be purchased in plastic coated bottles and stored in their polyfoam shipping container. Make certain that the polyfoam container is also labeled. It would be an unsafe practice to have to remove the bottle from the container to determine its contents.

Specially constructed acid storage cabinets are commercially available and provide an ideal, secure, corrosion-resistant facility for acids. Some cabinets are made of particle board laminated with a corrosion resistant laminate, with hinges and hardware Teflon® coated. Perchloric acid should not be stored in this type of cabinet. Another acceptable alternative is cabinets made entirely of polypropylene. Metal cabinets are not acceptable because of their tendency to corrode.

CAUSTICS

Caustics (bases) are compounds such as sodium hydroxide (lye) and calcium oxide (lime). They should never be stored next to acid because large amounts of heat can be generated if contact is made. Since most caustics are not volatile and do not create corrosive vapors, they do not need to be stored in special cabinets. However, caustics are extremely corrosive to human tissue and should be handled with as much care as concentrated acids. A spill control plan should also be available and posted.

OXIDIZERS

Oxidizers are a class of chemical compounds that can react violently with flammable and combustible materials. They should be stored separately from these materials, as well as reducing agents such as diborane, bisulfites, and hydrides. Perchloric acid is commonly found in laboratories and presents special problems. It must be used only in a water wash-down fume hood made of stainless steel. Do not use around wooden tables or benches. Store perchloric acid bottles on glass or ceramic trays that have enough volume to hold all of the acid if the bottle should break.

Common Oxidizers

Solids	
Ammonium dichromate	Nitrates, salts of
Ammonium perchlorate	Periodic acid
Ammonium persulfate	Permanganic acid
Benzoyl peroxide	Peroxides, salts of
Bromates, salts of	Potassium dichromate
Calcium hypochlorite	Potassium ferricyanide
Cerric sulfate	Potassium permanganate
Chlorates, salts of	Potassium persulfate
Chromium trioxide	Sodium bismuthate
Ferric chloride	Sodium chlorite
Iodates, salts of	Sodium dichromate
Iodine	Sodium nitrite
Magnesium Perchlorate	Sodium perborate
Manganese Dioxide	

Common Oxidizers (continued)

Liquids

Bromine	Nitric acid
Chromic acid	Perchloric acid
Hydrogen peroxide	Sulfuric acid

Gases

Chlorine	Nitrogen oxide
Chlorine dioxide	Oxygen
Fluorine	Ozone
Nitrogen dioxide	

TOXIC SUBSTANCES

Toxic chemicals can be harmful by contact, inhalation, and ingestion. Highly toxic chemicals and carcinogens should be stored in ventilated storage areas in unbreakable secondary containers. Such storage areas should have limited access and exhibit warning signs. Chemicals with a high chronic toxicity such as mutagens, teratogens and carcinogens, should be identified with a label. It is important that employees do not handle these materials without being made aware of their hazards and be given proper training. Typically known and probable carcinogens are:

Arsenic powder and arsenicals	Sodium arsenite
Arsenic pentoxide	Acrylonitrile
Arsenic trichloride	Cadmium powder
Arsenic trioxide	Cadmium chloride
Asbestos	Cadmium sulfate
Benzene	Carbon tetrachloride
Benzidine	Chloroform
Chromium powder	Ethylene oxide
Chromium (VI) oxide	Nickel powder
Lead arsenate	*o*-Toluidine
Sodium arsenate	

PEROXIDE FORMING COMPOUNDS

Many ethers and other similar compounds tend to react with oxygen in the air to form unstable peroxides, which may explode with extreme violence. Under the proper conditions and the absence of safety precautions, these chemicals can pose a serious hazard to personnel.

Ethers are highly susceptible to forming peroxides, even in unopened containers. Age alone can contribute to this condition. Processes which concentrate these chemicals by evaporation or distillation increase the likelihood of forming a detonatable mixture. Subjecting this mixture to unusual heat, shock, or friction may cause an explosion. Peroxide formation, as a result of evaporation around the cap of a container, appears as particles or a crusty conglomerate. Special attention should also be given to peroxidizable compounds whose age is unknown or that have physical characteristics different from those of the pure substance.

There are precautions that can be taken to make handling of these materials safer.

Additives are available for inhibiting peroxide formation, and there are also procedures for removal of peroxides. These practices, in themselves, can have a degree of danger and still may only postpone the problem.

A few simple storage precautions can help to greatly reduce the hazard of peroxide formation. These compounds should be stored in airtight containers in a dark, cool, and dry area. Containers should be labeled with the date when received, when opened, and the recommended disposal date. Purchase only small quantities of peroxidizable chemicals. This limits the amount exposed to the air and allows you to use up the reagent before the expected date of first peroxide formation. If not used, it should be disposed of in accordance with applicable regulations. A program should be followed to test for the presence of peroxides at designated intervals based upon the susceptibility of peroxide formation.

Knowledge of peroxide formation, implementation of safe storage and disposal procedures, and the use of proper personal protective equipment will help prevent injury and promote a safer working environment. If pure peroxides are found, however, do not attempt to move the container or remove its lid. Only a ''bomb squad'' should dispose of pure peroxides.

Examples of Peroxide Forming Chemicals

Isopropyl ether	Cyclohexene
Ethyl ether	Cyclooctene
p-Dioxane	Potassium metal
Tetrahydrofuran	Vinylidene chloride
Acetaldehyde	Acetal
Acrylaldehyde	Crotonaldehyde
Tetralin	Vinyl ether

PYROPHORIC SUBSTANCES

Any liquid or solid that will ignite spontaneously in air below 130°F (54.4°C) is considered pyrophoric. Phosphorus and titanium dichloride are examples of pyrophoric solids, and tributyl aluminum is a pyrophoric liquid. Potassium, sodium, lithium hydride are spontaneously flammable in moist air, as they react exothermically with water. Such materials must be stored in an inert atmosphere or under kerosene. Finely divided metals such as barium and nickel will spark when slight friction is applied; they are used as tips in pocket lighters.

LIGHT-SENSITIVE CHEMICALS

These chemicals will degrade upon exposure to light and should be stored in amber bottles. Examples are ethyl ether, silver salts, and mercury compounds.

WATER-REACTIVE CHEMICALS

These chemicals will react with water to yield flammable or toxic gases. Accordingly, they must be protected from contact with water or moisture in the air. Water reactive chemicals are particularly hazardous in a fire situation so the symbol should be posted outside the storeroom for the benefit of firemen. Hygroscopic chemicals absorb moisture from air, turning a dry powder into a wet crystaline mass that is difficult to remove from bottles. Storage in polyethylene bags and tightly sealed containers can minimize this problem.

Some Water-Reactive Chemicals

Solids

Aluminum chloride, anhydrous	Maleic anhydride
Calcium carbide	Phosphorus pentachloride
Calcium oxide	Phosphorus pentasulfide
Ferrous sulfide	Potassium
Lithium	Sodium
Magnesium	

Liquids

Acetyl chloride	Stannic chloride
Chlorosulfonic acid	Sulfur chloride
Phosphorus trichloride	Sulfuryl chloride
Silicon tetrachloride	Thionyl chloride

FLAMMABLES

Control of flammable liquid hazards starts with the proper selection and use of safety cans and flammable safety storage cabinets. Design specifications of safety cans and storage cabinets as set forth by the National Fire Protection Association (NFPA) must be followed in order to be in compliance with OSHA.

Flammable liquids are volatile by nature, and it is their vapors combined with air, not the liquids themselves, that ignite and burn. Flammable liquids are widely used throughout industry and are recognized as among the most hazardous materials handled in large volumes on the job.

Definitions

Important terms as defined by OSHA follow:

I. Flammable liquid means any liquid having a flashpoint below 100°F (37.8°C), except any mixture having components with flashpoints of 100°F (37.8°C) or higher, the total of which make up 99 percent or more of the total volume of the mixture. Flammable liquids known as Class I liquids are divided into three classes as follows:
   A. Class IA shall include liquids having flashpoints below 73°F (22.8°C), and having a boiling point below 100°F (37.8°C).
   B. Class IB shall include liquids having flashpoints below 73°F (22.8°C), and having a boiling point at or above 100°F (37.8°C).
   C. Class IC shall include liquids having flashpoints at or above 73°F.
II. Flashpoint means the minimum temperature at which a liquid gives vapor within a test vessel in sufficient concentration to form an ignitable mixture with air near the surface of the liquid.
III. Vapor pressure shall mean the pressure measured in pounds per square inch (psi) (absolute) exerted by a volatile liquid as determined by the "Standard Method of Test for Vapor Pressure of Petroleum Products (Reid Method)," American Society for Testing and Materials—ASTM D323–68.

Another important term used to indicate the relative hazard of the liquid is the explosive or flammable range. This is the percentage range of liquid vapor in air by volume, within which ignition can occur. Explosive range figures are based on normal atmospheric temperatures and pressures. Ethyl alcohol has an explosive range between 3.3% and 19%, indicating that any concentration of ethyl alcohol vapor in the air between these percentage limits will ignite at any temperature above 55°F when an ignition source is present.

Flammable liquids are hazardous because of the vapors that evolve. When vapors combine with air, there is the possibility of ignition and burning. As the temperature increases, a flammable liquid becomes more hazardous because of the increased rate at which its vapors are evolved. Fire and explosion danger presented by flammable liquids can usually be eliminated by safe storing, dispensing, and handling procedures. Ways to eliminate the hazards of flammable liquids can be accomplished by modifications which reduce the areas of exposed liquids, or by substituting a nonflammable or less flammable material, or a liquid with a higher flashpoint. Vapors must be controlled by confinement, local exhaust removal, or area ventilation so that they will not accumulate to form a flammable mixture.

Efforts should be made to know the location of all flammable liquids. Inspect the containers for proper labeling. If a cap or cover is lost, or if there is evidence of significant mechanical damage to the container, the contents should generally be transferred immediately. The transfer should be made to a clean, undamaged container of the same type or better. Flammable liquids should be transferred in an area that is adequately ventilated, has adequate electrical grounding facilities, and is free from all sources of ignition.

Flammable liquids may also present health hazards from both skin contact with the liquid and inhalation of its toxic vapors. Some flammable liquids are primary skin

Store flammable liquids in approved safety cans.

irritants that destroy tissue, and some are skin sensitizers. Dermatitis may result when the liquids dissolve natural skin oils. Inhalation hazards exist in all cases, varying in degree with the duration of the exposure, and the concentration and toxicity of the vapor.

Health hazards can often be controlled by measures such as substitution of a less hazardous material for those that are dangerous to health, minimizing worker contact by changing the process or by isolating, and by safe handling methods or containment. Respiratory protective devices should be used where there is intermittent hazardous exposure or when such exposures are impractical to fully control by other means.

When practical, flammable liquids should be stored in safety cans. Approved safety cans meet basic safety requirements and are available to hold up to 5 gallons. The cap mechanism is spring loaded and self-closing to provide a leaktight spout seal and pressure relief venting. Flame arrestor screens inside the spout effectively prevent flashback to the can contents. This holds the temperature below the ignition point of the vapor air mixture inside the can. Safety cans should be UL and/or FM listed.

Size restrictions of flammable liquid containers are established by NFPA and adopted by OSHA. The following table lists the maximum allowable size of container specifications for storage of flammable liquids.

Maximum Allowable Size of Containers and Portable Tanks.

	Class IA	Class IB	Class IC	Class II	Class III
Glass or approved plastic	1 pt	1 qt	1 gal	1 gal	1 gal
Metal (other than DOT drums)	1 gal	5 gal	5 gal	5 gal	5 gal
Safety cans	2 gal	5 gal	5 gal	5 gal	5 gal
Metal drums (DOT spec.)	60 gal	60 gal	60 gal	60 gal	60 gal
Approved portable tanks	660 gal	660 gal	660 gal	660 gal	660 gal

Container exemptions: (a) Medicines, beverages, foodstuffs, cosmetics, and other common consumer items, when packaged according to commonly accepted practices, shall be exempt from the requirements of 1910.106(d)(2)(i) and (ii).

Flammable liquid storage cabinets are found throughout industry for the safe storage of flammable liquids. The purpose of these cabinets is to insulate the flammable liquids they store, thereby delaying their involvement in a fire. Metal cabinets should have double wall construction and be made of 18 gauge steel with $1\frac{1}{2}$ inch air space on top, bottom, and sides. The door should have a continuous hinge with a three-point latch arrangement, and the door sill should be raised at least 2 inches above the bottom of the cabinet to retain spilled liquids. Venting of a storage cabinet is optional under OSHA; however, some jurisdictions mandate that the interior of cabinets be vented to minimize the accumulation of vapors inside the cabinet.

Both NFPA and OSHA accept wooden storage cabinets for flammables, provided they are made from high density plywood.

It may also be necessary to store 55 gallon drums in storage cabinets. Special consideration should be given to grounding the drum to the cabinet and bonding the drum to the dispensing container. Drum vents are also available for the safe removal of flammable liquids.

Moderate numbers of containers may be stored in a flammable liquid storage cabinet. However, OSHA says that not more than 60 gallons of highly or moderately flammable liquids (Classes I and II) and not more than 120 gallons of combustible liquids (Class III) shall be stored in a storage cabinet.

Not more than three flammable storage cabinets may be located in a single fire area, except in an industrial occupancy where additional cabinets may be located in the same fire area if the additional cabinet, or group of not more than three cabinets, is separated from other cabinets or group of cabinets by at least 100 feet. Most commercially available and approved storage cabinets are built to hold 60 gallons or less of flammable liquids.

Depending upon the particular occupancy of a building, certain lesser quantities of flammable liquids can be stored in a safe place, outside of a specific storage cabinet or room.

Large quantities are often required to be stored in a separate storage room or area, or possibly may require storage in a flammable liquid warehouse with an automatic sprinkling system. Provisions covering locations where flammable or combustible liquids may be stored are arranged in the NFPA 30 Code according to increasing size, from storage cabinets, inside rooms, cutoff rooms, and attached buildings to warehouses.

Always observe the following guidelines when handling flammable materials:

1. Prevent the build up of potentially explosive atmospheres with good ventilation. When working with appreciable quantities of flammable materials, use an efficient exhaust hood.
2. Handle flammable materials only in areas free of ignition sources, such as sparking devices and open flames.
3. Never heat flammable materials with an open flame. Instead use water baths, oil baths, or heating mantels.

Cabinet specially designed for storage of flammable materials.

REFRIGERATORS

Refrigerators for chemicals have become part of many laboratory and industrial settings. Their uses may be to preserve items, deactivate or reduce reaction rates, increase shelf life, or reduce volatility of materials. Safe storage of flammable liquids requires special modifications of the refrigerator. There are basically three types of refrigerator to choose from, depending on the articles to be stored and the area in which the unit is to be located. Care must be taken in choosing the appropriate model for its intended use. Proper consideration will provide safety in use and avoid needless expense.

Selecting the correct refrigerator begins with determining the type of commodity to be stored. Perishable goods and items that do not generate flammable vapors can be safely stored in standard commercial refrigerators. Storing flammables in a standard refrigerator could result in a serious explosion. It is a recommended procedure to label standard refrigerators as not suitable for storage of flammable liquids.

Flammable solvents and liquids require special refrigeration, but first there is the consideration whether refrigeration is even needed. For example, consider ethyl ether, with a flash point of $-49°F$. Refrigerated storage of ethyl ether would not eliminate its vapor generation because the cold storage temperature is still above its flash point. Since the coldest setting of most common refrigerators/freezers is above this flash point, refrigeration is not advantageous for storage of ethyl ether. Limited quantities of ethyl ether could be stored in safety cans in a flammable liquid storage cabinet. Refrigerated storage should only be provided if it protects the integrity of the product or will reduce or eliminate hazards associated with it.

During storage of flammable liquids in a refrigerator, leaking vapors from the containers can accumulate in the enclosed space. This concentration may reach the lower explosion limit (LEL) of that material. This is a potentially dangerous situation in that any spark will cause an explosion. Standard refrigerators do not provide protection against interior sparking so they cannot be used for storage of flammables. This type of storage requires specially designed refrigerators that protect against interior spark induced explosions.

The main feature that makes these refrigerators safer than commercial ones is the no-spark interior construction. There are no internal electrical components such as switches, thermostats, or lights which could possibly spark and ignite the flammable vapors inside the refrigerator. Electrical equipment is mounted on the outside of the refrigerator so that exposure to hazardous concentrations of vapors is minimal. A flammable materials storage refrigerator is prewired for easy installation. Simply plug it into a standard wall outlet. These refrigerators will provide safe storage of volatile flammable materials in areas which are not classified as hazardous locations.

Some storage areas have the potential for flammable vapors to be present in the ambient room air. These hazardous locations need refrigeration units which not only have no spark interior, but also have spark-free electrical components throughout the entire unit. Explosionproof refrigerators have all the same features as flammable materials refrigerators, and also allow the unit to be located in a hazardous environment where flammable vapors are present immediately in the surrounding area. Explosionproof refrigerators are UL listed for use in Class I, Group C and D hazardous locations. These units are generally characterized by having a totally enclosed thermostat and compressor, wiring through conduit and enclosed overload and relay switch.

It is important to realize that an explosionproof refrigerator is only needed if the entire area is classified as a hazardous location because of high concentrations of flammable vapors or gases that could be present. The need for such a unit is only justified if all the other electrical equipment in the area is also rated explosionproof. Most laboratories do not require an explosionproof unit because they are not classified as hazardous areas and the other electrical devices are not explosionproof. In this case, a flammable materials refrigerator will provide safe storage without the expense of purchasing and specially wiring an explosionproof unit.

Storage of perishable goods, process solutions, chemicals, and flammable liquids can be accomplished safely with the proper type of refrigerator. It is necessary to know the physical properties of the items being refrigerated in order to choose the correct unit. Food should never be stored in refrigerators designated for chemicals.

CHEMICAL STOREROOM

Rooms designated for storage of chemicals are recommended. Storeroom specifications include the proper placement of doorways, ventilation systems and location of safety equipment, such as fire and smoke alarms, fire extinguishers and sprinklers, eye/face/body wash stations, and warning signs.

Before proceeding with the actual storeroom design, however, several agencies should be consulted. These include OSHA (29 CFR 2910.35, 36, 37 Subpart E) and state and local fire and building codes. These agencies provide specific criteria for chemical storage rooms.

Proper placement of doorways within a chemical storeroom is critical not only for access, but for egress as well. There should be two clearly marked exits. Interior walls should be of masonry construction without false ceilings. A false ceiling may allow fire or toxic fumes to travel to other portions of the building.

Storerooms should be fully lit and contain no blind alleys or aisle obstructions. They should also be clearly identified and secured when not in use. Emergency procedures, telephone numbers, and evacuation routes should be posted for fast reference.

The storeroom should be well ventilated, exhausting the air to the outside in a manner that will not recirculate contaminated air. This area should be supplied with adequate independent air conditioning and/or dehumidifying systems to keep a constant cool, dry atmosphere below 25°C (77°F).

When an emergency occurs, properly marked and positioned emergency equipment may save lives. A warning system should be available in the event of an emergency,

and should include emergency and evacuation procedures for all key personnel.

Personal protective equipment (goggles, clothing, globes, etc.), self-contained breathing apparatus (SCBA), emergency spill control supplies, and fire extinguishers should be stored outside of the storeroom. In case of an emergency inside a storeroom, equipment stored within may not be accessible.

All fire and smoke alarms should be mounted with a backup power supply in fire prone areas and tested regularly. Automatic sprinkler systems may be installed in the storeroom; however, chemical reactivity hazards with water must be considered first. Consultation with a chemical safety specialist should be considered before installation.

An eye/face/body wash station in a storeroom should be located within 100 feet or reachable within 10 seconds. These units should be periodically checked and maintained.

A safe chemical storeroom is a clean room; therefore, good housekeeping is important. Housekeeping procedures should be followed on a daily basis; all chemicals that are contaminated, unlabeled, or no longer used or needed should be discarded routinely. Chemical contamination can be reduced by prohibiting the return of unused chemicals to the stock bottle or container.

Regular inspections for decomposition of the chemicals should be performed and recorded in an inspection log. Labels should also be checked for degradation and secure attachment to bottles to prevent identification and chemical disposal problems.

Storerooms should not be used as preparation areas because an accident could contaminate the rest of the chemicals or could lead to a major emergency involving stored chemicals. Preparation and repackaging should be done in a separate area. A storeroom is a place where chemicals are stored and inventoried in a safe and convenient manner. It should be accessible to laboratory or production workers, at convenient

An emergency eye wash should be a part of all active chemical storage areas.

times, to eliminate the necessity of duplicating storage facilities. Procedures should be established for the operation of a storeroom and responsibility for safety and inventory control assigned to one person. A distinction should be made between storerooms where a variety of chemicals in relative small amounts are stored and storage rooms where a reserve supply is stored. Safety procedures between these facilities could differ.

The storage area must be maintained in a clean and orderly fashion by promptly removing all packing material and empty cartons and by emptying waste receptacles. A special waste receptacle should be designated for broken glass. When filled, it should be sealed and identified for disposal.

When disposing of chemicals, one should follow all federal, state and local regulations to assure environmentally safe and effective disposal.

When either a small or large chemical spill occurs, time counts in cleaning up. The first line of defense in controlling a chemical spill is containing the initial spill. Chemically inert spill trays large enough to contain the spill if breakage or leakage should occur should be placed under the containers. Spill dams or dikes may also be considered.

If the spill should happen outside of the tray, then Spill Control Pillows or neutralizers, either acidic or caustic, should be used to clean up the spill. If the spill contains a flammable chemical, nonsparking tools should be used in the cleanup process.

The person in charge of cleanup should contact the proper agencies: Environmental Protection Agency (EPA), Department of Natural Resources (DNR), local police and fire departments or any other agency that might be required to be notified depending upon the severity of the spill. Once the chemical spill is completely cleaned up, proper disposal methods for the chemical and the material should be followed.

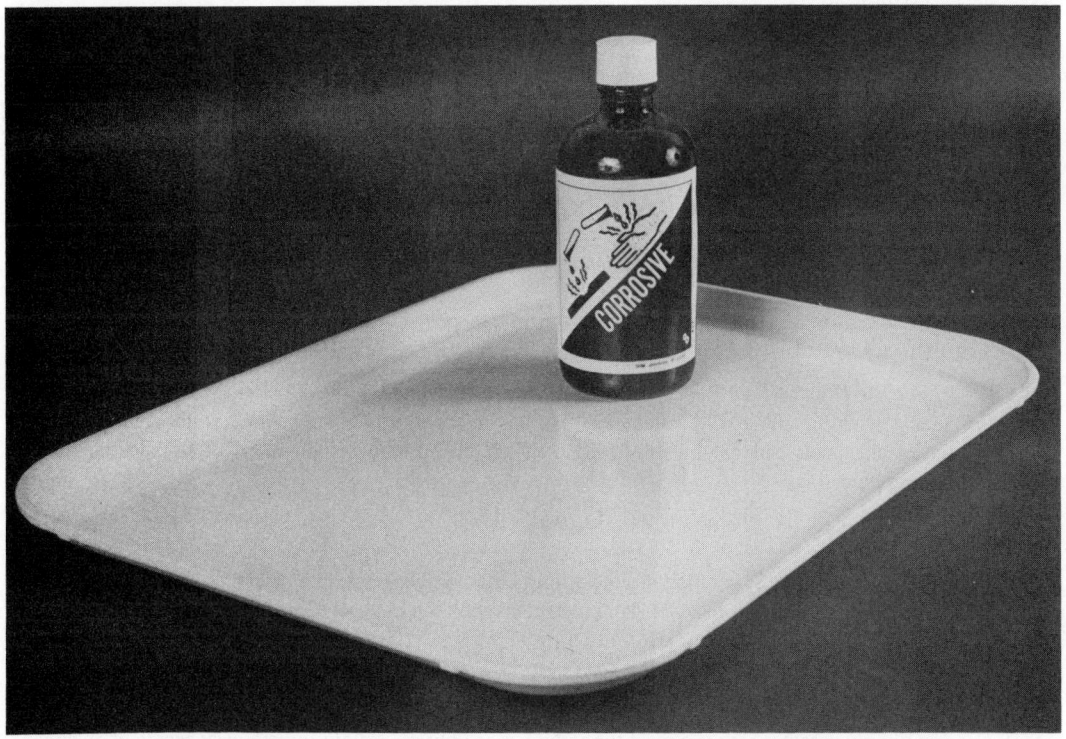

Trays can prevent spills from contaminating work surfaces.

COMPRESSED GASES

Gas cylinder storage can be very dangerous because of the extremely high pressure of the gas in the cylinder. All gas cylinders should be stored with valve covers on and should be tightly secured to prevent tipping. Gas cylinders should be stored in a dry, cool location away from fumes, direct and indirect heat or flames, or chemicals that are highly flammable or corrosive. Never attempt to move a gas cylinder by grasping its valve; use cylinder transfer carts. Storage considerations also must include separate storage areas and proper labeling for empty and full cylinders.

Use the minimum size compressed gas cylinder to do the job. When ordering cylinders containing hazardous gases, make certain they are stored and used in fume hoods or ventilated cabinets and that adequate personal protective equipment is available. Do not expose cylinders to temperatures higher than 50°C. Some rupture devices on cylinders will release at about 65°C. Lecture bottles may not be fitted with these rupture devices and may explode if exposed to high temperatures. Never use a cylinder that cannot be properly identified.

Hazardous Gases and Their Properties

	Flammable	*Oxidant*	*Corrosive*	*Toxic*
Acetylene	X	—	—	—
Arsine	X	—	—	X
Carbon monoxide	X	—	—	X
Chlorine	—	X	X	X
Diborane	X	—	—	X
Hydrogen	X	—	—	—
Fluorine	—	X	X	X
Phosgene	—	—	—	X
Sulfur dioxide	—	X	—	X

REFERENCES

American Chemical Society, *Safety in Academic Chemistry Laboratories,* American Chemical Society, 1985.

    A small, softcover book packed with a wealth of practical information on chemical safety in labs. Complete and easy to understand text includes guidelines on safe chemical storage.

Bretherick, L., *The Handbook of Reactive Chemical Hazards,* Butterworths, Boston, MA, 1979.

    Provides a wide sampling of documented information on the likely reaction/hazard potential for a broad range of chemicals and incompatibilities. Included in this information is select storage, handling, and transporting guidelines.

Green, M. E., and Turk, A., *Safety in Working with Chemicals,* Macmillan Publishers, New York, NY, 1978.

    Compiled by two college chemistry teachers, this softcover book serves as a safety outline for workers and/or students alike. Equipment hazards, lab precautions including housekeeping and hygiene, and storage tips are just a few of the topics discussed.

Sax, N. Irving and Richard J. Lewis, Sr., Eds., *Hawley's Condensed Chemical Dictionary,* Eleventh Edition, Van Nostrand Reinhold, New York, NY, 1987.

    Concise, thorough text which is a must for any reference library. Allows quick, easy access of chemical data. Includes not only chemical and physical information but also storage and transportation information.

National Research Council, *Prudent Practices for Handling Hazardous Chemicals in Laboratories,* National
Academy Press, 1981.

A practical reference guide concerned with safe practices in working with hazardous chemicals in
laboratories. Several chapters specifically cover chemical storerooms and storing chemicals in laboratories.
General considerations for safe lab storage and special storage guidelines for toxics, flammables, and
compressed gases are included.

Renfrew, M. M., "Safety in the Chemical Labatory," *Journal of Chemical Education,* 1981.

A collection of journal articles from the Safety Column of the Journal of Chemical Education. Several
articles on storage are included: storage in labs, storage of gas cylinders, storage stabilization, and
storeroom safety.

Sax, N. Irving, *Dangerous Properties of Industrial Materials,* Sixth Edition, Van Nostrand Reinhold,
New York, NY, 1984.

Over 19,000 industrial and laboratory chemicals are included in this exhaustive reference. Subjects
covered are storage and handling, toxicity data, and chemical incompatibilities, to name a few. A standard
in the workplace.

Steere, N. V., *Handbook of Laboratory Safety,* CRC Press, 1972.

Covers virtually every aspect of laboratory safety from ventilation and radiation hazards to electrical
and biological hazards. A very comprehensive reference filled with valuable information. Several reference
sections include chemical storeroom and storage guidelines.

2

RESPIRATORS

Sheldon H. Rabinovitz, Ph.D., C.I.H.

Cincinnati, Ohio

INTRODUCTION

Whenever a substance is used in the workplace, moral, liability, and regulatory considerations dictate that one should determine whether use of the substance will result in a health hazard to employees. This determination should also include the potential for suddenly occurring hazards caused by accidents and malfunctions. Often such an evaluation can be made based on the toxicity and physical properties of the material and a knowledge of how the material will be used, including amounts. When an inhalation hazard is indicated, air sampling may be required to complete the evaluation. Regardless of the methods used to identify hazards, action must be taken to eliminate any identified problems or deal with the potential for hazards created accidentally.

While substitution of a less hazardous substance for the hazardous material or the use of permanent engineering controls or work practices are the preferred methods of eliminating chemical hazards, respirators can be used in certain situations to protect exposed workers from airborne hazards or to permit escape and rescue during emergencies. In some circumstances respirators may be the only feasible means of protection. Respirators may also be used for protection in oxygen deficient atmospheres.

This chapter describes what respirators are and what types are commercially available, when they should be used, how they are selected, and the components of a respiratory program, including regulatory requirements. By learning how respirators work, the reader will gain an understanding of their limitations and why complete respirator programs are necessary if workers are to be adequately protected. The chapter concludes with a list of references which can provide more detailed information on the subject.

Only the basic concepts and problems associated with the use of respirators in the workplace are presented here. The reader will not be able to set up a respirator program solely on the information provided in this chapter. No attempt is made to select individual respirators for particular situations, or to identify all elements in an acceptable respirator program. Respirator programs are regulated mainly by the Occupational Health and Safety Administration (OSHA). The regulations change periodically; before a respiratory protection program is set up, the latest regulations should be obtained and studied.

Preferred Methods of Protecting Employees from Airborne Chemical Hazards

Good industrial hygiene practice dictates that engineering or work practices be considered first in protecting the employee from exposure to toxic chemicals. Good control mechanisms will reliably prevent exposure of employees to hazardous levels of a contaminant. Ideal controls require little or no maintenance, are not subject to breakdown or deterioration, and do not require positive action by workers or management to make them work. Substitution of a nontoxic material for the toxic material is an ideal way to protect the health of employees, since all potential for exposure is eliminated.

When substitution is not a viable option, isolation, use of closed systems, and ventilation are common protective measures. Closed systems prevent contaminants from becoming airborne by eliminating all openings. Ventilation systems function either by collecting airborne contaminants directly at the source of the emission (local exhaust ventilation), or by diluting the airborne contaminant by introducing fresh air into the general area (general dilution ventilation). The advantage of these controls is that they can be tested to ensure that they work effectively and, if maintained, will continue to work. The systems can be inspected periodically and when problems occur (for example, reduced airflows in ventilation systems or broken seals in closed systems) they can be remedied.

OSHA recognizes the reliability of engineering controls and requires such controls to protect workers from excessive exposure to regulated chemicals where feasible. OSHA does not identify specific methods of compliance for the majority of chemicals regulated by the agency. However, for about 23 chemicals where compliance methods are specified, engineering controls are identified. For some substances and operations (for example, asbestos, arsenic, lead, and coke oven emissions) very specific control requirements are stated; further, in these cases if engineering controls cannot reduce exposures below the limit, they must still be implemented in conjunction with the use of respirators. The controls are required because if the respirators fail, the employee will not be exposed to as great a hazard as would occur if no controls were in place. Also, the lower the ambient concentration, the greater the protection that will be offered by using a given respirator. This requirement indicates the relative importance OSHA attaches to engineering controls over respirators in protecting employees from excessive exposure to airborne contaminants.

Limitations of Respirators

The following list summarizes the major problems associated with respirators:

1. While air sampling tests can be made to determine the effectiveness of engineering controls, testing the effectiveness of the protection offered by respirators is very difficult.
2. Employees do not like to wear respirators because:

   · Many respirators are very uncomfortable to wear, especially in hot environments or when they have to be worn for long periods of time
   · Some employees have psychological problems wearing respirators (claustrophobia)
   · Employees often do not believe a hazard exists, especially when the chemicals have no odor or color and health effects take years to occur; consequently, they will not wear their respirators conscientiously

3. Obtaining a good seal with the respirator can be difficult for some workers because of facial contour or size, and they will not get the protection expected. Determining which employees will receive adequate protection is difficult, expensive, and time consuming.
4. If employees have facial hair, scars, or significant deformities, the likelihood of not achieving good face fits increases.
5. When more protective respirators are required, worker mobility may be lost. The weight of some respirators significantly reduces the ability of the worker to function and is a possible source of discomfort.

6. The ability of air purifying elements to filter certain hazardous substances is poor, reducing the effectiveness of the respirator.
7. Recent studies have suggested that the protection offered by certain respirators as determined in laboratory studies may not be realized in the workplace.
8. Although there is a government certification program for respirators, it has significant limitations. The certification program cannot be relied upon to ensure that a respirator will offer a certain degree of protection to every worker.

Situations Where Respirators Are Commonly Used

Even though there are difficulties in using respirators to achieve a healthful condition, a number of situations exist where they are the only practical solution. The employer should not hesitate to use respirators where they are indicated, so long as the proper precautions are followed and the expected level of protection is realized. Situations where employees must enter, leave, or work in hazardous atmospheres and should use respirators are as follows:

· While engineering controls are being installed or tested
· While engineering controls are being repaired or maintained
· During firefighting activities
· During escape from suddenly occurring hazardous atmospheres
· To rescue workers or eliminate hazardous conditions associated with emergencies
· For operations where other controls are not feasible
· For certain short-term operations where installing engineering controls would be economically impractical

The use of respirators for intermittent or short-term operations avoids some of the problems associated with long-term use. For example, the discomfort factor is not as significant when the respirator is only worn for a short time. Also, the likelihood of an employee breaking the seal while in the contaminated area probably increases as the wearing time increases.

Respirators are also used as permanent solutions for long-term operations when other controls are not feasible. Proper employee motivation and support programs to ensure that the respirator functions properly are critical in a respirator protection program. The last section of this chapter describes the elements necessary to make a respirator program successful.

DESCRIPTIONS OF RESPIRATORS

A respirator consists of an enclosure or facepiece that covers the mouth, the mouth and nose, or the entire face, or a helmet or hood which covers the head. The hood is connected by means of a hose to either a clean air supply or to an air filtering device which provides clean air for breathing in environments containing airborne hazards or lacking oxygen. Respirators connected to a clean air source are called *atmosphere-supplying respirators*. Respirators connected to a filtering device which removes contaminants from the ambient air to make it suitable for breathing are called *air-purifying respirators*. Atmosphere-supplying respirators include both air-line (supplied air) or self-contained breathing apparatus (SCBA) type respirators.

Different types of facepieces can be used with most respirators. They are divided into two types, tight fitting and loose fitting. Tight fitting facepieces form a seal between the facepiece and the skin and are the only types that can be used with negative pressure respirators. Tight fitting facepieces usually consist of half masks which cover the face from below the chin to over the nose and full facepieces which cover the face from the hairline to below the chin. There is also a quarter-mask facepiece which just covers the mouth and nose. It is not widely used, and is more easily dislodged than half-facepiece respirators. Loose fitting facepieces consist of helmets or hoods; these require a forced air supply, at some constant flow rate. Full facepieces, helmets, and hoods provide eye protection, while half facepieces do not. The major types of respirators are shown in Figure 1.

Positive and Negative Pressure Respirators

Both air-purifying and atmosphere-supplying respirators can be either positive pressure or negative pressure devices. A positive pressure respirator maintains a positive pressure with respect to ambient pressure inside the facepiece during both inhalation and exhalation. The positive pressure is maintained by forcing air into the facepiece from a hose connected to a pressurized tank, compressor, or blower motor (fan). Regulator valves are also used when the air supply comes from a high pressure source. A negative pressure respirator has negative pressure in the facepiece relative to ambient pressure during inhalation and positive pressure during exhalation. Air for breathing is drawn into the facepiece by the inhalation pressure. The air may be ambient air drawn through filters or it may come from an external source (tank or nearby clean air) through hoses.

Most positive and negative pressure tight fitting respirators contain an inhalation valve and an exhalation valve. The exhalation valve ensures that outside air does not enter the respirator during inhalation, and the inhalation valve prevents exhaled air in the facepiece from exiting through the intake valve.

Supplied Air and SCBA Respirators

Atmosphere-supplying respirators consist of two types: (1) self contained breathing apparatus (SCBA), where the air supply comes from a source carried by the user; and (2) supplied air respirators, where the air supply comes from a point at some distance from the user and is supplied to the facepiece by an air-line hose. These respirators are also called air-line respirators.

There are two types of SCBA: open circuit and closed circuit. In open circuit systems, breathing air is supplied to the facepiece and the exhaled air is vented directly out of the facepiece into the atmosphere. In closed circuit systems, the exhaled air is recycled by removing the carbon dioxide and replenishing the consumed oxygen with additional oxygen from a self contained solid, liquid, or gaseous source.

Today most commercially available atmosphere-supplying respirators are positive pressure devices. While negative pressure atmosphere-supplying respirators are still certified, they do not provide the protection offered by positive pressure devices and are rarely recommended. Advances in the technology of positive pressure atmosphere suppling devices have rendered most atmosphere-supplying negative pressure devices obsolete.

Atmosphere-supplying positive pressure respirators come in two modes. One is pressure demand, where a valve senses the pressure in the facepiece and provides the volume of air necessary to maintain a positive pressure in the facepiece under all conditions. Theoretically, no matter how hard the user is breathing, sufficient air will be supplied into the facepiece to maintain a positive pressure. The other type is a continuous flow positive pressure system, where a constant volume of air is continuously supplied into the facepiece. Depending on the volume of air supplied, the user could breathe hard enough to create a negative pressure (with respect to the ambient pressure) in the facepiece. The continuous flow system has another disadvantage in that it uses more air than a pressure demand system. For these reasons, pressure demand systems are usually preferred.

Air-Purifying Respirators

Most air-purifying respirators are negative pressure devices, although some contain a portable blower which draws air through a filter and then delivers it to the facepiece

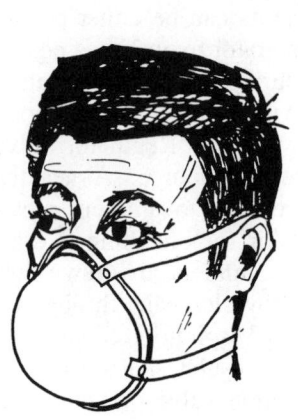

Single-Use Respirator

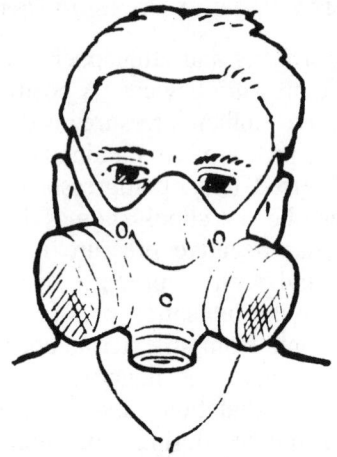

Half-Mask Respirator

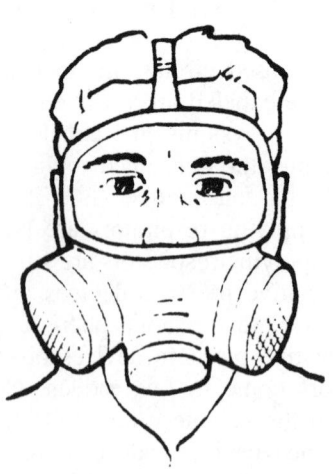

Full Facepiece Respirator

Powered Air-Purifying Respirator

Figure 1. Air-Purifying Respirators.

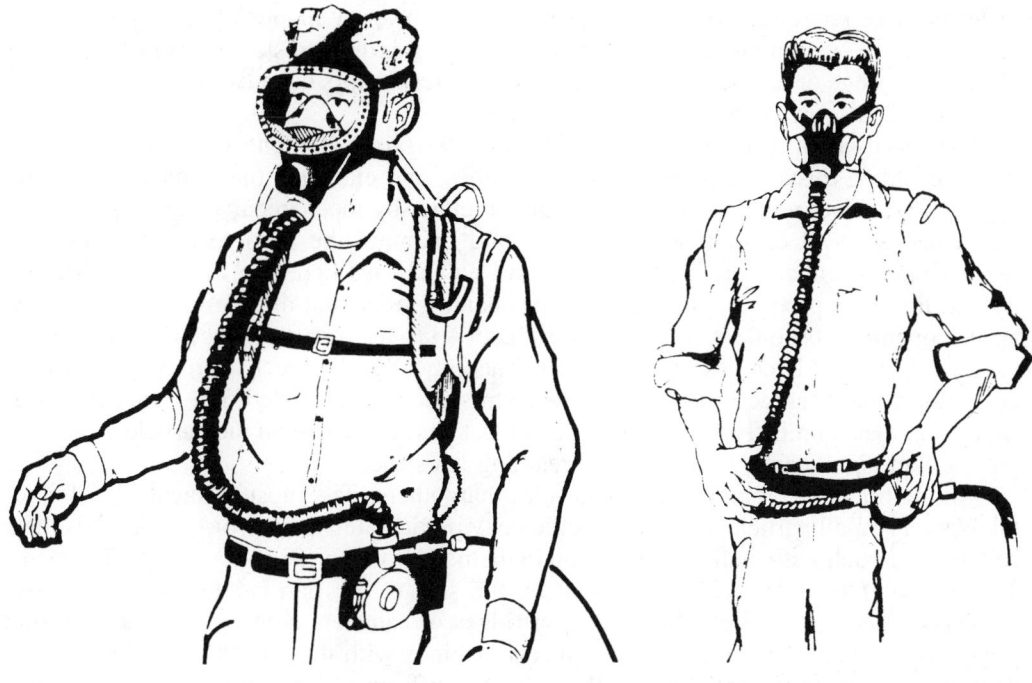

**Air Line Respirator
with Full Facepiece**

Air Line Respirator with Half Facepiece

**Full Facepiece,
Self-Contained Breathing Apparatus**

Oxygen-Generating SCBA (Closed Circuit)

Figure 1 (cont'd). Atmosphere-Supplying Respirators.

under positive pressure. (These devices are called powered air-purifying respirators.) If the filtering device on an air-purifying respirator is a large canister instead of a smaller filtering cartridge or cartridges (some air-purifying respirators have two cartridges instead of one), it is also called a gas mask.

Air-purifying filter elements are divided into two types: elements that remove particulates, including dusts, mists, aerosols, and fumes; and elements that remove gases and vapors. Particulate removing elements function either by mechanically trapping particles like a net or by electrostatically attracting and retaining the particles. These trapping mechanisms essentially are inexhaustible and will always continue to collect particles; in fact, the more particles collected on a mechanical filter, the more efficient it becomes as the openings become smaller. Unfortunately, as the openings become smaller or clogged, breathing becomes more difficult. Theoretically, the mechanism of electrostatic collection could become exhausted, if all electrostatic sites are filled, but from a practical standpoint that is unlikely to happen. Filters must be replaced when the particles collected on the filter significantly increase the breathing resistance.

Some filters can retain smaller particles than others. The most efficient filter for retaining very small particles is the high efficiency particulate air filter or HEPA. This filter can retain particles as small as 0.3 micron in diameter with an efficiency of 99.97 percent. The type of filter selected is dependent on the size of the particulate and its toxicity.

Gases and vapors are collected in cartridges or canisters containing a medium that physically adsorbs the contaminant or reacts or binds with the gas or vapor to retain it. This mechanism can be satiated and the collection efficency varies with the contaminant. Thus, unlike particulate filters, which filter more efficiently with time, gas and vapor cartridges have limited absorption capacity and will fail after a certain time depending on the contaminant and the concentration. The user can only know when the cartridge or canister is satiated and the contaminant is passing through if the contaminant has warning properties such as odor. When a contaminant passes through a spent cartridge, it is called breakthrough.

Single Use and Disposable Respirators

The facepiece on most air-purifying respirators consists of a molded rubber-type material to which disposable cartridges are attached. In recent years, a new type of air-purifying respirator has increased in popularity. It is called the single use or disposable respirator, which is made for protection against particulates and gases and vapors. These respirators are tight fitting half masks where the filtering medium essentially constitutes the whole facepiece. They are less expensive than moldable reuseable respirators, but can be more difficult to fit because of their construction.

The government certification program makes a distinction between single use and disposable respirators. Single use respirators are certified only for protection against pneumoconiosis and fibrosis-producing dusts, but not for use against contaminants causing systemic damage such as kidney or liver damage. Disposable respirators, which may look and be constructed just like single use respirators, can be certified for protection against certain systemic poisons including both dusts and gases and vapors. The next section describes the certification process.

When the filter medium collects particulates to the point where an increase in breathing resistance is noted or where the sorbent material has been satiated with the contaminant gas or vapor, the entire respirator is thrown away. These respirators can be reused until the filter media no longer functions properly.

Because they are not composed of a molded rubber type material, the shape of single use respirators is not a constant. This can cause difficulty in assuring a good, tight seal with each use. Workers often prefer these types of respirators over moldable respirators because of their light weight and feel. Because of the difficulty of assuring a good fit, considerable controversy exists regarding the effectiveness of these types of respirators as compared to reusable (moldable) respirators. Also, while some disposable or single use respirators have exhalation valves, they do not come in a HEPA model. Thus the protection they offer is not adequate for certain highly toxic particulates.

RESPIRATOR CERTIFICATION PROGRAM

The National Institute for Occupational Safety and Health (NIOSH) and the Mine Safety and Health Agency (MSHA), two federal government organizations, jointly certify respirators. The certification program consists of inspection, examination, and testing of respirators in accordance with Title 30 Code of Federal Regulations, Part 11 (30 CFR 11). A certified respirator has passed certain tests and is made according to specified construction standards, but this does not ensure that an employee wearing such a respirator will receive adequate protection. Adequate protection depends on proper selection, fitting, and use of a certified respirator.

Certification requirements vary by type of respirator and intended use. Minimum specifications have been established for each grouping or class of respirator. In addition to inspecting and testing respirators, the manufacturer must implement an adequate quality control program in order to obtain certification. The certified product must display a certification number, which identifies the type of certification given.

Certification indicates that a respirator meets the specifications of a particular class. Some certified respirators may just meet those specifications, while others certified in the same class may exceed the standards and offer greater protection. When a given class of respirators is recommended, the poorest acceptable performing respirator in that class is supposed to offer adequate protection. When selecting a respirator within a recommended class of respirators for a given situation, however, the user should determine which respirator in that class would be the best. That information can be obtained by comparing the manufacturers' literature for the respirators being considered.

Respirator certification classes and types include:

1. Self-contained breathing apparatus (SCBA) for entry into and escape from or only escape from hazardous atmospheres

   · Closed circuit
   · Open circuit
     —Demand type (negative pressure)
     —Pressure demand type (positive pressure)

2. Gas masks for entry into and escape from or only escape from hazardous atmospheres containing adequate oxygen to support life
3. Supplied air respirators during entry into and escape from hazardous atmospheres

   · Type A supplied air—hose mask respirators (large air-line hose attached to hand operated or motor driven blower where the wearer could breathe fresh air through the hose if the motor failed; these types of respirators are rarely used now
   · Type AE supplied air respirators—having a helmet or hood which offers head and neck protection
   · Type B supplied air respirator—hose mask respirator as described above but

without the blower motor, for use in atmospheres not immediately dangerous to life or health

· Type C supplied air respirator—an air-line respirator for entry into and escape from atmospheres not immediately dangerous to life or health which can operate either under negative or positive pressure

· Several other classes of supplied air respirators involving additional neck and head protection

4. Dust, fume, and mist respirators including both air-purifying and powered air-purifying respirators

· Dust respirators with replaceable or reusable filters designed to protect against dusts having an air contamination level not less than 0.05 mg/m^3 or dusts having an air contamination level not less than 2 million particles per cubic foot of air

· Dust, mist, and fume respirators with replaceable filter designed to protect against fumes of various metals having an air contamination level not less than 0.05 mg/m^3

· Dust and mist respirators with replaceable filters designed to protect against mists of materials having an air contamination level not less than 0.05 mg/m^3 or 2 million particles per cubic foot

· Dust, mist, and fume respirators with replaceable filters designed to protect against air contaminants having an air contamination level less than 0.05 mg/m^3 (these are HEPA filter type air-purifying respirators)

· Single use respirators designed as respiratory protection against pneumoconiosis- and fibrosis-producing dusts, or dusts and mists

5. Chemical cartridge respirators designed for use as respiratory protection during entry into or escape from atmospheres not immediately dangerous to life or health with adequate levels of oxygen to support life for the following substances:

Type of Chemical Cartridge	Maximum Use Concentration (ppm)
ammonia	300
chlorine	10
hydrogen chloride	50
methylamine	100
sulfur dioxide	50
organic vapor	1000

Organic cartridge respirators can only be used for those organics which have adequate warning properties and which do not generate high heats of reaction with the sorbent material in the cartridge. Also, the organic cartridge respirator cannot be used for any concentration exceeding the immediately dangerous to life or health (IDLH) concentration. The IDLH is discussed in the respirator selection section. There are certifications for other chemicals such as ethylene oxide, formaldehyde, hydrogen sulfide, and phosphine where manufacturers have submitted requests for such approvals. The user should obtain the NIOSH certification book to determine if an appropriate cartridge respirator exists for a specific chemical.

6. Pesticide respirators including those designed for use as respiratory protection during entry into and escape from atmospheres which contain pesticide hazards and include

gas masks and chemical cartridge air-purifying respirators (including powered air-purifying respirators)

As mentioned, the certification test includes inspection, examination, and testing of the respirator along with a review of the manufacturers quality control program. During the inspection, a determination is made if the respirator has the required components for the class in which certification is sought. Examples of components required (where applicable) include:

Type of Respirator	Examples of Components
SCBA	Pressure gages, safety relief valve, harness, flexible breathing tubes
gas masks	canister or cartridge, external check valve, face-piece or mouthpiece and noseclip
supplied-air respirator	air supply valve, air supply hose, detachable couplings, respirator harness
air-purifying respirators	filter unit, harness, face-piece, mouthpiece with noseclip, hood, or helmet

In addition to component requirements, there are performance requirements. Examples of some of the requirements (where applicable) include:

· Hose requirements including flexiblity and length
· Protection of valves
· Air velocity and noise level restriction
· Inhalation and exhalation resistance requirements
· Minimum requirements for breathing gas
· Types of gages that can be used
· Service life timer requirements for SCBA

After a respirator has met all the component and performance standard requirements, it undergoes testing. Examples of testing include filter media efficiency, leakage, air flow velocities, noise levels, breathing resistance tests, service time tests (SCBA), breakthrough test (chemical cartridge respirators), and manhandling tests. The testing varies for different classes. For classes offering more protection such as HEPA type air-purifying respirators, the filter efficiency tests are more restrictive than the tests for a dust and mist respirator. After a respirator passes all program requirements, it is certified.

While certification is extremely important in assuring the respirator user that the respirator meets certain specifications, it does not necessarily mean that it will offer adequate protection in a given situation. The reasons why no such guarantees come with certification include:

· The certification does not identify all use conditions
· The fit testing done in the certification is limited and does not define a minimum protection level that all users can expect to have

· The filtration efficiency testing conducted during certification is limited and does not ensure the efficiency or when breakthrough will occur for a given chemical with a cartridge respirator

For particulate filter efficiency testing, only two types of particulates are used: silica and lead fume. Thus while air-purifying respirators are used to protect against fibers, no filters are tested using fibers. For chemical cartridge respirators for protection against organic vapors, the cartridge is only tested with one chemical, carbon tetrachloride (CCl_4) and only at one concentration, 1000 ppm. There are a number of organics which will not be effectively collected in a typical charcoal cartridge even though charcoal will collect CCl_4. When a manufacturer applies for certification for a respirator that offers protection against a specific chemical, NIOSH will usually test the cartridge with that gas or vapor.

The testing of face seals is done qualitatively using a small panel of human volunteers. The testing is not extensive enough to determine what percentage of the population will receive a given level of protection. Actually, not all certifications include fit testing as part of the certification process.

To address some of the problems in certification, NIOSH is planning to make changes in the process. As the changes must also be approved of by the Mine Safety and Health Administration (MSHA) and submitted to the Office of Management and Budget (OMB) for review before promulgating the changes, the process is slow. At this time it is not known exactly what changes will be made. Some of the 30 CFR 11 regulations are clearly outdated. For example, the regulations currently permit certification of single use respirators for protection against asbestos. This occurred because at the time these regulations were made, asbestos was thought to only cause pneumoconiosis and fibrosis and was not considered to be a carcinogen. Because asbestos is now regulated as a carcinogen, OSHA does not permit the use of single use respirators.

PROTECTION FACTORS

To address the problem of determining the actual protection offered by a respirator, NIOSH funded a project at the Los Alamos National Testing Laboratory to conduct quantitative fit testing on human volunteers. The results of that study were to establish protection levels offered by various types of respirators.

The study would only measure respirator leakage. When testing air-purifying respirators, HEPA filter cartridges were used in all respirators as they are extremely efficient and the contamination levels measured in the facepiece would only be due to leakage around the faceseal, assuming the respirator was functioning properly. To determine what percentage of an average working population would receive a given level of protection, an anthropomorphic selected panel (a group of test subjects whose range of facial features are representative of the general working population) was carefully selected.

Each member of the panel, after familiarizing himself or herself with the respirator using a fixed protocol was then quantitatively tested to determine the protection received using that respirator. In quantitative fit testing, the subject enters a chamber containing a known level of a contaminant. A probe is connected into the facepiece of the respirator worn by the subject, which is connected to a continuous recording analytical device. The concentration inside the facepiece (which is the concentration presumably breathed by the test subject) is then compared to the concentration of the contaminant in the chamber. The result of dividing the concentration of contaminant in the chamber by the concentration of contaminant in the facepiece is called the protection factor. The greater the difference between the outside contaminant level and the level inside the facepiece, the greater the protection factor, which corresponds to adequacy of the faceseal and the pressure inside the facepiece. If the pressure inside the facepiece is positive with respect to the outside, leakage is not as important.

Because of the variability of facial contours and the limited sizes and shape of respirators, a certain percentage of the working population will always have some leakage with some or all respirators. Thus, for the test done at Los Alamos, the protection factor (FP) was set as that multiple of chamber concentration to facepiece concentration that was consistently attained by 95% of the test group. That meant that many of the test subjects would get considerably higher protection factors than the levels reported by Los Alamos to be attained by 95% of the group, and 5% would get less protection.

The results of the Los Alamos tests are listed in Table 1. These data are over ten years old, and the appropriateness of some of the protocols used in testing respirators has been questioned. In the table, for example, the protection factor for half-face air-purifying respirators is 10. That means that a person correctly wearing that respirator who has been properly fit-tested and using a cartridge that will quantitatively remove the contaminant in the atmosphere, can expect to be inhaling one tenth or less of the concentration of the contaminant in the environment.

If the cartridge fails, the seal is broken, or valves or any other part of the respirator fails, the rated protection factor can no longer be expected. One reason that the wearer must first be properly fit-tested is to ensure that the individual is not one of the 5 percent whose facial contour will not permit achieving a 10-fold protection factor.

Selecting a respirator with the proper protection factor is just one consideration in

Table 1. Respirator Protection Factors (as reported by the Los Alamos National Laboratories in 1975).

Respirator Class	Facepiece Pressure	Protection Factor
1. Air-purifying		
single-use, dust	−	5
quarter-mask, dust	−	5
half-mask, dust, fume, HEPA, and gas and vapor	−	10
full facepiece, HEPA, gas and vapor	−	50
2. Atmosphere-supplying		
A. Supplied air:		
demand half mask	−	10
demand full facepiece	−	50
pressure demand or continuous flow, half-mask	+	1000
pressure demand or continuous flow, full-facepiece	+	2000
B. Self Contained Breathing Apparatus (SCBA):		
open circuit, demand, full facepiece	−	50
open circuit, pressure demand, full facepiece	+	10,000+
closed circuit, oxygen tank type, full facepiece	−	50

Note: Other considerations may prevent using a given respirator up to its rated class protection factor. See the text for further information regarding respirator restrictions.

choosing a respirator. The following section describes how to select a respirator for a given situation.

RESPIRATOR SELECTION

After a decision is made that respirators are the only feasible method of adequately protecting employees for a given situation, the employer must select a suitable respirator. The selection process will depend on the situation where the respirator is used, the physical state and concentration of the contaminant, and the toxicity of the contaminant. The level of protection required will be dependent on the concentration and toxicity of the contaminant.

The selection process can be difficult because the work situation might indicate the use of a respirator that may not offer adequate protection. Usually the user will want to select a comfortable and inexpensive respirator but may not be able to because of protection factor requirements and other restrictions.

At times, the user may want to use a respirator that offers more protection than is required. Reasons for using a respirator having a higher protection factor than dictated by the use conditions include the possibility of encountering higher concentrations than anticipated or knowing that the concentration limit selected as being safe is based on limited data. Also, some employees may be hypersensitive and require additional protection. This section describes the process of selecting a respirator for a given situation. That process is based on several basic principles, which include:

- The level of protection required (protection factor) is based on the toxicity and concentration of the contaminant.
- The effectiveness of air-purifying respirators is dependent on how well and for how long the filtering medium will work.
- Full facepiece, tight fitting respirators leak less than half facepiece respirators.
- Leakage of ambient air into the facepiece due to poor face seals is not as likely with positive pressure respirators as it is with negative pressure respirators. (Thus positive pressure respirators have higher protection factors than negative pressure respirators.
- Leakage of ambient air into the facepiece from pressure demand atmosphere-supplying respirators will be less than for most continuous flow devices. (Continuous flow devices can create a negative pressure in the facepiece when the user inhales more air than supplied by the device. This usually will not happen in a pressure demand respirator.)
- Air-purifying respirators for gases and vapors are often not effective for high concentrations or for certain types of contaminants.
- Single use or disposable respirators may not offer as good a faceseal as moldable respirators. (A certain amount of controversy exists regarding this point.)
- SCBA respirators can offer the highest levels of protection but are cumbersome and expensive.
- Supplied air respirators can offer high levels of protection, but restrict mobility.

In the following discussion, the selection process will be described by the situations where the respirator will be used. Within each type of use situation, the different types of respirators that can be used depending on the physical and toxic properties of the contaminant will be identified. For some conditions, the correct respirator may not be dependent on the physical state of the contaminant (such as when atmosphere supplying respirators are used). The discussion will also mention the benefits of using each type of recommended respirator and when other types of suitable respirators may want to be considered.

Firefighting

A condition where only one type respirator is used regardless of the toxicity or physical condition of the contaminant is firefighting. In that situation only self-contained breathing apparatus (SCBA) with a full facepiece (FF) operated in pressure demand (PD) mode should be used. This type of SCBA is recommended because a very high protection factor is required due to the potential presence of high concentrations of toxic contaminants encountered in fires and the need for mobility.

SCBA are heavy and can be cumbersome. Further, they are expensive to purchase and they must be carefully maintained and inspected. However, they offer the highest degree of protection and mobility. Actually a pressure demand (PD), full facepiece (FF) air-line respirator theoretically offers the same protection factor as a full facepiece SCBA operated in pressure demand. However, the air-line respirator is not rated as high as a SCBA because of the possibility that the air-line could be cut. If the air-line respirator also has an auxiliary air supply tank known as an escape bottle or auxiliary SCBA, then the wearer would have an air supply to escape should the air-line be damaged. When an air-line respirator has such an escape bottle, then the protection factor of that respirator is the same as a SCBA (full facepiece and operated in pressure demand mode). But for firefighting where a high degree of mobility is required, only the SCBA, FF and PD is acceptable.

Whenever respiratory protection is required for work in oxygen deficient atmospheres, then only SCBA or an air-line respirator with an auxiliary SCBA can be used. NIOSH certifies air-line respirators without auxiliary SCBA and all air-purifying respirators only for atmospheres containing at least 19.5% oxygen.

Oxygen Deficiency

Air normally contains about 21% oxygen. At oxygen concentrations below 16%, significant effects begin to occur which end in death at a concentration at about 6% or less. Often when in an oxygen deficient atmosphere, a worker receives little warning that there is a problem. At an oxygen concentration of 19.5%, no effects of oxygen deficiency are seen and that level provides a margin of safety.

While NIOSH only certifies SCBA and air-line respirators with auxiliary bottles for oxygen deficient atmospheres, they need not be full facepiece or pressure demand types. These recommendations are only for oxygen deficient atmospheres containing no contaminants. If contaminants are present, in addition to the oxygen deficiency, FF and PD may also be required depending on the toxicity and concentration of those contaminants. See the discussion on determining the required protection level for the contaminant present to identify the proper respirator for such a situation. The physical state of the contaminant is not important because air-purifying respirators are not an option in oxygen deficient atmospheres.

Emergencies

If the respirator is to be used for emergencies such as rescue or to stop the cause of the emergency, then only two types of respirators are recommended. They include

SCBA with FF and operated in PD or an air-line respirator with FF operated in PD with an auxiliary SCBA or escape bottle.

Only respirators with the highest protection factors are recommended for entering emergency situations because these situations usually involve the presence of either very high or unknown concentrations of known contaminants or the presence of unidentified contaminants. If the concentration is unknown, then it must be assumed to be high and require respirators having the highest protection factors.

Emergency Escape Situations

The respirators used for emergency work are often completely different from respirators used only to escape from suddenly occurring hazardous atmospheres. While the respirators listed above for working in unknown conditions can also be used for escape, most escape equipment is characterized by the ease with which it can be put on, the mobility it affords, and the relatively short time for which it provides protection. Escape-type SCBA can be certified for as short a time as three minutes. While air-purifying gas masks cannot be used for emergency entry into unknown concentrations, they can be used for escape as long as the environment contains at least 19.5% oxygen.

For escape, speed in putting on the respirator and ability to leave the area without incurring any permanent health effects or incurring any escape-impeding conditions such as severe eye irritation are important. Where long distances are involved in an escape route and very high concentrations or lack of oxygen may occur, then escape-type SCBA would be recommended. For a particular application, the user must determine the conditions that might occur during an emergency and consult a vendor to determine the correct type of escape type respirator.

IDLH

One of the most important concentration use restrictions in selecting respirators is the concept of the IDLH, which is the Immediately Dangerous to Life or Health concentration. As defined by ANSI, the IDLH is any atmosphere that poses an immediate hazard to life or produces immediate, irreversible debilitating effects. The definition could also apply for any escape impairing symptoms such as eye irritation which prevents seeing while escaping. NIOSH, in addition to the ANSI definition, includes delayed irreversible effects in their definition of the IDLH. The difference is that delayed effects would include cancer from exposure to chemical carcinogens and also exposure to radiation such as radon daughters.

Because exposure to concentrations above the IDLH would result in permanent damage or prevent a worker from escaping if his respirator failed, only the most reliable respirators should be used in these situations. Thus whenever the contaminant concentration exceeds the IDLH, only the respirators used in emergency entry situations are recommended. These respirators include SCBA with FF and operated in PD and air-line respirators with FF operated in PD with an auxiliary SCBA.

IDLH values have been set for many chemicals regulated by OSHA and can be found in the NIOSH document entitled ''Pocket Guide to Chemical Hazards,'' which is contained in the list of references. The employer whose workers are using a chemical for which no IDLH value has been identified should evaluate available toxicological data to determine an IDLH. If insufficient data are available to determine an IDLH, any exposure should be considered life threatening and assumed to be above the IDLH.

Routine and Nonroutine Use of Respirators

The remaining conditions where respirators are to be used involve continuous or intermittent and routine or nonroutine planned work situations. One primary difference between the situations to be described and those previously described is that in these situations the type and concentration of the contaminants are known. If the type and concentration are not known, then the respirators described previously must be used (not including escape-only respirators). When conditions are not known only the most protective respirators can be used because very high concentrations may exist.

Thus, the first step in determining what type of respirator to use is to identify the highest exposure to which any employee might be exposed and base the respirator selection on providing protection for that level of contamination. For contaminants having ceiling limits (the limit concentration may not be exceeded at any time) the highest instantaneous concentration is the level which must be reduced to an acceptable level. If the limit has no ceiling, but is an average exposure limit, then the average exposure serves as the number which must be reduced. However, some respirators have concentration limits which cannot be exceeded. For example, organic cartridge respirators cannot be used for concentrations above 1000 ppm no matter what the limit of the organic contaminant. If the average exposure to a given organic is 800 ppm (assume the time weighted average limit is 500 ppm), but for a short period of time the exposure is 1200 ppm, organic cartridge air purifying respirators cannot be used.

These types of problems are most likely to be encountered in short term operations where the average full shift exposure concentration may not be much over the limit because the exposure time in the contaminated area is short. However, while the exposure time is short, the concentration may be high. In that situation, the overall required exposure reduction level might be misleadingly small. Thus, the user must be careful to consider all respirator concentration use restrictions.

After the user has identified the highest instantaneous or average concentration to which the employee will be exposed, the next step is to identify what a safe exposure would be. The user has some flexibility in this area. At a minimum, no exposure above OSHA limits for regulated chemicals is permitted. For a number of chemicals, the OSHA limit has proven to be quite reliable and based on current knowledge, an employer could be considered as having fulfilled his ethical, legal, and moral obligations in following the OSHA limit. However, for some OSHA regulated chemicals, the limits are based on data that may be more than 17 years old and possibly not protect the entire work population.

The American Conference of Governmental Industrial Hygienists (ACGIH) has been and is continuing to recommend exposure limits called TLVs which are designed to protect the majority of healthy workers. Actually, many of the OSHA permitted exposure limits (PEL) were taken from the 1968 TLV list published by the ACGIH. Notice that these limits are not designed to protect all workers. If the employer uses such limits, they should determine if any employees would not be protected under those standards. Because these limits are periodically updated and more current than many OSHA standards, many organizations use these limits as guidelines for protection. As long as the limitations of TLVs are known and dealt with appropriately, compliance with these standards can help ensure a healthful working environment.

NIOSH also makes recommendations for exposure limits called RELs (recommended exposure limits). These limits are designed to protect the entire working population with the possible exception of certain hypersensitive individuals. These limits are often but not always less than the TLVs. There are also other standards such as those published by the American Industrial Hygiene Association (AIHA). For many newer substances, no limit may exist and the user will have to determine one. Again, an evaluation of the available toxicity data must be completed as with the procedure for determining IDLHs.

In reviewing the limits from different organizations and the basis for the standards, the user may be uncomfortable in assuming that any exposure below that limit is safe. Because of human variability and the quality and quantity of data upon which many standards have been set, such discomfort can be well founded. Therefore the user may wish to choose a higher protection factor than is necessary to comply with the standard. When taking the problems of achieving good respiratory protection into consideration, the user is advised to seek as high a protection as is consistent with the situation.

The following statements summarize the problems in choosing an adequate level of protection:

· The response to toxic chemicals varies within the work population—there is no magic number which separates healthy conditions from unhealthy conditions for a given work population.
· The concentration of contaminants in the workplace often varies in time and space, making it difficult to really know what the employee is exposed to.
· The degree of accuracy used to set safe exposure limits varies from substance to substance.
· The effectiveness of the respirator will vary based on the quality of the respirator, the motivation of the wearer, and a number of other factors.

After the exposure associated with the operation and the desired exposure limit has been identified, the user is now ready to calculate the needed protection factor. Divide the actual exposure level by the exposure limit chosen. That number identifies the minimum protection factor needed.

Next, determine if the contaminant is an eye irritant or can cause eye damage. If it can, then either a full facepiece, helmet, or hood type respirator will be required. If not, half facepiece and single use respirators may still be an option. When continuing the selection process, the user must not forget to remember the use restriction for eye problems.

After the protection factor has been identified, the user is ready to review Table 1 to determine those respirators that meet the protection factor requirements. Except when very high protection factors are required, air-purifying respirators will probably be an option. Air-purifying respirators are used more than any other type of respirator. They are portable like the SCBA but without the weight or bulk. For many air-purifying respirators, they can be used for a full shift without changing the cartridges (varies for different contaminants and concentrations). Certainly, they are at least initially less expensive to purchase and are far easier to maintain.

Unfortunately, they also have the highest potential for abuse. They can be used with the wrong air-purifying elements. Because most air-purifying respirators are negative pressure devices, they require a tight facial seal to function properly. Unfortunately, because of the limited range of sizes and facial design of respirators, not everyone can get a good seal even when following the manufacturer's donning instructions. With positive pressure respirators, a perfect fit is not as critical, as air would always leak out of a positive pressure respirator. This assumes that the volume of air entering the facepiece always exceeds the volume of air being inhaled. This is not true for all continuous flow air-supplying respirators.

When a good seal is required for a negative pressure respirator, the fit can cause significant discomfort. If the respirator must be worn for long periods of time, the employee may periodically remove it or adjust it, causing exposures during that period. Even if that period is short, it may be long enough to cause an overexposure. Wearing a respirator for long periods in hot weather can cause the skin to become irritated, thereby increasing the level of discomfort.

Thus, the least expensive and restrictive respirators also usually provide the least protection. While the most protective respirators such as the SCBA provide the most protection, they are difficult to use for all but short term operations where movements through narrow openings and passageways are not required.

If the required PFs are low enough, air-purifying respirators are often selected because of their initial low cost and the few restrictions they impose on work performance. In selecting a suitable cartridge for an air-purifying respirator, the physical state of the contaminant must be known.

For dusts, mists, and fumes a suitable cartridge is usually available. For particulates (general grouping for dusts, mist, aerosols, and fumes) the type of filter is not only dependent on the physical state, but also on the exposure limit and the protection factor. For all contaminants with an exposure limit greater than 0.05 mg/m^3, mechanical or electrostatic filters are suitable. However, for particulates with limits less than 0.05 mg/m^3 or when protection factors of 50 or greater are required, then only HEPA filters are permitted based on NIOSH certification requirements.

For gases and vapors, NIOSH has some specific certification approvals. These approvals carry maximum concentration limitations and other requirements that must be consulted. The NIOSH certification manual (contained in the list of references) should be reviewed.

NIOSH approves organic cartridges as a class of chemicals. Because not all organics will be effectively captured by chemical cartridges and cartridges have a limited capture capacity, such respirators cannot be used for organics which do not have adequate odor warning properties. The user can consult the literature for information on odor detection levels. However, such data can be misleading, depending on how the tests were conducted. Sometimes, the reported odor detection limit will only apply to persons with very sensitive odor thresholds. The user must be careful in making a determination that all employees will be able to detect the presence of the chemical if the cartridge fails.

If the chemical is an irritant and not known to cause permanent damage at levels somewhat above the limit (but not in excess of 3 times the limit and only for chemicals not having a ceiling exposure limit), the joint NIOSH/OSHA Standards Completion Project allows some flexibility in setting odor limit detection levels. If the contaminant causes systemic damage or has a ceiling limit, then the chemical odor limit should be less than the exposure limit.

While organic chemical cartridges have use limits of 1,000 ppm, canisters in gas masks can be used for up to 5,000 ppm for face mounted canisters and 20,000 ppm for front/back mounted canisters. All other restrictions including odor warnings and protection factors still apply. In any case, the restriction with the lowest number will be the limiting factor whenever respirators are selected. A review of some of the restrictions for using air-purifying respirators include:

· Do not use any filter type respirator for unknown concentrations or IDLH environments.
· Only use HEPA filter respirators when a protection factor of 50 or greater is required

and for contaminants with a limit of less than 0.05 mg/m^3 (consult OSHA standards for exceptions, for example, the lead standard).

· Adequate oxygen must be available (at least 19.5%).
· Use a cartridge known to efficiently collect the contaminant present. This requires some data on service life.
· Do not use air-purifying chemical cartridge respirators for contaminants with poor warning properties.
· Do not use chemical cartridges for organic vapors at concentrations above 1000 ppm: higher concentrations are permissible for canisters.
· Use full facepieces for contaminants that are eye irritants.
· Consult OSHA standards and NIOSH certifications for other concentration limitations for specific gases and vapors.

ELEMENTS OF AN ACCEPTABLE RESPIRATOR PROGRAM

Respirators can offer adequate protection when used correctly. However, even when the correct respirator is used, the practice of handing an employee a respirator without having a respirator program in place will usually result in poor protection. Respirators have limitations which workers must know. The correct respirator must be selected for the situation and conscientously used. Assurances must be made that the actual protection afforded the employee is at least as good as the designated protection factor assigned for that respirator. And the respirator must be properly maintained and stored to continue providing adequate protection.

This section describes program elements required by OSHA whenever employees are protected from inhalation hazards (exposures above OSHA standards) through the use of respirators. In addition, other program elements not required by OSHA but which are considered good practices are also discussed. Also included are helpful suggestions for setting up a functional respirator program.

OSHA Requirements

OSHA requires the following program elements any time respirators are used to protect employees from airborne contaminants:

· The employer is responsible for providing the respirator.
· The employer will provide training for workers using respirators regarding their limitations, inspection, maintenance, how to wear the respirator, and how to test the effectiveness of the fit.
· The employer will inspect and maintain respirators used by employees.
· The employer shall provide a written program identifying how respirators are selected and used.
· The worker shall use the provided respiratory protection as per instructions and training received.
· Respirators shall be cleaned and disinfected regularly.
· Respirators shall be stored in a convenient, clean, and sanitary place.
· Conduct appropriate surveillance of the work area conditions and exposures and review the effectiveness of the respirator program routinely.
· Medical determination shall be made of the physical ability of the workers to wear respirators correctly and still perform their work.

- When available, use a respirator that is certified by NIOSH.
- Where practical, respirators should be assigned to workers for their exclusive use.

In addition to the OSHA requirements, other factors must be considered in setting up a successful respirator program. The following discussion describes program aspects that supplement OSHA requirements or discuss how to implement the OSHA respirator standard.

Management Commitment, Training, and Motivation of Employees

An important part of the respirator program is management commitment. This commitment should come from top management and be reflected through all levels of supervision down to the employee. If the company does not really care if employees are adequately protected, a respirator program is not likely to succeed even if the correct respirators are chosen.

This commitment is also necessary to fund the program properly. Often management considers respirators to be an inexpensive solution to their toxic atmosphere problem; however, the cost of a respirator program extends well beyond the initial purchase price of the respirators. Manpower costs of training, fit testing, maintenance, storage, supply, repair, and possibly reduced work output can be high. The cost of replacement filters for air-purifying respirators or charging tanks, of inspecting SCBA, and of maintaining compressor equipment including air filters and alarms for air-line respirators can be significant. When employees see that management is spending the money required to run the program effectively, they are more likely to appreciate the need for respiratory protection and wear their respirators conscientiously.

As mentioned, respirators are often uncomfortable, restrict vision, are bulky, and make oral communication difficult. Proper motivation of employees is critical to make a respirator program successful. The following suggestions should be considered:

- Provide the employee with several different types of respirators, possibly from different manufacturers. This will give the employee a sense of involvement and increase the likelihood of finding a more comfortable respirator.
- During training, educate the employee as to the harm that might occur if he does not conscientiously wear his or her respirator.
- Provide adequate time for inspection, fit testing, and cleaning. As mentioned, if the employee sees that the employer is taking the program seriously, then the employee is more likely to do the same. Make all levels of supervision responsible for the enforcement of respirator usage.
- Attempt to set up peer pressure for employees to work in a safe manner so that one employee does not jeopardise another. Not wearing a respirator could be the sign of a worker who does not care and could be pressured by other employees to wear their respirator.

One person knowledgeable in conducting a respirator program should be placed in charge. If someone who is not familiar with respirators is to be in charge, then they should do the following:

- If the person in charge does not wish to learn what is necessary to run a program, retain a consultant to set up the program and monitor the program on a periodic basis to ensure that adequate protection is maintained.

· Attend respirator training courses to become familiar with respirators.
· Have respirator salespeople visit and allow them to teach the person in charge about running a respirator program. Manufacturers often have knowledgeable salespeople who will spend time to obtain your business. However, they will often attempt to protect themselves from any liability associated with your respirator program. These sales representatives will probably not specifically tell you that a given respirator will ensure that your employees will not suffer any ill effects when working in a given environment. However, they will tell you that a given respirator is designed to offer protection against the contaminant in your plant and that it should reduce the employees exposure to levels below a given limit.
· Obtain a reference library and become more familiar with running respirator programs. A list of references can be found at the end of this chapter.

Fit Testing

A certified respirator will not necessarily provide the protection factor rating of its class for every employee. There is a high probability that if correctly worn, most employees will receive that level of protection, but not all. Therefore, each employee must be tested to determine if he or she is getting adequate protection. This is most commonly done through fit testing. The following types of fit testing should be done:

· Before using the respirator in the work atmosphere, workers should be given either a qualitative or quantitative fit test. These tests should be repeated routinely, based on need. However, they should be performed at least yearly.
· Every time the employee puts on a tight fitting respirator, he should perform a positive and negative pressure face fit test. Because the employee is to perform the test every time it is put on, it must be simple to do and require no special tools. The manufacturer usually provides instructions for such tests. These tests usually consist of covering the filter intake ports or exhalation valve with the hands and then inhaling or exhaling gently to see if a negative or positive pressure is maintained in the facepiece. For some respirators it is difficult to cover the openings involved in the positive and negative fit tests. For those respirators, the manufacturer may offer a simple method for conducting the tests or the wearer may only be able to perform the positive or the negative test but not both.
· Many of these test cannot be done with single use or disposable respirators which is one of the problems associated with their use.

Qualitative fit tests involve exposing an employee to an atmosphere containing a nontoxic airborne substance that can be smelled or tasted or that will cause irritation. If the employee senses the presence of the substance while wearing the respirator, he informs the tester that he senses the material and fails the test. Conversely if he does not sense the substance, he passes. It is a qualitative test because the exact concentration of the substance is not known. Even if the exact concentration were known, the test would still be qualitative because the sensitivity of the person being tested is not known. While specific instructions are provided to generate the substance in the exposure chamber, the amount actually generated is not exact. However, the nose is very sensitive and the test can be quite reliable.

The procedures for conducting qualitative fit tests are stated in the OSHA lead standard (29 CFR 1910.1025). The concentration of the contaminant does vary. If the tester makes it strong, the chances of an employee with a poor fit passing the test is reduced. However, some employees with an adequate fit may fail the test if they are very sensitive to the smell or taste of the contaminant and the generated concentration is too high. On the other hand, if the generated concentration is too low, then no employee with an adequate fit is likely to fail, but some employees with a poor fit who may not be extremely sensitive to the chemical may pass when they should have failed. While problems do exist in conducting qualitative tests, if done carefully, they can be effective in identifying employees who cannot achieve good fit.

Other problems associated with the use of qualitative fit tests include:

- The employee can say he cannot smell the chemical when he can. He might lie in fear of losing his job if he does not pass. If an irritant smoke is used as the test agent, the employee cannot lie because the smoke is sufficiently irritating to cause him to reflexively cough.
- The test is not conducted under actual use conditions. While the employee is told to make various motions and read a certain passage containing words that require a number of different facial movements, such simulation is not the same as testing under actual work conditions. Sometimes, a worker will pass the test while reciting a passage, but fail if he smiles. Smiling is not part of the current OSHA qualitative fit test protocol. This problem also occurs with quantitative fit testing.

As described earlier, quantitative fit testing involves generating a known concentration of a nontoxic substance and having the employee wear the respirator in that test environment. The facepiece of that respirator has a probe inserted into it so that the concentration of the contaminant in the facepiece can be measured. The concentration of the contaminant in the test atmosphere is divided by the concentration of the contaminant in the facepiece to determine the protection factor provided by the respirator.

A particulate is used as the contaminant because particulates can easily be continuously measured by a detection machine that is rugged, portable, easily calibrated, and not too expensive. Phthalates have commonly been used as the test particulates; however, they have now been identified as possible carcinogens. Even though no one would be exposed continuously to high levels of the chemical, alternatives are available. Corn oil is used successfully but it is important to keep the testing unit clean to avoid the oil from becoming rancid.

As with the qualitative test, the quantitative procedure is not conducted in the workplace and thus may underestimate the level of protection. As the testing apparatus is monitoring the facepiece concentration of the contaminant, the employee recites a speech (the rainbow passage given in the OSHA lead standard) and the average penetration concentration of the contaminant is determined. To pass the test, the calculated protection factor must be greater than some preset number. That number should be well above the protection factor suggested for the class of that respirator. The safety factor required takes into account the differences between the protection factor measured in the testing location and that actually provided in the workplace.

Because the facepiece of the respirator must be probed to measure the concentration inside the facepiece, quantitative testing of single use and disposable respirators is difficult. Also, the time and expense for conducting quantitative fit testing is significantly greater than for qualitative fit testing. While quantitative fit testing does not have the disadvantage of not providing a known test concentration and a known detection sensitivity, many professionals believe that qualitative fit testing can accurately detect those employees who cannot get a good fit. The program administrator must evaluate the pros and cons of each method and determine which one makes the most sense for the particular application. Perhaps a combination of the two would be a good compromise in certain situations. An outside company could conduct the quantitative fit tests while the employer supplements with qualitative fit tests.

Maintenance, Training, Medical Evaluation, and Inspections

After employees are fitted and tested with the correct respirator, the respirators must be properly maintained. For routine use, the employer should provide the employee with his or her own respirator. Time must be given for the employee to clean and inspect the respirator daily. The respirator must be stored in a location where it will not be damaged or come into contact with any toxic materials.

The employee must be adequately trained in how to inspect and clean the respirator if he is to be responsible for it. These respirators should be periodically checked by the program manager to determine if the employee is taking proper care of the respirator. The employee should be told that if at any time he suspects that the respirator is not functioning properly while being used in a contaminated environment, he should leave the area to determine the problem. In situations where abruptly leaving the work area could have a major impact on production, a mechanism should be available to replace the employee quickly.

Minor repairs should be made by someone who is proficient and knowledgeable. Having a good supply of parts that commonly go bad will help ensure that respirators are repaired quickly. Only the specified parts made by the respirators' manufacturer can be used for repairs. If parts from different manufacturers are used, the NIOSH certification is not valid. Major repairs, especially for air-line and SCBA respirators, should be made by an authorized representative of the manufacturer.

Employees should have ample opportunity to wear the respirator in a clean area before using it. This permits any employees who are apprehensive about wearing respirators a chance to familiarize themselves with the equipment in an open and nonstressful situation where they may be able to overcome their fear. In this situation, the employee can determine that wearing a respirator does not significantly increase the energy needed to breathe. If this is done properly, with the right respirator, employees can be motivated to want to wear their respirators. Once the employee wears his respirator long enough, some of the initial discomfort may cease.

While OSHA requires that a physician determine the conditions that must be met for having an employee wear a respirator, in most situations, the actual exam need not be extensive. Recent studies suggest that if an employee can do his job safely without a respirator, then in most situations he can also do that same job safely with a respirator. One exception is when SCBA is used because of the weight involved. For air-purifying respirators, the added physiological strain associated with wearing a respirator is not significant. Recent medical studies suggest that even employees with some pulmonary function problems can successfully wear respirators. However, the data are not complete and information should be treated carefully.

The physician should take a detailed history, including any previous respirator use, and confirm the history information with a physical exam. If the physician does not suspect any medical problems that would interfere with wearing a respirator no further tests are indicated. In addition, ANSI recommends a pulmonary function test be routinely given to employees who wear respirators.

The physician should be given the discretion to conduct tests in addition to the history and physical to determine the ability of an employee to wear a respirator. Special consideration for heat stress should be taken when respirators are worn in hot environments. This is especially true when rebreathing units are used (the recirculated air becomes quite hot) or encapsulating suits are worn.

Maintaining a quality respirator program is perhaps more difficult where the respirators are only used intermittently or for emergencies. In these situations a tendency exists not to inspect the stored respirators and not to continuously train employees in their proper use. Motivating employees to wear respirators in an emergency is usually not difficult because of the recognition of the need and the short time the respirators are usually worn.

When respirators are stored in the work area for emergencies, OSHA requires periodic inspections. OSHA requires that all stored SCBA gear be inspected monthly and that the inspections be documented. Employees may tamper with the respirators or they may deteriorate with time. Unfortunately, these respirators cannot be locked up or they will not be available when needed. Remember, even when not used often, respirators must be serviced and employees periodically trained.

In addition to the use of respirators, other special precautions are required for working in certain hazardous areas such as confined spaces. When working in an IDLH atmosphere, an observer stationed in a clean environment in contact with the worker in the hazardous environment is required. The reader should consult the reference list for working in confined spaces to learn all the requirements.

Respirators must be cleaned and disinfected routinely and any time a different employee is to wear one. Examples of disinfecting procedures include:

· 2 minute immersion in a hypochlorite solution (50 ppm chlorine)
· 2 minute immersion in an aqueous iodine solution (50 ppm iodine)
· 2 minute immersion in a quaternary ammonium solution (200 ppm quaternary ammonium in water)

When respirators are used at high or low temperatures, special considerations are required. At low temperatures, lenses on full facepieces, helmets, and hoods may fog and valves may freeze. The respirators user should carefully follow the manufacturer's instructions regarding the temperature limits in which the respirator can be safely worn. In hot weather, the use of respirators may increase the potential for heat stress. When heat stress is possible, employees must be carefully monitored.

With proper motivation, continued training, and frequent inspection, respirator programs can and do work. Such programs are not cheap, and they require management commitments and a decision regarding the level of protection the company wishes to provide its employees.

REFERENCES

Title 30 parts 11, 12, 13, and 14, *Federal Register,* Saturday, March 25, 1972, Volume 37, Number 59, Part II.

Title 29 part 1910.134 (Respirator Standard) and 1910.1001–1047 (Health Standards including specific respirator selection requirements for certain chemicals) *Federal Register,* Wednesday, October 18, 1972 Volume 37, number 202 Part II.

National Institute for Occupational Safety and Health, ''A Guide to Industrial Respiratory Protection,'' HEW Publication No. (NIOSH) 76–189, June 1976.

National Institute for Occupational Safety and Health, ''NIOSH Certified Equipment List as of October 1, 1986,'' DHHS (NIOSH) Publication No. 87–102, October 1986.

Walter E. Ruch and Bruce J. Held, *Respiratory Protection, OSHA, and the Small Businessman*, Ann Arbor Science Publishers Inc., Post Office Box 1425, Ann Arbor, Mich 48106, 1975.

American National Standard Practices for Respiratory Protection, Z88.2–1980, American National Standards Institute, 1430 Broadway, New York, NY 10018.

American National Standards for Respiratory Protection, ANSI Z88.6–1984, American National Standards Institute, 1430 Broadway New York, NY 10018.

Amoore, J. E., and Hautala, E., "Odor as an Aid to Chemical Safety: Odor Thresholds Compared with Threshold Limit Values and Volatilities for 214 Industrial Chemicals in Air and Water Dilution," *J. Appl. Toxicol.*, **3**(6): 272–290 (1983).

National Institute for Occupational Safety and Health, *NIOSH Pocket Guide to Chemical Hazards*, Fifth Printing, DHHS (NIOSH) Publication No. 85–114.

Pocket Guide to Industrial Respiratory Protection, National Safety Council, 444 North Michigan Ave., Chicago, Illinois, 60611.

Birkner, L. R., *Respiratory Protection: A Manual and Guideline*, 1980, American Industrial Hygiene Association, Akron, Ohio, 44311.

NIOSH/OSHA/USCG/EPA, *Occupational Safety and Health Guidance Manual for Hazard Waste Site Activities*, October 1985, DHHS (NIOSH) Publication No. 85–115.

Respiratory Protection, U.S. Department of Labor, OSHA, (OSHA—3079), 1984.

National Institute for Occupational Safety and Health, *Personal Protective Equipment for Hazardous Materials Incidents: A Selection Guide*, DHHS (NIOSH) Publication No. 84–114, 1984.

American National Standard Commodity Specification for Air, ANSI Z86.1–1973 (Compressed Gas Association Commodity Specification for Air, G-7.1, 1973), American National Standards Institute, Inc., 1430 Broadway, New York, NY 10018.

National Institute for Occupational Safety and Health, "Criteria for a Recommended Standard . . . Working in Confined Spaces," DHEW (NIOSH) Publication No. 80–106, December 1979.

3

SELECTION OF CHEMICAL PROTECTIVE CLOTHING[1]

Gerard C. Coletta

Tillinghast/TPF&C
San Francisco, California

INTRODUCTION AND BACKGROUND

Recent studies have shown that chemical protective clothing (CPC), even some traditionally thought to offer tough, all-purpose protection, actually can degrade, permit permeation, and leak under conditions of direct chemical contact.[2,3] In its 1985 review of personal protective equipment, the Office of Technology Assessment of the U.S. Congress reported that the effectiveness of protective clothing has not been demonstrated satisfactorily.[4] Many items have been tested only in laboratory situations that were not designed to simulate, or even approximate, true workplace conditions.

Perhaps the most publicized failures have occurred during emergency cleanup of waste chemicals and while fighting structural fires involving chemical storage. Nevertheless, even the most tranquil chemical operations in industry and research are not immune. Nondestructive permeation through clothing materials, including glove materials, can occur at such low rates that it is not easily noticed. Because of this, many people continue to select or recommend protective clothing that inadequately resists chemicals in the workplace. Some workers may be regularly exposed to toxic chemicals without even knowing it.

The uncertainty associated with its performance has caused CPC to be considered a "hidden monster" by much of the safety and health profession.

That's the bad news, but it is also the good news. The primary reason these deficiencies are now being identified is that tools needed to help make proper selections are being developed. These include test methods for measuring CPC performance under simulated workplace conditions, techniques for ensuring the effective use of available protection, and recommendations for the care of individual clothing items after their selection. As these tools continue to grow in number, they are confirming that protective clothing performance is a complex but manageable interaction of physical and chemical phenomena.

Every person who selects and distributes items of protective clothing to workers assumes a personal liability for exercising his best efforts in the selection process. This includes, but isn't limited to, safety and health professionals, department managers, first-line supervisors, and purchasing agents. Selection decisions should be based on a review of the best available information and data describing the work environment and CPC protection levels.

To help the user in his/her efforts, this chapter reviews the current state-of-the-art in chemical protective clothing selection procedures. It addresses issues like identifying chemical hazards in the workplace, procedures for determining needed protective clothing performance, test methods, and testing strategy for both the CPC user and manufacturer, the influence of regulatory agencies, and, lastly, the need for independent testing laboratories in certifying CPC performance. Together with information on the hazards of individual chemicals presented in Section II, the present chapter identifies a pathway for greatly improving protection made available to anyone exposed to chemicals in his work environment.

Chemical Hazards in the Workplace

Occupational exposure to hazardous or toxic chemicals is an important health problem in the United States. The magnitude of this problem is evidenced by the fact that chemically induced dermatitis has been cited as a major occupational disease.[5-8] The U.S. Bureau of Labor Statistics (BLS) reported that, in 1978, skin diseases accounted for 46% of all occupational illnesses reported by the private industrial sector.[9] In fact, the BLS report probably underestimated the true impact of this problem since reporting rules only require listing diseases that result either in lost work time or in medical treatment beyond first aid. Additionally, many workers may be hesitant to report minor rashes or skin irritations to their employers.

The most significant concern with skin contact by chemicals is the potential for systemic damage after absorption through the skin. The human body's tolerance for an absorbed chemical varies from person to person, but depends broadly on a combination of the amount absorbed and the chemical's toxicity. Deleterious effects can range from temporary intoxication to chronic sensitization to degenerative diseases like cancer. Substances which the body cannot break down or eliminate may build up through intermittent exposures over a long period until the accumulation is sufficient to cause symptoms. This is certainly more insidious than indicated by the outward appearance of a dermatitis. Surprisingly little research has been reported on this aspect of occupational health. Hence, there are no published standards or thresholds of harm for skin exposures to most toxic chemicals.

Interest in the performance of chemical protective clothing has grown tremendously during the last few years. This is due, in part, to the recognition that many American workers are exposed to toxic chemicals to some degree. More specifically, workers who handle chemicals on either a routine or an emergency basis are often subjected to direct contact by those chemicals. Examples of such situations include:

- Handling liquid chemicals during electronic component manufacture;
- Maintenance and quality assurance activities for chemical production;
- Application of pesticides and other agricultural chemicals;
- Chemical waste handling; and,
- Emergency chemical spill response by public agencies.

Also, interest in protective clothing has grown because of (1) a greater emphasis on the use of personal protective equipment to replace more costly engineering controls, (2) the development and aggressive marketing of new base materials for protective clothing with attendant claims of greater levels of worker protection, and (3) the need for "totally protective" suits for use at hazardous waste disposal sites and in certain governmental and industrial operations.

Incentives for an Aggressive CPC Program

Primary responsibility for ensuring workplace safety and health lies with the employer. In the U.S., regulatory requirements clearly assign accountability to management, with the expectation that they will determine and implement state-of-the-art or "common practice" administrative and technical programs for worker protection. This especially includes protection from chemical hazards.

Chemical-related injuries and illnesses directly impact an organization's financial health. Workers' compensation costs increase in proportion to the frequency and severity of accidents, regardless of whether a company is insured by a commercial carrier or is self-insured. Further, there is a growing tendency to use tort law for recovery of personal damages well above the levels of remuneration provided by workers' compensation. The use of the tort mechanism exposes an individual manager's personal assets and resources as well as the organization's.

Interruptions in manufacturing operations or services can be a direct consequence of lost time. This is particularly true if an injury or illness is severe enough to require a replacement worker. Training time and initial inefficiency can substantially lower overall output.

Also, frequent chemical accidents can impact employee morale. Productivity can be reduced by distractions caused by fear of chemical injury or illness.

Performance vs. Design Specifications

The primary objective of chemical protection is that all workers, to the greatest extent possible, be free from direct contact with toxic chemicals.

In an ideal world, such protection should be accomplished through a judicious mix of engineering controls, good work practices and, as a last resort, use of protective clothing. In the real world, however, much of the burden will continue to be borne by chemical protective clothing. CPC is often the first and only line of defense. Therefore, the most pragmatic approach to raising the level of protection is to ensure superior products as well as superior selection techniques. Superior products emanate from the development of performance specifications that are tied directly to workplace conditions and exposures.

For a few limited applications, design specifications may be acceptable. However, for general industry, they are not. The specification of design details for chemical protective clothing, including materials, closures, and assembly techniques, is self-limiting because it usually permits no room for new or improved products.

On the other hand, performance specifications leave the field wide open for the development of innovative and better-working protective clothing. As long as CPC users specify performance levels that are clearly defined, measurable, and realistic, the process will work and work well. Users will enjoy the ability to select from products that are constantly demonstrating greater cost effectiveness through improved protection levels and longer effective lifetimes.

Important Performance Parameters

As users of chemical protective clothing continue to become more aware of the value of measuring the performance of protective clothing under simulated workplace conditions, the traditional approach of basing selection decisions on the results of simplistic tests will no longer be accepted. Materials and designs not supported by quantitative performance data will succumb in the marketplace to those that are.

Current testing strategy focuses on evaluating a hierarchy of these performance parameters:

- *Resistance to Degradation:* the resistance to deterioration of physical properties during single-sided contact by a chemical. Significant properties that can be evaluated include changes in thickness, weight, elongation, tear strength, and puncture resistance.
- *Resistance to Penetration:* the resistance to flow of chemical on a non-molecular level through closures, seams, pinholes, or other imperfections in a protective clothing material.
- *Resistance to Permeation:* the resistance to chemical movement through a clothing material on a molecular level via absorption, diffusion, and then desorption.

Permeation is usually expressed in terms of both breakthrough time and a steady state permeation rate after breakthrough (e.g., the rate at which a chemical passes through a given area of clothing per unit time: common units are micrograms per square centimeter per minute).

In many cases (such as handling suspected carcinogens), breakthrough time may be the single most important criterion for selecting CPC. It is defined as the elapsed time from initial contact by a chemical on the outside surface of a clothing material to the first detection of that chemical on the inside surface. Measured breakthrough times are readily determined by permeation testing, and are dependent on the sensitivity of the analytical method used in the test.

Permeation rates will have more value in the future when and if acceptable levels of skin contact by chemicals are determined.

Testing that uses these three performance parameters presents clothing materials with challenges of increasing rigor. Only the best performers would be expected to undergo permeation testing, since this method represents the most expensive and time-consuming protocol. Permeation resistance is considered the single most important performance issue relating to worker protection.

The American Society for Testing and Materials (ASTM), through its Committee F-23 on Protective Clothing, has made significant progress in developing standardized methods for testing these parameters. Methods for permeation resistance (designated F-739) and penetration resistance (designated F-903) already have been promulgated.[10,11] A method for degradation is currently under development.

The past several years have seen the beginnings of comprehensive industrywide testing, especially of permeation testing (this was the first standard method to be developed and approved by ASTM). A number of CPC manufactuers, CPC users, and independent contract laboratories have developed testing expertise and are generating performace data.

Generally published data still are limited, however. Perhaps the best available source is a book based on recent quantitative test data entitled *Guidelines for the Selection of Chemical Protective Clothing,*[12] offered by the American Conference of Governmental Industrial Hygienists (ACGIH). The next major publication, expected in mid 1987, will be a monograph on chemical protective clothing now being developed by the American Industrial Hygiene Association (AIHA). In addition, several major protective clothing manufacturers have begun to include performance assessments based on quantitative testing in their product literature.

Regulatory Agencies

Until recently, most federal agencies chartered to address occupational or environmental health and safety problems have taken an arm's-length approach to regulating or otherwise impacting the selection and use of protective clothing.

To be fair, mention must be made of exceptions to this governmental approach. Researchers easily can find specific performance and design requirements in the literature. However, such information and data are limited and relate primarily to past military and other highly specialized applications that have little to do with the needs of general industry or public emergency response organizations.

But times are changing. Several governmental agencies in the U.S. now have become active, even leading participants in the process of developing test methods and other tools needed to help foster proper protective clothing use. The U.S. Coast Guard (the Department of Transportation, DOT), the Environmental Protection Agency (EPA), the National Institute of Occupational Safety and Health (NIOSH), and the U.S. Fire Administration (part of the Federal Emergency Management Agency, FEMA) are the key players. Actually, the past few years have seen a remarkable growth in the involvement of these agencies in better defining protective clothing performance. General industry would do well to keep up with these developments. Contacts at each agency are listed at the end of this chapter under important CPC information sources.

At the present time, the most important accomplishment at the federal level is the signing of a Memorandum of Understanding by these four agencies and the Occupational Safety and Health Administration (OSHA). This inter-agency agreement provides a mechanism for formally coordinating development activities for chemical protective clothing. It should serve to minimize overlap and duplication of effort over the long-term.

In addition, these agencies have begun to participate individually in developing standardized test methods, to generate and publish performance data, and to develop specifications for emergency response ensembles. Examples of these activities are summarized briefly in the following paragraphs:

NIOSH has established an objective of providing general industry with improved guidelines for selecting chemical protective clothing. These proposed guidelines are to be based on available information and data. NIOSH also hopes to develop a mathematical model for predicting performance without extensive laboratory testing.

The Coast Guard and the EPA's Release Control Branch have divided responsibility in the various federal regions for responding to chemical spills and problem chemical waste sites.

The Coast Guard's current research and development thrust is to improve protection for its response teams in the field. Several internal projects, as well as close cooperation with standards-writing organizations like ASTM, are currently underway in support of this effort.

The EPA is addressing chemical protective clothing under its Toxic Substances Control Act (TSCA). The agency is requiring that chemical manufacturers submit test data on protective clothing as part of the TSCA premanufacturing notification process. Also under the auspices of TSCA, the EPA is funding the development of modeling techniques for predicting chemical protective clothing performance with minimum test data. Further, the EPA will be supporting some protective clothing research with moneys available under its Pesticide Program and Superfund.

The Fire Administration has initiated a number of projects with the Coast Guard. The most pertinent is a subjective evaluation of preproduction prototypes of a new generation of chemical protective ensembles. The project will entail a joint field evaluation of these suits.

Also, the Fire Administration is working closely with the National Fire Protection Association's (NFPA's) Technical Committee on Protective Equipment for Firefighters. The thrust of this work is to improve turnout gear performance in fires involving chemicals.

As shown by this brief listing of activities and projects, the recent involvement of government agencies in the evolution of the state-of-the-art of chemical protective clothing has been positive.

TYPES OF CHEMICAL PROTECTIVE CLOTHING

Chemical protective clothing is made from a variety of rubber and plastic materials. Each material is resistant only to certain chemicals. Recent test data generated by ASTM's Committee F-23 on Protective Clothing have confirmed that no single material is resistant to all chemicals or chemical mixtures. Workplace exposures and protective clothing materials should be viewed as matched pairs. The notion held by some users that there are one or two all-purpose materials is a myth.

Available types of protection include full and partial body ensembles as well as individual components like gloves, arm guards, aprons, jackets, and coveralls. The type of CPC selected should provide covering for the part of the body at risk. For example, gloves may be sufficient in protecting against a minor splash or spill, while a complete ensemble would be indicated to protect against a continuous spray from a parting pipe joint.

CPC has been selected properly only when performance data indicate they have a minimum resistance that lasts for the duration of expected workplace exposures. When such a match cannot be made, the employer should go back into the workplace to either improve engineering controls or adjust work practices to reduce exposure in a manner that accommodates available levels of protection.

Gloves and Arm Guards

Hand and arm protection, as illustrated in Figure 1, is necessary to prevent injury from chemicals as well as associated heat and sharp or abrasive objects. Good practice dictates

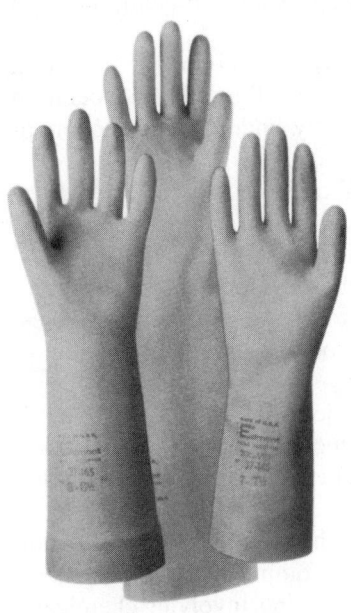

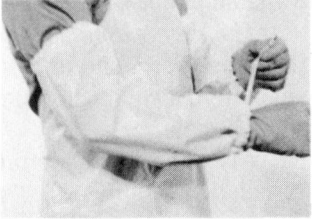

Figure 1. Gloves and Arm Guards.

that gloves and arm guards be used in handling all chemicals except water or clearly nonhazardous substances. They should have a demonstrated resistance to the specific chemicals they will be protecting against.

In selecting gloves or arm guards, fit is of utmost importance. Proper fit maintains dexterity and minimizes additional problems that can be caused by ill-fitting gloves.

Workers should check the integrity of gloves before each use. Leaks can be detected by inflating each glove with air or nitrogen and looking for weak spots, tears, or pinholes.

Gloves should be discarded if any problems are found or if gloves have become soiled with hard-to-remove substances. Regular cleaning and drying are as important as using the proper type.

In addition to gloves, arm guards (including glove sleeves or gauntlets) may be needed to protect the area between the wrist and elbow. Good practice requires operators working with any corrosive substance to wear resistant sleeves. This helps protect the arm area against splashes and spills. Gloves should be cuffed to the outside so that liquids do not drip inside and run onto the wrist and hand.

Ensembles

Depending on the parts of the body exposed to physical or chemical hazards, protection can range from full body to partial body encapsulation as shown in Figure 2.

For example, responding to a major chemical spill or entering a contaminated tank requires full body protection with an external air supply. On the other hand, pouring and mixing liquid chemicals or working in dusty, dirty environments may require only partial body protection. Partial body encapsulation can be accomplished with coveralls, aprons, bib overalls, coats, and jackets.

Ensembles should be worn by all workers who mix, pour, or work with mechanically agitated chemicals. This includes exposure to corrosives, solvents, oxidizing agents, and poisons.

Additionally, protective ensembles should be worn by workers handling radioactive materials, biological agents, dirt, grease, and toxic dusts. For employees working with radioactive materials or biological agents, the protective clothing should be sealed at all openings. Masking tape can be used for this purpose if necessary.

Protection may also be used in situations with no immediate danger of skin contact, but in which street clothes should not be contaminated.

HUMAN FACTORS

Before delving into detailed discussions on proper CPC selection, attention should be placed on several parameters that influence the ability of protective clothing to interface successfully with both its users and its external surroundings. Although long recognized as pertinent, these parameters—better expressed as human factors—have never been assigned their true level of importance. No matter how good the resistance to direct chemical contact, protective clothing is of little value if it is not worn by those it is intended to protect. User acceptance should be maximized through both comfort and function. Ideally, users should be able to forget that they are wearing protective clothing while in the midst of their day-to-day tasks or while responding to emergency conditions.

Inadequate sizing has evolved into a primary human factors issue in nearly every use of chemical protective clothing. In the United States, for example, dimensions have been drawn traditionally from the Anglo-Saxon male. The notion that "three sizes fit all" has not worked well in a country in which workforces are composed of men

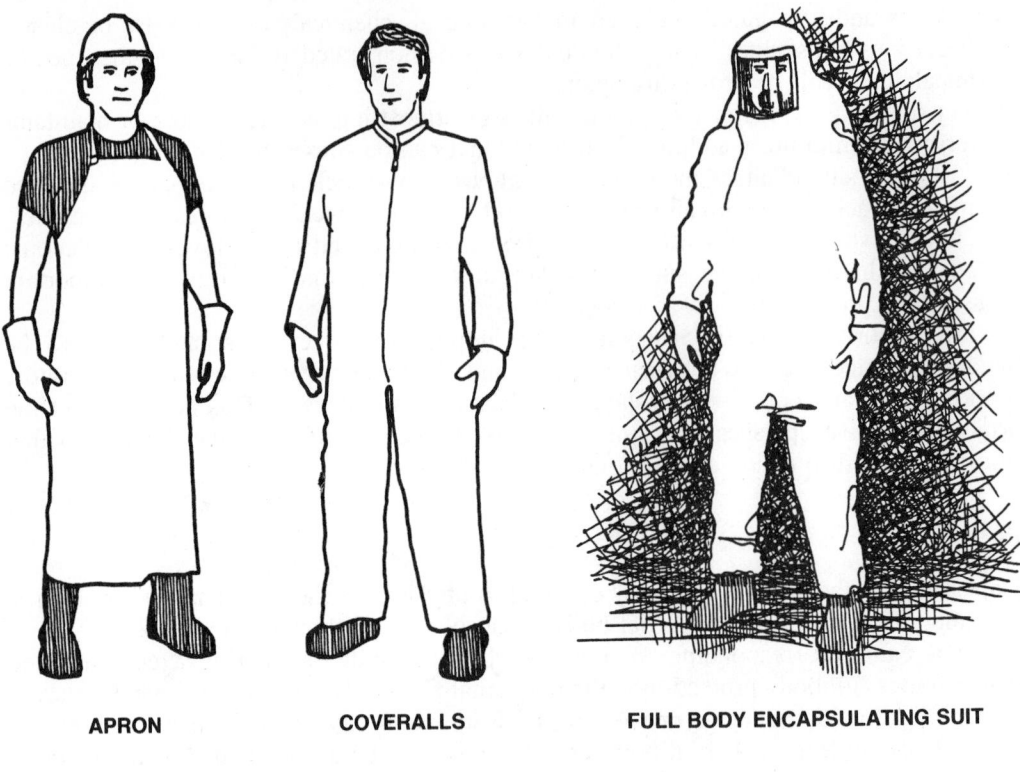

APRON   COVERALLS   FULL BODY ENCAPSULATING SUIT

PANTS   BIB OVERALL

Figure 2. Protective Clothing Ensembles.

and women of all nationalities. This has resulted in a need to clothe people of all sizes and shapes. A classic failing has been in developing gloves suitable for the large number of small Asian women working in West Coast industries. Too-large gloves have made handling tools and equipment so difficult and resulting productivity so low that they simply are not used in many cases. Often, supervisors just look the other way.

While sizing justifiably draws attention, it is not alone in impacting comfort and function. In fact, it is difficult to separate sizing from thermal comfort, moisture transmission, bulk, flexibility, and ability to integrate with other protective clothing items. Certainly the military and emergency services like firefighters have addressed these issues. However, their research and development thrusts have been aimed mainly at clothing items used in extreme environments. Only now are developments beginning to filter into the commercial sector.

Some protective clothing manufacturers, either alone or through industry groups like the Industrial Safey Equipment Association (ISEA), have begun addressing human factors problems. Further, ASTM's Committee F-23 has formed a subcommittee expressly to explore the development of standardized sizing criteria as well as guidelines for comfort and function.

SELECTING CHEMICAL PROTECTIVE CLOTHING

With recent efforts by ASTM and others to develop standardized test methods, employers have an opportunity to provide high levels of protection to their employees. To take full advantage of this opportunity, safety directors, industrial hygienists, and responsible department managers should follow a well thought out protective clothing selection process. The process should be based on balancing the risks of skin contact by specific chemicals against the levels of protection provided by available items of chemical protective clothing. Purchasing agents should never be permitted to order CPC without the benefit of a thorough selection process.

Successful selection results from planning and conscientious follow-through. Four key elements are important. As shown in Figure 3, they are (1) conducting a risk assessment, (2) specifying needed levels of protection, (3) matching needed protection with available products, and (4) compromising when needed protection is not available.

These elements should be considered the cornerstones of any effective selection process. If pursued with diligence, they will maximize benefits to be derived from the effective use of chemical protective clothing.

Conducting a Risk Assessment

First in selecting proper chemical protective clothing, and an activity usually treated lightly, is to evaluate protection needs or, better stated, to assess the level of risk to all affected individuals in the workplace. Conducting a risk assessment involves two primary steps. These are determining the likelihood of skin exposure and then identifying the consequences of direct skin contact:

Step 1: Determining the Likelihood of Skin Exposure. Identifying workplace hazards and then establishing the likelihood and duration of skin exposures is not an easy job. The process involves reviewing normal day-to-day operations as well as estimating the possibility of a spill or leak resulting from equipment failure or human failure. Identifying failure scenarios is particularly important when using highly toxic substances.

Assessing normal operations to determine the durations for possible exposures starts

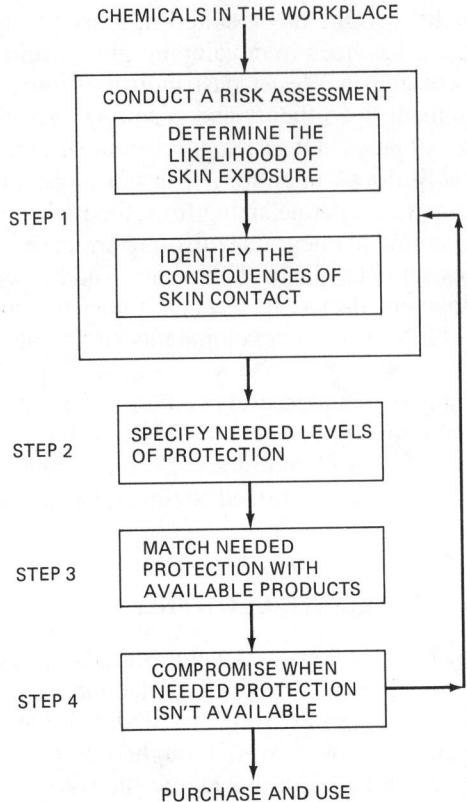

Figure 3. Key Elements in the CPC Selection Process.

with an inventory of chemicals used or produced, focuses on defining individual tasks involving those chemicals, and concludes with an evaluation of chemical handling procedures. It is essential to note here that skin exposure does not always equate to direct skin contact. Skin contact is the result of exposure if no effective barrier is present to isolate the skin.

Normal operations might include laboratory work, chemical-intensive manufacturing operations, reactor sampling or maintenance tasks. Maintenance tasks can be the most difficult to assess because they are often unpredictable and sporadic and can last from as little as five minutes to as much as eight hours or more.

Estimating the possibility of a chemical spill or leak requires an even more pragmatic review of equipment, conditions, and procedures associated with handling and containing chemicals of interest. For example, the integrity of engineering controls and the thoroughness of preventive maintenance are each important in minimizing accidental exposures. Also important are individual work practices and emergency response procedures.

In-house accident statistics, workers' compensation experience reports, and equipment maintenance/repair records provide excellent sources of information that can either confirm or refute initial assessments of operations and the subsequent estimates of spill or leak. By building such experience factors into the process as cross-checks and by modifying the first results accordingly, a realistic or ''reasonable man'' determination of potential skin exposure can be generated.

Step 2: Identifying the Consequences of Skin Contact. The second major step in this selection process is identifying the consequences of direct skin contact by individual chemicals.

In identifying the consequences of skin contact, the primary question to be answered is whether direct contact could result in injury or illness. Some information on the effects of skin contact is available in the published literature, including this handbook and toxicology texts, producers' documentation (material safety data sheets), and government research reports. This information is constantly changing and growing as the effects of even the most widely used chemicals are studied further and become better known. Whether chronic, low-level contact by certain substances is acceptable or not is an open question. This even includes common, frequently used chemicals like organic solvents.

Historically, a tendency has existed to protect against only those substances with known, well documented toxicity. However, a good rule of thumb for anyone using chemical protective clothing is to consider every chemical harmful until its degree of toxicity is clearly known.

The consequences of skin contact can be significant and many. The use of chemicals impacts nearly every aspect of our twentieth-century existence. In the occupational sector, the use of chemicals as raw materials, fuels, and cleaning agents is wide and deep. Most workers involved in the production, use, and transportation of chemicals can be exposed to numerous compounds capable of causing harm.

In investigating individual chemicals, surface effects on the skin should be considered, but so should chemical absorption through the skin. Published information should always be supplemented by available in-house experience. A ranking or matrix of effects can be developed to show the extent of possible injury or illness on contact.

Finally, combining the toxicological consequences of direct skin contact with the information on the likelihood and duration of skin exposure yields an overall assessment of the levels of risk facing individual employees.

Specifying Needed Levels of Protection

This risk assessment is now the foundation for selecting proper chemical protective clothing. It permits the development of target performance levels for chemical protective clothing.

Performance levels should be as specific as possible in defining the type and duration of protection required under both normal conditions and emergency conditions. These levels should include specifying such parameters as minimum breakthrough time and, when appropriate, steady-state permeation rates for acceptable levels of skin contact. Degradation resistance, including retention of physical properties like flexibility and puncture resistance, is important also.

Matching Needed Protection with Available Products

Needed levels of protection are then compared to protection provided by available clothing products.

The most widely available sources of information on chemical protective clothing are the product catalogs published by equipment manufacturers and vendors. These booklets usually contain descriptions of the types, sizes, and varieties of protective clothing produced by each manufacturer. This and supporting information should be collected prior to purchasing to permit satisfactory completion of this recommended CPC selection process.

In some cases, materials of construction are also included with the product descriptions. Many protective clothing manufacturers publish in-depth information on the chemical resistance of the materials from which their products are made. This information is often presented as resistance ratings (excellent, good, fair, poor) or use recommendations for individual chemicals.

Vendor Literature—Limitations. Test data provided for some items of personal protective equipment have been generated without the benefit of standardized test procedures.

For example, many of the traditional resistance tables for chemical protective clothing are based on simple immersion tests in which a material specimen was merely immersed and then observed for some time period. There is no standard time for immersion and the criteria for the performance ratings associated with any given test are likely to vary from observer to observer. Except in cases of complete failure, immersion tells little about a material's resistance to permeation by a chemical. This is true because permeation can occur without any visible degradation or change in the materials.

Several manufacturers have begun to provide information derived from ASTM's new, standardized laboratory tests. These new methods have been designed to simulate worst-case chemical contacts. Some also provide information on abrasion, tear, cut, puncture, and other strength characteristics.

Even so, a few manufacturers are now following an opposite approach and have elected to discontinue publishing any recommendations. They have taken this approach in order to reduce legal liabilities for the performance of their products once in the field. While this response may minimize their liability, it is counterproductive from the point of view of helping protective clothing users make informed decisions.

A compromise that may solve this problem appears to be on the horizon. Manufacturers may begin conducting more in-depth product testing using standardized laboratory tests in return for the ability to publish the numerical results without interpretation. Users then would be forced to decide what is "excellent," "good," "fair," or "poor" for individual applications. Those purchasing CPC should watch for this change.

In the meantime, all vendor literature should be reviewed with care and scrutiny. The more in-depth the background gathered on published performance ratings, the more likely users will make proper selection decisions.

Vendor Literature—Precautions. Even though today's product performance information should be improved, it's all that is available to most protective clothing users. Therefore, several precautions are in order when it is used:

- Only a few of the available product catalogs include information on the criteria for their resistance ratings. This information is necessary in any attempt to form conclusions regarding expected workplace performance and should be solicited from the vendor aggressively by users.
- As part of soliciting background for performance ratings, users should request information on test methods and testing protocols. This information is important in correlating expected workplace conditions with test conditions and in allowing products to be compared directly against each other.
- Many product information tables are outdated. Since the tables were first generated, actual materials of construction or their formulations likely have changed. Questions

should be raised about the appropriateness of test data to the products currently being marketed.

Through this process, potential candidates are identified for further scrutiny in terms of initial cost, ability to be cleaned or decontaminated, comfort, and durability. With all this information, a sound business decision can be made that should result in proper levels of protection for workers.[13]

Compromising When Needed Protection Is Not Available

It is possible that protective clothing currently available in the marketplace will not meet the initial in-house performance requirements, especially for breakthrough time. In such a situation, either process conditions or, more important, employee job requirements and procedures should be reviewed and modified to permit a successful match.

EMPLOYEE TRAINING

Once the selection process has been completed and chemical protective clothing purchased, a clothing management program should be set up by those work groups actually using the CPC. Ongoing attention to maintenance and upkeep is necessary for effective long-term CPC performance.

The single most important element in a chemical protective clothing management program is employee training. Emphasis must be placed on the proper use of protective clothing and, especially, on the limitations of the clothing. Training should be held for new employees, for transferring employees and during the introduction of new CPC products into the workplace. Refresher courses should be held periodically thereafter.

Pre-Use Inspection

The safety director, industrial hygienist, or department manager should visually inspect every item of chemical protective clothing as soon as it is removed from its package. The product received should be compared with the item actually ordered. Defective or incorrect items should be replaced.

Individual items should be checked for defects such as faulty seams, nonuniform coatings, pinholes, and malfunctioning closures or other assemblies. Some flexible materials may stiffen during extended storage—products should be flexed to expose surface cracks or other signs of shelf-life deterioration.

Full-body ensembles should be checked for integrity, including the operation of pressure relief valves and fittings at the wrists, ankles, and neck.

Putting on Clothing

Each worker should thoroughly inspect the clothing he/she is to use immediately before donning, especially older equipment. Of principal concern are cuts, tears, punctures, discoloration, or stiffness. All are indicative of wear or deterioration.

For full protective clothing ensembles, all closures should be secured and checked. The overall fit should be evaluated. An improper fit can represent a severe hazard. Clothing too small in size can restrict worker movements and accelerate fatigue. Clothing too large increases the possibility of snag and compromises dexterity and coordination.

Defective items should be discarded and replaced.

In-Use Inspection

During the course of a workday, each worker should periodically inspect his/her protective clothing. Of principal concern are tears, punctures, seam discontinuities, or closure failures that may have developed. Evidence of chemical attack, such as discoloration, swelling, or softening, should also be noted.

Any physically damaged or chemically degraded item should be removed and replaced as soon as safely possible.

Storage and Reuse

Once absorbed, many chemicals continuously permeate through clothing materials. Washing with soap and water removes surface contamination, but does not remove the absorbed chemical. After surface decontamination, some of the absorbed chemical will continue to permeate the material and it may ultimately appear on the inside surface. Clothing should be checked inside and out for discoloration and, if possible, wipe-tested for suspected chemicals prior to reuse. This precaution is particularly important for full ensembles, which often are reused because of the high costs of replacement.

Different types and materials of protective clothing should not be stored together. For example, virtually indistinguishable black gloves may be made from nitrile, neoprene, Viton, polyvinyl chloride, or butyl. Each material has unique chemical barrier properties. A mix up in gloves significantly increases the chance that a worker will wear the wrong material. Gloves should be separated, perhaps by using the manufacturer's stamped product number as a guide.

TEST METHODS

No item of chemical protective clothing can provide complete protection indefinitely. As pointed out earlier, the notion held by some users that there are all-purpose materials for CPC is a myth, at least for now. Workplace exposures and protective clothing materials should be viewed as matched pairs. End items, whether gloves, arm guards, aprons, hoods, boots, or full ensembles, have been selected properly only when performance data indicate a resistance to chemicals that lasts for the duration of anticipated exposures.

This requirement for matching performance directly with exposure opens the door to a need for rigorously tested protective clothing. Safety directors, industrial hygienists, and responsible department managers should insist on receiving product test data as part of the CPC selection process. Some large users may be able to conduct performance tests themselves, but most data will come from either CPC manufacturers or independent laboratories.

Test methods should simulate workplace conditions as closely as possible to help users determine what levels of protection are available. Temperatures and other appropriate environmental parameters should be duplicated; actual chemicals of interest should be used in proper form and concentrations; and, most important, exposure by the chemicals to protective clothing materials should be on the normal outside surface only.

One-sided exposure is an important criterion when testing chemical protective clothing. In actual use, spills, splashes, and leaks will wet or coat an item's outside surface. The chemicals rarely, if ever, come in contact with edges or inside surfaces. If edges or inside surfaces are contacted, the clothing has already failed in its mission of providing protection. To truly simulate end-use conditions, test method designs should reflect this fact.

History of CPC Testing

Traditionally, protective clothing manufacturers have depended on simple immersion tests as a primary source for performance information on chemical resistance. Because such tests permit edge and inside surface contact, they have caused data to be generated and published that can be inaccurate and misleading. Immersion tests are particularly inappropriate for evaluating multi-layered composites.

A number of protective clothing manufacturers developed their own test methods and data reporting schemes for inhouse use and, sometimes, external marketing efforts. These were based on simple criteria for convenient apparatus design and testing conditions. Frequently, criteria bore little resemblance to the work environment. Minimum emphasis was placed on standardization.

Furthermore, some manufacturers marketed their products on the basis of other crude test methods designed to measure resistance to physical hazards. Consequently, protective clothing items would be selected on their ability to withstand abrasion, cut, and puncture. There might or might not be an attempt to match physical strength with the results of chemical immersion testing.

Much of the information developed from such immersion and physical property tests still exists in available literature. This is one reason why many users are still selecting protective clothing items that do not represent a match between chemical exposures and materials.

With such test data, comparison of product performance from one data source to another is difficult, if not impossible. The resulting selection process is confusing, especially for those without laboratory facilities of their own to make direct comparisons.

This swirl of confusion has prompted the aggressive movement toward standardizing test methodology and reporting performance data for chemical protective clothing. In the United States, the principal forum for these activities is ASTM's Committee F-23 on Protective Clothing. In Europe, it is the International Standards Organization (ISO). In the Scandinavian countries, it is the Nordic Coordinating Group on Protective Clothing as a Technical Prophylactic Measure (NOKOBETEF).

Progress to date has been substantial. New test methods have been promulgated and several others are in various stages of development. A critical landmark was the early recognition that evaluating the performance of chemical protective clothing is a complex problem that must account for a multiplicity of variables, some of which are controllable during testing, some of which are not. Even more important, a workable strategy for efficiency in CPC testing has begun to emerge.

Testing Strategy

At this stage in the evolution of workplace-related test methods, each material that is a candidate for use in protective clothing must be tested against each chemical to which it will be exposed. This is a far-reaching statement, because the possible combinations of materials and chemicals is limitless. Later developments may well identify shortcuts and techniques for accurately predicting performance within material categories or within chemical groupings. However, at the present time, the evolution of such useful correlations is in its infancy.

Therefore, current testing is being driven by this one-on-one strategy broken into four parts:

- Begin testing with flat specimens of materials that are candidates for CPC. Specimens should consist of either a single layer or, when appropriate, a composite of multiple layers arranged in proper order.
- Screen as many candidates as possible using the simplest, least expensive test method. This test method will most probably evaluate resistance to degradation.
- Take only the best candidate materials through an entire testing protocol, presuming that such a protocol will involve testing both chemical resistance and physical hazard resistance. Complex, expensive test methods, like that for permeation resistance, should be reserved for the best performers.
- Upgrade from testing flat material specimens to testing assembled end items as the final confirmation of design and construction integrity. Air-supplied ensembles, for example, may require pressure testing to ensure there are no leaks.

Following this strategy, and as noted earlier in this chapter, flat material testing is based on a hierarchy of three parameters: resistance to degradation, resistance to penetration, and resistance to permeation.

Not only does this approach represent an organized plan for evaluating performance of both candidate materials and end items, but it also provides a route for exercising overall cost containment during what could be a lengthy and expensive testing process.

Chemical Resistance

To date, chemical resistance test method development efforts have been oriented toward bench-top laboratory techniques. With the availability of sophisticated analytical equipment, this focus is permitting an examination of many variables required to simulate workplace conditions.

A second generation of methods, still some time away, will evolve from these laboratory methods as field tests with significantly lower costs. For now, however, chemical resistance is being evaluated in the laboratory.

Degradation Test Method. This test method is currently under development by ASTM's Committee F-23. By this new method, the resistance of a protective clothing material to degradation by a liquid chemical is determined by (1) measuring the thickness, weight, and elongation of fresh specimens of the clothing material; (2) contacting additional, separate specimens of the material with the chemical of interest; and (3) measuring the thickness, weight, and elongation of this second set of specimens to identify changes resulting from contact with the chemical. The list of physical properties can be expanded to include resistance to abrasion, cut, puncture, and the like.

The method is expected to be used as an initial screen to provide an evaluation of flat specimens cut from materials that are candidates for items of protective clothing. Materials that demonstrate acceptable resistance can then be evaluated further to determine resistance to penetration and permeation.

Penetration Test Method. The penetration test method promulgated by ASTM is denoted as Standard Test Method F-903.[11]

The resistance of a protective clothing material to penetration by a liquid chemical is determined by subjecting one side of a material specimen to the chemical under a

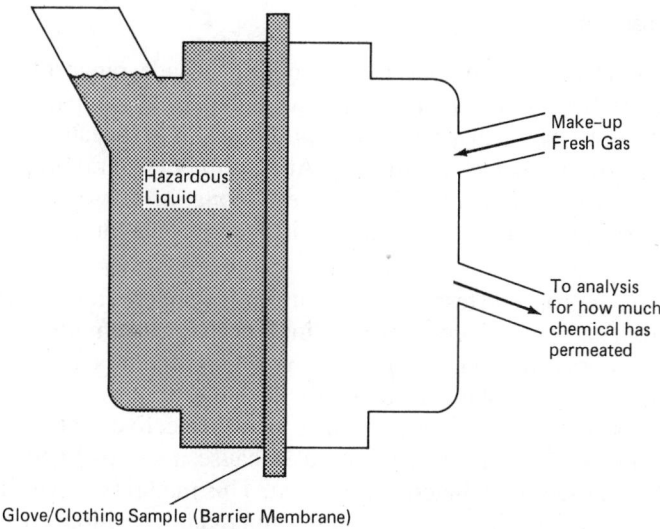

Glove/Clothing Sample (Barrier Membrane)

Figure 4. Schematic of Permeation Test Cell.

pressure of 2 psig and noting the time at which visible penetration occurs. In the test apparatus, the material specimen acts as a simple partition separating the liquid chemical from the viewing surface.

This test determines resistance to penetration using a simple pass/fail criterion. Penetration, or failure, indicates flow through seams, pinholes, or other imperfections.

Permeation Test Method. Designed as ASTM Standard Test Method F-739, this permeation test method was the first and also the most complex procedure so far developed by Committee F-23.[10] With its most recent revision, the method can be used to evaluate a material challenged by either liquid or gaseous chemicals. Since its promulgation in 1981, F-739 has become the standard by which permeation resistance is determined.

The resistance of a protective clothing material to low-level permeation is evaluated by measuring the initial breakthrough time of the chemical and then monitoring the rate of subsequent permeation through the material specimen. In the test apparatus (Figure 4), the material specimen acts as a barrier separating the chemical of interest from a medium that collects the chemical as it permeates.

The collecting medium, which can be either a liquid or a gas, is analyzed quantitatively for its concentration of permeated chemical. This analysis allows direct calculation of the amount of chemical that has permeated the material as a function of time after initial contact. The level of detectability of the chemical is dependent on the sensitivity of the analytical equipment used.

By measuring breakthrough time, the method is used to estimate the duration of maximum protection provided by a protective clothing material under the worst-case condition of continuous chemical contact. Then, by measuring the permeation rate, the method may be used to identify protective clothing materials that limit potential exposures to acceptable, steady-state dermal contact levels.*

---

* At the present time, such contact data are not available for most chemicals. They are yet to be developed by recognized occupational health authorities by derivation from threshold limit values, 50 percent lethal dose, or other similar indices.

Physical Properties

Although hundreds of test methods for evaluating a wide range of physical properties of materials have been developed by private and public groups, most have not been applicable to testing chemical protective clothing. This is because they do not make a proper accounting of workplace conditions. Accordingly, several proposals for correcting this deficiency have been made. The furthest along in the development process are those for measuring resistance to cut and resistance to puncture.

Cut Test Method. A proposed test method is currently under review by ASTM's Committee F-23. A thorough scrutiny of protective clothing performance requirements has indicated that an existing simple method now used by two CPC manufacturers might be adapted, with minor alteration, as a standard method.

By the proposed method, the resistance of a protective clothing material to cut is determined by measuring the force required to cause a sharp-edged, uniformly moving blade to score the surface of a material specimen. This method defines the blade configuration and the material specimen size, conditioning, and position on the test apparatus. The blade is intended to correspond to a common workplace cut hazard like the edge of a piece of broken glass.

Puncture Test Method. This draft test method is also under a review by ASTM's Committee F-23. An existing test method could be adaptable here as well.

By the current proposal, the resistance of clothing materials to puncture is determined by measuring the force required to cause a pointed, uniformly moving penetrometer to puncture a material specimen. The document defines the penetrometer configuration and the material specimen size, conditioning, and position in the test apparatus. The penetrometer is intended to correspond to a common workplace puncture hazard like a 4d nail.

As discussed earlier, methods for measuring performance parameters such as cut and puncture should be integrated within the degradation protocol to better evaluate changes that result from direct surface contact by chemicals.

Applying Data

A word of advice is appropriate for chemical protective clothing users who apply the results of these laboratory test methods to the selection process. Because the severity of worker exposures to chemicals in an occupational environment often changes significantly on a minute-by-minute basis, precise simulation is difficult. Therefore, a major premise in designing such test methods is that worst-case exposures would be paramount.

In testing chemical resistance, worst-case exposures are continuous, one-sided contacts. Splashes and other intermittent exposures pose risks that are less severe. Accordingly, performance data should be treated as approximations rather than absolutes. By adding safety factors to levels of required protection, and by reviewing performance of the materials relative to one another, proper selections can be made.

Furthermore, as test data continue to be generated, a master data base should evolve which will consolidate as much information as possible. Many professionals anticipate that such a data base will be accessible generally and will minimize the need for manufacturers and laboratories to conduct repetitive testing on frequently used material/ chemical combinations. Testing resources and funding could then be turned toward more unusual combinations or, more important, to developing new, superior products.

LABORATORY VS FIELD PERFORMANCE

As discussed in the previous section, laboratory tests can be time-consuming and expensive. They are in controlled environments with well-defined test procedures and exposures. The method for evaluating resistance to chemical permeation, for instance, requires sophisticated analytical equipment, continuous chemical contact on the material specimens and expert interpretation of results.

Applying costly laboratory data gathered in a controlled setting to the process of selecting chemical protective clothing for use under field conditions raises a number of important questions:

· Can quick and inexpensive field tests supplement, or even replace in some cases, the laboratory tests?
· Will predictive mathematical models be developed so that the need for laboratory tests can be reduced?
· What is the practicality of assembling a centralized data base that consolidates performance test data and makes these data available to the users in the field?

The answers to these questions suggest that within a few years, CPC users will have tools at their finger tips that will make the selection process a straightforward, routine task, even under field conditions.

Laboratory Methods

In an ideal world, laboratory testing would be conducted in all cases by protective clothing users as well as manufacturers to determine acceptable performance. The laboratory tests would be matched against the type and severity of hazards identified during a thorough risk analysis of the workplace.

However, laboratory tests take time and require evaluations of individual material specimens on a one-by-one basis with chemical and physical hazards under conditions of specified temperature and relative humidity. This situation is further exacerbated when determining performance of the same generic clothing material from different manufacturers. Variations in raw materials, additives and manufacturing techniques can result in significant changes in, say, resistance to permeation among the different products. Because the number of chemicals and protective clothing materials is so large, possible one-to-one combinations can easily reach the hundreds of thousands or more.

The real world, therefore, is seeking evaluation techniques that demand less commitment of resources, yet are equally as predictive as laboratory methods. There is no question that laboratory tests will continue, and even increase, in importance. However, they will become the basis for simplified, less costly field tests. They will also become the basis for performance predictions using mathematical models.

Field Methods

The most common request of field test methods is that they be "quick and dirty." In translation, this means simply that protective clothing performance can be evaluated by any user at any location with test methods that:

· Need very little time to set up and run;
· Make use of readily available equipment and supplies;

· Require only small amounts of material to be tested;
· Require only small amounts of challenge chemical.

Of course, these methods must generate data that are both reproducible and meaning-ful. Meaningful implies that the methods generate results that would be close to the results of parallel laboratory tests if they were conducted.

The most straightforward approach to developing field test methods is to simplify already developed laboratory methods following the four criteria cited above. A classic analogy for this process is the evolution of today's available, inexpensive personal computer systems for home and office use from the massive and costly mainframe computers of the early days of vacuum-tube systems.

Evolution into field test methods will occur. Some methods, like those for evaluating resistance to physical hazards, are already moving down this road. Although they are being developed as laboratory techniques, they are considered crossover methods because they already meet one or two of the criteria for field test methods. Additional refinements should carry them fully into the ranks of field methods.

Predictive Methods

Field test methods, however, still require evaluation of chemicals and clothing materials on a one-to-one basis under specified temperature and relative humidity. An alternative approach in the field selection of chemical protective clothing is the use of predictive mathematical models. In the long run, this will be a most fruitful course of action, since minimum testing will be required.

It will not be necessary to test every item of protective clothing against every chemical or mixture of chemicals. A few tests would be conducted and the results tied into a mathematical formulation for accurate prediction of performance, even at tempera-tures and relative humidities that are different from those of the original tests. The current candidates for modeling are the laboratory methods that evaluate complex phenom-ena, such as resistance to permeation.

A most promising model for determining permeation resistance is based on solubility data obtained from simple immersion tests. Surprisingly, these immersion tests are similar to the early resistance tests used by CPC manufacturers and now discounted. The major difference is that a numerical value for solubility (saturation concentration of chemical in a clothing material specimen) is the end result, rather than the earlier visual, subjective assessment of a material's integrity.

A high chemical solubility in a protective clothing material appears to correlate with short chemical breakthrough times and high permeation rates. Therefore, the expecta-tion is that the numerical solubilities can be used in conjunction with a mathematical model to predict overall permeation resistance. Evaluating composites of multiple layers may add some complexity to this concept.

Further, a series or battery of "standard" chemicals exhibiting a range of solubility characteristics may be all that is needed to test a material's resistance. Every material that is a candidate for use in chemical protective clothing would be put through an immersion test with each of the standard chemicals. The solubility would then be deter-mined and the number plugged into the appropriate mathematical model.

But again, the success of such a predictive model rests squarely on a strong data base generated by use of the laboratory test methods. A primary data base for chemical protective clothing is essential, regardless of whether field tests or mathematical models are used to reduce the resources committed to performance testing. Both field tests and mathematical models must refer back to solid laboratory test data.

PERFORMANCE CERTIFICATION

The year is 2000 and Milton X. Archibald, senior vice president for occupational health and safety, has just been informed by his manufacturing operations that his company will begin using a new chemical, Toxic ZZ, on a regular basis. Because of several unique properties, there are no alternatives to using this substance. Although his engineering controls represent the state of the art, some manual process intervention is necessary. Archibald has been asked, therefore, to prepare his workforce for handling Toxic ZZ on a day-to-day basis as well as in emergency situations.

A User's Dream

Archibald has activated his usual procedure for selecting chemical protective clothing. His staff is beginning to conduct a risk assessment of the planned process, including storage, on-site transportation, hands-on use and disposal of Toxic ZZ. Performance requirements for his protective clothing will be based on reliable estimates of the likelihood of skin exposure, its duration, and the known consequences of direct skin contact.

Once developed, the information will be sent to the Protective Clothing Institute (PCI). PCI is one of several independent, not-for-profit testing laboratories that have become the authority on chemical protective clothing performance. Based on risk assessment information from users such as Archibald, PCI issues recommendations on proper protective clothing materials. The best part is that PCI certifies the performance of its recommendations based on standardized test methods and conditions simulating worst-case conditions in Archibald's process.

PCI has made Archibald's life much easier during the last few years. All he needs to do after receiving PCI's recommendations is to make a selection within normal business constraints imposed by his company. His primary constraint will be cost effectiveness in considering disposable protective clothing versus reusable protective clothing that can be decontaminated after contact by Toxic ZZ.

Back to Reality

This snapshot of the future presents an interesting scenario. It suggests that most, if not all, CPC testing has been moved to a few central laboratories. Better yet, labs like PCI make a legal commitment by certifying their test results.

Returning to the present, Archibald realizes that he doesn't yet have the luxury of a PCI. He doesn't have a staff, and he isn't even a vice president. In fact, methodology for conducting a thorough risk assessment of chemical handling operations as a way to set CPC performance expectations is only now evolving.

Archibald indeed has a request from management to help bring Toxic ZZ into his workplace. He will conduct a risk assessment, will generate rudimentary performance requirements, and will evaluate a number of candidate protective clothing materials and designs. But because there is no PCI, Archibald faces a dilemma faced by most CPC users in gathering performance data on his candidate materials.

Archibald has several options, albeit not fully satisfactory. He can rely on manufacturers' brochures, a few of which are now based on quantitative performance data drawn from the recently developed ASTM standard test methods. But it is unlikely that any of these data include tests with Toxic ZZ. He can research literature for any available performance information based on Toxic ZZ. He can contract one of three or four private laboratories to evaluate his candidate protective clothing materials using the standard test methods. Or, as a final option, he could set up and conduct the tests in his own laboratory.

But like many other users, Archibald does not have resources or experience available to do his own performance testing. The development of reliable "quick and dirty" field tests is in its infancy. And since workable, predictive mathematical models are several years away, all performance testing must be laboratory based. His only real option then, as a conscientious CPC user, is to contract with a private lab for comparative performance data. He recognizes that these labs currently do not certify their work.

A Real Need

Archibald's company can be considered typical of most chemical protective clothing users. They have protective clothing needs that can be defined, at least in terms of worst-case exposures. Standard test methods that can translate these needs into quantified performance levels are quickly becoming available.

Archibald has concluded that a set of independent laboratories to certify CPC performance is needed. These labs would be capable of conducting tests for a wide range of users and manufacturers.

Such laboratories would be equipped with the latest analytical instrumentation and laboratory facilities to efficiently and safely work with hazardous chemicals. These labs would automate their test apparatus to reduce the cost of multiple analyses as required by standard test methods.

By offering these services, the central labs could reduce needless repetition of tests already conducted by others. The labs might be the custodians of a database built from the results of all previous testing. They could become the purveyors of applied modeling techniques. As experienced, not-for-profit organizations, the laboratories would become recognized for their consistency and credibility.

Conceptually, a certification laboratory is an excellent idea. The question, however, of who would fund such an endeavor is raised. Sponsors must be willing to accept some risk of liability if a clothing material fails to provide anticipated protection in the field, for whatever reasons.

But a precedent for certifying the performance of personal protective equipment already has been established. NIOSH certifies respirators for use in the workplace. The Safety Equipment Institute (SEI, a part of the ISEA) currently certifies hard hats, eyewear, and shower/eye wash units.

Whatever the structure, certifying labs would be cost-effective, reputable and professional groups entrusted to provide the best in available performance data. For their own benefit, CPC users should encourage the development of such a concept.

THE PROMISE OF THINGS TO COME

This chapter began by noting that uncertainty in performance has caused chemical protective clothing to emerge as a prime occupational safety and health concern. A growing understanding of the need to specify performance levels during the selection of CPC is leading to rapid changes in product design and manufacture. Protective clothing users should watch for these changes and, even more importantly, encourage their vendors to keep them aware of the changes.

Today's technological growth will result in a whole new generation of materials during the next few years. This already is evident from the entry of at least one new, highly superior protective clothing material into the marketplace for full ensembles.

Economic decisions surrounding the selection of chemical protective clothing will focus more heavily on comparisons between reusable and disposable items that offer comparable protection. Decontamination will assume a dominant role in this process.

Further, assembly techniques will improve and no longer be the "Achilles' heel" of protection provided by either individual clothing items or full ensembles. Overall performance specifications will result in the evolution of new designs for seams and closures.

Materials

Traditionally, most chemical protective clothing has been produced from natural or synthetic elastomeric materials. Examples include natural rubber, neoprene, butyl rubber, nitrile rubber, polyvinyl chloride, polyvinyl acetate, and styrene-butadiene rubber. A more contemporary material is Viton©, a proprietary fluoroelastomer that is highly chemical resistant, but expensive. These elastomers are used in both free-standing form and as coatings on support substrates like cotton, polyester, or nylon fabric. Tyvek©, a proprietary nonwoven fabric, is being used in disposable items in both coated and uncoated form.

Levels of protection afforded by each of these materials is specific to each chemical to be resisted. Aside from chemical reactions associated with degradation, and porosity that results in penetration, material/chemical compatibility is most important in the permeation process. Permeation resistance is highly dependent on the structural makeup of an elastomer. Density, polarity, molecular complexity, and additives all play an important role.

Recently, a New Hampshire company developed a line of new materials for use in chemical protective clothing, especially full ensembles. During its development work, this company relied for guidance on performance data from the ASTM Standard Test Method on Permeation Resistance. The new materials are Teflon© composites on woven and nonwoven Nomex©. Data from initial permeation testing indicate that performance is far superior to other commercially available materials.

Also, in an effort to improve resistance by blocking the physical and chemical mechanisms associated with permeation, at least one group is working to develop a process for depositing thin, flexible metallic coatings on a variety of traditional elastromeric materials.

These two examples show that a new generation of improved materials for chemical protective clothing is not only possible, but probable. The ability to quantify performance levels already is starting to have a major impact on new product design objectives.

Decontamination

Much like the pot of gold at the end of a rainbow, decontaminating chemical protective clothing after direct contact by a hazardous chemical has been an elusive goal.

The standard process for cleaning protective clothing has been a water and detergent wash. This method is effective only on the outside surfaces of clothing materials and for those chemicals that are water or detergent soluble. A water wash usually does little to extract a chemical that has begun to permeate into the thickness of an elastomeric matrix. Surface cleaning that does not reach in and pull out permeated chemicals leaves behind a reservoir that will subsequently migrate to both the inside and outside surfaces. A user could then be contacted by the hazardous chemical immediately upon donning the ''clean'' protective clothing.

The traditional solution to this problem has been to discard contaminated items rather than to attempt cleaning and reuse. Because of this, a strong market for inexpensive, disposable protective clothing has evolved. Such disposables are of limited application: usually one-time use. In other cases, more expensive, nondisposable materials must be used to provide adequate protection. However, even nondisposable clothing is often discarded if an actual chemical contact is confirmed. Response to a recent PCB spill cost the clean-up company more than a million dollars for protective clothing alone because workers' ensembles were replaced after each use.

The growing understanding of physical and chemical mechanisms associated with the performance of chemical protective clothing also is leading to a better understanding of requirements for cleaning methodology. ASTM's Committee F-23 has taken a first step in this arena by initiating the development of a standard test method for establishing levels of contamination as well as for determining the effectiveness, or efficiency, of cleaning processes.

Another company is attempting to modify for chemical protective clothing a Freon-wash system originally designed to clean clothing contaminated by radioactive particulates. With the development of new protective clothing materials that minimize or block the permeation process, such an inert solvent system might well be effective. If so, the cost-per-use of nondisposable protective clothing will fall dramatically.

As with many other aspects of chemical protective clothing, there is much yet to do. A strong economic incentive exists for the development of an effective decontamination system. Many people predict a close relationship will evolve between individual materials and decontamination research efforts.

Assembly Techniques

Techniques used in assembling items of chemical protective clothing from flat material stock have represented a weak link in protection. CPC users should ensure that seams, closures and other discontinuities are as resistant and as physically strong as the base materials used in the protective clothing. However, a recent study by the National Institute of Occupational Safety and Health in Morgantown, West Virginia, indicated that this is not always the case.[14] NIOSH has confirmed that seams and closures are often a limiting factor in protective clothing's ability to provide protection.

Seams most often are sewn or glued. Sewing puts needle holes through the material layers and, unless the holes are filled or covered, provides an opportunity for direct chemical penetration. Gluing should be continuous with no breaks that also could permit chemical penetration. The adhesive used in such seams should be resistant to the challenge chemical as the base materials. Historically, neither of these seaming techniques have performed as well as expected.

Heat sealing is the most recent advance in seaming techniques. Thermal energy is provided by either a direct heat source or a radio frequency (RF) source. This can be a relatively foolproof technique, as long as the base materials are capable of melting and then adhering upon rehardening. Close control of the sealing temperatures is important.

Designs for closures like zippers and exhalation valves for full ensembles seemingly have not changed in many years. Metal and nylon zippers, for example, remain the mainstay.

Besides leakage, another problem is that closures can be difficult to attach to many protective clothing materials. Because closures and base materials are frequently dissimilar, adhesives or heat sealing techniques that work with one may not work with the other. Then, the only recourse is sewing.

Improved assembly techniques provide an interesting challenge to those who are creative and opportunistic. Performance specifications should encourage renewed efforts to develop improved seams and closures as well as to develop improved base materials.

The Future

There is no doubt that chemical protective clothing is changing. Technological advances are encouraging development of an entirely new generation of materials and designs with markedly improved levels of protection. The era of accepting the status quo has ended; therefore, protective clothing users should confront performance issues head on. The standard test methods that are now becoming available are providing realistic quantitative measures of effectiveness.

IMPORTANT CPC INFORMATION SOURCES

Professional Groups

- American Society for Testing and Materials (ASTM)
  1916 Race St.
  Philadelphia, PA 19103
  Contact: Ann McKlindon (215)299-5400
- American Industrial Hygiene Association (AIHA)
  475 Wolf Ledges Parkway
  Akron, OH 44311
  Contact: Chairman, Personal Protective Devices Committee (216)762-7294
- National Fire Protection Association (NFPA)
  Battery March Park
  Quincy, MA 02269
  Contact: Bruce W. Teele (617)770-3000
- Industrial Safety Equipment Association (ISEA)
  1901 N. Moore St.
  Arlington, VA 22209
  Contact: Catherine J. Morin (703)525-1695

Government Agencies

- U.S. Coast Guard
  Washington R & D Office
  2100 2nd St., S.W.
  Washington, DC 20593
  Contact: Lt. Jeffrey O. Stull (202)426-1023
- U.S. Environmental Protection Agency (EPA)
  Pesticide Program (TS-769)
  401 M St., S.W.
  Washington, DC 20460
  Contact: Alan P. Nielsen (703)557-0267

- U.S. Environmental Protection Agency (EPA)
  Releases Control Branch HWERL
  Woodbridge Ave.
  Edison NJ 08837
  Contact: Michael D. Royer (201)321-6633
- U.S. Fire Administration
  10820 Bethel Rd.
  Frederick, MD 21701
  Contact: Robert T. McCarthy (301)447-6711
- National Institute for Occupational Safety and Health (NIOSH)
  Division of Safety Research
  944 Chestnut Ridge Rd.
  Morgantown, WV 26505
  Contact: Stephen R. Berardinelli

Testing Laboratories

- Arthur D. Little, Inc.
  15 Acorn Park
  Cambridge, MA 02140
  Contact: Arthur D. Schwope (617)864-5770
- E. I. Du Pont De Nemours & Co.
  Haskell Laboratory, CR&D
  Elkton Rd., P.O. Box 50
  Newark, DE 19711
  Contact: Norman W. Henry III (302)366-5250
- Radian Corporation
  8501 Mo-Pac Blvd.
  P.O. Box 9948
  Austin, TX
  Contact: Meridith Conolay (512)454-4797

REFERENCES

1. Excerpts taken from: Coletta, G. C., "Chemical Protective Clothing," a six-part series, *Occupational Health and Safety,* April–September, 1985.
2. Sansome, E. B., and Tewori, Y. B., "The Permeability of Laboratory Gloves to Selected Solvents," *American Industrial Hygiene Association Journal,* **39:**169–174, 1978.
3. Williams, J. R., "Permeation of Glove Materials by Physiologically Harmful Chemicals," *American Industrial Hygiene Association Journal,* **40:**877–882, 1979.
4. "Preventing Illness and Injury in the Workplace," U.S. Congress, Office of Technology Assessment, OTA-H-256, April 1985.
5. *The Prevention of Occupational Skin Diseases,* 3rd Printing, The Soap and Detergent Association, New York, 1981.
6. "A Summary of the NIOSH Open Meeting on Chemical Protective Clothing," held on June 3, 1981, NIOSH, Rockville, Maryland, 1981.
7. Mansdorf, S. Z., and Miles, B., "A Protective Dermal Film System," presented at the American Industrial Hygiene Association Conference, Los Angeles, California, May, 1978.
8. "Report of the Advisory Committee on Cutaneous Hazards," Report to the Assistant Secretary of Labor, OSHA, Washington, D.C., December 19, 1978.
9. "Occupational Injuries and Illness in 1978: Summary," Report 586, Bureau of Labor Statistics, Washington, D.C., 1980.
10. ASTM Standard Test Method F-739, "Resistance of Protective Clothing Materials to Permeation by Liquids or Gases," *Annual Book of ASTM Standards,* Section 15, Volume 15.07.
11. ASTM Standard Test Method F-903, "Resistance of Protective Clothing Materials to Penetration by Liquids," *Annual Book of ASTM Standards,* Section 15, Volume 15.07.
12. Schwope, A.D., et al., "Guidelines for the Selection of Chemical Protective Clothing," Publication 0460, American Conference of Governmental Industrial Hygienists, Cincinnati, Ohio 45211.
13. Coletta, G. C., and M. W. Spence, "Managing the Selection and Use of Chemical Protective Clothing," p. 235, ASTM Special Technical Publication 900, August, 1986.
14. Berardinelli, S. P., and L. Cottingham, "Evaluation of Chemical Protective Garment Seams and Closures for Resistance to Liquid Penetration," p. 263, ASTM Special Technical Publication 900, August, 1986.

4

FIRE PROTECTION

Arthur C. Smith

Manager
Corporate Environmental Safety and Health
AT&T Information Systems
Bernardsville, New Jersey

INTRODUCTION

Annual industrial fire losses constitute a high percentage of the national total from a property standpoint. The National Fire Protection Association (NFPA) estimated business and industry losses in 1984 of 133,000 structural fires, and property losses amounting to $1,821,000. These fires, as a general rule, did not result in extensive loss of life. The potential for loss of life from fire is always present, and is related to the fire hazards and risk of the operations, the number of people in the building and their operations, the type and severity of fire that might occur, and the design features of the building that could affect life safety. *America Burning,* a look at the U.S. fire record in 1973, cited the "twin tides" of "ignorance and indifference" as the major causes of this poor fire record. Historically, we hear the comment "I won't have a fire, it happens to others." This indifference often extends to meeting minimal standards set by building codes. Postmortems of large-loss fires indicate that confusion and lack of effective action at the time of the emergency contribute to the loss.

Purpose

As corporations retain greater risks through self insurance and recognize the greater investments in these facilities, there is an increasing need for those responsible for fire protection to have more than a superficial knowledge of fire protection. A management commitment to an aggressive program of loss prevention is the foundation of efforts to reduce loss. A manager is charged with the responsibility of producing a product or service. If a facility sustains a major emergency and is either disrupted or destroyed even the most effective management will fail to sustain profitability. Many managers find that in addition to their regular responsibilities they have been given the added responsibility of fire protection for their facility. It is often impossible for the manager to wade through the many and varied works that address fire protection. This chapter is designed to assist the manager responsible for fire protection.

As a manager, you must first recognize and establish the following priorities: life safety first, preservation of company assets next, and protection of the building last. Convincing management of the importance of these priorities can be difficult, since money is often more freely spent for aesthetics than on fire protection. It must be explained that while a dollar return on expenditures for loss control may be difficult to demonstrate, a loss prevention program is financially justified. Accomplishing this task requires interaction with many other people, including but not limited to upper management; engineers who design new laboratories, storage facilities, and other occupancies; security personnel responsible for the movement of people within the complex; insurance representatives who assess and evaluate recommendations; and building maintenance personnel responsible for servicing and maintaining fire safety equipment.

Scope

This section will guide you in addressing the life safety and property conservation needs of a company. It will not give you a fill-in-the-blank approach nor provide all the details or techniques that may be employed. It will help you start planning, which in turn helps to define your problem. To do this, examine the following definitions: *Hazard*—something with a potential for damaging life or property. *Vulnerability*—the susceptibility of life and/or property to damage, if a hazard manifests its potential. *Risk*—the probability that damage will occur if a hazard manifests itself. Ask yourself, "What if?" What loss expectancy can be tolerated? The range of hazards is wide. For example, the high-technology electronics industry utilizing such materials as solvents, which range from extreme fire hazards to health hazards, and compressed gases —for example, the highly toxic arsine and pyrophoric silane.

DETERMINING HAZARDS

Most facilities handle chemical substances of some kind. Even offices have duplicating fluids or powders. Awareness of the hazards of materials is an important step in loss prevention. If your industry utilizes anything you suspect might be dangerous to employees, begin with the following questions:

1. Can it form a flammable mixture in air?
2. Is it corrosive?
3. Can it aid or add to combustion?
4. Is it self-ignitable?
5. What are its toxic effects?
6. Is it an asphyxiant?
7. Can it react with other materials?

The answers to these questions will identify potential problems that must be considered and controlled. After determining that there is indeed a problem or at least a potential for one, you should consider the following:

1. Is there a substitute that is less hazardous?
2. Can a process change be instituted that would reduce the hazard?
3. Can a ventilation system assist in reducing the hazard?
4. Is monitoring available to detect the hazardous vapors or gases?
5. Training is the most important component to the overall safety and successful avoidance and conclusion of an emergency involving hazardous materials. Training should address handling, storage, disposal, and all other actions necessary in the event of a spill or release of material.

DESIGN AND CONSTRUCTION

The first item on your agenda when examining either your facility or changes to the facility is to determine what code is enforced in your area. The following codes are used in various parts of the country: the Building Officials and Code Administrators Code (BOCA); the National Fire Codes, published by the National Fire Protection Association; National Building Code, published by the American Insurance Association; Uniform Building Code, published by International Conference of Building Officials; and Standard Building Code, published by Southern Building Code Commerce International. The only deviation you will find exists in large urban areas such as New York, Chicago, etc., where a code to address specific problems has been developed. Building codes classify occupancies with regard to fire severity and classify construction categories. Codes address building area, height, construction, and materials as well as items important to evacuating people safely from a fire and retarding the spread of fire. Codes are often quite confusing to the uninitiated, and it is wise to look to your insurer for guidance when you are in doubt. You should look through their documents and become familiar with them.

Enclosure

Fire is bound by the laws of physics and predictions can be made as to how it may spread and impact on your facility. A fire can start whenever there are fuels and men, women, or children around. The idea is to limit the ability of a fire to spread to other areas. This is one of the purposes of building codes. Accepting this reasoning helps you understand the intent of the requirements established by the codes. The building provides the passive fire defenses and the fire containment which affords life safety to the occupants. One aspect that is often overlooked is the occupancy changes or construction changes that are frequently conducted in a facility. During renovations the building and occupants are more vulnerable to emergencies created by hazards from blocked fire exits, travel restrictions, accumulations of construction materials and trash. The other matter to be addressed is that possible physical changes to the structure may increase the occupancy of a floor, add a hazardous process, or alter egress from the area.

Egress

Codes establish the design, construction, and arrangement of elements required to provide a safe means of egress from structures. For every location in a building there must be a means of egress or path of travel which a person can move through to gain access to the outside. These exits must provide protection from fire along their entire length. This is accomplished by an enclosure with construction having a designated degree of fire resistance. These egress elements consist of doorways, passageways, corridors, stairways, escalators, ramps, and combinations of any of these. Exits must be arranged so as to be discernable and accessible with unobstructed access. They should also be arranged to lead directly to the exterior or an area of refuge with supplemental means of egress that will not be obstructed or impaired by fire or smoke. The elements of egress, their numbers and arrangement, are determined by (a) the type of occupancy and (b) the number of occupants. Basically there are more restrictive requirements where dangers exist from either what is being performed, stored, or used and/or the number of occupants. There are allowances made for the installation of automatic fire suppression systems. When sprinklers are present, travel distances to exits are allowed to be increased.

As you walk through your facility, begin to look for obstacles that would hinder the safety of people using the corridors, stairways, and doors. Items that may not have appeared to be important up to now will begin to take on a different value. The door that is usually held open with a chock or some other means is now recognized as a danger that would expose people to smoke and heat. Doors that are designed to be self closing to isolate people from dangers are often found inoperative. The storage of items in the halls, stairwells, and corridors is now recognized as a danger that impedes safe passage. There are devices available that will assist in eliminating problems such as these. There are electromagnetic devices that hold doors open and are activated or released by smoke detectors, pull stations or sprinklers. These provide protection to those using the means of egress that would otherwise not exist.

FIRE PROTECTION

Emergencies do not give warning before striking; therefore, it is essential to have active fire protection devices in place that can provide this warning. Active devices do something when a fire occurs. Detection devices can do one or all of the following: alert occupants, notify the fire department, control elevators, and activate equipment. Some devices detect the products of combustion from a fire and are best known as smoke detectors. There are also specialized devices for hazardous areas that detect the flicker of flame. Automatic suppression devices are used to control and suppress fires in special equipment, in general areas, and as a barrier defense. Detectors and automatic suppression devices can be used in conjunction with one another and provide some unique protection to occupants and building. Standpipes are internal hydrants that enable trained teams or the fire department to manually fight the fire with hose streams from the standpipes. The following section will describe how these systems operate and what protection they can provide.

Detection Systems

The primary function of a fire detector is to respond to a fire, and transform this information into a visual and audible signal which alerts building occupants. Detectors sense the beginnings of a fire so that actions by the occupants can be started. These actions take the form of evacuation and/or suppression of the fire. There are a number and variety of heat, smoke, and flame detectors in use today. The most common device a manager will encounter is the smoke detector. These systems are installed for general area detection often as an alternative to automatic sprinkler protection or where sprinklers are not required. The system may be as simple as a single detector, or it can be a more sophisticated microprocessesor system capable of providing addresses for each unit in the facility. Detector systems constantly check their circuitry to ensure against problems such as breaks in the wiring or removal of units. The environment and the type of occupancy to be protected will dictate the type of detector to be considered. The energy levels or heat release, the type of fire and quantity of smoke anticipated should also be considered. For example, where the ignition of a high energy release fire from a flammable liquid is probable the use of a thermal detector is advised. This same detector would be ineffective where a low energy fire in electronic equipment is likely. Detectors have limitations which must be recognized for proper application and effectiveness.

More often than not, improper system function is due to inadequate consideration to the aforementioned issues. A list of detectors in use today is given on p. 77.

Thermal Detectors:

- *Fixed temperature* detectors are in common use for limited applications. The most frequently seen type is the fusible link, which is used on sprinkler heads, in sliding fire doors, and in such special applications as deep fat fryers.
- *Rate of rise* detectors respond to temperature increases greater than 15°F per minute. These have limited application.

Other Types:

- *Photoelectric* detectors respond to the visible products of combustion. Environments containing dust or any suspended mists cause alarm. This type should be used in areas where visible products of combustion will be generated, e.g., where there are flammable liquids.
- *Ionization* detectors also sense products of combustion and are widely used. These units are particularly useful in areas with highly susceptible electronic equipment. These locations require fire be detected at the earliest stage to avoid irreparable damage.
- *Flame detectors* respond to visible flame and are used where volatile materials are present. They are extremely fast and can sense a flame in milliseconds. This sensitivity makes their use appropriate for explosion suppression systems where detection and release of a suppressant must be instantaneous. Today's technology has produced a detection system that utilizes fiber optics. The fiber optic light guide is placed around a room's ceiling with a light transmitter at one end and a receiver at the other end. When a fire occurs the light reception is altered and causes alarming of the system.

Extinguishing Systems

A sprinkler system employs automatic heat-activated heads attached to piping which is connected to a water supply; the system is designed to discharge the water on the fire from heads opened by heat. The most common types of system in use are described below.

- *Dry pipe sprinklers* are systems employing sprinkler heads attached to piping containing air or nitrogen under pressure. When the sprinkler head opens from the heat of a fire the air is released allowing the water to flow from a control valve. The water is then discharged from that head opened by the fire. These systems are used where the danger of freezing of liquids exists.
- *Wet pipe sprinkler* systems employ automatic sprinklers attached to a piping system containing water. This is probably the most common type of system in use.
- *Pre-action systems* are systems with automatic heads attached to the piping, with a supplemental fire detection system installed in the same areas as the sprinklers. Activation of the supplemental system by a fire opens a valve which permits water to flow through the opened head. These systems are utilized in areas where there is a fear of water unnecessarily damaging what is contained in the occupancy.
- *Deluge systems* consist of open sprinkler heads attached piping which is connected to a water supply through a valve that is opened by a detection system or manually. These systems are used where protection from possible exposure to fires is needed.

A recent development in sprinkler technology is the automatic on/off head. This type of head is activated by the heat of the fire which causes the head to open and allows the water to flow. Once the fire has been extinguished the head will close, shutting off the flow of water. If the fire should rekindle the head will recycle once again. This type of head limits water damage, which is a commonly cited drawback of sprinkler systems. There have been other developments that provide less time lag for a sprinkler head to respond to the heat of a fire. These heads are called quick response and have been found to be capable of extinguishing fires where the older design had only controlled the fire.

Managers advocating installation of sprinklers are often confronted with the fears of those who do not know just how successful sprinklers are. A 28-year study by the U.S. Department of Energy was published in 1982, showing how successful sprinklers were during this period. About one-third of all fires in the survey were completely extinguished by the operation of a single sprinkler. There had been no loss of life due to fire in a sprinklered building; the chance of any damage to, or from, a sprinkler system is about one per year for every 800 systems—and nearly half the incidents were so slight that the damage to, or from, the system was negligible. And finally, sprinkler systems were found to be 98 percent effective in controlling or extinguishing fires.

One common fear about sprinkler systems is water damage. However, if there is no sprinkler to extinguish the fire you may lose everything. This is usually more serious than water damage. As a practical matter, employees should know where the sprinkler shut-off is and what to do when a fire occurs. More often than not no one knows where the sprinkler shut-off is; as a result, water continues to flow unnecessarily, causing damage that could have been prevented with a little training. Another complaint is the effect sprinkler heads have aesthetically. This has been overcome with designs that can completely hide the presence of sprinkler heads.

Halon systems are used in high value electronic occupancies such as computer rooms where other extinguishing media cannot be used. These systems consist of a detection system which activates a control valve releasing the halon through a piping system and nozzle in the room. The room is flooded with the gas, which extinguishes the fire. A major advantage of this medium is that it requires no cleanup after a fire. Halon is also noncorrosive to electrical circuitry and considered nontoxic to living things.

Carbon dioxide systems in the past were used extensively for high value areas where water could not be used. This medium extinguishes by displacing oxygen to a point where there is not enough oxygen concentration left to support fire. This is important to remember since insufficient oxygen for a fire means that it is also insufficient to support life. These systems have been used in applications ranging from deep fat fryers in kitchens to dip tanks and fur vaults. Installation is similar to that of the halon system.

Alarm Systems

Signaling devices are installed to provide a warning or to alert the occupants of a facility that evacuation is necessary. They also can alert the emergency action team to take action. Some of these systems also notify the local fire service. The major differences that exist between these systems must be recognized and occupants must become familiar with exactly what they mean and what they do. The local signaling device only notifies the occupants. It can utilize anything from a pre-recorded message to a distinctive sound or a light to provide this warning. Whatever is used it must be capable of being understood above background noise levels. While these systems provide the important purpose of notification for evacuation, it should be recognized that this is only part of the benefit that can be derived. Notification of the fire department is another important benefit. This would assure that assistance is on the way immediately upon the recognition of a fire. There are manual switch boxes which can be "pulled" upon the detection of a fire. Detectors can transmit an alarm to occupants, trigger sprinklers, and activate special extinguishing devices.

Portable Extinguishers

Fire extinguishers are often placed in a facility in response to insurance company requests, or various laws. Little thought is given to type, distribution, or training. The ability of these devices to keep a small incident from becoming a major fire is often overlooked.

The effectiveness of fire extinguishers is stated in National Fire Protection Standard No. 10(1). They advise consideration be given to:

1. Proper location
2. Working condition
3. Proper type for specific hazard(s)
4. Detection of the fire while still incipient
5. Employee training

The type of extinguisher to use is determined by assessing the class of fire that would predominate in an area. Consideration must next be given to the amount of fuel present and the size of the fire that could be anticipated. This information also will help you determine the number of extinguishers needed to adequately protect an area.

The major types of extinguishers in use are:

1. *Pressurized.* This type contains $2\frac{1}{2}$ gallons of water with air pressure to force the water out. It is used on class A fires (ordinary combustibles). It has a reach of 30 feet and will completely discharge in about 60 seconds. One consideration when using pressurized water is the potential for water freezing. Anti-freeze added to the water reduces this possibility.

2. *Carbon Dioxide* (CO_2). This type is used on class B and C fires (flammable liquids and electrical). These extinguishers range in size from 5 lb hand extinguishers to large wheeled units weighing 50 lb. CO_2 units have a range of 3 to 6 feet and last about 10 to 30 seconds.

3. *Dry Chemical.* There are two basic types in use, regular dry chemical and multi-purpose. Regular dry chemical is suitable for class B and C fires (flammable liquids and electrical). Dry chemical units are also available from 3 lb hand type to 150 lb wheeled units. A multi-purpose dry chemical unit suitable for use on classes A, B, and C (ordinary combustibles, flammable liquids, and electrical) is also available. The multi-purpose extinguisher is available in 3 lb portables and wheeled units up to 125 lb. The multi-purpose units have a range of 5 to 20 feet and last from 10 to 30 seconds. A precaution that must be exercised when using either of these extinguishers is to avoid use in areas where electronic equipment of any kind is present. All dry chemicals used in fire extinguishers are very fine powders that are dispersed everywhere when expelled. When the powder is heated by fire or exposed to a humidity of 50 percent it forms a gummy deposit on surfaces. If left on components for any period of

time it can cause them to overheat. It is also capable of forming a weak solution of phosphoric acid, which can seriously damage circuit boards.

4. *Halon.* These extinguishers leave no residue and are most suitable for expensive electronic equipment areas. Units range in size from $2\frac{1}{2}$ to 10 lb, have a range of 4 to 8 feet, and last from 10 to 30 seconds.

An important point to establish is that extinguishers are first aid devices that are to be used only when the operator believes that the fire can be controlled or extinguished with one extinguisher. Personal safety must not be jeopardized while utilizing extinguishers.

Proficient use of any extinguisher is possible for all trained operators. Use the acronymn "PASS" to train your employees so that they will know how to use every extinguisher they encounter: *P*ull the pin (the locking pin); *A*im the nozzle; *S*queeze the handle; and *S*weep the nozzle from side to side.

EMERGENCY ACTION PLAN

Emergencies are unpredictable because they occur unexpectedly. A manager may not be able to predict what type of emergency may occur or when, but a good plan can prevent it from becoming a major problem to the company. An emergency action plan should be instituted with the objective of providing life safety for employees, protection of property, and the resumption of normal business as quickly as possible. Implementation of the program will achieve compliance with local, state, and federal laws. This outline provides sufficient information to develop a comprehensive plan but will not address every possible type of emergency that may occur. Emergencies can originate from fire, explosion, chemical spill or release and natural disasters, to name a few situations. A plan that would attempt to provide for every emergency would become much too complex and cumbersome to enact. It is advisable to meet with the local Fire Department to coordinate plans to maximize the efficiency of an evacuation plan. This is especially important in regard to high rise buildings, because of the many new techniques requiring movement to areas of refuge rather than total evacuation. Before you can start to assemble a plan you must recognize that you will probably need help, since you may not possess all the knowledge or information needed to go into a plan. When one individual has sole responsibility for planning, the range of the plan will be limited by that individual's knowledge. Comprehensive planning requires the labor of a few who can really make it work. The plan will then reflect the expertise of a variety of sources ensuring accuracy and completeness. The extent and implementation must be determined by evaluating the:

- · Probability of an event
- · Hazards present in actual processes or stored materials
- · Complexity of the building
- · Number of employees present
- · Automatic fire protection systems
- · Water supply for fire protection
- · Type of Fire Department available (paid, volunteer, etc.)

There are also mandatory minimum requirements spelled out in various codes and regulations. They include:

· A formal written plan available for review
· Posted egress routes
· Records of evacuation drills, which should be performed every 6 months
· Maintenance of monthly fire safety inspection reports.

Team Development

Emergencies can and do disrupt the best plans and operations. Preplanning can keep an incident from becoming a disaster. When a fire occurs, quick notification and evacuation may prevent fatalities or serious injuries. It is important to give serious consideration to developing an Emergency Action Team. Team members should be willing participants rather than appointed members. They should also normally be at their work location and not required to be at frequent meetings. They should possess the ability to communicate the urgency of the situation without becoming excited. One cannot use anyone who may shout ''fire'' in a crowded theatre. Previous experience and knowledge of fire extinguishers and/or first aid are additional criteria to be considered in selection of team members. The size of the team will vary with the size of the building and the complexity of the floors or exit routes. A team should consist of at least the following members:

· Coordinator
· Deputy coordinator
· Floor wardens (at least one per floor)
· Deputy floor wardens (OSHA recommends one per 20 occupants)
· Special wardens for handicapped employees who may require assistance during
  emergencies.

In the event of an alarm being sounded for an emergency the team would be activated and have the following responsibilities:

Coordinator: Ensure notification of the Fire Department; alert wardens that evacuation is required; confirm assistance of handicapped; follows status of evacuation with wardens; assure direction of local agencies to emergency site; sound ''all clear'' when it is safe to return to workplace: and notify appropriate management of emergency.

Deputy Coordinator: Aid the coordinator; take responsibility in the absence of the Coordinator.

Wardens: Determine if evacuation is necessary; personally direct actions to be taken; report activities to the coordinator. Wardens should be capable of suppressing small fires through the use of the proper extinguisher.

Deputy Wardens: Search assigned areas and assist wardens as needed; assume duties of wardens in their absence.

Special Wardens: Two Wardens should be assigned to each handicapped person to ensure safe evacuation during emergencies.

Note: Those responsible for life safety must recognize that some employees may become temporarily ''handicapped,'' that is they could be rendered mobility-impaired as the result of accidents involving broken legs, ankles etc.

Training

Training should be flexible, since each action plan has unique requirements. Testing the evacuation plan through drills usually reveals these unique conditions. Once the plan is reasonably stable (THERE SHOULD BE A DEADLINE SET FOR COMPLE- TION) the coordinator should formalize the training by documenting the procedures and responsibilities for each of the team members. Classroom training sessions should be followed by practice evacuation drills to assure understanding by all involved. The team members should be given copies of their responsibilities and procedures. Training sessions should be conducted at least twice a year to maintain familiarity with all aspects of the plan and address any changes that may have occurred since the last meeting. It is strongly recommended that local fire and police departments be called upon to join in the pre-planning procedures and to coordinate drills. Drills are simulated emergencies that allow testing of the plan. You must not ignore the need for continuing training via drills. Mismanagement of emergencies can be traced more often to ignorance of the plan and lack of training than to anything else. Drills are aimed at getting the answers to these questions: Is the plan adequate? Are personnel trained to function under the plan? Finally, the plan should never just be put on the shelf once completed. Plans become outdated, people change, processes change, and so should the plan. A regular review period on at least an annual basis is necessary. Insurance companies are an excellent source of assistance for training with resources of films and printed material.

OSHA

OSHA 29 CFR, Part 1910, Sub Part L, offers the employer of ten or more employees options that will be suited to their particular needs. The standards provide for:

- Employee emergency action and fire prevention plans
- Fire brigades
- Design, installation and testing of fire protection systems
- Use of fire protection systems on hazardous materials.

OSHA has provided options to the employer regarding the action of the employees during emergencies. Basically the overview is this:

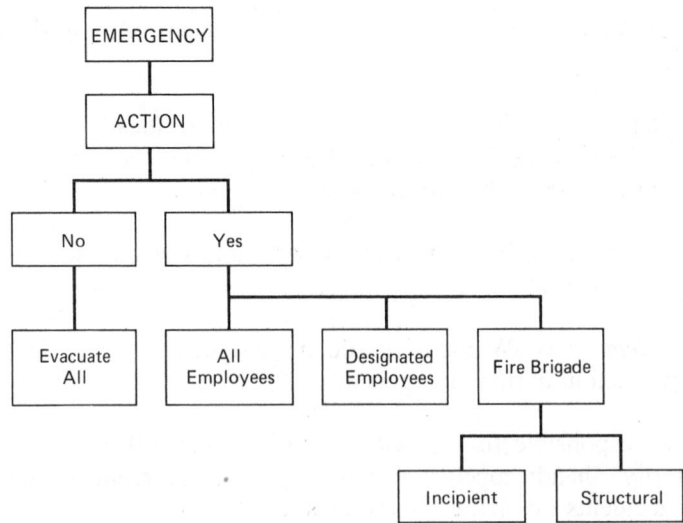

The plan provides four options if you decide to take action during an emergency, and total evacuation if you plan to do nothing.

The total evacuation plan does not allow for any activities to control or extinguish a fire. This option provides for the life safety of employees and totally neglects property conservation.

The next two options involve employee participation in the emergency. These action options can involve either (A) all employees (B) or only designated employees. Option A requires that all employees be trained and prepared to fight a fire. The law stipulates at least annual "hands on" training. Option B is selected by the employer who designated only certain employees to fight a fire in their immediate work area.

Another option is the engagement of the fire brigade, a formal group organized for the purpose of attacking a fire anywhere within the facility. Within this option is an incipient brigade which will fight incipient fires with portable extinguishers, class 2 standpipes, or small hose systems. No special protective clothing or breathing apparatus is required under this option.

The final choice is a structural brigade which is assigned to fight interior structural fires anywhere in the facility. This group is also responsible for rescue activities and requires full protective clothing and breathing apparatus. Participants should also be determined to be physically able to perform their duties during emergencies.

Decisions as to which selection to choose should be based on the following criteria:

· The local Fire Department's capabilities, considering volunteer versus paid, manpower available, and distance of the facility from the firehouse.
· Water supply available for firefighting and automatic suppression devices.
· Number of employees available for participation in the selected option.
· Types of hazard involved in the facility.
· Size of the facility.
· Availability of automatic detection and suppression systems.
· Presence of smoke exhaust systems.
· Availability of egress systems.

BIBLIOGRAPHY

Occupational Safety & Health Administration 1910.155 Sub Part L outlines the requirements for all aspects of fire protection and should be referred to for complete requirements and information.

1910.156	Fire Brigades
1910.157	Portable Fire Extinguishers
1910.158–163	Fixed Fire Suppression Systems
1910.164	Fire Detection Systems
1910.38	Sub Part E Evacuation and Fire Prevention Plans.

Planer, Robert. G. *Fire Loss Control: A Management Guide,* Marcel Dekker, Inc., 270 Madison Avenue, New York, NY 1979.

Building Officials & Code Administrators International, Inc., 4051 W Flossmoor Rd., Country Club Hills, Illinois, 60477, 1984.

National Fire Codes, National Fire Protection Association, Batterymarch Park, Quincy, MA 02269, 1987.

Bugbee, Percy. *Principles of Fire Protection,* National Fire Protection Association, Batterymarch Park, Quincy, MA 02269, 1978.

Bryan, John L., *Automatic Sprinkler and Standpipe Systems,* National Fire Protection Association, Batterymarch Park, Quincy, MA 02269, 1976.

5

FIRST AID IN THE WORKPLACE

Shirley Ann Conibear, M.D., M.P.H.
Associate Professor, College of Medicine
University of Illinois at Chicago
Vice President, Carnow, Conibear and Associates, Ltd.
Chicago, Illinois

Victoria Cooper Musselman, M.Ed.
Director, Educational Services
Carnow, Conibear and Associates, Ltd.
Chicago, Illinois

DEFINING THE SCOPE

The term first aid refers to assistance given to an injured or ill person in an emergency or in situations where regular professional medical treatment is not readily available. It is by definition provided by persons who are not health professionals. This distinction has blurred somewhat with the development of emergency medical systems and the training of emergency medical technicians (EMTs) who administer first aid and may also provide some form of more definitive treatment such as intubation and intravenous fluids. At a worksite, a variety of professionals may provide first aid, ranging from lay persons who have taken a first aid training course, to persons trained as emergency medical technicians (EMTs), to paramedics, to nurses, and finally to physicians, who may staff a medical department at a large industrial facility. Where formally trained health professionals such as physicians or nurses are employed, the design and implementation of emergency and routine treatment is usually left up to them. This chapter will deal primarily with how to set up a first aid program at the plant without a doctor or nurse on site. In circumstances where these health professionals are on site or available through phone consultation, the information in this chapter is still applicable, but the program may be extended and expanded to make full use of available medical expertise.

The first step in preventing health problems in the workplace is anticipating how, when, and where they might develop. A first aid program should be only one part of a general health and safety program in the workplace. The goal of the overall program is to prevent or minimize the health impact of workplace accidents and exposures. However, no preventive program is ever 100% effective and so every health and safety program must anticipate and plan for catastrophic events. The first aid program provides immediate care for both emergency and urgent problems and forms a bridge to a health care delivery system that can provide definitive diagnosis and treatment. That system may be an in-plant medical department, a clinic or doctor's office, or a hospital emergency room.

First aid provides secondary prevention, that is, preventing an injury or illness from getting worse, preventing permanent damage, or even curing. It is important to view first aid programs in this context. Unless it is a part of a primary prevention program of proper industrial hygiene and safety, a first aid program can become literally the provision of a "bandaid" for workers' injuries and illnesses. A good first aid program is necessary to but by no means sufficient as a plant-wide health and safety program.

Another important reason why the first aid program must be an integral part of the overall health and safety program is to provide feed-back on the type, circumstances and frequency of accidents and illnesses occuring at the plant. Frequently, a major incident will be preceeded by a number of small incidents or "near misses" involving a dangerous situation. Good communication between the first aid team and the hygienist and safety officer may help to pinpoint a problem area before a major problem occurs.

Much of the literature on first aid deals with incidents which are unlikely to occur in most workplaces, such as motor vehicle accidents, drownings, and ingestion of poisons. Conversely, the workplace presents the potential for accidents not covered in most first aid manuals, such as exposure to toxic gases or pesticides. The first step in designing an effective first aid program at a worksite is to assess the types of illness and injury likely to occur at that particular facility. In order to anticipate the problems, you must enumerate and analyze the hazards. For instance, light manufacturing facilities with a lot of grinding and polishing of metal may have problems with eye injuries, whereas high physical demand occupations may present more foot, lower extremity, and back injuries, as well as problems due to heat stress. Outdoor work may expose workers to contact with poisonous insects or plants. Table 1 lists some of the common physical hazards you should think about when assessing the potential first aid needs at a facility.

In addition to physical hazards, you must assess the chemicals that pose hazards at a facility in terms of their number, relative toxicity, quantity handled, potential for worker exposure, possible route of contact and absorption, and the nature of the injury they can cause. Since the implementation of OSHA's Hazard Communication Standard, most companys have already gathered material safety data sheets (MSDSs) on all chemicals used or made in the plant. You should start by collecting and reading all these MSDSs. This will make your job of assessing the chemical hazard much easier. Each Material Safety Data Sheet has a section which deals with acute and chronic health effects, including emergency treatment. In addition, the nature of the proper treatment and the

---

Table 1. Physical Hazards Leading to Problems Requiring First Aid Treatment.

1. Temperature extremes (frostbite, heat exhaustion)
2. Poisonous plants, animals and insects (allergic reactions, skin rash)
3. Electrical hazards (burns, respiratory and cardiac arrest)
4. Materials handling (leading to sprains, strains, crushing injuries, dislocations, fractures)
5. Working at heights (falls leading to fractures, etc)
6. Ultra-violet light (skin and eye burns)
7. Operating moving equipment (leading to cuts, crushing injuries, amputations)
8. Hot materials (burns)
9. Grinding, cutting, polishing (leading to eye injuries, splinters)

antidote, if one exists, is usually specified on the MSDS. This should serve as a guide in developing treatment protocols, selecting appropriate supplies and equipment, and designing and locating the first aid facility(s). In addition to the identity of each chemical, you should have a general idea of the amount of each chemical used and stored at the facility. Find out how, when, and where each chemical is handled in the workplace. This will provide insight into the likelihood of human overexposure and will identify which workers are at risk of overexposure. Knowing exactly what chemicals each worker handles can help in identifying the exposure in an emergency. For instance, knowing that an unconscious worker routinely handles cyanides can point the way toward rapid and appropriate treatment if the symptoms seem to indicate CN-exposure.

In addition to determining the hazards that exist at the facility, you should also characterize the employee population in terms of age, race, sex, and individual medical history. An older male population or any other group likely to have a high rate of heart attacks indicates a greater need for CPR (cardio-pulmonary resuscitation) trained individuals in each department on each shift as opposed to the need in a working population composed mainly of young females. Individuals who may have a more frequent or particular need for first aid such as persons with seizure disorders, allergies, cardiac conditions, or insulin dependent diabetes should be identified and this information should be readily available to those responsible for delivering first aid. In some situations, certain individuals may be especially susceptible or have conditions which contraindicate the use of the usual antidote or treatment. For example, persons with deficiency of G-6-PD, an enzyme present in red blood cells, should not be treated with methylene blue, the antidote used to treat methemoglobinemia, a condition caused by exposure to nitrates and amino compounds. Persons with a common medical condition such as hyper-tension may be at increased risk in some exposure situations such as heat stress. Persons with severe chronic lung disease can suddenly stop breathing if given oxygen. This type of information can be vitally important, not only to the first aider, but also to every succeeding level of medical personnel such as the EMT in the ambulance and the physician in the emergency room who must give definitive treatment to the patient.

Finally, you must anticipate what hazards may be generated by an accident, spill, leak, fire, explosion or improper mixing of chemicals. Find out what chemicals may form as a result of mixing chemicals normally kept separate in the workplace. For instance, if alkaline solutions used in tanning leather are mixed with acid, lowering the pH, highly dangerous hydrogen sulfide gas can be released. While hydrogen sulfide is not a raw material or end product of a tannery, the potential for its generation always exists and this must be taken into account when developing the first aid program. Likewise, pyrolysis or combustion products should also receive consideration. When chlorinated hydrocarbon solvents such as trichlorethylene are heated or subjected to ultraviolet light near a metal surface, phosgene, a highly toxic gas, can be formed. Certain types of plastic when burned can generate large quantities of cyanide vapors. These types of situations must be anticipated when designing a first aid program. This information can be gained by reading about the industry in question (see references), by looking back through a plant's accident reports, workman's compensation insurance claims, and by talking to long time employees or workers with technical expertise (chemical engineers, industrial hygienists, etc).

ESTABLISHING POLICIES AND PROCEDURES

As described in the preceeding section, the first step in establishing an effective first aid program at the worksite is to identify the potential physical and chemical hazards and to characterize the population at risk. The program must be viewed in the context of the overall health and safety program at the facility. The next step is to establish policies and procedures that will be followed by those who implement the first aid program. Policies delineate who has authority and responsibility in certain situations and both prohibit and prescribe actions by certain persons. They are especially needed in urgent situations where a conflict between persons at different organizational levels may occur. This is a common situation in health and safety matters where line and staff personnel interact.

In establishing policies, several legal and ethical issues must be considered. States may have different laws governing the level or type of care that can be administered by non-physicians, so it is important to review these state statutes. Find out what, if any, licensing or certifying procedures exist for the health care professional you plan to use, including emergency medical technician, paramedic, licensed practical nurse, nurse, and so forth. Investigate the legal liability and what type of insurance, if any, the company will provide for individuals administering first aid. While many states have so-called good samaritan laws which exempt individuals from suits in certain first aid situations, negligence in the form of improper or incomplete treatment is never exempt. While there is no guarantee, a formally trained person is less likely to make such errors. Whatever level of health care provider will be used, it is advisable to insist that that person either be licensed by the state, or certified by a professional organization, or through a group such as the American Red Cross if applicable. Their certificate or license should be prominently displayed in the first aid room. By requiring up-to-date licensing or certification, one can assure a certain level of training and expertise. Refresher courses, continuing education, or re-certification should be encouraged if not required.

Certain ethical considerations should be articulated in regard to the first-aid program. Medical ethics demands that the provider of health care always consider what is best for the patient and act in that person's best interest. This may mean exercising restraint by limiting treatment or procedures of questionable benefit to the patient. It may mean referring the patient further along in the medical care system in order to provide more sophisticated diagnostic tests or treatment. It may mean keeping the employee away from his job for additional time in order to assure that his condition is stable. At times, decisions like these may be in direct conflict with the immediate desire of the individual's supervisor, who may want the individual to return to work as soon as possible in order to meet production quotas, reduce injury statistics, or hold down medical expenditures. These are classic dilemmas in occupational medicine and are familiar to physicians and nurses who work in this field. However, they may be new and unexpected to first-aiders or emergency medical technicians who may more easily give in to pressure from persons in authority. It is important to articulate a policy stating who is in charge and just what the first-aider's responsibility and authority is in these circumstances. Company policy pertaining to those who provide first aid care should also insure that first-aiders cannot and will not be penalized in any way for the proper and timely performance of their duties of administering first aid care. For instance, an individual who leaves his job to administer cardiopulmonary resuscitation (CPR) must not be docked for failure to meet a production quota. A less obvious example is when the first aider stays with an injured or ill worker in the first aid room until the worker can be driven home or to a hospital.

Another basic policy pertaining to the first aid program should be that first aiders may seek advice from a more knowledgeable health professional if they are at all doubtful about how to proceed. First aiders should be supported in and encouraged to make this decision by insuring that referral to more highly trained medical personnel is readily available and is not systematically discouraged either by concerns about cost or lost time. The first aider should always have access to a nurse or physician as back up. This may involve only a phone call to discuss the situation or actually sending the patient to be seen. The first aider should have the authority to take this action on his or her own. In case of a conflict with production personnel about disposition of a patient, the first-aider should be able to call on this expertise to confirm their decision.

Two other essential policies concern informed consent by the person receiving treatment and confidentiality of the information gathered. Informed consent is recognized as a patient's basic right in medical care. It means that the individual being treated must be told what will be done, why it is recommended, and what the possible outcome will be. The patient must agree to the procedure before anything is done. In the case of someone who is unconscious or rendered mentally incompetent (intoxicated by solvent fumes, for instance) consent is considered to be implied. Some persons may refuse certain treatment on the basis of religious beliefs, fear, distrust, or other reasons. If their refusal creates a potentially dangerous situation in the opinion of the first-aider, this must be explained. If the patient still refuses, he should be referred to a physician or nurse for follow-up. His refusal should be documented in writing for the record.

Confidentiality of medical and personal information is absolutely essential for a successful first aid program and must be assured through a policy statement. Medical information can be revealed only with the prior written consent of the patient, or to another health care provider when a referral is made. When collecting initial information on pre-existing medical conditions, it should be made clear to the individual that this information will be kept on file and will be made available to designated individuals responsible for providing first aid and medical care. If a work-related health problem develops or if an illness results in work restrictions, the supervisor may be informed. Most workers are willing to agree to this type of system as long as you can assure them that no one at the plant, such as fellow workers, management or personnel staff will have casual access to the files. If this cannot be assured, workers may be reluctant to reveal pertinent medical information or to accept treatment by plant personnel. This situation can jeopardize the credibility and effectiveness of the entire first aid program. Therefore, policies regarding confidentiality should be clearly stated, widely publicized and strictly enforced. Each worker should sign a release of information form which explicitly indicates who will have access to the records. Workers always have the right to see their own medical records.

SETTING UP A FIRST AID PROGRAM

The number and type of personnel needed to provide first aid and emergency care must be individualized for each particular facility, taking into account the total number of employees, the shift schedule, particularly if a swing shift schedule is used, and the physical layout and size of the facility. A minimum of one first aid or health care professional should be available on each shift. Where the second or third shift is small, a person with less training may be used. For instance, a licensed practical nurse or an EMT may be employed on the largest shift with the most production while only a first aid trained person is available on second shift. The number and type of injuries at the plant will determine whether or not one needs a full-time person whose duties include first aid and emergency care or whether a first-aid trained individual with a regular job assignment is sufficient. Where communication between buildings or departments is poor or where distances are long, more than one first-aid trained individual is probably needed.

Some situations can be well served by designating an employee with a regular job assignment to also do first aid as needed. A person who was a military corpsman or who has had EMT training may already be employed in an unrelated job. If not, someone should be recruited and trained. When selecting this individual, look for someone with an interest as well as the intelligence and personality to perform the job. A mature individual with a calm, quiet manner who can perform well in an emergency is ideal. While these same qualities certainly apply when hiring a trained health professional such as an EMT or LPN, theoretically, they have already developed many of these characteristics through training and practice in their chosen profession. However, not every health professional is comfortable with or enthusiastic about dealing with emergency or urgent situations. Not everyone is comfortable working alone in a relatively independent situation. These are all areas which should be explored when hiring a professional for the position.

Once the number and type of professional is determined, lines of communication need to be established to handle emergency and routine situations. An extremely critical factor in most emergencies is the time it takes to get the injured or ill person into the hands of the first aider (or health professional). An essential part of setting up a first aid program is to deseminate simple, clear directions for where to go in an emergency and how to contact the first aider. If the first aid room is not staffed by a full-time person, the system is made more complex by the need to contact the first aider whererever he or she may be in the plant. The specific method will vary depending upon whether there is a voice paging system, pagers, radios, or readily available telephones. Everyone in the plant must know how to and be capable of initiating a speedy encounter with the first aider. This may mean simply going to the first aid room, but in the case of a badly injured or unconscious worker or someone who cannot be moved, plant personnel must know how to summon help.

Another part of the emergency communication system involves the manner in which the first aider calls for help from the local emergency medical system. A nearby emergency room and reliable ambulance company should be selected. Proximity and availability are two major concerns when selecting an ambulance company and a hospital emergency room. The qualifications of their personnel are also important. If available, an emergency room staffed by full-time emergency room physicians trained in that specialty is best. Secondary considerations have to do with the availability of other medical specialities which may be needed, such as orthopedics, ophthalmology, and general surgery. The ambulance should have EMTs on board who can communicate with a physician from the ambulance. The ambulance and emergency room staff should be informed about any unusual injuries or illness which might be expected to occur at the facility. If specific antidotes may be required, such as for cyanide, both the ambulance company and the hospital emergency room should be informed so that they can stock the drugs.

The second line of commnication deals with systematic information gathering and reporting. This involves setting up a procedure to ensure that all newly hired employees inform the first aid department about medical conditions they have which might necessitate emergency care or make them specially susceptible to certain exposures. A preplacement health questionnaire can serve this function.

The first aider should make written documentation of every encounter with an employee and keep this in the employee's confidential file. Many incidents will have to be reported to OSHA and some may require a report to the company's workmen compensation insurance carrier if outside medical services are used. In some cases, these reporting functions may be assigned to the personnel department. In-house reports detailing number and type of incident by department, time of day, day of the week, etc. should be compiled on a regular basis and given to management and other health and safety personnel with names removed.

Certain equipment and facilities must be provided. In general, a separate first aid room is needed in all but the very smallest of facilities (less than 25 workers). The first aid room should be clean and quiet and provide visual and voice privacy. It must have hot and cold running water, soap, and clean towels, and preferably be near bathroom facilities. If possible, it should be readily accessible from all areas of the plant and should be well identified with a sign. It must be well lit and have a telephone in addition to a bed, pillow, sheets, blanket, stretcher, and a chair with arms. Ideally, it should be divided by a partition with the bed in a separate room or a screen provided for privacy. There should be hooks or hangers to hang clothing and cabinets to store supplies and equipment.

Supplies must be individualized for the number and type of employees as well as the nature of physical and chemical hazards. "Unit" type first aid kits are available from many drug companies and a selection should be made among these kits. Unit kits are organized to deal with various levels of care and types of problems. For instance, a kit for dealing with a snake bite would have all the supplies needed in such an emergency. Likewise, a kit for lower limb fractures would have an appropriate size splint, elastic wrap and so forth. The American National Standards Institute has published minimum requirements for the contents of these kits and it is recommended that you consult this standard as well as a physician familiar with your facility in order to select from among the many available units. You may also want to consider special antidote kits containing prescription drugs for emergency treatment. For instance, there is a special antidote kit for cyanide exposure and a unit dose treatment kit for allergic reaction to insect stings. These contain pre-measured amounts of a specific antidote or drug antagonist in a ready to administer form. These must be prescribed by a physician and purchased only after consultation with and instruction by a physician familiar with your facility. Needless to say, specific training for persons with access to these kits is an absolute necessity. Many of the medications in the kits age and become useless after a period of time. It is necessary to have a routine for periodically inventorying all equipment and supplies to determine if any are out of date or missing. Access to these special items is always problematic in that limiting access through locked cabinets means risking a situation in which a needed drug is unavailable due to a lost or misplaced key whereas unlimited access risks unavailability due to theft, misplacement or prior use without replacement. Unless the item can be used as a recreational drug or has abuse potential, it is probably better to err on the side of greater access with frequent inventory. Where hazardous materials are handled in various parts of the plant, you may want to provide satellite kits made available to designated workers trained in their use.

TREATMENT PROTOCOLS

Many books have been published describing treatment protocols or detailed instructions for treating common medical emergencies. Some sources are listed in the bibliography and include the American Red Cross's first aid manuals, training manuals for emergency medical technicians, as well as several texts which deal specifically with emergencies caused by exposure to hazardous chemicals. While emergencies vary from plant to plant, those listed in Table 2 are the most common and should be provided for in a first aid program. Many of them such as burns, allergic reactions and cuts can range in

**Table 2. Common Emergencies or Minor
Complaints That a First Aid Program
Should Be Able to Handle.**

Emergencies
  Respiratory and/or Cardiac arrest
  Asthma
  Choking
  Seizure
  Hypoglycemia
  Fainting
  Allergic reaction

  Burns
  Cuts
  Foreign body in eye
  Acid or alkali burns
  Fractures
  Heat stroke
  Frost bite
  Inhalation of irritant or asphyxiant gas
  Bleeding
  Toxic narcosis
  Electrocution

Minor complaints
  Headache
  Upset stomach
  Muscle aches
  Menstrual discomfort
  Upper respiratory complaints
  Fever

severity from minor annoyances to major medical emergencies. A good first aid program should be able to cope with all levels of these problems. The fact that some may be directly caused by exposure in the workplace whereas others may be workplace related only in that they occur there is incidental to providing for them in a first aid program. It is advisable to establish certain treatment protocols which will be followed in each of these cases. Different books, agencies, or organizations publish slightly different protocols for dealing with each. One protocol should be selected, preferably in consultation with a physician. All equipment and supplies should be available for following that protocol and the first aider should be trained to use the selected protocol. To handle certain types of emergencies such as heart attacks or choking, it is advisable to encourage as many employees as possible to undergo cardiopulmonary resuscitation (CPR) training and certification. Such individuals can greatly supplement a first aid program. They should be identified to their fellow workers and management. By virtue of their training, they can be expected to follow a rigid protocol.

In regard to acute poisoning with chemicals, many standard text books deal with this, albeit largely in terms of ingestion which is more likely in the non-workplace setting. In the workplace, inhalation and skin contact are the two major routes by which absorption and intoxication takes place. This not only has implications in terms of estimating the dose that a person has received, but also in identifying the material and interrupting continued exposure. In general, the generic steps in dealing with exposure to a hazardous chemical are the following:

1. Recognize that a potentially dangerous exposure has occurred and that a problem exists. The problem may be urgent or non-urgent depending upon the material and the dose. The initial condition of the exposed person is not necessarily a good indication of how urgent the problem may be. Many inhalation exposures take up to 12 hours to produce their maximum effect, sometimes resulting ultimately in pulmonary edema and death.
2. The second, third and fourth steps ought to occur almost simultaneously. They involve first of all calling for help. This means help in containing the chemical as well as summoning or activating the emergency medical system.
3. Alert other workers to the hazard. This prevents exposure of other individuals.
4. Remove the exposed individual(s) from the exposure. It is absolutely essential that any rescuer first take measures to protect himself before attempting rescue. In a well designed health and safety program, appropriate respiratory protection will be readily available and rescuers will be trained in its use. There are many examples of cases involving multiple deaths because successive rescuers did not take proper precautions and died in the attempt.
5. Once the individual is removed from exposure, the basic tenents of first aid apply. The critical factor is to establish an airway and support respiration and cardiac function.
6. It is important at this point to interrupt further exposure by removing contaminated clothing and flushing the skin with clean water if possible. Remember that hair and shoes may be contaminated also. Take care in handling contaminated clothing.
7. The next step is to identify the agent. This may mean consulting the Material Safety Data Sheets, other workers in the area, or technical personnel.
8. If there is a specific antidote or antagonist available, it should be administered immediately if available and if the first aider in attendance is familiar with its proper use. Certain poisonous gases like cyanide and hydrogen sulfide do not respond to treatment with oxygen alone. If either is potentially present in the workplace, provision should made for stocking an appropriate antidote. Likewise, specific antidotes exist for organophosphate and carbamate pesticides. These may be life-saving when administered shortly after exposure.
9. It is important to have the exposed individual examined by a physician before allowing them to go home or return to work. Many chemicals have a latency period of several hours before their effects begin to show. Emergency room personnel should be informed of the suspected or actual agent of exposure and should have been briefed previously on its typical pattern of effect.

A well run first aid program is a source of satisfaction to those who participate in it and reassuring to its potential customers. It can create good will and improve morale by its very presence. A poorly run program can have just as strong a negative effect so it is very important to do it right. This section has dealt with some of the basics in a first aid program. The following section will tell you where and how you can get extra help and link up with other health service organizations.

INFORMATION FOR A FIRST AID PROGRAM

If you are responsible for developing a first aid program in a chemical or other type of manufacturing setting, there is some essential information that you need. This chapter has already suggested that following the guidelines established by the American Red Cross is a good place to start. However, when implementing a first aid program in an industrial setting, the information you need must directly address the potential hazards created by the substances in use in that facility. Individuals involved in implementing the first aid plan need to know an appropriate level of information about possible health hazards of chemicals used in the operations carried out at the facilities involved. As has been stated before, the ultimate objective is to prevent accidents. However, knowing the possible hazards of the substances and processes involved in your facility, as well as the requirements for handling accidents and emergency situations, is essential.

First of all, you should be aware of any laws and regulations that require you to take specific action in relationship to first aid and emergency preparedness. The most important one to recognize is the OSHA requirement 29 CFR Subpart K, 1910.15. In this regulation, OSHA requires that all facilities identify and train an individual to be responsible for first aid if there is no "infirmary, clinic or hospital in near proximity." Additionally, you may be required by your city or county, as well as the state and the federal government, to provide information to authorities involved in hazardous materials planning programs. The requirements of these laws will be described in full a little later.

Another government required program that parallels some of the requirements for establishing first aid programs are the right-to-know rules and laws. These laws were mentioned earlier in this chapter as a source of information. It is most likely that you have already established a hazard communication or employee right-to-know program at your facility. This existing program requires some of the basic information that you need to develop for your first aid and emergency preparedness plan. There are three parts of the right-to-know program that overlap with information needed for your industrial first aid program.

1. *Chemical Inventory.* You must have a complete inventory of all chemicals in your facility. This information must be evaluated to determine the potential health hazards involved. For example, there may be chemicals in your facility that represent a fire or explosion hazard, an inhalation hazard and/or may be corrosive. Without this information, it is impossible to determine what kind of first aid program is required and what kind of training your personnel should have that can prepare them to react appropriately to an accident that may occur as a result of your chemical inventory.

2. *Material Safety Data Sheets.* If you are a user of chemicals in your facility, you should have received and will have maintained material safety data sheets on all your chemicals as is required by the OSHA Hazard Communication Standard. These material safety data sheets are the documents used to transfer information about the hazards associated with the chemicals appropriate use, as well as the proper control procedures to use if accidents occur. It will be necessary to review these MSDSs on a regular basis to make sure that the first aid precautions that are provided on these documents do not change and that, if they do, appropriate notification is made. The technical accuracy of MSDSs vary widely. Each MSDS should be reviewed for completeness. If the information is either incomplete or inaccurate you can always contact the manufacturer directly.

If your company manufactures chemicals, it is required to develop and produce material safety data sheets to be supplied to the purchasers of your products. If this is the case, information about the hazards associated with the chemicals in your facility should be readily available. You can base your first aid program on the information available to you from internal sources, such as the toxicology and/or medical department, product liability department, or the manufacturing department which is charged with the responsibility for MSDS development. If you are a smaller company you may need to develop an appropriate program for writing MSDSs. Working with a consultant may be necessary in order to prepare appropriate MSDSs.

3. *Labels.* Under the Hazard Communication Standard, all products that enter the workplace must be labeled as well. These labels may or may not have first aid information on them since it is not required by law; however, the label must contain information about the product's name so that you can easily refer back to the material safety data sheet to find out information about its proper use. Having a proper label on a container is essential. Through the products' name, the label provides the link to additional information on the MSDS.

When establishing a first aid program, it is essential that the person in charge of this effort be in communication with other people involved in the safety and health prevention activities at the company. It is possible that in many small facilities one person is in charge of hazard communication, industrial hygiene, safety and injury prevention, first aid, and other areas of preventive health care. If this is the case, then it is essential that this individual understand and implement a full scope of hazard control programs, of which first aid is only one.

Another regulatory requirement you must address if you are a manufacturer of chemicals rather than just a consumer of chemicals is that it is your responsibility to provide first aid advice on the material safety data sheets you ship with your products. You are also responsible for providing first aid and emergency procedures advice when the chemical is shipped from your facility. There is not enough space here to delve into the detail of all the requirements for the transportation of chemicals and the hazardous materials prevention programs that are associated with it; however, you should be aware that you are responsible for supplying information about safety and health aspects of your products, including first aid and emergency preparedness, as they get into the marketplace. Transportation safety should not be ignored.

In addition to the documentation existing on Material Safety Data Sheets regarding the hazards associated with chemicals and the first aid procedures to be used in the case of overexposure, there are other avenues of possible information that may be useful to you. The Chemical Manufacturer's Association has established several telephone hot lines which can be used by anyone to gather information about chemicals on an emergency basis. A list of Resources at the end of this chapter provides the telephone numbers. It is important to note, however, that you must be aware of the name of the chemical or chemicals involved in an emergency in order to get this information. This is another reason why labeling all containers in the workplace is important.

Since the main objective of any health and safety program is to avoid as much as possible the need for first aid, it is prudent to review material safety data sheets and labels prior to any accident occurring. If this review reveals deficiencies in information on the material safety data sheets and labels, before an accident occurs is the time to contact poison control centers or the Chemical Manufacturers Association hot line to fill in deficient information. As extra protection, it is important that among the various postings and emergency information available to you, that your local Poison Control Center as well as the Chemical Manufacturers Association hot line be easily accessible to you in time of emergency.

When an accident occurs, it is unlikely that you will have the time or ability to refer to written materials available to you. However, having a small library of basic first aid books is a necessity. These can be used as references on a day to day basis, as text books and refresher information when needed. You also should consider subscribing to a magazine or periodical that can keep you informed on a monthly basis about state-of-the-art first aid information.

Another source of information to consider is on-line computer data bases that may provide needed information for you. This type of resource could be most useful to you if you were involved in developing first aid information for MSDSs or labels. A list of books, journals, and computer data bases appears at the end of this chapter for your convenience.

OUTSIDE RESOURCES FOR YOUR FIRST AID PROGRAM

Depending upon the size of your operation you may or may not have to rely on outside resources in the case of an emergency. If you are a large chemical manufacturing facility, it is likely that you have several nurses and a physician on staff or, at least, on call at all times. You may also have an ambulance available at your facility to transport employees to your own medical facility on your own premises. However, if you are a smaller company, you may not even have a registered nurse on duty 24 hours a day. In this case, you will definitely need to make arrangements with outside resources that can provide you services you will need in case of an emergency. As has been discussed previously, the level and amount of first aid care that is provided will depend upon the level of expertise of the people providing those services at your facility. In some situations, even simple cuts and bruises may be referred outside your facility while at other places a simple cut or bruise will be handled by a first aider using a first aid kit at the site of the accident. You will determine the kinds of outside resources you will need depending upon the first aid plan that you have established.

Even if your company has its own vehicles to transport an injured employee to a local facility, it is recommended that all firms develop relationships with an ambulance service or a "911" emergency facility that is connected to the police or fire department. In case of a major accident where on-site first aid alone is not adequate, it will be necessary to use other resources to move the disabled to hospitals, even if you have your own vehicles. Relationships with these ambulance services should be maintained on a regular basis with regular notification and communication with them, even if the resources are not used. It would be, at best, embarrassing, at worst, life threatening to call an ambulance service after 6 months to a year of not being in contact with them to find out that they no longer existed. If this were to occur at the time of an emergency, you would be hard pressed to find a substitute on the spot.

There are institutions that can provide backup when the first aid matters cannot be handled completely at your facility. Hospital emergency rooms and urgent care centers throughout the country are becoming more aware of the need to be prepared for occupational accidents. It is possible to shop around in your area to find the hospital close enough to you that may have facilities for emergency room admissions as well as pre-employment physicals at the best price. Proximity and price, however, are not always the only criteria to consider when choosing to make arrangements for the hospital emergency room or an urgent care center. It might make more sense in the case of a chemical manufacturing facility that works with or manufactures some acutely hazardous chemicals to go an additional mile or two and pay some additional dollars to be able to have occupational medicine expertise. Though in the short run your routine emergency room visits may be slightly more expensive, in the long run, the increased cost may allow the institution to employ a more skilled staff who can help you handle future problems.

When relationships are developed between hospital emergency rooms or the urgent care center in your area, it is important that the institution be supplied with information about the potential hazards of your facility. The best arrangement is one where the hospital facility is willing and able to accept and store a set of your material safety data sheets.

Maintaining an ongoing relationship with the hospital and its doctors is also important. For example, it is worthwhile to consider having the local emergency room physicians come and tour your plant to see what type of facility it is and better understand the possible accidents that may occur. It is important to remember that you want to make sure the doctors are aware of the potential hazards at your facility so they can be prepared to treat any injury.

Community right-to-know laws, as corollaries to worker right-to-know laws or employee right-to-know laws, are being enacted at the municipal, township, county, state, and federal levels. It is now quite common for industrial operations, particularly those involved in manufacturing or using chemicals, to be compelled by law to share information with government and quasi-government groups within their communities about the quantities and types of chemicals used and the preventive as well as emergency procedures developed. This information sharing process is designed to make communities safer and provides a method for communities to gather information and therefore better control the kind of chemical and other manufacturing operations that go on in their communities.

These laws are relevant to the first aid program because emergency preparedness is a central focus of this legislative initiative. These laws and regulations mandate that industry provide information about the hazards that exist at the workplace to a variety of community bodies including the fire department, local police, health departments, and emergency planning agencies. Some of this information may go beyond what would be considered part of a first aid program at an industrial facility; however, it is important to recognize that the first aid program is one element of an entire health and safety program that should be designed to control chemical and physical hazards from the time raw materials and formulated chemicals enter the door of a plant to the time they leave and reach their destinations in whatever form via the transportation system.

Cooperating with these emergency planning agencies as required by law will help you establish and codify your first aid policies and procedures. These coordination activities should be seen as fulfilling two different goals. First, you have emergency care needs which have been previously discussed, and your needs should be shared with the surrounding providers. If there is an accident or injury in your plant and your first aid facilities are incapable of handling it, you will want to be able to have access to outside resources to support you in managing an accident. The second important goal of these programs has to do with the long-term planning needs. By sharing information through the emergency planning agencies as defined by community right-to-know laws, you will have a better understanding of total emergency preparedness for your community. Expansion plans at your facility may require additional first aid and emergency preparedness resources. These activities may need to be coordinated with your community plans. For the long term, it is also sensible to use these regulatory initiatives to get you in touch with others in your community regarding future plans and needs regarding outside support for emergency response. Development of emergency response networks and teams may be advantageous to you and may, in the long run, cut costs as a result of sharing resources.

In summary, it is important to understand that, when considering first aid as a management responsibility in your facility, its scope goes beyond the band-aid and ace bandage that we think of as traditional first aid. In fact, first aid, in a sense, runs the gamut from injury prevention through to disaster planning. Those people involved in the first aid process must be aware and educated about possible implications of first aid intervention and accident occurrence. The training and information needs of first aiders and others involved in your facility who are potentially involved in accident situations are comprehensive and must be maintained and upgraded on a regular basis. The remaining section of this chapter will talk about these training and information needs.

TRAINING AND INFORMATION NEEDS

Establishing a first aid program requires a substantial amount of sharing of information within your organization. First of all, everybody must know something about how to handle a first aid emergency. Every employee at your facility must know that there is a first aid plan, and where to report an accident or injury. In order to insure that everyone at the facility is aware of these procedures, information about them must be part of the new employee orientation. This information should be reinforced using signs and placards that describe the procedures as well as verbal reinforcement at safety meetings.

The first person to receive information about an injury or an accident, often a supervisor, needs to know who to contact to treat the person involved if he or she is not a first aider. If it is only a minor injury, it may go no further than the supervisor or a first line first aider administering some simple first aid and sending the person back to work. Besides learning the basic first aid skills, the person providing the treatment must be familiar with the plant first aid plan and his particular role and responsibility in it. He or she must be aware of the steps to take if the accident or injury is more severe than the person can handle.

If the accident or injury is more severe than a simple cut, a whole series of events is set into motion. Even though it is the desire of all involved to focus on prevention of accidents and injuries, in the worst case situation, emergencies require that a well-trained staff be available to execute well-planned disaster control activities. Acquiring this level of expertise in the first aid area requires more than simple information exchange. Attending courses, refresher seminars, performing practice drills and reading journals regularly is required to ensure that state-of-the-art knowledge is maintained.

The more likely the possibility of an accident, for example if there are explosive chemicals in the work environment, the more necessary it is that there be formal plans and identified individuals who are responsible for planning and reviewing accidents and emergency procedures. The more complex the potential hazards and the related emergency preparations, the more training is required to prepare someone to handle the situation.

If serious hazards exist in larger operations, the facility is likely to have more staff trained and capable of providing first aid and emergency care. In the smaller facility, the most likely situation is that there will be no nurse or physician on staff. It is then essential that the company pay to have someone become an American Red Cross trained first aider. Since this is the minimum required by OSHA, every plant as stated before, should have such an individual identified and trained. If there is only one or two trained first aiders in the facility, relationships with outside organizations, such as ambulance services, fire departments and emergency medical technicians, must be very strong.

In a small facility, an additional type of trained individual to consider would be a person trained in CPR, or cardiopulmonary resuscitation. A CPR trained individual can help when there is a need to assist someone who is suffering from some kind of cardiopulmonary event. There may be some individuals who have on their own received CPR training who are working at your plant. You should be aware of them and include them in your first aid planning. However, a CPR trained individual is not a substitute for first aid training by the American Red Cross.

In larger facilities, the likelihood is that there is a nurse or physician's assistant on the premises. They can get American Red Cross instructor certificates with some additional training. This allows them to then train others in the facility. If your facility is large enough to employ a nurse, it is important that the first aiders who the nurse trains be aware of the special problems of industrial first aid, specifically the special hazards existing as a result of the workplace. This is also necessary for the lone first aider who is not supported by a nurse at the facility. This information about special industrial hazards of your plant may not be easily accessible for the lone first aider. Classes or conferences may be available in larger cities. Trade associations may offer information. Individuals may be sent to specific seminars. Also, this information may be available in a book about industrial hazards such as this one. However, awareness of the special hazards in the work environment are essential for the first aider and should go beyond the rudimentary training given all employees about the hazards in the workplace under worker hazard communication, right-to-know laws and rules.

In the larger facilities where there are possibly several nurses and a medical staff as well as safety professionals, industrial hygienists and others involved in accident and injury prevention strategies, it is important that at the time of a disastrous situation that they be aware of each other's presence and responsibility. The training program must include this information. Regular meetings of emergency response teams may be appropriate in these situations.

In order to insure that people respond appropriately in actual practice, whether they be a potential accident victim or trained first aider, it is essential that they be trained and retrained on a regular basis and that reinforcement of this training also occur regularly. For the most responsible parties involved, such as members of the first aid team or those responsible for CPR and accidents, monthly if not biweekly refresher information should be available. This may mean that the lone first aider reads a magazine or attends a seminar. In situations where several people are involved with first aid, more formal interaction should take place. This can be part of existing safety meetings, but sometimes meetings alone are inadequate. If respirators are required for confined space entry as part of a potential first aid response, those respirators must be checked on a regular basis and those involved must be confident that the respirators are functioning appropriately. Fire drills or accident simulation exercises may be a good way to see if your staff function properly.

We all understand that the safety process is one that must be continuous. Though formal monthly sessions for people with active responsibilities in case of an accident or injury may be appropriate, it is the whole atmosphere in the plant regarding these events that is important. In general, employees should be aware, on a daily basis, of first aid and emergency preparedness plans. This may mean posting of signs in local areas, giving emergency telephone numbers or instructions, and installing safety warning systems. Continuous information flow can make people aware of hazards and prevent accidents. Without this kind of continual activity, the likelihood is that there will be a low level of cooperation and preparedness at the time of an event when it is most needed. Therefore, a well-planned program will include regular training and reinforcement procedures that are carried out on a regular, consistent, and organization-wide basis.

SOURCES OF FIRST AID INFORMATION

A list of basic reference information about first aid follows. However, if you are responsible for first aid in your plant, it is important that you keep up with state-of-the-art current knowledge about these issues. This can be done through subscribing to magazines, some of which are listed below, and getting involved in professional organizations that will keep you informed of new developments in these areas. In addition, through community right-to-know initiatives, industrial networks should be developed so that you can share information with your colleagues at other facilities in your location and make sure that your program meets the standards developed for your particular locality.

In closing, it is necessary to point out once again that the key and primary objective of a first aid program is that it fit in and be a part of an entire company-wide health and safety prevention program. The idea is to prevent accidents and injuries as much as possible; however, a prevention program must acknowledge that accidents do happen. Because of this the best way to minimize the impact of such accidents and injuries is to have a well prepared first aid program that can prevent minor accidents from becoming major injuries in your facility.

BIBLIOGRAPHY

Books

American Red Cross, *Advanced First Aid and Emergency Care*, New York, Doubleday & Company, Inc., Garden City, 1979.

American Red Cross, *Standard First Aid and Personal Safety*, Doubleday & Company, Inc., Garden City, New York, 1979.

Campbell, John E., *Basic Trauma Life Support*, Brady Communications Company, Inc., Bowie, MD 20715, 1985.

Caroline, Nancy L., *Emergency Medical Treatment*, Little, Brown and Company, Boston, 1982.

Lefevre, Mark J., *First Aid Manual for Chemical Accidents*, Van Nostrand Reinhold Company, New York, 1980.

Sax, N. Irving, *Dangerous Properties of Industrial Materials*, 6th edition, Van Nostrand Reinhold Company, New York, 1984.

Windholtz, Martha, *The Merck Index*, Merck and Company Inc., Rahway, New Jersey, 1983.

Magazines

National Safety and Health News, National Safety Council, 444 N. Michigan Avenue, Chicago, IL 60611.

Industrial Fire World, P.O. Box 9161, College Station, TX 77840.

Dangerous Properties of Industrial Materials Report, Van Nostrand Reinhold Company, New York, 1980–.

Telephone Hot Lines

CHEM-TREC Chemical Transportation Emergency Center, Chemical Manufacturers Association
1-800-424-9300
Information about spills, leaks, or fires at the time of the emergency is provided here.

CRC—Chemical Referral Center, Chemical Manufactures Association
  1-800-CMA-8200
  1-202-887-1315 Collect, in Washington, DC and Alaska
  Routine information and referrals about chemicals.

Computer Data Bases

RTECS, US Department of Health and Human Services, Public Health Service, Centers for Disease Control,
  National Institute for Occupational Safety and Health, Cincinnati, Ohio 45226
TOXNET, MEDLARS, National Library of Medicine, Bethesda, Md. 20894

Audio-Visual Teaching Aids

National Fire Protection Association, Battery March Park, Quincy, Massachusetts 02269, 1-800-344-3555
  Films and videos for teaching
Film Communications, 11136 Neddington Street, North Hollywood, CA 91601
  First aid audio-visual materials

Reference Works

American National Standard Institute, "Minimum Requirements for Industrial Limit-Type First Aid Kits"
  (ANSIZ308.1), 1430 Broadway, NY, NY, 10018, 1978.
Parmeggiani, L. *Encyclopedia of Occupational Health and Safety,* International Labor Organization, Geneva,
  Switzerland, 1983.

SECTION II
General Chemicals

KEY TO ABBREVIATIONS

abs - absolute
ACGIH - American Conference of Governmental Industrial Hygienists
alc - alcohol
alk - alkaline
amorph - amorphous
anhy - anhydrous
approx - approximately
aq - aqueous
atm - atmosphere
autoign - autoignition
aw - atomic weight
af - atomic formula
bp - boiling point
b range - boiling range
CAS - Chemical Abstracts Service
cc - cubic centimeter
CC - closed cup
CL - ceiling limit
COC - Cleveland open cup
compd(s) - compound(s)
conc - concentration, concentrated
contg - containing
cryst - crystal(s), crystalline
d - density
D - day(s)
decomp, decomposes, decomposition
deliq - deliquescent
dil - dilute
DOT - U.S. Department of Transportation
EPA - U.S. Environmental Protection Agency
eth - ether
expl - explosive, explosion
(F) - Fahrenheit
flash p - flash point
flam - flammable
fp - freezing point
g, gm - gram
glac - glacial
gran - granular, granules
hygr - hygroscopic
H, hr - hour(s)
HR: - hazard rating
htd - heated
htg - heating
IARC - International Agency for Research on Cancer
incomp - incompatible
insol - insoluble
irr - irritant, irritating
kg - kilogram (one thousand grams)
L,l - liter

lel - lower explosive level
liq - liquid
M - minute(s)
m^3 - cubic meter
mg - milligram
misc - miscible
μ, u - micron
mf - molecular formula
mL, ml - milliliter
mg - milligrams
mm - millimeter
mod - moderate, moderately
mp - melting point
mppcf - million particles per cubic foot
mw - molecular weight
mumem - mucous membrane
mw - molecular weight
NIOSH - National Institute for Occupational Safety and Health
ng - nanogram
nonflam - nonflammable
NTP - National Toxicology Program
OC - open cup
org - organic
OSHA - Occupational Safety and Health Administration
PEL - permissible exposure level
petr - petroleum
pg - picogram (one trillionth of a gram)
Pk - peak concentration
pmole - picomole
powd - powder
ppb - parts per billion (v/v)
pph - parts per hundred (v/v)(percent)
ppm - parts per million (v/v)
ppt - parts per trillion (v/v)
prep - preparation
PROP - properties
refr - refractive
rhomb - rhombic
S,sec - second(s)
slt, sltly - slight, slightly
sol - soluble
soln - solution
solv(s) - solvent(s)
spont - spontaneous(ly)
subl - sublimes
TCC - Tag closed cup
tech - technical
temp - temperature
THR - toxic and hazard review

TLV - Threshold Limit Value
TOC - Tag open cup
TWA - time weighted average
U, unk - unknown, unreported
μ - micron
uel - upper explosive limits
μg - microgram
ULC, ulc - Underwriters Laboratory Classification
vac - vacuum
vap - vapor
vap d - vapor density
vap press - vapor pressure

visc - viscosity
vol - volume
W - week(s)
Y - year(s)
% - percent(age)
> - is greater than
< - is less than
$\geqslant$ - is greater than or equal to
$\leqslant$ - is less than or equal to
° - degrees of temperature in Celsius (Centigrade)
°(F),°F - temperature in degrees Fahrenheit

A

A. ANTRACTICUS (AUSTRALIA) VENOM   HR: 3
NIOSH: YX 4178200

SYNS: VENOM, AUSTRALIAN ELAPIDAE SNAKE, ACANTHOPHIS ANTARCTICUS * ACANTHOPHIS ANTARCTICUS (AUSTRALIA) VENOM * AUSTRALIAN DEATH ADDER SNAKE VENOM

THR: Poison via subcutaneous or intraperitoneal routes.

ABATE   HR: 3
CAS: 3383-96-8   NIOSH: TF 6890000
mf: $C_{16}H_{20}O_6P_2S_3$ mw: 466.48

PROP: White crystals. mp: 30°.

SYNS: O,O-DIMETHYL PHOSPHOROTHIOATE-O,O-DIESTER WITH 4,4'-THIODIPHENOL * ENT 27, 165 * O,O,O',O'-TETRAMETHYL-O,O'-THIODI-P-PHENYLENE PHOSPHOROTHIOATE * O,O'-(THIODI-4,1-PHENYLENE)BIS(O,O-DIMETHYL PHOSPHOROTHIOATE) * TEMOPHOS

THR: Poison by ingestion. Moderately toxic via skin contact. A cholinesterase inhibitor type of insecticide. See also parathion. When heated to decomposition, it emits toxic PO_x and SO_x fumes.

ABIES ALBA OIL   HR: 2
CAS: 8021-27-0   NIOSH: AA 4350000

PROP: The colorless to pale-yellow oil from the steam distillation of the crushed cones of *Abies Alba Mill.*

SYNS: OIL OF ABIES ALBA * OIL OF FUR * OIL OF SILVER FIR * OIL OF SILVER PINE * SILVER FIR NEEDLE OIL * SILVER FIR OIL * SILVER PINE OIL * TEMPLIN OIL

THR: Moderately toxic via skin contact. When heated to decomposition, it emits acrid smoke and irritating fumes.

ABRIN   HR: 3
PROP: A toxalbumin obtained from the seeds of jequirity, *Abrus precatorius L., Leguminosae*. White powder.

SYNS: ABRINS * AGGLUTININ * CRAB'S EYES * INDIAN LICORICE SEED * JUMBLE BEAD * PRAYER BEAD * TOXALBUMIN

THR: Poison to humans by ingestion. Poison experimentally by intraperitoneal and intravenous routes. See also ricin. When heated to decomposition, it emits acrid fumes and irritating smoke.

ABSINTHIUM   HR: 2
PROP: Dried leaves and flowering tops of *Artemisia absinthium L.*

SYNS: WORMWOOD * THUJONE

THR: Moderately toxic by ingestion. Habitual users develop "absinthism" with tremors, vertigo, vomiting, and hallucinations. An allergen. May cause dermatitis. See also beta-thujone.

ACACIA GUM   HR: 1
mw: 240,000, d: 1.35-1.49.

SYNS: GUM ARABIC * SPHERES OR "TEARS"

THR: A weak allergen via skin contact, inhalation, or ingestion. Slight fire hazard. For further information, see Vol. 1, No. 3 of *DPIM Report.*

ACECLIDINE   HR: 3
CAS: 827-61-2   NIOSH: VD 6195000
mf: $C_9H_{15}NO_2$ mw: 169.25

SYNS: 3-ACETOXYQUINUCLIDINE GLAUCOSTAT * 3-QUINUCLIDINOL ACETATE

THR: Poison by ingestion, subcutaneous, and intravenous routes. When heated to decomposition, it emits toxic fumes of NO_x.

ACENAPHTHANTHRACENE   HR: 3
CAS: 5779-79-3   NIOSH: CU 1575000
mf: $C_{20}H_{14}$ mw: 254.34

SYNS: BENZ(K)ACEPHENANTHRENE * 4,5-DIHYDROBENZ(K)ACEPHENANTHRYLENE * 3:4-DIMETHYLENE-1:2-BENZANTHRACENE

THR: Possible carcinogen. Poison by intravenous route. When heated to decomposition, it emits acrid smoke and irritating fumes.

ACEPROMAZINE MALEATE   HR: 3
CAS: 3598-37-6   NIOSH: OB 2450000
mf: $C_{19}H_{22}N_2OS \cdot C_4H_4O_4$ mw: 442.57

SYNS: ACETYLPROMAZINE MALEATE (1:1) * MALEATE ACIDE DE L'ACETYL-3-DIMETHYLAMINO-3-PROPYL-10-PHENOTHIAZINE (FRENCH)

THR: Poison by ingestion, subcutaneous, and intravenous routes. When heated to decomposition, it emits highly toxic fumes of NO_x and SO_x.

ACETAL HR: 2
CAS: 105-57-7 NIOSH: AB 2800000
DOT: 1088
mf: $C_6H_{14}O_2$ mw: 118.20

PROP: Colorless, volatile liquid, agreeable odor, nutty after-taste. bp: 102.7°, flash p: −5°F (CC), lel = 1.65%, uel = 10.4%, d: 0.831; autoign temp: 446°F, vap press: 10 mm @ 8.0°, vap d: 4.08, mp: −100°. Sl sol in water, misc in alc and ether.

SYNS: ACETAAL (DUTCH) * ACETAL DIETHYLIQUE (FRENCH) * ACETALE (ITALIAN) * 1,1-DIAETHOXY-AETHAN (GERMAN) * DIAETHYLACETAL (GERMAN) * 1,1-DIETHOXY-ETHAAN (DUTCH) * 1,1-DIETHOXYETHANE * DIETHYL ACETAL * 1,1-DIETOSSIETANO (ITALIAN) * ETHYLIDENE DIETHYL ETHER * USAF DO-45

DOT Classification: Label: Flammable Liquid

THR: Moderately toxic by ingestion or intraperitoneal routes. A skin and eye irritant. A narcotic. Dangerous fire hazard when exposed to heat or flame; can react vigorously with oxidizing materials.

ACETALDEHYDE HR: 2
CAS: 75-07-0 NIOSH: AB 1925000
DOT: 1089
mf: C_2H_4O mw: 44.06

PROP: Colorless, fuming liquid; pungent, fruity odor; mp−123.5°; bp 20.8°; lel = 4.0%; uel = 57%; flash p−36°F (CC); d 0.7827 @ 20°/20°; autoign temp 347°F; vap d 1.52. Misc in water, alcohol, and ether.

SYNS: ACETALDEHYD (GERMAN) * ACETIC ALDEHYDE * ALDEHYDE ACETIQUE (FRENCH) * ALDEIDE ACETICA (ITALIAN) * ETHANAL * ETHYL ALDEHYDE * NCI-C56326 * OCTOWY ALDEHYD (POLISH)
ACGIH TLV: TWA 100 ppm; STEL 150 ppm
OSHA PEL: TWA 200 ppm
DFG MAK: 50 ppm (90 mg/m^3)
DOT Classification: Label: Flammable Liquid

THR: Poison by intratracheal and intravenous routes. Mutagenic data, an experimental teratogen. A human systemic irritant; a narcotic. A common air contaminant. Dangerous when exposed to heat or flame; can react vigorously with acid anhydrides, alcohols, ketones, phenols, NH_3, HCN, H_2S, halogens, P, isocyanates and strong alkalies, amines. For further information, see Vol. 3, No. 3 of *DPIM Report*.

ACETALDEHYDE-N-METHYL-N-FORMYLHYDRAZONE HR: 3
CAS: 16568-02-8 NIOSH: LQ 8500000
mf: $C_4H_8N_2O$ mw: 100.14

SYNS: ETHYLIDENE GYROMITRIN * N-METHYL-N-FORMYL HYDRAZONE OF ACETALDEHYDE

THR: Experimental teratogen. Poison via ingestion and possibly other routes. When heated to decomposition, it emits toxic fumes of NO_x.

ACETALDEHYDE OXIME HR: 3
CAS: 107-29-9 NIOSH: AB 2975000
mf: C_2H_5NO mw: 59.08

PROP: Cryst material; sol in water, alc, ether; mp: (α) 46.5°; mp (β) 12°; d: 0.966; bp: 114.5°. flash p: ≤ 72°F

SYNS: ACETALDOXIME * ALDOXIME * ETHANAL OXIME * ETHYLIDENEHYDROXYLAMINE * USAF AM-5

THR: Poison via intraperitoneal route. When heated to decomposition, it emits toxic fumes of NO_x.

ACETALDOL HR: 3
CAS: 107-89-1 NIOSH: ES 3150000
mf: $C_4H_8O_2$ mw: 88.12

PROP: Clear, white-to-yellow syrupy liquid. bp: 83° @ 20 mm, flash p: 150°F (OC), d: 1.11, autoign temp: 482°F, vap d: 3.04.

SYNS: ALDOL * 3-HYDROXYBUTANAL * BETA-HYDROXYBUTYRALDEHYDE * 3-HYDROXYBUTYRALDEHYDE * OXYBUTYRIC ALDEHYDE

THR: Poison via skin contact. A skin and eye irritant. Moderate fire hazard when exposed to heat or flame; emits crotonaldehyde and water when heated. See crotonaldehyde. Can react with oxidizing materials.

ACETAMIDE HR: 3
CAS: 60-35-5 NIOSH: AB 4025000
mf: C_2H_5NO mw: 59.08

PROP: Colorless cryst, mousey odor; mp: 81°; bp: 221.2°; d: 1.159 @ 20°/4°; vap press: 1 mm @ 65°; decomp in hot water.

SYNS: ACETIC ACID AMIDE; ACETIMIDIC ACID * ETHANAMIDE * METHANECARBOXAMIDE * NCI-C02108

THR: Experimental carcinogen. When heated to decomposition, it emits toxic fumes of NO_x. For further information see Vol. 3, No. 6 of *DPIM Report.*

5-ACETAMIDE-1,3,4-THIADIAZOLE-2-SULFONAMIDE   HR: 3
CAS: 59-66-5   NIOSH: AC 8225000
mf: $C_4H_6N_4O_3S_2$ mw: 222.26

SYNS: 2-ACETAMIDO-5-SULFONAMIDO-1,3,4-THIADIAZOLE * ACETAMIDOTHIADIAZOLESULFONAMIDE * ACETAMOX * ACETAZOLAMID * ACETAZOLAMIDE * ACETAZOLEAMIDE * ACETOZALAMIDE * CARBONIC ANHYDRASE INHIBITOR NO. 6063

THR: Experimental teratogen. Moderate acute toxicity by various routes. When heated to decomposition, it emits very toxic fumes of NO_x and SO_x.

5-ACETAMIDO-3-(5-NITRO-2-FURYL)-6H-1,2,4-OXADIAZINE   HR: 3
CAS: 24143-08-6   NIOSH: AC 6492000
mf: $C_9H_8N_4O_5$ mw: 252.21

SYN: N-(3-(5-NITRO-2-FURYL)-6H-1,2,4-OXA-DIAZINYL)ACETAMIDE

THR: Experimental carcinogen. When heated to decomposition, it emits toxic fumes of NO_x.

2-ACETAMIDO-4-(5-NITRO-2-FURYL)THIAZOLE   HR: 3
CAS: 531-82-8   NIOSH: AC 6650000
mf: $C_9H_7N_3O_4S$ mw: 253.25

SYNS: 2-ACETAMINO-4-(5-NITRO-2-FURYL)THIAZOLE * 2-ACETYLAMINO-4-(5-NITRO-2-FURYL)THIAZOLE * N-(4-(5-NITRO-2-FURANYL)-2-THIAZOLYL)ACETAMIDE * N-(4-(5-NITRO-2-FURYL)-2-THIAZOLYL)ACETAMIDE * N-(4-(5-NITRO-2-FURYL)THIAZOL-2-YL)ACETAMIDE

THR: Experimental carcinogen. When heated to decomposition, it emits very toxic fumes of SO_x and NO_x.

2-ACETAMIDOPHENATHRENE   HR: 3
CAS: 4120-77-8   NIOSH: AC 7175000
mf: $C_{16}H_{13}NO$ mw: 235.30

SYNS: 2-ACETAMINOPHENANTHRENE * 2-ACE-TYLAMINOPHENANTHRENE * 2-PHENANTHRYL-ACETAMIDE * N-2-PHENANTHRYLACETAMIDE * N-(2-PHENANTHRYL)ACETAMIDE

THR: Experimental carcinogen and tumorigen. When heated to decomposition, it emits toxic fumes of NO_x.

ACETAMINE YELLOW CG   HR: 3
CAS: 2832-40-8   NIOSH: AC 3662000
mf: $C_{15}H_{15}N_3O_2$ mw: 269.33

SYNS: ACTIOQUINONE LIGHT YELLOW * C.I. DISPERSE YELLOW 3 * N-(4-((2-HYDROXY-5-METHYLPHENYL)AZO)PHENYL)ACETAMIDE * 4'-((6-HYDROXY-M-TOLYL)AZO)ACETANILIDE * INTRASPERSE YELLOW GBA EXTRA * MICROSETILE YELLOW GR * NACELAN FAST YELLOW CG * NCI-C53781

THR: An experimental carcinogen. An allergen. When heated to decomposition, it emits toxic fumes of NO_x. For further information, see C.I. Disperse Yellow 3, Vol. 1, No. 3 of *DPIM Report.*

ACETANILIDE   HR: 3
CAS: 103-84-4   NIOSH: AD 7350000
mf: C_8H_9NO

PROP: White, shining crystalline scales; mp: 113.5°; bp: 305°; flash p: 345°F (OC); d: 1.2105 @ 4°/4°; autoign temp: 1004°F; vap press: 1 mm @ 114.0°; vap d: 4.65. Somewhat sol in water, alc and ether.

SYNS: ACETAMIDOBENZENE * ACETANIL * ACETIC ACID ANILIDE * ACETOANILIDE * ACETYLAMINOBENZENE * ACETYLANILINE * N-ACETYLANILINE * ANTIFEBRIN * N-PHE-NYLACETAMIDE * USAF EK-3

THR: Poison by ingestion, intravenous, and possibly other routes. When heated to decomposition, it emits toxic NO_x fumes. Slight fire hazard when exposed to heat or flame. See aniline. For further information, see Vol. 3, No. 6 of *DPIM Report.*

(ACETATO)(DIETHOXYPHOSPHINYL)-MERCURY   HR: 3
CAS: 5421-48-7   NIOSH: OV 5950000
mf: $C_6H_{13}HgO_5P$ mw: 396.75

SYN: (DIETHOXY-PHOSPHINYL)MERCURY ACE-TATE

THR: Poison by intraperitoneal route. See also mercury compounds. When heated to decomposition, it emits very toxic fumes of Hg and PO_x.

(ACETATO)(2,3,5,6-TETRAMETHYL-PHENYL) MERCURY   HR: 3
CAS: 21450-81-7   NIOSH: OV 6680000
mf: $C_{12}H_{16}HgO_2$ mw: 392.87

THR: Poison by intravenous route. See also mercury compounds. When heated to decomposition, it emits toxic fumes of Hg.

ACETAZOLAMIDE SODIUM   HR: 3
CAS: 1424-27-7   NIOSH: AC 8400000
mf: $C_4H_5N_4O_3S_2 \cdot Na$ mw: 244.24

SYNS: ACETAZOLAMIDE SODIUM SALT
* SODIUM ACETAZOLAMIDE

THR: An experimental teratogen. When heated to decomposition, it emits very toxic fumes of NO_x and SO_x.

ACETIC ACID   HR: 2
CAS: 64-19-7   NIOSH: AF 1225000
DOT: 2789/2790
mf: $C_2H_4O_2$ mw: 60.06

PROP: Clear, colorless liquid, pungent odor. mp: 16.7°, bp: 118.1°, flash p: 109°F (CC), lel = 5.4%, uel = 16.0% @ 212°F, d: 1.049 @ 20°/4°, autoign temp: 869°F, vap press: 11.4 mm @ 20°, vap d: 2.07. Misc water, alc and ether.

SYNS: ACETIC ACID (AQUEOUS SOLUTION) (DOT)
* ACETIC ACID, GLACIAL (DOT) * ACIDE ACE-TIQUE (FRENCH) * ACIDO ACETICO (ITALIAN)
* AZIJNZUUR (DUTCH) * ESSIGSAEURE (GER-MAN) * ETHANOIC ACID * ETHYLIC ACID
* GLACIAL ACETIC ACID * METHANECARBOX-YLIC ACID * OCTOWY KWAS (POLISH)
* VINEGAR ACID
ACGIH TLV: TWA 10 ppm; STEL 15 ppm
OSHA PEL: TWA 10 ppm
DFG MAK: 10 ppm (25 mg/m³)
DOT Classification: Label: Corrosive

THR: Mutagenic data. Moderately toxic by various routes. Caustic, can cause burns, lachrymation, and conjunctivitis. Inhalation causes irritation of mucous membranes. A common air contaminant. Moderate fire and explosion haz-ard when exposed to heat or flame; can react vigorously with oxidizing materials. When heated to decomposition, it emits irritating fumes. Incompatible with chromic acid; 5-azidotetrazole; sodium peroxide; nitric acid; acetaldehyde; 2-amino-ethanol; NH_4NO_3; BrF_5; ClF_3; chlorosulfonic acid; $(O_3 + diallyl methyl carbinol)$; ethylenediamine; ethylene imine; H_2O_2; $(HNO_3 + acetone)$; potassium-*tert*-butoxide; oleum; $HClO_4$; permanganates; $P(OCN)_3$; PCl_3; KOH; NaOH; *n*-xylene. To fight fire, use CO_2, dry chemical, alcohol foam, foam and mist. For further information see Vol. 3, No. 6 of *DPIM Report*.

ACETIC ACID-*tert*-BUTYL ESTER   HR: 3
CAS: 540-88-5   NIOSH: AF 7400000
mf: $C_6H_{12}O_2$ mw: 116.18

SYNS: ACETIC ACID-1,1-DIMETHYLETHYL ESTER
* T-BUTYL ACETATE

OSHA PEL: TWA 200 ppm

THR: Poison by inhalation and ingestion. (See also esters.) When heated to decomposition, it emits acrid smoke and irritating fumes. For further information, see Vol. 3, No. 6 of *DPIM Report*.

ACETIC ACID-4,6-DINITRO-*o*-CRESYL ESTER   HR: 3
CAS: 18461-55-7   NIOSH: GP 0175000
mf: $C_9H_8N_2O_6$ mw: 240.19

SYN: 4,6-DINITRO-O-KRESYLESTER KYSELINY OCTOVE (CZECH)

THR: Poison by ingestion and intraperitoneal routes. Severe eye irritant. Moderate skin irritant. When heated to decomposition it emits toxic fumes, such as NO_x.

ACETIC ACID ISOPROPYL ESTER   HR: 2
CAS: 108-21-4   NIOSH: AI 4930000
DOT: 1220
mf: $C_5H_{10}O_2$ mw: 102.15

PROP: Colorless aromatic liquid. mp: −73°, bp: 88.4°, lel = 1.8%, uel = 7.8%, fp: −69.3°, flash p: 40°F, d: 0.874 @ 20°20°, autoign temp: 860°F, vap press: 40 mm @ 17.0°, d: 3.52. Somewhat sol in water, misc in alc, ether.

SYNS: ACETATE D'ISOPROPYLE (FRENCH)
* ACETIC ACID, 1-METHYLETHYL ESTER (9CI)

* 2-ACETOXYPROPANE * ISOPROPILE (ACETATO DI) (ITALIAN) * ISOPROPYLACETAAT (DUTCH) * ISOPROPYLACETAT (GERMAN) * ISOPROPYL ACETATE (DOT) * 2-PROPYL ACETATE

OSHA PEL: TWA 250 ppm
DOT Classification: Label: Flammable
 Liquid

THR: Moderately toxic by ingestion. A human systemic irritant; irritant to the eyes. Mildly toxic by inhalation. See also esters. Narcotic in high concentrations. Chronic exposure can cause liver damage. Dangerous fire hazard when exposed to heat, flame, or oxidizers. Moderately explosive when exposed to heat or flame. Dangerous; keep away from heat and open flame; incompatible with oxidizing materials. To fight fire, use foam, CO_2, dry chemical.

ACETIC ACID METHYL ESTER   HR: 2
CAS: 79-20-9   NIOSH: AI 9100000
DOT: 1231
mf: $C_3H_6O_2$ mw: 74.09

PROP: Colorless, volatile liquid. mp: $-98.7°$, lel = 3.1%, uel = 16%, bp: 57.8°, ulc: 85-90, flash p: 14°F, d: 0.92438, autoign temp: 935°F, vap press: 100 mm @ 9.4°, vap d: 2.55. Moderately sol in water, misc in alc, ether.

SYNS: ACETATE DE METHYLE (FRENCH) * METHYLACETAAT (DUTCH) * METHYLACE-TAT (GERMAN) * METHYL ACETATE (DOT) * METHYLE (ACETATE DE) (FRENCH) * METHYLESTER KISELINY OCTOVE (CZECH) * METILE (ACETATO DI) (ITALIAN) * OCTAN METYLU (POLISH)

OSHA PEL: TWA 200 ppm
DOT Classification: Label: Flammable
 Liquid

THR: Moderately toxic by several routes. A human systemic irritant. (See also esters.) Dangerous fire hazard when exposed to heat, flame, or oxidizers. Moderate explosion hazard when exposed to heat or flame.

ACETIC ACID METHYLNITROSAMINO-
METHYL ESTER   HR: 3
CAS: 56856-83-8   NIOSH: AI 9150000
mf: $C_4H_8N_2O_3$ mw:132.14

SYNS: N-ALPHA-ACETOXYMETHYL-N-METHYLNI-TROSAMINE * ALPHA-ACETOXY DIMETHYL-NITROSAMINE * ACETOXYMETHYL-METHYL-NITROSAMIN (GERMAN) * ACETOXYMETHYL

METHYLNITROSAMINE * 1-ACETOXY-N-NITRO-SODIMETHYLAMINE * N-NITROSO-N-METHYL-N-ACETOXYMETHYLAMINE

THR: Poison by ingestion, subcutaneous, intravenous, and intraperitoneal routes. Mutagenic data. An experimental carcinogen and teratogen. (See also esters.) When heated to decomposition, it emits toxic fumes of NO_x.

ACETIC ACID MYRCENYL
ESTER   HR: 1
CAS: 1118-39-4   NIOSH: RH 3585000
mf: $C_{12}H_{20}O_2$ mw: 196.32

SYNS: 2-METHYL-6-METHYLENE-7-OCTEN-2-OL ACETATE * ACETIC ACID, 2-METHYL-6-METHY-LENE-7-OCTEN-2-YL ESTER * 3-METHYLENE-7-METHYL-1-OCTEN-7-YL ACETATE * MYRCENYL ACETATE * 2-METHYL-6-METHYLENE-7-OCTEN-2-YL ACETATE

THR: A skin irritant. See also esters. When heated to decomposition it emits acrid smoke and irritating fumes.

ACETIC ACID VINYL ESTER   HR: 2
CAS: 108-05-4   NIOSH: AK 0875000
DOT: 1301
mf: $C_4H_6O_2$ mw: 86.10

PROP: Colorless mobile liquid, polymerizes to solid on exposure to light. mp: $-92.8°$, bp: 73°, flash p: 18°F, d: 0.9335 @ 20°, autoign temp: 800°F, vap press: 100 mm @ 21.5°, lel = 2.6%, uel = 13.4%, vap d: 3.0 Misc in alc, ether. Somewhat sol in water.

SYNS: ACETIC ACID ETHENYL ESTER * 1-ACE-TOXYETHYLENE * ETHENYL ACETATE * OCTAN WINYLU (POLISH) * VINILE (ACETATO DI) (ITALIAN) * VINYLACETAAT (DUTCH) * VINYLACETAT (GERMAN) * VINYL ACETATE (DOT) * VINYLE (ACETATE DE) (FRENCH) * VINYL A MONOMER

DOT Classification: Label: Flammable
 Liquid

THR: An experimental carcinogen. Moderately toxic by ingestion, inhalation, and intraperitoneal routes. A skin and eye irritant. Highly dangerous fire hazard when exposed to heat, flame, or oxidizers. Incompatible (explosive) with 2-amino ethanol; chlorosulfonic acid; ethylenediamine; ethyleneimine; HCl; HF; HNO_3; oleum; peroxides; H_2SO_4. For further information, see Vinyl Acetate Vol. 2, No. 2 of *DPIM Report*.

ACETIC ANHYDRIDE   HR: 2
CAS: 108-24-7   NIOSH: AK 1925000
DOT: 1715
mf: $C_4H_6O_3$ mw: 102.10

PROP: Colorless, very mobile, strongly refractive liquid; very strong acetic odor. mp: $-73.1°$, bp: 140°, flash p: 129°F (CC), d: 1.082 @ 20°/4°, lel = 2.9%, uel = 10.3%, autoign temp: 734°F, vap press: 10 mm @ 36.0°, vap d: 3.52. Somewhat sol in cold water, decomp in hot water, misc in alc, decomp in hot alc, misc in ether.

SYNS: ACETIC ACID, ANHYDRIDE * ACETIC OXIDE * ACETYL ANHYDRIDE * ACETYL ETHER * ACETYL OXIDE * ANHYDRIDE ACETIQUE (FRENCH) * ANIDRIDE ACETICA (ITALIAN) * AZIJNZUURANHYDRIDE (DUTCH) * ESSIGSAEUREANHYDRID (GERMAN) * ETHANOIC ANHYDRATE * OCTOWY BEZWODNIK (POLISH)
ACGIH TLV: TWA CL 5 ppm
DFG MAK: 5 ppm (20 mg/m^3)
OSHA PEL: TWA 5 ppm.
DOT Classification: Label: Corrosive

THR: Moderately toxic by inhalation, ingestion, and skin contact. Moderate fire and explosion hazard when exposed to heat or flame. Incompatible with 2-aminoethanol; aniline; boric acid; chlorosulfonic acid; CrO_3; (CrO_3 + acetic acid); ethylenediamine; ethyleneimine; glycerol; HCl; HNO_3; oleum; $HClO_4$; H_2O_2; HF; permanganates; NaOH; Na_2O_2; H_2SO_4; water; N_2O_2; 1,3-diphenyl triazine; (glycerol + phosphoryl chloride); metal nitrates; $KMnO_4$. When heated to decomposition it emits toxic fumes; can react vigorously with oxidizing materials, will react violently on contact with water or steam. To fight fire, use CO_2, dry chemical, water mist, alcohol foam. For further information, see Vol. 3, No. 3 of *DPIM Report*.

ACETOACETANILIDE   HR: 3
CAS: 102-01-2   NIOSH: AK 4200000
mf: $C_{10}H_{11}NO_2$ mw: 177.22

PROP: White crystalline solid. Mp: 85°, bp: decomp, flash p: 365°F (COC), d: 1.260 @ 20°, vap press: 0.01 mm @ 20°.

SYNS: ACETOACETAMIDOBENZENE * ACETOACETIC ACID ANILIDE * ACETOACETIC ANILIDE * ((ACETOACETYL)AMINO)BENZENE * ACETOACETYLANILINE * ACETYLACETANILIDE * ALPHA-ACETYLACETANILIDE * N-(ACETYL-

ACETYL)ANILINE * BETA-KETOBUTYRANILIDE * N-PHENYLACETOACETAMIDE * USAF EK-1239

THR: Poison by intraperitoneal route. A weak allergen. See also acetanilide. Slight fire hazard when exposed to heat or flame. Dangerous disaster hazard; see aniline and cyanide. When heated to decomposition it emits toxic NO_x fumes. To fight fire, use alcohol foam, water mist, CO_2, dry chemical.

ACETOACET-o-TOLUIDIDE   HR: 2
CAS: 93-68-5   NIOSH: AK 6550000
mf: $C_{11}H_{13}NO_2$ mw: 191.25

PROP: Cryst; mp: 106°; bp: decomp; d: 1.300 @ 20°; vap press: 0.01 mm @ 20°; flash p: 320°F (COC).

SYNS: ACETOACETYL-2-METHYLANILIDE * 2-ACETOACETYLAMINOTOLUENE * 2'-METHYLACETOACETANILIDE

THR: Moderately toxic by ingestion and intraperitoneal route. When heated to decomposition, it emits toxic fumes of NO_x.

ACETOACETYL-o-ANISIDINE   HR: 2
CAS: 92-15-9   NIOSH: BZ 5600000
mf: $C_{11}H_{13}NO_3$ mw: 207.25

PROP: Crystals. mp: 86.6°; flash p: 325°F (OC); d: 1.132 @ 86.6°/20°; vap d: 7.0.

SYNS: O-ACETOACETANISIDE * ACETOACETIC ACID-O-ANISIDIDE * 2-ACETOACETYLAMINOANISOLE * ACETOACETYL-O-ANISIDE * ACETOACETYL-O-ANISINE * ACETOACET-O-ANISIDIN (CZECH) * O-METHOXYACETOACETANILIDE * 2-METHOXYACETOACETANILIDE * 2'-METHOXYACETOACETANILIDE

THR: Moderately toxic by ingestion. A skin and eye irritant. When heated to decomposition, it emits toxic fumes of NO_x. Slight fire hazard when exposed to heat or flame or oxidizing materials. To fight fire, use CO_2, mist, dry chemicals.

ACETOIN   HR: 3
CAS: 513-86-0   NIOSH: EL 8790000
mf: $C_4H_8O_2$ mw: 88.12

PROP: Slightly yellow liquid or cryst solid, sol in alc, misc with water; d: 1.016, bp: 147-148°, mp: 15°.

SYNS: ACETYL METHYL CARBINOL * 2-BUTANOL-3-ONE * DIMETHYLKETOL * 3-HY-

DROXY-2-BUTANONE * 1-HYDROXYETHYL METHYLKETONE * GAMMA-HYDROXY-BETA-OXOBUTANE

THR: Experimental carcinogen and tumorigen. A moderate skin irritant. When heated to decomposition, it emits acrid smoke.

ACETOL (1)   HR: 3
CAS: 116-09-6   NIOSH: UC 2800000
mf: $C_3H_6O_2$ mw: 74.09

PROP: Colorless liquid: d: 1.084 @ 20°/4°; mp: −7°; bp: 145°-146° decomp; misc in water, alc and ether.

SYN: 1-HYDROXY-2-PROPANONE

THR: An experimental teratogen. Moderately toxic by ingestion. An allergen. Implicated in aplastic anemia. A 10 g dose may be fatal to an adult. Skin contact, inhalation, or ingestion can cause asthma, sneezing, irritation of eyes and nose, hives and eczema. Slight fire hazard when exposed to heat or flame. For further information, see Vol. 1, No. 3 of *DPIM Report*.

ACETOL (2)   HR: 3
CAS: 50-78-2   NIOSH: VO 0700000
mf: $C_9H_8O_4$ mw: 180.17

PROP: Colorless needles; mp: 135°. Very sl sol in alc, sol in benzene. Solubility in water = 1% @ 37°, in ether = 5% @ 20°.

SYNS: SALICYLIC ACID, ACETATE * ACETO-PHEN * ACETOSALIC ACID * O-ACETOXYBEN-ZOIC ACID * 2-ACETOXYBENZOIC ACID * ACETYLSALICYLIC ACID * ACETYLSALICYL-SAURE (GERMAN) * ACIDE ACETYLSALICYLIQUE (FRENCH) * ASPIRIN * EMPIRIN * O-CAR-BOXYPHENYL ACETATE

THR: Mutagenic data, an experimental teratogen. Poison by ingestion and possibly other routes. A human pulmonary and gastrointestinal irritant. Implicated in aplastic anemia. A 10 gram dose to an adult may be fatal. An allergen; skin contact, inhalation, or ingestion can cause asthma, sneezing, irritation of eyes and nose, hives and eczema. Slight fire hazard when exposed to heat or flame. For further information, see Vol. 1, No. 4 of *DPIM Report*.

ACETOMETHOXANE   HR: 3
CAS: 828-00-2   NIOSH: AH 1350000
mf: $C_8H_{14}O_4$ mw: 174.22

PROP: Yellow to amber clear liquid. Sol in water and org solvents. d: 1.068-1.075 @ 25°/25°; bp: 66°-68° @ 3 mm; fp: < −25°.

SYNS: ACETIC ACID, 2,6-DIMETHYL-M-DIOXAN-4-YL ESTER * ACETOMETHOXAN * 6-ACE-TOXY-2,4-DIMETHYL-M-DIOXANE * DIMETH-OXANE * 2,6-DIMETHYL-M-DIOXAN-4-OL ACETATE * 2,6-DIMETHYL-M-DIOXAN-4-YL ACETATE * DIOXIN (BACTERICIDE) (OBS.) * NCI-C56213

THR: An experimental carcinogen. Moderately toxic by ingestion. See also esters. When heated to decomposition, it emits acrid smoke.

ACETONE   HR: 2
CAS: 67-64-1   NIOSH: AL 3150000
DOT: 1090
mf: C_3H_6O mw: 58.09

PROP: Colorless liquid, fragrant mint-like odor. mp; −94.6°, bp: 56.48°, ulc = 90, flash p: 0°F (CC), lel = 2.6%, uel = 12.8%, d: 0.7972 @ 15°, autoign temp:(color) 869°F, vap press: 400 mm @ 39.5°, vap d: 2.00. Misc in water, alc, and ether.

SYNS: ACETON (GERMAN, DUTCH, POLISH) * DIMETHYLFORMALDEHYDE * DIMETHYLKE-TAL * DIMETHYL KETONE * KETONE PROPANE * BETA-KETOPROPANE * METHYL KETONE * PROPANONE * 2-PROPANONE * PYRO-ACETIC ACID * PYROACETIC ETHER

ACGIH TLV: TWA 750 ppm; STEL 1000 ppm
DFG MAK: 1000 ppm (2400 mg/m^3)
OSHA PEL: TWA 1000 ppm.
DOT Classification: Label: Flammable Liquid

THR: Moderately toxic by various routes. A skin and eye irritant @ 500 ppm. Narcotic in high concentration. In industry, no injurious effects have been reported other than skin irritation resulting from its defatting action, or headache from prolonged inhalation. A common air contaminant. Dangerous disaster hazard due to fire and explosion hazard, can react vigorously with oxidizing materials. Incompatible with ($CHCl_3$ + a base); CrO; Cr(OCl)$_2$; (nitric + acetic acid); (nitric + sulfuric acid); NOCl; nitrosyl perchlorate; nitryl perchlorate; permonosulfuric acid; potassium tert-butoxide; NaOBr; (sulfuric acid + potassium dichromate); (thiodiglycol + hydrogen peroxide); trichloromelamine; bromoform; air; HNO$_3$; activated C; chloroform; H_2SO_4; BF_3; Br_2; chromyl chloride; H_2O_2; F_2O_2; SCl_2; thiotrithiazyl perchlorate; H_2O_5S. To fight fire, use CO_2, dry chemical, alcohol foam. For further information, see Vol. 4, No. 3 of *DPIM Report*.

ACETONE CHLOROFORM   HR: 3
CAS: 57-15-8   NIOSH: UC 0175000
mf: $C_4H_7Cl_3O$ mw: 177.46

PROP: Crystals, camphor odor; mp: 97°, bp: 167°.

SYNS: ANHYDROUS CHLOROBUTANOL * CHLORBUTANOL * CHLORBUTOL * CHLO-ROBUTANOL * TRICHLORO-T-BUTYL ALCOHOL * T-TRICHLOROBUTYL ALCOHOL * 1,1,1-TRI-CHLORO-2-METHYL-2-PROPANOL

THR: Poison by ingestion. A narcotic. A skin and eye irritant. (See also chloral hydrate, which acts similarly.) Dangerous; can react with oxidizing materials (see phosgene). Slight fire hazard when exposed to heat or flame.

ACETONE OIL   HR: 3
NIOSH: AL 6700000
DOT: 1091

PROP: (a) Standard: light, lemon-yellow. (b) Refined: almost water white. (c) Heavy: dark, orange-yellow. bp: (a) 75°-160°, (c) 80°-225°, d: (a) 0.826-0.830, (b) 0.812, (c) 0.885-0.865. DOT Classification: Label: Flammable

THR: Some carcinogenic activity. Dangerous fire and explosion hazard when exposed to heat or flame. Can react vigorously with oxidizing materials. To fight fire, use CO_2, dry chemical.

ACETONE SEMICARBAZONE   HR: 3
CAS: 110-20-3   NIOSH: AL 7175000
mf: $C_4H_9N_3O$ mw: 115.16

PROP: mp: 190°-199° (decomp): sol in cold water, sl sol in cold alc; insol in ether.

THR: Poison by intravenous route. When heated to decomposition, it emits toxic fumes of NO_x.

ACETONITRILE   HR: 3
CAS: 75-05-8   NIOSH: AL 7700000
DOT: 1648
mf: C_2H_3N mw: 41.06

PROP: Colorless liquid, aromatic odor. mp: −45°, bp: 81.1°, flash p: 42°F(COC), d: 0.7868 @ 20°/20°, vap d: 1.42, vap press: 100 mm

@ 27°, lel = 4.4%, uel = 16%, autoign temp: 975°F. Misc in water, alc, and ether.

SYNS: ACETONITRIL (GERMAN, DUTCH) * CYANOMETHANE * CYANURE DE METHYL (FRENCH) * ETHANENITRILE * ETHYL NITRILE * METHANECARBONITRILE * METHYL CYANIDE * NCI-C60822 * USAF EK-488
ACGIH TLV: TWA 40 ppm; STEL 60 ppm
DFG MAK: 40 ppm (70 mg/m³)
OSHA PEL: TWA 40 ppm.
DOT Classification: Label: Flammable Liquid

THR: Poison by ingestion. Human central nervous system effects. A skin and eye irritant. (See also nitriles.) Dangerous fire hazard when exposed to heat, flame, or oxidizers. Explosion Hazard: See cyanide. When heated to decomposition it emits highly toxic fumes of cyanide, NO_x. Will react with water, steam, or acids to produce toxic and flammable vapors. Incompatible with oleum; chlorosulfonic acid; perchlorates; nitrating agents; indium; dinitrogen tetraoxide; N-fluoro compounds (i.e., perfluorourea + acetonitrile); HNO_3; H_2SO_4; SO_3. To fight fire, use foam, CO_2, dry chemical. For further information, see Methyl cyanide, Vol. 4, No. 1 of *DPIM Report*.

ACETONYL CHLORIDE   HR: 3
CAS: 78-95-5   NIOSH: UC 0700000
DOT: 1695
mf: C_3H_5ClO mw: 92.53

PROP: Colorless liquid, pungent odor. mp: −44.5°, bp: 119°, d: 1.162.

SYNS: CHLORACETONE (FRENCH) * CHLORO-ACETONE * CHLOROPROPANONE * 1-CHLORO-2-PROPANONE * MONOCHLOROACETONE

DOT Classification: Forbidden

THR: Experimental carcinogen, mutagenic data. Poison by inhalation, ingestion, and skin contact. A lachrymator poison gas. See chlorinated hydrocarbons, aliphatic and acetone. Moderate fire hazard when exposed to heat or flame, or oxidizers. Old material can explode. When heated to decomposition it emits highly toxic fumes of phosgene.

3-(alpha-ACETONYLFURFURYL)-4-HYDROXYCOUMARIN   HR: 3
CAS: 117-52-2   NIOSH: GN 4850000
mf: $C_{17}H_{14}O_5$ mw: 298.31

PROP: White powder; practically insol in water, sol in alcohols. mp: 124°.

SYNS: COUMAFURYL * FOUMARIN * 3-(1-FURYL-3-ACETYLETHYL)-4-HYDROXYCOUMARIN

THR: Poison by inhalation and ingestion. See also warfarin.

ACETOPHENAZINE   HR: 3
CAS: 5714-00-1   NIOSH: OB 4180000
mf: $C_{23}H_{29}N_3O_2S \cdot 2C_4H_4O_4$ mw: 643.77

SYNS: 10-(3-(4-(2-HYDROXYETHYL)-1-PIPER-AZINYL)PROPYL)PHENOTHIAZIN-2-YL METHYL KE-TONE DIMALEATE * SCH 6673 * TINDAL

THR: Poison by ingestion, intraperitoneal, and intravenous routes. Severe eye irritant.

p-ACETOPHENETIDIDE   HR: 3
CAS: 62-44-2   NIOSH: AM 4375000
mf: $C_{10}H_{13}NO_2$ mw: 179.24

SYNS: 1-ACETAMIDO-4-ETHOXYBENZENE * ACETO-PARA-PHENALIDE * ACETO-PARA-PHENETIDIDE * PARA-ACETOPHENETIDIDE * ACETPHENETIDIN * ACETYLPHENETIDIN * BROMO SELTZER * CORICIDIN * DARVON COMPOUND * EMPIRIN COMPOUND * PARA-ETHOXY-ACETANILID * N-PARA-ETHOXYPHEN-YLACETAMIDE * PARACETOPHENETIDIN * PERCODAN * PARA-PHENACETIN * P-ACE-TOPHENETIDE * ACETOPHENETIDIN * ACE-TOPHENETIDINE * P-ACETOPHENETIDINE * ACETO-4-PHENETIDINE * ACETOPHENETIN * ACET-P-PHENALIDE * P-ACETPHENETIDIN * ACET-P-PHENETIDIN * N-ACETYL-P-PHENETI-DINE * P-ETHOXYACETANILIDE * 4-ETHOXY-ACETANILIDE * 4'-ETHOXYACETANILIDE * N-(4-ETHOXYPHENYL)ACETAMIDE * SINUTAB

THR: Experimental carcinogen, mutagenic data. Poison by various routes. Human central nervous system effects. Chronic effects consist of weight loss, insomnia, shortness of breath, weakness and often aplastic anemia. When heated to decomposition, it emits toxic fumes of NO_x. For further information, see Phenacetin, Vol. 6, No. 1 of *DPIM Report*.

ACETOPHENONE   HR: 3
CAS: 98-86-2   NIOSH: AM 5250000
mf: C_8H_8O mw: 120.16

PROP: Colorless liquid or plates. mp: 19.7°, bp: 202.3°, flash p: 180°F (OC), d: 1.026 @ 20°/4°, vap d: 4.14, vap press: 1 mm @ 15°,

autoign temp: 1060°F. Insol in water, sol in alc, and ether.

SYNS: ACETYLBENZENE * BENZOYL METHIDE * HYPNONE * METHYL PHENYL KETONE * 1-PHENYLETHANONE * PHENYL METHYL KE-TONE * USAF EK-496

THR: Poison by intraperitoneal route. Moderately toxic by ingestion. Narcotic in high concentration. A hypnotic. See ketones. Moderate fire hazard when exposed to heat, flame or oxidizers. To fight fire, use foam, CO_2, dry chemical.

ACETOPROMAZINE   HR: 3
CAS: 61-00-7   NIOSH: OB 2275000
mf: $C_{19}H_{22}N_2OS$ mw: 326.49

SYNS: ACEPROMAZINA * ACEPROMAZINE * ACEPROMIZINA * ACETHYLPROMAZIN * ACETAZINE * 3-ACETYL-10-(3-DIMETHYLAMINOPROPYL)PHENOTHIAZINE * ACETYLPROMAZINE * ANATRAN * ANER-GAN * ATRAVET * ATSETOZIN * AY-57,062 * AZEPROMAZINE * 1522 CB * 10-(3-DIMETHYLAMINOPROPYL)PHENOTHIAZINE-3-ETH-YLONE * 1-(10-(3-(DIMETHYLAMINO)PROPYL)-10H-PHENOTHIAZIN-2-YL)ETHANONE * 10-(3-DIMETHYLAMINOPROPYL)PHENOTHIAZIN-3-YLMETHYL KETONE * LISERGAN * NOTEN-QUIL * NOTENSIL * NOTESIL * PLEGECYL * PLEGICIN * PLIVAPHEN * SOPRINTIN * SOPRONTIN * SOPROTIN * SV-1522 * VETRANQUIL * WY-1172

THR: Poison by ingestion, intravenous, and subcutaneous routes.

p-ACETOTOLUIDIDE   HR: 2
CAS: 103-89-9   NIOSH: AN 2930000
mf: $C_9H_{11}NO$ mw: 149.21

PROP: Cryst; bp: 307°; flash p: 335°F (CC) d: 1.212; vap d: 5.14; mp: 153°.

SYNS: P-ACETAMIDOTOLUENE * P-ACETO-TOLUIDE * 4-ACETOTOLUIDE * 4-(ACETYL-AMINO)TOLUENE * ACETYL-P-TOLUIDINE * N-ACETYL-P-TOLUIDIDE * P-METHYLACET-ANILIDE * 4-METHYLACETANILIDE * 4'-METHYLACETANILIDE

THR: Moderately toxic by ingestion. See also acetanilide. Slight fire hazard. When heated to decomposition, it emits toxic fumes of NO_x. To fight fire, use water, foam, CO_2, dry chemical.

N-ACETOXY-4-ACETAMIDO-BIPHENYL   HR: 3
CAS: 26541-56-0   NIOSH: AF 1600000
mf: $C_{16}H_{15}NO_3$ mw: 269.32

SYNS: ACETIC ACID (N-ACETYL-N-(4-BIPHENYL)-AMINO) ESTER * ACETIC ACID, ESTER WITH N-4-BIPHENYLYLACETOHYDROXAMIC ACID * N-ACETOXY-4-BIPHENYLACETAMIDE * N-(4-BIPHENYLYL)ACETOHYDROXAMIC ACETATE

THR: An experimental carcinogen. Mutagenic data. See also esters. When heated to decomposition, it emits toxic fumes of NO_x.

N-ACETOXY-N-ACETYL-2-AMINOFLUORENE   HR: 3
CAS: 6098-44-8   NIOSH: NC 3420000
mf: $C_{17}H_{15}NO_3$ mw: 281.33

SYNS: N-ACETOXY-2-ACETAMIDOFLUORENE * N-ACETOXY-2-ACETYLAMINOFLUORENE * N-ACETOXY-2-FLUORENYLACETAMIDE * N-(FLUOREN-2-YL)ACETOHYDROXAMIC ACETAMIDE

THR: An experimental carcinogen. Mutagenic data.

2-ACETOXYACRYLONITRILE   HR: 3
CAS: 3061-65-2   NIOSH: AG 4550000
mf: $C_5H_5NO_2$ mw: 111.11

SYNS: ALPHA-ACETOXYACRYLONITRILE * ALPHA-CYANOVINYL ACETATE

THR: Poison by inhalation, ingestion, and skin contact. A skin irritant. See also nitriles. When heated to decomposition, it emits toxic fumes of NO_x.

1-ACETOXY-1,3-BUTADIENE   HR: 2
CAS: 1515-76-0   NIOSH: EJ 1225000
mf: $C_6H_8O_2$ mw: 112.14

SYN: ACETIC ACID, 1,3-BUTADIENYL ESTER

THR: Poison by inhalation. Moderately toxic by other routes. Mutagenic data. When heated to decomposition, it emits acrid smoke.

ACETOXYCYCLOHEXIMIDE   HR: 3
CAS: 3326-96-3   NIOSH: MA 5075000
mf: $C_{17}H_{25}NO_6$ mw: 339.43

SYNS: 3-(2-(5-ACETOXY-3,5-DIMETHYL-2-OXO-CYCLOHEXYL)-2-HYDROXYETHYL)GLUTARIMIDE * E-73 ACETATE

THR: Poison by ingestion, intraperitoneal, and subcutaneous routes. When heated to decomposition, it emits toxic fumes, such as NO_x.

1-ACETOXY-1,4-DIHYDRO-4-(HYDROXYAMINO)QUINOLINE ACETATE (ESTER)   HR: 3
CAS: 38539-23-0   NIOSH: VB 4800000
mf: $C_{13}H_{13}N_2O_4$ mw: 261.28

SYN: 1-ACETOXY-1,4-DIHYDRO-4-HYDROXYAMINO QUINOLINE ACETATE (ESTER)

THR: An experimental carcinogen. Mutagenic data. See also esters. When heated to decomposition, it emits toxic fumes of NO_x.

2-ACETOXYETHYLTRIMETHYL-AMMONIUM CHLORIDE   HR: 3
CAS: 60-31-1   NIOSH: FZ 9800000
mf: $C_7H_{16}NO_2 \cdot Cl$ mw: 181.69

SYNS: ACETYLCHOLINE HYDROCHLORIDE * ACETYLCHOLINIUM CHLORIDE * (2-HYDROXYETHYL)TRIMETHYLAMMONIUM CHLORIDE ACETATE

THR: Poison by subcutaneous, intravenous, intraperitoneal, and parenteral routes. Moderately toxic by ingestion. When heated to decomposition, it emits very toxic fumes of NO_x and Cl^-.

N-ACETOXYFLUORENYL-ACETAMIDE   HR: 3
CAS: 38105-27-0   NIOSH: AF 1650000
mf: $C_{17}H_{15}NO_3$ mw: 281.33

SYNS: ACETIC ACID ESTER WITH N-(FLUOREN-3-YL)ACETOHYDROXAMIC ACID * N-ACETOXY-3-FLUORENYLACETAMIDE * N-(FLUOREN-3-YL)ACETOHYDROXAMIC ACETATE

THR: An experimental carcinogen. See also esters. When heated to decomposition, it emits toxic fumes of NO_x.

p-(ACETOXYMERCURI)ANILINE   HR: 3
CAS: 6283-24-5   NIOSH: OV 5550000
mf: $C_8H_9HgNO_2$ mw: 351.77

PROP: Colorless crystals, insol in water. mp: 167°

SYNS: (ACETATO)(P-AMINOPHENYL)MERCURY * P-AMINOPHENYLMERCURIC ACETATE

THR: Poison by intravenous routes. (See also mercury compounds and aniline.) When heated to decomposition, it emits very toxic fumes of NO$_x$ and Hg.

2-(ACETOXYMERCURI)-4-NITROANILINE   HR: 3
CAS: 54481-45-7   NIOSH: BW 6825000
mf: C$_8$H$_8$HgN$_2$O$_4$ mw: 396.77

SYN: ACETATO(2-AMINO-5-NITROPHENYL)MERCURY

THR: Poison by intraperitoneal route. See also mercury compounds and nitro compounds of aromatic hydrocarbons. When heated to decomposition, it emits very toxic fumes of Hg and NO$_x$.

7-ACETOXYMETHYL-12-METHYL-BENZ(a)ANTHRACENE   HR: 3
CAS: 2517-98-8   NIOSH: CW 9100000
mf: C$_{22}$H$_{18}$O$_2$ mw: 314.40

SYN: 12-METHYLBENZ(A)ANTHRACENE-7-METHANOL ACETATE (ESTER)

THR: An experimental carcinogen. See also esters. When heated to decomposition, it emits acrid smoke.

p-ACETOXYNITROBENZENE   HR: 3
CAS: 830-03-5   NIOSH: AJ 1150000
mf: C$_8$H$_7$NO$_4$ mw: 181.16

SYNS: P-NITROPHENOL ACETATE * P-NITROPHENYL ACETATE * 4-NITROPHENYL ACETATE

THR: Poison by intravenous route. When heated to decomposition, it emits toxic fumes of NO$_x$.

1-ACETOXY-N-NITROSODIPROPYL-AMINE   HR: 3
NIOSH: JL 9330000
mf: C$_8$H$_{16}$N$_2$O$_3$ mw: 188.26

SYNS: N-(ALPHA-ACETOXY)PROPYL-N-N-PROPYL-NITROSAMINE * ACETIC ACID, 1-(PROPYLNITROSAMINO)PROPYL ESTER * 1-(PROPYLNITROSAMINO)PROPYL ACETATE

THR: An experimental carcinogen. Mutagenic data. Moderately toxic by subcutaneous route. See also esters and amines. When heated to decomposition, it emits toxic fumes of NO$_x$.

17-ACETOXY-19-NOR-17-alpha-PREGN-4-EN-20-YN-3-ONE   HR: 3
CAS: 51-98-9   NIOSH: RC 8965000
mf: C$_{22}$H$_{28}$O$_3$ mw: 340.50

SYNS: 17-BETA-ACETOXY-19-NOR-17-ALPHA-PREGN-4-EN-20-YN-3-ONE * (17-ALPHA)-17-(ACETYLOXY)-19-NORPREGN-4-EN-20-YN-3-ONE * 17-ACETYLOXY(17-ALPHA)-19-NOR-PREGN-4-ESTREN-17-BETA-OL-ACETATE-3-ONE * 17-ENT * 17-ALPHA-ETHINYL-19-NORTESTOSTERONE ACETATE * 17-ALPHA-ETHINYL-19-NORTESTOSTERONE-17-BETA-ACETATE * 17-ALPHA-ETHYNYL-17-BETA-ACETOXY-19-NORANDROST-4-EN-3-ONE * 17-ALPHA-ETHYNYL-17-HYDROXYESTR-4-EN-3-ONE ACETATE * 17-ALPHA-ETHYNYL-19-NORTESTOSTERONE ACETATE * 17-HYDROXY-19-NOR-17-ALPHA-PREGN-4-EN-20-YN-3-ONE ACETATE * 17-BETA-HYDROXY-19-NOR-17-ALPHA-PREGN-4-EN-20-YN-3-ONE ACETATE * NORETHINDRONE-17-ACETATE * 19-NORETHISTERONE ACETATE * 19-NORETHYNYLTESTOSTERONE ACETATE * NORETHYSTERONE ACETATE

THR: An experimental carcinogen and teratogen. Mutagenic data. When heated to decomposition, it emits acrid smoke and irritating fumes.

ACETOXYPHENYLMERCURY   HR: 3
CAS: 62-38-4   NIOSH: OV 6475000
mf: C$_8$H$_8$HgO$_2$ mw: 336.75

PROP: Lustrous crystals, slt sol in water. mp: 149°.

SYNS: ACETATE PHENYLMERCURIQUE (FRENCH) * (ACETATO)PHENYLMERCURY * ACETIC ACID, PHENYLMERCURY DERIV. * (ACETOXYMERCURI)BENZENE * CERESAN * FENYLMERCURIACETAT (CZECH) * MERCURIPHENYL ACETATE * OCTAN FENYLRTUTNATY (CZECH) * PHENYLMERCURIACETATE * PHENYL MERCURIC ACETATE * PHENYLMERCURY ACETATE * PHENYLQUECKSILBERACETAT (GERMAN)

THR: Poison by ingestion, intravenous, intraperitoneal, subcutaneous, and possibly other routes. An experimental teratogen. Mutagenic data. A skin and eye irritant. See mercury compounds. When heated to decomposition, it emits toxic fumes of Hg.

3-ACETOXYPHENYLTRIMETHYL-AMMONIUM IODIDE   HR: 3
CAS: 17427-00-8   NIOSH: BO 2275000
mf: C$_{11}$H$_{16}$NO$_2$ • I   mw: 321.18

SYN: NU 2017

THR: A poison via subcutaneous and intravenous routes. When heated to decomposition, it emits very toxic fumes of NO$_x$ and I$^-$.

ACETOXYTRIETHYLSTANNANE   **HR: 3**
CAS: 1907-13-7   NIOSH: WH 5950000
mf: $C_8H_{18}O_2Sn$ mw: 264.95

SYNS: TRIAETHYLZINNACETAT (GERMAN)
* TRIETHYLTIN ACETATE

OSHA PEL: TWA 100 ug(Sn)/m³ (skin)

THR: Poison by ingestion and intravenous routes. (See also tin compounds.) When heated to decomposition, it emits acrid smoke and irritating fumes.

ACETOXYTRIHEXYLSTANNANE   **HR: 3**
CAS: 2897-46-3   NIOSH: WH 6125000
mf: $C_{20}H_{42}O_2Sn$ mw: 433.31

SYNS: TRIHEXYLTIN ACETATE * TRI-N-HEXYL-ZINNACETAT (GERMAN)

THR: Poison by intravenous route. Moderately toxic by ingestion. See also tin compounds. When heated to decomposition, it emits acrid smoke and fumes.

**ACETOXYTRIPHENYLSTAN-
NANE**   **HR: 3**
CAS: 900-95-8   NIOSH: WH 6650000
mf: $C_{20}H_{18}O_2Sn$ mw: 409.07

PROP: Practically insol, crystalline solid. mp: 120°.

SYNS: ACETATOTRIPHENYLSTANNANE * PHEN-TIN ACETATE * TRIPHENYLACETO STANNANE
* ACETATE DE TRIPHENYL-ETAIN (FRENCH)
* ACETATO DI STAGNO TRIFENILE (ITALIAN)
* ACETOXY-TRIPHENYL-STANNAN (GERMAN)
* ACETOXY-TRIPHENYLSTANNANE * BRESTAN
* ENT 25208 * FENTIN ACETAAT (DUTCH)
* FENTIN ACETAT (GERMAN) * FENTIN ACE-TATE * FENTINE ACETATE (FRENCH) * FINTIN ACETATO (ITALIAN) * PHENTINOACETATE
* TIN TRIPHENYL ACETATE * TRIFENYL-TIN-ACETAAT (DUTCH) * TRIPHENYLTIN ACETATE
* TRIPHENYL-ZINNACETAT (GERMAN)

OSHA PEL: TWA 100 ug(Sn)/m³ (skin)

THR: Poison by ingestion, skin contact, intraperitoneal, intravenous, and subcutaneous routes. An experimental carcinogen. A fungicide and algicide. See also tin compounds. When heated to decomposition, it emits acrid smoke and fumes.

ACETPHENARSINE   **HR: 3**
CAS: 97-44-9   NIOSH: CF 8400000
mf: $C_8H_{10}AsNO_5$ mw: 275.11

PROP: Crystalline material, slightly water sol. Decomp @ 240°-250°.

SYNS: M-ARSANILIC ACID, N-ACETYL-4-HY-DROXY- * 3-ACETAMIDO-4-HYDROXYBENZENE-ARSONIC ACID * 3-ACETAMIDO-4-HYDROXY-PHENYLARSONIC ACID * ACETARSOL
* ACETARSONE * ACETPHENARSINE
* 3-ACETYLAMINO-4-HYDROXYPHENYLARSONIC ACID * N-ACETYL-4-HYDROXY-M-ARSANILIC ACID * AMARSAN * AMOEBAL * ARSPHEN
* ARSONIC ACID * (3-(ACETYLAMINO)-4-HYDROXYPHENYL)- (9CI) * ARSONINE
* DEVEGAN * DISPARICIDA
* DYNARSAN * EHRLICH 594 * 190 F
* F 190 * FOURNEAU 190 * GINARSOL
* GOYL * GYNOPLIX * KHARO-PHEN * KUBARSOL * LIMARSOL MALAGRIDE
* MEXYL * MONARGAN * NILACID
* ORALCID * ORARSAN * OSARSAL
* OSARSOL * OSARSOLE * OSVARSAN
* OSVARSON * PALLICID * PAROXYL
* SPIROCID * SPIROZID * STOVARSAL
* STOVARSOL * STOVARSOLAN * SVC
* VAGISEPT * VAGOFLOR

OSHA PEL: TWA 500 ug(As)/m³

THR: Poison by ingestion and intravenous routes. Mutagenic data. When heated to decomposition, it emits As and NO_x fumes.

ACETYL ACETONE   **HR: 2**
CAS: 123-54-6   NIOSH: SA 1925000
mf: $C_5H_8O_2$ mw: 100.13

PROP: Colorless to sltly yellow, flammable liquid. Pleasant odor.Sol in water; misc in alc, ether, chloroform, acetone, glacial acetic acid. mp: −23.2°, bp: 139° @ 746 mm, flash p: 105°F (OC), d: 0.976, vap d: 3.45, autoign temp: 644°F.

SYNS: ACETOACETONE * DIACETYLMETHANE
* PENTANEDIONE * PENTANEDIONE-2,4

THR: Moderately toxic via oral, intraperitoneal, inhalation routes. A skin, eye irritant. Moderate fire hazard when exposed to heat or flame; can react with oxidizing materials. To fight fire, use alcohol foam, CO_2, dry chemical. Incompatible with oxidizing materials. For further information, see Vol. 1, No. 7 of *DPIM Report*.

4-ACETYLAMINOFLUORENE   **HR: 3**
CAS: 28322-02-3   NIOSH: AB 9530000
mf: $C_{15}H_{13}NO$ mw: 223.29

SYNS: 4-ACETYLAMINOFLUOREN (GERMAN) * N-FLUOREN-4-YLACETAMIDE * N-4-FLUOR-ENYLACETAMIDE

THR: An experimental carcinogen. Mutagenic data. When heated to decomposition, it emits toxic fumes of NO_x.

2-ACETYLAMINOFLUORENONE HR: 3
CAS: 3096-50-2   NIOSH: AC 7000000
mf: $C_{15}H_{11}NO_2$ mw: 237.27

SYNS: 2-ACETYLAMINO-9-FLUORENONE * 9-OXO-2-FLUORENYLACETAMIDE * N-(9-OXO-2-FLUORENYL)ACETAMIDE

THR: An experimental carcinogen. When heated to decomposition, it emits toxic fumes of NO_x.

1-ACETYLAZIRIDINE HR: 3
CAS: 460-07-1   NIOSH: CM 6300000
mf: C_4H_7NO mw: 85.12

SYN: ACETYLETHYLENEIMINE

THR: Poison by intraperitoneal route. An experimental carcinogen. When heated to decomposition, it emits toxic fumes of NO_x.

ACETYL BENZOYLPEROXIDE (SOLID) HR: 3
CAS: 644-31-5   NIOSH: SD 7860000
mf: $C_9H_8O_4$ mw: 180.17

PROP: White crystals. Sol in oils, alc, ether and chloroform. mp: 36°-37°, bp: 130° @ 19 mm.
DOT Classification: Forbidden, Label: Organic Peroxide

THR: Poison by inhalation and ingestion. Severe irritant. A powerful oxidizing agent which is corrosive to the skin and mucous membranes. See also peroxides, organic. Dangerous; shock or heat will cause detonation with evolution of toxic fumes; will react with water or steam to produce heat; can react vigorously with reducing materials. Moderate fire hazard by spontaneous chemical reaction. To fight fire, use CO_2 or dry chemical.

ACETYL BENZOYLPEROXIDE (SOLUTION) HR: 1
CAS: 644-31-5   NIOSH: SD 7861000

PROP: Solution contains not over 40% acetyl benzoyl peroxide
DOT Classification: Label: Organic Peroxide

THR: A highly irritating material to skin, eyes, and mucous membranes. See also acetyl benzoyl peroxide, solid.

N-ACETYL-4-BIPHENYL-HYDROXYLAMINE HR: 3
CAS: 4463-22-3   NIOSH: AK 8225000
mf: $C_{14}H_{13}NO_2$ mw: 227.28

SYNS: N-4-BIPHENYLYLACETOHYDROXAMIC ACID * 4-BIPHENYLACETHYDROXAMIC ACID * N-HYDROXY-4-ACETAMIDOBIPHENYL * N-4-(N-HYDROXYACETAMIDO)BIPHENYL * N-HYDROXY-4-ACETAMIDODIPHENYL * N-HYDROXY-4-ACETYLAMINOBIPHENYL

THR: An experimental carcinogen. Mutagenic data. When heated to decomposition, it emits toxic fumes of NO_x.

N-(N-ACETYL-3-(p-(BIS(2-CHLORO-ETHYL)AMINO)PHENYL)ALANYL-3-PHENYLALANINE ETHYL ESTER HR: 3
CAS: 3733-45-7   NIOSH: AY 3050000

SYN: ETHYL ESTER OF N-ACETYL-DL-SARCOLY-SYL-L-PHENYLALANINE

THR: Poison by ingestion and intramuscular routes. When heated to decomposition, it emits very toxic fumes of Cl^- and NO_x.

ACETYL BROMIDE HR: 3
CAS: 506-96-7   NIOSH: AO 5955000
DOT: 1716
mf: C_2H_3BrO mw: 122.96

PROP: Colorless fuming liquid; turns yellow in air. mp: −96.5°, bp: 76.7°, d: 1.52 @ 9.5°/4°. Decomp in water and alc; misc in benzene, ether and chloroform.
DOT Classification: Label: Corrosive

THR: Poison by inhalation, ingestion, skin contact, and intraperitoneal routes. See also hydrobromic acid, acetic acid. Great explosion hazard by spontaneous chemical reaction; decomposes violently upon contact with moisture or alcohols. When heated to decomposition, it emits highly corrosive and toxic fumes of carbonyl bromide and bromine; will react with water or steam to produce heat. To fight fire, use dry chemical, CO_2. For further information, see Vol. 1, No. 8 of *DPIM Report*.

1-ACETYL-3-(2-BROMO-2-ETHYLBUTYRYL)UREA   HR: 2
CAS: 77-66-7   NIOSH: YR 6475000
mf: $C_9H_{15}BrN_2O_3$ mw: 279.17

SYNS: ACETYLBROMODIETHYLACETYLCARB-
AMIDE * N-ACETYL-N-BROMODIETHYL-
ACETYLCARBAMIDE * N-ACETYL-N-
BROMODIETHYLACETYLUREA
* N-ACETYL-N'-ALPHA-BROMO-ALPHA-
ETHYLBUTYRYLCARBAMIDE * 1-ACETYL-3-
(ALPHA-BROMO-ALPHA-ETHYLBUTYRYL)UREA

THR: Psychotropic effect by ingestion in humans. Moderately toxic by injection. When
heated to decomposition, it emits very toxic
fumes of Br⁻ and NO_x.

o-ACETYL-2-sec-BUTYL-4,6-DINITROPHENOL   HR: 3
CAS: 2813-95-8   NIOSH: AF 7140000
mf: $C_{12}H_{14}N_2O_6$ mw: 282.28

SYNS: ACETIC ACID, (2,4-DINITRO-6-S-BUTYL-
PHENYL) ESTER * ACETIC ACID, (4,6-DINI-
TRO-2-S-BUTYLPHENYL) ESTER * 2-SEC-
BUTYL-4,6-DINITROPHENYLACETATE
* 6-SEC-BUTYL-2,4-DINITROPHENYLACETATE
* 2,4-DINITRO-6-S-BUTYLFENYLESTER KYSELINY
OCTOVE (CZECH) * 4,6-DINITRO-2-S-BUTYL-
PHENYL ACETATE * BETA-(2-HYDROXY-3,5-
DINITROPHENYL)BUTANE ACETATE * 2-(1-
METHYLPROPYL)-4,6-DINITROPHENYL ACETATE

THR: Poison by ingestion. A skin and eye irritant. See also esters. When heated to decomposition, it emits toxic fumes of NO_x.

3'-o-ACETYLCALOTROPIN   HR: 3
CAS: 36573-63-4   NIOSH: EW 8850000
mf: $C_{31}H_{42}O_{10}$ mw: 574.73

PROP: A glycoside isolated *Asclepius CUNS-
SUICA.*

SYN: ASCLEPIN

THR: Poison by ingestion and intraperitoneal
routes. When heated to decomposition, it emits
acrid smoke and irritating fumes.

ACETYL CHLORIDE   HR: 3
CAS: 75-36-5   NIOSH: AO 6390000
DOT: 1717
mf: C_2H_3ClO mw: 78.50

PROP: Colorless, fuming liquid. mp: −112°,
bp: 51°-52°, flash p: 40°F (CC), autoign temp:

734°F, d: 1.1051 @ 20°/4°, vap d: 2.70. lel
= 5%. Decomp in water and alc, misc in benzene, ether and chloroform.

SYNS: ACETIC ACID CHLORIDE * ACETIC
CHLORIDE * ETHANOYL CHLORIDE

DOT Classification: Label: Flammable
   Liquid

THR: Poison by inhalation and ingestion. A
human systemic irritant. Readily hydrolyzes violently to form HCl and acetic acid. Can decompose during preparation. Dangerous fire hazard
when exposed to heat or flame. Explosion hazard
by spontaneous chemical reaction, with dimethyl sulfoxide or ethanol. When heated to decomposition it emits highly toxic fumes of phosgene; reaction with water or steam produces
heat and toxic or corrosive fumes of chlorides.
To fight fire, use CO_2 or dry chemical. Incompatible with PCl_3; dimethyl sulfoxide; H_2O. For
further information, see Vol. 1, No. 8 and Vol.
3, No. 3 of *DPIM Report.*

N-ACETYL-L-CYSTEINE   HR: 3
CAS: 616-91-1   NIOSH: HA 1660000
mf: $C_5H_9NO_3S$ mw: 163.21

SYNS: ACETYLCYSTEINE * N-ACETYLCYSTEINE
* N-ACETYL-N-CYSTEINE * MERCAPTURIC ACID

THR: Poison by intraperitoneal route. Moderately toxic by other routes. When heated to
decomposition, it emits very toxic fumes of NO_x
and SO_x.

ACETYL DIGITOXIN-alpha   HR: 3
CAS: 1111-39-3   NIOSH: IH 2450000
mf: $C_{43}H_{66}O_{14}$ mw: 807.09

SYNS: ALPHA-ACETYLDIGITOXIN * ACYLANID

THR: Poison by ingestion and intravenous
routes. When heated to decomposition, it emits
acrid smoke and fumes.

ACETYL DIGITOXIN-beta   HR: 3
CAS: 1264-51-3   NIOSH: IH 2625000
mf: $C_{43}H_{66}O_{14}$ mw: 807.09

SYN: BETA-ACETYLDIGOXIN

THR: Poison by ingestion and intravenous
routes. When heated to decomposition, it emits
acrid smoke and fumes.

ACETYL DIGOXIN-alpha   HR: 3
CAS: 5511-98-8   NIOSH: IH 6300000
mf: $C_{43}H_{66}O_{15}$ mw: 823.09

SYNS: ALPHA-ACETYLDIGOXIN * DIGORID A

THR: Poison by ingestion, intravenous, and intraduodenal routes. When heated to decomposition, it emits acrid smoke and fumes.

ACETYL DIGOXIN-beta   HR: 3
CAS: 5355-48-6   NIOSH: IH 6330000
mf: $C_{43}H_{66}O_{15}$ mw: 823.09

SYN: BETA-ACETYLDIGOXIN

THR: Poison by ingestion, intravenous, and intraduodenal routes. Heat decomposition emits acrid smoke and fumes.

ACETYLENE   HR: 1
CAS: 74-86-2   NIOSH: AO 9600000
DOT: 1001
mf: C_2H_2 mw: 26.04

PROP: Colorless gas, garlic-like odor. Flammable. bp: $-84.0°$ (sublimes), lel = 2.5%, uel = 82%, mp: $-81.8°$, flash p: 0°F (CC), d: 1.173 g/liter @ 0°, autoign temp: 581°F, vap press: 40 atm @ 16.8°, vap d: 0.91. d: (liq) 0.613 @ $-80°$. d: (solid) 0.730 @ $-85°$. Quite sol in water, very sol in alc, almost misc in ether.

SYNS: ACETYLEN * ETHINE * ETHYNE

OSHA PEL: CL 2500 ppm
DOT Classification: Label: Flammable Gas

THR: Narcotic in high concentration, otherwise a simple asphyxiant. In general industrial practice, acetylene does not constitute a serious toxic hazard. It is a very dangerous fire hazard when exposed to heat, flame, or oxidizers. Moderate explosion hazard when exposed to heat or flame or by spontaneous chemical reaction. At high pressures and moderate temperatures, and in the absence of air, acetylene has been known to decompose explosively. Incompatible with copper; brass; copper salts; copper carbide; pyroforic Co; Hg; Hg salts; K; Ag; Ag salts; RbH; CsH; halogens; HNO_3; NaH; halogens. Acetylene + halide + UV can explode. Molten K ignites in C_2H_2 and then explodes. C_2H_2 reacts vigorously with trifluoromethyl hypofluorite. With O_2, C_2H_2 can detonate very powerfully. See acetylides. When ignited, it burns with an intensely hot flame; can react vigorously with oxidizing materials. To fight fire, use CO_2, water spray, or dry chemical. Stop flow of gas. For further information, see Vol. 1, No. 2 of *DPIM Report*.

ACETYLENE CHLORIDE   HR: 3
mf: CHCCl   mw: 60.5

PROP: A gas; bp: $-31°$, vap d: 2.0, mp: $-126°$.

SYN: CHLOROETHYNE

THR: Dangerous fire hazard by spontaneous chemical reaction. Spontaneously flammable in air. Shock will explode it. When heated to decomposition, it emits highly toxic fumes of phosgene; can react vigorously with oxidizing materials. See also chlorinated hydrocarbons, aliphatic.

ACETYLENEDICARBOXAMIDE   HR: 3
CAS: 543-21-5   NIOSH: AO 9900000
mf: $C_4H_4N_2O_2$ mw: 112.10

PROP: Produced by *Str. Reticuli var. Aqua myceticus* and is identical to Cellocidin

SYNS: ACETYLENEDICARBOXYLIC ACID DIAMIDE * AQUAMYCIN * 2-BUTYNEDIAMIDE * CELLOCIDIN * LENAMYCIN

THR: Poison by intravenous and intraperitoneal routes. When heated to decomposition, it emits toxic fumes of NO_x.

ACETYLENEDICARBOXYLIC ACID
MONOPOTASSIUM SALT   HR: 3
CAS: 928-04-1   NIOSH: AP 0700000
mf: $C_4HO_4 \cdot K$ mw: 152.15

SYN: MONOPOTASSIUM SALT OF ACETYLENEDICARBOXYLIC ACID

THR: Poison by ingestion, intravenous, and intraperitoneal routes. When heated to decomposition, it emits acrid smoke and fumes, such as KO_x.

ACETYLENE DICHLORIDE   HR: 2
CAS: 540-59-0   NIOSH: KV 9360000
mf: $C_2H_2Cl_2$ mw: 96.94

SYNS: 1,2-DICHLOR-AETHEN (GERMAN) * DICHLORO-1,2-ETHYLENE (FRENCH) * SYM-DICHLOROETHYLENE * 1,2-DICHLOROETHYLENE * NCI-C56031
ACGIH TLV: TWA 200 ppm; STEL 250 ppm
OSHA PEL: TWA 200 ppm

THR: Moderately toxic by various routes. When heated to decomposition, it emits highly toxic fumes of Cl^-.

trans-ACETYLENE DICHLORIDE   HR: 2
CAS: 156-60-5   NIOSH: KV 9400000
mf: $C_2H_2Cl_2$ mw: 96.94

PROP: Colorless liquid, pleasant odor. mp: $-50°$, bp: 48°, flash p: 36°F, autoign temp: 860°F, lel = 9.7%, uel = 12.8%, d: 1.2743 @ 25°/4°, vap press: 400 mm @ 30.8°, vap d: 3.34.

SYNS: TRANS-1,2-DICHLOROETHYLENE

THR: Moderately toxic by inhalation with human central nervous system effects. Exposure to high vapor concentration can cause nausea, vomiting, weakness, tremor and cramps. Recovery is usually prompt following removal from exposure. Dermatitis may result from de-fatting action on skin. Dangerous fire hazard when exposed to heat, flame or oxidizers. Moderate explosion hazard in the form of vapor when exposed to flame. See chlorides. To fight fire, use water, foam, CO_2, dry chemical.

ACETYLENE TETRABROMIDE   HR: 3
CAS: 79-27-6   NIOSH: KI 8225000
DOT: 2504
mf: $C_2H_2Br_4$ mw: 345.68

PROP: Colorless to yellow liquid. bp: 151° @ 54 mm, fp: $-1°$, d: 2.9638 @ 20°/4°, autoign temp: 635°F.

SYNS: MUTHMANN'S LIQUID * 1,1,2,2-TETRABROMAETHAN (GERMAN) * TETRABROMOACETYLENE * 1,1,2,2-TETRABROMOETANO (ITALIAN) * S-TETRABROMOETHANE * 1,1,2,2-TETRABROMOETHANE * 1,1,2,2-TETRABROOMETHAAN (DUTCH)

OSHA PEL: TWA 1 ppm
ACGIH TLV: TWA 1 ppm
DFG MAK: 1 ppm (14 mg/m^3)
DOT Classification: ORM-A, Label: None

THR: Poison by inhalation and ingestion. An experimental carcinogen. Mutagenic data. It is an irritant and a narcotic. When heated, it emits highly toxic fumes of carbonyl bromide.

ACETYLENE TETRACHLORIDE   HR: 3
CAS: 79-34-5   NIOSH: KI 8575000
mf: $C_2H_2Cl_4$ mw: 167.84

PROP: Heavy, colorless, mobile liquid, chloroform-like odor. mp: $-43.8°$, bp: 146.4°, d: 1.600 @ 20°/4°.

SYNS: 1,1,2,2-CZTEROCHLOROETAN (POLISH) * 1,1-DICHLORO-2,2-DICHLOROETHANE * NCI-C03554 * 1,1,2,2-TETRACHLOORETHAAN (DUTCH) * 1,1,2,2-TETRACHLORAETHAN (GERMAN) * 1,1,2,2-TETRACHLORETHANE (FRENCH) * SYM-TETRACHLOROETHANE * 1,1,2,2-TETRACHLOROETHANE * 1,1,2,2-TETRACLOROETANO (ITALIAN) * TETRACHLORURE D'ACETYLENE (FRENCH)

OSHA PEL: TWA 5 ppm (skin)
DFG MAK: 1 ppm (1 mg/m^3)

THR: Poison by inhalation and ingestion. An experimental carcinogen. Moderately toxic by skin contact. Considered the most toxic of the common chlorinated hydrocarbons. A concentration of 3 ppm produces an odor. An irritant to the eyes, upper respiratory tract, and mucous membranes. Can cause dermatitis. It can produce yellow atrophy and cirrhosis of the liver. Central nervous and peripheral nerve system effects. The initial symptoms resulting from exposure to the vapor are lacrimation, salivation and irritation of the nose and throat. Continued exposure to high concentrations results in restlessness, dizziness, nausea, vomiting and narcosis. Mutagenic data. Considered to be a very severe industrial hazard and its use has been restricted or even forbidden in certain countries. Reacts violently with N_2O_4, 2,4-dinitrophenyl disulfide and contact with sodium or potassium. When heated in contact with solid potassium hydroxide, a spontaneously flammable gas is evolved. Any water can cause appreciable hydrolysis, even at room temperature, and both hydrolysis and oxidation become comparatively rapid above 110°. When heated it emits highly toxic decomposition products. For further information, see Vol. 5, No. 4 of *DPIM Report*.

ACETYLENE TRICHLORIDE   HR: 3
CAS: 79-01-6   NIOSH: KX 4550000
DOT: 1710
mf: C_2HCl_3 mw: 131.38

PROP: Stable, colorless, heavy, mobile liquid, chloroform-like odor. mp: $-73°$, bp: 87.1°, fp: $-86.8°$, d: 1.45560 @ 25°/4°, autoign temp: 788°F; vap press: 100 mm @ 32°, vap d: 4.53, flash p: none, lel = 12.5%, uel = 90%.

SYNS: BLANCOSOLV * 1-CHLORO-2,2-DICHLOROETHYLENE * CIRCOSOLV * 1,1-DICHLORO-2-CHLOROETHYLENE * ETHINYL TRICHLORIDE

∗ ETHYLENE TRICHLORIDE ∗ NCI-C04546
∗ TRICHLOORETHEEN (DUTCH) ∗ TRICHLO-
RAETHEN (GERMAN) ∗ TRICHLORETHENE
(FRENCH) ∗ TRICHLORETHYLENE ∗ TRICHLO-
ROETHENE ∗ TRICHLOROETHYLENE ∗ 1,1,2-
TRICHLOROETHYLENE ∗ 1,2,2-TRICHLOROETH-
YLENE ∗ TRI-CLENE ∗ TRICLORETENE
(ITALIAN) ∗ TRICLOROETILENE (ITALIAN)
∗ TRIELINA (ITALIAN)

OSHA PEL: TWA 100 ppm; CL 200; Pk 300/
5M/2H
DOT Classification: ORM-A, Label: None

THR: Poison by intravenous route. Moderately toxic by other routes. An experimental carcinogen. Mutagenic data. Severe eye irritant. A form of addiction has been observed in exposed workers. A common air contaminant. Can react violently with Al; Ba; N_2O_4; Li; Mg; liquid O_2; O_2; KOH; KNO_3; Na; NaOH; Ti. Dangerous disaster hazard; see chlorides.

ACETYL ETHYL TETRAMETHYL
TETRALIN   HR: 3
CAS: 88-29-9   NIOSH: AL 3031000
mf: $C_{18}H_{26}O$ mw: 258.44

PROP: White crystals.

SYNS: 2′-ACETONAPHTHONE, 3′-ETHYL-5′,6′,
7′,8′-TETRAHYDRO-5′,5′,8′-TETRAMETHYL-
∗ 7-ACETYL-1,1,4,4-TETRAMETHYL-1,2,3,4-
TETRAHYDRONAPHTHALENE ∗ ACETYLETHYL
TETRAMETHYL TETRALIN ∗ 6-ACETYL-1,1,4,4-
TETRAMETHYL-7-ETHYL-1,2,3,4,-TETRALIN
∗ AETT ∗ ETHANONE, 1-(3-ETHYL-5,6,
7,8-TETRAHYDRO-5,5,8,8-TETRAMETHYL-2-
NAPHTHALENYL)(9CI) ∗ 3′-ETHYL-5′,6′,
7′,8′-TETRAHYDRO-5′,5′,8′-TETRAMETHYL-2′-
ACETONAPHTHONE ∗ 1-(3-ETHYL-5,6,7,8-
TETRAHYDRO-5,5,8,8-TETRAMETHYL-2-
NAPHTHALENYL)-ETHANONE ∗ MUSK 36A
∗ POLYCYCLIC MUSK ∗ VERSALIDE

THR: Poison by ingestion and intraperitoneal routes. Moderately toxic by other routes, especially by skin contact. A moderate skin irritant. Exposure causes central nervous system effects.

ACETYL FLUORIDE   HR: 3
CAS: 557-99-3   NIOSH: AP 2800000
mf: C_2H_3FO mw: 62.05

PROP: d: 1.002 @ 15°/4°; mp: −60°; bp: 20.8°. Sl sol in alc, ether, acetone and benzene.

SYN: METHYLCARBONYL FLUORIDE

THR: Very toxic by inhalation. See also fluorides. When heated to decomposition, it emits toxic fumes of F^-.

ACETYL HYDRAZIDE   HR: 3
CAS: 1068-57-1   NIOSH: AI 1225000
mf: $C_2H_6N_2O$ mw: 74.10

SYNS: ACETHYDRAZIDE ∗ ACETOHYDRAZIDE
∗ N-ACETYLHYDRAZINE ∗ ENT-61241
∗ ETHANEHYDRAZONIC ACID ∗ MONOACETYL-
HYDRAZINE

THR: Poison by intraperitoneal route. Exposure can cause hemolysis and liver damage. See also phenyl hydrazine. When heated to decomposition, it emits toxic fumes of NO_x.

N-ACETYL-4-HYDROXY-m-ARSANILIC
ACID   HR: 3
CAS: 97-44-9   NIOSH: CF 8400000
mf: $C_8H_{10}AsNO_5$ mw: 275.11

PROP: Crystals; sl sol in water. decomp @ 240°-250°.

SYNS: 3-ACETAMIDO-4-HYDROXYBENZENEAR-
SONIC ACID ∗ 3-ACETAMIDO-4-HYDROXY-PHE-
NYLARSONIC ACID ∗ ACETPHENARSINE
∗ 3-ACETYLAMINO-4-HYDROXYPHENYLARSONIC
ACID ∗ STOVARSOL

OSHA PEL: TWA 500 ug(As)/m³

THR: Poison by ingestion and intravenous routes. See also arsenic compounds. When heated to decomposition, it emits very toxic fumes of NO_x and As.

ACETYLIDES
THR: See individual compounds. Severe explosion hazard when shocked or exposed to heat. See also acetylene.

ACETYL IODIDE   HR: 3
CAS: 507-02-8   NIOSH: AP 4670000
DOT: 1898
mf: C_2H_3IO mw: 169.95

PROP: Brown transparent, fuming liquid; bp: 108°; d: 2.067 @ 20°/4°; decomp in water, alc; sol in ether.
DOT Classification: Label: Corrosive

THR: A toxic, corrosive material. Reacts with water or steam to produce toxic and corrosive fumes. Dangerous to use. When heated to decomposition, it emits toxic fumes of I^-.

ACETYLKIDAMYCIN HR: 3
CAS: 39393-24-8   NIOSH: CB 4584000
mf: $C_{46}H_{58}N_2O_{13}$ mw: 847.06

THR: Poison by intravenous and intraperitoneal routes. Moderately toxic by ingestion. When heated to decomposition, it emits toxic fumes of NO_x.

1-ACETYLLYSERGIC ACID DIETHYLAMIDE BITARTRATE HR: 3
CAS: 63938-24-9   NIOSH: KE 2100000
mf: $C_{22}H_{27}N_3O_2 \cdot 2C_4H_4O_6$ mw: 661.68

SYN: 1-ACETYL-9,10-DIDEHYDRO-N,N-DIETHYL-6-METHYLERGOLINE-8-BETA-CARBOXAMIDE BITARTRATE

THR: Poison by ingestion and intravenous routes. Very small amounts produce psychotropic effects. When heated to decomposition, it emits toxic fumes of NO_x.

d-1-ACETYL LYSERGIC ACID MONOETHYLAMIDE HR: 3
CAS: 50485-03-5   NIOSH: KE 2275000
mf: $C_{20}H_{23}N_3O_2$ mw: 337.46

SYNS: 1-ACETYL-9,10-DIDEHYDRO-N-ETHYL-6-METHYLERGOLINE-8-BETA-CARBOXAMIDE * 1-ACETYLLYSERGIC ACID ETHYLAMIDE

THR: Poison by ingestion and intravenous routes. Very small amounts produce psychotropic effects. When heated to decomposition, it emits toxic fumes of NO_x.

o-ACETYL-beta-METHYLCHOLINE CHLORIDE HR: 3
CAS: 62-51-1   NIOSH: BR 5250000
mf: $C_8H_{18}NO_2 \cdot Cl$ mw: 195.72

PROP: mp: 172°-173°. Very sol in water and alc; decomp in alkalies and ether.

SYNS: (2-HYDROXYPROPYL)TRIMETHYLAMMONIUMCHLORIDE ACETATE * METHACHOLINE CHLORIDE * METHACHOLINIUM CHLORIDE * METHYLACETYL CHOLINE * BETA-METHYLACETYLCHOLINE CHLORIDE * TRIMETHYL-BETA-ACETOXYPROPYLAMMONIUM CHLORIDE

THR: Poison by subcutaneous, intravenous, and intraperitoneal routes. Moderately toxic by other routes. When heated to decomposition, it emits very toxic fumes of Cl^- and NO_x.

N-ACETYL-N-(2-METHYL-4-((2-METHYLPHENYL)AZO)PHENYL)-ACETAMIDE HR: 3
CAS: 83-63-6   NIOSH: AB 4230000
mf: $C_{18}H_{19}N_3O_2$ mw: 309.40

SYNS: DIACETOTOLUIDE * DIACETYLAMINOAZOTOLUENE * N,N-DIACETYL-O-TOLYLAZO-O-TOLUIDINE * 4-O-TOLYLAZO-O-DIACETOTOLUIDE * 4'-(O-TOLYLAZO)-O-DIACETOTOLUIDIDE

THR: An experimental carcinogen. When heated to decomposition, it emits dangerous and toxic fumes of NO_x.

ACETYLMETHYLNITROSOUREA HR: 3
CAS: 28895-91-2   NIOSH: YR 7000000
mf: $C_4H_7N_3O_3$ mw: 145.14

SYNS: ACETYL-METHYL-NITROSO-HARNSTOFF (GERMAN) * N'-ACETYL-METHYLNITROSOUREA * N-METHYL-N-NITROSO-N'-ACETYLUREA * 1-METHYL-1-NITROSOACETYLUREA

THR: Poison by ingestion. Experimental carcinogen. Mutagenic data. When heated to decomposition, it emits toxic fumes of NO_x.

ACETYL NITRATE HR: 3
CAS: 591-09-3   NIOSH: AF 3925000
mf: $C_2H_3NO_4$ mw: 105.06

PROP: Colorless, fuming, mobile liquid; bp: 22° @ 70 mm; d: 1.24 @ 15°/4°.

SYN: ACETIC ACID, ANHYDRIDE WITH NITRIC ACID (1:1)

THR: A poison irritant. Corrosive to the eyes. Violently unstable. Incompatible with HgO and other active oxides. When heated to decomposition, it emits toxic fumes of NO_x and/or explodes.

4-(ACETYLOXY)-12,13-EPOXY-3,7,15-TRIHYDROXY TRICHOTHEC-9-EN-8-ONE-(3-alpha,4-beta,7-beta) HR: 3
CAS: 23255-69-8   NIOSH: YD 0160000
mf: $C_{17}H_{22}O_8$ mw: 354.39

PROP: Isolated from the culture filtrate of *fusarium nivale*

SYNS: 4-ACETYLOXY-12,13-EPOXY-3,7,15-TRIHYDROXY-(3-ALPHA,4-BETA,7-BETA)-TRICHOTHEC-9-EN-8-ONE * 3,7,15-TRIHYDROXY-4-ACETOXY-8-OXO-12,13-EPOXY-DELTA(SUP

9)-TRICHOTHECENE ∗ 3,7,15-TRIHYDROXY-SCIRP-4-ACETOXY-9-EN-8-ONE

THR: Poison by ingestion, intraperitoneal, subcutaneous, and intravenous routes. An experimental carcinogen. When heated to decomposition, it emits smoke and acrid fumes.

5-(1-ACETYLOXY-2-PROPENYL)-1,3-BENZODIOXOLE   HR: 3
CAS: 34627-78-6   NIOSH: DF 4921000
mf: $C_{12}H_{12}O_4$ mw: 220.24

SYN: 1'-ACETOXYSAFROLE

THR: An experimental carcinogen. Mutagenic data. When heated to decomposition, it emits acrid smoke.

ACETYL PEROXIDE   HR: 3
CAS: 110-22-5   NIOSH: AP 8500000
mf: $C_4H_6O_4$ mw: 118.04

PROP: Solid or colorless crystals or liquid. Slightly sol in cold water, decomp; d: 1.18. mp: 30°, bp: 63° @ 21 mm; can explode.

SYN: DIACETYL PEROXIDE

THR: Strong irritant. An experimental carcinogen. Dangerous fire hazard by spontaneous chemical reaction. A powerful oxidizing agent; can cause ignition of organic materials on contact. Severe explosion hazard when shocked or exposed to heat. It may explode spontaneously, possibly when more than 24 hours old; it should be used up as soon as prepared. It will react with water or steam to produce heat; can react vigorously with reducing materials; can emit toxic fumes on contact with acid or acid fumes. To fight fire, use CO_2, dry chemical.

ACETYL PEROXIDE SOLUTION   HR: 3
NIOSH: AP 8550000
DOT: 2084

PROP: Crystal clear liquid. mp: −7°; flash p: 113°F (OC); d: 1.18 @ 20°. Solution contains not over 25% acetyl peroxide .

SYN: DIACETYL PEROXIDE (SOLUTION)

DOT Classification: Label: Organic Peroxide

THR: Strong irritant to skin, eyes, and mucous membranes. See also acetyl peroxide, solid and peroxides, organic. Moderate fire hazard when exposed to heat or flame, or by spontaneous

chemical reaction. An oxidizing agent. When heated to decomposition, it emits toxic fumes; can react vigorously with oxidizing materials. To fight fire, use foam, CO_2.

1-ACETYL-2-PICOLINOLHYDRAZINE   HR: 3
CAS: 17433-31-7   NIOSH: MU 7700000
mf: $C_8H_{11}N_3O_2$ mw: 181.22

SYNS: 1-ACETYL-2-PICOLINOYLHYDRAZINE ∗ NCI-C04739 ∗ NSC-68626

THR: Poison by ingestion and intravenous routes. An experimental carcinogen. When heated to decomposition, it emits toxic fumes such as NO_x.

o-ACETYLSTERIGMATOCYSTIN   HR: 3
CAS: 58086-32-1   NIOSH: LV 1735000
mf: $C_{20}H_{14}O_7$ mw: 366.34

THR: An experimental carcinogen. Mutagenic data. When heated to decomposition, it emits acrid smoke.

1-ACETYL-2-THIOHYDANTOIN   HR: 3
CAS: 584-26-9   NIOSH: MT 8575000
mf: $C_5H_6N_2O_2S$ mw: 158.19

PROP: Insol in water and ether; sl sol in alc. mp: 175°-176°.

SYNS: USAF B-7 ∗ USAF BE-0405

THR: Poison by intravenous and intraperitoneal routes. When heated to decomposition, it emits very toxic fumes such as SO_x and NO_x.

ACETYL THIOUREA   HR: 3
CAS: 591-08-2   NIOSH: YR 7700000
mf: $C_3H_6N_2OS$ mw: 118.17

PROP: Sol in hot water and alc; sl sol in ether; mp: 166°-167°.

SYNS: 1-ACETYL-2-THIOUREA ∗ USAF EK-4890

THR: Poison by ingestion and intraperitoneal routes. See also sulfides. When heated to decomposition, it emits very toxic fumes of NO_x and SO_x.

N-ACETYL TRIMETHYLCOLCHICINIC ACID   HR: 3
CAS: 477-27-0   NIOSH: GH 0520000
mf: $C_{21}H_{23}NO_6$ mw: 385.45

SYN: 7-ACETAMIDO-10-HYDROXY-1,2,3-TRIME-
THOXY-6,7-DIHYDROBENZO(A)HEPTALEN-
9(5H)-ONE

THR: Poison by many routes. When heated to decomposition, it emits toxic fumes of NO_x.

ACID BLUE 1   HR: 3
CAS: 129-17-9   NIOSH: BP 6830000
mf: $C_{27}H_{31}N_2O_6S_2 \cdot Na$ mw: 566.71

SYNS: ANHYDRO-4,4'-BIS(DIETHYLAMINO)-
TRIPHENYLMETHANOL-2',4''-DISULPHONIC ACID,
MONOSODIUM SALT * C.I. ACID BLUE 1, SO-
DIUM SALT * COSMETIC GREEN BLUE R25396
* 4,4'-DI(DIETHYLAMINO)-4',6'-DISULPHOTRI-
PHENYLMETHANOL ANHYDRIDE, SODIUM SALT

THR: An experimental carcinogen. Moderately toxic by several routes. When heated to decomposition, it emits very toxic fumes of NO_x and SO_x. For further information, see Vol. 1, No. 4 of *DPIM Report*.

ACID BLUE 92   HR: 2
CAS: 3861-73-2   NIOSH: QJ 6300000
mf: $C_{26}N_{19}N_3O_{10}S_3 \cdot 3Na$ mw: 698.63

SYNS: ACID BLUE A * ACID LEATHER BLUE R
* ACID WOOL BLUE RL * AIREDALE BLUE RL
* CALCOCID FAST BLUE SR * C.I. 13390
* COLACID BLUE A * COOMASSIE BLUE
* CYANINE ACID BLUE R * FAST WOOL BLUE R
* FENAZO BLUE SR * PONTACYL FAST BLUE R
* SODIUM AMAZOLENE * SODIUM ANAZOLENE
* TERTRACID FAST BLUE SR * TRISODIUM-4'-
ANILINO-8-HYDROXY-1,1'-AZONAPHTHALENE-
3,6,5'-TRISULFONATE * WOOL FAST BLUE R

THR: Moderately toxic by various routes. When heated to decomposition, it emits very toxic fumes of SO_x, NO_x, and Na_2O.

ACID BRILLIANT GREEN BS   HR: 3
CAS: 3087-16-9   NIOSH: BQ 1150000
mf: $C_{27}H_{26}N_2O_7S_2 \cdot Na$ mw: 577.66

SYNS: C.I. 44090 * C.I. ACID GREEN 50,
MONOSODIUM SALT * C.I. FOOD GREEN 4
* WOOL GREEN S (BIOLOGICAL STAIN)

THR: An experimental carcinogen. Mutagenic data. When heated to decomposition, it emits very toxic fumes of SO_x and NO_x.

ACID BUTYL PHOSPHATE   HR: 3
CAS: 12788-93-1   NIOSH: TB 8490000

DOT: 1718
mf: $C_4H_{10}O_4P$ mw: 153.1

PROP: Water white liquid, sol in alc, acetone and toluene, insol in water, petroleum and naphtha. D: 1.120-1.125 @ 25°/40°; flash p: 230°F (COC)

SYNS: ACID BUTYL PHOSPHATE * n-BUTYL
ACID PHOSPHATE * BUTYL PHOSPHORIC ACID

DOT Classification: Label: Corrosive

THR: Toxic and corrosive. When heated to decomposition, it emits highly toxic fumes. (See also phosphoric acid.) Slight fire hazard when exposed to heat or flame.

ACID CARBOYS, EMPTY
PROP: *Warning:* These containers may contain concentrated vapors or even some liquid acid remaining from their original contents. Therefore, they can give rise to all the hazards of their original contents.

ACID GREEN 3   HR: 3
CAS: 4680-78-8   NIOSH: BQ 4375000
mf: $C_{37}H_{36}N_2O_6S_2 \cdot Na$ mw: 691.86

SYNS: ACIDAL GREEN G * ACID GREEN
* C.I. ACID GREEN 3, MONOSODIUM SALT
* C.I. FOOD GREEN 1 * FD AND C GREEN NO.
1 * FDC GREEN 1 * GUINEA GREEN B
* PONTACYL GREEN BL * SULFACID BRILLIANT
GREEN 1B

THR: An experimental carcinogen. Mutagenic data. When heated to decomposition, it emits very toxic fumes of SO_x and NO_x. For further information, see Vol. 1, No. 2 of *DPIM Report*.

ACONITE   HR: 3
CAS: 8063-12-5   NIOSH: AR 5570000

PROP: The dried tuberous root of *Aconitum Napellus*, composed of several alkaloids, the chief one being aconitine.

SYNS: ACONITUM * FRIARS COWL
* MONKSHOOD * MOUSEBANE * WOLFSBANE

THR: Poison by all routes including absorption through the skin. (The poisonous alkaloid can be absorbed through the skin sufficiently to cause death.) Slight fire hazard when exposed to heat or flame. When heated to decomposition, it emits highly toxic fumes.

ACONITINE   HR: 3
CAS: 302-27-2   NIOSH: AR 5960000
mf: $C_{34}H_{49}NO_{11}$ mw: 647.74

PROP: White crystalline alkaloid; feeble bitter taste. mp: 204°.

SYN: ACETYL BENZOYL ACONINE

THR: Poison by all routes including absorption through the skin. When heated to decomposition, it emits highly toxic fumes of NO_x. For further information, see Vol. 1, No. 3 of *DPIM Report*.

ACONITINE (CRYSTALLINE)   HR: 3
CAS: 302-27-2   NIOSH: AR 5960000
mf: $C_{34}H_{47}NO_{11}$ mw: 645.82

SYNS: ACONITANE * ACONITIN CRISTALLISAT (GERMAN)

THR: Poison by intravenous, subcutaneous, and intraperitoneal routes. When heated to decomposition, it emits toxic fumes of NO_x.

ACONITINE NITRATE   HR: 3
mf: $C_{34}H_{47}NO_{11}HNO_3 \cdot 5HOH$ mw: 798.8

PROP: Crystals. mp: 200° (decomp)

THR: Poison by various routes. (See aconitine.) When heated to decomposition, it emits highly toxic fumes of NO_x, can react vigorously with reducing materials. (See also nitrates). Moderate fire hazard by spontaneous chemical reaction; an oxidizing agent.

ACRIDINE   HR: 3
CAS: 260-94-6   NIOSH: AR 7175000
mf: $C_{13}H_9N$ mw: 179.23

PROP: Small colorless needles. mp: 110.5°, bp: 346°, d: 1.005 @ 19.7°/4°, vap press: 1 mm @ 129.4°. Slightly sol in hot water, sol in alc, ether and CS_2.

SYNS: 9-AZAANTHRACENE * 10-AZAANTHRACENE * BENZO(B)QUINOLINE * 2,3-BENZOQUINOLINE * DIBENZO(B,E)PYRIDINE

THR: Poison by subcutaneous and intravenous routes. Mutagenic data. A skin, eye, and mucous membrane irritant. When heated to decomposition, it emits toxic fumes of NO_x. For further information, see Vol. 1, No. 8 of *DPIM Report*.

4'-(9-ACRIDINYLAMINO)METHANESUL-
PHON-m-ANISIDIDE   HR: 3
CAS: 51264-14-3   NIOSH: PB 1080000
mf: $C_{21}H_{19}N_3O_3S$ mw: 393.49

SYNS: NSC 141549 * 4'-(9-ACRIDINYLAMINO) METHYLSULFONYL-M-ANISIDINE * NSC 249992

THR: Poison by intravenous route. Systemic blood and gastrointestinal effect in humans. Mutagenic data. When heated to decomposition, it emits very toxic fumes of NO_x and SO_x.

ACRIFLAVINIUM CHLORIDE   HR: 3
CAS: 8048-52-0   NIOSH: AR 9660000
mf: $C_{14}H_{14}ClN_3 \cdot Cl_3H_{11}N_3$ mw: 469.03

SYNS: ACRIFLAVINE MIXTURE WITH PROFLAVINE * 3,6-DIAMINO-10-METHYLACRIDINIUM CHLORIDE MIXTURE WITH 3,6-ACRIDINEDIAMINE * 3,6-DIAMINOACRIDINE MIXTURE WITH 3,6-DIAMINO-10-METHYLACRIDINIUM CHLORIDE * 2,8-DIAMINO-10-METHYLACRIDINIUM CHLORIDE MIXTURE WITH 2,8-DIAMINOACRIDINE

THR: Poison by subcutaneous route. An experimental carcinogen. Mutagenic data. When heated to decomposition it emits very toxic fumes of NO_x and Cl^-.

ACROLEIN   HR: 3
CAS: 107-02-8   NIOSH: AS 1050000
DOT: 1092
mf: C_3H_4O mw: 56.07

PROP: Colorless or yellowish liquid, disagreeable choking odor. Sol in water, alc and ether. mp: −87.7°, bp: 52.5°, flash p: <0°F, d: 0.841 @ 20°/4°, autoign temp: unstable (455°F), lel = 2.8%, uel = 31%, vap. d: 1.94.

SYNS: ACRALDEHYDE * ACROLEINA (ITALIAN) * ACROLEINE (DUTCH, FRENCH) * ACRYLALDEHYD (GERMAN) * ACRYLALDEHYDE * ACRYLIC ALDEHYDE * ALDEIDE ACRILICA (ITALIAN) * ALDEHYDE ACRYLIQUE (FRENCH) * ALLYL ALDEHYDE * AKROLEIN (CZECH) * AKROLEINA (POLISH) * ETHYLENE ALDEHYDE * NSC 8819 * PROPENAL (CZECH) * 2-PROPENAL * PROP-2-EN-1-AL * 2-PROPEN-1-ONE

OSHA PEL: TWA 0.1 ppm
ACGIH TLV: TWA 0.1 ppm; STEL 0.3 ppm
DFG MAK: 0.1 ppm (0.25 mg/m³)
DOT Classification: Flammable Liquid

THR: Poison by most routes. An experimental carcinogen. Mutagenic data. Dangerous fire hazard when exposed to heat, flame, or oxidizers. Explosion hazard: Incompatible with acids; alkalis; amines; SO_2; thiourea; metal salts; oxidants; (light + heat). When heated to decomposition, it emits highly toxic fumes; can react vigorously with oxidizing materials. To fight fire, use CO_2, dry chemical or alcohol foam. For further information, Vol. 3, No. 3 of *DPIM Report*.

ACROLEIN DIACETATE   HR: 3
CAS: 869-29-4   NIOSH: UC 9625000
mf: $C_7H_{10}O_4$ mw: 158.17

PROP: Liquid, mp: −36.6°, bp: 107° @ 50 mm, flash p: 180°F (OC), d: 1.0749 @ 20°/20°, vap d: 5.46.

SYNS: ALLYLIDENE DIACETATE * DIACETOXY-PROPENE * 1,1-DIACETOXYPROPENE-2 * 3,3-DIACETOXYPROPENE

THR: Poison by inhalation, ingestion, and skin contact. A skin and eye irritant. Moderate fire hazard when exposed to heat or flame; can react with oxidizing materials. To fight fire, water may be used to blanket the fire, foam, CO_2, dry chemical.

ACRONYCINE   HR: 3
CAS: 7008-42-6   NIOSH: UQ 0330000
mf: $C_{20}H_{19}NO_3$ mw: 321.40

SYNS: 3,12-DIHYDRO-6-METHOXY-3,3,12-TRI-METHYL-7-HPYRANO(2,3-C)ACRIDIN-7-ONE * NCI-C01536 * NSC 403169

THR: An experimental carcinogen. Mutagenic data. When heated to decomposition, it emits toxic fumes of NO_x.

ACRYLAMIDE   HR: 3
CAS: 79-06-1   NIOSH: AS 3325000
DOT: 2074
mf: C_3H_5NO mw: 71.09

PROP: White crystalline solid. Very sol in water, alc and ether; mp: 84.5 ± 0.3°, bp: 125° @ 25 mm, d: 1.122 @ 30°, vap press: 1.6 mm @ 84.5°, vap d: 2.45.

SYNS: ACRYLIC AMIDE * AKRYLAMID (CZECH) * ETHYLENECARBOXAMIDE * PROPENAMIDE * 2-PROPENAMIDE

OSHA PEL: TWA 300 ug/m³ (skin)
DFG MAK: 0.3 mg/m³
DOT Classification: Poison B

THR: Poison by ingestion, skin contact, and intraperitoneal routes. Mutagenic data. A skin and eye irritant. When heated to decomposition,

it emits acrid fumes and NO_x. For further information, see Vol. 2, No. 4 of *DPIM Report*.

ACRYLIC ACID   HR: 3
CAS: 79-10-7   NIOSH: AS 4375000
DOT: 2218
mf: $C_3H_4O_2$ mw: 72.07

PROP: Liquid, acrid odor. Misc in water, benzene, alc, chloroform, ether and acetone. mp: 13°, bp: 141°, d: 1.062, vap press: 10 mm @ 39.9°, flash p: 130°F (OC), vap. d: 2.45

SYNS: ACROLEIC ACID * ACRYLIC ACID, GLACIAL * ETHYLENECARBOXYLIC ACID * PROPENE ACID * 2-PROPENOIC ACID * VINYL-FORMIC ACID

ACGIH TLV: TWA 10 ppm
DOT Classification: Label: Corrosive

THR: Poison by ingestion, skin contact, and intraperitoneal routes. An experimental teratogen. A skin and eye irritant. Corrosive. For further information, see Vol. 1, No. 7 of *DPIM Report*.

ACRYLIC ACID-beta-CHLOROETHYL ESTER   HR: 3
CAS: 2206-89-5   NIOSH: AS 6300000
mf: $C_5H_7ClO_2$ mw: 134.57

SYNS: CHLOROETHYL ACRYLATE * BETA-CHLOROETHYL ACRYLATE * 2-CHLOROETHYL ACRYLATE * 2-CHLOROETHANOL ACRYLATE * 2-PROPENOIC ACID-2-CHLOROETHYL ESTER

THR: Poison by inhalation, ingestion, and intraperitoneal routes. A skin and eye irritant. (See also esters.) When heated to decomposition, it emits toxic fumes of Cl^-.

ACRYLIC ACID-2-ETHOXYETHYL ESTER   HR: 2
CAS: 106-74-1   NIOSH: AS 9800000
mf: $C_7H_{12}O_3$ mw: 144.19

SYNS: CELLOSOLVE ACRYLATE * ETHOXY-ETHYL ACRYLATE * 2-ETHOXYETHYL ACRYLATE * 2-ETHOXYETHYL-2-PROPENOATE * ETHYL-ENE GLYCOL MONOETHYL ETHER ACRYLATE * ETHYLENE GLYCOL MONOETHYL ETHER PRO-PENOATE * 2-PROPENOIC ACID-2-ETHOXYETHYL ESTER

THR: Moderately toxic by various routes. A skin and eye irritant. (See also esters.) When heated to decomposition, it emits acrid smoke and fumes.

ACRYLIC ACID ESTER WITH HYDRA-CRYLONITRILE   HR: 3
CAS: 106-71-8   NIOSH: AT 1500000
mf: $C_6H_7NO_2$ mw: 125.14

PROP: Liquid, sol in water; d: 1.069; bp: polymerizes; fp: −16.9°; flash p: 255°F (COC); vap. d: 4.3.

SYNS: ACRYLIC ACID-2-CYANOETHYL ESTER
* CYANOETHYL ACRYLATE * 2-CYANOETHYL ACRYLATE * 2-CYANOETHYL PROPENOATE
* HYDRACRYLONITRILE ACRYLATE * 2-PROPENOIC ACID, 2-CYANOETHYL ESTER

THR: Poison by ingestion. A skin and eye irritant. (See also esters and nitriles.) When heated to decomposition, it emits toxic fumes of NO_x and cyanide.

ACRYLIC ACID ETHYL ESTER (INHIBITED)   HR: 2
CAS: 140-88-5   NIOSH: AT 0710000
DOT: 1917
mf: $CH_2CHCOOC_2H_5$ mw: 100.11

PROP: Colorless liquid. Acrid, penetrating odor. Sl sol in water. fp: $<-72°$, lel = 1.8%, flash p: 60°F (OC), d: 0.941 @ 20°/4° vap press: 29.3 mm @ 20°, vap d: 3.45. bp: 100°-101°.

SYNS: ETHYL PROPENOATE * ETHYL ACRYLATE, INHIBITED

DFG MAK: 25 ppm (100 mg/m³)
DOT Classification: Flammable Liquid

THR: Moderately toxic by various routes. Irritant to mucous membranes. (See also esters.) Dangerous fire hazard when exposed to heat or flame; can react vigorously with oxidizing materials. Violent reaction with chlorosulfonic acid. To fight fire, use alcohol foam.

ACRYLIC ACID-2-ETHYLHEXYL ESTER   HR: 2
CAS: 103-11-7   NIOSH: AT 0855000
mf: $C_{11}H_{20}O_2$ mw: 184.31

PROP: bp: 130° @ 50 mm; fp: −90°; flash p: 180°F (OC); d: 0.8869 @ 20°/20°; vap press: 1 mm @ 50°; vap d: 6.35

SYNS: 2-ETHYLHEXYL ACRYLATE * 2-ETHYLHEXYL-2-PROPENOATE * OCTYL ACRYLATE
* 2-PROPENOIC ACID-2-ETHYLHEXYL ESTER

THR: Moderately toxic by various routes. A skin and eye irritant. (See also esters.) Moderate

fire hazard. When heated to decomposition, it emits acrid smoke. To fight fire, use alcohol foam, CO_2, dry chemical. For further information, see 2-Ethylhexyl Acrylate, Vol. 3, No. 2 of *DPIM Report*.

ACRYLIC ACID-2-HYDROXYPROPYL ESTER   HR: 3
CAS: 999-61-1   NIOSH: AT 1925000
mf: $C_6H_{10}O_3$ mw: 130.16

SYNS: BETA-HYDROXYPROPYL ACRYLATE
* 2-HYDROXYPROPYL ACRYLATE * 1,2-PROPANEDIOL-1-ACRYLATE * 2-PROPENOIC ACID-2-HYDROXYPROPYL ESTER * PROPYLENE GLYCOL MONOACRYLATE

THR: Poison by ingestion and subcutaneous routes. (See also esters.) When heated to decomposition, it emits acrid smoke and fumes.

ACRYLONITRILE   HR: 3
CAS: 107-13-1   NIOSH: AT 5250000
DOT: 1093
mf: C_3H_3N mw: 53.07

PROP: Colorless, mobile liquid; mild odor. Sol in water. mp: −82°, bp: 77.3°, fp: −83°, flash p: 30°F (TCC), lel = 3.1%, uel = 17%, d: 0.806 @ 20°/4°, autoign temp: 898°F, vap press: 100 mm @ 22.8°, vap d: 1.83, flash p: (of 5% aq soln): <50°F

SYNS: ACRYLNITRIL (GERMAN, DUTCH)
* ACRYLONITRILE MONOMER * AKRYLONITRYL (POLISH) * CIANURO DI VINILE (ITALIAN)
* CYANOETHYLENE * CYANURE DE VINYLE (FRENCH) * ENT 54 * NITRILE ACRILICO (ITALIAN) * NITRILE ACRYLIQUE (FRENCH)
* PROPENENITRILE * 2-PROPENENITRILE
* VINYL CYANIDE

OSHA PEL: TWA 2 ppm; CL 10 ppm/15M
ACGIH TLV: TWA 2 ppm
TRK: 3 ppm (7 mg/m³)
DOT Classification: Label: Flammable
  Liquid and Poison

THR: Poison by inhalation, ingestion, skin contact, and other routes. An experimental carcinogen and teratogen. Mutagenic data. Human blood and central nervous system effects. See also cyanide and nitriles. A systemic irritant. Dangerous fire hazard when exposed to heat, flame, or oxidizers. Moderate explosion hazard when exposed to flame. Can react vigorously with oxidizing materials (see cyanide). To fight fire, use CO_2, dry chemical or alcohol foam. Incompatible with $AgNO_3$, benzyltrimethyl ammonium hydroxide, strong acids, amines, strong alkalis, Br_2, 1,2,3,4-tetrahydrocarbazole. For further information, see Vol. 3, No. 3 of *DPIM Report*.

ACRYLONITRILE POLYMER WITH STYRENE HR: 2
CAS: 9003-54-7   NIOSH: AT 6978000
mf: $(C_8H_8 \cdot C_3H_3N)x$

SYNS: ACRYLONITRILE-STYRENE COPOLYMER * ACRYLONITRILE-STYRENEPOLYMER * ACRYLONITRILE-STYRENE RESIN * POLYSTYRENE-ACRYLONITRILE * 2-PROPENENITRILE POLYMER WITH ETHENYLBENZENE * STYRENE-ACRYLONITRILE COPOLYMER * STYREN-ACRYLONITRILE-POLYMER

THR: See nitriles. When heated to decomposition, it emits toxic fumes of NO_x.

ACRYLONITRILE POLYMER WITH 1,3-BUTADIENE AND STYRENE HR: 2
CAS: 9003-56-9   NIOSH: AT 6970000

SYNS: ACRYLONITRILE-BUTADIENE-STYRENE COPOLYMER * ACRYLONITRILE-1,3-BUTADIENE-STYRENE COPOLYMER * ACRYLONITRILE-BUTADIENE-STYRENE POLYMER * ACRYLONITRILE-1,3-BUTADIENE-STYRENE POLYMER * ACRYLONITRILE-BUTADIENE-STYRENE RESIN * ACRYLONITRILE-BUTADIENE-STYRENE TERPOLYMER * ACRYLONITRILE-STYRENE-BUTADIENE RESIN * BAKELITE * BUTADIENE-ACRYLONITRILE-STYRENE COPOLYMER * BUTADIENE-ACRYLONITRILE-STYRENE TERPOLYMER * BUTADIENE-STYRENE-ACRYLONITRILE COPOLYMER * 2-PROPENENITRILE POLYMER WITH 1,3-BUTADIENE AND ETHENYLBENZENE * POLYACRYLONITRILE WITH 1,3-BUTADIENE AND ETHENYLBENZENE * STYRENE-ACRYLONITRILE-BUTADIENE COPOLYMER * STYRENE-ACRYLONITRILE-BUTADIENE POLYMER * STYRENE-ACRYLONITRILE-BUTADIENE RESIN * STYRENE-ACRYLONITRILE-BUTADIENE TERPOLYMER * STYRENE-BUTADIENE-ACRYLONITRILE COPOLYMER

THR: See nitriles. When heated to decomposition, it emits toxic fumes of NO_x.

ACRYLONITRILE POLYMER WITH 1,3-BUTADIENE AND STYRENE, COMBUSTION PRODUCTS HR: 2
NIOSH: AT 6975000
mf: $(C_3H_3N)x$

SYNS: ABS(PYROLYSISPRODUCTS) * ACRYLONITRILE-BUTADIENE-STYRENE (PYROLYSIS PRODUCTS) * POLYACRYLONITRILE * 2-PROPENENITRILE HOMOPOLYMER (9CI)

THR: Moderately toxic by inhalation. When heated to decomposition, it emits toxic fumes of NO_x.

ACTINIC RADIATION
PROP: Outdoor workers such as fishermen, sailors, soldiers and farmers show a high incidence of skin cancer. The commonest acute manifestation of actinic radiation effects on skin is sunburn leading often to skin cancer.

ACTINOMYCIN D HR: 3
CAS: 50-76-0   NIOSH: AU 1575000
mf: $C_{62}H_{86}N_{12}O_{16}$ mw: 1255.60

SYNS: ACTINOMYCIN I * NCI-C04682 * NSC 3053 * ONCOSTATIN K

THR: Poison by ingestion, intravenous, subcutaneous, and intraperitoneal routes. An experimental carcinogen and teratogen. Mutagenic data. A systemic skin irritant. When heated to decomposition, it emits toxic fumes of NO_x. For further information, see Vol. 1, No. 3 of *DPIM Report*.

ACTINOMYCIN L HR: 3
NIOSH: AU 1830000

SYN: ACTINOMYCIN 2104L

THR: An experimental carcinogen.

ACTINOMYCIN S HR: 3
CAS: 12623-78-8   NIOSH: AU 1925000

SYN: ACTINOMYCIN 1048A

THR: An experimental carcinogen.

ACTIVATED CARBON HR: 2
at wt: 12.01

PROP: Black amorphous mass. mp: 3500°; bp: 4000°; d: 3.51

SYN: CHARCOAL, ACTIVATED

THR: Usually contains small amounts of irritating and possibly toxic impurities. (See also coal tar.) Moderate fire hazard when exposed to heat or flame. Moderate explosion hazard when dust is exposed to flame. See dust explosions. To fight fire, use water.

1-ADAMANTANAMINE HYDRO-
CHLORIDE   HR: 3
CAS: 665-66-7   NIOSH: AU 4375000
mf: $C_{10}H_{17}N \cdot ClH$ mw: 187.74

SYNS: ADAMANTANAMINE HYDROCHLORIDE
* ADAMANTINE HYDROCHLORIDE * ADAMAN-
TYLAMINE HYDROCHLORIDE * 1-ADAMANTYL-
AMINE HYDROCHLORIDE * AMANTADINE HY-
DROCHLORIDE * AMINOADAMANTANE
HYDROCHLORIDE * 1-AMINOADAMANTENE
HYDROCHLORIDE * NSC 83653

THR: Poison by ingestion, intraperitoneal, and
intravenous routes. An experimental teratogen.
When heated to decomposition, it emits very
toxic fumes of NO_x and HCl.

ADAPIN   HR: 3
CAS: 1229-29-4   NIOSH: HQ 4375000
mf: $C_{19}H_{21}NO \cdot ClH$ mw: 315.87

SYNS: CIDOXEPIN HYDROCHLORIDE * 11-DI-
METHYLAMINO PROPYLIDENE-6H-DI-
BENZ(B,E)OXEPIN

THR: Poison by ingestion, subcutaneous, intraperitoneal, and intravenous routes. Has central
nervous system and blood pressure effects.

ADENINE   HR: 3
CAS: 73-24-5   NIOSH: AU 6125000
mf: $C_5H_5N_5$ mw: 135.15

SYNS: ADENINIMINE * 6-AMINOPURINE
* 6-AMINO-1H-PURINE * 6-AMINO-3H-PURINE
* 6-AMINO-9H-PURINE * 1,6-DIHYDRO-6-IMI-
NOPURINE * 3,6-DIHYDRO-6-IMINOPURINE
* 1H-PURIN-6-AMINE * USAF CB-18
* VITAMIN B4

THR: Poison by intraperitoneal route. Moderately toxic by ingestion. An experimental teratogen. Mutagenic data. When heated to decomposition, it emits toxic fumes of NO_x .

ADIPAMIDE   HR: 2
CAS: 628-94-4   NIOSH: AU 7800000
mf: $C_6H_{12}N_2O_2$ mw: 144.20

PROP: Crystals. mp: 220°. Sol in alc.

SYNS: ADIPIC ACID DIAMIDE * ADIPIC DI-
AMIDE * 1,4-BUTANEDICARBOXAMIDE
* HEXANEDIAMIDE (9CI) * NCI-C02095

THR: Moderately toxic by ingestion. When
heated to decomposition, it emits toxic fumes
of NO_x .

ADIPIC ACID   HR: 3
CAS: 124-04-9   NIOSH: AU 8400000
mf: $C_6H_{10}O_4$ mw: 146.16

PROP: Monoclinic prisms. Mp: 152°, flash p:
385°F (CC), d: 1.360 @ 25°/4°, vap press: 1
mm @ 159.5°, vap d: 5.04, autoign temp 788°F,
bp: 265° @ 10 mm. Very sol in alc. Solubility
in water = 1.4% @15°; 0.6% @ 15° in ether.

SYNS: ADIPINIC ACID * 1,4-BUTANEDICAR-
BOXYLIC ACID * 1,6-HEXANEDIOIC ACID
* KYSELINA ADIPOVA (CZECH) * MOLTEN
ADIPIC ACID

THR: Poison by intraperitoneal route. Moderately toxic by other routes. An eye irritant. Slight
fire hazard when exposed to heat or flame; can
react with oxidizing materials. When heated to
decomposition, it emits acrid smoke and fumes.
For further information, see Vol. 3, No. 3 of
DPIM Report.

ADIPONITRILE   HR: 3
CAS: 111-69-3   NIOSH: AV 2625000
mf: $C_6H_8N_2$ mw: 108.16

PROP: Water white liquid, practically odorless.
mp: 2.3°, bp: 295°, flash p: 199.4°F (OC), d:
0.965 @ 20°/4°, vap d: 3.73.

SYNS: ADIPIC ACID DINITRILE * ADIPIC ACID
NITRILE * 1,4-DICYANOBUTANE * ADIPODINI-
TRILE * HEXANEDINITRILE * HEXANEDIOIC
ACID DINITRILE * NITRILE ADIPICO (ITALIAN)
* TETRAMETHYLENE CYANIDE

THR: Poison by inhalation, ingestion, subcutaneous, and intraperitoneal routes. (See also hydrocyanic acid and nitriles.) Moderate fire hazard when exposed to heat or flame. When heated
to decomposition, it emits highly toxic fumes;
can react with oxidizing materials. To fight fire,
use foam, CO_2, dry chemical.

d-ADRENALINE   HR: 3
CAS: 150-05-0   NIOSH: DO 2800000
mf: $C_9H_{13}NO_3$ mw: 183.23

PROP: Light brown or nearly white crystals.
mp: 211-212°. Very sltly sol water, alc, 1:1
chloroform + ether.

SYNS: L-(+)-ADRENALINE * D-EPINEPHRINE

THR: Poison by subcutaneous and intravenous
routes. Can cause contact dermatitis. Slight fire
hazard when heated.

1-ADRENALINE CHLORIDE   HR: 3
CAS: 55-31-2   NIOSH: DO 3150000
mf: $C_9H_{13}NO_3 \cdot ClH$ mw: 219.69

SYNS: (-)-ADRENALINE HYDROCHLORIDE
* 1,2-BENZENEDIOL, 4-(1-HYDROXY-2-
(METHYLAMINO)ETHYL)-, HYDROCHLORIDE, (R)-
(9CI) * (-)-EPINEPHRINE HYDROCHLORIDE
* 1-ADRENALINE HYDROCHLORIDE * ADREN-
ALIN HYDROCHLORIDE * 1-1-(3,4-DIHYDROXY-
PHENYL)-2-METHYLAMINO-1-ETHANOL HYDRO-
CHLORIDE * 1-EPINEPHRINE CHLORIDE
* 1-EPINEPHRINE HYDROCHLORIDE * GELATIN-
EPINEPHRINE * 1-METHYLAMINOETHANOL-
CATHECHOL HYDROCHLORIDE * NCI-C55663

THR: Poison by ingestion, subcutaneous, intraperitoneal, and intraduodenal routes. When
heated to decomposition, it emits very toxic
fumes of Cl^- and NO_x .

ADRIAMYCIN   HR: 3
CAS: 23214-92-8   NIOSH: AV 9800000
mf: $C_{27}H_{29}NO_{11} \cdot ClH$ mw: 580.03

PROP: Isolated from cultures of *Streptomyces
Peucetius var. Caesius*.

SYNS: ADRIAMYCIN-HCL * 14-HYDROXYDAU-
NOMYCIN * 14′-HYDROXYDAUNOMYCIN
* 14-HYDROXYDAUNORUBICINE * NCI-C01514
* NSC-123127

THR: Poison by intraperitoneal, subcutaneous,
and intravenous routes. An experimental carcinogen and teratogen. Mutagenic data. Systemic
blood and cardiovascular effects. When heated
to decomposition, it emits very toxic fumes of
NO_x and HCl. For further information, see Vol.
1, No. 3 of *DPIM Report*.

AFLATOXIN   HR: 3
CAS: 1402-68-2   NIOSH: AW 5950000

THR: Poison by ingestion. Moderately toxic
by other routes. An experimental carcinogen.
Mutagenic data.

AFLATOXIN 495   HR: 3
NIOSH: AW 6825000

PROP: Mixture of 15% of Aflatoxin B1, 9%
Aflatoxin G1 and less than 1% Aflatoxins B2
and C2.

THR: Poison by ingestion, intravenous, and subcutaneous routes. Mutagenic data.

AFLATOXIN B1   HR: 3
CAS: 1162-65-8   NIOSH: GY 1925000
mf: $C_{17}H_{12}O_6$ mw: 312.29

PROP: A metabolite of *Aspergillus Flavus Link
ex Fries*. A crystalline material. mp: 268°.

SYNS: AFBI * AFLATOXIN B

THR: Acute poison by ingestion and intraperitoneal routes. An experimental carcinogen and
teratogen. Mutagenic data. When heated to decomposition, it emits acrid smoke. For further
information, see Vol. 1, No. 4 of *DPIM Report*.

AFLATOXIN B1-2,3-DICHLORIDE HR: 3
CAS: 63976-04-5   NIOSH: GY 1700000
mf: $C_{17}H_{12}Cl_2O_6$ mw: 383.19

THR: An experimental carcinogen. When
heated to decomposition, it emits toxic fumes
of Cl^- .

AFLATOXIN B2   HR: 3
CAS: 7220-81-7   NIOSH: GY 1722000
mf: $C_{17}H_{14}O_6$ mw: 314.31

SYN: DIHYDROAFLATOXIN B1

THR: Poison by ingestion. An experimental carcinogen. Mutagenic data.

AFLATOXIN G1   HR: 3
CAS: 1165-39-5   NIOSH: LV 1720000
mf: $C_{17}H_{12}O_7$ mw: 328.29

PROP: A metabolite of *Aspergillus Flavus Link
ex Fries*.

THR: Poison by ingestion. An experimental and
suspected human carcinogen. When heated to
decomposition, it emits acrid smoke and irritating fumes. For further information, see Vol.
1, No. 4 of *DPIM Report*.

AFLATOXIN G1 mixed with AFLATOXIN
B1   HR: 3
NIOSH: AW 6825000

PROP: Metabolites of *Aspergillus Flavus Link
ex Fries* Aflatoxin G1, 56.4%; Alfatoxin B1,
37.7%.

THR: Poison by ingestion, subcutaneous, and
intraperitoneal routes. An experimental carcinogen.

AFLATOXIN G2   HR: 3
CAS: 7241-98-7   NIOSH: LV 1700000
mf: $C_{17}H_{14}O_7$ mw: 330.31

THR: Acute poison by ingestion. An experimental carcinogen. Mutagenic data. When heated to decomposition, it emits acrid smoke and irritating fumes.

AFLATOXIN MI   HR: 3
CAS: 6795-23-9   NIOSH: GY 1880000
mf: $C_{17}H_{12}O_7$ mw: 328.29

SYN: 4-HYDROXYAFLATOXIN Bl

THR: Poison by ingestion. An experimental carcinogen. Mutagenic data. When heated to decomposition, it emits acrid smoke and irritating fumes.

AFLATOXIN Ro   HR: 3
CAS: 29611-03-8   NIOSH: GY 1934000
mf: $C_{17}H_{14}O_6$ mw: 314.31

SYN: AFLATOXICOL

THR: An experimental carcinogen. Mutagenic data. When heated to decomposition, it emits acrid smoke and irritating fumes.

AGAR   HR: 1
CAS: 9002-18-0   NIOSH: AW 7950000

PROP: Unground, thin translucent membranous pieces; ground, pale, buff powder, sol in boiling H_2O, insol in cold H_2O and organic solvents.

SYNS: AGAR-AGAR * AGAR AGAR FLAKE * AGAR-AGAR GUM * BENGAL GELATIN * BENGAL ISINGLASS * CEYLON ISINGLASS * CHINESE ISINGLASS * DIGENEA SIMPLEX MUCILAGE * JAPAN AGAR * JAPAN ISINGLASS * LAYOR CARANG * NCI-C50475

THR: Low toxicity by ingestion.

AGENT ORANGE   HR: 2
CAS: 39277-47-9   NIOSH: AG 8100000

SYNS: 2,4-D, n-BUTYL ESTER MIXED WITH 2,4,5-T, n-BUTYL ESTER (1:1) * 2,4,5-T, n-BUTYL ESTER MIXED WITH 2,4-D, n-BUTYL ESTER

THR: Moderately toxic by ingestion. Contains toxic impurities. (See also esters.) When heated to decomposition, it emits acrid smoke and irritating fumes.

AGERITE   HR: 3
CAS: 103-16-2   NIOSH: SJ 7700000
mf: $C_{13}H_{12}O_2$ mw: 200.25

SYNS: AGERITE ALBA * BENZYL HYDROQUINONE * P-BENZYLOXYPHENOL * HYDROQUI-

NONE BENZYL ETHER * HYDROQUINONE MONOBENZYL ETHER * P-HYDROXYPHENYL BENZYL ETHER * MONOBENZYL ETHER HYDROQUINONE * MONOBENZYL HYDROQUINONE

THR: An experimental carcinogen. A skin irritant. (See also ethers.) When heated to decomposition, it emits acrid smoke and irritating fumes.

AHR-1680   HR: 3
CAS: 21820-82-6   NIOSH: RQ 2625000
mf: $C_{17}H_{22}N_2O_2$ mw: 286.41

SYN: 5-(2-(3,6-DIHYDRO-4-PHENYL-1(2H)-PYRIDYL)ETHYL)-3-METHYL-2-OXAZOL IDINONE

THR: Poison by ingestion, intraperitoneal, and intravenous routes. When heated to decomposition it emits toxic fumes of NO_x.

AIR (COMPRESSED) (DOT)   HR: 2
NIOSH: AX 5275000
DOT: 1002

PROP: Bluish, mobile liquid. O_2 + N_2; bp: $-189°$ (liq); flash p: none; autoign temp: none. DOT Classification: Label: Nonflammable Gas

THR: Liquid air can cause tissue damage due to low temperature. Personnel exposed to compressed air may develop caisson disease (the bends, the chokes) if decompression is too rapid. Moderate explosion hazard when containers under pressure are shocked or exposed to heat or flame or flammable materials; i.e., ethyl ether, hydrocarbons or charcoal, which have been in contact with liquid air may explode very easily. Ordinary oxidation is greatly accelerated in compressed air. Moderately dangerous disaster hazard; can react vigorously with reducing materials.

AIZEN MALACHITE GREEN   HR: 3
CAS: 569-64-2   NIOSH: BQ 1180000
mf: $C_{23}H_{25}N_2$ • Cl   mw: 364.95

SYNS: ADC MALACHITE GREEN CRYSTALS * ANILINE GREEN * BENZALDEHYDE GREEN * CHINA GREEN (BIOLOGICAL STAIN) * C.I. 42000 * C.I. BASIC GREEN 4 * DIABASIC MALACHITE GREEN * TETROPHENE GREEN M * TETRAMETHYL DIAPARA-AMIDO-TRIPHENYL CARBINOL

THR: Poison by ingestion and intraperitoneal routes. Mutagenic data. When heated to decomposition it emits very toxic fumes of NO_x and Cl^-.

AJMALICINE HR: 3
CAS: 483-04-5 NIOSH: AX 7875000
mf: $C_{21}H_{24}N_2O_3$ mw: 352.47

SYNS: ALKALOID II * HYDROSARPAN
* RAUBASINE * RAUMALINA * SUBSTANCE II
* TETRAHYDROSERPENTINE * PY-TETRAHY-
DROSERPENTINE * VINCAIN * VINCEINE
* DELTA-YOHIMBINE

THR: Poison by ingestion, intraperitoneal, and intravenous routes. Central nervous system effects. When heated to decomposition, it emits highly toxic NO_x .

AJMALINE HR: 3
CAS: 4360-12-7 NIOSH: AX 8050000
mf: $C_{20}H_{26}N_2O_2$ mw: 326.48

SYNS: CARDIORYTHMINE * RAUGALLINE
* RAUWOLFIN * RAUWOLFINE * RHYTMA-
TON * TAJMALIN * TAKYCOR

THR: Poison by intraperitoneal and intravenous routes. When heated to decomposition, it emits toxic fumes of NO_x .

ALCOHOL, DENATURED HR: 3
PROP: Liquid. Composed of alcohol and denaturants.

SYN: DENATURED SPIRITS

THR: Dependent on alcohols in question, generally ethanol with methanol as a denaturant. Dangerous fire hazard; can react vigorously with oxidizing materials. Moderate explosion hazard: see ethanol, methanol, propanol.

ALCOHOLS, N.O.S. (See also specific compound).
THR: No general statement can be made due to wide variations in toxic effects. Dangerous fire hazard when exposed to heat or flame. Can react violently in contact with (H_2O + H_2SO_4), HOCl, Cl_2, isocyanates, $LiAl_4$, N_2O_4, $HClO_4$, H_2SO_5 (Caro's acid), $Ba(ClO_4)_2$, $(CH_2)_2O$, acetaldehyde, diethyl aluminum bromide, hexamethylene diisocyanate, triisobutyl aluminum.

ALDACTAZIDE HR: 3
CAS: 52-01-7 NIOSH: TU 4725000
mf: $C_{24}H_{32}O_4S$ mw: 416.62

SYNS: 7-ALPHA-ACETYLTHIO-3-OXO-17-ALPHA-
PREGN-4-ENE-21,17-BETA-CARBOLACTONE

* 7-ALPHA-ACETYLTHIO-3-OXO-17-BETA-
PREGN-4-ENE-21,17-BETA-CARBOLACTONE
* ALDACTIDE * 3'-(3-OXO-7-ALPHA-ACETYL-
THIO-17-BETA-HYDROXYANDROST-4-EN-17-BETA-
YL)PROPIONIC ACID LACTONE * ALDACTONE
* 3-(3-KETO-7-ALPHA-ACETYLTHIO-17-BETA-
HYDROXY-4-ANDROSTEN-17-ALPHA-YL)PROPIONIC
ACID LACTONE * 17-ALPHA-PREGN-4-ENE-
21-CARBOXYLIC ACID, 1-HYDROXY-7-ALPHA-
MERCAPTO-3-OXO-ALPHA-LACTONE * SPIRO
(17H-CYCLOPENTA(A)PHENAUTHRENE-17,2'-
(3'H)-FURAN)

THR: Poison by intraperitoneal route. Glandular effects by ingestion. Experimental and suspected human carcinogen. When heated to decomposition, it emits toxic fumes of SO_x .

ALDEHYDE AMMONIA HR: 2
CAS: 75-39-8 NIOSH: AB 1950000
mf: $CH_3CH(NH_2)OH$ mw: 61.08

PROP: White crystalline solid. bp: 110°; mp: 97°.

SYNS: ACETALDEHYDE AMMONIA * ACETAL-
DEHYDE AMINE SALT * 1-AMINO ETHANOL
* α-AMINO ETHANOL

THR: Moderately irritant to skin, eyes, and mucous membranes by ingestion and inhalation. Moderate fire and explosion hazard when exposed to heat or flame; readily decomposes to acetaldehyde and ammonia when heated. When heated to decomposition, it emits toxic fumes of NO_x and NH_3. Incompatible with oxidizing materials.

ALDEHYDES. See also specific compounds.
THR: All the aldehydes possess anesthetic properties, but this is obscured by their highly irritating action on the eyes and mucous membranes of the respiratory tract. The lower aldehydes, very soluble in water, act chiefly on the eyes and tissues of the upper respiratory tract. The higher aldehydes, less soluble in water, tend to penetrate more deeply into the respiratory system and may affect the lungs. A toxicity hazard rating is more accurate for the lower molecular weight aldehydes. Some higher aldehydes and also the aromatic aldehydes may exhibit much lower toxicity.

ALDOMET HR: 3
CAS: 555-30-6 NIOSH: AY 5950000
mf: $C_{10}H_{13}NO_4$ mw: 211.24

SYNS: L-(−)-3-(3,4-DIHYDROXYPHENYL)-2-METHYLALANINE * L(−)-BETA-(3,4-DIHYDROXYPHENYL)-ALPHA-METHYLALANINE * L-ALPHA-METHYL-3,4-DIHYDROXYPHENYLALANINE * L-(−)-ALPHA-METHYL-BETA-(3,4-DIHYDROXYPHENYL)ALANINE * ALPHA-METHYL-L-3,4-DIHYDROXYPHENYLALANINE * ALPHA-METHYL-BETA-(3,4-DIHYDROXYPHENYL)-L-ALANINE * METHYLDOPA * L-ALPHA-METHYLDOPA * ALPHA-METHYL-L-DOPA * NCI-C55721

THR: Poison by intraperitoneal route. Moderately toxic by ingestion and other routes. When heated to decomposition it emits toxic fumes of NO_x.

ALDRIN   HR: 3
CAS: 309-00-2   NIOSH: LO 2100000
DOT: 2761
mf: $C_{12}H_8Cl_6$ mw: 364.90

PROP: Crystals, insol in water, sol in aromatics, esters, ketones, paraffins and halogenated solvents. mp: 104°-105°.

SYNS: ENT 15,949 * HEXACHLOROHEXAHYDRO-ENDO-EXO-DIMETHANONAPHTHALENE * 1,2,3,4,10,10-HEXACHLORO-1,4,4a,5,8,8a-HEXAHYDRO-1,4,5,8-DIMETHANONAPHTHALENE * 1,2,3,4,10,10-HEXACHLORO-1,4,4a,5,8,8a-HEXAHYDRO-EXO-1,4,-ENDO-5,8-DIMETHANONAPHTHALENE * 1,2,3,4,10,10-HEXACHLORO-1,4,4a,5,8,8a-HEXAHYDRO-1,4-ENDO-EXO-5,8-DIMETHANONAPHTHALENE * NCI-C00044

OSHA PEL: TWA 250 ug/m³ (skin)
ACGIH TLV: TWA 0.25 mg/m³
DFG MAK: 0.25 mg/m³
DOT Classification: Poison B ORM-A

THR: Poison by ingestion, skin contact, and other routes. An experimental carcinogen and teratogen. Continued acute exposure causes liver damage. (See also chlorinated hydrocarbons.) When heated to decomposition, it emits toxic fumes of Cl⁻. For further information, see Vol. 3, No. 5 of *DPIM Report*.

ALFALFA MEAL   HR: 1
THR: An allergen. Skin contact may cause dermatitis. Moderate fire hazard when exposed to heat or flame; by spontaneous chemical reaction. Avoid moisture content extremes. Fires may smolder for 72 hours before becoming noticeable.

ALIDINE DIHYDROCHLORIDE   HR: 3
CAS: 63905-54-4   NIOSH: NS 5425000
mf: $C_{23}H_{29}NO_4 \cdot ClH$ mw: 419.99

SYNS: 1-(p-AMINOPHENETHYL)-4-PHENYLISONIPECOTIC ACID, ETHYL ESTER, DIHYDROCHLORIDE * 1-(4-AMINOPHENETHYL)-4-PHENYLISONIPECOTIC ACID, ETHYL ESTER DIHYDROCHLORIDE * 4-ETHOXYCARBONYL-1-(2-HYDROXY-3-PHENOXYPROPYL) 4-PHENYLPIPERIDINE HYDROCHLORIDE * ETHYL- 1-(p-AMINOPHENETHYL)-4-PHENYLISONIPECOTATEDIHYDROCHLORIDE * ETHYL 1-(4-AMINOPHENYLETHYL)-4-PHENYLISONIPECOTATE DIHYDROCHLORIDE * ETHYL 1-(4-AMINOPHENYLETHYL)-4-PHENYLISONIPECOTATE DIHYDROCHLORIDE

THR: Poison by ingestion, intraperitoneal, subcutaneous, and intravenous routes. Psychotropic effects by ingestion. See also esters. When heated to decomposition, it emits very toxic fumes of HCl and NO_x.

ALIPHATIC AND AROMATIC EPOXIDES.   HR: 3
THR: Experimental carcinogen of skin, lung, and blood-forming tissues.

ALKALIES. (See also specific compound.)
PROP: A term loosely applied to the hydroxides and carbonates of the alkali metals and alkaline earth metals, as well as the bicarbonate and hydroxide of ammonium. They can neutralize acids, change the color of indicators and impart a soapy taste and feel to aqueous solutions.

THR: Variable toxicity. As a group, constitute the commonest causes of contact dermatitis. Systemically, ammonia is most troublesome.

ALKALOID POISONS, ALSO THEIR SALTS, LIQUID, NOS, OR ALKALOID POISONS, AND THEIR SALTS, SOLID, NOS   HR: 3
SYN: ALKALOIDS

THR: Nearly all alkaloid salts are poisonous. Some are also allergens. See specific alkaloid salt. Dangerous; heat decomp emits highly toxic fumes of NO_x and SO_x.

ALKANES
PROP: All colorless neutral liquids with light aromatic odors. See also individual alkanes as listed.

SYNS: N-PENTANE + 2 ISOMERS, N-HEXANE + 4 ISOMERS, N-HEPTANE + 8 ISOMERS, N-OCTANE + 17 ISOMERS

THR: Variable toxicity. Hexane can cause neuropathy upon chronic exposure.

ALKYLBENZENESULFONATE   HR: 3
NIOSH: DB 4370000

SYNS: BENZENESULFONIC ACID, ALKYL DERIV * ABS

THR: Poison by intravenous route. Moderately toxic by other routes. Very reactive with F_2. See also sulfonates. When heated to decomposition, it emits toxic fumes of SO_x.

ALKYL DIMETHYLBENZYL AMMONIUM CHLORIDE   HR: 3
CAS: 8001-54-5   NIOSH: BO 3150000

PROP: Alkyl group contains from C_8-C_{18}. d: (50%) 0.9884 @ 20°. Clear, mobile liquid.

SYNS: ALKYLDIMETHYL(PHENYLMETHYL)-QUATERNARY AMMONIUM CHLORIDES * BENZALKONIUM CHLORIDE * HYAMINE 3500 * PHENEENE GERMICIDAL SOLUTION AND TINCTURE * QUATERNARY AMMONIUM COMPOUNDS, ALKYLBENZYLDIMETHYL, CHLORIDES * TRITON K-60 * ZEPHIRAN CHLORIDE

THR: Poison by ingestion, intraperitoneal, intravenous, and parenteral routes. When heated to decomposition, it emits very toxic fumes of NO_x, NH_3 and Cl^-.

ALKYL DIMETHYL-3,4-DICHLOROBENZENE AMMONIUM CHLORIDE   HR: 3
NIOSH: BO 2975000

PROP: A cationic surfactant used at dilutions of 1:1000 as a germicide.

SYNS: ARAKONIUM CHLORIDE * DICHLORAN * TETRASAN

THR: Poison by ingestion. When heated to decomposition, it emits very toxic fumes of NO_x, NH_3, and Cl^-.

ALKYL(C_8C_{18})DIMETHYL-3,4-DICHLOROBENZYLAMMONIUM CHLORIDE   HR: 3
CAS: 8023-53-8   NIOSH: BO 3200000

SYNS: ALKYL(C_8H_{17} to $C_{18}H_{37}$) DIMETHYL-3,4-DICHLOROBENZYL AMMONIUM CHLORIDE * TETROSAN

THR: Poison by intraperitoneal route. Moderately toxic by all other routes. Can cause liver and kidney damage. A moderate allergen. Mutagenic data. See also esters. When heated to decomposition, it emits very toxic fumes of NO_x, NH_3, and Cl^-.

ALLETHRIN   HR: 3
NIOSH: GZ 1925000
mf: $C_{19}H_{26}O_3$ mw: 302.45

PROP: A viscous liquid.

SYNS: (+)-ALLELRETHONYL (+)-CIS,TRANS-CHRYSANTHEMATE * ALLYL CINERIN * ALLYL HOMOLOG OF CINERIN I * D,1-2-ALLYL-4-HYDROXY-3-METHYL-2-CYCLOPENTEN-1-ONE-D, 1-CHRYSANTHEMUMMONOCARBOXYLATE * 3-ALLYL-4-KETO-2-METHYLCYCLOPENTENYL CHRYSANTHEMUMMONOCARBOXYLATE * 3-ALLYL-2-METHYL-4-OXO-2-CYCLOPENTEN-1-YL CHRYSANTHEMATE * DL-3-ALLYL-2-METHYL-4-OXOCYCLOPENT-2-ENYL-DL-CIS TRANS CHRYSANTHEMATE * ENT 17,510 * SYNTHETIC PYRETHRINS

THR: Poison by intraperitoneal route. Moderately toxic by ingestion. An allergen. An insecticide. It can cause liver and kidney damage by all routes of entry into the body. Lung congestion may occur due to exposure. Local contact may cause contact dermatitis. Inhalation may cause asthma, coughing, wheezing, running nose and eyes. Mutagenic data. See also esters. Slight fire hazard. Heat decomposition emits acrid fumes.

(+)-cis-ALLETHRIN   HR: 3
NIOSH: GZ 1460000
mf: $C_{19}H_{26}O_3$ mw: 302.45

SYN: CYCLOPROPANECARBOXYLIC ACID, 2,2-DIMETHYL-3-(2-METHYLPROPENYL)-, ESTER WITH 2-ALLYL-4-HYDROXY-3-METHYL-2-CYCLOPENTEN-ONE, (+)-(Z)-

THR: Poison by ingestion and inhalation. See also esters. Heat decomposition emits acrid smoke and irritating fumes.

trans-(+)-ALLETHRIN   HR: 3
CAS: 28434-00-6   NIOSH: GZ 1472000
mf: $C_{19}H_{26}O_3$ mw: 302.45

SYNS: D-ALLETHROLONE CHRYSANTHEMUMATE * (+)-ALLETHRONYL (+)-TRANS-CHRYSANTHEMUMATE

THR: Poison by inhalation and ingestion. See allethrin and esters. When heated to decomposition it emits acrid and irritating fumes.

ALLETHRIN RACEMIC MIXTURE　　HR: 3
NIOSH: GZ 1476000
mf: $C_{19}H_{26}O_3 \cdot 4Cl_9H_{26}O_3$　　mw: 1512.25

SYNS: 4-HYDROXY-3-METHYL-2-CYCLOPENTEN-1-ONE, CIS- MIXED WITH TRANS-2,2-DIMETHYL-3-(2-METHYL-PROPENYL)CYCLOPROPANECAR-BOXYLIC ACID ESTER WITH 2-ALLYL-4-HYDROXY-3-METHYL-2-CYCLOPENTEN-1-ONE (1:4)

THR: Poison by inhalation. Moderately toxic by ingestion. When heated to decomposition it emits toxic fumes of Cl^-.

ALLOBARBITAL　　　　　　　　HR: 3
CAS: 52-43-7　　　　NIOSH: CQ 3325000
mf: $C_{10}H_{12}N_2O_3$　　mw: 208.24

SYNS: ALLYLBARBITURAL　*　2,4,-6(1H,3H,5H)-PYRIMIDINETRIONE, 5,5-DI-2-PRO-PENYL- (9CI)　*　ALLOBARBITONE　*　DIALLYL-BARBITAL　*　5,5-DIALLYLBARBITURIC ACID

THR: Poison by ingestion, intraperitoneal, subcutaneous, and intravenous routes. Psychotropic effects by ingestion. When heated to decomposition, it emits toxic fumes of NO_x.

ALLONAL　　　　　　　　　　　HR: 3
CAS: 77-02-1　　　　NIOSH: CP 8925000
mf: $C_{10}H_{14}N_2O_3$　　mw: 210.26

SYNS: ALLIONAL　*　5-ALLYL-5-ISOPROPYLBAR-BITURATE　*　ALLYLISOPROPYLBARBITURIC ACID　*　5-ALLYL-5-ISOPROPYLBARBITURIC ACID　*　ALLYLISOPROPYLMALONYLUREA　*　ALLYL-PROPYMAL　*　APROBARBITONE　*　ISOPROPYL-ALLYLBARBITURIC ACID

THR: Poison by ingestion, subcutaneous, and intraperitoneal routes. Psychotropic effects by ingestion. When heated to decomposition it emits toxic NO_x.

ALLORPHINE　　　　　　　　　HR: 3
CAS: 62-67-9　　　　NIOSH: QC 7700000
mf: $C_{19}H_{21}NO_3$　　mw: 311.41

SYNS: N-ALLYL-7,8-DEHYDRO-4,5-EPOXY-3,6-DIHYDROXYMORPHINAN　*　N-ALLYL-N-DES-METHYLMORPHINE　*　N-ALLYLNORMORPHINE

THR: Poison by intravenous and subcutaneous routes. Moderately toxic by many other routes. Psychotropic effects. When heated to decomposition it emits very toxic fumes of NO_x.

ALLOXAN　　　　　　　　　　　HR: 3
CAS: 50-71-5　　　　NIOSH: BA 4200000
mf: $C_4H_2N_2O_4$　　mw: 142.08

PROP: d: 1.70; mp: 256° (decomp); sol in water, alc, benzene and acetone.

SYNS: MESOXALYLCARBAMIDE　*　MESOXALYL-UREA　*　2,4,5,6(1H,3H)-PYRIMIDINETETRONE　*　2,4,5,6-PYRIMIDINTETRON (CZECH)　*　2,4,5,6-TETRAOXOHEXAHYDROPYRIMIDINE

THR: Poison by intraperitoneal, intravenous, subcutaneous, and rectal routes. Moderately toxic by ingestion. An experimental carcinogen. An eye irritant. When heated to decomposition, it emits toxic fumes of NO_x. For further information, see Vol. 1, No. 4 of *DPIM Report*.

ALLYL ACETATE　　　　　　　　HR: 3
CAS: 591-87-7　　　　NIOSH: AF 1750000
mf: $C_5H_8O_2$　　mw: 100.13

PROP: Liquid; vap d: 3.45; bp: 104°; d: 0.928. insol in water. flash p: 72°F.

SYNS: ACETIC ACID ALLYL ESTER　*　ACETIC ACID-2-PROPENYL ESTER　*　3-ACETOXYPROPENE

THR: Poison by ingestion. Moderately toxic by inhalation and skin contact. A skin and eye irritant. When heated to decomposition, it emits acrid smoke and irritating fumes. Dangerous fire hazard.

ALLYL ALCOHOL　　　　　　　　HR: 3
CAS: 107-18-6　　　　NIOSH: BA 5075000
DOT: 1098
mf: C_3H_6O　　mw: 58.09

PROP: Limpid liquid, pungent odor, sol in water, alc and ether. mp: −129°, bp: 96°-97°, lel = 2.5%, uel = 18%, flash p: 70°F (CC), d: 0.854 @ 20°/4°, autoign temp: 713°F, vap press: 10 mm @ 10.5°, vap d: 2.00. Misc in water, alc, and ether.

SYNS: ALCOOL ALLILCO (ITALIAN)　*　ALCOOL ALLYLIQUE (FRENCH)　*　ALLILOWY ALKOHOL (POLISH)　*　ALLYLALKOHOL (GERMAN)　*　ALLYLIC ALCOHOL　*　3-HYDROXYPROPENE　*　PROPENOL　*　PROPEN-1-OL-3　*　1-PROPEN-

3-OL * 2-PROPEN-1-OL * PROPENYL ALCO-
HOL * 2-PROPENYL ALCOHOL * VINYLCARBI-
NOL

OSHA PEL: TWA 2 ppm (skin)
ACGIH TLV: TWA 2 ppm; STEL 4 ppm
DFG MAK 2 ppm (5 mg/m^3)
DOT Classification: Label: Flammable
Liquid and Poison

THR: Poison by inhalation, ingestion, skin contact, and intraperitoneal routes. A skin, eye,
and systemic irritant. Dangerous fire and explosion hazard when exposed to heat or flame or
oxidizers; incompatible with CCl_4; chlorosulfonic acid; HNO_3; H_2SO_4; oleum; NaOH; diallyl
phosphite; PCl_3; tri-n-bromomelamine. When
heated to decomposition it emits toxic fumes;
to fight fire: CO_2, alcohol foam, dry chemical.
For further information, see Vol. 1, No. 7 of
DPIM Report.

ALLYLAMINE   HR: 3
CAS: 107-11-9   NIOSH: BA 5425000
mf: C_3H_7N mw: 57.11

PROP: Colorless liquid, burning taste, sharp
odor. bp: 56.5°; d: 0.761 @ 20°/4°; flash p:
−20°; autoign temp: 705°F; vap d: 2.00; lel
= 2.2%; uel = 22%. Misc in water, alc, and
ether.

SYNS: 3-AMINOPROPENE * 3-AMINOPROPY-
LENE * MONOALLYLAMINE * 2-PROPENAMINE
* 2-PROPEN-1-AMINE

THR: Poison by inhalation, ingestion, and skin
contact. A systemic irritant. Extraordinary precautions against fumes are advised. Dangerous
fire and explosion hazard when exposed to heat,
flame or oxidizers. Highly reactive. To fight
fire: alcohol foam, CO_2, dry chemical. For further information, see Vol. 2, No. 6 of *DPIM
Report*.

p-ALLYLANISOLE   HR: 3
CAS: 140-57-0   NIOSH: BZ 8225000
mf: $C_{10}H_{12}O$ mw: 148.22

PROP: Isolated from rind of Persea Gratissima
Garth, and from Oil of Estragon; found in oils
of Russian Anise, Basil, Fennel, Turpentine,
and others.

SYNS: TARRAGON * 4-ALLYL-1-METHOXYBEN-
ZENE * CHAVICOL METHYL ETHER * ISOAN-
ETHOLE * P-METHOXYALLYLBENZENE
* METHYL CHAVICOL

THR: An experimental carcinogen. Moderate
acute toxicity by many routes. A skin irritant.
When heated to decomposition, it emits acrid
smoke and irritating fumes. For further information, see Vol. 1, No. 3 of *DPIM Report*.

ALLYL BROMIDE   HR: 3
CAS: 106-95-6   NIOSH: UC 7090000
DOT: 1099
mf: C_3H_5Br mw: 120.99

PROP: Colorless liquid, pungent odor. mp:
−119°, bp: 71.3°, flash p: 30°F, d: 1.3980 @
20°/4°, autoign temp: 563°F, vap d: 4.17, lel
= 4.4%, uel = 7.3%. Insol in water.

SYNS: BROMALLYLENE * 3-BROMOPROPENE
* 3-BROMOPROPYLENE

DOT Classification: Label: Flammable
Liquid

THR: Poison by ingestion and intraperitoneal
routes. Mutagenic data. See also allyl chloride.
Dangerous fire and explosion hazard when exposed to heat, flame or oxidizers; To fight fire:
alcohol foam, water spray or mist, CO_2, dry
chemical.

5-ALLYL-5-sec-BUTYLBARBITURIC
ACID   HR: 3
CAS: 115-44-6   NIOSH: CP 8225000
mf: $C_{11}H_{16}N_2O_3$ mw: 224.29

SYNS: 5-ALLYL-5-(1-METHYLPROPYL) BARBI-
TURIC ACID * SEC-BUTYL ALLYL BARBITURIC
ACID * 2,4,6(1H,3H,5H)-PYRIMIDINETRIONE,
5-(1-METHYLPROPYL)-5-(2-PROPENYL)- (9CI)

THR: Poison by ingestion, intraperitoneal, and
intravenous routes. Psychotropic effects by ingestion. When heated to decomposition it emits
toxic NO_x.

ALLYL-sec-BUTYL THIOBARBITURIC
ACID   HR: 3
CAS: 2095-58-1   NIOSH: CP 8300000
mf: $C_{11}H_{16}N_2O_2S$ mw: 240.35

SYN: 5-ALLYL-5-SEC-BUTYL-2-THIOBARBITURIC
ACID

THR: Poison by intravenous and rectal routes.
When heated to decomposition it emits very
toxic fumes of SO_x and NO_x.

ALLYL CARBAMATE   HR: 3
CAS: 2114-11-6   NIOSH: EY 8512000
mf: $C_4H_7NO_2$ mw: 101.12

THR: An experimental carcinogen. See also carbamates. When heated to decomposition it emits toxic NO_x.

ALLYL CHLORIDE   HR: 3
CAS: 107-05-1   NIOSH: UC 7350000
DOT: 1100
mf: C_3H_5Cl mw: 76.53

PROP: Colorless liquid. Mp: $-136.4°$, bp: $44.6°$, d: 0.938 @ $20°/4°$, flash p: $-25°F$, lel = 2.9%, uel = 11.2%, autoign temp: $905°F$, vap d: 2.64. Solubility = <0.1 in water.

SYNS: ALLILE (CLORURO DI) (ITALIAN)
* ALLYLCHLORID (GERMAN) * ALLYLE (CHLORURE D') (FRENCH) * CHLORALLYLENE
* 3-CHLOROPRENE * 1-CHLORO PROPENE-2
* 3-CHLOROPROPENE * 1-CHLORO-2-PROPENE
* 3-CHLOROPROPYLENE * 3-CHLORPROPEN (GERMAN) * NCI-C04615

OSHA PEL: TWA 1 ppm
ACGIH TLV: TWA 1 ppm; STEL 2 ppm
DFG MAK 1 ppm (3 mg/m^3)
DOT Classification: Label: Flammable
  Liquid

THR: Poison by inhalation, ingestion, intraperitoneal, and intravenous routes. An experimental carcinogen. Mutagenic data. A skin, eye, and mucous membrane irritant. Chronic exposure may cause liver and kidney damage. The vapors of allyl chloride are quite irritating to the eyes, nose and throat and contact of the liquid with the skin, in addition to local vasoconstriction and numbness, may lead to rapid absorption and distribution through the body. If remedial measures are not taken promptly, such contact may result in burns and internal injuries. In low concentrations, its odor is irritating enough to furnish warning of its presence in most cases. Concentrations of the vapors high enough to cause serious effects, including damage to the lungs, especially on repeated exposure, may be tolerable. Consequently, the warning characteristics should never be disregarded. Dangerous fire and explosion hazard when exposed to heat, flame or oxidizers. Incompatible with HNO_3; H_2SO_4; ethylene imine; ethylenediamine; chlorosulfonic acid; oleum; NaOH; (benzene or toluene to $-70°$ in presence of trichlorotriethyl dialuminum or ethyl aluminum dichloride or diethyl aluminum chloride). To fight fire: CO_2, alcohol foam, dry chemical. For further information, see Vol. 1, No. 7 of *DPIM Report*.

ALLYL CHLOROCARBONATE   HR: 2
CAS: 2937-50-0   NIOSH: LQ 5775000
DOT: 1722
mf: $C_4H_5ClO_2$ mw: 120.54

PROP: Liquid. bp: $106°-114°$, flash p: $88°F$ (CC), d: 1.14, vap d: 4.2.

SYNS: ALLYL CHLOROCARBONATE (DOT)
* ALLYL CHLOROFORMATE

DOT Classification: Label: Flammable
  Liquid

THR: Moderately toxic by inhalation. Dangerous when exposed to heat, open flame (or sparks), or powerful oxidizers. Can react with oxidizing materials. To fight fire: Alcohol foam, spray or mist, dry chemical.

4-ALLYL-1,2-DIMETHOXYBENZENE   HR: 3
CAS: 93-15-2   NIOSH: CY 2450000
mf: $C_{11}H_{14}O_2$ mw: 178.25

SYNS: 1-ALLYL-3,4-DIMETHOXYBENZENE
* 4-ALLYLVERATROLE * 1,2-DIMETHOXY-4-ALLYLBENZENE * 1-(3,4-DIMETHOXYPHENYL)-2-PROPENE * 1,3,4-EUGENOL METHYL ETHER
* EUGENYL METHYL ETHER * METHYL EUGENOL * VERATROLE METHYL ETHER

THR: Poison by intravenous route; moderately toxic by ingestion and other routes. A skin irritant. When heated to decomposition, it emits acrid smoke and irritating fumes.

ALLYLESTRENOL   HR: 3
CAS: 432-60-0   NIOSH: KG 7960000
mf: $C_{21}H_{32}O$ mw: 300.53

SYNS: 17-ALPHA-ALLYLESTR-4-EN-17-BETA-OL
* 17-ALPHA-ALLYL-4-ESTREN-17-BETA-OL
* 17-ALPHA-ALLYL-3-DEOXY-19-NORTESTOSTERONE * 17-ALPHA-ALLYL-17-BETA-HYDROXY-DELTA(SUP 4)-ESTREN * 17-ALPHA-ALLYL-17-BETA-HYDROXY-4-ESTRENE
* 17A-ALLYL-17B-HYDROXY-4-ESTRENE
* 17-ALPHA-ALLYL-4-OESTRENE-17-BETA-OL

THR: An experimental teratogen. When heated to decomposition it emits acrid smoke and irritating fumes.

ALLYL FLUORIDE   HR: 3
mf: C_3H_5F mw: 60.07

PROP: Colorless gas. bp: $-10°$.

SYN: 3-FLUOROPROPENE

THR: Poison by inhalation and ingestion. A strong irritant. See also fluorides. When heated to decomposition it emits highly toxic fumes of F^-. Incompatible with water or steam to produce toxic and corrosive fumes.

ALLYL FORMATE HR: 3
CAS: 1838-59-1   NIOSH: LQ 9800000
mf: $C_4H_6O_2$ mw: 86.10

PROP: Liquid, slightly water sol, sol in organic solvents. d: 0.948 @ 18°/4°, bp: 83°, flash p: $< -50°F$.

SYN: FORMIC ACID, ALLYL ESTER

THR: Poison by ingestion. Very flammable and reactive. Dangerous fire hazard. See esters. When heated to decomposition it yields irritating smoke and fumes.

ALLYL HYDROPEROXIDE HR: 3
mf: $C_3H_6O_2$ mw: 74.1

THR: A potentially explosive liquid. Unstable to heat, light, and solid alkalies. Highly toxic. See also peroxides, organic.

ALLYL IODIDE HR: 3
mf: C_3H_5I mw: 168.0

PROP: Yellow liquid, pungent odor. mp: −99°; bp: 103.1°; d: 1.825 @ 20°/4°; vap d: 5.8.

THR: Poison by inhalation and ingestion. A powerful irritant. When heated to decomposition it emits highly toxic fumes of I^-. Moderately flammable. Incompatible with oxidizing materials. To fight fire: water, foam, CO_2, dry chemical.

ALLYLISOBUTYLBARBITURATE HR: 3
CAS: 77-26-9   NIOSH: CP 8750000
mf: $C_{11}H_{16}N_2O_3$ mw: 224.29

SYNS: ALLYLBARBITAL * ALLYLBARBITONE * ALLYLISOBUTYLBARBITAL * 5-ALLYL-5-ISO-BUTYLBARBITURIC ACID * 5-ALLYL-5-(2'-METHYL-N-PROPYL) BARBITURIC ACID * BUTAL-BARBITAL * ISOBUTYLALLYLBARTURIC ACID * 2,4,6(1H,3H,5H)-PYRIMIDINETRIONE, 5-(2-METHYLPROPYL)-5-(2-PROPENYL)- (9CI) * TETRALLOBARBITAL

THR: Poison by ingestion, subcutaneous, and intraperitoneal routes. Psychotropic effects by ingestion. When heated to decomposition it emits toxic NO_x.

ALLYL ISOTHIOCYANATE HR: 3
CAS: 57-06-7   NIOSH: NX 8225000
mf: C_4H_5NS mw: 99.16

PROP: Colorless to pale yellow liquid; irr odor. mp: −80°, bp: 150.7°, flash p: 115°F, d: 1.013-1.016 @ 25°/25°, vap press: 10 mm @ 38.3°, vap d: 3.41

SYNS: ALLYL ISOSULFOCYANATE * ALLYL MUSTARD OIL * ALLYLSENFOEL (GERMAN) * ISOTHIOCYANATE D'ALLYLE (FRENCH) * MUSTARD OIL * NCI-C50464 * OLEUM SINAPIS VOLATILE * 2-PROPENYL ISOTHIOCYA-NATE * SENF OEL (GERMAN) * SYNTHETIC MUSTARD OIL * VOLATILE OIL OF MUSTARD

THR: Poison by ingestion, skin contact, intravenous, subcutaneous, and intraperitoneal routes. An experimental carcinogen. An allergen. May cause contact dermatitis. Moderately flammable. Highly reactive. When heated to decomposition it or contact with acid or acid fumes emits highly toxic fumes of cyanide. To fight fire, use foam, CO_2, dry chemical. For further information, see Vol. 1, No. 1 of *DPIM Report*.

ALLYL MERCAPTAN HR: 3
mf: CH_2CHCH_2SH mw: 74.1

PROP: Water-white liquid with a strong garlic odor; darkens on standing. d: 0.925 @ 23°/4°; bp: 68°.

THR: Poison by inhalation and ingestion. Strong irritant to skin and mucous membranes. When heated to decomposition it emits highly toxic fumes of SO_x. Dangerous fire hazard. To fight fire: water mist or spray, alcohol foam, CO_2, or dry chemical.

2-ALLYLOXYETHANOL HR: 3
CAS: 111-45-5   NIOSH: KJ 5425000
mf: $C_5H_{10}O_2$ mw: 102.15

SYNS: 2-ALLOXYETHANOL (CZECH) * USAF DO-47

THR: Poison by intraperitoneal route. Moderately toxic by ingestion and parenteral routes. Severe eye irritant. Moderate skin irritant. When heated to decomposition, it emits acrid smoke and irritating fumes.

alpha-ALLYL PHENETHYLAMINEHY-DROCHLORIDE HR: 3
CAS: 4255-24-7   NIOSH: SG 9275000
mf: $C_{11}H_{15}N \cdot HCl$ mw: 174.29

SYNS: ALETAMINE HYDROCHLORIDE ✳ ETH-
YLAMINE, 1-ALLYL-2-PHENYL-, HYDROCHLORIDE

THR: Poison by ingestion, subcutaneous, and
intravenous routes. When heated to decomposition, it emits toxic fumes of HCl.

o-ALLYL PHENOL   HR: 3
CAS: 1745-8-1-9   NIOSH: SJ 3850000
mf: $C_9H_{10}O$ mw: 134.19

PROP: mp: 10°; bp: 230°; d: 1.033 @ 18°/4°.
Sol in water, alc, chloroform, ether.

SYN: 2-ALLYL PHENOL

THR: Poison by intraperitoneal route. An experimental carcinogen. When heated to decomposition, it emits acrid smoke and fumes. For
further information, see Vol. 1, No. 1 of *DPIM
Report*.

ALLYL PHENYL ETHER   HR: 3
CAS: 1746-13-0   NIOSH: DA 8575000
mf: $C_9H_{10}O$ mw: 134.19

PROP: d: 0.986; bp: 191.7. Insol in water.

SYNS: USAF DO-23 ✳ (2-PROPENYLOXY)BEN-
ZENE

THR: Poison by intravenous and intraperitoneal
routes. See ethers. When heated to decomposition, it emits acrid smoke and irritating fumes.

ALLYL PROPYL DISULFIDE   HR: 3
CAS: 2179-59-1   NIOSH: JO 0350000
mf: $C_6H_{12}S_2$ mw: 148.30

PROP: Liquid, pungent odor.

SYN: ONION OIL

OSHA PEL: TWA 2 ppm
ACGIH TLV: TWA 2 ppm; STEL 3 ppm
DFG MAK: 2 ppm (12 mg/m^3)

THR: Poison by inhalation and ingestion. A
powerful irritant. Moderately flammable by exposure to heat, flame or oxidizers. When heated
to decomposition, it emits highly toxic SO_x.
To fight fire: Foam, CO_2, dry chemical. For
further information, see Vol. 1, No. 5 of *DPIM
Report*.

ALLYL SULFIDE   HR: 3
CAS: 592-88-1   NIOSH: BC 4900000
mf: $C_6H_{10}S$ mw: 114.22

PROP: Colorless liquid, garlic odor; mp: −83°;
bp: 139°; d: 0.8881; vap d: 3.90.

SYNS: ALLYL MONOSULFIDE ✳ DIALLYL
MONOSULFIDE ✳ DIALLYL SULFIDE ✳ DIALLYL
THIOETHER ✳ OIL GARLIC ✳ THIOALLYL
ETHER

THR: Poison by intravenous route. Moderately
toxic by intraperitoneal route. An irritant to skin,
eyes, and mucous membranes. See sulfides.
When heated to decomposition, it emits toxic
SO_x.

1-ALLYLTHEOBROMINE   HR: 3
CAS: 2530-99-6   NIOSH: XH 2350000
mf: $C_{10}H_{12}N_4O_2$ mw: 220.26

SYN: ALLYLTHEOBROMINE

THR: Poison by subcutaneous and intravenous
routes. Central nervous system effects. When
heated to decomposition, it emits highly toxic
NO_x.

ALLYL THIOCYANATE   HR: 3
CAS: 764-49-8   NIOSH: XL 2000000
mf: C_4H_5NS mw: 99.16

PROP: Colorless oil. d: 1.024; mp: −102.5°
bp: 152°; sol in water, alc and ether.

SYNS: ALLYLRHODANID (GERMAN) ✳ ALLYL
SULFOCYANIDE ✳ THIOCYANIC ACID, ALLYL ES-
TER

THR: Poison by intraperitoneal and subcutaneous routes. See thiocyanates. When heated to
decomposition, it emits very toxic NO_x, SO_x,
and CN^-.

1-ALLYL-2-THIOUREA   HR: 3
                          NIOSH: YR 8050000
mf: $C_4H_8N_2S$ mw: 116.20

PROP: Colorless prisms; d: 1.219; mp: 77°-
78°; Sol hot water, insol in benzene, alc, very
sltly sol in ether.

SYNS: ALLYLTHIOCARBAMIDE ✳ N-ALLYL-
THIOUREA ✳ 1-ALLYLTHIOUREA

THR: Poison by ingestion, subcutaneous, and
intravenous routes. When heated to decomposition, it emits very toxic NO_x and SO_x.

ALLYL TRICHLORIDE   HR: 3
CAS: 96-18-4   NIOSH: TZ 9275000
mf: $C_3H_5Cl_3$ mw: 147.43

PROP: bp: 142°, d: 1.414 @ 20°/20°, flash p:
180°F (OC).

SYNS: GLYCEROL TRICHLOROHYDRIN * GLYC-
ERYL TRICHLOROHYDRIN * NCI-C60220
* 1,2,3-TRICHLOROPROPANE

OSHA PEL: TWA 50 ppm

THR: Poison by inhalation, ingestion, and possibly other routes. Moderately toxic by skin contact. Mutagen data. Moderately flammable by
heat, flames (sparks), or powerful oxidizers.
Heat decomposition yields highly toxic Cl^-.
To fight fire: water (as a blanket), spray, mist,
dry chemical.

ALLYL TRICHLOROSILANE   HR: 3
CAS: 107-37-9   NIOSH: VV 1530000
DOT: 1724
mf: $C_3H_5Cl_3Si$ mw: 175.52

PROP: Colorless liquid, pungent, irr odor; bp:
117.5°; d: 1.217 @ 27°; flash p: 95°F (COC).
DOT Classification: Label: Corrosive

THR: Poison by intravenous route. Corrosive.
See silanes. When heated to decomposition, it
emits toxic Cl^-. A dangerous fire hazard. To
fight fire: Foam, mist, spray, dry chemical.

ALLYLUREA   HR: 3
CAS: 557-11-9   NIOSH: YR 7875000
mf: $C_4H_8N_2O$ mw: 100.14

PROP: mp: 85°. Very sol in water and alc;
very sltly sol in ether.

SYNS: ALLYLCARBAMIDE * MONOALLYLUREA

THR: Poison by ingestion. When heated to decomposition, it emits toxic fumes of NO_x.

ALLYL VINYL ETHER   HR: 2
CAS: 3917-15-5   NIOSH: KM 9450000
mf: C_5H_8O mw: 84.13

PROP: Very slightly sol in water. d: 0.8, bp:
67°, flash p: <68°F (OC).

THR: Moderately toxic by inhalation and ingestion. See ethers. Dangerous fire and explosion
hazard from heat, sparks,or powerful oxidizers.
To fight fire: water may be ineffective; use alcohol foam, dry chemical, mist. Heat decomp
yields acrid, irritating fumes. Becomes shock
and heat sensitive on storage.

ALMOND OIL   HR: 1
PROP: Fixed, non-drying oil; oily liquid. Composition: oleic, linoleic, myristic, palmitic acids. d: 0.910-0.915 @ 25°/25°.

SYNS: ALMOND OIL EXPRESSED * ALMOND
OIL SWEET

THR: May cause contact dermatitis. Combustible. To fight fire: use alcohol foam, dry chemical, water, mist.

ALMOND OIL, BITTER   HR: 3
PROP: Composition: Chief known constituents
are benzaldehyde, hydrocyanic acid, benzaldehyde cyanhydrin. bp: 179°; d: 1.045-1.070 @
15°

THR: Poison if not separated from its hydrogen
cyanide. May cause contact dermatitis. Combustible.

ALUMINON   HR: 3
CAS: 569-58-4   NIOSH: GU 4800000
mf: $C_{22}H_{23}N_3O_9$ mw: 473.48

SYNS: AMMONIUM AURINTRICARBOXYLATE
* AURINTRICARBOXYLIC ACID AMMONIUM SALT

THR: Poison by intravenous route. When heated
to decomposition, it emits toxic fumes such as
NO_x.

ALUMINUM   HR: 3
CAS: 7429-90-5   NIOSH: BD 0330000
DOT: 1309/1383/1396
mf: Al   mw: 26.98

PROP: A silvery ductile metal. mp: 660°, bp:
2450°, d: 2.702, vap press 1 mm @ 1284°.
Sol in HCl, H_2SO_4 and alkalies.

SYNS: ALAUN (GERMAN) * ALUMINA FIBRE
* ALUMINIUM FLAKE * ALUMINUM DE-
HYDRATED * ALUMINUM, METALLIC,
POWDER (DOT) * ALUMINUM POWDER
* C.I. 77000
ACGIH TLV: (Metal and oxide) TWA 10 mg/
    m^3; (pyro powders and welding fumes) TWA
    5 mg/m^3; (soluble salts and alkyls) TWA 2
    mg/m^3
DOT Classification: Flammable solid

THR: Inhalation of finely divided powder has
been reported as a cause of pulmonary fibrosis.
May be implicated in Alzheimers disease. Dust
is moderately flammable/explosive by heat,
flame, or chemical reaction with powerful oxidizers. To fight fire: special mixtures of dry
chemical. For further information, see Vol. 4,
No. 5 of *DPIM Report*.

ALUMINUM BOROHYDRIDE　　HR: 3
mf: AlB_3H_{12}　　mw: 71.53

PROP: Liquid. bp: 44.5°; mp: −64.5°; vap press: 400 mm @ 28.1°.

SYN: ALUMINUM TETRAHYDROBORATE

THR: See hydrides and boron compounds. Dangerous by spontaneous chemical reaction; ignites spontaneously in air, particularly in moist air. Explodes in O_2 at temperatures as low as 20°. An explosive range of 5% to 90%. Incompatible with water, steam, oxidizing materials, acid or acid fumes; will react with water or steam to produce heat, H_2, or toxic fumes. To fight fire: CO_2, dry chemical.

ALUMINUM BROMIDE　　HR: 2
CAS: 7727-15-3　　NIOSH: BD 0350000
DOT: 1725/2580
mf: $AlBr_3$　　mw: 266.71

PROP: White to yellow-red lumps. mp: 97.5°; bp: 263.3° @ 748 mm; d: 3.2; vap press: 1 mm @ 81.3°.

SYNS: ALUMINUM BROMIDE (ANHYDROUS)
* ALUMINUM TRIBROMIDE * TRIBROMOALUMINUM

DOT Classification: Corrosive label

THR: A toxic, corrosive material. See also bromides. Mixtures with sodium or potassium explode violently upon impact. When heated to decomposition, it emits toxic fumes of Br^-. Do not add H_2O to anhydrous material. Hydrolysis can be violent.

ALUMINUM CHLORIDE　　HR: 2
CAS: 7446-70-0　　NIOSH: BD 0525000
mf: $AlCl_3$　　mw: 133.33

PROP: White hex deliquescent crystals. d: 2.44; mp: 194° @ 5.2 atm; bp: subl @ 181°; vap press: 1 mm @ 100.0°. Violently sol in water, sol in alc and ether.

SYNS: ALLUMINIO(CLORURO DI) (ITALIAN)
* ALUMINIUMCHLORID (GERMAN) * ALUMINUM CHLORIDE (1:3) * ALUMINUM TRICHLORIDE * CHLORURE D'ALUMINIUM (FRENCH)
* TRICHLOROALUMINUM

THR: Moderately toxic by ingestion. Mutagenic data. The dust is an irritant by ingestion, inhalation, and skin contact. Dangerous, see HCl; will react with water or steam to produce heat and toxic or corrosive fumes of Cl^-.

ALUMINUM CHLORIDE HEXAHYDRATE　　HR: 3
CAS: 7784-13-6　　NIOSH: BD 0530000
mf: $AlCl_3 \cdot 6H_2O$　　mw: 241.45

SYNS: ALUMINUM(III) CHLORIDE,HEXAHYDRATE
* ALUMINUM TRICHLORIDEHEXAHYDRATE

THR: Corrosive and irritating to tissue. Mutagenic data. When heated to decomposition, it emits toxic fumes of Cl^-.

ALUMINUM CHLORIDE HYDROXIDE　　HR: 3
CAS: 12042-91-0　　NIOSH: BD 0550000

SYNS: ALUMINUM CHLORHYDRATE * ALUMINUM CHLORHYDROL * ALUMINUM CHLORHYDROXIDE * ALUMINUM CHLOROHYDROXIDE
* ALUMINUM HYDROXIDE CHLORIDE * ALUMINUM HYDROXYCHLORIDE * BASIC ALUMINUM CHLORATE * CHLOROPENTAHYDROXYDIALUMINUM

THR: Poison by inhalation and ingestion. A skin irritant. See aluminum and sodium hydroxide. When heated to decomposition, it emits toxic fumes of Cl^-.

ALUMINUM ETHYLATE　　HR: 3
mf: $Al(OC_2H_5)_3$　　mw: 162.15

PROP: Liquid. Decomp by H_2O. bp: 200° @ 6-8 mm; mp: 140°.

THR: Strong irritant to skin, eyes, and mucous membranes by inhalation. See organo metals.

ALUMINUM FLUORIDE　　HR: 3
CAS: 7784-18-1　　NIOSH: BD 0725000
mf: AlF_3　　mw: 83.98

PROP: Solid; mp: 1291°; subl @ 1260°; d: 2.88; vap press: 1 mm @ 1238°; bp: 1537°.

SYNS: ALUMINIUM FLUORURE (FRENCH)
* ALUMINUM TRIFLUORIDE * FLUORID HLINITY (CZECH)

THR: A poison. Moderately toxic by subcutaneous routes. An eye irritant. Violently impact-sensitive when in contact with Na and K. When heated to decomposition, it emits toxic F^-. For further information, see Vol. 2, No. 1 of *DPIM Report*.

ALUMINUM HYDRIDE　　HR: 3
CAS: 7784-21-6　　NIOSH: BD 0930000
DOT: 2463
mf: AlH_3　　mw: 30.01

PROP: Colorless powder.

SYNS: ALUMINUM TRIHYDRIDE * ALPHA-ALU-
MINUM TRIHYDRIDE

DOT Classification: Label: Flammable Solid
and Dangerous When Wet

THR: Hydrides of some metals (such as AsH_3)
are extremely toxic; little is known about AlH_3.
See hydrides. Dangerous fire hazard. Spont
flammable in air or O_2. Evolves H_2 upon contact
with moisture. Severe explosion hazard by
chemical reaction wherein H_2 gas is produced,
also in contact with dimethyl ether contaminated
by CO_2. Will react with water or steam to produce heat and H_2; reacts with oxidizing materials. On contact with acid or acid fumes, it can
emit toxic fumes.

ALUMINUM HYDROXIDE HR: 3
CAS: 21645-51-2   NIOSH: BD 0940000
mf: AlH_3O_3 mw: 78.01

PROP: White crystalline powder, balls or granules, insol in water; sol in mineral acids and
caustic soda. d: 2.42; mp: $-H_2O$ @ 300°.

SYNS: ALUMIGEL * ALUMINA HYDRATE
* ALUMINA HYDRATED * ALUMINA TRIHY-
DRATE * ALPHA-ALUMINA TRIHYDRATE
* ALUMINIC ACID * ALUMINIUM HYDROXIDE
* ALUMINUM HYDRATE * ALUMINUM(III) HY-
DROXIDE * ALUMINUM HYDROXIDE GEL
* ALUMINUM OXIDE-3H₂O * ALUMINUM OXIDE
HYDRATE * ALUMINUM OXIDE TRIHYDRATE
* ALUMINUM TRIHYDRAT * ALUMINUM TRI-
HYDROXIDE * C.I. 77002

THR: Poison by intraperitoneal route. Systemic
gastrointestinal tract effect. When coprecipitated
with bismuth hydroxide and reduced by H_2, it
is violently flammable in air. Incompatible with
chlorinated rubber. For further information, see
Vol. 2, No. 1 of *DPIM Report*.

ALUMINUM IODIDE
mf: AlI_3 mw: 407.7

PROP: White leaflets. mp: 191°; bp: 360°; d:
3.98 @ 25°; vap press: 1 mm @ 178.0° (sublimes).

THR: See iodides. Incompatible with water.

**ALUMINUM MAGNESIUM PHOS-
PHIDE** HR: 3
                          NIOSH: BD 1000000
DOT: 1419
mf: Mg_3AlP_3 mw: 192.8

SYN: MAGNESIUM ALUMINUM PHOSPHIDE (DOT)

DOT Classification: Label: Flammable Solid
and Dangerous When Wet

THR: Poison by inhalation and ingestion. Dangerous fire hazard. Evolves spontaneously flammable PH_3 in contact with water. See phosphides, phosphine, and magnesium compounds.

ALUMINUM METHYL HR: 3
mf: $Al(CH_3)_3$ mw: 72.07

PROP: Colorless liquid; bp: 130°; mp: 0°.

THR: Related alkyl aluminum compounds are
poisonous and strong irritants. Very flammable
by spontaneous chemical reaction with air. Incompatible with water, halogenated hydrocarbons, and oxidizing materials. When heated to
decomposition, it emits toxic fumes. To fight
fire: do not use water, foam, or halogenated
extinguishing agents. Use dry chemical.

ALUMINUM (III) NITRATE (1:3) HR: 3
CAS: 13473-90-0   NIOSH: BD 1040000
DOT: 1438
mf: $N_3O_9 \cdot Al$ mw: 213.01

PROP: White crystals

SYNS: ALUMINUM NITRATE (DOT) * ALUMI-
NUM TRINITRATE * NITRIC ACID, ALUMIUM
SALT * NITRIC ACID, ALUMINUM(3+) SALT

DOT Classification: Label: Oxidizer

THR: See nitrates. A powerful oxidizer. When
heated to decomposition, it emits toxic NO_x.

ALUMINUM OXIDE (2:3) HR: 3
CAS: 1344-28-1   NIOSH: BD 1200000
mf: Al_2O_3 mw: 101.96

PROP: White powder; mp: 2050°; bp: 2977°,
d: 3.5-4.0; vap press: 1 mm @ 2158°.

SYNS: ACTIVATED ALUMINUM OXIDE * ALU-
MINA * ALPHA-ALUMINA * GAMMA-ALUMINA
* BETA-ALUMINA * ALUMINIUM OXIDE
* ALUMINUM OXIDE * ALPHA-ALUMINUM OX-
IDE * BETA-ALUMINUM OXIDE * GAMMA-
ALUMINUM OXIDE * ALUMINUM SESQUIOXIDE

THR: An experimental carcinogen by intrapleural route. Inhalation of finely divided particles may cause lung damage (Shaver's disease).
Incompatible with hot chlorinated rubber. For
further information, see Alumina, Vol. 1, No.
5 of *DPIM Report*.

ALUMINUM PHOSPHIDE   HR: 3
CAS: 20859-73-8   NIOSH: BD 1400000
DOT: 1397
mf: AlP   mw: 57.95

PROP: Dark gray or dark yellow crystals. d: 2.85 @ 25°/4°. mp: >1000°.

SYNS: ALUMINIUM FOSFIDE (DUTCH) * ALUMINUM MONOPHOSPHIDE * ALUMINIUM PHOSPHIDE * ALUMINUM PHOSPHIDE (DOT) * FOSFURI DI ALLUMINIO (ITALIAN) * PHOSPHURES D'ALUMIUM (FRENCH)

DOT Classification: Label: Flammable Solid and Dangerous When Wet

THR: Poison by inhalation and ingestion. Releases phosphine readily. See also phosphine. Dangerous; in contact with water or steam it yields PH_3, which is spont flammable in air. When heated to decomposition it yields toxic PO_x.

ALUMINUM PICRATE   HR: 3
mf: $Al(C_6H_2O(NO_2)_3)_3$ mw: 711.3

PROP: A solid.

THR: A poison. A powerful irritant. Very flammable by reaction with reducing materials. Severe explosion hazard when shocked or exposed to heat. See also explosives (high). When heated to decomposition, it emits highly toxic fumes of NO_x and explodes.

ALUMINUM(III) SILICATE (2:1)   HR: 3
CAS: 1302-76-7   NIOSH: BD 1450000
mf: $O_5Si•2Al$ mw: 162.05

SYNS: CERAMIC FIBRE * CYANITE * DISTHENE * KYANITE * SILICIC ACID ALUMINUM SALT

THR: An experimental carcinogen. For further information, see Vol. 1, No. 5 of *DPIM Report*.

ALUMINUM SULFATE (2:3)   HR: 3
CAS: 10043-01-3   NIOSH: BD 1700000
mf: $O_{12}S_3•2Al$ mw: 342.14

PROP: White powder. mp: decomp @ 770°, d: 2.71. Solubility in water = 36.4% @ 20°.

SYNS: ALUM * ALUMINUM SULFATE (2:3) * ALUMINUM TRISULFATE * CAKE ALUM * DIALUMINUM SULPHATE * DIALUMINUM TRISULFATE * SULFURIC ACID, ALUMINUM SALT (3:2)

THR: Poison by inhalation, ingestion, and intraperitoneal routes. Hydrolyzes to form sulfuric acid which irritates tissue, especially lungs. When heated to decomposition, it emits toxic fumes of SO_x. For further information, see Vol. 2, No. 1 of *DPIM Report*.

ALUMINUM THALLIUM SULFATE   HR: 3
mf: $AlTl(SO_4)_2•12H_2O$ mw: 639.6

PROP: Cubic, octagonal, colorless crystals. mp: 91°; d: 2.32 @ 20°/4°.

THR: A poison. See thallium compounds.

ALUMINUM TRIPROPYL   HR: 3
mf: $Al(C_3H_7)_3$ mw: 156.24

PROP: Liquid

SYN: TRIPROPYL ALUMINUM

THR: Related alkyl aluminum compounds are poisons. Very flammable by spont reaction with air. See diisobutyl aluminum chloride. Incompatible with halogenated hydrocarbons. Explosion Hazard: Hydrolyzes to evolve flammable vapor. To fight fire: do not use water, foam or halogenated extinguishing agents. Use dry chemical or a special powder extinguisher.

ALYPIN   HR: 3
CAS: 963-07-5   NIOSH: EL 2350000
mf: $C_{16}H_{26}N_2O_2$ mw: 278.44

PROP: White crystalline powder.

SYNS: ALYPINE * 2-BUTANOL, 1-(DIMETHYLAMINO)-2-((DIMETHYLAMINO)METHYL)-, BENZOATE,(ESTER)

THR: Poison by subcutaneous, intravenous, and intraperitoneal routes. An allergen. See esters. When heated to decomposition, it emits toxic fumes of NO_x.

AMATOL   HR: 2
PROP: A high explosive, Composition: NH_4NO_3; 80% and TNT; 20% d: 1.47.

THR: Moderately toxic by inhalation and oral routes. An allergen. May cause contact dermatitis. See nitrates. Dangerous fire hazard. An explosive by shock, spontaneous chemical reaction, or exposure to flame. Decomposition emits highly toxic fumes.

AMBERGRIS TINCTURE   HR: 2
CAS: 9000-02-6   NIOSH: BD 8650000

PROP: Concretion from intestine of sperm whale, composed mostly of cholesterol.

SYNS: AMBRA * AMBER * GRAY AMBER

THR: Moderately toxic by skin contact. When heated to decomposition, it emits acrid smoke and irritating fumes.

AMBUSH   HR: 2
CAS: 52645-53-1   NIOSH: GZ 1255000
mf: $C_{21}H_{20}Cl_2O_3$ mw: 391.31

SYNS: PERMETRINA (PORTUGUESE) * 3-PHE-NOXYBENZYL (+-)-3-(2,2-DICHLOROVINYL)-2,2-DIMETHYLCYCLOPROPANECARBOXYLATE * (3-PHENOXYPHENYL)METHYL 3-(2,2-DICHLOR-ETHENYL)-2,2-DIMETHYLCYCLOPROPANECAR-BOXYLATE * POUNCE

THR: Moderately toxic by ingestion. When heated to decomposition, it emits toxic fumes of Cl^-.

AMERICIUM   HR: 3
af: Am   at wt: 243

PROP: A silvery, somewhat malleable metal. mp: 994°; bp: 2607°; d: 13.67 @ 20°.

THR: A poison. Bone-seeking, long lived radio-active element. Flammable, see powdered metals. In a disaster, this highly toxic radioactive material can be disseminated over a wide area, causing a long-lived inhalation hazard which is difficult to remove from surfaces or from the body once it enters.

AMERICIUM TRICHLORIDE   HR: 3
mf: $AmCl_3$ mw: 349.4

THR: See americium. Due to its radioactivity it can cause radiolysis and build pressure in sealed containers and eventually explode. When heated to decomposition, it emits toxic fumes of Cl^-.

AMETYCIN   HR: 3
CAS: 50-07-7   NIOSH: CN 0700000
mf: $C_{15}H_{18}N_4O_5$ mw: 334.37

SYNS: MUTAMYCIN(MITOMYCIN FOR INJECTION) * MYTOMYCIN * NSC 26980 * NCI-C04706

THR: Poison by ingestion, intravenous, intra-peritoneal, and possibly other routes. An experimental carcinogen and teratogen. Mutagenic data. When heated to decomposition, it emits toxic fumes of NO_x.

AMICARDINE   HR: 3
CAS: 82-02-0   NIOSH: LV 1050000
mf: $C_{14}H_{12}O_5$ mw: 260.26

SYNS: 5,8-DIMETHOXY-2-METHYL-4',5'-FURANO-6,7-CHROMONE * 5,8-DIMETHOXY-2-METHYL-6,7-FURANOCHROMONE * 4,9-DIMETHOXY-7-METHYL-5H-FURO(3,2-G)(1)BENZOPYRAN-5-ONE * 5,8-DIMETHOXY-2-METHYL-4',5'-FURO-6,7-CHROMONE * 4,9-DIMETHOXY-7-METHYL-5-OXO-1,8-DIOXABENZ-(F)INDENE * 4,9-DIMETHOXY-7-METHYL-5-OXOFURO(3,2-G)(1)BENZOPYRAN * 4,9-DIMETHOXY-7-METHYL-5-OXOFURO (3,2-G)-1,2-CHROMENE * KELINCOR

THR: Poison by ingestion and intravenous routes. When heated to decomposition, it emits acrid smoke and irritating fumes.

AMIDES
PROP: Organic compounds containing the structural group $-CONH_2$, and closely related to the organic acids with the grouping COOH. Common examples are: acetamide (CH_3CONH_2) and urea ($CO(NH_2)_2$).

THR: Variable toxicity. Most of the saturated amides have low toxicity, but the unsaturated and N-substituted amides are irritants and may be absorbed via skin contact. Can cause injury to the liver, kidney, and brain.

AMIDITHION   HR: 3
CAS: 919-76-6   NIOSH: TE 1575000
mf: $C_7H_{16}NO_4PS_2$ mw: 273.33

SYNS: O,O-DIMETHYL-S-(2-METHOXYETHYL-CARBAMOYLMETHYL)DITHIOPHOSPHATE * O,O-DIMETHYL-S-(2-METHOXYETHYLCAR-BAMOYL METHYL)PHOSPHORODITHIOATE * ENT 27, 160 * S-(N-2-METHOXYETHYL-CARBAMOYLMETHYL)DIMETHYL PHOPHORO-THIOLOTHIONATE

THR: Poison by ingestion. Moderately toxic by other routes. When heated to decomposition, it emits very toxic fumes of SO_x, PO_x and NO_x.

AMINE 220   HR: 2
CAS: 95-38-5   NIOSH: NJ 2800000
mf: $C_{22}H_{42}N_2O$ mw: 350.66

PROP: Liquid. bp: 235° @ 1 mm, flash p: 465°F (OC), d: 0.9300 @ 20°/20°, vap d: 12.1.

SYNS: 2-(8-HEPTADECENYL)-2-IMIDAZOLINE-1-ETHANOL * 1-HYDROXYETHYL-2-HEPTADECE-

NYLGLYOXALIDINE * 1-(2-HYDROXYETHYL)-2-HEPTADECENYLGLYOXALIDINE * 1-(2-HYDROXYETHYL)-2-N-HEPTADECENYL-2-IMIDAZOLINE * 1-(2-HYDROXYETHYL)-2-HEPTADECENYL-2-IMIDAZOLINE

THR: Moderately toxic by ingestion. Combustible; can react with oxidizing materials. To fight fire: foam, CO_2, dry chemical. When heated to decomposition, it emits toxic fumes of NO_x.

AMINES See also specific compounds.
                                      HR: 3
PROP: A large group of organic compounds containing nitrogen and considered as derived from ammonia (NH_3) by replacement of one or more H atoms by an organic radical.

THR: Variable; some are poisons, some are only slightly toxic. Many are skin irritants and some are sensitizers. See aromatic amines and also fatty amines.

3'-AMINOACETANILIDE   **HR: 3**
CAS: 102-28-3   NIOSH: AD 8050000
mf: $C_8H_{10}N_2O$ mw: 150.20

PROP: mp: softens: bp: decomp @ 787°; sol in water, acetone, alc and ether, sltly sol in benzene, insol in ligroin.

SYNS: M-ACETAMINOANILINE * M-(ACETYL-AMINO)ANILINE * 3-ACETYLAMINOANILINE * N-ACETYL-M-FENYLENEDIAMIN (CZECH) * N-ACETYL-M-PHENYLENEDIAMINE * 3-AMINOACETANILID (CZECH) * M-AMINOACETANILIDE

THR: Poison by intravenous route. Moderately toxic by ingestion. An eye irritant. When heated to decomposition, it emits toxic fumes of NO_x.

AMINOACETONITRILE SULFATE   **HR: 3**
CAS: 5466-22-8   NIOSH: MC 2100000
mf: $C_4H_8N_4 \cdot H_2O_4S$ mw: 210.24

SYN: AMINOACETONITRILE BISULFATE

THR: Poison by ingestion. An experimental teratogen. See nitriles and sulfates. When heated to decomposition, it emits very toxic fumes of NO_x, SO_x and CN^-.

2-AMINOACETOPHENONE   **HR: 3**
CAS: 613-89-8   NIOSH: AM 5775000
mf: C_8H_9NO mw: 135.18

PROP: Yellow, oily liquid. bp: 251° (slt decomp); insol in water, sol in alc and ether.

SYNS: OMEGA-AMINOACETOPHENONE * PHENACYLAMINE

THR: An experimental carcinogen and teratogen. When heated to decomposition, it emits toxic fumes of NO_x.

3'-AMINOACETOPHENONE   **HR: 2**
CAS: 99-03-6   NIOSH: AM 5800000
mf: C_8H_9NO mw: 135.18

PROP: Yellow, oily liquid. bp: 251° (slt decomp); insol in water, sol in alc and ether.

SYNS: M-ACETYLANILINE * 3-ACETYLANILINE * BETA-AMINOACETOPHENONE * M-AMINOACETOPHENONE * M-AMINOACETYLBENZENE

THR: Moderately toxic by several routes. An eye irritant. When heated to decomposition, it emits toxic fumes of NO_x.

p-AMINO ACETOPHENONE   **HR: 3**
CAS: 99-92-3   NIOSH: AM 5500000
mf: C_8H_9NO mw: 135.18

PROP: Crystalline; mp: 106°; bp: 293°-295°. Sol in hot water, alc and ether.

SYNS: 4-ACETYLANILINE * P-AMINO ACETOPHENONE * 4'-AMINOACETOPHENONE * P-AMINOACETYLBENZENE

THR: Poison by intraperitoneal route. When heated to decomposition, it emits toxic fumes of NO_x.

9-AMINOACRIDINE   **HR: 3**
CAS: 90-45-9   NIOSH: AR 7300000
mf: $C_{13}H_{10}N_2$ mw: 194.25

SYNS: 9-ACRIDINAMINE * 5-AMINOACRIDINE * IZOACRIDINA

THR: Poison by intraperitoneal and subcutaneous routes. Mutagenic data. When heated to decomposition, it emits toxic fumes of NO_x.

6-AMINO-4-((3-AMINO-4-(((4-((1-METHYLPYRIDINIUM-4-YL)AMINO)PHENYL)-AMINO)CARBONYL)PHENYL)AMINO)-1-METHYLQUINOLINIUM),DIIODIDE   **HR: 3**
                                      NIOSH: VC 3511000
mf: $C_{29}H_{29}N_6O \cdot 2I$ mw: 731.44

THR: Poison by intraperitoneal route. Mutagenic data. When heated to decomposition, it emits very toxic fumes of NO_x and I^-.

1-AMINOANTHRAQUINONE HR:
CAS: 82-45-1   NIOSH: CB 5075000
mf: $C_{14}H_9NO_2$ mw: 223.24

PROP: Red needles. mp: 256°; bp: subl. Insol in water, sol in HCl, alc, benzene, ether and chloroform.

SYNS: 1-AMINO-9,10-ANTHRACENEDIONE * 1-AMINOANTHRACHINON (CZECH) * ALPHA-AMINOANTHRAQUINONE * 1-AMINO-9,10-AN-THRAQUINONE * ALPHA-ANTHRAQUINONYL-AMINE * C.I. 37275

THR: An experimental neoplastigen. Severe eye irritant; moderately toxic via intraperitoneal route. When heated to decomposition, it emits toxic NO_x.

2-AMINOANTHRAQUINONE
CAS: 117-79-3   NIOSH: CB 5120000
mf: $C_{14}H_9NO_2$ mw: 223.24

PROP: Red needles from alc. mp: 302°; bp: subl. Insol in water and ether; sol in alc and benzene.

SYNS: 2-AMINO-9,10-ANTHRACENEDIONE * BETA-AMINOANTHRAQUINONE * NCI-C01876

THR: An experimental carcinogen. Moderately toxic via intraperitoneal route. When heated to decomposition, it emits toxic NOx. For further information, see Vol. 4, No. 6 of *DPIM Report*.

2-AMINO-5-AZOTOLUENE HR: 3
CAS: 97-56-3   NIOSH: XU 8800000
mf: $C_{14}H_{15}N_3$ mw: 225.32

SYNS: C.I. 11160 * AMINOAZOTOLUENE (IN-DICATOR) * O-AMIDOAZOTOLUOL (GERMAN) * O-AMINOAZOTOLUENE * 4'-AMINO-2,3'-AZOTOLUENE * O-AMINOAZOTOLUENO (SPAN-ISH) * 4'-AMINO-2:3'-AZOTOLUENE * O-AMINOAZOTOLUOL * 4-AMINO-2',3-DI-METHYLAZOBENZENE * 4'-AMINO-2,3'-DIMETHYLAZOBENZENE * BUTTER YELLOW * C.I. 11160B * C.I. SOLVENT YELLOW 3 * 2',3-DIMETHYL-4-AMINOAZOBENZENE * 2-METHYL-4-((2-METHYLPHENYL)AZO)BEN-ZENAMINE * O-TOLUENEAZO-O-TOLUIDINE * O-TOLUOL-AZO-O-TOLUIDIN (GERMAN) * 5-(O-TOLYLAZO)-2-AMINOTOLUENE * 4-(O-TOLYLAZO)-O-TOLUIDINE

THR: An experimental carcinogen. Moderately toxic by several routes. Mutagenic data. When heated to decomposition, it emits toxic fumes of NO_x. For further information, see Vol. 6, No. 4 of *DPIM Report*.

m-AMINOBENZAL FLUORIDE HR: 3
CAS: 98-16-8   NIOSH: XU 9180000
mf: $C_7H_6F_3N$ mw: 161.14

PROP: Colorless liquid with aniline-like odor. mp: 3°, bp: 189°, d: 1.303 @ 15.5°/15.5°, vap d: 5.56.

SYNS: 3-AMINOBENZOTRIFLUORIDE * M-(TRI-FLUOROMETHYL)ANILINE * 3-(TRIFLUORO-METHYL)ANILINE * 3-(TRIFLUOROMETHYL)BEN-ZENAMINE * M-AMINOBENZOTRIFLUORIDE * USAF MA-4

THR: Poison by inhalation, ingestion, and intraperitoneal routes. See fluorides. When heated to decomposition, it emits very toxic fumes of fluorides and NO_x.

2-AMINOBENZENETHIOL HR: 3
CAS: 137-07-5   NIOSH: DC 0600000
mf: C_6H_7NS mw: 125.20

PROP: Liquid. mp: 23°, bp: 227.2°, flash p: 175°F, d: 1.168, vap d: 4.3.

SYNS: O-AMINOTHIOPHENOL * 2-AMINOTHIO-PHENOL * O-MERCAPTOANILINE * USAF EK-4376

THR: Poison by intraperitoneal route. Moderately toxic by ingestion. Moderately flammable. Can react with oxidizing materials. To fight fire: water, foam, CO_2, mist or spray, dry chemical.

2-AMINOBENZIMIDAZOLE HR: 3
CAS: 934-32-7   NIOSH: DD 5775000
mf: $C_7H_7N_3$ mw: 133.17

PROP: Aq leaflets; mp: 222°-224°; sol in water, alkalies, alc, acetone, very sltly sol in ether.

SYN: USAF EK-4037

THR: Poison by intravenous and intraperitoneal routes. Moderately toxic by ingestion. Mutagenic data. When heated to decomposition, it emits toxic NO_x.

p-AMINOBENZOIC ACID HR: 3
CAS: 150-13-0   NIOSH: DG 1400000
mf: $C_7H_7NO_2$ mw: 137.15

PROP: Yellowish to red crystals. mp: 187°. Sol in water, alc and ether.

SYNS: GAMMA-AMINOBENZOIC ACID * 4-AMI-NOBENZOIC ACID * 1-AMINO-4-CARBOXYBEN-ZENE * ANTI-CHROMOTRICHIA FACTOR * BACTERIAL VITAMIN H1 * 4-CARBOXYANI-LINE * P-CARBOXYPHENYLAMINE * CHROMO-TRICHIA FACTOR * PABA * TRICHOCHROMO-GENIC FACTOR * VITAMIN H

THR: An experimental carcinogen. Moderately toxic by ingestion and intravenous routes. Ingesting large doses can cause nausea, vomiting, skin rash, methemoglobinemia and possibly toxic hepatitis. Combustible. When heated to decomposition, it emits toxic fumes of NO_x.

p-AMINOBENZOIC ACID-3-(beta-DI-ETHYLAMINO)ETHOXY)PROPYL-ESTER HR: 3
CAS: 63917-76-0 NIOSH: DG 2000000
mf: $C_{16}H_{26}N_2O_3$ mw: 294.44

THR: Poison by intravenous and subcutaneous routes. See esters. When heated to decomposition, it emits toxic fumes of NO_x.

p-AMINOBENZOIC ACID-2-DIETHYLAM-INOETHYL ESTER HR: 3
CAS: 59-46-1 NIOSH: DG 2100000
mf: $C_{13}H_{20}N_2O_2$ mw: 236.35

SYNS: 4-AMINOBENZOIC ACID DIETHYLAMINO-ETHYL ESTER * P-AMINOBENZOYLDI-ETHYLAMINOETHANOL * DIETHYL-AMINOETHYL-P-AMINOBENZOATE * BETA-DIETHYLAMINOETHYL-4-AMINOBENZOATE * 2-DIETHYLAMINOETHYL-P-AMINOBENZOATE * NISSOCAINE * NOVOCAINE * PROCAINE

THR: Poison by ingestion, intraperitoneal, intravenous, and subcutaneous routes. Central nervous system effects. See esters. When heated to decomposition, it emits toxic fumes of NO_x.

p-AMINOBENZOPHENONE HR: 3
CAS: 1137-41-3 NIOSH: DJ 0175000
mf: $C_{13}H_{11}NO$ mw: 197.25

PROP: Leaflets from alc. mp: 124°; very sltly sol in cold water, very sol in alc.

SYN: USAF A-233.

THR: Poison by intraperitoneal route. See ketones. When heated to decomposition, it emits toxic fumes of NO_x.

AMINOBENZOYLDIBUTYLAMINOPRO-PANOL HYDROCHLORIDE HR: 3
NIOSH: DG 1650000
mf: $C_{18}H_{30}N_2O_2 \cdot ClH$ mw: 342.96

SYN: P-AMINOBENZOIC ACID-3-(DIBUTYLAMINO) PROPYL ESTER HYDROCHLORIDE

THR: Poison by subcutaneous, intraperitoneal, and possibly other routes. See esters. When heated to decomposition, it emits very toxic fumes of HCl and NO_x.

p-AMINOBENZOYLDIETHYLAMINOETH-ANOL HYDROCHLORIDE HR: 3
CAS: 51-05-8 NIOSH: DG 2275000
mf: $C_{13}H_{20}N_2O_2 \cdot ClH$ mw: 272.81

SYNS: P-AMINOBENZOIC ACID-2-DIETHYLAMINO-ETHYL ESTER HYDROCHLORIDE * ANESTHESOL * ATOXICOCAINE * PROCAINE HYDROCHLO-RIDE * 2-DIETHYLAMINOETHYL-P-AMINOBEN-ZOATE HYDROCHLORIDE * DIETHYLAMINOETHA-NOL-4-AMINOBENZOATE HYDROCHLORIDE * NOVOCAIN-CHLORHYDRAT (GERMAN) * NOVOCAIN HYDROCHLORID (GERMAN) * NOVOCAINE HYDROCHLORIDE

THR: Poison by intravenous, intraperitoneal, intraspinal, parenteral, and possibly other routes. Moderately toxic by subcutaneous routes. See esters. When heated to decomposition, it emits very toxic fumes of HCl and NO_x.

p-AMINOBENZOYLDIMETHYLAMINO-1,2-DIMETHYLPROPANOL HYDRO-CHLORIDE HR: 3
CAS: 532-62-7 NIOSH: EL 2625000
mf: $C_{14}H_{22}N_2O_2 \cdot ClH$ mw: 286.84

SYNS: 3-DIMETHYLAMINO-1,2-DIMETHYLPROPYL P-AMINOBENZOATE HYDROCHLORIDE * TOTO-CAINE HYDROCHLORIDE * TUTOCAINE HYDRO-CHLORIDE

THR: Poison by subcutaneous, intravenous, intraperitoneal, and intraspinal routes. When heated to decomposition, it emits very toxic fumes of HCl and NO_x.

2-AMINOBENZOXAZOLE HR: 3
CAS: 4570-41-6 NIOSH: DM 4500000
mf: $C_7H_6N_2O$ mw: 134.15

THR: Poison by intravenous and intraperitoneal routes. Moderately toxic by ingestion. When heated to decomposition, it emits toxic fumes of NO_x.

5-AMINO-1-BIS(DIMETHYLAMIDE)-PHOSPHORYL-3-PHENYL-1,2,4-TRIAZOLE HR: 3
CAS: 1031-47-6 NIOSH: TA 1400000
mf: $C_{12}H_{19}N_6OP$ mw: 294.34

SYNS: 5-amino-1-(bis(dimethylamino)phosphinyl)-3-phenyl-1,2,4-triazole * 5-amino-1-bis(dimethylamido)phosphoryl-3-phenyl-1,2,4-triazole * ent 27,223 * 5-amino-3-fenil-1-bis(-dimetilamino)-fosforil-1,2,4-triazolo (italian) * 5-amino-3-fenyl-1-bis-(dimethyl-amino)-fosforyl-1,2,4-triazool (dutch) * 5-amino-3-phenyl-1-bis (dimethyl-amino)-phosphoryle-1,2,4-triazole-(french) * 5-amino-3-phenyl-1-bis(dimethylamino)-phosphoryl-1h-1,2,4-triazol (german) * 5-amino-3-phenyl-1,2,4-triazole-1-yl-n,n,n',n'-tetramethylphosphodiamide * 5-amino-3-phenyl-1,2,4-triazolyl-1-bis(dimethylamido)pheosphate * 5-amino-3-phenyl-1,2,4-triazolyl-n,n,n',n'-tetramethyl-phosphonamide * bis(dimethylamino)-3-amino-5-phenyl-triazolyl phosphine oxide * triamifos (german, dutch, italian)

THR: Poison by ingestion, skin contact, and possibly other routes. Mutagenic data. When heated to decomposition, it emits very toxic fumes of PO_x and NO_x.

AMINOBIS(PROPYLAMINE) HR: 3
CAS: 56-18-8 NIOSH: JL 9450000
DOT: 2269
mf: $C_6H_{17}N_3$ mw: 131.26

SYNS: bis-(3-aminopropyl)amine * 3,3-diaminodipropylamine * 3,3'-diaminodipropylamine * dipropylenetriamine * iminobis(propylamine) * 3,3'-iminobis-(propylamine) * initiating explosive iminobispropylamine (dot)

DOT Classification: Label: Corrosive

THR: Poison by skin contact. Moderately toxic by ingestion. A skin and eye irritant. When heated to decomposition, it emits toxic fumes of NO_x. An explosive.

2-AMINOBUTAN-1-OL HR: 2
CAS: 96-20-8 NIOSH: EK 9625000
mf: $C_4H_{11}NO$ mw: 89.16

PROP: Water-white liquid. mp: $-2°$, bp: 178°, flash p: 165°F (OC), d: 0.944 @ 20°/20°, vap d: 3.06.

SYNS: 2-amino-1-butanol * butanol-2-amine

THR: Moderately toxic by ingestion and intraperitoneal routes. Moderately flammable when exposed to heat, flame or oxidizing materials. To fight fire: water spray, alcohol foam, dry chemical. When heated to decomposition, it yields NO_x.

3-AMINO-2-BUTOXYBENZOIC ACID-2-DIETHYLAMINOETHYL ESTER HYDROCHLORIDE HR: 3
CAS: 3624-87-1 NIOSH: DG 1500000
mf: $C_{17}H_{28}N_2O_3 \cdot ClH$ mw: 344.93

SYNS: 2-butoxy-3-aminobenzoic acid beta-diethylaminoethyl ester hydrochloride * 2'-diethylaminoethyl-3-amino-2-butoxybenzoate hydrochloride * beta-diethylaminoethyl-2-butoxy-3-aminobenzoate hydrochloride * methambucaine hydrochloride * primacaine hydrochloride

THR: Poison by subcutaneous, intravenous, and intraperitoneal routes. See esters. When heated to decomposition, it emits very toxic fumes of HCl and NO_x.

3-(2-AMINOBUTYL)INDOLE ACETATE HR: 3
CAS: 118-68-3 NIOSH: NL 3850000
mf: $C_{12}H_{16}N_2 \cdot C_2H_4O_2$ mw: 248.36

SYNS: dl-alpha-ethyltryptamine acetate * alpha-ethyltryptamine acetate * etryptamine acetate * indole-3-(2-aminobutyl) acetate

THR: Poison by ingestion, intraperitoneal, and intravenous routes. See esters. When heated to decomposition, it emits toxic fumes of NO_x.

gamma-AMINOBUTYRIC ACID CETYL ESTER HR: 3
CAS: 34562-99-7 NIOSH: EL 7713000
mf: $C_{20}H_{41}NO_2$ mw: 327.62

SYNS: cetyl gamma-aminobutyrate * cetyl gaba

THR: Poison by intravenous and intraperitoneal routes. See esters. When heated to decomposition, it emits toxic fumes of NO_x.

6-AMINOCAPROIC ACID HR: 2
CAS: 60-32-2 NIOSH: MO 6300000
mf: $C_6H_{13}NO_2$ mw: 131.20

SYNS: EPSILON-AMINOCAPROIC ACID * AMI-NOKAPRON * OMEGA-AMINOCAPROIC ACID * EPSILON-AMINOHEXANOIC ACID * EPSICA-PRON * EPSILON-LEUCINE * EPSILON-NORLEU-CINE * OMEGA-AMINOHEXANOIC ACID

THR: Moderately toxic by intravenous and intraperitoneal routes. An experimental teratogen. A moderate irritant. When heated to decomposition, it emits toxic fumes such as NO_x.

2-AMINO-5-CHLOROBENZOX-AZOLE   HR: 3
CAS: 61-80-3   NIOSH: DM 4550000
mf: $C_7H_5ClN_2O$ mw: 168.59

SYNS: MCN-485 * USAF MA-12

THR: Poison by ingestion, intraperitoneal, and intravenous routes. Central nervous system effects by ingestion. When heated to decomposition, it yields toxic fumes of Cl^- and NO_x.

4-AMINO-5-CHLORO-N-(2-(DIETHYL-AMINO)ETHYL)-n-ANISAMIDE   HR: 3
CAS: 364-62-5   NIOSH: BZ 3300000
mf: $C_{14}H_{22}ClN_3O_2$ mw: 299.84

SYNS: 4-AMINO-5-CHLORO-N-(2-(DIETHYL-AMINO)ETHYL)-2-METHOXYBENZAMIDE * 5-CHLORO-2-METHOXYPROCAINAMIDE * N-(DIETHYLAMINOETHYL)-2-METHOXY-4-AMINO-5-CHLOROBENZAMIDE * 2-METHOXY-5-CHLOROPROCAINAMIDE

THR: Poison by ingestion and intravenous routes. Central nervous system effects. When heated to decomposition, it emits very toxic fumes of NO_x and Cl^-.

2-AMINO-4-CHLOROTOLUENE   HR: 3
CAS: 95-79-4   NIOSH: XU 5075000
mf: C_7H_8ClN mw: 141.61

PROP: Solid; bp: 241°, mp: 29°.

SYNS: 1-AMINO-3-CHLORO-6-METHYLBENZENE * 4-CHLORO-2-AMINOTOLUENE * 3-CHLORO-6-METHYLANILINE * 5-CHLORO-o-TOLUIDINE * NCI-CO2051

THR: Moderately toxic by ingestion. An experimental carcinogen. When heated to decomposition, it emits very toxic fumes of Cl^- and NO_x.

1-AMINOCYCLOPENTANE-1-CARB-OXYLIC ACID   HR: 3
CAS: 52-52-8   NIOSH: GY 2625000
mf: $C_6H_{11}NO_2$ mw: 129.18

SYNS: 1-AMINO-1-CYCLOPENTANECARBOXYLIC ACID * CYCLOLEUCINE * NSC 1026

THR: Poison by ingestion and intravenous routes. Central nervous system effects by ingestion. When heated to decomposition, it emits toxic fumes of NO_x.

AMINODARONE   HR: 3
CAS: 1951-25-3   NIOSH: OB 1360000
mf: $C_{25}H_{29}INO_3$ mw: 518.45

SYNS: 2-BUTYL-3-BENZOFURANYL P-((2-DIETHYLAMINO)ETHOXY)-m,m-DIIODOPHENYL KETONE * 2-BUTYL-3-(3,5-DIIODO-4-(2-DIETHYLAMINO-ETHOXY)BENZOYL)BENZOFURAN * 2-N-BUTYL-3',5'-DIIODO-4'-N-DIETHYLAMINOETHOXY-3-BENZOYLBENZOFURAN

THR: Poison by intravenous and intraperitoneal routes. When heated to decomposition, it emits very toxic fumes of I^- and NO_x.

3-AMINO-2,5-DICHLOROBENZOIC ACID   HR: 3
CAS: 133-90-4   NIOSH: DG 1925000
mf: $C_7H_5Cl_2NO_2$ mw: 206.03

SYNS: ACP-M-728 * 2,5-DICHLORO-3-AMINO-BENZOIC ACID * NCI-C00055

THR: Moderately toxic by ingestion. An experimental carcinogen. When heated to decomposition, it emits highly toxic fumes such as Cl^- and NO_x. For further information, see Vol. 1, No. 3 of *DPIM Report*.

p-AMINO-n-(2-DIETHYLAMINOETHYL)-BENZAMIDE   HR: 3
CAS: 51-06-9   NIOSH: CV 2275000
mf: $C_{13}H_{21}N_3O$ mw: 235.37

SYNS: P-AMINOBENZOIC DIETHYLAMINOETHYL-AMIDE * BENZAMIDE, 4-AMINO-N-(2-(DIETHYL-AMINO)ETHYL)- (9CI) * NOVOCAINAMIDE * PROCAINAMIDE

THR: Poison by intravenous and intraperitoneal routes. Moderately toxic by ingestion causing pulmonary and systemic skin effects. When heated to decomposition, it emits toxic fumes of NO_x.

1-(4-AMINO-6,7-DIMETHOXY-2-QUINA-ZOLINYL-4-(2-FURANYLCARBONYL) PIPERAZINE   HR: 3
CAS: 19216-56-9   NIOSH: VA 1300000
mf: $C_{19}H_{21}N_5O_4$ mw: 383.45

SYNS: FURAZOSIN * 2-(4-(2-FUROYL)PIPERA-ZIN-1-YL)-4-AMINO-6,7-DIMETHOXYQUINAZOLINE

THR: Peripheral nervous system and pulmonary system effects. When heated to decomposition, it emits toxic fumes of NO_x.

1-AMINO-3-DIMETHYLAMINO-PROPANE   HR: 2

CAS: 109-55-7   NIOSH: TX 7525000
mf: $C_5H_{14}N_2$ mw: 102.21

PROP: Colorless liquid; mp: $<-70°$, bp: 123°, flash p: 100°F (OC), d: 0.8100 @ 30°, vap press: 10 mm @ 30°, vap d: 3.52.

SYNS: N,N-DIMETHYL-N-(3-AMINOPROPYL) AMINE * 3-(DIMETHYLAMINO)PROPYLAMINE * N,N-DIMETHYL-1,3-DIAMINOPROPANE * N,N-DIMETHYL-1,3-PROPANEDIAMINE * N,N-DIMETHYL-1,3-PROPYLENEDIAMINE

THR: Moderately toxic by ingestion. A skin and eye irritant. Very flammable when exposed to heat, flame or oxidizers. To fight fire: alcohol foam, CO_2, dry chemical. Emits toxic fumes of NO_x when heated.

3-AMINO-1,4-DIMETHYL-5H-PYRIDO-(4,3-b)INDOLE ACETATE   HR: 3

NIOSH: UU 9352000
mf: $C_{13}H_{13}N_3 \cdot C_2H_4O_2$ mw: 271.35

SYN: TRP-P-1 (ACETATE)

THR: An experimental carcinogen. Mutagenic data. When heated to decomposition, it emits toxic fumes of NO_x.

2-AMINOETHANETHIOL   HR: 3

CAS: 60-23-1   NIOSH: KJ 0175000
mf: C_2H_7NS mw: 77.16

SYNS: 2-AMINOETHYL MERCAPTAN * CYSTEAMIDE * CYSTEAMINE * DECARBOXYCYSTEINE * MERCAPTAMINE * (2-MERCAPTOETHYL)AMINE * THIOETHANOLAMINE

THR: Poison by intravenous, subcutaneous, and intraperitoneal routes. Moderately toxic by ingestion. An experimental teratogen. Mutagenic data. When heated to decomposition, it emits very toxic fumes of SO_x and NO_x.

3-AMINO-4-ETHOXYACET-ANILIDE   HR: 3

CAS: 17026-81-2   NIOSH: AD 8575000
mf: $C_{10}H_{14}N_2O_2$ mw: 194.26

SYNS: 2-AMINO-4-ACETAMINIFENETOL (CZECH) * NCI-C01887

THR: Moderately toxic by ingestion. An experimental carcinogen. An eye irritant. When heated to decomposition, it emits toxic fumes of NO_x.

2-AMINOETHOXYETHANOL   HR: 3

CAS: 929-06-6   NIOSH: KJ 6125000
DOT: 1760
mf: $C_4H_{11}NO_2$ mw: 105.16

SYNS: 2-(2-AMINOETHOXY)ETHANOL * DIGLYCOLAMINE

DOT Classification: Label: Corrosive

THR: Very toxic by skin contact. Corrosive and a powerful irritant. When heated to decomposition, it emits toxic fumes of NO_x.

3-AMINO-9-ETHYLCARBAZOLE   HR: 3

CAS: 132-32-1   NIOSH: FE 3590000
mf: $C_{14}H_{14}N_2$ mw: 210.30

SYN: 3-AMINO-N-ETHYLCARBAZOLE

THR: Poison by ingestion and intraperitoneal routes. An experimental carcinogen. When heated to decomposition, it emits toxic fumes of NO_x. For further information, see 2-amino ethyl ethanol amine, Vol. 4, No. 6 of *DPIM Report*.

3-AMINO-9-ETHYLCARBAZOLEHYDRO-CHLORIDE   HR: 3

CAS: 6109-97-3   NIOSH: FE 3675000
mf: $C_{14}H_{14}N_2 \cdot ClH$ mw: 246.76

SYN: NCI-C03043

THR: Poison by ingestion. An experimental carcinogen. When heated to decomposition, it emits very toxic fumes of NO_x and HCl.

N-AMINOETHYL ETHANOL-AMINE   HR: 2

CAS: 111-41-1   NIOSH: KJ 6300000
mf: $C_4H_{12}N_2O$ mw: 104.18

PROP: Colorless liquid; bp: 243.7°, flash p: 216°F, d: 1.0304 @ 20°/20°, autoign temp: 695°F, vap press: <0.01 mm @ 20°, vap d: 3.59.

SYNS: N-HYDROXYETHYL-1,2-ETHANEDIAMINE * N-(BETA-HYDROXYETHYL)ETHYLENEDIAMINE * N-(2-HYDROXYETHYL)ETHYLENEDIAMINE * MONOETHANOLETHYLENEDIAMINE

THR: Moderately toxic by ingestion and skin contact. A severe eye irritant and moderate skin irritant. Combustible. To fight fire: alcohol foam, mist, dry chemical. When heated to decomposition, it emits toxic fumes of NO_x. For further information, see 2-amino ethyl ethanol amine, Vol. 2, No. 3 of *DPIM Report*.

3-(2-AMINOETHYL)INDOLE HYDRO-CHLORIDE   HR: 3
CAS: 343-94-2   NIOSH: NL 4375000
mf: $C_{10}H_{12}N_2 \cdot ClH$ mw: 196.70

SYNS: BETA-INDOLAETHYLAMIN-CHLORHYDRAT (GERMAN) * BETA-3-INDOLYLETHYLAMINE HYDROCHLORIDE * INDOLE-3-ETHYLAMINE HYDROCHLORIDE * BETA-INDOLE-ETHYLAMINE HYDROCHLORIDE * TRYPTAMINE HYDROCHLORIDE

THR: Poison by intravenous and intraperitoneal routes. Moderately toxic by other routes. When heated to decomposition, it emits very toxic NO_x and HCl.

3-(2-AMINOETHYL)INDOL-5-OL   HR: 3
CAS: 50-67-9   NIOSH: NM 2450000
mf: $C_{10}H_{12}N_2O$ mw: 176.24

SYNS: 5-HYDROXY-3-(BETA-AMINOETHYL)INDOLE * 5-HYDROXYTRYPTAMINE * SEROTONIN

THR: Poison by intravenous, subcutaneous, and intraperitoneal routes. Moderately toxic by other routes. An experimental teratogen. When heated to decomposition, it emits toxic fumes of NO_x.

2-beta-AMINOETHYLISOTHIO-UREA   HR: 3
CAS: 56-10-0   NIOSH: UM 0175000
mf: $C_3H_9N_3S \cdot 2BrH$ mw: 281.05

SYNS: BETA-AMINOAETHYL-ISOTHIURONIUM DIHYDROBROMID(GERMAN) * S-2-AMINOETHYL-ISOTHIOURONIUM BROMIDE HYDROBROMIDE * 2-(BETA-AMINOETHYL)ISOTHIOURONIUM BROMIDE HYDROBROMIDE * BETA-AMINOETHYLISOTHIURONIUM BROMIDE HYDROBROMIDE * S-AMINOETHYLISOTHIURONIUM BROMIDE HYDROBROMIDE * S-(BETA-AMINOETHYL)ISOTHIURONIUM BROMIDE HYDROBROMIDE * S-(2-AMINOETHYL)ISOTHIURONIUM BROMIDE HYDROBROMIDE * BETA-AMINOETHYLISOTHIURONIUM BROMIDE HYDROBROMIDE * 2-AMINOETHYLISOTHIURONIUM BROMIDE HYDROBROMIDE * 2-AMINOETHYLISOTHIOURONIUM DIBROMIDE * 2-AMINOETHYLISOTHIURONIUM

DIHYDROBROMIDE * S-(2-AMINOETHYL)-PSEUDOTHIOUREA DIHYDROBROMIDE * 2-AMINOETHYLTHIOPSEUDOUREA DIHYDROBROMIDE * 2-(2-AMINOETHYL)-2-THIOPSEUDOUREA HYDROBROMIDE * USAF XR-31

THR: Poison by subcutaneous and intravenous routes. Moderately toxic by other routes. When heated to decomposition, it emits very toxic NO_x, SO_x, and HBr.

N-AMINOETHYLMORPHOLINE   HR: 3
CAS: 2038-03-1   NIOSH: QD 7350000
mf: $C_6H_{14}N_2O$ mw: 130.22

PROP: Liquid. mp: 25.6°; bp: 204.2°; flash p: 347°F (OC); d: 0.9915 @ 20°/20°; vap d: 4.49

SYNS: BETA-AMINOAETHYL-MORPHOLIN (GERMAN) * 4-MORPHOLINEETHANAMINE

THR: Poison by skin contact. Moderately toxic by other routes. A skin irritant. Moderately flammable when exposed to heat, flame or oxidizing materials. To fight fire: alcohol foam, dry chemical. When heated to decomposition, it emits toxic fumes of NO_x.

N-AMINOETHYLPIPERAZINE   HR: 3
CAS: 140-31-8   NIOSH: TK 8050000
DOT: 2815
mf: $C_6H_{15}N_3$ mw: 129.24

PROP: Light colored liquid; d: 0.9852 @ 20°/20°, mp: −19°, bp: 220.4°, flash p: 200°F (OC), vap d: 4.4.

SYNS: N-(BETA-AMINOETHYL)PIPERAZINE * N-(2-AMINOETHYL)PIPERAZINE * 1-(2-AMINOETHYL)PIPERAZINE * USAF DO-46

DOT Classification: Label: Corrosive

THR: Poison by intraperitoneal route, by ingestion and skin contact. A skin irritant. See amines. Moderately flammable when exposed to heat, flame or sparks, powerful oxidizers. To fight fire: alcohol foam. When heated to decomposition, it emits toxic fumes of NO_x.

2-AMINOETHYL-2-THIOPSEUDO-UREADICHLORIDE   HR: 3
CAS: 871-25-0   NIOSH: UM 0140000
mf: $C_3H_9N_3S \cdot 2Cl$ mw: 190.11

SYN: AET DICHLORIDE

THR: Poison by parenteral route. When heated to decomposition, it emits very toxic fumes of Cl^-, NO_x and SO_x.

4-AMINO-4'-FLUORODIPHENYL   HR: 3
CAS: 324-93-6   NIOSH: DU 9625000
mf: $C_{12}H_{10}FN$ mw: 187.23

SYNS: 4'-FLUORO-4-AMINODIPHENYL * 4'-FLUORO-4-BIPHENYLAMINE

THR: Very poisonous by ingestion. An experimental carcinogen. When heated to decomposition, it emits very toxic fumes of F^- and NO_x.

1-AMINO-4-HYDROXYANTHRAQUI-NONE   HR: 3
CAS: 116-85-8   NIOSH: CB 5600000
mf: $C_{14}H_9NO_3$ mw: 239.24

PROP: Red-violet powder. mp: 207°-208°. Sol in water, HCl, alc, ether and benzene.

SYNS: 1-AMINO-4-OXYANTHRAQUINONE (RUSSIAN) * C.I. 60710 * 1-HYDROXY-4-AMINOANTHRAQUINONE * 4-HYDROXY-1-ANTHRAQUINONYLAMINE

THR: Poison by intraperitoneal route. Mutagenic data. When heated to decomposition it emits toxic fumes of NO_x.

2-AMINO-3-HYDROXYBENZOIC ACID   HR: 3
CAS: 548-93-6   NIOSH: DG 2625000
mf: $C_7H_7NO_3$ mw: 153.15

SYNS: 3-HYDROXYANTHRANILIC ACID
* 3-HYDROXY-ANTHRANILSAEURE (GERMAN)
* 3-OXYANTHRANILIC ACID

THR: An experimental carcinogen. Mutagenic data. When heated to decomposition it emits toxic fumes such as NO_x.

3-AMINO-4-HYDROXYBENZOIC ACID METHYL ESTER   HR: 2
CAS: 536-25-4   NIOSH: DG 2750000
mf: $C_8H_9NO_3$ mw: 167.18

SYNS: AMINOBENZ * ORTHOCAINE * ORTHODERM * ORTHOFORM

THR: Moderately toxic by several routes. See esters. When heated to decomposition it emits toxic fumes of NO_x.

1-AMINO-2-METHYLANTHRA-QUINONE   HR: 3
CAS: 82-28-0   NIOSH: CB 5740000
mf: $C_{15}H_{11}NO_2$ mw: 237.27

SYNS: C.I. 60700 * 2-METHYL-1-ANTHRAQUINONYLAMINE * NCI-C01901

THR: An experimental carcinogen. When heated to decomposition it emits toxic fumes of NO_x.

alpha-(AMINOMETHYL)-m-HYDROXY-BENZYL ALCOHOL   HR: 3
CAS: 536-21-0   NIOSH: DN 7250000
mf: $C_8H_{11}NO_2$ mw: 153.20

SYNS: 1-(M-HYDROXYPHENYL)-2-AMINOETHANOL * 1-(3'-HYDROXYPHENYL)-2-AMINOETHANOL * M-HYDROXYPHENYLETHANOLAMINE * 1-(3-HYDROXYPHENYL)-1-HYDROXY-2-AMINOETHANE

THR: Poison by ingestion, subcutaneous, intravenous, and intraperitoneal routes. When heated to decomposition it emits toxic fumes of NO_x.

5-AMINOMETHYL-3-ISOXYZOLE   HR: 3
CAS: 2763-96-4   NIOSH: NY 3325000
mf: $C_4H_6N_2O_2$ mw: 114.12

SYNS: 3-HYDROXY-5-AMINOMETHYLISOXAZOLE-AGARIN * MUSCIMOL * PANTHERINE

THR: Poison by ingestion, subcutaneous, intravenous, and intraperitoneal routes. Central nervous system effects. When heated to decomposition it emits toxic fumes of NO_x. For further information, see Muscimol, Vol. 2, No. 3 of *DPIM Report*.

4-AMINO-2-METHYL-1-NAPH-THOL   HR: 3
CAS: 130-24-5   NIOSH: QL 3350000
mf: $C_{11}H_{11}NO$ mw: 173.23

SYNS: 2-METHYL-4-AMINO-1-NAPHTHOL
* VITAMIN K5

THR: Poison by intraperitoneal route. When heated to decomposition it emits toxic NO_x.

2-AMINO-6-(1'-METHYL-4'-NITRO-5'-IMIDAZOLYL)MERCAPTOPURINE   HR: 3
CAS: 5581-52-2   NIOSH: UO 7505000
mf: $C_9H_8N_8O_2S$ mw: 292.31

SYNS: 6-BENZYLAMINOPURINE * GUANERAN

THR: Poison by intraperitoneal route. Moderately toxic by ingestion. An experimental teratogen. When heated to decomposition it emits very toxic SO_x and NO_x.

2-AMINO-2-METHYL-1,3-PROPANE-DIOL   HR: 2
CAS: 115-69-5   NIOSH: TY 2975000
mf: $C_4H_{11}NO_2$ mw: 105.16

PROP: A clear liquid. mp: 110° bp: 151° @ 10 mm, vap d: 3.63.

SYNS: AMINOGLYCOL * ISOBUTANDIOL-2-AMINE * PENTAERYTHRITOL DICHLOROHYDRIN

THR: Moderately toxic by ingestion. Combustible. Can react with oxidizing materials. When heated to decomposition it emits toxic fumes of NO_x.

2-AMINO-4-METHYLPYRIDINE   HR: 3
CAS: 695-34-1   NIOSH: TJ 5150000
mf: $C_6H_8N_2$ mw: 108.16

PROP: Crystals. mp: 99°; bp: 230.9°; vap d: 3.73.

SYNS: 2-AMINO-4-PICOLINE * METHYL-4-AMINO-2-PYRIDINE * 4-METHYL-2-AMINOPYRIDINE * 4-PICOLYLAMINE

THR: Poison by subcutaneous and intravenous routes. Combustible. When heated to decomposition it emits toxic fumes of NO_x.

2-AMINO-6-METHYLPYRIDINE   HR: 3
CAS: 1824-81-3   NIOSH: US 1885000
mf: $C_6H_8N_2$ mw: 108.16

PROP: mp: 43.7°; bp: 214.4; vap d: 3.73.

THR: Poison by intravenous route. When heated to decomposition it emits toxic fumes of NO_x.

2-AMINO-1-METHYL-5H-PYRIDO-(4,3-b)INDOLE   HR: 2
NIOSH: UU 9354000
mf: $C_{12}H_{11}N_3$ mw: 197.2

SYN: TRP-P-2

THR: Mutagenic data. When heated to decomposition it emits toxic fumes of NO_x.

3-AMINO-1,5-NAPHTHALENEDISULFONIC ACID
CAS: 131-27-1   NIOSH: QJ 6140000
mf: $C_{10}H_9NO_6S_2$ mw: 303.32

SYNS: 2-AMINO-4,8-NAPHTHALENEDISULFONIC ACID * 7-AMINO-1,5-NAPHTHALENEDISULFONIC

ACID * C ACID * 4,8-DISULFO-2-NAPHTHALAMINE * KYSELINA-2-NAFTYLAMIN-4,8-DISULFONOVA (CZECH) * BETA-NAPHTHYLAMINEDISULFONIC ACID * BETA-NAPHTHYLAMINE-4,8-DISULFONIC ACID * 2-NAPHTHYLAMINE-4,8-DISULFONIC ACID

THR: Eye irritant. See also sulfonates. When heated to decomposition it emits very toxic fumes of NO_x and SO_x.

4-AMINO-1-NAPHTHALENESULFONIC ACID   HR: 3
CAS: 84-86-6   NIOSH: QK 1270000
mf: $C_{10}H_9NO_3S$ mw: 223.26

SYNS: 1-AMINONAPHTHALENE-4-SULFONIC ACID * 1-AMINO-4-SULFONAPHTHALENE * 1,4-NAPHTHIONIC ACID * ALPHA-NAPHTHYLAMINE-P-SULFONIC ACID * 1-NAPHTHYLAMINE-4-SULFONIC ACID * PIRIA'S ACID * USAF M-5

THR: Poison by intraperitoneal route. See sulfonates. When heated to decomposition it emits very toxic fumes of NO_x and SO_x.

5-AMINO-2-NAPHTHALENESULFONIC ACID   HR: 1
CAS: 119-79-9   NIOSH: QK 1285000
mf: $C_{10}H_9NO_3S$ mw: 223.26

SYNS: 1-AMINO-6-NAPHTHALENESULFONIC ACID * 1-AMINO-6-SULFONAPHTHALENE * CLEVE'S ACID-1,6 * CLEVE'S BETA-ACID * KYSELINA CLEVE (CZECH) * KYSELINA-1-NAFTYLAMIN-6-SULFONOVA (CZECH) * 1-NAPHTHYLAMINE-6-SULFONIC ACID * 5-NAPHTHYLAMINE-2-SULFONIC ACID

THR: An eye irritant. See sulfonates. When heated to decomposition it emits very toxic fumes of NO_x and SO_x.

8-AMINO-2-NAPHTHOL   HR: 3
CAS: 118-46-7   NIOSH: QL 3331000
mf: $C_{10}H_9NO$ mw: 159.20

PROP: Crystalizes in benzene or ligroin. mp: 95°-97° (decomp); sol in hot water, alkali and HCl.

THR: Poison by intravenous route. When heated to decomposition it emits toxic NO_x.

1-AMINO-2-NAPHTHOL HYDRO-CHLORIDE   HR: 3
CAS: 1198-27-2   NIOSH: QL 3335000
mf: $C_{10}H_9NO \cdot ClH$ mw: 195.66

PROP: Needles from alc. mp: 201°; slightly sol in water; sol in alc and ether.

SYN: 2-HYDROXY-1-NAPHTHYLAMINE HYDRO-CHLORIDE

THR: An experimental carcinogen. Mutagenic data. When heated to decomposition it emits very toxic fumes of HCl and NO_x.

2-AMINO-1-NAPHTHOL HYDRO-CHLORIDE   HR: 3
CAS: 41772-23-0   NIOSH: QL 3340000
mf: $C_{10}H_9NO \cdot ClH$ mw: 195.66

PROP: Needles. mp: 255° (decomp); sol in alc.

SYN: 1-HYDROXY-2-NAPHTHYLAMINE HYDRO-CHLORIDE

THR: An experimental carcinogen. When heated to decomposition it emits very toxic fumes of NO_x and HCl.

6-AMINONICOTINAMIDE   HR: 3
CAS: 329-89-5   NIOSH: US 4550000
mf: $C_6H_7N_3O$ mw: 137.16

SYNS: 6-AMINONIKOTINSAEUREAMID (GERMAN) * 6-AMINONICOTINIC ACID AMIDE * 6-AMINO-NICOTINSAEUREAMID (GERMAN) * NSC 21206

THR: Very poisonous by intraperitoneal route. An experimental teratogen. Mutagenic data. When heated to decomposition it emits toxic fumes of NO_x.

2-AMINO-4-NITROANILINE   HR: 3
CAS: 99-56-9   NIOSH: ST 2975000
mf: $C_6H_7N_3O_2$ mw: 153.16

SYNS: C.I. 76020 * 1,2-DIAMINO-4-NITRO-BENZENE * NCI-C03941 * 4-NITRO-1,2-BEN-ZENEDIAMINE * 4-NITRO-1,2-DIAMINOBENZENE * 4-NITRO-O-PHENYLENE-DIAMINE * 4-NITRO-1,2-PHENYLENEDIAMINE * P-NITRO-O-PHENYL-ENEDIAMINE

THR: An experimental carcinogen and terato-gen. Mutagenic data. Moderately toxic by inges-tion. When heated to decomposition it emits toxic fumes of NO_x.

4-AMINO-2-NITROANILINE   HR: 3
CAS: 5307-14-2   NIOSH: ST 3000000
mf: $C_6H_7N_3O_2$ mw: 153.16

SYNS: C.I. 76070 * 1,4-DIAMINO-2-NITRO-BENZENE * NCI-C02222 * NITRO-P-PHENYL-

ENEDIAMINE * 2-NITRO-1,4-BENZENEDIAMINE * 2-NITRO-1,4-DIAMINOBENZENE * O-NITRO-P-PHENYLENEDIAMINE * 2-NITRO-1,4-PHENY-LENEDIAMINE * 2-NITROL-P-PHENYLENEDIAMINE

THR: Poison by intraperitoneal route. Moder-ately toxic by ingestion. An experimental carci-nogen and teratogen. Mutagenic data. When heated to decomposition it emits toxic fumes of NO_x.

4-AMINO-2-NITROPHENOL   HR: 3
CAS: 119-34-6   NIOSH: SJ 6125000
mf: $C_6H_6N_2O_3$ mw: 154.14

SYNS: 4-HYDROXY-3-NITROANILINE * NCI-C03963 * O-NITRO-P-AMINOPHENOL * 2-NI-TRO-4-AMINOPHENOL * OXIDATION BASE 25

THR: Very poisonous by intraperitoneal route. Moderately toxic by ingestion. An experimental carcinogen. When heated to decomposition it emits toxic fumes of NO_x. For further informa-tion, see Vol. 1, No. 7 of *DPIM Report*.

2-((4-AMINO-2-NITROPHENYL)AMINO)-ETHANOL   HR: 3
CAS: 2871-01-4   NIOSH: KJ 6500000
mf: $C_8H_{11}N_3O_3$ mw: 197.22

SYNS: HC RED NO. 3 * NCI-C54922

THR: Experimental carcinogen. No acute toxic-ity data. When heated to decomposition it emits toxic fumes of NO_x.

2-AMINO-4-(p-NITROPHENYL)THIA-ZOLE   HR: 3
CAS: 2104-09-8   NIOSH: XJ 2860000
mf: $C_9H_7N_3O_2S$ mw: 221.25

THR: An experimental carcinogen. When heated to decomposition it emits toxic NO_x and SO_x.

2-AMINO-5-NITROTHIAZOLE   HR: 3
CAS: 121-66-4   NIOSH: XJ 2800000
mf: $C_3H_3N_3O_2S$ mw: 145.15

SYNS: AMINONITROTHIAZOLE * NCI-C03065 * USAF EK-6561

THR: Poison by intraperitoneal route. An ex-perimental carcinogen. Mutagenic data. When heated to decomposition it emits very toxic fumes of NO_x and SO_x. Incompatible with HNO_3 and H_2SO_4. A preparative hazard.

2-AMINOPHENOL   HR: 3
CAS: 95-55-6   NIOSH: SJ 4725000
mf: C_7H_7NO mw: 109.14

PROP: Colorless needles. mp: 173°; bp: subl.
Sol in water, alc; very sol in ether.

SYNS: 2-AMINO-1-HYDROXYBENZENE * O-AM-
INOPHENOL * BASF URSOL 3GA * C.I. 76520
* C.I. OXIDATION BASE 17

THR: Poison by intraperitoneal and subcutaneous routes. Moderately toxic by ingestion. Mutagenic data. When heated to decomposition it
emits toxic NO_x.

4-AMINOPHENOL   HR: 3
CAS: 123-30-8   NIOSH: SJ 5075000
mf: C_6H_7NO mw: 109.14

PROP: Colorless crystals, slightly sol in water,
alc and ether, insol in chloroform. mp: 189.6°-
190.2°, bp: 284° (decomp).

SYNS: 4-AMINO-1-HYDROXYBENZENE * P-AM-
INOPHENOL * BASF URSOL P BASE * P-AMI-
NOFENOL (CZECH) * C.I. OXIDATION BASE 6A
* FOURAMINE P * P-HYDROXYANILINE
* 4-HYDROXYANILINE * URSOL P BASE

THR: Poison by ingestion. Moderately toxic
by subcutaneous route. An allergen and skin
and eye irritant. Mutagenic data. Can cause contact dermatitis, bronchial asthma, and methemoglobinemia with cyanosis. When heated to decomposition it emits toxic fumes of NO_x.

m-AMINOPHENOL   HR: 3
CAS: 591-27-5   NIOSH: SJ 4900000
mf: C_6H_7NO mw: 109.14

PROP: Prisms from toluene. mp: 123°. Sol in
water, alc; sltly sol in ether.

SYNS: M-AMINOFENOL (CZECH) * 3-AMINO-
1-HYDROXYBENZENE * 3-AMINOPHENOL
* C.I. 76545 * C.I. OXIDATION BASE 7
* FOURAMINE EG * 3-HYDROXYANILINE

THR: Poison by subcutaneous and intraperitoneal routes. Moderately toxic by ingestion. A
skin and eye irritant. When heated to decomposition it emits toxic fumes of NO_x.

7-(D-alpha-AMINOPHENYLACETAMI-
DO)DESACETOXYCEPHALOSPORANIC
ACID   HR: 2
CAS: 15686-71-2   NIOSH: XI 0350000
mf: $C_{16}H_{17}N_3O_4S$ mw: 347.42

SYNS: 7-(D-2-AMINO-2-PHENYLACETAMIDO)-3-

METHYL-DELTA (SUP 3)-CEPHEM-4-CARBOXYLIC
ACID * CEPHALEXIN * CEPOREXINE

THR: Moderately toxic by several routes. When
heated to decomposition it emits very toxic
fumes of NO_x and SO_x.

N-beta-(p-AMINOPHENYL)ETHYLNOR-
MEPERIDINE   HR: 3
CAS: 144-14-9   NIOSH: NS 5250000
mf: $C_{22}H_{28}N_2O_2$ mw: 352.52

SYNS: 1-(P-AMINOPHENETHYL)-4-PHENYLISONI-
PECOTIC ACID, ETHYL ESTER * 1-(P-AMINO-
PHENETHYL)-4-PHENYLPIPERIDINE-4-CARBOXYLIC
ACID ETHYL ESTER * N-(BETA-(P-AMINOPHEN-
YL)ETHYL)-4-PHENYL-4-CARBETHOXYPIPERIDINE
* ETHYL-1-(P-AMINOPHENETHYL)-4-PHENYLISO-
NIPECOTATE

THR: Poison by ingestion, subcutaneous, intravenous, and intraperitoneal routes. When
heated to decomposition it emits toxic fumes
of NO_x.

2-AMINO-5-PHENYL-OXAZOLINE FOR-
MATE   HR: 3
CAS: 13425-22-4   NIOSH: RQ 4725000
mf: $C_9H_{10}N_2O \cdot C_4H_4O_4$ mw: 278.29

SYNS: AMINOREXFUMARATE * MENOCIL

THR: Poison by ingestion. Central nervous system effects. When heated to decomposition it
emits toxic fumes of NO_x.

AMINOPHON   HR: 2
CAS: 51249-05-9   NIOSH: SZ 6900000
mf: $C_{18}H_{37}NO_3P$ mw: 346.53

SYNS: 1-(BUTYLAMINO)CYCLOHEXYLPHOS-
PHONIC ACID DIBUTYL ESTER * O,O-DIBUTYL-
1-BUTYLAMINO-CYCLOHEXYLPHOSPHONATE

THR: Moderately toxic by several routes. When
heated to decomposition it emits very toxic
fumes of PO_x and NO_x.

1-AMINOPROPAN-2-OL   HR: 3
CAS: 78-96-6   NIOSH: UA 5775000
mf: C_3H_9NO mw: 75.13

PROP: Liquid, slight ammonia odor, sol in water. d: 0.969, mp: 1.4°, flash p: 171°F, vap d:
2.6.

SYNS: ALPHA-AMINOISOPROPYL ALCOHOL
* 1-AMINO-2-PROPANOL * 2-HYDROXYPRO-
PYLAMINE * ISOPROPANOLAMINE * MONO-
ISOPROPANOLAMINE

THR: Poison by intraperitoneal route. Moderately toxic by skin contact. A skin and eye irritant. Moderately flammable by heat, flame, sparks, powerful oxidizers. When heated to decomposition it emits toxic fumes of NO_x. Incompatible with cellulose nitrate. Use alcohol foam to fight fire.

3-AMINOPROPANOL HR: 2
mf: C_3H_9NO mw: 75.11

PROP: Colorless liquid, fishy odor. bp: 168° @ 500 mm; flash p: 175°F (TOC); fp: 12.4°; d: 0.9786 @ 30°; vap press: 2.1 mm @ 60°; vap d: 2.59.

THR: Moderately toxic by skin contact and ingestion. An irritant. See amines. Moderately flammable. Incompatible with oxidizing materials. Use foam, CO_2, dry chemical to fight fire.

AMINOPROPANOL PYROCATECHOL-HYDROCHLORIDE HR: 3
CAS: 138-61-4   NIOSH: DN 3850000
mf: $C_9H_{13}NO_3 \cdot ClH$ mw: 219.69

SYNS: 3,4-DIHYDROXYNOREPHEDRINE HYDRO-CHLORIDE * 3,4-DIHYDROXYPHENYLAMINOPRO-PANOL HYDROCHLORIDE * 3,4-DIHYDROXYPHE-NYLPROPANOLAMINE HYDROCHLORIDE * ISOADRENALINE HYDROCHLORIDE * ALPHA-METHYLNORADRENALINE HYDROCHLORIDE * NORHOMOEPINEPHRINE HYDROCHLORIDE

THR: Poison by subcutaneous and intravenous routes. When heated to decomposition it emits very toxic fumes of NO_x and Cl^-.

2-AMINOPROPIONITRILE HR: 3
mf: $C_3H_6N_2$ mw: 70.1

THR: A poison; dangerous fire hazard. Can explode in storage. See also nitriles.

3-AMINOPROPIONITRILE HR: 3
CAS: 151-18-8   NIOSH: UG 0350000
mf: $C_3H_6N_2$ mw: 70.11

PROP: Amine odor, liquid. bp: 185°.

SYNS: BETA-AMINOPROPIONITRILE * BETA-CYANOETHYLAMINE

THR: An experimental teratogen. Nitriles usually have cyanide-like effects. See cyanide. Easily oxidized and unstable. When heated to decomposition it emits toxic fumes of NO_x. For fire and explosion hazards see cyanides.

beta-AMINOPROPIONITRILE FUMA-RATE HR: 3
CAS: 2079-89-2   NIOSH: UG 0700000
mf: $C_3H_6N_2 \cdot 2C_4H_4O_4$ mw: 302.27

SYNS: BETA-APN * BAPN FUMARATE

THR: An experimental teratogen. When heated to decomposition it emits toxic fumes of NO_x.

p-AMINOPROPIOPHENONE HR: 3
CAS: 70-69-9   NIOSH: UG 7350000
mf: $C_9H_{11}NO$ mw: 149.21

PROP: mp: 140°

SYNS: PAPP * USAF UCTL-1856

THR: Poison by intraperitoneal and possibly other routes. Ingestion of large doses can cause cyanosis.

3-(2-AMINOPROPYL)INDOLE HR: 3
CAS: 299-26-3   NIOSH: NL 4550000
mf: $C_{11}H_{14}N_2$ mw: 174.27

SYNS: INDOPAN * ALPHA-METHYL-BETA-INDOL-AETHYLAMINE (GERMAN) * ALPHA-METHYL-BETA-INDOLEETHYLAMINE * ALPHA-METHYL-TRYPTAMINE

THR: Poison by ingestion and intraperitoneal routes. Moderately toxic by subcutaneous route. Psychotropic effects by ingestion. An experimental teratogen. When heated to decomposition it emits toxic fumes of NO_x.

3-(gamma-AMINOPROPYL)-INDOLEHY-DROCHLORIDE HR: 3
CAS: 18237-15-5   NIOSH: NL 4710000
mf: $C_{11}H_{14}N_2 \cdot ClH$ mw: 210.73

SYNS: 3-(GAMMA-AMINOPROPYL)-INDOLE HY-DROCHLORIDE * HOMOTRYPTAMINE HYDRO-CHLORIDE * GAMMA-3-INDOLYLPROPYLAMINE HYDROCHLORIDE * INDOLE-3-PROPYLAMINE HYDROCHLORIDE

THR: Poison by intravenous and intraperitoneal routes. When heated to decomposition it emits very toxic fumes of HCl and NO_x.

4-AMINOPROPYLMORPHOLINE HR: 2
CAS: 123-00-2   NIOSH: QD 7700000
DOT: 1760
mf: $C_7H_{16}N_2O$ mw: 144.25

PROP: Liquid; mp: −15°, bp: 224.7°, flash

p: 220°F (OC), d: 0.9872 @ 20°/20°, vap press: 0.06 mm @ 20°, vap d: 4.97.

SYN: N-AMINOPROPYLMORPHOLINE (DOT)

DOT Classification: Label: Corrosive

THR: A corrosive material. Moderately toxic by several routes. A skin and eye irritant. Combustible. Can react with oxidizing materials. use alcohol foam, dry chemical to fight fire. When heated to decomposition it emits toxic fumes of NO_x.

AMINOPTERIDINE   HR: 3
CAS: 54-62-6   NIOSH: MA 1050000
mf: $C_{19}H_{20}N_8O_5$ mw: 440.47

PROP: Yellow needles, sol in sodium hydroxide soln.

SYNS: 4-AMINO-4-DEOXYPTEROYLGLUTAMATE * 4-AMINO-PGA * AMINOPTERIN * 4-AMINOPTEROYLGLUTAMIC ACID * A-NINOPTERIN * APGA * ENT-26079 * NSC 739

THR: Poison by ingestion and intraperitoneal routes. A teratogen. Mutagenic data. Can cause blood problems and bone marrow depression. When heated to decomposition it emits toxic fumes of NO_x.

o-AMINOPYRIDINE   HR: 3
CAS: 504-29-0   NIOSH: US 1575000
DOT: 2671
mf: $C_5H_6N_2$ mw: 94.13

PROP: White powder or crystals. mp: 58.1, bp: 210.6°. Sol in water and ether, very sol in alc, sltly sol in ligroin.

SYNS: ALPHA-AMINOPYRIDINE * 2-AMINOPYRIDINE * AMINO-2-PYRIDINE
ACGIH TLV: TWA 0.5 ppm
DFG MAK: 0.5 ppm (2 mg/m³)
OSHA PEL: TWA 0.5 ppm
DOT Classification: Poison B

THR: Poison by inhalation, subcutaneous, intravenous, and intraperitoneal routes. Toxic effects resemble strychnine poisoning. When heated to decomposition it emits highly toxic fumes of NO_x.

4-AMINOPYRIDINE   HR: 3
CAS: 504-24-5   NIOSH: US 1750000
mf: $C_5H_6N_2$ mw: 94.13

PROP: Needles from benzene, mp: 158°; sol in water, sltly sol in benzene and ether.

SYNS: AMINO-4-PYRIDINE * 4-PYRIDINAMINE

THR: Poison by ingestion, subcutaneous, and intraperitoneal routes. When heated to decomposition it emits toxic fumes of NO_x.

AMINOPYRINE SODIUM SULFONATE   HR: 3
CAS: 68-89-3   NIOSH: PB 1300000
mf: $C_{13}H_{17}N_3O_4S \cdot Na$ mw: 334.38

SYNS: (ANTIPYRINYLMETHYLAMINO)METHANESULFONIC ACID SODIUM SALT * METHYLAMINOANTIPYRINE SODIUM METHANESULFONATE * 4-METHYLAMINO-1,5-DIMETHYL-2-PHENYL-3-PYRAZOLONE SODIUM METHANESULFONATE * METHYLAMINOPHENYLDIMETHYLPYRAZOLONE METHANESULFONATE SODIUM * 1-PHENYL-2,3-DIMETHYL-5-PYRAZOLONE-4-METHYLAMINOMETHANESULFONATESODIUM * 1-PHENYL-2,3-DIMETHYLPYRAZOLONE-(5)-4-METHYLAMINOMETHANESULFONICACID SODIUM * PHENYL DIMETHYL PYRAZOLON METHYL AMINOMETHANE SODIUM SULFONATE * 4-SODIUM METHANESULFONATE METHYLAMINE-ANTIPYRINE * SODIUM METHYLAMINOANTIPYRINE METHANESULFONATE * SODIUM-4-METHYLAMINO-1,5-DIMETHYL-2-PHENYL-3-PYRAZOLONE 4-METHANESULFONATE * SODIUM NORAMIDOPYRINE METHANESULFONATE * SODIUM-1-PHENYL-2,3-DIMETHYL-4-METHYLAMINOPYRAZOLON-n-METHANESULFONATE * SODIUM-1-PHENYL-2,3-DIMETHYL-5-PYRAZOLONE-4-METHYLAMINO METHANESULFONATE * SODIUM PHENYLDIMETHYLPYRAZOLON-METHYLAMINO-METHANE SULFONATE

THR: Poison by subcutaneous route. Moderately toxic by several other routes. An experimental teratogen. See sulfonates. When heated to decomposition it emits very toxic fumes of NO_x and SO_x.

4-AMINOSALICYLIC ACID   HR: 2
CAS: 65-49-6   NIOSH: VO 1225000
mf: $C_7H_7NO_3$ mw: 153.14

PROP: Minute cryst from alc. Sol in dil acid or base; mp: 150°. Very sol in water and alc; sltly sol in ether.

SYNS: 4-AMINO-2-HYDROXYBENZOIC ACID * AMINOSALICYLIC ACID * P-AMINOSALICYLIC ACID * 2-HYDROXY-4-AMINOBENZOIC ACID

* 3-HYDROXY-4-CARBOXYANILINE * KYSELI-NA-P-AMINOSALICYLOVA (CZECH)

THR: Moderately toxic by several routes. An eye irritant. Mutagenic data. When heated to decomposition it emits toxic fumes of NO_x.

5-AMINOSALICYLIC ACID   HR: 2
CAS: 89-57-6   NIOSH: VO 1400000
mf: $C_7H_7NO_3$ mw: 153.14

PROP: Needles; mp: decomp @ 260°-280°. Sltly sol in hot water, insol in alc, sol in HCl and CS_2.

SYNS: 5-AMINO-2-HYDROXYBENZOIC ACID
* M-AMINOSALICYLIC ACID

THR: Moderately toxic. When heated to decomposition it emits toxic fumes of NO_x.

p-AMINOSALICYLIC ACID-2-(DIETHYL-AMINO)ETHYL ESTER HYDRO-CHLORIDE   HR: 3
NIOSH: VO 1602000
mf: $C_{12}H_{18}N_2O_3 \cdot ClH$ mw: 274.78

SYNS: 4-AMINO-2-HYDROXYBENZOIC ACID, 2-(DIETHYLAMINO)ETHYL ESTER, HYDROCHLORIDE (9CI) * P-AMINOSALICYLSAEUREDIAETHYLAMI-NOAETHYLESTER-CHLORHYDRAT (GERMAN)
* HCL SALZ DES P-AMINO-SALICYLSAEURE-DIA-ETHYLAMINOAETHYLESTER (GERMAN)

THR: Poison by subcutaneous, intravenous, and intraperitoneal routes. When heated to decomposition it emits very toxic fumes of NO_x and HCl.

trans-4-AMINOSTILBENE   HR: 3
CAS: 4309-66-4   NIOSH: WJ 3540000
mf: $C_{14}H_{13}N$ mw: 195.28

SYNS: TRANS-4-STILBENE * TRANS-4-N-STIL-BENAMINE

THR: An experimental carcinogen. Mutagenic data. When heated to decomposition it emits toxic fumes of NO_x.

2-(p-AMINOSTYRYL)-6-(p-ACETYLAMI-NOBENZOYLAMINO)QUINOLINE METHOACETATE   HR: 3
CAS: 3432-10-8   NIOSH: VC 3500000
mf: $C_{27}H_{25}N_4O_2 \cdot C_2H_3O_2$ mw: 496.61

SYN: STYRYL 430

THR: An experimental carcinogen. Mutagenic data. When heated to decomposition it emits toxic fumes of NO_x.

4-AMINO-2,2,5,5-TETRAKIS(TRIFLUO-ROMETHYL)-3-IMIDAZOLINE   HR: 3
CAS: 23757-42-8   NIOSH: NJ 1750000
mf: $C_7H_3F_{12}N_3$ mw: 357.13

SYNS: 5-AMINO-2,2,4,4-TETRAKIS(TRIFLUORO-METHYL)IMIDAZOLIDINE * EXP 338

THR: Poison by intraperitoneal, ingestion, and intravenous routes. When heated to decomposition it emits very toxic fumes of F^- and NO_x.

2-AMINO-1,3,4-THIADIAZOLE   HR: 3
CAS: 4005-51-0   NIOSH: XI 3000000
mf: $C_2H_3N_3S$ mw: 101.14

THR: Poison by subcutaneous route. An experimental teratogen. When heated to decomposition it emits very toxic fumes of NO_x and SO_x.

2-AMINOTHIAZOLE   HR: 3
CAS: 96-50-4   NIOSH: XJ 2100000
mf: $C_3H_4N_2S$ mw: 100.15

PROP: Light brown crystals; mp: 90°; bp: decomp. Sltly sol in water, alc, ether; sol in hot alc.

SYNS: AMINOTHIAZOLE * 2-THIAZYLAMINE
* USAF EK-P-5501

THR: Poison by intraperitoneal route. Moderately toxic by ingestion. Spontaneous ignition occurs at 100°. Incompatible with HNO_3, H_2SO_4. When heated to decomposition it emits very toxic fumes of NO_x and SO_x.

AMINOTRIACETIC ACID   HR: 3
CAS: 139-13-9   NIOSH: AJ 0175000
mf: $C_6H_9NO_6$ mw: 191.16

SYNS: N,N-BIS(CARBOXYMETHYL)GLYSINE
* NCI-C02766 * NITRILOTRIACETIC ACID
* TRIGLYCINE * TRIGLYCOLLAMIC ACID
* VERSENE NTA ACID

THR: Poison by intraperitoneal route. Moderately toxic by ingestion. An experimental carcinogen. When heated to decomposition it emits toxic fumes of NO_x.

3-AMINOTRIAZOLE   HR: 3
CAS: 61-82-5   NIOSH: XZ 3850000
mf: $C_2H_4N_4$ mw: 84.10

SYNS: AMITROLE * 3-AMINO-S-TRIAZOLE
* 3-AMINO-1,2,4-TRIAZOLE * 2-AMINO-1,3,4-TRIAZOLE * 3-AMINO-1H-1,2,4-TRIAZOLE
* ENT 25445 * 1H-1,2,4-TRIAZOL-3-AMINE
* USAF XR-22

DFG MAK: 0.2 mg/m^3

THR: Poison by intraperitoneal route. Moderately toxic by ingestion. An experimental carcinogen. Mutagenic data. When heated to decomposition it emits toxic fumes of NO$_x$. For further information, see 3-Amino-1,3,4-Triazole Vol. 4, No. 2 of *DPIM Report*.

4-AMINO-3,5,6-TRICHLOROPICOLINIC ACID   HR: 3
CAS: 1918-02-1   NIOSH: TJ 7525000
mf: C$_6$H$_3$Cl$_3$N$_2$O$_2$ mw: 241.46; Crystals; mp: 218°.

SYNS: CHLORAMP (RUSSIAN) * 4-AMINO-3,5,6-TRICHLORO-2-PICOLINIC ACID * 4-AMINO-3,5,6-TRICHLORPICOLINSAEURE (GERMAN) * PICLORAM * NCI-C00237 * 3,5,6-TRICHLORO-4-AMINOPICOLINIC ACID

THR: An experimental carcinogen. Moderately toxic by ingestion. When heated to decomposition it emits very toxic fumes of Cl$^-$ and NO$_x$.

AMINOUNDECANOIC ACID   HR: 3
CAS: 2432-99-7   NIOSH: YQ 2285000
mf: C$_{11}$H$_{23}$NO$_2$ mw: 201.35

SYNS: 11-AMINOUNDECANOIC ACID * NCI-C50613

THR: An experimental carcinogen. See amines. When heated to decomposition it emits toxic fumes of NO$_x$.

AMITAL   HR: 3
CAS: 57-43-2   NIOSH: CQ 5075000
mf: C$_{11}$H$_{18}$N$_2$O$_3$ mw: 226.31

PROP: Slightly bitter crystals.

SYNS: AMYLBARBITONE * AMYLOBARBITAL * ETHYLISOPENTYLBARBITURIC ACID * ISO-AMYLETHYLBARBITURIC ACID * AMOBARBITAL * AMYLOBARBITONE * AMYTAL * 5-ETHYL-5-ISOAMYLBARBITURIC ACID * 5-ETHYL-5-ISO-AMYLMALONYL UREA * 5-ETHYL-5-ISOPENTYL-BARBITURIC ACID * 5-ETHYL-5-(3-METHYLBU-TYL)BARBITURIC ACID * 5-ISOAMYL-5-ETHYLBARBITURIC ACID * NSC 10815

THR: Poison by ingestion. See barbiturates. When heated to decomposition it emits toxic fumes of NO$_x$.

AMMONIA   HR: 3
CAS: 7664-41-7   NIOSH: BO 0875000

DOT: 1005
mf: H$_3$N   mw: 17.04

PROP: Colorless gas, extremely pungent odor, liquefied by compression. mp: −77.7°, bp: −33.35°, lel = 16%, uel = 25%, d: 0.771 g/liter @ 0°, 0.817 g/liter @ -79°, autoign temp: 1204°F, vap press: 10 atm @ 25.7°, vap d: 0.6. Very sol in water, mod sol in alc.

SYNS: AMMONIA ANHYDROUS * AMMONIAC (FRENCH) * AMMONIACA (ITALIAN) * AMMONIA GAS * AMMONIAK (GERMAN) * AMONIAK (POLISH) * SPIRIT OF HARTSHORN

DOT Classification: Label: Nonflammable Gas
ACGIH TLV: TWA 25 ppm; STEL 35 ppm
DFG MAK: 50 ppm (35 mg/m^3)
OSHA PEL: TWA 50 ppm

THR: Poison by inhalation, ingestion, and possibly other routes. An eye, mucous membrane, and systemic irritant. Mutagenic data. A common air contaminant. Difficult to ignite. Explosion hazard when exposed to flame or in a fire. NH$_3$ + air in a fire can detonate. Incompatible in contact with Ag; acetaldehyde; acrolein; B; BI$_3$; halogens; BrF$_5$; HClO$_3$; ClO; ClF$_3$; chlorites; chlorosilane; CrO$_3$; chromyl chloride; (ethylene dichloride + liquid ammonia); ethylene oxide; Au; hexachloromelamine; (hydrazine + alkali metals); HBr; HOCl; Mg(ClO$_4$)$_2$; Hg; HNO$_3$; NO$_2$; N$_2$O$_4$; NCl$_3$; NF$_3$; nitryl chloride; OF$_2$; P$_2$O$_5$; P$_2$O$_3$; picric acid; (K + AsH$_3$); (K + PH$_3$); (K + NaNO$_2$); KClO$_3$; potassium ferricyanide; potassium mercuric cyanide; AgCl; (Na + CO); Sb; S; SCl$_2$; tellurium hydropentachloride; trichloromelamine; boron halides; ClN$_3$; ethylene oxide; NO$_2$Cl; gold(III)chloride; CrO$_3$; dichlorine oxide; O$_3$F$_2$; ammonium peroxo disulfate; H$_2$O$_2$; (O$_2$ + Pt); AgNO$_3$; Ag$_2$O; SbH$_3$; TeCl$_4$; tetramethylammonium amide; SOCl$_2$; thiotrithiazylchloride. Emits toxic fumes of NH$_3$ and NO$_x$ when exposed to heat. Stop flow of gas is an effective way to fight fire. For further information, see Vol. 3, No. 3 of *DPIM Report*.

AMMONIA-D   HR: 3
mf: ND$_3$ mw: 20.05

PROP: Gas. mp: −74°; bp: −33.4°.

SYN: TRIDEUTERIO AMMONIA

THR: See ammonia. Incompatible with oxidizing materials.

AMMONIUM ACETATE HR: 3
CAS: 631-61-8 NIOSH: AF 3675000
mf: $C_2H_4O_2 \cdot H_3N$ mw: 77.10

PROP: Crystals; mp: 114°; d: 1.07.

SYN: ACETIC ACID, AMMONIUM SALT

THR: Poison by intravenous route. Moderately toxic by other routes. When heated to decomposition it emits toxic fumes of NO_x and NH_3. For further information, see Vol. 2, No. 3 of *DPIM Report*.

AMMONIUM AZIDE
mf: NH_4N_3 mw: 60.1

PROP: Colorless plates; mp: 160°; bp: explodes; d: 1.346; vap press: 1 mm @ 59.2° (sublimes).

THR: Poison by inhalation and ingestion. See azides. Moderately flammable. Unstable. Explosion hazard upon rapid heating.

AMMONIUM BICARBONATE (1:1)
HR: 3
CAS: 1066-33-7 NIOSH: BO 8600000
mf: $HCO_3 \cdot H_4N$ mw: 79.1

PROP: Hard, colorless to white crystals, water sol; faint ammonia odor, stable @ room temp; volatile, decomp @ 60°; mp: 107.5° (rapid heating); d: 1.586, insol alc.

SYNS: ACID AMMONIUM CARBONATE * AMMONIUM CARBONATE * AMMONIUM HYDROGEN CARBONATE * CARBONIC ACID, MONOAMMONIUM SALT * MONOAMMONIUM CARBONATE

THR: Poison by intravenous route. When heated to decomposition it emits toxic fumes of NO_x and NH_3. For further information, see Vol. 4, No. 2 of *DPIM Report*.

AMMONIUM BICHROMATE HR: 3
CAS: 7789-09-5 NIOSH: HX 7650000
DOT: 1439
mf: $Cr_2H_8N_2O_7$ mw: 252.10

PROP: Red crystals. mp: decomp, d: 2.936.

SYNS: AMMONIO (DICROMATO DI) (ITALIAN) * AMMONIUMBICHROMAAT (DUTCH) * AMMONIUMDICHROMAT (GERMAN) * AMMONIUMDICHROMAAT (DUTCH) * AMMONIUM DICHROMATE * AMMONIUM DICHROMATE (VI) * BICHROMATE D'AMMONIUM (FRENCH)

DOT Classification: Label: Oxidizer

THR: Poison by inhalation, ingestion, skin contact, and subcutaneous routes. See chromium compounds. An unstable oxidizer. Moderately flammable; reacts with reducing agents. For further information, see Ammonium Dichromate, Vol. 3, No. 5 of *DPIM Report*.

AMMONIUM BROMATE HR: 3
mf: NH_4BrO_3 mw: 145.96

PROP: Colorless crystals. Very sol in water; mp: explodes.

THR: An unstable, explosive oxidizing material. See bromates. Severe explosion hazard.

AMMONIUM BROMIDE HR: 2
CAS: 12124-97-9 NIOSH: BO 9170000
mf: BrH_4N mw: 97.96

PROP: Colorless, cubic, sltly hygroscopic cryst; mp: subl @ 452°; bp: 235° in vac; d: 2.429; vap press: 1 mm @ 198.3°.

SYN: HYDROBROMIC ACID MONOAMMONIATE

THR: Mutagenic data. See bromides. When heated to decomposition it emits very toxic fumes of NO_x, Br^- and NH_3. Incompatible with BrF_3, IF_7, K.

AMMONIUM CALCIUM ARSENATE HR: 3
mf: $NH_4CaAsO_4 \cdot 6H_2O$ mw: 305.1

PROP: Colorless crystals. mp: 140° (decomp), d: 1.905 @ 15°; sltly sol in cold water; sol in hot water; sol in NH_4Cl and NH_4OH.

THR: A poison. See arsenic compounds.

AMMONIUM CARBAMATE HR: 3
CAS: 1111-78-0 NIOSH: EY 8575000
mf: $CH_3NO_2 \cdot H_3N$ mw: 78.09

PROP: White, crystalline rhombic powder, sol in water and alc, ammonia odor. Sublimates at 60°.

SYN: AMMONIUM AMINOFORMATE

THR: Poison by intravenous route. See carbamates. For further information, see Vol. 2, No. 3 of *DPIM Report*.

AMMONIUM CARBONATE HR: 3
CAS: 506-87-6 NIOSH: BP 1925000
mf: $(NH_4)_2CO_3$ mw: 96.09

PROP: Colorless crystals.

SYNS: AMMONIUMCARBONAT (GERMAN) * CARBONIC ACID, AMMONIUM SALT * CARBONIC ACID, DIAMMONIUM SALT * DIAMMONIUM CARBONATE

THR: Poison by subcutaneous and intravenous routes. When heated to decomposition it emits toxic fumes of NO_x and NH_3. For further information, see Vol. 2, No. 3, of *DPIM Report*.

AMMONIUM CHLORATE　　　HR: 3
CAS: 10192-29-7　　　NIOSH: FN 9760000
DOT: 1445
mf: ClH_3NO_3　　mw: 100.49

PROP: White crystals or mass.
DOT Classification: Label: Oxidizer

THR: A powerful oxidizer. Moderately flammable due to spont chemical reaction. Explosion hazard due to shock, chemical reaction, or exposure to heat. When contaminated it is very sensitive. When heated to decomposition it emits highly toxic fumes of Cl^- and NO_x. Incompatible with reducing materials; BrF_3; BrF_5;

AMMONIUM CHLORIDE　　　HR: 3
CAS: 12125-02-9　　　NIOSH: BP 4550000
DOT: 9085
mf: $H_4N \cdot Cl$　　mw: 53.50

PROP: White crystals. bp: 520°, mp: 337.8°, d: 1.520, vap press: 1 mm @ 160.4° (sublimes).

SYNS: AMMONIUMCHLORID (GERMAN) * AMMONIUM MURIATE * CHLORID AMONNY (CZECH) * SAL AMMONIA * SAL AMMONIAC

DOT Classification: ORM-E
ACGIH TLV: TWA 10 mg/m³; STEL 20 mg/m³

THR: Poison by subcutaneous, intravenous, and intramuscular routes. Moderately toxic by other routes. An eye irritant. Can react violently with NH_4; NO_3; BrF_3; IF_7; $KClO_3$. When heated to decomposition it emits very toxic fumes of NO_x, Cl^-, and NH_3. For further information, see Vol. 2, No. 3 of *DPIM Report*.

AMMONIUM CHLOROPALLA-DATE (IV)　　　HR: 3
CAS: 19168-23-1　　　NIOSH: BP 5390000
mf: $Cl_6H_8N_2Pd$　　mw: 355.20

PROP: Red-brown crystals; d: 2.418; mp: decomposes.

SYNS: AMMONIUM HEXACHLOROPALLADATE * DIAMMONIUM HEXACHLOROPALLADATE

THR: A poison and a skin irritant. When heated to decomposition it emits very toxic fumes of NO_x, Cl^-, and NH_3.

AMMONIUM CHLOROPLATI-NATE　　　HR: 3
CAS: 16919-58-7　　　NIOSH: BP 5425000
mf: $Cl_6Pt \cdot 2H_4N$　　mw: 443.89

PROP: Cubic, yellow crystals; d: 3.065; mp: decomposes.

SYNS: AMMONIUM HEXACHLOROPLATINATE(IV) * AMMONIUM PLATINIC CHLORIDE * DIAMMONIUM HEXACHLOROPLATINATE (2-) * PLATINIC AMMONIUM CHLORIDE

OSHA PEL: TWA 2 ug(Pt)/m³

THR: Poison by inhalation and ingestion. Pulmonary system effects. See platinum compounds. When heated to decomposition it emits very toxic fumes of Cl^-, NO_x and NH_3.

AMMONIUM CHROMATE
mf: $(NH_4)_2CrO_4$　　mw: 152.1

PROP: Yellow, crystalline material; mp: decomp @ 180°; d: 1.91 @ 12°. sol in cold water.

THR: A poison. A carcinogen. See also chromium compounds. A powerful oxidizer. An explosion hazard when shocked or heated. When heated to decomposition it emits toxic fumes of NH_3 and NO_x. Incompatible with reducing agents. For further information see Vol. 2, No. 3 of *DPIM Report*.

AMMONIUM CYANIDE　　　HR: 3
mf: NH_4CN　　mw: 44.1

PROP: Solid, white powder or crystals, mp: 36° (decomp); bp: sublimes @ 40°; d: 1.002 @ 100°; vap press: 400 ppm @ 20.5°. Very sol water, alc; decomp in hot water.

THR: A poison. See cyanide. When heated to decomposition it emits toxic CN^-, NH_3, NO_x.

AMMONIUM DIFLUORIDE MIXED WITH HYDROCHLORIC ACID　　　HR: 3
　　　　　　　　　NIOSH: BP 8205000
DOT: 1760

SYN: WHITE ACID (DOT)

DOT Classification: Label: Corrosive

THR: A corrosive. Poison by inhalation, ingestion, and skin contact. See components. When heated to decomposition it emits very toxic fumes of F^-, HF and HCl.

AMMONIUM FLUOBORATE   HR: 3
CAS: 13826-83-0   NIOSH: BQ 6100000
mf: NH_4BF_4 mw: 104.9

PROP: White, rhombic crystals; d: 1.871 @ 15°; mp: subl; sol in NH_4OH, water.

SYNS: AMMONIUM BOROFLUORIDE * AMMONIUM FLUOROBORATE * AMMONIUM TETRA-FLUOROBORATE * AMMONIUM TETRAFLUOROBORATE(1-)

THR: A poison; strong irritant. See fluorides and boron compounds. When heated to decomposition it emits very toxic fumes of F^-, NO_x and NH_3.

AMMONIUM FLUORIDE   HR: 3
CAS: 12125-01-8   NIOSH: BQ 6300000
DOT: 2505
mf: $H_4N \cdot F$ mw: 37.05

PROP: Colorless crystals; mp: subl; d: 1.009 @ 25°.

SYNS: AMMONIUM FLUORURE (FRENCH) * NEUTRAL AMMONIUM FLUORIDE

OSHA PEL: TWA 2500 ug(F)/m^3
DOT Classification: ORM-B; Label: None

THR: Poison by subcutaneous and intraperitoneal routes. See fluorides. When heated to decomposition it emits very toxic fumes of F^-, NO_x and NH_3. Incompatible with ClF_3. For further information, see Vol. 3, No. 5 of *DPIM Report*.

AMMONIUM HEXAFLUOROTITA-NATE   HR: 3
CAS: 16962-40-6   NIOSH: XR 1500000
mf: $F_6Ti \cdot H_4N_2$ mw: 193.96

THR: Poison by intravenous route. See fluorides, ammonia, and titanium compounds. When heated to decomposition it emits very toxic fumes of F^- and NO_x.

AMMONIUM HEXAFLUOROVANA-DATE   HR: 3
CAS: 13815-31-1   NIOSH: YW 0750000
mf: $F_6H_{12}N_3V$ mw: 219.09

SYN: HEXAFLUORO VANADATE (3-) TRIAMMO-NIUM SALT

THR: Poison by intravenous route. See fluorides and vanadium compounds. When heated to decomposition it emits very toxic NH_3, NO_x, and fluorides.

AMMONIUM HYDROGEN FLUO-RIDE   HR: 3
CAS: 1341-49-7   NIOSH: BQ 9200000
DOT: 1727
mf: F_2H_5N mw: 57.06

PROP: White cryst; d: 1.51; mp: 124.6°. Will etch glass. Water sol.

SYNS: AMMONIUM BIFLUORIDE * AMMONIUM HYDROGEN FLUORIDE, SOLID

DOT Classification: ORM-B; Label: None

THR: Caustic poison and strong irritant by all routes. See HF. When heated to decomposition it emits very toxic fumes of F^-, NO_x and NH_3. For further information, see Vol. 3, No. 5 of *DPIM Report*.

AMMONIUM HYDROGEN FLUORIDE (SOLUTION)   HR: 3
CAS: 1341-49-7   NIOSH: BQ 9205000
DOT: 2817

SYN: AMMONIUM HYDROGEN FLUORIDE SOLUTION (DOT)

DOT Classification: Label: Corrosive

THR: Caustic poison and strong irritant by all routes. See also HF. When heated to decomposition it emits very toxic fumes of HF, F^- and NO_x.

AMMONIUM HYDROSULFIDE   HR: 3
mf: NH_4HS mw: 51.11

PROP: Powder or crystals. mp: 118° (150 atm); d: 1.17; vap press: 400 mm @ 21.8°.

THR: Poison by ingestion, intraperitoneal and skin contact routes. A strong irritant. Penetrates the skin readily. See sulfides. Fire hazard: see sulfides. When heated to decomposition it emits toxic fumes of SO_x, NH_3 and NO_x.

AMMONIUM HYDROSULFIDE (SOLU-TION)   HR: 3
CAS: 12124-99-1   NIOSH: BQ 9480000
DOT: 2683

SYN: AMMONIUM HYDROSULFIDE SOLUTION (DOT)

DOT Classification: ORM-A; Label: None

THR: Poison due to easily liberated H_2S. See ammonium sulfide. When heated to decomposition it emits very toxic fumes of NO_x, SO_x, NH_3, and H_2S.

AMMONIUM HYDROXIDE   HR: 3

CAS: 1336-21-6   NIOSH: BQ 9625000
DOT: 2672
mf: $H_4N \cdot HO$ mw: 35.06

PROP: Colorless liquid. Very pungent odor. mp: $-77°$. Sol in water. Soln contains not more than 44% ammonia.

SYNS: AMMONIA AQUEOUS * AMMONIA SOLUTION (DOT) * AQUA AMMONIA

DOT Classification: Label: Corrosive

THR: Poison by inhalation and ingestion. An eye and systemic irritant. Mutagenic data. Incompatible with acrolein; nitromethane; acrylic acid; chlorosulfonic acid; dimethyl sulfate; halogens; (Au + aqua regia); HCl; HF; HNO_3; oleum; β-propiolactone; propylene oxide; $AgNO_3$; Ag_2O; (Ag_2O + C_2H_5OH); $AgMnO_4$; H_2SO_4. Dangerous; emits irritating fumes, and liquid can inflict burns. Use with adequate ventilation. When heated to decomposition it emits NH_3 and NO_x. For further information, see Vol. 2, No. 3 of *DPIM Report*.

AMMONIUM IODATE

mf: H_4INO_3 mw: 192.94

PROP: Colorless crystals. d: 3.309 @ 21°; mp: 150° (decomp). Sltly sol in cold water, insol in hot water.

THR: No toxicity data. See iodates. A powerful, unstable oxidizer. When heated to decomposition it emits very toxic fumes of I^- and NO_x. Has detonated upon contact with a scoop, possibly due to contamination by ammonium periodate.

AMMONIUM IODIDE   HR: 2

mf: NH_4I mw: 145

PROP: Colorless, hygroscopic crystals. mp: subl @ 551°; bp: 220° (vacuo); d: 2.514 @ 25°; vap press: 1 mm @ 210.9°.

THR: See iodides. Incompatible with BrF_3; IF_7; K. When heated to decomposition it emits toxic fumes of I^-, NH_3, and NO_x.

AMMONIUM MAGNESIUM ARSENATE   HR: 3

mf: $NH_4MgAsO_4 \cdot 6H_2O$ mw: 289.4

PROP: Colorless crystals. mp: decomp: d: 1.932 @ 15°. Very sltly water sol.

THR: See arsenic compounds and magnesium compounds. When heated to decomposition it emits very toxic fumes of As, NH_3 and NO_x.

AMMONIUM MAGNESIUM CHROMATE   HR: 3

mf: $(NH_4)_2CrO_4 \cdot MgCrO_4 \cdot 6H_2O$ mw: 400.5

PROP: Yellow crystals. mp: decomp; d; 1.84; very water sol.

THR: See chromium compounds and magnesium compounds. Moderately flammable; can explode. Incompatible with reducing agents. When heated to decomposition it can emit toxic fumes of NH_3 and NO_x.

AMMONIUM MERCAPTOACETATE   HR: 3

CAS: 5421-46-5   NIOSH: AI 7525000
mf: $C_2H_3O_2S \cdot H_3N$ mw: 108.15

PROP: Colorless liquid; strong skunk-like odor.

SYNS: AMMONIUM THIOGLYCOLATE * AMMONIUM THIOGLYCOLLATE * THIOGLYCOLLIC ACID, AMMONIUM SALT * USAF MO-2

THR: Poison by intraperitoneal route. An allergen, can cause contact dermatitis. Emits hydrogen sulfide. See sulfides. When heated to decomposition it emits very toxic NO_x, SO_x and NH_3.

AMMONIUM MOLYBDATE   HR: 3

CAS: 13106-76-8   NIOSH: QA 4900000
mf: $MoO_4 \cdot 2H_4N$ mw: 196.04

SYNS: DIAMMONIUM MOLYBDATE * MOLYBDIC ACID DIAMMONIUM SALT * AMMONIUM PARAMOLYBDATE

OSHA PEL: TWA 5 mg(Mo)/m^3

THR: Poison by ingestion and intraperitoneal route. Moderately toxic by other routes. An irritant. See molybdenum compounds. When heated to decomposition it emits toxic fumes of NH_3 and NO_x.

AMMONIUM(I) NITRATE(1:1)   HR: 2

CAS: 6484-52-2   NIOSH: BR 9050000
DOT: 1942/0222/2426
mf: $HNO_3 \cdot H_3N$ mw: 80.06

PROP: Colorless cryst; mp: 169.6°; d: 1.725 @ 25°; bp: decomp >210°. Solubility: 192/100 @ 20°.

SYNS: AMMONIUM NITRATE * AMMONIUM NITRATE (DOT) * NITRIC ACID, AMMONIUM SALT

DOT Classification: Label: Oxidizer

THR: A powerful oxidizer and an allergen. See nitrates. Can ignite when mixed with acetic acid. Use water in large amounts to fight fire. It is important that the mass of materials be kept cool and that burning be extinguished promptly. Ventilate well. May explode under confinement and high temperatures. When heated to decomposition it emits highly toxic fumes of NO_x, can react vigorously with reducing materials. Incompatible with powdered metals; (NH_4Cl + heat); (C + heat); chlorides; organic matter; P; (K + $(NH_4)_2SO_4$); NaOCl; $NaClO_4$; (NaK + $(NH_4)_2SO_4$); S; acetic acid; $KMnO_4$; urea; NH_4NO_3; Al; Sb; Bi; Cd; Cr; Co; Cu; Fe; Pb; Mg; Mn; Ni; Sn; Brass; stainless steel; Zn; alkali metals @ <200°; glowing charcoal. Occasional explosions in presence of oil, $(NH_4)_2SO_4$ with K or Na will explode. When the nitrate is mixed with wax, oil, stearates, contact with fused nitrite causes incandescence. For further information, see Vol. 2, No. 3 of *DPIM Report*.

AMMONIUM-N-NITROSOPHENYLHYDROXYLAMINE　　　HR: 3
CAS: 135-20-6　　　NIOSH: NC 4725000
mf: $C_6H_6N_2O_2 \cdot H_4N$　　mw: 156.19

SYNS: CUPFERRON * N-HYDROXY-N-NITROSO-BENZENAMINE, AMMONIUM SALT * KUPFERRON (CZECH) * NCI-C03258 * N-NITROSOFENYL-HYDROXYLAMIN AMONNY (CZECH) * N-NITROSOPHENYLHYDROXYLAMIN AMMONIUM SALZ (GERMAN) * N-NITROSOPHENYLHYDROXYL-AMINE AMMONIUM SALT

THR: Poison by intravenous route. An experimental carcinogen. Powerful eye irritant. See amines. When heated to decomposition it emits very toxic NH_3 and NO_x.

AMMONIUM OXALATE　　　HR: 3
mf: $(NH_4)_2C_2O_4 \cdot H_2O$　　mw: 142.12

PROP: Colorless crystals. mp: decomp; d: 1.50; slightly sol in water.

THR: A poison. See oxalates. Can react violently with (NaOCl + ammonium acetate). When heated to decomposition it can emit toxic fumes of NH_3 and NO_x.

AMMONIUM PERCHLORATE　　　HR: 3
mf: NH_4ClO_4　　mw: 117.50

PROP: White crystals. mp: decomp; d: 1.95.

THR: See perchlorates. Easily ignited by friction. Can explode when mixed with sugar; charcoal; or on contact with hot copper pipes. Can be sensitized by nitryl perchlorate; KIO_4; $KMnO_4$; metals (as cocrystallized impurities). When contaminated by powdered carbon, ferrocene, S, organic matter, powdered metals, it becomes impact sensitive. When heated to decomposition it emits toxic fumes of NH_3, Cl^- and NO_x. For further information, see Vol. 2, No. 3 of *DPIM Report*.

AMMONIUM PERCHLORYL AMIDE　　　HR: 3
mf: $H_5N_2O_3Cl$　　mw: 116.6

PROP: mp: 80°

THR: A shock sensitive explosive. May detonate @ 80°. When heated to decomposition it emits very toxic fumes of NH_3, NO_x, and Cl^-.

AMMONIUM-m-PERIODATE　　　HR: 3
mf: NH_4IO_4　　mw: 209

PROP: Colorless crystals. mp: explodes; d: 3.056.

THR: A contact explosive. See iodates, iodides. When heated to decomposition it can emit toxic fumes of NH_3, NO_x and I^-. Heat, impact, and touch as from a scoop or an abrasive impact.

AMMONIUM PERMANGANATE　　　HR: 3
mf: NH_4MnO_4　　mw: 137.0

PROP: Crystalline solid. mp: explodes; d: 2.208 @ 10°.

THR: See manganese compounds. A powerful oxidizer. Moderately flammable by chemical reaction with reducing agents. Explosive when shocked or warmed to 60°. Can be exploded by percussion. When heated to decomposition it emits toxic fumes of NO_x and NH_3. Incompatible with reducing material, friction.

AMMONIUM PEROXO BORATE　　　HR: 3
mf: $BH_4NO_3 \cdot 1/2H_2O$　　mw: 85.86

PROP: White crystals. mp: decomp; slightly sol in water.

THR: Potentially explosive by heat, friction, or impact. See boron compounds. When heated to decomposition it emits toxic fumes of NO_x and NH_3.

AMMONIUM PEROXY CHRO-MATE   HR: 3

mf: $(NH_4)_3CrO_2$ mw: 234.1

PROP: Red-brown crystals. mp: decomp @ 40°, bp: explodes @ 50°.

THR: A poison. See chromium compounds. Moderately flammable by chemical reaction with reducing agents. A powerful oxidizer. Moderately explosive when heated. When heated to decomposition it emits toxic NO_x, NH_3.

AMMONIUM PERSULFATE   HR: 3

CAS: 7727-54-0   NIOSH: SE 0350000
DOT: 1444
mf: $O_8S_2 \cdot 2H_4N$ mw: 228.22

PROP: White crystals. mp: decomp @ 120°, d: 1.982.

SYNS: AMMONIUM PEROXYDISULFATE * PER-SULFATE D'AMMONIUM (FRENCH)

DOT Classification: Oxidizer

THR: Poison by intravenous and intraperitoneal routes. Moderately toxic by ingestion. A powerful oxidizer. Releases oxygen when heated. Can explode when mixed with Na_2O_2, powdered Al and H_2). Dangerous, emits SO_x, NH_3 and NO_x. Can react vigorously with reducing agents. For further information, see Ammonium Peroxydisulfate, Vol. 2, No. 3 of *DPIM Report*.

AMMONIUM PHOSPHATE DI-BASIC   HR: 2

CAS: 7783-28-0   NIOSH: TB 9375000
mf: $H_6N_2 \cdot H_3O_4P$ mw: 132.08

PROP: White crystals or powder, sol in water, insol in alcohol. d: 1.619, mp: 155° (decomp).

SYNS: AMMONIUM PHOSPHATE * DIAMMO-NIUM HYDROGEN PHOSPHATE * DIBASIC AMMO-NIUM PHOSPHATE * SECONDARY AMMONIUM PHOSPHATE

THR: Low to moderate toxicity. See phosphates. When heated to decomposition it emits very toxic fumes of PO_x, NO_x, and NH_3.

AMMONIUM PHOSPHIDE   HR: 3

mf: $P(NH_4)_3$ mw: 85.07

THR: Poison by inhalation and ingestion. See phosphine. When heated to decomposition it emits toxic fumes of PO_x, NO_x, and NH_3.

AMMONIUM PHOSPHITE   HR: 3

NIOSH: BS 3850000
mf: H_6NO_3P mw: 99.04

SYN: AMMONIUM ORTHOPHOSPHITE

THR: Poison by inhalation. See phosphites. When heated to decomposition it emits very toxic fumes of NO_x, NH_3, and PO_x.

AMMONIUM PICRATE   HR: 3

CAS: 131-74-8   NIOSH: BS 3855000
DOT: 0004
mf: $C_6H_3N_3O_7 \cdot H_3N$ mw: 246.16

PROP: Red or yellow rhombic crystals. d: 1.719, mp: decomp; bp: explodes @ 423°. Solubility: 1.1/100 @ 20°.

SYNS: AMMONIUM PICRONITRATE * OBELINE PICRATE * PICRATE OF AMMONIA (DOT) * PICRIC ACID, AMMONIUM SALT

DOT Classification: Label: Explosive A

THR: An allergen. Moderately irritating to skin, eyes, and mucous membranes. See picric acid, nitrates. Moderately flammable by spontaneous chemical reaction. A powerful oxidizer. Dangerous explosive when shocked or heated. See explosives, high. When heated to decomposition it emits highly toxic fumes of NO_x, etc; can react vigorously with reducing materials. For further information, see Vol. 2, No. 3 of *DPIM Report*.

AMMONIUM PICRATE (WET)   HR: 3

CAS: 131-74-8   NIOSH: BS 3856000
DOT: 1310

PROP: Bright yellow crystals; bitter taste. Compound contains 10% or more water.

SYNS: AMMONIUM PICRATE, WET (DOT) * EXPLOSIVE D * PICRATOL * PICRIC ACID, AMMONIUM SALT

DOT Classification: Label: Flammable Solid

THR: An explosive. A flammable material that is a powerful oxidizer. See picric acid. When heated to decomposition it emits fumes of NO_x; explodes.

AMMONIUM POTASSIUM SELENIDE MIXED WITH AMMONIUM POTASSIUM SULFIDE   HR: 3

CAS: 64046-00-0

NIOSH: BS 4030000
mf: $H_4KNSe + NH_4KS$ mw: 136.11 + 89.21 = 225.32

THR: An experimental carcinogen. See selenides, sulfides. When heated to decomposition it emits very toxic fumes of NO_x, NH_3, SO_x, and Se.

2,4-D AMMONIUM SALT HR: 3
CAS: 2307-55-3 NIOSH: AG 7075000
mf: $C_8H_6Cl_2O_3 \cdot H_3N$ mw: 238.08

THR: Poison. Central nervous system effects by inhalation. An experimental carcinogen. When heated to decomposition it emits very toxic fumes of Cl^-, NO_x, and NH_3.

AMMONIUM SILICO FLUORIDE HR: 3
CAS: 1309-32-6 NIOSH: GQ 9450000
mf: $SiF_6 \cdot 2NH_4$ mw: 178.19

SYNS: AMMONIUM FLUOSILICATE * AMMONIUM HEXAFLUOROSILICATE * DIAMMONIUM HEXAFLUOROSILICATE * FLUOSILICATE DE AMMONIUM (FRENCH)

THR: Poison by ingestion. See fluosilicates and fluorides. When heated to decomposition, it emits very toxic fumes of F^- and NO_x. For further information, see Vol. 4, No. 3 of *DPIM Report*.

AMMONIUM SULFAMATE HR: 2
CAS: 7773-06-0
DOT: 9089
mf: $H_6N_2O_3S$ mw: 114.1

PROP: Deliquescent crystalline material (white crystalline solid). bp: 160° (decomp); mp: 131°.

SYN: AMMATE

DOT Classification: ORM-E
ACGIH TLV: TWA 10 mg/m^3
DFG MAK: 15 mg/m^3

THR: Moderately toxic by ingestion and intraperitoneal routes. Somewhat explosive when heated or by spontaneous chemical reaction in a hot acid solution. A powerful oxidizer. See sulfonates. When heated to decomposition it emits toxic fumes of NO_x. For further information, see Vol. 2, No. 3 of *DPIM Report*.

AMMONIUM SULFATE (2:1) HR: 3
CAS: 7783-20-2 NIOSH: BS 4500000
DOT: 2506
mf: $O_4S \cdot (H_4N)_2$ mw: 132.16

PROP: Brown-gray to white cryst. mp: $> 280°$ (decomp); d: 1.77.

SYNS: AMMONIUM SULPHATE * DIAMMONIUM SULFATE * SULFURIC ACID, DIAMMONIUM SALT

DOT Classification: ORM-B; Label: None

THR: Moderately toxic by several routes. See sulfates. Incompatible with $(K + NH_4NO_3)$; $KClO_3$; KNO_2; $(NaK + NH_4NO_3)$. When heated to decomposition it emits very toxic fumes of NO_x, NH_3, and SO_x.

AMMONIUM SULFIDE HR: 3
CAS: 12124-99-1 NIOSH: BS 4900000
mf: $(H_4N)_2S$ mw: 68.2

PROP: Yellow, hygroscopic cryst. mp: decomp.

SYNS: AMMONIUM MERCAPTAN * MONOAMMONIUM SULFIDE * AMMONIUM BISULFIDE * AMMONIUM HYDROGEN SULFIDE * AMMONIUM HYDROSULFIDE * AMMONIUM SULFHYDRATE * SIRNIK AMONNY (CZECH)

THR: Poison by ingestion, skin contact, subcutaneous, intravenous, intradermal, parenteral, and intraperitoneal routes. Pyroforic in air. See sulfides. When heated to decomposition it emits very toxic fumes of SO_x, NO_x, and NH_3. Incompatible with zinc.

AMMONIUM SULFIDE (SOLUTION) HR: 3
CAS: 12135-76-1 NIOSH: BS 4920000
DOT: 2683
DOT Classification: Label: Flammable Liquid

THR: Poison due to emission of H_2S on contact with acid or acid fumes. Fatal poisoning has been reported from use in hair waving lotion. See ammonium sulfide and sulfides. For further information, see Vol. 2, No. 4 of *DPIM Report*.

AMMONIUM THIOCYANATE HR: 3
CAS: 1762-95-4 NIOSH: XK 7875000
mf: $CNS \cdot H_4N$ mw: 76.13

PROP: Colorless solid or deliquescent crystals. mp: 149.6°; bp: decomp @ 170°; d: 1.305.

SYNS: AMMONIUM RHODANIDE * AMMONIUM SULFOCYANATE * AMMONIUM SULFOCYANIDE * USAF EK-P-433

THR: Poison by ingestion. Moderately toxic by other routes. Gastrointestinal tract effects.

See thiocyanates. When heated to decomposition it emits toxic fumes of NH_3, NO_x, SO_x, and CN^-. Incompatible with $KClO_3$ and mixtures with $Pb(NO_3)_2$. For further information, see Vol. 2, No. 3 of *DPIM Report*.

AMMONIUM THIOGLYCOLATE   HR: 3
mf: $HSCH_2COONH_4$ mw: 109.1

PROP: Colorless liquid, strong skunk-like odor.

THR: Poison by inhalation, ingestion, and intraperitoneal routes. A strong allergen. Can cause contact dermatitis. See also thiocyanates. When heated to decomposition it or contact with acid or acid fumes emits highly toxic fumes of sulfides. See also hydrogen sulfide.

AMMONIUM TRICHLOROACE-
TATE   HR: 3
mf: $NH_4O_2CCCl_3$ mw: 180.6

THR: Poison by inhalation and ingestion. A powerful irritant. When heated to decomposition it or contact with acid or acid fumes emits toxic fumes of Cl^-, NH_3, and NO_x. Incompatible with water or steam.

AMMONIUM VANADATE   HR: 3
CAS: 7803-55-6   NIOSH: YW 0875000
mf: $O_3V \cdot H_4N$ mw: 116.99

PROP: Colorless to yellow crystals. mp: 200° (decomp); d: 2.326.

SYNS: VANADIC ACID, AMMONIUM SALT * AMMONIUM METAVANADATE

THR: Poison by ingestion, subcutaneous, intravenous, intratracheal, and intraperitoneal routes. Mutagenic data. See also vanadium compounds. When heated to decomposition it emits toxic fumes of NH_3 and NO_x.

AMMONIUM VANADI-ARSE-
NATE   HR: 3
NIOSH: BT 5100000
mf: $H_{16}N_4O_2 \cdot As_2O_5V_2$ mw: 515.92

THR: Poison by subcutaneous and intravenous routes. See arsenic and vanadium compounds. When heated to decomposition it emits very toxic fumes of NO_x and As.

AMMONIUM VANADO-ARSE-
NATE   HR: 3
NIOSH: BT 5180000
mf: $H_{40}N_{10}O_5 \cdot 3As_2O_5 \cdot 4O_4V_2$ mw: 1388.78

THR: Poison by subcutaneous and intravenous routes. See arsenic and vanadium compounds. When heated to decomposition it emits very toxic NO_x, NH_3, and As.

AMOSITE (See Also Asbestos)   HR: 3
CAS: 12172-73-5   NIOSH: BT 6825000

SYNS: AMOSITE ASBESTOS * MYSORITE

OSHA PEL: TWA 2 fb/mL

THR: A human carcinogen. Mutagenic data. See also asbestos.

AMPHETAMINE   HR: 3
CAS: 60-15-1   NIOSH: SH 9000000
mf: $C_9H_{13}N$ mw: 135.23

SYNS: BETA-AMINOPROPYLBENZENE * DES-OXYNOREPHEDRINE * ALPHA-METHYLPHENETH-YLAMINE * 1-PHENYL-2-AMINO-PROPAN (GER-MAN) * 1-PHENYL-2-AMINOPROPANE * BETA-PHENYLISOPROPYLAMIN (GERMAN) * (PHENYLISOPROPYL)AMINE * BETA-PHENYL-ISOPROPYLAMINE

THR: Poison by ingestion, subcutaneous, intravenous, and intraperitoneal routes. Central nervous system effects by ingestion. When heated to decomposition it emits very toxic fumes of NO_x.

D-AMPHETAMINE   HR: 3
CAS: 51-64-9   NIOSH: SH 9100000
mf: $C_9H_{13}N$ mw: 135.23

SYNS: D-2-AMINO-1-PHENYLPROPANE * (+)-AMPHETAMINE * DEXAMPHETAMINE * DEXE-DRINE * D-1-PHENYL-2-AMINOPROPAN (GER-MAN) * D-1-PHENYL-2-AMINOPROPANE

THR: Poison by ingestion. subcutaneous, intravenous, and intraperitoneal routes. Chronic exposure causes central nervous system damage and blood-pressure effects. When heated to decomposition it emits toxic NO_x.

AMPHETAMINE HYDROCHLOR-
IDE   HR: 3
CAS: 2706-50-5   NIOSH: SH 9798000
mf: $C_9H_{13}N \cdot ClH$ mw: 171.69

THR: Poison by subcutaneous, intramuscular, intravenous, and intraperitoneal routes. Can cause blood pressure and central nervous system effects by subcutaneous route. When heated to decomposition it emits very toxic fumes of HCl and NO_x.

DL-AMPHETAMINE SULFATE   HR: 3
CAS: 60-13-9   NIOSH: SI 1750000
mf: $C_{18}H_{26}N_2 \cdot H_2O_4S$ mw: 368.54

SYNS: AMFETAMINA * AMFETAMINE
* (+-)-2-AMINO-1-PHENYLPROPANE SULFATE
* (+-)-AMPHETAMINE SULFATE * ANFETA-
MINA * BENNIE * BENZAMPHETAMINE
* BENZEDRYNA * BENZIES * BETAFEN
* CARTWHEELS * DEOXYNOREPHEDRINE
* DESOXYNOREPHEDRINE * HEARTS *
IBIOZEDRINE * LINAMPHETA * ALPHA-
METHYLPHENETHYLAMINE SULFATE, (+-)-
* NCI-C55710 * NOREPHEDRANE * PEACHES
* PHARMEDRINE * PHENAMINE * PHENE-
DRINE * (+-)-PHENISOPROPYLAMINE SULFATE
* BETA-PHENYL ISOPROPYLAMINE SULFATE
* PSYCHEDRYNA * PSYCHEDRINUM * RACE-
PHEN * ROSES * STIMULAN

THR: Poison by ingestion, subcutaneous, intravenous, and intraperitoneal routes. Mutagenic data. Causes central nervous system effects by ingestion. When heated to
decomposition it emits very toxic NO_x and SO_x.

AMPHETANE PHOSPHATE   HR: 3
CAS: 139-10-6   NIOSH: SI 0875000
mf: $C_9H_{13}N \cdot H_3O_4P$ mw: 233.23

SYNS: DL-AMPHETAMINE PHOSPHATE * AL-
PHA-METHYLPHENETHYLAMINE PHOSPHATE, DL-
MIXTURE * MONOBASIC DL-ALPHA-METHYL-
PHENETHYLAMINE PHOSPHATE * MONOBASIC
RACEMIC AMPHETAMINE PHOSPHATE

THR: Poison by ingestion and intraperitoneal
routes. When heated to decomposition it emits
very toxic fumes of PO_x and NO_x.

AMPHIBOLE (SEE ALSO ASBES-
TOS)   HR: 3
NIOSH: BU 1390000

SYNS: MAGNESIAAREVEDSONITE * MAGNESIA-
ARPHVEDSONITE

THR: An experimental carcinogen. See also asbestos.

AMPHOTERICIN B   HR: 3
CAS: 1397-89-3   NIOSH: BU 2625000
mf: $C_{47}H_{73}NO_{17}$ mw: 924.21

SYNS: AMPHOMORONAL * AMPHOTERICINE B
* FUNGILIN * FUNGISONE

THR: Poison by intravenous and intraperitoneal
routes. Mutagenic data. When heated to decomposition it emits toxic fumes of NO_x.

AMYL ACETATE (MIXED ISO-
MERS)   HR: 1
NIOSH: AJ 2010000
mf: $C_7H_{14}O_2$ mw: 130.21

PROP: Colorless liquid, pear-like odor. mp:
$-78.5°$; bp: 148° @ 737 mm; ulc: 55-60; lel
= 1.1%; uel = 7.5%; flash p: 77°F (CC); d:
0.879 @ 20°/20°; autoign temp: 714°F; vap d:
4.5.

SYN: ACETIC ACID, AMYL ESTER

DFG MAK: 100 ppm (525 mg/m^3)

THR: A skin irritant. Low oral toxicity. See
esters, amyl alcohol, and acetic acid. When
heated to decomposition it emits acrid smoke
and irritating fumes.

AMYL ALCOHOL   HR: 2
mf: $C_5H_{12}O$ mw: 88.1

PROP: Clear liquid; mp: $-79°$; bp: 137.8°; flash
p: 91°F (CC); d: 0.8168 @ 20°/20°; ulc: 40;
lel = 1.2%; uel = 10% @ 212°F, vap press:
1 mm @ 13.6°; 10 mm @ 44.9°, vap d; 3.04
sol in water; misc in alc, ether.

SYNS: ALCOOL AMYLIQUE (FRENCH)
* N-AMYL ALCOHOL * AMYL ALCOHOL, NOR-
MAL * N-BUTYLCARBINOL * N-AMYLALKO-
HOL (CZECH) * PENTANOL-1 * N-PENTANOL
* PENTAN-1-OL * PENTASOL * PENTYL AL-
COHOL * PRIMARY AMYL ALCOHOL

THR: Moderately toxic by ingestion and skin
contact. An eye and upper respiratory irritant
by inhalation. Ingestion can cause headache,
nausea, vomiting, delirium, and methemoglobin
formation. Extremely flammable if exposed to
heat, flame or powerful oxidizers. To fight fire:
alcohol foam, dry chemical. Moderately explosive when exposed to flame. Incompatible with
oxidizing materials; hydrogen trisulfide. For further information, see Vol. 2, No. 3 of *DPIM
Report*.

tert-N-AMYL ALCOHOL, RE-
FINED   HR: 2
mf: $C_4H_9O(CH_3)$ mw: 88.15

PROP: Colorless liquid. mp: $-11.9°$; bp:
101.8°; flash p: 105°F (CC); d: 0.809; autoign
temp: 819°F; vap press: 10 mm @ 17.2°;
lel = 1.2%; uel = 9%; vap d: 3.03.

SYNS: TERT-AMYL ALCOHOL * AMYLENE HYDRATE * DIMETHYL ETHYL CARBINOL * 2-METHYL-2-BUTANOL * 2-METHYL BUTA-NOL-2 * 3-METHYLBUTAN-3-OL * TERT-PEN-TANOL

THR: Moderately toxic by ingestion, subcutaneous, and intraperitoneal routes. Moderately irritating to mucous membranes. It is a narcotic in high concentrations. Moderately flammable. To fight fire: alcohol foam, dry chemical. Incompatible with oxidizing materials.

5-n-AMYL-1:2-BENZANTHRA-CENE   HR: 3
CAS: 63018-99-5   NIOSH: CX 2800000
mf: $C_{23}H_{22}$ mw: 298.45

SYN: 8-PENTYLBENZ(a)ANTHRACENE

THR: An experimental carcinogen. When heated to decomposition it emits acrid smoke and irritating fumes.

4-AMYL-N-BENZOHYDRYLPYRIDINIUM BROMIDE   HR: 3
CAS: 63905-98-6   NIOSH: UU 4017000
mf: $C_{22}H_{26}N \cdot Br$ mw: 384.40

SYN: B-45

THR: Poison by intraperitoneal, subcutaneous, ingestion, and intravenous routes. See also bromides. When heated to decomposition it emits very toxic fumes of NO_x and Br^-.

d-AMYL BROMIDE   HR: 3
mf: $CH_3(CH_2)_4Br$ mw: 151.1

PROP: Colorless liquid; bp: 120°; flash p: 90°F; fp: < −30°; d: 1.211 @ 25°/25°.

THR: Poison by intraperitoneal route. It can cause liver damage, is narcotic in high concentrations, and is a local irritant. See also bromides. Extremely flammable. To fight fire: alcohol foam, water mist or spray, dry chemical. When heated to decomposition it emits very toxic bromides. Incompatible with oxidizing materials.

alpha-AMYL CINNAMALDEHYDE   HR: 2
CAS: 122-40-7   NIOSH: GD 6825000
mf: $C_{14}H_{18}O$ mw: 202.32

PROP: Yellow liquid. d: 0.971; bp: 174°-175° @ 20 mm.

SYNS: ALPHA-AMYL CINNAMIC ALDEHYDE

* ALPHA-AMYL-BETA-PHENYLACROLEIN
* JASMINALDEHYDE * ALPHA-PENTYLCINNAM-ALDEHYDE

THR: Moderately toxic by ingestion. A mild skin irritant. See aldehydes. When heated to decomposition it emits acrid smoke and irritating fumes.

α,η-AMYLENE   HR: 2
CAS: 513-35-9   NIOSH: EM 7650000
DOT: 2460
mf: C_5H_{10} mw: 70.15

PROP: Liquid, disagreeable odor; mp: −124° bp: 30.1°; lel = 1.6%; uel = 8.7%; flash p: 0°F (OC); d: 0.643; vap d: 2.42; autoign temp: 527°F.

SYNS: BETA-ISO-AMYLENE * 2-METHYL-2-BU-TENE * TRIMETHYLETHYLENE

DOT Classification: Label: Flammable Liquid

THR: Moderately toxic by ingestion and inhalation. Narcotic in high concentration. A simple asphyxiant. Extremely flammable. Moderately explosive when exposed to heat, flame, or powerful oxidizers. To fight fire: alcohol foam, spray, mist, dry chemical. When heated to decomposition it emits acrid smoke and irritating fumes.

AMYLENES, MIXED   HR: 2
mf: C_5H_{10} mw: 70.58

PROP: Water white liquid; bp: 32.2°; flash p: 0°F; d: 0.66 @ 20°.

THR: Moderately toxic. See also alpha-n-amylene. Very flammable; reacts with heat, flame and oxidizing materials. To fight fire: foam, CO_2, dry chemical.

AMYL ETHYL KETONE   HR: 2
CAS: 541-85-5   NIOSH: MJ 7350000
mf: $C_8H_{16}O$ mw: 128.24

PROP: Liquid with mild fruity odor, sol in many organic solvents. bp: 157°-162°, d: 0.822 @ 20°/20°, flash p: 138°F.

SYNS: ETHYL AMYL KETONE * 3-METHYL-5-HEPTANONE * 5-METHYL-3-HEPTANONE

OSHA PEL: TWA 25 ppm

THR: Moderately irritating to skin, eyes, and mucous membranes by inhalation and ingestion.

Narcotic in high concentration. See also ketones. When heated to decomposition it emits acrid smoke. Dangerously flammable from heat, flame, oxidizers. To fight fire: foam, CO_2, dry chemical.

N-AMYL FORMATE   HR: 2
CAS: 638-49-3   NIOSH: LQ 9370000
DOT: 1109
mf: $C_6H_{12}O_2$ mw: 116.18

PROP: Clear liquid; d: 0.902; 0.893 @ 15°/4°; mp: −73.5°; bp: 130.4°; flash p: 80°F; very sltly sol in water, misc in alc and ether.

SYNS: AMYL FORMATE (DOT) * PENTYL FORMATE * M-PENTYL FORMATE

DOT Classification: Flammable Liquid

THR: A moderate irritant by ingestion and skin contact. See also esters. Dangerously flammable; reacts vigorously with heat, flame, oxidizing materials. To fight fire: foam, CO_2, dry chemical.

AMYL LACTATE   HR: 1
mf: $C_8H_{16}O_3$ mw: 160.2

PROP: Colorless liquid. bp: 210°; flash p: 175°F; d: 0.960 @ 20°.

THR: An irritant by inhalation and ingestion. See also esters. Moderately flammable. Incompatible with heat; flame; oxidizing materials. To fight fire: foam, CO_2, dry chemical.

AMYL LAURATE   HR: 1
mf: $C_5H_{11}O_2C(CH_2)_{10}CH_3$ mw: 270.44

PROP: bp: 290°; flash p: 300°F; d: 0.86.

THR: Unknown toxicity. It may defat skin and cause contact dermatitis. Combustible. Incompatible with oxidizing materials. To fight fire: CO_2, dry chemical.

AMYL METHYL ALCOHOL   HR: 2
CAS: 105-30-6   NIOSH: SA 7175000
mf: $C_6H_{14}O$ mw: 102.20

PROP: Liquid. bp: 130°, flash p: 114°F (CC), d: 0.804, vap d: 3.52.

SYNS: 1,3-DIMETHYL BUTANOL * ISOHEXYL ALCOHOL * ISOPROPYL DIMETHYL CARBINOL * METHYLAMYL ALCOHOL * METHYL ISOBUTYL CARBINOL * 2-METHYLPENTANOL-1 * 2-METHYL-2-PROPYLETHANOL

THR: Moderately toxic by ingestion and skin

contact. A skin irritant, a systemic irritant by inhalation. Moderately flammable; can react with oxidizing materials. To fight fire: CO_2, dry chemical. When heated to decomposition it emits smoke and acrid fumes.

N-AMYL-N-METHYLNITROS-
AMINE   HR: 3
CAS: 13256-07-0   NIOSH: SC 1225000
mf: $C_6H_{14}N_2O$ mw: 130.22

SYNS: METHYLAMYLNITROSAMIN (GERMAN) * METHYLAMYLNITROSAMINE * METHYL-N-AMYLNITROSAMINE * N-METHYL-N-NITROSO-PENTYLAMINE * METHYL-N-PENTYLNITROSAMINE * NITROSOMETHYL-N-PENTYLAMINE

THR: Poison by ingestion, subcutaneous, and intraperitoneal routes. An experimental carcinogen. Mutagenic data. When heated to decomposition it emits toxic NO_x.

AMYL NITRATE   HR: 2
CAS: 1002-16-0   NIOSH: QV 0600000
mf: $C_5H_{11}NO_3$ mw: 133.17

PROP: Liquid; bp: 145°, flash p: 125°F (OC), d: 0.99.

SYN: NITRATE D'AMYLE (FRENCH)

THR: Moderately toxic by inhalation. Moderately flammable when exposed to heat or flame or by spontaneous chemical reaction. An oxidizing agent. When heated to decomposition it emits toxic fumes of NO_x.

N-AMYL NITRITE   HR: 2
CAS: 463-04-7   NIOSH: RA 1140000
DOT: 1113
mf: $C_5H_{11}NO_2$ mw: 117.17

PROP: Clear yellowish liquid, peculiar ethereal fruity odor and pungent aromatic taste. bp: 96°-99°, d: 0.8528 @ 20°/4°, autoign temp: 408°F, vap d: 4.0.

SYNS: NITROUS ACID, PENTYL ESTER * AMYL NITRITE (DOT) * 1-NITROPENTANE * PENTYL NITRITE

DOT Classification: Label: Flammable
   Liquid

THR: Moderately toxic by inhalation and ingestion. Causes flushing of skin, rapid pulse, headache, and fall in blood pressure. Mutagenic data. See also nitrites and esters. Moderately flammable when exposed to heat or flame or by spontaneous chemical reaction. To fight fire: alcohol foam. An oxidizing material. Vapors explode when heated. It will react with oxidizing or reducing materials. When heated to decomposition it emits toxic fumes of NO_x.

AMYLOCAINE   HR: 3
CAS: 644-26-8   NIOSH: EL 2800000
mf: $C_{14}H_{21}NO_2$ mw: 235.36

SYNS: AMYLEINE * 2-BUTANOL,1-(DIMETHYL-AMINO)-2-METHYL-, BENZOATE (ESTER) * STOVAINE

THR: Poison by intravenous, subcutaneous, and intraperitoneal routes. See also esters. When heated to decomposition it emits toxic fumes of NO_x.

4-n-AMYLPHENOL   HR: 3
CAS: 14938-35-3   NIOSH: SM 6750000
mf: $C_{11}H_{16}O$ mw: 164.27

PROP: Liquid; bp: 342°; vap d: 5.66; flash p: 219°F (OC); d: 0.966.

SYN: O-AMYL PHENOL

THR: An experimental carcinogen. Moderately flammable. To fight fire: foam, CO_2, dry chemical. When heated to decomposition it emits acrid smoke and irritating fumes.

2-sec-AMYLPHENOL   HR: 3
                          NIOSH: SM 6700000
mf: $C_{11}H_{16}O$ mw: 164.27

PROP: Clear, straw colored liquid; very sltly sol in water; sol in oils and organic solvents; d: 0.955-0.971 @ 30°/30°; bp: 235°-250°; flash p.: 200°F.

SYN: O-SEC AMYL PHENOL

THR: Poison by intravenous route. An experimental carcinogen. Moderately flammable when exposed to heat or flame. To fight fire: foam, fog, dry chemical, water mist or spray, multi-purpose dry chemical. When heated to decomposition it emits acrid smoke and irritating fumes.

4-sec-AMYLPHENOL   HR: 3
                          NIOSH: SM 6800000
mf: $C_{11}H_{16}O$ mw: 164.27

PROP: d: <1.0, bp: 482°-516°F, flash p: 270°F.

SYN: P-SEC AMYL PHENOL

THR: An experimental carcinogen. Combustible when exposed to heat or flame. To fight fire:

dry chemical, water mist, CO_2. When heated to decomposition it emits acrid smoke and fumes.

4-tert-AMYLPHENOL   HR: 2
CAS: 80-46-6   NIOSH: SM 6825000
mf: $C_{11}H_{16}O$ mw: 164.27

PROP: Colorless needles. bp: 250°, mp: 92°-93°, flash p: 232°F (OC).

SYNS: P-TERT-AMYLPHENOL * P-(ALPHA,AL-PHA-DIMETHYLPROPYL)PHENOL * P-(1,1-DI-METHYLPROPYL)PHENOL * P-t-PENTYLPHENOL

THR: Moderately toxic by ingestion and skin contact. Combustible. When heated to decomposition it emits toxic fumes. To fight fire: dry chemical, water mist, CO_2. Incompatible with oxidizing materials.

n-AMYL THIOCYANATE   HR: 3
CAS: 32446-40-5   NIOSH: XL 1750000
mf: $C_6H_{11}NS$ mw: 129.24

PROP: Pale yellow oil; d: 0.905; bp: 197°. Insol in water, sol in alc and ether.

SYN: THIOCYANIC ACID, AMYL ESTER

THR: Poison by subcutaneous and intraperitoneal route. See also thiocyanates, esters. When heated to decomposition it emits toxic fumes of NO_x and SO_x.

AMYTAL SODIUM   HR: 3
CAS: 64-43-7   NIOSH: CQ 5250000
mf: $C_{11}H_{17}N_2O_3 \cdot Na$ mw: 248.29

SYNS: 5-ETHYL-5-ISOPENTYLBARBITURIC ACID SODIUM SALT * 5-ETHYL-5-(3-METHYLBUTYL)-BARBITURIC ACID SODIUM DERIVATIVE * 5-ISOAMYL-5-ETHYLBARBITURIC ACID, SODIUM DERIV * SODIUM AMYLOBARBITONE * SODIUM ETHYLISOAMYLBARBITURATE * SODIUM ISOAMYLETHYL BARBITURATE

THR: Poison by ingestion, subcutaneous, in-travenous, parenteral, intraperitoneal, and rectal routes. When heated to decomposition it emits toxic NO_x.

ANAGESTONE ACETATE MIXED WITH MESTRANOL (10:1)   HR: 3
                          NIOSH: TU 5540000
mf: $C_{24}H_{36}O_3$ mw: 372.60

SYN: ANATROPIN MIXED WITH MESTRANOL (10:1)

THR: An experimental carcinogen. When heated to decomposition it emits acrid smoke and irritating fumes.

ANGEL DUST   HR: 3
CAS: 956-90-1   NIOSH: TN 2272600
mf: $C_{17}H_{25}N \cdot ClH$ mw: 279.89

PROP: Crystals. mp: 46°-46.5°, bp: 135°-137°.

SYNS: CI395 * DOA * ELEPHANT TRANQUIL-IZER * ELYSION * HOG * PCP * PEACE PILL * PHENCYCLIDINE HYDROCHLORIDE * 1-(1-PHENYLCYCLOHEXYL)PIPERIDINE HYDRO-CHLORIDE * SERNYL * SERNYLAN * SER-NYL HYDROCHLORIDE * TRANK

THR: Poison by ingestion, subcutaneous, intravenous, and intraperitoneal routes. Psychotropic and other central nervous system effects. An experimental teratogen. Often mixed with other drugs of abuse, yielding totally unpredictable effects. When heated to decomposition it emits very toxic fumes of HCl and NO_x.

ANG-STERANTHRENE   HR: 3
CAS: 517-85-1   NIOSH: HN 0350000
mf: $C_{23}H_{18}$ mw: 294.41

SYNS: 5,5a,6,7-TETRAHYDRO-4H-DI-BENZ(f,g,j)ACEANTHRYLENE * ANG.-STERAN-THREN (GERMAN)

THR: An experimental tumorigen by skin contact. When heated to decomposition it emits acrid smoke and irritating fumes.

ANGUIDIN   HR: 3
CAS: 2270-40-8   NIOSH: YD 0112000
mf: $C_{18}H_{26}O_7$ mw: 354.44

SYNS: ANG 66 * 4-BETA,15-DIACETOXY-3-ALPHA-HYDROXY-12,13-EPOXYTRICHOTHEC-9-ENE * DIACETOXYSCIRPENOL * 12,13-EPOXY-4-BETA,15-DIAZETOXY-3-ALPHA-HYDROXYTRI-CHOTHEC-9-ENE

THR: Poison by ingestion and intravenous routes. Mutagenic data. When heated to decomposition it emits acrid smoke and fumes.

ANHYDRIDES   HR: 2
PROP: Chemical compounds derived from acids by elimination of a molecule of water. Thus sulfur trioxide, SO_3, is the anhydride of sulfuric acid (H_2SO_4); carbon dioxide, CO_2, is the anhydride of carbonic acid, H_2CO_3; phthalic acid, $C_6H_4(CO_2H)_2$, minus water gives phthalic anhydride, $C_6H_4(CO_2)O$. This term should not be confused with anhydrous, meaning without water.

THR: Anhydrides are acidic and react with bases in tissue. Thus, they tend to attack and irritate tissue.

ANILINE   HR: 3
CAS: 62-53-3   NIOSH: BW 6650000
DOT: 1547
mf: C_6H_7N mw: 93.14

PROP: Colorless, oily liquid, characteristic odor; bp: 184.4°; lel = 1.3%; ulc: 20-25; flash p: 158°F (CC); fp: −6.2°; d: 1.02 @ 20°/4°; autoign temp: 1139°F; vap press: 1 mm @ 34.8°; vap d: 3.22.

SYNS: AMINOBENZENE * AMINOPHEN * ANILIN (CZECH) * ANILINA (ITALIAN, POL-ISH) * ANILINE OIL * BENZENAMINE * BLUE OIL * C.I. 76000 * HUILE D'ANI-LINE (FRENCH) * NCI-C03736 * PHENYL-AMINE

OSHA PEL: TWA 5 ppm (skin)
ACGIH TLV: TWA 2 ppm
DFG MAK: 2 ppm/8 mg/m³
DOT Classification: Label: Poison

THR: Poison by most routes including inhalation and ingestion. An experimental carcinogen. A skin and eye irritant, and a mild sensitizer. In the body, aniline causes formation of methemoglobin, resulting in prolonged anoxemia and depression of the central nervous system; less acute exposure causes hemolysis of the red blood cells, followed by stimulation of the bone marrow. The liver may be affected with resulting jaundice. Long-term exposure to aniline dye manufacture has been associated with malignant bladder growths. A common air contaminant. Moderately flammable when exposed to heat or flame. To fight fire: alcohol foam, CO_2, dry chemical. It can react vigorously with oxidizing materials. When heated to decomposition it emits highly toxic fumes of NO_x. Violent reactions caused by mixing with acetic anhydride; chlorosulfonic acid; hexachloro melamine; HNO_3; (HNO_3 + N_2O_4 + H_2SO_4); (nitrobenzene + glycerine); oleum; O_3; (HCHO + $HClO_4$); perchromates; performic acid; K_2O_2; β-propiolactone; $AgClO_4$; Na_2O_2; H_2SO_4; trichloromelamine; BCl_3; nitromethane; acids; peroxy disulfuric-acid; peroxo mono sulfuric-acid; F_2; FNO_3; FO_3Cl; diisoopropyl peroxy-dicarbonate; n-haloimides; trichloronitro methane. For further information, see Vol. 3, No. 5 of *DPIM Report*.

ANILINE ANTIMONYL TARTRATE   HR: 3
CAS: 1300-14-7   NIOSH: BY 7875000
mf: $C_6H_8N \cdot C_4H_4O_7Sb$ mw: 379.98

PROP: White crystals.

SYN: ANTIMONYL ANILINE TARTRATE

THR: Poison by intraperitoneal route. See antimony compounds and aniline. When heated to decomposition it emits very toxic fumes of Sb and NO_x.

ANILINE DYES   HR: variable
THR: The finished dyes are generally very much less toxic than many of the intermediates occurring or used in the manufacture of the dyes. Some of the aniline dyes cause local irritating effects to the eyes, mucous membranes and skin; the basic dyes are believed to be more irritating than the acid dyes. Allergic responses to aniline dyes have been known to occur. See also specific compounds. When heated to decomposition it emits toxic fumes of NO_x, possibly SO_x.

ANILINE HYDROCHLORIDE   HR: 3
CAS: 142-04-1   NIOSH: CY 0875000
mf: $C_6H_7N \cdot HCl$ mw: 129.59

PROP: Crystals; vap d: 4.46; d: 1.22; mp: 198°; bp: 245°; flash p: 380°F (CC).

SYN: ANILINE CHLORIDE

THR: See aniline. A dangerous disaster hazard. When heated to decomposition or when in contact with acid or acid fumes it emits highly toxic fumes of aniline and chlorine compounds. Combustible. Incompatible with oxidizers. To fight fire: water, CO_2, water mist or spray, dry chemical. For further information, see Vol. 4, No. 4 of *DPIM Report*.

ANILINE OIL DRUMS, EMPTY   HR: 3
THR: See aniline. Combustible if full of vapors, such drums may ignite under the proper conditions. A dangerous disaster hazard if many drums are involved. They emit highly toxic fumes of aniline.

ANILINE VIOLET   HR: 3
CAS: 548-62-9   NIOSH: BO 9000000
mf: $C_{25}H_{30}N_3 \cdot Cl$ mw: 408.03

SYNS: AIZEN CRYSTAL VIOLET EXTRA PURE * GENTIAN VIOLET * HEXAMETHYL-P-ROSANILINE HYDROCHLORIDE * HEXAMETHYL VIOLET * METHYLROSANILINE CHLORIDE * NCI-c55969

THR: Poison by ingestion, intravenous, intraperitoneal, and intraduodenal routes. A skin irritant. Mutagenic data. When heated to decomposition it emits very toxic fumes of NO_x and Cl^-.

2-ANILINOETHANOL   HR: 3
CAS: 122-98-5   NIOSH: KJ 7175000
mf: $C_8H_{11}NO$ mw: 137.20

PROP: d: 1.1, bp: 268°, flash p: 305°F (OC).

SYNS: N-(2-HYDROXYETHYL)PHENYLAMINE * PHENYL ETHANOLAMINE * N-PHENYLETHANOLAMINE * 2-(PHENYLAMINO)ETHANOL

THR: Poison by skin contact and intravenous routes. Moderately toxic by ingestion. A skin and eye irritant. Combustible when exposed to heat or flame. To fight fire: dry chemical, water mist. When heated to decomposition it emits toxic fumes of NO_x.

p-ANILINOPHENOL   HR: 2
CAS: 122-37-2   NIOSH: SJ 6950000
mf: $C_{12}H_{11}NO$ mw: 185.24

PROP: Gray solid leaflets. mp: 70°, bp: 330°.

SYNS: 4-ANILINOPHENOL * PARA-HYDROXYDIFENYLAMIN (CZECH) * P-HYDROXYDIFENYLAMIN (CZECH) * P-HYDROXYDIPHENYLAMINE * 4-HYDROXYDIPHENYLAMINE * P-OXYDIPHENYLAMINE * N-PHENYL-P-AMINOPHENOL

THR: Moderately toxic by ingestion. An eye irritant. See also amines. When heated to decomposition it emits toxic fumes of NO_x.

ANILITE   HR: 3
A high explosive mixture composed of liquid NO_2 and carbon disulfide or gasoline. Extremely sensitive to shock.

p-ANISALDEHYDE   HR: 2
CAS: 123-11-5   NIOSH: BZ 2625000
mf: $C_8H_8O_2$ mw: 136.15

PROP: Colorless oil; d: 1.123 @ 20°/4°; mp: 2.5°; bp: 247°-248°. Very sltly sol in water, misc in alc and ether.

SYNS: ANISIC ALDEHYDE * P-METHOXYBENZ-ALDEHYDE * 4-METHOXYBENZALDEHYDE

THR: Moderately toxic by ingestion. A skin irritant. Mutagenic data. When heated to decomposition it emits acrid smoke and irritating fumes.

ANISEED OIL   HR: 2
PROP: Colorless or pale yellow liquid. Composition: 80%-90% anethol, methyl chavicol, anisaldehyde. d: 0.978-0.988 @ 25°/25°.

THR: Moderately toxic by ingestion. A weak sensitizer. May cause contact dermatitis. Combustible.

ANISE OIL   HR: 2
CAS: 8007-70-3   NIOSH: BZ 4200000

PROP: Consists of (80-90%) of Anethole. Small quantities of methyl chavicol, p-methoxyaceto-phenone and other materials also. Found in the dried ripe fruit of *Impinella Anisum L;* d: 0.978-0.988 @ 25°/25°.

SYNS: ANISEED OIL * ANIS OEL (GERMAN) * OIL OF ANISE * STAR ANISE OIL

THR: Moderately toxic by ingestion. A weak sensitizer. May cause contact dermatitis. Combustible. When heated to decomposition it emits acrid smoke and irritating fumes.

m-ANISIC ACID   HR: 3
CAS: 586-38-9   NIOSH: BZ 4375000
mf: $C_8H_8O_3$ mw: 152.15

PROP: Needles from aq solns; mp: 107°-109°; bp: 170°-172° @ 10 mm. Sol in hot water, alc and ether.

SYNS: M-METHOXYBENZOIC ACID * 3-ME-THOXYBENZOIC ACID

THR: Poison by intraperitoneal route. When heated to decomposition it emits acrid smoke and irritating fumes.

p-ANISIC ACID, ETHYL ESTER   HR: 2
CAS: 94-30-4   NIOSH: BZ 4697000
mf: $C_{10}H_{12}O_3$ mw: 180.21

PROP: Liquid; d: 1.103 @ 25°/25°; mp: 7°-8°; bp: 269°-270°. Insol in water, sol in alc and ether.

SYNS: ETHYL ANISATE * ETHYL-P-ANISATE * ETHYL-P-METHOXYBENZOATE * ETHYL-4-METHOXYBENZOATE

THR: Moderately toxic by ingestion. See also esters. When heated to decomposition it emits acrid smoke and irritating fumes.

o-ANISIC ACID, HYDRAZIDE   HR: 3
CAS: 7466-54-8   NIOSH: BZ 4700000
mf: $C_8H_{10}N_2O_2$ mw: 166.20

SYNS: O-METHOXYBENZOHYDRAZIDE * O-ME-THOXYBENZOIC ACID HYDRAZIDE * 2-ME-THOXYBENZOIC ACID HYDRAZIDE * O-METHOX-YBENZOYLHYDRAZIDE * 2-METHOXYBENZOYL-HYDRAZINE * 2-METHOXYBENZOYL HYDRAZIDE

THR: An experimental carcinogen. When heated to decomposition it emits toxic fumes of NO_x.

p-ANISIC ACID, HYDRAZIDE   HR: 3
CAS: 3290-99-1   NIOSH: BZ 4800000
mf: $C_8H_{10}N_2O_2$ mw: 166.20

SYNS: ANISIC ACID HYDRAZIDE * ANISIC HY-DRAZIDE * ANISOYLHYDRAZINE * P-ANISOYL-HYDRAZINE * P-METHOXYBENZOIC ACID HY-DRAZIDE * 4-METHOXYBENZOIC ACID HYDRAZIDE * P-METHOXYBENZOIC HYDRAZIDE * 4-METHOXYBENZOYL HYDRAZIDE * (P-ME-THOXYBENZOYL)HYDRAZINE * 4-METHOXYBEN-ZOYLHYDRAZINE

THR: Poison by intravenous route. An experimental carcinogen. When heated to decomposition it emits toxic fumes of NO_x.

p-ANISIDINE   HR: 2
CAS: 104-94-9   NIOSH: BZ 5450000
mf: C_7H_9NO mw: 123.16

PROP: Plates from aq soln; d: 1.089 @ 55°/55°; mp: 57.2°; bp: 243°; vap d: 4.28. Sol in hot water, alc and ether.

SYNS: P-AMINOANISOLE * 4-AMINOANISOLE * 1-AMINO-4-METHOXYBENZENE * 4-ANISI-DINE * P-ANISYLAMINE * P-METHOXYANILINE * 4-METHOXYANILINE * 4-METHOXYBENZEN-AMINE * 4-METHOXYBENZENEAMINE * P-ME-THOXYPHENYLAMINE

DFG MAK: 0.1 ppm/0.5 mg/m³

THR: Moderately toxic by several routes. A mild sensitizer. May cause a contact dermatitis. See also aniline. When heated to decomposition it evolves toxic fumes, NO_x.

m-ANISIDINE ANTIMONYL TARTRATE HR: 3
CAS: 64090-82-0   NIOSH: BZ 5775000

THR: Poison by intraperitoneal route. See also antimony compounds. When heated to decomposition it emits very toxic fumes of NO_x and Sb.

o-ANISIDINE HYDROCHLORIDE HR: 3
CAS: 134-29-2   NIOSH: BZ 6500000
mf: $C_7H_9NO \cdot ClH$ mw: 159.63

SYNS: c.i. 37115 * 2-methoxyaniline hydrochloride * nci-c03747

THR: An experimental carcinogen. Mutagenic data. When heated to decomposition it emits very toxic fumes of NO_x and HCl. For further information, see Vol. 6, No. 5 of *DPIM Report*.

p-ANISIDINE HYDROCHLORIDE HR: 3
CAS: 20265-97-8   NIOSH: BZ 6600000
mf: $C_7H_9NO \cdot ClH$ mw: 159.63

SYN: nci-c03758

THR: An experimental carcinogen. When heated to decomposition it emits very toxic fumes of NO_x and HCl.

ANISOLE HR: 2
CAS: 100-66-3   NIOSH: BZ 8050000
mf: C_7H_8O mw: 108.15

PROP: Mobile liquid, clear straw color; vap d: 3.72, mp: $-37.3°$, bp: 153.8°, flash p: 125°F (COC), d: 0.996 @ 18°/4°, vap press: 10 mm @ 42.2°, autoign temp: 887°F. Insol in water, sol in alc and ether.

SYNS: methoxybenzene * methyl phenyl ether * phenyl methyl ether

THR: Moderately toxic by ingestion. A skin irritant. Moderately flammable when exposed to heat, flame, or oxidizing materials. To fight fire, use foam, CO_2, dry chemical. When heated to decomposition it emits acrid fumes.

ANISOYL CHLORIDE HR: 3
CAS: 100-07-2   NIOSH: CA 0270000
DOT: 1729
mf: $C_8H_7ClO_2$ mw: 170.60

PROP: Needle-like crystals, insol in water, sol in ether and acetone. Mp: 22°, bp: 262°-263°, (slight decomp).

SYNS: methoxybenzoyl chloride * p-anisyl chloride

DOT Classification: Label: Corrosive

THR: Corrosive to skin, eyes, mucous membranes, and other tissue. Evolves HCl by hydrolysis. See hydrochloric acid. Can explode spontaneously at room temperature. When heated to decomposition it emits toxic fumes of Cl^- and may explode. See also chlorides.

ANTHANTHRENE HR: 3
CAS: 191-26-4   NIOSH: HO 5900000
mf: $C_{22}H_{12}$ mw: 276.34

SYNS: anthanthren (german) * anthranthrene * dibenzo(cd, mk)pyrene * dibenzo-(def,mno)chrysene

THR: An experimental carcinogen. Mutagenic data. A polycyclic hydrocarbon found in polluted air. When heated to decomposition it emits acrid fumes.

ANTHOPHYLITE (SEE ALSO ASBESTOS) HR: 3
CAS: 17068-78-9   NIOSH: CA 8400000

SYNS: anthophyllite * azbolen asbestos * ferroanthophyllite

THR: A human carcinogen. See also asbestos.

2-ANTHRACENAMINE HR: 3
CAS: 613-13-8   NIOSH: CA 9275000
mf: $C_{14}H_{11}N$ mw: 193.26

PROP: Yellow leaflets from alc. Mp: 238°; bp: subl @ 93° @ 9 mm. Insol in water, sltly sol in alc and ether.

SYNS: beta-aminoanthracene * 2-aminoanthracene * 2-anthracylamine * 2-anthrylamine * 2-anthramine

THR: An experimental carcinogen. Mutagenic data. See also amines. When heated to decomposition it emits toxic fumes of NO_x.

ANTHRACENE HR: 3
CAS: 120-12-7   NIOSH: CA 9350000
mf: $C_{14}H_{10}$ mw: 178.24

PROP: Colorless crystals, violet fluorescence. Mp: 217°, lel = 0.6%, flash p: 250°F (CC), d: 1.24 @ 27°/4°, autoign temp: 1004°F, vap press: 1 mm @ 145.0°, (sublimes), vap d: 6.15, bp: 339.9°. Insol in water. Sol in alc @ 1.9/100 @ 20°; in ether = 12.2/100 @ 20°.

SYNS: ANTHRACEN (GERMAN) * ANTHRACIN * GREEN OIL * PARANAPHTHALENE * TETRA OLIVE N2G

THR: An experimental carcinogen. A skin irritant and allergen. Mutagenic data. Combustible when exposed to heat, flame, or oxidizing materials. Moderately explosive when exposed to flame, Ca(OCl)$_2$, chromic acid. To fight fire, use water, foam, CO$_2$, water spray or mist, dry chemical. Incompatible with fluorine. For further information, see Vol. 4, No. 6 of *DPIM Report*.

1,8,9-ANTHRACENETRIOL HR: 3
CAS: 480-22-8 NIOSH: CB 1225000
mf: C$_{14}$H$_{10}$O$_3$ mw: 226.24

PROP: Yellow powder. Mp: 178°-180°. Insol in water, sol in fat, hot alc, benzene and dilute alkalies.

SYNS: ANTHRALIN * 1,8,9-ANTHRATRIOL * DIHYDROXY-ANTHRANOL * 1,8-DIHYDROXYANTHRANOL * 1,8-DIHYDROXY-9-ANTHRANOL * 1,8-DIHYDROXY-9-ANTHRONE * DIOXYANTHRANOL * 1,8,9-TRIHYDROXYANTHRACENE

THR: An experimental carcinogen. Mutagenic data. Skin contact can cause folliculitis. Absorption can cause kidney damage and intestinal disturbances. Combustible when heated. When heated to decomposition it emits acrid smoke and irritating fumes.

ANTHRACITE PARTICLES
PROP: Black powder or dust.

SYN: COAL DUST

THR: Variable toxicity depending upon SiO$_2$ content. Moderately flammable when exposed to heat, flame, or chemical reaction with oxidizers. Slightly explosive when exposed to flame.

ANTHRANILIC ACID HR: 3
CAS: 118-92-3 NIOSH: CB 2450000
mf: C$_7$H$_7$NO$_2$ mw: 137.15

PROP: Needle-like crystals. mp: 146°, bp: subl, d: 1.412 @ 20°. Solubility: in water = 0.35/100 @ 14°, in 90% alc = 10.7/100 @ 10°, in ether = 16/100 @ 70°.

SYNS: o-AMINOBENZOIC ACID * 2-AMINOBENZOIC ACID * 1-AMINO-2-CARBOXYBENZENE * O-CARBOXYANILINE * 2-CARBOXYANILINE * CARBOXYANILINE * NCI-C01730 * ORTHO-AMIDOBENZOIC ACID * ORTHO-AMINOBENZOIC ACID * VITAMIN L

THR: An experimental carcinogen. Moderately toxic by intraperitoneal route. Combustible. When heated to decomposition it emits toxic fumes of NO$_x$.

ANTHRANILIC ACID, CINNAMYL ESTER HR: 3
CAS: 87-29-6 NIOSH: CB 2725000
mf: C$_{16}$H$_{15}$NO$_2$ mw: 253.32

SYNS: 2-AMINOBENZOIC ACID-3-PHENYL-2-PROPENYL ESTER * CINNAMYL ALCOHOL ANTHRANILATE * CINNAMYL-2-AMINOBENZOATE * CINNAMYL-o-AMINOBENZOATE * CINNAMYL ANTHRANILATE * NCI-C03510 * 3-PHENYL-2-PROPENYLANTHRANILATE * 3-PHENYL-2-PROPEN-1-YL ANTHRANILATE

THR: An experimental carcinogen. See also esters. When heated to decomposition it emits toxic fumes of NO$_x$. For further information, see Cinnamyl Anthranilate, Vol. 1, No. 5 of *DPIM Report*.

ANTHRANILIC ACID, METHYL ESTER HR: 2
CAS: 134-20-3 NIOSH: CB 3325000
mf: C$_8$H$_9$NO$_2$ mw: 151.18

PROP: Plates from alc; d: 1.23; mp: 226°-228°; bp: 225°-230° @ 15 mm. Very sol in water, and hot abs alc = 23/100; insol in ether and chloroform.

SYNS: o-AMINOBENZOIC ACID METHYL ESTER * 2-AMINOBENZOIC ACID METHYL ESTER * O-CARBOMETHOXYANILINE * 2-CARBOMETHOXYANILINE * 2-(METHOXYCARBONYL)ANILINE * METHYL-O-AMINOBENZOATE * METHYL-2-AMINOBENZOATE * METHYL ANTHRANILATE

THR: Moderately toxic by ingestion. A skin irritant. See also esters. When heated to decomposition it emits toxic fumes of NO$_x$.

ANTHRANILONITRILE HR: 3
CAS: 1885-29-6 NIOSH: CB 4575000
mf: C$_7$H$_6$N$_2$ mw: 118.15

PROP: Needles from CS_2; mp: 51°; bp: 267°-268° @ 777 mm. Very sltly sol in water, sol in alc and ether.

SYNS: o-AMINOBENZONITRILE * 2-AMINOBENZONITRILE * O-CYANOANILINE * 2-CYANOANILINE

THR: Poison by intravenous route. See also nitriles. When heated to decomposition it emits toxic fumes of NO_x and CN^-.

ANTHRAQUINONE   HR: 2
CAS: 84-65-1   NIOSH: CB 4725000
mf: $C_{14}H_8O_2$ mw: 208.22

PROP: Yellow cryst; mp: 286°, bp: 376.9°, flash p: 365°F (CC), d: 1.438, vap press: 1 mm @ 190.0°, vap d: 7.16. Insol in water, very sltly sol in ether, in alc = 0.05/100 @ 18°, in hot alc = 2.25/100.

SYNS: 9,10-ANTHRACENEDIONE * ANTHRADIONE * 9,10-ANTHRAQUINONE * 9,10-DIOXOANTHRACENE

THR: Moderately toxic by several routes. A mild allergen. Combustible when exposed to heat or flame. To fight fire, use water, foam, CO_2, water spray or mist, dry chemical. When heated to decomposition it emits acrid smoke and irritating fumes.

ANTIMONY   HR: 3
CAS: 7440-36-0   NIOSH: CC 4025000
DOT: 2871
mf: Sb   mw: 121.75

PROP: Silvery or gray lustrous metal. Mp: 630°, bp: 1635°; d: 6.684 @ 25°, vap press: 1 mm @ 886°. Insol in water, sol in hot conc H_2SO_4.

SYNS: ANTIMONY BLACK * ANTIMONY REGULUS * ANTYMON (POLISH) * C.I. 77050 * STIBIUM

OSHA PEL: TWA 0.5 mg/m³
DFG MAK: 0.5 mg/m³
DOT Classification: Poison B

THR: Poison by intraperitoneal and other unspecified routes. See also antimony compounds. Moderate fire and explosion hazard in the forms of dust and vapor, when exposed to heat or flame. See also powdered metals. For further information, see Vol. 2, No. 1 of *DPIM Report*. When heated or on contact with acid, emits toxic fumes of SbH_3, electrolysis of acid sulfides and stirred Sb halide yields explosive Sb. It can react violently with NH_4NO_3; halogens; BrN_3; BrF_3; $HClO_3$; ClO; ClF_3; HNO_3; KNO_3; $KMnO_4$; K_2O_2; $NaNO_3$; oxidants.

ANTIMONY AMMONIA TRIACETIC ACID   HR: 3
CAS: 72017-60-8   NIOSH: AF 5810000
mf: $C_{12}H_{14}N_2O_{12}Sb \cdot 2H_2O$ mw: 536.07

SYN: ATA-sb

THR: Poison by intraperitoneal route. Systemic effects. See also antimony compounds. When heated to decomposition it emits toxic fumes of NO_x, Sb, and NH_3.

ANTIMONY (III) CHLORIDE   HR: 3
CAS: 10025-91-9   NIOSH: CC 4900000
DOT: 1733
mf: Cl_3Sb mw: 228.10

PROP: Colorless rhombic deliq crystals. d: 3.06; mp: 73.4°; bp: 220°; vap press: 1 mm @ 49.2° (subl). Sol in water @ 20°, sol in alc, benzene and chloroform.

SYNS: ANTIMONOUS CHLORIDE * ANTIMONY CHLORIDE * ANTIMONIO (TRICLORURO DI) (ITALIAN) * ANTIMONTRICHLORID (GERMAN) * ANTIMONY TRICHLORIDE, SOLID (DOT) * ANTIMOONTRICHLRIDE (DUTCH) * BUTTER OF ANTIMONY * CHLORID ANTIMONITY (CZECH) * CHLORURE ANTIMONIEUX (FRENCH) * C.I. 77056 * TRICHLOROSTIBINE

OSHA PEL: TWA 500 ug(Sb)/m³
DOT Classification: Label: Corrosive

THR: Moderately toxic by ingestion. Pulmonary and gastrointestinal systemic effects. Corrosive by vigorous reaction with moisture, generating heat and hydrogen chloride gas (a strong irritant) which can cause pulmonary edema when inhaled. Systemic effects can be caused by the antimony. See also antimony compounds. Mutagenic data. When heated to decomposition it emits very toxic fumes of chlorine and antimony. It can react violently with aluminum, potassium, and sodium. For further information, see Ammonium Trichloride, Vol. 2, No. 1 of *DPIM Report*.

ANTIMONY (III) CHLORIDE (SOLUTION)   HR: 3
CAS: 10025-91-9   NIOSH: CC 4910000
DOT: 1733
DOT Classification: Label: Corrosive

THR: A Poison, corrosive, and an irritant. See also antimony (III) chloride and antimony compounds. When heated to decomposition it emits very toxic fumes of Cl^- and Sb. $(Sb^{+++} +$ hot $HClO_3)$ can form an explosive mixture.

ANTIMONY (V) CHLORIDE　　　　HR: 3
CAS: 7647-18-9　　　NIOSH: CC 5075000
DOT: 1730
mf: Cl_5Sb　　mw: 299.01

PROP: Red-yellow oil, liquid; offensive odor; mp: 2.8°; bp: 140°; d: 2.336; vap press: 1 mm @ 22.7°. Decomp in water, sol in HCl, HBr and CS_2.

SYNS: ANTIMONIC CHLORIDE * ANTIMONIO (PENTACLORURO DI) (ITALIAN) * ANTIMONPENTACHLORID (GERMAN) * ANTIMONY PENTA-CHLORIDE * ANTIMONY PENTACHLORIDE (DOT) * ANTIMOONPENTACHLORIDE (DUTCH) * ANTIMONY PERCHLORIDE * BUTTER OF ANTIMONY * PENTACHLOROANTIMONY * PENTACHLORURE D'ANTIMOINE (FRENCH) * PERCHLORURE D'ANTIMOINE (FRENCH)

OSHA PEL: TWA 500 ug(Sb)/m^3
DOT Classification: Label: Corrosive

THR: Poison by ingestion. Corrosive. Mutagenic data. See antimony compounds and antimony (III) chloride. When heated to decomposition it emits very toxic fumes of Cl^- and Sb.

ANTIMONY (V) CHLORIDE (SOLUTION)　　　　HR: 3
　　　　　　NIOSH: CC 5080000
DOT: 1733

SYNS: ANTIMONIC CHLORIDE * ANTIMONY PERCHLORIDE * ANTIMONY PENTACHLORIDE SOLUTION (DOT)

DOT Classification: Label: Corrosive

THR: Poisonous, corrosive, and an irritant. See also antimony (V) chloride. When heated to decomposition it emits very toxic fumes of Sb and Cl^-.

ANTIMONY COMPOUNDS　　　　HR: 3
THR: Most antimony compounds are poisons by ingestion, inhalation, and intraperitoneal routes. See also antimony. Locally antimony compounds irritate the skin and mucous membranes. $(Sb^{+++}$ and hot $HClO_3)$ can form an explosive mixture.

ANTIMONY (III) FLUORIDE (1:3)　HR: 3
CAS: 7783-56-4　　　NIOSH: CC 5150000
mf: F_3Sb　　mw: 178.75

PROP: Colorless rhombic deliq crystals. mp: 292°; bp: 376° (subl); d: 4.379 @ 20.9°. Sol in water @ 20°.

SYNS: ANTIMOINE FLUORURE (FRENCH) * ANTIMONOUS FLUORIDE * ANTIMONY TRIFLUORIDE * TRIFLUOROANTIMONY

THR: Poison by subcutaneous route. See also fluorides and antimony compounds. When heated to decomposition it emits very toxic fumes of F^- and Sb. For further information, see Vol. 3, No. 5 of *DPIM Report*.

ANTIMONY NITRIDE　　　　HR: 3
mf: NSb　　mw: 135.76

THR: See also antimony compounds, nitrides. Explosively decomposes upon warming (in vacuo). When heated to decomposition it emits very toxic fumes of Sb, NO_x, and NH_3.

ANTIMONY OXIDE　　　　HR: 3
CAS: 1327-33-9　　　NIOSH: CC 5650000
mf: O_3Sb_2　　mw: 291.50

PROP: White cubes; d: 5.2; mp: 573°; bp: 656°. Very sltly sol in water, sol in KOH, HCl.

SYNS: ANTIMONIOUS OXIDE * ANTIMONY PEROXIDE * ANTIMONY SESQUIOXIDE * ANTIMONY WHITE * ANTIMONY TRIOXIDE * C.I. PIGMENT WHITE 11 * DECHLORANE-A-O * DIANTIMONY TRIOXIDE * FLOWERS OF ANTIMONY * NCI-C55152
ACGIH TLV: TWA 500 ug/m^3

THR: Poison by intravenous and subcutaneous routes. Moderately toxic by other routes. An experimental carcinogen. Mutagenic data. See also antimony compounds. When heated to decomposition it emits toxic Sb fumes. Incompatible with chlorinated rubber and heat of 216°; BrF_3. For further information, see Vol. 2, No. 1 of *DPIM Report*.

ANTIMONY (V) PENTAFLUORIDE　　　　HR: 3
CAS: 7783-70-2　　　NIOSH: CC 5800000
DOT: 1732
mf: F_5Sb　　mw: 216.75

PROP: Oily, colorless liquid. Very reactive. mp: 7.0°, bp: 149.5°, d: (liq) 2.99 @ 23°. Sol in water and KF.

SYNS: ANTIMONY FLUORIDE * ANTIMONY (V) FLUORIDE * PENTAFLUOROANTIMONY

DOT Classification: Label: Corrosive

THR: A very reactive, corrosive liquid to skin, eyes, mucous membranes. See also fluorides and antimony compounds. Violent reaction with phosphates. When heated to decomposition it emits very toxic fumes of F^- and Sb.

ANTIMONY PENTASULFIDE   HR: 3
CAS: 1315-04-4   NIOSH: CC 6125000
mf: S_5Sb_2 mw: 403.80

PROP: Orange-yellow powder; mp: (decomp); d: 4.120.

SYNS: ANTIMONIAL SAFFRON * ANTIMONIC SULFIDE * ANTIMONY RED * ANTIMONY SULFIDE * C.I. 77061 * GOLDEN ANTIMONY SULFIDE

OSHA PEL: TWA 500 ug(Sb)/m^3

THR: Toxic by intraperitoneal route. See also antimony, sulfides. Flammable when exposed to heat or by chemical reaction with powerful oxidizers. Use water to fight fire. Moderately explosive when shocked or by spont chemical reaction in contact with powerful oxidizers. When heated to decomposition it or contact with acid or acid fumes emits highly toxic fumes of oxides of sulfur and antimony. Incompatible with water or steam to produce toxic and flammable vapors, with oxidizers (i.e., $Ag(ClO_3)_2$; $HClO_3$; ClO_2; $Mg(ClO_3)_2$; TlO; $Zn(ClO_3)_2$).

ANTIMONY PENTOXIDE   HR: 3
CAS: 1314-60-9   NIOSH: CC 6300000
mf: O_5Sb_2 mw: 323.50

PROP: Yellowish-white powder. d: 3.78; mp: decomp @ 380°. Very sltly sol in water, sltly sol in KOH.

SYNS: ANTIMONIC "ACID" * ANTIMONIC OXIDE * ANTIMONY PENTAOXIDE * DIANTIMONY PENTOXIDE * STIBIC ANHYDRIDE

OSHA PEL: TWA 500 ug(Sb)/m^3

THR: Toxic by intraperitoneal route. See also antimony compounds. Flammable by chemical reaction with reducing compounds. An oxidizer. See also antimony.

ANTIMONY POTASSIUM TARTRATE   HR: 3
CAS: 28300-74-5   NIOSH: CC 6825000
DOT: 1551
mf: $C_4H_4O_7Sb \cdot K$ mw: 324.93

PROP: Colorless cryst to white powder; d: 2.607; mp: $-H_2O$ @ 100°.

SYNS: ANTIMONYL POTASSIUM TARTRATE * POTASSIUM ANTIMONYL-d-TARTRATE * POTASSIUM ANTIMONY TARTRATE * EMETIQUE (FRENCH) * ENT 50,434 * POTASSIUM ANTIMONYL TARTRATE * TARTAR EMETIC * TARTARIZED ANTIMONY * TARTRATE ANTIMONIO-POTASSIQUE (FRENCH) * TARTRATED ANTIMONY

OSHA PEL: TWA 500 ug(Sb)/m^3
DOT Classification: ORM-A; Label: None

THR: Poison by ingestion, subcutaneous, intravenous, intramuscular, and intraperitoneal routes. Large doses cause severe liver damage. Used medicinally, the therapeutic dose is close to the toxic dose. For disaster hazard, see also antimony compounds. For further information, see Vol. 1, No. 8 of *DPIM Report*.

dl-ANTIMONY POTASSIUM TARTRATE   HR: 3
CAS: 64070-12-8   NIOSH: CC 7175000
mf: $C_4H_4O_7Sb \cdot K$ mw: 324.93

SYNS: POTASSIUM ANTIMONYL-d,l-TARTRATE * dl-TARTARIC ACID, ANTIMONY POTASSIUM SALT

THR: Poison by intraperitoneal route. See also antimony compounds. When heated to decomposition it emits toxic fumes of Sb.

meso-ANTIMONY POTASSIUM TARTRATE   HR: 3
CAS: 64070-10-6   NIOSH: CC 7525000
mf: $C_4H_4O_7Sb \cdot K$ mw: 324.93

SYN: POTASSIUM ANTIMONYL-meso-TARTRATE

THR: Poison by intraperitoneal route. See also antimony compounds. When heated to decomposition it emits toxic fumes of Sb.

ANTIMONY SODIUM DIMETHYL CYSTEINO TARTRATE   HR: 3
                             NIOSH: CC 7875000

PROP: Made up of 5.8 parts of sodium antimony tartrate and 10 parts of dimethyl cysteine.

SYN: SODIUM ANTIMONYL DIMETHYLCYSTEINE TARTRATE

OSHA PEL: TWA 500 ug(Sb)/m^3

THR: Poison by subcutaneous and intramuscular routes. Moderately toxic by other routes. See also antimony compounds. When heated to decomposition it emits very toxic Sb fumes and Na_2O.

ANTIMONY (III) SODIUM GLUCO-NATE   HR: 3
CAS: 12550-17-3   NIOSH: CC 7900000
mf: $C_6H_8O_7Sb \cdot Na$ mw: 336.88

SYNS: SODIUM ANTIMONY GLUCONATE
* SODIUM ANTIMONY (III) GLUCONATE
* TRIVALENT SODIUM ANTIMONYL GLUCONATE

THR: Poison by intravenous and intraperitoneal routes. See also antimony compounds. When heated to decomposition it emits toxic fumes of Sb and Na_2O.

ANTIMONY (V) SODIUM GLUCO-NATE   HR: 3
CAS: 12001-86-4   NIOSH: CC 7930000

SYN: SODIUM ANTIMONY (V) GLUCONATE

THR: Poison by intraperitoneal route. See also antimony compounds. When heated to decomposition it emits toxic fumes of Sb.

ANTIMONY SODIUM TARTRATE   HR: 3
CAS: 34521-09-0   NIOSH: CC 8050000
mf: $C_4H_4O_7Sb \cdot Na$ mw: 308.82

SYNS: SODIUM ANTIMONYL TARTRATE
* ANTIMONY SODIUM OXIDE l-(+)-TARTRATE
* NATRIUMANTIMONYLTARTRAT (GERMAN)
* SODIUM ANTIMONY TARTRATE

OSHA PEL: TWA 500 ug(Sb)/m^3

THR: Poison by subcutaneous, intravenous, and intraperitoneal routes. Mutagenic data. See also antimony compounds. When heated to decomposition it emits toxic fumes of Sb.

ANTIMONY (V) SODIUM TAR-TRATE   HR: 3
CAS: 34521-09-0   NIOSH: CC 8075000

SYN: SODIUM ANTIMONY (V) TARTRATE

THR: Poison by intravenous route. Mutagenic data. See also antimony compounds. When heated to decomposition it emits toxic fumes of Sb.

ANTIMONY TRIBROMIDE   HR: 3
CAS: 7789-61-9
mf: $SbBr_3$ mw: 361.51

PROP: Yellow deliquescent, crystalline mass. Decomp by water. mp: 96.6°; bp: 280°; d: 4.145; vap press: 1 mm @ 93.9°.

SYNS: ANTIMONY BROMIDE * TRIBROMO STIB-INE

THR: Corrosive to skin, eyes, and mucous membranes. Reaction with water liberates HBr and antimony trioxide. Can cause severe burns. For further information, see Vol. 3, No. 5 of *DPIM Report*.

ANTIMONY TRIETHYL   HR: 3
mf: $Sb(C_2H_5)_3$ mw: 209.0

PROP: Liquid, water insol. d: 1.324 @ 16°; mp: −29°; bp: 159.5°.

THR: Alkyl metal compounds are often highly toxic. See also antimony compounds. Dangerous fire hazard by spontaneous chemical reaction. Explodes in air, water, carbon tetrachloride, other halogenated hydrocarbons, dimethyl formamide, triethyl borine. When heated to decomposition it emits highly toxic fumes of Sb.

ANTIMONY TRIIODIDE   HR: 3
mf: SbI_3 mw: 502.5

PROP: Red-to-yellow crystals. mp: 170°; bp: 401°; d: 4.768 @ 22°; vap press: 1 mm @ 163.6°.

THR: Poison by ingestion. See also iodides and antimony compounds. Incompatible with sodium, potassium. When heated to decomposition it emits highly toxic Sb fumes and I^-.

ANTIMONY TRIMETHYL   HR: 3
mf: $Sb(CH_3)_3$ mw: 166.9

PROP: Liquid, sltly sol in water. bp: 80.6°; d: 1.523 @ 15°

SYN: TRIMETHYL STIBINE

THR: Moderately toxic by subcutaneous route. See also antimony triethyl and antimony compounds. Dangerous fire hazard by spontaneous reaction in air. Explodes in water. When heated to decomposition it emits highly toxic fumes of antimony. Incompatible with oxidizing materials; halogenated hydrocarbons.

ANTIMONY TRISULFIDE　　　HR: 3
CAS: 1345-04-6　　　NIOSH: CC 9450000
DOT: 1325
mf: S_3Sb_2　　mw: 339.68

PROP: Red-to-black crystals. mp: 546°; d: 4.64; bp: about 1150°. Sol in H_2SO_4, solubility in water = 0.002/100 @ 20° (decomp).

SYNS: ANTIMONOUS SULFIDE * ANTIMONY GLANCE * ANTIMONY ORANGE * ANTIMONY SULFIDE * C.I. 77060 * CRIMSON ANTIMONY * NEEDLE ANTIMONY

OSHA PEL: TWA 500 ug(Sb)/m^3
DOT Classification: ORM-A; Label: None

THR: Toxic by intraperitoneal route. Blood and gastrointestinal system effects by inhalation. See also antimony compounds and sulfides. Spontaneously flammable when exposed to strong oxidizers. Flammable when exposed to heat or flame. Moderately explosive by spontaneous reaction with chlorates, perchlorates, ClO, thallic oxide. When heated to decomposition it or contact with acid or acid fumes emits highly toxic fumes of oxides of sulfur and antimony; will react with water or steam to produce toxic and flammable vapors.

ANTIMONY TRITELLURIDE　　　HR: 3
mf: Sb_2Te_3　　mw: 626.4

PROP: Gray powder; mp: 629°; d: 6.50 @ 13°

SYN: ANTIMONY TELLURIDE

THR: No toxicity data, but probably a poison. See also antimony and tellurium compounds. Flammable by spontaneous reaction with strong oxidizers. Moderately explosive by chemical reaction in contact with chlorates and perchlorates. When heated to decomposition it or contact with acid or acid fumes emits highly toxic fumes of Sb and tellurium. Incompatible with water or steam and oxidizing materials.

ANTIPYRINE　　　HR: 2
CAS: 60-80-0　　　NIOSH: CD 2450000
mf: $C_{11}H_{12}N_2O$　　mw: 188.23

PROP: Fine, white cryst; mp: 113°; bp: 319° @ 174 mm; d: 1.19. Very sol in water and alc, sltly sol in ether.

SYNS: DIMETHYLOXYQUINAZINE * 2,3-DI-METHYL-1-PHENYL-3-PYRAZOLIN-5-ONE * 2,3-DIMETHYL-1-PHENYL-5-PYRAZOLONE * OXYDIMETHYLQUINAZINE * PHENAZONE

(PHARMACEUTICAL) * 1-PHENYL-2,3-DI-METHYLPYRAZOLE-5-ONE * 1-PHENYL-2,3-DI-METHYL-5-PYRAZOLONE

THR: Moderately toxic via oral, subcutaneous, and intravenous routes. When heated to decomposition it emits toxic fumes of NO_x.

APHOLATE　　　HR: 3
CAS: 52-46-0　　　NIOSH: XX 9450000
mf: $C_{12}H_{24}N_9P_3$　　mw: 387.36

SYNS: AZIRIDINE-1,3,5,2,4,6-TRIAZATRIPHOS-PHORINE DERIVATIVE * 1-AZIRIDINYLPHOS-PHONITRILE TRIMER * ENT 26,316 * HEXA-(1-AZIRIDINYL)TRIPHOSPHOTRIAZINE * 2,2,4,4,6,6-HEXAHYDRO-2,2,4,4,6,6-HEXAKIS(1-AZIRI-DINYL)-1,3,5,2, 4,6-TRIAZATRIPHOSPHORINE * 2,2,4,4,6,6-HEXAKIS(1-AZIRIDINYL)CYCLO-TRIPHOSPHAZA-1,3,5-TRIENE * 2,2,4,4,6,6-HEXAKIS(1-AZIRIDINYL)-2,2,4,4,6,6-HEXAHY-DRO-1,3,5,2,4,6-TRIAZATRIPHOSPHORINE * HEXAKIS(AZIRIDINYL)PHOSPHOTRIAZINE * NSC-26812

THR: Poison by ingestion and intramuscular routes. An experimental carcinogen and teratogen. Mutagenic data. When heated to decomposition it emits very toxic fumes of NO_x and PO_x.

APOATROPINE　　　HR: 3
CAS: 500-55-0　　　NIOSH: YM 4025000
mf: $C_{17}H_{21}NO_2$　　mw: 271.39

SYNS: 1-ALPHA-H,5-ALPHA-H-TROPAN-3-AL-PHA-OL, ATROPATE (ESTER) * ATROPYLTRO-PEINE * ATROPAMINE

THR: Poison by ingestion and intraperitoneal routes. See also esters. When heated to decomposition it emits toxic fumes of NO_x.

APOCODEINE　　　HR: 3
mf: $C_{18}H_{19}NO_2$　　mw: 281.34

PROP: White crystalline solid. mp: 124°.

THR: Poison by inhalation and ingestion. A weak sensitizer and may cause contact dermatitis. See also codeine. When heated to decomposition it emits highly toxic fumes of NO_x.

APOMINE BLACK GX　　　HR: 3
CAS: 1937-37-7　　　NIOSH: QJ 6160000
mf: $C_{34}H_{25}N_9O_7S_2$ • 2Na　　mw: 781.78

SYNS: CHLORAMINE BLACK C * C.I. 30235 * MITSUI DIRECT BLACK GX * NCI-C54557

THR: An experimental carcinogen. Mutagenic data. When heated to decomposition it emits very toxic fumes of NO_x and SO_x.

APORMORPHINE HR: 3
CAS: 58-00-4   NIOSH: CE 0700000
mf: $C_{17}H_{17}NO_2$ mw: 267.35

PROP: White crystalline alkaloid. mp: 195° (decomp).

SYNS: 6A-beta-APORPHINE-10,11-DIOL * APOMORFIN * APOMORPHINE

THR: Poison by subcutaneous, intravenous, intraperitoneal, and rectal routes. Central nervous system effects. A powerful emetic. A weak sensitizer and may cause contact dermatitis. When heated to decomposition it emits highly toxic fumes of NO_x.

ARABIC GUM HR: 2
CAS: 9000-01-5   NIOSH: CE 5945000
mw: 240,000

SYNS: ACACIA * ACACIA DEALBATA GUM * ACACIA GUM * ACACIA SENEGAL * ACACIA SYRUP * AUSTRALIAN GUM * GUM OVALINE * GUM SENEGAL * INDIAN GUM * SENEGAL GUM * STARSOL NO. 1 * WATTLE GUM * GUM ARABIC * NCI-c50748

THR: Hives, eczema, and angiodema may result from inhalation or ingestion. A weak allergen. Combustible. When heated to decomposition it emits acrid smoke.

ARABINOCYTIDINE HR: 3
CAS: 147-94-4   NIOSH: HA 5425000
mf: $C_9H_{13}N_3O_5$ mw: 243.25

SYNS: 4-AMINO-1-ARABINOFURANOSYL-2-OXO-1,2-DIHYDROPYRIMIDINE * 1-beta-d-ARABINOFURANOSYL-4-AMINO-2(1H)PYRIMIDINONE * 1-ARABINOFURANOSYLCYTOSINE * 1-beta-ARABINOFURANOSYLCYTOSINE * 1-(beta-d-ARABINOFURANOSYL)CYTOSINE * BETA-d-ARABINOSYLCYTOSINE * CYTOSINE-beta-ARABINOSIDE * CYTOSINE beta-d-ARABINOSIDE * NCI-c04728 * NSC 63878 * 2(1H)-PYRIMIDINONE, 4-AMINO-1-beta-d-ARABINOFURANOSYL-(9CI)

THR: An experimental teratogen. Central nervous system effects by the intravenous route. An eye irritant. Mutagenic data. When heated to decomposition it emits toxic fumes of NO_x.

9-beta-d-ARABINO FURANOSYL ADENINE HR: 3
CAS: 5536-17-4   NIOSH: AU 6200000
mf: $C_{10}H_{13}N_5O_4$ mw: 267.28

SYNS: ADENINE ARABINOSIDE * ARABINOSYLADENINE * beta-d-ARABINOSYLADENINE * 9-ARABINOSYLADENINE

THR: An experimental teratogen. Central nervous system, blood, and other systemic effects by the intravenous route. When heated to decomposition it emits toxic fumes of NO_x.

ARACHIDONIC ACID HR: 3
CAS: 506-32-1   NIOSH: CE 6675000
mf: $C_{20}H_{32}O_2$ mw: 304.52

SYNS: ARCHIDONATE * (ALL-Z)-5,8,11,14-EICOSATETRAENOIC ACID

THR: Poison by intravenous route. When heated to decomposition it emits acrid smoke and irritating fumes.

ARATHANE HR: 3
CAS: 6119-92-2   NIOSH: GQ 5775000
mf: $C_{18}H_{24}N_2O_6$ mw: 364.44

PROP: Liquid.

SYNS: CAPRYLDINITROPHENYL CROTONATE * 2-CAPRYL-4,6-DINITROPHENYL CROTONATE * CROTONATE DE 2,4-DINITRO 6-(1-METHYL-HEPTYL)-PHENYLE (FRENCH) * 4,6-DINITRO-2-CAPRYLPHENYL CROTONATE * 4,6-DINITRO-2-(2-CAPRYL)PHENYL CROTONATE * DINITRO(1-METHYLHEPTYL)PHENYL CROTONATE * 2,4-DINITRO-6-(1-METHYLHEPTYL)-PHENYL CROTONATE * 2,4-DINITRO-6-(2-OCTYL)PHENYL CROTONATE * ENT 24727 * (6-(1-METHYL-HEPTYL)-2,4-DINITRO-FENYL)-CROTONAAT (DUTCH) * (6-(1-METHYL-HEPTYL)-2,3-DINITRO-PHENYL)-CROTONAT (GERMAN) * 2-(1-METHYLHEPTYL)-4,6-DINITROPHENYL CROTONATE * (6-(1-METIL-EPITL)-2,4-DINITRO-FENIL)-CROTONATO (ITALIAN)

THR: Poison by ingestion and intravenous routes. An experimental carcinogen. See nitrates. When heated to decomposition it emits toxic fumes of NO_x.

ARECOLINE HR: 3
CAS: 63-75-2   NIOSH: QT 2100000
mf: $C_8H_{13}NO_2$ mw: 155.22

PROP: Oily liquid. bp: 209°.

SYNS: ARECAIDINE METHYL ESTER * ARECO-LINE BASE * METHYL-1,2,5,6-TETRAHYDRO-1-METHYLNICOTINATE * N-METHYL-DELTA-TET-RAHYDRONICOTINIC ACID METHYL ESTER * N-METHYLTETRAHYDROPYRIDINE-BETA-CAR-BOXYLIC ACID METHYL ESTER * 1,2,5,6-TET-RAHYDRO-1-METHYLNICOTINIC ACID, METHYL ES-TER

THR: Poison by inhalation, ingestion, and subcutaneous routes. An experimental carcinogen. See also esters. Combustible, can react with oxidizing materials. When heated to decomposition it emits highly toxic fumes of NO_x.

ARGON HR: 1
CAS: 7440-37-1   NIOSH: CF 2300000
DOT: 1006

PROP: Colorless, inert gas. Af: A, at wt: 39.94, mp: $-189.2°$, bp: $-185.7°$, d: 1.784 g/L @ $0°$, 1.40 @ $-186°$, 1.65 @ $-233°$. Solubility in water 3.36 mL/100 g @ 20°.

SYN: ARGON-40

DOT Classification: Label: Nonflammable Gas

THR: A simple asphyxiant gas. As an inert gas, it has no specific inherent dangerous properties. For further information, see Vol. 1, No. 5 of *DPIM Report*. Classified as a simple asphyxiant gas. Gases of this type have no specific toxicity effect, but they act by excluding O_2 from the lungs. The effect of simple asphyxiant gases is proportional to the extent to which they diminish the amount (partial pressure) of O_2 in the air that is breathed. The oxygen may be diminished to 2/3 of its normal percentage in air before appreciable symptoms develop, and this in turn requires the presence of a simple asphyxiant in a conc of 33% in the mixture of air and gas. When the simple asphyxiant reaches a conc of 50%, marked symptoms can be produced. A conc of 75% is fatal in a matter of minutes. The first symptoms produced by simple asphyxiant gases such as argon are rapid respirations and air hunger. Mental alertness is diminished and muscular coordination is impaired. Later, judgment becomes faulty and all sensations are depressed. Emotional instability often results and fatigue occurs rapidly. As the asphyxia progresses, there may be nausea and vomiting, prostration and loss of consciousness, and finally, convulsions, deep coma and death.

ARGYROL HR: 2
PROP: Brown to black crystals. Composition: 19-23% silver.

SYN: MILD SILVER PROTEIN

THR: An allergen. May cause contact dermatitis. Continued application can cause argyria. See also silver and silver compounds.

ARISTOCORT HR: 3
CAS: 124-94-7   NIOSH: TU 3850000
mf: $C_{21}H_{27}FO_6$ mw: 394.48

SYNS: 9-ALPHA-FLUORO-16-ALPHA-HYDROXY-PREDNISOLONE * 9-ALPHA-FLUORO-11-BETA,16-ALPHA-17,21-TETRAHYDROXYPREGNA-1,4-DIENE-3,20-DIONE * 9-ALPHA-FLUORO-11-BETA,16-ALPHA-17,21-TETRAHYDROXY-1,4-PREGNADIENE-3,20-DIONE * 9-ALPHA-FLUORO-11-BETA,16-ALPHA,17-ALPHA,21-TETRAHY-DROXYPREGNA-1,4-DIENE-3,20-DIONE FLUOXY-PREDNISOLONE * 11-BETA,16-ALPHA,17-AL-PHA,21-TETRAHYDROXY-9-ALPHA-FLUORO-1,4-PREGNADIENE-3,20-DIONE TRIAMCINOLONE

THR: Poison by intramuscular and intravenous routes. An experimental teratogen. When heated to decomposition it emits toxic fumes of F^-.

ARISTOCORT ACETONIDE HR: 3
CAS: 76-25-5   NIOSH: TU 3920000
mf: $C_{24}H_{31}O_6F$ mw: 434.55

SYNS: TRIAMCINOLONE-16,17-ACETONIDE * 9-ALPHA-FLUORO-16-ALPHA-17-ALPHA-ISO-PROPYLEDENE DIOXY PREDNISOLONE * 9-AL-PHA-FLUORO-16-HYDROXYPREDNISOLONE ACETO-NIDE

THR: Poison by subcutaneous route. An experimental teratogen. When heated to decomposition it emits acrid smoke and toxic fumes of F^-.

ARISTOLOCHINE HR: 3
CAS: 1398-06-7   NIOSH: CF 3325000
mf: $C_{17}H_{11}NO_7$ mw: 341.29

From alcoholic extract of *Aristolochia Indico*.

SYNS: ARISTOLOCHIC ACID * BIRTHWORT * 8-METHOXY-6-NITROPHENANTHOL (3,4-d) 1,3-DIOXOLE-5-CARBOXYLIC ACID * NSC-50413

THR: Poison by intravenous route. A potent experimental carcinogen. When heated to decomposition it emits toxic fumes of NO_x.

From International Register of Potentially Toxic Chemicals: April 1982. Vol 5 No. 1: Under the provision of World Health Organization Resolutions WHA 16.36, WHA 23.61 of the Sixteenth, Twenty-third and Twenty-sixth World Health Assemblies on Clinical and Pharmacological Evaluation of Drugs/Drug Efficacy/ Quality, and Safety of Drugs, the Ministry of Health of the Federal Republic of Germany has informed the Pharmaceuticals Unit of the World Health Organization of the circumstances in which drugs containing aristolochic acid have been withdrawn from the national market. The decision, which was implemented with immediate effect, resulted from the demonstration of a carcinogenic potential in a three-month oral toxicity study undertaken in rats. The findings in the various dosage groups were as follows:

10 mg/kg/day: each of 18 animals sacrificed at the end of the dosage period had developed malignant papillomatous lesions of the stomach. Other animals sacrificed after a four-month treatment-free interval were found to have multiple carcinomatous lesions. Sites most frequently involved included the duodenum, mesentery, liver, kidney, bladder, lungs, thoracic cavity, diaphragm, and skin.

1 mg/kg/day: 16 of 18 animals sacrificed immediately had developed papillomatous changes in the stomach. Evidence of carcinomatous change within these lesions was demonstrated in several animals maintained for a further period of three months without further exposure.

0.1 mg/kg/day: No pathological changes were detected in the control group among animals sacrificed immediately, but papillomatous changes developed in 8 of 12 animals sacrificed eight months after a three-month period of exposure.

The Federal Health Office regards these findings as indicative that aristolochic acid has a particularly potent carcinogenic potential, since; an unusually short period of exposure is sufficient to induce malignant changes; malignant changes continue to develop rapidly following withdrawal of the stimulus; malignant changes are apparently induced in many tissues; a clear dose-dependent effect is evident.

Aristolochic acid is claimed to promote phagocytosis and to have immunostimulant activity. A growth-inhibiting effect on experimentally-induced tumours has been described, but this effect has not been shown to have any clinical relevance. Extracts of species of *Aristolochiacea* have traditionally been used as a bitter, and a broad range of therapeutic effects has been claimed.

The Federal Health Office considers that these substances have no justifiable use having regard to their apparent risks. The regulatory decision consequently relates not only to branded drugs containing aristolochic acid but to the sale of herbal preparations or extracts prepared from plants of the *Aristolochiaceae* family. Only homeopathic preparations prepared in a dilution of at least 10^{-11} will remain available. (Source: WHO Drug Information Circular # 197 26 October 1981).

ARNICA   HR: 3
PROP: An alcoholic infusion

SYNS: WOLFSBANE * MOUNTAIN TOBACCO

THR: Poison by inhalation and ingestion. A moderate irritant and allergen. It can cause gastroenteritis, nervous disturbances, and collapse. May cause contact dermatitis. Combustible when exposed to heat or flame. Incompatible with oxidizing materials.

AROMATIC AMINES   HR: 3
Amines which contain one or more rings of unsaturated or cyclic HC, such as benzene. There are a vast number of such amines. The term is largely due to the characteristic odor. Many of these aromatic amines are recognized as carcinogenic to the human bladder, ureter, and renal pelvis, and carcinogenic to the intestines, lung, liver, and prostate. See also amines.

AROMATIC SPIRITS OF AMMONIA
PROP: Colorless liquid, suffocating odor of ammonia. Composition: 10% by weight of NH_3 in alcohol.

THR: See also ammonia. A dangerous fire hazard due to its alcohol content. Moderately explosive. When heated, emits toxic fumes of ammonia. Incompatible with oxidizing materials.

ARSANILIC ACID   HR: 3
CAS: 98-50-0   NIOSH: CF 7875000
mf: $C_6H_8AsNO_3$ mw: 217.06

PROP: Needles from aq solns; mp: 232°; bp: decomp, $-H_2O$ @ 15°. Very sol in hot water, alc.; insol in ether and benzene.

SYNS: P-AMINOBENZENEARSONIC ACID
* 4-AMINOBENZENEARSONIC ACID * AMINO-
PHENYLARSINE ACID * P-AMINOPHENYLARSINE
ACID * P-AMINOPHENYLARSINIC ACID
* P-AMINOPHENYLARSONIC ACID * 4-AMINO-
PHENYLARSONIC ACID * P-ANILINEARSONIC
ACID * ANTOXYLIC ACID * P-ARSANILIC ACID
* 4-ARSANILIC ACID * ATOXYLIC ACID

OSHA PEL: TWA 500 ug(As)/m^3

THR: Poison by ingestion, intravenous, and intraperitoneal routes. See also arsenic compounds
and aniline. A human carcinogen. Flammable,
decomposes with heat to yield flammable vapors. When heated to decomposition it or contact
with acid or acid fumes emits highly toxic fumes
of As and NO$_x$.

ARSANILIC ACID, MONOSODIUM
SALT   HR: 3
CAS: 127-85-5   NIOSH: CF 9625000
mf: C$_6$H$_7$AsNO$_3$ • Na   mw: 239.05

PROP: Tetra hydrate; white odorless cryst powder, faint salty taste. Sol in water, somewhat
sol in alc.

SYNS: NCI-C61176 * (4-AMINOPHENYL)AR-
SONIC ACID SODIUM SALT * ANHYDROUS SO-
DIUM ARSANILATE * ARSANILIC ACID SODIUM
SALT * ATOXYL * SODIUM AMINARSONATE
* SODIUM-P-AMINOBENZENEARSONATE
* SODIUM AMINOPHENOL ARSONATE * SO-
DIUM-P-AMINOPHENYLARSONATE * SODIUM-AN-
ILINE ARSONATE * SODIUM ANILARSONATE
* SODIUM ARSANILATE * SODIUM-P-ARSANI-
LATE * SODIUM ARSONILATE

OSHA PEL: TWA 500 ug(AS)/m^3

THR: Poison by subcutaneous route. Can cause
blindness. When heated to decomposition it
emits very toxic fumes of As and NO$_x$.

ARSENIC   HR: 3
CAS: 7440-38-2   NIOSH: CG 0525000
DOT: 1558
mf: As   mw: 74.92

PROP: Silvery to black, brittle, crystalline and
amorphous metalloid. mp: 814° @ 36 atm, bp:
subl @ 612°, d: black crystals 5.724 @ 14°;
black amorphous 4.7, vap press: 1 mm @ 372°
(sublimes). Insol in water; sol in HNO$_3$. See
also arsenic vapor.

SYNS: ARSENICALS * ARSENIC-75 * AR-

SENIC BLACK * ARSEN (GERMAN, POLISH)
* COLLOIDAL ARSENIC * GREY ARSENIC
* METALLIC ARSENIC

OSHA PEL: TWA 500 ug/m^3
ACGIH TLV: TWA 0.2 mg/m^3
TRK: 0.2 mg/m^3 calculated as As in that portion
of dust that can possibly be inhaled.
DOT Classification: Label: Poison

THR: Poison by subcutaneous, intramuscular,
and intraperitoneal routes. Systemic, skin, and
gastrointestinal effects. A human carcinogen.
An experimental teratogen. Mutagenic data.
Flammable in the form of dust when exposed
to heat or flame or by chemical reaction with
powerful oxidizers such as bromates; chlorates;
iodates; peroxides; lithium; NCl$_3$; KNO$_3$;
KMnO$_4$; Rb$_2$C$_2$; AgNO$_3$; NOCl; IF$_5$; CrO$_3$;
ClF$_3$; ClO; BrF$_3$; BrF$_5$; BrN$_3$; RbC≡CH;
CsC≡CH. Slightly explosive in the form of
dust when exposed to flame. When heated or
on contact with acid or acid fumes, emits highly
toxic fumes; can react vigorously on contact
with oxidizing materials. Incompatible with bromine azide; dirubidium acetylide; halogens; palladium; zinc; platinum; NCl$_3$; AgNO$_3$; CrO$_3$;
Na$_2$O$_2$; hexafluoro isopropylideneamino lithium. For further information, see Vol. 4, No.
1 of *DPIM Report*.

m-ARSENIC ACID   HR: 3
CAS: 10102-53-1   NIOSH: CG 0760000
mf: AsHO$_3$ mw: 123.93

SYN: METAARSENIC ACID

THR: See also arsenic compounds. When heated
to decomposition it emits toxic fumes of As.

o-ARSENIC ACID   HR: 3
CAS: 7778-39-4   NIOSH: CG 7000000
mf: AsH$_3$O$_4$ mw: 141.95

SYNS: ACIDE ARSENIQUE LIQUIDE (FRENCH)
* ORTHOARSENIC ACID

OSHA PEL: TWA 500 ug(As)/m^3

THR: Poison by ingestion. See also arsenic compounds. When heated to decomposition it emits
toxic fumes of As. For further information, see
Vol. 2, No. 3 (as Arsenic Acid) in *DPIM Report*.

ARSENIC ACID (SOLUTION)   HR: 3
CAS: 7778-39-4   NIOSH: CG 0765000

SYN: ORTHOARSENIC ACID

DOT Classification: Label: Poison

THR: A poison. See also o-arsenic acid. For further information, see Vol. 2, No. 3 of *DPIM Report*.

ARSENIC ACID, CALCIUM SALT (2:3)   HR: 3
CAS: 7778-44-1   NIOSH: CG 0830000
DOT: 1573
mf: $As_2O_8 \cdot 3Ca$ mw: 398.08

PROP: Colorless amorphous powder. d: 3.620. Solubility: in water = 0.013/100 @ 25°.

SYNS: ARSENIATE DE CALCIUM (FRENCH)
* CALCIUMARSENAT * CALCIUM ARSENATE
* CALCIUM ORTHOARSENATE * KALZIUMARSE-
NIAT (GERMAN) * TRICALCIUMARSENAT (GER-
MAN) * TRICALCIUM ARSENATE

OSHA PEL: TWA 1 mg/m^3
DOT Classification: Label: Poison

THR: Poison by ingestion. Moderately toxic by skin contact. A human carcinogen. When heated to decomposition it emits toxic fumes of As. For further information, see Calcium Arsenate, Vol. 2, No. 1 of *DPIM Report*.

ARSENIC ACID, DISODIUM SALT   HR: 3
CAS: 7778-43-0   NIOSH: CG 0875000
mf: $Na_2HAsO_4 \cdot 7H_2O$ mw: 312.01

PROP: Colorless powder, effloresces. d: 1.88; mp: $-7H_2O$ @ 130°; bp: decomp @ 150°. Solubility in water = 61/100 @ 15°; sol in glycerol.

SYNS: DISODIUM ARSENIC ACID * DISODIUM HYDROGEN ARSENATE * DISODIUM HYDROGEN ORTHOARSENATE * DISODIUM MONOHYDROGEN ARSENATE * SODIUM ARSENATE * DISODIUM ARSENIC ACID * DISODIUM ARSENATE * DISODIUM HYDROGEN ARSENATE * DISO-DIUM HYDROGEN ORTHOARSENATE * DISODIUM MONOHYDROGEN ARSENATE * SODIUM ACID ARSENATE * SODIUM ARSENATE DIBASIC, AN-HYDROUS

OSHA PEL: TWA 500 ug(As)/m^3

THR: Poison by intraperitoneal route. Mutagenic data. See arsenic compounds. When heated to decomposition it emits toxic fumes of As.

ARSENIC ACID, DISODIUM SALT, HEPTAHYDRATE   HR: 3
CAS: 10048-95-0   NIOSH: CG 0900000
mf: $AsHO_4 \cdot 2Na \cdot 7H_2O$ mw: 312.05

SYNS: DISODIUM ARSENATE, HEPTAHYDRATE
* SODIUM ACID ARSENATE, HEPTAHYDRATE
* SODIUM ARSENATE, DIBASIC, HEPTAHYDRATE
* SODIUM ARSENATE HEPTAHYDRATE

THR: Poison by subcutaneous route. An experimental teratogen. See also arsenic compounds. When heated to decomposition it emits toxic fumes of As.

o-ARSENIC ACID, HEMIHYDRATE   HR: 3
CAS: 7774-41-6   NIOSH: CG 0950000
mf: $AsH_3O_4 \cdot 1/2H_2O$ mw: 150.96

PROP: White translucent crystals. mp: 35.5°; bp: $-H_2O$ @ 160°; d: 2.0-2.5.

SYNS: ARSENIC ACID, SOLID (DOT) * OR-THOARSENIC ACID HEMIHYDRATE

DOT Classification: Label: Poison

THR: Poison by intravenous route. See also arsenic compounds. When heated to decomposition it emits toxic fumes of arsenic.

ARSENIC ACID, LEAD SALT   HR: 3
CAS: 7645-25-2   NIOSH: CG 1000000
mf: $AsH_3O_4 \cdot 7Pb$ mw: 1592.28

SYNS: ARSENIATE DE PLOMB (FRENCH)
* LEAD ARSENATE

OSHA PEL: TWA 150 ug/m^3

THR: Poison by ingestion. See also lead compounds and arsenic compounds. When heated to decomposition it emits very toxic fumes of Pb and As.

ARSENIC ACID, MAGNESIUM SALT   HR: 3
CAS: 10103-50-1   NIOSH: CG 1050000
DOT: 1622
mf: $AsH_3O_4 \cdot 7Mg$ mw: 312.12

PROP: Monoclinic, white cryst. d: 2.60-2.61.

SYNS: ARSENIATE DE MAGNESIUM (FRENCH)
* MAGNESIUM ARSENATE * MAGNESIUM AR-SENATE PHOSPHOR

OSHA PEL: TWA 500 ug(As)/m^3
DOT Classification: Label: Poison

THR: Poison by ingestion. See also arsenic compounds. When heated to decomposition it emits toxic fumes of As.

ARSENIC ACID, MONOPOTASSIUM SALT   HR: 3
CAS: 7784-41-0   NIOSH: CG 1100000
DOT: 1677
mf: $AsH_2O_4 \cdot K$ mw: 180.04

SYNS: MONOPOTASSIUM ARSENATE * MONO-POTASSIUM DIHYDROGEN ARSENATE * MAC-QUER'S SALT * MONOPOTASSIUM ARSENATE * POTASSIUM ACID ARSENATE * POTASSIUM ARSENATE * POTASSIUM DIHYDROGEN ARSE-NATE * POTASSIUM HYDROGEN ARSENATE

DOT Classification: Label: Poison

THR: A human carcinogen. Mutagenic data. See also arsenic compounds. When heated to decomposition it emits toxic fumes of As. For further information, see Vol. 3, No. 4 (as Potassium Arsenate) of *DPIM Report*.

ARSENIC ACID, SODIUM SALT   HR: 3
CAS: 7631-89-2   NIOSH: CG 1225000
DOT: 1685
mf: $AsH_3O_4 \cdot xNa$

SYNS: SODIUM ARSENATE (DOT) * SODIUM ORTHOARSENATE

OSHA PEL: TWA 500 $ug(As)/m^3$
DOT Classification: Label: Poison

THR: Poison by ingestion, intravenous, and intraperitoneal routes. A human carcinogen. An experimental teratogen. Mutagenic data. See also arsenic compounds. When heated to decomposition it emits toxic fumes of As. For further information, see Sodium Arsenate, Vol. 2, No. 6 of *DPIM Report*.

ARSENIC(V) ACID, TRISODIUM SALT, HEPTAHYDRATE (1:3:7)   HR: 3
CAS: 64070-83-3   NIOSH: CG 1300000
mf: $AsO_4 \cdot 3Na \cdot 7H_2O$ mw: 334.03

SYN: TRISODIUM ARSENATE, HEPTAHYDRATE

THR: Poison by intraperitoneal route. An experimental carcinogen. See also arsenic compounds. When heated to decomposition it emits toxic fumes of As.

ARSENICAL DIP   HR: 3
                              NIOSH: CG 1330000
DOT: 1557

SYNS: ARSENICAL DIP, LIQUID (DOT) * SHEEP DIP

DOT Classification: Label: Poison

THR: A poison. See also arsenic compounds.

ARSENICAL DUST   HR: 3
CAS: 8028-73-7   NIOSH: CG 1335000
DOT: 1562

SYNS: ARSENICAL FLUE DUST * FLUE DUST, ARSENIC-CONTG.

DOT Classification: Label: Poison

THR: A poison. See also arsenic compounds.

ARSENICAL FLUE DUST (DOT)   HR: 3
                              NIOSH: CG 1340000
DOT Classification: Label: Poison

THR: Poison by inhalation and ingestion. See also arsenic compounds.

ARSENIC BISULFIDE   HR: 3
mf: As_2S_2 mw: 214

PROP: Red-brown crystals. bp: 565°; mp: beta = 307°; d: alpha = 3.506 @ 19°; beta = 3.254 @ 19°.

SYNS: ARSENIC SULFIDE * REALGAR.

THR: A poison. See also arsenic compounds and sulfides. Flammable in the form of dust when exposed to heat or flame. Explosion hazard when intimately mixed with powerful oxidizers such as Cl_2, KNO_3 and chlorates. It will react with water or steam to produce toxic and flammable vapors.

ARSENIC(III) BROMIDE   HR: 3
CAS: 7784-33-0   NIOSH: CG 1375000
DOT: 1555
mf: $AsBr_3$ mw: 314.65

PROP: Colorless rhombic crystals. mp: 32.8°; bp: 220.0°; vap press: 1 mm @ 41.8°; d: 3.3972 @ 25°, (liq), 3.3282.

SYNS: ARSENOUS TRIBROMIDE * TRIBRO-MOARSINE * ARSENIC TRIBROMIDE * ARSE-NOUS BROMIDE

DOT Classification: Label: Poison

THR: A poison. See also arsenic compounds and bromides. When heated to decomposition it emits very toxic fumes of As and Br^-. For further information, see Arsenic Tribromide, Vol. 2, No. 3. of *DPIM Report*.

ARSENIC CHLORIDE HR: 3
CAS: 7784-34-1   NIOSH: CG 1750000
DOT: 1560
mf: $AsCl_3$ mw: 181.28

PROP: Colorless oily liquid. d: 2.15 @ 25°; mp: −16°; bp: 130°. Decomp in water and by UV light; misc in chloroform, CCl_4, ether, iodine, P, S, alkalic iodides, oils and fats. Vap d: 6.25; vap press: 10 mm @ 23.5°.

SYNS: ARSENOUS TRICHLORIDE (9CI) * TRICHLOROARSINE * ARSENIC BUTTER * ARSENIC(III) CHLORIDE * ARSENIOUS CHLORIDE * ARSENOUS CHLORIDE * CHLORURE ARSENIEUX (FRENCH) * CHLORURE D'ARSENIC (FRENCH) * FUMING LIQUID ARSENIC * TRICHLORURE D'ARSENIC (FRENCH)

OSHA PEL: TWA 500 ug(As)/m^3
DOT Classification: Label: Poison

THR: A poison via inhalation. Mutagenic test data. See also arsenic compounds and chlorides. Very poisonous; fumes in air. When heated to decomposition it emits very toxic fumes of As and Cl^-. Highly reactive. Explodes with Na; K; Al on impact.

ARSENIC COMPOUNDS HR: 3
SYN: ARSENICALS

THR: Used as insecticides, herbicides, silvicides, defoliants, desiccants and rodenticides. Poisoning from arsenic compounds may be acute or chronic. Acute poisoning usually results from swallowing arsenic compounds; chronic poisoning from either swallowing or inhaling. Acute allergic reactions to arsenic compounds used in medical therapy have been fairly common. The type and severity of reaction depending upon the compound of arsenic. Inorganic arsenicals are more toxic than organics. Trivalent is more toxic than pentavalent. Acute arsenic poisoning (from ingestion) results in marked irritation of the stomach and intestines with nausea, vomiting, and diarrhea. In severe cases, the vomitus and stools are bloody and the patient goes into collapse and shock with weak, rapid pulse, cold sweats, coma, and death. Chronic arsenic poisoning, whether through ingestion or inhalation, may manifest itself in many different ways. There may be disturbances of the digestive system such as loss of appetite, cramps, nausea, constipation, or diarrhea. Liver damage may occur, resulting in jaundice. Disturbances of the blood, kidneys, and nervous system are not infrequent. Arsenic can cause a variety of skin abnormalities including itching, pigmentation, and even cancerous changes. A characteristic of arsenic poisoning is the great variety of symptoms that can be produced. A recognized carcinogen of the skin, lungs, liver. An experimental carcinogen of the mouth, esophagus, larynx, bladder and para nasal sinus. Dangerous; When heated to decomposition or when metallic arsenic contacts acids or acid fumes, or when water solutions of arsenicals are in contact with active metals such as Fe, Al, Zn, they emit highly toxic fumes of arsenic. For further information, see Vol. 1, No. 3 of *DPIM Report*.

In treating acute poisoning from ingestion BAL (dimercaptol) is of questionable effectiveness for acute and chronic poisoning with trivalent arsenicals, such as arsenic trioxide, arsine and arsenites. It is of no value for pentavalent arsenicals, such as cacodylic acid, methanearsonic acid, sodium cacodylate, MSMA, DSMA, arsanilic acid, arsenic acid, and arsenates. Vomiting and gastric lavage are the preferred emergency treatments for acute arsenical poisoning. Modern medical treatment of arsenical poisoning uses exchange transfusion and dialysis (A. E. De Palma, *J. Occup Med.*, Vol. 11,582-587 (1969). Note: Arsenic compounds are common air contaminants.

ARSENIC DIETHYL HR: 3
mf: $[As(C_2H_5)_2]_2$ mw: 266.2.

PROP: Liquid or oil. bp: 185°-190°, d: about 1.

THR: See also arsenic compounds. A dangerous fire hazard by spont chemical reaction. Dangerous when heated; see also arsenic. Incompatible with oxidizing materials.

ARSENIC DIMETHYL HR: 3
mf: $[As(CH_3)_2]_2$ mw: 210.0

PROP: Colorless to yellow oily liquid. mp: −6°; bp: 186°; d: 1.15.

THR: Poison by inhalation and ingestion. See also arsenic compounds. Flammable. Dangerous when heated; see also arsenic.

ARSENIC HEMISELENIDE HR: 3
mf: As_2Se mw: 228.78

THR: See also arsenic and selenium compounds. When heated to decomposition it emits As fumes and Se fumes. Incompatible with oxidizing materials.

ARSENIC IODIDE   HR: 3
CAS: 7784-45-4   NIOSH: CG 1950000
DOT: 1557
mf: AsI_3 mw: 455.62

PROP: Red hexagonal crystals. mp: 141.8°; bp: 403°; d: 4.38 @ 13°. Solubility: in water = 6/100 @ 25°, in CS_2 = 5.2/100.

SYNS: ARSENIC TRIIODIDE * ARSENOUS IODIDE * ARSENOUS TRIIODIDE (9CI) * TRIIODOARSINE

DOT Classification: Label: Poison

THR: A poison. See also arsenic compounds and iodides. Can form a shock sensitive compound with sodium, potassium. When heated to decomposition it emits very toxic fumes of I^- and As.

ARSENIC IODIDE MIXED WITH MERCU-RIC IODIDE   HR: 3
CAS: 8012-54-2   NIOSH: CG 1955000

PROP: A 100 mL solution contains 0.9 to 1.0 g arsenic iodide and 1.05 g mercuric iodide.

SYN: DONOVAN'S SOLUTION

THR: Poison by inhalation and ingestion. See also arsenic compounds, iodides and mercury compounds. When heated to decomposition it emits very toxic fumes of Hg, As, and I^-.

ARSENIC PENTASULFIDE   HR: 3
mf: As_2S_5 mw: 310.2

PROP: Brownish-yellow glassy, amorphous, highly refractive mass. mp: 500° (subl).

THR: See also arsenic compounds and sulfides. Flammable in the form of dust when exposed to heat or flame. Explosive when intimately mixed with powerful oxidizers, such as Cl_2, KNO_3, and chlorates. Will react with water and steam to produce toxic and flammable vapors. Incompatible with water, steam, and strong oxidizers.

ARSENIC PENTOXIDE   HR: 3
CAS: 1303-28-2   NIOSH: CG 2275000
DOT: 1559
mf: As_2O_5 mw: 229.84

PROP: White, amorphous, deliquescent solid. mp: decomp @ 800°; d: 4.32. Sol in alc, in water = 65.8/100 @ 20°.

SYNS: ARSENIC ACID * ANHYDRIDE ARSENI-QUE (FRENCH) * ARSENIC ACID ANHYDRIDE * ARSENIC ANHYDRIDE * ARSENIC OXIDE * ARSENIC (V) OXIDE * DIARSENIC PENTOXIDE

OSHA PEL: TWA 500 ug(As)/m^3
DOT Classification: Label: Poison

THR: Poison by ingestion and intravenous routes. A human carcinogen. Mutagenic data. See also arsenic compounds. Reacts vigorously with Rb_2C_2. When heated to decomposition it emits toxic fumes of As. For further information, see Vol. 2, No. 3 of *DPIM Report*.

ARSENIC PHOSPHIDE   HR: 3
mf: AsP   mw: 105.9

PROP: Brown to red powder. mp: subl with decomp.

THR: See also arsenic compounds and phosphine for toxicity data. Flammable by spontaneous chemical reaction. Phosphine is liberated upon contact with moisture. Dangerous when heated, see also phosphorous and arsenic compounds. Incompatible with water or steam, oxidizing materials.

ARSENIC SULFIDE   HR: 3
CAS: 1303-33-9   NIOSH: CG 2638000
DOT: 1557
mf: As_2S_3 mw: 246.04

PROP: Yellow or red cryst. bp: 707°; d: 3.43; mp: 312°. Insol in water, sol in alkalies.

SYNS: ARSENIC SESQUISULFIDE * ARSENIC SULFIDE YELLOW * ARSENIC TERSULPHIDE * ARSENIC TRISULFIDE * ARSENIC YELLOW * ARSENIOUS SULPHIDE * ARSENOUS SULFIDE * C.I. 77086 * DIARSENIC TRISULFIDE * KING'S YELLOW * ORPIMENT

DOT Classification: Label: Poison

THR: A poison. A human carcinogen. See also As compounds, sulfides, and arsenic bisulfide. Reacts violently with H_2O_2; (KNO_3 + S). When heated to decomposition it or contact with acid or acid fumes emits highly toxic fumes of SO_2, H_2S, and As; reacts with water or steam to emit toxic and flammable vapors. For further information, see Vol. 3, No. 5 of *DPIM Report*.

ARSENIC TRIFLUORIDE   HR: 3
CAS: 7784-35-2   NIOSH: CG 5775000
mf: AsF_3 mw: 131.92

PROP: Colorless liquid. d: 3.01; mp: −5.95; bp: 51°; vap press: 100 mm @ 13.2°, 400 mm @ 41.5°. Insol in water, sol in alc, benzene and Hg.

SYNS: ARSENIC FLUORIDE * TRIFLUOROARSINE * ARSENOUS FLUORIDE

OSHA PEL: TWA 500 ug(As)/m^3

THR: A poison by inhalation. See also fluorides and As compounds. Strong reaction with P_2O_3. When heated to decomposition it emits very toxic fumes of As and F$^-$.

ARSENIC TRIIODIDE MIXED WITH MERCURIC IODIDE HR: 3
CAS: 8012-54-2   NIOSH: CG 3200000
DOT: 2810

SYN: ARSENIOUS AND MERCURIC IODIDE SOLUTION (DOT)

DOT Classification: Label: Poison

THR: A poison. See also As compounds, Hg compounds, and iodides. When heated to decomposition it emits very toxic fumes of Hg, As, and I$^-$.

ARSENIC TRIOXIDE HR: 3
CAS: 1327-53-3   NIOSH: CG 3325000
DOT: 1561
mf: As_4O_6 mw: 395.68

PROP: Colorless rhombic crystals (dimer, claudetite). d: 4.15; mp: 278°; bp: 460°. Solubility in water = 1.82/100 @ 20°, sol in alc. Colorless cubes. d: 3.865; mp: 309°; solubility in water = 1.2/100 @ 20°.

SYNS: ACIDE ARSENIEUX (FRENCH) * ANHYDRIDE ARSENIEUX (FRENCH) * ARSENIC BLANC (FRENCH) * ARSENIC OXIDE * ARSENIC (III) OXIDE * ARSENIC SESQUIOXIDE * ARSENIGEN SAURE (GERMAN) * ARSENIOUS ACID * ARSENIOUS OXIDE * ARSENIOUS TRIOXIDE * ARSENOUS ACID * ARSENOUS ACID ANHYDRIDE * ARSENOUS ANHYDRIDE * ARSENOUS OXIDE * ARSENOUS OXIDE ANHYDRIDE * CRUDE ARSENIC * DIARSENIC TRIOXIDE * WHITE ARSENIC

OSHA PEL: TWA 500 ug(As)/m^3
DOT Classification: Label: Poison

THR: Poison by ingestion, subcutaneous, intradermal, intravenous, and possibly other routes. Systemic effects. A human carcinogen. Muta-

genic data. Reacts vigorously with Rb_2C_2; CIF_3; F_2; Hg; OF_2; $NaClO_3$. See also As compounds. For further information, see Vol. 3, No. 5 of *DPIM Report*.

ARSENIDES HR: 3
Compounds of As and H or metals, (i.e., transitional, alkaline earth, or rare-earth). These materials are dangerous because they readily emit very toxic arsine, arsenic fumes when exposed to heat, moisture, acids, acid fumes.

ARSENIOUS ACID SODIUM SALT HR: 3
CAS: 14060-38-9   NIOSH: CG 3850000
mf: $AsH_3O_3 \cdot 7Na$ mw: 286.88

PROP: Colorless or grayish-white powder. d: 1.87.

SYNS: ARSONIC ACID, SODIUM SALT (9CI) * ARSENIOUS ACID, SODIUM SALT POLYMERS * NATRIUMARSENIT (GERMAN) * SODIUM ORTHOARSENITE

OSHA PEL: TWA 500 ug(As)/m^3

THR: Poison by intraperitoneal and subcutaneous routes. Moderately toxic by ingestion. When heated to decomposition it emits toxic fumes of As.

ARSENOPYRITE HR: 3
CAS: 1303-18-0   NIOSH: CG 4340000
mf: AsFeS   mw: 162.83

SYNS: ARSENOMARCASITE * MISPICKEL

THR: Poison by intravenous route. When heated to decomposition it emits very toxic fumes of As and SO_x.

ARSINE HR: 3
CAS: 7784-42-1   NIOSH: CG 6475000
DOT: 2188
mf: AsH_3 mw: 77.95

PROP: Colorless gas, mild garlic odor. d: 2.695 g/L; bp: −62.5°. Solubility in water = 28 mg/100 @ 20°, sol in benzene, chloroform; vap d: 2.66; mp: −116°.

SYNS: ARSENIC HYDRIDE * ARSENIC TRIHYDRIDE * ARSENIURETTED HYDROGEN * ARSENOUS HYDRIDE * ARSENOWODOR (POLISH) * ARSENWASSERSTOFF (GERMAN) * HYDROGEN ARSENIDE

OSHA PEL: TWA 0.05 ppm
DFG MAK: 0.05 ppm/0.2 mg/m^3
DOT Classification: Label: Poison and
Flammable Gas

THR: Poison by inhalation. A human carcinogen. Has effects on red blood cells, gastrointestinal system, central nervous system, and other systemic effects. See also As and As compounds. Flammable when exposed to flame. Moderately explosive when exposed to Cl_2, HNO_3, $(K + NH_3)$ or open flame. Dangerous, more toxic than its oxidation product; When heated to decomposition it emits highly toxic fumes.

ARSINE BORON TRIBROMIDE HR: 3
mf: $AsH_3 \cdot BBr_3$ mw: 328.6

THR: See also bromides, As compounds, B compounds. A highly unstable compound. Ignites in air. When heated to decomposition it emits very toxic fumes of As and Br^-.

ARSINE-TRI-1-PIPERIDINIUM CHLORIDE HR: 3
CAS: 67360-94-5   NIOSH: TN 4430000
mf: $C_{15}H_{33}AsN_3 \cdot 3Cl$ mw: 436.78

SYN: ARSINOTRIS PIPERIDINIUM TRICHLORIDE

THR: Poison by intravenous route. When heated to decomposition it emits very toxic fumes of As, NO_x and Cl^-.

ARSPHENOXIDE HR: 3
CAS: 538-03-4   NIOSH: SJ 5600000
mf: $C_6H_6AsNO_2 \cdot ClH$ mw: 235.51

SYNS: 2-AMINO-4-ARSENOSOPHENOL HYDROCHLORIDE * 3-AMINO-4-HYDROXY-PHENARSINE HYDROCHLORIDE * 3-AMINO-4-HYDROXYPHENYLARSINE OXIDE HYDROCHLORIDE * 3-AMINO-4-HYDROXYPHENYL ARSINOXIDE HYDROCHLORIDE * OXOPHENARSINE HYDROCHLORIDE

THR: Poison by ingestion, intravenous, and intraperitoneal routes. See also As compounds. When heated to decomposition it emits very toxic fumes of As, NO_x, and HCl.

ASALIN HR: 3
CAS: 13425-94-0   NIOSH: YV 9365000

SYN: ETHYL ESTER OF N-ACETYL-DL-SARCOSYL-LYL-DL-VALINE

THR: Poison by ingestion, intramuscular, rectal, and intraperitoneal routes. See also esters.

ASBESTOS HR: 3
CAS: 1332-21-4   NIOSH: CI 6475000

Generic name for naturally occurring mineral silicate fibers of the Serpentine and Amphibole series.

DOT: 2212/2590

SYNS: ACTINOLITE * AMIANTHUS * AMOSITE * AMPHIBOLE * ANTHOPHYLLITE * ASBESTOSE (GERMAN) * ASBESTOS FIBER * ASCARITE * CHRYSOTILE * CROCIDOLITE * TREMOLITE

OSHA PEL: TWA 2 fb/cc
TRK: (Fine dust particles which are able to reach the alveolar area of the lung) 0.05 resp. 1×10^6 fibers/m^3 (definition of fiber: length greater than 5 μm; diameter less than 3 μm; length/diameter greater than 3:1, equivalent to 1 fiber/cc); 2.0 mg/m^3 applicable when there is less than or equal to 2.5 wt % asbestos in fine dust (additionally to TRK value).
DOT Classification: ORM-C

THR: A human carcinogen. Pulmonary system effects by inhalation. Usually at least 4 to 7 years of exposure are required before serious lung damage (fibrosis) results. A common air contaminant. For further information, see Vol. 1, No. 1 of *DPIM Report*.

ASCARIDOLE HR: 3
CAS: 512-85-6   NIOSH: OT 0175000
mf: $C_{10}H_{16}O_2$ mw: 168.26

PROP: Colorless unstable liquid. mp: 3.3°, bp: 40° @ 2 mm; 115° @ 15 mm, d: 1.011 @ 13°/15°.

SYNS: ASCARISIN * 1,4-PEROXIDO-P-MENTHENE-2

THR: Poison by ingestion. An experimental carcinogen. See also chenopodium oil and peroxides, organic. Flammable spontaneous chemical reaction. An oxidizer. Explodes when heated > 130° or when exposed to organic acids. Dangerous; heating emits toxic fumes and may explode; reacts with reducing materials.

L-ASCORBIC ACID HR: 2
CAS: 50-81-7   NIOSH: CI 7650000
mf: $C_6H_8O_6$ mw: 176.14

PROP: White crystals, sol in water, slightly

sol in alcohol, insol in ether, chloroform, benzene, petroleum ether; oils and fats. mp: 192°.

SYNS: ASCORBIC ACID * L(+)-ASCORBIC ACID * ASCORBUTINA * CEVITAMIC ACID * CEVITAMIN * 3-KETO-L-GULOFURANOLACTONE * L-3-KETOTHREOHEXURONIC ACID LACTONE * NATRASCORB INJECTABLE * NCI-C54808 * 3-OXO-L-GULOFURANOLACTONE * VITACIN * VITAMIN C * VITAMISIN * VITASCORBOL * XITIX * L-XYLOASCORBIC ACID

THR: Moderately toxic. Blood systemic effects. Mutagenic data. When heated to decomposition it emits acrid smoke and irritating fumes. For further information, see Vol. 1, No. 4 of *DPIM Report*.

L-ASPARAGINASE HR: 3
CAS: 9015-68-3   NIOSH: CI 9000000

SYNS: ASPARAGINASE * l-ASPARAGINASE X * L-ASPARAGINASI (ITALIAN) * L-ASPARAGINE AMIDOHYDROLASE * LEUCOGEN * NSC-109229

THR: An experimental carcinogen. Systemic effects by intramuscular route.

ASPHALT HR: 3
CAS: 8052-42-4   NIOSH: CI 9900000

PROP: Black or dark brown mass. bp: <470°, flash p: 400+°F (CC), d: 0.95−1.1, autoign temp: 905°F.

DOT: 1999

SYNS: ASPHALTUM * BITUMEN * JUDEAN PITCH * MINERAL PITCH * PETROLEUM PITCH * ROAD ASPHALT (DOT) * ROAD TAR (DOT)
ACGIH TLV: TWA 5 mg/m^3
DOT Classification: ORM-C; Label: None

THR: A moderate irritant. May contain carcinogenic components. Combustible when exposed to heat or flame. Use foam, CO_2, dry chemical to fight fire. For further information, see Vol. 2, No. 1 of *DPIM Report*.

ASPHALT (CUT BACK) HR: 3
NIOSH: CI 9905000
DOT: 1999

PROP: A liquid petroleum product, solubility of residue from distillation in carbon tetrachloride = 99.5%, flash p: < 50°F.

SYNS: ROAD ASPHALT (DOT) * ROAD TAR, LIQUID (DOT)

DOT Classification: Label: Flammable Liquid

THR: Contains carcinogenic components. A dangerous fire hazard when exposed to heat or flame. To fight fire use dry chemical, water mist, fog. When heated to decomposition it emits smoke and irritating, acrid fumes.

ASPIRIN, PHENACETIN AND CAFFEINE HR: 2
CAS: 8003-03-0   NIOSH: DG 0935000

Composed of 50% Aspirin, 46% Phenacetin, and 4% Caffeine.

SYNS: 2-(ACETYLOXY)BENZOIC ACID, MIXED WITH 3,7-DIHYDRO-1,3,7-TRIMETHYL-1H-PURINE-2,6-DIONE AND N-(4-ETHOXYPHENYL)ACETAMIDE * NCI-C02697

THR: An experimental carcinogen. Moderately toxic by ingestion. See caffeine, p-acetophenetidide, and 2-(acetyloxy)benzoic acid. When heated to decomposition it emits toxic fumes of NO_x.

ATABRINE HR: 3
CAS: 83-89-6   NIOSH: AR 7700000
mf: $C_{23}H_{30}ClN_3O$ mw: 400.01

PROP: Bright yellow crystals. mp: decomp @ 248°.

SYNS: 6-CHLORO-9-((4-(DIETHYL AMINO)-1-METHYL BUTYL)AMINO)-2-METHOXYACRIDINE * 3-CHLORO-7-METHOXY-9-(1-METHYL-4-DIETHYLAMINOBUTYLAMINO)ACRIDINE * 2-METHOXY-6-CHLORO-9-DIETHYLAMINOPENTYL-AMINOACRIDINE * QUINACRINE

THR: Poison by subcutaneous route. Moderately toxic by ingestion. Mutagenic data. Has been implicated in aplastic anemia. When heated to decomposition, it emits very toxic fumes of Cl^- and NO_x.

ATROMID S HR: 3
CAS: 637-07-0   NIOSH: UE 9480000
mf: $C_{12}H_{15}ClO_3$ mw: 242.72

SYNS: 2-(4-CHLOROPHENOXY)-2-METHYLPROPANOIC ACID ETHYL ESTER * ETHYL CHLOROPHENOXYISOBUTYRATE * ETHYL-PARA-CHLOROPHENOXYISOBUTYRATE * SKLERO-TABLINEN * AMOTRIL * ALPHA-P-CHLOROPHENOXYISO-

BUTYRYL ETHYL ESTER * 2-(P-CHLOROPHE-NOXY)-2-METHYLPROPIONIC ACID ETHYL ESTER * ETHYL-ALPHA-P-CHLOROPHENOXYISOBUTY-RATE * ETHYL-ALPHA-(4-CHLOROPHENOXY) ISOBUTYRATE * ETHYL-2-(P-CHLOROPHENOXY) ISOBUTYRATE * ETHYL-ALPHA-(P-CHLOROPHE-NOXY)-ALPHA-METHYLPROPIONATE * ETHYL-ALPHA-(4-CHLOROPHENOXY)-ALPHA-METHYLPRO-PIONATE * ETHYL 2-(P-CHLOROPHENOXY)-2-METHYLPROPIONATE * ETHYL 2-(4-CHLOROPHE-NOXY)-2-METHYLPROPIONATE * ETHYL CLOFIB-RATE * ICI 28257

THR: An experimental carcinogen. Moderately toxic by ingestion and other routes. Central nervous system effects. When heated to decomposition it emits toxic fumes of Cl^-.

ATROPINE   HR: 3
CAS: 51-55-8   NIOSH: CK 0700000
mf: $C_{17}H_{23}NO_3$ mw: 289.41

PROP: Colorless cryst alkaloid.

SYNS: ATROPIN (GERMAN) * DL-HYOSCYA-MINE * 2-PHENYLHYDRACRYLIC ACID-3-ALPHA-TROPANYL ESTER * BETA-PHENYL-GAMMA-OXYPROPIONSAEURE-TROPYL-ESTER (GERMAN) * 1-ALPHA-H,5-ALPHA-H-TROPAN-3-ALPHA-OL (+-)-TROPATE (ESTER) * DL-TROPANYL-2-HYDROXY-1-PHENYLPROPIONATE * TROPIC ACID-3-ALPHA-TROPANYL ESTER * TROPIC ACID, ESTER WITH TROPINE * TRO-PINE TROPATE * DL-TROPYLTROPATE * (+,-)-TROPYL TROPATE

THR: Poison by ingestion, subcutaneous, intravenous, and intraperitoneal routes. An alkaloid. When heated to decomposition it emits toxic fumes of NO_x.

ATROPINE SULFATE (1:1)   HR: 3
CAS: 2472-17-5   NIOSH: CK 1925000
mf: $C_{17}H_{23}NO_3 \cdot H_2O_4S$ mw: 387.49

THR: Poison by intravenous route. Mutagenic data. See also atropine and sulfates. When heated to decomposition it emits very toxic fumes of NO_x and SO_x.

ATROPINE SULFATE (2:1)   HR: 3
CAS: 55-48-1   NIOSH: CK 2450000
mf: $C_{34}H_{46}N_2O_6 \cdot H_2O_4S$ mw:676.90

SYNS: ATROPIN SIRAN (CZECH) * ATROPIN-SULFAT (GERMAN) * SULFATE D'ATROPINE (FRENCH) * 1-ALPHA-H,5-ALPHA-H-TROPAN-3-

ALPHA-OL (+-)-TROPATE (ESTER), SULFATE (2: 1) SALT * DL-TROPANYL-2-HYDROXY-1-HENYL-PROPIONATE SULFATE

THR: Poison by subcutaneous, intravenous, and intraperitoneal routes. Moderately toxic by ingestion. Pulmonary system effects. See also atropine. When heated to decomposition it emits very toxic fumes of NO_x and SO_x.

ATX II   HR: 3
CAS: 60748-45-0   NIOSH: CK 4438200

PROP: A polypeptide isolated from the sea anemone, *Anemonia Sulcata*.

SYN: SEA ANEMONE TOXIN II

THR: A deadly poison by intravenous and parenteral routes.

AUREINE   HR: 3
CAS: 130-01-8   NIOSH: VT 5710000
mf: $C_{18}H_{25}NO_5$ mw: 335.44

SYNS: 12-HYDROXYSENECIONAN-11,16-DIONE * SENECIONINE

THR: Poison by intravenous, intraperitoneal, and possibly other routes. When heated to decomposition it emits toxic fumes of NO_x.

AUREMETINE   HR: 3
PROP: Percentage composition = 28% emetine, 16% auramine and 56% iodine.

THR: Poison by ingestion. When heated to decomposition it emits very toxic fumes of I^- and NO_x.

1-AUROTHIO-D-GLUCOPYRA-NOSE   HR: 3
CAS: 12192-57-3   NIOSH: MD 6475000
mf: $C_6H_{11}O_5S \cdot Au$ mw: 392.20

SYNS: AUROMYOSE * AUROTAN * AURO-THIOGLUCOSE * AURUMINE * AUTHRON * BRENOL * (D-GLUCOPYRANOSYLTHIO)GOLD * (1-D-GLUCOSYLTHIO)GOLD * GLYSANOL B * GOLD THIOGLUCOSE * GTG * ORONOL * ROMOSOL * SOLGANAL * SOLGANAL B * (1-THIO-D-GLUCOPYRANOSATO)GOLD * THIOGLUCOSE D'OR (FRENCH)

THR: An experimental carcinogen. Moderate acute toxicity. See also gold compounds. When heated to decomposition it emits very toxic fumes of SO_x.

AVERTIN HR: 3
CAS: 75-80-9 NIOSH: KM 3675000
mf: $C_2H_3Br_3O$ mw: 282.78

PROP: Crystals, ethereal odor, aromatic taste, slightly water-sol, sol in alcohol and organic solvents. mp: 70°-82°, bp: 92°-93° @ 10 mm.

SYNS: 2,2,2-TRIBROMOETHANOL * 2,2,2-TRI-BROMOETHYL ALCOHOL

THR: Poison by intravenous route. Moderately toxic by ingestion and other routes. Dangerous when heated; see also bromides.

AVICOL HR: 3
CAS: 151-06-4 NIOSH: SH 1225000
mf: $C_{10}H_{14}ClN \cdot ClH$ mw: 220.16

SYNS: P-CHLORO-ALPHA,ALPHA-DIMETHYLPHEN-ETHYLAMINE HYDROCHLORIDE * 4-CHLORO-ALPHA,ALPHA-DIMETHYLPHENETHYLAMINE HYDROCHLORIDE * 1-(P-CHLOROPHENYL)-2-METHYL-2-AMINOPROPANE HYDROCHLORIDE * ALPHA,ALPHA-DIMETHYL-P-CHLORO-PHENETHYLAMINE HYDROCHLORIDE

THR: Poison by ingestion, intraperitoneal, and intravenous routes. When heated to decomposition it emits very toxic fumes of HCl and NO_x.

AZACYTIDINE HR: 3
CAS: 320-67-2 NIOSH: XZ 30175000
mf: $C_8H_{12}N_4O_5$ mw: 244.24

SYNS: AZACITIDINE * 5′-AZACYTIDINE * 4-AMINO-1-BETA-D-RIBOFURANOSYL-D-TRIAZIN-2(1H)-ONE * ANTIBIOTIC U 18496 * 5-AZACYTIDINE * NCI-C01569 * NSC-102816

THR: Poison by intravenous and intraperitoneal routes. Moderately toxic by ingestion. Blood and gastrointestinal system effects by intravenous route. An experimental carcinogen and teratogen. Mutagenic data. When heated to decomposition it emits toxic fumes of NO_x.

AZATADINE MALEATE HR: 3
CAS: 3978-86-7 NIOSH: DE 8025500
mf: $C_{20}H_{22}N_2 \cdot 2C_4H_4O_4$ mw:522.60

SYNS: 6,11-DIHYDRO-11-(1-METHYL-4-PIPERIDYLIDENE)5H-BENZO(5,6)CYCLOHEPTA (1,2-b) PYRIDINE DIMALEATE * SCH 10649

THR: Poison by ingestion, subcutaneous, and intraperitoneal route. When heated to decomposition it emits toxic fumes of NO_x.

AZATHIOPRINE HR: 3
CAS: 446-86-6 NIOSH: UO 8925000
mf: $C_9H_7N_7O_2S$ mw: 277.29

SYNS: AZOTHIOPRINE * METHYLNITROIMIDA-ZOLYLMERCAPTOPURINE * 6-(METHYL-p-NITRO-5-IMIDAZOLYL)-THIOPURINE * 6-((1-METHYL-4-NITROIMIDAZOL-5-YL)THIO)-PURINE * 6-(1-METHYL-p-NITRO-5-IMID-AZOLYL)-THIOPURINE * 6-(1-METHYL-4-NI-TROIMIDAZOL-5-YLTHIO)PURINE * NCI-C03474

THR: Poison by intradermal and intraperitoneal routes. An experimental carcinogen. Mutagenic data. When heated to decomposition it emits very toxic fumes of NO_x and SO_x. For further information, see Vol. 1, No. 4 of *DPIM Report*.

AZIDES (See also specific compound).
THR: Variable toxicity. Many azides are poisonous, cause fall in blood pressure and some inhibit enzyme action, thus resembling nitrites and cyanides. An azide is a compound of H or a metal ion and the monovalent $-N_3$ radical. All of its salts and the acid are unstable and some decompose explosively; although lead azide, which is one of the most important azides, is not very sensitive. Dangerous; shock and heat will explode it; When heated to decomposition it emits highly toxic fumes; if exposed to CS_2, it forms violently explosive salts. Organic azides are sensitized by metal salts or traces of strong acid.

AZIDITHION HR: 2
CAS: 78-57-9 NIOSH: TD 5600000
mf: $C_6H_{12}N_5O_2PS_2$ mw: 281.32

SYNS: 2,4-DIAMINO-6-DIMETHOXYPHOS-PHINOTHIONYLTHIOMETHYL-S-TRIAZINE * S-(4,6-DIAMINO-1,3,5-TRIAZIN-2-YL)-METHYL)-O,O-DIMETHYL-DITHIOFOSFAAT (DUTCH) * S-(4,6-DIAMINO-1,3,5-TRIAZIN-2-YL)-METHYL)-O,O-DIMETHYL-DITHIOPHOSPHAT (GER-MAN) * 4,6-DIAMINO-1,3,5-TRIAZIN-2-YL-METHYL-O,O-DIMETHYL PHOSPHORODITHIOATE * S-((4,6-DIAMINO-S-TRIAZIN-2-YL)METHYL)-O,O-DIMETHYL PHOSPHORODITHIOATE * S-(4,6-DIAMINO-1,3,5-TRIAZIN-2-YLMETHYL)-O,O-DIMETHYL PHOSPHORODITHIOATE * S-(4,6-DIAMINO-1,3,5-TRIAZIN-2-YLMETHYL) DIMETHYL PHOSPHOROTHIOLOTHIONATE * O,O-DIMETHYL-S-(4,6-DIAMINO-1,3,5-TRI-AZINYL-2-METHYL) DITHIOPHOSPHATE

* O,O-DIMETHYL- S-(4,6-DIAMINO-S-TRI-
AZIN-2-YLMETHYL)PHOSPHORODITHIOATE
* O,O-DIMETHYL-S-(4,6-DIAMINO-1,3,5-
TRIAZIN-2-YL)METHYL PHOSPHORODITHIOATE
* O,O-DIMETHYL-S-(4,6-DIAMINO-1,3,5-TRI-
AZIN-2-YL)METHYL PHOSPHOROTHIOLOTHIONATE
* DITHIOPHOSPHATE DE O,O-DIMETHYLE ET
DE S-(4,6-DIAMINO-1,3,5-TRIAZINE-2-YL)-
METHYLE) (FRENCH) * ENT 25,760

THR: Moderately toxic by ingestion and possibly other routes. When heated to decomposition
it emits very toxic fumes of NO_x, PO_x, and
SO_x.

AZINPHOS METHYL   HR: 3
CAS: 86-50-0   NIOSH: TE 1925000
mf: $C_{10}H_{12}N_3O_3PS_2$ mw: 317.34

PROP: Crystals, slightly water-sol but sol in
organic solvents, brown, waxy solid. d: 1.44,
mp: 74°.

DOT: 2783

SYNS: BENZOTRIAZINEDITHIOPHOSPHORIC ACID
DIMETHOXY ESTER * S-(3,4-DIHYDRO-4-OXO-
BENZO(ALPHA)(1,2,3)TRIAZIN-3-YLMETHYL)-O,
O-DIMETHYL PHOSPHORODITHIOATE * S-(3,4-
DIHYDRO-4-OXO-1,2,3-BENZOTRIAZIN-3-YL-
METHYL)- O,O-DIMETHYL PHOSPHORODITHIOATE
* O,O-DIMETHYL-S-(BENZAZIMINOMETHYL)
DITHIOPHOSPHATE * O,O-DIMETHYL-S-(1,2,3-
BENZOTRIAZINYL-4-KETO)METHYL PHOSPHORO-
DITHIOATE * O,O-DIMETHYL-S-(3,4-DIHYDRO-
4-KETO-1,2,3-BENZOTRIAZINYL-3-METHYL)
DITHIOPHOSPHATE * DIMETHYLDITHIO-
PHOSPHORIC-ACID N-METHYLBENZAZIMIDE
ESTER * O,O-DIMETHYL-S-(4-OXO-3H-1,2,3-
BENZOTRIZIANE-3-METHYL)PHOSPHORODITHIOATE
* O,O-DIMETHYL-S-(4-OXOBENZOTRIAZINO-
3-METHYL)PHOSPHORODITHIOATE * O,O-
DIMETHYL-S-(4-OXO-1,2,3-BENZOTRI-
AZINO(3)-METHYL) THIOTHIONOPHOSPHATE
* O,O-DIMETHYL-S-((4-OXO-3H-1,2,3-BENZO-
TRIAZIN-3-YL)-METHYL)-DITHIOFOSFAAT (DUTCH)
* O,O-DIMETHYL-S-((4-OXO-3H-1,2,3-BENZO-
TRIAZIN-3-YL)-METHYL)-DITHIOPHOSPHAT (GER-
MAN) * O,O-DIMETHYL-S-4-OXO-1,2,3-BENZO-
TRIAZIN-3(4H)-YLMETHYL PHOSPHORODITHIOATE
* O,O-DIMETIL-S-((4-OXO-3H-1,2,3-BENZO-
TRIAZIN-3-IL)-METIL)-DITIOFOSFATO (ITALIAN)
* ENT 23,233 * 3-(MERCAPTOMETHYL)-
1,2,3-BENZOTRIAZIN-4(3H)-ONE-O,O-DIMETHYL
PHOSPHORODITHIOATE-S-ESTER * METHYL GU-
THION * METILTRIAZOTION * NCI-C00066

OSHA PEL: TWA 200 ug/m^3
ACGIH TLV: TWA 0.2 mg/m^3
DFG MAK: 0.2 mg/m^3
DOT Classification: Poison B

THR: Poison by inhalation, ingestion, skin contact, intravenous, intraperitoneal, and possibly
other routes. An experimental carcinogen. See
also parathion and esters. When heated to decomposition it emits very toxic fumes of PO_x,
SO_x, and NO_x. For further information, see Vol.
3, No. 4 (as Guthion) of *DPIM Report*.

AZIRIDINE CARBOXYLIC ACID ETHYL
ESTER   HR: 3
CAS: 671-51-2   NIOSH: CM 6800000
mf: $C_5H_9NO_2$ mw: 115.15

SYNS: N-CARBETHOXYETHYLENIMINE
* N-(ETHOXYCARBONYL)AZIRIDINE * N-ETH-
OXYCARBONYLETHYLENEIMINE * ETHOXY-
CARBONYL-1-ETHYLENIMINE * ETHYL
AZIRIDINECARBOXYLATE * ETHYL-1-AZIRI-
DINECARBOXYLATE * ETHYL AZIRIDINOCAR-
BOXYLATE * ETHYL-1-AZIRIDINYLCARBOXY-
LATE * ETHYL AZIRIDINYLFORMATE

THR: Poison by intravenous and intraperitoneal
routes. Mutagenic data. See also esters. When
heated to decomposition it emits toxic fumes
of NO_x.

1-AZIRIDINE ETHANOL   HR: 3
CAS: 1072-52-2   NIOSH: CM 7000000
mf: C_4H_9NO mw: 87.14

SYNS: 2-(1-AZIRIDINYL)ETHANOL * BETA-HY-
DROXY-1-ETHYLAZIRIDINE * 2-HYDROXY-1-
ETHYLAZIRIDINE * N-(BETA-HYDROXYETHYL)-
AZIRIDINE * N-(2-HYDROXYETHYL)AZIRIDINE
* N-HYDROXYETHYL ETHYLENE IMINE * N-(2-
HYDROXYETHYL)ETHYLENIMINE * 1-(2-HY-
DROXYETHYL)ETHYLENIMINE

THR: Poison by ingestion, skin contact, and
intravenous routes. An experimental carcinogen. A skin and eye irritant. When heated to
decomposition it emits toxic fumes of NO_x.

AZOBENZENE   HR: 3
CAS: 103-33-3   NIOSH: CN 1400000
mf: $C_{12}H_{10}N_2$ mw: 182.23

PROP: Orange monoclinic crystals. mp: 68°;
bp: 297°; d: 1.203 @ 20°/4°; vap press: 1 mm
@ 103.5°. Insol in water. Sol in alc = 4.2/

100 @ 20 °, in ether (ligroin) = 12/100 @ 20°.

SYNS: AZODIBENZENEAZOFUME * AZOBEN-ZEEN (DUTCH) * AZOBENZIDE * AZOBENZOL * AZOBISBENZENE * AZODIBENZENE * BEN-ZENEAZOBENZENE * DIAZOBENZENE * DIPHE-NYLDIAZENE * 1,2-DIPHENYLDIAZENE * DIPHENYLDIIMIDE * ENT 14,611 * NCI-c02926 * USAF EK-704

THR: An experimental carcinogen. Moderately toxic by ingestion and possibly other routes. When heated to decomposition it emits toxic fumes of NO_x. For further information, see Vol. 1, No. 3 of *DPIM Report*.

AZODICARBAMIDE
CAS: 123-77-3   NIOSH: LQ 1040000
mf: $C_2H_4N_4O_2$ mw: 116.10

PROP: Orange-red powder. Decomp @ 180°-200°. Very sltly sol in hot water; insol in alc. Decomp in hot HCl.

SYNS: 1,1'-AZOBISCARBAMIDE * AZOBISCAR-BONAMIDE * AZOBISCARBOXAMIDE * 1,1'-AZOBIS(FORMAMIDE) * AZODICARBOAMIDE * AZODICARBONAMIDE * AZODICARBOXAMIDE * AZODICARBOXYLIC ACID DIAMIDE * DELTA-(1,1')-BIUREA * DIAZENEDICARBOXAMIDE * NCI-C55981

THR: When heated to decomposition it emits toxic fumes of NO_x. Suspected experimental carcinogen.

"AZODRIN"   HR: 3
mf: $C_6H_{14}O_5NP$ mw: 211.2

PROP: Reddish-brown solid, mild ester odor. bp: 125°

SYN: MONOCROTOPHOS

THR: Poison by ingestion and skin contact. See also esters. A dangerous fire hazard. When heated to decomposition it evolves highly toxic fumes of NO_x and PO_x.

AZO ETHANE   HR: 3
CAS: 821-14-7   NIOSH: KH 4025000
mf: $C_4H_{10}N_2$ mw: 86.16

SYN: AZOAETHAN (GERMAN)

THR: An experimental teratogen and carcino-gen, and transplacental carcinogen. Moderate acute toxicity. When heated to decomposition it emits toxic fumes of NO_x. An unstable, dan-gerously explosive material in concentrated state. For further information, see Vol. 1, No. 4 of *DPIM Report*.

1,1'-AZONAPHTHALENE   HR: 3
CAS: 487-10-5   NIOSH: CO 1400000
mf: $C_{20}H_{14}N_2$ mw: 282.35

PROP: Red needles from acetic acid. mp: 190°; bp: subl > 190°; Insol in water, sol in acetic acid, very slightly sol in alcohol, very sol in benzene.

THR: An experimental carcinogen. When heated to decomposition it emits toxic fumes of NO_x.

AZOTOMYCIN   HR: 3
CAS: 7644-67-9   NIOSH: MO 7260000
mf: $C_{17}H_{23}N_7O_8$ mw: 453.47

SYNS: DUAZOMYCIN B * NSC-56654

THR: Poison by unspecified route. When heated to decomposition it emits toxic fumes such as NO_x.

AZOXYBENZENE   HR: 3
CAS: 495-48-7   NIOSH: CO 4025000
mf: $C_{12}H_{10}N_2O$ mw: 198.23

PROP: Yellow, rhombic crystals. d: 1.248 @ 20°/20°; mp: 36°; bp: decomp. Insol in water, sol in alc = 11.4/100 @ 15°, sol in ether (ligroin) = 43.5/100 @ 15°.

SYNS: AZOBENZENE OXIDE * AZOSSIBENZENE (ITALIAN) * AZOXYBENZEEN (DUTCH) * AZOXYBENZIDE * AZOXYBENZOL (GERMAN) * AZOXYDIBENZENE * ORDINARY AZOXYBEN-ZENE

THR: Poison by subcutaneous route. Moderately toxic by ingestion, skin contact, and other routes. A skin and eye irritant. Combustible. When heated to decomposition it emits toxic fumes of NO_x.

AZOXYETHANE   HR: 3
                      NIOSH: HM 2970000
mf: $C_4H_{10}N_2O$ mw: 102.16

SYNS: AZOXYAETHAN (GERMAN) * DIETHYL-DIAZENE-1-OXIDE

THR: Poison by subcutaneous and intravenous routes. An experimental transplacental teratogen and carcinogen. When heated to decomposition it emits toxic fumes of NO_x.

AZOXYMETHANE   **HR: 3**
CAS: 25843-45-2   NIOSH: PA 2975000
mf: $C_2H_6N_2O$ mw: 74.10
SYN: AOM

THR: Poison by subcutaneous route. An experimental carcinogen. Mutagenic data. When heated to decomposition it emits toxic fumes of NO_x.

B

BACILLUS SUBTILIS BPN   **HR: 3**
CAS: 1395-21-7   NIOSH: CO 9450000

PROP: A commercial raw Proteolytic Enzyme used in laundry detergents

SYNS: BACILLOMYCIN (8CI,9CI) * BACILLO-MYCIN R * FUNGOCIN * SUBTILISIN BPN

THR: A poison via intraperitoneal route. Severe eye irritant.

BACILLUS SUBTILIS CARLS-BERG   **HR: 2**
CAS: 9014-01-1   NIOSH: CO 9550000

PROP: A commercial raw Proteolytic Enzyme used in laundry detergents.

SYNS: BACILLOPEPTIDASE A * BACILLOPEPTI-DASE B * SUBTILISIN (9CI) * SUBTILOPEPTI-DASE BPN′

THR: Moderately toxic by ingestion; an eye irritant.

BACITRACIN   **HR: 3**
CAS: 1405-87-4   NIOSH: CP 0175000

PROP: An antibiotic; white to pale buff, hygroscopic powder, odorless or slight odor, freely sol in water, sol in alcohol, methanol and glacial acetic acid, insol in acetone, chloroform and ether.

SYNS: PARENTRACIN * TOPITRACIN * USAF CB-7

THR: A poison by ingestion, intraperitoneal, subcutaneous, and intravenous routes. A food additive permitted in feed and drinking water of animals; also permitted in food for human consumption.

BACTERIODES FRAGILIS ENDO-TOXIN,EXTRACTS   **HR: 3**
NIOSH: CP 0925000

PROP: Phenol-water extracts from *Bacteriodes Fragilis* 62/73 strain.

THR: A poison via intravenous and intradermal routes.

BAGASSE DUST   **HR: 1**
THR: A nuisance dust from the fibrous residue of cane sugar manufacture. Inhalation can cause bronchial asthma, sneezing, rhinorrhea, pneumonitis, etc. See also cotton dust. Fire and explosion hazard when exposed to heat, flame, or oxidizers. See also dust explosions.

BAKELITE   **HR: 3**
CAS: 9002-86-2   NIOSH: KV 0350000
mf: $(C_2H_3Cl)n$

PROP: Polymers with molecular weights ranging from 60,000-150,000. White powder; d: 1.406.

SYNS: ATACTIC POLY(VINYL CHLORIDE) * CHLOROETHENE HOMOPOLYMER * POLYVI-NYLCHLORID (GERMAN) * POLYVINYL CHLO-RIDE * NCI-C60797 * POLY(CHLOROETH-YLENE) * POLY(VINYL CHLORIDE) * VINYL CHLORIDE HOMOPOLYMER * VINYL CHLORIDE POLYMER

THR: A suspected human and experimental carcinogen. An inhalant poison. Can cause allergic dermatitis. Reacts violently with F_2. When heated to decomposition, it emits toxic fumes of Cl^-.

BAL   **HR: 3**
CAS: 59-52-9   NIOSH: UB 2625000
mf: $C_3H_8OS_2$ mw: 124.23

PROP: Viscous, oily liquid, pungent odor. bp: 140° @ 40 mm, vap d: 4.3, d: 1.2385 @ 25°/4°.

SYNS: BRITISH ANTILEWISITE * DIMERCAPTOL PROPANOL * DIMERCAPTOL * 2,3-DIMERCAP-TOPROPANOL * 2,3-DIMERCAPTOPROPAN-1-OL * 2,3-DIMERCAPTOL-1-PROPANOL * 1,2-DI-THIOGLYCEROL * 2,3-DITHIOPROPANOL * USAF ME-1

THR: Poison via intramuscular, intraperitoneal, and intravenous routes. An experimental teratogen. Affects the human central nervous system. It causes redness and swelling when applied locally to the skin, but does not produce blisters or ulcers. Intensely irritating to eyes and mucous membranes. Systemic symptoms are caused by injection. When heated to decomposition, it emits toxic fumes of SO_x.

BARBITAL   **HR: 3**
CAS: 57-44-3   NIOSH: CQ 3500000
mf: $C_8H_{12}N_2O_3$ mw: 184.22

SYNS: DIETHYLBARBITONE * 5,5-DIETHYL-
BARBITURIC ACID * DIETHYLMALONYLUREA
* SEDEVAL * VERONAL

THR: Moderately toxic by intraperitoneal and subcutaneous routes. Ingestion causes psychological effects in humans. Mutagenic data. When heated to decomposition, it emits toxic fumes of NO_x.

BARBITAL SODIUM   HR: 3
CAS: 144-02-5   NIOSH: CQ 3850000
mf: $C_8H_{11}N_2O_3 \cdot Na$ mw: 206.20

PROP: Bitter crystals or powder.

SYNS: BARBITONE SODIUM * NATRIUMBARBI-
TALS (GERMAN) * SODIUM DIETHYLBARBITU-
RATE * SODIUM-5,5-DIETHYLBARBITURATE
* SODIUM ETHYLBARBITAL * SODIUM MALO-
NYLUREA * SODIUM VERONAL * SOLUBLE
BARBITAL * VERONAL SODIUM

THR: Moderately toxic by ingestion. Large doses cause marked depression (sometimes preceded by excitation), prolonged coma, and death. Allergic skin reactions may occur on contact. Implicated in development of aplastic anemia. A truly habit forming drug. An experimental teratogen. Mutagenic data. Combustible. When heated to decomposition, it emits toxic fumes of NO_x.

BARBITURATES
SYNS: DERIVATIVES OF BARBITURIC ACID;
* BARBITAL * BARBITONE * BARBITAL SO-
DIUM

THR: See barbital sodium.

BARIUM   HR: 3
CAS: 7440-39-3   NIOSH: CA 8370000
DOT: 1399/1400/1854
af: Ba   at wt: 137.36

PROP: Silver-white, slightly lustrous, somewhat malleable metal. mp: 725°, bp: 1640°, d: 3.5 @ 20°, vap press: 10 mm @ 1049°.
ACGIH TLV: TWA 0.5 mg/m^3
DFG MAK: 0.5 mg/m^3
DOT Classification: Some are flammable or explosive

THR: See also barium compounds. Dust is dangerous and explosive when exposed to heat, flame, or chemical reaction. Incompatible with acids; CCl_4; $C_2Cl_3F_3$; $C_2H_2FCl_3$; C_2Cl_4; C_2HCl_3;

water; 1,1,2-trichloro trifluoro ethane; fluorotrichloroethane; fluorotrichloromethane; trichloroethylene (can detonate on contact). For further information, see Vol. 3, No. 4 of *DPIM Report*.

BARIUM ACETATE   HR: 3
CAS: 543-80-6   NIOSH: AF 4550000
mf: $C_4H_6O_4 \cdot Ba$ mw: 255.44

PROP: White cryst. Water sol.

SYNS: ACETIC ACID, BARIUM SALT * BARIUM
DIACETATE * OCTAN BARNATY (CZECH)

OSHA PEL: TWA 500 ppm

THR: Poison via intravenous and subcutaneous routes. When heated to decomposition, it emits acrid smoke. See also barium compounds.

BARIUM AZIDE   HR: 3
CAS: 18810-58-7   NIOSH: CQ 8500000
mf: BaN_6 mw: 221.40

PROP: Monoclinic prisms. mp: $-N_2$ @ about 120°, bp: explodes, d: 2.936.

THR: See also barium compounds (soluble) and azides. Moderate explosion hazard when shocked or exposed to around 275° heat. Spontaneously flammable in air. Very unstable.

BARIUM AZIDE (WET)   HR: 3
CAS: 18810-58-7   NIOSH: CQ 8510000
DOT: 1571

PROP: Compound contains 50% or more water.
DOT Classification: Label: Flammable Solid

THR: A poison, flammable and possibly explosive. See also barium compounds and azides.

BARIUM BENZOATE   HR: 3
mf: $Ba(C_7H_5O_2)_2 \cdot 2H_2O$ mw: 415.61

PROP: White nacreous leaflets. mp: $-2H_2O$ @ 100°.

THR: Deadly poison! See barium compounds (soluble).

BARIUM BROMATE   HR: 3
mf: $Ba(BrO_3)_2 \cdot H_2O$ mw: 411.21

PROP: White crystals or crystalline powder. mp: decomp @ 260°; d: 3.99 @ 18°.

THR: A violent poison. See also barium compounds (soluble) and bromine. Fire hazard by chemical reaction with easily oxidized materials. At 300°, it decomposes with violence. Incompatible with Al; As; C; Cu; metal sulfides; organic matter; P; S; reducing materials. When heated to decomposition, it emits toxic bromides.

BARIUM CARBIDE
mf: BaC_2 mw: 161.4

PROP: Gray crystals. d: 3.75.

THR: See also barium compounds (soluble). A fire and explosion hazard by chemical reaction with moisture to form acetylene. Incompatible with Se; S; H_2O. To fight fire, use CO_2, dry chemical.

BARIUM CARBONATE (1:1)   HR: 3
CAS: 513-77-9   NIOSH: CQ 8600000
mf: $CO_3 \cdot Ba$ mw: 197.35

PROP: White powder. mp: 1740 @ 90 atm; bp: decomp; d: 4.43.

SYNS: CARBONIC ACID, BARIUM SALT (1:1) * C.I. 77099

OSHA PEL: TWA 500 ug(Ba)/m^3

THR: Poison to humans by ingestion. See also barium compounds (soluble). Incompatible with BrF_3; 2-furanpercarboxylic acid.

BARIUM CHLORATE   HR: 3
CAS: 13477-00-4   NIOSH: FN 9770000
DOT: 1445
mf: $Cl_2O_6 \cdot Ba$ mw: 304.24

PROP: Colorless prisms or white powder. mp: $-H_2O$ @ 414°, d: 3.18.

SYN: CHLORIC ACID, BARIUM SALT

DOT Classification: Label: Oxidizer

THR: A poison. See also barium compounds (soluble) and chlorates. For fire and explosion hazards, see chlorates. Incompatible with Al; As; C; charcoal; Cu; MnO_2; metal sulfides; S_4N_4; organic matter; P; S.

BARIUM CHLORATE, WET (DOT)   HR: 3
CAS: 13477-00-4   NIOSH: FN 9775000
DOT: 1445

SYN: CHLORIC ACID, BARIUM SALT (WET)

DOT Classification: Label: Oxidizer

THR: A poison! See also barium compounds (soluble) and chlorates. A powerful oxidizer.

When heated to decomposition, it emits toxic fumes of Cl$^-$.

BARIUM CHLORIDE   HR: 3
CAS: 10361-37-2   NIOSH: CQ 8750000
mf: $BaCl_2$ mw: 208.24

PROP: Colorless flat crystals. mp: transition @ 925° to cubic crystals, bp: 1560°, d: 3.856 @ 24°.

SYNS: SBa 0108E * BARIUM DICHLORIDE * NCI-C61074

OSHA PEL: TWA 500 ug(Ba)/m^3

THR: Poison to humans by ingestion, subcutaneous, intravenous, and intraperitoneal routes. See also barium compounds (soluble). When heated to decomposition, it emits toxic fumes of Cl$^-$.

BARIUM CHROMATE (VI)   HR: 3
CAS: 10294-40-3   NIOSH: CQ 8760000
mf: $Ba \cdot CrH_2O_4$ mw: 255.36

PROP: Heavy yellow crystalline powder. d: 4.498 @ 15°.

SYNS: BARIUM CHROMATE (1:1) * BARIUM CHROMATE OXIDE * BARYTA YELLOW * C.I. 77103 * LEMON YELLOW * STEINBUHL YELLOW

THR: A poison. See also barium compounds (soluble) and chromium compounds. For fire hazard, see chromates. It reacts vigorously with reducing materials.

BARIUM COMPOUNDS (SOLUBLE)   HR: 3
THR: The soluble barium salts, such as the chloride and sulfide, are poisonous when ingested. The insoluble sulfate used in radiography is not acutely toxic. See also barium sulfate. Few cases of industrial systemic poisoning have been reported, but one investigator describes a fatal case of poisoning attributed to barium oxide, the symptoms being severe abdominal pain with vomiting, dyspnoea, rapid pulse, paralysis of the arm and leg, and eventually cyanosis and death. The same investigator produced paralysis in animals with barium oxide and carbonate. The usual result of exposure to the sulfide, oxide, and carbonate is irritation of the eyes, nose, and throat, and of the skin, producing dermatitis. The salts mentioned are somewhat caustic.

BARIUM CYANIDE   **HR: 3**
CAS: 542-62-1   NIOSH: CQ 8785000
DOT: 1565
mf: C_2BaN_2 mw: 189.38

PROP: White crystalline powder.

SYN: BARIUM DICYANIDE

DOT Classification: Label: Poison

THR: A deadly poison. See also cyanide and barium compounds (soluble). When heated to decomposition, it emits toxic fumes of CN^-. For further information, see Vol. 3, No. 4 of *DPIM Reports.*

BARIUM CYANOPLATINITE   **HR: 3**
mf: $BaPt(CN)_4 \cdot 4H_2O$ mw: 508.6

PROP: (a) monoclinic yellow crystals; (b) rhombic crystals; mp: $-2H_2O$ @ 100°; d: (a) 2.076, (b) 2.085.

THR: A poison. See also barium compounds (soluble), cyanide, and platinum compounds. When heated to decomposition, it emits highly toxic fumes of CN^- and NO_x.

BARIUM DICHROMATE   **HR: 3**
mf: $BaCr_2O_7$ mw: 353.38

PROP: Brownish-red crystalline masses.

SYN: BARIUM BICHROMATE

THR: A poison. See also barium compounds (soluble) and chromium compounds. Some chromates are carcinogenic. A moderate fire hazard by chemical reaction with easily oxidized materials. A powerful oxidizer. Incompatible with reducing materials.

BARIUM FLUOBORATE   **HR: 3**
CAS: 13862-62-9   NIOSH: CQ 8925000
mf: $B_2F_8 \cdot Ba$ mw: 310.96

SYNS: BARIUM BIS(TETRAFLUOROBORATE)
* BARIUM TETRAFLUOROBORATE.

THR: Poison by ingestion. See also barium compounds, boron compounds, and fluorides. When heated to decomposition, it emits toxic fumes of F^-.

BARIUM FLUORIDE   **HR: 3**
CAS: 7787-32-8   NIOSH: CQ 9100000
mf: BaF_2 mw: 175.34

PROP: White powder; mp: 1280°, bp: 2137°, d: 4.89.

SYN: BARYUM FLUORURE (FRENCH)

OSHA PEL: TWA 500 ug(Ba)/m^3

THR: A poison by ingestion and intraperitoneal routes. See also fluorides and barium compounds (soluble). When heated to decomposition, it emits toxic fumes of F^-.

BARIUM HYDRIDE   **HR: 3**
mf: BaH_2 mw: 139.38

PROP: Gray crystals or lumps; mp: decomp @ 675°; bp: 1400°; d: 4.21 @ 0°.

THR: A poison. Rapidly decomposed by water and acids. See also barium compounds (soluble) and hydrides. In powder form, it ignites spontaneously in air and reacts vigorously with water. It is incompatible with water; acids; and metal halogenates. A dangerous fire hazard because moisture may cause it to ignite. To fight fire, use dry chemical, graphite, CO_2.

BARIUM HYPOPHOSPHITE   **HR: 3**
mf: $Ba(H_2PO_2)_2 \cdot H_2O$ mw: 285.38

PROP: Crystalline powder. mp: decomp; d: 2.90 @ 17°.

THR: A poison. See also barium compounds (soluble) and hypophosphites. When heated to decomposition, it emits highly toxic fumes of PO_x. Incompatible with $KClO_3$.

BARIUM IODATE   **HR: 3**
mf: $Ba(IO_3)_2$ mw: 487.20

PROP: White crystalline powder. mp: decomp; d: 4.998.

THR: A poison. See also barium compounds (soluble) and iodates. A powerful oxidizer. When heated to decomposition, it emits toxic fumes of I^-. Incompatible with Al; As; C; Cu; metal sulfides; organic matter.

BARIUM(II) NITRATE (1:2)   **HR: 3**
CAS: 10022-31-8   NIOSH: CQ 9625000
DOT: 1446
mf: $N_2O_6 \cdot Ba$ mw: 261.36

PROP: Lustrous crystals; mp: 592°, bp: decomp, d: 3.24 @ 23°.

SYNS: BARIUM DINITRATE * DUSICNAN BARNATY (CZECH) * NITRATE DE BARYUM (FRENCH)

OSHA PEL: TWA 500 ug(Ba)/m^3
DOT Classification: Label: Oxidizer

THR: A poison via intravenous and oral routes. An irritant to skin and eyes. See also barium compounds (soluble) and nitrates. When heated to decomposition, it emits very toxic fumes of NO_x. Incompatible with $(Mg+BaO_2+Zn)$; Al; and Mg alloys.

BARIUM NITRIDE   **HR: 3**
mf: Ba_3N_2 mw: 440.10

PROP: Colorless crystals. bp: 1000° (vacuo); d: 4.783 @ 25°/4°.

THR: A poison. See also barium compounds (soluble) and ammonia. A moderate fire hazard by spontaneous chemical reaction with water to liberate flammable vapor; see also ammonia. Dangerous; explodes upon heating and by spontaneous chemical reaction to liberate vapor NH_3 which can form explosive mixtures with air. See also ammonia. Incompatible with air and moisture.

BARIUM OXIDE   **HR: 3**
CAS: 1304-28-5   NIOSH: CQ 9800000
DOT: 1884
mf: BaO   mw: 153.34

PROP: White to yellowish-white powder. mp: 1923°, bp: 2000° (approx), d: 5.72.

SYNS: BARIUM MONOXIDE * BARIUM PROTOXIDE * BARYTA * CALCINED BARYTA * OXYDE DE BARYUM (FRENCH)

OSHA PEL: TWA 500 ug(Ba)/m^3
DOT Classification: ORM-B, Label: None

THR: A poison via subcutaneous route. See also barium compounds (soluble). Slight fire hazard by spontaneous chemical reaction; produces heat on contact with water or steam. Incompatible with H_2S; hydroxylamine; N_2O_4; triuranium octaoxide; SO_3.

BARIUM PEROXIDE   **HR: 3**
CAS: 1304-29-6   NIOSH: CR 0175000
DOT: 1449
mf: BaO_2 mw: 169.34

PROP: Grayish-white powder. mp: 450°, bp: loses O @ 800°, d: 4.96.

SYNS: BARIO (PEROSSIDO DI) (ITALIAN) * BARIUM BINOXIDE * BARIUM DIOXIDE * BARIUMPEROXID (GERMAN) * BARIUMPEROXYDE (DUTCH) * BARIUM SUPEROXIDE * DIOXYDE DE BARYUM (FRENCH)

DOT Classification: Label: Oxidizer

THR: A poison via subcutaneous route. See also barium compounds (soluble) and peroxides, inorganic. Powerful oxidizer. Incompatible with H_2S; water; peroxy formic acid; hydroxylamine solution; mixture of $(Mg+Zn+Ba(NO_3)_2)$; organic matter.

BARIUM SILICOFLUORIDE   **HR: 3**
CAS: 17125-80-3   NIOSH: CR 0525000
mf: $F_6Si \cdot Ba$ mw: 279.43

PROP: White crystalline powder. d: 4.29 @ 21°/4°, mp: 300° (decomp).

SYNS: BARIUM FLUOROSILICATE * BARIUM HEXAFLUOROSILICATE(2-) * BARIUM SILICON FLUORIDE * BARIUM FLUOSILICATE * BARIUM HEXAFLUOROSILICATE

THR: A poison by ingestion. See also barium compounds (soluble) and fluosilicates.

BARIUM SULFATE   **HR: 3**
CAS: 7727-43-7   NIOSH: CR 0600000
mf: $O_4S \cdot Ba$ mw: 233.40

PROP: White, heavy, odorless powder, not sol in water or dilute acids. d: 4.50 @ 15°, mp: 1580°.

SYNS: BAYRITES * ARTIFICIAL BARITE * ARTIFICIAL HEAVY SPAR * BARYTES * BLANC FIXE * C.I. 77120 * ENAMEL WHITE * PRECIPITATED BARIUM SULPHATE * SULFURIC ACID, BARIUM SALT (1:1)

THR: An experimental carcinogen via intrapleural route. A relatively insoluble salt used as an opaque medium in radiography. Soluble impurities can lead to toxic reactions. See also barium compounds (soluble). Heating with aluminum can produce an explosion. When heated to decomposition, it emits toxic fumes of SO_x. Incompatible with Al; P. For further information see Vol. 1, No. 1 of *DPIM Report*.

BARIUM SULFIDE   **HR: 3**
mf: BaS   mw: 169.4

PROP: Cubic colorless crystals. d: 4.25 @ 15°, mp: 1200°.

THR: A poison. See also barium compounds (soluble) and sulfides. A moderate fire hazard by spontaneous chemical reaction; air, moisture, or acid fumes may cause it to ignite. For explosion and disaster hazards, see sulfides. To fight fire, use CO_2, dry chemical. Incompatible with Cl_2O; PbO_2; $KClO_3$; KNO_3; $Ca(NO_3)_2$; $SrNO_3$; $CaClO_3$; $SrClO_3$; ClO.

BARIUM ZIRCONIUM(IV) OXIDE   HR: 2
CAS: 12009-21-1   NIOSH: CR 0875000
mf: $O_4Zr_4 \cdot Ba$ mw: 566.22

PROP: Light gray-buff powder, insol in water and alkalies, sltly sol in acid. d: 5.52, mp: 2510°.

SYNS: BARIUM ZIRCONIUM TRIOXIDE * BARIUM ZIRCONATE

THR: Moderately toxic by ingestion and intraperitoneal routes. Inhalation produces interstitial pneumonitis. See also zirconium compounds and barium compounds.

BASIC PARAFUCHSINE   HR: 3
CAS: 569-61-9   NIOSH: CX 9850000
mf: $C_{19}H_{17}N_3 \cdot ClH$ mw: 323.85

SYNS: C.I. BASIC RED 9, MONOHYDROCHLORIDE * 4,4′-((4-IMINO-2,5-CYCLOHEXADIEN-1-YLIDENE)METHYLENE) DIANILINE MONOHYDROCHLORIDE * NCI-C54739 * PARAFUCHSIN (GERMAN) * PARAFUCHSINE * PARA-MAGENTA * PARAROSANILINE HYDROCHLORIDE * 4,4′4″-TRIAMINOTRIPHENYLMETHAN-HYDROCHLORID (GERMAN)

THR: An experimental carcinogen. Mutagenic data. When heated to decomposition it emits very toxic fumes of HCl and NO_x.

BASIC ZINC CHROMATE   HR: 3
CAS: 13530-65-9   NIOSH: GB 3290000
mf: $CrO_4Zn \cdot Zn(OH)_2$ mw: 280.76

SYNS: CHROMIC ACID, ZINC SALT * BUTTERCUP YELLOW * PRIMROSE YELLOW * PURE ZINC CHROME * ZINC CHROMATE * ZINC TETRAOXYCHROMATE 76A * ZINC YELLOW

THR: A poison via intravenous route. A human and and experimental carcinogen. Mutagenic data. See also chromium compounds.

BAYGON   HR: 3
CAS: 114-26-1   NIOSH: FC 3150000
mf: $C_{11}H_{15}NO_3$ mw: 209.27

PROP: A white to tan crystalline solid, slightly sol in water, sol in all polar organic solvents.

SYNS: BAYER 39007 * ENT 25,671 * O-ISOPROPOXYPHENYL METHYLCARBAMATE * O-ISOPROPOXYPHENYL-N-METHYLCARBAMATE * 2-ISOPROPOXYPHENYL-N-METHYLCARBAMATE * 2-(1-METHYLETHOXY)PHENOL METHYLCARBAMATE * N-METHYL-2-ISOPROPOXYPHENYLCARBAMATE

THR: A poison via oral, intraperitoneal, intravenous, and intramuscular routes. Mutagenic data. Moderately irritating to skin. See also carbamates. When heated to decomposition, it emits toxic fumes of NO_x.

BAY OIL   HR: 2
NIOSH: CT 0350000

PROP: Consists mainly of eugenol and chavicol (55-65%), major portion of balance consists of terpenes (alpha pinene, myrcene and dipentene) small quantities of citrol, nerol, cineol and other terpenoids have also been found.

SYNS: BAY LEAF OIL * BOIS D'INDE * LAUREL LEAF OIL * MYRCIA OIL * OIL OF MYRCIA * MYRICIA OIL * OIL OF BAY

THR: Moderately toxic by ingestion. When heated to decomposition, it emits acrid smoke.

BAYTHION   HR: 3
CAS: 14816-18-3   NIOSH: MD 4740000
mf: $C_{12}H_{15}N_2O_3PS$ mw: 298.32

SYNS: BENZOYL CYANIDE-O-(DIETHOXYPHOSPHINOTHIOYL)OXIME * O,O-DIAETHYL-O-(ALPHA-CYANO-BENZYLIDENAMINO)-MONOTHIOPHOSPHAT (GERMAN) * ALPHA-(((DIETHOXYPHOSPHINOTHIOYL)OXY)IMINO)BENZENEACETONITRILE * (DIETHOXY-THIOPHOSPHORYLOXYIMINO)-PHENYL ACETONITRILE * PHENYLGLYOXYLONITRILE OXIME-O,O-DIETHYL PHOSPHOROTHIOATE

THR: Poison by ingestion. When heated to decomposition, it emits very toxic fumes of NO_x, PO_x, and SO_x.

BELLADONNA   HR: 3
SYN: DEADLY NIGHTSHADE, FROM WHICH THE ALKALOIDS, ATROPINE AND BELLADONNINE ARE DERIVED.

THR: A deadly poison. See also hyoscyamine and atropine. Local contact may cause a contact dermatitis. A poisonous constituent of some berries and plants, and of some folk remedies.

BENADRYL HYDROCHLORIDE   HR: 3
CAS: 147-24-0   NIOSH: KR 7000000
mf: $C_{17}H_{21}NO \cdot ClH$ mw: 291.85

SYNS: BENZHYDRAMINE HYDROCHLORIDE
* 2-(BENZHYDRYLOXY)-N,N-DIMETHYLETHYL-
AMINEHYDROCHLORIDE * DIMETHYLAMINE
BENZHYDRYL ESTER HYDROCHLORIDE * BETA-
DIMETHYLAMINOETHYL BENZHYDRYL ETHER
HYDROCHLORIDE * DIPHENYLHYDRAMINE HY-
DROCHLORIDE * 2-DIPHENYLMETHOXY-N,N-
DIMETHYLETHYLAMINE HYDROCHLORIDE
* ALPHA-HYDROXYDIPHENYLMETHANE-BETA-
DIMETHYLAMINOETHYL ETHER HYDROCHLORIDE
* NCI-C56075

THR: Poison by ingestion, subcutaneous, intravenous, and intraperitoneal routes. A suspected carcinogen. See also esters, ethers. When
heated to decomposition, it emits very toxic
fumes of NO_x and HCl.

BENZONITE HR: 3
CAS: 1302-78-9 NIOSH: CT 9450000

PROP: A clay containing appreciable amounts
of the clay mineral montmorillonite; light yellow
or green, cream, pink, gray to black, plastic,
insol in water and common organic solvents.

SYNS: ALBAGEL PREMIUM USP 4444 * MAG-
BOND * MONTMORILLONITE * PANTHER
CREEK BENTONITE * SOUTHERN BENTONITE

THR: Poison by intravenous route. A general
purpose food additive.

BENZ(e)ACEPHENANTHRYLENE HR: 3
CAS: 205-99-2 NIOSH: CU 1400000
mf: $C_{20}H_{12}$ mw: 252.32

PROP: mp: 168°.

SYNS: 3,4-BENZ(E)ACEPHENANTHRYLENE
* 2,3-BENZFLUORANTHENE * 3,4-BENZFLU-
ORANTHENE * BENZO(B)FLUORANTHENE
* 2,3-BENZOFLUORANTHENE

THR: An experimental carcinogen. Mutagenic
data. When heated to decomposition, it emits
acrid smoke and irritating fumes.

BENZACINE HYDROCHLORIDE HR: 3
CAS: 71-79-4 NIOSH: DD 3150000
mf: $C_{18}H_{21}O_3ClH$ mw: 403.28

SYNS: DIMETHYLAMINOETHYLDIPHENYLHY-
DROXYACETATEHYDROCHLORIDE * DIMETHYL-
N-ETHYL BENZILATE, HYDROCHLORIDE
* BETA-DIMETHYLAMINOETHYL BENZILATE
HYDROCHLORIDE * 2-(DIMETHYLAMINO)ETHYL
BENZILATE HYDROCHLORIDE * DIMETHYL-
AMINOETHYL BENZYLATE HYDROCHLORIDE

THR: Poison by ingestion, intravenous, and intraperitoneal routes. When heated to decomposition it emits toxic fumes of HCl.

BENZ(c)ACRIDINE
CAS: 225-51-4 NIOSH: CU 2975000
mf: $C_{17}H_{11}N$ mw: 229.29

PROP: mp: 108°.

SYNS: 12-AZABENZ(A)ANTHRACENE * 3,4-
BENZACRIDINE * 7,8-BENZACRIDINE (FRENCH)
* 3,4-BENZOACRIDINE * ALPHA-CHRYSIDINE
* ALPHA-NAPHTHACRIDINE

THR: An experimental carcinogen. Mutagenic
data. When heated to decomposition, it emits
toxic fumes of NO_x.

BENZ(c)ACRIDINE-7-CARBOXALDE-
HYDE HR: 3
CAS: 3301-75-5 NIOSH: CU 3100000
mf: $C_{18}H_{11}NO$ mw: 257.30

SYNS: 3,4-BENZACRIDINE-9-ALDEHYDE
* 7-FORMYLBENZ(C)ACRIDINE * 7-FORMYL-
BENZO(C)ACRIDINE

THR: An experimental carcinogen. When
heated to decomposition, it emits toxic fumes
of NO_x.

BENZALDEHYDE HR: 3
CAS: 100-52-7 NIOSH: CU 4375000
DOT: 1989
mf: C_7H_6O mw: 106.13

PROP: Refractive liquid. mp: −26°, bp: 179°,
flash p: 148°F, d: 1.050 @ 15°/4 °, autoign
temp: 377°F, vap press: 1 mm @ 26.2°, vap
d: 3.65.

SYNS: ALMOND ARTIFICIAL ESSENTIAL OIL
* ARTIFICIAL ALMOND OIL * BENZENECAR-
BONAL * BENZOIC ALDEHYDE * NCI-C56133

DOT Classification: Combustible Liquid,
Label: None

THR: Poison by ingestion and intraperitoneal
routes. An allergen. Acts as a feeble local anesthetic. See also aldehydes. Local contact may
cause contact dermatitis. Causes central nervous
system depression in small doses and convulsions in larger doses. A skin irritant. Mutagenic
data. Moderate fire hazard when exposed to
heat or flame. To fight fire, use water (may
be used as a blanket), alcohol, foam, dry chemical. Incompatible with oxidizers such as peroxy-

formic acid. For further information, see Vol. 1, No. 8 of *DPIM Report*.

BENZALDEHYDE GREEN   HR: 3
CAS: 633-03-4   NIOSH: BP 6825000
mf: $C_{27}H_{33}N_2 \cdot HO_4S$ mw: 482.69

PROP: Bright green cryst.

SYNS: ANILINE GREEN * C.I. 42040
* C.I. BASIC GREEN 1, SULFATE (1:1)

THR: Poison by ingestion, intraperitoneal, and intravenous routes. A skin irritant. See also sulfates. When heated to decomposition, it emits very toxic fumes of NO_x and SO_x.

BENZALIN   HR: 3
CAS: 146-22-5   NIOSH: DF 2450000
mf: $C_{15}H_{11}N_3O_3$ mw: 281.29

SYNS: 1,3-DIHYDRO-7-NITRO-5-PHENYL-2H-1,4-BENZODIAZEPIN-2-ONE * EPIBENZALIN
* NITRAZEPAM * 7-NITRO-5-PHENYL-2,3-DIHYDRO-1H-1,4-BENZODIAZEPIN-2-ONE

THR: Poison by intraperitoneal and intravenous routes. Moderately toxic by ingestion. Mutagenic data. When heated to decomposition it emits toxic fumes of NO_x.

BENZ(a)ANTHRACENE   HR: 3
CAS: 56-55-3   NIOSH: CV 9275000
mf: $C_{18}H_{12}$ mw: 228.30

PROP: Colorless leaflets or plates; bp: 400°, mp: 160°.

SYNS: BENZANTHRACENE * 1,2-BENZ(A)AN-THRACENE * 1,2-BENZANTHRENE * BEN-ZOANTHRACENE * 1,2-BENZOANTHRACENE
* BA * 1,2-BENZANTHRACENE * 1,2-BENZ-ANTHRAZEN (GERMAN) * BENZANTHRENE
* BENZO(A)ANTHRACENE * BENZO(A)PHENAN-THRENE * BENZO(B)PHENANTHRENE * 2,3-BENZOPHENANTHRENE * 2,3-BENZPHENAN-THRENE * NAPHTHANTHRACENE
* TETRAPHENE

THR: An experimental carcinogen. Mutagenic data. It is found in oils, waxes, smoke, food, drugs. When heated to decomposition, it emits acrid smoke and irritating fumes.

BENZ(a)ANTHRACENE-1,2-DIHYDRO-DIOL   HR: 3
CAS: 60967-88-6   NIOSH: CW 1780000
mf: $C_{18}H_{14}O_2$ mw: 262.32

SYNS: BA-1,2-DIHYDRODIOL * TRANS-1,2-DIHYDROXY-1,2-DIHYDROBENZ(A)ANTHRACENE

THR: An experimental carcinogen. Mutagenic data. When heated to decomposition, it emits acrid smoke and irritating fumes.

BENZ(a)ANTHRACENE-3,4-DIHYDRO-DIOL   HR: 3
CAS: 60967-89-7   NIOSH: CW 1783000
mf: $C_{18}H_{14}O_2$ mw: 262.32

SYNS: TRANS-3,4-DIHYDRO-3,4-DIHYDROXY-BENZO(A)ANTHRACENE * TRANS-3,4-DIHY-DROXY-3,4-DIHYDROBENZ(A)ANTHRACENE

THR: An experimental carcinogen. Mutagenic data. When heated to decomposition, it emits acrid smoke and irritating fumes.

BENZ(a)ANTHRACENE-5,6-DIHYDRO-DIOL   HR: 3
CAS: 3719-37-7   NIOSH: CW 1786000
mf: $C_{18}H_{14}O_2$ mw: 262.32

SYNS: BA-5,6-DIHYDRODIOL * TRANS-5,6-DIHYDROXY-5,6-DIHYDROBENZ(A)ANTHRACENE

THR: An experimental carcinogen. Mutagenic data. When heated to decomposition, it emits acrid smoke and irritating fumes.

trans-BENZ(a)ANTHRACENE-8,9-DIHY-DRODIOL   HR: 3
CAS: 34501-24-1   NIOSH: CW 1790000
mf: $C_{18}H_{14}O_2$ mw: 262.32

SYNS: BA-8,9-DIHYDRODIOL * TRANS-8,9-DIHYDROXY-8,9-DIHYDROBENZ(A)ANTHRACENE

THR: An experimental carcinogen. Mutagenic data. When heated to decomposition, it emits acrid smoke and irritating fumes.

BENZ(a)ANTHRACENE-10,11-DIHYDRO-DIOL   HR: 3
CAS: 60967-90-0   NIOSH: CW 1795000
mf: $C_{18}H_{14}O_2$ mw: 262.32

SYNS: BA-10,11-DIHYDRODIOL * TRANS-10,11-DIHYDROXY-10,11-DIHYDROBENZ(A)AN-THRACENE

THR: An experimental carcinogen. When heated to decomposition, it emits acrid smoke and irritating fumes.

BENZ(a)ANTHRACENE-7,12-DIMETHA-NOLDIACETATE   HR: 3
CAS: 63018-62-2   NIOSH: CW 2100000
mf: $C_{24}H_{20}O_4$ mw: 372.44

SYNS: ACETIC ACID, BENZ(A)ANTHRACENE-7,12-DIMETHANOL DIESTER * 9,10-BISACE-TOXYMETHYL-1,2-BENZANTHRACENE

THR: An experimental carcinogen. See also esters. When heated to decomposition, it emits acrid smoke and irritating fumes.

BENZ(a)ANTHRACENE-7-METH-ANOL   HR: 3
CAS: 16110-13-7   NIOSH: CW 8400000
mf: $C_{19}H_{14}O$ mw: 258.33

SYN: 10-HYDROXYMETHYL-1,2-BENZANTHRA-CENE

THR: An experimental carcinogen. When heated to decomposition, it emits acrid smoke and irritating fumes.

BENZ(a)ANTHRACENE-7-METHANOL ACETATE   HR: 3
CAS: 17526-24-8   NIOSH: CW 8575000
mf: $C_{21}H_{16}O_2$ mw: 300.37

SYNS: ACETIC ACID, BENZ(A)ANTHRACENE-7-METHANOL ESTER * 10-ACETOXYMETHYL-1,2-BENZANTHRACENE

THR: An experimental carcinogen. See also esters. When heated to decomposition, it emits acrid smoke and irritating fumes.

BENZAZOLINE HYDROCHLO-RIDE   HR: 3
CAS: 59-97-2   NIOSH: NJ 2350000
mf: $C_{10}H_{12}N_2 \cdot ClH$ mw: 196.70

SYNS: 2-BENZYL-2-IMIDAZOLINE MONOHYDRO-CHLORIDE * BENZYLIMIDAZOLINE HYDROCHLO-RIDE

THR: Poison by intravenous route. Human cardiovascular effects. When heated to decomposition, it emits very toxic fumes of NO_x and HCl.

BENZEDRINE   HR: 3
CAS: 300-62-9   NIOSH: SH 9450000
mf: $C_9H_{13}N$ mw: 135.23

PROP: Liquid; bp: 200°, flash p: < 212°F (OC), d: 0.931, vap d: 4.65.

SYNS: DL-AMPHETAMINE * (+-)-BENZEDRINE * DL-BENZEDRINE * (+-)-DESOXYNOREPHED-RINE * (+-)-ALPHA-METHYLPHENETHYLAMINE * DL-1-PHENYL-2-AMINOPROPANE * PRO-FAMINA * PSYCHEDRINE * SYMPATEDRINE

THR: Poison by ingestion, subcutaneous, intraperitoneal, and intravenous routes. A stimulant. Overdoses cause hyperactivity, restlessness, insomnia, rapid pulse, rise in blood pressure, dilated pupils, dryness of the throat. A slight fire hazard when exposed to heat flame or oxidizers. When heated to decomposition, it emits NO_x. To fight fire, use CO_2, dry chemical, alcohol foam, water mist, fog.

BENZEDRINE SULFATE   HR: 3
CAS: 156-31-0   NIOSH: SI 1225000
mf: $C_{18}H_{26}N_2 \cdot H_2O_4S$ mw: 368.54

SYNS: PHENETHYLAMINE, ALPHAMETHYL-, SUL-FATE (2:1) * DIAMPHETAMINE SULFATE * DL-ALPHA-METHYLPHENETHYLAMINE SULFATE * 1-PHENYL-2-AMINOPROPANE SULFATE

THR: A poison via intraperitoneal and subcutaneous routes. See also sulfates. When heated to decomposition, it emits very toxic fumes of SO_x and NO_x. See also benzedrine.

d-BENZEDRINE SULFATE   HR: 3
CAS: 51-63-8   NIOSH: SI 1400000
mf: $C_{18}H_{26}N_2 \cdot H_2O_4S$ mw: 368.54

SYNS: AMPHEDRINE * AMPHEREX * (+)-AMPHETAMINE SULFATE * D-AMPHETAMINE SULFATE * DEXAMPHETAMINE SULFATE * DEXAMYL * DEXEDRINA * DEXEDRINE SULFATE * DEXIES * D-ALPHA-METHYLPHENE-THYLAMINE SULFATE * OBESEDRIN * FAST-BALLS * HEARTS * DEXTROAMPHETAMINE SULFATE * DEXTRO-ALPHA-METHYLPHENETHY-LAMINE SULFATE * ORANGES * PHENEDRINE * PHENOPROMIN * D-1-PHENYL-2-AMINOPRO-PANE SULFATE * DEXTRO-1-PHENYL-2-AMINO-PROPANE SULFATE * D-BETA-PHENYLISOPROPY-LAMINE SULFATE * DEXTRO-BETA-PHENYLISO-PROPYLAMINE SULFATE

THR: Poison by ingestion, intraperitoneal, subcutaneous, and intravenous routes. An experimental teratogen. A habit-forming stimulant. See also sulfates. When heated to decomposition, it emits very toxic fumes of SO_x and NO_x.

l-BENZEDRINE SULFATE   HR: 3
CAS: 51-62-7   NIOSH: SI 1575000
mf: $C_{18}H_{26}N_2 \cdot H_2O_4S$ mw: 368.54

SYNS: (-)-AMPHETAMINE SULFATE * L-AM-
PHETAMINE SULFATE * LEVEDRINE * L-1-
PHENYL-2-AMINOPROPANE SULFATE

THR: A poison via subcutaneous and intraperitoneal routes. See also sulfates. When heated to
decomposition, it emits very toxic fumes of SO_x
and NO_x.

BENZENAMINE HYDROCHLO-
RIDE   HR: 3
CAS: 142-04-1   NIOSH: CY 0875000
mf: $C_6H_7N \cdot ClH$ mw: 129.60

PROP: Crystals; vap d: 4.46, d: 1.22, mp: 198°,
bp: 245°, flash p: 380°F (OC).

SYNS: ANILINE HYDROCHLORIDE * "ANILINE
SALT" * CHLORHYDRATE D'ANILINE (FRENCH)
* CHLORID ANILINU (CZECH) * NCI-CO3736
* USAF EK-442

THR: Poison by intraperitoneal route. An experimental carcinogen. Moderate skin irritant,
severe eye irritant. See also aniline. Slight fire
hazard when exposed to heat or flame. When
heated to decomposition or on contact with acid
or acid fumes, it emits highly toxic fumes of
aniline and chlorine compounds; can react vigorously with oxidizing materials. To fight fire,
use water, CO_2, water mist or spray, dry chemical.

BENZENE   HR: 3
CAS: 71-43-2   NIOSH: CY 1400000
DOT: 1114
mf: C_6H_6 mw: 78.12

PROP: Clear colorless liquid. mp: 5.51°, bp:
80.093°-80.094°, flash p: 12°F (CC), d: 0.8794
@ 20°, autoign temp: 1044°F, lel: 1.4%, uel:
8.0%, vap press: 100 mm @ 26.1°, vap d:
2.77, ulc: 95-100.

SYNS: (6)ANNULENE * BENZEEN (DUTCH)
* BENZEN (POLISH) * BENZOL * BENZOLENE
* BENZOLO (ITALIAN) * BICARBURET OF HY-
DROGEN * CARBON OIL * COAL NAPHTHA
* CYCLOHEXATRIENE * FENZEN (CZECH)
* MINERAL NAPHTHA * MOTOR BENZOL
* NCI-C55276 * PHENYL HYDRIDE * PYRO-
BENZOLE

OSHA PEL: TWA 10 ppm; CL 25 ppm; Pk
50 ppm/10M/8H

ACGIH TLV: TWA 10 ppm; BEI (total phenol
in urine) 50 mg/L
TRK: 8 ppm; 26 mg/m^3
DOT Classification: Label: Flammable
Liquid

THR: Poison by intravenous and possibly other
routes. Moderately toxic by inhalation, ingestion, subcutaneous, and intraperitoneal routes.
A strong eye and mild skin irritant. Central
nervous system and blood system effects by
inhalation and ingestion. A human carcinogen
(myeloid leukemia). An experimental teratogen
and tumorigen. Mutagenic data. A narcotic.
Chronic benzene poisoning by inhalation is important in industry, although poisoning by skin
contact has been reported. Elimination is chiefly
through the lungs. A common air contaminant.

Poisoning occurs most commonly via inhalation of the vapor, though benzene can penetrate the skin and poison in that way. Locally,
benzene has a comparatively strong irritating
effect, producing erythema and burning, and,
in more severe cases, edema and even blistering.
Exposure to high concentrations of the vapor
(3000 ppm or higher) may result from failure
of equipment or spillage. Such exposure, while
rare in industry, may result in acute poisoning,
characterized by the narcotic action of benzene
on the central nervous system.. The anesthetic
action of benzene is similar to that of other
anesthetic gases, consisting of a preliminary
stage of excitation followed by depression and,
if exposure is continued, death through respiratory failure. The chronic, rather than the acute
form, of benzene poisoning is important in industry. It is a recognized leukemogen. There
is no specific blood picture occurring in cases
of chronic benzol poisoning. The bone marrow
may be hypoplastic, normal, or hyperplastic,
the changes reflected in the peripheral blood.
Anemia, leucopenia, macrocytosis, reticulocy-
tosis, thrombocytopenia, high color index, and
prolonged bleeding time may be present. Cases
of myeloid leukemia have been reported. For
the supervision of the worker, repeated blood
examinations are necessary, including hemoglobin determinations, white and red cell counts
and differential smears. Where a worker shows
a progressive drop in either red or white cells,
or where the white count remains low, 5,000
per cu mm or the red count <4.0 million per
cu mm, on two successive monthly examina-

tions, he should be immediately removed from exposure. Elimination is chiefly through the lungs, when fresh air is breathed. The portion absorbed is oxidized, and the oxidation products are combined with sulfuric and glycuronic acids and eliminated in the urine. This may be used as a diagnostic sign. Benzene has a definite cumulative action, and exposure to relatively high concentration is not serious from the point of view of causing damage to the blood-forming system, provided the exposure is not repeated. On the other hand, daily exposure to concentrations of 100 ppm or less will usually cause damage if continued over a protracted period of time. In acute poisoning, the worker becomes confused and dizzy, complains of tightening of the leg muscles and of pressure over the forehead, then passes into a stage of excitement. If allowed to remain in exposure, he quickly becomes stupefied and lapses into coma. In nonfatal cases, recovery is usually complete and no permanent disability occurs. In chronic poisoning the onset is slow, with the symptoms vague; fatigue, headache, dizziness, nausea and loss of appetite, loss of weight and weakness are common complaints in early cases. Later, pallor, nosebleeds, bleeding gums, menorrhagia, petechiae and purpura may develop. There is great individual variation in the signs and symptoms of chronic benzene poisoning. Benzene is a common air contaminant. It is an experimental mutagen, carcinogen, teratogen. A dangerous fire hazard when exposed to heat or flame; can react vigorously with oxidizing materials, such as BrF_5; Cl_2; CrO_3; O_2N; ClO_4; O_2; O_3; perchlorates; $(AlCl_3 + FClO_4)$; $(H_2SO_4 + permanganates)$; K_2O_2; $(AgClO_4 + acetic\ acid)$; Na_2O_2. Moderate explosion hazard when exposed to heat or flame. Use with adequate ventilation. Highly flammable. To Fight Fire: Foam, CO_2, dry chemical. Incompatible with diborane. For further information see Vol. 2, No. 4 and Vol. 3, No. 3 of *DPIM Report*.

BENZENEARSONIC ACID   HR: 3
CAS: 98-05-5   NIOSH: CY 3150000
mf: $C_6H_7AsO_3$ mw: 202.05

PROP: Colorless crystals, water-sol; d: 1.760, mp: 160° decomp.

SYNS: PHENYL ARSENIC ACID * PHENYLAR-SONIC ACID

THR: A deadly poison by ingestion and intravenous routes. See also arsenic compounds. When heated to decomposition, it emits toxic fumes of As.

BENZENECARBOXALDEHYDE   HR: 3
CAS: 63021-32-9   NIOSH: CU 3750000
mf: $C_{19}H_{15}N$ mw: 257.35

SYNS: 7-ETHYLBENZ(C)ACRIDINE * 9-ETHYL-3,4-BENZACRIDINE * PHENYLMETHANAL

THR: An experimental carcinogen. See also aldehydes. When heated to decomposition, it emits toxic fumes of NO_x.

BENZENE CHLORIDE   HR: 2
CAS: 108-90-7   NIOSH: CZ 0175000
DOT: 1134
mf: C_6H_5Cl mw: 112.56

PROP: Clear, colorless liquid. bp: 131.7°, lel = 1.3%, uel = 7.1%, @ 150°, mp: −45°, flash p: 85°F (CC), d: 1.113 @ 15.5°/15.5°, autoign temp: 1180°F, vap press: 10 mm @ 22.2°, vap d: 3.88.

SYNS: CHLOORBENZEEN (DUTCH) * CHLOR-BENZENE * CHLORBENZOL * CHLOROBENZEN (POLISH) * CHLOROBENZENE * CLOROBEN-ZENE (ITALIAN) * MONOCHLOORBENZEEN (DUTCH) * MONOCHLORBENZENE * MONO-CHLORBENZOL (GERMAN) * MONOCHLOROBEN-ZENE * MONOCLOROBENZENE (ITALIAN) * NCI-C54886 * PHENYL CHLORIDE

OSHA PEL: TWA 75 ppm
DOT Classification: Label: Flammable
   Liquid

THR: Moderately toxic by ingestion. Strong narcotic with slight irritant qualities. Dichlorobenzols are strongly narcotic. Little is known of the effects of repeated exposures at lower concentrations, but it may cause kidney and liver damage. The industrial illnesses reported may possibly be due to nitrobenzol. Dangerous fire hazard when exposed to heat or flame. Also violent reaction with $AgClO_4$, dimethyl sulfoxide. Moderate explosion hazard when exposed to heat or flame. Reacts vigorously with oxidizers. See also chlorine compounds. To fight fire, use foam, CO_2, dry chemical, water to blanket fire. For further information, see Chlorobenzene, Vol. 2, No. 4 of *DPIM Report*.

1,000 ppm causing narcosis in guinea pigs followed by death after 20 H exposure. Some of the symptoms described (methemoglobinemia) suggest that other substances, such as nitrobenzol, may have been partially responsible for the few cases of industrial illness reported. It is possible that prolonged exposure to chlorobenzol may cause kidney and liver damage. Also somnolence, loss of consciousness, twitchings of the extremities, cyanosis, deep, rapid respirations and a small, irregular pulse are the chief symptoms occurring in acute exposures. The urine may be burgundy red, and the red blood cells show degenerative and regenerative changes. Fire Hazard: Dangerous, when exposed to heat or flame. Also violent reaction with $AgClO_4$, dimethyl sulfoxide.

1,3-BENZENEDICARBONITRILE   HR: 3
CAS: 626-17-5   NIOSH: CZ 1900000
mf: $C_8H_4N_2$ mw: 128.14

PROP: Colorless crystals; water insol, sol in benzene, acetone; vap d: 4.42; mp: 138°; bp: subl.

SYNS: M-PHTHALODINITRILE * M-DICYANO-BENZENE * 1,3-DICYANOBENZENE * ISOFTA-LODINITRIL (CZECH) * ISOPHTHALODINITRILE * ISOPHTHALONITRILE * NITRIL KYSELINY ISOFTALOVE (CZECH)

THR: Poison by ingestion. See also nitriles. When heated to decomposition, it emits toxic fumes of NO_x and CN^-.

p-BENZENEDINITRILE   HR: 2
CAS: 623-26-7   NIOSH: CZ 1925000
mf: $C_8H_4N_2$ mw: 128.14

PROP: Crystals. vap d: 4.42.

SYNS: 4-CYANOBENZONITRILE * P-DICYANO-BENZENE * 1,4-DICYANOBENZENE * NITRIL KYSELINY TEREFTALOVE (CZECH) * P-PHTHALO-DINITRILE * TEREFTALODINITRIL (CZECH) * TEREPHTHALONITRILE

THR: Moderately toxic by intraperitoneal route. Slightly toxic by ingestion. See also nitriles. A moderate skin and eye irritant. See also cyanide, NO_x.

BENZENE HEXACHLORIDE   HR: 3
CAS: 608-73-1   NIOSH: GV 3150000
mf: $C_6H_6Cl_6$ mw: 290.82

PROP: Technical grade contained 68.7% alpha-BHC, 6.5% beta-BHC and 13.5% gamma-BHC. White crystalline powder. mp: 113°, vap press: 0.0317 mm @ 20°.

SYNS: BHC * COMPOUND-666 * ENT 8,601 * GAMMEXANE * HEXACHLOROCYCLOHEXANE * 1,2,3,4,5,6-HEXACHLOROCYCLOHEXANE

THR: Poison by ingestion, skin contact, and subcutaneous routes. An experimental carcinogen, neoplastigen, and tumorigen by ingestion and skin contact. Implicated in aplastic anemia. The isomers have different actions; the gamma and alpha isomers are central nervous system stimulants. The beta and delta isomers are central nervous system depressants. The use of thermal vaporizers with lindane has caused acute poisoning by inhalation. Lindane is more toxic than DDT or dieldrin. When heated to decomposition, it emits highly toxic fumes of phosgene, HCl and Cl^-.

A toxic organochlorine pesticide which is persistent in the environment and accumulates in mammalian tissue. For cattle the oral LD_{50} = < 100 mg/kg. The several isomers of hexachlorocyclohexane have different actions. The γ and α isomers are central nervous system stimulants, the principal symptom being convulsions. The β and Δ isomers are central nervous system depressants. The dangerous acute dose of the technical mixture has been estimated at about 30 g and the dangerous dose of lindane at about 7 to 15 g. However, as already mentioned, a single dose of 45 mg (or approximately 0.65 mg/kg) of lindane caused convulsions. Lindane shows a marked difference in toxicity to different species. Its toxic effect on laboratory animals compares favorably with that of DDT, but for several domestic animals, notably calves, lindane is more toxic than DDT or dieldrin. On a chronic systemic basis the α, β and γ isomers are experimental carcinogens. Human carcinogenicity not definite. Has been implicated in aplastic anemia.

Dermatitis, and perhaps other manifestations based on sensitivity represent a sort of chronic, though probably not systemic intoxication, which has been observed in human beings.

The signs and symptoms of confirmed acute poisoning in man have paralleled those of experimental animals. These signs and symptoms are: excitation, hyperirritability, loss of equilibrium, clonic-tonic convulsions, and later depression. There is some evidence that the pulmonary edema and vascular collapse may be of neurogenic origin also. The symptoms in animals systemically poisoned by the γ isomer alone are essentially similar to those caused by mixtures, although the onset may be earlier. Men acutely exposed to high air concentrations of lindane and its decomposition products show headache, nausea, and irritation of eyes, nose and throat.

Urticaria has followed exposure to lindane vapor in rare instances. Unlike the signs and symptoms already mentioned, this allergic manifestation occurs only in susceptible individuals, and usually only after a period of sensitization.

BENZENE HEXACHLORIDE-alpha-isomer   HR: 3
CAS: 319-84-6   NIOSH: GV 3500000
mf: $C_6H_6Cl_6$ mw: 290.82

SYNS: ALPHA-BENZENEHEXACHLORIDE * ALPHA-BHC * ENT 9,232 * ALPHA-HEXA-CHLOROCYCLOHEXANE * ALPHA-1,2,3,4,5,6-HEXACHLOROCYCLOHEXANE * 1-ALPHA,2-ALPHA,3-BETA,4-ALPHA,5-BETA,6-BETA-HEXACHLOROCYCLOHEXANE * ALPHA-LINDANE

THR: An experimental carcinogen. Mutagenic data. When heated to decomposition, it emits toxic fumes of Cl⁻.

delta-BENZENE HEXACHLOR-IDE   HR: 2
CAS: 319-8-6-8   NIOSH: GV 4550000
mf: $C_6H_6Cl_6$ mw: 290.82

SYNS: 1,2,3,4,5,6-HEXACHLORO CYCLOHEX-ANE-DELTA ISOMER * DELTA BHC * ENT 9,236 * DELTA HCH * DELTA LINDANE

THR: Moderately toxic by ingestion. See also benzene hexachloride. When heated to decomposition, it emits highly toxic fumes of Cl⁻, HCl, and phosgene.

BENZENE HEXACHLORIDE-gamma isomer   HR: 3
CAS: 58-89-9   NIOSH: GV 4900000
DOT: 2761
mf: $C_6H_6Cl_6$ mw: 290.82

SYNS: GAMMA-BENZENE HEXACHLORIDE * LINDANE * GAMMA-BHC * ENT 7,796 * GAMMAHEXANE * HEXACHLORAN * GAMMA-HEXACHLORAN * GAMMA-HEXACHLORANE * GAMMA-HEXACHLOROBENZENE * 1-AL-PHA,2-ALPHA,3-BETA,4-ALPHA,5-ALPHA,BETA-HEXACHLOROCYCLOHEXANE * GAMMA-

HEXACHLOROCYCLOHEXANE * 1,2,3,4,5,6-HEXACHLOROCYCLOHEXANE, GAMMA-ISOMER

OSHA PEL: TWA 500 ug/m³ (skin)
DOT Classification: ORM-A, Label: None

THR: A human systemic poison by ingestion. Also a poison by intraperitoneal, intravenous, dermal, and intramuscular routes. An experimental carcinogen. Mutagenic data. See also benzene hexachloride. When heated to decomposition, it emits toxic fumes of Cl⁻, HCl, and phosgene. For further information, see Lindane in Vol. 3, No. 1 of *DPIM Report*.

BENZENE HEXACHLORIDE (MIXED ISOMERS)   HR: 3
CAS: 608-73-1   NIOSH: GV 4950000
mf: $C_6H_6Cl_6$ mw: 290.82

PROP: Technical BHC contains about 64% alpha, 10% beta, 13% gamma, 9% delta and 1% epsilon isomers of 1,2,3,4,5,6-hexachlorocyclohexane.

SYNS: 1,2,3,4,5,6-HEXACHLOROCYCLOHEXANE (MIXTURE OF ISOMERS) * BENZAHEX * TECHNICAL BHC * TECHNICAL HCH

THR: An experimental carcinogen. See also benzene hexachloride. Potentially dangerous reaction with DMF in presence of Fe, also CCl₄. When heated to decomposition, it emits highly toxic fumes of Cl⁻, HCl, and phosgene. For further information, see 1,2,3,4,5,6-Hexachlorocyclohexane, gamma, Vol. 1, No. 4 of *DPIM Report*.

trans-alpha-BENZENEHEXA-CHLORIDE   HR: 3
CAS: 319-85-7   NIOSH: GV 4375000
mf: $C_6H_6Cl_6$ mw: 290.82

SYNS: BETA-BENZENEHEXACHLORIDE * BETA-BHC * BETA-HCH * BETA-HEXACHLOROBEN-ZENE * 1-ALPHA,2-BETA,3-ALPHA,4-BETA,5-ALPHA,6-BETA-HEXACHLOROCYCLOHEXANE * BETA-HEXACHLOROCYCLOHEXANE * BETA-1,2,3,4,5,6-HEXACHLOROCYCLOHEXANE * BETA-LINDANE

THR: An experimental carcinogen. When heated to decomposition, it emits very toxic fumes of Cl⁻, HCl and phosgene.

BENZENESULFONIC ACID   HR: 3
CAS: 98-11-3   NIOSH: DB 4200000
mf: $C_6H_6O_3S$ mw: 158.18

PROP: Deliquescent plates or tablets. mp: 43°-44°.

SYN: PHENYLSULFONIC ACID

THR: Poison by ingestion and probably inhalation. A skin irritant. See also sulfates and sulfonates.

BENZENESULFONYL CHLORIDE   HR: 3
CAS: 98-09-9   NIOSH: DB 8750000
mf: $C_6H_5ClO_2S$ mw: 176.62

SYNS: BENZENE SULFONECHLORIDE * BENZENESULFONIC (ACID) CHLORIDE * BENZENOSULFOCHLOREK (POLISH) * BENZENOSULPHOCHLORIDE

THR: Poison by inhalation and intraperitoneal routes. Reacts vigorously with dimethyl sulfoxide, methyl formamide. See also sulfonates. When heated to decomposition, it emits toxic fumes of Cl^- and SO_x. Incompatible with dimethyl sulfoxide.

BENZENESULPHONYL FLUORIDE   HR: 3
CAS: 368-43-4   NIOSH: DB 8940000
mf: $C_6H_5FO_2S$ mw: 160.17

PROP: Clear liquid; bp: 209°, fp: −5°, flash p: 196°F, d: 1.329, vap press: 8 mm @ 80°, vap d: 5.52.

THR: It appears to be a poison by vapor inhalation and intraperitoneal routes. Slightly irritant to skin. A moderate fire hazard when exposed to heat or flame. See also fluorides, chlorides, and sulfates; can react vigorously with oxidizing materials. To fight fire, use water, foam, CO_2, water spray or mist, dry chemical.

BENZENETHIOL   HR: 3
CAS: 108-98-5   NIOSH: DC 0525000
mf: C_6H_6S mw: 110.18

PROP: Liquid, repulsive odor; bp: 168.3°, d: 1.0728 @ 25°/4°.

SYNS: PHENYL MERCAPTAN * THIOPHENOL
ACGIH TLV: TWA 0.5 ppm

THR: Poison by ingestion, inhalation, intraperitoneal, and dermal routes. A severe eye irritant. Can cause severe dermatitis and acute exposure is capable of causing headache and dizziness. See also mercaptans. When heated to decomposition it or contact with acids emits toxic fumes of SO_x.

1,2,4-BENZENETRIOL   HR: 3
CAS: 533-73-3   NIOSH: DC 4200000
mf: $C_6H_6O_3$ mw: 126.12

SYNS: HYDROXYHYDROQUINONE * HYDROXYQUINOL * 1,2,4-TRIHYDROXYBENZENE

THR: Poison by subcutaneous and intraperitoneal routes. When heated to decomposition, it emits acrid smoke and irritating fumes.

BENZETHONIUM CHLORIDE   HR: 3
CAS: 121-54-0   NIOSH: BO 7175000
mf: $C_{27}H_{42}NO_2 \cdot Cl$ mw: 448.15

PROP: Colorless, sol crystals.

SYNS: BENZYLDIMETHYL(2-(2-(p-(1,1,3,3-TETRAMETHYLBUTYL)PHENOXY)ETHOXY)-ETHYL)AMMONIUM CHLORIDE * BENZYLDIMETHYL-P-(1,1,3,3-TETRAMETHYLBUTYL)PHENOXYETHOXY-ETHYLAMMONIUM CHLORIDE * DIISOBUTYLPHENOXYETHOXYETHYLDIMETHYL BENZYL AMMONIUM CHLORIDE * HYAMINE 1622 * P-TERT-OCTYLPHENOXYETHOXYETHYL-DIMETHYLBENZYLAMMONIUM CHLORIDE

THR: Poison by ingestion, subcutaneous, intraperitoneal, and intravenous routes. An experimental carcinogen. An eye irritant. When heated to decomposition, it emits very toxic fumes of Cl^- and NO_x. For further information, see Benzethonium Chloride, Vol. 1, No. 1 of *DPIM Report*.

BENZHEXOL HYDROCHLORIDE   HR: 3
CAS: 52-49-3   NIOSH: TN 2625000
mf: $C_{20}H_{31}NO \cdot ClH$ mw: 337.98

SYNS: ARTANE HYDROCHLORIDE * BENZHEXOL CHLORIDE * ALPHA-CYCLOHEXYL-ALPHA-PHENYL-1-PIPERIDINEPROPANOL HYDROCHLORIDE * 1-PHENYL-1-CYCLOHEXYL-3-PIPERIDYL-1-PROPANOL HYDROCHLORIDE * 3-(1-PIPERIDYL)-1-CYCLOHEXYL-1-PHENYL-1-PROPANOL HYDROCHLORIDE * TRIHEXYLPHENIDYL HYDROCHLORIDE

THR: Poison by ingestion, intraperitoneal, intravenous, and subcutaneous routes. Causes human psychotropic effects. When heated to decomposition, it emits very toxic fumes of NO_x and HCl.

BENZHYDRAZIDE   HR: 3
CAS: 613-94-5   NIOSH: DH 1575000
mf: $C_7H_8N_2O$ mw: 136.17

SYN: BENZOYL HYDRAZIDE

THR: Poison by intraperitoneal route. An experimental carcinogen. When heated to decomposition, it emits toxic fumes of NO_x.

BENZHYDRYL **HR: 3**
CAS: 58-73-1   NIOSH: KR 6825000
mf: $C_{17}H_{21}NO$ mw: 255.39

SYNS: BENADRYL * BENZHYDRAMINE
* BENZHYDRIL * O-BENZHYDRYLDIMETHYL-
AMINOETHANOL * 2-(BENZHYDRYLOXY)-
N,N-DIMETHYLETHYLAMINE * 2-(BENZOHY-
DRYLOXY)-N,N-DIMETHYLETHYLAMINE
* BETA-DIMETHYLAMINOETHANOL DIPHENYL-
METHYL ETHER * ALPHA-(2-DIMETHYLAMINO-
ETHOXY)DIPHENYLMETHANE * BETA-DI-
METHYLAMINO-AETHYL-BENZHYDRYL-AETHER
(GERMAN) * BETA-DIMETHYLAMINOETHYL-
BENZHYDRYLETHER * DIPHENYLHYDRAMINE
* 2-(DIPHENYLMETHOXY)-N,N-DIMETHYL-
ETHYLAMINE

THR: A human poison. Poison by ingestion, intravenous, intraperitoneal, and subcutaneous routes. See also ethers. Mutagenic data. When heated to decomposition, it emits toxic fumes of NO_x.

BENZIDAMINE HYDROCHLO-
RIDE **HR: 3**
CAS: 132-69-4   NIOSH: NK 7875000
mf: $C_{19}H_{23}N_3O \cdot ClH$ mw: 345.91

SYNS: BENZINDAMINE HYDROCHLORIDE
* BENZYDAMINE HYDROCHLORIDE * 1-BEN-
ZYL-3-GAMMA-DIMETHYLAMINOPROPOXY-1H-IN-
DAZOLE HYDROCHLORIDE * 1-BENZYL-3-(3-(DI-
METHYLAMINO)PROPOXY)-1H-INDAZOLE
HYDROCHLORIDE

THR: Poison by intraperitoneal, subcutaneous, and intravenous routes. An experimental teratogen. When heated to decomposition, it emits very toxic fumes of HCl and NO_x.

BENZIDINE **HR: 3**
CAS: 92-87-5   NIOSH: DC 9625000
DOT: 1885
mf: $C_{12}H_{12}N_2$ mw: 184.26

PROP: Grayish-yellow crystalline powder; white or slightly reddish crystals, powder or leaf; mp: 127.5°-128.7° @ 740 mm, bp: 401.7°, d: 1.250 @ 20°/4°.

SYNS: BENZIDIN (CZECH) * BENZIDINA (ITALIAN) * BENZYDYNA (POLISH) * 4,4'-BIPHE-
NYLDIAMINE * 4,4'-DIAMINOBIPHENYL
* P-DIAMINODIPHENYL * 4,4'-DIAMINODIPHE-
NYL * P,P'-DIANILINE

OSHA PEL: Carcinogen
DOT Classification: Poison B

THR: Poison by ingestion, inhalation, and subcutaneous routes. A human and experimental carcinogen. Mutagenic data. Can cause damage to blood, including hemolysis and bone marrow depression. On ingestion causes nausea and vomiting which may be followed by liver and kidney damage. See also aromatic amines. Any exposure is considered extremely hazardous. When heated to decomposition, it emits highly toxic fumes of NO_x. For further information, see Vol. 3, No. 4 of *DPIM Report*.

BENZIDINE HYDROCHLORIDE **HR: 3**
CAS: 531-85-1   NIOSH: DD 0600000
mf: $C_{12}H_{12}N_2 \cdot 2ClH$ mw: 257.18

SYN: DIHIDROCLORURO DE BENZIDINA (SPANISH)

THR: Poison by ingestion. An experimental carcinogen. Mutagenic data. When heated to decomposition, it emits very toxic fumes of HCl and NO_x.

BENZIDINE SULFATE **HR: 3**
CAS: 631-86-2   NIOSH: DD 1575000
mf: $C_{12}H_{12}N_2 \cdot H_2O_4S$ mw: 282.34
OSHA PEL: Carcinogen

THR: An experimental carcinogen. See also benzedine and sulfates. When heated to decomposition, it emits toxic fumes of SO_x and NO_x.

BENZILIC ACID-beta-DIETHYLAMINO-
ETHYLESTER HYDROCHLORIDE
HR: 3
CAS: 57-37-4   NIOSH: DD 2800000
mf: $C_{20}H_{25}NO_3 \cdot ClH$ mw: 363.92

SYNS: BENACTIZINE HYDROCHLORIDE * BEN-
ZILATE DU DIETHYLAMINO-ETHANOL CHLORHY-
DRATE (FRENCH) * BETA-DIETHYLAMINOETHYL
BENZILATE HYDROCHLORIDE * 2-DIETHYLAMI-
NOETHYL BENZILATE HYDROCHLORIDE * 2-DI-
ETHYLAMINOETHYL DIPHENYLGLYCOLATE
HYDROCHLORIDE * 2-(DIFENYL-HYDROXY-
ACETOXY)ETHYL-DIETHYLAMMONIUMCHLORID
(CZECH) * DIPHENYLGLYCOLLIC ACID-2-
(DIETHYLAMINO)ETHYL ESTER HYDROCHLORIDE

THR: Poison by ingestion, intraperitoneal, subcutaneous, intradermal, and intravenous routes.

Causes human psychotropic effects. When heated to decomposition, it emits very toxic fumes of NO$_x$ and HCl.

BENZIMIDAZOLE   HR: 3
CAS: 51-17-2   NIOSH: DD 5425000
mf: C$_7$H$_6$N$_2$ mw: 118.15

PROP: Tabular crystals sol in alcohol, sparingly sol in water. mp: 170.5°, bp: >360°.

SYNS: 1H-BENZIMIDAZOLE (9CI) * BENZOIMI-DAZOLE * O-BENZIMIDAZOLE * 1,3-BENZO-DIAZOLE * 1,3-DIAZAINDENE * N,N′-METHE-NYL-O-PHENYLENEDIAMINE * NSC 759

THR: Poison by intravenous route. When heated to decomposition, it emits highly toxic fumes of NO$_x$.

BENZIMIDAZOLE METHYLENE MUS-TARD   HR: 3
CAS: 6898-43-7   NIOSH: DD 6300000
mf: C$_{14}$H$_{19}$Cl$_2$N$_3$ • ClH   mw: 336.72

SYNS: BENZIMIDAZOLE MUSTARD * 2-(BIS-(2-CHLOROETHYL)AMINOMETHYL)-5,5-DI-METHYLBENZIMIDAZOLE HYDROCHLORIDE * 2-(DI-2-CHLOROETHYL)AMINOMETHYL-5,6-DIMETHYLBENZIMIDAZOLE * NSC-23892

THR: An experimental carcinogen. When heated to decomposition, it emits very toxic fumes HCl and NO$_x$.

2-BENZIMIDAZOLETHIOL   HR: 3
CAS: 583-39-1   NIOSH: DE 1050000
mf: C$_7$H$_6$N$_2$S   mw: 150.21

SYNS: 2-MERCAPTOBENZIMIDAZOLE * MER-CAPTOBENZOIMIDAZOLE * MERKAPTOBENZIMI-DAZOL (CZECH) * 2-MERCAPTOBENZOIMIDA-ZOLE * NCI-C60980 * O-PHENYLENETHIO-UREA * USAF EK-6540 * USAF XF-21

THR: Poison by intraperitoneal and intravenous routes. Skin and eye irritant. When heated to decomposition, it emits toxic fumes of SO$_x$ and NO$_x$.

BENZIN   HR: 3
CAS: 8030-30-6   NIOSH: DE 3030000
DOT: 1255/1256/1271/2553

PROP: Made from American coal oil and consists chiefly of pentane, hexane and heptane.

SYNS: AROMATIC SOLVENT * HI-FLASH NAPH-THAETHYLEN * NAPHTHA * PETROLEUM BEN-ZIN * PETROLEUM DISTILLATES (NAPHTHA) * PETROLEUM ETHER

OSHA PEL: TWA 500 ppm (skin)
DOT Classification: Combustible Liquid, Label: None

THR: A poison via intravenous route; See also naphtha. Flammable liquid. See also pentane, hexane, heptane.

1,2-BENZISOTHIAZOL-3(2H)-ONE-1,1-DIOXIDE   HR: 3
CAS: 81-07-2   NIOSH: DE 4200000
mf: C$_7$H$_5$NI$_3$S   mw: 183.19

PROP: mp: 228° (decomp); bp: sublimes. Crystals or powder.

SYNS: ANHYDRO-O-SULFAMINE BENZOIC ACID * 3-BENZISOTHIAZOLINONE-1,1-DIOXIDE * O-BENZOSULFIMIDE * BENZO-2-SULPHIMIDE * BENZO-SULPHIMIDE * O-BENZOYL SULFIMIDE * 1,2-DIHYDRO-2-KETOBENZISOSULPHONAZOLE * 2,3-DIHYDRO-3-OXOBENZISOSULFONAZOLE * 3-HYDROXY BENZISOTHIAZOL-S,S-DIOXIDE * INSOLUBLE SACCHARINE * SACCAHARIMIDE * SACCHARINA * SACCHARIN ACID * SAC-CHARINE * SACCHARINOL * SACCHARINOSE * SACCHAROL * SUCRE EDULCOR * O-SUL-FOBENZIMIDE * O-SULFOBENZOIC ACID IMIDE

THR: An experimental carcinogen. Mutagenic data. A non-nutritive sweetener food additive. When heated to decomposition, it emits toxic NO$_x$ and SO$_x$.

BENZO(b)CHRYSENE   HR: 3
CAS: 214-17-5   NIOSH: DE 7340000
mf: C$_{22}$H$_{14}$ mw: 278.36

SYNS: 2,3-BENZOCHRYSENE * 3,4-BENZOTET-RACENE * 3,4-BENZOTETRAPHENE * DI-BENZO-2,3,7,8-PHENANTHRENE

THR: An experimental neoplastigen. When heated to decomposition, it emits acrid smoke and irritating fumes.

BENZO(c)CHRYSENE   HR: 3
CAS: 194-69-4   NIOSH: DE 7350000
mf: C$_{22}$H$_{14}$ mw: 278.36

SYN: 1,2,5,6-DIBENZPHENANTHRENE

THR: An experimental carcinogen. When heated to decomposition, it emits acrid smoke and irritating fumes.

BENZO(g)CHRYSENE HR: 3
CAS: 196-78-1 NIOSH: DE 7525000
mf: $C_{22}H_{14}$ mw: 278.36

SYNS: 1,2,3,4-DIBENZOPHENANTHRENE
* 1,2,3,4-DIBENZPHENANTHRENE

THR: An experimental carcinogen. When heated to decomposition, it emits acrid smoke and irritating fumes.

N-6-(3,4-BENZOCOUMARINYL)ACET-AMIDE HR: 3
CAS: 5096-19-5 NIOSH: HP 8750000
mf: $C_{15}H_{10}NO_3$ mw: 252.26

SYN: N-(6-OXO-6H-DIBENZO(B,D)PYRAN-1-YL)-ACETAMIDE

THR: An experimental carcinogen. When heated to decomposition, it emits very toxic fumes of NO_x.

BENZOCTAMINE HR: 3
CAS: 17243-39-9 NIOSH: KJ 4375000
mf: $C_{18}H_{19}N$ mw: 249.38

SYNS: TACITIN * N-METHYL ETHANOANTHRA-CENE-9(10H)-METHYLAMINE

THR: Poison by ingestion, intraperitoneal, subcutaneous, and intravenous routes. When heated to decomposition, it emits toxic fumes of NO_x.

BENZODIOXANE HYDROCHLO-RIDE HR: 3
CAS: 135-87-5 NIOSH: TM 4550000
mf: $C_{14}H_{19}NO_2 \cdot ClH$ mw: 269.80

SYNS: BENODAINE HYDROCHLORIDE * 1-(1,4-BENZODIOXAN-2-YLMETHYL)PIPERIDINEHYDRO-CHLORIDE * 2-PIPERIDINOMETHYL-1,4-BENZODIOXAN HYDROCHLORIDE
* 2-(1-PIPERIDYLMETHYL)-1,4-BENZODIOXAN HYDROCHLORIDE * PIPEROXANE HYDRO-CHLORIDE

THR: Poison by intraperitoneal, intramuscular, and intravenous routes. When heated to decomposition, it emits very toxic fumes of NO_x and HCl.

1,3-BENZODIOXOLE-5-(2-PROPEN-1-OL) HR: 3
CAS: 5208-87-7 NIOSH: DF 4920000
mf: $C_{10}H_{10}O_3$ mw: 178.20

SYNS: 1'-HYDROXYSAFROLE * 1,2-METHYL-ENEDIOXY-4-(1-HYDROXYALLYL)BENZENE
* ALPHA-VINYLPIPERONYL ALCOHOL

THR: An experimental carcinogen of the esophagus via oral route. Mutagenic data. When heated to decomposition, it emits acrid smoke and irritating fumes.

BENZOEPIN HR: 3
CAS: 115-29-7 NIOSH: RB 9275000
mf: $C_9H_6Cl_6O_3S$ mw: 406.91

PROP: A mixture of 2 isomers, brown crystals, nearly insol in water, sol in most organic solvents. mp(α): 106°, mp(β): 212°, d: 1.745 @ 20°/20°.

SYNS: ENT 23,979 * ALPHA,BETA-1,2,3,4,7,7-HEXACHLOROBICYCLO(2.2.1)-2-HEPTENE-5,6-BISOXYMETHYLENE SULFITE * 1,2,3,4,7,7-HEXACHLOROBICYCLO(2.2.1)HEPTEN-5,6-BI-OXYMETHYLENESULFITE * HEXACHLOROHEXA-HYDROMETHANO * 2,4,3-BENZODIOXATHIEPIN-3OXIDE * 6,7,8,9,10,10-HEXACHLORO-1,5,5a,6,9,9a-HEXAHYDRO-6,9-METHANO-2,4,3-BENZODIOXATHIEPIN-3-OXIDE * 1,4,5,6,7,7-HEXACHLORO-5-NORBORNENE-2,3-DIMETHANOL CYCLIC SULFITE * NCI-C00566

THR: Very poisonous by ingestion and dermal route. An experimental teratogen and carcinogen. A central nervous system stimulant producing convulsions. A highly toxic organochlorine pesticide which does not accumulate significantly in human tissue. Absorption is normally slow, but is increased by alcohols, oil, emulsifiers. Dangerous; see also chlorides and sulfur compounds.

BENZO(j)FLUORANTHENE HR: 3
CAS: 205-82-3 NIOSH: DF 6300000
mf: $C_{20}H_{12}$ mw: 252.32

SYNS: 10,11-BENZFLUORANTHENE * 7,8-BEN-ZOFLUORANTHENE * DIBENZO(A,J,K)FLUORENE

THR: An experimental carcinogen. When heated to decomposition, it emits acrid smoke and irritating fumes.

BENZOFUROLINE HR: 3
CAS: 10453-86-8 NIOSH: GZ 1310000
mf: $C_{22}H_{26}O_3$ mw: 338.48

SYNS: 5-BENZYL-3-FURYL METHYL(+)-cis,transCHRYSANTHEMATE * (5-BENZYL-3-FURYL) METHYL-2,2-DIMETHYL-3-(2-METHYL-PROPENYL)-CYCLOPROPANECARBOXYLATE
* DIMETHYL-3-(2-METHYL-1-PROPENYL)CYCLO-PROPANECARBOXYLATE * ENT 27474
* RESMETRINA (PORTUGUESE)

THR: Poison by inhalation and intravenous routes. Moderately toxic by ingestion and skin route. When heated to decomposition, it emits acrid and irritating fumes.

BENZOGUANAMINE   HR: 3
CAS: 23844-24-8   NIOSH: DK 1575000
mf: $C_{22}H_{32}N_2O_5$ mw: 404.56

PROP: Crystals. mp: 227°, d: 1.4.

SYNS: 2-ACETOXY-3-DIETHYLCARBAMYL-9,10-DIMETHOXY-1,2,3,4,6,7-HEXAHYDRO -11B-BEN-ZO(A)QUINOLIZINE * BENZOQUINAMIDE

THR: Poison by ingestion and intravenous routes. When heated to decomposition, it emits toxic fumes of NO_x.

BENZOIC ACID   HR: 3
CAS: 65-85-0   NIOSH: DG 0875000
mf: $C_7H_6O_2$ mw: 122.13

PROP: White powder; mp: 121.7°, bp: 249°, flash p: 250°F (CC), d: 1.316, autoign temp: 1060°F, vap press: 1 mm @ 96.0° (sublimes), vap d: 4.21.

SYNS: ACIDE BENZOIQUE (FRENCH) * BEN-ZENECARBOXYLIC ACID * KYSELINA BEN-ZOOVA (CZECH) * PHENYL CARBOXYLIC ACID * PHENYLFORMIC ACID

THR: Poison by vapor inhalation. Moderately irritating to human skin. Severe eye irritant in rabbits. A slight fire hazard when exposed to heat or flame; can react with oxidizing materials. To fight fire, use water, CO_2, water spray or mist, dry chemical. When heated to decomposition, it emits acrid smoke and irritating fumes. For further information, see Vol. 3, No. 4 of *DPIM Report.*

BENZOIC ACID, BENZYL ESTER   HR: 2
CAS: 120-51-4   NIOSH: DG 4200000
mf: $C_{14}H_{12}O_2$ mw: 212.26

PROP: Found in Peru and Tolu Balsams, in Ylang-Ylang and in about 20 other essential oils. Liquid. mp. 21°, bp: 324°, flash p: 298°F (CC), d: 1.114, vap d: 7.3, autoign temp: 898°F.

SYNS: BENZYL ALCOHOL BENZOIC ESTER * BENZYL BENZENECARBOXYLATE * BENZYL BENZOATE * BENZYL PHENYLFORMATE

THR: Moderately toxic by ingestion and dermal routes. No data on chronic effects. Slight fire hazard when exposed to heat or flame; can react

with oxidizing materials. To fight fire, use CO_2, water spray or mist, dry chemical. When heated to decomposition, it emits acrid and irritating fumes and smoke. For further information, see Benzyl benzoate, Vol. 2, No. 3 of *DPIM Report.*

BENZONITRILE   HR: 2
CAS: 100-47-0   NIOSH: DI 2450000
mf: C_7H_5N mw: 103.13

PROP: Transparent, colorless oil, almond-like odor; d: 1.246 @ 20°/4°, bp: 191°, mp:-12.8°.

SYNS: BENZOIC ACID NITRILE * CYANOBEN-ZENE * PHENYL CYANIDE

THR: Moderately toxic by ingestion, inhalation, and dermal routes. See also nitriles. A skin irritant. When heated to decomposition, it emits toxic fumes of CN^- and NO_x. For further information, see Vol. 3, No. 4 of *DPIM Report.*

BENZO(rst)PENTAPHENE   HR: 3
CAS: 189-55-9   NIOSH: DI 5775000
mf: $C_{24}H_{14}$ mw: 302.38

PROP: Green-yellow needles; mp: 280°-282°.

SYNS: DIBENZO(A,I)PYRENE * 1,2,7,8-DIBEN-ZOPYRENE * 3,4:9,10-DIBENZOPYRENE

THR: An experimental carcinogen. Mutagenic data. When heated to decomposition, it emits acrid smoke and irritating fumes.

BENZO(c)PHENANTHRENE   HR: 3
CAS: 195-19-7   NIOSH: DI 8225000
mf: $C_{18}H_{12}$ mw: 228.30

SYNS: 3,4-BENZOPHENANTHRENE * 3,4-BENZPHENANTHRENE

THR: An experimental carcinogen. When heated to decomposition, it emits acrid and irritating fumes.

BENZOPHENONE   HR: 2
CAS: 119-61-9   NIOSH: DI 9950000
mf: $C_{13}H_{10}O$ mw: 182.23

PROP: Rhombic, white crystals, persistent rose-like odor; mp (α): 49°, mp (β): 26°, mp (γ): 47°, bp: 305.4°, d (a): 1.0976 @ 50/50°, d (β): 1.108 @ 23°/40°, vap press: 1 mm @ 108.2

SYNS: BENZOYLBENZENE * DIPHENYL KETONE * DIPHENYLMETHANONE * ALPHA-OXODIPHE-NYLMETHANE * PHENYL KETONE

THR: Moderately toxic by ingestion and intra-peritoneal routes. See also ketones. Slight fire hazard when heated. Incompatible with oxidizers. When heated to decomposition, it emits acrid and irritating fumes. For further information, see Vol. 2, No. 1 of *DPIM Report*.

BENZO(a)PYRENE HR: 3
CAS: 50-32-8   NIOSH: DJ 3675000
mf: $C_{20}H_{12}$ mw: 252.32

PROP: Yellow crystals insol in water, sol in benzene, toluene, xylene; mp: 179°, bp: 312° @ 10 mm.

SYNS: BENZO(D,E,F)CHRYSENE * 3,4-BENZO-PIRENE (ITALIAN) * 3,4-BENZOPYRENE * 6,7-BENZOPYRENE * 3,4-BENZPYREN (GER-MAN) * BENZ(A)PYRENE * 3,4-BENZ(A)PY-RENE * 3,4-BENZYPYRENE

THR: A poison via subcutaneous route. An experimental carcinogen. A common air contaminant of water, food, and smoke. Mutagenic data. When heated to decomposition, it emits acrid smoke. For further information, see Vol. 5, No. 1 of *DPIM Report*.

BENZO(e)PYRENE HR: 3
CAS: 192-97-2   NIOSH: DJ 4200000
mf: $C_{20}H_{12}$ mw: 252.32

SYNS: 1,2-BENZOPYRENE * 1,2-BENZPYRENE * 4,5-BENZOPYRENE

THR: An experimental carcinogen. Mutagenic data. When heated to decomposition, it emits acrid smoke and irritating fumes.

BENZO(a)PYRENE-6-CARBOXYALDE-HYDE HR: 3
CAS: 13312-42-0   NIOSH: DJ 4910000
mf: $C_{21}H_{12}O$ mw: 280.33

SYNS: 3,4-BENZPYRENE-5-ALDEHYDE * 6-FORMYLBENZO(A)PYRENE

THR: An experimental carcinogen. When heated to decomposition, it emits irritating fumes.

anti-BENZO(a)PYRENE-7,8-DIHYDRO-DIOL-9,10-OXIDE HR: 3
NIOSH: DJ 5050000
mf: $C_{20}H_{14}O_3$ mw: 302.34

SYNS: ANTI-BP-7,8-DIHYDRODIOL-9,10-OXIDE * BENZO(A)PYRENE-7,8-DIHYDRODIOL-9,10-EXPOXIDE (ANTI) * BP-7,8-DIHYDRODIOL-9,10-EPOXIDE(ANTI)

THR: An experimental carcinogen. Mutagenic data. When heated to decomposition, it emits acrid smoke and irritating fumes.

BENZO(a)PYRENE-6-METHA-NOL HR: 3
CAS: 21247-98-3   NIOSH: DJ 6410000
mf: $C_{21}H_{14}O$ mw: 282.35

SYN: 6-HYDROXYMETHYLBENZO(A)PYRENE

THR: An experimental carcinogen. Mutagenic data. When heated to decomposition, it emits irritating fumes.

BENZO(a)PYRENE-4,5-OXIDE HR: 3
CAS: 37574-47-3   NIOSH: DJ 7900000
mf: $C_{20}H_{12}O$ mw: 268.32

SYNS: BENZO(A)PYRENE-4,5-EPOXIDE * BP-4,5-EPOXIDE

THR: An experimental carcinogen. Mutagenic data. When heated to decomposition, it emits acrid and irritating fumes.

BENZO(a)PYRENE-7,8-OXIDE HR: 3
CAS: 36504-65-1   NIOSH: DJ 7920000
mf: $C_{20}H_{12}O$ mw: 268.32

SYNS: BENZO(A)PYRENE-7,8-EPOXIDE * 6-BETA,7-ALPHA-DIHYDROBENZO(10,11)-CHRYSENO(1,2-B)OXIRENE

THR: An experimental carcinogen. Mutagenic data. When heated to decomposition, it emits irritating fumes.

BENZO(a)PYRENE-9,10-OXIDE HR: 3
CAS: 36504-66-2   NIOSH: DJ 7929000
mf: $C_{20}H_{12}O$ mw: 268.32

SYN: BP-9,10-OXIDE

THR: An experimental carcinogen. Mutagenic data. When heated to decomposition, it emits acrid smoke and irritating fumes.

BENZO(a)PYRENE-11,12-OXIDE HR: 3
CAS: 60448-19-3   NIOSH: DJ 7940000
mf: $C_{20}H_{12}O$ mw: 268.32

SYN: BP-11,12-OXIDE

THR: An experimental carcinogen. Mutagenic data. When heated to decomposition, it emits acrid smoke and irritating fumes.

BENZO(a)PYREN-2-OL   HR: 3
CAS: 56892-30-9   NIOSH: DJ 8200000
mf: $C_{20}H_{12}O$ mw: 268.32

SYN: 2-HYDROXYBENZO(A)PYRENE

THR: An experimental carcinogen. Mutagenic
data. When heated to decomposition, it emits
acrid smoke and irritating fumes.

BENZO(a)PYREN-3-OL   HR: 3
CAS: 13345-21-6   NIOSH: DJ 8225000
mf: $C_{20}H_{12}O$ mw: 268.32

SYNS: 3-HYDROXYBENZO(A)PYRENE * 8-HY-
DROXY-3,4-BENZPYRENE

THR: An experimental carcinogen. Mutagenic
data. When heated to decomposition, it emits
acrid smoke and irritating fumes.

BENZO(a)PYREN-6-OL   HR: 3
CAS: 33953-73-0   NIOSH: DJ 8260000
mf: $C_{20}H_{12}O$ mw: 268.32

SYN: 6-HYDROXYBENZO(A)PYRENE

THR: An experimental carcinogen. Mutagenic
data. When heated to decomposition, it emits
acrid smoke and fumes.

BENZO(a)PYREN-7-OL   HR: 3
CAS: 37994-82-4   NIOSH: DJ 8270000
mf: $C_{20}H_{12}O$ mw: 268.32

SYN: 7-HYDROXYBENZO(A)PYRENE

THR: An experimental carcinogen. Mutagenic
data. When heated to decomposition, it emits
acrid smoke and fumes.

p-BENZOQUINONE   HR: 3
CAS: 106-51-4   NIOSH: DK 2625000
mf: $C_6H_4O_2$ mw: 108.10

PROP: Yellow crystals, characteristic irr odor;
mp: 115.7°, bp: sublimes, d: 1.318 @ 20°/4°.

SYNS: BENZO-CHINON (GERMAN) * 1,4-BEN-
ZOQUINE * 1,4-BENZOQUINONE * CHINON
(DUTCH, GERMAN) * 1,4-CYCLOHEXADIENE-
DIONE * 2,5-CYCLOHEXADIENE-1,4-DIONE
* 1,4-CYCLOHEXADIENE DIOXIDE * 1,4-DIOS-
SIBENZENE (ITALIAN) * 1,4-DIOXYBENZENE
* 1,4-DIOXY-BENZOL (GERMAN) * NCI-
C55845 * QUINONE * P-QUINONE * USAF
P-220
ACGIH TLV: TWA 0.1 ppm
DFG MAK: 0.1 ppm; 0.4 mg/m^3

THR: Poison by ingestion, inhalation, intraperitoneal, and intravenous routes. An experimental
carcinogen. Quinone has a characteristic, irritating odor. Causes severe damage to the skin
and mucous membranes by contact with it in
the solid state, in solution, or in the form of
condensed vapors. Locally, it causes discoloration, severe irritation, erythema, swelling, and
the formation of papules and vesicles, whereas
prolonged contact may lead to necrosis. When
the eyes become involved, it causes dangerous
disturbances of vision.

BENZOQUINONE AZIRIDINE   HR: 3
CAS: 800-24-8   NIOSH: DK 3325000
mf: $C_{16}H_{22}N_2O_6$ mw: 338.40

SYNS: AZIRIDYL BENZOQUINONE * 2,5-
BIS(1-AZIRIDINYL)-3,6-BIS(2-METHOXYETH-
OXY)-2,5-CYCLOHEXADIENE-1,4-DIONE
* 2,5-BIS(1-AZIRIDINYL)-3,6-BIS(2-METH-
OXYETHOXY)-P-BENZOQUINONE
* 3,6-BIS(BETA-METHOXYETHOXY)-
2,5-BIS(ETHYLENIMINO)-P-BENZOQUINONE
* 2,5-BISMETHOXYETHOXY-3,6-BISETHYLENE-
IMINO-1,4-BENZOQUINONE * 3,6-BIS (BETA-
METHOXYETHOXY)-2,5-BIS(ETHYLENEIMINO)-
P-BENZOQUINONE * NSC-17262

THR: Poison by intravenous route. An experimental carcinogen. Mutagenic data. When
heated to decomposition, it emits toxic fumes
of NO$_x$.

1,4-BENZOQUINONE-N'-BENZOYLHY-
DRAZONE OXIME   HR: 3
CAS: 495-73-8   NIOSH: DH 6125000
mf: $C_{13}H_{11}N_3O_2$ mw: 241.27

SYNS: P-BENZOQUINONE OXIME BENZOYLHY-
DRAZONE * CHINONOXIMEBENZOYLHYDRAZONE
* QUINONE OXIME BENZOYLHYDRAZONE

THR: Poison by ingestion. When heated to decomposition, it emits toxic NO$_x$.

BENZOTHIAZIDE   HR: 3
CAS: 91-33-8   NIOSH: DK 8400000
mf: $C_{15}H_{14}ClN_3O_4S_3$ mw: 431.95

SYNS: AQUATAG * 3-((BENZYLTHIO)METHYL)-
6-CHLORO-1,2,4-BENZOTHIADIAZINE-7-SUL-
FONAMIDE-1,1-DIOXIDE * 3-BENZYLTHIO-
METHYL-6-CHLORO-2H-1,2,4-BENZOTHIADIA-
ZINE-7-SULFONAMIDE-1,1-DIOXIDE
* 3-BENZYLTHIOMETHYL-6-CHLORO-7-SULFA-

MOYL-1,2,4-BENZOTHIADIAZINE-1,1-DIOXIDE
* 3-BENZYLTHIOMETHYL-6-CHLORO-
7-SULFAMYL-1,2,4-BENZOTHIADIAZINE-
1,1-DIOXIDE * 3-BENZYLTHIOMETHYL-6-
CHLORO-7-SULFAMYL-2H-1,2,4-BENZOTHIADIA-
ZINE-1,1-DIOXIDE

THR: Poison by intravenous route. When heated to decomposition, it emits very toxic fumes of SO_x, NO_x, and Cl^-.

BENZOTHIAZOLE   HR: 3
CAS: 95-16-9   NIOSH: DL 0875000
mf: C_7H_5NS mw: 135.19

PROP: Liquid, odor of quinoline, slightly water sol. d: 1.246 @ 20°/4°, bp: 228° @ 765 mm.

SYNS: BENZOSULFONAZOLE * 1-THIA-3-
AZAINDENE * USAF EK-4812

THR: Poison by intraperitoneal and intravenous routes. When heated to decomposition, it emits very toxic sulfides, cyanides, NO_x.

BENZOTHIAZOLE DISULFIDE   HR: 3
CAS: 120-78-5   NIOSH: DL 4550000
mf: $C_{14}H_8N_2S_4$ mw: 332.48

PROP: Cream to light yellow powder; mp: 175°, d: 1.5.

SYNS: 2-BENZOTHIAZOLYL DISULFIDE * BIS-
(BENZOTHIAZOLYL)DISULFIDE * BIS(2-BENZO-
THIAZYL) DISULFIDE * DI-2-BENZOTHIAZOLYL-
DISULFIDE * 2,2'-DIBENZOTHIAZYLDISULFIDE
* DIBENZOYLTHIAZYL DISULFIDE * DIBENZTHI-
AZYL DISULFIDE * 2,2'-DITHIOBIS(BENZOTHIA-
ZOLE) * 2-MERCAPTOBENZOTHIAZOLEDISULFIDE
* 2-MERCAPTOBENZOTHIAZYLDISULFIDE

THR: Poison by intravenous route. An experimental carcinogen. When heated to decomposition, it emits very toxic fumes of SO_x and NO_x.

2-BENZOTHIAZOLETHIOL   HR: 3
CAS: 149-30-4   NIOSH: DL 6475000
mf: $C_7H_5NS_2$ mw: 167.25

PROP: Light yellow powder; mp: 170°, d: 1.42 @ 25°.

SYNS: 2-MERCAPTOBENZOTHIAZOLE * NCI-
C56519 * USAF GY-3 * USAF XR-29

THR: Poison by ingestion and intraperitoneal routes. An experimental carcinogen. When heated to decomposition it or contact with acids

or acid fumes emits toxic SO_x and NO_x. Incompatible with oxidizers.

2-BENZOTHIAZOLYL-N-MORPHOLINO-SULFIDE   HR: 3
CAS: 102-77-2   NIOSH: DL 5950000
mf: $C_{11}H_{12}N_2OS_2$ mw: 252.37

SYNS: 2-BENZOTHIAZOLYLSULFENYL MORPHO-
LINE * 4-(2-BENZOTHIAZOLYLTHIO)MORPHOLINE
* 2-(MORPHOLINOTHIO)BENZOTHIAZOLE
* MORPHOLINYLMERCAPTOBENZOTHIAZOLE
* 2-(4-MORPHOLINYLTHIO)BENZOTHIAZOLE
* N-(OXYDIETHYLENE)BENZOTHIAZOLE-2-SUL-
FENAMIDE * USAF CY-7

THR: Poison by intraperitoneal route. An experimental carcinogen. See also sulfides. When heated to decomposition, it emits very toxic fumes of NO_x and SO_x.

1H-BENZOTRIAZOLE   HR: 3
CAS: 95-14-7   NIOSH: DM 1225000
mf: $C_6H_5N_3$ mw: 119.14

PROP: Needle-like crystals; mp: 100°, bp: 204° @ 15 mm.

SYNS: 1,2,-AMINOZOPHENYLENE * AZIMIDO-
BENZENE * AZIMINOBENZENE * BENZISOTRIA-
ZOLE * 1,2,3-BENZOTRIAZOLE * 2,3-DIA-
ZAINDOLE * NCI-C03521 * NSC-3058
* 1,2,3-TRIAZAINDENE

THR: Poison by intravenous route. An experimental carcinogen. Mutagenic data. When heated to decomposition, it emits toxic fumes. Has detonated during vacuum distillation.

BENZOTRIFLUORIDE   HR: 3
CAS: 98-08-8   NIOSH: XT 9450000
mf: $C_7H_5F_3$ mw: 146.12

PROP: Water white liquid, aromatic odor; mp: −29.1°, bp: 104°, flash p: 54°F (CC), d: 1.197 @ 15.5°/15.5°, vap d: 5.04, vap press: 11 mm @ 0°.

SYNS: (TRIFLUOROMETHYL)BENZENE * BEN-
ZENYL FLUORIDE * BENZYLIDYNE FLUORIDE
* PHENYLFLUOROFORM * ALPHA,ALPHA,
ALPHA-TRIFLUOROTOLUENE * OMEGA-
TRIFLUOROTOLUENE * USAF MA-16

THR: Poison by intraperitoneal route. See also fluorides. Dangerous fire hazard. To fight fire, use water, foam, CO_2, spray mist, dry chemical. When heated to decomposition, it emits toxic fumes of F^-. Incompatible with oxidizing materials.

BENZO(b)TRIPHENYLENE HR: 3
CAS: 215-58-7   NIOSH: DM 1925000
mf: $C_{22}H_{14}$ mw: 278.36

PROP: Clear plates or leaflets; mp: 267°.

SYNS: DIBENZ(A,C)ANTHRACENE * 1,2:3,4-DIBENZANTHRACENE

THR: An experimental carcinogen. Mutagenic data. When heated to decomposition, it emits acrid smoke and irritating fumes.

S-((3-BENZOXAZOLINYL-6-CHLORO-2-OXO)METHYL) O,O-DIETHYLPHOSPHO-RODITHIOATE HR: 3
CAS: 2310-17-0   NIOSH: TD 5175000
mf: $C_{12}H_{15}ClNO_4PS_2$ mw: 367.82

SYNS: S-(6-CHLORO-3-(MERCAPTOMETHYL)-2-BENZOXAZOLINONE)- O,O-DIETHYL PHOSPHORO-DITHIOATE * 3-(6-CHLORO-2-OXOBENZOXAZO-LIN-3-YL)METHYL-O,O-DIETHYL PHOSPHORO-THIOLOTHIONATE * O,O-DIAETHYL-S-(6-CHLOR-2-OXO-BEN(B)-1,3-OXALIN-3-YL)-METHYL-DIT HIOPHOSPHAT (GERMAN) * O,O-DIETHYL-S-((6-CHLOOR-2-OXO-BENZOXAZOLIN-3-YL)-METHYL)-DITHIO FOSFAAT (DUTCH) * O,O-DIETHYL- S-(6-CHLOROBENZOXAZOLINYL-3-METHYL)DITHIOPHOSPHATE * O,O-DI-ETHYL- S-((6-CHLORO-2-OXOBENZOXAZOLIN-3-YL)METHYL) PHOSPHORODITHIOATE * O,O-DIETHYL-S-(6-CHLORO-2-OXO-BENZOXAZOLIN-3-YL)METHYL-PHOSPHORO THIOLOTHIONATE * 3-DIETHYLDITHIOPHOSPHO-RYLMETHYL-6-CHLOROBENZOXAZOLONE-2 * O,O-DIETIL-S-((6-CLORO-2-OXO-BENZOSSAZO-LIN-3-IL)-METIL)-DITIOFOSFATO (ITALIAN) * ENT 27,163

THR: Poison by ingestion, skin, and other routes. A cholinesterase inhibitor. See also parathion. When heated to decomposition, it emits very toxic fumes of Cl^-, NO_x, PO_x, and SO_x.

BENZOYL CHLORIDE HR: 3
CAS: 98-88-4   NIOSH: DM 6600000
DOT: 1736
mf: C_7H_5ClO mw: 140.57

PROP: Colorless, fuming pungent liquid, decomp in water; mp: $-0.5°$, bp: 197°, flash p: 162°F (CC), d: 1.2187 @ 15°/15°, vap press: 1 mm @ 32.1°, vap d: 4.88.

SYN: BENZENECARBONYL CHLORIDE

DOT Classification: Label: Corrosive

THR: Implicated as a carcinogen. Powerful irritant to skin, eyes, and mucous membranes, and by ingestion and inhalation. Moderate fire hazard when exposed to heat or flame. See also chlorides. Will react with water or steam to produce heat and toxic and corrosive fumes. To fight fire, use alcohol foam, CO_2, dry chemical. Incompatible with dimethyl sulfoxide; $(NaN_3 + KOH)$; water; steam; oxidizers. For further information, see Vol. 2, No. 1 of *DPIM Report*.

BENZOYLOXYTRIBUTYLSTAN-NANE HR: 3
CAS: 4342-36-3   NIOSH: WH 6710000
mf: $C_{19}H_{32}O_2Sn$ mw: 411.20

SYNS: TRIBUTYLTIN BENZOATE * TRI-N-BU-TYL-ZINN BENZOATE (GERMAN)

THR: Poison by ingestion and intravenous routes. See also tin compounds. When heated to decomposition, it emits acrid smoke and irritating fumes.

BENZOYL PEROXIDE HR: 3
CAS: 94-36-0   NIOSH: DM 8575000
DOT: 2085
mf: $C_{14}H_{10}O_4$ mw: 242.24

PROP: White, granular, tasteless, odorless powder, insol in water, sol in benzene, acetone, chloroform; mp: 103°-105° (decomp), bp: decomps explosively, autoign temp: 176°F.

SYNS: BENZOYLPEROXID (GERMAN) * BEN-ZOYLPEROXYDE (DUTCH) * BENZOYL SUPEROX-IDE * DIBENZOYLPEROXID (GERMAN) * DI-BENZOYLPEROXYDE (DUTCH) * LUCIDOL * PEROSSIDO DI BENZOILE(ITALIAN) * PEROX-YDE DE BENZOYLE (FRENCH)

OSHA PEL: TWA 5 mg/m^3
ACGIH TLV: TWA 5 mg/m^3
DFG MAK: 5 mg/m^3
DOT Classification: Label: Organic Peroxide

THR: Poison by inhalation and intraperitoneal routes. Moderately toxic by ingestion and dermal routes. Can cause dermatitis and asthmatic effects, testicular atrophy, and vasodilation. An allergen. Moderate fire hazard by spontaneous chemical reaction in contact with reducing agents; a powerful oxidizer. Dangerous explosion hazard; may explode spontaneously, when heated to above melting point, or when overheated under confinement. Reacts violently in contact with various organic or inorganic acids; alcohols; amines; metallic naphthenates; as well as with polymerization accelerators; i.e., dimethylaniline; $(CCl_4 + C_2H_4)$; methyl methacrylate. Decomposition yields dense white smoke of benzoic acid, phenyl benzoate, terphenyls, biphenyls, benzene and carbon dioxide. To fight fire, use water spray, foam. All precautions must be taken to guard against fire and explosion hazards. Keep in a cool place; out of the direct rays of the sun; away from sparks, open flames and other sources of heat; away from shock, rough handling, friction from grinding, etc. Isolated storage is required; keep away from possible contact with acids, alcohols, ethers or other reducing agents or polymerization catalysts such as dimethylaniline. Complete instructions on storage and handling available from manufacturer. For further information, see Vol. 2, No. 1 of *DPIM Report*.

BENZOYL PEROXIDE, WET

PROP: A paste or wetted granular material containing at least 30% water. Autoign temp 176°F.

THR: See also benzoyl peroxide. Moderate fire hazard by chemical reaction with reducing agents; a powerful oxidizer. Mixed with a large surplus of water (i.e., 30%), this material is relatively safe. It is most dangerous when it contains very little water (1% or less). To fight fire, use water, foam or spray. Care must be taken to prevent drying out of wet material.

BENZYL ACETATE   HR: 3
CAS: 140-11-4   NIOSH: AF 5075000
mf: $C_9H_{10}O_2$ mw: 150.19

PROP: Liquid; mp: −51.5°, bp: 213.5°, flash p: 216°F (CC), d: 1.06, autoign temp: 862°F, vap press: 1 mm @ 45°, vap d: 5.1.

SYNS: ACETIC ACID BENZYL ESTER * ACETIC ACID PHENYLMETHYL ESTER * ALPHA-ACETOXYTOLUENE * BENZYL ETHANOATE * NCI-c06508

THR: Poison by inhalation. A human systemic irritant by inhalation. See also esters. Slight fire hazard when exposed to heat or flame; can react with oxidizing materials. To fight fire, use alcohol foam, CO_2. When heated to decomposition, it emits irritating fumes.

BENZYL ALCOHOL   HR: 3
CAS: 100-51-6   NIOSH: DN 3150000
mf: C_7H_8O mw: 108.15

PROP: Found in Jasmine, Hyacinth, Ylang-Ylang Oils and at least two dozen other essential oils. Water white liquid, faint aromatic odor; mp: −15.3°, bp: 205.7°, flash p: 213°F (CC), d: 1.050 @ 15°/15°, autoign temp: 817°F, vap press: 1 mm @ 58.0°, vap d: 3.72.

SYNS: BENZAL ALCOHOL * BENZENECARBINOL * BENZENEMETHANOL * BENZOYL ALCOHOL * ALPHA-HYDROXYTOLUENE * NCI-c06111 * PHENOLCARBINOL * PHENYLCARBINOL * PHENYLMETHANOL * PHENYLMETHYL ALCOHOL * ALPHA-TOLUENOL

THR: A species dependent oral poison. A skin and eye irritant. Slight fire hazard when exposed to heat or flame; can react with oxidizing materials and acids. To fight fire, use alcohol foam, CO_2, dry chemical. For further information, see Vol. 4, No. 6 of *DPIM Report*.

BENZYL-6-AMINOPENICILLINIC ACID   HR: 3
CAS: 61-33-6   NIOSH: XH 9400000
mf: $C_{16}H_{18}N_2O_4S$ mw: 334.42

SYNS: BENZOPENICILLIN * BENZYLPENICILLIN * BENZYLPENICILLIN G * BENZYLPENICILLINIC ACID * FREE BENZYLPENICILLIN * PENICILLIN G * PHENYLACETAMIDOPENICILLANIC ACID * (PHENYLMETHYL) PENICILLINIC ACID

THR: Poison by ingestion, intravenous, intracerebral, intraspinal, and subcutaneous routes. An experimental carcinogen. When heated to decomposition, it emits very toxic fumes of NO_x and SO_x.

BENZYLBARBITAL   HR: 3
CAS: 36226-64-9   NIOSH: CQ 0504500
mf: $C_{13}H_{14}N_2O_3$ mw: 246.29

SYNS: ETHYLBENZYLBARBITURIC ACID * 2,4,6(1H,3H,5H)-PYRIMIDINETRIONE, 5-ETHYL-5-(PHENYLMETHYL)- (9CI) * 5-BENZYL-5-ETHYLBARBITURICACID

THR: Poison by intraperitoneal and subcutaneous routes. When heated to decomposition, it emits toxic fumes of NO_x.

BENZYL BROMIDE   HR: 2
CAS: 100-39-0   NIOSH: XS 7965000
DOT: 1737
mf: C_7H_7Br mw: 171.05

PROP: Clear refractive liquid, pleasant odor, lachrymator, insol in water; mp: $-4.0°$, bp: 198°, d: 1.438 @ 22°/0°, vap d: 5.8.

SYNS: (BROMOMETHYL)BENZENE * BROMO-PHENYLMETHANE * OMEGA-BROMOTOLUENE * ALPHA-BROMOTOLUENE

DOT Classification: Label: Corrosive

THR: Intensely irritating and corrosive to skin, eyes, and mucous membranes. Large doses cause central nervous system depression. Mutagenic data. See also bromides. For further information, see Vol. 2, No. 3 in *DPIM Report*.

BENZYL BUTYL PHTHALATE   HR: 2
CAS: 85-68-7   NIOSH: TH 9990000
mf: $C_{19}H_{20}O_4$ mw: 312.39

PROP: Clear, oily liquid; mp: $<-35°$, bp: 370°, flash p: 390°F, d: 1.116 @ 25°/25°, vap d: 10.8.

SYNS: BUTYL BENZYL PHTHALATE * NCI-c54375

THR: Moderately toxic by inhalation and intraperitoneal routes. See also esters. Slight fire hazard when exposed to heat or flame; can react with oxidizers. To fight fire, use spray or mist, CO_2, dry chemical. When heated to decomposition, it emits acrid smoke and irritating fumes. For further information, see Butyl Benzyl Phthalate, Vol. 2, No. 2 of *DPIM Report*.

BENZYL CHLOROFORMATE   HR: 3
CAS: 501-53-1   NIOSH: LQ 5860000
DOT: 1739
mf: $C_8H_7ClO_2$ mw: 170.60

PROP: Colorless to pale yellow liquid, odor of phosgene.

SYNS: BENZYLCARBONYL CHLORIDE * BENZYL CHLOROCARBONATE * BENZYLOXYCARBONYL CHLORIDE * CARBOBENZYLOXY CHLORIDE * CHLOROFORMIC ACID BENZYL ESTER

DOT Classification: Label: Corrosive

THR: Poison by ingestion and inhalation routes. A powerfully corrosive irritant. See also esters and chlorides. Will react with water or steam to produce heat and toxic and corrosive fumes of Cl^-.

BENZYL CINNAMATE   HR: 2
CAS: 103-41-3   NIOSH: GD 8400000
mf: $C_{16}H_{14}O_2$ mw: 238.30

PROP: Found in balsams of Peru, Tolu, Styrax, Copaiba and others. White crystals, aromatic odor; mp: 39°, bp: 350.0°, vap press: 1 mm @ 173.8°.

SYNS: BENZYL ALCOHOL, CINNAMIC ESTER * BENZYL GAMMA-PHENYLACRYLATE * CIN-NAMEIN

THR: Moderately toxic by ingestion. A mild allergen and skin irritant. See also esters. When heated to decomposition, it emits acrid smoke and irritating fumes.

1-BENZYL-2,5-DIMETHYL SEROTONIN
HYDROCHLORIDE   HR: 3
CAS: 525-02-0   NIOSH: NL 4025000
mf: $C_{19}H_{22}N_2O \cdot ClH$ mw: 330.89

SYNS: 3-(2-AMINOETHYL)-1-BENZYL-5-ME-THOXY-2-METHYLINDOLE HYDROCHLORIDE * 1-BENZYL-2-METHYL-3-(2-AMINOETHYL)-5-METHOXYINDOLE HYDROCHLORIDE * 1-BENZYL-2-METHYL-5-METHOXYTRYPTAMINE HYDROCHLO-RIDE

THR: Poison by intraperitoneal route. Causes psychotropic effects in humans. When heated to decomposition, it emits very toxic fumes of HCl and NO_x.

BENZYL ETHER   HR: 2
CAS: 103-50-4   NIOSH: DQ 6125000
mf: $C_{14}H_{14}O$ mw: 198.28

PROP: Liquid; mp: 5°, bp: 298°, flash p: 275°F (CC), d: 1.036, vap d: 6.84.

SYNS: BENZYL OXIDE (CZECH) * DIBENZYL-ETHER (CZECH)

THR: Moderately toxic by ingestion. Vapors are probably narcotic in high concentration. See also ethers. A skin and eye irritant. Slight fire hazard when exposed to heat or flame; can react with oxidizing materials. Moderate explosion hazard by spontaneous chemical reaction. See also ethers. To fight fire, use CO_2, dry chemical.

BENZYL FORMATE   HR: 2
CAS: 104-57-4   NIOSH: LQ 5400000
mf: $C_8H_8O_2$ mw: 136.16

SYN: BENZYL METHANOATE

THR: Moderately toxic by ingestion and dermal route. Probably narcotic in high concentrations. See also esters. When heated to decomposition, it emits acrid, irritating fumes.

BENZYL ISOTHIOCYANATE   HR: 3
CAS: 622-78-6   NIOSH: NX 8250000
mf: C_8H_7NS mw: 149.22

PROP: Orange-red crystalline solid; mp: 41°,
bp: 230°, d: 1.125

SYNS: ISOTHIOCYANIC ACID, BENZYL ESTER
* BENZYL MUSTARD OIL * BENZYLSENFOEL
(GERMAN)

THR: Poison by intraperitoneal and subcutaneous routes. See also esters. Intensely irritating.
See also thiocyanates. Moderate fire hazard via
heat, flame, and oxidizers. To fight fire, use
water, spray, foam, dry chemical. When heated
to decomposition, it emits very toxic NO_x and
SO_x.

BENZYL ISOTHIOUREA HYDROCHLO-
RIDE   HR: 3
CAS: 538-28-3   NIOSH: UM 0738000
mf: $C_8H_{10}N_2S \cdot ClH$ mw: 202.72

SYNS: 2-BENZYLISOTHIOURONIUM CHLORIDE
* BENZYL THIOPSEUDOUREA HYDROCHLORIDE
* S-BENZYLTHIURONIUM CHLORIDE * USAF
EK-2124

THR: Poison by ingestion, intraperitoneal, subcutaneous, and intravenous routes. When heated
to decomposition, it emits very toxic fumes of
HCl, SO_x, and NO_x.

BENZYL PENICILLINIC ACID SODIUM
SALT   HR: 3
CAS: 69-57-8   NIOSH: XH 9800000
mf: $C_{16}H_{17}N_2O_4S \cdot Na$ mw: 356.40

SYNS: AMERICAN PENICILLIN * SODIUM BEN-
ZYLPENICILLIN G * SODIUM BENZYLPENICILLI-
NATE * SODIUM PENICILLIN G * SODIUM PEN-
ICILLIN II

THR: Poison by intramuscular route. An experimental carcinogen. Moderately toxic via intravenous route. When heated to decomposition,
it emits very toxic fumes of NO_x and SO_x.

BENZYL THIOCYANATE   HR: 3
CAS: 3012-37-1   NIOSH: XK 8155000
mf: C_8H_7NS mw: 149.22

PROP: Orange-red crystalline solid; mp: 41°;
bp: 230°; d: 1.125

SYN: BENZYL MUSTARD OIL

THR: Poison by intraperitoneal route. See also
thiocyanates. When heated to decomposition,
it emits very toxic NO_x, SO_x, and CN^-.

BENZYL TRICHLORIDE   HR: 3
CAS: 98-07-7   NIOSH: XT 9275000
mf: $C_7H_5Cl_3$ mw: 195.47

PROP: Clear, colorless to yellowish liquid; penetrating odor; mp: −5°, bp: 221°, d: 1.38 @
15.5°/15.5°, vap d: 6.77.

SYNS: BENZENYL CHLORIDE * BENZENYL TRI-
CHLORIDE * BENZYLIDYNE CHLORIDE
* PHENYLTRICHLOROMETHANE * TOLUENE TRI-
CHLORIDE * 1-(TRICHLOROMETHYL)BENZENE
* TRICHLOROPHENYLMETHANE * AL-
PHA,ALPHA,ALPHA-TRICHLOROTOLUENE
* OMEGA,OMEGA,OMEGA-TRICHLOROTOLUENE
* BENZOTRICHLORIDE * CHLORURE DE BEN-
ZENYLE (FRENCH) * PHENYL CHLOROFORM
* TRICHLOORMETHYLBENZEEN (DUTCH)
* TRICHLORMETHYLBENZOL (GERMAN)
* TRICHLOROMETHYLBENZENE * TRICLOROME-
TILBENZENE (ITALIAN) * TRICLORQTOLUENE
(ITALIAN)

THR: Poison by inhalation. Skin, eye, and mucous membrane irritant. Large doses can cause
central nervous system depression. Mutagenic
data. See also HCl. For further information,
see Vol. 6, No. 1 of DPIM Report.

BENZYL TRIMETHYL AMMONIUM IO-
DIDE   HR: 3
CAS: 4525-46-6   NIOSH: BO 8585000
mf: $C_{10}H_{16}N \cdot I$ mw: 277.17

SYNS: BENZYLDIMETHYLAMINE METHIODIDE
* PHENMETHYL-TRIMETHYLAMMONIUM IODIDE

THR: Poison by intraperitoneal and intravenous
routes. See also iodides. When heated to decomposition, it emits very toxic fumes of
NO_x and I^-.

BERBERINE   HR: 3
CAS: 2086-83-1   NIOSH: DR 9870000
mf: $C_{20}H_{18}NO_4$ mw: 336.39

PROP: White to yellow crystals. mp (anhyd):
145°.

SYNS: BERBERIN * 9,10-DIMETHOXY-2,3-
(METHYLENEDIOXY)-7,8,13,13A-TETRAHYDRO-
BERBINIUM

THR: An alkaloid poison by subcutaneous route.
In humans, toxic doses lower the temperature,
increase peristalsis, and cause death by central
paralysis. Should carry a poison label. Should
never be ingested without the advice of a physi-

cian. Should not be handled excessively, since it may be absorbed through the skin and have a toxic effect upon the body. Mutagenic data. When heated to decomposition, it emits highly toxic fumes of NO_x.

BERBERINE SULFATE TRIHY-
DRATE   HR: 3
CAS: 69352-97-2   NIOSH: DR 9880000
mf: $C_{40}H_{36}N_2O_8 \cdot O_4S \cdot 3H_2O$ mw: 822.90

SYNS: 5,6-DIHYDRO-9,10-DIMETHOXYBEN-ZO(G)-1,3-BENZODIOXOLO(5,6-A)QUINOLIZINIUM SULFATE TRIHYDRATE * 7,8,13,13A-TETRA-DEHYDRO-9,10-DIMETHOXY-2,3-(METHYLENEDI-OXY)BERBINIUM SULFATE TRIHYDRATE * UM-BELLATINE SULFATE TRIHYDRATE

THR: Poison by intraperitoneal and subcutaneous routes. See also berberine and sulfates. When heated to decomposition, it emits very toxic SO_x and NO_x.

BERTRANDITE   HR: 3
CAS: 12161-82-9   NIOSH: DS 1225000
mf: $H_{10}O_9Si_2 \cdot H_2O \cdot Be_4$ mw: 264.34

SYN: BERYLLIUM SILICATE HYDRATE

THR: An experimental and possibly human carcinogen. See also beryllium and beryllium compounds. When heated to decomposition, it emits very toxic BeO.

BERYL   HR: 3
CAS: 1302-52-9   NIOSH: DS 1400000
mf: $Al_2O_{18}Si_6 \cdot 3Be$ mw: 537.53

PROP: Green, blue, yellow or white crystals. d: 2.63-2.91.

SYNS: BERYL ORE * BERYLLIUM ALUMINIUM SILICATE * BERYLLIUM ALUMINOSILICATE

OSHA PEL: TWA 2 ug/m³; CL 5; Pk 25/30M/8H

THR: An experimental and suspected human carcinogen. See also beryllium compounds.

BERYLLIUM   HR: 3
CAS: 7440-41-7   NIOSH: DS 1750000
DOT: 1567
Af: Be   Aw: 9.01

PROP: A grayish-white, hard light metal; mp: 1278°, bp: 2970°, d: 1.85.

SYNS: BERYLLIUM-9 * GLUCINUM

OSHA PEL: TWA 2 ug/m³; CL 5; Pk 25/30M/8H
ACGIH TLV: TWA 0.002 mg/m³
TRK: Grinding of beryllium 0.005 mg/m³ calculated as Be in that portion of dust that can possibly inhaled; metal and metal-alloys other 0.002 mg/m³ calculated as Be in that portion of dust that can possibly be inhaled
DOT Classification: Poison B, Flammable Solid Powder and Poison (metal)

THR: An experimental and suspected human carcinogen and cause of pulmonary damage. See also beryllium compounds. A moderate fire hazard in the form of dust or powder, or when exposed to flame or by spontaneous chemical reaction. Slight explosion hazard in the form of powder or dust. When heated to decomposition it emits very toxic fumes of BeO. Incompatible with halocarbons, i.e., CCl_4; C_2HCl_3. It will flash or spark on impact. Reacts with Li; P. For further information, see Vol. 1, No. 3 of *DPIM Report*.

BERYLLIUM ACETATE   HR: 3
CAS: 543-81-7   NIOSH: AF 5250000
mf: $C_4H_6O_4 \cdot Be$ mw: 127.11

PROP: Plates; mp: decomp @ 300°.

SYN: BERYLLIUM ACETATE, NORMAL

OSHA PEL: TWA 2 ug/m³; CL 5; Pk 25/30M/8H

THR: An intraperitoneal poison. An experimental carcinogen. See also beryllium compounds. When heated to decomposition, it emits toxic BeO dust.

BERYLLIUM ALUMINUM ALLOY   HR: 3
CAS: 12770-50-2   NIOSH: DS 2200000

PROP: Alloy is 62% Beryllium and 38% Aluminum

SYNS: ALUMINIUM ALLOY, AL,Be * ALUMINUM BERYLLIUM ALLOY * BERYLLIUM-ALLUMINIUM ALLOY

THR: An experimental carcinogen. See also beryllium compounds. When heated to decomposition, it emits very toxic BeO.

BERYLLIUM CARBONATE   HR: 3
CAS: 66104-24-3   NIOSH: DS 2350000
mf: $C_2H_2Be_3O_8$ mw: 181.07

SYNS: BERYLLIUM CARBONATE, BASIC
* BERYLLIUMOXIDE CARBONATE * BIS(CAR-
BONATO(2-))DIHYDROXYTRIBERYLLIUM

THR: An experimental and suspected human carcinogen. See also beryllium compounds. When heated to decomposition, it emits toxic BeO fume.

BERYLLIUM CARBONATE (1:1)   HR: 3
CAS: 13106-47-3   NIOSH: DS 2400000
mf: $CO_3 \cdot Be$ mw: 69.02

SYN: CARBONIC ACID BERYLLIUM SALT (1:1)

OSHA PEL: TWA 2 ug/m^3; CL 5; Pk 25/30M/ 8H

THR: Poison by intraperitoneal route. See also beryllium compounds. When heated to decomposition, it emits highly toxic fumes of BeO.

BERYLLIUM CHLORIDE   HR: 3
CAS: 7787-47-5   NIOSH: DS 2625000
mf: $BeCl_2$ mw: 79.91

PROP: Colorless deliquescent needles; mp: 440°, bp: 520°, d: 1.899 @ 25°, vap press: 1 mm @ 291° (sublimes).

SYN: BERYLLIUM DICHLORIDE

OSHA PEL: TWA 2 ug/m^3; CL 5; Pk 25/30M/ 8H

THR: Poison by ingestion and intraperitoneal route. Mutagenic data; suspected human carcinogen, and experimental carcinogen. When heated to decomposition, it emits very toxic fumes of BeO and Cl$^-$. For further information, see Vol. 3, No. 5 of *DPIM Report*.

BERYLLIUM CHLORIDE TETRAHY-
DRATE   HR: 3
CAS: 13466-27-8   NIOSH: DS 2675000
mf: $BeCl_2 \cdot 4H_2O$ mw: 151.99

THR: An intraperitoneal poison. See also beryllium compounds and chlorides. When heated to decomposition, it emits very toxic Cl$^-$ and BeO.

BERYLLIUM COMPOUNDS   HR: 3
THR: Beryllium and its compounds are considered to be experimental carcinogens, tumorigens, and neoplastigens. Beryllium compounds can enter the body through inhalation of dusts and fumes, and may act locally on the skin. Even alloys of low beryllium content have been

shown to be dangerous. In industry, inhalation of the dust can cause severe lung damage with symptoms appearing within months. A delayed form of lung damage can occur up to five years after the work exposure. Lesions and tumors may result from penetrating wounds caused by broken fluorescent tubes. Effects have been reported in persons living near processing plants and families of beryllium workers. The fluoride, ammonium fluoride, sulfate, oxide, and hydroxide occur during extraction from beryllium ore. Exposure to the oxide may occur in processing of beryllium alloys and beryllium ceramics. There may be exposure to the carbonate and more complex salts such as ZnMnBe silicate during fluorescent tube manufacture.

The extraction of Be from its ore is attended by exposure to acid salts of the metal, particularly the fluoride (BeF_2), the ammonium fluoride and the sulfate ($BeSO_4$) and also to beryllium oxide (BeO), and hydroxide [$Be(OH)_2$]. Exposure to the oxide also occurs in the casting of Be alloys and in operations with beryllia ceramics. In the manufacture of fluorescent powders, lamps and sign tubes there may be exposure to Be carbonate and to more complex salts, such as ZnMnBe silicate. Even alloys of low Be content have been shown to be dangerous. Be compounds can enter the body through inhalation of the dusts and fumes and they may act locally on the skin. Exposure to Be compounds encountered in the extraction of the metal or its oxide from the ore, particularly the halide salts, has been attended, in certain individuals, by the development of dermatitis of an edematous and papulovesicular type, chronic skin ulcers, rhinitis, nasopharyngitis, epistaxis, bronchitis and in severe cases, by the development of an acute pneumonitis, with cough, scanty sputum, low-grade fever, rales, dyspnea and substernal pain. Radiographs show diffuse haziness throughout both lungs, followed by the appearance of soft, ill-defined opacities. The condition occurs while the worker is exposed, sometimes within 1 or 2 months of starting work, and recovery occurs within 2 months, as a rule, though radiographic changes sometimes persist for longer periods. Certain investigators have reported occasional failure of complete resolution, followed by fibrosis. In severe cases of pneumonitis the patient may die. Necropsies have revealed diffuse pulmonary edema, hemorrhagic extravasation, large numbers of plasma cells and a relative absence of polymorphonuclear infiltration. On the basis of experimental work with animals, certain investigators are of the opinion that the acute upper and lower respiratory effects are due chiefly to the acid radical present in the dust or fume, but this view has little support. A delayed form of lung disease, characterized by the occurrence of granulomatous areas in the lung tissue, has been reported in workers manufacturing fluorescent powders, lamps and sign tubes, casting beryllium master alloys, and in the production of beryllium from beryl ore. Symptoms can start during exposure, but they might be delayed up to 5 years or more after leaving work. The commonest symptoms are coughing, shortness of breath, loss of appetite, loss of weight, and fatigue. Rales are usually present in the bases and axillae, and the red cell count is frequently elevated. Cyanosis is common and the pulse and respiratory rates are often increased. Radiographically, three stages of the disease are described: (1) a diffuse, uniform granular shadowing extending throughout both lung fields; (2) a diffuse reticular pattern on the granular background; (3) the appearance of distinct nodules scattered through the lungs, with some enlargement and blurring of the hilar shadows. The intensity of the shadowing is usually greater in the middle third of the lung fields. The prognosis is poor. Clinical improvement may occur gradually over a period of several years, but there appears to be little tendency for the radiographic shadowing to clear. In certain cases, the disease has progressed gradually for some months or years, with death resulting from respiratory and cardiac failure. In several instances necropsies have shown the presence of a diffuse fibrosis with coarse strands of hyalinized collagen between the alveoli and, in some places, replacing them. The hyalinized areas contained granulomatous foci, the alveolar walls are thickened and fibrosed, the blood vessels being engorged and dilated. In some cases the hilar lymph nodes show granulomatous change and fibrosis. Granulomatous change has also been noted in the liver and hyaline fibrosis in the spleen. Two cases of delayed lung disease not coming to autopsy have presented papular lesions on the dorsum of the hands; on the biopsy these showed "sarcoid-like" lesions with central necrosis.

Several cases have been reported in which localized granulomatous lesions developed following penetrating wounds caused by splinters of glass from broken fluorescent light tubes. Several weeks or months following the accident, swellings were noted in the injured areas and excision revealed granulomatous tumors, which in one case was shown to contain beryllium. Several cases of beryllium granuloma have been reported in persons residing near processing plants and in families of beryllium workers. Be and its compounds are considered as experimental and suspected human carcinogens. There is no specific treatment, but temporary remissions have been produced by ACTH and cortisone.

BERYLLIUM, COMPOUND WITH NIOBIUM (12:1)   HR: 3
NIOSH: DS 3600000
mf: $Be_{12}Nb$ mw: 201.03

THR: An experimental carcinogen. See also beryllium compounds. When heated to decomposition in air it emits very toxic fumes of beryllium oxide.

BERYLLIUM, COMPOUND WITH TITANIUM (12:1)   HR: 3
NIOSH: DS 5030000
mf: $Be_{12}Ti$ mw: 156.02

THR: An experimental carcinogen. See also beryllium compounds, titanium compounds. When heated to decomposition, it emits very toxic fumes of BeO.

BERYLLIUM, COMPOUND WITH VANADIUM (12:1)   HR: 3
CAS: 12400-16-7   NIOSH: DS 5050000
mf: $Be_{12}V$ mw: 159.06

THR: An experimental carcinogen. See also beryllium compounds, vanadium compounds. When heated to decomposition, it emits very toxic fumes of BeO.

BERYLLIUM FLUORIDE   HR: 3
CAS: 7787-49-7   NIOSH: DS 2800000
mf: BeF_2 mw: 47.01

PROP: Amorphous, colorless mass; mp: 800°, d: 1.986 @ 25°.

SYN: BERYLLIUM DIFLUORIDE

OSHA PEL: TWA 2 ug/m³; CL 5; Pk 25/30M/ 8H

THR: Poison by ingestion, subcutaneous, and intraperitoneal routes. An experimental and suspected human carcinogen. See also beryllium compounds and fluorides. When heated to decomposition, it emits very toxic fumes of BeO and F^-. Incompatible with Mg. For further information, see Vol. 3, No. 5 of *DPIM Report*.

BERYLLIUM HYDROGEN PHOSPHATE (1:1)   HR: 3
CAS: 13598-15-7   NIOSH: DS 2975000
mf: $BeHO_4P$ mw: 104.99

SYNS: BERYLLIUM PHOSPHATE * PHOSPHORIC ACID, BERYLLIUM SALT (1:1) * PHOSPHOROUS ACID, BERYLLIUM SALT

OSHA PEL: TWA 2 ug/m^3; CL 5; Pk 25/30M/8H

THR: An intravenous poison. An experimental and suspected human carcinogen. See also Be compounds and phosphates. When heated to decomposition, it emits very toxic fumes of BeO and PO_x.

BERYLLIUM HYDROXIDE   HR: 3
CAS: 13327-32-7   NIOSH: DS 3150000
mf: $H_2O_2 \cdot Be$ mw: 43.03

PROP: Amorphous powder or crystals; mp: decomp @ 138°, d(cr): 1.909.

SYNS: BERYLLIUM DIHYDROXIDE * BERYLLIUM HYDRATE

OSHA PEL: TWA 2 ug/m^3; CL 5; Pk 25/30M/8H

THR: An experimental and suspected human carcinogen. See also beryllium compounds. When heated to decomposition, it emits very toxic fumes of BeO.

BERYLLIUM MANGANESE ZINC SILICATE
        NIOSH: DS 3550000
mf: $BeMnO_4SiZn$ mw: 221.41

SYNS: MANGANESE ZINC BERYLLIUM SILICATE * ZINC MANGANESE BERYLLIUM SILICATE

THR: An experimental carcinogen via inhalation and intravenous routes. See also beryllium compounds, manganese compounds, and zinc compounds. When heated to decomposition, it emits very toxic fumes of BeO.

BERYLLIUM NITRATE   HR: 3
CAS: 7787-55-5   NIOSH: DS 3675000
mf: $N_2O_6 \cdot Be \cdot 2H_2O$ mw: 169.07

PROP: White, yellowish crystals, deliquescent; mp: 60°, bp: decomp @ 100°-200°.
OSHA PEL: TWA 2 ug/m^3; CL 5; Pk 25/30M/8H

THR: An intraperitoneal poison. See also beryllium compounds and nitrates. When heated to decomposition, it emits very toxic fumes of BeO and nitrates. For further information, see Vol. 2, No. 1 of *DPIM Report*.

BERYLLIUM OXIDE   HR: 3
CAS: 1304-56-9   NIOSH: DS 4025000
mf: BeO   mw: 25.01

PROP: White amorphous powder; mp: 2530°±30°, bp: 3900° (approx), d: 3.025.

SYNS: BERYLLIA * BERYLLIUM MONOXIDE

OSHA PEL: TWA 2 ug/m^3; CL 5; Pk 25/30M/8H

THR: An experimental and suspected human carcinogen. See also beryllium compounds. Incompatible with (Mg + heat). When heated to decomposition, it emits very toxic fumes of BeO. For further information, see Vol. 1, No. 1 of *DPIM Report*.

BERYLLIUM OXYACETATE   HR: 3
CAS: 19049-40-2   NIOSH: DS 2900000
mf: $C_{12}H_{18}Be_4O_{13}$ mw: 406.34

SYNS: BERYLLIUM ACETATE, BASIC * BERYLLIUM OXIDE ACETATE * HEXAKIS(MU-ACETATO-O:O'))-MU(SUP 4)-OXOTETRABERYLLIUM * HEXAKIS(MU-ACETATO)-MU(SUP 4)-OXOTETRABERYLLIUM

THR: A poison. See also beryllium compounds. When heated to decomposition, it emits toxic BeO.

BERYLLIUM OXYFLUORIDE   HR: 3
CAS: 63990-88-5   NIOSH: DS 4200000
mf: BeF_2O_2 mw: 79.01
OSHA PEL: TWA 2 ug/m^3; CL 5 ug/m^3; Pk 25 ug/m^3/30M/8H

THR: Poison by ingestion, subcutaneous, intravenous, and intraperitoneal routes. See also beryllium compounds and fluorides. When heated to decomposition, it emits very toxic fumes of BeO and F^-.

BERYLLIUM SULFATE (1:1)   HR: 3
CAS: 13510-49-1   NIOSH: DS 4800000
mf: $O_4S \cdot Be$ mw: 105.07

PROP: Crystals; mp: 550°-600° (decomp), d: 2.443.

SYN: SULFURIC ACID, BERYLLIUM SALT (1:1)

OSHA PEL: TWA 2 ug/m³; CL 5; Pk 25/30M/ 8H

THR: Acute poison by ingestion, intraperitoneal, subcutaneous, intravenous, and intratracheal routes. An experimental and suspected human carcinogen. See also sulfates. Mutagenic data. When heated to decomposition, it emits very toxic SO_x and BeO. For further information, see Vol. 2, No. 1 of *DPIM Report*.

BERYLLIUM SULFATE TETRAHYDRATE (1:1:4)   HR: 3
CAS: 7787-56-6   NIOSH: DS 5000000
mf: $O_4S \cdot Be \cdot 4H_2O$ mw: 177.15
OSHA PEL: TWA 2 ug/m³; CL 5; Pk 25/30M/ 8H

THR: Poison by intravenous route. An experimental and suspected human carcinogen. See also beryllium compounds and sulfates. When heated to decomposition, it emits very toxic fumes of BeO and SO_x. For further information, see Vol. 1, No. 1 of *DPIM Report*.

BERYLLIUM ZINC SILICATE   HR: 3
CAS: 39413-47-3   NIOSH: VV 9100000
mf: $O_2Si \cdot Zn \cdot Be$ mw: 134.47

SYN: ZINC BERYLLIUM SILICATE

THR: An experimental and suspected human carcinogen. See also beryllium compounds, zinc compounds and silicates.

BETAMETHASONE   HR: 3
CAS: 378-44-9   NIOSH: TU 4000000
mf: $C_{22}H_{29}FO_5$ mw: 392.51

SYNS: 9-ALPHA-FLUORO-16-BETA-METHYL-PREDNISOLONE * 9-ALPHA-FLUORO-16-BETA-METHYL-1,4-PREGNADIENE-11-BETA, 17-ALPHA,21-TRIOL-3,20-DIONE * 9-FLUORO-11-BETA,17,21-TRIHYDROXY-16-BETA-METHYLPREGNA-1,4-DIENE-3,20-DIONE * 9-ALPHA-FLUORO-11-BETA,17,21-TRI-HYDROXY-16-BETA-METHYLPREGNA-1,4-DIENE-3,20-DIONE * 9-ALPHA-FLUORO-11-BETA, 17-ALPHA,21-TRIHYDROXY-16-BETA-METHYLPREGNA-1,4-DIENE-3,20-DIONE * 16-BETA-METHYL-1,4-PREGNADIENE-9-

ALPHA-FLUORO-11-BETA,17-ALPHA,21-TRIOL-3,20-DIONE

THR: An experimental teratogen. When heated to decomposition, it emits toxic fumes of F⁻.

BETEL NUT   HR: 3
                              NIOSH: DS 8150000

PROP: Mottled brown, with fawn color. Extract of 50 g sun dried Betel Nut in 100 mL boiling water.

SYNS: ARECA NUT * PINANG * ARECA CATECHU

THR: An experimental carcinogen.

delta-BHC   HR: 2
CAS: 319-86-8   NIOSH: GV 4550000
mf: $C_6H_6Cl_6$ mw: 290.82

SYNS: DELTA-BENZENEHEXACHLORIDE * ENT 9,234 * 1-ALPHA,2-ALPHA,3-ALPHA,4-BETA,5-ALPHA,6-BETA-HEXACHLOROCYCLOHEX-ANE * DELTA-HEXACHLOROCYCLOHEXANE * DELTA-1,2,3,4,5,6-HEXACHLOROCYCLOHEX-ANE * DELTA-LINDANE

THR: Moderately toxic by ingestion. When heated to decomposition, it emits toxic fumes of Cl⁻.

BHT (FOOD GRADE)   HR: 3
CAS: 128-37-0   NIOSH: GO 7875000
mf: $C_{15}H_{24}O$ mw: 220.39

PROP: White, crystalline solid; bp: 265°, fp: 68°, flash p: 260°F (TOC), d: 1.048 @ 20°/ 4°, vap d: 7.6.

SYNS: BUTYLATED HYDROXYTOLUENE * BUTYLHYDROXYTOLUENE * 2,6-DI-TERT-BU-TYL-P-CRESOL * 2,6-DI-TERC. BUTYL-P-KRESOL (CZECH) * 2,6-DI-TERT-BUTYL-1-HYDROXY-4-METHYLBENZENE * 3,5-DI-TERT-BUTYL-4-HYDROXYTOLUENE * 2,6-DI-TERT-BUTYL-P-METHYLPHENOL * 2,6-DI-TERT-BUTYL-4-METHYLPHENOL * 4-HYDROXY-3, 5-DI-TERT-BUTYLTOLUENE * METHYLDI-TERT-BUTYLPHENOL * 4-METHYL-2,6-DI-TERC. BUTYLFENOL (CZECH) * 4-METHYL-2,6-DI-TERT-BUTYLPHENOL * NCI-C03598
ACGIH TLV: TWA 10 mg/m³

THR: An experimental teratogen and carcinogen. Moderately toxic by ingestion. Used as a food additive. It may be a tumorigen. A severe eye irritant and a mild skin irritant to humans. Slight fire hazard when exposed to heat or flame. When heated to decomposition, it emits toxic fumes; can react with oxidizing materials. To fight fire, use CO_2, dry chemical.

4',4'''-BIACETANILIDE   HR: 3
CAS: 613-35-4   NIOSH: DT 2800000
mf: $C_{16}H_{16}N_2O_2$ mw: 268.34

PROP: mp: 329°.

SYNS: N,N'-(1,1'-BIPHENYL)-4,4'-DIYLBIS-ACETAMIDE 4',4'''-BIACETANILIDE * N,N'-4,4'-BIPHENYLYLENEBISACETAMIDE * 4,4'-DIACETYLAMINOBIPHENYL * N,N'-DIACETYL BENZIDINE * 4,4'-DIACETYLBENZIDINE

THR: An experimental carcinogen. Mutagenic data. When heated to decomposition, it emits toxic fumes of NO_x.

5,5'-BIANTHRANILIC ACID   HR: 3
CAS: 2130-56-5   NIOSH: DV 3325000
mf: $C_{14}H_{12}N_2O_4$ mw: 272.28

SYNS: 3,3'-BENZIDINEDICARBOXYLIC ACID * 4,4'-DIAMINO-3,3'-BIPHENYLDICARBOXYLIC ACID * 4,4'-DIAMINOBIPHENYL-3,3'-DICAR-BOXYLIC ACID * 3,3'-DICARBOXYBENZIDINE

THR: An experimental tumorigen. When heated to decomposition, it emits toxic fumes of NO_x.

BIBENZYL   HR: 3
CAS: 103-29-7   NIOSH: DT 4375000
mf: $C_{14}H_{14}$ mw: 182.28

PROP: Flash p: 264°F, autoign temp: 896°F, d: 1.0, vap d: 6.29, bp: 285°.

SYNS: DIBENZYL * 1,2-DIPHENYLETHANE

THR: Poison by intravenous route; moderately toxic by intraperitoneal route. Slight fire hazard. To fight fire, use water, spray, mist, alcohol foam, dry chemical.

BICYCLOPENTADIENE   HR: 3
CAS: 77-73-6   NIOSH: PC 1050000
mf: $C_{10}H_{12}$ mw: 132.22

PROP: Colorless crystals; mp: 32.9°, bp: 166.6°, d: 0.976 @ 35°, vap press: 10 mm @ 47.6°, vap d: 4.55, flash p: 90°F (OC).

SYNS: BISCYCLOPENTADIENE * 1,3-CYCLO-PENTADIENE, DIMER * DICYCLOPENTADIENE * DIMER CYKLOPENTADIENU (CZECH) * 3A,4,7,7A-TETRAHYDRO-4,7-METHANOINDENE

ACGIH TLV: TWA 5 ppm

THR: Poison by ingestion and intraperitoneal routes. Moderately toxic by inhalation. A skin and eye irritant. Dangerous fire hazard when exposed to heat or flame; can react with oxidizing materials. To fight fire, use alcohol foam. When heated to decomposition, it emits acrid fumes.

1,1'-BI(ETHYLENE OXIDE)   HR: 3
CAS: 1464-53-5   NIOSH: EJ 8225000
mf: $C_4H_6O_2$ mw: 86.10

PROP: Colorless liquid. bp: 142°, mp: 19°, d: 1.113 @ 18°/4°.

SYNS: 2,2'-BIOXIRANE * BUTADIENE DIOXIDE * 2,4-DIEPOXYBUTANE * 1,2:3,4-DIEPOXYBU-TANE * ERYTHRITOL ANHYDRIDE

THR: Poison by ingestion, inhalation, and dermal exposure. An experimental tumorigen. A severe skin and eye irritant. Mutagenic data. When heated to decomposition, it emits acrid smoke and irritating fumes.

BINAPACRYL   HR: 3
CAS: 485-31-4   NIOSH: GQ 5600000
mf: $C_{15}H_{18}N_2O_6$ mw: 322.35

SYNS: 2-SEC-BUTYL-4,6-DINITROPHENYL-3,3-DIMETHYLACRYLATE * 2-SEC-BUTYL-4,6-DINI-TROPHENYL-3-METHYL-2-BUTENOATE * 2-SEC-BUTYL-4,6-DINITROPHENYL-3-METHYLCROTONATE * 3,3-DIMETHYL-ACRYLATE DE 2,4-DINITRO-6-(1-METHYLPROPYLE) PHENYLE (FRENCH) * 2,4-DINITRO-6-SEC-BUTYLPHENYL-2-METHYL-CROTONATE * 4,6-DINITROPHENYL-2-SEC-BUTYL-3-METHYL-2-BUTENONATE * (6-(1-METHYL-PROPYL)-2,4-DINITRO-FENYL)-3,3-DIMETHYL-ACRYLAAT (DUTCH) * 2-(1-METHYLPROPYL)-4,6-DINITROPHENYL BETA,BETA-DIMETHACRYLATE * (6-(1-METHYL-PROPYL)-2,4-DINITRO-PHENYL)-3,3-DIMETHYL-ACRYLAT (GERMAN) * (6-(1-METIL-PROPIL)-2,4-DINITRO-FENIL)-3,3-DIMETIL-ACRILATO (ITALIAN)

THR: Very poisonous by all routes. A cholines-terase inhibitor. See also parathion. Dangerous; see phosphorus compounds. For further information, see Vol. 2, No. 4 of *DPIM Report*.

BINDON ETHYL ETHER   HR: 3
CAS: 69382-20-3   NIOSH: NK 5650000
mf: $C_{20}H_{14}O_3$ mw: 302.34

SYNS: BINDON ATHYLATHER * 2-(3-ETHOXY-1-INDANYLIDENE)-1,3-DINDANDIONE

THR: Poison by intraperitoneal route. An experimental teratogen. When heated to decomposition, it emits acrid smoke and irritating fumes.

BIOALLETHRIN **HR: 3**
CAS: 584-79-2 NIOSH: GZ 1950000

SYNS: D-TRANS ALLETHRIN * BIOALETRINA (PORTUGUESE) * (+)-TRANS-CHRYSANTHE-MUMIC ACID ESTER OF (+−)-ALLETHROLONE * ENT 16275

THR: Poison by ingestion and intravenous routes. When heated to decomposition, it emits acrid and irritating fumes.

BIPERIDEN **HR: 3**
CAS: 514-65-8 NIOSH: TN 3675000
mf: $C_{21}H_{29}NO$ mw: 311.51

SYNS: ALPHA-(BICYCLO(2.2.1)HEPT-5-EN-2-YL)-ALPHA-PHENYL-1-PIPERIDINO PROPANOL * 1-BICYCLOHEPTENYL-1-PHENYL-3-PIPERI-DINO-PROPANOL-1 * 3-PIPERIDINO-1-PHENYL-1-BICYCLOHEPTENYL-1-PROPANOL * 3-PIPERIDINO-1-PHENYL-1-BICYCLO(2.2.1)-HEPTEN-(5)-YL-PROPANOL-(1)(GERMAN) * 3-PIPERIDINO-1-PHENYL-1-BICYCLOHEPTENYL-1-PROPANOL

THR: Poison by subcutaneous and intravenous routes. Less toxic by ingestion. When heated to decomposition, it emits toxic fumes of NO_x.

BIPHENYL **HR: 3**
CAS: 92-52-4 NIOSH: DU 8050000
mf: $C_{12}H_{10}$ mw: 154.22

PROP: White scales, pleasant odor; mp: 70°, bp: 255°, flash p: 235°F (CC), d: 0.991 @ 75°/4°, autoign temp: 1004°F, vap d: 5.31, lel = 0.6% @ 232°, uel = 5.8% @ 331°F.

SYNS: BIBENZENE * 1,1′-BIPHENYL * DI-PHENYL * LEMONENE * PHENYLBENZENE

OSHA PEL: TWA 0.2 ppm
ACGIH TLV: TWA 0.2 ppm
DFG MAK: 0.2 ppm; 1 mg/m³

THR: Powerful, irritating poison by inhalation. Moderately toxic by ingestion. An experimental tumorigen and carcinogen. Mutagenic data. Slight fire hazard when exposed to heat or flame; can react with oxidizing materials. To fight fire, use CO₂, dry chemical, water spray, mist, fog. For further information, see Vol. 1, No. 5 of *DPIM Report.*

2-BIPHENYLAMINE **HR: 3**
CAS: 90-41-5 NIOSH: DU 8850000
mf: $C_{12}H_{11}N$ mw: 169.24

SYNS: O-AMINOBIPHENYL * 2-AMINOBIPHENYL * O-AMINODIPHENYL * 2-AMINODIPHENYL * O-BIPHENYLAMINE * NCI-C50282 * O-PHENYLANILINE

THR: A possible carcinogen. Moderately toxic by ingestion. Mutagenic data. When heated to decomposition, it emits toxic fumes of NO_x.

4-BIPHENYLAMINE **HR: 3**
CAS: 92-67-1 NIOSH: DU 8925000
mf: $C_{12}H_{11}N$ mw: 169.24

PROP: Colorless crystals; mp: 53°, bp: 302°, d: 1.160 @ 20°/20°, autoign temp: 842°F.

SYNS: p-AMINOBIPHENYL * 4-AMINOBIPHENYL * 4-AMINODIFENIL (SPANISH) * P-AMINODI-PHENYL * 4-AMINODIPHENYL * BIPHENYL-AMINE * (1,1′-BIPHENYL)-4-AMINE * P-BI-PHENYLAMINE * PARAAMINODIPHENYL * P-PHENYLANILINE * XENYLAMIN (CZECH) * XENYLAMINE

THR: Poison by ingestion and inhalation. A human carcinogen. Experimental tumorigen. An irritant. Effects resemble those of benzidine. See also benzidine. Mutagenic data. When heated to decomposition, it emits toxic fumes of NO_x. Slight to moderate fire hazard when exposed to heat, flames (sparks), or powerful oxidizers. To fight fire, use water spray, mist, dry chemical.

2,4′-BIPHENYLDIAMINE **HR: 3**
CAS: 492-17-1 NIOSH: DV 2100000
mf: $C_{12}H_{12}N_2$ mw: 184.26

PROP: Needles, very slightly sol in alcohol and ether; mp: 54.4°, bp: 363°.

SYNS: O,P′-BIANILINE * (1,1′-BIPHENYL)-2,4′-DIAMINE * O,P′-DIAMINOBIPHENYL * 2,4′-DIAMINODIPHENYL * 2,4′-DIPHENYL-DIAMINE * DIPHENYLINE

THR: An oral poison. An experimental tumorigen and suspected carcinogen. When heated to decomposition, it emits toxic fumes of NO_x.

4-BIPHENYLHYDROXYLAMINE HR: 3
CAS: 6810-26-0 NIOSH: NC 3150000
mf: $C_{12}H_{11}NO$ mw: 185.24

SYNS: N-4-BIPHENYLYLHYDROXYLAMINE
* 4-HYDROXYLAMINOBIPHENYL

THR: An experimental carcinogen. See also amines. Mutagenic data. When heated to decomposition, it emits highly toxic fumes of NO_x.

BIPHENYL, mixed with BIPHENYL OXIDE (3:7) HR: 2
CAS: 8004-13-5 NIOSH: DV 1500000

PROP: Eutectic mixture 73.5% phenylether and 26.5% biphenyl by weight

SYNS: PHENYL ETHER-DIPHENYL MIXTURE
* DIPHENYL MIXED WITH DIPHENYL OXIDE
* DOWTHERM A

OSHA PEL: TWA 1 ppm

THR: A powerful inhalation irritant in humans. When heated to decomposition it emits acrid smoke and irritating fumes.

2-BIPHENYLOL HR: 3
CAS: 90-43-7 NIOSH: DV 5775000
mf: $C_{12}H_{10}O$ mw: 170.22

SYNS: (1,1'-BIPHENYL)-2-OL * O-BIPHENYLOL
* 2-HYDROXYBIFENYL (CZECH) * O-HYDROXY-
BIPHENYL * 2-HYDROXYBIPHENYL * O-HY-
DROXYDIPHENYL * 2-HYDROXYDIPHENYL
* NCI-C50351 * ORTHOHYDROXYDIPHENYL
* ORTHOPHENYLPHENOL * ORTHOXENOL
* O-PHENYLPHENOL * 2-PHENYLPHENOL
* USAF EK-2219 * O-XENOL

THR: A poison via intraperitoneal route; moderately toxic by ingestion. Severe irritant to skin and eyes. Mutagenic data. When heated to decomposition, it emits acrid smoke and irritating fumes.

4-BIPHENYLOL HR: 3
CAS: 92-69-3 NIOSH: DV 5850000
mf: $C_{12}H_{10}O$ mw: 170.22

SYNS: P-HYDROXYBIPHENYL * 4-HYDROXYBI-
PHENYL * P-HYDROXYDIPHENYL * 4-HY-
DROXYDIPHENYL * PARAXENOL * P-PHENYL-
PHENOL * 4-PHENYLPHENOL

THR: Acute poison by intraperitoneal route. An experimental carcinogen and tumorigen. When heated to decomposition, it emits acrid, irritating fumes.

2-BIPHENYLOL, SODIUM SALT HR: 3
CAS: 132-27-4 NIOSH: DV 7700000
mf: $C_{12}H_9O \cdot Na$ mw: 192.20

SYNS: DOWICIDE * 2-HYDROXYDIPHENYL SO-
DIUM * MIL-DU-RID * SODIUM-2-HYDROXYDI-
PHENYL * SODIUM-O-PHENYLPHENATE
* SODIUM-2-PHENYLPHENATE * SODIUM-O-
PHENYLPHENOLATE * SODIUM-O-PHENYL-
PHENOXIDE

THR: An experimental carcinogen. Moderately toxic by ingestion; mild human skin irritant. When heated to decomposition, it emits Na_2O, acrid fumes.

2,2'-BIPYRIDINE HR: 3
CAS: 366-18-7 NIOSH: DW 1750000
mf: $C_{10}H_8N_2$ mw: 156.20

PROP: White crystals, sol in 2200 parts water, very sol in alcohol, ether, benzene, chloroform and petroleum ether; mp: 69.7°, bp: 272°-273°.

SYNS: 2,2'-BIPYRIDIN * ALPHA,ALPHA'-BI-
PYRIDINE * ALPHA,ALPHA'-BIPYRIDYL
* 2,2'-BIPYRIDYL * ALPHA,ALPHA'-DIPYRI-
DYL * 2,2'-DIPYRIDYL

THR: Poison by ingestion and intraperitoneal route. An experimental teratogen and tumorigen. Mutagenic data. When heated to decomposition, it emits toxic fumes of NO_x.

BIRCH TAR OIL HR: 2
CAS: 8001-88-5 NIOSH: DW 3000000

PROP: Brown liquid; leather-like odor. d: 0.886-0.950. Found in the tar of the bark and wood of *Betula Pendula Roth* (Fam. *Betulaceae*) and prepared by steam distillation of the tar obtained by dry distillation of the bark and wood.

THR: Moderately irritating to eyes and mucous membranes. A mild allergen. Slight fire hazard when exposed to heat or flame; can react with oxidizing materials.

2,7-BIS(ACETAMIDO)FLUORENE HR: 3
CAS: 304-28-9 NIOSH: AC 0700000
mf: $C_{17}H_{16}N_2O_2$ mw: 280.35

SYNS: 2,7-DIACETAMIDOFLUORENE * 2,7-DI-
ACETYLAMINOFLUORENE * 2,7-FLUORENYLBIS-
ACETAMIDE * N,N'-2,7-FLUORENYLENEDIACET-
AMIDE * N,N'-FLUOREN-2,7-YLBISACETAMIDE

* N,N'-FLUOREN-2,7-YLENEBISACETAMIDE
* N,N'-2,7-FLUORENYLENEBISACETAMIDE
* N,N'-(FLUOREN-2,7-YLENE)BIS(ACETYLAMINE)

THR: An experimental carcinogen and tumorigen. Mutagenic data. When heated to decomposition, it emits toxic fumes of NO_x.

BIS(ACETYLACETONATO) TITANIUM OXIDE   HR: 3
CAS: 14024-64-7   NIOSH: XR 2330000
mf: $C_{10}H_{14}O_5Ti$ mw: 262.14

SYNS: BIS(2,4-PENTANEDIONATO)TITANIUM OXIDE * TITANIUM ACETONYL ACETONATE * TITANIUM OXIDE BIS(ACETYLACETONATE) * TITANIUM, OXOBIS(2,4-PENTANEDIONATO-O,O') * TITANYL BIS(ACETYLACETONATE)

THR: An experimental tumorigen. Moderate intraperitoneal toxicity. When heated to decomposition, it emits acrid smoke and irritating fumes.

BIS(ACETYL ACETONE)COPPER   HR: 3
CAS: 13395-16-9   NIOSH: GL 6520000
mf: $C_{10}H_{14}O_4 \cdot Cu$ mw: 261.78

SYNS: BIS(2,4-PENTANEDIONATO)COPPER * COPPER(II) ACETYLACETONATE * COPPER BIS(ACETYLACETONATE) * COPPER BIS(ACETYLACETONE) * COPPER BIS(2,4-PENTANEDIONATE) * COPPER DIACETYLACETONATE * CUPRIC ACETYLACETONATE

THR: Poison by intravenous route. See also copper compounds. When heated to decomposition, it emits acrid smoke + Cu fumes.

1,3-BIS(4-ALDOX IMINOPYRIDINIUM)DIMETHYL ETHER BICHLORIDE   HR: 3
CAS: 114-90-9   NIOSH: UU 6825000
mf: $C_{14}H_{16}N_4O_3 \cdot Cl_2$ mw: 359.24

SYNS:
BIS(4-HYDROXYIMINOMETHYLPYRIDINIUM-1-METHYL)ETHER DICHLORIDE * OBIDOXIME CHLORIDE

THR: Poison by intraperitoneal, intravenous, intramuscular, and other routes. Moderately toxic by ingestion. When heated to decomposition, it emits toxic fumes of Cl^- and NO_x.

BIS(4-AMINO-3-CHLOROPHENYL) ETHER   HR: 3
CAS: 28434-86-8   NIOSH: KM 9625000
mf: $C_{12}H_{10}Cl_2N_2O$ mw: 269.14

SYNS: 3,3'-DICHLOR-4,4'-DIAMINO-DIPHENYLAETHER (GERMAN) * 3,3'-DICHLORO-4,4'-DIAMINODIPHENYL ETHER * 4,4'-OXYBIS(2-CHLOROANILINE) * 4,4'-OXYBIS(2-CHLORO-BENZENAMINE)

THR: An experimental carcinogen. When heated to decomposition, it emits toxic fumes of Cl^- and NO_x.

4,4'-BIS(1-AMINO-8-HYDROXY-2,4-DISULFO-7-NAPHTHYLAZO)-3,3'-BITOLYL, TETRASODIUM SALT   HR: 3
CAS: 314-13-6   NIOSH: QJ 6440000
mf: $C_{34}H_{24}N_6O_{14}S_4 \cdot 4Na$ mw: 960.84

SYNS: 4,4'-BIS(7-(1-AMINO-8-HYDROXY-2,4-DISULFO)NAPHTHYLAZO)-3,3'-BITOLYL, TETRASODIUM SALT * 4,4'-BIS(1-AMINO-8-HYDROXY-2,4-DISULPHO-7-NAPHTHYLAZO)-3,3'-BITOLYL, TETRASODIUM SALT * C.I. 23860

THR: An experimental carcinogen, teratogen, and tumorigen. When heated to decomposition, it emits very toxic fumes of SO_x and NO_x.

BIS(gamma-AMINOPROPYL)METHYLAMINE   HR: 3
CAS: 105-83-9   NIOSH: JL 9625000
mf: $C_7H_{19}N_3$ mw: 145.29

PROP: Liquid, completely miscible in water; d: 0.9307 @ 20°/20°, bp: 240.6°, fp: −29.6°, flash p: 220°F.

SYNS: BIS(OMEGA-AMINOPROPYL)METHYLAMINE * BIS(3-AMINOPROPYL)METHYLAMINE * N,N-BIS(GAMMA-AMINOPROPYL)METHYLAMINE * N,N-BIS(3-AMINOPROPYL)METHYLAMINE * 3,7'-DIAMINO-N-METHYLDIPROPYLAMINE * METHYLBIS(3-AMINOPROPYL)AMINE

THR: Poison by inhalation and dermal routes. Moderately toxic by ingestion. A skin and eye irritant. See also amines. A moderate fire hazard when exposed to heat or flame. To fight fire, use foam, fog, dry chemical. When heated to decomposition, it emits toxic fumes of NO_x.

2,5-BIS(1-AZIRIDINYL)-3-(2-CARBAMOYLOXY-1-METHOXYETHYL)-6-METHYL-1,4-BENZOQUINONE   HR: 3
CAS: 24279-91-2   NIOSH: GU 5300000
mf: $C_{15}H_{19}N_3O_5$ mw: 321.37

SYN: 2,5-BIS(1-AZIRIDINYL)-3-(2-HYDROXY-1-METHOXYETHYL)-6-METHYL-P-BENZOQUINONE CARBAMATE (ESTER)

THR: A poison via oral, intraperitoneal, subcutaneous, and intravenous routes. Mutagenic data. When heated to decomposition, it emits toxic NO_x.

BIS(2-BENZOTHIAZOLYLTHIO) ZINC   HR: 3
CAS: 155-04-4   NIOSH: DL 7000000
mf: $C_{14}H_8N_2S_4 \cdot Zn$ mw: 397.85

SYNS: 2-BENZOTHIAZOLETHIOL, ZINC SALT (2: 1) * BIS(MERCAPTOBENZOTHIAZOLATO)ZINC * 2-MERCAPTOBENZOTHIAZOLE ZINC SALT * USAF GY-7 * ZINC BENZOTHIAZOLYL MERCAPTIDE * ZINC-2-BENZOTHIAZOLETHIOLATE * ZINC BENZOTHIAZYL-2-MERCAPTIDE * ZINC BENZOTHIAZOL-2-YLTHIOLATE * ZINC MERCAPTOBENZOTHIAZOLATE * ZINC-2-MERCAPTOBENZOTHIAZOLE * ZINC MERCAPTOBENZOTHIAZOLE SALT

THR: An experimental carcinogen. See also zinc compounds and mercaptans. A fungicide. When heated to decomposition, it emits very toxic fumes of SO_x, NO_x and Zn.

2,5-BIS(BIS-(2-CHLOROETHYL)AMINO-METHYL)HYDROQUINONE   HR: 3
CAS: 4420-79-5   NIOSH: MX 4200000
mf: $C_{16}H_{24}Cl_4N_2O_2$ mw: 418.22

SYNS: HYDROQUINONE MUSTARD * NSC 18321 * WEATHERBEE MUSTARD

THR: An intravenous poison. An experimental carcinogen. A powerful irritant. When heated to decomposition, it emits highly toxic fumes of NO_x and Cl^-.

1,2-BIS(BROMOACETOXY)-ETHANE   HR: 3
CAS: 3785-34-0   NIOSH: AF 5960000
mf: $C_6H_8Br_2O_4$ mw: 303.96

SYNS: BROMOACETIC ACID ETHYLENE ESTER * ETHYLENE BIS(BROMOACETATE) * ETHYLENE BROMOACETATE * ETHYLENE GLYCOL, BIS-(BROMOACETATE)

THR: An intraperitoneal and intravenous poison. See also esters and bromides. When heated to decomposition, it emits toxic fumes of Br^-.

2,2-BIS(2-BROMOETHYL)-1,3-PRO-PANEDIOL   HR: 3
CAS: 3296-90-0   NIOSH: TY 3195000
mf: $C_7H_{14}Br_2O_2$ mw: 290.03

SYN: NCI-C55516

THR: A suspected carcinogen. When heated to decomposition, it emits toxic Br^-.

BIS(BUTOXYMALEOYLOXY)DIBUTYL-STANNANE   HR: 3
CAS: 15546-16-4   NIOSH: WH 6712000
mf: $C_{24}H_{40}O_8Sn$ mw: 575.33

SYNS: DI-N-BUTYLTIN DI(MONOBUTYL)MALEATE * DI-N-BUTYL-ZINN-DI(MONOBUTYL)MALEINAT (GERMAN)

THR: An oral poison. See also tin compounds. When heated to decomposition, it emits acrid smoke and irritating fumes.

BIS(BUTOXYMALEOYLOXY)DIOCTYL-STANNANE   HR: 2
CAS: 29575-02-8   NIOSH: WH 6714000
mf: $C_{32}H_{56}O_8Sn$ mw: 687.57

SYNS: DI-N-OCTYLTIN BIS(BUTYL MALEATE) * DI-N-OCTYLTIN DIMONOBUTYLMALEATE * DI-N-OCTYLZINN-DIMONOBUTYLMALEINAT (GERMAN)

THR: Moderately toxic by ingestion. See also tin compounds. When heated to decomposition, it emits acrid smoke and irritating fumes.

BIS(3-tert-BUTYL-4-HYDROXY-6-METHYLPHENYL) SULFIDE   HR: 3
CAS: 96-69-5   NIOSH: GP 3150000
mf: $C_{22}H_{30}O_2S$ mw: 358.58

PROP: Light gray to tan powder; mp: 150°, d: 1.10.

SYNS: BIS(4-HYDROXY-5-TERT-BUTYL-2-METHYLPHENYL) SULFIDE * 4,4'-THIOBIS-(6-TERT-BUTYL-M-CRESOL) * 4,4'-THIOBIS-(2-TERT-BUTYL-5-METHYLPHENOL) * 4,4'-THIOBIS(6-TERT-BUTYL-3-METHYLPHENOL) * 4,4'-THIOBIS(3-METHYL-6-TERT-BUTYLPHENOL) * 1,1'-THIOBIS(2-METHYL-4-HYDROXY-5-TERT-BUTYLBENZENE) * USAF B-15

THR: Poison by intraperitoneal route and probably ingestion and inhalation. See also sulfides. When heated to decomposition, it emits highly toxic fumes of SO_x.

BIS(2-CARBOXYETHYL) SULFIDE   HR: 3
CAS: 111-17-1   NIOSH: UF 7875000
mf: $C_6H_{10}O_4S$ mw: 178.22

PROP: Very sol in alc, hot water, acetate. Sltly sol in water; mp: 134°.

SYNS: 4-THIAHEPTANEDIOIC ACID * THIODI-HYDRACRYLIC ACID * THIODIPROPIONIC ACID * BETA,BETA′-THIODIPROPIONIC ACID * 3,3′-THIODIPROPIONIC ACID

THR: A poison by intraperitoneal and intravenous routes. Moderately toxic by ingestion. Mild skin irritant; severe eye irritant. When heated to decomposition, it emits toxic fumes of SO_x.

BIS(p-CHLOROBENZOYL) PEROXIDE   HR: 3
CAS: 94-17-7   NIOSH: SD 7875000
DOT: 2113
mf: $C_{14}H_8Cl_2O_4$ mw: 311.12

PROP: A white granular material, insol in water, sol in organic solvents.

SYNS: P-CHLOROBENZOYL PEROXIDE * P,P′-DICHLOROBENZOYL PEROXIDE

DOT Classification: Label: Organic Peroxide

THR: Poison by ingestion and intraperitoneal route. Probably an irritant to skin and mucous membranes. See also peroxides, organic. Dangerous fire hazard; a powerful oxidizer. Store in a cool place away from fire hazards, sparks, open flames, and out of the direct rays of the sun. Dangerous explosion hazard; this material may be caused to explode by heat (over 38°) or contamination. Any contaminant which acts as an accelerator to the polymerization or decomposition of this material can cause an explosion. Heat or contact with certain fumes or mists can cause it to explode. To fight small fires, use Cl^-, CO_2, or foam extinguishers may be used. Water spray or mist may also be used. Dry chemical is effective.

trans-N,N′-BIS(2-CHLOROBENZYL)-1,4-CYCLOHEXANEBIS(METHYLAMINE) DIHYDROCHLORIDE   HR: 3
CAS: 366-93-8   NIOSH: GU 7025000
mf: $C_{22}H_{28}Cl_2N_2 \cdot 2ClH$ mw: 464.34

SYNS: TRANS- 1,4-BIS(2-DICHLOROBENZYLAMINOETHYL)CYCLOHEXANE DICHLORHYDRATE (FRENCH) * TRANS-N,N′-(1,4-CYCLOHEXYLENEDIMETHYLENE)BIS(2-CHLOROBENZYLAMINE) DIHYDROCHLORIDE

THR: An experimental teratogen. Moderately

toxic by ingestion. When heated to decomposition, it emits very toxic fumes of HCl, NO_x, and Cl^-.

N,N-BIS(beta-CHLOROETHYL)-D,L-ALANINE HYDROCHLORIDE   HR: 3
CAS: 3374-04-7   NIOSH: AY 4200000
mf: $C_7H_{13}Cl_2NO_2 \cdot ClH$ mw: 250.57

SYNS: ALANINE MUSTARD * NSC 17663

THR: Poison by intracerebral and intravenous routes. A nonspecific effect in man. When heated to decomposition, it emits very toxic Cl^-, NO_x, and HCl.

BIS-beta-CHLOROETHYLAMINE   HR: 3
CAS: 334-22-5   NIOSH: IA 0175000
mf: $C_4H_9Cl_2N$ mw: 142.04

SYNS: N,N-BIS-(BETA-CHLORAETHYL)-AMIN (GERMAN) * NH-LOST * NOR-NITROGEN MUSTARD * NSC-10873

THR: Poison by intraperitoneal, subcutaneous, and intravenous routes. Mutagenic data. When heated to decomposition, it emits very toxic NO_x and Cl^-.

BIS(2-CHLOROETHYL)AMINE HYDROCHLORIDE   HR: 3
CAS: 821-48-7   NIOSH: IA 1225000
mf: $C_4H_9Cl_2N \cdot ClH$ mw: 178.50

SYNS: BIS(BETA-CHLOROETHYL)-AMINE HYDROCHLORIDE * NOR-LOST HYDROCHLORID (GERMAN) * NORNITROGEN MUSTARD HYDROCHLORIDE * 2,2′-DICHLORO DIETHYLAMINE HYDROCHLORIDE

THR: A poison by inhalation, intraperitoneal, intramuscular, and subcutaneous routes. An experimental teratogen. Mutagenic data. When heated to decomposition, it emits toxic fumes of HCl, NO_x and Cl^-.

4′-(BIS(2-CHLOROETHYL)AMINO)-ACETANILIDE   HR: 3
CAS: 1215-16-3   NIOSH: AD 9300000
mf: $C_{12}H_{16}Cl_2N_2O$ mw: 275.20

SYN: P-ACETYLAMINOPHENYL DERIVATIVE OF NITROGEN MUSTARD

THR: Poison via intraperitoneal route. An experimental teratogen. When heated to decomposition, it emits very toxic fumes of Cl^- and NO_x.

4'-(BIS(2-CHLOROETHYL)AMINO)-2-FLUORO ACETANILIDE   HR: 3
CAS: 1492-93-9   NIOSH: AD 9310000
mf: $C_{12}H_{15}Cl_2FN_2O$ mw: 293.19

SYN: P-FLUOROACETYLAMINOPHENYL DERIVA-
TIVE OF NITROGEN MUSTARD

THR: Poison by intraperitoneal route. An experimental teratogen. When heated to decomposition, it emits very toxic fumes of Cl^-, F^-,
and NO_x.

3-(p-(BIS(beta-CHLOROETHYL)AMINO)-PHENYL)-D,L-ALANINE HYDROCHLO-RIDE   HR: 3
CAS: 1465-26-5   NIOSH: AY 3980000
mf: $C_{13}H_{18}Cl_2N_2O_2 \cdot ClH$ mw: 341.69

SYNS: 4-(BIS(2-CHLOROETHYL)AMINO)DL-
PHENYLALANINE MONOHYDROCHLORIDE
∗ MELPHALAN (RUSSIAN) ∗ NCS-14210
∗ DL-SARCOLYSINE HYDROCHLORIDE
∗ SARCOLYSIN HYDROCHLORIDE
∗ SARKOKLORIN

THR: A poison via intravenous route. Mutagenic
data. When heated to decomposition it emits
very toxic fumes of Cl^-, NO_x, and HCl.

L-3-(p-(BIS(2-CHLOROETHYL)AMINO)-PHENYL)ALANINE MONOHYDROCHLO-RIDE   HR: 3
CAS: 13469-52-8   NIOSH: AY 4025000
mf: $C_{13}H_{18}Cl_2N_2O_2 \cdot ClH$ mw: 341.69

SYNS: NSC-8806 ∗ L-PHENYLALANINE MUS-
TARD HYDROCHLORIDE

THR: Poison by intravenous route. An experimental carcinogen. Affects the human gastrointestinal tract. When heated to decomposition,
it emits very toxic fumes of Cl^-, NO_x, and
HCl.

L-3-(p-(BIS(2-CHLOROETHYL)AMINO)-PHENYL)-N-FORMYLALANINE   HR: 3
CAS: 35849-41-3   NIOSH: AY 3850000
mf: $C_{14}H_{18}Cl_2N_2O_3$ mw: 333.24

SYN: N-FORMYL-L-P-SARCOLYSIN

THR: An intraperitoneal poison. Moderately
toxic by ingestion. Affects the human gastrointestinal tract. When heated to decomposition, it emits very toxic fumes of Cl^- andBM
NO_x.

0-(4-(BIS(2-CHLOROETHYL)AMINO)-PHENYL-DL-TYROSINE   HR: 3
CAS: 857-95-4   NIOSH: AY 3340000
mf: $C_{19}H_{22}Cl_2N_2O_3$

SYN: PHENTYRIN

THR: A poison by ingestion, intravenous, and
intraperitoneal routes. When heated to decomposition, it emits very toxic fumes of Cl^- and
NO_x.

o-(4-BIS(beta-CHLOROETHYL)AMINO-o-TOLYLAZO)BENZOIC ACID   HR: 3
CAS: 4213-40-5   NIOSH: DG 4425000
mf: $C_{18}H_{19}Cl_2N_3O_2$ mw: 380.30

SYN: NSC-16498

THR: Poison by intraperitoneal and intravenous
routes. Mutagenic data. When heated to decomposition, it emits very toxic fumes of Cl^- and
NO_x.

5-(BIS(2-CHLOROETHYL)AMINO)URA-CIL   HR: 3
CAS: 66-75-1   NIOSH: YQ 8925000
mf: $C_8H_{11}Cl_2N_3O_2$ mw: 252.12

SYNS: CHLORETHAMINACIL ∗ AMINOURACIL
MUSTARD ∗ 5-(BIS(2-CHLOROETHYL)AMINO)-
2,4(1H,3H)PYRIMIDINEDIONE ∗ 5-N,N-
BIS(2-CHLOROETHYL)AMINOURACIL
∗ 5-(DI-(beta-CHLOROETHYL)AMINO)URACIL
∗ 5-(DI-2-CHLOROETHYL)AMINOURACIL
∗ 2,6-DIHYDROXY-5-BIS(2-CHLOROETHYL)-
AMINOPYRAMIDINE ∗ ENT 50439 ∗ NCI-CO-
4820 ∗ NSC-34462 ∗ URACIL MUSTARD
∗ URAMUSTINE

THR: An acute poison by ingestion and intraperitoneal route. Mutagenic data. An experimental carcinogen, teratogen, and neoplastigen.
When heated to decomposition, it emits very
toxic fumes of Cl^- and NO_x.

N,N-BIS(2-CHLOROETHYL)BENZYL-AMINE   HR: 3
CAS: 55-51-6   NIOSH: DP 1575000
mf: $C_{11}H_{15}Cl_2N$ mw: 232.17

SYNS: BENZYL NOR-MECHLORETHAMINE
∗ BENZYLBIS(BETA-CHLOROETHYL)AMINE
∗ N,N-BIS(2-CHLOROETHYL)BENZENEMETHAN-
AMINE

THR: An inhalant and subcutaneous poison.
When heated to decomposition, it emits very
toxic fumes as Cl^- and NO_x.

N,N-BIS(2-CHLOROETHYL)-3-CHLORO-6-ETHOXYBENZYLAMINE HYDRO-CHLORIDE   HR: 2
NIOSH: DP 1579000
mf: $C_{13}H_{18}Cl_3NO \cdot ClH$ mw: 347

SYN: 2-ETHOXY-5-CHLOROBENZYL-BIS-(BETA-CHLOROETHYL) AMINE HYDROCHLO-RIDE

THR: Mutagenic data. When heated to decomposition, it emits very toxic fumes such as Cl^- and NO_x.

BIS(2-CHLOROETHYL) ETHER   HR: 3
CAS: 111-44-4   NIOSH: KN 0875000
mf: $C_4H_8Cl_2O$ mw: 143.02

PROP: Colorless, stable liquid. bp: 178.5°, fp: −51.9°, flash p: 131°F (CC), d: 1.2220 @ 20°/20°, autoign temp: 696°F, vap press: 0.7 mm @ 20°, vap d: 4.93.

SYNS: BIS(BETA-CHLOROETHYL) ETHER
* 1-CHLORO-2-(BETA-CHLOROETHOXY)ETHANE
* 2,2′-DICHLOORETHYLETHER (DUTCH)
* 2,2′-DICHLOR-DIAETHYLAETHER (GERMAN)
* 2,2′-DICHLORETHYL ETHER * BETA,BETA-DICHLORODIETHYL ETHER * DICHLOROETHYL ETHER * DI(BETA-CHLOROETHYL)ETHER
* BETA,BETA′-DICHLOROETHYL ETHER
* SYM-DICHLOROETHYL ETHER * 2,2′-DICHLOROETHYL ETHER * DICHLOROETHYL OXIDE
* 2,2′-DICLOROETILETERE (ITALIAN)
* DWUCHLORODWUETYLOWY ETER (POLISH)
* ENT 4,504 * ETHER DICHLORE (FRENCH)
* 1,1′-OXYBIS(2-CHLORO)ETHANE * OXYDE DE CHLORETHYLE (FRENCH)

OSHA PEL: TWA CL 15 ppm (skin)
DFG MAK: 10 ppm (60 mg/m³)

THR: A poison by ingestion and inhalation. Moderately toxic via skin contact. A skin, eye, and mucous membrane irritant. Exposure to 1000 ppm for 30 to 60 minutes may result in death within days. The odor is easily detectable at 35 ppm which causes only slight irritation. Flammable when exposed to heat, flame or oxidants. Dangerous explosion hazard; reacts vigorously with oleum; chlorosulfonic acid. See ethers. When heated to decomposition, it emits highly toxic fumes; reacts with water or steam to evolve toxic and corrosive fumes; can react vigorously with oxidizing materials. To fight fire, use water, foam, mist, fog, spray, dry chemical.

BIS(2-CHLOROETHYL)ETHYL-AMINE   HR: 3
CAS: 538-07-8   NIOSH: YE 1225000
mf: $C_6H_{13}Cl_2N$ mw: 170.10

SYNS: ETHYLBIS(BETA-CHLOROETHYL)AMINE
* ETHYLBIS(2-CHLOROETHYL)AMINE

THR: Poison by inhalation, intravenous, skin, subcutaneous, and intraperitoneal routes. When heated to decomposition, it emits very toxic fumes of Cl^- and NO_x.

BIS(beta-CHLOROETHYL)FOR-MAL   HR: 3
CAS: 111-91-1   NIOSH: PA 3675000
mf: $C_5H_{10}Cl_2O_2$ mw: 173.05

PROP: Liquid; bp: 217.5°, flash p: 230°F (OC), d: 1.23, vap d: 5.9.

SYNS: BIS(2-CHLOROETHYL)FORMAL * DI-CHLOROETHYL FORMAL * DI-2-CHLOROETHYL FORMAL * FORMALDEHYDE BIS(BETA-CHLORO-ETHYL) ACETAL

THR: Poison by ingestion, inhalation, and skin absorption. A skin and eye irritant. Slight fire hazard when exposed to heat or flame. Dangerous; see also chlorides. Incompatible with oxidizers. To fight fire, use alcohol foam, foam, CO_2, dry chemical. When heated to decomposition, it emits toxic fumes of Cl^-. For further information, see Vol. 6, No. 3 of *DPIM Report*.

BIS(beta-CHLOROETHYL)METHYL-AMINE   HR: 3
CAS: 51-75-2   NIOSH: IA 1750000
mf: $C_5H_{11}Cl_2N$ mw: 156.07

PROP: Dark liquid; mp: 1° @ 10 mm, d: 1.09 @ 25°, vap press: 0.17 mm @ 25°, vap d: 5.9.

SYNS: BIS(2-CHLOROETHYL)METHYLAMINE
* N,N-BIS(2-CHLOROETHYL)METHYLAMINE
* CHLORAMINE * CHLORAMINE (THE NITRO-GEN MUSTARD) * DICHLOREN (GERMAN)
* BETA,BETA′-DICHLORODIETHYL-N-METHYL-AMINE * DI(2-CHLOROETHYL)METHYLAMINE
* 2,2′-DICHLORO-N-METHYLDIETHYLAMINE
* METHYLBIS(BETA-CHLOROETHYL)AMINE
* N-METHYL-BIS-CHLORAETHYLAMIN (GERMAN)
* N-METHYL-BIS(BETA-CHLOROETHYL)AMINE
* N-METHYL-2,2′-DICHLORODIETHYLAMINE
* METHYLDI(2-CHLOROETHYL)AMINE * N-METHYL-LOST * MUSTINE * NITROGEN MUSTARD * NSC 762

THR: A deadly poison by unspecified routes in humans. An experimental carcinogen, teratogen, tumorigen, neoplastigen. Powerful irritant via eyes, dermal, inhalation, and intravenous routes. Mutagenic data. When heated to decomposition, it emits very toxic fumes of Cl$^-$ and NO$_x$.

BIS(2-CHLOROETHYL)METHYLAMINE HYDROCHLORIDE   HR: 3
CAS: 55-86-7   NIOSH: IA 2100000
mf: C$_5$H$_{11}$Cl$_2$N • ClH   mw: 192.53

SYNS: CHLORAMINE * 2-CHLORO-N-(2-CHLO-ROETHYL)-N-METHYLETHANAMINE HYDROCHLO-RIDE * BETA,BETA'-DICHLORODIETHYL-N-METHYLAMINE HYDROCHLORIDE * DI(2-CHLOROETHYL)METHYLAMINE HYDROCHLORIDE * 2,2'-DICHLORO-N-METHYLDIETHYLAMINE HY-DROCHLORIDE * N-LOST * N-METHYL-BIS-BETA-CHLORETHYLAMINE HYDROCHLORIDE * METHYLBIS(2-CHLOROETHYL)AMINE HYDRO-CHLORIDE * N-METHYL-2,2'-DICHLORODIETH-YLAMINE HYDROCHLORIDE * N-METHYL-DI-2-CHLOROETHYLAMINE HYDROCHLORIDE * METHYLDI(BETA-CHLOROETHYL)AMINE HY-DROCHLORIDE * METHYLDI(2-CHLOROETHYL)-AMINE HYDROCHLORIDE * NCI-C56382 * NITROGEN MUSTARD HYDROCHLORIDE * NSC-762 HYDROCHLORIDE

THR: Acute poison by ingestion, intravenous, subcutaneous, intraperitoneal, and parenteral routes. An experimental teratogen, carcinogen, neoplastigen, and tumorigen. Mutagenic data.

N,N-BIS(2-CHLOROETHYL)-2-NAPHTHYLAMINE   HR: 3
CAS: 494-03-1   NIOSH: QM 2450000
mf: C$_{14}$H$_{15}$Cl$_2$N   mw: 268.20

SYNS: 2-BIS(2-CHLOROETHYL)AMINONAPHTHA-LENE * BIS(2-CHLOROETHYL)-BETA-NAPHTHYL-AMINE * DICHLOROETHYL-BETA-NAPHTHYL-AMINE * DI(2-CHLOROETHYL)-BETA-NAPHTHYL-AMINE * N,N-DI(2-CHLOROETHYL)-BETA-NAPHTHYLAMINE * 2-N,N-DI(2-CHLORO-ETHYL)NAPHTHYLAMINE * NAPHTHYLAMINE MUSTARD * BETA-NAPHTHYL-BIS-(BETA-CHLOROETHYL)AMINE * 2-NAPHTHYLBIS(2-CHLOROETHYL)AMINE * BETA-NAPHTHYL-DI-(2-CHLOROETHYL)AMINE * NSC-62209

THR: A human carcinogen. Moderately toxic by intraperitoneal route. When heated to decomposition, it emits very toxic fumes of Cl$^-$ and NO$_x$.

N,N'-BIS(2-CHLOROETHYL)-N-NITRO-SOUREA   HR: 3
CAS: 154-93-8   NIOSH: YS 2625000
mf: C$_5$H$_9$Cl$_2$N$_3$O$_2$ mw: 214.07

SYNS: BIS(2-CHLOROETHYL)NITROSOUREA * 1,3-BIS(BETA-CHLOROETHYL)-1-NITROSOUREA * 1,3-BIS-(2-CHLOROETHYL)-1-NITROSOUREA * NCI-C04773 * NSC-409962

THR: Poison by ingestion, intravenous, intraperitoneal, and subcutaneous routes. An experimental teratogen and tumorigen, and suspected human carcinogen. Mutagenic data. Human blood, gastrointestinal tract, and other unspecified effects. When heated to decomposition, it emits very toxic fumes of Cl$^-$ and NO$_x$.

BIS(2-CHLOROETHYL)SULFIDE   HR: 3
CAS: 505-60-2   NIOSH: WQ 0900000
mf: C$_4$H$_8$Cl$_2$S   mw: 159.08

PROP: Colorless (if pure), to light yellow, oily liquid; bp: 228°, fp: 14.4°, flash p: 221°F, d: 1.2741 @ 20°/4°, vap d: 5.4, vap press: 0.09 mm @ 30°.

SYNS: BIS(BETA-CHLOROETHYL)SULFIDE * BIS(2-CHLOROETHYL)SULPHIDE * 1-CHLORO-2-(BETA-CHLOROETHYLTHIO)ETHANE * 2,2'-DI-CHLORODIETHYL SULFIDE * DI-2-CHLOROETHYL SULFIDE * BETA,BETA'-DICHLOROETHYL SUL-FIDE * BETA,BETA-DICHLOR-ETHYL-SULPHIDE * 2,2'-DICHLOROETHYL SULPHIDE * DISTILLED MUSTARD * KAMPSTOFF "LOST" * MUSTARD HD * MUSTARD GAS * MUSTARD VAPOR * SCHWEFEL-LOST * S-LOST * SULFUR MUS-TARD GAS * SULFUR MUSTARD * SULPHUR MUSTARD GAS * 1,1'-THIOBIS(2-CHLOROETH-ANE) * YELLOW CROSS LIQUID * YPERITE

THR: Poison by inhalation, skin contact, subcutaneous, and intravenous routes. A severe skin and moderate eye irritant. An experimental neoplastigen, carcinogen, and tumorigen by several routes. Mutagenic data. A military blistering gas. Strongly effects the skin, eyes, lungs, and gastric system. Pulmonary lesions are often fatal. it penetrates the skin deeply to injure blood vessels. Minute amounts can cause inflammation. Secondary infections are common. Combustible when exposed to heat or flame; can be ignited by a large explosive charge. Dangerous; when heated to decomposition or on contact with acid or acid fumes, it emits highly toxic fumes of oxides of sulfur and chlorides; will react with water or steam to produce toxic and corrosive fumes; can react vigorously with oxidizing materials. To fight fire, use water, foam, CO_2, dry chemical. Incompatible with bleaching powder.

Treatment of local lesions is mainly by cleanliness and emollients, similar to that of burns. Oils protect the skin only slightly. Immediately after exposure, the poison may be partly removed by scrubbing the victim with kerosene, but penetration is so rapid that this treatment is not successful if delayed for 15 to 30 minutes.

BIS(2-CHLOROETHYL)SULFONE   HR: 3
CAS: 471-03-4   NIOSH: WR 3325000
mf: $C_4H_8Cl_2O_2S$ mw: 191.08

SYNS: BIS(BETA-CHLOROETHYL)SULFONE
* MUSTARD GAS SULFONE * MUSTARD SUL-
FONE * YPERITE SULFONE

THR: A poison via intravenous and subcutaneous routes. Moderately toxic via inhalation. See also sulfonates. When heated to decomposition, it emits very toxic fumes of Cl^- and SO_x.

BIS(2-CHLOROISOPROPYL) ETHER   HR: 3
CAS: 108-60-1   NIOSH: KN 1750000
DOT: 2490
mf: $C_6H_{12}Cl_2O$ mw: 171.08

PROP: Colorless liquid. bp: 187.8°, fp: $-20°$, flash p: 185°F (OC), d: 1.11 @ 25°/25°, vap d: 6.0, vap press: 0.10 mm @ 20°.

SYNS: BIS(2-CHLORO-1-METHYLETHYL) ETHER
* (2-CHLORO-1-METHYLETHYL) ETHER
* DICHLORODIISOPROPYL ETHER * 2,2'-DI-
CHLOROISOPROPYL ETHER * NCI-C50044

DOT Classification: Label: Corrosive

THR: Poison by ingestion; moderately toxic by skin absorption and inhalation. See also ethers. Moderate fire hazard when exposed to heat, flame, or powerful oxidizers. See ethers for explosion hazard. Dangerous; when heated to decomposition, emits highly toxic fumes of chlorides. To fight fire, use water to blanket fire; foam, CO_2, dry chemical. Incompatible with oxidizing materials. For further information, see Vol. 6, No. 3 of DPIM Report.

BIS-1,2-(CHLOROMETHOXY)-ETHANE   HR: 3
CAS: 13483-18-6   NIOSH: KH 5450000
mf: $C_4H_8Cl_2O_2$ mw: 159.02

PROP: Viscous liquid; bp: 99°-100° @ 22 mm, d: 1.2879 @ 14°/15°.

SYN: ETHYLENE GLYCOL BIS(CHLOROMETHYL)-
ETHER

THR: An experimental neoplastigen and carcinogen. See also ethers. When heated to decomposition, it emits toxic fumes of Cl^-. For further information see Vol. 1, No. 5 of DPIM Report.

1,4-BIS(CHLOROMETHOXYMETHYL)-BENZENE   HR: 3
CAS: 56894-91-8   NIOSH: CY 8420000
mf: $C_{10}H_{12}Cl_2O_2$ mw: 235.12

SYN: BIS-1,4-(CHLOROMETHOXY)-P-XYLENE

THR: An experimental carcinogen, neoplastigen, and tumorigen. When heated to decomposition, it emits toxic fumes of Cl^-.

BIS(CHLOROMETHYL) ETHER   HR: 3
CAS: 542-88-1   NIOSH: KN 1575000
mf: $C_2H_4Cl_2O$ mw: 114.96

PROP: Volatile liquid; bp: 105°, d: 1.315 @ 20°, vap d: 4.0. flash p: <19°.

SYNS: CHLORO(CHLOROMETHOXY)METHANE
* DICHLORDIMETHYLAETHER (GERMAN)
* SYM-DICHLORO-DIMETHYL ETHER * DI-
METHYL-1,1'-DICHLOROETHER * OXYBIS-
(CHLOROMETHANE)
ACGIH TLV: TWA 0.001 ppm

THR: Poison by inhalation, ingestion, and skin absorption. See also ethers. An experimental and human carcinogen, neoplastigen, and tumorigen. Mutagenic data. When heated to decomposition, it emits very toxic fumes of Cl^-. For further information, see Vol. 6, No. 3 of DPIM Report.

1,1-BIS(4-CHLOROPHENYL)-2,2-DI-CHLOROETHANE   HR: 3
CAS: 72-54-8   NIOSH: KI 0700000
mf: $C_{14}H_{10}Cl_4$ mw: 320.04

PROP: Crystalline solid; mp: 110°, vap d: 11.

SYNS: 1,1-BIS(P-CHLOROPHENYL)-2,2-DICHLO-
ROETHANE * 2,2-BIS(P-CHLOROPHENYL)-1,1-
DICHLOROETHANE * 2,2-BIS(4-CHLOROPHE-
NYL)-1,1-DICHLOROETHANE * p,p'-DDD
* 1,1-DICHLOOR-2,2-BIS(4-CHLOOR FENYL)-
ETHAAN (DUTCH) * 1,1-DICHLOR-2,2-BIS(4-

CHLOR-PHENYL)-AETHAN (GERMAN) * 1,1-DI-
CHLORO-2,2-BIS(P-CHLOROPHENYL)ETHANE
* 1,1-DICHLORO-2,2-BIS(4-CHLOROPHENYL)-
ETHANE (FRENCH) * 1,1-DICHLORO-2,2-BIS-
(PARACHLOROPHENYL)ETHANE * 1,1-DI-
CHLORO-2,2-DI(4-CHLOROPHENYL)ETHANE
* P,P'-DICHLORODIPHENYLDICHLOROETHANE
* 1,1-DICLORO-2,2-BIS(4-CLORO-FENIL)-ETANO
(ITALIAN) * ENT 4,225 * NCI-C00475
* TETRACHLORODIPHENYLETHANE

THR: Poison by ingestion. Experimental tumorigen, neoplastigen, carcinogen. Mutagenic data.
Moderate skin toxicity. See also DDT. When
heated to decomposition, it emits toxic fumes
of Cl⁻. For further information, see Vol. 5,
No. 3 of *DPIM Report*.

2,2-BIS(p-CHLOROPHENYL)-1,1-DI-
CHLOROETHYLENE   HR: 3
CAS: 72-55-9   NIOSH: KV 9450000
mf: $C_{14}H_8Cl_4$ mw: 318.02

SYNS: 1,1-DICHLORO-2,2-BIS(P-CHLOROPHE-
NYL)ETHYLENE * P,P'-DICHLORODIPHENYL
DICHLOROETHYLENE * 1,1'-DICHLORO-
ETHENYLIDENE)BIS(4-CHLOROBENZENE)
* NCI-C00555

THR: An experimental carcinogen and neoplastigen. Moderately toxic by ingestion. Mutagenic
data. When heated to decomposition it emits
very toxic fumes of Cl⁻.

1,1-BIS(p-CHLOROPHENYL)-2,2,2-TRI-
CHLOROETHANOL   HR: 3
CAS: 115-32-2   NIOSH: DC 8400000
mf: $C_{14}H_9Cl_5O$ mw: 370.48

PROP: Material used in cancer bioassay was
40-60% pure

SYNS: 1,1-BIS(4-CHLOROPHENYL)-2,2,2-TRI-
CHLOROETHANOL * 4-CHLORO-ALPHA-
(4-CHLOROPHENYL)-ALPHA-(TRI-
CHLOROMETHYL)BENZENEMETHANOL
* DI-(P-CHLOROPHENYL)TRICHLOROMETHYL-
CARBINOL * 4,4'-DICHLORO-ALPHA-(TRICHLO-
ROMETHYL)BENZHYDROL * ENT 23,648
* KELTHANETHANOL * NCI-C00486
* 2,2,2-TRICHLOOR-1,1-BIS(4-CHLOOR
FENYL)-ETHANOL (DUTCH) * 2,2,2-TRICHLOR-
1,1-BIS(4-CHLOR-PHENYL)-AETHANOL (GERMAN)
* 2,2,2-TRICHLORO-1,1-BIS(4-CHLOROPHENYL)-
ETHANOL (FRENCH) * 2,2,2-TRICHLORO-1,1-
BIS(4-CLORO-FENIL)-ETANOLO (ITALIAN)

* 2,2,2-TRICHLORO-1,1-DI-(4-CHLOROPHE-
NYL)ETHANOL

THR: Acute poison by ingestion and skin absorption. An experimental carcinogen. Moderate acute toxicity via intraperitoneal route. When
heated to decomposition, it emits toxic fumes
of Cl⁻.

BIS(2-CHLOROVINYL)CHLORO
ARSINE   HR: 3
CAS: 40334-69-8   NIOSH: CG 8350000
mf: $C_4H_4AsCl_3$ mw: 233.35

SYN: LEWISITE II

THR: A dermal and subcutaneous poison. See
also arsenic compounds. When heated to decomposition, it emits very toxic fumes of As and
Cl⁻.

BIS(beta-CYANOETHYL)AMINE   HR: 3
CAS: 111-94-4   NIOSH: UG 2975000
mf: $C_6H_9N_3$ mw: 123.18

PROP: Liquid; mp: −5.5°, bp: 173° @ 10 mm,
d: 1.0165 @ 30°, vap d: 3.3.

SYNS: BIS-(2-CYANOETHYL)AMINE * N,N-
BIS(2-CYANOETHYL)AMINE * DI-(2-CYANO-
ETHYL)AMINE * IMINO-BETA,BETA'-DIPROPIONI-
TRILE * BETA,BETA-IMINODIPROPIONITRILE
* BETA,BETA'-IMINODIPROPIONITRILE * 3,3'-
IMINODIPROPIONITRILE * USAF A-8564

THR: A poison via intraperitoneal route. Moderately toxic by ingestion and skin absorption.
An eye irritant. See also nitriles and amines.
When heated to decomposition, it emits toxic
fumes of NO_x.

cis-BIS(CYCLOPENTYLAMMINE)-
PLATINUM(II)   HR: 3
CAS: 38780-36-8   NIOSH: TP 2210000
mf: $C_{10}H_{22}Cl_2N_2Pt$ mw: 436.33

SYNS: CIS-DICHLOROBIS(CYCLOPENTYLAMMINE)-
PLATINUM(II) * CIS-DICYCLOPENTYLAMMINE-
DICHLOROPLATINUM (II)

THR: An experimental carcinogen. Mutagenic
data. See also platinum compounds. When
heated to decomposition, it emits very toxic
fumes of Cl⁻ and NO_x.

BIS(DECANOYLOXY)DI-n-BUTYLSTAN-
NANE   HR: 3
CAS: 3465-75-6   NIOSH: WH 6715310
mf: $C_{28}H_{56}O_4Sn$ mw: 575.53

SYN: BIS(DECANOYLOXY)DI-N-BUTYLTIN

THR: Poison by ingestion. See also tin compounds. When heated to decomposition, it emits acrid and irritating fumes.

BIS-DEHYDRO ISYNOLIC ACID
METHYL ESTER   HR: 3
CAS: 5684-13-9   NIOSH: SF 7300000
mf: $C_{19}H_{22}O_3$ mw: 298.41

SYNS: DEHYDROFOLLICULINIC ACID * 1-
ETHYL-2-METHYL-7-METHOXY-1,2,3,4-TETRAHY-
DROPHENANTHRYL-2-CARBOXYLIC ACID
* 7-METHYLBISDEHYDRODOISYNOLIC ACID
* METILESTER DEL ACIDO BISDEHIDROISYNOLICO
(SPANISH) * 16,17-SECO-13-ALPHA-ESTRA-1,
3,5,6,7,9-PENTAEN-17-OIC ACID, METHYL ESTER
* TETRADEHYDRODOISYNOLIC ACID METHYL
ETHER

THR: An experimental tumorigen. See also ethers and esters. When heated to decomposition, it emits acrid and irritating fumes.

BIS(DIBUTYL DITHIO CARBAMATO)-
NICKEL   HR: 3
CAS: 13927-77-0   NIOSH: QR 6140000
mf: $C_{18}H_{36}N_2S_4 \cdot Ni$ mw: 467.51

SYNS: DIBUTYLDITHIOCARBAMIC ACID, NICKEL
SALT * NICKEL DIBUTYLDITHIOCARBAMATE

THR: An experimental tumorigen. See also nickel compounds. When heated to decomposition, it emits very toxic fumes of SO_x and NO_x.

BIS(DIBUTYL DITHIO CARBAMATO)-
ZINC   HR: 3
CAS: 136-23-2   NIOSH: ZH 0175000
mf: $C_{18}H_{38}N_2S_4Zn$ mw: 476.19

PROP: White powder; mp: 104°-108°; d: 1.24
@ 20°/20°.

SYNS: DIBUTYLDITHIOCARBAMIC ACID ZINC
SALT * USAF GY-5 * ZINC-BIBUTYLDITHIO-
CARBAMATE * ZINC-DIBUTYLDITHIOCARBAMATE
* ZINC-N,N-DIBUTYLDITHIOCARBAMATE

THR: An intraperitoneal poison. An experimental tumorigen. See also zinc compounds and carbamates. When heated to decomposition, it emits very toxic fumes of NO_x and SO_x.

BIS(DIETHYL DITHIO CARBAMATO)-
CADMIUM   HR: 3
CAS: 14239-68-0   NIOSH: EU 9850000
mf: $C_{10}H_{20}CdN_2S_4$ mw: 408.96

SYNS: CADMIUM DIETHYL DITHIOCARBAMATE
* ETHYL CADMATE

THR: An experimental tumorigen. Mutagenic data. See also cadmium compounds and carbamates. When heated to decomposition, it emits very toxic fumes of NO_x and SO_x.

BIS(DIETHYL DITHIO CARBAMATO)-
MERCURY   HR: 3
CAS: 14239-51-1   NIOSH: OV 7360000
mf: $C_{10}H_{20}HgN_2S_4$ mw: 497.15

THR: An intraperitoneal and intravenous poison. See also mercury compounds and carbamates. When heated to decomposition, it emits very toxic fumes of NO_x, SO_x, and Hg.

BIS(DIETHYL DITHIO CARBAMATO)-
ZINC   HR: 3
CAS: 14324-55-1   NIOSH: ZH 0350000
mf: $C_{10}H_{22}N_2S_4 \cdot Zn$ mw: 363.95

PROP: White powder; d: 1.47 @ 20°/20°.

SYNS: DIETHYLDITHIOCARBAMIC ACID ZINC
SALT * ZINC DIETHYLDITHIOCARBAMATE
* ZINC-N,N-DIETHYLDITHIOCARBAMATE

THR: An experimental tumorigen and carcinogen. Mutagenic data. See also zinc compounds and carbamates. Severe irritant to eyes, nose, and throat. When heated to decomposition, it emits very toxic fumes of NO_x and SO_x.

1,4-BIS(N,N'-DIETHYLENE PHOSPHA-
MIDE)PIPERAZINE   HR: 3
CAS: 738-99-8   NIOSH: SZ 1300000
mf: $C_{12}H_{24}N_6O_2P_2$ mw: 346.36

SYN: 1,4-PIPERAZINEDIYLBIS(BIS(1-AZIRIDINYL)-
PHOSPHINE OXIDE

THR: Poison by ingestion and subcutaneous route. Mutagenic data. When heated to decomposition, it emits very toxic fumes of PO_x and NO_x.

BIS(DIETHYL THIO CARBAMOYL) DI-
SULFIDE   HR: 3
CAS: 97-77-8   NIOSH: JO 1225000
mf: $C_{10}H_{20}N_2S_4$ mw: 296.56

PROP: Yellow-white cryst; mp: 72°.

SYNS: ANTABUSE * AVERSAN
* BIS(DIETHYLAMINO)THIOXOMETHYL)DISULPHIDE
* BIS(N,N-DIETHYLTHIOCARBAMOYL) DISULFIDE
* BIS(DIETHYLTHIOCARBAMOYL)DISULPHIDE

* BIS(N,N-DIETHYLTHIOCARBAMOYL)DISULPHIDE
* DISULFIRAM * 1,1'-DITHIOBIS(N,N-DIETHYL-
THIOFORMAMIDE) * NCI-C02959 * TETRA-
ETHYLTHIOPEROXYDICARBONIC DIAMIDE
* TETRAETHYLTHIRAM DISULPHIDE * N,N,
N',N'-TETRAETHYLTHIURAM DISULPHIDE
ACGIH TLV: TWA 2 mg/m^3

THR: Poison in humans by ingestion. Also poison by intraperitoneal routes. An experimental neoplastigen and carcinogen. Toxic symptoms when accompanied by ingestion of alcohol. See also bis(dimethyl thio carbamyl)disulfide. Dangerous for further information, see Vol. 1, No. 5 of *DPIM Report*.

BIS(DIETHYL THIO)CHLORO METHYL PHOSPHONATE   HR: 3
CAS: 34491-12-8   NIOSH: TA 4250000
mf: $C_5H_{12}ClOPS_2$ mw: 218.71

SYN: ENT 27,267

THR: Poison by ingestion, skin contact, and intraperitoneal routes. When heated to decomposition, it emits very toxic fumes of SO_x, PO_x and Cl^-.

BIS(DIMETHYL AMIDO)FLUORO PHOSPHATE   HR: 3
CAS: 115-26-4   NIOSH: TD 4025000
mf: $C_4H_{12}FN_2OP$ mw: 154.15

SYNS: BIS(DIMETHYLAMIDO)-PHOSPHORYL
FLUORIDE * BIS(DIMETHYLAMINO)FLUORO-
PHOSPHATE * BISDIMETHYLAMINOFLUORO-
PHOSPHINE OXIDE * ENT 19,109 * FLUO-
PHOSPHORIC ACID DI(DIMETHYLAMIDE)
* FLUORURE DE N,N,N',N'-TETRAMETHYLE
PHOSPHORO-DIAMIDE (FRENCH) * N,N,N',N'-
TETRAMETHYL-DIAMIDO-FOSFORZUUR-FLUORIDE
(DUTCH) * TETRAMETHYLDIAMIDOPHOS-
PHORIC FLUORIDE * N,N,N',N'-TETRA-
METHYL-DIAMIDO-PHOSPHORSAEURE-FLUORID
(GERMAN) * TETRAMETHYLPHOSPHORODIAMIDIC
FLUORIDE * N,N,N,N-TETRAMETHYLPHOSPHO-
RODIAMIDIC FLUORIDE * N,N,N',N'-
TETRAMETIL-FOSFORODIAMMIDO-FLUORURO
(ITALIAN)

THR: Poison by ingestion, dermal, intraperitoneal, subcutaneous, and intravenous routes. When heated to decomposition, it emits very toxic fumes of F^-, NO_x and PO_x.

3,6-BIS(DIMETHYL AMINO)ACRIDINE   HR: 3
CAS: 494-38-2   NIOSH: AR 7600000
mf: $C_{17}H_{19}N_3$ mw: 265.39

SYNS: C.I. 46005 * ACRIDINE ORANGE
* 2,8-BISDIMETHYLAMINOACRIDINE * C.I. NO.
46005:1 * 3,6-DI(DIMETHYLAMINO)ACRIDINE
* N,N,N'-TETRAMETHYL-3,6-ACRIDINEDIAMINE

THR: Poison by subcutaneous route. An experimental tumorigen and carcinogen. Mutagenic data. When heated to decomposition, it emits toxic fumes of NO_x. For further information, see Acridine Orange, Vol. 1, No. 3 of *DPIM Report*.

BIS(beta-DIMETHYL AMINO ETHYL)-SUCCINATE BIS(METHYL IODIDE)   HR: 3
CAS: 541-19-5   NIOSH: GA 5250000
mf: $C_{14}H_{30}N_2O_4 \cdot 2I$ mw: 517.92

SYNS: CHOLINE, IODIDE, SUCCINATE (2:1)
* DIACETYLCHOLINE DIIODIDE * SUCCINIC
ACID BIS(BETA-DIMETHYLAMINOETHYL) ESTER
BISMETHIODIDE * SUCCINIC ACID, DIESTER
WITH CHOLINE IODIDE * SUCCINYLDICHOLINE
IODIDE * O-O-SUCCINYLDICHOLINE IODIDE

THR: An intravenous poison. When heated to decomposition, it emits very toxic fumes of NO_x and I^-.

3,7-BIS(DIMETHYL AMINO)PHENAZA THIONIUM CHLORIDE   HR: 3
CAS: 61-73-4   NIOSH: SO 5600000
mf: $C_{16}H_{18}N_3S \cdot Cl$ mw: 319.88

SYNS: METHYLENE BLUE USP XII (MEDICINAL)
* C.I. 52 015 (CZECH) * 3,7-BIS(DIMETHYL-
AMINO)PHENOTHIAZIN-5-IUM CHLORIDE
* METHYLTHIONINE CHLORIDE * METHYL-
THIONIUM CHLORIDE * TETRAMETHYLTHIONINE
CHLORIDE

THR: Poison by ingestion, intraperitoneal, intravenous, and subcutaneous routes. Mutagenic data. When heated to decomposition, it emits very toxic fumes of NO_x, SO_x, and Cl^-.

BIS(DIMETHYL DITHIO CARBAMATO)-ZINC   HR: 3
CAS: 137-30-4   NIOSH: ZH 0525000
mf: $C_6H_{12}N_2S_4 \cdot Zn$ mw: 305.81

PROP: White powder; mp: 248°-250°; d: 1.65 @ 20°/20°.

SYNS: DIMETHYLCARBAMODITHIOIC ACID, ZINC SALT * DIMETHYLDITHIOCARBAMATE ZINC SALT * ZINC BIS(DIMETHYLDITHIOCARBAMATE) * ENT 988 * TSIRAM (RUSSIAN) * USAF P-2 * NCI-C50442 * BIS(DIMETHYLCARBAMODITHIOATO-S,S')ZINC * BIS-DIMETHYLDITHIOCARBAMATE DE ZINC (FRENCH) * BIS(N,N-DIMETIL-DITIOCARBAMMATO) DI ZINCO (ITALIAN) * ZINC BIS(DIMETHYLDITHIOCARBAMOYL)DISULPHIDE * ZINC BIS(DIMETHYLTHIOCARBAMOYL)-DISULFIDE * ZINC DIMETHYLDITHIOCARBAMATE * ZINC N,N-DIMETHYLDITHIOCARBAMATE * ZINK-BIS(N,N-DIMETHYL-DITHIOCARBAMAAT) (DUTCH) * ZINK-BIS(N,N-DIMETHYL-DITHIO-CARBAMAT) (GERMAN)

THR: Poison by ingestion, intraperitoneal, and intravenous routes. Moderately toxic by inhalation. Mutagenic data. An experimental carcinogen and tumorigen. See also zinc compounds and carbamates. Severe irritant to eyes, nose, and throat. When heated to decomposition, it emits very toxic fumes of NO_x and SO_x.

1,1'-BIS(3,5-DIMETHYL MORPHOLINO CARBONYL METHYL)-4,4'-BIPY-RIDYNIUM DICHLORIDE   HR: 3
CAS: 4636-83-3   NIOSH: UU 2592000
mf: $C_{26}H_{36}N_4O_4 \cdot 2Cl$ mw: 539.56

SYNS: 1,1'-BIS(2-(3,5-DIMETHYL-4-MORPHO-LINYL)-2-OXOETHYL)-4,4'-BIPYRIDINIUM DICHLORIDE * MORPHANQUAT DICHLORIDE

THR: Poison by ingestion. When heated to decomposition, it emits very toxic fumes of Cl^- and NO_x.

BIS(DIMETHYL THIOCARBAMYL)DI-SULFIDE   HR: 3
CAS: 137-26-8   NIOSH: JO 1400000
DOT: 2771
mf: $C_6H_{12}N_2S_4$ mw: 240.44

PROP: Crystals, insol in water, sol in alcohol, ether, acetone, chloroform. mp: 156°, d: 1.30, bp: 129° @ 20 mm.

SYNS: BIS((DIMETHYLAMINO)CARBONOTHIOYL) DISULPHIDE * BIS(DIMETHYLTHIOCARBAMOYL) DISULFIDE * DISOLFURO DI TETRAMETIL-TIOURAME (ITALIAN) * DISULFURE DE TETRA-METHYLTHIOURAME (FRENCH) * ALPHA,AL-PHA'-DITHIOBIS(DIMETHYLTHIO)FORMAMIDE * N,N'-(DITHIODICARBONOTHIOYL)BIS(N-METHYLMETHANAMINE) * METHYL THIURAMDI-

SULFIDE * TERAMETHYL THIURAM DISULFIDE * TETRAMETHYLDIURANE SULPHITE * TET-RAMETHYLENETHIURAM DISULPHIDE * TETRAMETHYLTHIOCARBAMOYLDISULPHIDE * TETRAMETHYLTHIOPEROXYDICARBONIC DI-AMIDE * TETRAMETHYLTHIORAMDISULFIDE (DUTCH) * TETRAMETHYL-THIRAM DISULFID (GERMAN) * TETRAMETHYLTHIURAM BISULFIDE * TETRAMETHYLTHIURAM DISULFIDE * N,N,N',N'-TETRAMETHYLTHIURAM DISULFIDE * TETRAMETHYL THIURANE DISULFIDE * TET-RAMETHYLTHIURUM DISULFIDE * THIRAM (DOT) * TIURAM (POLISH) * USAF B-30 * USAF EK-2089 * USAF P-5

OSHA PEL: TWA 5 mg/m^3
DOT Classification: ORM-A, Label: None

THR: Poison by ingestion and intraperitoneal routes. An experimental teratogen, tumorigen, and carcinogen. Mutagenic data. Affects human pulmonary system. A mild allergen and irritant. Acute poisoning in experimental animals produced liver, kidney, and brain damage. Dangerous in a fire; see NO_x and SO_x. For further information, see Vol. 1, No. 5 of *DPIM Report*.

BIS(DIMETHYL THIO CARBAMOYL)-SULFIDE   HR: 3
CAS: 97-74-5   NIOSH: WQ 1750000
mf: $C_6H_{12}N_2S_3$ mw: 208.38

SYNS: BIS(DIMETHYLTHIOCARBAMYL) MONOSUL-FIDE * TETRAMETHYLTHIURAMMONIUM SULFIDE * TETRAMETHYLTHIURAM MONOSULFIDE * TETRAMETHYLTHIURAMONOSULFIDE * TET-RAMETHYLTHIURAM SULFIDE * USAF B-32 * USAF EK-P-6255

THR: Poison by ingestion and intraperitoneal routes. Mutagenic data. An experimental tumorigen. When heated to decomposition, it emits very toxic fumes of NO_x and SO_x.

BIS(2,3-EPOXY CYCLOPENTYL) ETHER   HR: 3
CAS: 2386-90-5   NIOSH: RN 9100000
mf: $C_{10}H_{14}O_3$ mw: 182.24

SYNS: 2,2'-OXYBIS-6-OXABICYCLO-(3.1.0)-HEXANE * BIS(2,3-EPOXYCYCLOPENTYL) ETHER

THR: An experimental neoplastigen. A skin and oral irritant. See also ethers. When heated to decomposition, it emits acrid smoke and irritating fumes.

2,4-BIS(ETHYLAMINO)-6-CHLORO-s-TRIAZINE   HR: 3
CAS: 122-34-9   NIOSH: XY 5250000
mf: $C_7H_{12}ClN_5$ mw: 201.69

SYNS: 1-CHLORO, 3,5-BISETHYLAMINO-2,4,6-TRIAZINE * 2-CHLORO-4,6-BIS(ETHYLAMINO)-s-TRIAZINE * 2-CHLORO-4,6-BIS(ETHYLAMINO)-1,3,5-TRIAZINE

THR: An intravenous poison. Mutagenic data. An experimental tumorigen. A skin and eye irritant. When heated to decomposition, it emits very toxic fumes of Cl^- and NO_x.

2,6-BIS(ETHYLEN IMINO)-4-AMINO-s-TRIAZINE   HR: 3
CAS: 6708-69-6   NIOSH: XY 2975000
mf: $C_7H_{10}N_6$ mw: 178.23

THR: A poison via intraperitoneal and intravenous routes. When heated to decomposition, it emits toxic fumes of NO_x.

BIS(2-ETHYL HEXANOYL OXY)DIBUTYL STANNANE   HR: 3
CAS: 2781-10-4   NIOSH: WH 6714500
mf: $C_{24}H_{48}O_4Sn$ mw: 519.41

SYN: DI-N-BUTYLTIN DI-2-ETHYLHEXANOATE

THR: Poison by ingestion and intravenous route. See also tin compounds. When heated to decomposition, it emits acrid smoke and irritating and toxic fumes.

BIS(2-ETHYLHEXYL)PHOSPHATE   HR: 3
CAS: 298-07-7   NIOSH: TB 7875000
DOT: 1902
mf: $C_{16}H_{35}O_4P$ mw: 322.48

SYNS: BIS(2-ETHYLHEXYL)HYDROGEN PHOSPHATE * BIS(2-ETHYLHEXYL)ORTHOPHOSPHORIC ACID * BIS(2-ETHYLHEXYL)PHOSPHORIC ACID * DI(2-ETHYLHEXYL)PHOSPHATE * 2-ETHYL-1-HEXANOL HYDROGEN PHOSPHATE

DOT Classification: Label: Corrosive

THR: An intraperitoneal poison. A corrosive material. A skin and eye irritant. When heated to decomposition, it emits toxic fumes of PO_x.

BIS(2-ETHYLHEXYL)PHTHALATE   HR: 3
CAS: 117-81-7   NIOSH: TI 0350000
mf: $C_{24}H_{38}O_4$ mw: 390.62

SYNS: BIS(2-ETHYLHEXYL)-1,2-BENZENEDICARBOXYLATE * DI(2-ETHYLHEXYL)ORTHOPHTHALATE * DI(2-ETHYLHEXYL)PHTHALATE * DI-SEC-OCTYL PHTHALATE * DOP * 2-ETHYLHEXYL PHTHALATE * NCI-C52733 * OCTOIL

OSHA PEL: TWA 5 mg/m^3

THR: Poison by intravenous route. Suspected human carcinogen and an experimental teratogen. Affects the human gastrointestinal tract. A mild skin and eye irritant. When heated to decomposition, it emits acrid smoke. For further information, see Di-(2-Ethylhexyl)Phthalate, Vol. 2, No. 2 of *DPIM Report*.

BIS(2-ETHYLHEXYL)SEBACATE   HR: 2
CAS: 122-62-3   NIOSH: VS 1000000
mf: $C_{26}H_{50}O_4$ mw: 426.76

PROP: Light, clear liquid, mild odor; bp: 248° @ 9 mm, fp: −55°, flash p: 410°F, d: 0.913 @ 25°/25°, vap d: 14.7.

SYNS: DECANEDIOIC ACID, BIS(2-ETHYLHEXYL) ESTER * DI(2-ETHYLHEXYL)SEBACATE * DIOCTYL SEBACATE * 2-ETHYLHEXYL SEBACATE * OCTYL SEBACATE

THR: Moderately toxic by ingestion and intravenous route. See also esters. Slight fire hazard when exposed to heat or flame; can react with oxidizing materials. To fight fire, use foam, CO_2, dry chemical. When heated to decomposition, it emits acrid and irritating fumes.

BIS(ETHYL MERCURI)PHOSPHATE   HR: 3
CAS: 2440-45-1   NIOSH: OW 4375000
mf: $C_4H_{11}Hg_2O_4P$ mw: 555.30

PROP: Solid

SYNS: ETHYLMERCURIC PHOSPHATE * ETHYLMERCURY PHOSPHATE * LIGNASAN FUNGICIDE

THR: Poison by ingestion and subcutaneous route. See also mercury compounds, organic. When heated to decomposition, it emits very toxic fumes of Hg and PO_x.

BISETHYL XANTHOGEN DISULFIDE   HR: 3
CAS: 502-55-6   NIOSH: LQ 7700000
mf: $C_6H_{10}O_2S_4$ mw: 242.40

SYNS: BIETHYLXANTHOGENTRISULFIDE * BIS(ETHYLXANTHIC)DISULFIDE * DIETHYLDI-

THIO BIS(THIONOFORMATE) * DIETHYLXANTHO-GEN DISULFIDE * DITHIOBIS(THIOFORMIC ACID)-O,O-DIETHYL ESTER * ETHYL XANTHO-GEN DISULFIDE

THR: Poison by ingestion and intraperitoneal route. Moderately toxic by skin absorption. See also esters and sulfides. Dangerous; when heated to decomposition, it emits highly toxic fumes of SO_x.

BIS(L-HISTIDINE)COBALT   HR: 3
CAS: 14873-10-0   NIOSH: BJ 9675000
mf: $C_{12}H_{14}N_6O_5 \cdot Co$ mw: 365.25

SYNS: ALPHA-AMINOIMIDAZOLE-4-PROPIONIC ACID, COBALT(2+) SALT * BIS(L-HISTIDI-NATO)COBALT * KOBALT HISTIDIN (GERMAN)

THR: An intraperitoneal and intravenous poison. See also cobalt compounds. When heated to decomposition, it emits toxic fumes of NO_x.

BISHYDROXY COUMARIN   HR: 3
CAS: 66-76-2   NIOSH: GN 7875000
mf: $C_{19}H_{12}O_6$ mw: 336.31

PROP: Very small crystals, slight pleasant odor, bitter taste, sol in alkali; mp: 287°-293°.

SYNS: BIS(4-HYDROXYCOUMARIN-3-YL)METH-ANE * DI-(4-HYDROXY-3-COUMARINYL)-METHANE * DI-4-HYDROXY-3,3'-METHYL-ENEDICOUMARIN * 3,3'-METHYLEEN-BIS(4-HYDROXY-CUMARINE) (DUTCH) * 3,3'-METHYLEN-BIS(4-HYDROXY-CUMARIN) (GERMAN) * 3,3'-METHYLENEBIS(4-HY-DROXY-1,2-BENZOPYRONE) * 3,3'-METHYL-ENEBIS(4-HYDROXYCOUMARIN) * 3,3'-METHYLENE-BIS(4-HYDROXYCOUMARINE) (FRENCH) * 3,3'-METILEN-BIS(4-IDROSSI-CU-MARINA) (ITALIAN)

THR: Poison by ingestion, intravenous, and intraperitoneal routes. An anticoagulant. Excessive doses can cause hemorrhages.

BIS(4-HYDROXY-3-COUMARIN) ACETIC ACID ETHYL ESTER   HR: 3
CAS: 548-00-5   NIOSH: AF 5775000
mf: $C_{22}H_{16}O_8$ mw: 408.38

SYNS: BIS-(4-HYDROXY-3-COUMARINYL)ETHYL ACETATE * BIS-3,3'-(4-HYDROXYCOUMARI-NYL)ACETIC ACID ETHYL ESTER * BIS(4-HY-DROXY-2-OXO-2H-1-BENZOPYRAN-3-YL)ACETIC ACID ETHYL ESTER * 3,3'-(CARBOXYMETHYL-ENE)BIS(4-HYDROXYCOUMARIN) ETHYL ESTER * ETHYL BIS(4-HYDROXYCOUMARINYL)ACETATE * ETHYL BIS(4-HYDROXY-3-COUMARINYL)-ACETATE * ETHYLDICOUMAROL ACETATE * ETHYL 4,4'-DIHYDROXYDICOUMARINYL-3,3'-ACETATE

THR: An intraperitoneal poison. Moderately toxic by ingestion and subcutaneous routes. See also esters. When heated to decomposition, it emits acrid and irritating fumes.

BIS(2-HYDROXY ETHYL)AMINE   HR: 2
CAS: 111-42-2   NIOSH: KL 2975000
mf: $C_4H_{11}NO_2$ mw: 105.16

PROP: A faintly colored, viscous liquid; mp: 28°, bp: 269.1° (decomp), flash p: 305°F (OC), d: 1.0919 @ 30°/20°, autoign temp: 1224°F, vap press: 5 mm @ 138°, vap d: 3.65.

SYNS: DIAETHANOLAMIN (GERMAN) * DIE-THANOLAMIN (CZECH) * DIETHANOLAMINE * 2,2'-DIHYDROXYDIETHYLAMINE * DI(2-HY-DROXYETHYL)AMINE * 2,2'-IMINODIETHANOL * NCI-C55174

THR: Moderately toxic by ingestion, intraperitoneal, and subcutaneous routes. A mild skin and severe eye irritant. Slight fire hazard when exposed to heat or flame; can react with oxidizing materials. To fight fire, use alcohol foam, water, CO_2, dry chemical. When heated to decomposition, it emits toxic fumes such as NO_x.

BIS(2-HYDROXYETHYL)-2-(2-CHLORO ETHYL THIO)ETHYL SULFONIUM, CHLORIDE   HR: 3
CAS: 64036-91-5   NIOSH: WR 7470000
mf: $C_8H_{18}ClO_2S_2 \cdot Cl$ mw: 281.28

SYN: BETA-CHLOROETHYL BETA-(BIS(BETA-HY-DROXYETHYL)SULFONIUM)ETHYL SULFIDE CHLORIDE

THR: A poison by skin absorption and intravenous routes. See also sulfonates. When heated to decomposition, it emits very toxic fumes of Cl^- and SO_x.

N,N-BIS(2-HYDROXY ETHYL)DODECAN AMIDE   HR: 3
CAS: 120-40-1   NIOSH: JR 1925000
mf: $C_{16}H_{33}NO_3$ mw: 287.50

SYNS: N,N-BIS(HYDROXYETHYL)LAURAMIDE * N,N-BIS(BETA-HYDROXYETHYL)LAURAMIDE

* BIS(2-HYDROXYETHYL)LAURAMIDE * N,N-
BIS(2-HYDROXYETHYL)LAURAMIDE * COCONUT
OIL AMIDE OF DIETHANOLAMINE * DIETHANOL-
LAURAMIDE * N,N-DIETHANOLLAURAMIDE
* N,N-DIETHANOLLAURIC ACID AMIDE
* LAURIC ACID DIETHANOLAMIDE * LAUROYL
DIETHANOLAMIDE * LAURYL DIETHANOLAMIDE
* NCI-C55323

THR: Suspected experimental carcinogen. Moderately toxic by ingestion. When heated to decomposition, it emits toxic fumes of NO_x.

BIS(HYDROXY METHYL)FURATRIAZINE   HR: 3
CAS: 794-93-4   NIOSH: PC 3200000
mf: $C_{11}H_{11}N_5O_5$ mw: 293.27

SYNS: 3-BIS(HYDROXYMETHYL)AMINO-6-(5-NI-
TRO-2-FURYLETHENYL)-1,2,4-TRIAZINE
* N-(6-(5-NITROFURFURYLIDENEMETHYL)-
1,2,4-TRIAZIN-3-YL)IMINODIMETHANOL
* ((6-(2-(5-NITRO-2-FURYL)VINYL)-AS-TRIA-
ZIN-3-YL)IMINO)DIMETHANOL * 3-DI(HY-
DROXYMETHYL)AMINO-6-(5-NITRO-2-FURYL-
ETHENYL)-1,2,4-TRIAZINE

THR: An experimental carcinogen. When heated to decomposition, it emits toxic fumes of NO_x.

BIS(ISOOCTYL OXYCARBONYL METHYL THIO)DIOCTYL STANNANE   HR: 3
CAS: 26401-97-8   NIOSH: WH 6723000
mf: $C_{36}H_{72}O_4S_2Sn$ mw: 751.89

SYNS: DIISOOCTYL ((DIOCTYLSTANNYLENE)-
DITHIO)DIACETATE * DIOCTYLTIN BIS(ISOOCTYL
MERCAPTOACETATE) * DIOCTYLTIN-S,S'-BIS-
(ISOOCTYL MERCAPTOACETATE) * DIOCTYLTIN
BIS(ISOOCTYL THIOGLYCOLATE) * DI-N-OCTYL-
TIN DIISOOCTYL THIOGLYCOLATE * DI-N-OC-
TYL-ZINN-DI-ISOOCTYLTHIOGLYKOLAT (GERMAN)

OSHA PEL: TWA 100 ug(Sn)/m^3 (skin)

THR: Moderately toxic via oral route. See also tin compounds. When heated to decomposition, it emits toxic fumes of SO_x.

1,3-BISMALEIMIDO BENZENE   HR: 3
CAS: 3006-93-7   NIOSH: ON 6125000
mf: $C_{14}H_8N_2O_4$ mw: 268.24

SYNS: 1,3-DIMALEIMIDOBENZENE * N,N'-
(M-PHENYLENE)BISMALEIMIDE * N,N'-(M-
PHENYLENEDIMALEIMIDE)

THR: Poison by ingestion and intraperitoneal route. When heated to decomposition, it emits toxic fumes of NO_x.

BIS(METHANE SULFONYL)-D-MANNITOL   HR: 3
CAS: 1187-00-4   NIOSH: OP 2975000
mf: $C_8H_{18}O_{10}S_2$ mw: 338.38

SYNS: 1,6-BIS-O-METHYLSULFONYL-d-MANNI-
TOL * 1,6-DIMESYL-d-MANNITOL * 1,6-DI-
METHANESULFONATE-d-MANNITOL * 1,6-DI-
METHANE-SULFONOXY-d-MANNITOL * 1,6-
DIMETHANESULPHONOXY-1,6-DIDEOXY-d-
MANNITOL * D-MANNITOL BUSULFAN
* MANNITOL MYLERAN * NSC-37538

THR: Poison by intravenous route. Moderate intraperitoneal toxicity. Mutagenic data. An experimental neoplastigen. When heated to decomposition, it emits toxic fumes of SO_x.

N,N'-BISMORPHOLINE DISULFIDE   HR: 3
CAS: 103-34-4   NIOSH: QE 3325000
mf: $C_8H_{16}N_2O_2S_2$ mw: 236.38

PROP: Tan to gray powder; mp: 122° min; d: 1.36 @ 25°.

SYNS: BISMORPHOLINO DISULFIDE * DIMOR-
PHOLINE DISULFIDE * DIMORPHOLINO DISULFIDE
* DITHIOBISMORPHOLINE * 4,4'-DITHIOBIS-
(MORPHOLINE) * N,N-DITHIODIMORPHOLINE
* 4,4'-DITHIODIMORPHOLINE * 4,4'-DITHIO-
MORPHOLINE * MORPHOLINODISULFIDE
* MORPHOLINE DISULFIDE * USAF B-17
* USAF EK-T-6645

THR: Poison by intraperitoneal and intravenous routes. Moderate oral toxicity. See also morpholine. When heated to decomposition it emits very toxic fumes of NO_x and SO_x.

BISMUTH   HR: 3
CAS: 7440-69-9   NIOSH: EB 2600000
Af: Bi   Aw: 208.98

PROP: Hexagonal silver-white or reddish metallic crystals; mp: 271.3°, bp: 1420°-1560°, d: 9.80, vap press: 1 mm @ 1021°.

SYN: BISMUTH-209

THR: Poisonous to man. See also bismuth compounds. Flammable when exposed to flame and by chemical reaction with [Bi(OH)$_3$ + Al(OH)$_3$]; coprecipitated and H_2 reduced yields a spontaneously flammable product. Moderately dangerous; can react with acid or acid fumes to emit toxic fumes. Incompatible with Al; BrF_3; acids; NOF; NH_4NO_3; $HClO_3$; Cl_2; IF_5; HNO_3; $HClO_4$. For further information see Vol. 3, No. 5 of *DPIM Report*.

BISMUTH ARSPHENAMINE SULFO-NATE   HR: 3
CAS: 12001-47-7   NIOSH: EB 2625000
mf: $C_{21}H_{24}As_3Bi_2N_3O_{12}S_3 \cdot 3Na$ mw: 1318.35

SYNS: BISMARSEN * SULFARSPHENAMINE BISMUTH

THR: A poison via intraperitoneal, intramuscular routes. See also arsenic compounds and bismuth compounds. When heated to decomposition it emits very toxic fumes of Na_2O, NO_x, SO_x, As and Bi.

BISMUTH COMPOUNDS   HR: 3
THR: Bi and its salts can cause kidney damage, although the degree of such damage is usually mild. Large doses can be fatal. Industrially it is considered one of the less toxic of the heavy metals, although intoxication has occurred from its use in medicine. The similarity between the pharmacologic and toxic behaviors of lead and Bi has been pointed out in the literature. Like lead, Bi may be liberated from tissue deposits during periods of acidosis. Serious and sometimes fatal poisoning may occur from the injection of large doses into closed cavities and from extensive application to burns. Death of animals from Bi nephritis following injections of soluble salts occurs within several hours to 24 days, the time being generally inversely proportional to the dose, and it appears to be in the order of 5-10 times higher than the dose by slow intravenous injection for rabbits. It is stated that the administration of Bi should be stopped when gingivitis appears, for otherwise serious ulcerative stomatitis is likely to result. Other toxic results may develop, such as malaise, albuminuria, diarrhea, skin reactions and sometimes serious exodermatitis. Industrial Bi poisoning has not been reported, although Bi absorbed in industrial cases may complicate a diagnosis of plumbism, since the dark line in the gums, which is often present in lead poisoning, is also produced by bismuth. All Bi compounds do not have equal toxicity. See also individual entries.

    Treatment and Antidotes: Personnel showing some of the symptoms noted above which might indicate that they were absorbing too much Bi into the body should be removed from exposure as soon as possible. Get medical advice. Personnel should be cautioned against careless handling of these materials.

BISMUTH DIMETHYL DITHIOCARBA-MATE   HR: 3
CAS: 21260-46-8   NIOSH: EB 3400000
mf: $C_9H_{18}N_3S_6 \cdot Bi$ mw: 569.64

SYNS: BISMATE * TRIS(DIMETHYLDITHIO-CARBAMATO)BISMUTH

THR: An experimental tumorigen. See also bismuth compounds and thiocarbamic acid. When heated to decomposition, it emits very toxic fumes of SO_x and NO_x.

BISMUTH NITRATE   HR: 3
CAS: 10361-44-1   NIOSH: EB 2984400
mf: BiN_3O_9 mw: 395.01

PROP: Triclinic, colorless, slightly hygroscopic crystals. bp: $-5H_2O$ @ 80°, d: 2.83, mp: 30° (decomp).

SYN: NITRIC ACID, BISMUTH(3+) SALT

THR: An intravenous poison. Moderate intraperitoneal toxicity. See also Bi compounds and nitrates.

BISMUTH PENTAFLUORIDE   HR: 3
mf: BiF_5 mw: 303.98

PROP: Crystals. Sublimes @ 550°.

THR: An irritant poison via oral and inhalation routes. Decomposes readily on contact with moisture to yield O_3 and bismuth trifluoride. See fluorides and ozone. Very dangerous. When heated to decomposition it emits highly toxic fumes of F^-. Incompatible with water and petrolatum @ > 50°; moisture; acids. Reacts violently liberating much heat and ozone.

BISMUTH SODIUM THIOGLYCOL-LATE   HR: 3
CAS: 150-49-2   NIOSH: EB 2984000
mf: $C_6H_6BiNa_3O_6S_3$ mw: 548.25

SYNS: SODIUM BISMUTH THIOGLYCOLATE
* SODIUM BISMUTH THIOGLYCOLLATE
* THIOBISMOL

THR: A poison via intraperitoneal, intravenous, and intramuscular routes. A systemic toxicant in children. See also Bi compounds. When heated to decomposition, it emits very toxic fumes of SO_x and Na_2O.

BIS(OCTANOYLOXY)DI-n-BUTYL STANNANE   HR: 3
NIOSH: WH 6733100
mf: $C_{24}H_{48}O_4Sn$ mw: 519.41

SYN: KAPRYLAN DI-N-BUTYLCINICITY (CZECH)

THR: Poison by ingestion. A skin and eye irritant. See also tin compounds. When heated to decomposition, it emits acrid and irritating fumes.

BIS(8-OXYQUINOLINE)COPPER   HR: 3
CAS: 10380-28-6   NIOSH: VC 5250000
mf: $C_{18}H_{12}CuN_2O_2$ mw: 351.86

PROP: Yellow-green powder.

SYNS: BIS(8-QUINOLINATO)COPPER * BIS(8-QUINOLINOLATO)COPPER * COPPER-8-HYDROXYQUINOLATE * COPPER-8-HYDROXYQUINOLINATE * COPPER-8-HYDROXYQUINOLINE * COPPER (2+) OXINATE * COPPER OXYQUINOLATE * COPPER OXYQUINOLINE * COPPER QUINOLATE * COPPER-8-QUINOLATE * COPPER-8-QUINOLINOLATE * COPPER-8-QUINOLINOL * COPPER QUINOLINOLATE * CUPRIC-8-HYDROXYQUINOLATE * CUPRIC-8-QUINOLINOLATE * 8-HYDROXYQUINOLINE COPPER COMPLEX * OXYQUINOLINOLEATE DE CUIVRE (FRENCH)

THR: An intraperitoneal poison. A carcinogen. An experimental tumorigen. See also copper compounds. When heated to decomposition, it emits toxic fumes of NO_x.

BISPENTA FLUORO SULFUR OXIDE   HR: 3
CAS: 42310-84-9   NIOSH: WS 4920000
mf: $F_{10}OS_2$ mw: 270.12

SYN: SULFUR FLUORIDE OXIDE

THR: Poison by inhalation. See also fluorides. When heated to decomposition, it emits very toxic fumes of F^- and SO_x.

BISPHENOL A   HR: 3
CAS: 80-05-7   NIOSH: SL 6300000
mf: $C_{15}H_{16}O_2$ mw: 228.31

PROP: White flakes, mild phenolic odor, insol in water, sol in alcohol and dilute alkalies, slightly sol in CCl_4.

SYNS: 2,2-BIS-4'-HYDROXYFENYLPROPAN (CZECH) * BIS(4-HYDROXYPHENYL) DIMETHYLMETHANE * BIS(4-HYDROXYPHENYL)PROPANE * 2,2-BIS(p-HYDROXYPHENYL)PROPANE * 2,2-BIS(4-HYDROXYPHENYL)PROPANE * P,P'-DIHYDROXYDIPHENYLDIMETHYLMETHANE * 4,4'-DIHYDROXYDIPHENYLDIMETHYLMETHANE * P,P'-DIHYDROXYDIPHENYLPROPANE * 2,2-(4,4'-DIHYDROXYDIPHENYL)PROPANE * 4,4'-DIHYDROXYDIPHENYLPROPANE * 4,4'-DIHYDROXYDIPHENYL-2,2-PROPANE * 4,4'-DIHYDROXY-2,2-DIPHENYLPROPANE * DIMETHYLMETHYLENE-p,p'-DIPHENOL * BETA-DI-p-HYDROXYPHENYLPROPANE * 2,2-DI(4-HYDROXYPHENYL)PROPANE * DIMETHYL BIS(p-HYDROXYPHENYL)METHANE * 2,2-DI(4-PHENYLOL)PROPANE * P,P'-ISOPROPYLIDENEBISPHENOL * 4,4'-ISOPROPYLIDENEBISPHENOL * P,P'-ISOPROPYLIDENEDIPHENOL * NCI-C50635

THR: An intraperitoneal poison. Moderately toxic by ingestion and skin contact. A skin and eye irritant. When heated to decomposition, it emits acrid and irritating fumes.

BISPHENOL A DIGLYCIDYL ETHER   HR: 3
CAS: 1675-54-3   NIOSH: TX 3800000
mf: $C_{21}H_{24}O_4$ mw: 340.45

SYNS: 2,2-BIS(4-(2,3-EPOXYPROPYLOXY)PHENYL)PROPANE * BIS(4-GLYCIDYLOXYPHENYL)DIMETHYAMETHANE * 2,2-BIS(p-GLYCIDYLOXYPHENYL)PROPANE * 2,2-BIS(4-HYDROXYPHENYL)PROPANE, DIGLYCIDYL ETHER * BIS(4-HYDROXYPHENYL)DIMETHYLMETHANE DIGLYCIDYL ETHER * 2,2-BIS(p-HYDROXYPHENYL)PROPANE, DIGLYCIDYL ETHER * DIGLYCIDYL BISPHENOL A ETHER * DIGLYCIDYL ETHER OF 2,2-BIS(p-HYDROXYPHENYL)PROPANE * DIGLYCIDYL ETHER OF 2,2-BIS(4-HYDROXYPHENYL)PROPANE * DIGLYCIDYL ETHER OF BISPHENOL A * DIGLYCIDYL ETHER OF 4,4'-ISOPROPYLIDENEDIPHENOL * 4,4'-DIHYDROXYDIPHENYLDIMETHYLMETHANE DIGLYCIDYL ETHER * P,P'-DIHYDROXYDIPHENYLDIMETHYLMETHANE DIGLYCIDYL ETHER * 4,4'-ISOPROPYLIDENEDIPHENOL DIGLYCIDYL ETHER

THR: An experimental carcinogen. Mutagenic data. Skin and eye irritant. See also ethers. When heated to decomposition, it emits acrid and irritating fumes.

1,4-BIS(PHENYL AMINO)BEN-ZENE HR: 3

CAS: 74-31-7 NIOSH: ST 2275000
mf: $C_{18}H_{16}N_2$ mw: 260.36

PROP: A solid; d: 1.20, vap d: 9.0.

SYNS: AGERITE * N,N′-DIFENYL-p-FENYLENDI-AMIN (CZECH) * DIPHENYL-p-PHENYLENEDIA-MINE * N,N′-DIPHENYL-p-PHENYLENEDIAMINE * P-PHENYLAMINODIPHENYLAMINE * 4-PHE-NYLAMINODIPHENYLAMINE * USAF GY-2

THR: An intraperitoneal poison. Moderately toxic by ingestion. A weak allergen. An experimental tumorigen. Severe eye irritant. Slight fire hazard when exposed to heat or flame. Moderately dangerous; when heated to decomposition, it emits toxic fumes of NO_x; can react with oxidizing materials.

trans-1,2-BIS(n-PROPYL SULFONYL)-ETHYLENE HR: 3

CAS: 1113-14-0 NIOSH: KU 8085000
mf: $C_8H_{16}O_4S_2$ mw: 240.36

SYNS: CHEMAGRO B-1843 * VANCIDE PA

THR: Poison by ingestion. Moderately toxic via intraperitoneal route. See also sulfonates. When heated to decomposition, it emits toxic fumes of SO_x.

1,4-BIS(p-TOLYAMINO)ANTHRAQUI-NONE HR: 2

CAS: 128-80-3 NIOSH: CB 5775000
mf: $C_{28}H_{22}N_2O_2$ mw: 418.52

SYNS: BIS-1,4-p-TOLYLAMINOANTHRCHINON (CZECH) * C.I. 61565 * C.I. SOLVENT GREEN 3 * D AND C GREEN NO. 6

THR: Moderately toxic by ingestion. A severe eye irritant. When heated to decomposition, it emits toxic fumes of NO_x.

BIS(TRIBUTYL TIN)OXIDE HR: 3

CAS: 56-35-9 NIOSH: JN 8750000
mf: $C_{24}H_{54}OSn_2$ mw: 596.16

SYNS: BIS-(TRI-N-BUTYLCIN)OXID (CZECH) * BIS(TRI-N-BUTYLZINN)-OXYD (GERMAN) * ENT 24,979 * HEXABUTYLDISTANNOXANE * HEXABUTYLDITIN * KYSLICNIK TRI-N-BUTYL-CINICITY (CZECH) * OTBE (FRENCH) * OXY-BIS(TRIBUTYLTIN) * OXYDE DE TRIBUTYLETAIN * TRIBUTYLTIN OXIDE

OSHA PEL: TWA 100 ug(Sn)/m^3

THR: A poison by ingestion, intraperitoneal, and intravenous routes. Moderate dermal toxicity. An eye irritant. See also tin compounds. When heated to decomposition, it emits acrid and irritating fumes. For further information, see Tri-n-butyltinoxide, Vol. 1, No. 5 of *DPIM Report*.

BIS(TRICHLORO METHYL)TRISUL-FIDE HR: 3

CAS: 2532-50-5 NIOSH: YL 7700000
mf: $C_2Cl_6S_3$ mw: 332.90

SYNS: BIS-TRICHLOROMETHYL-TRISULFID (CZECH) * TRITHIOBIS(TRICHLOROMETHANE)

THR: An intravenous poison. Moderately toxic by ingestion. A skin and eye irritant. See also sulfides. When heated to decomposition, it emits very toxic fumes of Cl^- and SO_x.

BIS(TRIETHYL TIN) SULFATE HR: 3

CAS: 57-52-3 NIOSH: XQ 7175000
mf: $C_6H_{16}O_4SSn$ mw: 302.97

SYNS: STANNANE, TRIETHYLHYDROXY-, SUL-FATE (2:1) (8CI) * TRIETHYLHYDROXYTIN SUL-FATE

OSHA PEL: TWA 100 ug(Sn)/m^3 (skin)

THR: Poison by ingestion, intraperitoneal, subcutaneous, intravenous, and parenteral routes. See also tin compounds and sulfates. When heated to decomposition, it emits toxic fumes of SO_x.

BIS(TRINITROPHENYL)SULFIDE HR: 2

CAS: 28930-30-5 NIOSH: WQ 2850000
mf: $C_{12}H_4N_6O_{12}S$ mw: 456.28

SYNS: HEXANITRODIPHENYLSULFIDE * PICRYL SULFIDE

THR: Moderately toxic by ingestion. See also sulfides and nitro compounds of aromatic hydrocarbons. See nitrates for fire and explosion hazard. This material is a powerful explosive and has an added military advantage in that its explosive gases contain irritating and very toxic SO_x. See also explosives, high.

BIS(TRIPHENYL PHOSPHINE)NICKEL DITHIO CYANATE HR: 3

CAS: 15709-62-3 NIOSH: QR 6525000
mf: $C_{38}H_{30}N_2NiP_2S_2$ mw: 699.47

SYNS: NICKEL, BIS(TRIPHENYLPHOSPHINE)-, DI-THIOCYANATE * PHOSPHINE, NICKEL BIS(TRI-PHENYL-, DITHIOCYANATE

THR: An intravenous poison. See also nickel compounds and thiocyanates. When heated to decomposition, it emits very toxic fumes of SO_x, PO_x, NO_x, and CN^-.

BIS(TRIPHENYL SILYL)CHRO-MATE   HR: 2
CAS: 1624-02-8   NIOSH: GB 2685000
mf: $C_{36}H_{30}CrO_4Si_2$ mw: 634.84

SYN: CHROMIC ACID, BIS(TRIPHENYLSILYL) ES-TER

THR: Moderately toxic by ingestion and skin contact. See also chromium compounds and es-ters. When heated to decomposition, it emits toxic fumes of CrO_3 particulates.

BIS(TRIPHENYL TIN)SULFIDE   HR: 3
CAS: 77-80-5   NIOSH: JN 8850000
mf: $C_{36}H_{30}SSn_2$ mw: 732.10

SYN: 1,1,1,3,3,3-HEXAPHENYLDISTANNTHIANE

THR: A poison via intravenous route. See also tin compounds and sulfides. When heated to decomposition, it emits toxic fumes of SO_x.

BITIODIN   HR: 3
CAS: 5169-78-8   NIOSH: TM 7870000
mf: $C_{15}H_{17}NS_2$ mw: 275.45

SYNS: 3-(DI-2-THIENYLMETHYLENE)-1-METHYL-PIPERIDINE ∗ 1-METHYL-3-PIPERIDYLIDENEDI(2-THIENYL)METHANE

THR: A poison via intraperitoneal, intravenous, and intramuscular routes. Moderately toxic by ingestion. When heated to decomposition, it emits very toxic fumes of NO_x and SO_x.

(m,o'-BITOLYL)-4-AMINE   HR: 3
CAS: 13394-86-0   NIOSH: DU 9100000
mf: $C_{14}H_{15}N$ mw: 197.30

SYNS: 3,2'-DIMETHYL-4-AMINODIPHENYL
∗ 3,2'-DIMETHYL-4-BIPHENYLAMINE

THR: An experimental carcinogen and tumori-gen. Mutagenic data. When heated to decompo-sition it emits toxic fumes of NO_x.

BITTER ALMOND OIL   HR: 3
PROP: Volatile oil from dried ripe kernels of bitter almonds or from other kernels containing amygdalin, such as apricots, cherries, plums, and especially peaches. Colorless to yellow, very refractive liquid; characteristic odor and taste of benzaldehyde. d: 1.028 − 1.060 @ 25°/25°. Sltly sol in water, miscible with alco-hol, ether, oils. Keep cool and protected from light.

THR: Very poisonous. Human toxicity due to HCN component.

BITTER ORANGE OIL   HR: 2
                          NIOSH: RJ 3392000

PROP: Main constituent is d-Limonene. Pale yellow liquid, bitter taste. d: 0.842-0.848 @ 25°/25°. Very sltly sol in water; miscible with absolute alc; sol in 4 vols alc, in 1 vol glacial acetic acid. Keep well closed, cool and protected from light.

THR: A skin irritant. See also d-limonene. When heated to decomposition, it emits acrid and irri-tating fumes.

BLADEX   HR: 3
CAS: 21725-46-2   NIOSH: UG 1490000
mf: $C_9H_{13}ClN_6$ mw: 240.73

PROP: A white, crystalline material; mp: 167°.

SYNS: 2-CHLORO-4-(1-CYANO-1-METHYLETH-YLAMINO)-6-ETHYLAMINO-1,3,5-TRIAZINE
∗ 2-(4-CHLORO-6-ETHYLAMINO-s-TRIAZINE-2-YLAMINO)-2-METHYL-PROPIONITRILE
∗ 2-((4-CHLORO-6-(ETHYLAMINO)-s-TRIAZIN-2-YL)AMINO)-2-METHYLPROPIONITRILE
∗ 2-(4-CHLORO-6-ETHYLAMINO-1,3,5-TRIA-ZINE-2-YLAMINO)-2-METHYLPROPIONITRILE
∗ CYANAZINE

THR: Poison by ingestion. Moderately toxic by skin contact. Mutagenic data. See also ni-triles. When heated to decomposition, it emits very toxic fumes of Cl^-, NO_x, and CN^-. For further information, see Cyanazine Vol. 3, No. 1 of *DPIM Report*.

BLEOMYCIN A2   HR: 3
CAS: 11116-31-7   NIOSH: EC 5988500

THR: Mutagenic data. Noted for adverse pulmo-nary effects (in man). When heated to decompo-sition, it emits toxic fumes of NO_x.

BONE OIL   HR: 2
CAS: 8001-85-2   NIOSH: ED 0780000

PROP: Product of destructive distillation of bones in preparation of bone charcoal containing nitrogenous compounds such as pyridine, ani-line, methylamine, and pyrrole.

SYNS: ANIMAL OIL * BONE OIL (DOT)
* DIPPEL'S OIL * OIL OF HARTSHORN

DOT Classification: ORM-A, Label: None

THR: Moderately toxic by ingestion.

BORAZINE   **HR: 1**
mf: $B_3H_6N_3$ mw: 80.5

SYN: BORAZOLE

PROP: Colorless liquid; mp: $-58°$, bp: $53°$,
d: 0.824 @ $0°$.

THR: Powerful irritant to skin, eyes, and mucous membranes. See also boron compounds.
Dangerous fire hazard by chemical reaction to
produce flammable or even spontaneously flammable gases. Hydrolyzes in water to evolve boron hydrides. Sealed ampoules exposed to light
explode. Dangerous; when heated to decomposition, it emits toxic fumes; upon reaction with
water, can evolve toxic and flammable gases.

BORDEAUX ARSENITE   **HR: 3**
                                 NIOSH: ED 3880000
DOT: 2759
DOT Classification: Label: Poison

THR: A poison. See also arsenic compounds
and copper compounds. When heated to decomposition, it emits toxic fumes of As.

BORIC ACID   **HR: 3**
CAS: 10043-35-3   NIOSH: ED 4550000
mf: BH_3O_3 mw: 61.84

PROP: White crystals or powder; mp: $185°$ (decomp), $-$ 1.5HOH @ $300°$, d: 1.435 @ $15°$.

SYNS: BORACIC ACID * BORSAURE (GERMAN)
* NCI-C56417 * ORTHOBORIC ACID

THR: Poison by ingestion and subcutaneous
routes. Moderately toxic via skin contact to infants and children. An experimental teratogen;
causes gastrointestinal effects. Mutagenic data.
See also boron compounds. Incompatible with
K; $(CH_3CO)_2O$. For further information, see
Vol. 1, No. 8 of *DPIM Report*.

BORIC ACID, ETHYL ESTER   **HR: 3**
CAS: 34099-73-5   NIOSH: ED 4590000
DOT: 1176
mf: $C_2H_7BO_3$ mw: 89.90

PROP: Colorless liquid, mild odor, decomp
in water. bp: $120°$, flash p: $52°F(CC)$, d: 0.864
@ $26.5°$, vap d: 5.04.

SYN: ETHYLBORATE (DOT)

DOT Classification: Label: Flammable
Liquid

THR: A poison; a severe eye irritant. See also
boron compounds and esters. Dangerous fire
hazard when exposed to heat or flame; will react
with water or steam to produce flammable vapors. To fight fire, use CO_2, dry chemical. Incompatible with oxidizers; heat; open flame.

BORNEOL   **HR: 2**
CAS: 507-70-0   NIOSH: ED 7000000
mf: $C_{10}H_{18}O$ mw: 154.28

PROP: Hexagonal crystals, peppery odor and
burning taste; mp: $208°$, bp: $212°$, flash p: $150°F$,
d: 1.01 @ $20°/4°$, vap d: 5.31.

SYNS: BAROS CAMPHOR * BHIMSAIM CAM-
PHOR * BORNEO CAMPHOR * TRANS-BOR-
NEOL * BORNYL ALCOHOL * 2-CAMPHANOL
* DRYOBALANOPS CAMPHOR * 2-HYDROXY-
CAMPHANE * MALAYAN CAMPHOR * SUMA-
TRA CAMPHOR

THR: Moderately toxic by ingestion. A mild
irritant. Moderate fire hazard when exposed to
heat or flame; can react with oxidizing materials.
To fight fire, use water, CO_2, water spray, dry
chemical. When heated to decomposition, it
emits acrid smoke and fumes.

BORON   **HR: 3**
CAS: 7440-42-8   NIOSH: ED 7350000
mf: B   mw: 10.81

PROP: Monoclinic crystals, yellow or brown
amorphous powder. mp: $2300°$, bp: $2550°$, d:
3.33 @ $20°$.

THR: A poison by ingestion. See also boron
compounds. Moderate fire hazard in the form
of dust when exposed to air or by chemical
reaction. An explosion hazard in the form of
dust which ignites on contact with air. See also
iron dust. Incompatible with NH_3; Br_2; BrF_3;
Cs_2C_2; Cl_2; CuO; F_2; HIO_3; PbO_2; HNO_3; NO;
NOF; N_2O; $KClO_3$; KNO_3; Rb_2C_2; AgF; S;
BrF_5; IF_5; metal fluorides; inter halogens; nitryl
fluoride (FNO_2); OF_2; KNO_2; NO_x; Na_2O_2; PbO;
air. For further information see Vol. 3, No. 5
of *DPIM Report*.

BORON BROMIDE   **HR: 3**
CAS: 10294-33-4   NIOSH: ED 7400000

DOT: 2692

mf: BBr_3 mw: 250.54

PROP: Colorless, fuming liquid; mp: $-45°$, bp: $91.7°$, d: 2.650 @ $0°$, vap press: 40 mm @ $14.0°$, 100 mm @ $33.5°$.

SYN: BORON TRIBROMIDE (DOT)

DOT Classification: Label: Corrosive

THR: A poison. A skin, eye, and mucous membrane irritant. See also boron compounds and HBr. Dangerous; when heated to decomposition, it can explode; emits toxic fumes of bromides and will react with water or steam to produce toxic and corrosive fumes and possibly explode. Incompatible with K; Na; H_2O.

BORON COMPOUNDS HR: 3

THR: Very toxic and therefore considered an industrial poison. Used in medicine as sodium borate, boric acid or borax, which is a common cleanser. Fatal poisoning of children has been caused in some instances by the accidental substitution of boric acid for powdered milk. The medical literature reveals instances of accidental poisoning due to boric acid; oral ingestion of borates or boric acid; and presumably absorption of boric acid from wounds and burns. The fatal dose of orally ingested boric acid for an adult is somewhat $>$ 15 to 20 g and for an infant from 5-6 g. Boron is one of a group of elements, such as Pb, Mn, As, which affects the central nervous system. Boron poisoning causes depression of the circulation, persistent vomiting and diarrhea, followed by profound shock and coma. The temperature becomes subnormal and a scarletina-form rash may cover the entire body. Containers of boric acid should be plainly labeled and should differ radically from those which contain powdered milk, particularly in institutions such as hospitals.

BORON FLUORIDE HR: 3

CAS: 7637-07-2 NIOSH: ED 2275000

DOT: 1008

mf: BF_3 mw: 67.81

PROP: Colorless gas. Pungent, irritating odor; mp: $-126.8°$, bp: $-99.9°$, d: 2.99 g/L.

SYNS: BORON TRIFLUORIDE * FLUORURE DE BORE (FRENCH)

OSHA PEL: CL 1 ppm

DOT Classification: Label: Nonflammable Gas and Poison

THR: A poison by inhalation. A strong irritant. See also boron compounds, fluorides. Dangerous; when heated to decomposition or upon contact with water or steam, will produce toxic and corrosive fumes of F^-. Incompatible with alkali metals; alkaline earth metals (except Mg); alkyl nitrates; CaO.

BORON HYDRIDE HR: 3

CAS: 19287-45-7 NIOSH: HQ 9275000

mf: B_2H_6 mw: 27.68

PROP: Colorless gas; sickly sweet odor; mp: $-165.5°$; bp: $-92.5°$; d: 0.447 (liquid @ $-112°$); 0.577 (solid @ $-183°$); vap press: 224 mm @ $-112°$; autoign temp: $38°-52°$; lel = 0.9%; uel = 98%; flash p: $-90°F$.

SYNS: DIBORANE(6) * DIBORANE * DIBORON HEXAHYDRIDE * BOROETHANE

OSHA PEL: TWA 0.1 ppm

THR: Poison by skin contact. An irritant poison to skin, eyes, and mucous membranes comparable with chlorine, fluorine, arsine, and phosgene. The liquid causes local inflammation, blisters, redness, and swelling. Injuries to central nervous system, liver, and kidneys have also been produced in experimental animals. Similar observations have been reported in humans resulting at times in a reaction resembling metal fume fever. Human exposure to pentaborane has produced signs of severe central nervous system irritation such as drowsiness, dizziness, visual disturbances, muscle twitching and, in severe cases, painful muscle spasm. Dangerously flammable when exposed to heat or flame or by chemical reaction. Pentaborane (stable) is spontaneously flammable in air. On contact with moisture, hydrogen is usually evolved. Highly explosive when exposed to heat; flame; air; HNO_3; O_2. Diborane reacts explosively with Cl_2. Other boron hydrides evolve H_2 upon contact with moisture or can propagate a flame rapidly enough to cause an explosion. Heat can cause these materials to decompose violently or at least to evolve H_2; they also react with water or steam to evolve hydrogen. Powerful oxidizing agents such as chlorine gas, etc., can react violently with boron hydrides. For further information, see diborane, Vol. 2, No. 1 of *DPIM Report*.

BORON OXIDE HR: 3

CAS: 1303-86-2 NIOSH: ED 7900000

mf: B_2O_3 mw: 69.62

PROP: Vitreous, colorless crystals; mp: 450° (approx), bp: 1860°, d: 2.46.

SYNS: BORIC ANHYDRIDE * BORON SESQUIOXIDE * BORON TRIOXIDE * FUSED BORIC ACID

OSHA PEL: TWA 15 mg/m^3
ACGIH TLV: TWA 10 mg/m^3
DFG MAK: 15 mg/m^3

THR: See also boron compounds. Mixed with CaO and put into fused $CaCl_2$, the mixture incandesces.

BORON PHOSPHIDE   HR: 3
mf: BP   mw: 41.79

PROP: Maroon powder; mp: 200°.

THR: A poison. See also boron compounds and phosphides. Fire hazard; ignites @ 200°. Deflagrates with fused alkali nitrates. See phosphides for explosion hazard. Incompatible with HNO_3; oxidants, i.e., nitrates.

BORON TRICHLORIDE   HR: 3
CAS: 10294-34-5   NIOSH: ED 1925000
DOT: 1741
mf: BCl_3 mw: 117.16

PROP: Colorless, fuming liquid. Pungent, irr odor. mp: −107°, bp: 12.5°, d: 1.434 @ 0°, vap press: 1 atm @ 12.7°, vap d: 4.03.

SYNS: BORON CHLORIDE * CHLORURE DE BORE (FRENCH)

DOT Classification: Label: Corrosive

THR: A poison. An irritant to skin, eyes and mucous membranes. See also boron compounds and hydrochloric acid. Dangerous; when heated to decomposition, it emits toxic fumes of chlorides; will react with water or steam to produce heat, toxic and corrosive fumes. Incompatible with aniline; hexafluorisopropylidene amino lithium; NO_2; PH_3; grease; organic matter; O_2.

BORON TRIIODIDE   HR: 3
mf: BI_3 mw: 391.52

PROP: Colorless, hygroscopic plates; mp: 43°, bp: 210°, d: 3.35 @ 50°.

THR: A poison. See also boron compounds and iodides. Incompatible with NH_3; P; water; ethers; carbohydrates; POCl.

BP-7,8-DIHYDRODIOL   HR: 3
CAS: 61443-57-0   NIOSH: DJ 5000000
mf: $C_{20}H_{14}O_2$ mw: 286.34

SYN: (+,−)-trans-7,8-DIHYDROXY-7,8-DIHYDROBENZO(a)PYRENE

THR: An experimental carcinogen, neoplastigen, and tumorigen. Mutagenic data. When heated to decomposition it emits acrid smoke and fumes.

BRACKEN FERN, CHLOROFORM
FRACTION   HR: 3
                    NIOSH: EE 1520500

PROP: Chloroform fraction of tannin isolated from Bracken Fern (*Pteridium Aquilinum*)

THR: An experimental carcinogen.

BRACKEN FERN, DRIED   HR: 3
                    NIOSH: EE 1520000

SYNS: PTERIS AQUALINA * PTERIDIUM AQUILINUM * S. EGRELTRI ATUNUN (TURKISH)

THR: An experimental carcinogen, tumorigen, and neoplastigen. Mutagenic data.

BRACKEN FERN TANNIN   HR: 3
                    NIOSH: EE 1521000

SYNS: PTERIDIUM AQUILINUM TANNIN
* TANNIN FROM BRACKEN FERN

THR: An intraperitoneal poison. An experimental neoplastigen. Mutagenic data.

BRADYKININ   HR: 3
CAS: 58-82-2   NIOSH: EE 1530000
mf: $C_{50}H_{73}N_{15}O_{11}$ mw: 1060.38

SYNS: BRADYKININ (SYNTHETIC) * KALLIDIN
* SYNTHETIC BRADYKININ

THR: An experimental teratogen via intravenous route. When heated to decomposition, it emits toxic fumes of NO_x.

BRILLIANT BLUE R   HR: 3
CAS: 2580-78-1   NIOSH: CB 1050000
mf: $C_{22}H_{16}N_2O_{11}S_3$ • 2Na   mw: 626.56

SYNS: C.I. 61200 * REMAZOL BRILLIANT BLUE R

THR: An experimental tumorigen. When heated to decomposition, it emits very toxic fumes of NO_x and SO_x.

BRISTAMIN HYDROCHLORIDE   HR: 3
CAS: 6152-43-8   NIOSH: KQ 7600000
mf: $C_{17}H_{21}NO$ • ClH   mw: 291.85

SYNS: N,N-DIMETHYL-2-(alpha-PHENYL-O-TOLOXY)-ETHYLAMINE HYDROCHLORIDE * PHENYLTOLOXAMINE HYDROCHLORIDE

THR: Poison by ingestion, intravenous, and intraperitoneal routes. When heated to decomposition it emits very toxic fumes of HCl and NO_x.

BROMADRYL   HR: 3
CAS: 13977-28-1   NIOSH: KQ 8400000
mf: $C_{18}H_{22}BrNO \cdot ClH$ mw: 384.78

SYNS: 2-(1-(4-BROMODIPHENYL)ETHOXY)-N,N-DIMETHYLETHYLAMINE HYDROCHLORIDE * P-BROMO-ALPHA-METHYLBENZHYDRYL-2-DIMETHYLAMINOETHYL ETHER HYDROCHLORIDE * 2-((P-BROMO-ALPHA-METHYL-ALPHA-PHENYL-BENZYL)OXY)-N,N-DIMETHYLETHYLAMINE HYDROCHLORIDE * 2-(1-(4-BROMOPHENYL)-1-PHENYLETHOXY)-N,N-DIMETHYLETHANAMINE HYDROCHLORIDE * 1-(P-BROMOPHENYL)-1-PHENYL-1-(2-DIMETHYLAMINOETHOXY)ETHANE HYDROCHLORIDE * (2-(1-P-BROMOPHENYL-1-PHENYLETHOXY)ETHYL)DIMETHYLETHYLAMINE HYDROCHLORIDE * BETA-DIMETHYLAMINOETHYL-P-BROMO-ALPHA-METHYLBENZHYDRYL ETHER HYDROCHLORIDE

THR: Poison by ingestion and intravenous route. See also ethers. When heated to decomposition, it emits very toxic fumes of HCl, Br^-, and NO_x.

BROMATES   HR: 2
THR: Generally considered to be more toxic than chlorates causing central nervous system paralysis. They may form methemoglobin, but less actively than chlorates. See also specific compounds as listed. Moderate fire hazard in the form of gas, vapor, or dust by chemical reaction with (powdered metals + acids); Al; As; CaH_2; C; Cu; powdered metals; metal sulfides; organic matter; PH_4I; P; SrH; S; (H_2SO_4 + metals). Dangerous; when heated to decomposition they emit toxic fumes of bromides; they can react with reducing materials.

BROMELAIN   HR: 3
CAS: 9001-00-7   NIOSH: EF 8575000

SYNS: BROMELIN * PLANT PROTEASE CONCENTRATE

THR: A poison via intraperitoneal and intravenous routes. When heated to decomposition, it emits toxic fumes.

BROMIC ACID, POTASSIUM SALT   HR: 2
CAS: 7758-01-2   NIOSH: EF 8725000
DOT: 1484
mf: $BrO_3 \cdot K$ mw: 167.01

PROP: White crystals or crystalline powder; mp: 350° (approx), decomp @ 370°, d: 3.27 @ 17.5°. Sltly sol in cold water; sol in hot water.

SYN: POTASSIUM BROMATE (DOT)

DOT Classification: Label: Oxidizer

THR: A powerful oxidizer. An irritant to skin, eyes, and mucous membranes. A food additive permitted in food for human consumption. Violent reaction with Al; As; C; Cu; $Pb(C_2H_3O_2)_2$; metal sulfides; organic matter; P; Se; S. Mutagenic data. When heated to decomposition, it emits very toxic fumes of Br^- and K_2O.

BROMIDES   HR: 3
THR: The most common inorganic bromides are Na, K, NH_4, Ca and Mg bromides. Methyl and ethyl bromides are among the most common organic bromides. Exposure to the inorganic bromides produces depression, emaciation, and, in severe cases, psychoses and mental deterioration. Bromide rashes (bromoderma), especially of the face and resembling acne and furunculosis, often occur when bromide inhalation or administration is prolonged. Organic bromides, such as methyl bromide and ethyl bromide, are volatile liquids of relatively high toxicity. See also specific compounds. When strongly heated, they emit highly toxic fumes of Br^-. For further information, see Vol. 1, No. 4 of *DPIM Report*.

BROMINE   HR: 3
CAS: 7726-95-6   NIOSH: EF 9100000
DOT: 1744
mf: Br_2 mw: 159.82

PROP: Rhombic crystals or dark red liquid; fp: −7.3°, bp: 58.73°, d: 2.928 @ 59°, 3.12 @ 20°, vap press: 175 mm @ 21°, 1 atm @ 58.2°, vap d: 5.5.

SYNS: BROM (GERMAN) * BROME (FRENCH) * BROMO (ITALIAN) * BROOM (DUTCH)

OSHA PEL: TWA 0.1 ppm
ACGIH TLV: TWA 0.1 ppm; STEL 0.3 ppm
DFG MAK: 0.1 ppm (0.7 mg/m^3)
DOT Classification: Label: Corrosive

THR: A poison by ingestion and inhalation. The action of bromine is essentially the same as that of chlorine, being an irritant to the mucous membranes of the eyes and upper respiratory tract. Severe exposures may result in pulmonary edema. Usually, however, the irritant qualities of the chemical force the workman to leave the exposure before serious poisoning can result. Chronic exposure is similar to the therapeutic ingestion of excessive bromides. See also bromides. Regular physical examinations should be made upon people who work with bromine or bromides. Moderate fire hazard in the form of liquid or vapor by spontaneous chemical reaction with reducing materials. A very powerful oxidizer. Highly dangerous; when heated, it emits highly toxic fumes; will react with water or steam to produce toxic and corrosive fumes. Incompatible with acetaldehyde; C_2H_2; acrylonitrile; Al; NH_3; Sb; B; Ca_3N_2; Cs_2O; Cs_2C_2; CsC_2H; ClF_3C_2; CuH_2; Cu_2C_2; dimethyl formamide; ethyl phosphine; F_2; Ge; H_2; Fe_2C; isobutyrophenone; Li; Li_2C_2; Li_2Si_2; Mg_3P_2; CH_3OH; $Ni(Co)_4$; NI_3; olefins; OF_2; O_3; PH_3; P; PO_x; K; Rb_2C_2; RbC_2H; AgN_3; Na; Na_2C_2; NaC_2H; Sr_3P; Sn; UC_2; ZrC2; reducing materials. For further information see Vol. 3, No. 5 of *DPIM Report*.

BROMINE AZIDE   HR: 3
mf: BrN_3 mw: 121.93

PROP: Crystals or red liquid; mp: 45°, bp: explodes.

SYN: BROMOAZIDE

THR: A poison. See also Br and azides. Can self detonate (even in solution). Moderate fire hazard in the form of vapor by chemical reaction. A powerful oxidant. See also Br. Can explode spontaneously. Moderately explosive when exposed to heat. Dangerous; when heated to decomposition, emits highly toxic fumes of Br and explodes. Incompatible with Sb; As; ethyl ether; P; Ag; Na; metals.

BROMINE DIOXIDE   HR: 3
mf: BrO_2 mw: 111.91

PROP: Light yellow crystals; mp: 0° (decomp).

THR: See bromine. Very unstable material. Moderate fire hazard in the form of vapor by chemical reaction with reducing agents. A strong oxidant. Dangerous; when heated to de-

composition, emits highly toxic fumes of Br. Incompatible with water or steam to produce toxic and corrosive fumes; reducing materials. Unstable unless stored at low temperatures. Rapid heating may explode it.

BROMINE FLUORIDE   HR: 3
mf: BrF   mw: 98.91

THR: A poison. Powerful irritant. Very reactive. See also Br, F. Incompatible with organic matter; water.

BROMINE PENTAFLUORIDE   HR: 3
CAS: 7789-30-2   NIOSH: EF 9350000
DOT: 1745
mf: BrF_5 mw: 174.91

PROP: Colorless fuming liquid; mp: −61.3, bp: 40.5, d: 2.466 @ 25°, vap d: 6.05.
ACGIH TLV: TWA 0.1 ppm
DOT Classification: Label: Oxidizer

THR: Poisonous and corrosive. See also Br and fluorides. Dangerous; will react with water or steam to produce toxic and corrosive fumes. Liquefied gas reacts violently with many organic compounds and some inorganic compounds. It is a powerful oxidizer. When heated to decomposition, it emits very toxic fumes of F^- and Br^-. Incompatible with acids; halogens; nonmetals; oxides; H_2; H_2S; cork; grease; paper; wax; $ClCH_3$; water; acetic acid; NH_3; As; C_6H_6; H_2S; cellulose; charcoal; C_2H_5OH; I; alkaline halides; metallic halides; metal oxides; metals; CH_4; HNO_3; organic matter; Se; $S;H_2SO_4$.

BROMINE TRIFLUORIDE   HR: 3
CAS: 7787-71-5   NIOSH: EF 9360000
DOT: 1746
mf: BrF_3 mw: 136.91

PROP: Colorless, fuming liquid. Mp: 8.8°, bp: 127°, d: 2.84.
DOT Classification: Label: Oxidizer and Poison

THR: Poisonous and corrosive. See also bromine pentafluoride, fluorides, bromine. Very dangerous. Very reactive; a powerful oxidizer. Incompatible with NH_4Br; NH_4Cl; NH_4I; Sb; SbOCl; Sb_2O_3; As; $BaCl_2$; Bi_2O_5; B; Br; $CdCl_2$; $CaCl_2$; Co; CCl_4; CI_4; CsCl; I; LiCl; $MnIO_3$; metals; Mo; Nb; Nb_2O_5; organic matter; P; $PtBr_4$; $PtCl_4$; (Pt + KFO); KBr; KCl; KI; $RhBr_4$; RbCl; AgCl; NaBr; NaCl; NaI; $SnCl_2$; S; Ta;

Ta_2O_5; Sn; Ti; W; UO_x; V; H_2O; rubber; plastics; organic materials. Explosive reaction with ammonium halides; $SbCl_3O$; Co; halogens; C; CI_4; CH_3Cl; C_6H_6; ether; 2-pentanone; pyridine; silicone grease; solvents (acetone, ether, and toluene @ $-80°$); U; UF_6.

alpha-BROMOACETIC ACID   HR: 3
CAS: 79-08-3   NIOSH: AF 5955000
mf: $C_2H_3O_2Br$ mw: 158.96

PROP: Hygroscopic cryst, sol in water and alc.; d: 1.93; mp: 50°; bp: 208°.

SYNS: BROMOETHANIOC ACID   *   MONOBRO-
MOACETIC ACID

DOT Classification: Label: Corrosive

THR: A powerful irritant and poison. Corrosive. See also bromides. When heated to decomposition, it emits toxic fumes of Br^-.

BROMOACETYLENE
mf: CHCBr   mw: 104.9

PROP: Gas; bp: $-2°$, vap d: 4.684

SYNS: BROMOETHYNE   *   BROMACETYLENE

THR: Variable toxicity. Probably similar to dibromoacetylene. Dangerous fire hazard by spontaneous chemical reaction. A spontaneously flammable gas. Highly explosive. When heated to decomposition, it burns and emits toxic fumes. Incompatible with oxidizing materials in contact with air even when solid at $-196°$ or upon distillation.

(o-BROMO BENZYL)ETHYL DIMETHYL AMMONIUM-p-TOLUENE SULFO-
NATE   HR: 3
CAS: 61-75-6   NIOSH: BO 9450000
mf: $C_{11}H_{17}BrN \cdot C_7H_7O_3S$ mw: 414.40

SYNS: BRETYLIUM-P-TOLUENESULFONATE
*   BRETYLIUM TOSYLATE

THR: A poison by ingestion, intraperitoneal, subcutaneous, intravenous, and intramuscular routes. See also sulfonates. When heated to decomposition, it emits very toxic fumes of SO_x, NO_x and Br^-.

BROMOBENZYLNITRILE   HR: 3
CAS: 5798-79-8   NIOSH: AL 8050000
mf: C_8H_6BrN mw: 196.06

PROP: Pure: yellowish-white crystals; tech: brown oily liquid with pungent odor of sour fruit; mp: 29°; bp: 242°; fp: 25.5°; flash p: none; d: 1.5160 @ 20°; vap d: 6.8; vap press: 0.011 mm @ 20°.

SYNS: BROMBENZYL CYANIDE   *   ALPHA-BRO-
MOBENZYL CYANIDE   *   ALPHA-BROMOBENZYL-
NITRILE   *   ALPHA-BROMOPHENYLACETONITRILE
*   ALPHA-BROMO-ALPHA-TOLUNITRILE

THR: Poison by ingestion. Moderately toxic by inhalation. See also nitriles. When heated to decomposition, it emits very toxic fumes of NO_x, Br^-, and CN^-. For further information, see Bromobenzyl Cyanide, Vol. 2, No. 3 of *DPIM Report*.

2-BROMOBUTANE   HR: 3
CAS: 78-76-2   NIOSH: EJ 6228000
mf: C_4H_9Br mw: 137.04

PROP: Colorless liquid; fp: $< -50°$, bp: 91.4°, flash p: 70°F, d: 1.257 @ 25°/25°.

SYNS: SEC-BUTYL BROMIDE   *   METHYLETHYL-
BROMOMETHANE

THR: An experimental neoplastigen. Narcotic in high concentrations. See also chlorinated hydrocarbons, aliphatic. Dangerous fire hazard when exposed to heat or flame. When heated to decomposition, it emits toxic fumes (Br^-); can react with oxidizing materials. To fight fire, use water, spray or mist, foam, CO_2, dry chemical. For further information, see sec-Butyl Bromide, Vol. 1, No. 1 of *DPIM Report*.

BROMOCHLOROMETHANE   HR: 3
CAS: 74-97-5   NIOSH: PA 5250000
mf: CH_2BrCl mw: 129.39

PROP: Clear, colorless liquid, sweet odor. Bp: 67.8°, fp: $-88°$, flash p: none: d: 1.930 @ 25°/25°, vap d: 4.46.

SYNS: CHLOROBROMOMETHANE   *   HALON
1011   *   METHYLENE CHLOROBROMIDE
*   MONO-CHLORO-MONO-BROMO-METHANE

OSHA PEL: TWA 200 ppm
ACGIH TLV: TWA 200 ppm; STEL 250 ppm

THR: Poison by ingestion and inhalation. This material has a narcotic action of moderate intensity, although of prolonged duration. Animals exposed for several weeks to 1000 ppm of this substance had blood bromide levels as high as 350 mg/100 g of blood. Therefore, until further data are available, it should be considered at least as toxic as carbon tetrachloride and more than minimal exposure to its vapors should be avoided. Dangerous; when heated to decomposition, it emits highly toxic fumes of halides, Br^- and Cl^-.

O-(4-BROMO-2-CHLOROPHENYL)-O-ETHYL-S-PROPYL PHOSPHORO-THIOATE   HR: 3
CAS: 41198-08-7   NIOSH: TE 6975000
mf: $C_{11}H_{15}BrClO_3PS$ mw: 373.65

SYNS: CGA 15324 * POLYCRON

THR: Poison by ingestion and skin absorption. When heated to decomposition, it emits very toxic SO_x, PO_x, Br^-, and Cl^-.

BROMOCRIPTINE   HR: 3
CAS: 25614-03-3   NIOSH: KE 8250000
mf: $C_{32}H_{40}BrN_5O_5$ mw: 654.68

SYNS: 2-BROMOERGOCRYPTINE * 2-BROMO-ALPHA-ERGOKRYPTIN * 2-BROMO-12'-HY-DROXY-2'-(1-METHYLETHYL)-5'-ALPHA-(2-METHYLPROPYL)ERGOTAMIN-3',6',18-TRIONE

THR: A poison via intravenous route. An experimental carcinogen. When heated to decomposition, it emits very toxic fumes such as Br^- and NO_x.

5-BROMO-2'-DEOXY URIDINE   HR: 3
CAS: 59-14-3   NIOSH: YU 7350000
mf: $C_9H_{11}BrN_2O_5$ mw: 307.13

SYNS: 5-BROMODEOXYURIDINE * 5-BROMO-2-DEOXYURIDINE * 5-BROMOURACIL-2-DEOXYRI-BOSIDE * 5-BROMODESOXYURIDINE * 5-BRO-MOURACIL DEOXYRIBOSIDE

THR: An experimental teratogen. Mutagenic data. Moderate intraperitoneal toxicity. When heated to decomposition, it emits very toxic fumes of Br^- and NO_x.

BROMO DICHLORO METHANE   HR: 3
CAS: 75-27-4   NIOSH: PA 5310000
mf: $CHBrCl_2$ mw: 163.83

PROP: Colorless liquid. bp: 89.2°–90.6°, d: 1.971 @ 25°/25°.

SYNS: DICHLOROBROMOMETHANE * NCI-c55243

THR: Suspected carcinogen. Moderately toxic by ingestion. Probably narcotic in high concentration. When heated to decomposition, it emits

very toxic fumes of Br^- and Cl^-. For further information, see Vol. 6, No. 3 of *DPIM Report.*

BROMO DIPHENYL METHANE   HR: 3
CAS: 776-74-9   NIOSH: PA 5350000
DOT: 1770
mf: $C_{13}H_{11}Br$ mw: 247.15

PROP: Solid, decomp in hot water, sol in alc, very sol in benzene. Mp: 45°; bp: 193° @ 26 mm.

SYN: DIPHENYL METHYL BROMIDE, SOLID (DOT)

DOT Classification: Label: Corrosive

THR: A corrosive poison. See also bromides. Dangerous in a fire.

BROMO EOSINE   HR: 3
CAS: 17372-87-1   NIOSH: LM 5850000
mf: $C_{20}H_8Br_4O_5 \cdot 2Na$ mw: 693.90

SYNS: BROMOFLUORESCEIC ACID * C.I. 45380 * D&C RED NO. 22 * DISODIUM EOSIN * EOSINE SODIUM SALT * EOSIN GELBLICH (GERMAN) * SODIUM EOSINATE * 2,4,5,7-TETRABROMO-9-O-CARBOXYPHENYL-6-HYDROXY-3-ISOXANTHONE, DISODIUM SALT * 2',4',5',7'-TETRABROMOFLUORESCEIN DI-SODIUM SALT * TETRABROMOFLUORESCEIN SOLUBLE * 2-(2,4,5,7-TETRABROMO-6-HY-DROXY-3-OXO-3H-XANTHENE-9-YL)BENZOIC ACID, DISODIUM SALT

THR: An experimental carcinogen. Moderate intravenous toxicity. When heated to decomposition, it emits very toxic fumes of Br^-.

3-BROMO-1,2-EPOXYPROPANE   HR: 3
CAS: 3132-64-7   NIOSH: TX 4115000
mf: C_3H_5BrO mw: 136.99

PROP: Flash p: < 22°.

SYN: EPIBROMOHYDRIN

THR: An intraperitoneal poison. Mutagenic data. See also bromides. When heated to decomposition, it emits toxic fumes of Br^-.

BROMOETHANE   HR: 2
CAS: 74-96-4   NIOSH: KH 6475000
DOT: 1891
mf: C_2H_5Br mw: 108.98

PROP: Colorless, volatile liquid. Mp: −119°, bp: 38.4°, lel = 6.7%, uel = 11.3%, flash p:

$< -4°F$,d: 1.451 @ 20°/4°, autoign temp: 952°F, vap press: 400 mm @ 21°, vap d 3.76.

SYNS: BROMURE D'ETHYLE * ETHYL BROMIDE * ETYLU BROMEK (POLISH) * MONOBROMO-ETHANE * NCI-C55481

OSHA PEL: TWA 200 ppm
DFG MAK: 200 ppm (890 mg/m³)
DOT Classification: Poison B

THR: Moderately irritating via inhalation; irritant to eyes and mucous membranes. It is readily decomposed into volatile toxic products, such as HBr and Br, particularly in the presence of hot surfaces or open flame. Physiologically, it is an anesthetic and narcotic. Its vapors are markedly irritating to the lungs on inhalation for even short periods. It can produce acute congestion and edema. Liver and kidney damage in humans has been reported. It is much less toxic than methyl bromide, but more toxic than ethyl chloride. It is a preparative hazard. Dangerously flammable by heat, open flame (sparks), oxidizers. Moderately explosive when exposed to flame. When heated to decomposition, it emits highly toxic fumes of bromine; will react with water or steam to produce toxic and corrosive fumes; can react vigorously with oxidizing materials. To fight fire, use CO_2, dry chemical.

2-BROMO ETHANOL   HR: 3
CAS: 540-51-2   NIOSH: KJ 8225000
mf: C_2H_5BrO mw: 124.98

SYNS: ETHYLENEBROMOHYDRIN * GLYCOL BROMOHYDRIN

THR: An acute intraperitoneal poison. An experimental neoplastigen. Mutagenic data. When heated to decomposition, it emits toxic fumes of Br^-.

2-BROMO-2-ETHYL BUTYRYL UREA   HR: 3
CAS: 77-65-6   NIOSH: YS 2975000
mf: $C_7H_{13}BrN_2O_2$ mw: 237.13

SYNS: BROMODIETHYLACETYLCARBAMIDE * BROMODIETHYLACETYLUREA * (ALPHA-BROMO-ALPHA-ETHYLBUTYRYL)CARBAMIDE * (ALPHA-BROMO-ALPHA-ETHYLBUTYRYL)UREA * 1-BROMO-ETHYL-BUTYRYL-UREA * NCI-C03805

THR: Poison by ingestion; moderately toxic via intravenous route. When heated to decomposition, it emits very toxic fumes of NO_x and Br^-.

2-BROMO ETHYL ETHYL ETHER
mf: C_4H_9BrO mw: 155

PROP: Liquid. Vap d: 5.25, flash p: 5°C.

THR: An insecticide. See also ethers for discussion of fire, explosion, and disaster hazards.

BROMOFORM   HR: 3
CAS: 75-25-2   NIOSH: PB 5600000
DOT: 2515
mf: $CHBr_3$ mw: 252.75

PROP: Colorless liquid or hexagonal crystals. Mp: 6°-7°, bp: 149.5°, flash p: none, d: 2.890 @ 20°/4°.

SYNS: BROMOFORME (FRENCH) * BROMOFOR-MIO (ITALIAN) * METHENYL TRIBROMIDE * NCI-C55130 * TRIBROMMETHAAN (DUTCH) * TRIBROMMETHAN (GERMAN) * TRIBROMO-METAN (ITALIAN) * TRIBROMOMETHANE

OSHA PEL: TWA 0.5 ppm (skin)
DOT Classification: Poison B

THR: An experimental neoplastigen. Metabolic poison. Moderate toxicity via ingestion and subcutaneous route. This material causes lachrymation. It can damage the liver to a serious degree and cause death. It has been said that its medicinal application has resulted in numerous poisonings. It has anesthetic properties similar to those of chloroform, but it is not sufficiently volatile for inhalation purposes and is far too toxic to be recommended. Inhalation of small amounts of this material causes irritation, provoking the flow of tears and saliva, and reddening of the face. Abuse can lead to addiction and serious consequences. Dangerous; when heated to decomposition, it emits highly toxic fumes of Br^-. Incompatible with Li; NaK alloy; acetone; potassium hydroxide. For further information, see Vol. 2, No. 6 of *DPIM Report*.

2-BROMO-D-LYSERGIC ACID DI-ETHYLAMIDE   HR: 3
CAS: 478-84-2   NIOSH: KE 2625000
mf: $C_{20}H_{26}BrN_3O$ mw: 404.40

SYNS: BROM LSD * 2-BROMO-9,10-DIDEHY-DRO-N,N-DIETHYL-6-METHYLERGOLINE-8-BETA-CARBOXAMIDE * 2-BROM-D-LYSERGIC ACID DIETHYLAMINE * BROMLYSERGAMIDE * 9,10-DIDEHYDRO-N,N-DIETHYL-2-BROMO-

6-METHYLERGOLINE-8-BETA-CARBOXAMIDE
* USAF SZ-1

THR: Poison by intraperitoneal and intravenous routes. An experimental teratogen. Causes central nervous system effects in humans. When heated to decomposition, it emits very toxic fumes such as Br^- and NO_x.

BROMO METHANE   HR: 3
CAS: 74-83-9   NIOSH: PA 4900000
DOT: 1062
mf: CH_3Br mw: 94.95

PROP: Colorless, transparent, volatile liquid or gas, burning taste, chloroform-like odor; bp: 3.56°, lel = 13.5%, uel = 14.5%, fp: −93°, flash p: none, d: 1.732 @ 0°/0°, autoign temp: 998°F, vap d: 3.27, vap press: 1824 mm @ 25°.

SYNS: BROM-METHAN (GERMAN) * BROMO-METANO (ITALIAN) * BROMURE DE METHYLE (FRENCH) * BROMURO DI METILE (ITALIAN) * BROOMMETHAAN (DUTCH) * METHYL-BROMID (GERMAN) * METHYL BROMIDE * METYLU BROMEK (POLISH) * MONOBROMO-METHANE

OSHA PEL: TWA CL 20 ppm (skin)
DOT Classification: Label: Poison

THR: A poison via inhalation in children. Affects the human gastrointestinal tract. Extremely irritating to skin and can produce severe burns. A powerful fumigant gas which is one of the most toxic of the common organic halides. It is hemotoxic and narcotic with delayed action; it is cumulative and damaging to nervous system, kidneys, lung. Central nervous system effects include blurred vision, mental confusion, numbness, tremors, speech defects. Methyl bromide is reported to be 8 times more toxic on inhalation than ethyl bromide. Moreover, because of its greater volatility, methyl bromide is a much more frequent cause of poisoning. Death following acute poisoning is usually caused by its irritant effect on the lungs. In chronic poisoning, death is due to injury to the central nervous system. Fatal poisoning has always resulted from exposures to relatively high concs of methyl bromide vapors (from 8,600 to 60,000 ppm). Nonfatal poisoning has resulted from exposure to concentrations as low as 100-500 ppm. In addition to the lung and central nervous system injury mentioned, the kidneys may be damaged with development of albuminuria and, in fatal cases, cloudy swelling and/or tubular degeneration. The liver may be enlarged. There are no characteristic blood changes. Moderately flammable when exposed to heat or flame. Moderately explosive when exposed to sparks or flame. When heated to decomposition, it emits highly toxic bromides. To fight fire, use foam, water, CO_2, dry chemical. Incompatible with metals (Al); dimethyl sulfoxide; ethylene oxide. Forms explosive mixtures with air within narrow limits at atmospheric pressure, but wider at higher pressure. For further information, see Vol. 5, No. 6 of *DPIM Report*.

7-BROMO METHYL BENZ(a)ANTHRACENE   HR: 3
CAS: 24961-39-5   NIOSH: CW 0380000
mf: $C_{19}H_{13}Br$ mw: 321.23

SYNS: 7-BMBA * ICR 498

THR: An acute intravenous poison. An experimental carcinogen and neoplastigen. Mutagenic data. When heated to decomposition, it emits toxic fumes of Br^-.

2-BROMO-3-METHYL BUTYRYL UREA   HR: 2
CAS: 496-67-3   NIOSH: YS 3150000
mf: $C_6H_{11}BrN_2O_2$ mw: 223.10

SYNS: BROMCARBAMIDE * ALPHA-BRO-MISOVALERYLUREA * BROMOCARBAMIDE * ALPHA-BROMOISOVALERIC ACID UREIDE * BROMOVALEROCARBAMIDE * ALPHA-BROMO-BETA-DIMETHYLPROPANOYLUREA * ALPHA-BROMOISOVALEROYLUREA * (ALPHA-BROMOISOVALERYL)UREA * MONOBRO-MOISOVALERYLUREA

THR: Moderately toxic by ingestion. Affects the central nervous system in women. When heated to decomposition, it emits very toxic fumes of Br^- and NO_x.

7-BROMO METHYL-12-METHYL BENZ(a)ANTHRACENE   HR: 3
CAS: 16238-56-5   NIOSH: CW 0525000
mf: $C_{20}H_{15}Br$ mw: 335.26

SYN: ICR 502

THR: An experimental neoplastigen and carcinogen. When heated to decomposition, it emits toxic fumes of Br^-.

1-BROMO-2-METHYL PROPANE   HR: 3
CAS: 78-77-3   NIOSH: TX 4140000
mf: C_4H_9Br mw: 137.04

PROP: Flash p: 22°C.

SYNS: I-BUTYL BROMIDE * ISOBUTYL BRO-
MIDE

THR: An experimental neoplastigen. Dangerous
fire hazard. When heated to decomposition, it
emits toxic fumes of Br^-.

2-BROMO-2-NITRO-1,3-PROPANE-
DIOL   HR: 3
CAS: 52-51-7   NIOSH: TY 3385000
mf: $C_3H_6BrNO_4$ mw: 200.01

SYN: 2-BROMO-2-NITROPROPAN-1,3-DIOL

THR: Poison by ingestion and intraperitoneal
route. A skin and eye irritant. When heated to
decomposition, it emits very toxic fumes of NO_x
and Br^-.

4-BROMOPHENOL   HR: 3
CAS: 106-41-2   NIOSH: SJ 7960000
mf: C_6H_5BrO mw: 173.02

SYN: P-BROMOPHENOL

THR: An experimental tumorigen. See also bro-
mophenols. When heated to decomposition, it
emits toxic fumes of Br^-.

BROMO PHENOLS (m,p,o).   HR: 2
PROP: (m) Crystals, insol in water, sol in alcohol, ether and alkalis. (p) crystals, sltly sol in
water, sol in alcohol, ether, chloroform and
glacial acetic acid. (o) yellow to oily red liquid,
unpleasant odor, insol in water, sol in alcohol,
ether and chloroform; $HO(C_6H_4)Br$, mw: 173
(m,p,o), d: (p) 1.840 (15°), 1.5875 (80°), d:
(o) 1.5, mp: (m) 33°, mp: (p) 64°, mp: (o) 6°,
bp: (m) 236°, bp: (p) 238°, bp: (o) 194°.

THR: Moderately toxic by subcutaneous route.
Dangerous in a fire. See also bromides.

BROMO PHENYL HYDRAMINE HYDRO-
CHLORIDE   HR: 3
CAS: 1808-12-4   NIOSH: KQ 8420000
mf: $C_{17}H_{20}BrNO \cdot ClH$ mw: 370.75

SYNS: BETA-(P-BROMOBENZHYDRYLOXY)-
ETHYLDIMETHYLAMINE HYDROCHLORIDE
* 2-(4-BROMOBENZOHY-
DRYLOXY)-ETHYL-DIMETHYLAMINE HYDRO-
CHLORIDE

THR: Poison by ingestion, intravenous, and intraperitoneal routes. When heated to decomposition, it emits very toxic fumes of Br^-, NO_x,
and HCl.

1-BROMOPROPANE   HR: 2
CAS: 106-94-5   NIOSH: TX 4110000
mf: C_3H_7Br mw: 123.01

PROP: Liquid. Mp: $-110°$, bp: 70.9°, d: 1.353
@ 20°/4°, autoign temp: 914°F. flash p: <22°,
lel = 4.6%.

SYN: PROPYL BROMIDE

THR: Mildly toxic by inhalation. Mutagenic
data. Dangerous fire hazard when heated or exposed to flame or oxidizers. Dangerous; when
heated to decomposition, it emits highly toxic
fumes of bromides; can react with oxidizing
materials. To fight fire, use water, foam, CO_2,
dry chemical.

BROMO-2-PROPANONE   HR: 2
CAS: 598-31-2   NIOSH: UC 0525000
DOT: 1569
mf: C_3H_5BrO mw: 136.99

SYN: BROMOACETONE

DOT Classification: Label: Poison Gas

THR: Moderately toxic by inhalation. When
heated to decomposition, it emits toxic fumes
of Br^-. For further information, see Bromoacetone, Vol. 2, No. 2 of DPIM Report.

3-BROMOPROPIONIC ACID   HR: 3
CAS: 590-92-1   NIOSH: UE 7875000
mf: $C_3H_5BrO_2$ mw: 152.99

SYN: BETA-BROMOPROPIONIC ACID

THR: An experimental carcinogen. Moderate
intraperitoneal toxicity. When heated to decomposition, it emits toxic fumes of Br^-.

3-BROMO PROPIONITRILE   HR: 3
CAS: 2417-90-5   NIOSH: UG 1050000
mf: C_3H_4BrN mw: 133.99

SYN: USAF DO-51

THR: An intraperitoneal and parenteral poison.
See also nitriles. When heated to decomposition,
it emits very toxic fumes of NO_x, CN^-, and
Br^-.

2-((6-(5-BROMO-2-PYRIDYL OXY)-HEXYL)AMINO)ETHANE THIOL HYDROCHLORIDE HR: 3
CAS: 41287-56-3 NIOSH: KJ 0280000
mf: $C_{13}H_{21}BrN_2OS \cdot ClH$ mw: 369.79

THR: Poison by ingestion and intraperitoneal route. When heated to decomposition, it emits very toxic Br^-, Cl^-, SO_x, and NO_x.

N-BROMO SUCCINIMIDE HR: 3
CAS: 128-08-5 NIOSH: WN 2275000
mf: $C_4H_4BrNO_2$ mw: 178.00

PROP: White to pale buff, fine crystalline powder with faint odor of bromine; mp: 173°-175°, d: 2.098.

SYNS: N-BROMOSUCCIMIDE * SUCCINBROMIMIDE * SUCCINIBROMIMIDE

THR: An intraperitoneal poison. An irritant poison to skin, eyes and mucous membranes. Dangerous in a fire; see bromides, NO_x. Incompatible with aniline; diallyl sulfide; hydrazine hydrate.

BROMOTRICHLOROMETHANE HR: 3
CAS: 75-62-7 NIOSH: PA 5400000
mf: $CBrCl_3$ mw: 198.27

PROP: Colorless liquid; bp: 103.8°-105.1°; d: 1.997 @ 25°/25°.

THR: Poison by ingestion. Narcotic in high concentration. See also chloroform. When heated to decomposition, it emits very toxic fumes of Cl^- and Br^-. Incompatible with ethylene.

BROMO TRIFLUOROETHYLENE HR: 3
mf: BrF_3C_2 mw: 160.94

THR: A poison. Flammable gas or liquid. When heated to decomposition, it emits highly toxic fumes of Br^-, F^- and $COCF_2$. Incompatible with powerful oxidizers; gas ($>-3°$); O_2; ignites in air.

BROMO TRIFLUORO METHANE HR: 2
CAS: 75-63-8 NIOSH: PA 5425000
DOT: 1009
mf: $CBrF_3$ mw: 148.92

SYNS: BROMOFLUOROFORM * HALON 1301 * MONOBROMOTRIFLUORMETHANE (DOT) * TRIFLUOROMONOBROMOMETHANE

OSHA PEL: TWA 1000 ppm
DOT Classification: Label: Nonflammable Gas

THR: Moderately toxic by inhalation. Dangerous in a fire; see also bromides and fluorides. Incompatible with aluminum.

BRONCHOSPASMIN HR: 3
CAS: 13055-82-8 NIOSH: XH 5099000

SYNS: 7-(3-(2-(3,5-DIHYDROXYPHENYL-2-HYDROXY-ETHYLAMINO)PROPYL)THEOPHYLLINE HYDROCHLORIDE * REPROTEROL-HCL

THR: Poison by intravenous route. When heated to decomposition it emits very toxic fumes of HCl and NO_x.

BRUCINE HR: 3
CAS: 357-57-3 NIOSH: EH 8925000
DOT: 1570
mf: $C_{23}H_{26}N_2O_4$ mw: 394.51

PROP: Monoclinic prisms; mp: 178°. An alkaloid extracted from Strychnos seeds.

SYNS: BRUCIN (GERMAN) * BRUCINA (ITALIAN) * BRUCINE ALKALOID * DIMETHOXY STRYCHNINE * 10,11-DIMETHYSTRYCHNINE

DOT Classification: Label: Poison

THR: A deadly poison by ingestion and intraperitoneal routes. An alkaloid-like strychnine, but 1/6 as toxic. Dangerous; when heated it emits toxic fumes of NO_x. For further information, see Vol. 3, No. 5 of *DPIM Report*.

BRUCINE METHIODIDE HR: 3
CAS: 60723-51-5 NIOSH: EH 9105000
mf: $C_{23}H_{26}N_2O_4 \cdot CH_3I$ mw: 536.45

SYNS: BRUCINE IODOMETHYLATE * BRUCINE IODOMETHYLE (FRENCH)

THR: A poison via intravenous route. See also brucine. When heated to decomposition, it emits very toxic fumes of NO_x and HI.

BUFORMIN HYDROCHLORIDE HR: 3
CAS: 15537-73-2 NIOSH: NJ 7020000
mf: $C_6H_{15}N_5 \cdot 7ClH$ mw: 412.48

SYNS: BUFONAMIN * DIABRIN * INSULAMIN

THR: A poison by ingestion. When heated to decomposition, it emits very toxic fumes of HCl and NO_x.

BUFOTALINE HR: 3
CAS: 471-95-4 NIOSH: EI 3150000
mf: $C_{26}H_{36}O_6$ mw: 444.62

SYNS: 3-BETA-14,16-BETA-TRIHYDROXY-5-BETA-BUFA-20,22-DIENOLIDE, 16-ACETATE * BUFOTALIN

THR: A poison by ingestion, subcutaneous, and intravenous routes. When heated to decomposition, it emits acrid and irritating fumes.

BUPICAINE HYDROCHLORIDE (±)   HR: 3
CAS: 27262-48-2   NIOSH: TK 6120000
mf: $C_{18}H_{28}N_2O \cdot ClH$ mw: 324.94

SYN: 1-BUTYL-2',6'-PIPECOLOXYLIDIDE, HYDROCHLORIDE (±)

THR: A poison by ingestion, subcutaneous, intravenous, and parenteral routes. When heated to decomposition, it emits very toxic HCl and NO_x.

BUTACAINE   HR: 3
CAS: 149-16-6   NIOSH: UB 0875000
mf: $C_{18}H_{30}N_2O_2$ mw: 306.50

PROP: Colorless, odorless powder. Mp: 98°-100°.

SYNS: 3-(P-AMINOBENZOXY)-1-DI-N-BUTYLAMINOPROPANE * P-AMINOBENZOYLDIBUTYL-AMINOPROPANOL * BUTYN * 3-(DIBUTYL-AMINO)-1-PROPANOL-P-AMINOBENZOATE * 3-DIBUTYLAMINOPROPYL-P-AMINOBENZOATE

THR: A poison via subcutaneous and intravenous routes. A weak allergen. Slight fire hazard. When heated to decomposition, it emits toxic fumes of NO_x.

BUTACAINE SULFATE   HR: 3
CAS: 149-15-5   NIOSH: UB 1050000
mf: $C_{36}H_{60}N_4O_4 \cdot H_2O_4S$ mw: 711.08

SYNS: 3-(P-AMINOBENZOXY)-1-DI-N-BUTYLAMINOPROPANE SULFATE * P-AMINOBENZOYLDI-BUTYLAMINOPROPANOL SULFATE * BUTYN SULFATE * DIBUTYLAMINOPROPYL-P-AMINO-BENZOATE SULFATE * 3'-DIBUTYLAMINO-PROPYL-4-AMINOBENZOATE SULFATE

THR: A poison by ingestion and intraperitoneal routes. See also sulfates. When heated to decomposition, it emits very toxic fumes of SO_x and NO_x.

1,3-BUTADIENE   HR: 3
CAS: 106-99-0   NIOSH: EI 9275000

DOT: 1010
mf: C_4H_6 mw: 54.10

PROP: Colorless gas, mild aromatic odor. Very reactive. Bp: -4.5°, mp: -113°, fp: -108.9°, flash p: -105°F, lel = 2.0%, uel = 11.5%, d: 0.621 @ 20°/4°, autoign temp: 788°F, vap d: 1.87, vap press: 1840 mm @ 21°.

SYNS: BIETHYLENE * BIVINYL * BUTADIEEN (DUTCH) * BUTA-1,3-DIEEN (DUTCH) * BUTADIEN (POLISH) * BUTA-1,3-DIEN (GER-MAN) * BUTA-1,3-DIENE * ALPHA-GAMMA-BUTADIENE * NCI-C50602 * PYRROLYLENE * VINYLETHYLENE

OSHA PEL: TWA 1000 ppm
ACGIH TLV: TWA 1000 ppm
DOT Classification: Flammable Gas

THR: An experimental carcinogen. Inhalation of high concentrations can cause unconsciousness and death. The vapors are irritating to eyes and mucous membranes. If spilled on skin or clothing, can cause burns or frost bite (due to rapid vaporization). Chronic systemic poisoning in humans has not been reported. Mutagenic data. Dangerous fire hazard when exposed to heat, flame, or powerful oxidizers. Dangerously explosive if heated under pressure, in air, mixed with phenol, ClO_2, crotonaldehyde. May form explosive peroxides upon exposure to air. Moderately dangerous when heated; it emits acrid fumes; can react with oxidizing materials. To fight fire, stop flow of gas.

L-BUTADIENE DIEPOXIDE   HR: 3
CAS: 30031-64-2   NIOSH: EJ 8575000
mf: $C_4H_6O_2$ mw: 86.10

SYNS: (2s,3s)-DIEPOXYBUTANE * L-DIEPOXY-BUTANE * L-1,2:3,4-DIEPOXYBUTANE * (2s,3s)-1,2:3,4-DIEPOXYBUTANE * NSC-32606

THR: An intraperitoneal poison. An experimental neoplastigen and carcinogen. When heated to decomposition, it emits acrid and irritating fumes.

BUTALBITAL SODIUM   HR: 3
CAS: 125-88-2   NIOSH: CP 9300000
mf: $C_{10}H_{14}N_2O_3 \cdot Na$ mw: 233.25

SYNS: APROBARBITAL SODIUM * APROBARBI-TONE SODIUM * SODIUM-5-ALLYL-5-ISOPROPYL-BARBITURATE

THR: A poison via intraperitoneal and subcutaneous routes. When heated to decomposition, it emits toxic fumes of NO_x and Na_2O.

BUTALLYLONAL   HR: 3
CAS: 1142-70-7   NIOSH: CQ 0700000
mf: $C_{11}H_{15}BrN_2O_3$ mw: 303.19

SYNS: 5-(2-BROMOALLYL)-5-SEC-BUTYLBARBI-
TURIC ACID * 5-(2'-BROMOALLYL)-5-(1'-
METHYL-N-PROPYL)BARBITURIC ACID
* 5-SEC-BUTYL-5-(BETA-BROMOALLYL)BARBI-
TURIC ACID * 2,4,6(1H,3H,5H)-PYRIMIDINE-
TRIONE, 5-(2-BROMO-2-PROPENYL)-5-(1-
METHYLPROPYL)-(9CI)

THR: Poison by ingestion, intraperitoneal, and subcutaneous routes. Human psychotropic effects by ingestion. When heated to decomposition, it emits very toxic fumes of Br^-, Na_2O and NO_x.

BUTALLYLONAL SODIUM   HR: 3
CAS: 3486-86-0   NIOSH: CQ 0875000
mf: $C_{11}H_{14}BrN_2O_3 \cdot Na$ mw: 325.17

SYNS: SEC-BUTYL-BROM-ALLYL BARBITURIC
ACID SODIUM SALT * SODIUM-5-(2-BROMOAL-
LYL)-5-SEC-BUTYLBARBITURATE

THR: Poison by ingestion and intraperitoneal route. When heated to decomposition, it emits very toxic fumes of Br^- and NO_x.

N-BUTANE   HR: 2
CAS: 106-97-8   NIOSH: EJ 4200000
DOT: 1011/1075/1969
mf: C_4H_{10} mw: 58.14

PROP: Colorless gas, faint disagreeable odor. bp: $-0.5°$, fp: $-138°$, lel = 1.9%, uel = 8.5%, flash p: $-76°F$ (CC), d: 0.599, autoign temp: 761°F, vap press: 2 atm @ 18.8°, vap d: 2.046.

SYNS: BUTANEN (DUTCH) * BUTANI (ITALIAN)
* DIETHYL * METHYLETHYLMETHANE

DFG MAK: 1000 ppm (2350 mg/m^3)
DOT Classification: Flammable Gas

THR: Moderately toxic via inhalation. Causes drowsiness. An asphyxiant. A general purpose food additive. Very dangerous fire hazard when exposed to heat, flame, or oxidizers. Highly explosive when exposed to flame, also when mixed with [$Ni(CO)_4$ + O_2]. To fight fire, stop flow of gas. Incompatible with flame; [$Ni(CO)_4$ + O_2]; oxidizing materials.

1,3-BUTANE DIAMINE   HR: 3
CAS: 590-88-5   NIOSH: EJ 6700000
mf: $C_4H_{12}N_2$ mw: 88.18

PROP: Liquid. Bp: $142°-150°$, flash p: 125°F, d: 0.85, vap d: 3.04.

SYN: 1,3-DIAMINOBUTANE

THR: A poison via dermal route. Severe skin and eye irritant. Moderate fire hazard when exposed to heat or flame. To fight fire, use alcohol foam, foam, CO_2, dry chemical. Incompatible with oxidizing materials. When heated to decomposition, it emits toxic fumes of NO_x.

1,2-BUTANEDIOL   HR: 2
CAS: 584-03-2   NIOSH: EK 0380000
mf: $C_4H_{10}O_2$ mw: 90.14

PROP: d: 1.0, vap d: 3.1, bp: 194°, flash p: 194°F.

SYN: 1,2-BUTYLENE GLYCOL

THR: Moderately toxic by ingestion. Moderate fire hazard when exposed to heat or flame. To fight fire, use alcohol foam. When heated to decomposition, it emits acrid and irritating fumes.

1,3-BUTANEDIOL   HR: 1
CAS: 107-88-0   NIOSH: EK 0440000
mf: $C_4H_{10}O_2$ mw: 90.14

PROP: Viscous liquid. Bp: 207.5°, fp: $< -50°$, flash p: 250°F, d: 1.006 @ 20°/20°, autoign temp: 741°F, vap press: 0.06 mm @ 20°, vap d: 3.2.

SYNS: BUTANE-1,3-DIOL * BETA-BUTYLENE
GLYCOL * 1,3-BUTYLENE GLYCOL * 1,3-
DIHYDROXYBUTANE

THR: Mildly toxic by ingestion and subcutaneous routes. See also ethylene glycol. A food additive permitted in food for human consumption. Eye irritant. Slight fire hazard when exposed to heat or flame. To fight fire, use foam, alcohol foam, CO_2, dry chemical. Incompatible with oxidizing materials. When heated to decomposition, it emits acrid fumes. For further information, see 1,3-Butylene glycol, Vol. 3, No. 2 of *DPIM Report*.

1,4-BUTANEDIOL   HR: 2
CAS: 110-63-4   NIOSH: EK 0525000
mf: $C_4H_{10}O_2$ mw: 90.14

PROP: Nearly odorless, colorless, viscid liquid; bp: 228°, fp: 20.9°, flash p: 250°F (OC), d: 1.0154 @ 25°/4°, vap d: 3.1.

SYNS: BUTANE-1,4-DIOL * 1,4-BUTYLENE GLYCOL * 1,4-DIHYDROXYBUTANE * 1,4-TETRAMETHYLENE GLYCOL

THR: Moderately toxic by ingestion and intraperitoneal routes. Slight fire hazard when exposed to heat or flame. To fight fire, use alcohol foam, mist, foam, CO_2, dry chemical. When heated to decomposition, it emits acrid smoke and fumes. Incompatible with oxidizing materials.

2,3-BUTANEDIOL   HR: 1
CAS: 513-85-9   NIOSH: EK 0532000
mf: $C_4H_{10}O_2$ mw: 90.14

PROP: Colorless liquid or solid. Bp: 180°, fp: 19°, flash p: 185°F (TOC), d: 1.0095 @ 20°/20°, autoign temp: 756°F, vap press: 0.17 mm @ 20°, vap d: 3.1.

SYNS: 2,3-BUTYLENE GLYCOL * 2,3-DIHYDROXYBUTANE * DIMETHYLENE GLYCOL

THR: Mildly toxic by ingestion. See also ethylene glycol. Moderate fire hazard when exposed to heat or flame. To fight fire, use alcohol foam, CO_2, dry chemical. Incompatible with oxidizing materials. When heated to decomposition, it emits acrid smoke and fumes.

1,4-BUTANEDIOL DIMETHYL SULFONATE   HR: 3
CAS: 55-98-1   NIOSH: EK 1750000
mf: $C_6H_{14}O_6S_2$ mw: 246.32

PROP: White crystals; mp: 114°-118°.

SYNS: 1,4-BIS(METHANESULFONOXY)BUTANE * (1,4-BIS(METHANESULFONYLOXY)BUTANE) * 1,4-BUTANEDIOL DIMETHANESULPHONATE * 1,4-DIMETHANESULFONOXYBUTANE * 1,4-DI(METHANESULFONYLOXY)BUTANE * 1,4-DIMETHANESULPHONYLOXYBUTANE * 1,4-DIMETHYLSULFONOXYBUTANE * MYLERAN * NCI-C01592 * NSC-750 * TETRAMETHYLENE BIS(METHANESULFONATE) * TETRAMETHYLENE DIMETHANE SULFONATE

THR: Poison by ingestion, intraperitoneal, and intravenous routes. An experimental teratogen, neoplastigen, and carcinogen. Mutagenic data. See also sulfonates. When heated to decomposition, it emits toxic fumes of SO_x.

2,3-BUTANEDIONE   HR: 3
CAS: 431-03-8   NIOSH: EK 2625000
DOT: 2346
mf: $C_4H_6O_2$ mw: 86.10

PROP: Greenish-yellow liquid, strong odor; bp: 88°, flash p: 80°F, d: 0.9904 @ 15°/15°, vap d: 3.00.

SYNS: BIACETYL * DIACETYL * 2,3-DIKETOBUTANE * DIMETHYL DIKETONE * DIMETHYLGLYOXAL

DOT Classification: Label: Flammable Liquid

THR: A poison via intraperitoneal route and moderately toxic by ingestion. Used as a synthetic flavoring substance and adjuvant. Dangerous fire hazard when exposed to heat or flame. To fight fire, use alcohol foam, CO_2, dry chemical.

BUTANESULTONE   HR: 3
CAS: 1633-83-6   NIOSH: RP 4300000
mf: $C_4H_8O_3S$ mw: 136.18

SYNS: DELTA-BUTANE SULTONE * 1,4-BUTANESULTONE * 1,4-BUTYLENE SULFONE * DELTA-VALEROSULTONE

THR: A subcutaneous and intravenous poison. An experimental tumorigen. Moderately toxic by ingestion. Mutagenic data. See also sulfonates. When heated to decomposition, it emits toxic fumes of SO_x.

n-BUTANETHIOL   HR: 3
CAS: 109-79-5   NIOSH: EK 6300000
DOT: 2347
mf: $C_4H_{10}S$ mw: 90.20

PROP: Colorless liquid, skunk-like odor; mp: −116°, bp: 98°, d: 0.8365 @ 25°/4°, flash p: 35°F, vap d: 3.1.

SYNS: BUTYL MERCAPTAN * NCI-C60866

OSHA PEL: TWA 10 ppm
DFG MAK: 0.5 ppm (1.5 mg/m^3)
DOT Classification: Label: Flammable Liquid

THR: An intraperitoneal poison. Moderately toxic by inhalation and ingestion. Reacts violently with HNO_3. An eye irritant. Dangerous fire hazard by exposure to heat, flame, sparks, or powerful oxidizers. To fight fire, use alcohol foam. When heated to decomposition it emits toxic SO_x. Incompatible with acid; acid fumes; oxidizing materials; heat; flame; sparks.

BUTANOL-4-BUTYL NITROSAMINE   HR: 3
CAS: 3817-11-6   NIOSH: EL 1225000
mf: $C_8H_{18}N_2O_2$ mw: 174.28

SYNS:BUTYL-BUTANOL-NITROSAMINE * N-NITROSO-n-BUTYL-(4-HYDROXYBUTYL)AMINE * BUTYL-BUTANOL(4)-NITROSAMIN (GERMAN) * N-BUTYL-N-(4-HYDROXYBUTYL)NITROSAMINE * N-BUTYL-(4-HYDROXYBUTYL)NITROSAMINE * 4-(BUTYLNITROSAMINO)-1-BUTANOL * 4-(N-BUTYLNITROSAMINO)-1-BUTANOL * 4-HYDROXYBUTYLBUTYLNITROSAMINE

THR: An experimental tumorigen, neoplastigen, and carcinogen. Mutagenic data. Moderately toxic by ingestion. See also nitrates. When heated to decomposition it emits toxic fumes of NO_x.

4-BUTANOLIDE   HR: 3
CAS: 96-48-0   NIOSH: LU 3500000
mf: $C_4H_6O_2$ mw: 86.10

PROP: Colorless liquid, mild odor. Mp: $-44°$, bp: 206°, flash p: 209°F (OC), d: 1.124 @ 25°/4°, vap d: 3.0.

SYNS: 4-DEOXYTETRONIC ACID * GAMMA-HYDROXYBUTYROLACTONE * TETRAHYDRO-2-FURANONE * BUTYRIC ACID LACTONE * GAMMA-BUTYROLACTONE * BUTYRYL LACTONE * ALPHA-BUTYROLACTONE * DIHYDRO-2(3H)-FURANONE * 4-HYDROXYBUTANOIC ACID LACTONE * GAMMA-HYDROXYBUTYRIC ACID CYCLIC ESTER * 4-HYDROXYBUTYRIC ACID, GAMMA-LACTONE * NCI-C55878

THR: An experimental tumorigen. Less acutely toxic than beta-propiolactone. Slight fire hazard when exposed to heat or flame; can react with oxidizing materials. To fight fire, use foam, alcohol foam, CO_2, dry chemical. When heated to decomposition, it emits acrid and irritating fumes. For further information, see gamma Butyrolactone, Vol. 1, No. 3 of *DPIM Report*.

2-BUTANONE   HR: 3
CAS: 78-93-3   NIOSH: EL 6475000
DOT: 1193/1232
mf: C_4H_8O mw: 72.12

PROP: Colorless liquid, acetone-like odor. Bp: 79.57°, fp: $-85.9°$, lel = 1.8%, uel = 11.5%, flash p: 22°F (TOC), d: 0.80615 @ 20°/20°,

vap press: 71.2 mm @ 20°, autoign temp: 960°F, vap d: 2.42, ULC: 85-90.

SYNS: AETHYLMETHYLKETON (GERMAN) * BUTANONE-2 (FRENCH) * ETHYL METHYL CETONE (FRENCH) * ETHYLMETHYLKETON (DUTCH) * ETHYL METHYL KETONE * MEK * METHYL ACETONE * METHYL ETHYL KETONE * METHYL ETHYL KETONE (DOT) * METILETILCHETONE (ITALIAN) * METYLOETYLOKETON (POLISH)

OSHA PEL: TWA 200 ppm
ACGIH TLV: TWA 200 ppm; STEL 300 ppm
DFG MAK: 200 ppm (590 mg/m³)
DOT Classification: Label: Flammable Liquid

THR: Experimental teratogen. Moderately toxic by ingestion and dermal routes. A strong irritant. Affects peripheral nervous system and central nervous system. See also ketones. Eye irritation @ 350 ppm. Dangerous fire hazard when exposed to heat or flame. Moderately explosive when exposed to flame. To fight fire, use alcohol foam, CO_2, dry chemical. Incompatible with chlorosulfonic acid; oleum; potassium-tert-butoxide; heat or flame; chloroform; hydrogen peroxide; nitric acid.

BUTAZOLIDINE SODIUM   HR: 3
CAS: 129-18-0   NIOSH: UQ 8300000
mf: $C_{19}H_{20}N_2O_2 \cdot Na$ mw: 331.40

SYNS: 3,5-DIOXO-1,2-DIPHENYL-4-N-BUTYLPYRAZOLIDIN SODIUM * DIPHENYLDIOXOBUTYL-PYRAZOLIDINE-BUTAZOLIDINE-SODIUM * 4-BUTYL-1,2-DIPHENYL-3,5-PYRAZOLIDINEDIONE SODIUM SALT * SODIUM PHENYLBUTAZONE

THR: A poison by ingestion in humans. A poison via subcutaneous, intravenous, and intraperitoneal routes. When heated to decomposition, it emits toxic fumes of NO_x, Na_2O.

trans-2-BUTENAL   HR: 3
CAS: 123-73-9   NIOSH: GP 9625000
mf: C_4H_6O mw: 70.10

PROP: Water white mobile liquid, pungent, suffocating odor; bp: 104°, fp: $-76.0°$, lel = 2.1%, uel = 15.5%, flash p: 55°F, d: 0.853 @ 20°/20°, vap d: 2.41, autoign temp: 450°F.

SYNS: ALDEHYDE CROTONIQUE (FRENCH) * 2-BUTENAL, (E)- * CROTENALDEHYDE * CROTONALDEHYDE * CROTONIC ALDEHYDE

* BETA-METHYL ACROLEIN * NCI-C56279
* PROPYLENE ALDEHYDE

OSHA PEL: TWA 2 ppm
DOT Classification: Label: Flammable
Liquid

THR: A poison by ingestion, inhalation, and dermal routes. A lachrymating material which is very dangerous to the eyes. Can cause corneal burns and is irritating to the skin. See also aldehydes. Dangerous fire hazard when exposed to heat or flame. To fight fire, use alcohol foam, CO_2, dry chemical. Treatment and Antidotes: In case of contact, immediately flush the skin or eyes with water for at least 15 minutes. Get medical attention. Incompatible with 1,3-butadiene; oxidizing materials.

1-BUTENE   **HR: 2**
mf: C_4H_8 mw: 56.11

PROP: A colorless flammable gas, slightly aromatic odor. Bp: $-6.3°$, fp: $-185.3°$, lel = 1.6%, uel = 9.3% flash p: $-80°$ ($-112°F$), d: 0.668 @ $0°/1°$, vap d: 1.93, vap press: 3480 mm @ $21°$, autoign temp: $723°F$.

SYN: ALPHA-BUTYLENE

THR: An asphyxiant. Very dangerous fire hazard when exposed to heat, flame, or oxidizers. To fight fire, stop flow of gas. Moderately explosive when exposed to flame. Incompatible with oxidizing materials; aluminum tris-tetrahydroborate.

cis-2-BUTENE   **HR: 2**
mf: C_4H_8 mw: 56.11

PROP: Colorless flammable gas, slightly aromatic odor. Bp: $1°$, fp: $-139°$, flash p: $-100°F$, d: 0.627 @ $15.5°/15.5°$, vap press: 1410 mm @ $21°$, autoign temp: $615°F$, lel = 1.7%, uel = 9.0%, vap d: 1.9.

SYNS: DIMETHYLETHYLENE * PSEUDO-BUTYL-ENE

THR: A simple asphyxiant. Very dangerous fire hazard when exposed to heat or flame. Very likely to explode. Incompatible with oxidizing materials. To fight fire, stop flow of gas.

trans-2-BUTENE   **HR: 2**
mf: C_4H_8 mw: 56.11

PROP: A colorless, flammable gas, slightly aro-

matic odor. Bp: $2.5°$, fp: $-105.6°$, flash p: $-100°F$, d: 0.613 @ $15.5°/15.5°$ vap d: 1.95, vap press: 1592 mm @ $21°$, autoign temp: 615 F, lel = 1.8%, uel = 9.7%, vap d: 1.9.

THR: A simple asphyxiant. Very dangerous fire hazard when exposed to heat or flame. Very likely to explode. To fight fire, stop flow of gas. Incompatible with oxidizing materials.

cis-BUTENEDIOIC ACID   **HR: 2**
CAS: 110-16-7   NIOSH: OM 9625000
mf: $C_4H_4O_4$ mw: 116.08

PROP: White crystals, faint acidulous odor. Mp: $130.5°$, bp: $135°$ decomp, d: 1.590 @ $20°/4°$, vap d: 4.0.

SYNS: MALEIC ACID * BUTENEDIOIC ACID, (Z)- * CIS-1,2-ETHYLENEDICARBOXYLIC ACID
* MALEINIC ACID * MALENIC ACID
* TOXILIC ACID

THR: Moderately toxic by ingestion and skin absorption. A skin and eye irritant. Believed to be more toxic than its isomer, fumaric acid. Slight fire hazard. When heated to decomposition, it emits acrid smoke and fumes.

3-BUTENE NITRILE   **HR: 3**
CAS: 109-75-1   NIOSH: EM 8050000
mf: C_4H_5N mw: 67.10

PROP: Colorless liquid, onion-like odor; bp: $116°-119°$, d: 0.8341 @ $20°/4°$, mp: $-87°$.

SYNS: ALLYL CYANIDE * 1-BUTENE-4-NITRILE
* BETA-BUTENONITRILE * VINYLACETONITRILE

THR: A poison by ingestion. Moderately toxic by inhalation and dermal routes. A skin irritant. See also nitriles. Dangerous; emits highly toxic fumes of NO_x and CN^- when heated to decomposition or on contact with acids or acid fumes. To fight fire, use alcohol foam, mist.

1-BUTENE OXIDE   **HR: 2**
CAS: 106-88-7   NIOSH: EK 3675000
mf: C_4H_8O mw: 72.12

PROP: Colorless liquid, sol in water, miscible with most organic solvents. d: 0.8312 @ $20°/20°$, bp: $63°$, flash p: $5°F$, lel = 1.5%, uel = 18.3%.

SYNS: 1,2-BUTENE OXIDE * 1,2-BUTYLENE OXIDE * 1,2-EPOXYBUTANE * NCI-C55527

THR: Moderately irritating by ingestion, inhalation, and dermal routes. Mutagenic data. Dangerous fire hazard when exposed to heat, flame, or powerful oxidizers. To fight fire, use dry chemical, water spray, mist or fog, alcohol foam.

2-BUTEN-1-OL   HR: 2
CAS: 6117-91-5   NIOSH: EM 9275000
mf: C_4H_8O mw: 72.12

PROP: Colorless liquid. Mp: <30°, bp: 118°, flash p: 92°F, d: 0.8726 @ 0°/4°, vap d: 2.49.

SYNS: 2-BUTENOL * 2-BUTENYL ALCOHOL * CROTONYL ALCOHOL * CROTYL ALCOHOL

THR: Moderately toxic by ingestion, inhalation, and dermal routes. Dangerous fire hazard when exposed to heat or flame; can react with oxidizing materials. To fight fire, use alcohol foam, CO_2, dry chemical. When heated to decomposition, it emits acrid smoke and fumes.

3-BUTEN-2-ONE   HR: 3
CAS: 78-94-4   NIOSH: EM 9800000
mf: C_4H_6O mw: 70.10

PROP: Colorless liquid, powerfully irritating odor; bp: 81.4°, flash p: 20°F (CC), d: 0.8393 @ 25°/4°, vap d: 2.41.

SYNS: 3-BUTENE-2-ONE * METHYL-VINYL-CETONE (FRENCH) * METHYLVINYLKETON (GERMAN) * METHYLVINYL KETONE * VINYL METHYL KETONE

THR: An irritant poison to skin, eyes, and mucous membranes, and via intraperitoneal routes. A lachrymator. See also ketones. Dangerous fire hazard when exposed to heat, flame or oxidizers. Upon exposure to heat or flame, it emits toxic and irritating fumes; can react with oxidizing materials. To fight fire, use CO_2, dry chemical.

BUTEN-3-YNE
mf: C_4H_4 mw: 52.05

PROP: lel = 2%, uel = 100% d: 0.68 @ 1.7 atm, vap d: 1.8, bp: 11°, flash p: < −5°

SYN: VINYL ACETYLENE

THR: No toxicity data. Very dangerous fire hazard spontaneous combustion, heat, flame, or oxidizers. Incompatible with air. Store out of contact with air to avoid formation of explosive compounds.

BUTISOL   HR: 3
CAS: 125-40-6   NIOSH: CQ 1750000
mf: $C_{10}H_{16}N_2O_3$ mw: 212.28

SYNS: BUTABARBITAL * 5-SEC-BUTYL-5-ETHYLBARBITURIC ACID * 5-SEC-BUTYL-5-ETHYLMALONYL UREA * 5-ETHYL-5-(1-METHYLPROPYL)BARBITURATE * 5-ETHYL-5-(1-METHYLPROPYL)BARBITURIC ACID * 2,4,6(1H,3H,5H)-PYRIMIDINETRIONE, 5-ETHYL-5-(1-METHYLPROPYL)-(9CI) * SEC-BUTABARBITAL

THR: Poison via ingestion, intraperitoneal, and subcutaneous routes. Human psychotropic effects by ingestion. When heated to decomposition, it emits toxic fumes of NO_x.

BUTISOL SODIUM   HR: 3
CAS: 143-81-7   NIOSH: CQ 2275000
mf: $C_{10}H_{15}N_2O_3 \cdot Na$ mw: 234.26

SYNS: BUTABARBITAL SODIUM * 5-SEC-BU-TYL-5-ETHYLBARBITURIC ACID SODIUM SALT * 5-ETHYL-5-(1-METHYLPROPYL)BARBITURIC ACID SODIUM SALT * 2,4,6(1H,3H,5H)-PYRIMI-DINETRIONE, 5-ETHYL-5-(1-METHYLPROPYL)-, MONOSODIUM SALT * SECBUBARBITAL SODIUM * SODIUM BUTABARBITAL * SODIUM-5-SEC-BUTYL-5-ETHYLBARBITURATE * SODIUM-5-ETHYL-5-SEC-BUTYLBARBITURATE * SODIUM-5-ETHYL-5-(1-METHYLPROPYL)BARBITURATE

THR: A poison by ingestion, intraperitoneal, and intravenous routes. Causes central nervous system and psychotropic effects in humans. When heated to decomposition, it emits toxic fumes of NO_x.

BUTOBARBITAL   HR: 3
CAS: 77-28-1   NIOSH: CQ 1575000
mf: $C_{10}H_{16}N_2O_3$ mw: 212.28

SYNS: BUTOBARBITURAL * BUTOBARBITONE * 5-BUTYL-5-ETHYLBARBITURIC ACID * 5-ETHYL-5-N-BUTYLBARBITURIC ACID * 2,4, 6(1H,3H,5H)-PYRIMIDINETRIONE, 5-BUTYL-5-ETHYL- (9CI)

THR: A poison by ingestion, intraperitoneal, subcutaneous, and intravenous effects. Psychotropic and central nervous system effects in humans by ingestion. When heated to decomposition, it emits toxic fumes of NO_x.

BUTONATE   HR: 2
CAS: 126-22-7   NIOSH: ET 0175000
mf: $C_8H_{14}Cl_3O_5P$ mw: 327.54

SYNS: DIMETHOXY-2,2,2-TRICHLORO-1-N-BUTYRYLOXY-ETHYLPHOSPHINE OXIDE

* O,O-DIMETHYL-(1-BUTYRYLOXY-2,2,2-TRI-
CHLOROETHYL) PHOSPHONATE * 0,0-DIMETHYL
2,2,2-TRICHLORO-1-(N-BUTYRYLOXY)-
ETHYLPHOSPHONATE * ENT 20,852

THR: Moderately toxic by ingestion and low
toxicity by dermal route. When heated to decomposition, it emits highly toxic fumes of PO_x
and Cl^-.

BUTOPHEN   **HR: 3**
CAS: 6365-83-9   NIOSH: SK 0525000
mf: $C_{10}H_{12}N_2O_5 \cdot H_3N$ mw: 257.28

SYNS: 2-SEC-BUTYL-4,6-DINITROPHENOL, AM-
MONIUM SALT * 4,6-DINITRO-2-SEC.BUTYL-
FENOLATE AMMONY (CZECH) * 4,6-DINI-
TRO-O-SEC-BUTYLPHENOL AMMONIUM SALT
* 4,6-DINITRO-2-SEC-BUTYLPHENOL AM-
MONIUM SALT * 2-(1-METHYL-N-PROPYL)
4,6-DINITROPHENOL, AMMONIUM SALT

THR: A poison by ingestion and skin absorption.
An eye irritant. When heated to decomposition,
it emits very toxic fumes of NH_3 and NO_x.

**2-BUTOXYETHANOL PHOS-
PHATE**   **HR: 3**
CAS: 78-51-3   NIOSH: KJ 9800000
mf: $C_{18}H_{39}O_7P$ mw: 398.54

PROP: Light-colored liquid, butyl-like odor.
mp: −70°; bp: 200°-230° @ 4 mm, flash p:
435°F, d: 1.02 @ 20°/20°, vap press: 0.03 mm
@ 150°, vap d: 13.8.

SYNS: TRI(2-BUTOXYETHYL) PHOSPHATE
* TRIBUTYL CELLOSOLVE PHOSPHATE
* TRIS(2-BUTOXYETHYL) PHOSPHATE

THR: A poison via intravenous route. Moderately toxic by ingestion. Slight fire hazard when
exposed to heat or flame. Dangerous; see also
phosphates; can react with oxidizing materials.
To fight fire, use water, foam, CO_2, dry chemical. When heated to decomposition, it emits
toxic fumes of PO_x.

**2-(2-BUTOXY ETHOXY)ETHYL THIO-
CYANATE**   **HR: 3**
CAS: 112-56-1   NIOSH: XK 8400000
mf: $C_9H_{17}NO_2S$ mw: 203.33

SYNS: 2-(2-(BUTOXY)ETHOXY)ETHYL
THIOCYANIC ACID ESTER * BUTOXYRHO-
DANODIETHYL ETHER * BETA-BUTOXY-
BETA′-THIOCYANODIETHYL ETHER * 2-

BUTOXY-2′-THIOCYANODIETHYL ETHER
* 1-BUTOXY-2-(2-THIOCYANOETHOXY)ETHANE
* BUTYL CARBITOL RHODANATE * BUTYL
CARBITOL THIOCYANATE * ENT 6

THR: A poison by ingestion, skin contact, and
intravenous routes. Irritating to skin and mucous
membranes. High concentrations can cause central nervous system depression. An insecticide.
See also thiocyanates, esters, ethers. When
heated to decomposition, it emits very toxic
fumes of SO_x, NO_x and CN^-.

**1-BUTOXY ETHOXY-2-PRO-
PANOL**   **HR: 2**
CAS: 124-16-3   NIOSH: UA 8050000
mf: $C_9H_{20}O_3$ mw: 176.29

PROP: Sol in water. D: 0.9310 @ 20°/20°,
bp: 230.3°, fp: −90°, flash p: 250°F (OC).

SYN: 1-(2-BUTOXYETHOXY)-2-PROPANOL

THR: Moderately toxic by ingestion and dermal
routes. Slight fire hazard when exposed to heat
or flame. To fight fire, use alcohol foam, dry
chemical, spray, or mist. When heated to decomposition, it emits acrid and irritating fumes.

2-BUTOXYETHYL ACETATE   **HR: 2**
CAS: 112-07-2   NIOSH: KJ 8925000
mf: $C_8H_{16}O_3$ mw: 160.24

PROP: Colorless liquid, fruity odor, sol in hydrocarbons and organic solvents, insol in water.
Bp: 192.3°, d: 0.9424 @ 20°/20°, fp: −63.5°,
flash p: 190°F.

SYNS: 2-BUTOXYETHANOL ACETATE * BUTYL
CELLOSOLVE ACETATE * ETHYLENE GLYCOL
MONOBUTYL ETHER ACETATE * GLYCOL MONO-
BUTYL ETHERACETATE

THR: Moderately toxic by ingestion and dermal
route. Mild skin irritant. Moderate fire hazard
when exposed to heat, flame, or oxidizers. To
fight fire, use alcohol foam. When heated to
decomposition, it emits acrid smoke and irritating fumes.

**p-BUTOXY PHENYL ACETOHYDROX-
AMIC ACID**   **HR: 3**
CAS: 2438-72-4   NIOSH: AK 8280000
mf: $C_{12}H_{17}NO_3$ mw: 223.30

SYN: 4-BUTOXYPHENYLACETOHYDROXAMIC ACID

THR: An experimental teratogen. Mutagenic
data. When heated to decomposition, it emits
toxic fumes of NO_x.

4'-BUTOXY-3-PIPERIDINO PROPIOPHE-NONE HYDROCHLORIDE   HR: 3
CAS: 536-43-6   NIOSH: UG 8750000
mf: $C_{18}H_{27}NO_2 \cdot ClH$ mw: 325.92

SYNS: 4-N-BUTOXY-BETA-(1-PIPERIDYL)PROPIO-
PHENONE HYDROCHLORIDE * DYCLONINE HY-
DROCLORIDE * PIPERIDINE, 1-(2-(4-BUTOXY-
BENZOYL)ETHYL), HYDROCHLORIDE

THR: A poison via intraperitoneal, subcutaneous, and intravenous routes. When heated to
decomposition, it emits very toxic fumes of HCl
and NO_x.

BUTRIZOL   HR: 3
CAS: 16227-10-4   NIOSH: XZ 4150000
mf: $C_6H_{11}N_3$ mw: 125.20

SYN: 4-N-BUTYL-4H-1,2,4-TRIAZOLE

THR: A poison by ingestion and skin routes.
When heated to decomposition, it emits toxic
fumes of NO_x.

N-BUTYL ACETATE   HR: 2
CAS: 123-86-4   NIOSH: AF 7350000
DOT: 1123
mf: $C_6H_{12}O_2$ mw: 116.18

PROP: Colorless liquid; bp: 126°, fp: −73.5°,
ulc: 50-60, lel = 1.4%, uel = 7.5%, flash p:
72°F, d: 0.88 @ 20°/20°, autoign temp: 797°F,
vap press: 15 mm @ 25°.

SYNS: ACETATE DE BUTYLE (FRENCH)
* BUTILE (ACETATI DI) (ITALIAN) * BUTY-
LACETAT (GERMAN) * 1-BUTYL ACETATE
* BUTYLACETATEN (DUTCH) * BUTYLE (ACE-
TATE DE) (FRENCH) * BUTYL ETHANOATE
* OCTAN N-BUTYLU (POLISH)

OSHA PEL: TWA 150 ppm
ACGIH TLV: TWA 150 ppm; STEL 200 ppm
DFG MAK: 200 ppm (950 mg/m^3)
DOT Classification: Label: Flammable
   Liquid

THR: Moderately toxic by inhalation and intraperitoneal routes. A skin and eye irritant. A
systemic irritant in humans. See also esters. A
mild allergen. High concentrations are irritating
to eyes and respiratory tract and cause narcosis.
Evidence of chronic systemic toxicity is incon-
clusive. Dangerous fire hazard when exposed
to heat or flame; can react with oxidizing materials. Moderately explosive when exposed to
flame. To fight fire, use alcohol foam, CO_2,

dry chemical. When heated to decomposition,
it emits acrid and irritating fumes. Incompatible
with potassium-tert-butoxide. For further information, see Vol. 4, No. 3 of *DPIM Report*.

sec-BUTYL ACETATE   HR: 2
CAS: 105-46-4   NIOSH: AF 7380000
DOT: 1123
mf: $C_6H_{12}O_2$ mw: 116.18

PROP: Colorless liquid; mild odor. Bp: 112°;
flash p: 18°; d: 0.862-0.866 @ 20°/20°; vap d:
4.00. lel = 1.3%, uel = 7.5%.

SYNS: ACETATE DE BUTYLE SECONDAIRE
(FRENCH) * ACETIC ACID-2-BUTOXY ESTER
* ACETIC ACID-1-METHYLPROPYL ESTER (9CI)
* 2-BUTYL ACETATE * SEC-BUTYL ALCOHOL
ACETATE

OSHA PEL: TWA 200 ppm
ACGIH TLV: TWA 200 ppm
DFG MAK: 200 ppm (950 mg/m^3)
DOT Classification: Flammable Liquid

THR: An irritant and allergen. See also esters.
Moderate fire hazard. To fight fire, use alcohol
foam, CO_2, dry chemical. When heated to de-
composition, it emits acrid and irritating fumes.
For further information, see Vol. 4, No. 6 of
DPIM Report.

BUTYL ACETOACETATE   HR: 2
CAS: 591-60-6   NIOSH: AK 5100000
mf: $C_8H_{14}O_3$ mw: 158.22

PROP: bp: 214°; flash p: 185°F; d: 0.96; vap
d: 5.55.

THR: An eye irritant. See also esters; acetoacetic
acid, 3,7-dimethyl-2,6-octadienyl ester; n-butyl
alcohol. Moderate fire hazard. To fight fire,
use alcohol foam, CO_2, dry chemical. When
heated to decomposition, it emits acrid and irritating fumes.

N-BUTYL ALCOHOL   HR: 3
CAS: 71-36-3   NIOSH: EO 1400000
DOT: 1120
mf: $C_4H_{10}O$ mw: 74.14

PROP: Colorless liquid; bp: 117.5°, ulc: 40,
lel = 1.4%, uel = 11.2%, fp: −88.9°, flash
p: 95°-100°F, d: 0.80978 @ 20°/4°, autoign
temp: 689°F, vap press: 5.5 mm @ 20°, vap
d: 2.55.

SYNS: ALCOOL BUTYLIQUE (FRENCH) * BUTA-
NOL (FRENCH) * 1-BUTANOL * N-BUTANOL

∗ BUTANOLEN (DUTCH) ∗ BUTANOLO (ITALIAN) ∗ BUTAN-1-OL ∗ BUTYL HYDROXIDE ∗ BUTYLOWY ALKOHOL (POLISH) ∗ 1-HYDROXYBUTANE ∗ METHYLOLPROPANE ∗ PROPYLCARBINOL ∗ PROPYLMETHANOL

OSHA PEL: TWA 100 ppm
ACGIH TLV: TWA CL 50 ppm
DFG MAK: 100 ppm (300 mg/m^3)
DOT Classification: Label: Flammable
  Liquid

THR: A poison by ingestion. Moderately toxic via dermal route. Moderately irritating via inhalation route to humans. Though animal experiments have shown the butyl alcohols to possess toxic properties, they have produced few cases of poisoning in industry probably because of their low volatility. The use of normal butyl alcohol is reported to have resulted in irritation of the eyes, with corneal inflammation, slight headache and dizziness, slight irritation of the nose and throat, and dermatitis about the fingernails and along the side of the fingers. Keratitis has also been reported. See also alcohols. Dangerous fire hazard when exposed to heat, flame, or oxidizers. Moderately explosive when exposed to flame. When heated to decomposition, it emits toxic fumes. To fight fire, use water spray, alcohol foam, CO_2, dry chemical. Incompatible with Al; chromium trioxide; oxidizing materials.

sec-BUTYL ALCOHOL   HR: 2
CAS: 78-92-2   NIOSH: EO 1750000
mf: $C_4H_{10}O$ mw: 74.14

PROP: Colorless liquid. Mp: −89°, bp: 99.5°, flash p: 14°, d: 0.808 @ 20°/4°, autoign temp: 763°F, vap press: 10 mm @ 20°, vap d: 2.55, lel = 1.7% @ 212°F, uel = 9.8% @ 212°F.

SYNS: ALCOOL BUTYLIQUE SECONDAIRE (FRENCH) ∗ BUTAN-2-OL ∗ SEC-BUTANOL ∗ 2-BUTANOL ∗ 2-BUTYL ALCOHOL ∗ BUTYLENE HYDRATE ∗ ETHYLMETHYL CARBINOL ∗ 2-HYDROXYBUTANE ∗ METHYLETHYLCARBINOL

OSHA PEL: TWA 150 ppm
ACGIH TLV: TWA 100 ppm
DFG MAK: 100 ppm (300 mg/m^3)

THR: Moderately toxic by ingestion and inhalation. See also n-butyl alcohol. An eye irritant. Dangerous fire hazard when exposed to heat

or flame. To Fight Fire, use water spray, alcohol foam, CO_2, dry chemical. Incompatible with oxidizing materials.

tert-BUTYL ALCOHOL   HR: 2
CAS: 75-65-0   NIOSH: EO 1925000
mf: $C_4H_{10}O$ mw: 74.14

PROP: Colorless liquid or rhombic prisms or plates; mp: 25.3°, bp: 82.8°, flash p: 50°F (CC), d: 0.7887 @ 20°/4°, autoign temp: 896°F, vap press: 40 mm @ 24.5°, vap d: 2.55, lel = 2.4%, uel = 8.0%.

SYNS: ALCOOL BUTYLIQUE TERTIAIRE (FRENCH) ∗ T-BUTANOL ∗ T-BUTYL HYDROXIDE ∗ 1,1-DIMETHYLETHANOL ∗ NCI-C55367 ∗ TRIMETHYLCARBINOL

OSHA PEL: TWA 100 ppm
ACGIH TLV: TWA 100 ppm; STEL 150 ppm
DFG MAK: 100 ppm (300 mg/m^3)

THR: Moderately toxic by ingestion and intraperitoneal routes. Dangerous fire hazard when exposed to heat or flame. Moderately explosive in the form of vapor when exposed to flame. To fight fire, use alcohol foam, CO_2, dry chemical. Incompatible with oxidizing materials; H_2O_2.

N-BUTYLAMINE   HR: 3
CAS: 109-73-9   NIOSH: EO 2975000
DOT: 1125
mf: $C_4H_{11}N$ mw: 73.16

PROP: Liquid, ammonia-like odor. Mp: −50°, bp: 77°, flash p: 10°F (OC), 10°F (CC), d: 0.74-0.76 @ 20°/20°, autoign temp: 594°F, vap d: 2.52, lel = 1.7%, uel = 9.8%.

SYNS: 1-AMINOBUTAN (GERMAN) ∗ 1-AMINOBUTANE ∗ 1-AMINO-BUTAAN (DUTCH) ∗ 1-BUTANAMINE ∗ N-BUTILAMINA (ITALIAN) ∗ N-BUTYLAMIN (GERMAN) ∗ MONO-N-BUTYLAMINE

OSHA PEL: TWA CL 5 ppm (skin)
ACGIH TLV: TWA CL 5 ppm
DFG MAK: 5 ppm (15 mg/m^3)
DOT Classification: Label: Flammable
  Liquid

THR: An experimental tumorigen. Moderately toxic by ingestion, inhalation, and by skin absorption. Mutagenic data. Dangerous fire hazard when exposed to heat or flame. To fight fire, use alcohol foam, CO_2, dry chemical. When heated to decomposition, it emits toxic NO_x. Incompatible with oxidizing materials; perchloryl fluoride. For further information, see Vol. 6, No. 2 of *DPIM Report*.

sec-BUTYLAMINE   HR: 3
CAS: 13952-84-6   NIOSH: EO 3325000
DOT: 1125
mf: $C_4H_{11}N$ mw: 73.16

PROP: Liquid; mp: −104°, bp: 63°, flash p: 15°F, d: 0.724 @ 20°.

SYNS: 2-AMINOBUTANE * TUTANE

DFG MAK: 5 ppm (15 mg/m³)
DOT Classification: Flammable Liquid

THR: A poison by ingestion. A powerful irritant. See also n-butyl amine, and amines. Moderately irritating to skin. Dangerous fire hazard when exposed to heat or flame. To fight fire, use alcohol foam, water spray or mist, dry chemical. Incompatible with oxidizing materials. For further information, see Vol. 3, No. 6 of *DPIM Report*.

tert-BUTYLAMINE   HR: 2
CAS: 75-64-9   NIOSH: EO 3330000
DOT: 1125
mf: $C_4H_{11}N$ mw: 73.16

PROP: Colorless liquid; mp: −67.5°, bp: 44°-46°, d: 0.700 @ 15°, lel = 1.7% @ 212°F, uel = 8.9% @ 212°F, vap d: 2.5, autoign temp: 716°F.

SYNS: 2-AMINOISOBUTANE * 2-AMINO-2-METHYLPROPANE * 1,1-DIMETHYLETHYLAMINE * TRIMETHYLAMINOMETHANE

DOT Classification: Flammable Liquid

THR: Moderately toxic by ingestion. See also butylamine and amines. Very dangerous fire hazard when exposed to heat or flame. To fight fire, use alcohol foam. When heated to decomposition it emits toxic fumes of NO_x. For further information, see Vol. 5, No. 6 of *DPIM Report*.

p-(BUTYL AMINO)BENZOIC ACID-2(DIMETHYL AMINO)ETHYL ESTER   HR: 3
CAS: 94-24-6   NIOSH: DG 4725000
mf: $C_{15}H_{24}N_2O_2$ mw: 264.41

SYNS: P-(BUTYLAMINO)BENZOIC ACID, 2-(DI-METHYLAMINO)ETHYL ESTER * P-BUTYLAMINO-BENZOYL-2-DIMETHYLAMINOETHANOL * DI-METHYLAMINOETHYL-P-BUTYL-AMINOBENZOATE * 2-DIMETHYLAMINOETHYL-P-BUTYLAMINO-BENZOATE * TETRACAINE

THR: A poison via intravenous, intraperitoneal, and subcutaneous routes. See also esters. When heated to decomposition, it emits toxic fumes of NO_x.

2-BUTYL AMINO ETHANOL   HR: 2
CAS: 111-75-1   NIOSH: KK 0175000
mf: $C_6H_{15}NO$ mw: 117.22

PROP: Liquid. Bp: 192°, flash p: 170°F (OC), d: 0.89, vap d: 4.03.

THR: Moderately toxic by ingestion and intraperitoneal route. A skin and severe eye irritant. See also amines. Moderate fire hazard when exposed to heat or flame. To fight fire, use alcohol foam, foam, CO_2, dry chemical. Incompatible with oxidizing materials. When heated to decomposition, it emits toxic fumes of NO_x.

tert-BUTYL AMINO ETHYL METHACRYLATE   HR: 3
CAS: 3775-90-4   NIOSH: OZ 3500000
mf: $C_{10}H_{19}NO_2$ mw: 185.30

PROP: Liquid; bp: 100°-105°; d: 0.914. flash p: 205°F (OC).

SYN: 2-(TERT-BUTYLAMINO)ETHYL METHACRY-LATE

THR: An intraperitoneal poison. See also esters. Slight fire hazard when exposed to heat or flame. To fight fire, use alcohol foam, water spray or mist, dry chemical. When heated to decomposition, it emits toxic fumes of NO_x.

alpha'-((tert-BUTYL AMINO)METHYL)-4-HYDROXY-m-XYLENE-alpha, alpha'-DIOL   HR: 3
CAS: 18559-94-9   NIOSH: ZE 4400000
mf: $C_{13}H_{21}NO_3$ mw: 239.35

SYNS: 2-(TERT-BUTYLAMINO)-1-(4-HYDROXY-3-HYDROXYMETHYLPHENYL)ETHANOL * 4-HY-DROXY-3-HYDROXYMETHYL-ALPHA-((TERT-BUTYLAMINO)METHYL)BENZYL ALCOHOL

THR: An intravenous poison. Affects the human cardiovascular system. When heated to decomposition, it emits toxic fumes of NO_x.

p-BUTYLAMINO SALICYLIC ACID, 2-(DI-ETHYLAMINO)ETHYL ESTER HYDRO-CHLORIDE HR: 3
NIOSH: VO 1925000
mf: $C_{17}H_{28}N_2O_3 \cdot ClH$ mw: 344.93

SYN: HCl SALZ DES P-N-N-BUTYLAMINOSA-LICYLSAEUREDIAETHYLAMINOAETHYLESTER (GERMAN)

THR: A poison via intraperitoneal, subcutaneous, and intravenous routes. An eye irritant. See also esters. When heated to decomposition, it emits very toxic fumes of NO_x and HCl.

BUTYLATED HYDROXY ANISOLE HR: 3
CAS: 25013-16-5 NIOSH: SL 1945000
mf: $C_{11}H_{16}O_2$ mw: 180.27

SYNS: BUTYLHYDROXYANISOLE * TERT-BU-TYLHYDROXYANISOLE * TERT-BUTYL-4-HY-DROXYANISOLE

THR: An experimental tumorigen. Moderately toxic by ingestion. Used as an antioxidant in foods. Mutagenic data. When heated to decomposition, it emits acrid and irritating fumes.

n-BUTYLBENZENE HR: 1
CAS: 104-51-8 NIOSH: CY 9070000
mf: $C_{10}H_{14}$ mw: 134.24

PROP: Colorless liquid. Mp: $-81.2°$, bp: 182.1°, fp: $-88.2°$, flash p: 160°F (TOC), d: 0.8601 @ 20°/4°, vap press: 1 mm @ 22.7°, autoign temp: 774°F, lel = 0.8%, uel = 5.8%, vap d: 4.6.

SYN: 1-PHENYLBUTANE

THR: Mildly toxic by ingestion. Moderate fire hazard when exposed to heat or flame. To fight fire, use alcohol foam, CO_2, dry chemical. Incompatible with oxidizing materials. When heated to decomposition, it emits acrid and irritating fumes.

sec-BUTYLBENZENE HR: 2
CAS: 135-98-8 NIOSH: CY 9100000
mf: $C_{10}H_{14}$ mw: 134.24

PROP: Colorless liquid. Mp: $-82.7°$, bp: 173.5°, fp: $-75.8°$, flash p: 126°F (TOC), d: 0.8621 @ 20°, vap press: 1 mm @ 18.6°, vap d: 4.62, autoign temp: 788°F, lel = 0.8%, uel = 6.9%.

SYN: 2-PHENYLBUTANE

THR: Moderately toxic by ingestion. Moderate fire hazard when exposed to heat or flame. To fight fire, use foam, CO_2, dry chemical, water spray or mist. Incompatible with oxidizing materials.

tert-BUTYLBENZENE HR: 2
CAS: 98-06-6 NIOSH: CY 9120000
mf: $C_{10}H_{14}$ mw: 134.24

PROP: Colorless liquid. Bp: 168.2°, fp: $-58°$, flash p: 140°F (TOC), d: 0.8665 @ 20°, vap press: 1 mm @ 13.0°, vap d: 4.62, autoign temp: 842°F, lel = 0.7% @ 212°F, uel = 5.7% @ 212°F.

SYNS: 2-METHYL-2-PHENYLPROPANE * PSEU-DOBUTYLBENZENE * TRIMETHYLPHENYLMETH-ANE

THR: Mildly toxic by ingestion. Moderate fire hazard when exposed to heat or flame. To fight fire, use foam, CO_2, dry chemical, water spray, fog, mist. Incompatible with oxidizing materials. When heated to decomposition, it emits acrid and irritating fumes.

5-BUTYL-2-BENZIMID AZOLECAR-BAMIC ACID METHYL ESTER HR: 3
CAS: 14255-87-9 NIOSH: DD 6495000
mf: $C_{13}H_{17}N_3O_2$ mw: 247.33

SYNS: PARBENDAZOLE * WORM GUARD

THR: An experimental teratogen. Moderate acute toxicity by ingestion. When heated to decomposition, it emits toxic fumes of NO_x.

BUTYL BENZOATE HR: 2
CAS: 136-60-7 NIOSH: DG 4925000
mf: $C_{11}H_{14}O_2$ mw: 178.25

PROP: Liquid. Mp: $-21.5°$, bp: 250°, flash p: 225°F (OC), d: 1.0073 @ 20°/20°, vap press: <0.01 mm @ 20°, vap d: 6.15.

THR: Moderately toxic by skin absorption. Severe skin irritant. An eye irritant. Slight fire hazard when exposed to heat or flame; can react with oxidizing materials. To fight fire, use CO_2, dry chemical, water mist, fog, spray. When heated to decomposition, it emits acrid and irritating fumes.

p-tert-BUTYL BENZOIC ACID HR: 2
CAS: 98-73-7 NIOSH: DG 4708000
mf: $C_{11}H_{14}O_2$ mw: 178.25

PROP: Colorless, fine crystalline powder. Mp: 166.3°, d: 1.142 @ 20°/4°.

SYN: TBBA

THR: Moderately toxic by ingestion. An irritant. Slight fire hazard when exposed to heat or flame. To fight fire, use foam, CO_2, dry chemical. When heated to decomposition, it emits acrid smoke and irritating fumes. Incompatible with oxidizing materials.

N-BUTYLBIGUANIDE HYDROCHLORIDE   HR: 3
CAS: 1190-53-0   NIOSH: DU 1135000
mf: $C_6H_{15}N_5 \cdot ClH$ mw: 193.72

SYNS: 1-BUTYLBIGUANIDE HYDROCHLORIDE * N-BUTYL-IMIDODICARBONIMIDIC DIAMIDE MONOHYDROCHLORIDE (9CI) * 1-BUTYLDIGUANIDE HYDROCHLORIDE

THR: A poison via intravenous and intraperitoneal routes. When heated to decomposition, it emits very toxic fumes of HCl and NO_x.

N-BUTYL-N,N-BIS(HYDROXY ETHYL)-AMINE   HR: 1
CAS: 102-79-4   NIOSH: KK 0525000
mf: $C_8H_{19}NO_2$ mw: 161.28

PROP: Liquid; bp: 262°, flash p: 245°F (OC), d: 0.97, vap d: 5.55.

SYNS: N-BUTYLDIETHANOLAMINE * N-BUTYL-2,2′-IMINODIETHANOL

THR: Mildly toxic via ingestion. No chronic effects data. Mild skin irritant. Severe eye irritant. Slight fire hazard when exposed to heat or flame. To fight fire, use alcohol foam, foam, CO_2, dry chemical. Incompatible with oxidizing materials. When heated to decomposition it, emits toxic fumes of NO_x.

tert-BUTYL BROMIDE   HR: 3
                        NIOSH: TX 4150000
mf: $(CH_3)_3CBr$ mw: 137.0

PROP: Colorless liquid; mp: −20°; bp: 73.3°; fp: −18°; d: 1.215 @ 25°/25°.

SYNS: 2-BROMOISOBUTANE * 2-BROMO-2-METHYLPROPANE * TRIMETHYLBROMOMETHANE

THR: An experimental neoplastigen. When heated to decomposition, it emits toxic fumes of Br^-.

n-BUTYL-n-BUTANOATE   HR: 2
CAS: 109-21-7   NIOSH: ES 8120000
mf: $C_8H_{16}O_2$ mw: 144.24

PROP: Liquid. Bp: 166°, flash p: 128°F (OC), d: 0.874, vap d: 5.0.

SYN: N-BUTYL-N-BUTYRATE

THR: Moderately toxic via intraperitoneal route. Mildly toxic by ingestion. Moderately irritating to eyes and mucous membranes and via inhalation. Narcotic in high concentrations. Moderate fire hazard when exposed to heat or flame. To fight fire, use alcohol foam, foam, CO_2, dry chemical. Incompatible with oxidizing materials. When heated to decomposition, it emits acrid and irritating fumes.

BUTYL CARBAMATE   HR: 3
CAS: 592-35-8   NIOSH: EZ 0175000
mf: $C_5H_{11}NO_2$ mw: 117.17

SYNS: USAF FO-1 * USAF EL-101

THR: A poison via intraperitoneal route; moderately toxic via subcutaneous route. See also carbamates. Mutagenic data. An experimental neoplastigen and teratogen. When heated to decomposition, it emits toxic fumes of NO_x.

BUTYL CARBITOL ACETATE   HR: 2
CAS: 124-17-4   NIOSH: KJ 9275000
mf: $C_{10}H_{20}O_4$ mw: 204.30

PROP: Colorless liquid. Fp: −32.2°, bp: 247°, flash p: 240°F (OC), d: 0.981 @ 20°/20°, autoign temp: 570°F, vap press: 0.01 mm @ 20°.

SYNS: 2-(2-BUTOXYETHOXY)ETHANOL ACETATE * 2-(2-BUTOXYETHOXY)ETHYL ACETATE * DIGLYCOL MONOBUTYL ETHER ACETATE

THR: Moderately toxic by ingestion. Mild skin and eye irritant. Slight fire hazard when exposed to heat or flame; emits decomposition products. To fight fire, use foam, CO_2, dry chemical. Incompatible with oxidizing materials; heat; flame. When heated to decomposition, it emits acrid and irritating fumes.

BUTYL CARBO BUTOXY METHYL PHTHALATE   HR: 3
CAS: 85-70-1   NIOSH: TI 0535000
mf: $C_{18}H_{24}O_6$ mw: 336.42

SYNS: BUTYL PHTHALATE BUTYL GLYCOLATE * BUTYL PHTHALYL BUTYL GLYCOLATE * DIBUTYL-O-(O-CARBOXYBENZOYL) GLYCOLATE * DIBUTYL-O-CARBOXYBENZOYLOXYACETATE

THR: An experimental teratogen. Mutagenic data. An eye irritant. Mildly toxic via intraperitoneal route. When heated to decomposition, it emits acrid and irritating fumes.

N-BUTYL-(3-CARBOXY PROPYL)NITRO-SAMINE   HR: 3
CAS: 38252-74-3   NIOSH: ES 8100000
mf: $C_8H_{16}N_2O_3$ mw: 188.26

SYNS: 4-(BUTYLNITROSOAMINO)BUTANOIC ACID * N-NITROSO-N-BUTYL-N-(3-CARBOXYPROPYL)-AMINE

THR: An experimental carcinogen. Mutagenic data. When heated to decomposition, it emits toxic fumes of NO_x.

BUTYL CELLOSOLVE   HR: 3
CAS: 111-76-2   NIOSH: KJ 8575000
mf: $C_6H_{14}O_2$ mw: 118.20

PROP: Clear, mobile liquid, pleasant odor. bp: 168.4°-170.2°, fp: -74.8°, flash p: 160°F (COC), d: 0.9012 @ 20°/20°, vap press: 300 mm @ 140°. Much used in industry.

SYNS: BUTOKSYETYLOWY ALKOHOL (POLISH) * 2-BUTOSSI-ETANOLO (ITALIAN) * 2-BUTOXY-AETHANOL (GERMAN) * 2-BUTOXY-1-ETHANOL * O-BUTYL ETHYLENE GLYCOL * BUTYLGLY-COL (FRENCH, GERMAN) * DOWANOL EB * ETHYLENE GLYCOL-N-BUTYL ETHER * GLY-COL MONOBUTYL ETHER

OSHA PEL: TWA 50 ppm (skin)

THR: A human poison via inhalation. A poison via ingestion, intravenous, and skin routes. Moderately toxic via intraperitoneal and subcutaneous routes. Mild skin and eye irritant. Moderate fire hazard when exposed to heat or flame. To fight fire, use foam, CO_2, dry chemical. Incompatible with oxidizing materials; heat; flame. When heated to decomposition, it emits acrid and irritating fumes. For further information, see ethylene glycol monobutyl ether, Vol. 4, No. 2 of *DPIM Report*.

N-BUTYL CHLORIDE   HR: 2
CAS: 109-69-3   NIOSH: EJ 6300000
DOT: 1127
mf: C_4H_9Cl mw: 92.58

PROP: Colorless liquid; mp: -123.1°, bp: 78°, lel = 1.9%, uel = 10.1%, flash p: 15°F (OC), d: 0.884, autoign temp: 860°F, vap d: 3.20.

SYNS: 1-CHLOROBUTANE * CHLORURE DE BU-TYLE (FRENCH) * NCI-C06155 * N-PROPYL-CARBINYL CHLORIDE

DOT Classification: Label: Flammable Liquid

THR: Moderately toxic by ingestion, inhalation, and other routes. See chlorinated hydrocarbons, aliphatic. Skin and eye irritant. Dangerous fire hazard when exposed to heat or flame. Moderately explosive when exposed to flame. When heated to decomposition, it emits highly toxic fumes of phosgene. To fight fire, use foam, CO_2, dry chemical. Incompatible with oxidizing materials.

tert-BUTYL CHLORIDE   HR: 3
CAS: 507-20-0   NIOSH: TX 5040000
mf: C_4H_9Cl mw: 92.58

PROP: Flash p: 32°F, d: 0.87, vap d: 3.2, bp: 51°.

SYNS: 2-CHLOROISOBUTANE * 2-CHLORO-2-METHYLPROPANE * TRIMETHYLCHLOROMETH-ANE

THR: An experimental neoplastigen. Dangerous fire hazard via heat, flame (sparks), and oxidizers. To fight fire, use water, spray, fog, alcohol foam, dry chemical. When heated to decomposition, it emits toxic fumes of Cl^-.

4-tert-BUTYL-2-CHLORO PHENYL METHYL METHYL PHOSPHORAMI-DATE   HR: 3
CAS: 299-86-5   NIOSH: TB 3850000
mf: $C_{12}H_{19}ClNO_3P$ mw: 291.74

SYNS: ENT 25,602-X * O-METHYL-O-2-CHLORO-4-TERT-BUTYLPHENYL-N-METHYLAMIDO-PHOSPHATE * O-(4-TERT BUTYL-2-CHLOOR-FE-NYL)-O-METHYL-FOSFORZUUR-N-METHYL-AMIDE (DUTCH) * 4-TERT. BUTYL 2-CHLOROPHENYL METHYLPHOSPHORAMIDATE DE METHYLE (FRENCH) * O-(4-TERT-BUTYL-2-CHLOR-PHE-NYL)-O-METHYL-PHOSPHORSAEURE-N-METHYL AMID (GERMAN) * O-(4-TERZ.-BUTIL-2-CLO-RO-FENIL)-O-METIL-FOSFORAMMIDE (ITALIAN)

THR: A poison by ingestion. Moderately toxic via skin and other routes. When heated to decomposition, it emits very toxic fumes of PO_x, NO_x, and Cl^-.

tert-BUTYL CHROMATE   HR: 3
CAS: 1189-85-1   NIOSH: GB 2900000
mf: $C_8H_{18}CrO_4$ mw: 230.26

SYN: CHROMIC ACID, DI-t-BUTYL ESTER

OSHA PEL: TWA CL 100 ug(CrO3)/m^3

THR: No toxicity data. See also chromium compounds and esters. A very flammable mixture. When heated to decomposition, it emits acrid and irritating fumes.

2-tert-BUTYL-p-CRESOL   HR: 3
CAS: 2409-55-4   NIOSH: GO 7000000
mf: $C_{11}H_{16}O$ mw: 164.27

PROP: Clear liquid, sol in organic solvents and aqueous potassium hydroxide. fp: 23.1°, bp: 244°, d: 0.922; flash p: 116°F.

SYNS: 2-t-BUTYL-p-CRESOL * 2-tert-BUTYL-p-KRESOL (CZECH) * 2-T-BUTYL-4-METHYLPHE-NOL

THR: A poison via intravenous route. Moderately toxic by ingestion and dermal routes. A severe skin and eye irritant. Moderate fire hazard when exposed to heat, flame, or oxidizers. To fight fire, use alcohol foam, foam, water spray, fog, dry chemical. When heated to decomposition, it emits acrid and irritating fumes.

N-BUTYL CYCLOHEXYL AMINE   HR: 3
CAS: 10108-56-2   NIOSH: GX 1050000
mf: $C_{10}H_{21}N$ mw: 155.32

PROP: flash p: 200°F (OC), d: 0.8, bp: 210°.

THR: A poison by ingestion. Moderately toxic via dermal route. See also amines. A skin irritant. Slight fire hazard when exposed to heat or flame. To fight fire, use alcohol foam. When heated to decomposition, it emits toxic fumes of NO_x.

N-(4-tert-BUTYL CYCLOHEXYL)-3,3-DI-PHENYL PROPYLAMINE HYDROCHLO-RIDE   HR: 3
CAS: 61925-70-0   NIOSH: UH 9950000
mf: $C_{25}H_{25}N \cdot ClH$ mw: 375.97

SYN: MG 18037

THR: A poison via intraperitoneal route; moderately toxic by ingestion. When heated to decomposition, it emits very toxic fumes of HCl and NO_x.

BUTYL DICHLORO PHENOXY ACE-TATE   HR: 2
CAS: 94-80-4   NIOSH: AG 8050000
mf: $C_{12}H_{14}Cl_2O_3$ mw: 277.16

SYNS: BUTYL 2,4-D * BUTYL (2,4-DICHLORO-PHENOXY)ACETATE * 2,4-D BUTYL ESTER * (2,4-DICHLOROPHENOXY)ACETIC ACID, BUTYL ESTER

THR: Moderately toxic by ingestion and other routes. Human central nervous system effects. See also esters. When heated to decomposition, it emits toxic fumes of Cl$^-$.

BUTYL DICHLORO PHENOXY ACE-TATE (Mixed Isomer)   HR: 2
CAS: 64047-35-4   NIOSH: AG 7875000
mf: $C_{12}H_{14}Cl_2O_3$ mw: 277.16

SYN: BUTYL ESTER OF DICHLOROPHENOXY-ACETIC ACID

THR: Moderate acute toxicity by ingestion. When heated to decomposition, it emits toxic fumes of Cl$^-$.

2-n-BUTYL-3-DIMETHYLAMINO-5,6-METHYLENE DIOXY INDENE HY-DROCHLORIDE   HR: 3
          NIOSH: NK 8926000
mf: $C_{16}H_{21}NO_2 \cdot ClH$ mw: 295.84

SYNS: BU-MDI * 6-BUTYL-5-DIMETHYLAMINO-5H-INDENO(5,6-D)-1,3-DIOXOLE HYDROCHLORIDE

THR: A poison via intraperitoneal and intravenous routes. When heated to decomposition, it emits very toxic fumes of NO_x and HCl.

2-sec-BUTYL-4,6-DINITRO PHE-NOL   HR: 3
CAS: 88-85-7   NIOSH: SJ 9800000
mf: $C_{10}H_{12}N_2O_5$ mw: 240.24

PROP: Crystals. Vap d: 7.73.

SYNS: 4,6-DINITRO-2-SEC.BUTYLFENOL (CZECH) * 2,4-DINITRO-6-SEC-BUTYLPHENOL * 4,6-DI-NITRO-O-SEC-BUTYLPHENOL * 4,6-DINITRO-2-SEC-BUTYLPHENOL * DINITROBUTYLPHENOL * 4,6-DINITRO-2-(1-METHYL-N-PROPYL)PHENOL * 2,4-DINITRO-6-(1-METHYL-PROPYL)PHENOL (FRENCH) * DINOSEBE (FRENCH) * ENT 1,122 * 6-(1-METHYL-PROPYL)-2,4-DINITROFENOL (DUTCH) * 2-(1-METHYLPROPYL)-4,6-DINITRO-PHENOL * 6-(1-METIL-PROPIL)-2,4-DINITRO-FENOLO (ITALIAN)

THR: A poison by ingestion, skin, intraperitoneal, and other routes. Mutagenic data. An eye irritant. An experimental teratogen. When heated to decomposition, it emits toxic fumes of NO_x.

4-BUTYL-1,2-DIPHENYL-3,5-DIOXO PYRAZOLIDINE   HR: 3
CAS: 50-33-9   NIOSH: UQ 8225000
mf: $C_{19}H_{20}N_2O_2$ mw: 308.41

SYNS: BUTIWAS-SIMPLE * 4-BUTYL-1,2-DI-
PHENYLPYRAZOLIDINE-3,5-DIONE * 4-BUTYL-
1,2-DIPHENYL-3,5-PYRAZOLIDINEDIONE
* 3,5-DIOXO-1,2-DIPHENYL-4-N-BUTYLPYRA-
ZOLIDENE * 3,5-DIOXO-1,2-DIPHENYL-4-
N-BUTYL-PYRAZOLIDIN * 3,5-DIOXO-
1,2-DIPHENYL-4-M-BUTYL-PYRAZOLIDINE
* DIPHENYLBUTAZONE * 1,2-DIPHENYL-4-
BUTYL-3,5-DIOXOPYRAZOLIDINE * 1,2-DI-
PHENYL-4-BUTYL-3,5-PYRAZOLIDINEDIONE
* 1,2-DIPHENYL-3,5-DIOXO-4-BUTYLPYRAZOLI-
DINE * 1,2-DIPHENYL-2,3-DIOXO-4-N-BUTYL-
PYRAZOLINE * NCI-C56531

THR: A poison by ingestion, intraperitoneal,
subcutaneous, intravenous, and intramuscular
routes. A carcinogen in women. An experimental teratogen. A systemic poison in women. A
moderate eye irritant. When heated to decomposition, it emits toxic fumes of NO_x.

2-BUTYLENE DICHLORIDE   HR: 3
CAS: 110-57-6   NIOSH: EM 4903000
mf: $C_4H_6Cl_2$ mw: 125.00

PROP: Colorless liquid. Mp: 1°-3°, bp: 156°,
d: 1.183 @ 25°/4°.

SYNS: 1,4-DICHLOROBUTENE-2 (TRANS)
* 1,4-DICHLORO-2-BUTENE

THR: A poison by ingestion and inhalation.
Moderately toxic via dermal route. An experimental neoplastigen and carcinogen. When
heated to decomposition, it emits toxic fumes
of Cl^-.

BUTYL ESTER PHOSPHORIC ACID   HR: 3
CAS: 12788-93-1   NIOSH: TB 8490000
DOT: 1718
mf: $C_4H_{10}O_4P$ mw: 153.11

SYNS: ACID BUTYL PHOSPHATE (DOT)
* N-BUTYL ACID PHOSPHATE (DOT) * BUTYL
PHOSPHORIC ACID (DOT)

DOT Classification: Label: Corrosive

THR: A corrosive poison and irritant to skin,
eyes, mucous membranes. See also esters and
phosphoric acid. When heated to decomposition, it emits toxic fumes of PO_x.

n-BUTYL ETHER   HR: 2
CAS: 142-96-1   NIOSH: EK 5425000
DOT: 1149
mf: $C_8H_{18}O$ mw: 130.26

PROP: Colorless liquid. Mp: −95°, bp: 142°,
flash p: 77°F, d: 0.769 @ 20°/20°, autoign temp:
382°F, vap d: 4.48, lel = 1.5%, uel = 7.6%.

SYNS: 1-BUTOXYBUTANE * DI-N-BUTYL ETHER
* DIBUTYL OXIDE * ETHER BUTYLIQUE
(FRENCH)

DOT Classification: Label: Flammable
   Liquid

THR: Moderately toxic via inhalation in humans. Mildly toxic by ingestion and dermal
routes. A skin and eye irritant. Dangerous fire
hazard; Moderately explosive; see also ethers.
Incompatible with NCl_3 and oxidizing materials.
When heated, it emits acrid fumes. To fight
fire, use alcohol foam, dry chemical.

BUTYL ETHYL ACETALDEHYDE   HR: 2
CAS: 123-05-7   NIOSH: MN 7525000
DOT: 1191
mf: $C_8H_{16}O$ mw: 128.24

PROP: Bp: 163.4°, flash p: 125°F (OC). autoign
temp: 387°F, d: 0.8205, vap press: 1.8 mm
@ 20°, vap d: 4.42.

SYNS: ALPHA-ETHYLCAPROALDEHYDE
* ETHYLHEXALDEHYDE (DOT) * 2-ETHYLHEX-
ALDEHYDE * 2-ETHYLHEXANAL

DOT Classification: Combustible Liquid,
   Label: None

THR: Moderately toxic by ingestion and inhalation. A skin and eye irritant. See also aldehydes.
Dangerous fire hazard; spontaneously flammable in air. To fight fire, use foam, CO_2, dry
chemical, water spray, mist, fog. Incompatible
with oxidizing materials. When heated to decomposition, it emits acrid and irritating fumes.
For further information, see 2-ethyl hexaldehyde, Vol. 3, No. 2 of *DPIM Report*.

BUTYL ETHYL ACETIC ACID   HR: 2
CAS: 149-57-5   NIOSH: MO 7700000
mf: $C_8H_{16}O_2$ mw: 144.24

PROP: Flash p: 260°F(OC).

SYNS: ALPHA-ETHYLCAPROIC ACID * 2-
ETHYLHEXANOIC ACID * 2-ETHYLHEXOIC ACID

THR: Moderately toxic by ingestion and dermal
route. Mild skin irritant. Severe eye irritant.

Slight fire hazard (combustible). When heated to decomposition, it emits acrid and irritating fumes.

tert-BUTYL FORMAMIDE   HR: 3
CAS: 2425-74-3   NIOSH: LQ 1450000
mf: $C_5H_{11}NO$ mw: 101.17

THR: A poison via intravenous route. When heated to decomposition, it emits toxic fumes of NO_x.

N-BUTYL FORMATE   HR: 2
CAS: 592-84-7   NIOSH: LQ 5500000
DOT: 1128
mf: $C_5H_{10}O_2$ mw: 102.15

PROP: Colorless liquid; mp: $-90°$, bp: $106.0°$, flash p: 64°F (CC), d: 0.911, autoign temp: 612°F, vap press: 40 mm @ 31.6°, vap d: 3.52, lel = 1.7%, uel = 8%.

SYN: BUTYL FORMATE (DOT)

DOT Classification: Label: Flammable
  Liquid

THR: Moderately toxic by ingestion and inhalation. An irritant and narcotic in high concentrations. See also esters, butyl alcohol, and formic acid. Dangerous fire hazard when exposed to heat or flame. To fight fire, use alcohol foam, foam, CO_2, dry chemical. Incompatible with oxidizing materials. When heated to decomposition, it emits acrid and irritating fumes.

BUTYL GLYCIDYL ETHER   HR: 2
CAS: 2426-08-6   NIOSH: TX 4200000
mf: $C_7H_{14}O_2$ mw: 130.21

SYNS: N-BUTYL GLYCIDYL ETHER * 2,3-EP-OXYPROPYL BUTYL ETHER * GLYCIDYL BUTYL ETHER

OSHA PEL: TWA 50 ppm
ACGIH TLV: TWA 25 ppm
DFG MAK: 50 ppm (270 mg/m^3)

THR: Moderately toxic via ingestion, intraperitoneal, and skin routes. Mutagenic data. A skin and eye irritant. See also ethers. When heated to decomposition, it emits acrid and irritating fumes.

6-BUTYL HYDROPEROXIDE   HR: 3
CAS: 75-91-2   NIOSH: EQ 4900000
DOT: 2093/2094
mf: $C_4H_{10}O_2$ mw: 90.14

PROP: Water white liquid, slightly sol in water, very sol in esters and alcohols, flash p: 80°F or above, fp: $-35°$, d: 0.860, vap d: 2.07.

SYNS: TERC. BUTYLHYDROPEROXID (CZECH)
* HYDROPEROXYDE DE BUTYLE TERTIAIRE (FRENCH) * 1,1-DIMETHYL ETHYL HYDRO PER-OXIDE.

DOT Classification: Organic peroxide

THR: A poison via ingestion and inhalation. Severe skin and eye irritant. Mutagenic data. At highest dosage levels, symptoms noted were severe depression, incoordination, and cyanosis. Death was due to respiratory arrest. Very dangerous fire hazard when exposed to heat or flame, or by spontaneous chemical reaction such as with reducing materials. Moderately explosive. To fight fire, use alcohol foam, CO_2, dry chemical.

n-BUTYL IODIDE   HR: 3
CAS: 542-69-8   NIOSH: EK 4400000
mf: C_4H_9I mw: 184.03

SYN: 1-IODOBUTANE

THR: An experimental neoplastigen. Mild acute toxicity via inhalation. See also iodides. When heated to decomposition, it emits toxic fumes of I^-.

n-BUTYL ISOCYANATE   HR: 3
CAS: 111-36-4   NIOSH: NQ 8250000
DOT: 2485
mf: C_5H_9NO mw: 99.15

PROP: Colorless liquid. Bp: 115°, d: 0.880 @ 20°/4°.

SYN: ISOCYANIC ACID, BUTYL ESTER

DOT Classification: Label: Flammable
  Liquid and Poison

THR: A poison via ingestion and intravenous routes. Moderate toxicity via inhalation. A powerful irritant to eyes, skin, and mucous membranes. Flammable liquid. See also cyanates, NO_x.

tert-BUTYL ISOPROPYL BENZENE HY-DROPEROXIDE   HR: 3
CAS: 30026-92-7   NIOSH: MX 2430000
DOT: 2091
mf: $C_{13}H_{20}O_2$ mw: 208.33

PROP: Crystals.

SYN: TERTIARY BUTYL ISOPROPYL BENZENE HYDROPEROXIDE (DOT)

DOT Classification: Label: Organic Peroxide

THR: Powerful irritant. See also peroxides, organic. Dangerous fire hazard when exposed to heat or flame or by chemical reaction. Incompatible with oxidizing or reducing materials.

BUTYL LITHIUM
mf: C_4H_9Li　　mw: 63.94

THR: Unknown, but probably very toxic. Extremely flammable. Incompatible with air; water; CO_2 can cause it to ignite and burn rapidly. To fight fire, use dry chemical; see special instructions of manufacturer.

n-BUTYLMERCURIC CHLORIDE　HR: 3
CAS: 543-63-5　　　NIOSH: OV 7700000
mf: C_4H_9ClHg　　mw: 293.17

SYN: BMC

OSHA PEL: TWA 10 ug(Hg)/m^3; CL 40

THR: A poison via subcutaneous route. Mutagenic data. See also mercury compounds, organic. When heated to decomposition, it emits very toxic fumes of Cl$^-$ and Hg.

n-BUTYL MESITYL OXIDE OXALATE　HR: 2
CAS: 532-34-3　　　NIOSH: UP 7000000
mf: $C_{12}H_{18}O_4$　　mw: 226.30

PROP: Yellow to reddish liquid; bp: 113°, d: 1.052-1.060 @ 25°/25°, flash p: 315°F.

SYNS: BUTYL-3,4-DIHYDRO-2,2-DIMETHYL-4-OXO-2H-PYRAN-6-CARBOXYLATE ∗ N-BUTYL ESTER OF 3,4-DIHYDRO-2,2-DIMETHYL-4-OXO-2H-PYRAN-6-CARBOXYLIC ACID ∗ N-BUTYL-MESITYLOXID OXALATE ∗ 2-CARBO-N-BUTOXY-6,6-DIMETHYL-5,6-DIHYDRO-1,4-PYRONE ∗ 3,4-DIHYDRO-2,2-DIMETHYL-4-OXO-2H-PYRAN-6-CARBOXYLIC ACID-N-BUTYL ESTER ∗ ALPHA,ALPHA-DIMETHYL-ALPHA′-CARBO-BUTOXY-DIHYDRO-GAMMA-PYRONE ∗ 2,2-DIMETHYL-6-CARBOBUTOXY-2,3-DIHYDRO-4-PYRONE ∗ ENT 9

THR: Moderately toxic by ingestion. Can produce an experimental liver necrosis. A mild skin irritant. See also esters. Slight fire hazard. When heated to decomposition it emits acrid and irritant fumes.

6-tert-BUTYL-3-METHYL-2,4-DINITRO ANISOLE　HR: 3
CAS: 83-66-9　　　NIOSH: BZ 8575000
mf: $C_{12}H_{16}N_2O_5$　　mw: 268.30

SYNS: 2,6-DINITRO-3-METHOXY-4-tert-BUTYL-TOLUENE ∗ MUSK AMBRETTE

THR: A poison via ingestion. A skin irritant. When heated to decomposition, it emits toxic fumes of NO$_x$.

n-BUTYL NITRITE　HR: 3
CAS: 544-16-1　　　NIOSH: RA 0780000
mf: $C_4H_9NO_2$　　mw: 103.14

PROP: Oily liquid, characteristic odor, miscible in alcohol and ether; bp: 75°, d: 0.9114 @ 0°/4°, vap d: 3.5, flash p: 10°.

SYNS: BUTYL NITRITE ∗ NCI-C56553 ∗ NITROUS ACID, N-BUTYL ESTER

THR: A poison via ingestion and inhalation. An irritant. Resembles amyl nitrite in causing fall in blood pressure, headache, pulse throbbing, and weakness. See also nitrites, butyl alcohol, and esters. Mutagenic data. Moderate fire hazard when exposed to heat or flame or by spontaneous chemical reaction. When heated to decomposition, it emits toxic fumes of NO$_x$.

sec-BUTYL NITRITE　HR: 2
CAS: 924-43-6　　　NIOSH: RA 0800000
mf: $C_4H_9NO_2$　　mw: 103.14

PROP: Liquid; bp: 68°, d: 0.8981 @ 0°/4°, vap d: 3.5.

SYNS: NITROUS ACID, sec-BUTYL ESTER ∗ NITROUS ACID, 1-METHYL PROPYL ESTER

THR: Moderately toxic via intraperitoneal route. See also esters. Moderately flammable when exposed to heat or flame or by spontaneous chemical reaction. An oxidizer. See nitrites for explosion hazard. When heated to decomposition it emits toxic NO$_x$. To fight fire, use water, spray, foam, dry chemical.

BUTYL NITROS AMINO METHYL ACETATE　HR: 3
CAS: 56986-36-8　　　NIOSH: AF 7477000
mf: $C_7H_{14}N_2O_3$　　mw: 174.23

SYNS: N-BUTYL-N-(ACETOXYMETHYL)NITROSAMINE ∗ ACETOXYMETHYLBUTYLNITROSAMINE ∗ N-(ACETOXY)METHYL-N-N-BUTYLNITROSAMINE ∗ BUTYL ACETOXYMETHYLNITROSAMINE

＊ N-NITROSO-N-(1-ACETOXYMETHYL)BUTYL-
AMINE

THR: Mutagenic data. Moderately toxic by ingestion. When heated to decomposition, it emits
toxic fumes of NO_x.

N-BUTYL-N-NITROSO AMYL AMINE　　　　　　　　　　　　HR: 2
CAS: 16339-05-2　　　　NIOSH: SC 0450000
mf: $C_9H_{20}N_2O$　　mw: 172.31

SYNS: BUTYLAMYLNITROSAMIN (GERMAN)
＊ N-BUTYL-N-NITROSOPENTYLAMINE ＊ N-BU-
TYL-N-PENTYLINITROSAMINE ＊ N-NITROSO-N-
BUTYLPENTYLAMINE ＊ N-NITROSO-N-BUTYL-N-
PENTYLAMINE

THR: Moderately toxic via subcutaneous route.
See also nitrosamines. When heated to decomposition, it emits toxic fumes of NO_x.

N-BUTYL-N-NITROSO-1-BUTA-MINE　　　　　　　　　　　　HR: 3
CAS: 924-16-3　　　　NIOSH: EJ 4025000
mf: $C_8H_{18}N_2O$　　mw: 158.28

PROP: Pale yellow liquid; bp: 235°.

SYNS: DI-N-BUTYLNITROSAMIN (GERMAN)
＊ DI-N-BUTYLNITROSAMINE ＊ N,N-DI-N-BU-
TYLNITROSAMINE ＊ N,N-DIBUTYLNITROSOAMINE
＊ N-NITROSO-DI-N-BUTYLAMINE

THR: An experimental carcinogen and neoplastigen. Mutagenic data. Moderately toxic via ingestion, subcutaneous, and intraperitoneal routes.
A suspected human carcinogen. When heated
to decomposition, it emits toxic fumes of NO_x.
For further information, see N-nitroso dibutyl
amine, Vol. 2, No. 5 of *DPIM Report*.

N-BUTYL-N-NITROSO ETHYL CARBA-MATE　　　　　　　　　　　　HR: 3
CAS: 6558-78-7　　　　NIOSH: EZ 1275000
mf: $C_7H_{14}N_2O_3$　　mw: 174.23

SYN: N-BUTYL-N-NITROSOURETHAN

THR: A poison via inhalation. Moderately toxic
by ingestion. Mutagenic data. An experimental
neoplastigen. See also carbamates. When heated
to decomposition, it emits toxic fumes of NO_x.

n-BUTYL NITROSO UREA　　　　　HR: 3
CAS: 869-01-2　　　　NIOSH: YS 3850000
mf: $C_5H_{11}N_3O_2$　　mw: 145.19

SYNS: BUTYLNITROSOHARNSTOFF (GERMAN)
＊ N-BUTYL-N-NITROSOUREA ＊ 1-BUTYL-1-NI-
TROSOUREA ＊ N-NITROSOBUTYLUREA

THR: Mutagenic data. An experimental brain
carcinogen. An acute poison by ingestion. Moderately toxic via acute subcutaneous route.
When heated to decomposition, it emits toxic
fumes of NO_x.

2-BUTYL-1-OCTANOL　　　　　　HR: 2
CAS: 3913-02-8　　　　NIOSH: RH 0885000
mf: $C_{12}H_{26}O$　　mw: 186.38

PROP: Liquid. Mp: −80°, flash p: 230°F(OC),
bp: 253.3°, d: 0.8355 @ 20°/20°, vap d: 6.42.

SYN: 2-BUTYLOCTYL ALCOHOL

THR: Mild toxicity via ingestion. A skin and
eye irritant. See also alcohols. Slightly flammable when exposed to heat or flame. To fight
fire, use CO_2, dry chemical. When heated to
decomposition, it emits acrid and irritating
fumes. Incompatible with oxidizing materials.

BUTYL OLEATE　　　　　　　　HR: 2
　　　　　　　　　　　NIOSH: RG 4123500
mf: $C_{22}H_{42}O_2$　　mw: 338.64

PROP: Liquid. Bp: 173°, flash p: 356°F(OC),
d: 0.873, vap d: 11.3.

SYN: (Z)-9-OCTADECENOIC ACID, BUTYL ESTER

THR: A skin irritant. See also esters, butyl alcohol, and oleic acid. Slightly flammable when
exposed to heat or flame. To fight fire, use
CO_2, dry chemical. When heated to decomposition, it emits acrid and irritating fumes. Incompatible with oxidizing materials.

N-BUTYL-N-(2-OXOBUTYL)NITROSA-MINE　　　　　　　　　　　　HR: 2
CAS: 61734-89-2　　　　NIOSH: EL 7030000
mf: $C_8H_{16}N_2O_2$　　mw: 172.26

SYN: N-NITROSO-N-(2-OXOBUTYL)BUTYLAMINE

THR: Moderately toxic. Mutagenic data. When
heated to decomposition, it emits toxic fumes
of NO_x.

tert-BUTYL PERACETATE　　　　HR: 3
CAS: 107-71-1　　　　NIOSH: SD 8925000
mf: $C_6H_{12}O_3$　　mw: 132.18

PROP: Clear, colorless benzene solution, insol
in water, sol in organic solvents; d: 0.923, vap
press: 50 mm @ 26°, flash p: <80°F (COC).

SYN: T-BUTYL PEROXYACETATE

THR: A poison via ingestion. Moderate toxicity
via inhalation. An eye irritant. See also perox-

ides, organic. Sensitive to shock and heat; can explode in contact with organic matter. Dangerous fire hazard via heat, flame, reducers. To fight fire, use dry chemical, alcohol foam, spray and mist. Pure ester is shock sensitive and detonates. Explodes with great violence when rapidly heated to critical temperature.

tert-BUTYL PERBENZOATE   HR: 2
CAS: 614-45-9   NIOSH: SD 9450000
mf: $C_{11}H_{14}O_3$ mw: 194.25

PROP: Colorless to slight yellow liquid, mild aromatic odor. Insol in water, sol in organic solvents; bp: 112° (decomp), flash p: 19°, fp: 8°, vap press: 0.33 mm @ 50°, d: 1.0.

SYNS: TERT-BUTYLPERBENZOAN (CZECH)
* T-BUTYL PEROXY BENZOATE * PERBENZOATE DE BUTYLE TERTIAIRE (FRENCH)

THR: Moderate toxicity via ingestion. A powerful skin and eye irritant. See also peroxides, organic for fire hazard. Dangerous explosive in contact with organic matter. To fight fire, see peroxides, organic.

tert-BUTYL PEROXIDE   HR: 3
CAS: 110-05-4   NIOSH: ER 2450000
mf: $C_8H_{18}O_2$ mw: 146.26

PROP: Clear, water white liquid. Mp: −40°, bp: 80° @ 284 mm, flash p: 65°F (OC), d: 0.79, vap press: 19.51 mm @ 20°, vap d: 5.03.

SYNS: DI-tert-BUTYLPEROXID (GERMAN) ·
* DI-t-BUTYL PEROXIDE * DI-tert-BUTYL PEROXYDE (DUTCH) * PEROSSIDO DI BUTILE TERZIARIO (ITALIAN) * PEROXYDE DE BUTYLE TERTIAIRE (FRENCH)

THR: Moderate toxicity via intraperitoneal route. Powerful irritant via ingestion and inhalation. A skin and eye irritant. See peroxides, organic, for fire and explosion hazards. To fight fire: Warning, water may not work.

tert-BUTYL PEROXY PIVALATE   HR: 3
CAS: 927-07-1   NIOSH: SE 0950000
mf: $C_9H_{18}O_3$ mw: 174.27

PROP: Colorless liquid, insol in water and ethylene glycol, sol in most organic solvents. d: 0.854 @ 25°/25°, fp: < 19°, flash p: > 155°F (OC), rapid decomp @ 21°.

THR: Poison. Mild toxicity via ingestion. See also peroxides, organic. Moderately flammable

via heat, flame (sparks), oxidizers. Can explode on heating. To fight fire, use water, fog, mist, alcohol foam, dry chemical.

o-sec-BUTYLPHENOL   HR: 3
CAS: 89-72-5   NIOSH: SJ 8920000
mf: $C_{10}H_{14}O$ mw: 150.24

PROP: Colorless liquid. Bp: 226°-228° @ 25 mm, fp: 12°, flash p: 225°F, d: 0.981 @ 25°/25°.

SYN: 2-sec-BUTYLFENOL (CZECH)

ACGIH TLV: TWA 5 ppm

THR: A poison via intravenous route. Moderate toxicity via ingestion. A skin and eye irritant. Slight fire hazard. To fight fire, use foam, spray, CO_2, dry chemical. When heated to decomposition, it emits acrid and irritating fumes.

p-sec-BUTYL PHENOL   HR: 2
mf: $(CH_3CHC_2H_5)C_6H_4OH$ mw: 150.2

PROP: Nearly white flakes. Bp: 135.4°-136.5° @ 25 mm, fp: 51°, flash p: 240°F, d: 0.963 @ 60°/60°.

THR: Moderate toxicity via ingestion. Slightly flammable when exposed to heat or flame. When heated to decomposition, it emits toxic fumes. To fight fire, use foam, CO_2, dry chemical. Incompatible with oxidizing materials.

4-tert-BUTYLPHENOL   HR: 2
CAS: 98-54-4   NIOSH: SJ 8925000
mf: $C_{10}H_{14}O$ mw: 150.24

PROP: Crystals or practically white flakes; bp: 238°, fp: 97°, d: 0.9081 @ 114°/4°, vap press: 1 mm @ 70.0°, vap d: 5.1.

SYNS: p-tert-BUTYLFENOL (CZECH) * P-TERT-BUTYLPHENOL * 1-HYDROXY-4-TERT-BUTYL-BENZENE

DFG MAK: 0.08 ppm (0.5 mg/m³)

THR: Moderate toxicity via skin route. A skin and eye irritant. Slight fire hazard when exposed to heat or flame; can react with oxidizing materials. To fight fire, use foam, CO_2, dry chemical. When heated to decomposition, it emits acrid and irritating fumes.

o-sec-BUTYLPHENYL CARBAMATE   HR: 3
NIOSH: EZ 1310000
mf: $C_{12}H_{17}NO_2$ mw: 207.30

THR: A poison via ingestion. See also carbamates. When heated to decomposition, it emits toxic fumes of NO_x.

BUTYL PHENYL ETHER   HR: 2
CAS: 1126-79-0   NIOSH: KN 5300000
mf: $C_{10}H_{14}O$ mw: 150.24

PROP: flash p: 180°F (OC); d: 0.9; vap d: 5.2; bp: 210°.

SYN: BUTOXYPHENYL

THR: Moderate toxicity by ingestion. See also ethers. When heated to decomposition, it emits acrid and irritating fumes.

BUTYL PHOSPHORO TRITHIO-
ATE   HR: 3
CAS: 78-48-8   NIOSH: TG 5425000
mf: $C_{12}H_{27}OPS_3$ mw: 314.54

PROP: Liquid, insol in water, sol in aliphatic, aromatic and chlorinated hydrocarbons. Bp: 150° @ 0.3 mm.

SYNS: S,S,S-TRIBUTYL PHOSPHOROTRITHIOATE * S,S,S-TRIBUTYL TRITHIOPHOSPHATE

THR: A poison via ingestion, dermal, and intraperitoneal routes. Animal experiments show an anti-cholinesterase effect. See also parathion. Dangerous; see phosphates and sulfates.

5-BUTYL PICOLINIC ACID   HR: 3
CAS: 536-69-6   NIOSH: US 5625000
mf: $C_{10}H_{13}NO_2$ mw: 179.24

SYNS: 5-BUTYL-2-PYRIDINECARBOXYLIC ACID * FUSARIC ACID

THR: A poison via ingestion, intraperitoneal, and intravenous routes. When heated to decomposition, it emits toxic fumes of NO_x.

1-BUTYL-2',6'-PIPECOLOXY-
LIDIDE   HR: 3
CAS: 2180-92-9   NIOSH: TK 6060000
mf: $C_{18}H_{28}N_2O$ mw: 288.48

SYN: DL-BUPIVACAINE

THR: A poison via subcutaneous and intravenous routes. Affects the human central nervous system via intravenous route. When heated to decomposition, it emits toxic fumes of NO_x.

BUTYL PROPANOATE   HR: 2
CAS: 590-01-2   NIOSH: UE 8245000
mf: $C_7H_{14}O_2$ mw: 130.2

PROP: Water white liquid, apple-like odor; mp: −89.6°, bp: 145.4°, flash p: 90°F, d: 0.875 @ 20°, autoign temp: 800°F, vap d: 4.49.

SYNS: BUTYL PROPIONATE * N-BUTYL PROPIONATE * PROPANOIC ACID, BUTYLESTER (9CI)

THR: Mildly toxic by ingestion. A mild irritant. See also esters, n-butyl alcohol and propionic acid. Dangerously flammable when exposed to heat or flame. To fight fire, use foam, CO_2, dry chemical. Incompatible with oxidizing materials.

BUTYL-2-PROPENOATE   HR: 3
CAS: 141-32-2   NIOSH: UD 3150000
mf: $C_7H_{12}O_2$ mw: 128.19

PROP: Water white, extremely reactive monomer; bp: 69° @ 50 mm, fp: −64.6°, flash p: 120°F (OC), d: 0.89 @ 25°/25°, vap press: 10 mm @ 35.5°, vap d: 4.42.

SYNS: ACRYLIC ACID-N-BUTYL ESTER * N-BUTYL ACRYLATE
ACGIH TLV: TWA 10 ppm

THR: Poison. Moderate toxicity via ingestion, inhalation, intraperitoneal, and skin routes. A skin and eye irritant. Moderately flammable when exposed to heat or flame. To fight fire, use foam, CO_2, dry chemical. When heated to decomposition, it emits acrid and irritating fumes. Incompatible with oxidizing materials.

4-tert-BUTYLPYROCATECHOL   HR: 3
CAS: 98-29-3   NIOSH: UX 1400000
mf: $C_{10}H_{14}O_2$ mw: 166.24

PROP: fp: 52°; flash p: 265°F; bp: 285°; d: 1.049 @ 60°/25°.

SYN: 4-6-BUTYLCATECHOL

THR: A poison via intravenous route. Moderately toxic by ingestion and skin absorption. A human skin irritant. A skin and eye irritant. Slightly flammable. To fight fires, use CO_2, dry chemical, fog, mist. When heated to decomposition, it emits acrid and irritating fumes.

BUTYL STANNOIC ACID   HR: 3
CAS: 2273-43-0   NIOSH: WH 6770000
mf: $C_4H_{10}O_2Sn$ mw: 208.83

SYN: BUTYLHYDROXYOXOSTANNANE

THR: A poison via intravenous route. See also tin compounds. When heated to decomposition, it emits acrid and irritating fumes.

1-BUTYL-3-SULFANILYL UREA   HR: 3
CAS: 339-43-5   NIOSH: YS 4200000
mf: $C_{11}H_{17}N_3O_3S$ mw: 271.37

SYNS: N-(4-AMINOBENZENESULFONYL)-N'-
BUTYLUREA * N'-(BUTYLCARBAMOYL)SULFA-
NILAMIDE * N(SUP 1)-SULFANILYL-N(SUP
2)-BUTYLCARBAMIDE * N(SUP 1)-SULFA-
NILYL-N(SUP 2)-BUTYLUREA

THR: A poison via intraperitoneal route. An
experimental teratogen. When heated to decomposition, it emits very toxic fumes of NO_x and
SO_x.

n-BUTYL THIOCYANATE   HR: 3
CAS: 628-83-1   NIOSH: XK 8500000
mf: C_5H_9NS mw: 115.21

SYNS: 1-THIOCYANOBUTANE * N-BUTYL RHO-
DANATE * BUTYRHODANID (GERMAN)

THR: A poison via ingestion and subcutaneous
route. When heated to decomposition, it emits
very toxic fumes of NO_x and SO_x.

n-BUTYL THIOUREA   HR: 3
CAS: 1516-32-1   NIOSH: YS 4375000
mf: $C_5H_{12}N_2S$ mw: 132.25

SYN: USAF D-5

THR: A poison via ingestion and intraperitoneal
routes. When heated to decomposition, it emits
very toxic fumes of NO_x and SO_x.

BUTYL TITANATE   HR: 3
CAS: 5593-70-4   NIOSH: XR 1585000
mf: $C_{16}H_{36}O_4 \cdot Ti$ mw: 340.42

PROP: Colorless to light yellow liquid, odor
of butanol; mp: $-55°$, bp: $312°$, flash p: $170°F$,
vap d: 11.5.

SYN: TETRABUTYLTITANATE (CZECH)

THR: A poison via intravenous route. Moderately toxic via ingestion. See butyl alcohol and
titanium compounds. Moderately flammable
when exposed to heat or flame. To fight fire,
use water, spray, foam, dry chemical. When
heated to decomposition, it emits acrid and irritating fumes. Incompatible with oxidizing materials.

p-tert-BUTYLTOLUENE   HR: 3
CAS: 98-51-1   NIOSH: XS 8400000
mf: $C_{11}H_{16}$ mw: 148.27

PROP: Colorless liquid.

SYNS: p-METHYL-tert-BUTYLBENZENE
* 1-METHYL-4-tert-BUTYLBENZENE

OSHA PEL: TWA 10 ppm
ACGIH TLV: TWA 10 ppm; STEL 20 ppm
DFG MAK: 10 ppm (60 mg/m^3)

THR: A poison via inhalation. Moderately toxic
via ingestion. A skin and eye irritant. A human
irritant which affects the central nervous system.
Inhalation of vapors causes irritation of lungs
and depression of central nervous system. Prolonged exposure may result in damage to liver
and kidneys. Moderately flammable when exposed to heat or flame. Dangerous; when heated,
emits highly toxic, acrid, and irritating fumes;
can react with oxidizing materials.

1-BUTYL-3-(p-TOLYL SULFONYL)-
UREA   HR: 3
CAS: 64-77-7   NIOSH: YS 4550000
mf: $C_{12}H_{18}N_2O_3S$ mw: 270.38

SYNS: 1-BUTYL-3-(P-METHYLPHENYLSULFO-
NYL)UREA * N-BUTYL-N'-P-TOLUENESULFO-
NYLUREA * N-N-BUTYL-N'-TOSYLUREA
* 1-BUTYL-3-TOSYLUREA * NCI-CO1763
* ORINASE * N-(SULFONYL-P-METHYLBEN-
ZENE)-N'-N-BUTYLUREA * TOLBUTAMIDE
* 1-P-TOLUENESULFONYL-3-BUTYLUREA
* N-(P-TOLYLSULFONYL)-N'-BUTYLCARBAMIDE
* 3-(P-TOLYL-4-SULFONYL)-1-BUTYLUREA

THR: An experimental teratogen in women. Mutagenic data. Moderate toxicity by ingestion,
intraperitoneal, and subcutaneous routes. Implicated in aplastic anemia. When heated to decomposition, it emits very toxic fumes of NO_x and
SO_x.

BUTYL-2,4,5-TRICHLORO PHENOXY
ACETATE   HR: 3
CAS: 93-79-8   NIOSH: AJ 8485000
mf: $C_{12}H_{13}Cl_3O_3$ mw: 311.60

SYNS: N-BUTYLESTER KYSELINI 2,4,5-TRI-
CHLORFENOXYOCTOVE (CZECH) * N-BUTYL
(2,4,5-TRICHLOROPHENOXY)ACETATE * 2,4,5-
T-N-BUTYL ESTER * 2,4,5-TRICHLOROPHEN-
OXYACETIC ACID, BUTYL ESTER

THR: An experimental teratogen. A skin and
eye irritant. Moderately toxic via ingestion. See
also esters. When heated to decomposition, it
emits toxic fumes of Cl^-.

BUTYL TRICHLORO SILANE   HR: 3
CAS: 7521-80-4   NIOSH: VV 2080000

DOT: 1747
mf: $C_4H_9Cl_3Si$　　mw: 191.57

PROP: Liquid; vap d: 6.4; flash p: 130°F (OC); d: 1.2.
DOT Classification: Label: Corrosive

THR: A corrosive poison. See also chlorosilanes. Moderately flammable via heat, flame (sparks), oxidizers. To fight fire, use water to blanket fire, fog, mist, dry chemical, alcohol foam. Dangerous; when heated to decomposition, it emits highly toxic fumes of chlorides. Incompatible with water or steam to produce heat and toxic and corrosive fumes.

BUTYL TRICHLORO STANNANE　HR: 2
CAS: 1118-46-3　　　NIOSH: WH 6780000
mf: $C_4H_9Cl_3Sn$　　mw: 282.17

SYN: CHLORID-N-BUTYLCINICITY (CZECH)

THR: Moderately toxic via ingestion. A skin and eye irritant. See also tin compounds. When heated to decomposition, it emits toxic fumes of Cl^-.

N-BUTYLUREA　　　　　　　　　HR: 2
CAS: 592-31-4　　　NIOSH: YS 3675000
mf: $C_5H_{12}N_2O$　　mw: 116.19

SYN: NCI-CO2131

THR: Moderately toxic via parenteral route. Mutagenic data. When heated to decomposition, it emits toxic fumes of NO_x.

BUTYL VINYL ETHER
mf: $C_6H_{12}O$　　mw: 100.2

PROP: Liquid; mp: −92°, bp: 93.3°, flash p: −1°, d: 0.77, vap d: 3.4.

THR: Unknown toxicity. See also ethers. Moderately explosive by spontaneous chemical reaction. See also ethers. Dangerously flammable. To fight fire, use alcohol foam.

2-BUTYNE-1,4-DIOL　　　　　　HR: 3
CAS: 110-65-6　　　NIOSH: ES 0525000
mf: $C_4H_6O_2$　　mw: 86.10

PROP: Straw to amber crystals; mp: 57.5°, bp: 194° @ 100 mm.

SYN: 1,4-BUTYNEDIOL

THR: A poison via ingestion. Skin sensitizer upon long or repeated contact. Moderately explosive. When heated to decomposition, it emits

acrid fumes and may explode. Incompatible with heat or spontaneous chemical reaction in contact with certain materials; i.e., mercury salts; strong acids and alkali earth hydroxides; halides at high temperatures.

N-BUTYRALDEHYDE　　　　　　HR: 2
CAS: 123-72-8　　　NIOSH: ES 2275000
DOT: 1129
mf: C_4H_8O　　mw: 72.12

PROP: Colorless liquid. Mp: −100°, bp: 74.7°, flash p: 20°F (CC), (−6°C), d: .902 @ 20°/4°, autoign temp: 446°F, lel = 2.5%, uel = 12.5%, vap d: 2.5.

SYNS: ALDEHYDE BUTYRIQUE (FRENCH) * ALDEIDE BUTIRRICA (ITALIAN) * N-BUTANAL (CZECH) * BUTYRALDEHYD (GERMAN) * BUTYRALDEHYDE (CZECH) * N-BUTYL ALDEHYDE * BUTYRAL * BUTYRIC ALDEHYDE * NCI-C56291

DOT Classification: Label: Flammable Liquid

THR: Moderate toxicity via ingestion and dermal subcutaneous routes. Severe skin and eye irritant. Powerful inhalant irritant in humans. See also aldehydes. Dangerously flammable when exposed to heat or flame. To fight fire, use foam, CO_2, dry chemical. Incompatible with oxidizing materials; reacts vigorously with chlorosulfonic acid; HNO_3; oleum; H_2SO_4.

m-BUTYRALDEHYDE OXIME　　HR: 3
CAS: 110-69-0　　　NIOSH: ES 3500000
mf: C_4H_9NO　　mw: 87.14

PROP: Liquid; mp: −29.5°, bp: 152°, flash p: 136°F (CC), d: 0.923, vap d: 3.01.

SYNS: BUTANAL OXIME * N-BUTYRALDOXIME * USAF AM-6

THR: A poison via intraperitoneal route. Moderately flammable when exposed to heat or flame. To fight fire, use alcohol foam, dry chemical. When heated to decomposition, it emits toxic fumes of NO_x. Incompatible with oxidizing materials.

BUTYRALDOXIME　　　　　　　HR: 3
mf: C_4H_9NO　　mw: 87.1

PROP: Liquid. Mp: −29.5°, bp: 152°, flash p: 136°F (CC), d: 0.923, vap d: 3.01.

THR: A poison via intraperitoneal route. Moderately flammable when exposed to heat or flame.

Highly explosive. Can explode during vacuum distillation. Incompatible with oxidizing materials; metallic impurities. To fight fire, use alcohol foam, dry chemical.

N-BUTYRIC ACID   HR: 2
CAS: 107-92-6   NIOSH: ES 5425000
DOT: 2820
mf: $C_4H_8O_2$ mw: 88.12

PROP: Liquid. Mp: $-7.9°$, bp: $163.5°$, flash p: $161°F$, fp: $-5.5°$, d: 0.9590 @ $20°/20°$, autoign temp: $846°F$, vap press: 0.43 mm @ $20°$, vap d: 3.04, lel = 2.0%, uel = 10.0%.

SYNS: BUTANOIC ACID * BUTTERSAEURE (GERMAN) * ETHYLACETIC ACID * 1-PROPANECARBOXYLIC ACID * PROPYLFORMIC ACID

DOT Classification: Label: Corrosive

THR: Moderately toxic by ingestion, dermal, subcutaneous, intraperitoneal, and intravenous routes. Mutagenic data. Severe skin and eye irritant. A corrosive material. A synthetic flavoring substance and adjuvant. Incompatible with chromium trioxide. Moderately flammable when exposed to heat or flame; can react with oxidizing materials. To fight fire, use alcohol foam, CO_2, dry chemical. When heated to decomposition, it emits acrid and irritating fumes. For further information, see Vol. 2, No. 3 of *DPIM Report*.

beta-BUTYROLACTONE   HR: 3
CAS: 3068-88-0   NIOSH: RQ 8050000
mf: $C_4H_6O_2$ mw: 86.10

SYNS: 3-HYDROXYBUTANOIC ACID, BETA-LACTONE * HYDROXYBUTYRIC ACID LACTONE * 4-METHYL-2-OXETANONE

THR: An experimental carcinogen and neoplastigen. A moderate skin irritant. Low acute oral toxicity. Mutagenic data. When heated to decomposition, it emits acrid and irritating fumes.

BUTYRONITRILE   HR: 3
CAS: 109-74-0   NIOSH: ET 8750000
mf: C_4H_7N mw: 69.12

PROP: Colorless liquid, slightly sol in water, sol in alcohol and ether. d: 0.796 @ $15°$, mp: $-112.6°$, bp: $117°$, flash p: $79°F$ (OC).

SYNS: BUTANENITRILE * PROPYL CYANIDE

THR: A poison via ingestion and dermal routes. See nitriles and cyanide. A skin irritant. Dangerously flammable when exposed to heat, flame, or oxidizers. See also nitriles. To fight fire, use alcohol foam. Incompatible with heat; flame; oxidizers.

1-n-BUTYRYLAZIRIDINE   HR: 2
CAS: 10431-86-4   NIOSH: CM 6475000
mf: $C_6H_{11}NO$ mw: 113.18

SYNS: BUTYRYLETHYLENIMINE * BUTYRYLETHYLENEIMINE

THR: Moderate toxicity via intraperitoneal route. Mutagenic data. When heated to decomposition it emits toxic fumes of NO_x.

BUTYRYL CHLORIDE   HR: 3
mf: C_4H_7ClO mw: 106.51

PROP: Clear, colorless liquid with sharp odor. Mp: $-89°$, bp: $101°$, d: 1.028 @ $20°/20°$, vap d: 3.67, flash p: $<21°$.

THR: An irritant poison to skin, eyes, and mucous membranes. Dangerous; when heated to decomposition, it emits highly toxic fumes of Cl^-. Incompatible with water or steam to produce toxic and corrosive fumes; oxidizing materials.

BUX-TEN   HR: 3
CAS: 8065-36-9   NIOSH: FC 3510000

PROP: A low-melting amber solid, very sol in xylene, ethanol, nearly insol in water; mp: $26.4°$.

SYNS: METHYLCARBAMIC ACID-m-(1-METHYL)-BUTYL)PHENYL ESTER MIXED WITH CARBAMIC ACID, METHYL-m-(1-ETHYLPROPYL)PHENYL ESTER (3:1) * ORTHO 5353

THR: A poison via ingestion. Moderate toxicity via skin contact. See also esters. When heated to decomposition, it emits toxic fumes of NO_x.

C

CACODYL   HR: 3
mf: $(CH_3)_2As\text{-}As(CH_3)_2$ mw: 210.0

PROP: bp: 165°; fp: −6°; d: 1.15.

SYN: TETRAMETHYL DIARSYL

THR: Poison by ingestion and inhalation. See also arsenic compounds. Dangerous fire hazard by spontaneous chemical reaction. Ignites spontaneously in dry air. Dangerous; see also arsenic, can react vigorously with oxidizing materials, i.e., air; Cl_2.

CACODYL SULFIDE   HR: 3
mf: $((CH_3)_2As)_2S$ mw: 242

PROP: Oily liquid; slightly sol in water. bp: 211°.

SYN: DICACODYL SULFIDE

THR: Poison by ingestion and inhalation. See also arsenic compounds and sulfides. Dangerous fire hazard when exposed to heat or by spontaneous chemical reaction, i.e. in air. Dangerous; see also arsenic and oxides of sulfur; can react vigorously with oxidizing materials.

CADIA DEL PERRO   HR: 3
                          NIOSH: EU 9500000

PROP: Aqueous extract from the dried leaves of the plant (JNCIAM 46,1131,71)

SYN: KRAMERIA IXINA

THR: An experimental carcinogen and neoplastigen. When heated to decomposition it emits acrid fumes.

CADMIUM   HR: 3
CAS: 7440-43-9   NIOSH: EU 9800000
mf: Cd   mw: 112.40

PROP: Hexagonal crystals, silver-white malleable metal mp: 320.9°, bp: 767 ±2°, d: 8.642, vap press: 1 mm @ 394°.

SYNS: C.I. 77180 * KADMIUM (GERMAN)

OSHA PEL: TWA 200 ug/m³; CL 600
ACGIH TLV: TWA (Dusts and salts) 0.05 mg/m³; BAT: Blood 1.5 ug/dl; Urine 15 ug/L

THR: Poison to humans by inhalation and other routes. Poison by ingestion, intraperitoneal, subcutaneous, intramuscular, and intravenous routes. See also cadmium compounds. Dust is moderately flammable and explosive when exposed to heat, flame, or by chemical reaction with oxidizing agents; metals; HN_3; Zn; Se; Te. For further information, see Vol. 3, No. 6 of *DPIM Report*.

CADMIUM (II) ACETATE   HR: 3
CAS: 543-90-8   NIOSH: EU 9810000
mf: $C_2H_4O_2 \cdot 1/2Cd$ mw: 116.25

PROP: Monoclinic colorless crystals, odor of acetic acid; mp: 256°, bp: decomp, d: 2.341.

SYNS: BIS(ACETOXY)CADMIUM * CADMIUM DIACETATE * C.I. 77185

THR: Poison by intraperitoneal route. Mutagenic data. See cadmium compounds. When heated to decomposition it emits toxic fumes of cadmium. For further information, see Vol. 4, No. 4 of *DPIM Report*.

CADMIUM CAPRYLATE   HR: 3
CAS: 2191-10-8   NIOSH: RH 0370000
mf: $C_{16}H_{30}O_4 \cdot Cd$ mw: 398.86

SYN: OCTANOIC ACID, CADMIUM SALT (2:1)

THR: Poison by ingestion and intratracheal routes. See cadmium compounds. When heated to decomposition it emits toxic fumes of cadmium.

CADMIUM CHLORATE   HR: 3
mf: $CdCl_2O_6$ mw: 279.31

PROP: Colorless, deliquescent prisms; mp: 80°; d: 2.28 @ 18°.

THR: See cadmium compounds and chlorates for toxicity data. Moderate fire hazard by chemical reaction with reducing agents. A powerful oxidizing agent. Reacts violently with Sb_2S_3; As_2S_3; CuS; SnS_2; SnS. Moderate explosion hazard when shocked or exposed to heat. See also chlorates.

CADMIUM CHLORIDE   HR: 3
CAS: 10108-64-2   NIOSH: EV 0175000
mf: $CdCl_2$ mw: 183.30

PROP: Hexagonal, colorless crystals; mp: 568°, d: 4.047 @ 25°, vap press: 10 mm @ 656°, bp: 960°.

SYNS: CADDY * CADMIUM DICHLORIDE * KADMIUMCHLORID (GERMAN)

OSHA PEL: TWA 200 ug(Cd)/m^3; CL 600

THR: Poison by ingestion, intraperitoneal, sub-cutaneous, inhalation, and intravenous routes. An experimental carcinogen and teratogen. Mutagenic data. See also cadmium compounds. Reacts violently with BrF$_3$; K. When heated to decomposition it emits very toxic fumes of Cd and Cl$^-$. For further information, see Vol. 2, No. 3 of *DPIM Report*.

CADMIUM COMPOUNDS   HR: 3
NIOSH: EV 0260000

THR: Poison by ingestion, however the irritating and emetic action is so violent that little of the cadmium has time to be absorbed, and fatal poisoning rarely ensues. Cases of human poisoning have been reported from ingestion of food or beverages prepared or stored in cadmium-plated containers. Inhalation of fumes or dusts affects the respiratory tract and the kidneys. Brief exposure to high concentrations may result in pulmonary edema and death. Fatal concentrations may be breathed without sufficient discomfort to warn a workman to leave the exposure. Cadmium oxide fumes can cause metal fume fever resembling that caused by zinc oxide fumes. A teratogen. An experimental carcinogen.

CADMIUM FLUOBORATE   HR: 3
CAS: 14486-19-2   NIOSH: EV 0525000
mf: B$_2$CdF$_8$ mw: 286.02

SYN: FLUOROBORATE

THR: Poison by ingestion. Moderately toxic by inhalation. See tetrafluoborate. When heated to decomposition it emits very toxic fumes of Cd and F$^-$. For further information, see Fluoroborate Vol. 2, No. 3 of *DPIM Report*.

CADMIUM FLUORIDE   HR: 3
CAS: 7790-79-6   NIOSH: EV 0700000
mf: CdF$_2$ mw: 150.40

PROP: Cubic white crystals. Mp: 1100°, bp: 1758°, d: 6.64, vap press: 1 mm @ 1112°.

SYN: CADMIUM FLUORURE (FRENCH)

OSHA PEL: TWA 200 ug(Cd)/m^3; CL 600

THR: Poison by subcutaneous route. Violent reaction with K. See also fluorides and cadmium

compounds. When heated to decomposition it emits very toxic fumes of Cd and F$^-$.

CADMIUM FLUOSILICATE   HR: 3
CAS: 17010-21-8   NIOSH: EV 0875000
mf: CdF$_6$Si   mw: 254.49

PROP: Hexagonal, colorless crystals.

SYN: TL 1070

THR: Poison by ingestion and inhalation. See also cadmium compounds and hexafluorosilicate(2-)dihydrogen. When heated to decomposition it emits very toxic fumes of Cd and F$^-$.

CADMIUM (II) NITRATE TETRAHY-
DRATE (1:2:4)   HR: 3
CAS: 10022-68-1   NIOSH: EV 1850000
mf: N$_2$O$_6$ • Cd • 4H$_2$O   mw: 308.50

SYN: DUSICNAN KADEMNATY (CZECH)

THR: Poison by ingestion. A skin and eye irritant. See also cadmium compounds and nitrates. When heated to decomposition it emits very toxic fumes of Cd and NO$_x$. For further information see Vol. 2, No. 4 of *DPIM Report*.

CADMIUM OXIDE   HR: 3
CAS: 1306-19-0   NIOSH: EV 1925000
mf: CdO   mw: 128.40

PROP: (1) amorphous, brown crystals; (2) cubic, brown crystals. Mp (1): <1426°, mp(2): decomp @ 950°, bp: 1559°, d(1): 6.95, d(2): 8.15, vap press: 1 mm @ 1000°.

SYNS: KADMU TLENEK (POLISH) * NCI-c02551

OSHA PEL: TWA 200 ug(Cd)/m^3; CL 600
ACGIH TLV: TWA Production 0.05 mg/m^3

THR: Poison by ingestion and inhalation. An experimental carcinogen. Systemic poison. When heated to decomposition it emits toxic fumes of Cd. Incompatible with magnesium.

CADMIUM OXIDE FUME   HR: 3
CAS: 1306-19-0   NIOSH: EV 1930000

SYN: CADMIUM FUME

OSHA PEL: TWA 110 ug/m^3
ACGIH TLV: (fume) TWA ceiling limit 0.05 mg/m^3

THR: Poison by inhalation. Pulmonary system effects. Strong irritant via inhalation. When heated to decomposition it emits toxic fumes of Cd. For further information see Vol. 4, No. 4 of *DPIM Report*.

CADMIUM SULFATE (1:1)   HR: 3
CAS: 10124-36-4   NIOSH: EV 2700000
mf: $O_4S \cdot Cd$ mw: 208.46

PROP: Rhombic white crystals. mp: 1000°, d: 4.691.

SYN: SULPHURIC ACID, CADMIUM SALT (1:1)

THR: An experimental carcinogen and teratogen. Mutagenic data. See also cadmium compounds. When heated to decomposition it emits very toxic fumes of Cd and SO_x. For further information, see Vol. 2, No. 4 of *DPIM Report*.

CADMIUM SULFATE (1:1) HYDRATE (3:8)   HR: 3
CAS: 7790-84-3   NIOSH: EV 2850000
mf: $O_4S \cdot Cd \cdot 2.66H_2O$ mw: 256.51

SYNS: CADMIUM SULFATE OCTAHYDRATE * CADMIUM SULPHATE

THR: Poison by subcutaneous route. An experimental carcinogen. Mutagenic data. See also cadmium compounds. When heated to decomposition it emits very toxic fumes of Cd and SO_x.

CADMIUM SULFATE TETRAHYDRATE   HR: 3
CAS: 13477-21-9   NIOSH: EV 2900000
mf: $O_4S \cdot Cd \cdot 4H_2O$ mw: 280.54

THR: An experimental carcinogen. See also cadmium compounds. When heated to decomposition it emits very toxic fumes of Cd and SO_x.

CADMIUM SULFIDE   HR: 3
CAS: 1306-23-6   NIOSH: EV 3150000
mf: CdS   mw: 144.46

PROP: Hexagonal, yellow-orange crystals. Mp: 1750 @ 100 atm, bp: subl in N_2, d: 4.82.

SYNS: CADMIUM SULPHIDE * C.I. 77199 * NCI-C02711

THR: An experimental carcinogen. Mutagenic data. See also cadmium compounds and sulfides. When heated to decomposition it emits very toxic fumes of Cd and SO_x.

CAFFEINE   HR: 3
CAS: 58-08-2   NIOSH: EV 6475000
mf: $C_8H_{10}N_4O_2$ mw: 194.22

PROP: White, fleecy masses; mp: 236.8°.

SYNS: CAFFEIN * COFFEINE * METHYLTHEOBROMIDE * NCI-C02733 * THEINE * 1,3,7-TRIMETHYL-2,6-DIOXOPURINE * 1,3,7-TRIMETHYLXANTHINE

THR: Poison by ingestion and intravenous routes. An experimental teratogen and carcinogen. Central nervous system effects. Mutagenic data. Implicated in increased fetal losses. A general purpose food additive, large doses (above 1.0 g) cause palpitation, excitement, insomnia, dizziness, headache, and vomiting. Continued excessive use of caffeine in tea or coffee may lead to digestive disturbances, constipation, palpitations, shortness of breath, and depressed mental states. It is also implicated in cardiac disorders under those conditions. When heated to decomposition it emits toxic fumes of NO_x. For further information, see Vol. 1, No. 1 of *DPIM Report*.

CAFFEINE HYDROBROMIDE   HR: 3
CAS: 5743-18-0   NIOSH: EV 6599500
mf: $C_8H_{10}N_4O_2 \cdot BrH$ mw: 275.14

SYNS: CAFFEINE BROMIDE * 3,7-DIHYDRO-1,3,7-TRIMETHYL-1H-PURINE-2,6-DIONE, MONOHYDROBROMIDE

THR: Poison by ingestion, subcutaneous, and intravenous routes. See also caffeine and bromides. When heated to decomposition it emits very toxic fumes of NO_x and HBr.

CAJEPUTOL   HR: 3
CAS: 470-82-6   NIOSH: OS 9275000
mf: $C_{10}H_{18}O$ mw: 154.28

SYNS: 1,8-CINEOLE * 1,8-EPOXY-P-MENTHANE * EUCALYPTOL * LIMONENE OXIDE * NCI-C56575 * 1,8-OXIDO-P-MENTHANE

THR: Poison by subcutaneous and intramuscular routes. An experimental carcinogen. When heated to decomposition it emits acrid smoke.

CALCIUM   HR: 2
CAS: 7440-70-2   NIOSH: EV 8040000
DOT: 1401/1855
af: Ca   aw: 40.08

PROP: Silver-white, soft metal. Mp: 842°, bp: 1484°, d: 1.54 @ 20°, vap press: 10 mm @ 983°.

SYNS: CALCIUM, METAL, CRYSTALLINE (DOT) * CALCIUM, METAL (DOT)

DOT Classification: Label: Flammable Solid and Dangerous When Wet

THR: See also calcium compounds. Moderate fire hazard when heated or in intimate contact with moisture or acids, evolves hydrogen. See also hydrogen. Moderate explosion hazard in intimate contact with very powerful oxidizing agents; i.e., Cl_2; ClF_3; F_2; O_2; Si; S; V_2O_3. Reacts with moisture or acids to liberate large quantities of hydrogen; can develop explosive pressure in containers. To fight fire, use special mixtures of dry chemical. Incompatible with air; asbestos cement; halogens; lead dichloride; phosphorus (V) oxide; silicon; sodium; mixed oxides; sulfur; water.

CALCIUM ACETARSONE   HR: 3
CAS: 64046-96-4   NIOSH: CF 8575000
mf: $C_8H_{10}AsNO_5 \cdot 7Ca$ mw: 555.67

SYN: N-ACETYL-4-HYDROXY-M-ARSANILIC ACID, CALCIUM SALT

OSHA PEL: TWA 500 ug(As)/m^3

THR: Poison by ingestion. See also arsenic compounds. When heated to decomposition it emits very toxic fumes of As and NO_x.

CALCIUM ACETATE   HR: 3
CAS: 62-54-4   NIOSH: AF 7525000
mf: $C_4H_6O_4 \cdot Ca$ mw: 158.18

SYNS: BROWN ACETATE * CALCIUM DIACE-TATE * GRAY ACETATE * LIME ACETATE * LIME PYROLIGNITE * SORBO-CALCIAN * VINEGAR SALTS

THR: Poison by intravenous route. See also calcium compounds. When heated to decomposition it emits acrid fumes.

CALCIUM ARSENITE   HR: 3
CAS: 27152-57-4   NIOSH: CG 3380000
DOT: 1574
mf: $AsCaHO_3$ mw: 164.01

PROP: White, granular powder.

SYNS: CALCIUM ARSENITE, SOLID (DOT) * MONOCALCIUM ARSENITE

DOT Classification: Poison B, Label: Poison

THR: A poison via inhalation and ingestion. See also arsenic compounds.

CALCIUM-o-BENZOSULFIMIDE   HR: 2
CAS: 6485-34-3   NIOSH: DE 4250000
mf: $C_{14}H_{10}N_2O_6S_2 \cdot Ca$ mw: 406.46

PROP: White crystalline powder, odorless or faint aromatic odor, sol in water.

SYNS: 1,2-BENZISOTHIAZOL-3(2H)-ONE, 1,1-DI-OXIDE, CALCIUM SALT * CALCIUM-O-BENZO-SULPHIMIDE * CALCIUM SACCHARIN * CAL-CIUM SACCHARINA * CALCIUM SACCHARINATE * SULPHOBENZOIC IMIDE CALCIUM SALT

THR: See saccharin. Dangerous disaster hazard; see also sulfates, NO_x. An experimental carcinogen.

CALCIUM BISULFITE (solution)   HR: 3
CAS: 13780-03-5   NIOSH: EV 9290000
DOT: 2693/1923

PROP: Colorless or slightly yellowish liquid, strong sulfur dioxide odor, d: 1.06.

SYNS: CALCIUM BISULFITE SOLUTION (DOT) * CALCIUM HYDROGEN SULFITE SOLUTION (DOT)

DOT Classification: Label: Corrosive

THR: A poison by ingestion. Strong irritant via skin contact, ingestion, and inhalation routes. See also sulfites and sulfurous acid.

CALCIUM CARBIDE   HR: 3
CAS: 75-20-7   NIOSH: EV 9400000
DOT: 1402
mf: C_2Ca mw: 64.10

PROP: Rhombic, gray crystals. Mp: approx 2300°, d: 2.222.
DOT Classification: Label: Flammable Solid and Dangerous When Wet

THR: A poison via ingestion, inhalation routes. Acetylene is evolved when calcium carbide is in contact with moisture. See also calcium hydroxide and acetylene. Moderate fire hazard on contact with moisture, acid or acid fumes, evolves heat or flammable vapors. It incandesces with HCl gas; PbF_2; Mg. Incompatible with Se; (KOH + Cl_2); $AgNO_3$; Na_2O_2; $SnCl_2$; S; water. Moderate explosion hazard, see acety-lene. For further information, see Vol. 2, No. 1 of *DPIM* Report.

CALCIUM (II) CARBONATE (1:1)   HR: 1
CAS: 1317-65-3   NIOSH: EV 9580000
mf: $CO_3 \cdot Ca$ mw: 100.09

PROP: White powder; mp: $825°(\alpha)$; $1339°$ (β) @ 102.5 atm, d: 2.7-2.95.

SYNS: CHALK * DOLOMITE * LIMESTONE * MARBLE * NATURAL CALCIUM CARBONATE * PORTLAND STONE * SOHNHOFEN STONE ACGIH TLV: TWA (marble) 10 mg/m^3

THR: See also calcium compounds. A nutrient and/or dietary supplement food additive, a general-purpose food additive. Calcium carbonate is a common air contaminant. Incompatible with F_2; $(Mg + H_2)$; fluorine.

CALCIUM CHLORATE   HR: 2
CAS: 10137-74-3   NIOSH: FN 9800000
DOT: 1452/2429
mf: $Cl_2O_6 \cdot Ca \cdot 2H_2O$ mw: 243.03

PROP: Monoclinic, white-yellowish, deliquescent crystals. mp: $-H_2O$ @ $>100°$, d: 2.711.

SYNS: CALCIUM CHLORATE (DOT) * CHLORATE DE CALCIUM (FRENCH)

DOT Classification: Label: Oxidizer

THR: Moderately toxic by ingestion and intraperitoneal routes. A powerful oxidant. See chlorates for fire, disaster, and explosion hazard. Incompatible with Al; As; C; Cu; charcoal; MnO_2; metal sulfides; S; dibasic organic acids; organic matter; P.

CALCIUM CHLORIDE   HR: 3
CAS: 10043-52-4   NIOSH: EV 9800000
mf: $CaCl_2$ mw: 110.98

PROP: Cubic, colorless, deliquescent crystals. Mp: 772°, bp: $>1600°$, d: 2.512 @ 25°.

THR: Poison by intravenous, intramuscular, intraperitoneal, and subcutaneous routes. Moderately toxic by ingestion. See also calcium compounds. Reacts violently with $(B_2O_3 + CaO)$; BrF_3. When heated to decomposition it emits toxic fumes of Cl^-. Incompatible with methyl vinyl ether; water; zinc. For further information, see Vol. 2, No. 1 of *DPIM Report*.

CALCIUM CHLORITE   HR: 3
CAS: 14674-72-7   NIOSH: EV 9850000
DOT: 1453
mf: $CaCl_2O_4$ mw: 174.98

PROP: White solid.
DOT Classification: Label: Oxidizer

THR: See chlorites for toxicity and hazard data. With Cl_2 yields explosive ClO_2.

CALCIUM CHROMATE   HR: 3
CAS: 13765-19-0   NIOSH: GB 2750000
DOT: 9096
mf: $CrO_4 \cdot Ca$ mw: 156.08

PROP: Monoclinic prisms, yellow color.

SYNS: CHROMIC ACID, CALCIUM SALT (1:1) * CALCIUM CHROMATE ($CaCrO_4$) * CALCIUM CHROME YELLOW * CALCIUM CHROMIUM OXIDE ($CaCrO_4$) * CALCIUM MONOCHROMATE * C.I. PIGMENT YELLOW 33 * GELBIN

TRK: 0.1 mg/m^3 calculated as CrO_3 in that portion of dust that can possibly be inhaled.
DOT Classification: ORM-E oxidizer

THR: A human carcinogen and neoplastigen. Mutagenic data. See also chromium compounds. A powerful oxidizer.

CALCIUM CHROMATE (VI) DIHYDRATE   HR: 3
CAS: 8012-75-7   NIOSH: GB 2800000
mf: $CrO_4 \cdot Ca \cdot 2H_2O$ mw: 192.12

SYNS: CHROMIC ACID, CALCIUM SALT (1:1), DIHYDRATE * CALCIUM CHROME YELLOW * C.I. 77223 * C.I. PIGMENT YELLOW 33 * GELBIN YELLOW ULTRAMARINE * PIGMENT YELLOW 33 * STEINBUHL YELLOW

OSHA PEL: TWA CL 100 ug(CrO_3)/m^3 (skin)

THR: Poison by ingestion and implant routes. An experimental carcinogen. See also chromium compounds. A powerful oxidizer.

CALCIUM COMPOUNDS
THR: Variable toxicity. The fumes evolved by burning calcium in air are composed of calcium oxide (quick lime) which are irritant to the skin, eyes, and mucous membranes. Generally speaking, calcium compounds should be considered toxic only when they contain toxic components (such as arsenic, etc.) or as calcium oxide or hydroxide. Calcium compounds are common air contaminants.

CALCIUM CYANAMIDE   HR: 3
CAS: 156-62-7   NIOSH: GS 6000000
DOT: 1403

PROP: Hexagonal, rhombohedral, colorless crystals; mp: 1300°, subl > 1500°. Compound not hydrated; compound contains more than 0.1% calcium (FEREAC 41,15972,76)

SYNS: CYANAMIDE, CALCIUM SALT (1:1) * CALCIUM CARBIMIDE * CALCIUM CYANAMID * CYANAMIDE * CYANAMIDE CALCIQUE (FRENCH) * LIME-NITROGEN * NCI-C02937 * NITROLIME * USAF CY-2

ACGIH TLV: TWA 0.5 mg/m^3
DOT Classification: Flammable solid; dangerous when wet

THR: A possible carcinogen. Moderately toxic by ingestion. A skin, eye, and mucous membrane irritant. Systemic effects. Calcium cyanamide is not believed to have a cumulative action. The fatal dose, by ingestion, is probably around 20 to 30 g for an adult. It does not have a cyanide effect. See also amides and cyanide for disaster hazard. For further information, see Vol. 2, No. 6 of *DPIM Report*.

CALCIUM CYANIDE   HR: 3
CAS: 592-01-8   NIOSH: EW 0700000
DOT: 1575
mf: C_2CaN_2 mw: 92.12

PROP: Rhombohedral crystals, white powder; mp: decomp > 350°.

SYNS: CALCIUM CYANIDE, SOLID (DOT) * CALCYANIDE * CYANOGAS * CYANURE DE CALCIUM (FRENCH)

DOT Classification: Poison B, Label: Poison

THR: A deadly poison by ingestion and inhalation. See also cyanides. When heated to decomposition it emits toxic fumes of NO_x, CN^-. For further information, see Vol. 2, No. 1 of *DPIM Report*.

CALCIUM CYANIDE (mixture)   HR: 3
CAS: 592-01-8   NIOSH: EW 0710000
DOT: 1575

SYN: CALCIUM CYANIDE MIXTURE, SOLID (DOT)

DOT Classification: Poison B, Label: Poison

THR: A poison. See also cyanides and calcium compounds. When heated to decomposition it emits toxic fumes of NO_x, CN^-.

CALCIUM CYCLOHEXYLSULPHA-
MATE   HR: 3
CAS: 139-06-0   NIOSH: GV 7100000
mf: $C_{12}H_{24}N_2O_6S_2 \cdot Ca$ mw: 396.58

PROP: White crystalline powder, almost odorless, freely sol in water, practically insol in alcohol, benzene, chloroform and ether.

SYNS: CALCIUM CYCLOHEXANESULFAMATE * CALCIUM CYCLOHEXYLSULFAMATE * CYCLAMATE CALCIUM * CYCLOHEXANESULFAMIC ACID, CALCIUM SALT * KALZIUMZYKLAMATE (GERMAN) * SUCARYL CALCIUM

THR: Poison by ingestion and intravenous routes. An experimental carcinogen, and neoplastigen. Mutagenic data. When heated to decomposition it emits very toxic fumes of SO_x and NO_x.

CALCIUM FLUORIDE   HR: 3
CAS: 7789-75-5   NIOSH: EW 1760000
mf: CaF_2 mw: 78.08

PROP: Cubic, colorless crystals, luminous with heat; mp: 1360°, d: 3.180.

SYN: FLUORSPAR

THR: An experimental teratogen. See also fluorides and calcium compounds. Moderately toxic by intraperitoneal route. When heated to decomposition it emits toxic fumes of F^-. For further information, see Vol. 1, No. 8 of *DPIM Report*.

CALCIUM FORMATE   HR: 3
CAS: 544-17-2   NIOSH: LQ 5600000
mf: $C_2H_2O_4 \cdot Ca$ mw: 130.12

SYN: MRAVENCAN VAPENATY (CZECH)

THR: Poison by intravenous route. Moderately toxic by ingestion. Severe eye irritant. When heated to decomposition it emits acrid smoke.

CALCIUM GLUCONATE   HR: 2
CAS: 299-28-5   NIOSH: EW 2100000
mf: $C_{12}H_{22}O_{14} \cdot Ca$ mw: 430.42

PROP: White fluffy powder or granules, odorless, sol in hot water, less sol in cold water, insol in alcohol, acetic acid and other organic solvents. mp: $-H_2O$ @ 120°.

SYN: GLUCONATE DE CALCIUM (FRENCH)

THR: Moderately toxic by intraperitoneal and intravenous routes. Systemic effects due to skin exposure. See also calcium compounds.

CALCIUM HYDROXIDE   HR: 1
CAS: 1305-62-0   NIOSH: EW 2800000
mf: CaH_2O_2 mw: 74.10

PROP: Rhombic, trigonal, colorless crystals.
mp: $-H_2O$ @ 580°, bp: decomp, d: 2.343.

SYNS: CALCIUM HYDRATE * HYDRATED LIME
* LIME WATER * SLAKED LIME

THR: Low toxicity by ingestion. Irritant to skin,
eyes, mucous membranes, and respiratory system. Causes dermatitis. Dust is considered to
be an important industrial hazard. A common
air contaminant. Violent reaction with maleic
anhydride; nitroethane; nitromethane; nitroparaffins; nitropropane; P. See also calcium
compounds. For further information, see Vol.
1, No. 8 of *DPIM Report*.

CALCIUM (II) NITRATE (1:2)   HR: 1
CAS: 10124-37-5   NIOSH: EW 2985000
DOT: 1454
mf: $N_2O_6 \cdot Ca$ mw: 164.1

SYN: CALCIUM NITRATE (DOT)

DOT Classification: Label: Oxidizer

THR: An irritant. See also nitrates and calcium
compounds.

CALCIUM OXIDE   HR: 3
CAS: 1305-78-8   NIOSH: EW 3100000
DOT: 1910
mf: CaO   mw: 56.08

PROP: Cubic, colorless crystals; mp: 2580°,
d: 3.37, bp: 2850°.

SYNS: BURNT LIME * CALCIA * CALX
* LIME * LIME, BURNED * LIME, UNSLAKED
(DOT) * OXYDE DE CALCIUM (FRENCH)
* QUICKLIME * WAPNIOWY TLENEK (POLISH)

OSHA PEL: TWA 5 mg/m³
DFG MAK: 5 mg/m³
DOT Classification: ORM-B

THR: A caustic and irritating material. See also
calcium compounds. A common air contaminant. A powerful caustic to living tissue. Violent
reaction with (B_2O_3 + $CaCl_2$); BF_3; ClF_3; F_2;
HF; P_2O_5; water. Incompatible with hydrogen
fluoride; interhalogens; phosphorus (V) oxide;
water. For further information, see Vol. 2, No.
1 of *DPIM Report*.

CALCIUM-d-PANTOTHENATE   HR: 2
CAS: 137-08-6   NIOSH: RU 4375000
mf: $C_{19}H_{34}N_2O_{10} \cdot Ca$ mw: 490.63

PROP: White, slightly hygroscopic powder,
odorless, sol in water and glycerol, insol in
alcohol, chloroform and ether. mp: 170°-172°,
decomp @ 195°-196°.

SYNS: CALCIUM D(+)-N-(ALPHA,gamma-DIHY-
DROXY-beta,beta-DIMETHYLBUTYRYL)-beta-
ALANINATE * CALCIUM PANTOTHENATE
* D-CALCIUM PANTOTHENATE * DEXTRO CAL-
CIUM PANTOTHENATE * N-(2,4-DIHYDROXY-
3,3-DIMETHYLBUTYRYL)-BETA-ALANINE CALCIUM
* PANTOTHENATE CALCIUM * (+)-PANTO-
THENIC ACID CALCIUM SALT * VITAMIN B-5

THR: Moderately toxic by intraperitoneal, subcutaneous, and intravenous routes. See also calcium compounds. When heated to decomposition it emits toxic fumes of NO_x.

CALCIUM PENTOBARBITAL   HR: 3
CAS: 7563-42-0   NIOSH: CQ 5950000
mf: $C_{11}H_{18}N_2O_3 \cdot xCa$ mw: 506.87

SYNS: 2,4,6(1H,3H,5H)-PYRIMIDINETRIONE, 5-
ETHYL-5-(1-METHYLBUTYL)-,CALCIUM SALT
* NEMBUTAL CALCIUM * PENTOBARBITAL
CALCIUM

THR: Poison by ingestion and intravenous
routes. When heated to decomposition it emits
toxic fumes of NO_x.

CALCIUM PERMANGANATE   HR: 3
CAS: 10118-76-0   NIOSH: EW 3860000
DOT: 1456
mf: $Mn_2O_8 \cdot Ca$ mw: 277.96

PROP: Violet, deliquescent crystals; mp: decomp, d: 2.4.
DOT Classification: Label: Oxidizer

THR: A poison. See also calcium compounds
and permanganates. Incompatible with acetic
acid; acetic anhydride; hydrogen peroxide.

CALCIUM PEROXIDE   HR: 3
CAS: 1305-79-9   NIOSH: EW 3865000
DOT: 1457
mf: CaO_2 mw: 72.08

PROP: Yellow crystals or powder or white crystals; mp: decomp @ 275°.
DOT Classification: Label: Oxidizer

THR: Irritant in concentrated form, will react
with moisture to form slaked lime. See also
calcium compounds, calcium hydroxide, and
peroxides, inorganic. Moderate fire hazard if

hot and mixed with finely divided combustible material. A strong alkali. An oxidizer. An explosion hazard when intimately mixed with finely divided reducing agents such as organic matter. Incompatible with oxidizable materials.

CALCIUM PHOSPHIDE   HR: 3
CAS: 1305-99-3   NIOSH: EW 3870000
DOT: 1360
mf: Ca_3P_2 mw: 182.18

PROP: Red crystals; mp: >1600°, d: 2.238 @ 25°.
DOT Classification: Label: Flammable Solid and Dangerous When Wet

THR: Incompatible with Cl_2; ClO; HCl; O_2; S; water. See also calcium compounds and phosphides. It liberates highly toxic and spontaneously flammable PH_3 in contact with moisture, acid, or acid fumes. When heated to decomposition it emits toxic fumes of PO_x. For further information, see Vol. 2, No. 1 of *DPIM* Report.

CALCIUM RESINATE   HR: 1
CAS: 9007-13-0   NIOSH: EW 3974000
DOT: 1313/1314
mf: $Ca(C_{44}H_{62}O_4)_2$ mw: 1349.50

PROP: Yellowish white amorphous powder or lumps.

SYN: LIMED ROSIN

DOT Classification: Label: Flammable Solid

THR: Moderate fire hazard when heated; can react with oxidizing materials. When heated to decomposition it emits acrid smoke.

CALCIUM RESINATE (FUSED)   HR: 1
                      NIOSH: EW 3971000

SYN: LIMED ROSIN, FUSED

DOT Classification: Label: Flammable Solid

THR: When heated to decomposition it emits acrid smoke. A flammable material and a moderate fire hazard when exposed to heat, flame, oxidizers.

CALCIUM SILICOFLUORIDE   HR: 3
CAS: 16925-39-6   NIOSH: EW 4025000
mf: CaF_6Si mw: 182.17

PROP: White, crystalline powder. d: 2.662 @ 17.5°.

SYNS: CALCIUM FLUOSILICATE * CALCIUM HEXAFLUOROSILICATE

THR: Poison by ingestion and subcutanous routes. See also hexafluoro silicate(2-)dihydrogen and calcium compounds. When heated to decomposition it emits toxic fumes of F^-.

CALCIUM (II) SULFATE DIHYDRATE (1:1:2)   HR: 3
CAS: 10101-41-4   NIOSH: EW 4150000
mf: $O_4S \cdot Ca \cdot 2H_2O$ mw: 172.18

PROP: Colorless crystals. d: 2.32, mp: 128° ($-1.5H_2O$), bp: 163° ($-2H_2O$).

SYNS: ALABASTER * C.I. 77231 * GYPSUM * GYPSUM STONE * LAND PLASTER * LIGHT SPAR * MINERAL WHITE * NATIVE CALCIUM SULFATE * PRECIPITATED CALCIUM SULFATE * SULFURIC ACID, CALCIUM(2+) SALT, DIHYDRATE * TERRA ALBA

THR: An experimental carcinogen mainly via the human pulmonary system. Long considered a nuisance dust (depending on silica content). When heated to decomposition it emits toxic fumes of SO_x.

CALCIUM SULFIDE   HR: 3
mf: CaS   mw: 72.14

PROP: Cubic, colorless crystals; bp: decomp, d: 218 @ 15°.

SYNS: OLDHAMITE * HEPAR CALCIS * CALCIC LIVER OF SULFUR

THR: A poison via inhalation route. See also sulfides. Reacts violently with PbO_2; $KClO_3$; KNO_3. Incompatible with oxidants.

CALCIUM THIOCYANATE   HR: 3
CAS: 2092-16-2   NIOSH: XK 8540000
mf: $C_2N_2S_2 \cdot Ca$ mw: 156.24

PROP: White crystals; deliquescent.

SYNS: CALCIUM RHODANID (GERMAN) * THIOCYANIC ACID, CALCIUM SALT (2:1)

THR: Poison by ingestion and intravenous routes. See also thiocyanates. When heated to decomposition it emits toxic fumes of NO_x and SO_x.

CALCIUM TRISODIUM DIETHYLENE TRIAMINE PENTAACETATE   HR: 3
CAS: 2531-75-1   NIOSH: MB 8210000
mf: $C_{14}H_{18}N_3O_{10} \cdot CaNa_3$ mw: 497.40

SYNS: CALCIUM SALT OF DIETHYLENETRIAMINE-PENTAACETIC ACID * Ca-DTPA * CALCIUM TRISODIUM SALT OF DIETHYLENE TRIAMINE PENTA ACETIC ACID

THR: An experimental teratogen. Moderate acute toxicity by intraperitoneal route. When heated to decomposition it emits toxic fumes of Na_2O and NO_x.

CAMPHENE   HR: 2
CAS: 79-92-5   NIOSH: EX 1055000
DOT: 9011
mf: $C_{10}H_{16}$ mw: 136.26

PROP: Cubic crystals; mp: 50°-51°, bp: 159°, d: 0.842 @ 54°/4°.
DOT Classification: ORM-A, Label: None

THR: No toxicity data. Slight fire hazard; yields flammable vapors when heated, can react with oxidizing materials. To fight fire, use water spray, foam, fog, CO_2. When heated to decomposition it emits acrid smoke and irritating fumes.

CAMPHOR   HR: 3
CAS: 76-22-2   NIOSH: EX 1225000
DOT: 2717
mf: $C_{10}H_{16}O$ mw: 152.26

PROP: White, transparent, crystalline masses, penetrating odor, pungent, aromatic taste; mp: 180°, bp: 204°, lel = 0.6%, uel = 3.5%, flash p: 150°F (CC), d: 0.992 @ 25°/4°, autoign temp: 871°F, vap d: 5.24.

SYNS: 2-BORNANONE * 2-CAMPHANONE * CAMPHOR-NATURAL * FORMOSA CAMPHOR * GUM CAMPHOR * HUILE DE CAMPHRE (FRENCH) * JAPAN CAMPHOR * KAMPFER (GERMAN) * 2-KETO-1,7,7-TRIMETHYLNOR-CAMPHANE * LAUREL CAMPHOR * MATRI-CARIA CAMPHOR * 1,7,7-TRIMETHYLBICYCLO-(2.2.1)-2-HEPTANONE

OSHA PEL: TWA 2 ppm
ACGIH TLV: TWA 2 ppm; STEL 3 ppm
DOT Classification: Flammable solid

THR: Poison by ingestion and other routes to humans. Poison by subcutaneous and intraperitoneal routes. A local irritant. Swallowing it causes nausea, vomiting, dizziness, excitation, and convulsions. A moderate fire hazard when exposed to heat or flame; can react with oxidizing materials. Vapor is explosive when exposed to heat or flame or CrO_3. To fight fire, use foam, carbon dioxide, dry chemical. For further information, see Vol. 1, No. 8 of *DPIM Report*.

CAMPHOR, (1R,4R)-(+)-   HR: 2
CAS: 464-49-3   NIOSH: EX 1260000
mf: $C_{10}H_{16}O$ mw: 152.26

SYNS: BICYCLO(2.2.1)HEPTAN-2-ONE, 1,7,7-TRIMETHYL-, (1R)- * (+)-2-BORNANONE * D-2-BORNANONE * D-2-CAMPHANONE * CAMPHOR, (+)- * (+)-CAMPHOR * D-CAMPHOR * D-(+)-CAMPHOR * CAMPHOR USP * JAPANESE CAMPHOR

THR: Moderate toxicity by ingestion, subcutaneous, intraperitoneal, and intravenous routes. A skin irritant. When heated to decomposition it emits acrid and irritating fumes. For further information, see Vol. 1, No. 8 of *DPIM Report*.

CAMPHOR OIL   HR: 3
CAS: 8008-51-3   NIOSH: EX 1490000
DOT: 1130

PROP: Colorless or yellowish, oily, fragrant liquid, bp: 175°-200° flash p: 117°F (CC), d: 0.875-0.900 @ 20°/20°. Insol in water; sol in chloroform, ether, oils, in approx 3 vols alc. Keep well closed, cool and protected from light. Found in the trees and bark of *Cinnamomum Carphora Sieb* (Fam. *Lauraceae*) and prepared by fractional distillation of crude camphor oil after the camphor has been crystallized out; a white, viscous liquid with Cineole as the principal ingredient along with monoterpenes.

SYNS: CAMPHOR OIL, RECTIFIED * CAMPHOR OIL YELLOW * FORMOSA CAMPHOR OIL * FORMOSE OIL OF CAMPHOR * JAPANESE CAMPHOR OIL * JAPANESE, OIL OF CAMPHOR * LIGHT CAMPHOR OIL * OIL OF CAMPHOR WHITE * LIGHT OIL OF CAMPHOR * LIQUID CAMPHOR * OIL OF CAMPHOR RECTIFIED * OIL CAMPHOR SASSAFRASSY * WHITE OIL OF CAMPHOR

DOT Classification: Label: Combustible Liquid

THR: Poison by ingestion. Central nervous system effects. A skin irritant. See also safrol and camphor. Moderate fire hazard when exposed to heat or flame; can react with oxidizing materials. To fight fire, use foam, CO_2, dry chemical, mist, fog.

CANDICIDIN   HR: 3
CAS: 1403-17-4   NIOSH: EX 4300000
mf: $C_{63}H_{85}N_{21}O_{19}$ mw: 1440.69

SYNS: CANDEPTIN * VANOBID

THR: Poison by intraperitoneal, intravenous, and subcutaneous routes. When heated to decomposition it emits toxic fumes of NO_x.

CANNABIS   HR: 3
CAS: 8063-14-7   NIOSH: EX 6300000

PROP: A resinous, bitter substance from *Cannabis Sativa,* greenish black mass.

SYNS: BHANG * CHARAS * CME * GANJA * HASACH * HASHISH * INDIAN CANNABIS * INDIAN HEMP * MARIHUANA

THR: An experimental teratogen. Moderately toxic by ingestion and other possible routes. Blood pressure effects. Mutagenic data. An allergen. When ingested or inhaled as smoke, it can cause euphoria, delirium, hallucinations, drowsiness, weakness, and hyporeflexia. An overdose can cause coma and death. Dried material can burn, can react with oxidizing materials.

CANTHARIDES   HR: 3
mf: $C_{10}H_{12}O_4$ mw: 196.15

PROP: Brown to black powder and scales. mp: 218°, bp: subl @ 90°.

SYNS: BLISTERING FLIES * BLISTERING BEETLES * SPANISH FLY

THR: Strong irritant via dermal, oral, inhalation, and contact with eyes. An allergen. Can cause conjunctivitis, keratitis, blepharitis, slight swelling of cornea and inflammation of iris. It is often mistakenly used as an aphrodisiac, but it is much too dangerous and irritating a material for this purpose. On decomposition by heat it, emits toxic fumes.

CANTHARIDES CAMPHOR   HR: 3
CAS: 56-2-5-7   NIOSH: RN 8575000
mf: $C_{10}H_{12}O_4$ mw: 196.22

PROP: Orthorhombic plates, scales; mp: 218°. Sublimes at about 110°. Insol in cold water, sol in hot water; sol in oils.

SYNS: CANTHARIDIN * CANTHARONE * 2,3-DIMETHYL-7-OXABICYCLO(2.2.1)-HEPTANE-2,3-DICARBOXYLIC ANHYDRIDE

* HEXAHYDRO-3A,7A-DIMETHYL-4,7-EPOXYISO-BENZOFURAN-1,3-DIONE

THR: Poison by ingestion in humans. Poison by subcutaneous route. An experimental carcinogen. See cantharides. When heated to decomposition it emits acrid smoke and fumes.

CANTHARIDINE   HR: 3
CAS: 56-25-7   NIOSH: RN 8575000
mf: $C_{10}H_{12}O_4$ mw: 196.22

SYNS: CANTHARIDES CAMPHOR * CANTHARIDIN * CANTHARONE * EXO-1,2-CIS-DI-METHYL-3,6-EPOXYHEXAHYDROPHTHALIC ANHYDRIDE * 2,3-DIMETHYL-7-OXABICYCLO-(2.2.1)HEPTANE-2,3-DICARBOXYLIC ANHYDRIDE * HEXAHYDRO-3A,7A-DIMETHYL-4,7-EPOXYISO-BENZOFURAN-1,3-DIONE

THR: Poison by ingestion and subcutaneous route. A suspected carcinogen and neoplastigen. See also cantharides. When heated to decomposition it emits acrid and irritating fumes. For further information, see Vol. 1, No. 2 of *DPIM Report.*

CAP   HR: 3
CAS: 302-22-7   NIOSH: TU 3750000
mf: $C_{23}H_{29}ClO_4$ mw: 404.97

SYNS: 17-ALPHA-ACETOXY-6-CHLORO-6-DEHY-DROPROGESTERONE * 17-ALPHA-ACETOXY-6-CHLORO-6,7-DEHYDROPROGESTERONE * 17-ALPHA-ACETOXY-6-CHLORO-4,6-PREGNA-DIENE-3,20-DIONE * 17-ALPHA-ACETOXY-6-CHLOROPREGNA-4,6-DIENE-3,20-DIONE * 17-(ACETYLOXY)-6-CHLOROPREGNA-4,6-DIENE-3,20-DIONE * 6-CHLORO-17-ALPHA-ACETOXY-4,6-PREGNADIENE-3,20-DIONE * DELTA(SUP 6)-6-CHLORO-17-ALPHA-ACETOXY-PROGESTERONE * 6-CHLORO-DELTA(SUP 6)-17-ACETOXYPROGESTERONE * 6-CHLORO-DELTA(SUP 6)-(17-ALPHA)ACETOXYPROGESTER-ONE * 6-CHLORO-DELTA(SUP 6)-DEHYDRO-17-ACETOXYPROGESTERONE * 6-CHLORO-6-DEHY-DRO-17-ALPHA-ACETOXYPROGESTERONE * 6-CHLORO-6-DEHYDRO-17-ALPHA-HYDROXY-PROGESTERONE ACETATE * 6-CHLORO-17-AL-PHA-HYDROXYPREGNA-4,6-DIENE-3,20-DIONE ACETATE * 6-CHLORO-17-ALPHA-HYDROXY-DELTA(SUP 6)-PROGESTERONE ACETATE * 6-CHLORO-DELTA(SUP 4,6)-PREGNADIENE-17-ALPHA-OL-3,20-DIONE-17-ACETATE * 6-CHLO-RO-PREGNA-4,6-DIEN-17-ALPHA-OL-3,20-DIONE ACETATE * 6-DEHYDRO-6-CHLORO-17-ALPHA-ACETOXYPROGESTERONE

THR: An experimental carcinogen and teratogen. When heated to decomposition it emits toxic fumes of Cl⁻.

CAPSAICIN   HR: 3
CAS: 404-86-4   NIOSH: RA 8530000
mf: $CH_2NHCO(CH_2)_4CH=CHCH(CH_3)_2$
$C_6H_4(OH)(OCH_3)$ mw: 305.46

PROP: Monoclinic, rectangular plates and scales. Mp: 65°C, bp: 210-220°C. Freely soluble in ethanol, ether, benzene, chloroform; slightly soluble in carbon disulfide; insoluble in water. Highly volatile with a pungent odor.

SYNS: CAPSAICINE * N-((4-HYDROXY-3-METHOXYPHENYL)METHYL)-8-METHYL-6-NONENAMIDE * TRANS-N-((4-HYDROXY-3METHOXYPHENYL)METHYL)-8-METHYL-6-NONEAMIDE * TRANS-8-METHYL-N-VANILLYL-6-NONEAMIDE * NCI-C56564

THR: Poison by intravenous route. Irritating to mucous membranes. When heated to decomposition it emits toxic fumes of NO$_x$. Capsaicin produced erythema and burning without blistering the human skin. Capsicum is considered a moderate irritant to human skin and a high irritant to gastric mucosa. Irritating to mucous membranes, produces severe gastritis and diarrhea. Intragastric infusion of capsaicin in humans increased the DNA content of the gastric aspirate. Capsaicin inhibits transplanted tumors in mice. Capsicum chilles fed to rats produced tumors in 15 of 26 animals. For further information see Vol. 1, No. 4 of *DPIM Report*.

CAPTAN   HR: 3
CAS: 133-06-2   NIOSH: GW 5075000
DOT: 9099
mf: $C_9H_8Cl_3NO_2S$ mw: 300.59

PROP: Odorless crystals, insol in water, sol in benzene and chloroform.

SYNS: ENT 26,538 * LE CAPTANE (FRENCH) * NCI-C00077 * N-TRICHLOROMETHYLMERCAPTO-4-CYCLOHEXENE-1,2-DICARBOXIMIDE * N-(TRICHLOROMETHYLMERCAPTO)-DELTA(SUP 4)-TETRAHYDROPHTHALIMIDE * N-TRICHLOROMETHYLTHIOCYCLOHEX-4-ENE-1,2-DICARBOXIMIDE * N-TRICHLOROMETHYLTHIO-CIS-DELTA(SUP 4)-CYCLOHEXENE-1,2-DICARBOXIMIDE * N-(TRICHLOROMETHYLTHIO)-4-CYCLOHEXENE-1,2-DICARBOXIMIDE * N-((TRICHLOROMETHYL)THIO)-4-CYCLOHEXENE-1,2-DICARBOXIMIDE * N-TRICHLOROMETHYLTHIOTETRAHYDROPHTHALIMIDE * N-((TRICHLOROMETHYL)THIO)-TETRAHYDROPHTHALIMIDE * N-TRICHLOROMETHYLTHIO-3A,4,7,7A-TETRAHYDROPHTHALIMIDE

ACGIH TLV: TWA 5 mg/m³
DOT Classification: ORM-E

THR: An experimental carcinogen and teratogen. Moderate toxicity by inhalation; low toxicity by ingestion. Dangerous when decomposed by heat; see chlorides and sulfates. For further information, see cis-N-((Trichloromethyl)-thio)-4-Cyclohexene-1,2,-Dicarboximide, Vol. 1, No. 4 of *DPIM Report*, also see Captan, Vol. 3, No. 5.

CARAWAY OIL   HR: 2
CAS: 8000-42-8   NIOSH: EY 6300000

PROP: The main constituent of Caraway Oil is 1-Carvone, found in the fruits of *Carum Carvi L.*

SYNS: KUEMMEL OIL (GERMAN) * OIL OF CARAWAY

THR: Moderately toxic by ingestion and dermal routes. A skin irritant. See also 1-6,8(9)-p-menthadien-2-one. When heated to decomposition it emits acrid smoke and irritating fumes.

CARBACHOL CHLORIDE   HR: 3
CAS: 51-83-2   NIOSH: GA 0875000
mf: $C_6H_{15}N_2O_2 \cdot Cl$ mw: 182.68

SYNS: CARBACHOLINE CHLORIDE * CARBAMINOYLCHOLINE CHLORIDE * CARBAMOYLCHOLINE CHLORIDE * GAMMA-CARBAMOYL CHOLINE CHLORIDE * CARBAMYLCHOLINE CHLORIDE * CHOLINE CARBAMATE CHLORIDE * CHOLINE CHLORINE CARBAMATE * (2-HYDROXYETHYL)TRIMETHYL AMMONIUM CHLORIDE CARBAMATE * LENTINE (FRENCH)

THR: Poison by ingestion, subcutaneous, intravenous and intraperitoneal routes. Systemic skin effects. When heated to decomposition it emits very toxic fumes of Cl⁻ and NO$_x$. For further information, see Carbachol, Vol. 1, No. 7 of *DPIM Report*.

CARBAMATES   HR: 3
PROP: Compounds based upon carbamic acid, NH_2COOH. Used only in the form of its numerous salts and derivatives. Many carbamates are from moderately to highly toxic. Some carba-

mates appear to be carcinogenic, teratogenic and/or mutagenic.

4-CARBAMIDOPHENYL BIS(CARBOXY METHYL THIO)ARSENITE   HR: 3
CAS: 120-02-5   NIOSH: AI 6500000
mf: $C_{11}H_{13}AsN_2O_5S_2$ mw: 392.30

SYNS: MERCAPTOACETIC ACID DIESTER WITH DI-THIO-P-UREIDOBENZENEARSONOUS ACID * BIS(CARBOXYMETHYLMERCAPTO)(P-UREIDO-PHENYL)ARSINE * CHEMOTHERAPY CENTER NO. 914 * MERCAPTOACETIC ACID, DIESTER WITH DITHIO-P-UREIDOBENZENEARSONOUS ACID * PHENYL UREA-P-DI(CARBOXYMETHYL) THIOAR-SENITE * THIOCARBARSONE * (P-UREIDO-PHENYLARSYLENEDITHIO)DIACETIC ACID

THR: Poison by intraperitoneal and intravenous routes. Moderately toxic by ingestion. See also esters. When heated to decomposition it emits very toxic fumes of As and SO_x.

N-CARBAMOYL ARSANILIC ACID   HR: 3
CAS: 121-59-5   NIOSH: CF 9275000
mf: $C_7H_9AsN_2O_4$ mw: 260.10

PROP: White, nearly odorless powder, slt acid taste, sol in alc and water. mp: 174°.

SYNS: P-ARSONOPHENYLUREA * P-CARBA-MIDOBENZENEARSONIC ACID * P-CARBAMINO PHENYL ARSONIC ACID * CARBAMINOPHENYL-P-ARSONIC ACID * 4-CARBAMYLAMINOPHENYL-ARSONIC ACID * N-CARBAMYL ARSANILIC ACID * P-UREIDOBENZENEARSONIC ACID * 4-UREIDO-1-PHENYLARSONIC ACID

OSHA PEL: TWA 500 ug(As)/m^3

THR: Poison by ingestion. Moderately toxic by intraperitoneal route. An experimental carcinogen. See also arsenic compounds. When heated to decomposition it emits very toxic fumes of As and NO_x.

N-(CARBAMOYL METHYL)ARSANILIC ACID   HR: 3
CAS: 618-25-7   NIOSH: CF 9450000
mf: $C_8H_{11}AsN_2O_4$ mw: 274.13

PROP: White crystalline powder.

SYNS: 4-ARSONOPHENYLGLYCINAMIDE * SODIUM-N-PHENYLGLYCINAMIDE-P-ARSONATE * TRYPARSAMIDE

OSHA PEL: TWA 500 ug(As)/m^3

THR: Poison by ingestion and intramuscular route. Moderately toxic by intravenous route. See also arsenic compounds. When heated to decomposition it emits very toxic fumes of As and NO_x.

N-(CARBAMOYLMETHYL)-2-DIAZO-ACETAMIDE   HR: 3
CAS: 817-99-2   NIOSH: AB 4900000
mf: $C_4H_6N_4O_2$ mw: 142.14

SYNS: N-DIAZOACETILGLICINA-AMIDE (ITALIAN) * N-(DIAZOACETYL)GLYCINAMIDE * DIAZO-ACETYLGLYCINAMIDE * N-DIAZOACETYLGLY-CINE AMIDE

THR: An experimental carcinogen. Mutagenic data. When heated to decomposition it emits toxic fumes of NO_x.

1-CARBAMYL-2-PHENYL HYDRA-ZINE   HR: 3
CAS: 103-03-7   NIOSH: FD 0575000
mf: $C_7H_9N_3O$ mw: 151.19

PROP: Crystals. mp: 172°.

SYNS: 1-CARBAMOYL-2-PHENYLHYDRAZINE * 2-PHENYLDIAZENECARBOXAMIDE * 2-PHE-NYLHYDRAZIDE, CARBAMIC ACID * 1-PHENYL-HYDRAZINE CARBOXAMIDE * 2-PHENYLHYDRA-ZINECARBOXAMIDE * 1-PHENYLSEMICARBAZIDE

THR: Poison by intraperitoneal route. An experimental carcinogen. When heated to decomposition it emits toxic fumes of NO_x.

CARBANILIC ACID ETHYL ESTER   HR: 3
CAS: 101-99-5   NIOSH: FD 8925000
mf: $C_9H_{11}NO_2$ mw: 165.21

PROP: Crystals; mp: 53°, bp: 238° (slt decomp), d: 1.106.

SYNS: ETHYL CARBANILATE * ETHYL-N-PHE-NYLCARBAMATE * PHENYLETHYL CARBAMATE * N-PHENYLURETHANE

THR: Poison by intraperitoneal and intravenous routes. An experimental carcinogen and neoplastigen. See also esters. When heated to decomposition it emits toxic fumes of NO_x.

CARBANILIC ACID ISOPROPYL ESTER   HR: 3
CAS: 122-42-9   NIOSH: FD 9100000
mf: $C_{10}H_{13}NO_2$ mw: 179.24

PROP: A white crystalline solid, sol in acetone and benzene; mp: 90°.

SYNS: ISOPROPIL-N-FENIL-CARBAMMATO (ITAL-IAN) * ISOPROPYL CARBANILATE * ISOPROPYL CARBANILIC ACID ESTER * ISOPROPYL-N-FENYL-CARBAMAAT (DUTCH) * ISOPROPYL-N-PHENYL-CARBAMAT (GERMAN) * ISOPROPYL PHENYL-CARBAMATE * ISOPROPYL-N-PHENYLCARBA-MATE * O-ISOPROPYL N-PHENYL CARBAMATE * ISOPROPYL-N-PHENYLURETHAN (GERMAN) * N-PHENYLCARBAMATE D'ISOPROPYLE (FRENCH) * N-PHENYL ISOPROPYL CARBAMATE * USAF D-9

THR: Poison by intraperitoneal route. An experimental carcinogen and neoplastigen. See also esters. When heated to decomposition it emits toxic fumes of NO_x.

CARBANOLATE   HR: 3
CAS: 116-06-3   NIOSH: UE 2275000
mf: $C_7H_{14}N_2O_2S$ mw: 190.29

PROP: A solid material.

SYNS: ALDICARBE (FRENCH) * ENT 27,093 * 2-METHYL-2-(METHYLTHIO)PROPANAL, O-((METHYLAMINO)CARBONYL)OXIME * 2-METHYL-2-(METHYLTHIO)PROPIONALDE-HYDE O-(METHYLCARBAMOYL)OXIME * 2-METHYL-2-METHYLTHIO-PROPIONALDEHYD-O-(N-METHYL-CARBAMOYL)-OXIM(GERMAN) * 2-METIL-2-TIOMETIL-PROPIONALDEID-O-(N-METIL-CARBAMOIL)-OSSIMA (ITALIAN) * NCI-C08640

THR: Poison by ingestion, subcutaneous, and skin routes. Powerful systemic poison, pesticide, nematocide, acaricide. When heated to decomposition it emits very toxic fumes of NO_x and SO_x. For further information see aldicarb, Vol. 4, No. 2 of *DPIM Report*.

CARBARYL   HR: 3
CAS: 63-25-2   NIOSH: FC 5950000
mf: $C_{12}H_{11}NO_2$ mw: 201.24

PROP: White crystals; mp: 142°, d: 1.232 @ 20°/20°.

SYNS: METHYLCARBAMIC ACID-1-NAPHTHYL-ESTER * CARBATOX-60 * CRAG SEVIN * ENT 23,969 * EXPERIMENTAL INSECTICIDE 7744 * KARBARYL (POLISH) * N-METHYL-CARBAMATE DE 1-NAPHTYLE (FRENCH) * METHYLCARBAMATE-1-NAPHTHALENOL

* METHYLCARBAMATE-1-NAPHTHOL
* METHYLCARBAMIC ACID-1-NAPHTHYL ESTER
* N-METHYL-1-NAFTYL-CARBAMAAT (DUTCH)
* N-METHYL-1-NAPHTHYL-CARBAMAT (GERMAN)
* N-METHYL-ALPHA-NAPHTHYLCARBAMATE
* N-METHYL-1-NAPHTHYL CARBAMATE
* N-METHYL-ALPHA-NAPHTHYLURETHAN
* N-METIL-1-NAFTIL-CARBAMMATO (ITALIAN)
* ALPHA-NAFTYL-N-METHYLKARBAMAT (CZECH)
* 1-NAPHTHOL-N-METHYLCARBAMATE
* 1-NAPHTHYL METHYLCARBAMATE * ALPHA-NAPHTHYL N-METHYLCARBAMATE * 1-NAPH-THYL-N-METHYLCARBAMATE * SEVIN

DOT Classification: ORM-A, Label: None

THR: Poison by ingestion, intravenous, intraperitoneal, and possibly other routes. Mutagenic data. An experimental carcinogen, teratogen, and tumorigen. An eye and severe skin irritant. Absorbed by all routes, although skin absorption is slow. No accumulation in tissue. Symptoms include blurred vision, headache, stomach ache, vomiting. Symptoms similar to but less severe than those due to parathion. A reversible cholinesterase inhibitor. See also carbamates and esters. When heated to decomposition it emits toxic fumes of NO_x. For further information see Carbaryl, Vol. 1, No. 5 of *DPIM Report*.

CARBAZOLE   HR: 3
CAS: 86-74-8   NIOSH: FE 3150000
mf: $C_{12}H_9N$ mw: 167.22

PROP: White crystals; mp: 244.8°, bp: 354.8°, d: 1.10 @ 18°/4°, vap press: 400 mm @ 323.0°.

SYNS: 9-AZAFLUORENE * DIBENZO(B,D)PYR-ROLE * DIPHENYLENEIMINE * DIPHENYLENI-MIDE * DIPHENYLENIMINE * USAF EK-600

THR: Poison by intraperitoneal route. Moderately toxic by ingestion. When heated to decomposition it emits toxic fumes of NO_x.

CARBETHOXY MALATHION   HR: 3
CAS: 121-75-5   NIOSH: WM 8400000
DOT: 2783
mf: $C_{10}H_{19}O_6PS_2$ mw: 330.38

PROP: Brown to yellow liquid, characteristic odor, miscible in organic solvents, slightly water-sol; d: 1.23 @ 25°/4°, mp: 2.9°, bp: 156° @ 0.7 mm.

SYNS: COMPOUND 4049 * CYTHION * DICARBOETHOXYETHYL-O,O-DIMETHYL PHOS-

PHORODITHIOATE * 1,2-DI(ETHOXYCARBON-YL)ETHYL-O,O-DIMETHYL PHOSPHORODITHIOATE * S-(1,2-DI(ETHOXYCARBONYL)ETHYL DIMETHYL PHOSPHOROTHIOLOTHIONATE * DIETHYL MER-CAPTOSUCCINATE, O,O-DIMETHYL DITHIOPHOS-PHATE, S-ESTER * DIETHYL MERCAPTOSUCCI-NATE, O,O-DIMETHYL PHOSPHORODITHIOATE * DIETHYL MERCAPTOSUCCINATE, O,O-DIMETHYL THIOPHOSPHATE * O,O-DIMETHYL-S-(1,2-BIS(ETHOXYCARBONYL)ETHYL)DITHIOPHOSPHATE * O,O-DIMETHYL-S-(1,2-DICARBETHOXYETHYL) DITHIOPHOSPHATE * O,O-DIMETHYL-S-(1,2-DICARBETHOXYETHYL)PHOSPHORODITHIOATE * O,O-DIMETHYL-S-(1,2-DICARBETHOXYETHYL) THIOTHIONOPHOSPHATE * O,O-DIMETHYL-S-1,2-DI(ETHOXYCARBAMYL)ETHYL PHOSPHORODITHIO-ATE * O,O-DIMETHYL-S-1,2-DIKARBETOXYLE-THYLDITIOFOSFAT (CZECH) * O,O-DIMETHYLDI-THIOPHOSPHATE DIETHYLMERCAPTOSUCCINATE * DITHIOPHOSPHATE DE O,O-DIMETHYLE ET DE S-(1,2-DICARBOETHOXYETHYLE) (FRENCH) * ENT 17,034 * EXPERIMENTAL INSECTICIDE 4049 * S-(1,2-BIS(AETHOXY-CARBONYL)-AETHYL)-O,O-DIMETHYL-DITHIOPHASPHAT-(GERMAN) * S-(1,2-BIS(ETHOXY-CARBONYL)-O,O-DIMETHYL-DITHIOFOSFAAT (DUTCH) * S-(1,2-BIS(ETHOXYCARBONYL)ETHYL) O,O-DIMETHYL PHOSPHORODITHIOATE * S-1,2-BIS(ETHOXYCARBONYL)ETHYL-O,O-DIMETHYL THIOPHOSPHATE * S-(1,2-BIS(ETOSSI-CAR-BONIL)-ETIL)-O,O-DIMETIL-DITIOFOSFATO (ITAL-IAN) * NCI-C00215 * OLEOPHOSPHOTHION * PHOSPHORODITHIOIC ACID, O,O-DIMETHYL ES-TER, S-ESTER WITH DIETHYL MERCAPTOSUCCI-NATE * PHOSPHOTHION * MALATHION * MALATHION (DOT) * MALATHION LV CON-CENTRATE

OSHA PEL: TWA 15 mg/m^3 (skin)
DOT Classification: ORM-A, Label: None

THR: Poison by ingestion in humans. Poison by ingestion, intraperitoneal, intravenous, and subcutaneous routes. An experimental carcino-gen. Central nervous system effects. Mutagenic data. Has caused allergic sensitization of the skin. An organic phosphate cholinesterase inhib-itor. Dangerous when decomposed by heat; see also phosphates and parathion. For further infor-mation see malathion, Vol. 1, No. 6 of *DPIM Report*.

CARBITOL ACETATE   HR: 2
CAS: 112-15-2   NIOSH: KK 8925000
mf: C$_8$H$_{16}$O$_4$ mw: 176.24

PROP: Liquid; bp: 217.4°, fp: −25°, flash p: 230°F (OC), d: 1.0114 @ 20°/20°, vap press: 0.05 mm @ 20°, vap d: 6.07.

SYNS: DIETHYLENE GLYCOL MONOETHYL ETHER ACETATE * DIGLYCOL MONOETHYL ETHER ACE-TATE

THR: Moderately toxic by ingestion. Mild skin and eye irritant. See also glycols. Slight fire hazard when exposed to heat; can react with oxidizing materials. To fight fire, use alcohol foam, water, CO$_2$, dry chemical.

CARBITOL CELLOSOLVE   HR: 2
CAS: 111-90-0   NIOSH: KK 8750000
mf: C$_6$H$_{14}$O$_3$ mw: 134.20

PROP: Colorless liquid, mild pleasant odor. bp: 201.9°, flash p: 201°F (OC), d: 0.9902 @ 20°/4°, vap d: 4.62.

SYNS: CARBITOL * CARBITOL SOLVENT * DIETHYLENE GLYCOL MONOETHYL ETHER * DIOXITOL * DOWANOL * ETHOXY DIGLY-COL * 2-(2-ETHOXYETHOXY)ETHANOL * ETHYL CARBITOL * ETHYL DIETHYLENE GLY-COL * LOSUNGSMITTEL APV * MONOETHYL ETHER OF DIETHYLENE GLYCOL * POLY-SOLV

THR: Moderately toxic by intravenous route. Mild skin and eye irritant. Slight fire hazard when exposed to heat; can react with oxidizing materials. To fight fire, use alcohol foam, CO$_2$, dry chemical. When heated to decomposition it emits acrid smoke and irritating fumes.

CARBON   HR: 2
CAS: 7440-44-0   NIOSH: FF 5250000
mf: C   mw: 12.01

PROP: Black crystals, powder or diamond form. mp: 3652°-3697° (subl), bp: approx 4200°, d(amorphous): 1.8-2.1, d(graphite): 2.25, d(diamond): 3.51, vap press: 1 mm @ 3586°.

SYNS: BLACK PEARLS * COLUMBIAN CARBON * CARBONE (ITALIAN) * CHARCOAL BLACK * C.I. 77266 * PURIFIED CHARCOAL

OSHA PEL: TWA 3500 ug/m^3

THR: Moderately toxic by intravenous route. It can cause a dust irritation, particularly to the eyes and mucous membranes. See also car-bon black, soot. Slight fire hazard when exposed to heat. Dust is explosive when exposed to heat or flame or (NH$_4$NO$_3$ + heat); (NH$_4$ClO$_4$ @

240°); bromates; $Ca(OCl)_2$; chlorates; Cl_2; (Cl_2 + $Cr(OCl)_2$); ClO; F_2; iodates; IO_5; $(Pb(NO_3)_2$; $HgNO_3$; HNO_3; (oils + air); (K + air); Na_2S; $Zn(NO_3)_2$. Incompatible with air; metals; oxidants; unsaturated oils.

CARBON BLACK   HR: 1
PROP: A generic term applied to a family of high-purity colloidal carbons commercially produced by carefully controlled pyrolysis of gaseous or liquid hydrocarbons. Carbon blacks, including commercial colloidal carbons such as furnace blacks, lamp blacks and acetylene blacks, usually contain less than several tenths percent of extractible organic matter and less than one percent ash.

SYNS: LAMP BLACK * ACETYLENE BLACK
* FURNACE BLACK
ACGIH TLV: TWA 3.5 mg/m^3

THR: Low toxicity by ingestion, inhalation, and skin contact. See also carbon. A nuisance dust in high concentrations. While it is true that the tiny particulates of carbon black contain some molecules of carcinogenic materials, the carcinogens are apparently held tightly and are not eluted by hot or cold water, gastric juices or blood plasma. For further information see Vol. 3, No. 2 of *DPIM Report*.
Refs: Nau, C. A., Taylor, G. T., Lawrence, C. H., "Properties and Physiological Effects of Thermal Carbon Black." *Journal of Occupational Medicine*. Nov. 1976, Vol 18, No. 11, pp. 732-734. Nau, C. A., Neal, J., Stembridge, V. A., "A Study of the Physiological Effects of Carbon Black." *Archives of Environmental Health*, Dec. 1960, Vol. 1, pp. 512-533, American Medical Association.

CARBON DIOXIDE   HR: 3
CAS: 124-38-9   NIOSH: FF 6400000
DOT: 1013/1845/2187
mf: CO_2 mw: 44.01

PROP: Colorless, odorless gas; mp: subl @ −78.5°, (−56.6° @ 5.2 atm), vap d: 1.53.

SYNS: ANHYDRIDE CARBONIQUE (FRENCH)
* CARBONIC ACID GAS * CARBONIC ANHYDRIDE * DRY ICE * KOHLENSAURE (GERMAN)

OSHA PEL: TWA 5000 ppm
ACGIH TLV: TWA 5000 ppm
DFG MAK: 5000 ppm (9000 mg/m^3)
DOT Classification: Nonflammable gas

THR: An experimental teratogen. Asphyxiant. An eye irritant. Contact of carbon dioxide snow with the skin may cause a "burn." See discussion of simple asphyxiants under Argon. Reacts vigorously with (Al + Na_2O_2); Cs_2O; $Mg(C_2H_5)_2$; Li; (Mg + Na_2O_2); K; KHC; Na; Na_2C_2; NaK; Ti. Incompatible with acrylaldehyde; aziridine; dicaesium oxide; metal acetylides; metals; sodium peroxide.

CARBON DIOXIDE (liquefied)   HR: 1
CAS: 124-38-9   NIOSH: FF 6425000
DOT: 1013/1845/2187
DOT Classification: Label Nonflammable Gas

THR: See carbon dioxide.

CARBON DIOXIDE MIXED WITH NITROUS OXIDE
CAS: 53569-62-3   NIOSH: FF 6480000
DOT: 1015

PROP: Gas. Composition: CO_2 + N_2O.

SYN: CARBON DIOXIDE-NITROUS OXIDE MIXTURE (DOT)

DOT Classification: Label: Nonflammable Gas

THR: No toxicity data. See components as listed. Slight fire hazard. An oxidizing mixture. Can react with reducing materials.

CARBON DIOXIDE MIXED WITH OXYGEN   HR: 1
CAS: 8063-77-2   NIOSH: FF 6485000
DOT: 1014

SYN: CARBON DIOXIDE-OXYGEN MIXTURE (DOT)

DOT Classification: Label: Nonflammable Gas

THR: No toxicity data. See components as listed.

CARBON DISULFIDE   HR: 3
CAS: 75-15-0   NIOSH: FF 6650000
DOT: 1131
mf: CS_2 mw: 76.13

PROP: Clear, colorless liquid, nearly odorless when pure; mp: −110.8°, bp: 46.5°, lel = 1.3%, uel = 50%, flash p: −22°F (CC), d: 1.261 @ 20°/20°, autoign temp: 257°F, vap press: 400 mm @ 28°, vap d: 2.64.

SYNS: CARBON BISULFIDE * CARBONE (SU-
FURE DE) (FRENCH) * CARBONIO (SOLFURO DI)
(ITALIAN) * CARBON SULFIDE * DITHIOCAR-
BONIC ANHYDRIDE * KOHLENDISULFID (SCHWE-
FELKOHLENSTOFF) (GERMAN) * KOOLSTOFDI-
SULFIDE (ZWAVELKOOLSTOF) (DUTCH)
* NCI-C04591 * SCHWEFELKOHLENSTOFF
(GERMAN) * SULPHOCARBONIC ANHYDRIDE
* WEGLA DWUSIARCZEK (POLISH)

OSHA PEL: TWA 20 ppm; CL 30
ACGIH TLV: TWA 10 ppm (skin)
DFG MAK: 10 ppm (30 mg/m^3)
DOT Classification: Flammable liquid

THR: Poison by ingestion and intraperitoneal
route. Moderately toxic by inhalation. An experimental teratogen. Mutagenic data. The chief
toxic effect is on the central nervous system,
acting as a narcotic and anesthetic in acute poisoning with death following from respiratory
failure. In chronic poisoning, the effect on the
nervous system is one of central and peripheral
damage, which may be permanent if the damage
has been severe. A dangerous fire hazard when
exposed to heat, flame, sparks, or friction. Severe explosion hazard when exposed to heat
or flame, reacts violently with Al; Cl_2; azides;
CsN_3; ClO; ethylamine diamine; ethylene imine;
F_2; $Pb(N_3)_2$; LiN_3; NO; N_2O_4; (H_2SO_4 + permanganates); K; KN_3; RbN_3; NaN_3Zn. When
heated to decomposition it emits highly toxic
fumes of SO_x; can react vigorously with oxidizing materials. To fight fire, use water, CO_2,
dry chemical, fog, mist. Incompatible with air,
rust; halogens; metal azides; metals; oxidants.
For further information, see Vol. 3, No. 5 of
DPIM Report.

CARBON MONOXIDE HR: 3
CAS: 630-08-0 NIOSH: FG 3500000
DOT: 1016
mf: CO mw: 28.01

PROP: Colorless, odorless gas; mp: $-207°$,
bp: $-191.3°$, lel = 12.5%, uel = 74.2%, d:
(gas) 1.250 g/L @ 0°, (liq) 0.793, autoign temp:
1128°F.

SYNS: CARBONE (OXYDE DE) (FRENCH)
* CARBONIC OXIDE * CARBONIO (OSSIDO DI)
(ITALIAN) * EXHAUST GAS * FLUE GAS
* KOHLENMONOXID (GERMAN) * KOOL-
MONOXYDE (DUTCH) * OXYDE DE CARBONE
(FRENCH) * WEGLA TLENEK (POLISH)

OSHA PEL: TWA 50 ppm
ACGIH TLV: TWA 50 ppm; STEL 400 ppm;
BEI *carboxyhemoglobin in blood less than
8% (*CO in end-exhaled air less than 40
ppm);
DFG MAK: 30 ppm (33 mg/m^3); BAT blood
5%
DOT Classification: Label: Flammable Gas

THR: Poison by inhalation in humans. An experimental teratogen. Human central nervous system effects. Can cause asphyxiations by preventing hemoglobin from binding oxygen. After being removed from exposure, the half-life in
blood is one hour. Chronic exposure effects can
occur at lower concentrations. A common air
contaminant. A dangerous fire hazard when exposed to flame. Severe explosion hazard when
exposed to heat or flame. Severe reaction with
BrF_3; Cs_2O; ClF_3; IF_7; (Li + H_2O); NF_3; O_2;
OF_2; (K + O_2); Ag_2O; (Na + NH_3). To fight
fire, stop flow of gas. Incompatible with fluorine
and oxygen; interhalogens; metal oxides; metals;
oxidants. For further information, see Vol. 3,
No. 5 of *DPIM Report*. Very dangerous when
exposed to flame. Acute cases of poisoning resulting from brief exposures to high concentrations seldom results in any permanent disability
if recovery takes place. Chronic effects as the
result of repeated exposure to lower concentrations have been described, particularly in the
Scandinavian literature. Auditory disturbances
and contraction of the visual fields have been
demonstrated. Glycosuria does occur, and heart
irregularities have been reported. Other workers
have found that where the poisoning has been
relatively long and severe, cerebral congestion
and edema may occur, resulting in long-lasting
mental or nervous damage. Repeated exposure
to low conc of the gas, up to 100 ppm in air,
is generally believed to cause no signs of poisoning or permanent damage. Industrially, sequelae
are rare, as exposure, though often severe, is
usually brief. It is a common air contaminant.

CARBON REMOVER (LIQUID) HR: 2
NIOSH: FG 4400000
DOT: 1132

PROP: Flash p: <80°F.
DOT Classification: Label: Flammable
Liquid

THR: No toxicity data. Dangerous fire hazard
when exposed to heat or flame; can react with

oxidizing materials. To fight fire, use CO_2, dry chemical.

CARBON TETRABROMIDE   HR: 3
CAS: 558-13-4   NIOSH: FG 4725000
DOT: 2516
mf: CBr_4 mw: 331.65

PROP: Colorless, monoclinic tablets; mp: (α) 48.4°, (β) 90.1°, bp: 189.5°, d: 3.42, vap press: 40 mm @ 96.3°.

SYNS: CARBON BROMIDE * TETRABROMO-METHANE
ACGIH TLV: TWA 0.1 ppm; STEL 0.3 ppm
DOT Classification: Poison B

THR: Poison by subcutaneous and intravenous routes. Moderately toxic by ingestion. Severe reaction with Li. Narcotic in high concentration. See chlorinated hydrocarbons, aliphatic, for disaster hazard. When heated to decomposition it emits toxic fumes of Br^-.

CARBON TETRACHLORIDE   HR: 3
CAS: 56-23-5   NIOSH: FG 4900000
DOT: 1846
mf: CCl_4 mw: 153.81

PROP: Colorless liquid, heavy, ethereal odor; mp: −22.6°, bp: 76.8°, fp: −22.9°, flash p: none, d: 1.597 @ 20°, vap press: 100 mm @ 23.0°.

SYNS: BENZINOFORM * CARBONA * CARBON CHLORIDE * CARBON TET * CZTEROCHLOREK WEGLA (POLISH) * ENT 4,705 * FREON 10 * METHANE TETRACHLORIDE * PERCHLORO-METHANE * TETRACHLOORKOOLSTOF (DUTCH) * TETRACHLOORMETAAN (DUTCH) * TETRA-CHLORKOHLENSTOFF, TETRA (GERMAN) * TETRACHLORMETHAN (GERMAN) * TETRA-CHLOROCARBON * TETRACLOROMETANO (ITALIAN) * TETRACHLOROMETHANE * TETRACHLORURE DE CARBONE (FRENCH) * TETRACLORURO DI CARBONIO (ITALIAN)

OSHA PEL: TWA 10 ppm; CL 25; Pk 200/5M/4H
ACGIH TLV: TWA 5 ppm
DFG MAK: 10 ppm (65 mg/m^3)
DOT Classification: ORM-A

THR: Poison by ingestion. Moderately toxic by inhalation. An experimental carcinogen and teratogen, and suspected human carcinogen. Central nervous system, pulmonary, gastrointestinal and other systemic effects in humans. A severe eye and mild skin irritant. Damages liver, kidneys, and lungs. A narcotic. Individual susceptibility varies widely. Contact dermatitis can result from skin contact. Dangerous when heated to decomposition, emits highly toxic fumes of phosgene. Incompatible with aluminum trichloride; calcium disilicide; chlorine trifluoride; decarborane (14); dibenzoyl peroxide; N,N-dimethylformamide; 1,2,3,4,5,6-hexachlorocyclohexane; dinitrogen tetraoxide; fluorine; metals; potassium-tert-butoxide. Severe reaction with allyl alcohol; Al; $Al(C_2H_5)_3$; Ba; (benzoyl peroxide + C_2H_4); Be; BrF_3; $Ca(OCl)_2$; diborane; C_2H_4; dimethyl formamide; disilane; F_2; Li; Mg; liquid O_2; Pu; K; ($AgClO_4$ + HCl); potassium-*tert*-butoxide; Na; NaK; tetrasilane; trisilane; U; Zr; burning wax. See chlorinated HC. It has been banned from household use by FDA. For further information, see Vol. 3, No. 5 of *DPIM Report*.

Carbon tetrachloride has a narcotic action resembling that of chloroform, though not as strong. Following exposures to high conc, the victim may become unconscious, and if exposure is not terminated, death can follow from respiratory failure. In cases of narcosis that recover, the after-effects are more serious than those of delayed chloroform poisoning, usually taking the form of damage to the kidneys, liver and lungs. Exposure to lower concentrations, insufficient to produce unconsciousness, usually results in severe gastro-intestinal upset, and may progress to serious kidney and hepatic damage. The kidney lesion is an acute nephrosis; the liver involvement consists of an acute degeneration of the central portions of the lobules. Where recovery takes place, there may be no permanent disability. Marked variation in individual susceptibility to carbon tetrachloride exists, some persons appear to be unaffected by exposures which seriously poison their fellow-workers. Alcoholism and previous liver and kidney damage seem to render the individual more susceptible. Concentrations of the order of 1,000 to 1,500 ppm are sufficient to cause symptoms if exposure continues for several hours. Repeated daily exposure to such conc may result in poisoning.

Though the common form of poisoning following industrial exposure is usually one of gastrointestinal upset, which may be followed by renal damage, other cases have been reported in which the CNS has been affected with production of polyneuritis, narrowing of the visual fields, and other neurological changes. Prolonged exposure to small amounts of carbon tetrachloride has also been reported as causing cirrhosis of the liver.

Locally, a dermatitis may be produced following long or repeated contact with the liquid. The skin oils are removed, and the skin becomes red, cracked and dry. The effect of carbon tetrachloride on the eyes either as a vapor or as a liquid, is one of irr with lacrimation and burning.

Industrial poisoning is usually acute, with malaise, headache, nausea, dizziness, and confusion, which may be followed by stupor and sometimes loss of consciousness. Symptoms of liver and kidney damage may follow later, with development of dark urine, sometimes jaundice and liver enlargement, followed by scanty urine, albumenuria and renal casts, uremia may develop and cause death. Where the exposure has been less acute, the picture is usually one of headache, dizziness, nausea, vomiting epigastric distress, loss of appetite, and fatigue. Visual disturbances (blind spots, spots before the eyes, a visual "haze" and restriction of the visual fields), secondary anemia, and occasionally a slight jaundice may occur. Dermatitis may be noticed on the exposed parts.

CARBON TETRAFLUORIDE   HR: 1
CAS: 75-73-0   NIOSH: FG 4920000
mf: CF_4 mw: 88.01

PROP: Colorless gas; mp: $-184°$, bp: $-127.7°$, d: 1.96 @ $-184°$.

SYNS: FREON 14 * HALON 14 * TETRA-FLUOROMETHANE

THR: Low toxicity by inhalation. Less chronically toxic than carbon tetrachloride. See halogenated hydrocarbons. Violent reaction with Al. See also halogenated hydrocarbons, aliphatic. When heated to decomposition it emits toxic fumes of F^-.

CARBON TRIFLUORIDE   HR: 2
CAS: 75-46-7   NIOSH: PB 6900000
mf: CHF_3 mw: 70.02

PROP: Colorless, odorless gas. mp: $-163°$, bp: $-82.2°$, d: 1.52 (liq) @ $-100°$.

SYNS: FLUOROFORM * FREON F-23 * METHYL TRIFLUORIDE * TRIFLUOROMETH-ANE

THR: Narcotic in high concentration. A mild

respiratory irritant. Mutagenic data. See also fluorides. When heated to decomposition it emits toxic fumes of F^-.

CARBONYL FLUORIDE   HR: 3
CAS: 353-50-4   NIOSH: FG 6125000
DOT: 2417
mf: CF_2O mw: 66.01

PROP: Colorless gas, pungent, hygroscopic; mp: $-114°$, bp: $-83°$, d: 1.139 @ $-114°$.

SYNS: CARBON OXYFLUORIDE * CARBONYL DIFLUORIDE * FLUOPHOSGENE * FLUORO-PHOSGENE

ACGIH TLV: TWA 2 ppm; STEL 5 ppm
DOT Classification: Poison A

THR: Poison by inhalation. A powerful irritant. See also hydrofluoric acid and fluorine. Hydrolyzes instantly upon contact with moisture. See carbon monoxide for fire, disaster, and explosion hazards. See also fluorides. Incompatible with hexafluoroisopropylideneamino-lithium.

CARBONYLS   HR: 3
PROP: The (CO) group attached to a metal.

THR: Most carbonyls are highly toxic. The toxicity of carbonyls depends in part; but not always entirely, on their ready decomposition which releases carbon monoxide. Symptoms are due in part to carbon monoxide and in part to the direct irritating action of the carbonyl. See specific carbonyl in question. Moderate fire and explosion hazard when exposed to heat or flame. More or less readily evolves carbon monoxide. See also carbon monoxide and powdered metals. When heated to decomposition it emits highly toxic fumes of carbon monoxide; they react with water or steam to produce toxic and flammable vapors; they can react vigorously with oxidizing materials.

CARBONYL SULFIDE   HR: 3
CAS: 463-58-1   NIOSH: FG 6400000
mf: COS   mw: 60.07

PROP: Gas or liquid. mp: $-138°$, bp: $49.9°$, lel = 12%, uel = 28.5%, d: liq 1.24 @ $-87°$, vap d: 2.1.

SYNS: CARBON OXIDE SULFIDE * CARBON-OXYSULFIDE

THR: Poison by inhalation. Narcotic in high concentration. An irritant. May liberate highly toxic hydrogen sulfide upon decomposition. Very dangerous fire hazard when exposed to heat or flame. Moderate explosion hazard when exposed to heat or flame. See also sulfides. Can react vigorously with oxidizing materials. To fight fire, stop flow of gas or use CO_2, dry chemical or water spray. See also sulfides.

2-CARBOXY-4′-(DIMETHYL AMINO)-AZOBENZENE   HR: 3
CAS: 493-52-7   NIOSH: DG 8960000
mf: $C_{15}H_{15}N_3O_2$ mw: 269.33

PROP: Shiny violet crystals.

SYNS: C.I. ACID RED 2 * 4′-DIMETHYLAMI-NOAZOBENZENE-2-CARBOXYLIC ACID * 2-((4-DIMETHYLAMINO)PHENYLAZO)BENZOIC ACID

THR: Mutagenic data. An experimental carcinogen. When heated to decomposition it emits toxic fumes of NO_x.

(4-(CARBOXY METHOXY)-3-CHLORO-PHENYL)(5,5-DIETHYL-2,4,6(1H,3H,5H)-PYRIMIDINETRIONATO-O(sup 2)MER-CURY, MONOSODIUM SALT   HR: 3
CAS: 36568-91-9   NIOSH: OV 7900000
mf: $C_{16}H_{18}ClHgN_2O_6$ • Na   mw: 593.39

SYNS: MERBAPHEN * NOVASUROL

THR: Poison by intravenous route. See also mercury compounds. When heated to decomposition it emits very toxic fumes of Cl^-, NO_x, and Hg vapors.

(9-(o-CARBOXYPHENYL)-6-(DIETHYL-AMINO)-3H-XANTHEN-3-YLIDENE) DI-ETHYLAMMONIUM CHLORIDE   HR: 3
CAS: 81-88-9   NIOSH: BP 3675000
mf: $C_{28}H_{31}N_2O_3$ • Cl   mw: 379.06

SYNS: 9-o-CARBOXYPHENYL-6-DIETHYLAMINO-3-ETHYLIMINO-3-ISOXANTHENE, 3-ETHOCHLORIDE * C.I. FOOD RED 15 * DIETHYL-M-AMINO-PHE-NOLPHTHALEIN HYDROCHLORIDE * FD AND C RED NO. 19 * TETRAETHYLDIAMINO-O-CAR-BOXY-PHENYL-XANTHENYL CHLORIDE * TETRA-ETHYLRHODAMINE

THR: Poison by intraperitoneal and intravenous routes. An experimental carcinogen. Mutagenic data. Moderately toxic by ingestion. When heated to decomposition it emits very toxic fumes of NO_x and Cl^-.

CARDIO-GREEN   HR: 3
CAS: 3599-32-4   NIOSH: DE 3150000
mf: $C_{43}H_{48}N_2O_6S_2$ • Na   mw: 776.04

SYN: ICG

THR: Poison by intraperitoneal and intravenous routes. When heated to decomposition it emits very toxic fumes of SO_x and NO_x.

beta-CARDONE   HR: 3
CAS: 959-24-0   NIOSH: PB 0600000
mf: $C_{12}H_{20}N_2O_3S$ • ClH   mw: 308.86

SYNS: 4′-(1-HYDROXY-2-(ISOPROPYLAMINO)-ETHYL)-METHANESULFOANILIDE HYDROCHLORIDE * 4-(2-ISOPROPYLAMINE-1-HYDROXYETHYL)-METHANESULFOANILIDE HYDROCHLORIDE * ISOPROPYLAMINOHYDROXYETHYLMETHANE-SULFONALIDE HYDROCHLORIDE * N-ISOPROPYL-BETA-(4-METHANESULFONAMIDOPHENYL)-ETHANOLAMINE HYDROCHLORIDE

THR: Poison by ingestion in humans. Poison by intravenous and intraperitoneal routes. When heated to decomposition it emits very toxic fumes of HCl, SO_x, and NO_x.

CARRAGEEN   HR: 3
CAS: 9000-07-1   NIOSH: FI 0700000

PROP: A sulfated polysaccharide. Dried plant of seaweed *Chondrus Crispus,* yellow-white when powdered, insol in organic solvents. Dried, bleached *Chondrus Crispus* containing salts of sulfated polygalactose esters.

SYNS: 3,6-ANHYDRO-D-GALACTAN * CARRA-GEENAN GUM * CARRAGHEANIN * CARRA-GHEEN * CHONDRUS EXTRACT * GUM CHON 2 * GUM CHROND * IRISH MOSS GELOSE * PEARLPUSS * PELLUGEL * PIG-WRACK * SELF ROCK MOSS

THR: Poison by intravenous route. An experimental carcinogen and neoplastigen. When heated to decomposition it emits acrid smoke.

CARRAGEENAN, DEGRADED   HR: 3
                       NIOSH: FI 0706000

PROP: Carrageenan derived from *Eucheuma spinosum,* degraded by acid hydrolysis; average molecular weight 20,000-40,000.

THR: An experimental carcinogen. See also carrageen. When heated to decomposition it emits acrid smoke and irritating fumes.

CARVACROL   HR: 3
CAS: 499-75-2   NIOSH: FI 1225000
mf: $C_{10}H_{14}O$ mw: 150.24

SYNS: 2-P-CYMENOL * 2-HYDROXY-P-CYMENE
* ISOPROPYL-O-CRESOL * ISOTHYMOL
* 2-METHYL-5-ISOPROPYLPHENOL * 5-ISOPRO-
PYL-2-METHYLPHENOL * O-THYMOL

THR: Poison by ingestion and subcutaneous
route. When heated to decomposition it emits
acrid smoke and irritating fumes.

CARVON   HR: 3
CAS: 469-62-5   NIOSH: EL 2900000
mf: $C_{22}H_{29}NO_2$ mw: 339.52

SYNS: DEXTROPROPOXYPHENE * ALPHA-(+)-
4-DIMETHYLAMINO-1,2-DIPHENYL-3-METHYL-2-
BUTANOL PROPIONATE ESTER * PROPOXYPHENE,
(+)-

THR: Poison by ingestion, intraperitoneal, and
intravenous routes. Human pulmonary system
and blood pressure effects. When heated to decomposition it emits toxic fumes of NO_x.

CARYOPHYLLENE   HR: 2
CAS: 87-44-5   NIOSH: DT 8400000
mf: $C_{15}H_{26}$ mw: 206.41

PROP: Found in oil of Clove, Cinnamon Leaves
and Copaiba Balsam and in minor quantities
in various other essential oils, especially laven-
der; prepared by isolation from clove leaf oil,
clove stem oil, cinnamon leaf oil or pine oil
fractions.

SYNS: BETA-CARYOPHYLLENE * 8-METHY-
LENE-4,11,11-(TRIMETHYL)BICYCLO(7.2.0)-
UNDEC-4-ENE

THR: A skin irritant. Mutagenic data. When
heated to decomposition it emits acrid smoke
and irritating fumes.

CASSIA OIL   HR: 3
CAS: 8007-80-5   NIOSH: FI 4050000

PROP: Chief constituent is Cinnamic Aldehyde,
found in the leaves and twigs of *Cinnamomum
Cassia Blume*.

SYNS: KASSIA OEL (GERMAN) * OIL OF CAS-
SIA

THR: Poison by skin contact. A human skin
irritant. See cinnamaldehyde, aldehydes. When
heated to decomposition it emits acrid smoke
and irritating fumes.

CASTOR BEAN   HR: 3
NIOSH: FI 4095000

SYNS: CASTOR BEANS (DOT) * CASTOR
POMACE (DOT)

THR: Poison by ingestion in humans. It contains
the very toxic materials ricin and ricinine. Potent
allergen. See also ricin. When heated to decomposition it emits toxic fumes of NO_x.

CASTOR OIL   HR: 2
CAS: 8001-79-4   NIOSH: FI 4100000

PROP: A colorless to pale yellow viscous liquid, characteristic odor. mp: $-12°$, bp: $313°$,
flash p: 445°F (CC), d: 0.96, autoign temp:
840°F.

SYNS: AROMATIC CASTOR OIL * CASTOR OIL
AROMATIC * NCI-c55163 * NEOLOID
* OIL OF PALMA CHRISTI * PHORBYOL
* RICINUS OIL * RICIRUS OIL * TANGANTAN-
GAN OIL

THR: Moderately toxic by ingestion. An allergen. A food additive permitted in food for human consumption. An eye irritant. A purgative.
Slight fire hazard when exposed to heat. Spontaneous heating: Yes. To fight fire, use CO_2, dry
chemical, fog, mist.

CASTRIX   HR: 3
CAS: 535-89-7   NIOSH: UV 8050000
mf: $C_7H_{10}ClN_3$ mw: 171.65

PROP: Sltly water sol crystals.

SYNS: 2-CHLOOR-4-DIMETHYLAMINO-6-
METHYL-PYRIMIDINE (DUTCH) * 2-CHLORO-
4-DIMETHYLAMINO-6-METHYL-PYRIMIDINE
* 2-CHLORO-4-METHYL-6-DIMETHYLAMINO-
PYRIMIDINE * 2-CLORO-4-DIMETILAMINO-6-
METIL-PIRIMIDINA (ITALIAN) * CRIMIDIN (GER-
MAN) * CRIMIDINA (ITALIAN) * CRIMIDINE

THR: Poison by ingestion and other routes. Can
cause central nervous system damage and convulsions. Intensely poisonous to mammals.
When heated to decomposition, it emits very
toxic fumes of Cl^- and NO_x.

CEDAR LEAF OIL   HR: 2
CAS: 8007-20-3   NIOSH: FJ 1520000

PROP: Yellowish volatile oil; d: 0.910-0.920.
Constituent is d-alpha-Thujone, found in leaves
of *Thuja Occidentalis L*.

SYN: OIL THUJA

THR: Moderately toxic via ingestion route. A skin irritant. Ingestion of large quantities causes hypertension, bradycardia, tachypnea, convulsions, death. When heated to decomposition, it emits acrid smoke.

CEFAZEDONE   HR: 3
NIOSH: RR 5873000

mf: $C_{18}H_{15}Cl_2N_5O_5S_3$ mw: 548.46

SYN: REFOSPOREN

THR: A teratogen. Slightly toxic by intravenous route. When heated to decomposition, it emits very toxic fumes of Cl^-, NO_x, and SO_x.

CELLOPHANE   HR: 3
CAS: 9005-81-6   NIOSH: FJ 4100000
mf: $(C_6H_{10}O_5)_n$

THR: An experimental tumorigen. See also polymers. When heated to decomposition, it emits acrid smoke and irritating fumes.

"CELLOSOLVE" ACETATE   HR: 2
CAS: 111-15-9   NIOSH: KK 8225000
mf: $C_6H_{12}O_3$ mw: 132.18

PROP: Colorless liquid with a mild, pleasant ester-like odor; bp: 156.4°, flash p: 117°F (COC), lel = 1.7%, fp: −61.7°, d: 0.9748 @ 20°/20°, autoign temp: 715°F, vap press: 1.2 mm @ 20°, vap d: 4.72.

SYNS: ACETATE D'ETHYLGLYCOL (FRENCH) * ACETATO DI CELLOSOLVE (ITALIAN) * 2-AETHOXY-AETHYLACETAT (GERMAN) * ETHOXY ACETATE * 2-ETHOXYETHANOL, ESTER WITH ACETIC ACID * 2-ETHOXY-ETHYLACETAAT (DUTCH) * ETHOXYETHYL ACETATE * 2-ETHOXYETHYLE, ACETATE DE (FRENCH) * BETA-ETHOXYETHYL ACETATE * ETHYL CELLOSOLVE ACETAAT (DUTCH) * ETHYLENE GLYCOL ETHYL ETHER ACETATE * ETHYLENE GLYCOL MONOETHYL ETHER ACETATE * ETHYLENE GLYCOL MONOETHYL ETHER ACETATE (DOT) * ETHYLGLYKOLACETAT (GERMAN) * 2-ETOSSIETIL-ACETATO (ITALIAN) * GLYCOL MONOETHYL ETHER ACETATE * OCTAN ETOKSYETYLU (POLISH) * POLY-SOLV EE ACETATE

OSHA PEL: TWA 100 ppm (skin)

THR: Moderately toxic by ingestion and dermal route. Mild skin and eye irritant. See also glycols. Moderate fire hazard when exposed to heat or flame; can react with oxidizing materials.

Moderate explosion hazard in the form of vapor when heated. To fight fire, use alcohol foam, CO_2, dry chemical. When heated to decomposition, it emits acrid smoke and irritating fumes. For further information see ethylene glycol monoethyl ether acetate, Vol. 4, No. 2 of *DPIM Report*.

CELLOSOLVE SOLVENT   HR: 2
CAS: 110-80-5   NIOSH: KK 8050000
mf: $C_4H_{10}O_2$ mw: 90.14

PROP: Colorless liquid, practically odorless; bp: 135.1°, lel = 1.8%, uel = 14%, flash p: 202°F(CC), d: 0.9360 @ 15°/15°, autoign temp: 455°F, vap press: 3.8 mm @ 20°, vap d: 3.10.

SYNS: CELLOSOLVE * DOWANOL EE * ETHER MONOETHYLIQUE DE L'ETHYLENE-GLYCOL (FRENCH) * 2-ETHOXYETHANOL * ETHYL CELLOSOLVE * ETHYLENE GLYCOL MONOETHYL ETHER (DOT) * ETOKSYETYLOWY ALKOHOL (POLISH) * GLYCOL MONOETHYL ETHER * NCI-C54853 * POLY-SOLV EE

OSHA PEL: TWA 200 ppm (skin)

THR: Moderately toxic by ingestion and dermal routes. See also glycols. Slight fire hazard when exposed to heat or flame; can react with oxidizing materials. Moderate explosion hazard in the form of vapor when exposed to heat or flame. To fight fire, use alcohol foam, dry chemical. For further information see ethylene glycol monoethyl ether, Vol. 4, No. 2 of *DPIM Report*.

CELLULOSE TETRANITRATE
CAS: 9004-70-0   NIOSH: FJ 6000000

PROP: White amorphous solid.
$C_{12}H_{16}(ONO_2)_4O_6$, mw: 504.3, d: 1.66, flash p: 55°F.

SYNS: CELLOIDIN * CELLULOSE NITRATE * COLLODION COTTON * COLLOXYLIN * GUNCOTTON * NITROCELLULOSE * NITROCOTTON * PYRALIN * PYROXYLIN * PYROXYLIN PLASTICS (DOT) * PYROXYLIN PLASTIC SCRAP (DOT) * SOLUBLE GUN COTTON * XYLOIDIN

THR: Highly dangerous in the dry state when exposed to heat, flame, or powerful oxidizers. When wet with 35% of denatured ethanol it is about as hazardous as ethanol alone or gasoline. Dry cellulose tetranitrate burns rapidly with intense heat and ignites easily. Moderately dangerous explosion hazard. See explosives, high. To fight fire, use copious volumes of water; alcohol foam. CO_2 is effective in extinguishing fires of nitrocellulose solvents.

CEMENT (LEATHER)   HR: 2
NIOSH: FJ 8000000

SYN: CEMENT, LEATHER (DOT)

DOT Classification: Label: Flammable Liquid

THR: No toxicity data. A flammable liquid.

CEMENT (LIQUID)   HR: 2
NIOSH: FJ 8010000

THR: No toxicity data. A combustible liquid.

CEMENT, PORTLAND   HR: 1
PROP: Fine gray powder composed of compounds of lime, aluminum, silica and iron oxide as $(4CaO \cdot Al_2O_3 \cdot Fe_2)_3$, $(3CaOAl_2O_3)$, $(3CaO \cdot SiO_2)$, and $(2CaOSiO_2)$. Small amounts of magnesia, sodium, potassium, chromium and sulfur are also present in combined form.

THR: Slightly toxic by inhalation. A moderate irritant. An allergen. Cement dust is a common air contaminant.

CEMENT (PYROXYLIN)   HR: 2
NIOSH: FJ 8020000

DOT Classification: Label: Flammable Liquid

THR: Dangerous fire hazard when exposed to heat or flame; can react with oxidizing materials.

CEMENT (ROOFING LIQUID)   HR: 2
NIOSH: FJ 8030000

DOT Classification: Label: Flammable Liquid

THR: Dangerous fire hazard when exposed to heat or flame; can react with oxidizing materials.

CEMENT (RUBBER)
NIOSH: FJ 8040000

PROP: Flash p: 50°F or less.
DOT Classification: Label: Flammable Liquid

THR: Often contains benzene or other toxic solvents. See specific constituent. Dangerous fire

hazard when exposed to heat or flame; can react with oxidizing materials.

CEPACOL CHLORIDE   HR: 3
CAS: 123-03-5   NIOSH: UU 4900000
mf: $C_{21}H_{38}N \cdot Cl$ mw: 340.05

SYNS: CEEPRYN CHLORIDE * CETYLPYRIDINIUM CHLORIDE * N-HEXADECYLPYRIDINIUM CHLORIDE * 1-HEXADECYLPYRIDINIUM CHLORIDE

THR: Poison by ingestion, intraperitoneal, subcutaneous, and intravenous routes. Moderate skin irritant. When heated to decomposition, it emits very toxic fumes of NO_x and Cl^-. For further information, see Cetylpyridinium Chloride, Vol. 2, No. 4 of *DPIM Report*.

CERIC OXIDE   HR: 2
CAS: 1306-38-3   NIOSH: FK 4550000
mf: CeO_2 mw: 172.12

SYN: CERIUM DIOXIDE

THR: Moderately toxic by ingestion. See also cerium compounds.

CERIUM   HR: 2
PROP: Cubic or hexagonal, steel gray crystals. Ce, at wt: 140.13, mp: 815°, bp: 3257°, d: (cubic form): 6.90; hexagonal form 6.75.

THR: Cerium resembles aluminum in its pharmacological action as well as in its chemical properties. The insoluble salts such as the oxalates are stated to be non-toxic even in large doses. It is used to prevent vomiting in pregnancy. The average dose is from 0.05 to 0.5 g. Cerium tartrate has been found to produce a direct injurious action on the hearts of small animals. The effect on the central nervous system of the rare-earth metals following inhalation may preclude welding operations with these materials to any large extent. Cerium is stated to produce polycythemia but is useless in the treatment of anemia owing to its toxic effects. The salts of cerium increase the blood coagulation rate. See also rare earths. Moderate fire hazard; ignites spontaneously in air at 150°-180°. A strong reducing agent. See also iron dust. Severe reaction with halogens; P. Moderate explosion hazard in the form of dust when exposed to flame. Incompatibles: self-explodes; metals;

halogens; phosphorus; silicon. For further information, see Vol. 1, No. 8 of *DPIM Report*.

CERIUM ACETATE   HR: 3
CAS: 537-00-8   NIOSH: FK 4900000
mf: $C_6H_9O_6 \cdot Ce$ mw: 317.27

SYNS: CEROUS ACETATE ∗ CERIUM TRIACETATE

THR: Human central nervous system effects. See also cerium compounds. When heated to decomposition, it emits acrid and irritating fumes.

CERIUM CHLORIDE   HR: 3
CAS: 7790-86-5   NIOSH: FK 5075000
mf: $CeCl_3$ mw: 246.47

PROP: Colorless, deliquescent crystals; mp: 848°, bp: 1727°, d: 3.92.

SYNS: CERIUM TRICHLORIDE ∗ CEROUS CHLORIDE

THR: Poison by intravenous, intraperitoneal, and subcutaneous routes. Moderately toxic by ingestion. See also cerium compounds. When heated to decomposition, it emits toxic fumes of Cl^-.

CERIUM CITRATE   HR: 3
CAS: 512-24-3   NIOSH: FK 5425000
mf: $C_6H_8O_6 \cdot Ce$ mw: 332.26

THR: Poison by intraperitoneal route. See also cerium compounds. When heated to decomposition, it emits acrid and irritating fumes.

CERIUM COMPOUNDS
PROP: The toxicity of cerium compounds may be taken to be that of cerium, except when the anion has a toxicity of its own. See also cerium.

CERIUM EDETATE   HR: 3
CAS: 15158-67-5   NIOSH: FK 5950000

THR: Poison by intraperitoneal route. See also cerium compounds. When heated to decomposition, it emits acrid and irritating fumes.

CERIUM FLUORIDE   HR: 3
CAS: 7758-88-5   NIOSH: FK 6125000
mf: CeF_3 mw: 197.12

PROP: White, hexagonal crystals; d: 6.16; mp: 1430°; bp: 2327°. Insol in water, sol in H_2SO_4.

SYNS: CERIUM FLUORURE (FRENCH) ∗ CERIUM TRIFLUORIDE

THR: Poison. See fluorides and cerium compounds. When heated to decomposition, it emits toxic fumes of F^-.

CERIUM (III) NITRATE, HEXAHYDRATE (1:3:6)   HR: 3
CAS: 10294-41-4   NIOSH: FK 6300000
mf: $N_3O_9 \cdot Ce \cdot 6H_2O$ mw: 434.27

SYNS: CERIUM NITRATE, HEXAHYDRATE ∗ CEROUS NITRATE HEXAHYDRATE

THR: Poison by intraperitoneal and intravenous routes. Moderately toxic by ingestion. See also cerium compounds and nitrates. When heated to decomposition, it emits toxic fumes of NO_x.

CERIUM (III) TETRAHYDROALUMINATE
mf: Al_3CeH_{12} mw: 236.46

PROP: Unstable; decomp @ $-80°$; self-ignites in air.

THR: No toxicity data. See also cerium compounds and aluminum compounds. Since it self ignites in air it is a dangerous fire hazard.

CESIUM   HR: 3
CAS: 7440-46-2   NIOSH: FK 9225000
mf: Cs   mw: 132.91

PROP: Hexagonal crystals, silver-white, ductile metal or possibly a silvery liquid. Mp: 28.5°, bp: 705°, d: 1.873, vap press: 1 mm @ 279°.

THR: Cesium is quite similar to potassium in its elemental state. It has been shown, however, to have pronounced physiological action in experimentation with animals. Hyperirritability, including marked spasms, has been shown to follow the administration of cesium in amounts equal to the potassium content of the diet. It has been found that replacing the potassium in the diet of rats with cesium, caused death after 10-17 days. Dangerous fire hazard by chemical reaction; reacts with Cl_2; O_2; P. Can ignite spontaneously in moist air. See also sodium. Moderate explosion hazard by chemical reaction; reacts with moisture to liberate hydrogen. Moderately dangerous disaster hazard; will react with water or steam to produce heat and hydrogen; on contact with oxidizing materials, it can react vigorously. Incompatible with air; oxygen; water.

CESIUM BROMIDE   HR: 2
CAS: 7787-69-1   NIOSH: FK 9275000
mf: BrCs   mw: 212.82

THR: Moderately toxic by intraperitoneal route. See also cesium and bromides. When heated to decomposition, it emits toxic fumes of Br^-.

CESIUM CARBONATE HR: 2
CAS: 534-17-8 NIOSH: FK 9400000
mf: $CO_3 \cdot 2Cs$ mw: 325.83

SYN: DICESIUM CARBONATE

THR: Mutagenic data. See also cesium.

CESIUM CHLORIDE HR: 2
CAS: 7647-17-8 NIOSH: FK 9625000
mf: $ClCs$ mw: 168.36

PROP: d: 3.99, mp: 646°, bp: 1303°.

THR: Moderately toxic by ingestion and intraperitoneal routes. Mutagenic data. Reacts violently with BF_3. See also cesium. When heated to decomposition, it emits toxic fumes of Cl^-.

CESIUM FLUORIDE HR: 3
CAS: 13400-13-0 NIOSH: FK 9650000
mf: CsF mw: 151.91

THR: Poison by unspecified routes. See also fluorides and cesium compounds. When heated to decomposition, it emits toxic fumes of F^-. Incompatible with benzenediazonium tetrafluoroborate; difluoroamine.

CESIUM HYDROXIDE HR: 3
CAS: 21351-79-1 NIOSH: FK 9800000
mf: $CsHO$ mw: 149.92

PROP: Colorless to yellowish, very deliquescent crystals; mp: 272.3°, d: 3.675.

DOT: 2681/2682

SYN: CESIUM HYDRATE

DOT Classification: Corrosive

THR: Poison by intraperitoneal route. Moderately toxic by ingestion. See also cesium. A powerful caustic. A skin and eye irritant.

CESIUM IODIDE HR: 2
CAS: 7789-17-5 NIOSH: FL 0350000
mf: CsI mw: 259.81

THR: Moderately toxic by ingestion and intraperitoneal route. See also cesium and iodides. When heated to decomposition, it emits toxic fumes of I^-.

CESIUM (I) NITRATE (1:1) HR: 2
CAS: 7789-18-6 NIOSH: FL 0700000
mf: $NO_3 \cdot Cs$ mw: 194.92

PROP: Colorless, hexagonal or cubic, glittering crystalline powder. Mp: 414°, bp: decomp, d: 3.685; 2.71 @ 500° (liq).

THR: Moderately toxic by intraperitoneal route. See also cesium and nitrates. Mutagenic data. When heated to decomposition, it emits toxic fumes of NO_x.

CETYL PYRIDINIUM CHLORIDE MONOHYDRATE HR: 3
CAS: 6004-24-6 NIOSH: UU 5075000
mf: $C_{21}H_{38}N \cdot Cl \cdot H_2O$ mw: 358.07

SYN: CEEPRYN

THR: Poison by ingestion, intraperitoneal, intravenous, and subcutaneous routes. When heated to decomposition, it emits very toxic fumes of HCl and NO_x.

CHAMOMILE HR: 2
CAS: 520-36-5 NIOSH: LK 9276000
mf: $C_{15}H_{10}O_5$ mw: 270.25

PROP: Blue liquid, turning brownish-yellow. Composed of amyl and butyl esters of angelic and tiglic acids, butyric acid, etc. d: 0.905-0.915 @ 15°/15°.

SYNS: APIGENOL * C.I. NATURAL YELLOW 1 * 4′,5,7-TRIHYDROXYFLAVONE

THR: Mutagenic data. Also a mild allergen. When heated to decomposition, it emits acrid and irritating fumes.

CHAMOMILE OIL (GERMAN) HR: 2
CAS: 8002-66-2 NIOSH: FL 7180000

PROP: Blue-yellowish-brown liquid. Composition: amyl and butyl esters of angelic, tiglic acids, butyric acid; d: 0.905-0.915 @ 15°/15°. By steam distillation of the flowers and stalks of *Matrilaria Chamomilla L.*

THR: A mild allergen; a skin irritant. See also esters. When heated to decomposition it emits acrid and irritating fumes.

CHAMOMILE OIL (ROMAN) HR: 2
CAS: 8002-66-2 NIOSH: FL 7181000

PROP: Blue liquid, turning brownish-yellow. Composition: amyl and butyl esters of angelic and tiglic acids, butyric acid, etc.; d: 0.905-0.915 @ 15°/15°. By the steam distillation of the dried flowers of *Anthemis Nobilis L.*

THR: A mild allergen, skin irritant. See also esters. Slight fire hazard when heated. When heated to decomposition, it emits acrid smoke and irritating fumes.

CHARCOAL (ACTIVATED)

CAS: 64365-11-3　　　　NIOSH: FL 7242500

THR: No toxicity data. A flammable solid. See also charcoal (briquettes).

CHARCOAL (BRIQUETTES)

CAS: 16291-96-6　　　　NIOSH: FL 7243500

PROP: Black amorphous solid. Composition: C + impurities; mw: 12.0, mp: > 3500°, bp: 4200°, d: 3.51.

SYN: CHARCOAL BRIQUETTES (DOT)

THR: Carbon itself has no toxic action, but it contains impurities and these may be toxic. Fire hazard; reacts with liquid air; $Ba(ClO_3)_2$; BrF_5; ClO; $Ca(ClO_3)_2$; ClF_2; F_2; H_2O_2; $Mg(ClO_3)_2$; $(O_2 + wood)$; perchlorates; peroxides; $(P + air)$; K; $KClO_3$; KNO_3; RuO_4; $AgNO_3$; $NaClO_3$; $(AgCl + NaO_2)$; S; $(S + NaNO_3)$; $Zn(ClO_3)_2$. Heats spontaneously, particularly when wet, freshly calcined, or tightly packed; it can ignite and burn. Slight explosion hazard when exposed to heat or flame. To fight fire, use water, mist, foam or dry chemical.

CHARCOAL SCREENINGS, MADE FROM "PINON" WOOD (DOT)

NIOSH: FL 7286000

DOT Classification: Label: Flammable Solid

THR: No toxicity data. See also charcoal (briquettes).

CHARCOAL (SHELL)

NIOSH: FL 7244000

DOT Classification: Label: Flammable Solid

THR: No toxicity data. A flammable solid. See also charcoal (briquettes).

CHARCOAL (WOOD, GROUND, CRUSHED, GRANULATED OR PULVERIZED)

NIOSH: FL 7248000

DOT Classification: Label: Flammable Solid

THR: No toxicity data. See also charcoal (briquettes).

CHARCOAL (WOOD, LUMP)

NIOSH: FL 7248500

DOT Classification: Label: Flammable Solid

THR: No toxicity data. See also charcoal (briquettes).

CHARCOAL WOOD SCREENINGS, OTHER THAN "PINON" WOOD SCREENINGS (DOT)

NIOSH: FL 7288000

DOT Classification: Label: Flammable Solid

THR: No toxicity data. See also charcoal (briquettes).

CHELIDONINE　　　　　　　　　　HR: 3

CAS: 476-32-4　　　　NIOSH: FL 9450000
mf: $C_{20}H_{19}NO_5$　　mw: 353.40

PROP: White crystalline powder. mp: 135°-136°.

THR: Poison by intravenous and subcutaneous routes. A central nervous system depressant, causing sleepiness, depression, slowing of the pulse, and in large doses, coma and circulatory failures. Slight fire hazard. When heated to decomposition, it emits toxic fumes of NO_x.

CHENOPODIUM OIL　　　　　　　HR: 3

CAS: 8006-99-3　　　　NIOSH: FM 2997000

PROP: American wormseed; ingredients are ascaridol, cymene, camphor and saponins.

PROP: Colorless or pale yellow liquid, characteristic disagreeable odor and taste, not water sol. Composition: 60-70% ascaridol. D: 0.950-0.980 @ 25°/25°. Sol in 8 vols 70% alc; sltly sol in glacial acetic acid. Keep well closed, cool and protected from light.

THR: Poison by ingestion and dermal routes. A skin irritant. See also ascaridol. When heated to decomposition, it emits acrid smoke and irritating fumes.

CHERRY BARK OAK　　　　　　　HR: 3

NIOSH: FM 3005000

PROP: Tannin containing fraction of bark used (JNCIAM 57,207,76)

SYNS: QUERCUS FALCATA PAGODAEFOLIA
* TANNIN FROM CHERRY BARK OAK

THR: An experimental neoplastigen. See also tannin. When heated to decomposition it emits acrid and irritating fumes.

CHERRY LAUREL OIL

PROP: Volatile oil from leaves of *Prunus laurocerasus L., Rosacene.* Pale yellow liquid; odor and taste similar to oil of bitter almond. d: 1.054-1.066 @ 20°/20°. Sltly sol in water, sol in 2 vols 70% alc, benzene, chloroform, ether. Keep well closed, cool and protected from light.

THR: Very poisonous. Hydrogen cyanide component is responsible for highly toxic properties. See also cyanides. When heated, it emits highly toxic fumes of cyanide.

CHESTNUT TANNIN HR: 3
NIOSH: FM 3015000

SYNS: CASTANEA SATIVA MILL TANNIN * TANNIN FROM CHESTNUT

THR: An experimental carcinogen. See also tannin. When heated to decomposition, it emits acrid and irritating fumes.

CHLODITHANE HR: 3
CAS: 53-19-0   NIOSH: KH 7880000
mf: $C_{14}H_{10}Cl_4$ mw: 320.04

SYNS: 2-(o-CHLOROPHENYL)-2-(p-CHLOROPHE-NYL)-1,1-DICHLOROETHANE * 1,1-DICHLORO-2,2-BIS(2,4'-DICHLOROPHENYL)ETHANE * 1,1-DICHLORO-2-(o-CHLOROPHENYL)-2-(p-CHLOROPHENYL)ETHANE * o,p'-DICHLORODIPHENYLDICHLOROETHANE * 2,4'-DICHLOROPHENYLDICHLOROETHANE * NCI-C04933 * NSC 38721

THR: An experimental tumorigen. It causes central nervous system, gastrointestinal system, systemic skin effects, and blood effects in humans. When heated to decomposition, it emits toxic fumes of Cl^-.

CHLORACETONITRILE HR: 3
CAS: 107-14-2   NIOSH: AL 8225000
mf: C_2H_2ClN mw: 75.50

SYNS: ALPHA-CHLOROACETONITRILE * 2-CHLOROACETONITRILE * CHLOROMETHYL CYANIDE * MONOCHLOROACETONITRILE * MONOCHLOROMETHYL CYANIDE * USAF KF-5

THR: Poison by ingestion, inhalation, and dermal routes. A skin irritant. See also nitriles. When heated to decomposition it emits very toxic fumes of Cl^-, NO_x, and CN^-.

CHLORAL HYDRATE HR: 3
CAS: 302-17-0   NIOSH: FM 8750000
mf: $C_2HCl_3O \cdot H_2O$ mw: 165.40

PROP: Transparent, colorless crystals, aromatic, penetrating, slightly acrid odor and slightly bitter, caustic taste; mp: 52°, bp: 97.5°, d: 1.9.

SYNS: AQUACHLORAL * TRICHLORACETAL-DEHYD-HYDRAT (GERMAN) * TRICHLOROACET-ALDEHYDEMONOHYDRATE * 2,2,2-TRICHLORO-1,1-ETHANEDIOL

THR: Poison to humans by ingestion, intravenous and rectal routes. Mutagenic data. An experimental tumorigen. Human central nervous system effects. Slight fire hazard when heated.
Treatment and Antidotes: A physician should be called at once. Wash out the stomach. Administer strychnine hypodermically, also give caffeine, and caffeine with sodium benzoate. Maintain the temperature of the patient with the use of electric pads, hot water bottles and blankets if necessary.

2-CHLORALLYL DIETHYLDITHIOCAR-BAMATE HR: 3
CAS: 95-06-7   NIOSH: EZ 5075000
mf: $C_8H_{14}ClNS_2$ mw: 223.80

PROP: amber liquid; bp: 129° @ 1 mm.

SYNS: 2-CHLOROALLYL-N,N-DIETHYLDITHIOCAR-BAMATE * DIETHYLDITHIOCARBAMIC ACID-2-CHLOROALLYL ESTER * NCI-C00453 * VEGA-DEX

THR: An experimental carcinogen. Mutagenic data. Mild skin irritant. See also carbamates, esters. When heated to decomposition, it emits very toxic fumes of Cl^-, NO_x, and SO_x.

CHLORAMBUCIL HR: 3
CAS: 305-03-3   NIOSH: ES 7525000
mf: $C_{14}H_{19}Cl_2NO_2$ mw: 304.24

SYNS: 4-(BIS(2-CHLOROETHYL)AMINO)BEN-ZENEBUTANOIC ACID * GAMMA-(P-BIS(2-CHLOROETHYL)AMINOPHENYL)BUTYRIC ACID * 4-(P-(BIS(2-CHLOROETHYL)AMINO)PHENYL)-BUTYRIC ACID * 4(P-BIS(BETA-CHLOROETHYL)-AMINOPHENYL)BUTYRIC ACID * CHLORAMINO-PHENE * CHLOROBUTINE * N,N-DI-2-CHLORO-ETHYL-GAMMA-P-AMINOPHENYLBUTYRIC ACID * P-(N,N-DI-2-CHLOROETHYL)AMINOPHENYL BUTYRIC ACID * P-N,N-DI-(BETA-CHLORO-ETHYL)AMINOPHENYL BUTYRIC ACID * GAM-MA-(P-DI(2-CHLOROETHYL)AMINOPHENYL)-BUTYRIC ACID * NCI-C03485 * NSC-3088 * PHENYLBUTYRIC ACID NITROGEN MUSTARD

THR: Poison by intravenous, intraperitoneal, and subcutaneous routes. An experimental carcinogen, teratogen, and neoplastigen. Mutagenic data. When heated to decomposition, it emits very toxic fumes of Cl^- and NO_x. For further information see Vol. 5, No. 1 of *DPIM Report*.

CHLORAMINE-T   HR: 2
mf: $C_7H_7ClNNaO_2S \cdot 3H_2O$ mw: 281.

PROP: White or faintly yellow crystals. Slight chlorine odor, water sol.

SYNS: CHLORAMINE * SODIUM-P-TOLUENE SULFON CHLORAMIDE

THR: Inhalation of vapors can cause vasomotor rhinitis and asthma. A mild irritant and allergen. Dangerous disaster hazard; see also sulfonates and chlorides. A water disinfectant.

CHLORAMPHENICOL   HR: 3
CAS: 56-75-7   NIOSH: AB 6825000
mf: $C_{11}H_{12}Cl_2N_2O_5$ mw: 323.15

PROP: Cryst; sl sol H_2O; mp: 151°.

SYNS: CHLOROMYCETIN * CHLORONITRIN * D-CHLORAMPHENICOL * D-THREO-CHLORAMPHENICOL * D-(-)-THREO-CHLORAMPHENICOL * DEXTROMYCETIN * D-(-)-THREO-2-DICHLOROACETAMIDO-1-P-NITROPHENYL-1,3-PROPANEDIOL * D-THREO-N-DICHLOROACETYL-1-P-NITROPHENYL-2-AMINO-1,3-PROPANEDIOL * D-(-)-THREO-2,2-DICHLORO-N-(BETA-HYDROXY-ALPHA-(HYDROXYMETHYL))-P-NITROPHENETHYLACETAMIDE * D-THREO-N-(1,1'-DIHYDROXY-1-P-NITROPHENYLISOPROPYL)-DICHLOROACETAMIDE * NCI-C55709 * D-(-)-THREO-1-P-NITROPHENYL-2-DICHLORACETAMIDO-1,3-PROPANEDIOL * D-THREO-1-(P-NITROPHENYL)-2-(DICHLOROACETYLAMINO)-1,3-PROPANEDIOL * NSC 3069

THR: Poison by intraperitoneal, intravenous, and subcutaneous routes. Mutagenic data. An oral carcinogen in women. Human central nervous system effects and blood effects. An experimental teratogen. When heated to decomposition, it emits very toxic fumes of NO_x and Cl^-.

CHLORATES
PROP: Chlorates are a combination of a metal or hydrogen and ClO_3^- monovalent radical. They are crystalline and somewhat deliquescent.

THR: The principal toxic effects of chlorates are the production of methemoglobin in the blood and destruction of red blood corpuscles. The latter may lead to irritation of the kidneys. Damage to heart muscle has been reported. Dangerous fire hazard in contact with flammable matter. When contaminated with oxidizable materials, they are particularly sensitive to friction, heat, and shock; they are powerful oxidizing agents. Dangerous explosion hazard when shocked, exposed to heat, or rubbed, particularly when contaminated with sugar; charcoal; shellac; sulfur; starch; sawdust; sulfuric acid; ammonium compounds; cyanides; phosphorous or antimony sulfide; Al; (metals + acids); As_2S_3; CaH_2; MnO_2; metal sulfides; organic acids; powdered metals; Hg_3P_4; PHI_4; SCN; (S + Cu); Se; NaH_2PO_2; SrH; SO_2. Chlorates when mixed with combustible materials may form explosive mixtures. For instance, potassium chlorate, when mixed with sulfur or with other combustible substances explodes on friction. Pure chlorates which have been spilled on the floor, or mixed with small amounts of impurities, become very sensitive to shock and friction. Water is considered the best agent for fighting fires involving chlorates. When heated to decomposition, they can emit toxic fumes and explode; can react with reducing materials.

CHLORCYCLIZINE DIHYDROCHLORIDE   HR: 3
CAS: 129-71-5   NIOSH: TL 2100000
mf: $C_{18}H_{21}ClN_2 \cdot 2ClH$ mw: 373.78

SYNS: 1-(4-CHLOROBENZHYDRYL)-4-METHYLPIPERAZINE DIHYDROCHLORIDE * DI-PARALENE-2-HYDROCHLORIDE * N-METHYL-N'-(4-CHLOROBENZHYDRYL)PIPERAZINE DIHYDROCHLORIDE * PERAZIL DIHYDROCHLORIDE

THR: Poison by ingestion, subcutaneous, and intraperitoneal routes. When heated to decomposition, it emits very toxic fumes of Cl^-, NO_x, and HCl.

CHLORCYCLIZINE HYDROCHLORIDE   HR: 3
CAS: 14362-31-3   NIOSH: TL 2200000
mf: $C_{18}H_{21}ClN_2 \cdot 7ClH$ mw: 556.08

SYNS: CHLORCYCLIZINIUM CHLORIDE . * 1-(P-CHLOROBENZHYDRYL)-4-METHYLPIPERAZINE HYDROCHLORIDE * DIPARALENE HYDROCHLORIDE

THR: Poison by ingestion, subcutaneous, and intraperitoneal routes. An experimental teratogen. When heated to decomposition, it emits very toxic fumes of HCl and NO_x.

CHLORDANE   HR: 3
CAS: 57-74-9   NIOSH: PB 9800000
mf: $C_{10}H_6Cl_8$ mw: 409.76

PROP: Colorless to amber, odorless, viscous liquid. bp: 175°, d: 1.57-1.63 @ 15.5°/15.5°.

DOT: 2762

SYNS: BELT * CHLOORDAAN (DUTCH) * GAMMA-CHLORDAN * CHLORODANE * CLORDAN (ITALIAN) * ENT 9,932 * ENT 25,552-X * NCI-C00099 * OCTACHLOR * OCTACHLORODIHYDRODICYCLOPENTADIENE * 1,2,4,5,6,7,8,8-OCTACHLORO-2,3, 3A,4,7,7A-HEXAHYDRO-4,7-METHANOINDENE * 1,2,4,5,6,7,8,8-OCTACHLORO-2,3, 3A,4,7,7A-HEXAHYDRO-4,7-METHANO-1H-IN-DENE * 1,2,4,5,6,7,8,8-OCTACHLORO-3A,4,7,7A-HEXAHYDRO-4,7-METHYLENE INDANE * OCTACHLORO-4,7-METHANOHYDROINDANE * OCTACHLORO-4,7-METHANOTETRAHYDROIN-DANE * 1,2,4,5,6,7,8,8-OCTACHLORO-4,7-METHANO-3A,4,7,7A-TETRAHYDROINDANE * 1,2,4,5,6,7,8,8-OCTACHLOOR-3A,4,7,7A-TETRAHYDRO-4,7-ENDO-METHANO-INDAAN (DUTCH) * 1,2,4,5,6,7,8,8-OCTACHLORO-3A,4,7,7A-TETRAHYDRO-4,7-METHANOINDANE * 1,2,4,5,6,7,8,8-OCTACHLORO-3A,4,7,7A-TETRAHYDRO-4,7-METHANOINDANE * 1,2,4,5, 6,7,10,10-OCTACHLORO-4,7,8,9-TETRAHYDRO-4,7-METHYLENEINDANE * 1,2,4,5,6,7,8,8-OC-TACHLOR-3A,4,7,7A-TETRAHYDRO-4,7-ENDO-METHANO-INDAN (GERMAN) * 1,2,4,5,6,7,8, 8-OTTOCHLORO-3A,4,7,7A-TETRAIDRO-4,7-ENDO-METANO-INDANO (ITALIAN)
ACGIH TLV: TWA 0.5 mg/m³; STEL 2 mg/m³
DFG MAK: 0.5 mg/m³
DOT Classification: Flammable Liquid/ Combustible Liquid

THR: Poison to humans by ingestion and other routes, such as inhalation, intravenous, and intraperitoneal routes. Mutagenic data. An experimental carcinogen. When heated to decomposition, it emits toxic fumes of Cl⁻. For further information, see Vol. 3, No. 5 of *DPIM Report*.

A central nervous system stimulant whose exact mode of action is unknown, but may involve microsomal enzyme stimulation. Animals poisoned by this and related compounds show an extremely marked loss of appetite and neurological symptoms. The fatal dose to man is unknown. It has been estimated to be between 6 to 60 g (1/5 to 2 ounces). One person receiving an accidental skin application of 25% solution (amounting to something over 30 g of technical chlordane) developed symptoms within about 40 minutes and died before medical attention was obtained. In two patients, death followed exposure to low oral doses of chlordane (2-4 g); on microscopic examination both patients showed severe chronic fatty degeneration of the liver, characteristic of chronic alcoholism. Although these two fatalities cannot be attributed exclusively to chlordane, they are entirely consistent with previous observations that the toxicity of other chlorinated hydrocarbons is much enhanced in the presence of chronic liver damage. The dangerous chronic dose in man is unknown.

One person poisoned by chlordane developed convulsions within 40 minutes of gross skin contamination and died, apparently of respiratory failure, before medical aid could be obtained.

Acutely poisoned experimental animals show similar signs. Experimental animals exposed to repeated small doses exhibit hyperexcitability, tremors, and convulsions, and those which survive long enough show marked anorexia and loss of weight. Symptoms in animals frequently occur within an hour of the administration of a large dose, but death often is delayed for several days depending on the dosage and route of administration. In any event, symptoms are of longer duration with chlordane than with DDT under similar conditions.

Laboratory findings are essentially normal, except that the insecticide may be demonstrated in tissues of poisoned animals by means of bioassay. A method for specific, quantitative chemical analysis for chlordane is now available using small amounts of subcutaneous fat. Chronically poisoned animals show degenerative changes in the liver and kidney tubules.

Treatment and Antidotes: Removal of the poison from the skin or the alimentary tract should be attempted. Oil laxatives should be avoided. The nervous symptoms may be combatted with pentobarbital or phenobarbital.

CHLORENDIC ACID   HR: 2
CAS: 115-28-6   NIOSH: RB 9000000

SYNS: 1,4,5,6,7,7-HEXACHLORO-5-NORBOR-
NENE-2,3-DICARBOXYLIC ACID * KYSELINA
3,6-ENDOMETHYLEN-3,4,5,6,7,7-HEXACHLOR-
DELTA(SUP 4)-TETRAHYDROFTALOVA (CZECH)
* KYSELINA HET (CZECH) * NCI-C55072

THR: Moderately toxic by ingestion. When
heated to decomposition, it emits toxic fumes
of Cl^-.

CHLORFENVINFOS   HR: 3
CAS: 470-90-6   NIOSH: TB 8750000
mf: $C_{12}H_{14}Cl_3O_4P$ mw: 359.58

SYNS: O-2-CHLOOR-1-(2,4-DICHLOOR-FENYL)-
VINYL-O,O-DIETHYLFOSFAAT (DUTCH) * O-2-
CHLOR-1-(2,4-DICHLOR-PHENYL)-VINYL-O,O-DIA-
ETHYLPHOSPHAT (GERMAN) * DIETHYL-1-(2,4-
DICHLOROPHENYL)-2-CHLOROVINYL PHOSPHATE
* CHLORFENVINPHOS * 2-CHLORO-1-(2,4-DI-
CHLOROPHENYL)VINYL DIETHYL PHOSPHATE
* BETA-2-CHLORO-1-(2',4'-DICHLOROPHENYL)
VINYL DIETHYLPHOSPHATE * CHLOROFENVIN-
PHOS * CHLORPHENVINFOS * CHLORPHENVIN-
PHOS * O-2-CLORO-1-(2,4-DICLORO-FENIL)-
VINYL-O,O-DIETILFOSFATO (ITALIAN)
* O,O-DIETHYL O-(2-CHLORO-1-(2',4'-
DICHLOROPHENYL)VINYL) PHOSPHATE * ENT
24969 * PHOSPHATE DE O,O-DIETHYLE ET DE
O-2-CHLORO-1-(2,4-DICHLOROPHENYL) VINYLE
(FRENCH)

THR: Poison by ingestion, skin contact, intraperitoneal, subcutaneous, and intravenous
routes. Human blood system effects by skin
contact. A cholinesterase inhibitor. See also
parathion. When heated to decomposition it
emits very toxic fumes of Cl^- and PO_x. For
further information see Vol. 2, No. 4 of *DPIM
Report*.

CHLORHYDRIN   HR: 3
CAS: 96-24-2   NIOSH: TY 4025000
mf: $C_3H_7ClO_2$ mw: 110.55

PROP: Colorless liquid. bp: 213° decomp, d:
1.326.

SYNS: CHLORODEOXYGLYCEROL * 1-CHLORO-
2,3-DIHYDROXYPROPANE * 3-CHLORO-1,2-DI-
HYDROXYPROPANE * ALPHA-CHLOROHYDRIN
* 1-CHLOROPROPANE-2,3-DIOL * 3-CHLORO-
PROPANE-1,2-DIOL * 3-CHLORO-1,2-PROPANE-
DIOL * 3-CHLOROPROPYLENE GYLCOL

* 2,3-DIHYDROXYPROPYL CHLORIDE * GLYC-
ERIN-ALPHA-MONOCHLORHYDRIN * GLYCEROL
CHLOROHYDRIN * GLYCEROL-ALPHA-CHLORO-
HYDRIN * GLYCEROL-ALPHA-MONOCHLOROHY-
DRIN * GLYCERYL-ALPHA-CHLOROHYDRIN
* MONOCHLOROHYDRIN * ALPHA-MONOCHLO-
ROHYDRIN

THR: Poison by ingestion, inhalation, and intraperitoneal route. Mutagenic data. An eye irritant. Slight fire hazard when exposed to heat
or flame. When heated to decomposition, it
emits toxic fumes of Cl^-.

CHLORIC ACID   HR: 3
CAS: 7790-93-4   NIOSH: FN 9750000
DOT: 2626
mf: $ClHO_3$ mw: 84.46

PROP: Colorless solution; mp: $< -20°$, bp: decomp @ 40°, d: 1.282 @ 14.2°.

SYN: CHLORINE DIOXIDE HYDRATE, FROZEN
(DOT)

DOT Classification: Label: Oxidizer and
Poison

THR: Poison. Strongly irritating by ingestion
and inhalation. See also chlorates, chlorine.
Dangerous fire hazard; ignites organic matter
upon contact; a very powerful oxidizing agent.
Greater than 40% is unstable, reacts violently
with NH_3; Sb; Sb_2S_3; As_2S_3; Bi; CuS; PHI_4;
SnS_2; SnS. Reacts vigorously with reducing materials. Incompatible with cellulose; copper sulphide; oxidizable materials. For further information see Vol. 4, No. 1 of *DPIM Report*.

CHLORIDES
THR: Varies widely. Sodium chloride (table
salt) has very low toxicity, while carbonyl chloride (phosgene) is lethal in small doses. See
therefore specific entries. When heated to decomposition or on contact with acids or acid
fumes, they evolve highly toxic chloride fumes.
Some organic chlorides decompose to yield
phosgene.

CHLORINATED HYDROCARBONS, AL-
IPHATIC
SYN: CHLORINATED HC, ALIPHATIC

THR: Variable toxicity. The substitution of a
Cl (or other halogen) atom for a hydrogen greatly
increases the anesthetic action of the aliphatic
hydrocarbons and increases the range of their

systemic effects. In many cases, the chlorine derivative is quite toxic. In general, the unsaturated chlorine derivatives are more narcotic but less toxic than the saturated derivatives. In the saturated group, the narcotic effect is proportional to the number of chlorine atoms. This relationship is not true for toxicity. The toxic action may result from repeated low exposures. Individual susceptibility varies widely. It often causes experimental carcinogen of liver, lung, skin, and blood forming tissues. When heated to decomposition, they emit highly toxic fumes of phosgene; they react violently with Al; liquid O_2; K; Na. In dealing with these chlorinated hydrocarbons, it must be remembered that a toxic action may result from repeated exposure to concentrations which are too low to produce a narcotic effect, and which, consequently, are too low to give warning of danger. Individual susceptibility is also important when poisoning by this group of solvents is being considered. Certain workmen may be seriously affected by concentrations that seem to have no effect on fellow employees with the same exposure. A carcinogen of the liver, lung, skin and blood forming tissues.

CHLORINATED HYDROCARBONS, AROMATIC

SYN: CHLORINATED HC AROMATIC

THR: In most instances, it is difficult to predict the toxicity of these compounds. However, in the case of most aromatic chlorine compounds, their toxicity is usually no greater, and frequently is less, than that of the corresponding aliphatic hydrocarbons, with the notable exception of naphthalene. React violently with Al; liquid O_2; K; Na. When heated to decomposition, they emit toxic fumes of Cl^-; they can react with oxidizing materials.

CHLORINE   HR: 3

CAS: 7782-50-5   NIOSH: FO 2100000
mf: Cl_2 mw: 70.90

PROP: Greenish-yellow gas, liquid, or rhombic crystals. mp: $-101°$, bp: $-34.5°$, d: (liq) 1.47 @ 0° (3.65 atm), vap press: 4800 mm @ 20°, vap d: 2.49.

DOT: 1017

SYNS: BERTHOLITE ∗ CHLOOR (DUTCH) ∗ CHLOR (GERMAN) ∗ CHLORE (FRENCH) ∗ CLORO (ITALIAN)

OSHA PEL: TWA 1 ppm
ACGIH TLV: TWA 1 ppm; STEL 3 ppm
DFG MAK: 0.5 ppm (1.5 mg/m^3)
DOT Classification: Nonflammable Gas; Poison

THR: Poison to humans by inhalation. A strong irritant to eyes and mucous membranes. Chlorine is extremely irritating to the mucous membranes of the eyes @ 3 ppm and the respiratory tract. Combines with moisture to liberate nascent oxygen and form HCl. Both these substances, if present in quantity, cause inflammation of the tissues with which they come in contact. If the lung tissues are attacked, pulmonary edema may result. A concentration of 3.5 ppm produces a detectable odor; 15 ppm causes immediate irritation of the throat. Concentrations of 50 ppm are dangerous for even short exposures, 1,000 ppm may be fatal, even where the exposure is brief. Because of its intensely irritating properties, severe industrial exposure seldom occurs, as the worker is forced to leave exposure before he can be seriously affected. In cases where this is impossible, the initial irritation of the eyes and mucous membranes of the nose and throat is followed by cough, a feeling of suffocation, and later, pain and a feeling of constriction in the chest. If exposure has been severe, pulmonary edema may follow, with rales being heard over the chest. It is a common air contaminant. Can react to cause fires or explosions upon contact with turpentine; ether; ammonia gas; illuminating gas; hydrocarbons; hydrogen and metal dusts; polydimethyl siloxane; polypropylene; drawing wax; rubber; sulfamic acid; $As_2(CH_3)_4$; UC_2; acetaldehyde; C_2H_2; alcohols; alkylisothiourea salts; alkyl phosphines; Al; Sb; As; AsS_2; AsH_3; Ba_3P_2; C_6H_6; Bi; B; BPI_2; B_2S_3; brass BrF_5; Ca; (CaC_2 + KOH); $Ca(ClO_2)_2$; Ca_3N_2 Ca_3P_2; C; CS_2; Cs; $CsHC_2$; Co_2O; Cs_3N; (C + $Cr(OCl)_2$)); Cu; CuH_2; CuC_2; dialklyl phosphines; diborane; dibutyl phthalate; $Zn(C_2H_5)_2$; C_2H_6; C_2H_4; ethylene imine; $C_2H_5PH_2$; F_2; Ge; glycerol; $(NH_2)_2$; $(H_2O + KOH)$; I_2; hydroxylamine; Fe; FeC_2; Li; Li_2C_2; Li_6C_2; Mg; Mg_2P_3; Mn; Mn_3P_2; HgO; HgS; Hg; Hg_3P_2; CH_4; Nb; NI_3; OF_2; H_2SiO; (OF_2 + Cu); PH_3; P; $P(SNC)_3$; P_2O_3; PCB's; K; KHC_2; KH; Ru; $RuHC_2$; Si; SiH_2; Ag_2O; Na; $NaHC_2$; Na_2C_2; SnF_2; SbH_3; Sr_3P; Te; Th; Sn; WO_2; U; V; Zn; ZrC_2. Will react with water or steam to produce toxic and corrosive fumes of HCl. Incompatible with alcohols;

aluminum; bromine pentafluoride; tert butanol; carbon disulphide; dibutyl phthalate; dicaesium oxide; diethyl ether; dioxygen difluoride; fluorine; glycerol; hexachlorodisilane; hydrocarbons; hydrogen; metal acetylides and carbides; metal hydrides; metal phosphides; metals; nitrogen compounds; non-metal hydrides; non-metals; oxygen difluoride; phosphorus compounds; polychlorobiphenyl; silicones; steel; sulphides; synthetic rubber; tetraselenium tetranitride; trialkylboranes; tungsten dioxide. For further information, see Vol. 1, No. 3 of *DPIM Report*.

CHLORINE AZIDE   HR: 1
mf: ClN_3 mw: 77.47

PROP: An explosive gas.

SYN: CHLOR(O)AZIDE

THR: Strong irritant by inhalation. Severe explosion hazard when shocked; exposed to heat, flame, or 1,3-butadiene; C_2H_6; C_2H_4; CH_4; C_3H_8. When heated to decomposition, it emits highly toxic fumes of chlorine and NO_x; will react with water or steam to produce toxic and corrosive fumes of HCl. Self-explodes or is incompatible with ammonia; phosphorus; silver azide; sodium.

CHLORINE OXIDE   HR: 3
CAS: 10049-04-4   NIOSH: FO 3000000
mf: ClO_2 mw: 67.45

PROP: Red-yellow gas or orange-red crystals; mp: $-59°$; bp: 9.9° @ 731 mm explodes; d: 3.09 g/L @ 11°.

SYNS: CHLORINE DIOXIDE * CHLORINE PEROXIDE * CHLORINE(IV) OXIDE

OSHA PEL: TWA 0.1 ppm
ACGIH TLV: TWA 0.1 ppm; STEL 0.3 ppm
DFG MAK: 0.1 ppm (0.3 mg/m³)
DOT Classification: Forbidden (not hydrated); Oxidizer (hydrated)

THR: A poison via strong irritant by inhalation. A powerful oxidizer. Reacts violently with P; KOH; S; concentrations from 0.1 to 1.0 atmosphere of > 10% in air, explodes; also F_2Hg; organic matter; NHF_2. When heated to decomposition, it emits highly toxic fumes of chlorine; will react with water or steam to produce toxic and corrosive fumes of HCl. Incompatible with chlorine dioxide; carbon monoxide; hydrogen; mercury; non-metals; phosphorus pentachloride.

CHLORINE PENTAFLUORIDE   HR: 3
CAS: 13637-63-3   NIOSH: FO 2975000
mf: ClF_5 mw: 130.34

THR: Poison by inhalation. See also chlorine, fluorine, fluorides and chlorine trifluoride. When heated to decomposition, it emits very toxic fumes of Cl^- and F^-. Incompatible with nitric acid; water.

CHLORINE TRIFLUORIDE   HR: 3
CAS: 7990-91-2   NIOSH: FO 2800000
mf: ClF_3 mw: 92.45

PROP: Colorless gas to yellow liquid, sweet odor, mp: $-83°$, bp: 11.8°, d: 1.77 @ 13°.

DOT: 1749

SYN: TIRFLUORURE DE CHLORE (FRENCH)

OSHA PEL: TWA CL 0.1 ppm
ACGIH TLV: TWA 0.1 ppm
DFG MAK: 0.1 ppm (0.4 mg/m³)
DOT Classification: Oxidizer; Poison; Corrosive

THR: Poison by inhalation. An eye irritant. See also fluorides, chlorine, and fluorine. Spontaneously flammable. Reacts violently with organic matter; glass wool; acetic acid; Al; Sb; As; Cu; H_2; I_2; Ir; Fe; Pb; Mg; Mo; Os; P; K; Rh; Se; Si; Ag; Na; S; Te; Sn; W; Zn; oxides; NH_3; benzene; CO; CrO_3; ether; graphite; H_2S; HgI_2; HNO_3; K_2CO_3; KI; rubber; $AgNO_3$; NaOH; H_2SO_4; V_2P_5; water; WO_3. When heated to decomposition or on contact with acid or acid fumes, emits highly toxic fumes; will react with water or steam to produce much heat and toxic and corrosive fumes; reacts vigorously with reducing materials. Incompatible with acids; ammonium fluoride; carbon tetrachloride; fluorinated polymers; fuels; hydrogen-containing materials; iodine; metals, metal oxides, metal salts, non-metals, or non-metal oxides; nitrocompounds; organic materials; polychlorotrifluoroethylene; refractory materials.

CHLORITES
THR: See individual chlorites. Reacts violently with NH_3; organic matter; metals.

2-CHLOROACETALDEHYDE   HR: 3
CAS: 107-20-0   NIOSH: AB 2450000
DOT: 2232
mf: C_2H_3ClO mw: 78.50

PROP: Clear, colorless liquid, pungent odor; bp: 90.0°-100.1° (40% soln), fp: −16.3° (40% soln), flash p: 190°F, d: 1.19 @ 25°/25° (40% soln), vap press: 100 mm @ 45° (40% soln).

SYNS: CHLOROACETALDEHYDE MONOMER * 2-CHLORO-1-ETHANAL * MONOCHLOROAC-ETALDEHYDE

OSHA PEL: TWA CL 1 ppm
DOT Classification: Poison B

THR: Poison by ingestion, intraperitoneal, and dermal routes. Mutagenic data. See also aldehydes. Moderate fire hazard when exposed to heat or flame. To fight fire, use water, foam, CO_2, dry chemical. Dangerous, when heated to decomposition it emits very toxic chlorides; reacts with oxidizing materials. For further information, see Vol. 2, No. 4 of *DPIM Report*.

2-CHLORO ACETAMIDE   HR: 3
CAS: 79-07-2   NIOSH: AB 5075000
mf: C_2H_4ClNO mw: 93.52

PROP: Cryst, mod water sol; mp: 120°; bp: 225° (decomp).

SYNS: CHLORACETAMID (GERMAN) * CHLO-ROACETAMIDE * ALPHA-CHLOROACETAMIDE * 2-CHLOROETHANAMIDE * USAF DO-29

THR: Poison by ingestion, intravenous, and intraperitoneal routes. When heated to decomposition, it emits very toxic Cl^- and NO_x.

CHLOROACETIC ACID   HR: 3
CAS: 79-11-8   NIOSH: AF 8575000
mf: $C_2H_3ClO_2$ mw: 94.50

PROP: Colorless cryst; mp: (α) 63°; (β) 56°; (τ) 50°; bp: 189°; flash p: 259°F; d: 1.58 @ 20°/20°; vap d: 3.26.

SYNS: ACIDE MONOCHLORACETIQUE (FRENCH) * ACIDOMONOCLOROACETICO (ITALIAN) * CHLORACETIC ACID * ALPHA-CHLOROACETIC ACID * CHLOROETHANOIC ACID * MONO-CHLOORAZIJNZUUR (DUTCH) * MONOCHLOR-ACETIC ACID * MONOCHLORESSIGSAEURE (GERMAN) * MONOCHLOROACETIC ACID * MONOCHLOROETHANOIC ACID * NCI-C60231

THR: Poison by ingestion and subcutaneous route. An experimental tumorigen. A powerful skin, eye, and mucous membrane irritant. Slight fire hazard. To fight fire, use water spray, fog, mist, dry chemical, foam. When heated to de-

composition, it emits toxic fumes of Cl^-. For further information, see Vol. 3, No. 5 of *DPIM Report*.

p-CHLORO ACETO ACETANI-LIDE   HR: 2
CAS: 101-92-8   NIOSH: AK 4375000
mf: $C_{10}H_{10}ClNO_2$ mw: 211.66

PROP: Crystals; mp: 133°, bp: decomp, flash p: 320°F (CC), d: 1.348 @ 20°, vap press: <0.011 mm @ 20°, vap d: 7.31.

SYNS: ACETOACET-P-CHLOROANILIDE * ACE-TOACETYL-4-CHLOROANILIDE * 4'-CHLOROACE-TOACETANILIDE

THR: Moderately toxic by intraperitoneal route. Slight fire hazard when exposed to heat or flame. See also aniline, phosgene and cyanides; can react vigorously with oxidizing materials. To fight fire, use water, foam, CO_2, dry chemical. When heated to decomposition, it emits very toxic fumes of Cl^- and NO_x.

CHLOROACETONE   HR: 3
CAS: 78-95-5   NIOSH: UC 0700000
mf: C_3H_5ClO mw: 92.53

PROP: Colorless liq; pungent odor; mp: −44.5°; bp: 119°; d: 1.162.

SYNS: ACETONYL CHLORIDE * CHLORACE-TONE (FRENCH) * CHLOROPROPANONE * 1-CHLORO-2-PROPANONE * MONOCHLOR-ACETONE * MONOCHLOROACETONE

THR: Poison by ingestion. An experimental tumorigen. A lachrymator poison gas. See also chlorinated hydrocarbons, aliphatic, and acetone. Mutagenic data. Moderate fire hazard when exposed to heat or flame. Old material can explode. When heated to decomposition, it emits highly toxic fumes of phosgene; can react vigorously with oxidizing materials. Polymerization explosion has occurred.

2-CHLOROACETOPHENONE   HR: 3
CAS: 532-27-4   NIOSH: AM 6300000
mf: C_8H_7ClO mw: 154.60
DOT: 1697

SYNS: ALPHA-CHLOROACETOPHENONE * OMEGA-CHLOROACETOPHENONE * 1-CHLO-ROACETOPHENONE * CHLOROMETHYL PHENYL KETONE * MACE (LACRYMATOR) * NCI-C55107 * PHENACYL CHLORIDE * PHENYL-

CHLOROMETHYLKETONE * PHENYL CHLORO-
METHYL KETONE

OSHA PEL: TWA CL 0.05 ppm
DOT Classification: Irritant (IMO:Poison B)

THR: Poison in humans by inhalation. Poison
by ingestion, intraperitoneal, and intravenous
routes. A lacrymator. See also ketones. When
heated to decomposition, it emits toxic fumes
of Cl⁻. For further information see Vol. 4, No.
1 of *DPIM Report*.

p-CHLOROACETOPHENONE   HR: 3
CAS: 99-91-2   NIOSH: KM 5600000
mf: C_8H_7ClO mw: 154.60

PROP: Pale straw-colored liquid or white crystals, fragrant, non-persistent odor. mp: 56°, bp:
237°-247°, fp: 59°, d: 1.19 @ 25°/25°, vap press:
0.012 mm @ 0°, vap d: 5.2, flash p: 244°F.

SYNS: USAF DO-1 * PHENACYLCHLORIDE
* PHENYLCHLOROMETHYLKETONE

THR: Poison by ingestion and inhalation. A
powerful irritant and lachrymator. When heated
to decomposition, it emits toxic fumes, will
react with water or steam to produce toxic and
corrosive fumes of Cl⁻. Slight fire hazard. To
fight fire, use water, foam, alcohol foam, dry
chemical.

4'-CHLOROACETYL ACETANI-
LIDE   HR: 2
CAS: 140-49-8   NIOSH: AE 1050000
mf: $C_{10}H_{10}ClNO_2$ mw: 211.66

SYNS: P-ACETAMIDOPHENACYL CHLORIDE
* P-(ACETYLAMINO)PHENACYL CHLORIDE
* NCI-C03770

THR: Moderately toxic by ingestion. When
heated to decomposition, it emits very toxic
fumes of Cl⁻ and NO$_x$.

CHLOROACETYL CHLORIDE   HR: 3
CAS: 79-04-9   NIOSH: AO 6475000
DOT: 1752
mf: $C_2H_2Cl_2O$ mw: 112.94

PROP: Water white or sltly yellow liquid; bp:
105°-106°, fp: −22.5°, flash p: none, d: 1.495
@ 0°.

SYNS: CHLOROACETIC ACID CHLORIDE
* CHLOROACETIC CHLORIDE * CHLORURE DE
CHLORACETYLE (FRENCH) * MONOCHLOROACE-
TYL CHLORIDE

DOT Classification: Corrosive

THR: Poison by ingestion, inhalation, and intravenous routes. A lacrymator. When heated
to decomposition, it emits toxic fumes of
Cl.

alpha-CHLOROALLYL CHLOR-
IDE   HR: 3
CAS: 542-75-6   NIOSH: UC 8310000
mf: $C_3H_4Cl_2$ mw: 110.97

PROP: Liquid; bp: 103°-110°, flash p: 95°F,
d: 1.22, vap d: 3.8.

SYNS: gamma-CHLOROALLYL CHLORIDE
* DICHLOROPROPENE * 1,3-DICHLOROPROPENE
* 1,3-DICHLOROPROPENE-1 * alpha,gamma-
DICHLOROPROPYLENE * 1,3-DICHLOROPRO-
PYLENE * NCI-C03985

THR: Poison by ingestion and inhalation. Moderately toxic by dermal route. A strong irritant.
Has produced liver and kidney injury in experimental animals. Mutagenic data. Dangerous fire
hazard when exposed to heat, flame or oxidizers.
See also chlorides; can react vigorously with
oxidizing materials. To fight fire, use water,
foam, CO_2, dry chemical.

1-CHLOROANTHRAQUINONE   HR: 3
CAS: 82-44-0   NIOSH: CB 6150000
mf: $C_{14}H_8ClNO_2$ mw: 257.68

SYNS: 1-CHLORANTHRACHINON (CZECH)
* alpha-CHLOROANTHRAQUINONE *
1-CHLORO-9,10-ANTHRAQUINONE * alpha-
MONOCHLOROANTHRAQUINONE

THR: Poison by intravenous route. A skin and
eye irritant. When heated to decomposition, it
emits very toxic fumes of Cl⁻ and NO$_x$.

o-CHLOROBENZALDEHYDE   HR: 3
CAS: 89-98-5   NIOSH: CU 5075000
mf: C_7H_5ClO mw: 140.57

SYNS: 2-CHLOORBENZALDEHYDE (DUTCH)
* 2-CHLORBENZALDEHYD (GERMAN) * 2-
CHLOROBENZALDEHYDE * 2-CLOROBENZAL-
DEIDE (ITALIAN) * O-CHLOROBENZENECARBOX-
ALDEHYDE * USAF M-7

THR: Poison by intraperitoneal and intravenous
routes. When heated to decomposition, it emits
toxic fumes of Cl⁻.

(o-CHLORO BENZAL)MALONO NITRILE   HR: 3
CAS: 2698-41-1   NIOSH: OO 3675000
mf: $C_{10}H_5ClN_2$ mw: 188.62

PROP: White crystals, solid; mp: 95°, bp: 313°.

SYNS: 2-CHLOROBENZALMALONONITRILE
* o-CHLOROBENZYLIDENE MALONITRILE
* 2-CHLOROBENZYLIDENE MALONONITRILE
* beta,beta-DICYANO-o-CHLOROSTYRENE
* NCI-C55118 * USAF KF-11 * PROPANEDI-
NITRILE, ((2-CHLOROPHENYL)METHYLENE)

OSHA PEL: TWA 0.05 ppm

THR: Poison by ingestion, intraperitoneal, and intravenous routes. Moderately toxic by inhalation. Human central nervous system effects. A human skin and eye irritant. See also nitriles. Human exposure data suggest relatively low systematic toxicity, but intense irritation of eyes, skin, and mucous membranes. When heated to decomposition, it emits very toxic fumes of Cl^-, NO_x, and CN^-.

o-CHLOROBENZONITRILE   HR: 3
CAS: 873-32-5   NIOSH: DI 2625000
mf: C_7H_4ClN mw: 137.57

PROP: (o and p). Crystals.

SYNS: o-CHLORBENZONITRIL (CZECH) * NITRIL KYSELINY-o-CHLORBENZOOVE (CZECH)

THR: Poison by ingestion and intraperitoneal route. A severe eye irritant. Dangerous hazard when heated to decomposition; see also cyanides and chlorides; they will react with water, steam, acid, or acid fumes to produce toxic fumes.

p-CHLOROBENZYL-p-CHLOROPHENYL SULFIDE   HR: 2
CAS: 103-17-3   NIOSH: WQ 2975000
mf: $C_{13}H_{10}Cl_2S$ mw: 269.19

PROP: Crystals, almond-like odor, insol in water, sol in most organic solvents; mp: 75°-76°, d: 1.4210 @ 25°/4°, vap press: 1.21 ×10⁻⁵ mm @ 30°.

SYNS: (4-CHLOOR-BENZYL)-(4-CHLOOR-FENYL)-SULFIDE (DUTCH) * CHLORBENSID (GERMAN) * CHLORBENZIDE * (4-CHLOR-BENZYL)-(4-CHLOR-PHENYL)-SULFID (GERMAN) * P-CHLO-ROBENZYL-P-CHLOROPHENYL SULPHIDE
* 4-CHLOROBENZYL-4-CHLOROPHENYL SULPHIDE
* 1-CHLORO-4-(((4-CHLOROPHENYL)METHYL)-THIO)BENZENE * 4-CHLOROPHENYL-4'-CHLORO-

BENZYL SULFIDE * (4-CLORO-BENZIL)-(4-CLORO-FENIL)-SOLFURO (ITALIAN) * P,P'-DI-CHLORODIPHENYL SULFIDE * ENT 20,696
* SULFURE DE 4-CHLOROBENZYLE ET DE 4-CHLOROPHENYLE (FRENCH)

THR: Moderately toxic by ingestion. Has caused liver and kidney injury and skin irritation in experimental animals. When heated to decomposition it emits Cl^-, SO_x. For hazards, see also chlorides and sulfur compounds.

trans-CHLORO(2-(3-BROMO PROPION AMIDO)CYCLO HEXYL)MERCURY   HR: 3
CAS: 73926-87-1   NIOSH: OV 9400000
mf: $C_9H_{15}BrClHgNO$ mw: 469.20

SYNS: CHLORO(2-(3-BROMOPROPIONAMIDO)CY-CLOHEXYL), MERCURY (E)- * 3-BROMO-N-(2-CHLOROMERCURICYCLOHEXYL)PROPIONAMIDE

THR: Poison by intravenous route. See also mercury compounds. When heated to decomposition, it emits very toxic fumes of Br^-, Cl^-, NO_x, and Hg.

1-CHLOROBUTADIENE   HR: 2
                              NIOSH: EI 9623000
mf: C_4H_5Cl mw: 88.54

PROP: Colorless liquid; bp: 59.4°, d: 0.9583, flash p: −4°F, lel = 4.0%, uel = 20.0%, vap d: 3.0.

SYN: 1-CHLORO-1,3-BUTADIENE

THR: Moderately toxic by ingestion, inhalation, and subcutaneous routes. Exposure to the vapor first causes irritation of the respiratory tract, followed by depression of respiration and, if exposure is continued, asphyxia. The vapor is a central nervous system depressant; blood pressure is lowered. Lung changes accompany exposure to the higher concs. Humans exposed to chloroprene have been reported to develop dermatitis, conjunctivitis, corneal necrosis, anemia, temporary loss of hair, nervousness and irritability. Mutagenic data. Dangerous when heated to decomposition; see also chlorides. Dangerous fire hazard when exposed to heat or flame. To fight fire, use alcohol foam.

2-CHLORO-1,3-BUTADIENE POLYMER   HR: 2
CAS: 9010-98-4   NIOSH: EI 9660000
mf: $(C_4H_5Cl)n$

SYNS: 2-CHLORO-1,3-BUTADIENE HOMOPOLY-MER (9CI) * CHLOROBUTADIENE POLYMER * NEOPRENE * POLY(2-CHLOROBUTADIENE) * POLY(2-CHLORO-1,3-BUTADIENE) * POLY-CHLOROPRENE

THR: Suspected carcinogen. When heated to decomposition, it emits toxic fumes of Cl⁻.

2-CHLOROBUTANE   HR: 3
CAS: 78-86-4   NIOSH: EJ 6475000
mf: C₄H₉Cl   mw: 92.58

PROP: Flash p: 14°F, d: 0.87, vap d: 3.2, bp: 68.50.

SYN: sec-BUTYL CHLORIDE

THR: An experimental neoplastigen. Moderately toxic by inhalation. Dangerous fire hazard from heat, open flame (sparks), or oxidizers. To fight fire, use water, water spray, fog, mist, dry chemical, alcohol foam. When heated to decomposition, it emits toxic fumes of Cl⁻.

m-CHLORO CARBANILIC ACID-4-CHLORO-2-BUTYNYL ESTER   HR: 3
CAS: 101-27-9   NIOSH: FD 7700000
mf: C₁₁H₉Cl₂NO₂   mw: 258.11

SYNS: 2-BUTYNYL-4-CHLORO-M-CHLOROCARBA-NILATE * CARBYNE * CBN * (4-CHLOOR-BUT-2-YN-YL)-N-(3-CHLOOR-FENYL)-CARBAMAAT (DUTCH) * (4-CHLOR-BUT-2-IN-YL)-N-(3-CHLOR-PHENYL)-CARBAMAT (GERMAN) * (4-CLORO-BUT-2-IN-IL)-N-(3-CLORO-FENIL)-CARBAMMATO (ITALIAN) * 4-CHLORO-2-BUTY-NYL-M-CHLOROCARBANILATE * 4-CHLORO-BUT-2-YNYL-3-CHLOROPHENYLCARBAMATE * 4-CHLORO-2-BUTYNYL-N-(3-CHLOROPHENYL)-CARBAMATE * M-CHLOROCARBANILIC ACID, 4-CHLORO-2-BUTYNYL ESTER * N-(3-CHLORO PHENYL) CARBAMATE DE 4-CHLORO 2-BUTYNYLE (FRENCH)

THR: Poison by ingestion and other routes. Moderately toxic by inhalation. Mutagenic data. See also carbamates, esters. When heated to decomposition, it emits very toxic fumes of Cl⁻ and NOₓ.

2-CHLORO-N-(2-CHLOROETHYL)-N-METHYL ETHANAMINE-N-OXIDE
                                        HR: 3
CAS: 126-85-2   NIOSH: IZ 2200000
mf: C₅H₁₁Cl₂NO   mw: 172.07

SYNS: 2,2′-DICHLORO-N-METHYLDIETHYL-AMINE-N-OXIDE * MECHLORETHAMINE OXIDE

* METHYLBIS(BETA-CHLOROETHYL)AMINE-N-OX-IDE * N-METHYL-DI-2-CHLOROETHYLAMINE-N-OXIDE * NITROGEN MUSTARD AMINE OXIDE * NITROGEN MUSTARD-N-OXIDE * N-OXYD-LOST (GERMAN) * N-OXYD-MUSTARD (GER-MAN) * NSC 10107

THR: Poison by ingestion, intraperitoneal, and subcutaneous routes. An experimental carcinogen and tumorigen. Mutagenic data.

2-CHLORO-N-(2-CHLORO ETHYL)-N-METHYL ETHANAMINE-N-OXIDE HYDROCHLORIDE   HR: 3
CAS: 302-70-5   NIOSH: IA 2275000
mf: C₅H₁₁Cl₂NO • ClH   mw: 208.53

SYNS: 2,2′-DICHLORO-N-METHYLDIETHYL-AMINE-N-OXIDE HYDROCHLORIDE * ME-CHLORETHAMINE OXIDE HYDROCHLORIDE * METHYLBIS(beta-CHLOROETHYL)AMINE-N-OXIDE HYDROCHLORIDE * N-METHYLBIS(2-CHLOROETHYL)AMINE-N-OXIDE HYDROCHLORIDE * N-METHYL-2,2′-DICHLORODIETHYLAMINE-N-OXIDE HYDROCHLORIDE * METHYLDI(2-CHLOROETHYL)AMINE-N-OXIDE HYDROCHLORIDE * NITROGEN MUSTARD-N-OXIDE HYDROCHLO-RIDE * N-OXYD-LOST * NSC-10107

THR: Poison by intravenous and other routes. An experimental carcinogen. Mutagenic data.

4-CHLORO-m-CRESOL   HR: 3
CAS: 59-50-7   NIOSH: GO 7100000
mf: C₇H₇ClO   mw: 142.59

PROP: Odorless crystals (when pure). Somewhat sol in water, very sol in organic solvents; mp: 66°, bp: 235°.

SYNS: p-CHLOR-m-CRESOL * p-CHLOROCRE-SOL * 4-CHLORO-M-CRESOL * 2-CHLORO-HY-DROXYTOLUENE * 6-CHLORO-3-HYDROXY-TOLUENE * 4-CHLORO-3-METHYLPHENOL * 3-METHYL-4-CHLOROPHENOL

THR: Poison by ingestion and subcutaneous routes. An allergen. See also cresol. Dangerous; when heated to decomposition it emits very toxic Cl⁻. For hazard, see also phosgene. Incompatible with sodium hydroxide. For further information see Vol. 6, No. 1 of *DPIM Report*.

CHLOROCYCLINE   HR: 3
CAS: 82-93-9   NIOSH: TL 1925000
mf: C₁₈H₂₁ClN₂   mw: 300.86

SYNS: CHLORCYCLINE * CHLORCYCLIZINE * 1-(4-CHLOROBENZHYDRYL)-4-METHYLPIPERA-

ZINE * 1-(p-CHLORO-alpha-PHENYLBENZYL)-4-
METHYLPIPERAZINE

THR: Poison by intraperitoneal route. An experimental teratogen. When heated to decomposition, it emits very toxic fumes of Cl⁻ and NOₓ.

2-CHLORO-N,N-DIALLYLACET-AMIDE   HR: 3

CAS: 93-71-0   NIOSH: AB 5250000
mf: $C_8H_{12}ClNO$ mw: 173.66

PROP: Amber liq, sltly sol in H_2O, sol in alc, hexane, xylene. bp: 74° @ 0.3 mm.

SYNS: N,N-DIALLYLCHLOROACETAMIDE
* N,N-DIALLYL-alpha-CHLOROACETAMIDE
* N,N-DIALLYL-2-CHLOROACETAMIDE * NCI-
CO4035

THR: Poison by dermal route. Moderately toxic by ingestion. When heated to decomposition, it emits very toxic fumes of Cl⁻ and NOₓ.

CHLORODIBROMOMETHANE   HR: 2

CAS: 124-48-1   NIOSH: PA 6360000
mf: $CHBr_2Cl$ mw: 208.29

PROP: Colorless to pale yellow, heavy liquid; bp: 118°-122°, fp: < −20°, d: 2.440 @ 25°/25°.

SYNS: DIBROMOCHLOROMETHANE * NCI-
C55254

THR: Moderately toxic by ingestion. Compounds of this type are generally irritating and narcotic. See also bromoform and chloroform. When heated to decomposition, it emits toxic fumes of Cl⁻ and Br⁻. For further information see Vol. 5, No. 2 of *DPIM Report*.

2-CHLORO-2′,6′-DIETHYL-N-(METHOXY METHYL)ACETANILIDE   HR: 2

CAS: 15972-60-8   NIOSH: AE 1225000
mf: $C_{14}H_{20}ClNO_2$ mw: 269.80

SYNS: ALOCHLOR * 2-CHLORO-N-(2,6-DI-
ETHYL)PHENYL-N-METHOXYMETHYLACETAMIDE
* LASSO * METHACHLOR

THR: Moderately toxic by ingestion, dermal, and other routes. When heated to decomposition, it emits very toxic fumes of Cl⁻ and NOₓ.

6-CHLORO-3,4-DIHYDRO-2H-1,2,4-BEN-ZOTHIADIAZINE-7-SULFONAMIDE- 1,1-DIOXIDE   HR: 3

CAS: 58-93-5   NIOSH: DK 9100000
mf: $C_7H_8ClN_3O_4S_2$ mw: 297.75

SYNS: DIHYDROCHLOROTHIAZIDE * 6-
CHLORO-3,4-DIHYDRO-7-SULFAMOYL-2H-1,
2,4-BENZOTHIADIAZINE-1,1-DIOXIDE * 6-
CHLORO-7-SULFAMOYL-3,4-DIHYDRO-2H-1,2,
4-BENZOTHIADIAZINE-1,1-DIOXIDE * 3,4-
DIHYDRO-6-CHLORO-7-SULFAMYL-1,2,4-BENZO-
THIADIAZINE-1,1-DIOXIDE * 3,4-DIHYDRO-
CHLOROTHIAZIDE * HYDRO-DIURIL
* HYPOTHIAZIDE * NCI-C55925

THR: Poison by intraperitoneal and intravenous routes. Moderately toxic by ingestion, subcutaneous, and other routes. When heated to decomposition, it emits very toxic fumes of SOₓ, Cl⁻, and NOₓ.

7-CHLORO-1,3-DIHYDRO-3-HYDROXY-5-PHENYL-2H-1,4-BENZODIAZEPINE-2-ONE   HR: 3

CAS: 604-75-1   NIOSH: DF 1400000
mf: $C_{15}H_{11}ClN_2O_2$ mw: 286.73

SYNS: 7-CHLORO-3-HYDROXY-5-PHENYL-1,3-
DIHYDRO-2H-1,4-BENZODIAZEPIN-2-ONE
* TAZEPAM

THR: An experimental neoplastigen. Mutagenic data. Moderately toxic by ingestion and intraperitoneal routes. Central nervous system effects in children by ingestion. When heated to decomposition, it emits very toxic fumes of Cl⁻ and NOₓ.

CHLORO DIISOBUTYL ALUMI-NUM   HR: 2

CAS: 1779-25-5   NIOSH: BD 0560000
mf: $C_8H_{18}AlCl$ mw: 176.69

SYNS: ALUMINIO DIISOBUTIL-MONOCLORURO
(ITALIAN) * BIS(ISOBUTYL)ALUMINUM CHLO-
RIDE * DIISOBUTYLALUMINUM CHLORIDE
* DIISOBUTYLALUMINUM MONOCHLORIDE
* DIISOBUTYLCHLOROALUMINUM

THR: Moderately toxic by inhalation. See also aluminum. Self-ignites in air is thus a fire hazard. When heated to decomposition, it emits toxic fumes of Cl⁻.

P-CHLORODIMETHYLAMINOAZOBEN-ZENE   HR: 3

CAS: 2491-76-1   NIOSH: BX 5060000
mf: $C_{14}H_{14}ClN_3$ mw: 259.76

SYNS: 4′-CHLORO-4-DIMETHYLAMINOAZOBEN-
ZENE * N,N-DIMETHYL-p-((p-CHLOROPHENYL)-
AZO)ANILINE

THR: An experimental teratogen and neoplastigen. Moderately toxic by subcutaneous route. When heated to decomposition it emits very toxic fumes of Cl$^-$ and NO$_x$.

2'-CHLORO-N,N-DIMETHYL-4-STIL-BENAMINE   HR: 3
CAS: 63020-91-7   NIOSH: WJ 3675000
mf: C$_{16}$H$_{16}$ClN   mw: 257.78

SYNS: 2'-CHLORO-4-DIMETHYLAMINOSTILBENE * 2'-CHLORO-4-STILBENYL-N,N-DIMETHYL-AMINE

THR: An experimental tumorigen and carcinogen. When heated to decomposition, it emits very toxic fumes of Cl$^-$ and NO$_x$.

1-CHLORO-2,4-DINITROBEN-ZENE   HR: 3
CAS: 97-00-7   NIOSH: CZ 0525000
mf: C$_6$H$_3$ClN$_2$O$_4$ mw: 202.56

PROP: Yellow rhombic crystals, insol in water. mp(α): 53.4°, mp(β): 43°, mp(γ): 27°, bp: 315°, lel = 2.0%, uel = 22%, flash p: 382°F (CC), d(α): 1.687 @ 22°, d(β):1.680 @ 20°/4°, vap d: 6.98.

SYNS: 1-CHLOOR-2,4-DINITROBENZEEN (DUTCH) * 4-CHLORO-1,3-DINITROBENZENE * 1-CHLORO-2,4-DINITROBENZOL (GERMAN) * 1-CLORO-2,4-DINITROBENZENE (ITALIAN) * 2,4-DINITROCHLOROBENZENE * 1,3-DINITRO-4-CHLOROBENZENE * 2,4-DINITRO-1-CHLOROBENZENE *

DOT Classification: Label: Poison

THR: Poison by intraperitoneal and dermal routes. Moderately toxic by ingestion. A human skin irritant and sensitizer. An allergen. Mutagenic data. Slight fire hazard when exposed to heat or flame. Explosion hazard when exposed to flame, sparks, heated to 150°, or, if confined, by shock. Reacts violently with hydrazine hydrate. See nitrates for hazard. To fight fire, use CO$_2$, dry chemical.

CHLORO DIPHENYL ARSINE   HR: 3
CAS: 712-48-1   NIOSH: CG 9900000
mf: C$_{12}$H$_{10}$AsCl   mw: 264.59

PROP: Colorless crystals when pure, technical product is dark brown liquid. bp: 333° (decomp), fp: 44°, d: 1.363 @ 40° (solid): 1.358 @ 45°

(liquid), vap press: 0.00049 mm @ 20°, vap d: 9.15.

SYNS: BLUE CROSS * DIPHENYLCHLOORARSINE (DUTCH) * DIPHENYLCHLOROARSINE * SNEEZING GAS

THR: Poison by inhalation and dermal routes. A powerful irritant. See also arsenic compounds and chloride for hazards. When heated to decomposition it evolves As and Cl$^-$.

1-CHLORO-2,3-EPOXY PRO-PANE   HR: 3
CAS: 106-89-8   NIOSH: TX 4900000
mf: C$_3$H$_5$ClO   mw: 92.53

PROP: Colorless, mobile liquid, irr chloroform-like odor; bp: 117.9°, flash p: 105.1°F (OC) (40°C), mp -25.6°C, d: 1.1761 @ 20°/20°, vap press: 10 mm @ 16.6°, vap d: 3.29. Insol in H$_2$O; misc in alc, ether, chloroform, carbon tetrachloride, trichloroethylene; immisc in petr hydrocarbons.

SYNS: 1-CHLOOR-2,3-EPOXY-PROPAAN (DUTCH) * 1-CHLOR-2,3-EPOXY-PROPAN (GERMAN) * 3-CHLORO-1,2-EPOXYPROPANE * (CHLOROMETHYL)ETHYLENE OXIDE * 2-(CHLOROMETHYL)OXIRANE * gamma-CHLOROPROPYLENE OXIDE * 3-CHLORO-1,2-PROPYLENE OXIDE * 1-CLORO-2,3-EPOSSIPROPANO (ITALIAN) * EPICHLOORHYDRINE (DUTCH) * EPICHLORHYDRIN (GERMAN) * EPICHLORHYDRINE (FRENCH) * EPICHLOROHYDRIN * EPICHLOROHYDRIN (DOT) * alpha-EPICHLOROHYDRIN * (DL)-alpha-EPICHLOROHYDRIN * EPICHLOROHYDRYNA (POLISH) * EPICLORIDRINA (ITALIAN) * 1,2-EPOXY-3-CHLOROPROPANE * 2,3-EPOXYPROPYL CHLORIDE * GLYCEROL EPICHLORHYDRIN

OSHA PEL: TWA 5 ppm (skin)
DOT Classification: Label: Flammable Liquid and Poison

THR: Poison by ingestion, inhalation, intraperitoneal, and subcutaneous routes. An experimental carcinogen. Moderately toxic by skin absorption. Mutagenic data. Human eye effects (systemic). Skin irritant and sensitizer. Moderate fire hazard when exposed to heat or flame. When heated to decomposition, it emits toxic fumes of Cl$^-$. Incompatible with isopropylamine; potassium tert-butoxide; trichloroethylene. For further information, see Epichlorhydrin, Vol. 3, No. 3 of *DPIM Report*.

2-CHLORO ETHANOL PHOS-PHATE   HR: 3
CAS: 115-96-8   NIOSH: KK 2450000
mf: $C_6H_{12}Cl_3O_4P$ mw: 285.50

PROP: Flash p: 421°F (COC); boil range: 210°-220° @ 20 mm d: 1.425 @ 20°/20°; autoign temp: 1115°F; vap press: 0.5 mm @ 145°.

SYNS: NCI-C60128 * NIAX FLAME RETAR-DANT 3 CF * TRI(2-CHLOROETHYL)PHOSPHATE * TRIS(BETA-CHLOROETHYL) PHOSPHATE * TRIS(2-CHLOROETHYL) PHOSPHATE

THR: Poison by intraperitoneal route. Moderately toxic by ingestion. Mild skin and eye irritant. See also esters. Slight fire hazard when exposed to heat or flame. When heated to decomposition, it emits very toxic fumes of PO_x and Cl^-.

2-CHLOROETHYL ALCOHOL   HR: 3
CAS: 107-07-3   NIOSH: KK 0875000
mf: C_2H_5ClO mw: 80.52

PROP: Colorless liquid, faint ethereal odor; mp: −69°, bp: 128.8°, flash p: 140°F (OC), d: 1.197 @ 20°/4°, autoign temp: 797°F, vap press: 10 mm @ 30.3°, vap d: 2.78, lel = 4.9%, uel = 15.9%.

SYNS: AETHYLENECHLORHYDRIN (GERMAN) * 2-CHLOORETHANOL (DUTCH) * 2-CHLORA-ETHANOL (GERMAN) * beta-CHLOROETHYL ALCOHOL * delta-CHLOROETHANOL * CHLO-ROETHYLOWY ALKOHOL (POLISH) * 2-CLORO-ETHANOLO (ITALIAN) * ETHYLEEN-CHLOORHY-DRINE (DUTCH) * ETHYLENE CHLOROHYDRIN (DOT) * GLICOL MONOCLORIDRINA (ITALIAN) * GLYCOL CHLOROHYDRIN * GLYCOLMONO-CHLOORHYDRINE (DUTCH) * GLYCOL MONO-CHLOROHYDRIN * MONOCHLORHYDRINE DU-GLYCOL (FRENCH) * 2-MONOCHLOROETHANOL * NCI-C50135

OSHA PEL: TWA 5 ppm (skin)
DOT Classification: Label: Poison

THR: A narcotic poison by ingestion, inhalation, skin absorption, intraperitoneal, and subcutaneous routes affecting the nervous system and the liver, spleen, and lungs. Mutagenic data. Mild skin irritant. Severe eye irritant. Moderate fire hazard when exposed to heat, flame or oxidizers. When heated to decomposition, it emits highly toxic fumes of phosgene; will react with water or steam to produce toxic and corrosive fumes; can react with oxidizing materials. To fight fire, use alcohol foam, CO_2, dry chemical. Violent reaction with chlorosulfonic acid; ethylene diamine; sodium hydroxide.

1-(2-CHLOROETHYL)-3-CYCLOHEXYL-1-NITROSOUREA   HR: 3
CAS: 13010-47-4   NIOSH: YS 4900000
mf: $C_9H_{16}ClN_3O_2$ mw: 233.73

SYNS: CCNU * NCI-C04740 * NSC-79037

THR: Poison by ingestion, intraperitoneal, subcutaneous, intravenous and parenteral routes. Mutagenic data. Human gastrointestinal tract effects, blood effects. An experimental teratogen and tumorigen. When heated to decomposition, it emits very toxic fumes of Cl^- and NO_x.

N-(2-CHLORO ETHYL)DIETHYL AMINE   HR: 3
CAS: 100-35-6   NIOSH: YE 0700000
mf: $C_6H_{14}ClN$ mw: 135.66

SYNS: (2-CHLOROETHYL)DIETHYLAMINE * beta-CHLOROTRIETHYLAMINE * 2-CHLORO-TRIETHYLAMINE * 2-(DIETHYLAMINO)CHLORO-ETHANE * DIETHYLAMINOETHYL CHLORIDE * beta-(DIETHYLAMINO)ETHYL CHLORIDE * 2-(DIETHYLAMINO)ETHYL CHLORIDE * DIETHYL(2-CHLOROETHYL)AMINE * N-DI-ETHYLAMINOETHYL CHLORIDE

THR: Poison by ingestion and skin absorption. A skin and eye irritant. See also amines. When heated to decomposition, it emits very toxic fumes of Cl^- and NO_x.

N-(2-CHLORO ETHYL)DIMETHYL AMINE   HR: 2
CAS: 107-99-3   NIOSH: KQ 9015000
mf: $C_4H_{10}ClN$ mw: 107.6

PROP: Liquid; vap d: 3.72.

SYNS: CHLORO(DIMETHYLAMINO)ETHANE * (2-CHLOROETHYL)DIMETHYLAMINE * DI-METHYLAMINOETHYL CHLORIDE * beta-(DI-METHYLAMINO)ETHYL CHLORIDE * 2-DIMETH-YLAMINOETHYLCHLORIDE * DIMETHYL(2-CHLO-ROETHYL)AMINE * NITROGEN HALF MUSTARD

THR: Probably an irritant poison. Mutagenic data. When heated to decomposition, it emits highly toxic fumes of chlorides and NO_x.

CHLORO ETHYLENE-1,1-DICHLORO ETHYLENE POLYMER   HR: 3
CAS: 9011-06-7   NIOSH: KV 9750000
mf: $(C_2H_3Cl \cdot C_2H_2Cl_2)x$

SYNS: 1,1-DICHLOROETHENE POLYMER WITH CHLOROETHENE * 1,1-DICHLOROETHYLENE-MONOCHLOROETHYLENE POLYMER * 1,1-DICHLOROETHYLENEPOLYMER WITH CHLOROETHYLENE * VINYL CHLORIDE COPOLYMER WITH VINYLIDENE CHLORIDE * VINYL CHLORIDE-1,1-DICHLOROETHYLENE COPOLYMER * VINYLIDENE CHLORIDE-VINYL CHLORIDE POLYMER

THR: An experimental carcinogen. When heated to decomposition, it emits very toxic fumes of Cl$^-$.

CHLOROETHYL ETHYL SULFIDE   HR: 3
CAS: 693-07-2   NIOSH: WQ 3250000
mf: C_4H_9ClS mw: 124.64

SYNS: 2-CHLOROETHYL ETHYL THIOETHER * 2-ETHYLTHIOETHYL CHLORIDE * HALF-MUSTARD GAS * 2-CHLOROETHYL ETHYL SULFIDE * 1-CHLORO-2-(ETHYLTHIO)ETHANE * ETHYL-2-CHLOROETHYL SULFIDE * BETA-ETHYLMERKAPTOETHYLCHLORID (CZECH) * 2-(ETHYLTHIO)CHLOROETHANE

THR: Poison by ingestion. Mutagenic data. A skin and eye irritant. See also ethers, sulfides. When heated to decomposition, it emits very toxic fumes of Cl$^-$ and SO$_x$.

CHLORO ETHYL MERCURY   HR: 3
CAS: 107-27-7   NIOSH: OV 9800000
mf: C_2H_5ClHg mw: 265.11

PROP: Silvery, irridescent leaflets. mp: 192.5°.

SYNS: CERESAN * ETHYLMERCURIC CHLORIDE * ETHYLMERCURY CHLORIDE

OSHA PEL: TWA 10 ug(Hg)/m^3; CL 40

THR: Poison by ingestion, skin absorption, subcutaneous, inhalation, intraperitoneal, and other routes. Mutagenic data. See also mercury compounds, organic. When heated to decomposition, it emits very toxic fumes of Cl$^-$ and Hg.

2-CHLORO ETHYL METHANE SULFONATE   HR: 3
CAS: 3570-58-9   NIOSH: KK 1960000
mf: $C_3H_7ClO_3S$ mw: 158.61

SYNS: beta-CHLOROETHYLMETHANESULFONATE * NSC 18016

THR: Poison by intravenous and intraperitoneal routes. Mutagenic data. See also sulfonates.

When heated to decomposition, it emits very toxic fumes of Cl$^-$ and SO$_x$.

1-(2-CHLOROETHYL)-3-(4-METHYL-CYCLOHEXYL)-1-NITROSOUREA   HR: 3
CAS: 13909-09-6   NIOSH: YS 5000000
mf: $C_{10}H_{18}ClN_3O_2$ mw: 247.76

SYNS: N-(2-CHLOROETHYL)-N'-(TRANS-4-METHYLCYCLOHEXYL)-N-NITROSOUREA * NCI-c04955 * NSC-95441 * SEMUSTINE

THR: Poison by ingestion, intraperitoneal, intravenous, and parenteral routes. Mutagenic data. An experimental tumorigen. Human blood effects and gastrointestinal system effects. When heated to decomposition, it emits very toxic fumes of Cl$^-$ and NO$_x$.

2-CHLORO ETHYL-N-NITROSO URETHANE   HR: 3
CAS: 6296-45-3   NIOSH: EZ 2100000
mf: $C_5H_9ClN_2O_3$ mw: 180.61

SYNS: N-(beta-CHLOROETHYL)-N-NITROSOURETHAN * ETHYL-N-(beta-CHLOROETHYL)-N-NITROSOCARBAMATE

THR: Poison by inhalation, ingestion, and intraperitoneal routes. An experimental tumorigen. See also carbamates. When heated to decomposition, it emits very toxic fumes of Cl$^-$ and NO$_x$.

1-CHLORO-3-ETHYL-1-PENTEN-4-YN-3-OL   HR: 3
CAS: 113-18-8   NIOSH: SB 4725000
mf: C_7H_9ClO mw: 144.61

SYNS: AETHYL-CHLORVYNOL * beta-CHLOROVINYL ETHYLETHYNYL CARBINOL * 3-(beta-CHLOROVINYL)-1-PENTYN-3-OL * ETHYL-beta-CHLOROVINYLETHYNYL CARBINOL

THR: Poison by ingestion, subcutaneous, intraperitoneal, and intravenous routes. Human central nervous system effects. When heated to decomposition, it emits toxic fumes of Cl$^-$.

2-CHLOROETHYL VINYL ETHER   HR: 3
CAS: 110-75-8   NIOSH: KN 6300000
mf: C_4H_7ClO mw: 106.56

PROP: Liquid. bp: 109° @ 740 mm, d: 1.0525, flash p: 80°F (OC), mp: −70.3°.

SYNS: VINYL-beta-CHLOROETHYL ETHER * VINYL 2-CHLOROETHYL ETHER

THR: Poison by ingestion and inhalation. A severe eye and skin irritant. See also ethers.

Dangerous fire hazard when exposed to heat, flame, or oxidizers. See ethers for explosion hazard. See also chlorides for When heated to decomposition it emits Cl^-, phosgene; it; can react with oxidizing materials. To fight fire, use alcohol foam, dry chemical.

CHLOROFORM   HR: 3
CAS: 67-66-3   NIOSH: FS 9100000
mf: $CHCl_3$ mw: 119.37

PROP: Colorless liquid, heavy, ethereal odor; mp: $-63.5°$, bp: $61.26°$, fp: $-63.5°$, flash p: none, d: 1.49845 @ $15°$, vap press: 100 mm @ $10.4°$, vap d: 4.12.

DOT: 1888

SYNS: CHLOROFORME (FRENCH) * CLORO-FORMIO (ITALIAN) * FORMYL TRICHLORIDE * FREON 20 * METHANE TRICHLORIDE * METHENYL TRICHLORIDE * METHYL TRI-CHLORIDE * NCI-C02686 * TRICHLOORME-THAAN (DUTCH) * TRICHLORMETHAN (CZECH) * TRICHLOROFORM * TRICHLOROMETHANE * TRICLOROMETANO (ITALIAN)
ACGIH TLV: TWA 10 ppm
DFG MAK: 10 ppm (50 mg/m^3)
DOT Classification: ORM-A (IMO: Poison B)

THR: Poison to humans by ingestion, inhalation. Moderately toxic by intraperitoneal and subcutaneous routes. A suspected human carcinogen. An experimental teratogen and carcinogen. Human central nervous system effects and systemic effects. It has been widely used as an anesthetic. However, due to its toxic effects, this use is being abandoned. 68,000-82,000 ppm in air kill most animals in a few minutes. 14,000 ppm is dangerous to life after an exposure of from 30 to 60 minutes. 5,000-6,000 ppm can be tolerated by animals for 1 hour without serious disturbances. The maximum concentration tolerated for several hours or for prolonged exposure with slight symptoms is 2,000-2,500 ppm. The harmful effects are narcosis, and damage to the liver and heart. Prolonged administration as an anesthetic may lead to such serious effects as profound toxemia and damage to the liver, heart and kidneys. Chloroform causes irritation of the conjunctiva. Upon inhalation, it causes dilation of the pupils with reduced reaction to light, as well as reduced intraocular pressure (experimental). The material is well known as an an-esthetic. In the initial stages there is a feeling of warmth of the face and body, then an irritation of the mucous membranes and skin followed by nervous aberration. Prolonged inhalation will bring on paralysis accompanied by cardiac respiratory failure and finally death. Experimental prolonged but light anesthesia in dogs produces a typical hepatitis. Inhalation of the concentrated chloroform vapor results in irritation of the mucous surfaces exposed to it. The narcosis is ordinarily preceded by a stage of excitation which is followed by loss of reflexes, sensation, and consciousness. See also chlorinated hydrocarbons. Reacts violently with (acetone + a base); Al; disilane; Li; Mg; nitrogen tetroxide; K; (perchloric acid + phosphorus pentoxide); (KOH + methanol); K-*tert*-butoxide; Na; (NaOH + methanol); sodium methylate; NaK. A skin and eye irritant. Slight fire hazard when exposed to high heat. When heated to decomposition, it emits toxic fumes of Cl^-. Incompatible with acetone; alkali; dinitrogen tetroxide; fluorine; metals; potassium tert-butoxide; sodium, sodium hydroxide, methanol; sodium methoxide; triisopropylphosphine. For further information, see Vol. 3, No. 5 of *DPIM Report*.

Treatment and Antidotes: If it has been ingested, or there has been great overexposure, the following antidotes may be applied: emetics, stomach siphon, friction, cold douche, fresh air, strychnine (hypodermically; from 1/120 to 1/60 grain), rubefacients, artificial respiration, etc. If during exposure to unknown amounts of chloroform vapor, the patient should feel any of the symptoms noted above, he should immediately be moved to fresh air and kept under observation until the symptoms disappear.

4-CHLORO-N-FURFURYL-5-SULFA-MOYL ANTHRANILIC ACID   HR: 3
CAS: 54-31-9   NIOSH: CB 2625000
mf: $C_{12}H_{11}ClN_2O_5S$ mw: 330.76

SYNS: 4-CHLORO-N-(2-FURYLMETHYL)-5-SULFA-MOYLANTHRANILIC ACID * CHLOR-N-(2-FURYL-METHYL)-5-SULFAMYLANTHRANILSAEURE (GER-MAN) * FUROSEMIDE * LASEX * LASIX * NCI-C55936

THR: Poison by intravenous route. Moderately toxic by ingestion. Mutagenic data. Human central nervous system effects; a diuretic. Ingestion can damage liver and, rarely, affects hearing. When heated to decomposition, it emits very toxic fumes of Cl^-, NO_x, and SO_x.

CHLORO HYDROQUINONE   HR: 3
CAS: 615-67-8   NIOSH: MX 4800000
mf: $C_6H_5ClO_2$ mw: 144.56

THR: Poison by ingestion and intraperitoneal route. Moderately toxic by skin absorption. When heated to decomposition, it emits toxic fumes of Cl^-.

2-CHLORO-4-(HYDROXY MERCURI)-PHENOL   HR: 3
CAS: 538-04-5   NIOSH: OW 0650000
mf: $C_6H_5ClHgO_2$ mw: 345.15

PROP: Insol solid. Contains 20% Mercury.

SYNS: (3-CHLORO-4-HYDROXYPHENYL)HY-DROXYMERCURY * SEMESAN

THR: Poison by ingestion, inhalation, and intravenous route. See also mercury compounds. When heated to decomposition, it emits very toxic fumes of Cl^- and Hg.

(-)-N-((5-CHLORO-8-HYDROXY-3-METHYL-1-OXO-7-ISOCHROMANYL)-CARBONYL)-3-PHENYLALANINE   HR: 3
CAS: 303-47-9   NIOSH: AY 4375000
mf: $C_{29}H_{18}ClNO_6$ mw: 403.84

SYNS: (R)N-((5-CHLORO-3,4-DIHYDRO-8-HY-DROXY-3-METHYL-1-OXO-1H-2-BENZOPYRAN-7-YL)PHENYLALANINE * NCI-C56586
* OCHRATOXIN A

THR: Poison by ingestion, intraperitoneal, intravenous, and subcutaneous routes. An experimental carcinogen and teratogen. When heated to decomposition, it emits very toxic fumes of Cl^- and NO_x.

5-CHLORO-2-((2-HYDROXY-1-NAPH-THYL)AZO)-p-TOLUENESULFONIC ACID, BARIUM SALT   HR: 3
CAS: 5160-02-1   NIOSH: DB 5500000
mf: $C_{17}H_{12}ClN_2O_4S \cdot 1/2Ba$ mw: 444.49

SYNS: 5-CHLORO-2-((2-HYDROXY-1-NAPHTHAL-ENYL)AZO)-4-METHYLBENZENESULFONICACID, BARIUM SALT (2:1) * 5-CHLORO-2-((2-HY-DROXY-1-NAPHTHALENYL)AZO)-4-METHYL-BENZENESULPHONICACID, BARIUM SALT * 5-CHLORO-2-((2-HYDROXY-1-NAPHTHYL)-AZO)-p-TOLUENESULPHONIC ACID, BARIUM SALT * 1-(4-CHLORO-o-SULFO-5-TOLYLAZO)-2-NAPHTHOL,BARIUM SALT * NCI-C53792

THR: A carcinogen. When heated to decomposition, it emits very toxic fumes of SO_x, NO_x, and Cl^-.

5-CHLORO-7-IODO-8-QUINOLI-NOL   HR: 3
CAS: 130-26-7   NIOSH: VC 5075000
mf: C_9H_5ClINO mw: 305.50

SYNS: 5-CHLORO-8-HYDROXY-7-IODOQUINOLINE
* 5-CHLORO-7-IODO-8-HYDROXYQUINOLINE
* IODOCHLORHYDROXYQUIN * IODOCHLORHY-DROXYQUINOL * IODOCHLORHYDROXYQUINO-LINE * 7-IODO-5-CHLORO-8-HYDROXYQUINO-LINE

THR: Poison by ingestion. Human eye (systemic) and central nervous system effects. When heated to decomposition, it emits very toxic fumes of Cl^-, I^-, and NO_x.

2-CHLORO-N-ISOPROPYLACETANI-LIDE   HR: 3
CAS: 1918-16-7   NIOSH: AE 1575000
mf: $C_{11}H_{14}ClNO$ mw: 211.71

SYNS: alpha-CHLORO-N-ISOPROPYLACETANILIDE
* 2-CHLORO-N-ISOPROPYL-N-PHENYLACETAMIDE
* N-ISOPROPYL-alpha-CHLOROACETANILIDE
* N-ISOPROPYL-2-CHLOROACETANILIDE

THR: Poison by ingestion and dermal route. A pre-emergence, selective herbicide. When heated to decomposition, it emits very toxic fumes of Cl^- and NO_x.

3-CHLOROLACTONITRILE   HR: 3
CAS: 33965-80-9   NIOSH: OD 8575000
mf: C_3H_4ClNO mw: 105.53

THR: Poison by intravenous, intraperitoneal, and dermal routes. When heated to decomposition, it emits very toxic fumes of Cl^- and NO_x.

p-CHLORO-MERCURIC BENZOIC ACID   HR: 3
CAS: 59-85-8   NIOSH: OV 8050000
mf: $C_7H_5ClHgO_2$ mw: 357.16

SYNS: BENZOIC ACID-P-(CHLOROMERCURI)
* USAF D-3

THR: Poison by intraperitoneal route. See also mercury compounds. When heated to decomposition, it emits very toxic fumes of Cl^- and Hg.

CHLOROMETHANE   HR: 3
CAS: 74-8-7-3   NIOSH: PA 6300000
DOT: 1063
mf: CH_3Cl mw: 50.49

PROP: Colorless gas, ethereal odor and sweet taste; d: 0.918 @ 20°/4°; mp: −97°; bp: −23.7°;

flash p: <32°F; lel = 8.1%; uel = 17%; autoign temp: 1170°F; vap d: 1.78. Sltly sol in water, miscible with chloroform, ether, glacial acetic acid; sol in alcohol.

SYNS: ARTIC * METHYL CHLORIDE * MONO-CHLOROMETHANE

OSHA PEL: TWA 100 ppm; CL 200; Pk 300/ 5M/3H
ACGIH TLV: TWA 50 ppm; STEL 100 ppm
DFG MAK: 50 ppm (105 mg/m³)
DOT Classification: Label: Flammable Gas

THR: Chloromethane has very slight irritating properties and may be inhaled without noticeable discomfort. It has some narcotic action, but this effect is weaker than that of chloroform. Acute poisoning, characterized by the narcotic effect, is rare in industry. Repeated exposure to low concentrations causes damage to the central nervous system and, less frequently, to the liver, kidneys, bone marrow and cardiovascular system. Hemorrhages into the lungs, intestinal tract and dura have been reported. Sprayed on the skin, chloromethane produced anesthesia through freezing of the tissues as it evaporates. In exposures to high concentrations, dizziness, drowsiness, incoordination, confusion, nausea and vomiting, abdominal pains, hiccoughs, diplopia and dimness of vision are followed by delirium, convulsions and coma. Death may be immediate, and if the exposure is not fatal, recovery is usually slow, and degenerative changes in the central nervous system are not uncommon. The liver, kidneys and bone marrow may be affected, with resulting acute nephritis and anemia. Death may occur several days after exposure, resulting from degenerative changes in the heart, liver and especially the kidneys. Used as a food additive permitted in food for human consumption. Very dangerous fire hazard when exposed to heat, flame or powerful oxidizers. Moderate explosion hazard when exposed to flame (sparks). Incompatibles: Al; Mg; K; Na; NaK; aluminum trichloride; ethylene; inter-halogens; metals. Dangerous; when heated to decomposition it emits highly toxic fumes of Cl⁻. To Fight fire: Stop flow of gas; CO_2, dry chemical or water spray. For further information see Vol. 2, No. 4 of *DPIM Report*.

CHLOROMETHANE SULFONYL CHLO-RIDE   HR: 3
CAS: 3518-65-8   NIOSH: PB 2800000
mf: $CH_2Cl_2O_2S$ mw: 148.99

SYNS: CHLORID KYSELINY CHLORMETHANSULFO-NOVE (CZECH) * CHLORMETHANSULFOCHLORID (CZECH)

THR: Poison by ingestion. A skin and eye irritant. When heated to decomposition, it emits very toxic fumes of Cl⁻ and SO_x.

CHLOROMETHAPYRILENE   HR: 3
CAS: 148-65-2   NIOSH: US 7350000
mf: $C_{14}H_{18}ClN_3S$ mw: 295.86

SYNS: 2-((5-CHLORO-2-THENYL)(2-DIMETHYLAMINOETHYL)AMINO)PYRIDINE * CHLOROTHENYLPYRAMINE * N,N-DIMETHYL-N'-(2-PYRIDYL)-N'-(5-CHLORO-2-THENYL)ETHYL-ENEDIAMINE * NCI-C60559

THR: Poison by intraperitoneal route. When heated to decomposition, it emits very toxic fumes of Cl⁻, NO_x, and SO_x.

p-CHLORO-N-METHYLAMPHET-AMINE   HR: 3
CAS: 1199-85-5   NIOSH: SH 0875000
mf: $C_{10}H_{14}ClN$ mw: 183.70

SYN: P-CHLORO-N-alpha-DIMETHYLPHENETHYL-AMINE

THR: Poison by intraperitoneal and subcutaneous routes. When heated to decomposition it emits very toxic fumes of Cl⁻ and NO_x.

7-CHLOROMETHYL BENZ(a)ANTHRA-CENE   HR: 3
CAS: 6325-54-8   NIOSH: CW 0960000
mf: $C_{19}H_{13}Cl$ mw: 276.77

SYN: ICR 451

THR: Poison by intravenous route. Mutagenic data. An experimental neoplastigen. When heated to decomposition it emits toxic Cl⁻.

6-CHLOROMETHYL BENZO(a)-PYRENE   HR: 3
CAS: 49852-84-8   NIOSH: DJ 4920000
mf: $C_{21}H_{13}Cl$ mw: 300.79

THR: An experimental carcinogen and neoplastigen. When heated to decomposition it emits very toxic fumes of Cl⁻.

2-CHLORO-6-METHYLCARBANILIC ACID-2-(PYRROLIDINYL)ETHYL ESTER HYDROCHLORIDE   HR: 3
NIOSH: FD 8580000
mf: $C_{14}H_{19}ClN_2O_2$ • ClH   mw: 319.26

SYNS: C 3067 ∗ 2-(PYRROLIDINYL)ETHYL-2-CHLORO-6-METHYLCARBANILATE HYDROCHLORIDE

THR: Poison by intraperitoneal, subcutaneous, and intravenous routes. An eye irritant. See also esters. When heated to decomposition it emits very toxic fumes of HCl, NO$_x$, and Cl$^-$.

3-CHLORO-4-METHYL-7-COUMARINYL DIETHYL PHOSPHATE   HR: 3
CAS: 321-54-0   NIOSH: GN 5950000
mf: C$_{14}$H$_{16}$ClO$_6$P   mw: 346.72

SYNS: COUMAPHOS-O-ANALOG ∗ O,O-DI(2-CHLOROETHYL)-7-(3-CHLORO-4-METHYLCOUMARINYL)PHOSPHATE ∗ DIETHYL-3-CHLORO-4-METHYL-7-COUMARINYL PHOSPHATE ∗ O,O-DIETHYL-O-(3-CHLORO-4-METHYLCOUMARIN-7-YL) PHOSPHATE

THR: Poison by ingestion. When heated to decomposition it emits very toxic fumes of PO$_x$ and Cl$^-$.

CHLOROMETHYL ETHYL ETHER   HR: 3
CAS: 3188-13-4   NIOSH: KN 6430000
DOT: 2354
mf: C$_3$H$_7$ClO   mw: 94.54

PROP: Flash p: <−2.2°F.

SYNS: CHLOROMETHOXY ETHANE ∗ ETHOXY CHLOROMETHANE ∗ ETHOXY METHYL CHLORIDE ∗ UN 2354(DOT)

DOT Classification: Label: Flammable Liquid and Poison

THR: A poison via inhalation and ingestion. Possibly a carcinogen. A very dangerous fire and explosion hazard. See also ethers.

7-CHLOROMETHYL-12-METHYL BENZ(a)ANTHRACENE   HR: 3
CAS: 13345-62-5   NIOSH: CW 1000000
mf: C$_{20}$H$_{15}$Cl   mw: 290.80

SYN: IRC 453

THR: Poison by intravenous route. An experimental neoplastigen. Mutagenic data. When heated to decomposition it emits toxic Cl$^-$.

CHLOROMETHYL METHYL ETHER   HR: 3
CAS: 107-30-2   NIOSH: KN 6650000
DOT: 1239
mf: C$_2$H$_5$ClO   mw: 80.52

PROP: Flash p: <73.4°F.

SYNS: CHLORDIMETHYLETHER (CZECH) ∗ DIMETHYLCHLOROETHER ∗ ETHER METHYLIQUE MONOCHLORE (FRENCH) ∗ METHYL CHLOROMETHYL ETHER, ANHYDROUS (DOT)

OSHA PEL: Carcinogen
DOT Classification: Label: Flammable Liquid and Poison

THR: Poison by inhalation. Moderately toxic by ingestion. A suspected human carcinogen. An experimental carcinogen, tumorigen, and neoplastigen. See also ethers. When heated to decomposition, it emits toxic fumes of Cl$^-$.

4-CHLORO-N-METHYL-3-(METHYLSULFAMOYL)BENZAMIDE   HR: 2
CAS: 3688-85-5   NIOSH: CV 2450000
mf: C$_9$H$_{11}$ClN$_2$O$_3$S   mw: 262.73

SYNS: C.I. 456 ∗ DIAPAMIDE ∗ THIAMIZIDE

THR: Moderately toxic by ingestion and intraperitoneal route. When heated to decomposition it emits very toxic SO$_x$, NO$_x$, and Cl$^-$.

2-CHLORO-N-METHYL-N-NITROSO-ETHYLAMINE   HR: 3
CAS: 16339-16-5   NIOSH: KR 1970000
mf: C$_3$H$_7$ClN$_2$O   mw: 122.57

SYNS: METHYL-2-CHLORAETHYLNITROSAMIN (GERMAN) ∗ N-NITROSOMETHYL-2-CHLOROETHYLAMINE

THR: Poison by ingestion and intravenous route. An experimental tumorigen. When heated to decomposition it emits very toxic fumes of Cl$^-$ and NO$_x$.

7-CHLORO-1-METHYL-5-PHENYL-1H-1,5-BENZODIAZEPINE-2,4(3H,5H)-DIONE   HR: 3
CAS: 22316-47-8   NIOSH: DE 9600000
mf: C$_{16}$H$_{13}$ClN$_2$O$_2$ mw: 300.76

SYNS: CLOREPIN ∗ 1-PHENYL-5-METHYL-8-CHLORO-1,2,4,5-TETRAHYDRO-2,4-DIOXO-3H-1,5-BENZODIAZEPINE

THR: Poison by ingestion and intraperitoneal route. Moderately toxic by subcutaneous route. When heated to decomposition it emits very toxic fumes of NO$_x$ and Cl$^-$.

2-CHLORO-11-(4-METHYLPIPERAZINO)-DIBENZO(b,f)(1,4)THIAZEPINE   HR: 3
CAS: 2058-52-8   NIOSH: HQ 2100000
mf: $C_{18}H_{18}ClN_3S$ mw: 343.90

SYNS: 2-chloro-11-(4-methyl-1-pipera-zinyl)dibenzo(b,f)(1,4)thiazepine
* dibenzothiazepine

THR: Poison by ingestion. When heated to decomposition it emits very toxic fumes of Cl^-, NO_x, and SO_x.

1-CHLORO-2-METHYLPROPANE   HR: 3
mf: C_4H_9Cl mw: 92.57

PROP: Flash p: 21.2°F; lel = 2.0%; uel = 8.7%.

THR: A poison via ingestion and inhalation routes. A very dangerous fire and explosion hazard.

3-CHLORO-2-METHYLPROPENE   HR: 2
CAS: 563-47-3   NIOSH: UC 8050000
mf: C_4H_7Cl mw: 90.56

PROP: Colorless, volatile liquid, disagreeable odor; bp: 72.17°, lel = 2.3%, uel = 9.3%, fp: < −80°, d: 0.9257 @ 20°/4°, vap press: 101.7 mm @ 20°, vap d: 3.12, flash p: −2.2°F.

SYNS: gamma-chloroisobutylene * 3-chlor-2-methyl-prop-1-en (german)
* 3-chloro-2-methyl-1-propene * 3-cloro-2-metil-prop-1-ene (italian) * chlorure de methallyle (french) * cloruro di metallile (italian) * isobutenyl chloride * methallyl chloride * 2-methyl-allyl-chlorid (german) * beta-methylallyl chloride * 2-methylallyl chloride * nci-c54820

THR: Mildly toxic by inhalation. An irritant. Mutagenic data. Dangerous fire hazard when exposed to heat, flame, or oxidizers. Moderately explosive when exposed to heat or flame. On decomposition it emits highly toxic fumes of chlorides; can react vigorously with oxidizing materials. To fight fire, use alcohol foam, CO_2, dry chemical.

3-(CHLOROMETHYL) PYRIDINE HYDROCHLORIDE   HR: 3
CAS: 6959-48-4   NIOSH: US 7000000
mf: $C_6H_6ClN • ClH$ mw: 164.04

SYN: nci-c03838

THR: Poison by ingestion. An experimental carcinogen. When heated to decomposition it emits very toxic fumes of NO_x, HCl, and Cl^-.

5'-CHLORO-2-(METHYL(2-(PYRROLIDI-NYL)ETHYL)AMINO)-o-ACETOTOLUI-DIDE DIHYDROCHLORIDE   HR: 3
NIOSH: AN 3440600
mf: $C_{16}H_{24}ClN_3O • 2ClH$ mw: 382.80

SYNS: c 5420 * 5'-chloro-2-(methyl(2-(pyrrolidinyl)ethyl)amino)-o-acetotolui-dide dihydrochloride

THR: Poison by intraperitoneal and subcutaneous routes. An eye irritant. When heated to decomposition it emits very toxic fumes of Cl^-, NO_x, and HCl.

1-CHLORO-3-NITROBENZENE   HR: 3
CAS: 121-73-3   NIOSH: CZ 0940000
DOT: 1578
mf: $C_6H_4ClNO_2$ mw: 157.56

PROP: Yellowish crystals, mp: 46°, bp: 236°, d: 1.534 @ 20°/4°.

SYNS: chloro-m-nitrobenzene * m-chloronitrobenzene * m-nitrochlorobenzene

DOT Classification: Label: Poison

THR: Poison by inhalation and ingestion. When absorbed, it forms methemoglobin and gives rise to cyanosis and blood changes. Its effects are analogous to those of nitrobenzene. It can cause poisoning by the pulmonary route and its effects are cumulative. Chemically, it is probably reduced in the body to chloroaniline, which is also poisonous. The p- compound is thought to be somewhat less toxic than the o- compound. In industry, it is the dust of this material that is most often the source of intoxication. Dangerous fire hazard; see nitrates and phosgene; can react with oxidizing materials.

1-CHLORO-4-NITROBENZENE   HR: 3
CAS: 100-00-5   NIOSH: CZ 1050000
DOT: 1578
mf: $C_6H_4ClNO_2$ mw: 157.56

SYNS: 1-chloor-4-nitrobenzeen (dutch)
* 1-chlor-4-nitrobenzol (german)
* p-chloronitrobenzene * 4-chloro-1-nitrobenzene * 1-cloro-4-nitrobenzene (italian) * p-nitrochlorobenzol (german)
* p-nitrochloorbenzeen (dutch) * p-ni-

TROCHLOROBENZENE ∗ P-NITROCLOROBENZENE (ITALIAN)

OSHA PEL: TWA 1 mg/m³ (skin)
DOT Classification: Label: Poison

THR: A poison. An experimental carcinogen. When heated to decomposition it emits very toxic fumes of NO_x and Cl^-.

CHLORO-o-NITROBENZENE   HR: 3
CAS: 88-73-3   NIOSH: CZ 0875000
DOT: 1578
mf: $C_6H_4ClNO_2$ mw: 157.56

PROP: Yellow crystals, mp: 32°-33°, bp: 245°-246°, d: 1.305, flash p: 261°F.

SYNS: O-CHLORONITROBENZENE ∗ 1-CHLORO-2-NITROBENZENE ∗ 2-CHLORO-1-NITROBENZENE ∗ O-NITROCHLOROBENZENE

DOT Classification: Label: Poison

THR: Poison by ingestion and inhalation. See also 1-chloro-3-nitrobenzene. An experimental carcinogen and neoplastigen. To fight fire, use water, foam.

CHLORONITROPROPANE   HR: 3
CAS: 2425-66-3   NIOSH: TX 5250000
mf: $C_3H_6ClNO_2$ mw: 123.55

SYNS: CHLORONITROPROPAN (POLISH) ∗ 1-CHLORO-2-NITROPROPANE

THR: Poison by ingestion, inhalation, and skin absorption. When heated to decomposition it emits very toxic fumes of Cl^- and NO_x.

1-CHLORO-1-NITROPROPANE   HR: 3
CAS: 600-25-9   NIOSH: TX 5075000
mf: $C_3H_6ClNO_2$ mw: 123.55

PROP: Liquid, bp: 139.5°, flash p: 144°F (OC), d: 1.209 @ 20°/20°, vap d: 4.26.

SYN: KORAX

OSHA PEL: TWA 20 ppm

THR: Poison by ingestion, inhalation, and subcutaneous routes. Causes injury to kidneys, liver, and cardiovascular system. Moderate fire hazard when exposed to heat, flame (sparks), and oxidizers. Moderately explosive when exposed to heat. Dangerous; see chlorides; can react with oxidizing materials. To fight fire, use alcohol foam, water, CO_2, dry chemical.

2-CHLORO-2-NITROPROPANE   HR: 3
CAS: 594-71-8   NIOSH: TX 5425000
mf: $C_3H_6ClNO_2$ mw: 123.55

PROP: Liquid, bp: 134°, flash p: 135°F (OC), d: 1.197 @ 20°/20°, vap d: 4.26.

THR: Poison by subcutaneous route. Moderately toxic by ingestion. Moderate fire hazard via heat, flame, and oxidizers. Explodes on rapid heating. See also nitrates and Cl^- for disaster hazard.

CHLOROPENTAFLUORO ETH-ANE   HR: 2
CAS: 76-15-3   NIOSH: KH 7877500
DOT: 1020
mf: C_2ClF_5 mw: 154.47

PROP: Colorless gas, insol in water, sol in alc and ether; bp: −39.3°, mp: −77°.

SYNS: FREON 115 ∗ MONOCHLOROPENTA-FLUOROETHANE

DOT Classification: Nonflammable gas

THR: Moderately toxic by inhalation. A food additive permitted in food for human consumption. A nonflammable gas. Dangerous on heating to decomposition, it emits chlorides and fluorides.

CHLOROPHACINONE   HR: 3
CAS: 3691-35-8   NIOSH: NK 5335000
mf: $C_{23}H_{15}ClO_3$ mw: 374.83

SYNS: 2-((P-CHLOROPHENYL)PHENYLACETYL)-1,3-INDANDIONE ∗ CHLOORFACINON (DUTCH) ∗ 2(2-(4-CHLOOR-FENYL-2-FENYL)-ACETYL)-IN-DAAN-1,3-DION (DUTCH) ∗ CHLORFACINON (GERMAN) ∗ 2-(ALPHA-P-CHLOROPHENYLACE-TYL)INDANE-1,3-DIONE ∗ ((4-CHLORPHENYL)-1-PHENYL)-ACETYL-1,3-INDANDION (GERMAN) ∗ 2(2-(4-CHLOROPHENYL)-2-PHENYLACETYL)IN-DAN-1,3-DIONE ∗ CHLORPHACINON (ITALIAN) ∗ 2(2-(4-CHLOR-PHENYL-2-PHENYL)ACETYL)IN-DAN-1,3-DION (GERMAN) ∗ 2(2-(4-CLORO-FENIL-2FENIL)-ACETIL)INDAN-1,3-DIONE (ITAL-IAN) ∗ 2-(2-PHENYL-2-(4-CHLOROPHENYL)-ACETYL)-1,3-INDANDIONE

THR: Poison by ingestion and skin absorption. When heated to decomposition it emits toxic fumes of Cl^-.

CHLOROPHENAMIDINE   HR: 3
CAS: 6164-98-3   NIOSH: LQ 4375000
mf: $C_{10}H_{13}ClN_2$ mf: 196.70

SYNS: CHLORDIMEFORM * N'-(4-CHLORO-O-TOLYL)-N,N-DIMETHYLFORMAMIDINE * N'-(4-CHLORO-2-METHYLPHENYL)-N,N-DIMETHYLMETH-ANIMIDAMIDE * N'-(4-CHLOR-O-TOLYL)-N, N-DIMETHYLFORMAMIDIN (GERMAN) * N,N-DIMETHYL-N'-(2-METHYL-4-CHLOROPHENYL)-FORMAMIDINE * METHANIMIDAMIDE, N'-(4-CHLORO-2-METHYLPHENYL)-N,N-DI-METHYL-N'-(2-METHYL-4-CHLORPHENYL)-FOR-MAMIDIN-HYDROCHLORID (GERMAN)

THR: Poison by ingestion, skin absorption, and intraperitoneal routes. When heated to decomposition it emits very toxic fumes of NO_x and Cl^-. For further information, see Chlordimeform, Vol. 2, No. 6 of *DPIM Report*.

2-CHLOROPHENOL   HR: 3
CAS: 95-57-8   NIOSH: SK 2625000
mf: C_6H_5ClO mw: 128.56

PROP: Light amber liquid, bp: 174.5°, fp: 7°, d: 1.256 @ 25°/25°, flash p: 147°F, vap press: 1 mm @ 12.1°.

SYNS: O-CHLOROPHENOL

THR: Poison by intraperitoneal, intravenous, and subcutaneous routes. Moderately toxic by ingestion. An experimental tumorigen. Moderate fire hazard when exposed to heat, flame, or oxidizers. Dangerous when decomposed by heat; see also phenol and chlorides. Can react with oxidizers. To fight fire, use alcohol foam. For further information, see o-chlorophenol, Vol. 4, No. 6 of *DPIM Report*.

3-CHLOROPHENOL   HR: 3
CAS: 108-43-0   NIOSH: SK 2450000
mf: C_6H_5ClO mw: 128.56

PROP: Crystals, mp: 32.5°, bp: 214°, d: 1.245, vap press: 1 mm @ 44.2°.

SYN: M-CHLOROPHENOL

THR: Poison by intraperitoneal route. An experimental tumorigen. Moderately toxic by ingestion. When heated to decomposition, it emits toxic fumes of Cl^-. For further information, see Vol. 6, No. 5 of *DPIM Report*.

4-CHLOROPHENOL   HR: 3
CAS: 106-48-9   NIOSH: SK 2800000
mf: C_6H_5ClO mw: 128.56

PROP: Needle-like, white to straw colored crys-

tals, unpleasant odor; fp: 42.8°, flash p: 250°F, d: 1.246 @ 60°/25°, vap press: 1 mm @ 49.8°.

SYNS: P-CHLORFENOL (CZECH) * P-CHLORO-PHENOL * PARACHLOROPHENOL

THR: Poison by ingestion and intraperitoneal routes. Moderately toxic by subcutaneous route. A skin and eye irritant. Slight fire hazard when exposed to heat or flame. Dangerous when heated to decomposition it emits chlorides and phenol. To fight fire, use water, spray, mist, fog, foam, dry chemical. For further information, see Vol. 2, No. 6 of *DPIM Report*.

4-(3-(2-CHLOROPHENOTHIAZIN-10-YL)PROPYL)-1-PIPERAZINEETHA-NOL   HR: 3
CAS: 58-39-9   NIOSH: TL 7175000
mf: $C_{21}H_{26}ClN_3OS$ mw: 404.01

SYNS: 2-CHLORO-10-3-(1-(2-HYDROXYETHYL)-4-PIPERAZINYL)PROPYL PHENOTHIAZINE * 1-(2-HYDROXYETHYL)-4-(3-(2-CHLORO-10-PHENOTHIAZINYL)PROPYL)PIPERAZINE * gamma-(4-(beta-HYDROXYETHYL)PIPERAZIN-1-YL)PROPYL-2-CHLOROPHENOTHIAZINE

THR: Poison by ingestion, intravenous, subcutaneous, and intramuscular routes. When heated to decomposition it emits very toxic fumes of SO_x, NO_x, and Cl^-.

p-CHLORO PHENOXY ACETIC ACID   HR: 2
CAS: 122-88-3   NIOSH: AG 0175000
mf: $C_8H_7ClO_3$ mw: 186.60

SYNS: (4-CHLOROPHENOXY)ACETIC ACID * TOMATOTONE

THR: Moderately toxic by ingestion and intraperitoneal route. Mutagenic data. When heated to decomposition it emits toxic fumes of Cl^-.

4-CHLOROPHENYL BENZENESULFO-NATE   HR: 3
CAS: 80-38-6   NIOSH: DB 5600000
mf: $C_{12}H_9ClO_3S$ mw: 268.72

PROP: Colorless crystals, insol in water, sol in organic solvents. mp: 62°.

SYNS: BENZENESULFONATE DE 4-CHLOROPHE-NYLE (FRENCH) * BENZENESULFONIC ACID, 4-CHLOROPHENYL ESTER * (4-CHLOOR-FENYL)-BENZEEN-SULFONAAT (DUTCH) * P-CHLOROFE-NYLESTER KYSELINY BENZENSULFONOVE (CZECH)

* P-CHLOROPHENYL BENZENESULFONATE
* 4-CHLOROPHENYL BENZENESULPHONATE
* (4-CHLOR-PHENYL)-BENZOLSULFONAT (GERMAN) * (4-CLORO-FENIL)-BENZOL-SOLFONATO (ITALIAN) * FENIZON (FRENCH) * MURVESCO * TRIFENSON

THR: Moderately toxic by ingestion. A severe eye and moderate skin irritant. See also esters. An acaricide. Dangerous; see also chlorides and sulfonates for disaster hazard.

1-(p-CHLORO-alpha-PHENYLBENZYL)-HEXAHYDRO-4-METHYL-1H-1,4-DIAZEPINEDIHYDROCHLORIDE　　HR: 3
CAS: 1982-36-1　　　NIOSH: HM 3675000
mf: $C_{19}H_{23}ClN_2 \cdot 2ClH$　　mw: 387.81

SYNS: HOMOCHLORCYCLIZINE DIHYDROCHLORIDE * HOMOCHLOROCYCLIZINE DIHYDROCHLORIDE

THR: Poison by a variety of routes. When heated to decomposition it emits very toxic fumes of HCl, Cl$^-$, and NO$_x$.

1-(p-CHLORO-alpha-PHENYLBENZYL)-4-(2-((2-HYDROXYETHOXY)ETHYL)PIPERAZINE　　HR: 3
CAS: 68-88-2　　　NIOSH: KK 2275000
mf: $C_{21}H_{27}ClN_2O_2$　　mw: 374.95

SYNS: ATARAX * ATERAX * 1-(P-CHLOROBENZHYDRYL)-4-(2-(2-HYDROXYETHOXY)-ETHYL)DIETHYLENEDIAMINE * 1-(P-CHLOROBENZHYDRYL)-4-(2-(2-HYDROXYETHOXY)ETHYL)PIPERAZINE * N-(4-CHLOROBENZHYDRYL)-N'-(HYDROXYETHOXYETHYL)PIPERAZINE * 1-(P-CHLORODIPHENYLMETHYL)-4-(2-(2-HYDROXYETHOXY)ETHYL)PIPERAZINE * 2-(2-(4-(P-CHLORO-ALPHA-PHENYLBENZYL)-1-PIPERAZINYL)ETHOXY)ETHANOL * VESPARAZWIRKSTOFF

THR: An acute poison by ingestion, intraperitoneal, and intravenous routes. An experimental teratogen. When heated to decomposition it emits very toxic fumes of Cl$^-$ and NO$_x$.

4-CHLOROPHENYL 4-CHLOROBENZENESULFONATE　　HR: 2
CAS: 80-33-1　　　NIOSH: DB 5250000
mf: $C_{12}H_8Cl_2O_3S$　　mw: 303.16

SYNS: CHLOORFENSON (DUTCH) * (4-CHLOOR-FENYL)-4-CHLOOR-BENZEEN-SULFONAAT (DUTCH) * CHLOREFENIZON (FRENCH)

* CHLORFENSON * 4-CHLOROBENZENESULFONATE DE 4-CHLOROPHENYLE (FRENCH) * P-CHLOROBENZENESULFONIC ACID-P-CHLOROPHENYL ESTER * P-CHLOROPHENYL-P-CHLOROBENZENESULFONATE * P-CHLOROPHENYL-P-CHLOROBENZENESULPHONATE * 4-CHLOROPHENYL-4-CHLOROBENZENESULPHONATE * (4-CHLOR-PHENYL)-4-CHLOR-BENZOL-SULFONATE (GERMAN) * (4-CLORO-FENIL)-4-CLORO-VENZOL-SOLFONATO (ITALIAN) * ENT 16,358 * TRICHLORFENSON

THR: Moderately toxic by ingestion. A pesticide. When heated to decomposition it emits very toxic fumes of Cl$^-$ and SO$_x$.

3-(p-CHLOROPHENYL)-1,1-DIMETHYLUREA　　HR: 3
CAS: 150-68-5　　　NIOSH: YS 6300000
mf: $C_9H_{11}ClN_2O$　　mw: 198.67

PROP: Crystals, nearly water-insol, slight odor; mp: 171°, vap press: 2×10^{-3} mm @ 100°.

SYNS: 3-(4-CHLOOR-FENYL)-1,1-DIMETHYL-UREUM (DUTCH) * N-(P-CHLOROPHENYL)-N',N'-DIMETHYLUREA * 1-(P-CHLOROPHENYL)-3,3-DIMETHYLUREA * 3-(4-CHLOROPHENYL)-1,1-DIMETHYLUREA * 1-(4-CHLORO PHENYL)-3,3-DIMETHYLUREE (FRENCH) * 3-(4-CHLOR-PHENYL)-1,1-DIMETHYL-HARNSTOFF (GERMAN) * 3-(4-CLORO-FENIL)-1,1-DIMETIL-UREA (ITALIAN) * N,N-DIMETHYL-N'-(4-CHLOROPHENYL)UREA * 1,1-DIMETHYL-3-(P-CHLOROPHENYL)UREA * MONURON * NCI-C02846 * USAF P-8 * USAF XR-41

THR: An experimental carcinogen. Moderately toxic by ingestion and intraperitoneal route. Has produced anemia and methemoglobinemia in experimental animals. When heated to decomposition it emits very toxic fumes of NO$_x$ and Cl$^-$.

6-(o-CHLOROPHENYL)-8-ETHYL-1-METHYL-4H-s-TRIAZOLO(3,4-c)-THIENO(2,3-e)(1,4)-DIAZEPINE　　HR: 3
　　　　　　NIOSH: XJ 7152000
mf: $C_{17}H_{15}ClN_4S$　　mw: 342.87

SYN: ETIZOLAM

THR: An experimental teratogen. When heated to decomposition it emits very toxic fumes of Cl$^-$, SO$_x$, and NO$_x$.

p-CHLOROPHENYL ISOCYA-
NATE   HR: 3
CAS: 104-12-1   NIOSH: NQ 8575000
mf: C_7H_4ClNO mw: 153.57

PROP: White solid, sol in organic solvents, mp: 28°, bp: 106.5° @ 30 mm, flash p: 230°F.

SYNS: ISOCYANIC ACID-P-CHLOROPHENYL ESTER * P-CHLORFENYLISOKYANAT (CZECH)

THR: Poison by inhalation and ocular routes. Moderately toxic by ingestion. A severe eye and moderate skin irritant. Dangerous, can explode on distillation. See also chlorides and cyanates.

1-(p-CHLOROPHENYL)-5-ISOPROPYL-
BIGUANIDE   HR: 3
CAS: 500-92-5   NIOSH: DU 1225000
mf: $C_{11}H_{16}ClN_5$ mw: 253.77

PROP: White powder, mp: 244°.

SYNS: BIGUMAL * CHLOROGUANIDE

THR: Poison by ingestion, intravenous, and intraperitoneal routes. Dangerous. See also chlorides and nitrogen oxide for disaster hazard.

N-3-CHLOROPHENYL ISOPROPYL
CARBAMATE   HR: 3
CAS: 101-21-3   NIOSH: FD 8050000
mf: $C_{10}H_{12}ClNO_2$ mw: 213.68

PROP: Light brown crystalline solid, faint characteristic odor, mp: 41°, bp: 247° (decomp).

SYNS: N-(3-CHLOOR-FENYL)-ISOPROPYL CARBA-MAAT (DUTCH) * N-(3-CLORO-FENIL)-ISOPRO-PIL-CARBAMMATO (ITALIAN) * N-(3-CHLORO PHENYL) CARBAMATE D'ISOPROPYLE (FRENCH) * N-(3-CHLOROPHENYL)CARBAMIC ACID,-ISOPROPYL ESTER * (3-CHLOROPHENYL)-CARBAMIC ACID,-1-METHYLETHYL ESTER * N-(3-CHLOR-PHENYL)-ISOPROPYL-CARBAMAT (GERMAN) * CHLORPROPHAME (FRENCH) * ENT 18,060 * ISOPROPYL META-CHLORO-CARBANILATE * ISOPROPYL-3-CHLOROCARBANI-LATE * ISOPROPYL-N-(3-CHLOROPHENYL)CAR-BAMATE * O-ISOPROPYL N-(3-CHLOROPHENYL)-CARBAMATE * ISOPROPYL-3-CHLOROPHENYL-CARBAMATE

THR: An experimental carcinogen and neoplastigen. Moderately toxic by ingestion and inhalation. See also carbamates. An insecticide and an herbicide. Dangerous, when heated to decomposition it emits highly toxic fumes of phosgene.

3-(4-CHLOROPHENYL)-1-METHOXY-1-
METHYLUREA   HR: 3
CAS: 1746-81-2   NIOSH: YS 6425000
mf: $C_9H_{11}ClN_2O_2$ mw: 214.67

SYNS: 3-(4-CHLORPHENYL)-1-METHOXY-1-METHYLHARNSTOFF (GERMAN) * MONOLI-NURON

THR: An experimental teratogen. Moderately toxic by ingestion. When heated to decomposition it emits very toxic fumes of Cl^- and NO_x.

2-(o-CHLOROPHENYL)-2-(METHYLAMI-
NO)CYCLOHEXANONE HYDROCHLO-
RIDE   HR: 3
CAS: 1867-66-9   NIOSH: GW 1400000
mf: $C_{13}H_{16}ClNO \cdot ClH$ mw: 274.21

SYNS: CI 581 * KETAMINE HYDROCHLORIDE

THR: Poison by intraperitoneal and intravenous routes. Human central nervous system and psychotropic effects. When heated to decomposition it emits very toxic fumes of Cl^-, NO_x, and HCl.

o-CHLOROPHENYL METHYLCARBA-
MATE   HR: 3
CAS: 3942-54-9   NIOSH: FB 6350000
mf: $C_8H_8ClNO_2$ mw: 185.62

SYNS: 2-CHLOROPHENYL-N-METHYLCARBAMATE * HOPCIDE

THR: A poison by ingestion. See also carbamates. When heated to decomposition it emits very toxic fumes of Cl^- and NO_x.

3-(p-CHLOROPHENYL)-1-METHYL-1-(1-
METHYL-2-PROPYNYL)UREA   HR: 3
CAS: 3766-60-7   NIOSH: YS 6475000
mf: $C_{12}H_{13}ClN_2O$ mw: 236.72

SYNS: N'-(4-CHLOROPHENYL)-N-ISOBUTINYL-N-METHYLUREA * N-(4-CHLORPHENYL)-N'-METHYL-N'-ISOBUTINYLHARNSTOFF (GERMAN)

THR: An experimental teratogen. Moderately toxic by ingestion and intraperitoneal routes. When heated to decomposition it emits very toxic fumes of Cl^- and NO_x.

2-(3-(4-(3-CHLOROPHENYL)-1-PIPERA-
ZINYL)PROPYL)-1,2,4-TRIZOLO(4,3-
a)PYRIDIN-3(2H)-ONE HYDROCHLO-
RIDE   HR: 3
                       NIOSH: YM 1431000
mf: $C_{19}H_{22}ClN_5O \cdot ClH$ mw: 408.37

SYNS: TRAZODONE HYDROCHLORIDE * TRIT-TICO

THR: Poison by intravenous route. When heated to decomposition it emits very toxic fumes of Cl^-, NO_x, and HCl.

1-(p-CHLOROPHENYLSULFONYL)-3-PROPYLUREA   HR: 3
CAS: 94-20-2   NIOSH: YS 6650000
mf: $C_{10}H_{13}ClN_2O_3S$ mw: 276.76

SYNS: 1-(P-CHLOROBENZENESULFONYL)-3-PRO-PYLUREA * NCI-CO1752 * N-PROPYL-N'-(P-CHLOROBENZENESULFONYL)UREA * 1-PROPYL-3-(P-CHLOROBENZENESULFONYL)UREA * N-PRO-PYL-N'-P-CHLORPHENYLSULFONYLCARBAMIDE

THR: An experimental teratogen. Moderately toxic by ingestion. Mutagenic data. Human systemic and gastrointestinal tract effects. When heated to decomposition it emits very toxic fumes of Cl^-, NO_x, and SO_x.

p-CHLOROPHENYL-2,4,5-TRICHLORO-PHENYL SULFONE   HR: 3
CAS: 116-29-0   NIOSH: WR 5850000
mf: $C_{12}H_6Cl_4O_2S$ mw: 356.04

PROP: Crystals, nearly water-insol; mp: 147°.

SYNS: 4-CHLOROPHENYL, 2,4,5-TRICHLOROPHE-NYL SULFONE * ENT 23,737 * 2,4,4',5-TET-RACHLOOR-DIFENYL-SULFON (DUTCH) * 2,4,-4',5-TETRACHLOR-DIPHENYL-SULFON (GERMAN) * 2,4,4',5-TETRACHLORODIPHENYL SULFONE * 2,4,5,4'-TETRACHLORODIPHENYLSULPHONE * 2,4,4',5-TETRACLORO-DIFENIL-SOLFONE (ITALIAN)

THR: Moderately toxic by ingestion. A food additive permitted in food for human consumption. Dangerous, when heated to decomposition it emits highly toxic fumes of chlorides and SO_x.

CHLOROPHENYLTRICHLOROSI-LANE   HR: 3
CAS: 26571-79-9   NIOSH: VV 2650000
DOT: 1753
mf: $C_6H_4Cl_4Si$ mw: 245.99

PROP: Colorless to pale yellow liquid, readily hydrolyzed by moisture with the liberation of HCl (a mixture of 3 isomers); bp: 230°, d: 1.439 @ 25°/25°, flash p: 255°F (COC).
DOT Classification: Label: Corrosive

THR: A poison irritant by ingestion and inhalation. See also chlorosilanes. Very irritating to skin, eyes, and mucous membranes. Slight fire hazard when exposed to heat or flame. Dangerous; see also chlorides. In contact with water it readily hydrolyzes to HCl and evolves heat.

CHLOROPHYLL   HR: 3
CAS: 1406-65-1   NIOSH: FW 6420000

THR: Poison by intravenous and intraperitoneal route.

CHLOROPLATINIC ACID   HR: 3
CAS: 16941-12-1   NIOSH: TP 1500000
mf: $Cl_6Pt \cdot 2H$ mw: 409.81

PROP: Brownish-yellow, very deliquescent, crystalline mass, easily sol in water and alcohol; d: 2.431, mp: 60°.

SYNS: CHLOROPLATINIC (IV) ACID * DIHY-DROGEN HEXACHLOROPLATINATE * HEXA-CHLOROPLATINIC (IV) ACID * HYDROGEN HEXACHLOROPLATINATE(4+) * PLATINIC CHLORIDE

THR: Poison by intravenous and intraperitoneal routes. See platinum compounds and chlorides. Incompatible with BrF_3. Mutagenic data. When heated to decomposition it emits toxic fumes of Cl^-.

CHLOROPROMAZINE   HR: 3
CAS: 50-53-3   NIOSH: SN 8925000
mf: $C_{17}H_{19}ClN_2S$ mw: 318.89

SYNS: CHLORO-3-(DIMETHYLAMINO-3-PROPYL)-10 PHENOTHIAZINE (FRENCH) * 2-CHLORO-10-(3-(DIMETHYLAMINO)PROPYL)PHENOTHIAZINE * CHLORPROMAZIN * SKF-2601 * THORA-ZINE

THR: Poison by ingestion, inhalation, intravenous, intraperitoneal, and subcutaneous routes. Mutagenic data. A human teratogen; human central nervous system and systemic effects. Has been implicated in aplastic anemia. When heated to decomposition it emits very toxic fumes of Cl^-, NO_x, and SO_x.

CHLOROPROMAZINE HYDROCHLO-RIDE   HR: 3
CAS: 69-09-0   NIOSH: SO 1750000
mf: $C_{17}H_{19}ClN_2S \cdot ClH$ mw: 355.35

SYNS: 2-CHLORO-10-(3-DIMETHYLAMINOPRO-PYL) PHENOTHIAZINE MONOHYDROCHLORIDE

✳ 10-(3-DIMETHYLAMINOPROPYL)-2-CHLORO-
PHENOTHIAZINE MONOHYDROCHLORIDE
✳ NCI-C05210 ✳ PHENOTHIAZINE HYDROCHLO-
RIDE ✳ THORAZINE HYDROCHLORIDE

THR: Poison by ingestion, intraperitoneal, intravenous, and subcutaneous routes. See also
chlorpromazine. When heated to decomposition
it emits very toxic fumes of Cl^-, NO_x, SO_x,
and HCl.

1-CHLOROPROPANE   HR: 2
CAS: 540-54-5   NIOSH: TX 4400000
DOT: 1278
mf: C_3H_7Cl mw: 78.55

PROP: Colorless liquid, chloroform-like odor.
mp: $-122.8°$, bp: $47.2°$, lel = 2.6%, uel =
11.1%, flash p: <0°F, d: 0.890, vap d: 2.71,
autoign temp: 968°F.

SYN: N-PROPYL CHLORIDE

DOT Classification: Label: Flammable
  Liquid

THR: A moderately poisonous irritant to skin,
eyes, and mucous membranes. Narcotic in high
concentrations. See also chlorinated hydrocarbons, aliphatic. Dangerous fire hazard when exposed to heat, flame, or oxidizers. Moderately
explosive when exposed to flame. Keep away
from heat and open flame; can react vigorously
with oxidizing materials. See chlorides. To fight
fire use CO_2, dry chemical.

2-CHLORO-1-PROPANOL   HR: 3
CAS: 78-89-7   NIOSH: UA 8925000
mf: C_3H_7ClO mw: 94.55

PROP: Colorless liquid, mild non-residual odor;
bp: 133.5°, flash p: 125°F (CC), d: 1.103 @
20°, vap d: 3.26.

SYNS: 2-CHLOROPROPYL ALCOHOL ✳ PRO-
PYLENECHLOROHYDRIN

THR: Poison by ingestion. Moderately toxic
by inhalation. A skin and eye irritant. Moderate
fire hazard when exposed to heat, flame, or
powerful oxidizers. When heated to decomposition it emits highly toxic fumes of chlorides;
can react with oxidizing materials. To fight fire,
use alcohol foam, CO_2, dry chemical.

2-CHLORO-1-PROPENE   HR: 3
CAS: 557-98-2   NIOSH: UC 7200000
DOT: 2456
mf: C_3H_5Cl mw: 76.53

PROP: Colorless liq; bp: 22.65°, fp: $-137.4°$,
d: 0.918 @ 9°, flash p: $-4°$. lel = 4.5%; uel
= 16%.

SYN: 2-CHLOROPROPENE (DOT)

DOT Classification: Label: Flammable
  Liquid

THR: Poison by ingestion and inhalation. Mutagenic data. Very dangerous fire hazard via heat,
flame, sparks, or powerful oxidizers. Dangerous; reacts with powerful oxidizers; see also
chlorides. To fight fire, use water, spray, mist,
fog, dry chemical, alcohol foam.

alpha-CHLOROPROPIONIC
ACID   HR: 2
CAS: 598-78-7   NIOSH: UE 8575000
mf: $C_3H_5ClO_2$ mw: 108.53

PROP: Sol in water; d: 1.260-1.268 @ 20°,
bp: 183-187°, flash p: 225°F.

THR: Moderately toxic by ingestion. Moderate
fire hazard when exposed to heat or flame. Dangerous. See chlorides for disaster hazard. To
fight fire, use water, foam, alcohol foam.

3-CHLOROPROPIONITRILE   HR: 3
CAS: 542-76-7   NIOSH: UG 1400000
mf: C_3H_4ClN mw: 89.53

PROP: Colorless liquid; mp: $-51°$, bp: 176°
decomp, flash p: 168°F (CC), d: 1.1363 @ 25°,
vap press: 6 mm @ 50°, vap d: 3.09.

SYNS: 3-CHLOROPROPANONITRILE ✳ beta-
CHLOROPROPIONITRILE ✳ USAF A-8798

THR: Poison by ingestion, intravenous, and intraperitoneal routes. See also nitriles. Moderate
fire hazard in its liquid form when exposed to
heat or flame. To fight fire, use alcohol foam,
water, foam, CO_2, or dry chemical. When
heated to decomposition it emits very toxic
fumes of Cl^- and NO_x.

p-CHLOROPROPIOPHENONE   HR: 3
CAS: 6285-05-8   NIOSH: UG 9275000
mf: C_9H_9ClO mw: 168.63

SYN: USAF EK-5296

THR: Poison by intraperitoneal and intravenous
routes. When heated to decomposition it emits
toxic fumes of Cl^-.

2-CHLOROPYRIDINE   HR: 3
CAS: 109-09-1   NIOSH: US 5950000
mf: C_5H_4ClN mw: 113.55

PROP: Colorless oily liquid; bp: 170°, d: 1.205 @ 25°, vap press: 1 mm @ 13.3°, vap d: 3.93.

THR: Poison by ingestion, inhalation, intraperitoneal, and dermal routes. Slight fire hazard when exposed to heat or flame. Dangerous; see phosgene; can react with oxidizing materials. When heated to decomposition it emits very toxic fumes of Cl^- and NO_x.

CHLOROQUINE   HR: 3
CAS: 54-05-7   NIOSH: VB 2360000
mf: $C_{18}H_{26}ClN_3$ mw: 319.92

SYNS: 7-CHLORO-4-(4-DIETHYLAMINO-1-METHYLBUTYLAMINO)QUINOLINE * N(SUP 4)-(7-CHLORO-4-QUINOLINYL)-N(SUP 1),N(SUP 1)-DIETHYL-1,4-PENTANEDIAMINE * WIN 244

THR: Poison by ingestion, intraperitoneal, intravenous, and subcutaneous routes. Mutagenic data. A human teratogen. When heated to decomposition it emits very toxic fumes of Cl^- and NO_x. For further information see Vol. 6, No. 3 of *DPIM Report*.

CHLOROQUINE DIPHOSPHATE   HR: 3
CAS: 50-63-5   NIOSH: VB 2450000
mf: $C_{18}H_{26}ClN_3 \cdot 2H_3O_4P$ mw: 515.92

SYNS: 7-CHLOR-4-(4-(DIAETHYLAMINO)-1-METHYLBUTYLAMINO)-CHINOLINDIPHOSPHAT (GERMAN) * 7-CHLORO-4-((4'-DIETHYLAMINO-1-METHYLBUTYL)AMINO)QUINOLINE DIPHOSPHATE

THR: Poison by intravenous and intraperitoneal routes. Mutagenic data. Human musculo-skeletal effects. When heated to decomposition it emits very toxic fumes of Cl^-, NO_x, and PO_x.

CHLOROQUINE MUSTARD   HR: 3
CAS: 4213-44-9   NIOSH: VB 0175000
mf: $C_{18}H_{24}Cl_3N_3 \cdot 2ClH$ mw: 461.72

SYNS: 4-((4-(BIS(2-CHLOROETHYL)AMINO)-1-METHYLBUTYL)AMINO-7-CHLOROQUINOLINE, DI-HYDROCHLORIDE * ICR-25A * NSC-17118

THR: Poison by intravenous route. An experimental carcinogen. When heated to decomposition it emits very toxic fumes of Cl^-, NO_x, and HCl.

CHLOROSILANES   HR: 3
PROP: Compounds of Si, Cl and H where the total number of atoms of Cl and H add up to 4. SiH_xCl_{4-x}.

THR: Poison by ingestion and inhalation routes, and a poisonous irritant to skin, eyes, and mucous membranes. Toxicity based on HCl which is formed upon hydrolysis of a chlorosilane. Self-ignites in air; with a little ammonia, it forms a self-igniting product. Dangerous; when heated to decomposition they emit highly toxic fumes of chlorides; react with water or steam to produce heat and toxic and corrosive fumes of HCl.

CHLOROSULFURIC ACID   HR: 3
CAS: 7790-94-5   NIOSH: FX 5730000
mf: $ClHO_3S$ mw: 116.52

PROP: Clear to cloudy, colorless to pale yellow liquid, sharp odor; mp: −80°; bp: 151.0°; d: 1.766 @ 18°; vap press: 1 mm @ 32°; vap d: 4.02.

SYNS: CHLOROSULFONIC ACID * SULFONIC ACID, MONOCHLORIDE * SULFURIC CHLORO-HYDRIN

THR: See also sulfuric acid. A poison via irritation. Chlorosulfonic acid can cause severe acid burns and is very irritating to the eyes, lungs and mucous membranes. It can cause acute toxic effects either in the liquid or vapor state. Inhalation of concentrated vapor may cause loss of consciousness with serious damage to lung tissue. Contact of liquid with the eyes can cause severe burns if not immediately and completely removed. It also causes severe skin burns due to its highly corrosive action. Upon ingestion it will irritate the mouth, esophagus and stomach to a serious degree and on contact with skin cause dermatitis. It may cause conjunctivitis even in the vapor form. If spilled on a person remove all contaminated clothing, wash contaminated skin with a lot of water, followed by baking soda solution. Irrigate eyes with warm water for 15 M. Consult a physician. Vent stored drums 2 times per month to control pressure of H_2 produced by action of acid on metal of drum. Incompatible with acetic acid; acetic anhydride; acetonitrile; acrolein; acrylic acid; acrylonitrile; allyl alcohol; allyl chloride; 2-amino ethanol; ammonium hydroxide; aniline; *n*-butyraldehyde; creosote oil; cresol; cumene; dichloroethyl ether; diethylene glycol monomethyl ether; diisobutylene; diisopropyl ether; epichloro hydrin; ethyl acetate; ethyl acrylate; ethylene chlorohydrin; ethylene cyanohydrin; ethylene diamine; ethylene glycol; ethylene glycol monoethyl ether acetate; ethylene imine; glyoxal; HCl; HF; H_2O_2; isoprene; mesityl oxide; metal powders; methyl ethyl ketone; HNO_3; 2-nitropropane; P; β-propiolactone; propylene oxide; pyridene; NaOH; H_2SO_4; sulfolane; styrene monomer; vinyl acetate; vinylidene chloride; silver nitrate; water; organic matter; combustibles. Dangerous; see sulfuric acid and HCl and sulfonates. Decomposes explosively on contact with water, alcohol, acids. To Fight fire: Avoid water; use dry chemicals. For further information see Vol. 3, No. 5 of *DPIM Report*.

CHLOROTHIOFORMIC ACID ETHYL ESTER   HR: 3
CAS: 2812-73-9   NIOSH: LQ 6950000
DOT: 2826
mf: C_3H_5ClOS mw: 124.59

SYN: ETHYL CHLOROTHIOFORMATE (DOT)

DOT Classification: Label: Corrosive

THR: Poison by ingestion and inhalation, and a poisonous irritant to skin, eyes, and mucous membranes. See also esters. When heated to decomposition it emits very toxic fumes of Cl^- and SO_x.

2-(2-CHLORO-p-TOLUIDINO)-2-IMID-AZOLIDINE   HR: 3
NIOSH: NI 8938000
mf: $C_{10}H_{12}ClN_3$ mw: 209.70

SYN: 2-(2-CHLORO-4-METHYLPHENYL)AMINO-1,3-DIAZACYCLOPENT-2-ENE

THR: Poison by ingestion, intravenous, and intraperitoneal routes. When heated to decomposition it emits very toxic fumes of Cl^- and NO_x.

CHLOROTRIANISENE   HR: 3
CAS: 569-57-3   NIOSH: KV 0600000
mf: $C_{23}H_{21}ClO_3$ mw: 380.89

SYNS: 1,1',1''-(1-CHLORO-1-ETHENYL-2-YLI-DENE)-TRIS(4-METHOXYBENZENE) * CHLORO-TRIS(P-METHOXYPHENYL)ETHYLENE * NSC-10108 * TRI-P-ANISYLCHLOROETHYLENE * TRIS(P-METHOXYPHENYL)CHLOROETHYLENE

THR: An experimental tumorigen. Possible carcinogen. When heated to decomposition, it emits very toxic fumes of Cl^-.

CHLOROTRIBUTYL STANNANE   HR: 3
CAS: 1461-22-9   NIOSH: WH 6820000
mf: $C_{12}H_{27}ClSn$ mw: 325.53

SYN: CHLORID TRI-N-BUTYLCINICITY (CZECH)

THR: Poison by ingestion and skin absorption. An eye irritant. See also tin compounds. When heated to decomposition it emits toxic fumes of Cl^-.

2-CHLORO-6-(TRICHLOROMETHYL)-PYRIDINE   HR: 3
CAS: 1929-82-4   NIOSH: US 7525000
mf: $C_6H_3Cl_4N$ mw: 230.90

SYNS: DOWCO-163 * NITRAPYRIN * N-SERVE NITROGEN STABILIZER

THR: Poison by ingestion. When heated to decomposition it emits very toxic fumes of Cl^- and NO_x.

CHLORO(TRIETHYLPHOSPHINE)-GOLD   HR: 3
CAS: 15529-90-5   NIOSH: MD 5431000
mf: $C_6H_{15}AuClP$ mw: 350.60

SYNS: SK&F 36914 * TRIETHYLPHOSPHINE-AUROUS CHLORIDE

THR: Poison by ingestion. See also gold. When heated to decomposition it emits very toxic fumes of Cl^- and PO_x.

CHLOROTRIFLUOROETHYLENE   HR: 3
CAS: 79-38-9   NIOSH: KV 0525000
DOT: 1082
mf: C_2ClF_3 mw: 116.47

PROP: A gas; lel = 24%, uel = 40.3%, flash p: $-18°F$.

SYNS: CHLORTRIFLUORAETHYLEN (GERMAN) * GENETRON 1113 * TRIFLUOROCHLORETH-YLENE (DOT) * 1,1,2-TRIFLUORO-2-CHLORO-ETHYLENE * TRIFLUOROMONOCHLOROETHYL-ENE * TRIFLUOROVINYL CHLORIDE

DOT Classification: Label: Flammable Gas

THR: Poison by ingestion and intraperitoneal route. Moderately toxic by inhalation. Violent reaction when mixed with ($Br_2 + O_2$) or (ClF_3 + water). Very dangerous fire hazard via heat, flames (sparks), or oxidizers. Dangerous; see chlorides and fluorides for disaster hazard. To fight fire, stop flow of gas. Incompatible with 1,1-dichloro-ethylene; oxygen.

CHLORO TRIFLUORO METHANE   HR: 1
CAS: 75-72-9   NIOSH: PA 6410000
mf: $CClF_3$ mw: 104.46

PROP: Colorless gas, ethereal odor; mp: −181°, bp: −80°.

SYNS: FREON 13 * MONOCHLOROTRIFLUORO-METHANE (DOT) * TRIFLUOROCHLOROMETHANE * TRIFLUOROMETHYL CHLORIDE * TRIFLUORO-MONOCHLOROCARBON

DOT Classification: Label: Nonflammable Gas

THR: A mild irritant. Narcotic in high concentrations. Reacts violently with Al. Dangerous; when heated to decomposition it emits highly toxic fumes of chlorides and fluorides.

4-(4-(4-CHLORO-alpha,alpha,alpha-TRI-FLUORO-m-TOLYL)-4-HYDROXY-PIPERIDINO)BUTYROPHENONE-4′-FLUORO-,HYDROCHLORIDE   HR: 3
CAS: 17230-87-4   NIOSH: EU 2100000
mf: $C_{22}H_{22}ClF_4NO_2 \cdot ClH$ mw: 480.36

SYNS: CLOFLUPEROL HYDROCHLORIDE * R 9298

THR: Poison by ingestion, subcutaneous, and intravenous routes. When heated to decomposition it emits very toxic Cl^-, F^-, NO_x, and HCl.

CHLOROTRIMETHYLSTANNANE   HR: 3
CAS: 1066-45-1   NIOSH: WH 6850000
mf: C_3H_9ClSn mw: 199.26

PROP: Mp: 37°.

SYNS: CHLOROTRIMETHYLTIN * TRIMETHYL-CHLOROSTANNANE * TRIMETHYLCHLOROTIN * TRIMETHYLSTANNYL CHLORIDE * TRI-METHYLTIN CHLORIDE

THR: Poison by intravenous route. See also tin compounds. When heated to decomposition it emits toxic fumes of Cl^-.

CHLOROTRINITROMETHANE   HR: 3
CAS: 1943-16-4   NIOSH: PA 6430000
mf: $CClN_3O_6$ mw: 185.49

THR: Poison by intraperitoneal route. When heated to decomposition it emits very toxic fumes of Cl^- and NO_x.

CHLOROTRIPHENYLSTANNANE   HR: 3
CAS: 639-58-7   NIOSH: WH 6860000
mf: $C_{18}H_{15}ClSn$ mw: 385.47

PROP: Colorless crystals, insol in water, sol in organic solvents; mp: 106°, bp: 240° @ 13.5 mm.

SYNS: TRIPHENYLCHLOROSTANNANE * TRIPHE-NYLCHLOROTIN * CHLOROTRIPHENYLTIN * FENTIN CHLORIDE * TRIPHENYLTIN CHLO-RIDE

THR: Poison by ingestion and intravenous routes. See also tin compounds. When heated to decomposition it emits toxic fumes of Cl^-.

CHLOROTRIPROPYLSTANNANE   HR: 3
CAS: 2279-76-7   NIOSH: WH 6870000
mf: $C_9H_{21}ClSn$ mw: 283.44

PROP: Colorless liquid. Sol in organic solvents; d: 1.2678 @ 28°; mp: −23.5°.

SYN: TRI-N-PROPYLTIN CHLORIDE

THR: Poison by intravenous route. See also tin compounds. When heated to decomposition it emits toxic fumes of Cl^-.

CHLORO(TRIVINYL)STANNANE   HR: 3
CAS: 10008-90-9   NIOSH: WH 6870500
mf: C_6H_9ClSn mw: 235.29

SYN: TRIVINYLTIN CHLORIDE

THR: Poison by intravenous route. See also tin compounds and chlorides. When heated to decomposition, it emits toxic fumes of Cl^-.

CHLOROVINYLARSINE DICHLORIDE   HR: 3
CAS: 541-25-3   NIOSH: CH 2975000
mf: $C_2H_2AsCl_3$ mw: 207.31

PROP: Liquid, faint odor of geranium; bp: 190° decomp, fp: −13°, d: 1.888 @ 20°/4°, vap press: 0.4 mm @ 20°, vap d: 7.15.

SYNS: 2-CHLOROVINYLDICHLOROARSINE * beta-CHLOROVINYLBICHLOROARSINE * DICHLORO(2-CHLOROVINYL)ARSINE * LEWIS-ITE (ARSENIC COMPOUND)

THR: Poison by ingestion, inhalation, dermal, subcutaneous, and intravenous routes. See also arsenic compounds. A blistering type military poison. Has a delayed action similar to distilled mustard gas. This gas exhibits a systemic poisoning effect on humans. Lewisite is absorbed through skin; as little as 2 mL on the skin can cause death of an adult. Dangerous; see also arsenic and chlorides.

CHLOROWAX 500C
CAS: 63449-39-8   NIOSH: FY 2290000

SYN: NCI-C53587

THR: No acute toxicity data. When heated to decomposition it emits toxic fumes of Cl⁻.

4-CHLORO-3,5-XYLENOL
CAS: 88-04-0   NIOSH: ZE 6850000
mf: C_8H_9ClO mw: 156.62

PROP: Crystals, phenolic odor; slightly water sol; mp: 115.5°; bp: 246°.

SYNS: CHLORO XYLENOL * p-CHLORO-m-XYLENOL

THR: See also chlorinated aromatic hydrocarbons. When heated to decomposition it emits toxic fumes of Cl⁻.

(4-CHLORO-6-(2,3-XYLIDINO)-2-PYRIMIDINYLTHIO)ACETIC ACID   HR: 3
CAS: 50892-23-4   NIOSH: AG 2915000
mf: $C_{14}H_{14}ClN_3O_2S$ mw: 323.82

SYN: WY-14,643

THR: An experimental carcinogen. When heated to decomposition it emits very toxic fumes of Cl⁻, NO_x, and SO_x.

CHLOROZOTOCIN   HR: 3
CAS: 54749-90-5   NIOSH: LZ 5758000
mf: $C_9H_{16}ClN_3O_7$ mw: 313.73

SYNS: 2-(((((2-CHLOROETHYL)NITROSOAMINO)CARBONYL)AMINO)-2-DEOXY-D-GLUCOSE * 2-(3-(2-CHLORO-ETHYL)-3-NITROSOUREIDO)-2-DEOXY-D-GLUCOSO-PYRANOSE * NSC 178248

THR: Poison by intravenous route. An experimental carcinogen. Mutagenic data. When heated to decomposition it emits very toxic fumes of Cl⁻ and NO_x.

CHLORPHENIRAMINE MALEATE   HR: 3
CAS: 1102-47-2   NIOSH: US 6500000
mf: $C_{16}H_{19}ClN_2 \cdot 7C_4H_4O_4$ mw: 1087.38

SYN: CHLORPROPHENPYRIDAMINE MALEATE

THR: Poison by ingestion, intraperitoneal, intravenous, and subcutaneous routes. When

heated to decomposition it emits very toxic fumes of Cl⁻ and NO_x.

CHLORPHENTERMINE   HR: 3
CAS: 461-78-9   NIOSH: SH 1050000
mf: $C_{10}H_{14}ClN$ mw: 183.70

SYNS: p-CHLORO-alpha,alpha-DIMETHYLPHEN-ETHYLAMINE * beta-(p-CHLOROPHENYL)-alpha,alpha-DIMETHYLETHYLAMINE * CHLOR-PHENTERAMINE

THR: Poison by ingestion and intravenous routes. When heated to decomposition it emits very toxic fumes of Cl⁻ and NO_x.

gamma-(4-(p-CHLORPHENYL)-4-HYDROXPIPERIDINO)-p-FLUORBUTYRO-PHENONE   HR: 3
CAS: 52-86-8   NIOSH: EU 1575000
mf: $C_{21}H_{23}ClFNO_2$ mw: 375.90

SYNS: 1-(3-P-FLUOROBENZOYLPROPYL)-4-P-CHLOROPHENYL-4-HYDROXYPIPERIDINE * 4'-FLUORO-4-(4-HYDROXY-4-(4'-CHLOROPHENYL)PIPERIDINO)BUTYROPHENONE * 4-(4-HYDROXY-4'-CHLORO-4-PHENYLPIPERI-DINO)-4'-FLUOROBUTYROPHENONE

THR: Poison by ingestion, intraperitoneal, intravenous, and subcutaneous routes. Mutagenic data. A human teratogen. Human central nervous system effects. When heated to decomposition it emits very toxic fumes of F⁻, Cl⁻, and NO_x.

CHLORPROETHAZINE HYDRO-CHLORIDE   HR: 3
CAS: 4611-02-3   NIOSH: SN 8225000
mf: $C_{19}H_{23}ClN_2S \cdot ClH$ mw: 383.41

SYN: 2-CHLORO-10-(3'-DIETHYLAMINOPROPYL)PHENOTHIAZINE HYDRO-CHLORIDE

THR: Poison by ingestion, intravenous, intraperitoneal, and subcutaneous routes. When heated to decomposition it emits very toxic fumes of Cl⁻, NO_x, SO_x, and HCl.

CHLORQUINOX   HR: 2
CAS: 3495-42-9   NIOSH: VD 3550000
mf: $C_8H_2Cl_4N_2$ mw: 267.92

SYNS: LUCEL * 5,6,7,8-TETRACHLORO-QUINOXALINE

THR: Moderately toxic by ingestion. When heated to decomposition it emits very toxic fumes of Cl⁻ and NO_x.

CHLORTETRACYCLINE   HR: 3
CAS: 57-62-5   NIOSH: QI 7750000
mf: $C_{22}H_{23}ClN_2O_8$ mw: 478.92

PROP: Golden yellow crystals, slightly sol in water, very sol in aqueous soln pH 7.65, freely sol in the "cellosolves," dioxane, "Carbitol," sol in methanol, ethanol, butanol, acetone, ethyl acetate, and benzene, insol in ether and petroleum ether; mp: 168°-169°.

SYNS: AUREOMYCIN * 7-CHLOROTETRACY-CLINE

THR: Poison by intravenous and intraperitoneal routes. Moderately toxic by ingestion and inhalation. A food additive permitted in the feed and drinking water of animals and/or for the treatment of food-producing animals. Also a food additive permitted in food for human consumption. Dangerous; see also chlorides, NO_x.

CHOLANTHRENE   HR: 3
CAS: 479-23-2   NIOSH: FZ 2625000
mf: $C_{20}H_{14}$ mw: 254.34

SYNS: BENZ(J)ACEANTHRYLENE * 7,8-DI-METHYLENEBENZ(A)ANTHRACENE

THR: An experimental tumorigen and carcinogen. When heated to decomposition it emits acrid smoke and irritating fumes.

CHOLECALCIFEROL   HR: 3
CAS: 67-97-0   NIOSH: VS 2900000
mf: $C_{27}H_{44}O$ mw: 384.71

SYNS: 9,10-SECOCHOLESTA-5,7,10(19)-TRIEN-3-BETA-OL * D3-VIGANTOL * OLEOVITAMIN D3 * VITAMIN D3

THR: Poison by ingestion. When heated to decomposition it emits acrid smoke and irritating fumes.

CHOLEST-5-EN-3-ONE   HR: 3
CAS: 601-54-7   NIOSH: FZ 7875000
mf: $C_{27}H_{44}O$ mw: 384.71

SYNS: DELTA(SUP 5)-CHOLESTENONE * CHO-LESTENONE * 5-CHOLESTEN-3-ONE

THR: An experimental carcinogen. When heated to decomposition it emits acrid smoke and irritating fumes.

CHOLESTEROL   HR: 3
CAS: 57-88-5   NIOSH: FZ 8400000
mf: $C_{27}H_{46}O$ mw: 386.73

PROP: White or faint yellow, pearly leaflets; mp: 148.5°, bp: 360° decomp.

SYNS: CHOLEST-5-EN-3-beta-OL * 5-CHOLES-TEN-3-beta-OL * 5:6-CHOLESTEN-3-beta-OL * CHOLESTERIN * 3-beta-HYDROXYCHOLEST-5-ENE * PROVITAMIN D

THR: An experimental teratogen, tumorigen, and carcinogen. When heated to decomposition it emits acrid smoke and irritating fumes. For further information, see Vol. 1, No. 7 of *DPIM Report*.

CHOLESTERYL-p-BIS(2-CHLOROETH-YL)AMINO PHENYLACETATE   HR: 3
CAS: 3546-10-9   NIOSH: AF 5457000
mf: $C_{39}H_{59}Cl_2NO_2$ mw: 644.89

SYNS: (P-(BIS(2-CHLOROETHYL)AMINO)PHE-NYL)ACETIC ACID CHOLESTEROL ESTER * CHO-LESTEROL, (P-(BIS(2-CHLOROETHYL)AMINO)PHE-NYL) ACETATE * FENESTERIN * NCI-C01558 * NSC 104469

THR: An experimental carcinogen and neoplastigen. When heated to decomposition it emits very toxic fumes of Cl^- and NO_x.

CHOLIC ACID (HYDRATE)   HR: 3
CAS: 81-25-4   NIOSH: FZ 9350000
mf: $C_{24}H_{40}O_5 \cdot H_2O$ mw: 526.67

PROP: The most abundant bile acid, the monohydrate crystallizes in plates from dilute acetic acid, sol in glacial acetic acid, acetone, and alcohol. Slightly sol in chloroform, practically insol in water and benzene.

SYNS: CHOLSAEURE (GERMAN) * 3-alpha,7-alpha,12-alpha-TRIHYDROXY-5-beta-CHOLAN-24-OIC ACID * 3,7,12-TRIHYDROXY-CHOLAN-24-OIC ACID (3-alpha,5-beta,7-alpha, 12-alpha

THR: Poison by intraperitoneal route. Moderately toxic by subcutaneous route.

CHOLINE SUCCINATE (2:1) (ESTER)   HR: 3
CAS: 306-40-1   NIOSH: GA 7000000
mf: $C_{14}H_{30}N_2O_4$ mw: 290.46

SYNS: CHOLINE SUCCINATE (ESTER) * DIACE-TYLCHOLINE * SUCCINIC ACID DIESTER WITH CHOLINE * SUCCINOYLCHOLINE * SUCCINYL-BISCHOLINE * SUCCINYLDICHOLINE

THR: Poison by ingestion, intraperitoneal, intravenous, and subcutaneous routes. When heated to decomposition it emits toxic fumes of NO_x.

CHROMIC ACETATE   HR: 3
CAS: 1066-30-4   NIOSH: AG 2975000
mf: $C_6H_9O_6 \cdot Cr$ mw: 229.15

PROP: Gray, green powdered or bluish green pasty mass.

SYNS: CHROMIUM(III) ACETATE * CHROMIUM TRIACETATE

THR: An experimental carcinogen and tumorigen. Moderately toxic by intravenous route. Mutagenic data. See also chromium compounds. When heated to decomposition it emits acrid smoke and irritating fumes. For further information see Vol. 5, No. 6 of *DPIM Report*.

CHROMIC ACID   HR: 3
CAS: 7738-94-5   NIOSH: GB 2450000
mf: CrH_2O_4 mw: 118.02

SYNS: ACIDE CHROMIQUE (FRENCH)
* CHROMIC (VI) ACID

OSHA PEL: TWA CL 100 ug(CrO_3/m^3

THR: A poison and a carcinogen. Mutagenic data. See also chromium compounds. When heated to decomposition it emits smoke and irritating fumes. Incompatible with acetone. For further information, see Vol. 3, No. 3 of *DPIM Report*.

CHROMIC ACID (MIXTURE)   HR: 3
                        NIOSH: GB 2650000
DOT: 1463
mf: CrO_3 mw: 100.01

PROP: mp: 196°; d: 2.70; dark red cryst; decomp @ 250° to $Cr_2O_3 + O_2$; a powerful oxidizer. Water sol.

SYNS: CHROMIUM TRIOXIDE * CHROMIC AN-HYDRIDE

DOT Classification: Label: Oxidizer

THR: A poison. See also chromium compounds and chromates. A powerful irritant of skin, eyes, and mucous membranes; can cause a dermatitis, bronchoasthma, "chrome holes," damage to the eyes. May explode in a fire. Incompatible with acetic acid; acetic anhydride; tetrahydronaphthalene; acetone; alcohols; alkali metals; ammonia; arsenic; bromine penta fluoride; butyric acid; n,n-dimethylformamide; hydrogen sulfide; peroxyformic acid; phosphorus; potassium hex-

acyanoferrate; pyridine; selenium; sodium; sulfur.

CHROMIC ACID (SOLUTION)   HR: 3
                        NIOSH: GB 2670000

SYN: CHROMIC ACID SOLUTION (DOT)

DOT Classification: Label: Corrosive

THR: A poison via many routes. See chromic acid, dry. See also chromium compounds.

CHROMIC CHROMATE   HR: 3
CAS: 24613-89-6   NIOSH: GB 2850000
mf: $Cr_3O_{12} \cdot 2Cr$ mw: 452.00

SYNS: CHROMIC ACID, CHROMIUM (3+)SALT (3:2) * CHROMIUM CHROMATE

THR: An experimental neoplastigen and carcinogen. See also chromium compounds. Very powerful oxidizer.

CHROMITE (MINERAL)   HR: 3
CAS: 1308-31-2   NIOSH: GB 4000000
mf: Cr_2FeO_4 mw: 223.85

SYNS: CHROME ORE * CHROMITE ORE
* IRON CHROMITE
ACGIH TLV: TWA 0.05 mg/m^3

THR: An experimental carcinogen. See also chromium compounds and iron.

CHROMIUM   HR: 3
CAS: 7440-47-3   NIOSH: GB 4200000
Af: Cr   Aw: 52.0

SYN: CHROME

OSHA PEL: TWA 1 mg/m^3
ACGIH TLV: TWA 0.5 mg/m^3

THR: An experimental carcinogen and tumorigen. Powder will ignite spontaneously in air. See also chromium compounds. Incompatible with oxidants. For further information, see Vol. 3, No. 3 of *DPIM Report*.

CHROMIUM (III) CHLORIDE   HR: 3
CAS: 10025-73-7   NIOSH: GB 5425000
mf: Cl_3Cr mw: 158.36

PROP: Bp: 1300° (subl).

SYNS: CHROMIC CHLORIDE * CHROMIUM CHLORIDE, ANHYDROUS * CHROMIUM CHLO-RIDE * CHROMIUM TRICHLORIDE * C.I. 77295 * PURATRONIC CHROMIUM CHLORIDE * TRICHLOROCHROMIUM

THR: Poison by skin absorption. Moderately toxic by ingestion. Violent reaction with Li; nitrogen. When heated to decomposition it emits toxic fumes of Cl^-.

CHROMIUM COMPOUNDS   HR: 3

THR: Chromic acid and its salts have a corrosive action on the skin and mucous membranes. The lesions are confined to the exposed parts, affecting chiefly the skin of the hands and forearms and the mumem of the nasal septum. The characteristic lesion is a deep, penetrating ulcer, which, for the most part, does not tend to suppurate, and which is slow in healing. Small ulcers, about the size of a matchhead or end of a lead pencil may be found, chiefly around the base of the nails, on the knuckles, dorsum of the hands and forearms. These ulcers tend to be clean, and progress slowly. They are frequently painless, even though quite deep. They heal slowly, and leave scars. On the mucous membranes of the nasal septum the ulcers are usually accompanied by purulent discharge and crusting. If exposure continues, perforation of the nasal septum may result, but produces no deformity of the nose. Chromate salts are experimental and human carcinogens of the lungs, nasal cavity and paranasal sinus, also experimental carcinogens of the stomach and larynx. Hexavalent compounds are more toxic than the trivalent. Eczematous dermatitis due to trivalent chromium compounds has been reported.

CHROMIUM (VI) OXIDE (1:3)   HR: 3
CAS: 1333-82-0   NIOSH: GB 6650000
DOT: 1463/1755
mf: CrO_3 mw: 100.00

PROP: Red rhomb, deliq cryst; d: 2.70; mp: 196°; bp: decomp, sol = 61.7 g/100 cc @ 0°; 67.45 g/100 cc @ 100°.

SYNS: ANHYDRIDE CHROMIQUE (FRENCH) * ANIDRIDE CROMICA (ITALIAN) * CHROME (TRIOXYDE DE) (FRENCH) * CHROMIC ACID; CHROMIC (VI) ACID * CHROMIC ACID, SOLID; CHROMIC ACID, SOLID (DOT) * CHROMIC ANHYDRIDE * CHROMIC ANHYDRIDE (DOT) * CHROMIC TRIOXIDE; CHROMIC TRIOXIDE (DOT) * CHROMIUM OXIDE; CHROMIUM (VI) OXIDE * CHROMIUM TRIOXIDE * CHROMIUM (6+) TRIOXID * CHROMSAEUREANHYDRID (GERMAN) * CHROMTRIOXID (GERMAN) * CHROOMTRIOXYDE (DUTCH) * CHROOMZUURANHYDRIDE (DUTCH) * CHROMO (TRIOSSIDO DI) (ITALIAN)

* MONOCHROMIUM OXIDE) * MONOCHROMIUM TRIOXIDE * PURATRONIC CHROMIUM TRIOXIDE

OSHA PEL: CL 100 ug(CrO_3)/m^3
DFG MAK: 0.1 mg/m^3
DOT Classification: Label: Oxidizer

THR: Poison by subcutaneous route. An experimental teratogen and carcinogen. Mutagenic data. See also chromium compounds. Powerful oxidizer. Incompatible with acetic acid; acetic anhydride; acetic anhydride + tetrahydronaphthalene; acetone; alcohols; alkali metals; ammonia; arsenic; bromine pentafluoride; butyric acid; N,N-dimethylformamide; hydrogen sulfide; peroxyformic acid; phosphorus; potassium hexacyanoferrate; pyridine; selenium; sodium; sulfur.

CHROMIUM OXYCHLORIDE   HR: 3
CAS: 14977-61-8   NIOSH: GB 5775000
DOT: 1758
mf: Cl_2CrO_2 mw: 154.90

PROP: Dark red liquid, musty burning odor; mp: −96.5°; bp: 115.7°; d: 1.9145 @ 25°/4°; vap press: 20 mm @ 20°.

SYNS: CHROMYL CHLORIDE * CHLORURE DE CHROMYLE (FRENCH) * CHROMIC OXYCHLORIDE * CHROMIUM CHLORIDE OXIDE * CHROMIUM DICHLORIDE DIOXIDE * CHROMIUM DIOXIDE DICHLORIDE * CHROMIUM (VI) DIOXYCHLORIDE * CHROMYLCHLORID (GERMAN) * CHROMOXYLCHLORIDE (DUTCH) * CROMILE, CLORURO DI (ITALIAN) * CROMO, OSSICLORURO DI (ITALIAN) * DICHLORODIOXOCHROMIUM * DIOXODICHLOROCHROMIUM * OXYCHLORURE CHROMIQUE (FRENCH)

DOT Classification: Label: Corrosive

THR: Poison by inhalation and subcutaneous routes. A strong irritant. Hydrolyzes to form chromic and hydrochloric acids. See chromium compounds. Reacts violently with alcohol; ether; acetone; turpentine; NH_3; (Cl_2 + C); F_2; P; PCl_3; NaN_3; S; SCl. Dangerous; see chlorides for disaster hazard. During preparation can violently explode. Incompatible with ammonia; disulfur dichloride; organic solvents; phosphorus; phosphorus trichloride; sodium azide; sulfur.

CHROMIUM(6+)ZINC OXIDE HYDRATE (1:2:6:1)   HR: 3
CAS: 15930-94-6   NIOSH: GB 3260000
mf: $CrO_4 \cdot H_2O_2 \cdot Zn_2 \cdot H_2O$ mw: 298.78

SYNS: BUTTERCUP YELLOW * ZINC CHROMATE HYDROXIDE * ZINC CHROMATE (VI) HYDROXIDE * ZINC HYDROXYCHROMATE * ZINC YELLOW

THR: A human and experimental carcinogen. See also chromium and zinc compounds.

CICUTOXIN   HR: 3
CAS: 505-75-9   NIOSH: MI 3950000
mf: $C_{17}H_{22}O_2$ mw: 258.39

THR: Poison by ingestion and intravenous routes. When heated to decomposition it emits acrid smoke and irritating fumes.

C.I. DIRECT BLUE 1, TETRASODIUM SALT   HR: 3
CAS: 2610-05-1   NIOSH: QJ 6430000
mf: $C_{34}H_{28}N_6O_{16}S_4 \cdot 4Na$ mw: 996.88

SYNS: NCI-C61109 * AIREDALE BLUE FFD * AMANIL SKY BLUE 6B * CHROME LEATHER SKY BLUE * C.I. 24410 * C.I. DIRECT BLUE 1

THR: An experimental teratogen. When heated to decomposition it emits very toxic fumes of NO_x and SO_x.

C.I. DIRECT BLUE 6, TETRASODIUM SALT   HR: 3
CAS: 2602-46-2   NIOSH: QJ 6400000
mf: $C_{32}H_{20}N_6O_4S_4 \cdot Na$ mw: 936.82

PROP: A dye.

SYNS: AIREDALE BLUE 2BD * AIZEN DIRECT BLUE 2BH * C.I. 22610 * DIPHENYL BLUE 2B * DIRECT BLUE 6 * INDIGO BLUE 2B * NAPHTAMINE BLUE 2B * NCI-C54579 * NIAGARA BLUE 2B * SODIUM DIPHENYL-4,4'-BIS-AZO-2''-8''-AMINO-1''-NAPHTHOL-3'', 6''-DISULPHONATE

THR: An experimental carcinogen, teratogen, tumorigen, and neoplastigen. When heated to decomposition it emits very toxic fumes of NO_x and SO_x.

C.I. DIRECT BLUE 14, TETRASODIUM SALT   HR: 3
CAS: 72-57-1   NIOSH: QJ 6475000
mf: $C_{34}H_{28}N_6O_{14}S_4 \cdot 4Na$ mw: 964.88

SYNS: CHLORAZOL BLUE 3B * CHROME LEATHER BLUE 3B * C.I. 23850 * C.I. DIRECT BLUE 14 * SODIUM DITOLYLDIAZOBIS-8-AMINO-1-NAPHTHOL-3,6-DISULFONATE

THR: Poison by intraperitoneal, intravenous, and subcutaneous routes. An experimental carcinogen, teratogen, tumorigen, and neoplastigen. Mutagenic data. When heated to decomposition it emits very toxic fumes of NO_x and SO_x.

C.I. DIRECT BROWN   HR: 3
CAS: 16071-86-6   NIOSH: GL 7375000
mf: $C_{31}H_{20}N_6O_9S \cdot Cu \cdot 2Na$ mw: 762.15

SYNS: C.I. 30145 * DIPHENYL FAST BROWN BRL * NCI-C 54568 * SATURN BROWN LBR

THR: An experimental carcinogen and neoplastigen. When heated to decomposition it emits very toxic fumes of Na_2O, SO_x, and NO_x.

C.I. DIRECT VIOLET 1, DISODIUM SALT   HR: 3
CAS: 2586-60-9   NIOSH: QK 1420000
mf: $C_{32}H_{22}N_6O_8S_2 \cdot 2Na$ mw: 728.70

SYNS: CHLORAZOL VIOLET N * C.I. 22570

THR: An experimental tumorigen. When heated to decomposition it emits very toxic fumes of NO_x and SO_x.

C.I. FOOD RED 3   HR: 2
CAS: 3567-69-9   NIOSH: QK 1925000
mf: $C_{20}H_{12}N_2O_7S_2 \cdot 2Na$ mw: 502.44

SYNS: C.I. ACID RED 14, DISODIUM SALT * 4-HYDROXY-3,4'-AZODI-1-NAPHTHALENESUL-FONIC ACID, DISODIUM SALT * DISODIUM SALT OF 2-(4-SULPHO-1-NAPHTHYLAZO)-1-NAPHTHOL-4-SULPHONIC ACID * DISODIUM 2-(4-SULFO-1-NAPHTHYLAZO)-1-NAPHTHOL-4-SULFONATE * DISODIUM 2-(4-SULPHO-1-NAPHTHYLAZO)-1-NAPHTHOL-4-SULPHONATE * EXTRACT D AND C RED NO. 10 * FOOD RED 5 * 4-HYDROXY-3,4'-AZODI-1-NAPHTHALENESULPHONIC ACID, DI-SODIUM SALT * 4-HYDROXY-3-((4-SULFO-1-NAPHTHALENYL)AZO)-1-NAPHTHALENESULFONIC ACID, DISODIUM SALT * NCI-C53849 * 2-(4-SULFO-1-NAPHTHYLAZO)-1-NAPHTHOL-4-SULFONIC ACID, DISODIUM SALT

THR: Mutagenic data. Moderately toxic by intraperitoneal route. An experimental carcinogen. When heated to decomposition it emits very toxic fumes of SO_x and NO_x.

CINNAMIC ACID, NICKEL(II) SALT   HR: 3
CAS: 63938-16-9   NIOSH: GE 0350000
mf: $C_{18}H_{14}O_4 \cdot Ni$ mw: 353.03

THR: Poison by parenteral route. See also nickel compounds. When heated to decomposition it emits acrid smoke and irritating fumes.

CINNAMMALDEHYDE   HR: 3
mf: C_9H_8O mw: 132.16

PROP: Yellowish oil; cinnamic odor; sol in 5 volumes of 60% alcohol, very sltly sol in H_2O; d: 1.048-1.052 @ 25°/25°; mp: −8°; bp: 246°.

SYNS: CINNAMIC ALDEHYDE * 3-PHENYL PROPENAL * CINNAMYL ALDEHYDE

THR: Poison by intraperitoneal route. Moderately toxic by ingestion and inhalation. Incompatible with NaOH.

d-CINNAMYLEPHEDRINE HYDROCHLORIDE   HR: 3
CAS: 64043-53-4   NIOSH: DO 0350000
mf: $C_{19}H_{23}NO \cdot ClH$ mw: 317.89

SYN: CINNAMYLEPHEDRINE HYDROCHLORIDE, DEXTRO

THR: Poison by subcutaneous route. Poison to humans by intradermal route. When heated to decomposition it emits very toxic fumes of HCl and NO_x.

CITRIC ACID   HR: 2
mf: $C_6H_8O_7$ mw: 192.13

PROP: Colorless, odorless crystals; mp: 153°; bp: decomp; d: 1.542.

SYN: β-HYDROXYTRICARBALLYLIC ACID

THR: Moderately toxic by ingestion and inhalation. An irritating organic acid, some allergenic properties. Incompatible with metal nitrates. For further information, see Vol. 1, No. 8 of *DPIM Report*.

CITRONELLOL   HR: 3
CAS: 106-22-9   NIOSH: RH 3400000
mf: $C_{10}H_{20}O$ mw: 156.30

SYNS: 2,6-DIMETHYL-2-OCTEN-8-OL * 3,7-DIMETHYL-6-OCTEN-1-OL * RODINOL

THR: Poison by intravenous route. Moderately toxic by ingestion, skin absorption, and intramuscular routes. When heated to decomposition it emits acrid smoke and irritating fumes.

CLAVACIN   HR: 3
CAS: 149-29-1   NIOSH: LV 2625000
mf: $C_7H_6O_4$ mw: 154.13

PROP: Colorless crystals; mp: 111°.

SYNS: * PATULIN * 2,4-DIHYDROXY-2H-PYRAN-delta-3(6H),alpha-ACETIC ACID-3,4-LACTONE * (2,4-DIHYDROXY-2H-PYRAN-3(6H)-YLIDENE)ACETIC ACID-3,4-LACTONE * EXPANSIN * 4-HYDROXY-4H-FURO(3,2-C)PYRAN-2(6H)-ONE

THR: Poison by ingestion, intraperitoneal, intravenous, and other routes. An experimental neoplastigen. Mutagenic data. When heated to decomposition it emits acrid smoke and irritating fumes. For further information, see Expansin, Vol. 1, No. 3 of *DPIM Report*.

CLEP   HR: 3
CAS: 1532-19-0   NIOSH: US 8650000
mf: $C_7H_7Cl_2N \cdot ClH$ mw: 212.51

SYNS: 2-(alpha,beta-DICHLORETHYL)-PYRIDINE HYDROCHLORIDE * 2-(1,2-DICHLOROETHYL)PYRIDINE HYDROCHLORIDE

THR: An experimental carcinogen, tumorigen, neoplastigen. When heated to decomposition it emits very toxic fumes of HCl and NO_x.

CLOAZEPAM   HR: 3
CAS: 1622-61-3   NIOSH: DF 2100000
mf: $C_{15}H_{10}ClN_3O_3$ mw: 315.73

SYNS: 1,3-DIHYDRO-7-NITRO-5-(2-CHLOROPHENYL)-2H-1,4-BENZODIAZEPIN-2-ONE * CLONAZEPAM

THR: Poison by ingestion. When heated to decomposition it emits very toxic fumes of Cl^- and NO_x.

CLORGYLINE HYDROCHLORIDE   HR: 3
CAS: 17780-75-5   NIOSH: UI 0675000
mf: $C_{13}H_{15}Cl_2NO \cdot ClH$ mw: 308.65

SYN: N-METHYL-N-PROPARGYL-3-(2,4-DICHLOROPHENOXY)PROPYLAMINE HYDROCHLORIDE

THR: Poison by ingestion, intraperitoneal, intravenous, and subcutaneous routes. When heated to decomposition it emits very toxic fumes of HCl and NO_x.

COAL, GROUND BITUMINOUS (DOT)   HR: 1
                              NIOSH: GF 8300000
DOT: 1361

PROP: Black powder or chunks.

SYNS: COAL FACINGS * SEA COAL

DOT Classification: Label: Flammable Solid

THR: Depends upon content of SiO_2. See also silica. Moderate fire hazard when exposed to heat; can react with oxidizing materials. Moderate spontaneous heating. Sltly explosive when exposed to flame.

COAL TAR DYE, LIQUID (DOT)   HR: 3
NIOSH: GF 8620000
DOT: 2801
DOT Classification: Label: Corrosive

THR: Many of the coal tar dyes are quite harmless and are permitted for foods, drugs, and cosmetics. Some of them may be allergens or carcinogens.

COAL TAR NAPHTHA (DOT)   HR: 2
CAS: 65996-79-4   NIOSH: GF 8635000
DOT: 2552

PROP: Dark straw colored to colorless liquid. bp: 149°-216°; flash p: 107°F (CC); d: 0.862-0.892; autoign temp: 531°F.

SYNS: NAPHTHA, COAL TAR * NAPHTHA SOLVENT

DOT Classification: Combustible Liquid, Label: None

THR: Moderately toxic by inhalation. Can cause unconsciousness which may go to coma, stentorious breathing and bluish tint to the skin. Recovery follows removal from exposure. In mild form, intoxication resembles drunkeness. On a chronic basis, no true poisoning. A common air contaminant. See also mineral oils.

COBALT   HR: 3
mf: Co   mw: 58.93

PROP: Silver-gray metal; mp: 1495°; bp: 2000°; d: 8.9.
ACGIH TLV: TWA (metal, dust, and fume) 0.05 mg/m³ (trial limit)
TRK: 0.5 mg/m³ calculated as cobalt in that portion of dust that can possibly be inhaled in the production of cobalt powder and catalysts; hard metal (tungsten carbide) and magnet production (processing of powder, machine pressing, and mechanical processing of unsintered articles); others 0.1 mg/m³ calculated as cobalt in that portion of dust that can possibly be inhaled.

THR: An experimental carcinogen. See also specific cobalt compounds. Incompatible with acetylene; hydrazinium nitrate; oxidants. Moderate fire hazard when exposed to heat, flame, or spontaneous chemical reaction; see also powdered metals. Pyroforic cobalt reacts violently with acetylene; air; NH_4NO_3. For further information, see Vol. 1, No. 3 of *DPIM Report*.

COBALT ALLOY, Co,Cr   HR: 3
CAS: 11114-92-4   NIOSH: GF 9180000

SYNS: COBALT-CHROMIUM ALLOY * HASTELLOY C * HAYNES STELLITE 21 * VITALLIUM

THR: An experimental carcinogen. See also cobalt and chromium compounds and alloys. Violent reaction with molten Li.

COBALT CARBONYL   HR: 3
CAS: 10210-6-8-1   NIOSH: GG 0300000
mf: $C_8Co_2O_8$ mw: 341.94

PROP: Orange platelets. d: 1.87; mp: 51°; decomp above 52°. Decomp on exposure to air. Insol in water; sol in organic solvents.

SYNS: COBALT OCTACARBONYL * COBALT TETRACARBONYL * DICOBALT CARBONYL

THR: Poison by ingestion. See also CO; carbonyls and Co compounds. See also carbonyls and CO for fire hazard. When heated to decomposition it emits CO.

COBALT DIACETATE   HR: 3
CAS: 71-48-7   NIOSH: AG 3150000
mf: $C_4H_6O_4 \cdot Co$ mw: 177.03

SYNS: ACETIC ACID, COBALT(2+) SALT * COBALT ACETATE * COBALT(2+) ACETATE * COBALT(II) ACETATE * COBALTOUS DIACETATE

THR: Poison by intravenous route. Mutagenic data. See also cobalt compounds. When heated to decomposition it emits acrid smoke and irritating fumes.

COBALT NITROPRUSSIDE   HR: 3
CAS: 63919-21-1   NIOSH: LJ 8400000

SYN: COBALT NITROSOPENTACYANOFERRATE(3)

THR: Poison by ingestion and intraperitoneal routes. When heated to decomposition it emits very toxic fumes of CN^- and NO_x.

COBALTOUS CHLORIDE   HR: 3
CAS: 7646-7-9-9   NIOSH: GF 9800000
mf: Cl_2Co mw: 129.83

PROP: Blue powder; mp: 724°; bp: 1049°; d: 3.348.

SYNS: COBALT CHLORIDE * COBALT DICHLO-RIDE * COBALT MURIATE * KOBALT CHLORID (GERMAN)

THR: Poisonous to human children by ingestion. Poison by ingestion, skin absorption, intraperi-toneal, intravenous, and subcutaneous routes. An experimental mutagen and teratogen. Incom-patible with metals. See also cobalt. When heated to decomposition it emits toxic fumes of Cl^-. For further information, see Vol. 2, No. 5 of DPIM Report.

COBALTOUS NITRATE   HR: 3
CAS: 10141-0-5-6   NIOSH: GG 1109000
mf: $N_2O_6 \cdot Co$ mw: 182.95

PROP: Mp: 55°; d: 1.87.

SYNS: COBALT (II) NITRATE (1:2) * NITRIC ACID, COBALT (2+) SALT * COBALT DINITRATE

THR: Poison by ingestion, intramuscular, and subcutaneous routes. An experimental teratogen and tumorigen. When heated to decomposition it emits toxic fumes of NO_x. Incompatible with carbon. For further information, see Vol. 2, No. 5 of DPIM Report.

COCAINE   HR: 3
CAS: 50-36-2   NIOSH: YM 2800000
mf: $C_{17}H_{21}NO_4$ mw: 303.39

PROP: Colorless to white crystals; mp: 98°.

SYNS: 1-alpha-H,5-alpha-H-TROPANE-2-beta-CARBOXYLIC ACID, 3-beta-HYDROXY-,METHYL ESTER, BENZOATE * L-COCAINE * BENZOYL-METHYLECGONINE * BERNICE * BERNIES * BURESE * 2-beta-CARBOMETHOXY-3-beta-BENZOXYTROPANE * "C" CARRIE * CECIL * CHOLLY * (-)-COCAINE * beta-COCAINE * 1-COCAINE * COKE * CORINE * ECGO-NINE, METHYL ESTER, BENZOATE (ESTER) * ERITROXILINA * ERYTROXYLIN * GIRL * GOLD DUST * HAPPY DUST * KOKAIN * KOKAN * KOKAYEEN * NEUROCAINE * STAR DUST * 3-TROPANYLBENZOATE-2-CAR-BOXYLIC ACID METHYL ESTER

THR: Poison by ingestion, intraperitoneal, in-travenous, subcutaneous, and parenteral routes.

Human central nervous system effects. See also esters. When heated to decomposition it emits highly toxic fumes. Treatment and Antidotes: Wash stomach out immediately with water and sodium bicarbonate (60 g to 1 pint). If a solution of permanganate is handy (1 crystal to 8 ounces of water), use it to wash out stomach. Give inhalations of ammonia. Allay convulsions with chloral or chloroform. If breathing is disturbed, give artificial respiration and cardiac massage after respiration fails. Inhalation of oxygen plus 5% carbon dioxide should be given. Get medical advice.

COCAINE CHLORIDE   HR: 3
CAS: 53-21-4   NIOSH: YM 3050000
mf: $C_{17}H_{21}NO_4 \cdot ClH$ mw: 339.85

SYNS: 1-alpha-H,5-alpha-H-TROPANE-2-beta-CARBOXYLIC ACID, 3-beta-HYDROXY-,METHYL ESTER, BENZOATE (ESTER), HYDROCHLORIDE * (-)-COCAINE HYDROCHLORIDE * L-COCAINE HYDROCHLORIDE * COCAIN-CHLORHYDRAT (GERMAN) * COCAINE HYDROCHLORIDE * COCAINE MURIATE * SAL DE MERCK

THR: Poison by ingestion, intravenous, intra-peritoneal, and subcutaneous routes. An eye irritant. See also cocaine. When heated to de-composition it emits very toxic fumes of NO_x and HCl.

CODEINE   HR: 3
CAS: 76-57-3   NIOSH: QD 0893000
mf: $C_{18}H_{21}NO_3$ mw: 299.40

PROP: Cryst from H_2O or dil alc; mp: 155°; d: 1.32 @ 20°/4°.

SYNS: METHYLMORPHINE * MORPHINE MONO-METHYL ETHER * MORPHINE-3-METHYL ETHER

THR: Poison by ingestion, intraperitoneal, in-travenous, and subcutaneous routes. See also ethers. An addictive drug. Moderate fire hazard when exposed to heat or flame. To fight fire, use alcohol foam. When heated to decomposi-tion it emits toxic fumes of NO_x.

CODEINE HYDROCHLORIDE   HR: 3
CAS: 1422-07-7   NIOSH: QD 1050000
mf: $C_{18}H_{21}NO_3 \cdot ClH$ mw: 335.86

THR: Poison by subcutaneous and parenteral routes. See also codeine. An addictive drug. When heated to decomposition it emits very toxic fumes of NO_x and HCl.

CODEINE NICOTINATE (ester)   HR: 3
CAS: 3688-66-2   NIOSH: QD 1225000
mf: $C_{24}H_{24}N_2O_4$ mw: 404.50

SYNS: NICOCODINE * NICOTINIC ACID, ESTER
WITH CODEINE * NICOTINIC ACID, 7,8-DIDEHY-
DRO-4,5-alpha-EPOXY-3-METHOXY-17-METHYL-
MORPHINAN-6-alpha-YL ESTER

THR: Poison by ingestion and intraperitoneal
routes. See also codeine, and esters. An addictive drug. When heated to decomposition it
emits toxic fumes of NO_x.

CODEINE PHOSPHATE   HR: 3
CAS: 52-28-8   NIOSH: QD 1310000
mf: $C_{18}H_{21}NO_3 \cdot H_3O_4P$ mw: 397.40

THR: Poison by ingestion, intravenous, intramuscular, intraperitoneal, and subcutaneous
routes. See also codeine. When heated to decomposition it emits very toxic fumes of NO_x
and PO_x.

CODEINE SULFATE   HR: 3
CAS: 1420-53-7   NIOSH: QD 1400000
mf: $C_{36}H_{42}N_2O_6 \cdot O_4S$ mw: 694.86

THR: Poison by ingestion, intravenous, and subcutaneous routes. An experimental teratogen.
See also codeine. An addictive drug. When
heated to decomposition it emits very toxic
fumes of NO_x and SO_x.

COLLODION   HR: 3
CAS: 9004-70-0   NIOSH: GH 2110000

PROP: Soln of nitrated cellulose in (ether +
alcohol). $C_{12}H_{16}O_6(NO_3)_4C_{13}H_{17}O_7(NO_3)_3$,
mw: 975, flash p: <0°F.
DOT Classification: Label: Flammable
   Liquid

THR: Very dangerous fire hazard when exposed
to heat or flame. To fight fire, use alcohol foam.
When heated to decomposition it emits
toxic fumes of NO_x.

COLTSFOOT   HR: 3
NIOSH: GJ 9880000

PROP: It is herb of the tribe *Senecione* and
from family *Compositae*.

SYNS: KAN-TO-KA (JAPANESE) * TUSSILAGO
FARFARA L

THR: An experimental carcinogen by ingestion.

CONIUM MACULATUM   HR: 3
NIOSH: GL 1223600

PROP: Colorless, oily liquid with mousy odor;
bp: 166.5°, fp: −2.5°, d: 0.844-0.848 @ 20°/
4°. Lupine Plant whose toxic agent is Coniine,
fed as green or dried plant.

THR: Toxic principle of poison hemlock. Slight
fire hazard when heated. An experimental teratogen.

COPPER   HR: 3
CAS: 7440-50-8   NIOSH: GL 5325000
Af: Cu   Aw: 63.54

PROP: A metal with a distinct reddish color;
mp: 1083°, bp: 2324°, d: 8.92, vap press: 1
mm @ 1628°.

SYNS: BRONZE POWDER * C.I. 77400
* COPPER BRONZE * GOLD BRONZE
ACGIH TLV: TWA 0.2 mg/m³ (fume)
DFG MAK: (dust) 1 mg/m³; (fume) 0.1 mg/m³

THR: Poison to humans by ingestion. See also
copper compounds. Reacts violently with C_2H_2;
NH_4NO_3; bromates; chlorates; iodates; Cl_2;
ClF_3; (Cl_2 + OF_2); ethylene oxide; F_2; H_2O_2;
hydrazine mononitrate; hydrazoic acid; H_2S;
$Pb(N_3)_2$; K_2O_2; NaN_3; Na_2O_2. Incompatible
with 1-bromo-2-propyne. For further information, see Vol. 1, No. 5 of *DPIM Report*.

COPPER ACETATE   HR: 2
CAS: 142-71-2   NIOSH: AG 3480000
mf: $C_4H_6O_4 \cdot Cu$ mw: 181.64

PROP: Greenish blue powd or small crystals.

SYNS: ACETIC ACID, CUPRIC SALT * COP-
PER(2+) ACETATE * COPPER(II) ACETATE
* COPPER DIACETATE * COPPER(2+) DIACE-
TATE * CRYSTALLIZED VERDIGRIS * CRYS-
TALS OF VENUS * CUPRIC ACETATE * CUPRIC
DIACETATE * NEUTRAL VERDIGRIS * OCTAN
MEDNATY (CZECH)

THR: Moderately toxic by ingestion. When
heated to decomposition it emits acrid smoke
and irritating fumes.

COPPER ALLOY, Cu,Be   HR: 3
CAS: 11133-98-5   NIOSH: GL 5830000

SYN: BERYLLIUM-COPPER ALLOY

THR: An experimental carcinogen. Cases of
berylliosis have been reported from exposure

to so called low beryllium alloys. See also beryllium compounds and copper compounds. When heated to decomposition it emits very toxic fumes of BeO.

COPPER ARSENATE HYDROXIDE   HR: 3
CAS: 16102-92-4   NIOSH: GL 5900000
mf: AsCu$_2$HO$_5$ mw: 283.01

PROP: A green solid.

SYNS: CUPROUS ARSENATE, BASIC * COPPER ARSENATE (BASIC)

THR: A poison via various routes. See also arsenic compounds and copper compounds. When heated to decomposition it emits toxic As.

COPPER (II) CARBONATE HYDROXIDE (2:1:2)   HR: 3
CAS: 12069-69-1   NIOSH: GL 6910000
mf: CO$_3$ • H$_2$O$_2$ • 2Cu   mw: 221.11

PROP: Green powder; mp: decomp @ 200°, d: 4.0.

SYNS: BASIC COPPER CARBONATE * (CARBONATO)DIHYDROXYDICOPPER * COPPER CARBONATE HYDROXIDE * CUPRIC CARBONATE * MALACHITE

THR: Poison by ingestion. See also copper compounds.

COPPER (I) CHLORIDE   HR: 3
CAS: 7768-89-6   NIOSH: GL 6990000
mf: ClCu   mw: 98.99

PROP: Cubic white crystals; mp: 422°, bp: 1366°, d: 3.53, vap press: 1 mm @ 546°.

SYNS: CHLORID MEDNY (CZECH) * CUPROUS CHLORIDE

THR: Poison by ingestion. See also chlorides and copper compounds. Reacts violently with K. When heated to decomposition it emits toxic fumes of Cl$^-$. For further information, see Vol. 1, No. 8 of *DPIM Report*.

COPPER (II) CHLORIDE (1:2)   HR: 3
CAS: 1344-67-8   NIOSH: GL 7000000
mf: Cl$_2$Cu   mw: 134.44

PROP: Yellowish-brown hygroscopic powder; mp: 498°, d: 3.054.

SYNS: COPPER CHLORIDE (DOT) * CUPRIC CHLORIDE

OSHA PEL: TWA 1 mg(Cu)/m^3

THR: Poison by ingestion, inhalation, and intraperitoneal routes. Mutagenic data. See also copper compounds and chlorides. Can react violently with K and Na. When heated to decomposition it emits toxic fumes of Cl$^-$. For further information, see Vol. 1, No. 8 of *DPIM Report*.

COPPER COMPOUNDS
THR: As the sublimed oxide, copper may be responsible for one form of metal fume fever. Inhalation of Cu dust has caused, in animals, hemolysis of the red blood cells, deposition of hemofuscin in the liver and pancreas, and injury to the lung cells; injection of the dust has caused cirrhosis of the liver and pancreas, and a condition closely resembling hemochromatosis, or bronzed diabetes. However, considerable trial exposure to Cu compounds has not resulted in such disease. As regards local effect, copper chloride and sulfate have been reported as causing irritation of the skin and conjunctivae which may be on an allergic basis. Cuprous oxide is irritating to the eyes and upper respiratory tract. Discoloration of the skin is often seen in persons handling Cu, but this does not indicate any actual injury from Cu. There is an excess of cancer cases in the Cu smelting industry. In man the ingestion of a large quantity of copper sulfate has caused vomiting, gastric pain, dizziness, exhaustion, anemia, cramps, convulsions, shock, coma and death. Symptoms attributed to damage to the nervous system and kidney have been recorded, jaundice has been observed and, in some cases, the liver has been enlarged. Deaths have been reported to have occurred following the ingestion of as little as 27 g of the salt, while other victims have recovered after having taken up to 120 g. Many Cu-containing compounds are used as fungicides. Many Cu salts form highly unstable acetylides. Those formed in basic solutions from (Cu$^+$ salts + C$_2$H$_2$) are less stable than those formed from Cu^{++} salts. (Cu salts + hydrazine) react strongly, and with nitro-methane are explosive.

COPPER CYANIDE   HR: 3
CAS: 644-92-3   NIOSH: GL 7150000
mf: CCuN   mw: 89.56

PROP: Monoclinic white prisms; mp: 473° in N$_2$, bp: decomp, d: 2.92.

SYNS: CUPRICIN * CUPROUS CYANIDE

THR: A poison. See also cyanides and Cu compounds. Reacts violently with Mg. When heated to decomposition it emits very toxic CN^- and NO_x.

COPPER (II) CYANIDE   HR: 3
CAS: 14763-77-0   NIOSH: GL 7175000
DOT: 1587
mf: C_2CuN_2 mw: 115.58

PROP: Yellowish-green powder; mp: decomp before melting.

SYNS: COPPER CYANIDE (DOT) * CUPRIC CYANIDE (DOT) * CYANURE DE CUIVRE (FRENCH)

DOT Classification: Label: Poison

THR: Poison by intraperitoneal route. See also cyanides and Cu compounds. When heated to decomposition it emits toxic fumes of NO_x and CN^-. Incompatible with magnesium.

COPPER DIMETHYLDITHIOCARBA-MATE   HR: 3
CAS: 137-29-1   NIOSH: FA 0175000
mf: $C_6H_{12}N_2S_4 \cdot Cu$ mw: 303.98

SYNS: COMPOUND-4018 * CUMATE * DI-METHYLDITHIOCARBAMIC ACID COPPER SALT

THR: Poison by intraperitoneal route. Mutagenic data. See also copper compounds and carbamates. When heated to decomposition it emits very toxic fumes of NO_x and SO_x.

COPPER EDTA COMPLEX   HR: 3
CAS: 54453-03-1   NIOSH: AH 4280000

SYN: (ETHYLENEDINITRILO)TETRA-ACETIC ACID COPPER(II) COMPLEX

THR: Poison by intraperitoneal route. See also copper compounds. When heated to decomposition it emits toxic fumes of NO_x.

COPPER FUME   HR: 2
                            NIOSH: GL 7525000
Af: Cu

SYN: MIEDZ (POLISH)

OSHA PEL: TWA 100 ug/m^3

THR: Human irritant via systemic effects. See also copper.

COPPER HYDROXIDE   HR: 3
CAS: 20427-59-2   NIOSH: GL 7600000
mf: $H_2O_2 \cdot Cu$ mw: 97.56

PROP: Blue, gelatinous or amorphous powder; d: 3.368.

SYNS: CUPRAVIT BLUE * CUPRIC HYDROXIDE * KOCIDE

THR: Poison by ingestion and inhalation. See also copper compounds.

COPPER ORTHOARSENITE   HR: 3
CAS: 10290-12-7   NIOSH: CG 3385000
DOT: 1586
mf: $AsCuHO_3$ mw: 187.47

PROP: Yellowish-green powder; mp: decomp.

SYNS: ACID COPPER ARSENITE * COPPER AR-SENITE, SOLID (DOT) * CUPRIC ARSENITE * CUPRIC GREEN * SCHEELES GREEN * SCHEELE'S MINERAL * SWEDISH GREEN

DOT Classification: Label: Poison

THR: Poison. See also arsenic compounds and Cu compounds.

COPPER (I) OXIDE   HR: 2
CAS: 1317-39-1   NIOSH: GL 8050000
mf: Cu_2O mw: 143.08

PROP: Octahedral, cubic red crystals; mp: 1235°, bp: $-O_2$ @ 1800°, d: 6.0.

SYNS: BROWN COPPER OXIDE * CUPROUS OX-IDE * OLEOCUIVRE * OLEO NORDOX * YELLOW CUPROCIDE

OSHA PEL: TWA 1 mg(Cu)/m^3

THR: Moderately toxic by ingestion. See also Cu compounds. Incompatible with peroxyformic acid.

COPPER (II) SULFATE (1:1)   HR: 3
CAS: 7758-98-7   NIOSH: GL 8800000
mf: $O_4S \cdot Cu$ mw: 159.60

PROP: Blue crystals or blue, crystalline granules or powder; d: 2.284.

SYNS: CUPRIC SULFATE * ROMAN VITRIOL * SULFATE DE CUIVRE (FRENCH) * SULFURIC ACID, COPPER(2+) SALT (1:1)

OSHA PEL: TWA 1 mg(Cu)/m^3

THR: Poison by intraperitoneal route. Moderately toxic by ingestion and inhalation. An experimental tumorigen. Human systemic and gastrointestinal tract effects. Mutagenic data. Reacts violently with hydroxylamine; Mg. See also copper compounds and sulfuric acid. When heated to decomposition it emits toxic fumes of SO_x. Incompatible with hydroxylamine.

COPPER (II) SULFATE PENTAHYDRATE (1:1:5)   HR: 3
CAS: 7768-99-8   NIOSH: GL 8900000
mf: $O_4S \cdot Cu \cdot 5H_2O$ mw: 249.70

PROP: Mp: $-4H_2O$ @ 110°.

SYNS: BLUE COPPERRAS * BLUESTONE * BLUE VITRIOL * CUPRIC SULFATE PENTAHYDRATE * KUPFERSULFAT-PENTAHYDRAT (GERMAN) * ROMAN VITRIOL * SALZBURG VITRIOL * SULFURIC ACID, COPPER(2+) SALT, PENTAHYDRATE

THR: Poison by intraperitoneal route. Moderately toxic to humans by ingestion. Human systemic toxicity. When heated to decomposition it emits toxic fumes of SO_x.

COPPER (I) SULFIDE   HR: 3
CAS: 22205-45-4   NIOSH: GL 8910000
mf: Cu_2S mw: 159.14

PROP: Rhombic, black crystals. mp: 1100°, d: 5.6.

SYNS: COPPER SULFIDE * CUPRASULFIDE * CUPROUS SULFIDE * DICOPPER MONOSULFIDE * DICOPPER SULFIDE

THR: Mutagenic data. See also sulfides and copper compounds. When heated to decomposition it emits toxic fumes of SO_x.

CORN OIL   HR: 2
CAS: 8001-30-7   NIOSH: GM 4800000

PROP: Light yellow, clear, oily liquid, faint characteristic odor; mp: $-10°$, flash p: 490°F (CC), d: 0.92, autoign temp: 740°F. From wet milling of *Zea mays*.

THR: Human skin irritant. May be an allergen. Slight fire hazard when exposed to heat or flame. Dangerous when stored if leakage impregnates rags, waste, etc. Moderate spontaneous heating. To fight fire, use CO_2, dry chemical.

CORROSIVE SUBLIMATE   HR: 3
CAS: 7487-94-7   NIOSH: OV 9100000
mf: Cl_2Hg mw: 271.49

PROP: White crystals or powder; mp: 276°, bp: 302°, d: 5.440 @ 25°, vap press: 1 mm @ 136.2°.

SYNS: BICHLORIDE OF MERCURY * BICHLORURE DE MERCURE (FRENCH) * CHLORID RTUTNATY (CZECH) * CORROSIVE MERCURY CHLORIDE * MERCURIC BICHLORIDE * MERCURY BICHLORIDE * MERCURY PERCHLORIDE * NCI-C60173 * QUECKSILBER CHLORID (GERMAN) * PERCHLORIDE OF MERCURY * SUBLIMAT (CZECH)

DOT Classification: Label: Poison

THR: Poison to humans by ingestion. Poison by inhalation, subcutaneous, intravenous, intraperitoneal, intramuscular, and dermal routes. Mutagenic data. A skin and eye irritant. Human gastrointestinal tract and musculo-skeletal effects. See also mercury compounds. Reacts violently with K; Na. When heated to decomposition it emits very toxic fumes of Hg and Cl^-.

CORTISOL   HR: 3
CAS: 50-23-7   NIOSH: GM 8925000
mf: $C_{21}H_{30}N_{40}O_5S$ mw: 954.97

SYNS: ANTI-INFLAMMATORY HORMONE * CORTISOL ALCOHOL * CORTISPRAY * HIDRO-COLISONA * 11-BETA-HYDROCORTISONE * HYDROCORTISONE FREE ALCOHOL * HYDROCORTISYL * HYDROCORTONE * 17-HYDROXYCORTICOSTERONE * 11-beta-HYDROXYCORTISONE * NSC 10483 * 4-PREGNENE-11-beta,17-alpha,21-TRIOL 3,20-DIONE * REICHSTEIN'S SUBSTANCE M * 11-beta,17,21-TRIHYDROXYPREGN-4-ENE-3,20-DIONE * 11-beta,17-alpha-21-TRIHYDROXY-4-PREGNENE-3,20-DIONE

THR: An experimental teratogen. Moderately toxic by subcutaneous route. Mutagenic data. When heated to decomposition it emits very toxic fumes of SO_x and NO_x.

CORUNDUM   HR: 3
CAS: 1302-74-5   NIOSH: GN 0231000
mf: Al_2O_3 mw: 101.96

PROP: A varicolored mineral; d: 3.95-4.10.

SYNS: KORUND * EMERY

THR: An experimental carcinogen by intraperitoneal route. It is mainly a nuisance dust. Low acute toxicity.

CORUNDUM FUME   HR: 3
PROP: Half finely divided alumina, half silica.

THR: Poison by intratracheal route. See also aluminum oxide (2:3) and silica.

COTTON DUST   HR: 2
NIOSH: GN 2275000
OSHA PEL: TWA 1 mg/m^3
ACGIH TLV: TWA 0.2 mg/m^3
DFG MAK: 1.5 mg/m^3 (raw cotton)

THR: Human pulmonary effects. Causes a mild febrile condition of the lungs resembling metal fume fever. Coarser grades of cotton contain more dust than the finer varieties, and therefore constitute a greater hazard. It is considered an inert dust and indeed it is, within the meaning of the term. However, it can cause some illness, due to the allergens or fungi in the cotton or on the dust. Workers in processing rooms may develop conjunctivitis or blepharitis from the burned products of the gassing of the double yarn. It is a mild allergen. Inhalation may produce bronchial asthma, sneezing and eczema in sensitized persons. Moderate fire and explosion hazard when exposed to heat or flame; can react with oxidizing materials.

COTTONSEED OIL (DEODORIZED WINTERIZED)   HR: 3
NIOSH: GN 2815000

PROP: Oily, pale yellow, nearly odorless liquid from seeds of species of *gossypium*. Flash p: 486°F (CC), fp: 0° to 5°, d: 0.915-0.921 @ 25°/25°, autoign temp: 650°F.

SYNS: NCI-C50168 * DEODORIZED WINTERIZED COTTONSEED OIL

THR: An experimental teratogen. An allergen. Slight fire hazard when exposed to heat or flame. However, if allowed to impregnate rags or oily waste, it can cause a dangerous fire hazard. Moderate spontaneous heating. To fight fire, use CO$_2$, dry chemical. For further information, see Vol. 1, No 3 of *DPIM Report*.

COUMADIN   HR: 3
CAS: 81-81-2   NIOSH: GN 4550000
mf: C$_{19}$H$_{16}$O$_4$ mw: 308.35

PROP: Colorless, odorless, tasteless crystals; mp: 161°. Sol in acetone, dioxane; slightly sol in methanol, ethanol; very sol in alkaline aqusol; insol in water, benzene.

SYNS: 3-(alpha-ACETONYLBENZYL)-4-HYDROXYCOUMARIN * COUMAFENE * 4-HYDROXY-3-(3-OXO-1-FENYL-BUTYL) CUMARINE (DUTCH) * 4-HYDROXY-3-(3-OXO-1-PHENYL-BUTYL)-CUMARIN (GERMAN) * 4-IDROSSI-3-(3-OXO-)-FENIL-BUTIL)-CUMARINE (ITALIAN) * 3-(alpha-PHENYL-beta-ACETYLETHYL)-4-HYDROXYCOUMARIN * 3-(1'-PHENYL-2'-ACETYLETHYL)-4-HYDROXYCOUMARIN * (PHENYL-1 ACETYL-2 ETHYL)-3-HYDROXY-4 COUMARINE (FRENCH) * WARFARIN * WARFARINE (FRENCH)

OSHA PEL: TWA 100 ug/m^3

THR: Strongly poisonous by ingestion. Poison by inhalation, intravenous, intraperitoneal, and intramuscular routes. An experimental teratogen. Moderately toxic by ingestion and dermal routes. The possibility of human poisoning by warfarin must be kept in mind, although the safety factors make it appear unlikely that poisoning will occur except with suicidal intent or as a result of gross carelessness and ignorance.

COUMADIN SODIUM   HR: 3
CAS: 129-06-6   NIOSH: GN 4725000
mf: C$_{19}$H$_{15}$O$_4$ • Na   mw: 330.33

SYNS: 3-(alpha-ACETONYLBENZYL)-4-HYDROXY-COUMARIN SODIUM SALT * MAREVAN (SODIUM SALT) * PANWARFIN * PROTHROMBIN * SODIUM COUMADIN * SODIUM WARFARIN * WARFARIN SODIUM

THR: Poison to humans by ingestion. Poison by intravenous route. When heated to decomposition it emits toxic fumes of Na$_2$O.

COUMAPHOS   HR: 3
CAS: 56-72-4   NIOSH: GN 6300000
mf: C$_{14}$H$_{16}$ClO$_5$PS   mw: 362.78

SYNS: 3-CHLORO-7-HYDROXY-4-METHYL-COUMARIN-O,O-DIETHYL PHOSPHOROTHIOATE * 3-CHLORO-7-HYDROXY-4-METHYL-COUMARIN-O-ESTER WITH-O,O-DIETHYL PHOSPHOROTHIOATE * O-3-CHLORO-4-METHYL-7-COUMARINYL-O,O-DIETHYL PHOSPHOROTHIOATE * 3-CHLORO-4-METHYL-7-COUMARINYL DIETHYL PHOSPHOROTHIOATE * 3-CHLORO-4-METHYL-7-HYDROXYCOUMARIN DIETHYL THIOPHOSPHORIC ACID ESTER * 3-CHLORO-4-METHYLUMBELLIFERONE-O-ESTER WITH-O,O-DIETHYL PHOSPHOROTHIOATE * CUMAFOS (DUTCH) * O,O-DIAETHYL-O-(3-CHLOR-4-METHYL-CUMARIN-7-YL)-MONOTHIOPHOSPHAT (GERMAN) * O,O-DIETHYL-O-(3-CHLOOR-4-METHYL-CUMARIN-7-YL)MONOTHIOFOSFAAT (DUTCH) * O,O-DIETHYL O-(3-CHLORO-4-METHYL-7-COU-

MARINYL)PHOSPHOROTHIOATE * O,O-DIETHYL O-(3-CHLORO-4-METHYLCOUMARINYL-7) THIOPHOSPHATE * O,O-DIETHYL O-(3-CHLORO-4-METHYL-2-OXO-2H-BENZOPYRAN-7-YL)PHOSPHOROTHIOATE * O,O-DIETHYL 3-CHLORO-4-METHYL-7-UMBELLIFERONE THIOPHOSPHATE * O,O-DIETHYL O-(3-CHLORO-4-METHYLUMBELLIFERYL)PHOSPHOROTHIOATE * DIETHYL 3-CHLORO-4-METHYLUMBELLIFERYL THIONOPHOSPHATE * DIETHYL THIOPHOSPHORIC ACIDESTER OF 3-CHLORO-4-METHYL-7-HYDROXY-COUMARIN * O,O-DIETIL-O-(3-CLORO-4-METIL-CUMARIN-7-IL-MONOTIOFOSFATO) (ITALIAN) * NCI-C08662

THR: Poison by ingestion, intraperitoneal, dermal, ocular, and other routes. When heated to decomposition, it emits very toxic fumes of SO_x, PO_x, and Cl^-. For further information see coumaphos, Vol. 4, No. 1 of *DPIM Report*.

COUMARIN   HR: 3
CAS: 91-64-5   NIOSH: GN 4200000
mf: $C_9H_6O_2$ mw: 146.15

PROP: Crystals, fragrant, pleasant odor, burning taste;ᐧ mp: 70°, bp: 291.0°, vap press: 1 mm @ 106.0°.

SYNS: 2H-1-BENZOPYRAN-2-ONE * 1,2-BENZOPYRONE * cis-O-COUMARINIC ACID LACTONE * COUMARINIC ANHYDRIDE * O-HYDROXYCINNAMIC ACID LACTONE * NCI-C60297 * 2-OXO-1,2-BENZOPYRAN * TONKA BEAN CAMPHOR

THR: Poison by ingestion, intraperitoneal, and subcutaneous routes. An experimental carcinogen and tumorigen. Mutagenic data. Slight fire hazard when exposed to heat or flame.

C-QUENS   HR: 3
CAS: 8065-91-6   NIOSH: TU 3775000

SYNS: CHLORMADINONE ACETATE MIXED WITH MESTRANOL * 6-CHLORO-6-DEHYDRO-17-alpha-ACETOXYPROGESTERONE MIXED WITH MESTRENOL * MESTRANOL MIXED WITH 6-CHLORO-6-DEHYDRO-17-alpha-ACETOXYPROGESTERONE * LUTESTRAL (FRENCH) * MESTRENOL MIXED WITH 6-CHLORO-6-DEHYDRO-17-alpha-ACETOXYPROGESTERONE

THR: A carcinogen in women by ingestion. An experimental neoplastigen. When heated to decomposition it emits toxic fumes of Cl^-.

CRAG HERBICIDE   HR: 2
CAS: 136-78-7   NIOSH: KK 4900000
mf: $C_8H_7Cl_2O_5S \cdot Na$ mw: 309.10

SYNS: 2-(2,4-DICHLOROPHENOXY)ETHANOL HYDROGEN SULFATE SODIUM SALT * 2,4-DICHLOROPHENOXYETHYL SULFATE, SODIUM SALT * NATRIUM-2,4-DICHLORPHENOXYATHYLSULFAT (GERMAN) * SODIUM-2-(2,4-DICHLOROPHENOXY)ETHYL SULFATE * SODIUM-2,4-DICHLOROPHENOXYETHYL SULPHATE * SODIUM-2,4-DICHLOROPHENYL CELLOSOLVE SULFATE

OSHA PEL: TWA 15 mg/m³

THR: Moderately toxic by ingestion. Produces experimental liver and kidney damage. Strong solutions are irritant to skin. When heated to decomposition it emits very toxic fumes of Cl^- and SO_x.

CRESOL   HR: 3
CAS: 1319-77-3   NIOSH: GO 5950000
DOT: 2022/2076
mf: C_7H_8O mw: 108.15

PROP: Description (U.S.P. XVI): mixture of isomeric cresols obtained from coal tar, colorless or yellowish to brown-yellow or pinkish liquid, phenolic odor; mp: 10.9°-35.5°, bp: 191°-203°, flash p: 178°F, d: 1.030-1.038 @ 25°/25°, vap press: 1 mm @ 38-53°, vap d: 3.72.

SYNS: ACEDE CRESYLIQUE (FRENCH) * CRESOLI (ITALIAN) * CRESYLIC ACID * HYDROXYTOLUOLE (GERMAN) * KRESOLE (GERMAN) * KRESOLEN (DUTCH) * KREZOL (POLISH)

OSHA PEL: TWA 5 ppm (skin)
ACGIH TLV: TWA 5 ppm
DFG MAK: (all isomers) 5 ppm (22 mg/m³)
DOT Classification: Corrosive (IMO: Poison B)

THR: Moderately toxic by ingestion and inhalation. Corrosive to skin and mucous membranes. Systemic poisoning has rarely been reported, but it is possible that absorption may result in damage to the kidneys, liver, and nervous system. The main hazard accompanying its use in industry lies in severe chemical burns and dermatitis. Moderate fire hazard when exposed to heat or flame. Slightly explosive in the form of vapor when exposed to heat or flame. Reacts violently with HNO_3; oleum; chlorosulfonic acid. Explosive Range: 1.35% @ 300°F. When heated to decomposition it emits highly toxic fumes; can react vigorously with oxidizing materials. To fight fire, use foam, CO_2, dry chemical.

m-CRESOL   HR: 3
CAS: 108-39-4   NIOSH: GO 6125000
mf: C_7H_8O mw: 108.15

PROP: Colorless to yellowish liquid, phenolic odor; mp: 10.9° bp: 202.8°, lel: 1.1% @ 302°F, flash p: 202°F, d: 1.034 @ 20°/4°, autoign temp: 1038°F, vap press: 1 mm @ 52.0°, vap d: 3.72.

SYNS: 3-CRESOL * M-CRESYLIC ACID * 1-HYDROXY-3-METHYLBENZENE * M-HYDROXYTOLUENE * M-KRESOL * M-METHYLPHENOL * 3-METHYLPHENOL * M-OXYTOLUENE

OSHA PEL: TWA 5 ppm (skin)

THR: Poison by ingestion, dermal, intravenous, intraperitoneal, and subcutaneous routes. An experimental neoplastigen. Severe eye and skin irritant. See cresol for fire and disaster hazard. Moderately explosive in the form of vapor when exposed to heat or flame. For further information see Vol. 6, No. 1 of *DPIM Report*.

o-CRESOL   HR: 3
CAS: 95-48-7   NIOSH: GO 6300000
mf: C_7H_8O mw: 108.15

PROP: Crystals or liquid darkening with exposure to air and light. mp: 30.8°, bp: 190.8°, flash p: 178°F, d: 1.047 @ 20°/4°, autoign temp: 1110°F, vap press: 1 mm @ 38.2°, vap d: 3.72, lel = 1.4% @ 300°F.

SYNS: 2-CRESOL * O-CRESYLIC ACID * ORTHOCRESOL * 1-HYDROXY-2-METHYLBENZENE * O-HYDROXYTOLUENE * O-KRESOL (GERMAN) * O-METHYLPHENOL * 2-METHYLPHENOL * O-OXYTOLUENE

OSHA PEL: TWA 5 ppm (skin)

THR: Poison by ingestion. Moderately toxic by dermal route. See also cresol. An experimental neoplastigen. Moderate fire hazard via heat, flame, oxidants. See also cresol for explosion and disaster hazards. To fight fire, water may be used to blanket fire; foam, fog, mist, dry chemical. For further information see Vol. 5, No. 3 of *DPIM Report*.

p-CRESOL   HR: 3
CAS: 106-44-5   NIOSH: GO 6475000
mf: C_7H_8O mw: 108.15

PROP: Found in a score of essential oils, including Ylang-Ylang and Oil of Jasmine. Crystals, phenolic odor; mp: 35.5°, bp: 201.8°, lel = 1.1% @ 302°F, flash p: 202°F, d: 1.0341 @ 20°/4°,

PROP: autoign temp: 1038°F, vap press 1 mm @ 53.0°, vap d: 3.72.

SYNS: 4-CRESOL * P-CRESYLIC ACID * 1-HYDROXY-4-METHYLBENZENE * P-HYDROXYTOLUENE * 4-HYDROXYTOLUENE * P-KRESOL * 1-METHYL-4-HYDROXYBENZENE * P-METHYLPHENOL * 4-METHYLPHENOL * P-OXYTOLUENE * para-CRESOL * PARA-METHYL PHENOL

OSHA PEL: TWA 5 ppm (skin)

THR: Poison by ingestion. Moderately toxic by dermal route. A severe skin and eye irritant. With 7,12-dimethyl benz(a)anthracene, it is an experimental neoplastigen. See also cresol. Slight fire hazard when exposed to heat or flame. Moderately explosive in the form of vapor when exposed to heat or flame. See also cresol for disaster hazard. To fight fire, use CO_2, dry chemical, alcohol foam.

CROCIDOLITE (see ASBESTOS)   HR: 3
CAS: 12001-28-4   NIOSH: GP 8225000

SYNS: BLUE ASBESTOS * CROCIODOLITE * KROKYDOLITH (GERMAN)

THR: A human carcinogen. An experimental tumorigen and neoplastigen. Mutagenic data. Considered to be the most carcinogenic of all asbestos.

CROTONIC ACID   HR: 3
CAS: 3724-65-0   NIOSH: GQ 2800000
mf: $C_4H_6O_2$ mw: 86.10

PROP: Colorless needle-like crystals; bp: 185°, mp: 72°, flash p: 190°F (COC), d: 1.018 @ 15°/4°, vap press: 0.19 mm @ 20°, vap d: 2.97.

SYNS: alpha-BUTENOIC ACID * 2-BUTENOIC ACID * alpha-CROTONIC ACID * 3-METHYLACRYLIC ACID * beta-METHYLACRYLIC ACID * SOLID CROTONIC ACID

DOT Classification: Label: Corrosive

THR: Poison by intraperitoneal route. Moderately toxic by ingestion, subcutaneous, and dermal routes. Powerful irritant and corrosive. Moderate fire hazard when exposed to heat or flame; can react with oxidizing materials. To fight fire, use alcohol Foam, CO_2, dry chemical. When heated to decomposition it emits acrid smoke and irritating fumes.

CROTON OIL   HR: 3
PROP: Oil from the seeds of *Croton Tiglium.*

CAS: 8001-28-3   NIOSH: GQ 6300000

PROP: Brownish-yellow, viscid oil, slight offensive odor. Composition: croton resin, glycerides of fatty acids and crotin; d: 0.935 @ 25°/25°.

SYNS: CROTONOEL (GERMAN) * CROTON RESIN * CROTON TIGLIUM L. OIL * OLEUM TIGLII * OLIO DI CROTON (ITALIAN)

THR: Poison by parenteral and intraperitoneal routes. An experimental neoplastigen and tumorigen by dermal route. A skin and eye irritant. An allergen. Mutagenic data. When heated to decomposition it emits toxic fumes.

CROTONYLENE (DOT)   HR: 1
CAS: 503-17-3   NIOSH: GQ 7210000
mf: CH_3CCCH_3 mw: 54.09

PROP: Liquid; bp: 27°, flash p: < −4°F, lel = 1.4%, d: 0.688 @ 25°, vap d: 1.91.

SYN: 2-BUTYNE

DOT Classification: Label: Flammable Liquid

THR: A simple asphyxiant. See also argon. Very dangerous fire hazard when exposed to heat or flame; can react with oxidizing materials. Moderately explosive in the form of vapor when exposed to heat or flame. To fight fire, use foam, CO_2, dry chemicals.

CROTOXYPHOS   HR: 3
CAS: 7700-17-6   NIOSH: GQ 5075000
mf: $C_{14}H_{19}O_6P$ mw: 314.30

SYNS: CROTONIC ACID, 3-HYDROXY-, alpha-METHYLBENZYL ESTER, DIMETHYL PHOSPHATE, (E) - * 2-BUTENOIC ACID, 3-((DIMETHOXY-PHOSPHINYL)OXY)-, 1-PHENYLETHYL ESTER, (E) - (9CI) * CIODRIN VINYL PHOSPHATE; DIMETHYL PHOSPHATE OF alpha-METHYLBENZYL-3-HYDROXY-cis-CROTONATE * ENT 24,717 * 1-METHYLBENZYL-3-(DIMETHOXYPHOSPHINYLOXO) ISOCROTONATE * alpha-METHYL BENZYL-3-(DIMETHOXY-PHOSPHINYLOXY)-cis-CROTO-

NATE * alpha-METHYLBENZYL-3-HYDROXY-CROTONATE DIMETHYL PHOSPHATE * PANTOZOL 1 * cis-2-(1-PHENYLETHOXY) CARBONYL-1-METHYLVINYL DIMETHYLPHOSPHATE * SD 4294 * SHELL SD 4294 * VOLFAZOL

THR: Poison by ingestion, dermal, intraperitoneal, and subcutaneous routes. Mutagenic data. When heated to decomposition it emits highly toxic PO_x. For further information see Vol. 2, No. 5 of *DPIM Report.*

CRYPTOHALITE   HR: 3
CAS: 1309-32-6   NIOSH: GQ 9450000
mf: $F_6Si \cdot 2H_4N$ mw: 178.19

PROP: mp: subl, d: 2.01.

SYNS: AMMONIUM FLUOSILICATE * AMMONIUM HEXAFLUOROSILICATE * AMMONIUM SILICOFLUORIDE * DIAMMONIUM HEXAFLUOROSILICATE * FLUOSILICATE DE AMMONIUM (FRENCH)

THR: Poison by ingestion. See also hexafluosilicate (2-) dihydrogen and fluorides. When heated to decomposition it emits very toxic fumes of F^- and NO_x.

CUMALDEHYDE   HR: 3
CAS: 122-03-2   NIOSH: CU 7000000
mf: $C_{10}H_{12}O$ mw: 148.22

PROP: Found in at least 50 essential oils such as Cumin, Eucalyptus species, Cinnamon, Boldo and Rue, and as main constituent of Oil of Pectis Papposa Harn and Gray.

SYNS: p-CUMIC ALDEHYDE * CUMINALDEHYDE * CUMINIC ALDEHYDE * CUMINYL ALDEHYDE * p-ISOPROPYLBENZALDEHYDE * 4-ISOPROPYLBENZALDEHYDE * p-ISOPROPYLBENZENECARBOXALDEHYDE

THR: Moderately toxic by ingestion and dermal routes. A skin irritant. When heated to decomposition it emits acrid smoke and irritating fumes.

CUMENE   HR: 2
CAS: 98-82-8   NIOSH: GR 8575000
DOT: 1221
mf: C_9H_{12} mw: 120.21

PROP: Colorless liquid; mp: −96.0°, bp: 152°, flash p: 111°F, d: 0.864 @ 20°/4°, vap press: 10 mm @ 38.3°, autoign temp: 795°F, lel = 0.9%, uel = 6.5%, vap d: 4.1.

SYNS: CUMEEN (DUTCH) * CUMOL * 2-FENILPROPANO (ITALIAN) * 2-FENYL-PROPAAN

(DUTCH) * ISOPROPYLBENZEEN (DUTCH) * ISOPROPILBENZENE (ITALIAN) * ISOPROPYL BENZENE * ISOPROPYLBENZOL * ISOPROPYL-BENZOL (GERMAN) * 2-PHENYLPROPANE

OSHA PEL: TWA 50 ppm (skin)
ACGIH TLV: TWA 50 ppm
DFG MAK: 50 ppm (245 mg/m^3)
DOT Classification: Flammable Liquid

THR: Very irritating in humans via inhalation route. Moderately toxic by ingestion and inhalation. Potential narcotic action. Central nervous system depressant. There is no apparent difference between the toxicity of natural cumene or that derived from petroleum. Skin and eye irritant. See also benzene and toluene. Moderate fire hazard when exposed to flame; can react with oxidizing materials. Violent reaction with HNO_3; oleum; chlorosulfonic acid. To fight fire, use foam, CO_2, dry chemical. For further information, see Vol. 4, No. 1 of *DPIM Report*.

m-CUMENOL METHYLCARBA-MATE   HR: 3
CAS: 64-00-6   NIOSH: FB 7875000
mf: $C_{11}H_{15}NO_2$ mw: 193.27

SYNS: m-CUMENYL METHYLCARBAMATE * ENT 25,500 * ENT 25,543 * m-ISOPRO-PYLPHENOL-n-METHYLCARBAMATE * m-ISO-PROPYLPHENYL METHYLCARBAMATE * m-ISO-PROPYLPHENYL-n-METHYLCARBAMATE * 3-ISOPROPYLPHENYL METHYLCARBAMATE * n-METHYL-m-ISOPROPYLPHENYL CARBAMATE * n-METHYL-3-ISOPROPYLPHENYL CARBAMATE

THR: Poison by ingestion, intraperitoneal, intravenous, intramuscular, and dermal routes. See also carbamates. When heated to decomposition it emits toxic fumes of NO_x.

CUPRIC ACETOARSENITE   HR: 3
CAS: 13002-03-8   NIOSH: GL 6475000
mf: $C_4H_6As_6Cu_4O_{16}$ mw: 1013.78

PROP: Emerald green powder.

SYNS: (ACETATO-O)(TRIMETAARSENITO)DICOP-PER * ACETOARSENITE DE CUIVRE (FRENCH) * COPPER ACETO ARSENITE * EMERALD GREEN * FRENCH GREEN * IMPERIAL GREEN * KING'S GREEN * MEADOW GREEN * MIN-ERAL GREEN * MOSS GREEN * PARIS GREEN * PARROT GREEN * SCHWEINFURTERGRUN * SCHWEINFURT GREEN * VIENNA GREEN

DOT Classification: Label: Poison

THR: Poison by ingestion and inhalation. See also arsenic compounds. When heated to decomposition it emits very toxic fumes of As.

CURARE   HR: 3
CAS: 8063-06-7   NIOSH: GS 3325000

PROP: Brown, brittle, resinous mass. Rendering of an Indian name given to the unstandardized extracts derived mainly from the bark of various species of *Strychnos* and *Chondodendron*.

SYNS: OURARI * INTOCOSTRINE * URARI * WOORALI * WOORARI * WOURARA

THR: Poison by intraperitoneal, intravenous, and subcutaneous routes. When heated to decomposition it emits highly toxic fumes.

CURETARD   HR: 3
CAS: 29929-77-9   NIOSH: VC 2300000
mf: $(C_{12}H_{14}N_2O)n$

SYNS: n-NITROSO-2,2,4-TRIMETHYL-1,2-DIHY-DROQUINOLINE,POLYMER * 1-NITROSO-2,2,4-TRIMETHYL-1(2H)-QUINOLINE, POLYMER * QUINOLINE, 1-NITROSO-2,2,4-TRIMETHYL-1,2,-DIHYDRO-,(POLYMER)

THR: An experimental carcinogen, tumorigen, and neoplastigen. See also polymers. When heated to decomposition it emits toxic fumes of NO_x.

CUTTING OILS   HR: 3
THR: Often carcinogenic. Can cause dermatitis. See also mineral oils. Slight fire hazard when exposed to heat or flame. This oil is the cause of "cutting oil" dermatitis. Although it is generally caused by an insoluble oil, it can occasionally be caused by a soluble one. Many have looked for a causative factor other than the oil itself. Bacteria have frequently been blamed, although insoluble oils are usually sterile while the soluble oils may contain bacteria. The metal slivers which occur in these oils after use have also been blamed as well as the sulfur, chlorine and inhibitors which they contain. The oil itself can plug the pores, forming boils. They are often carcinogenic. See also mineral oils.

Treatment and Antidotes: Disinfectant agents have been added to the oils to cut down the infectious effects. Filters have been installed in the cutting oil lines to filter out the tiny metal slivers which may irritate and cause infection of the skin. General housecleaning measures are also thought advisable to promptly remove any excess oil which gets onto the hands and skin. It has been found that occasionally changing from one oil to another may bring relief from the irritation. This would also indicate that the specific irritant is not present in all oils.

CYANACETIC ACID HYDRA-ZIDE   HR: 3
CAS: 140-87-4   NIOSH: AG 4200000
mf: $C_3H_5N_3O$ mw: 99.11

PROP: Mp: 115°; a solid.

SYNS: CYANOACETHYDRAZIDE * CYANO-ACETIC ACID HYDRAZIDE * CYANOACETOHY-DRAZIDE * CYANOACETYLHYDRAZIDE * CYANOETHYDRAZIDE * CYANACETHYDRA-ZIDE * CYANACETOHYDRAZIDE * ALPHA-CYANOACETOHYDRAZIDE * CYANACETYL-HYDRAZIDE * (CYANOACETYL)HYDRAZIDE * USAF KF-18 * MALONITRILE HYDRAZIDE * MALONONITRILE HYDRAZIDE

THR: Poison by ingestion and intraperitoneal route. When heated to decomposition it emits toxic fumes of NO_x and CN^-.

CYANAMIDE   HR: 3
CAS: 420-04-2   NIOSH: GS 5950000
mf: CH_2N_2 mw: 42.05

PROP: Deliquescent crystals; mp: 45°, bp: 260°, flash p: 285°F, d: 1.282, vap d: 1.45.

SYNS: AMIDOCYANOGEN * CARBAMONITRILE * CARBIMIDE * CYANOGEN NITRIDE * CYAN-OGENAMIDE * HYDROGEN CYANAMIDE * USAF EK-1995

THR: Poison by ingestion, inhalation, and intra-peritoneal route. Slight fire hazard when exposed to heat or flame. When heated to decomposition or on contact with acid or acid fumes, it emits toxic fumes of CN^- and NO_x. To fight fire, use CO_2, dry chemical. Incompatible with moisture (water); acids; alkalies.

CYANATES
THR: Variable. See individual entry. When heated to decomposition or on contact with acid or acid fumes, they emit toxic fumes.

CYANATOTRIBUTYLSTAN-NANE   HR: 3
CAS: 4027-17-2   NIOSH: WH 6871000
mf: $C_{13}H_{27}NOSn$ mw: 332.10

SYN: TRIBUTYLTIN CYANATE

THR: Poison by intravenous route. See also cyanates and tin compounds. When heated to decomposition it emits toxic fumes of NO_x.

CYANIDES   HR: 3
CAS: 57-12-5   NIOSH: GS 7175000
mf: CN^- mw: 26.02

SYN: CYANURE (FRENCH)

ACGIH TLV: TWA 5 mg/m^3
DFG MAK: 5 mg/m^3

THR: Very poisonous. Death can occur within seconds of inhalation or ingestion. Cyanide directly stimulates the chemoreceptors of the carotid and aortic bodies with a resultant hyperpnea. Cardiac irregularities are often noted, but the heart invariably outlasts the respirations. Death is due to respiratory arrest of central origin. It can occur within seconds or minutes of the inhalation of high concs of HCN gas. Because of slower absorption, death may be more delayed after the ingestion of cyanide salts, but the critical events still occur within the first hour. Two other sources of cyanide have been responsible for human poisoning. One of these is amygdalin, a cyanogenic glycoside found in apricot, peach, and similar fruit pits and in sweet almonds. Amygdalin is a chemical combination of glucose, benzaldehyde, and cyanide from which the latter can be released by the action of β-glucosidase or emulsion. Although these enzymes are not found in mammalian tissues, the human intestinal microflora appears to possess these or similar enzymes capable of effecting cyanide release resulting in human poisoning. For this reason amygdalin may be as much as 40 times more toxic by the oral route as compared with intravenous injection. Amygdalin is the major ingredient of Laetrile, and this alleged anticancer drug has also been responsible for human cyanide poisoning. An ethical drug that may also cause cyanide poisoning in overdose is the potent vascular smooth muscle relaxant sodium nitroprusside. Although nitroprusside is related chemically to ferricyanide, unlike the latter it penetrates into erythrocytes and reacts with hemoglobin to release its cyanide (Smith and Kruszyna, 1974). Fortunately, the therapeutic margin for nitroprusside appears to be quite large. Cyanide is commonly found in certain rat and pest poisons, silver and metal polishes, photographic solutions, and fumigating products. Compounds such as potassium cyanide can also be readily purchased from chemical stores. Cyanide is readily absorbed from all routes, including the skin, mumem, and by inhalation, although alkali salts of cyanide are toxic only when ingested. Death may occur with ingestion of even small amounts of sodium or potassium cyanide and can occur within minutes or hours depending on route of exposure. Inhalation tion of toxic fumes represents a potentially rapidly fatal type of exposure. Sodium nitroprusside (Smith and Kruszyna, 1974) and apricot seeds (Sayre and Kaymakcalan, 1941) have also caused cyanide poisoning. A blood cyanide level of greater than 0.2 μg/mL is considered toxic. Lethal cases have usually had levels above 1 μg/mL. Clinically, cyanide poisoning is reported to produce a bitter, almond odor on the breath of the patient; however, only a small proportion of the population is genetically able to discern this characteristic odor. Typically, cyanide has a bitter, burning taste, and following poisoning, symptoms of salivation, nausea without vomiting, anxiety, confusion, vertigo, giddiness, lower jaw stiffness, convulsions, opisthotonos, paralysis, coma, cardiac arrhythmias, and transient respiratory stimulation followed by respiratory failure may occur. Bradycardia is a common finding, but in most cases heartbeat usually outlasts respiration (Wexler et al., 1947). A prolonged expiratory phase is considered to be characteristic of cyanide poisoning. (Casarett and Doull's, "Toxicology, the Basic Science of Poisons" 2nd ed. Doull, Klaassen and Amdur (eds). Macmillan Pub. Co. Inc. New York, N.Y.) The volatile cyanides resemble HCN physiologically, inhibiting tissue oxidation and causing death through asphyxia. Cyanogen is probably as toxic as HCN; the nitriles are generally considered somewhat less toxic, probably because of their lower volatility. The non-volatile cyanide salts appear to be relatively nontoxic systemically, so long as they are not ingested and care is taken to prevent the formation of HCN. Workers, such as electroplaters and picklers who are daily exposed to cyanide solns may develop a "cyanide" rash, characterized by itching, and by macular, papular, and vesicular eruptions. Frequently there is secondary infection. Exposure to small amounts of cyanide compounds over long periods of time is reported to cause loss of appetite, headache, weakness, nausea, dizziness, and symptoms of irritation of the upper respiratory tract and eyes. See also specific compounds. Moderate fire hazard by chemical reaction with heat; moisture; acid. Many cyanides evolve HCN rather easily. This is a flammable gas and is highly toxic. Carbon dioxide from the air is sufficiently acidic to liberate HCN from cyanide solns. See also hydrocyanic acid. See hydrocyanic acid for explosion hazard. Explodes if melted with nitrite or chlorate @ about 450°. Violent reaction with F_2; Mg; nitrates; HNO_3; nitrites. Dangerous; on contact with acid, acid fumes, water or steam, they will produce toxic and flammable vapors.

CYANOACETIC ACID   HR: 3
CAS: 372-09-8   NIOSH: AG 3675000
mf: $C_3H_3NO_2$ mw: 85.07

PROP: Solid; mp: 66°; bp: 108° @ 15 mm.

SYNS: ACIDE CYANACETIQUE (FRENCH) * CYANESSIGSAEURE (GERMAN) * MALONIC MONONITRILE * MONOCYANOACETIC ACID * USAF KF-17

THR: Poison by intraperitoneal route. Moderately toxic by ingestion and subcutaneous route. See also nitriles. Reacts violently with furfuryl alcohol. When heated to decomposition it emits toxic fumes of NO_x and CN^-.

CYANODIMETHYLARSINE   HR: 3
CAS: 683-45-4   NIOSH: CH 2100000
mf: C_3H_6AsN mw: 131.02

SYN: DIMETHYLCYANOARSINE

THR: Poison by ingestion and inhalation. Self ignites in air. See also arsenic compounds and cyanides. When heated to decomposition it emits very toxic fumes of As and NO_x.

2-CYANO-2'-FLUORODIETHYL ETHER   HR: 3
CAS: 353-18-4   NIOSH: UG 2330000
mf: C_5H_8FNO mw: 117.14

SYNS: 2-CYANOETHYL-2'-FLUOROETHYLETHER * 2-FLUORO-2'-CYANODIETHYL ETHER

THR: Poison by intraperitoneal and subcutaneous routes. See also ethers. When heated to decomposition it emits toxic F^-, NO_x and CN^-.

CYANOGEN   **HR: 3**
CAS: 460-19-5   NIOSH: GT 1925000
DOT: 1026
mf: C_2N_2 mw: 52.04

PROP: Colorless gas, pungent odor; mp: −34.4°, bp: −21.0°, d: 0.866 @ 17°/4°, lel = 6.6%, uel = 32%, vap d: 1.8.

SYNS: DICYANOGEN * CYANOGENE (FRENCH) * CYANOGEN GAS (DOT) * ETHANEDINITRILE * OXALONITRILE
ACGIH TLV: TWA 10 ppm
DFG MAK: 10 ppm/22 mg/m^3
DOT Classification: Label: Flammable and Poison Gas (Poison A)

THR: Poison. Irritant by inhalation and subcutaneous routes. Very toxic via human eye. See also cyanides. Violent reaction with F_2; O_2. Very dangerous fire hazard via heat, flames (sparks), oxidizers. When heated to decomposition or on contact with acid, acid fumes, water or steam, will react to produce highly toxic fumes of NO_x and CN^-. To fight fire, stop flow of gas. For further information, see Vol. 2, No. 1 of *DPIM Report*.

CYANOGEN BROMIDE   **HR: 3**
CAS: 506-68-3   NIOSH: GT 2100000
mf: CBrN   mw: 105.93

PROP: Colorless needles; mp: 52°, bp: 61.6°, d: 2.015 @ 20°/4°, vap press: 100 mm @ 22.6°.

SYNS: BROMINE CYANIDE * BROMOCYANOGEN * BROMURE DE CYANOGEN (FRENCH)

THR: A poison via inhalation route. See also Br, and cyanogen. When heated to decomposition it emits very toxic CN^- and Br^-. For further information see Vol. 1, No. 8 of *DPIM Report*.

CYANOGEN CHLORIDE   **HR: 3**
CAS: 506-77-4   NIOSH: GT 2275000
DOT: 1589
mf: CClN   mw: 61.47

PROP: Colorless liquid or gas, lachrymatory and irr odor; mp: −6.5°, bp: 13.1°, d: 1.218 @ 4°/4°, vap press: 1010 mm @ 20°, vap d: 1.98.

SYNS: CHLORCYAN * CHLORINE CYANIDE * CHLOROCYANOGEN * CHLORURE DE CYANO-GENE (FRENCH)
ACGIH TLV: TWA (CL) 0.3 ppm

DOT Classification: Nonflammable Gas and Poison A Gas

THR: Very toxic via human eyes. A poisonous irritant via inhalation and ocular routes. An insecticide. See also cyanides and HCl. Highly dangerous; when heated to decomposition or on contact with water or steam, will react to produce highly toxic and corrosive fumes such as Cl^-, CN^- and NO_x. For further information see Vol. 6, No. 1 of *DPIM Report*.

CYANOGEN IODIDE   **HR: 3**
CAS: 506-78-5   NIOSH: NN 1750000
mf: CIN   mw: 152.92

PROP: Colorless; mp: 146.5°, vap press: 1 mm @ 25.2°.

SYNS: IODINE CYANIDE * NCI * JODCYAN

THR: A poison via subcutaneous route. Violent reaction with P. See cyanides and iodides. When heated to decomposition it emits very toxic fumes of NO_x, CN^- and I^-.

CYANOMETHYL ACETATE   **HR: 3**
CAS: 1001-55-4   NIOSH: MC 7700000
mf: $C_4H_5NO_2$ mw: 99.10

PROP: Colorless liquid; mp: −22.5°, bp: 200°, d: 1.123 @ 15°.

THR: A poison via skin, oral, and inhalation routes. When heated to decomposition it emits very toxic fumes of NO_x and CN^-.

CYANOTRIMEPRAZINE MALE-ATE   **HR: 3**
CAS: 63833-98-7   NIOSH: SN 7000000
mf: $C_{19}H_{21}N_3S \cdot C_4H_4O_4$ mw:439.57

SYNS: CYAMEPROMAZINE MALEATE * CYANO-3 (DIMETHYLAMINO-3 METHYL-2 PROPYL)-10PHENOTHIAZINE MALEATE * 10-(3-(DIMETHYLAMINO)-2-METHYLPROPYL)-PHENO-THIAZINE-2-CARBONITRILE MALEATE

THR: Poison by intravenous and intraperitoneal routes. Moderately toxic by ingestion and subcutaneous routes. See also nitriles. When heated to decomposition it emits very toxic fumes of NO_x and SO_x.

CYCAD HUSK   **HR: 3**
                          NIOSH: GT 8300000

PROP: The active substance in the cycad meal is aglycone of *Cycasin*, a methylazoxymethanol.

SYN: CYCAS CIRCINALIS HUSK

THR: An experimental tumorigen. When heated to decomposition it emits toxic and irritating fumes.

CYCAD MEAL   HR: 3
NIOSH: GT 8310000

PROP: Obtained from the nut of *Cycas Circinalis L.*

THR: An experimental neoplastigen. When heated to decomposition it emits toxic fumes.

CYCASIN   HR: 3
CAS: 14901-08-7   NIOSH: LZ 5950000
mf: $C_8H_{16}N_2O_7$ mw: 252.26

SYNS: METHYLAZOXYMETHANOL GLUCOSIDE
* CYCAS REVOLUTA GLUCOSIDE * CYKAZINE
* beta-D-GLUCOSYLOXYAZOXYMETHANE
* METHYLAZOXYMETHANOL-beta-d-GLUCOSIDE

THR: An experimental tumorigen, and brain carcinogen. Mutagenic data. A poison via oral route. When heated to decomposition it emits toxic fumes of NO_x. For further information see Vol. 1, No. 3 of *DPIM Report.*

CYCLAZOCINE   HR: 3
CAS: 3572-80-3   NIOSH: PB 8585000
mf: $C_{18}H_{25}NO$ mw: 271.44

SYNS: 2-CYCLOPROPYLMETHYL-5,9-DIMETHYL-2′-HYDROXY-6,7-BENEOMORPHAN * 3-CYCLO-PROPYLMETHYL-6(eq),11(ax)-DIMETHYL-2,6-METHANO-3-BENZAZOCIN-8-OL * 3-(CYCLOPROPYLMETHYL)1-1,2,3,4,5,6-HEXAHY-DRO-6,11-DIMETHYL-2,6-METHANO-3-BENZAZO-CIN-8-OL * 2-CYCLOPROPYLMETHYL-2′-HY-DROXY-5,9-DIMETHYL-6,7-BENZOMORPHAN

THR: A poison via subcutaneous, intravenous, and intraperitoneal routes. When heated to decomposition it emits toxic fumes of NO_x.

CYCLIC ETHYLENE SULFITE   HR: 3
CAS: 3741-38-6   NIOSH: KW 3900000
mf: $C_2H_4O_3S$ mw: 108.12

SYNS: ETHYLENE SULFITE * 1,2-ETHYLENE SULFITE * GLYCOL SULFITE

THR: A poison via acute intraperitoneal route. An experimental tumorigen. When heated to decomposition it emits toxic fumes of SO_x.

CYCLOBUTANE   HR: 1
mf: C_4H_8 mw: 56.11

PROP: A gas; mp: −50°; bp: 12.9°; flash p: <50°F(CC); d: 0.708 @ 11°; vap d: 1.93; lel = 1.8%.

SYN: TETRAMETHYLENE

THR: May be a simple asphyxiant. See also cyclohexane. Very dangerous fire hazard when exposed to heat or flame; can react with oxidizing materials. To fight fire, stop flow of gas; CO_2, dry chemicals or water spray.

CYCLOBUTENE   HR: 1
mf: C_4H_6 mw: 54.09

PROP: Gas; bp: 2.4°; d: 0.733 @ 0°/4°; flash p: <15°F.

SYN: CYCLOBUTYLENE

THR: May be a simple asphyxiant. Dangerous fire hazard when exposed to heat or flame; can react with oxidizing materials.

CYCLOCHLOROTINE   HR: 3
CAS: 12663-46-6   NIOSH: GU 2100000
mf: $C_{24}H_{30}Cl_2N_5O_7$ mw: 571.49

PROP: White needles, mp: 251°, decomp; Chlorine containing peptide produced by *P. Islandicum.*

SYN: ISLANDITOXIN

THR: Very poisonous via oral, subcutaneous, and intravenous routes. An experimental carcinogen. When heated to decomposition it emits very toxic fumes of Cl^- and NO_x.

CYCLOHEPTANE   HR: 2
mf: C_7H_{14} mw: 98.19

PROP: An oil; mp: −12°; bp: 117°; flash p: 59°F; d: 0.8099 @ 20°/4°; vap d: 3.3.

SYN: SUBERANE

THR: A flammable material. See also cyclohexane. Dangerous fire hazard when exposed to heat or flame; can react with oxidizing materials. To fight fire: Foam, CO_2, dry chemicals.

CYCLOHEPTANONE   HR: 2
CAS: 502-42-1   NIOSH: GU 3325000
mf: $C_7H_{12}O$ mw: 112.19

PROP: Liquid, nearly insol in water; bp: 181°, d: 0.9490 @ 20°/4°.

SYNS: KETOCYCLOHEPTANE * KETOHEPTA-METHYLENE * SUBERONE

THR: Moderate toxicity via intraperitoneal route causing mainly central nervous system depression.

CYCLOHEPTENYL ETHYLBARBITURIC ACID   HR: 3
CAS: 509-86-4   NIOSH: CQ 2450000
mf: $C_{13}H_{18}N_2O_3$ mw: 250.33

SYNS: 5-(1-CYCLOHEPTEN-1-YL)-5-ETHYLBARBI-TURIC ACID * CYCLOHEPTENYLETHYLMALONY-LUREA * 5-ETHYL-5-(1'-CYCLOHEPTENYL)-BAR-BITURIC ACID * HEPTABARBITAL * HEPTABARBITONE * 2,4,6(1H,3H,5H)-PYRIMIDINETRIONE, 5-(1-CYCLOHEPTEN-1-YL)-5-ETHYL-(9CI)

THR: A poison via intraperitoneal, and intravenous routes. An oral psychotropic material in humans. When heated to decomposition it emits toxic fumes of NO_x.

CYCLOHEXANE   HR: 3
CAS: 110-82-7   NIOSH: GU 6300000
DOT: 1145
mf: C_6H_{12} mw: 84.18

PROP: Colorless mobile liquid, pungent odor; mp: 6.5°, bp: 80.7°, fp: 4.6°, flash: p: 1.4°F, ulc: 90-95, lel = 1.3%, uel = 8.4%, d: 0.7791 @ 20°/4°, autoign temp: 473°F, vap press: 100 mm @ 60.8°, vap d: 2.90.

SYNS: CYCLOHEXAAN (DUTCH) * CYCLO-HEXAN (GERMAN) * CYKLOHEKSAN (POLISH) * HEXAHYDROBENZENE * HEXAMETHYLENE * HEXANAPHTHENE * CICLOESANO (ITALIAN)

OSHA PEL: TWA 300 ppm
ACGIH TLV: TWA 300 ppm
DOT Classification: Flammable Liquid

THR: Moderately irritating via inhalation and oral routes. Irritating to skin. Eye irritant in humans. Dangerous fire hazard when exposed to heat or flame; can react with oxidizing materials. Moderate explosion hazard in the form of vapor when exposed to flame. When mixed hot with liquid N_2O_4 an explosion resulted. To fight fire, use foam, CO_2, dry chemical, spray, fog. Incompatible with dinitrogen tetroxide.

CYCLOHEXANOL   HR: 2
mf: $C_6H_{12}O$ mw: 100.16

PROP: Colorless needles or viscous liquid, hygroscopic, camphor-like odor; mp: 24°; bp:

161.5°; flash p: 154°F (CC); d: 0.9449 @ 25°/4°; vap press: 1 mm @ 21.0°; vap d: 3.45; autoign temp: 572°F.

SYN: HEXAHYDROPHENOL

THR: Moderate toxicity via oral and inhalation routes. Narcotic in high concentrations. Has caused damage to kidneys, liver and blood vessels in experimental animals. Moderate fire hazard when exposed to heat or flame; can react with oxidizing materials. Violent reaction with HNO_3. To fight fire, use alcohol foam, foam, CO_2, dry chemical. Incompatible with oxidants.

CYCLOHEXANONE   HR: 2
CAS: 108-94-1   NIOSH: GW 1050000
DOT: 1915
mf: $C_6H_{10}O$ mw: 98.16

PROP: Colorless liquid, acetone-like odor; mp: −45.0°, bp: 115.6°. ulc: 35-40, lel = 1.1% @ 100°, flash p: 111°F, d: 0.9478 @ 20°/4°, autoign temp: 788°F, vap press: 10 mm @ 38.7°, vap d: 3.4.

SYNS: CICLOESANONE (ITALIAN) * CYCLOHEX-ANON (DUTCH) * CYKLOHEKSANON (POLISH) * KETOHEXAMETHYLENE * NCI-C55005 * PIMELIC KETONE

OSHA PEL: TWA 50 ppm
ACGIH TLV: TWA 25 ppm
DFG MAK: 50 ppm (200 mg/m^3)
DOT Classification: Flammable Liquid

THR: Moderately toxic via oral, inhalation, subcutaneous, and intraperitoneal routes. Skin and eye irritant. Human inhalation irritant. Mild narcotic properties have also been ascribed to it. See also cyclohexane. Moderate fire hazard when exposed to heat or flame; can react vigorously with oxidizing materials such as HNO_3. Slight explosion hazard in its vapor form, when exposed to flame. To fight fire, use alcohol foam, dry chemical or CO_2. When heated to decomposition it emits acrid smoke and irritating fumes. Incompatibles: Hydrogen peroxide; nitric acid. For further information see Vol. 5, No. 6 of *DPIM Report*.

CYCLOHEXANONE-Δ   HR: 2
mf: C_6H_8O mw: 96.12

PROP: Liquid; bp: 155.5°, flash p: 93°F (CC), vap d: 3.31, vap press: 4 mm @ 20°.

THR: Skin contact can cause a dermatitis. Irritating to eyes, skin and mucous membranes. Can damage the liver and kidneys. Dangerous fire hazard when exposed to flame and heat; can react with oxidizing materials. To fight fire, use CO_2, dry chemical.

CYCLOHEXENE HR: 2
CAS: 110-83-8
DOT: 2256
mf: C_6H_{10} mw: 82.15

PROP: Colorless liquid; bp: 83°; fp: −103.7°; flash p: < 21.2°F; d: 0.8102 @ 20°/4°; vap press: 160 mm @ 38°; autoign temp: 590°F; vap d: 2.8; lel = 1.2%.

SYN: 1,2,3,4-TETRAHYDROBENZENE

DOT Classification: Flammable Liquid

THR: Moderate toxicity via inhalation route. Dangerous fire hazard when exposed to flame; can react with oxidizers. Dangerous; keep away from heat and open flame. To fight fire, use foam, CO_2, dry chemical.

CYCLOHEXENE OXIDE HR: 3
CAS: 286-20-4   NIOSH: RN 7175000
mf: $C_6H_{10}O$ mw: 98.16

PROP: Clear liquid; 98.14, bp: 129.5°, flash p: 81°F, d: 0.9678 @ 25°/4°, vap d: 3.5.

SYNS: 7-OXABICYCLO(4.1.0)HEPTANE * CYCLOHEXENE-1-OXIDE * CYCLOHEXENE EPOXIDE * 1,2-CYCLOHEXENE OXIDE * CYCLOHEXYLENE OXIDE * 1,2-EPOXYCYCLOHEXANE * TETRAMETHYLENEOXIRANE

THR: Mutagenic data. An experimental tumorigen. Moderately toxic via oral, inhalation, intraperitoneal, intramuscular, and skin routes. When heated to decomposition it emits acrid smoke and irritating fumes.

2-CYCLOHEXEN-1-ONE HR: 3
CAS: 930-69-7   NIOSH: GW 7000000
mf: C_6H_8O mw: 96.14

SYN: CYCLOHEXENONE

THR: A poison via oral, inhalation, intraperitoneal, and skin routes. When heated to decomposition it emits acrid smoke and irritating fumes.

CYCLOHEXENYLETHYLENE HR: 3
CAS: 100-40-3   NIOSH: GW 6650000
mf: C_8H_{12} mw: 108.20

PROP: Liquid; bp: 128°, fp: −109°, flash p:

60°F (TOC), d: 0.832 @ 20°/4°, autoign temp: 517°F, vap press: 25.8 mm @ 38°, vap d: 3.76.

SYNS: BUTADIENE DIMER * 4-ETHENYL-1-CYCLOHEXENE * NCI-C54999 * 1,2,3,4-TETRAHYDROSTYRENE * 1-VINYLCYCLOHEXENE-3 * 1-VINYLCYCLOHEX-3-ENE * 4-VINYLCYCLOHEXENE-1 * 4-VINYL-1-CYCLOHEXENE

THR: An experimental carcinogen. Moderately toxic via oral, inhalation routes. Low toxicity via skin route. Dangerous fire hazard when exposed to heat, flame or oxidizers. Dangerous; when exposed to heat or open flame, can react with oxidizers. To fight fire, use foam, CO_2, dry chemical.

CYCLOHEXENYL TRICHLOROSILANE HR: 2
CAS: 10137-69-6   NIOSH: VV 2800000
mf: $C_6H_9Cl_3Si$ mw: 215.59

PROP: Colorless fuming liquid; HCl odor; bp: 202°; d: 1.263 @ 25°/25°; flash p: 200°F (COC).

THR: Moderately toxic via oral and skin routes. A skin irritant. A corrosive material; fumes in moist air releasing HCl. When heated to decomposition it emits toxic fumes of Cl^-.

CYCLOHEXIMIDE HR: 3
CAS: 66-81-9   NIOSH: MA 4375000
mf: $C_{15}H_{23}NO_4$ mw: 281.39

PROP: Crystals, mod sol in water, sol in chloroform, ether and acetone; mp: 116°.

SYNS: ACTI-AID * ACTIDIONE * ACTIDONE * 3-(2-(3,5-DIMETHYL-2-OXOCYCLOHEXYL)-2-HYDROXYETHYL)GLUTARIMIDE * NEOCYCLOHEXIMIDE * NSC-185 * U-4527

THR: A poison via acute toxicity. An experimental teratogen. When exposed to heat it emits toxic fumes, including NO_x. For further information see Vol. 2, No. 5 of *DPIM Report*.

CYCLOHEXYL ACETATE HR: 2
CAS: 622-45-7   NIOSH: AG 5075000
mf: $C_8H_{15}O_2$ mw: 142.22

PROP: Pale yellow liquid; bp: 177°; d: 0.996; vap d: 4.9; flash p: 136°F; autoign temp 633°F.

SYNS: CYCLOHEXANOL ACETATE * CYCLOHEXANOLAZETAT (GERMAN) * CYCLOHEXANYL ACETATE

THR: A systemic irritant to humans. Moderate toxicity via subcutaneous route. When heated to decomposition it emits acrid smoke and irritating fumes.

CYCLOHEXYLAMINE   HR: 3
CAS: 108-91-8   NIOSH: GX 0700000
DOT: 2357
mf: $C_6H_{13}N$ mw: 99.20

PROP: Liquid, strong fishy odor; mp: $-17.7°$, bp: $134.5°$, flash p: $69.8°F$, d: 0.865 @ $25°/25°$, autoign temp: $560°F$, vap d. 3.42.

SYNS: AMINOCYCLOHEXANE * AMINOHEXAHYDROBENZENE * CYCLOHEXANAMINE * HEXAHYDROANILINE * HEXAHYDROBENZENAMINE

DFG MAK: 10 ppm (40 mg/m^3)
DOT Classification: Flammable Liquid

THR: A poison via intraperitoneal and dermal routes. Mutagenic data. Moderately toxic by oral, subcutaneous, and parenteral routes. Can cause dermatitis, convulsions. Dangerous fire hazard when exposed to heat or flame. Incompatible with nitric acid. Dangerous; when heated to decomp, emits highly toxic fumes; can react vigorously with oxidizing materials. To fight fire, use alcohol foam, CO_2, dry chemical.

N-CYCLOHEXYL-2-BENZOTHIAZOLE-SULFENAMIDE   HR: 3
CAS: 95-33-0   NIOSH: DL 6250000
mf: $C_{13}H_{16}N_2S_2$ mw: 264.43

PROP: Light tan or buff powder; mp: $94°$, d: 1.27 @ $25°$.

SYNS: DURAX * SULFENAMIDE TS

THR: A poison via intravenous route. An experimental tumorigen. Dangerous; see also sulfonates.

2-CYCLOHEXYL-4,6-DINITRO-PHENOL   HR: 3
CAS: 131-89-5   NIOSH: SK 6650000
mf: $C_{12}H_{14}N_2O_5$ mw: 266.28

PROP: Crystals.

SYNS: 6-CICLOESIL-2,4-DINITR-FENOLO (ITALIAN) * 2-CYCLOHEXYL-4,6-DINITROFENOL (DUTCH) * 6-CYCLOHEXYL-2,4-DINITROPHENOL * DINITROCYCLOHEXYLPHENOL * DINITRO-O-CYCLOHEXYLPHENOL * 2,4-DINITRO-6-CYCLOHEXYLPHENOL * 4,6-DINITRO-O-CYCLOHEXYLPHENOL * DINITROCYCLOHEXYLPHENOL (DOT) * ENT 157

THR: A poison via oral, intraperitoneal, intravenous, subcutaneous, and other routes. Moder-

ately toxic via skin contact. A skin irritant. Fire hazard: See nitrates. Can react with oxidizers. When heated to decomposition it emits toxic fumes of NO_x.

CYCLOHEXYLMETHANE   HR: 2
CAS: 108-87-2   NIOSH: GV 6125000
mf: C_7H_{14} mw: 98.21

PROP: Colorless liquid; mp: $-126.4°$, lel = 1.2%, uel = 6.7%, bp: $100.3°$, flash p: $25°F$ (CC), d: 0.7864 @ $0°/4°$, 0.769 @ $20°/4°$, vap press: 40 mm @ $22.0°$, vap d: 3.39, autoign temp: $482°F$.

SYNS: HEXAHYDROTOLUENE * METHYLCYCLOHEXANE * METYLOCYKLOHEKSAN (POLISH) * TOLUENE HEXAHYDRIDE

OSHA PEL: TWA 500 ppm

THR: Moderate toxicity via oral route. This material does not cause irritation to the eyes and nose, and even at the level of 500 ppm, exhibits only a very faint odor. Therefore, it cannot be said to have any warning properties. It is believed to be about 3 times as toxic as hexane and can cause death by tetanic spasm as has been noted in animals. In sublethal concentration it causes narcosis and anesthesia. Dangerous fire hazard when exposed to heat, flame or oxidizers. Moderately explosive when exposed to heat or flame. Dangerous; upon exposure to heat or flame, can react vigorously with oxidizing materials. To fight fire, use foam, CO_2, dry chemical.

1-CYCLOHEXYL-1-PHENYL-3-PYRROLIDINO-1-PROPANOL   HR: 3
CAS: 77-37-2   NIOSH: UY 1925000
mf: $C_{19}H_{29}NO$ mw: 287.49

SYNS: 1-CYCLOHEXYL-1-PHENYL-3-(1-PYRROLIDINYL)-1-PROPANOL * PROCYCLIDINE * TRICYCLAMOL

THR: A poison via intraperitoneal, and intravenous routes. When heated to decomposition it emits toxic fumes of NO_x.

CYCLOHEXYLTRICHLORO-SILANE   HR: 3
CAS: 98-12-4   NIOSH: VV 2890000
mf: $C_6H_{11}Cl_3Si$ mw: 217.61

THR: A highly toxic and corrosive material. When heated to decomposition it emits toxic fumes of Cl^-.

CYCLOOCTAFLUOROBUTANE   HR: 1
CAS: 115-25-3   NIOSH: GU 1779500
mf: C_4F_8 mw: 200.03

PROP: Colorless, nonflammable gas, odorless; bp: $-6.04°$, mp: $-41.4°$, d(liquid): 1.513 @ $-70°F$.

SYNS: FREON C-318 * OCTAFLUOROCYCLOBU-TANE * PERFLUOROCYCLOBUTANE * PROPEL-LANT C318

THR: Low toxicity via oral and inhalation routes. Can cause slight transient effects at high concentrations. No anesthesia or central nervous system effects. A food additive permitted in food for human consumption. Mutagenic data. When heated to decomposition emits highly toxic fluorides.

1,3,5,7-CYCLOOCTATETRAENE   HR: 2
mf: C_8H_8 mw: 104.15

PROP: Liquid; mp: $-7°$; bp: $140.6°$; fp: $-4.7°$; vap press: 7.9 mm @ $25°$; flash p: $<71.6°F$.

THR: May be a simple asphyxiant. Very flammable. Moderate fire hazard when exposed to heat or flame; can react with oxidizing materials. To fight fire, spray, mist, fog, foam, dry chemicals.

CYCLOPAMINE   HR: 3
CAS: 4449-51-8   NIOSH: GY 0750000
mf: $C_{27}H_{41}NO_2$ mw: 411.69

PROP: Derived from *Veratrum Californicum*.

SYNS: ALKALOID V * 11-DEOXOJERVINE

THR: An experimental teratogen. A poison via oral route. When heated to decomposition it emits toxic fumes of NO_x.

1,3-CYCLOPENTADIENE   HR: 2
CAS: 542-92-7   NIOSH: GY 1000000
mf: C_5H_6 mw: 66.11

PROP: Colorless liquid; mp: $-85°$, bp: $42.5°$, d: 0.80475 @ $19°/4°$. flash p: $77°F$.

SYNS: CYCLOPENTADIENE * PYROPENTYLENE

OSHA PEL: TWA 75 ppm
DFG MAK: 75 ppm (200 mg/m^3)

THR: Moderate toxicity via inhalation route. An insecticide and fungicide. Moderate fire hazard when exposed to heat or flame; can react

with oxidizing materials. Moderately explosive in the form of gas when exposed to heat or by chemical reaction. It decomposes violently at high temperature and pressure. Dimerisation is highly exothermic. Incompatible with nitric acid; oxides of nitrogen; oxygen; sulfuric acid.

CYCLOPENTADIENYLMANGANESE TRICARBONYL   HR: 3
CAS: 12079-65-1   NIOSH: OO 9720000
mf: $C_8H_5MnO_3$ mw: 204.07

SYN: MANGANESE CYCLOPENTADIENYLTRICAR-BONYL

THR: A poison via oral, inhalation, and intravenous routes. See also manganese compounds and carbon monoxide. A mild narcotic; damages kidneys. When heated to decomposition it emits acrid smoke and irritating fumes.

CYCLOPENTANE   HR: 2
CAS: 287-92-3   NIOSH: GY 2390000
DOT: 1146
mf: C_5H_{10} mw: 70.15

PROP: Colorless liquid; bp: $49.3°$, fp: $-93.7°$, flash p: $19.4°F$, autoign temp: $716°F$, d: 0.745 @ $20°/4°$, vap press: 400 mm @ $31.0°$, vap d: 2.42.

SYNS: CYCLOPENTANE (DOT) * PENTAMETHY-LENE

DOT Classification: Flammable Liquid

THR: Low toxicity via oral and inhalation routes. High concentrations have narcotic action. Dangerous fire hazard when exposed to flame; can react with oxidizers. To fight fire, use foam, CO_2, dry chemical.

CYCLOPENTANONE   HR: 2
CAS: 120-92-3   NIOSH: GY 4725000
mf: C_5H_8O mw: 84.13

PROP: Liquid; mp: $-58.2°$, bp: $130.6°$, flash p: $79°F$, d: 0.9509 @ $18°/4°$, vap d: 2.3.

SYNS: ADIPIC KETONE * KETOCYCLOPENTANE * KETOPENTAMETHYLENE

THR: Moderate toxicity via intraperitoneal, oral, inhalation and subcutaneous routes. A skin irritant. See also ketones. Dangerous fire hazard when exposed to flame; can react with oxidizers. To fight fire, use alcohol foam, foam, CO_2, dry chemical. Incompatibles: Hydrogen peroxide; nitric acid.

CYCLOPENTA(cd)PYRENE   HR: 3
CAS: 27208-37-3   NIOSH: GY 5850000
mf: $C_{18}H_{10}$ mw: 226.28

THR: Mutagenic data. An experimental neoplastigen. When heated to decomposition it emits acrid smoke and irritating fumes.

CYCLOPENTENE   HR: 2
CAS: 142-29-0   NIOSH: GY 5950000
mf: C_5H_8 mw: 68.13

PROP: Liquid; mp: −93.3°; bp: 44.242°; fp: −135.2°; flash p: −20°F; d: 0.77199 @ 20°.

THR: Moderate toxicity via inhalation, oral and dermal routes. Dangerous fire hazard when exposed to flame or heat; can react with oxidizing materials. Dangerous. Keep away from heat and open flame. To fight fire, use foam, CO_2, dry chemical.

CYCLOPHOSPHAMIDE AND MNU
(1 : 2)   HR: 3
                    NIOSH: RP 5965000
mf: $C_7H_{15}Cl_2N_2O_2P \cdot C_4H_5N_6O_2$ mw: 430.26

PROP: A combination of these two drugs is used in chemotherapy to combat far advanced malignant tumors.

SYN: MNU AND CYCLOPHOSPHAMIDE (2 : 1)

THR: In humans it affects the gastrointestinal tract and the blood picture. When heated to decomposition it emits very toxic fumes of Cl^-, NO_x and PO_x.

CYCLOPROPANE   HR: 2
CAS: 75-19-4   NIOSH: GZ 0690000
mf: C_3H_6 mw: 42.09

PROP: Colorless gas; mp: −126.6°, bp: −33.5°, lel = 2.4%, uel = 10.4%, d: 1.879 g/L @ 0°, autoign temp: 932°F.

SYN: TRIMETHYLENE

THR: Moderately toxic via inhalation route. High concentrations are narcotic. Used as a surgical anesthetic. Very dangerous fire hazard when exposed to heat or flame; can react with oxidizing materials. Moderately explosive in the form of vapor when exposed to heat or flame. To fight fire, stop flow of gas. Use CO_2, dry chemical or water spray.

CYCLOTETRAMETHYLENETETRA-
NITRAMINE   HR: 3
CAS: 2691-41-0   NIOSH: XF 7450000
mf: $C_4H_8N_8O_8$ mw: 296.20

SYN: OCTOGEN

THR: A poison via oral, and intravenous routes. Moderately toxic via oral route. When heated to decomposition it emits toxic fumes of NO_x.

p-CYMENE   HR: 1
CAS: 99-87-6   NIOSH: GZ 5950000
mf: $C_{10}H_{14}$ mw: 134.24

PROP: Liquid; mp: −68.2°, bp: 176°, lel = 0.7%, @ 100°, ulc: 30-35, flash p: 117°F (CC), d: 0.86, autoign temp: 817°F, vap d: 4.62, vap press: 1 mm @ 17.3°, flash p: (technical) 127°F, uel (technical) = 5.6%. Found in nearly 100 volatile oils including Lemongrass, sage, thyme, coriander, star anise, and cinnamon. Sol in alc, ether, acetone, benzene.

SYNS: CAMPHOGEN * CYMOL * 4-ISOPRO-PYL-1-METHYLBENZENE * P-ISOPROPYLTOLUENE * P-METHYLISOPROPYL BENZENE * 1-METHYL-4-ISOPROPYLBENZENE

THR: Low toxicity via oral and inhalation routes, although humans sustain central nervous system effects at low dose rates. Moderate fire hazard when exposed to heat, flame or oxidizers. Slight explosion hazard in the form of vapor. To fight fire, use foam, CO_2, dry chemical.

CYPRESS OIL   HR: 2
CAS: 8013-86-3   NIOSH: GZ 9090000

PROP: The constituents include furfural, d-alpha-pinene, d-camphene, cymene, d-terpineol, l-cadinene, sylvestrene, cypress camphor and cedrol.

THR: A moderate skin irritant. When heated to decomposition it emits acrid smoke and irritating fumes.

CYPROMID   HR: 3
CAS: 2759-71-9   NIOSH: GZ 1050000
mf: $C_{10}H_9Cl_2NO$ mw: 230.10

SYNS: CIPROMID * CLOBBER * 3,4′-DICHLOROCYCLOPROPANECARBOXANILIDE * N-(3,4-DICHLOROPHENYL)-CYCLOPROPANECARBOXAMIDE

THR: A poison via oral and inhalation routes. An herbicide. Moderately toxic via skin route.

When heated to decomposition it emits very toxic fumes of HCl and NO$_x$.

CYPROSTERONE ACETATE   HR: 3
CAS: 427-51-0   NIOSH: GZ 2230000
mf: C$_{24}$H$_{29}$ClO$_4$ mw: 416.98

SYNS: 6-CHLORO-1,2-alpha-METHYLENE-6-DE-HYDRO-17-alpha-HYDROXYPROGESTERONE ACETATE * 6-CHLORO-delta(SUP 6)-1,2-alpha-METHYLENE-17-alpha-HYDROXYPROGESTERONE ACETATE * 6-CHLORO-1,2-alpha-METHYLENE-17-alpha-HYDROXY-delta(SUP 6)-PROGESTERONE ACETATE * 1,2-alpha-METHYLENE-6-CHLORO-delta(SUP 6)-17-alpha-HYDROXYPROGESTERONE ACETATE * 1,2-alpha-METHYLENE-6-CHLORO-PREGNA-4,6-DIENE-3,20-DIONE 17-alpha-ACE-TATE * 1,2-alpha-METHYLENE-6-CHLORO-(SUP 4,6)-PREGNADIENE-17-alpha-OL-3,20-DIONE 17-alpha-ACETATE * 1,2-alpha-METHYLENE-6-CHLORO-delta-(SUP 4,6)-PREGNADIENE-17-al-pha-OL-3,20-DIONE ACETATE

THR: An experimental teratogen and tumorigen. When heated to decomposition it emits toxic fumes of Cl$^-$.

L-CYSTEINE   HR: 2
CAS: 52-90-4   NIOSH: HA 1600000
mf: C$_3$H$_7$NO$_2$S   mw: 121.17

PROP: An amino acid derived from cystine, occurring naturally in the *l*-form, which will be considered here. Colorless crystals, sol in water, ammonium hydroxide and acetic acid, insol in ether, acetone, benzene, carbon disulfide and carbon tetrachloride.

SYNS: CYSTEIN * CYSTEINE * L-(+)-CY-STEINE * HALF-CYSTEINE * BETA-MERCAP-TOALANINE * THIOSERINE

THR: Mutagenic data. Moderately toxic via oral route. A nutrient and/or dietary supplement food additive. When heated to decomposition it emits very toxic fumes of SO$_x$ and NO$_x$. For further information see Vol. 3, No. 1 of *DPIM Report*.

CYTOCHALASIN E   HR: 3
CAS: 36011-19-5   NIOSH: HA 5360000
mf: C$_{28}$H$_{32}$NO$_7$ mw: 494.61

PROP: Food storage mold metabolite of *aspergillus clavatus*.

THR: A poison via oral, intraperitoneal, and parenteral routes. When heated to decomposition it emits toxic fumes of NO$_x$.

CYTOSTASAN   HR: 3
CAS: 3543-75-7   NIOSH: DE 1590000

SYNS: gamma(1-METHYL-5-BIS(beta-CHLORAETHYL)AMINOBENZIMIDAZOLYL-)BUTTER-SAEUREHYDROCHLORID(GERMAN) * gamma-(1-METHYL-5-BIS(beta-CHLOROAETHYL) AMINO-BENZIMIDAZOYL) BUTTERSAUERHYDROCHLORID-(GERMAN)

THR: An experimental teratogen, and carcinogen. A poison via oral and intraperitoneal routes. When heated to decomposition it emits very toxic fumes of HCl and NO$_x$.

CYTOXAL ALCOHOL   HR: 3
CAS: 4465-94-5   NIOSH: RP 5945000
mf: C$_7$H$_{17}$Cl$_2$N$_2$O$_3$P • C$_6$H$_{13}$N   mw: 378.33

SYNS: CYTOXYL ALCOHOL CYCLOHEXYLAMMO-NIUM SALT * 2-(BIS(2-CHLOROETHYL)AMINO)-TETRAHYDROOXAZAPHOSPHORINE CYCLOHEXYL-AMINE SALT * N,N-BIS(2-CHLOROETHYL)-N'-(3-HYDROXYPROPYL)PHOSPHORODIAMIDATE, CYCLOHEXYLAMMONIUM SALT * NCI-C04922 * NSC-52695 * PHOSPHORODIAMIDIC ACID, N,N-BIS(2-CHLOROETHYL)-N'-3-HYDROXYLPRO-PYLCYCLOHEXYLAMINE SALT

THR: An experimental teratogen, carcinogen, and tumorigen. When heated to decomposition it emits very toxic fumes of NO$_x$, PO$_x$ and Cl$^-$.

D

D&C RED NO. 5   HR: 3
CAS: 3761-53-3   NIOSH: QJ 6825000
mf: $C_{18}H_{14}N_2O_7S_2$ • 2Na   mw: 480.44

SYNS: C.I. ACID RED 26 ∗ C.I. FOOD RED 5
∗ 3-HYDROXY-4-(2,4-XYLYLAZO)-2,7-
NAPHTHALENEDISULFONIC ACID, DISODIUM SALT
∗ 4-((2,4-DIMETHYLPHENYL)AZO)-3-HYDROXY-
2,7-NAPHTHALENEDISULFONIC ACID, DISODIUM
SALT ∗ DISODIUM (2,4-DIMETHYLPHENYLAZO)-
2-HYDROXYNAPHTHALENE-3,6-DISULFONATE
∗ DISODIUM (2,4-DIMETHYLPHENYLAZO)-2-HY-
DROXYNAPHTHALENE-3,6-DISULPHONATE
∗ DISODIUM SALT OF 1-(2,4-XYLYLAZO)-2-
NAPHTHOL-3,6-DISULFONIC ACID ∗ DISODIUM
SALT OF 1-(2,4-XYLYLAZO)-2-NAPHTHOL-3,6-DI-
SULPHONIC ACID ∗ 1-XYLYLAZO-2-NAPHTHOL-
3,6-DISULFONIC ACID, DISODIUM SALT ∗ 1-(2,
4-XYLYLAZO)-2-NAPHTHOL-3,6-DISULPHONIC
ACID, DISODIUM SALT

THR: An experimental carcinogen and tumorigen. Moderately toxic by intraperitoneal route.
When heated to decomposition it emits very
toxic NO_x and SO_x.

DACARBAZINE   HR: 3
CAS: 4342-03-4   NIOSH: NI 3950000
mf: $C_6H_{10}N_6O$ mw: 182.22

SYNS: 4-(3,3-DIMETHYL-1-TRIAZENO)IMIDA-
ZOLE-5-CARBOXAMIDE ∗ 4-(5)-(3,3-DIMETHYL-
1-TRIAZENO)IMIDAZOLE-5(4)-CARBOXAMIDE
∗ 5-(3,3-DIMETHYL-1-TRIAZENO)IMIDAZOLE-4-
CARBOXAMIDE ∗ NCI-C04717 ∗ NSC-45388

THR: An experimental carcinogen. Moderately
toxic. Human blood and gastrointestinal system
effects by intravenous route. When heated to
decomposition it emits toxic fumes of NO_x.

DALMANE   HR: 3
CAS: 1172-18-5   NIOSH: DF 0875000
mf: $C_{21}H_{23}ClFN_3O$ • 2ClH   mw: 460.84

SYNS: 7-CHLORO-1-(2-(DIETHYLAMINO)ETHYL)-
5-(O-FLUOROPHENYL)-2H-1,4-BENZODIAZEPIN-2-
ONE ∗ DALMADORM HYDROCHLORIDE
∗ FLURAZEPAM HYDROCHLORIDE

THR: Poison by intravenous and intraperitoneal
routes. Moderately toxic by ingestion and other
routes. Habituating and possibly addictive.
When heated to decomposition it emits very
toxic fumes of HCl, F^-, NO_x, and Cl^-.

DANIFOS   HR: 3
CAS: 7173-84-4   NIOSH: TE 8400000
mf: $C_{11}H_{16}ClO_3PS_2$ mw: 326.81

SYNS: S-((P-CHLOROPHENYLTHIO)METHYL) O,O-
DIETHYL PHOSPHORODITHIOATE ∗ O,O-DI-
ETHYL-S-p-CHLOROPHENYL THIOMETHYLPHOS-
PHOROTHIOATE

THR: Poison by ingestion and skin contact.
When heated to decomposition it emits very
toxic fumes of Cl^-, PO_x and SO_x.

DAUNOMYCIN   HR: 3
PROP: Isolated from cultures of a *Streptomyces*.

CAS: 20830-81-3   NIOSH: HB 7875000
mf: $C_{27}H_{29}NO_{10}$ mw: 527.57

PROP: Thin red needles. mp: 190° (decomp).

SYNS: DAUNORUBICIN ∗ LEUKAEMOMYCIN C
∗ NCI-C04693 ∗ NSC-82151 ∗ STREPTO-
MYCES PEUCETIUS

THR: Poison by ingestion, subcutaneous, intravenous, and intraperitoneal routes. An experimental carcinogen and teratogen. Mutagenic
data. When heated to decomposition it emits
toxic fumes of NO_x. For further information,
see Vol. 1, No. 3 of *DPIM Report*.

DCDD   HR: 3
CAS: 33857-26-0   NIOSH: HP 3100000
mf: $C_{12}H_6Cl_2O_2$ mw: 253.08

PROP: Colorless cryst; mp: 210°.

SYNS: 2,7-DICHLORODIBENZO-p-DIOXIN
∗ 2,7-DICHLORODIBENZODIOXIN ∗ NCI-C03667

THR: An experimental carcinogen and teratogen. An eye irritant. When heated to decomposition it emits toxic fumes of Cl^-.

DDT   HR: 3
CAS: 50-29-3   NIOSH: KJ 3325000
DOT: 2761
mf: $C_{14}H_9Cl_5$ mw: 354.48

PROP: Colorless crystals or white to slightly
off-white powder. Odorless or with slight aromatic odor. Mp: 108.5°-109°.

SYNS: alpha,alpha-BIS(P-CHLOROPHENYL)-
beta,beta,beta-TRICHLORETHANE * 2,2-BIS(p-
CHLOROPHENYL)-1,1,1-TRICHLOROETHANE
* CHLOROPHENOTHANE * p,p'-DDT * p,p'-
DICHLORODIPHENYLTRICHLOROETHANE * 4,4'
DICHLORODIPHENYLTRICHLOROETHANE * DI-
PHENYLTRICHLOROETHANE * ENT 1,506
* NCI-C00464 * 1,1,1-TRICHLOOR-2,2-BIS(4-
CHLOOR FENYL)-ETHAAN (DUTCH) * 1,1,1-TRI-
CHLOR-2,2-BIS(4-CHLOR-PHENYL)-AETHAN
(GERMAN) * 1,1,1-TRICHLORO-2,2-DI(4-
CHLOROPHENYL)-ETHANE * 1,1,1-TRICLORO-
2,2-BIS(4-CLORO-FENIL)-ETANO (ITALIAN)

OSHA PEL: TWA 1 mg/m^3 (skin)
ACGIH TLV: TWA: 1 mg/m^3
DFG MAK: 1 mg/m^3
DOT Classification: ORM-A

THR: Poison by ingestion, skin contact, subcutaneous, intravenous, and intraperitoneal routes.
Central nervous system effects. An experimental
carcinogen, teratogen, and mutagen. See also
chlorinated hydrocarbons. DDT and certain of
its degradation products, particularly DDE, are
stored in fat. This storage effect leads to a concentration of DDT at higher levels of the food
chain. A single oral dose of 20g in man produced
highly dangerous long term effects, but was
not fatal. Common solution formulations of
DDT are much more dangerous. Less is known
about chronic doses. A common air contaminant. DDT stored in the fat is at least largely
inactive since a greater total dose may be stored
in an experimental animal than is sufficient as
a lethal dose for that same animal if given at
one time. A study based on 75 human cases
reported an average of 5.3 ppm of DDT stored
in the fat. A higher content of DDT and its
derivatives (up to 434 ppm of DDE and 648
ppm of DDT) was found in workers who had
very extensive exposure. Without exception, the
samples were taken from persons who were either asymptomatic or suffering from some disease completely unrelated to DDT. Careful hospital examination of workers, who had been
very extensively exposed and who had volun-
teered for examination revealed no abnormality
which could be attributed to DDT. Much higher
levels than have been found in man have been
observed in the fat of experimental animals
which were apparently asymptomatic. DDT
stored in the fat is eliminated only very gradually
when further dosage is discontinued. After a
single dose, the secretion of DDT in the milk

and its excretion in the urine reach their height
within a day or two and continue at a lower
level thereafter. Dangerous acute dose in man:
A dose of 20 g has proved highly dangerous
though not fatal to man. This dose was taken
by 5 persons who vomited an unknown portion
of the material and even so recovered only in-
completely after 5 weeks. Smaller doses produced less important symptoms with relatively
rapid recovery. Experimental ingestion of 1.5
g resulted in great discomfort and moderate neurological changes including paraesthesia,
tremor, moderate ataxia, exaggeration of part
of the reflexes, headache, and fatigue. Vomiting
followed only after 11 hours. Recovery was
complete on the following day. The fatal dose
of DDT for man is not known. Judging from
the literature, no one has ever been killed by
DDT in the absence of other insecticides and/
or a variety of toxic solvents. However, these
common solvent formulations are highly fatal
when taken in small doses, partly because of
the toxicity of the solvent, and perhaps because
of the increased absorbability of the DDT; several fatal cases in man have been reported. Acute
oral toxicity for man = 250 mg/kg. Acute oral
LD$_{50}$ (rat) = 113 mg/kg (technical grade). Federal fruit and vegetable tolerance = 7 ppm.
Dangerous chronic dose in man: Even less is
known of the hazard of chronic DDT poisoning.
It is known that certain experimental animals
fed diets containing one part of DDT per million store the compound in their fat. The
storage of DDT in man has been mentioned
above. The exact significance of these findings is not known and their further investigation is of the greatest importance. Human volunteers have ingested up to 35 mg/day for
21 months with no noticed ill effects. For
further information see Vol. 5, No. 1 of *DPIM
Report*.

DECABORANE(14)   HR: 3
CAS: 17702-41-9   NIOSH: HD 1400000
DOT: 1868
mf: B$_{10}$H$_{14}$ mw: 122.24

PROP: Colorless needles. Mp: 99.7°, d: 0.94.
(solid), d: 0.78 (liquid @ 100°), vap press: 19
mm @ 100°.

SYNS: DECABORANE (DOT) * BORON HYDRIDE

OSHA PEL: TWA 300 ug/m^3 (skin)
ACGIH TLV: TWA: 0.05 ppm; STEL 0.15 ppm

DFG MAK: 0.05 ppm (0.3 mg/m^3)

DOT Classification: Label: Flammable Solid and Poison

THR: Poison by inhalation, ingestion, skin contact, and intraperitoneal routes. Spontaneous ignition in O_2. When heated to decomposition it emits toxic fumes of boron oxides. Incompatible with ethers; halocarbons; O_2 @ 100°; dimethyl sulfoxide. See also boron compounds and boron hydride. For further information see Vol. 1, No. 8 of *DPIM Report*.

DECAFENTIN   HR: 3
CAS: 15652-38-7   NIOSH: TA 2100000

mf: $C_{28}H_{36}P \cdot C_{18}H_{15}BrClSn$ mw: 868.99

SYNS: DECYLTRIPHENYLPHOSPHONIUM BROMO-CHLOROTRIPHENYLSTANNATE * STANNORAM

THR: Poison by skin contact. Moderately toxic by ingestion. See also tin compounds. When heated to decomposition it emits very toxic fumes of PO_x, Br^-, and Cl^-.

DECAHYDRONAPHTHALENE   HR: 2
CAS: 91-17-8   NIOSH: QJ 3150000

mf: $C_{10}H_{18}$ mw: 138.28

PROP: Water white liquid. Mp (*cis*): −43.3°, mp (*trans*): −30.7°, bp: (*cis*): 194.6°, flash p: 136°F (CC), autoign temp: 482°F, vap press: (*cis*) 1 mm @ 22.5°, (*trans*) 10 mm @ 47.2°, d: 0.8963, vap d: 4.76, lel = 0.7% @ 212°F, uel = 4.9% @ 212°F.

SYNS: BICYCLO(4.4.0)DECANE * DECALIN * DEKALINA (POLISH) * PERHYDRONAPHTHA-LENE

DOT Classification: Combustible Liquid; Label: None

THR: Moderately toxic by inhalation and ingestion. Strong irritant to skin, eyes, and mucous membranes. A systemic irritant by inhalation. Can cause kidney damage. Flammable when exposed to heat or flame; can react with oxidizing materials. To fight fire: Foam, CO_2, dry chemical.

1-DECANAL   HR: 2
PROP: Found in over 50 sources, including Citrus Oils, Citronella and Lemongrass.

CAS: 112-31-2   NIOSH: HD 6000000

mf: $C_{10}H_{20}O$ mw: 156.30

PROP: Colorless to light yellow liquid, floral fatty odor, sol in 80% alcohol, fixed oils, volatile oils and mineral oils, insol in water and glycerol. d: 0.831-0.838 @ 15°.

SYNS: ALDEHYDE C-10 * CAPRALDEHYDE * 1-DECYL ALDEHYDE

THR: Moderately toxic by ingestion. A skin irritant. See aldehydes. When heated to decomposition it emits acrid smoke and irritating fumes.

DECANE   HR: 3
CAS: 124-18-5   NIOSH: HD 6550000

mf: $C_{10}H_{22}$ mw: 142.32

PROP: Liquid. Mp: −29.7°, bp: 174.1°, lel = 0.8%, uel = 5.4%, flash p: 115°F (CC), d: 0.730 @ 20°/4°, autoign temp: 410°F, vap press: 1 mm @ 16.5°, vap d: 4.90.

THR: An experimental carcinogen. A simple asphyxiant. Narcotic in high concentrations. See also argon. Flammable when exposed to heat or flame; can react with oxidizing materials. Moderately explosive in its vapor form. To fight fire: Foam, CO_2, dry chemical.

1-DECANEAMINE   HR: 3
CAS: 2016-57-1   NIOSH: HD 6475000

mf: $C_{10}H_{23}N$ mw: 157.34

PROP: Liquid. Mp: 17°, bp: 95° @ 10 mm, flash p: 210°F, d: 0.79 @ 20°, vap d: 5.5.

SYNS: DECYLAMINE * 1-AMINODECANE

THR: Poison by ingestion and skin contact. See amines and amines, fatty. A skin irritant. Flammable when exposed to heat or flame; can react with oxidizing materials. To fight fire: Alcohol foam, foam, dry chemical. When heated to decomposition it emits toxic fumes of NO_x.

DECANOIC ACID   HR: 3
CAS: 334-48-5   NIOSH: HD 9100000

mf: $C_{10}H_{20}O_2$ mw: 172.30

PROP: White crystals, unpleasant odor, sol in most organic solvents and in dilute nitric acid, insol in water. D: 0.8858, bp: 270°, mp: 31.4°.

SYNS: N-CAPRIC ACID * N-DECYLIC ACID * NEO-FAT 10 * 1-NONANECARBOXYLIC ACID

THR: Poison by intravenous route. When heated to decomposition it emits acrid smoke and irritating fumes.

DECYL ALCOHOL HR: 3

PROP: Found in Sweet Orange and a few other essential oils.

CAS: 112-30-1   NIOSH: HE 4375000
mf: $C_{10}H_{22}O$ mw: 158.32

PROP: Viscous, refractive liquid. Mp: 7°, bp: 232.9°, flash p: 180°F (OC), d: 0.8297 @ 20°/4°, vap press: 1 mm @ 69.5°, vap d: 5.3.

SYNS: ALCOHOL C-10 * CAPRIC ALCOHOL * CAPRINIC ALCOHOL * N-DECANOL * 1-DECANOL * N-DECYL ALCOHOL * DECYLIC ALCOHOL * NONYLCARBINOL * PRIMARY DECYL ALCOHOL

THR: Moderately toxic by skin contact. A human skin and eye irritant. An experimental carcinogen. See also alcohols. Flammable when exposed to heat or flame; can react with oxidizing materials. To fight fire: Foam, CO_2, dry chemical. When heated to decomposition it emits acrid smoke and irritating fumes.

DEERTONGUE INCOLORE HR: 2

CAS: 8024-14-4   NIOSH: HE 7200000

PROP: Found in leaves of Liatris odoratissima, contains Coumarin.

SYNS: DEER'S TONGUE * LIATRIX OLEORESIN * VANILLA PLANT

THR: Moderately toxic by ingestion and skin contact. See also coumarin. When heated to decomposition it emits toxic fumes of NO_x.

1,2-DEHYDRO-3-METHYLCHOLANTHRENE HR: 3

CAS: 3343-10-0   NIOSH: CU 1225000
mf: $C_{21}H_{14}$ mw: 266.35

SYNS: 3-METHYLBENZ(J)ACEANTHRYLENE * 3-METHYLCHOLANTHRYLENE * 20-METHYLCHOLANTHRYLENE

THR: An experimental carcinogen. When heated to decomposition it emits acrid smoke and irritating fumes.

DEHYDRORETRONECINE HR: 3

CAS: 23107-12-2   NIOSH: VH 7260000
mf: $C_8H_{11}NO_2$ mw: 153.20

SYN: 1H-PYRROLIZINE-7-METHANOL 2,3-DIHYDRO-1-HYDROXY- (R)-

THR: Poison by intraperitoneal route. An experimental carcinogen. Mutagenic data. When heated to decomposition it emits toxic fumes of NO_x.

DEHYDROSTILBESTROL HR: 3

CAS: 84-17-3   NIOSH: SL 0580000
mf: $C_{18}H_{18}O_2$ mw: 266.36

SYNS: 3,4-BIS(P-HYDROXYPHENYL)-2,4-HEXADIENE * DIENESTROL * 4,4'-(1,2-DIETHYLIDENE-1,2-ETHANEDIYL)BISPHENOL * P,P'-(DIETHYLIDENEETHYLENE)DIPHENOL * 4,4'-(DIETHYLIDENEETHYLENE)DIPHENOL * DI(P-OXYPHENYL)-2,4-HEXADIENE * 4,4'-HYDROXY-gamma,delta-DIPHENYL-beta,delta-HEXADIENE

THR: An experimental carcinogen. When heated to decomposition it emits acrid smoke and irritating fumes.

DEMAROL HR: 3

CAS: 57-42-1   NIOSH: NS 5775000
mf: $C_{15}H_{21}NO_2$ mw: 247.37

SYNS: ETHYL-1-METHYL-4-PHENYLISONIPECOTATE * ETHYL-1-METHYL-4-PHENYLPIPERIDINE-4-CARBOXYLATE * N-METHYL-4-PHENYL-4-CARBETHOXYPIPERIDINE * 1-METHYL-4-PHENYLISONIPECOTIC ACID, ETHYL ESTER * 1-METHYL-4-PHENYL-PIPERIDIN-4-CARBONSAEURE-AETHYLESTER-HYDROCHLORID (GERMAN) * 1-METHYL-4-PHENYLPIPERIDINE-4-CARBOXYLIC ACID ETHYL ESTER

THR: Poison by ingestion, subcutaneous, intravenous, intradermal, parenteral, intraperitoneal, and possibly other routes. A pharmaceutical pain killer. See also esters. When heated to decomposition it emits toxic fumes of NO_x.

4-DEMETHOXYDAUNOMYCIN HR: 3

CAS: 58957-92-9   NIOSH: HB 7876000
mf: $C_{26}H_{27}NO_9$ mw: 497.54

SYNS: 4-DEMETHOXYDAUNORUBICIN * NSC-256439

THR: Poison by ingestion and intravenous routes. Mutagenic data. When heated to decomposition it emits toxic fumes of NO_x.

DEMETHYLIMIPRAMINE HYDROCHLORIDE HR: 3

CAS: 58-28-6   NIOSH: HO 0525000
mf: $C_{18}H_{22}N_2 \cdot ClH$ mw: 302.88

SYNS: 10,11-DIHYDRO-5-(3-(METHYLAMINO)-PROPYL)-5H-DIBENZ(b,f)AZEPINE HYDROCHLOR-

IDE ∗ DESMETHYLIMIPRAMINE HYDROCHLORIDE ∗ N-(GAMMA-METHYLAMINO-PROPYL)IMINODIBENZYL HYDROCHLORIDE

THR: Poison by ingestion, intraperitoneal, subcutaneous, and intravenous routes. When heated to decomposition it emits very toxic fumes of NO$_x$ and HCl.

DEMETON-O + DEMETON-S HR: 3
CAS: 8065-48-3 NIOSH: TF 3150000

PROP: A light brown liquid, sulfur compound odor.

SYNS: PHOSPHOROTHIOIC ACID, O,O-DIETHYL O-(2-(ETHYLTHIO)ETHYL) ESTER, MIXED WITH O,O-DIETHYL S-(2-(ETHYLTHIO)ETHYL) ESTER (7:3) ∗ DEMETON ∗ DIETHOXY THIOPHOSPHORIC ACID ESTER OF 2-ETHYLMERCAPTOETHANOL ∗ O,O-DIETHYL 2-ETHYLMERCAPTOETHYL THIOPHOSPHATE ∗ O,O-DIETHYL O(AND S)-2-(ETHYLTHIO)ETHYL PHOSPHOROTHIOATE MIXTURE ∗ ENT 17,295 ∗ MERCAPTOPHOS ∗ SYSTOX

OSHA PEL: TWA 100 mg/m^3 (skin)
ACGIH TLV: TWA 0.01 ppm
DFG MAK: 0.01 ppm (0.1 mg/m^3)

THR: Poison by ingestion, skin contact, intramuscular, intravenous, and intraperitoneal routes. An experimental teratogen. Mutagenic data. An insecticide which inhibits cholinesterase which causes the buildup of acetylcholine. Doses are cumulative. If illness occurs, it is acute in nature whether caused by a single large dose or by repeated exposure. Persons poisoned with demeton may be expected to show the following symptoms: headache, giddiness, blurred vision, weakness, nausea, diarrhea, and discomfort in the chest. When heated to decomposition it emits very toxic fumes of PO$_x$ and SO$_x$. See parathion.

Treatment of Poisoning: Keep the patient fully atropinized, using 1 to 2 mg of atropine sulfate per dose and up to 10 to 20 mg per day. Remove any remaining poison. Administer oxygen and other supportive measures early and be prepared for the immediate use of mechanical artificial respiration. Watch the patient continuously. See also parathion.

DEMETON-O-METHYL HR: 3
CAS: 867-27-6 NIOSH: TG 1650000
mf: C$_6$H$_{15}$O$_3$PS$_2$ mw: 230.30

SYNS: DEMETON-O-METILE (ITALIAN) ∗ O,O-

DIMETHYL-O-(2-AETHYLTHIO-AETHYL MONOTHIO-PHOSPHAT (GERMAN) ∗ O,O-DIMETHYL-O-(2-ETHYL-THIO-ETHYL)-MONOTHIOFOSFAAT (DUTCH) ∗ O,O-DIMETHYL-O-ETHYLMERCAPTOETHYL THIOPHOSPHATE ∗ O,O-DIMETHYL-O-2-(ETHYL-THIO)ETHYL PHOSPHOROTHIOATE ∗ O,O-DIMETIL-O-(2-ETILTIO-ETIL)-MONOTIOFOSFATO (ITALIAN) ∗ ENT 18,862 ∗ beta-ETHYL-MERCAPTOETHYL DIMETHYL THIONOPHOSPHATE ∗ O-(2-(ETHYLTHIO)ETHYL)-O,O-DIMETHYL PHOSPHOROTHIOATE ∗ 2-(ETHYLTHIO)ETHYL DIMETHYL PHOSPHOROTHIONATE ∗ O-METHYL-DEMETON ∗ METHYLMERCAPTOPHOS ∗ METHYLSYSTOX ∗ THIOPHOSPHATE DE O,O-DIETHYLE ET DE O-2-ETHYLTHIO-ETHYLE (FRENCH)

THR: Poison by ingestion and intravenous routes. An experimental teratogen. See also demeton-O + demeton-S. When heated to decomposition it emits very toxic fumes of PO$_x$ and SO$_x$.

DEMETON-O-METHYL SULFOXIDE HR: 3
CAS: 301-12-2 NIOSH: TG 1420000
mf: C$_6$H$_{15}$O$_3$PS$_2$ mw: 230.30

SYNS: DEMETON-S-METHYL SULFOXIDE ∗ O,O-DIMETHYL S-(2-ETHTHIONYLETHYL) PHOSPHOROTHIOATE ∗ DIMETHYL-S-(2-ETHTHIONYL-ETHYL) THIOPHOSPHATE ∗ O,O-DIMETHYL-S-(2-(ETHYLSULFINYL)ETHYL)-O,O-DIMETHYL ESTER ∗ O,O-DIMETHYL-S-(2-ETHYLSULFINYL-ETHYL)-MONOTHIOFOSFAAT (DUTCH) ∗ O,O-DIMETHYL S-(2-ETHYLSULFINYL)ETHYL THIOPHOSPHATE ∗ O,O-DIMETHYL S-ETHYLSULPHINYLETHYL PHOSPHOROTHIOLATE ∗ O,O-DIMETHYL-S-(3-OXO-3-THIA-PENTYL)-MONOTHIOPHOSPHAT (GERMAN) ∗ O,O-DIMETIL-S-(2-ETIL-SOLFINIL-ETIL)-MONOTIOFOSFATO (ITALIAN) ∗ ENT 24,964 ∗ S-(2-(ETHYLSULFINYL)ETHYL)-O,O-DIMETHYL PHOSPHOROTHIOATE ∗ ISOMETHYLSYSTOX SULFOXIDE ∗ METAISOSYSTOXSULFOXIDE ∗ OXYDEMETON-METILE (ITALIAN) ∗ THIOPHOSPHATE DE O,O-DIMETHYLE ET DE S-2-ETHYL-SULFINYLETHYLE (FRENCH)

THR: Poison by ingestion, skin contact, intravenous, intraperitoneal, and possibly other routes. Moderately toxic by inhalation. Mutagenic data. When heated to decomposition it emits very toxic fumes of PO$_x$ and SO$_x$.

DEMETON-S HR: 3
CAS: 126-75-0 NIOSH: TF 3130000
mf: C$_8$H$_{19}$O$_3$PS$_2$ mw: 258.36

SYNS: O,O-DIAETHYL-S-(2-AETHYLTHIO-AETH-YL)-MONOTHIOPHOSPHAT (GERMAN) * DIETHYL S-(2-ETHIOETHYL)THIOPHOSPHATE * O,O-DI-ETHYL S-(2-ETHTHIOETHYL)PHOSPHORO-THIOATE * O,O-DIETHYL S-ETHYL-2-ETHYLMERCAPTOPHOSPHOROTHIOLATE * O,O-DIETHYL S-2-(ETHYLTHIO)ETHYL PHOSPHORO-THIOATE * O,O-DIETHYL S-(2-(ETHYLTHIO)-ETHYL) PHOSPHOROTHIOLATE * O,O-DIETHYL-S-(2-ETHYLTHIO-ETHYL)-MONOTHIOFOSFAAT (DUTCH) * O,O-DIETIL-S-(2-ETILTIO-ETIL)-MONOTIOFOSFATO (ITALIAN) * O,O-DIETYL-S-2-ETYLMERKAPTOETYLTIOFOSFAT (CZECH) * ISODEMETON * THIOPHOSPHATE DE O,O-DI-ETHYLE ET DE S-(2-ETHYLTHIO-ETHYLE) (FRENCH)

THR: Poison by intraperitoneal route. See also esters and Demeton-O + Demeton-S. When heated to decomposition it emits very toxic fumes of PO_x and SO_x.

DEMETON-S-METHYL   HR: 3
CAS: 919-86-8   NIOSH: TG 1750000
mf: $C_6H_{15}O_3PS_2$ mw: 230.30

SYNS: DEMETON-S-METILE (ITALIAN) * O,O-DIMETHYL-S-(2-AETHYLTHIO-AETHYL)-MONO-THIOPHOSPHAT (GERMAN) * O,O-DIMETHYL S-(2-ETHTHIOETHYL)PHOSPHOROTHIOATE * DIMETHYL-S-(2-ETHTHIOETHYL)THIOPHOS-PHATE * O,O-DIMETHYL S-ETHYLMERCAPTO-ETHYL THIOPHOSPHATE * O,O-DIMETHYL-S-(2-ETHYLTHIO-ETHYL)-MONOTHIOFOSFAAT (DUTCH) * O,O-DIMETHYL S-(2-(ETHYLTHIO)ETHYL) PHOSPHOROTHIOATE * ISOMETASYSTOX * ISOMETHYLSYSTOX * O,O-DIMETHYL-S-(3-THIAPENTYL)-MONOTHIOPHOSPHAT (GERMAN) * O,O-DIMETIL-S-(2-ETILITIO-ETIL)-MONOTIO FOSFATO (ITALIAN) * S-(2-(ETHYLTHIO)ETHYL)-O,O-DIMETHYL PHOSPHOROTHIOATE * S-(2-(ETHYLTHIO)ETHYL) DIMETHYL PHOSPHOROTHIO-LATE * S-(2-(ETHYLTHIO)ETHYL)-O,O-DI-METHYL THIOPHOSPHATE * METAISOSEPTOX * METAISOSYSTOX * METASYSTOX FORTE * METHYL DEMETON THIOESTER * METHYL ISOSYSTOX * METHYL-MERCAPTOFOS TEOLOVY * THIOPHOSPHATE DE O,O-DIMETHYLE ET DE S-2-ETHYLTHIOETHYLE (FRENCH)

THR: Poison by ingestion, inhalation, skin contact, intraperitoneal, and other routes. Mutagenic data. See also demeton-O + demeton-S. When heated to decomposition it emits very toxic fumes of PO_x and SO_x.

DEMETON-S-METHYL SULPHONE   HR: 3
CAS: 17040-19-6   NIOSH: TF 9050000
mf: $C_6H_{15}O_5PS_2$ mw: 262.30

SYNS: DEMETON-S-METHYLSULFON (GER-MAN) * O,O-DIMETHYL S-(2-ETHSUL-FONYLETHYL)PHOSPHOROTHIOATE * DIMETHYL S-(2-ETHSULFONYLETHYL)THIOPHOS-PHATE * O,O-DIMETHYL S-ETHYL-2-SULFO-NYLETHYL PHOSPHOROTHIOLATE * O,O-DI-METHYL S-ETHYLSULPHONYLETHYL PHOSPHORO-THIOLATE * DIOXYDEMETON-S-METHYL * ISOMETASYSTOX SULFONE * ISOMETHYL-SYSTOX SULFONE

THR: Poison by ingestion, inhalation, intraperitoneal, intravenous, and other routes. Mutagenic data. See also demeton-O + demeton-S. When heated to decomposition it emits very toxic fumes of PO_x and SO_x.

DEOXYCHOLATIC ACID   HR: 3
CAS: 83-44-3   NIOSH: FZ 2100000
mf: $C_{24}H_{40}O_4$ mw: 392.64

PROP: A solid; mp: 178°.

SYNS: 3-alpha,12-alpha-DIHYDROXY-5-beta-CHOLAN-24-OIC ACID * CHOLEIC ACID * DEOXYCHOLIC ACID * DESOXYCHOLIC ACID * DESOXYCHOLSAEURE (GERMAN) * 3,12-DI-HYDROXYCHOLANIC ACID * 3-alpha,12-alpha-DIHYDROXYCHOLANIC ACID * 3-alpha,12-al-pha-DIHYDROXY-5-beta-CHOLANOIC ACID * 3-alpha,12-alpha-DIHYDROXYCHOLANSAEURE (GERMAN) * 17-beta-(1-METHYL-3-CARBOXY-PROPYL)-ETIOCHOLANE-3-alpha,12-alpha-DIOL

THR: Poison by intraperitoneal route. Moderately toxic by intravenous route. An experimental carcinogen. When heated to decomposition it emits acrid smoke and irritating fumes.

2'-DEOXY-5-FLUORO URIDINE   HR: 3
CAS: 50-91-9   NIOSH: YU 7525000
mf: $C_9H_{11}FN_2O_5$ mw: 246.22

SYNS: 5-FLUOROURACIL DEOXYRIBOSIDE * DEOXYFLUOROURIDINE * 1-beta-D-2'-DEOX-YRIBOFURANOSYL-5-FLUROURACIL * FLUORO-DEOXYURIDINE * 5-FLUORODEOXYURIDINE * 5-FLUORO-2-DEOXYURIDINE * 5-FLUORO-2'-DEOXYURIDINE * 5-FLUOROURACIL-2'-DEOXY-RIBOSIDE * NSC-27640

THR: Poison by ingestion and possibly other routes. Mutagenic data. An experimental teratogen. Toxic human gastrointestinal tract effects. When heated to decomposition it emits very toxic fumes of F^- and NO_x.

2-DEOXYGLUCOSE   HR: 3
CAS: 154-17-6   NIOSH: MO 3325000
mf: $C_6H_{12}O_5$ mw: 164.18

SYNS: D-2-DEOXYGLUCOSE * 2-DEOXY-D-ARABINO-HEXOSE * NSC 15193

THR: An experimental teratogen. Moderately toxic by intraperitoneal route. When heated to decomposition it emits acrid smoke and irritating fumes.

2'-DEOXY-5-IODO URIDINE   HR: 3
CAS: 54-42-2   NIOSH: YU 7700000
mf: $C_9H_{11}IN_2O_5$ mw: 354.12

SYNS: 1-beta-D-2'-DEOXYRIBOFURANOSYL-5-IODOURACIL * 5-IODODEOXYURIDINE * 5-IODO-2'-DEOXYURIDINE * 5-IODOURACIL DEOXYRIBOSIDE * NSC 39661

THR: An experimental teratogen. Mutagenic data. Moderately toxic by intraperitoneal route. When heated to decomposition it emits very toxic fumes of I^- and NO_x.

DERRIS   HR: 3
                          NIOSH: HG 4725000

PROP: The constituents of Derris are Rotenone, Dequelin, Tephrosin and Toxicarol.

SYNS: DERRIS RESINS * DERRIS ROOT

THR: Poison by ingestion and inhalation. See also rotenone.

DES DISODIUM SALT   HR: 3
                          NIOSH: WJ 5835000
mf: $C_{18}H_{18}O_2 \cdot 2Na$ mw: 312.34

SYN: alpha,alpha'-DIETHYL-4,4'-STILBENEDIOL DISODIUM SALT

THR: An experimental neoplastigen. When heated to decomposition it emits acrid smoke and irritating fumes.

DESMETHYLDOXEPIN   HR: 3
CAS: 5626-16-4   NIOSH: HQ 4260000
mf: $C_{17}H_{17}NO$ mw: 251.35

SYN: DIBENZ(B,E)OXEPIN-delta(SUP 11(6H)),gamma-PROPYLAMINE

THR: Poison by ingestion, intravenous, and in-

traperitoneal routes. When heated to decomposition it emits toxic fumes of NO_x.

dl-DESOXY EPHEDRINE HYDRO-CHLORIDE   HR: 3
CAS: 300-42-5   NIOSH: SH 5075000
mf: $C_{10}H_{15}N \cdot ClH$ mw: 185.72

SYNS: N,alpha-DIMETHYLPHENETHYLAMINE HYDROCHLORIDE * N-METHYL-beta-PHENYL-ISOPROPYLAMINHYDROCHLORID (GERMAN) * METHYLAMPHETAMINE HYDROCHLORIDE * SPEED

THR: Poison by ingestion, intravenous, intraperitoneal, subcutaneous, and intramuscular routes. Human central nervous system effects. An eye irritant. When heated to decomposition it emits very toxic fumes of NO_x and HCl.

DESOXYN   HR: 3
CAS: 7632-10-2   NIOSH: SH 4900000
mf: $C_{10}H_{15}N$ mw: 149.26

SYNS: DESOXYEPHEDRINE * METHEDRINE * N-METHYLAMPHETAMINE * N-METHYL-beta-PHENYLISOPROPYLAMIN (GERMAN) * N-METHYL-beta-PHENYLISOPROPYLAMINE * 1-PHENYL-2-METHYLAMINOPROPAN (GERMAN) * 1-PHENYL-2-METHYLAMINOPROPANE * alpha-PHENYL-beta-METHYLAMINOPROPANE

THR: Poison by ingestion, intraperitoneal, subcutaneous, and intravenous routes. An experimental teratogen. When heated to decomposition it emits toxic fumes of NO_x.

2-DESOXY PHENOBARBITAL   HR: 3
CAS: 125-33-7   NIOSH: UV 9100000
mf: $C_{12}H_{14}N_2O_2$ mw: 218.28

SYNS: 2-DEOXYPHENOBARBITAL * DESOXYPHENOBARBITONE * 5-ETHYLDIHYDRO-5-PHENYL-4,6(1H,5H)-PYRIMIDINEDIONE * 5-ETHYLHEXAHYDRO-4,6-DIOXO-5-PHENYLPYRIMIDINE * 5-ETHYLHEXAHYDRO-5-PHENYLPYRIMIDINE-4,6-DIONE * 5-ETHYL-5-PHENYLHEXAHYDROPYRIMIDINE-4,6-DIONE * HEXAMIDINE (THE ANTISPASMODIC) * NCI-C56360 * 5-PHENYL-5-ETHYL-HEXAHYDRO-PYRIMIDINE-4,6-DIONE

THR: Poison by ingestion and possibly other routes. An experimental teratogen. When heated to decomposition it emits toxic fumes of NO_x.

DEUTERIOMORPHINE   HR: 3
CAS: 67293-88-3   NIOSH: QC 8800000
mf: $C_{17}H_{16}D_3NO_3$ mw: 288.37

THR: Poison by subcutaneous and intracerebral routes. See also neoprene. When heated to decomposition it emits toxic fumes of NO_x.

DEUTERIUM
mf: D_2 mw: 4

PROP: A gas; lel = 5%; uel = 75%.

SYN: D_2

THR: See also hydrogen. Very dangerous fire hazard via ignition sources, oxidizers. To fight fire, stop flow of gas.

DEUTERIUM FLUORIDE   HR: 3
CAS: 14333-26-7   NIOSH: HH 3000000
mf: DF   mw: 21.01

THR: Poison by inhalation. See also fluorides and hydrogen fluoride. When heated to decomposition it emits toxic fumes of F^-.

DEUTERIUM OXIDE
mf: D_2O mw: 20

THR: Incompatible with pentafluorophenyllithium. See also H_2O (water).

DEXEPIN   HR: 3
CAS: 1668-19-5   NIOSH: HQ 4300000
mf: $C_{19}H_{21}NO$ mw: 279.41

SYN: 11-(3-DIMETHYLAMINOPROPYLIDENE)-6,11-DIHYDRODIBENZ(B,E)OXIPIN

THR: Poison to humans by ingestion; poison by intravenous and intraperitoneal routes. When heated to decomposition it emits toxic fumes of NO_x.

DEXTRAN 1   HR: 3
CAS: 9004-54-0   NIOSH: HH 9230000

PROP: A linear water-soluble polymer of average molecular weight 200,000.

THR: An experimental neoplastigen and carcinogen by subcutaneous and intraperitoneal routes.

DEXTRAN 2   HR: 3
CAS: 9004-54-0   NIOSH: HH 9232500

PROP: A linear, water soluble polymer of average molecular weight 100,000.

THR: An experimental carcinogen by subcutaneous and intraperitoneal routes.

DEXTRAN 5   HR: 3
CAS: 9004-54-0   NIOSH: HH 9235000

PROP: A highly branched, water soluble polymer.

THR: An experimental neoplastigen by subcutaneous route.

DEXTRAN 10   HR: 3
CAS: 9004-54-0   NIOSH: HH 9240000

PROP: A branched, water-soluble polymer of average molecular weight 89,400.

THR: An experimental carcinogen by subcutaneous, intravenous, and intraperitoneal routes.

DEXTRAN 11   HR: 3
CAS: 9004-54-0   NIOSH: HH 9245000

PROP: A highly branched, water soluble polymer of average molecular weight 71,400.

THR: An experimental neoplastigen by subcutaneous, intravenous, and intraperitoneal routes.

DEXTRINS   HR: 1
CAS: 9004-53-9   NIOSH: HH 9450000
mf: $(C_6H_{10}O_5)n \cdot xH_2O$

PROP: An intermediate product formed by the hydrolysis of starches, yellow or white powder or granules, sol in water, insol in alcohol and ether, colloidal in properties and describes a class of substances (hence it has no formula).

SYNS: STARCH GUM * ARTIFICIAL GUM * VEGETABLE GUM * TAPIOCA * DEXTRAN

THR: Mildly toxic by intravenous route. A substance migrating to food from packaging material. When heated to decomposition it emits acrid smoke and irritating fumes.

2-DIACETAMIDO FLUORENE   HR: 3
CAS: 642-65-9   NIOSH: AC 0350000
mf: $C_{17}H_{15}NO_2$ mw: 265.33

SYNS: N-DIACETYL-2-AMINOFLUORENE * N,N-DIACETYL-2-AMINOFLUORENE * N,N-DIACETYL-2-FLUORENAMINE * 2-DIACETYLAMINOFLUORENE * 2-FLUORENYLDIACETAMIDE * N-FLUOREN-2-YLDIACETAMIDE * N-2-FLUORENYLDIACETAMIDE

THR: An experimental tumorigen, carcinogen and neoplastigen by ingestion and skin contact. When heated to decomposition it emits toxic fumes of NO_x.

1,3-DIACETIN   HR: 2
CAS: 25395-31-7   NIOSH: AK 3325000
mf: $C_7H_{12}O_5$ mw: 176.19

PROP: Crystals; d: 1.178 @ 15°/15°; bp: 280°; mp: 40°.

SYNS: DIACETYLGLYCEROL * GLYCERIN DI-ACETATE * GLYCEROL DIACETATE * 1,2,3-PROPANETRIOL, DIACETATE * 2,3-DIACETIN * DIACETYL GLYCERINE * GLYCERYL-1,3-DI-ACETATE * (HYDROXYMETHYL)ETHYLENE ACE-TATE

THR: Moderately toxic by subcutaneous and intravenous routes. Mildly toxic by ingestion. When heated to decomposition it emits acrid smoke and irritating fumes.

DIACETOXY DIBUTYL STAN-NANE   HR: 3
CAS: 1067-33-0   NIOSH: WH 6880000
mf: $C_{12}H_{24}O_4Sn$ mw: 351.05

PROP: Clear, colorless liquid with a slight acetic acid odor, bp: decomp, fp: 5°-10°, flash p: 290°F (OC), d: 1.31 @ 25°, vap d: 12.1.

SYNS: BIS(ACETYLOXY)DIBUTYLSTANNANE * DIACETOXYBUTYLTIN * DIACETOXYDIBUTYL-TIN * DIBUTYL TIN DIACETATE * NCI-C02028

OSHA PEL: TWA 100 ug(Sn)/m³ (skin)

THR: Poison by ingestion and intravenous routes. See also tin compounds. Combustible when exposed to heat or flame; can react with oxidizing materials. To fight fire: water, foam, CO_2, dry chemical. When heated to decomposition it emits acrid smoke and irritating fumes.

DIACETOXYSCIRPENOL   HR: 3
CAS: 2270-40-8   NIOSH: YD 0112000
mf: $C_{18}H_{26}O_7$ mw: 354.44

SYNS: 4-beta,15-DIACETOXY-3-alpha-HY-DROXY-12,13-EPOXYTRICHOTHEC-9-ENE * 12,13-EPOXY-TRICHOTHEC-9-ENE-3-alpha,4-beta,15-TRIOL-4,15-DIACETATE * TRICHOTHEC-9-ENE, 12,13-EPOXY-4-beta,15-DIAZETOXY-3-al-pha-HYDROXY-

THR: Poison by intravenous and ingestion routes. Mutagenic data. When heated to decomposition it emits acrid smoke and fumes.

DIACETYL DIOXIME   HR: 3
CAS: 95-45-4   NIOSH: EK 2975000
mf: $C_4H_8N_2O_2$ mw: 116.14

SYNS: 2,3-DIISONITROSOBUTANE * DIMETHYL-GLYOXIME

THR: Poison by ingestion. Mutagenic data. When heated to decomposition it emits toxic fumes of NO_x.

3,4-DI(ACETYLTHIOMETHYL)-5-HY-DROXY-6-METHYLPYRIDINE HYDROBROMIDE   HR: 3
CAS: 73622-67-0   NIOSH: AJ 6835000
mf: $C_{12}H_{15}NO_3S_2$ • BrH   mw: 366.32

SYNS: 4,5-DI(MERCAPTOMETHYL)-2-METHYL-3-PYRIDINOL DITHIOACETATE HYDROBROMIDE * 4,5-DIMERCAPTOPYRIDOXINDI-THIOACETAT HYDROBROMID (GERMAN)

THR: Poison by ingestion, subcutaneous, and intravenous routes. See also bromides. When heated to decomposition it emits very toxic fumes of NO_x, SO_x, and HBr.

DIALICOR   HR: 3
CAS: 2192-21-4   NIOSH: UH 0350000
mf: $C_{21}H_{27}NO_2$ • ClH   mw: 361.95

SYNS: 1-(2-(2-(DIETHYLAMINO)ETHOXY)-PHENYL)-3-PHENYL-1-PROPANONE * (O-BETA-DIETHYLAMINOETHOXY)-PHENYL PROPIOPHENONE HYDROCHLORIDE

THR: Poison by ingestion, intravenous, and intraperitoneal routes. When heated to decomposition it emits very toxic fumes of NO_x and HCl.

DIALLATE   HR: 3
CAS: 2303-16-4   NIOSH: EZ 8225000
mf: $C_{10}H_{17}Cl_2NOS$ mw: 270.24

PROP: Brown liquid, slightly sol in water, sol in organic solvents, bp: 150° @ 9 mm, mp: 25°-30°.

SYNS: DIALLAAT (DUTCH) * DIALLAT (GER-MAN) * 5-2,3-DICHLOROALLYLDIISOPROPYL-THIOCARBAMATE * S-(2,3-DICHLOR-ALLYL)-N,N-DIISOPROPYL-MONOTHIOCARBAMAAT (DUTCH) * S-(2,3-DICHLORO-ALLIL)-N,N-DIISOPROPIL-MONOTIOCARBAMMATO (ITALIAN) * 2,3-DI-CHLOROALLYL N,N-DIISOPROPYLTHIOLCARBA-MATE * 2,3-DICHLORO-2-PROPENE-1-THIOL, DIISOPROPYLCARBAMATE * DI-ISOPROPYLTHIO-LOCARBAMATE DE S-(2,3-DICHLOROALLYLE) (FRENCH)

THR: Poison by ingestion and possibly other routes. Moderately toxic by skin contact. An experimental carcinogen. Mutagenic data. When heated to decomposition it emits very toxic fumes of Cl⁻, NO$_x$ and SO$_x$. For further information see Vol. 3, No. 1. of *DPIM Report*.

DIALLYLAMINE HR: 3
CAS: 124-02-7   NIOSH: UC 6650000
mf: $C_6H_{11}N$ mw: 97.18

PROP: Liquid, sol in water, d: 0.7889 @ 20°, bp: 112°, fp: −100°. flash p.: 69.8°F.

SYN: DI-2-PROPENYLAMINE

THR: Poison by intraperitoneal and dermal routes. Moderately toxic by ingestion and inhalation. Human pulmonary system effects. A skin irritant. See also amines. A dangerous fire hazard when exposed to heat or flame. When heated to decomposition it emits toxic fumes of NO$_x$.

DIALLYL DIBROMO STANNANE HR: 3
CAS: 17381-88-3   NIOSH: WH 6881000
mf: $C_6H_{10}Br_2Sn$ mw: 360.67

SYN: DIALLYLTIN DIBROMIDE

THR: Poison by intravenous route. See bromides and tin compounds. When heated to decomposition it emits toxic fumes of Br⁻.

DIALLYL ETHER HR: 3
CAS: 557-40-4   NIOSH: KN 7525000
mf: $C_6H_{10}O$ mw: 98.16

PROP: Liquid, odor of radishes, bp: 94.3°, d: 0.805, vap d: 3.38, flash p: 20°F (OC).

SYNS: ALLYLETHER * PROPENYL ETHER

THR: Poison by ingestion. A skin and eye irritant. A dangerous fire hazard when exposed to heat, flame, or oxidizing materials. See also ethers. To fight fire, use alcohol foam. Violent explosion has occurred during distillation.

DIALLYL MALEATE HR: 3
CAS: 999-21-3   NIOSH: ON 0700000
mf: $C_{10}H_{12}O_4$ mw: 196.22

PROP: Liquid; vap d: 6.6.

THR: Poison by ingestion and intraperitoneal routes. Moderately toxic by skin contact. A skin and eye irritant. When heated to decomposition it emits acrid smoke and irritating fumes.

N,N'-DIALLYLNORTOXIFERINIUM DICHLORIDE HR: 3
CAS: 15180-03-7   NIOSH: RC 9868000
mf: $C_{44}H_{50}N_4O_2 \cdot 2Cl$ mw: 737.88

SYNS: DIALLYLNORTOXIFERINE DICHLORIDE

* 4,4'-DIDEMETHYL-4,4'-DI-2-PROPENYLTOXIFE-RINE I DICHLORIDE

THR: Poison by ingestion, intraperitoneal, subcutaneous, and intravenous routes. When heated to decomposition it emits very toxic fumes of NO$_x$ and Cl⁻.

DIAMIDINE HR: 3
CAS: 140-64-7   NIOSH: CV 6500000
mf: $C_{19}H_{24}N_4O_2 \cdot C_4H_{12}O_8S_2$ mw: 592.75

SYNS: 4,4'-DIAMIDINODIPHENOXYPENTANE DI(BETA-HYDROXYETHANESULFONATE * 4,4'-DIAMIDINO-alpha,omega-DIPHENOXYPENTANE ISETHIONATE * P,P'-(PENTAMETHYLENEDIOXY)-DIBENZAMIDINE BIS(beta-HYDROXYETHANESUL-FONATE) * PENTAMIDINE DIISETHIONATE * PENTAMIDINE ISETHIONATE * USAF XR-10

THR: Poison by intraperitoneal route. Blood effects in women by intramuscular route. When heated to decomposition it emits very toxic fumes of NO$_x$ and SO$_x$.

DIAMINIDE MALEATE HR: 3
CAS: 59-33-6   NIOSH: UT 1225000
mf: $C_{17}H_{23}N_3O \cdot C_4H_4O_4$ mw: 401.51

SYNS: N-DIMETHYLAMINOETHYL-N-P-METHOXY-alpha-AMINOPYRIDINE MALEATE * 2-((2-(DIMETHYLAMINO)ETHYL)(P-METHOXY-BENZYL)AMINO)PYRIDINE BIMALEATE * 2-((2-(DIMETHYLAMINO)ETHYL)(P-METH-OXYBENZYL)AMINO)PYRIDINE MALEATE * N-P-METHOXYBENZYL-N'-N'-DIMETHYL-N-al-pha-PYRIDYLETHYLENEDIAMINE MALEATE * PYRANISAMINE MALEATE

THR: Poison by ingestion, subcutaneous, and intraperitoneal routes. When heated to decomposition it emits toxic fumes of NO$_x$.

2,6-DIAMINOACRIDINE HR: 3
CAS: 3407-94-1   NIOSH: AR 8675000
mf: $C_{13}H_{11}N_3$ mw: 209.27

SYNS: 3,7-DIAMINOACRIDINE * ACRAMINE RED * 2,6-ACRIDINEDIAMINE * DIFLAVINE (ACRIDINE)

THR: Poison by intraperitoneal and subcutaneous routes. When heated to decomposition it emits toxic fumes of NO$_x$.

3,6-DIAMINOACRIDINE SULPHATE (1:1) HR: 3
CAS: 553-30-0   NIOSH: AR 9065000
mf: $C_{13}H_{11}N_3 \cdot H_2O_4S$ mw: 307.35

SYNS: 3,6-ACRIDINEDIAMINE SULFATE (2:1)
* 3,6-DIAMINOACRIDINE BISULPHATE * 3,6-
DIAMINOACRIDINIUMMONOHYDROGEN SUPHATE
* 2,8-DIAMINOACRIDINIUM SULPHATE * 3,6-
ACRIDINEDIAMINE SULFATE (1:1) * PROFLA-
VINE HEMISULFATE

THR: An experimental carcinogen. Mutagenic
data. When heated to decomposition it emits
very toxic fumes of NO_x and SO_x.

3,6-DIAMINOACRIDINIUM   HR: 3
CAS: 92-62-6   NIOSH: AR 8670000
mf: $C_{13}H_{11}N_3$ mw: 209.27

SYNS: 2,8-DIAMINOACRIDINIUM * 3,7-DI-
AMINO-5-AZAANTHRACENE * 3,6-ACRIDINEDI-
AMINE * 2,8-DIAMINOACRIDINE (EUROPEAN)
* 3,6-DIAMINOACRIDINE * PROFLAVINE

THR: Poison by intravenous and subcutaneous
routes. An experimental carcinogen. Mutagenic
data. When heated to decomposition it emits
toxic fumes of NO_x.

2,4-DIAMINOANISOLE SUL-
PHATE   HR: 3
CAS: 39156-41-7   NIOSH: ST 2705000
mf: $C_7H_{10}N_2O \cdot xH_2O_4S$ mw: 824.75

SYNS: 2,4-DIAMINOANISOLE SULPHATE
* 2,4-DIAMINOANISOLE SULFATE
* 2,4-DIAMINO-ANISOL SULPHATE * 2,4-DI-
AMINO-1-METHOXYBENZENE * 4-METHOXY-1,
3-BENZENEDIAMINE SULFATE * 4-METHOXY-1,
3-BENZENEDIAMINE SULPHATE * 4-METHOXY-
M-PHENYLENEDIAMINE SULFATE * 4-METHOXY-
M-PHENYLENEDIAMINE SULPHATE * NCI-C01989

THR: Poison by intraperitoneal route. An experimental carcinogen. Mutagenic data. When
heated to decomposition it emits very toxic
fumes of NO_x and SO_x.

1,5-DIAMINOANTHRARUFIN   HR: 3
CAS: 145-49-3   NIOSH: CB 6480000
mf: $C_{14}H_{10}N_2O_4$ mw: 270.26

SYNS: 1,5-DIAMINO-4,8-DIHYDROXYANTHRA-
QUINONE * 4,8-DIAMINOANTHRARUFIN
* 4,8-DIAMINO-1,5-DIHYDROXYANTHRAQUINONE
* LEUCO-1,5-DIAMINO-4,8-DIHYDROXYANTHRA-
QUINONE * 1,5-DIHYDROXY-4,8-DIAMINOAN-
THRACHINON (CZECH) * 1,5-DIHYDROXY-4,8-
DIAMINOANTHRAQUINONE

THR: Poison by intravenous route. An eye irritant. Mutagenic data. When heated to decomposition it emits toxic fumes of NO_x.

3,6-DIAMINO-2,7-DIMETHYL-
ACRIDINE   HR: 3
CAS: 92-26-2   NIOSH: AR 8750000
mf: $C_{15}H_{15}N_3$ mw: 237.33

SYNS: ACRIDINE YELLOW BASE * 2,8-DI-
AMINO-3,7-DIMETHYLACRIDINE

THR: Poison by subcutaneous route. Mutagenic
data. When heated to decomposition it emits
toxic fumes of NO_x.

2,7-DIAMINO-10-ETHYL-9-PHENYLPHE-
NANTHRIDINIUM BROMIDE   HR: 3
CAS: 1239-45-8   NIOSH: SF 7950000
mf: $C_{21}H_{20}N_3 \cdot Br$ mw: 394.35

SYNS: 2,7-DIAMINO-9-PHENYL-10-ETHYLPHE-
NANTHRIDINIUM BROMIDE * 2,7-DIAMINO-9-
PHENYLPHENANTHRIDINE ETHOBROMIDE

THR: Poison by intraperitoneal route. Mutagenic data. See bromides. When heated to decomposition it emits very toxic fumes of NO_x
and Br^-.

4,6-DIAMINO-2-(5-NITRO-2-FURYL)-s-
TRIAZINE   HR: 3
CAS: 720-69-4   NIOSH: XY 6895000
mf: $C_7H_6N_6O_3$ mw: 222.19

THR: An experimental carcinogen. Mutagenic
data. When heated to decomposition it emits
toxic fumes of NO_x.

2,4-DIAMINO-6-PIPERIDINOPYRIMI-
DINE-3-OXIDE   HR: 3
CAS: 38304-91-5   NIOSH: UV 8200000
mf: $C_9H_{15}N_5O$ mw: 209.29

SYN: 6-AMINO-1,2-DIHYDRO-1-HYDROXY-2-
IMINO-4-PIPERIDINOPYRIMIDINE

THR: Poison by intravenous route. Moderately
toxic by ingestion and intraperitoneal routes.
When heated to decomposition it emits toxic
fumes of NO_x.

2,6-DIAMINOPYRIDINE   HR: 3
CAS: 141-86-6   NIOSH: US 7570000
mf: $C_5H_7N_3$ mw: 109.15

PROP: Crystals. Mp: 120.8°, bp: 285°.

THR: Poison by intravenous and intraperitoneal
routes. When heated to decomposition it emits
toxic fumes of NO_x.

2,4-DIAMINOTOLUENE DIHYDRO-CHLORIDE   HR: 3
CAS: 636-23-7   NIOSH: XT 0175000
mf: $C_7H_{10}N_2 \cdot 2ClH$ mw: 195.11

SYNS: METATOLYLENEDIAMINE DIHYDROCHLORIDE * 2,4-TOLUENEDIAMINE DIHYDROCHLORIDE

THR: Poison by intraperitoneal route. Moderately toxic by ingestion. An experimental neoplastigen. See also chlorides. When heated to decomposition it emits very toxic fumes of NO_x and HCl.

2,5-DIAMINOTOLUENE DIHYDRO-CHLORIDE   HR: 3
CAS: 615-45-2   NIOSH: XT 0350000
mf: $C_7H_{10}N_2 \cdot 2ClH$ mw: 195.11

SYNS: 1,4-BENZENEDIAMINE, 2-METHYL-, DIHYDROCHLORIDE (9CI) * P-TOLUENEDIAMINE DIHYDROCHLORIDE

THR: Poison by ingestion. An experimental teratogen. See also chlorides. When heated to decomposition it emits very toxic fumes of NO_x and HCl.

2,6-DIAMINOTOLUENE DIHYDRO-CHLORIDE   HR: 3
CAS: 15481-70-6   NIOSH: XT 0370000
mf: $C_7H_{10}N_2 \cdot 2ClH$ mw: 195.11

SYN: NCI-C50317

THR: An experimental tumorigen. See also chlorides. When heated to decomposition it emits very toxic fumes of NO_x and HCl.

2,5-DIAMINOTOLUENE SULFATE   HR: 3
CAS: 615-50-9   NIOSH: XT 0525000
mf: $C_7H_{10}N_2 \cdot H_2O_4S$ mw: 220.27

SYNS: 1,4-BENZENEDIAMINE, 2METHYL-, SULFATE (1:1) (9CI) * TOLUENE-2,5-DIAMINE, SULFATE (1:1) (8CI) * C.I. 76043 * P-DIAMINOTOLUENE SULFATE * 2,5-DIAMINOTOLUENE SULPHATE * 2-METHYL-1,4-BENZENEDIAMINE SULFATE * 2-METHYL-p-PHENYLENEDIAMINE SULPHATE * NCI-C01832 * P-TOLUENEDIAMINE SULFATE * 2,5-TOLUENEDIAMINE SULFATE * TOLUENE-2,5-DIAMINE SULPHATE * P-TOLUENEDIAMINE SULPHATE * TOLUYLENE-2,5-DIAMINE SULPHATE * P-TOLUYLENEDIAMINE SULPHATE * P-TOLYLENEDIAMINE SULPHATE

THR: Poison by ingestion and intraperitoneal routes. An experimental carcinogen by ingestion. See also sulfates. When heated to decomposition it emits very toxic fumes of NO_x and SO_x.

3,5-DIAMINO-s-TRIAZOLE   HR: 3
CAS: 1455-77-2   NIOSH: XZ 4535000
mf: $C_2H_5N_5$ mw: 99.12

SYNS: GUANAZOLE * NCI-C04819 * NSC 1895

THR: An experimental carcinogen by intraperitoneal route. Human white blood cell effects by intravenous route. When heated to decomposition it emits toxic fumes of NO_x.

DIAMMINEMALONATOPLATINUM (II)   HR: 3
CAS: 38780-43-7   NIOSH: TP 2485000
mf: $C_3H_8N_2O_4Pt$ mw: 331.22

THR: Poison by intraperitoneal route. Mutagenic data. See also platinum compounds. When heated to decomposition it emits toxic fumes of NO_x.

DIAMMONIUM HYDROGEN ARSENATE   HR: 3
CAS: 7784-44-3   NIOSH: CG 0850000
DOT: 9080
mf: $AsH_3O_4 \cdot 2H_3N$ mw: 176.03

PROP: White powder or crystal; mp: decomp to yield NH_3.

SYNS: DIAMMONIUM MONOHYDROGEN ARSENATE * AMMONIUM ACID ARSENATE * AMMONIUM ARSENATE, SOLID (DOT) * DIAMMONIUM ARSENATE * DIBASIC AMMONIUM ARSENATE * SECONDARY AMMONIUM ARSENATE

DOT Classification: Label: Poison

THR: A poison. See also arsenic. When heated to decomposition it emits very toxic fumes of As and NO_x.

DIAMYL AMINE   HR: 3
CAS: 2050-92-2   NIOSH: RZ 9100000
mf: $C_{10}H_{23}N$ mw: 157.34

PROP: Water white liquid, bp: 202°, flash p: 124°F, d: 0.777 @ 20°/20°, vap d: 5.42.

SYNS: DI-N-AMYLAMINE * DIPENTYLAMINE * PENTYL PENTYLAMINE

THR: Poison by inhalation, ingestion, and skin contact. A skin irritant. See also amines. Flammable when exposed to heat or flame; can react with oxidizing materials. To fight fire, use alcohol foam, foam, CO_2, dry chemical. When heated to decomposition it emits toxic fumes of NO_x.

DI-n-AMYLNITROSAMINE   HR: 3
CAS: 13256-06-9   NIOSH: JJ 1575000
mf: $C_{10}H_{22}N_2O$ mw: 186.34

SYNS: DIAMYLNITROSAMIN (GERMAN)
* DI-N-PENTYLNITROSAMINE * DIPENTYLNI-
TROSAMINE * N-NITROSODI-N-PENTYLAMINE

THR: An experimental tumorigen by ingestion and subcutaneous routes. Mutagenic data. Moderately toxic by ingestion and other routes. When heated to decomposition it emits toxic fumes of NO_x. For further information see N-Nitroso Dipentylamine Vol. 1, No. 5 of *DPIM Report*.

DIANEMYCIN   HR: 3
CAS: 35865-33-9   NIOSH: HL 552100
mf: $C_{47}H_{78}O_{14}$ mw: 867.25

THR: Poison by ingestion, intraperitoneal, and subcutaneous routes. When heated to decomposition it emits acrid smoke and irritating fumes.

DIANHYDROGALACTITOL   HR: 3
CAS: 23261-20-3   NIOSH: LW 5320000
mf: $C_6H_{10}O_4$ mw: 146.16

SYNS: 1,2:5,6-DIANHYDROGALACTITOL
* DULCITOLDIEPOXIDE * NSC 132313

THR: Poison by ingestion, intravenous, and intraperitoneal routes. Mutagenic data. When heated to decomposition it emits acrid smoke and irritating fumes.

o-DIANISIDINE   HR: 3
CAS: 119-90-4   NIOSH: DD 0875000
mf: $C_{14}H_{16}N_2O_2$ mw: 244.32

PROP: Colorless crystals; mp: 137°-138°; flash p: 403°F; vap d: 8.5.

SYNS: O-DIANISIDIN (CZECH, GERMAN)
* O-DIANISIDINA (ITALIAN) * O,O'DIANISIDINE
* 3,3'-DIMETHOXYBENZIDIN (CZECH) * 3,3'-
DIMETHOXYBENZIDINE * 3,3'-DIMETOSSIBEN-
ZODINA (ITALIAN)

THR: An experimental carcinogen. Mutagenic data. Moderately toxic by ingestion. Combustible when exposed to heat or flame. When heated to decomposition it emits toxic fumes of NO_x.

DIANISIDINE DIISOCYANATE   HR: 3
CAS: 91-93-0   NIOSH: NQ 8800000
mf: $C_{16}H_{12}N_2O_4$ mw: 296.30

SYNS: 4,4'-DIISOCYANATO-3,3'-DIMETHOXY-
1,1'-BIPHENYL * 3,3'-DIMETHOXYBENZIDINE-
4,4'-DIISOCYANATE * 3,3'-DIMETHOXY-4,4'-
BIPHENYLENE DIISOCYANATE * NCI-C02175

THR: Poison by intravenous route. An experimental carcinogen by ingestion. When heated to decomposition it emits toxic fumes of NO_x.

3,4-DIANISYL-3-HEXENE   HR: 3
CAS: 7773-34-4   NIOSH: WJ 5400000
mf: $C_{20}H_{24}O_2$ mw: 296.44

SYNS: 3,4-BIS(P-METHOXYPHENYL)-3-HEXENE
* alpha,alpha'-DIETHYL-4,4'-DIMETHOXYSTIL-
BENE * TRANS-ALPHA,ALPHA'-DIETHYL-4,4'-
DIMETHOXYSTILBENE * (E)-1,1'-(1,2-DIETH-
YL-1,2-ETHENE-DIYL)BIS(4-METHOXYBENZENE)
* DIETHYLSTILBESTROL DIMETHYL ETHER
* 4,4'-DIMETHOXY-ALPHA,BETA-DIETHYLSTIL-
BENE * STILBESTROL DIMETHYL ETHER

THR: An experimental carcinogen. When heated to decomposition it emits acrid and irritating fumes.

DIAQUODIAMMINEPLATINUM
DINITRATE   HR: 3
CAS: 13601-02-0   NIOSH: TP 2310000
mf: $H_{10}N_2O_2Pt \cdot N_2O_6$ mw: 389.23

SYN: CIS-DIAQUODIAMMINEPLATINUM(II) DINI-
TRATE

THR: Poison by intraperitoneal route. Systemic skin effects. Mutagenic data. When heated to decomposition it emits toxic fumes of NO_x.

DIATOMACEOUS EARTH   HR: 3
                         NIOSH: VV 7309000

PROP: Composed of skeletons of small aquatic plants related to algae and contains as much as 88% amorphous silica.

SYNS: DIATOMITE * INFUSORIAL EARTH
* KIESELGUHR

THR: Poison by inhalation and ingestion. The dust may cause fibrosis of the lungs. Roasting or calcining produces cristobalite, and tridymite, thus increasing the fibrogenicity of the material. See silica (various forms).

DIAZIDE   HR: 3
CAS: 333-41-5   NIOSH: TF 3325000

DOT: 2783
mf: $C_{12}H_{21}N_2O_3PS$ mw: 304.38

PROP: Liquid with faint ester-like odor, miscible in organic solvents, bp: 84° @ 0.002 mm, d: 1.116 @ 20°/4°.

SYNS: O,O-DIAETHYL-O-(2-ISOPROPYL-4-METHYL-PYRIMIDIN-6-YL)-MONOTHIOPHOSPHAT (GERMAN) * DIAZINON * O,O-DIETHYL-O-(2-ISOPROPYL-4-METHYL-PYRIMIDIN-6-YL)MONO-THIOFOSFAAT (DUTCH) * O,O-DIETHYL-O-(2-ISOPROPYL-4-METHYL-6-PYRIMIDINYL)-PHOSPHOROTHIOATE * O,O-DIETHYL O-(2-ISOPROPYL-6-METHYL-4-PYRIMIDINYL) PHOSPHOROTHIOATE * O,O-DIETHYL-O-(2-ISO-PROPYL-4-METHYL-6-PYRIMIDYL)PHOSPHORO-THIOATE * O,O-DIETHYL-O-(2-ISOPROPYL-4-METHYL-6-PYRIMIDYL) THIONOPHOSPHATE * O,O-DIETHYL 2-ISOPROPYL-4-METHYLPYRI-MIDYL-6-THIOPHOSPHATE * DIETHYL 4-(2-ISO-PROPYL-6-METHYLPYRIMIDINYL)PHOSPHOROTHIO-NATE * O,O-DIETIL-O-(2-ISOPROPIL-4-METIL-PIRIMIDIN-6-IL)-MONOTIOFOSFATO(ITALIAN) * O,O-DIETHYL O-6-METHYL-2-ISOPROPYL-4-PYRIMIDINYL PHOSPHOROTHIOATE * ENT 19,507 * O-2-ISOPROPYL-4-METHYLPYRIMI-DYL-O,O-DIETHYL PHOSPHOROTHIOATE * ISO-PROPYLMETHYLPYRIMIDYL DIETHYL THIOPHOS-PHATE * NCI-C08673 * THIOPHOSPHATE DE O,O-DIETHYLE ET DE O-2-ISOPROPYL-4-METHYL-6-PYRIMIDYLE (FRENCH)

DOT Classification: ORM-A, Label: None

THR: Poison by ingestion, skin contact, subcutaneous, intravenous, intraperitoneal, and possibly other routes. Central nervous system effects. A skin and eye irritant. Mutagenic data. When heated to decomposition it emits very toxic fumes of NO_x, PO_x and SO_x.

DIAZOACETIC ESTER   HR: 3
CAS: 623-73-4   NIOSH: AG 5775000
mf: $C_4H_6N_2O_2$ mw: 114.12

SYNS: DIAZOACETIC ACID, ETHYL ESTER * DIAZOESSIGSAEURE-AETHYLESTER (GERMAN) * ETHOXYCARBONYLDIAZO-METHANE * ETHYL DIAZOACETATE

THR: Poison by ingestion and intravenous routes. An experimental tumorigen by ingestion and intravenous routes. See also esters. When heated to decomposition it emits toxic fumes of NO_x. Can explode.

N-(DIAZOACETYL)GLYCINE HYDRAZINE   HR: 3
CAS: 820-75-7   NIOSH: MB 9450000
mf: $C_4H_7N_5O_2$ mw: 157.16

SYNS: N-DIAZOACETILGLICINA-IDRAZIDE (ITAL-IAN) * DIAZOACETYLGLYCINE HYDRAZIDE * N-DIAZOACETYL GLYCYLHYDRAZIDE * NSC-58404

THR: An experimental carcinogen and neoplastigen. Mutagenic data. Moderately toxic (acute). When heated to decomposition it emits toxic fumes including NO_x.

DIAZOMETHANE   HR: 3
CAS: 334-88-3   NIOSH: PA 7000000
mf: CH_2N_2 mw: 42.05

PROP: Yellow gas at ordinary temp; mp: −145°, bp: −23°, d: 1.45.

SYNS: DIAZIRINE * AZIMETHYLENE

OSHA PEL: TWA 0.2 ppm

THR: An experimental carcinogen. A poison irritant by inhalation. A powerful allergen. It can cause pulmonary edema and frequently causes hypersensitivity leading to asthmatic symptoms. Highly explosive when shocked, exposed to heat or by chemical reaction. Undiluted liquid or gas may explode on contact with alkali metals; rough surfaces; heat (200°). When heated to decomposition or on contact with acid or acid fumes, emits highly toxic fumes of NO_x. May explode. Incompatible with alkali metals; calcium sulfate. For further information see Vol. 1, No. 3 of *DPIM Report*.

DIAZOURACIL   HR: 3
CAS: 2435-76-9   NIOSH: YQ 9630000
mf: $C_4H_2N_4O_2$ mw: 138.10

SYN: 2,4-DIOSSI-5-DIAZOPIRIMIDINA (ITALIAN)

THR: Poison by subcutaneous route. When heated to decomposition it emits toxic fumes of NO_x.

DIBEKACIN   HR: 3
CAS: 34493-98-6   NIOSH: WK 1985000
mf: $C_{18}H_{37}N_5O_8$ mw: 451.60

SYN: 3′,4′-DIDEOXYKANAMYCIN B

THR: Poison by intramuscular and intravenous route. When heated to decomposition it emits toxic fumes of NO_x.

DIBENAMINE HYDROCHLOR-IDE   HR: 3
CAS: 55-43-6   NIOSH: HQ 6825000
mf: $C_{16}H_{18}ClN \cdot ClH$ mw: 296.26

SYNS: N-(2-CHLOROETHYL)DIBENZYLAMINE HY-DROCHLORIDE * N,N-DIBENZYLAMINOETHYL CHLORIDE HYDROCHLORIDE * DIBENZYLCHLOR-ETHYLAMINE HYDROCHLORIDE * N,N-DIBEN-ZYL-beta-CHLOROETHYLAMINE HYDROCHLORIDE * N,N-DIBENZYL-2-CHLOROETHYLAMINE HYDRO-CHLORIDE

THR: Poison by intraperitoneal route. Psycho-tropic effects by intravenous route. When heated to decomposition it emits very toxic fumes of Cl^-, NO_x, and HCl.

DIBENZ(a,j)ACEANTHRYLENE   HR: 3
CAS: 203-20-3   NIOSH: HM 9800000
mf: $C_{24}H_{14}$ mw: 302.38

SYN: 15,16-BENZDEHYDROCHOLANTHRENE

THR: Poison by intravenous route. An experi-mental tumorigen. When heated to decomposi-tion it emits acrid smoke and irritating fumes.

13H-DIBENZ(bc,j)ACEANTHRY-LENE   HR: 3
CAS: 201-42-3   NIOSH: HN 0175000
mf: $C_{23}H_{14}$ mw: 290.37

SYNS: 13H-ACENAPHTHO(1,8-AB)PHENAN-THRENE * 1′,9-METHYLENE-1,2:5,6-DIBENZ-ANTHRACENE

THR: An experimental carcinogen. When heated to decomposition it emits acrid smoke and irritating fumes.

DIBENZACEPIN   HR: 3
CAS: 1977-10-2   NIOSH: HQ 4025000
mf: $C_{18}H_{18}ClN_3O$ mw: 327.84

SYNS: 2-CHLORO-11-(4-METHYL--1-PIPERAZINYL)-DIBENZO(B,F)(1,4)OXOAZEPINE * DIBENZOAZEPINE

THR: Poison by ingestion, intraperitoneal, sub-cutaneous, and intravenous routes. When heated to decomposition it emits very toxic fumes of Cl^- and NO_x.

DIBENZ(a,h)ACRIDINE   HR: 3
CAS: 226-36-8   NIOSH: HN 0875000
mf: $C_{21}H_{13}N$ mw: 279.35

SYNS: 1,2,5,6-DIBENZACRIDINE * 1,2,5,6-DI-NAPHTHACRIDINE

THR: An experimental carcinogen and tumori-gen. When heated to decomposition it emits toxic fumes of NO_x.

DIBENZ(a,j)ACRIDINE   HR: 3
CAS: 224-42-0   NIOSH: HN 1050000
mf: $C_{21}H_{13}N$ mw: 279.35

SYNS: 7-AZADIBENZ(A,J)ANTHRACENE * 1,2,7,8-DIBENZACRIDINE * 3,4,5,6-DIBEN-ZACRIDINE * 3,4,6,7-DINAPHTHACRIDINE

THR: An experimental carcinogen and tumori-gen. Mutagenic data. When heated to decompo-sition it emits toxic fumes of NO_x.

DIBENZ(c,h)ACRIDINE   HR: 3
CAS: 224-53-3   NIOSH: HN 1225000
mf: $C_{21}H_{13}N$ mw: 279.35

SYNS: 14-AZADIBENZ(A,J)ANTHRACENE * 1,2,7,8-DIBENZACRIDINE (FRENCH) * 3,4:5,6-DIBENZACRIDINE

THR: An experimental tumorigen. Mutagenic data. When heated to decomposition it emits toxic fumes of NO_x.

DIBENZ(a,j)ACRIDINE METHO-SULFATE   HR: 3
CAS: 63918-83-2   NIOSH: HN 1925000
mf: $C_{21}H_{13}N \cdot C_2H_6O_4S$ mw: 405.49

SYN: 3,4:5,6-DIBENZACRIDINE METHOSULFATE

THR: An experimental carcinogen. See also sul-fonates. When heated to decomposition it emits very toxic fumes of SO_x and NO_x.

DIBENZ(a,h)ANTHRACENE   HR: 3
CAS: 53-70-3   NIOSH: HN 2625000
mf: $C_{22}H_{14}$ mw: 278.36

SYNS: 1,2,5,6-DIBENZANTHRACEEN (DUTCH) * 1,2:5,6-DIBENZANTHRACENE * DIBEN-ZO(A,H)ANTHRACENE

THR: Poison by intravenous route. An experi-mental carcinogen and neoplastigen by most routes. Mutagenic data. When heated to decom-position it emits acrid smoke and irritating fumes. For further information see Vol. 4, No. 6 of *DPIM Report*.

DIBENZ(a,j)ANTHRACENE   HR: 3
CAS: 224-41-9   NIOSH: HN 2800000
mf: $C_{22}H_{14}$ mw: 278.36

SYN: 1,2:7,8-DIBENZANTHRACENE

THR: An experimental tumorigen. Mutagenic data. When heated to decomposition it emits acrid smoke and irritating fumes.

5H-DIBENZ(b,f)AZEPINE-5-CARBOXAMIDE   HR: 3
CAS: 298-46-4   NIOSH: HN 8225000
mf: $C_{15}H_{12}N_2O$ mw: 236.29

SYNS: 5-CARBAMOYL-5H-DIBENZ(B,F)AZEPINE
* 5-CARBAMOYL-5H-DIBENZO(B,F)AZEPINE
* 5-CARBAMOYLDIBENZO(B,F)AZEPINE
* 5-CARBAMYLDIBENZO(B,F)AZEPINE
* 5-CARBAMYL-5H-DIBENZO(B,F)AZEPINE

THR: Poison by intraperitoneal route. Moderately toxic by ingestion. Skin, red blood cell, and other systemic effects by ingestion. When heated to decomposition it emits toxic fumes of NO_x.

DIBENZEPIN   HR: 3
CAS: 4498-32-2   NIOSH: HP 2500000
mf: $C_{18}H_{21}N_3O$ mw: 295.42

SYN: 10-(2-(DIMETHYLAMINO)ETHYL)-5,10-DI-HYDRO-5-METHYL-11H-DIBENZO(B,E)(1,4)-DIAZEPIN-11-ONE

THR: Poison to humans by ingestion. Poison by intraperitoneal, intravenous, and subcutaneous routes. When heated to decomposition it emits toxic fumes of NO_x.

DIBENZEPINE HYDROCHLORIDE   HR: 3
CAS: 315-80-0   NIOSH: HP 2625000
mf: $C_{18}H_{21}N_3O \cdot ClH$ mw: 331.88

SYNS: DIBENZEPIN HYDROCHLORIDE * 5-METHYL-10-BETA-DIMETHYLAMINOAETHYL-10,11-DIHYDRO-11-OXO-5-DIBENZO(B,E)(1,4)DIAZEPIN

THR: Poison by ingestion, intravenous, and intraperitoneal routes. When heated to decomposition it emits very toxic fumes of HCl and NO_x.

7H-DIBENZO(c,g)CARBAZOLE   HR: 3
CAS: 194-59-2   NIOSH: HO 5600000
mf: $C_{20}H_{13}N$ mw: 267.34

PROP: Needles, mp: 158°.

SYNS: 3,4,5,6-DIBENZCARBAZOLE * 3,4,5,6-DINAPHTHACARBAZOLE

THR: Poison by intraperitoneal route. An experimental neoplastigen and tumorigen by most

routes. When heated to decomposition it emits toxic fumes of NO_x.

DIBENZO(b,def)CHRYSENE   HR: 3
CAS: 189-64-0   NIOSH: HO 5775000
mf: $C_{24}H_{14}$ mw: 302.38

SYNS: DIBENZO(A,H)PYRENE * 1,2,6,7-DI-BENZOPYRENE * 3,4,8,9-DIBENZOPYRENE
* 3,4,8,9-DIBENZPYRENE

THR: An experimental carcinogen by skin contact. When heated to decomposition it emits acrid smoke and irritating fumes. For further information see dibenzo(a,h)pyrene, Vol. 5, No. 2 of *DPIM Report*.

DIBENZO(def,p)CHRYSENE   HR: 3
CAS: 191-30-0   NIOSH: HO 6125000
mf: $C_{24}H_{14}$ mw: 302.38

SYNS: DIBENZO(A,D)PYRENE * DIBENZO(A,L)PYRENE * 1,2:3,4-DIBENZOPYRENE
* 1,2,9,10-DIBENZOPYRENE * 1,2,3,4-DI-BENZPYRENE

THR: An experimental carcinogen by skin contact. When heated to decomposition it emits acrid smoke and irritating fumes.

DIBENZO-p-DIOXIN   HR: 3
CAS: 262-12-4   NIOSH: HP 3090000
mf: $C_{12}H_8O_2$ mw: 184.20

PROP: Crystals, mp: 123°.

SYNS: DIBENZODIOXIN * DIBENZO(1,4)DIOXIN
* DIBENZO(B,E)(1,4)DIOXIN * DIPHENYLENE DIOXIDE * NCI-C03656 * OXANTHRENE
* PHENODIOXIN

THR: An experimental carcinogen by skin contact. When heated to decomposition it emits acrid smoke and irritating fumes.

13H-DIBENZO(a,g)FLUORENE   HR: 3
CAS: 207-83-0   NIOSH: HP 3850000
mf: $C_{21}H_{14}$ mw: 266.35

SYN: 1,2,5,6-DIBENZOFLUORENE

THR: An experimental carcinogen by skin contact. When heated to decomposition it emits acrid smoke and irritating fumes.

2-DIBENZOFURANAMINE   HR: 3
CAS: 3693-22-9   NIOSH: HP 4550000
mf: $C_{12}H_9NO$ mw: 183.22

SYNS: 3-AMINODIBENZOFURAN * 2-AMINODI-PHENYLENE OXIDE

THR: An experimental carcinogen by ingestion. When heated to decomposition it emits toxic fumes of NO_x.

DIBENZOSUBERONE OXIME   HR: 3
CAS: 1785-74-6   NIOSH: HP 1300000
mf: $C_{15}H_{13}NO$ mw: 223.29

SYN: 10,11-DIHYDRO-5H-DIBENZO(A,D)-CYCLOHEPTEN-5-ONE OXIME

THR: Poison by intraperitoneal and intravenous routes. When heated to decomposition it emits toxic fumes of NO_x.

DIBENZ(b,f)(1,4)OXAZEPINE   HR: 3
CAS: 257-07-8   NIOSH: HQ 3950000
mf: $C_{13}H_9NO$ mw: 195.23

SYN: EA 3547

THR: Poison by inhalation and intravenous routes. Moderately toxic by ingestion. A human skin and eye irritant. When heated to decomposition it emits toxic fumes of NO_x.

DIBENZYLINE HYDROCHLOR-IDE   HR: 3
CAS: 63-92-3   NIOSH: DP 3750000
mf: $C_{18}H_{22}ClNO \cdot ClH$ mw: 340.32

SYNS: N-PHENOXYISOPROPYL-N-BENZYL-beta-CHLOROETHYLAMINE HYDROCHLORIDE * 2-(N-BENZYL-2-CHLOROETHYLAMINO)-1-PHENOXYPRO-PANE HYDROCHLORIDE * BENZYL(2-CHLORO-ETHYL)(1-METHYL-2-PHENOXYETHYL)AMINE HYDROCHLORIDE * N-(2-CHLOROETHYL)-N-(1-METHYL-2-PHENOXYETHYL)BENZENE-METHANAMINE HYDROCHLORIDE * N-(2-CHLO-ROETHYL)-N-(1-METHYL-2-PHENOXYETHYL) BENZYLAMINE, HYDROCHLORIDE * NCI-C01661 * N-2-PHENOXYISOPROPYL-N-BENZYL-CHLORO-ETHYLAMINE HYDROCHLORIDE

THR: Poison by intraperitoneal and subcutaneous routes. An experimental carcinogen. When heated to decomposition it emits very toxic fumes of NO_x and Cl^-.

DIBENZYLMERCURY   HR: 3
CAS: 780-24-5   NIOSH: OW 2070000
mf: $C_{14}H_{14}Hg$ mw: 382.87

PROP: Colorless crystals, sol in organic solvents.

THR: Poison by intravenous route. See also mercury compounds, organic. When heated to decomposition it emits toxic fumes of Hg.

DIBROMDULCITOL   HR: 3
CAS: 10318-26-0   NIOSH: LW 5425000
mf: $C_6H_{12}Br_2O_4$ mw: 308.00

SYNS: 1,6-DIBROMO-1,6-DIDEOXYDULCITOL
* 1,6-DIBROMO-1,6-DIDEOXYGALACTITOL
* 1,6-DIBROMODULCITOL * NCI-C04795
* NSC-104800

THR: Poison by ingestion and intraperitoneal routes. Gastrointestinal system effects by ingestion. An experimental carcinogen. When heated to decomposition it emits very toxic fumes of Br^-.

2,2'-DIBROMOBIACETYL   HR: 3
CAS: 6305-43-7   NIOSH: EK 2850000
mf: $C_4H_4Br_2O_2$ mw: 243.90

SYN: alpha,alpha'-DIBROMOBIACETYL

THR: Poison by intravenous and intraperitoneal routes. When heated to decomposition it emits toxic fumes of Br^-.

1,4-DIBROMO-2-BUTENE   HR: 3
CAS: 6974-12-5   NIOSH: EM 4550000
mf: $C_4H_6Br_2$ mw: 213.92

SYN: TL 80

THR: Poison by ingestion, inhalation, and intraperitoneal routes. A skin and eye irritant. When heated to decomposition it emits toxic fumes of Br^-.

DIBROMOCHLOROPROPANE   HR: 3
CAS: 96-12-8   NIOSH: TX 8750000
mf: $C_3H_5Br_2Cl$ mw: 236.35

PROP: Bp: 196°, flash p: 170°F (TOC).

SYNS: 1-CHLORO-2,3-DIBROMOPROPANE
* 3-CHLORO-1,2-DIBROMOPROPANE * NEMA-GON * DIBROMCHLORPROPAN (GERMAN)
* 1,2-DIBROM-3-CHLOR-PROPAN (GERMAN)
* 1,2-DIBROMO-3-CHLOROPROPANE * 1,2-DI-BROMO-3-CLORO-PROPANO (ITALIAN) * 1,2-DIBROOM-3-CHLOORPROPAAN (DUTCH) * NCI-C00500 * NEMATOCIDE

OSHA PEL: TWA 1 ppb

THR: Poison by ingestion and inhalation. Moderately toxic by skin contact. Mutagenic data. A skin and eye irritant. An experimental carcinogen and tumorigen. Narcotic in high concentrations. Has been implicated in causing human male sterility in factory workers. Dangerous;

see also bromides and chlorides for disaster hazard. For further information see 1, 2-Dibromo-3-chloropropane, Vol. 1, No. 3 of *DPIM Report*.

2,6-DIBROMO-4-CYANOPHENYL OCTANOATE   HR: 3
CAS: 1689-99-2   NIOSH: DI 3325000
mf: C$_{15}$H$_{17}$Br$_2$NO$_2$ mw: 403.15

SYNS: NCR CE EE DOV7 * BROMOXYNIL OCTANOATE * 3,5-DIBROMO-4-OCTANOYLOXY-BENZONITRILE

THR: Poison by ingestion and possibly other routes. When heated to decomposition it emits very toxic fumes of NO$_x$ and Br$^-$.

DIBROMODIBUTYLSTANNANE   HR: 3
CAS: 996-08-7   NIOSH: WH 6882000
mf: C$_8$H$_{18}$Br$_2$Sn   mw: 392.77

PROP: mp: 20°

SYNS: DIBROMODIBUTYLTIN * DIBUTYL TIN DIBROMIDE

THR: Poison by ingestion. Moderately toxic by skin contact. See tin compounds. When heated to decomposition it emits toxic fumes of Br$^-$.

3,5-DIBROMO-4-HYDROXY-BENZONITRILE   HR: 3
CAS: 1689-84-5   NIOSH: DI 3150000
mf: C$_7$H$_3$Br$_2$NO   mw: 276.93

SYNS: BROMOXYNIL * 4-HYDROXY-3,5-DI-BROMOBENZONITRILE

THR: Poison by ingestion, intravenous, and possibly other routes. An herbicide. When heated to decomposition it emits highly toxic fumes of NO$_x$ and Br$^-$. For further information see bromoxynil, Vol. 2, No. 4 of *DPIM Report*.

1,6-DIBROMOMANNITOL   HR: 3
CAS: 488-41-5   NIOSH: OP 2800000
mf: C$_6$H$_{12}$Br$_2$O$_4$ mw: 308.00

SYNS: 1,6-DIBROMO-1,6-DIDEOXY-D-MANNITOL * 1,6-DIBROMO-1,6-D-DIDESOXYMANNITOL * NCI-C04762 * NSC-94100

THR: An experimental neoplastigen and teratogen. Moderately toxic by several routes. When heated to decomposition it emits toxic fumes of Br$^-$.

DIBROMOMETHANE   HR: 2
CAS: 74-95-3   NIOSH: PA 7350000
mf: CH$_2$Br$_2$ mw: 173.85

PROP: Colorless, heavy liquid, bp: 95.6°-97.4°, fp: <50°, d: 2.485 @ 25°/25°, vap d: 6.05.

SYNS: METHYLENE BROMIDE * METHYLENE DIBROMIDE

THR: Moderately toxic by subcutaneous route. Mutagenic data. When heated to decomposition it emits toxic fumes of Br$^-$.

3,4-DIBROMO NITROSO PIPERIDINE   HR: 3
CAS: 57541-73-8   NIOSH: TM 7054000
mf: C$_5$H$_8$Br$_2$N$_2$O   mw: 271.97

SYN: N-NITROSO-3,4-DIBROMOPIPERIDINE

THR: An experimental carcinogen. Mutagenic data. When heated to decomposition it emits very toxic fumes of Br$^-$ and NO$_x$.

DIBROMOPHENYLARSINE   HR: 3
CAS: 696-24-2   NIOSH: CH 2700000
mf: C$_6$H$_5$AsBr$_2$ mw: 311.85

SYN: PHENYLDIBROMOARSINE

THR: Poison by intravenous and dermal routes. See also arsenic compounds. When heated to decomposition it emits very toxic fumes of As and Br$^-$.

2,3-DIBROMOPROPION-ALDEHYDE   HR: 3
CAS: 5221-17-0   NIOSH: UE 0800000
mf: C$_3$H$_4$Br$_2$O   mw: 215.89

THR: Poison by intravenous route. Mutagenic data. When heated to decomposition it emits toxic fumes of Br$^-$.

DIBUCAINE   HR: 3
CAS: 85-79-0   NIOSH: GD 3150000
mf: C$_{20}$H$_{29}$N$_3$O$_2$ mw: 343.52

SYNS: 2-BUTOXY-N-(beta-DIETHYLAMINO-ETHYL)CINCHONINAMIDE * 2-BUTOXY-N-(2-(DIETHYLAMINO)ETHYL)CINCHONINAMIDE * alpha-BUTYLOXYCINCHONINIC ACID DIETHYL ETHYLENEDIAMINE * 2-BUTOXYQUINOLINE-4-CARBOXYLIC ACID DIETHYLAMINOETHYLAMIDE * N-(2-(DIETHYLAMINO)ETHYL)-2-BUTOXYCIN-CHONINAMIDE * NUPERCAINE

THR: Poison to humans by ingestion. Poison by subcutaneous, intravenous, intraperitoneal, and other routes. When heated to decomposition it emits toxic fumes of NO_x.

N-DIBUTYLAMINE   HR: 3
CAS: 111-92-2   NIOSH: HR 7780000
mf: $C_8H_{19}N$ mw: 129.28

PROP: Liquid. Mp: $-59°$; bp: 159°, flash p: 125°F (OC), d: 0.76, vap d: 4.46, vap press: 2 mm @ 20°.

SYN: DI-N-BUTYLAMINE

THR: Poison by inhalation and ingestion. Moderately toxic by skin contact. A skin and eye irritant. Mutagenic data. Flammable when exposed to heat or flame; can react with oxidizing materials. To fight fire, use alcohol foam, foam, CO_2, dry chemical. When heated to decomposition it emits toxic fumes of NO_x.

DIBUTYLAMINOETHANOL   HR: 3
CAS: 102-81-8   NIOSH: KK 3850000
DOT: 2873
mf: $C_{10}H_{23}NO$ mw: 173.34

PROP: Liquid, bp: 222°, flash p: 220°F (OC), d: 0.85, vap d: 6.0.

SYNS: 2-N-DIBUTYLAMINOETHANOL * BETA-N-DIBUTYLAMINOETHYL ALCOHOL * N,N-DIBU-TYLETHANOLAMINE * N,N-DIBUTYL-N-(2-HY-DROXYETHYL)AMINE

DOT Classification: Poison B

THR: Poison by intraperitoneal route. Moderately toxic by ingestion and skin contact. A severe eye and skin irritant. Flammable when exposed to heat or flame; can react with oxidizing materials. To fight fire, use CO_2, dry chemical. When heated to decomposition it emits toxic fumes of NO_x.

DIBUTYLBIS(LAUROYLOXY)-STANNANE   HR: 3
CAS: 77-58-7   NIOSH: WH 7000000
mf: $C_{32}H_{64}O_4Sn$ mw: 631.65

PROP: Pale yellow liquid to colorless solid (when pure). mp: 23°, bp: non-distillable @ 10 mm, flash p: 455°F (OC), d: 1.066 @ 20°/20°, vap d: 21.8.

SYNS: BIS(LAUROYLOXY)DI(N-BUTYL)STANNANE * DIBUTYLBIS(LAUROYLOXY)TIN * DIBUTYLTIN DILAURATE * DIBUTYL-ZINN-DILAURAT (GER-MAN) * LAUDRAN DI-N-BUTYLCINICITY (CZECH) * LAURIC ACID, DIBUTYLSTANNYLENE SALT * TIN DIBUTYL DILAURATE

THR: Poison by ingestion and intraperitoneal routes. A skin and eye irritant. Avoid the vapor produced by heating. Combustible when exposed to heat or flame; reacts with oxidizers.

DIBUTYL BUTANEPHOSPHO-NATE   HR: 3
CAS: 78-46-6   NIOSH: SZ 7000000
mf: $C_{12}H_{27}O_3P$ mw: 250.36

PROP: Colorless liquid, mild odor, bp: 128° @ 2.5 mm, flash p: 311° (COC), d: 8.62.

SYN: DIBUTYL BUTYLPHOSPHONATE

THR: Poison by intraperitoneal and intravenous routes. Flammable when exposed to heat or flame. When heated to decomposition it emits highly toxic fumes of oxides of phosphorus; it can react vigorously with oxidizing materials. To fight fire, use foam, CO_2, dry chemical.

DI-n-BUTYL-CARBAMYLCHOLINE SULPHATE   HR: 3
CAS: 532-49-0   NIOSH: BQ 5775000
mf: $C_{30}H_{66}N_4O_4 \cdot O_4S$ mw: 643.06

SYNS: (2-DIBUTYLCARBAMYLOXYETHYL)-DI-METHYLETHYLAMMONIUM SULFATE * DIMETH-YL-ETHYL-BETA-HYDROXYETHYL-AMMONIUM-SULFATE-DI-N-BUTYL-CARBAMATE * DIMETHYLETHYL-BETA-HYDROXYETHYLAMMO-NIUM SULFATE DIBUTYLURETHAN

THR: Poison by intraperitoneal and subcutaneous routes. See also sulfates. When heated to decomposition it emits very toxic fumes of NO_x and SO_x.

DI-tert-BUTYL CHROMATE   HR: 3
mf: $C_8H_{18}CrO_4$ mw: 231.67

THR: A poison. Incompatible with tert-butanol; chromium trioxide. See also chromium compounds.

DIBUTYLDICHLOROGERMANE   HR: 3
CAS: 4593-81-1   NIOSH: LY 5020000
mf: $C_8H_{18}Cl_2Ge$ mw: 257.75

THR: Poison by intraperitoneal route. See also germanium compounds. When heated to decomposition it emits very toxic fumes of Cl^-.

DIBUTYLDICHLOROSTANNANE HR: 3
CAS: 683-18-1 NIOSH: WH 7100000
mf: $C_8H_{18}Cl_2Sn$ mw: 303.85

PROP: White, crystalline solid. Mp: 43°, bp: 135° @ 10 mm, flash p: 335°F (OC), d: 1.36 @ 50°, vap press: 2 mm @ 100°, vap d: 10.5.

SYNS: DIBUTYLDICHLOROTIN * DI-N-BUTYLTIN DICHLORIDE * DICHLORODIBUTYLSTANNANE * DICHLORODIBUTYLTIN * CHLORID DI-N-BUTYLCINICITY (CZECH) * DIBUTYLTIN DICHLORIDE * DI-N-BUTYL-ZINN-DICHLORID (GERMAN)

OSHA PEL: TWA 100 ug(Sn)/m^3 (skin)

THR: Poison by ingestion, intravenous, intraperitoneal, and possibly other routes. Moderately toxic by skin contact. See also tin compounds. Combustible when exposed to heat or flame. A dangerous material; emits highly toxic fumes of HCl; will react with water or steam to produce heat and toxic fumes; can react vigorously with oxidizing materials. To fight fire, use water, foam, CO_2, dry chemical.

DIBUTYL(DIFORMYLOXY)-STANNANE HR: 3
CAS: 7392-96-3 NIOSH: WH 7125500
mf: $C_{10}H_{20}O_4Sn$ mw: 322.99

SYN: MRAVENCAN DI-N-BUTYLCINICITY (CZECH)

THR: Poison by ingestion. A severe skin and eye irritant. When heated to decomposition it emits acrid, toxic and irritating fumes.

DIBUTYL DIPENTANOYLOXY STANNANE HR: 3
CAS: 3465-74-5 NIOSH: WH 7133000
mf: $C_{18}H_{36}O_4Sn$ mw: 435.23

SYNS: DI(PENTANOYLOXY)DIBUTYLSTANNANE * VALERAN DI-N-BUTYLCINICITY (CZECH)

THR: Poison by ingestion. A severe skin and eye irritant. See also tin compounds. When heated to decomposition it emits acrid, toxic and irritating fumes.

N,N-DI-sec-BUTYL DITHIOOX-AMIDE HR: 3
CAS: 26818-53-1 NIOSH: RO 9275000
mf: $C_{10}H_{20}N_2S_2$ mw: 232.44

THR: Poison by intravenous route. When heated to decomposition it emits very toxic fumes of NO_x and SO_x.

DIBUTYL ESTER SULFURIC ACID HR: 3
CAS: 625-22-9 NIOSH: WS 7700000
mf: $C_8H_{18}O_4S$ mw: 210.32

SYNS: DIBUTYL SULFATE * DI-N-BUTYLSULFAT (GERMAN)

THR: Poison by ingestion. An experimental tumorigen. See esters and sulfates. When heated to decomposition it emits toxic fumes of SO_x.

DI-sec-BUTYL FLUOROPHOSPHO-NATE HR: 3
CAS: 625-17-2 NIOSH: TE 5430000
mf: $C_8H_{18}FO_3P$ mw: 212.23

SYNS: DI-sec-BUTYLFLUOROPHOSPHATE * TL 1266

THR: Poison by inhalation. Pulmonary system effects. When heated to decomposition it emits very toxic fumes of F^- and PO_x.

N,N-DI-n-BUTYLFORMAMIDE HR: 3
CAS: 761-65-9 NIOSH: LQ 1833000
mf: $C_9H_{19}NO$ mw: 157.29

SYN: DBF

THR: Poison by intraperitoneal route. When heated to decomposition it emits toxic fumes of NO_x.

DIBUTYL FUMARATE HR: 3
CAS: 105-75-9 NIOSH: LT 1225000
mf: $C_{12}H_{20}O_4$ mw: 228.32

PROP: Colorless, clear, mobile, liquid, typical odor. Bp: 285.1°, fp: −19°, flash p: 300°F (OC), d: 0.986 @ 20°/20°, vap d: 7.88.

THR: Poison by intraperitoneal route. An eye and mucous membrane irritant. Combustible when exposed to heat or flame; can react with oxidizing materials. To fight fire, use foam, CO_2, dry chemical.

DIBUTYL LEAD DIACETATE HR: 3
CAS: 2587-84-0 NIOSH: TP 4400000
mf: $C_{12}H_{24}O_4Pb$ mw: 439.55

THR: Poison by ingestion, intraperitoneal, and intravenous routes. See also lead compounds. When heated to decomposition it emits toxic fumes of Pb.

DIBUTYL MALEATE HR: 3
CAS: 105-76-0 NIOSH: ON 0875000
mf: $C_{12}H_{20}O_4$ mw: 228.32

PROP: Liquid, mp: −85° (sets to a glass), bp: 281°, flash p: 285°F (OC), d: 0.9964 @ 20°/20°, vap d: 7.9.

SYN: 2-BUTENEDIOIC ACID, DIBUTYL ESTER

THR: Poison by intraperitoneal route. Moderately toxic by oral route. An eye and mild skin irritant. See also esters and butyl alcohol. Combustible when exposed to heat or flame; can react with oxidizing materials. To fight fire, use foam, CO_2, dry chemical, alcohol foam. When heated to decomposition it emits acrid and irritating fumes.

DIBUTYLMERCURY   HR: 3
CAS: 629-35-6   NIOSH: OW 2100000
mf: $C_8H_{18}Hg$ mw: 314.85

PROP: Liquid, bp: 105° @ 10 mm, d: 1.779, vap d: 10.8.
OSHA PEL: TWA 10 ug(Hg)/m^3; CL 40

THR: Poison by intraperitoneal route. See also mercury compounds, organic. Flammable when exposed to heat or flame. When heated to decomposition or on contact with acid or acid fumes it emits highly toxic fumes of mercury; can react vigorously with oxidizing materials.

DI-sec-BUTYLMERCURY   HR: 3
CAS: 691-88-3   NIOSH: OW 2275000
mf: $C_8H_{18}Hg$ mw: 314.85
OSHA PEL: TWA 10 ug(Hg)/m^3; CL 40

THR: Poison by intraperitoneal route. See also mercury compounds, organic. When heated to decomposition it emits toxic fumes of Hg.

DIBUTYLOXOSTANNANE   HR: 3
CAS: 818-08-6   NIOSH: WH 7175000
mf: $C_8H_{18}OSn$ mw: 248.95

PROP: White amorphous powder, mp: decomp without melting, d: 0.5, vap d: 8.6.

SYNS: DIBUTYLOXOTIN * DIBUTYLSTANNANE OXIDE * DIBUTYLTIN OXIDE * DI-N-BUTYLTIN OXIDE * DI-N-BUTYL-ZINN-OXYD (GERMAN) * KYSLICNIK DI-N-BUTYLCINICITY (CZECH)

OSHA PEL: TWA 100 ug(Sn)/m^3 (skin)

THR: Poison by ingestion and intraperitoneal routes. A skin and severe eye irritant. Flammable when exposed to flame; can react with oxidizing materials. To fight fire use dry chemical, fog, CO_2. When heated to decomposition it emits acrid, toxic and irritating fumes.

2,6-DI-sec-BUTYLPHENOL   HR: 3
CAS: 5510-99-6   NIOSH: SK 8225000
mf: $C_{14}H_{22}O$ mw: 206.36

PROP: Amber liquid; bp: 152°-165° @ 25 mm; fp: −50°; flash p: 280°F; d: 0.936 @ 25°/4°.

SYN: 2,6-DI-SEC-BUTYLFENOL (CZECH)

THR: Poison by intravenous route. Moderately toxic by ingestion. A severe skin and eye irritant. Combustible when exposed to heat or flame; can react with oxidizing materials. To fight fire, use foam, CO_2, dry chemical. When heated to decomposition it emits acrid and irritating fumes.

N,N′-DI-S-BUTYL-p-PHENYLENEDI-AMINE   HR: 3
CAS: 101-96-2   NIOSH: SS 9040000
mf: $C_{14}H_{24}N_2$ mw: 220.40

PROP: Liquid, mp: 17.8°, flash p: 285°F (OC), d: 0.94-0.95 @ 24°/24°.

SYN: TENAMENE 2

THR: Poison by ingestion and inhalation. Moderately toxic by skin contact. Corrosive to skin. A mild allergen. Symptoms are sweating, flushing, shortness of breath and slow pulse. Combustible when exposed to heat or flame; can react with oxidizing materials. To fight fire, use foam, CO_2, dry chemical. Heat decomposition emits toxic fumes of NO_x.

DI-n-BUTYL PHTHALATE   HR: 3
CAS: 84-74-2   NIOSH: TI 0875000
mf: $C_{16}H_{22}O_4$ mw: 278.38

PROP: Oily liquid, mild odor, bp: 340°, fp: −35°, flash p: 315°F (CC), d: 1.047-1.049 @ 20°/20°, autoign temp: 757°F, vap d: 9.58.

SYNS: O-BENZENEDICARBOXYLIC ACID, DIBUTYL ESTER * BENZENE-O-DICARBOXYLIC ACID DI-N-BUTYL ESTER * DIBUTYL-1,2-BENZENEDICARBOXYLATE * DIBUTYL PHTHALATE

OSHA PEL: TWA 5 mg/m^3
ACGIH TLV: TWA 5 mg/m^3
DOT Classification: ORM-E

THR: An experimental teratogen. Mutagenic data. Human systemic eye effects by ingestion. See also esters, phthalic acid and butyl alcohol. Combustible when exposed to heat or flame; can react with oxidizing materials. Violent reaction with Cl_2. To fight fire, use CO_2, dry chemical. Incompatible with chlorine. For further information see Vol. 5, No. 4 of *DPIM Report*.

DICHLORACETIC ACID   HR: 2
CAS: 79-43-6   NIOSH: AG 6125000
mf: $C_2H_2Cl_2O_2$ mw: 128.94

PROP: Colorless corrosive liquid, pungent odor, mp (a): 10°, mp (b): −4°, bp: 194°, d: 1.5634 @ 20°/4°, vap press: 1 mm @ 44.0°, vap d: 4.45.

SYNS: DICHLORETHANOIC ACID * 2,2-DICHLO-ROACETIC ACID * DICHLOROETHANOIC ACID

DOT Classification: Label: Corrosive

THR: Moderately toxic by skin contact and ingestion. It is corrosive to the skin, eyes, and mucous membranes. Dangerous; when heated to decomposition it emits highly toxic fumes of chlorides; will react with water or steam to produce toxic and corrosive fumes.

1,2-DICHLORETHANE   HR: 3
CAS: 107-06-2   NIOSH: KI 0525000
mf: $C_2H_4Cl_2$ mw: 98.96

PROP: Colorless liquid, pleasant odor, sweet taste, bp: 83.5°, ulc: 60-70, lel = 6.2%, uel = 15.9%, fp: −35.7°, flash p: 56°F, d: 1.257 @ 20°/4°, autoign temp: 775°F, vap press: 100 mm @ 29.4°, vap d: 3.35.

SYNS: AETHYLENCHLORID (GERMAN) * 1,2-BICHLOROETHANE * BICHLORURE D'ETHYLENE (FRENCH) * CHLORURE D'ETHYLENE (FRENCH) * CLORURO DI ETHENE (ITALIAN) * 1,2-DICHLOORETHAAN (DUTCH) * 1,2-DICHLOR-AETHAN (GERMAN) * 1,2-DICHLORETHANE * DICHLORO-1,2-ETHANE (FRENCH) * SYM.-DICHLOROETHANE * ALPHA,BETA-DICHLORO-ETHANE * 1,2-DICHLOROETHANE * 1,2-DICLOROETANO(ITALIAN) * ENT1,656 * ETH-YLEENDICHLORIDE (DUTCH) * ETHYLENE DICHLORIDE * FREON 150 * GLYCOL DI-CHLORIDE * NCI-C00511

OSHA PEL: TWA 50 ppm; CL 100; Pk 200/5M/3H
DOT Classification: Label: Flammable Liquid

THR: Poison by intravenous route. Moderately toxic by inhalation, ingestion, and other routes. Central nervous system and gastrointestinal system effects. It may also cause dermatitis, edema of the lungs, toxic effects on the liver and kidneys, and severe corneal effects. A strong narcotic. An experimental transplacental carcinogen and teratogen. A strong local irritant. Its smell and irritant effects warn of its presence at relatively safe concentrations. A dangerous fire hazard if exposed to heat, flame or oxidizers. Violent reaction with Al; N_2O_4; NH_3; dimethylamino-propylamine. Moderately explosive in the form of vapor when exposed to flame. Dangerous; when heated to decomposition it emits highly toxic fumes of phosgene; can react vigorously with oxidizing materials and emit vinyl chloride and HCl. To fight fire, use water, foam, CO_2, dry chemicals.

DICHLORMETHAZANONE   HR: 3
CAS: 5571-97-1   NIOSH: XJ 1225000
mf: $C_{11}H_{11}Cl_2NO_3S$ mw: 308.19

SYNS: 2-(3,4-DICHLOROPHENYL)TETRAHYDRO-3-METHYL-4H-1,3-THIAZIN-4-ONE-1,1-DIOXIDE * 2-(3,4-DICHLOROPHENYL)-3-METHYL-4-META-THIAZANONE-1,1-DIOXIDE

THR: Poison by ingestion and intraperitoneal routes. When heated to decomposition it emits very toxic fumes of Cl^-, SO_x and NO_x.

DICHLOROACETYLENE   HR: 3
CAS: 7572-29-4   NIOSH: AP 1080000
mf: C_2Cl_2 mw: 94.92

SYN: DICHLOROETHYNE

ACGIH TLV: TWA CL 0.1 ppm
DOT Classification: Forbidden

THR: Poison by inhalation. Central nervous system effects. Can be formed by thermal decomposition (>70°) from trichloroethylene. Symptoms include a disabling nausea and intense jaw pain. See chlorinated hydrocarbons, aliphatic. Strong explosive when shocked or exposed to heat or air. Highly dangerous; shock and heat will explode it; when heated to decomposition or on contact with acid or acid fumes it emits highly toxic fumes of chlorides; can react vigorously with oxidizing materials.

o-DICHLOROBENZENE   HR: 3
CAS: 955-50-1   NIOSH: CZ 4500000
mf: $C_6H_4Cl_2$ mw: 147.00

PROP: Clear liquid, mp: −17.5°, bp: 180°-183°, fp: −22°, flash p: 151°F, d: 1.307 @ 20°/20°, vap d: 5.05, autoign temp: 1198°F, lel = 2.2%, uel = 9.2%.

SYNS: o-DICHLOR BENZOL * O-DICHLOROBEN-ZENE * 1,2-DICHLOROBENZENE * NCI-c54944

OSHA PEL: TWA CL 50 ppm
ACGIH TLV: TWA CL 50 ppm
DFG MAK: 50 ppm (300 mg/m³)
DOT Classification: ORM-A, Label: None

THR: Poison by ingestion and intravenous routes. Moderately toxic by inhalation and other routes. An eye, skin, and mucous membrane irritant. Causes liver and kidney injury. An experimental carcinogen. See also benzene chloride. Flammable when exposed to heat or flame. Dangerous; see also chlorides; can react vigorously with oxidizing materials. To fight fire, use water, foam, CO₂ dry chemical. Incompatible with aluminum.

p-DICHLOROBENZENE   HR: 3
CAS: 106-46-7   NIOSH: CZ 4550000
DOT: 1592
mf: $C_6H_4Cl_2$ mw: 147.00

PROP: White crystals, penetrating odor, mp: 53°, bp: 173.4°, flash p: 150°F (CC), d: 1.4581 @ 20.5°/4°, vap press: 10 mm @ 54.8°, vap d: 5.08.

SYNS: 1,4-DICHLOORBENZEEN (DUTCH) * P-DICHLORBENZOL (GERMAN) * P-DICHLOROBENZOL * 1,4-DICHLOROBENZENE * 1,4-DICLOROBENZENE (ITALIAN) * NCI-C54955 * PARADICHLOROBENZOL

OSHA PEL: TWA 75 ppm
ACGIH TLV: TWA 75 ppm; STEL 110 ppm
DFG MAK: 75 ppm (450 mg/m³)
DOT Classification: ORM-A (IMO: Poison B)

THR: Poison by ingestion and possibly other routes. Moderately toxic by inhalation. Can cause liver injury in humans. An experimental carcinogen. Mutagenic data. Flammable when exposed to heat, flame, or oxidizers. Dangerous; can react vigorously with oxidizing materials. See also chlorides. To fight fire, use water, foam, CO₂, dry chemical. For further information see 1,4 dichlorobenzene, Vol. 4, No. 2 of *DPIM Report*.

3,4-DICHLOROBENZENEDIAZO-THIOUREA   HR: 3
CAS: 5836-73-7   NIOSH: XX 9625000
mf: $C_7H_6Cl_2N_4S$ mw: 249.13

SYNS: (3,4-DICHLOOR-FENYL-AZO)-THIOUREUM (DUTCH) * 3,4-DICHLOROBENZENE DIAZOTHIO-CARBAMID * 3,4-DICHLOROPHENYLAZO-THIOUREA * 3,4-DICHLOROPHENYL-AZO-THIOUREE (FRENCH) * (3,4-DICHLOR-PHENYL-AZO)-THIOHARNSTOFF (GERMAN) * (3,4-DICLORO-FENIL-AZO)-TIOUREA (ITALIAN)

THR: Poison by ingestion, intraperitoneal, and possibly other routes. When heated to decomposition it emits very toxic fumes of Cl^-, NO_x and SO_x.

DICHLOROBENZIDINE BASE   HR: 3
CAS: 91-94-1   NIOSH: DD 0525000
mf: $C_{12}H_{10}Cl_2N_2$ mw: 253.14

PROP: Crystals, insol in water; sol in alcohol, benzene, and glacial acetic acid, mp: 133°.

SYNS: 4,4'-DIAMINO-3,3'-DICHLOROBIPHENYL * 3,3'-DICHLORBENZIDIN (CZECH) * 3,3'-DICHLOROBENZIDINA (SPANISH) * O,O'-DICHLOROBENZIDINE * 3,3'-DICHLOROBENZIDENE * 3,3'-DICHLORO-4,4'-BIPHENYLDIAMINE * 3,3'-DICHLOROBIPHENYL-4,4'-DIAMINE * 3,3'-DICHLORO-4,4'-DIAMINOBIPHENYL

TRK: 0.1 mg/m³

THR: An experimental carcinogen and tumorigen. Mild acute toxicity by ingestion. Mutagenic data. When heated to decomposition it emits very toxic fumes of Cl^- and NO_x. For further information see 3,3'-Dichlorobenzidine, Vol. 3, No. 2 of *DPIM Report*.

3,4-DICHLOROBENZOIC ACID   HR: 3
CAS: 51-44-5   NIOSH: DG 7175000
mf: $C_7H_4Cl_2O_2$ mw: 191.01

SYNS: SYNSTIGMINE * SYNTOSTIGMIN * VAGOSTIGMIN

THR: Poison by subcutaneous route. When heated to decomposition it emits toxic fumes of Cl^-.

cis-DICHLOROBIS(PYRROLIDINE)-PLATINUM (II)   HR: 3
CAS: 38780-42-6   NIOSH: TP 2501000
mf: $C_8H_{18}Cl_2N_2Pt$ mw: 408.27

SYN: CIS-DIPYRROLIDINEDICHLOROPLATINUM (II)

THR: Poison by intraperitoneal route. An experimental carcinogen. See also platinum compounds. When heated to decomposition it emits very toxic fumes of Cl^- and NO_x.

1,4-DICHLORO-2-BUTYNE   HR: 3
CAS: 821-10-3   NIOSH: ER 9600000
mf: $C_4H_4Cl_2$ mw: 122.98

SYN: 1,4-DICHLOROBUTYNE

THR: Poison by intravenous route. When heated to decomposition it emits toxic fumes of Cl^-. Hazardous.

DICHLORO(2-CHLOROVINYL)ARSINE OXIDE   HR: 3
CAS: 333-25-5   NIOSH: CH 7300000

SYN: LEWISITE 1 OXIDE

THR: Poison by ingestion, intravenous, and subcutaneous routes. See also arsenic compounds. When heated to decomposition it emits very toxic fumes of Cl^- and As.

2,6-DICHLORO-N-CYCLOPROPYL-N-ETHYL ISONICOTINAMIDE   HR: 3
CAS: 20373-56-2   NIOSH: NR 9800000
mf: $C_{11}H_{12}Cl_2N_2O$ mw: 259.15

SYN: ABBOTT-28440

THR: Poison by ingestion and intraperitoneal route. When heated to decomposition it emits very toxic fumes of Cl^- and NO_x.

trans-DICHLORO DIAMMINE PLATINUM (II)   HR: 3
CAS: 14913-33-8   NIOSH: TP 2455000
mf: $C_{12}H_6N_2Pt$ mw: 300.07

SYN: TRANS-DIAMMINEDICHLOROPLATINUM (II)

THR: Poison by intraperitoneal route. Mutagenic data. See also platinum compounds. When heated to decomposition it emits toxic fumes of NO_x and Cl^-.

DICHLORODIETHYLSILANE   HR: 3
CAS: 1719-53-5   NIOSH: VV 3060000
mf: $C_4H_{10}Cl_2Si$ mw: 157.13

PROP: Liquid, mp: $-96°$; bp: $131.0°$, d: 1.05, vap d: 5.41; flash p: 75.2°F.

SYN: DIETHYLDICHLOROSILANE (DOT)

DOT Classification: Label: Flammable Liquid

THR: Poison by intraperitoneal route. Moderately toxic by ingestion. Corrosive to tissue. See also chlorosilanes. Dangerous fire hazard when exposed to heat, flame or oxidizers. Dangerous; will react with water or steam to produce heat and toxic and corrosive fumes; can react vigorously with oxidizing materials. See also chlorides. To fight fire, use foam, CO_2, dry chemical.

DICHLORODIETHYLSTANNANE   HR: 3
CAS: 866-55-7   NIOSH: WH 7200000
mf: $C_4H_{10}Cl_2Sn$ mw: 247.73

PROP: Water white crystal, sol; mp: 85°; bp: 220°.

SYNS: DICHLORODIETHYLTIN * DIETHYLDI-CHLOROSTANNANE * DIETHYLSTANNYL DI-CHLORIDE * DIETHYLTIN DICHLORIDE

OSHA PEL: TWA 100 ug(Sn)/m^3 (skin)

THR: Poison by ingestion and intravenous routes. See also tin compounds and chlorides. When heated to decomposition it emits toxic fumes of tin and Cl^-.

DICHLORODIFLUORO-METHANE   HR: 2
CAS: 75-71-8   NIOSH: PA 8200000
DOT: 1028
mf: CCl_2F_2 mw: 120.91

PROP: Colorless, almost odorless gas; mp: $-158°$, bp: $-29°$, vap press: 5 atm @ 16.1°.

SYNS: DICHLORODIFLUOROMETHANE (DOT) * DIFLUORODICHLOROMETHANE * DWUCHLO-RODWUFLUOROMETAN (POLISH) * FLUOROCAR-BON-12 * FREON F-12 * GENETRON 12 * HALON

OSHA PEL: TWA 1000 ppm
ACGIH TLV: TWA 1000 ppm
DFG MAK: 1000 ppm (4950 mg/m^3)
DOT Classification: Label: Nonflammable Gas

THR: Central nervous system and systemic eye effects by inhalation. Narcotic in high concentrations. Dangerous; when heated to decomposition it emits highly toxic fumes of phosgene and fluorides. Can react violently with Al.

DICHLORODIHEXYLSTANNANE   HR: 3
CAS: 2767-41-1   NIOSH: WH 7220000
mf: $C_{12}H_{26}Cl_2Sn$ mw: 359.97

SYN: DIHEXYLTIN DICHLORIDE

THR: Poison by ingestion and intravenous routes. See also tin compounds. When heated to decomposition it emits toxic fumes of tin and Cl$^-$.

(2,5-DICHLORO-3,6-DIHYDROXY-p-BEN-ZOQUINOLATO)MERCURY   HR: 3
                    NIOSH: OW 2300000
mf: $C_6Cl_2HgO_4$ mw: 407.55

SYNS: P-BENZOQUINONE, 2,5-DICHLORO-3,6-DIHYDROXY-, MERCURY SALT * (2,5-DI-CHLORO-3,6-DIHYDROXY-P-BENZOQUINONE), MERCURY SALT

THR: Poison by intravenous route. See also mercury. When heated to decomposition it emits very toxic fumes of Cl$^-$ and Hg.

5,5'-DICHLORO-2,2'-DIHYDROXY-3,3'-DINITROBIPHENYL   HR: 3
CAS: 15595-24-1   NIOSH: DV 5075000
mf: $C_{12}H_6Cl_2N_2O_6$ mw: 345.10

SYNS: 4,4'-DICHLORO-6,6'-DINITRO-O,O'-BI-PHENOL * 5,5'-DICHLORO-3,3'-DINITRO(1,1'-BIPHENYL)-2,2'-DIOL

THR: An acute poison by ingestion. An experimental teratogen. When heated to decomposition it emits very toxic fumes of Cl$^-$ and NO$_x$.

DICHLORODIMETHYLSILANE   HR: 3
CAS: 75-78-5   NIOSH: VV 3150000
mf: $C_2H_6Cl_2Si$ mw: 129.07

PROP: Liquid, mp: below −86°, bp: 70.5°, d: 1.07 @ 25°/25°, vap d: 4.45, flash p: 15.8°F; lel = 3.4%; uel = 9.5%.

SYNS: DIMETHYLDICHLOROSILANE (DOT) * DIMETHYL-DICHLORSILAN (CZECH)

DOT Classification: Label: Flammable Liquid

THR: Poison by inhalation and intraperitoneal routes. Moderately toxic by ingestion. A severe eye irritant. A dangerous fire hazard when exposed to heat, flames (sparks), or oxidants. Dangerous; see chlorides; reacts violently with water or steam to produce heat and toxic and corrosive fumes. To fight fire, use dry chemical, CO$_2$, alcohol foam.

DICHLORODIMETHYL-STANNANE   HR: 3
CAS: 753-73-1   NIOSH: WH 7245000
mf: $C_2H_6Cl_2Sn$ mw: 219.67

PROP: Solid, water-sol, mp: 90°, bp: 190°.

SYNS: DICHLORODIMETHYLTIN * DIMETHYLDI-CHLOROSTANNANE * DIMETHYLDICHLOROTIN * DIMETHYLTIN DICHLORIDE

OSHA PEL: TWA 100 ug(Sn)/m^3 (skin)

THR: Poison by ingestion, intravenous,and possibly other routes. See also tin compounds. When heated to decomposition it emits toxic fumes of tin and Cl$^-$.

DICHLORODIOCTYLSTANNANE   HR: 3
CAS: 3542-36-7   NIOSH: WH 7247000
mf: $C_{16}H_{34}Cl_2Sn$ mw: 416.09

SYNS: DIOCTYLTIN DICHLORIDE * DI-N-OC-TYLTINDICHLORIDE * DI-N-OCTYL-ZINN DI-CHLORID (GERMAN)

THR: Poison by ingestion and intravenous routes. See also tin compounds. When heated to decomposition it emits toxic fumes of Cl$^-$.

trans-2,3-DICHLORO-1,4-DIOX-ANE   HR: 3
CAS: 3883-43-0   NIOSH: JG 9800000
mf: $C_4H_6Cl_2O_2$ mw: 157.00

THR: An experimental neoplastigen. Moderately toxic by ingestion. When heated to decomposition it emits toxic fumes of Cl$^-$.

DICHLORODIPROPYL-STANNANE   HR: 3
CAS: 867-36-7   NIOSH: WH 7255000
mf: $C_6H_{14}Cl_2Sn$ mw: 275.79

PROP: Colorless crystals. Sol in organic solvents, mp: 81°.

SYNS: DICHLORODIPROPYLTIN * DIPROPYLTIN CHLORIDE * DI-N-PROPYLTIN DICHLORIDE * DIPROPYLTIN DICHLORIDE

OSHA PEL: TWA 100 ug(Sn)/m^3 (skin)

THR: Poison by ingestion. See also tin compounds and chlorides. When heated to decomposition it emits toxic fumes of Cl$^-$.

DICHLOROETHYLARSINE   HR: 3
CAS: 598-14-1   NIOSH: CH 3500000
mf: $C_2H_5AsCl_2$ mw: 174.89

PROP: Colorless liquid, fruity, biting, irr odor, mp: −65°, bp: 156° decomp, d: 1.742 @ 14°, vap press: 2.29 mm @ 21.5°, vap d: 6.03.

SYNS: ARSENIC DICHLOROETHANE * DICK (GERMAN) * ETHYLDICHLOROARSINE

THR: Deadly poison by inhalation and intravenous routes, probably by ingestion also. Severe irritant. Used as a military poison gas. See also arsenic compounds. Dangerous; on contact with acid or acid fumes it emits highly toxic fumes of arsenic and phosgene; will react with water or steam to produce toxic and corrosive fumes. Can react with oxidizing materials.

1,1-DICHLOROETHYLENE   HR: 3
CAS: 75-35-4   NIOSH: KV 9275000
mf: $C_2H_2Cl_2$ mw: 96.94

PROP: Colorless volatile liquid, bp: 31.6°, lel = 7.3%, uel = 16.0%, fp: −122°, flash p: 0°F (OC), d: 1.213 @ 20°/4°, autoign temp: 1058°F.

SYNS: CHLORURE DE VINYLIDENE (FRENCH) * 1,1-DICHLOROETHENE * 1-1-DCE * NCI-C54262 * VINYLIDENE DICHLORIDE

THR: Poison by inhalation and ingestion. An experimental carcinogen and mutagen by skin contact, inhalation, and other routes. See also vinyl chloride. A dangerous fire hazard when exposed to heat or flame. Moderately explosive in the form of gas, when exposed to heat or flame. Also can explode spontaneously; reacts violently with chlorosulfonic acid; HNO_3; oleum. Can react vigorously with oxidizing materials. To fight fire, use alcohol foam, CO_2, dry chemical. Incompatible with air; chlorotrifluoroethylene; ozone; perchloryl fluoride.

DICHLORO(ETHYLENEDIAMMINE)-
PLATINUM(II)   HR: 3
CAS: 14096-51-6   NIOSH: TP 2497100
mf: $C_2H_8Cl_2N_2Pt$ mw: 326.11

THR: Poison by intraperitoneal route. Mutagenic data. See also platinum compounds. When heated to decomposition it emits very toxic fumes of Cl^- and NO_x.

DI-2-CHLOROETHYL MALEATE   HR: 3
CAS: 63917-06-6   NIOSH: ON 1050000
mf: $C_8H_{10}Cl_2O_4$ mw: 241.08

THR: Poison by ingestion and skin contact. When heated to decomposition it emits toxic fumes of Cl^-.

2,3-DICHLORO-N-ETHYLMALEINI-
MIDE   HR: 3
CAS: 20198-77-0   NIOSH: ON 5175000
mf: $C_6H_5Cl_2NO_2$ mw: 194.02

SYN: N-ETHYLDICHLOROMALEINIMIDE

THR: Poison by intraperitoneal and intravenous routes. An experimental teratogen When heated to decomposition it emits very toxic fumes of Cl^- and NO_x.

DICHLOROETHYLPHENYL-
SILANE   HR: 3
CAS: 1125-27-5   NIOSH: VV 3270000
mf: $C_8H_{10}Cl_2Si$ mw: 205.17

PROP: Liquid.

SYN: ETHYL PHENYL DICHLOROSILANE (DOT)

DOT Classification: Label: Corrosive

THR: Poison irritant to skin, eyes, and mucous membranes. Poison by ingestion and inhalation. Dangerous; when heated to decomposition it emits highly toxic fumes of chlorides and phenol; will react with water or steam to produce toxic and corrosive fumes; can react with oxidizing materials.

DICHLOROETHYLSILANE   HR: 3
CAS: 1789-58-8   NIOSH: VV 3230000
mf: $C_2H_6Cl_2Si$ mw: 129.07

PROP: Liquid, vap d: 4.45, flash p: < 73.4°F.

SYN: ETHYL DICHLOROSILANE (DOT)

DOT Classification: Label: Flammable Liquid

THR: Poison irritant to skin, eyes, mucous membranes and via ingestion and inhalation. Dangerous fire hazard if exposed to heat, open flames or powerful oxidizers. Dangerous; when heated it emits highly toxic fumes of phosgene; will react with water or steam to produce heat and toxic and corrosive fumes. To fight fire, use foam, dry chemical, mist, spray.

DICHLOROFENTHION   HR: 3
CAS: 97-17-6   NIOSH: TF 0350000
mf: $C_{10}H_{13}Cl_2O_3PS$ mw: 315.16

PROP: A nonvolatile, residual organic phosphate nematocide and insecticide, insol in water, sol in most organic solvents, bp: 166° @ 0.1 mm, d: 1.3.

SYNS: O,O-DIAETHYL-O-2,4-DICHLOR-PHENYL-MONOTHIOPHOSPHAT (GERMAN) * DICHLOFENTHION * O-2,4-DICHLOROPHENYL-O,O-DIETHYL PHOSPHOROTHIOATE * 2,4-DICHLOR-PHENYL DIETHYL PHOSPHOROTHIONATE * O,O-DI-

ETHYL-O-(2,4-DICHLOOR-FENYL)-MONOTHIOFOS-
FAAT (DUTCH) * O,O-DIETHYL O-(2,4-DICHLO-
ROPHENYL) PHOSPHOROTHIOATE * DIETHYL
2,4-DICHLOROPHENYL PHOSPHOROTHIONATE
* O,O-DIETHYL O-2,4-DICHLOROPHENYL THIO-
PHOSPHATE * O,O-DIETIL-O-(2,4-DICLORO-
FENIL)-MONOTIOFOSFATO (ITALIAN) * ENT
17470

THR: Poison by ingestion. Mildly toxic by skin
contact. Very toxic insecticide. See also esters,
parathion. When heated to decomposition it
emits very toxic fumes of PO_x, SO_x and Cl^-.

2,3-DICHLOROHEXAFLUORO-
BUTENE-2   HR: 3
CAS: 303-04-8   NIOSH: EM 4910000
mf: $C_4Cl_2F_6$ mw: 232.94

SYN: 2,3-DICHLORO-1,1,1,4,4,4-HEXAFLUORO-
BUTENE-2

THR: Poison by inhalation. Moderately toxic
by ingestion. When heated to decomposition
it emits very toxic fumes of Cl^- and F^-.

6,7-DICHLORO-10-(3-(N-(2-HYDROXY-
ETHYL)METHYLAMINO)PROPYL)
ISOALLOXAZINE SULFATE   HR: 3
     NIOSH: NP 2315000
mf: $C_{16}H_{17}Cl_2N_5O_3 \cdot H_2O_4S$ mw: 496.36

THR: Poison by intraperitoneal, intravenous,
intramuscular, and subcutaneous routes. When
heated to decomposition it emits very toxic
fumes of SO_x, NO_x and Cl^-.

DICHLOROMALEIMIDE   HR: 3
CAS: 1193-54-0   NIOSH: ON 5150000
mf: $C_4HCl_2NO_2$ mw: 165.96

SYN: DICHLOROMALEINIMIDE

THR: Poison by intraperitoneal route. An experimental teratogen. When heated to decomposition it emits very toxic fumes of Cl^- and NO_x.

3'5'-DICHLOROMETHOTREX-
ATE   HR: 3
CAS: 528-74-5   NIOSH: MA 1250000
mf: $C_{20}H_{20}Cl_2N_8O_5$ mw: 523.38

SYNS: 3',5'-DICHLOROAMETHOPTERIN
* 3',5'-DICHLORO-4-AMINO-4-DEOXY-N(SUP
10)-METHYLPTEROGLUTAMIC ACID * N-(3,5-DI-
CHLORO-4-((2,4-DIAMINO-6-PTERIDINYL-
METHYL)METHYLAMINO)BENZOYL)GLUTAMIC
ACID * NCI-C04875 * NSC-29630

THR: An experimental carcinogen. When
heated to decomposition it emits very toxic
fumes of NO_x and Cl^-.

DICHLOROMETHYLARSINE   HR: 3
CAS: 593-89-5   NIOSH: CH 4375000
mf: CH_3AsCl_2 mw: 160.86

PROP: Colorless liquid, bp: 134.5°, fp: −59°,
flash p: >221°F, d: 1.838 @ 20°/4°, vap press:
10 mm @ 24.3°, vap d: 5.40.

SYNS: METHYLARSINE DICHLORIDE * METH-
YLDICHLOROARSINE

DOT Classification: Label: Poison Gas

THR: Poison irritant to skin, eyes, and mucous
membranes and via ingestion and inhalation.
A blistering type of military poison. It is rapidly
detoxified in the body. A moderately persistent
gas. See also chloro-vinyl arsine dichloride and
arsenic compounds. Combustible when exposed
to heat or flame. Dangerous; when heated to
decomposition or on contact with acid or acid
fumes it emits highly toxic fumes of chlorides
and arsenic; can react vigorously with oxidizing
materials. To fight fire, use water, foam, CO_2,
dry chemical.

DICHLORO-N-METHYLMALEI-
MIDE   HR: 3
CAS: 1123-61-1   NIOSH: ON 5200000
mf: $C_5H_3Cl_2NO_2$ mw: 179.99

SYNS: 2,3-DICHLORO-N-METHYLMALEIMIDE
* N-METHYLDICHLOROMALEINIMIDE

THR: Poison by intraperitoneal and intravenous
routes. An experimental teratogen. When heated
to decomposition it emits very toxic fumes of
Cl^- and NO_x.

DICHLOROMETHYLSILANE   HR: 3
CAS: 75-54-7   NIOSH: VV 3500000
mf: CH_4Cl_2Si mw: 115.04

PROP: Colorless liquid; sol in benzene, ether,
and heptane. Bp: 41°, d: 1.10 @ 27°, flash p:
−26°F.

SYNS: METHYL DICHLOROSILANE (DOT)
* METHYL-DICHLORSILAN (CZECH)

DOT Classification: Label: Flammable
     Liquid

THR: Poison by inhalation to skin, eyes, and
mucous membranes. Dangerous fire hazard

when exposed to heat or flame. To fight fire, use water, foam, CO_2, mist. When heated to decomposition it emits toxic fumes of Cl^-. Ignites spontaneously in air.

2,3-DICHLORO-1,4-NAPHTHO-QUINONE HR: 3
CAS: 117-80-6   NIOSH: QL 7525000
mf: $C_{10}H_4Cl_2O_2$ mw: 227.04

PROP: Golden yellow crystals, insol in water, mod sol in organic solvents, mp: 193°, vap d: 7.8.

SYNS: 2,3-DICHLORO-1,4-NAPHTHAQUINONE * 2,3-DICHLOR-1,4-NAPHTHOCHINON (GERMAN) * 2,3-DICHLORONAPHTHOQUINONE * 2,3-DI-CHLORO-ALPHA-NAPHTHOQUINONE * 2,3-DI-CHLORONAPHTHOQUINONE-1,4 * ENT 3,776

THR: An experimental carcinogen. Moderately toxic by ingestion. A skin, eye, and mucous membrane irritant. Large doses can cause central nervous system depression. A dangerous hazard when heated; see also chlorides.

2,4-DICHLORO-4'-NITRODIPHENYL ETHER HR: 3
CAS: 1836-75-5   NIOSH: KN 8400000
mf: $C_{12}H_7Cl_2NO_3$ mw: 284.10

SYNS: 2,4-DICHLORO-1-(4-NITROPHENOXY)BEN-ZENE * 2,4-DICHLOROPHENYL-P-NITROPHENYL ETHER * 2,4-DICHLOROPHENYL-4-NITROPHE-NYL ETHER * NCI-C00420 * NITROFENE (FRENCH)

THR: An experimental carcinogen and teratogen. Moderate acute toxicity by ingestion, inhalation, and skin contact. Severe eye irritant; skin irritant. See also ethers. When heated to decomposition it emits very toxic fumes of Cl^- and NO_x.

1,1-DICHLORO-1-NITROETHANE HR: 3
CAS: 594-72-9   NIOSH: KI 1050000
DOT: 2650
mf: $C_2H_3Cl_2NO_2$ mw: 143.96

PROP: Liquid, bp: 124°, flash p: 168°F(OC), d: 1.4153 @ 20°/20°, vap d: 4.97.

SYNS: 1,1-DICHLOOR-1-NITROEITHAAN (DUTCH) * 1,1-DICHLOR-1-NITROAETHAN (GERMAN) * DICHLORONITROETHANE * 1,1-DICLORO-1-NITROETANO (ITALIAN)

OSHA PEL: TWA CL 10 ppm
DFG MAK: 10 ppm (60 mg/m^3)
DOT Classification: ORM: Poison B

THR: Poison by ingestion. Strong irritant. Inhalation causes pulmonary edema. Flammable when exposed to heat, flame or oxidizers. Dangerous; when heated to decomposition it emits highly toxic fumes of Cl^- and NO_x; can react vigorously with oxidizing materials. To fight fire, use water, CO_2, dry chemical.

1,2-DICHLORO-3-NITRONAPHTHA-LENE HR: 3
CAS: 6240-55-7   NIOSH: QJ 3675000
mf: $C_{10}H_5Cl_2NO_2$ mw: 242.06

THR: An experimental carcinogen. When heated to decomposition it emits very toxic fumes of Cl^- and NO_x.

2,4-DICHLORO-6-NITROPHENOL HR: 3
CAS: 609-89-2   NIOSH: SL 0350000
mf: $C_6H_3Cl_2NO_3$ mw: 208.00

SYN: 2,4-DICHLOR-6-NITROFENOL (CZECH)

THR: Poison by ingestion. Severe eye irritant. When heated to decomposition it emits very toxic fumes of Cl^- and NO_x.

2',5-DICHLORO-4'-NITRO-SALICYLANILIDE HR: 3
CAS: 50-65-7   NIOSH: VN 8400000
mf: $C_{13}H_8Cl_2N_2O_4$ mw: 327.13

SYNS: 5-CHLORO-2'-CHLORO-4'-NITROSALICY-LANILIDE * 2-CHLORO-4-NITROPHENYLAMIDE-6-CHLOROSALICYLIC ACID * N-(2-CHLORO-4-NITROPHENYL)-5-CHLOROSALICYLAMIDE * 2',5-DICHLOR-4'-NITRO-SALIZYLSAEUREANI-LID (GERMAN) * 2-HYDROXY-5-CHLORO-N-(2-CHLORO-4-NITROPHENYL)BENZAMIDE

THR: Poison by intraperitoneal and intravenous routes. Moderately toxic by ingestion. When heated to decomposition it emits very toxic fumes of Cl^- and NO_x.

2',5-DICHLORO-4'-NITROSALICYLANI-LIDE, 2-AMINOETHANOL SALT HR: 3
CAS: 1420-04-8   NIOSH: VN 8575000
mf: $C_{13}H_8Cl_2N_2O_4 \cdot C_2H_7NO$ mw: 388.23

SYNS: 5-CHLORO-N-(2-CHLORO-4-NITROPHE-NYL)-2-HYDROXYBENZAMIDE COMPOUND WITH 2-

AMINOETHANOL (1:1) * 5,2'-DICHLORO-4'-NI-TROSALICYLANILIDE, ETHANOLAMINE SALT * 5,2-DICHLORO-4-NITROSALICYLIC ANILIDE-2-AMINOETHANOL SALT * 2',5-DICHLORO-4'-NI-TROSALICYLOYLANILIDE ETHANOLAMINE SALT * NCI-C00431

THR: Poison by ingestion and intraperitoneal routes. An experimental carcinogen. When heated to decomposition it emits very toxic fumes of NO_x and Cl^-.

2,2'-DICHLORO-N-NITROSODI-PROPYLAMINE   HR: 3
NIOSH: JL 9635000
mf: $C_6H_{12}Cl_2N_2O$ mw: 199.10

SYN: NITROSOBIS(2-CHLOROPROPYL)AMINE

THR: An experimental carcinogen. Mutagenic data. When heated to decomposition it emits very toxic fumes of Cl^- and NO_x.

3,4-DICHLORONITROSOPIPERI-DINE   HR: 3
CAS: 57541-72-7   NIOSH: TM 7067000
mf: $C_5H_8Cl_2N_2O$ mw: 183.05

SYN: N-NITROSO-3,4-DICHLOROPIPERIDINE

THR: An experimental tumorigen. Moderately toxic by ingestion and other routes. Mutagenic data. See also chlorinated phenols. When heated to decomposition it emits very toxic fumes of Cl^- and NO_x.

3,4-DICHLORO-N-NITROSO-PYRROLIDINE   HR: 3
CAS: 59863-59-1   NIOSH: UY 0800000
mf: $C_4H_6Cl_2N_2O$ mw: 169.02

THR: An experimental carcinogen. When heated to decomposition it emits very toxic fumes of Cl^- and NO_x.

DICHLOROPHENARSINE HYDRO-CHLORIDE   HR: 3
CAS: 536-29-8   NIOSH: SJ 5775000
mf: $C_6H_6AsCl_2NO \cdot ClH$ mw: 290.41

SYNS: 2-AMINO-4-DICHLOROARSINOPHENOL HY-DROCHLORIDE * 3-AMINO-4-HYDROXYPHENYL DICHLORARSINE HYDROCHLORIDE * DICHLORO-MAPHARSEN

THR: Poison by intravenous and other routes. Human gastrointestinal tract effects. See also arsenic compounds. When heated to decomposi-

tion it emits very toxic fumes of As, NO_x, HCl and Cl^-.

2,4-DICHLOROPHENOL   HR: 3
CAS: 120-83-2   NIOSH: SK 8575000
mf: $C_6H_4Cl_2O$ mw: 163.00

PROP: Colorless crystals, mp: 45°, bp: 210°, flash p: 237°F, d: 1.383 @ 60°/25°, vap d: 5.62, vap press: 1 mm @ 53.0°.

SYN: NCI-C55345

THR: An experimental carcinogen. Moderately toxic by ingestion, intraperitoneal, and subcutaneous routes. Combustible when exposed to heat or flame. Dangerous; when heated to decomposition or on contact with acid or acid fumes, it emits highly toxic fumes of chlorides; can react vigorously with oxidizing materials. To fight fire, use alcohol foam, foam, CO_2, dry chemical. For further information see Vol. 6, No. 4 of *DPIM Report*.

2,6-DICHLOROPHENOL   HR: 2
CAS: 87-65-0   NIOSH: SK 8750000
mf: $C_6H_4Cl_2O$ mw: 163.00

SYN: 2,6-DICHLORFENOL (CZECH)

THR: Moderately toxic by ingestion, intraperitoneal, and subcutaneous routes. A skin and eye irritant. When heated to decomposition it emits toxic fumes of Cl^-. For further information see Vol. 4, No. 5 of *DPIM Report*.

2,4-DICHLOROPHENOL BENZENE-SULFONATE   HR: 3
CAS: 97-16-5   NIOSH: SK 9100000
mf: $C_{12}H_8Cl_2O_3S$ mw: 303.16

SYNS: BENZENESULPHONIC ACID, 2,4-DICHLO-ROPHENYL ESTER * 2,4-DICHLOROPHENYL BENZENESULFONATE * 2,4-DICHLOROPHENYL BENZENESULPHONATE * 2,4-DICHLOROPHE-NYL ESTER OF BENZENESULFONIC ACID

THR: Poison by intravenous route. An experimental carcinogen. Moderately toxic by ingestion and other routes. An irritant. See esters. When heated to decomposition it emits very toxic fumes of Cl^- and SO_x.

DICHLOROPHENOXYACETIC ACID   HR: 3
CAS: 94-75-7   NIOSH: AG 6825000
DOT: 2765
mf: $C_8H_6Cl_2O_3$ mw: 221.04

PROP: White powder, mp: 141°; bp: 160° @ 0.4 mm; vap d: 7.63.

SYNS: ACIDE 2,4-DICHLORO PHENOXYACETIQUE (FRENCH) * ACIDO (2,4-DICLORO-FENOSSI)-ACETICO (ITALIAN) * 2,4-D ACID * (2,4-DICHLOOR-FENOXY)-AZIJNZUUR (DUTCH) * 2,4-DICHLORPHENOXYACETIC ACID * (2,4-DICHLOR-PHENOXY)-ESSIGSAEURE (GERMAN)

OSHA PEL: TWA 10 mg/m^3
DFG MAK: 10 mg/m^3
DOT Classification: ORM-A

THR: Poison by ingestion, subcutaneous, intravenous, and intraperitoneal routes. Moderately toxic by skin contact. Central nervous system and gastrointestinal system effects. An experimental carcinogen and teratogen. Mutagenic data. Can cause liver and kidney injury. Ingestion can cause nausea, vomiting, and central nervous system depression. When heated to decomposition it emits toxic fumes of Cl$^-$.

5,6-DICHLORO-1-PHENOXYCARBONYL-2-TRIFLUOROMETHYLBENZIMID-AZOLE   HR: 3
CAS: 14255-88-0   NIOSH: DD 6650000
mf: C$_{15}$H$_7$Cl$_2$F$_2$N$_2$O$_2$ mw: 375.14

SYNS: 5,6-DICHLORO-2-TRIFLUOROMETHYL-BENZIMIDAZOLE-1-CARBOXYLATE * ENT 27438 * PHENYL-5,6-DICHLORO-2-TRIFLUOROMETHYL-BENZIMIDAZOLE-1-CARBOXYLATE

THR: Poison by ingestion, intraperitoneal, and other routes. When heated to decomposition it emits very toxic fumes of F$^-$, Cl$^-$ and NO$_x$.

DICHLOROPHENYLARSINE   HR: 3
CAS: 696-28-6   NIOSH: CH 5425000
mf: C$_6$H$_5$AsCl$_2$ mw: 222.93

PROP: Colorless gas or liquid, changes to yellow, bp: 255°-275°, fp: −15.6°, d: 1.654 @ 20°, vap press: 0.021 mm @ 20°, vap d: 7.7.

SYNS: PHENYLARSINEDICHLORIDE * PHENYL DICHLOROARSINE

DOT Classification: Label: Poison

THR: Poison by inhalation, ingestion, and skin contact. See also arsenic. A lachrymator type of military poison gas. Dangerous; when heated to decomposition, it emits highly toxic fumes of arsenic; will react with heat, water or steam to produce corrosive fumes of Cl$^-$.

3-(3,4-DICHLOROPHENYL)-1,1-DI-METHYLUREA   HR: 3
CAS: 330-54-1   NIOSH: YS 8925000
mf: C$_9$H$_{10}$Cl$_2$N$_2$O   mw: 233.11

PROP: Crystals, low sol in water and hydrocarbon solvents, mp: 159°.

SYNS: 3-(3,4-DICHLOROPHENOL)-1,1-DIMETHYLUREA * N'-(3,4-DICHLOROPHENYL)-N,N-DIMETHYLUREA * 1-(3,4-DICHLORO-PHENYL)-3,3-DIMETHYLUREE (FRENCH) * 3-(3,4-DICHLOR-PHENYL)-1,1-DIMETHYL-HARNSTOFF (GERMAN) * 3-(3,4-DICHLOOR-FENYL)-1,1-DIMETHYLUREUM (DUTCH) * 3-(3,4-DICLORO-FENYL)-1,1-DIMETIL-UREA (ITALIAN) * 1,1-DIMETHYL-3-(3,4-DICHLORO-PHENYL)UREA * USAF P-7 * USAF XR-42

THR: An experimental teratogen and carcinogen. Moderately toxic by ingestion and intraperitoneal routes. Dangerous; when heated to decomposition emits highly toxic fumes of Cl$^-$ and NO$_x$.

cis-DICHLORO(o-PHENYLENEDI-AMINE)PLATINUM (II)   HR: 3
CAS: 38780-39-1   NIOSH: TP 2497300
mf: C$_6$H$_8$Cl$_2$N$_2$Pt   mw: 374.15

SYN: DICHLORO(1,2-PHENYLENEDIAMMINE)-PLATINUM (II)

THR: Poison by intraperitoneal route. Mutagenic data. See platinum compounds. When heated to decomposition it emits very toxic fumes of Cl$^-$ and NO$_x$.

3-(3,4-DICHLOROPHENYL)-1-METHOXY-METHYLUREA   HR: 2
CAS: 330-55-2   NIOSH: YS 9100000
mf: C$_9$H$_{10}$Cl$_2$N$_2$O$_2$ mw: 249.11

PROP: Solid, slightly sol in water, partially sol in acetone and alcohol. mp: 93°-94°.

SYNS: 3-(3,4-DICHLOOR-FENYL)-1-METHOXY-1-METHYLUREUM(DUTCH) * 3-(3,4-DICHLORO-FENIL)-1-METOSSI-1-METIL-UREA (ITALIAN) * 3-(3,4-DICHLOROPHENYL)-1-METHOXY-1-METHYLUREA * N'-(3,4-DICHLOROPHE-NYL)-N-METHOXY-N-METHYLUREA * 1-(3,4-DICHLOROPHENYL)3-METHOXY-3-METHYLUREE(FRENCH) * N-(3,4-DICHLORO-PHENYL)-N'-METHYL-N'-METHOXYUREA * 3-(3,4-DICHLOR-PHENYL)-1-METHOXY-1-

METHYL-HARNSTOFF (GERMAN) * 1-METHOXY-
1-METHYL-3-(3,4-DICHLOROPHENYL)UREA

THR: Moderately toxic by ingestion. See also
3-(p-chlorophenyl)-1,1-dimethylurea. When
heated to decomposition it emits very toxic
fumes of Cl⁻ and NO$_x$.

o-(2,4-DICHLOROPHENYL)-o-METHYL-ISOPROPYLPHOSPHORAMIDO-THIOATE   HR: 3
CAS: 299-85-4   NIOSH: TB 5075000
mf: $C_{10}H_{14}Cl_2NO_2PS$ mw: 314.18

SYNS: o-(2,4-DICHLOROPHENYL)-O-METHYL
N-ISOPROPYLPHOSPHORAMIDOTHIOATE * DOW
1329 * ENT 25647 * ISOPROPYLPHOS-
PHORAMIDOTHIOIC ACID, O-2,4-DICHLOROPHE-
NYL-O-METHYL ESTER

THR: Poison by ingestion. Moderately toxic
by skin contact. When heated to decomposition
it emits very toxic fumes of Cl⁻, NO$_x$, PO$_x$
and SO$_x$.

N-(3,5-DICHLOROPHENYL)SUCCIN-IMIDE   HR: 3
CAS: 24096-53-5   NIOSH: WN 2500000
mf: $C_{10}H_7Cl_2NO_2$ mw: 244.08

SYN: DIMETHACHLON

THR: An experimental carcinogen. Moderately
toxic by ingestion. When heated to decomposition it emits very toxic fumes of Cl⁻ and NO$_x$.

1,2-DICHLOROPROPANE   HR: 2
CAS: 78-87-5   NIOSH: TX 9625000
mf: $C_3H_6Cl_2$ mw: 112.99

PROP: Colorless liquid, bp: 96.8°, flash p:
60°F, d: 1.1593 @ 20°/20°, vap press: 40 mm
@ 19.4°, vap d: 3.9, autoign temp: 1035°F,
lel = 3.4%, uel = 14.5%.

SYNS: BICHLORURE DE PROPYLENE (FRENCH)
* alpha,beta-DICHLOROPROPANE * DWUCHLO-
ROPROPAN (POLISH) * ENT 15,406 * NCI-
C55141 * PROPYLENE DICHLORIDE * alpha,beta-PROPYLENE DICHLORIDE

OSHA PEL: TWA 75 ppm
ACGIH TLV: TWA 75 ppm; STEL 110 ppm
DFG MAK: 75 ppm (350 mg/m³)
DOT Classification: Label: Flammable
    Liquid

THR: Moderately toxic by inhalation and ingestion. Severe skin and eye irritant. Mutagenic

data. Can cause liver, kidney, and heart damage.
Can cause dermatitis. One of the more toxic
chlorinated hydrocarbons. A suggested order
of increasing toxicity is dichloromethane, trichloroethylene, carbon tetrachloride, dichloro-
propane, dichloroethane. Animals exposed to
high concentrations often showed marked visceral congestion, fatty degeneration of the liver,
kidney, and, less frequently, of the heart. They
also showed areas of coagulation and necrosis
of the liver. There was found to be a heavy
mortality among mice exposed to 400 ppm concentrations. Dangerous, when exposed to heat
or flame. Reacts violently with Al. Dangerous;
see chlorides; can react vigorously with oxidizing materials. To fight fire, use water, foam,
CO_2, dry chemical. Incompatible with aluminum; o-dichlorobenzene; 1,2-dichloroethane.

1,3-DICHLOROPROPANE   HR: 2
CAS: 142-28-9   NIOSH: TX 9660000
mf: $C_3H_6Cl_2$ mw: 112.99

PROP: Colorless liquid, bp: 125°, d: 1.201 @
15°, vap d: 3.90, flash p: 69.8°F.

SYN: TRIMETHYLENE DICHLORIDE

THR: Moderately toxic by ingestion. Mutagenic
data. See also chlorinated hydrocarbons, aliphatic, and 1,2-dichloropropane. Dangerous;
when heated to decomposition it emits highly
toxic fumes of phosgene.

1,2-DICHLOROPROPANE MIXTURE   HR: 3
CAS: 8003-19-8   NIOSH: TX 9800000

PROP: D-D Soil fumigant consists of chlorinated C_3 hydrocarbons (100%); 1,3-dichloro-
propene, 3,3-dichloropropene, 1,2-dichloropro-
pane, 2,3-dichloropropene; and related C_3
chlorinated hydrocarbons.

SYNS: 1,3-DICHLOROPROPENE * DICHLORO-
PROPANE-DICHLOROPROPENE MIXTURE * DOW-
FUME N * ENT 8,420

THR: Poison by ingestion and inhalation. Mutagenic data. Severe skin and eye irritant. When
heated to decomposition it emits toxic fumes
of Cl⁻.

1,3-DICHLORO-2-PROPANOL   HR: 3
CAS: 96-23-1   NIOSH: UB 1400000
mf: $C_3H_6Cl_2O$ mw: 128.99

PROP: Colorless liquid, ether-like odor, bp:
174°, d: 1.367 @ 20°/4°, vap press: 1 mm @

28.0°, vap d: 4.45, flash p: 165°F (OC), mp: −4°.

SYNS: DICHLOROISOPROPYL ALCOHOL
* α-DICHLOROHYDRIN

THR: Poison by inhalation. Moderately toxic by ingestion and skin contact. A skin irritant. Action may be similar to carbon tetrachloride but more irritating to mucous membranes. Flammable when exposed to heat, flame or oxidizers. Dangerous; when heated to decomposition it emits highly toxic fumes of phosgene. To fight fire, use alcohol foam, dry chemical, fog, mist, spray.

2,3-DICHLOROPROPANOL   HR: 3
CAS: 616-23-9   NIOSH: UB 1225000
mf: $C_3H_6Cl_2O$ mw: 128.99

SYNS: 1,2-DICHLORO-3-PROPANOL * 1,2-DI-CHLOROPROPANOL-3 * 2,3-DICHLORO-1-PROPA-NOL * GLYCEROL alpha,beta-DICHLOROHYDRIN

THR: Poison by intravenous route. Mutagenic data. Skin and eye irritant. See also chlorinated hydrocarbons, aromatic. When heated to decomposition it emits toxic fumes of Cl^-.

cis-1,3-DICHLOROPROPENE   HR: 3
CAS: 10061-01-5   NIOSH: UC 8325000
mf: $C_3H_4Cl_2$ mw: 110.97

SYNS: (Z)-1,3-DICHLOROPROPENE * CIS-1,3-DICHLOROPROPYLENE * 1,3-DICHLORO-1-PRO-PENE, (Z)- (9CI)

THR: An experimental neoplastigen. Mutagenic data. When heated to decomposition it emits toxic fumes of Cl^-. For further information see Vol. 6, No. 5 of *DPIM Report*.

2,3-DICHLOROPROPENE   HR: 3
CAS: 78-88-6   NIOSH: UC 8400000
mf: $C_3H_4Cl_2$ mw: 110.97

PROP: Flash p: 50°F.

SYN: 2,3-DICHLORO-1-PROPENE

THR: Poison by ingestion. Moderately toxic by inhalation and skin contact. Mutagenic data. Severe skin irritant. When heated to decomposition it emits toxic fumes of Cl^-. For further information see Vol. 6, No. 4 of *DPIM Report*.

2,3-DICHLOROPROPIONAL-DEHYDE   HR: 3
CAS: 10140-89-3   NIOSH: UE 0875000
mf: $C_3H_4Cl_2O$ mw: 126.97

PROP: Liquid: vap d: 4.4.

SYN: alpha,beta-DICHLOROPROPIONALDEHYDE

THR: Poison by ingestion, inhalation, and skin contact. Mutagenic data. Severe skin and eye irritant. When heated to decomposition it emits toxic fumes of Cl^-.

3,5-DICHLOROSALICYLIC ACID   HR: 3
CAS: 320-72-9   NIOSH: VO 2450000
mf: $C_7H_4Cl_2O_3$ mw: 207.01

SYN: USAF DO-68

THR: Poison by intraperitoneal route. When heated to decomposition it emits toxic fumes of Cl^-.

3,4-DICHLOROSULFOLANE   HR: 3
CAS: 3001-57-8   NIOSH: XN 0480000
mf: $C_4H_6Cl_2O_2S$ mw: 189.06

SYNS: 3,4-DICHLOROTETRAHYDROTHIOPHENE-1,1-DIOXIDE * DICHLOROTHIOLANE DIOXIDE

THR: Poison by intravenous and intraperitoneal routes. Moderately toxic by ingestion and skin contact. When heated to decomposition it emits very toxic fumes of Cl^- and SO_x.

DICHLOROTETRAFLUOROACE-TONE   HR: 3
CAS: 127-21-9   NIOSH: UC 1575000
mf: $C_3Cl_2F_4O$ mw: 198.93

PROP: A colorless liquid. Miscible with water and many organic solvents. bp: 45.2°, fp: $<-100°$.

SYNS: SYM-DICHLOROTETRAFLUOROACETONE * 1,3-DICHLORO-1,1,3,3-TETRAFLUORO-2-PRO-PANONE

THR: Poison by ingestion, skin contact, inhalation, and intravenous routes. When heated to decomposition it emits very toxic fumes of Cl^- and F^-.

DICHLOROTETRAFLUORO-ETHANE   HR: 2
CAS: 1320-37-2   NIOSH: KI 1100000
mf: $C_2Cl_2F_4$ mw: 170.92

PROP: Colorless gas; bp: 3.5°.

SYNS: DWUCHLOROCZTEROFLUOROETAN (POL-ISH) * FLUORANE 114 * FREON 114

OSHA PEL: TWA 1000 ppm

THR: A mildly toxic irritant; narcotic in high concentrations. An asphyxiant. Reacts violently with Al. Dangerous; see fluorides and chlorides for disaster hazard.

2,3-DICHLOROTETRAHYDRO-FURAN   HR: 3
CAS: 3511-19-1   NIOSH: LU 0525000
mf: $C_4H_6Cl_2O$ mw: 141.00

THR: An experimental carcinogen. When heated to decomposition it emits toxic fumes of Cl^-.

2,6-DICHLOROTHIOBENZAMIDE   HR: 3
CAS: 1918-13-4   NIOSH: CV3850000
mf: $C_7H_5Cl_2NS$ mw: 206.09

SYNS: CHLOROTHIAMIDE * 2,6-DICHLOROBEN-ZENECARBOTHIOAMIDE

THR: Poison by ingestion and intraperitoneal route. Moderately toxic by skin contact. When heated to decomposition it emits very toxic fumes of Cl^-, NO_x and SO_x.

DICHLOROTITANOCENE   HR: 3
CAS: 1271-19-8   NIOSH: XR 2050000
mf: $C_{10}H_{10}Cl_2Ti$ mw: 249.00

SYNS: DICHLOROBIS(ETA(SUP 5)-2,4-CYCLO-PENTADIEN-1-YLTITANIUM * DICHLORODICY-CLOPENTADIENYLTITANIUM
* DICHLORODI-PI-CYCLOPENTADIENYLTITANIUM
* DICYCLOPENTADIENYLTITANIUMDICHLORIDE
* DICYCLOPENTADIENYLDICHLOROTITANIUM
* NCI-C04502 * TITANIUM, DICHLOROBIS-(ETA(SUP 5)-2,4-CYCLOPENTADIEN-1-YL- (9CI)

THR: Poison by intravenous and intraperitoneal routes. An experimental neoplastigen and carcinogen. See also titanium compounds. When heated to decomposition it emits toxic fumes of Cl^-.

DICHLOROTRIPHENYLANTI-MONY   HR: 3
CAS: 594-31-0   NIOSH: CC 5095000
mf: $C_{18}H_{15}Cl_2Sb$ mw: 423.98

SYNS: ANTIMONY, TRIPHENYLDICHLORIDE
* DICHLOROTRIPHENYLSTIBINE * TRIPHENYL-ANTIMONY DICHLORIDE

THR: Poison by ingestion. See also antimony compounds. When heated to decomposition it emits very toxic fumes of Cl^- and Sb.

S-DICHLOROVINYL-L-CYSTEINE   HR: 3
CAS: 627-72-5   NIOSH: AY 5075000
mf: $C_5H_7Cl_2NO_2S$ mw: 216.09

SYNS: S-(1,2-DICHLOROETHYLENEYL)-L-CYSTEINE * L-3-((1,2-DICHLOROVINYL)THIO)-ALANINE

THR: Poison by intravenous and intraperitoneal routes. When heated to decomposition it emits very toxic fumes of Cl^-, NO_x and SO_x.

alpha,alpha'-DICHLORO-o-XYLENE   HR: 3
CAS: 612-12-4   NIOSH: ZE 4085000
mf: $C_8H_8Cl_2$ mw: 175.06

SYNS: 1,2-BIS(CHLOROMETHYL)BENZENE
* O-XYLYLENE DICHLORIDE

THR: Poison by intravenous route. See also chlorinated aromatic hydrocarbons. When heated to decomposition it emits toxic Cl^-.

DICOBALT EDETATE   HR: 3
CAS: 36599-65-7   NIOSH: GF 9275000
mf: $C_{10}H_{12}CoN_2O_8 \cdot Co$ mw: 406.10

SYNS: COBALTATE(2−), ((ETHYLENE-DINITRILO)TETRAACETATO(2−))-, COBALT(2+) SALT * DICOBALT EDTA * KOBALT-EDTA (GERMAN)

THR: Poison by intravenous and intraperitoneal routes. See also cobalt compounds. When heated to decomposition it emits toxic fumes of NO_x.

DICUMENE CHROMIUM   HR: 3
CAS: 12001-89-7   NIOSH: GB 5800000
mf: $C_{18}H_{24} \cdot Cr$ mw: 292.42

SYN: DICUMENYLCHROMIUM

THR: Poison by skin contact and intravenous route. Moderately toxic by ingestion. A skin and eye irritant. See also chromium compounds. When heated to decomposition it emits acrid and irritating fumes.

DI(2-CYANOETHYL)SULFIDE   HR: 3
CAS: 111-97-7   NIOSH: UG 4375000
mf: $C_6H_8N_2S$ mw: 140.22

PROP: Mp: (α) 28.65°; (β): 22.10°; d: 1.1095 @ 30°. White crystals.

SYNS: beta,beta'-DICYANODIETHYL SULFIDE
* NITRIL KYSELINY BETA,BETA'-THIODIPROPIO-

NOVE (CZECH) * BETA,BETA'-THIODIPROPIONI-
TRILE * USAF HA-5

THR: Poison by intraperitoneal route. Moderately toxic by ingestion. A skin and eye irritant. When heated to decomposition it emits very toxic fumes of NO_x, SO_x and CN^-.

cis-DICYCLOBUTYLAMMINE-DICHLOROPLATINUM (II)   HR: 3
CAS: 38780-37-9   NIOSH: TP 2195000
mf: $C_8H_{18}Cl_2N_2Pt$ mw: 408.27

SYNS: CIS-BIS(CYCLOBUTYLAMMINE)DICHLORO-PLATINUM(II) * CIS-DICHLOROBIS(CYCLOBUTYL-AMMINE)PLATINUM(II)

THR: Poison by intraperitoneal route. Mutagenic data. See also platinum compounds. When heated to decomposition it emits very toxic fumes of Cl^- and NO_x.

N,N-DICYCLOHEXYLAMINE   HR: 3
CAS: 101-83-7   NIOSH: HY 4025000
mf: $C_{12}H_{23}N$ mw: 181.36

PROP: Liquid, fishy odor, mp: $-1°$, bp: $256°$, flash p: >210°F (OC), d: 0.910, vap d: 6.27.

SYNS: N-CYCLOHEXYLCYCLOHEXANAMINE
* DICYKLOHEXYLAMIN (CZECH) * DODECAHY-DRODIPHENYLAMINE

THR: Poison by ingestion and subcutaneous routes. An experimental carcinogen. Severe skin and eye irritant. See also cyclohexylamine. When heated to decomposition it emits toxic fumes of NO_x. Combustible when exposed to heat or flame; can react with oxidizing materials. To fight fire, use alcohol foam, CO_2, dry chemical.

DICYCLOHEXYLAMINE NITRITE   HR: 3
CAS: 3129-91-7   NIOSH: HY 4200000
mf: $C_{12}H_{23}N \cdot HNO_2$ mw: 228.38

SYNS: DICYCLOHEXYLAMINONITRITE * DICY-CLOHEXYLAMMONIUM NITRITE * DICYKLOHEX-YLAMIN NITRIT (CZECH) * DODECAHYDROPHE-NYLAMINE NITRITE * DUSITAN DICYKLOHEXYLAMINU (CZECH)

THR: Poison by ingestion and subcutaneous route. An experimental tumorigen. When heated to decomposition it emits very toxic fumes of HNO_2 and NO_x.

(2-(DICYCLOPENTYLACETOXY)ETHYL)-TRIETHYLAMMONIUM BRO-MIDE   HR: 3
CAS: 2001-81-2   NIOSH: BR 1210000
mf: $C_{20}H_{38}NO_2 \cdot Br$ mw: 404.50

SYNS: AETHOBROMID DES ALPHA,ALPHA-DICY-CLOPENTYLESSIGSAEURE-BETA'-DIAETHYLAMINO AETHYLESTER (GERMAN) * 2-((DICYCLOPEN-TYLACETYL)OXY)-N,N,N-TRIETHYL-ETHANAMI-NIUM BROMIDE * TRIETHYL(2-HYDROXY-ETHYL)-AMMONIUM BROMIDE, DICYCLOPENTYL-ACETATE

THR: Poison by intravenous and intraperitoneal routes. Moderately toxic by ingestion. See also bromides. When heated to decomposition it emits very toxic fumes of Br^- and NO_x.

3',4'-DIDEOXYKANAMYCIN B SUL-FATE   HR: 3
CAS: 64070-13-9   NIOSH: NZ 3160000
mf: $C_{18}H_{37}N_5O_8 \cdot O_4S$ mw: 547.66

SYN: DKB SULFATE

THR: Poison by intramuscular and intravenous routes. Moderately toxic by other routes. When heated to decomposition it emits very toxic fumes of NO_x and SO_x.

DIELDRIN   HR: 3
CAS: 60-57-1   NIOSH: IO 1750000
DOT: 2761
mf: $C_{12}H_8Cl_6O$ mw: 380.90

PROP: White crystals, odorless, insol in water, sol in common organic solvents, mp: 150°, vap d: 13.2.

SYNS: DIELDRINE (FRENCH) * ENT 16,225
* HEXACHLOROEPOXYOCTAHYDRO-ENDO,EXO-DI-METHANONAPHTHALENE * NCI-C00124

OSHA PEL: TWA 250 ug/m³ (skin)
ACGIH TLV: TWA 0.25 mg/m³
DFG MAK: 0.25 mg/m³
DOT Classification: ORM-A

THR: Poison by inhalation, ingestion, skin contact, intravenous, and intraperitoneal routes. Absorbed readily through the skin and by other routes. It is a central nervous system stimulant. See also chlorinated hydrocarbons; see also aldrin. Dieldrin is considerably more toxic than DDT by ingestion and skin contact. Dieldrin or its derivatives may accumulate in the body from chronic low dosages. Dangerous; when heated to decomp, emits highly toxic fumes of chlorides. For further information see Vol. 6, No. 1 of *DPIM Report*.

d,l-DIEPOXYBUTANE   HR: 3
CAS: 298-18-0   NIOSH: EJ 8400000
mf: $C_4H_6O_2$ mw: 86.10

PROP: Colorless liquid, bp: 138°, mp: 4°, d: 1.112 @ 18°/4°.

SYNS: DL-BUTADIENE DIOXIDE * 1,2:3,4-DIANHYDRO-DL-THREITOL * (±)-1,2:3,4-DI-EPOXYBUTANE * DL-1,2:3,4-DIEPOXYBUTANE

THR: Poison by ingestion, inhalation and skin contact. A skin irritant. An experimental carcinogen, neoplastigen, and tumorigen. When heated to decomposition it emits acrid smoke and irritating fumes. For further information see Vol. 4, No. 3 of *DPIM Report*.

meso-1,2,3,4-DIEPOXYBUTANE   HR: 3
CAS: 564-00-1   NIOSH: EJ 8750000
mf: $C_4H_6O_2$ mw: 86.10

SYNS: (R*,S*)-2,2'-BIOXIRANE * 1,2:3,4-DIANHYDROERYTHRITOL * MESO-DIEPOXYBUTANE * (R*,S*)-DIEPOXYBUTANE * ERYTHRITOL ANHYDRIDE

THR: An acute poison by intraperitoneal route. An experimental carcinogen and neoplastigen. When heated to decomposition it emits acrid smoke and irritating fumes. For further information see Vol. 4, No. 3 of *DPIM Report*.

1,2,9,10-DIEPOXYDECANE   HR: 3
                        NIOSH: HD 7800000
mf: $C_{10}H_{18}O_2$ mw: 170.28

THR: An experimental carcinogen. When heated to decomposition it emits acrid smoke and irritating fumes.

1,2:5,6-DIEPOXYHEXANE   HR: 3
CAS: 1888-89-7   NIOSH: MO 1575000
mf: $C_6H_{10}O_2$ mw: 114.16

THR: An experimental tumorigen and neoplastigen. When heated to decomposition it emits acrid smoke and irritating fumes.

1,2:8,9-DIEPOXYNONANE   HR: 3
CAS: 24829-11-6   NIOSH: RA 6475000
mf: $C_9H_{16}O_2$ mw: 156.25

THR: An experimental carcinogen. When heated to decomposition it emits acrid smoke and irritating fumes.

1,2,7,8-DIEPOXYOCTANE   HR: 3
CAS: 2426-07-5   NIOSH: RG 9450000
mf: $C_8H_{14}O_2$ mw: 142.22

SYN: 1,2-EPOXY-7,8-EPOXYOCTANE

THR: Poison by skin contact. Moderately toxic by ingestion. An experimental carcinogen. Mutagenic data. When heated to decomposition it emits acrid and irritating fumes.

DIESEL EXHAUST PARTICLES   HR: 2
                        NIOSH: HZ 1760000

PROP: Particulate samples collected from the exhaust of a 1979 2.3 L diesel powered automobile running on No. 2 diesel fuel.

THR: Mutagenic data. Suspected carcinogen.

DIETHANOLAMMONIUM MALEIC HYDRAZIDE   HR: 3
CAS: 5716-15-4   NIOSH: UR 6000000
mf: $C_4H_{11}NO_2 \cdot C_4H_4N_2O_2$ mw: 217.26

SYNS: ETHANOL, 2,2'-IMINODI-, COMPOUND WITH 1,2-DIHYDRO-3,6-PYRIDAZINEDIONE (1:1) * 6-HYDROXY-3-(2H)-PYRIDAZINONE DIETHANOLAMINE * MALEIC HYDRAZIDE DIETHANOLAMINE SALT * NCI-C54660

THR: An experimental carcinogen. Moderately toxic by ingestion. Mutagenic data. When heated to decomposition it emits toxic fumes of NO_x.

N,N-DIETHYLACETAMIDE   HR: 3
CAS: 685-91-6   NIOSH: AB 7000000
mf: $C_6H_{13}NO$ mw: 115.20

PROP: Liquid; mp: <65°; bp: 180°; flash p: 170°F; d: 0.92; vap d: 4.0.

THR: An experimental teratogen and carcinogen. Moderately toxic by intravenous and intraperitoneal routes. Flammable when exposed to heat or flame. To fight fire, use foam, mist, CO_2, dry chemical. When heated to decomposition it emits toxic fumes of NO_x. For further information see Vol. 1, No. 1 of *DPIM Report*.

DIETHYLACETIC ACID   HR: 2
CAS: 88-09-5   NIOSH: ET 1400000
mf: $C_6H_{12}O_2$ mw: 116.18

PROP: Colorless, volatile liquid, pineapple-like odor, mp: $-93°$, bp: $121.0°$, flash p: $78°F$ (CC), d: 0.8788, vap press: 10 mm @ $15.3°$, vap d: 4.0, autoign temp: $865°F$.

SYNS: 2-ETHYL BUTANOIC ACID * ALPHA-ETHYLBUTYRIC ACID * 2-ETHYLBUTYRIC ACID * 3-PENTANECARBOXYLIC ACID

THR: Moderately toxic by ingestion and skin contact. An irritant to skin, eyes, and mucous membranes. See also esters. Narcotic in high concentrations. Dangerous when exposed to heat or flame; can react vigorously with oxidizing materials. To fight fire, use CO_2, dry chemical, alcohol foam.

1-DIETHYLACETYLAZIRIDINE   HR: 3
CAS: 62019-57-8   NIOSH: CM 6875000
mf: $C_8H_{15}NO$ mw: 141.24

SYN: DIETHYLACETYLETHYLENEIMINE

THR: An experimental neoplastigen. Mutagenic data. When heated to decomposition it emits toxic fumes of NO_x.

DIETHYLAMINE   HR: 2
CAS: 109-89-7   NIOSH: HZ 8750000
DOT: 1154
mf: $C_4H_{11}N$ mw: 73.16

PROP: Colorless liquid, ammoniacal odor, mp: $-38.9°$, bp: $55.5°$, flash p: $-0.4°F$, d: 0.7108 @ $20°/20°$, autoign temp: $594°F$, vap press: 400 mm @ $38.0°$, vap d: 2.53, lel = 1.8%, uel = 10.1%.

SYNS: DIAETHYLAMIN (GERMAN) * DIETILA-MINA (ITALIAN) * DWUETYLOAMINA (POLISH)

OSHA PEL: TWA 25 ppm
DFG MAK: 10 ppm
DOT Classification: Label: Flammable Liquid

THR: Moderately toxic by ingestion, inhalation, and skin contact. A skin and eye irritant. See also amines. Dangerous when exposed to heat or flame; can react with oxidizing materials. To fight fire, use alcohol foam, CO_2, dry chemical. When heated to decomposition it emits toxic fumes of NO_x. Incompatible with cellulose nitrate.

N,N-DIETHYLAMINOACETONI-TRILE   HR: 3
CAS: 3010-02-4   NIOSH: AL 8575000
mf: $C_6H_{12}N_2$ mw: 112.20

SYN: NITRIL KISELINY DIETHYLAMINOOCTOVE (CZECH)

THR: Poison by ingestion, inhalation, and skin contact. A skin and eye irritant. When heated to decomposition it emits toxic fumes of NO_x.

o-(DIETHYLAMINOETHOXY)BENZ-ANILIDE   HR: 3
CAS: 6376-26-7   NIOSH: CV 7525000
mf: $C_{19}H_{24}N_2O_2$ mw: 312.45

SYNS: o-DIAETHYLAMINOAETHOXY-BENZANILID (GERMAN) * 2-(2-(DIETHYLAMINO)ETHOXY)-BENZANILIDE

THR: Poison by ingestion, subcutaneous, and intravenous (acute) routes. An experimental teratogen. When heated to decomposition it emits toxic fumes of NO_x.

2-(2-(DIETHYLAMINO)ETHOXY)-5-METHYLBENZANILIDE   HR: 3
CAS: 17822-74-1   NIOSH: CV 8575000
mf: $C_{20}H_{25}N_2O_2$ mw: 325.47

THR: Poison by subcutaneous route. An experimental teratogen and tumorigen. When heated to decomposition it emits toxic fumes of NO_x.

2-(DIETHYLAMINO)ETHYLCHLORO-DIPHENYL ACETATEHYDRO-CHLORIDE   HR: 3
CAS: 902-83-0   NIOSH: AF 9100000
mf: $C_{20}H_{24}ClNO_2 \cdot ClH$ mw: 382.36

SYNS: 2-CHLORO-2,2-DIPHENYLACETIC ACID 2--(DIETHYLAMINO)ETHYL ESTER HYDROCHLORIDE * DIAMINOPHEN * DIAPHEN (NEURO-PLEGIC)

THR: Poison by subcutaneous and intraperitoneal routes. When heated to decomposition it emits very toxic fumes of Cl^-, NO_x and HCl.

2-(2-(DIETHYLAMINO)ETHYL)-2-PHE-NYL-4-PENTENOIC ACID ETHYL ESTER   HR: 3
CAS: 14557-50-7   NIOSH: SB 2975000
mf: $C_{19}H_{29}NO_2$ mw: 303.49

SYN: UCB 6249

THR: Poison by ingestion, intraperitoneal, and intravenous routes. See also esters. When heated to decomposition it emits toxic fumes of NO_x.

2-DIETHYLAMINOETHYLPROPYLDI-PHENYLACETATE   HR: 3
CAS: 302-33-0   NIOSH: AJ 3400000
mf: $C_{23}H_{31}NO_2$ mw: 353.55

SYN: SKF-525-A

THR: Poison by intraperitoneal and intravenous routes. Moderately toxic by ingestion. When heated to decomposition it emits toxic fumes of NO_x.

N-(2-(DIETHYLAMINO)ETHYL)-2,4,6-TRI-METHYLBENZAMIDE, HYDRO-CHLORIDE   HR: 3
                    NIOSH: CV 4205000
mf: $C_{16}H_{26}N_2O \cdot ClH$ mw: 298.90

SYNS: C 3235 * BETA-ISODURYLAMIDE, N-(2-(DIETHYLAMINO)ETHYL)-,HYDROCHLORIDE

THR: Poison by intraperitoneal and subcutaneous routes. Eye irritant. When heated to decomposition it emits very toxic fumes of HCl and NO_x.

7-DIETHYLAMINO-5-METHYL-s-TRIAZOLO(1,5-a)PYRIMIDINE   HR: 3
CAS: 15421-84-8   NIOSH: XZ 5700000
mf: $N_5C_{10}H_{15}$ mw: 205.30

SYN: N,N-DIETHYL-5-METHYL-(1,2,-4)TRIAZOLO(1,5-A)PYRIMIDINE-7-AMINE

THR: Poison by ingestion, intraperitoneal, subcutaneous, and intravenous routes. See also amines. When heated to decomposition it emits toxic fumes of NO_x.

2-DIETHYLAMINOPROPIOPHENONE-HYDROCHLORIDE   HR: 3
CAS: 134-80-5   NIOSH: UG 9625000
mf: $C_{13}H_{19}NO \cdot ClH$ mw: 241.79

SYNS: alpha-BENZOYLTRIETHYLAMINE HYDROCHLORIDE * DIETHYLPROPIONE HYDROCHLORIDE * DIETHYLPROPION HYDROCHLORIDE * 1-PHENYL-2-DIETHYLAMINOPROPANONE-1 HYDROCHLORIDE * 1-PHENYL-2-DIETHYLAMINO-1-PROPANONE HYDROCHLORIDE

THR: Poison by intraperitoneal and intravenous routes. Moderately toxic by ingestion. When heated to decomposition it emits very toxic fumes of HCl and NO_x.

6,8-DIETHYLBENZ(a)ANTHRA-CENE   HR: 3
CAS: 36911-94-1   NIOSH: CW 1725000
mf: $C_{22}H_{20}$ mw: 284.42

THR: An experimental carcinogen. When heated to decomposition it emits acrid smoke and irritating fumes.

9,10-DIETHYL-1,2-BENZANTHRA-CENE   HR: 3
CAS: 16354-52-2   NIOSH: CW 1750000
mf: $C_{22}H_{20}$ mw: 284.42

SYN: BENZ(A)ANTHRACENE, 7,12-DIETHYL

THR: An experimental carcinogen. When heated to decomposition it emits acrid smoke.

DIETHYL BENZENE   HR: 2
CAS: 25340-17-4   NIOSH: CZ 5600000
mf: $C_{10}H_{14}$ mw: 134.24

PROP: Colorless, mobile liquid, bp: 183.8°, flash p: 134°F, d: 0.868 @ 25°/25°, autoign temp: 743°-842°F, vap press: 1 mm @ 20.7°, vap d: 4.62.

THR: Moderately toxic by ingestion. See also ethyl benzene. A skin and eye irritant. Flammable when exposed to heat or flame; can react with oxidizing materials. To fight fire, use CO_2, dry chemical.

m-DIETHYLBENZENE   HR: 2
CAS: 141-93-5   NIOSH: CZ 5620000
mf: $C_{10}H_{14}$ mw: 134.24

THR: Moderately toxic by ingestion. When heated to decomposition it emits acrid and irritating fumes.

N,N-DIETHYLBENZENESULFON-AMIDE   HR: 3
CAS: 1709-50-8   NIOSH: DB 1925000
mf: $C_{10}H_{15}NO_2S$ mw: 213.32

THR: An experimental teratogen. Moderately toxic by ingestion. When heated to decomposition it emits very toxic fumes of SO_x and NO_x.

DIETHYLBERYLLIUM   HR: 3
mf: $C_4H_{10}Be$ mw: 67.13

PROP: Colorless liquid. mp: 12°; bp: 110° @ 15 mm; vap d: 2.3.

THR: Very poisonous. See also beryllium compounds. Dangerous fire hazard when exposed to heat or flame. Spontaneously flammable in air. When heated to decomposition it emits poisonous fumes of beryllium; can react vigorously with oxidizing materials. To fight fire, use special extinguishing agents, dry chemical. Explodes with water.

DIETHYLBIS(OCTANOYLOXY)-STANNANE   HR: 3
CAS: 2641-56-7   NIOSH: WH 7268000
mf: $C_{20}H_{40}O_4Sn$ mw: 463.29

SYNS: DIETHYLTIN DICAPRYLATE * DIETHYL-
TIN DIOCTANOATE

THR: Poison by ingestion. See also tin compounds. When heated to decomposition it emits
acrid and irritating fumes.

DIETHYLCADMIUM   HR: 3
mf: $C_4H_{10}Cd$ mw: 170.5

PROP: An oil; decomp by moisture; d: 1.6562;
mp: $-21°$; bp: 64°.

THR: A poison. See also cadmium compounds.
A dangerous fire and explosion hazard. Its fumes
explode in air or when exposed to heat or flame.
When heated to decomposition it emits highly
toxic fumes of cadmium.

DIETHYLCARBAMAZINE ACID
CITRATE   HR: 3
CAS: 1642-54-2   NIOSH: TL 1225000
mf: $C_{10}H_{21}N_3O \cdot C_6H_8O_7$ mw:391.48

SYNS: DIETHYLCARBAMAZANE CITRATE
* DIETHYLCARBAMAZINE CITRATE * DIETHYL-
CARBAMAZINE HYDROGEN CITRATE * 1-DI-
ETHYLCARBAMOYL-4-METHYLPIPERAZINE DIHY-
DROGEN CITRATE * N,N-DIETHYL-4-METHYL-
1-PIPERAZINE CARBOXAMIDE CITRATE
* N,N-DIETHYL-4-METHYL-1-PIPERAZINECAR-
BOXAMIDE DIHYDROGEN CITRATE

THR: Poison by intravenous route. Moderately
toxic by ingestion. When heated to decomposition it emits toxic fumes of NO_x.

O,O-DIETHYL S-(CARBETHOXY)-
METHYL PHOSPHORO-
THIOLATE   HR: 3
CAS: 2425-25-4   NIOSH: AI 7000000
mf: $C_8H_{17}O_5PS$ mw: 256.28

SYNS: ACETOPHOS * ACETOXON * O,O-DI-
ETHYL-S-CARBOETHOXYMETHYL PHOSPHORO-
THIOATE * O,O-DIETHYL-S-CARBOETHOXY-
METHYL THIOPHOSPHATE * PHOSPHOROTHIOIC
ACID,-O,O-DIETHYL ESTER, S-ESTER WITH ETHYL
MERCAPTOACETATE

THR: Poison by ingestion, intraperitoneal, and
other routes. When heated to decomposition it
emits very toxic fumes of PO_x and SO_x.

DIETHYL CARBONATE   HR: 3
CAS: 105-58-8   NIOSH: FF 9800000
mf: $C_5H_{10}O_3$ mw: 118.15

PROP: Colorless liquid, mild odor, mp: 43°,
bp: 125.8°, flash p: 77°F (OC), d: 0.975 @
20°/4°, vap press: 10 mm @ 23.8°, vap d: 4.07.

SYNS: DIAETHYLCARBONAT (GERMAN)
* ETHOXYFORMIC ANHYDRIDE * ETHYL CAR-
BONATE * NCI-C60899

THR: An experimental tumorigen. Moderately
toxic by intraperitoneal route. Flammable when
exposed to heat or flame; can react with oxidizing materials. To fight fire, use foam, CO_2,
dry chemical. When heated to decomposition
it emits acrid smoke.

o,o-DIETHYL-o-(2-CHLORO-1,2,5-DI-
CHLOROPHENYLVINYL)PHOSPHORO-
THIOATE   HR: 3
CAS: 1757-18-2   NIOSH: TE 7525000
mf: $C_{12}H_{14}Cl_3O_3PS$ mw: 375.64

SYNS: O-(2-CHLORO-1-(2,5-DICHLOROPHENYL)-
VINYL)O,O-DIETHYL PHOSPHOROTHIOATE
* ENT 27,102

THR: Poison by ingestion and skin contact.
When heated to decomposition it emits very
toxic fumes of Cl^-, PO_x and SO_x.

DIETHYL CHLOROPHOSPHATE   HR: 3
CAS: 814-49-3   NIOSH: TD 1400000
mf: $C_4H_{10}ClO_3P$ mw: 172.56

PROP: Water white liq, bp: 60° @ 2 mm, d:
1.1915 @ 25°/25°, vap d: 5.94.

SYNS: CHLOROPHOSPHORIC ACID, DIETHYL ES-
TER * DIETHOXYPHOSPHORUS OXYCHLORIDE

THR: Acute poison by skin contact. Poison by
ingestion. A cholinesterase inhibitor. See also
parathion and esters. When heated to decomposition it emits very toxic fumes of Cl^- and PO_x.

N,N-DIETHYL CYSTEAMINE   HR: 3
CAS: 100-38-9   NIOSH: KJ 0525000
mf: $C_6H_{15}NS$ mw: 133.28

SYNS: N-DIAETHYL CYSTEAMIN (GERMAN)
* N-DIETHYL CYSTEAMINE

THR: Poison by intraperitoneal and subcutaneous routes. When heated to decomposition it
emits very toxic fumes of NO_x and SO_x.

o,o-DIETHYL-S-(3,4-DICHLOROPHE-
NYLTHIO)METHYL PHOSPHORO-
THIOATE   HR: 3
CAS: 3152-41-8   NIOSH: TD 5950000
mf: $C_{11}H_{13}Cl_2O_2PS_3$ mw: 375.29

SYNS: S-((3,4-DICHLOROPHENYLTHIO)-METHYL)-O,O-DIETHYL PHOSPHORODITHIOATE * ENT 25,555-X

THR: Poison by ingestion. When heated to decomposition it emits very toxic fumes of Cl^-, PO_x and SO_x.

o,o-DIETHYL-S-(2-DIETHYLAMINO-ETHYL) THIOPHOSPHATE   HR: 3
CAS: 78-53-5   NIOSH: TF 0525000
mf: $C_{10}H_{24}NO_3PS$ mw: 269.38

PROP: Liquid; bp: 110° @ 0.2 mm; mp: 98°.

SYNS: S-(DIETHYLAMINOETHYL)-O,O-DIETHYL PHOSPHOROTHIOATE * DIETHYL-S-2-DIETHYLAMINOETHYL PHOSPHOROTHIOATE * (2-DIETHYLAMINO)ETHYLPHOSPHOROTHIOIC ACID-O,O-DIETHYL ESTER * O,O-DIETHYL-S-2-DIETHYLAMINOETHYL PHOSPHOROTHIOATE * O,O-DIETHYL-S-DIETHYLAMINOETHYL PHOSPHOROTHIOLATE * O,O-DIETHYL-S-2-DIETHYLAMINOETHYL PHOSPHOROTHIOLATE * O,O-DIETHYL-S-(BETA-DIETHYLAMINO)ETHYL PHOSPHOROTHIOLATE * ENT 24,980-X

THR: Poison by ingestion and intraperitoneal route. A cholinesterase inhibitor. See also parathion. When heated to decomposition it emits very toxic fumes of NO_x, PO_x, and SO_x.

DIETHYLDIIODOSTANNANE   HR: 3
CAS: 2767-55-7   NIOSH: WH 7270000
mf: $C_4H_{10}I_2Sn$ mw: 430.63

PROP: Very sltly sol white crystals, mp: 45°, bp: 240-245° (decomp).

SYN: DIETHYLTIN DIIODIDE

THR: Poison by ingestion and intraperitoneal route. See also tin compounds and iodides. When heated to decomposition it emits toxic fumes of I^-.

p-DIETHYL p′-DIMETHYLTHIO-PYROPHOSPHATE   HR: 3
CAS: 64048-13-1   NIOSH: XN 4300000
mf: $C_6H_{16}O_5P_2S_2$ mw: 294.28

SYN: P′-DIETHYL P-DIMETHYL THIOPYROPHOSPHATE

THR: Poison by intramuscular and intraperitoneal routes. When heated to decomposition it emits very toxic fumes of SO_x and PO_x.

N,N′-DIETHYL-N,N′-DINITROSO-ETHYLENEDIAMINE   HR: 3
CAS: 7346-14-7   NIOSH: KV 3675000
mf: $C_6H_{14}N_4O_2$ mw: 174.24

SYNS: N,N′-DINITROSO-N,N′-DIETHYLETHYLENEDIAMINE * NSC 62579

THR: An experimental carcinogen. When heated to decomposition it emits toxic fumes of NO_x.

DIETHYL DIPHENYL DICHLORO-ETHANE   HR: 3
CAS: 72-56-0   NIOSH: KH 5790000
mf: $C_{18}H_{20}Cl_2$ mw: 307.28

SYNS: 1,1-BIS(P-ETHYLPHENYL)-2,2-DICHLOROETHANE * 2,2-BIS(P-ETHYLPHENYL)-1,1-DICHLOROETHANE * 1,1-DICHLORO-2,2-BIS-(P-ETHYLPHENYL)ETHANE * 2,2-DICHLORO-1,1-BIS(P-ETHYLPHENYL)ETHANE * DI(P-ETHYLPHENYL)DICHLOROETHANE * NCI-C02868

THR: An experimental tumorigen and carcinogen by ingestion. When heated to decomposition it emits toxic fumes of Cl^-.

1,3-DIETHYL-1,3-DIPHENYL-UREA   HR: 3
CAS: 85-98-3   NIOSH: FE 0350000
mf: $C_{17}H_{20}N_2O$ mw: 268.39

PROP: Colorless crystals, mp: 73°, d: 1.12, bp: 326°, flash p: 302°F (CC), vap d: 9.3.

SYNS: BIS(N-ETHYL-N-PHENYL)UREA * N,N-DIETHYLCARBANILIDE * N,N′-DIETHYL-N,N′-DIPHENYLUREA * SYM-DIETHYLDIPHENYL-UREA * USAF EK-1047

THR: Poison by intraperitoneal route. Combustible when exposed to heat or flame. Probably a low explosion hazard although it is a component of smokeless explosive mixtures. Highly dangerous; when heated to decomposition it burns and emits very toxic fumes of NO_x. Very slightly flammable. To fight fire, use dry chemical, CO_2, spray or mist.

DIETHYLDITHIOCARBAMIC ACID SELENIUM (II) SALT   HR: 3
CAS: 136-92-5   NIOSH: EZ 6300000
mf: $C_{10}H_{20}N_2S_4 \cdot Se$ mw: 375.52

THR: Poison by ingestion and intraperitoneal route. See also selenium compounds and carbamates. When heated to decomposition it emits very toxic fumes of NO_x, SO_x, and Se.

DIETHYLDITHIOCARBAMIC ANHYDROSULFIDE   HR: 3

CAS: 5827-03-2   NIOSH: EZ 4725000
mf: $C_{11}H_{24}NO_2PS_3$ mw: 329.51

SYNS: DIETHYLDITHIOCARBAMIC ANHYDRIDE OF O,O-DIISOPROPYL THIONOPHOSPHORIC ACID * N,N-DIETHYLTHIOCARBAMYL-O,O-DIISOPRO-PYLDITHIOPHOSPHATE * O,O-DIISOPROPYL-S-DIETHYLDITHIOCARBAMOYLPHOSPHORODITHIOATE * O,O-DIIOSPROPYL DITHIOPHOSPHORIC ACID ESTER OF-N,N-S-DIETHYLTHIOCARBAMOYL-O,O-DI-ISOPROPYL PHOSPHOROTHIOATE * DIISOPROPYL ESTER OF DITHIOCARBAMYL PHOSPHOROTHIOIC ACID * ENT 24,725

THR: Poison by ingestion. See also esters. When heated to decomposition it emits very toxic fumes of PO_x, NO_x, and SO_x.

DIETHYLENE GLYCOL   HR: 3

CAS: 111-46-6   NIOSH: ID 5950000
mf: $C_4H_{10}O_3$ mw: 106.14

PROP: Clear, colorless, practically odorless, syrupy liquid, bp: 245.8°, fp: −8°, flash p: 255°F, d: 1.1184 @ 20°/20°, autoign temp: 444°F, vap press: 1 mm @ 91.8°, vap d: 3.66.

SYNS: BIS(2-HYDROXYETHYL) ETHER * DI-GLYCOL * BETA,BETA′-DIHYDROXYDIETHYL ETHER * ETHYLENE DIGLYCOL * GLYCOL ETHYL ETHER * 3-OXAPENTANE-1,5-DIOL * 3-OXA-1,5-PENTANEDIOL * 2,2′-OXY-BISETHANOL * 2,2′-OXYDIETHANOL

THR: Poison by inhalation and subcutaneous route. A mild eye and skin irritant. An experimental carcinogen. See also ethers. Combustible when exposed to heat or flame; can react with oxidizing materials. To fight fire, use alcohol foam, water, CO_2, dry chemical. When heated to decomposition it emits acrid smoke and irritating fumes.

DIETHYLENE GLYCOL DINITRATE   HR: 2

CAS: 693-21-0   NIOSH: ID 6825000
DOT: 0075
mf: $C_4H_8N_2O_7$ mw: 196.14

PROP: Liquid, vap d: 6.76.

SYNS: BIS(HYDROXYAETHYL)-AETHER-DINITRAT (GERMAN) * DIETHYLENGLYKOLDINITRATE (CZECH) * DIGLYCOLDINITRAAT (DUTCH) * DIGLYKOLDINITRAT (GERMAN) * DI(HY-DROXYETHYL) ETHER DINITRATE * DINITRATE DE DIETHYLENE-GLYCOL (FRENCH) * DINITRO-DIGLICOL (ITALIAN) * DINITRODIGLYKOL (CZECH)

DOT Classification: Forbidden

THR: Moderately toxic by ingestion. Ingestion can cause a drop in blood pressure and cardiac disturbances. See also nitrates. A dangerous fire hazard. A severe explosion hazard when shocked or exposed to heat. Used in low freezing dynamites and some permissible explosives. See also explosives, high. Dangerous; shock and heat will explode it; when heated, emits toxic fumes of NO_x; can react vigorously with oxidizing or reducing materials.

DIETHYLENETRIAMINE   HR: 3

CAS: 111-40-0   NIOSH: IE 1225000
DOT: 2079
mf: $C_4H_{13}N_3$ mw: 103.20

PROP: Yellow, viscous liquid, mild ammonia-cal odor, mp: −39°, bp: 207°, flash p: 215°F (OC), d: 0.9586 @ 20°/20°, autoign temp: 750°F, vap press: 0.22 mm @ 20°, vap d: 3.48.

SYNS: AMINOETHYLETHANDIAMINE * 3-AZA-PENTANE-1,5-DIAMINE * BIS(2-AMINOETHYL)-AMINE * BIS(BETA-AMINOETHYL)AMINE * 2,2′-DIAMINODIETHYLAMINE

DOT Classification: Corrosive

THR: Poison by skin contact and intraperitoneal routes. Moderately toxic by ingestion. A severe skin and eye irritant. High concentration of vapors causes irritation of respiratory tract, nausea, and vomiting. Repeated exposures can cause asthma and sensitization of skin. Combustible when exposed to heat or flame; can react with oxidizing materials. To fight fire, use alcohol foam. When heated to decomposition it emits toxic fumes of NO_x.

DIETHYL ESTER SULFURIC ACID   HR: 3

CAS: 64-67-5   NIOSH: WS 7876000
mf: $C_4H_{10}O_4S$ mw: 154.20

PROP: Colorless oily liquid, faint ethereal odor,. mp: −25°; bp: decomp to ethyl ether, flash p: 220°F (CC); d: 1.172 @ 25°/4°; autoign temp: 817°F; vap press: 1 mm @ 47.0°; vap d: 5.31.

SYNS: DIAETHYLSULFAT (GERMAN) * DIETHYL SULFATE * ETHYL SULFATE

THR: Poison by inhalation and subcutaneous routes. Moderately toxic by ingestion and skin contact. A severe skin irritant. Mutagenic data. An experimental tumorigen and carcinogen. Combustible when exposed to heat or flame. Dangerous; see also sulfates, can react with oxidizing materials; moisture causes liberation of H_2SO_4; violent reaction with potassium tert-butoxide. To fight fire, use alcohol foam, H_2O foam, CO_2, dry chemicals.

N,N-DIETHYLETHANOLAMINE HR: 3

CAS: 100-37-8   NIOSH: KK 5075000
mf: $C_6H_{15}NO$ mw: 117.22

PROP: Colorless, hygroscopic liquid, bp: 162°, flash p: 140°F (OC), d: 0.8851 @ 20°/20°, vap press: 1.4 mm @ 20°, vap d: 4.03.

SYNS: DIAETHYLAMINOAETHANOL (GERMAN) * DIETHYLAMINOETHANOL * N-DIETHYLAMI-NOETHANOL * BETA-DIETHYLAMINOETHANOL * 2-(DIETHYLAMINO)ETHANOL * N,N-DI-ETHYL-N-(BETA-HYDROXYETHYL)AMINE * 2-HYDROXYTRIETHYLAMINE

OSHA PEL: TWA 10 ppm (skin)

THR: Poison by intraperitoneal, intravenous, and intramuscular routes. Moderately toxic by ingestion, skin contact, and other routes. A severe eye and moderate skin irritant. Flammable when exposed to heat or flame; can react with oxidizing materials. To fight fire, use alcohol foam, CO_2, dry chemical. When heated to decomposition it emits toxic fumes of NO_x.

o,o-DIETHYL-S-(N-ETHOXYCARBO-NYL-N-METHYLCARBAMOYLMETHYL) PHOSPHORO DITHIOATE HR: 3

CAS: 2595-54-2   NIOSH: FB 3850000
mf: $C_{10}H_{20}NO_5PS_2$ mw: 329.40

SYNS: o,o-DIAETHYL-S-(3-METHYL-2,4-DIOXO-5-OXA-3-AZA-HEPTYL)-DITHIOPHOSPHAT (GERMAN) * o,o-DIETHYL S-(N-ETHOXYCAR-BONYL-N-METHYLCARBAMOYLMETHYL) PHOS-PHOROTHIOLOTHIONATE * o,o-DIETHYL S-(N-METHYL-N-CARBOETHOXYCARBAMOYLMETHYL) DITHIOPHOSPHATE * o,o-DIETHYL-S-(3-METHYL-2,4-DIOXO-5-OXA-3-AZA-HEPTYL)-DI-THIOFOSFAAT (DUTCH) * o,o-DIETIL-S-(N-ETOSSI-CARBONIL-N-METIL-CARBAMOIL-METIL)-

DITIOFOSFATO (ITALIAN) * DITHIOPHOSPHATE DE O,O-DIETHYLE ET DE S-N-METHYL N-CARBO-ETHOXY CARBAMOYLMETHYLE (FRENCH) * N-ETHOXYCARBONYL-N-METHYLCARBAMOYL-METHYL-O,O-DIETHYL PHOSPHORODITHIOATE * S-((ETHOXYCARBONYL)METHYLCARBAMOYL)-METHYL-O,O-DIETHYL PHOSPHORODITHIOATE

THR: Poison by ingestion, skin contact, and subcutaneous routes. When heated to decomposition it emits very toxic fumes of SO_x, PO_x and NO_x.

DIETHYL FLUOROPHOSPHATE HR: 3

CAS: 358-74-7   NIOSH: TE 5600000
mf: $C_4H_{10}FO_3P$ mw: 156.11

PROP: A liquid with a sweet or fruity odor, mp: low, bp: 170°, d: 1.15 (approx), vap d: 5.38.

SYNS: T-1036 * TL 345

THR: Poison by inhalation and skin contact. See also fluorides. Dangerous; when heated to decomposition or on contact with acid or acid fumes, emits highly toxic fumes of fluorides and oxides of phosphorus.

1,2-DIETHYLHYDRAZINE HR: 3

CAS: 1615-80-1   NIOSH: MV 2275000
mf: $C_4H_{12}N_2$ mw: 88.18

PROP: Sol in alcohol and ether; bp: 86°, d: 0.797 @ 26°.

SYNS: N-N'-DIETHYLHYDRAZINE * 1,2-DIA-ETHYLHYDRAZINE (GERMAN) * SYM-DIETHYL-HYDRAZINE * HYDRAZOETHANE

THR: An experimental carcinogen, teratogen, and tumorigen. It is also a transplacental carcinogen. When heated to decomposition it emits toxic fumes of NO_x.

1,2-DIETHYLHYDRAZINE DIHYDROCHLORIDE HR: 3

CAS: 7699-31-2   NIOSH: MV 2295000
mf: $C_4H_{12}N_2 \cdot 2ClH$ mw: 161.10

THR: An experimental teratogen. When heated to decomposition it emits very toxic fumes of HCl and NO_x.

o,o-DIETHYL-S-2-ISOPROPYLMERCAP-TOMETHYLDITHIOPHOSPHATE HR: 3

CAS: 78-52-4   NIOSH: TD 9625000
mf: $C_8H_{19}O_2PS_3$ mw: 274.42

SYNS: o,o-DIETHYL S-(ISOPROPYLMERCAPTO-
METHYL) PHOSPHORODITHIOATE ✶ O,O-DIETHYL
S-(ISOPROPYLTHIOMETHYL) PHOSPHORODITHIOATE
✶ ENT 22,865

THR: Poison by ingestion and subcutaneous
routes. When heated to decomposition it emits
very toxic fumes of PO_x and SO_x.

DIETHYL LEAD DIACETATE   HR: 3
CAS: 15773-47-4   NIOSH: OG 0470000
mf: $C_8H_{16}O_4Pb$ mw: 383.43

THR: Poison by ingestion. See also lead. When
heated to decomposition it emits toxic fumes
of Pb.

N,N-DIETHYLLYSERGAMIDE   HR: 3
CAS: 50-37-3   NIOSH: KE 4100000
mf: $C_{20}H_{25}N_4O$ mw: 323.48

SYNS: ACID ✶ CUBES ✶ DELYSID ✶ 9,10-
DIDEHYDRO-N,N-DIETHYL-6-METHYL-ERGOLINE-8-
beta-CARBOXAMIDE ✶ D-LSD ✶ HEAVENLY
BLUE ✶ LSD ✶ LSD-25 ✶ LYSERGAMID
✶ LYSERGAMIDE, N,N-DIETHYL- ✶ LYSERG-
AURE DIETHYLAMID ✶ D-LYSERGIC ACID DI-
ETHYLAMIDE ✶ LYSERGIC ACID DIETHYLAMIDE-
25 ✶ LYSERGIDE ✶ LYSERGSAUEREDIAETH-
YLAMID ✶ PEARLY GATES ✶ ROYAL BLUE
✶ WEDDING BELLS

THR: Poison by ingestion and intravenous route.
Mutagenic data. Human central nervous system
and psychotropic effects. An experimental teratogen. A much abused hallucinogen. When
heated to decomposition it emits toxic fumes
of NO_x.

DIETHYL MERCURY   HR: 3
CAS: 627-44-1   NIOSH: OW 2350000
mf: $C_4H_{10}Hg$ mw: 258.73

PROP: Colorless liquid, hazel-like odor, bp:
159°, d: 2.4660 @ 20°.

THR: A deadly poison by inhalation. See also
mercury compounds, organic. Flammable when
exposed to heat or flame. Dangerous; when
heated to decomposition or on contact with acid
or acid fumes it emits highly toxic fumes of
mercury; can react with oxidizing materials.

1,1-DIETHYL-3-METHYL-3-
NITROSOUREA   HR: 3
CAS: 50285-72-8   NIOSH: YS 9380000
mf: $C_6H_{13}N_3O_2$ mw: 159.22

SYNS: NITROSO-1,1-DIETHYL-3-METHYLUREA

✶ NITROSOMETHYLDIAETHYLHARNSTOFF
✶ NITROSOMETHYLDIETHYLUREA ✶ 1-NITROSO-
1-METHYL-3,3-DIETHYLUREA

THR: An experimental tumorigen. When heated
to decomposition it emits toxic fumes of NO_x.

o,o-DIETHYL-o, p-(METHYLSULFINYL)-
PHENYL THIOPHOSPHATE   HR: 3
CAS: 115-90-2   NIOSH: TF 3850000
mf: $C_{11}H_{17}O_4PS_2$ mw: 308.37

SYNS: O,O-DIAETHYL-O-4-METHYLSULFINYL-
PHENYL-MONOTHIOPHOSPHAT (GERMAN)
✶ O,O-DIETHYL O-(P-(METHYLSULFINYL)PHENYL)
PHOSPHOROTHIOATE ✶ ENT 24,945

THR: Poison by ingestion, skin contact, intraperitoneal, and other routes. When heated to
decomposition it emits very toxic fumes of SO_x
and PO_x.

DIETHYL (4-METHYLUMBELLIFERYL)
THIONOPHOSPHATE   HR: 3
CAS: 299-45-6   NIOSH: GN 7525000
mf: $C_{14}H_{17}O_5PS$ mw: 328.34

SYNS: O,O-DIAETHYL-O-(4-METHYL-COUMARIN-
7-YL)-MONOTHIOPHOSPHAT (GERMAN) ✶ DIETH-
OXY THIOPHOSPHORIC ACID ESTER OF
7-HYDROXY-4-METHYL COUMARIN ✶ O,O-DI-
ETHYL-O-(2-KETO-4-METHYL-7-ALPHA',BETA'-
BENZO-ALPHA'-PYRANYL) THIOPHOSPHATE
✶ O,O-DIETHYL-O-(4-METHYLCOUMARIN-7-YL)-
MONOTHIOFOSFAAT (DUTCH) ✶ O,O-
DIETHYL-O-(4-METHYL-7-COUMARINYL)
PHOSPHOROTHIOATE ✶ O,O-DIETHYL-O-(4-
METHYL-7-COUMARINYL) THIONOPHOSPHATE
✶ O,O-DIETHYL-O-(4-METHYLCOUMARINYL-7)
THIOPHOSPHATE ✶ O,O-DIETHYL-O-(4-METHYL-
7-KUMARINYL) ESTER KYSELINY THIOFOSFOR-
ESCNE (CZECH) ✶ O,O-DIETHYL-O-(4-METHYL-
UMBELLIFERONE) ESTER OF THIOPHOSPHORIC
ACID ✶ O,O-DIETHYL-O-(4-METHYLUMBELLI-
FERONE) PHOSPHOROTHIOATE ✶ O,O-DIETIL-O-
(4-METILCUMARIN-7-IL)-MONOTIOFOSFATO
(ITALIAN) ✶ O,O-DIETYL-O-4-METHYLKUMARI-
NYL(7)TIOFOSFAT (CZECH) ✶ 4-METHYL-7-
HYDROXYCOUMARIN DIETHOXYTHIOPHOSPHATE
✶ 4-METHYLUMBELLIFERONE-O,O-DIETHYL
THIOPHOSPHATE ✶ THIOPHOSPHATE DE O,O-DI-
ETHYLE ET DE O-(4-METHYL-7-COUMARINYLE)
(FRENCH)

THR: Poison by ingestion, skin contact, intraperitoneal, and subcutaneous routes. When
heated to decomposition it emits very toxic
fumes of SO_x and PO_x.

N,N-DIETHYLNICOTINAMIDE   HR: 3
CAS: 59-26-7   NIOSH: QS 4375000
mf: $C_{10}H_{14}N_2O$ mw: 178.26

SYNS: DIAETHYL-NICOTINAMID (GERMAN)
* DIETHYL-NICOTAMIDE * CORNOTONE
* COROTONIN * PYRIDINE-3-CARBOXYDIETH-
YLAMIDE * PYRIDINE-3-CARBOXYLIC ACID DI-
ETHYLAMIDE * NICOTINIC ACID DIETHYLAMIDE

THR: Poison by ingestion, intravenous, intraperitoneal, and subcutaneous routes. When
heated to decomposition it emits toxic fumes
of NO_x.

o,S-DIETHYL-o-(4-NITROPHENYL)THIO-
PHOSPHATE   HR: 3
CAS: 597-88-6   NIOSH: TF 4900000
mf: $C_{10}H_{14}NO_5PS$ mw: 291.28

SYNS: O,S-DIETHYL-O-(P-NITROPHENYL) PHOS-
PHOROTHIOATE * O,S-DIETHYL-O-(4-NITROPHE-
NYL)PHOSPHOROTHIOATE

THR: Poison by ingestion, skin contact, intraperitoneal, subcutaneous, and intramuscular
routes. When heated to decomposition it emits
very toxic fumes of NO_x.

DIETHYL OXALATE   HR: 3
CAS: 95-92-1   NIOSH: RO 2800000
mf: $C_6H_{10}O_4$ mw: 146.16

PROP: Colorless, oily aromatic liquid, decomp
in water, mp: $-40.6°$, bp: $185.4°$, flash p: $168°F$
(OC), d: 1.08426 @ 15°, 1.0785 @ 20°/4°,
vap d: 5.04.

SYNS: DIETHYL ETHANEDIOATE * ETHYL OXA-
LATE

THR: Poison by ingestion. See also esters. Flammable when exposed to heat or flame; can react
with oxidizing materials. Will give off irritating
fumes on combustion. To fight fire, use foam,
CO_2, dry chemical.

5,5-DIETHYL-1,3-OXAZIN-2,4-
DIONE   HR: 3
CAS: 702-54-5   NIOSH: RP 6300000
mf: $C_8H_{13}NO_3$ mw: 171.22

SYNS: 5,5-DIETHYLDIHYDRO-2H-1,3-OXAZINE-
2,4(3H)-DIONE * 5,5-DIETHYL-1,3-OXAZINE-
2,4-DIONE * 5,5-DIETHYLTETRAHYDRO-2H-1,
3-OXAZINE-2,4(3H)-DIONE

THR: Poison by ingestion, intravenous, intra-

peritoneal, subcutaneous, and intramuscular
routes. When heated to decomposition it emits
toxic fumes of NO_x.

2,2-DIETHYL-4-PENTENAMIDE   HR: 3
CAS: 512-48-1   NIOSH: SB 2145000
mf: $C_9H_{17}NO$ mw: 155.27

SYNS: DIAETHYLALLYLACETAMIDE (GERMAN)
* EPINOVAL

THR: Poison to humans by ingestion. Pulmonary and cardiovascular effects by ingestion.
When heated to decomposition it emits toxic
fumes of NO_x.

DIETHYL-para-PHENYLENEDI-
AMINE   HR: 3
CAS: 93-05-0   NIOSH: SS 9275000
mf: $C_{10}H_{16}N_2$ mw: 164.28

THR: Poison by ingestion, subcutaneous, and
intravenous routes. Human systemic skin effects
by skin contact. When heated to decomposition
it emits toxic fumes of NO_x.

DIETHYL PHENYLTIN ACETATE   HR: 3
CAS: 64036-46-0   NIOSH: WH 5700000
mf: $C_{12}H_{18}O_2Sn$ mw: 312.99

SYN: ACETOXYDIETHYLPHENYLSTANNANE

THR: Poison by ingestion. See also tin compounds. When heated to decomposition it emits
acrid and irritating fumes.

DIETHYL PHOSPHOROCYANI-
DATE   HR: 3
CAS: 2942-58-7   NIOSH: TD 2500000
mf: $C_5H_{10}NO_3P$ mw: 163.13

SYNS: DIETHOXY-PHOSPHORYL CYANIDE
* DIETHYLCYANOPHOSPHATE * DIETHYL CYA-
NOPHOSPHONATE

THR: Poison by intravenous, intraperitoneal,
and subcutaneous routes. When heated to decomposition it emits very toxic fumes of NO_x
and PO_x.

DIETHYL-o-PHTHALATE   HR: 3
CAS: 84-66-2   NIOSH: TI 1050000
mf: $C_{12}H_{14}O_4$ mw: 222.26

PROP: Clear, colorless liquid; mp: $-40.5°$, bp:
302°, flash p: $325°F$ (OC), d: 1.110, vap d:
7.66.

SYNS: 1,2-BENZENEDICARBOXYLIC ACID, DI-

ETHYL ESTER * ETHYL PHTHALATE * NCI-c60048

THR: Poison by intravenous route. An experimental teratogen. Moderately toxic by ingestion and other routes. An eye irritant and systemic irritant by inhalation. Narcotic in high concentrations. When heated to decomposition it emits acrid smoke. Combustible when exposed to heat or flame. To fight fire, use water spray, mist, foam. For further information see ethyl phthalate, Vol. 4, No. 3 of *DPIM Report*.

3,3-DIETHYL-1-(m-PYRIDYL)TRIAZENE   HR: 3
CAS: 21600-43-1   NIOSH: XY 0495000
mf: $C_9H_{14}N_4$ mw: 178.27

SYNS: 1-(PYRIDYL-3-)-3,3-DIAETHYL-TRIAZEN (GERMAN) * 1-PYRIDYL-3,3-DIETHYLTRIAZENE * 1-(PYRIDYL-3)-3,3-DIETHYLTRIAZENE * M-PYRIDYL-DIETHYL-TRIAZENE * 1-(3-PYRIDYL)-3,3-DIETHYLTRIAZENE

THR: Poison by ingestion and subcutaneous routes. Mutagenic data. An experimental teratogen, neoplastigen, and tumorigen. When heated to decomposition it emits toxic fumes of NO_x.

o,o-DIETHYL-o-2-QUINOXALYL THIOPHOSPHATE   HR: 3
CAS: 13593-03-8   NIOSH: TF 6125000
mf: $C_{12}H_{15}N_2O_3PS$ mw: 298.32

SYNS: O,O-DIAETHYL-O-(CHINOXALYL-(2))-MONOTHIOPHOSPHAT (GERMAN) * O,O-DIETHYL O-QUINOXALIN-2-YL PHOSPHOROTHIOATE * O,O-DIETHYL-O-(2-QUINOXALINYL) PHOSPHOROTHIOATE * O,O-DIETHYL-O-(2-QUINOXALYL) PHOSPHOROTHIOATE * ENT 27394

THR: Poison by ingestion, inhalation, and skin contact. When heated to decomposition it emits very toxic fumes of NO_x, PO_x, and SO_x.

N,N-DIETHYLSELENOUREA   HR: 3
CAS: 5117-17-9   NIOSH: YS 9450000
mf: $C_5H_{12}N_2Se$ mw: 179.15

SYN: USAF B-100

THR: Poison by intraperitoneal route. See also selenium compounds. When heated to decomposition it emits very toxic fumes of NO_x and Se.

N,N-DIETHYL-4-STILBENAMINE   HR: 3
CAS: 40193-47-3   NIOSH: WJ 3825000
mf: $C_{18}H_{21}N$ mw: 251.40

SYNS: DIETHYLAMINO STILBENE * 4-STIL-BENYL-N,N-DIETHYLAMINE

THR: An experimental neoplastigen. When heated to decomposition it emits toxic fumes of NO_x.

DIETHYLSTILBESTEROL   HR: 3
CAS: 56-53-1   NIOSH: WJ 5600000
mf: $C_{18}H_{20}O_2$ mw: 268.38

PROP: Small cryst; mp: 171°.

SYNS: 4,4'-(1,2-DIETHYL-1,2-ETHENEDIYL)BIS-PHENOL * DIETILESTILBESTROL (SPANISH) * TRANS-4,4'-(1,2-DIETHYL-1,2-ETHENEDIYL)-BISPHENOL * ALPHA,ALPHA'-DIETHYLSTILBEN-EDIOL * TRANS-ALPHA,ALPHA'-DIETHYL-4,4'-STILBENEDIOL * ALPHA,ALPHA'-DIETHYL-4,4'-STILBENEDIOL * TRANS-DIETHYLSTILBES-TEROL * DIETHYLSTILBESTROL * TRANS-DI-ETHYLSTILBESTROL * DIETHYLSTILBOESTEROL * TRANS-DIETHYLSTILBOESTEROL * 4,4'-DI-HYDROXYDIETHYLSTILBENE * 4,4'-DIHY-DROXY-ALPHA,BETA-DIETHYLSTILBENE * 3,4' (4,4'-DIHYDROXYPHENYL)HEX-3-ENE * 3,4-BIS(P-HYDROXYPHENYL)-3-HEXENE * DES (SYNTHETIC ESTROGEN) * STILBESTROL * NEO-OESTRANOL 1 * NSC-3070

THR: Poison by intraperitoneal and subcutaneous routes. Moderately toxic by ingestion and other routes. A human teratogen, carcinogen, tumorigen, and experimental neoplastigen by various routes. A suspected transplacental carcinogen. Glandular system effects by skin contact. Mutagenic data. Implicated in male impotence and enlargement of male breasts. See also 19-nor-17-alpha-pregna-1,3,5(10)-trien-2-yne-3,17-diol (estrogen). Flammable. For further information see Vol. 6, No. 2 of *DPIM Report*.

DIETHYLSTILBESTROL DIPALMITATE   HR: 3
CAS: 63019-08-9   NIOSH: WJ 5700000
mf: $C_{50}H_{80}O_4$ mw: 745.30

SYNS: alpha,alpha'-DIETHYL-4,4'-STILBENE-DIOL DIPALMITATE * 4,4'-DIHYDROXY-al-pha,beta-DIETHYLSTILBENE PALMITATE

THR: An experimental carcinogen. When heated to decomposition it emits acrid smoke and irritating fumes.

DIETHYLSTILBESTROL DIPRO-
PIONATE   HR: 3
CAS: 130-80-3   NIOSH: WJ 5775000
mf: $C_{24}H_{28}O_4$ mw: 380.52

PROP: Crystals. mp: 104°.

SYNS: TRANS-4,4'-(1,2-DIETHYL-1,2-
ETHENEDIYL)BISPHENOL DIPROPIONATE
* ALPHA,ALPHA'-DIETHYL-4,4'-STILBENEDIOL,
DIPROPIONATE * ALPHA,ALPHA'-DIETHYL-4,4'-
STILBENEDIOL TRANS-DIPROPIONATE * TRANS-
ALPHA,ALPHA'-DIETHYL-4,4'-STILBENEDIOL DI-
PROPIONATE * DIETHYLSTILBENE DIPROPIONATE
* ALPHA,ALPHA'-DIETHYL-4,4'-STILBENEDIOL
DIPROPIONYL ESTER * DIETHYLSTILBESTEROL
DIPROPIONATE * DIETHYLSTILBESTROL PROPIO-
NATE * DIHYDROXYDIETHYLSTILBENE DIPRO-
PIONATE * 4,4'-DIHYDROXY-ALPHA,BETA-DI-
ETHYLSTILBENE DIPROPIONATE * DIPROPIONATO
DE ESTILBENE (SPANISH) * P,P'-DIPROPIONOXY-
TRANS-ALPHA,BETA-DIETHYLSTILBENE * STIL-
BESTROL DIPROPIONATE

THR: A human carcinogen. An experimental
tumorigen and neoplastigen. When heated to
decomposition it emits acrid smoke and irritating
fumes.

2,2-DIETHYL-3-THIOMORPHO-
LINONE   HR: 3
CAS: 69226-06-8   NIOSH: XM 3054000
mf: $C_8H_{15}NOS$ mw: 173.30

THR: Poison by intravenous route. Moderately
toxic by ingestion and intraperitoneal route.
When heated to decomposition it emits very
toxic fumes of NO_x and SO_x.

1,3-DIETHYLTHIOUREA   HR: 3
CAS: 105-55-5   NIOSH: YS 9800000
mf: $C_5H_{12}N_2S$ mw: 132.25

SYNS: N,N'-DIETHYLTHIOCARBAMIDE * N,N'-
DIETHYLTHIOUREA * 1,3-DIETHYL-2-THIOUREA
* NCI-C03816 * USAF EK-1803

THR: Poison by ingestion. An experimental carcinogen. When heated to decomposition it emits
very toxic fumes of NO_x and SO_x.

DIETHYL-m-TOLUAMIDE   HR: 3
CAS: 134-62-3   NIOSH: XS 3675000
mf: $C_{12}H_{17}NO$ mw: 191.30

PROP: A liquid, sol in water, alcohol and ether,
bp: 160° @ 19 mm, d: 0.996 @ 20°/4°.

SYNS: DIETHYL-m-TOLUAMIDE * ENT 20218
* ENT 22542 * M-TOLUIC ACID DIETHYLAM-
IDE * DIETHYLTOLUAMIDE * N,N-DIETHYL-M-
TOLUAMIDE * 3-METHYL-N,N-DIETHYLBENZAM-
IDE

THR: Poison by intravenous route. Moderately
toxic by ingestion and skin contact. A systemic
irritant by skin contact. Mutagenic data. Can
cause central nervous system disturbances.
When heated to decomposition it emits toxic
fumes of NO_x.

DIETHYL TRIAZENE   HR: 3
CAS: 63980-20-1   NIOSH: XY 0350000
mf: $C_4H_{11}N_3$ mw: 101.18

THR: An experimental teratogen. When heated
to decomposition it emits toxic fumes of NO_x.

N,N-DIETHYLVANILLAMIDE   HR: 3
CAS: 304-84-7   NIOSH: YW 5250000
mf: $C_{12}H_{17}NO_3$ mw: 223.30

SYNS: VANILLIC ACID-N,N-DIETHYLAMIDE
* DIETHYLAMIDE DE VANILLIQUE * 3-ME-
THOXY-4-HYDROXYBENZOIC ACID DIETHYLAMIDE
* VANILLIC ACID DIETHYLAMIDE * VANILLIN-
SAEURE-DIAETHYLAMID (GERMAN)

THR: Poison by ingestion, intravenous, subcutaneous, and intraperitoneal routes. When heated
to decomposition it emits toxic fumes of NO_x.

DI(ETHYLXANTHOGEN)TRISUL-
FIDE   HR: 3
CAS: 1851-77-0   NIOSH: LR 0440000
mf: $C_6H_{10}O_2S_5$ mw: 274.46

SYNS: BIS(ETHYLXANTHOGEN) TRISULFIDE
* DEFOLIANT 713

THR: Poison by ingestion. When heated to decomposition it emits toxic fumes of SO_x.

DIETHYLZINC   HR: 3
CAS: 557-20-0   NIOSH: ZH 2070000
mf: $C_4H_{10}Zn$ mw: 123.51

PROP: Liquid; mp: −28°; bp: 118°; d: 1.2065
@ 20°/4°.

SYNS: ZINC ETHIDE * ZINC ETHYL (DOT)

DOT Classification: Label: Flammable
Liquid

THR: No toxicity data available, but it is pre-
sumed to be a poison. See also zinc compounds.

Dangerously flammable by spontaneous chemical reaction in air, or with oxidizing materials. A dangerous explosion hazard. Reacts violently with Cl_2; water; air; hydrazine. Dangerous fire hazard; can react vigorously with oxidizing materials. To fight fire, Do not use water, foam or halogenated extinguishing agents. Use dry materials, such as graphite, sand, etc. Incompatible with water and air.

DIETROL HR: 3
CAS: 63868-62-2   NIOSH: QE 2450000
mf: $C_{12}H_{17}NO \cdot 2C_4H_6O_6$ mw: 491.50

SYNS: 3,4-DIMETHYL-2-PHENYLMORPHOLINE BITARTRATE * PHENDIMETRAZINE BITARTRATE

THR: Poison by ingestion, intraperitoneal, and intravenous routes. When heated to decomposition it emits toxic fumes of NO_x.

2,10-DIFLUOROBENZO(rst)PENTA-PHENE HR: 3
CAS: 61735-78-2   NIOSH: DI 6140000
mf: $C_{24}H_{12}F_2$ mw: 338.36

THR: An experimental carcinogen. When heated to decomposition it emits toxic fumes of F^-.

DIFLUORODIMETHYLSTAN-NANE HR: 3
CAS: 3582-17-0   NIOSH: WH 7285000
mf: $C_2H_6F_2Sn$ mw: 186.77

PROP: White crystals. Water sol. Bp: decomp < 360°.

SYNS: DIMETHYLTIN FLUORIDE * DIMETHYL-TIN DIFLUORIDE

THR: Poison by intravenous route. See also tin compounds and fluorides. When heated to decomposition it emits toxic fumes of F^-.

DIFLUOROPHENYLARSINE HR: 3
CAS: 368-97-8   NIOSH: CH 5970000
mf: $C_6H_5AsF_2$ mw: 190.03

SYN: PHENYLDIFLUOROARSINE

THR: Poison by skin contact and intravenous route. See also fluorides, arsenic compounds. When heated to decomposition it emits very toxic fumes of As and F^-.

5-(2,4-DIFLUOROPHENYL)SALICYLIC ACID HR: 3
CAS: 22494-42-4   NIOSH: DV 2030000
mf: $C_{13}H_8F_2O_3$ mw: 250.21

SYNS: 2',4'-DIFLUORO-4-HYDROXY-(1,1'-BI-PHENYL)-3-CARBOXYLIC ACID * 2',4'-DI-FLUORO-4-HYDROXY-3-BIPHENYLCARBOXYLIC ACID * 2',4'-DIFLUORO-4-HYDROXY-(1',1-DI-PHENYL)-3-CARBOXYLIC ACID * 2-(HYDROXY)-5-(2,4-DIFLUOROPHENYL)BENZOIC ACID

THR: Poison by ingestion. Human systemic effects by ingestion. When heated to decomposition it emits toxic fumes of F^-. See also fluorides.

DIFOLATAN HR: 3
CAS: 2425-06-1   NIOSH: GW 4900000
mf: $C_{10}H_9Cl_4NO_2S$ mw: 349.06

SYNS: ORTHO 5865 * SULFONIMIDE * N-(1,1,2,2-TETRACHLORAETHYLTHIO)-CYCLO-HEX-4-EN-1,4-DIACARBOXIMID(GERMAN) * N-1,1,2,2-TETRACHLOROETHYLMERCAPTO-4-CYCLOHEXENE-1,2-CARBOXIMIDE * N((1,1,2,2-TETRACHLOROETHYL)SULFENYL)-CIS-4-CY-CLOHEXENE-1,2-DICARBOXIMIDE * N-(1,1,2,2-TETRACHLOROETHYLTHIO)-4-CYCLOHEXENE-1,2-DICARBOXIMIDE

THR: Poison by intraperitoneal route. Moderately toxic by ingestion. Mutagenic data. An experimental teratogen. When heated to decomposition it emits very toxic fumes of Cl^-, NO_x and SO_x.

DIFUMARATE HR: 3
NIOSH: TL 7712000
mf: $C_{31}H_{36}ClN_3O_5S \cdot 2C_4H_4O_4$ mw: 830.37

SYN: 4-(3-(2-CHLOROPHENOTHIAZIN-10-YL)PRO-PYL-1-PIPERAZINEETHANOL, 3,4,5-TRIMETHOXYBENZOATE DIFUMARATE

THR: An experimental teratogen. Moderately toxic by ingestion. When heated to decomposition it emits very toxic fumes of Cl^-, NO_x and SO_x.

DIGAMMACAINE HR: 3
CAS: 63906-88-7   NIOSH: CV 5788000
mf: $C_{21}H_{26}N_2O \cdot ClH$ mw: 358.95

SYNS: 1-BENZAMIDO-1-PHENYL-3-PIPERIDINO-PROPANE HYDROCHLORIDE * N-(3-BENZAMIDO-3-PHENYL)PROPYL PIPERIDINE HYDROCHLORIDE

THR: Poison by ingestion, intramuscular, sub-cutaneous, and intravenous routes. When heated to decomposition it emits very toxic fumes of HCl and NO_x.

DIGITALIS   HR: 3
NIOSH: IH 1990000

PROP: Dried whole leaf of *digitalis purpurea*.
Composition: digitoxin (0.2-0.4%), etc.

SYN: FOXGLOVE

THR: Poison by intraperitoneal and possibly
other routes. 2.5 g or 30 cc of the tincture is
a toxic dose. An overdose can be fatal. It has
been implicated in aplastic anemia. It contains
digitalin, digitalein, digitonin, and digitoxin (the
most toxic component).

DIGITONIN   HR: 3
CAS: 11024-24-1   NIOSH: IH 2050050
mf: $C_{56}H_{92}O_{29}$ mw: 1229.48

SYN: DIGITIN

THR: Poison by ingestion, intravenous, intraperitoneal, and subcutaneous routes. See also
digitalis. When heated to decomposition it emits
acrid smoke and irritating fumes.

DIGITOXIN   HR: 3
CAS: 71-63-6   NIOSH: IH 2275000
mf: $C_{41}H_{64}O_{13}$ mw: 765.05

SYNS: DIGITALIN * DIGITALINE CRISTALLISEE
* DIGITALINE NATIVELLE * DIGITALINUM VE-
RUM * DIGITOPHYLLIN * DIGITOXIGENIN TRI-
DIGITOXOSIDE

THR: Poison by ingestion, intradermal, intravenous, intraperitoneal, and possibly other routes.
Central nervous system and cardiovascular system effects. An eye irritant. See also digitalis.
When heated to decomposition it emits acrid
smoke and irritating fumes.

DIGLYCIDYL ETHER   HR: 3
CAS: 2238-07-5   NIOSH: KN 2350000
mf: $C_6H_{10}O_3$ mw: 130.16

PROP: Liquid.

SYNS: DGE * DI(2,3-EPOXY)PROPYL ETHER

OSHA PEL: TWA CL 0.5 ppm
DFG MAK: 0.1 ppm

THR: Poison by ingestion, inhalation, and intravenous routes. Moderately toxic by skin contact. Severe eye and skin irritant. An experimental tumorigen. Chronic exposure causes bone
marrow depression. Suspected carcinogen. See
also ethers for hazard information.

DIGOXIN   HR: 3
CAS: 20830-75-5   NIOSH: IH 6125000
mf: $C_{41}H_{64}O_{14}$ mw: 781.05

PROP: White crystalline powder; mp: 265°.
Glycoside isolated from *Digitalis Lanata*.

SYNS: DIGOXINE * LANOXIN

THR: Poison by ingestion, intraperitoneal, intravenous, intraduodenal, subcutaneous, and intramuscular routes. See also digitalis. When
heated to decomposition it emits acrid and irritating fumes.

epsilon-DIGOXIN ACETATE   HR: 3
NIOSH: IH 6335000
mf: $C_{43}H_{66}O_{15}$ mw: 823.09

SYN: EPSILON-ACETYLDIGOXIN (GERMAN)

THR: Poison by ingestion and intravenous
routes. See also digitalis. When heated to decomposition it emits acrid and irritating fumes.

DIHEPTYLMERCURY   HR: 3
CAS: 51622-02-7   NIOSH: OW 2450000
mf: $C_{14}H_{30}Hg$ mw: 399.03
OSHA PEL: TWA 10 ug(Hg)/m^3; CL 40

THR: Poison by intraperitoneal route. See also
mercury. When heated to decomposition it emits
toxic fumes of Hg.

DIHEXYLAMINE   HR: 3
CAS: 143-16-8   NIOSH: IH 6600000
mf: $C_{12}H_{27}N$ mw: 185.40

PROP: Liquid; bp: 233-243°, flash p: 220°F
(OC), d: 0.78, vap d: 6.38.

SYN: DI-N-HEXYLAMINE

THR: Poison by ingestion, skin contact, and
intravenous routes. A mild skin irritant. Flammable when exposed to heat or flame; can react
with oxidizing materials. To fight fire, use CO_2,
dry chemical. When heated to decomposition
it emits toxic fumes of NO_x.

DIHYDANTOIN   HR: 3
CAS: 57-41-0   NIOSH: MU 1050000
mf: $C_{15}H_{12}N_2O_2$ mw: 252.29

SYNS: DIFENILHIDANTOINA (SPANISH) * DI-
LANTIN * DIPHENYLHYDANTOIN * 5,5-DIPHE-
NYLHYDANTOIN * 5,5-DIPHENYLIMIDAZOLIDIN-
2,4-DIONE * NCI-C55765

THR: Poison by ingestion, subcutaneous, intravenous, and intraperitoneal routes. Psycho-

tropic, central nervous system, glandular, skin, and other systemic effects by ingestion. An experimental carcinogen, neoplastigen, and teratogen by several routes. When heated to decomposition it emits toxic fumes of NO_x. For further information see Diphenyl Hydantoin, Vol. 1, No. 5 of *DPIM Report*.

1,2-DIHYDROBENZ(e)ACEANTHRYLENE   HR: 3
CAS: 3697-25-4   NIOSH: CU 0700000
mf: $C_{20}H_{14}$ mw: 254.34

SYN: 4,10-ACE-1,2-BENZANTHRACENE; BENZ(E)-ACEANTHRENE

THR: An experimental tumorigen. When heated to decomposition it emits acrid smoke and irritating fumes.

1,2-DIHYDROBENZO(a)ANTHRACENE   HR: 3
CAS: 60968-08-3   NIOSH: CW 1777000
mf: $C_{18}H_{14}$ mw: 230.32

SYN: 1,2-DIHYDROBENZ(A)ANTHRACENE

THR: An experimental neoplastigen by skin contact. When heated to decomposition it emits acrid smoke and irritating fumes.

3,4-DIHYDROBENZO(a)ANTHRACENE   HR: 3
CAS: 60968-01-6   NIOSH: CW 1778000
mf: $C_{18}H_{14}$ mw: 230.32

SYN: 3,4-DIHYDROBENZ(A)ANTHRACENE

THR: An experimental neoplastigen by skin contact. When heated to decomposition it emits acrid smoke and irritating fumes.

6,13-DIHYDROBENZO(e)(1)BENZOTHIO-PYRANO(4,3-b)INDOLE   HR: 3
CAS: 10023-25-3   NIOSH: DE 6300000
mf: $C_{19}H_{13}NS$ mw: 287.39

SYN: BENZO(E)(1)BENZOTHIOPYRANO(4,3-b)IN-DOLE,6,13-DIHYDRO-

THR: An experimental neoplastigen. Mutagenic data. When heated to decomposition it emits very toxic fumes of SO_x and NO_x.

7,8-DIHYDROBENZO(a)PYRENE   HR: 3
CAS: 17673-23-8   NIOSH: DJ 4965000
mf: $C_{20}H_{14}$ mw: 254.34

THR: An experimental neoplastigen. Mutagenic data. When heated to decomposition it emits acrid smoke and irritating fumes.

9,10-DIHYDROBENZO(e)PYRENE   HR: 3
NIOSH: DJ 4965500
mf: $C_{20}H_{14}$ mw: 254.34

SYN: 9,10-H2 B(E)P

THR: An experimental neoplastigen. Mutagenic data. When heated to decomposition it emits acrid smoke and irritating fumes.

meso-DIHYDROCHOLAN-THRENE   HR: 3
NIOSH: FZ 2800000
mf: $C_{20}H_{16}$ mw: 256.36

SYN: 6,12B-DIHYDROCHOLANTHRENE

THR: An experimental tumorigen. When heated to decomposition it emits acrid smoke and irritating fumes.

DIHYDROCHOLESTEROL   HR: 3
CAS: 360-68-9   NIOSH: FZ 6350000
mf: $C_{27}H_{48}O$ mw: 388.75

SYNS: 3-BETA-CHOLESTANOL * 3-BETA-HY-DROXYCHOLESTANE * KOPROSTERIN (GERMAN)

THR: An experimental neoplastigen. When heated to decomposition it emits acrid smoke and irritating fumes.

DIHYDROCODEINE   HR: 3
CAS: 125-28-0   NIOSH: QD 1680000
mf: $C_{18}H_{23}NO_3$ mw: 301.42

SYNS: DIHYDRIN * 7,8-DIHYDROCODEINE * 6-HYDROXY-3-METHOXY-N-METHYL-4,5-EPOX-YMORPHINAN

THR: Poison by ingestion and subcutaneous routes. Can cause drug dependency with repeated doses. When heated to decomposition it emits toxic fumes of NO_x.

DIHYDROCODEINE BITAR-TRATE   HR: 3
CAS: 5965-13-9   NIOSH: QD 1729000
mf: $C_{18}H_{23}NO_3 \cdot C_4H_6O_6$ mw: 451.52

SYNS: DIHYDRO-CODEINE TARTRATE (1:1) * DIHYDROCODEINE ACID TARTRATE

THR: Poison by ingestion, subcutaneous, and parenteral routes. When heated to decomposition it emits toxic fumes of NO_x.

9,10-DIHYDRO-1,2,5,6-DIBENZ-ANTHRACENE   HR: 3

CAS: 57816-08-7   NIOSH: HN 3675000

mf: $C_{22}H_{16}$ mw: 280.38

SYN: 7,14-DIHYDRODIBENZ(A,B)ANTHRACENE

THR: An experimental tumorigen by skin contact. When heated to decomposition it emits acrid smoke and irritating fumes.

9,10-DIHYDRO-9,10-DIHYDROXY-BENZO(a)PYRENE   HR: 3

CAS: 24909-09-9   NIOSH: DJ 5010000

mf: $C_{20}H_{14}O_2$ mw: 286.34

SYN: 9,10-DIHYDROBENZO(A)PYRENE-9,10-DIOL

THR: An experimental neoplastigen and tumorigen by skin contact. When heated to decomposition it emits acrid smoke and irritating fumes.

trans-4,5-DIHYDRO-4,5-DIHYDROXY-BENZO(a)PYRENE   HR: 3

CAS: 37571-88-3   NIOSH: DJ 4990000

mf: $C_{20}H_{14}O_2$ mw: 286.34

SYN: trans-4,5-DIHYDROBENZO(A)PYRENE-4,5-DIOL

THR: An experimental neoplastigen by skin contact. Mutagenic data. When heated to decomposition it emits acrid smoke and irritating fumes.

trans-3,4-DIHYDRO-3,4-DIHYDROXY-DMBA   HR: 3

NIOSH: CW 1795050

mf: $C_{20}H_{18}O_2$ mw: 290.38

SYN: trans-3,4-DIHYDRO-3,4-DIHYDROXY-7,12-DIMETHYLBENZ(a)ANTHRACENE

THR: An experimental neoplastigen by skin contact. Mutagenic data. When heated to decomposition it emits acrid smoke and irritating fumes.

trans-8,9-DIHYDRO-8,9-DIHYDROXY-DMBA   HR: 3

NIOSH: CW 1795060

mf: $C_{20}H_{18}O_2$ mw: 290.38

SYN: trans-8,9-DIHYDRO-8,9-DIHYDROXY-7,12-DIMETHYLBENZ(a)ANTHRACENE

THR: An experimental tumorigen by skin contact. Mutagenic data. When heated to decomposition it emits acrid smoke and irritating fumes.

trans-8,9-DIHYDRO-8,9-DIHYDROXY-7-METHYLBENZ(a)ANTHRACENE   HR: 3

CAS: 64521-15-9   NIOSH: CW 1797000

mf: $C_{18}H_{16}O_2$ mw: 264.34

THR: An experimental tumorigen by skin contact. Mutagenic data. When heated to decomposition it emits acrid smoke and irritating fumes.

7,8-DIHYDRO-7,8-DIHYDROXY-5-METHYLCHRYSENE   HR: 3

NIOSH: GC 1175000

mf: $C_{19}H_{16}O_2$ mw: 276.35

SYN: 7,8-DIHYDRO-5-METHYL-7,8-CHRYSENE-DIOL

THR: An experimental neoplastigen by skin contact. Mutagenic data. When heated to decomposition it emits acrid and irritating fumes.

5,6-DIHYDRO-7,12-DIMETHYL-BENZ(a)-ANTHRACENE   HR: 3

CAS: 35281-29-9   NIOSH: CW 1800000

mf: $C_{20}H_{18}$ mw: 258.38

THR: An experimental carcinogen by skin contact. When heated to decomposition it emits acrid smoke and irritating fumes.

3,4-DIHYDRO-1,11-DIMETHYL-CHRYSENE   HR: 3

CAS: 52171-93-4   NIOSH: GC 0800000

mf: $C_{20}H_{18}$ mw: 258.38

THR: An experimental neoplastigen by skin contact. Mutagenic data. When heated to decomposition it emits acrid smoke and irritating fumes.

16,17-DIHYDRO-11,17-DIMETHYLCY-CLOPENTA(a)PHENANTHRENE   HR: 3

CAS: 5831-16-3   NIOSH: GY 5680000

mf: $C_{19}H_{17}$ mw: 245.36

SYN: 11,17-DIMETHYL-16,17-DIHYDRO-15H-CYCLOPENTA(A)PHENANTHRENE

THR: An experimental tumorigen by skin contact. When heated to decomposition it emits acrid smoke and irritating fumes.

(±)-9,10-DIHYDRO-N,N-DIMETHYL-2-TRIFLUOROMETHYL)-9-ANTHRACENE PROPANAMINE   HR: 3

CAS: 35764-71-7   NIOSH: CB 0730000

mf: $C_{20}H_{22}F_3N$ mw: 333.43

SYNS: (±)-9,10,DIHYDRO-N,N-DIMETHYL2-
(TRIFLUORMETHYL)-9ENTHRACENPROPANAMIN
(GERMAN) * SKF 25971

THR: Poison by ingestion. When heated to decomposition it emits very toxic fumes of F⁻ and NO_x.

BP-7,8-DIHYDRODIOL   HR: 3
CAS: 61443-57-0   NIOSH: DJ 5000000
mf: $C_{20}H_{14}O_2$ mw: 286.34

SYN: (+,−)-trans-7,8-DIHYDROXY-7,8-DIHY-
DROBENZO(a)PYRENE

THR: An experimental carcinogen, neoplastigen, and tumorigen. Mutagenic data. When heated to decomposition it emits acrid smoke and fumes.

BP-7,8-DIHYDRODIOL   HR: 3
CAS: 60864-95-1   NIOSH: DJ 5003000
mf: $C_{20}H_{14}O_2$ mw: 286.34

SYN: (−)-trans-7,8-DIHYDROXY-7,8-DIHYDRO-
BENZO(a)PYRENE

THR: An experimental neoplastigen by skin contact. Mutagenic data. When heated to decomposition it emits acrid smoke and fumes.

5,10-DIHYDRO-5,10-DIOX-ONAPHTHO(2,3-b)-p-DITHIIN-2,3-DICARBONITRILE   HR: 3
CAS: 3347-22-6   NIOSH: QL 0700000
mf: $C_{14}H_4N_2O_2S_2$ mw: 296.32

SYNS: 2,3-DICARBONITRILO-1,4-DIATHIAAN-
THRACHINON (GERMAN) * 2,3-DICYANO-1,4-
DITHIA-ANTHRAQUINONE * 2,3-DINITRILO-1,4-
DITHIOANTHRACHINON (GERMAN) * 2,3-DINI-
TRILO-1,4-DITHIA-ANTHRAQUINONE
* 1,4-DITHIAANTHRAQUINONE-2,3-DINITRILE

THR: Poison by ingestion. Moderately toxic by other routes. See also nitriles. When heated to decomposition it emits very toxic fumes of NO_x, SO_x and CN⁻.

DIHYDROERGOTAMINE   HR: 3
CAS: 511-12-6   NIOSH: II 3675000
mf: $C_{33}H_{37}N_5O_5$ mw: 583.65

SYN: DEHYDROERGOTAMINE

THR: Poison by intravenous and subcutaneous routes. When heated to decomposition it emits toxic fumes of NO_x.

DIHYDROERGOTAMINE TARTRATE (2:1)   HR: 3
CAS: 5989-77-5   NIOSH: KE 8000000
mf: $C_{66}H_{74}N_{10}O_{10} \cdot C_4H_6O_6$ mw: 1317.60

THR: Poison by intravenous, intraperitoneal, and subcutaneous routes. When heated to decomposition it emits toxic fumes of NO_x.

DIHYDRO-beta-ERYTHROIDINE HYDROBROMIDE   HR: 3
CAS: 29734-68-7   NIOSH: KF 3325000
mf: $C_{16}H_{21}NO_3 \cdot BrH$ mw: 356.30

THR: Poison by intravenous route. When heated to decomposition it emits very toxic fumes of Br⁻ and NO_x.

DIHYDRO-beta-ERYTHROIDINE HYDROCHLORIDE   HR: 3
CAS: 1328-90-1   NIOSH: KF 3500000
mf: $C_{16}H_{21}NO_3 \cdot ClH$ mw: 311.84

THR: Poison by ingestion and subcutaneous routes. When heated to decomposition it emits very toxic fumes of HCl and NO_x.

6,13-DIHYDRO-4-FLUORO-BENZO(g)(1)-BENZOTHIOPYRANO(4,3-b)IN-DOLE   HR: 3
NIOSH: DE 6330000
mf: $C_{19}H_{12}FNS$ mw: 305.38

THR: Mutagenic data. When heated to decomposition it emits very toxic fumes of F⁻, NO_x and SO_x.

6,11-DIHYDRO-4-FLUORO(1)BENZO-THIOPYRANO(4,3-b)INDOLE   HR: 3
NIOSH: DL 9940000
mf: $C_{15}H_{10}FNS$ mw: 255.32

THR: An experimental neoplastigen. Mutagenic data. When heated to decomposition it emits very toxic fumes of F⁻, SO_x, and NO_x.

6,7-DIHYDRO-6-(2-HYDROXYETHYL)-5H-DIBENZ(c,e)AZEPINE   HR: 3
CAS: 63918-74-1   NIOSH: HN 9625000
mf: $C_{16}H_{17}NO$ mw: 239.34

SYN: RO 2-3599

THR: Poison by intravenous and intraperitoneal routes. When heated to decomposition it emits toxic fumes of NO_x.

DIHYDROISOCODEINE ACID TARTRATE   HR: 3
CAS: 6414-38-6   NIOSH: QD 1727000
mf: $C_{18}H_{23}NO_3 \cdot C_4H_6O_6$ mw: 451.52

SYNS: DIHYDROISOCODEINE TARTRATE
* DIHYDRO ISOCODEINE TARTRATE (1:1)

THR: Poison by subcutaneous route. See also codeine. When heated to decomposition it emits toxic fumes of NO_x.

2,3-DIHYDRO-9H-ISOXAZOLO(3,2-b)-QUINAZOLIN-9-ONE   HR: 3
CAS: 37795-69-0   NIOSH: NY 3450000
mf: $C_{10}H_8N_2O_2$ mw: 188.20

SYN: w-2429

THR: Poison by ingestion and intravenous routes. When heated to decomposition it emits toxic fumes of NO_x.

12-beta-13-alpha-DIHYDROJERVINE   HR: 3
CAS: 21842-58-0   NIOSH: NY 9050000
mf: $C_{26}H_{39}NO_3$ mw: 413.66

THR: Poison by ingestion. An experimental teratogen by ingestion. When heated to decomposition it emits toxic fumes of NO_x.

15,16-DIHYDRO-11-METHOXYCYCLOPENTA(a)PHENANTHREN-17-ONE   HR: 3
CAS: 5836-85-1   NIOSH: GY 5745000
mf: $C_{18}H_{14}O_2$ mw: 262.32

THR: An experimental carcinogen by skin contact. When heated to decomposition it emits acrid smoke and irritating fumes.

10,11-DIHYDRO-5-(3-METHYL-AMINOPROPYL)-5H-DIBENZ(b,f)-AZEPINE   HR: 3
CAS: 50-47-5   NIOSH: HO 0350000
mf: $C_{18}H_{22}N_2$ mw: 266.42

SYNS: DEMETHYLIMIPRAMINE * DIMETHYL-IMIPRAMINE
* METHYLAMINOPROPYLIMINODIBENZYL
* MONODEMETHYLIMIPRAMINE

THR: Poison by ingestion, intraperitoneal, subcutaneous, and intravenous, and possibly other routes. An experimental teratogen. When heated to decomposition it emits toxic fumes of NO_x.

2,3-DIHYDRO-2-METHYLBENZOPYRANYL-7,N-METHYLCARBAMATE   HR: 3
CAS: 1563-67-3   NIOSH: FB 9625000
mf: $C_{11}H_{13}NO_3$ mw: 207.25

SYN: ENT 27,324

THR: Poison by ingestion and subcutaneous routes. See also carbamates. When heated to decomposition it emits toxic fumes of NO_x.

DIHYDROMORPHINE HYDROCHLORIDE   HR: 3
CAS: 1421-28-9   NIOSH: QD 4200000
mf: $C_{17}H_{21}NO_3 \cdot ClH$ mw: 323.85

SYNS: DILAUDID HYDROCHLORIDE * PARA-MORFAN

THR: Poison by subcutaneous and intravenous routes. Moderately toxic by ingestion. See also morphine. When heated to decomposition it emits very toxic fumes of NO_x and HCl.

DIHYDROMORPHINONE   HR: 3
CAS: 466-99-9   NIOSH: QD 2500000
mf: $C_{17}H_{19}NO_3$ mw: 285.37

SYNS: HYDROMORPHONE * HYMORPHAN
* LAUDICON

THR: Poison by ingestion and subcutaneous route. See also (-)morphine. When heated to decomposition it emits toxic fumes of NO_x.

DIHYDRONE HYDROCHLORIDE   HR: 3
CAS: 124-90-3   NIOSH: QD 2470000
mf: $C_{18}H_{21}NO_4 \cdot ClH$ mw: 351.86

SYNS: DIHYDROOXYCODEINONE HYDROCHLORIDE * DIHYDROXYCODEINONE HYDROCHLORIDE * 4,5-EPOXY-14-HYDROXY-3-METHOXY-17-METHYL-MORPHINAN-6-ONE, HYDROCHLORIDE, (5-ALPHA)- (9CI)

THR: Poison by intravenous and subcutaneous routes. When heated to decomposition it emits very toxic fumes of NO_x and HCl. See also (-)morphine.

2,3-DIHYDROPHORBOL MYRISTATE ACETATE   HR: 3
                 NIOSH: GZ 0670000
mf: $C_{36}H_{58}O_8$ mw: 618.94

SYNS: DPMA * 2,3-DIHYDROPHORBOL ACE-TATE MYRISTATE

THR: An experimental neoplastigen. When heated to decomposition it emits acrid smoke and irritating fumes.

DIHYDROPYRAN HR: 3
CAS: 110-87-2 NIOSH: UP 7700000
mf: C_5H_8O mw: 84.13

PROP: Colorless, mobile liquid, ethereal odor, bp: 85.6°, flash p: 0°F, d: 0.923 @ 20°/4°, vap d: 2.90.
DOT Classification: Label: Flammable
Liquid

THR: Poison by intraperitoneal route. Dangerous fire hazard when exposed to heat or flame; can react vigorously with oxidizing materials. Dangerous! Keep away from heat and open flame! To fight fire, use alcohol foam, CO_2, dry chemical. When heated to decomposition it emits acrid smoke and irritating fumes.

DIHYDROSAFROLE HR: 3
CAS: 94-58-6 NIOSH: DA 6125000
mf: $C_{10}H_{12}O_2$ mw: 164.22

PROP: An oily liquid, bp: 228°, d: 1.0695 @ 20°.

SYNS: 1,2-(METHYLENEDIOXY)-4-PROPYLBEN-ZENE * 5-PROPYL-1,3-BENZODIOXOLE
* 4-PROPYL-1,2-METHYLENEDIOXYBENZENE

THR: An experimental carcinogen. Moderately toxic by ingestion and intraperitoneal route. A skin irritant. When heated to decomposition it emits acrid smoke and irritating fumes.

DIHYDROSTILBESTROL HR: 3
CAS: 84-16-2 NIOSH: SL 0600000
mf: $C_{18}H_{22}O_2$ mw: 270.40

SYNS: MESO-3,4-BIS(P-HYDROXYPHENYL)-N-HEXANE * 3,4-BIS(P-HYDROXYPHENYL)HEXANE
* 4,4'-(1,2-DIETHYLETHYLENE)DIPHENOL
* DIHYDRODIETHYLSTILBESTROL * 4,4'-DIHY-DROXY-ALPHA,BETA-DIETHYLDIPHENYLETHANE
* 4,4'-DIHYDROXY-GAMMA,DELTA-DIPHENYL-HEXANE * GAMMA,DELTA-DI(P-HYDROXYPHE-NYL)-HEXANE * MESO-3,4-DI(P-HYDROXYPHE-NYL)-N-HEXANE

THR: An experimental carcinogen and neoplastigen. When heated to decomposition it emits acrid smoke and irritating fumes.

1,4-DIHYDROXYANTHRA-QUINONE HR: 3
CAS: 81-64-1 NIOSH: CB 6600000
mf: $C_{14}H_8O_4$ mw: 240.22

PROP: Crystals, mp: 194°, bp: 450.0°, vap press: 1 mm @ 196.7°, vap d: 8.3.

SYNS: 1,4-DIHYDROXYANTHRACHINON (CZECH)
* 1,4-DIOXYANTHRAQUINONE (RUSSIAN)
* 1,4-DIHYDROXY-9,10-ANTHRAQUINONE
* QUINIZARIN

THR: Poison by intravenous route. Mutagenic data. An eye irritant. A weak allergen. When heated to decomposition it emits acrid smoke and irritating fumes.

3,3'-DIHYDROXYBENZIDINE HR: 3
CAS: 2373-98-0 NIOSH: DV 4900000
mf: $C_{12}H_{12}N_2O_2$ mw: 216.26

SYN: 3,3'-DIOXYBENZIDINE

THR: An experimental carcinogen and neoplastigen by skin contact, ingestion, and other routes. When heated to decomposition it emits toxic fumes of NO_x.

2,4-DIHYDROXYBENZOPHE-NONE HR: 3
CAS: 131-56-6 NIOSH: DJ 0700000
mf: $C_{13}H_{10}O_3$ mw: 214.23

SYNS: EASTMAN INHIBITOR DHPB * UF 1

THR: Poison by intravenous and intraperitoneal routes. Severe eye irritant. When heated to decomposition it emits acrid smoke and irritating fumes.

DI(2-HYDROXY-n-PROPYL)-AMINE HR: 3
CAS: 53609-64-6 NIOSH: JL 9650000
mf: $H_{14}N_2O_6$ mw: 138.16

SYNS: 2,2'-BISHYDROXYPROPYLNITROSAMINE
* N-BIS(2-HYDROXYPROPYL)NITROSAMINE
* 2,2'-DIHYDROXY-DI-N-PROPYLNITROSOAMINE
* DIISOPROPANOLNITROSAMINE
* N-NITROSOBIS(2-HYDROXYPROPYL)AMINE

THR: An experimental carcinogen, neoplastigen, and tumorigen. Mutagenic data. See also nitrosamines. When heated to decomposition it emits toxic fumes of NO_x.

4,8-DIHYDROXYQUINALDIC ACID HR: 3
CAS: 59-00-7 NIOSH: UZ 9275000
mf: $C_{10}H_7NO_4$ mw: 205.18

PROP: Sulfur-yellow crystals; mp: 286°; insol in water; sol in aq alkali, hydroxides, hot dil HCl.

SYNS: 4,8-DIHYDROXYQUINALDINIC ACID
* 4,8-DIHYDROXYQUINOLINE-2-CARBOXYLIC
ACID * XANTHURENIC ACID

THR: An experimental neoplastigen. When heated to decomposition it emits toxic fumes of NO_x.

2,3-DIHYDROXYTOLUENE   HR: 3
CAS: 488-17-5   NIOSH: UX 1910000
mf: $C_7H_8O_2$ mw: 124.15

SYNS: 2,3-TOLUENEDIOL * 1,2-BENZENE-DIOL, 3-METHYL- * 3-METHYLCATECHOL * 3-METHYLPYROCATECHOL

THR: Poison by intravenous route. Mutagenic data. When heated to decomposition it emits acrid smoke and fumes.

DIHYDROXYVITAMIN D3   HR: 3
CAS: 32222-06-3   NIOSH: FZ 4645000

SYNS: CALCITRIOL * 1a,25-DIHYDROXYCHO-LECALCIFEROL

THR: Poison by ingestion. When heated to decomposition it emits acrid smoke and fumes.

1,3-DIIMINOISOINDOLINE   HR: 3
CAS: 3468-11-9   NIOSH: NR 3500000
mf: $C_8H_7N_3$ mw: 145.18

SYNS: 1,3-DIIMINOISOINDOLIN (CZECH) * 1,3-DIIMINOISOINDOLINE * PHTHALIMIDI-MIDE

THR: Poison by ingestion (acute). An experimental carcinogen. Severe eye and skin irritant. When heated to decomposition it emits toxic fumes of NO_x.

DIIODOHYDROXYQUIN   HR: 3
CAS: 83-73-8   NIOSH: VC 5775000
mf: $C_9H_5I_2NO$ mw: 396.95

SYNS: DIIODOHYDROXYQUINOLINE * 5,7-DI-IODO-8-HYDROXYQUINOLINE * 5,7-DIIODO-OX-INE * 5,7-DIIODO-8-QUINOLINOL * DIODOHY-DROXYQUIN * 8-HYDROXY-5,7-DIIODOQUINOLINE

THR: Poison by ingestion, intravenous, and intraperitoneal routes. Systemic eye effects by ingestion in humans. When heated to decomposition it emits very toxic fumes of I^- and NO_x.

2,6-DIIODO-4-NITROPHENOL   HR: 3
CAS: 305-85-1   NIOSH: SL 0750000
mf: $C_6H_3I_2NO_3$ mw: 390.90

SYN: DIISOPHENOL

THR: Poison by ingestion, intraperitoneal, sub-cutaneous, intravenous, and parenteral routes. When heated to decomposition it emits very toxic fumes of I^- and NO_x.

DIISOBUTYLAMINE   HR: 3
CAS: 110-96-3   NIOSH: TX 1750000
mf: $C_8H_{19}N$ mw: 129.28

PROP: Water white liquid, amine odor; mp: $-70°$, bp: 139°, flash p: 69.8°F, d: 0.745 @ 20°/4°, vap press: 10 mm @ 30.6°, vap d: 4.46.

THR: Poison by ingestion. Dangerous, when exposed to heat or flame; can react vigorously with oxidizing materials. To fight fire, use alcohol foam, CO_2, dry chemical. When heated to decomposition it emits toxic fumes of NO_x.

DIISOBUTYL CARBINOL   HR: 2
CAS: 108-82-7   NIOSH: MJ 3325000
mf: $C_9H_{20}O$ mw: 144.29

PROP: Colorless liquid, bp: 173.3°, fp: $-65°$, flash p: 165°F, d: 0.8121 @ 20°/20°, vap press: 0.3 mm @ 20°, vap d: 4.98, lel = 0.8% @ 212°F, uel = 6.1% @ 212°F.

SYNS: 2,6-DIMETHYL-4-HEPTANOL * 2,6-DI-METHYL HEPTANOL-4

THR: Moderately toxic by ingestion. A powerful systemic irritant by inhalation. A skin and eye irritant. Can cause central nervous system and liver damage when ingested. Flammable when exposed to heat or flame; can react with oxidizing materials. To fight fire, use alcohol foam, foam, CO_2, dry chemical. When heated to decomposition it emits acrid smoke and fumes. For further information see Vol. 1, No. 8 of *DPIM Report*.

DIISOBUTYL KETONE   HR: 2
CAS: 108-83-8   NIOSH: MJ 5775000
DOT: 1157
mf: $C_9H_{18}O$ mw: 142.27

PROP: Liquid, bp: 166°, flash p: 140°F, d: 0.81, vap d: 4.9, lel = 0.8% @ 212°F, uel = 6.2% @ 212°F.

SYNS: DIISOBUTILCHETONE (ITALIAN) * DI-ISOBUTYLCETONE (FRENCH) * DIISOBUTYLKE-TON (DUTCH, GERMAN) * DIISOBUTYL KETONE (DOT) * S-DIISOPROPYLACETONE * 2,6-DIME-THYL-HEPTAN-4-ON (DUTCH, GERMAN) * 2,6-DIMETHYLHEPTAN-4-ONE * 2,6-DI-

METHYL-4-HEPTANONE * 2,6-DIMETIL-EPTAN-4-ONE (ITALIAN) * ISOBUTYL KETONE * ISOVALERONE

OSHA PEL: TWA 50 ppm
DFG MAK: 50 ppm
DOT Classification: Combustible Liquid, Label: None

THR: Moderately toxic by ingestion and inhalation. A mild irritant. Narcotic in high concentrations. See also ketones. Flammable when exposed to heat or flame; can react with oxidizing materials. To fight fire, use CO_2, dry chemical, water spray, mist or fog.

DIISOBUTYLOXOSTANNANE   HR: 3
CAS: 61947-30-6   NIOSH: WH 7310000
mf: $C_8H_{18}OSn$ mw: 248.95

SYNS: KYSLICNIK DIISOBUTYLCINICITY (CZECH) * DIISOBUTYLTIN OXIDE

THR: Poison by ingestion. A severe eye and skin irritant. See also tin compounds. When heated to decomposition it emits acrid smoke and fumes.

DIISOBUTYL PHTHALATE   HR: 3
CAS: 84-69-5   NIOSH: TI 1225000
mf: $C_{16}H_{22}O_4$ mw: 278.38

PROP: Liquid; mp: −64°, flash p: 385°F, d: 1.039-1.043, vap d: 9.59.

SYN: HEXAPLAS M/1B

THR: An experimental teratogen. Moderately toxic by intraperitoneal route. Combustible when exposed to heat or flame. To fight fire, use foam, CO_2, dry chemical. When heated to decomposition it emits acrid smoke and fumes.

1,6-DIISOCYANATOHEXANE   HR: 3
CAS: 822-06-0   NIOSH: MO 1740000
mf: $C_8H_{12}N_2O_2$ mw: 168.22

SYNS: HEXAMETHYLENE-1,6-DIISOCYANATE * 1,6-HEXAMETHYLENE DIISOCYANATE * 1,6-HEXANEDIOL DIISOCYANATE * METHYLENO-BIS-FENYLOIZOCYJANIAN (POLISH) * SZESCIOMETYLENODWUIZOCYJANIAN (POLISH)

THR: Poison by intravenous route. Moderately toxic by ingestion and skin contact. When heated to decomposition it emits toxic fumes of NO_x.

DI-ISO-CYANATOLUENE   HR: 3
CAS: 584-84-9   NIOSH: CZ 6300000
mf: $C_9H_6N_2O_2$ mw: 174.17

PROP: Clear, faintly yellow liquid, bp: 118°-120° @ 10 mm, fp: 20°, d: 1.22 @ 20°/4°, flash p: 270°F (OC), vap d: 6.0, lel = 0.9%, uel = 9.5%.

SYNS: DI-ISOCYANATE DE TOLUYLENE (FRENCH) * 2,4-DIISOCYANATOTOLUENE * DIISOCYANAT-TOLUOL (GERMAN) * 4-METHYL-PHENYLENE DIISOCYANATE * 4-METHYL-PHENYLENE ISOCYANATE * NCI-C50533 * TOLUEEN-DIISOCYANAAT (DUTCH) * TOLUEN-DISOCIANATO (ITALIAN) * TOLUENE DIISOCYANATE * TOLUENE-2,4-DIISOCYANATE * TOLUILENODWUIZOCYJANIAN (POLISH) * TULUYLENDIISOCYANAT (GERMAN) * TOLUYLENE-2,4-DIISOCYANATE * TOLYLENE-2,4-DIISOCYANATE * 2,4-TOLYLENEDIISOCYANATE

OSHA PEL: TWA CL 0.02 ppm
DOT Classification: Label: Poison

THR: Poison by inhalation and intravenous route. Moderately toxic by ingestion. Pulmonary system effects and strong systemic irritation by inhalation. A severe skin and eye irritant. Capable of producing severe dermatitis and bronchial spasm. A common air contaminant. Combustible when exposed to heat or flame. Dangerous; when heated to decomposition it emits highly toxic fumes of NO_x. To fight fire, use dry chemical, CO_2.

DIISOPROPYLAMINE   HR: 2
CAS: 108-18-9   NIOSH: IM 4025000
DOT: 1158
mf: $C_6H_{15}N$ mw: 101.22

PROP: Colorless liquid, bp: 83-84°, flash p: 19.4°F. d: 0.722 @ 220.0°, vap d: 3.5.
OSHA PEL: TWA 5 ppm (skin)
DOT Classification: Label: Flammable Liquid

THR: Moderately toxic by inhalation, ingestion, and several other routes. A severe eye irritant. Inhalation of fumes can cause pulmonary edema. See also amines. Dangerous when exposed to heat or flame; can react vigorously with oxidizing materials. Dangerous! Keep away from heat and open flame! Emits toxic fumes of NO_x. To fight fire, use alcohol foam, foam, CO_2, dry chemical.

DIISOPROPYL ESTER SULFURIC ACID   HR: 3
CAS: 2973-10-6   NIOSH: WS 8050000
mf: $C_6H_{14}O_4S$ mw: 182.26

SYNS: DI-ISOPROPYLSULFAT (GERMAN)
* DI-ISOPROPYLSULFATE * ISOPROPYL SULFATE

THR: An experimental carcinogen. Moderately toxic by ingestion and skin contact. See also esters and sulfates. When heated to decomposition it emits toxic fumes of SO_x.

DIISOPROPYL ETHER   HR: 2
CAS: 108-20-3   NIOSH: TZ 5425000
mf: $C_6H_{14}O$ mw: 102.20

PROP: Colorless liquid, ethereal odor, miscible in water, mp: $-60°$, bp: 68.5°, lel = 1.4%, uel = 7.9%, flash p: $-18°F$ (CC), d: 0.719 @ 25°, autoign temp: 830°F, vap press: 150 mm @ 25°, vap d: 3.52.

SYNS: DIISOPROPYL OXIDE * ETHER ISOPRO-
PYLIQUE (FRENCH) * ISOPROPYL ETHER
* 2-ISOPROPOXYPROPANE * IZOPROPYLOWY
ETER (POLISH)

OSHA PEL: TWA 500 ppm
DOT Classification: Label: Flammable
  Liquid

THR: Moderately toxic. A systemic irritant by inhalation. A skin irritant. An extremely dangerous fire hazard when exposed to heat, flame, sparks, or oxidizers. A severe explosive when exposed to heat or flame. Can form peroxides and explode upon shaking unless treated with sodium sulfite. Violent reaction with chlorosulfonic acid; HNO_3. Dangerous; keep away from heat, sparks or open flame; under some conditions shock will explode it; emits highly toxic fumes; reacts vigorously with oxidizing materials. To fight fire, use alcohol foam, CO_2, foam, dry chemical.

DIISOPROPYLMERCURY   HR: 3
CAS: 1071-39-2   NIOSH: OW 2975000
mf: $C_6H_{14}Hg$ mw: 286.79

PROP: Liquid; bp: 63° @ 10 mm, d: 2.0024, vap d: 9.9.
OSHA PEL: TWA 10 ug(Hg)/m³; CL 40

THR: Poison. See also mercury compounds, organic. When heated to decomposition it emits toxic fumes of Hg.

DIISOPROPYLOXOSTANNANE   HR: 3
CAS: 23668-76-0   NIOSH: WH 7525000
mf: $C_6H_{14}OSn$ mw: 220.89

PROP: Solid. Insol in water.

SYNS: KYSLICNIK DIISOPROPYLCINICITY (CZECH)
* DIISOPROPYLTIN OXIDE

THR: Poison by ingestion. Severe eye and skin irritant. See also tin compounds. When heated to decomposition it emits acrid smoke and irritating fumes.

DIISOPROPYL PARAOXON   HR: 3
CAS: 3254-66-8   NIOSH: TC 2450000
mf: $C_{12}H_{18}NO_6P$ mw: 303.28

SYNS: DIISOPROPYL-P-NITROPHENYL PHOSPHATE
* O,O-DIISOPROPYL-O,p-NITROPHENYL PHOS-
PHATE

THR: Poison by ingestion and intraperitoneal routes. When heated to decomposition it emits very toxic fumes of NO_x and PO_x.

2,6-DIISOPROPYLPHENOL   HR: 3
                              NIOSH: SL 0810000
mf: $C_{12}H_{18}O$ mw: 178.3

PROP: A colorless liquid or solid; bp: 242.4°, fp: 17.9°, flash p: 235°F (CC), d: 0.955 @ 20°/4°.

THR: Poison by intravenous route. Combustible when exposed to heat or flame; can react with oxidizing materials. To fight fire, use foam, CO_2, dry chemical. When heated to decomposition it emits acrid smoke, fumes.

DIISOPROPYLTIN DICHLORIDE   HR: 3
CAS: 38802-82-3   NIOSH: WH 7240000
mf: $C_6H_{14}Cl_2Sn$ mw: 275.79

PROP: Colorless crystals. Sol in water. Mp: 84°

SYN: DICHLORODIISOPROPYLSTANNANE

OSHA PEL: TWA 100 ug(Sn)/m³ (skin)

THR: Poison by intravenous route. See also tin compounds and chlorides. When heated to decomposition it emits toxic fumes of Cl^-.

DILANTIN   HR: 3
CAS: 630-93-3   NIOSH: MU 1400000
mf: $C_{15}H_{11}N_2O_2 \cdot Na$ mw: 274.27

SYNS: DILANTIN SODIUM * DIPHENYLHYDAN-
TOIN SODIUM * 5,5-DIPHENYLHYDANTOIN SO-
DIUM * HYDANTOIN * PHENYTOIN SODIUM
* SODIUM-5,5-DIPHENYL-2,4-IMIDAZOLIDINE-
DIONE * SODIUM DIPHENYLHYDANTOIN
* SODIUM DIPHENYL HYDANTOINATE * SO-
DIUM-5,5-DIPHENYLHYDANTOINATE

THR: Poison by ingestion, subcutaneous, intravenous, and possibly other routes. Glandular system effects by ingestion. An experimental carcinogen and teratogen by several routes. When heated to decomposition it emits very toxic fumes of NO_x.

DILATOL HYDROCHLORIDE HR: 3
CAS: 849-55-8 NIOSH: DO 8300000
mf: $C_{19}H_{25}NO_2 \cdot ClH$ mw: 335.91

SYNS: P-HYDROXY-ALPHA-(1-((1-METHYL-3-
-PHENYLPROPYL)AMINO)ETHYL)BENZYL ALCOHOL
HYDROCHLORIDE * 1-P-HYDROXYPHENYL-
2-(1'-METHYL-3'-PHENYLPROPYLAMINO)-1-PRO-
PANOL HYDROCHLORIDE

THR: Poison by ingestion, intraperitoneal, and intravenous routes. When heated to decomp, it emits very toxic fumes of Cl^- and NO_x.

DILAUDID HR: 3
CAS: 71-68-1 NIOSH: QD 2625000
mf: $C_{17}H_{19}NO_3 \cdot ClH$ mw: 321.83

SYNS: DIHYDROMORPHINONE HYDROCHLORIDE
* HYDROMORPHONE HYDROCHLORIDE

THR: Poison by subcutaneous and intravenous routes. See also morphine. When heated to decomposition it emits very toxic fumes of NO_x and HCl.

DIMECRON HR: 3
CAS: 13171-21-6 NIOSH: TC 2800000
mf: $C_{10}H_{19}ClNO_5P$ mw: 299.72

PROP: Colorless liquid, sol in water and organic solvents.

SYNS: (2-CHLOOR-3-DIETHYLAMINO-1-METHYL-
3-OXO-PROP-1-EN-YL)-DIMETHYL-FOSFAAT
(DUTCH) * (2-CHLOR-3-DIAETHYLAMINO-1-
METHYL-3-OXO-PROP-1-EN-YL)-DIMETHYL-PHOS-
PHAT (GERMAN) * 2-CHLORO-2-DIETHYLCAR-
BAMOYL-1-METHYLVINYL DIMETHYLPHOSPHATE
* 1-CHLORO-DIETHYLCARBAMOYL-1-PROPEN-2-
YL DIMETHYL PHOSPHATE * (2-CLORO-3-
DIETIL-AMINO-1-METIL-3-OXO-PROP-1-EN-IL)-DI-
METIL-FOSFATO (ITALIAN) * DIMETHYL-2-
CHLORO-2-DIETHYLCARBAMOYL-1-METHYLVINYL
PHOSPHATE * O,O-DIMETHYL-O-(2-CHLORO-
2(N,N-DIETHYLCARBAMOYL)-1-METHYLVINYL)-
PHOSPHATE * DIMETHYL DIETHYLAMIDO-
1-CHLOROCROTONYL (2) PHOSPHATE
* DIMETHYL PHOSPHATE OF 2-CHLORO-N,N-
DIETHYL-3-HYDROXYCROTONAMIDE * ENT

25515 * FOSFAMIDON (DUTCH) * FOSFA-
MIDONE (ITALIAN) * NCI-C00588 * PHOS-
PHAMIDON * PHOSPHATE DE DIMETHYLE ET
DE (2-CHLORO-2-DIETHYLCARBAMOYL-1-
METHYL-VINYLE) (FRENCH)

THR: Poison by ingestion, skin contact, subcutaneous, and intravenous routes. Mutagenic data. An experimental carcinogen. When heated to decomposition it emits very toxic fumes of Cl^-, NO_x and PO_x.

DIMEFLINE HR: 3
CAS: 1165-48-6 NIOSH: LK 8400000
mf: $C_{20}H_{21}NO_3$ mw: 323.42

SYNS: 8-(DIMETHYLAMINOMETHYL)-7-ME-
THOXY-3-METHYLFLAVONE * MALIVAN
* N-(7-METHOXY-3-METHYL-4-OXO-2-PHENYL-
4H-CHROMEN-8-YL)METHYL-N,N-DIMETHYLAMINE
* REANIMIL * REMEFLIN

THR: Poison by ingestion, intraperitoneal, intravenous, rectal, and subcutaneous routes. When heated to decomposition it emits very toxic fumes of NO_x.

DIMERCUROUS METHANE ARSO-
NATE HR: 3
CAS: 63869-15-8 NIOSH: PA 2000000
mf: $CH_3AsO_3 \cdot 2Hg$ mw: 539.14

SYN: METHANEARSONIC ACID, DIMERCURY SALT

THR: Poison by intraperitoneal route. See also mercury and arsenic. When heated to decomposition it emits very toxic fumes of As and Hg.

DIMETHISOQUIN HYDROCHLOR-
IDE HR: 3
CAS: 2773-92-4 NIOSH: NW 6950000
mf: $C_{17}H_{24}N_2O \cdot ClH$ mw: 308.89

SYNS: 3-BUTYL-1-(2-(DIMETHYLAMINO)
ETHOXY)ISOQUINOLINE HYDROCHLORIDE
* 1-(BETA-DIMETHYLAMINOETHOXY)-3-N-
BUTYLISOQUINOLINE MONOHYDROCHLORIDE
* QUOTANE HYDROCHLORIDE

THR: Poison by intraperitoneal and intravenous routes. When heated to decomposition it emits very toxic fumes of HCl and NO_x.

DIMETHOATE-ETHYL HR: 3
CAS: 116-01-8 NIOSH: TE 0960000
mf: $C_6H_{14}NO_3PS_2$ mw: 243.30

SYNS: O,O-DIMETHYL S-(N-ETHYLCARBAMOYL-
METHYL) DITHIOPHOSPHATE ∗ O,O-DIMETHYL
S-(N-ETHYLCARBAMOYLMETHYL) PHOSPHORO-
DITHIOATE ∗ ENT-25,506 ∗ S-(N-ETHYLCAR-
BAMOYLMETHYL) DIMETHYL PHOSPHORODI-
THIOATE ∗ N-MONOETHYLAMIDE OF O,O-
DIMETHYLDITHIOPHOSPHORYLACETIC ACID

THR: Poison by ingestion. Moderately toxic by
skin contact. When heated to decomposition it
emits very toxic fumes of NO_x, PO_x, and SO_x.

DIMETHOATE OXYGEN ANALOG   HR: 3
CAS: 1113-02-6   NIOSH: TF 8050000
mf: $C_5H_{12}NO_4PS$ mw: 213.21

SYNS: O,O-DIMETHYL-S-((N-METHYL-CARBA-
MOYL)-METHYL)-MONOTHIOFOSFAAT (DUTCH)
∗ O,O-DIMETHYL-S-(N-METHYL-CARBAMOYL)-
METHYL-MONOTHIOPHOSPHAT (GERMAN)
∗ O,O-DIMETHYL S-((METHYLCARBA-
MOYL)METHYL)PHOSPHOROTHIOATE
∗ O,O-DIMETHYL-S-(N-METHYLCARBAMOYL-
METHYL)PHOSPHOROTHIOATE ∗ O,O-DIMETHYL
S-(N-METHYLCARBAMOYLMETHYL)
PHOSPHOROTHIOLATE ∗ DIMETHYL-S-
(N-METHYL-CARBAMOYL-METHYL)PHOSPHORO-
THIOLATE ∗ O,O-DIMETHYL S-(N-METHYL-
CARBAMOYLMETHYL) THIOPHOSPHATE
∗ O,O-DIMETHYL-S-(2-OXO-3-AZABUTYL)-
MONOTHIOPHOSPHATE ∗ O,O-DIMETIL-S-(N-ME-
TIL-CARBAMOIL)-METIL-MONOTIOFOSFATO (ITAL-
IAN) ∗ ENT 25,776 ∗ PHOSPHOROTHIOIC
ACID, O,O-DIMETHYL S-(2-(METHYLAMINO)-2-
OXOETHYL) ESTER

THR: Poison by ingestion and possibly other
routes. Moderately toxic by skin contact. Mutagenic data. When heated to decomposition it
emits very toxic fumes of NO_x, PO_x, and SO_x.

7,12-DIMETHOXYBENZ(a)ANTHRACENE   HR: 3
CAS: 16354-53-3   NIOSH: CW 2275000
mf: $C_{20}H_{16}O_2$ mw: 288.36

THR: An experimental neoplastigen. When
heated to decomposition it emits acrid smoke
and fumes.

3,3'-DIMETHOXYBENZIDINE DIHYDROCHLORIDE   HR: 3
CAS: 20325-40-0   NIOSH: DD 1050000
mf: $C_{14}H_{15}N_2O_2$ • 2ClH   mw: 317.24

SYN: C.I. DISPERSE BLACK 6 DIHYDROCHLORIDE

THR: An experimental carcinogen by ingestion.
When heated to decomposition it emits very
toxic fumes of NO_x and HCl.

DIMETHOXY-DDT   HR: 3
CAS: 72-43-5   NIOSH: KJ 3675000
mf: $C_{16}H_{15}Cl_3O_2$ mw: 345.66

PROP: Crystals, mp: 78°, vap d: 12.

SYNS: 1,1,1-TRICHLORO-2,2-BIS(P-METHOXY-
PHENOL)ETHANOL ∗ 2,2-BIS(P-ANISYL)-1,1,1-
TRICHLOROETHANE ∗ 1,1-BIS(P-METHOXYPHE-
NYL)-2,2,2-TRICHLOROETHANE ∗ 2,2-BIS(P-
METHOXYPHENYL)-1,1,1-TRICHLOROETHANE
∗ 2,2-DI-P-ANISYL-1,1,1-TRICHLOROETHANE
∗ P,P'-DIMETHOXYDIPHENYLTRICHLOROETHANE
∗ DI(P-METHOXYPHENYL)-TRICHLOROMETHYL
METHANE ∗ ENT 1,716 ∗ P,P'-METHOXY-
CHLOR ∗ METOKSYCHLOR (POLISH) ∗ NCI-
C00497 ∗ 1,1,1-TRICHLORO-2,2-BIS(P-ANI-
SYL)ETHANE ∗ 1,1'-(2,2,2-
TRICHLOROETHYLIDENE)BIS(4-METHOXYBENZENE)
∗ 1,1,1-TRICHLORO-2,2-BIS(P-METHOXYPHE-
NYL)ETHANE ∗ 1,1,1-TRICHLORO-2,2-DI(4-
METHOXYPHENYL)ETHANE

OSHA PEL: TWA 15 mg/m^3

THR: An experimental tumorigen and carcinogen. Moderately toxic by ingestion, intraperitoneal and dermal routes. An irritant and an allergen. An insecticide. Prolonged exposure may
cause kidney injury. See also DDT. Mutagenic
data. Dangerous when heated to decomposition
emits highly toxic chlorides.

3,4-DIMETHOXYDOPAMINE   HR: 3
CAS: 120-20-7   NIOSH: SH 2300000
mf: $C_{10}H_{15}NO_2$ mw: 181.26

PROP: Colorless to pale yellow liquid, mp:
15°, bp: 156° @ 10 mm, d: 1.08 @ 28°/4°,
vap d: 6.25.

SYNS: 3,4-DIMETHOXYPHENETHYLAMINE
∗ 3,4-DIMETHOXY-BETA-PHENETHYLAMINE
∗ DIMETHOXYPHENYLETHYLAMINE ∗ 3,4-DI-
METHOXYPHENYLETHYLAMINE ∗ 3,4-DIME-
THOXY-BETA-PHENYLETHYLAMINE ∗ BETA-(3,
4-DIMETHOXYPHENYL)ETHYLAMINE ∗ 2-(3,4-
DIMETHOXYPHENYL)ETHYLAMINE ∗ 3,4-DIME-
THOXYPHENYLETHYLAMINE (BASE) ∗ DIME-
THYLMESCALINE ∗ HOMOVERATRYLAMINE

THR: A poison via intravenous route; moderately toxic by intraperitoneal route. When heated
to decomposition it emits toxic fumes of NO_x.

DIMETHOXY ETHYL PHTHALATE HR: 3
CAS: 117-82-8   NIOSH: TI 1400000
mf: $C_{14}H_{18}O_6$ mw: 282.32

PROP: Light colored, clear liquid, mild aromatic odor, mp: $-40°$ (forms gel), bp: $190°$-$210°$ @ 4 mm, flash p: $360°F$, d: 1.171 @ $20°/20°$, vap press: 0.3 mm @ $150°$, vap d: 9.75.

THR: An experimental teratogen. Moderately toxic by ingestion, inhalation, and intraperitoneal routes. Combustible when exposed to heat or flame; can react with oxidizing materials. To fight fire, use water, foam, CO_2, dry chemical. When heated to decomposition it emits acrid smoke, fumes.

DIMETHOXYMETHANE HR: 2
CAS: 109-87-5   NIOSH: PA 8750000
DOT: 1234
mf: $C_3H_8O_2$ mw: 76.11

PROP: Colorless liquid, pungent odor, mp: $-104.8°$, bp: $42.3°$, d: 0.864 @ $20°/4°$, vap press: 330 mm @ $20°$, vap d: 2.63, autoign temp: $459°F$, flash p: $-0.4°F$.

SYNS: FORMAL * FORMALDEHYDE DIMETHYLACETAL * METHYLENE DIMETHYL ETHER * METHYLAL (DOT) * METYLAL (POLISH)

OSHA PEL: TWA 1000 ppm
ACGIH TLV: TWA 1000 ppm
DFG MAK: 1000 ppm
DOT Classification: Label: Flammable
  Liquid

THR: Can cause injury to lungs, liver, kidneys, and the heart. A narcotic and anesthetic in high concentrations. Flammable when exposed to heat, flame or oxidizers. Moderately explosive when exposed to heat or flame. Dangerous upon exposure to heat or flame; can react vigorously with oxidizing materials. To fight fire, use foam, CO_2, dry chemical. When heated to decomposition it emits acrid smoke, fumes.

2,5-DIMETHOXY-4-METHYLAMPHETAMINE
CAS: 15588-95-1   NIOSH: SH 2975000
mf: $C_{13}H_{21}NO_2$ mw: 223.35

SYNS: DOM * STP

THR: A poison via ingestion and intraperitoneal routes. When heated to decomposition it emits toxic fumes of NO_x.

2,5-DIMETHOXY-4-METHYLAMPHETAMINE HYDROCHLORIDE HR: 3
CAS: 15589-00-1   NIOSH: SH 2445000
mf: $C_{12}H_{19}NO_2 \cdot ClH$ mw: 245.78

SYNS: AMPHETAMINE, 2,5-DIMETHOXY-4-METHYL-, HYDROCHLORIDE * 2,5-DIMETHOXY-ALPHA,4-DIMETHYLPHENETHYLAMINE HYDROCHLORIDE

THR: A poison by ingestion, intraperitoneal, and intravenous routes. When heated to decomposition it emits very toxic NO_x and HCl.

3,4-DIMETHOXYPHENETHYLAMINE HYDROCHLORIDE HR: 3
CAS: 635-85-8   NIOSH: SH 3000000
mf: $C_{10}H_{15}NO_2 \cdot ClH$ mw: 217.72

SYN: 3,4-DIMETHOXY-BETA-PHENYLETHYLAMINE HYDROCHLORIDE

THR: A poison via intraperitoneal and intravenous routes. When heated to decomposition it emits very toxic NO_x and HCl.

2,6-DIMETHOXYPHENOL
CAS: 91-10-1   NIOSH: SL 0900000
mf: $C_8H_{10}O_3$ mw: 154.18

SYNS: 1,3-DIMETHYL PYROGALLATE * PYROGALLOL DIMETHYLETHER * PYROGALLOL-1,3-DIMETHYL ETHER

THR: A poison via intravenous route. See also ethers. When heated to decomposition it emits acrid smoke and fumes.

1-(2,5-DIMETHOXYPHENYL)-2-AMINOPROPANE HR: 3
CAS: 24973-25-9   NIOSH: SH 3145000
mf: $C_{11}H_{17}NO_2 \cdot ClH$ mw: 231.75

SYNS: 2,5-DIMETHOXYAMPHETAMINE HYDROCHLORIDE * 2,5-DIMETHOXY-alpha-METHYL-beta-PHENYLETHYLAMINEHYDROCHLORIDE

THR: A poison via intraperitoneal and intravenous routes. Mutagenic data. When heated to decomposition it emits very toxic NO_x and HCl.

1-(3,4-DIMETHOXYPHENYL)-2-AMINOPROPANE HR: 3
CAS: 13078-75-6   NIOSH: SH 3235000
mf: $C_{11}H_{17}NO_2 \cdot ClH$ mw: 231.75

SYNS: 3,4-DIMETHOXYAMPHETAMINE HYDROCHLORIDE * 3,4-DIMETHOXY-ALPHAMETHYL-BETA-PHENYLETHYLAMINEHYDROCHLORIDE

THR: A poison via intraperitoneal and intravenous routes. When heated to decomposition it emits very toxic fumes of NO_x and HCl.

3-(DIMETHOXYPHOSPHINYLOXY)N-METHYL-cis-CROTONAMIDE   HR:3
CAS: 6923-22-4   NIOSH: TC 4375000
mf: $C_7H_{14}NO_5P$ mw: 223.19

PROP: A reddish-brown solid, mild ester odor, bp: 125°.

SYNS: (E)-DIMETHYL 1-METHYL-3-(METHYLAMINO)-3-OXO-1-PROPENYL PHOSPHATE * O,O-DIMETHYL-O-(2-N-METHYLCARBAMOYL-1-METHYL-VINYL)-FOSFAAT (DUTCH) * O,O-DIMETHYL-O-(2-N-METHYLCARBAMOYL-1-METHYL)-VINYL-PHOSPHAT (GERMAN) * O,O-DIMETHYL-O-(2-N-METHYLCARBAMOYL-1-METHYL-VINYL) PHOSPHATE * DIMETHYL-1-METHYL-2-(METHYLCARBAMOYL)VINYLPHOSPHATE, CIS * DIMETHYL PHOSPHATE ESTER OF 3-HYDROXY-N-METHYL-CIS-CROTONAMIDE * DIMETHYL PHOSPHATE OF 3-HYDROXY-N-METHYL-CIS-CROTONAMINE * O,O-DIMETIL-O-(2-N-METILCARBAMOIL-1-METIL-VINIL)-FOSFATO (ITALIAN) * ENT 27,129 * 3-HYDROXY-N-METHYL-CIS-CROTONAMIDE DIMETHYL PHOSPHATE * CIS-1-METHYL-2-METHYL CARBAMOYL VINYL PHOSPHATE * PHOSPHATE DE DIMETHYLE ET DE 2-METHYLCARBAMOYL1-METHYL VINYLE (FRENCH) * PHOSPHORIC ACID, DIMETHYL ESTER, ESTER WITH CIS-3-HYDROXY-N-METHYLCROTONAMIDE

THR: A poison by ingestion, skin contact, inhalation, intraperitoneal, subcutaneous, and intravenous routes. Mutagenic data. When heated to decomposition it emits very toxic NO_x and PO_x.

3-(DIMETHOXYPHOSPHINYLOXY)-N-METHYL-N-METHOXY-cis-CROTONAMIDE   HR: 3
CAS: 25601-84-7   NIOSH: TC 4025000
mf: $C_8H_{16}NO_6P$ mw: 253.22

SYNS: ENT 27625 * 3-HYDROXY-N-METHOXY-N-METHYL-CIS-CROTONAMIDE,DIMETHYL PHOSPHATE

THR: A poison via ingestion and skin contact. When heated to decomposition it emits very toxic fumes of NO_x and PO_x.

2′,5′-DIMETHOXYSTILBEN-AMINE   HR: 3
CAS: 23435-31-6   NIOSH: WJ 3833000
mf: $C_{16}H_{17}NO_2$ mw: 255.34

SYNS: 4-(2,5-DIMETHOXYPHENETHYL)ANILINE * 2,5-DIMETHOXY-4′-AMINOSTILBENE(TRANS) * 4-(2,5-DIMETHOXY)STILBENAMINE

THR: An experimental carcinogen by ingestion. When heated to decomposition it emits toxic fumes of NO_x.

N,N-DIMETHYLACETAMIDE   HR: 3
CAS: 127-19-5   NIOSH: AB 7700000
mf: C_4H_9NO mw: 87.14

PROP: Liquid, mp: −20°, bp: 165°, d: 0.9448 @ 15.5°, vap d: 3.01, vap press: 1.3mm @ 25°, flash p: 171°F (TOC), lel = 2.0%, uel = 11.5% @ 740 mm and 160°.

SYNS: ACETDIMETHYLAMIDE * ACETIC ACID, DIMETHYLAMIDE * DIMETHYLACETONE AMIDE * DIMETHYLAMIDE ACETATE * HALLUCINOGEN * NSC 3138

OSHA PEL: TWA 10 ppm (skin)
ACGIH TLV: TWA 10 ppm
DFG MAK: 10 ppm

THR: An experimental teratogen by several routes. Moderately toxic by skin contact, intravenous, and intraperitoneal routes. Systemic effects by inhalation. A skin irritant. Less toxic than dimethylformamide. Flammable when exposed to heat and flame. A moderate explosion hazard. When heated to decomposition it emits toxic fumes of NO_x. For further information see Vol. 1, No. 5 of *DPIM Report*.

o,o-DIMETHYL-S-(2-(ACETYLAMINO)-ETHYL) DITHIOPHOSPHATE   HR: 3
CAS: 13265-60-6   NIOSH: TE 0350000
mf: $C_6H_{14}NO_3PS_2$ mw: 243.30

SYNS: O,O-DIMETHYL S-(2-ACETYLAMINO-ETHYL) PHOSPHORODITHIOATE * N-((O,O-DIMETHYLPHOSPHORODITHIOYL)ETHYL)ACETAMIDE * ENT 27,346 * NSC 190945 * PHOSPHORODITHIOIC ACID, O,O-DIMETHYL ESTER, S-ESTER WITH N-(2-MERCAPTOETHYL)ACETAMIDE

THR: Poison by ingestion, skin contact, intraperitoneal, subcutaneous, and other routes. See also esters. When heated to decomposition it emits very toxic NO_x, PO_x and SO_x.

o,S-DIMETHYLACETYLPHOSPHORO-AMIDOTHIOATE   HR: 3
CAS: 30560-19-1   NIOSH: TB 4760000
mf: $C_4H_{10}NO_3PS$ mw: 183.18

SYNS: ACEPHAT (GERMAN) * ACETYLPHOS-PHORAMIDOTHIOIC ACID-O,S-DIMETHYL ESTER * ENT 27822

THR: Poison by ingestion. Moderately toxic by skin contact. See also esters. When heated to decomposition it emits very toxic fumes of NO_x, PO_x and SO_x.

2-(3,3-DIMETHYLALLYL)CYCLAZO-CINE   HR: 3
CAS: 359-83-1   NIOSH: PB 8750000
mf: $C_{19}H_{27}NO$ mw: 285.47

SYNS: 2-(3,3-DIMETHYLALLYL)-2′,2′-HY-DROXY-5,9-DIMETHYL-6,7-BENZOMORPHAN * 1,2,3,4,5,6-HEXAHYDRO-3-(3-METHYL-2-BUTENYL)-6,11-DIMETHYL-8-HYDROXY-2,6-METHANO-3-BENZAZOCINE * 2′-HYDROXY-5,9-DIMETHYL-2-(3,3-DIMETHYLALLYL)-6,7-BENZOMORPHAN * DL-2′-HYDROXY-5,9-DI-METHYL-2-(3,3-DIMETHYLALLYL)-6,7-BENZO-MORPHAN * 3-(3-METHYL-2-BUTENYL)-1,2,3,4,5,6-HEXAHYDRO-6,11-DIMETHYL-2,6-METH-ANO-3-BENZAZOCIN-8-OL * 3-(3-METHYL-2-BUTENYL)-1,2,3,4,5,6-HEXAHYDRO-8-HYDROXY-6,11-DIMETHYL-2,6-METHANO-3-BENZAZOCINE

THR: Poison by ingestion, subcutaneous, intramuscular, and intravenous routes. Human musculo-skeletal effects. Can cause drug dependency and other central nervous system effects. When heated to decomposition it emits toxic fumes of NO_x.

2-(3,3-DIMETHYLALLYL)-5-ETHYL-2′-HYDROXY-9-METHYL-6,7-BENZOMORPHAN   HR: 3
CAS: 3639-66-5   NIOSH: PB 9000000
mf: $C_{20}H_{29}NO$ mw: 299.50

SYN: 5-ETHYL-2′-HYDROXY-2(N)-(3-METHYL-2-BUTENYL)-9-METHYL-6,7-BENZOMORPHAN

THR: A poison via subcutaneous and intravenous routes. When heated to decomposition it emits toxic fumes of NO_x.

DIMETHYLAMINE   HR: 3
CAS: 124-40-3   NIOSH: IP 8750000
DOT: 1032/1160
mf: C_2H_7N mw: 45.10

SYNS: DIMETHYLAMINE AQUEOUS SOLUTION (DOT) * DMA

OSHA PEL: TWA 10 ppm

DFG MAK: 10 ppm
DOT Classification: Label: Flammable Liquid

THR: Poison by ingestion. Corrosive to the eyes, skin, and mucous membranes. When heated to decomposition it emits toxic fumes of NO_x. Incompatible with acrylaldehyde; fluorine; maleic anhydride.

DIMETHYLAMINE (ANHY-DROUS)   HR: 3
CAS: 124-40-3   NIOSH: IP 8755000
mf: C_2H_7N mw: 45.10

PROP: Colorless gas, bp: 6.88°, flash p: 0°F, fp: −92.19°, d: 0.6804 @ 0°/4°, autoign temp: 752°F, vap d: 1.55, lel = 2.8%, uel = 14.4%. Sol in water, ether, alc.
DOT Classification: Label: Flammable Gas

THR: Poison by ingestion. An irritant. See also dimethylamine. Very dangerous fire hazard when exposed to heat or flame; keep away from heat and open flame! Can react vigorously with oxidizing materials. Moderately explosive when exposed to flame. When heated to decomposition it emits toxic fumes of NO_x. To fight fire, stop flow of gas, foam, CO_2, dry chemical.

DIMETHYLAMINE BORANE   HR: 3
CAS: 74-94-2   NIOSH: IP 9450000
mf: $C_2H_7N \cdot BH_3$ mw: 58.94

SYN: BORANE, COMPOUND WITH DIMETHYL-AMINE (1:1)

THR: Poison by ingestion, intraperitoneal, and intravenous routes. See also amines. A skin and eye irritant. When heated to decomposition it emits toxic fumes of NO_x.

4-DIMETHYLAMINE-m-CRESYLMETH-YLCARBAMATE   HR: 3
CAS: 2032-59-9   NIOSH: FC 0175000
mf: $C_{11}H_{16}N_2O_2$ mw: 208.29

SYNS: 4-DIMETHYLAMINO-3-CRESYL METHYL-CARBAMATE * (4-DIMETHYLAMINO-3-METHYL-PHENYL)N-METHYL-CARBAMAAT (DUTCH) * (4-DIMETHYLAMINO-3-METHYL-PHENYL)N-METHYL-CARBAMAT (GERMAN) * (4-DIMETH-YLAMINO-3-METHYL-PHENYL)N-METHYL-CARBA-MATE * (4-DIMETILAMINO-3-METIL-FENIL)-N-METIL-CARBAMMATO (ITALIAN) * 4-(DIMETH-YLAMINO)-M-TOLYL METHYLCARBAMATE * ENT 25,784 * N-METHYLCARBAMATE DE

4-DIMETHYLAMINO-3-METHYL PHENYLE (FRENCH)

THR: A poison via ingestion, skin contact, intraperitoneal, subcutaneous and other routes. See also carbamates. When heated to decomposition it emits toxic fumes of NO$_x$.

4-(DIMETHYLAMINE)-3,5-XYLYL-N-METHYLCARBAMATE   HR: 3
CAS: 315-18-4   NIOSH: FC 0700000
mf: C$_{12}$H$_{18}$N$_2$O$_2$ mw: 222.32

PROP: Crystals, mp: 85°; vap press: <0.1 mm @ 139°.

SYNS: 4-(DIMETHYLAMINO)-3,5-DIMETHYLPHENOL METHYLCARBAMATE (ESTER) * 4-(DIMETHYLAMINO)-3,5-DIMETHYLPHENYL N-METHYLCARBAMATE * 4-(DIMETHYLAMINO)-3, 5-XYLENOL, METHYLCARBAMATE (ESTER) * 4-(N,N-DIMETHYLAMINO)-3,5-XYLYL N-METHYLCARBAMATE * 4-DIMETHYLAMINO-3,5-XYLYL METHYLCARBAMATE * 4-DIMETHYLAMINO-3,5-XYLYL N-METHYLCARBAMATE * ENT 25766 * METHYLCARBAMIC ACID, 4-(DIMETHYLAMINO)-3,5-XYLYL ESTER * METHYL-4-DIMETHYLAMINO-3,5-XYLYL CARBAMATE * METHYL-4-DIMETHYLAMINO-3, 5-XYLYL ESTER OF CARBAMIC ACID * NCI-C00544

THR: A poison by ingestion, intraperitoneal, and other routes. An experimental carcinogen, neoplastigen, and tumorigen. Moderate skin toxicity. See also esters and carbamates. When heated to decomposition it emits toxic fumes of NO$_x$. For further information see Vol. 5, No. 3 of *DPIM Report*.

DIMETHYLAMINOANTIPYRINE   HR: 3
CAS: 58-15-1   NIOSH: CD 2625000
mf: C$_{13}$H$_{17}$N$_3$O   mw: 231.33

PROP: Colorless leaflets, somewhat sol in water; mp: 107°-109°.

SYNS: AMINOPYRINE * 4-(DIMETHYLAMINO)ANTIPYRINE * DIMETHYLAMINOAZOPHENE * 4-DIMETHYLAMINO-2,3-DIMETHYL-1-PHENYL-3-PYRAZOLIN-5-ONE * 4-DIMETHYLAMINO-2,3-DIMETHYL-1-PHENYL-5-PYRAZOLONE * DIMETHYLAMINOPHENAZON (GERMAN) * DIMETHYLAMINOPHENAZONE * 4-DIMETHYLAMINOPHENAZONE * 4-DIMETHYLAMINO-1-PHENYL-2,3-DIMETHYLPYRAZOLONE * 1,5-DIMETHYL-4-DIMETHYLAMINO-2-PHENYL-3-PYRAZOLONE

* 2,3-DIMETHYL-4-DIMETHYLAMINO-1-PHENYL-5-PYRAZOLONE * 3-KETO-1,5-DIMETHYL-4-DIMETHYLAMINO-2-PHENYL-2,3-DIHYDROPYRAZOLE * 1-PHENYL-2,3-DIMETHYL-4-DIMETHYLAMINO-PYRAZOL-5-ONE * 1-PHENYL-2,3-DIMETHYL-4-DIMETHYLAMINOPYRAZOLONE-5 * PYRAMIDONE

THR: Poison by ingestion, subcutaneous, intramuscular, intravenous, intraperitoneal, and possibly other routes. Mutagenic data. Can cause bone marrow depression resulting in leucopenia. Has been implicated in development of aplastic anemia. Mixed with NaNO$_2$ (1:1) it is an experimental carcinogen. When heated to decomposition it emits toxic fumes of NO$_x$.

1-(4-DIMETHYLAMINOBENZAL)INDENE   HR: 3
CAS: 443-30-1   NIOSH: NK 8350000
mf: C$_{18}$H$_{17}$N   mw: 247.36

SYNS: (4-DIMETHYLAMINOBENZYLIDENE)INDENE * N,N-DIMETHYL-ALPHA-INDOLYLIDENE-P-TOLUIDINE * 4-(1H-INDEN-1-YLIDENEMETHYL)-N,N-DIMETHYLBENZENAMINE * NSC-80087 * P-TOLUIDINE, N,N-DIMETHYL-ALPHA-INDOLYLIDENE

THR: An experimental tumorigen. Moderately toxic by intraperitoneal route. When heated to decomposition it emits toxic fumes of NO$_x$.

p-DIMETHYLAMINOBENZALRHODANINE   HR: 3
CAS: 536-17-4   NIOSH: VI 8090000
mf: C$_{12}$H$_{12}$N$_2$OS$_2$ mw: 264.38

SYNS: P-(DIMETHYLAMINO)BENZAL-5-RHODANINE * 5-(P-DIMETHYLAMINOBENZAL)RHODANINE * P-DIMETHYLAMINOBENZYLIDENE RHODAMINE * 5-(P-DIMETHYLAMINOBENZOYLIDENE)RHODANINE * USAF PD-20

THR: A poison via intraperitoneal route. When heated to decomposition it emits very toxic fumes of NO$_x$ and SO$_x$.

(DIMETHYLAMINO)CARBONYL CHLORIDE   HR: 3
CAS: 79-44-7   NIOSH: FD 4200000
mf: C$_3$H$_6$ClNO   mw: 107.55

PROP: Liquid; mp: −33°, bp: 165°-167°, d: 1.678 @ 20°/4°, vap d: 3.73.

SYNS: CHLOROFORMIC ACID DIMETHYLAMIDE * DIMETHYLCARBAMIC ACID CHLORIDE

* DIMETHYLCARBAMIC CHLORIDE * DI-METHYLCARBAMIDOYL CHLORIDE * DI-METHYLCARBAMOYL CHLORIDE * N,N-DIMETHYLCARBAMOYL CHLORIDE
* DIMETHYLCARBAMYL CHLORIDE * N,N-DIMETHYLCARBAMYL CHLORIDE

THR: Poison by intraperitoneal route. Moderately toxic by inhalation and ingestion. An experimental carcinogen, tumorigen, and neoplastigen. Mutagenic data. Can cause skin and papillary tumors by skin contact, and squamous cell carcinoma by inhalation. Dangerous; see chlorides. Will react with water or steam to produce toxic and corrosive fumes. When heated to decomposition it emits very toxic fumes of Cl^- and NO_x. A powerful lachrymator.

4-(DIMETHYLAMINO)-2,2-DIPHENYLVALERAMIDE HR: 3
CAS: 60-56-8   NIOSH: YV 4400000
mf: $C_{19}H_{24}N_2O$ mw: 296.45

SYNS: AMINOPENTAMIDE * ALPHA,ALPHA-DI-PHENYL-GAMMA-DIMETHYLAMINOVALERAMIDE
* 3-METHYL-4-DIMETHYLAMINO-2,2-DIPHENYL-BUTYRAMIDE

THR: Poison by ingestion, intraperitoneal, intravenous, and possibly other routes. When heated to decomposition it emits toxic fumes of NO_x.

3-(2-(DIMETHYLAMINO)ETHYL)-INDOLE HR: 3
CAS: 61-50-7   NIOSH: NL 7350000
mf: $C_{12}H_{16}N_2$ mw: 188.30

SYN: N,N-DIMETHYLTRYPTAMINE

THR: Poison by intravenous and intraperitoneal routes. Human central nervous system effects. When heated to decomposition it emits toxic fumes of NO_x.

3-(2-DIMETHYLAMINOETHYL)-5-INDOLOL HR: 3
CAS: 487-93-4   NIOSH: NM 2800000
mf: $C_{12}H_{16}N_2O$ mw: 204.30

SYNS: BUFOTENIN * 3-(BETA-DIMETHYLAMI-NOETHYL)-5-HYDROXYINDOLE * N,N-DI-METHYL-5-HYDROXYTRYPTAMINE * N,N-DIMETHYLSEROTONIN * 5-HYDROXY-N,N-DIMETHYLTRYPTAMINE

THR: Poison by intraperitoneal route. Human psychotropic effects by intravenous route. When

heated to decomposition it emits toxic fumes of NO_x.

DIMETHYLAMINOETHYL METHACRYLATE HR: 3
CAS: 2867-47-2   NIOSH: OZ 4200000
mf: $C_8H_{15}NO_2$ mw: 157.24

PROP: Liquid, sol in water and organic solvents; d: 0.933 @ 25°, bp: 182°-190°, flash p: 165°F (TOC), vap d: 5.4.

SYNS: METHACRYLIC ACID, 2-(DIMETHYLAMI-NO)ETHYL ESTER * N,N-DIMETHYLAMINOETHYL METHACRYLATE * BETA-DIMETHYLAMINO-ETHYL METHACRYLATE * 2-DIMETHYLAMIN-OETHYL METHACRYLATE * 2-(DIMETHYLAMI-NO)ETHYL METHACRYLATE * USAF RH-3

THR: Poison by intraperitoneal route. Skin, eye, and mucous membrane irritant. A powerful lachrymator, an ester. Flammable when exposed to sparks, heat, open flame or oxidizers. To fight fire, use alcohol foam, dry chemical, spray. When heated to decomposition it emits toxic fumes of NO_x.

5-DIMETHYLAMINOETHYLOXYIMINO-5H-DIBENZO(a,d)CYCLOHEPTA-1,4-DIENE HYDROCHLORIDE HR: 3
CAS: 4985-15-3   NIOSH: HP 1150000
mf: $C_{19}H_{22}N_2O \cdot ClH$ mw: 330.89

SYNS: 5-(DIMETHYLAMINOAETHYL-OXYIMINO)-5H-DIBENZO(A,D)CYCLOHEPTA-1,4-DIENHYDRO-CHLORID (GERMAN) * 5-(DIMETILAMINOETILOX-IMINO-5H-DIBENZO(A,D)CICLOEPTA-1,4-DIENE) CLORIDRATO(ITALIAN)

THR: Poison by ingestion, subcutaneous, intravenous, intramuscular, and intraperitoneal routes. When heated to decomposition it emits very toxic fumes of NO_x and HCl.

2-DIMETHYLAMINOFLUORENE HR: 3
CAS: 13261-62-6   NIOSH: LL 5250000
mf: $C_{15}H_{15}N$ mw: 209.31

SYNS: 2-DIMETHYLAMINO-FLUOREN (GERMAN)
* N,N-DIMETHYL-2-AMINOFLUORENE * 2-FLUORENYLDIMETHYLAMINE

THR: An experimental tumorigen. When heated to decomposition it emits toxic fumes of NO_x.

anti-8-(N,N-DIMETHYLAMINOMETHYL)-DIBENZOBICYCLO(3.2.1)OCTADIENE HYDROCHLORIDE HR: 3
NIOSH: PB 9250000

SYN: 10,11-DIHYDRO-N,N-DIMETHYL-5,10-METHANO-5H-DIBENZO(A,D)CYCLOHEPTENE-12-METHANAMINE HCl

THR: Poison by ingestion and intravenous routes. When heated to decomposition it emits very toxic fumes of NO_x and HCl.

trans-2-((DIMETHYLAMINO)METHYL-IMINO)-5-(2-(5-NITRO-2-FURYL)VINYL)-1,3,4-OXADIAZOLE   HR: 3
CAS: 55738-54-0   NIOSH: RO 0800000
mf: $C_{11}H_{12}N_5O_4$ mw: 278.28

THR: An experimental carcinogen. Mutagenic data. When heated to decomposition it emits toxic fumes of NO_x.

2-DIMETHYLAMINOMETHYL-1-(m-METHOXYPHENYL)CYCLOHEXA-NOL   HR: 3
CAS: 2914-77-4   NIOSH: GV 8885000
mf: $C_{16}H_{25}NO_2$ mw: 263.42

THR: Poison by ingestion, subcutaneous, and intravenous routes. When heated to decomposition it emits toxic fumes of NO_x.

5-(DIMETHYLAMINO)-1-NAPHTHALENE-SULFONYL CHLORIDE   HR: 3
CAS: 605-65-2   NIOSH: QK 3688000
mf: $C_{12}H_{12}ClNO_2$ mw: 237.70

SYNS: 1-CHLOROSULFONYL-5-DIMETHYLAMINO-NAPHTHALENE * DIMETHYLAMINONAPHTHA-LENESULFONYL CHLORIDE * 1-DIMETHYLAMI-NONAPHTHALENE-5-SULFONYL CHLORIDE * 1-(DIMETHYLAMINO)-5-NAPHTHALENESULFON-YLCHLORIDE * 5-DIMETHYLAMINONAPHTHYL-5-SULFONYL CHLORIDE

THR: Poison by intravenous route. When heated to decomposition it emits very toxic fumes of Cl^- and NO_x.

5-((p-(DIMETHYLAMINO)PHENYL)AZO)-ISOQUINOLINE   HR: 3
CAS: 63040-64-2   NIOSH: NW 9625000
mf: $C_{17}H_{16}N_4$ mw: 276.37

SYN: N,N'-DIMETHYL-4-(5'-ISOQUINOLINYLAZO)-ANILINE

THR: An experimental tumorigen. When heated to decomposition it emits toxic fumes of NO_x.

4-((4-(DIMETHYLAMINO)PHENYL)AZO)-2,6-LUTIDINE-1-OXIDE   HR: 3
CAS: 7349-99-7   NIOSH: OL 0350000
mf: $C_{15}H_{18}N_4O$ mw: 270.37

SYNS: N,N-DIMETHYL-4-(4'-(2',6'-DIMETHYL-PYRIDYL-1'-OXIDE)AZO)ANILINE * 2,6-DI-METHYLPYRIDINE-1-OXIDE-4-AZO-P-DIMETHYL-ANILINE

THR: An experimental neoplastigen. When heated to decomposition it emits toxic fumes of NO_x.

4-((p-(DIMETHYLAMINO)PHENYL)AZO)-N-METHYLACETANILIDE   HR: 3
CAS: 33804-48-7   NIOSH: AE 2250000
mf: $C_{17}H_{20}N_4O$ mw: 296.41

SYN: N',N'-DIMETHYL-4'-AMINO-N-ACETYL-N-MONOMETHYL-4-AMINOAZOBENZENE

THR: Poison by intraperitoneal route. An experimental carcinogen. When heated to decomposition it emits toxic fumes of NO_x.

5-((p-(DIMETHYLAMINO)PHENYL)AZO)-7-METHYLQUINOLINE   HR: 3
CAS: 17400-65-6   NIOSH: VB 6125000
mf: $C_{18}H_{18}N_4$ mw: 290.40

SYNS: N,N-DIMETHYL-4-(5'-(7'-METHYLQUINO-LYL)AZO)ANILINE * 7'-METHYL-5'-(P-DIMETHYLAMINOPHENYLAZO)QUINOLINE

THR: An experimental carcinogen. When heated to decomposition it emits toxic fumes of NO_x.

5-((p-DIMETHYLAMINO)PHENYL)AZO)-QUINOLINE   HR: 3
CAS: 17416-17-0   NIOSH: VB 5425000
mf: $C_{17}H_{16}N_4$ mw: 276.37

SYNS: N,N-DIMETHYL-P-(5'-QUINOLYLAZO)ANI-LINE * N,N-DIMETHYL-4-(5'-QUINOLYLAZO)-ANILINE

THR: An experimental carcinogen and tumorigen. When heated to decomposition it emits toxic fumes of NO_x.

6-((p-(DIMETHYLAMINO)PHENYL)AZO)-QUINOLINE   HR: 3
CAS: 30041-69-1   NIOSH: VB 5600000
mf: $C_{17}H_{16}N_4$ mw: 276.37

SYNS: N,N-DIMETHYL-4-(6'-QUINOLYLAZO)ANI-LINE * QUINOLINE-6-AZO-P-DIMETHYLANILINE

THR: An experimental carcinogen. When heated to decomposition it emits toxic fumes of NO_x.

2-(p-DIMETHYLAMINOPHENYL)-1,6-DIMETHYLQUINOLINIUM CHLORIDE   HR: 3

CAS: 24220-18-6   NIOSH: VC 3606000
mf: $C_{19}H_{21}N_2 \cdot Cl$ mw: 312.87

THR: Poison by ingestion and intraperitoneal route. When heated to decomposition it emits very toxic fumes of NO_x and Cl^-.

3-(DIMETHYLAMINO)PROPIONITRILE   HR: 3

CAS: 1738-25-6   NIOSH: UG 1575000
mf: $C_5H_{10}N_2$ mw: 98.17

PROP: Liquid; mp: $-43°$, bp: $170°$, d: 0.8617, vap d: 3.35, flash p: $<71.6°F$.

SYN: BETA-DIMETHYLAMINOPROPIONITRILE

THR: Poison by intravenous route. Moderately toxic by ingestion and skin contact. See also nitriles. Flammable when exposed to heat, flame or oxidizers. Dangerous; when heated to decomposition it emits highly toxic fumes of NO_x; can react with oxidizing materials. To fight fire, use foam, CO_2, dry chemical.

5-(3-(DIMETHYLAMINO)PROPYL)-5H-DIBENZ(b,f)AZEPINE   HR: 3

CAS: 303-54-8   NIOSH: HO 1480000
mf: $C_{19}H_{22}N_2$ mw: 278.43

THR: Poison by intravenous route. When heated to decomposition it emits toxic fumes of NO_x.

N-(gamma-DIMETHYLAMINOPROPYL)-IMINODIBENZYL HYDROCHLORIDE   HR: 3

CAS: 13-52-0   NIOSH: HO 1925000
mf: $C_{19}H_{24}N_2 \cdot ClH$ mw: 316.91

SYNS: 5-(3-DIMETHYLAMINOPROPYL)-10,11-DIHYDRO-5H-DIBENZ(B,F)AZEPINEHYDROCHLORIDE * N-(3-DIMETHYLAMINOPROPYL)IMINODIBENZYL HYDROCHLORIDE

THR: Poison to humans by ingestion. Poison by intravenous and intraperitoneal routes. Human central nervous system effects by ingestion. When heated to decomposition it emits very toxic fumes of NO_x and HCl.

5-DIMETHYLAMINO-6-PROPYL-5H-INDENO(5,6-d)-1,3-DIOXOLE HYDROCHLORIDE   HR: 3

NIOSH: NK 8927000
mf: $C_{15}H_{19}NO_2 \cdot ClH$ mw: 281.81

SYNS: 2-PROPYL-3-DIMETHYLAMINO-5,6-METHYLENEDIOXYINDENE HYDROCHLORIDE
* 2-N-PROPYL-3-DIMETHYLAMINO-5,6-METHYLENEDIOXYINDENE HYDROCHLORIDE

THR: Poison by intraperitoneal and intravenous route. When heated to decomposition it emits very toxic fumes of NO_x and HCl.

10-(2-(DIMETHYLAMINO)PROPYL)-PHENOTHIAZINE   HR: 3

CAS: 60-87-7   NIOSH: SO 6825000
mf: $C_{17}H_{20}N_2S$ mw: 284.45

SYNS: (2-DIMETHYLAMINO-2-METHYL)ETHYL-N-DIBENZOPARATHIAZINE * N-(2'-DIMETHYLAMINO-2'-METHYL)ETHYLPHENOTHIAZINE * 10-(2-(DIMETHYLAMINO)-2-METHYLETHYL)PHENOTHIAZINE * N-DIMETHYLAMINO-2-METHYLETHYL THIODIPHENYLAMINE * DIMETHYLAMINO-ISOPROPYL-PHENTHIAZIN (GERMAN) * (DIMETHYLAMINO-2-PROPYL-10-PHENOTHIAZINE HYDROCHLORIDE (FRENCH) * NCI-C60673

THR: Poison by ingestion, intravenous, intramuscular, intraperitoneal, and subcutaneous routes. When heated to decomposition it emits very toxic fumes of NO_x and SO_x.

2-(DIMETHYLAMINO) RESERPILINATE   HR: 3

CAS: 5585-67-1   NIOSH: VG 6570000
mf: $C_{26}H_{35}N_3O_5$ mw: 469.64

SYN: RESERPILIN-24-OIC ACID, 2-(DIMETHYLAMINO) ETHYL ESTER

THR: Poison by intraperitoneal route. Moderately toxic by ingestion and subcutaneous routes. When heated to decomposition it emits toxic fumes of NO_x.

DIMETHYLAMINOSUCCINAMIC ACID   HR: 3

CAS: 1596-84-5   NIOSH: WM 9625000
mf: $C_6H_{12}N_2O_3$ mw: 160.20

SYNS: BUTANEDIOIC ACID MONO(2,2-DIMETHYLHYDRAZIDE) * N-DIMETHYL AMINO-BETA-CARBAMYL PROPIONIC ACID * N-DIMETHYLAMINOSUCCINAMIDSAEURE (GERMAN) * N-(DIMETHYLAMINO)SUCCINAMIC ACID * NCI-C03827 * SUCCINIC ACID-2,2-DIMETHYLHYDRAZIDE * SUCCINIC-1,1-DIMETHYL HYDRAZIDE

THR: Poison by intraperitoneal route. An experimental carcinogen. When heated to decomposition it emits toxic fumes of NO_x.

**4-((4-(DIMETHYLAMINO)-o-TOLYL)-
AZO)-2-PICOLINE-1-OXIDE   HR: 3**
CAS: 7347-48-0   NIOSH: TJ 6125000
mf: $C_{15}H_{18}N_4O$ mw: 270.37

SYNS: N,N-DIMETHYL-4-(4'-
(2'METHYLPYRIDYL-1OXIDE)AZO)-O-TOLUIDINE
* N,N'DIMETHYL-3-METHYL-4(4'-(2'-METHYL-
PYRIDYL-1'OXIDE)AZO)ANILINE

THR: An experimental neoplastigen. When
heated to decomposition it emits toxic fumes
of NO_x.

N,N-DIMETHYLANILINE   HR: 3
CAS: 121-69-7   NIOSH: BX 4725000
DOT: 2253
mf: $C_8H_{11}N$ mw: 121.20

PROP: Liquid; mp: 2.5°, bp: 193.1°, flash p:
145°F (CC), d: 0.9557 @ 20°/4°, ulc: 20-25,
autoign temp: 700°F, vap press: 1 mm @ 29.5°,
vap d: 4.17.

SYNS: (DIMETHYLAMINO)BENZENE * N,N-DI-
METHYLBENZENEAMINE * N,N-DIMETHYLPHE-
NYLAMINE * DWUMETYLOANILINA (POLISH)
* NCI-C56428

OSHA PEL: TWA 5 ppm (skin)
ACGIH TLV: TWA 5 ppm
DFG MAK: 5 ppm
DOT Classification: IMO: Poison B

THR: Poison to humans by ingestion. Moderately toxic by skin contact. A skin irritant. Physiological action is similar to but less toxic than
aniline. A central nervous system depressant.
Flammable when exposed to heat, flame or oxidizers. Dangerous; when heated to decomposition it emits highly toxic fumes of aniline; can
react with oxidizing materials such as benzoyl
peroxide. To fight fire, use foam, CO_2, dry
chemical. Incompatible with dibenzoyl peroxide; diisopropyl peroxydicarbonate. For further
information see Vol. 5, No. 3 of *DPIM Report*.

9,10-DIMETHYLANTHRACENE   HR: 3
CAS: 781-43-1   NIOSH: CA 9685000
mf: $C_{16}H_{14}$ mw: 206.30

THR: An experimental carcinogen. Mutagenic
data. When heated to decomposition it emits
acrid smoke and fumes.

DIMETHYLARSINE   HR: 3
mf: C_2H_7As mw: 106.07

PROP: Colorless liquid; bp: 36°; d: 1.213 @
29°/4°; vap d: 3.65.

SYN: CACODYL HYDRIDE

THR: A poison. See also arsine and aresenic
compounds. A dangerous fire hazard by spontaneous ignition in air. When heated to decomposition it emits toxic fumes of As. It is more
toxic than its oxidation products; reacts vigorously with oxidizing agents. To fight fire, ex-
clude O_2 or allow fire to burn or apply water,
foam, dry chemical, water spray, CO_2.

3,6'-DIMETHYLAZOBENZENE   HR: 3
CAS: 28842-05-9   NIOSH: CN 2400000
mf: $C_{14}H_{14}N_2$ mw: 210.30

SYNS: 2:3'-AZOTOLUENE * 2,3'-DIMETHYL-
AZOBENZENE

THR: An experimental neoplastigen and tumorigen. When heated to decomposition it emits
toxic fumes of NO_x.

7,9-DIMETHYLBENZ(c)ACRIDINE   HR: 3
CAS: 963-89-3   NIOSH: CU 3480000
mf: $C_{19}H_{15}N$ mw: 257.35

SYN: 3,10-DIMETHYL-7,8-BENZACRIDINE
(FRENCH)

THR: An experimental tumorigen. Mutagenic
data. When heated to decomposition it emits
toxic fumes of NO_x.

**5,7-DIMETHYL-1,2-BENZACRI-
DINE   HR: 3**
CAS: 53-69-0   NIOSH: CU 3500000
mf: $C_{19}H_{15}N$ mw: 257.35

SYN: 8,10-DIMETHYLBENZ(A)ACRIDINE

THR: Poison by intravenous route. An experimental carcinogen. Mutagenic data. When
heated to decomposition it emits toxic fumes
of NO_x.

**6,9-DIMETHYL-1,2-BENZACRI-
DINE   HR: 3**
CAS: 2381-40-0   NIOSH: CU 3485000
mf: $C_{19}H_{15}N$ mw: 257.35

SYNS: 2,10-DIMETHYL-7,8-BENZACRIDINE
(FRENCH) * 7,10-DIMETHYLBENZ(C)ACRIDINE

THR: An experimental tumorigen. Mutagenic
data. When heated to decomposition it emits
toxic fumes of NO_x.

DIMETHYLBENZANTHRACENE HR: 3
CAS: 57-97-6 NIOSH: CW 3850000
mf: $C_{20}H_{16}$ mw: 256.36

SYNS: 7,12-DIMETHYLBENZANTHRACENE
* 9,10-DIMETHYLBENZ(A)ANTHRACENE
* 7,12-DIMETHYLBENZO(A)ANTHRACENE
* 7,12-DIMETHYLBENZ(A)ANTHRACENE
* 9,10-DIMETHYL-1,2-BENZANTHRACENE
* 9,10-DIMETHYL-1,2-BENZANTHRAZEN (GERMAN) * 1,4-DIMETHYL-2,3-BENZPHENANTHRENE * NCI-C03918

THR: Poison by ingestion, intravenous, and subcutaneous routes. An experimental teratogen, transplacental brain carcinogen, tumorigen, and neoplastigen. Mutagenic data. When heated to decomposition it emits acrid smoke and fumes.

6,8-DIMETHYLBENZ(a)ANTHRACENE HR: 3
CAS: 317-64-6 NIOSH: CW 2975000
mf: $C_{20}H_{16}$ mw: 256.36

SYN: 6,8-DIMETHYL-1,2-BENZANTHRACENE

THR: An experimental neoplastigen. When heated to decomposition it emits acrid smoke and fumes.

6,7-DIMETHYL-1,2-BENZANTHRACENE HR: 3
CAS: 57-97-6 NIOSH: CW 4375000
mf: $C_{20}H_{16}$ mw: 256.36

SYN: 9,10-DIMETHYLBENZ(A)ANTHRACENE

THR: An experimental carcinogen. When heated to decomposition it emits acrid smoke and fumes.

DIMETHYLBENZYLAMINE HYDROCHLORIDE HR: 3
CAS: 1875-92-9 NIOSH: DP 4550000
mf: $C_9H_{13}N \cdot ClH$ mw: 171.69

SYNS: DIMETHYLBENZYLAMMONIUM CHLORIDE
* USAF EL-78

THR: Poison by intraperitoneal route. When heated to decomposition it emits toxic fumes of NO_x and HCl.

N,N-DIMETHYL-N'-BENZYL-N'-(2-PYRIDYL)ETHYLENEDIAMINE HR: 3
NIOSH: KV 1950000
mf: $C_{16}H_{21}N_3$ mw: 255.40

SYNS: 2-(N-BENZYL-N-(2-DIMETHYLAMINO-ETHYL)AMINO)PYRIDINE * N-BENZYL-N',N'-DIMETHYL-N-(2-PYRIDYL)ETHYLENEDIAMINE
* BENZYL-ALPHA-PYRIDYL-DIMETHYL-AETHYL-ENDIAMIN (GERMAN) * N-BENZYL-N-(2-PYRIDYL)-N',N'-DIMETHYL ETHYLENEDIAMINE

THR: Poison by subcutaneous and intravenous route. When heated to decomposition it emits toxic fumes of NO_x.

DIMETHYLBERYLLIUM HR: 3
mf: C_2H_6Be mw: 39.09

PROP: White needles; bp: sublimes @ 200°.

THR: Poison. See also beryllium compounds. Flammable when exposed to heat or flame. Dangerous; when heated to decomposition it emits highly toxic fumes of beryllium oxides; can react with oxidizing materials. Explosive with water. Incompatible with moist air; carbon dioxide.

1,1-DIMETHYLBIGUANIDE HR: 3
CAS: 657-24-9 NIOSH: DU 1790000
mf: $C_4H_{11}N_5$ mw: 129.20

SYNS: N,N-DIMETHYLBIGUANIDE * N,N-DI-METHYLDIGUANIDE * METFORMIN

THR: Poison by subcutaneous and intraperitoneal routes. An experimental teratogen. When heated to decomposition it emits toxic fumes of NO_x.

2,2-DIMETHYLBUTANE HR: 2
CAS: 75-83-2 NIOSH: EJ 9300000
mf: C_6H_{14} mw: 86.20

PROP: Liquid; bp: 49.7°, mp: −98.2°, flash p: −54°F, fp: −101.9°, d: 0.649, autoign temp: 797°F; vap press: 400 mm @ 31.0°, vap d: 3.00, lel = 1.2%, uel = 7.0%.

SYN: NEOHEXANE (DOT)

DOT Classification: Label: Flammable Liquid

THR: Unknown toxicity. Probably an irritant and narcotic in high concentration. An extremely dangerous fire hazard when exposed to heat or flame! Can react vigorously with oxidizing materials. An explosion hazard. To fight fire, use foam, CO_2, dry chemical.

5-(1,3-DIMETHYL-2-BUTENYL)-5-ETHYL-BARBITURIC ACID HR: 3
CAS: 3625-18-1 NIOSH: CQ 4025000
mf: $C_{12}H_{18}N_2O_3$ mw: 238.32

SYNS: MCNEIL 481 ∗ 5-(1,3-DIMETHYL-2-BU-TENYL)-5ETHYL-2,4,6(1H,3H,5H)-PYRIMIDINETRIONE

THR: Poison by ingestion, intravenous, and intraperitoneal routes. When heated to decomposition it emits toxic fumes of NO_x.

1,3-DIMETHYL BUTYLAMINE   HR: 3
CAS: 108-09-8   NIOSH: EO 4460000
mf: $C_6H_{15}N$ mw: 101.22

PROP: A liquid; bp: 106°-109°, flash p: 55°F (OC), d: 0.750 @ 20°/20°.

THR: Poison by ingestion and intravenous routes. Moderately toxic by inhalation and skin contact. See also amines. Dangerous when exposed to heat or flame; can react vigorously with oxidizing materials. When heated to decomposition it emits toxic fumes of NO_x. An explosion hazard. To fight fire, use foam, CO_2, dry chemical.

DIMETHYLCARBAMIC ESTER of 8-OXYMETHYLQUINOLINIUM METHYLSULFATE   HR: 3
CAS: 63680-76-2   NIOSH: VC 3600000
mf: $C_{13}H_{15}N_2O_2 \cdot CH_3O_4S$ mw: 342.40

SYNS: CARBAMIC ACID, N,N-DIMETHYL-, 8-QUINOLINYL ESTER, METHOSULFATE ∗ QUINO-LINIUM, 8-HYDROXY-1-METHYL-, METHYL-SULFATE, DIMETHYLCARBAMATE

THR: Poison by ingestion and intravenous route. See also esters, carbamates. When heated to decomposition it emits very toxic fumes of NO_x and SO_x.

DIMETHYL(2-CHLOROETHYL)AMINE HYDROCHLORIDE   HR: 3
CAS: 4584-46-7   NIOSH: KQ 9020000
mf: $C_4H_{10}ClN \cdot ClH$ mw: 144.06

SYN: 2-CHLORO-N,N-DIMETHYLETHYLAMINE HY-DROCHLORIDE

THR: Poison by intraperitoneal and subcutaneous routes. An experimental neoplastigen. Mutagenic data. When heated to decomposition it emits very toxic fumes of Cl^- and NO_x.

1, 2-DIMETHYLCHRYSENE   HR: 3
CAS: 15914-23-5   NIOSH: GC 0875000
mf: $C_{20}H_{16}$ mw: 256.36

THR: An experimental carcinogen and tumorigen. When heated to decomposition it emits acrid smoke and fumes.

1,11-DIMETHYLCHRYSENE   HR: 3
CAS: 52171-92-3   NIOSH: GC 0895000
mf: $C_{20}H_{16}$ mw: 256.36

SYN: 5,7-DIMETHYLCHRYSENE

THR: An experimental neoplastigen. Mutagenic data. When heated to decomposition it emits acrid smoke and fumes.

4,5-DIMETHYLCHRYSENE   HR: 3
                NIOSH: GC 0970000
mf: $C_{20}H_{16}$ mw: 256.36

THR: An experimental carcinogen. When heated to decomposition it emits acrid smoke and fumes.

DIMETHYLCYANAMIDE   HR: 3
CAS: 1467-79-4   NIOSH: GS 6475000
mf: $C_3H_6N_2$ mw: 70.11

PROP: Colorless, mobile liquid; mp: −41.0°, bp: 160°, flash p: 160°F (TCC), d: 0.8767 @ 30°, vap press: 40 mm @ 80°, vap d: 2.55.

THR: Poison by ingestion, subcutaneous, and intraperitoneal routes. Flammable when exposed to heat, flame or oxidizers. Dangerous; see also cyanide; will react with water or steam to produce toxic NO_x and CN^- and flammable vapors; can react with oxidizing materials. To fight fire, use foam, CO_2, dry chemical. For further information see Vol. 1, No. 7 of *DPIM Report*.

DIMETHYL-1,2-DIBROMO-2,2-DICHLO-ROETHYL PHOSPHATE   HR: 3
CAS: 300-76-5   NIOSH: TB 9450000
mf: $C_4H_7Br_2Cl_2O_4P$ mw: 380.80

PROP: Slightly sol in aliphatic hydrocarbons, very sol in aromatic hydrocarbons; mp: 27.0°.

SYNS: O-(1,2-DIBROM-2,2-DICHLORAETHYL)-O,O-DIMETHYL-PHOSPHAT (GERMAN) ∗ 1,2-DI-BROMO-2,2-DICHLOROETHYL DIMETHYL PHOS-PHATE ∗ O-(1,2-DIBROMO-2,2-DICLORO-ETIL)-O,O-DIMETIL-FOSTATO (ITALIAN) ∗ O-(1,2-DIBROOM-2,2-DICHLOOR-ETHYL)-O,O-DIMETHYL-FOSFAAT (DUTCH) ∗ O,O-DIMETHYL-O-(1,2-DIBROMO-2,2-DICHLOROETHYL)-PHOS-PHATE ∗ O,O-DIMETHYL O-2,2-DICHLORO-1, 2-DIBROMOETHYL PHOSPHATE ∗ ENT 24988 ∗ PHOSPHATE DE O,O-DIMETHLE ET DE O-(1,2-DIBROMO-2,2-DICHLORETHYLE) (FRENCH)

OSHA PEL: TWA 3 mg/m^3
ACGIH TLV: TWA 3 mg/m^3

THR: Poison by inhalation and ingestion. Moderately toxic by skin contact. A severe skin irritant. Mutagenic data. An insecticide of the cholinesterase inhibitor type, and a nonsystemic acaricide. See also parathion and phosphorous compounds, inorganic. Dangerous; see also chlorides, bromides. When heated to decomposition it emits very toxic fumes of Br⁻, Cl⁻ and PO$_x$. For further information see Vol. 5, No. 3 of *DPIM Report*.

DIMETHYL DICHLOROVINYL PHOSPHATE   HR: 3
CAS: 62-73-7   NIOSH: TC 0350000
mf: $C_4H_7Cl_2O_4P$ mw: 220.98

PROP: Liquid, slightly sol in water and glycerine; miscible with aromatic and chlorinated hydrocarbon solvents and alcohol; bp: 120° @ 14 mm, bp: 77° @ 1 mm.

SYNS: (2,2-DICHLOOR-VINYL)-DIMETHYL-FOSFAAT (DUTCH) * DICHLOORVO (DUTCH) * DICHLORFOS (POLISH) * 2,2-DICHLOROETHENYL DIMETHYL PHOSPHATE * 2,2-DICHLOROETHENYL PHOSPHORIC ACID DIMETHYL ESTER * DICHLOROPHOS * (2,2-DICHLORO-VINIL)DIMETILFOSFATO (ITALIAN) * 2,2-DICHLOROVINYL DIMETHYL PHOSPHATE * 2,2-DICHLOROVINYL DIMETHYL PHOSPHORIC ACID ESTER * DICHLOROVOS * (2,2-DICHLOR-VINYL)-DIMETHYL-PHOSPHAT (GERMAN) * O-(2,2-DICHLORVINYL)-O,O-DIMETHYLPHOSPHAT (GERMAN) * DIMETHYL 2,2-DICHLOROETHENYL PHOSPHATE * DIMETHYL 2,2-DICHLOROVINYL PHOSPHATE * O,O-DIMETHYL DICHLOROVINYL PHOSPHATE * O,O-DIMETHYL O-2,2-DICHLOROVINYL PHOSPHATE * O,O-DIMETHYL-O-(2,2-DICHLOR-VINYL)-PHOSPHAT (GERMAN) * ENT 20738 * NCI-C00113 * PHOSPHATE DE DIMETHYLE ET DE 2,2-DICHLOROVINYLE (FRENCH) * VAPONA * VAPONITE

OSHA PEL: TWA 1 mg/m³ (skin)

THR: Poison by inhalation, skin contact, ingestion, subcutaneous, intravenous, intraperitoneal, and possibly other routes. Human blood effects by inhalation. An experimental carcinogen. Mutagenic data. A cholinesterase inhibitor, it is used in flea (pest) collars for pets. No neurotoxicity has been observed. It is very rapidly metabolized and excreted. When heated to decomposition it emits very toxic fumes of Cl⁻ and PO$_x$. For further information see Dichlorovos, Vol. 1 No. 3, of *DPIM Report*.

2,6-DIMETHYL-1,1-DIETHYLPIPERIDINIUM BROMIDE   HR: 3
CAS: 19072-57-2   NIOSH: TN 4900000
mf: $C_{11}H_{24}BrN$ mw: 250.27

SYNS: AGILENE * SC-1950

THR: Poison by ingestion, intraperitoneal, and intravenous routes. When heated to decomposition it emits very toxic fumes of Br⁻ and NO$_x$.

5,5-DIMETHYLDIHYDRORESORCINOL DIMETHYLCARBAMATE   HR: 3
CAS: 122-15-6   NIOSH: FA 1500000
mf: $C_{11}H_{17}NO_3$ mw: 211.29

SYNS: DIMETHYLCARBAMATE DE 5,5-DIMETHYL DIHYDRORESORCINOL (FRENCH) * 5,5-DIMETHYL-4,5-DIHYDRO-3RESORCYL-DIMETHYL-CARBAMAT (GERMAN) * (5,5-DIMETHYL-3-OXO-CYCLOHEX-1-EN-YL)-N,N-DIMETHYL-CARBAMAAT (DUTCH) * (5,5-DIMETHYL-3-OXO-CYCLOHEX-1-EN-YL)-N,N-DIMETHYL-CARBAMAT (GERMAN) * 5,5-DIMETHYL-3-OXO-1-CYCLOHEXEN-1-YL DIMETHYLCARBAMATE * 5,5-DIMETHYL-3-OXOCYCLOHEX-1-ENYL DIMETHYL-CARBAMATE * (5,5-DIMETIL-3-OXO-CICLOES-1-EN-IL)-N,N-DIMETIL-CARBAMMATO (ITALIAN) * ENT 24,738

THR: Poison by ingestion and possibly other routes. See also carbamates. When heated to decomposition it emits toxic fumes of NO$_x$.

2,3-DIMETHYL-8-(DIMETHYLAMINO-METHYL)-7-METHOXYCHROMONE HYDROCHLORIDE   HR: 3
CAS: 38035-28-8   NIOSH: GB 8075000
mf: $C_{15}H_{19}NO_3 \cdot ClH$ mw: 297.81

THR: Poison by ingestion, intraperitoneal, and subcutaneous routes. When heated to decomposition it emits very toxic fumes of NO$_x$ and HCl.

9,9-DIMETHYL-10-DIMETHYLAMINO-PROPYLACRIDAN HYDROGEN TARTRATE   HR: 3
CAS: 3759-07-7   NIOSH: AR 9275000
mf: $C_{20}H_{26}N_2 \cdot C_4H_4O_6$ mw: 442.56

SYNS: DIMETHACRINE TARTRATE * 10-(3-(DIMETHYLAMINO)PROPYL)-9,9-DIMETHYL-ACRIDAN TARTRATE (1:1) * 9,9-DIMETHYL-10-(3-DIMETHYLAMINO)PROPYLACRIDINE TARTRATE * DIMETACRINE BITARTRATE * DIMETACRIN HYDROGENTARTRATE

THR: Poison by ingestion, intravenous, and intraperitoneal routes. When heated to decomposition it emits toxic fumes of NO_x.

o,o-DIMETHYL-o-(p-(N,N-DIMETHYL-SULFAMOYL) PHENYL)-PHOSPHOROTHIOATE   HR: 3

CAS: 52-85-7   NIOSH: TF 7650000
mf: $C_{10}H_{16}NO_5PS_2$ mw: 325.36

PROP: Crystalline powder, very sol in chloroform and carbon tetrachloride, sltly sol in water, mp: 55°.

SYNS: o-(4-((DIMETHYLAMINO)SULFONYL)PHE-NYL) o,o-DIMETHYL PHOSPHOROTHIOATE * ENT 25,644

THR: Poison by ingestion and intramuscular routes. Moderately toxic by skin contact. A cholinesterase inhibitor. See also parathion. Dangerous; see phosphates, NO_x and SO_x.

2,5-DIMETHYLDINITROSOPIPERA-ZINE   HR: 3

CAS: 55556-88-2   NIOSH: TL 6205000
mf: $C_6H_{14}N_4O_2$ mw: 174.24

PROP: Mixture approximately 25% cis and 75% trans conformers.

SYN: 2,5-DIMETHYL-DNPZ

THR: An experimental carcinogen. Mutagenic data. When heated to decomposition it emits toxic fumes of NO_x.

2,6-DIMETHYLDINITROSOPIPERA-ZINE   HR: 3

CAS: 55380-34-2   NIOSH: TL 6210000
mf: $C_6H_{14}N_4O_2$ mw: 174.24

SYN: 2,6-DIMETHYL-DNPZ

THR: An experimental tumorigen. Mutagenic data. When heated to decomposition it emits toxic fumes of NO_x.

N,N-DIMETHYL-2,2-DIPHENYLACET-AMIDE   HR: 2

CAS: 957-51-7   NIOSH: AB 8050000
mf: $C_{16}H_{17}NO$ mw: 239.34

PROP: White solid, very slightly sol in water; mod sol in acetone, dimethyl formamide, and phenyl cellosolve. Mp: 134.5°-135.5°.

SYNS: N,N,DIMETHYL-1,2-DIPHENYLACETAMIDE * 2,2-DIPHENYL-N,N-DIMETHYLACETAMIDE

THR: Moderately toxic by ingestion, intraperitoneal, subcutaneous, and possibly other routes. When heated to decomposition it emits toxic fumes of NO_x.

2,2'-DIMETHYLDIPROPYLINITROSO-AMINE   HR: 3

CAS: 997-95-5   NIOSH: JL 9655000
mf: $C_8H_{18}N_2O$ mw: 158.28

SYN: N-NITROSO-2,2'-DIMETHYLDI-N-PROPYLA-MINE

THR: An experimental tumorigen and neoplastigen. Mildly toxic by subcutaneous route. When heated to decomposition it emits toxic fumes of NO_x.

o,o-DIMETHYL DITHIOPHOSPHORYL-ACETIC ACID-N-METHYL-N-FORMYL-AMIDE   HR: 3

CAS: 2540-82-1   NIOSH: TE 1050000
mf: $C_6H_{12}NO_4PS_2$ mw: 257.28

SYNS: ENT 27,257 * o,o-DIMETHYL S-(N-FORMYL-N-METHYLCARBAMOYLMETHYL) PHOS-PHORODITHIOATE * o,o-DIMETHYL-S-(3-METHYL-2,4-DIOXO-3-AZA-BUTYL)-DITHIOFOS-FAAT (DUTCH) * o,o-DIMETHYL-S-(3-METHYL-2,4-DIOXO-3-AZA-BUTYL)-DITHIOPHOSPHAT (GERMAN) * o,o-DIMETHYL S-(N-METHYL-N-FORMYLCARBAMOYLMETHYL)PHOSPHORODITHIO-ATE * o,o-DIMETHYL PHOSPHORODITHIOATE N-FORMYL-2-MERCAPTO-N-METHYLACETAMIDE S-ESTER * o,o-DIMETIL-S-(N-FORMIL-N-METIL-CARBAMOIL-METIL)-DITIOFOSFATO (ITALIAN) * S-(2-(FORMYLMETHYLAMINO)-2-OXOETHYL) o,o-DIMETHYLPHOSPHORODITHIOATE * N-FORMYL-N-METHYLCARBAMOYLMETHYL o,o-DIMETHYL PHOSPHORODITHIOATE * S-(N-FORMYL-N-METHYLCARBAMOYLMETHYL) o,o-DI-METHYL PHOSPHORODITHIOATE * S-(N-FORMYL-N-METHYLCARBAMOYLMETHYL) DIMETHYL PHOSPHOROTHIOLOTHIONATE * o,o-DIMETHYL-S-(N-METHYL-N-FORMYL-CARBAMOYLMETHYL)-DITHIOPHOSPHAT (GERMAN)

THR: Poison by ingestion, skin contact, and intravenous routes. See also esters. When heated to decomposition it emits very toxic fumes of NO_x, PO_x, and SO_x.

(o,o-DIMETHYLDITHIOPHOSPHORYL-PHENYL)ACETIC ACID ETHYL ESTER   HR: 3

NIOSH: AI 7875000
mf: $C_{12}H_{17}O_4PS_2$ mw: 320.38

SYNS: O,O-DIMETHYL S-(1-CARBOETHOXYBEN-
ZYL) DITHIOPHOSPHATE * O,O-DIMETHYL S-AL-
PHAETHOXY-CARBONYLBENZYL PHOSPHORODI-
THIOATE * O,O-DIMETHYL-S-(PHENYLACETIC
ACID ETHYL ESTER) PHOSPHORODITHIOATE
* O,O-DIMETHYL S-(PHENYL)(CARBOETHOXY)-
METHYL PHOSPHORODITHIOATE * (DIMETHYL
S-(PHENYLETHOXYCARBONYLMETHYL)PHOS-
PHOROTHIOLOTHIONATE) * ENT 27386GC
* S-alpha-ETHOXYCARBONYLBENZYL O,O-DI-
METHYL PHOSPHORODITHIOATE * S-alpha-ETH-
OXYCARBONYLBENZYL DIMETHYL PHOSPHOROTHI-
OLOTHIONATE * ETHYL alpha-((DIMETHOXY-
PHOSPHENOTHIOYL)THIO)BENZENEACETATE
* ETHYL-O,O-DIMETHYL PHOSPHORODITHIOYL-
PHENYL ACETATE * ETHYL ESTER OF O,O-DI-
METHYLDITHIOPHOSPHORYL alpha-PHENYL
ACETATE ACID * ETHYL MERCAPTOPHENYL-
ACETATE O,O-DIMETHYL PHOSPHOROCITHIOATE

THR: Poison by ingestion and other unspecified
routes. See also esters. When heated to decomposition it emits very toxic fumes of PO_x and
SO_x.

6,7-DIMETHYLESCULETIN   HR: 3
CAS: 120-08-1   NIOSH: GN 6550000
mf: $C_{11}H_{10}O_4$ mw: 206.21

SYNS: AESCULETIN DIMETHYL ETHER * 6,7-
DIMETHOXYCOUMARIN * ESCULETIN DIMETHYL
ETHER * SCOPARONE

THR: Poison by ingestion and intraperitoneal
routes. When heated to decomposition it emits
acrid smoke and fumes.

O,O-DIMETHYL-S-ETHYLSULPHONIO-
ETHYLMETHYL PHOSPHOROTHIO-
LATE   HR: 3
          NIOSH: TF 8800000
mf: $C_7H_{18}O_3PS_2$ mw: 245.34

SYN: PHOSPHOROTHIOIC ACID, O,O-DIMETHYL
S-ETHYLMETHYLSULFONIOETHYL ESTER, INNER
SALT

THR: Poison by ingestion and intravenous
routes. When heated to decomposition it emits
very toxic fumes of PO_x and SO_x.

7,8-DIMETHYLFLUORANTHENE   HR: 3
CAS: 38048-87-2   NIOSH: LL 4375000
mf: $C_{18}H_{14}$ mw: 230.32

THR: An experimental neoplastigen. When
heated to decomposition it emits acrid smoke
and fumes.

7,12-DIMETHYL-5-FLUOROBENZ(a)-
ANTHRACENE   HR: 3
          NIOSH: CW 4410000
mf: $C_{20}H_{15}F$ mw: 274.35

SYN: 5-FLUORO-7,12-DIMETHYLBENZ(A)AN-
THRACENE

THR: An experimental carcinogen. When
heated to decomposition it emits toxic fumes
of F^-.

7,12-DIMETHYL-11-FLUOROBENZ(a)-
ANTHRACENE   HR: 3
CAS: 2023-61-2   NIOSH: CW 4460000
mf: $C_{20}H_{15}F$ mw: 274.35

SYN: 11-FLUORO-7,12-DIMETHYLBENZ(A)AN-
THRACENE

THR: An experimental neoplastigen and carcinogen. When heated to decomposition it emits
toxic fumes of F^-.

N,N-DIMETHYL-p-((p-FLUOROPHE-
NYL)AZO)ANILINE   HR: 3
CAS: 150-74-3   NIOSH: BX 5595000
mf: $C_{14}H_{14}FN_3$ mw: 243.31

SYNS: 4'-FLUORO-N,N-DIMETHYL-4-
AMINOAZOBENZENE * 4'-FLUORO-P-DIMETH-
YLAMINOAZOBENZENE * 4'-FLUORO-4-DI-
METHYLAMINOAZOBENZENE
* 4'-FLUORO-N,N-DIMETHYL-P-PHENYLAZOANI-
LINE * P-((P-FLUOROPHENYL)AZO)-N,N-DI-
METHYLANILINE * 4-((4-FLUOROPHENYL)AZO)-
N,N-DIMETHYLBENZENAMINE

THR: An experimental tumorigen. When heated
to decomposition it emits very toxic fumes of
F^- and NO_x.

N,N-DIMETHYL-p-(3-FLUOROPHENYL-
AZO)ANILINE   HR: 3
CAS: 332-54-7   NIOSH: BX 5950000
mf: $C_{14}H_{14}FN_3$ mw: 243.31

SYN: 3'-FLUORO-4-DIMETHYLAMINOAZOBENZENE

THR: An experimental teratogen, neoplastigen,
tumorigen. Moderately toxic by subcutaneous
route. When heated to decomposition it emits
very toxic fumes of F^- and NO_x.

DIMETHYLFLUOROPHOSPHATE HR: 3
CAS: 5954-50-7   NIOSH: TE 6125000
mf: $C_2H_6FO_3P$ mw: 128.05

PROP: Liquid; mp: low, bp: 149°, d: 1.28, vap d: 4.42.

SYN: FLUOPHOSPHORIC ACID, DIMETHYL ESTER

THR: Poison by inhalation and intravenous routes. See also esters. Dangerous; see fluorides and phosphates.

DIMETHYL FORMAMIDE   HR: 3
CAS: 68-12-2   NIOSH: LQ 2100000
DOT: 2265
mf: C_3H_7NO mw: 73.11

PROP: Colorless, mobile liquid, bp: 152.8°, lel = 2.2% @ 100°, uel = 15.2% @ 100°, flash p: 136°, fp: −61°, d: 0.9445 @ 25°/4°, autoign temp: 833°F, vap press: 3.7 mm @ 25°, vap d: 2.51.

SYNS: N-FORMYLDIMETHYLAMINE * DIMETH-YLFORMAMID (GERMAN) * N,N-DIMETHYL FOR-MAMIDE * DIMETILFORMAMIDE (ITALIAN) * DWUMETHYLOFORMAMID (POLISH) * NCI-C60913 * NSC 5356

OSHA PEL: TWA 10 ppm (skin)
DOT Classification: Flammable or Combustible Liquid

THR: Poison by skin contact, intravenous, and intraperitoneal routes. Moderately toxic by several other routes. A skin and eye irritant. Central nervous system effects by inhalation. An experimental teratogen. Flammable when exposed to heat or flame; can react with oxidizing materials. Explosion hazard when exposed to flame. Avoid contact with halogenated hydrocarbons; inorganic nitrates; Br_2; CCl_4; CrO_3; (2,5-dimethyl pyrrole + $P(OCl)_3$); C_6Cl_6; organic nitrates; methylene diisocyanates; P_2O_3; $Al(C_2H_5)_3$. To fight fire, use foam, CO_2, dry chemical. For further information see Vol. 1, No. 3 of *DPIM Report*.

DIMETHYLFORMOCARBOTHIAL-DINE   HR: 3
CAS: 533-74-4   NIOSH: XI 2800000
mf: $C_5H_{10}N_2S_2$ mw: 162.29

PROP: Crystals, sol in alc, mp: 107°.

SYNS: CRAG FUNGICIDE 974 * 3,5-DIMETHYL-PERHYDRO-1,3,5-THIADIAZIN-2-THION (CZECH, GERMAN) * 3,5-DIMETHYLTETRAHYDRO-1,3,5-THIADIAZINE-2-THIONE * 3,5-DIMETHYL-1,-2,3,5-TETRAHYDRO-1,3,5-THIADIAZINETHIONE-2 * 3,5-DIMETHYLTETRAHYDRO-1,3,5-2H-THIA-DIAZINE-2-THIONE * 3,5-DIMETHYL-1,3,

5-2H-TETRAHYDROTHIADIAZINE-2-THIONE * 3,5-DIMETHYLTETRAHYDRO-2H-1,3,5-THIA-DIAZINE-2-THIONE * 3,5-DIMETHYL-2-THIONO-TETRAHYDRO-1,3,5-THIADIAZINE * 3,5-DIMETIL-PERIDRO-1,3,5-TIHADIAZIN-2-TIONE (ITALIAN) * MYLON (CZECH) * TETRAHY-DRO-2H-3,5-DIMETHYL-1,3,5-THIADIAZINE-2-THIONE * TETRAHYDRO-3,5-DIMETHYL-2H-1,3,5-THIADIAZINE-2-THIONE * 2-THIO-3,5-DIMETHYLTETRAHYDRO-1,3,5-THIADIAZINE

THR: Poison by ingestion and intraperitoneal routes. Moderately toxic by other routes. A skin and severe eye irritant. A mild primary skin irritant and sensitizer. When heated to decomposition it emits very toxic fumes of NO_x and SO_x.

1,1-DIMETHYLHYDRAZINE   HR: 3
CAS: 57-14-7   NIOSH: MV 2450000
DOT: 1163
mf: $C_2H_8N_2$ mw: 60.12

PROP: Colorless liquid, ammonia-like odor. Hygroscopic, water-miscible, bp: 63.3°, fp: −58°, flash p: 5°F, d: 0.782 @ 25°/4°, vap press: 157 mm @ 25°, vap d: 1.94, autoign temp: 480°F, lel = 2%, uel = 95%.

SYNS: 1,1-DIMETHYLHYDRAZINE (GERMAN) * DIMETHYLHYDRAZINE * ASYMMETRIC DI-METHYLHYDRAZINE * N,N-DIMETHYLHYDRA-ZINE * UNS-DIMETHYLHYDRAZINE * UNSYM-DIMETHYLHYDRAZINE * 1,1-DIMETHYL HYDRAZINE * DIMETHYLHYDRAZINE UNSYMMETRICAL (DOT) * UDMH

OSHA PEL: TWA 1 mg/m^3 (skin)
ACGIH TLV: TWA 0.5 ppm
DOT Classification: Label: Flammable Liquid and Poison

THR: Poison by inhalation, intraperitoneal, and subcutaneous routes. An experimental carcinogen and tumorigen. Dangerous when exposed to heat, flame or oxidizers. Highly dangerous; when heated to decomposition it emits highly toxic fumes of NO_x; can react vigorously with oxidizing materials such as air; H_2O_2; HNO_3; fuming HNO_3; (HNO_3 + N_2O_4); NO. To fight fire, use alcohol foam, CO_2, dry chemical. For further information see Vol. 4, No. 3 of *DPIM Report*.

1,2-DIMETHYL HYDRAZINE   HR: 3
CAS: 540-73-8   NIOSH: MV 2625000

DOT: 2382
mf: $C_2H_8N_2$ mw: 60.12

PROP: Clear, colorless, flammable, hygroscopic liquid, fishy ammonia odor, flash p: < 73.4°F, bp: 81°, mp: −9°, d: 0.8274 @ 20°/4°.

SYNS: N,N'DIMETHYLHYDRAZINE * 1,2-DI-METHYLHYDRAZIN (GERMAN) * SYM-DI-METHYLHYDRAZINE

DOT Classification: Flammable Liquid

THR: Poison by ingestion, inhalation, subcutaneous,and intramuscular routes. An experimental carcinogen and tumorigen. When heated to decomposition it emits toxic fumes of NO_x. For further information see Vol. 4, No. 3 of *DPIM Report*.

1,2-DIMETHYLHYDRAZINE
DIHYDROCHLORIDE   HR: 3
CAS: 306-37-6   NIOSH: MV 2800000
mf: $C_2H_8N_2 \cdot 2ClH$ mw: 133.04

SYNS: N,N'-DIMETHYLHYDRAZINE DIHYDRO-CHLORIDE * SYM-DIMETHYLHYDRAZINE DIHY-DROCHLORIDE

THR: Poison by ingestion and subcutaneous routes. Experimental tumorigen, neoplastigen, and carcinogen. When heated to decomposition it emits very toxic fumes of HCl and NO_x.

1,1-DIMETHYLHYDRAZINE
HYDROCHLORIDE   HR: 3
CAS: 593-82-8   NIOSH: MV 2900000
mf: $C_2H_8N_2 \cdot ClH$ mw: 96.58

THR: Poison by ingestion, intraperitoneal, and intravenous routes. An experimental carcinogen. When heated to decomposition it emits very toxic fumes of HCl and NO_x.

1,2-DIMETHYLHYDRAZINE
HYDROCHLORIDE   HR: 3
CAS: 540-73-8   NIOSH: MV 2940000
mf: $C_2H_8N_2 \cdot ClH$ mw: 96.58

SYN: SYM-DIMETHYLHYDRAZINE HYDROCHLO-RIDE (DMH)

THR: Poison by ingestion, intraperitoneal, and intravenous routes. An experimental neoplastigen, and tumorigen. Mutagenic data. When heated to decomposition it emits very toxic fumes of HCl and NO_x.

DIMETHYL-5-(1-ISOPROPYL-3-METHYL-PYRAZOLYL)-CARBAMATE   HR: 3
CAS: 119-38-0   NIOSH: FA 2100000
mf: $C_{10}H_{17}N_3O_2$ mw: 211.30

SYNS: DIMETHYLCARBAMATE-D'L-ISOPROPYL-3-METHYL-5-PYRAZOLYLE (FRENCH) * ENT 19,060 * ISOLANE (FRENCH) * (1-ISOPROPIL-3-METIL-1H-PIRAZOL-5-IL)-N,N-DIMETIL-CARBAM-MATO (ITALIAN) * (1-ISOPROPYL-3-METHYL-1H-PYRAZOL-5-YL)-N,N-DIMETHYLCARBAMAAT (DUTCH) * (1-ISOPROPYL-3-METHYL-1H-PYRA-ZOL-5-YL)-N,N-DIMETHYL-CARBAMAT (GERMAN) * 1'-ISOPROPYL-3-METHYL-5-PYRAZOLYL DI-METHYLCARBAMATE * 5-METHYL-2-ISOPRO-PYL-3-PYRAZOLYL DIMETHYLCARBAMATE

THR: Poison by skin contact, ingestion, intraperitoneal, and possibly other routes. An experimental carcinogen. When heated to decomposition it emits toxic fumes of NO_x.

o,o-DIMETHYL-S-ISOPROPYL-2-SULFI-NYLETHYLPHOSPHOROTHIO-ATE   HR: 3
CAS: 2674-91-1   NIOSH: TF 8800000
mf: $C_7H_{17}O_4PS_2$ mw: 260.33

SYNS: O,O-DIMETHYL-S-(2-(ETHYLSULFINYL)-ISOPROPYL) PHOSPHOROTHIOATE- * S-(2-(ETH-YLSULFINYL)ISOPROPYL) PHOSPHOROTHIOIC ACID-O,O-DIMETHYL ESTER

THR: Poison by ingestion and intraperitoneal routes. When heated to decomposition it emits very toxic fumes of PO_x and SO_x.

DIMETHYL LEAD DIACETATE   HR: 3
CAS: 20917-34-4   NIOSH: TP 4450000
mf: $C_6H_{12}O_4Pb$ mw: 355.37

SYN: DIACETOXYDIMETHYLPUMBANE

THR: Poison by ingestion. See also lead. When heated to decomposition it emits toxic fumes of Pb.

alpha,beta-DIMETHYLMALEIC
ANHYDRIDE   HR: 3
CAS: 766-39-2   NIOSH: ON 4025000
mf: $C_6H_6O_3$ mw: 126.12

THR: An experimental carcinogen. See also anhydrides. When heated to decomposition it emits acrid smoke and fumes.

3,3-DIMETHYL-1,p-METHOXY-
PHENYLTRIAZENE   HR: 3
CAS: 7203-92-1   NIOSH: XY 1225000
mf: $C_9H_{13}N_3O$ mw: 179.25

SYNS: 1-P-METHOXYFENYL-3,3-DIMETHYL-TRIAZEN (CZECH) * 1-(P-METHOXYPHENYL)-3,3-DIMETHYL-TRIAZENE * 1-(4-METHYLOXY-PHENYL)-3,3-DIMETHYLTRIAZINE

THR: Poison by ingestion and subcutaneous routes. An experimental carcinogen. Mutagenic data. When heated to decomposition it emits toxic fumes of NO_x.

o,o-DIMETHYL-S-(5-METHOXY-1,3,4-THIADIAZOLINYL-3-METHYL)DITHIO PHOSPHATE   HR: 3
CAS: 950-37-8   NIOSH: TE 2100000
mf: $C_6H_{11}N_2O_4PS_3$ mw: 302.34

SYNS: (O,O-DIMETHYL)-S-(-2-METHOXY-delta(SUP 2)-1,3,4-THIADIAZOLIN-5-ON-4-YLMETHYL)DITHIOPHOSPHATE * S-(2,3-DIHY-DRO-5-METHOXY-2-OXO-1,3,4-THIADIAZOL-3-METHYL) DIMETHYL PHOSPHOROTHIOLOTHIONATE * O,O-DIMETHYL-S-(2-METHOXY-1,3,4-THIADIA-ZOL-5(4H)-ONYL-(4)-METHYL) PHOSPHORODI-THIOATE * O,O-DIMETHYL-S-(2-METHOXY-1,3,4-THIADIAZOL-5-(4H)-ONYL-(4)-METHYL)-DI-THIOPHOSPHAT (GERMAN) * O,O-DIMETHYL-S-((2-METHOXY-1,3,4 (4H)-THIODIAZOL-5-ON-4-YL)-METHYL)-DITHIOFOSFAAT (DUTCH) * O,O-DIMETIL-S-((2-METOSSI-1,3,4-(4H)-TIADIZAOL-5-ON-4-IL)-METIL)-DITIFOSFATO (ITALIAN) * ENT 27193 * S-((5-METHOXY-2-OXO-1,3,4-THIADIAZOL-3(2H)-YL)METHYL) O,O-DIMETHYL PHOSPHORODITHIOATE

THR: Poison by ingestion, skin contact, and other routes. Moderately toxic by inhalation. Severe eye irritant. When heated to decomposition it emits very toxic fumes of NO_x, PO_x, and SO_x.

N,N-DIMETHYL-N'-(((METHYLAMINO)-CARBONYL)OXY)PHENYLMETHANI-MIDAMIDE MONOHYDRO-CHLORIDE   HR: 3
CAS: 23422-53-9   NIOSH: FC 2800000
mf: $C_{11}H_{15}N_3O_2 \cdot ClH$ mw: 257.75

SYNS: M-(((DIMETHYLAMINO)METHYLENE)-AMINO)PHENYLMETHYL CARBAMATE,HYDRO-CHLORIDE * 3-DIMETHYLAMINOMETHYLENE-IMINOPHENYL-N-METHYLCARBAMATE, HYDROCHLORIDE * ENT-27566

THR: Poison by ingestion, inhalation, and skin contact. When heated to decomposition it emits very toxic fumes of NO_x and HCl.

o,o-DIMETHYL METHYLCAR-BAMOYLMETHYL PHOSPHORO-DITHIOATE   HR: 3
CAS: 60-51-5   NIOSH: TE 1750000
mf: $C_5H_{12}NO_3PS_2$ mw: 229.27

SYNS: DIMETHOAAT (DUTCH) * DIMETHOATE * DIMETHOAT (GERMAN) * DIMETHOAT TECHNISCH 95% * O,O-DIMETHYL-DITHIOPHOSPHORYLACETIC ACID, N-MONOMETH-YLAMIDE SALT * O,O-DIMETHYL S-(2-(METH-YLAMINO)-2-OXOETHYL) PHOSPHORODITHIOATE * O,O-DIMETHYL-S-(N-METHYL-CAR-BAMOYL)-METHYL-DITHIOFOSFAAT (DUTCH) * (O,O-DIMETHYL-S-(N-METHYL-CARBAMOYL-METHYL)-DITHIOPHOSPHAT) (GERMAN) * O,O-DIMETHYL S-(N-METHYLCARBAMOYLMETHYL) DI-THIOPHOSPHATE * O,O-DIMETHYL S-(N-METH-YLCARBAMOYLMETHYL) PHOSPHORODITHIOATE * O,O-DIMETHYL S-(N-METHYLCARBAMYL-METHYL) THIOTHIONOPHOSPHATE * O,O-DI-METHYL-S-(N-MONOMETHYL)-CARBAMYL METH-YLDITHIOPHOSPHATE * O,O-DIMETHYL-S-(2-OXO-3-AZA-BUTYL)-DITHIOPHOSPHAT (GERMAN) * O,O-DIMETIL-S-(N-METIL-CARBAMOIL-METIL)-DITIOFOSFATO (ITALIAN) * DITHIOPHOSPHATE DE-O,O-DIMETHYLE ET DE S(-N-METHYLCARBA-MOYL-METHYLE)(FRENCH) * ENT 24,650 * S-METHYLCARBAMOYLMETHYL O,O-DIMETHYL PHOSPHORODITHIOATE * N-MONOMETHYLAMIDE OF O,O-DIMETHYLDITHIOPHOSPHORYLACETIC ACID * NC-262 * NCI-C00135 * PHOSPHAMID * PHOSPHORODITHIOIC ACID, O,O-DIMETHYL-S-(2-(METHYLAMINO)-2-OXOETHYL) ESTER (9CI)

THR: Poison by ingestion, skin contact, intrave-nous, intraperitoneal, subcutaneous, and possi-bly other routes. An experimental carcinogen and teratogen. Mutagenic data. See also esters. When heated to decomposition it emits very toxic fumes of NO_x, PO_x, and SO_x.

N',N'-DIMETHYL-N-((METHYLCARBA-MOYL)OXY)-1-METHYLTHIOOXAMI-MIDIC ACID   HR: 3
CAS: 23135-22-0   NIOSH: RP 2300000
mf: $C_7H_{13}N_3O_3S$ mw: 219.29

SYNS: 2-DIMETHYLAMINO-1-(METHYLTHIO)GLY-OXAL-O-METHYLCARBAMOYLMONOXIME * METHYL-2-(DIMETHYLAMINO)-N-(((METHYL-AMINO)CARBONYL)OXY)-2-OXOETHANIMIDOTH-IOATE * S-METHYL-1-(DIMETHYLCARBAMOYL)-N-((METHYLCARBAMOYL)OXY)THIOFORMIMIDATE * METHYL-N',N'-DIMETHYL-N-((METHYLCARBA-MOYL)OXY)-1-THIOOXAMIMIDATE

THR: Poison by ingestion and inhalation. Moderately toxic by skin contact. When heated to decomposition it emits very toxic fumes of NO_x and SO_x.

3,3-DIMETHYL-1-(m-METHYLPHENYL)-TRIAZENE   HR: 3
CAS: 20241-03-6   NIOSH: XY 1300000
mf: $C_9H_{13}N_3$ mw: 163.25

SYNS: 3,3-DIMETHYL-1-(M-TOLYL)TRIAZENE * 1-(M-METHYLPHENYL)-3,3-DIMETHYL-TRIAZENE * 1-(3-METHYLPHENYL)-3,3-DIMETHYL-TRIAZENE

THR: Poison by ingestion and intraperitoneal routes. Moderately toxic by subcutaneous route. An experimental carcinogen. When heated to decomposition it emits toxic fumes of NO_x.

o,o-DIMETHYL-o-(3-METHYL) PHOSPHOROTHIOATE   HR: 3
CAS: 122-14-5   NIOSH: TG 0350000
mf: $C_9H_{12}NO_5PS$ mw: 277.25

SYNS: O,O-DIMETHYL-O-(3-METHYL-4-NITROFENYL)-MONOTHIOFOSFAAT (DUTCH) * O,O-DIMETHYL-O-(3-METHYL-4-NITRO-PHENYL)-MONOTHIOPHOSPHAT (GERMAN) * O,O-DIMETHYL-O-(3-METHYL-4-NITROPHENYL) PHOSPHOROTHIOATE * O,O-DIMETHYL-O-(3-METHYL-4-NITROPHENYL) THIOPHOSPHATE * DIMETHYL-3-METHYL-4-NITROPHENYLPHOSPHOROTHIONATE * O,O-DIMETHYL-O-(4-NITRO-3-METHYLPHENYL)THIOPHOSPHATE * O,O-DIMETHYL-O-4-NITRO-M-TOLYL PHOSPHOROTHIOATE * O,O-DIMETIL-O-(3-METIL-4-NITRO-FENIL)-MONOTIOFOSFATO (ITALIAN) * ENT 25,715 * THIOPHOSPHATE DE O,O-DIMETHYLE ET DE O-(3-METHYL-4-NITROPHENYLE) (FRENCH)

THR: Poison by ingestion, intravenous, intraperitoneal, and possibly other routes. Moderately toxic by skin contact and subcutaneous routes. When heated to decomposition it emits very toxic fumes of NO_x, PO_x and SO_x. For further information see Fenitrothion, Vol. 2, No. 4 of *DPIM Report*.

N,N-DIMETHYL-4-(2-METHYL-4-PYRIDYLAZO)ANILINE-N-OXIDE   HR: 3
CAS: 7347-46-8   NIOSH: BX 6650000
mf: $C_{14}H_{16}N_4O$ mw: 256.34

SYNS: 4-((4-(DIMETHYLAMINO)PHENYL)AZO)-2-PICOLINE 1-OXIDE * N,N-DIMETHYL-4-(4'-(2'-METHYLPYRIDYL-1'-OXIDE)AZO)ANILINE * 2-METHYLPYRIDINE-1-OXIDE-4-AZO-p-DIMETHYLANILINE

THR: An experimental neoplastigen. When heated to decomposition it emits toxic fumes of NO_x.

o,o-DIMETHYL-o-(4-(METHYLSULFONYL)-m-TOLYL) PHOSPHOROTHIOATE   HR: 3
CAS: 3761-42-0   NIOSH: TF 9420000
mf: $C_{10}H_{15}O_5PS_2$ mw: 310.34

SYN: O,O-DIMETHYL O-((4-METHYLTHIO)-M-TOLYL) PHOSPHOROTHIOATE SULFONE

THR: Poison by ingestion and intraperitoneal routes. When heated to decomposition it emits very toxic fumes of PO_x and SO_x.

o,o-DIMETHYL-o-4-(METHYLTHIO)-3,5-XYLYL PHOSPHOROTHIOATE   HR: 3
CAS: 55-37-8   NIOSH: TF 9800000
mf: $C_{11}H_{17}O_3PS_2$ mw: 292.37

SYNS: BAY 37342 * O,O-DIMETHYL O-(3,5-DIMETHYL-4-METHYLTHIOPHENYL) PHOSPHOROTHIOATE * ENT 25684

THR: Poison by ingestion. When heated to decomposition it emits very toxic fumes of PO_x and SO_x.

DIMETHYLMYLERAN   HR: 3
CAS: 55-93-6   NIOSH: MO 2400000
mf: $C_8H_{16}O_6S_2$ mw: 272.36

SYNS: 2,5-DIMETHANESULFOMYLOXYHEXANE * 1,4-DIMETHANESULFONOXY-1,4-DIMETHYLBUTANE * NSC-23890

THR: Poison by intravenous and intraperitoneal routes. Mutagenic data. When heated to decomposition it emits very toxic fumes of SO_x.

N,N-DIMETHYL-1-NAPHTHYLAMINE   HR: 3
CAS: 86-56-6   NIOSH: QM 2825000
mf: $C_{12}H_{13}N$ mw: 171.26

SYNS: 1-DIMETHYLAMINONAPHTHALENE * DIMETHYL-ALPHA-NAPHTHYLAMINE * ALPHA-DIMETHYLNAPHTHYLAMINE * N,N-DIMETHYL-ALPHA-NAPHTHYLAMINE

THR: Poison by intraperitoneal route. Moderately toxic by ingestion. When heated to decomposition it emits toxic fumes of NO_x.

N,N-DIMETHYL-p-(1-NAPHTHYLAZO) ANILINE   HR: 3
CAS: 607-59-0   NIOSH: BX 7000000
mf: $C_{18}H_{17}N_3$ mw: 275.38

SYNS: P-DIMETHYLAMINOBENZENEAZO-1-
NAPHTHALENE * P-DIMETHYLAMINOBENZENE-1-
AZO-1-NAPHTHALENE

THR: An experimental carcinogen. When heated to decomposition it emits toxic fumes of NO$_x$.

DIMETHYLNITRAMINE   HR: 3
CAS: 4164-28-7   NIOSH: IQ 0450000
mf: $C_2H_6N_2O_2$ mw: 90.10

SYNS: DIMETHYLNITRAMIN (GERMAN)
* DIMETHYLNITROAMINE * N-NITRODIMETH-
YLAMINE

THR: Poison by intraperitoneal route. Moderately toxic by ingestion. An experimental tumorigen. When heated to decomposition it emits toxic fumes of NO$_x$.

1,2-DIMETHYL-5-NITROIMID-AZOLE   HR: 3
CAS: 551-92-8   NIOSH: NI 5075000
mf: $C_5H_7N_3O_2$ mw: 141.15

SYN: DIMETRIDAZOLE

THR: An experimental neoplastigen. Mutagenic data. When heated to decomposition it emits toxic fumes of NO$_x$.

3,3-DIMETHYL-1-(p-NITROPHENYL) TRIAZENE   HR: 3
CAS: 7227-92-1   NIOSH: XY 1450000
mf: $C_8H_{10}N_4O_2$ mw: 194.22

SYNS: 1-P-NITROFENYL-3,3-DIMETHYLTRIAZEN
(CZECH) * 1-(P-NITROPHENYL-3,3-DIMETHYL-
TRIAZEN (GERMAN) * 1-(P-NITROPHENYL)-3,3-
DIMETHYL-TRIAZENE * 1-(4-NITROPHENYL)-
3,3-DIMETHYLTRIAZENE

THR: Poison by subcutaneous route. Moderately toxic by ingestion. An experimental neoplastigen and carcinogen. When heated to decomposition it emits toxic fumes of NO$_x$.

2,3-DIMETHYL-4-NITROPYRIDINE-1-OXIDE   HR: 3
                       NIOSH: UT 2807000
mf: $C_7H_8N_2O_3$ mw: 168.17

THR: An experimental carcinogen. Mutagenic data. When heated to decomposition it emits toxic fumes of NO$_x$.

DIMETHYLNITROSAMINE   HR: 3
CAS: 62-75-9   NIOSH: IQ 0525000
mf: $C_2H_6N_2O$ mw: 74.10

PROP: Yellow liquid, sol in water, alcohol and ether. Bp: 152°, d: 1.005 @ 20°/4°.

SYNS: DIMETHYLNITROSAMIN (GERMAN)
* N,N-DIMETHYLNITROSAMINE * DIMETHYLNI-
TROSOAMINE * N-METHYL-N-NITROSOMETHANA-
MINE * N-NITROSODIMETHYLAMINE

THR: Poison by ingestion, inhalation, intraperitoneal, subcutaneous, and possibly other routes. An experimental tumorigen, neoplastigen, and carcinogen. Mutagenic data. See also nitrosamine. Has caused fatal liver disease in humans. A transplacental carcinogen and teratogen. When heated to decomposition it emits toxic fumes of NO$_x$. For further information see N-Nitrosodimethylamine, Vol. 2. No. 6 of *DPIM Report*.

N,N-DIMETHYL-p-NITROSOANI-LINE   HR: 3
CAS: 138-89-6   NIOSH: BX 7175000
mf: $C_8H_{10}N_2O$ mw: 150.20

SYNS: P-(DIMETHYLAMINO)NITROSOBENZENE
* 4-(DIMETHYLAMINO)NITROSOBENZENE
* DIMETHYL(P-NITROSOPHENYL)AMINE
* NCI-C01821 * P-NITROSO-N,N-DIMETHYL-
ANILINE * 4-NITROSODIMETHYLANILINE
* PARANITROSODIMETHYLANILIDE

THR: Poison by ingestion. An experimental carcinogen. When heated to decomposition it emits toxic fumes of NO$_x$. Incompatible with acetic anhydride; acetic acid.

2,6-DIMETHYLNITROSOMORPHO-LINE   HR: 3
CAS: 1456-28-6   NIOSH: QE 2150000
mf: $C_6H_{12}N_2O_2$ mw: 144.20

SYN: NITROSO-2,6-DIMETHYLMORPHOLINE

THR: Poison by ingestion and subcutaneous routes. An experimental carcinogen, tumorigen, and neoplastigen. Mutagenic data. When heated to decomposition it emits toxic fumes of NO$_x$.

1,3-DIMETHYLNITROSOUREA   HR: 3
CAS: 13256-32-1   NIOSH: YT 0350000
mf: $C_3H_7N_3O_2$ mw: 117.13

SYNS: DIMETHYLNITROSOHARNSTOFF (GERMAN)
* N,N'-DIMETHYLNITROSOUREA * 1,3-DI-
METHYL-N-NITROSOUREA

THR: Poison by ingestion and intravenous routes. An experimental tumorigen, teratogen, and carcinogen. When heated to decomposition it emits toxic fumes of NO_x.

DIMETHYLOCTADECYLBENZYL-AMMONIUM CHLORIDE   HR: 3
CAS: 122-19-0   NIOSH: BO 7000000
mf: $C_{27}H_{50}N \cdot Cl$ mw: 424.23

SYNS: BENZYLDIMETHYLSTEARYLAMMONIUM CHLORIDE * BENZYLSTEARYLDIMETHYLAMMONIUM CHLORIDE * DIMETHYLBENZYLOCTADECYLAMMONIUM CHLORIDE * N-OCTADECYL-N-BENZYL-N,N-DIMETHYLAMMONIUMCHLORIDE * OCTADECYLDIMETHYLBENZYLAMMONIUM CHLORIDE * STEARYLDIMETHYLBENZYLAMMONIUM CHLORIDE * TALLOW BENZYL DIMETHYLAMMONIUM CHLORIDE

THR: Poison by intraperitoneal route. Moderately toxic by ingestion. A human skin irritant and severe eye irritant. When heated to decomposition it emits very toxic fumes of NO_x and Cl^-.

3,3-DIMETHYL-2-OXETHANONE   HR: 3
CAS: 1955-45-9   NIOSH: RQ 7525000
mf: $C_5H_8O_2$ mw: 100.13

SYNS: 3,3-DIMETHYL-2-OXETANONE * DIMETHYL PROPIOLACTONE * 3,3-DIMETHYL-BETA-PROPIOLACTONE * NCI-C04126 * PIVALIC ACID LACTONE

THR: Poison by ingestion. An experimental carcinogen. When heated to decomposition it emits acrid smoke and fumes.

DIMETHYLOXOSTANNANE   HR: 3
CAS: 2273-45-2   NIOSH: WH 7526500
mf: C_2H_6OSn mw: 164.77

PROP: White powder. Insol in water.

SYN: DIMETHYLTIN OXIDE

THR: Poison by intravenous route. See also tin compounds. When heated to decomposition it emits smoke and irritating fumes.

DIMETHYL PARANITROPHENYL THIONOPHOSPHATE   HR: 3
CAS: 3820-53-9   NIOSH: TG 0240000
mf: $C_8H_{10}NO_5PS$ mw: 263.22

PROP: Cryst; vap d: 9.1; mp: 38°; d: 1.235 @ 20°/4°.

SYNS: O,O-DIMETHYL-S-P-NITROFENYL ESTER KYSELINY THIOFOSFORECNE (CZECH) * O,O-DIMETHYL S-(P-NITROPHENYL) PHOSPHOROTHIOATE * O,O-DIMETHYL S-(4-NITROPHENYL)THIOPHOSPHATE

THR: Poison by ingestion and subcutaneous routes. When heated to decomposition it emits very toxic fumes of NO_x, PO_x, and SO_x.

alpha,alpha-DIMETHYLPHENETHYL-AMINE   HR: 3
CAS: 122-09-8   NIOSH: SH 4025000
mf: $C_{10}H_{15}N$ mw: 149.26

SYNS: alpha,alpha-DIMETHYL-beta-PHENYLETHYLAMINE * 2-PHENYL-tert-BUTYLAMINE

THR: Poison by ingestion, intravenous, and intraperitoneal routes. When heated to decomposition it emits toxic fumes of NO_x.

N,N-DIMETHYL-p-PHENYLAZO-ANILINE   HR: 3
CAS: 60-11-7   NIOSH: BX 7350000
mf: $C_{14}H_{15}N_3$ mw: 225.32

PROP: Yellow crystalline tablets, insol in water, sol in strong mineral acids and oils.

SYNS: C.I. 11020 * C.I. SOLVENT YELLOW 2 * USAF EK-338 * P-DIMETHYLAMINOAZOBENZEN (CZECH) * DIMETHYLAMINOAZOBENZENE * N,N-DIMETHYL-4-AMINOAZOBENZENE * N,N-DIMETHYL-P-AMINOAZOBENZENE * P-DIMETHYLAMINOAZOBENZENE * 4-DIMETHYLAMINOAZOBENZENE * 4-(N,N-DIMETHYLAMINO)AZOBENZENE * DIMETHYLAMINOAZOBENZOL * P-DIMETHYLAMINO-AZOBENZOL (GERMAN) * 4-DIMETHYLAMINOAZOBENZOL * 4-DIMETHYLAMINOPHENYLAZOBENZENE * N,N-DIMETHYL-P-AZOANILINE * N,N-DIMETHYL-4-(PHENYLAZO)BENZAMINE * N,N-DIMETHYL-4-(PHENYLAZO)BENZENAMINE

THR: Poison by ingestion and intraperitoneal routes. An experimental teratogen, carcinogen, neoplastigen, and tumorigen. Mutagenic data. When heated to decomposition it emits toxic fumes of NO_x. For further information see Vol. 5, No. 3 of *DPIM Report*.

N,N-DIMETHYL-p-PHENYLAZOANILINE-N-OXIDE   HR: 3
CAS: 2747-31-1   NIOSH: BX 7525000
mf: $C_{14}H_{15}N_3O$ mw: 241.32

SYNS: 4-DIMETHYLAMINOAZOBENZENE AMINE-N-OXIDE * N,N-DIMETHYLAMINOAZOBENZENE-N-OXIDE

THR: Poison by intraperitoneal route. Moderately toxic by ingestion. An experimental tumorigen. When heated to decomposition it emits toxic fumes of NO_x.

2-DI(N-METHYL-N-PHENYL-tert-BUTYL CARBAMOYLMETHYL)AMINO-ETHANOL   HR: 3

CAS: 126-27-2   NIOSH: AC 3325000
mf: $C_{28}H_{41}N_3O_3$ mw: 467.72

SYNS: N,N-BIS(N-METHYL-N-PHENYL-TERT-BUTYLACETAMIDO)-BETA-HYDROXYETHYLAMINE * 2,2'-((2-HYDROXYETHYL)IMINO BIS(N-(ALPHA,ALPHA-DIMETHYLPHENETHYL)-N-METHYL-ACETAMIDE

THR: Poison by ingestion, intratracheal, intravenous, intraperitoneal, intramuscular, subcutaneous, and implant routes. When heated to decomposition it emits toxic fumes of NO_x.

N,N-DIMETHYL-p-PHENYLENE-DIAMINE   HR: 3

CAS: 99-98-9   NIOSH: ST 1050000
mf: $C_8H_{12}N_2$ mw: 136.22

THR: Poison by ingestion, inhalation, skin contact, subcutaneous, and intravenous routes. Toxic via human skin. When heated to decomposition it emits toxic fumes of NO_x.

N,N-DIMETHYL-p-PHENYLENE-DIAMINE   HR: 3

CAS: 105-10-2   NIOSH: ST 0875000
mf: $C_8H_{12}N_2$ mw: 136.22

THR: Poison by intraperitoneal route. When heated to decomposition it emits toxic fumes of NO_x.

o,p-DIMETHYL-beta-PHENYLETHYL-HYDRAZINE DIHYDROGENSULFATE   HR: 3

CAS: 154-99-4   NIOSH: MV 3200000

SYNS: BETA-(2,4-DIMETHYLPHENYL)ETHYLHYDRAZINE DIHYDROGEN SULPHATE * LON 41

THR: Poison by ingestion and subcutaneous routes. When heated to decomposition it emits very toxic fumes of SO_x and NO_x.

3,5-DIMETHYLPHENYL-N-METHYLCARBAMATE   HR: 3

CAS: 2655-14-3   NIOSH: FC 8925000
mf: $C_{10}H_{13}NO_2$ mw: 179.24

SYNS: 3,5-XYLENOL METHYLCARBAMATE * 3,5-XYLENYL-N-METHYLCARBAMATE * 3,5-XYLYL-N-METHYLCARBAMATE

THR: Poison by ingestion and possibly other routes. See also carbamates. When heated to decomposition it emits toxic fumes of NO_x.

3,4-DIMETHYL-2-PHENYLMORPHO-LINEHYDROCHLORIDE   HR: 3

CAS: 7635-51-0   NIOSH: QE 2625000
mf: $C_{12}H_{17}NO \cdot ClH$ mw: 227.76

SYNS: PHENDIMETRAZINE HYDROCHLORIDE * D-2-PHENYL-3,4-DIMETHYLMORPHOLINE HYDROCHLORIDE

THR: Poison by ingestion, intraperitoneal, subcutaneous, and intravenous routes. When heated to decomposition it emits very toxic fumes of NO_x and HCl.

1,1-DIMETHYL-4-PHENYLPIPERAZINE IODIDE   HR: 3

CAS: 54-77-3   NIOSH: TM 3385000
mf: $C_{12}H_{19}N_2 \cdot I$ mw: 318.23

SYN: 1,1-DIMETHYL-4-PHENYLPIPERAZINIUM IODIDE

THR: Poison by intravenous and intramuscular routes. When heated to decomposition it emits very toxic fumes of NO_x and I^-.

1,2-DIMETHYL-3-PHENYL-3-PYRROLIDYLPROPIONATE   HR: 3

CAS: 3734-17-6   NIOSH: UY 5640000
mf: $C_{15}H_{21}NO_2$ mw: 247.37

SYN: PRODILIDINE

THR: Poison by ingestion, intravenous, and subcutaneous routes. When heated to decomposition it emits toxic fumes of NO_x.

3,3-DIMETHYL-1-PHENYLTRIAZENE   HR: 3

CAS: 7227-91-0   NIOSH: XY 2100000
mf: $C_8H_{11}N_3$ mw: 149.22

SYNS: 1-FENYL-3,3-DIMETHYLTRIAZIN * 1-PHENYL-3,3-DIMETHYLTRIAZENE

THR: Poison by ingestion, intraperitoneal, and subcutaneous routes. Mutagenic data. An experimental carcinogen, tumorigen, and teratogen. When heated to decomposition it emits toxic fumes of NO_x. Decomposes explosively on attempted distillation at atmospheric pressure.

DIMETHYL PHOSPHATE ESTER WITH 2-CHLORO-N-METHYL-3-HYDROXY-CROTONAMIDE   HR: 3

CAS: 34491-04-8   NIOSH: TC 2870000
mf: $C_7H_{13}ClNO_5P$ mw: 257.63

SYNS: CIBA C-768 * ENT 27,357 * NSC 190955

THR: Poison by ingestion and subcutaneous routes. See also esters. When heated to decomposition it emits very toxic Cl^-, NO_x and PO_x.

o,S-DIMETHYL PHOSPHORAMIDO-THIOATE   HR: 3

CAS: 10265-92-6   NIOSH: TB 4970000
mf: $C_2H_8NO_2PS$ mw: 141.14

PROP: Mp: 40°. Crystals slightly water sol; sol in alcohol.

SYNS: O,S-DIMETHYL ESTER AMIDE OF AMIDOTHIOATE * ENT 27,396 * TAMARON * THIOPHOSPHORSAEURE-O,S-DIMETHYLESTER-AMID (GERMAN)

THR: Poison by ingestion, inhalation, skin contact, subcutaneous, and intraperitoneal routes. A cholinesterase inhibitor type of insecticide. See also parathion. When heated to decomposition it emits very toxic fumes of NO_x, PO_x and SO_x.

o,o-DIMETHYLPHOSPHOROTHIOIC ACID-o-(4-METHYLTHIO)-m-TOLYLESTER   HR: 3

CAS: 55-38-9   NIOSH: TF 9625000
mf: $C_{10}H_{15}O_3PS_2$ mw: 278.34

SYNS: O,O-DIMETHYL O-4-(METHYLMERCAPTO)-3-METHYLPHENYL PHOSPHOROTHIOATE * O,O-DIMETHYL-O-4-(METHYLMERCAPTO)-3-METHYL-PHENYL THIOPHOSPHATE * O,O-DIMETHYL O-(3-METHYL-4-METHYLMERCAPTOPHENYL)-PHOSPHOROTHIOATE * O,O-DIMETHYL-O-(3-METHYL-4-METHYLTHIO-FENYL)-MONOTHIOFOS-FAAT (DUTCH) * O,O-DIMETHYL-O-(3-METH-YL-4-METHYLTHIOPHENYL)-MONOTHIOPHOSPHAT (GERMAN) * O,O-DIMETHYL-O-3-METHYL-4-METHYLTHIOPHENYL PHOSPHOROTHIOATE * O,O-DIMETHYL-O-(4-METHYLTHIO-3-METHYL-PHENYL) PHOSPHOROTHIOATE * O,O-DI-

METHYL-O-(4-(METHYLTHIO)-m-TOLYL) PHOSPHOROTHIOATE * O,O-DIMETIL-O-(3-METIL-4-METILTIO-FENIL)-MONOTIOFOSFATO (ITALIAN) * ENT 25,540 * NCI-C08651 * THIOPHOSPHATE DE O,O-DIMETHYLE ET DE O-(3-METHYL-4-METHYLTHIOPHENYLE) (FRENCH)

THR: Poison by ingestion, skin contact, intraperitoneal, intravenous, intramuscular, and possibly other routes. An experimental teratogen. A carcinogen. See also esters. When heated to decomposition it emits very toxic fumes of PO_x and SO_x.

DIMETHYL PHTHALATE   HR: 3

CAS: 131-11-3   NIOSH: TI 1575000
mf: $C_{10}H_{10}O_4$ mw: 194.20

PROP: Colorless, odorless liquid, bp: 283.7°, flash p: 295°F (CC), d: 1.189 @ 25°/25°, autoign temp: 1032°F, vap d: 6.69, vap press: 1 mm @ 100.3°.

SYNS: 1,2-BENZENEDICARBOXYLIC ACID, DIMETHYL ESTER * DIMETHYL-1,2-BENZENEDICARBOXYLATE * DIMETHYL BENZENEORTHODICARBOXYLATE * DMP * ENT 262 * METHYL PHTHALATE * PHTHALSAEUREDIMETHYLESTER (GERMAN) * PHTHALIC ACID METHYL ESTER

OSHA PEL: TWA 5 mg/m^3
ACGIH TLV: TWA 5 mg/m^3

THR: An experimental teratogen. Moderately toxic by ingestion and intraperitoneal routes. See also esters. An eye irritant. Combustible when exposed to heat or flame; can react with oxidizing materials. To fight fire, use CO_2, dry chemical. When heated to decomposition it emits acrid smoke and fumes. For further information see DMP, Vol. 2, No. 4 of *DPIM Report*.

2,2-DIMETHYLPROPANOYL CHLORIDE   HR: 2

CAS: 3282-30-2   NIOSH: AO 7200000
mf: C_5H_9ClO mw: 120.59

SYNS: 2,2-DIMETHYLPROPIONYL CHLORIDE * NEOPANTANOYL CHLORIDE * PIVALIC ACID CHLORIDE * PIVALOLYL CHLORIDE * PIVALOYL CHLORIDE (8CI) * PIVALYL CHLORIDE * TRIMETHYL ACETYL CHLORIDE (DOT)

DOT Classification: Label: Corrosive

THR: An irritant to skin, eyes, and mucous membranes. When heated to decomposition it emits toxic fumes of Cl^-.

(3,3-DIMETHYL-1-(m-PYRIDYL-N-OXIDE))TRIAZENE   HR: 3
CAS: 21600-52-0   NIOSH: XY 2300000
mf: $C_7H_{10}N_4O$ mw: 166.21

SYNS: 3-(3',3'-DIMETHYLTRIAZENO)-PYRIDIN-N-OXID (GERMAN) * 3-(3',3'-DIMETHYLTRIAZENO)-PYRIDINE-N-OXIDE * 1-(PYRIDYL-3-N-OXID)-3,3-DIMETHYL-TRIAZEN (GERMAN) * 1-(PYRIDYL-3-N-OXIDE)-3,3-DIMETHYLTRIAZENE

THR: Poison by intravenous and subcutaneous routes. Mutagenic data. An experimental carcinogen. When heated to decomposition it emits toxic fumes of NO_x.

S-(4,6-DIMETHYL-2-PYRIMIDINYL)-o,o-DIETHYL PHOSPHORODITHIOATE   HR: 3
CAS: 333-40-4   NIOSH: TE 3150000
mf: $C_{10}H_{17}N_2O_2PS_2$ mw: 292.38

SYNS: ENT 25,737 * STAUFFER R-3413

THR: Poison by ingestion. When heated to decomposition it emits very toxic fumes of NO_x, PO_x, and SO_x.

N,N-DIMETHYL-4-STILBEN-AMINE   HR: 3
CAS: 1145-73-9   NIOSH: WJ 3850000
mf: $C_{16}H_{17}N$ mw: 223.34

SYNS: 4-DIMETHYLAMINOSTILBEN (GERMAN) * N,N-DIMETHYL-P-STYRYLANILINE * STILBENYL-N,N-DIMETHYLAMINE

THR: Poison by ingestion. An experimental tumorigen and carcinogen. When heated to decomposition it emits toxic fumes of NO_x.

(E)-N,N-DIMETHYL-4-STILBEN-AMINE   HR: 3
CAS: 838-95-9   NIOSH: WJ 4000000
mf: $C_{16}H_{17}N$ mw: 223.34

SYNS: TRANS-P-(DIMETHYLAMINO)STILBENE * TRANS-4-DIMETHYLAMINOSTILBENE * TRANS-N,N-DIMETHYL-4-STILBENAMINE

THR: Poison by ingestion and possibly other routes. Mutagenic data. An experimental carcinogen and tumorigen. When heated to decomposition it emits toxic fumes of NO_x.

2,4-DIMETHYL SULFOLANE   HR: 3
CAS: 1003-78-7   NIOSH: XN 0525000
mf: $C_6H_{12}O_2S$ mw: 148.24

PROP: Solid, bp: 280°, flash p: 290°F (OC), d: 1.1362 @ 20°/4°, vap press: 0.006 mm @ 20°.

SYN: TETRAHYDRO-2,4-DIMETHYLTHIOPHENE 1,1-DIOXIDE

THR: Poison by ingestion, intraperitoneal, and intravenous routes. Combustible when exposed to heat or flame. Dangerous; see also sulfates; can react with oxidizing materials. To fight fire, use water, foam, CO_2, dry chemical. When heated to decomposition it emits toxic fumes of SO_x.

DIMETHYL SULFOXIDE   HR: 3
CAS: 67-68-5   NIOSH: PV 6210000
mf: C_2H_6OS mw: 78.14

PROP: Clear, water white, hygroscopic liquid; mp: 18.5°, bp: 189°, flash p: 203°F (OC), d: 1.100 @ 20°, vap press: 0.37 mm @ 20°, lel = 2.6%, uel = 28.5%, autoign temp: 419°F.

SYNS: METHYL SULFOXIDE * DIMETHYL SULPHOXIDE * DMSO * GAMASOL 90

THR: An experimental teratogen. Moderately toxic by several routes. A skin and eye irritant and allergen. Mutagenic data. Can cause an anaphylactic reaction, and corneal opacity. It freely penetrates the skin and may carry dissolved chemicals with it into the body. Combustible when exposed to heat or flame. Reacts violently with many acyl and aryl halides; bromobenzoyl acetanilide; cyanuric chloride; CH_3Br; NIO_4; P_2O_3; AgF. When heated to decomposition it emits toxic fumes of SO_x; can react with oxidizing materials. To fight fire, use foam, alcohol foam, CO_2, dry chemical. Incompatible with boron compounds; dinitrogen tetraoxide; iodine pentafluoride; magnesium perchlorate; metal oxosalts; non-metal halides; perchloric acid; periodic acid; silver difluoride; sodium hydride; sulfur trioxide. For further information see Vol. 1, No. 1 of *DPIM Report*.

DIMETHYL TEREPHTHALATE   HR: 3
CAS: 120-61-6   NIOSH: WZ 1225000
mf: $C_{10}H_{10}O_4$ mw: 194.20

SYNS: 1,4-BENZENE DICARBOXYLIC ACID, DIMETHYL ESTER (9CI) * DIMETHYL-1,4-BENZENE DICARBOXYLATE * METHYL-4-CARBOMETHOXY BENZOATE * NCI-C50055 * TEREPHTHALIC ACID METHYL ESTER

THR: An experimental carcinogen. A skin and eye irritant. Moderately toxic by ingestion. When heated to decomposition it emits acrid smoke and fumes.

DIMETHYLTIN DIBROMIDE   HR: 3
CAS: 2767-47-7   NIOSH: WH 6883000
mf: $C_2H_6Br_2Sn$ mw: 308.59

PROP: Colorless crystals. Sol in water and organic solvents. Mp: 76°; bp: 208°-213°.

SYN: DIBROMODIMETHYL STANNANE

THR: Poison by intravenous route. See also tin compounds and bromides. When heated to decomposition it emits toxic fumes of Br^-.

N,N-DIMETHYL-p-(m-TOLYLAZO)ANILINE   HR: 3
CAS: 55-80-1   NIOSH: BX 8250000
mf: $C_{15}H_{17}N_3$ mw: 239.35

SYNS: 4-(N,N-DIMETHYLAMINO)-3'-METHYL-AZOBENZENE * 3'-METHYLBUTTERGELB (GERMAN) * 3'-METHYL-4-DIMETHYLAMINO-AZOBENZEN (CZECH) * M'-METHYL-P-DIMETHYLAMINOAZOBENZENE * 3'-METHYL-4-DIMETHYLAMINOAZOBENZENE * 3'-METHYL-N,N-DIMETHYL-4-AMINOAZOBENZENE * 3'-METHYLDIMETHYLAMINOAZOBENZOL (GERMAN)

THR: An experimental tumorigen. Mutagenic data. Moderately toxic by ingestion. When heated to decomposition it emits toxic fumes of NO_x.

5-(3,3-DIMETHYL-1-TRIAZENO) IMIDAZOLE-4-CARBOXAMIDE CITRATE   HR: 3
CAS: 64038-56-8   NIOSH: NI 3975000
mf: $C_6H_{10}N_6O \cdot C_6H_8O_7$ mw: 374.36

SYN: DTIC CITRATE

THR: Poison by intraperitoneal route. When heated to decomposition it emits toxic fumes of NO_x.

1,1-DIMETHYL-3-(alpha,alpha,alpha-TRIFLUORO-m-TOLYL) UREA   HR: 3
CAS: 2164-17-2   NIOSH: YT 1575000
mf: $C_{10}H_{11}F_3N_2O$ mw: 232.23

SYNS: NCI-C08695 * N-(M-TRIFLUOROMETH-YLPHENYL)-N',N'-DIMETHYLUREA * N-(3-TRI-FLUOROMETHYLPHENYL)-N'-N'-DIMETHYL-UREA * 3-(M-TRIFLUOROMETHYLPHENYL)-1, 1-DIMETHYLUREA

THR: Poison by ingestion. Possibly toxic by other routes. An experimental carcinogen. Mutagenic data. When heated to decomposition it emits very toxic fumes of F^- and NO_x.

o,o'-DIMETHYLTUBOCURARINE   HR: 3
CAS: 5152-30-7   NIOSH: YO 5250000
mf: $C_{40}H_{48}N_2O_6$ mw: 652.90

SYNS: DIMETHYL TUBOCURARINE * D,O-DI-METHYLTUBOCURARINE * N,N',O,O-TETRA-METHYL-(+)-TUBOCURINE

THR: Poison by intravenous route. When heated to decomposition it emits toxic fumes of NO_x.

1,3-DIMETHYLUREA   HR: 3
CAS: 96-31-1   NIOSH: YS 9868000
mf: $C_3H_8N_2O$ mw: 88.13

PROP: Colorless crystals, water and alcohol sol, d: 1.14, mp: 106°, bp: 270°.

SYNS: SYM-DIMETHYLUREA * N,N'-DIMETHYL-HARNSTOFF (GERMAN) * N,N'-DIMETHYLUREA

THR: An experimental teratogen. Moderately toxic by ingestion. When heated to decomposition it emits toxic fumes of NO_x.

N,N-DIMETHYL-p-(3,4-XYLYLAZO)-ANILINE   HR: 3
CAS: 3025-73-8   NIOSH: BX 8925000
mf: $C_{16}H_{19}N_3$ mw: 253.38

SYN: 3',4'-DIMETHYL-4-DIMETHYLAMINOZOBEN-ZENE

THR: An experimental neoplastigen and tumorigen. When heated to decomposition it emits toxic fumes of NO_x.

DIMORPHOLAMINE   HR: 3
CAS: 119-48-2   NIOSH: QD 9800000
mf: $C_{20}H_{38}N_4O_4$ mw: 298.62

SYNS: N,N'-DIBUTYL-N,N'-DICARBOXYMORPHO-LIDE-ETHYLENEDIAMINE * N,N'-ETHYLENE-BIS(N-BUTYL-4-MORPHOLINECARBOXAMIDE)

THR: Poison by ingestion, intraperitoneal, intravenous and subcutaneous routes. See also amines. When heated to decomposition it emits toxic fumes of NO_x.

2,4-DINITROANILINE   HR: 3
CAS: 97-02-9   NIOSH: BX 9100000
mf: $C_6H_5N_3O_4$ mw: 183.14.

PROP: Yellow, needle-like crystals, insol in water, mp: 188°, flash p: 435°F (CC), d: 1.615, vap d: 6.31.

SYNS: 2,4-DINITRANILINE * 2,4-DINITROANI-LIN (GERMAN) * 2,4-DINITROANILINA (ITALIAN) * NCI-C60753

THR: Poison by ingestion and intraperitoneal routes. A severe eye irritant. Mutagenic data. See also nitroanilines. Combustible when exposed to heat or flame. Dangerous; when heated to decomposition it emits highly toxic fumes; can react with oxidizing materials. To fight fire, use CO_2, dry chemical.

2,4-DINITROANISOL   HR: 3
CAS: 119-27-7   NIOSH: DA 5250000
mf: $C_7H_6N_2O_5$ mw: 198.15

PROP: Colorless to yellow crystals. mp: 89°, bp: sublimes, d: 1.341 @ 20°/4°, vap d: 6.83.

SYNS: 2,4-DINITROANISOLE * ALPHA-DINI-TROANISOLE * 2,4-DINITROPHENYLMETHYL ETHER * 1-METHOXY-2,4-DINITROBENZENE

THR: Poison by ingestion. Mutagenic data. See also nitro compounds of aromatic hydrocarbons and nitrates. When heated to decomposition it emits toxic fumes of NO_x.

m-DINITROBENZENE   HR: 3
CAS: 99-65-0   NIOSH: CZ 7350000
mf: $C_6H_4N_2O_4$ mw: 168.12

PROP: Yellowish crystals, mp: 89°, bp: 301°.

SYNS: BINITROBENZENE * 1,3-DINITROBEN-ZENE * 2,4-DINITROBENZENE * 1,3-DINITRO-BENZOL * DWUNITROBENZEN (POLISH)

OSHA PEL: TWA 1 mg/m³ (skin)

THR: Poison by ingestion and intravenous routes. Human blood effects by skin contact. Mutagenic data. When heated to decomposition it emits toxic fumes of NO_x. Incompatible with tetranitromethane; nitric acid. See also o- and p-dinitrobenzene. For further information see Vol. 6, No. 1 of *DPIM Report*.

o-DINITROBENZENE   HR: 3
CAS: 528-29-0   NIOSH: CZ 7450000
mf: $C_6H_4N_2O_4$ mw: 168.12

PROP: Colorless needles or plates, mp: 118°, bp: 319°, flash p: 302°F (CC), d: 1.571 @ 0°/4°, vap d: 5.79.

SYN: o-DINITROBENZOL

ACGIH TLV: TWA 0.15 ppm

THR: Poison by inhalation and ingestion. Moderately toxic by skin contact. Can cause liver, kidney, and central nervous system injury. See also m- and p-dinitrobenzene. Combustible when exposed to heat or flame. A severe explosion hazard when shocked or exposed to heat or flame. It is used in bursting charges and to fill artillery shells. Dangerous; when heated to decomposition it emits highly toxic fumes of NO_x and explodes; can react vigorously with oxidizing materials. To fight fire, use water, CO_2, dry chemical. Incompatible with nitric acid. For further information see Vol. 5, No. 3 of *DPIM Report*.

p-DINITROBENZENE   HR: 3
CAS: 100-25-4   NIOSH: CZ 7525000
mf: $C_6H_4N_2O_4$ mw: 168.12

PROP: White crystals, mp: 173°, bp: 299°. Volatile with steam.

SYN: DITHANE A-4

OSHA PEL: TWA 1 mg/m³ (skin)

THR: Poison by ingestion. Mutagenic data. See also o- and m-dinitrobenzene. When heated to decomposition it emits toxic fumes of NO_x. For further information see Vol. 3, No. 3, *DPIM Report*.

4,4'-DINITROBIPHENYL   HR: 3
CAS: 1528-74-1   NIOSH: DV 4000000
mf: $C_{12}H_8N_2O_4$ mw: 244.22

SYN: 4,4'-DINITROBIFENYL (CZECH)

THR: An experimental carcinogen. Eye irritant. When heated to decomposition it emits toxic fumes of NO_x.

2,4-DINITRO-1-CHLORO-NAPHTHA-LENE   HR: 3
CAS: 2401-85-6   NIOSH: QJ 2450000
mf: $C_{10}H_5ClN_2O_4$ mw: 252.62

SYN: 1-CHLORO-2,4-DINITRONAPHTHALENE

THR: A poison. An experimental carcinogen and neoplastigen. See also 2,4-dinitroaniline. When heated to decomposition it emits very toxic fumes of Cl^- and NO_x.

3,5-DINITRO-p-CRESOL   HR: 3
CAS: 63989-82-2   NIOSH: GO 9600000
mf: $C_7H_6N_2O_5$ mw: 198.15

PROP: Crystals.

THR: Poison by intraperitoneal and possibly other routes. Strong irritant to eyes, skin, and mucous membranes. Can cause brain, liver, and kidney damage. See also 4,6-dinitro-*o*-cresol. When heated to decomposition it emits toxic fumes of NO_x.

4,6-DINITRO-o-CRESOL   HR: 3
CAS: 534-52-1   NIOSH: GO 9625000
mf: $C_7H_6N_2O_5$ mw: 198.15

PROP: Yellow prismatic crystals, mp: 85.8°, vap d: 6.82.

SYNS: 2,4-DINITRO-O-CRESOL * 4,6-DINI-TRO-O-CRESOLO (ITALIAN) * 3,5-DINITRO-2-HYDROXYTOLUENE * 4,6-DINITROKRESOL (DUTCH) * 4,6-DINITRO-O-KRESOL (CZECH) * DINITROMETHYL CYCLOHEXYLTRIENOL * 2,4-DINITRO-6-METHYLPHENOL * DNOK (CZECH) * DWUNITRO-O-KREZOL (POLISH) * LE DINITROCRESOL-4,6 (FRENCH) * 2-METHYL-4,6-DINITROPHENOL * ZAHLREICHE BEZEICHNUNGEN (GERMAN)

OSHA PEL: TWA 200 ug/m³ (skin)
ACGIH TLV: TWA 0.2 mg/m³

THR: Poison by ingestion, inhalation, dermal, intraperitoneal, and intravenous routes. Less toxic than the p- form, but is still highly toxic. Severe eye and mild skin irritant. For fire and disaster hazards, see nitrates. For further information see Vol. 4, No. 1 of *DPIM Report*.

2,6-DINITRO-p-CRESOL   HR: 3
CAS: 609-93-8   NIOSH: GO 9800000
mf: $C_7H_6N_2O_5$ mw: 198.15

SYNS: DNPC * DINITRO-P-CRESOL * VICTORIA ORANGE * VICTORIA YELLOW

THR: Poison by intraperitoneal route. When heated to decomposition it emits toxic fumes of NO_x.

3,5-DINITRO-o-CRESOL SODIUM SALT   HR: 3
CAS: 2312-76-7   NIOSH: GP 1050000
mf: $C_7H_5N_2O_5 \cdot Na$ mw: 220.13

PROP: Brilliant orange-yellow dye.

SYNS: 4,6-DINITRO-O-CRESOL SODIUM SALT * 2,4-DINITRO-6-METHYLPHENOL SODIUM SALT * SODIUM-4,6-DINITRO-O-CRESOXIDE

THR: Poison by ingestion, skin contact, and subcutaneous routes. See also dinitrocresol. Flammable; see nitrates for fire and disaster hazards.

2,4-DINITRO-1-FLUOROBENZENE   HR: 3
CAS: 70-34-8   NIOSH: CZ 7800000
mf: $C_6H_3FN_2O_4$ mw: 186.11

PROP: Crystals; sol in ether, benzene, propylene glycol. Mp: 26°, bp: 137° @ 20 mm.

SYNS: 1,2,4-FLUORODINITROBENZENE * 1-FLUORO-2,4-DINITROBENZENE

THR: Poison by ingestion, skin contact, and subcutaneous routes. An experimental carcinogen and neoplastigen. A powerful irritant and vesicant. See also fluorides. Mutagenic data. Dangerous. When heated to decomposition it emits highly toxic fumes of NO_x and fluorides.

2,4-DINITRO-1-NAPHTHOL   HR: 3
CAS: 605-69-6   NIOSH: QL 3850000
mf: $C_{10}H_6N_2O_5$ mw: 234.18

PROP: Yellow needles or leaflets, mp: 138°, vap d: 8.08.

SYNS: 2-4 DINITRO-alpha-NAPHTOL (FRENCH) * GOLDEN YELLOW * MANCHESTER YELLOW * MARITUS YELLOW * NAPHTHOL YELLOW * NAPHTHYLENE YELLOW * SAFFRON YELLOW

THR: Poison by skin contact, subcutaneous, intramuscular, intravenous, and intraperitoneal routes. Toxic to the skin. See nitrates for fire, disaster, and explosion hazards.

DINITROPHENOL   HR: 3
CAS: 25550-58-7   NIOSH: SL 2625000
mf: $C_6H_4N_2O_5$ mw: 184.12

THR: Poison by ingestion and subcutaneous routes. When heated to decomposition it emits toxic fumes of NO_x.

2,3-DINITROPHENOL   HR: 3
CAS: 66-56-8   NIOSH: SL 2700000
mf: $C_6H_4N_2O_5$ mw: 184.12

PROP: Yellow needles, mp: 144°, d: 1.681 @ 20°, vap d: 6.35.

THR: Poison by unspecified routes. Can cause fatal liver damage and induced fever and kidney damage. Inhalation of dust can be fatal. A skin irritant and an allergen. A powerful stimulant of the metabolism by excessive oxidation. For fire hazard, see nitrates. Highly explosive when exposed to heat. It is used as a component of some shell and bomb charges. See also explosives, high.

2,4-DINITROPHENOL   HR: 3
CAS: 51-28-5   NIOSH: SL 2800000
mf: $C_6H_4N_2O_5$ mw: 184.12

PROP: Yellow crystals, mp: 112°, d: 1.683 @ 24°, vap d: 6.35.

SYNS: 2,4-DINITROFENOL (DUTCH) * DINITRO-FENOLO (ITALIAN) * ALPHA-DINITROPHENOL * 1-HYDROXY-2,4-DINITROBENZENE * NSC 1532

THR: Poison by ingestion, intravenous, intraperitoneal, subcutaneous, intramuscular, and possibly other routes. Mutagenic data. A skin irritant. Phytotoxic. See also nitrates. When heated to decomposition it emits toxic fumes of NO_x. For further information see Vol. 3, No. 2 of *DPIM Report*.

2,5-DINITROPHENOL   HR: 3
CAS: 329-71-5   NIOSH: SL 2900000
mf: $C_6H_4N_2O_5$ mw: 184.12

SYNS: gamma-DINITROPHENOL * 2,5-DNP

THR: A poison. When heated to decomposition it emits toxic fumes of NO_x.

2,6-DINITROPHENOL   HR: 3
CAS: 573-56-8   NIOSH: SL 2975000
mf: $C_6H_4N_2O_5$ mw: 184.12

PROP: Yellow crystals, mp: 63°, vap d: 6.35.

SYN: beta-DINITROPHENOL

THR: Poison by intramuscular and other routes. For fire and disaster hazards, see nitrates. Moderately explosive when exposed to heat. For further information see Vol. 3, No. 2 of *DPIM Report*.

3,4-DINITROPHENOL   HR: 3
CAS: 577-71-9   NIOSH: SL 3000000
mf: $C_6H_4N_2O_5$ mw: 184.12

THR: A poison. When heated to decomposition it emits toxic fumes of NO_x.

3,5-DINITROPHENOL   HR: 3
CAS: 586-11-8   NIOSH: SL 3050000
mf: $C_6H_4N_2O_5$ mw: 184.12

THR: A poison. When heated to decomposition it emits toxic fumes of NO_x.

2,4-DINITROPHENYLMORPHINE-HYDROCHLORIDE   HR: 3
CAS: 63732-56-9   NIOSH: QD 0800000
mf: $C_{23}H_{21}N_3O_7 \cdot ClH$ mw: 487.93

SYN: 2,4-DINITROPHENYL ETHER OF MORPHINE

THR: Poison by intravenous, intraperitoneal, and subcutaneous routes. See also ethers. When heated to decomposition it emits very toxic fumes of HCl and NO_x.

4,6-DINITROQUINOLINE-1-OXIDE   HR: 3
CAS: 1596-52-7   NIOSH: VB 7875000
mf: $C_9H_5N_3O_5$ mw: 235.17

THR: An experimental carcinogen. Mutagenic data. When heated to decomposition it emits toxic fumes of NO_x.

DINITROSODIMETHYLPROPANE DIAMINE   HR: 3
CAS: 6972-76-5   NIOSH: TX 7700000
mf: $C_5H_{12}N_4O_2$ mw: 160.21

SYNS: N,N'-DIMETHYL-N,N'-DINITROSO-1,3-PROPANEDIAMINE * N,N'-DINITROSO-N,N'-DIMETHYL-1,3-PROPANEDIAMINE * NSC 62580

THR: An experimental neoplastigen. When heated to decomposition it emits toxic fumes of NO_x.

DINITROSOPENTAMETHYLENE-TETRAMINE   HR: 3
CAS: 101-25-7   NIOSH: XA 5250000
mf: $C_5H_{10}N_6O_2$ mw: 186.21

SYNS: N,N-DINITROSOPENTAMETHYLENETETRAMINE * N(SUP 1),N(SUP 3)-DINITROSOPENTAMETHYLENETETRAMINE * N,N'-DINITROSOPENTAMETHYLENETETRAMINE * 3,4-DI-N-NITROSOPENTAMETHYLENETETRAMINE * 3,7-DI-N-NITROSOPENTAMETHYLENETETRAMINE * 3,7-DINITROSO-1,3,5,7-TETRAAZABICYCLO-(3,3,1)-NONANE * 1,5-ENDOMETHYLENE-3,7-DINITROSO-1,3,5,7-TETRAAZACYCLOOCTANE * 1,5-METHYLENE-3,7-DINITROSO-1,3,5,7-TETRAAZACYCLOOCTANE * NSC 73599

THR: Poison by intravenous, subcutaneous, and intraperitoneal routes. Moderately toxic by ingestion. See also amines. When heated to decomposition it emits toxic fumes of NO_x.

DINITROSOPIPERAZINE HR: 3
CAS: 140-79-4   NIOSH: TL 6300000
mf: $C_4H_8N_4O_2$ mw: 144.16

PROP: White crystals, mp: 158°, vap d: 4.97.

SYNS: DINITROSOPIPERAZIN (GERMAN)
* N,N'-DINITROSOPIPERAZINE * 1,4-DINITRO-
SOPIPERAZINE * NSC 339 * USAF DO-36

THR: Poison by ingestion, subcutaneous, and intraperitoneal routes. Mutagenic data. An experimental carcinogen, tumorigen, and neoplastigen. A stomach insecticide. See also nitrates. When heated to decomposition it emits toxic fumes of NO_x.

3,5-DINITRO-o-TOLUAMIDE HR: 3
CAS: 148-01-6   NIOSH: XS 4200000
mf: $C_8H_7N_3O_5$ mw: 225.18

PROP: Yellowish solid, very slightly sol in water, sol in acetone, acetonitrile and dimethyl formamide, mp: 177°.

SYNS: 2-METHYL-3,5-DINITROBENZAMIDE
* ZOALENE
ACGIH TLV: TWA 5 mg/m^3; STEL 10 mg/m^3

THR: Poison by intravenous route. Mutagenic data. Moderately toxic by ingestion. For disaster hazard, see nitrates.

2,4-DINITROTOLUENE HR: 3
CAS: 121-14-2   NIOSH: XT 1575000
mf: $C_7H_6N_2O_4$ mw: 182.15

PROP: Yellow needles, mp: 69.5°, bp: 300°, d: 1.521 @ 15°, vap d: 6.27, flash p: 404°F.

SYNS: 2,4-DINITROTOLUOL * 1-METHYL-2,4-
DINITROBENZENE * NCI-C01865

OSHA PEL: TWA 1500 ug/m^3 (skin)
ACGIH TLV: TWA 1.5 mg/m^3

THR: Poison by ingestion and subcutaneous route. Mutagenic data. An experimental carcinogen and neoplastigen. An irritant and an allergen. Can cause anemia, methemoglobinemia, cyanosis and liver damage. See also trinitrotoluene. Slight fire hazard. To fight fire, use water spray or mist, dry chemical. When heated to decomposition it emits toxic fumes of NO_x.

Incompatible with nitric acid. For further information see Vol. 3, No. 2 of *DPIM Report*.

4,6-DINITRO-1,2,3-TRICHLORO-BENZENE HR: 3
CAS: 6379-46-0   NIOSH: CZ 7900000
mf: $C_6HCl_3N_2O_4$ mw: 271.44

SYNS: 1,2,3-TRICHLORO-4,6-DINITROBENZENE
* VINCIDE PB

THR: An experimental carcinogen. When heated to decomposition it emits very toxic fumes of HCl and NO_x, Cl$^-$.

DINOPROSTONE HR: 3
CAS: 363-24-6   NIOSH: UK 8000000
mf: $C_{20}H_{32}O_5$ mw: 352.52

SYNS: (5Z,11-ALPHA,13E,15S)-11,15-DIHY-
DROXY-9-OXOPROSTA-5,13-DIEN-1-OIC ACID
* 7-(3-HYDROXY-2-(3-HYDROXY-1-OCTENYL)-
5-OXOCYCLOPENTYL)-5-HEPTENOIC ACID
* PROSTAGLANDIN E2

THR: Poison by subcutaneous route. Mutagenic data. When heated to decomposition it emits acrid smoke and fumes.

DIOCTYL FUMARATE HR: 3
NIOSH: LT 0525000
mf: $C_{20}H_{36}O_4$ mw: 340.56

PROP: Clear mobile liquid, mild odor; bp: 211°-220°, flash p: 365°F (COC), d: 0.942 @ 20°/20°.

SYNS: BIS(2-ETHYLHEXYL) FUMARATE
* DI(2-ETHYLHEXYL) FUMARATE * 2-ETHYL-
HEXYL FUMARATE

THR: Poison by intraperitoneal route. Severe skin irritant and eye irritant. See also esters and fumaric acid. Combustible; can react with oxidizing materials, heat, flame. To fight fire, use foam, CO_2, dry chemical.

n-DIOCTYL PHTHALATE HR: 3
CAS: 117-84-0   NIOSH: TI 1925000
mf: $C_{24}H_{38}O_4$ mw: 390.62

SYNS: o-BENZENEDICARBOXYLIC ACID, DIOCTYL
ESTER * DIOCTYL-o-BENZENEDICARBOXYLATE
* DIOCTYL PHTHALATE * OCTYL PHTHALATE

THR: An experimental teratogen. A skin and severe eye irritant. See also esters. When heated to decomposition it emits acrid smoke and fumes.

DIOCTYL(1,2-PROPYLENEDIOXYBIS-(MALEOYLDIOXY))STANNANE
HR: 3

CAS: 69226-45-5   NIOSH: WH 7640000
mf: $C_{27}H_{42}O_8Sn$ mw: 613.38

SYNS: DI-N-OCTYLTIN DI(1,2-PROPYLENEGLY-
COLMALEATE) * DI-N-OCTYL-ZINN-
DI-(1,2-PROPYLENGLYKOLMALEINAT)(GERMAN)

THR: Poison by intraperitoneal route. See also
tin compounds. When heated to decomposition
it emits acrid smoke and fumes.

DIOCTYL SODIUM SULFOSUCCI-NATE
HR: 3

CAS: 577-11-7   NIOSH: WN 0525000
mf: $C_{20}H_{38}O_7S \cdot Na$ mw: 445.63

SYNS: BIS(ETHYLHEXYL) ESTER OF SODIUM SUL-
FOSUCCINIC ACID * BIS(2-ETHYLHEXYL)SODIUM
SULFOSUCCINATE * BIS(2-ETHYLHEXYL)-S-SO-
DIUM SULFOSUCCINATE * 1,4-BIS(2-ETHYL-
HEXYL) SODIUM SULFOSUCCINATE * DIOCTYL
ESTER OF SODIUM SULFOSUCCINATE * DI-(2-
ETHYLHEXYL) SODIUM SULFOSUCCINATE
* SODIUM BIS(2-ETHYLHEXYL) SULFOSUCCINATE
* SODIUM DI-(2-ETHYLHEXYL) SULFOSUCCINATE
* SODIUM DIOCTYL SULFOSUCCINATE * SO-
DIUM SULFODI-(2-ETHYLHEXYL) SULFOSUCCINATE

THR: Poison by intravenous route. Moderately
toxic by ingestion. A skin and eye irritant. See
also esters. When heated to decomposition it
emits toxic fumes of SO_x.

DIOCTYLTIN-3,3'-THIODIPROPIO-NATE
HR: 3

CAS: 3594-15-8   NIOSH: JH 4760000
mf: $C_{22}H_{42}O_4SSn$ mw: 521.39

SYN: 2,2-DIOCTYL-1,3-DIOXA-2-STANNA-7-
THIADECAN-4,10-DIONE

THR: Poison by intravenous route. See also
tin compounds. When heated to decomposition
it emits toxic fumes of SO_x.

sym-DIOLEPOXIDE
HR: 3
NIOSH: DJ 5055850
mf: $C_{20}H_{14}O_3$ mw: 302.34

SYNS: (−)-BP 7-ALPHA,8-BETA-DIOL-9-
BETA,10-BETA-EPOXIDE 2 * (−)-(E)-7-AL-
PHA,8-BETA-DIHYDROXY-9-BETA,10-BETA-
EPOXY-7,8,9,10-TETRAHYDROBENZO(A)PYRENE

THR: An experimental carcinogen. Mutagenic
data. When heated to decomposition it emits
acrid smoke and fumes.

p-DIOXANE
HR: 3
CAS: 123-91-1   NIOSH: JG 8225000
DOT: 1165
mf: $C_4H_8O_2$ mw: 88.12

PROP: Colorless liquid, pleasant odor, mp: 12°,
bp: 101.1°, lel = 2.0%, uel = 22.2%, flash
p: 54°F (CC), d: 1.0353 @ 20°/4°, autoign temp:
356°F, vap press: 40 mm @ 25.2°, vap d: 3.03.

SYNS: DIETHYLENE DIOXIDE * 1,4-DIETHY-
LENE DIOXIDE * DIETHYLENE ETHER * DI-
(ETHYLENE OXIDE) * DIOKSAN (POLISH)
* DIOSSANO-1,4 (ITALIAN) * DIOXAAN-1,4
(DUTCH) * 1,4-DIOXACYCLOHEXANE * DI-
OXAN-1,4 (GERMAN) * P-DIOXAN (CZECH)
* 1,4-DIOXANE * DIOXANNE (FRENCH)
* DIOXYETHYLENE ETHER * GLYCOL ETHYL-
ENE ETHER * NCI-C03689 * TETRAHYDRO-P-
DIOXIN * TETRAHYDRO-1,4-DIOXIN

OSHA PEL: TWA 100 ppm (skin)
DOT Classification: Label: Flammable
  Liquid

THR: Poison by inhalation. Moderately toxic
by ingestion and several other routes. Eye and
possibly other systemic effects by inhalation.
An experimental carcinogen and tumorigen. The
irritant effects probably provide sufficient warning, in acute exposures, to enable the workman
to leave exposure before he is seriously affected.
Repeated exposure to low concentrations has
resulted in human fatalities, the organs chiefly
affected being the liver and kidneys. Dangerous
fire hazard when exposed to heat or flame; can
react vigorously with oxidizing materials. Explosion hazard when exposed to flame or by
chemical reaction with oxidizers. Violent reaction with (H_2 + Raney Ni); $AgClO_4$. To fight
fire, use alcohol foam, CO_2, dry chemical.

DIOXIME-p-BENZOQUINONE
HR: 3
CAS: 105-11-3   NIOSH: DK 4900000
mf: $C_6H_6N_2O_2$ mw: 138.14

SYNS: DIOXIME-1,4-CYCLOHEXADIENEDIONE
* DIOXIME-2,5-CYCLOHEXADIENE-1,4-DIONE
* NCI-C03850 * P-QUINONE DIOXIME

THR: An experimental carcinogen and neoplastigen. Moderately toxic by ingestion. When
heated to decomposition it emits toxic fumes
of NO_x.

DIOXOLAN
HR: 2
CAS: 100-79-8   NIOSH: JI 0400000
mf: $C_6H_{12}O_3$ mw: 132.18

PROP: Water white liquid, mp: −26.4°, bp: 75°, flash p: 35°F (OC), d: 1.065, vap press: 70 mm @ 20°, vap d: 2.6.

SYNS: ACETONE, CYCLIC (HYDROXYMETHYL) ETHYLENE ACETAL * 2,2-DIMETHYL-1,3-DIOXOLANE-4-METHANOL * 2,2-DIMETHYL-5-HYDROXYMETHYL-1,3-DIOXOLANE * 2,2-DIMETHYL-4-OXYMETHYL-1,3-DIOXOLANE * DIOXOLANE (DOT) * GLYCEROLACETONE * GLYCEROL DIMETHYLKETAL * 4-HYDROXYMETHYL-2,2-DIMETHYL-1,3-DIOXOLANE * ISOPROPYLIDENE GLYCEROL

DOT Classification: Label: Flammable Liquid

THR: Moderately toxic by ingestion. An eye irritant. Dangerous, when exposed to heat or flame; can react vigorously with oxidizing materials. Dangerous! Keep away from heat and open flame! To fight fire, use alcohol foam, CO_2, dry chemical.

o-(1,3-DIOXOLAN-2-YL)PHENYL METHYLCARBAMATE   HR: 3
CAS: 6988-21-2   NIOSH: FC 1925000
mf: $C_{11}H_{13}NO_4$ mw: 223.25

SYNS: 2-(1,3-DIOXOLAN-2-YL)PHENYL-N-METHYLCARBAMAT * 2-(1,3-DIOXOLANE-2-YL) PHENYL N-METHYLCARBAMATE * ELOCRON * ENT 27,389

THR: Poison by ingestion, skin contact, and possibly other routes. A toxic contact and systemic insecticide. When heated to decomposition it emits toxic fumes of NO_x.

DIOXOPROMETHAZINE HYDROCHLORIDE   HR: 3
CAS: 13754-56-8   NIOSH: SO 7350000
mf: $C_{17}H_{20}N_2O_2S \cdot ClH$ mw: 352.91

SYN: 5,5-DIOXO-10-(2-(DIMETHYLAMINO) PROPYL)PHENOTHIAZINE HYDROCHLORIDE

THR: Poison by ingestion, intraperitoneal, intravenous, and subcutaneous routes. When heated to decomposition it emits very toxic fumes of NO_x, SO_x and HCl.

DIPENTYL LEAD DIACETATE   HR: 3
CAS: 18279-20-4   NIOSH: OG 0750000
mf: $C_{14}H_{28}O_4Pb$ mw: 467.61

THR: Poison by ingestion. See also lead compounds. When heated to decomposition it emits toxic fumes of Pb.

DIPENTYLTIN DICHLORIDE   HR: 3
CAS: 1118-42-9   NIOSH: WH 7250000
mf: $C_{10}H_{22}Cl_2Sn$ mw: 331.91

SYN: DICHLORODIPENTYLSTANNANE

THR: Poison by intravenous route. See also tin compounds and chlorides. When heated to decomposition it emits toxic fumes of Cl^-.

DIPHENADIONE   HR: 3
CAS: 82-66-6   NIOSH: NK 5600000
mf: $C_{23}H_{16}O_3$ mw: 340.39

PROP: Pale yellow crystals, sol in acetone and acetic acid, mp: 147°.

SYNS: 2-DIPHENYLACETYL-1,3-DIKETOHYDRINDENE * 2-DIPHENYLACETYL-1,3-INDANDIONE * 2-(DIPHENYLACETYL)INDAN-1,3-DIONE * 2-(DIPHENYLACETYL)-1H-INDENE-1,3(2H)-DIONE

THR: Poison by ingestion. Inhibits blood clotting, leading to hemorrhages. Action similar to coumadin (warfarin). When heated to decomposition it emits acrid smoke and fumes.

DIPHENYLACETONITRILE   HR: 3
CAS: 86-29-3   NIOSH: AL 9800000
mf: $C_{14}H_{11}N$ mw: 193.26

SYNS: BENZYHYDRYLCYANIDE * DIPHENYL-ALPHA-CYANOMETHANE * DIPHENYLMETHYL-CYANIDE * ALPHA-PHENYLBENZYLCYANIDE * ALPHA-PHENYLPHENYLACETONITRILE * USAF KF-13

THR: Poison by intraperitoneal and intravenous routes. An experimental carcinogen. See also nitriles. When heated to decomposition it emits toxic fumes of NO_x and CN^-.

2,3-DIPHENYLACRYLONITRILE   HR: 3
CAS: 2510-95-4   NIOSH: AT 6225000
mf: $C_{15}H_{11}N$ mw: 205.27

SYNS: BENZYLIDENEPHENYLACETONITRILE * ALPHA-CYANOSTILBENE * 2,3-DIPHENYL-ACRYLONITRILE * ALPHA,BETA-DIPHENYLACRYLONITRILE * ALPHA-PHENYLCINNAMONITRILE * ALPHA-STILBENECARBONITRILE * USAF A-9789

THR: Poison by intraperitoneal route. See also nitriles. When heated to decomposition it emits toxic fumes of NO_x.

DIPHENYLAMINE   HR: 3
CAS: 122-39-4   NIOSH: JJ 7800000
mf: $C_{12}H_{11}N$ mw: 169.24

PROP: Crystals, floral odor. Sol in benzene, ether and carbon disulfide, mp: 52.9°, bp: 302.0°, flash p: 307°F (CC), d: 1.16, autoign temp: 1173°F, vap press: 1 mm @ 108.3°, vap d: 5.82.

SYNS: ANILINOBENZENE * BIG DIPPER * C.I. 10355 * N-PHENYLANILINE
ACGIH TLV: TWA 10 mg/m³

THR: Poison by ingestion. An experimental teratogen. Action similar to aniline but less severe. See also aniline, amines, and aromatic amines. Combustible when exposed to heat or flame. Can react violently with hexachloromelamine, trichloromelamine. Dangerous; when heated to decomp, emits highly toxic fumes of NO_x; can react with oxidizing materials. To fight fire, use CO_2, dry chemical. For further information see Vol. 2, No. 5 of *DPIM Report*.

DIPHENYL CARBONATE   HR: 3
CAS: 102-09-0   NIOSH: FG 0500000
mf: $C_{13}H_{10}O_3$ mw: 214.23

SYN: PHENYL CARBONATE

THR: An experimental neoplastigen and carcinogen. When heated to decomposition it emits acrid smoke and fumes.

1,2-DIPHENYL-1-(DIMETHYLAMINO)-ETHANE   HR: 3
CAS: 14148-99-3   NIOSH: KR 5920000
mf: $C_{16}H_{19}N \cdot ClH$ mw: 261.82

SYN: N,N-DIMETHYL-1,2-DIPHENYLETHYLAMINE HYDROCHLORIDE, (R) (−)-

THR: Poison by ingestion, intravenous, and subcutaneous routes. When heated to decomposition it emits very toxic fumes of HCl and NO_x.

DIPHENYL ETHER   HR: 2
CAS: 101-84-8   NIOSH: KN 8970000
mf: $C_{12}H_{10}O$ mw: 170.22

PROP: Colorless crystals, geranium odor, mp: 28°, bp: 257°, flash p: 239°F, d: 1.0728 @ 20°, vap d: 5.86, autoign temp: 1148°F, lel = 0.8%, uel = 1.5%.

SYNS: BIPHENYL OXIDE * GERANIUM CRYSTALS * PHENOXYBENZENE * PHENYL ETHER

OSHA PEL: TWA 1 ppm

THR: Moderately toxic by ingestion and inhalation. Prolonged exposure damages liver, spleen, kidneys and thyroids and upsets gastrointestinal tract. A skin and eye irritant. Flammable when exposed to heat or flame; can react with oxidizing materials. For explosion hazard, see also ethers. To fight fire, use water, foam, CO_2, dry chemical.

DIPHENYLGUANIDINE   HR: 3
CAS: 102-06-7   NIOSH: MF 0875000
mf: $C_{13}H_{13}N_3$ mw: 211.29

PROP: White powder, mp: 145°; d: 1.115 @ 25°.

SYNS: N,N'-DIPHENYLGUANIDINE * 1,3-DIPHENYLGUANIDINE * DPG * MELANILINE * NCI-C60924 * USAF EK-1270 * VULKAZIT

THR: Poison by ingestion and intraperitoneal and possibly other routes. Dangerous; when heated to decomposition it emits toxic fumes of NO_x.

DIPHENYLMERCURY   HR: 3
CAS: 587-85-9   NIOSH: OW 3150000
mf: $C_{12}H_{10}Hg$ mw: 354.81

PROP: White crystals, insol in water, d: 2.318, mp: 122° (sublimes), bp: 204° @ 10.5 mm.

THR: Poison by intraperitoneal route. Moderately toxic by ingestion. See also mercury compounds. When heated to decomposition it emits toxic fumes of Hg. Incompatible with non-metal oxides.

DIPHENYL METHANE DIISOCYANATE   HR: 2
CAS: 101-68-8   NIOSH: NQ 9350000
DOT: 2489
mf: $C_{15}H_{10}N_2O_2$ mw: 250.27

PROP: Crystals or yellow fused solid, mp: 37.2°, bp: 194°-199° @ 5 mm, d: 1.19 @ 50°, vap press: 0.001 mm @ 40°.

SYNS: BIS(P-ISOCYANATOPHENYL)METHANE * BIS(1,4-ISOCYANATOPHENYL)METHANE * BIS(4-ISOCYANATOPHENYL)METHANE * DIFENYLMETHAAN-DISSOCYANAAT (DUTCH) * DIFENIL-METAN-DIISOCIANATO (ITALIAN) * 4-4'-DIISOCYANATE DE DIPHENYLMETHANE (FRENCH) * 4,4'-DIISOCYANATODIPHENYL-METHANE * DIPHENYLMETHAN-4,4'-DIISOCYA-

NAT (GERMAN) * P,P'-DIPHENYLMETHANE DI-ISOCYANATE * 4,4'-DIPHENYLMETHANE DIISO-CYANATE * DIPHENYLMETHANE 4,4'-DI-ISOCYANATE * METHYLENEBIS(4-ISOCYANATO-BENZENE) * 1,1-METHYLENEBIS(4-ISOCYANA-TOBENZENE) * METHYLENEBIS(P-PHENYLENE ISOCYANATE) * P,P'-METHYLENEBIS(PHENYL ISOCYANATE) * METHYLENEBIS(P-PHENYL ISO-CYANATE) * METHYLENEBIS(4-PHENYL ISOCYA-NATE) * 4,4'-METHYLENEBIS(PHENYL ISOCYA-NATE) * 4,4'-METHYLENEDIPHENYL DIISOCYANATE * METHYLENEDI-P-PHENYLENE DIISOCYANATE * 4,4'-METHYLENEDIPHENYLENE ISOCYANATE * NACCONATE 300 * NCI-c50668

OSHA PEL: TWA CL 20 ppb
ACGIH TLV: TWA CL 0.02 ppm
DOT Classification: IMO: Poison B

THR: Moderately toxic. An irritant and allergic sensitizer. When heated to decomposition it emits toxic fumes of NO_x and SO_x.

4-(DIPHENYLMETHOXY)-1-METHYLPI-PERIDINE CHLOROTHEO-PHYLLINE   HR: 3
NIOSH: TM 7823000
mf: $C_{19}H_{23}NO \cdot C_7H_7ClN_4O_2$ mw: 496.06

SYN: P 284

THR: Poison by ingestion and intravenous routes. When heated to decomposition it emits very toxic fumes of NO_x and Cl^-.

DIPHENYLNITROSAMINE   HR: 3
CAS: 86-30-6   NIOSH: JJ 9800000
mf: $C_{12}H_{10}N_2O$ mw: 198.24

PROP: Green crystals, mp: 144°.

SYNS: DIPHENYLNITROSAMIN (GERMAN) * NCI-C02880 * N-NITROSODIFENYLAMIN (CZECH) * N-NITROSODIPHENYLAMINE

THR: An experimental carcinogen. Moderately toxic by ingestion. Mutagenic data. An eye irritant. Dangerous fire hazard when exposed to heat, flame, or oxidizing materials. Dangerous; when heated to decomposition it emits highly toxic fumes of NO_x; can react vigorously with oxidizing materials. For further information see Vol. 5, No. 4 of *DPIM Report*.

3,3-DIPHENYL-2-OXETANONE   HR: 3
CAS: 16230-71-0   NIOSH: RQ 7700000
mf: $C_{15}H_{12}O_2$ mw: 224.27

SYNS: 2,2-DIPHENYL-3-HYDROXYPROPIONIC ACID LACTONE * alpha,alpha-DIPHENYL-beta-PROPIOLACTONE

THR: An experimental carcinogen. When heated to decomposition it emits acrid smoke and fumes.

alpha,alpha-DIPHENYL-2-PIPERIDINEMETHANOL   HR: 3
CAS: 467-60-7   NIOSH: TN 0420000
mf: $C_{18}H_{21}NO$ mw: 267.40

SYNS: ALPHA-(2-PIPERIDYL)BENZHYDROL * ALPHA-PIPRADOL * PYRIDROL

THR: Poison by ingestion. When heated to decomposition it emits toxic fumes of NO_x.

1,1-DIPHENYL-2-PROPYNYL-N-CYCLOHEXYLCARBAMATE   HR: 3
CAS: 10087-89-5   NIOSH: GU 7175000
mf: $C_{22}H_{23}NO_2$ mw: 333.46

SYN: CYCLOHEXANECARBAMIC ACID, 1,1-DIPHE-NYL-2-PROPYNYL ESTER

THR: An experimental tumorigen. Mutagenic data. When heated to decomposition it emits toxic fumes of NO_x.

DIPHENYLPYRAZONE   HR: 3
CAS: 57-96-5   NIOSH: UQ 8575000
mf: $C_{23}H_{20}N_2O_3S$ mw: 404.51

SYNS: 1,2-DIPHENYL-4-(2'-PHENYLSULFIN-ETHYL)-3,5-PYRAZOLIDINEDIONE * 4-(PHENYL-SULFOXYETHYL)-1,2-DIPHENYL-3,5-PYRAZOLIDI-NEDIONE * SULFINPYRAZINE * SULFOXYPHE-NYLPYRAZOLIDINE * USAF GE-13

THR: Poison by ingestion, intravenous, and intraperitoneal routes. When heated to decomposition it emits very toxic fumes of NO_x and SO_x.

DIPHENYLTHIOCARBAZONE   HR: 3
CAS: 60-10-6   NIOSH: LQ 9450000
mf: $C_{13}H_{12}N_4S$ mw: 256.35

PROP: Bluish-black crystalline powder, insol in water, sparingly sol in alcohol; freely sol in carbon tetrachloride and chloroform.

SYNS: DITHIZON * DITHIZONE * USAF EK-3092

THR: Poison by intraperitoneal route. Can cause eye injury and glycosuria. Dangerous; when heated to decomposition it emits highly toxic fumes of NO_x and SO_x.

DIPHENYLTHIOUREA HR: 2

CAS: 102-08-9 NIOSH: FE 1225000
mf: $C_{13}H_{12}N_2S$ mw: 228.33

PROP: White to faint gray powder, mp: 154°, bp: decomp, d: 1.32 @ 25°.

SYNS: s-DIPHENYLTHIOCARBAMIDE * N,N'-DI-PHENYLTHIOUREA * SYM-DIPHENYLTHIOUREA * 1,3-DIPHENYLTHIOUREA * 1,3-DIPHENYL-2-THIOUREA * SULFOCARBANILIDE * THIOCAR-BANILIDE * USAF EK-245

THR: Moderately toxic by ingestion and intraperitoneal routes. Dangerous; when heated to decomposition it emits highly toxic fumes of SO_x and NO_x.

1,3-DIPHENYLTRIAZENE HR: 3

CAS: 136-35-6 NIOSH: XY 2625000
mf: $C_{12}H_{11}N_3$ mw: 197.26

PROP: Golden yellow crystals, mp: 98°-99°, bp: explodes, vap d: 6.8.

SYNS: DIAZOAMINOBENZEN (CZECH) * DI-AZOAMINOBENZENE * P-DIAZOAMINOBENZENE * DIAZOAMINOBENZOL (GERMAN)

THR: Poison by ingestion. An experimental tumorigen. Strongly explosive when shocked or exposed to heat. When heated to decomposition it emits toxic fumes of NO_x. Incompatible with acetic anhydride.

DIPOTASSIUM PERSULFATE HR: 2

CAS: 7727-21-1 NIOSH: SE 0400000
mf: $H_2O_8S_2 \cdot 2K$ mw: 272.34

PROP: White, odorless crystals, mp: decomp @ 100°, d: 2.477.

SYNS: POTASSIUM PERSULFATE * POTASSIUM PEROXYDISULFATE * POTASSIUM PEROXYDI-SULPHATE

THR: Moderately toxic. An irritant and allergen. Flammable when exposed to heat or by chemical reaction. It liberates oxygen above 100° when dry or @ about 50° when in solution. Dangerous; when heated to decomposition it emits highly toxic fumes of SO_x; can react with reducing materials.

DIPROPYL MERCURY HR: 3

CAS: 628-85-3 NIOSH: OW 3325000
mf: $C_6H_{14}Hg$ mw: 286.79

PROP: Colorless liquid, immisc in water; d: 2.0208; bp: 190°.

OSHA PEL: TWA 10 ug(Hg)/m³; CL 40

THR: Poison by intraperitoneal route. See also mercury compounds, organic. When heated to decomposition it emits toxic fumes of Hg.

DI-n-PROPYLNITROSAMINE HR: 3

CAS: 621-64-7 NIOSH: JL 9700000
mf: $C_6H_{14}N_2O$ mw: 130.22

SYNS: N-NITROSODI-N-PROPYLAMINE * N-NI-TROSO-N-PROPYL-1-PROPANAMINE

THR: Poison by ingestion and subcutaneous routes. An experimental tumorigen, carcinogen, and neoplastigen. Mutagenic data. See also nitrosamines. When heated to decomposition it emits toxic fumes of NO_x.

DIPROPYLOXOSTANNANE HR: 3

CAS: 7664-98-4 NIOSH: WH 8225000
mf: $C_6H_{14}OSn$ mw: 220.89

SYNS: KYSLICNIK DI-N-PROPYLCINICITY (CZECH) * DIPROPYLTIN OXIDE

THR: Poison by ingestion. A skin and eye irritant. See also tin compounds. When heated to decomposition it emits acrid smoke and fumes.

p-(DIPROPYLSULFAMOYL)BENZOIC ACID SODIUM SALT HR: 3

CAS: 23795-03-1 NIOSH: DG 9450000
mf: $C_{13}H_{18}NO_4S \cdot Na$ mw: 307.37

SYNS: P-(DI-N-PROPYLSULFAMYL)BENZOIC ACID, SODIUM SALT * P-(DIPROPYLSULFAMOYL)BEN-ZOIC ACID, SODIUM SALT * PROBENECID SO-DIUM SALT

THR: Poison by intraperitoneal and intravenous routes. Moderately toxic by ingestion and subcutaneous routes. Human glandular effects by ingestion. When heated to decomposition it emits very toxic fumes of NO_x and SO_x.

DIPYRIDYL HYDROGEN PHOS-PHATE HR: 3

CAS: 21000-42-0 NIOSH: DW 1970000
mf: $C_{12}H_{14}N_2 \cdot 2C_2H_6O_4P$ mw: 436.2

SYN: DIPYRIDYL PHOSPHATE

THR: Poison by ingestion, skin contact and possibly other routes. When heated to decomposition it emits very toxic fumes of PO_x and NO_x.

DI-3-PYRIDYLMERCURY HR: 3

CAS: 20738-78-7 NIOSH: OW 3330000
mf: $C_{10}H_8HgN_2$ mw: 356.79

THR: Poison by intravenous route. See also mercury compounds, organic. When heated to decomposition it emits very toxic fumes of NO_x and Hg.

DIQUAT DIBROMIDE　　　　HR: 3
CAS: 85-00-7　　　　NIOSH: JM 5690000
mf: $C_{12}H_{12}N_2 \cdot 2Br$　　mw: 344.08

PROP: Yellow cryst, water sol; mp: 355°.

SYNS: 9,10-DIHYDRO-8A,10,-DIAZONIAPHENAN-THRENE DIBROMIDE ∗ 9,10-DIHYDRO-8A, 10A-DIAZONIAPHENANTHRENE(1,1'-ETHYLENE-2, 2'-BIPYRIDYLIUM)DIBROMIDE ∗ 5,6-DIHYDRO-DIPYRIDO(1,2A;2,1C)PYRAZINIUM DIBROMIDE ∗ 6,7-DIHYDROPYRIDO(1,2-A;2',1'-C)PYRA-ZINEDIUM DIBROMIDE ∗ 1,1'-ETHYLENE-2, 2'-BIPYRIDYLIUM DIBROMIDE ∗ ETHYLENE DIPYRIDYLIUM DIBROMIDE ∗ 1,1-ETHYLENE 2,2-DIPYRIDYLIUM DIBROMIDE ∗ 1,1'-ETHYL-ENE-2,2'-DIPYRIDYLIUM DIBROMIDE

THR: Poison by ingestion, subcutaneous, intravenous, intraperitoneal, and possibly other routes. A skin and eye irritant. An experimental teratogen. Mutagenic data. A herbicide, see also paraquat. When heated to decomposition it emits very toxic fumes of NO_x and Br^-.

DIQUAT DICHLORIDE　　　　HR: 3
CAS: 4032-26-2　　　　NIOSH: JM 5750000
mf: $C_{12}H_{12}N_2 \cdot 2Cl$　　mw: 255.16

SYN: 1,1'-ETHYLENE-2,2'-DIPYRIDINIUM DI-CHLORIDE

THR: Poison by ingestion, intravenous, and subcutaneous routes. When heated to decomposition it emits very toxic fumes of NO_x and Cl^-.

3,3'-DISELENODIALANINE　　　HR: 2
CAS: 1464-43-3　　　　NIOSH: AY 6030000
mf: $C_6H_{12}N_2O_4Se_2$　　mw: 334.1

SYNS: SELENIUM CYSTINE ∗ SELENOCYSTINE

THR: Mutagenic data. See also selenium. When heated to decomposition it emits very toxic fumes of NO_x and Se.

DISILANE　　　　HR: 3
mf: H_6Si_2　　mw: 62.22

PROP: Gas, repulsive odor; mp: −132.5°; bp: −14.5°; d: 0.686 @ −25°/4°.

SYN: SILICOETHANE

THR: Poison by inhalation. See also hydrides and silanes. Dangerous, when exposed to heat or flame or by chemical reaction; can react with oxidizing materials. Ignites spontaneously in air. Reacts violently with CCl_4; $CHCl_3$; O_2; SF_6.

DISODIUM ARSENITE　　　　HR: 3
CAS: 15120-17-9　　　　NIOSH: CG 3400000
mf: $AsO_2 \cdot Na_2$　　mw: 152.90

SYN: ARSENIOUS ACID, DISODIUM SALT

THR: Poison by ingestion, skin contact, and intraperitoneal routes. Mutagenic data. See also arsenic compounds. When heated to decomposition it emits toxic fumes of As.

DISODIUM-N-(3-(CARBOXYMETHYL-THIOMERCURI)-2-METHOXYPROPYL)-alpha- CAMPHORAMATE　　　　HR: 3
CAS: 21259-76-7　　　　NIOSH: OV 8700000
mf: $C_{16}H_{27}HgNO_6S \cdot 2Na$　　mw: 608.07

SYNS: DIUCARDYN SODIUM ∗ SODIUM MER-CAPTOMERIN ∗ THIOMERIN SODIUM

THR: Poison by subcutaneous, intramuscular, and intravenous routes. See also mercury. When heated to decomposition it emits very toxic fumes of Hg, NO_x, and SO_x.

DISODIUM CHROMATE　　　　HR: 3
CAS: 7775-11-3　　　　NIOSH: GB 2955000
mf: $CrO_4 \cdot 2Na$　　mw: 161.98

SYNS: CHROMATE OF SODA ∗ CHROMIUM DI-SODIUM OXIDE ∗ NEUTRAL SODIUM CHROMATE ∗ SODIUM CHROMATE

OSHA PEL: TWA CL 100 ug(CrO3)/m^3

THR: Poison by skin contact, intraperitoneal, intravenous, subcutaneous, and intradermal routes. A human carcinogen by inhalation. Mutagenic data. A powerful oxidizer. For further information see Sodium Chromate Vol. 1, No. 8 of *DPIM Report*.

DISODIUM-3,6-ENDOXOHEXA-HYDROPHTHALATE　　　　HR: 3
CAS: 129-67-9　　　　NIOSH: RN 8225000
mf: $C_8H_8O_5 \cdot 2Na$　　mw: 230.14

PROP: Water sol solid; mp: 144°.

SYNS: DINATRIUM-(3,6-EPOXY-CYCLOHEXAAN-1,2-DICARBOXYLAAT) (DUTCH) ∗ DINATRI-UM-(3,6-EPOXY-CYCLOHEXAN-1,2-DICARBOXY-LAT) (GERMAN) ∗ DISODIUM 3,6-EPOXYCYCLO-HEXANE-1,2-DICARBOXYLATE ∗ DISODIUM 7-OXABICYCLO(2.2.1)HEPTANE-2,3-

DICARBOXYLATE * DISODIUM SALT OF ENDOT-HALL * DISODIUM SALT OF 7-OXABICYCLO(2.2.1)HEPTANE-2,3-DICARBOXYLIC ACID * 3,6-ENDOXOHEXAHYDROPHTHALIC ACID DISODIUM SALT * (3,6-EPOSSI-CICLOESAN-1,2-DICARBOSSILATO) DISODICO (ITALIAN) * 3,6-EPOXY-CYCLOHEXANE 1,2-CARBOXYLATE DISO-DIQUE (FRENCH)

THR: Poison by ingestion and skin contact. Very irritating to eyes, skin, and mucous membranes. A defoliant and an herbicide. When heated to decomposition it emits acrid smoke and fumes.

DISODIUM FUMARATE HR: 3
CAS: 17013-01-3 NIOSH: LT 1830000
mf: $C_4H_2O_4 \cdot 2Na$ mw: 160.64

SYN: SODIUM FUMARATE

THR: Moderately toxic by ingestion and other routes. Causes gastrointestinal system problems by ingestion. When heated to decomposition it emits toxic fumes of Na_2O.

DISODIUM HEXAFLUOROSILI-CATE HR: 3
CAS: 16893-85-9 NIOSH: VV 8410000
mf: $F_6Si \cdot 2Na$ mw: 188.07

SYNS: DISODIUM HEXAFLUOROSILICATE (2-) * FLUOSILICATE DE SODIUM * SILICON SODIUM FLUORIDE * SODIUM FLUOROSILICATE * SODIUM FLUOSILICATE * SODIUM HEXAFLUO-ROSILICATE * SODIUM SILICOFLUORIDE

OSHA PEL: TWA 2500 ug(F)/m^3

THR: Poison by ingestion and subcutaneous route. When heated to decomposition it emits very toxic fumes of F^- and Na_2O.

DISODIUM INDIGO-5,5-DISULFO-NATE HR: 3
CAS: 860-22-0 NIOSH: DU 3000000
mf: $C_{16}H_{10}N_2O_8S_2 \cdot 2Na$ mw: 468.38

SYNS: CARMINE BLUE (BIOLOGICAL STAIN) * C.I. 73015 * DISODIUM SALT OF 1-INDIGO-TIN-S,S'-DISULPHONIC ACID * FD AND C BLUE NO. 2 * INDIGO CARMINE * INDIGO CARMINE (BIOLOGICAL STAIN) * INDIGO CARMINE DISO-DIUM SALT * INDIGO EXTRACT * 5,5'-INDI-GOTINDISULFONIC ACID * INDIGOTINE DISODIUM SALT * INTENSE BLUE * MAPLE INDIGO CAR-MINE * SACHSISCHBLAU * SODIUM 5,5'-INDI-GOTIDISULFONATE * SOLUBLE INDIGO

THR: Poison by subcutaneous route. An experimental neoplastigen. When heated to decomposition it emits very toxic fumes of SO_x, NO_x, and Na_2O.

DISODIUM METHANEARSE-NATE HR: 3
CAS: 144-21-8 NIOSH: PA 2275000
mf: $CH_3AsO_3 \cdot 2Na$ mw: 183.94

PROP: Crystals, water-sol hydrate, mp: 132°-139°.

SYNS: DISODIUM METHANEARSONATE * DISO-DIUM METHYLARSENATE * DISODIUM METHYL-ARSONATE * DISODIUM MONOMETHYLARSO-NATE * SODIUM METHANEARSONATE * SODIUM METHARSONATE * SODIUM METH-YLARSONATE

THR: A human carcinogen. Moderately toxic by ingestion. See also arsenic compounds. When heated to decomposition it emits toxic fumes of As.

DISODIUM MOLYBDATE HR: 3
CAS: 7631-95-0 NIOSH: QA 5075000
mf: $MoO_4 \cdot 2Na$ mw: 205.92

SYNS: NATRIUMMOLYBDAT (GERMAN) * SODIUM MOLYBDATE(VI)

OSHA PEL: TWA 5 mg(Mo)/m^3

THR: Poison by intraperitoneal route. Moderately toxic by several other routes. See also molybdenum compounds.

DISODIUM-2-(p-(gamma-PHENYLPRO-PYLAMINO)BENZENESULFONAMIDO) PYRIDINE HR: 3
CAS: 53778-51-1 NIOSH: TZ 2575000
mf: $C_{20}H_{19}N_3O_8S_3 \cdot 2Na$ mw: 571.58

SYN: 1-PHENYL-3-(P-2-PYRIDYLSULFAMOYLANI-LINO)-1,3-PROPANEDISULFONIC ACID DISODIUM SALT

THR: Poison by ingestion, intravenous, and subcutaneous route. When heated to decomposition it emits very toxic fumes of NO_x and SO_x.

DISODIUM PYROPHOSPHATE HR: 3
CAS: 7758-16-9 NIOSH: UX 6475000
mf: $H_2O_7P_2 \cdot Na_2$ mw: 221.94

PROP: White crystalline powder, sol in water, d: 1.862, mp: 220° (decomp).

SYNS: DINATRIUMPYROPHOSPHAT (GERMAN)
∗ DIPHOSPHORIC ACID, DISODIUM SALT
∗ DISODIUM DIHYDROGEN PYROPHOSPHATE
∗ DISODIUM DIPHOSPHATE ∗ SODIUM ACID PY-
ROPHOSPHATE ∗ SODIUM PYROPHOSPHATE

THR: Poison by intravenous route. Moderately toxic by ingestion and subcutaneous route. An irritant to skin, eyes, and mucous membranes. When heated to decomposition it emits toxic fumes of PO_x.

DISODIUM SELENATE HR: 3
CAS: 13410-01-0 NIOSH: VS 6650000
mf: $O_4Se \cdot 2Na$ mw: 188.94

PROP: Colorless rhombic crystals, d: 3.098.

SYNS: SELENIC ACID, DISODIUM SALT
∗ NATRIUMSELENIAT (GERMAN) ∗ SODIUM
SELENATE

OSHA PEL: TWA 200 ug(Se)/m^3

THR: Poison by ingestion, intravenous, subcutaneous, and intraperitoneal routes. Mutagenic data. An experimental carcinogen by ingestion. Effects similar to arsenic. Human systemic effects by ingestion including liver and kidney damage. See also selenium compounds, arsenic. When heated to decomposition it emits toxic fumes of Se and Na_2O.

DISTIGMINE BROMIDE HR: 3
CAS: 15876-67-2 NIOSH: UU 5277000
mf: $C_{22}H_{32}N_4O_4 \cdot 2Br$ mw: 576.40

SYNS: 3,3′-(1,6-HEXANEDIYLBIS-((METHYL-
IMINO)CARBONYL)OXY)BIS(1-METHYLPYRI-
DINIUMDIBROMIDE) ∗ 3-HYDROXY-1-
METHYLPYRIDINIUM BROMIDE
HEXAMETHYLENEBIS(METHYLCARBAMATE)

THR: Poison by ingestion, intravenous, and intraperitoneal routes. When heated to decomposition it emits very toxic fumes of NO_x and Br^-.

DITHIAZANINE HR: 3
CAS: 7187-55-5 NIOSH: DL 7050000
mf: $C_{23}H_{23}N_2S_2$ mw: 391.60

THR: Poison by ingestion, intravenous, and intraperitoneal routes. When heated to decomposition it emits very toxic fumes of NO_x and SO_x.

DITHIAZANINE IODIDE HR: 3
CAS: 514-73-8 NIOSH: DL 7060000
mf: $C_{23}H_{24}N_2S_2 \cdot I$ mw: 519.51

SYNS: 3,3′-DIETHYLPENTAMETHINETHIACYANINE
IODIDE ∗ 3,3′-DIETHYLTHIADICARBOCYANINE
IODIDE ∗ 3-ETHYL-2-(5-(3-ETHYL-2-BENZO-
THIAZOLINYLIDENE)-1,3-PENTADIENYL)BENZO-
THIAZOLIUM IODIDE

THR: Poison by ingestion, intraperitoneal, and intravenous routes. When heated to decomposition it emits very toxic fumes of I^-, SO_x and NO_x.

2,2′-DITHIOBISANILINE HR: 3
CAS: 1141-88-4 NIOSH: BX 9540000
mf: $C_{12}H_{12}N_2S_2$ mw: 248.38

SYNS: BIS(2-AMINOPHENYL)DISULFIDE
∗ O,O′-DIAMINO DIPHENYL DISULFIDE ∗ O,O-
DITHIO-BIS-ANILINE ∗ USAF AB-315

THR: Poison by intravenous and intraperitoneal routes. Moderately toxic by ingestion. Severe eye irritant. When heated to decomposition it emits very toxic fumes of NO_x and SO_x.

2,2′-DITHIOBIS(PYRIDINE-1-OXIDE)-MAGNESIUM SULFATE TRIHY-DRATE HR: 3
NIOSH: UT 6900000
mf: $C_{10}H_8N_2O_2S_2 \cdot MgOS \cdot 3H_2O$ mw: 378.75

SYN: MDS

THR: An experimental teratogen. When heated to decomposition it emits very toxic fumes of SO_x and NO_x.

DITHIOBIURET HR: 3
CAS: 541-53-7 NIOSH: EC 1575000
mf: $C_2H_5N_3S_2$ mw: 135.22

PROP: Crystals, mp: 181°, bp: decomp, d: 1.522 @ 30°.

SYN: 2-THIO-1-(THIOCARBAMOYL)-UREA

THR: Poison by ingestion and inhalation causing respiratory paralysis. Dangerous; when heated to decomposition it emits highly toxic fumes of SO_x and NO_x.

DITHIODIGLYCOL HR: 3
CAS: 1892-29-1 NIOSH: KK 7525000
mf: $C_4H_{10}O_2S_2$ mw: 154.26

SYN: USAF TH-9

THR: Poison by intraperitoneal route. When heated to decomposition it emits toxic fumes of SO_x.

DITHIOLANE IMINOPHOSPHATE   HR: 3
CAS: 333-29-9   NIOSH: NJ 6400000
mf: $C_7H_{14}NO_2OS_3$ mw: 271.37

SYNS: CYCLIC ETHYLENE (DIETHOXYPHOSPHINO-
THIOYL)DITHIOIMIDOCARBONATE * CYCLIC
ETHYLENE ESTER OF (DIETHOXYPHOSPHINO-
THIOYL)DITHIOIMIDOCARBONIC ACID * DI-
ETHYL-N-1,3-DITHIOLANYL-2-IMINO PHOSPHATE
* ENT 25,809

THR: Poison by ingestion and skin contact.
When heated to decomposition it emits very
toxic fumes of NO_x and SO_x.

DITHION   HR: 3
CAS: 572-48-5   NIOSH: GW 4200000
mf: $C_{17}H_{21}O_5PS$ mw: 368.41

PROP: Crystals nearly insol in water, mp: 88°.

SYNS: O,O-DIETHYL 7-HYDROXY-3,4-TETRA-
METHYLENE COUMARINYL PHOSPHOROTHIOATE
* O,O-DIETHYL O-(7,8,9,10-TETRAHYDRO-6-OX-
OBENZO(C)CHROMAN-3-YL)PHOSPHOROTHIOATE
* O,O-DIETHYL O-(7,8,9,10-TETRAHYDRO-6-
OXO-6H-DIBENZO(B,D)PYRAN-3-YL)PHOSPHO-
ROTHIOATE * O,O-DIETHYL O-(3,4-TETRA-
METHYLENECOUMARINYL-7) THIOPHOSPHATE
* DITHIONE * ENT-24986 * 7-HYDROXY-
3,4-TETRAMETHYLENECOUMARIN-O,O-DIETHYL
THIOPHOSPHATE

THR: Poison by ingestion and possibly other
routes. See also parathion. When heated to decomposition it emits very toxic fumes of PO_x
and SO_x.

DITHIOOXAMIDE   HR: 3
CAS: 79-40-3   NIOSH: RP 1575000
mf: $C_2H_4N_2S_2$ mw: 120.20

SYNS: DITHIOXAMIDE * ETHANEDITHIOAMIDE
* HYDRORUBEANIC ACID * RUBEANIC ACID
* USAF B-43 * USAF EK-4394 * USAF MK-6

THR: Poison by ingestion, intraperitoneal, and
intravenous routes. When heated to decomposition it emits very toxic fumes of NO_x and SO_x.

DITHIOSYSTOX   HR: 3
CAS: 298-04-4   NIOSH: TD 9275000
mf: $C_8H_{19}O_2PS_3$ mw: 274.42

SYNS: O,O-DIAETHYL-S-(3-THIA-PENTYL)-DI-
THIOPHOSPHAT (GERMAN) * O,O-DIAETHYL-
S-(2-AETHYLTHIO-AETHYL)-DITHIOPHOSPHAT

(GERMAN) * O,O-DIETHYL-S-(2-ETHTHIOETHYL)
PHOSPHORODITHIOATE * O,O-DIETHYL-S-(2-
ETHTHIOETHYL) THIOTHIONOPHOSPHATE
* O,O-DIETHYL-S-(2-ETHYLMERCAPTOETHYL)
DITHIOPHOSPHATE * O,O-DIETHYL-S-(2-
ETHYLTHIO-ETHYL)-DITHIOFOSFAAT (DUTCH)
* O,O-DIETHYL-2-ETHYLTHIOETHYL PHOSPHORO-
DITHIOATE * O,O-DIETHYL-S-2-(ETHYLTHIO)-
ETHYL PHOSPHORODITHIOATE * O,O-DIETIL-S-
(2-ETILTIO-ETIL)-DITIOFOSFATO (ITALIAN)
* DITHIODEMETON * DITHIOPHOSPHATE DE
O,O-DIETHYLE ET DE S-(2-ETHYLTHIO-ETHYLE)
(FRENCH) * O,O-ETHYL-S-2(ETHYL THIO)ETHYL
PHOSPHORODITHIOATE * ENT23,437 * S-2-
(ETHYLTHIO)ETHYL O,O-DIETHYL ESTER OF
PHOSPHORODITHIOIC ACID

THR: A poison by ingestion, skin, intraperitoneal, intravenous, and other routes. Mutagenic
data. See also esters. When heated to decomposition it emits very toxic SO_x and PO_x.

DITOLYLETHANE   HR: 3
CAS: 27755-15-3   NIOSH: KI 3421000
mf: $C_{16}H_{18}$ mw: 210.34

THR: Poison by subcutaneous route. Moderately
toxic by ingestion. An experimental tumorigen
by ingestion. When heated to decomposition
it emits acrid smoke and fumes.

DI-o-TOLYLGUANIDINE   HR: 3
CAS: 97-39-2   NIOSH: MF 1400000
mf: $C_{15}H_{17}N_3$ mw: 239.35

PROP: White crystals, mp: 179°, d: 1.10 @
20°/4°, vap d: 8.24.

SYNS: DIORTHOTOLYLGUANIDINE * USAF A-
6598

THR: Poison by ingestion and intraperitoneal
routes. When heated to decomposition it emits
toxic fumes of NO_x.

DI-o-TOLYLTHIOUREA   HR: 3
CAS: 137-97-3   NIOSH: FE 0700000
mf: $C_{15}H_{16}N_2S$ mw: 256.39

PROP: Cryst; mp: 178°; vap d: 8.85.

SYNS: 2,2'-DIMETHYLTHIOCARBANILIDE
* DI-O-TOLUYLTHIOUREA * USAF EK-1651

THR: Poison by intraperitoneal route. Moderately toxic by ingestion. When heated to decomposition it emits very toxic fumes of NO_x and
SO_x.

DITRAN   **HR: 3**
CAS: 8015-54-1   NIOSH: OO 7175000

SYN: JB 329

THR: Poison by intravenous and intraperitoneal routes. Human psychotropic effects by ingestion.

DIVINYL SULFONE   **HR: 3**
CAS: 77-77-0   NIOSH: KM 7175000
mf: $C_4H_6O_2S$ mw: 118.16

SYNS: TL 797 * VINYL SULFONE

THR: Poison by ingestion, skin contact, intravenous, subcutaneous, and intraperitoneal routes. Skin and eye irritant. Dangerous; see sulfonates.

**DIXYRAZINE DIHYDROCHLOR-
IDE**   **HR: 3**
CAS: 60539-20-0   NIOSH: KL 8050000
mf: $C_{24}H_{33}N_3O_2S \cdot 2ClH$ mw: 500.58

SYN: 2-(2-(4-(2-METHYL-3-PHENOTHIAZIN-10-YLPROPYL)-1-PIPERAZINYL)ETHOXY)ETHANOL DIHYDROCHLORIDE

THR: Poison by ingestion, intravenous, and intraperitoneal routes. When heated to decomposition it emits very toxic fumes of Cl^-, SO_x, and NO_x.

DMA   **HR: 3**
CAS: 38222-35-4   NIOSH: GY 7200000
mf: $C_8H_{13}NO_3$ mw: 171.22

SYNS: 2-CYCLOPENTEN-1-ONE, 2,5-DIHYDROXY-3-DIMETHYLAMINO-5-METHYL- * DIMETHYLAMINO HEXOSE REDUCTIONE

THR: Poison by ingestion and intraperitoneal routes. When heated to decomposition it emits toxic fumes of NO_x.

DMNT   **HR: 3**
CAS: 26049-69-4   NIOSH: XJ 4600000
mf: $C_9H_{10}N_4O_3S$ mw: 254.29

SYN: 2-(2,2-DIMETHYLHYDRAZINO)-4-(5-NITRO-2-FURYL)THIAZOLE

THR: An experimental carcinogen. Mutagenic data. When heated to decomposition it emits very toxic fumes of NO_x and SO_x.

DODECANE   **HR: 3**
CAS: 112-40-3   NIOSH: JR 2125000
mf: $C_{12}H_{26}$ mw: 170.38

PROP: Liquid, mp: $-12°$, bp: 216.2°, lel = 0.6%, flash p: 165°F, d: 0.750, vap press: 1 mm @ 47.8°, vap d: 5.96, autoign temp: 401°F.

SYNS: BIHEXYL * DIHEXYL * N-DODECAN (GERMAN) * DUODECANE

THR: An experimental carcinogen. Irritating and narcotic in high concentrations. Flammable when exposed to heat, flame or oxidizers. Moderately explosive in the form of vapor when exposed to flame. Dangerous; keep away from heat and open flame. To fight fire, use foam, CO_2, dry chemical.

t-DODECANETHIOL   **HR: 3**
CAS: 25103-58-6   NIOSH: JR 3150000
mf: $C_{12}H_{26}S$ mw: 202.44

PROP: White to light yellow liquid, bp: 200°-235°, flash p: 205°F (OC), d: 0.85 @ 25°/25°, vap d: 6.98.

SYNS: TERC.DODECYLMERKAPTAN (CZECH) * TERT-DODECYLMERCAPTAN * TERT-DODECYLTHIOL * 2,3,3,4,4,5-HEXAMETHYL-2-HEXANETHIOL

THR: Poison by ingestion. See also mercaptans. Dangerous; see also sulfates. Can react vigorously with oxidizing materials. Combustible when exposed to heat or flame. To fight fire, use foam, CO_2, dry chemical.

**DODECATRIETHYLAMMONIUM BRO-
MIDE**   **HR: 3**
CAS: 18186-71-5   NIOSH: BQ 2800000
mf: $C_{18}H_{40}N \cdot Br$ mw: 350.50

THR: Poison by ingestion and intraperitoneal routes. See also bromides. When heated to decomposition it emits very toxic fumes of NO_x and Br^-.

DODECENE EPOXIDE   **HR: 3**
CAS: 2855-19-8   NIOSH: JR 2450000
mf: $C_{12}H_{24}O$ mw: 184.36

SYN: 1,2-EPOXYDODECANE

THR: An experimental tumorigen. When heated to decomposition it emits acrid smoke and fumes.

**DODECENYLSUCCINIC ANHY-
DRIDE**   **HR: 3**
CAS: 25377-73-5   NIOSH: WN 1225000
mf: $C_{16}H_{27}NO_3$ mw: 281.44

PROP: Light yellow, clear viscous oil, bp: 180°-182° @ 5 mm, flash p: 352°F (COC), d: 1.002 @ 25°/4°.

THR: Poison by intraperitoneal route. An irritant and sensitizer. Combustible when exposed to heat or flame; can react with oxidizing materials. To fight fire, use foam, CO_2, dry chemical. When heated to decomposition it emits toxic fumes of NO_x.

DODECYL ALCOHOL   HR: 3
CAS: 112-53-8   NIOSH: JR 5775000
mf: $C_{12}H_{26}O$ mw: 186.38

PROP: Colorless solid, floral odor, sol in 2 parts of 70% alcohol, insol in water, d: 0.8201 @ 24°/4°, mp: 24°, bp: 259°, flash p: 260°F, autoign temp: 527°F.

SYNS: ALCOHOL C-12 * 1-DODECANOL * N-DODECANOL * N-DODECYL ALCOHOL * DUODECYL ALCOHOL * LAURIC ALCOHOL * LAURINIC ALCOHOL * LAURYL ALCOHOL * N-LAURYL ALCOHOL, PRIMARY * LAURYL 24

THR: An experimental tumorigen by skin contact. Moderately toxic. A severe skin irritant. Combustible when exposed to heat or flame. To fight fire, use dry chemical, CO_2. When heated to decomposition it emits acrid smoke and fumes.

DODECYLAMINE   HR: 3
CAS: 124-22-1   NIOSH: JR 6475000
mf: $C_{12}H_{27}N$ mw: 185.40

PROP: Oil, amine odor; fp: 28.3°; vap press: 64 mm @ 170°.

SYN: LAURYLAMINE

THR: Poison by intraperitoneal route. Moderately toxic by ingestion. Severe eye and skin irritant. See also amines. When heated to decomposition it emits toxic fumes of NO_x.

DODECYL BENZENE SODIUM SULFONATE   HR: 3
CAS: 25155-30-0   NIOSH: DB 6825000
mf: $C_{18}H_{29}O_3S \cdot Na$ mw: 348.52

PROP: White to light yellow flakes, granules or powder.

SYNS: DODECYLBENZENSULFONAN SODNY (CZECH) * NACCANOL NR * SODIUM DODE-CYLBENZENESULFONATE * SODIUM LAURYLBEN-ZENESULFONATE * SOL SODOWA KWASU LAURYLOBENZENOSULFONOWEGO(POLISH) * SULFAPOLU (POLISH) * P-1',1',4',4'-TETRAMETHYLOKTYLBENZENSULFONAN SODNY (CZECH) * ULTRAWET K

THR: Poison by intravenous route. Moderately toxic by ingestion. Severe eye irritant. See also sulfonates. For further information see Sodium dodecyl benzene sulfonate, Vol. 3, No. 1 of *DPIM Report*.

N-DODECYLGUANIDINE ACE-TATE   HR: 3
CAS: 2439-10-3   NIOSH: MF 1750000
mf: $C_{13}H_{29}N_3 \cdot C_2H_4O_2$ mw: 287.51

PROP: Crystals, sol in hot water and alc, mp: 136°.

SYNS: DODECYLGUANIDINE ACETATE * DO-DINE ACETATE * DODINE, MIXTURE WITH GLYODIN * ENT 16,436 * LAURYLGUANIDINE ACETATE

THR: An experimental carcinogen. Moderately toxic by ingestion, skin contact, and possibly other routes. When heated to decomposition it emits very toxic fumes of NO_x.

DODECYLPHENOL   HR: 2
CAS: 27193-86-8   NIOSH: SL 3675000
mf: $C_{18}H_{30}O$ mw: 262.48

PROP: Straw-colored liquid, phenolic odor, bp: 154°-168°, flash p: 325°F (OC), d: 0.93 @ 20°/20°, vap d: 9.04.

THR: Moderately toxic by ingestion. Combustible when exposed to heat or flame. When heated to decomposition it emits toxic fumes; can react with oxidizing materials. To fight fire, use CO_2, dry chemical.

DODECYL-p-TOLYL TRIMETHYL AM-MONIUM CHLORIDE   HR: 3
CAS: 1399-80-0   NIOSH: BO 3675000

PROP: Aqueous preparation containing approximately 40% methyl dodecylbenzyl trimethyl ammonium chloride and 10% methyl dodecyl-xylene bis(trimethyl ammonium chloride)

SYNS: ALKYL(C9-15)TOLYL METHYLTRIMETHYL AMMONIUM CHLORIDE * HYAMINE 2389 * METHYL DODECYL BENZYL AMMONIUM CHLO-RIDE

THR: Poison by ingestion, intraperitoneal, and intravenous routes. A severe eye irritant. When heated to decomposition it emits very toxic fumes of NO_x and Cl^-.

DODECYLTRICHLOROSILANE   HR: 3
CAS: 4484-72-4   NIOSH: VV 3940000
mf: $C_{12}H_{25}Cl_3Si$ mw: 303.81

PROP: Colorless to yellow liquid, readily hydrolyzed by moisture with the production of hydrochloric acid, bp: 288°, d: 1.026 @ 25°/25°.

SYN: TRICHLORODODECYLSILANE

THR: Poison. An irritant; corrosive. See also silane. Dangerous; see also chlorides.

DOLANTIN HYDROCHLORIDE   HR: 3
CAS: 50-13-5   NIOSH: NS 5950000
mf: $C_{15}H_{21}NO_2 \cdot ClH$ mw: 283.83

SYNS: CHLORBICYCLENE (FRENCH) * ETHYL-1-METHYL-4-PHENYLISONIPECOTATE HYDROCHLORIDE * ETHYL-1-METHYL-4-PHENYLPIPERIDINE-4-CARBOXYLATE HYDROCHLORIDE * ETHYL-1-METHYL-4-PHENYLPIPERIDYL-4-CARBOXYLATE HYDROCHLORIDE * 1-METHYL-4-CARBETHOXY-4-PHENYLPIPERIDINE HYDROCHLORIDE * N-METHYL-4-PHENYL-4-CARBETHOXYPIPERIDINE HYDROCHLORIDE * 1-METHYL-4-PHENYL-4-CARBOETHOXYPIPERIDINE HYDROCHLORIDE * 1-METHYL-4-PHENYLISONIPECOTIC ACID ETHYL ESTER HYDROCHLORIDE

THR: Poison by ingestion, subcutaneous, intravenous, and intraperitoneal routes. When heated to decomposition it emits very toxic fumes of HCl and NO_x.

L-DOPA   HR: 2
CAS: 59-92-7   NIOSH: AY 5600000
mf: $C_9H_{11}NO_4$ mw: 197.21

SYNS: DIHYDROXY-L-PHENYLALANINE * L-DIHYDROXYPHENYLALANINE * L-ALPHA-DIHYDROXYPHENYLALANINE * L-BETA-(3,4-DIHYDROXYPHENYL)ALANINE * L-3,4-DIHYDROXYPHENYL-ALPHA-ALANINE * BETA-(3,4-DIHYDROXYPHENYL)-L-ALANINE * 3-(3,4-DIHYDROXYPHENYL)-L-ALANINE * 3,4-DIHYDROXYPHENYLALANINE * (−)-3,4-DIHYDROXYPHENYLALANINE * 3,4-DIHYDROXYPHENYL-L-ALANINE * 3,4-DIHYDROXY-L-PHENYLALANINE * L-3,4-DIHYDROXYPHENYLALANINE * L-O-HYDROXYTYROSINE

THR: Moderately toxic by several routes. Psychotropic and other central nervous system ef-

fects in humans. When heated to decomposition it emits toxic fumes of NO_x.

L-DOPA HYDROCHLORIDE   HR: 3
CAS: 5796-14-5   NIOSH: AY 5775000
mf: $C_9H_{11}NO_4 \cdot ClH$ mw: 233.67

SYNS: L-3-(3,4-DIHYDROXYPHENYL)ALANINE * 3-HYDROXY-L-TYROSINE HYDROCHLORIDE

THR: Poison by intravenous route. Moderately toxic by ingestion. When heated to decomposition it emits very toxic fumes of NO_x and HCl.

DOPAMINE   HR: 3
CAS: 51-61-6   NIOSH: UX 1088000
mf: $C_8H_{11}NO_2$ mw: 153.20

SYN: 4-(2-AMINOETHYL)PYROCATECHOL

THR: Poison by subcutaneous and intraperitoneal routes. When heated to decomposition it emits toxic fumes of NO_x.

DORIDEN   HR: 3
CAS: 77-21-4   NIOSH: MA 5725000
mf: $C_{13}H_{15}NO_2$ mw: 217.29

SYNS: 3-ETHYL-3-PHENYL-2,6-DIKETOPIPERIDINE * 3-ETHYL-3-PHENYL-2,6-DIOXOPIPERIDINE * alpha-ETHYL-alpha-PHENYLGLUTARIMIDE * 2-ETHYL-2-PHENYLGLUTARIMIDE * 3-ETHYL-3-PHENYL-2,6-PIPERIDINEDIONE * 3-PHENYL-3-ETHYL-2,6-DIKETOPIPERIDINE * 3-PHENYL-3-ETHYL-2,6-DIOXOPIPERIDINE * 2-PHENYL-2-ETHYLGLUTARIC ACID IMIDE * alpha-PHENYL-alpha-ETHYLGLUTARIMIDE

THR: Poison to humans by ingestion. Affects the central nervous system. When heated to decomposition it emits toxic fumes of NO_x.

DOWCO 179   HR: 3
CAS: 2921-88-2   NIOSH: TF 6300000
mf: $C_9H_{11}Cl_3NO_3PS$ mw: 350.59

SYNS: O,O-DIETHYL-O-3,5,6-TRICHLORO-2-PYRIDYL PHOSPHOROTHIOATE * O,O-DIAETHYL-O-3,5,6-TRICHLOR-2-PYRIDYLMONOTHIOPHOSPHAT (GERMAN) * ENT 27311

THR: Poison by ingestion, skin contact, inhalation, subcutaneous, and other routes. When heated to decomposition it emits very toxic fumes of Cl^-, NO_x, PO_x, and SO_x.

DOWFUME EB-5   HR: 3
CAS: 53908-27-3   NIOSH: JT 7525000

PROP: Contains ethylene dichloride (30%), carbon tetrachloride (63%) and ethylene dibromide (7%).

THR: Poison by ingestion. See also components as listed. When heated to decomposition it emits very toxic fumes of Cl^- and Br^-.

DRAMAMINE
CAS: 523-87-5   NIOSH: XH 5082000
mf: $C_{17}H_{21}NO \cdot C_7H_7ClN_4O_2$ mw: 470.02

SYNS: o-BENZHYDRYLDIMETHYLAMINOETHANOL-8-CHLOROTHEOPHYLLINATE * 2-(BENZHYDRYLOXY)-N,N-DIMETHYLETHYLAMINE COMPOUND WITH 8-CHLOROTHEOPHYLLINE * ETHYLAMINE-2-(DIPHENYLMETHOXY)-N,N-DIMETHYL-, COMPOUND WITH 8-CHLOROTHEOPHYLLINE (1:1) * NCI-c60639

THR: A poison via intravenous route. See also amines. A drug much used for motion sickness. When heated to decomposition it emits very toxic fumes of NO_x and Cl^-.

DROPERIDOL   HR: 3
CAS: 548-73-2   NIOSH: DE 2100000
mf: $C_{22}H_{22}FN_3O_2$ mw: 379.47

SYNS: DEHYDROBENZPERIDOL * DEIDRO-BENZPERIDOLO * DIHIDROBENZPERIDOL * 1-(1-(3-(P-FLUOROBENZOYL)PROPYL)-1,2,3,6-TETRAHYDRO-4-PYRIDYL)-2 -BENZIMIDAZOLINONE * 1-(1-(4-(P-FLUOROPHENYL-4-OXOBUTYL)-1,2,3,6-TETRAHYDRO-4-PYRIDYL)-2-BENZIMIDAZOLINONE

THR: Poison by subcutaneous, intravenous, and intraperitoneal routes. Moderately toxic by ingestion. When heated to decomposition it emits very toxic fumes of F^- and NO_x.

DYNAMITE
PROP: Major constituent is nitroglycerin

SYNS: DYNAMITE (DOT) * GELATINE DYNAMITE

DOT Classification: Class A Explosive, Label: Explosive

A high explosive used industrially in construction and mining. The name generally refers to a mixture containing as its principal explosive ingredient either glyceryl trinitrate (nitroglycerin) or ammonium nitrate, suitably sensitized. It does not apply to black blasting powders, chlorate powders, and other deflagrating mixtures.

An ordinary blasting cap or an electric blasting cap is used for detonating a charge of dynamite. The various classes and grades of dynamites are made from mixtures composed of an explosive compound or a mixture of explosive compounds, a dope, and an antacid. If any of the explosive ingredients are in a liquid state they are referred to as the "explosive oil," which is usually composed of glyceryl trinitrate (nitroglycerin) and about 25-30% of ethylene glycol dinitrate.

When the explosive oil is gelatinized the explosive is known as a gelatin or an ammonia gelatin dynamite. See also dynamite, ammonia gelatin and dynamite, gelatin.

Blasting gelatin is a gelatinized mass of an elastic nature obtained by incorporating nitrocotton with an explosive oil into which is mixed about 1% of antacid. See also dynamite, blasting gelatin.

Dynamites may be in bulk form (bag powder) or put up in cartridge form, the most common size being 1¼ in. in diameter and 8 in. long, although for holes of small diameter, cartridges as small as ⅞ in. in diameter are also used. In large diameter well-drill holes for quarry blasting, cartridge diameters up to 10 in. and lengths up to 30 inches may be used.

An integral part of a stick of dynamite is the paraffined paper wrapper that not only holds the ingredients together but enters into the explosive reaction.

The wrapper also affords some measure of protection from moderate exposure to dampness. For blasting in wet operations, a gelatinized dynamite which resists the absorption of water is used.

The strength of straight dynamite is graded by its explosive oil content (% by weight), while for any other class of dynamite, the strength is determined experimentally in comparison with the various grades of the straight dynamites. For example, a 40% straight dynamite is one which contains 40% of explosive oil; a 40% strength ammonia dynamite, as determined by tests, equals a 40% straight dynamite in strength. In other words a 40% strength ammonia dynamite will release the same energy as an equivalent weight of a 40% straight dynamite.

THR: See also nitrates. For fire hazard, see explosives, high, and nitrates. While this material is a powerful explosive when detonated by shock or heat, it is only moderately hazardous. See also explosives, high and nitrates. Dangerous; shock and heat will explode it; when heated to decomposition it emits highly toxic fumes of NO_x and CO, etc. It can react vigorously with oxidizing materials.

DYPHONATE **HR: 3**
CAS: 944-22-9   NIOSH: TA 5950000
mf: $C_{10}H_{15}OPS_2$ mw: 246.34

SYNS: o-aethyl-s-phenyl-aethyl-dithio-phosphonat (german) * ent 25,796 * o-ethyl-s-phenyl ethylphosphonodi-thioate * o-ethyl-s-phenyl ethyldithio-phosphonate

THR: Poison by ingestion and skin contact and possibly other routes. When heated to decomposition it emits very toxic fumes of PO_x and SO_x.

DYSPROSIUM **HR: 2**
af: Dy   at wt: 162.5

PROP: Bright, lustrous, silvery metal, mp: 1409°; bp: 2335°; d: 8.540 @ 25°.

THR: It may exhibit an anti-coagulant effect. See also rare earths. Flammable; an active reducing agent. Reacts violently in air and to halogens.

DYSPROSIUM CHLORIDE **HR: 3**
CAS: 10025-74-8   NIOSH: JW 0700000
mf: Cl_3Dy mw: 268.85

PROP: Shiny yellow crystals, d: 3.67 @ 0°/4°, mp: 718°, bp: 1500°. A sol salt.

THR: Poison by intraperitoneal route. See also chlorides and rare earths.

DYSPROSIUM CITRATE **HR: 3**
CAS: 13074-91-4   NIOSH: GE 7650000
mf: $C_6H_5O_7 \cdot Dy$ mw: 351.61

SYN: citric acid, dysprosium(3+) salt (1:1)

THR: Poison by intraperitoneal route. See also rare earths. When heated to decomposition it emits acrid smoke and fumes.

DYSPROSIUM (III) NITRATE HEXAHY-DRATE (1:3:6) **HR: 3**
CAS: 35725-30-5   NIOSH: JW 1050000
mf: $N_3O_9 \cdot Oy \cdot 6H_2O$ mw: 456.65

SYN: nitric acid dysprosium(3+) salt hex-ahydrate

THR: Poison by intraperitoneal route. Moderately toxic by ingestion. See also rare earths. When heated to decomposition it emits very toxic fumes of NO_x. See also nitrates.

E

EDATHAMIL DISODIUM HR: 3
CAS: 139-33-3 NIOSH: AH 4375000
mf: $C_{10}H_{14}N_2O_8 \cdot 2Na$ mw: 336.24

SYNS: D'E.D.T.A. DISODIQUE (FRENCH)
* DISODIUM (ETHYLENEDINITRILO)TETRAACE-
TATE * DISODIUM (ETHYLENEDINITRILO)TET-
RAACETIC ACID * DISODIUM DIACID ETHYLENE-
DIAMINETETRAACETATE * DISODIUM
DIHYDROGEN ETHYLENEDIAMINETETRAACETATE
* DISODIUM DIHYDROGEN(ETHYLENEDINITRILO)
TETRAACETATE * DISODIUM EDETATE * DI-
SODIUM EDTA * DISODIUM ETHYLENEDIAMINE-
TETRAACETATE * DISODIUM ETHYLENEDI-
AMINETETRAACETIC ACID * DISODIUM SEQUES-
TRENE * DISODIUM TETRACEMATE * DISO-
DIUM VERSENATE * DISODIUM VERSENE
* ETHYLENE DIAMINE TETRAACETIC ACID DISO-
DIUM SALT * EDETATE DISODIUM * EDTA,
DISODIUM SALT * ENDRATE DISODIUM
* ETHYLENEDIAMINETETRAACETATE DISODIUM
SALT

THR: A poison by ingestion, intraperitoneal,
and intravenous routes. An experimental teratogen. See disodium ethylene diamine tetra-ace-
tate. The calcium disodium salt of EDTA is
used as a chelating agent in treating lead poisoning. When heated to decomposition it emits toxic
fumes of NO_x and Na_2O.

EDATHANIL TETRASODIUM HR: 3
CAS: 64-02-8 NIOSH: AH 5075000
mf: $C_{10}H_{12}N_2O_8 \cdot 4Na$ mw: 380.20

SYNS: N,N'-ETHYLENE DIAMINE DIACETIC ACID
TETRASODIUM SALT * EDETIC ACID TETRASO-
DIUM SALT * EDTA TETRASODIUM SALT
* ENDRATE TETRASODIUM * ETHYLENEBIS-
(IMINODIACETIC ACID) TETRASODIUM SALT
* SODIUM ETHYLENEDIAMINETETRAACETATE
* SODIUM ETHYLENEDIAMINETETRAACETIC ACID
* SODIUM SALT OF ETHYLENEDIAMINETETRA-
ACETIC ACID TETRASODIUM EDETATE * TETRA-
SODIUM EDTA * TETRASODIUM ETHYLENEDI-
AMINETETRAACETATE * TETRASODIUM
ETHYLENEDIAMINETETRACETATE * TETRASO-
DIUM (ETHYLENEDINITRILO)TETRAACETATE
* TETRASODIUM SALT EDTA * TETRASODIUM
SALT OF EDTA * TETRASODIUM SALT OF ETH-
YLENEDIAMINETETRACETICACID

THR: A poison by intraperitoneal route. An
eye irritant. See also ethylene diaminetetra acetic acid, disodium salt. When heated to decomposition it emits toxic fumes of NO_x and Na_2O.

EICOSANOIC ACID HR: 3
CAS: 506-30-9 NIOSH: JX 3780000
mf: $C_{20}H_{40}O_2$ mw: 312.60

PROP: White leaflets; mp: 77°; bp: 328° slt
decomp in water; very sol in hot alc absolute;
very sol in ether.

SYNS: ARACHIC ACID * ARACHIDIC ACID

THR: An experimental neoplastigen. When
heated to decomposition it emits acrid smoke
and fumes.

ELAIOMYCIN HR 3
CAS: 23315-05-1 NIOSH: EL 5075000
mf: $C_{13}H_{26}N_2O_3$ mw: 258.41

PROP: Metabolite of *Streptomyces Hepaticus*.

SYN: D-THREO-METHOXY-3-(1-OCTENYL-ONN-
AZOXY)-2-BUTANOL

THR: An experimental tumorigen. Poison by
subcutaneous and intravenous routes. A brain
carcinogen. When heated to decomposition it
emits toxic fumes of NO_x.

ELAVIL HR: 3
CAS: 50-48-6 NIOSH: HO 9275000
mf: $C_{20}H_{23}N$ mw: 277.44

SYNS: AMITRIPTYLIN (GERMAN) * AMITRIPTY-
LINE * 10,11-DIHYDRO-N,N-DIMETHYL-5H-DI-
BENZO(A,D)HEPTALENE-DELTA(SUP
5),GAMMA-PROPYLAMINE * 5-(3-DIMETHYLAM-
INOPROPYLIDENE)-10,11-DIHYDRO-5H-DIBEN-
ZO(A,D)CYCLOHEPTENE * 5-(GAMMA-DIMETH-
YLAMINOPROPYLIDENE)-10,11-DIHYDRO-5H-
DIBENZO(A,D)CYCLOHEPTENE * 5-(3-DIMETH-
YLPROPYLIDENE)DIBENZO(A,D)(1,4)CYCLOHEPTA-
DIENE * ELANIL

THR: Poison by ingestion, intraperitoneal, subcutaneous and intravenous routes. Affects human central nervous system and cardiovascular
systems. When heated to decomposition it emits
toxic fumes such as NO_x.

ELAVIL HYDROCHLORIDE HR: 3
CAS: 549-18-8 NIOSH: HO 9450000
mf: $C_{20}H_{23}N \cdot ClH$ mw: 313.90

SYNS: AMITRIPTYLINE CHLORIDE * AMITRYP-
TYLINE HYDROCHLORIDE * DAMILEN HYDRO-
CHLORIDE * 3-(3-DIMETHYLAMINOPROPYLI-
DENE)-1:2-4:5-DIBENZOCYCLOHEPTA-1:4-DIENE
* 5-(3-DIMETHYLAMINOPROPYLIDENE)DI-
BENZO(A,D)(1,4)CYCLOHEPTADIENE HYDRO-
CHLORIDE * PROHEPTADIEN MONOHYDROCHLO-
RIDE * TRYPTIZOL HYDROCHLORIDE

THR: Poison by intraperitoneal, intravenous,
ingestion and subcutaneous routes. Mutagenic
data. Affects central nervous and cardiovascular
systems in humans by ingestion. When heated
to decomposition it emits very toxic fumes such
as HCl and NO_x.

ELYMOCLAVINE   HR: 3
CAS: 548-43-6   NIOSH: KE 6340000

PROP: A close chemical relative of LSD found
in ergot fungi and bindweeks of the genus *Ipo-
moea*, family *Convolvulaceae*.

THR: Poison by intraperitoneal route. An experimental teratogen. When heated to decomposition it emits acrid smoke and fumes. For
further information see Vol. 1, No. 3 of *DPIM
Report*.

EMAZOL RED B   HR: 3
CAS: 19526-81-9   NIOSH: QK 2100000
mf: $C_{18}H_{16}N_2O_{10}S_3 \cdot 2Na$ mw: 562.52

THR: An experimental tumorigen. When heated
to decomposition it emits very toxic fumes of
NO_x and SO_x.

EMETINE   HR: 3
CAS: 483-18-1   NIOSH: DK 1750000
mf: $C_{29}H_{40}N_2O_4$ mw: 480.71

PROP: White powder or lumps, bitter taste,
darkens on exposure mp: 74°.

SYNS: (−)-EMETINE * NSC-33669

THR: Poison by ingestion, inhalation, intraperitoneal, and subcutaneous routes. Mutagenic
data. An experimental tumorigen. It is one of
the two potent alkaloids obtained from the Brazilian plant ipecac. The therapeutic use of various ipecac preparations caused many cases of
poisoning, in some instances with fatal results.
The toxic effects are particularly prominent if
the drug is given intravenously. Special care
should therefore be exercised when administering it in this manner. The symptoms of intoxication are gastrointestinal irritation and salivation,

as well as general edema, which follows renal
insufficiency, hemoptysis, flaccid paralysis, peripheral neuritis, aphonia, difficulties in swallowing, delirium, coma and failure of the heart.
A severe eye irritant. The fatal dose is considered to be approximately 2 g, whether administered over a short or relatively long period. The
drug seems to have a cumulative effect. Dangerous; when heated to decomposition it emits
highly toxic fumes of NO_x.

EMETINE ANTIMONY IODIDE   HR: 3
NIOSH: JY 5000000

PROP: Percentage composition = 34% Emetine
and 14% Antimony.

SYN: ANTIMONY EMETINE IODIDE

THR: Poison by gastrointestinal tract in humans
by ingestion route. When heated to decomposition it emits very toxic fumes such as I^-, NO_x
and Sb.

EMETINE BISMUTH IODIDE   HR: 3
CAS: 8001-15-8   NIOSH: JY 5075000

PROP: Composition = 25% Emetine and 17%
Bismuth.

SYN: BISMUTH EMETINE IODIDE

THR: Poison by ingestion route. When heated
to decomposition it emits very toxic fumes such
as I^- and NO_x.

1-EMETINE DIHYDROCHLORIDE   HR: 3
CAS: 316-42-7   NIOSH: JY 5250000
mf: $C_{29}H_{40}N_2O_4 \cdot 2ClH$ mw: 553.63

SYNS: AMEBICIDE * (−)-EMETINE DIHYDRO-
CHLORIDE * NSC-33669

THR: Poison by ingestion, intraperitoneal, subcutaneous, and intravenous routes. A human
eye irritant. When heated to decomposition it
emits very toxic fumes such as Cl^-, NO_x and
HCl.

ENAVID   HR: 3
CAS: 8015-30-3   NIOSH: RC 9050000

PROP: Mixture of 98.5% (17-alpha)-19-norp-
regn-4-en-20-yn-3-one, 17-hydroxy- and 1.5%
(17-alpha)-19-norpregna-1,3,5(10)-trien-20-
yn-17-ol

SYNS: ENOVID * MESTRANOL MIXED WITH
NORETHYNODREL * NORETHYNODREL MIXED
WITH MESTRANOL

THR: A neoplastigen in women by ingestion. An experimental carcinogen, neoplastigen, and tumorigen.

ENDOTHION HR: 3
CAS: 2778-04-3 NIOSH: TF 8225000
mf: $C_9H_{13}O_6PS$ mw: 280.25

SYNS: O,O-DIMETHYL-S-(5-METHOXY-4-OXO-4H-PYRAN-2-YL)PHOSPHOROTHIOATE * O,O-DI-METHYL-S-(5-METHOXYPYRONYL-2-METHYL) THIOPHOSPHATE * ENT 24,653 * 5-METHOXY-2-(DIMETHOXYPHOSPHINYLTHIO-METHYL)PYRONE-4 * S-5-METHOXY-4-OXOPY-RAN-2-YLMETHYL DIMETHYL PHOSPHOROTHIOATE * S-((5-METHOXY-4H-PYRON-2-YL)-METHYL)-O,O-DIMETHYL-MONOTHIOFOSFAAT (DUTCH) * S-((5-METHOXY-4H-PYRON-2-YL)-METHYL)-O,O-DIMETHYL-MONOTHIOPHOSPHAT(GERMAN) * S-(5-METHOXY-4-PYRON-2-YLMETHYL) DIMETHYL PHOSPHOROTHIOATE * S-((5-ME-TOSSI-4H-PIRON-2-IL)-METIL)-O,O-DIMETIL-MONO-TIOFOSFATO (ITALIAN) * THIOPHOSPHATE DE O,O-DIMETHYLE ET DE S-((5-METHOXY-4-PYRONYL)-METHYLE)(FRENCH)

THR: Poison by ingestion, dermal, and other routes. When heated to decomposition it emits very toxic fumes of PO_x and SO_x.

ENDRIN HR: 3
CAS: 72-20-8 NIOSH: IO 1575000
DOT: 2761
mf: $C_{12}H_8Cl_6O$ mw: 380.90

PROP: White crystals, mp: decomp @ 200°.

SYNS: ENDRINE (FRENCH) * ENT 17,251 * EXPERIMENTAL INSECTICIDE 269 * HEXA-CHLORO EPOXY OCTAHYDRO-ENDO,ENDO-DIMETH-ANO NAPHTHALENE * NCI-C00157

OSHA PEL: TWA 100 ug/m^3 (skin)
ACGIH TLV: TWA 0.1 mg/m^3 (skin)
DFG MAK: 0.1 mg/m^3
DOT Classification: Poison B

THR: EXTREMELY poisonous by ingestion and intravenous routes. Poison by dermal route. An experimental teratogen and carcinogen. Mutagenic data. A central nervous system stimulant. Highly toxic to birds, fish and man. Many cases of fatal poisoning attributed to it. Does not accumulate in human tissue. In man, ingestion of 1 mg/kg caused symptoms. Dangerous in a fire. See also aldrin. Has reacted violently

with parathion. For further information see Vol. 5, No. 2 of *DPIM Report*.

EPE HR: 3
CAS: 33419-42-0 NIOSH: KC 0190000
mf: $C_{29}H_{32}O_{14}$ mw: 604.61

SYNS: 4'-DEMETHYLEPI PODOPHYLLO TOXIN 9-(4,6-O-ETHYLIDENE-beta-D-GLUCOPYRANOSIDE * NSC 141540

THR: An experimental teratogen. Causes blood problems in humans by ingestion and intravenous routes. Mutagenic data. When heated to decomposition it emits acrid smoke and fumes.

EPHEDRINE HR: 3
CAS: 299-42-3 NIOSH: KB 0700000
mf: $C_{10}H_{15}NO$ mw: 165.26

PROP: White granules, mp: 79° (dl), bp: 255° (decomp). Sol in ether and chloroform.

SYNS: EPHEDRIN * L(-)-EPHEDRINE * 1-EPHEDRINE * alpha-HYDROXY-beta-METHYL AMINE PROPYLBENZENE * 1-HYDROXY-2-METHYLAMINO-1-PHENYLPROPANE * 1-alpha-(1-METHYLAMINOETHYL)BENZYL ALCOHOL * (−)-alpha-(1-METHYLAMINOETHYL)BENZYL ALCOHOL * 1-2-METHYLAMINO-1-PHENYLPRO-PANOL * 1-PHENYL-2-METHYLAMINOPROPANOL

THR: VERY poisonous by intravenous route. Poison by ingestion, subcutaneous, and intramuscular routes. Causes rapid pulse, rise in blood pressure, and other actions similar to epinephrine. Has been known to cause allergic sensitization. When heated to decomposition it emits toxic fumes of NO_x. For further information see Vol. 1, No. 4 of *DPIM Report*.

EPHEDRINE HYDROCHLORIDE HR: 3
CAS: 50-98-6 NIOSH: KB 1400000
mf: $C_{10}H_{15}NO \cdot ClH$ mw: 201.72

THR: Poison to humans by skin exposure and intradermal route. Poison by subcutaneous and intravenous routes. When heated to decomposition it emits very toxic fumes such as Cl$^-$ and NO_x.

d-EPHEDRINE HYDROCHLO-RIDE HR: 3
CAS: 24221-86-1 NIOSH: KB 1925000
mf: $C_{10}H_{15}NO \cdot ClH$ mw: 201.72

THR: Poison by intraperitoneal, subcutaneous and intravenous routes. When heated to decomposition it emits very toxic fumes such as HCl and NO$_x$.

dl-EPHEDRINE HYDROCHLO-RIDE   HR: 3
CAS: 134-71-4   NIOSH: KB 1575000
mf: C$_{10}$H$_{15}$NO•ClH   mw: 201.72

SYNS: dl-alpha-(1-(METHYLAMINO)ETHYL) BEN-ZYL ALCOHOL HYDROCHLORIDE * 1-PHENYL-2-METHYLAMINOPROPANOL-1 * RACEPHEDRINE HYDROCHLORIDE

THR: Poison by subcutaneous, intravenous and intraperitoneal routes. When heated to decomposition it emits very toxic fumes such as HCl and NO$_x$.

1-EPHEDRINE SULFATE   HR: 3
CAS: 134-72-5   NIOSH: KB 2625000
mf: C$_{20}$H$_{30}$N$_2$O$_2$•H$_2$O$_4$S   mw: 428.60

SYNS: ISOFEDROL * 1-ALPHA-(1-(METHYLAMI-NO)ETHYL)BENZYL ALCOHOL SULFATE * NCI-c55652 * 1-PHENYL-2-METHYLAMINE-PROPA-NOL-1-SULFATE

THR: A human poison by intravenous route causing cardiovascular problems. Poison by intraperitoneal route. Moderately toxic by ingestion routes. When heated to decomposition it emits very toxic fumes such as NO$_x$, SO$_x$, and H$_2$SO$_4$.

EPHININE HYDROCHLORIDE   HR: 3
CAS: 62-32-8   NIOSH: UX 1925000
mf: C$_9$H$_{13}$NO$_2$•ClH   mw: 203.69

SYNS: 3,4-DIHYDROXYPHENYLETHYLMETHYLAMINE HYDRO-CHLORIDE * 3,4-DIHYDROXYPHENYL-1-METH-YLAMINO-2-ETHANE HYDROCHLORIDE * 4-(2-METHYLAMINOETHYL)PYROCATECHOL HYDRO-CHLORIDE * METHYL-(beta-(3,4-DIHYDROXY PHENYL ETHYL) AMINE HYDROCHLORIDE * N-METHYLDOPAMINE HYDROCHLORIDE

THR: Poison by subcutaneous and intraperitoneal routes. When heated to decomposition it emits very toxic fumes of NO$_x$ and HCl.

dl-EPINEPHRINE   HR: 3
CAS: 329-65-7   NIOSH: DO 2975000
mf: C$_9$H$_{13}$NO$_3$ mw: 183.23

SYN: EPINEPHRINE RACEMIC

THR: VERY poisonous by intravenous and intraperitoneal routes. When heated to decomposition it emits toxic fumes such as NO$_x$.

3,4-EPOXY-1-BUTENE   HR: 3
CAS: 930-22-3   NIOSH: EM 7350000
mf: C$_4$H$_6$O   mw: 70.10

PROP: Liquid, mp: $-135°$, bp: 67°, flash p: $<-58°F$ (CC), d: 0.869, autoign temp: 806°F, vap d: 2.41.

SYNS: BUTADIENE MONOXIDE * 1,2-EPOXY-BUTENE-3

THR: Poison by intraperitoneal route. An experimental tumorigen. Mutagenic data. Dangerous when exposed to heat or flame. Moderately dangerous; when heated to decomposition it emits acrid fumes; can react with oxidizing materials. To fight fire, use CO$_2$, dry chemical, water spray.

EPOXYCHOLESTEROL   HR: 3
CAS: 1250-95-9   NIOSH: FZ 6360000
mf: C$_{27}$H$_{46}$O$_2$ mw: 402.73

SYNS: CHOLESTEROL-alpha-EPOXIDE * CHO-LESTEROL 5-alpha,6-alpha-EPOXIDE * CHOLES-TEROL alpha-OXIDE * 5-alpha,6-alpha-EPOXY-CHOLESTANOL

THR: Mutagenic data. An experimental carcinogen. When heated to decomposition it emits acrid smoke and fumes.

1,2-EPOXYETHYLBENZENE   HR: 3
CAS: 96-09-3   NIOSH: CZ 9625000
mf: C$_8$H$_8$O   mw: 120.16

PROP: Colorless liquid, bp: 194.2, flash p: 165°F (OC), fp: $-36.7°$, d: 1.0469 @ 25°/4°, vap d: 4.14.

SYNS: EPOXYETHYLBENZENE (8CI) * EPOXY-STYRENE * ALPHA,BETA-EPOXYSTYRENE * NCI-c54977 * 1-PHENYL-1,2-EPOXYETHANE * PHENYLETHYLENE OXIDE * 2-PHENYLOXI-RANE * STYRENE EPOXIDE * STYRENE OXIDE * STYRYL OXIDE

THR: An experimental carcinogen. A skin and eye irritant. Moderately toxic by ingestion, inhalation, intraperitoneal, and skin routes. Mutagenic data. Moderate hazard when exposed to heat, flame or oxidizers. When heated it emits acrid fumes; can react with oxidizing materials. To fight fire, use foam, CO$_2$, dry chemical.

EPOXYHEPTACHLOR   HR: 3
CAS: 1024-57-3   NIOSH: PB 9450000
mf: C$_{10}$H$_5$Cl$_7$O   mw: 399.30

SYNS: ENT 25,584 * HEPTACHLOR EPOXIDE
* 1,4,5,6,7,8,8-HEPTACHLORO-2,3-EPOXY-2,
3,3A,4,7,7A-HEXAHYDRO-4,7-METHANOINDENE
* 1,4,5,6,7,8,8-HEPTACHLORO-2,3-EPOXY-
3A,4,7,7A-TETRAHYDRO-4,7-METHANOINDAN
* 2,3,4,5,6,7,7-HEPTACHLORO-1A,1B,5,5A,
6,6A-HEXAHYDRO-2,5-METHANO-2H-INDENO-
(1,2-B)OXIRENE

THR: Poison by ingestion and intravenous routes. An experimental carcinogen. Mutagenic data. When heated to decomposition it emits toxic fumes of Cl$^-$. For further information see Vol. 5, No. 1 of *DPIM Report*.

1,2-EPOXYHEXADECANE   HR: 3
CAS: 7320-37-8   NIOSH: ML 9450000
mf: $C_{16}H_{32}O$ mw: 240.48

SYNS: HEXADECENE EPOXIDE * NCI-C55538

THR: An experimental tumorigen. When heated to decomposition it emits acrid smoke and fumes.

cis-9,10-EPOXYOCTADECANOIC ACID   HR: 3
CAS: 2443-39-2   NIOSH: RG 1300000
mf: $C_{18}H_{34}O_3$ mw: 298.52

SYNS: cis-9,10-EPOXYOCTADECANOATE
* EPOXYOLEIC ACID * 9,10-EPOXYSTEARIC
ACID * cis-9,10-EPOXYSTEARIC ACID
* cis-3-OCTYL-OXIRANEOCTANOIC ACID

THR: An experimental carcinogen. When heated to decomposition it emits acrid smoke and fumes.

EPOXYPROPANE   HR: 3
CAS: 75-56-9   NIOSH: TZ 2975000
mf: C_3H_6O mw: 58.09

PROP: Colorless liquid, ethereal odor, sol in water, alcohol and ether, bp: 33.9°, lel = 2.8%, uel = 37%, fp: −104.4°, flash p: −35°F (TOC), d: 0.8304 @ 20°/20°, vap press: 400 mm @ 17.8°, vap d: 2.0.

SYNS: 1,2-EPOXYPROPANE * METHYL ETHYL-
ENE OXIDE * METHYL OXIRANE * NCI-
C50099 * OXYDE DE PROPYLENE (FRENCH)
* PROPENE OXIDE * 1,2-PROPYLENE OXIDE

OSHA PEL: TWA 100 ppm

THR: Poison by intraperitoneal route. Moderately toxic by ingestion, inhalation, and skin contact. Mutagenic data. A skin and eye irritant.

An experimental neoplastigen and carcinogen. A food additive permitted in food for human consumption. See also aliphatic and aromatic epoxides. An insecticidal fumigant. Highly dangerous when exposed to heat or flame. Severe explosion hazard when exposed to flame. Incompatible with NH$_4$OH; chlorosulfonic acid; HCl; HF; HNO$_3$; oleum; H$_2$SO$_4$. Dangerous; can react vigorously with oxidizing materials. To fight fire, use alcohol foam, CO$_2$, dry chemical.

2,3-EPOXY-1-PROPANOL   HR: 3
CAS: 556-52-5   NIOSH: UB 4375000
mf: $C_3H_6O_2$ mw: 74.09

PROP: Colorless liquid, entirely sol in water, alcohol and ether, d: 1.165 @ 0°/4°, bp: 167° (decomp).

SYNS: EPIHYDRIN ALCOHOL * GLYCIDOL
* GLYCIDYL ALCOHOL * 3-HYDROXY-1,2-
EPOXYPROPANE * NCI-C55549

OSHA PEL: TWA 50 ppm

THR: Poison by intraperitoneal route. Moderately toxic by ingestion, inhalation, intraperitoneal, and dermal routes. Mutagenic data. A skin irritant. See also diglycidyl ether. Animal experiments suggest somewhat lower toxicity than related epoxy compounds. Readily absorbed through the skin. Causes nervous excitation followed by depression.

2,3-EPOXYPROPYL ACRYLATE   HR: 3
CAS: 106-90-1   NIOSH: AS 9275000
mf: $C_6H_8O_3$ mw: 128.14

PROP: Insol in water, bp: 57.2° @ 2 mm, flash p: 141°F (OC), d: 1.1, vap d: 4.4.

SYNS: ACRYLIC ACID, 2,3-EPOXYPROPYL ESTER
* GLYCIDYL ACRYLATE * GLYCIDYL PROPEN-
ATE * 2-PROPENOIC ACID, OXIRANYLMETHYL
ESTER

THR: Poison by ingestion, inhalation, and skin contact. Mutagenic data. A skin and eye irritant. See also esters. Moderate hazard by heat, flames, oxidizers. To fight fire, use foam, dry chemical, CO$_2$. When heated to decomposition it emits acrid smoke and fumes.

2,3-EPOXYPROPYL OLEATE   HR: 3
CAS: 5431-33-4   NIOSH: RK 0700000
mf: $C_{21}H_{38}O_3$ mw: 338.59

SYNS: 2,3-epoxy-1-propanol oleate
* 2,3-epoxypropyl ester of oleic acid
* glycidyl oleate * glycidyl octadecenoate * oleic acid glycidyl ester
* oxiranylmethyl ester of 9-octadecenoic
acid

THR: Poison by intravenous route. Moderately toxic by ingestion. Mutagenic data. An experimental carcinogen. See also esters. When heated to decomposition it emits acrid smoke and fumes.

EPOXY RESINS, CURED

THR: Most cured resins have little or no toxicity. If curing is incomplete there may be residues of high toxicity curing agents such as the organic amines: *m*-phenylene diamine, diethylene triamine, tetraethylene pentamine, and hexamethylene tetramine, as well as phthalic anhydride and related compounds. See also polymers. When heated to decomposition they emit highly toxic fumes.

EPOXY RESINS, UNCURED   HR: 3
SYN: polymers of epichlorohydrin and 2,2-bis(4-hydroxy phenyl)piperazine

THR: Animal experiments have shown disturbed blood formation. The degree of toxicity of uncured epoxy resins varies, and is partly dependent on the extent of unreacted curing agents. See also epoxy resins, cured, and specific agents. See also polymers. Dangerous; When heated to decomposition they emit highly toxic fumes.

EQUILENIN   HR: 3
CAS: 517-09-9   NIOSH: KG 6125000
mf: $C_{18}H_{18}O_2$ mw: 266.36

PROP: Leaflets from acetone and ethanol; very sltly sol in water.

SYNS: equilenina (spanish); equilenine
* 3-hydroxyestra-1,3,5(10),6,8-penaten-17-one

THR: An experimental tumorigen. When heated to decomposition it emits acrid smoke and fumes.

EQUILIN   HR: 3
CAS: 474-86-2   NIOSH: KG 6650000
mf: $C_{18}H_{20}O_2$ mw: 268.38

SYNS: 1,3,5,7-estratetraen-3-ol-17-one
* 3-hydroxyestra-1,3,5(10),7-tetraen-17-one

THR: An experimental neoplastigen. When heated to decomposition it emits acrid smoke and fumes.

ERBIUM CHLORIDE   HR: 3
CAS: 10138-41-7   NIOSH: KD 8575000
mf: Cl_3Er mw: 273.61

SYN: erbium trichloride

THR: Poison by intraperitoneal and subcutaneous routes. Moderately toxic by ingestion. When heated to decomposition it emits toxic fumes of Cl^-.

ERBIUM (III) NITRATE (1:3)   HR: 3
CAS: 10168-80-6   NIOSH: KD 9050000
mf: $N_3O_9 \cdot Er$ mw: 353.29

PROP: Reddish crystals.

SYN: nitric acid, erbium (3+) salt

THR: Poison by intravenous and intraperitoneal routes. See also nitrates. When heated to decomposition it emits toxic fumes of NO_x.

ERBIUM (III) NITRATE, HEXAHYDRATE (1:::::3:6)   HR: 3
CAS: 13476-05-6   NIOSH: KD 9100000
mf: $N_3O_9 \cdot Er \cdot 6H_2O$ mw: 461.41

PROP: mp: $-4H_2O$ @ 130°.

SYN: nitric acid, erbium (3+) salt, hexahydrate

THR: Poison by intraperitoneal and intravenous routes. See also nitrates. When heated to decomposition it emits toxic fumes of NO_x.

ERGOCALCIFEROL   HR: 3
CAS: 50-14-6   NIOSH: KE 1050000
mf: $C_{28}H_{44}O$ mw: 396.72

PROP: White crystals, odorless, insol in water, sol in alcohol, chloroform, ether and fatty acids, mp: 115°-118°.

SYNS: calciferol * ergosterol activated
* ergosterol, irradiated * irradiated ergosta-5,7,22-trien-3-beta-ol * oleovitamin d * 9,10,secoergosta-5,7,10(19),22-tetraen 3-beta-ol * vitamin d2

THR: Poison to women by ingestion, intraperitoneal, intravenous and intramuscular routes. A nutrient and/or dietary supplement food additive.

ERGOT   HR: 3
CAS: 129-51-1   NIOSH: KE 5250000
mf: $C_{19}H_{23}N_3O_2 \cdot C_4H_4O_4$ mw: 441.53

PROP: Composition: ergot amine, ergosine, ergocristine, ergocryptine, ergocornine, ergosinine, ergocristinine, ergocryptinine, ergotaminine, etc.

SYNS: ERGOMETRINE ACID MALEATE * ERGOMETRINE MALEATE * ERGONOVINE, MALEATE (1:1) (SALT) * ERGOTRATE * ERGOTRATE MALEATE * SPURRED RYE

THR: Poison by intravenous and other routes causing vomiting, diarrhea, thirst, tachycardia, confusion, coma, central nervous system symptoms, gastrointestinal disturbances, gangrene; circulatory changes can follow ingestion. When heated to decomposition it emits toxic fumes of NO_x.

ERGOTAMINE   HR: 3
CAS: 113-15-5   NIOSH: KE 7700000
mf: $C_{33}H_{35}N_5O_5$ mw: 581.63

PROP: A specific alkaloid present in rye ergot.

THR: Poison by intravenous and subcutaneous routes. Mutagenic data. When heated to decomposition it emits toxic fumes such as NO_x.

ERGOTAMINE TARTRATE   HR: 3
CAS: 379-79-3   NIOSH: KE 8225000
mf: $C_{66}H_{70}N_{10}O_{10} \cdot C_4H_6O_6$ mw: 1313.56

SYNS: ERGOTAMINE BITARTRATE * GOTAMINE TARTRATE * GYNERGEN * NEO-ERGOTIN

THR: Poison by ingestion, intravenous, and subcutaneous routes. An experimental teratogen. Affects the cardiovascular system in humans. See also ergot. When heated to decomposition it emits toxic fumes of NO_x. For further information see Vol. 1, No. 3 of *DPIM Report*.

ERYTHROMYCIN   HR: 3
CAS: 114-07-8   NIOSH: KF 4375000
mf: $C_{37}H_{67}NO_{13}$ mw: 734.05

PROP: An antibiotic, white or slightly yellow crystalline powder, odorless, freely sol in alcohol, chloroform and ether, very sltly sol in water, mp: 133°-138°.

SYNS: ERYTHROCIN * ERYTHROGRAN * ERYTHROMYCIN A * PEDIAMYCIN * PROPIOCINE

THR: Poison by intravenous, intramuscular, and intraperitoneal routes. Moderately toxic by subcutaneous route. Mutagenic data. When heated to decomposition it emits toxic fumes of NO_x.

ESCIN, SODIUM SALT   HR: 3
                          NIOSH: KF 6298600
mf: $C_{55}H_{85}O_{24} \cdot Na$ mw: 1153.39

PROP: A mixture of SAPONINS occurring in the seed of the horse chestnut tree.

SYNS: AESCIN SODIUM SALT * AESCUSAN SODIUM SALT * Na-AESCINAT * REPARIL SODIUM SALT

THR: Poison by intravenous route. When heated to decomposition it emits toxic fumes such as Na_2O.

ESTERS   HR: Variable
PROP: A large group of organic compounds which correspond structurally to salts in inorganic chemistry. They are considered as being derived from acids by the replacement of hydrogen by an organic alkyl radical.

THR: No general statement can be made as to the toxicity of esters. Many may be highly volatile and hence can act as asphyxiants or narcotics. Skin absorption, as well as inhalation, may be important routes of absorption for those esters which are volatile and have a high solvent action. The degree of toxicity covers from slight to poison. Esters generally hydrolyze upon contact with moisture; hence a rough guide to the toxicity of a given ester may be the sum of the toxicities of the products of hydrolysis. Incompatible with nitrates.

ESTRADIOL (β)   HR: 3
CAS: 50-28-2   NIOSH: KG 2975000
mf: $C_{18}H_{24}O_2$ mw: 272.42

PROP: Needles out of benzene, acetone; mp: 223°(175°); bp: decomp; Sol in dioxone, alc and ether.

SYNS: DIHYDROFOLLICULAR HORMONE * DIHYDROFOLLICULIN * DIHYDROMENFORMON * DIHYDROTHEELIN * 3,17-beta-DIHYDROXYESTRA-1,3,5(10)-TRIENE * 3,17-beta-DIHYDROXY-1,3,5(10)-ESTRATRIENE * 3,17-beta-DIHYDROXYOESTRA-1,3,5-TRIENE * 3,17-beta-DIHYDROXY-1,3,5(10)-OESTRATRIENE * DIHYDROXYOESTRIN * DIMENFORMON PROLONGATUM * DIOGYNETS * 3,17-EPIDI-

HYDROXYESTRATRIENE * 3,17-EPIDIHYDROX-
YOESTRATRIENE * ESTRADIOL-17-BETA
* alpha-ESTRADIOL * beta-ESTRADIOL
* 3,17-beta-ESTRADIOL * 17-beta-ESTRADIOL
* D-3,17-beta-ESTRADIOL * cis-ESTRADIOL
* D-ESTRADIOL * ESTRA-1,3,5(10)-TRIENE-
3,17-beta-DIOL * 17-beta-ESTRA-1,3,5(10)-
TRIENE-3,17-DIOL * 1,3,5-ESTRATRIENE-3,17-
beta-DIOL * NSC-9895 * OESTRADIOL
* alpha-OESTRADIOL * beta-OESTRADIOL
* 3,17-beta-OESTRADIOL * cis-OESTRADIOL
* D-OESTRADIOL * D-3,17-beta-OESTRADIOL
* OESTRADIOL-17-beta * OESTRA-1,3,5(10)-
TRIENE-3,17-beta-DIOL * 17-beta-OESTRA-
1,3,5(10)-TRIENE-3,17-DIOL * 17-beta-OH-ES-
TRADIOL * 17-beta-OH-OESTRADIOL

THR: An experimental teratogen, tumorigen,
neoplastigen, and carcinogen. A hormone much
used in medicine. When heated to decomposition it emits acrid smoke and fumes. For further
information see Vol. 1, No. 4 of *DPIM Report*.

ESTRADIOL-3-BENZOATE   HR: 3
CAS: 50-50-0   NIOSH: KG 4050000
mf: $C_{25}H_{28}O_3$ mw: 376.53

PROP: White or slightly yellow to brownish
crystalline powder; odorless. Almost insol in
water; sol in alcohol, acetone, and dioxane; sparingly sol in vegetable oils; sltly sol in ether,
mp: 191°-196°

SYNS: BENZOIC ACID ESTRADIOL * BENZOATE
D'OESTRADIOL (FRENCH) * DE GRAAFINA
* DIHYDROESTRIN BENZOATE * DIHYDROFOL-
LICULIN BENZOATE * DIMENFORMON BENZOATE
* ESTRADIOL-17-beta-BENZOATE * ESTRA-
DIOL-17-beta-3-BENZOATE * beta-ESTRADIOL
BENZOATE * beta-ESTRADIOL-3-BENZOATE
* 17-beta-ESTRADIOL BENZOATE * 17-beta-
ESTRADIOL-3-BENZOATE * ESTRADIOL MONO-
BENZOATE * 17-beta-ESTRADIOL MONOBEN-
ZOATE * ESTRA-1,3,5(10)-TRIENE-3,17-DIOL
(17-beta)-3-BENZOATE * ESTRA-1,3,5(10)-
TRIENE-3,17-beta-DIOL, 3-BENZOATE
* 1,3,5(10)-ESTRATRIENE-3,17-beta-DIOL
3-BENZOATE * HYDROXYESTRIN BENZOATE
* OESTRADIOL-3-BENZOATE * beta-OESTRA-
DIOL BENZOATE * beta-OESTRADIOL 3-BEN-
ZOATE * 17-beta-OESTRADIOL 3-BENZOATE
* OESTRADIOL MONOBENZOATE * 1,3,5(10)-
OESTRATRIENE-3,17-beta-DIOL 3-BENZOATE
* OVAHORMON BENZOATE * OVOCYCLIN BEN-
ZOATE

THR: An experimental teratogen, tumorigen,
and carcinogen. A food additive permitted in
the feed and drinking water and/or for treatment
of food-producing animals. See also estradiol
(β). When heated to decomposition it emits acrid
smoke and fumes. For further information see
Vol. 1, No. 4 of *DPIM Report*.

ESTRADIOL DIPROPIONATE   HR: 3
CAS: 113-38-2   NIOSH: KG 4725000
mf: $C_{24}H_{32}O_4$ mw: 384.56

SYNS: DIMENFORMON DIPROPIONATE * ESTRA-
DIOL-3,17-DIPROPIONATE * beta-ESTRADIOL DI-
PROPIONATE * beta-ESTRADIOL-3,17-DIPRO-
PIONATE * 3,17-beta-ESTRADIOL DIPROPIONATE
* 17-beta-ESTRADIOL DIPROPIONATE * ES-
TRAL-3,5(10)-TRIENE-3,17-DIOL (17-beta)-DI-
PROPIONATE * 1,3,5(10)-ESTRATRIENE-3,17-
beta-DIOL DIPROPIONATE * OESTRADIOL
DIPROPIONATE * OESTRADIOL-3,17-DIPROPIO-
NATE * beta-OESTRADIOL DIPROPIONATE
* 3,17-beta-OESTRADIOL DIPROPIONATE
* 17-beta-OESTRADIOL DIPROPIONATE

THR: Poison by intravenous and parenteral
routes. An experimental carcinogen and tumorigen. See also estradiol (β). When heated to
decomposition it emits acrid smoke. For further information see Vol. 1, No. 4 of *DPIM
Report*.

ESTRADIOL MUSTARD   HR: 3
CAS: 22966-79-6   NIOSH: KG 7300000
mf: $C_{42}H_{50}Cl_4N_2O_4$ mw: 788.74

PROP: mp: 40-65° (freeze dried).

SYNS: BIS((4-(BIS(2-CHLOROETHYL)AMINO)-
BENZENE)ACETATE)ESTRA-1,3,5(10)-TRIENE-3,
17-DIOL(17-beta) * BIS((4-(BIS(2-CHLORO-
ETHYL)AMINO)BENZENE)ACETATE)OESTRA-
1,3,5(10)-TRIENE-3,17-DIOL(17-beta) * BIS-
((P-(BIS(2-CHLOROETHYL)AMINO)PHENYL)ACE-
TATE)ESTRADIOL * BIS((P-(BIS(2-CHLO-
ROETHYL)AMINO)PHENYL)ACETATE)ESTRA-1,3,
5(10)-TRIENE-3,17-beta-DIOL * BIS((P- (BIS(2-
CHLOROETHYL)AMINO)PHENYL)ACETATE)OES-
TRADIOL * BIS((P-BIS(2-CHLOROETHYL)AMI-
NOPHENYL)ACETATE)OESTRA-1,3,5(10)-TRIENE-
3,17-beta-DIOL * NCI-C01570 * NSC
112259 * OESTRADIOL MUSTARD

THR: An experimental carcinogen and neoplastigen. When heated to decomposition it emits
very toxic fumes of Cl^- and NO_x.

ESTRADIOL POLYESTER WITH PHOSPHORIC ACID   HR: 3
CAS: 28014-46-2   NIOSH: KG 5792000
mf: $(C_{18}H_{24}O_2 \cdot H_3O_4P)x$ mw: about 26,000

SYNS: ESTRADIOL PHOSPHATE POLYMER * ESTRA-1,3,5(10)-TRIENE-3,17 DIOL (17-beta)-, POLYMER WITH PHOSPHORIC ACID * (17-beta)-ESTRA-1,3,5(10)-TRIENE-3,17-DIOL POLYMER WITH PHOSPHORIC ACID * OESTRADIOL PHOSPHATE POLYMER * OESTRADIOL POLYESTER WITH PHOSPHORIC ACID * POLY-(ESTRADIOL PHOSPHATE)

THR: An experimental carcinogen. See also esters and polymers and phosphoric acid. When heated to decomposition it emits toxic fumes of PO_x.

ESTRIOL   HR: 3
CAS: 50-27-1   NIOSH: KG 8225000
mf: $C_{18}H_{24}O_3$ mw: 288.42

PROP: Small white crystals, d: 0.965; bp: 214.6°

SYNS: ESTRA-1,3,5(10)-TRIENE-3,16-alpha,17-beta-TRIOL * 1,3,5-ESTRATRIENE-3-beta,16-alpha,17-beta-TRIOL * (16-alpha,17-beta)-ESTRA-1,3,5(10)-TRIENE-3,16,17-TRIOL * ESTRATRIOL * 3,16-alpha,17-beta-ESTRIOL * 16-alpha,17-beta-ESTRIOL * FOLLICULAR HORMONE HYDRATE * 16-alpha-HYDROXY ESTRADIOL * NSC-12169 * OESTRA-1,3,5(10)-TRIENE-3,16-alpha,17-beta-TRIOL * 1,3,5-OESTRATRIENE-3-beta-3,16-alpha,17-beta-TRIOL * (16-alpha,17-beta)-OESTRA-1,3,5(10)-TRIENE-3,16,17-TRIOL * 3,16-alpha,17-beta-OESTRIOL * 16-alpha,17-beta-OESTRIOL * 3,16-alpha,17-beta-TRIHYDROXY ESTRA-1,3,5(10)-TRIENE * 3,16-alpha,17-beta-TRIHYDROXY-delta-1,3,5-ESTRATRIENE * 3,16-alpha,17-beta-TRIHYDROXY-delta-1,3,5-OESTRATRIENE * 3,16-alpha,17-beta-TRIHYDROXYOESTRA-1,3,5(10)-TRIENE

THR: Poison by ingestion and subcutaneous routes. An experimental carcinogen, neoplastigen, and tumorigen. When decomposed by heat emits irritant fumes.

ESTRONE   HR: 3
CAS: 53-16-7   NIOSH: KG 8575000
mf: $C_{18}H_{22}O_2$ mw: 270.40

PROP: White crystals; mp: 254°; Insol in water; sol in alcohol, benzene, and ether.

SYNS: ESTERONE * 1,3,5-ESTRATRIEN-3-OL-17-ONE * ENDOFOLLICULINA * 1,3,5(10)-ESTRATRIEN-3-OL-17-ONE * delta-1,3,5-ESTRATRIEN-3-beta-OL-17-ONE * ESTRONA (SPANISH) * FOLLICULAR HORMONE * 3-HYDROXYESTRA-1,3,5(10)-TRIEN-17-ONE * 3-HYDROXY-17-KETO-ESTRA-1,3,5-TRIENE * 3-HYDROXY-17-KETO-OESTRA-1,3,5-TRIENE * 3-HYDROXY-1,3,5(10)-OESTRATRIEN-17-ONE * 3-HYDROXY-OESTRA-1,3,5(10)-TRIEN-17-ONE * delta-1,3,5-OESTRATRIEN-3-beta-OL-17-ONE * 1,3,5-OESTRATRIEN-3-OL-17-ONE * 1,3,5(10)-OESTRATRIEN-3-OL-17-ONE

THR: Poison by intraperitoneal and subcutaneous routes. An experimental teratogen, carcinogen, tumorigen, neoplastigen. When heated to decomposition it emits acrid smoke and fumes. For further information see Vol. 1, No. 4 of *DPIM Report*.

ESTRONE BENZOATE   HR: 3
CAS: 2393-53-5   NIOSH: KG 8750000
mf: $C_{25}H_{26}O_3$ mw: 374.51

SYNS: BENZOATE D'OESTRONE (FRENCH) * 3-(BENZOYLOXY)ESTRA-1,3,5(10)-TRIEN-17-ONE * 3-HYDROXYESTRA-1,3,5(10)-TRIEN-17-ONE BENZOATE * KETOHYDROXYESTRIN BENZOATE * OESTRONBENZOAT (GERMAN)

THR: An experimental carcinogen, tumorigen, neoplastigen. When heated to decomposition it emits acrid smoke and fumes.

ETHACRYNIC ACID   HR: 3
CAS: 58-54-8   NIOSH: AG 6600000
mf: $C_{13}H_{12}Cl_2O_4$ mw: 303.15

SYNS: (2,3-DICHLORO-4-(2-METHYL-ENEBUTYRYL)PHENOXY)ACETIC ACID * 2,3-DICHLORO-4-(2-METHYLENEBUTYL)PHENOXY ACETIC ACID * (4-(2-METHYLENEBUTYRYL)-2,3-DICHLORO-PHENOXY)ACETIC ACID * METHYLENEBUTYRYL PHENOXYACETIC ACID

THR: Poison by intravenous route. Moderately toxic by ingestion. Human peripheral nervous system effects by ingestion. When heated to decomposition it emits toxic fumes of Cl^-.

ETHANAMINE   HR: 3
CAS: 75-04-7   NIOSH: KH 2100000
DOT: 1036
mf: C_2H_7N mw: 45.10

PROP: Colorless gas or liquid, strong ammoniacal odor, bp: 16.6°; flammable, lel = 4.95%, uel = 20.75%, fp: −80.6°, flash p: −0.4°F; d: 0.662 @ 20°/4°, autoign temp: 725°F, vap d: 1.56, vap press: 400 mm @ 20°. Miscible with water, alc, ether.

SYNS: AETHYLAMINE (GERMAN) * AMINO-ETHANE * ETHYLAMINE * ETILAMINA (ITALIAN) * ETYLOAMINA (POLISH) * MONOETHYLAMINE

OSHA PEL: TWA 10 ppm
DOT Classification: Label: Flammable
  Liquid

THR: Poison by ingestion, inhalation, skin, and intravenous routes. Moderately irritating to eyes and mucous membranes. Very dangerous fire hazard when exposed to heat or flame. Moderate explosion hazard when exposed to spark or flame. Dangerous; keep away from heat and open flame; can react vigorously with oxidizing materials. Decomposes to emit NO_x. To fight fire, stop flow of gas; use alcohol foam, dry chemical. Incompatible with cellulose nitrate; oxidizers. For further information see Vol. 5, No. 5 of *DPIM Report*.

ETHANAMINE (ANHYDROUS)   HR: 3
CAS: 75-04-7   NIOSH: KH 2150000

SYN: MONOMETHYLAMINE, ANHYDROUS (DOT)

DOT Classification: Label: Flammable Gas

THR: A flammable gas. See also ethanamine. When heated to decomposition it emits toxic fumes such as NO_x.

ETHANAMINE (AQUEOUS SOLU-TION)   HR: 3
CAS: 75-04-7   NIOSH: KH 2155000

SYN: MONOMETHYLAMINE, AQUEOUS SOLUTION (DOT)

DOT Classification: Label: Flammable
  Liquid

THR: A flammable liquid. See also ethanamine. When heated to decomposition it emits toxic fumes such as NO_x.

ETHANE   HR: 2
CAS: 74-84-0   NIOSH: KH 3800000
mf: C_2H_6 mw: 30.08

PROP: Colorless, odorless gas, mp: −172°, bp: −88.6°, lel = 3.0%, uel = 12.5%, fp:

−183.2°, d: 0.446 @ 0° (liquid), autoign temp: 959°F, vap d: 1.04; flash p: −202°F.

SYNS: BIMETHYL * DIMETHYL * ETHYL HYDRIDE * METHYLMETHANE

DOT Classification: Label: Flammable
  Gas

THR: A simple, very flammable asphyxiant gas. See argon for properties of simple asphyxiants. Very dangerous fire hazard when exposed to heat or flame. Moderately explosive when exposed to flame. To fight fire, stop flow of gas. Incompatible with chlorine; dioxygenyl tetrafluoroborate; oxidizing materials; heat or flame.

1,2-ETHANEDIAMINE   HR: 3
CAS: 107-15-3   NIOSH: KH 8575000
mf: $C_2H_8N_2$ mw: 60.12

PROP: Volatile, colorless, hygroscopic liquid, ammonia-like odor, mp: 8.5°, bp: 117.2°, flash p: 110°F (CC), d: 0.8994 @ 20°/4°, vap press: 10.7 mm @ 20°, vap d: 2.07, autoign temp: 725°F.

SYNS: AETHYLENEDIAMIN (GERMAN) * AE-THALDIAMIN (GERMAN) * 1,2-DIAMINOAETHAN (GERMAN) * 1,2-DIAMINO-ETHAAN (DUTCH) * 1,2-DIAMINOETHANE * 1,2-DIAMINO-ETHANO (ITALIAN) * DIMETHYLENEDIAMINE * ETHYLENEDIAMINE * 1,2-ETHYLENEDIAMINE * ETHYLEENDIAMINE (DUTCH) * ETHYLENE-DIAMINE (FRENCH) * NCI-C60402

OSHA PEL: TWA 10 ppm
DOT Classification: Label: Flammable
  Liquid

THR: An irritant poison in humans by inhalation. Poison by intraperitoneal and subcutaneous routes. Moderately toxic by ingestion and skin contact. Severe eye irritant. An allergen and sensitizer. Used as a food additive in human food. Flammable when exposed to heat, flame or oxidizers. Can react violently with acetic acid; acetic anhydride; acrolein; acrylic acid; acrylonitrile; allyl chloride; CS_2; chlorosulfonic acid; epichlorohydrin; ethylene chlorohydrin; HCl; mesityl oxide; HNO_3; oleum; $AgClO_4$; H_2SO_4; β-propiolactone; vinyl acetate. To fight fire, use CO_2, dry chemical, alcohol foam. When heated to decomposition it emits toxic fumes of NO_x. For further information see ethylene diamine, Vol. 4, No. 2 of *DPIM Report*.

1,2-ETHANEDITHIOL   HR: 3
CAS: 540-63-6   NIOSH: KI 3325000
mf: $C_2H_6S_2$ mw: 94.20

THR: Poison by ingestion, subcutaneous, and intravenous routes. When heated to decomposition it emits very toxic fumes of SO_x.

ETHANETHIOL   HR: 2
CAS: 75-08-1   NIOSH: KI 9625000
DOT: 2363
mf: C_2H_6S mw: 62.14

PROP: Colorless liquid, penetrating garlic-like odor, mp: $-147°$, bp: $36.2°$, lel = 2.8%, uel = 18.2%, d: 0.83907 @ $20°/4°$, autoign temp: 570°F, vap d: 2.14, flash p: $<-0.4°F$.

SYNS: AETHANETHIOL (GERMAN) * AETHYL-MERCAPTAN (GERMAN) * ETANTIOLO (ITALIAN) * ETHAANTHIOL (DUTCH) * ETHYL HYDROSUL-FIDE * ETHYLMERCAPTAAN (DUTCH) * ETHYL MERCAPTAN * ETHYL MERCAPTAN (DOT) * ETHYLMERKAPTAN (CZECH) * ETHYL SULF-HYDRATE * ETHYL THIOALCOHOL * ETILMER-CAPTANO (ITALIAN) * THIOETHANOL * THIO-ETHYL ALCOHOL

OSHA PEL: TWA CL 10 ppm
DOT Classification: Label: Flammable
    Liquid

THR: Moderately toxic by ingestion and intra-peritoneal routes. A severe eye irritant. Moderate skin irritant. Inhalation causes central nervous system effects in humans. Very dangerous fire hazard when exposed to heat, flame or oxidizers. Moderate explosion hazard when exposed to spark or flame. Violent reaction with $Ca(OCl)_2$. Dangerous when heated to decomposition or on contact with acid or acid fumes it emits highly toxic fumes of oxides of sulfur; will react with water or steam to produce toxic and flammable vapors; can react vigorously with oxidizing materials. To fight fire, use CO_2, dry chemical.

ETHANOL   HR: 3
CAS: 64-17-5   NIOSH: KQ 6300000
DOT: 1170
mf: C_2H_6O mw: 46.08

PROP: Clear, colorless, fragrant liquid, burning taste, bp: $78.32°$, ulc: 70, lel = 3.3%, uel = 19% @ 60°, fp: $<-130°$, flash p: 55.6°F, d: 0.7893 @ $20°/4°$, autoign temp: 793°F, vap press: 40 mm @ 19°, vap d: 1.59. Misc in water, alcohol, chloroform and ether.

SYNS: ABSOLUTE ETHANOL * AETHANOL (GERMAN) * AETHYLALKOHOL (GERMAN) * ALCOHOL * ALCOHOL, ANHYDROUS * ALCOHOL DEHYDRATED * ALCOOL ETHYLI-QUE (FRENCH) * ALCOOL ETILICO (ITALIAN) * ALKOHOL (GERMAN) * ANHYDROL * CO-LOGNE SPIRIT * COLOGNE SPIRITS (DOT) * ETHYL ALCOHOL * ETANOLO (ITALIAN) * ETHANOL 200 PROOF * ETHYLALCOHOL (DUTCH) * ETHYL ALCOHOL ANHYDROUS * ETHYL HYDRATE * ETHYL HYDROXIDE * ETYLOWY ALKOHOL (POLISH) * FERMEN-TATION ALCOHOL * GRAIN ALCOHOL * METHYLCARBINOL * MOLASSES AL-COHOL * NCI-C03134 * POTATO ALCO-HOL * SPIRITS OF WINE

OSHA PEL: TWA 1000 ppm
ACGIH TLV: TWA 1,000 ppm
DFG MAK: 1,000 ppm ($1,900$ mg/m^3)
DOT Classification: Label: Flammable
    Liquid

THR: Moderate toxicity by ingestion, intravenous and dermal routes. Probably also by inhalation route. Mutagenic data. The systemic effect of ethanol differs from that of methanol. Ethanol is rapidly oxidized in the body to carbon dioxide and water, and in contrast to methanol, no cumulative effect occurs. Though ethanol possesses narcotic properties, concentrations sufficient to produce this effect are not reached in industry. Exposure to concentrations of 5,000-10,000 ppm results in irritation of the eyes and mucous membranes of the upper respiratory tract. If continued for an hour, stupor and drowsiness may result. Concentrations below 1,000 ppm usually produce no signs of intoxication. There is no concrete evidence that repeated exposure to ethanol vapor results in cirrhosis of the liver. Large doses can cause alcohol poisoning. Repeated ingestions can lead to alcoholism. It is a central nervous system depressant, a teratogen, and a tumorigen. Causes gastrointestinal tract and glandular effects in humans. Exposure to concentrations of over 1,000 ppm may cause headache, irritation of the eyes, nose and throat, and, if long continued, drowsiness and lassitude, loss of appetite and inability to concentrate. Dangerous fire hazard when exposed to heat or flame. Incompatible with acetyl chloride; (Ag_2O + NH_4OH); BrF_5; $Ca(OCl)_2$; ClO_3; CrO_3; $Cr(OCl)_2$; (cyanuric acid + H_2O); H_2O_2; HNO_3; (H_2O_2 + H_2SO_4); (I + CH_3OH + HgO); disulfuryl difluoride; oxidants; platinum; potassium; potassium-tert-butoxide; silver nitrate; silver oxide; [Mn(ClO$_4$)$_2$ + 2,2-dimethoxy propane]; Hg(NO$_3$)$_2$; HClO$_4$; perchlorates; (H$_2$SO$_4$ + permanganates); HMnO$_4$; KO$_2$; KOC(CH$_3$)$_3$; (Ag + HNO$_3$); AgNO$_3$; AgClO$_4$; NaH$_3$N$_2$; UO$_2$(ClO$_4$)$_2$. Moderately explosive when exposed to flame. To fight fire, use alcohol foam, CO$_2$, dry chemical. For further information see Vol. 1, No. 7 of *DPIM Report.*

ETHANOXYTRIPHETOL   HR: 3
CAS: 67-98-1   NIOSH: KK 5550000
mf: C$_{27}$H$_{33}$NO$_3$ mw: 419.61

SYNS: 1-(4-(2-DIETHYLAMINOETHOXY)PHENYL)-1-PHENYL-2-(P-ANISYL)ETHANOL * 1-(P-(2-(DIETHYLAMINO)ETHOXY)PHENYL)-1-PHENYL-2-(P-METHOXYPHENYL)ETHANOL * (p-2-DIETHYLAMINOETHOXYPHENYL)-1-PHENYL-2-p-ANISYLETHANOL

THR: An experimental teratogen. When heated to decomposition it emits toxic fumes of NO$_x$.

ETHENE TETRA CARBO NITRILE   HR: 3
CAS: 670-54-2   NIOSH: KM 7300000
mf: C$_6$N$_4$ mw: 128.10

PROP: Colorless crystals, sublimes >120°, mp: 198°-200°, bp: 223°.

SYN: TETRACYANOETHYLENE

THR: Poison by ingestion and intravenous routes. See also nitriles. When heated to decomposition it emits toxic fumes of NO$_x$ and CN$^-$.

1-ETHENYL PYRENE   HR: 3
CAS: 17088-21-0   NIOSH: UR 2730000
mf: C$_{18}$H$_{12}$ mw: 228.2

SYNS: 1-VINYLPYRENE * 3-VINYLPYRENE

THR: An experimental tumorigen. Mutagenic data. When heated to decomposition it emits acrid smoke and fumes.

4-ETHENYL PYRENE   HR: 3
NIOSH: UR 2740000
mf: C$_{18}$H$_{12}$ mw: 228.2

SYN: 4-VINYLPYRENE

THR: An experimental tumorigen. When heated to decomposition it emits acrid smoke and fumes.

1-ETHENYL-2-PYRROLIDINONE   HR: 2
CAS: 88-12-0   NIOSH: UY 6107000
mf: C$_6$H$_9$NO   mw: 111.16

PROP: Colorless liquid, water-sol, bp: 148° @ 100 mm, fp: 13.5°, flash p: 209°F (OC), d: 1.04 @ 25°, autoign temp: 687°F, vap d: 3.8; fire p: 213°F.

SYNS: VINYLBUTYROLACTAM * N-VINYLPYRROLIDINONE * N-VINYL-2-PYRROLIDINONE * 1-VINYL-2-PYRROLIDINONE * VINYLPYRROLIDONE * N-VINYLPYRROLIDONE * N-VINYL-2-PYRROLIDONE * 1-VINYL-2-PYRROLIDONE

THR: Moderately toxic by ingestion, inhalation, and skin contact. A severe eye irritant. Probably irritating and narcotic in high concentrations. Combustible when exposed to heat or flame. To fight fire, use alcohol foam, CO$_2$, dry chemical. Dangerous; when heated to decomposition it emits highly toxic fumes of NO$_x$; can react vigorously with oxidizing materials.

ETHERS   HR: 2
PROP: Organic compounds in which an oxygen atom is interposed between two carbon atoms in the structure of the molecule.

THR: The simpler ethers such as ethyl ether, isopropyl ether, etc., are powerful narcotics which in large doses can cause death. The danger from ethers is usually acute seldom chronic. There are seldom after-effects to ether intoxication although continued exposure to small concentrations (not enough to cause an overt symptom) has been known to cause loss of appetite, excessive thirst and fatigue. The most common ethers such as ethyl and methyl are particularly dangerous fire hazards. The common ones are easily ignited and have low flash points. It is necessary to control smoking, open flames or even the use of hot plates in areas where low molecular weight ethers are apt to reach 1% concentration or more in air. Only electrical equipment of explosion-proof type (Group C Classification) is permitted to be operated in ether areas. Ethers should not be stored near powerful oxidizers or in areas of high fire hazard. They should be kept cool and the containers electrically grounded to avoid sparks. Dangerous explosion hazard when heated or exposed to flame or sparks. Besides the risk of explosion from air mixtures of ether vapors, ethers tend to form peroxides upon standing. When ethers containing peroxides are heated they can detonate. See ethyl ether. Dangerous; shock or heat can cause gaseous ethers to escape from their containers and create flammable or even explosive conditions. Incompatible with oxidizing materials; BI_3.

ETHINYL ESTRADIOL

CAS: 57-63-6 NIOSH: RC 8925000
mf: $C_{29}H_{24}O_2$ mw: 296.44

SYNS: 3,17-BETA-DIHYDROXY-17-ALPHA-ETHY-NYL-1,3,5(10)-OESTRATRIENE ∗ 3,17-BETA-DIHYDROXY-17-ALPHA-ETHYNYL-1,3,5(10)-ES-TRATRIENE ∗ 17-ALPHA-ETHINYL-3,17-DIHYDROXY-DELTA(SUP 1,3,5)-ESTRATRIENE ∗ 17-ALPHA-ETHINYL-3,17-DIHYDROXY-DELTA(SUP. 1,3,5)OESTRATRIENE ∗ 19-NOR-17-alpha-PREGNA-1,3,5(10)-TRIEN-2-YNE-3,17-DIOL ∗ 17-ETHINYLESTRADIOL ∗ 17-ETHINYL-3,17-ESTRADIOL ∗ 17-ALPHA-ETHINYLESTRADIOL ∗ 17-ALPHA-ETHINYL-17-BETA-ESTRADIOL ∗ 17-ALPHA-ETHINYLES-TRA-1,3,5(10)-TRIENE-3,17-BETA-DIOL ∗ ETHINYLESTRIOL ∗ ETHINYLOESTRADIOL ∗ 17-ETHINYL-3,17-OESTRADIOL ∗ ETHINYL-OESTRANOL ∗ 17-ALPHA-ETHINYLOESTRA-1,3,5(10)-TRIENE-3,17-BETA-DIOL ∗ 17-ALPHA-INYL-DELTA(SUP 1,3,5(10))OESTRATRIENE-3,17-BETA-DIOL ∗ ETHINYLOESTRIOL ∗ 17-ETHYNYL-3,17-DIHYDROXY-1,3,5-OESTRATRIENE ∗ ESTROGEN ∗ ETHYNYLESTRADIOL ∗ 17-ALPHA-ETHYNYLESTRADIOL ∗ 17-ALPHA-ETH-YNYLESTRADIOL-17-BETA ∗ 17-ALPHA-ETH-YNYL-1,3,5(10)-ESTRATRIENE-3,17-BETA-DIOL ∗ 17-ALPHA-ETHYNYLESTRA-1,3,5(10)-TRIENE-3,17-BETA-DIOL ∗ ETHYNYLOESTRADIOL ∗ 17-ETHYNYLOESTRADIOL ∗ 17-ALPHA-ETH-YNYLOESTRADIOL ∗ 17-ALPHA-ETHYNYL-17-BETA-OESTRADIOL ∗ 17-ALPHA-ETHYNYLOES-TRADIOL-17-BETA ∗ 17-ETHYNYLOESTRA-1,3,5(10)-TRIENE-3,17-BETA-DIOL ∗ 17-ALPHA-ETHYNYL-1,3,5-OESTRATRIENE-3,17-BETA-DIOL ∗ 17-ALPHA-ETHYNYL-1,3,5(10)-OESTRA-TRIENE-3,17-BETA-DIOL ∗ 17-ALPHA-ETHYNYL-OESTRA-1,3,5(10)-TRIENE-3,17-BETA-DIOL ∗ (17-ALPHA)-19-NORPREGNA-1,3,5(10)-TRIEN-20-YNE-3,17,DIOL

THR: An experimental tumorigen, carcinogen, and neoplastigen. Human glandular effects. Moderately toxic by ingestion. When heated to decomposition it emits acrid smoke and fumes.

17-alpha-ETHINYL-5,10-ESTRENO-LONE HR: 3

CAS: 68-23-5 NIOSH: RC 8980000
mf: $C_{20}H_{26}O_2$ mw: 298.46

SYNS: 17-alpha-ETHINYL-delta(SUPER)-5,10-19-NORTESTOSTERONE ∗ 17-HYDROXY-19-NOR-17-alpha-PREGN-5(10)-EN-20-YN-3-ONE ∗ 17-alpha-ETHINYL-ESTRA(5,10)ENEOLONE ∗ 17-alpha-ETHYNYL-5(10)-ESTREN-17-OL-3-ONE ∗ 17-ALPHA-ETHYNYLESTR-5-(10)-EN-17-BETA-OL-3-ONE ∗ 17-ALPHA-ETHYNYL-ESTR-5(10)-EN-3-ON-17-BETA-OL ∗ 17-ALPHA-ETHYNYL-17-HYDROXY-5(10)-ES-TREN-3-ONE ∗ 17-ALPHA-ETHYNYL-17-BETA-HYDROXYESTER-5(10)-EN-3-ONE ∗ 17-ALPHA-ETHYNYL-17-BETA-HYDROXY-DELTA(SUP 5(10))-ESTREN-3-ONE ∗ 17-ALPHA-ETHYNYL-17-BETA-HYDROXY-DELTA(SUP. 5(10))-ESTREN-3-ONE ∗ 17-ALPHA-ETHYNYL-17-BETA-HYDROXY-3-OXO-DELTA(SUP5(10))-ESTRENE ∗ 17-ALPHA-ETHYNYL-19-NOR-5(10)-ANDROSTEN-17-BETA-OL-3-ONE ∗ (17-ALPHA)-17-HYDROXY-19-NORPREGN-5-(10)-EN-20-YN-3-ONE ∗ 17-HYDROXY(17-ALPHA)-19-NORPREGN-5(10)-EN-20-YN-3-ONE ∗ 19-NOR-ETHINYL-5,10-TESTOSTERONE

THR: An experimental teratogen, neoplastigen, and carcinogen. When heated to decomposition it emits acrid smoke and fumes.

dl-ETHIONINE HR: 3

CAS: 67-21-0 NIOSH: ES 7000000
mf: $C_6H_{13}NO_2S$ mw: 163.26

PROP: Crystals; decomp @ 273°.

SYNS: (+ × −)-ETHIONINE ∗ NSC 751

THR: An experimental teratogen and tumorigen. Moderately toxic by intraperitoneal route. When heated to decomposition it emits very toxic fumes of NO_x and SO_x.

ETHOMEEN C/15 HR: 2

CAS: 61791-14-8 NIOSH: KO 9100000

PROP: Polyoxyethylene(5)cocoa amine alkyl links C8-C18 which consists of dodecyl(47%), undecyl(18%), decyl (9%), octyl(8%), hexadecyl(10%) and octadecyl(5%).

THR: Moderately toxic by ingestion route. When heated to decomposition it emits acrid smoke and fumes.

ETHOXY ACETYLENE HR: 2

mf: C_4H_6O mw: 70.09

PROP: Insol in water, flash p: 19.4°F; d: 0.8; vap d: 2.4; bp: 61°.

No toxicity data available. Very dangerous fire hazard when exposed to heat, flame or oxidizers. To fight fire, use foam, dry chemical, CO_2. Incompatible with ethyl magnesium iodide.

ETHOXY CARBONYL DIGOXIN   HR: 3
CAS: 73987-52-7   NIOSH: IH 6342000

SYN: CARBAETHOXYDIGOXIN (GERMAN)

THR: Poison by ingestion and intravenous routes.

2-ETHOXY-3,4-DIHYDRO-2H-PYRAN   HR: 2
CAS: 103-75-3   NIOSH: UP 8925000
mf: $C_7H_{12}O_2$ mw: 128.19

PROP: d: 1.0; bp: 143°; flash p: 111°F (OC)

SYNS: 2-ETHOXY-2,3-DIHYDRO-gamma-PYRAN * 2-ETHOXY-3,4-DIHYDRO-1,2-PYRAN * 2-ETHOXY DIHYDROPYRAN, IN PREGNANCY DIAGNOSIS

THR: Moderately toxic by skin contact and inhalation. Skin and eye irritant. Flammable when exposed to flame, sparks, oxidizers. When heated to decomposition it emits acrid smoke and fumes. To fight fire, use dry chemical, foam, fog.

N-ETHOXYMORPHOLINO DIAZENIUM FLUOROBORATE   HR: 3
CAS: 22960-71-0   NIOSH: HM 3150000
mf: $C_6H_{13}N_2O_2 \cdot BF_4$ mw: 232.02

THR: An experimental carcinogen. When heated to decomposition it emits very toxic NO_x and F^-.

ETHOXY-4-NITRO PHENOXY PHENYL PHOSPHINE SULFIDE   HR: 3
CAS: 2104-64-5   NIOSH: TB 1925000
mf: $C_{14}H_{14}NO_4PS$ mw: 323.32

PROP: Liquid or pale yellow crystals with an aromatic odor, nearly insol in water, sol in organic solvents, d: 1.268 @ 25°, mp: 36°.

SYNS: O-AETHYL-O-(4-NITRO-PHENYL)-PHENYL-MONOTHIOPHOSPHONAT (GERMAN) * ENT 17,798 * EPN * O-ETHYL-O-((4-NITRO-FENYL)-FENYL)-MONOTHIOFOSFONAAT(DUTCH) * O-ETHYL O-(4-NITROPHENYL)BENZENE-THIONOPHOSPHONATE * ETHYL P-NITROPHENYL BENZENETHIONOPHOSPHONATE * ETHYL P-NITROPHENYL BENZENETHIOPHOSPHATE * ETHYL P-NITROPHENYL BENZENETHIOPHOS-PHONATE * ETHYL P-NITROPHENYL PHENYL-PHOSPHONOTHIOATE * O-ETHYL O-P-NITRO-PHENYL PHENYLPHOSPHONOTHIOLATE * O-ETHYL O-(4-NITROPHENYL) PHENYLPHOS-PHONOTHIOATE * O-ETHYL O-P-NITROPHENYL PHENYLPHOSPHOROTHIOATE * ETHYL P-NITROPHENYL THIONOBENZENEPHOSPHATE * ETHYL P-NITROPHENYL THIONOBENZENEPHOS-PHONATE * O-ETHYL PHENYL P-NITROPHENYL THIOPHOSPHONATE * O-ETIL-O-((4-NITRO-FENIL)-FENIL)-MONOTIOFOSFONATO (ITALIAN) * PHENYLTHIOPHOSPHONATE DE O-ETHYLE ET O-4-NITROPHENYLE (FRENCH)

OSHA PEL: TWA 0.5 ug/m^3

THR: VERY poisonous by ingestion, dermal, and intraperitoneal routes. A highly toxic insecticide. A cholinesterase inhibitor. See related compound, parathion. This material is extremely hazardous on contact with skin, inhalation or ingestion. Dangerous; when heated to decomposition, emits highly toxic fumes of SO_x and PO_x and NO_x.

4-ETHOXY PHENYL UREA   HR: 3
CAS: 150-69-6   NIOSH: YT 2275000
mf: $C_9H_{12}N_2O_2$ mw: 180.23

PROP: Needle-like crystals; mp: 174°.

SYNS: N-(4-ETHOXYPHENYL)UREA * p-PHENE-THYLUREA * p-AETHOXYPHENYLHARNSTOFF (GERMAN) * DULCINE * NCI-C02073 * p-PHENETOLCARBAMID (GERMAN) * p-PHEN-ETOLCARBAMIDE * PHENETHYLCARBAMID (GERMAN)

THR: Poison by ingestion in children. Moderately toxic by ingestion route. An experimental tumorigen. Acts on the central nervous system in women. An experimental carcinogen. In adults 20 g to 40 g produces dizziness, nausea, methemoglobinemia, cyanosis, hypotension. When heated to decomposition it emits toxic fumes of NO_x.

4-ETHOXY-beta-(1-PIPERIDYL)PROPIO PHENONE HYDROCHLORIDE   HR: 3
CAS: 63815-42-9   NIOSH: UH 1660000
mf: $C_{16}H_{23}NO_2 \cdot ClH$ mw: 297.86

THR: Poison by intraperitoneal, subcutaneous, and intravenous routes. When heated to decomposition it emits very toxic fumes of HCl and NO_x.

3-ETHOXY PROPIONALDE-
HYDE   HR: 3
CAS: 63918-98-9   NIOSH: UE 1575000
mf: $C_5H_{10}O_2$ mw: 102.15

PROP: Liquid, mp: $-69.4°$, bp: $135.2°$, flash
p: $100°F$ (OC), d: 0.918 @ $20°/20°$, vap d:
3.63, vap press: 5.5 mm @ $20°$.

THR: Poison by ingestion, inhalation and dermal
routes. See also aldehydes. Skin and eye irritant.
Dangerous fire hazard when exposed to heat
or flame; can react with oxidizing materials.
To fight fire, use alcohol foam, CO_2, dry chemical.

3-ETHOXY PROPIONIC ACID   HR: 2
CAS: 1331-11-9   NIOSH: UF 2800000
mf: $C_5H_{10}O_3$ mw: 118.15

PROP: Liquid, mp: $-10.7°$, bp: $219°$, flash
p: $225°F$ (OC), d: 1.0474, vap d: 4.08.

THR: Moderately toxic by skin contact. A skin
and eye irritant. Combustible when exposed to
heat or flame; can react with oxidizing materials.
To fight fire, use alcohol foam, CO_2, dry chemical. When heated to decomposition it emits acrid
smoke and fumes.

ETHOXYTRIGLYCOL   HR: 1
CAS: 112-50-5   NIOSH: KK 8950000
mf: $C_8H_{18}O_4$ mw: 178.26

PROP: Bp: $255.4°$, flash p: $275°F$ (OC), d:
1.0208 @ $20°/20°$, vap press: 0.01 mm @ $20°$.

SYNS: 2-(2-(2-ETHOXYETHOXY)ETHOXY)ETHA-
NOL * ETHOXYTRIETHYLENE GLYCOL * TRI-
ETHYLENE GLYCOL ETHYL ETHER

THR: Slightly toxic by ingestion and dermal
routes. An eye irritant. Combustible when exposed to heat or flame; can react with oxidizing
materials. To fight fire, use foam, alcohol foam,
CO_2, dry chemical.

N-ETHYLACETANILIDE   HR: 3
CAS: 529-65-7   NIOSH: AE 2830000
mf: $C_{10}H_{13}NO$ mw: 163.24

PROP: White crystals, faint odor, mp: $54°$, bp:
$258°$, flash p: $126°F$, d: 0.994, vap d: 5.62.

SYNS: ACETETHYLANILIDE * ETHYLACETANI-
LIDE

THR: Poison by ingestion. Flammable when
exposed to heat or flame; can react with oxidizing materials. To fight fire, use foam, CO_2,
dry chemical. When heated to decomposition
it emits toxic fumes of NO_x.

ETHYL ACETATE   HR: 2
CAS: 141-78-6   NIOSH: AH 5425000
DOT: 1173
mf: $C_4H_8O_2$ mw: 88.12

PROP: Colorless liquid, fragrant odor, mp:
$-83.6°$, bp: $77.15°$, ulc: 85-90, lel = 2.2%,
uel = 11%, flash p: $24°F$, d: 0.8946 @ $25°$,
autoign temp: $800°F$, vap press: 100 mm @
$27.0°$, vap d: 3.04.

SYNS: ACETIC ETHER * ACETIDIN * ACE-
TOXYETHANE * AETHYLACETAT (GERMAN)
* ESSIGESTER (GERMAN) * ETHYLACETAAT
(DUTCH) * ETHYL ACETIC ESTER * ETHYLE
(ACETATE D') (FRENCH) * ETHYL ETHANOATE
* ETILE (ACETATO DI) (ITALIAN) * OCTAN
ETYLU (POLISH) * VINEGAR NAPHTHA

OSHA PEL: TWA 400 ppm
ACGIH TLV: TWA 400 ppm
DFG MAK: 400 ppm ($1400 mg/m^3$)
DOT Classification: Flammable Liquid

THR: Moderately toxic by inhalation, intraperitoneal, and subcutaneous routes. Human eye
irritant. Human systemic irritant. See also esters. Ethyl acetate is irritating to mucous surfaces, particularly the eyes, gums and respiratory passages, and is also mildly narcotic. On
repeated or prolonged exposures, it causes conjunctival irritation and corneal clouding. It can
cause dermatitis. High concentrations have a
narcotic effect and can cause congestion of the
liver and kidneys. Chronic poisoning has been
described as producing secondary anemia, leucocytosis and cloudy swelling, and fatty degeneration of the viscera. Used as a synthetic flavoring substance and adjuvant. Dangerous fire
hazard when exposed to heat or flame; can react
vigorously with chlorosulfonic acid; ($LiAlH_2$
+ 2-chloromethyl furan); oleum; K-*tert*-butox-
ide. Moderate explosion hazard when exposed
to flame. To fight fire, use CO_2, dry chemical
or alcohol foam. For further information see
Vol. 4, No. 1 of *DPIM Report*.

ETHYL ACETYL ACETATE   HR: 2
CAS: 141-97-9   NIOSH: AK 5250000
mf: $C_6H_{10}O_3$ mw: 130.16

PROP: Colorless liquid, fruity odor, bp: 180.8°, fp: −45°, flash p: 185°F (COC), autoign temp: 563°F, d: 1.0261 @ 20°/20°, vap press: 1 mm @ 28.5°, vap d: 4.48.

SYNS: ACETOACETIC ESTER * ACTIVE ACETYL ACETATE * DIACETIC ETHER * ETHYL ACETOACETATE * ETHYL ACETYLACETONATE * ETHYL-3-OXOBUTANOATE * ETHYL-3-OXOBUTYRATE

THR: Moderately toxic by ingestion. A skin and eye irritant. See also esters. Flammable when exposed to heat or flame; can react with oxidizing materials. Violent reaction with (Zn + tribromoneopentyl alcohol). To fight fire, use alcohol foam, CO_2, dry chemical.

ETHYL ACETYLENE   HR: 2
mf: C_4H_6 mw: 54

PROP: A colorless, highly flammable gas, bp: 8.3°; d: 0.669 @ 0°/0°; mp: −130°; flash p: <30°F (TOC); <7°C (Gas >8°C).

SYN: 1-BUTYNE

THR: No toxicity data. Probably an asphyxiant; see also acetylene. Dangerous fire hazard when exposed to heat, open flame or powerful oxidizers. Dangerous explosion hazard; details not known. To fight fire, stop flow of gas.

ETHYL ACRYLATE   HR: 3
CAS: 140-88-5   NIOSH: AT 0700000
mf: $C_5H_8O_2$ mw: 100.13

PROP: Colorless liquid, acrid penetrating odor, bp: 99.8°; fp: <−72°: lel = 1.8%; flash p: 60°F (OC); 48.2°F d: 0.941 @ 20°/4°; vap press; 29.3 mm @ 20°; vap d: 3.45.

SYNS: ACRYLIC ACID ETHYL ESTER * ACRYLATE D'ETHYLE (FRENCH) * ACRYLSAEUREAETHYLESTER (GERMAN) * AETHYLACRYLAT (GERMAN) * ETHOXYCARBONYLETHYLENE * ETHYLACRYLAAT (DUTCH) * ETHYLAKRYLAT (CZECH) * ETHYL-2-PROPENOATE * ETIL ACRILATO (ITALIAN) * ETILACRILATULUI (ROUMANIAN) * NCI-C50384

OSHA PEL: TWA 25 ppm (skin)
ACGIH TLV: TWA 5 ppm (skin)

THR: Poison by inhalation and ingestion routes. Moderately toxic by intraperitoneal route. A skin and eye irritant. An experimental carcinogen. A human systemic irritant. See also esters.

Ingestion administration of 0.42 g or more per kg of body weight in rabbits resulted in fatal poisoning. Characterized in its terminal stages by dyspnea, cyanosis, and convulsive movements. It caused severe local irritation of the gastro-enteric tract; and toxic degenerative changes of cardiac, hepatic, renal, and splenic tissues were observed. It gave no evidence of cumulative effects. When applied to the intact skin of rabbits, the ethyl ester caused marked local irritation, erythema, edema, thickening, and vascular damage. Animals subjected to a fairly high concentrations of these esters suffered irritation of the mucous membranes of the eyes, nose and mouth as well as lethargy, dyspnea, and convulsive movements. A substance migrating to food from packaging materials. Dangerous fire hazard when exposed to heat or flame; can react vigorously with oxidizing materials. Violent reaction with chlorosulfonic acid. When heated to decomposition it emits smoke and acrid fumes. To fight fire, use CO_2, dry chemical or alcohol foam. For further information see Vol. 1, No. 2 of DPIM Report.

ETHYL-4-AMINOBENZOATE   HR: 3
CAS: 94-09-7   NIOSH: DG 2450000
mf: $C_9H_{11}NO_2$ mw: 165.21

PROP: Crystals, mp: 88°-90°.

SYNS: P-AMINOBENZOIC ACID ETHYL ESTER * ANESTHESIN * BENZOCAINE * ETHYL-p-AMINOBENZOATE * NORCAIN * PARATHESIN

THR: A poison by ingestion. A skin irritant and a mild sensitizer. Local contact may cause contact dermatitis. Used medicinally. See also esters, ethyl alcohol and p-amino benzoic acid. Moderately dangerous; when heated to decomposition it emits highly toxic fumes of NO_x.

2-ETHYL AMINO ETHANOL   HR: 3
CAS: 110-73-6   NIOSH: KK 9100000
mf: $C_4H_{11}NO$ mw: 89.16

PROP: Flash p: 160°F (OC), d: 0.92, vap d: 3.06, bp: 161°.

SYN: 2-N-MONOETHYLAMINOETHANOL

THR: A poison by dermal route; moderate toxicity by ingestion and intraperitoneal routes. A skin and eye irritant. Flammable by heat, flames and oxidizers. To fight fire, use alcohol foam, dry chemical, CO_2. When heated to decomposition it emits toxic fumes such as NO_x.

2-ETHYLAMINO-1,3,4-THIADIAZOLE   HR: 3
CAS: 13275-68-8   NIOSH: XI 3900000
mf: $C_4H_7N_3S$ mw: 129.20

SYN: NSC 4730

THR: A poison by intraperitoneal and subcutaneous routes. An experimental teratogen. When heated to decomposition it emits very toxic fumes of NO_x and SO_x.

N-ETHYL ANILINE   HR: 3
CAS: 103-69-5   NIOSH: BX 9780000
mf: $C_8H_{11}N$ mw: 121.20

PROP: Clear, yellow-brown oil, mp: $-63.5°$, bp: 204°, d: 0.958 @ 25°/25°, vap press: 1 mm @ 38.5°, vap d: 4.18, flash p: 185°F (OC).

SYNS: AETHYLANILIN (GERMAN) * ANILINO-ETHANE * N-ETHYLAMINOBENZENE * ETHYL-ANILINE * N-ETHYLBENZENAMINO * ETHYL-PHENYLAMINE

THR: A poison by ingestion and intraperitoneal routes. Slightly toxic by skin contact. An allergen. Flammable by heat, flame, oxidizers. Highly dangerous; on decomposition or on contact with acid or acid fumes it emits highly toxic fumes of aniline and NO_x; can react with oxidizing materials. To fight fire, use dry chemical, CO_2, foam. Incompatible with nitric acid.

2-ETHYL ANILINE   HR: 2
CAS: 578-54-1   NIOSH: BX 9800000
mf: $C_8H_{11}N$ mw: 121.20

PROP: Yellow liquid, darkens upon standing, mp: $-63.5°$, bp: 215°, flash p: 185°F (OC), d: 0.98 @ 25°/25°, vap d: 4.17.

SYNS: O-AMINOETHYLBENZENE * O-ETHYLANI-LINE * 2-ETHYLBENZENAMINE

THR: Moderate toxicity by ingestion. See also N-ethylaniline. Flammable when exposed to heat or flame. Dangerous; when heated to decomposition it emits highly toxic fumes of aniline and NO_x; can react with oxidizing materials. To fight fire, use foam, CO_2, dry chemical.

4-ETHYL ANILINE   HR: 3
CAS: 589-16-2   NIOSH: BX 9900000
mf: $C_8H_{11}N$ mw: 121.20

PROP: D: 0.963; mp: 65.8°; bp: 205.5°; Insol in water; misc in alc and eth.

SYNS: 1-AMINO-4-ETHYLBENZENE * P-ETH-YLANILINE

THR: A poison by intravenous and other routes. See also o-ethylaniline. When heated to decomposition it emits toxic fumes of NO_x.

ETHYL BENZENE   HR: 2
CAS: 100-41-4   NIOSH: DA 0700000
DOT: 1175
mf: C_8H_{10} mw: 106.18

PROP: Colorless liquid, aromatic odor. Misc in alcohol and ether, insol in NH_3; sol in SO_2, bp: 136.2°, fp: $-94.9°$, flash p: 59°F, d: 0.8669 @ 20°/4°, autoign temp: 810°F, vap press: 10 mm @ 25.9°, vap d: 3.66, lel = 1.2%, uel = 6.8%.

SYNS: AETHYLBENZOL (GERMAN) * ETHYL-BENZEEN (DUTCH) * ETHYLBENZOL * ETIL-BENZENE (ITALIAN) * ETYLOBENZEN (POLISH) * NCI-C56393 * PHENYLETHANE

OSHA PEL: TWA 100 ppm (skin)
ACGIH TLV: TWA 100 ppm; STEL 125 ppm; BEI 2 g/L (mandelic acid in urine at end of shift)
DFG MAK: 100 ppm (440 mg/m³)
DOT Classification: Label: Flammable Liquid

THR: Moderate toxicity by irritation to skin, eyes, mucous membranes and by ingestion and inhalation routes. The liquid is an irritant to the skin and mucous membranes. A concentration of 0.1% of the vapor in air is an irritant to human eyes, and a concentration of 0.2% is extremely irritating at first, then causes dizziness, irritation of the nose and throat and a sense of constriction in the chest. Exposure of guinea pigs to 1% concentration has been reported as causing ataxia, loss of consciousness, tremor of the extremities and finally death through respiratory failure. The pathological findings were congestion of the brain and lungs, with edema. No data are available regarding the effect of chronic exposure. An experimental teratogen. Dangerous fire hazard when exposed to heat or flame; can react vigorously with oxidizing materials. To fight fire, use foam, CO_2, dry chemical. For further information see Vol. 2, No. 6 of *DPIM Report*.

ETHYL BENZOATE   HR: 2
CAS: 93-89-0   NIOSH: DH 0200000
mf: $C_9H_{10}O_2$ mw: 150.19

PROP: Colorless, aromatic liquid, mp: $-34.6°$, bp: 213.4°, flash p: >204°F, d: 1.048 @ 20°/

20°, vap press: 1 mm @ 44.0°, vap d: 5.17, autoign temp: 914°F. Insol in water; misc in petroleum, alcohol, chloroform, and ether.

SYNS: BENZOIC ETHER * ESSENCE OF NIOBE

THR: Moderately toxic by skin contact and ingestion. See also esters. Combustible when exposed to heat or flame; can react with oxidizing materials. To fight fire, use foam, CO_2, dry chemical. When heated to decomposition it emits acrid smoke and fumes.

ETHYL BROMACETATE HR: 3
CAS: 105-36-2 NIOSH: AF 6000000
mf: $C_4H_7BrO_2$ mw: 167.02

PROP: Colorless to straw-colored liquid, bp: 158.8°, fp: <−20°, flash p: 118°F, d: 1.514 @ 13°/4°, vap d: 5.8. insol in water; misc in alcohol and ether.

SYNS: BROMOACETIC ACID, ETHYL ESTER * ETHOXYCARBONYLMETHYL BROMIDE * ETHYL BROMOACETATE * ETHYL-alpha-BROMOACETATE * ETHYL MONOBROMOACE-TATE

THR: An experimental neoplastigen. An irritant to skin, eyes and mucous membranes. Flammable by heat, flame, oxidizers. Dangerous; when heated to decomposition or on contact with acid or acid fumes, it emits highly toxic fumes of bromides; will react with water or steam to produce toxic and corrosive fumes. To fight fire, use water as a fire blanket.

ETHYL BROMOPHOS HR: 3
CAS: 4824-78-6 NIOSH: TE 7000000
mf: $C_{10}H_{12}BrCl_2O_3PS$ mw: 394.06

SYNS: O-(4-BROMO-2,5-DICHLOROPHENYL)-O,O-DIETHYL PHOSPHOROTHIOATE * O-(4-BROMO-2,5 DICHLOROPHENYL)-O,O-DIETHYLPHOSPHORO-THIONATE * O,O-DIAETHYL-O-(4-BROM-2,5-DICHLOR)-PHENYL-MONOTHIOPHOSPHAT (GER-MAN) * O,O-DIETHYL-O-(4-BROOM-2,5-DI-CHLOOR-FENYL)-MONOTHIOFOSFAAT (DUTCH) * O,O-DIETHYL O-(2,5-DICHLORO-4-BROMOPHE-NYL) THIOPHOSPHATE * O,O-DIETHYL O-2,5-DI-CHLORO-4-BROMOPHENYL-PHOSPHOROTHIOATE * O,O-DIETIL-O-(4-BROMO-2,5 DICLORO-FENIL)-MONOTIOFOSFATO (ITALIAN) * ENT 27,258 * THIOPHOSPHATE DE O,O-DIETHYLE ET DE O-(2,5-DICHLORO-4-BROMO) PHENYLE (FRENCH)

THR: A poison by ingestion and other routes. Moderate toxicity by skin contact. When heated

to decomposition it emits very toxic fumes of Br^-, PO_x, SO_x and Cl^-.

2-ETHYLBUTANOL HR: 2
CAS: 97-95-0 NIOSH: EL 3850000
mf: $C_6H_{14}O$ mw: 102.20

PROP: Clear liquid, bp: 148.9°, flash p: 135°F (COC), d: 0.8328, vap press: 0.9 mm @ 20°, vap d: 3.4.

SYNS: 2-ETHYLBUTANOL-1 * 2-ETHYL-1-BU-TANOL * 2-ETHYLBUTYL ALCOHOL * sec-HEXYL ALCOHOL * 3-METHYLOLPENTANE * sec-PENTYLCARBINOL * 3-PENTYLCARBINOL * PSEUDOHEXYL ALCOHOL

THR: Moderate toxicity by ingestion, dermal contact and inhalation. A skin and eye irritant. Flammable when exposed to heat or flame; can react with oxidizing materials. To fight fire, use dry chemical, CO_2, foam, fog.

2-ETHYL-1-BUTENE HR: 2
CAS: 760-21-4 NIOSH: EM 7400000
mf: C_6H_{12} mw: 84.18

PROP: Flash p: <−4°, autoign temp: 599°F, d: 0.69, vap d: 2.9, bp: 62°.

THR: A human eye irritant. Dangerous fire hazard by heat, flames, oxidizers. To fight fire, use dry chemical, CO_2, foam, spray.

2-ETHYL BUTYL ACRYLATE HR: 2
CAS: 3953-10-4 NIOSH: AT 0300000
mf: $C_9H_{16}O_2$ mw: 156.25

PROP: Clear, colorless liquid, bp: 82° @ 10 mm, fp: −70°, flash p: 125°F (OC), d: 0.8964 @ 20°/20°, vap press: 1.7 mm @ 20°.

SYN: 2-PROPENOIC ACID, 2-ETHYLBUTYL ESTER

THR: Moderately toxic by skin contact. Slightly toxic by ingestion. A skin and eye irritant. See also esters. Flammable when exposed to heat or flame; can react with oxidizing materials. To fight fire, use foam, CO_2, dry chemical. When heated to decomposition it emits acrid smoke and fumes.

2-ETHYL BUTYLAMINE HR: 3
CAS: 617-79-8 NIOSH: EO 4900000
mf: $C_6H_{15}N$ mw: 101.22

PROP: Water-white liquid, bp: 110°-113°, flash p: 64°F (OC), d: 0.739 @ 20°/20°, vap d: 3.5.

SYN: 2-ETHYL-1-BUTANAMINE

THR: A poison by ingestion. A skin and eye irritant. Moderately toxic by dermal route. Dangerous fire hazard when exposed to heat or flame; can react vigorously with oxidizing materials. Decomposes into NO_x. To fight fire, use dry chemical, CO_2, foam.

ETHYL BUTYL ETHER HR: 2
CAS: 628-81-9 NIOSH: KN 4725000
mf: $C_6H_{14}O$ mw: 102.20

PROP: Colorless liquid, insol in water; misc in alcohol and ether, bp: 92°, mp: −124°, flash p: 40°F, d: 0.7528 @ 20°/20°, vap d: 3.52.

SYN: ETHER ETHYLBUTYLIQUE (FRENCH)

THR: Moderately toxic by ingestion route. A skin and eye irritant. See also ethers. Dangerous fire hazard when exposed to heat or flame; can react vigorously with oxidizing materials. See also ethers. To fight fire, use alcohol foam, CO_2, dry chemical.

ETHYL-n-BUTYL NITROS AMINE HR: 3
CAS: 4549-44-4 NIOSH: EO 5075000
mf: $C_6H_{14}N_2O$ mw: 130.22

SYNS: AETHYL-N-BUTYL-NITROSOAMIN (GERMAN) * N-ETHYL-N-NITROSOBUTYLAMINE * N-NITROSO-N-BUTYLETHYLAMINE * N-NITROSOETHYL-N-BUTYLAMINE

THR: A poison by ingestion and intravenous routes. An experimental carcinogen. Mutagenic data. When heated to decomposition it emits toxic fumes of NO_x.

ETHYL CAPROATE HR: 2
CAS: 123-66-0 NIOSH: MO 7735000
mf: $C_8H_{16}O_2$ mw: 144.24

PROP: Colorless liquid, mild odor, bp: 163°, flash p: 130°F (OC), d: 0.875-0.881 @ 20°/20°, vap d: 5.0.

SYNS: ETHYL BUTYLACETATE (DOT) * ETHYL HEXANOATE

THR: Moderate skin irritant. See also esters. When heated to decomposition it emits acrid smoke and fumes. Flammable when exposed to heat or flame; can react with oxidizing materials. To fight fire, use CO_2, foam, dry chemical.

2-(N-ETHYL CARBAMOYL HYDROXY METHYL)FURAN HR: 3
CAS: 63833-90-9 NIOSH: LU 9170000
mf: $C_8H_{11}NO_3$ mw: 169.20

THR: A poison by intraperitoneal route. Mutagenic data. When heated to decomposition it emits toxic fumes such as NO_x.

ETHYL CHLORACETATE HR: 3
CAS: 105-39-5 NIOSH: AF 9110000
mf: $C_4H_7ClO_2$ mw: 122.56

PROP: Colorless liquid, fruity pungent odor, bp: 143.6°, fp: −26.6° flash p: 100°F, d: 1.159 @ 20°/4°, vap press: 10 mm @ 37.5°, vap d: 4.3. Insol in water; misc in alcohol and ether.

SYNS: CHLOROACETIC ACID, ETHYL ESTER * ETHYL CHLOROACETATE * ETHYL-ALPHA-CHLOROACETATE * ETHYL CHLOROETHANOATE * ETHYL MONOCHLORACETATE * ETHYL MONOCHLOROACETATE

THR: A poison by skin contact. An experimental neoplastigen. An eye irritant. Dangerous fire hazard when exposed to heat or flame. When heated to decomposition it emits highly toxic fumes of chlorides; will react with water or steam to produce toxic and corrosive fumes; can react with oxidizing materials. To fight fire, use water, foam, CO_2, dry chemical.

ETHYL CHLORIDE HR: 2
CAS: 75-00-3 NIOSH: KH 7525000
mf: C_2H_5Cl mw: 64.52

PROP: Colorless liquid or gas; ether-like odor, sol in water; misc in alcohol and ether, burning taste, bp: 12.3°, lel = 3.8%, uel = 15.4%, fp: −139°, flash p: −58°F (CC), d: 0.9214 @ 0°/4°, autoign temp: 966°F, vap press: 1000 mm @ 20°, vap d: 2.22.

SYNS: AETHYLCHLORID (GERMAN) * AETHYLIS CHLORIDUM * CHLOORETHAAN (DUTCH) * CHLOROAETHAN (GERMAN) * CHLOROETHANE * CHLORURE D'ETHYLE (FRENCH) * CHLORYL ANESTHETIC * CLOROETANO (ITALIAN) * CLORURO DI ETILE (ITALIAN) * ETHER HYDROCHLORIC * ETHER MURIATIC * ETYLU CHLOREK (POLISH) * KELENE * MONOCHLORETHANE * MURIATIC ETHER * NCI-C06224

OSHA PEL: TWA 1000 ppm

THR: Moderately toxic by ingestion and inhalation routes. An irritant to skin, eyes and mucous membranes and harmful to the eyes. In the case of guinea pigs, the symptoms attending exposure are similar to those caused by methyl chloride, except that the signs of lung irritation are not as pronounced. It gives some warning of its presence because it is irritating, but it is possible to tolerate exposure to it until one becomes unconscious. It is the least toxic of all the chlorinated hydrocarbons. It can cause narcosis, although the effects are usually transient. Highly dangerous fire hazard when exposed to heat, flame or oxidizers. Severe explosion hazard when exposed to flame. Highly dangerous! Keep away from heat and open flame; forms phosgene on combustion; reacts with water or steam to produce toxic and corrosive fumes; can react vigorously with oxidizing materials. To fight fire, use carbon dioxide. For further information see Vol. 1, No. 4 of *DPIM Report*.

ETHYL CHLORO BENZENE   HR: 1
CAS: 1331-31-3   NIOSH: CZ 0700000
mf: C_8H_9Cl mw: 140.62

PROP: Clear, colorless liquid, mp: $-62.6°$, bp: 184.3°, flash p: 147°F, d: 1.05 @ 25°/25°, vap press: 1 mm @ 19.2°, vap d: 4.86.

SYN: CHLOROETHYLBENZENE

THR: Skin and eye irritant. See also chlorinated hydrocarbons, aromatic, and benzene chloride. Flammable when exposed to heat or flame; reacts with oxidizing materials. Dangerous; see chlorides. To fight fire, use foam, CO_2, dry chemical.

ETHYL CHLORO FORMATE   HR: 3
CAS: 541-41-3   NIOSH: LQ 6125000
mf: $C_3H_5ClO_2$ mw: 108.53

PROP: Colorless liquid, decomp in water, mp: $-80.6°$, bp: 94°, flash p: 35.6°F, d: 1.138 @ 20°/4°, vap d: 3.74, autoign temp: 932°F. Misc in alcohol, benzene, ether, chloroform.

SYNS: CHLOROCARBONATE D'ETHYLE (FRENCH) * CLHORAMEISENSAEUREAETHYLESTER (GERMAN) * ETHYLCHLOORFORMIAAT (DUTCH) * ETHYL CHLOROCARBONATE * ETHYLE, CHLOROFORMIAT D' (FRENCH) * ETIL CLOROCARBONATO (ITALIAN) * ETIL CLOROFORMIATO (ITALIAN)

DOT Classification: Label: Flammable Liquid and Poison

THR: A poison by intraperitoneal, ingestion and inhalation routes. Moderate skin toxicity. Dangerous fire hazard when exposed to heat, flame,

oxidizers. Dangerous; when heated to decomposition, emits highly toxic fumes of chlorides; will react with water or steam to produce toxic and corrosive fumes; can react vigorously with oxidizing materials. To fight fire, use CO_2, dry chemical.

ETHYL-trans-CINNAMATE   HR: 2
CAS: 103-36-6   NIOSH: GD 9010000
mf: $C_{11}H_{12}O_2$ mw: 176.23

PROP: Nearly colorless, oily liquid; faint cinnamon odor, d: 1.049 @ 20°/4°; bp: 271°; mp: 9°; Insol in water; misc in alcohol and ether.

SYNS: ETHYL-beta-PHENYLACRYLATE * ETHYL-3-PHENYLPROPENOATE * ETHYL-CINNAMATE

THR: Moderate toxicity by ingestion. See also esters. When heated to decomposition it emits acrid smoke and fumes.

cis-(2-ETHYLCROTONYL) UREA   HR: 2
CAS: 95-04-5   NIOSH: YT 2800000
mf: $C_7H_{12}N_2O_2$ mw: 156.21

SYNS: ECTYLCARBAMIDE * ECTYLUREA * EKTYLCARBAMID * 2-ETHYL-cis-CROTONYLUREA * (alpha-ETHYL-cis-CROTONYL)CARBAMIDE * 2-ETHYLCROTONYLUREA

THR: Moderate toxicity by ingestion and intraperitoneal routes. When heated to decomposition it emits toxic fumes of NO_x.

ETHYL CYANOACETATE   HR: 2
CAS: 105-56-6   NIOSH: AG 4110000
mf: $C_5H_7NO_2$ mw: 113.13

PROP: Colorless to pale straw-colored liquid, mp: $-22.5°$, bp: 206°, flash p: 230°F, d: 1.06 @ 25°/25°, vap press: 1 mm @ 67.8°, vap d: 3.9.

SYNS: CYANACETATE ETHYLE (GERMAN) * CYANOACETIC ACID ETHYL ESTER * CYANOACETIC ESTER * ETHYL CYANOETHANOATE * MALONIC ACID ETHYL ESTER NITRILE * USAF KF-25

THR: Moderate toxicity by intraperitoneal and subcutaneous routes. Combustible when exposed to heat or flame. Dangerous; when heated to decomposition or on contact with acid or acid fumes it emits highly toxic fumes of cyanide; will react with water or steam to produce toxic and flammable vapors; can react with oxidizing materials. To fight fire, use CO_2, dry chemical.

N-ETHYL(CYCLOHEXYL)AMINE   HR: 2
CAS: 5459-93-8   NIOSH: GX 1225000
mf: $C_8H_{17}N$ mw: 127.26

PROP: Slightly sol in water; flash p: 86°F (OC), d: 0.8, vap d: 4.4.

SYN: N-ETHYL-CYCLOHEXYLAMINE

THR: Moderately toxic by ingestion, inhalation and dermal contact. A skin irritant. Dangerous fire hazard when exposed to heat, flame or oxidizers. To fight fire, use alcohol foam, mist, spray, dry chemical.

ETHYL DECABORANE   HR: 3
CAS: 26747-87-5   NIOSH: HD 1925000
mf: $C_2H_{18}B_{10}$ mw: 150.30

THR: A poison by inhalation. See also boron compounds.

ETHYL(DI-(1-AZIRIDINYL)PHOSPHI-NYL)CARBAMATE   HR: 3
CAS: 302-49-8   NIOSH: EY 8800000
mf: $C_7H_{14}N_3O_3P$ mw: 219.21

SYNS: (BIS(1-AZIRIDINYL)PHOSPHINYL)CAR-BAMIC ACID, ETHYL ESTER ∗ BIS(ETHYL-ENIMIDO)PHOSPHORYLURETHAN ∗ ETHYL (BIS(1-AZIRIDINYL)PHOSPHINYL)CARBAMATE ∗ NSC 37095

THR: A poison by intraperitoneal route. See also esters. When heated to decomposition it emits very toxic fumes of PO_x and NO_x.

ETHYL-4,4'-DICHLOROBENZI-LATE   HR: 3
CAS: 510-15-6   NIOSH: DD 2275000
mf: $C_{16}H_{14}Cl_2O_3$ mw: 325.20

PROP: Viscous liquid, sometimes yellow, sltly sol in water, bp: 156°-158°, vap press: 2.2 × 10^{-6} mm @ 20°.

SYNS: 4,4'-DICHLORBENZILSAEUREAETHYLES-TER (GERMAN) ∗ ETHYL-p,p'-DICHLOROBENZI-LATE ∗ 4,4'-DICHLOROBENZILIC ACID ETHYL ESTER ∗ ETHYL-4,4'-DICHLORODIPHENYL GLYCOLLATE ∗ ETHYL-4,4'-DICHLOROPHENYL GLYCOLLATE ∗ ETHYL-2-HYDROXY-2,2-BIS(4-CHLOROPHENYL)ACETATE ∗ NCI-C00408 ∗ NCI-C60413

THR: An experimental carcinogen. Moderately toxic by inhalation, ingestion, and skin contact. A skin and eye irritant. When heated to decomposition it emits toxic fumes of Cl^-. For further information see chlorobenzilate, Vol. 5, No. 1 of *DPIM Report*.

1-ETHYL-1,4-DIHYDRO-7-METHYL-4-OXO-1,8-NAPHTHYRIDINE-3-CAR-BOXYLIC ACID   HR: 3
CAS: 389-08-2   NIOSH: QN 2885000
mf: $C_{12}H_{12}N_2O_3$ mw: 232.26

SYNS: 3-CARBOXY-1-ETHYL-7-METHYL-1,8-NAPHTHIDIN-4-ONE ∗ 1,4-DIHYDRO-1-ETHYL-7-METHYL-4-OXO-1,8-NAPHTHYRIDINE-3-CAR-BOXYLIC ACID ∗ 1-ETHYL-7-METHYL-1,4-DIHYDRO-1,8-NAPHTHYRIDIN-4-ONE-3-CAR-BOXYLIC ACID ∗ 1-ETHYL-7-METHYL-4-OXO-1,4-DIHYDRO-1,8-NAPHTHYRIDINE-3-CARBOX-YLIC ACID ∗ NCI-C56199

THR: A poison by intravenous route. Moderately toxic by ingestion, intraperitoneal, and subcutaneous routes. Damages central nervous system by ingestion in humans. Mutagenic data. When heated to decomposition it emits toxic fumes of NO_x.

ETHYL DIMETHYL AMIDO CYANO PHOSPHATE   HR: 3
CAS: 77-81-6   NIOSH: TB 4550000
mf: $C_5H_{11}N_2O_2P$ mw: 162.15

PROP: A colorless to brownish liquid, bp: decomp @ 238°, fp: −49.4°, flash p: 172°F, d: 1.073 @ 25°, vap press: 0.07 mm @ 25°, vap d: 5.63.

SYNS: DIMETHYLAMIDOETHOXYPHOSPHORYL CYANIDE ∗ DIMETHYLPHOSPHORAMIDOCYANI-DIC ACID, ETHYL ESTER ∗ ETHYL-N,N-DIMETH-YLPHOSPHORAMIDOCYANIDATE ∗ TABUN

THR: VERY poisonous by all routes. LD (man) = 0.01 mg/kg. A nerve gas. Vapor does not penetrate skin; liquid does so rapidly. The primary physiological action is on the sympathetic nervous system, causing a vasoparesis. Vapors when inhaled can cause nausea, vomiting and diarrhea, which can be followed by muscular twitchings and convulsions. See also parathion. Flammable when exposed to heat or flame. Highly dangerous; emits highly toxic fumes; can react with oxidizing materials. When heated to decomposition it emits very toxic fumes of

PO_x and NO_x. For further information see Tabun, Vol. 1, No. 2, of *DPIM Report*.

o-ETHYL-S-(2-DIMETHYL AMINO ETHYL)-METHYL PHOSPHONO THIOATE   HR: 3

NIOSH: TB 1092000

mf: $C_7H_{18}NO_2PS$ mw: 211.29

SYN: o-AETHYL-S-(2-DIMETHYLAMINOAETHYL)-METHYLPHOSPHONOTHIOATE (GERMAN)

THR: A poison by ingestion, intraperitoneal, intravenous, and intramuscular routes. When heated to decomposition it emits very toxic fumes of NO_x, PO_x and SO_x.

ETHYL DIMETHYL METHANE   HR: 2
CAS: 78-78-4   NIOSH: EK 4430000
mf: C_5H_{12} mw: 72.17

PROP: Colorless liquid with pleasant odor, bp: 27.8°, fp: $-160.5°$, flash p: $< -60°F$ (CC), vap press: 595 mm @ 21.1°, vap d: 2.48, lel = 1.4%, uel = 7.6%.

SYNS: ISOAMYLHYDRIDE * ISOPENTANE
* 2-METHYLBUTANE

THR: See pentane for toxicity data. Highly dangerous fire hazard when exposed to heat, flame or oxidizers; can react with oxidizing materials. To fight fire, use foam, CO_2, dry chemical.

o-ETHYL-S,S-DIPHENYL DITHIO PHOSPHATE   HR: 3
CAS: 17109-49-8   NIOSH: TE 3850000
mf: $C_{14}H_{15}O_2PS_2$ mw: 310.38

PROP: A clear yellow to light brown liquid, d: 1.23.

SYNS: O-AETHYL-S,S-DIPHENYL-DITHIOPHOSPHAT (GERMAN) * DITHIOPHOSPHORSAEURE-O-AETHYL-S,S-DIPHENYLESTER (GERMAN)
* EDIFENPHOS * O-ETHYL-S,S-DIPHENYL PHOSPHORODITHIOATE

THR: A poison by ingestion and intraperitoneal routes. A cholinesterase inhibitor. See also parathion. When heated to decomposition it emits very toxic fumes of SO_x and PO_x. For further information see Edifenfos, Vol. 2, No. 4. of *DPIM Report*.

o-ETHYL-S,S-DIPROPYL PHOSPHORO DITHIOATE   HR: 3
CAS: 13194-48-4   NIOSH: TE 4025000
mf: $C_8H_{19}O_2PS_2$ mw: 242.36

SYNS: ENT 27,318 * ETHOPROP * ETHO-PROPHOS

THR: A poison by ingestion and dermal routes. A cholinesterase inhibitor. See also parathion. When heated to decomposition it emits very toxic fumes of PO_x and SO_x. For further information see Ethoprop, Vol. 2, No. 4 of *DPIM Report*.

s-ETHYL-N,N-DI-n-PROPYL-THIOCARBAMATE   HR: 3
CAS: 759-94-4   NIOSH: FA 4550000
mf: $C_9H_{19}NOS$ mw: 189.35

SYNS: S-AETHYL-N,N-DIPROPYLTHIOLCARBAMAT (GERMAN) * N,N-DIPROPYLTHIOCARBAMIC ACID S-ETHYL ESTER * S-ETHYL-N,N-DIPROPYLTHIO-CARBAMATE * ETHYL DI-N-PROPYLTHIOLCAR-BAMATE * ETHYL-N,N-DI-N-PROPYLTHIOLCAR-BAMATE

THR: A poison by inhalation, intravenous, and ingestion routes. Moderate toxicity by dermal route. When heated to decomposition it emits very toxic fumes of NO_x and SO_x.

ETHYLENE   HR: 1
CAS: 74-85-1   NIOSH: KU 5340000
mf: C_2H_4 mw: 28.06

PROP: Soluble in water, alcohol; ether; d: 0.975, colorless gas, sweet odor and taste, bp: $-103.9°$, mp: $-169.4°$, lel = 2.7%, uel = 36%, d: 0.610 @ 0°, autoign temp: 914°F, vap d: 0.98, fp: $-181°$.

SYNS: ACETENE * ATHYLEN (GERMAN)
* ELAYL * ETHENE * LIQUID ETHYENE
* OLEFIANT GAS

DOT Classification: Label: Flammable Gas

THR: Mildly toxic by inhalation. High concentrations cause anesthesia. A simple asphyxiant. A common air contaminant. It is phytotoxic. Very dangerous fire hazard when exposed to heat or flame. Moderate explosion hazard when exposed to flame. Flammable gas! To fight fire, stop flow of gas, CO_2, dry chemical or fine water spray. Incompatible with $AlCl_3$; (CCl_4 + benzoyl peroxide); (bromtrichloro methane + $AlCl_3$); O_3; CCl_4; Cl_2; NO_x; ($AlCl_3$ + nitromethane); (bromotrichloromethane); tetrafluoroethylene; trifluorohypofluorite. For further information see Vol. 4, No. 1 of *DPIM Report*.

ETHYLENE ACRYLATE   HR: 3
CAS: 2274-11-5   NIOSH: AT 0350000
mf: $C_8H_{10}O_4$ mw: 170.18

SYNS: ACRYLIC ACID, ETHYLENE ESTER
* ETHYLDIOL ACRILATE (RUSSIAN) * ACRYLIC
ACID, ETHYLENE GLYCOL DIESTER * ETHYLENE
DIACRYLATE * ETHYLENE GLYCOL DIACRYLATE
* 2-PROPENOIC ACID-1,2-ETHANEDIYL ESTER

THR: A poison by ingestion and inhalation
routes. Moderate toxicity by dermal route. See
also esters. When heated to decomposition it
emits acrid fumes.

ETHYLENEBIS(DITHIOCARBAMATO)
ZINC   HR: 3
CAS: 12122-67-7   NIOSH: ZH 3325000
mf: $C_4H_6N_2S_4 \cdot Zn$ mw: 275.73

PROP: Light-colored powder insol in water.

SYNS: ENT 14,874 * 1,2-ETHANEDIYLBIS-
(CARBAMODITHIOATO) (2-)-S,S'-ZINC * 1,2-
ETHANEDIYLBISCARBAMODITHIOIC ACID, ZINC
COMPLEX * 1,2-ETHANEDIYLBISCARBAMO-
THIOIC ACID, ZINC SALT * ETHYLENEBIS(DI-
THIOCARBAMIC ACID), ZINC SALT * ETHYL
ZIMATE * ZINK-(N,N'-AETHYLEN-BIS(DITHIO-
CARBAMAT)) (GERMAN) * ZINC ETHYLENE-
1,2-BISDITHIOCARBAMATE * ZINEB

THR: Moderate toxicity by ingestion, inhalation, and other routes. Mutagenic data. An experimental teratogen and carcinogen. See also
zinc compounds and carbamates. When heated
to decomposition it emits very toxic fumes of
NO_x and SO_x.

ETHYLENE DIACETATE   HR: 2
CAS: 111-55-7   NIOSH: KW 4025000
mf: $C_6H_{10}O_4$ mw: 146.16

PROP: Colorless liquid or crystals, mp: $-31°$,
bp: 191°, flash p: 205°F (OC), d: 1.128 @ 0°/
4°, vap press: 1 mm @ 38.3°, vap d: 5.04.

SYNS: ETHYLENE ACETATE * ETHYLENE GLY-
COL ACETATE * GLYCOL DIACETATE

THR: Moderate toxicity by intraperitoneal route.
Slight toxicity by ingestion and dermal routes.
A mild eye irritant. Combustible when exposed
to heat or flame; can react with oxidizing materials. To fight fire, use alcohol foam, CO_2, dry
chemical. When heated to decomposition it
emits acrid smoke and fumes.

ETHYLENE DIAMINE TETRAACETIC
ACID   HR: 3
CAS: 60-00-4   NIOSH: AH 4025000
mf: $C_{10}H_{16}N_2O_8$ mw: 292.28

PROP: Colorless crystals, slightly water-sol,
insol in common organic solvents, mp: decomp
@ 240°.

SYNS: ACIDE ETHYLENEDIAMINETETRACETIQUE
(FRENCH) * EDETIC ACID * EDTA (CHELAT-
ING AGENT) * EDTA ACID * ETHYLENEDI-
AMINETETRAACETATE * ETHYLENEDIAMINE-N,
N,N',N'-TETRAACETIC ACID * ETHYLENEDINI-
TRILOTETRAACETIC ACID * GLYCINE, N,N'-
1,2-ETHANEDIYLBIS(N-(CARBOXYMETHYL)-
(9CI)

THR: A poison by intraperitoneal route. Mutagenic data. An experimental teratogen. A com-
plexing agent. When heated to decomposition
it emits toxic fumes of NO_x. For further information see Vol. 1, No. 4 of DPIM Report.

1,2-ETHYLENE DIBROMIDE   HR: 3
CAS: 106-93-4   NIOSH: KH 9275000
mf: $C_2H_4Br_2$ mw: 187.88

PROP: Colorless, heavy, liquid, sweet odor,
bp: 131.4°, fp: 9.3°, flash p: none, d: 2.172
@ 25°/25°, 2.1707 @ 25°/4°, vap press: 17.4
mm @ 30°, vap d: 6.48.

SYNS: AETHYLENBROMID (GERMAN) * BRO-
MOFUME * BROMURO DI ETILE (ITALIAN)
* 1,2-DIBROMAETHAN (GERMAN) * 1,2-DI-
BROMOETANO (ITALIAN) * SYM-DIBROMO-
ETHANE * ALPHA,BETA-DIBROMOETHANE
* 1,2-DIBROMOETHANE * DIBROMURE
D'ETHYLENE (FRENCH) * 1,2-DIBROOMETHAAN
(DUTCH) * DOWFUME 40 * DWUBROMOETAN
(POLISH) * ENT 15,349 * ETHYLENE BRO-
MIDE GLYCOL DIBROMIDE NCI-C00522

OSHA PEL: TWA 20 ppm; CL 30; Pk 50/5M/
8H

THR: Mutagenic data. Experimental teratogen,
carcinogen, and neoplastigen. An insecticide.
Implicated in worker sterility. See also ethylene
dichloride. When heated to decomposition it
emits toxic fumes of Br^-. For further information see Vol. 1, No. 5 of DPIM Report.

ETHYLENE DICHLORIDE   HR: 3
CAS: 107-06-2   NIOSH: KI 0525000

DOT: 1184
mf: $C_2H_4Cl_2$ mw: 98.96

SYNS: 1,2-DICHLOROETHANE * AETHYLEN-
CHLORID (GERMAN) * 1,2-BICHLOROETHANE
* BICHLORURE D'ETHYLENE (FRENCH)
* BROCIDE * CHLORURE E'ETHYLENE
(FRENCH) * CLORURO DI ETHENE (ITALIAN)
* 1,2-DICHLORETHANE * 1,2-DICHLOORE-
THAAN (DUTCH) * 1,2-DICHLOR-AETHAN (GER-
MAN) * DICHLOREMULSION * sym-DICHLO-
ROETHANE * DICHLOROETHYLENE * 1,2-
DICLOROETANO (ITALIAN) * DUTCH LIQUID
* DUTCH OIL * EDC * ENT 1,656
* ETHYLEENDICHLORIDE (DUTCH) * ETHYL-
ENE CHLORIDE * 1,2-ETHYLENE DICHLORIDE
* GLYCOL DICHLORIDE * NCI-C00511

OSHA PEL: TWA 50 ppm; CL 100 ppm; Pk
   200 ppm/5M/3H
ACGIH TLV: TWA 10 ppm
DFG MAK: 20 ppm (80 mg/m^3)
DOT Classification: Label: Flammable
   Liquid and Poison

THR: Poison by ingestion, inhalation, and intravenous routes. Moderately toxic by skin contact, intraperitoneal, and subcutaneous routes.
Vapors produce irritation of respiratory tract
and conjunctiva, corneal clouding, equilibrium
disturbances, narcosis, abdominal cramps. This
substance has been listed as a carcinogen by
the EPA. Flammable liquid. For further information see Vol. 5, No. 1 of *DPIM Report*.

ETHYLENE DIISO THIOURONIUM
DIBROMIDE   HR: 3
CAS: 6943-65-3   NIOSH: UM 6790000
mf: $C_4H_{10}N_4S_2 \cdot 2BrH$ mw: 340.14

SYNS: 2,2'-ETHYLENE-BIS-(2-THIOPSEU-
DOUREA), DIHYDROBROMIDE * PSEUDOUREA,
2,2'-ETHYLENEBIS(2-THIO-, DIHYDROBROMIDE

THR: A poison by intraperitoneal, intravenous,
and other routes. See also bromides. When
heated to decomposition it emits very toxic NO$_x$,
SO$_x$ and HBr.

ETHYLENE GLYCOL   HR: 2
CAS: 107-21-1   NIOSH: KW 2975000
mf: $C_2H_6O_2$ mw: 62.08

PROP: Colorless, sweet-tasting liquid. Hygroscopic, bp: 197.5°, lel = 3.2%, fp: −13°, flash
p: 232°F (CC), d: 1.113 @ 25°/25°, autoign

temp: 752°F, vap d: 2.14, vap press: 0.05 mm
@ 20°.

SYNS: ATHYLENGLYKOL (GERMAN) * 1,2-DI-
HYDROXYETHANE * 1,2-ETHANEDIOL * ETH-
YLENE ALCOHOL * GLYCOL * GLYCOL
ALCOHOL * MONOETHYLENE GLYCOL
* NCI-C00920

THR: Moderately irritating by skin, eyes and
mucous membranes, and by ingestion, intravenous and intraperitoneal routes. (Lethal dose
for man reported to be 100 mL.) If ingested it
causes initial central nervous system stimulation
followed by depression. Later, it causes kidney
damage which can terminate fatally. Very toxic
in particulate form upon inhalation. Combustible when exposed to heat or flame; can react
violently with chlorosulfonic acid; oleum;
H_2SO_4; $HClO_4$; P_2S_5. Moderate explosion hazard when exposed to flame. To fight fire, use
alcohol foam, water, foam, CO_2, dry chemical.
For further information see Vol. 4, No. 3 of
DPIM Report.

ETHYLENE GLYCOL DIFORM-
ATE   HR: 3
CAS: 629-15-2   NIOSH: KW 5250000
mf: $C_4H_6O_4$ mw: 118.10

PROP: Liquid, mp: −10°, bp: 177°, flash p:
200°F (OC), d: 1.2277 @ 20°/20°, vap d: 4.07.

SYNS: ETHYLENE FORMATE * GLYCOL DIFOR-
MATE

THR: A poison by ingestion. Severe eye irritant.
Flammable when exposed to heat or flame; can
react with oxidizing materials. To fight fire,
use CO_2, dry chemical. When heated to decomposition it emits acrid smoke and fumes.

ETHYLENE GLYCOL DINITRATE   HR: 2
CAS: 628-96-6   NIOSH: KW 5600000
mf: $C_2H_4N_2O_6$ mw: 152.08

PROP: Yellow liquid, mp: −20°, bp: explodes
@ 114°, d: 1.483 @ 8°, vap d: 5.25.

SYNS: DINITROGLICOL (ITALIAN) * DINITRO-
GLYCOL * ETHANEDIOL DINITRATE * ETHYL-
ENE DINITRATE * ETHYLENGLYKOLDINITRAT
(CZECH) * GLYCOL DINITRATE * GLYCOL (DI-
NITRATE DE) (FRENCH) * GLYCOLDINITRAAT
(DUTCH) * GLYKOLDINITRAT (GERMAN)
* NITROGLYKOL (CZECH)

OSHA PEL: TWA CL 1 mg/m^3 (skin)

ACGIH TLV: TWA 0.05 ppm (skin)
DFG MAK: 0.05 ppm (0.3 mg/m^3)

THR: Moderately toxic by ingestion, inhalation and dermal routes. Can cause lowered blood pressure, leading to headache, dizziness and weakness. See nitrates for fire and explosion hazards.

ETHYLENE GLYCOL METHYL
ETHER   HR: 2
CAS: 109-86-4   NIOSH: KL 5775000
mf: $C_3H_8O_2$ mw: 76.11

PROP: Colorless liquid, mild, agreeable odor. Misc in water; alcohol, ether, benzene. Bp: 124.5°, fp: −86.5°, flash p: 115°F (OC), lel = 2.5%, uel = 14%, d: 0.9660 @ 20°/4°, autoign temp: 545°F, vap press: 6.2 mm @ 20°, vap d: 2.62.

SYNS: AETHYLENGLYKOL-MONOMETHYLAETHER (GERMAN) * DOWANOL EM * ETHER MONO-METHYLIQUE DE L'ETHYLENE-GLYCOL (FRENCH) * GLYCOLMETHYL ETHER * GLYCOL MONO-METHYL ETHER * 2-METHOXY-AETHANOL (GERMAN) * 2-METHOXYETHANOL * METHOXYHYDROXYETHANE * METHYL CELLOSOLVE (DOT) * METHYL GLYCOL * METHYLGLYKOL (GERMAN) * 2-METOS-SIETANOLO (ITALIAN) * METOKSYETYLOWY ALKOHOL (POLISH) * METIL CELLOSOLVE (ITALIAN)

OSHA PEL: TWA 25 ppm (skin)

THR: Moderately toxic in humans by ingestion and inhalation routes. Moderate via skin, ingestion, inhalation, intraperitoneal, and intravenous routes. Skin and eye irritants. When used under conditions which do not require the application of heat, this material probably presents little hazard to health. However, in the manufacture of fused collars, which require pressing with a hot iron, cases have been reported showing disturbance of the hemopoietic system with or without neurological signs and symptoms. The blood picture may resemble that produced by exposure to benzene. Two cases reported had severe aplastic anemia with tremors and marked mental dullness. One case had multiple neuritis and four others had normal reflexes. The commonest change in the blood picture was the finding of immature neutrophils (shift to the left); in other cases there were a reduction in number of the blood platelets and a macrocytic anemia. The

persons affected had been exposed to vapors of methyl "Cellosolve", ethanol and methanol, ethyl acetate and petroleum naphtha. Flammable when exposed to heat or flame; can react with oxidizing materials to form explosive peroxides. To fight fire, use alcohol foam, CO_2, dry chemical. For further information see ethylene glycol monomethyl ether, Vol. 4, No. 4 of *DPIM Report*.

ETHYLENE GLYCOL MONOMETHYL
ETHER   HR: 2
mf: $CH_3OCH_2CH_2OH$ mw: 76.09

PROP: Colorless liquid, mild agreeable odor, bp: 124.5, fp -86.5, flash p: 115 F (OC), lel = 2.5%, uel = 14%, d: 0.9660 @ 20/4, autoign temp: 545 F, vap press: 6.2 mm @ 20, vap d: 2.62.

SYNS: 2-METHOXYETHANOL * METHYL "CEL-LOSOLVE"

THR: Moderately toxic by ingestion, inhalation, and intravenous routes. Central nervous system effects by inhalation. Blood effects by exposure to vapor similar to effects from benzene. Severe damage probably takes the form of aplastic anemia. Unheated material is of little hazard to health. Flammable when exposure to heat or flame; can react with oxidizing materials to form explosive peroxides. To fight fire, use alcohol foam, CO_2, dry chemical.

ETHYLENE GLYCOL MONOMETHYL
ETHER ACETATE   HR: 2
CAS: 110-49-6   NIOSH: KL 5950000
mf: $C_5H_{10}O_3$ mw: 118.15

PROP: Colorless liquid, bp: 143°, fp: −70°, flash p: 111°F (CC), d: 1.005 @ 20°/20°, vap d: 4.07, lel = 1.7%, uel = 8.2%.

SYNS: ACETATE DE L'ETHER MONOMETHYLIQUE DE L'ETHYLENE-GLYCOL (FRENCH) * ACETATE DE METHYLE GLYCOL (FRENCH) * ACETATO DI-METIL CELLOSOLVE (ITALIAN) * AETHYL-ENGLYKOLMETHYLAETHERACETAT (GERMAN) * ETHYLENE GLYCOL MONOMETHYL ETHER ACETATE(DOT) * GLYCOL MONOMETHYL ETHER ACETATE * 2-METHOXYAETHYLACETAT (GERMAN) * 2-METHOXY-ETHYL ACETAAT (DUTCH) * 2-METHOXYETHYL ACETATE * 2-METHOXYETHYLE, ACETATE DE (FRENCH) * METHYL CELLOSOLYE ACETAAT (DUTCH) * METHYL CELLOSOLVE ACETATE * METHYL-

GLYKOLACETAT(GERMAN) * METHYL GLYCOL ACETATE * METHYL GLYCOL MONOACETATE * 2-METOSSIETILACETATO(ITALIAN)

OSHA PEL: TWA 25 ppm (skin)

THR: Moderately toxic by ingestion, intraperitoneal, and subcutaneous routes. An inhalation irritant in humans. A mild eye irritant. Flammable when exposed to heat or flame; can react with oxidizing materials. To fight fire, use CO_2, dry chemical. When heated to decomposition it emits acrid smoke and fumes.

ETHYLENE GLYCOL STEARATE   HR: 3
CAS: 111-60-4   NIOSH: WI 4100000
mf: $C_{20}H_{40}O_3$ mw: 328.60

SYNS: STEARIC ACID, 2-HYDROXYETHYL ESTER * ETHYLENE GLYCOL, MONOSTEARATE * GLYCOL MONOSTEARATE * GLYCOL STEARATE * STEARIC ACID, MONOESTER WITH ETHYLENE GLYCOL * USAF KE-11

THR: A poison by intraperitoneal route. See also esters. When heated to decomposition it emits acrid smoke and fumes.

ETHYLENE OXIDE   HR: 3
CAS: 75-21-8   NIOSH: KX 2450000
mf: C_2H_4O mw: 44.06

PROP: Colorless gas at room temperature, mp: $-111.3°$, bp: $10.7°$, ulc: 100, lel = 3.0%, uel = 100%, flash p: $-4°F$, d: 0.8711 @ $20°/20°$, autoign temp: $804°F$, vap press: 1095 mm @ $20°$, vap d: 1.52. Misc in water and alc. very sol in ether.

SYNS: AETHYLENOXID (GERMAN) * DIHYDROOXIRENE * DIMETHYLENE OXIDE * 1,2-EPOXYAETHAN (GERMAN) * 1,2-EPOXYETHANE * ETHYLENE (OXYDE D') (FRENCH) * ETHYLEENOXIDE (DUTCH) * ETHYLENE OXIDE (DOT) * ETILENE (OSSIDO DI) (ITALIAN) * ETYLENU TLENEK (POLISH) * NCI-C50088 * OXACYCLOPROPANE * ALPHA,BETA-OXIDOETHANE * OXIRAAN (DUTCH) * OXIRANE

OSHA PEL: TWA 50 ppm
DOT Classification: Label: Flammable Liquid

THR: A poison by ingestion, inhalation, intraperitoneal, and intravenous routes. An experimental teratogen and neoplastigen. Mutagenic data. A powerful systemic irritant. An irritant to skin and eyes and mucous membranes of respiratory tract. High concentrations can cause pulmonary edema. See aliphatic and aromatic epoxies. Very dangerous fire hazard when exposed to heat or flame. Severe explosion hazard when exposed to flame. To fight fire, use alcohol foam, CO_2, dry chemical. Incompatible with acids and bases; alcohols; air; m-nitroaniline; aluminum chloride; aluminum oxide; ammonia; trimethyl amine; copper; iron chlorides; iron oxides; magnesium perchlorate; mercaptans; potassium; tin chlorides; contaminants; alkane thiols; bromoethane. For further information see Vol. 4, No. 2 of *DPIM Report*.

ETHYLENE SULFATE   HR: 3
CAS: 1072-53-3   NIOSH: JH 4800000
mf: $C_2H_4O_4S$ mw: 124.12

SYNS: ETHYLENE GLYCOL, CYCLIC SULFATE * GLYCOL SULFATE * SULFURIC ACID, CYCLIC ETHYLENE ESTER

THR: Mutagenic data. An experimental neoplastigen. See also sulfates, esters. When heated to decomposition it emits toxic fumes of SO_x.

ETHYLENE SULFIDE   HR: 3
CAS: 420-12-2   NIOSH: KX 3500000
mf: C_2H_4S mw: 60.12

PROP: Colorless liquid, bp: $55°-56°$ decomp, d: 1.0368 @ $0°/4°$, vap d: 2.07.

SYNS: 2,3-DIHYDROTHIIRENE * AETHYLENSULFID (GERMAN) * ETHYLENE EPISULFIDE * THIACYCLOPROPANE

THR: A poison by irritation of skin, eyes and mucous membranes, and by ingestion, inhalation, intraperitoneal and subcutaneous routes. Dangerous; when heated to decomposition, or on contact with acid or acid fumes it emits highly toxic fumes of SO_x; can react with oxidizing materials.

ETHYLENIMINE   HR: 3
CAS: 151-56-4   NIOSH: KX 5075000
mf: C_2H_5N mw: 43.08

PROP: Oil, pungent ammoniacal odor, water-white liquid, bp: $55°-56°$, fp: $-71.5°$, flash p: $12°F$, d: 0.832 @ $20°/4°$, autoign temp: $608°F$, vap press: 160 mm @ $20°$, vap d: 1.48, lel = 3.6%, uel = 46%.

SYNS: AETHYLENIMIN (GERMAN) * AMINOETHYLENE * AZACYCLOPROPANE * AZIRIDIN

(GERMAN) * AZIRIDINE * DIHYDRO-1H-AZI-
RINE * DIMETHYLENEIMINE * DIMETHYLENI-
MINE * ETHYLEENIMINE (DUTCH) * ETHYL-
ENEIMINE * ETHYLIMINE * ETILENIMINA
(ITALIAN)

OSHA PEL: TWA 1 mg/m^3 (skin)

THR: A poison by irritation of skin, eyes, mucous membranes and ingestion, dermal, intraperitoneal and inhalation routes. An experimental carcinogen and neoplastigen. Mutagenic
data. An allergic sensitizer of skin. Causes
opaque cornea, keratoconus and necrosis of cornea (experimental). Has been known to cause
severe human eye injury. Drinking of carbon-
ated beverages is recommended as antidote to
this material in stomach. Dangerous fire hazard
when exposed to heat, flame or oxidizers. Reacts
violently with acetic acid; acetic anhydride;
acrolein; acrylic acid; allyl chloride; CS_2; Cl_2;
chlorosulfonic acid; epichlorohydrin; glyoxal;
HCl; HF; HNO_3; oleum; β-propiolactone; Ag;
NaOCl; H_2SO_4; vinyl acetate. Dangerous; heat
and/or the presence of catalytically active metals
or chloride ions can cause a violent exothermic
reaction. This material should be handled as
per instructions of the manufacturer. To fight
fire, use alcohol foam, CO_2, dry chemical.

ETHYL ETHER HR: 2
CAS: 60-29-7   NIOSH: KI 5775000
mf: $C_4H_{10}O$ mw: 74.14

PROP: A clear, volatile liquid, sweet, pungent
odor, Sol in water; misc in alcohol and ether;
sol in chloroform. mp: −116.2°, bp: 34.6°, ulc:
100, lel = 1.85%, uel = 36%, flash p: −49°F,
d: 0.7135 @ 20°/4°, autoign temp: 320°F, vap
press: 442 mm @ 20°, vap d: 2.56.

SYNS: AETHER * ANAESTHETIC ETHER
* ANESTHETIC ETHER * DIAETHYLAETHER
(GERMAN) * DIETHYL ETHER * DIETHYL OX-
IDE * DWUETYLOWY ETER (POLISH) * ETERE
ETILICO (ITALIAN) * ETHER ETHYLIQUE
(FRENCH) * ETHOXYETHANE * 1,1′-OXYBI-
SETHANE * OXYDE D'ETHYLE (FRENCH)
* SOLVENT ETHER

OSHA PEL: TWA 400 ppm
ACGIH TLV: TWA 400 ppm; STEL 500 ppm
DOT Classification: Label: Flammable
  Liquid

THR: Moderately toxic by ingestion and inhalation routes. Mutagenic data. An irritant. Ether
is not corrosive or dangerously reactive. However, it must not be considered safe for individuals to inhale or ingest. It is not toxic in the
sense of being a poison. It is, however, a depressant of the central nervous system and is capable
of producing intoxication, drowsiness, stupor,
and unconsciousness. Death due to respiratory
failure may result from severe and continued
exposure. Human eye irritant. Very dangerous
fire hazard when exposed to heat or flame; can
react vigorously with acetyl peroxide; air; liquid
air; bromoazide; Cl_2; ClF_3; CrO_3; $Cr(OCl)_2$;
$LiAlH_2$; $NOClO_4$; O_2; $NClO_2$; O_3; $HClO_4$;
(H_2SO_4 + permanganates); K_2O_2; Na_2O_2;
[$(C_2H_5)_3Al$ + air]; [$(CH_3)_3Al$ + air]. Severe
explosion hazard when exposed to heat or flame.
See also ethers. To fight fire, use alcohol foam,
CO_2, dry chemical. For further information see
Vol. 4, No. 1 of *DPIM Report*.

5-ETHYL-5-(1-ETHYLPROPYL)BARBI-
TURIC ACID HR: 3
CAS: 17013-37-5   NIOSH: CQ 4790000
mf: $C_{11}H_{18}N_3O_3$ mw: 240.32

SYNS: ISOMEBUMAL * 5-ETHYL-5-(1-ETHYL-
PROPYL)2,4,6(1H,3H,5H)-PYRIMIDINETRIONE

THR: A poison by intraperitoneal route. When
heated to decomposition it emits toxic fumes
of NO_x.

ETHYL-10-FLUORODECANO-
ATE HR: 3
CAS: 353-03-7   NIOSH: HD 9500000
mf: $C_{12}H_{23}FO_2$ mw: 218.35

SYNS: ETHYL-OMEGA-FLUORODECANOATE
* ETHYL-9-FLUORONONANECARBOXYLATE

THR: A poison by intraperitoneal and parenteral
routes. When heated to decomposition it emits
toxic fumes of F^-.

ETHYL-8-FLUORO OCTANOATE HR: 3
CAS: 332-97-8   NIOSH: RH 0720000
mf: $C_{10}H_{19}FO_2$ mw: 190.29

SYNS: 8-FLUOROOCTANOIC ACID, ETHYL ESTER
* ETHYL-OMEGA-FLUOROOCTANOATE

THR: A poison by intraperitoneal, subcutaneous, parenteral, and other routes. See also esters
and fluorides. When heated to decomposition
it emits toxic fumes of F^-.

ETHYL FORMATE   HR: 3
CAS: 109-94-4   NIOSH: LQ 8400000
mf: $C_3H_6O_2$ mw: 74.09

PROP: Water white liquid, pleasant aromatic odor, mp: $-79°$, bp: $54.3°$, lel = 2.7%, uel = 13.5%, flash p: $-4°F$ (CC), d: 0.9236 @ $20°/20°$, autoign temp: $851°F$, vap press: 100 mm @ $5.4°$, vap d: 2.55.

SYNS: AETHYLFORMIAT (GERMAN) * ETHYLE (FORMIATE D') (FRENCH) * ETHYLFORMIAAT (DUTCH) * ETHYL FORMIC ESTER * ETHYL METHANOATE * ETILE (FORMIATO DI) (ITAL-IAN) * FORMIC ETHER * MROWCZAN ETYLU (POLISH)

OSHA PEL: TWA 100 ppm
ACGIH TLV: TWA 100 ppm
DOT Classification: Label: Flammable
  Liquid

THR: An experimental tumorigen; a powerful inhalation irritant in humans; Moderate toxicity by ingestion and inhalation. See also esters. Dangerous fire hazard when exposed to heat, flame or oxidizers. To fight fire, use alcohol foam, spray, mist, dry chemical.

ETHYL GUTHION   HR: 3
CAS: 2642-71-9   NIOSH: TD 8400000
mf: $C_{12}H_{16}N_3O_3PS_2$ mw: 245.40

SYNS: AZINFOS-ETHYL (DUTCH) * AZINPHOS-AETHYL (GERMAN) * AZINPHOS ETHYL * AZINPHOS-ETILE (ITALIAN) * BENZOTRIAZINE DERIVATIVE OF AN ETHYL DITHIOPHOSPHATE * O,O-DIAETHYL-S-((4-OXO-3H-1,2,3-BENZO-TRIAZIN-3-YL)-METHYL)-DITHIOPHOSPHAT (GERMAN) * O,O-DIETHYL S-(4-OXOBENZO-TRIAZINO-3-METHYL)PHOSPHORODITHIOATE * O,O-DIETHYL-S-(4-OXO-3H-1,2,3-BENZO-TRIAZINE-3-YL)-METHYL-DITHIOPHOSPHATE * O,O-DIETHYL-S-((4-OXO-3H-1,2,3-BENZO-TRIAZIN-3-YL)-METHYL)-DITHIO FOSFAAT (DUTCH) * O,O-DIETHYL PHOSPHORODITHIO-ATE S-ESTER WITH 3-(MERCAPTOMETHYL)-1,2,3-BENZOTRIAZIN-4(3H)-ONE * O,O-DIETIL-S-((4-OXO-3H-1,2,3-BENZOTRIAZIN-3-IL)-METIL)-DITIOFOSFATO (ITALIAN) * 3,4-DIHY-DRO-4-OXO-3-BENZOTRIAZINYLMETHYL O,O-DIETHYL PHOSPHORODITHIOATE * S-(3,4-DIHYDRO-4-OXO-1,2,3-BENZOTRIAZIN-3-YLMETHYL)-O,O-DIETHYL PHOSPHORODITHIOATE * ENT 22,014 * TRIAZOTION (RUSSIAN)

THR: VERY poisonous by ingestion and intra-peritoneal routes. A poison by dermal routes. A cholinesterase inhibitor. See also parathion. When heated to decomposition it emits very toxic SO_x, PO_x and NO_x.

2-ETHYL-1-HEXENE   HR: 3
CAS: 1632-16-2   NIOSH: MP 6825000
mf: C_8H_{16} mw: 112.24

PROP: Colorless liquid, bp: $120°$, d: 0.7270 @ $20°/20°$, vap d: 3.87.

SYNS: 2-ETHYL HEXENE-1 * USAF DO-21

THR: A poison by inhalation. Combustible; can react with oxidizing materials. When heated to decomposition it emits acrid smoke and fumes.

ETHYL HEXYLENE GLYCOL   HR: 2
CAS: 94-96-2   NIOSH: MO 2625000
mf: $C_8H_{18}O_2$ mw: 146.26

PROP: Practically colorless, somewhat viscous, odorless liquid, bp: $243.1°$, flash p: $260°F$ (OC), fp: $-40°$, d: 0.9422 @ $20°/20°$, vap press: <0.01 mm @ $20°$, vap d: 5.03.

SYNS: CARBIDE 6-12 * COMPOUND 6-12 IN-SECT REPELLENT * ETHOHEXADIOL * ETHYL HEXANEDIOL * 2-ETHYL-1,3-HEXANEDIOL * 2-ETHYL-3-PROPYL-1,3-PROPANEDIOL * 3-HYDROXYMETHYL-N-HEPTAN-4-OL * OCTYLENE GLYCOL * REPELLENT 612 * RUTGERS 612

THR: Moderate toxicity by ingestion route. A skin irritant. An experimental tumorigen. Com-bustible when exposed to heat or flame; can react with oxidizing materials. To fight fire, use alcohol foam, foam, dry chemical.

ETHYL HYDROCUPREINE HYDROCHLORIDE   HR: 3
CAS: 3413-58-9   NIOSH: MW 6300000
mf: $C_{21}H_{28}N_2O_2 \cdot ClH$ mw: 376.97

SYNS: NUMOQUIN HYDROCHLORIDE * OPTO-QUINHYDROCHLORIDE

THR: A poison by intravenous route in man causing cardiovascular problems. When heated to decomposition it emits very toxic fumes such as Cl^- and NO_x.

ETHYL-2-HYDROXY ETHYL NITROSAMINE   HR: 3
CAS: 13147-25-6   NIOSH: KL 0350000
mf: $C_4H_{10}N_2O_2$ mw: 118.16

SYNS: AETHYL-AETHANOL-NITROSOAMIN (GER-MAN) * 2-(ETHYLNITROSAMINO)ETHANOL * N-ETHYL-N-HYDROXYETHYLNITROSAMINE * N-NITROSOAETHYLAETHANOLAMIN (GERMAN) * N-NITROSOETHYLETHANOLAMINE * N-NITROSO-N-ETHYL-N-(2-HYDROXYETHYL)AMINE

THR: An experimental carcinogen. When heated to decomposition it emits toxic fumes of NO_x.

ETHYL-p-HYDROXY PHENYL KETONE   HR: 3
CAS: 70-70-2   NIOSH: UH 1925000
mf: $C_9H_{10}O_2$ mw: 150.19

SYNS: P-HYDROXYPHENYL-1-PROPANONE * 1-(4-HYDROXYPHENYL)-1-PROPANONE * HYDROXYPROPIOPHENONE * P-HYDROXY-PROPIOPHENONE * 4-HYDROXYPROPIOPHENONE * P-OXYPROPIOPHENONE * PAROXYPROPIONE * P-PROPIONYLPHENOL * USAF EK-3302

THR: A poison by intraperitoneal route. Moderate toxicity by subcutaneous route. When heated to decomposition it emits acrid smoke and fumes.

ETHYLIDENE DICHLORIDE   HR: 3
CAS: 75-34-3   NIOSH: KI 0175000
mf: $C_2H_4Cl_2$ mw: 98.96

PROP: Colorless liquid, aromatic, ethereal odor, hot saccharine taste, mp: $-97.7°$, lel = 5.6%, bp: 57.3°, flash p: 22°F (TOC), d: 1.174 @ 20°/4°, vap press: 230 mm @ 25°, vap d: 3.44, autoign temp: 856°F.

SYNS: AETHYLIDENCHLORID (GERMAN) * CHLORINATED HYDROCHLORIC ETHER * CHLORURE D'ETHYLIDENE (FRENCH) * 1,1-DICHLOORETHAAN (DUTCH) * CLORURO DI ETILIDENE (ITALIAN) * 1,1-DICHLORAETHAN (GERMAN) * 1,1-DICHLOROETHANE * 1,1-DICLOROETANO (ITALIAN) * ETHYLIDENE CHLORIDE * NCI-C04535

OSHA PEL: TWA 100 ppm

THR: An experimental teratogen and tumorigen. Moderate toxicity by ingestion. Liver damage reported in experimental animals. Dangerous fire hazard when exposed to heat or flame. Moderately explosive when exposed to heat or flame. Dangerous; when heated to decomposition it emits highly toxic fumes of phosgene; can react vigorously with oxidizing materials. To fight fire, use alcohol foam, water, foam, CO_2, dry chemical. For further information see 1,1 dichloroethane, Vol. 4, No. 3 of *DPIM Report*.

ETHYL ISOBUTYRATE   HR: 3
CAS: 97-62-1   NIOSH: NQ 4675000
mf: $C_6H_{12}O_2$ mw: 116.18

PROP: Colorless, volatile liquid, fruity, aromatic odor, mp: $-88°$, bp: 110°-111°, d: 0.870, vap press: 40 mm @ 33.8°, vap d: 4.01, flash p: <64.4°F.

SYNS: ISOBUTYRIC ACID, ETHYL ESTER * ETHYL ISOBUTANOATE * ETHYL-2-METHYL-PROPANOATE * ETHYL-2-METHYLPROPIONATE

THR: Moderately toxic by intraperitoneal route. See also esters. Dangerous fire hazard when exposed to heat or flame; can react vigorously with oxidizing materials. To fight fire, use foam, CO_2, dry chemical. When heated to decomposition it emits acrid smoke and fumes.

ETHYL ISOCYANATE   HR: 3
CAS: 109-90-0   NIOSH: NQ 8825000
mf: C_3H_5NO mw: 71.09

PROP: Bp: 60°, d: 0.90 @ 20°/4°, vap d: 2.45.

SYNS: ISOCYANIC ACID, ETHYL ESTER * ISOCYANATOETHANE

THR: A poison by intravenous route. When heated to decomposition it emits toxic fumes of NO_x.

ETHYL ISONICOTINATE   HR: 3
CAS: 1570-45-2   NIOSH: NS 1450000
mf: $C_8H_9NO_2$ mw: 151.18

SYNS: 4-PYRIDINECARBOXYLIC ACID, ETHYL ESTER * ISONICOTINIC ACID, ETHYL ESTER

THR: A poison by intravenous route. When heated to decomposition it emits toxic fumes of NO_x.

ETHYL ISOPROPYL BARBITURIC ACID   HR: 3
CAS: 76-76-6   NIOSH: CQ 5425000
mf: $C_9H_{14}N_2O_3$ mw: 198.25

SYNS: 5-ETHYL-5-ISOPROPYLBARBITURIC ACID * PROBARBITAL * PROBARBITONE * 2,4,5(1H,3H,5H)-PYRIMIDINETRIONE, 5-ETHYL-5-(1-METHYLETHYL)- (9CI)

THR: A poison by intraperitoneal, subcutaneous, ingestion, and intravenous routes. A human psychotropic agent by ingestion. When heated to decomposition it emits toxic fumes of NO_x.

ETHYL ISOPROPYL NITROSO-AMINE   HR: 3
CAS: 16339-04-1   NIOSH: IA 3400000
mf: $C_5H_{12}N_2O$ mw: 116.19

SYNS: AETHYL-ISOPROPYL-NITROSOAMIN (GERMAN) * 1-METHYL-N-NITROSODIETHYLAMINE * N-NITROSOETHYLISOPROPYLAMINE

THR: An experimental carcinogen and tumorigen. Moderately toxic by ingestion route. See also nitrosoamines. When heated to decomposition it emits toxic fumes of NO_x.

ETHYL MALONATE   HR: 3
CAS: 105-53-3   NIOSH: OO 0700000
mf: $C_7H_{12}O_4$ mw: 160.19

PROP: Clear, colorless liquid, bp: 198.9°, fp: −49.8°, flash p: 200°F (OC), d: 1.055 @ 25°/25°, vap press: 1 mm @ 40.0°, vap d: 5.52.

SYNS: CARBETHOXYACETIC ESTER * DICARBETHOXYMETHANE * DIETHYL MALONATE * DIETHYL PROPANEDIOATE * MALONIC ESTER * METHANEDICARBOXYLIC ACID, DIETHYL ESTER * PROPANEDIOIC ACID, DIETHYL ESTER

THR: An experimental teratogen. A skin irritant. See also esters. Combustible when exposed to heat or flame; can react with oxidizing materials. To fight fire, use water to blanket fire; foam, CO_2, dry chemical. When heated to decomposition it emits acrid smoke and fumes.

ETHYL MERCURI CHLORENDI-MIDE   HR: 3
CAS: 2597-93-5   NIOSH: OW 3362000
mf: $C_{11}H_7Cl_6HgNO_2$ mw: 598.48

SYNS: N-(ETHYLMERCURI)-1,4,5,6,7,7-HEXACHLOROBICYCLO(2.2.1)HEPT-5-ENE- 2,3-DICARBOXIMIDE * N-ETHYLMERCURI-3,4,5,6,7,7-HEXACHLORO-3,6-ENDOMETHYLENE-1,2,3,6- TETRAHYDROPHTHALIMIDE * N-ETHYL-MERCURI-1,2,3,6-TETRAHYDRO-3,6-ENDOMETH-ANO- 3,4,5,6,7,7-HEXACHLOROPHTHALIMIDE

THR: A poison by ingestion and other routes. See also mercury compounds. When heated to decomposition it emits very toxic fumes of Cl^-, Hg and NO_x.

ETHYLMERCURY-p-TOLUENE SULFONAMIDE   HR: 3
CAS: 517-16-8   NIOSH: OW 3850000
mf: $C_{15}H_{17}HgNO_2S$ mw: 475.98

PROP: Crystals, pungent, garlic-like odor, water-insol.

SYNS: N-ETHYLMERCURI-N-PHENYL-P-TOLUENE-SULFONAMIDE * N-(ETHYLMERCURI)-P-TOLU-ENESULFONANILIDE * N-(ETHYLMERCURI)-P-TOLUENESULPHONANILIDE * ETHYLMERCURY P-TOLUENESULFANILIDE * ETHYLMERCURY-P-TOLUENESULFONANILIDE * (N-PHENYL-P-TOLUENESULFONAMIDO)ETHYLMERCURY

THR: A poison by ingestion and other routes. Mutagenic data. See also mercury compounds and sulfonamides. When heated to decomposition it emits very toxic fumes of Hg, NO_x and SO_x.

ETHYL METHACRYLATE   HR: 3
CAS: 97-63-2   NIOSH: OZ 4550000
mf: $C_6H_{10}O_2$ mw: 114.16

PROP: A liquid, mp: $<-75°$, bp: 119°, lel = 1.8%, uel = saturation, flash p: 68°F (OC), d: 0.911 @ 25°/25°, vap d: 3.94.

SYNS: ETHYL-ALPHA-METHYL ACRYLATE * ETHYL-2-METHYLACRYLATE * ETHYL-2-METHYL-2-PROPENOATE

THR: A poison by intraperitoneal route. Moderate toxicity by ingestion. A skin irritant. Dangerous fire and explosion hazard when exposed to heat or flame; can react with oxidizing materials. To fight fire, use CO_2, dry chemical. When heated to decomposition it emits acrid smoke and fumes.

ETHYL METHANESULFONATE   HR: 3
CAS: 62-50-0   NIOSH: PB 2100000
mf: $C_3H_8O_3S$ mw: 124.17

SYNS: ETHYL ESTER OF METHANESULFONIC ACID * ETHYL ESTER OF METHYLSULFONIC ACID * ETHYL METHANSULFONATE * HALF-MYLERAN * METHANESULPHONIC ACID ETHYL ESTER * METHYLSULFONIC ACID, ETHYL ESTER * NSC 26805

THR: A poison by intraperitoneal route. Mutagenic data. An experimental teratogen and brain carcinogen. See also sulfonates and esters. When heated to decomposition it emits toxic fumes of SO_x.

ETHYLMETHIAMBUTENE HYDROCHLORIDE   HR: 3

CAS: 64037-50-9   NIOSH: BA 6345000
mf: $C_{15}H_{19}NS_2 \cdot ClH$ mw: 313.93

SYNS: N-ETHYL-N-1-DIMETHYL-3,3-DI-2-THIENYLALLYLAMINE HYDROCHLORIDE * N-ETHYL-N-1-DIMETHYL-3,3-DI-2-THIENYL-2-PROPENAMINE HYDROCHLORIDE * 3-ETHYL-METHYLAMINO-1,1-DI(2'-THIENYL)BUT-1-ENE HYDROCHLORIDE * ETHYLMETHYLTHIAMBU-TENE HYDROCHLORIDE

THR: A poison by ingestion and subcutaneous routes. When heated to decomposition it emits very toxic fumes of NO_x, SO_x and HCl.

5-ETHYL-5-(1-METHYL-1-BUTENYL)-BARBITURATE   HR: 3

CAS: 125-42-8   NIOSH: CQ 5595000
mf: $C_{11}H_{16}N_2O_3$ mw: 224.29

SYNS: 5-ETHYL-5-(1-METHYL-1-BUTENYL)BAR-BITURIC ACID * 2,4,6(1H,3H,5H)-PYRIMIDINE-TRIONE, 5-ETHYL-5-(1-METHYL-1-BUTENYL)-(9CI)

THR: A poison by ingestion and intraperitoneal route. When heated to decomposition it emits toxic fumes of NO_x.

ETHYL-N-METHYL CARBAMATE   HR: 3

CAS: 105-40-8   NIOSH: FC 2625000
mf: $C_4H_9NO_2$ mw: 103.14

PROP: Needles, mp: 54°, bp: 177°.

SYN: N-METHYL URETHAN

THR: An experimental tumorigen. Moderately toxic by intraperitoneal and subcutaneous routes. When heated to decomposition it emits toxic fumes of NO_x.

ETHYL METHYLENE PHOSPHORO DITHIOATE   HR: 3

CAS: 563-12-2   NIOSH: TE 4550000
mf: $C_9H_{22}O_4P_2S_4$ mw: 384.49

PROP: Liquid, sltly sol in water, sol in xylene, chloroform, acetone. Mp: −13°, d: 1.220 @ 20°/4°.

SYNS: BIS(S-(DIETHOXYPHOSPHINOTHIOYL)-MERCAPTO)METHANE * BIS(DITHIOPHOSPHATE DE O,O-DIETHYLE) DE S,S'-METHYLENE (FRENCH) * ENT 24,105 * METHANEDITHIOL-S,S-DIESTER WITH O,O-DIETHYL PHOSPHORO-DITHIOATE * METHYLEEN-S,S'-BIS(O,O-DIETHYL-DITHIOFOSFAAT) (DUTCH) * METHYLENE-S,S'-BIS(O,O-DIAETHYL-DITHIO-PHOSPHAT) (GERMAN) * S,S'-METHYLENE O,O,O',O'-TETRAETHYL PHOSPHORODITHIOATE * METILEN-S,S'-BIS(O,O-DIETIL-DITIOFOSFATO) (ITALIAN) * NIAGARA 1240 * PHOSPHORO-DITHIOIC ACID, O,O-DIETHYL ESTER, S,S-DIESTER WITH METHANEDITHIOL * O,O,O',O'-TETRA-AETHYL-BIS(DITHIOPHOSPHAT) (GERMAN) * O,O,O,O-TETRAETHYL S,S'-METHYLENEBIS-(DITHIOPHOSPHATE) * O,O,O',O'-TETRAETHYL S,S'-METHYLENEBISPHOSPHORDITHIOATE * O,O,O',O'-TETRAETHYL-S,S'-METHYLENE-BISPHOSPHORODITHIOATE * TETRAETHYL S,S'-METHYLENE BIS(PHOSPHOROTHIOLOTHIO-NATE) * O,O,O',O'-TETRAETHYL S,S'-METH-YLENE DI(PHOSPHORODITHIOATE)

THR: VERY poisonous by ingestion and intra-peritoneal routes. A poison by dermal route. Effects the human central nervous system and blood picture. An insecticide. Used also as a food additive permitted in the feed and drinking water of animals and/or for the treatment of food-producing animals. Also permitted in food for human consumption. Dangerous; when heated to decomposition emits highly toxic fumes of oxides of sulfur and phosphorus. See also parathion. For further information see eth-ion, Vol. 4, No. 1 of *DPIM Report*.

N-ETHYL-N-METHYL-p-(PHENYLAZO)-ANILINE   HR: 3

CAS: 2058-66-4   NIOSH: BY 0430000
mf: $C_{15}H_{17}N_3$ mw: 239.35

SYNS: P-ETHYLMETHYLAMINOAZOBENZENE * 4-ETHYLMETHYLAMINOAZOBENZENE * 4-(METHYLETHYL)AMINOAZOBENZENE * N-METHYL-N-ETHYL-P-AMINOAZOBENZENE

THR: An experimental carcinogen. When heated to decomposition it emits toxic fumes of NO_x.

5-ETHYL-N-METHYL-5-PHENYL BARBI-TURIC ACID   HR: 3

CAS: 115-38-8   NIOSH: CQ 6650000
mf: $C_{13}H_{14}N_2O_3$ mw: 246.29

SYNS: MEPHOBARBITAL * MEPHOBARBITONE * N-ETHYLMETHYLPHENYLBARBITURIC ACID * 1-METHYL-5-ETHYL-5-PHENYLBARBITURIC ACID * METHYLPHENOBARBITAL * N-METH-YLPHENOBARBITAL * 1-METHYLPHENOBAR-

BITAL * N-METHYLPHENOLBARBITOL
* METHYLPHENYLBARBITURIC ACID * N-
METHYL-5-PHENYL-5-ETHYLBARBITAL
* 1-METHYL-5-PHENYL-5-ETHYLBARBITURIC
ACID * N-METHYLPHENOBARBITAL
* 5-PHENYL-5-ETHYL-3-METHYLBARBITURIC
ACID

THR: A poison by intraperitoneal and ingestion.
A teratogen in humans by ingestion. When
heated to decomposition it emits toxic NO_x.

3-(5-ETHYL-1-METHYL-2-PYRROLIDINYL)INDOLE   HR: 3
CAS: 19369-07-4   NIOSH: NL 8760000
mf: $C_{15}H_{20}N_2$ mw: 228.37

THR: A poison by intraperitoneal route. When
heated to decomposition it emits toxic fumes
of NO_x.

2-ETHYL-2-METHYLSUCCINIMIDE   HR: 2
CAS: 77-67-8   NIOSH: WN 2800000
mf: $C_7H_{11}NO_2$ mw: 141.19

SYNS: EPILEO PETIT MAL * ETHOSUCCIMIDE
* ETHOSUCCINIMIDE * ETHOSUXIMIDE
* alpha-ETHYL-alpha-METHYLSUCCINIMIDE
* 3-ETHYL-3-METHYLPYRROLIDINE-2,5-DIONE
* gamma-METHYL-gamma-ETHYL-SUCCINIMIDE

THR: Moderate toxicity by ingestion and intraperitoneal route. When heated to decomposition
it emits toxic fumes of NO_x.

N-ETHYL-alpha-METHYL-m-(TRIFLUOROMETHYL)PHENETHYLAMINE   HR: 3
CAS: 458-24-2   NIOSH: SH 6820000
mf: $C_{12}H_{16}F_3N$ mw: 231.29

SYNS: FENFLURAMINE * 3-(TRIFLUORO-
METHYL)-N-ETHYL-alpha-METHYL PHENETHYL
AMINE

THR: A human poison by ingestion. A poison
by ingestion and intraperitoneal routes. See also
amines. When heated to decomposition it emits
very toxic fumes of F^- and NO_x.

ETHYL MORPHINE   HR: 3
CAS: 76-58-4   NIOSH: QD 0850000
mf: $C_{19}H_{23}NO_3$ mw: 313.43

THR: A poison by intraperitoneal and subcutaneous routes. See also (-)morphine. When

heated to decomposition it emits toxic fumes
of NO_x.

ETHYL MORPHINE HYDROCHLORIDE DIHYDRATE   HR: 3
CAS: 6746-59-4   NIOSH: QD 0875000
mf: $C_{19}H_{23}NO_3 \cdot ClH \cdot 2H_2O$ mw: 385.93

PROP: White, microscopic crystalline powder,
mp: 125° decomp, vap d: 13.3.

SYNS: 7,8-DIDEHYDRO-4,5-alpha-EPOXY-3-
ETHOXY-17-METHYLMORPHINAN-6-alpha-OL
HYDROCHLORIDE DIHYDRATE * DIONIN
* ETHYLMORPHINE HYDROCHLORIDE

THR: A poison by subcutaneous route. Can
cause habituation. See also codeine. Dangerous;
when heated to decomposition it emits very toxic
NO_x and HCl.

4-ETHYLMORPHOLINE   HR: 3
CAS: 100-74-3   NIOSH: QE 4025000
mf: $C_6H_{13}NO$ mw: 115.20

PROP: Colorless liquid, bp: 138°, flash p:
89.6°F (OC), d: 0.916 @ 20°/20°, vap d: 4.00.

SYN: N-ETHYL MORPHOLINE

OSHA PEL: TWA 20 ppm (skin)
ACGIH TLV: TWA 5 ppm (skin)

THR: A poison by intravenous route. Moderately toxic by ingestion route. A systemic irritant
in humans. A skin and eye irritant. Dangerous
fire hazard when exposed to heat or flame; incompatible with oxidizing materials. To fight
fire, use alcohol foam, foam, CO_2, dry chemical. When heated to decomposition it emits toxic
fumes of NO_x.

ETHYL NITRATE   HR: 3
CAS: 625-58-1   NIOSH: QU 7900000
mf: $C_2H_5NO_3$ mw: 91.08

PROP: Colorless liquid, pleasant odor, sweet
taste, mp: $-112°$, bp: 88.7°, explodes @ 185°F,
lel = 3.8%, flash p: 50°F (CC), d: 1.105 @
20°/4°, vap d: 3.14.

SYN: NITRIC ETHER (DOT)

DOT Classification: Label: Flammable
   Liquid

THR: Moderately toxic by ingestion and inhalation routes. See also nitrates. Dangerous fire
hazard when exposed to heat or flame. Moder-

ately explosive when exposed to heat. See nitrates. To fight fire, use foam, CO_2, dry chemical, water to blanket fire. Incompatible with Lewis acids.

ETHYL NITRITE   HR: 3
CAS: 109-95-5   NIOSH: RA 0810000
mf: $C_2H_5NO_2$ mw: 75.08

PROP: Colorless or yellowish liquid, highly aromatic, ethereal odor. Decomp on standing. Very sltly sol in water; misc in alc and eth, bp: 16.4°, lel = 3.0%, uel = 50%, explodes at 194°F, flash p: −31°F (CC), d: 0.900 @ 15.5°, autoign temp: 194°F, vap d: 2.59. Can explode > 90°C.

SYNS: NITROUS ACID ETHYL ESTER * ETHYL NITRITE (DOT) * NITROSYL ETHOXIDE * NITROUS ETHER * NITROUS ETHER (DOT) * NITROUS ETHYL ETHER

DOT Classification: Label: Flammable Liquid

THR: A poison by inhalation in children. Narcotic in high concentrations; lowers blood pressure. Methemoglobinemia has been reported. Very dangerous fire hazard when exposed to heat or flame. A powerful oxidizer. Severe explosion hazard especially @ >90°. Highly dangerous when heated to decomposition or on contact with acid or acid fumes, emits highly toxic fumes of NO_x; reacts vigorously with oxidizers. To fight fire, use foam, CO_2, dry chemical or water spray.

(ETHYL NITROSAMINO)METHYL
ACETATE   HR: 3
CAS: 65986-80-3   NIOSH: AH 5601500
mf: $C_5H_{10}N_2O_3$ mw: 146.17

SYNS: N-ETHYL-N-(ACETOXYMETHYL)NITROSAMINE * N-ACETOXYMETHYL-N-NITROSOETHYLAMINE * ACETOXYMETHYLETHYLNITROSAMINE * AETHYL ACETOXYMETHYLNITROSAMIN (GERMAN) * ETHYL ACETOXYMETHYLNITROSAMINE

THR: Mutagenic data. An experimental tumorigen. Moderate toxicity by ingestion. When heated to decomposition it emits toxic fumes of NO_x.

N-ETHYL-N-NITROSOBIURET   HR: 3
CAS: 32976-88-8   NIOSH: EC 1750000
mf: $C_4H_8N_4O_3$ mw: 160.16

SYNS: ETHYL NITROSO BIURET * N-NITROSO-N-ETHYL BIURET

THR: An experimental transplacental brain carcinogen and tumorigen by ingestion. Moderate acute toxicity by ingestion. When heated to decomposition it emits toxic fumes of NO_x.

N-ETHYL-N-NITROSO-N'-
NITROGUANIDINE   HR: 3
CAS: 63885-23-4   NIOSH: MF 2275000
mf: $C_3H_7N_5O_3$ mw: 161.15

SYNS: N-AETHYL-N'-NITRO-N-NITROSOGUANIDIN (GERMAN) * N-ETHYL-N'-NITRO-N-NITROSO-GUANIDINE

THR: An experimental neoplastigen, carcinogen and tumorigen. Mutagenic data. When heated to decomposition it emits very toxic fumes of NO_x.

1-ETHYL-1-NITROSOUREA
CAS: 759-73-9   NIOSH: YT 3150000
mf: $C_3H_7N_3O_2$ mw: 117.13

PROP: Pale yellow crystals, mp: 103° (decomp) 1% water soln @ 20°.

SYNS: AETHYLNITROSO-HARNSTOFF (GERMAN) * ENU * ETHYLNITROSOUREA * N-ETHYL-N-NITROSO-UREA * NITROSOETHYLUREA * NSC 45403

THR: A poison by ingestion, subcutaneous, and intravenous routes. Mutagenic data. A transplacental brain carcinogen, teratogen, neoplastigen, and tumorigen. A suspected human carcinogen. When heated to decomposition it emits toxic fumes of NO_x. For further information see Vol. 5, No. 3 of *DPIM Report*.

ETHYL NORADRENALINE
HYDROCHLORIDE   HR: 3
CAS: 3198-07-0   NIOSH: DN 7875000
mf: $C_{10}H_{15}NO_3 \cdot ClH$ mw: 233.72

SYNS: alpha-(1-AMINOPROPYL)PROTOCATE-CHUYL ALCOHOL HYDROCHLORIDE * 1-(3,4-DIHYDROXYPHENYL)-2-AMINO-1-BUTANOL HYDROCHLORIDE * 1-(3,4-DIHYDROXYPHE-NYL)-1-HYDROXY-2-AMINOBUTANE HYDRO-CHLORIDE * ETHYL NOREPINEPHRINE HYDRO-CHLORIDE * alpha-ETHYLNOREPINEPHRINE HYDROCHLORIDE

THR: A poison by intravenous and subcutaneous routes. Causes pulmonary problems in humans.

When heated to decomposition it emits very toxic fumes of Cl^- and NO_x.

ETHYL ORTHOFORMATE　　HR: 2
CAS: 122-51-0　　　NIOSH: RM 6475000
mf: $C_7H_{16}O_3$　　mw: 148.23

PROP: Clear liquid, pungent odor, bp: 145.9°, flash p: 86°F (CC), d: 0.895 @ 20°/20°, vap press: 10 mm @ 40.5°, vap d: 5.11.

SYNS: ORTHOFORMIC ACID, TRIETHYL ESTER * ETHYLESTER KYSELINY ORTHOMRAVENCI (CZECH) * ORTHOMRAVENCAN ETHYLNATY (CZECH) * TRIETHOXYMETHANE * TRIETHYL ORTHOFORMATE

THR: Moderate toxicity by ingestion and inhalation. An irritant to skin and eyes. See also esters. Dangerous fire hazard when exposed to heat, flame or oxidizers; can react with oxidizing materials. To fight fire, use foam, CO_2, dry chemical.

p-((m-ETHYLPHENYL)AZO)-N,N-DI-METHYLANILINE　　HR: 3
CAS: 17010-65-0　　　NIOSH: BY 0525000
mf: $C_{16}H_{19}N_3$　　mw: 253.38

SYNS: N,N-DIMETHYL-P((M-ETHYLPHENYL) AZO)ANILINE * 3'-ETHYL-4-DIMETHYL AMINOAZOBENZENE

THR: An experimental carcinogen. When heated to decomposition it emits toxic fumes of NO_x.

p-((p-ETHYLPHENYL)AZO)-N,N-DI-METHYLANILINE　　HR: 3
CAS: 5302-41-0　　　NIOSH: BY 0700000
mf: $C_{16}H_{19}N_3$　　mw: 253.38

SYNS: N,N-DIMETHYL-4'-ETHYL-4-AMINOAZO-BENZENE * N,N-DIMETHYL-P-((4-ETHYLPHE-NYL)AZO)ANILINE * 4'-ETHYL-4-DIMETHYL-AMINOAZOBENZENE

THR: An experimental carcinogen. When heated to decomposition it emits toxic fumes of NO_x.

p-(4-ETHYLPHENYLAZO)-N-METHYL-ANILINE　　HR: 3
CAS: 55398-27-1　　　NIOSH: BY 0800000
mf: $C_{15}H_{17}N_3$　　mw: 239.35

SYNS: 4'-ETHYL-N-METHYL-4-AMINOAZOBEN-ZENE * N-METHYL-4'-ETHYL-P-AMINOAZOBEN-ZENE

THR: An experimental tumorigen. When heated to decomposition it emits toxic fumes of NO_x.

N-ETHYL-1-((p-(PHENYLAZO)PHE-NYL)AZO)-2-NAPHTHYLAMINE　HR: 3
CAS: 6368-72-5　　　NIOSH: QM 2850000
mf: $C_{24}H_{21}N_5$　　mw: 379.50

SYNS: C.I. 26050 * N-ETHYL-1-((P-(PHE-NYLAZO)PHENYL)AZO)-2-NAPHTHALENAMINE * N-ETHYL-1-((4-(PHENYLAZO)PHENYL)AZO)-2-NAPHTHYLAMINE * FAT RED 7B * OIL VIO-LET * (PHENYLAZO-4-PHENYLAZO)-1-ETHYL-AMINO-2-NAPHTHALENE * 1-(4-PHENYLAZO-PHENYLAZO)-2-ETHYLAMINONAPHTHALENE

THR: An experimental carcinogen. When heated to decomposition it emits toxic fumes of NO_x.

5-ETHYL-5-PHENYLBARBITURIC ACID　　HR: 3
CAS: 50-06-6　　　NIOSH: CQ 6825000
mf: $C_{12}H_{12}N_2O_3$　　mw: 232.26

SYNS: BARBIPHENYL * PHENYLETHYLBARBI-TURATE * PHENYL-ETHYL-BARBITURIC ACID * 5-PHENYL-5-ETHYLBARBITURIC ACID * PHENYLETHYLMALONYLUREA * 5-ETHYL-5-PHENYL-2,4,6-(1H,3H,5H)PYRIMIDINETRIONE

THR: A poison by ingestion, subcutaneous, intravenous, and rectal routes. An experimental teratogen and neoplastigen; affects the central nervous system, a systemic skin hazard, and causes systemic psychotropic effects. Mutagenic data. When heated to decomposition it emits toxic fumes of NO_x. For further information see Phenobarbital, Vol. 1, No. 2 in *DPIM Report*.

ETHYL PHENYL HYDANTOIN　　HR: 3
CAS: 631-07-2　　　NIOSH: MU 2452000
mf: $C_{11}H_{12}N_2O_2$　　mw: 204.25

PROP: Colorless, odorless crystalline powder, mp: 199°-200°.

SYNS: 5-ETHYL-5-PHENYL-2,4-IMIDAZOLIDINE-DIONE * NIRVANOL

THR: A poison by subcutaneous route. Moderately irritating to skin, eyes and mucous membranes. Combustible. When heated to decomposition it emits toxic fumes such as NO_x.

ETHYL PHOSPHORODICHLORI-DATE   HR: 3
CAS: 1498-51-7   NIOSH: TD 4390000
mf: $C_2H_5Cl_2O_2P$ mw: 162.94

SYN: DICHLOROPHOSPHORIC ACID, ETHYL ESTER

THR: A corrosive material which is very toxic to tissue. When heated to decomposition it emits very toxic fumes of Cl^- and PO_x.

5-ETHYL-alpha-PICOLINE   HR: 2
CAS: 104-90-5   NIOSH: TJ 6825000
mf: $C_8H_{11}N$ mw: 121.20

PROP: Liquid, bp: 174°, d: 0.9184 @ 23°/4°, flash p: 165° (OC).

SYNS: ALDEHYDE COLLIDINE * 3-ETHYL-6-METHYLPYRIDINE * 5-ETHYL-2-METHYLPYRI-DINE * 5-ETHYL-2-PICOLINE * 2-METHYL-5-ETHYLPYRIDINE * 6-METHYL-3-ETHYLPYRIDINE * METHYL ETHYL PYRIDINE (DOT)

THR: Moderately toxic by ingestion and dermal routes. A skin and eye irritant. Dangerous; when heated to decomposition it emits toxic fumes; incompatible with HNO_3 and other oxidizers. Flammable by heat, open flames, oxidizers. To fight fire, use alcohol foam. For further information see 2-methyl-5-ethyl pyridine, Vol. 3, No. 6 in *DPIM Report*.

1-ETHYLPIPERIDINE   HR: 3
CAS: 766-09-6   NIOSH: TN 0250000
mf: $C_7H_{15}N$ mw: 113.23

PROP: Flash p: 66.2°F.

SYN: N-AETHYLPIPERIDIN (GERMAN)

THR: A poison by intravenous and subcutaneous routes. An eye irritant. When heated to decomposition it emits toxic fumes of NO_x.

ETHYL PROPIONATE   HR: 2
CAS: 105-37-3   NIOSH: UF 3675000
mf: $C_5H_{10}O_2$ mw: 102.15

PROP: Water-white liquid, pineapple-like odor. Misc in alc and ether, bp: 99°, mp: −72.6°, flash p: 54°F (CC), d: 0.895 @ 15.5°, autoign temp: 824°F, vap press: 40 mm @ 27.2°, vap d: 3.52, lel = 1.9%, uel = 11%.

SYNS: PROPIONATE D'ETHYLE (FRENCH) * PROPIONIC ETHER

DOT Classification: Label: Flammable Liquid

THR: Moderate toxicity by ingestion and intra-peritoneal routes. A skin irritant. Dangerous fire hazard when exposed to heat or flame; can react vigorously with oxidizing materials. See also ethers. To fight fire, use foam, CO2, dry chemical. When heated to decomposition it emits acrid smoke and fumes.

ETHYL PYRAZINYL PHOSPHORO-THIOATE   HR: 3
CAS: 297-97-2   NIOSH: TF 5775000
mf: $C_8H_{13}N_2O_3PS$ mw: 248.26

SYNS: O,O-DIAETHYL-O-(PYRAZIN-2YL)-MONO-THIOPHOSPHAT (GERMAN) * O,O-DIETHYL O,2-PYRAZINYL PHOSPHOROTHIOATE * DI-ETHYL O-2-PYRAZINYL PHOSPHOROTHIONATE * O,O-DIETHYL O-2-PYRAZINYL PHOSPHOTHIO-NATE * O,O-DIETHYL O-PYRAZINYL THIO-PHOSPHATE * ENT 25,580 * NEMATOCIDE * PHOSPHOROTHIOIC ACID, O,O-DIETHYL O-2-PYRAZINYL ESTER

THR: A poison by ingestion and dermal routes. A cholinesterase inhibitor. See also parathion. When heated to decomposition, it emits highly toxic fumes of NO_x, PO_x, and SO_x.

ETHYL SULFATE   HR: 3
CAS: 64-6-7-5   NIOSH: WS 7875000
mf: $C_4H_{10}O_4S$ mw: 154.20

PROP: Colorless, oily liquid, peppermint odor, bp: 209.5° (decomp); mp: −25°; d: 1.172 @ 15°/4°; flash p: 220°F; vap d: 5.31; vap press: 1 mm @ 47°. Insol in water, decomp by hot water, misc with alcohol, ether, autoign temp: 817°F.

SYNS: DIETHYL SULFATE * SULFURIC ACID DI-ETHYL ESTER

THR: A poison by subcutaneous and inhalation routes. Moderate acute toxicity by ingestion and skin routes. Mutagenic data. An experimental tumorigen, transplacental brain carcinogen. Combustible when exposed to heat or flame. When heated to decomposition it emits toxic fumes of SO_x. Incompatible with 2,7-dinitro-9-phenylphenanthridine; water; oxidizing materials. To fight fire, use alcohol foam, water, foam, CO_2.

ETHYL SULFIDE   HR: 2
CAS: 352-93-2   NIOSH: LC 7200000
mf: $C_4H_{10}S$ mw: 90.20

PROP: Liquid, garlic-like odor, mp: −102°, bp: 92-93°, d: 0.837 @ 20°/4°, vap d: 3.11. Flash p: 14°F.

SYNS: DIETHYL SULFIDE * DIETHYLSULFID (CZECH) * DIETHYLTHIOETHER * ETHYL MONOSULFIDE * ETHYLTHIOETHANE * ETHYL THIOETHER * SULFODOR (CZECH) * THIOETHYL ETHER

THR: Slight toxicity by ingestion. Skin and eye irritant. See also sulfides. Moderate fire hazard by heat, flame (sparks) or oxidizers. When heated to decomposition it yields highly toxic fumes of SO_x; also from the action of acids or acid fumes. Can react with water or steam to produce toxic and flammable vapors; can react with oxidizing materials. To fight fire, use water spray or mist, dry chemical, CO_2, foam.

4-(ETHYLSULFONYL)-1-NAPHTHALENESULFONAMIDE HR: 3
CAS: 842-00-2 NIOSH: QK 0810000
mf: $C_{12}H_{13}NO_4S_2$ mw: 299.38

SYN: 4-ETHYLSULPHONYLNAPHTHALENE-1-SULFONAMIDE

THR: An experimental carcinogen and neoplastigen. When heated to decomposition it emits very toxic fumes of NO_x and SO_x.

ETHYL TELLURAC HR: 3
CAS: 20941-65-5 NIOSH: WY 2950000
mf: $C_{20}H_{40}N_4S_8 \cdot Te$ mw: 720.72

PROP: Orange, yellow powder, d: 1.44, mp: 108-118°

SYNS: DIETHYLDITHIO CARBAMIC ACID TELLURIUM SALT * NCI-C02857 * TELLURIUM DIETHYLDITHIOCARBAMATE * TETRAKIS(DIETHYLCARBAMODITHIOATO-S,S')TELLURIUM * TETRAKIS(DIETHYLDITHIOCARBAMATO)-TELLURIUM

THR: An experimental tumorigen and carcinogen. A suspected human carcinogen. See also tellurium. When heated to decomposition it emits very toxic fumes of NO_x, SO_x and Te.

3'-ETHYL-5',6',7',8'-TETRAHYDRO-5',5',8',8'-TETRAMETHYL-2'-ACETONAPHTHONE HR: 3
CAS: 88-29-9 NIOSH: AL 3031000
mf: $C_{18}H_{26}O$ mw: 258.44

SYNS: 7-ACETYL-6-ETHYL-1,1,4,4-TETRAMETHYL-1,2,3,4-TETRAHYDRONAPHTHALENE

* ACETYLETHYLTETRAMETHYLTETRALIN
* 6-ACETYL-1,1,4,4-TETRAMETHYL-7-ETHYL-1,2,3,4-TETRALIN * 3'-ETHYL-5',6',7',8'-TETRAHYDRO-5',5',8',8'-TETRAMETHYL-2'-ACETONAPHTHONE * 1-(3-ETHYL-5,6,7,8-TETRAHYDRO-5,5,8,8-TETRAMETHYL-2-NAPHTHALENYL)-ETHANONE * POLYCYCLIC MUSK

THR: A poison by ingestion and intraperitoneal routes. Moderately toxic by subcutaneous route. A skin irritant. When heated to decomposition it emits acrid smoke and fumes.

ETHYL THIOCYANATE HR: 3
CAS: 542-90-5 NIOSH: XK 9900000
mf: C_3H_5NS mw: 87.15

PROP: D: 0.996; mp: −85.5; bp: 145°; insol in water; misc in alc and ether.

SYNS: AETHYLRHODANID (GERMAN) * ETHYL RHODANATE * ETHYL SULFOCYANATE

THR: A poison by ingestion, subcutaneous, and intravenous routes. See also thiocyanates. When heated to decomposition it emits very toxic fumes of NO_x and SO_x.

2-ETHYL THIO ISONICOTINAMIDE HR: 3
CAS: 536-33-4 NIOSH: NS 0350000
mf: $C_8H_{10}N_2S$ mw: 166.26

SYNS: alpha-ETHYLISONICOTINIC ACID THIOAMIDE * 2-ETHYLISONICOTINIC ACID THIOAMIDE * 2-ETHYLISONICOTINIC THIOAMIDE * alpha-ETHYLISONICOTINOYLTHIOAMIDE * ETHYLISOTHIAMIDE * ALPHA-ETHYLISOTHIONICOTINAMIDE * 2-ETHYLISOTHIONICOTINAMIDE * 2-ETHYL-4-PYRIDINECARBOTHIOAMIDE * 2-ETHYL-4-THIOAMIDYLPYRIDINE * 2-ETHYL-4-THIOCARBAMOYLPYRIDINE * ALPHA-ETHYLTHIOISONICOTINAMIDE * ETHYONOMIDE * NCI-C01694

THR: An experimental carcinogen and teratogen. It affects the human peripheral nervous system and is a systemic poison in humans. Moderate acute toxicity ingestion and intraperitoneal routes. When heated to decomposition it emits very toxic fumes of SO_x and NO_x.

ETHYL THIO PYROPHOSPHATE HR: 3
CAS: 3689-24-5 NIOSH: XN 4375000
mf: $C_8H_{20}O_5P_2S_2$ mw: 322.34

PROP: A liquid almost insol in water.

SYNS: BIS-O,O-DIETHYLPHOSPHOROTHIONIC AN-
HYDRIDE * DITHIODIPHOSPHORIC ACID, TETRA-
ETHYL ESTER * DI(THIOPHOSPHORIC) ACID,
TETRAETHYL ESTER * DITHIOPYROPHOSPHATE
DE TETRAETHYLE (FRENCH) * ENT 16,273
* PYROPHOSPHORODITHIOIC ACID, TETRAETHYL
ESTER * O,O,O,O-TETRAETHYL-DITHIO-DIFOS-
FAAT (DUTCH) * TETRAETHYL DITHIONOPYRO-
PHOSPHATE * TETRAETHYL DITHIOPYRO-
PHOSPHATE * O,O,O,O-TETRAETHYL
DITHIOPYROPHOSPHATE * TETRAETHYL DITHIO
PYROPHOSPHATE, LIQUID (DOT) * O,O,O,O-
TETRAETIL-DITIO-PIROFOSFATO (ITALIAN)

OSHA PEL: TWA 200 ug/m^3 (skin)
DOT Classification: Label: Poison

THR: A poison by ingestion, skin contact, intramuscular, intraperitoneal, subcutaneous, and intravenous routes. A cholinesterase inhibitor. See
also parathion.

ETHYL TOSYLATE   HR: 3
CAS: 80-40-0   NIOSH: XT 6825000
mf: $C_9H_{12}O_3S$ mw: 200.27

PROP: Liquid, mp: 33°; bp: 221.3°; flash p:
316°F(CC); d: 1.17; vap d: 6.98.

SYNS: ETHYL-P-METHYL BENZENESULFONATE
* ETHYL-P-TOLUENESULFONATE * ETHYL-P-
TOSYLATE * P-TOLUOLSULFONSAEUREAETHYL
ESTER (GERMAN)

THR: Mutagenic data. An experimental tumorigen. Moderately toxic by subcutaneous route.
See also esters. Combustible when exposed to
heat or flame. When heated to decomposition
it emits highly toxic fumes of SO$_x$; can react
with oxidizing materials. To fight fire, use CO_2,
dry chemical.

ETHYL TRICHLORO PHENYL ETHYL
PHOSPHONOTHIOATE   HR: 3
CAS: 327-98-0   NIOSH: TB 0700000
mf: $C_{10}H_{12}Cl_3O_2PS$ mw: 333.60

SYNS: ENT 25,712 * O-ETHYL-O-2,4,5-TRI-
CHLOROPHENYL ETHYLPHOSPHONOTHIOATE
* 2,4,5-TRICHLORO-PHENOL-O-ESTER WITH O-
ETHYL ETHYLPHOSPHONOTHIOATE

THR: Poison by ingestion, skin contact, and
other unspecified routes. When heated to decomposition it emits very toxic fumes of Cl$^-$, PO$_x$
and SO$_x$.

ETHYL TRICHLOROSILANE   HR: 3
CAS: 115-21-9   NIOSH: VV 4200000
mf: $C_2H_5Cl_3Si$ mw: 163.51

PROP: Liquid, mp: −105.6°, bp: 99.5°, flash
p: 72°F (OC), d: 1.24 @ 25°/25°, vap d: 5.6.

SYNS: ETHYL SILICON TRICHLORIDE * TRI-
CHLOROETHYLSILANE * TRICHLOROETHYLSILI-
CANE

THR: A poison by inhalation and intraperitoneal
routes. Moderately toxic by ingestion. Dangerous fire hazard when exposed to heat, flame
or oxidizers. Dangerous when heated to decomposition it emits highly toxic phosgene; will
react with water or steam to produce heat and
toxic and corrosive fumes; can react vigorously
with oxidizing materials. To fight fire, use CO_2,
dry chemical.

ETHYLUREA and SODIUM NITRITE
(2:1)   HR: 3
                       NIOSH: YT 3650000

SYNS: AETHYLHARNSTOFF UND NATRIUMNITRIT
(GERMAN) * AETHYLHARNSTOFF UND NITRIT
(GERMAN)

THR: An experimental teratogen and neoplastigen. When heated to decomposition it emits
toxic fumes of NO$_x$.

ETHYL VANILLIN   HR: 3
CAS: 121-32-4   NIOSH: CU 6125000
mf: $C_9H_{10}O_3$ mw: 166.19

PROP: Fine, crystalline needles, mp: 76.5°.

SYNS: 3-ETHOXY-4-HYDROXYBENZALDEHYDE
* 4-HYDROXY-3-ETHOXYBENZALDEHYDE
* PROTOCATECHUIC ALDEHYDE ETHYL ETHER
* QUANTROVANIL * VANILLAL

THR: An experimental tumorigen. Moderately
toxic by ingestion, intraperitoneal and intravenous routes. Used as a synthetic flavoring substance and adjuvant.

ETHYNODIOL ACETATE   HR: 3
CAS: 297-76-7   NIOSH: RC 8963000
mf: $C_{24}H_{32}O_4$ mw: 384.56

SYNS: 3-BETA, 17-BETA-DIACETOXY-17-ALPHA-
ETHYNYL-4-OESTRENE * 3-BETA,17-BETA-DI-
ACETOXY-19-NOR-17-ALPHA-PREGN-4-EN-20-YNE
* 19-NOR-17-alpha-PREGN-4-EN-20-YNE-3-
beta,17-DIOL DIACETATE * ETHYNODIOL DI-

ACETATE * BETA-ETHYNODIOL DIACETATE
* 17-ALPHA-ETHYNYL-3,17-DIHYDROXY-4-
ESTRENE DIACETATE * 17-ALPHA-ETHYNYL-
ESTR-4-ENE-3-BETA,17-BETA-DIOL ACETATE
* 17-ALPHA-ETHYNYL-4-ESTRENE-3-BETA,17-
BETA-DIOL DIACETATE * 17-ALPHA-ETHYNYL-
4-ESTRENE-3-BETA,17-DIOL DIACETATE
* 17-ALPHA-ETHYNYL-19-NORANDROST-4-
ENE-3-BETA,17-BETA-DIOL DIACETATE
* (3-BETA,17-ALPHA)-19-NORPREGN-4-EN-20-
YNE-3,17-DIOL DIACETATE

THR: An experimental carcinogen and neoplastigen. When heated to decomposition it emits acrid smoke and fumes.

ETP   HR: 3
CAS: 29767-20-2   NIOSH: KC 0180000
mf: $C_{32}H_{32}O_{13}S$ mw: 656.70

SYNS: 4'-DEMETHYL-EPIPODOPHYLLOTOXIN-
beta-D-THENYLIDENE-GLUCOSIDE * 4'-DE-
METHYL 1-O-(4,6-O,O-(2-THENYLIDENE)-beta-
D-GLUCOPYRANOSYL)EPIPODOPHYLLOTOXIN
* NSC-122819

THR: A poison by intraperitoneal route. Causes gastrointestinal tract problems in humans by intravenous route; also causes blood dyscrasia problems. Mutagenic data. When heated to decomposition it emits very toxic fumes of SO_x. For further information see Vol. 1, No. 5 of *DPIM Report*.

EUCALYPTUS OIL   HR: 2
CAS: 8000-48-4   NIOSH: LE 2530000

PROP: Chief constituent is Eucalyptol. Colorless to pale-yellow liquid. Spicy odor. Composition: eucalyptol, aldehydes, d-pinene, mp: −15.4° (approx), d: 0.905-0.925 @ 25°/25°.

SYNS: DINKUM OIL * EUKALYPTUS OEL (GER-MAN) * OIL OF EUCALYPTUS

THR: A poison in children by ingestion route affecting central nervous system. A moderate skin irritant. Moderately toxic in man by ingestion. When heated to decomposition it emits acrid smoke and fumes.

EUGENOL   HR: 2
CAS: 97-53-0   NIOSH: SJ 4375000
mf: $C_{10}H_{12}O_2$ mw: 164.22

PROP: Colorless or yellowish liquid, oily, spicy odor, sol in alcohol, chloroform, ether and vola-

tile oils, very sltly sol in water, d: 1.064-1.070, bp: 253.5°.

SYNS: 4-ALLYLGUAIACOL * 4-ALLYL-1-HY-
DROXY-2-METHOXYBENZENE * 4-ALLYL-2-
METHOXYPHENOL * CARYOPHYLLIC ACID
* EUGENIC ACID * 1-HYDROXY-2-METHOXY-
4-ALLYLBENZENE * 1-HYDROXY-2-METHOXY-
4-PROP-2-ENYLBENZENE * 2-METHOXY-4-
PROP-2-ENYLPHENOL * 2-METHOXY-4-(2-PRO-
PENYL)PHENOL * NCI-C50453 * SYNTHETIC
EUGENOL

THR: Mutagenic data. Moderately toxic by ingestion, intraperitoneal and subcutaneous routes. A skin irritant. A synthetic flavoring and adjuvant.

EUROPIUM CHLORIDE   HR: 3
CAS: 10025-76-0   NIOSH: LE 7525000
mf: $EuCl_3$ mw: 258.31

THR: A poison by intraperitoneal route. Moderately toxic by ingestion route. See also rare earths. When heated to decomposition it emits very toxic fumes of Cl^-.

EUROPIUM (III) NITRATE, HEXAHYDRATE (1:3:6)   HR: 3
CAS: 10031-53-5   NIOSH: LE 8050000
mf: $N_3O_9 \cdot Eu \cdot 6H_2O$ mw: 446.11

SYN: NITRIC ACID, EUROPIUM(3+) SALT, HEXAHYDRATE

THR: A poison by intraperitoneal route. Slight toxicity by ingestion route. See also rare earths. When heated to decomposition it emits very toxic fumes. See also nitrates.

EVIPAL   HR: 3
CAS: 56-29-1   NIOSH: CQ 2625000
mf: $C_{12}H_{16}N_2O_3$ mw: 236.30

SYNS: HEXENAL (BARBITURATE) * 5-(1-CY-
CLOHEXEN-1-YL)-1,5-DIMETHYLBARBITURIC ACID
* 5-(1-CYCLOHEXENYL-1)-1-METHYL-5-METH-
YLBARBITURIC ACID * 5-(DELTA-1,2-CYCLO-
HEXENYL)-5-METHYL-N-METHYL-BARBITUR-
SAEURE (GERMAN) * CYCLONAL * 1,5-DI-
METHYL-5-(1-CYCLOHEXENYL)BARBITURIC ACID
* EVIPAN * HEXABARBITAL * HEXOBARBI-
TAL * HEXOBARBITONE * N-METHYL-5-CY-
CLOHEXENYL-5-METHYLBARBITURIC ACID

THR: A poison by subcutaneous, intraperitoneal, intravenous, and rectal routes. Moderately toxic by ingestion route. See also barbiturates. When heated to decomposition it emits toxic fumes of NO_x.

EXPLOSIVES, LOW

SYNS: BLACK BLASTING POWDER * GUNPOW-DER * ''A'' BLASTING POWDER * ''B'' BLASTING POWDER

Composition: Black powder is composed of saltpeter, charcoal and sulfur in the approximate proportions of 6:1:1. (''A'' blasting powder uses KNO_3 and ''B'' blasting powder uses $NaNO_3$). Low explosives are explosives which deflagrate; this differentiates them both in composition and properties from high explosives, which detonate. A deflagrating explosive is one that burns progressively over a relatively sustained period of time in comparison with a detonating explosive, which decomposes almost instantaneously.

THR: Dangerous fire hazard when exposed to heat or flame or by chemical reaction. Although most safety men now look upon black blasting powder with disfavor, it is one of the oldest and most generally used explosives in commercial work. It burns with extreme rapidity instead of detonating as high explosives do; it is highly sensitive to flame or sparks or friction and gives off much flame, which is hot and of great length of duration. These properties make it extremely hazardous for use in mines (especially coal mines) and quarries. The gases given off in detonation are not only very hot but frequently contain harmful constituents. Notwithstanding its numerous deficiencies, from a safety standpoint it has action characteristics that make it valuable in both coal mining and quarrying, though it has relatively little utility in metal mining. It is difficult to use effectively in wet places and this is its main disadvantage from an efficiency standpoint. Pellet powder is black blasting powder in consolidated (pellet or stick) form rather than in grains or granules, and it has few if any real advantages over black blasting powder, notwithstanding the fairly prevalent idea that it is a ''safe'' explosive.

Smokeless powders have a somewhat different composition from that of black blasting powder and are used chiefly for sportsmen's ammunition and, more widely, for military purposes. They are decidedly sensitive to flame and impact but ordinarily are so packaged that if reasonable judgement is used they are relatively harmless.

Most black powder fires start from sparks. Ignition results in an explosion so quickly that no attempt can be made to fight the fire. Every effort should be made to prevent fires from reaching stores of black powder; but if this fails, fire-fighting forces should be withdrawn at least 800 ft from the fire and should protect themselves against an explosion by seeking any cover available or by lying flat on the ground. If an explosion does occur, every effort should be made to prevent flames from spreading to neighboring magazines. Fire-fighting forces should be cautioned against approaching a fire which may involve black powder to avoid being trapped or injured by an explosion. Most explosions from black powder originate from sparks and the following safety rules should be strictly enforced and obeyed.

1) Open no containers in a magazine to which explosives or ammunitions are stored. This should be done only in a building free from all other explosives or ammunitions; or in suitable weather in the open, at least 100 ft from the nearest magazine. The quantity at or near such an operation should be limited to 100 lbs. Safety tools only should be used in opening or closing containers or in other operations involving black powder. Safety shoes (non-insulating) should be worn in all rooms where black powder is handled and by all persons engaged in handling black powder. The wearing of all non-conductive shoes, such as rubber is prohibited. If the handling of black powder is carried on over a concrete floor; the floor should be covered with a tarpaulin or other suitable material.

Loose black powder is extremely dangerous. Whenever it is necessary to handle loose black powder, not over 50 lbs should be permitted at or near such operations. If black powder is spilled on benches or floors, all work should be stopped until it has been removed and the explosive hazard of any remaining dust or particles has been neutralized with water. Rooms or buildings in which black powder is handled should be inspected frequently for dust, and all such dust should be immediately removed with water.

Explosive Hazard: Severe when exposed to heat or flame or by chemical reaction. Black powder is the slowest acting of all explosives. It has a shearing and heaving action, tending to blast materials in large, firm fragments. The action derives from a relatively slow development of gas pressure so that it must be carefully loaded and closely confined. The application is limited because it disintegrates in water and therefore cannot be used in wet work, except with special precautions. It also produces more smoke and fumes than most dynamites.

Precautions in Handling: Black powder is the most treacherous explosive material used today and it is regarded as one of the worst known explosive hazards. When ignited unconfined it burns with explosive violence and will explode if ignited under even slight confinement. It can be ignited easily by very small sparks, heat and friction. It is subject to rapid deterioration in the presence of moisture, but if kept dry it retains its explosive properties for many years. It is used to ignite smokeless powder, propelling charges, airplane flares and bursting charges of hand grenades, as a bursting charge in shrapnel, practice bombs, practice trench-mortar shells, in saluting charges, smoke-puff charges, time and percussion fuses, pellets, primers and primer detonators, and in expelling charges of pyrotechnic signals.

When black powder is shipped or received, each container should be inspected for holes and weak spots. It should be examined particularly for small holes, such as those made by nails which are visible only upon close examination. Damaged containers should not be repaired. The contents should instead be transferred to new or serviceable containers. Metal containers for export shipment should be crated. Usually two containers are packed in each crate. Repainting of containers and repacking of black powder, contained in damaged or unserviceable containers, constitutes the principal maintenance activity for such stocks. Black powder containers are also subjected to sweating, which rusts metal drums or kegs, so that repainting is necessary to keep containers serviceable. Repainting should not be done in a magazine in which explosives or ammunition are stored. It may be done in a nearly empty magazine, or in clear weather in the open, at least 100 ft from the nearest magazine. The quantity of black powder, at or near such operations should be limited to 100 lbs.

The marking on repainted containers should be checked carefully to see that it is a facsimile of the old. Furthermore, the metal caps on certain types of black powder containers deteriorate in storage. Replacement of these caps may be allowed, but the same safety precautions as outlined above for repainting containers should be followed. Operations, such as removal of black powder from containers and its transfer from unserviceable to serviceable drums should be conducted in strict compliance with applicable portions of the above outlined safety regulations. Black powder operations of any kind should be conducted in special buildings which are not used for other purposes at the same time. The floors of such buildings should be covered with suitable materials. Absolute cleanliness should be maintained at all times in and around each operation. Non-insulating safety shoes should be worn by personnel in all assembly operations. All equipment should be electrically grounded and this should be determined by tests. Empty metal containers which have held black powder should be thoroughly washed inside with water before they are disposed of. Serious explosions have occurred from supposedly empty cans. Wooden containers should be destroyed by burning. Safety tools only should be used in opening or closing containers or in handling black powder. Processes should be so laid out as to bring about frequent grounding of all operators handling this material.

Destruction: Black powder is best destroyed by dumping it into water, preferably a stream or large body of water, but most states have anti-stream pollution laws that may be violated in this manner. It is well to be informed on this subject and to be certain that no damage will be done to water supplies.

If it is inconvenient or inadvisable to destroy black powder thus, it may be burned. If that is done the contents of only one container holding 25 lbs or less should be burned at a time. The powder should be poured on the ground in a trail not wider than 2 inches and no part of such a trail should parallel another part at a distance of less than 10 ft. A train of combustible material should be laid to the explosive for igniting it, as with dynamite.

"Pellet powder" (black powder in compressed stick form) may be destroyed either by throwing it into water after removing the wrapper, or as described for dynamite. If dry and in good condition, black powder burns rapidly, especially in small grain size, with a yellow or pinkish-blue flame and dense smoke.

The empty powder containers should be

washed out, as explosions are said to have occurred from "empty" containers.

EXPLOSIVES, PERMITTED

"Permissible" explosives are essentially high explosives (dynamite) modified by the introduction of "dopes." The function of the dopes in general is to decrease flame temperature and to a smaller extent, the length and duration of flame, when the explosive is converted from a solid into a gas; in other words, when it is fired or detonated. The designation "permissible" is given to an explosive of modified dynamite type after it has passed certain tests made by the Federal Bureau of Mines. The permissible character of such explosives depends not only upon the ingredients in the explosive, but also on certain well-defined specifications as to handling and use. As with the dynamites, there are several different types and grades; "permissibles," hydrated "permissibles," organic nitrate "permissibles," nitroglycerin "permissibles," ammonium perchlorate "permissibles," and gelatin "permissibles." Essentially all of those now used to any extent are in either the ammonium nitrate or the gelatin classes. See also dynamite.

The ammonium nitrate "permissible" explosives contain relatively little nitroglycerin and relatively large proportions of ammonium nitrate. The latter is an explosive but one less sensitive to impact, sparks, and flames than nitroglycerin. This type of permissible explosive is now used extensively, as it has a rather wide range of strength, rate of detonation, density, size of cartridge, etc., and can be utilized not only in dry but also to some extent in fairly wet holes if charged carefully and fired promptly. Gelatin "permissibles," explosives are more suitable than ammonium nitrate "permissibles," for wet holes, and in general are stronger and more violent than the ammonium nitrate types.

All "permissible" explosives are strong, must be used in relatively small quantities (less than 1.5 lbs) per hole to retain their permissibility, give off considerable quantities of toxic gases on detonation, and, while much safer than black blasting powder or dynamite, must be stored, handled, and used with care.

Classification Upon Basis of Toxic Gases: All "permissible" explosives, when detonated, emit some toxic gases and a much larger volume of nontoxic gases. In order that the toxic products may not become a menace to the life or health of miners, no explosive is now or can become "permissible" if upon detonation it evolves more than 158 liters (5.5 cu ft) of toxic gases per 1.5 lb charge as determined by tests in the Bichel pressure gage.

Classification upon the basis of the volume of toxic gases produced by 580 g (1.5 lb) of explosive is as follows: *Class A,* not more than 53 liters; *Class B,* between 53 and 106 liters; and *Class C,* between 106 and 158 liters. (These classifications are not to be confused with the I.C.C. Classification of explosives)

Field tests were made with a 1.5 lb charge of a "permissible" explosive that produced, in the Bichel gage, the maximum allowable quantity of poisonous gases (158 liters per 1.5 lbs); these tests indicated that in a narrow entry, without artificial ventilation. 1800 ppm of carbon monoxide (the only poisonous gas present) was produced, as shown by analysis of an air sample taken 2 minutes after the shot. Another sample of the air taken 2 minutes later contained 800 ppm of carbon monoxide. Under no conditions should miners or shot firers return to the place until the poisonous gases have been removed by adequate ventilation.

It is provided further that, in accordance with the provisions and conditions, explosives enumerated on the "permissible" lists of the Bureau of Mines are "permissible" in use only when they satisfy the following requirements:

1. That the explosive be in all respects similar to the sample submitted by the manufacturer for test, and that the diameters of the cartridges used must be those that have been approved.

2. That electric detonators (not fuse and detonators) be used of not less efficiency than No. 6, the detonation charge of which shall consist of a 1 g mixture of 80 parts of mercury fulminate and 20 parts of potassium chlorate (or their equivalents), and that the required electric firing must be done by means of a "permissible" type blasting unit.

3. That the explosive be stored in surface magazines under proper conditions, so that it will not undergo change in character, and that after taking underground it be used in less than 36 hours.

4. That the coal to be blasted be undercut or equivalently relieved; that, to prevent blow-through, all portions of the borehole must be at least 18 inches from relief in any direction;

that, to prevent blowouts, the charge be properly confined with not less than 2 ft of clay (if the length of the hole will not permit the charge desired and 2 ft of stemming, at least half the length of the hole shall be filled with stemming) or other incombustible stemming and not be on the solid; that, to prevent the hole being on the solid; it shall be at least 6 in. shorter than the depth of the undercut or equivalent relief, and, when placed adjacent to the roofs, ribs, or floor, all but 12 in at the rear of the hole must be at least 6 inches from the adjacent surface as projected into the coal to be blasted, and all parts of the hole shall be free from the adjacent surface as projected into the coal to be blasted; that the shot be not a dependent shot; and that the shot hole be cleaned before charging.

5. That the quantity used for a shot (1) be not in excess of 680 g (1.5 lbs) when fired in accordance with these requirements and (2) when used under certain additional requirements or restrictions be not in excess of 1,361 g (3 lbs). The use of charges over 1.5 lbs and not exceeding 3 lbs is approved tentatively pending further investigation. For charges of over 1.5 lbs, the following additional requirements must be observed, (a) Shot holes must be 6 ft or more in length. (b) Explosives must be charged in a continuous train, with no cartridges deliberately deformed or crushed, with all cartridges in contact with each other, and with the end cartridges touching the rear of the hole and the stemming, respectively, (c) Examination for gas must be made in the blasting area before and after a shot is fired. (d) The "permissible" explosive must be one showing toxic gas emission that will place it either in Class A or Class B.

6. That the region in which the blasting is done be kept well protected by rock dust or otherwise in accordance with Bureau of Mines inspection standards.

7. That the shot not be fired in the presence of a dangerous percentage of firedamp. Examination for firedamp to be made at the blasting area before shooting in a gassy mine.

See also articles on dynamite, nitroglycerin, trinitrotoluene, pentaerythritol tetranitrate, nitrates, ammonium nitrates, picrates, azides, fulminates.

F

FD AND C BLUE 1   HR: 3
CAS: 38444-45-9   NIOSH: BQ 4725000
mf: $C_{37}H_{36}N_2O_9S_3 \cdot 2Na$ mw: 794.91

SYNS: BLUE DYE NUMBER 1 FOOD ADDITIVE
* BRILLIANT BLUE FCD NO. 1 * C.I. 42090
* C.I. ACID BLUE 9, DISODIUM SALT * C.I.
FOOD BLUE 2 * COSMETIC BLUE LAKE
* D AND C BLUE NO. 1 * D AND C BLUE NO.
4 * FOOD BLUE 1 * FOOD BLUE 2 * FOOD
BLUE DYE NO. 1

THR: An experimental neoplastigen and carcinogen. When heated to decomposition it emits
very toxic NO_x, Na_2O, and SO_x.

FD AND C BLUE NO. 1   HR: 3
CAS: 2650-18-2   NIOSH: BQ 4550000
mf: $C_{37}H_{36}N_2O_9S_3 \cdot 2H_3N$ mw: 783.01

SYNS: C.I. 42090 * C.I. ACID BLUE 9, DIAM-
MONIUM SALT * C.I. DIRECT BROWN 78, DIAM-
MONIUM SALT * C.I. FOOD BLUE 2 * D AND
C BLUE NO. 4 * FD AND C BLUE NO. 1-ALUMI-
NUM LAKE * F D + C BLUE 1 * FOOD
BLUE 1

THR: An experimental neoplastigen and tumorigen. Slightly toxic by subcutaneous route. When
heated to decomposition it emits very toxic NH_3,
NO_x and SO_x.

FD AND C GREEN NO. 2   HR: 3
CAS: 5141-20-8   NIOSH: BQ 4900000
mf: $C_{37}H_{36}N_2O_9S_3 \cdot 2Na$ mw: 794.91

SYNS: C.I. FOOD GREEN 2 * A.F. GREEN NO.
2 * C.I. 42095 * C.I. ACID GREEN 5
* C.I. ACID GREEN 5, DISODIUM SALT
* D AND C GREEN NO. 4 * FAST ACID
GREEN N

THR: An experimental carcinogen and neoplastigen. Moderately toxic by intravenous route.
When heated to decomposition it emits very
toxic NO_x, Na_2O and SO_x.

FD AND C GREEN NO. 3   HR: 3
CAS: 2353-45-9   NIOSH: BQ 4425000
mf: $C_{37}H_{36}N_2O_{10}S_3 \cdot 2Na$ mw: 810.91

SYNS: AIZEN FOOD GREEN NO. 3 * C.I.
42053 * C.I. FOOD GREEN 3 * FAST GREEN
FCF

THR: An experimental neoplastigen and carcinogen. When heated to decomposition it emits
very toxic NO_x, Na_2O and SO_x.

FD AND C RED NO. 1   HR: 3
CAS: 3564-09-8   NIOSH: QJ 6650000
mf: $C_{19}H_{16}N_2O_7S_2 \cdot 2Na$ mw: 494.47

SYNS: A.F. RED NO. 1 * C.I. 16155
* C.I. FOOD RED 6, DISODIUM SALT * DISO-
DIUM-3-HYDROXY-4-((2,4,5-TRIMETHYLPHE-
NYL)AZO)-2,7-NAPHTHALENEDISULFONATE
* DISODIUM-3-HYDROXY-4-((2,4,5-TRIMETHYL-
PHENYL)AZO)-2,7-NAPHTHALENEDISULFONIC ACID
* DISODIUM-3-HYDROXY-4-((2,4,5-TRIMETHYL-
PHENYL)AZO)-2,7-NAPHTHALENEDISULPHONATE
* DISODIUM-3-HYDROXY-4-((2,4,5-TRIMETHYL-
PHENYL)AZO)-2,7-NAPHTHALENEDISULPHONIC
ACID * EXT. D AND C RED NO. 15 * 3-HY-
DROXY-4-((2,4,5-TRIMETHYLPHENYL)AZO)-2,7-
NAPHTHALENEDISULFONIC ACID,DISODIUM SALT
* SODIUM CUMENEAZO-BETA-NAPHTHOL DISUL-
FONATE * SODIUM CUMENEAZO-BETA-NAPH-
THOL DISULPHONATE * USACERT RED NO. 1

THR: An experimental carcinogen and neoplastigen. When heated to decomposition it emits
very toxic fumes of SO_x and NO_x.

FENAMIPHOS   HR: 3
CAS: 22224-92-6   NIOSH: TB 3675000
mf: $C_{13}H_{22}NO_3PS$ mw: 303.39

SYNS: O-AETHYL-O-(3-METHYL-4-METHYLTHIO-
PHENYL)-ISOPROPYLAMIDO-PHOSPHOR SAEURE
ESTER (GERMAN) * BAY 68138 * ENT 27572
* ETHYL 3-METHYL-4-(METHYLTHIO)PHENYL(1-
METHYLETHYL)PHOSPHORAMIDATE * ETHYL-4-
(METHYLTHIO)-m-TOLYL ISOPROPYL PHOSPHOR
AMIDATE * ISOPROPYLAMINO-O-ETHYL-(4-
METHYLMERCAPTO-3-METHYLPHENYL)PHOS-
PHATE * 1-(METHYLETHYL)-ETHYL 3-METHYL-
4-(METHYLTHIO)PHENYL PHOSPHORAMIDATE

THR: A poison by ingestion, inhalation, and
skin routes. When heated to decomposition it
emits very toxic fumes of NO_x, PO_x and SO_x.
For further information see Vol. 3, No. 1 of
DPIM Report.

FENAZOXINE   HR: 3
CAS: 13669-70-0   NIOSH: DM 4210000
mf: $C_{17}H_{19}NO$ mw: 253.2

SYNS: 3,4,5,6,7-TETRAHYDRO-5-METHYL-1-PHENYL 1H-2,5-BENZOXAZOCINE * NEFOPAM

THR: A poison by ingestion, intravenous and intramuscular routes. When heated to decomposition it emits toxic fumes such as NO_x.

FERBAM   HR: 3
CAS: 14484-64-1   NIOSH: NO 8750000
mf: $C_9H_{18}N_3S_6 \cdot Fe$ mw: 416.51

PROP: Black solid, sltly sol in water, mp: decomp 180°.

SYNS: DIMETHYLCARBAMODITHIOIC ACID, IRON(3+) SALT * DIMETHYLDITHIOCARBAMIC ACID, IRON SALT * DIMETHYLDITHIOCARBAMIC ACID, IRON(3+) SALT * EISENDIMETHYLDITHIOCARBAMAT (GERMAN) * EISEN(III)-TRIS-(N,N-DIMETHYLDITHIOCARBAMAT) (GERMAN) * FERMATE FERBAM FUNGICIDE * FERMOCIDE * FERRADOW * FERRIC DIMETHYLDITHIOCARBAMATE * IRON DIMETHYLDITHIOCARBAMATE * TRIS(DIMETHYLCARBAMODITHIOATO-S,S')IRON * TRIS)DIMETHYLDITHIOCARBAMATO)IRON * TRIS(N,N-DIMETHYLDITHIOCARBAMATO) IRON(111) * VANCIDE FE95

OSHA PEL: TWA 15 mg/m³

THR: A poison by intraperitoneal route. Moderately toxic by ingestion. An experimental carcinogen. A fungicide. Mutagenic data. When heated to decomposition it emits very toxic fumes of NO_x and SO_x.

FERRIC CHLORIDE   HR: 3
CAS: 7705-08-0   NIOSH: LJ 9100000
mf: $C_{13}Fe$ mw: 162.20

PROP: Black-brown solid, mp: 292°, bp: 319.0°, d: 2.90 @ 25°, vap press: 1 mm @ 194.0°.

SYNS: CHLORURE FERRIQUE (FRENCH) * FLORES MARTIS * IRON (III) CHLORIDE * IRON TRICHLORIDE * PERCHLORURE DE FER (FRENCH)

THR: A poison by ingestion and intraperitoneal routes. Used as a trace mineral added to animal feeds. Violent reaction with allyl chloride; Na; K. Dangerous; when heated to decomposition emits highly toxic fumes of HCl; will react with water to produce toxic and corrosive fumes. For further information see Vol. 3, No. 4 of *DPIM Report*.

FERRIC FLUORIDE   HR: 3
CAS: 7783-50-8   NIOSH: NO 6865000
mf: F_3Fe mw: 112.85

PROP: Green crystals, d: 3.87.

SYNS: IRON FLUORIDE * IRON TRIFLUORIDE

THR: A poison by intravenous route. See also fluorides. When heated to decomposition it emits toxic fumes of F^-.

FERRIC SULFATE   HR: 3
CAS: 10028-22-5   NIOSH: NO 8505000
mf: $Fe_2O_{12}S_3$ mw: 399.88

SYNS: DIIRON TRISULFATE * IRON PERSULFATE * IRON SESQUISULFATE * IRON SULFATE (2:3) * IRON TERSULFATE * SULFURIC ACID, IRON (3^+) SALT $(3:2)$

THR: See also other ferric salts. When heated to decomposition it emits toxic fumes of SO_x.

FERROCENE   HR: 3
CAS: 102-54-5   NIOSH: LK 0700000
mf: $C_{10}H_{10}Fe$ mw: 186.05

PROP: Orange crystals, camphor odor, insol in water, sol in alcohol and ether, mp: 174°, sublimes @ >100°, volatile in steam.

SYNS: BIS CYCLOPENTADIENYL IRON * DI-2,4-CYCLOPENTADIEN-1-YL IRON * IRON DICYCLOPENTADIENYL * IRON BIS(CYCLOPENTADIENE)

THR: A poison by intraperitoneal and intravenous routes. Moderate toxicity by ingestion. An experimental tumorigen. Flammable. Reacts violently with NH_4ClO_4. Dangerous; when heated to decomposition it emits toxic fumes. For further information see Vol 1, No. 4 of *DPIM Report*.

FERROCHROME (EXOTHERMIC)   HR: 3
CAS: 11114-46-8   NIOSH: LK 0750000

SYNS: CARBON FERROCHROMIUM * CHROME FERROALLOY * CHROMIUM ALLOY * CHROMIUM ALLOY, BASE, Cr,C,Fe,N,Si (FERROCHROMIUM) * EXOTHERMIC FERROCHROME (DOT) * FERROCHROMIUM

THR: A poison by inhalation route. An experimental carcinogen. See also chromium compounds.

FERROMANGANESE (EXOTHERMIC)   HR: 3
CAS: 12604-53-4   NIOSH: LK 0900000

SYN: EXOTHERMIC FERROMANGANESE (DOT)

THR: See also manganese compounds. The dust will burn violently and give off toxic fumes of MnO_2.

FERROUS CHLORIDE HR: 3
CAS: 7758-94-3   NIOSH: NO 5400000
mf: Cl_2Fe mw: 126.75

PROP: Green to yellow, deliquescent crystals, mp: 614°-670°, bp: 1026°, d: 3.16, vap press: 10 mm @ 700°.

SYNS: IRON(II) CHLORIDE (1:2) * IRON DI-CHLORIDE * IRON PROTOCHLORIDE

THR: A poison by intraperitoneal route. See also chlorides. Can react violently with ethylene oxide; K; Na. Mutagenic data. Dangerous; see also chlorides.

FERROUS GLUCONATE HR: 3
CAS: 299-29-6   NIOSH: LZ 5150000
mf: $C_{12}H_{22}O_{14} \cdot Fe$ mw: 446.19

PROP: Yellowish-gray or pale greenish-yellow, fine powder or granules with slight odor, sol in water and glycerin, insol in alcohol.

SYNS: FERGON PREPARATIONS * FERLUCON * FERRONICUM * GLUCO-FERRUM * IRON GLUCONATE * IROX (GADOR) * NIONATE

THR: An experimental tumorigen. Causes central nervous system problems in children by ingestion route. A nutrient food additive. A trace mineral added to animal feeds. When heated to decomposition it emits acrid smoke and fumes.

FERROUS SULFATE HR: 3
CAS: 7782-63-0   NIOSH: NO 8510000
mf: $O_4S \cdot Fe \cdot 7H_2O$ mw: 278.05

PROP: Monoclinic crystals, d: 2.99-3.08.

SYN: IRON(II) SULFATE (1:1), HEPTAHYDRATE

THR: A poison by intravenous route. Moderate toxicity by ingestion. Mutagenic data. Used as a nutrient and/or dietary supplement food as well as a trace mineral added to animal feed; it is a substance migrating to foods from packaging materials. Decomposition emits toxic fumes of SO_x.

FERROVANADIUM DUST HR: 3
CAS: 12604-58-9   NIOSH: LK 2900000
mf: FeV   mw: 106.8

PROP: A gray to black dust.
OSHA PEL: TWA 1 mg/m^3

THR: Can cause pulmonary damage. See also vanadium and iron dust. Flammable when exposed to heat or flame. Slight explosion hazard when dust is exposed to flame.

FIBROUS GLASS HR: 2
The possibility of lung problems due to inhalation of fine particles or flakes or fibers of fiberglas on workers or users has often been raised. It now appears that the extensive medical research so far reported has shown no consistent evidence of chronic health effects in men who are exposed to man-made vitreous fibers. In laboratory animals exposed by inhalation no evidence of chronic ill effects were noted even after massive dosing with highly respirable particles. In some studies where massive doses of fine diameter fibers were implanted into mice, cancer development in the pleura was noted. Also some animal studies involving injection of fibers into the tracheaes resulted in a minimal fibrosis. However, these experiments are not directly extrapolatible to man. Exposure to glass fibers sometimes causes irritation of the skin, and, less frequently, irritation of the eyes, nose or throat. This is not an allergic reaction, but simply a mechanical irritation. Skin irritation typically is experienced by individuals who are newly exposed to fibrous glass and it usually diminishes after several days of exposure. Good personal and industrial hygiene practices minimize the amount of discomfort experienced.

FIREMASTER BP-6 HR: 3
CAS: 59536-65-1   NIOSH: LK 5060000

PROP: Consists mainly of penta-, hexa-, and heptabromobiphenyl, with lesser amounts of tetra- and other brominated biphenyls

SYNS: HEXABROMOBIPHENYL (TECHNICAL GRADE) * PBB * POLYBROMINATED BIPHE-NYLS

THR: An experimental carcinogen and teratogen by ingestion route. When heated to decomposition it emits very toxic Br^-.

FIREMASTER FFI HR: 3
CAS: 67774-32-7   NIOSH: LK 5065000

PROP: 2,4,5,2',4',5'-hexabromobiphenyl is the predominant isomer

SYNS: 2,4,5,2',4',5'-HEXABROMOBIPHENYL * POLYBROMINATED BIPHENYL

THR: An experimental neoplastigen. When heated to decomposition it emits very toxic fumes of Br⁻.

FLUANISONE HYDROCHLORIDE HR: 3
CAS: 17160-71-3 NIOSH: EU 3675000
mf: $C_{21}H_{25}FN_2O_2 \cdot ClH$ mw: 392.94

SYNS: 4'-FLUORO-4-(4-(o-METHOXYPHENYL)-1-PIPERAZINYL)BUTYROPHENONE,HYDROCHLORIDE * ANTI-PICA * HALOANISONE COMPOSITUM

THR: A poison by subcutaneous, intraperitoneal, and intravenous routes. When heated to decomposition it emits very toxic HCl, NO_x and F⁻.

FLUOMINE DUST HR: 3
CAS: 62207-76-5 NIOSH: GG 0575000
mf: $C_{16}H_{12}CoF_2N_2O_2$ mw: 361.23

SYNS: N,N'-ETHYLENEBIS(3-FLUOROSALICYLIDENEIMINATO)COBALT (II) * N,N'-ETHYLENE BIS(3-FLUORO SALICYLIDENEIMINATO)COBALT (II)

THR: A poison by ingestion and inhalation. See also cobalt compounds. When heated to decomposition it emits very toxic fumes of F⁻ and NO_x.

FLUORACIZINE HR: 3
CAS: 30223-48-4 NIOSH: SO 4700000
mf: $C_{20}H_{21}F_3N_2OS$ mw: 394.49

SYN: 10-DIETHYLAMINOPROPIONYL-3-TRIFLUOROMETHYL PHENOTHIAZINE HYDROCHLORIDE

THR: An experimental teratogen. When heated to decomposition it emits very toxic SO_x, NO_x and F⁻.

FLUORANTHENE HR: 3
CAS: 206-44-0 NIOSH: LL 4025000
mf: $C_{16}H_{10}$ mw: 202.26

PROP: A polycyclic hydrocarbon. Colorless solid, mp: 120°, bp: 367°, vap press: 0.01 mm @ 20°.

SYNS: BENZO(JK)FLUORENE * IDRYL * 1,2-(1,8-NAPHTHYLENE)BENZENE

THR: A poison by intravenous route. Moderately toxic by ingestion and skin contact. An experimental tumorigen. Mutagenic data. Combustible when exposed to heat or flame. When heated to decomposition it emits acrid smoke and fumes.

FLUOREN-2-AMINE HR: 3
CAS: 153-78-6 NIOSH: LL 5075000
mf: $C_{13}H_{11}N$ mw: 181.25

SYNS: AMINOFLUOREN (GERMAN) * 2-AMINOFLUORENE * 2-FLUORENAMINE * 2-FLUORENEAMINE

THR: An experimental carcinogen and neoplastigen. Mutagenic data. When heated to decomposition it emits toxic fumes of NO_x.

N-FLUOREN-2-YL ACETAMIDE HR: 3
CAS: 53-96-3 NIOSH: AB 9450000
mf: $C_{15}H_{13}NO$ mw: 223.29

SYNS: 2-ACETAMIDOFLUORENE * N-ACETYL-2-AMINOFLUORENE

THR: Mutagenic data. An experimental brain carcinogen, neoplastigen, and teratogen. Moderately toxic by ingestion. When heated to decomposition it emits toxic fumes of NO_x. For further information see Vol. 5, No. 5 of *DPIM Report*.

N-FLUOREN-2-YL BENZO HYDROXAMIC ACID HR: 3
CAS: 3671-71-4 NIOSH: DG 0350000
mf: $C_{20}H_{15}NO_2$ mw: 301.36

SYN: N-HYDROXY-N-2-FLUORENYLBENZAMIDE

THR: An experimental carcinogen. When heated to decomposition it emits toxic fumes such as NO_x.

N-FLUOREN-2-YL-2,2,2-TRIFLUORO ACETAMIDE HR: 3
CAS: 363-17-7 NIOSH: AC 1050000
mf: $C_{15}H_{10}F_3NO$ mw: 277.26

SYNS: N-(2-FLUORENYL)-2,2,2-TRIFLUOROACETAMIDE * 2-TRIFLUOROACETYLAMINOFLUORENE * 2,2,2-TRIFLUORO-N-(FLUOREN-2-YL)ACETAMIDE

THR: An experimental carcinogen. See also fluorides. When heated to decomposition it emits very toxic F⁻ and NO_x.

FLUORESCEIN HR: 3
CAS: 2321-07-5   NIOSH: LM 5075000
mf: $C_{20}H_{14}O_5$ mw: 334.34

PROP: Orange-red crystalline powder, mp: 314°-316° with decomp.

SYNS: 9-(o-CARBOXYPHENYL)-6-HYDROXY-3-I-SOXANTHENONE ∗ C.I. 45350 (FREE ACID) ∗ D AND C YELLOW NO. 7 ∗ 3',6'-DIHYDROX-YFLUORAN ∗ DIHYDROXYFLUORANE ∗ 3,6-FLUORANDIOL ∗ HIDACID FLUORESCEIN ∗ RESORCINOLPHTHALEIN ∗ SOAP YELLOW F

THR: A poison by intravenous route. Moderately toxic by ingestion route. When heated to decomposition it emits acrid smoke and fumes.

FLUORESCEIN SODIUM HR: 3
CAS: 518-47-8   NIOSH: LM 5425000
mf: $C_{20}H_{10}O_5 \cdot 2Na$ mw: 376.28

PROP: Orange-red powder; sol in water; sltly sol in alcohol.

SYNS: AIZEN URANINE ∗ CALCOCID URANINE B4315 ∗ CERTIQUAL FLUORESCEINE ∗ D&C YELLOW NO. 8 ∗ DISODIUM-6-HYDROXY-3-OXO-9-XANTHENE-O-BENZOATE ∗ FLUORESCEIN SODIUM B.P ∗ HIDACID URANINE ∗ NCI-c54706 ∗ RESORCINOL PHTHALEIN SODIUM ∗ SODIUM FLUORESCEIN ∗ SODIUM SALT OF HYDROXY-O-CARBOXY PHENYL FLUORONE ∗ SOLUBLE FLUORESCEIN

THR: An experimental tumorigen. Moderate toxicity by intraperitoneal route. When heated to decomposition it emits toxic fumes, including Na_2O fumes. For further information see Vol. 1, No. 5 of *DPIM Report.*

2-FLUORETHYL (1,1'-BIPHENYL)-4-ACETATE HR: 3
CAS: 4242-33-5   NIOSH: AH 3050000
mf: $C_{16}H_{15}FO_2$ mw: 258.31

PROP: A brown solid, sol in organic solvents. An insecticide, mp: 60.6°.

SYNS: 2-FLUOROETHYL (4-BIPHENYLYL)ACETATE ∗ BETA-FLUOROETHYLIC ESTER OF XENYLACETIC ACID

THR: A poison by ingestion and skin contact. See also fluorides. When heated to decomposition it emits toxic fumes of F^-.

FLUORIDES HR: 3
BAT: 7 mg/g creatinine end of exposure; 4 mg/g creatinine ca 16 hours after end of exposure

THR: Inorganic fluorides are generally highly irritating and toxic. Acute effects resulting from exposure to fluorine compounds are due to HF. Chronic fluorine poisoning, or "fluorosis," occurs among miners of cryolite, and consists of a sclerosis of the bones, caused by fixation of the calcium by the fluorine. There may also be some calcification of the ligaments. The teeth are mottled, and there is osteosclerosis and ostemalacia. The bony and ligamentous changes are demonstrable by x-ray. Estimated LD (man) = 2.5 to 5.9 g of F^-. Large doses can cause very severe nausea, vomiting, diarrhea, abdominal burning and cramp-like pains. It is not taken up by the thyroid and does not interfere with iodine uptake. Can cause or aggravate attacks of asthma. Can cause severe bone changes, making normal movements painful. Some signs of pulmonary fibrosis are noted. Some enzyme systems effects are reported. An irritant to the eyes, skin and mucous membranes. Also loss of weight, anorexia, anemia, wasting and cachexia, and dental defects are among the common findings in chronic fluorine poisoning. There may be an eosinophilia, and impairment of growth in young workers. Symptoms of intoxication include gastric, intestinal, circulatory, respiratory and nervous complaints and skin rashes. Organic fluorides are generally less toxic than other halogenated hydrocarbons. Common air contaminants. Dangerous; when heated to decomposition, or on contact with acid or acid fumes, they emit highly toxic fumes. See also fluorine.

FLUORINE HR: 3
CAS: 7782-41-4   NIOSH: LM 6475000
mf: F_2 mw: 38.00

PROP: Pale yellow gas, mp: -218°, bp: -187°, d: 1.14 @ -200°, 1.108 @ -188°, vap d: 1.695.

SYNS: BIFLUORIDEN (DUTCH) ∗ FLUOR (DUTCH, FRENCH, GERMAN, POLISH) ∗ FLUO-RINE (DOT) ∗ FLUORO (ITALIAN) ∗ FLUO-RURES ACIDE (FRENCH) ∗ FLUORURI ACIDI (ITALIAN) ∗ SAEURE FLUORIDE (GER-MAN)

OSHA PEL: TWA 0.1 ppm
DOT Classification: Label: Poison and Oxidizer

THR: A poison by irritation to skin, eyes, mucous membranes, and by ingestion and inhalation. See also fluorides. A most powerful caustic irritant to tissue. Dangerous fire and explosion hazard. Reacts violently with many materials. Incompatible with ammonia; cesium heptafluoro propoxide; covalent halides; cyanoguanidine; halocarbons; hexalithium disilicide; hydrocarbons; HF; hydrogen; seleninyl fluoride; hydrogen sulfide; ice; nitric acid; non-metal oxides; non-metals; oxygen; sodium acetate; sodium bromate; sodium dicyanamides; water; most organic matter; H-containing molecules; oxides of S, N, P; alkali metals; alkaline earths. It reacts violently with halogens; halogen acids; P; S; hydrazine; ClO_2; C; coke; charcoal; cyanamide; cyanides; KNO_3; (PbO + glycerol); CCl_4; silicides; silicates; alkenes; alkyl benzenes; CS_2; $Cr(OCl)_2$; B; Al; Tl; Sn; Sb; Te; Se; S; P; As; natural gas; liquid air; perfluoropropionyl fluoride; phenol formaldehyde resin; polyamides; polychloroprene; polyethylene; polyvinyl chloride acetate; polyurethane. Many reactions go on even at $< -160°$. Highly dangerous; when heated it emits highly toxic fumes; will react with water or steam to produce heat and toxic and corrosive fumes. For further information see Vol. 3, No. 4 of *DPIM Report*.

FLUORINE NITRATE   HR: 3
mf: FNO_3 mw: 81.01

PROP: Colorless gas, acrid odor, d: (liquid) 1.507 @ −45.9°; (solid) 1.951 @ −193.2°; mp: −175°; bp: −45.9°.

SYN: NITROGEN TRIOXYFLUORIDE

THR: VERY powerful irritant to tissues of skin, eyes and mucous membranes and by ingestion and inhalation routes. See also fluorine. Dangerous fire hazard; very powerful oxidizer. Conflagrates upon contact with reducers such as alcohol, ether, aniline. Very dangerous explosion hazard. Liquid explodes upon slight shock. Upon warming, emits highly toxic fumes of F^- and NO_x. Incompatible with gases; organic materials.

FLUORINE PERCHLORATE   HR: 3
mf: $ClFO_4$ mw: 118.45

PROP: Colorless gas, pungent, acrid odor, mp: −167.3°; bp: −15.9°.

SYN: CHLORINE TETROXYFLUORIDE

THR: A poison by irritation to skin, eyes and mucous membranes and by inhalation and ingestion. Very unstable. Known to be very corrosive and a powerful oxidizer. See also fluorine. Very powerful oxidizer. See perchlorates. Very dangerous; explodes on slightest provocation, and when warm, emits highly toxic fumes of F^- and Cl^-. Incompatible with rough surfaces; dirt; grease; heating; melting; KI; rubber tubing. Self explodes on exposure to hydrogen.

FLUOROACETALDEHYDE   HR: 3
CAS: 1544-46-3   NIOSH: AB 2900000
mf: C_2H_3FO mw: 62.05

THR: A poison by intraperitoneal and subcutaneous routes. See also fluorides and aldehydes. When heated to decomposition it emits toxic fumes of F^-.

FLUOROACETAMIDE   HR: 3
CAS: 640-19-7   NIOSH: AC 1225000
mf: C_2H_4FNO mw: 77.07

SYNS: COMPOUND 1081 ∗ 2-FLUOROACETAMIDE ∗ FLUOROACETIC ACID AMIDE ∗ MONOFLUOROACETAMIDE

THR: A poison by ingestion, skin, intraperitoneal, subcutaneous, and other routes. See also fluorides. When heated to decomposition it emits very toxic fumes of F^- and NO_x.

7-FLUORO-2-ACETAMIDO-FLUORENE   HR: 3
CAS: 343-89-5   NIOSH: AC 2275000
mf: $C_{15}H_{12}FNO$ mw: 241.28

SYNS: 7-FLUORO-2-ACETYLAMINOFLUORENE ∗ N-(7-FLUOROFLUORENE-2-YL)ACETAMIDE

THR: An experimental carcinogen. See also fluorides. When heated to decomposition it emits toxic fumes of F^- and NO_x.

FLUOROACETANILIDE   HR: 3
CAS: 330-68-7   NIOSH: AE 2975000
mf: C_8H_8FNO mw: 153.17

SYNS: AFL 1082 ∗ TL 1312

THR: A poison by ingestion, inhalation, and other routes. See also fluorides. When heated to decomposition it emits very toxic fumes of F^- and NO_x.

FLUOROACETONITRILE   HR: 3
CAS: 503-20-8   NIOSH: AM 0175000
mf: C_2H_2FN mw: 59.05

SYN: FLUOROMETHYL CYANIDE

THR: A poison by intraperitoneal and subcutaneous routes. See also cyanides and fluorides. When heated to decomposition it emits very toxic fumes of F^- and NO_x.

1-FLUORO-2-ACETYL AMINO FLUORENE   HR: 3
CAS: 2824-10-4   NIOSH: AC 1400000
mf: $C_{15}H_{12}FNO$ mw: 241.28

SYNS: 1-FLUORO-2-FAA * N-(1-FLUORO-FLUOREN-2-YL)ACETAMIDE

THR: An experimental carcinogen. ·See also amines and fluorides. When heated to decomposition it emits very toxic fumes of F^- and NO_x.

FLUOROACETYL CHLORIDE   HR: 3
CAS: 359-06-8   NIOSH: AO 6825000
mf: C_2H_2ClFO mw: 96.49

THR: A poison by inhalation route. See also fluorides and chlorides. When heated to decomposition it emits very toxic fumes of Cl^- and F^-.

o-(FLUOROACETYL)SALICYLIC ACID   HR: 3
CAS: 364-71-6   NIOSH: VO 3325000
mf: $C_9H_7FO_4$ mw: 198.16

SYN: SALICYLIC ACID, FLUOROACETATE

THR: A poison by ingestion and subcutaneous routes. See also fluorides. When heated to decomposition it emits toxic fumes of F^-.

5-FLUORO AMYLAMINE   HR: 3
CAS: 592-79-0   NIOSH: SC 0900000
mf: $C_5H_{12}FN$ mw: 105.18

SYN: 5-FLUOROPENTYLAMINE

THR: A poison by intraperitoneal and subcutaneous routes. When heated to decomposition it emits very toxic fumes of F^- and NO_x.

5-FLUOROAMYL THIOCYANATE   HR: 3
CAS: 661-18-7   NIOSH: XL 0950000
mf: $C_6H_{10}FNS$ mw: 147.23

SYN: 5-FLUOROPENTYL THIOCYANATE

THR: A poison by intraperitoneal and subcutaneous routes. See also thiocyanates and fluorides. When heated to decomposition it emits very toxic F^-, NO_x and SO_x.

4-FLUOROANILINE   HR: 3
CAS: 371-40-4   NIOSH: BY 1575000
mf: C_6H_6FN mw: 111.13

PROP: D: 1.1724, bp: 187.4°, mp: −1.9°

SYN: 4-FLUORANILIN (CZECH)

THR: A poison by ingestion. A skin and eye irritant. Mutagenic data. When heated to decomposition it emits very toxic fumes of NO_x and F^-.

4-FLUORO BENZANTHRACENE   HR: 3
CAS: 388-72-7   NIOSH: CW 6475000
mf: $C_{18}H_{11}F$ mw: 246.29

SYNS: 4-FLUOROBENZ(A)ANTHRACENE * 4'-FLUORO-1,2-BENZANTHRACENE

THR: An experimental neoplastigen. When heated to decomposition it emits toxic fumes of F^-.

1-(3-(4-FLUORO BENZOYL)PROPYL)-4-PIPERIDYL-N-ISOPROPYL CARBAMATE   HR: 3
CAS: 20977-50-8   NIOSH: FB 3235000
mf: $C_{19}H_{27}FN_2O_3$ mw: 350.48

SYN: AL-1021

THR: A poison by ingestion and intravenous routes. Causes psychotropic effects in humans by ingestion. When heated to decomposition it emits very toxic F^- and NO_x.

1-(4'-FLUORO BENZOYL)-3-PYRROLIDINYL PROPANE MALEATE   HR: 3
          NIOSH: EU 4730000
mf: $C_{14}H_{18}FNO • C_4H_4O_4$ mw: 351.41

SYNS: 1-(4'-FLUOROBENZOIL)-3-PIRROLIDINO-PROPANO MALEATO (ITALIAN) * 4'-FLUORO-4-(1-PYRROLIDINYL)BUTYROPHENONE MALEATE

THR: A poison by ingestion and intravenous routes. When heated to decomposition it emits very toxic fumes of F^- and NO_x.

4-FLUOROBUTYL THIOCYANATE   HR: 3
CAS: 353-17-3   NIOSH: XL 0525000
mf: C_5H_8FNS mw: 133.20

THR: A poison by intraperitoneal and subcutaneous route. See also thiocyanates and fluorides. When heated to decomposition it emits very toxic F^-, NO_x and SO_x.

FLUOROBUTYROPHENONE HR: 3

CAS: 1893-33-0   NIOSH: DW 1050000
mf: $C_{21}H_{30}FN_3O_2$ mw: 375.54

SYNS: DIPIPERONE * 1'-(3-(P-
FLUOROBENZOYL)PROPYL)(1,4'-BIPIPERIDINE 1-
4'-CARBOXAMIDE * 4'-FLUORO-4-(4-N-PIPERI-
DINO-4-CARBAMIDOPIPERIDINO)BUTYROPHENONE
* P-FLUORO-GAMMA-(4-PIPERIDINO-4-
CARBAMOYLPIPERIDINO)BUTYROPHENONE
* PIPERONYL

THR: A poison by ingestion, subcutaneous, and
intravenous routes. When heated to decomposition it emits very toxic F^- and NO_x.

5-FLUORO-2'-DEOXYCYTIDINE HR: 3

CAS: 10356-76-0   NIOSH: HA 3850000
mf: $C_9H_{12}FN_3O_4$ mw: 245.24

SYN: 5-FLUORODEOXYCYTIDINE

THR: An experimental teratogen. Moderate toxicity by intraperitoneal route. When heated to
decomposition it emits very toxic fumes of F^-
and NO_x.

10-FLUORO-9,12-DIMETHYLBENZ(a) ACRIDINE HR: 3

CAS: 386-38-9   NIOSH: CU 3600000
mf: $C_{19}H_{14}FN$ mw: 275.34

SYNS: 2,10-DIMETHYL-3-FLUORO-5,6-BENZA-
CRIDINE * 3-FLUORO-2,10-DIMETHYL-5,6-BEN-
ZACRIDINE

THR: An experimental tumorigen. When heated
to decomposition it emits toxic F^- and NO_x.

FLUORO ETHANOIC ACID HR: 3

CAS: 144-49-0   NIOSH: AH 5950000
mf: $C_2H_3FO_2$ mw: 78.05

PROP: Colorless solid, water sol, mp: 33°; bp:
165°.

SYNS: ACIDE-MONOFLUORACETIQUE (FRENCH)
* ACIDO MONOFLUOROACETIO (ITALIAN)
* CYMONIC ACID * 2-FLUOROACETIC ACID
* FLUOROACETIC ACID * GIFBLAAR POISON
* MONOFLUORAZIJNZUUR (DUTCH) * MONO-
FLUORESSIGSAURE (GERMAN) * MONOFLUO-
ROACETATE * MONOFLUOROACETIC ACID

THR: A poison by subcutaneous, intraperitoneal, intravenous, and other routes. Effects the
human central nervous system, causing convulsions and ventricular fibrillation. See also sodium fluoroacetate. When heated to decomposition it emits toxic fumes of F^-.

FLUOROETHANOL HR: 3

CAS: 63919-01-7   NIOSH: KL 1400000
mf: C_2H_5FO mw: 64.07

SYNS: MONOFLUORETHANOL * MONOFLUO-
ROETHANOL

THR: A poison by intraperitoneal, inhalation,
and subcutaneous routes. When heated to decomposition it emits very toxic fumes of F^-.

2-FLUOROETHANOL HR: 3

CAS: 371-62-0   NIOSH: KL 1575000
mf: C_2H_5FO mw: 64.07

SYN: beta-FLUOROETHANOL

THR: A poison by inhalation, intraperitoneal,
subcutaneous, and intravenous routes. When
heated to decomposition it emits very toxic
fumes of F^-.

beta-FLUOROETHYL FLUOROACE-TATE HR: 3

CAS: 459-99-4   NIOSH: AH 7525000
mf: $C_4H_6F_2O_2$ mw: 124.10

SYN: 2-FLUOROETHYL FLUOROACETATE

THR: A poison by inhalation, subcutaneous,
and parenteral routes. See also esters and fluorides. When heated to decomposition it emits
toxic fumes of F^-.

2-FLUORO ETHYL-gamma-FLUORO BUTYRATE HR: 3

CAS: 371-29-9   NIOSH: ET 2275000
mf: $C_6H_{10}F_2O_2$ mw: 152.16

SYN: beta-FLUOROETHYL-gamma-FLUOROBUTY-
RATE

THR: A poison by inhalation route. When heated
to decomposition it emits toxic fumes of F^-.

2-FLUOROETHYL-5-FLUOROHEX-OATE HR: 3

CAS: 63765-78-6   NIOSH: MO 8330000
mf: $C_8H_{14}F_2O_2$ mw: 180.22

THR: A poison by inhalation, intramuscular,
subcutaneous, and intravenous routes. When
heated to decomposition it emits very toxic F^-.

4'-FLUORO-4-(8-FLUORO-2,3,4,5-TET-RAHYDRO-1H-PYRIDO(4,3-b)INDOL-2-YL)BUTYROPHENONE
HYDROCHLORIDE   HR: 3
CAS: 31540-62-2   NIOSH: EU 2625000
mf: $C_{21}H_{20}F_2N_2O \cdot ClH$ mw: 390.89

SYN: ABBOTT-30360

THR: A poison by ingestion, intramuscular, and intravenous routes. When heated to decomposition it emits very toxic HCl, NO_x and F^-.

FLUOROMETHANE   HR: 2
mf: CH_3F mw: 34.03

PROP: Colorless gas, d: (liquid) 0.8774 @ $-78°$; (gas) 1.1951 (air = 1); (gas) 1.0813 (oxygen = 1); mp: $-141.8°$; bp: $-75.7°$ @ 872 mm; $-78.2°$ @ 760 mm. Freely sol in alc and ether. Agreeable ether like odor.

THR: Narcotic in high concentrations. May act as a simple asphyxiant. Burns with evolution of HF. Flame being about as colorless as that of alcohol. When heated to decomposition or burned it emits toxic fumes of F^-.

3-FLUORO-7-METHYLBENZ(a)ANTHRA-CENE   HR: 3
CAS: 2606-87-3   NIOSH: CW 6825000
mf: $C_{19}H_{13}F$ mw: 260.32

SYN: 3'-FLUORO-10-METHYL-1,2-BENZANTHRA-CENE

THR: An experimental neoplastigen. When heated to decomposition it emits toxic fumes of F^-.

7-FLUORO-10-METHYL-1,2-BENZ-ANTHRACENE   HR: 3
CAS: 1881-76-1   NIOSH: CW 7525000
mf: $C_{19}H_{13}F$ mw: 260.32

SYN: 10-FLUORO-7-METHYLBENZ(A)ANTHRA-CENE

THR: An experimental neoplastigen. When heated to decomposition it emits toxic fumes of F^-.

4'-FLUORO-4-(4-METHYL PIPERIDINO) BUTYRO PHENONE HYDROCHLO-RIDE   HR: 3
CAS: 1622-79-3   NIOSH: EU 4025000
mf: $C_{16}H_{22}FNO \cdot ClH$ mw: 299.85

SYN: METHYLPERONE HYDROCHLORIDE

THR: A poison by ingestion, subcutaneous, and intravenous routes. When heated to decomposition it emits very toxic F^-, NO_x and HCl.

8-(4-p-FLUORO PHENYL-4-OXOBUTYL)-2-METHYL-2,8-DIAZASPIRO(4.5)-DECANE-1,3-DIONE   HR: 3
CAS: 2804-00-4   NIOSH: HM 2650000
mf: $C_{19}H_{23}FN_2O_3$ mw: 346.44

SYNS: F-33 * FR-33 * R 7158

THR: A poison by ingestion and intravenous routes. When heated to decomposition it emits very toxic fumes of F^- and NO_x.

2-FLUORO-2-PROPEN-1-OL   HR: 3
CAS: 5675-31-0   NIOSH: UD 5075000
mf: C_3H_5FO mw: 76.08

THR: A poison by ingestion, skin contact, and inhalation. When heated to decomposition it emits toxic fumes of F^-.

4'-FLUORO-4-(4-(2-PYRIDYL)-1-PIPERA-ZINYL)BUTYROPHENONE   HR: 3
CAS: 1649-18-9   NIOSH: EU 4550000
mf: $C_{19}H_{22}FN_3O$ mw: 327.44

SYNS: AZEPERONE * 4'-FLUORO-4-(4-(2-PYRI-DYL)-1-PIPERAZINYL)BUTYROPHENONE

THR: A poison by intravenous and subcutaneous routes. When heated to decomposition it emits very toxic fumes of F^- and NO_x.

4'-FLUORO-4-(n-(4-PYRROLIDINAMIDO-4-m-TOLYPIPERIDINO)-BUTYROPHE-NONE   HR: 3
CAS: 2266-22-0   NIOSH: EU 4725000
mf: $C_{27}H_{33}FN_2O_2$ mw: 436.62

SYNS: MEPERIDIDE * METHYLPERIDIDE

THR: A poison by subcutaneous and intravenous routes. When heated to decomposition it emits very toxic fumes of F^- and NO_x.

FLUOROSULFURIC ACID   HR: 3
CAS: 7789-21-1   NIOSH: LP 0715000
mf: FHO_3S mw: 100.07

PROP: Colorless, fuming, highly corrosive liquid, mp: $-87.3°$, bp: 163.5°, d: 1.743 @ 15°.

SYNS: FLUOSULFONIC ACID * FLUOROSUFONIC ACID

THR: A poison irritant to skin, eyes, and mucous membranes, and by ingestion and inhalation.

See also fluorides, sulfuric acid and fluosulfonates.

3-FLUOROTYROSIN   HR: 3
CAS: 139-26-4   NIOSH: YP 2625000
mf: $C_9H_{10}FNO_3$ mw: 199.20

SYNS: M-FLUOROTYROSINE * FLUORTHYRIN * 3-FLUOROTYROSINE * 3-FLUORTYROSIN (GERMAN)

THR: A poison by ingestion, subcutaneous, and skin routes. See also fluorides. When heated to decomposition it emits very toxic fumes of F^- and NO_x.

FLUOROURACIL   HR: 3
CAS: 51-21-8   NIOSH: YR 0350000
mf: $C_4H_3FN_2O_2$ mw: 130.09

SYNS: NSC-19893 * 5-FLUORACIL (GERMAN) * 5-FLUORPROPYRIMIDINE-2,4-DIONE * 5-FLUOROURACIL

THR: A poison by intraperitoneal, ingestion and other routes. Moderate toxicity by intravenous, and parenteral routes. Mutagenic data. An experimental teratogen. Causes gastrointestinal tract, cardiovascular system, white blood cell, and other unspecified effects in humans. When heated to decomposition it emits very toxic fumes of F^- and NO_x.

5-FLUOROURIDINE   HR: 3
CAS: 316-46-1   NIOSH: YU 8050000
mf: $C_9H_{11}FN_2O_6$ mw: 262.22

SYN: 5-FUR

THR: A poison by intraperitoneal and subcutaneous routes. Mutagenic data. An experimental teratogen. When heated to decomposition it emits very toxic fumes of F^- and NO_x.

FLUOSULFONATES   HR: 3
THR: No toxicity data. These are salts of a strong acid. Probably poison. See also fluorides. Dangerous; when heated to decomposition they emits highly toxic fumes of fluorides and oxides of sulfur. Will react with water or steam to produce toxic and corrosive fumes.

FLURAZEPAM   HR: 3
CAS: 17617-23-1   NIOSH: DF 0820000
mf: $C_{21}H_{23}ClFN_3O$ mw: 387.92

THR: Has effects on human central nervous systems by ingestion route. Moderate acute toxicity

by ingestion. When heated to decomposition it emits very toxic Cl^-, F^- and NO_x.

cis-(Z)-FLURENTHIXOL   HR: 3
CAS: 2709-56-0   NIOSH: TL 9900000
mf: $C_{23}H_{25}F_3N_2OS$ mw: 434.56

SYNS: 2-TRIFLUOROMETHYL-9-(3-(4- (beta-HYDROXYETHYL-1-PIPERAZINYL)PROPYLIDENE)THIOXANTHENE * 4-(3-(2-(TRIFLUOROMETHYL)-9H-THIOXANTHEN-9-YLIDENE)PROPYL)-1-PIPERAZINEETHANOL

THR: A poison by ingestion, intraperitoneal, and intravenous routes. When heated to decomposition it emits very toxic F^-, NO_x and SO_x.

FOLIC ACID   HR: 3
CAS: 59-30-3   NIOSH: LP 5425000
mf: $C_{19}H_{19}N_7O_6$ mw: 441.45

PROP: A member of the vitamin B complex. Orange yellow needles or platelets, odorless, slightly sol in water, insol in lipid solvents, sol in dilute alkali hydroxide and carbonate solns.

SYNS: FOLACIN * FOLCYSTEINE * NSC 3073 * PTEGLU * PTEROYL-L-GLUTAMIC ACID * USAF CB-13 * VITAMIN BC * VITAMIN M

THR: A poison by intraperitoneal and intravenous routes. A food additive permitted in food for human consumption. When heated to decomposition it emits toxic fumes, including NO_x.

FOOD RED 2   HR: 3
CAS: 915-67-3   NIOSH: QJ 6550000
mf: $C_{20}H_{11}N_2O_{10}S_3 \cdot 3Na$ mw: 604.48

SYNS: D AND C RED 2 * ACID AMARANTH * AMARANTH * AMARANTHE USP (BIOLOGICAL STAIN) * AZO RED R * AZO RUBY S * BORDEAUX * C.I. 16185 * DYE FDC RED 2 * FAST RED * FD AND C RED NO. 2 * FD AND C RED NO. 2-aluminum LAKE * 3-HYDROXY-4-((4-SULFO-1-NAPHTHYL)AZO)-2,7-NAPHTHALENEDISULFONIC ACID, TRISODIUM SALT * FRUIT RED A GEIGY * 2-HYDROXY-1,1′-AZONAPHTHALENE-3,6,4′-TRISULFONIC ACID TRISODIUM SALT * 3-HYDROXY-4-((4-SULFO-1-NAPHTHALENYL)AZO)-2,7-NAPHTHLENEDISULFONICACID, TRISODIUM SALT * 3-HYDROXY-4-((4-SULPHO-1-NAPHTHALENYL)AZO)-2,7-NAPHTHALENEDISULPHONICACID, TRISODIUM SALT * MAPLE AMARANTH * RASPBERRY RED FOR JELLIES * RED DYE

NO. 2 * 1-(4-SULFO-NAPHTHYLAZO)-2-NAPH-THOL-3,6-DISULFONIC ACID TRISODIUM SALT * 1-(4-SULPHO-1-NAPHTHYLAZO)-2-NAPHTHOL-3,6-DISULPHONIC ACID, TRISODIUM SALT * TRISODIUM SALT OF 1-(4-SULFO-1-NAPHTH-YLAZO)-2-NAPHTHOL-3,6-DISULFONIC ACID * TRISODIUM SALT OF 1-(4-SULPHO-1-NAPHTH-YLAZO)-2-NAPHTHOL-3,6-DISULPHONICACID

THR: Mutagenic data. An experimental carcinogen and teratogen. Moderately toxic by intraperitoneal route. When heated to decomposition it emits very toxic fumes of SO_x and NO_x. For further information see Amaranth, Vol. 1, No. 3 of *DPIM Report*.

FOOD RED 4   HR: 3
CAS: 4548-53-2   NIOSH: QK 2705000
mf: $C_{18}H_{14}N_2O_7S_2 \cdot 2Na$ mw: 480.44

SYNS: C.I. 14700 * C.I. 16045 * C.I. FOOD RED 1, DISODIUM SALT * 3-((2,4-DI-METHYL-5-SULFOPHENYL)AZO)-4-HYDROXY-1-NAPHTHALENESULFONICACID, DISODIUM SALT * DYE FD & C RED NO. 4 * 4-HYDROXY-3-((6-SULFO-2,4-XYLYL)AZO)-1-NAPHTHALENE-SULFONIC ACID, DISODIUM SALT * 1306 RED * 12101 RED * RED NO. 1 * RED NO. 4 * 2-(6-SULFO-2,4-XYLYLAZO)-1-NAPHTHOL-4-SULFONIC ACID,DISODIUM SALT * USACERT FD & C RED NO. 4

THR: An experimental carcinogen. When heated to decomposition it emits very toxic fumes of SO_x and NO_x.

FOOD RED 7   HR: 3
CAS: 2611-82-7   NIOSH: QJ 6530000
mf: $C_{20}H_{14}N_2O_{10}S_3 \cdot 3Na$ mw: 607.51

SYNS: ACID BRILLIANT SCARLET 3R * BRILLIANT PONCEAU 3R * BRILLIANT SCARLET * COCHINEAL RED A SPECIALLY PURE * C.I. 16255 * 7-HYDROXY-8-((4-SULFO-1-NAPH-THYL)AZO)-1,3-NAPHTHALENEDISULFONIC ACID,TRISODIUM SALT * FOOD RED NO. 102 * JAVA SCARLET 3R * NEW COCCINE EXTRA PURE A * STRAWBERRY RED A GEIGY * 1-(4-SULPHO-1-NAPHTHYLAZO)-2-NAPHTHOL-6,8-DISULPHONIC ACID, TRISODIUM SALT * VICTORIA SCARLET RED

THR: An experimental tumorigen. Moderately toxic by intraperitoneal route. When heated to decomposition it emits very toxic fumes of NO_x and SO_x.

FOOD YELLOW 3   HR: 3
CAS: 2783-94-0   NIOSH: QK 2450000
mf: $C_{16}H_{10}N_2O_7S_2 \cdot 2Na$ mw: 452.38

SYNS: C.I. 15985 * C.I. FOOD YELLOW 3, DISODIUM SALT * DISODIUM SALT OF 1-P-SUL-PHOPHENYLAZO-2-NAPHTHOL-6-SULPHONIC ACID * DYE FD & C YELLOW NO. 6 * FD & C NO. 6 * FD & C YELLOW 6 * FD & C YELLOW NO. 6 aluminum LAKE * 6-HYDROXY-5-((p-SULFOPHENYL)AZO)-2-NAPHTHALENESULFONIC ACID, DISODIUM SALT * FOOD YELLOW 6 * 6-HYDROXY-5-((4-SULFOPHENYL)AZO)-2-NAPHTHALENESULFONIC ACID, DISODIUM SALT * NCI-C53907 * 1-P-SULFOPHENYLAZO-2-HY-DROXYNAPHTHALENE-6-SULFONATE, DISODIUM SALT * 1-P-SULFOPHENYLAZO-2-NAPHTHOL-6-SULFONIC ACID,DISODIUM SALT * SUN YELLOW EXTRA PURE A * USACERT FD & C YELLOW NO. 6 * YELLOW ORANGE SPECIALLY PURE 85 * YELLOW SY FOR FOOD

THR: Possible experimental carcinogen. Moderate intraperitoneal toxicity. When heated to decomposition it emits very toxic fumes of SO_x and NO_x.

FORMALDEHYDE   HR: 3
CAS: 50-00-0   NIOSH: LP 8925000
mf: CH_2O mw: 30.03

PROP: Clear, water-white, very sltly acid, gas or liquid, pungent odor. Pure formaldehyde is not available commercially because of its tendency to polymerize. It is sold as aqueous solns containing from 37% to 50% formaldehyde by weight and varying amounts of methanol. Some alcoholic solns are used industrially and the physical properties and hazards may be greatly influenced by the solvent; lel = 7.0%, uel = 73.0%, autoign temp: 806°F, d: 1.0, bp: −3°F, flash p: (37% methanol-free): 185°F, flash p: (15% methanol-free): 122°F.

SYNS: ALDEHYDE FORMIQUE (FRENCH) * ALDEIDE FORMICA (ITALIAN) * FANNOFORM * FORMALDEHYD (CZECH, POLISH) * FORMALDEHYDE, AS FORMALIN SOLUTION (DOT) * FORMALIN * FORMALINA (ITALIAN) * FORMALINE (GERMAN) * FORMALIN-LOE-SUNGEN (GERMAN) * FORMALITH * FORMIC ALDEHYDE * FORMOL * FYDE * HOCH * KARSAN * METHANAL * METHYL ALDE-HYDE * METHYLENE OXIDE * NCI-C02799 * OPLOSSINGEN (DUTCH) * OXOMETHANE * OXYMETHYLENE

OSHA PEL: TWA 3 ppm; CL 5; Pk 10/30M/ 8H

THR: A poison irritant to skin, eyes, and mucous membranes. If swallowed it causes violent vomiting and diarrhea which can lead to collapse. A fungicide. A common air contaminant. Mutagenic data and possible tumorigen of the lung. Frequent or prolonged exposure can cause hypersensitivity leading to contact dermatitis, possibly of an eczematoid nature. An air concentration of 20 ppm is quickly irritating to eyes. Very dangerous fire hazard for gas, moderate for vapors. Will burn above flash point if exposed to flame, sparks, etc. Should formaldehyde be involved in a fire, irritant gaseous formaldehyde may be evolved. When aqueous formaldehyde solutions are heated above their flash points, a potential for explosion hazard exists. High formaldehyde concentration or methanol content lowers flash point. Reacts with NO_x @ about 180°; the reaction becomes explosive. Also reacts violently with ($HClO_4$-aniline) and performic acid; nitromethane; magnesium carbonate; H_2O_2. Moderately dangerous because of irritating vapor which may exist in toxic concentrations locally if storage tank is ruptured. To fight fire, stop flow of gas (for pure form); alcohol foam for 37% methanol-free form. For further information see Vol. 3, No. 3 of *DPIM Report*.

FORMAL-gamma-TRIMETHYL AMMO-NIUM PROPANEDIOL HR: 3
CAS: 541-66-2　　　NIOSH: BQ 2700000
mf: $C_7H_{16}NO_2 \cdot I$　　mw: 273.14

SYNS: ((1,3-DIOXOLAN-4-YL)METHYL)TRI-METHYLAMMONIUM IODIDE ∗ OXAPROPANIUM IODIDE

THR: A poison by subcutaneous, ingestion, and intravenous routes. When heated to decomposition it emits very toxic fumes of NO_x and I^-.

FORMAMIDE HR: 3
CAS: 75-12-7　　　NIOSH: LQ 0525000
mf: CH_3NO　　mw: 45.05

PROP: Colorless, hygroscopic and oily liquid, misc in water and alc; very sltly sol in ether, mp: 2.5°, fp: 2.6°, vap press: 29.7 mm @ 129.4°, flash p: 310°F (COC), bp: 210° decomp, d: 1.134 @ 20°/40°; 1.1292 @ 25°/4°.

SYNS: CARBAMALDEHYDE ∗ METHANAMIDE

THR: Moderate toxicity by intramuscular and ingestion routes. An irritant to skin, eyes and mucous membranes. An experimental teratogen. Flammable. Vapors will burn in air above 310°F. Incompatible with I_2; pyridine; SO_3. When heated to decomposition emits toxic fumes of NO_x. Has exploded while in storage. For further information see Vol. 1, No. 1 of *DPIM Report*.

FORMIC ACID HR: 2
CAS: 64-18-6　　　NIOSH: LQ 4900000
mf: CH_2O_2　　mw: 46.03

PROP: Colorless, fuming liquid, pungent penetrating odor. Misc in water and alc, bp: 100.8°, fp: 8.2°, flash p: 156°F (OC), d: 1.2267 @ 15°/4°, 1.220 @ 20°/4°, autoign temp: 1114°F, vap press: 40 mm @ 24.0°, vap d: 1.59, flash p: (90% soln): 122°F, autoign temp (90% soln): 813°F, lel (90% soln) = 18%, uel (90% soln) = 57%.

SYNS: ACIDE FORMIQUE (FRENCH) ∗ ACIDO FORMICO (ITALIAN) ∗ AMEISENSAEURE (GERMAN) ∗ FORMYLIC ACID ∗ HYDROGEN CARBOXYLIC ACID ∗ METHANOIC ACID ∗ MIERENZUUR (DUTCH)

OSHA PEL: TWA 5 ppm

THR: Moderate toxicity by ingestion, intraperitoneal and intravenous routes. A severe eye irritant. Mild skin toxicity. A substance migrating to food from packaging materials. Flammable when exposed to heat or flame; can explode with furfuryl alcohol; H_2O_2; $Tl(NO_3)_3 \cdot 3H_2O$; nitromethane; P_2O_5. To fight fire, use CO_2, dry chemical, alcohol foam. For further information see Vol. 3, No. 4 of *DPIM Report*.

N-FORMYL-N-HYDROXYGLY-CINE HR: 3
CAS: 689-13-4　　　NIOSH: MC 0550000
mf: $C_3H_5NO_4$　　mw: 119.09

SYNS: ASYMMETRIN ∗ N-FORMYL HYDROX-YAMINOACETIC ACID ∗ HADACIDINE ∗ HADACIN ∗ NSC 521778

THR: An experimental teratogen. Moderate toxicity by intraperitoneal route. When heated to decomposition it emits toxic fumes of NO_x.

N-FORMYL-N-METHYLHYDRA-ZINE HR: 3
CAS: 758-17-8　　　NIOSH: LQ 8940000
mf: $C_2H_6N_2O$　　mw: 74.10

SYN: N-METHYL-N-FORMLYHYDRAZINE

THR: A poison by ingestion and other routes. An experimental teratogen and carcinogen. When heated to decomposition it emits toxic fumes of NO_x.

2-FORMYL-1-METHYLPYRIDINIUM CHLORIDE OXIME   HR: 3
CAS: 51-15-0   NIOSH: UU 4200000
mf: $C_7H_9N_2O \cdot Cl$ mw: 172.63

SYNS: 2-FORMYL-N-METHYLPYRIDINIUM OXIME CHLORIDE * 2-(HYDROXYIMINOMETHYL)-1-METHYLPYRIDINIUM CHLORIDE * 1-METHYL-2-ALDOXIMINOPYRIDINIUM CHLORIDE * 1-METHYL-2-PYRIDINIUM ALDOXIME CHLORIDE * N-METHYLPYRIDINIUM CHLORIDE-2-ALDOXIME * 2-PYRIDINE-ALDOXIME CHLORIDE * PYRIDINE-2-ALDOXIME METHOCHLORIDE * 2-PYRIDINE ALDOXIME METHYL CHLORIDE * PYRIDINIUM ALDOXIME METHOCHLORIDE

THR: A poison by intravenous, intramuscular, and intraperitoneal routes. Moderate toxicity by ingestion route. Causes central nervous system effects in man by intravenous route. When heated to decomposition it emits very toxic fumes of Cl^- and NO_x.

FREON 113   HR: 2
CAS: 354-5-8-5   NIOSH: KJ 3975000
mf: $C_2Cl_3F_3$ mw: 187.37

PROP: Colorless gas, mp: 13.2°, bp: 45.8°, d: 1.5702, autoign temp: 1256°F.

SYNS: FLUOROCARBON 113 * 1,1,2-TRI-CHLORO-1,2,2-TRIFLUOROETHANE * 1,1,2-TRI-FLUORO-1,2,2-TRICHLOROETHANE * UCON FLUOROCARBON 113

OSHA PEL: TWA 1000 ppm

THR: Affects the central nervous system in humans. Mild skin irritation. Low toxicity by inhalation and ingestion. Slightly combustible when exposed to heat or flame. Violent reaction with Al; Ba; Li; Sm; NaK alloy; Ti. Dangerous, see also chlorides and fluorides.

FRUCTOSE   HR: 3
CAS: 7660-25-5   NIOSH: LS 7000000
mf: $C_6H_{12}O_6$ mw: 180.18

SYNS: LEVULOSE * FRUIT SUGAR * FRU-TABS * LAEVORAL * LAEVOSAN * LEVU-GEN

THR: An experimental tumorigen. When heated to decomposition it emits acrid smoke and fumes. For further information see Vol. 1, No. 2 of *DPIM Report*.

FTBG   HR: 3
CAS: 19982-87-7   NIOSH: DJ 1400000
mf: $C_{15}H_{12}F_4N_4 \cdot ClH$ mw: 360.77

SYNS: 4-FLUORO-4'-TRIFLUOROMETHYLBENZO-PHENONE GUANYLHYDRAZONEHYDROCHLORIDE * WR 09792

THR: A poison by ingestion and intraperitoneal routes. When heated to decomposition it emits very toxic HCl, F^- and NO_x.

FUEL OIL   HR: 2
                         NIOSH: LS 8950000

PROP: A petroleum fraction consisting of a complex mixture of hydrocarbons (C_9 and higher) boiling over the range 325 to 650 degrees F. Brown, slightly viscous, liquid; flash p: 100°F; d: <1; autoign temp: 494°F.

SYNS: DISTILLATE * GAS OIL * HOME HEAT-ING OIL NO. 2

THR: See also diesel oil; mineral oils; kerosene. To fight fire, use CO_2, dry chemical. When heated to decomposition it emits acrid smoke and fumes. For further information see Fuel Oils, Vol. 1, No. 7 of *DPIM Report*.

FUGU POISON   HR: 3
CAS: 4368-28-9   NIOSH: IO 1450000
mf: $C_{11}H_{17}N_3O_8$ mw: 319.31

SYNS: SPHEROIDINE * TARICHATOXIN * TETRODONTOXIN * TETRODOTOXIN * TETRODOXIN

THR: A poison by intraperitoneal, subcutaneous, intravenous and other routes. When heated to decomposition it emits toxic NO_x. For further information see Tetradoxin, Vol. 1, No. 5 in *DPIM Report*.

FUJITHION   HR: 3
CAS: 3309-87-3   NIOSH: TE 8225000
mf: $C_8H_{10}ClO_3PS$ mw: 252.66

SYNS: S-(P-CHLOROPHENYL)-O,O-DIMETHYL PHOSPHOTHIOATE * O,O-DIMETHYL-S-P-CHLOROPHENYL PHOSPHOROTHIOATE

THR: Poison by ingestion. Moderately toxic by skin contact. When heated to decomposition it emits very toxic fumes of Cl^-, PO_x and SO_x.

FULMINATES

THR: Variable toxicity. Dangerous fire hazard. Keep away from heat and open flame! Severe explosion hazard when shocked or exposed to heat or flame (see also Explosives, High). Storage and Handling: The fulminates are a group of explosives which are very sensitive to heat, impact, and friction when dry. They should be kept moist till ready for use. If compressed beyond 25,000 psi they become what is known as "dead-pressed," i.e., not capable of being exploded by flame. Fulminates are subject to deterioration when stored in hot climates. They decompose completely and violently when detonated. They can be ignited with a flame or "spit," with a fuse, or with an electrically heated wire. They are widely used as initiators or primers for bringing about the detonation of high explosives or the ignition of powder. They are commonly used in combination with substances which provide a more prolonged blow and a bigger flame than fulminates alone. In the reinforced type of detonator, fulminates are made more effective by the addition of a more sensitive and powerful high explosive such as tetryl. This material is generally used in the manufacture of caps and detonators for initiating explosions for military, industrial, and sporting purposes. All precautions required for protection of magazines apply to storage of these materials. They should not be handled when frozen. Wet fulminate of mercury or wet floor coverings containing small quantities of fulminates may be burned on windrows of flammable material. Nonexplosive products are formed by neutralizing fulminates with cold sodium thiosulfate. All floors, tables and walls where the dry fulminates have been used should be washed with this solution.

In the manufacture of mercury fulminate, the fumes given off are toxic and flammable. Care is required to prevent fulminate dust from being carried off in the exhaust system: deposits thus made have caused explosions. Careful attention should be given to cleanliness as foreign or gritty materials in the product may cause an unexpected explosion. The floors on which fulminates are used, should be covered with 16-inch cloth inserted rubber packing or its equal. All cracks and crevices should be covered. The walls of these rooms should be covered with glazed, water-proof material. Frequent washing with neutralizing solution is necessary. In manufacture, the fulminate is dried on muslin squares on a drying table. Drying tables may be heated with hot water or the dry house may be heated with an air blower system to between 50 and 60°. Primer caps and detonators loaded with fulminate of mercury are less sensitive than the dry bulk material but must be handled with great care. Fires involving these assemblies should be treated the same as for the bulk material. They will explode as soon as fire reaches them. Stocks in an assembly or loading room should be kept as small as possible. Examples of fulminates commonly used in the explosive industry are mercury fulminate, copper fulminate and silver fulminate.

FULVINE   HR: 3

CAS: 6029-87-4   NIOSH: RC 1300000
mf: $C_{16}H_{23}NO_5$ mw: 309.40

SYN: CRISPATINE

THR: A poison by intraperitoneal and other routes. An experimental teratogen. Mutagenic data. When heated to decomposition it emits toxic fumes of NO_x.

FUMARINE   HR: 3

CAS: 130-86-9   NIOSH: VS 2800000
mf: $C_{20}H_{19}NO_5$ mw: 353.40

SYNS: CORYDININE * PROTOPINE

THR: A poison by ingestion and intraperitoneal routes. When heated to decomposition it emits toxic fumes of NO_x.

FUMARYL CHLORIDE   HR: 3

CAS: 627-63-4   NIOSH: LT 2800000
mf: $C_4H_2Cl_2O_2$ mw: 152.96

PROP: Clear, straw-colored liquid, mp: 160°, d: 1.408 @ 20°/4°.

SYNS: CHLORURE DE FUMARYLE (FRENCH) * DICHLORID KYSELINY FUMAROVE (CZECH) * FUMARYLCHLORID (CZECH) * FUMARYL CHLORIDE (DOT)

THR: A poison irritant to skin, eyes and mucous membranes; also by inhalation route. See also HCl. Moderately toxic by ingestion and skin routes. A corrosive agent. Dangerous; when heated to decomposition it emits highly toxic fumes of phosgene and HCl; will react with water or steam to produce toxic and corrosive fumes.

FUNICOLOSIN   HR: 3

CAS: 11055-06-4   NIOSH: LT 5270000
mf: $C_{27}H_{41}NO$ mw: 395.69

THR: A poison by ingestion and intraperitoneal routes. When heated to decomposition it emits toxic fumes such as NO_x.

FURADAN   HR: 3
CAS: 1563-66-2   NIOSH: FB 9450000
mf: $C_{12}H_{15}NO_3$ mw: 221.28

PROP: An odorless, white crystalline solid, sltly water-sol, mp: 105°-152°, d: 1.180 @ 20°/20°, vap press: 2×10^{-5} mm @ 33°.

SYNS: 2,3-DIHYDRO-2,2-DIMETHYLBENZOFURA-NYL-7-N-METHYLCARBAMATE * 2,3-DIHYDRO-2,2-DIMETHYL-7-BENZOFURANYL METHYLCARBA-MATE * ENT 27,164 * 2,2-DIMETHYL-2,2-DIHYDROBENZOFURANYL-7-N-METHYLCAR-BAMATE

THR: Poison by inhalation, ingestion, skin contact, and possibly other routes. See also carbamates. When heated to decomposition it emits toxic fumes of NO_x.

2-FURALDEHYDE   HR: 3
CAS: 98-01-1   NIOSH: LT 7000000
mf: $C_5H_4O_2$ mw: 96.09

PROP: Colorless-yellowish liquid, almond-like odor, bp: 161.7° @ 764 mm, lel = 2.1%, uel = 19.3%, flash p: 140°F (CC), d: 1.161 @ 20°/20°, autoign temp: 600°F, vap d: 3.31.

SYNS: ARTIFICIAL ANT OIL * FURAL * 2-FURANALDEHYDE * 2-FURANCARBONAL * 2-FURANCARBOXALDEHYDE * FURFURAL * FURFURAL (DOT) * FURFURALDEHYDE * FURFURALE (ITALIAN) * 2-FURIL-METANALE (ITALIAN) * 2-FURYL-METHANAL * NCI-C56177 * PYROMUCIC ALDEHYDE

OSHA PEL: TWA 5 ppm (skin)

THR: A poison by inhalation in humans. A severe eye irritant. An acute poison by ingestion, inhalation and intraperitoneal routes. Mutagenic data. Moderate toxicity by dermal route to skin, eyes and mucous membranes. The liquid is dangerous to the eyes. The vapor is irritating to mucous membranes and is a central nervous system poison. However, its low volatility reduces its toxicity effect. Furfural which has been ingested has produced cirrhosis of the liver in rats. In industry there is a tendency to minimize the danger of acute effects resulting from exposure to it. This is true, particularly, because of its low volatility. Little is known concerning the possibility of chronic effects, such as nervous disturbances, following prolonged or severe exposure to this material. Flammable when exposed to heat or flame; can react with oxidizing materials. Moderately explosive when exposed to heat or flame or by chemical reaction. An exothermic resinification of almost explosive violence can occur upon contact with strong mineral acids or alkalies. Moderately dangerous. Keep away from heat and open flames. To fight fire, use alcohol foam, CO_2, dry chemical. For further information see Furfural, Vol. 1, No. 2 of *DPIM Report*.

FURAN   HR: 3
CAS: 110-00-9   NIOSH: LT 8524000
mf: C_4H_4O mw: 68.08

PROP: Water white liquid, mp: −85.65°, bp: 31.36°, lel = 2.3%, uel = 14.3%, flash p: −32°F, d: 0.937 @ 20°/4°, vap d: 2.35.

SYNS: DIVINYLENE OXIDE * FURFURAN * NCI-C56202 * OXACYCLOPENTADIENE * OXOLE * TETROLE

THR: A poison by inhalation and intraperitoneal routes. A narcotic. Absorbed also by dermal route. Probably at least moderately toxic by ingestion and dermal routes. The exposure concentration limit of 10 ppm together with its low boiling point requires that adequate ventilation be provided in areas handling this chemical. Contact with liquid must be avoided since this chemical can be absorbed through the skin. Thorough washing with soap and water followed by prolonged rinsing should be done immediately after accidental contact. Highly dangerous fire hazard when exposed to heat or flame; can react with oxidizing materials. Unstabilized, it may form unstable peroxides on exposure to air and should always be tested before distillation. Washing with an aqueous solution of ferrous sulfate slightly acidified with sodium bisulfate will remove these peroxides. Confirm by test. Contact with acids can initiate a violent exothermic reaction. Moderately explosive when exposed to flame. The low boiling point of this material makes it easy to obtain explosive concentrations of the vapor in inadequately ventilated areas. Highly dangerous upon exposure to heat or flame; can react vigorously with oxidizing materials. To fight fire, use CO_2, dry chemical.

FURFURAMIDE   HR: 3
CAS: 494-47-3   NIOSH: LU 1925000
mf: $C_{15}H_{12}N_2O_3$ mw: 268.29

PROP: Needles from alcohol; mp: 117-21°; bp: 250° decomp; insol in water; decomp in acid; sol in alc and ether.

SYNS:
2-(BIS(FURFURYLIDENAMINO))METHYLFURAN
* HYDROFURAMIDE

THR: A poison irritant to skin, eyes and mucous membranes and by ingestion. A component of fungicides. Causes intense pulmonary irritation and reported to cause liver and kidney damage. See also amines and amides. When heated to decomposition it emits toxic fumes of NO_x.

FURFURYL ALCOHOL   HR: 3
CAS: 98-00-0   NIOSH: LU 9100000
mf: $C_5H_6O_2$ mw: 98.11

PROP: Clear, colorless, mobile liquid, mp: −31°, lel = 1.8%, uel = 16.3%, both between 72°-122°, bp: 171° @ 750 mm, flash p: 167°F (OC), d: 1.129 @ 20°/4°, autoign temp: 915°F, vap press: 1 mm @ 31.8°, vap d: 3.37.

SYNS: FURFURAL ALCOHOL * FURYL ALCO-HOL * 2-FURANCARBINOL * 2-FURAN-METHANOL * 2-FURFURYLALKOHOL (CZECH) * 2-FURYLCARBINOL * ALPHA-FURYLCAR-BINOL * 2-HYDROXYMETHYLFURAN * NCI-c56224

OSHA PEL: TWA 50 ppm

THR: A poison by ingestion and inhalation routes. Moderate toxicity by intraperitoneal and intravenous routes. Flammable when exposed to heat; can react with oxidizing materials. Moderately explosive when exposed to heat or flame. Incompatible with acids; H_2O_2. To fight fire, use alcohol foam, CO_2, dry chemical.

2-(2-FURYL)-3-(5-NITRO-2-FURYL) ACRYLAMIDE   HR: 3
CAS: 3688-53-7   NIOSH: AS 3500000
mf: $C_{11}H_8N_2O_5$ mw: 248.21

SYNS: FURYLAMIDE * FURYLFURAMIDE

THR: Mutagenic data. An experimental carcinogen and neoplastigen. Moderate acute toxicity by ingestion. When heated to decomposition it emits toxic fumes of NO_x.

FUSARENONE X   HR: 3
CAS: 23255-6-9-8   NIOSH: YD 0160000
mf: $C_{17}H_{22}O_8$ mw: 354.39

PROP: Isolated from the culture filtrate of *Fusarium Nivale*

SYNS: 4-ACETYLOXY-12,13-EPOXY-3,7,15-TRIHYDROXY-(3-alpha,4-beta,7-beta)-TRICHO-THEC-9-EN-8-ONE * NIVALENOL-4-O-ACETATE * 3,7,15-TRIHYDROXYSCIRP-4-ACETOXY-9-EN-8-ONE * 3,7,15-TRIHYDROXY-4-ACETOXY-8-OXO-12,13-EPOXY-DELTA(SUP 9)-TRICHOTHE-CENE

THR: A poison by ingestion, intraperitoneal, subcutaneous, intravenous, and other routes. Mutagenic data. An experimental teratogen and carcinogen. When heated to decomposition it emits acrid smoke and fumes.

FUSARIOTOXIN T 2   HR: 3
CAS: 21259-20-1   NIOSH: YD 0100000
mf: $C_{24}H_{34}O_9$ mw: 466.58

PROP: A strain of *F. tricinctum* isolated from infected corn

SYNS: 4,15-DIACETOXY-8-(3-METHYLBUTYRY-LOXY)-12,13-EPOXY-delta-9-TRICHOTHECEN-3-OL * 3-HYDROXY-4,15-DIACETOXY-8-(3-METHYL-BUTYRYLOXY)-12,13-EPOXY-delta (SUP 9)-TRI-CHOTHECENE * T-2 MYCOTOXIN

THR: A poison by ingestion and intravenous routes. An experimental teratogen. When heated to decomposition it emits acrid smoke and fumes.

FUSEL OIL   HR: 3
CAS: 123-51-3   NIOSH: LV 5600000

PROP: Composition of grain fusel oil is methanol, ethanol, acetaldehyde and other alcohols

SYNS: FUSELOEL (GERMAN) * HUILE DE FUSEL (FRENCH)

THR: Mutagenic data. Suspected of containing carcinogens. See also individual components. When heated to decomposition it emits acrid smoke and fumes.

G

GADOLINIUM   HR: 3
CAS: 7440-54-2   NIOSH: LW 3850000
af: Gd   aw: 157.25

PROP: A yellow-white malleable and ductile
metallic element, a rare earth, stable in dry air,
reacts slowly with H_2O, mp: 1312°, bp: 3233°,
d: 7.898 @ 25°.

THR: An experimental tumorigen. It may act
as an anticoagulant. It can react violently with
air; halogens. See also rare earths.

GADOLINIUM CHLORIDE   HR: 3
CAS: 10138-52-0   NIOSH: LW 4025000
mf: Cl_3Gd mw: 263.60

PROP: White monoclinic crystals; d: 4.52 @
0°; mp: approx 609°; sol in H_2O.

SYN: GADOLINIUM TRICHLORIDE

THR: A poison via intraperitoneal route. See
also chlorides and gadolinium. A skin and eye
irritant. See also rare earths. When heated to
decomposition it emits very toxic fumes of Cl^-.

GADOLINIUM CITRATE   HR: 3
CAS: 3088-53-7   NIOSH: LW 4200000

THR: A poison via intraperitoneal route. See
also rare earths. When heated to decomposition
it emits acrid smoke and fumes.

GADOLINIUM (III) NITRATE
(1:3)   HR: 3
CAS: 10168-81-7   NIOSH: LW 4550000
mf: $N_3O_9 \cdot Gd$ mw: 343.28

THR: A poison by intraperitoneal route. Moderate toxicity by ingestion. See also rare earths,
nitrates. When heated to decomposition it emits
toxic fumes of NO_x.

GALACTOSE   HR: 3
CAS: 59-23-4   NIOSH: LW 5490000
mf: $C_6H_{12}O_6$ mw: 180.18

PROP: α form prisms from water or ethanol;
mp: 167°. Freely sol in hot H_2O, sol in pyridine,
sltly sol in alc. β form crystals; mp: 167°.

SYNS: CEREBROSE * BRAIN SUGAR

THR: An experimental teratogen. When heated
to decomposition it emits acrid smoke and
fumes.

GALLIC ACID   HR: 3
CAS: 149-91-7   NIOSH: LW 7525000
mf: $C_7H_6O_5$ mw: 170.13

PROP: White- to pale fawn-colored odorless
crystals, somewhat water-sol, d: 1.694, mp:
225°-250° (decomp), $-H_2O$ @ 100°-120°.

SYN: 3,4,5-TRIHYDROXYBENZOIC ACID

THR: A poison by intravenous route. Mildly
toxic by ingestion. When heated to decomposition it emits acrid smoke and fumes. For further
information see Vol. 3, No. 4 of *DPIM Report*.

GALLIUM (3+) CHLORIDE   HR: 3
CAS: 13450-90-3   NIOSH: LW 9100000
mf: $GaCl_3$ mw: 176.07

PROP: Colorless needles; mp: 78°

THR: A poison by inhalation, subcutaneous,
intravenous, and intraperitoneal routes. When
heated to decomposition it emits very toxic
fumes of Cl^-.

GALLIUM CITRATE and COM-
POUNDS   HR: 3
CAS: 27905-02-8   NIOSH: GE 7675000
mf: $C_6H_5O_7 \cdot Ga$ mw: 258.83

SYN: CITRIC ACID, GALLIUM SALT (1:1)

THR: A poison by subcutaneous route. Preliminary investigations were done with the oxide,
tartrate, benzoate, and anthranilate, which were
used by some investigators in the experimental
treatment of syphilis. Amounts up to 15 mg/
kg of body weight were injected intravenously
and were tolerated without harm by laboratory
animals. Larger doses produce hemorrhagic nephritis. In the case of gallium lactate, work
done at the Naval Medical Research Institute
showed that intravenous injections of about 40
mg of gallium per kg of body weight in rats
or rabbits was lethal. Metallic gallium as well
as the nitrate produced no skin injury and subcutaneous injection of relatively large amounts
could be tolerated both by rabbits and rats without evidence of injury. It has, however, been
demonstrated that gallium remains in the tissues
for long periods of time following intramuscular
injection of soluble gallium salts. Tissue distribution experiments indicate that it behaves like

bismuth and mercury in that one respect. When heated to decomposition it emits acrid smoke and fumes.

GALLIUM (III) NITRATE (1:3)   HR: 3
CAS: 13494-90-1   NIOSH: LW 9625000
mf: $N_3O_9 \cdot Ga$ mw: 255.75

PROP: White, deliquescent cryst, mp: decomp @ 110°, bp: converts to Ga_2O_3 @ 200°.

THR: A poison by intraperitoneal and subcutaneous routes. In humans it causes central nervous system problems, gastrointestinal tract effects and damage to blood. See also gallium. When heated to decomposition it emits toxic fumes of NO_x.

GASOLINE (from 50-100 octane)   HR: 3
CAS: 8006-61-9   NIOSH: LX 3300000

PROP: Clear, aromatic, volatile liquid, a mixture of aliphatic HC, flash p: −50°F, d: <1.0, vap d: 3.0-4.0, ulc: 95-100, lel = 1.3%, uel = 6.0%, autoign temp: 536°-853°F, bp: initially 39°; after 10% distilled = 60°; after 50% = 110°; after 90% = 170°; final bp: 204°. Insol in H_2O, freely sol in absolute alc, ether, chloroform, benzene.

SYN: PETROL

THR: A poison by inhalation. Repeated or prolonged dermal exposure causes dermatitis. Can cause blistering of skin. Inhalation and by ingestion routes causes central nervous system depression. Pulmonary aspiration can cause severe pneumonitis. Some addiction has been reported from inhalation of fumes. Even brief inhalations of high concentrations can cause a fatal pulmonary edema. It can cause hyperemia of the conjunctiva and other disturbances of the eyes. The vapors are considered to be moderately poisonous. If its concentration in air is sufficiently high to reduce the oxygen content below that needed to maintain life, it acts as a simple asphyxiant. Gasoline is a common air contaminant. See also mineral oils. Dangerous fire and explosion hazard when exposed to heat or flame; can react vigorously with oxidizing materials. To fight fire, use foam, CO_2, dry chemical.

GASOLINE ENGINE EXHAUST "TAR"   HR: 3
NIOSH: LX 3350000

THR: An experimental carcinogen.

GELSEMINE   HR: 3
CAS: 509-15-9   NIOSH: LX 9100000
mf: $C_{20}H_{22}N_2O_2$ mw: 322.44

PROP: An alkaloid, mp: 178°. Sltly sol in H_2O; sol in alc, benzene, chloroform, ether, acetone, dilute acids.

SYN: GELSEMIN

THR: A poison by subcutaneous and intraperitoneal routes. A poisonous alkaloid. A central nervous system stimulant, can cause muscular weakness, respiratory arrest. When heated to decomposition it emits toxic fumes of NO_x.

GENTAMICIN   HR: 3
CAS: 1403-66-3   NIOSH: LY 2450000

SYNS: GARAMYCIN * GENTAMYCIN * GENTAMYCIN-CREME (GERMAN) * UROMYCINE

THR: A poison by intravenous, intramuscular, and subcutaneous routes. Affects the peripheral nervous system by intravenous route. When heated to decomposition it emits acrid smoke and fumes.

GENTAMYCIN SULFATE   HR: 3
CAS: 1405-41-0   NIOSH: LY 2625000

SYN: GM SULFATE

THR: A poison by intravenous and intramuscular routes. Moderate intraperitoneal toxicity. When heated to decomposition it emits very toxic fumes of SO_x.

GERANIOL   HR: 3
CAS: 106-24-0   NIOSH: RG 5830000
mf: $C_{10}H_{18}O$ mw: 154.28

PROP: Colorless to pale yellow, liquid oil, pleasant geranium odor, d: 0.870-0.890 @ 15°, mp: 15°, bp: 230°.

SYNS: 2,6-DIMETHYL-TRANS-2,6-OCTADIEN-8-OL * 3,7-DIMETHYL-TRANS-2,6-OCTADIEN-1-OL * 3,7-DIMETHYL-(E)-2,6-OCTADIEN-1-OL * GERANIOL ALCOHOL * GERANYL ALCOHOL * GUANIOL * LEMONOL

THR: Poison by intravenous route. Moderately toxic by ingestion and intramuscular routes. When heated to decomposition it emits acrid smoke and fumes.

GERMANE   HR: 3
CAS: 7782-65-2   NIOSH: LY 4900000

DOT: 2192

mf: GeH_4 mw: 76.63

PROP: Colorless gas; mp: $-165°$; bp: $-90°$; d: 1.523 @ $-142°/4°$; sltly sol in hot HCl, decomp in nitric acid.

SYNS: GERMANIUM TETRAHYDRIDE * GERMA-NIUM HYDRIDE * MONOGERMANE

DOT Classification: Label: Poison Gas and Flammable Gas

THR: A poison by inhalation. A hemolytic gas. See hydrides and germanium compounds and germanium. Self-ignites in air. Incompatible with Br_2.

GERMANIC OXIDE (CRYSTAL-LINE)   HR: 3
CAS: 1310-53-8   NIOSH: LY 5205000

THR: A poison by intraperitoneal route. When heated to decomposition it emits acrid smoke and fumes.

GERMANIUM BROMIDE   HR: 3
CAS: 13450-92-5   NIOSH: LY 5220000
mf: Br_4Ge mw: 392.23

PROP: Gray-white crystals, mp: 26.1°, bp: 186.5°, d: 3.232 @ 29°/29°.

SYN: GERMANIUM TETRABROMIDE

THR: A poison by intravenous route. See also germanium compounds. When heated to decomposition it emits very toxic fumes of bromides.

GERMANIUM CHLORIDE   HR: 3
CAS: 10038-98-9   NIOSH: LY 5225000
mf: Cl_4Ge mw: 214.39

PROP: Colorless mobil liquid. Fumes in air. Peculiar acidic odor but can be distinguished from that of concentrated HCl. Volatile @ room temp. Sol in benzene, ether, other organic solvents, mp: $-49.5°$, bp: 83.1°, d: 1.879 @ 20°/20°.

SYN: GERMANIUM TETRACHLORIDE

THR: A poison by intravenous route. Irritating to skin, eyes and mucous membranes of respiratory tract. See also germanium compounds. Dangerous; when heated to decomposition emits toxic fumes of chlorides; will react violently with water or steam to produce toxic and corrosive fumes.

GERMANIUM COMPOUNDS   HR: 2
THR: Little is known about the toxicity of organic germanium compounds but they may resemble other organometals in having higher toxicity than inorganic forms. Germanium compounds are considered to be of a low order of toxicity, but rare instances of poisoning have been reported in the literature. Interest is high in this material because of its close chemical relationship to arsenic. It has been found that the dioxide stimulates the generation of red blood cells, but it is believed to be relatively nontoxic. When germanium is given in sublethal amounts, it causes a pronounced tolerance to be exhibited. Germanium compounds are considered much less toxic than the corresponding lead and tin compounds. Buffered germanium dioxides in solution have been found to be nonirritating to the skin. Germanium hydride is a hemolytic gas and has been shown to have toxic properties @ a concentration of 100 ppm. It can cause death @ a concentration of 150 ppm.

GIBBERELLIC ACID   HR: 3
CAS: 77-06-5   NIOSH: LY 8990000
mf: $C_{19}H_{22}O_6$ mw: 346.41

PROP: A plant growth-promoting hormone, crystals, sltly sol in water, ether; sol in methanol, ethanol, acetone, aqueous solns of sodium bicarbonate and sodium acetate. Mod sol in ethyl acetate, mp: 233°-235°.

SYNS: FLORALTONE * GIBBERELLIN * NCI-C55823 * 2,4A,7-TRIHYDROXY-1-METHYL-8-METHYLENEGIBB-3-ENE-1,10-CARBOXYLICACID 1-4-LACTONE

THR: An experimental tumorigen and carcinogen. Mutagenic data. A food additive permitted in food for human consumption. When heated to decomposition it emits acrid smoke and fumes.

GILSONITE   HR: 2
CAS: 12002-43-6   NIOSH: LY 9148000

PROP: A black solid hydrocarbon mineral formed from petroleum millions of years ago by geologic processes.

SYNS: NCI-C55185 * UINTAITE

THR: Moderately irritating to skin, eyes and mucous membranes. An allergen. Has been known to cause photosensitization of skin. Flammable when exposed to heat or open flame.

Moderately hazardous; when heated to decomposition it emits toxic fumes. To fight fire, use water, foam, dry chemical and CO_2.

GITOXIN HR: 3
CAS: 4562-36-1   NIOSH: LZ 0525000
mf: $C_{41}H_{64}O_{14}$ mw: 781.05

PROP: Stout prisms from chloroform and methanol. Decomp @ 285° (rapid heating). Almost insol in chloroform, ethyl acetate and acetone. Dissolves in a mixture of chloroform and alc or pyridine or dil alc.

SYNS: ANHYDROGITALIN * BIGITALIN * PSEUDODIGITOXIN

THR: A poison by ingestion and intravenous routes. When heated to decomposition it emits acrid smoke and fumes.

GITOXIN PENTA ACETATE HR: 3
CAS: 7242-04-8   NIOSH: LZ 1225000
mf: $C_{51}H_{74}O_{19}$ mw: 991.25

PROP: Rhombic crystals; mp: 151°-155°.

SYN: PENTAACETYLGITOXIN

THR: A poison by ingestion and intravenous routes. When heated to decomposition it emits acrid smoke and fumes.

alpha-d-GLUCOCHLORALOSE HR: 3
CAS: 16879-93-3   NIOSH: FM 9450000
mf: $C_8H_{11}Cl_3O_6$ mw: 309.54

SYNS: ANHYDROGLUCOCHLORAL * MONO-TRICHLOR-AETHYLIDEN-alpha-GLUCOSE (GERMAN) * 1,2-O-(2,2,2-TRICHLOROETHYLIDENE)-alpha-d-GLUCOFURANOSE

THR: An experimental tumorigen. A poison by intraperitoneal and ingestion. When heated to decomposition it emits toxic fumes of Cl^-.

GLUCO PROSCILLARIDIN A HR: 3
CAS: 124-99-2   NIOSH: EI 3325000
mf: $C_{36}H_{52}O_{13}$ mw: 692.88

SYNS: 14-beta-HYDROXY-3-beta-SCIL-LOBIOSIDOBUFA-4,20,22-TRIENOLIDE * SCILLAGLYKOSID A (GERMAN) * 3-beta-SCILLOBIOSIDO-14-beta-HYDROXY-delta-4,20, 22-BUFATRIENOLID(GERMAN) * TRANSVAALIN

THR: A poison by intravenous route. When heated to decomposition it emits acrid smoke and fumes.

d-GLUCOSE HR: 1
CAS: 50-99-7   NIOSH: LZ 6600000
mf: $C_6H_{12}O_6$ mw: 180.18

PROP: Colorless crystals or white crystalline or granular powder, odorless, sol in water, sltly sol in alcohol, d: 1.544, mp: 146°. α form monohydrate, crystals from water, mp: 83°. α form anhydrous crystals from hot ethanol or water; mp: 146°. Very sparingly sol in absolute alc, ether, acetone; sol in hot glacial acetic acid, pyridine, aniline. β form crystals from hot H_2O + ethanol, from dil acetic acid or from pyridine; mp: 148°-155°.

SYNS: ANHYDROUS DEXTROSE * CORN SUGAR * DEXTROSE * BLOOD SUGAR * DEXTROSOL * GLUCOLIN * GLUCOSE * D-GLUCOSE, AN-HYDROUS * GLUCOSE LIQUID * GRAPE SUGAR * SIRUP

THR: Very low acute toxicity. Substance migrating to food from packaging materials. Reacts violently with (Na_2O_2 + KNO_3). When heated to decomposition it emits acrid smoke and fumes. For further information see Vol. 2, No. 1 of *DPIM Report*.

l-GLUTAMIC ACID HR: 3
CAS: 56-86-0   NIOSH: LZ 9700000
mf: $C_5H_9NO_4$ mw: 147.15

PROP: A nonessential amino acid present in all complete proteins. Crystals, mp (*dl* form): 194°, d (*dl* form): 1.4601 @ 20°/4°, mp (1 form): 224°-225°, d (1 form): 1.538 @ 20°/4°.

SYNS: L-2-AMINOGLUTARIC ACID * alpha-AMINOGLUTARIC ACID * 2-AMINOPENTANEDI-OIC ACID * 1-AMINOPROPANE-1,3-DICARBOX-YLIC ACID * GLUTACID * alpha-GLUTAMIC ACID * L-GLUTAMINIC ACID * GLUTAMINOL * GLUTATON

THR: A poison to humans by intravenous route. A general-purpose food additive. When heated to decomposition it emits toxic fumes of NO_x.

GLYCERINE HR: 1
CAS: 56-81-7   NIOSH: MA 8050000
mf: $C_3H_8O_3$ mw: 92.11

PROP: Colorless or pale yellow liquid, odorless, syrupy, sweet and warm taste, mp: 17.9 (solidifies @ a much lower temp), bp: 290°, ulc: 10-20, flash p: 320°F, d: 1.260 @ 20°/4°, autoign temp: 698°F, vap press: 0.0025 mm @ 50°, vap d: 3.17.

SYNS: GLYCERIN, ANHYDROUS * GLYCERITOL * GLYCYL ALCOHOL * 1,2,3-PROPANETRIOL * SYNTHETIC GLYCERIN * 90 TECHNICAL GLYCERINE * 1,2,3-TRIHYDROXYPROPANE

THR: Low toxicity by ingestion, subcutaneous and intravenous routes. In the form of mist it is an inhalation irritant. A general-purpose food additive. It migrates to food from packaging materials. A skin and eye irritant. Combustible when exposed to heat, flame or powerful oxidizers. Can react violently with acetic anhydride; (aniline + nitrobenzene); $Ca(OCl)_2$; CrO_3; Cr_2O_3; (F_2 + PbO); ($HClO_4$ + PbO); $KMnO_4$; K_2O_2; $AgClO_4$; Na_2O_2; NaH. To fight fire, use alcohol foam, CO_2, dry chemical. For further information see glycerol, Vol. 3, No. 4 of *DPIM Report*.

GLYCEROL (TRI(CHLOROMETHYL))-ETHER   HR: 3
CAS: 38571-73-2   NIOSH: UA 1850000
mf: $C_6H_{11}Cl_3O_3$ mw: 237.52

SYN: TRIS-1,2,3-(CHLOROMETHOXY)PROPANE

THR: An experimental carcinogen and neoplastigen. When heated to decomposition it emits toxic fumes of Cl^-.

GLYCERYL-o-TOLYL ETHER   HR: 3
CAS: 59-47-2   NIOSH: TZ 1225000
mf: $C_{10}H_{14}O_3$ mw: 182.24

SYNS: O-CRESOL GLYCERYL ETHER * CRESOSSIDIOLO * CRESOSSIPROPANDIOLO * CRESOXYDIOL * CRESOXYPROPANEDIOL * O-CRESYL ALPHA-GLYCERYL ETHER * ALPHA,BETA-DIHYDROXY-GAMMA-(2-METHYLPHENOXY)PROPANE * 1,2-DIHYDROXY-3-(2-METHYLPHENOXY)PROPANE * 3-(2-METHYLPHENOXY)-1,2-PROPANEDIOL * 3-O-TOLOXY-1,2-PROPANEDIOL * ALPHA-(O-TOLYL)GLYCERYL ETHER * 1-ORTHO-TOLYLGLYCEROL ETHER * 3-(O-TOLYLOXY)PROPANE-1,2-DIOL

THR: A poison by intraperitoneal, intravenous, and subcutaneous routes. Moderately toxic by ingestion. See also ethers. When heated to decomposition it emits acrid smoke and fumes.

GLYCIDALDEHYDE   HR: 3
CAS: 765-34-4   NIOSH: MB 3150000
mf: $C_3H_4O_2$ mw: 72.1

PROP: Colorless liquid, bp: 113°, d: 1.1403 @ 20°/4°.

SYNS: EPIHYDRINALDEHYDE * EPIHYDRINE ALDEHYDE * 2,3-EPOXY-1-PROPANAL * 2,3-EPOXYPROPIONALDEHYDE * GLYCIDAL * OXIRANE-CARBOXALDEHYDE

THR: An experimental carcinogen and neoplastigen. Mutagenic data. Powerful skin sensitizer and irritant. Irritating also to eyes and mucous membranes by inhalation. When heated to decomposition it emits acrid smoke and fumes.

GLYCINE NITROGEN MUSTARD   HR: 3
CAS: 2619-97-8   NIOSH: MB 8480000
mf: $C_6H_{11}Cl_2NO_2 \cdot ClH$ mw: 236.54

SYNS: N,N-BIS(beta-CHLOROETHYL)GLYCINE HYDROCHLORIDE * GLYCINE MUSTARD * NSC 17661

THR: Very poisonous by intraperitoneal, intravenous, and intracerebral routes. When heated to decomposition it emits very toxic fumes of Cl^- and NO_x.

GLYCOPYRRONIUM BROMIDE   HR: 3
CAS: 596-51-0   NIOSH: UY 4455000
mf: $C_{19}H_{28}NO_3 \cdot Br$ mw: 398.39

SYNS: 1,1-DIMETHYL-3-HYDROXYPYRROLIDINIUM BROMIDE-alpha-CYCLOPENTYLMANDELATE * 3-HYDROXY-1,1-DIMETHYLPYRROLIDINIUM BROMIDE-alpha-CYCLOPENTYLMANDELATE * 1-METHYL-3-PYRROLIDINYL alpha-PHENYLCYCLOPENTANEGLYCOLATE METHOBROMIDE * beta-1-METHYL-3-PYRROLIDYL-alpha-CYCLOPENTYLMANDELATE METHOBROMIDE

THR: A poison by intravenous route. Moderate toxicity by ingestion. See also bromides. When heated to decomposition it emits very toxic fumes of NO_x and Br^-.

GLYOXAL   HR: 3
CAS: 107-22-2   NIOSH: MD 2625000
mf: $C_2H_2O_2$ mw: 58.04

PROP: Yellow crystals or light yellow liquid, mp: 15°, bp: 50.4°, d: 1.14 @ 20°/4°. Sol in anhyd solvents. Very sol absolute alc and ether.

SYNS: BIFORMAL * DIFORMAL * ETHANEDIAL * 1,2-ETHANEDIONE * GLYOXYLALDEHYDE * OXAL * OXALALDEHYDE

THR: A poison by intraperitoneal route. Moderate toxicity by ingestion. Mutagenic data. A mild to severe skin and eye irritant. Violent reactions with water; air; chlorosulfonic acid; ethylene imine; HNO_3; oleum; NaOH. Can explode during manufacture.

GOLD   HR: 3
CAS: 7440-57-5   NIOSH: MD 5435000
af: Au   aw: 196.97

PROP: Cubic, yellow, ductile, metallic crystals, mp: 1064.76°, bp: 2700°, d: 19.3 (liquid); 17.0 @ 1063°, vap press: 1 mm @ 1869°.

SYNS: BURNISH GOLD * COLLOIDAL GOLD * GOLD FLAKE * GOLD LEAF * GOLD POWDER * MAGNESIUM GOLD PURPLE * SHELL GOLD

THR: An experimental tumorigen. Can form explosive compounds with NH_3; (NH_4OH, aqua regia); H_2O_2. Incompatible with aqua regia; mixtures containing chlorides, bromides, or iodides if they can generate nascent halogens; with many oxidizing mixtures, especially those containing halogens. Also with alkali cyanides; solutions of thiocyanates and double cyanides.

GOLD SODIUM THIOMALATE   HR: 3
CAS: 12244-57-4   NIOSH: MD 5435000
mf: $C_4H_3AuO_4S \cdot 2Na$ mw: 390.08

SYNS: ((1,2-DICARBOXYETHYL)THIO)GOLD DISODIUM SALT * (DIHYDROGEN MERCAPTOSUCCINATO)GOLD DISODIUM SALT * DISODIUM AUROTHIOMALATE * (MERCAPTOBUTANEDIOATO(1-))GOLD DISODIUM SALT * TAURE(O)DON

THR: A poison to humans. It affects the skin, blood picture, the central nervous system and is a systemic toxin. When heated to decomposition it emits very toxic Na_2O and SO_x. For further information see Vol. 2, No. 2 of *DPIM Report*.

GOLD SODIUM THIOSULFATE DIHYDRATE   HR: 3
CAS: 10210-36-3   NIOSH: MD 6300000
mf: $O_6S_4 \cdot Au \cdot 3Na \cdot 2H_2O$ mw: 526.22

SYNS: AURICIDINE * AUROLIN * AUROPEX * AUROPIN * AUROSAN * AUROTHION * SODIUM AUROTHIOSULPHATE DIHYDRATE

THR: An experimental poison in many animals. Causes gastrointestinal problems in humans by intramuscular routes. When heated to decomposition it emits very toxic SO_x and Na_2O.

GOSSYPOL   HR: 2
CAS: 303-45-7   NIOSH: DU 3100000

PROP: A polyphenolic yellow pigment isolated from cottonseed pigment glands
mf: $C_{30}H_{30}O_8$ mw: 518.60

PROP: Mp: 180° from ether; mp: 199° from chloroform, 214° from ligroin; very sltly sol in methanol, ethanol, ether, chloroform, DMF, freely sol (with slow decomp) in dilute solns of ammonia, sodium carbonate. Insol in H_2O.

SYNS: 2,2'-BIS(1,6,7-TRIHYDROXY-3-METHYL-5-ISOPROPYL-8-ALDEHYDONAPHTHALENE * 8-FORMYL-1,6,7-TRIHYDROXY-5-ISOPROPYL-3-METHYL-2,2'- BISNAPHTHALENE

THR: Moderately toxic by ingestion. May be irritating to gastrointestinal tract. In experimental animals large doses cause edema of lungs, shortness of breath, paralysis. When heated to decomposition it emits acrid smoke and fumes. For further information see Vol. 2, No. 2 of *DPIM Report*.

GRAMICIDIN   HR: 3
CAS: 1405-97-6   NIOSH: MD 8225000
mf: $C_{148}H_{210}N_{30}O_{26}$ mw: 2825.88

PROP: Spear-shaped or lenticular platelets; mp: 229°-230°; almost insol in H_2O. Sol in lower alc, acetic acid, pyridine. Practically insol in ether, hydrocarbons.

THR: VERY poisonous by intravenous, intraperitoneal, and parenteral routes. When heated to decomposition it emits very toxic fumes of NO_x.

GRAPEFRUIT OIL   HR: 3
CAS: 8016-20-4   NIOSH: RI 7199000

THR: An experimental tumorigen. A skin irritant.

GRISEOVIRIDIN   HR: 3
CAS: 53216-90-3   NIOSH: ME 4200000
mf: $C_{22}H_{27}N_3O_7S$ mw: 477.58

PROP: Polymorphic crystals decomp @ 158°-166°; 194°-200° or 230°-240° depending on the crystal modification. Sol in pyridine; mod sol in lower alcohols, very slightly sol in H_2O and nonpolar solvents.

THR: A poison by intraperitoneal, subcutaneous, and intravenous routes. When heated to decomposition it emits very toxic SO_x and NO_x.

GRISOFULVIN   HR: 3
CAS: 126-07-8   NIOSH: WG 9800000
mf: $C_{17}H_{17}ClO_6$ mw: 352.79

SYNS: GRISCOFULVIN * GRISEOFULVIN * (+)-GRISEOFULVIN * GRISEOFULVIN-FORTE * GRISEOFULVINUM * NSC 34533 * SPIROFULVIN * USAF SC-2

THR: A poison by intravenous and intraperitoneal routes. Mutagenic data. An experimental teratogen, neoplastigen, and carcinogen.

GUAIACOL   HR: 2
CAS: 90-05-1   NIOSH: SL 7525000
mf: $C_7H_8O_2$ mw: 124.15

PROP: Clear, pale yellow liquid or solid. Characteristic odor, darkens on exposure to air and light; d (crystals): 1.129; d (liq): about 1.112. Misc with alc, chloroform, ether, oils, glacial acetic acid; Sltly sol in petroleum ether; sol in NaOH soln, mp: 28°, bp: 202°-209°, flash p: 180°F (OC), d: 1.097 @ 25°/25°.

SYNS: GUAICOL * O-HYDROXYANISOLE * 2-HYDROXYANISOLE * 1-HYDROXY-2-METHOXYBENZENE * O-METHOXYPHENOL * 2-METHOXYPHENOL * PYROGUAIAC ACID * 2-METHOXY-4-PROPENYLPHENOL * PROPENYLGUAIACOL * 4-PROPENYLGUAIACOL

THR: Mutagenic data. Protect from light. Moderate toxicity by ingestion route. Ingestion produces burning in mouth and throat, gastrointestinal distress, tremors and collapse. See also phenol. Flammable when exposed to heat or flame; can react with oxidizing materials. To fight fire, use foam, CO_2, dry chemical. When heated to decomposition it emits acrid smoke and fumes.

GUANETHIDINE MONOSULFATE   HR: 3
CAS: 645-43-2   NIOSH: MF 3150000
mf: $C_{10}H_{22}N_4 \cdot H_2O_4S$ mw: 296.44

SYNS: N-(2-GUANIDINO ETHYL)HEPTAMETHYLENIMINE SULFATE * (2-(HEXAHYDRO-1(2H)-AZOCINYL)ETHYL) GUANIDINE HYDROGEN SULFATE * 2-(OCTAHYDRO-1-AZOCINYL)ETHYL GUANIDINE SULPHATE

THR: A poison by intraperitoneal route. An eye irritant in humans. When heated to decomposition it emits very toxic fumes of SO_x and NO_x.

GUANIDINE   HR: 3
CAS: 113-00-8   NIOSH: ME 7750000
mf: CH_5N_3 mw: 59.09

PROP: Hygroscopic colorless crystals; mp: $50 \pm °$; very sol in water; sol in alc. On heating to 160° it converts to melamine and NH_3. Keep well closed.

SYNS: AMINOFORMAMIDINE * AMINOMETHANAMIDINE * CARBAMAMIDINE * CARBAMIDINE * IMINOUREA

THR: A poison by intraperitoneal and subcutaneous routes. When heated to decomposition it emits toxic fumes of NO_x.

GYPSOGENIN   HR: 3
CAS: 639-14-5   NIOSH: RK 0178000
mf: $C_{30}H_{46}O_4$ mw: 470.76

PROP: Needles or leaflets from methanol, mp: 274°-276°.

SYNS: 3-beta-HYDROXY-23-OXO-OLEAN-12-EN-28-OIC ACID * ALBSAPOGENIN * GYPSOPHILASAPOGENIN * GYPSOPHILASAPONIN

THR: A poison by subcutaneous and intravenous routes. Moderate ingestion toxicity. When heated to decomposition it emits acrid smoke and fumes.

H

HAFNIUM   HR: 3
CAS: 7440-58-6   NIOSH: MG 4600000
DOT: 1326/2545
af: Hf   aw: 178.49

PROP: A silvery ductile, lustrous metal element, mp: 2227°, bp: 4602°, d: 13.31 @ 20°.

SYN: HAFNIUM METAL, DRY (DOT)

OSHA PEL: TWA 500 ug/m^3
DOT Classification: Label: Flammable Solid

THR: A poison by unspecified routes. See also iron dust. It can self-explode in powder form. Incompatible with nonmetals; oxidants.

HAFNIUM OXYCHLORIDE   HR: 3
CAS: 13759-17-6   NIOSH: MG 4725000
mf: Cl_2HfO mw: 265.39

SYN: HAFNIUM CHLORIDE OXIDE

THR: A poison by ingestion and intraperitoneal routes. See also hafnium. When heated to decomposition it emits very toxic fumes of Cl$^-$.

HAFNOCENE DICHLORIDE   HR: 3
CAS: 12116-66-4   NIOSH: MG 4815000
mf: $C_{10}H_{10}Cl_2Hf$ mw: 379.59

SYNS: DICHLOROBIS(eta-CYCLOPENTADIENYL)-HAFNIUM * DICHLORODICYCLOPENTADIENYL-HAFNIUM * DICHLORODI-pi-CYCLOPENTA-DIENYLHAFNIUM * DICYCLOPENTADIENYLHAF-NIUM DICHLORIDE * HAFNIUM DICYCLOPEN-TADIENE DICHLORIDE

THR: A poison via intravenous route. See also hafnium. When heated to decomposition it emits very toxic fumes such as chlorides.

HALOTHANE   HR: 3
CAS: 151-67-7   NIOSH: KH 6550000
mf: $C_2HBrClF_3$ mw: 197.39

PROP: Nonflammable, highly volatile liquid, characteristic sweetish, not unpleasant odor; d: 1.871 @ 20°/4°; bp: 50.2°; 20° @ 243 mm; Sensitive to light, miscible with petr ether, other fat solvents; sltly sol in H_2O.

SYNS: FTOROTAN (RUSSIAN) * 2-BROMO-2-CHLORO-1,1,1-TRIFLUOROETHANE
* FLUOTHANE * 1,1,1-TRIFLUORO-2-BROMO-2-CHLOROETHANE * 1,1,1-TRIFLUORO-2-CHLORO-2-BROMOETHANE * 2,2,2-TRIFLUORO-1-CHLORO-1-BROMOETHANE

BAT: Blood (end of work week, end of shift) 250 ug/dl; urine 10 mg/L

THR: An experimental teratogen. Moderately toxic by ingestion and inhalation in humans. When heated to decomposition it emits very toxic fumes of F$^-$, Cl$^-$ and Br$^-$. For further information see Vol. 1, No. 5 of *DPIM Report*.

HALVISOL   HR: 3
CAS: 13382-33-7   NIOSH: TL 0200000
mf: $C_{21}H_{27}FN_2O_2$ mw: 358.50

SYNS: ANISOPIROL * (+−)-alpha-(P-FLUORO-PHENYL)-4-(O-METHOXYPHENYL)-1-PIPERAZINE-BUTANOL * DL-1-(4-FLUOROPHENYL)-4-(1-(4-(2-METHOXY-PHENYL))-PIPERAZINYL)BUTANOL

THR: A poison via subcutaneous and intravenous routes. When heated to decomposition it emits very toxic fumes of F$^-$ and NO$_x$.

HAMAMELIS   HR: 2
                        NIOSH: MG 8325000

SYNS: NCI-C50544 * SNAPPING HAZEL
* SPOTTED ALDER * STRIPED ALDER
* TOBACCO WOOD * WINTER BLOOM
* WITCH HAZEL

THR: A mild irritant. Combustible when exposed to heat or flame; can react with oxidizing materials.

HARMINE   HR: 3
CAS: 442-51-3   NIOSH: UV 0175000
mf: $C_{13}H_{12}N_2O$ mw: 212.27

PROP: An alkaloid isolated from *Banisteria Caapi sp.*, a South American narcotic agent.

SYNS: 7-METHOXY-1-METHYL-9H-PYRIDO(3,4-b)INDOLE * LEUCOHARMINE * 6-ME-THOXYHARMAN * 1-METHYL-7-METHOXY-BETA-CARBOLINE * TELEPATHINE * YAGEINE
* YAJEINE

THR: Poison by intravenous and subcutaneous routes. Human central nervous system effects by intramuscular route. When heated to decomposition it emits toxic fumes of NO$_x$.

HCDD HR: 3
CAS: 34465-46-8   NIOSH: HP 3200000
mf: $C_{12}H_2Cl_6O_2$ mw: 390.84

PROP: Colorless solid, mp: 239°.

SYN: HEXACHLORODIBENZO-p-DIOXIN

THR: Deadly poison by ingestion. An experimental teratogen.

HEF-2 JET FUEL HR: 3
PROP: Mixture of alkylpentaborane derivatives

THR: A poison by inhalation.

HELIOTRINE HR: 3
CAS: 303-33-3   NIOSH: MH 6125000
mf: $C_{16}H_{27}NO_5$ mw: 313.44

SYN: HELIOTRON

THR: An acute poison in rats and mice and mammals. An experimental teratogen and tumorigen. When heated to decomposition it emits toxic fumes of NO_x.

HELVETICOSIDE DIHYDRATE HR: 3
CAS: 13495-01-7   NIOSH: HY 6117000
mf: $C_{29}H_{42}O_9 \cdot 2H_2O$ mw: 570.75

PROP: Needles from oil, methanol; mp: 153°-157°.

THR: A deadly poison via intravenous route.

HEMATITE HR: 3
CAS: 1317-60-8   NIOSH: MH 7600000

PROP: Consists mainly of Fe_2O_3

SYNS: BLOOD STONE * HAEMATITE * IRON ORE * RED IRON ORE

THR: An experimental carcinogen via inhalation and ingestion.

HEMICHOLINIUM-3-DIBROMIDE HR: 3
CAS: 312-45-8   NIOSH: QF 2450000
mf: $C_{24}H_{34}N_2O_4 \cdot 2Br$ mw: 574.42

SYNS: 2,2'-(1,1'-BIPHENYL-4,4'-DIYLBIS-(2-HYDROXY-4,4-DIMETHYL-MORPHOLINIUM DIBROMIDE * HEMICHOLINIUM BROMIDE * HEMICHOLINIUM-3-BROMIDE * HEMICHOLINIUM DIBROMIDE

THR: A poison via intraperitoneal and intravenous routes. See also bromides. When heated to decomposition it emits very toxic fumes of NO_x and Br^-.

HEPARIN HR: 2
CAS: 9005-49-6   NIOSH: MI 0700000

SYNS: HED-HEPARIN * HEPARINATE * LIQUAEMIN SODIUM * SODIUM HEPARIN * THROMBOLIQUINE

THR: Moderately toxic via intravenous and intraperitoneal routes. Mutagenic data.

HEPTACHLOR HR: 3
CAS: 76-44-8   NIOSH: 0700000
mf: $C_{10}H_5Cl_7$ mw: 373.30

PROP: Crystals, nearly insol in water; sol in organic solvents, mp: 96°.

SYNS: 1,4,5,6,7,8,8A-HEPTACHLORO-3A,4,7,7A-TETRAHYDRO-4,7-METHANOINDANE * 3-CHLOROCHLORDENE * NCI-c00180 * ENT 15,152 * EPTACLORO (ITALIAN) * HEPTACHLOOR (DUTCH) * 1,4,5,6,7,8,8-HEPTACHLOOR-3A,4,7,7A-TETRAHYDRO-4,7-ENDO-METHANO-INDEEN (DUTCH) * HEPTA-CHLOR * HEPTACHLORE (FRENCH) * 3,4,5,6,7,8,8-HEPTACHLORODICYCLOPENTADIENE * 3,4,5,6,7,8,8A-HEPTACHLORODICYCLO-PENTADIENE * 1,4,5,6,7,8,8-HEPTACHLORO-3A,4,7,7A-TETRAHYDRO-4,7-ENDOMETHANOIN-DENE * 1,4,5,6,7,8,8-HEPTACHLORO-3A,4,7,7A-TETRAHYDRO-4,7-METHANOINDENE * 1(3A),4,5,6,7,8,8-HEPTACHLORO-3A(1),4,7,7A-TETRAHYDRO-4,7-METHANOINDENE * 3A,4,5,6,7,8,8-HEPTACHLORO-3A,4,7,7A-TETRAHYDRO-4,7-METHANOINDENE * 1,4,5,6,7,8,8-HEPTACHLORO-3A,4,7,7A-TETRAHYDRO-4,7-METHANOL-1H-INDENE * 1,4,5,6,7,8,8-HEPTACHLORO-3A,4,7,7A-TETRAHYDRO-4,7-METHYLENE INDENE * 1,4,5,6,7,10,10-HEPTACHLORO-4,7,8,9-TETRAHYDRO-4,7-METHYLENEINDENE * 1,4,5,6,7,10,10-HEPTACHLORO-4,7,8,9-TETRAHYDRO-4,7-ENDOMETHYLENEINDENE

OSHA PEL: TWA 500 ug/m^3 (skin)

THR: A poison by ingestion, skin contact, intraperitoneal, intravenous, and other routes. Mutagenic data. An experimental carcinogen. Acute exposure causes liver damage. Chronic doses have caused liver damage in experimental animals. See closely related chlordane. In man, a dose of 1-3 g can cause serious symptoms, especially where liver impairment is the case. Acute symptoms include tremors, convulsions, kidney damage, respiratory collapse and death. Dangerous; when heated to decomposition it emits highly toxic fumes of Cl^-. For further information see Vol. 6, No. 5 of *DPIM Report*. *NOTE:* The EPA has cancelled registration of pesticides containing this with the exception of its use through subsurface ground insertion for termite control.

HEPTAFLUOROBUTYRIC ACID   HR: 3
CAS: 375-22-4   NIOSH: ET 4025000
mf: $C_4HF_7O_2$ mw: 214.05

PROP: Colorless liquid; sharp, butyric acid odor; bp: 210° @ 735 mm.

THR: A poison via intraperitoneal route. When heated to decomposition it emits toxic fumes of F^-. Will react with H_2O or steam to produce corrosive fumes.

N-HEPTANE   HR: 3
CAS: 142-82-5   NIOSH: MI 7700000
DOT: 1206
mf: C_7H_{16} mw: 100.23

PROP: Colorless liquid, bp: 98.52, lel = 1.05%, uel = 6.7%, fp: −90.5°, flash p: 25°F (CC), d: 0.684 @ 20°/4°, autoign temp: 433.4° F, vap press: 40 mm @ 22.3°, vap d: 3.45. Sltly sol in alcohol; misc in ether and chlorform. Insol in H_2O.

SYNS: DIPROPYL METHANE * EPTANI (ITALIAN) * HEPTAN (POLISH) * HEPTANEN (DUTCH) * HEPTYL HYDRIDE

OSHA PEL: TWA 500 ppm
DOT Classification: Label: Flammable
   Liquid

THR: A poison by inhalation and intravenous routes in humans. Narcotic in high concentrations. A central nervous system irritant. A volatile flammable liquid. Dangerous when exposed to heat or flame. Moderately explosive when exposed to heat or flame. Violent reaction with (P + Cl). Can react vigorously with oxidizing materials. To fight fire, use foam, CO_2, dry chemical.

1-HEPTANETHIOL   HR: 3
CAS: 1639-09-4   NIOSH: MJ 1400000
mf: $C_7H_{16}S$ mw: 132.29

SYNS: HEPTYL MERCAPTAN * N-HEPTYLMERCAPTAN * USAF EK-2122

THR: A poison via intraperitoneal route. When heated to decomposition it emits very toxic fumes of SO_x.

2-HEPTANOL   HR: 2
CAS: 543-49-7   NIOSH: MJ 2975000
mf: $C_7H_{16}O$ mw: 116.23

PROP: Liquid, bp: 160.4°, flash p: 160°F (OC), d: 0.8344 @ 0°, vap press: 1 mm @ 14.6°, vap d: 4.01. Insol in H_2O, sol in alc, ether, benzene.

SYNS: AMYL METHYL CARBINOL * HEPTANOL-2 * 2-HYDROXYHEPTANE * METHYL AMYL CARBINOL

THR: Moderately toxic by ingestion and skin contact. An irritant to skin and eyes. A moderate fire hazard from heat and oxidizers. To fight fire, use foam, CO_2, dry chemical.

2-HEPTANONE   HR: 2
CAS: 110-43-0   NIOSH: MJ 5075000
mf: $C_7H_{14}O$ mw: 114.18

PROP: Liquid, penetrating fruity odor, bp: 151.5°, flash p: 120°F (OC), autoign temp: 991°F, vap d: 3.94, d: 0.8197 @ 15°/4°; very sltly sol in water; sol in alcohol and ether.

SYNS: AMYL-METHYL-CETONE (FRENCH) * N-AMYL METHYL KETONE * METHYL-AMYL-CETONE (FRENCH) * METHYL-N-AMYL KETONE * METHYL AMYL KETONE (DOT)

OSHA PEL: TWA 100 ppm

THR: Moderate toxicity by ingestion and inhalation route. Flammable when exposed to heat or flame; can react with oxidizing materials. To fight fire, use foam, CO_2, dry chemical.

3-HEPTANONE   HR: 2
CAS: 106-35-4   NIOSH: MJ 5250000
mf: $C_7H_{14}O$ mw: 114.21

PROP: Clear liquid, mp: −36.7°, bp: 148°, flash p: 115°F (OC), d: 0.8198 @ 20°/20°, vap d: 3.93.

SYNS: AETHYLBUTYLKETON (GERMAN) * N-BUTYL ETHYL KETONE * EPTAN-3-ONE (ITALIAN) * ETHYLBUTYLCETONE (FRENCH) * ETHYLBUTYLKETON (DUTCH) * ETILBUTILCHETONE (ITALIAN) * HEPTAN-3-ON (DUTCH, GERMAN) * HEPTAN-3-ONE

OSHA PEL: TWA 50 ppm

THR: Moderately toxic by ingestion and inhalation. Flammable when exposed to heat or flame; can react with oxidizing materials. To fight fire, use foam, CO_2, dry chemical.

4-HEPTANONE   HR: 2
CAS: 123-19-3   NIOSH: MJ 5600000
mf: $C_7H_{14}O$ mw: 114.21

PROP: Colorless, refractive liquid, bp: 144°, mp: $-32.6°$, vap press: 5.2 mm @ 20°, flash p: 120°F (CC), d: 0.815, vap d: 3.93.

SYNS: BUTYRONE * DIPROPYL KETONE * GBL * HEPTAN-4-ONE * PROPYL KETONE

THR: Moderately toxic by ingestion and inhalation. See also ketones. Flammable when exposed to heat or flame; can react with oxidizing materials. To fight fire, use CO_2, dry chemical, alcohol foam, fog and mist.

HEPTYLMETHYLINITROSAMINE   HR: 3
CAS: 16338-99-1   NIOSH: MK 1260000
mf: $C_8H_{18}N_2O$ mw: 158.28

SYNS: METHYLHEPTYLNITROSAMIN (GERMAN) * N-METHYL-N-NITROSOHEPTYLAMINE * N-NITROSO-N-METHYLHEPTYLAMINE

THR: A poison via subcutaneous route. An experimental tumorigen. When heated to decomposition it emits toxic fumes of NO_x.

HEROIN   HR: 3
CAS: 561-27-3   NIOSH: QC 8050000
mf: $C_{21}H_{23}NO_5$ mw: 369.45

PROP: White, odorless, bitter crystals or crystalline powder, mp: 173°, bp: 273° @ 12 mm.

SYNS: ACETOMORFINE * ACETOMORPHINE * ASPRON * BOY * DIACEPHIN * DIACETYLMORFIN * DIACETYLMORPHINE * DIAMORFINA * DIAMORPHINE * DIAPHORM * DIASETIELMORFIEN * DIASETILMORFIN * DIASETYLMORFIIMI * DIAZETYLMORPHINE * 7,8-DIHYDRO-4,5-alpha-EPOXY-17-METHYL-MORPHINAN-3,6-alpha-DIOL DIACETATE * DOOJE * ECLORION * EROINA * ''H'' * HAIRY * HARRY * HEROIEN * HEROIIN * HEROLAN * HORSE * IEROIN * IROINI * JOY POWDER * MORPHACETIN * MORPHINE DIACETATE * PREZA * SCOT * WHITE STUFF

THR: A poison by ingestion and subcutaneous routes. Fatal dose between $\frac{1}{6}$ and 2 grains. Resembles morphine in its general results, but acts more strongly on the respiration and is therefore more poisonous. Its depressant effects on the cerebrum appear to be greater than that of co-deine. Large doses cause excitement and convulsions in animals and man. The more common symptoms are headache, disturbance of vision, slow, small, regular pulse, restlessness, cramps in the extremities, slight cyanosis, respiration slow and deep and death from respiratory paralysis. A poisonous habit-forming drug. See also morphine. Dangerous; when heated to decomposition it emits toxic fumes of NO_x. For further information see Vol. 1, No. 7 of *DPIM Report*. Treatment and Antidotes: Use the same treatment as for morphine, with special attention to control of the respiratory paralysis.

HEXACHLOROBENZENE   HR: 3
CAS: 118-74-1   NIOSH: DA 2975000
mf: C_6Cl_6 mw: 284.76

PROP: Monoclinic prisms, mp: 231°, bp: 323°-326°, flash p: 468°F, vap press: 1 mm @ 114.4°, vap d: 9.8. D: 2.44; insol in water; sol in benzene; very sltly sol in hot alcohol; sol in hot ether, and chloroform.

SYNS: HEXACHLORBENZOL (GERMAN) * JULIN'S CARBON CHLORIDE * PENTACHLOROPHENYL CHLORIDE * PERCHLOROBENZENE * PHENYL PERCHLORYL

MAK BAT: 15 ug/dL

THR: An experimental neoplastigen, carcinogen, tumorigen, and teratogen. A suspected human carcinogen. Moderately toxic by ingestion. Combustible when exposed to heat or flame. Violent reaction with dimethylformamide. Dangerous; when heated to decomposition emits highly toxic chlorides. To fight fire: CO_2, dry chemical. For further information see Vol. 4, No. 1 of *DPIM Report*.

HEXACHLORO-2,5-CYCLOHEXADIEN-1-ONE   HR: 3
CAS: 599-52-0   NIOSH: GU 5600000
mf: C_6Cl_6O mw: 300.76

SYNS: 2,3,4,4,5,6-HEXACHLORCYKLOHEXA-2,5-DIEN-1-ON (CZECH) * HEXACHLORFENOL (CZECH) * USAF DO-65

THR: A poison by ingestion and intraperitoneal routes. A severe skin and eye irritant. When heated to decomposition it emits toxic fumes of Cl^-.

1,2,3,4,7,8-HEXACHLORODIBENZO-p-DIOXIN   HR: 3

CAS: 39227-28-6   NIOSH: HP 3280000
mf: $C_{12}H_2Cl_6O_2$ mw: 390.84

THR: A poison by ingestion. See also chlorinated dibenzo-p-dioxins. When heated to decomposition it emits toxic fumes of Cl^-.

1,1,1,2,2,2-HEXACHLORO-ETHANE   HR: 3

CAS: 67-72-1   NIOSH: KI 4025000
mf: C_2Cl_6 mw: 236.72

PROP: Rhombic triclinic or cubic crystals, colorless, camphor-like odor, mp: 186.6° (sublimes), d: 2.091, vap press: 1 mm @ 32.7°, bp: 186.8° (triple point). Sol in alc, benzene, chloroform, ether, oils. Insol in H_2O.

SYNS: CARBON HEXACHLORIDE * ETHANE HEXACHLORIDE * ETHYLENE HEXACHLORIDE * HEXACHLOR-AETHAN (GERMAN) * HEXACHLOROETHANE (DOT) * HEXACHLOROETHYLENE * NCI-C04604 * PERCHLOROETHANE

OSHA PEL: TWA 1 ppm (skin)

THR: A poison via intravenous route. Moderately toxic by ingestion, intraperitoneal, and dermal routes. An experimental carcinogen. Liver injury has been described from exposure to this material. See also chlorinated hydrocarbons. Slight explosion hazard by spontaneous chemical reaction. Dehalogenation of this material by reaction with alkalies, metals, etc., will produce spontaneously explosive chloroacetylenes. Dangerous; when heated to decomposition emits highly toxic phosgene. For further information see Vol. 6, No. 4 of *DPIM Report*.

HEXACHLORONAPHTHALENE   HR: 3

CAS: 1335-87-1   NIOSH: QJ 7350000
mf: $C_{10}H_2Cl_6$ mw: 334.9

PROP: White solid.

SYN: HALOWAX 1014

OSHA PEL: TWA 200 ug/m³ (skin)

THR: A poison by ingestion, dermal and inhalation routes. Causes severe acne-form eruptions and toxic narcosis of liver. Absorbed by skin. When heated to decomposition it emits toxic fumes of Cl^-. For further information see Vol. 5, No. 1 of *DPIM Report*.

HEXACHLOROPHENE   HR: 3

CAS: 70-30-4   NIOSH: SM 0700000
mf: $C_{13}H_6Cl_6O_2$ mw: 406.89

PROP: Cryst; water insol; mp: 165°. Sol in alc, acetone, ether, chloroform, propylene glycol, polyethylene glycols, olive oil, cottonseed oil, dil solns of alkalies.

SYNS: 2,2'-DIHYDROXY-3,3',5,5',6,6'-HEXACHLORODIPHENYLMETHANE * 2,2'-DIHYDROXY-3,5,6,3',5',6'-HEXACHLORODIPHENYLMETHANE * G-11 * G-ELEVEN * 2,2',3,3',5,5'-HEXACHLORO-6,6'-DIHYDROXYDIPHENYLMETHANE * HEXACHLOROFEN (CZECH) * HEXACHLOROPHANE * HEXACHLOROPHEN * HEXOPHENE * METHANE, BIS(2,3,5-TRICHLORO-6-HYDROXYPHENYL) * 2,2'-METHYLENEBIS(3,4,6-TRICHLOROPHENOL) * NCI-C02653 * BIS(2-HYDROXY-3,5,6-TRICHLOROPHENYL)METHANE * BIS-2,3,5-TRICHLOR-6-HYDROXYFENYLMETHAN (CZECH) * BIS(3,5,6-TRICHLORO-2-HYDROXYPHENYL)-METHANE * COMPOUND G-11 * PHISOHEX * TRICHLOROPHENE

THR: Poison to humans by ingestion. Also poison by intraperitoneal and intravenous routes. An experimental teratogen and carcinogen. A skin and eye irritant. Affects the human central nervous system. Strong concentrations may be irritating but ordinary use of 1-2% is not. Used as a food additive permitted in the feed and drinking water of animals and/or for the treatment of food-producing animals; also permitted in food for human consumption.

For many years, the toxicologic hazard of hexachlorophene was unrecognized and the compound had a wide and virtually unrestricted use. However, studies by FDA scientists have shown that brain lesions occur from exposure to it in both rats and monkeys treated with it at levels only slightly higher than those of persons using soaps, toothpaste, shampoos, and a variety of other household products and cosmetics containing it. The FDA has now restricted sale of hexachlorophene and most preparations containing higher levels of the compound are available only through prescription. In the FDA studies, it was found that 2 weeks after onset of exposure, rats fed 500 ppm (25 mg/kg/day) hexachlorophene in their diet showed weakness in their hindquarters, which progressed to paralysis. Microscopic examination of the brain and spinal cord of these rats revealed a particular edema of the white matter resembling spongy degeneration of it in infants. When the animals were removed from the poisoned diet, they recovered gradually over a period of weeks. A similar symptom picture was noted in the monkey. Following ingestion of hexachlorophene, early symptoms are primarily gastrointestinal in nature, and include anorexia, nausea, vomiting, abdominal cramps, and diarrhea. Dehydration is sometimes severe and may be associated with shock. Dangerous; when heated to decomposition it emits highly toxic chlorides. For further information see Vol. 6, No. 2 of *DPIM Report*.

HEXACHLOROPROPENE   HR: 3
CAS: 1888-71-7   NIOSH: UD 0175000
mf: C_3Cl_6 mw: 248.73

SYN: HEXACHLOROPROPYLENE

THR: A poison by ingestion, inhalation and intraperitoneal routes. A powerful irritant. Dangerous, see also chlorides.

1-HEXADECANAMINE   HR: 3
CAS: 143-27-1   NIOSH: ML 9100000
mf: $C_{16}H_{35}N$ mw: 241.52

SYNS: ARMEEN 16D * CETYLAMINE * N-HEXADECYLAMINE * PALMITYLAMINE

THR: A poison by intraperitoneal route. When heated to decomposition it emits toxic fumes of NO_x.

HEXADECYL TRIMETHYL AMMONIUM BROMIDE   HR: 3
CAS: 57-09-0   NIOSH: BQ 7875000
mf: $C_{19}H_{42}N \cdot Br$ mw: 364.53

SYNS: CETYLTRIMETHYLAMMONIUM BROMIDE * N-CETYLTRIMETHYLAMMONIUM BROMIDE * N-HEXADECYLTRIMETHYLAMMONIUM BROMIDE * N-HEXADECYL-N,N,N-TRIMETHYLAMMONIUM BROMIDE * (1-HEXADECYL)TRIMETHYLAMMONIUM BROMIDE * TRIMETHYLCETYLAMMONIUM BROMIDE * TRIMETHYLHEXADECYLAMMONIUM BROMIDE

THR: A poison by ingestion, intraperitoneal, and subcutaneous routes. An experimental teratogen. A skin and eye irritant. See also bromides. When heated to decomposition it emits very toxic NO_x and Br⁻. For further information see N-Cetyltrimethyl Ammonium Bromide, Vol. 2, No. 4 of *DPIM Report*.

HEXADIMETHRINE BROMIDE   HR: 3
CAS: 9011-04-5   NIOSH: MO 1400000
mf: $C_{10}H_{24}N_2 \cdot 7C_3H_6Br_2$ mw: 1585.73

PROP: White hygroscopic, amorphous polymer; sol in H_2O up to 10%.

SYNS: 1,5-DIMETHYL-1,5-D-DIAZAUNDECAMETHYLENE POLYMETHOBROMIDE; * POLYBREME

THR: A poison via intravenous and intraperitoneal routes. When heated to decomposition it emits very toxic fumes of NO_x and Br⁻.

HEXAETHYL TETRAPHOSPHATE   HR: 3
CAS: 757-58-4   NIOSH: XF 1575000
DOT: 1611
mf: $C_{12}H_{30}O_{13}P_4$ mw: 506.30

PROP: Liquid, mp: −40°; bp: decomp above 150°.

SYNS: BLADAN BASE * ETHYL TETRAPHOSPHATE * HEXAETHYL TETRAPHOSPHATE, LIQUID (DOT) * TETRAPHOSPHATE HEXAETHYLIQUE (FRENCH)

DOT Classification: Poison B, Label: Poison

THR: A poison by ingestion, skin, intraperitoneal, subcutaneous, intravenous, and intramuscular routes. See also tetraethyl pyrophosphate. When heated to decomposition it emits toxic fumes of PO_x.

HEXAFLUOROACETONE   HR: 3
CAS: 684-16-2   NIOSH: UC 2450000
mf: C_3F_6O mw: 166.03

PROP: A colorless, non-flammable solvent liquid, d: 1.65 @ 25°.

SYN: NCI-C56440

THR: A poison via irritation to skin, eyes, and mucous membranes, and by ingestion, inhalation and possibly dermal routes. See also fluorine and fluorides. An experimental teratogen. When heated to decomposition it emits toxic fumes of F⁻. For further information see Vol. 1, No. 4 of *DPIM Report*.

HEXAFLUORO ACETONE TRIHYDRATE   HR: 3
CAS: 34202-69-2   NIOSH: UC 2700000
mf: $C_3F_6O \cdot 3H_2O$ mw: 220.09

SYN: GC 7787

THR: A poison by ingestion and dermal routes. When heated to decomposition it emits toxic fumes of F^-.

HEXAFLUORODICHLOROBU-
TENE HR: 3
CAS: 11111-49-2 NIOSH: EM 4905000

SYN: HFCB

THR: A poison via irritation to skin, eyes, and mucous membranes and inhalation. When heated to decomposition it emits very toxic fumes of Cl^- and F^-.

HEXAFLUORODIETHYL ETHER HR: 3
CAS: 333-36-8 NIOSH: KN 3675000
mf: $C_4H_4F_6O$ mw: 182.08

SYNS: BIS(TRIFLUOROETHYL)ETHER
* BIS(2,2,2-TRIFLUOROETHYL)ETHER
* 2-(2,2,2-TRIFLUOROETHOXY)-1,1,1-TRI-
FLUOROETHANE

THR: A poison via intravenous route; moderately toxic via ingestion, inhalation, and intraperitoneal routes. See also ethers. When heated to decomposition it emits toxic fumes of F^-.

HEXAFLUOROISOPROPANOL HR: 3
CAS: 920-66-1 NIOSH: UB 6450000
mf: $C_3H_2F_6O$ mw: 168.05

SYN: 1,1,1,3,3,3-HEXAFLUORO-2-PROPANOL

THR: A poison via inhalation, intravenous, and intraperitoneal routes. Moderately toxic by ingestion. When heated to decomposition it emits toxic fumes of F^-.

HEXAFLUOROPHOSPHORIC
ACID HR: 3
CAS: 16940-81-1 NIOSH: SY 7115000
DOT: 1782
mf: F_6HP mw: 145.98

PROP: Corrosive, colorless, clear liquid. mp: 31°, d: 1.65.

SYN: HYDROGEN HEXAFLUOROPHOSPHATE

DOT Classification: Label: Corrosive

THR: A poison by all routes with irritation to skin, eyes, and mucous membranes. See also hydrofluoric and phosphoric acids. Dangerous; when heated to decomposition it emits highly toxic F^- and PO_x.

HEXAFLUORO-2-PROPANONE SES-
QUIHYDRATE HR: 3
CAS: 13098-39-0 NIOSH: UC 2660000
mf: $C_3F_6O \cdot 3/2H_2O$ mw: 193.06

SYN: HEXAFLUOROACETONE SESQUIHYDRATE

THR: A poison by ingestion, intraperitoneal, and intravenous routes. When heated to decomposition it emits toxic fumes of F^-.

HEXAFLURONIUM BROMIDE HR: 3
CAS: 317-52-2 NIOSH: BQ 8225000
mf: $C_{36}H_{42}N_2 \cdot Br$ mw: 582.71

SYNS: 1,6-BIS(9 FLUORENYLDIMETHYL-AMMO-NIUM)HEXANE BROMIDE * HEXAFLUORENIUM DIBROMIDE * HEXAMETHYLENE BIS(9-FLUO-RENYLDIMETHYLAMMONIUM)DIBROMIDE * HEXAMETHYLENEBIS(DIMETHYL-9-FLUORENY-LAMMONIUM BROMIDE) * HEXAMETHYLENEBIS-(FLUOREN-9-YLDIMETHYLAMMONIUM BROMIDE)

THR: A poison via ingestion, intraperitoneal, subcutaneous, and intravenous routes. See also bromides. When heated to decomposition it emits very toxic fumes of NO_x and Br^-.

HEXAHYDRO-2H-AZEPIN-2-
ONE HR: 3
CAS: 105-60-2 NIOSH: CM 3675000
mf: $C_6H_{11}NO$ mw: 113.18

PROP: White crystals, mp: 69°, vap press: 6 mm @ 120°.

SYNS: CAPROLATTAME (FRENCH) * AMINOCA-PROIC LACTAM * 6-AMINOHEXANOIC ACID CYCLIC LACTAM * 2-AZACYCLOHEPTANONE * CAPROLACTAM * 6-CAPROLACTAM * EPSILON-CAPROLACTAM * OMEGA-CAPRO-LACTAM * CYCLOHEXANONE ISO-OXIME * EPSILON KAPROLAKTAM (POLISH) * HEXA-HYDRO-2-AZEPINONE * HEXANONISOXIM (GER-MAN) * HEXANONE ISOXIME * 1,6-HEXOLAC-TAM * E-KAPROLAKTAM (CZECH) * 2-KETO-HEXAMETHYLENIMINE * NCI-C50646 * 2-OXOHEXAMETHYLENIMINE * 2-PERHYDROAZEPINONE

THR: Moderate toxicity by ingestion, skin contact, intraperitoneal, and subcutaneous routes. A skin and eye irritant. A systemic human irritant. Mutagenic data. When heated to decomposition it emits toxic fumes of NO_x. For further information see ϵ-Caprolactam Vol. 1, No. 3 of *DPIM Report*.

1,2,3,4,5,6-HEXAHYDRO-6-METHYL-AZEPINO(4,5-b)INDOLE HYDRO-CHLORIDE   HR: 3
CAS: 15923-42-9   NIOSH: CM 3500000
mf: $C_{13}H_{16}N_2 \cdot ClH$ mw: 236.77

SYN: U-22394A

THR: A poison by ingestion and intraperitoneal routes. When heated to decomposition it emits very toxic fumes of NO_x and HCl.

HEXAHYDRO-1,3,5-s-TRIAZINE   HR: 3
CAS: 13980-04-6   NIOSH: XY 9470000
mf: $C_3H_6N_5O_3$ mw: 174.15

SYNS: HEXAHYDRO-1,3,5-TRINITROSO-s-TRIAZINE * HEXAHYDRO-1,3,5-TRINITROSO-1,3,5-TRIAZINE * 1,3,5-TRINITROSO-1,3,5-TRIAZA-CYCLOHEXANE * TRINITROSOTRIMETHYLENE-TRIAMINE * TRINITROSOTRIMETHYLENTRIAMIN (GERMAN)

THR: A poison by ingestion. An experimental tumorigen. When heated to decomposition it emits toxic fumes of NO_x.

HEXA(HYDROXYMETHYL)MEL-AMINE   HR: 3
CAS: 531-18-0   NIOSH: PC 3700000
mf: $C_9H_{18}N_6O_6$ mw: 306.33

SYNS: HEXAKIS(HYDROXYMETHYL)MELAMINE * HEXAMETHYLOLMELAMIN (CZECH) * HEX-AMETHYLOLMELAMINE * (s-TRIAZINE-2,4,6-TRIYLTRINITRILO)HEXAMETHANOL * 2,4,6-TRIS(DI(HYDROXYMETHYL)AMINO)-1,3,5-TRIAZINE

THR: A poison via intravenous route. A skin and eye irritant. When heated to decomposition it emits toxic fumes of NO_x.

HEXAMETHONIUM BROMIDE   HR: 3
CAS: 55-97-0   NIOSH: BQ 8575000
mf: $C_{12}H_{30}N_2 \cdot 2Br$ mw: 362.26

PROP: Crystals; mp: 274°-276°; sol in water, alc, acid to litmus. Aq solns are stable.

SYNS: alpha, omega-BIS(TRIMETHYL AMMONI-UM)HEXANE DIBROMIDE * HEXAMETHIONIUM BROMIDE * HEXAMETHONIUM DIBROMIDE * HEXAMETHYLENEBIS(TRIMETHYLAMMONIUM) BROMIDE

THR: A poison via intraperitoneal and intrave-nous routes. Moderately toxic by ingestion. An

experimental teratogen. See also bromides. When heated to decomposition it emits very toxic NO_x and Br^-.

HEXAMETHONIUM DIIODIDE   HR: 3
CAS: 870-62-2   NIOSH: BQ 8750000
mf: $C_{12}H_{30}N_2 \cdot 2I$ mw: 456.24

SYNS: HEXAMETHONIUM IODIDE * HEXA-METHYLENEBIS(TRIMETHYLAMMONIUM IODIDE)

THR: A poison via intravenous route. See also iodides. When heated to decomposition it emits very toxic fumes of NO_x and I^-.

HEXAMETHYLBENZENE
CAS: 87-85-4   NIOSH: DA 3200000
mf: $C_{12}H_{18}$ mw: 162.30

PROP: Plates from ethanol; mp: 165.5°; bp: 265°; insol in water; very sol in ether.

THR: An experimental neoplastigen. When heated to decomposition it emits acrid smoke and fumes. Incompatible with nitromethane (can cause it to explode).

HEXAMETHYLDITIN   HR: 3
CAS: 661-69-8   NIOSH: WH 8280000
mf: $C_6H_{18}Sn_2$ mw: 327.62

SYN: HEXAMETHYLDISTANNANE

THR: A poison by ingestion. See also tin com-pounds. When heated to decomposition it emits acrid smoke and fumes.

1-(2-HEXAMETHYLENEIMINOETHYL)-2-OXOCYCLOHEXANECARBOXYLIC ACID BENZYL ESTER HYDROCHLO-RIDE   HR: 3
CAS: 1169-26-2   NIOSH: GU 8560000
mf: $C_{22}H_{31}NO_3 \cdot ClH$ mw: 394.00

SYN: 2-(beta-HEXAMETHYLENIMINOAETHYL)-CYCLOHEXANON-2-CARBONSAUREBENZYLESTER-HYDROCHLORIDE (GERMAN)

THR: A poison via ingestion, intraperitoneal, subcutaneous, and intravenous routes. See also esters. When heated to decomposition it emits very toxic fumes of HCl and NO_x.

HEXAMETHYLENETETRAMINE   HR: 3
CAS: 100-97-0   NIOSH: MN 4725000
mf: $C_6H_{12}N_4$ mw: 140.22

PROP: Odorless rhombic crystals from alcohol, mp: 280° sublimes, flash p: 482°F, d: 1.33 @ −5°. Very sltly sol in hot ether.

SYNS: AMMONIOFORMALDEHYDE * FORMA-MINE * HEXAFORM * HEXAMETHYLENAMINE * HEXAMETHYLENTETRAMIN (GERMAN) * HEXAMINE * HEXILMETHYLENAMINE * METHENAMINE * 1,3,5,7-TETRAAZAADAM-ANTANE * UROTROPINE

THR: A poison via subcutaneous route. Low toxicity via intravenous route. A moderate irritant to skin, eyes and mucous membranes. An experimental neoplastigen. Some persons suffer a skin rash if they come in contact with this material or the fumes evolved when it is heated. Pure hexamethylenetetramine may be taken internally in small amounts and is used in medicine as a urinary antiseptic. Its major industrial use is in the manufacture of phenolic resins. It is combustible and can be readily ignited when a flame is applied directly to its surface. It liberates formaldehyde on decomposition. Flammable when exposed to heat or flame; can react with oxidizing materials. Reacts violently with Na_2O_2.

HEXAMETHYLMELAMINE   HR: 3
CAS: 645-05-6   NIOSH: OS 1050000
mf: $C_9H_{18}N_6$ mw: 210.33

PROP: A solid material, insol in water, sol in acetone.

SYNS: ENT 50852 * NCI-C50259 * NSC 13875 * 2,4,6-TRIS(DIMETHYLAMINO)-S-TRIAZINE

THR: A poison by ingestion, intraperitoneal, and intravenous routes. Mutagenic data. An experimental neoplastigen. Affects the human gastrointestinal tract. When heated to decomposition it emits toxic fumes of NO_x.

HEXAMETHYL PHOSPHOR-AMIDE   HR: 3
CAS: 680-31-9   NIOSH: TD 0875000
mf: $C_6H_{18}N_3OP$ mw: 179.24

PROP: Clear, colorless, mobile liquid, spicy odor, bp: 233°, fp: 6°, d: 1.024 @ 25°/25°, vap d: 6.18.

SYNS: TRI(DIMETHYLAMINO)PHOSPHINEOXIDE * ENT 50882 * HEXAMETHYLPHOSPHORIC ACID TRIAMIDE * HEXAMETHYLPHOSPHORIC TRIAMIDE * N,N,N,N,N,N-HEXAMETHYL-PHOSPHORIC TRIAMIDE * HEXAMETHYLPHOS-PHOROTRIAMIDE * HEXAMETHYLPHOSPHO-TRIAMIDE * HEXMETHYLPHOSPHORAMIDE * PHOSPHORIC TRIS(DIMETHYLAMIDE) * PHOSPHORYL HEXAMETHYLTRIAMIDE * TRIS(DIMETHYLAMINO)PHOSPHINE OXIDE * TRIS(DIMETHYLAMINO)PHOSPHORUS OXIDE

THR: Mutagenic data. An experimental carcinogen. Moderately toxic via ingestion, skin, and intravenous routes. When heated to decomposition it emits very toxic fumes of PO_x and NO_x.

HEXAMID   HR: 3
NIOSH: CQ 3501000
mf: $C_{18}H_{25}N_3O_3 \cdot ClH$ mw: 367.92

SYNS: 3-(2-(DIETHYLAMINO)ETHYL)-5-ETHYL-5-PHENYLBARBITURIC ACID HYDROCHLORIDE * 5,5-PHENYL-AETHYL-3-(BETA-DIAETHYLAM-INO-AETHYL)-2,4,6-TRIOXO-HEXAHYDROPYRIMI-DIN-HCl (GERMAN)

THR: A poison by ingestion and intravenous routes. Moderately toxic via subcutaneous route. When heated to decomposition it emits very toxic fumes of HCl and NO_x.

1-HEXANAL   HR: 3
CAS: 66-25-1   NIOSH: MN 7175000
DOT: 1207
mf: $C_6H_{12}O$ mw: 100.18

PROP: Liquid, mp: −56.3°, bp: 128.7°, flash p: 90°F (OC), d: 0.8156 @ 20°/20°, vap press: 8.6 mm @ 20°, vap d: 3.45. Colorless liquid with powerful fatty-green odor; reported in about a dozen essential oils.

SYNS: ALDEHYDE C-6 * CAPROALDEHYDE * CAPROIC ALDEHYDE * CAPRONALDEHYDE * N-CAPROYLALDEHYDE * HEXALDEHYDE (DOT)

DOT Classification: Label: Flammable Liquid

THR: Moderately toxic by ingestion and inhalation. A mild irritant to skin and eyes. When heated to decomposition it emits acrid smoke and fumes.

HEXANAMIDE   HR: 3
CAS: 628-02-4   NIOSH: MN 7875000
mf: $C_6H_{13}NO$ mw: 115.20

SYN: NCI-C02142

THR: Moderately toxic by ingestion. An experimental carcinogen. When heated to decomposition it emits toxic fumes such as NO_x.

N-HEXANE   HR: 2
CAS: 110-54-3   NIOSH: MN 9275000
DOT: 1208
mf: C_6H_{14} mw: 86.17

PROP: Colorless liquid, faint odor, bp: 69°, ulc: 90-95, lel = 1.2%, uel = 7.5%, fp: −95.6°, flash p: −9.4°F, d: 0.6603 @ 20°/4°, autoign temp: 437°F, vap press: 100 mm @ 15.8°, vap d: 2.97. Insol in water; misc in chloroform, ether, alcohol. Very volatile liquid.

SYNS: ESANI (ITALIAN) * HEKSAN (POLISH) * HEXANE (DOT) * HEXANEN (DUTCH) * NCI-C60571

OSHA PEL: TWA 500 ppm
ACGIH BEI: 2,5-hexanedione in urine, end of shift 5 mg/L
DOT Classification: Label: Flammable Liquid

THR: Low toxicity by ingestion. Used as a food additive permitted in food for human consumption. Can cause a motor neuropathy in exposed workers. May be irritating to respiratory tract and narcotic in high concentrations. Inhalation of 5000 ppm for 1/6 hour produces marked vertigo. 2500-1000 ppm for 12 hours/day produces drowsiness, fatigue, loss of appetite, paresthesia in distal extremities. 2500-500 ppm produces muscle weakness, cold pulsation in extremities, blurred vision, headache, anorexia and onset of polyneuropathy. 2000 ppm for 1/6 hour gives no symptoms. 1000-500 ppm for 3-6 months produces fatigue, loss of appetite, distal paresthesia. Dermal exposure resulted in no anesthesia, blister formation, irritation, itching, erythema, pigmentation, pain. Dangerous if abused. An eye irritant. Dangerous fire hazard when exposed to heat or flame. Moderate explosion hazard when exposed to heat or flame. Dangerous when heated or exposed to flame, can react vigorously with oxidizing materials. To fight fire, use CO_2, dry chemical. Incompatible with dinitrogen tetraoxide.

HEXANENITRILE   HR: 3
CAS: 628-73-9   NIOSH: MO 3900000
mf: $C_6H_{11}N$ mw: 97.18

SYNS: NC5 * CAPRONITRILE

THR: A poison by ingestion and subcutaneous routes. See also nitriles. When heated to decomposition it emits toxic fumes of NO_x and CN^-.

1-HEXANETHIOL   HR: 3
CAS: 111-31-9   NIOSH: MO 4550000
mf: $C_6H_{14}S$ mw: 118.26

SYNS: HEXYL MERCAPTAN * USAF EK-4628

THR: A poison via inhalation and intraperitoneal routes. Moderately toxic by ingestion. When heated to decomposition it emits very toxic fumes of SO_x.

2-HEXANONE   HR: 2
CAS: 591-78-6   NIOSH: MP 1400000
mf: $C_6H_{12}O$ mw: 100.18

PROP: Clear liquid, mp: −56.9°, bp: 127.2°, lel = 1.22%, uel = 8.0%, flash p: 95°F (OC), d: 0.830 @ 0°/4°, vap press: 10 mm @ 38.8°, vap d: 3.45, autoign temp: 991°F. Slightly sol in H_2O; sol in alc, ether.

SYNS: N-BUTYL METHYL KETONE * METHYL N-BUTYL KETONE

OSHA PEL: TWA 100 ppm

THR: Moderately toxic by ingestion, inhalation and intraperitoneal routes. Moderately irritating via skin and eye exposure. Dangerous fire hazard when exposed to heat or flame; can react with oxidizing materials. Moderately explosive when exposed to heat or flame. To fight fire, use alcohol foam, CO_2, dry chemical.

1-HEXANOYLAZIRIDINE   HR: 3
CAS: 45776-10-1   NIOSH: CM 7890000
mf: $C_8H_{15}NO$ mw: 141.24

SYNS: 1-CAPROYLAZIRIDINE * CAPROYL-ETHYLENEIMINE * HEXANOYLETHYLENEIMINE

THR: Mutagenic data. An experimental neoplastigen. When heated to decomposition it emits toxic fumes of NO_x.

2-HEXENAL   HR: 3
CAS: 505-57-7   NIOSH: MP 5880000
mf: $C_6H_{10}O$ mw: 98.16

SYNS: HEX-2-ENAL * HEX-2-EN-1-AL * HEXYLENIC ALDEHYDE * LEAF ALDEHYDE

THR: A poison via intraperitoneal route. When heated to decomposition it emits acrid smoke and fumes.

4-HEXEN-1-YN-3-OL HR: 3
CAS: 10138-60-0 NIOSH: MQ 0175000
mf: C_6H_8O mw: 96.14

THR: A poison by ingestion, inhalation, and skin contact. When heated to decomposition it emits acrid smoke and fumes.

4-HEXEN-1-YN-3-ONE HR: 3
CAS: 13061-80-8 NIOSH: MQ 0350000
mf: C_6H_6O mw: 94.12

THR: A poison by ingestion, inhalation, and skin contact. When heated to decomposition it emits acrid smoke and fumes.

HEXONE HR: 3
CAS: 108-10-1 NIOSH: SA 9275000
mf: $C_6H_{12}O$ mw: 100.18

PROP: Clear liquid, bp: 118°, lel = 1.4%, uel = 7.5%, flash p: 62.6°F, d: 0.803, fp: −80.2°, autoign temp: 858°F, vap press: 16 mm @ 20°, d: 3.45.

SYNS: ISOBUTYL-METHYLKETON (CZECH) * ISOBUTYL METHYL KETONE * ISOPROPYL-ACETONE * KETONE, ISOBUTYL METHYL * METHYL-ISOBUTYL-CETONE (FRENCH) * METHYLISOBUTYLKETON (DUTCH, GERMAN) * METHYL ISOBUTYL KETONE * METYLOIZO-BUTYLOKETON (POLISH) * 4-METHYL-PENTAN-2-ON (DUTCH, GERMAN) * 2-METHYL-4-PENTA-NONE * 4-METHYL-2-PENTANON (CZECH) * 4-METHYL-2-PENTANONE * METILISOBUTIL-CHETONE (ITALIAN) * 4-METILPENTAN-2-ONE (ITALIAN)

OSHA PEL: TWA 100 ppm

THR: A poison via intraperitoneal route. Moderately toxic by ingestion and inhalation. A skin and eye irritant. A human systemic irritant by inhalation. Very irritating to eyes and mucous membranes. Narcotic in high concentrations. See also ketones. Dangerous fire hazard when exposed to heat, flame or oxidizers. Violent reaction with potassium *tert*-butoxide. Moderate explosion hazard in the form of vapor when exposed to heat or flame. Dangerous; keep away from heat and open flame; can react vigorously with reducing materials. To fight fire, use alcohol foam, CO_2, dry chemical. Incompatible with air; potassium-tert-butoxide.

N-HEXYL ALCOHOL HR: 2
CAS: 111-27-3 NIOSH: MQ 4025000
mf: $C_6H_{14}O$ mw: 102.20

PROP: Colorless liquid, bp: 157.2°, fp: −44.6°, flash p: 145°F, d: 0.8186 @ 20°/4°, vap press: 1 mm @ 24.4°, vap d: 3.52. misc in alc and ether.

SYNS: AMYLCARBINOL * CAPROYL ALCOHOL * N-HEXANOL * 1-HYDROXYHEXANE * PENTYLCARBINOL

THR: Moderately toxic by ingestion and dermal contact. A skin and eye irritant. Flammable when exposed to heat or flame; can react with oxidizing materials. To fight fire, use alcohol foam, CO_2, dry chemical. For further information see 1-Hexanol Vol. 2, No. 2 of *DPIM Report*.

HEXYLAMINE HR: 2
CAS: 111-26-2 NIOSH: MQ 4540000
mf: $C_6H_{15}N$ mw: 101.22

PROP: Liquid, mp: −22.9°, bp: 131.4°, flash p: 85°F (OC), d: 0.7675 @ 20°/20°, vap d: 3.49.

SYNS: 1-AMINOHEXANE * 1-HEXANAMINE * N-HEXYLAMINE * MONO-N-HEXYLAMINE

THR: A poison via dermal contact and inhalation. Moderate oral toxicity. Dangerous fire hazard when exposed to heat or flame; can react with oxidizing materials. To fight fire, use alcohol foam, CO_2, dry chemical.

HEXYLRESORCINOL HR: 3
CAS: 136-77-6 NIOSH: VH 1575000
mf: $C_{12}H_{18}O_2$ mw: 194.30

PROP: Colorless liquid; bp: 179°; very sol in water; sol in benzene, eth and acetone. Pale yellow, heavy liquid becoming solid on standing @ room temp. Needles from benzene or petr ether, mp: 67.5°-69°. Pungent odor, sharp astringent taste; sol in ether, chloroform, acetone, alc, vegetable oils; slightly sol in petr ether.

SYNS: 4-HEXYL-1,3-DIHYDROXYBENZENE * P-HEXYLRESORCINOL * 4-HEXYLRESORCINOL * 4-N-HEXYLRESORCINOL * NCI-C55787 * SUCRETS

THR: A poison via ingestion, intraperitoneal, and other routes. Moderately toxic via subcutaneous route. An eye irritant. Concentrated solutions can cause burns on skin and mucous membranes on humans. When heated to decomposition it emits acrid smoke and fumes.

HEXYLTRICHLOROSILANE   HR: 3
CAS: 928-65-4   NIOSH: VV 4320000
DOT: 1784
mf: $C_6H_{13}Cl_3Si$ mw: 219.63
DOT Classification: Label: Corrosive

THR: A poison via irritation to skin, eyes and mucous membranes and by ingestion and inhalation. Dangerous; see chlorides; will react with water or steam to produce toxic and corrosive fumes.

HISPACID FAST ORANGE 2G   HR: 3
CAS: 1936-15-8   NIOSH: QJ 6500000
mf: $C_{16}H_{10}N_2O_7S_2 \cdot 2Na$ mw: 452.38

SYNS: ACID FAST ORANGE EGG * ACID LIGHT ORANGE G * ACID ORANGE 10 * BUCACID FAST ORANGE G * CALCOCID FAST LIGHT ORANGE 2G * C.I. 27 * 7-HYDROXY-8-(PHENYLAZO)-1,3-NAPHTHALENEDISULFONIC ACID, DISODIUM SALT * 7-HYDROXY-8-(PHENYLAZO)-1,3-NAPHTHALENEDISULPHONIC ACID, DISODIUM SALT * JAVA ORANGE 2G * INK ORANGE JSN * NCI-C53838 * ORANGE G DYE * 1-PHENYLAZO-2-NAPHTHOL-6,8-DISULFONIC ACID, DISODIUM SALT

THR: An experimental carcinogen. When heated to decomposition it emits very toxic SO_x and NO_x.

HISTAMETHIZINE   HR: 3
CAS: 569-65-3   NIOSH: TL 1970000
mf: $C_{25}H_{27}ClN_2$ mw: 390.99

SYNS: 1-(P-CHLOROBENZHYDRYL)-4-(M-METHYLBENZYL)DIETHYLENEDIAMINE * 1-P-CHLOROBENZHYDRYL-4-M-METHYLBENZYLPIPERAZINE * 1-(P-CHLORO-alpha-PHENYLBENZYL)-4-(M-METHYLBENZYL)PIPERAZINE * HISTAMETHINE * HISTAMETIZINE * HISTAMETIZYNE

THR: A poison via intravenous route. Moderately toxic by ingestion, and intramuscular routes. When heated to decomposition it emits very toxic fumes of Cl^- and NO_x.

HISTAMINE   HR: 2
CAS: 51-45-6   NIOSH: MS 1050000
mf: $C_5H_9N_3$ mw: 111.17

PROP: Colorless hygroscopic crystals or aqueous, very sol in water; sol in hot chloroform sol in alc; insol in ether. Deliquescent needles, mp: 84°, bp: 210° @ 18 mm.

SYNS: ERAMIN * ERGAMINE * ERGOTIDINE * FREE HISTAMINE * IMIDAZOLE-4-ETHYL-AMINE * 4-IMIDAZOLEETHYLAMINE * 5-IMIDAZOLEETHYLAMINE * beta-IMIDAZOLYL-4-ETHYLAMINE * 2-(4-IMIDAZOLYL)ETHYL-AMINE * THERAMINE

THR: A poison via intravenous and subcutaneous routes. Mutagenic data. The most potent capillary dilator known. Ingestion or inhalation of this material produces the following effects: flushing followed by pallor, dizziness, fainting, fall in blood pressure, headache, rapid, weak pulse. Allergic effects on skin (hives) may occur. Dangerous; when heated to decomposition it emits highly toxic fumes. Treatment and Antidotes: If swallowed, wash out stomach. Administer stimulants (epinephrine or ephedrine). Call a physician.

HOLMIUM
mf: Ho   mw: 164.93

PROP: Bp: 2720°, d: 8.78 @ 25°, vap press 2 mm @ 1630°. Bright metallic luster, soft, malleable metal, stable in dry air, oxidizes rapidly in moist air.

THR: No toxicity data. It may be an anticoagulant like the lanthanides. Can react violently with air; halogens.

HOLMIUM CHLORIDE   HR: 3
CAS: 10138-62-2   NIOSH: MS 7350000
mf: $HoCl_3$ mw: 271.28

PROP: Bright yellow cryst solid; mp: 718°

THR: A poison by intraperitoneal route. Slightly toxic by ingestion. See also rare earths. Dangerous; when heated to decomposition it emits highly toxic fumes of Cl^-.

HOLMIUM CITRATE   HR: 3
CAS: 13455-50-0   NIOSH: GE 7690000
mf: $C_6H_5O_7 \cdot Ho$ mw: 354.04

THR: A poison by intraperitoneal route. See also holmium, rare earths. When heated to decomposition it emits acrid smoke and fumes.

HOLMIUM(III) NITRATE, HEXAHYDRATE (1:::::3:6)   HR: 3
CAS: 35725-31-6   NIOSH: MS 8050000
mf: $N_3O_9 \cdot Ho \cdot 6H_2O$ mw: 459.08

SYN: NITRIC ACID, HOLMIUM(3+) SALT, HEXAHYDRATE

THR: A poison by intraperitoneal route. Moderately toxic by ingestion. See also holmium. See also rare earths. When heated to decomposition it emits toxic fumes of NO_x.

3-HOMOTETRA HYDRO CANNI-BINOL HR: 3
CAS: 117-51-1   NIOSH: HP 7875000
mf: $C_{22}H_{32}O_2$ mw: 328.54

SYNS: 3-HEXYL-7,8,9,10-TETRAHYDRO-6,6,9-TRIMETHYL-6H-DIBENZO(B,D)PYRAN-1-OL * 1-HYDROXY-3-N-HEXYL-6,6,9-TRIMETHYL-7,8,9,10-TETRAHYDRO-6-DIBENZOPYRAN * PARAHEXYL * PYRAHEXYL * SYNHEXYL

THR: A poison by intravenous route. Moderately toxic by ingestion and intraperitoneal routes.

HOSDON GRANULE HR: 3
CAS: 36614-38-7   NIOSH: TE 4465000
mf: $C_7H_{17}O_2PS_3$ mw: 260.39

SYNS: O,O-DIMETHYL-S-2-(ISOPROPYLTHIO)-ETHYLPHOSPHORODITHIOATE * ISOTHIOATE

THR: Poison by ingestion and skin contact. When heated to decomposition it emits very toxic fumes of POx and SOx.

HUMAN SPERM HR: 3
NIOSH: WG 8270000

THR: An experimental carcinogen. Low intraperitoneal toxicity.

HYCANTHONE METHANESUL-FONATE HR: 3
CAS: 23255-93-8   NIOSH: XO 1600000
mf: $C_{20}H_{24}N_2O_2S \cdot CH_4O_3S$ mw: 452.63

SYNS: 1-((2-(DIETHYLAMINO)ETHYL)AMINO)-4-(HYDROXYMETHYL)-9H-THIOXANTHEN-9-ONE MONOMETHANE-SULFONATE (SALT) * HYCANTHONE MESYLATE * HYCANTHONE MONOMETH-ANESULPHONATE

THR: A poison by intraperitoneal, subcutaneous, intravenous, and intramuscular routes. Mutagenic data. An experimental teratogen, carcinogen. A systemic toxin in humans. See also sulfonates. When heated to decomposition it emits very toxic fumes of SO_x.

HYDRACRYLONITRILE HR: 2
CAS: 109-78-4   NIOSH: MU 5250000
mf: C_3H_5NO mw: 71.09

PROP: Colorless to straw-colored liquid, bp: 228° decomp, fp: −46°, flash p: 265°F (OC), d: 1.0404 @ 25°, vap press: 0.08 mm @ 25°, vap d: 2.45. Misc with H_2O, acetone, methyl ethyl ketone, ethanol. Sltly sol in ether. Insol in benzene, petr ether, carbon disulfide, carbon tetrachloride.

SYNS: 2-CYANOETHANOL * 2-CYANOETHYL ALCOHOL * ETHYLENE CYANOHYDRIN * GLYCOL CYANOHYDRIN * BETA-HPN * 3-HYDROXYPROPANENITRILE * BETA-HYDROXYPROPIONITRILE * 3-HYDROXYPROPIONITRILE * METHANOLACETONITRILE * USAF RH-7

THR: Moderately toxic by ingestion and intraperitoneal routes. A mild irritant by skin and eye routes. See also nitriles. Combustible when exposed to heat or flame. Dangerous; when heated or on contact with acid or acid fumes, it emits highly toxic fumes of cyanide; will react with water or steam to produce toxic and flammable vapors; can react vigorously with chlorosulfonic acid; oleum; NaOH; H_2SO_4. See also cyanides. To fight fire, use CO_2, dry chemical, alcohol foam. For further information see ethylene cyanohydrin, Vol. 4, No. 2 of *DPIM Report*.

HYDRALAZINE HYDROCHLO-RIDE HR: 3
CAS: 304-20-1   NIOSH: TH 9000000
mf: $C_8H_8N_4 \cdot ClH$ mw: 196.66

PROP: Yellow crysts; decomp @ 273°; very sltly sol in ether.

SYNS: HYDRALAZINE CHLORIDE * HYDRALAZINE MONOHYDROCHLORIDE * 1-HYDRAZINO-PHTHALAZINE HYDROCHLORIDE * 1-HYDRAZINOPHTHLAZINE MONOHYDROCHLORIDE * 1(2H)-PHTHALAZINONE HYDRAZONE HYDROCHLORIDE

THR: A poison by intraperitoneal route. Mutagenic data. An experimental neoplastigen and suspected carcinogen. When heated to decomposition it emits very toxic NO_x and HCl.

HYDRAZINE HR: 3
CAS: 302-01-2   NIOSH: MU 7175000
DOT: 2029, 2030
mf: H_4N_2 mw: 32.06

PROP: Colorless, oily fuming liquid, or white crystals, mp: 1.4°, bp: 113.5°, flash p: 100°F (OC), d: 1.1011 @ 15° (liq), autoign temp:

can vary from 74°F in contact with iron rust to 270°F in contact with black iron to 313°F in contact with stainless steel to 518°F in contact with glass, vap d: 1.1; lel = 4.7%, uel = 100%.

SYNS: DIAMIDE * DIAMINE * HYDRAZINE BASE * HYDRAZINE, ANHYDROUS (DOT)

OSHA PEL: TWA 1 ppm (skin)
DFG MAK: 0.1 ppm
DOT Classification: Label: Flammable
  Liquid and Poison

THR: A poison by ingestion, intravenous and dermal routes. May cause skin sensitization as well as systemic poisoning. Hydrazine and some of its derivatives may cause damage to the liver and destruction of red blood cells. See phenyl hydrazine. An experimental neoplastigen, and carcinogen of lung, nervous system, liver, kidney, hematopoietic organs, breast, subcutaneous tissue. Mutagenic data. Dangerous fire hazard when exposed to heat, flame or oxidizing agents; a powerful reducing agent. Severe explosion when exposed to heat or flame or by chemical reaction with alkali metals; BaO; CaO; K; Na; NH_3; Cl_2; chromates; CuO; Cu^{++} salts; F_2; H_2O_2; iron rust; metallic oxides; Ni; $Ni(ClO_4)_2$; HNO_3; N_2O; O_2; liquid O_2; $K_2Cr_2O_7$; $Na_2Cr_2O_7$; tetryl; zinc diamide; $Zn(C_2H_5)_2$. Can self-ignite when absorbed on earth, asbestos, cloth, wood. It is a powerful explosive, much used in rocket fuels. It is very sensitive and must not be used without full and complete instructions from the manufacturer as to handling, storage and disposal. Dangerous; when heated to decomposition it emits highly toxic nitrogen compounds; may explode by heat or chemical reaction. For further information see Vol. 3, No. 4 of *DPIM Report*.

HYDRAZINECARBOXAMIDE   HR: 3
CAS: 57-56-7   NIOSH: VT 2975000
mf: CH_5N_3O mw: 75.09

SYNS: SEMICARBAZIDE * AMINOUREA * CARBAMYLHYDRAZINE * CARBAMIC ACID HYDRAZIDE * CARBAZAMIDE

THR: A poison by ingestion, intraperitoneal, subcutaneous, and intravenous routes. Mutagenic data. Affects the human central nervous system. An experimental tumorigen. When heated to decomposition it emits toxic fumes of NO_x. For further information see Vol. 4, No. 4 of *DPIM Report*.

HYDRAZINE HYDRATE   HR: 3
CAS: 7803-57-8   NIOSH: MV 8050000
mf: $H_4N_2 \cdot H_2O$ mw: 50.08

PROP: Colorless fuming, refractive liquid; mp: −51.7°; bp: 118.5° @ 740 mm. d: 1.03 @ 21°. Faint characteristic odor. Violent poison causes delayed eye irritation. A strong base, very corrosive, attacks glass, rubber, cork. Very powerful reducing agent. Misc with H_2O and alcohol. Insol in chloroform, ether.

SYN: IDRAZINA IDRATA (ITALIAN)

THR: A poison by ingestion. An experimental carcinogen. Mutagenic data. Incompatibles: HgO; Na; $SnCl_2$; 2,4-dinitrochlorobenzene. When heated to decomposition it emits toxic fumes of NO_x. For further information see Vol. 1, No. 5 of *DPIM Report*.

HYDRAZINE HYDROCHLORIDE   HR: 3
CAS: 2644-70-4   NIOSH: MV 4625000
mf: $H_4N_2 \cdot ClH$ mw: 68.52

THR: A poison by ingestion, intravenous, and intraperitoneal routes. When heated to decomposition it emits very toxic fumes of Cl^- and NO_x.

HYDRAZINE SULFATE (1:1)   HR: 3
CAS: 10034-93-2   NIOSH: MV 9625000
mf: $H_4N_2 \cdot H_2O_4S$ mw: 130.14

PROP: Colorless crystals, water-sol, insol in alcohol. Very sol in hot water. d: 1.378, mp: 85°.

SYNS: HYDRAZINE SULPHATE * IDRAZINA SOLFATO (ITALIAN) * NSC-150014 * SIRAN HYDRAZINU (CZECH)

THR: A poison by acute ingestion and intraperitoneal routes. An experimental carcinogen, tumorigen, and neoplastigen. Mutagenic data. In humans it causes gastrointestinal and central nervous system problems. When heated to decomposition it emits very toxic fumes of SO_x and NO_x. For further information see Vol. 1 No. 5 of *DPIM Report*.

2-HYDRAZINOETHANOL   HR: 3
CAS: 109-84-2   NIOSH: KL 2800000
mf: $C_2H_8N_2O$ mw: 76.12

PROP: Colorless, slightly viscous liquid, mp: −70°, bp: 145°-153° @ 25 mm, flash p: 224°F, vap d: 2.63; d: 1.11. Misc with H_2O, sol in lower alcohols; sltly sol in ether.

SYNS: beta-HYDROXYETHYLHYDRAZINE ∗ N-(2-HYDROXYETHYL)HYDRAZINE

THR: An experimental carcinogen. Mutagenic data. Flammable when exposed to heat or flame; can react with oxidizing materials. To fight fire, use foam, CO_2, dry chemical. When heated to decomposition it emits toxic fumes such as NO_x.

2-HYDRAZINO-4-(5-NITRO-2-FURYL)-THIAZOLE   HR: 3
CAS: 26049-68-3   NIOSH: XJ 4900000
mf: $C_7H_6N_4O_3S$ mw: 226.23

SYN: HNT

THR: An experimental carcinogen, neoplastigen, and tumorigen. When heated to decomposition it emits very toxic fumes of NO_x and SO_x.

2-HYDRAZINO-4-(4-NITROPHENYL)-THIAZOLE   HR: 3
CAS: 26049-70-7   NIOSH: XJ 5075000
mf: $C_9H_8N_4O_2S$ mw: 236.27

THR: An experimental carcinogen. When heated to decomposition it emits very toxic fumes of NO_x and SO_x.

4-HYDRAZINO-2-THIOURACIL   HR: 3
CAS: 63981-09-9   NIOSH: YR 0500000
mf: $C_4H_6N_4S$ mw: 142.20

SYN: 2-THIO-4-HYDRAZINOURACIL

THR: A poison by intraperitoneal route. When heated to decomposition it emits very toxic NO_x and SO_x.

HYDRAZOBENZENE   HR: 3
CAS: 122-66-7   NIOSH: MW 2625000
mf: $C_{12}H_{12}N_2$ mw: 184.26

PROP: Light or yellow crystals from ethanol; d: 1.58; mp: 131°; bp: decomp; very sltly sol in water; insol in acetylene.

SYNS: 1,2-DIPHENYLHYDRAZINE ∗ (SYM)-DIPHENYLHYDRAZINE ∗ HYDRAZOBENZEN (CZECH) ∗ NCI-c01854

THR: An experimental carcinogen. When heated to decomposition it emits toxic fumes of NO_x. For further information see 1,2-diphenyl hydrazine, Vol 6, No. 1 of *DPIM Report*.

HYDRAZOIC ACID   HR: 3
CAS: 7782-79-8   NIOSH: MW 2800000
mf: HN_3 mw: 43.04

PROP: Colorless liquid, very sol in water, intolerable pungent odor, mp: −80°, bp: 37°, d: 1.09 @ 25°/4°.

SYNS: AZOIMIDE ∗ DIAZOIMIDE ∗ HYDROGEN AZIDE ∗ HYDRONITRIC ACID ∗ TRIAZOIC ACID

THR: A poison by irritation to skin, eyes and mucous membranes and by ingestion and inhalation. Continued inhalation causes cough, headache, fall in blood pressure, collapse, chills and fever. High concentrations can cause fatal convulsions. Chronic exposure has been reported as causing injury to kidneys and spleen, hypotension, palpitation, ataxia, weakness. Dangerous explosion hazard when shocked or exposed to heat. Reacts violently with Cd; Cu; Ni; HNO_3; F_2.

2,2′-HYDRAZONODIETHANOL   HR: 3
CAS: 13529-51-6   NIOSH: KL 2850000
mf: $C_4H_{12}N_2O_2$ mw: 120.18

SYN: 1,1-DIETHANOLHYDRAZINE

THR: A poison by subcutaneous route. An experimental tumorigen. When heated to decomposition it emits toxic fumes such as NO_x.

HYDRIDES   HR: VARIABLE
THR: Variable toxicity. The hydrides of phosphorus, arsenic, sulfur, selenium, tellurium and boron which are of high toxicity produce local irritations and destroy red blood cells. They are particularly dangerous because of their volatility and ease of entry into the body. The hydrides of the alkali metals, alkaline earths, aluminum, zirconium and titanium react with moisture to evolve hydrogen and leave behind the hydroxide of the metallic element. This hydroxide is usually caustic. See also sodium hydroxide. Hydrides, metallic, primary type. This group includes the hydrides of calcium, lithium, magnesium, potassium, sodium and strontium. In the presence of moisture they are readily converted to hydroxides which are highly irritating to the skin by caustic and thermal action. Similar effects can occur on contact with eyes and respiratory mucous membranes. The volatile hydrides are flammable, some spontaneously so in air. All hydrides react violently on contact with powerful oxidizing agents. When heated or on contact with moisture or acids an exothermic reaction evolving hydrogen occurs. Often enough heat is evolved to cause ignition.

Hydrides require special handling instructions which should be obtained from the manufacturers. The volatile hydrides (such as hydrides of boron, arsenic, phosphorus, selenium, tellurium) form explosive mixtures with air. The nonvolatile hydrides (such as sodium, lithium, calcium) readily liberate hydrogen when heated or on contact with moisture or acids. Furthermore, hydrides form dust clouds which can explode due to contact with flames, sparks, heat or oxidizers. Highly dangerous; when heated, they can ignite at once or liberate hydrogen: they react with moisture or acids to evolve heat and hydrogen; on contact with powerful oxidizers violent reactions can occur.

HYDRIODIC ACID   HR: 3
CAS: 10034-85-2   NIOSH: MW 3760000
mf: HI   mw: 127.91

PROP: Colorless when freshly made, but rapidly turns yellowish or brown on exposure to light or air. Keep protected from light and air, preferably not above 3°. Misc with water and alc, mp: $-50.8°$, bp: $-35.38°$ @ 5 atm, d: 5.66g/liter @ 0°.

SYNS: ANHYDROUS HYDRIODIC ACID * HYDRIODIC ACID (DOT) * HYDROGEN IODIDE * HYDROGEN IODIDE SOLUTION (DOT)

THR: A poison by irritation to skin, eyes and mucous membranes and by ingestion and inhalation. Violent reaction with F_2; ($HClO_4$ + Mg); HNO_3; O_3; K; $KClO_3$; metals. Dangerous; when heated to decomposition it emits highly toxic iodides; will react with water or steam to produce toxic and corrosive fumes.

HYDROBROMIC ACID   HR: 3
CAS: 10035-10-6   NIOSH: MW 3850000
DOT: 1048/1788
mf: BrH   mw: 80.92

PROP: Colorless gas or pale yellow liquid, mp: $-87°$, bp: $-66.5°$, d: 3.50 g/L @ 0°. Misc with water, alc. Keep protected from light.

SYNS: ACIDE BROMHYDRIQUE (FRENCH) * ACIDO BROMIDRICO (ITALIAN) * ANHYDROUS HYDROBROMIC ACID * BROMOWODOR (POLISH) * BROMWASSERSTOFF (GERMAN) * BROOMWATERSTOF (DUTCH) * HYDROGEN BROMIDE

OSHA PEL: TWA 3 ppm
DOT Classification: Label: Corrosive

THR: A poison by ingestion, inhalation, and tissue irritation. Dangerous; see also bromides; reacts with water or steam to produce toxic and corrosive fumes. Reacts violently with F_2; Fe_2O_3; NH_3; O_3.

HYDROCARBON GAS (NONLIQUEFIED)   HR: 3
NIOSH: MW 3860000

PROP: Contains hydrogen, methane, carbon monoxide, lel = 5.3%, uel = 31%, autoign temp: 1200°F.

SYN: COAL GAS

DOT Classification: Label: Flammable Gas

THR: A poison by inhalation. See also carbon monoxide. Very dangerous fire hazard; see hydrogen and carbon monoxide. Moderate explosion hazard when exposed to heat or flame. Dangerous; see methane. To fight fire, stop flow of gas; CO_2, dry chemical or water spray.

HYDROCHLORIC ACID   HR: 3
CAS: 7647-01-0   NIOSH: MW 4025000
DOT: 1050/1789/2186
mf: HCl   mw: 36.46

PROP: Colorless gas or colorless fuming liquid, strongly corrosive, mp: $-114.3°$, bp: $-84.8°$, d: 1.639 g/L(gas) @ 0°, 1.194 @ $-26°$ (liquid), vap press: 4.0 atm @ 17.8°.

SYNS: ACIDE CHLORHYDRIQUE (FRENCH) * ACIDO CLORIDRICO (ITALIAN) * CHLOORWATERSTOF (DUTCH) * CHLOROHYDRIC ACID * CHLOROWODOR (POLISH) * CHLORWASSERSTOFF (GERMAN) * HYDROCHLORIC ACID (DOT) * HYDROCHLORIDE * MURIATIC ACID (DOT) * SPIRITS OF SALT (DOT)

OSHA PEL: TWA CL 5 ppm
DOT Classification: Label: Corrosive

THR: Mutagenic data. Irritating to skin, eyes and mucous membranes and by ingestion and inhalation. In general, hydrochloric acid causes little trouble in industry, other than from accidental splashes and burns. It is used as a general purpose food additive. It is a common air contaminant. Violent reactions with acetic anhydride; 2-amino ethanol; NH_4OH; Ca_3P_2; chlorosulfonic acid; ethylene diamine; ethylene imine; oleum; $HClO_4$; β-propiolactone; propylene oxide; ($AgClO_4$ + CCl_4); NaOH; H_2SO_4; U_3P_4; vinyl acetate. Also CaC_2; CsC_2H; Cs_2C_2; Li_6Si;

Mg_3B_2; $HgSO_4$; RbC_2H; Rb_2C_2; Na. Dangerous; see chlorides; will react with water or steam to produce toxic and corrosive fumes.

HYDROCHLORIC ACID, MIXED WITH NITRIC ACID (3:1)   HR: 3
CAS: 8007-56-5   NIOSH: MW 4100000
DOT: 1798

SYNS: AQUA REGIA * NITROHYDROCHLORIC ACID * NITROMURIATIC ACID

DOT Classification: Label: Corrosive

THR: See also hydrochloric acid and nitric acid and nitrosyl chloride. When heated to decomposition it emits very toxic HCl, HNO_3, Cl^- and NO_x.

HYDROCINNAMALDEHYDE   HR: 3
CAS: 104-53-0   NIOSH: MW 4890000
mf: $C_9H_{10}O$ mw: 134.19

PROP: Liquid; bp: 221-224°; insol in water.

SYNS: BENZENEPROPANAL * BENZYLACETAL-DEHYDE * DIHYDROCINNAMALDEHYDE * HYDROCINNAMIC ALDEHYDE * 3-PHENYL-1-PROPANAL * 3-PHENYLPROPIONALDEHYDE * 3-PHENYLPROPYL ALDEHYDE

THR: A poison by skin route. A skin irritant in humans. When heated to decomposition it emits acrid smoke and fumes.

HYDROCORTISONE SODIUM SUCCINATE   HR: 3
CAS: 125-04-2   NIOSH: GM 9015000
mf: $C_{25}H_{35}O_9 \cdot Na$ mw: 502.59

PROP: White, odorless, hygroscopic, amorphous solid, very sol in water and alcohol, insol in chloroform, very sltly sol in acetone. mp: 169°-171°.

SYN: CORTISOL SUCCINATE, SODIUM SALT

THR: An experimental teratogen. Used as a food additive permitted in food for human consumption. When heated to decomposition it emits toxic fumes of Na_2O.

HYDROCOUMARIN   HR: 3
CAS: 119-84-6   NIOSH: MW 5775000
mf: $C_9H_8O_2$ mw: 148.17

SYNS: 1,2-BENZODIHYDROPYRONE * DIHY-DROCOUMARIN * 3,4-DIHYDROCOUMARIN * MELILOTIN * NCI-C55890 * USAF DO-12

THR: A poison by intraperitoneal route. Moderately toxic by ingestion. A moderate skin irritant. When heated to decomposition it emits acrid smoke and fumes.

HYDROCYANIC ACID   HR: 3
CAS: 74-90-8   NIOSH: MW 6825000
DOT: 1614
mf: CHN   mw: 27.03

PROP: Odor of bitter almonds, mp: -13.2°, bp: 25.7°, lel = 5.6%, uel = 40%, flash p: 0°F (CC), d: 0.6876 @ 20°/4°, autoign temp: 1000°F, vap press: 400 mm @ 9.8°, vap d: 0.932. Misc in water, alc and ether.

SYNS: ACIDE CYANHYDRIQUE (FRENCH) * ACIDO CIANIDRICO (ITALIAN) * AERO LIQUID HCN * BLAUSAEURE (GERMAN) * BLAUW-ZUUR (DUTCH) * CYAANWATERSTOF (DUTCH) * CYANWASSERSTOFF (GERMAN) * CYCLON * CYCLONE B * CYJANOWODOR (POLISH) * HYDROCYANIC ACID, LIQUEFIED (DOT) * HYDROGEN CYANIDE * PRUSSIC ACID * ZACLON DISCOIDS

OSHA PEL: TWA 10 ppm (skin)
DOT Classification: Poison A, Label: Poison Gas and Flammable Gas

THR: VERY poisonous by ingestion, dermal, inhalation, intravenous, intraperitoneal and implantation routes. Hydrocyanic acid and the cyanides are true protoplasmic poisons, combining in the tissues with the enzymes associated with cellular oxidation. They thereby render the oxygen unavailable to the tissues, and cause death through asphyxia. The suspension of tissue oxidation lasts only while the cyanide is present; upon its removal, normal function is restored provided death has not already occurred. HCN does not combine easily with hemoglobin, but it does combine readily with met-hemoglobin to form cyanmethemoglobin. This fact is utilized in the treatment of cyanide poisoning, when an attempt is made to induce met-hemoglobin formation. The presence of cherry-red venous blood in cases of cyanide poisoning is due to the inability of the tissues to remove the oxygen from the blood. Exposure to concentrations of 100-200 ppm for periods of 30-60 minutes can cause death. In cases of acute cyanide poisoning, death is extremely rapid; though sometimes breathing may continue for a few minutes. In less acute cases, there is headache, dizziness, unsteadiness of gait, a feeling of suffocation, and nausea. Where the patient recovers, there is rarely any disability. An insecticide. Very dangerous fire hazard when exposed to heat, flame or oxidizers. Can polymerize @ 50°-60°, or catalyze with traces of alkali. Reacts violently with acetaldehyde. Severe explosion hazard when exposed to heat or flame or by chemical reaction with oxidizers. Under certain conditions, particularly contact with alkaline materials, HCN can polymerize or decompose explosively. The liquid is commonly stabilized by addition of acids. Highly dangerous; the gas forms explosive mixtures with air; will react with water, steam, acid or acid fumes to produce highly toxic fumes of cyanide. To fight fire, use CO_2, non-alkaline dry chemical, foam.

HYDROCYANIC ACID (SOLUTION)   HR: 3
CAS: 74-90-8   NIOSH: MW 6850000

SYNS: HYDROCYANIC ACID, SOLUTION (DOT) * PRUSSIC ACID SOLUTION

DOT Classification: Poison A, Label: Poison Gas and Flammable Gas

THR: A deadly poison to all living things. See also hydrocyanic acid.

HYDROFLUORIC ACID   HR: 3
CAS: 7664-39-3   NIOSH: MW 7875000
DOT: 1052
mf: FH   mw: 20.01

PROP: Clear, colorless, fuming corrosive liquid or gas, mp: −83.1°, bp: 19.54°, d: 0.901 g/L(gas); 0.699 @ 22° (liquid), vap press: 400 mm @ 2.5°.

SYNS: ACIDE FLUORHYDRIQUE (FRENCH) * ACIDO FLUORIDRICO (ITALIAN) * FLUORO-WODOR (POLISH) * FLUORWASSERSTOFF (GERMAN) * FLUORWATERSTOF (DUTCH) * HYDROFLUORIC ACID, ANHYDROUS (DOT) * HYDROGEN FLUORIDE (DOT)

OSHA PEL: TWA 3 ppm
DOT Classification: Label: Corrosive

THR: A poison by irritation to skin, eyes (@ 0.05 mg/L), mucous membranes and by ingestion. It is extremely irritating and corrosive to the skin and mucous membranes. Inhalation of the vapor may cause ulcers of the upper respiratory tract. Concentrations of 50-250 ppm are dangerous, even for brief exposures. Hydro-

fluoric acid produces severe skin burns which are slow in healing. The subcutaneous tissues may be affected becoming blanched and bloodless. Gangrene of the affected areas may follow. See also fluorides. Mutagenic data. It is a common air contaminant. Violent reaction with As_2O_3; P_2O_5; acetic anhydride; 2-amino ethanol; NH_4OH; $HBiO_3$; CaO; chlorosulfonic acid; ethylene diamine; ethylene imine; F_2; (HNO_3 + lactic acid); oleum; β-propiolactone; propylene oxide; Na; NaOH; H_2SO_4; vinyl acetate; HgO; sodium tetrafluoro silicate; n-phenyl azo piperidine. Dangerous; when heated it emits highly corrosive fumes of fluorides; will react with water or steam to produce toxic and corrosive fumes. For further information see Vol. 5, No. 6 of *DPIM Report*.

HYDROFLUORIC ACID (SOLUTION)   HR: 3
CAS: 7664-39-3   NIOSH: MW 7890000

SYNS: FLUORIC ACID * HYDROFLUORIC ACID SOLUTION

DOT Classification: Label: Corrosive

THR: A very corrosive solution. See also hydrofluoric acid.

HYDROGEN   HR: 2
CAS: 1333-74-0   NIOSH: MW 8900000
DOT: 1049/1966
mf: H_2 mw: 2.02

PROP: Colorless, odorless, tasteless gas, mp: −259.18°, bp: −252.8°, lel = 4.1%, uel = 74.2%, d: 0.0899 g/L, autoign temp: 752°F, vap d: 0.069.
DOT Classification: Label: Flammable Gas

THR: Practically no toxicity except that it may asphyxiate. Highly dangerous fire hazard when exposed to heat, flame or oxidizers. Flammable or explosive when mixed with air, O_2, chlorine. To fight fire, stop flow of gas. Severe explosion hazard when exposed to heat or flame. Violent reaction with (air + Pt); Br_2; Cl_2; I_2; ClF_3; (dioxane + Ni); F_2; Li; (Mg + $CaCO_3$); o-nitroanisole; NF_3; OF_2; (Pd + isopropyl alcohol); 3-methyl-2-penten-4-yn-1-ol; PbF_3; oxidants. Dangerous; can react vigorously with oxidizing materials.

HYDROGEN CHLORIDE   HR: 3
CAS: 7647-01-0   NIOSH: MW 9610000
mf: ClH   mw: 36.47

PROP: Colorless, corrosive, nonflammable gas. Pungent odor; fumes in air; d: 1.639 @ −137.77°; bp: −154.37° @ 1.0 mm.

SYN: HYDROCHLORIC ACID, ANHYDROUS (DOT)

DOT Classification: Label: Nonflammable Gas

THR: An acute poison by all routes. See also hydrogen chloride (vapor) and hydrogen chloride (aerosol), and hydrochloric acid. When heated it emits toxic fumes of HCl. Incompatible with Al; F_2; hexalithium disilicide; metal acetylides; metal carbides; $KMnO_4$; Na; H_2SO_4; tetraselenium tetranitride. For further information see Vol. 1, No. 7 of *DPIM Report*.

HYDROGEN CHLORIDE (AEROSOL)   HR: 3
CAS: 7647-01-0   NIOSH: MW 9620000

PROP: Saturated water aerosol mist

THR: High acute toxicity by inhalation. A very powerful irritant by inhalation, skin, eyes, and mucous membranes in humans. See also hydrogen chloride.

HYDROGEN CHLORIDE (VAPOR) HR: 3
CAS: 7647-01-0   NIOSH: MW 9625000
mf: ClH   mw: 36.46
OSHA PEL: TWA CL 5 ppm

THR: High acute toxicity to humans and experimentally by inhalation. See also hydrogen chloride.

HYDROGEN PEROXIDE, 90%   HR: 3
CAS: 7722-84-1   NIOSH: MX 0900000
DOT: 2015
mf: H_2O_2 mw: 34.02

PROP: Colorless heavy liquid, or, at low temp, a crystalline solid, d: 1.71 @ −20°, 1.46 @ 0°, vap press: 1mm @ 15.3°, unstable. Bitter taste; mp: −0.43°; bp: 152°. Misc with H_2O; sol in ether; insol in petr ether. Decomp by many organic solvents.

SYNS: ALBONE * HYDROGEN PEROXIDE SOLUTION (DOT) * INHIBINE * PERHYDROL * PEROSSIDO DI IDROGENO (ITALIAN) * PEROXAN * PEROXYDE D'HYDROGENE (FRENCH) * WASSERSTOFFPEROXID (GERMAN) * WATERSTOFPEROXYDE (DUTCH)

OSHA PEL: TWA 1 ppm

DOT Classification: Label: Oxidizer

THR: A poison by irritation to skin, eyes and mucous membranes and by ingestion and inhalation. Mutagenic data. A very powerful oxidizer. Pure H_2O_2, its solutions, vapors and mists are very irritating to body tissue. This irritation can vary from mild to severe depending upon the concentration of H_2O_2. For instance solutions of H_2O_2 of 35 wt% and over can easily cause blistering of the skin. Irritation caused by H_2O_2 which does not subside upon flushing the affected part with water should be treated by a physician. The eyes are particularly sensitive to this material. It is used as a general-purpose food additive; it is a substance which migrates to food from packaging materials. It is a common air contaminant. Dangerous fire hazard by chemical reaction with flammable materials. H_2O_2 is a powerful oxidizer, particularly in the concentrated state. It is important to keep containers of this material covered because (1) uncovered containers are much more prone to react with flammable vapors, gases, etc.; (2) because if uncovered, the water from an H_2O_2 solution can evaporate, concentrating the material and thus increasing the fire hazard of the remainder. For instance, solutions of H_2O_2 in concentration in excess of 65 wt% heat up spontaneously when decomposed to $H_2O + \frac{1}{2}O_2$. Thus 90 wt% solutions, when caused to decompose rapidly due to the introduction of a catalytic decomposition agent, can get quite hot and perhaps start fires. Severe explosion hazard when highly concentrated or pure H_2O_2 is exposed to heat, mechanical impact, detonation of a blasting cap, or caused to decomposition catalytically by metals, or on contact with acetic acid; acetic anhydride; acetone; (alcohols + H_2SO_4); acetal; alcohols; vinyl acetate; alcohols + tin (II) chloride; P_2O_5; P; H_2O; HNO_3; Sb_2S_3; As_2S_3; *tert*-butyl alcohol; cellulose; charcoal; (Cl_2 + KOH); + chlorosulfonic acid; CuS; ethanol; FeS; (formic acid + organic matter); H_2Se; hydrazine; (ketones + HNO_3); PbO_2; PbO; PbS; oxygenated compounds; nitrogenous bases; MnO_2; HgO; Hg_2O; MoS_2; organic matter; (2-methyl-1-phenyl-2-propanol + sulfuric acid); $KMnO_4$; $NaIO_3$; thiodiglycol; uns-dimethyl hydrazine. Carboxylic acids; ($FeSO_4$ + 2-methylpyridine + H_2SO_4); (HgO + HNO_3). BEWARE: Although many mixtures of H_2O_2 and organic materials do not explode upon contact, the resultant combination is detonatable either upon catching fire or by impact. The detonation velocity of aqueous solutions of H_2O_2 has been found to be about 6500 m/second for solutions of between 96 wt% and 100 wt% H_2O_2. Another source of H_2O_2 explosions is from sealing the material in strong containers. Under such conditions even gradual decomposition of H_2O_2 to $H_2O + \frac{1}{2}O_2$ can cause large pressures to build up in the containers which may then burst explosively. Highly dangerous because when heated, or shocked or contaminated, the concentrated material can explode or start fires.

HYDROGEN SELENIDE   HR: 3
CAS: 7783-07-5   NIOSH: MX 1050000
DOT: 2202
mf: H_2Se mw: 80.98

PROP: Colorless gas, mp: $-64°$, bp: $-41.4°$, d: 3.614 g/L (gas); 2.12 @ $-42°$ (liquid), vap press: 10 atm @ 23.4°. Flammable. Disagreeable odor. Sol in carbonyl chloride and carbon disulfide.

SYN: SELENIUM HYDRIDE

OSHA PEL: TWA 0.05 ppm
DOT Classification: Label: Flammable Gas and Poison

THR: Severe irritant to skin, eyes and mucous membranes and by inhalation. An allergen. This material is a hazardous compound of Se which can cause damage to the lungs and liver as well as conjunctivitis. It has been found that repeated 8-hour exposures to concentrations of 0.3 ppm prove experimentally fatal by causing a pneumonitis, as well as injury to the liver and spleen. Causes garlic odor of breath, dizziness, nausea. Concentrations of 0.3 ppm are readily detected by odor, but there is no noticeable irritant effect at that level. Concentrations of 1.5 ppm or higher are strongly irritating to the eyes and nasal passages. As in the case of hydrogen sulfide, the odor of hydrogen selenide in concentrations below 1 ppm disappears rapidly because of olfactory fatigue. Although the odor and irritating effects are both useful to an experienced investigator for estimating the concentration, they do not offer a dependable warning to workmen who may be exposed to gradually increasing amounts and therefore become used to it. Due to its extreme toxicity and irritating effects, it seldom is allowed to reach a concentration in which it is flammable in air. Very little data are available on possible chronic effects of this material, but it is logical to assume that when the concentration of this gas is low enough to avoid the irritant effects, only the systemic effects will be noticeable. Dangerous fire and explosion hazard; will react vigorously with powerful oxidizing agents, such as H_2O_2; HNO_3. Dangerous; forms explosive mixtures with air. See also hydrides. Highly dangerous; keep away from heat and open flame.

HYDROGEN SULFIDE   HR: 3
CAS: 7783-06-4   NIOSH: MX 1225000
mf: H_2S mw: 34.08

PROP: Colorless, flammable gas, offensive odor, mp: $-85.5°$, bp: $-60.4°$, lel = 4%, uel = 46%, autoign temp: 500°F, d: 1.539 g/L @ 0°, vap press: 20 atm @ 25.5°, vap d: 1.189.

SYNS: ACIDE SULFHYDRIQUE (FRENCH) * HYDROGENE SULFURE (FRENCH) * HYDROGEN SULFIDE (DOT) * IDROGENO SOLFORATO (ITALIAN) * SCHWEFELWASSERSTOFF (GERMAN) * SIARKOWODOR (POLISH) * STINK DAMP * SULFURETED HYDROGEN * SULFUR HYDRIDE * ZWAVELWATERSTOF (DUTCH)

OSHA PEL: TWA CL 20 ppm; Pk 50/10M
DOT Classification: Label: Flammable Gas and Poison

THR: A poison by irritation to eyes, mucous membranes and inhalation. It is both an irritant and an asphyxiant. Low concentrations of 20-150 ppm cause irritation of the eyes; slightly higher concentrations may cause irritation of the upper respiratory tract, and if exposure is prolonged, pulmonary edema may result. The irritant action has been explained on the basis that H_2S combines with the alkali present in moist surface tissues to form sodium sulfide, a caustic. With higher concentration the action of the gas on the nervous system becomes more prominent, and a 30-minute exposure to 500 ppm results in headache, dizziness, excitement, staggering gait, diarrhea and dysuria, followed sometimes by bronchitis or bronchopneumonia. The action on the nervous system is, with small amounts, one of depression; in larger amounts, it stimulates, and with very high amounts the respiratory center is paralyzed. Exposures of 800-1000 ppm may be fatal in 30 minutes, and high concentrations are instantly fatal. Fatal hydrogen sulfide poisoning may occur even more rapidly than that following exposure to a similar concentration of HCN. H_2S does not combine with the hemoglobin of the blood; its asphyxiant action is due to paralysis of the respiratory center. With repeated exposures to low concentrations, conjunctivitis, photophobia, corneal bullae, tearing, pain and blurred vision are the commonest findings. High concentrations may cause rhinitis, bronchitis, and occasionally pulmonary edema. Exposure to very high concentrations results in immediate death. Chronic poisoning results in headache, inflammation of the conjunctivae and eyelids, digestive disturbances, loss of weight and general debility. It is a common air contaminant. It is an insidious poison since sense of smell may be fatigued and fail to give warning of high concentrations. Very dangerous fire hazard when exposed to heat, flame or oxidizers. Moderate explosion hazard when exposed to heat or flame. Reacts violently with Na_2O_2; NI_3; NCl_3; NF_3; p-bromobenzenediazonium chloride; OF_2; HNO_3; PbO_2; F_2; Cu; CrO_3; ClF_3; ClO; BrF_5; acetaldehyde; (BaO + Hg_2O + air); (BaO + NiO + air); hydrated iron oxide, phenyl diazonium + chloride; (NaOH + CaO + air); Na; metal oxides; metals; rust; soda lime; oxidants. Highly dangerous; when heated to decomposition it emits highly toxic fumes of oxides of sulfur; can react vigorously with oxidizing materials. To fight fire, stop flow of gas. For further information see Vol. 3, No. 4 of *DPIM Report*.

HYDROQUINIDINE   HR: 3
CAS: 1435-55-8   NIOSH: MX 3016000
mf: $C_{20}H_{26}N_2O_2$ mw: 326.48

PROP: Plates from ether, needles from alc; mp: 169°; very sol in hot alc; sltly sol in H_2O and ether.

THR: A poison by ingestion and intravenous routes. When heated to decomposition it emits toxic fumes of NO_x.

HYDROQUINONE   HR: 3
CAS: 123-31-9   NIOSH: MX 3500000
mf: $C_6H_6O_2$ mw: 110.12

PROP: Colorless hexagonal prisms, mp: 170.5°, bp: 286.2°, flash p: 329°F (CC), d: 1.358 @ 20°/4°, autoign temp: 960°F (CC), vap press: 1 mm @ 132.4°, vap d: 3.81. Very sol in alc, ether. Sltly sol in benzene. Keep well closed and protected from light.

SYNS: ARCTUVIN * P-BENZENEDIOL * 1,4-BENZENEDIOL * BENZOHYDROQUINONE * BENZOQUINOL * 1,4-DIHYDROXY-BENZEEN (DUTCH) * 1,4-DIHYDROXYBENZEN (CZECH) * DIHYDROXYBENZENE * P-DIHYDROXYBENZENE * 1,4-DIHYDROXYBENZENE * 1,4-DIHYDROXY-BENZOL (GERMAN) * 1,4-DIIDRO-BENZENE (ITALIAN) * P-DIOXOBENZENE * HYDROCHINON (CZECH, POLISH) * HYDROQUINOL * ALPHA-HYDROQUINONE * P-HYDROQUINONE * P-HYDROXYPHENOL * IDROCHINONE (ITALIAN) * NCI-C55834 * BETA-QUINOL * USAF EK-356

OSHA PEL: TWA 2 mg/m³

THR: A poison by ingestion, intraperitoneal, intravenous and subcutaneous routes. An active allergen and a strong irritant. An experimental neoplastigen by implantation route. Absorption of this material by tissues can cause symptoms of illness, which resemble those induced by its *o* or *m* isomers. For instance, the ingestion of 1 g by an adult or a smaller quantity by a child may induce tinnitus, nausea, dizziness, a sensation of suffocation, an increased rate of respiration, vomiting, pallor, muscular twitchings, headache dyspnea, cyanosis, delirium, and collapse. The literature contains reports of fatal cases which have been caused by the ingestion of 5-12 g. Cases of dermatitis have resulted from skin contact with this material, and have also followed the application of an antiseptic oil which apparently contained traces of hydroquinone added as an antioxidant. The report also contains cases of keratitis and discoloration of the conjunctiva among personnel exposed to this material in concentrations ranging from 10 to 30 mg of the vapor or dust per cubic meter of air. It is considered to be more toxic than phenol. The inhalation of vapors of this material, particularly when liberated at high temperatures, must be avoided. Mutagenic data. Combustible when exposed to heat or flame; can react with oxidizing materials. Violent reaction with NaOH. Slight explosion hazard when exposed to heat. To fight fire, use water, CO_2, dry chemical. For further information see Vol. 2, No. 2 of *DPIM Report*.

Treatment and Antidotes: When personnel working with this material exhibit some of the symptoms listed above, they should immediately be removed to fresh air. If the symptoms do not subside quickly, consult a physician. In cases of dermatitis due to this material, removal from exposure will quickly clear up the symptoms. If this material accidentally comes into contact with the skin, it should be removed at once and the affected area washed with plenty of soap and water.

4'-HYDROXYACETANILIDE   HR: 3
CAS: 103-90-2   NIOSH: AE 4200000
mf: C₈H₉NO₂   mw: 151.18

SYNS: P-ACETAMIDOPHENOL * 4-ACETAMIDO-PHENOL * ACETAMINOPHEN * P-ACETAMINO-PHENOL * N-ACETYL-P-AMINOPHENOL * P-ACETYLAMINOPHENOL * P-(4-HYDROXY-PHENYL)ACETAMIDE * P-HYDROXYACETANILIDE * 4-HYDROXYACETANILIDE * NCI-C55801

THR: A human poison by ingestion and other routes. A systemic human toxin which affects the central nervous system. Experimentally, a poison by ingestion, and intraperitoneal routes. Moderate intravenous and subcutaneous toxicity. When heated to decomposition it emits toxic fumes of NO$_x$. For further information see Vol. 1, No. 4 of *DPIM Report*.

HYDROXYACETONITRILE   HR: 3
CAS: 107-16-4   NIOSH: AM 0350000
mf: C₂H₃NO   mw: 57.06

SYNS: CYANOMETHANOL * FORMALDEHYDE CYANOHYDRIN * GLYCOLIC NITRILE * GLY-COLONITRILE (8CI) * GLYCONITRILE * 2-HY-DROXYACETONITRILE * HYDROXYMETHYLINI-TRILE * USAF A-8565

THR: A poison by ingestion, inhalation, intraperitoneal, subcutaneous, and skin routes. See also nitriles. When heated to decomposition it emits toxic fumes of NO$_x$ and CN⁻.

N-HYDROXY-N-ACETYL-2-AMINO-FLUORENE   HR: 3
CAS: 53-95-2   NIOSH: AK 8575000
mf: C₁₅H₁₃NO₂   mw: 239.29

SYNS: FLUORENYL-2-ACETHYDROXAMIC ACID * N-FLUOREN-2-YLACETOHYDROXAMIC ACID * N-2-FLUORENYLACETOHYDROXAMIC ACID * N-HYDROXY-2-ACETAMIDOFLUORENE * 2-(N-HYDROXYACETAMIDO)FLUORENE * N-HYDROXY-2-ACETYLAMINOFLUORENE * N-HYDROXY-N-2-FLUORENYLACETAMIDE

THR: A poison by intraperitoneal route. Mutagenic data. An experimental carcinogen, neoplastigen, and tumorigen. When heated to decomposition it emits toxic fumes of NO$_x$.

4-(HYDROXYAMINO)QUINOLINE-1-OXIDE   HR: 3
CAS: 4637-56-3   NIOSH: VB 9800000
mf: C₉H₈N₂O₂   mw: 176.19

SYN: HYDROXYLAMINE, N-(4-QUINOLYL)-, 1'-OXIDE

THR: A poison by intravenous route. Mutagenic data. An experimental tumorigen and neoplastigen. When heated to decomposition it emits toxic fumes of NO$_x$.

4-HYDROXYAZOBENZENE   HR: 3
CAS: 1689-82-3   NIOSH: SM 8300000
mf: C₁₂H₁₀N₂O   mw: 198.24

PROP: Orange rhombic crystals from ethanol; mp: 155-6°; bp: 220-30°; very sol in ether.

SYNS: P-BENZENEAZOPHENOL * C.I. SOLVENT YELLOW 7 * P-HYDROXYAZOBENZENE * P-PHENYLAZOPHENOL * 4-PHENYLAZOPHE-NOL

THR: A poison by intraperitoneal route. When heated to decomposition it emits toxic fumes of NO$_x$.

p-HYDROXYBENZOIC ACID METHYL ESTER   HR: 2
CAS: 99-76-3   NIOSH: DH 2450000
mf: C₈H₈O₃   mw: 152.16

SYNS: METHYL ESTER OF P-HYDROXYBENZOIC ACID * METHYL P-HYDROXYBENZOATE * METHYLPARABEN * METHYL PARAHY-DROXYBENZOATE * METHYL-P-OXYBENZOATE

THR: Mutagenic data. Moderate acute ingestion and intraperitoneal toxicity. When heated to decomposition it emits acrid smoke and fumes.

(p-HYDROXYBENZYL)TARTARIC ACID   HR: 3
CAS: 469-65-8   NIOSH: WW 8200000
mf: C₁₁H₁₂O₇   mw: 256.23

PROP: From the bark of the Jamaica Dogwood.

SYNS: 2,3-DIHYDROXY-2-((4-HYDROXY-PHENYL)METHYL)BUTANEDIOIC ACID * PISCIDEIN * PISCIDIC ACID

THR: A poison by intraperitoneal, subcutaneous, and parenteral routes. When heated to decomposition it emits acrid smoke and fumes.

2-HYDROXY-3-BUTENENITRILE   HR: 3
CAS: 5809-59-6   NIOSH: EM 8225000
mf: C_4H_5NO mw: 83.10

SYN: ACROLEIN CYANOHYDRIN

THR: A poison by ingestion, skin contact, and inhalation. A skin and eye irritant. See also nitriles. When heated to decomposition it emits toxic fumes of NO_x and CN^-.

1-HYDROXYCHOLECALCI-
FEROL   HR: 3
CAS: 57651-82-8   NIOSH: FZ 4650000
mf: $C_{27}H_{44}O_2$ mw: 400.65

PROP: Mp: 134°-136° or 138°-139.5°.

SYNS: 1-ALPHA-DIHYDROXYVITAMIN D3
* 1-ALPHA-HYDROXYCHOLECALCIFEROL
* HYDROXYCHOLECALCIFEROL * 1-ALPHA-HY-DROXYVITAMIN D3 * 9,10-SECOCHOLESTA-5,7,10(19)-TRIENE-1-ALPHA,3-BETA-DIOL

THR: A poison by ingestion and intravenous routes. When heated to decomposition it emits acrid smoke and fumes.

4-HYDROXY-3,5-DIIODOBENZO-
NITRILE   HR: 3
CAS: 1689-83-4   NIOSH: DI 4025000
mf: $C_7H_3I_2NO$ mw: 370.91

PROP: Colorless solid, mp: 213°, sltly water-sol.

SYNS: 4-CYANO-2,6-DIIODOPHENOL * 4-CYANO-2,6-DIJODPHENOL (GERMAN) * 3,5-DI-IODO-4-HYDROXYBENZONITRILE * 3,5-DIJOD-4-HYDROXY-BENZONITRIL (GERMAN)

THR: Very poisonous by dermal route and a poison by ingestion. An herbicide. See also nitriles and iodides.

HYDROXYDIMETHYLARSINE
OXIDE   HR: 3
CAS: 75-60-5   NIOSH: CH 7525000
mf: $C_2H_7AsO_2$ mw: 138.01

PROP: Colorless crystals, odorless and sol in water, mp: 192°.

SYNS: ACIDE CACODYLIQUE (FRENCH)
* ACIDE DIMETHYLARSINIQUE (FRENCH)
* DIMETHYLARSENIC ACID * DIMETHYLARSI-NIC ACID * CACODYLIC ACID

THR: A poison by several routes. Moderately toxic by ingestion, subcutaneous, and intraperitoneal routes. A skin and eye irritant. A human carcinogen. An experimental tumorigen. See also arsenic compounds. Used as an herbicide, defoliant and silvicide. Hazardous when water solution is in contact with active metals; i.e., Fe; Al; Zn; or when heated to decomposition. When heated to decomposition it emits toxic fumes of As. For further information see caco-dylic acid, Vol. 6, No. 1 of *DPIM Report*.

3-HYDROXYDIMETHYL CROTONAMIDE
DIMETHYL PHOSPHATE   HR: 3
CAS: 141-66-2   NIOSH: TC 3850000
mf: $C_8H_{16}NO_5P$ mw: 237.22

SYNS: 3-(DIMETHOXYPHOSPHINYLOXY)-N,N-DI-METHYL-CIS-CROTONAMIDE * 3-(DIMETHOXY-PHOSPHINYLOXY)-N,N DIMETHYLISOCROTON-AMIDE * 3-(DIMETHYLAMINO)-1-METHYL-3-OXO-1-PROPENYL DIMETHYL PHOSPHATE
* CIS-2-DIMETHYLCARBAMOYL-1-METHYLVINYL DIMETHYLPHOSPHATE * O,O-DIMETHYL-O-(2-DIMETHYL-CARBAMOYL-1-METHYL-VINYL)PHOS-PHAT (GERMAN) * O,O-DIMETHYL O-(N,N-DIMETHYLCARBAMOYL-1METHYLVINYL)
PHOSPHATE * O,O-DIMETHYL-O-(1,4-DI-METHYL-3-OXO-4-AZA-PENT-1-ENYL)FOSFAAT (DUTCH) * O,O-DIMETHYL-O-(1,4-DIMETHYL-3-OXO-4-AZA-PENT-1-ENYL)PHOSPHATE
* DIMETHYL PHOSPHATE OF 3-HYDROXY-N,N-DIMETHYL-CIS-CROTONAMIDE * DIMETHYL PHOSPHATE ESTER WITH 3-HYDROXY-N,N-DI-METHYL-CIS-CROTONAMIDE * O,O-DI-METIL-O-(1,4-DIMETIL-3-OXO-4-AZA-PENT-1-ENIL)-FOSFATO (ITALIAN) * ENT 24,482
* DIMETHYLPHOSPHATE ESTER WITH (E)-3-HY-DROXY-N,N-DIMETHYLCROTONAMIDE * 3-HY-DROXY-N,N-DIMETHYL-CIS-CROTONAMIDE DI-METHYL PHOSPHATE * PHOSPHATE DE DIMETHYLE ET DE 2-DIMETHYLCARBAMOYL 1-METHYL VINYLE (FRENCH)

THR: Poison by ingestion, skin contact, subcu-taneous, intravenous, intraperitoneal, and possi-bly other routes. Mutagenic data. See also es-ters. When heated to decomposition it emits very toxic fumes of NOx and POx. For further information see Vol. 1, No. 5 in *DPIM Report*.

4-HYDROXY-N,N-DIMETHYLTRYPT-
AMINE   HR: 3
CAS: 520-53-6   NIOSH: NM 2625000
mf: $C_{12}H_{16}N_2O$ mw: 204.30

SYNS: 3-(2-(DIMETHYLAMINO)ETHYL)INDOL-4-OL * PSILOCINE * PSILOTSIN

THR: A poison by intravenous route. When heated to decomposition it emits toxic fumes of NO_x.

p-HYDROXYEPHEDRINE HR: 3
CAS: 365-26-4 NIOSH: DO 6600000
mf: $C_{10}H_{15}NO_2$ mw: 181.23

PROP: Cryst powder; mp: 152°-154°; very sltly sol in H_2O; alc, ether. Very sol in NaOH solns and dil acids.

SYNS:
P-HYDROXYPHENYLMETHYLAMINOPROPANOL
* 1-(4-HYDROXYPHENYL)-2-METHYLAMINOPRO-
PANOL * ALPHA-(1-METHYLAMINOETHYL)-
P-HYDROXYBENZYL ALCOHOL

THR: Highly toxic by intravenous route. When heated to decomposition it emits toxic fumes such as NO_x.

2-(2-HYDROXYETHOXY)ETHYL ESTER STEARIC ACID HR: 3
CAS: 106-11-6 NIOSH: WI 4025000
mf: $C_{22}H_{44}O_4$ mw: 372.66

SYNS: DIETHYLENE GLYCOL MONOSTEARATE
* DIETHYLENE GLYCOL STEARATE * DIETH-
YLENE GLYCOL, MONOESTER WITH STEARIC ACID
* DIGLYCOL MONOSTEARATE * DIGLYCOL
STEARATE * USAF KE-8

THR: A poison by intraperitoneal route. See also esters. When heated to decomposition it emits acrid smoke and fumes.

3-(2-HYDROXYETHYL)-3-METHYL-1-PHENYLTRIAZENE HR: 3
CAS: 21600-45-3 NIOSH: XY 2775000
mf: $C_9H_{13}N_3O$ mw: 179.25

SYNS: 1-PHENYL-3-METHYL-3-(2-HYDROXY-
AETHYL)-TRIAZEN (GERMAN) * 1-PHENYL-3-
METHYL-3-(2-HYDROXYETHYL)TRIAZENE

THR: A poison by subcutaneous route. An experimental carcinogen. When heated to decomposition it emits toxic fumes of NO_x.

1-(2-HYDROXYETHYL)-1-NITRO-SOUREA HR: 3
CAS: 13743-07-2 NIOSH: YT 4915000
mf: $C_3H_7N_3O_3$ mw: 133.13

SYN: HENU

THR: A poison by intraperitoneal route. Mutagenic data. An experimental carcinogen. When

heated to decomposition it emits toxic fumes of NO_x.

N-HYDROXY-2-FLUORENYLBENZENE-SULFONAMIDE HR: 3
CAS: 26630-60-4 NIOSH: DB 2625000
mf: $C_{19}H_{15}NO_3S$ mw: 337.41

SYN: N-HYDROXY-N-FLUORENYLBENZENE-SUL-
FONAMIDE

THR: An experimental carcinogen. When heated to decomposition it emits very toxic fumes of NO_x and SO_x.

2-HYDROXYINIPRAMINE HR: 3
NIOSH: HO 2975500
mf: $C_{19}H_{24}N_2O$ mw: 296.45

SYNS: GP 33679 * 5-(3-(DIMETHYLAMINO)-
PROPYL)-2-HYDROXY-10,11-DIHYDRO-5H-DI-
BENZ(b,f)AZEPINE

THR: Poison by intravenous route. Moderately toxic by ingestion. An experimental teratogen. When heated to decomposition it emits toxic fumes of NOx.

4-HYDROXY-3-IODO-5-NITROBENZONI-TRILE HR: 3
CAS: 1689-89-0 NIOSH: DI 4600000
mf: $C_7H_3IN_2O_3$ mw: 290.02

SYNS: DOVENIX * NITROXYNIL * TRODAX

THR: A poison by ingestion and parenteral routes. See also nitriles. When heated to decomposition it emits very toxic I^-, NO_x and CN^-.

HYDROXYLAMINE HR: 3
CAS: 7803-49-8 NIOSH: NC 2975000
mf: H_3NO mw: 33.04

PROP: Colorless liquid or white needles. Unstable, hygroscopic; decomp rapidly @ room temp. Decomp in hot H_2O; very sol in liquid ammonia and methanol; very slightly sol in ether, benzene, carbon disulfide and chloroform. mp: 34.0°, bp: 110.0°, flash p: explodes at 265°F, d: 1.227, vap press: 10 mm @ 47.2°.

SYN: OXAMMONIUM

THR: A poison by intraperitoneal and subcutaneous routes. Moderately irritating to eyes and mucous membranes. Corrosive to skin. Locally it is irritating, and systemically it can cause methemoglobinemia. An experimental teratogen. Mutagenic data. Dangerous fire and explosion hazards when exposed to heat or open flame. Incompatible with BaO_2; BaO; Cl_2; $CuSO_4$; PbO_2; PCl_5; PCl_3; $K_2Cr_2O_7$; $KMnO_4$; Na; Zn; Carbonyls; pyridine; metals; oxidizers. For further information see Vol. 2, No. 2 of *DPIM Report*.

o-(2-HYDROXY-4-METHOXYBENZOYL)-BENZOIC ACID   HR: 3
CAS: 4756-45-0   NIOSH: DH 2395000
mf: $C_{15}H_{12}O_5$ mw: 272.27

SYN: 2′-CARBOXY-2-HYDROXY-4-METHOXYBEN-ZOPHENONE(O-(2-HYDROXY-P-ANISOYL)BENZOIC ACID)

THR: A poison by intravenous route. When heated to decomposition it emits acrid smoke and fumes.

N-HYDROXY-N-METHYL-4-AMINOAZO-BENZENE   HR: 3
NIOSH: NC 4010000
mf: $C_{13}H_{13}N_3O$ mw: 227.29

SYNS: N-HYDROXY-MAB * N-METHYL-N-(P-(PHENYLAZO)PHENYL)HYDROXYLAMINE

THR: Mutagenic data. An experimental carcinogen. When heated to decomposition it emits toxic fumes of NO_x.

2-HYDROXY-3-METHYLCHOLAN-THRENE   HR: 3
CAS: 3308-64-3   NIOSH: FZ 4200000
mf: $C_{21}H_{16}O$ mw: 284.37

SYN: 3-METHYLCHOLANTHREN-2-OL

THR: An experimental carcinogen and neoplastigen. Mutagenic data. When heated to decomposition it emits acrid smoke and fumes.

7-HYDROXY-4-METHYLCOUMARIN SODIUM   HR: 3
CAS: 5980-33-6   NIOSH: GN 7570000
mf: $C_{10}H_7O_3 \cdot Na$ mw: 198.16

SYNS: HYMECROMONE SODIUM * METHYL-4-OMBELLIFERONE SODEE (FRENCH) * METHYL-4-UMBELLIFERONE SODIUM

THR: A poison by intraperitoneal and intravenous routes. When heated to decomposition it emits acrid smoke and fumes.

3-HYDROXY-9-METHYLGUA-NINE   HR: 3
CAS: 30345-28-9   NIOSH: MF 8414000
mf: $C_6H_7N_5O_2$ mw: 181.18

SYN: 2-AMINO-3-HYDROXY-1,7-DIHYDRO-8-METHYL-6H-PURIN-6-ONE

THR: An experimental tumorigen. When heated to decomposition it emits toxic fumes of NO_x.

7-HYDROXYMETHYL-12-METHYLBENZ-(a)ANTHRACENE   HR: 3
CAS: 568-75-2   NIOSH: CW 8750000
mf: $C_{20}H_{16}O$ mw: 272.36

SYNS: 12-METHYBENZ(A)ANTHRACENE-7-METH-ANOL * 7-OHM-MBA * 7-OHM-12-MBA

THR: Acute poison by intravenous route. An experimental teratogen, neoplastigen, and carcinogen. Mutagenic data. When heated to decomposition it emits acrid smoke and fumes.

12-HYDROXYMETHYL-7-METHYLBENZ-(a)ANTHRACENE   HR: 3
CAS: 568-70-7   NIOSH: CW 8925000
mf: $C_{20}H_{16}O$ mw: 272.36

SYN: 7-METHYLBENZ(A)ANTHRACENE-12-METH-ANOL

THR: An experimental carcinogen. Mutagenic data. When heated to decomposition it emits acrid smoke and fumes.

1-HYDROXYMETHYL-2-METHYLDIAM-IDE-2-OXIDE   HR: 3
CAS: 590-96-5   NIOSH: PC 2625000
mf: $C_2H_6N_2O_2$ mw: 90.10

SYN: METHYLAZOXYMETHANOL

THR: Mutagenic data. An experimental teratogen and tumorigen. When heated to decomposition it emits toxic fumes of NO_x.

d-N,N′-(1-HYDROXYMETHYLPROPYL)-ETHYLENEDINITROSAMINE   HR: 3
NIOSH: EL 3852000

SYNS: 2,2′-(ETHYLENEBIS(NITROSOIMINO))-BISBUTANOL * d-N,N′-(1-IDROSSIMETIL PRO-PIL)-ETILENDINITROSAMINA (ITALIAN)

THR: An experimental carcinogen. When heated to decomposition it emits toxic fumes of NO_x.

N-HYDROXYNAPHTHALIMIDE, DIETHYL PHOSPHATE   HR: 3
CAS: 1491-41-4   NIOSH: QK 5775000
mf: $C_{16}H_{16}NO_6P$ mw: 349.30

PROP: Tan crystalline powder, sol in methylene chloride, difficultly sol in most organic solvents, mp: 177.0°.

SYNS: BAYER 25820 ∗ O,O-DIETHYL N-HY-DROXYNAPHTHALIMIDE PHOSPHATE ∗ ENT 25,567

THR: A poison by ingestion and skin contact. A ruminant antihelminitic; a cholinesterase inhibitor. Dangerous; see also parathion. When heated to decomposition it emits toxic NO_x.

HYDROXYNOREPHEDRINE   HR: 3
CAS: 54-49-9   NIOSH: DN 4300000
mf: $C_9H_{13}NO_2$ mw: 167.23

SYNS: 1-ALPHA-(1-AMINOETHYL)-M-HYDROXY-BENZYL ALCOHOL ∗ M-HYDROXY NOREPHED-RINE ∗ M-HYDROXYPROPADRINE

THR: A poison by ingestion, intraperitoneal, subcutaneous, and intravenous routes. When heated to decomposition it emits toxic fumes of NO_x.

p-HYDROXYPHENYLBUTAZONE   HR: 3
CAS: 129-20-4   NIOSH: UQ 8400000
mf: $C_{19}H_{20}N_2O_3$ mw: 324.41

SYNS: 4-BUTYL-1-(P-HYDROXYPHENYL)-2-PHE-NYL-3,5-PYRAZOLIDINEDIONE ∗ 4-BUTYL-1-(4-HYDROXYPHENYL)-2-PHENYL-3,5-PYRAZOLIDINE-DIONE ∗ 4-BUTYL-2-(P-HYDROXYPHENYL)-1-PHENYL-3,5-PYRAZOLIDINEDIONE ∗ 4-BUTYL-2-(4-HYDROXYPHENYL)-1-PHENYL-3, 5-DIOXOPYRAZOLIDIDE ∗ 3,5-DIOXO-1-PHENYL-2-(P-HYDROXYPHENYL)-4-N-BUTYLPYRA-ZOLIDENE ∗ 1-(P-HYDROXYPHENYL)-2-PHENYL-4-BUTYL-3,5-PYRAZOLIDINEDIONE ∗ 1-P-HYDROXYPHENYL-2-PHENYL-3,5-DI-OXO-4-N-BUTYLPYRAZOLIDINE ∗ OXYPHEN-BUTAZONE ∗ OXYPHENYLBUTAZONE ∗ 1-PHENYL-2-(P-HYDROXYPHENYL)-3,5-DI-OXO-4-BUTYLPYRAZOLIDINE ∗ USAF GE-14

THR: A poison by intraperitoneal and intravenous routes. Moderately toxic by ingestion. This material affects the human gastrointestinal tract, the blood picture, and the glandular system. A suspected human carcinogen. When heated to decomposition it emits toxic fumes of NO_x.

p-HYDROXYPHENYLLACTIC ACID   HR: 3
CAS: 306-23-0   NIOSH: OD 5500000
mf: $C_9H_{10}O_4$ mw: 182.19

THR: An experimental carcinogen. When heated to decomposition it emits acrid smoke and fumes.

(m-HYDROXYPHENYL)TRIMETHYL-AMMONIUM IODIDE, METHYLCAR-BAMATE   HR: 3
CAS: 3983-39-9   NIOSH: BR 4200000
mf: $C_{11}H_{17}N_2O_2 \cdot I$ mw: 336.20

SYNS: CARBAMIC ACID, 3-DIMETHYLAMINOPHE-NYL ESTER, METHIODIDE ∗ CARBAMIC ACID, N-METHYL-3-DIMETHYLAMINOPHENYL ESTER METHIODIDE ∗ CARBAMIC ACID, (3-TRIMETH-YLAMMONIOPHENYL) ESTER, IODIDE

THR: A poison by subcutaneous and intravenous routes. See also iodides and carbamates and esters. When heated to decomposition it emits very toxic fumes of NO_x and I^-.

(m-HYDROXYPHENYL)TRIMETHYL-AMMONIUM METHYLSULFATE METH-YLPHENYLCARBAMATE   HR: 3
CAS: 64050-79-9   NIOSH: BR 4815000
mf: $C_{17}H_{21}N_2O_2 \cdot CH_3O_4S$ mw: 396.50

SYN: METHYLPHENYLCARBAMIC ESTER OF 3-OXYPHENYLTRIMETHYLAMMONIUM METHYLSUL-FATE

THR: A poison by ingestion, subcutaneous, and intravenous routes. See also sulfates, esters and carbamates. When heated to decomposition it emits very toxic fumes of NO_x and SO_x.

HYDROXYPROGESTERONE CAP-ROATE   HR: 3
CAS: 630-56-8   NIOSH: TU 5085000
mf: $C_{27}H_{40}O_4$ mw: 428.59

PROP: Dense needles; mp: 119°-121°.

SYNS: 17-alpha-HYDROXYPROGESTERONE CAPROATE ∗ 17-alpha-HYDROXY PROGESTER-ONE-N-CAPROATE ∗ 17-alpha-HYDROXYPRO-GESTERONE HEXANOATE ∗ 17-alpha-HEXAN-OYLOXYPREGN-4-ENE-3,20-DIONE ∗ 17-((1-OXOHEXYL)OXY)PREGN-4-ENE-3,20-DIONE ∗ PROGESTERONE CAPROATE

THR: An experimental carcinogen. When heated to decomposition it emits acrid smoke and fumes.

1-((2-HYDROXYPROPYL)-NITROSO)-AMINO)ACETONE HR: 3

NIOSH: AL 5890000

mf: $C_6H_{12}N_2O_3$ mw: 160.20

SYN: N-NITROSO(2-HYDROXYPROPYL)(2-OXO-PROPYL)AMINE

THR: A poison by subcutaneous route. Mutagenic data. An experimental carcinogen and neoplastigen. See also ketones. When heated to decomposition it emits toxic fumes of NO_x.

1-(3-HYDROXYPROPYL)THEOBROMINE HR: 2

CAS: 59413-14-8   NIOSH: XH 2388000

mf: $C_{10}H_{14}N_4O_3$ mw: 238.28

SYNS: 3,7-DIHYDRO-1-(3-HYDROXYPROPYL)-3,7-DIMETHYL-1H-PURINE-2,6-DIONE ∗ 1-(3-HYDROXYPROPYL)-3,7-DIMETHYLXANTHINE ∗ 1-(3-HYDROXYPROPYL)THEOBROMINE ∗ gamma-OXYPROPYLTHEOBROMIN (GERMAN) ∗ gamma-(gamma-OXYPROPYL)-THEOBROMIN (GERMAN)

THR: Moderately toxic by intraperitoneal and subcutaneous routes. When heated to decomposition it emits toxic fumes of NO_x.

HYDROXYSENKIRKINE HR: 3

CAS: 26782-43-4   NIOSH: VT 5970000

mf: $C_{19}H_{27}NO_7$ mw: 381.47

PROP: Isolated from the plant *Crotalaria Laburnifolia*

SYN: 8,12,18-TRIHYDROXY-4-METHYL-11,16-DIOXOSENECIONANIUM

THR: An experimental carcinogen. When heated to decomposition it emits toxic fumes of NO_x.

5-HYDROXYTETRACYCLINE HR: 3

CAS: 79-57-2   NIOSH: QI 7875000

mf: $C_{22}H_{24}N_2O_9$ mw: 460.48

SYNS: NCI-C56473 ∗ OXITETRACYCLIN ∗ OXYTETRACYCLINE AMPHOTERIC ∗ TERRAMYCIN

THR: A poison by intravenous route. Moderately toxic by ingestion and intraperitoneal routes. Mutagenic data. An experimental teratogen. In humans it is a cumulative toxin, and systemically affects the skin. When heated to decomposition it emits toxic fumes of NO_x.

(3-HYDROXY-p-TOLYL)TRIMETHYLAMMONIUM CHLORIDE,METHYLCARBAMATE HR: 3

CAS: 64050-03-9   NIOSH: BR 5915000

mf: $C_{12}H_{19}N_2O_2 \cdot Cl$ mw: 258.78

SYN: METHYLCARBAMIC ACID, 5-(TRIMETHYL-AMMONIO)-O-TOLYL ESTER, CHLORIDE

THR: A poison by ingestion, intraperitoneal, subcutaneous, intravenous, and implantation routes. See also esters, chlorides and carbamates. When heated to decomposition it emits very toxic NO_x and Cl^-.

HYDROXY TRIPHENYL STANNANE HR: 3

CAS: 76-87-9   NIOSH: WH 8575000

mf: $C_{18}H_{16}OSn$ mw: 367.03

PROP: Mp: 122°.

SYNS: ENT 28009 ∗ FINTINE HYDROXYDE (FRENCH) ∗ FINTIN HYDROXID (GERMAN) ∗ FENTIN HYDROXIDE ∗ FINTIN HYDROXYDE (DUTCH) ∗ FINTIN IDROSSIDO (ITALIAN) ∗ HYDROXYDE DE TRIPHENYL-ETAIN (FRENCH) ∗ HYDROXYTRIPHENYLTIN ∗ IDROSSIDO DI STAGNO TRIFENILE (ITALIAN) ∗ NCI-C00260 ∗ TRIFENYL-TINHYDROXYDE (DUTCH) ∗ TRIPHENYLTIN HYDROXIDE ∗ TRIPHENYLTIN OXIDE ∗ TRIPHENYL-ZINNHYDROXID (GERMAN)

OSHA PEL: TWA 100 ug(Sn)/m³ (skin)

THR: A poison by ingestion and intraperitoneal routes. Moderately toxic by other routes. A severe eye irritant. An experimental teratogen. See also tin compounds. When heated to decomposition it emits acrid smoke and fumes. For further information see fentin hydroxide in Vol. 6, No. 2 of *DPIM Report*.

HYDROXYUREA HR: 3

CAS: 127-07-1   NIOSH: YT 4900000

mf: $CH_4N_2O_2$ mw: 76.07

PROP: Needles from ethanol; mp: 133°-136°; bp: decomp; very sol in water; sol in hot alcohol.

SYNS: CARBAMOHYDROXAMIC ACID ∗ CARBAMOHYDROXIMIC ACID ∗ CARBAMOHYDROXYAMIC ACID ∗ CARBAMOYL OXIME ∗ CARBAMYL HYDROXAMATE ∗ HYDROXYLUREA ∗ NCI-C04831 ∗ NSC 32065

THR: An experimental teratogen and tumorigen. In humans it affects the blood picture and is toxic by unspecified routes. When heated to decomposition it emits toxic fumes of NO_x.

3-HYDROXYXANTHINE   HR: 3
CAS: 13479-29-3   NIOSH: ZD 8100000
mf: $C_5H_4N_4O_3$ mw: 168.13

SYN: XANTHINE-3-N-OXIDE

THR: A poison by intraperitoneal route. Moderately toxic by subcutaneous route. An experimental carcinogen and neoplastigen. When heated to decomposition it emits toxic fumes of NO_x. For further information see Vol. 1, No. 5 of *DPIM Report*.

7-HYDROXYXANTHINE   HR: 3
CAS: 16870-90-9   NIOSH: ZD 9150000
mf: $C_5H_4N_4O_3$ mw: 168.13

SYN: XANTHINE-7-N-OXIDE

THR: An experimental carcinogen and neoplastigen. When heated to decomposition it emits toxic fumes of NO_x.

HYOSCINE HYDROBROMIDE   HR: 3
CAS: 114-49-8   NIOSH: YM 4550000
mf: $C_{17}H_{21}NO_4 \cdot BrH$ mw: 384.31

SYNS: SCOPOLAMINE BROMIDE * (−)-SCOPO-LAMINE BROMIDE * (−)-SCOPOLAMINE HYDRO-BROMIDE * HYOSCINE BROMIDE * HYDRO-SCINE HYDROBROMIDE * (−)-HYOSCINE HYDROBROMIDE * l-HYOSCINE HYDROBROMIDE * HYOCINE F HYDROBROMIDE * HYOSCYINE HYDROBROMIDE

THR: A poison by intravenous route. Moderately toxic by ingestion, subcutaneous, intraduodenal, and intraperitoneal routes. Mutagenic data. See also bromides and esters. When heated to decomposition it emits very toxic NO_x and HBr.

(−)-HYOSCYAMINE   HR: 3
CAS: 101-31-5   NIOSH: NH 0875000
mf: $C_{17}H_{23}NO_3$ mw: 289.41

PROP: Very sol in alc, dil acids. White crystalline alkaloid, mp: 106°-108°.

SYNS: (−)-ATROPINE * DATURINE * HY-OSCYAMINE * l-HYOSCYAMINE * TROPIC ACID, (−)-, ESTER WITH TROPINE

THR: A poison by intravenous route. This is one of the atropine alkaloids and is very toxic, acting very much like atropine. It has the same

effect on the central nervous system but twice the effect on the peripheral nerves. The symptoms of poisoning from this are dryness of the throat and mouth, marked difficulty in swallowing, and a sensation of burning and thirst. The vision becomes impaired through dilation and loss of accommodation, and the eyes present a rather prominent, brilliant, staring appearance. The voice is husky and the tongue is red. Dangerous; when heated to decomposition it emits highly toxic NO_x.

HYPOCHLORITES   HR: 2
PROP: Salts of hypochlorous acid.

THR: Moderately toxic by ingestion and inhalation, and irritating to skin, eyes and mucous membranes. Flammable by chemical reaction with reducing agents. These are powerful oxidizers particularly at higher temperatures, even when chlorine and then oxygen are evolved or in the presence of moisture or carbon dioxide. With urea, it forms the highly explosive NCl_3. Dangerous; when heated or on contact with acid or acid fumes, emits highly toxic fumes of chlorine and chlorides; will react with water or steam to produce toxic and corrosive fumes; can react vigorously with oxidizers.

HYPOCHLOROUS ACID, CALCIUM SALT   HR: 2
CAS: 7778-54-3   NIOSH: NH 3485000
mf: $Cl_2O_2 \cdot Ca$ mw: 142.98

PROP: White powder. Compound contains 39% or less available chlorine

SYNS: BLEACHING POWDER (DOT) * CALCIUM CHLOROHYDROCHLORITE * CALCIUM HYPO-CHLORIDE * CALCIUM HYPOCHLORITE * CALCIUM OXYCHLORIDE * CHLORIDE OF LIME (DOT) * CHLORINATED LIME (DOT) * HTH * HY-CHLOR * LIME CHLORIDE

THR: Moderately toxic by ingestion and inhalation. Can cause severe irritation of skin and mucous membranes and emit fumes capable of causing pulmonary edema. Flammable by chemical reaction with combustible materials; i.e., anthracene; C; charcoal; C_2H_5OH; glycerol; grease; oil; mercaptans; methyl carbitol; nitromethane; organic matter; organic sulfides; phenol, 1-propanethiol; propylmercaptan; S; turpentine. Can explode with CCl_4; amines. A powerful oxidizer. Deflagration occurs in contact with combustible substances. Moderate explosion hazard in its solid form when heated. Explosive Range: When heated suddenly above 212°F. Dangerous when heated to decomposition or on contact with acid or acid fumes, it emits highly toxic fumes and explodes; will react with water or steam to produce toxic and corrosive fumes; can react vigorously with reducing materials. For further information see calcium hypochlorite in Vol. 4, No. 3 of *DPIM Report*.

I

IBYLCAINE HYDROCHLORIDE   HR: 3
CAS: 553-68-4   NIOSH: KL 4375000
mf: $C_{13}H_{20}N_2O_2 \cdot ClH$ mw: 272.81

SYNS: BUTETHAMINE HYDROCHLORIDE
* 2-(ISOBUTYLAMINO)ETHYL-p-AMINOBENZOATE
HYDROCHLORIDE * MONOCAINE HYDROCHLO-
RIDE

THR: A poison by intraperitoneal, intravenous,
and subcutaneous routes. When heated to decomposition it emits very toxic fumes of Cl^-
and NO_x.

IMIDAZOLE   HR: 3
CAS: 288-32-4   NIOSH: NI 3325000
mf: $C_3H_4N_2$ mw: 68.09

PROP: Prisms; mp: 90°-91°; bp: 257°, sol in
water; very sol in alc; sol in ether. chloroform,
pyridine; sltly sol in benzene.

SYNS: 1,3-DIAZA-2,4-CYCLOPENTADIENE
* 1,3-DIAZOLE * GLYOXALIN * IMINAZOLE
* PYRRO(B)MONAZOLE * USAF EK-4733

THR: A poison by subcutaneous and intraperitoneal routes. Moderately toxic by ingestion.
When heated to decomposition it emits toxic
fumes of NO_x.

IMIDAZOLE MUSTARD   HR: 3
CAS: 5034-77-5   NIOSH: NI 3930000
mf: $C_8H_{12}Cl_2N_6O$ mw: 279.16

SYNS: 5-(3,3-BIS(2-CHLOROETHYL)-1-TRI-
AZENO)IMIDAZOLE-4-CARBOXAMIDE * NCI-
C01616 * NSC-82196

THR: An experimental neoplastigen. Affects human blood picture. When heated to decomposition it emits very toxic Cl^- and NO_x.

2-IMIDAZOLIDINETHIONE   HR: 3
CAS: 96-45-7   NIOSH: NI 9625000
mf: $C_3H_6N_2S$ mw: 102.17

PROP: White crystals, water-soly = 9 g/100
mL @ 30°. Often occurs as a main degradation
product of the metal salts of ethylene bis-dithio-
carbamic acid.

SYNS: 4,5-DIHYDROIMIDAZOLE-2(3H)-THIONE
* ETHYLENE THIOUREA * N,N'-ETHYLENE-
THIOUREA * 1,3-ETHYLENE-2-THIOUREA

* L'ETHYLENE THIOUREE (FRENCH) * NCI-
C03372 * 2-THIOL-DIHYDROGLYOXALINE

THR: Mutagenic data. An experimental teratogen and carcinogen. Moderately toxic by ingestion. When heated to decomposition it emits
very toxic fumes of NO_x and SO_x. For further
information see Ethylene Thiourea Vol. 1, No.
2 of *DPIM Report*.

2-IMIDAZOLINETHIOL   HR: 3
          NIOSH: NJ 4600000
mf: $C_3H_6N_2S$ mw: 102.17

SYNS: 2-MERCAPTOIMIDAZOLINE * 2-MERKAP-
TOIMIDAZOLIN (CZECH) * RODANIN S-62
(CZECH) * USAF EL-62

THR: A poison by ingestion and intraperitoneal
routes. An eye irritant. When heated to decomposition it emits very toxic fumes of NO_x and
SO_x.

2-IMINO-5-PHENYL-4-OXAZOLIDI-
NONE   HR: 3
CAS: 2152-34-3   NIOSH: RQ 2975000
mf: $C_9H_8N_2O_2$ mw: 176.19

SYNS: PHENOXAZOLE * 5-PHENYL-2-IMINO-4-
OXAZOLIDINONE * 5-PHENYL-2-IMINO-4-OXO-
OXAZOLIDINE * PHENYL ISOHYDANTOIN
* PHENYLPSEUDOHYDANTOIN

THR: A poison by ingestion. Moderately toxic
by intraperitoneal route. When heated to decomposition it emits toxic fumes of NO_x.

IMIPHOS   HR: 3
CAS: 1078-7-9-1   NIOSH: SY 9625000
mf: $C_8H_{16}N_3OPS$ mw: 233.30

SYNS: BIS(1-AZIRIDINYL)(2-METHYL-3-THIAZO-
LIDINYL)PHOSPHINE OXIDE * MARCOPHANE
* MARKOFANE

THR: Poison by ingestion and intraperitoneal
route. When heated to decomposition it emits
very toxic fumes of SO_x, PO_x, and NO_x.

IMPIRAMINE-N-OXIDE   HR: 3
CAS: 2207-85-4   NIOSH: HO 2100000
mf: $C_{19}H_{24}N_2O$ mw: 296.45

PROP: White needle-like cryst; mp: 120°-123°
(decomp); sol in methanol, ether, acetone and
benzene. Hygroscopic.

SYNS: 5-(3-(DIMETHYLAMINO)PROPYL)-10,11-DIHYDRO-5H-DIBENZ(B,F)AZEPINE-5-OXIDE ∗ GP 38383 ∗ IPNO

THR: A poison by ingestion and intraperitoneal routes. When heated to decomposition it emits toxic fumes of NO_x.

1,3-INDANDIONE HR: 3
CAS: 606-23-5 NIOSH: NK 5070000
mf: $C_9H_6O_2$ mw: 146.15

PROP: Crystal or liquid; mp: 129-131° decomp; very sltly sol in cold water; sol in hot alc; sol in benzene.

SYNS: 1,3-DIKETOHYDRINDENE ∗ 1H-INDENE-1,3(2H)-DIONE

THR: A poison by intraperitoneal route. An experimental teratogen. When heated to decomposition it emits acrid smoke and fumes.

INDENE HR: 2
CAS: 95-13-6 NIOSH: NK 8225000
mf: C_9H_8 mw: 116.17

PROP: Liquid from coal tars. Water-insol, but miscible in organic solvents. D: 0.9968 @ 20°/4°, mp: −1.8°, bp: 181.6°.

SYN: INDONAPHTHENE

THR: Moderately toxic by ingestion, inhalation, and subcutaneous routes. Irritant to skin, eyes and mucous membranes. During nitration with $(H_2SO_4 + HNO_3)$ it has exploded.

INDERAL HR: 3
CAS: 525-66-6 NIOSH: UB 7500000
mf: $C_{16}H_{21}NO_2$ mw: 259.38

SYNS: ICI 45520 ∗ 1-ISOPROPYLAMINO-3-(1NAPHTHYLOXY)-2-PROPANOL ∗ PROPANALOL

THR: Poison by ingestion in humans acting as a psychotrope. A poison by intraperitoneal and intravenous routes. When heated to decomposition it emits toxic fumes of NO_x.

INDIUM HR: 3
CAS: 7440-74-6 NIOSH: NL 1050000
af: In aw: 114.82

PROP: Soft, silvery-white metal, mp: 156.61°, bp: 2080°, d: 7.31 @ 20°.

THR: A poison by subcutaneous route. Moderately toxic by ingestion. When this material is injected into animals by intraperitoneal or in-travenous routes it is found to be highly toxic. It affects the liver, heart, kidneys and the blood. However, available data are scanty. Toxicity is based on animal experiments. Flammable in the form of dust when exposed to heat or flame. Incandesces. See also iron dust. Incompatible with acetonitrile; dinitrogen tetroxide; mercury (II)bromide; S.

INDIUM CITRATE HR: 2
CAS: 4194-69-8 NIOSH: NL 1575000
mf: $C_{18}H_{15}O_{21}$ • In mw: 682.15

THR: Moderately toxic by subcutaneous route. See also indium. When heated to decomposition it emits acrid smoke and fumes.

INDIUM NITRATE HR: 3
CAS: 13770-61-1 NIOSH: NL 1750000
mf: InN_3O_9 mw: 300.85

THR: An experimental teratogen and severe skin irritant. See also nitrates.

INDIUM SULFATE HR: 3
CAS: 13464-82-9 NIOSH: NL 1925000
mf: $O_{12}S_3$ • In_2 mw: 517.82

PROP: Grayish white hygroscopic powders; d: 3.44; sol in H_2O. Keep well-closed.

SYNS: INDISULFAT (GERMAN) ∗ SULFURIC ACID, INDIUM SALT

THR: A poison by intravenous and subcutaneous routes. Moderately toxic by ingestion. See also indium and sulfates. When heated to decomposition it emits toxic fumes of SO_x.

INDIUM TRICHLORIDE HR: 3
CAS: 10025-82-8 NIOSH: NL 1400000
mf: Cl_3In mw: 221.17

PROP: Yellowish, deliquescent crystals; d: 4.0; mp: 586°; sublimes @ 500°; very sol in H_2O; keep tightly closed. Bp: volatile @ 600°.

SYN: INDIUM CHLORIDE

THR: A poison by subcutaneous, intraperitoneal, and intravenous routes. See also indium. When heated to decomposition it emits toxic fumes of Cl^-.

INDOLE HR: 3
CAS: 120-72-9 NIOSH: NL 2450000
mf: C_8H_7N mw: 117.16

PROP: Colorless to yellowish scales, intense fecal odor, sol in hot water, alcohol, ether and petroleum ether, insol in mineral oil and glycerol, mp: 52°, bp: 253°. Volatile with steam. Sol in hot water, hot alc, ether, benzene.

SYNS: 1-AZAINDENE * 1-BENZAZOLE * BENZOPYRROLE * 2,3-BENZOPYRROLE * INDOL (GERMAN) * KETOLE

THR: A poison by intraperitoneal and subcutaneous routes. An experimental carcinogen. Moderately toxic by ingestion and skin routes. When heated to decomposition it emits toxic fumes of NO_x.

1H-INDOLE-3-ACETIC ACID   HR: 3
CAS: 87-51-4   NIOSH: NL 3150000
mf: $C_{10}H_9NO_2$ mw: 175.20

PROP: Colorless leaves from benzene; mp: 165-168°; very sltly sol in cold water; sol in alc, ether and acetic acid; insol in chloroform.

SYNS: HETEROAUXIN * beta-INDOLEACETIC ACID * beta-INDOLE-3-ACETIC ACID * 3-INDOLEACETIC ACID * alpha-INDOL-3-YL-ACETIC ACID * beta-INDOLYLACETIC ACID * INDOLYL-3-ACETIC ACID * 3-INDOLYLACETIC ACID * omega-SKATOLE CARBOXYLIC ACID

THR: A poison by intraperitoneal route. Mutagenic data. An experimental teratogen and tumorigen. When heated to decomposition it emits toxic fumes of NO_x.

INDOLE-3-ACRYLIC ACID   HR: 3
CAS: 1204-06-4   NIOSH: NL 3680000
mf: $C_{11}H_9NO_2$ mw: 187.21

SYNS: 3-INDOLYLACRYLIC ACID * 3-(1-H-INDOL-3-YL)-2-PROPENOIC ACID

THR: An experimental carcinogen and neoplastigen. When heated to decomposition it emits toxic fumes of NO_x.

1H-INDOLE-3-BUTANOIC ACID   HR: 3
CAS: 133-32-4   NIOSH: NL 5250000
mf: $C_{12}H_{13}NO_2$ mw: 203.26

PROP: White crystals or powder, insol in water and chloroform, mp: 124°. Sol in acetone and ether.

SYNS: HORMEX ROOTING POWDER * HORMODIN * BETA-INDOLEBUTYRIC ACID * GAMMA-(INDOLE-3)-BUTYRIC ACID * 3-INDOLEBUTYRIC ACID * 3-INDOLYL-GAMMA-BUTYRIC ACID * GAMMA-(3-INDOLYL)BUTYRIC ACID

* GAMMA-(INDOL-3-YL)BUTYRIC ACID * INDOLYL-3-BUTYRIC ACID * 4-(INDOL-3-YL)BUTYRIC ACID

THR: A poison by ingestion and intraperitoneal routes. When heated to decomposition it emits toxic fumes of NO_x.

3-INDOLYLACETONITRILE   HR: 3
CAS: 771-51-7   NIOSH: AM 0700000
mf: $C_{10}H_8N_2$ mw: 156.20

SYNS: 3-(CYANOMETHYL)INDOLE * 3-INDOLACETONITRILE * INDOLEACETONITRILE * INDOLE-3-ACETONITRILE * 1H-INDOLE-3-ACETONITRILE * INDOLYLACETONITRILE * USAF CB-29

THR: A poison by intraperitoneal route. See also nitriles. When heated to decomposition it emits toxic NO_x and CN^-.

INDOMETHACIN   HR: 3
CAS: 53-86-1   NIOSH: NL 3500000
mf: $C_{19}H_{16}ClNO_4$ mw: 357.81

PROP: Crystals. One form mp: 155°-another 162°. Sol in ethanol, ether, acetone, castor oil; insol in water.

SYNS: N-P-CHLORBENZOYL-5-METHOXY-2-METHYLINDOLE-3-ACETIC ACID * 1-(P-CHLOROBENZOYL)-5-METHOXY-2-METHYLINDOLE-3-ACETIC ACID * 1-(P-CHLOROBENZOYL)-2-METHYL-5-METHOXYINDOLE-3-ACETIC ACID * 1-(P-CHLOROBENZOYL)-2-METHYL-5-METHOXY-3-INDOLE-ACETIC ACID * alpha-(1-(P-CHLOROBENZOYL)-2-METHYL-5-METHOXY-3-INDOLYL)ACETIC ACID * INDOMETHAZINE * NCI-C56144

THR: A poison by ingestion, intraperitoneal, and subcutaneous routes. A human psychotrope. When heated to decomposition it emits very toxic Cl^- and NO_x.

INSIDON DIHYDROCHLORIDE   HR: 3
CAS: 909-39-7   NIOSH: TL 9100000
mf: $C_{23}H_{29}N_3O \cdot 2ClH$ mw: 436.47

PROP: Long rectangular plates from water; mp: 90°.

SYNS: 4-(3-(5H-DIBENZ(B,F)AZEPIN-5-YL)PROPYL)-1-PIPERAZINEETHANOL DIHYDROCHLORIDE * 5-(gamma-(beta-HYDROXYETHYLPIPERAZINO)-PROPYL)-5H-DIBENZO(B,F)AZEPINE DIHYDROCHLORIDE * OPIPRAMOL DIHYDROCHLORIDE

THR: A poison by intraperitoneal, intravenous, and subcutaneous routes. Moderately toxic by ingestion. A human psychotrope. When heated to decomposition it emits very toxic NO_x and HCl. When heated to decomposition it emits toxic fumes of I^-.

IODATES
THR: Variable toxicity; similar to bromates and chlorates. See also specific compound.

IODIDES   HR: 3
THR: Similar in toxicity to bromides. Prolonged absorption of iodides may produce "iodism" which is manifested by skin rash, running nose, headache and irritation of mumem. In severe cases, the skin may show pimples, boils, redness, black and blue spots, hives and blisters. Weakness, anemia, loss of weight and general depression may occur. When heated to decomposition they can emit highly toxic iodine and iodine compounds.

IODINE   HR: 3
CAS: 7553-56-2   NIOSH: NN 1575000
mf: I_2 mw: 253.80

PROP: Rhombic, violet-black crystals, metallic luster, mp: 113.5°, bp: 185.24°, d: 4.93 (solid @ 25°), vap press: 1 mm @ 38.7°. Characteristic odor; sharp acrid taste; vap press (solid) = 0.030 mm @ 0°; very sol in aq solns of HI and iodides.

SYNS: IODE (FRENCH)   *   IODINE CRYSTALS
*   IODINE SUBLIMED   *   IODIO (ITALIAN)
*   JOD (GERMAN, POLISH)   *   JOOD (DUTCH)

OSHA PEL: TWA CL 0.1 ppm

THR: A poison irritant to skin, eyes and mucous membranes. The effect of iodine vapor upon the body is similar to that of chlorine and bromine, but it is more irritating to the lungs. Serious exposures are seldom encountered in industry, due to the low volatility of the solid at ordinary room temperatures. Signs and symptoms are irritation and burning of the eyes, lachrymation, coughing and irritation of the nose and throat. See also iodides. Ingestion of large quantities causes abdominal pain, nausea, vomiting, diarrhea. In severe cases, purging, excessive thirst and circulatory failure may develop. Doses of 2-3 g have been fatal. Incompatible with acetaldehyde; C_2H_2; Al; NH_3; NH_4OH; Sb; BrF_5; $CsHC_2$; Cs_2C_2; Cs_2O; Cl_2; ClF_3;

Cu_2C_2; ethanol; HgO; O_2; O_3F_2; (ethanol + methanol + HgO); F_2; Li; Li_2C_2; Li_6C; metals; Mg; OF_2; K; $RbHC_2$; Rb_2C_2; AgN_3; NaH; ZrC; SO_3; formamide; pyridine; P; ethanol + butadiene. Dangerous; when heated it emits highly toxic fumes of iodine and iodine compounds; can react vigorously with reducing materials. For further information see Vol. 1, No. 5 of *DPIM Report*.

IODINE (V) CHLORIDE   HR: 3
                          NIOSH: NN 1655000
DOT: 1792
mf: Cl_5I mw: 304.15

SYN: IODINE PENTACHLORIDE (DOT)

DOT Classification: Label: Oxidizer and Poison

THR: A poison and powerful oxidizer. Very irritating to skin, eyes, mucous membranes. When heated to decomposition it emits very toxic Cl^- and I^-.

IODINE MONOCHLORIDE   HR: 3
CAS: 7790-99-0   NIOSH: NN 1650000
DOT: 1792
mf: ClI   mw: 162.38

PROP: Black crystals or reddish-brown liquid. Exists in α, β forms; crystals α form (stable) black needles; sol in water, alc, ether, CS_2, acetic acid. Red-brown cryst or oily liquid, mp (α): 27°, (β): 14°, bp: 97.4 decomp @ 100°; d(α): 3.1822 @ 0°, (β): 3.24 @ 34°.

SYNS: IODINE CHLORIDE   *   PROTOCHLORURE D'IODE (FRENCH)   *   WIJS' CHLORIDE

DOT Classification: Label: Corrosive

THR: A poison irritant to skin, eyes and mucous membranes; also poison by ingestion and dermal routes. See also iodine and chlorine. Moderate explosion hazard when exposed to heat. Reacts violently with Al foil; CdS; PbS; organic matter; P; PCl_3; K; rubber; Ag_2S; Na; ZnS; metals. Dangerous; when heated to decomposition it emits highly toxic fumes of chlorine and iodine and may explode; will react with water or steam to produce toxic and corrosive fumes.

IODOACETAMIDE   HR: 3
CAS: 144-48-9   NIOSH: AC 4200000
mf: C_2H_4INO mw: 184.97

SYNS: alpha-IODOACETAMIDE   *   2-IODOACETAMIDE   *   USAF D-1   *   MONOIODOACETAMIDE

THR: A poison by intraperitoneal and intravenous routes. Mutagenic data. An experimental tumorigen. When heated to decomposition it emits very toxic fumes of I^- and NO_x.

IODOACETIC ACID   HR: 3
CAS: 64-69-7   NIOSH: AI 3500000
mf: $C_2H_3IO_2$ mw: 185.95

PROP: Colorless or white crystals. Sol in water and alc, mp: 82°-83° very sltly sol in ether.

SYNS: IODOACETATE * MONOIODOACETATE *˙ MONOIODOACETIC ACID

THR: A poison by subcutaneous, ingestion, and intravenous routes. An experimental teratogen, neoplastigen, and tumorigen. When heated to decomposition it emits toxic fumes of I^-. See also iodine.

IODOFORM   HR: 3
CAS: 75-47-8   NIOSH: PB 7000000
mf: CHI_3 mw: 393.72

PROP: Yellow powder or crystals; disagreeable odor; d: 4.1; mp: 120° (approx); bp: subl. Decomp @ high temp evolving iodine; unctuous touch, volatile with steam; very sol in H_2O; benzene, acetone; slightly sol in petr ether.

SYNS: NCI-C04568 * TRIIODOMETHANE

THR: A poison by subcutaneous route. Moderately toxic by ingestion. See also iodides. Incompatible with mercuric oxide; calomel; silver nitrate; tannin; Balsam Peru directly mixed; Li; acetone. Heated to decomposition it emits toxic fumes of I^-.

IODOMETHANE   HR: 3
CAS: 74-88-4   NIOSH: PA 9450000
mf: CH_3I mw: 141.94

PROP: Colorless liquid, turns brown on exposure to light, mp: −66.4°, bp: 42.5°, d: 2.279 @ 20°/4°, vap press: 400 mm @ 25.3°, vap d: 4.89. Sol in water @ 15°; misc in alc and ether.

SYNS: IODOMETANO (ITALIAN) * IODURE DE METHYLE (FRENCH) * JOD-METHAN (GERMAN) * JOODMETHAAN (DUTCH) * METHYL IODIDE * METHYLJODID (GERMAN) * METHYLJODIDE (DUTCH) * METYLU JODEK (POLISH) * MONOIODURO DI METILE (ITALIAN)

OSHA PEL: TWA 5 ppm (skin)

THR: A poison by ingestion and subcutaneous routes. Moderately toxic by inhalation and dermal routes. A human skin irritant. Mutagenic data. An experimental tumorigen, neoplastigen, and carcinogen. A strong narcotic and anesthetic. See also iodides. Incompatible with silver chlorite; sodium. When heated to decomposition it emits toxic fumes of I^-. For further information see Vol. 5, No. 6 of *DPIM Report*.

3-IODOPROPIONIC ACID   HR: 3
CAS: 141-76-4   NIOSH: UF 4900000
mf: $C_3H_5IO_2$ mw: 199.98

PROP: (a) Needles from water; d: 1.857; mp: 93-94°; bp: decomp; sltly sol in water; very sol in alc and ether; (b) needles; mp: 44.5°-45.5°; bp: 105°; very sltly sol in water; sol in alc, ether.

THR: An experimental tumorigen. Moderately toxic by skin contact. When heated to decomposition it emits toxic fumes of I^-.

3-IODO-2-PROPYNYL-2,4,5-TRICHLOROPHENYL ETHER   HR: 3
CAS: 777-11-7   NIOSH: KO 1225000
mf: $C_9H_4Cl_3IO$ mw: 361.38

SYNS: 2,4,5-TRICHLOROPHENYL IODOPROPARGYL ETHER * 2,4,5-TRICHLOROPHENYL-gamma-IODOPROPARGIL ETHER

THR: A poison by intraperitoneal route. Moderately toxic by ingestion. Skin and eye irritant. When heated to decomposition it emits very toxic Cl^- and I^-.

3-IODOTETRAHYDROTHIOPHENE-1,1-DIOXIDE   HR: 3
CAS: 17236-22-5   NIOSH: XN 0750000
mf: $C_4H_7IO_2S$ mw: 246.07

SYN: TETRAHYDRO-3-IODOTHIOPHENE-1,1-DIOXIDE

THR: A poison by ingestion, intraperitoneal, and intravenous routes. When heated to decomposition it emits very toxic I^- and SO_x.

IODOTRIMETHYLTIN   HR: 3
CAS: 811-73-4   NIOSH: WH 8581000
mf: C_3H_9ISn mw: 290.71

PROP: White powder; insol in water and organic solvents.

SYNS: TRIMETHYLSTANNYL IODINE * IODOTRIMETHYLSTANNANE * TRIMETHYLTIN IODIDE

THR: A poison by intravenous route. See also tin compounds. When heated to decomposition it emits toxic fumes of I^-.

IOXYNIL OCTANOATE HR: 3
CAS: 3861-47-0   NIOSH: RH 0410000
mf: $C_{15}H_{17}I_2NO_2$ mw: 497.13

SYNS: 4-cyano-2,6-diiodphenol capryl-saeureester (german) * 3,5-diiodo-4-hydroxybenzonitrile octanoate * 3,5-dijod-4-hydroxy-benzonitril caprysaeureester (german) * 3,5-diiodo-4-octanoyloxybenzonitrile

THR: Poison by ingestion. Mutagenic data. See also nitriles. When heated to decomposition it emits very toxic fumes of I^-, NO_x and CN^-.

IPECAC SYRUP HR: 3
NIOSH: NO 1700000

PROP: Dried rhizome and roots of Rio or Brazilian ipecac. Emetine, cephaline, emetamine, ipecacuanic acid, psychotrine, methyl psychotaine, resin.

SYN: ipecacuanha

THR: A poison by ingestion. A centrally acting emetic. Has caused fatalities. Symptoms include retention of urine, fever, diarrhea, violent abdominal pain, dehydration and cardiac irregularities. Can cause conjunctivitis with opacity of the cornea. See also emetine.

IPRATROPIUM BROMIDE HR: 3
CAS: 22254-24-6   NIOSH: YM 3700000

SYNS: (8r)-3-alpha-hydroxy-8-isopropyl-1-alpha-h,5-alpha-h-tropiumbromide-($\pm$)-tropate * ipratropiumbromid (german)

THR: A poison by intravenous and subcutaneous routes. Moderately toxic by ingestion. See also bromides. When heated to decomposition it emits toxic fumes of Br^-.

IRIDIUM HR: 2
af: Ir   aw: 192.2

PROP: Silver-white very hard element; metal; mp: 2450°; bp: approx 4500°; d: 22.65 @ 20°/4°. Highest specific gravity of all elements.

THR: No available toxicological data. Catalytic metal, may spontaneously ignite in air. Incompatible with interhalogens. Is attacked by F_2; Cl_2 @ red heat; by potassium sulfate or a mixture of potassium hydroxide and nitrate; on fusion; lead; zinc; tin.

IRIMONT HR: 3
NIOSH: NO 3700000

PROP: An iron complex for intramuscular application

SYN: fe 158/c

THR: A poison by subcutaneous and intravenous routes.

IRON (II) ARSENATE (3:2) HR: 3
CAS: 10102-50-8   NIOSH: NO 4580000
DOT: 1608
mf: $As_2O_8 \cdot 3Fe$ mw: 445.39

SYNS: arsenate of iron, ferrous * ferrous arsenate * iron arsenate

DOT Classification: Poison B, Label: Poison

THR: A deadly poison. See also arsenic compounds. When heated to decomposition it emits toxic fumes of As.

IRON (III) ARSENATE (1:1) HR: 3
CAS: 10102-49-5   NIOSH: NO 4585000
DOT: 1606
mf: $AsO_4 \cdot Fe$ mw: 194.77

SYNS: arsenate of iron, ferric * ferric arsenate

DOT Classification: Poison B, Label: Poison

THR: A deadly poison. See also arsenic compounds and iron compounds. When heated to decomposition it emits toxic fumes of As.

IRON (III)-o-ARSENITE PENTAHYDRATE HR: 3
CAS: 63989-69-5   NIOSH: NO 4600000
DOT: 1607
mf: $As_2Fe_2O_6 \cdot Fe_2O_3 \cdot 5H_2O$ mw: 607.34

PROP: Brown-yellow powder.

SYNS: ferric arsenite, solid (dot) * ferric arsenite * ferric arsenite, basic

DOT Classification: Poison B, Label: Poison

THR: A deadly poison. See also arsenic compounds. When heated to decomposition it emits toxic fumes of As.

IRON CARBONYL HR: 3
CAS: 13463-40-6   NIOSH: NO 4900000
mf: C_5FeO_5 mw: 195.90

PROP: Yellow to dark red viscous liquid, mp: −25°, bp: 103.0°, flash p: 5°F, d: 1.453 @ 25°/4°, vap press: 40 mm @ 30.3°.

SYNS: fer pentacarbonyle (french) * pentacarbonyliron

THR: A poison by inhalation, dermal, ingestion and intravenous routes. Inhalation of this material causes dizziness, nausea and vomiting. If continued, unconsciousness follows. Often there is a delayed reaction of chest pain, cough and difficult breathing. There may be cyanosis and circulatory collapse. In fatal cases death occurs from the fourth to eleventh day with pneumonitis and injury to kidneys, liver and brain. Iron carbonyl is less toxic than nickel carbonyl. Dangerous fire, disaster, and explosion hazard; see carbonyls. Pyrophoric in air! To fight fire, use water, foam, CO_2, dry chemical.

IRON-DEXTRAN COMPLEX   HR: 3
CAS: 9004-66-4   NIOSH: NO 5950000

PROP: For human use, it is a sterile dark brown colloidal solution, water-soluble. Approximate molecular weight is 180,000.

SYNS: DEXTRAN IRON COMPLEX * EISENDEXTRAN (GERMAN) * IMFERON * IRON DEXTRAN INJECTION * IRONORM INJECTION * URSOFERRAN

THR: A neoplastigen by intramuscular route in women. A suspected human carcinogen. An experimental tumorigen. Moderate acute toxicity by intravenous and intraperitoneal routes.

IRON-DEXTRIN COMPLEX   HR: 3
CAS: 9004-51-7   NIOSH: NO 6210000

PROP: For human use, it is a clear, brown colloidal solution. Approximate molecular weight is 230,000.

SYNS: ASTRAFER * DEXTRIFERRON * DEXTRIFERRON INJECTION * FERRIGEN * IRON CARBOHYDRATE COMPLEX * IRON DEXTRIN INJECTION

THR: A poison by intravenous route. An experimental neoplastigen and tumorigen.

IRON DISULFIDE   HR: 3
CAS: 12068-85-8   NIOSH: NO 6835000
mf: FeS_2 mw: 119.97

SYNS: IRON PYRITES * IRON SULFIDE

THR: A poison by inhalation and ingestion. See also sulfides and H_2S. When heated to decomposition it emits very toxic fumes due to contact with acids, acid fumes, high temperatures. Incompatible with water; powdered pyrites. Heats up spontaneously and ignites with combustibles.

IRON DUST   HR: 3
NIOSH: NO 6851000
af: Fe   at wt: 55.8

PROP: Silvery-white, tenacious, lustrous, ductile metal. mp: 1535°, bp: 3000°, d: 7.86, vap press: 1 mm @ 1787°. Iron dust from open hearth furnace; dust contained 52% iron.

THR: A poison by intraperitoneal route. Iron dust can cause conjunctivitis, choroiditis, retinitis and siderosis of tissues if iron contacts and remains in these tissues. Iron ore dust can cause palpebral conjunctivitis, massive pulmonary fibrosis and an increased incidence of lung cancer. An iron oxide fume is generated in welding operations and continued exposure to concentrations > 30 mg/m^3 of air can cause chronic bronchitis. Fresh iron oxide fume can cause metal fume fever. Iron compounds are suspected carcinogens of the lung, liver, connective tissue and reticuloendothelial tissue. Flammable in the form of dust when exposed to heat or flame. See also powdered metals. Reacts violently with Cl_2; ClF_3; F_2; H_2O_2; NO_2; P; Na_2C_2; H_2SO_4; air; water; polystyrene. Moderate explosion hazard in the form of dust when exposed to heat or flame. To fight fire, use special mixtures of dry chemical.

IRON (III) OXIDE   HR: 3
CAS: 1309-37-1   NIOSH: NO 7400000
mf: Fe_2O_3 mw: 159.70

SYNS: ANHYDROUS IRON OXIDE * BAUXITE RESIDUE * BLACK OXIDE OF IRON * BLENDED RED OXIDES OF IRON * BURNT SIENNA * BURNT UMBER * CAPUT MORTUUM * COLLOIDAL FERRIC OXIDE * C.I. 77491 * FERRIC OXIDE * IRON OXIDE RED * IRON SESQUIOXIDE * JEWELER'S ROUGE * MANUFACTURED IRON OXIDES * NATURAL IRON OXIDES * NATURAL RED OXIDE * OCHRE * RED IRON OXIDE * SIENNA * SPECULAR IRON * SYNTHETIC IRON OXIDE * VITRIOL RED * YELLOW FERRIC OXIDE * YELLOW OXIDE OF IRON

THR: A poison by subcutaneous route. An experimental tumorigen and possible carcinogen. Reacts violently with Al; $Ca(OCl)_2$; N_2H_4; ethylene oxide.

IRON OXIDE FUME   HR: 3
NIOSH: NO 7525000

SYN: ZELAZA TLENKI (POLISH)

OSHA PEL: TWA 10 mg/m^3

THR: See also iron dust.

IRON OXIDE, SACCHARATED   HR: 3
CAS: 8047-67-4   NIOSH: NO 7700000

PROP: Saccharated oxide of iron

SYNS: FEOJECTIN * FERRIC OXIDE, SACCHA-
RATED * FERRIC SACCHARATE . . . IRON OX-
IDE (MIX.) * FERRIVENIN * IRON OXIDE SAC-
CHARATED * IRON SACCHARATE * IRON
SUGAR * NEO-FERRUM * PROFERRIN
* SACCHARATED IRON

THR: An experimental tumorigen, carcinogen,
and musculoskeletal toxin.

IRON-POLYSACCHARIDE COM-PLEX   HR: 3
NIOSH: NO 8195000

PROP: Solution of iron and synthetically prepared polysaccharide of mean molecular weight
of about 20,000.

SYN: MUSCULARON

THR: An experimental tumorigen.

IRON SORBITOL CITRATE   HR: 3
CAS: 1338-16-5   NIOSH: NO 8350000
mf: $C_6H_{14}O_6 \cdot C_6H_8O_7 \cdot 7Fe$ mw: 765.29

SYN: IRON SORBITEX

THR: A poison by subcutaneous and intravenous
routes. When heated to decomposition it emits
acrid smoke and fumes.

IRON (II) SULFATE (1:1)   HR: 3
CAS: 7720-78-7   NIOSH: NO 8500000
mf: $O_4S \cdot Fe$ mw: 151.91

SYNS: COPPERAS * EXSICCATED FERROUS
SULFATE * FERROSULFAT (GERMAN) * FER-
ROUS SULFATE * GREEN VITRIOL * IRON
PROTOSULFATE * IRON VITRIOL * SULFURIC
ACID, IRON(2+) SALT (1:1)

ACGIH TLV: TWA 1 mg/m^3; STEL 2 mg/m^3

THR: A poison by ingestion to humans. An
experimental tumorigen, affects the central nervous system, the gastrointestinal tract, and is
a systemic toxin. Mutagenic data. When heated
to decomposition it emits toxic fumes of SO_x.

ISOACETOPHORONE   HR: 2
CAS: 78-59-1   NIOSH: GW 7700000
mf: $C_9H_{14}O$ mw: 138.23

PROP: Practically water-white liquid, bp:
215.2°, flash p: 184°F (OC), d: 0.9229, autoign
temp: 864°F, vap press: 1 mm @ 38.0°, vap
d: 4.77, lel = 0.8%, uel = 3.8%.

SYNS: ISOFORONE (ITALIAN) * ISOPHORONE
* IZOFORON (POLISH) * NCI-C55618
* 1,1,3-TRIMETHYL-3-CYCLOHEXENE-5-ONE
* 3,5,5-TRIMETHYL-2-CYCLOHEXEN-1-ON (GER-
MAN, DUTCH) * 3,5,5-TRIMETHYL-2-CYCLO-
HEXENE-1-ONE * 3,5,5-TRIMETIL-2-CICLOESEN-
1-ONE (ITALIAN)

OSHA PEL: TWA 25 ppm

THR: Moderately toxic by ingestion, inhalation,
and skin contact. A skin and eye irritant. A
human systemic irritant. See also ketones. Considered to be more toxic than mesityl oxide.
However, due to its low volatility, it is not a
dangerous industrial hazard. The response of
guinea pigs and rats to repeated inhalation of
the vapors indicates that it is one of the most
toxic of the ketones. It is chiefly a kidney poison.
It can cause irritation, lachrymation, possible
opacity of the cornea and necrosis of the cornea
(experimental). It is irritating at the level of
25 ppm to humans. In animal experiments death
during exposure was usually due to narcosis,
but occasionally to irritation of the lungs. Flammable when exposed to heat or flame; can react
with oxidizing materials. To fight fire, use foam,
CO_2, dry chemical. For further information see
isophorone, Vol. 2, No. 1 of *DPIM Report*.

ISOAMYL ALCOHOL   HR:
CAS: 123-51-3   NIOSH: EL 5425000
mf: $C_5H_{12}O$ mw: 88.17

PROP: Clear liquid, pungent, repulsive taste,
bp: 132°, ulc: 35-40, lel = 1.2%, uel = 9.0%
@ 212°F, flash p: 109°F (CC), d: 0.813, autoign
temp: 662°F, vap d: 3.04, mp: −117.2°. Sol
in water @ 14°; misc in alc and ether.

SYNS: ALCOOL AMILICO (ITALIAN) * ALCOOL
ISOAMYLIQUE (FRENCH) * AMYLOWY ALKOHOL
(POLISH) * FERMENTATION AMYL ALCOHOL
* ISOAMYL ALKOHOL (CZECH) * ISOAMYLOL
* ISOBUTYLCARBINOL * ISO-AMYLALKOHOL

(GERMAN) * ISOPENTANOL * ISOPENTYL AL-COHOL * 3-METHYL BUTANOL * 2-METHYL-4-BUTANOL * 3-METHYLBUTAN-1-OL * 3-METHYL-1-BUTANOL (CZECH) * 3-METIL-BUTANOLO (ITALIAN)

OSHA PEL: TWA 100 ppm

THR: A poison; moderately toxic by ingestion, intraperitoneal, skin, and intravenous routes. A skin and eye irritant. An experimental carcinogen. A human systemic irritant. Flammable when exposed to heat or flame; can react vigorously with reducing materials. Slight explosion hazard when exposed to flame. To fight fire, use alcohol foam, CO_2, dry chemical. When heated to decomposition it emits acrid smoke and fumes.

ISOBORNYL THIOCYANATO ACE-TATE   HR: 3
CAS: 115-31-1   NIOSH: AJ 6300000
mf: $C_{13}H_{19}NO_2S$ mw: 253.39

PROP: Yellow, oily liquid, d: 1.1465 @ 25°/4°, terpene-like odor; bp: 95° @ 0.06 mm; flash p: 82°C (180°F); very sol in alc, benzene, chloroform, ether. Insol in H_2O.

SYNS: ENT 92 * ISOBORNEOL THIOCYANATO-ACETATE * ISOBORNYL THIOCYANOACETATE * TERPINYL THIOCYANOACETATE * THIOCYAN-ATOACETIC ACID ISOBORNYL ESTER

THR: A poison by ingestion. Low toxicity by skin route. See also thiocyanates and esters. Very irritating to eyes, mucous membranes and skin. An insecticide and fly spray. When heated to decomposition it emits very toxic fumes of NO_x and SO_x. See also thiocyanates.

ISOBUTENE   HR: 1
CAS: 115-11-7   NIOSH: UD 0890000
DOT: 1055/1075
mf: C_4H_8 mw: 56.12

PROP: Volatile liquid or easily liquefied gas, bp: −6.9°, fp: −140.3°, flash p: <14°F, d: 0.600, autoign temp: 869°F, lel = 1.8%, uel = 9.6%. Insol in water, very sol in alc, ether, sulfuric acid.

SYNS: ISOBUTYLENE * METHYL PROPENE

DOT Classification: Label: Flammable Gas

THR: A simple asphyxiant; may have narcotizing action. Very dangerous fire hazard when exposed to heat or flame. Can react vigorously

with oxidizing materials. To fight fire, stop flow of gas.

ISOBUTYL ACETATE   HR: 1
CAS: 110-19-0   NIOSH: AI 4025000
DOT: 1213
mf: $C_6H_{12}O_2$ mw: 116.18

PROP: Colorless, neutral liquid, fruitlike odor, mp: −98.9°, bp: 118°, flash p: 64°F (CC) (18°C), d: 0.8685 @ 15°, vap press: 10 mm @ 12.8°, autoign temp: 793°F, vap d: 4.0, lel = 2.4%, uel = 10.5%. Very sol in alc; sltly sol in H_2O. Misc in alc + ether.

SYNS: ACETIC ACID, ISOBUTYL ESTER * ACETATE D'ISOBUTYLE (FRENCH) * ACETIC ACID-2-METHYLPROPYL ESTER * 2-METHYL-PROPYL ACETATE * 2-METHYL-1-PROPYL ACE-TATE * BETA-METHYLPROPYL ETHANOATE

OSHA PEL: TWA 150 ppm
DFG MAK: 200 ppm (950 mg/m³)
DOT Classification: Label: Flammable Liquid

THR: A skin and eye irritant. Low toxicity by ingestion and inhalation. See also esters. Upon absorption by the body it can hydrolyze to acetic acid and isobutanol. See also *n*-butyl acetate. Dangerous fire hazard when exposed to heat, flame or oxidizers. To fight fire, use alcohol foam, CO_2, dry chemical. For further information see Vol. 2, No. 2 of *DPIM Report*.

ISOBUTYL ALCOHOL   HR: 3
CAS: 78-83-1   NIOSH: NP 9625000
mf: $C_4H_{10}O$ mw: 74.14

PROP: Clear liquid, sweet odor. Flammable, bp: 107.90°, flash p: 82°F, ulc: 40-45, lel = 1.2%, uel = 10.9% @ 212°F, fp: −108°, d: 0.805 @ 20°/4°, autoign temp: 800°F, vap press: 10 mm @ 21.7°, vap d: 2.55. Sol in H_2O. Misc with alc, ether.

SYNS: 2-METHYL PROPANOL * ALCOOL ISO-BUTYLIQUE (FRENCH) * FERMENTATION BUTYL ALCOHOL * 1-HYDROXYMETHYLPROPANE * ISOBUTANOL * ISOBUTYLALKOHOL (CZECH) * ISOPROPYLCARBINOL * 2-METHYL-1-PROPA-NOL * 2-METHYLPROPAN-1-OL * 2-METHYL-PROPYL ALCOHOL

OSHA PEL: TWA 100 ppm
DFG MAK: 100 ppm (300 mg/m³)

THR: An experimental carcinogen. Moderately toxic by ingestion, inhalation and dermal routes.

May be mildly irritating to skin and mucous membranes, and in high concentrations, narcotic. Dangerous fire hazard when exposed to heat or flame. Moderate explosion hazard in the form of vapor when exposed to heat, flame or oxidizers. Emits toxic fumes when heated. Keep away from heat and open flame. To fight fire, use alcohol foam, CO_2, dry chemical. For further information see Vol. 2, No. 2 of *DPIM Report*.

ISOBUTYLAMINE HR: 2
CAS: 78-81-9 NIOSH: NP 9900000
DOT: 1214
mf: $C_4H_{11}N$ mw: 73.16

PROP: Colorless liquid, mp: $-85.5°$, bp: $68.6°$, flash p: 15°F, d: 0.731 @ 20°/20°, vap press: 100 mm @ 18.8°, autoign temp: 712°F, vap d: 2.5. Misc with H_2O, alc and ether.

SYNS: 1-AMINO-2-METHYLPROPANE * MONO-ISOBUTYLAMINE

DFG MAK: 5 ppm (15 mg/m^3)
DOT Classification: Label: Flammable
  Liquid

THR: A powerful irritant to skin, eyes and mucous membranes. Skin contact can cause blistering. Inhalation can cause headache and dryness of nose and throat. See also amines. Dangerous fire hazard when exposed to heat or flame. Can react vigorously with oxidizing materials. Can emit NO_x on burning. To fight fire, use dry chemical, foam, CO_2, alcohol foam.

ISOBUTYL CELLOSOLVE HR: 2
CAS: 4439-24-1 NIOSH: KL 3850000
mf: $C_6H_{14}O_2$ mw: 118.20

PROP: Colorless liquid; d: 0.903 @ 20°/4°; bp: 171.2°; misc in water, alc, ether.

SYNS: EKTASOLVE EIB * ETHYLENE GLYCOL MONOISOBUTYL ETHER

THR: Moderately toxic by ingestion, inhalation and skin contact. A mild skin irritant. When heated to decomposition it emits acrid smoke and fumes.

ISOBUTYLIDENEDIUREA HR: 3
CAS: 6104-30-9 NIOSH: YT 5300000
mf: $C_6H_{14}N_4O_2$ mw: 174.24

SYNS: DIUREIDOISOBUTANE * 1,1-DIUREID-ISOBUTANE * ISOBUTYLDIUREA * 1,1'-ISO-BUTYLIDENEBISUREA

THR: An experimental tumorigen. When heated to decomposition it emits toxic fumes of NO_x.

ISOBUTYL ISOBUTYRATE HR: 2
CAS: 97-85-8 NIOSH: NQ 5250000
mf: $C_8H_{16}O_2$ mw: 144.24

PROP: Liquid with fruity odor, mp: $-81°$, bp: 147.5°, d: 0.850-0.860 @ 20°/20°, vap press: 10 mm @ 39.9°. Insol in H_2O; misc with alc.

SYNS: ISOBUTYRIC ACID, ISOBUTYL ESTER
* 2-METHYLPROPYL ISOBUTYRATE

THR: Moderately toxic by ingestion and inhalation. An insect repellent. See also isobutyl alcohol. Combustible when exposed to heat or flame. Can react with oxidizing materials.

ISOBUTYL METHACRYLATE HR: 3
CAS: 97-86-9 NIOSH: OZ 4900000
mf: $C_8H_{14}O_2$ mw: 142.22

SYNS: ISOBUTYL-alpha-METHACRYLATE
* 2-METHYLPROPYL METHACRYLATE

THR: An experimental teratogen. Moderate intraperitoneal toxicity. See also esters. When heated to decomposition it emits acrid smoke and fumes.

ISOBUTYLMETHYLCARBINOL HR: 2
CAS: 108-11-2 NIOSH: SA 7350000
mf: $C_6H_{14}O$ mw: 102.20

PROP: Clear liquid, bp: 131.8°, fp: $<-90°$ (sets to a glass), flash p: 106°F, d: 0.8079 @ 20°/20°, vap press: 2.8 mm @ 20°, vap d: 3.53, lel = 1.0%, uel = 5.5%.

SYNS: ALCOOL METHYL AMYLIQUE (FRENCH)
* ISOBUTYLMETHYLMETHANOL * METHYL AMYL ALCOHOL * METHYLISOBUTYL CARBINOL
* 2-METHYL-4-PENTANOL * 4-METHYLPENTA-NOL-2 * 4-METHYL-2-PENTANOL * METI-LAMIL ALCOHOL (ITALIAN) * 4-METILPENTAN-2-OLO (ITALIAN)

OSHA PEL: TWA 25 ppm (skin)

THR: Moderately toxic by ingestion, inhalation, skin contact, and intraperitoneal routes. A strong irritant. Inhalation of high concentrations cause anesthesia. Flammable when exposed to heat or flame; can react with oxidizing materials. To fight fire, use alcohol foam.

ISOBUTYL NITRITE HR: 3
CAS: 542-56-3 NIOSH: RA 0805000
mf: $C_4H_9NO_2$ mw: 103.14

PROP: Liquid; d: 0.870 @ 22°/4°; bp: 67-68°; slightly sol in and decomp in water; misc in alc.

SYNS: NITROUS ACID, ISOBUTYL ESTER * NCI-C61052

THR: A poison by intraperitoneal route. Human blood effects by ingestion. See also esters and nitrites. When heated to decomposition it emits toxic fumes of NO_x.

p-ISOBUTYOXYBENZOIC ACID-3-(2'-METHYLPIPERIDINO)PROPYL ESTER   HR: 3
CAS: 63916-90-5   NIOSH: DH 3070000
mf: $C_{20}H_{31}NO_3$ mw: 333.52

THR: A poison by subcutaneous and intravenous routes. When heated to decomposition it emits toxic fumes of NO_x.

ISOBUTYRALDEHYDE   HR: 2
CAS: 78-84-2   NIOSH: NQ 4025000
mf: C_4H_8O mw: 72.12

PROP: Transparent, colorless, highly refractive liquid, pungent odor, mp: −65°, bp: 64°, flash p: −40°F (CC), d: 0.7938 @ 20°/4°, autoign temp: 434°F, lel = 1.6%, uel = 10.6%, vap d: 2.5. Sol in water; misc in alc, ether, benzene, carbon disulfide, acetone, toluene, chloroform.

SYNS: ISOBUTYRALDEHYD (CZECH) * ISOBU-TYLALDEHYDE * ISOBUTYRIC ALDEHYDE * 2-METHYLPROPANAL * 2-METHYL-1-PRO-PANAL * 2-METHYLPROPIONALDEHYDE * NCI-C60968

THR: Moderately toxic by ingestion and inhalation. See also aldehydes. Skin and eye irritant. Dangerous fire hazard when exposed to heat, flame or oxidizers. Moderate explosion hazard in the form of vapor when exposed to heat or flame. Dangerous; keep away from heat and open flame; can react vigorously with reducing materials. To fight fire, use dry chemical, CO_2, mist, foam. For further information see Vol. 2, No. 2 of *DPIM Report*.

ISOBUTYRIC ACID   HR: 3
CAS: 79-31-2   NIOSH: NQ 4375000
DOT: 2529
mf: $C_4H_8O_2$ mw: 88.12

PROP: Colorless liquid, pungent odor, mp: −47°, bp: 154.5°, flash p: 132°F (TOC), d:

0.949 @ 20°/4°, vap press: 1 mm @ 14.7°, vap d: 3.04, autoign temp: 935°F. Misc with alc, chloroform and ether. Sol in 6 parts H_2O.

SYNS: DIMETHYLACETIC ACID * ISOPROPYL-FORMIC ACID * alpha-METHYLPROPIONIC ACID * 2-METHYLPROPANOIC ACID * 2-METHYL-PROPIONIC ACID

DOT Classification: Label: Corrosive

THR: A poison by ingestion. Moderately toxic by dermal route. A skin irritant. Flammable when exposed to heat or flame; can react with oxidizing materials. To fight fire, use alcohol foam, CO_2, dry chemical.

ISOBUTYRONITRILE   HR: 3
CAS: 78-82-0   NIOSH: TZ 4900000
mf: C_4H_7N mw: 69.12

PROP: Colorless liquid, slightly sol in water, very sol in alcohol and ether, flash p: 46.4°F. d: 0.773 @ 20°/20°, bp: 107°, mp: −75°.

SYNS: ISOPROPYL CYANIDE * 2-METHYLPRO-PIONITRILE

THR: A poison by ingestion, skin contact, and subcutaneous routes. Moderate inhalation toxicity. See also nitriles. When heated to decomposition it emits toxic fumes of NO_x.

ISOCAINE   HR: 3
CAS: 533-28-8   NIOSH: TN 3325000
mf: $C_{16}N_{23}NO_2 \cdot ClH$ mw: 297.86

SYNS: O-AMINOBENZOYL DI(ISOPROPYLAMINO)-ETHANOL HYDROCHLORIDE * 3-BENZOXY-1-(2-METHYLPIPERIDINO)PROPANE HYDROCHLORIDE * DL-3-BENZOXY-1-(2-METHYLPIPERIDINO)PRO-PANE HYDROCHLORIDE * BENZOYL-gamma-(2-METHYLPIPERIDINE)-PROPANOL HYDROCHLORIDE * 3-(2-METHYLPIPERIDINO)PROPYL BENZOATE HYDROCHLORIDE * DL-(2-METHYLPIPERIDINO)-PROPYL BENZOATE HYDROCHLORIDE * gamma-(2-METHYLPIPERIDINO)PROPYL BEN-ZOATE HYDROCHLORIDE * (+×−)-gamma-(2-METHYLPIPERIDYL)PROPYL BENZOATE HYDRO-CHLORIDE * METHYCAINE HYDROCHLORIDE * NEOTHESIN HYDROCHLORIDE * PIPEROCAINE HYDROCHLORIDE

THR: A poison by intraperitoneal, intravenous, subcutaneous, and intraspinal routes. When heated to decomposition it emits very toxic fumes of NO_x and HCl.

ISOCARBOXAZID HR: 3
CAS: 59-63-2 NIOSH: NY 2625000
mf: $C_{12}H_{13}N_3O_2$ mw: 231.28

PROP: Crystals from methanol, practically tasteless; mp: 106°; very sltly sol in hot H_2O; sltly sol in alc, glycerol, propylene glycol.

SYNS: N'-BENZYL N-METHYL-5-ISOXAZOLECAR-BOXYLHYDRAZIDE-3 * 1-BENZYL-2-(5-METH-YL-3-ISOXAZOIYL-CARBONYL)HYDRAZINE * 1-BENZYL-1-(5-METHYL-3-ISOXAZOIYLCAR-BONYL)HYDRAZINE * ISOCARBONAZID * ISOCARBOSSAZIDE * ISOCARBOXAZIDE * 5-METHYL-3-ISOXAZOLECARBOXYLIC ACID-2-BENZYLHYDRAZIDE

THR: A poison by ingestion, intraperitoneal, and subcutaneous routes. When heated to decomposition it emits toxic fumes of NO_x.

ISODRIN HR: 3
CAS: 465-73-6 NIOSH: IO 1925000
mf: $C_{12}H_8Cl_6$ mw: 364.90

PROP: Crystals; mp: 240°-242°.

SYNS: ENT 19,244 * 1,2,3,4,10,10-HEXA-CHLORO-1,4,4a,5,8,8a-HEXAHYDRO-1,4,5,8-ENDO,ENDO-DIMETHANONAPHTHALENE

THR: A poison by ingestion, skin contact, and other routes. Causes liver injury, acne, and skin rashes. See also chlorinated phenols and aldrin. Dangerous, see also chlorides.

ISOFENPHOS HR: 3
NIOSH: DH 2255000
mf: $C_{15}H_{24}NO_4PS$ mw: 345.43

SYNS: 2-(O-AETHYL-N-ISOPROPYLAMIDO-THIOPHOSPHORYLOXY)-BENZOSAEURE-ISOPRO-PYLESTER(GERMAN) * O-ETHYL-O-(2-ISO-PROPOXY-CARBONYL)-PHENYL ISOPROPYL-PHOSPHORAMIDOTHIOATE * ISOPROPYL SALICYLATE-O-ESTERWITH O-ETHYLISOPRO-PYLPHOSPHORAMIDOTHIOATE * 1-METHYL-ETHYL-2-((ETHOXY((1-METHYLETHYL)AMINO)-PHOSPHINOTHIOYL)OXY)BENZOATE

THR: A poison by ingestion and dermal contact. When heated to decomposition it emits very toxic fumes of PO_x, NO_x and SO_x.

ISOMETHADONE HR: 3
CAS: 466-40-0 NIOSH: MP 1925000
mf: $C_{21}H_{27}NO$ mw: 309.49

SYNS: 6-DIMETHYLAMINO-5-METHYL-4,4-DI-PHENYL-3-HEXANONE * 1,1-DIPHENYL-1-(DIMETHYLAMINOISOPROPYL)BUTANONE-2 * ISOAMIDONE II

THR: A poison by ingestion and intravenous routes. When heated to decomposition it emits toxic fumes of NO_x.

ISONICOTINIC ACID HYDRA-ZIDE HR: 3
CAS: 54-85-3 NIOSH: NS 1750000
mf: $C_6H_7N_3O$ mw: 137.16

PROP: Consists of 12% W/V each of dodecyla-mine hydrochloride, trimethyl alkyl ammonium chloride, and methyl alkyl dipolyoxypropylene ammonium methyl sulfate.

SYNS: HYDRAZIDE * USAF CB-2 * 4-PYRIDI-NECARBOXYLIC ACID, HYDRAZIDE * NSC 9659 * ISONICOTINHYDRAZID * ISONICOTINOYL HY-DRAZID * ISONICOTINYL HYDRAZIDE

THR: A poison by ingestion, subcutaneous, in-traperitoneal, and intramuscular routes. An ex-perimental carcinogen, teratogen, neoplastigen, and tumorigen. In humans it is a poison by ingestion to the central nervous system, the pe-ripheral nervous system, and the glandular sys-tem. Mutagenic data. When heated to decompo-sition it emits toxic fumes of NO_x.

ISONICOTINIC ACID-2-ISOPROPYLHY-DRAZIDE HR: 3
CAS: 54-92-2 NIOSH: NS 1925000
mf: $C_9H_{13}N_3O$ mw: 179.25

SYNS: IPRONIAZID * 1-ISONICOTINYL-2-ISO-PROPYLHYDRAZINE * 1-ISONICOTINOYL-2-ISO-PROPYLHYDRAZINE * N-ISOPROPYL ISONICOTIN-HYDRAZIDE

THR: A poison by ingestion, intravenous and other routes. Moderately toxic by ingestion, skin contact, intramuscular, subcutaneous, and intra-peritoneal routes. An experimental tumorigen. When heated to decomposition it emits toxic fumes of NO_x.

ISONYL HR: 3
CAS: 503-01-5 NIOSH: MP 9100000
mf: $C_9H_{19}N$ mw: 141.29

PROP: Colorless oily liquid, characteristic amine odor, water-insol. See also 2-methylhex-ane.

SYNS: ISOMETHEPTENE * 6-METHYLAMINO-2-METHYLHEPTENE * METHYLISOOCTENYLAMINE

* 2-METHYL-6-METHYLAMINO-2HEPTENE
* METHYLOCTENYLAMINE * N-1,5-TRI-
METHYL-4-HEXENYLAMINE

THR: A poison by ingestion. Can cause head-ache, nausea, dizziness in humans. When heated to decomposition it emits toxic fumes of NO_x.

ISOOCTYL ALCOHOL   HR: 2
CAS: 26952-21-6   NIOSH: NS 7700000
mf: $C_8H_{18}O$ mw: 130.26

SYN: ISOOCTANOL

THR: Moderately toxic by ingestion and skin contact. Severe eye irritant. When heated to decomposition it emits acrid smoke and fumes. For further information see Vol. 2, No. 2 of *DPIM Report*.

2-ISOOCTYL AMINE   HR: 3
CAS: 543-82-8   NIOSH: MQ 4725000
mf: $C_8H_{19}N$ mw: 129.28

SYNS: 2-AMINO-6-METHYLHEPTANE * 6-AMINO-2-METHYLHEPTANE * ALPHA, EPSILON-DIMETHYLHEXYLAMINE * 1,5-DIMETHYLHEX-YLAMINE * 2-METHYL-6-AMINOHEPTANE * 6-METHYL-2-HEPTYLAMINE * 2-METIL-6-AMINO-EPTANO (ITALIAN)

THR: A poison by intraperitoneal and intramus-cular routes. When heated to decomposition it emits toxic fumes of NO_x.

ISOOCTYL-2,4-DICHLOROPHENOXY
ACETATE   HR: 3
CAS: 25168-26-7   NIOSH: AG 8575000
mf: $C_{16}H_{22}Cl_2O_3$ mw: 333.28

SYNS: 2,4-DICHLOROPHENOXYACETIC ACID ISO-OCTYL ESTER * 2,4-D ISOOCTYL ESTER * ISOOCTYL ALCOHOL (2,4-DICHLOROPHE-NOXY)ACETATE

THR: An experimental carcinogen and tumori-gen. See also esters. When heated to decomposi-tion it emits toxic fumes of Cl^-.

ISOPENTYL ALCOHOL ACE-
TATE   HR: 2
CAS: 123-92-2   NIOSH: NS 9800000
mf: $C_7H_{14}O_2$ mw: 130.21

PROP: Liquid, banana-like odor, bp: 142.0°, ulc: 55-60, lel = 1% @ 212°F, uel = 7.5%, flash p: 77°F, d: 0.876, autoign temp: 680°F, vap d: 4.49.

SYNS: BANANA OIL * ISOAMYL ACETATE * ISOAMYL ETHANOATE * ISOPENTYL ACE-TATE * 3-METHYLBUTYL ACETATE * 3-METHYL-1-BUTYL ACETATE * 3-METHYLBUTYL ETHANOATE * PEAR OIL

OSHA PEL: TWA 100 ppm

THR: Moderately irritating by ingestion and in-halation. Exposure to concentrations of about 1,000 ppm for 1 hour can cause headache, fa-tigue, pulmonary irritation and serious toxicity effects. Dangerous fire hazard when exposed to heat or flame; can react vigorously with reduc-ing materials. Moderate explosion hazard when exposed to heat or flame. To fight fire, use alcohol foam, CO_2, dry chemical. For further information see Vol. 2, No. 2 of *DPIM Report*.

ISOPHOSPHAMIDE   HR: 3
CAS: 3778-73-2   NIOSH: RP 6050000
mf: $C_7H_{15}Cl_2N_2O_2P$ mw: 261.11

SYNS: 3-(2-CHLOROETHYL)-2-((2-CHLOROETH-YL)AMINO) TETRAHYDRO-1,3,2-OXAZAPHOS-PHORINE-2-OXIDE * 2,3-(N,N)sup 1)-BIS(2-CHLOROETHYL)DIAMIDO-1,3,2-OXAZAPHOS-PHORIDINOXY * DIPHOSPHAMIDE * NCI-C01638 * NSC-109724

THR: Mutagenic data. An experimental carcino-gen and neoplastigen. A human central nervous system, blood picture, and white blood cell toxin. Moderate acute toxicity by intraperitoneal route. When heated to decomposition it emits very toxic Cl^-, NO_x and PO_x.

ISOPRAL   HR: 3
CAS: 76-00-6   NIOSH: UB 9800000
mf: $C_3H_5Cl_3O$ mw: 163.43

PROP: Crystals, camphor-like odor, pungent taste, water-sol, mp: 50°, bp: 162°.

SYNS: 1,1,1-TRICHLOROISOPROPYL ALCOHOL * 1,1,1-TRICHLORO-2-PROPANOL

THR: A poison by unspecified routes. Moder-ately toxic by ingestion. See also chlorinated hydrocarbons, aliphatic. Dangerous; see also chlorides.

ISOPRENE   HR: 3
CAS: 78-79-5   NIOSH: NT 4037000
mf: C_5H_8 mw: 68.11

PROP: Colorless, volatile liquid, mp: −146.7°, bp: 34°, flash p: −65°F, d: 0.6806 @ 20°/4°, autoign temp: 428°F, vap press: 400 mm @

15.4°, vap d: 2.35; fp: −145.95°. Insol in water; misc in alc and ether.

SYNS: beta-METHYLBIVINYL * 2-METHYLBU-TADIENE * 2-METHYL-1,3-BUTADIENE

DOT Classification: Label: Flammable Liquid

THR: Moderately irritating to skin, eyes and mucous membranes. Concentration of 2% in air not narcotic to mice but produces bronchial irritation. Concentration of 5% fatal to mice. No data on human exposures. Highly dangerous fire hazard when exposed to heat, flame or oxidizers; i.e., violent reaction with chlorosulfonic acid; HNO_3; oleum; H_2SO_4. Dangerous; can react vigorously with reducing materials. To fight fire: CO_2, dry chemical.

ISOPROPENYL METHYL KE-TONE   HR: 3
CAS: 814-78-8   NIOSH: EN 0175000
mf: C_5H_8O mw: 84.13

PROP: Bp: 98°; vap d: 2.9; lel = 1.8%; uel = 9.0%.

SYNS: 3-METHYL-3-BUTEN-2-ON (GERMAN) * 2-METHYL-1-BUTEN-3-ONE * METHYL ISO-PROPENYL KETONE

THR: A poison by ingestion, inhalation, and skin contact. Moderately toxic by intraperitoneal route. See also ketones. When heated to decomposition it emits acrid smoke and fumes.

ISOPROPYL ACETATE   HR: 3
CAS: 108-21-4   NIOSH: AI 4930000
mf: $C_5H_{10}O_2$ mw: 102.14

PROP: Flash p: 39.2°F; lel = 1.7%; uel = 7.8%; bp: 58°-60° @ 200 mm. Colorless aromatic liquid; mp: −73°; fp: −69.3°; d: 0.874 @ 20°/20°; autoign temp: 860°F; vap press: 40 mm @ 17.0°; d: 3.52.
OSHA PEL: TWA 250 ppm
DOT Classification: Label: Flammable Liquid

THR: Moderately toxic by ingestion. Low toxicity by inhalation. Narcotic in high concentrations. Chronic exposure can cause liver damage. Dangerous fire hazard when exposed to heat, flame or oxidizers. Moderate explosion hazard when exposed to heat or flame. Dangerous; keep away from heat and open flame; can react vigorously with oxidizing materials. To fight fire:

Foam, CO_2, dry chemical. For further information see Vol. 1, No. 3 of *DPIM Report*.

ISOPROPYLADRENALINE   HR: 3
CAS: 7683-59-2   NIOSH: DO 1750000
mf: $C_{11}H_{17}NO_3$ mw: 211.29

SYNS: 3,4-DIHYDROXY-alpha-((ISOPROPYLAM-INO)METHYL)BENZYLALCOHOL * 3,4-DIHY-DROXY-alpha-((ISOPROPYLAMINO)METHYL)-BENZYL ALCOHOL * ISOPROPYLAMINO-METHYL-3,4-DIHYDROXYPHENYL CARBINOL * ISOPROPYL NORADRENALINE * 1-ISOPRO-PYLNORADRENALINE

THR: Poison by ingestion, subcutaneous, intravenous, intraperitoneal, and possibly other routes. Cardiovascular system effects. Mutagenic data. When heated to decomposition it emits toxic fumes of NO_x.

ISOPROPYL ALCOHOL   HR: 2
CAS: 67-63-0   NIOSH: NT 8050000
mf: C_3H_8O mw: 60.11

PROP: Clear colorless liquid, slt odor, mp: −88.5°-−89.5°, bp: 82.5°, lel = 2.5%, uel = 12%, flash p: 53°F (CC), d: 0.7854 @ 20°/4°, vap d: 2.07, ulc: 70. Slightly bitter taste; fp: −89.5°; autoign temp: 852°F. Misc with water, alc, ether, chloroform. Insol in salt solns.

SYNS: ALCOOL ISOPROPILICO (ITALIAN) * ALCOOL ISOPROPYLIQUE (FRENCH) * DI-METHYLCARBINOL * ISOHOL * ISOPROPANOL * ISO-PROPYLALKOHOL (GERMAN) * PETRO-HOL * PROPAN-2-OL * 2-PROPANOL * 1-PROPANOL (GERMAN) * SEC-PROPYL ALCO-HOL * 1-PROPYLALKOHOL (GERMAN)

OSHA PEL: TWA 400 ppm

THR: Moderately toxic by ingestion and intraperitoneal routes. Low toxicity by dermal route. The single LD for a human adult = about 250 mL. An irritant to the eyes. Acts as a local irritant and in high concentration as a narcotic. It can cause corneal burns and often eye damage. It has good warning properties because it causes a mild irritation of the eyes, nose and throat at concentration levels of 400 ppm. It may induce a mild narcosis, the effects of which are usually transient, and it is somewhat less toxic than the normal isomer, but twice as volatile. It is not considered an important toxic hazard. There is some evidence that personnel can acquire a slight tolerance to this material, and single or repeated applications of it on the skin of rats, rabbits, dogs or human beings induced no untoward effects. It acts very much like ethanol in regard to absorption, metabolism and elimination but with a stronger narcotic action. Chronic injuries due to it have been detected in animals. Workers producing isopropanol show an excess of sinus cancers and laryngeal cancers. This may all or in part be due to the by-product, isopropyl oil. Humans have ingested up to 20 mL diluted with water and noticed only a sensation of heat and slight lowering of the blood pressure. There are, however, reports of serious illness from as little as 10 mL taken internally. A food additive permitted in food for human consumption. A common air contaminant. Absorbed by skin. Human Toxicology: Ingestion or inhalation of large quantities of the vapor may cause flushing, headache, dizziness, mental depression, nausea, vomiting, narcosis, anesthesia, coma; 100 mL can be fatal. Dangerous fire hazard when exposed to heat, flame or oxidizers. Moderate explosion hazard when exposed to heat or flame. Reacts violently with (H_2 + Pd); nitroform; oleum; $COCl_2$; potassium-*tert*-butoxide; Al; Al tri isopropoxide; crotonaldehyde; oxidants. Dangerous; keep away from heat and open flame; can react vigorously with oxidizing materials. To fight fire: CO_2, dry chemical, alcohol foam. For further information see Vol. 2, No. 2 of *DPIM Report*.

ISOPROPYLAMINE   HR: 2
CAS: 75-31-0   NIOSH: NT 8400000
mf: C_3H_9N mw: 59.08

PROP: Colorless liquid, amino odor, mp: −101.2°, flash p: −35°F (OC), d: 0.694 @ 15°/4°, autoign temp: 756°F, d: 2.03, bp: 33°-34°; lel = 2.3%; uel = 10.4%; misc with H_2O, alcohol, ether.

SYNS: 2-AMINO-PROPAAN (DUTCH) ∗ 2-AMINOPROPAN (GERMAN) ∗ 2-AMINO-PROPANO (ITALIAN) ∗ 2-AMINOPROPANE ∗ ISOPROPILAMINA (ITALIAN) ∗ MONOISOPROPYLAMINE

OSHA PEL: TWA 5 ppm
DOT Classification: Label: Flammable
  Liquid

THR: Moderately toxic by ingestion, inhalation and dermal routes. A strong irritant. Occasional contact causes sensitization. Narcotic in high concentrations. Incompatible with 1-chloro-1,3-epoxypropane; perchloryl fluoride. Very dan-

gerous fire hazard when exposed to heat, flame or oxidizers. Can react vigorously with oxidizing materials. When heated to decomposition it emits toxic fumes of NO_x. To fight fire, use alcohol foam, foam, CO_2, dry chemical.

alpha-((ISOPROPYLAMINO)METHYL)-NAPHTHALENEMETHANOL HYDROCHLORIDE   HR: 3
CAS: 51-02-5   NIOSH: QJ 9275000
mf: $C_{15}H_{19}NO_2$ • ClH   mw: 281.81

SYNS: ICI 38174 ∗ 2-ISOPROPYLAMINO-1-(2NAPHTHYL)ETHANOL HYDROCHLORIDE ∗ ALPHA-(((1-METHYLETHYL)AMINO)METHYL)-2-NAPHTHALENEMETHANOL, HYDROCHLORIDE ∗ NAPHTHYLISOPROTERENOL HYDROCHLORIDE

THR: A poison by intravenous and intraperitoneal routes. Moderate ingestion toxicity. An experimental carcinogen. When heated to decomposition it emits very toxic fumes of NO_x and HCl.

9-((3-(ISOPROPYLAMINO)PROPYL)-AMINO)-1-NITROACRIDINE DIHYDROCHLORIDE   HR: 3
CAS: 63710-43-0   NIOSH: TX 8257000
mf: $C_{19}H_{22}N_4O_2$ • 2ClH   mw: 411.37

SYN: N-(1-METHYLETHYL)-N'-(1-NITRO-9-ACRIDINYL)-1,3-PROPANEDIAMINE DIHYDROCHLORIDE

THR: A poison by ingestion and intravenous routes. Mutagenic data. When heated to decomposition it emits very toxic fumes of NO_x and HCl.

4-ISOPROPYLANTIPYRINE   HR: 3
CAS: 479-92-5   NIOSH: CD 2800000
mf: $C_{14}H_{18}N_2O$ mw: 230.34

SYNS: ISOPROPYL ANTIPYRINE ∗ 4-ISOPROPYL-2,3-DIMETHYL-1-PHENYL-3-PYRAZOLIN-5-ONE ∗ ISOPROPYLPHENAZONE ∗ 1-PHENYL-2,3-DIMETHYL-4-ISOPROPYL-3-PYRAZOLIN-5-ONE ∗ 1-PHENYL-2,3-DIMETHYL-4-ISOPROPYLPYRAZOL-5-ONE

THR: A poison by intraperitoneal route. Moderate ingestion toxicity. When heated to decomposition it emits toxic fumes of NO_x.

5-ISOPROPYL-1:2-BENZANTHRACENE   HR: 3
CAS: 63020-47-3   NIOSH: CW 7875000
mf: $C_{21}H_{18}$ mw: 270.39

SYN: 8-ISOPROPYLBENZ(A)ANTHRACENE

THR: An experimental tumorigen. When heated to decomposition it emits acrid smoke and fumes.

ISOPROPYLBENZENE HYDRO-PEROXIDE HR: 3
CAS: 80-15-9 NIOSH: MX 2450000
mf: $C_9H_{12}O_2$ mw: 152.21

PROP: Bp: 153°, flash p: 175°F; d: 1.05.

SYNS: CUMEENHYDROPEROXYDE (DUTCH) * CUMENE HYDROPEROXIDE * CUMENE HYDROPEROXIDE (DOT) * alpha-CUMYL HYDROPEROXIDE * CUMENYL HYDROPEROXIDE * CUMOLHYDROPEROXID (GERMAN) * CUMYL HYDROPEROXIDE * alpha-CUMYL HYDROPEROXIDE * alpha,alpha-DIMETHYLBENZYL HYDROPEROXIDE * HYDROPEROXYDE DE CUMENE (FRENCH) * HYDROPEROXYDE DE CUMYLE (FRENCH) * IDROPEROSSIDO DI CUMENE (ITALIAN) * IDROPEROSSIDO DI CUMOLO (ITALIAN)

DOT Classification: Label: Organic Peroxide

THR: A poison by ingestion, inhalation, intraperitoneal, and subcutaneous routes. Mutagenic data. A powerful irritant to skin and eyes. An experimental tumorigen. Flammable when exposed to heat or flame; can react with reducing materials. Can decompose violently @ 150° in concentrations of 91%, 95%. A strong oxidizing agent. See also peroxides. To fight fire, use foam, CO_2, dry chemical. For further information see Vol. 5, No. 6 of *DPIM Report*.

ISOPROPYL BORATE HR: 3
CAS: 5419-55-6 NIOSH: ED 5950000
mf: $C_9H_{21}BO_3$ mw: 188.11

PROP: Colorless liquid, mp: −59°, bp: 141.0°-142.4°, flash p: 82°F (TCC), d: 0.8138 @ 25°.

SYN: TRIISOPROPYL BORATE

THR: A poison by intravenous route. Moderately toxic by ingestion. A mild eye irritant. See esters and boron compounds. Dangerous fire hazard when exposed to heat, flame or oxidizers. Dangerous; can react vigorously with oxidizing materials. To fight fire, use foam, CO_2, dry chemical.

ISOPROPYL CARBAMATE HR: 3
CAS: 1746-77-6 NIOSH: FB 2975000
mf: $C_4H_9NO_2$ mw: 103.14

PROP: Prisms; mp: 60-61°; bp: 200°C; very sol in water, alc and ether.

SYN: CARBAMIC ACID-1-METHYLETHYL ESTER

THR: Mutagenic data. An experimental neoplastigen. Moderate subcutaneous toxicity. See also esters. When heated to decomposition it emits toxic fumes of NO_x.

ISOPROPYL CHLOROCARBO-NATE HR: 3
CAS: 108-23-6 NIOSH: LQ 6475000
mf: $C_4H_7ClO_2$ mw: 122.56

PROP: A clear, colorless liquid. A phosgene derivative. D: 1.090 @ 20°/4°; bp: 114-115° @ 768 mm; insol in water and alc; slt decomp; misc in ether and benzene. Flash p: < 73°F.

SYNS: ISOPROPYL CHLOROFORMATE * ISOPROPYL CHLOROMETHANOATE * CHLOROFORMIC ACID ISOPROPYL ESTER

THR: A poison by inhalation, skin contact, and ingestion. Severe eye irritant. Can cause pulmonary edema. Dangerous; see also chlorides. Has exploded while stored in a refrigerator. Reacts violently with phosgene.

ISOPROPYL DIETHYL DITHIOPHOSPHORYL ACETAMIDE HR: 3
CAS: 2275-18-5 NIOSH: TD 8225000
mf: $C_9H_{20}NO_3PS_2$ mw: 285.39

SYNS: O,O-DIETHYL S-(N-ISOPROPYLCARBAMOYLMETHYL) DITHIOPHOSPHATE * O,O-DIETHYLDITHIOPHOSPHORYLACETIC ACID, N-MONOISOPROPYLAMIDE * O,O-DIETHYL S-ISOPROPYLCARBAMOYLMETHYL PHOSPHORODITHIOATE * O,O-DIETHYL S-(N-ISOPROPYLCARBAMOYLMETHYL) PHOSPHORODITHIOATE * ENT 24,652 * N-MONOISOPROPYLAMIDE OF O,O-DIETHYLDITHIOPHOSPHORYLACETIC ACID

THR: A poison by ingestion, skin, and other routes. When heated to decomposition it emits very toxic NO_x, PO_x and SO_x.

ISOPROPYL-2,4-D ESTER HR: 3
CAS: 94-11-1 NIOSH: AG 8750000
mf: $C_{11}H_{12}Cl_2O_3$ mw: 263.13

SYNS: 2,4-D, ISOPROPYL ESTER * WEEDONE 128

THR: An experimental teratogen, tumorigen and carcinogen. Moderate acute toxicity by ingestion. See also esters. When heated to decomposition it emits toxic fumes of Cl^-.

ISOPROPYL FORMATE   HR: 3
CAS: 625-55-8   NIOSH: LQ 8750000
mf: $C_4H_8O_2$ mw: 88.12

PROP: Clear liquid, bp: 68.3°, flash p: 22°F
(CC), d: 0.873, autoign temp: 905°F, vap press:
100 mm @ 17.8°, vap d: 3.03.

SYN: NCI-C60106

THR: A poison by ingestion. A toxic fumigant.
See also esters. Dangerous fire hazard when
exposed to heat or flame. Can react vigorously
with oxidizing materials. To fight fire, use alcohol foam, foam, CO_2, dry chemical.

ISOPROPYL GLYCIDYL ETHER   HR: 3
CAS: 4016-14-2   NIOSH: TZ 3500000
mf: $C_6H_{12}O_2$ mw: 116.18

PROP: A liquid.

SYN: NCI-C56439

OSHA PEL: TWA 50 ppm

THR: Moderate inhalation and ingestion toxicity. A skin and eye irritant. See also ethers.
When heated to decomposition it emits acrid
smoke and fumes.

ISOPROPYL GLYCOL   HR: 2
CAS: 109-59-1   NIOSH: KL 5075000
mf: $C_5H_{12}O_2$ mw: 104.17

SYNS: ETHYLENE GLYCOL ISOPROPYL ETHER
* 2-ISOPROPOXYETHANOL * beta-HYDROXY-
ETHYL ISOPROPYL ETHER * ISOPROPYL CELLO-
SOLVE

THR: Moderate skin and inhalation toxicity.
Low ingestion toxicity. When heated to decomposition it emits acrid smoke and fumes.

ISOPROPYL-S-HYDROCHLO-
RIDE   HR: 3
CAS: 24426-36-6   NIOSH: NT 8575000
mf: $C_7H_{15}Cl_2N \cdot ClH$ mw: 220.59

SYNS: N,N-BIS(2-CHLOROETHYL)ISOPROPYLA-
MINE HYDROCHLORIDE * ISOPROPYL BIS(beta-
CHLOROETHYL)AMINE HYDROCHLORIDE

THR: A poison by ingestion, subcutaneous, and
intravenous routes. When heated to decomposition it emits very toxic fumes of HCl and
NO_x.

ISOPROPYL MEPROBAMATE   HR: 3
CAS: 78-44-4   NIOSH: FB 3325000
mf: $C_{12}H_{24}N_2O_4$ mw: 260.38

SYNS: ISOMEPROBAMATE * ISOPROPYLCAR-
BAMIC ACID, ESTER WITH 2-(HYDROXYMETHYL)-
2-METHYLPENTYL CARBAMATE * N-ISOPROPYL-
2-METHYL-2-PROPYL-1,3-PROPANEDIOL
DICARBAMATE * 2-METHYL-2-PROPYL-1,3-
PROPANEDIOL CARBAMATE ISOPROPYLCARBA-
MATE * NCI-C56235

THR: A poison by intravenous route. Moderate
ingestion and intraperitoneal toxicity. When
heated to decomposition it emits toxic fumes
of NO_x.

ISOPROPYL METHANEFLUORO-
PHOSPHONATE   HR: 3
CAS: 107-44-8   NIOSH: TA 8400000
mf: $C_4H_{10}FO_2P$ mw: 140.11

PROP: Bp: 147°; fp: −58°; d: 1.100 @ 20°;
vap press: 1.57 mm @ 20°; vap d: 4.86.

SYNS: ISOPROPYL METHYLFLUOROPHOSPHATE
* ISOPROPYL METHYLPHOSPHONOFLUORIDATE
* O-ISOPROPYL METHYLPHOSPHONOFLUORIDATE
* ISOPROPYL-METHYL-PHOSPHORYL FLUORIDE
* METHYLPHOSPHONOFLUORIDIC ACID ISOPROPYL
ESTER * SARIN II * SARIN

THR: A human poison by skin (a small drop
on skin can kill a man) and inhalation. A poison
by experimental ingestion, subcutaneous, intravenous, intramuscular, and intraperitoneal
routes. Affects human blood and central nervous
system. A human eye irritant. A so called
"nerve gas." See also ethyl dimethyl amido
cyanophosphate (or tabun), parathion. When
heated to decomposition or reacted with steam,
it emits very toxic fumes of F^- and PO_x. To
fight fire, use foam, CO_2, dry chemical.

1-ISOPROPYL-4-METHYLCYCLOHEX-
ANE HYDROPEROXIDE   HR: 3
CAS: 52061-60-6   NIOSH: MX 2500000
mf: $C_{10}H_{20}O_2$ mw: 172.30

SYNS: HEXAHYDRO-P-CYMENE HYDROPEROXIDE
* PARAMENTHANE HYDROPEROXIDE

THR: A poison irritant to skin, eyes, and mucous
membranes. See also peroxides, organic. A very
powerful oxidizer. When heated to decomposition it emits acrid smoke and fumes.

ISOPROPYL MYRISTATE   HR: 3
CAS: 110-27-0   NIOSH: XB 8600000
mf: $C_{17}H_{34}O_2$ mw: 270.44

PROP: Liquid of low viscosity, odorless, bp:
192.6° @ 20 mm; decomp @ 208°; d: 0.8532

@ 20°. Sol in castor oil, cottonseed oil, acetone, chloroform, ethyl acetate, ethanol, toluene and mineral oil. Insol in water, glycerol, propylene glycol. Dissolves many waxes.

SYNS: KESSCOMIR * ISOMYST * TETRADE-CANOIC ACID, ISOPROPYL

THR: An experimental tumorigen. A mild skin irritant. When heated to decomposition it emits toxic smoke and fumes.

ISOPROPYL NITRATE   HR: 2
CAS: 1712-64-7   NIOSH: QU 8930000
mf: $C_3H_7NO_3$ mw: 105.11

PROP: Liquid, bp: 102°, flash p: 51.8°F; uel = 100%.

SYNS: NITRIC ACID, ISOPROPYL ESTER

DOT Classification: Label: Flammable Liquid

THR: Moderately toxic. See also nitrates. Dangerous. When heated to decomposition it emits toxic fumes of NO_x. Incompatible with Lewis acids.

ISOPROPYL NITRITE   HR: 3
mf: $C_3H_7NO_2$ mw: 89.10

PROP: Flash p: < 50°F.

THR: Very toxic. Can cause vasodilation with fall in blood pressure, tachycardia, headache. Large doses can cause methemoglobinuria and cyanosis. Severe poisoning results in shock which can be fatal. When heated to decomposition it emits toxic fumes of NO_x. Quite flammable. See also nitrites.

ISOPROPYL OILS   HR: 3
                        NIOSH: NV 7900000

PROP: Isopropyl oils are a mixture containing trimeric and tetrameric polypropylenes.

THR: A by-product of isopropyl alcohol manufacture composed of trimeric and tetrameric polypropylene + small amounts of benzene, toluene, alkyl benzenes, polyaromatic ring compounds, hexane, heptane, acetone, ethanol, isopropyl ether and isopropyl alcohol. A human neoplastigen of para nasal sinuses, larynx and lung.

p-ISOPROPYLPHENOL   HR: 3
CAS: 99-89-8   NIOSH: SL 5950000
mf: $C_9H_{12}O$ mw: 136.21

PROP: Needles; d: 0.990 @ 20°; mp: 61°; bp: 228.2-229.2°; very sltly sol in water; sol in alc @ 25°; sol in ether @ 25°.

THR: A poison by intraperitoneal and intravenous routes. When heated to decomposition it emits acrid smoke and fumes.

ISOPROPYL PHOSPHOROFLUORIDATE   HR: 3
CAS: 55-91-4   NIOSH: TE 5075000
mf: $C_6H_{14}FO_3P$ mw: 184.17

PROP: Oily liquid, mp: −82°, bp: 46° @ 5 mm, d: 1.07 (approx), vap d: 5.24.

SYNS: DIISOPROPOXYPHOSPHORYL FLUORIDE * DIISOPROPYL FLUOROPHOSPHATE * O,O-DI-ISOPROPYL FLUOROPHOSPHATE * DIISOPROPYL FLUOROPHOSPHONATE * DIISOPROPYLFLUORO-PHOSPHORICACID ESTER * DIISOPROPYL PHOSPHOFLUORIDATE * DIISOPROPYL PHOS-PHOROFLUORIDATE * O,O'-DIISOPROPYL PHOS-PHORYL FLUORIDE * FLUOPHOSPHORIC ACID, DIISOPROPYL ESTER * FLUORODIISOPROPYL PHOSPHATE * ISOPROPYL FLUOPHOSPHATE * PHOSPHOROFLUORIDIC ACID,DIISOPROPYL ESTER

THR: A poison by ingestion, inhalation, skin contact, intraperitoneal, subcutaneous, intramuscular, ocular, and intravenous routes. Human eye irritant. Used in Germany as a basis for "nerve gases." An insecticide. Ingestion can cause damage to eyes, nausea, vomiting, diarrhea and central nervous system disturbances. See parathion. Dangerous; see fluorides and phosphorus compounds.

(±)-ISOPROTERENOL SULFATE   HR: 3
CAS: 114-45-4   NIOSH: DO 2100000
mf: $C_{22}H_{34}N_2O_6 \cdot H_2O_4S$ mw: 520.66

PROP: dl form: crystals from (acetone + methanol); mp: 128° (some decomp); sltly sol in alc. Insol in chloroform, ether, benzene. I form: crystals; mp: 164°-165°.

SYNS: DL-alpha-3,4-DIHYDROXYPHENYL-beta-I-SOPROPYLAMINOETHANOL SULFATE * DL-I-SOPRENALINE SULFATE * (±)-ISOPRENALINE SULFATE * DL-ISOPROTERENOL SULFATE

THR: A poison by ingestion, intravenous, intraperitoneal, and subcutaneous routes. Moderately toxic by inhalation. When heated to decomposition it emits very toxic fumes of NO_x and SO_x.

ISOPTIN   HR: 3
CAS: 52-53-9   NIOSH: YV 8300000
mf: $C_{27}H_{38}N_2O_4$ mw: 454.67

SYNS: 5-((3,4-DIMETHOXYPHENETHYL)METH-
YLAMINO)-2-(3,4-DIMETHOXYPHENYL)-2-ISOPRO-
PYLVALERONITRILE * IPROVERATRIL

THR: A poison by intraperitoneal and intravenous routes. Moderate ingestion toxicity. Affects human cardiovascular system and blood
pressure. When heated to decomposition it emits
toxic fumes of NO_x.

ISOQUINOLINE   HR: 3
CAS: 119-65-3   NIOSH: NW 6825000
mf: C_9H_7N mw: 129.17

PROP: Liquid, pungent odor, almost insol in
water, miscible with many organic solvents,
sol in dilute acids. Hygroscopic platelets when
solid, d: 1.09101 @ 30°/4°, mp: 26.48°, bp:
243°.

SYNS: 2-AZANAPHTHALENE * 2-BENZAZINE
* BENZO(C)PYRIDINE * LEUCOLINE

THR: A poison by ingestion and intraperitoneal
routes. Moderate skin toxicity. A severe skin
and eye irritant. When heated to decomposition
it emits toxic fumes of NO_x.

ISOSAFROLE   HR: 3
CAS: 120-58-1   NIOSH: DA 5950000
mf: $C_{10}H_{10}O_2$ mw: 162.20

PROP: Liquid, odor of anise, bp: 253°, mp:
8.2°.

SYNS: 1,2-METHYLENEDIOXY-4-PROPENYLBEN-
ZENE * 3,4-METHYLENEDIOXY-1-PROPENYL
BENZENE * 5-(1-PROPENYL)-1,3-BENZODI-
OXOLE * 4-PROPENYLCATECHOL METHYLENE
ETHER * 4-PROPENYL-1,2-METHYLENEDIOXY-
BENZENE

THR: An experimental carcinogen. A skin irritant. Moderately toxic by ingestion. When
heated to decomposition it emits acrid smoke
and fumes. For further information see Vol.
5, No. 5 of *DPIM Report*.

ISOSAFROLE-n-OCTYLSULF-
OXIDE   HR: 3
CAS: 120-62-7   NIOSH: DA 5775000
mf: $C_{18}H_{28}O_3S$ mw: 324.52

PROP: Water-insol, slightly sol in petroleum
oils, sol in most organic solvents.

SYNS: 1,2-(METHYLENEDIOXY)-4-(2-(OCTYL-
SULFINYL)PROPYL)BENZENE * 1-METHYL-2-
(3,4-METHYLENEDIOXYPHENYL)ETHYL OCTYLSUL-
FOXIDE * NCI-C02824 * N-OCTYLISOSAFROLE
SULFOXIDE * PIPERONYL SULFOXIDE

THR: Moderate acute toxicity by ingestion. An
experimental carcinogen and neoplastigen by
ingestion and subcutaneous routes. Reacts violently with $HClO_4$. Dangerous; when heated to
decomposition it emits highly toxic fumes of
SO_x.

ISO SYSTOX SULFOXIDE   HR: 3
CAS: 2496-92-6   NIOSH: TF 2850000
mf: $C_8H_{19}O_4PS_2$ mw: 274.36

SYNS: O,O-DIETHYL-S-(2-ETHTHIONYLETHYL)
PHOSPHOROTHIOATE * DIETHYL-S-(2-ETHTHIO-
NYLETHYL) THIOPHOSPHATE * O,O-DIETHYL-S-
ETHYL-2-ETHYLMERCAPTO PHOSPHOROTHIOLATE
SULFOXIDE

THR: A poison by ingestion and intraperitoneal
routes. When heated to decomposition it emits
very toxic fumes of PO_x and SO_x.

ISOTHIOCYANATOMETHANE   HR: 3
CAS: 556-61-6   NIOSH: PA 9625000
mf: C_2H_3NS mw: 73.12

PROP: Crystalline, sltly sol in water; bp: 119°;
d: 1.069; very sltly sol in water; misc in alc
and ether.

SYNS: ISOTHIOCYANATE DE METHYLE (FRENCH)
* ISOTHIOCYANIC ACID, METHYL ESTER
* ISOTIOCIANATO DI METILE (ITALIAN)
* METHYLISOTHIOCYANAAT (DUTCH)
* METHYL-ISOTHIOCYANAT (GERMAN)
* METHYL ISOTHIOCYANATE * METHYL
MUSTARD OIL * METHYLSENFOEL (GERMAN)

THR: A poison by ingestion, skin contact, and
subcutaneous routes. Very irritating to skin,
eyes, mucous membranes. A military poison.
When heated to decomposition it emits very
toxic fumes of NO_x and SO_x.

ISOTHIOCYANIC ACID, ETHYLENE
ESTER   HR: 3
CAS: 3688-08-2   NIOSH: NX 8580000
mf: $C_4H_4N_2S_2$ mw: 144.22

SYNS: ETHYLENE BIS ISOTHIOCYANATE
* AETHYLEN-BIS-THIURAMMONOSULFID (GER-
MAN) * AETHYLSENFOEL (GERMAN) * ETHYL-
ENE-BIS-THIURAMMONO-SULFIDE

THR: A poison by subcutaneous route. Moderate ingestion toxicity. When heated to decomposition it emits very toxic fumes of NO_x and SO_x.

ISOTHIOCYANIC ACID, PHENYL ESTER   HR: 3
CAS: 103-72-0   NIOSH: NX 9275000
mf: C_7H_5NS mw: 135.19

PROP: Pale yellow liquid. Insol in water; sol in alc and ether, mp: $-21°$, bp: $221°$, d: 1.1282.

SYNS: BENZENE-1-ISOTHIOCYANATE * PHENYL ISOTHIOCYANATE * PHENYL MUSTARD OIL * PHENYLSENFOEL (GERMAN) * THIOCARBANIL * USAF M-4

THR: A poison by ingestion, intraperitoneal and subcutaneous routes. Dangerous; when heated to decomposition or on contact with acid or acid fumes it emits highly toxic fumes of cyanide and SO_x.

ISOTHIOUREA   HR: 3
CAS: 62-56-6   NIOSH: YU 2800000
mf: CH_4N_2S mw: 76.13

PROP: White powder or crystals, mp: $177°$, bp: decomp, d: 1.405.

SYNS: PSEUDOTHIOUREA * beta-THIOPSEUDOUREA * THIOCARBAMIDE * 2-THIOUREA * USAF EK-497

THR: A human poison by ingestion, intraperitoneal, and other routes. Moderate subcutaneous toxicity. Mutagenic data. An experimental carcinogen, neoplastigen, and tumorigen. Affects womens' blood. A dose of 1 mg has proved fatal to a rat. It is said to cause depression of bone marrow with anemia, leukopenia and thrombocytopenia. May also cause allergic skin eruptions. A goitrogen. A cause of hepatic tumors upon chronic administration. May react violently with acrolein. Incompatibles: Acrylaldehyde; H_2O_2; HNO_3. When heated to decomposition it emits very toxic fumes of NO_x and SO_x.

ISOVALERIC ACID   HR: 2
CAS: 503-74-2   NIOSH: NY 1400000
mf: $C_5H_{10}O_2$ mw: 102.15

PROP: Colorless liquid; acid taste, disagreeable rancid-cheese odor. Solidifies @ $-37°$; d: 0.931 @ $20°/4°$; mp: $-34.5°$ $(-50°)$; sol in water @ $16°$; misc in alc and ether, bp: $175°-177°$; Keep tightly closed.

SYNS: DELPHINIC ACID * ISOPROPYLACETIC ACID * 3-METHYLBUTANOIC ACID * BETA-METHYLBUTYRIC ACID * ISOPENTANOIC ACID

DOT Classification: Label: Corrosive

THR: A poison by skin contact. Moderate ingestion and intravenous toxicity. A skin and eye irritant. When heated to decomposition it emits acrid smoke and fumes.

IVORY   HR: 3
NIOSH: NY 5615000

THR: An experimental tumorigen by implant.

J

JACOBINE   HR: 3
CAS: 6870-67-3   NIOSH: NY 5775000
mf: $C_{18}H_{25}NO_6$ mw: 351.44

PROP: An alkaloid isolated from *S. Jacobaea*

SYNS: 15,20-EPOXY-15,30-DIHYDRO-12-HY-
DROXY SENECIONAN-11,16-DIONE * NSC
89936

THR: A poison by intravenous and other routes.
Mutagenic data. An experimental carcinogen.
When heated to decomposition it emits toxic
fumes of NO_x.

JADE GREEN BASE   HR: 2
CAS: 128-58-5   NIOSH: JE 2315500
mf: $C_{36}H_{20}O_4$ mw: 516.56

SYNS: C.I. 59825 * DIMETHOXYVIOLAN-
THRONE * 16,17-DIMETHOXYVIOLANTHRONE
* ZELEN OSTANTHRENOVA BRILANTNI FFB
(CZECH)

THR: An eye irritant. Moderately toxic. When
heated to decomposition it emits acrid smoke,
fumes.

JAPACONITINE B   HR: 3
CAS: 1400-05-1   NIOSH: NY 6150100

SYN: JAPACONITINA B (ITALIAN)

THR: A poison.

JERVINE   HR: 3
CAS: 469-59-0   NIOSH: WG 9700000
mf: $C_{27}H_{39}NO_3$ mw: 425.59

PROP: Needles from (methanol + water); mp:
243.5°-244.5°. An alkamine isolated from
Veratrum album.

THR: A poison by ingestion and intravenous
routes. An experimental teratogen. When heated
to decomposition it emits toxic fumes of
NO_x.

JUNIPER BERRY OIL   HR: 2
CAS: 8012-91-7   NIOSH: NY 9870000

PROP: A volatile oil. Principal constituents include d-pinene, camphene, 1-terpineol-4 and
other oxygenated constituents; found in the fruit
of *juniperus communis L. (Fam. cupressaceae)*;
prepared by steam distillation of the dried ripe
fruit.

SYNS: OIL OF JUNIPER BERRY * WACHOLDER-
BEER OEL (GERMAN)

THR: Moderately toxic to humans by dermal
route. Low ingestion toxicity. Moderately irritant to skin, eyes and mucous membranes. An
allergen. A systemic irritant. If taken internally,
a severe kidney irritation similar to that caused
by turpentine may result. When heated to decomposition it emits acrid smoke and fumes.

K

KANAMYCIN A   **HR: 3**
CAS: 59-01-8   NIOSH: WK 1962000
mf: $C_{18}H_{36}N_4O_{11}$ mw: 484.58

SYNS: 4,6-DIAMINO-2-HYDROXY-1,3-CYCLO-
HEXANE 3,6'-DIAMINO-3,6'-DIDEOXYDI-alpha-
D-GLUCOSIDE ∗ KM (THE ANTIBIOTIC)

THR: Poison by intravenous route. Moderately
toxic by subcutaneous route. When heated to
decomposition it emits toxic fumes of NO_x.

KANAMYCIN SULFATE   **HR: 3**
CAS: 25389-94-0   NIOSH: 3225000
mf: $C_{18}H_{36}N_4O_{11} \cdot O_4S$ mw: 580.64

PROP: Irregular prisms, decomp over wide
range above 250°. Very sol in water; insol in
common alc and nonpolar solvents.

SYNS: KANNASYN ∗ KANTREX SULFATE

THR: A poison by intravenous route. Moderately toxic by intraperitoneal and intravenous
routes. When heated to decomposition it emits
very toxic NO_x and SO_x.

KARMINOMYCIN   **HR: 3**
CAS: 39472-31-6   NIOSH: NZ 7200000

SYN: CARMINOMYCIN

THR: Mutagenic data. An acute poison in rats,
mice and dogs. Affects the blood picture in
humans by intravenous route.

KELEVAN   **HR: 3**
CAS: 4234-79-1   NIOSH: PC 8400000
mf: $C_{17}H_{12}Cl_{10}O_4$ mw: 634.79

SYNS: GC-9160 ∗ 1,1a,3,3a,4,5,5a,5b,6-DE-
CACHLOROOCTAHYDRO-2-HYDROXY-1,3,4-METH-
ENO-1H-CYCLOBUTA(c,d)-PENTALENE-2-LEVU-
LINIC ACID, ETHYL ESTER

THR: Poison by ingestion and skin contact. See
also esters. When heated to decomposition it
emits toxic fumes of Cl^-.

KEPONE   **HR: 3**
CAS: 143-50-0   NIOSH: PC 8575000
mf: $C_{10}Cl_{10}O$ mw: 490.60

PROP: A chlorinated polycyclic ketone, a crystalline material, sltly water-sol, but sol in alcohols, ketones and acetic acid. Mp: decomp @

350°. Readily hydrates on exposure to room
temperature and humidity, and is normally used
as a mono- to trihydrate.

SYNS: CHLORDECONE ∗ 1,2,3,5,6,7,8,9,
10,10-DECA-CHLORO(5.2.1.0(sup 2,6).0(sup 3,
9).0(sup 5,8))DECANO-4-ONE ∗ DECACHLORO-
1,3,4-METHENO-2H-CYCLOBUTA(CD)PENTALEN-
2-ONE ∗ DECACHLOROOCTAHYDRO-1,3,4-
METHENO-2H-CYCLOBUTA(CD)PENTALEN-2-ONE
∗ 1,1A,3,3A,4,5,5,5A,5B,6-DECACHLOROOC-
TAHYDRO-1,3,4-METHENO-2H-CYCLOBUTA-
(CD)PENTALEN-2-ONE ∗ DECACHLORO-
PENTACYCLO(5.2.1.0(sup 2,6).0(sup 3,9).0(sup
5,8))DECAN-4-ONE ∗ DECACHLOROPENTACY-
CLO(5.3.0.0(sup 2,6).0(sup 4,10).0(sup 5,9)
DECAN-3-ONE ∗ DECACHLOROTETRACYCLO-
DECANONE ∗ DECACHLOROTETRAHYDRO-4,
7-METHANOINDENEONE ∗ ENT 16,391
∗ NCI-C00191

THR: A poison by ingestion, skin contact, and
other routes. An experimental carcinogen; a suspected human carcinogen. In man, inhalation,
absorption or ingestion can lead to central nervous system, liver and kidney damage, including bizarre symptoms caused by damage to the
nervous system. Usually, the symptoms are
tremors, ataxia, skin changes, hyperexcitability,
hyperactivity, muscle spasms, testicular atrophy, low sperm count, estrogenic effects, sterility, breast enlargement, liver lesions and cancer.
An insecticide and a fungicide. See also chlorides. For further information see Vol. 4, No.
4 of *DPIM Report*.

KERB   **HR: 3**
CAS: 23950-58-5   NIOSH: CV 3460000
mf: $C_{12}H_{11}Cl_2NO$ mw: 256.14

SYNS: N-(1,1-DIMETHYLPROPYNYL)-3,5-DI-
CHLOROBENZAMIDE ∗ PROMAMIDE ∗ PRO-
NAMIDE ∗ PROPYZAMIDE

THR: An experimental carcinogen. When
heated to decomposition it emits very toxic
fumes of Cl^- and NOx.

KEROSENE   **HR: 3**
CAS: 8008-20-6   NIOSH: OA 5500000

PROP: A pale yellow to water-white oily liquid,
bp: 175°-325°, ulc: 40, flash p: 150°-185°F, d:

0.80 to <1.0, lel = 0.7%, uel = 5.0%, autoign temp: 410°F, vap d: 4.5. Insol in H_2O; misc with other petr solvents. A mixture of petroleum hydrocarbons, chiefly of the methane series having from 10-16 carbon atoms per molecule.

SYNS: COAL OIL * KEROSINE

THR: A poison by intravenous route. Slightly toxic by ingestion causing irritation of the stomach and intestines with nausea and vomiting. Aspiration of vomitus can cause serious pneumonitis, particularly in young children. Inhalation of high concentration of vapor can cause headache and stupor. See also mineral oils. Flammable when exposed to heat or flame; can react with oxidizing materials. Moderate explosion hazard when exposed to heat or flame. To fight fire, use foam, CO_2, dry chemical.

KETENE   HR: 3
CAS: 463-51-4   NIOSH: OA 7700000
mf: C_2H_2O mw: 42.04

PROP: Colorless gas of disagreeable taste. Decomp in water, mp: −150°, bp: −56°, vap d: 1.45. Decomp in alc; sol in ether and acetone.

SYNS: CARBOMETHENE * ETHENONE
* KETO-ETHYLENE

OSHA PEL: TWA 0.5 ppm

THR: A poison by inhalation. Moderately toxic by ingestion. Can cause pulmonary edema.

KETOBEMIDONE HYDROCHLOR-
IDE   HR: 3
CAS: 5965-49-1   NIOSH: UC 2980000
mf: $C_{15}H_{21}NO_2$ mw: 247.37

SYNS: CLIRADON HYDROCHLORIDE
* 1-METHYL-4-(m-HYDROXYPHENYL)PIPER-
IDINE-4-ETHYLKETONE HYDROCHLORIDE

THR: Poison by ingestion, intravenous, and subcutaneous routes. When heated to decomposition it emits toxic fumes of NO_x.

KETONES
PROP: Organic compounds containing the chemical group =CO derived from secondary alcohols by oxidation. Acetone, which is dimethyl ketone, is the most familiar of this group of compounds. See also acetone. No general statement can be made as to the toxicity of ketones. Some are highly volatile and hence may have narcotic or anesthetic effects. Skin absorption as well as inhalation may be an important route of entry into the body. None of the ketones has been shown to have a high degree of chronic toxicity. Some are dangerous fire hazards. They react violently with aldehydes; HNO_3; (HNO_3 + H_2O_2); $HClO_4$. Common air contaminants.

L

LACTIC ACID   HR: 2
CAS: 50-21-5   NIOSH: OD 2800000
mf: $C_3H_6O_3$ mw: 90.09

PROP: Yellow to colorless crystals, mp: 16.8°, bp: 122° @ 15 mm, d: 1.249 @ 15°. Volatile with superheated steam; sol in H_2O, alc, furfurol; sltly sol in ether, insol in chloroform, petr ether, carbon disulfide. Misc in water, alc + ether.

SYNS: ACETONIC ACID * DL-LACTIC ACID * ETHYLIDENELACTIC ACID * 1-HYDROXYETHANECARBOXYLIC ACID * 2-HYDROXYPROPANOIC ACID * 2-HYDROXYPROPIONIC ACID * ALPHA-HYDROXYPROPIONIC ACID * KYSELINA MLECNA (CZECH) * MILCHSAURE (GERMAN) * MILK ACID * RACEMIC LACTIC ACID

THR: Mutagenic data. A skin and eye irritant. Moderately toxic by ingestion, rectal routes. Used as a general-purpose food additive. Incompatibles: HF; nitric acid (NHO_3 + HF). Emits acrid fumes and smoke when heated to decomposition.

LACTIC ACID, BERYLLIUM
SALT   HR: 3
CAS: 64059-26-3   NIOSH: OD 3730000

SYN: BERYLLIUM LACTATE

OSHA PEL: TWA 2 ug/m^3; CL 5; Pk 25/30M/8H

THR: A poison by intravenous route. See also beryllium compounds. When heated to decomposition it emits very toxic fumes of Be.

LACTIC ACID, IRON(2+) SALT
(2:1)   HR: 3
CAS: 5905-52-2   NIOSH: OD 5525000
mf: $C_6H_{10}O_6 \cdot Fe$ mw: 234.01

PROP: Greenish-white crystals, slight peculiar odor, mod sol in water, sltly sol in alcohol.

SYNS: FERROUS LACTATE * IRON(2+) LACTATE

THR: A poison by intravenous route. Moderately toxic by subcutaneous route. An experimental tumorigen. Nutrient and/or dietary supplement food additive. When heated to decomposition it emits acrid smoke and fumes.

LACTONITRILE   HR: 3
CAS: 78-97-7   NIOSH: OD 8225000
mf: C_3H_5NO mw: 71.09

PROP: Straw-colored liquid, mp: −40°, bp: 103° @ 50 mm, fp: −34°, flash p: 170°F (TCC), d: 0.9834 @ 25°, vap d: 2.45.

THR: A poison by ingestion, inhalation, skin contact, and subcutaneous routes. See also nitriles. In the presence of alkali, evolves HCN. Flammable when exposed to heat, flame or oxidizers. When heated to decomposition it evolves CN^-, NO_x. Incompatible with oxidizing materials. To fight fire, use foam, CO_2, dry chemical.

LACTOSE   HR: 3
CAS: 63-42-3   NIOSH: OD 9625000
mf: $C_{12}H_{22}O_{11}$ mw: 342.34

PROP: Colorless rhombic; d: 1.525 @ 20°; mp: 202° (anhydrous); bp: decomp; sol in water 17° cold; 40° hot; insol in alc and ether. Faintly sweet taste.

SYNS: 4-(beta-D-GALACTOSIDO)-D-GLUCOSE * D-LACTOSE * LACTOBIOSE * MILK SUGAR * SACCHARUM LACTIN

THR: An experimental teratogen and tumorigen. Moderately toxic by intravenous route. When heated to decomposition it emits acrid smoke and fumes.

LAETRILE   HR: 3
CAS: 1332-94-1   NIOSH: OD 9910000
mf: $C_{14}H_{15}NO_7$ mw: 309.30

PROP: Solid: mp: 214°-216°

SYNS: CYANOPHENYLMETHYL-beta-D-GLUCOPYRANOSIDURONIC ACID * 1-MANDELONITRILE-beta-GLUCURONIC ACID

THR: A human poison by ingestion. A poison by intraperitoneal route. Mutagenic data. Affects human central nervous system and gastrointestinal tract. A controversial treatment for cancer. When heated to decomposition it emits toxic fumes of NO_x.

LANATOSIDE C   HR: 3
CAS: 17575-22-3   NIOSH: OE 2100000
mf: $C_{49}H_{76}O_{20}$ mw: 985.25

PROP: Long, flat prisms from alc. Decomp @ 248°-250° after drying in high vacuum @ 150°. Very sol in pyridine and dioxane. Insol in ether, petr ether.

SYNS: DIGILANID C * ISOLANID

THR: A poison by intraperitoneal, intravenous, and ingestion. When heated to decomposition it emits acrid smoke and fumes.

LANTHANUM   HR: 3
af: La   aw: 138.91

PROP: Silvery-white, malleable and ductile metal element soft enough to cut with a knife. Very reactive rare earth metal; mp: 920°; bp: 3454°; d: 6.166 @ 25°.

THR: A poison by intravenous route. Lanthanum and other lanthanons can cause delayed blood clotting, leading to hemorrhages. Has caused liver injury in experimental animals. See also rare earths. Dangerous fire hazard in the form of dust when exposed to flame; can react vigorously with oxidizing materials. See also iron dust. Reacts violently with air; P; halogens. Moderate explosion hazard in the form of dust when exposed to flame or by chemical reaction. Incompatible with nitric acid; P; H_2O; C; N; B; Se; Si; S; halogens.

LANTHANUM ACETATE   HR: 3
CAS: 917-70-4   NIOSH: AI 5075000
mf: $C_2H_4O_2 \cdot 7La$ mw: 1032.43

SYNS: LANTHANACETAT (GERMAN) * LANTHANUM TRIACETATE

THR: A poison by intraperitoneal route. Moderately toxic by subcutaneous route; See also lanthanum and rare earths. When heated to decomposition it emits acrid smoke and fumes.

LANTHANUM CHLORIDE   HR: 3
CAS: 10099-58-8   NIOSH: OE 4375000
mf: Cl_3La mw: 245.26

PROP: Heptahydrate: triclinic crystals; sol in H_2O and alc.

THR: A poison by intraperitoneal and intravenous routes. Moderate toxicity by ingestion and subcutaneous routes. See also lanthanum and chlorides. See also rare earths. When heated to decomposition it emits toxic fumes of Cl^-.

LANTHANUM EDETATE   HR: 3
CAS: 11138-87-7   NIOSH: OE 4900000

THR: A poison by intraperitoneal route. See also lanthanum, rare earths. When heated to decomposition it emits acrid smoke and fumes.

LANTHANUM NITRATE   HR: 2
CAS: 10099-59-9   NIOSH: OE 5075000
mf: $N_3O_9 \cdot La$ mw: 324.94

PROP: Hexahydrate; white deliquescent crystals; mp: approx 40°; bp: 126°. Very sol in H_2O, alc. Keep well stoppered.

THR: Moderately toxic by intraperitoneal route. See also nitrates and lanthanum, rare earths. When heated to decomposition it emits toxic fumes of NO_x.

LAPPACONITINE   HR: 3
CAS: 32854-75-4   NIOSH: OE 5785000
mf: $C_{32}H_{44}N_2O_8$ mw: 584.69

PROP: Bitter crystals; mp: 217°-218°; sol in benzene, sltly sol in alc, ether, insol in H_2O. Chief alkaloid in aconitum septentrionale.

THR: A poison by ingestion and intravenous routes. When heated to decomposition it emits toxic fumes of NO_x.

LASIOCARPINE   HR: 3
CAS: 303-34-4   NIOSH: OE 7875000
mf: $C_{21}H_{33}NO_7$ mw: 411.55

PROP: An alkaloid isolated from *H. Lasiocarpum*.

SYNS: HELIOTRIDINE ESTER WITH LASIOCARPUM AND ANGELIC ACID * NCI-C01478

THR: A poison by intravenous, intraperitoneal, ingestion, and other routes. Mutagenic data. An experimental carcinogen. See also esters. When heated to decomposition it emits toxic fumes of NO_x.

LAURIC ACID   HR: 3
CAS: 143-07-7   NIOSH: OE 9800000
mf: $C_{12}H_{24}O_2$ mw: 200.36

PROP: Colorless, needle-like crystals, mp: 48°; bp: 225° @ 100 mm; d: 0.883; vap press: 1 mm @ 121.0°. Insol in water; sol in benzene, alc, ether and petroleum ether. Slight odor of bay oil.

SYNS: DODECANOIC ACID * DUODECYLIC ACID * LAUROSTEARIC ACID * NEO-FAT 12 * 1-UNDECANECARBOXYLIC ACID

THR: A poison by intravenous route. Low toxicity by ingestion. An experimental neoplastigen.

Combustible when exposed to heat or flame; can react with oxidizing materials. When heated to decomposition it emits acrid smoke and fumes.

LAUROYL PEROXIDE HR: 3
CAS: 105-74-8 NIOSH: OF 2625000
DOT: 2124
mf: $C_{24}H_{46}O_4$ mw: 398.70

PROP: White, tasteless, coarse powder, faint odor, mp: 53°-55°.

SYNS: DILAUROYL PEROXIDE * DODECANOYL PEROXIDE * PEROXYDE DE LAUROYLE (FRENCH)

DOT Classification: Label: Organic Peroxide

THR: A poison by irritation to skin, eyes, mucous membranes. See also peroxides, organic. An experimental tumorigen. It can irritate and cause burns upon the skin and mucous membranes; it is a powerful oxidizing agent. See also peroxides, organic.

LAURYLISOQUINOLINIUM BROMIDE HR: 3
CAS: 93-23-2 NIOSH: NX 5250000
mf: $C_{21}H_{32}N \cdot Br$ mw: 378.45

PROP: Deep amber, water-sol liquid, pleasant characteristic odor.

SYNS: 2-DODECYLISOQUINOLINIUM BROMIDE * INTEXSAN LQ75 * ISOTHAN

THR: A poison by ingestion. Severe eye irritant. Combustible when exposed to heat or flame. Dangerous; when heated to decomposition emits toxic fumes of Br^- and NO_x. Incompatible with oxidizing materials.

LEAD HR: 3
CAS: 7439-92-1 NIOSH: OF 7525000
af: Pb aw: 207.19

PROP: Bluish-gray, soft metal, mp: 327.43°, bp: 1740°, d: 11.34 @ 20°/4°. vap press: 1 mm @ 973°.

SYNS: C.I. 77575 * LEAD FLAKE * LEAD S2 * OLOW (POLISH)

OSHA PEL: TWA 200 ug/m³
ACGIH BEI: Lead in blood 50 ug/100 ml
BAT: 70 ug/dl; 45 ug/dl (women under 45 years)

THR: See also lead compounds. Affects human central nervous system. A poison by ingestion; moderately irritating. A common air contaminant. It is a carcinogen of the lungs and kidneys and an experimental teratogen. Flammable in the form of dust when exposed to heat or flame. See also powdered metals. Moderate explosion hazard in the form of dust when exposed to heat or flame. Incompatible with NH_4NO_3; ClF_3; H_2O_2; NaN_3; Na_2C_2; Zr; disodium acetylide; oxidants. Dangerous; when heated it emits highly toxic fumes; can react vigorously with oxidizing materials. For further information see Vol. 1, No. 1 of *DPIM Report*.

LEAD ACETATE HR: 3
CAS: 301-04-2 NIOSH: AI 5250000
mf: $C_4H_6O_4 \cdot Pb$ mw: 325.29

PROP: Trihydrate; colorless crystals or white granules or powder. Sltly acetic odor; slowly effloresces; d: 2.55; mp: 75° when rapidly heated. Decomp above 200°; very sol in glycerol. Keep well stoppered.

SYNS: ACETIC ACID LEAD (2+) SALT * ACETATE DE PLOMB (FRENCH) * BLEIACETAT (GERMAN) * LEAD (2+) ACETATE * LEAD(II) ACETATE * LEAD DIACETATE * LEAD DIBASIC ACETATE * NORMAL LEAD ACETATE * PLUMBOUS ACETATE * SALT OF SATURN * SUGAR OF LEAD

OSHA PEL: TWA 200 ug(Pb)/m³

THR: A poison by intraperitoneal, ingestion, subcutaneous, and intravenous routes. Mutagenic data. An experimental carcinogen, teratogen, and tumorigen. A suspected human carcinogen. See also lead compounds. An insecticide. When heated to decomposition it emits toxic fumes of lead. Incompatible with $KBrO_3$; acids; soluble sulfates; citrates; tartrates; chlorides; carbonates; alkalies; tannin phosphates; resorcinol; salicylic acid; phenol; chloral hydrate; sulfites; vegetable infusions; tinctures. For further information see Vol. 6, No. 2 of *DPIM Report*.

LEAD ACETATE, BASIC HR: 3
CAS: 1335-32-6 NIOSH: OF 8750000
mf: $C_4H_{10}O_8Pb_3$ mw: 807.71

PROP: White powder.

SYNS: BASIC LEAD ACETATE * BIS(ACETO)-DIHYDROXYTRILEAD * BIS(ACETATO)TETRAHYDROXYTRILEAD * LEAD MONOSUBACETATE

* LEAD SUBACETATE * MONOBASIC LEAD ACE-TATE

THR: An experimental carcinogen, tumorigen, and neoplastigen. A suspected human carcinogen. Mutagenic data. See also lead and lead compounds. When heated to decomposition it emits toxic fumes of lead.

LEAD ACETATE (II), TRIHYDRATE   HR: 3
CAS: 6080-56-4   NIOSH: OF 8050000
mf: $C_4H_6O_4 \cdot Pb \cdot 3H_2O$ mw: 379.35

SYNS: ACETIC ACID, LEAD(+2) SALT TRIHYDRATE * BIS(ACETATO)TRIHYDROXYTRILEAD * LEAD ACETATE TRIHYDRATE * LEAD DIACETATE TRIHYDRATE

THR: An experimental carcinogen. Moderately toxic by subcutaneous route. See also lead and lead compounds. When heated to decomposition it emits toxic fumes of lead. For further information see Vol. 1 No. 7 of *DPIM Report*.

LEAD ACID ARSENATE   HR: 3
CAS: 7784-40-9   NIOSH: CG 0980000
DOT: 1617
mf: $AsHO_4 \cdot Pb$ mw: 347.12

PROP: White crystals.

SYNS: ACID LEAD ARSENATE * ACID LEAD ORTHOARSENATE * DIBASIC LEAD ARSENATE * LEAD ARSENATE, SOLID (DOT) * LEAD ARSENATE (STANDARD)

OSHA PEL: TWA 150 ug/m^3
DOT Classification: Label: Poison

THR: A poison by ingestion. Moderately toxic by other routes. A human carcinogen and an experimental carcinogen; See also arsenic compounds and lead compounds. When heated to decomposition it emits very toxic fumes of As and Pb.

LEAD(II) ARSENITE   HR: 3
CAS: 10031-13-7   NIOSH: OF 8600000
DOT: 1618
mf: $As_2O_4 \cdot Pb$ mw: 421.03

PROP: White powder, d: 5.85. Insol in H_2O; sol in dil HNO_3.

SYN: LEAD ARSENITE, SOLID (DOT)

DOT Classification: Label: Poison

THR: A poison. See also lead compounds and arsenic compounds. When heated to decomposition it emits very toxic fumes of Pb and As.

LEAD AZIDE (II)   HR: 3
CAS: 13424-46-9   NIOSH: OF 8650000
mf: N_6Pb mw: 291.25

PROP: Colorless needles or white powder; explodes @ 350° or when shocked; very sol in acetic acid; insol in NH_4OH.

SYN: INITIATING EXPLOSIVE LEAD AZIDE (DOT)

DOT Classification: Label: Explosive

THR: A deadly poison. See also lead compounds and azides. Severe explosion hazard when shocked or exposed to heat or flame. Explodes at 250°. Will explode spontaneously during crystallization. Highly dangerous; when heated, emits highly toxic fumes of lead, NO_x. Incompatible with calcium stearate; copper or zinc; brass; CS_2.

LEAD CARBONATE   HR: 3
CAS: 598-63-0   NIOSH: OF 9275000
mf: $CO_3 \cdot Pb$ mw: 267.20

PROP: White, heavy powder. Decomp @ 400° leaving residue of PbO. Insol in water, alc; sol in acetic acid, dil HNO_3 (effervescence) d: 6.61.

SYNS: CARBONIC ACID, LEAD(2+) SALT (1:1) * DIBASIC LEAD CARBONATE * LEAD(2+) CARBONATE * WHITE LEAD

OSHA PEL: TWA 200 ug(Pb)/m^3

THR: A deadly poison. Moderately toxic by ingestion. An experimental carcinogen. See also lead, lead compounds. When heated to decomposition it emits toxic fumes of Pb. Incompatibles: F_2.

LEAD CHLORIDE   HR: 3
CAS: 7758-95-4   NIOSH: OF 9450000
DOT: 2291
mf: Cl_2Pb mw: 278.09

PROP: White crystals, mp: 501°, bp: 950°, d: 5.85, vap press: 1 mm @ 547°. Somewhat sol in cold water, more sol in hot H_2O. Very sol in ammonium chloride, NH_4NO_3, alkali hydroxides.

SYNS: LEAD (2+) CHLORIDE * LEAD (II) CHLORIDE * LEAD DICHLORIDE * PLUMBOUS CHLORIDE

OSHA PEL: TWA 200 ug(Pb)/m^3
DOT Classification: ORM-B, Label: None

THR: An experimental teratogen. Moderately toxic by ingestion. A poison. See also lead and lead compounds. Mutagenic data. When heated to decomposition it emits very toxic fumes of Pb and Cl$^-$. For further information see Vol. 6, No. 2 of *DPIM Report*.

LEAD CHROMATE   HR: 3
CAS: 7758-97-6   NIOSH: GB 2975000
mf: CrO$_4$ • Pb   mw: 323.19

PROP: Yellow or orange yellow powder. One of the most insol salts. Insol in acetic acid; sol in solns of fixed alkali hydroxides, dil HNO$_3$. mp: 844°. bp: decomp, d: 6.3.

SYNS: CHROMIC ACID, LEAD(2+) (1:1)
* CANARY CHROME YELLOW 40-2250
* CHROMATE DE PLOMB (FRENCH) * CHROME GREEN * C.I. 77600 * COLOGNE YELLOW
* CROCOITE * GIALLO CROMO (ITALIAN)
* KING'S YELLOW * LEAD CHROMATE
* LEAD CHROMATE(VI) * LEIPZIG YELLOW
* LEMON YELLOW * PARIS YELLOW
* PLUMBOUS CHROMATE

OSHA PEL: TWA CL 100 ug(Cr03)/m^3

THR: Mutagenic data. An experimental neoplastigen, tumorigen, carcinogen. A human carcinogen. See also lead, lead compounds, and chromium compounds. When heated to decomposition it emits toxic fumes of Pb. Incompatible with iron hexacyanoferrate (4$^-$). For further information see Vol. 1, No. 7 of *DPIM Report*.

LEAD CHROMATE, BASIC   HR: 3
CAS: 18454-12-1   NIOSH: OF 9800000
mf: CrO$_4$Pb • OPb   mw: 546.38

PROP: Red amorphous or crystal; mp: 920°.

SYNS: ARANCIO CROMO (ITALIAN) * AUSTRIAN CINNABAR * BASIC LEAD CHROMATE
* CHROMIUM LEAD OXIDE * C.I. 77601
* LEAD CHROMATE OXIDE * LEAD CHROMATE, RED

OSHA PEL: TWA 200 ug(Pb)/m^3

THR: Mutagenic data. An experimental carcinogen, tumorigen, and neoplastigen. A human carcinogen. See also lead and lead compounds. See also chromium compounds. When heated to decomposition it emits very toxic fumes of Pb.

LEAD COMPOUNDS   HR: 3
THR: Poisons. Lead poisoning is one of the commonest of occupational diseases. The presence of lead-bearing materials or lead compounds in an industrial plant does not necessarily result in exposure on the part of the worker. The lead must be in such form, and so distributed, as to gain entrance into the body or tissues of the worker in measurable quantity, otherwise no exposure can be said to exist. Some are carcinogens of the lungs and kidneys. Others are experimental neoplastigens and tumorigens. Mode of entry into body:

1. By inhalation of the dusts, fumes, mists or vapors. (Common air contaminants).

2. By ingestion of lead compounds trapped in the upper respiratory tract or introduced into the mouth on food, tobacco, fingers or other objects.

3. Through the skin; this route is of special importance in the case of organic compounds of lead, as lead tetraethyl. In the case of the inorganic forms of lead, this route is of no practical importance.

When lead is ingested, much of it passes through the body unabsorbed, and is eliminated in the feces. The greater portion of the lead that is absorbed is caught by the liver and excreted, in part, in the bile. For this reason, larger amounts of lead are necessary to cause poisoning if absorption is by this route, and a longer period of exposure is usually necessary to produce symptoms. On the other hand, upon inhalation, absorption takes place easily from the respiratory tract and symptoms tend to develop more quickly. From the point of view of industrial poisoning, inhalation of lead is much more important than is ingestion. Lead is a cumulative poison. Increasing amounts build up in the body and eventually a point is reached where symptoms and disability occur. Lead produces a brittleness of the red blood cells so that they hemolyze with but slight trauma; the hemoglobin is not affected. Due to their increased fragility, the red cells are destroyed more rapidly in the body than normally, producing an anemia which is rarely severe. The loss of circulating red cells stimulates the production of new young cells which, on entering the blood stream, are acted upon by the circulating lead, with resultant coagulation of their basophilic material. These cells after suitable staining, are recognized as "stippled cells." As regards the effect of lead on the white blood cells, there is no uniformity of opinion. In addition to its effect on the red cells of the blood, lead produces a damaging effect on the organs or tissues with which it comes in contact. No specific or characteristic lesion is produced. Autopsies of deaths attributed to lead poisoning and experimental work on animals have shown pathological lesions of the kidneys, liver, male gonads, nervous system, blood vessels and other tissues. None of these changes, however, have been found consistently. In cases of lead poisoning, the amount of lead found in the blood is frequently in excess of 0.07 mg per 100 cc of whole blood. The urinary lead excretion generally exceeds 0.1 mg per liter of urine. The toxicity of the various lead compounds appears to depend upon several factors: (1) the solubility of the compound in the body fluids; (2) the fineness of the particles of the compound; solubility is greater, of course, in proportion to the fineness of the particles; (3) conditions under which the compound is being used; where a lead compound is used as a powder; contamination of the atmosphere will be much less where the powder is kept damp. Of the various lead compounds, the carbonate, the monoxide and sulfate are considered to be more toxic than metallic lead or other lead compounds. Lead arsenate is very toxic, due to the presence of the arsenic radical. See also specific compound.

Diagnosis: A diagnosis of lead poisoning should not be made on the basis of any single clinical or laboratory finding. There must be a history of significant exposure, signs, and symptoms (as described above) compatible with the diagnosis, and confirmatory laboratory tests. Increase of stippled red blood cells, mild anemia, and elevated lead in blood and urine, i.e., more than 0.07 mg/100 mL. Blood and similar values per liter of urine. An increase of coproporphyrins and certain amino acids in urine may be present. Diagnostic mobilization of lead with calcium EDTA may be useful in questionable cases.

LEAD (II) CYANIDE   HR: 3
CAS: 592-05-2   NIOSH: OG 0175000
DOT: 1620
mf: C_2N_2Pb mw: 259.23

PROP: White powder.

SYNS: C.I. 77610 * C.I. PIGMENT YELLOW 48 * CYANURE DE PLOMB (FRENCH) * LEAD CYANIDE (DOT)

OSHA PEL: TWA 200 ug(Pb)/m^3
DOT Classification: Poison B, Label: Poison

THR: A poison by intraperitoneal route. See also lead compounds and cyanides. Violent reaction with Mg. A fire hazard as a powerful oxidizer. When heated to decomposition it emits very toxic fumes of Pb, CN$^-$ and NO$_x$. Incompatible with magnesium.

LEAD DIMETHYLDITHOCARBA-
MATE   HR: 3
CAS: 19010-66-3   NIOSH: OF 8850000
mf: $C_6H_{12}N_2S_4 \cdot Pb$ mw: 447.63

PROP: Solid, mp: 258°, d: 2.5

SYNS: BIS(DIMETHYLCARBAMODITHIOATO-S,S')LEAD * BIS(DIMETHYLDITHIOCARBA-MIATO)LEAD * LEDATE * NCI-C02891

THR: Poison. An experimental tumorigen and suspected carcinogen. See also lead compounds. Combustible. When heated to decomposition it emits very toxic fumes of Pb, NO$_x$ and SO$_x$.

LEAD DIOXIDE   HR: 3
CAS: 1309-60-0   NIOSH: OG 0700000
mf: O_2Pb mw: 239.19

PROP: Brown hexagonal crystals or dark-brown powder. Liberates O_2 when heated; Insol in H_2O; sol in HCl evolving Cl, sol in alkali iodide solns liberating iodine, sol in hot caustic alkali soln; mp: decomp @ 290°, d: 9.375.

SYNS: BIOXYDE DE PLOMB (FRENCH) * C.I. 77580 * LEAD BROWN * LEAD OXIDE BROWN * LEAD PEROXIDE * LEAD PEROXIDE (DOT) * LEAD SUPEROXIDE * PEROXYDE DE PLOMB (FRENCH)

OSHA PEL: TWA 200 ug(Pb)/m^3

THR: A poison by intraperitoneal route. See also lead compounds. Violent reactions with Al_4C_3; BaS; B; CaS; CsHC$_2$; ClF$_3$; H_2O_2; H_2S; hydroxylamine; Mo; performic acid; phenyl hydrazine; P; PCl$_3$; S; S(OCl)$_2$; W; Zr. Dangerous; see lead; can react with reducing material.

LEAD (II) FLUORIDE   HR: 3
CAS: 7783-46-2   NIOSH: OG 1225000
mf: F_2Pb mw: 245.21

PROP: Colorless solid, d (orthorhombic): 8.445; d (cubic): 7.750; mp: 824° bp: 1293°, vap press: 10 mm @ 904°, d: 8.24. Low solubility in H_2O; Solubility increases in presence of HNO$_3$ or nitrates.

SYNS: LEAD DIFLUORIDE * PLOMB FLUORURE (FRENCH) * PLUMBOUS FLUORIDE

OSHA PEL: TWA 200 ug(Pb)/m^3

THR: Poisonous! Moderate ingestion and subcutaneous toxicity. See also lead compounds and fluorides. Incompatible with CaC_2; F_2. When heated to decomposition it emits very toxic fumes of Pb and F^-. For further information see Vol. 6, No. 2 of *DPIM Report*.

LEAD(II) FLUOROSILICATE   HR: 3
CAS: 25808-74-6   NIOSH: VV 8450000
mf: $F_6Si \cdot Pb \cdot 2H_2O$ mw: 385.32

PROP: Monoclinic, colorless powder, mp: decomp.

SYN: HEXAFLUOROSILICATE (2-1) LEAD (II) SALT DIHYDRATE

OSHA PEL: TWA 200 ug(Pb)/m^3

THR: A poison by ingestion. When heated to decomposition it emits very toxic fumes of F^- and Pb.

LEAD-MOLYBDENUM CHRO-MATE   HR: 3
NIOSH: OG 1625000

SYNS: CHROMIC ACID, LEAD AND MOLYBDENUM SALT * CHROMIC ACID LEAD SALT WITH LEAD MOLYBDATE * CI PIGMENT RED 104 * LEAD CHROMATE, SULPHATE AND MOLYBDATE * MOLYBDENUM-LEAD CHROMATE * MOLYBDENUM ORANGE

THR: An experimental neoplastigen and tumorigen. See also lead, molybdenum and chromium compounds. A powerful oxidizer. When heated to decomposition it emits toxic fumes of lead.

LEAD MONOXIDE   HR: 3
CAS: 1317-36-8   NIOSH: OG 1750000
mf: OPb   mw: 223.19

PROP: Exists in 2 forms: red to reddish-yellow tetragonal crystals, stable at ordinary temperatures; yellow orthorhombic crystals stable >489°; d: 9.53; mp: 888°. Insol in H_2O, alc; sol in acetic acid, dil HNO_3, in warm solns of fixed alkali hydroxides.

SYNS: C.I. 77577 * LEAD OXIDE * LEAD PROTOXIDE * LEAD(II) OXIDE * LITHARGE * PLUMBOUS OXIDE * YELLOW LEAD OCHER

OSHA PEL: TWA 200 ug(Pb)/m^3

THR: A poison by inhalation and ingestion routes. A skin irritant. Mutagenic data. Avoid breathing dust. Wash thoroughly after contact with the material and before eating and smoking. An experimental carcinogen. See also lead compounds. Incompatible with chlorinated rubber; chlorine; ethylene; fluorine; glycerol; perchloric acid; hydrogen trisulfide; metal acetylides; metals; non-metals; peroxy-formic acid; seleninyl chloride. When heated to decomposition it emits toxic fumes of Pb.

LEAD (II) NITRATE (1:2)   HR: 3
CAS: 10099-74-8   NIOSH: OG 2100000
DOT: 1469
mf: $Pb(NO_3)_2$ mw: 331.23

PROP: White crystals, mp: decomp @ 470°, d: 4.53 @ 20°.

SYNS: LEAD DINITRATE * LEAD NITRATE * LEAD (2+) NITRATE * LEAD (II) NITRATE * NITRIC ACID, LEAD (2+) SALT * NITRATE DE PLOMB (FRENCH)

OSHA PEL: TWA 200 ug(Pb)/m^3
DOT Classification: Label: Oxidizer

THR: A poison by intraperitoneal route. Moderately toxic by ingestion. An experimental carcinogen and teratogen. When heated to decomposition it emits very toxic fumes of Pb and NO_x. Reacts violently with ammonium thiocyanate; carbon; lead hypophosphite; potassium acetate. See also lead compounds and nitrates. For further information see Vol. 6, No. 2 of *DPIM Report*.

LEAD NITRORESORCINATE   HR: 3
CAS: 51317-24-9   NIOSH: OG 2120000

SYNS: INITIATING EXPLOSIVE LEAD MONONITRORESORCINATE (DOT) * EXPLOSIVE HIGH * LEAD MONONITRORESORCINATE

DOT Classification: Class A Explosive, Label: Explosive

THR: A poison by ingestion and inhalation. See also lead and nitro compounds of aromatic hydrocarbons. See also explosives, high. When heated to decomposition it emits very toxic fumes of NO_x and Pb.

LEAD(IV)OXIDE BROWN   HR: 3
CAS: 1309-60-0   NIOSH: OG 0700000
DOT: 1872
mf: O_2Pb mw: 239.19

PROP: Brown hexagonal crystals, mp: decomp @ 290°; d: 9.375.

SYNS: BIOXYDE DE PLOMB (FRENCH) * C.I. 77580 * LEAD BROWN * LEAD PEROXIDE * LEAD DIOXIDE * LEAD SUPEROXIDE

OSHA PEL: TWA 200 ug(Pb)/m^3
DOT Classification: Label: Oxidizer

THR: A poison by intraperitoneal route. See also lead compounds. Incompatible with chlorine trifluoride; hydrogen sulfides; metal acetylides; metals; metal sulfides; nitroalkanes; nitrogen compounds; non-metals; peroxyformic acid; non-metal halides; potassium; sulfur dioxide. When heated to decomposition it emits toxic fumes of Pb.

LEAD OXIDE RED   **HR: 3**
CAS: 1314-41-6   NIOSH: OG 5425000
mf: O_4Pb_3 mw: 685.57

PROP: Bright red powder, mp: 890° (decomp), bp: 1472°, d: 8.32-9.16, vap press: 1 mm @ 943°.

SYNS: C.I. 77578 * LEAD ORTHOPLUMBATE * LEAD TETRAOXIDE * MINERAL ORANGE * MINERAL RED * MINIUM * ORANGE LEAD * PARIS RED * PLUMBOPLUMBIC OXIDE * RED LEAD * RED LEAD OXIDE * TRILEAD TETROXIDE

OSHA PEL: TWA 200 ug(Pb)/m^3

THR: A poison by intraperitoneal route. See also lead compounds. Moderately toxic by ingestion. Combustible by chemical reaction with reducing agents. An oxidizing agent. Incompatible with Al; $CsHC_2$; (F_2 + glycerol); H_2S_3; (glycerine + $HClO_4$); $RbHC_2$; (Si + Al); Na; SO_3; Ti; Zr. See also lead fumes (for decomposition hazard).

LEAD (II) PHOSPHATE (3:2)   **HR: 3**
CAS: 7446-27-7   NIOSH: OG 2675000
mf: O_8P_2 • 3Pb   mw: 811.51

PROP: Insol in H_2O, alc; sol in HNO_3, fixed alkali hydroxides. Hexagonal, colorless crystals or white powder, mp: 1014, d: 6.9-7.3.

SYNS: BLEIPHOSPHAT (GERMAN) * C.I. 77622 * LEAD ORTHOPHOSPHATE * LEAD PHOSPHATE * LEAD PHOSPHATE (3:2) * LEAD (2+) PHOSPHATE * NORMAL LEAD ORTHOPHOSPHATE * PHOSPHORIC ACID, LEAD (2+) SALT (2:3) * PLUMBOUS PHOSPHATE * TRILEAD PHOSPHATE

OSHA PEL: TWA 200 ug(Pb)/m^3

THR: Poison. An experimental tumorigen and carcinogen. A suspected human carcinogen. See also lead compounds. When heated to decomposition it emits very toxic fumes of Pb and PO_x.

LEAD (II) SULFATE (1:1)   **HR: 3**
CAS: 7446-14-2   NIOSH: OG 4375000
DOT: 2291
mf: O_4S • Pb   mw: 303.25

PROP: White cryst; mp: decomp @ 1000°; d: 6.2. Insol in alc, sol in NaOH, ammonium acetate or tartrate soln, concd HI. Practically insol in H_2O; somewhat more sol in dil HCl or HNO_3.

SYNS: ANGLISLITE * BLEISULFAT (GERMAN) * C.I. 77630 * FREEMANS WHITE LEAD * LEAD BOTTOMS * SULFATE DE PLOMB (FRENCH) * SULFURIC ACID, LEAD (2+) SALT (1:1)

OSHA PEL: TWA 200 ug(Pb)/m^3
DOT Classification: Label: Corrosive

THR: A poison by intraperitoneal route. Moderately toxic by ingestion. Mutagenic data. A strong irritant to skin, eyes and mucous membranes. See also lead compounds. Violent reaction with K. When heated to decomposition it emits very toxic fumes of Pb and SO_x.

LEAD SULFIDE   **HR: 3**
CAS: 1314-87-0   NIOSH: OG 4550000
mf: PbS   mw: 239.25

PROP: Silvery, metallic crystals or black powder. Insol in H_2O; sol in HNO_3; hot dil HCl, mp: 1114°, bp: 1281° (sublimes), d: 7.5, vap press: 1 mm @ 852°.

SYNS: G.I. 77640 * GALENA * NATURAL LEAD SULFIDE * PLUMBOUS SULFIDE

OSHA PEL: TWA 200 ug(Pb)/m^3

THR: Hazardous because of its Pb content. Moderately toxic by intraperitoneal route. See also sulfides and lead compounds. Violent reaction with ICl; H_2O_2. When heated to decomposition it emits very toxic fumes of Pb and SO_x. For further information see Vol. 6, No. 2 of *DPIM Report*.

LEAD TRINITRORESORCINATE   **HR: 3**
CAS: 15245-44-0   NIOSH: OG 6425000
mf: $C_6HN_3O_8Pb$ mw: 450.29

PROP: Orange-yellow, monoclinic crystals, mp: explodes @ 311°, d: 3.1-2.9.

SYNS: INITIATING EXPLOSIVE LEAD STYPHNATE (DOT) * LEAD STYPHNATE

DOT Classification: Label: Explosive A

THR: A poisonous material. See also lead compounds and nitrates. Very sensitive explosive. It is shock-sensitive and has detonated spontaneously when dry. See nitrates and explosives, high. Dangerous; explodes at 311°. Emits very toxic fumes of NO_x, Pb when heated to decomposition.

LEMON OIL   HR: 3
CAS: 8008-56-8   NIOSH: OG 8300000

SYNS: CEDRO OIL * OIL OF LEMON * ZITRO-NEN OEL (GERMAN)

THR: An experimental tumorigen. Moderate ingestion toxicity. A skin irritant. When heated to decomposition it emits acrid smoke and fumes.

LENTE INSULIN   HR: 3
CAS: 8049-62-5   NIOSH: NM 8900000

SYNS: INSULIN NOVO LENTE * INSULIN ZINC SUSPENSION * LENTE ILETIN

THR: An experimental carcinogen. When heated to decomposition it emits toxic fumes.

LEPTOPHOS   HR: 3
CAS: 21609-90-5   NIOSH: TB 1720000
mf: $C_{13}H_{10}BrCl_2O_2PS$ mw: 412.07

SYNS: O-(4-BROMO-2,5-DICHLOROPHENYL)-O-METHYL PHENYLPHOSPHONOTHIOATE * O-(2,5-DICHLORO-4-BROMOPHENYL)-O-METHYL PHENYL-THIOPHOSPHONATE * FOSVEL * PHOSVEL * VELSICOL VCS 506

THR: A poison by ingestion and subcutaneous route. Moderate skin toxicity. A pesticide. When heated to decomposition it emits very toxic SO_x, PO_x, Br^- and Cl^-.

LEPTRYL   HR: 3
CAS: 13093-88-4   NIOSH: TN 7700000
mf: $C_{22}H_{28}N_2O_2S$ mw: 384.58

SYNS: 2-METHOXY-10-(3-(4-HYDROXYPIPERI-DINO)-2-METHYLPROPYL)PHENOTHIAZINE * 3-METHOXY-10-(3-(4-HYDROXYPIPERIDYL)-2-METHYLPROPYL)PHENOTHIAZINE * 2-METHOXY-10-(2-METHYL-3-(4-HYDROXYPIPERIDINO)PRO-PYL)PHENOTHIAZINE * 1-(3-(2-METHOXY-PHENOTHIAZIN-10-YL)-2-METHYLPROPYL)-4-PIPERIDINOL

THR: A poison by ingestion, intraperitoneal, subcutaneous, and intravenous routes. When heated to decomposition it emits very toxic fumes of NO_x and SO_x.

LERGOTRILE MESYLATE   HR: 3
CAS: 51473-23-5   NIOSH: KE 6320000
mf: $C_{17}H_{20}ClN_3 \cdot CH_4O_3S$ mw: 397.96

SYN: D-2-CHLORO-6-METHYLERGOLINE-8-beta-ACETONITRILE METHANESULFONIC ACID SALT

THR: A poison by ingestion and intraperitoneal routes. Heated to decomposition it emits toxic fumes of SO_x, Cl^- and NO_x.

LETHANE (SPECIAL)   HR: 3
CAS: 63917-01-1   NIOSH: OH 2275000

PROP: A liquid; bp: 160°-190° @ 0.1 mm; bp: 120°-125° @ 0.28 mm. Insol in water, miscible with hydrocarbons and most organic solvents. A mixture of Lethane 60 (3 parts) and Lethane 384 (1 part).

THR: A poison by ingestion. Moderately toxic by skin. See also thiocyanates. Insecticides with n-butyl carbitolthiocyanate, etc., in a light petroleum base. Accidental and suicidal poisonings have occurred. Symptoms include drowsiness, followed by coma, the limbs becoming flaccid and the appearance of twitching and convulsions. The pupils may dilate and respiration may become labored. Cyanosis and vomiting occur. The lethanes are mild irritants and in higher doses narcotic. Can be absorbed by intact skin. When heated to decomposition it emits very toxic fumes of SO_x and NO_x. For further information see Vol. 2, No. 4 of *DPIM Report*.

I-LEUCINE   HR: 3
CAS: 61-90-5   NIOSH: OH 2850000
mf: $C_6H_{13}NO_2$ mw: 131.20

PROP: An essential amino acid; occurs in isomeric forms. Below we consider the *l* and *dl* forms. White crystals, sol in water, slightly sol in alcohol, insol in ether, mp (*dl*): 332° with decomp, mp (*l*): 295°, d: 1.239 @ 18°/4°. L form (natural): glistening hexagonal plates from aq alc; d: 1.291 @ 18°; sublimes @ 145°-148°; decomp @ 293°-295°.

SYNS: alpha-AMINOISOCAPROIC ACID * 2-AMINO-4-METHYLPENTANOIC ACID * alpha-AMINO-gamma-METHYLVALERIC ACID * LEUCINE

THR: An experimental teratogen. A nutrient and/or dietary supplement food additive. When heated to decomposition it emits toxic fumes of NO_x.

LEUCINOCAINE   **HR: 3**
CAS: 92-23-9   NIOSH: SA 6125000
mf: $C_{17}H_{28}N_2O_2$ mw: 292.47

SYN: 2-(DIETHYLAMINO)-4-METHYL-1-PENTANOL-P-AMINOBENZOATE (ESTER)

THR: A poison by subcutaneous and intravenous routes. When heated to decomposition it emits toxic fumes of NO_x.

LEUROCRISTINE   **HR: 3**
CAS: 57-22-7   NIOSH: OH 6300000
mf: $C_{46}H_{56}N_4O_{10}$ mw: 825.06

SYNS: NCI-C04864 * NSC-67574 * VINCRISTINE

THR: A poison by parenteral, intraperitoneal, and intravenous routes. Mutagenic data. An experimental teratogen. Affects human glandular system, central nervous system, and other unspecified systems. When heated to decomposition it emits toxic fumes of NO_x.

**LEUROCRISTINE SULFATE
(1:1)**   **HR: 3**
CAS: 2068-78-2   NIOSH: OH 6340000
mf: $C_{46}H_{56}N_4O_{10} \cdot SO_4$ mw: 921.16

SYNS: VINCRISTINE SULFATE * ONCORIN

THR: An experimental teratogen. When heated to decomposition it emits very toxic fumes of NO_x and SO_x.

LEVAMISOLE   **HR: 3**
CAS: 6649-23-6   NIOSH: NJ 5935000
mf: $C_{11}H_{12}N_2S$ mw: 204.31

SYN: 6-PHENYL-2,3,5,6-TETRAHYDROIMIDAZO(2,1-B)THIAZOLE

THR: Affects the human skin and blood picture. When heated to decomposition it emits very toxic fumes of NO_x and SO_x.

LEVORIN   **HR: 3**
CAS: 1403-17-4   NIOSH: OI 0250000

PROP: Polyene from *Actinomyces*.

THR: A poison by intraperitoneal route. Moderately toxic by ingestion and subcutaneous route.

LEVORPHANOL   **HR: 3**
CAS: 77-07-6   NIOSH: QD 1830000
mf: $C_{17}H_{23}NO$ mw: 257.41

SYNS: LEVO-DROMORAN * (−)-3-HYDROXY-N-METHYLMORPHINAN * LEVORPHAN

THR: A poison by ingestion, subcutaneous, and intravenous routes. When heated to decomposition it emits toxic fumes of NO_x.

LIBRIUM   **HR: 3**
CAS: 58-25-3   NIOSH: DE 9275000
mf: $C_{16}H_{14}ClN_3O$ mw: 299.78

SYNS: CHLORIDIAZEPOXIDE * CHLORODIAZEPOXIDE * 7-CHLORO-2-METHYLAMINO-5-PHENYL-3H-1,4-BENZODIAZEPINE 4-OXIDE * LIBRAX * LIBRININ

THR: A poison by ingestion to women causing central nervous system effect. Has been implicated in development of asplastic anemia. An experimental teratogen. When heated to decomposition it emits very toxic fumes such as NO_x.

LIBRIUM HYDROCHLORIDE   **HR: 3**
CAS: 438-41-5   NIOSH: DE 9450000
mf: $C_{16}H_{14}ClN_3O \cdot ClH$ mw: 336.24

SYNS: CHLORDIAZEPOXIDE MONOHYDROCHLORIDE * CHLORIDEAZEPOXIDE HYDROCHLORIDE * METHAMINODIAZEPOXIDE HYDROCHLORIDE * METHAMINODIAZEPINE HYDROCHLORIDE

THR: Poison by intraperitoneal and intravenous routes. An experimental teratogen. Moderately toxic by ingestion and subcutaneous routes. Mutagenic data. When heated to decomposition it emits very toxic fumes of HCl and NO_x.

LIDEPRAN HYDROCHLORIDE   **HR: 3**
CAS: 23257-56-9   NIOSH: TN 1050000
mf: $C_{14}H_{19}NO_2 \cdot ClH$ mw: 269.80

SYNS: LEVOPHACETOPERAN HYDROCHLORIDE * alpha-PHENYL-2-PIPERIDINEMETHANOL ACETATE HYDROCHLORIDE * 1-PHENYL-1-(2-PIPERIDYL)-1-ACETOXYMETHANE HYDROCHLORIDE * PHENYL-(2-PIPERIDYL)METHYL ACETATE HYDROCHLORIDE

THR: A poison by ingestion, intraperitoneal, subcutaneous, and intravenous routes. When heated to decomposition it emits very toxic fumes of NO_x and HCl.

d-LIMONENE
HR: 3

CAS: 5989-27-5 NIOSH: GW 6360000
mf: $C_{10}H_{16}$ mw: 136.26

PROP: Liquid, bp: 175.5°-176°; d: 0.8402 @ 25°/4°

SYNS: 1-METHYL-4-(1-METHYLETHENYL)-CY-CLOHEXENE, (R)- * D-(+)-LIMONENE * (+)-R-LIMONENE * NCI-C55572

THR: An experimental tumorigen. Reacts violently with (IF_3 + tetrafluoroethylene). See also p-mentha-1,8-diene. When heated to decomposition it emits acrid smoke and fumes. For further information see Limonene; Vol. 2, No. 1 of *DPIM Report*.

LIMONENE DIOXIDE
HR: 3

CAS: 96-08-2 NIOSH: OS 9100000
mf: $C_{10}H_{16}O_2$ mw: 168.26

SYNS: 1,2,8,9-DIEPOXYLIMONENE * DIPEN-TENE DIOXIDE * EPOXIDE 269 * 4-(1,2-EPOXY-1-METHYLETHYL)-1-METHYL-7-OXABICYCLO(4.1.0)HEPTANE

THR: An experimental tumorigen. Moderate skin and intramuscular toxicity. A skin irritant. When heated to decomposition it emits acrid smoke and fumes.

LIMONITE
HR: 3

NIOSH: OI 7710000

PROP: Consists mainly of hydrated sesquioxide of iron.

SYNS: BROWN HEMATITE * BROWN IRON ORE * BROWN IRONSTONE CLAY * IRON SESQUIOX-IDE HYDRATED

THR: An experimental carcinogen. See also iron.

LINOLEIC ACID
HR: 3

CAS: 60-33-3 NIOSH: RF 9990000
mf: $C_{18}H_{32}O_2$ mw: 280.50

PROP: Colorless oil, easily oxidized by air, sol in ether and ethanol. Misc with dimethyl formamide, fat solvents, oils. d: 0.9038 @ 18°/4°, mp: −12°, bp: 230° @ 16 mm.

SYNS: LEINOLEIC ACID * 9,12-LINOLEIC ACID * CIS,CIS-9,12-OCTADECADIENOIC ACID

* CIS-9,CIS-12-OCTADECADIENOIC ACID
* 9,12-OCTADECADIENOIC ACID

THR: Ingestion can cause nausea and vomiting. A nutrient and/or dietary supplement food additive. A human skin irritant. When heated to decomposition it emits acrid smoke and fumes. For further information see Vol. 1, No. 8 of *DPIM Report*.

LIQUEFIED CARBON DIOXIDE
HR: 2

mf: CO_2 mw: 44.0

PROP: Heavy gas or liquid under pressure, mp: −56.6° @ 3952 mm; bp: −78.5° (sublimes); d: 1.977g/L @ 0°, (liq) 1.101 @ −37°.

SYN: LIQUID CARBONIC GAS

THR: This material is stable when very cold and can damage tissue exposed to it. Solid CO_2 goes directly to gaseous CO_2 (sublimes) which is mainly an asphyxiant. See also carbon dioxide. Moderately dangerous.

LITHIUM
HR: 3

CAS: 7439-93-2 NIOSH: OJ 5540000
DOT: 1415
af: Li aw: 6.94

PROP: Silver-colored light metal, mixture of isotopes Li^6 and Li^7, mp: 179°, bp: 1317°, d: 0.534 @ 25°, vap press: 1 mm @ 723°. Sol in liq ammonia. Keep under mineral oil or other liq free from O_2 or water.

SYN: LITHIUM METAL

DOT Classification: Label: Flammable Solid and Dangerous When Wet

THR: See also lithium compounds for a discussion of the toxicity of the lithium ion. See also sodium for a discussion which applies to the toxicity of metallic lithium. Dangerous fire hazard when exposed to heat or flame, or contact with As; Be; Br_2; $CHBr_3$; maleic anhydride; carbides; CO_2; (CO + H_2O); CBr_4; CCl_4; Cl_2; $CHCl_3$; CrO_3; Cr; $CrCl_3$; cobalt alloys; FeS; diborane; Mn alloys; CH_2Cl_2; CH_2I_2; Mo_2O_3; Ni alloys; Nb_2O_5; $CFCl_3$; HNO_3; N_2; organic matter; O_2; P; Pt; rubber; silicates; NaCl; $NaNO_2$; S; Ta_2O_5; TiO_2;trichloroethylene; tetrachloroethylene; Fe alloys; trichlorotrifluoroethane; WO_3; V; V_2P_5; H_2O; $ZrCl_4$; CHI_3; I; H_2. Dangerous; when burned it emits toxic fumes of lithium oxide and hydroxide, will react with water or steam to produce heat and hydrogen; can react vigorously with oxidizing materials. Reacts with nitrogen at high temperatures. To fight fire, use special mixtures of dry chemical, soda ash, graphite. Incompatible with atmospheric gases; bromine pentafluoride; diazomethane; halocarbons; halogens; mercury; metal chlorides; nitrogen; metal oxides; nitric acid; nitryl fluoride; non-metal oxides; platinum; poly-1,1-difluoroethylene-hexafluoropropylene; sodium carbonate or sodium chloride; sulphur; trifluoromethylhypofluorite; water.

LITHIUM AMIDE   HR: 3
CAS: 7782-89-0   NIOSH: OJ 5590000
DOT: 1412
mf: H_2LiN mw: 22.97

PROP: White crystalline solid or powder. Sublimes in NH_3 current. Insol in anhydrous ether, benzene, toluene, mp: 380°-400°, d: 1.178 @ 17.50.

SYNS: LITHAMIDE * LITHIUM AMIDE, POWDER

DOT Classification: Label: Flammable
  Solid

THR: Ammonia is liberated and lithium hydroxide is formed when this compound is exposed to moisture. See also ammonia and lithium hydroxide. A powerful irritant to skin, eyes and mucous membranes. See ammonia for fire and explosion hazards. Reacts violently with H_2O. Moderately dangerous; will react with water or steam to produce toxic and flammable vapors; on contact with oxidizing materials, can react vigorously; on contact with acid or acid fumes, can evolve much heat. When heated to decomposition it emits very toxic fumes of Li_2O and NO_x.

LITHIUM ANTIMONY THIOMALATE   HR: 3
CAS: 305-97-5   NIOSH: WJ 1625000
mf: $C_{12}H_9O_{12}S_3Sb \cdot 6Li$ mw: 604.78

SYNS: ANTHIOLIMINE * ANTHIOMALINE
* LITHIUM ANTIMONIOTHIOMALATE * MERCAPTOSUCCINIC ACID ANTIMONATE (III) HEXALITHIUM SALT * MERCAPTOSUCCINIC ACID-S-ANTIMONY DERIVATIVE LITHIUM SALT

THR: A poison by intraperitoneal and intravenous routes. Affects the human gastrointestinal tract. See also antimony and lithium. When heated to decomposition it emits very toxic SO_x and Sb, Li_2O.

LITHIUM CARBONATE (2:1)   HR: 3
CAS: 554-13-2   NIOSH: OJ 5800000
mf: $CO_3 \cdot 2Li$ mw: 73.89

PROP: White, light alkaline powder. d: 2.11 @ 17.5°; mp: 618°. Insol in alc. @ 17.5°, white cryst.

SYNS: CARBONIC ACID, DILITHIUM SALT
* CARBONIC ACID LITHIUM SALT * DILITHIUM CARBONATE

THR: An experimental teratogen. A human cardiovascular toxin and glandular system toxin. Moderately toxic by ingestion, intraperitoneal, and subcutaneous routes. See also lithium compounds. Incompatible with fluorine.

LITHIUM CHLORIDE   HR: 3
CAS: 7447-41-8   NIOSH: OJ 5950000
mf: ClLi   mw: 42.39

PROP: Cubic, white deliquescent crystals, mp: 605°, bp: 1350°, d: 2.068 @ 25°, vap press: 1 mm @ 547°.

THR: A poison by ingestion. Moderately toxic by subcutaneous and intraperitoneal routes. An experimental teratogen and neoplastigen. Affects human central nervous system. This material has been recommended and used as a substitute for sodium chloride in ''salt-free'' diets, but cases have been reported in which the ingestion of lithium chloride has produced dizziness, ringing in the ears, visual disturbances, tremors and mental confusion. In most cases, the symptoms disappeared when use was discontinued. Prolonged absorption may cause disturbed electrolyte balance, impaired renal function. Reaction is violent with BrF_3. When heated to decomposition it emits toxic fumes of Cl^-.

LITHIUM COMPOUNDS
THR: Lithium compounds have been implicated in development of aplastic anemia. Lithium oxide, hydroxide, carbonate, etc., are strong bases and these solutions in water are very caustic. See also potassium compounds. Lithium ion has central nervous system toxicity. Large doses of lithium compounds have caused dizziness and prostration. Can cause kidney damage, anorexia, nausea, apathy, coma and death.

LITHIUM FLUORIDE   HR: 3
CAS: 7789-24-4   NIOSH: OJ 6125000
mf: FLi   mw: 25.94

PROP: Fine white powder, mp: 845°, bp: 1681°, d: 2.635 @ 20°, vap press: 1 mm @ 1047°. Sol in acids.

SYN: LITHIUM FLUORURE (FRENCH)

OSHA PEL: TWA 2500 ug(F)/m^3

THR: A poison by ingestion and subcutaneous routes. See also fluorides and lithium compounds. When heated to decomposition it emits toxic fumes of F$^-$.

LITHIUM HYDRIDE   HR: 3
CAS: 7580-67-8   NIOSH: OJ 6300000
DOT: 1414/2805
mf: HLi   mw: 7.95

PROP: White, translucent, crystals, mp: 680°; d: 0.76-0.77. Darkens rapidly on exposure to light. No solvent known. Decomp in H$_2$O liberating Li hydroxide and H$_2$.

SYN: HYDRURE DE LITHIUM (FRENCH)

OSHA PEL: TWA 25 ug/m^3
DOT Classification: Label: Flammable Solid and Dangerous When Wet

THR: A poison by inhalation. An eye irritant. In contact with moisture, lithium hydroxide is formed. Can ignite spontaneously in moist air. See also lithium compounds. The LiOH formed is very caustic and therefore highly toxic, particularly to lungs and respiratory tract, skin and mucous membranes. See also hydrides and Li$_2$O for fire and explosion hazards. Incompatible with oxygen; N$_2$O; liquid O$_2$; air. To fight fire, use special mixtures of dry chemical.

LOBELINE HYDROCHLORIDE   HR: 3
CAS: 64990-84-1   NIOSH: AM 9150000
mf: C$_{22}$H$_{27}$NO$_2$ • ClH   mw: 373.96

SYN: 2-(6-(beta-HYDROXYPHENETHYL)-1-METHYL-2-PIPERIDYL)ACETOPHENONEHYDRO-CHLORIDE

THR: A poison by intraperitoneal, subcutaneous, and intravenous routes. When heated to decomposition it emits very toxic fumes of NO$_x$ and HCl.

LOCUST BEAN GUM   HR: 1
CAS: 9000-40-2   NIOSH: OJ 8690000

PROP: Yellow-green color, odorless and tasteless but acquires a leguminous taste when boiled in H$_2$O. A galactomazannan polysaccharide; in powdered form, nearly pure white, insol in most organic solvents, mw: 310,000 (approx).

SYNS: ALGAROBA * CAROB BEAN GUM * CAROB FLOUR * NCI-C50419 * ST. JOHN'S BREAD

THR: Low toxicity by ingestion. A stabilizer food additive. Also a substance migrating to food from packaging materials. When heated to decomposition it emits acrid smoke and fumes.

LOMOTIL   HR: 3
CAS: 3810-80-8   NIOSH: NS 5300000
mf: C$_{30}$H$_{32}$N$_2$O$_2$ • ClH   mw: 489.10

SYNS: 1-(3-CYANO-3,3-DIPHENYLPROPYL)-4-PHENYLISONIPECOTIC ACID ETHYL ESTER HYDRO-CHLORIDE * DIPHENOXYLATE HYDROCHLORIDE

THR: Poison by ingestion. Human systemic pulmonary and central nervous system effect by ingestion. When heated to decomposition it emits very toxic fumes of HCl and NOx.

LONDON PURPLE   HR: 3
CAS: 8012-74-6   NIOSH: OJ 9280000
DOT: 1621

SYN: LONDON PURPLE, SOLID (DOT)

DOT Classification: Poison B, Label: Poison

THR: A poison. See also arsenic and aniline. When heated to decomposition it emits very toxic fumes of As and NO$_x$.

LUCANTHONE   HR: 3
CAS: 479-50-5   NIOSH: XO 1605500
mf: C$_{20}$H$_{24}$N$_2$OS   mw: 340.52

SYNS: 1-(2-'-DIETHYLAMINO)ETHYLAMINO-4-METHYLTHIOXANTHENONE * 9H-THIOXANTHEN-9-ONE, 1-((2->(DIETHYLAMINO)ETHYL)AMINO)4-METHYL- (9CI)

THR: Poison by intravenous route. Mutagenic data. When heated to decomposition it emits very toxic fumes of NOx and SOx.

LUCANTHONE METABOLITE   HR: 3
CAS: 3105-97-3   NIOSH: XO 1590000
mf: C$_{20}$H$_{24}$N$_2$O$_2$S   mw: 356.52

SYNS: 1-((2-(DIETHYLAMINO)ETHYL)AMINO)-4-(HYDROXYMETHYL)THIOXANTHEN-9-ONE * 1-((2-(DIETHYLAMINO)ETHYL)AMINO)-4-(HYDROXYMETHYL)9H-THIOXANTHEN-9-ONE

THR: A poison by subcutaneous, intravenous, and intramuscular routes. Mutagenic data. An experimental teratogen and carcinogen. Moderate ingestion toxicity. When heated to decomposition it emits very toxic fumes of NO_x and SO_x.

LUTEOSKYRIN   HR: 3
CAS: 21884-44-6   NIOSH: EK 6125000
mf: $C_{30}H_{22}O_{12}$ mw: 574.52

PROP: Yellow rectangular crystals, mp: 278° (decomp). Anthraquinoid hepatotoxin of *Penicillium Islandicum sopp*.

SYNS: 5H,6H-6,5A,13A,14-(1,2,3,4)-BUTANETETRAYCYCLOOCTA(1,2-B:5,6-B')-DINAPHTHALENE * 8,8'-DIHYDROXY-RUGULOSIN * FLAVOMYCELIN

THR: A poison by ingestion, intraperitoneal, subcutaneous, and intravenous routes. Mutagenic data. An experimental neoplastigen and carcinogen. When heated to decomposition it emits acrid smoke and fumes.

LUTETIUM CHLORIDE   HR: 3
CAS: 10099-66-8   NIOSH: OK 8400000
mf: Cl_3Lu mw: 281.32

PROP: Colorless crystals; sublimes above 750°; mp: 892° ± 2°; sol in H_2O.

THR: A poison by intraperitoneal route. See also rare earths. When heated to decomposition it emits toxic fumes of Cl^-.

LUTETIUM CITRATE   HR: 3
CAS: 63917-04-4   NIOSH: OK 8575000

THR: A poison by intraperitoneal route. See also rare earths. When heated to decomposition it emits acrid smoke and fumes.

LUTETIUM (III) NITRATE (1:3)   HR: 3
CAS: 10099-67-9   NIOSH: OK 8925000
mf: $N_3O_9 \cdot Lu$ mw: 361.00

SYN: NITRIC ACID, LUTETIUM(3+) SALT

THR: A poison by intraperitoneal route. See also nitrates, rare earths. Moderately dangerous; when heated to decomposition emits toxic fumes of NO_x.

3,4-LUTIDINE   HR: 3
CAS: 583-58-4   NIOSH: OK 9800000
mf: C_7H_9N mw: 107.17

PROP: Liquid; bp: 163.5°-164.5°.

SYN: 3,4-DIMETHYLPYRIDINE

THR: A poison by inhalation and skin contact. Moderate ingestion toxicity. When heated to decomposition it emits toxic fumes of NO_x.

d-LYSERGIC ACID DIETHYL
AMIDE   HR: 3
                        NIOSH: KE 4200000
mf: $C_{20}H_{25}N_3O$ mw: 323.48

SYNS: LYSERGIC ACID DIETHYLAMIDE, 1-ISOMER * USAF SZ-2

THR: A poison by intraperitoneal and intravenous routes. When heated to decomposition it emits toxic fumes of NO_x.

d-LYSERGIC ACID DIETHYLAMIDE
TARTRATE   HR: 3
                        NIOSH: KE 4390000
mf: $C_{40}H_{50}N_6O_2 \cdot C_4H_6O_6 \cdot 2CH_4O$ mw: 861.16

SYN: N-DIETHYL-6-METHYL-ERGOLINE-8-beta-CARBOXAMIDE-9-10-DIDEHYDRO-N,D-, TARTRATE WITH METHANOL (1:2)

THR: A poison by subcutaneous and intraperitoneal routes. Mutagenic data. When heated to decomposition it emits toxic fumes of NO_x.

d-LYSERGIC ACID DIMETHYL-
AMIDE   HR: 3
CAS: 4238-84-0   NIOSH: KE 6160000
mf: $C_{18}H_{21}N_3O$ mw: 295.42

SYNS: DAM-57 * 9,10-DIDEHYDRO-N,N,6-TRIMETHYLERGOLINE-8-beta-CARBOXAMIDE

THR: A poison by intravenous route. A psychotropic material by ingestion in humans. When heated to decomposition it emits toxic fumes of NO_x.

LYSERGIC ACID ETHYLAMIDE   HR:
CAS: 478-99-9   NIOSH: KE 4550000
mf: $C_{18}H_{21}N_3O$ mw: 295.42

SYNS: 9,10-DIDEHYDRO-N-ETHYL-6-METHYLERGOLINE-8-beta-CARBOXAMIDE; N-ETHYLLYSERGAMIDE * LAE-32 * D-LYSERGIC ACID MONOETHYLAMIDE

THR: A poison by intravenous route. Toxic in humans by ingestion causing psychotropic reactions. When heated to decomposition it emits toxic fumes of NO_x.

d-LYSERGIC ACID-L,2-PROPANOLAMIDE   HR: 3
CAS: 60-79-7   NIOSH: KE 5075000
mf: $C_{19}H_{23}N_3O_2$ mw: 325.45

SYNS: CORNOCENTIN * 9,10-DIDEHYDRO-N-(alpha-(HYDROXYMETHYL)ETHYL)-6METHYL-ERGOLINE-8-beta-CARBOXAMIDE * ERGO-BASINE * ERGOMETRINE * ERGONOVINE * ERGOTRATE * ERMETRINE * N-(alpha-(HYDROXYMETHYL)ETHYL)-D-LYSERGOMIDE * N-(1-(HYDROXYMETHYL)ETHYL)-D-LYSER-GOMIDE * D-LYSERGIC ACID-1-HYDROXYMETH-YLETHYLAMIDE * LYSERGIC ACID PROPANOL-AMIDE * D-LYSERGIC ACID-L,2-PROPANOL-AMIDE * NEOFEMERGEN * SECACORNIN * SYNTOMETRINE

THR: A poison by intravenous route. When heated to decomposition it emits toxic fumes of NO_x.

LYSERGIC ACID PYROLIDATE   HR: 3
CAS: 2385-87-7   NIOSH: KE 6314000
mf: $C_{20}H_{23}N_3O$ mw: 321.46

SYNS: LSD-25-PYRROLIDATE * D-LYSERGIC ACID PYRROLIDIDE

THR: A poison by intravenous route. Human psychotropic effects by ingestion. When heated to decomposition it emits toxic fumes of NO_x.

LYSSIPOLL   HR: 3
CAS: 147-20-6   NIOSH: TM 7810000
mf: $C_{19}H_{23}NO$ mw: 281.43

SYNS: 4-(BENZHYDRYLOXY)-1-METHYLPIPERI-DINE * DIPHENYLPYRILENE * 4-(DIPHENYL-METHOXY)-1-METHYLPIPERIDINE * N-METHYL-PIPERIDYL-(4)-BENZHYDRYLAETHER SALZSAUREN SALZE (GERMAN)

THR: A poison by ingestion and intravenous routes. An eye irritant. When heated to decomposition it emits toxic fumes of NO_x.

M

MAGENTA BASE
HR: 3

CAS: 3248-93-9   NIOSH: ZE 9705000

mf: $C_{20}H_{19}N_3$ mw: 301.42

SYNS: C.I. SOLVENT RED 41 * FUCHSINE BASE * ROSANILINE BASE

THR: Poison by ingestion. Human skin irritant. When heated to decomposition it emits toxic fumes of NO_x.

MAGNESIUM
HR: 3

CAS: 7439-95-4   NIOSH: OM 2100000

DOT: 1418/1869/2950

af: Mg   aw: 24.31

PROP: Hexagonal, silvery-white crystals of metal, mp: 651°, bp: 1100°, d: 1.74 @ 5°, d: 1.738 @ 20°, vap press: 1 mm @ 621°.

SYNS: MAGNESIO (ITALIAN) * MAGNESIUM METAL (DOT) * MAGNESIUM PELLETS * MAGNESIUM POWDERED * MAGNESIUM RIBBONS * MAGNESIUM TURNINGS

DOT Classification: Label: Flammable Solid and Dangerous When Wet

THR: Poison by ingestion. Inhalation of dust and irritating fumes can cause metal fume fever. Particles embedded in the skin can produce gaseous blebs and a gas 'gangrene' with a protracted course. A dangerous fire hazard in the form of dust or flakes when exposed to flame or oxidizing agents. In solid form, magnesium is difficult to ignite because heat is conducted rapidly away from the source of ignition; it must be heated above its melting point before it will burn. However, in finely divided form it may be ignited by a spark or the flame of a match. Magnesium fires do not flare up violently unless there is moisture present. Therefore, it must be kept away from water, moisture, etc. It may be ignited by a spark, match flame, or even spontaneously when the material is finely divided and damp, particularly with water-oil emulsion. Also, magnesium reacts with moisture, acids, etc., to evolve hydrogen, which is a highly dangerous fire and explosion hazard. Moderately explosive in the form of dust, when exposed to flame. Incompatible with air; beryllium fluoride; boron diiodophosphide; ethylene oxide; halocarbons; halogens or interhalogens; hydrogen iodide; metal cyanides; metal oxides; metal oxosalts; methanol; oxidants; polytetrafluoroethylene; potassium carbonate; silicon dioxide; sulfur; tellurium; (Al + $KClO_4$); NH_4NO_3; BeO; [$Ba(NO_3)_2$ + BaO_2 + Zn]; BPI_2; bromobenzyl trifluoride; $Cd(CN)_2$; CdO; CaC; carbonates; CCl_4; Cl_2; ClF_3; $CHCl_3$; $Co(CN)_2$; $Cu(CN)_2$; CuO; [$CuSO_4$(anhydrous) + NH_4NO_3 + $KClO_3$ + H_2O]; $CuSO_4$; F_2; AuCN; (H_2 + $CaCO_3$); HI; H_2O_2; I; $Pb(CN)_2$; HgO; $Hg(CN)_2$; CH_3Cl; MoO_3; $Ni(CN)_2$; HNO_3; NO_2; O_2 liquid; performic acid; phosphates; $KClO_3$; $KClO_4$; $AgNO_3$; Ag_2O; $NaClO_4$; (Na_2O_2 + CO_2); SnO_2; sulfates; trichloroethylene; $Zn(CN)_2$; ZnO; Na_2O_2. To fight fire, operators and fire fighters can approach a magnesium fire to within a few feet if no moisture is present. Water and ordinary extinguishers, such as CO_2, carbon tetrachloride, etc., should not be used on magnesium fires. G-1 powder or powdered talc should be used on open fires. Dangerous when heated, burns violently in air and emits fumes; will react with water or steam to produce hydrogen. For further information see Vol. 4, No. 2 of *DPIM Report*.

MAGNESIUM ACETATE
HR: 3

CAS: 142-72-3   NIOSH: AI 5600000

mf: $C_4H_6O_4 \cdot Mg$ mw: 142.41

PROP: Tetrahydrate, colorless or white deliquescent crystals; d: 1.45; mp: approx 80°; very sol in H_2O and alc. Keep container well closed.

SYNS: ACETIC ACID, MAGNESIUM SALT * MAGNESIUM DIACETATE

THR: Poison by intravenous route. See also magnesium. When heated to decomposition it emits acrid smoke and fumes.

MAGNESIUM CHLORIDE
HR: 3

CAS: 7786-30-3   NIOSH: OM 2800000

mf: Cl_2Mg mw: 95.21

PROP: Mp: 712° (rapid heating). Thin white to opaque gray granules and/or flakes, mp: 708°; bp: 1412°; d: 2.325. Sol in H_2O evolving much heat.

THR: Poison by intraperitoneal and intravenous routes. Moderately toxic by ingestion and subcutaneous routes. Mutagenic data. See also magnesium. When heated to decomposition it emits toxic fumes of Cl^-.

MAGNESIUM COMPOUNDS
HR: variable

THR: The inhalation of fumes of freshly sublimed magnesium oxide may cause metal fume fever. There is no evidence that magnesium produces true systemic poisoning. Particles of metallic magnesium or magnesium alloy which perforate the skin or gain entry through cuts and scratches may produce a severe local lesion characterized by the evolution of gas and acute inflammatory reaction, frequently with necrosis. The condition has been called a "chemical gas gangrene." Gaseous blebs may develop within 24 hours of the injury. The inflammatory response is marked at the site of injury and there may be signs of lymphangitis. The lesion is very slow to heal. The most serious hazard presented by magnesium is the danger from burns. Protection necessary for personnel handling and processing magnesium is usually no different from that which is necessary for other metals. The toxicity of magnesium compounds is usually that of the anion. Refer to magnesium and anion. See also specific compounds.

MAGNESIUM HEXAFLUOROSILI-CATE
HR: 3

CAS: 18972-56-0   NIOSH: VV 8575000
mf: $F_6Si \cdot Mg$ mw: 166.40

SYNS: FLUOSILICATE DE MAGNESIUM (FRENCH) * MAGNESIUM FLUOSILICATE * MAGNESIUM SILICOFLUORIDE

THR: Poison by ingestion and subcutaneous route. See also silicofluoric acid and magnesium compounds. When heated to decomposition it emits toxic fumes of F^-.

MAGNESIUM OXIDE
HR: 3

CAS: 1309-48-4   NIOSH: OM 3840000
mf: MgO   mw: 40.31

PROP: White, very fine powder, odorless. Very slightly sol in pure H_2O; sol in dil acids; insol in alc, mp: $2500°-2800°$; d: 3.65-3.75.

SYNS: CALCINED BRUCITE * CALCINED MAGNESIA * CALCINED MAGNESITE * MAGNESIA * MAGNESIA USTA * MAGNEZU TLENEK (POLISH) * SEAWATER MAGNESIA

OSHA PEL: TWA 15 mg/m^3

ACGIH TLV: TWA 10 mg/m^3
DFG MAK: (fume) 6 mg/m^3

THR: An experimental carcinogen. Human toxic effects of unspecified source. See also magnesium compounds. Inhalation of the fumes can produce in man a febrile reaction and a leukocytosis. Incompatible with interhalogens; phosphorus pentachloride; ClF_3.

MAGNESIUM PEROXIDE

CAS: 14452-57-4   NIOSH: OM 4100000
DOT: 1476
mf: MgO_2 mw: 56.31

PROP: White, tasteless, odorless powder. Insol in H_2O and slowly decomp evolving O_2. Sol in dil acids. Keep container closed.

SYN: MAGNESIUM PEROXIDE, SOLID (DOT)

DOT Classification: Label: Oxidizer

THR: See also peroxides, inorganic, and magnesium compounds. Flammable by chemical reaction with acidic materials and moisture; an oxidizing agent. Dangerous; reacts with reducing agents; will decompose violently in or near a fire.

MAGNESIUM PHOSPHIDE
HR: 3

CAS: 12057-74-8   NIOSH: OM 4200000
mf: Mg_3P_2 mw: 134.87

SYNS: FOSFURI DI MAGNESIO (ITALIAN) * MAGNESIUMFOSFIDE (DUTCH) * PHOSPHURE DE MAGNESIUM (FRENCH)

THR: Poison by inhalation. See also magnesium and phosphides. Reacts violently with Br_2; Cl_2; HNO_3; H_2O. When heated to decomposition it emits toxic fumes of PO_x and PH_3.

MAGNESIUM SHAVINGS

CAS: 7439-95-4   NIOSH: OM 4300000

SYNS: MAGNESIUM BORINGS * MAGNESIUM CLIPPINGS * MAGNESIUM SCALPINGS * MAGNESIUM SCRAP (DOT) * MAGNESIUM SHEET * MAGNESIUM TURNINGS

DOT Classification: Label: Flammable Solid and Dangerous When Wet

THR: See also magnesium and magnesium compounds.

MAGNESIUM SULFATE HEPTA-HYDRATE
HR: 2

mf: $MgO_4S \cdot 7H_2O$ mw: 246.48

PROP: Efflorescent crystals or powder, bitter, mp: $-7H_2O$ @ 200°; d: 1.68. Slightly sol in alc.

SYNS: EPSOM SALTS * BITTER SALTS

THR: Moderately toxic by several routes. Parenteral use or use in presence of renal insufficiency may lead to magnesium intoxication. See also magnesium compounds. An anticonvulsant and purgative. Incompatible with ethoxyethynyl alcohols.

MALEIC ACID-N-ETHYLIMIDE   HR: 3
CAS: 128-53-0   NIOSH: UX 9625000
mf: $C_6H_7NO_2$ mw: 125.14

PROP: Crystals, irr odor; mp: 45°.

SYNS: N-ETHYLMALEIMIDE * USAF B-121

THR: Poison by intraperitoneal route. Highly irritating vapors. When heated to decomposition it emits toxic fumes of NO_x.

MALEIC ANHYDRIDE   HR: 3
CAS: 108-31-6   NIOSH: ON 3675000
DOT: 2215
mf: $C_4H_2O_3$ mw: 98.06

PROP: Fused black or white crystals; mp: 52.8°, bp: 202°, flash p: 215°F (CC), d: 1.48 @ 20°/4°, autoign temp: 890°F, vap press: 1 mm @ 44.0°, vap d: 3.4, lel = 1.4%, uel = 7.1%. Sol in water @ 30°, forming maleic acid; very sltly sol in alc; sol in dioxane.

SYNS: CIS-BUTENEDIOIC ANHYDRIDE * 2,5-FURANDIONE * MALEIC ACID ANHYDRIDE * TOXILIC ANHYDRIDE

OSHA PEL: TWA 0.25 ppm
ACGIH TLV: TWA 0.25 ppm
DFG MAK: 0.2 ppm (0.8 mg/m³)
DOT Classification: ORM-A (IMO: Corrosive)

THR: An experimental carcinogen. Moderately toxic by ingestion and skin contact. Severe irritant to eyes, skin, and mucous membranes. Can cause pulmonary edema. Corrosive to skin and eyes. Combustible when exposed to heat or flame; will react with water or steam to produce heat; emits toxic fumes when heated; can react on contact with oxidizing materials. Violent reaction with alkali metals; amines; Ca(OH)$_2$; Li; K; KOH; Na; NaOH; pyridine. To fight fire, use alcohol foam. Incompatible with cations

or bases. For further information see Vol. 2, No. 3 of *DPIM Report*.

MALEIC HYDRAZIDE   HR: 3
CAS: 123-33-1   NIOSH: UR 5950000
mf: $C_4H_4N_2O_2$ mw: 112.10

PROP: Crystals, somewhat sol in water and alcohol, mp: > 300°.

SYNS: 1,2-DIHYDRO-3,6-PYRIDAZINEDIONE * ENT 18,870 * 6-HYDROXY-3(2H)-PYRIDAZINONE * MALEIC ACID HYDRAZIDE * 1,2-DIHYDROPYRIDAZINE-3,6-DIONE * N,N-MALEOYLHYDRAZINE * 1,2,3,6-TETRAHYDRO-3,6-DIOXOPYRIDAZINE

THR: An experimental carcinogen. Mutagenic data. Moderately toxic by ingestion. Can cause chronic liver damage and acute central nervous system effects. See also hydrazine. Dangerous; when heated to decomposition emits highly toxic fumes of NO_x. For further information see 1,2-dihydropyridazine-3-6 dione, Vol. 5, No. 5 of *DPIM Report*.

MALEIMIDE   HR: 3
CAS: 541-59-3   NIOSH: ON 4800000
mf: $C_4H_3NO_2$ mw: 97.08

SYNS: MALEINIMIDE * PYRROLE-2,5-DIONE * 3-PYRROLINE-2,5-DIONE

THR: An experimental teratogen. Poison by intraperitoneal and intravenous routes. When heated to decomposition it emits toxic fumes of NO_x.

MALONONITRILE   HR: 3
CAS: 109-77-3   NIOSH: OO 3150000
mf: $C_3H_2N_2$ mw: 66.07

PROP: White powder, d: 1.049 @ 34°/4°; mp: 30.5°; bp: 220°; flash p: 266°F (TOC). Sol in water, alc, ether.

SYNS: CYANOACETONITRILE * DICYANOMETHANE * MALONIC DINITRILE * NITRIL KYSELINY MALONOVE (CZECH) * USAF A-4600

THR: Poison by ingestion, subcutaneous, intravenous, intraperitoneal, and possibly other routes. Severe eye irritant. Combustible when exposed to heat or flame. To fight fire, use water, fog, spray, foam. Can spontaneously explode. When heated to decomposition it emits toxic fumes of NO_x and CN^-. Incompatible with bases.

MALTOSE
HR: 3
CAS: 69-79-4   NIOSH: OO 5250000
mf: $C_{12}H_{22}O_{11}$ mw: 342.31

PROP: Colorless needles; d: 1.540 @ 17°; mp: decomp; very sol in water; very sltly sol in cold alc; insol in ether.

SYNS: 4-(ALPHA-D-GLUCOPYRANOSIDO)-ALPHA-GLUCOPYRANOSE * 4-(ALPHA-D-GLUCOSIDO)-D-GLUCOSE * MALTOBIOSE * D-MALTOSE * MALT SUGAR * ALPHA-MALT SUGAR

THR: An experimental carcinogen. When heated to decomposition it emits acrid smoke and fumes.

MANDELIC ACID
HR: 3
CAS: 90-64-2   NIOSH: OO 6300000
mf: $C_8H_8O_3$ mw: 152.16

PROP: Large white crystals or powder, faint odor, bp: decomp, d: 1.30, mp: 117°-119°. Sol in water, alc and ether. Darkens and decomposes on prolonged exposure to light.

SYNS: AMYGDALIC ACID * AMYGDALINIC ACID * ALPHA-HYDROXY-ALPHA-TOLUIC ACID * ALPHA-HYDROXYPHENYLACETIC ACID * PARAMANDELIC ACID * PHENYLGLYCOLIC ACID * PHENYLHYDROXYACETIC ACID * RACEMIC MANDELIC ACID

THR: Poison by intramuscular route and moderately toxic by ingestion. Severe eye irritant. Continued absorption can cause kidney irritation.

MANDELIC ACID NITRILE
HR: 3
CAS: 532-28-5   NIOSH: OO 8400000
mf: C_8H_7NO mw: 133.16

PROP: Yellow viscous liquid, mp: −10°; bp: 170° decomp; d: 1.124.

SYNS: AMYGDALONITRILE * BENZALDEHYDE CYANOHYDRIN * BENZALDEHYDKYANHYDRIN (CZECH) * NITRIL KYSELINY MANDLOVE (CZECH)

THR: Poison by ingestion, intravenous, and subcutaneous routes. Mutagenic data. An eye irritant. See also nitriles. When heated to decomposition it emits toxic fumes of NO_x and CN^-.

MANGANESE
HR: 3
CAS: 7439-96-5   NIOSH: OO 9275000
af: Mn   aw: 54.94

PROP: Reddish-grey or silvery, brittle, metallic element, mp: 1260°, bp: 1900°, d: 7.20, vap press: 1 mm @ 1292°.

SYNS: COLLOIDAL MANGANESE * MANGAN (POLISH)

OSHA PEL: TWA CL 5 mg/m^3
ACGIH TWA: 5 mg/m^3 (dust)
DFG MAK: 5 mg/m^3

THR: An experimental carcinogen. Mutagenic data. Human toxicity caused by inhalation of the dust or fumes. Symptoms: languor, sleepiness, weakness, emotional disturbances, spastic gait, paralysis. See also manganese compounds. Flammable in the form of dust or powder, when exposed to flame. Moderately explosive in the form of dust, when exposed to flame. See also powdered metals. Violent reaction with (Al + air); Cl_2; F_2; H_2O_2; HNO_3; NO_2; P; SO_2. Moderately dangerous; will react with water or steam to produce hydrogen; can react with oxidizing materials. To fight fire, use special dry chemical. For further information see Vol. 1, No. 2 of *DPIM Report*.

MANGANESE ACETATE
HR: 2
CAS: 638-38-0   NIOSH: AI 5770000
mf: $C_4H_6O_4 \cdot Mn$ mw: 173.04

PROP: Pale red crystals, very sol in water and alc.

SYNS: ACETIC ACID MANGANESE(II) SALT (2:1) * DIACETYLMANGANESE * MANGANESE(2+) ACETATE * MANGANESE(II) ACETATE * MANGANESE DIACETATE * MANGANOUS ACETATE * OCTAN MANGANATY (CZECH)

THR: Moderately toxic by ingestion. See also manganese. When heated to decomposition it emits acrid smoke and fumes.

MANGANESE (II) CHLORIDE (1:2)
HR: 3
CAS: 7773-01-5   NIOSH: OO 9625000
mf: Cl_2Mn mw: 125.84

PROP: Cubic, deliquescent, pink crystals; mp: 650°, bp: 1190°, d: 2.977 @ 25°.

SYNS: MANGANESE DICHLORIDE * MANGANOUS CHLORIDE

THR: Poison by intraperitoneal, subcutaneous, intramuscular, intravenous, and parenteral routes. Mutagenic data. An experimental carcinogen. See also manganese compounds and chlorides. Reacts violently with K; Na; Zn. When heated to decomposition it emits toxic fumes of Cl^-.

MANGANESE COMPOUNDS   HR: 3

THR: Can cause central nervous system and pulmonary system damage by inhalation of fumes and dust. An experimental tumorigen. Chronic manganese poisoning is a clearly characterized disease which results from the inhalation of fumes or dusts of manganese. Exposure to heavy concentrations of dusts or fumes for as little as three months may produce the condition, but usually cases develop after 1-3 years of exposure. The central nervous system is the chief site of damage. If cases are removed from exposure shortly after the appearance of symptoms, some improvement in the patient's condition frequently occurs, though there may be some residual disturbances in gait and speech. When well established, however, the disease results in permanent disability. Exposure to dusts and fumes can possibly increase the incidence of upper respiratory infections and pneumonia. Chronic manganese poisoning begins usually with complaints of languor and sleepiness. This is followed by weakness in the legs and the development of a stolid, mask-like face, and the patient speaks with a slow monotonous voice. Then muscular twitchings appear, varying from a fine tremor of the hands to coarse, rhythmical movements of the arms, legs and trunk. Nocturnal cramps of the legs appear about the same time. There is a slight increase in tendon reflexes, ankle and patellar clonus, and a typical Parkinsonian slapping gait. The handwriting may be quite minute. The symptoms may simulate progressive bulbar paralysis, postencephalitic Parkinsonism, multiple sclerosis, amyotrophic lateral sclerosis and progressive lenticular degeneration (Wilson's Disease). Often a history of exposure is the only aid in establishing the diagnosis. Manganese compounds are common air contaminants.

MANGANESE DIOXIDE   HR: 3
CAS: 1313-1-3-9   NIOSH: OP 0350000
mf: MnO_2 mw: 86.94

PROP: Tetragonal crystals, mp: $-O_2$ @ 535°; d: 5.0. Insol in water, nitric or cold sulfuric acid.

SYNS: C.I. PIGMENT BLACK 14 * C.I. PIGMENT BROWN 8 * C.I. 77728 * MANGANESE

BLACK * CEMENT BLACK * PYROLUSITE BROWN

DOT Classification: ORM-B, Label: None

THR: Poison by inhalation, intravenous, and subcutaneous routes. A powerful oxidizer. See also manganese compounds. Flammable by chemical reaction. It must not be heated or rubbed in contact with easily oxidizable matter. Incompatible with aluminum; dirubidium acetylide; hydrogen sulfide; oxidants; potassium azide; ClF_3; H_2O_2; H_2SO_5; Na_2O_2. Moderately dangerous; keep away from heat and flammable materials.

MANGANESE OXIDE
CAS: 1317-35-7   NIOSH: OP 0895000
mf: Mn_3O_4 mw: 228.82

PROP: Brownish-black powder; d: 4.7. Insol in H_2O; sol in HCl, liberating chlorine.

SYNS: MANGANESE TETROXIDE * MANGANO-MANGANIC OXIDE * TRIMANGANESE TETRAOXIDE * TRIMANGANESE TETROXIDE

DFG MAK: 1 mg/m^3

THR: No toxicity data. See also manganese. Reacts violently @ <100°.

MANGANESE (II) OXIDE   HR: 3
CAS: 1344-43-0   NIOSH: OP 0900000
mf: MnO   mw: 70.94

PROP: Grass green powder, sol in acids, insol in water, d: 5.45, mp: 1650°, converted to Mn_3O_4 if heated in air.

SYNS: CASSEL GREEN * C.I. 77726 * MANGANESE MONOXIDE * MANGANOUS OXIDE

THR: Moderately to highly toxic by subcutaneous route. See also manganese compounds. Reacts violently with $Ca(OCl)_2$; F_2; H_2O_2.

MANGANESE (III) OXIDE   HR: 3
CAS: 1317-34-6   NIOSH: OP 0915000
mf: Mn_2O_3 mw: 157.88

PROP: Black fine powder; d: 4.50; insol in H_2O; sol in HCl evolving chlorine.

SYNS: C.I. 77727 * CASSEL BROWN * COLOGNE EARTH * COLOGNE UMBER * CULLEN EARTH * DIMANGANESE TRIOXIDE * MANGANESE MANGANATE * MANGANESE TRIOXIDE * MANGANIC OXIDE * RUBENS BROWN * WALNUT STAIN

THR: Moderately to highly toxic by subcutaneous route. See also manganese compounds.

MANGANESE (II) SULFATE
(1:1)   HR: 3
CAS: 7785-87-7   NIOSH: OP 1050000
mf: $O_4S \cdot Mn$ mw: 151.00

PROP: Reddish crystals, mp: 700°; bp: decomp @ 850°; d: 3.25. Very sol in H_2O, more so in boiling H_2O. Insol in alc.

SYNS: SULFURIC ACID, MANGANESE(2+) SALT
* MANGANOUS SULFATE

THR: Poison by ingestion and intraperitoneal routes. Mutagenic data. See also manganese compounds and sulfates. When heated to decomposition it emits toxic fumes of SO_x.

MANGANESE TRICARBONYL METHYL-
CYCLOPENTADIENYL   HR: 3
CAS: 12108-13-3   NIOSH: OP 1450000
mf: $C_9H_7MnO_3$ mw: 218.10

SYNS: COMBUSTION IMPROVER -2 * METHYL CYCLOPENTADIENYL MANGANESE TRICARBONYL * 2-METHYL CYCLOPENTADIENYL MANGANESE TRICARBONYL

THR: Poison by ingestion, inhalation and intraperitoneal routes. Moderately toxic by skin contact. See also manganese compounds and carbonyls. A skin irritant. When heated to decomposition it emits toxic fumes of CO.

MANNITOL HEXANITRATE   HR: 2
CAS: 15825-70-4   NIOSH: OP 3000000
mf: $C_6H_8N_6O_{18}$ mw: 452.17

PROP: Colorless crystals, bp: explodes @ 120°; d: 1.603 @ O°, mp: 106°-108°. Long needles in regular clusters from alc. Sol in alc, ether, insol in H_2O.

SYNS: HEXANITROL * NITRO MANNITE * NITROMANNITOL * D-MANNITOL HEXANITRATE

DOT Classification: Label: Explosive A

THR: Moderately toxic by ingestion and inhalation yielding a fall in blood pressure, which may result in weakness, headache and dizziness. Chronic exposure may produce methemoglobinemia with cyanosis. See nitrates for fire hazard. Strongly explosive when shocked or exposed to heat. Highly dangerous! See also explosives, high, and nitrates.

MANNOMUSTINE   HR: 3
CAS: 576-68-1   NIOSH: OP 2100000
mf: $C_{10}H_{22}Cl_2N_2O_4$ mw: 305.24

SYNS: 1,6-BIS(CHLOROETHYLAMINO)-1,6-BIS-DEOXY-D-MANNITOL * 1,6-BIS(CHLORO-ETHYLAMINO)-1,6-DIDEOXY-D-MANNITE * 1,6-BIS((BETA-CHLOROETHYL)AMINO)-1,6-DIDEOXY-D-MANNITOL * 1,6-BIS((2-CHLORO-ETHYL)AMINO)-1,6-DIDEOXY-D-MANNITOL * MANNIT-LOST (GERMAN) * MANNIT-MUSTARD (GERMAN) * MANNITOL NITROGEN MUSTARD

THR: Poison by intraperitoneal and intravenous routes. An experimental neoplastigen and carcinogen. When heated to decomposition it emits very toxic fumes of Cl^- and NO_x.

MANNOMUSTINE DIHYDROCHLO-
RIDE   HR: 3
CAS: 551-74-6   NIOSH: OP 2275000
mf: $C_{10}H_{23}Cl_2N_2O_4 \cdot 2ClH$ mw: 378.13

PROP: Crystals from 80% ethanol. Decomp @ 239°-241°; sol in H_2O; slightly sol in ethanol.

SYNS: 1,6-BIS-(CHLOROETHYLAMINO)-1,6-DESOXY-D-MANNITOLDIHYDROCHLORIDE * 1,6-BIS-(CHLOROETHYLAMINO)-1,6-DIDEOXY-D-MANNITEDIHYDROCHLORIDE * 1,6-DIDE-OXY-1,6-DI(2-CHLOROETHYLAMINO)-D-MANNI-TOLDIHYDROCHLORIDE * MANNITOL MUSTARD DIHYDROCHLORIDE * NSC-9698

THR: Poison by intravenous, subcutaneous, and parenteral routes. Mutagenic data. An experimental carcinogen. See also mannomustine. When heated to decomposition it emits very toxic fumes of Cl^-, NO_x, and HCl.

MAYTANSINE   HR: 3
CAS: 35846-53-8   NIOSH: OQ 2290000
mf: $C_{34}H_{46}ClN_3O_{10}$ mw: 692.21

PROP: Mp: 171°-172°. Active principle found in *Maytenus Serrata*.

SYNS: MAYT * NSC-153858

THR: An experimental teratogen. Moderately toxic by subcutaneous route. Human gastrointestinal tract effects. When heated to decomposition it emits very toxic fumes of Cl^- and NO_x.

7-MBA-3,4-DIHYDRODIOL   HR: 3
CAS: 64521-1-4-8   NIOSH: CW 1796000
mf: $C_{18}H_{16}O_2$ mw: 264.34

SYN: TRANS-3,4-DIHYDRO-3,4-DIHYDROXY-7-METHYLBENZ(a)ANTHRACENE

THR: An experimental neoplastigen by skin contact. Mutagenic data. When heated to decomposition it emits acrid smoke and irritating fumes.

MECLIZINE HYDROCHLORIDE   HR: 3
CAS: 36236-67-6   NIOSH: TL 2050000
mf: $C_{25}H_{27}ClN_2 \cdot ClH$ mw: 427.45

SYN: 1-(P-CHLORO-ALPHA-PHENYLBENZYL)-4-(M-METHYLBENZYL)PIPERAZINE HYDROCHLORIDE

THR: An experimental teratogen. When heated to decomposition it emits very toxic fumes of HCl, NO_x, and Cl^-.

MEDROXYPROGESTERONE ACETATE   HR: 3
CAS: 71-58-9   NIOSH: TU 5010000
mf: $C_{24}H_{34}O_4$ mw: 386.58

PROP: White to off-white, odorless crystalline powder, insol in water, freely sol in chloroform, sparingly sol in alcohol. Melting range 207°-209°.

SYNS: (6-ALPHA)-17-(ACETYLOXY)-6-METHYL-PREG-4-ENE-3,20-DIONE * 17-ALPHA-ACE-TOXY-6-ALPHA-METHYLPREGN-4-ENE-3,20-DIONE * 17-ACETOXY-6-ALPHA-METHYLPROGESTERONE * DEPO-PROVERA * 17-ALPHA-HYDROXY-6-ALPHA-METHYLPREGN-4-ENE-3,20-DIONE ACETATE * 17-ALPHA-HYDROXY-6-ALPHA-METHYL-PROGESTERONE ACETATE * 17-HYDROXY-6-AL-PHA-METHYLPREGN-4-ENE-3,20-DIONE ACETATE * 6-ALPHA-METHYL-17-ALPHA-ACETOXYPREGN-4-ENE-3,20-DIONE * 6-ALPHA-METHYL-17-AL-PHA-ACETOXYPROGESTERONE * 6-ALPHA-METHYL-17-ALPHA-HYDROXYPROGESTERONE ACETATE * 6-ALPHA-METHYL-4-PREGNENE-3,20-DION-17-ALPHA-OL ACETATE

THR: An experimental carcinogen, teratogen, and neoplastigen. Permitted in food for human consumption. When heated to decomposition it emits acrid smoke.

MELAMINE   HR: 2
CAS: 108-78-1   NIOSH: OS 0700000
mf: $C_3H_6N_6$ mw: 126.15

PROP: Monoclinic, colorless prisms, mp: <250°; bp: sublimes; d: 1.573 @ 250°; vap press: 50 mm @ 315°; vap d: 4.34. Sltly sol in water; very sltly sol in hot alc; insol in ether.

SYNS: CYANURAMIDE * CYANUROTRIAMIDE * CYANUROTRIAMINE * NCI-C50715 * 2,4,6-TRIAMINO-S-TRIAZINE

THR: Moderately toxic by ingestion. An eye, skin, and mucous membrane irritant. Causes dermatitis in humans. When heated to decomposition it emits toxic fumes of NO_x and CN^-.

MELIPAN   HR: 3
CAS: 3771-19-5   NIOSH: UF 6125000
mf: $C_{20}H_{22}O_3$ mw: 310.42

SYNS: 2-METHYL-2-(4-(1,2,3,4-TETRAHYDRO-1-NAPHTHALENYL)PHENOXY)PROPANOIC ACID * ALPHA-METHYL-ALPHA-(P-1,2,3,4-TETRAHY-DRONAPHTH-1-YLPHENOXY)PROPIONIC ACID * 2-METHYL-2-(4-(1,2,3,4-TETRAHYDRO-1-NAPHTHYL)PHENOXY)PROPANOIC ACID * 2-METHYL-2-(P-(1,2,3,4-TETRAHYDRO-1-NAPHTHYL)PHENOXY)PROPIONIC ACID

THR: A suspected human carcinogen. An experimental carcinogen. When heated to decomposition it emits acrid smoke and fumes.

p-MENTHANE-8-HYDROPEROXIDE   HR: 3
CAS: 80-47-7   NIOSH: OS 9450000
mf: $C_{10}H_{20}O_2$ mw: 172.30

PROP: Clear, pale yellow liquid, d: 0.910-0.925 @ 15.5°/4°.

SYN: P-MENTHANE HYDROPEROXIDE

THR: An experimental carcinogen. See also peroxides, organic. When heated to decomposition it emits acrid smoke and fumes. An irritant and powerful oxidizer.

MENTHOL   HR: 3
CAS: 89-78-1   NIOSH: OT 0350000
mf: $C_{10}H_{20}O$ mw: 156.26

PROP: Crystals or granules; peppermint taste and odor; d: 0.890 @ 15°/15°; vap press: 1 mm @ 56.0°; vap d: 5.38; mp: 41°-43°; bp: 212°; sltly sol in H_2O; very sol in alc, chloroform, ether, petroleum ether; glacial acetic acid, liquid petrolatum.

SYNS: P-MENTHAN-3-OL * NCI-C50000

THR: Poison by intravenous route. Moderately toxic by ingestion, subcutaneous, and intraperitoneal routes. An eye irritant. When heated to decomposition it emits acrid smoke and fumes. Incompatible with (medical) butylchloral hydrate + camphor; phenol; chloral hydrate; exalgine; betanaphthol; resorcinol or thymol in trituration, potassium permanganate; chromium trioxide; pyrogallol.

dl-MENTHOL   HR: 2
CAS: 15356-70-4   NIOSH: OT 0525000
mf: $C_{10}H_{20}O$ mw: 156.30

SYNS: 4-ISOPROPYL-1-METHYLCYCLOHEXAN-3-OL * 3-P-MENTHOL * DL-3-P-MENTHANOL * MENTHOL RACEMIC

THR: Moderately toxic by ingestion, intraperitoneal, and subcutaneous routes. A skin irritant. See also menthol and 1-menthol. When heated to decomposition it emits acrid smoke and fumes.

1-MENTHOL   HR: 3
CAS: 2216-51-5   NIOSH: OT 0700000
mf: $C_{10}H_{20}O$ mw: 156.30

PROP: Found in high concentrations in oils of Peppermint (*Mentha Piperita*) and Japanese Mint Oil (*Mentha Arvensis*), and in lower concentrations in Reunion Geranium Oil and in a large number of essential oils; prepared by isolation from *Mentha Arvensis* Oils.

SYN: U.S.P. MENTHOL

THR: Poison by intravenous route. Moderately toxic by ingestion, intraperitoneal, and subcutaneous routes. When heated to decomposition it emits acrid smoke and fumes.

MEPHEXAMIDE   HR: 3
CAS: 1227-6-1-8   NIOSH: AB 7175000
mf: $C_{15}H_{24}N_2O_3$ mw: 280.41

SYNS: N-(2-(DIETHYLAMINO)ETHYL)-2-(P-METHOXYPHENOXY)ACETAMIDE * 2-(P-METHOXYPHENOXY)-N-(2-(DIETHYLAMINO)ETHYL)ACETAMIDE * MEXEPHENAMIDE

THR: Poison by ingestion and intravenous routes. When heated to decomposition it emits toxic fumes of NO_x.

MEPHOSFOLAN   HR: 3
CAS: 950-1-0-7   NIOSH: JP 1050000
mf: $C_8H_{16}NO_3PS_2$ mw: 269.34

SYNS: CYCLIC PROPYLENE (DIETHOXYPHOSPHINYL)DITHIOIMIDOCARBONATE * P,P-DI-

ETHYL CYCLIC PROPYLENE ESTER OF PHOSPHONODITHIOIMIDOCARBONIC ACID * DIETHYL (4-METHYL-1,3-DITHIOLAN-2-YLIDENE)-PHOSPHOROAMIDATE * ENT-25,991 * 2-(DIETHOXYPHOSPHINYLIMINO)-4-METHYL-1,3-DITHIOLANE

THR: Poison by ingestion and skin contact. When heated to decomposition it emits very toxic fumes of NO_x, PO_x and SO_x. For further information see Vol. 3, No. 1 of *DPIM Report*.

MEPROSCILLARIN   HR: 3
NIOSH: EI 3450000
mf: $C_{29}H_{44}O_{10}$ mw: 552.73

SYN: CLIFT

THR: Poison by ingestion and intravenous routes. When heated to decomposition it emits acrid smoke and fumes.

MERCAPTANS   HR: 3
THR: Generally have a very offensive odor which may cause nausea and headache. High concentrations can produce unconsciousness with cyanosis, cold extremities, and rapid pulse. A common air contaminant. Dangerous; when heated to decomposition, they almost always emit highly toxic fumes of SO_x; they will react with water, steam or acids to produce toxic and flammable vapors; they can react violently with powerful oxidizers such as $Ca(OCl)_2$.

alpha-MERCAPTOACETANILIDE   HR: 3
CAS: 4822-44-0   NIOSH: AE 4375000
mf: C_8H_8NOS mw: 166.23

SYNS: 2-MERCAPTOACETANILIDE * THIOGLYCOLANILIDE * THIOGLYCOLIC ACID ANILIDE * USAF EK-6583

THR: Poison by intraperitoneal route. When heated to decomposition it emits very toxic fumes of NO_x and SO_x.

2-MERCAPTOACETIC ACID   HR: 3
CAS: 68-11-1   NIOSH: AI 5950000
DOT: 1940
mf: $C_2H_4O_2S$ mw: 92.12

PROP: Liquid, strong odor, mp: $-16.5°$, bp: 108° @ 15 mm. Misc with water alc, ether, chloroform, benzene.

SYNS: ACIDE THIOGLYCOLIQUE (FRENCH) * MERCAPTOACETIC ACID * ALPHA-MERCAPTOACETIC ACID * 2-THIOGLYCOLIC ACID

* THIOGLYCOLIC ACID (DOT) * THIOVANIC ACID * USAF CB-35

DOT Classification: Label: Corrosive

THR: Poison by ingestion, intraperitoneal, intravenous, and dermal routes. A corrosive and irritant to skin, eyes, and mucous membranes. See also hydrogen sulfide which is readily evolved by this compound. When heated to decomposition it emits toxic fumes of SO_x.

MERCAPTODIACETIC ACID   HR: 3
CAS: 123-93-3   NIOSH: AJ 6475000
mf: $C_4H_6O_4S$ mw: 150.16

PROP: A white powder, mp: 128°.

SYNS: (CARBOXYMETHYLTHIO)ACETIC ACID * DIMETHYLSULFIDE-ALPHA,ALPHA'-DICARBOXYLIC ACID * THIODIGLYCOLIC ACID * BETA,BETA'-THIODIGLYCOLIC ACID * 2,2'-THIODIGLYCOLIC ACID * THIODIGLYCOLLIC ACID * USAF CB-36 * USAF E-2

THR: Poison by intraperitoneal route. See also 2-mercapto acid. Dangerous; when heated to decomposition or on contact with acid or acid fumes it emits toxic fumes of SO_x.

2-MERCAPTOETHANOL   HR: 3
CAS: 60-24-2   NIOSH: KL 5600000
mf: C_2H_6OS mw: 78.14

PROP: Water white, mobile liquid, bp: 157°-158° (decomp) @ 742 mm, flash p: 165°F (COC), d: 1.1168 @ 20°/20°, vap press: 1.0 mm @ 20°, vap d: 2.69. Pure liquid is misc with H_2O, alc, ether and benzene.

SYNS: 1-ETHANOL-2-THIOL * 2-HYDROXYETHYL MERCAPTAN * BETA-MERCAPTOETHANOL * 2-THIOETHANOL * THIOGLYCOL * USAF EK-4196

THR: Poison by ingestion, skin contact, intravenous, and intraperitoneal routes. A severe eye and skin irritant. Mutagenic data. Flammable when exposed to heat, flame or oxidizers. Dangerous; when heated to decomposition it emits highly toxic fumes of oxides of sulfur; can react with oxidizing materials. To fight fire, use alcohol foam, CO_2, dry chemical.

beta-MERCAPTOETHYLAMINE DISULFIDE   HR: 3
CAS: 51-85-4   NIOSH: KR 7175000
mf: $C_4H_{12}N_2S_2$ mw: 152.30

SYNS: BIS(BETA-AMINOETHYL)DISULFIDE * CYSTAMINE * CYSTEINAMINE DISULFIDE * CYSTINAMIN (GERMAN) * CYSTINEAMINE * DECARBOXYCYSTINE * BETA,BETA'-DIAMINODIETHYL DISULFIDE * 2,2'-DITHIOBIS(ETHYLAMINE) * 2-MERCAPTOETHYLAMINE (OXIDIZED)

THR: Poison by intraperitoneal and subcutaneous routes. Mutagenic data. When heated to decomposition it emits very toxic fumes of NO_x and SO_x.

2-MERCAPTO-1-METHYLIMIDAZOLE   HR: 3
CAS: 60-56-0   NIOSH: NI 8615000
mf: $C_4H_6N_2S$ mw: 114.18

SYNS: MERCAPTAZOLE * METHIAMAZOLE * 1-METHYLIMIDAZOLE-2-THIOL * 1-METHYL-2-MERCAPTOIMIDAZOLE * THIAMAZOLE * USAF EL-30

THR: An experimental neoplastigen. Moderately toxic (acute) by intraperitoneal route. When heated to decomposition it emits very toxic fumes of NO_x and SO_x.

MERCAPTOPURINE RIBONUCLEOSIDE   HR: 3
CAS: 4988-64-1   NIOSH: UP 0700000
mf: $C_{10}H_{12}N_4O_4S$ mw: 284.32

SYNS: 6-MERCAPTOPURINE RIBOSIDE * NSC 4911 * RIBOFURANOSIDE, 9H-PURINE-6-THIOL-9

THR: An experimental teratogen. Moderately toxic by intraperitoneal route. When heated to decomposition it emits very toxic fumes of SO_x and NO_x.

p-MERCAPTO SULFADIAZINE   HR: 3
CAS: 67479-03-2   NIOSH: OU 7700000

SYN: USAF LO-3

THR: Poison by intraperitoneal route. When heated to decomposition it emits very toxic fumes of SO_x and NO_x.

d,3-MERCAPTOVALINE   HR: 3
CAS: 52-67-5   NIOSH: YV 9425000
mf: $C_5H_{11}NO_2S$ mw: 149.23

SYNS: (S)-PENICILLAMIN * REDUCED-D-PENICILLAMINE * DIMETHYLCYSTEINE * PENICILLAMINE * D-PENICILLAMINE

THR: An experimental teratogen and carcinogen. Human systemic and skin (systemic) effects. When heated to decomposition it emits very toxic fumes of NO_x and SO_x.

MERCUMATILIN SODIUM HR: 3
CAS: 8018-15-3   NIOSH: DJ 2360000
mf: $C_{14}H_{13}HgO_6 \cdot C_7H_8N_4O_2 \cdot Na$ mw: 681.04

SYNS: 2H-1-BENZOPYRAN-3-CARBOXYLIC ACID, 8-(3-(HYDROXYMERCURI)-2-METHOXYPROPYL)-2-OXO-, SODIUM SALT, COMPOUND WITH THEOPHYLLINE (1:1) * CUMERTILIN SODIUM * 8-(GAMMA-HYDROXYMERCURI-BETA-METHOXYPROPYL)-3-COUMARINCARBOXYLICACID THEOPHYLLINE SODIUM

THR: A poison by intravenous, intramuscular, and subcutaneous routes. Moderately toxic by ingestion. See also mercury compounds. When heated to decomposition it emits very toxic fumes of Hg, Na_2O and NO_x.

MERCURIC ACETATE HR: 3
CAS: 1600-27-7   NIOSH: AI 8575000
DOT: 1629
mf: $C_4H_6O_4 \cdot Hg$ mw: 318.69

PROP: White cryst or powder, d: 3.280. Slight acetic odor, sensitive to light; mp: 178°-180° (overheating causes decomp). Sol in alc. Keep stoppered and protected from light.

SYNS: ACETIC ACID, MERCURY(2+) SALT * BIS(ACETYLOXY)MERCURY * DIACETOXYMERCURY * MERCURIACETATE * MERCURIC DIACETATE * MERCURY ACETATE * MERCURY(2+) ACETATE * MERCURY (II) ACETATE * MERCURY DIACETATE * MERCURYL ACETATE

DOT Classification: Label: Poison

THR: Poison by ingestion, intravenous, and subcutaneous routes. An experimental teratogen. See also mercury compounds. When heated to decomposition it emits toxic fumes of Hg. For further information see Vol. 1, No. 3 of *DPIM Report*.

MERCURIC OXIDE HR: 3
CAS: 21908-53-2   NIOSH: OW 8750000
DOT: 1641
mf: HgO   mw: 216.59

PROP: Heavy, bright, orange-red or orange-yellow powder, mp: decomp @ 500°, d: 11.14.

Practically insol in H_2O; sol in dil HCl or HNO_3. Protect from light.

SYNS: MERCURY (II) OXIDE * C.I. 77760 * MERCURIC OXIDE, RED * MERCURIC OXIDE, SOLID (DOT) * MERCURIC OXIDE, YELLOW * RED OXIDE OF MERCURY * RED PRECIPITATE * OXYDE DE MERCURE (FRENCH) * YELLOW MERCURIC OXIDE * YELLOW OXIDE OF MERCURY * YELLOW PRECIPITATE

DOT Classification: Label: Poison

THR: Poison by ingestion and intramuscular routes. See mercury compounds, inorganic. Flammable by chemical reactions; an oxidizer. Reacts violently with Cl_2; hydrazine hydrate; H_2O_2; hypophosphorus acid; (I + CH_3OH + C_2H_5OH); Mg; P; phospham; NaK; S; acetyl nitrate; butadiene; hydrocarbons; S_2Cl_2; H_2S_3; methanethiol; reductants, non-metals. Dangerous; when heated to decomposition, emits highly toxic fumes of mercury; can react with reducing materials.

MERCURIC SULFOCYANATE HR: 3
CAS: 592-85-8   NIOSH: XL 1550000
DOT: 1646
mf: $C_2HgN_2S_2$ mw: 316.79

PROP: White odorless powder; slightly sol in cold H_2O; more sol in boiling H_2O (decomp); sol in dil HCl. Protect from light.

SYNS: MERCURIC SULFOCYANIDE * MERCURIC THIOCYANATE

DOT Classification: Label: Poison

THR: A poison. See also mercury compounds and thiocyanates. When heated to decomposition it emits very toxic fumes of Hg, NO_x, SO_x, and CN^-.

MERCURIPHENYL NITRATE HR: 3
CAS: 55-68-5   NIOSH: OW 8400000
mf: $C_6H_5HgNO_3$ mw: 339.71

PROP: Crystals; insol in cold water; mp: 176°-186°.

SYNS: NITRIC ACID, PHENYLMERCURY SALT * PHENYLMERCURIC NITRATE * PHENYLMERCURY NITRATE

THR: Poison by intravenous and subcutaneous routes. See also mercury compounds and nitrates. When heated to decomposition it emits very toxic fumes of Hg and NO_x.

MERCUROCHROME   HR: 3
CAS: 129-16-8   NIOSH: LM 5250000
mf: $C_{20}H_{10}Br_2HgO_6 \cdot 2Na$ mw: 752.69

SYNS: ASEPTICHROME * 2,7-DIBROMO-4-HY-
DROXYMERCURIFLUORESCEINE DISODIUM SALT
* DISODIUM 2,7-DIBROM-4-HYDROXY-MERCURI-
FLUORESCEIN * DISODIUM 2',7'-DIBROMO-4'-
(HYDROXYMERCURY)FLUORESCEIN * FLAVUROL
* FLUOROCHROME * GALLOCHROME
* GYNOCHROME * MERBROMIN * MERCURA-
NINE * MERCUROCHROME-220 SOLUBLE
* MERCUROCOL * MERCUROME * MERCURO-
PHAGE * PLANOCHROME

THR: Poison by ingestion, intravenous, and subcutaneous routes. Relatively non-irritating and
nontoxic to damaged skin or tissue. An antiseptic. See also mercury compounds, organic.
When heated to decomposition it emits very
toxic fumes including fumes of Na_2O, Br^-, and
Hg.

MERCUROUS CHLORIDE   HR: 3
CAS: 7546-30-7   NIOSH: OV 8750000
mf: Cl_2Hg_2 mw: 472.09

PROP: White, odorless, tasteless, heavy powder or crystals. Sunlight causes decomposition
into mercuric chloride and metallic Hg. Insol
in H_2O, alc and ether. Protect from light. Subl
@ 400°; d: 7.150.

SYNS: MERCURY (I) CHLORIDE * C.I. 77764
* CALOMEL * CALOMELANO (ITALIAN)
* CHLORURE MERCUREUX (FRENCH) * CLO-
RURO MERCUROSO (ITALIAN) * KALOMEL (GER-
MAN) * MERCUROCHLORIDE (DUTCH)
* MERCURY MONOCHLORIDE * MERCURY PRO-
TOCHLORIDE * MILD MERCURY CHLORIDE
* QUECKSILBER(I)-CHLORID (GERMAN)
* SUBCHLORIDE OF MERCURY

THR: Poison by ingestion. Mutagenic data. See
also mercury compounds. When heated to decomposition it emits very toxic fumes of Cl^-
and Hg. Medically incompatible with bromides;
iodides; alkali chlorides; sulfates; sulfites; carbonates; hydroxides; lime water; acacia; ammonia; golden antimony sulfide; cocaine; cyanides;
copper salts; hydrogen peroxide; iodine; iodoform; lead salts; silver salts; soap; sulfides.

MERCURY   HR: 3
CAS: 7439-97-6   NIOSH: OV 4550000
DOT: 2809
af: Hg   aw: 200.59

PROP: Silvery liquid, metallic element; mp:
−38.89°, bp: 356.9°, d: 13.546, vap press: 1
mm @ 126.2°, vap press: @ 25° = 2×10^{-3}
mm.

SYNS: COLLOIDAL MERCURY * KWIK (DUTCH)
* MERCURE (FRENCH) * MERCURIO (ITALIAN)
* MERCURY, METALLIC (DOT) * NCI-C60399
* QUECKSILBER (GERMAN) * QUICK SILVER
* RTEC (POLISH)

OSHA PEL: TWA CL 1 mg/$10m^3$
ACGIH TLV: TWA 0.05 mg/m^3
DFG MAK: 0.01 ppm (0.1 mg/m^3); BAT: (inor-
ganic and metallic) blood 5 ug/dl, urine 200
ug/l; (organic) blood 10 ug/dl
DOT Classification: ORM-B (IMO:
Corrosive), Label: None

THR: Poison by inhalation. An experimental
carcinogen. Human gastrointestinal tract and
central nervous system effects. See also mercury
compounds. Reacts violently with acetylene;
NH_3; BPI_2; Cl_2; ClO_2; CH_3N_3; Na_2C_2; nitromethane; (butyne diol + acid). Incompatible
with acetylenic compounds; ammonia; boron
diiodophosphide; ethylene oxide; metals; methyl
azide; methylsilane, oxygen; oxidants; tetracarbonylnickel, oxygen. For further information
see Vol. 1, No. 3 of *DPIM Report*.

MERCURY AMIDE CHLORIDE   HR: 3
CAS: 10124-48-8   NIOSH: OV 7020000
DOT: 1630
mf: ClH_2HgN mw: 252.07

PROP: White lumps or powder.

SYNS: AMINOMERCURIC CHLORIDE * MERCU-
RIC AMMONIUM CHLORIDE, SOLID * MERCURIC
CHLORIDE, AMMONIATED * MERCURY AMINE
CHLORIDE * MERCURY AMMONIATED
* WHITE MERCURY PRECIPITATED * WHITE
PRECIPITATE

DOT Classification: Label: Poison

THR: A poison. See also mercury compounds.
When heated to decomposition it emits very
toxic fumes of Cl^-, NO_x, and Hg.

MERCURY(II)-orthoARSENATE   HR: 3
CAS: 7784-37-4   NIOSH: OV 7040000
mf: $AsHO_4 \cdot Hg$ mw: 340.52

PROP: Yellow powder; mp: decomp. Insol in
H_2O, sol in HCl or HNO_3.

SYN: MERCURIC ARSENATE

THR: A poison. See also mercury and arsenic compounds. When heated to decomposition it emits very toxic fumes of Hg and As.

MERCURY (II) BENZOATE   HR: 3
CAS: 583-15-3   NIOSH: OV 7060000
DOT: 1631
mf: $C_{14}H_{10}O_4 \cdot Hg$ mw: 442.83

PROP: White crystalline powder; odorless, mp: 165°. Very sol in NaCl soln; sltly sol in alc. Protect from light.

SYNS: MERCURIC BENZOATE * MERCURIC BENZOATE, SOLID (DOT)

DOT Classification: Label: Poison

THR: A poison. See also mercury compounds. When heated to decomposition it emits toxic fumes of Hg.

MERCURY (I) BROMIDE (1:1)   HR: 3
CAS: 10031-18-2   NIOSH: OV 7410000
DOT: 1634
mf: BrHg   mw: 280.50

PROP: White-yellow tetragonal cryst or powder; odorless; d: 7.307; vap d: 19.3. Darkens on exposure to light. Sublimes @ approx 390° (decomp). Insol in H_2O, alc, ether; decomp by hot HCl or alkali bromides. Protect from light.

SYN: MERCUROUS BROMIDE, SOLID (DOT)

DOT Classification: Label: Poison

THR: A poison. See also mercury compounds and bromides. When heated to decomposition it emits very toxic fumes of Br^- and Hg.

MERCURY (II) BROMIDE (1:2)   HR: 3
CAS: 7789-47-1   NIOSH: OV 7415000
DOT: 1634
mf: Br_2Hg mw: 360.41

PROP: White crystals or cryst powder. Sensitive to light; mp: 237°; bp: 322° (sublimes); d: 6.109 @ 25°; vap press: 1 mm @ 136.5°; sublimes @ higher temp; very sol in hot alc, methanol, HCl, HBr, alkali bromide solns; slightly sol in chloroform.

SYNS: MERCURIC BROMIDE * MERCURIC BROMIDE, SOLID (DOT)

DOT Classification: Label: Poison

THR: A poison. See also mercury compounds and bromides. When heated to decomposition it emits very toxic fumes of Br^- and Hg. Incompatible with indium; Na; K.

MERCURY COMPOUNDS, INORGANIC   HR: 3
THR: Mercury is a general protoplasmic poison; after absorption it circulates in the blood and is stored in the liver, kidneys, spleen and bone. In industrial poisoning, the principal effect is upon the central nervous system and upon the mouth and gums. The cardinal symptoms of industrial mercury poisoning are stomatitis, tremors, and psychic disturbances. Usually the first complaints are of excessive salivation and pain on chewing; in severe cases there may be gingivitis, with loosening of the teeth, and a dark line on the gum margins, resembling the ''lead line.'' The psychic disturbance (so called ''erethism'') includes such changes as loss of memory, insomnia, lack of confidence, irritability, vague fears and depression. The dermatitis produced by fulminate of mercury takes the form of small, discrete ulcers on the exposed parts, and is usually accompanied by conjunctivitis and inflammation of the mucous membranes of the nose and throat. In humans it is readily absorbed by respiratory tract (elemental mercury vapor, mercury composed dusts) intact skin, and gastrointestinal tract, although occasional incidental swallowing of metallic mercury is without harm. Spilled and heated elemental mercury is particularly hazardous. A number of mercury compounds, in addition to the fulminate, can cause skin irritation and be absorbed through the skin. They are strong allergens and common air contaminants. Acute Toxicity: Soluble salts have violent corrosive effects on skin and mucous membranes; severe nausea, vomiting, abdominal pain, bloody diarrhea, kidney damage; death usually within 10 days. Dangerous; when heated to decomposition emits toxic fumes of Hg.

MERCURY COMPOUNDS, ORGANIC   HR: 3
DFG MAK: 0.01 mg/m^3

THR: The customary grouping of all organic mercurials in a single category is not fully justified by the toxicity of the compounds. Alkyl mercurials have very high toxicity; aryl compounds, particularly the phenyls, are much less toxic, and the organomercurials used in therapeutics are less toxic. The alkyls and aryls commonly cause skin burns and other forms of irritation, and both can be absorbed through the skin. Fatal poisoning has occurred due to exposure to alkyl mercurials and permanent damage to the brain has been reported. Phenyl mercurials appear to be no more toxic than metallic mercury. Organic mercury compounds, like organic lead compounds, seem to have an affinity for lipoid-containing organs, resulting in central nervous system disturbances such as from tetraethyl lead. These are common air contaminants. Dangerous; when heated to decomposition they emit highly toxic fumes of mercury.

MERCURY (II) CYANIDE   HR: 3
CAS: 592-04-1   NIOSH: OW 1515000
mf: C_2HgN_2 mw: 252.63

PROP: Colorless, odorless, transparent prisms, darkened by light; decomp @ 320°; d: 3.996. Slightly sol in ether.

SYNS: CYANURE DE MERCURE (FRENCH) * MERCURIC CYANIDE, SOLID (DOT)

DOT Classification: Label: Poison

THR: Poison by ingestion, intravenous, and intraperitoneal routes. Human central nervous system and systemic effects. See also cyanides and mercury compounds. Hydrolyzes to toxic fumes. When heated to decomposition it emits very toxic fumes of Hg, NO_x, and CN^-. Incompatible with fluorine; hydrogen cyanide; magnesium; sodium nitrite. For further information see Vol. 6, No. 1 of *DPIM Report*.

MERCURY FULMINATE (DRY)   HR: 3
CAS: 628-86-4   NIOSH: OW 4050000
mf: $C_2HgN_2O_2$ mw: 284.63

PROP: White solid; mp: explodes; d: 4.42.

SYNS: FULMINATE OF MERCURY, DRY * MERCURY FULMINATE (DOT)

DOT Classification: Forbidden

THR: See also mercury compounds. Dangerously flammable; should be kept moist until used. Dangerous explosion hazards; see also fulminates. When heated to decomposition it emits very toxic fumes of Hg and NO_x. Self-explodes. Incompatible with sulfuric acid.

MERCURY FULMINATE (WET)   HR: 3
CAS: 628-86-4   NIOSH: OW 4055000
DOT: 0135
mf: $C_2HgN_2O_2$ mw: 284.63

SYNS: FULMINATE OF MERCURY, WET * INITIATING EXPLOSIVE FULMINATE OF MERCURY (DOT)

DOT Classification: Label: Explosive A

THR: An explosive. It can be kept more safely in wet form for use. See also mercury fulminate (dry). When heated to decomposition it emits very toxic fumes of Hg and NO_x.

MERCURY (I) GLUCONATE   HR: 3
CAS: 63937-14-4   NIOSH: OW 4060000
DOT: 1637
mf: $C_6H_{11}O_7 \cdot Hg$ mw: 395.76

PROP: White solid.

SYNS: MERCUROUS GLUCONATE * MERCUROUS GLUCONATE, SOLID (DOT)

DOT Classification: Poison B, Label: Poison

THR: A poison. See also mercury. When heated to decomposition it emits toxic fumes of Hg.

MERCURY (I) IODIDE   HR: 3
CAS: 7783-30-4   NIOSH: OW 5075000
DOT: 1638
mf: HgI   mw: 327.49

PROP: Yellow tetragonal crystals or amorphous powder; heavy, odorless, mp: sublimes @ 140°; d: 7.70. mp: 290° when rapidly heated (partial decomp); insol in H_2O, alc, ether; sol in solns of mercurous or mercuric nitrates.

SYNS: IODURE DE MERCURE (FRENCH) * MERCUROUS IODIDE * MERCUROUS IODIDE, SOLID (DOT) * MERCURY PROTOIODIDE * YELLOW MERCURY IODIDE

DOT Classification: Poison B, Label: Poison

THR: Poison by ingestion and intraperitoneal routes. See also mercury and iodides. When heated to decomposition it emits very toxic fumes of Hg and I^-. Medically incompatible with soluble iodides.

MERCURY (II) IODIDE   HR: 3
CAS: 7774-29-0   NIOSH: OW 5250000
DOT: 1638
mf: HgI_2 mw: 454.39

PROP: Scarlet red, heavy, odorless, almost tasteless powder. Sensitive to light; d: 6.28; mp: 259°; bp: approx 350° (sublimes); very sol in alkali iodides, $HgCl_2$, $Na_2S_2O_3$.

SYNS: HYDRARGYRUM BIJODATUM (GERMAN) * MERCURIC IODIDE * MERCURIC IODIDE, RED

* MERCURIC IODIDE, SOLID (DOT) * MERCURY BINIODIDE * RED MERCURIC IODIDE

DOT Classification: Poison B, Label: Poison

THR: Poison by ingestion and intraperitoneal routes. See also mercury and iodides. When heated to decomposition it emits very toxic fumes of Hg and I⁻. Incompatible with chlorine trifluoride.

MERCURYMETHYLCHLORIDE   HR: 3
CAS: 115-09-3   NIOSH: OW 1225000
mf: CH₃ClHg   mw: 251.08

PROP: White crystals, characteristic odor, d: 4.063; mp: 170°.

SYNS: CHLOROMETHYLMERCURY * METHYL-MERCURIC CHLORIDE * METHYLMERCURY CHLORIDE * MONOMETHYL MERCURY CHLORIDE

OSHA PEL: TWA 10 ug(Hg)/m³; CL 40

THR: Poison by ingestion, intramuscular, intravenous, and intraperitoneal routes. Mutagenic data. An experimental carcinogen and teratogen. See also mercury. When heated to decomposition it emits very toxic fumes of Cl⁻ and Hg.

MERCURY MONOACETATE   HR: 3
CAS: 631-60-7   NIOSH: AI 8570000
DOT: 1629
mf: C₂H₃O₂ • Hg   mw: 259.64

PROP: Colorless scales or plates, mp: decomp. Sol in dil acetic acid; insol in alc, ether.

SYNS: MERCUROUS ACETATE * MERCUROUS ACETATE, SOLID (DOT) * MERCURY ACETATE

DOT Classification: Poison B, Label: Poison

THR: A poison. See also mercury compounds. When heated to decomposition it emits toxic fumes of Hg.

MERCURY (I) NITRATE (1:1)   HR: 3
CAS: 10415-75-5   NIOSH: OW 8000000
DOT: 1627
mf: NO₃ • Hg   mw: 262.60

SYNS: MERCUROUS NITRATE, SOLID (DOT) * NITRATE MERCUREUX (FRENCH) * NITRIC ACID, MERCURY(I) SALT

DOT Classification: Label: Oxidizer

THR: Poison by ingestion and intraperitoneal routes. See also mercury compounds. When

heated to decomposition it emits very toxic fumes of Hg and NO$_x$. Incompatible with phosphorus. A powerful oxidizer. For further information see mercurous nitrate, Vol. 6, No. 3 of *DPIM Report*.

MERCURY (II) NITRATE (1:2)   HR: 3
CAS: 10045-94-0   NIOSH: OW 8225000
mf: N₂O₆ • Hg   mw: 324.61

PROP: White-yellowish, deliquescent powder, mp: 79°; bp: decomp; d: 4.39.

SYNS: MERCURIC NITRATE * MERCURY NITRATE * NITRIC ACID, MERCURY(II) SALT * MERCURY PERNITRATE * NITRATE MERCURIQUE (FRENCH)

THR: Poison by intraperitoneal and subcutaneous routes. See also mercury compounds, inorganic, and nitrates. Forms a sensitive explosive product with acetylene; ethanol; PH₃; S; reacts violently with hypophosphoric acid; unsaturates; aromatics; phosphine. When heated to decomposition it emits very toxic fumes of Hg and NO$_x$.

MERCURY (I) OXIDE   HR: 3
CAS: 21908-53-2   NIOSH: OW 8700000
DOT: 1641
mf: Hg₂O   mw: 417.22

PROP: Black to grayish-black powder, mp: decomp @ 100°; d: 9.8. Insol in H₂O, sol in HNO₃; protect from light.

SYN: MERCUROUS OXIDE, BLACK, SOLID (DOT)

DOT Classification: Poison B, Label: Poison

THR: A poison. See also mercury compounds, inorganic. Fire Hazard: Moderate by chemical reaction; an oxidizer. Reacts violently with H₂O₂; K; Na; S; (H₂S + BaO + air). Dangerous; when heated to decomposition emits highly toxic fumes of mercury; can react with reducing materials. Incompatible with alkali metals; chlorine; ethylene; hydrogen peroxide; non-metals.

MERCURY (I) SULFATE   HR: 3
CAS: 7783-36-0   NIOSH: OX 0480000
DOT: 1628
mf: O₄S • 2Hg   mw: 497.24

PROP: White crystalline powder, mp: decomp; d: 7.56. Sltly sol in H₂O; sol in dil HNO₃.

SYN: MERCUROUS SULFATE, SOLID (DOT)

DOT Classification: Poison B, Label: Poison

THR: A poison. See also mercury compounds. When heated to decomposition it emits very toxic fumes of Hg and SO_x.

MERCURY(II) SULFATE (1:1)   HR: 3
CAS: 7783-35-9   NIOSH: OX 0500000
DOT: 1633
mf: $O_4S \cdot Hg$ mw: 296.65

PROP: White crystalline powder. Odorless; mp: decomp; d: 6.47. Sol in HCl, hot dilute H_2SO_4, concd solns of NaCl. Protect from light.

SYNS: MERCURIC SULFATE, SOLID (DOT) * MERCURY BISULFATE * MERCURY PERSULFATE * SULFATE MERCURIQUE (FRENCH) * SULFURIC ACID, MERCURY(2+) SALT (1:1)

DOT Classification: Poison B, Label: Poison

THR: Poison by ingestion. See also mercury compounds. When heated to decomposition it emits very toxic fumes of Hg and SO_x. For further information see mercury (II) sulfate, Vol. 6, No. 1 of *DPIM Report*.

MERTHIOLATE SODIUM   HR: 3
CAS: 54-64-8   NIOSH: OV 8400000
mf: $C_9H_9HgO_2S \cdot Na$ mw: 404.82

SYNS: ((O-CARBOXYPHENYL)THIO)ETHYLMERCURY SODIUM SALT * O-(ETHYLMERCURITHIO)BENZOIC ACID SODIUM SALT * ETHYLMERCURITHIOSALICYLIC ACID SODIUM SALT * MERCUROTHIOLATE * MERTHIOLATE SALT * SODIUM ETHYLMERCURIC THIOSALICYLATE * SODIUM-O-(ETHYLMERCURITHIO)BENZOATE * SODIUM ETHYLMERCURITHIOSALICYLATE * SODIUM MERTHIOLATE

THR: Poison by ingestion, subcutaneous, intravenous, and possibly other routes. An experimental neoplastigen. An eye irritant. See also mercury compounds. When heated to decomposition it emits very toxic fumes of Hg and SO_x.

MESCALINE   HR: 3
CAS: 54-04-6   NIOSH: SI 2625000
mf: $C_{11}H_{17}NO_3$ mw: 211.29

PROP: Crystals; mp: 35°-36°; bp: 180° @ 11 mm; Mod sol in H_2O, sol in alc, chloroform, benzene, practically insol in ether, petroleum ether.

SYNS: MEZCALINE * 3,4,5-TRIMETHOXYPHENETHYLAMINE

THR: Poison by intraperitoneal, parenteral, and possibly other routes. An experimental terato-

gen. Human central nervous system effects by ingestion, intravenous, and intramuscular routes. A drug of abuse. When heated to decomposition it emits toxic fumes of NO_x.

MESCALINE HYDROCHLORIDE   HR: 3
CAS: 832-92-8   NIOSH: SI 2800000
mf: $C_{11}H_{17}NO_3 \cdot ClH$ mw: 247.75

PROP: Needles; mp: 181°; sol in H_2O, alc.

SYNS: 3,4,5-TRIMETHOXYPHENETHYLAMINE HYDROCHLORIDE * 3,4,5-TRIMETHOXY-BETA-PHENYLETHYLAMINE HYDROCHLORIDE

THR: Poison by intravenous, intraperitoneal, and subcutaneous routes. Moderately toxic by ingestion. See also mescaline. When heated to decomposition it emits very toxic fumes of HCl and NO_x.

MESITYLENE   HR: 3
CAS: 108-67-8   NIOSH: OX 6825000
mf: C_9H_{12} mw: 120.21

PROP: A liquid, peculiar odor. Insol in water; misc in alc, benzene and ether, mp: −44.8°; d: 0.8637 @ 20°/4°; bp: 164.7°; autoign temp: 1022°F.

SYNS: SYM-TRIMETHYLBENZENE * 1,3,5-TRIMETHYLBENZENE * TRIMETHYLBENZOL

THR: Moderately toxic by inhalation and intraperitoneal routes. Human central nervous system effects. Reports of leucopenia, thrombocytopenia on experimental animals. Violent reaction with HNO_3. Flammable. To fight fire, use water spray, fog, foam, CO_2. When heated to decomposition it emits acrid smoke and fumes.

MESITYL OXIDE   HR: 3
CAS: 141-79-7   NIOSH: SB 4200000
DOT: 1229
mf: $C_6H_{10}O$ mw: 98.16

PROP: Oily colorless liquid, strong odor, mp: −59°, bp: 130.0°, flash p: 87°F (CC), d: 0.8539 @ 20°/4°, autoign temp: 652°F, vap press: 10 mm @ 26.0°, vap d: 3.38. Solidifies @ 41.5°; somewhat sol in water @ 20°. Misc in alc and ether and with most organic liquids.

SYNS: ISOBUTENYL METHYL KETONE * ISOPROPYLIDENEACETONE * MESITYLOXID (GERMAN) * MESITYLOXYDE (DUTCH) * METHYL ISOBUTENYL KETONE * 4-METHYL-3-PENTEN-

2-ON (DUTCH, GERMAN) * 4-METHYL-3-
PENTENE-2-ONE * 2-METHYL-2-PENTEN-4-ONE
* 4-METHYL-3-PENTEN-2-ONE * 4-METIL-3-
PENTEN-2-ONE (ITALIAN) * OSSIDO DI MESITILE
(ITALIAN) * OXYDE DE MESITYLE (FRENCH)

OSHA PEL: TWA 25 ppm
ACGIH TLV: TWA 15 ppm; STEL 25 ppm
DFG MAK: 25 ppm (100 mg/m^3)
DOT Classification: Label: Flammable
  Liquid

THR: Poison by intraperitoneal route. Moderately toxic by inhalation, ingestion, and skin contact. A systemic irritant by inhalation. Single exposures tend to indicate that this ketone has greater acute and narcotic action than isophorone. It can have harmful effects upon the kidneys and liver, and may damage the eyes and lungs to a serious degree. It can cause opaque cornea, keratoconus, and extensive necrosis of cornea. This compound is highly irritating to all tissues on contact; its vapors also are irritating. High concentrations are narcotic. Prolonged exposure can injure liver, kidneys and lungs. It is readily absorbed through intact skin. Dangerous when exposed to heat or flame; can react with oxidizing materials. Reacts violently with 2-amino ethanol; chlorosulfonic acid; ethylene diamine; HNO$_3$; oleum; H$_2$SO$_4$. To fight fire, use alcohol foam, CO$_2$, dry chemical.

MESOXALYLUREA MONO-
HYDRATE   HR: 3
CAS: 3237-50-1   NIOSH: UW 0492000
mf: C$_4$H$_2$N$_2$O$_4$ • H$_2$O   mw: 160.10

PROP: Triclinic, aqueous; mp: decomp @ 170°; very sol in water and alc; insol in ether.

SYNS: ALLOXAN MONOHYDRATE * MESOXA-
LYLCARBAMIDE MONOHYDRATE * 2,4,5,
6(1H,3H)-PYRIMIDINETETRONE HYDRATE
* 2,4,5,6-TETRAOXOHEXAHYDROPYRIMIDINE
HYDRATE

THR: Poison by intravenous route. An experimental teratogen. When heated to decomposition it emits toxic fumes of NO$_x$.

METAHYDROXYPROCAINE   HR: 3
CAS: 487-5-3-6   NIOSH: VO 1575000
mf: C$_{13}$H$_{20}$N$_2$O$_3$ mw: 252.35

SYNS: DIETHYLAMINOETHYL-P-AMINOSALICY-
LATE * 2-DIETHYLAMINOETHYL-P-AMINOSA-

LICYLATE * 2-DIETHYLAMINOETHYL-4-AMI-
NOSALICYLATE * DIETHYLAMINOETHYL-3-HY-
DROXY-4-AMINOBENZOATE
* DIETHYLAMINOETHANOL-P-AMINOSALICYLATE

THR: Poison by intravenous and intraperitoneal routes. When heated to decomposition it emits toxic fumes of NO$_x$.

METASYSTOX   HR: 3
CAS: 8022-00-2   NIOSH: TG 1760000

PROP: An oily liquid, slightly sol in water, d: 1.20.

SYNS: S(AND O)-2-(ETHYLTHIO)ETHYL-O,O-
DIMETHYL PHOSPHOROTHIOATE * METHYL
DEMETON * METHYL-MERCAPTOPHOS
* METHYL SYSTOX

DFG MAK: 0.5 ppm (5 mg/m^3)

THR: Poison by ingestion, skin contact, and possibly other routes. A cholinesterase inhibitor. See also parathion. An insecticide and acaricide.

METASYSTOX-S   HR: 3
CAS: 2674-91-1   NIOSH: TG 1575000
mf: C$_7$H$_{17}$O$_4$PS$_2$ mw: 260.33

SYNS: S-2-AETHYLSULFINYL-1-METHYL
AETHYL-O,O DIMETHYL-MONOTHIOPHOSPHAT
(GERMAN) * ENT 25,674 * S-2-ETHYL-SUL-
FINYL-1-METHYL-ETHYL-O,O-DIMETHYL-MONO-
THIOFOSFAAT (DUTCH) * S-2-ETHYL-SUL-
PHINYL-1-METHYL-ETHYL-O,O-DIMETHYL
PHOSPHOROTHIOLATE * S-2-ETIL-SULFINIL-1-
METIL-ETIL-O,O-DIMETIL-MONOTIOFOSFATO
(ITALIAN) * O,O-DIMETHYL-S-(ETHYLSULFI-
NYL-(2-ISOPROPYL)) PHOSPHOROTHIOATE
* PHOSPHOROTHIOIC ACID, O,O-DIMETHYL
S-(ETHYLSULFINYL-(2-ISOPROPYL)) ESTER
* THIOPHOSPHATE DE O,O-DIMETHYLE ET
DE S-2-(ISOPROPYLSULFINYL)-ETHYLE (FRENCH)

THR: Poison by ingestion, intraperitoneal, and other routes. Moderately toxic by skin contact. See also esters. When heated to decomposition it emits very toxic fumes of PO$_x$ and SO$_x$.

METHACRYLIC ACID   HR: 3
CAS: 79-41-4   NIOSH: OZ 2975000
mf: C$_4$H$_6$O$_2$ mw: 86.10

PROP: Corrosive liquid or colorless crystals, mp: 16°; bp: 163°; flash p: 171°F (COC); d: 1.014 @ 25° (glacial); vap press: 1 mm @ 25.5°; repulsive odor. Sol in warm H$_2$O; misc with alc, ether.

SYNS: ALPHA-METHYLACRYLIC ACID * 2-METHYLPROPENOIC ACID

ACGIH TLV: TWA 20 ppm

THR: Poison by intraperitoneal route. Flammable when exposed to heat, flame, or oxidizers. To fight fire, use alcohol foam, spray, mist, dry chemical. When heated to decomposition it emits acrid smoke and fumes.

METHACRYLIC ACID AMIDE   HR: 3
CAS: 79-39-0   NIOSH: UC 6475000
mf: C_4H_7NO mw: 85.12

SYNS: METHACRYLIC AMIDE * 2-METHYL-ACRYLAMIDE * ALPHA-METHYL ACRYLIC AMIDE * 2-METHYLPROPENAMIDE * USAF RH-1

THR: Poison by intraperitoneal route. Moderately toxic by ingestion and inhalation. When heated to decomposition it emits toxic fumes of NO_x.

METHACYCLINE HYDROCHLOR-IDE   HR: 3
CAS: 3963-95-9   NIOSH: OZ 6475000
mf: $C_{22}H_{22}N_2O_8 \cdot 7ClH$ mw: 697.68

PROP: Yellow crystalline powder; decomp @ approx 205°; bitter taste; sol in H_2O; slightly sol in alc; practically insol in ether, chloroform.

THR: Poison by intraperitoneal route. When heated to decomposition it emits very toxic fumes of NO_x and HCl.

METHADONE   HR: 3
CAS: 76-99-3   NIOSH: MJ 5950000
mf: $C_{21}H_{27}NO$ mw: 309.49

SYNS: ADANON * AMIDONE * DIAMINON * DOLOPHINE * HEPTADONE * HEPTANON * KETALGIN * MECODIN * PHENADONE * POLAMIDONE

THR: Poison by ingestion, intraperitoneal, intravenous, subcutaneous, and intraduodenal routes. An experimental teratogen. When heated to decomposition it emits toxic fumes of NO_x.

METHADONE HYDROCHLOR-IDE   HR: 3
CAS: 1095-90-5   NIOSH: MJ 6300000
mf: $C_{21}H_{27}NO \cdot ClH$ mw: 345.95

SYNS: ADANON HYDROCHLORIDE * AMIDONE HYDROCHLORIDE * DIAMINON HYDROCHLORIDE * DIASONE HYDROCHLORIDE * 6-DI-METHYLAMINO-4,4-DIPHENYL-3-HEPTANONE HYDROCHLORIDE * 1,1-DIPHENYL-1-(BETA-DIMETHYLAMINOPROPYL)BUTANONE-2 HYDRO-CHLORIDE

THR: Poison by ingestion, subcutaneous, intravenous, parenteral, and intraperitoneal routes. An experimental teratogen. Mutagenic data. When heated to decomposition it emits very toxic fumes of Cl^- and NO_x.

l-METHADONE HYDROCHLOR-IDE   HR: 3
CAS: 5967-73-7   NIOSH: MJ 6475000
mf: $C_{21}H_{27}NO \cdot ClH$ mw: 345.95

PROP: dl form crystals; mp: 241°.

SYNS: 1-6-DIMETHYLAMINO-4,4-DIPHENYL-3-HEPTANONE HYDROCHLORIDE * LEVADONE * LEVOTHYL * POLAMIDON

THR: Poison by intravenous, subcutaneous, and intraperitoneal routes. Human central nervous system effects by ingestion. *Caution:* Abuse leads to habituation or addiction. When heated to decomposition it emits very toxic fumes of Cl^- and NO_x.

d-METHADONE HYDROCHLOR-IDE   HR: 3
CAS: 15284-15-8   NIOSH: MJ 6650000
mf: $C_{21}H_{27}NO \cdot ClH$ mw: 345.95

SYN: D-DOLOPHINE HYDROCHLORIDE

THR: Poison by intravenous and intraperitoneal routes. Human central nervous system effects by ingestion. When heated to decomposition it emits toxic fumes of Cl^- and NO_x.

dl-METHADONE HYDROCHLOR-IDE   HR: 3
CAS: 125-56-4   NIOSH: MJ 6825000
mf: $C_{21}H_{27}NO \cdot ClH$ mw: 345.95

PROP: dl form: platelets from (alc + ether); mp: 235°; bitter taste; practically insol in ether, glycerol.

SYNS: ALGIDON * ALGOLYSIN * BUTALGIN * DEPRIDOL * 4,4-DIPHENYL-6-DIMETHYL-AMINO-3-HEPTANONE HYDROCHLORIDE (±)- * 1,1-DIPHENYL-1-(2-DIMETHYLAMINOPROPYL)-2-BUTANONE HYDROCHLORIDE, (±)- * DOLOPHIN HYDROCHLORIDE * FENADONE * KETALGIN HYDROCHLORIDE * MECODIN * MIADONE * PHENADONE HYDROCHLORIDE

THR: Poison by intravenous, intraperitoneal, and subcutaneous routes. When heated to decomposition it emits very toxic fumes of Cl$^-$ and NO$_x$.

METHANE HR: 1
CAS: 74-82-8   NIOSH: PA 1490000
DOT: 1972
mf: CH$_4$ mw: 16.05

PROP: Colorless, odorless, tasteless gas, bp: $-161.5°$, lel = 5.3%, uel = 15%, fp: $-183.2°$, d: 0.415 @ $-164°$, 0.7168 g/L, autoign temp: 650°, vap d: 0.6, flash p: $-368.6°$F. Sol in water, alc and ether. D: 0.554 @ 0°/4° (air = 1); or 0.7168 g/L. Mp: $-182.6°$; sol in alc, ether, organic solvents.

SYNS: FIRE DAMP * MARSH GAS * METHYL HYDRIDE

DOT Classification: Label: Flammable Gas

THR: A simple asphyxiant. See also argon. Very dangerous fire hazard when exposed to heat or flame. Reacts violently with BrF$_5$; Cl$_2$; ClO$_2$; NF$_3$; liquid O$_2$; OF$_2$. Dangerous explosion hazard when exposed to heat or flame. To fight fire, stop flow of gas, CO$_2$ or dry chemical. Incompatible with halogens or interhalogens; oxidants, air (forms explosive mixtures).

METHANE DICHLORIDE HR: 3
CAS: 75-09-2   NIOSH: PA 8050000
mf: CH$_2$Cl$_2$ mw: 84.93

PROP: Colorless volatile liquid, bp: 39.8°, lel = 15.5% in O$_2$, uel = 66.4% in O$_2$, fp: $-96.7°$, d: 1.326 @ 20°/4°, autoign temp: 1139°F, vap press: 380 mm @ 22°, vap d: 2.93.

SYNS: CHLORURE DE METHYLENE (FRENCH) * DICHLOROMETHANE (DOT) * FREON 30 * METHYLENE BICHLORIDE * METHYLENE CHLORIDE (DOT) * METHYLENE DICHLORIDE * METYLENU CHLOREK (POLISH) * NCI-c50102

OSHA PEL: TWA 500 ppm; CL 1000; Pk 2000/5M/2H
DOT Classification: ORM-A, Label: None

THR: Poison by ingestion and intravenous routes. Moderately toxic by inhalation and other routes. Blood and central nervous system effects by inhalation. A severe skin and moderate eye irritant. Mutagenic data. A strong narcotic but has few other acute toxicity effects. It will not form explosive mixtures with air at ordinary temperatures. However, it can be decomposed by contact with hot surfaces and open flame, and it can then yield toxic fumes which are irritating and will thus give warning of their presence. Reacts violently with Li; NaK; potassium-*tert*-butoxide; (KOH + *n*-methyl-*n*-nitrosourea). It will form explosive mixture in an atmosphere having high oxygen content; liquid O$_2$; N$_2$O$_4$; K; Na; NaK. Dangerous; when heated to decomposition emits highly toxic fumes of phosgene.

METHANESULFONIC ACID HR: 3
CAS: 75-75-2   NIOSH: PB 1140000
mf: CH$_4$O$_3$S   mw: 96.11

PROP: Solid. Sol in water, alc and ether; d: 1.4812 @ 18°/4°; mp: 20°; bp: 167° @ 10 mm. Corrosive to iron, steel, brass, copper and lead.

SYN: WSQ 1

THR: Poison by ingestion and intraperitoneal routes. See also sulfonates. When heated to decomposition it emits toxic fumes of SO$_x$. Incompatible with ethyl vinyl ether.

METHANETHIOL HR: 3
CAS: 74-93-1   NIOSH: PB 4375000
DOT: 1064
mf: CH$_4$S   mw: 48.11

PROP: Flammable gas. Odor of rotten cabbage; mp: $-123.1°$, vap d: 1.66, lel = 3.9%, uel = 21.8%. Bp: 5.95°; d: 0.8665 @ 20°/4°; solidifies @ $-123°$; flash p: $-0.4°$F.

SYNS: MERCAPTAN METHYLIQUE (FRENCH) * METHYTIOLO (ITALIAN) * METHAANTHIOL (DUTCH) * METHANTHIOL (GERMAN) * METHYLMERCAPTAAN (DUTCH) * METILMERCAPTANO (ITALIAN) * METHYL MERCAPTAN (DOT)

OSHA PEL: CL 10 ppm
DOT Classification: Label: Flammable Gas

THR: Poison by inhalation and subcutaneous routes. Mutagenic data. A common air contaminant. Very dangerous fire hazard when exposed to heat, flame or oxidizers. Dangerous; on decomposition, it emits highly toxic fumes of SO$_x$; will react with water, steam or acids to produce toxic and flammable vapors; can react vigorously with oxidizing materials. To fight fire, use alcohol foam, CO$_2$, dry chemical. Incompatible with mercury (II) oxide.

METHANOL   HR: 3
CAS: 67-56-1   NIOSH: PC 1400000
DOT: 1230
mf: CH_4O mw: 32.05

PROP: Clear, colorless, very mobile liquid; bp: 64.8° lel = 6.0%; uel = 36.5%; ulc: 70; mp: −97.8° flash p: 54°; d: 0.7915 @ 20°/4°; autoign temp: 878°F; vap press: 100 mm @ 21.2°; vap d: 1.11. Slight alcoholic odor when pure; crude material may have a repulsive pungent odor; misc in water, ethanol, ether, benzene, ketones and most other organic solvents.

SYNS: ALCOOL METHYLIQUE (FRENCH) * ALCOOL METILICO (ITALIAN) * CARBINOL * COLONIAL SPIRIT * COLUMBIAN SPIRITS (DOT) * METHANOL (DOT) * METANOLO (ITALIAN) * METHYL ALCOHOL (DOT) * METHYLALKOHOL (GERMAN) * METHYL HYDROXIDE * METYLOWY ALKOHOL (POLISH) * MONOHYDROXYMETHANE * PYROXYLIC SPIRIT * WOOD ALCOHOL * WOOD NAPHTHA * WOOD SPIRIT * COLUMBIAN SPIRIT * METHYL ALCOHOL

OSHA PEL: TWA 200 ppm
ACGIH TLV: TWA 200 ppm; STEL 250 ppm
DFG MAK: 200 ppm (260 mg/m³); BAT: urine 30 mg/L end of work week, last 4 hours of exposure
DOT Classification: Flammable Liquid

THR: Poison by ingestion, skin contact, intravenous, and intraperitoneal routes. Moderately toxic by inhalation and other routes. An eye and skin irritant. A systemic irritant by inhalation. A narcotic. Its main toxic effect is exerted upon the nervous system, particularly the optic nerves and possibly the retinae and can progress to permanent blindness. Once absorbed, methanol is only very slowly eliminated. Coma resulting from massive exposures may last as long as 2-4 days. In the body, the products formed by its oxidation are formaldehyde and formic acid, both of which are toxic. Because of the slowness with which it is eliminated, methanol should be regarded as a cumulative poison. Though single exposures to fumes may cause no harmful effect, daily exposure may result in the accumulation of sufficient methanol in the body to cause illness. Death from ingestion of less than 30 mL has been reported. A common air contaminant. It is used as a food additive permitted in foods for human consumption.

Dangerous fire hazard when exposed to heat, flame, or oxidizers. Moderate explosion hazard when exposed to flame. Violent reaction with CrO_3; (I + ethanol + HgO); $Pb(ClO_4)_2$; $HClO_4$; P_2O_3; (KOH + $CHCl_3$); (NaOH + $CHCl_3$). Dangerous; can react vigorously with oxidizing materials. To fight fire, use alcohol foam. Incompatible with beryllium dihydride; chloroform; cyanuric chloride; metals; oxidants; potassium tert-butoxide. For further information see Vol. 5, No. 5 of *DPIM Report*.

d-METHAPHETAMINE HYDRO-CHLORIDE   HR: 3
CAS: 51-57-0   NIOSH: SH 5425000
mf: $C_{10}H_{15}N \cdot ClH$ mw: 185.72

SYNS: D-DESOXYEPHEDRINE HYDROCHLORIDE * D-N,ALPHA-DIMETHYLPHENETHYLAMINE HYDROCHLORIDE * (+)-N,ALPHA-DIMETHYLPHENETHYLAMINE HYDROCHLORIDE * (+)-METHAPHETAMINE CHLORIDE * (+)-METHAPHETAMINE HYDROCHLORIDE * (+)-2-METHYLAMINO-2-PHENYLPROPANE HYDROCHLORIDE * D-N-METHYL-BETA-PHENYLISOPROPYLAMINE HYDROCHLORIDE * (+)-N-METHYLAMPHETAMINE HYDROCHLORIDE * USAF EL-36

THR: Poison by ingestion, intravenous, subcutaneous, and intraperitoneal routes. When heated to decomposition it emits very toxic fumes of HCl and NO_x.

METHAQUALONE HYDRO-CHLORIDE   HR: 3
CAS: 340-56-7   NIOSH: VA 4025000
mf: $C_{16}H_{14}N_2O \cdot ClH$ mw: 286.78

PROP: Crystals; mp: 255°-265°. Sol in ether, ethanol, almost insol in H_2O.

SYNS: METHYLQUINAZOLONE HYDROCHLORIDE * 2-METHYL-3-TOLYLCHINAZOLON-4 HYDROCHLORIDE (GERMAN) * 2-METHYL-3-O-TOLYL-4(3H)-QUINAZOLINONE HYDROCHLORIDE * 2-METHYL-3-(O-TOLYL)-4-QUINAZOLONE HYDROCHLORIDE

THR: Poison by ingestion, intraperitoneal, intravenous, and possibly other routes. When heated to decomposition it emits very toxic fumes of NO_x and HCl.

METHOHEXITAL SODIUM   HR: 3
CAS: 309-36-4   NIOSH: CQ 0400000
mf: $C_{14}H_{17}N_2O_3 \cdot Na$ mw: 284.32

PROP: Minute crystals; sol in H_2O;

SYNS: 5-ALLYL-1-METHYL-5-(1-METHYL-2-PENTYNYL)BARBITURIC ACID SODIUM SALT * BREVITAL SODIUM * ENALLYNYMAL SODIUM * METHOHEXITONE SODIUM * SODIUM-DL-5-ALLYL-1-METHYL-5-(1-METHYL-2-PENTYNYL)BARBITURATE * SODIUM A-DL-1-METHYL-5-ALLYL-5-(1-METHYL-2-PENTYNYL)BARBITURATE

THR: Poison by implant route. Excessive use may lead to addiction or habituation. Allergenic effects by intravenous route. When heated to decomposition it emits toxic fumes of NO_x and Na_2O.

METHOPHENAZINE DIFUMARATE HR: 3
CAS: 522-23-6   NIOSH: DH 9800000
mf: $C_{31}H_{36}ClN_3O_5S \cdot C_8H_4O_8$ mw: 826.33

SYN: BENZOIC ACID, 3,4,5-TRIMETHOXY-, 2-(4-(3-(2-CHLOROPHENOTHIAZIN-10-YL)PROPYL)-1-PIPERAZINYL)ETHYL ESTER, DIFUMARATE

THR: Poison (acute) by intraperitoneal, subcutaneous, and intravenous routes. Moderately toxic by ingestion. An experimental teratogen. See also esters. When heated to decomposition it emits very toxic fumes of Cl^-, SO_x, and NO_x.

METHORPHINAN HYDROBROMIDE HR: 3
CAS: 5985-35-3   NIOSH: QC 9625000
mf: $C_{17}H_{23}NO \cdot BrH$ mw: 338.33

SYNS: DROMORAN HYDROBROMIDE * DL-3-HYDROXY-N-METHYLMORPHINAN HYDROBROMIDE * RACEMORPHAN HYDROBROMIDE

THR: Poison by subcutaneous, intraperitoneal, and intravenous routes. When heated to decomposition it emits very toxic fumes of NO_x and HBr.

METHOTREXATE HR: 3
CAS: 59-05-2   NIOSH: MA 1225000
mf: $C_{20}H_{22}N_8O_5$ mw: 454.50

SYNS: AMETHOPTERIN * 4-AMINO-4-DEOXY-N(SUP 10)-METHYLPTEROYLGLUTAMATE * 4-AMINO-4-DEOXY-N(SUP 10)-METHYLPTEROYLGLUTAMIC ACID * 4-AMINO-10-METHYLFOLIC ACID * 4-AMINO-N(SUP 10)-METHYLPTEROYLGLUTAMIC ACID * N-BISMETHYLPTEROYLGLUTAMIC ACID * METHOPTERIN * METHOTEXTRATE * METHYLAMINOPTERIN * NCI-C04671 * NSC-740

THR: Poison by ingestion, intravenous, intraspinal, and intraperitoneal routes. Can cause pulmonary, blood, glandular, and other systemic effects by several routes. An experimental teratogen and tumorigen by several routes. An insect chemosterilant. When heated to decomposition it emits toxic fumes, including NO_x.

METHOXAMINE HYDROCHLORIDE HR: 3
CAS: 61-16-5   NIOSH: DN 4025000
mf: $C_{11}H_{17}NO_3 \cdot ClH$ mw: 247.71

PROP: Crystals; mp: 212°-216°; very sol in H_2O. Practically insol in ether, benzene, chloroform.

SYNS: 2-AMINO-1-(2,5-DIMETHOXYPHENYL)-1-PROPANOL HYDROCHLORIDE * ALPHA-(1-AMINOETHYL)-2,5-DIMETHOXYBENZYL ALCOHOL HYDROCHLORIDE * PRESSOMIN HYDROCHLORIDE * VASOXYL HYDROCHLORIDE

THR: Poison by ingestion, intravenous, and other routes. When heated to decomposition it emits very toxic fumes of HCl and NO_x.

dl,4-METHOXYAMPHETAMINE HYDROCHLORIDE HR: 3
CAS: 52740-56-4   NIOSH: SH 8125000
mf: $C_{10}H_{15}NO \cdot ClH$ mw: 201.72

SYN: 4-METHOXYAMPHETAMINE HYDROCHLORIDE

THR: Poison by ingestion, intravenous, and intraperitoneal routes. When heated to decomposition it emits very toxic fumes of HCl and NO_x.

6-METHOXYBENZO(a)PYRENE HR: 3
CAS: 52351-96-9   NIOSH: DJ 6900000
mf: $C_{21}H_{14}O$ mw: 282.35

THR: An experimental tumorigen. When heated to decomposition it emits acrid smoke and fumes.

5-METHOXYCHRYSENE HR: 3
CAS: 61413-39-6   NIOSH: GC 1300000
mf: $C_{19}H_{14}O$ mw: 258.33

THR: An experimental neoplastigen and tumorigen by skin contact. When heated to decomposition it emits acrid smoke and fumes.

5-METHOXYDIBENZ(a,h)ANTHRACENE HR: 3
CAS: 63019-72-7   NIOSH: HN 5250000
mf: $C_{23}H_{16}O$ mw: 308.39

SYNS: 3-METHOXY-DBA * 3-METHOXY-1,2: 5,6-DIBENZANTHRACENE

THR: An experimental tumorigen. When heated to decomposition it emits acrid smoke and fumes.

7-METHOXYDIBENZ(a,h)ANTHRA-CENE   HR: 3
CAS: 63041-72-5   NIOSH: HN 5425000
mf: $C_{23}H_{16}O$ mw: 308.39

SYNS: 9-METHOXY-DBA * 9-METHOXY-1, 2,5,6-DIBENZANTHRACENE

THR: An experimental neoplastigen and tumorigen. When heated to decomposition it emits acrid smoke and fumes.

4-METHOXY-9-(3-(ETHYL-2-CHLORO-ETHYL)   HR: 3
CAS: 38915-14-9   NIOSH: AR 8200000
mf: $C_{21}H_{26}ClN_3O \cdot 2ClH$ mw: 444.87

SYNS: ICR 377 * 9-(3-ETHYL-2-CHLORO-ETHYL)AMINO PROPYLAMINO)-4-METHOXY ACRIDINE DIHYDROCHLORIDE * AMINOPRO-PYLAMINO)ACRIDINE DIHYDROCHLORIDE

THR: A poison by intravenous route. An experimental neoplastigen. When heated to decomposition it emits very toxic fumes of Cl^-, NO_x and HCl.

1-METHOXY ETHYL ETHYLNITROSA-MINE   HR: 3
CAS: 61738-03-2   NIOSH: IA 3270000
mf: $C_5H_{12}N_2O_2$ mw: 132.19

SYN: 1-METHOXY-AETHYL-AETHYLNITROSAMIN (GERMAN)

THR: An experimental carcinogen. Moderately toxic by ingestion. See also nitrosamines. When heated to decomposition it emits toxic fumes of NO_x.

METHOXYETHYL MERCURIC ACE-TATE   HR: 3
CAS: 151-38-2   NIOSH: OV 6300000
mf: $C_5H_{10}HgO_3$ mw: 318.74

PROP: Crystals; water sol.

SYNS: ACETATO(2-METHOXYETHYL)MERCURY * MERCURAN * RADOSAN

OSHA PEL: TWA 10 ug(Hg)/m³; CL 40

THR: Poison by ingestion and possibly other routes. Mutagenic data. See also mercury com-

pounds. When heated to decomposition it emits toxic fumes of Hg.

2-METHOXYETHYLMERCURY CHLORIDE   HR: 3
CAS: 123-88-6   NIOSH: OW 0875000
mf: C_3H_7ClHgO mw: 295.14

PROP: Crystals.

SYNS: CERESAN UNIVERSAL NAZBEIZE * CHLORO(2-METHOXYETHYL)MERCURY * METHOXYAETHYLQUECKSILBERCHLORID (GER-MAN) * (BETA-METHOXYETHYL)MERCURIC CHLORIDE * METHOXYETHYL MERCURIC CHLO-RIDE * 2-METHOXYETHYLMERCURIC CHLORIDE * BETA-METHOXYETHYLMERCURY CHLORIDE * METHOXYETHYLMERCURY CHLORIDE

OSHA PEL: TWA 10 ug(Hg)/m³; CL 40

THR: Poison by ingestion, subcutaneous, and possibly other routes. Mutagenic data. See also mercury compounds and chlorides. When heated to decomposition it emits very toxic fumes of Hg and Cl^-.

1-METHOXY ETHYL METHYLNITROSA-MINE   HR: 3
CAS: 61738-05-4   NIOSH: KR 8880000
mf: $C_4H_{10}N_2O_2$ mw: 118.16

SYN: 1-METHOXY-AETHYL-METHYLNITROSAMIN (GERMAN)

THR: Poison (acute) by ingestion. An experimental carcinogen. See also nitrosamines. When heated to decomposition it emits toxic fumes of NO_x.

7-METHOXY-2-FAA   HR: 3
CAS: 16690-44-1   NIOSH: AC 5950000
mf: $C_{16}H_{15}NO_2$ mw: 253.32

SYNS: N-(7-METHOXY-2-FLUORENYL)ACET-AMIDE * N-(7-METHOXYFLUOREN-2-YL)ACET-AMIDE * 7-METHOXY-N-2-FLUORENYLACET-AMIDE

THR: An experimental carcinogen. When heated to decomposition it emits toxic fumes of NO_x.

1-METHOXY-2-FLUORENAMINE HY-DROCHLORIDE   HR: 3
CAS: 6893-24-9   NIOSH: LL 5425000
mf: $C_{14}H_{13}NO \cdot ClH$ mw: 247.74

SYN: 1-METHOXYFLUOREN-2-AMINE HYDRO-CHLORIDE

THR: An experimental carcinogen. When heated to decomposition it emits very toxic fumes of Cl$^-$ and NO$_x$.

METHOXYMETHYL ETHYL NITROSA-MINE   HR: 3
CAS: 61738-04-3   NIOSH: KR 8850000
mf: C$_4$H$_{10}$N$_2$O$_2$ mw: 118.16

SYN: METHOXYMETHYL-AETHYLNITROSAMINE (GERMAN)

THR: An experimental carcinogen. Moderately toxic (acute) by ingestion. See also nitrosamines. When heated to decomposition it emits toxic fumes of NO$_x$.

7-METHOXYMETHYL-12-METHYL-BENZ(a)ANTHRACENE   HR: 3
CAS: 013345-60-3   NIOSH: CX 0600000
mf: C$_{21}$H$_{18}$O   mw: 286.39

SYN: 9-METHYL-10-METHOXYMETHYL-1,2-BENZANTHRACENE

THR: An experimental tumorigen. When heated to decomposition it emits acrid smoke and fumes.

METHOXYMETHYL METHYLNITROSA-MINE   HR: 3
CAS: 39885-14-8   NIOSH: IQ 0250000
mf: C$_3$H$_8$N$_2$O$_2$ mw: 104.13

SYNS: METHOXYMETHYL-METHYLNITROSAMIN (GERMAN) * N-NITROSO-N-METHOXYMETHYL-METHYLAMINE

THR: An experimental carcinogen. Moderately toxic by ingestion. See also amines and nitrosamines. When heated to decomposition it emits toxic fumes of NO$_x$.

4-METHOXYMETHYLPYRIDOXINE HY-DROCHLORIDE   HR: 3
CAS: 3131-27-9   NIOSH: UT 5010000
mf: C$_9$H$_{13}$NO$_3$ • ClH   mw: 219.69

SYNS: 4-METHOXYMETHYL-5-HYDROXY-6-METHYL-3-PYRIDINEMETHANOL HYDROCHLORIDE * 4-METHOXYMETHYLPYRIDOXOL HYDROCHLORIDE

THR: Poison by ingestion, subcutaneous, and intravenous routes. When heated to decomposition it emits very toxic fumes of HCl and NO$_x$.

METHOXYOXIMERCURIPROPYLSUC-CINYL UREA   HR: 3
CAS: 140-20-5   NIOSH: OV 8575000
mf: C$_9$H$_{16}$HgN$_2$O$_6$ mw: 448.86

SYNS: N-((3-(HYDROXYMERCURI)-2-METHOXY-PROPYL)-CARBAMOYL)SUCCINAMIC ACID
* METHOXYHYDROXYMERCURIPROPYLSUCCINYL-UREA

THR: Poison by intramuscular and subcutaneous routes. See also mercury compounds. When heated to decomposition it emits very toxic fumes of Hg and NO$_x$.

p-METHOXYPHENYLACETIC ACID   HR: 3
CAS: 104-01-8   NIOSH: AI 8960000
mf: C$_9$H$_{10}$O$_3$ mw: 166.19

SYNS: ANISYL FORMATE * 2-(P-ANISYL)-ACETIC ACID * HOMOANISIC ACID * 4-METHOXYBENZENEACETIC ACID * P-METHOXYBEN-ZYL FORMATE * 4-METHOXYPHENYLACETIC ACID

THR: An experimental neoplastigen. Moderately toxic by ingestion and intraperitoneal routes. When heated to decomposition it emits acrid smoke and fumes.

N-(p-METHOXYPHENYL)-1-AZIRIDINE CARBOXAMIDE   HR: 3
CAS: 3647-17-4   NIOSH: CM 6770000
mf: C$_{10}$H$_{12}$N$_2$O$_2$ mw: 192.24

SYNS: 1-(1-AZIRIDINYL)-N-(P-METHOXYPHE-NYL)FORMAMIDE * P-METHOXYPHENYL-N-CAR-BAMOYLAZIRIDINE

THR: Poison by intravenous route. An experimental neoplastigen. When heated to decomposition it emits toxic fumes of NO$_x$.

2-METHOXYPROMAZINE   HR: 3
CAS: 61-01-8   NIOSH: SO 7600000
mf: C$_{18}$H$_{22}$N$_2$OS   mw: 314.48

PROP: Crystals. mp: 44°-48°.

SYNS: 10-(3-DIMETHYLAMINOPROPYL)-2-ME-THOXYPHENOTHIAZINE * 2-METHOXY-10-(3'-DIMETHYLAMINOPROPYL)PHENOTHIAZINE

THR: Poison by ingestion and intraperitoneal routes. When heated to decomposition it emits very toxic fumes of NO$_x$ and SO$_x$.

METHOXYPROMAZINE MALE-ATE   HR: 3
CAS: 3403-42-7   NIOSH: SO 7700000
mf: C$_{18}$H$_{22}$N$_2$OS • C$_4$H$_4$O$_4$ mw: 430.56

SYNS: 10-(3-(DIMETHYLAMINO)PROPYL)-2-ME-THOXY)PHENOTHIAZINE, MALEATE * METHO-PROMAZINE MALEATE * TENTONE MALEATE

THR: Poison by intraperitoneal route. Moderately toxic by ingestion. When heated to decomposition it emits very toxic fumes of NO_x and SO_x.

3-METHOXYPROPYLAMINE   HR: 3
CAS: 5332-73-0   NIOSH: UI 3335000
mf: $C_4H_{11}NO$ mw: 89.16

PROP: Colorless liquid, mp: $-75.7°$, bp: $116°$, flash p: $90°F$ (TOC), d: 0.8615 @ $30°$, vap press: 20 mm @ $30°$, vap d: 3.07.

SYN: 3-MPA

THR: Poison by intravenous route. Irritating to skin, eyes, and mucous membranes. Dangerous fire hazard when exposed to heat or flame; can react with oxidizing materials. To fight fire, use CO_2, dry chemical. When heated to decomposition it emits toxic fumes of NO_x.

5-METHOXY-3-(2-PYRROLIDINO-ETHYL)INDOLE   HR: 3
CAS: 3949-14-2   NIOSH: NM 0175000
mf: $C_{15}H_{20}N_2O$ mw: 244.37

SYNS: CT 4436 * METHOXY-5-PYRROLIDINO-2'-ETHYL-3-INDOLE

THR: Poison by intraperitoneal and intravenous routes. When heated to decomposition it emits toxic fumes of NO_x.

6-METHOXYTRYPTAMINE   HR: 3
CAS: 3610-36-4   NIOSH: NL 4125000
mf: $C_{11}H_{14}N_2O \cdot HCl$ mw: 226.73

SYN: 3-(2-AMINOETHYL)-6-METHOXYINDOLE HYDROCHLORIDE

THR: Poison by intravenous route. Moderately toxic by subcutaneous route. When heated to decomposition it emits very toxic fumes of HCl and NO_x.

METHYLACETAMIDE   HR: 3
CAS: 79-16-3   NIOSH: AC 5960000
mf: C_3H_7NO mw: 73.11

SYNS: N-METHYLACETAMIDE * MONOMETH-YLACETAMIDE

THR: Poison by ingestion and intravenous routes. Mutagenic data. An experimental teratogen. When heated to decomposition it emits toxic fumes of NO_x.

METHYLACETOPYRONONE   HR: 3
CAS: 520-45-6   NIOSH: UP 8050000
mf: $C_8H_8O_4$ mw: 168.16

PROP: Crystals, mod sol in water and organic solvents, mp: $109°$, bp: $269.0°$, vap press: 1 mm @ $91.7°$, vap d: 5.8.

SYNS: 3-ACETYL-6-METHYL-2,4-PYRANDIONE * 3-ACETYL-6-METHYLPYRANDIONE-2,4 * 3-ACETYL-6-METHYL-2H-PYRAN-2,4(3H)-DIONE * DEHYDRACETIC ACID * DEHYDRO-ACETIC ACID

THR: Poison by ingestion and intravenous routes. Moderately toxic by several other routes. An experimental carcinogen. Combustible.

METHYL-beta-ACETOXYETHYL-beta-CHLOROETHYLAMINE   HR: 3
CAS: 36375-30-1   NIOSH: KK 1575000
mf: $C_7H_{14}ClNO_2$ mw: 179.67

SYN: TL 1428

THR: Poison by intravenous and subcutaneous routes. When heated to decomposition it emits very toxic fumes of Cl^- and NO_x.

METHYLACETOXYMALONO-NITRILE   HR: 3
CAS: 7790-01-4   NIOSH: WW 9100000
mf: $C_6H_6N_2O_2$ mw: 138.14

SYN: 2-ACETOXYISOSUCCINODINITRILE

THR: Poison by ingestion, skin contact, and inhalation. See also nitriles. When heated to decomposition it emits toxic fumes of NO_x.

METHYL ACETYLACETATE   HR: 2
CAS: 105-45-3   NIOSH: AK 5775000
mf: $C_5H_8O_3$ mw: 116.13

PROP: Colorless liquid; mp: $27.5°$; bp: $170°$; flash p: $170°F$; autoign temp: $536°F$; d: 1.077; vap d: 4.00.

SYNS: ACETOACETIC METHYL ESTER * METH-YLACETOACETATE * METHYL ACETYLACETO-NATE * METHYL-3-OXOBUTYRATE

THR: Moderately toxic by ingestion. A skin and severe eye irritant. See also esters. Flammable. To fight fire, use foam, CO_2, dry chemical. When heated to decomposition it emits acrid smoke and fumes.

METHYLACRYLALDEHYDE   HR: 3
CAS: 78-85-3   NIOSH: OZ 2625000
mf: C_4H_6O mw: 70.10

PROP: Colorless liquid; mp: $-81°C$, bp: $73.5°$, flash p: $35°F$ (OC), d: 0.830 @ $20°/4°$, vap press: 120 mm @ 20°, vap d: 2.42. Sol in water.

SYNS: ISOBUTENAL * METHACROLEIN * METHACRYLIC ALDEHYDE * ALPHA-METH-YLACROLEIN * 2-METHYLACROLEIN * 2-METHYLPROPENAL (CZECH)

THR: Poison by ingestion, inhalation, and skin contact. Severe eye and skin irritant. See also aldehydes. Dangerously flammable when exposed to heat, flame or oxidizers. Dangerous; on decomposition it emits toxic fumes; can react vigorously with oxidizing materials. To fight fire, use CO_2, alcohol foam, foam, dry chemical.

METHYL ACRYLATE HR: 3
CAS: 96-33-3 NIOSH: AT 2800000
mf: $C_4H_6O_2$ mw: 86.10

PROP: Colorless liquid, acrid odor, lel = 2.8%, uel = 25%, fp: −75°, flash p: 27°F (OC), vap press: 100 mm @ 28°, vap d: 2.97; sol in alc and ether, d: 0.9561 @ 20°/4°; mp: -76.5°; bp: 70° @ 608 mm.

SYNS: ACRYLIC ACID METHYL ESTER * ACRYLATE DE METHYLE (FRENCH) * ACRYLSAE-UREMETHYLESTER (GERMAN) * METHYLACRY-LAAT (DUTCH) * METHYL PROPENATE * METHYL-ACRYLAT (GERMAN) * METHOXY-CARBONYLETHYLENE * METHYL PROPENOATE * METHYL-2-PROPENOATE * METILACRILATO (ITALIAN) * PROPENOIC ACID METHYL ESTER * 2-PROPENOIC ACID METHYL ESTER

OSHA PEL: TWA 10 ppm (skin)

THR: Poison by ingestion and intraperitoneal routes. Moderately toxic by inhalation and skin contact. A skin and eye irritant. Human irritant effects (systemic). See also esters. Chronic exposure has produced injury to lungs, liver, and kidneys in experimental animals. Dangerously flammable when exposed to heat, flame or oxidizers. Dangerous explosion hazard when exposed to heat, sparks, or flame. When heated to decomposition it emits toxic fumes; can react vigorously with oxidizing materials. To fight fire, use foam, CO_2, dry chemical.

METHYLACRYLONITRILE HR: 3
CAS: 126-98-7 NIOSH: UD 1400000
mf: C_4H_5N mw: 67.10

PROP: Mp: −36°, bp: 90.3°, d: 0.805, vap press: 40 mm @ 12.8°, flash p: 55°F.

SYNS: 2-CYANOPROPENE-1 * ISOPROPENE CYANIDE * ISOPROPENYLNITRILE * ALPHA-METHYLACRYLONITRILE * 2-METHYLPRO-PENENITRILE * USAF ST-40

THR: Poison by ingestion, inhalation, skin contact, and intraperitoneal routes. A skin and eye irritant See also nitriles. When heated to decomposition it emits toxic fumes of NO_x. For further information see Vol. 6, No. 1 of *DPIM Report*.

METHYLAMINE HR: 3
CAS: 74-89-5 NIOSH: PF 6300000
DOT: 1061/1235
mf: CH_5N mw: 31.07

PROP: Colorless gas or liquid. Powerful ammoniacal odor, bp: 6.3°; lel: 4.95% ; uel: 20.75%; mp: −93.5°; flash p: 32°F (CC); d: 0.662 @ 20°/4°, autoign temp: 806°F; vap d: 1.07. Flammable gas @ ordinary temp and pressure. Fuming liquid when liquefied; d: 0.699 @ −10.8°/4°. Sol in alc; misc with ether.

SYNS: AMINOMETHANE * METHYLAMINEN (DUTCH) * METILAMINE (ITALIAN) * METY-LOAMINA (POLISH) * MONOMETHYLAMINE

OSHA PEL: TWA 10 ppm
ACGIH TLV: TWA 10 ppm
DFG MAK: 10 ppm (12 mg/m³)
DOT Classification: Label: Flammable Gas/ Flammable Liquid

THR: Poison by subcutaneous route. A skin irritant. Very dangerous fire hazard. Moderately explosive from sparks or flame. To fight fire, stop flow of gas. When heated to decomposition it emits toxic fumes of NO_x. Incompatible with nitromethane. For further information see Vol. 5, No. 4 of *DPIM Report*.

METHYLAMINOCOLCHICIDE HR: 3
CAS: 63917-71-5 NIOSH: DF 9460000
mf: $C_{22}H_{26}N_2O_5$ mw: 398.50

SYN: METHYLCOLCHAMINONE

THR: Poison by intravenous and intraperitoneal routes. When heated to decomposition it emits toxic fumes of NO_x.

2-METHYLAMINOETHANOL HR: 3
CAS: 109-83-1 NIOSH: KL 6650000
mf: C_3H_9NO mw: 75.11

PROP: Viscous liquid, fishy odor. Corrosive to skin, cork and metals. Strong base; d: 0.9, vap d: 2.9, bp: 156°, flash p: 165°F (OC). Miscible with water, alc and ether. *Caution:* Irr to skin, eyes, mumem.

SYNS: BETA-(METHYLAMINO)ETHANOL * METHYLETHANOLAMINE * N-METHYLETHA-NOLAMINE * MONOMETHYLAMINOETHANOL * METHYL(BETA-HYDROXYETHYL)AMINE * MONOMETHYL-AMINOAETHANOL (GERMAN) * N-MONOMETHYLAMINOETHANOL * MONO-ETHYLETHANOLAMINE * USAF DO-50

THR: Poison by intraperitoneal route. Moderately toxic by ingestion and subcutaneous route. Severe eye irritant. A skin irritant. Moderately flammable when exposed to heat, flame, or oxidizers. To fight fire, use alcohol foam. When heated to decomposition it emits toxic fumes such as NO_x.

1-(3-METHYLAMINOPROPYL)-2-ADA-MANTANOL HYDROCHLORIDE   HR: 3
CAS: 52777-39-6   NIOSH: AU 5000000
mf: $C_{14}H_{25}NO \cdot ClH$ mw: 259.86

SYNS: 2-HYDROXY-1-(3-METHYLAMINOPRO-PYL)ADAMANTANE HYDROCHLORIDE * 2-HY-DROXY-N-METHYL-1-ADAMANTANEPROPANAMINE HYDROCHLORIDE * 3-(2-HYDROXY-1-ADAMAN-TYL)-N-METHYLPROPLYAMINE HYDROCHLORIDE

THR: Poison by ingestion and intraperitoneal routes. When heated to decomposition it emits very toxic fumes of NO_x and HCl.

METHYLANILINE   HR: 3
CAS: 100-61-8   NIOSH: BY 4550000
DOT: 2294
mf: C_7H_9N mw: 107.17

PROP: Colorless or slightly yellow liquid; becomes brown on exposure to air. Mp: −57°, d: 0.989 @ 20°/4°; bp: 194°-197°; sol in alc, ether; sltly sol in water.

SYNS: ANILINOMETHANE * (METHYLAMINO)-BENZENE * N-METHYLAMINOBENZENE * N-METHYLANILINE * N-METHYLBENZEN-AMINE * METHYLPHENYLAMINE * N-METH-YLPHENYLAMINE * MONOMETHYLANILINE * N-MONOMETHYLANILINE * N-PHENYLMETH-YLAMINE

OSHA PEL: TWA 2 ppm (skin)
ACGIH TLV: TWA 0.5 ppm
DFG MAK: 2 ppm (9 mg/m³)
DOT Classification: Poison B

THR: Poison by ingestion and intravenous routes. Moderately toxic by subcutaneous route. When heated to decomposition it emits toxic fumes of NO_x.

2-METHYL-p-ANISIDINE   HR: 3
CAS: 102-50-1   NIOSH: BZ 6730000
mf: $C_8H_{11}NO$ mw: 137.20

SYNS: M-CRESIDINE * 4-METHOXY-2-METH-YLANILINE * NCI-C02993

THR: An experimental carcinogen and tumorigen by ingestion. When heated to decomposition it emits toxic fumes of NO_x.

5-METHYL-o-ANISIDINE   HR: 3
CAS: 120-71-8   NIOSH: BZ 6825000
mf: $C_8H_{11}NO$ mw: 137.20

SYNS: 3-AMINO-P-CRESOL METHYL ETHER * 1-AMINO-2-METHOXY-5-METHYLBENZENE * 3-AMINO-4-METHOXYTOLUENE * 2-AMINO-4-METHYLANISOLE * 2-METHOXY-5-METHYL-ANILINE * 2-METHOXY-5-METHYLBENZEN-AMINE * 4-METHOXY-M-TOLUIDINE * NCI-C02982

THR: An experimental carcinogen and neoplastigen. Moderately toxic by ingestion. See also ethers. When heated to decomposition it emits toxic fumes of NO_x.

9-METHYLANTHRACENE   HR: 3
CAS: 779-02-2   NIOSH: CB 0700000
mf: $C_{15}H_{12}$ mw: 192.27

THR: An experimental carcinogen. Mutagenic data. When heated to decomposition it emits acrid smoke and fumes.

METHYLARSENIC SULFIDE   HR: 3
CAS: 2533-82-6   NIOSH: CH 7000000
mf: CH_3AsS mw: 122.02

SYNS: METHYLTHIOXOARSINE * METHYLAR-SINE SULFIDE * METHYLARSINIC SULFIDE

THR: Poison by ingestion. Moderately toxic by skin contact. Mutagenic data. See also arsenic compounds and sulfides. When heated to decomposition it emits very toxic fumes of As and SO_x.

2-METHYLAZIRIDINE   HR: 3
CAS: 75-55-8   NIOSH: CM 8050000
mf: C_3H_7N mw: 57.11

PROP: Liquid; vap d: 2.0, flash p.: 14°F.

SYNS: 2-METHYLAZACYCLOPROPANE * METH-YLETHYLENIMINE * 2-METHYLETHYLENIMINE * 1,2-PROPYLENEIMINE

OSHA PEL: TWA 2 ppm (skin)
DOT Classification: Label: Flammable Liquid

THR: Poison by ingestion, inhalation, and skin contact. Mutagenic data. Severe eye irritant. An experimental carcinogen by ingestion. Implicated as a brain carcinogen. A dangerous fire hazard when exposed to heat or flame. When heated to decomposition it emits toxic fumes of NO_x.

METHYL AZOXYMETHYL ACETATE   HR: 3
CAS: 592-62-1   NIOSH: PC 2800000
mf: $C_4H_8N_2O_3$ mw: 132.14

SYNS: METHYLAZOXYMETHANOL ACETATE * (METHYL-ONN-AZOXY)METHANOL, ACETATE (ESTER)

THR: Poison by ingestion and intravenous routes. Mutagenic data. An experimental teratogen, tumorigen, carcinogen, and neoplastigen. See also esters. When heated to decomposition it emits toxic fumes of NO_x.

4-METHYLBENZ(a)ANTHRA-CENE   HR: 3
CAS: 316-49-4   NIOSH: CX 1050000
mf: $C_{19}H_{14}$ mw: 242.33

SYN: 4'-METHYL-1:2-BENZANTHRACENE

THR: An experimental tumorigen. Mutagenic data. When heated to decomposition it emits acrid smoke and fumes.

5-METHYLBENZ(a)ANTHRA-CENE   HR: 3
CAS: 2319-96-2   NIOSH: CX 1225000
mf: $C_{19}H_{14}$ mw: 242.33

SYN: 3-METHYL-1,2-BENZANTHRACENE

THR: Poison by skin contact and subcutaneous routes. An experimental tumorigen. Mutagenic data. When heated to decomposition it emits acrid smoke and fumes.

6-METHYLBENZ(a)ANTHRA-CENE   HR: 3
CAS: 316-14-3   NIOSH: CX 1400000
mf: $C_{19}H_{14}$ mw: 242.33

SYN: 4-METHYL-1,2-BENZANTHRACENE

THR: Poison by skin contact and subcutaneous route. An experimental tumorigen. Mutagenic data. When heated to decomposition it emits acrid smoke and fumes.

8-METHYLBENZ(a)ANTHRA-CENE   HR: 3
CAS: 2381-31-9   NIOSH: CX 1750000
mf: $C_{19}H_{14}$ mw: 242.33

SYN: 5-METHYL-1,2-BENZANTHRACENE

THR: Poison by skin contact and subcutaneous route. An experimental tumorigen. Mutagenic data. When heated to decomposition it emits acrid smoke and fumes.

9-METHYLBENZ(a)ANTHRA-CENE   HR: 3
CAS: 2381-16-0   NIOSH: CX 1925000
mf: $C_{19}H_{14}$ mw: 242.33

SYN: 6-METHYL-1,2-BENZANTHRACENE

THR: An experimental tumorigen. Mutagenic data. When heated to decomposition it emits acrid smoke and fumes.

10-METHYL-1,2-BENZANTHRA-CENE   HR: 3
CAS: 2541-69-7   NIOSH: CX 1575000
mf: $C_{19}H_{14}$ mw: 242.33

SYNS: 7-MBA * 10-METHYL-1,2-BENZAN-THRACEN (GERMAN) * 7-METHYLBENZ(a)AN-THRACENE

THR: An experimental tumorigen and neoplastigen. Mutagenic data. When heated to decomposition it emits acrid smoke and fumes.

10-METHYLBENZ(a)ANTHRA-CENE   HR: 3
CAS: 2381-15-9   NIOSH: CX 2100000
mf: $C_{19}H_{14}$ mw: 242.33

SYN: 7-METHYL-1,2-BENZANTHRACENE

THR: An experimental tumorigen. Mutagenic data. When heated to decomposition it emits acrid smoke and fumes.

12-METHYLBENZ(a)ANTHRA-CENE   HR: 3
CAS: 2422-79-9   NIOSH: CX 2450000
mf: $C_{19}H_{14}$ mw: 242.33

SYN: 9-METHYL-1,2-BENZANTHRACENE

THR: An experimental tumorigen. Mutagenic data. When heated to decomposition it emits acrid smoke and fumes.

7-METHYLBENZ(a)ANTHRACENE-5,6-OXIDE HR: 3
CAS: 1155-38-0 NIOSH: CW 5075000
mf: $C_{19}H_{14}O$ mw: 258.33

SYN: 5,6-EPOXY-5,6-DIHYDRO-7-METHYL-BENZ(A)ANTHRACENE

THR: An experimental tumorigen and neoplastigen. Mutagenic data. When heated to decomposition it emits acrid smoke and fumes.

METHYL-2-BENZIMIDAZOLE HR: 3
CAS: 615-15-6 NIOSH: DD 9100000
mf: $C_8H_8N_2$ mw: 132.18

PROP: Needles from water; mp: 175°-176°; sol in hot water, NaOH; sltly sol in alc and ether.

SYN: 2-METHYLBENZIMIDAZOLE

THR: Poison by intravenous route. Moderately toxic by ingestion. Mutagenic data. When heated to decomposition it emits toxic fumes of NO_x.

METHYL BENZOATE HR: 2
CAS: 93-58-3 NIOSH: DH 3850000
mf: $C_8H_8O_2$ mw: 136.16

PROP: Colorless, liquid, fragrant odor, mp: −12.5°, bp: 199.6°, flash p: 181°F, d: 1.0937, vap press: 1 mm @ 39.0°, vap d: 4.69. Sol in water @ 30°; misc in alc and ether.

SYNS: METHYL BENZENECARBOXYLATE * NIOBE OIL * OIL OF NIOBE

THR: Moderately toxic by ingestion. A skin and eye irritant. Flammable when exposed to heat or flame; can react with oxidizing materials. To fight fire, use foam, CO_2, dry chemical, water to blanket fire.

5-METHYLBENZO(c)PHENAN-THRENE HR: 3
CAS: 652-04-0 NIOSH: DI 9450000
mf: $C_{19}H_{14}$ mw: 242.33

SYN: 2-METHYL-3,4-BENZPHENANTHRENE

THR: An experimental neoplastigen and tumorigen.

4-METHYL-p-BENZOPHENONE HR: 3
CAS: 134-84-9 NIOSH: DJ 1750000
mf: $C_{14}H_{12}O$ mw: 196.26

SYNS: PHENYL-P-TOLYL KETONE * USAF DO-54

THR: Poison by intraperitoneal route. When heated to decomposition it emits acrid smoke and fumes.

4'-METHYLBENZO(a)PYRENE HR: 3
CAS: 63041-77-0 NIOSH: DJ 7440000
mf: $C_{21}H_{14}$ mw: 266.35

SYNS: 7-METHYLBENZO(A)PYRENE * 4'-METHYL-3:4-BENZPYRENE

THR: An experimental neoplastigen and carcinogen. When heated to decomposition it emits acrid smoke and fumes.

2-METHYL-p-BENZOQUINONE HR: 3
CAS: 553-97-9 NIOSH: DK 6300000
mf: $C_7H_6O_2$ mw: 122.13

SYNS: METHYL-P-BENZOQUINONE * METHYL-1,4-BENZOQUINONE * 2-METHYLBENZOQUINONE-1,4 * 2-METHYL-1,4-QUINONE * P-TOLUQUINONE * 1,4-TOLUQUINONE

THR: Poison by ingestion. When heated to decomposition it emits acrid smoke and fumes.

2-((2-METHYLBENZO(b)THIEN-3-YL)-METHYL)-2-IMIDAZOLINE HYDRO-CHLORIDE HR: 3
CAS: 5090-37-9 NIOSH: NI 4805000
mf: $C_{13}H_{14}N_2S \cdot ClH$ mw: 266.81

SYNS: 4,5-DIHYDRO-2-((2-METHYLBENZO-(B)THIEN-3-YL)METHYL)-1H-IMIDAZOLE HYDRO-CHLORIDE * 2-METHYL-3-(DELTA(SUP 2)-IMIDAZOLINYLMETHYL)BENZO(B)THIOPHENE HYDROCHLORIDE

THR: Poison by ingestion, intravenous, and intraperitoneal routes. When heated to decomposition it emits very toxic fumes of NO_x, SO_x, and HCl.

8-METHYL-3:4-BENZPHENAN-THRENE HR: 3
CAS: 4076-40-8 NIOSH: DI 9275000
mf: $C_{19}H_{14}$ mw: 242.33

SYN: 4-METHYLBENZO(C)PHENANTHRENE

THR: An experimental tumorigen. When heated to decomposition it emits acrid smoke and fumes.

5-METHYL-3,4-BENZPYRENE HR: 3
CAS: 2381-39-7 NIOSH: DJ 7375000
mf: $C_{21}H_{14}$ mw: 266.35

SYNS: 6-METHYLBENZO(A)PYRENE * 5-METHYL-3,4-BENZOPYRENE

THR: An experimental neoplastigen, tumorigen, and carcinogen. Mutagenic data. When heated to decomposition it emits acrid smoke and fumes.

1-METHYL-2-BENZYLHYDRA-ZINE   HR: 3
CAS: 10309-79-2   NIOSH: MU 8925000
mf: $C_8H_{12}N_2$ mw: 136.22

SYN: 1-BENZYL-2-METHYLHYDRAZINE

THR: Poison by subcutaneous route. An experimental teratogen and tumorigen. When heated to decomposition it emits toxic fumes of NO_x.

N-METHYL-N-BENZYLNITROS-AMINE   HR: 3
CAS: 937-40-6   NIOSH: DP 6300000
mf: $C_8H_{10}N_2O$ mw: 150.20

SYNS: METHYL-BENZYL-NITROSOAMIN (GERMAN) * N-METHYL-N-NITROSOBENZYLAMINE * N-NITROSOBENZYLMETHYLAMINE * N-NITROSOMETHYLBENZYLAMINE

THR: Poison by ingestion. An experimental tumorigen. Mutagenic data. When heated to decomposition it emits toxic fumes of NO_x.

2-METHYL BUTANOL-1   HR: 2
CAS: 137-32-6   NIOSH: EL 5250000
mf: $C_5H_{12}O$ mw: 88.15

PROP: Colorless liquid, sltly sol in water, miscible with alcohol and ether, d: 0.81-0.82 @ 20°, fp: $< -70°$, bp: 128°, flash p: 122°F (OC), vap d: 3.0, lel = 1.4%; uel = 9.0%.

SYNS: 2-METHYLBUTANOL * SEC-BUTYLCARBINOL

THR: Moderately toxic by ingestion and skin contact. An eye, skin, and mucous membrane irritant. Can cause deafness, delerium, headache, nausea, and vomiting. Flammable when exposed to heat, flame, or oxidizers. Incompatible with H_2S_3. To fight fire, use alcohol foam, spray, mist, dry chemical.

2-METHYL BUTYLACRYLATE   HR: 3
CAS: 97-88-1   NIOSH: OZ 3675000
mf: $C_8H_{14}O_2$ mw: 142.22

PROP: Colorless liquid; ester odor; bp: 163°; flash p: 126°F (TOC); lel = 2%; uel = 8%;

autoign temp: 562°F; vap press: 4.9 mm @ 20°; d: 0.895 @ 20°/4°; vap d: 4.8.

SYNS: BUTIL METACRILATO (ITALIAN) * BUTYLMETHACRYLAAT (DUTCH) * N-BUTYL METHACRYLATE * BUTYL-2-METHACRYLATE * BUTYL-2-METHYL-2-PROPENOATE * METHACRYLATE DE BUTYLE (FRENCH) * METHACRYLSAEUREBUTYLESTER (GERMAN) * 2-METHYL-BUTYLACRYLAAT (DUTCH) * 2-METHYL-BUTYLACRYLAT (GERMAN)

THR: Poison by intraperitoneal route. Moderately toxic by other routes. An experimental teratogen. A skin irritant. Flammable. An explosion hazard. Violent polymerization can be caused by heat, moisture, oxidizers. To fight fire, use foam, dry chemical, CO_2. When heated to decomposition it emits acrid smoke and fumes.

METHYL-1-(BUTYLCARBAMOYL)-2-BENZIMIDAZOLYLCARBAMATE   HR: 3
CAS: 17804-35-2   NIOSH: DD 6475000
mf: $C_{14}H_{18}N_4O_3$ mw: 290.36

SYNS: 1-(BUTYLCARBAMOYL)-2-BENZIMIDAZOLECARBAMIC ACID, METHYL ESTER * 1-(N-BUTYLCARBAMOYL)-2-(METHOXY-CARBOXAMIDO)-BENZIMIDAZOL (GERMAN) * FUNGICIDE 1991

THR: Poison by ingestion. An experimental teratogen. Mutagenic data. When heated to decomposition it emits toxic fumes of NO_x.

METHYLBUTYLNITROSAMINE   HR: 3
CAS: 7068-83-9   NIOSH: EO 5425000
mf: $C_5H_{12}N_2O$ mw: 116.19

SYNS: METHYL-BUTYL-NITROSAMIN (GERMAN) * METHYL-N-BUTYLNITROSAMINE * N-METHYL-N-NITROSOBUTYLAMINE * N-NITROSO-N-BUTYLMETHYLAMINE * N-NITROSOMETHYL-N-BUTYLAMINE

THR: Poison (acute) by ingestion, inhalation, and subcutaneous routes. Mutagenic data. An experimental carcinogen and tumorigen. See also nitrosamines. When heated to decomposition it emits toxic fumes of NO_x.

8-(3-METHYLBUTYRYLOXY) DIACE-TOXYSCIRPENOL   HR: 3
CAS: 21259-20-1   NIOSH: VR 3545000
mf: $C_{24}H_{34}O_9$ mw: 466.58

SYN: 3-ALPHA-HYDROXY-4-BETA,15-DIACETOXY-8-ALPHA(3-METHYLBUTYRYLOXY)-12,13-EPOXY-TRICHLOTHEC-9-ENE

THR: An experimental carcinogen. A skin irritant. When heated to decomposition it emits acrid smoke and fumes.

METHYL CARBAMATE   HR: 3
CAS: 598-55-0   NIOSH: FC 2450000
mf: $C_2H_5NO_2$ mw: 75.07

PROP: Needles; bp: 177°; mp: 52°-54°; very sol in water, alc.

SYNS: METHYLURETHAN * METHYLURETHANE * NCI-C55594 * URETHYLANE

THR: Poison by intraperitoneal route. Mutagenic data. An experimental carcinogen. When heated to decomposition it emits toxic fumes of NO_x.

METHYLCARBAMIC ACID-o-CUMENYL ESTER   HR: 3
CAS: 2631-40-5   NIOSH: FB 7880000
mf: $C_{11}H_{15}NO_2$ mw: 193.27

SYNS: BAY 105807 * ENT 25670 * ISO-PROPYLPHENOL METHYLCARBAMATE * O-ISO-PROPYLPHENYL-N-METHYLCARBAMATE * 2-ISO-PROPYL-PHENYL-N-METHYLCARBAMATE

THR: Poison by ingestion, intravenous, and intraperitoneal routes. Moderately toxic by skin contact. When heated to decomposition it emits toxic fumes of NO_x.

METHYL CELLOSOLVE ACRYLATE   HR: 3
CAS: 3121-61-7   NIOSH: KL 6000000
mf: $C_6H_{10}O_3$ mw: 130.16

PROP: Liquid, bp: 61° @ 17 mm, flash p: 180°F (OC), d: 1.0134 @ 20°, vap d: 4.49.

SYNS: ACRYLIC ACID, 2-METHOXYETHYL ESTER * ETHYLENE GLYCOL MONOMETHYL ETHER ACRYLATE * GLYCOL MONOMETHYL ETHER ACRYLATE * 2-METHOXYETHANOL, ACRYLATE

THR: Poison by skin contact. Moderately toxic by ingestion and inhalation. Skin irritant. Flammable when exposed to heat or flame; can react with oxidizing materials. To fight fire, use foam, CO_2, dry chemical. When heated to decomposition it emits acrid smoke and fumes.

METHYL CHLOROCARBONATE   HR: 3
CAS: 79-22-1   NIOSH: FG 3675000
mf: $C_2H_3ClO_2$ mw: 94.50

PROP: Colorless liquid, bp: 71.4°, d: 1.223 @ 20°/4°, vap d: 3.26, flash p: 54°F, autoign temp: 940°F. Sltly sol in H_2O with gradual decomp; misc with alc, benzene, chloroform, ether. Caution: Vapors strongly irr to eyes.

SYNS: CHLORAMEISENSAEURE METHYLESTER (GERMAN) * CHLOROCARBONATE DE METHYLE (FRENCH) * METHYLCHLOORFORMIAT (DUTCH) * METHYL CHLOROFORMATE * METHYL CHLOROFORMATE (DOT) * METILCLOROFORMIATO (ITALIAN)

DOT Classification: Label: Organic Peroxide and Flammable Liquid

THR: Poison by ingestion, inhalation, and intraperitoneal routes. Human irritant (systemic) effects. Dangerous; when heated to decomposition it emits highly toxic fumes of methyl chloroformate and phosgene; will react with water or steam to produce toxic and corrosive fumes. Very dangerous fire hazard when exposed to heat sources, sparks, flame or oxidizers. When heated to decomposition it emits toxic fumes of Cl^-.

METHYL-beta-CHLOROETHYL-beta-HYDROXYETHYLAMINE HYDROCHLORIDE   HR: 3
CAS: 63905-05-5   NIOSH: PF 7350000
mf: $C_5H_{12}ClNO \cdot ClH$ mw: 174.09

THR: Poison by ingestion, subcutaneous, intraperitoneal, and intravenous routes. When heated to decomposition it emits very toxic fumes of Cl^-, NO_x, and HCl.

METHYLCHLOROTHION   HR: 2
CAS: 500-28-7   NIOSH: TE 8050000
mf: $C_8H_9ClNO_5PS$ mw: 297.66

SYNS: O-(3-CHLOOR-4-NITRO-FENYL)-O,O-DIMETHYL-MONOTHIOFOSFAAT (DUTCH) * CHLOOR-THION (DUTCH) * O-(3-CLORO-4-NITRO-FENIL)-O,O-DIMETIL-MONOTIOFOSFATO (ITALIAN) * O-(3-CHLOR-4-NITRO-PHENYL)-O,O-DIMETHYL-MONOTHIOPHOSPHAT (GERMAN) * O-(3-CHLORO-4-NITROPHENYL) O,O-DIMETHYL PHOSPHOROTHIOATE * CHLORTHION METHYL * CHLORTION (CZECH) * O,O-DIMETHYL-O-3-CHLOR-4-NITROFENYLTIOFOSFAT (CZECH) * O,O-DIMETHYL-O-(3-CHLOR-4-NITROPHENYL)-MONOTHIOPHOSPHAT (GERMAN) * O,O-DI-METHYL-O-(3-CHLORO-4-NITROPHENYL) PHOSPHOROTHIOATE * DIMETHYL-3-CHLORO-4-NITROPHENYL THIONOPHOSPHATE * O,O-DIMETHYL-O-(3-CHLORO-4-NITROPHENYL)

THIOPHOSPHATE * O,O-DIMETHYL-P-NITRO-M-CHLOROPHENYL THIOPHOSPHATE * O,O-DIMETHYL-O-4-NITRO-3-CHLOROPHENYL THIOPHOSPHATE * ENT 18,861 * P-NITRO-M-CHLOROPHENYL DIMETHYL THIONOPHOSPHATE * THIOPHOSPHATE DE O,O-DIMETHYLE ET DE O-3-CHLORO-4-NITROPHENYLE (FRENCH)

THR: Moderately toxic by ingestion, skin contact, intraperitoneal, and possibly other routes. When heated to decomposition it emits very toxic fumes of Cl^-, SO_x, PO_x, and NO_x. For further information see chlorothion, Vol. 2, No. 2 of *DPIM Report*.

METHYLCHLORTETRACYCLINE   HR: 3
CAS: 127-33-3   NIOSH: QI 7650000
mf: $C_{21}H_{21}ClN_2O_8$ mw: 464.89

SYNS: 7-CHLORO-6-DEMETHYLTETRACYCLINE DEMETHYLCHLOROTETRACYCLINE * 6-DEMETHYLCHLOROTETRACYCLINE * 6-DEMETHYL-7-CHLOROTETRACYCLINE DEMETHYLCHLORTETRACYCLINE * 6-DEMETHYLCHLORTETRACYCLINE * 6-DEMETHYL-7-CHLORTETRACYCLINE

THR: Poison by intravenous route. An experimental teratogen. Human systemic and skin (systemic) effects. When heated to decomposition it emits very toxic fumes of Cl^- and NO_x.

3-METHYLCHOLANTHRENE   HR: 3
CAS: 56-49-5   NIOSH: FZ 3675000
mf: $C_{21}H_{16}$ mw: 268.37

PROP: Pale yellow needles from benzene; mp: 176.5°, bp: 280° @ 80 mm, d: 1.28 @ 20°; sol in benzene, xylene, toluene; sltly sol in amyl alc; insol in H_2O.

SYNS: 1,2-DIHYDRO-3-METHYL-BENZ(J)ACEANTHRYLENE * METHYLCHOLANTHRENE * 20-METHYLCHOLANTHRENE

THR: Poison by intravenous and intraperitoneal routes. An experimental teratogen, tumorigen, carcinogen, and neoplastigen. Mutagenic data. When heated to decomposition it emits acrid smoke and fumes. For further information see Vol. 6, No. 1 of *DPIM Report*.

3-METHYLCHOLANTHRENE-2-ONE   HR: 3
CAS: 3343-08-6   NIOSH: FZ 4550000
mf: $C_{21}H_{14}O$ mw: 282.35

SYNS: 2-CHOLANTHRENONE, 3-METHYL- * 3-METHYLCHOLANTHREN-2-ONE

THR: Mutagenic data. An experimental carcinogen and neoplastigen. When heated to decomposition it emits acrid smoke and fumes.

20-METHYLCHOLANTHRENE PICRATE   HR: 3
CAS: 63041-80-5   NIOSH: CU 0950000
mf: $C_{21}H_{16} \cdot C_6H_3N_3O_7$ mw: 497.49

SYNS: 1,2-DIHYDRO-3-METHYLBENZ(J)ACEANTHRYLENE COMPOUND WITH 2,4,6-TRINITROPHENOL (1:1) * 3-METHYLCHOLANTHRENE COMPOUND WITH PICRIC ACID (1:1) * 2,4,6-TRINITROPHENOL COMPOUND WITH 1,2-DIHYDRO-3-METHYLBENZ(J)ACEANTHRYLENE

THR: An experimental carcinogen. When heated to decomposition it emits toxic fumes of NO_x.

20-METHYLCHOLANTHREN-15-ONE   HR: 3
CAS: 3343-07-5   NIOSH: FZ 4375000
mf: $C_{21}H_{14}O$ mw: 282.35

SYNS: 15-KETO-20-METHYLCHOLANTHRENE * 3-METHYLCHOLANTHREN-1-ONE

THR: An experimental neoplastigen and carcinogen. Mutagenic data. When heated to decomposition it emits acrid smoke and fumes.

3-METHYLCHRYSENE   HR: 3
CAS: 3351-31-3   NIOSH: GC 1380000
mf: $C_{19}H_{14}$ mw: 242.33

THR: An experimental carcinogen and neoplastigen. Mutagenic data. When heated to decomposition it emits acrid smoke and fumes.

4-METHYL CHRYSENE   HR: 3
CAS: 3351-30-2   NIOSH: GC 1400000
mf: $C_{19}H_{14}$ mw: 242.33

THR: An experimental carcinogen. Mutagenic data. When heated to decomposition it emits acrid smoke and fumes.

5-METHYLCHRYSENE   HR: 3
CAS: 3697-24-3   NIOSH: GC 1575000
mf: $C_{19}H_{14}$ mw: 242.33

THR: An experimental carcinogen, neoplastigen, and tumorigen. Mutagenic data. When heated to decomposition it emits acrid smoke and fumes.

6-METHYLCHRYSENE   HR: 3
CAS: 1705-85-7   NIOSH: GC 1750000
mf: $C_{19}H_{14}$ mw: 242.33

THR: An experimental tumorigen. Mutagenic data. When heated to decomposition it emits acrid smoke and fumes.

9-METHYL-10-CYANO-1,2-BENZAN-THRACENE   HR: 3
CAS: 63020-25-7   NIOSH: CW 1400000
mf: $C_{20}H_{14}N$ mw: 268.35

SYN: BENZ(A)ANTHRACENE, 7-CYANO-12-METHYL-

THR: An experimental tumorigen. When heated to decomposition it emits toxic fumes of NO_x.

METHYLCYCLOHEXANONE   HR: 2
CAS: 1331-22-2   NIOSH: GW 1575000
mf: $C_7H_{12}O$ mw: 112.19

PROP: Water-white to pale yellow liquid, acetone-like odor, mp: $-14°C$; bp: 160°-170°, flash p: 118°F (CC), d: 0.925 @ 15°/5°, vap d: 3.86; insol in water; sol in ether and alc.

SYN: METYLOCYKLOHEKSANON (POLISH)

THR: Moderately toxic by ingestion and skin contact. A toxic compound which can damage the kidneys and the liver. It is similar to cyclohexanol in its toxic action, although it is somewhat less active. Harmful exposures in industry are rare. Experimental animals can withstand prolonged exposures to from 0.02-0.05% by volume in air. Flammable when exposed to heat or flame. Can react violently with HNO_3 and other oxidizers. No spontaneous heating. To fight fire, use foam, CO_2, dry chemical.

2-METHYLCYCLOHEXANONE   HR: 3
CAS: 583-60-8   NIOSH: GW 1750000
mf: $C_7H_{12}O$ mw: 112.19

PROP: Liquid; d: 0.925 @ 20/4°; mp: $-14°$; bp: 165.1°; insol in water; sol in alc and ether.

SYNS: 2-METHYL-CYCLOHEXANON (GERMAN, DUTCH) * O-METHYLCYCLOHEXANONE * 2-METILCICLOESANONE (ITALIAN)

OSHA PEL: TWA 100 ppm (skin)

THR: Poison by intraperitoneal route. Moderately toxic by ingestion and skin contact. When heated to decomposition it emits acrid smoke and fumes.

N-METHYL-4-CYCLOHEXENE-1,2-DI-CARBOXIMIDE   HR: 3
NIOSH: GW 4820000
mf: $C_9H_{11}NO_2$ mw: 165.21

SYN: N-METHYL-1,2,3,6-TETRAHYDROPHTHALI-MIDE

THR: An experimental teratogen. When heated to decomposition it emits toxic fumes of NO_x.

METHYL DEMETON METHYL   HR: 3
CAS: 2587-90-8   NIOSH: TF 9450000
mf: $C_5H_{13}O_3PS_2$ mw: 216.27

PROP: Pale yellow oil; bp: 89° @ 0.15 mm; d: 1.207 @ 20°/4°; sol in H_2O @ room temp; sol in organic solvents.

SYNS: ETHANETHIOL, 2-(METHYLTHIO)-, O,O-DIMETHYL PHOSPHOROTHIOATE * ETHAN-ETHIOL, 2-(METHYLTHIO)-, S-ESTER WITH O,O-DIMETHYL PHOSPHOROTHIOATE * DEMETHIIN S.

THR: Poison by ingestion and skin contact. *Caution:* It is a cholinesterase inhibitor. When heated to decomposition it emits very toxic fumes of PO_x and SO_x.

N-METHYL-N-DESACETYLCOL-CHICINE   HR: 3
CAS: 477-30-5   NIOSH: GH 0800000
mf: $C_{21}H_{25}NO_5$ mw: 371.47

SYNS: COLCHINE, N-DEACETYL-NMETHYL * DEACETYLMETHYLCOLCHICINE * DEACETYL-N-METHYLCOLCHICINE * N-DEACETYL-N-METHYLCOLCHICINE * N-DESACETYL-N-METHYLCOLCHICINE * 6,7-DIHYDRO-1,2,3, 10-TETRAMETHOXY-7-(METHYLAMINO)-BENZO(ALPHA)HEPTALEN-9(5H)-ONE * NSC 3096 * SANTAVY'S SUBSTANCE F

THR: Poison by intraperitoneal, parenteral, intravenous, and intramuscular routes. Mutagenic data. An experimental teratogen. When heated to decomposition it emits toxic fumes of NO_x.

3-METHYLDIBENZ(a,h)ANTHRA-CENE   HR: 3
CAS: 63041-84-9   NIOSH: HN 5475000
mf: $C_{23}H_{16}$ mw: 292.39

SYN: 3'-METHYL-1:2:5:6-DIBENZANTHRACENE

THR: An experimental tumorigen. When heated to decomposition it emits acrid smoke and fumes.

4-METHYL-1,2,5,6-DIBENZANTHRA-CENE   HR: 3
CAS: 63041-85-0   NIOSH: HN 5500000
mf: $C_{23}H_{16}$ mw: 292.39

SYN: 6-METHYL DIBENZ(A,H)ANTHRACENE

THR: An experimental carcinogen. When heated to decomposition it emits acrid smoke and fumes.

METHYL DICHLORO ACETATE HR: 3
CAS: 116-5-4-1 NIOSH: AG 6625000
mf: $C_3H_4Cl_2O_2$ mw: 142.97

PROP: Colorless liquid, ethereal odor, bp: 143.0°; d: 1.3809 @ 19.2°/19.2°; vap d: 4.93.

SYNS: DICHLOROACETIC ACID METHYL ESTER * METHYL DICHLORO-ENTHANOATE

DOT Classification: Label: Corrosive

THR: Poison by inhalation. Severe irritant to eyes, skin, and mucous membranes. Hydrolyzes upon contact with moisture to form a product corrosive to tissue. See also dichloracetic acid and also esters. Dangerous; when heated to decomposition it emits highly toxic fumes of phosgene.

beta-METHYLDIGOXIN HR: 3
CAS: 30685-43-9 NIOSH: FH 5310000
mf: $C_{42}H_{66}O_{14} \cdot 1/2C_3H_6O$ mw: 824.12

SYN: 3BETA,12BETA,14BETA-TRIHYDROXY-5-BETA-CARD-20(22) ENOLIDE-3-(4'')-O-METHYL-TRIDIGITOXOSIDE)

THR: Poison by ingestion, intravenous, and subcutaneous routes. When heated to decomposition it emits acrid smoke and fumes.

11-METHYL-15,16-DIHYDRO-17-OXOCYCLOPENTA(a)PHENANTHRENE HR: 3
CAS: 892-17-1 NIOSH: GY 5775000
mf: $C_{18}H_{14}O$ mw: 246.32

SYNS: 15,16-DIHYDRO-11-METHYL-17H-CYCLOPENTA(A)PHENANTHREN-17-ONE * 11-METHYL-15,16-DIHYDRO-17H-CYCLOPENTA(A)PHENANTHREN-17-ONE

THR: An experimental carcinogen, neoplastigen, and tumorigen. When heated to decomposition it emits acrid smoke and fumes.

METHYLDIHYDROPYRAN HR: 3
NIOSH: UP 7950000
mf: $C_6H_{10}O$ mw: 98.16

THR: An experimental carcinogen. Moderate acute toxicity by ingestion and subcutaneous routes. When heated to decomposition it emits acrid smoke and fumes.

N-METHYL-O,O-DIMETHYLTHIOLO-PHOSPHORYL-5-THIA-3-METHYL-2-VALERAMIDE HR: 3
CAS: 2275-23-2 NIOSH: TF 7900000
mf: $C_8H_{18}NO_4PS_2$ mw: 287.36

SYNS: AMERICAN CYANAMID-43073 * O,O-DIMETHYL S-2-(1-N-METHYLCARBAMOYLETHYL-MERCAPTO)ETHYL THIOPHOSPHATE * O,O-DIMETHYL S-(2-(1-METHYLCARBA-MOYLETHYLTHIO)ETHYL) PHOSPHOROTHIOATE * DIMETHYL S-(2-(1-METHYLCARBAMOYL-ETHYLTHIO ETHYL) PHOSPHOROTHIOLATE * ENT 26,613 * N-METHYL-3-THIA-2-METHYL-VALERAMID DER O,O-DIMETHYL-THIOLPHOSPHORSAEURE (GERMAN)

THR: Poison by ingestion, skin contact, and possibly other routes. Mutagenic data. When heated to decomposition it emits very toxic fumes of PO_x, SO_x, and NO_x.

((2-METHYL-1,3-DIOXALAN-4-YL)-METHYL)TRIMETHYLAMMONIUM IODIDE HR: 3
CAS: 4386-79-2 NIOSH: BR 7875000
mf: $C_8H_{18}NO_2 \cdot I$ mw: 287.17

SYN: ETHYL-GAMMA-TRIMETHYLAMMONIUM PROPANEDIOL IODIDE

THR: Poison by ingestion, intravenous, and subcutaneous routes. See also iodides. When heated to decomposition it emits very toxic fumes of NO_x and I^-.

2,2'-METHYLENEBIS(4-CHLOROPHE-NOL) HR: 3
CAS: 97-23-4 NIOSH: SM 0175000
mf: $C_{13}H_{10}Cl_2O_2$ mw: 269.13

PROP: Crystals, nearly insol in water; mp: 178°, vap press: 10^{-4} mm @ 100°.

SYNS: BIS(5-CHLOR-2-HYDROXYPHENYL)-METHAN (GERMAN) * BIS(5-CHLORO-2-HY-DROXYPHENYL)METHANE * BIS-2-HYDROXY-5-CHLORFENYLMETHAN (CZECH) * BIS(2-HYDROXY-5-CHLOROPHENYL)METHANE * DICHLOORFEEN (DUTCH) * 5,5'-DI-CHLORO-2,2'-DIHYDROXYDIPHENYLMETHANE * DI-(5-CHLORO-2-HYDROXYPHENYL)-METHANE * 2,2'-DIHYDROXY-5,5'-DICHLO-RODIPHENYLMETHANE * O,O-METHYLEEN-BIS(4-CHLOORFENOL) (DUTCH) * O,O-METILEN-BIS(4-CLOROFENOLO) (ITALIAN)

THR: Poison by intravenous route. Moderately toxic by ingestion and other routes. A skin and severe eye irritant. Can cause cramps and diarrhea. Possibly similar to DDT. See also DDT. When heated to decomposition it emits toxic fumes of Cl^-.

2-METHYLENECYCLOPROPANYLALANINE   HR: 3
CAS: 156-56-9   NIOSH: GZ 0790000
mf: $C_7H_{11}NO_2$ mw: 141.19

PROP: Principal toxin of the *Blighia Sapida* tree in the West Indies.

SYNS: L-ALPHA-AMINO-BETA-METHYLENE-CYCLOPROPANEPROPIONIC ACID * ALPHA-AMINOMETHYLENECYCLOPROPANEPROPIONIC ACID * HYPOGLYCINE * HYPOGLYCINE A

THR: When heated to decomposition it emits toxic fumes of NO_x.

4,4'-METHYLENEDIANILINE   HR: 3
CAS: 101-77-9   NIOSH: BY 5425000
mf: $C_{13}H_{14}N_2$ mw: 198.29

PROP: Tan flakes or lumps, faint amine-like odor; mp: 90°; flash p: 440°F.

SYNS: 4-(4-AMINOBENZYL)ANILINE * BIS-P-AMINOFENYLMETHAN (CZECH) * BIS(P-AMINOPHENYL)METHANE * BIS(4-AMINOPHENYL)-METHANE * P,P'-DIAMINODIFENYLMETHAN (CZECH) * 4,4'-DIAMINODIPHENYLMETHAN (GERMAN) * DIAMINODIPHENYLMETHANE * P,P'-DIAMINODIPHENYLMETHANE * 4,4'-DIAMINODIPHENYLMETHANE * DI-(4-AMINOPHENYL)METHANE * DIANALINEMETHANE * 4,4'-DIPHENYLMETHANEDIAMINE * METHYLENEBIS(ANILINE) * 4,4'-METHYLENEBISANILINE * METHYLENEDIANILINE * P,P'-METHYLENEDIANILINE * NCI-C54604

THR: Poison by ingestion, subcutaneous, and intraperitoneal routes. Systemic effects by ingestion. An experimental tumorigen. A severe eye irritant. Mutagenic data. It is not rapidly absorbed through the skin in dangerous quantities. Combustible when exposed to heat or flame. When heated to decomposition it emits highly toxic fumes of aniline and NO_x.

METHYLENE DIMETHANESULFONATE   HR: 3
CAS: 156-72-9   NIOSH: PA 8925000
mf: $C_3H_8O_6S_2$ mw: 204.23

SYN: METHYLENE BIS(METHANESULFONATE)

THR: Mutagenic data. An experimental teratogen. See also sulfonates. When heated to decomposition it emits toxic fumes of SO_x.

4,4'-METHYLENEDIMORPHOLINE   HR: 3
CAS: 5625-90-1   NIOSH: QE 6125000
mf: $C_9H_{18}N_2O_2$ mw: 186.29

SYNS: BIS(MORPHOLINO-)METHAN (GERMAN) * BISMORPHOLINO METHANE

THR: An experimental tumorigen. Moderately toxic by subcutaneous route. When heated to decomposition it emits toxic fumes of NO_x.

3,4-METHYLENEDIOXY-alpha-ETHYL-beta-PHENYLETHYLAMINE   HR: 3
CAS: 42542-07-4   NIOSH: SH 6815000
mf: $C_{11}H_{15}NO_2 \cdot ClH$ mw: 229.73

THR: Poison by intravenous and intraperitoneal routes. When heated to decomposition it emits very toxic fumes of HCl and NO_x.

1-(3,4-METHYLENEDIOXYPHENYL)-2-AMINOPROPANE   HR: 3
CAS: 6292-91-7   NIOSH: SI 0610000
mf: $C_{10}H_{13}NO_2 \cdot ClH$ mw: 215.70

SYNS: 3,4-METHYLENEDIOXY-ALPHA-METHYL-BETA-PHENYLETHYLAMINE HYDROCHLORIDE * ALPHA-METHYL-3,4-METHYLENEDIOXYPHENETHYLAMINE HYDROCHLORIDE

THR: Poison by intravenous and intraperitoneal routes. Mutagenic data. When heated to decomposition it emits very toxic fumes of HCl and NO_x.

3,4-METHYLENEDIOXY-beta-PHENYLETHYLAMINE HYDROCHLORIDE   HR: 3
CAS: 1653-64-1   NIOSH: SH 9710000
mf: $C_9H_{11}NO_2 \cdot ClH$ mw: 201.67

THR: Poison by intravenous and intraperitoneal routes. When heated to decomposition it emits very toxic fumes of HCl and NO_x.

METHYLENE DITHIOCYANATE   HR: 3
CAS: 6317-18-6   NIOSH: XL 1560000
mf: $C_3H_2N_2S_2$ mw: 130.19

SYN: METHYLENDIRHODANID (CZECH) (GERMAN)

THR: Poison by ingestion, intravenous, and subcutaneous routes. See also thiocyanates. When heated to decomposition it emits very toxic fumes of NO_x and SO_x.

4,4'-METHYLENE DI-o-TOLUIDINE HR: 3
CAS: 838-88-0   NIOSH: BY 5300000
mf: $C_{15}H_{18}N_2$ mw: 226.35

PROP: mp: 149°.

SYNS: BIS-4-AMINO-3-METHYLFENYLMETHAN (CZECH) * 3,3'-DIMETHYL-4,4'-DIAMINODIPHENYLMETHANE * 4,4'-METHYLENEBIS(2-METHYLBENZENAMINE) * 4,4'-METHYLENEBIS(2-METHYLANILINE)

THR: An experimental carcinogen. Severe eye irritant. Moderately toxic by ingestion. When heated to decomposition it emits toxic fumes of NO_x.

METHYLENE DIURETHAN HR: 3
CAS: 3693-53-6   NIOSH: FC 2200000
mf: $C_7H_{14}N_2O_4$ mw: 190.23

SYN: N,N-METHYLENE-BIS(ETHYL CARBAMATE)

THR: An experimental carcinogen. When heated to decomposition it emits toxic fumes of NO_x.

METHYLENE GREEN HR: 3
CAS: 2679-01-8   NIOSH: SP 5775000
mf: $C_{17}H_{17}N_4O_2S \cdot Cl$ mw: 364.88

SYN: 3,7-BIS(DIMETHYLAMINO)-4-NITRO-PHENOTHIAZIN-5-IUM CHLORIDE

THR: Poison by intravenous and intraperitoneal route. Moderately toxic by ingestion. When heated to decomposition it emits very toxic fumes of Cl^-, SO_x, and NO_x.

METHYLEPHEDRINE HR: 3
CAS: 17605-71-9   NIOSH: DO 4620000
mf: $C_{11}H_{17}NO$ mw: 179.25

PROP: dl form: crystals from petroleum ether or methanol; mp: 63.5°-64.5°; very sol in usual solvents. d form: crystals; mp: 87°-87.5°, l form: crystals from petroleum ether; mp: 87°-88°.

SYNS: METHYLEPHEDRIN (GERMAN) * 1-PHENYL-2-DIMETHYLAMINOPROPANOL

THR: A poison by intraperitoneal route. Moderately toxic by ingestion. When heated to decomposition it emits toxic fumes of NO_x.

N-METHYLEPINEPHRINE
CAS: 554-99-4   NIOSH: DO 5425000
mf: $C_{10}H_{15}NO_3$ mw: 197.23

SYNS: 3,4-DIHYDROXY-ALPHA-(DIMETHYLAMINOMETHYL) BENZYL ALCOHOL * ALPHA-(3,4-DIHYDROXYPHENYL)-BETA-DIMETHYLAMINOETHANOL * ALPHA-(3,4-DIHYDROXYPHENYL)-ALPHA-HYDROXY-BETA-DIMETHYLAMINOETHANE * ALPHA-(DIMETHYLAMINOMETHYL)PROTOCATECHUYL ALCOHOL * N-METHYLADRENALINE

THR: Poison by intravenous, intraperitoneal, and subcutaneous routes. When heated to decomposition it emits toxic fumes of NO_x.

METHYLERGONOVINE MALEATE HR: 3
CAS: 57432-61-8   NIOSH: KE 5950000
mf: $C_{24}H_{29}N_3O_6$ mw: 455.56

PROP: White to pinkish-tan microcryst powder, odorless, bitter taste; sltly sol in H_2O, alc; very sltly sol in chloroform, ether.

SYNS: MALEIC ACID, METHYL ERGONOVINE * USAF UCTL-8 * BASOFORTINA * METHEGRIN

THR: Poison by intraperitoneal route. When heated to decomposition it emits toxic fumes of NO_x.

METHYL ETHER HR: 3
CAS: 115-10-6   NIOSH: PM 4780000
DOT: 1033
mf: C_2H_6O mw: 46.08

PROP: Colorless gas, ether odor, mp: −138.5°, bp: −23.7°, lel = 3.4%, uel = 27%, flash p: −42°F (CC), autoign temp: 662°F, vap d: 1.617. d: 0.661 (air = 1), sol in alc, water, ether.

SYNS: DIMETHYL ETHER * WOOD ETHER

DOT Classification: Label: Flammable Gas

THR: Poison by inhalation. See also ethyl ether. Very dangerous fire hazard when exposed to heat, flame, or oxidizers. Dangerous explosion hazard when exposed to flame, sparks, etc. Violent reaction with AlH_3; $LiAlH_2$. See also ethers. Highly dangerous. Keep in closed container away from heat and open flame. To fight fire, stop flow of gas.

3-METHYL-3-ETHYLGLUTARIMIDE HR: 3
CAS: 64-65-3   NIOSH: MA 4550000
mf: $C_8H_{13}NO_2$ mw: 155.22

SYNS: 2,6-DIOXO-4-METHYL-4-ETHYLPIPERIDINE * 4-ETHYL-4-METHYL-2,6-DIOXOPIPERIDINE

* BETA-ETHYL-BETA-METHYLGLUTARIMIDE
* 3-ETHYL-3-METHYLGLUTARIMIDE * 4-ETHYL-4-METHYL-2,6-PIPERIDINEDIONE
* 4-METHYL-4-ETHYL-2,6-DIOXOPIPERIDINE
* BETA-METHYL-BETA-ETHYLGLUTARIMIDE

THR: Poison by ingestion, intravenous, intraperitoneal, subcutaneous, and parenteral routes. When heated to decomposition it emits toxic fumes of NO_x.

METHYL ETHYL KETONE PEROXIDE HR: 3
CAS: 1338-23-4   NIOSH: EL 9450000
mf: $C_8H_{16}O_4$ mw: 176.24

SYNS: METHYLETHYLKETONHYDROPEROXIDE
* NCI-C55447
ACGIH TLV: TWA CL 0.2 ppm

THR: Poison by inhalation, ingestion, and intraperitoneal routes. A skin and eye irritant. Gastrointestinal tract effects by ingestion. When heated to decomposition it emits acrid smoke and fumes. For further information see Vol. 5, No. 4 of *DPIM Report*.

METHYL ETHYL KETONE SEMICARBAZONE HR: 3
CAS: 624-46-4   NIOSH: EM 0175000
mf: $C_5H_{11}N_3O$ mw: 129.19

SYN: 2-BUTANONE, SEMICARBAZONE

THR: Poison by intravenous and intraperitoneal routes. When heated to decomposition it emits toxic fumes of NO_x.

N,N-METHYLETHYLNITROSAMINE HR: 3
CAS: 10595-95-6   NIOSH: KR 9200000
mf: $C_3H_8N_2O$ mw: 88.13

SYNS: ETHYLMETHYLNITROSAMINE * METHYLAETHYLNITROSAMIN (GERMAN) * N-METHYL-N-NITROSO-ETHAMINE * N-METHYL-N-NITROSOETHYLAMINE * N-NITROSOETHYLMETHYLAMINE * N-NITROSOMETHYLETHYLAMINE

THR: Poison by ingestion. An experimental tumorigen. Mutagenic data. When heated to decomposition it emits toxic fumes of NO_x. For further information see N-nitrosomethyl ethylamine, Vol. 6, No. 3 of *DPIM Report*.

3-METHYL-5-ETHYL-5-PHENYLHYDANTOIN HR: 3
CAS: 50-12-4   NIOSH: MU 2275000
mf: $C_{12}H_{14}N_2O_2$ mw: 218.28

SYNS: 5-ETHYL-3-METHYL-5-PHENYLHYDANTOIN * 5-ETHYL-3-METHYL-5-PHENYL-2,4(3H, 5H)-IMIDAZOLEDIONE * 5-ETHYL-3-METHYL-5-PHENYLIMIDAZOLIDIN-2,4-DIONE * 3-METHYL-5,5-PHENYLETHYLHYDANTOIN
* PHENYLETHYLMETHYLHYDANTOIN

THR: Poison by ingestion and intraperitoneal routes. Systemic blood effects by ingestion. When heated to decomposition it emits toxic fumes of NO_x.

3-METHYLETHYNYLESTRADIOL HR: 3
CAS: 72-33-3   NIOSH: RC 8960000
mf: $C_{21}H_{26}O_2$ mw: 310.47

SYNS: 3-METHOXY-17-ALPHA-19-NORPREGNA-K,3,5(10)-TRIEN-20-YN-17-OL * 17-ALPHA-ETHINYL ESTRADIOL 3-METHYL ETHER * ETHINYLESTRADIOL-3-METHYL ETHER * 17-ALPHA-ETHINYL OESTRADIOL-3-METHYL ETHER
* ETHINYLOESTRADIOL-3-METHYL ETHER
* ETHYNYLESTRADIOL-3-METHYL ETHER
* 17-ETHYNYLESTRADIOL-3METHYL ETHER
* 17-ALPHA-ETHYNYLESTRADIOL-3-METHYL ETHER * 17-ALPHA-ETHYNYL-3-METHOXY-1, 3,5(10)-ESTRATRIEN-17-BETA-OL * (+)-17-ALPHA-ETHYNYL-17-BETA-HYDROXY-3-METHOXY-1,3,5(10)-ESTRATRIENE * (+)-17-ALPHA-ETHYNYL-17-BETA-HYDROXY-3-METHOXY-1,3,5(10)-OESTRATRIENE
* 17-ETHYNYL-3-METHOXY1,3,5(10)-ESTRATRIEN-17-BETA-OL * 17-ALPHA-ETHYNYL-3-METHOXY-17-BETA-HYDROXY-DELTA-1,3, 5(10)-ESTRATRIENE * 17-ALPHA-ETHYNYL-3-METHOXY-17-BETA-HYDROXY-DELTA-1,3, 5(10)-OESTRATRIENE * 17-ETHYNYL-3-METHOXY-1,3,5(10)-OESTRATIEN-17-BETA-OL
* 17-ETHYNYLOESTRADIOL 3METHYL ETHER
* 17-ALPHA-ETHYNYLOESTRADIOL 3-METHYL ETHER * 17-ALPHA-ETHYNYLOESTRADIOL METHYL ETHER * 3-METHOXY-17-ALPHA-ETHINYLESTRADIOL * 3-METHOXY-17-ALPHA-ETHINYLOESTRADIOL * 3-METHOXYETHYNYLESTRADIOL * 3-METHOXY-17-ALPHA-ETHYNYLESTRADIOL * 3-METHOXY-17-ALPHA-ETHYNYL-1, 3,5(10)-ESTRATRIEN-17-BETA-OL * 3-METHOXYETHYNYLOESTRADIOL * 3-METHOXY-17-ALPHA-ETHYNYL-1,3,5(10)-OESTRATRIEN-17-BETA-OL * 3-METHOXY-19-NOR-17-ALPHAPREGNA-1,3,5(10)-TRIEN-10-YN-17-OL
* (17-ALPHA)-3-METHOXY-19-NORPREGN-1, 3,5(10)-TRIEN-20-YN-17-OL * 3-METHYL-ETHYNYLOESTRADIOL

THR: An experimental neoplastigen and carcinogen by ingestion.

METHYL FLUOROACETATE   HR: 3
CAS: 453-18-9   NIOSH: AH 8400000
mf: $C_3H_5FO_2$ mw: 92.08

SYN: FLUOROACETIC ACID METHYL ESTER

THR: Poison by ingestion, inhalation, skin contact, subcutaneous, intramuscular, intraperitoneal, parenteral, and intravenous routes. See also esters and fluorides. When heated to decomposition it emits toxic fumes of F^-.

7-METHYL-9-FLUOROBENZ(c)ACRI-DINE   HR: 3
CAS: 482-41-7   NIOSH: CU 3852500
mf: $C_{18}H_{12}FN$ mw: 261.31

SYN: 3-FLUORO-10-METHYL-7,8-BENZACRIDINE (FRENCH)

THR: An experimental tumorigen. When heated to decomposition it emits very toxic fumes of F^- and NO_x.

METHYL-4-FLUOROBUTYRATE   HR: 3
CAS: 406-20-2   NIOSH: ET 3500000
mf: $C_5H_9FO_2$ mw: 120.14

SYNS: METHYL-GAMMA-FLUOROBUTYRATE * METHYL-OMEGA-FLUOROBUTYRATE

THR: Poison by ingestion, inhalation, intravenous, and subcutaneous routes. When heated to decomposition it emits toxic fumes of F^-.

METHYL-gamma-FLUOROCROTO-NATE   HR: 3
CAS: 2367-25-1   NIOSH: GQ 4025000
mf: $C_5H_7FO_2$ mw: 118.12

SYNS: CROTONIC ACID, 4-FLUORO-, METHYL ESTER * TL 1183

THR: Poison by inhalation. When heated to decomposition it emits toxic fumes of F^-.

METHYL-gamma-FLUORO-beta-HY-DROXYBUTYRATE   HR: 3
CAS: 63904-99-4   NIOSH: ET 2800000
mf: $C_5H_9FO_3$ mw: 136.14

SYNS: BUTYRIC ACID, GAMMA-FLUORO-BETA-HYDROXY-, METHYL ESTER * TL 1333

THR: Poison by ingestion and inhalation. See also esters. When heated to decomposition it emits toxic fumes of F^-.

METHYL-gamma-FLUORO-beta-HY-DROXYTHIOLBUTYRATE   HR: 3
CAS: 63732-23-0   NIOSH: EK 6150000
mf: $C_5H_9FO_2S$ mw: 152.20

SYN: BUTANETHIOIC ACID, 4-FLUORO-3-HYDROXY-, METHYL ESTER

THR: Poison by inhalation. When heated to decomposition it emits very toxic fumes of SO_x and F^-.

METHYL FLUOROSULFATE   HR: 3
CAS: 421-20-5   NIOSH: LP 0720000
mf: CH_3FO_3S mw: 114.10

PROP: Liquid, ethereal odor, bp: 92°, d: 1.427 @ 16°, vap d: 3.94.

SYNS: MAGIC METHYL * METHYL FLUORO-SULFONATE

THR: Poison by inhalation, ingestion, and skin contact. A skin, mucous membrane, and severe eye irritant. Mutagenic data. Dangerous; when heated to decomposition emits highly toxic fumes of fluorides and oxides of sulfur; will react with water, steam or acids to produce toxic and corrosive fumes.

N-METHYLFORMAMIDE   HR: 3
CAS: 123-39-7   NIOSH: LQ 3000000
mf: C_2H_5NO mw: 59.08

PROP: Flash p: <71.6°F.

SYNS: MONOMETHYLFORMAMIDE * NSC 3051

THR: An experimental teratogen by several routes. Violent reaction with benzene sulfonyl chloride. An eye irritant.

METHYL FORMATE   HR: 2
CAS: 107-31-3   NIOSH: LQ 8925000
DOT: 1243
mf: $C_2H_4O_2$ mw: 60.06

PROP: Colorless flammable liquid, agreeable odor, mp: −99.8°, bp: 31.5°, lel = 5.9%, uel = 20%, flash p: −2.2°F, d: 0.98149 @ 15°/4°, 0.975 @ 20°/4°, autoign temp: 869°F, vap press: 400 mm @ 16°/0°, vap d: 2.07. solidifies @ about 100°; Mod sol in water, methanol; misc in alc.

SYNS: FORMIATE DE METHYLE (FRENCH) * METHYL FORMATE (DOT) * METHYLFOR-MIAAT (DUTCH) * METHYLFORMIAT (GERMAN) * METHYL METHANOATE * METIL (FORMIATO DI) (ITALIAN)

OSHA PEL: TWA 100 ppm
ACGIH TLV: TWA 100 ppm; STEL 150 ppm
DOT Classification: Label: Flammable
  Liquid

THR: Moderately toxic by inhalation and ingestion. Inhalation of vapor can cause irritation
to nasal passages and conjunctiva, optic neuritis,
narcosis, retching, and death from pulmonary
irritation. Industrial fatalities have occurred only
with exposure to high concentrations. Very dangerous fire hazard when exposed to heat, flame,
or oxidizers. Moderately explosive when exposed to heat or flame. Dangerous; upon exposure to heat or flame it emits toxic fumes and
can react vigorously with oxidizing materials.
To fight fire, use alcohol foam, CO_2, dry chemical.

2-METHYLFURAN   HR: 3
CAS: 534-22-5   NIOSH: LU 2625000
DOT: 2301
mf: C_5H_6O mw: 82.11

PROP: Colorless, mobile liquid, ether-like
odor, bp: 63.7°, fp: −88.7°, flash p: −22°F,
d: 0.914 @ 20°/4°, vap press: 139 mm @ 20°,
vap d: 2.8.

SYNS: METHYLFURAN ∗ SILVAN (CZECH)

DOT Classification: Label: Flammable
  Liquid

THR: Poison by ingestion, inhalation, and possibly other routes. A severe eye irritant. Dangerous fire hazard when exposed to heat, flame,
or oxidizers. See also ethers for explosion hazard. Dangerous; upon exposure to heat or flame
it will emit toxic fumes and can react vigorously
with oxidizing materials. To fight fire, use CO_2,
dry chemical.

METHYL GAG   HR: 3
CAS: 31959-87-2   NIOSH: MF 3880000

SYNS: 1,1′-(METHYLETHANEDILIDENEDINI-
TRILO)BIGUANIDINE DIHYDROCHLORIDE DIHY-
DRATE ∗ NSC 32946

THR: Poison by unspecified route. Can cause
unspecified systemic effects. An experimental
neoplastigen by ingestion. When heated to decomposition it emits very toxic fumes of HCl
and NO_x.

METHYLGUANIDINE   HR: 3
CAS: 471-29-4   NIOSH: MF 3683000
mf: $C_2H_7N_3$ mw: 73.10

PROP: Colorless deliquescent; mp: decomp;
very sol in water; sol in alc. Strongly alkaline
mass.

SYNS: METHYLGUANIDIN (GERMAN) ∗ MONO-
METHYL GUANIDIN (GERMAN) ∗ MONOMETHYL-
GUANIDINE

THR: Poison by intravenous and intraperitoneal
routes. Mutagenic data. When heated to decomposition it emits toxic fumes of NO_x.

6-METHYL-2-HEPTYLHYDRA-
ZINE   HR: 3
                NIOSH: MV 2850000
mf: $C_8H_{20}N_2$ mw: 144.30

THR: Poison by ingestion, intravenous, and subcutaneous routes. When heated to decomposition it emits toxic fumes of NO_x.

METHYL HEXAFLUOROISOBUTY-
RATE   HR: 3
CAS: 360-54-3   NIOSH: NQ 4690000
mf: $C_5H_4F_6O_2$ mw: 210.09

SYNS: HEXAFLUOROISOBUTYRIC ACID METHYL
ESTER ∗ METHYL-3,3,3-TRIFLUORO-2-(TRI-
FLUOROMETHYL)PROPIONATE

THR: Poison by ingestion, intravenous, and intraperitoneal routes. When heated to decomposition it emits toxic fumes of $F^−$.

METHYLHYDRAZINE   HR: 3
CAS: 60-34-4   NIOSH: MV 5600000
DOT: 1244
mf: CH_6N_2 mw: 46.09

PROP: Colorless, hydroscopic liquid, ammonia-like odor; slightly sol in water, sol in alc
and ether, d: 0.874 @ 25°; mp: −20.9°; bp:
87.8°; vap d: 1.6; flash p: 73.4°F; fp: −52.4°;
autoign temp: 196°; lel = 2.5%; uel = 97 ±2%;
misc with H_2O, hydrazine; sol in hydrocarbons,
strong reducing agent.

SYNS: HYDRAZOMETHANE ∗ 1,-METHYLHY-
DRAZINE ∗ METHYLHYDRAZINE (DOT)
∗ MONOMETHYLHYDRAZINE ∗ MMH

OSHA PEL: TWA CL 350 ug/m³ (skin)
ACGIH TLV: TWA CL 0.2 ppm
DOT Classification: Label: Flammable
  Liquid and Poison

THR: Poison by inhalation, ingestion, skin contact, intraperitoneal, subcutaneous, and intrave-

nous routes. An experimental teratogen, neoplastigen, and carcinogen. May self-ignite in air. See also hydrazine. Dangerous fire hazard when exposed to heat or flame. To fight fire, use alcohol foam, CO_2, dry chemical. For further information see Vol 5, No. 4 of *DPIM Report*.

N-METHYL-3-HYDROXYMOR-PHINAN   HR: 3
CAS: 297-90-5   NIOSH: QC 9240000
mf: $C_{17}H_{23}NO$ mw: 257.41

SYNS: 1,3,4,9,10,10a-hexahydro-11-methyl-2h-10,4a-iminoethanophenanthren-6-ol, DL-MIXTURE * DL-3-HYDROXY-N-METHYL-MORPHINAN * (+-)-3-HYDROXY-N-METHYL-MORPHINAN * 2H-10, 4A-IMINOETHANOPHEN-ANTHREN-6-OL, 1,3,4,9,10,10A-HEXAHYDRO-11METHYL-, DL * METHORPHINAN

THR: Poison by ingestion, intravenous, and subcutaneous route. When heated to decomposition it emits toxic fumes of NO_x.

((METHYLIMINO)DIETHYLENE)BIS)-DIMETHYLETHYLAMMONIUM DIBROMIDE   HR: 3
CAS: 64049-83-8   NIOSH: BR 8065000
mf: $C_{13}H_{33}N_3Br_2$ mw: 391.31

SYNS: N,N,N',N'-3-PENTAMETHYL-N,N'-DIA-ETHYL-3-AZA-PENTAN-1,5-DIAMMONIUM-DIBRO-MID (GERMAN) * N,N,N',N'-PENTAMETHYL-N, N'-DIETHYL-3-AZA-PENTANE-1,5-DIAMMONIUM DIBROMIDE

THR: Poison by intravenous and subcutaneous routes. See also bromides. When heated to decomposition it emits very toxic fumes of NO_x and Br^-.

2-METHYL-1,3-INDANDIONE   HR: 3
CAS: 876-83-5   NIOSH: NK 6000000
mf: $C_{10}H_8O_2$ mw: 160.18

THR: Poison by intraperitoneal route. An experimental teratogen. When heated to decomposition it emits acrid smoke and fumes.

beta-METHYLINDOLE   HR: 2
CAS: 83-34-1   NIOSH: NM 0350000
mf: C_9H_9N mw: 131.19

PROP: Leaves from ligroin; mp: 95°; bp: 265° @ 755 mm; sol in cold water, alc, chloroform, ether and benzene.

SYNS: 3-METHYLINDOLE * SCATOLE * SKATOLE

THR: Moderately toxic by subcutaneous route. A very offensive odor. When heated to decomposition it emits toxic fumes of NO_x.

METHYL ISOCYANATE   HR: 3
CAS: 624-83-9   NIOSH: NQ 9450000
DOT: 2480
mf: C_2H_3NO mw: 57.06

PROP: Liquid, reacts with water; d: 0.9599 @ 20°/20°, bp: 39.1°, flash p: <5°F.

SYNS: ISOCYANIC ACID, METHYL ESTER * ISOCYANATE DE METHYLE (FRENCH) * ISO-CYANATOMETHANE * METHYLISOCYA-NAAT (DUTCH) * METHYL ISOCYANAT (GERMAN) * METIL ISOCIANATO (ITALIAN)

OSHA PEL: TWA 0.02 ppm (skin)
ACGIH TLV: TWA 0.02 ppm
DFG MAK: 0.01 ppm (0.025 mg/m³)
DOT Classification: Flammable Liquid and Poison

THR: Poison by inhalation, ingestion, and skin contact. A severe eye, skin, and mucous membrane irritant and a sensitizer. It can be absorbed through the skin. Exposure to high concentrations of the vapor can cause blindness; lung damage, including edema, permanent fibrosis, emphysema, and bronchitis; and gynecological effects. Most deaths are a result of lung tissue damage. Effects of cyanide poisoning have been noted but this may be due to impurities. A dangerous fire hazard when exposed to heat, flame, or oxidizers. To fight fire, use spray, foam, CO_2, dry chemical. When heated to decomposition it emits toxic fumes of NO_x. For further information see Vol. 5, No. 2 of *DPIM Report*.

1-METHYL-2-(p-(ISOPROPYLCARBA-MOYL)BENZYL)HYDRAZINE   HR: 3
CAS: 671-16-9   NIOSH: XS 4550000
mf: $C_{12}H_{19}N_3O$ mw: 221.34

SYNS: BENZAMIDE, N-(1-METHYLETHYL)-4-((2-METHYLHYDRAZINO)METHYL)- (9CI) * 2-(P-ISOPROPYLCARBAMOYLBENZYL)-1-METHYL HYDRAZINE * N-ISOPROPYL-ALPHA-(2-METH-YLHYDRAZINO)-P-TOLUAMIDE * 4-((2-METHYLHYDRAZINO)METHYL)-N-ISOPROPYLBEN-ZAMIDE * NSC-77213 * PROCARBAZIN (GERMAN)

THR: Poison by intravenous route. Mutagenic data. An experimental teratogen, carcinogen, and neoplastigen. Moderately toxic by intraperitoneal route. When heated to decomposition it emits toxic fumes of NO_x.

METHYL ISOPROPYL KETONE HR: 3
CAS: 563-80-4 NIOSH: EL 9100000
DOT: 2459
mf: $C_5H_{10}O$ mw: 86.15

SYNS: MIPK * ISOPROPYL METHYL KETONE * 3-METHYL-2-BUTANONE
ACGIH TLV: TWA 200 ppm
DOT Classification: Flammable Liquid

THR: Poison by ingestion. Moderately toxic by inhalation. A skin and eye irritant. When heated to decomposition it emits acrid smoke and fumes.

3-METHYL-4,5-ISOXAZOLEDIONE-4-((o-CHLOROPHENYL) HYDRA-ZONE) HR: 3
CAS: 5707-69-7 NIOSH: NY 2800000
mf: $C_{10}H_8ClN_3O_2$ mw: 237.66

SYNS: 4-(2-CHLOROPHENYLHYDRAZONE)-3-METHYL-5-ISOXAZOLONE * 3-METHYL-4,5-ISOXAZOLEDIONE 4-((2-CHLOROPHENYL)HYDRA-ZONE)

THR: Poison by ingestion and intraperitoneal route. When heated to decomposition it emits very toxic fumes of Cl^- and NO_x.

2-METHYLLACTONITRILE HR: 3
CAS: 75-86-5 NIOSH: OD 9275000
mf: C_4H_7NO mw: 85.12

PROP: Mp: $-20°$, bp: $82°$ @ 23 mm, d: 0.932 @ $19°$, autoign temp: $1270°F$, flash p: $165°F$, vap d: 2.93.

SYNS: ACETONCIANIDRINA (ITALIAN) * ACE-TONCIANHIDRINEI (ROUMANIAN) * ACETON-CYAANHYDRINE (DUTCH) * ACETONCYANHY-DRIN (GERMAN) * ACETONECYANHYDRINE (FRENCH) * ACETONE CYANOHYDRIN * ACE-TONE CYANOHYDRIN (DOT) * ACETONKYAN-HYDRIN (CZECH) * CYANHYDRINE D'ACETONE (FRENCH) * ALPHA-HYDROXYISOBUTYRONI-TRILE * 2-HYDROXY-2-METHYLPROPIONITRILE * USAF RH-8

DOT Classification: Poison B, Label: Poison

THR: Poison by intraperitoneal and subcutaneous routes. Readily decomposes to HCN and acetone. See also HCN, cyanides, NO_x and acetone. Do not store for long periods and keep cool. Combustible when exposed to heat or flame. Dangerous. To fight fire, use CO_2, dry chemical, alcohol foam. Can explode upon contact with H_2SO_4. For further information see acetone cyanohydrin, Vol. 4, No. 1 of *DPIM Report*.

1-METHYLLYSERGIC ACID ETHYLAM-IDE HR: 3
CAS: 7240-57-5 NIOSH: KE 4450000
mf: $C_{19}H_{23}N_3O$ mw: 309.45

SYNS: 9,10-DIDEHYDRO-N-ETHYL-1,6-DIMETH-YLERGOLINE-8-BETA-CARBOXAMIDE * D-1-METHYL LYSERGIC ACID MONOETHYLAMIDE * MLA-74

THR: Poison by ingestion and intravenous routes. Psychotropic effects by ingestion. When heated to decomposition it emits toxic fumes of NO_x.

4-METHYLMERCAPTO-3,5-XYLYL METHYLCARBAMATE HR: 3
CAS: 2032-65-7 NIOSH: FC 5775000
mf: $C_{11}H_{15}NO_2S$ mw: 225.33

SYNS: 3,5-DIMETHYL-4-METHYLTHIOPHENYL N-METHYLCARBAMATE * ENT 25,726 * METHYL CARBAMIC ACID-4-(METHYLTHIO)-3,5-XYLYL ESTER * 4-METHYLMERCAPTO-3,5-DIMETHYLPHENYL N-METHYLCARBAMATE * 3,5-DIMETHYL-4-(METHYLTHIO)PHENOL-METHYLCARBAMATE * 4-METHYLTHIO-3,5-DIMETHYLPHENYL METHYLCARBAMATE * 4-(METHYLTHIO)-3,5-XYLYL ISOMETHYL-CARBAMATE

THR: Poison by ingestion, skin contact, intraperitoneal, and possibly other routes. Moderately toxic by inhalation. See also esters and carbamates. When heated to decomposition it emits very toxic fumes of NO_x and SO_x.

METHYLMERCURIC DICYANDI-AMIDE HR: 3
CAS: 502-39-6 NIOSH: OW 1750000
mf: $C_3H_6HgN_4$ mw: 298.72

SYNS: CYANO(METHYLMERCURI)GUANIDINE * GUANIDINE, CYANO-, METHYLMERCURY DE-RIV. * METHYLMERCURIC CYANOGUANIDINE * METHYLMERCURY DICYANDIAMIDE

THR: Poison by ingestion, intraperitoneal, and possibly other routes. An experimental teratogen. See also mercury compounds. When heated to decomposition it emits very toxic fumes of Hg and NO$_x$.

METHYLMERCURICHLORENDI-MIDE   HR: 3
CAS: 5902-79-4   NIOSH: OW 4200000
mf: C$_{10}$H$_5$Cl$_6$HgNO$_2$ mw: 584.45

SYNS: N-(METHYLMERCURI)-1,4,5,6,7,7-HEXACHLOROBICYCLO(2.2.1)HEPT-5-ENE-2,3-DI-CARBOXIMIDE * N-METHYLMERCURI-1,2,3,6-TETRAHYDRO-3,6-ENDOMETHANO-3,4,5,6,7,7-HEXACHLOROPHTHALIMIDE * N-METHYL-MERCURI-1,2,3,6-TETRAHYDRO-3,6-METHANO-3,4,5,6,7,7-HEXACHLOROPHTHALIMIDE

THR: Poison by ingestion and other routes. See also mercury compounds, chlorides. When heated to decomposition it emits very toxic fumes of Cl$^-$, Hg, and NO$_x$.

METHYLMERCURY HYDROXIDE   HR: 3
CAS: 1184-57-2   NIOSH: OW 4900000
mf: CH$_4$HgO   mw: 232.64
OSHA PEL: TWA 10 ug(Hg)/m^3; CL 40

THR: Poison by intraperitoneal route. Mutagenic data. See also mercury compounds. When heated to decomposition it emits toxic fumes of Hg.

METHYLMERCURY QUINOLINO-LATE   HR: 3
CAS: 86-85-1   NIOSH: OW 7000000
mf: C$_{10}$H$_9$HgNO   mw: 359.79

SYNS: 8-(METHYLMERCURIOXY)QUINOLINE * METHYLMERCURY 8-HYDROXYQUINOLINATE * METHYLMERCURY BETA-HYDROXYQUINOLATE * METHYLMERCURY OXINATE * METHYLMER-CURY OXYQUINOLINATE * 8-(QUINOLINO-LATO)METHYL MERCURY * 8-QUINOLINOL, MERCURY COMPLEX

THR: Poison by ingestion. See also mercury compounds. When heated to decomposition it emits very toxic fumes of Hg and NO$_x$.

alpha-METHYLMESCALINE   HR: 3
CAS: 1082-88-8   NIOSH: SI 3325000
mf: C$_{12}$H$_{19}$NO$_3$ mw: 225.32

SYNS: 3,4,5-TRIMETHOXYAMPHETAMINE * TRIMETHOXYPHENYL-BETA-AMINOPROPANE * 3,4,5-TRIMETHOXYPHENYL-BETA-AMINOPRO-PANE

THR: Poison by intraperitoneal and other routes. Human psychotropic effects by ingestion. When heated to decomposition it emits toxic fumes of NO$_x$.

METHYL MESYLATE   HR: 3
CAS: 66-27-3   NIOSH: PB 2625000
mf: C$_2$H$_6$O$_3$S   mw: 110.14

PROP: Liquid; d: 1.046 @ 16°/4°, bp: 126.5° @ 756 mm; decomp in water; sol in alc and ether.

SYNS: AS-DIMETHYL SULPHATE * METHANE-SULPHONIC ACID METHYL ESTER * METHYL-METHANSULFONAT (GERMAN) * METHYL METH-ANESULFONATE * METHYL METHANSULFONATE * METHYL METHANSULPHONATE * NSC-50256

THR: Poison by intraperitoneal, intravenous, and subcutaneous routes. Mutagenic data. An experimental tumorigen and neoplastigen. See also esters. When heated to decomposition it emits toxic fumes of SO$_x$.

METHYL METHACRYLATE   HR: 2
CAS: 80-62-6   NIOSH: OZ 5075000
DOT: 1247
mf: C$_5$H$_8$O$_2$ mw: 100.13

PROP: Colorless liquid, very sltly sol in water, mp: −50°, bp: 101.0°, flash p: 50°F (OC), d: 0.936 @ 20°/4°, vap press: 40 mm @ 25.5°, vap d: 3.45, lel = 2.1%; uel = 12.5%.

SYNS: METHACRYLATE DE METHYLE (FRENCH) * METHACRYLSAEUREMETHYL ESTER (GERMAN) * METAKRYLAN METYLU (POLISH) * METHYL-METHACRYLAAT (DUTCH) * METHYL-METHA-CRYLAT (GERMAN) * METHYL METHACRYLATE MONOMER, UNINHIBITED (DOT) * METHYL-AL-PHA-METHYLACRYLATE * METHYL-2-METHYL-2-PROPENOATE * METIL METACRILATO (ITAL-IAN) * ''MONOCITE'' METHACRYLATE MONOMER * NCI-C50680

OSHA PEL: TWA 100 ppm
DOT Classification: Label: Flammable Liquid

THR: Moderately toxic by inhalation. A skin and eye irritant. A systemic irritant by inhalation. Central nervous system effects by inhalation. Mutagenic data. A common air contaminant. See also esters. Dangerous fire hazard when exposed to heat or flame; can react with oxidizing materials. Moderately explosive when exposed to heat, sparks, or flame. Reacts violently with benzoyl peroxide. To fight fire, use foam, CO_2, dry chemical. Incompatible with air; dibenzoyl peroxide.For further information see Vol. 6, No. 1 of *DPIM Report*.

METHYL METHANE SULFO-
NATE   HR: 3
CAS: 66-2-7-3   NIOSH: PB 2625000
mf: $C_2H_6O_3S$ mw: 110.14

SYNS: METHANESULFONIC ACID METHYL ESTER * MMS * NSC-50256

THR: Poison by ingestion, intraperitoneal, intravenous, and subcutaneous routes. Mutagenic data. An experimental neoplastigen, teratogen, and brain carcinogen. When heated to decomposition emits toxic fumes of SO_x.

1-METHYL-6-(1-METHYLALLYL)-2,5-DI-
THIOBIUREA   HR: 3
CAS: 926-93-2   NIOSH: EC 1490000
mf: $C_7H_{14}N_4S_2$ mw: 218.37

SYNS: I.C.I. 33,828 * 1-ALPHA-METHYLALLYLTHIOCARBAMOYL-2-METHYLTHIOCARBAMOYL-HYDRAZINE * N-((1-METHYLALLYL)THIOCARBAMOYL)-N′ -(METHYLTHIOCARBAMOYL)HYDRAZINE * 1-METHYL-6-(1-METHYLALLYL)DITHIOBIUREA

THR: An experimental teratogen. When heated to decomposition it emits very toxic fumes of SO_x and NO_x.

N-METHYL-N-(5-(N′-METHYLANILINO)-
2,4-PENTADIENYLIDENE) ANILINIUM
CHLORIDE   HR: 3
CAS: 13984-07-1   NIOSH: BZ 1500000
mf: $C_{19}H_{21}N_2 \cdot Cl$ mw: 312.87

THR: Poison by intraperitoneal route. Moderately toxic by ingestion. When heated to decomposition it emits very toxic fumes of NO_x and Cl^-.

METHYL-p-METHYLBENZENESULFO-
NATE   HR: 3
CAS: 80-48-8   NIOSH: XT 7000000
mf: $C_8H_{10}O_3S$ mw: 186.24

PROP: Light brown crystals; crystals of ethyl ligroin; d: 1.230-1.238 @ 25°/25°; vap d: 6.45; mp: 28°; insol in water; sol in benzene; very sol in alc and ether.

SYNS: METHYL-4-METHYLBENZENESULFONATE * METHYL TOLUENE-4-SULFONATE * METH-

YL-P-TOSYLATE * METHYLESTER KYSELINY P-TOLUENSULFONOVE (CZECH) * METHYL-P-TOLUENESULFONATE * METHYL TOSYLATE * P-TOLUOLSULFONSAEURE METHYL ESTER (GERMAN)

THR: Poison by subcutaneous routes. Moderately toxic by ingestion. An eye and severe skin irritant. A vesicant and skin sensitizer. An experimental tumorigen. See also esters and sulfonates. When heated to decomposition it emits toxic fumes of SO_x.

S-METHYL N-(METHYLCARBA-
MOYLOXY)THIOACETIMI-
DATE   HR: 3
CAS: 16752-77-5   NIOSH: AK 2975000
mf: $C_5H_{10}N_2O_2S$ mw: 162.23

PROP: White crystal solid; slight sulfurous odor; mod water sol; mp: 79°.

SYNS: METHYL N-((METHYLAMINO)CARBONYL)-OXY)ETHANIMIDO)THIOATE * METHYL-N-((METHYLCARBAMOYL)OXY)THIOACETIMIDATE * 2-METHYLTHIO-PROPIONALDEHYD-O-(METHYLCARBAMOYL)-OXIM (GERMAN)

THR: Poison by ingestion, inhalation, and subcutaneous routes. When heated to decomposition it emits very toxic fumes of NO_x and SO_x. For further information see under methomyl in Vol. 2, No. 5 of *DPIM Report*.

d-3-METHYL-N-METHYLMORPHINAN
PHOSPHATE   HR: 3
                        NIOSH: QC 7696000
mf: $C_{18}H_{25}N \cdot H_3O_4P$ mw: 353.44

SYNS: DIMEMORFAN PHOSPHATE * 3,17-DI-METHYL-9-ALPHA,13-ALPHA,14-ALPHA-MOR-PHINAN PHOSPHATE

THR: Poison by intravenous route. Moderately toxic by ingestion. See also phosphates. When heated to decomposition it emits very toxic fumes of NO_x and PO_x.

N-METHYLMITOMYCIN C   HR: 3
CAS: 801-52-5   NIOSH: CN 0525000
mf: $C_{16}H_{20}N_4O_5$ mw: 348.40

PROP: Antibiotic from *Streptomyces ARDUS* and *Streptomyces Verticillatus*.

SYNS: ENT-50825 * NSC-56410 * PORPHYROMYCIN

THR: Poison by ingestion, intravenous, and intraperitoneal routes. Human blood effects. Mutagenic data. When heated to decomposition it emits toxic fumes of NO_x.

2-METHYL-1,4-NAPHTHOQUINONE
HR: 3
CAS: 58-27-5 NIOSH: QL 9100000
mf: $C_{11}H_8O_2$ mw: 172.19

SYNS: KIPCA, OIL SOLUBLE * 2-METHYL-1,4-NAPHTHOCHINON (GERMAN) * 3-METHYL-1,4-NAPHTHOQUINONE * THYLOQUINONE * USAF EK-5185 * VITAMIN K3

THR: Poison by ingestion, intraperitoneal, and subcutaneous routes. Mutagenic data. An experimental tumorigen. When heated to decomposition it emits acrid smoke and fumes.

3-METHYL-2-NAPHTHYLAMINE **HR: 3**
CAS: 10546-24-4 NIOSH: QM 4025000
mf: $C_{11}H_{11}N$ mw: 157.23

THR: An experimental carcinogen and tumorigen. When heated to decomposition it emits toxic fumes of NO_x.

N-METHYL-N-(1-NAPHTHYL)FLUORO-ACETAMIDE **HR: 3**
CAS: 5903-13-9 NIOSH: AC 2700000
mf: $C_{13}H_{12}FNO$ mw: 217.26

SYNS: 1-(N-ACETAMIDOFLUOROMETHYL)-NAPHTHALENE * 2-FLUORO-N-METHYL-N-1-NAPHTHYLACETAMIDE * MNFA * N-METHYL-N-(1-NAPHTHYL)MONOFLUOROACETAMIDE

THR: Poison by ingestion, skin contact, subcutaneous, intravenous, and intraperitoneal routes. When heated to decomposition it emits very toxic fumes of F^- and NO_x.

METHYL NITRATE **HR: 3**
CAS: 598-58-3 NIOSH: QV 0100000
mf: CH_3NO_3 mw: 77.05

PROP: Colorless liquid; bp: 65° (explodes), d: 1.208 @ 20°/4°, vap d: 2.66, mp: −83°. Slightly sol in water; sol in alc, ether.

SYN: NITRIC ACID METHYL ESTER

THR: Poison by ingestion. Moderately toxic by inhalation. A severe eye, skin, and mucous membrane irritant by ingestion, skin contact, subcutaneous, and intraperitoneal routes. A dangerous fire and explosion hazard by spontaneous chemical reaction. It does not require external O_2 for combustion and can explode when shocked or heated. it is used as a rocket fuel. See also nitrates.

METHYL NITRITE **HR: 3**
CAS: 624-91-9 NIOSH: RA 1050000
mf: CH_3NO_2 mw: 61.05

PROP: Gas @ > 10.4°F; mp: −17°, bp: −12°, d: 0.991 @ 15°. Sol in alc, ether.

SYN: NITROUS ACID, METHYL ESTER

THR: Poison by inhalation. Narcotic in high concentration. See also nitrites and amyl nitrite. Dangerous fire and explosion hazards when exposed to heat or flame. Highly dangerous; when heated, emits highly toxic fumes of NO_x; can react vigorously with oxidizing materials.

2-METHYL-1-NITROANTHRAQUINONE **HR: 3**
CAS: 129-15-7 NIOSH: CB 7920000
mf: $C_{15}H_9NO_4$ mw: 267.25

SYNS: 2-METHYL-1-NITRO-9,10-ANTHRACENE-DIONE * NCI-C01923 * 1-NITRO-2-METHYL-ANTHRAQUINONE * 1-N-2-MA (RUSSIAN)

THR: An experimental carcinogen and neoplastigen. Mutagenic data. Moderately toxic by intraperitoneal route. See also xylene. When heated to decomposition it emits toxic fumes of NO_x.

m-METHYLNITROBENZENE **HR: 3**
CAS: 99-08-1 NIOSH: XT 2975000
mf: $C_7H_7NO_2$ mw: 137.15

PROP: Liquid, mp: 15.1°; flash p: 233°F (CC); d: 1.1630 @ 15°/4°; vap press: 1 mm @ 50.2°; vap d: 4.72; bp: 231.9°. Misc with alc, ether; sol in benzene; sol in water @ 30°.

SYNS: 3-METHYLNITROBENZENE * M-NITRO-TOLUENE * 3-NITROTOLUENE * 3-NITRO-TOLUOL

OSHA PEL: TWA 5 ppm (skin)

THR: Poison by ingestion. See also xylene. Combustible when exposed to heat, flame, or oxidizers. See also nitrates for disaster hazard. To fight fire, use water, CO_2, dry chemical. For further information see Vol. 6, No. 3 of *DPIM Report*.

o-METHYLNITROBENZENE   HR: 2
CAS: 88-72-2   NIOSH: XT 3150000
mf: $C_7H_7NO_2$ mw: 137.15

PROP: Yellowish liquid; mp: $-10°$; bp: 222.3°;
flash p: 223°F (CC); d: 1.1622 @ 19°/15°; vap
press: 1 mm @ 50°; vap d: 4.72. Insol in water;
sol SO_2 and petroleum ether; misc in alc, benzene and ether. Slightly sol in NH_3.

SYNS: 2-NITROTOLUENE * 2-METHYLNITRO-
BENZENE * O-NITROTOLUENE

OSHA PEL: TWA 5 ppm (skin)

THR: Moderately toxic by ingestion. Mucous
membrane effects by inhalation. See also xylene. Combustible when exposed to heat or open
flame. See also nitrates for disaster hazard. To
fight fire, use water spray, fog, foam, CO_2.
Incompatible with sodium hydroxide. For further information see Vol. 5, No. 3 of *DPIM
Report*.

p-METHYL NITROBENZENE   HR: 3
CAS: 99-99-0   NIOSH: XT 3325000
mf: $C_7H_7NO_2$ mw: 137.1

PROP: Yellowish crystals; bp: 238.3°; flash p:
223°F (CC); d: 1.286; vap press: 1 mm @ 53.7°;
vap d: 4.72; mp: 53°-54°. Insol in H_2O; sol in
alc, benzene, ether, chloroform, acetone.

SYNS: P-METHYLNITROBENZENE * 4-NITRO-
TOLUENE * 4-METHYLNITROBENZENE * NCI-
c60537 * P-NITROTOLUENE

OSHA PEL: TWA 5 ppm (skin)

THR: Moderately toxic by ingestion and intraperitoneal route. An experimental carcinogen.
Combustible. Can explode with H_2SO_4. To fight
fire, use CO_2, dry chemical, foam. When heated
to decomposition it emits toxic fumes of NO_x.
Explosive on standing. Incompatible with sulfuric acid; tetranitromethane. For further information see p-nitrotoluene, Vol. 3, No. 3 of
DPIM Report.

2-METHYL-5-NITROIMIDAZOLE-1-
ETHANOL   HR: 3
CAS: 443-48-1   NIOSH: NI 5600000
mf: $C_6H_9N_3O_3$ mw: 171.18

SYNS: 1-(BETA-ETHYLOL)-2-METHYL-5-NITRO-
3-AZAPYRROLE * 1-(BETA-HYDROXYETHYL)-2-
METHYL-5-NITROIMIDAZOLE * 1-(2-HYDROXY-
ETHYL)-2-METHYL-5-NITROIMIDAZOLE
* 1-HYDROXYETHYL-2-METHYL-5-NITROIMIDA-

ZOLE * 1-(2-HYDROXY-1-ETHYL)-2-METHYL-
5-NITROIMIDAZOLE * 2-METHYL-1-(2-
HYDROXYETHYL)-5-NITROIMIDAZOLE
* 2-METHYL-3-(2-HYDROXYETHYL)-4-NITRO-
IMIDAZOLE * 2-METHYL-5-NITROIMIDAZOLE-
1-ETHANOL

THR: Moderately toxic by ingestion. An experimental carcinogen. Mutagenic data. Human
peripheral nervous system and other systemic
effects by ingestion. When heated to decomposition it emits toxic fumes of NO_x.

N-METHYL-N'-NITRO-N-NITROSO-
GUANIDINE   HR: 3
CAS: 70-25-7   NIOSH: MF 4200000
mf: $C_2H_5N_5O_3$ mw: 147.12

PROP: Crystals.

SYNS: 1-METHYL-3-NITRO-1-NITROSOGUANIDINE
* N-METHYL-N-NITROSONITROGUANIDIN (GER-
MAN) * N-METHYL-N-NITROSO-N'-NITROGUANI-
DINE * N'-NITRO-N-NITROSO-N-METHYLGUANI-
DINE

THR: Poison by ingestion, subcutaneous, and
intravenous routes. An experimental teratogen,
carcinogen, and neoplastigen. Mutagenic data.
When heated to decomposition it emits very
toxic fumes of NO_x. Will explode under high
impact. For further information see Vol. 5, No.
4 of *DPIM Report*.

3-METHYL-4-NITROPYRIDINE-1-
OXIDE   HR: 3
CAS: 1074-98-2   NIOSH: UT 5775000
mf: $C_6H_6N_2O_3$ mw: 154.14

THR: An experimental tumorigen. Mutagenic
data. When heated to decomposition it emits
toxic fumes of NO_x.

2-METHYL-4-NITROQUINOLINE-1-
OXIDE   HR: 3
CAS: 4831-62-3   NIOSH: UZ 9850000
mf: $C_{10}H_8N_2O_3$ mw: 204.20

SYNS: 4-NITROQUINALDINE-N-OXIDE * 2-
METHYL-4-NITROQUINOLINE-1-OXIDE

THR: An experimental tumorigen. Mutagenic
data. When heated to decomposition it emits
toxic fumes of NO_x.

4-(N-METHYL-N-NITROSAMINO)-
1-(3-PYRIDYL)-1-BUTANONE   HR: 3
                         NIOSH: OB 6465000
mf: $C_{10}H_{13}N_3O_2$ mw: 207.26

SYNS: 4-(N-METHYL-N-NITROSOAMINO)-4-(3-PYRIDYL)-1-BUTANONE * N-METHYL-N-NITROSO-4-OXO-4-(3-PYRIDYL)BUTYL AMINE * 4-(NITROSOAMINO-N-METHYL)-1-(3-PYRIDYL)-1-BUTANONE * 4-(N-NITROSO-N-METHYLAMINO)-1-(3-PYRIDYL)-1-BUTANONE

THR: An experimental neoplastigen and carcinogen. See also amines. When heated to decomposition it emits toxic fumes of NO_x.

4-(4-N-METHYL-N-NITROSAMINO-STYRYL)QUINOLINE   HR: 3
CAS: 16699-10-8   NIOSH: VC 1650000
mf: $C_{18}H_{15}N_3O$ mw: 289.36

SYNS: N-METHYL-N-NITROSO-4-(2-(4-QUINOLINYL)ETHENYL)BENZENAMINE * NSC-101984

THR: An experimental carcinogen. When heated to decomposition it emits toxic fumes of NO_x.

METHYLNITROSOACETAMIDE   HR: 3
CAS: 7417-67-6   NIOSH: AC 6300000
mf: $C_3H_6N_2O_2$ mw: 102.11

SYNS: METHYLNITROSOACETAMID (GERMAN) * METHYLNITROSO-ACETAMIDE * N-METHYL-N-NITROSOACETAMIDE * N-NITROSO-N-METHYLACETAMIDE

THR: Poison (acute) by ingestion. An experimental tumorigen. Mutagenic data. When heated to decomposition it emits toxic fumes of NO_x.

N-METHYL-N-NITROSOALLYL AMINE   HR: 3
CAS: 4549-43-3   NIOSH: BA 7000000
mf: $C_4H_8N_2O$ mw: 100.14

SYNS: METHYLALLYLNITROSAMIN (GERMAN) * METHYLALLYLNITROSAMINE * N-NITROSOALLYLMETHYLAMINE * NITROSOMETHYLALLYLAMINE * N-NITROSOMETHYLALLYLAMINE

THR: Poison by ingestion and intravenous routes. Mutagenic data. An experimental tumorigen. When heated to decomposition it emits toxic fumes of NO_x.

N-METHYL-N-NITROSOANILINE   HR: 3
CAS: 614-00-6   NIOSH: BY 5775000
mf: $C_7H_8N_2O$ mw: 136.17

SYNS: PHENYLMETHYLNITROSAMINE * N-METHYL-N-NITROSOANILINE * N-METHYL-N-NITROSOBENZENAMINE * METHYLPHENYLNITROSAMINE * NITROSOMETHYLANILINE * N-NITROSO-N-METHYLANILINE * N-NITROSOMETHYLPHENYLAMINE * PHENYLMETHYLNITROSAMINE

THR: Poison by ingestion and intraperitoneal routes. Mutagenic data. An experimental teratogen, carcinogen, tumorigen, and neoplastigen. When heated to decomposition it emits toxic fumes of NO_x.

N-METHYL-N-NITROSOBIURET   HR: 3
CAS: 13860-69-0   NIOSH: EC 1925000
mf: $C_3H_6N_4O_3$ mw: 146.13

SYNS: 1-METHYL-1-NITROSOBIURET * N-METHYL-N-NITROSO-N'-CARBAMOYLUREA * N-NITROSO-N-METHYLBIURET

THR: Poison by ingestion. An experimental tumorigen by ingestion. Mutagenic data. When heated to decomposition it emits toxic fumes of NO_x.

METHYLNITROSOCYANAMIDE   HR: 3
CAS: 33868-17-6   NIOSH: GS 6480000
mf: $C_2H_3N_3O$ mw: 85.08

SYN: MNC

THR: Poison by ingestion. An experimental tumorigen by ingestion. Mutagenic data. When heated to decomposition it emits toxic fumes of NO_x.

1-METHYL-1-NITROSO-3-PHENYL-UREA   HR: 3
CAS: 21561-99-9   NIOSH: YT 8100000
mf: $C_8H_9N_3O_2$ mw: 179.20

SYNS: NITROSOMETHYLPHENYLUREA * 3-PHENYL-1-METHYL-1-NITROSOHARNSTOFF (GERMAN)

THR: An experimental tumorigen and carcinogen. When heated to decomposition it emits toxic fumes of NO_x.

N-METHYL-N-NITROSOPROPIONA-MIDE   HR: 3
CAS: 16395-80-5   NIOSH: UE 4525000
mf: $C_4H_8N_2O_2$ mw: 116.14

SYNS: METHYLNITROSO-PROPIONAMIDE * METHYL-NITROSOPROPIONSAEUREAMID (GERMAN) * METHYLNITROSOPROPIONYLAMIDE

THR: An experimental tumorigen. When heated to decomposition it emits toxic fumes of NO_x.

N-METHYL-N-NITROSOUREA   HR: 3
CAS: 684-93-5   NIOSH: YT 7875000
mf: $C_2H_5N_3O_2$ mw: 103.10

SYNS: N-METHYL-N-NITROSO-HARNSTOFF (GER-
MAN) * METHYLNITROSO-HARNSTOFF (GER-
MAN) * METHYLNITROSOUREA * METHYL-
NITROSOUREE (FRENCH) * N-NITROSO-N-
METHYLCARBAMIDE * NITROSOMETHYLUREA
* N-NITROSO-N-METHYLUREA * NSC 23909

THR: Poison (acute) by ingestion, implant, intraperitoneal, and intravenous routes. Mutagenic data. An experimental teratogen, tumorigen, neoplastigen, and transplacental brain
carcinogen. A suspected human carcinogen.
Can detonate with (KOH + CH_2Cl_2). When
heated to decomposition it emits toxic fumes
of NO_x. For further information see Vol. 5,
No. 4 of *DPIM Report*.

METHYL ORANGE   HR: 3
CAS: 140-56-7   NIOSH: CZ 1750000
mf: $C_8H_{10}N_3O_3S \cdot$ ^{Na}mw: 251.26

PROP: Yellow-brown crystals.

SYNS: P-DIMETHYLAMINOBENZENEDIAZOSODIUM
SULPHONATE * P-(DIMETHYLAMINO)BENZENE-
DIAZOSULFONATE * P-DIMETHYLAMINOBEN-
ZOLDIAZOSULFONAT (NATRIUMSALZ) (GERMAN)
* P-(DIMETHYLAMINO)BENZENEDIAZOSULPHO-
NATE * P-DIMETHYLAMINOBENZENE DIAZO
SODIUM SULFONATE * NCI-C03010 * SO-
DIUM P-(DIMETHYLAMINO)BENZENEDIAZOSUL-
FONATE * SODIUM 4-(DIMETHYLAMINO)-
BENZENEDIAZOSULFONATE * SODIUM
P-(DIMETHYLAMINO)BENZENEDIAZOSULPHONATE
* SODIUM 4-(DIMETHYLAMINO)BENZENEDIAZO-
SULPHONATE * SODIUM (4-(DIMETHYLAMINO)-
PHENYL)DIAZENESULFONATE

THR: Poison by ingestion, intravenous, intraperitoneal, and possibly other routes. Mutagenic
data. An experimental carcinogen. When heated
to decomposition it emits very toxic fumes of
NO_x and SO_x.

METHYL PARATHION   HR: 3
CAS: 298-00-0   NIOSH: TG 0175000
DOT: 2783
mf: $C_8H_{10}NO_5PS$ mw: 263.22

PROP: Crystals; vap d: 9.1; mp: 37°-38°; d:
1.358 @ 20°/4°. Sol in most organic solvents.

SYNS: O,O-DIMETHYL-O-P-NITROFENYLESTER
KYSELINY THIOFOSFORECNE (CZECH) * O,O-DI-
METHYL-O-(4-NITROFENYL)-MONOTHIOFOSFAAT
(DUTCH) * DIMETHYL P-NITROPHENYL MONO-
THIOPHOSPHATE' * O,O-DIMETHYL-O-(4-NITRO-
PHENYL)-MONOTHIOPHOSPHAT (GERMAN)
* O,O-DIMETHYL O-(P-NITROPHENYL) PHOSPHO-
ROTHIOATE * O,O-DIMETHYL-O-(4-NITROPHE-
NYL) PHOSPHOROTHIOATE * O,O-DIMETHYL
O-(P-NITROPHENYL) THIONOPHOSPHATE
* O,O-DIMETHYL-O-(P-NITROPHENYL)-THIONO-
PHOSPHAT (GERMAN) * DIMETHYL-P-NITROPHE-
NYL THIONPHOSPHATE * O,O-DIMETHYL O-P-NI-
TROPHENYLTHIOPHOSPHATE * DIMETHYL
PARATHION * DIMETHYLP-NITROPHENYLTHIO-
PHOSPHATE * O,O-DIMETIL-O-(4-NITRO-FENIL)-
MONOTIOFOSFATO(ITALIAN) * ENT17,292
* METYLPARATION(CZECH) * NCI-C02971
* P-NITROPHENYLDIMETHYLTHIONOPHOSPHATE-
* PARATHION METHYL * PARATHION-METILE
(ITALIAN) * THIOPHOSPHATE DE O,O-DIMETH-
YLE ET DE O-(4-NITROPHENYLE) (FRENCH)

ACGIH TLV: TWA 0.2 mg/m^3
DOT Classification: Poison B, Label: Poison

THR: Poison by inhalation, ingestion, skin contact, subcutaneous, intravenous, intraperitoneal,
and possibly other routes. Mutagenic data. A
cholinesterase inhibitor. See also parathion.
When heated to decomposition it emits very
toxic fumes of NO_x, PO_x, and SO_x. For further
information see Vol. 6, No. 1 of *DPIM Report*.

2-METHYL-2-PENTANOL-4-ONE   HR: 2
CAS: 123-42-2   NIOSH: SA 9100000
DOT: 1148'
mf: $C_6H_{12}O_2$ mw: 116.18

PROP: Liquid, faint pleasant odor, mp: −47°
to −54°, bp: 167.9°, flash p: 148°F, d: 0.9306
@ 25°/4°, autoign temp: 1118°F, vap d: 4.00,
vap press: 1.1 mm @ 20°, lel = 1.8%, uel =
6.9%, flash p: (acetone free): 136°F.

SYNS: DIACETONALCOHOL (DUTCH) * DIACE-
TONALCOOL (ITALIAN) * DIACETONALKOHOL
(GERMAN) * DIACETONE ALCOHOL (DOT)
* DIACETONE-ALCOOL (FRENCH) * DIKETONE
ALCOHOL * 4-HYDROXY-2-KETO-4-METHYL-
PENTANE * 4-HYDROXY-4-METHYL-2-PENTA-
NONE * 4-HYDROXY-4-METHYL-PENTAN-2-ON
(GERMAN, DUTCH) * 4-HYDROXY-4-METHYL
PENTAN-2-ONE * 4-IDROSSI-4-METIL-PENTAN-
2-ONE (ITALIAN)

OSHA PEL: TWA 50 ppm
DOT Classification: Label: Flammable
    Liquid

THR: Moderately toxic by ingestion and other routes. A skin, mucous membrane, and severe eye irritant. A systemic irritant by inhalation. Can cause anemia and damage to liver and kidneys. Narcotic in high concentration. Flammable when exposed to heat or flame; can react with oxidizing materials. To fight fire, use alcohol foam, foam, CO_2, dry chemical. When heated to decomposition it emits acrid smoke and fumes.

METHYLPENTYNOL CARBAMATE   HR: 3
CAS: 302-66-9   NIOSH: SC 5075000
mf: $C_7H_{11}NO_2$ mw: 141.19

PROP: Sltly water-sol crystals; mp: 57°; bp: 121° @ 16 mm.

SYNS: 3-CARBAMOYLOXY-3-METHYL-4-PENTYNE * 1-ETHYL-1-METHYL-2-PROPYNYL CARBAMATE * 2-ETHYNYL-2-BUTYL CARBAMATE * METHYLPARAFYNOL CARBAMATE * 3-METHYL-PENTIN-(1)-OL-(3) (GERMAN) * 3-METHYL-1-PENTYN-3-OL CARBAMATE * USAF EL-108

THR: Poison by ingestion, subcutaneous, and intraperitoneal routes. Human psychotropic effects by ingestion. A tranquilizer which can cause central nervous system depression and death by overdose. Toxic effects are enhanced with the use of alcohol and barbiturates. See also carbamates. When heated to decomposition it emits toxic fumes of NO_x.

METHYL PERFLUOROMETHACRYLATE   HR: 3
CAS: 685-09-6   NIOSH: AS 8575000
mf: $C_5H_3F_5O_2$ mw: 190.08

SYN: 3,3-DIFLUORO-2-(TRIFLUOROMETHYL)ACRYLIC ACID, METHYL ESTER

THR: Poison by ingestion, intravenous, and intraperitoneal routes. See also fluorides. When heated to decomposition it emits toxic fumes of F^-.

METHYLPERIDOL HYDROCHLORIDE   HR: 3
CAS: 3871-82-7   NIOSH: EU 2975000
mf: $C_{22}H_{26}FNO_2 \cdot ClH$ mw: 391.95

SYNS: 4'-FLUORO-4-(4-HYDROXY-4-P-TOLYLPIPERIDINO)BUTYROPHENONE, HYDROCHLORIDE * METHYLPERIDOL

THR: Poison by subcutaneous and intravenous routes. When heated to decomposition it emits very toxic fumes of F^-, HCl and NO_x.

METHYL PHENIDYL ACETATE   HR: 3
CAS: 113-45-1   NIOSH: TM 3675000
mf: $C_{14}H_{19}NO_2$ mw: 233.34

SYNS: METHYL ALPHA-PHENYL-ALPHA-(2-PIPERIDYL)ACETATE * NCI-C56280 * ALPHA-PHENYL-2-PIPERIDINEACETICACIDMETHYLESTER * ALPHA-PHENYL-2-PIPERODINEACETIC ACID METHYL ESTER

THR: Poison by ingestion, intraperitoneal, and subcutaneous routes. When heated to decomposition it emits toxic fumes of NO_x.

4-METHYLPHENYL ACETATE   HR: 2
CAS: 140-39-6   NIOSH: AJ 7570000
mf: $C_9H_{10}O_2$ mw: 150.19

PROP: Colorless liquid, d: 1.044 @ 16°; bp: decomp @ 360°; mp: 220°; vap d: 5.18; insol in water; misc in alc and ether.

SYNS: ACETIC ACID, 4-METHYLPHENYL ESTER * P-ACETOXYTOLUENE * 4-ACETOXYTOLUENE * P-CRESOL ACETATE * P-CRESYL ACETATE * 4-METHYLBENZOIC ACID METHYL ESTER * P-METHYLPHENYL ACETATE * PARACRESYL ACETATE * P-TOLYL ACETATE * P-TOLYL ETHANOATE

THR: Moderately toxic by ingestion and skin contact. See also esters. Moderately dangerous; when heated to decomposition it emits toxic smoke and fumes.

N-METHYL-p-(PHENYLAZO)ANILINE   HR: 3
CAS: 621-90-9   NIOSH: BY 5950000
mf: $C_{13}H_{13}N_3$ mw: 211.29

SYNS: 4-(METHYLAMINO)AZOBENZENE * N-METHYL-4-AMINOAZOBENZENE * N-METHYL-P-AMINOAZOBENZENE * P-MONOMETHYLAMINOAZOBENZENE * 4-MONOMETHYLAMINOAZOBENZENE

THR: An experimental tumorigen, neoplastigen, and teratogen. Mutagenic data. Moderately toxic by subcutaneous route. When heated to decomposition it emits toxic fumes of NO_x.

N-METHYL-4-(PHENYLAZO)-o-ANISIDINE   HR: 3
CAS: 10121-94-5   NIOSH: BZ 7000000
mf: $C_{14}H_{15}N_3O$ mw: 241.32

SYNS: 3-METHOXY-4-MONOMETHYLAMINOAZO-
BENZENE * 3-METHOXYMETHYLAMINOAZOBEN-
ZENE

THR: An experimental tumorigen and neoplastigen. When heated to decomposition it emits
toxic fumes of NO_x.

7-METHYL-9-PHENYLBENZ(C)ACRI-
DINE   HR: 3
NIOSH: CU 4100000
mf: $C_{24}H_{17}N$ mw: 319.42

SYNS: 5-METHYL-7-PHENYL-1:2-BENZACRIDINE
* 3 PHENYL, 10-METHYL,7:8 BENZACRIDINE
(FRENCH)

THR: An experimental tumorigen. When heated
to decomposition it emits toxic fumes of NO_x.

2-METHYL-N-PHENYLBENZA-
MIDE   HR: 3
CAS: 7055-03-0   NIOSH: CV 5582500
mf: $C_{14}H_{13}NO$ mw: 211.28

SYN: BAS-3050

THR: Poison by ingestion and intraperitoneal
routes. An experimental teratogen by ingestion.
Mutagenic data. When heated to decomposition
it emits toxic fumes of NO_x.

1-(o-METHYLPHENYL)-3,3-DIMETHYL-
TRIAZENE   HR: 3
CAS: 20240-98-6   NIOSH: XY 1350000
mf: $C_9H_{13}N_3$ mw: 163.25

SYNS: 3,3-DIMETHYL-1-(O-METHYLPHENYL)-
TRIAZENE * 3,3-DIMETHYL-1-(O-TOLYL)TRIA-
ZENE * 1-(O-METHYLPHENYL)-3,3-DIMETHYL-
TRIAZEN (GERMAN) * 1-(2-METHYLPHENYL)-
3,3-DIMETHYLTRIAZENE

THR: Poison by ingestion. Moderately toxic
by subcutaneous route. An experimental carcinogen. When heated to decomposition it emits
toxic fumes of NO_x.

METHYL PHENYLETHYL NITROSA-
MINE   HR: 3
CAS: 13256-11-6   NIOSH: SI 0770000
mf: $C_9H_{12}N_2O$ mw: 164.23

SYNS: N-METHYL-N-NITROSOPHENETHYLAMINE
* METHYL(2-PHENYLAETHYL)NITROSAMIN (GER-
MAN) * N-NITROSO-N-METHYL-2-PHENYLETH-
YLAMINE

THR: Poison (acute) by ingestion. Mutagenic
data. An experimental tumorigen by ingestion.
See also nitrosamines. When heated to decomposition it emits toxic fumes of NO_x.

5-METHYL-1-PHENYL-2-(PYRROLI-
DINYL)IMIDAZOLE   HR: 3
NIOSH: NI 7730000
mf: $C_{14}H_{17}N_3$ mw: 227.34

SYN: METHYL-5 PHENYL-1-(PYRROLIDINYL-1)-2
IMIDAZOLE (FRENCH)

THR: Poison by ingestion, intravenous, and intraperitoneal routes. When heated to decomposition it emits toxic fumes of NO_x.

1-METHYL-4-(PHENYLTHIO)PYRI-
DINIUM IODIDE   HR: 3
CAS: 73840-42-3   NIOSH: UU 6650500
mf: $C_{12}H_{12}NS \cdot I$ mw: 329.21

THR: Poison by intravenous and intraperitoneal
routes. When heated to decomposition it emits
very toxic fumes of NO_x, SO_x, and I^-.

METHYLPHOSPHODITHIOIC ACID-S-
(((p-CHLOROPHENYL)THIO)METHYL)-o-
METHYL ESTER   HR: 3
CAS: 18466-11-0   NIOSH: TA 7480000
mf: $C_9H_{12}ClOPS_2$ mw: 266.75

SYNS: ENT 27,180 * N 4548 * STAUFFER
N-4548

THR: Poison by ingestion and subcutaneous
routes. When heated to decomposition it emits
very toxic fumes of Cl^-, PO_x, and SO_x.

N-METHYLPIPERAZINE   HR: 3
CAS: 109-01-3   NIOSH: TM 1225000
mf: $C_5H_{12}N_2$ mw: 100.19

PROP: A hygroscopic solid, typical amine-like
odor, d: 0.9031 20°/20°, mp: 65.5°, bp: 139°,
flash p: 108°F (OC), vap d: 3.5.

THR: Poison by intraperitoneal route. Moderately toxic by inhalation, ingestion, and skin
contact. Flammable when exposed to heat or
flame; can react with oxidizing materials. To
fight fire, use alcohol foam, CO_2, dry chemical.

METHYL-4-(3-PIPERIDINOPROPIONYL-
AMINO)SALICYLATE, METHIODIDE
HR: 3
CAS: 73790-27-9   NIOSH: TN 4700500
mf: $C_{17}H_{25}N_2O_4 \cdot I$ mw: 448.34

SYN: 4-(3-PIPERIDINOPROPIONAMIDO) SALI-
CYCLIC ACID METHYL ESTER, METHIODIDE

THR: A poison by intraperitoneal, and intravenous routes. When heated to decomposition it
emits very toxic fumes of NO$_x$ and I$^-$.

N-METHYL-3-PIPERIDYL BENZI-
LATE   **HR: 3**
CAS: 3321-80-0   NIOSH: DD 3850000
mf: C$_{20}$H$_{23}$NO$_3$ mw: 325.44

SYNS: JB 336 * BENZILIC ACID-1-METHYL-3-
PIPERIDYL ESTER

THR: Poison by intravenous route. See also
esters. When heated to decomposition it emits
toxic fumes of NO$_x$.

10-(2-(1-METHYL-2-PIPERIDYL)ETHYL)-
2-(METHYLTHIO)PHENOTHIAZINE
                                          HR: 3
CAS: 50-52-2   NIOSH: SP 2100000
mf: C$_{21}$H$_{26}$N$_2$S$_2$ mw: 370.61

SYNS: MELLERIL * 2-METHYLMERCAPTO-
10-(2-N-METHYL-2-PIPERIDYL)-
ETHYL)PHENOTHIAZINE * THIORIDAZINE

THR: Poison by ingestion and intraperitoneal
routes. Human central nervous system effects
by ingestion. Mutagenic data. When heated to
decomposition it emits very toxic fumes of NO$_x$
and SO$_x$.

10-(2-(1-METHYL-2-PIPERIDYL)ETHYL)-
2-METHYLTHIOPHENOTHIAZINE HY-
DROCHLORIDE   **HR: 3**
CAS: 130-61-0   NIOSH: SP 2275000
mf: C$_{21}$H$_{26}$N$_2$S$_2$ • ClH   mw: 407.07

SYNS: MELLARIL HYDROCHLORIDE * 2-
METHYLMERCAPTO-10-(2-(N-METHYL-2-PI-
PERIDYL)ETHYLPHENOTHIAZINE HYDROCHLORIDE
* THIORIDAZINE HYDROCHLORIDE * USAF
SZ-3 * USAF SZ-B

THR: Poison by ingestion, intravenous, and intraperitoneal routes. Mutagenic data. When
heated to decomposition it emits very toxic
fumes of NO$_x$, SO$_x$, and HCl.

9-(1-METHYL-4-PIPERIDYLIDENE)THI-
OXANTHENE   **HR: 3**
CAS: 314-03-4   NIOSH: TN 1575000
mf: C$_{19}$H$_{19}$NS   mw: 293.45

SYNS: BP 400 * CALMIXENE

THR: Poison by ingestion and intravenous
routes. When heated to decomposition it emits
very toxic fumes of NO$_x$ and SO$_x$.

9-(METHYL-2-PIPERIDYL)METHYLCAR-
BAZOLE   **HR: 3**
CAS: 60706-49-2   NIOSH: FE 6274500
mf: C$_{19}$H$_{22}$N$_2$ mw: 278.43

SYN: 9-(1-METHYL-PIPERIDYL-(2)-METHYL)-
CARBAZOL (GERMAN)

THR: Poison by intravenous and intraperitoneal
routes. Moderately toxic by ingestion and subcutaneous routes. When heated to decomposition
it emits toxic fumes of NO$_x$.

METHYLPOTASSIUM   **HR: 3**
mf: CH$_3$K   mw: 54.14

THR: No toxicity data. However, since moisture
(as in all living tissue) is incompatible with
CH$_3$K, it must be considered a poison. (Dry)
Ignites spontaneously in air. Very dangerous.
When heated to decomposition it emits toxic
fumes of K$_2$O.

METHYLPREDNISOLONE   **HR: 3**
CAS: 83-43-2   NIOSH: TU 4146000
mf: C$_{22}$H$_{30}$O$_5$ mw: 374.46

PROP: Crystals; mp: 228°-237°.

SYNS: DELTA(SUP 1)-6-ALPHA-METHYLHYDRO-
CORTISONE * 6-ALPHA-METHYLPREDNISOLONE
* 11-BETA,17,21-TRIHYDROXY-6-ALPHA-
METHYLPREGNA-1,4-DIENE-3,20-DIONE
* 11-BETA,17-ALPHA,21-TRIHYDROXY-6-AL-
PHA-METHYL-1,4-PREGNADIENE-3,20-DIONE

THR: An experimental teratogen. When heated
to decomposition it emits acrid smoke and
fumes.

2-METHYLPROPANE   **HR: 1**
CAS: 75-28-5   NIOSH: TZ 4300000
DOT: 1075/1969
mf: C$_4$H$_{10}$ mw: 58.14

PROP: Colorless gas; bp: −11.7°, lel = 1.9%,
uel = 8.5%, fp: −160°, d: 0.5572 @ 20°, autoign temp: 864°F, vap d: 2.01.

SYN: ISOBUTANE

DOT Classification: Label: Flammable Gas

THR: An asphyxiant. A common air contaminant. Very dangerous fire and explosion hazards
when exposed to heat, flame, or oxidizers. To
fight fire, stop flow of gas.

2-METHYL-2-PROPANETHIOL   HR: 2
CAS: 75-66-1   NIOSH: TZ 2766000
mf: $C_4H_{10}S$ mw: 90.20

PROP: Liquid, skunk-like odor; mp: $-0.5°$,
d: 0.79-0.82 @ 15.5°/15.5°, flash p: $<-20°F$,
bp: 62-67°, vap d: 3.1.

SYNS: TERT-BUTANETHIOL * TERT-BUTYL
MERCAPTAN

THR: Moderately toxic by intraperitoneal route.
An eye irritant. Dangerous fire hazard when
exposed to heat or flame. Dangerous; when
heated to decomposition or on contact with acid
or acid fumes it emits highly toxic SO_x fumes;
can react vigorously with oxidizing materials.
To fight fire, use alcohol foam, dry chemical,
mist, fog.

METHYL PROPIONATE   HR: 2
CAS: 554-12-1   NIOSH: UF 5970000
DOT: 1248
mf: $C_4H_8O_2$ mw: 88.10

PROP: Colorless liquid, mp: $-87.0°$, bp: 79.8°,
flash p: 28°F (CC) $-2°C$, d: 0.937 @ 4°, autoign
temp: 876°F, vap press: 40 mm @ 11.0°, vap
d: 3.03, lel = 2.50%, uel = 13%, d: 0.915
@ 20°/4°. Sol in water @ 20°; misc in alc
and ethanol.

SYNS: PROPANOIC ACID, METHYL ESTER
* PROPIONATE DE METHYLE (FRENCH)

DOT Classification: Label: Flammable
  Liquid

THR: Moderately toxic by ingestion. Dangerous
(fire hazard) when exposed to heat, flame, or
oxidizers. To fight fire, use foam, CO_2, dry
chemical.

3-(1-METHYLPROPYL)-6-CHLOROPHE-
NYL METHYLCARBAMATE   HR: 3
CAS: 2917-19-3   NIOSH: FB 5075000
mf: $C_{12}H_{16}ClNO_2$ mw: 241.74

SYNS: METHYLCARBAMIC ACID-3-SEC-BUTYL-6-
CHLOROPHENYL ESTER * ENT 27,128

THR: Poison by ingestion. When heated to decomposition it emits very toxic fumes of Cl^-
and NO_x.

2-(1-METHYLPROPYL)PHENYL
METHYLCARBAMATE   HR: 3
CAS: 3766-81-2   NIOSH: FB 5425000
mf: $C_{12}H_{17}NO_2$ mw: 207.30

SYNS: O-SEC-BUTYLPHENYL METHYLCARBA-
MATE * 2-SEC-BUTYLPHENYL-N-METHYLCAR-
BAMATE

THR: Poison by ingestion and skin contact.
When heated to decomposition it emits toxic
fumes of NO_x.

2-METHYL-2-PROPYLTRIMETHYLENE
BUTYLCARBAMATE CARBA-
MATE   HR: 3
CAS: 4268-36-4   NIOSH: EZ 0875000
mf: $C_{13}H_{26}N_2O_4$ mw: 274.41

SYNS: N-N-BUTYL-2-METHYL-2-PROPYL-
1,3-PROPANEDIOL DICARBAMATE * N-
BUTYL-2-METHYL-2-PROPYL-1,3-PROPANEDIOL
DICARBAMATE * 2-(HYDROXYMETHYL)-2-
(METHYLPENTYL) BUTYLCARBAMATE CARBA-
MATE * 2-METHYL-2-PROPYL-1,3-PROPANE-
DIOL BUTYLCARBAMATE CARBAMATE

THR: Poison by intraperitoneal and intravenous
routes. Human central nervous system effects
by ingestion. When heated to decomposition
it emits toxic fumes of NO_x.

METHYLPYRAZOLYL DIETHYLPHOS-
PHATE   HR: 3
CAS: 108-34-9   NIOSH: TC 1750000
mf: $C_8H_{15}N_2O_4P$ mw: 234.22

SYNS: O,O-DIAETHYL-O-(3-METHYL-1H-PYRA-
ZOL-5-YL)-PHOSPHAT (GERMAN) * O,O-DIETH-
YL-O-(3-METHYL-1H-PYRAZOL-5-YL)-FOSFAAT
(DUTCH) * DIETHYL 3-METHYL-5-PYRAZOLYL
PHOSPHATE * O,O-DIETHYL O-(3-METHYL-5-
PYRAZOLYL) PHOSPHATE * O,O-DIETIL-O-(3-
METIL-1H-PIRAZOL-5-IL)-FOSFATO (ITALIAN)
* ENT 24,723 * 3-METHYLPYRAZOLYL-5-DI-
ETHYLPHOSPHATE * PHOSPHATE DE DIETHYLE
ET DE 3-METHYL-5-PYRAZOLYLE (FRENCH)
* PHOSPHORIC ACID, DIETHYL-(3-METHYL-
5-PYRAZOLYL) ESTER * PIRAZOXON
(ITALIAN)

THR: Poison by ingestion and subcutaneous
routes. When heated to decomposition it emits
very toxic fumes of NO_x and PO_x.

2-METHYL PYRIDINE   HR: 3
CAS: 109-06-8   NIOSH: TJ 4900000
mf: C_6H_7N mw: 93.14

PROP: Colorless liquid, strong unpleasant odor;
mp: $-70°$; bp: 129°; flash p: 102°F (OC); d:
0.95 @ 15°/4°; autoign temp: 1000°F; vap press:

10 mm @ 24.4°; vap d: 3.2; very sol in water; misc in alcohol and ether.

SYNS: ALPHA-METHYLPYRIDINE * 2-PICOLINE * ALPHA-PICOLINE

THR: Poison by skin contact. Moderately toxic by ingestion and inhalation. An eye and skin irritant. Flammable when exposed to heat and flame. To fight fire, use CO_2, dry chemical. When heated to decomposition it emits toxic fumes of NO_x. Incompatible with hydrogen peroxide; iron (II) sulfate; sulfuric acid.

4-METHYLPYRIDINE   HR: 3
CAS: 108-89-4   NIOSH: UT 5425000
mf: C_6H_7N mw: 93.14

PROP: Colorless liquid, disagreeable odor; bp: 145°, fp: 3.7°, d: 0.9571 @ 15°/4°, vap d: 3.21, flash p: 134°F (OC).

SYNS: GAMMA-PICOLINE * 4-PICOLINE

THR: Poison by skin contact. Moderately toxic by ingestion and inhalation. See also 2-methyl pyridine. A skin and severe eye irritant. Flammable by heat, flames, oxidizers. See also α-picoline for disaster hazard. To fight fire, use alcohol foam.

1-METHYLPYRROLIDINE   HR: 1
mf: $C_5H_{11}N$ mw: 85.15

PROP: Colorless to yellow liquid with penetrating amine-like odor; bp: 80.5°; fp: −90°; d: 0.8054 @ 20°/20°; flash p: 37.4°F; vap d: 2.9.

THR: This material is strongly alkaline. Liquid and vapors are corrosive to the skin, eyes, or mucous membranes. See also ammonia. Dangerous fire hazard; keep away from sparks, heat sources, and powerful oxidizers. Keep in closed containers. Dangerous; when heated to decomposition it emits highly toxic fumes. To fight fire, use alcohol foam.

3-(1-METHYL-2-PYRROLIDINYL)IN-DOLE   HR: 3
CAS: 7236-83-1   NIOSH: NM 1080000
mf: $C_{13}H_{16}N_2$ mw: 200.31

THR: Poison by intraperitoneal route. When heated to decomposition it emits toxic fumes of NO_x.

3-(1-METHYL-3-PYRROLIDINYL)IN-DOLE   HR: 3
CAS: 3671-00-9   NIOSH: NM 1090000
mf: $C_{13}H_{16}N_2$ mw: 200.31

THR: Poison by ingestion and intraperitoneal routes. When heated to decomposition it emits toxic fumes of NO_x.

10-((1-METHYL-3-PYRROLIDINYL)-METHYL)-PHENOTHIAZINE   HR: 3
CAS: 1982-37-2   NIOSH: SP 2975000
mf: $C_{18}H_{20}N_2S$ mw: 296.46

SYNS: METHDILAZINE * NCI-C60720

THR: Poison by ingestion and intraperitoneal routes. When heated to decomposition it emits very toxic fumes of SO_x and NO_x.

8-(METHYLQUINOLYL)-N-METHYL CAR-BAMATE   HR: 3
CAS: 14628-06-9   NIOSH: FC 4900000
mf: $C_{12}H_{12}N_2O_2$ mw: 216.26

SYNS: ENT 27,407 * GIEGY GS-13798 * GS-13,798

THR: Poison by ingestion, intraperitoneal, and subcutaneous routes. When heated to decomposition it emits toxic fumes of NO_x.

5-METHYLRESORCINOL   HR: 3
CAS: 504-15-4   NIOSH: VH 2100000
mf: $C_7H_8O_2$ mw: 124.15

SYNS: 1,3-DIHYDROXY-5-METHYLBENZENE * 3,5-DIHYDROXYTOLUENE * ORCINOL

THR: Poison by subcutaneous and intraperitoneal routes. Moderately toxic by ingestion and possibly other routes. When heated to decomposition it emits acrid smoke and fumes.

METHYL SALICYLATE   HR: 3
CAS: 119-36-8   NIOSH: VO 4725000
mf: $C_8H_8O_3$ mw: 152.14

PROP: Colorless, yellowish or reddish oily liquid. Odor and taste of gaultheria; mp: −8.6°; bp: 223.3°, ulc: 20-25, flash p: 214°F (CC), fp: −1.2°, d: 1.1840 @ 25°/25°; autoign temp: 850°F, vap press: 1mm @ 54.0°, vap d: 5.24. Sltly sol in water; sol in chloroform, ether, alc, glacial acetic acid.

SYNS: BETULA OIL * GAULTHERIA OIL, ARTIFICIAL * O-HYDROXYBENZOIC ACID, METHYL ESTER * METYLESTER KYSELINY SALICYLOVE (CZECH) * METHYL-O-HYDROXYBENZOATE * NATURAL WINTERGREEN OIL * SWEET BIRCH OIL * SYNTHETIC WINTERGREEN OIL * TEA-BERRY OIL * WINTERGREEN OIL, SYNTHETIC

THR: Poison to humans by ingestion. Moderately toxic by several other routes. A skin and severe eye irritant. Human systemic effects by ingestion. See also esters. Can cause kidney irritation, vomiting, and convulsions. Ingestion of relatively small amounts may cause severe poisoning and death. Combustible when exposed to heat or flame; can react with oxidizing materials. To fight fire, use CO_2, dry chemical. When heated to decomposition it emits acrid smoke and fumes.

METHYL-o-SILICATE   HR: 3
CAS: 681-84-5   NIOSH: VV 9800000
DOT: 2606
mf: $C_4H_{12}O_4Si$ mw: 152.25

PROP: Clear liquid; vap d: 5.25.

SYNS: METHYL SILICATE * TETRAMETHOXY SILANE
ACGIH TLV: TWA 1 ppm
DOT Classification: Flammable Liquid and Poison

THR: Poison by inhalation, intravenous, and intraperitoneal routes. Moderately toxic by ingestion. Severe eye irritant. See also silicates. This material can cause extensive necrosis (experimental), keratoconus, and opaque cornea; it also causes severe human eye injuries, as well as necrosis of corneal cells, which progresses long after the exposure has ceased. It is destructive and its effects resist treatment. Permanent blindness is possible from exposure to it. The kidney seems to be most subject to injury regardless of the mode of exposure. Pulmonary edema has also occurred. This material is more toxic than either ethyl silicate or silicic acid, although it has been thought that the injury caused is largely due to the action of the silicic acid.

METHYLSODIUM   HR: 3
mf: CH_3Na mw: 38.03

THR: Corrosive and irritating material. Self ignites in air immediately. Very dangerous. When heated to decomposition it emits toxic fumes of Na_2O. Incompatible with p-chloronitrobenzene. See also sodium and methanol. A dangerous fire hazard.

METHYL STEARATE   HR: 3
CAS: 112-61-8   NIOSH: WI 4460000
mf: $C_{19}H_{38}O_2$ mw: 298.57

PROP: Liquid to semi-solid; mp: 38°; bp: 215° @ 15 mm.; flash p. 307°F (CC); d: 0.860; sol in water and ether.

SYNS: METHOLENE 2218 * METHYL OCTADECANOATE * METHYL ESTER STEARIC ACID * OCTADECANOIC ACID, METHYL ESTER

THR: An experimental tumorigen. See also esters. Combustible when exposed to heat or flame, can react with oxidizing materials. To fight fire, use CO_2, dry chemical. When heated to decomposition it emits acrid smoke and fumes.

alpha-METHYLSTYRENE   HR: 3
CAS: 98-83-9   NIOSH: WL 5250000
DOT: 2303
mf: C_9H_{10} mw: 118.19

PROP: Colorless liquid; d: 0.862 @ 20°/4°; mp: −96.0°; bp: 152.4°; insol in water; misc in alc and ether.

SYNS: ISOPROPENIL-BENZOLO (ITALIAN) * ISOPROPENYLBENZENE * ISOPROPENYL-BENZEEN (DUTCH) * ISOPROPENYL-BENZOL (GERMAN) * ALPHA-METHYLSTYREEN (DUTCH) * ALPHA-METHYL-STYROL (GERMAN) * ALPHA-METIL-STIROLO (ITALIAN) * AS-METHYL-PHENYLETHYLENE * 2-PHENYLPROPENE * BETA-PHENYLPROPENE * 2-PHENYLPROPYLENE * BETA-PHENYLPROPYLENE

OSHA PEL: TWA CL 100 ppm
ACGIH TLV: TWA 50 ppm; STEL 100 ppm
DFG MAK: 100 ppm (480 mg/m³)
DOT Classification: Flammable Liquid

THR: Moderately toxic by inhalation. Human irritant (systemic) effects. A skin and eye irritant. When heated to decomposition it emits acrid smoke and fumes.

17-METHYL TESTOSTERONE   HR: 3
CAS: 58-18-4   NIOSH: BV 8400000
mf: $C_{20}H_{30}O_2$ mw: 302.50

SYNS: ANDROMETH * ANDROSAN * ANDROSAN (TABLETS) * ANDROST-4-EN-3-ONE, 17-HYDROXY-17-METHYL-, (17-BETA)-(9CI) * ANDROSTEN * ANERTAN * ANERTAN (TABLETS) * DELATESTRYL * DIANABOL * 17-BETA-HYDROXY-17-METHYLANDROST-4-EN-3-ONE * METHYLTESTOSTERONE * 17-ALPHA-METHYLTESTOSTERONE * TESTHORMONE

THR: Poison by intraperitoneal route. A human neoplastigen, carcinogen, and teratogen by ingestion and several other routes. A synthetic androgenic steroid. For further information see Vol. 1, No. 3 of *DPIM Report*.

METHYL THIOCYANATE   HR: 3
CAS: 556-64-9   NIOSH: XL 1575000
mf: C_2H_3NS mw: 73.12

PROP: Liquid; d: 1.068 @ 20°; mp: −51°; bp: 130-133°; very sltly sol in water; misc in alc and ether.

SYNS: METHYLRHODANID (GERMAN)
* METHYL SULFOCYANATE

THR: Poison by ingestion, intravenous, and subcutaneous routes. See also thiocyanates. When heated to decomposition it emits very toxic fumes of NO_x and SO_x.

METHYLTHIOINOSINE   HR: 3
CAS: 342-69-8   NIOSH: UO 8985000
mf: $C_{11}H_{14}N_4O_4S$ mw: 298.35

SYNS: 6-METHYLMERCAPTOPURINE RIBOSIDE
* NCI-C04784 * NSC 40774

THR: Poison by intraperitoneal and other routes. An experimental carcinogen. When heated to decomposition it emits very toxic fumes of SO_x and NO_x.

4-METHYLTHIOPHENYLDIMETHYL PHOSPHATE   HR: 3
CAS: 3254-63-5   NIOSH: TC 5075000
mf: $C_9H_{13}O_4PS$ mw: 248.25

SYNS: O,O-DIMETHYL O-(4-METHYLMERCAPTO-PHENYL) PHOSPHATE * DIMETHYL P-(METHYL-THIO)PHENYL PHOSPHATE * ENT 25734
* PHOSPHORIC ACID DIMETHYL-P-(METHYL-THIO)PHENYL ESTER

THR: Poison by ingestion, skin contact, and subcutaneous routes. See also esters. When heated to decomposition it emits very toxic fumes of SO_x and PO_x.

6-METHYLTHIOURACIL   HR: 3
CAS: 56-04-2   NIOSH: YR 0875000
mf: $C_5H_6N_2OS$ mw: 142.19

PROP: Bitter crystals or colorless liquid, decomp @ 326°-331°; sublimes readily; very sltly sol in ether, cold water. Very sol in alkaline hydroxides, NH_3. Odor of onions; slightly sol in alc, acetone. Almost insol in benzene, chloroform.

SYNS: 2,3-DIHYDRO-6-METHYL-2-THIOXO-4(1H)-PYRIMIDINONE * 2-MERCAPTO-6-METHYLPYRIMID-4-ONE * 2-MERCAPTO-4-HYDROXY-6METHYLPYRIMIDINE
* 2-MERCAPTO-6-METHYL-4-PYRIMIDONE * 6-METHYL-2-THIO-2,4-(1H3H)PYRIMIDINEDIONE
* METHYLTHIOURACIL * 4-METHYL-2-THIOURACIL * 6-METHYL-2-THIOURACIL
* 4-METHYLURACIL * 2-THIO-6-METHYL-1,3-PYRIMIDIN-4-ONE * 6-THIO-4-METHYLURACIL
* 2-THIO-4-OXO-6-METHYL-1,3-PYRIMIDINE
* USAF EK-6454

THR: Poison by intraperitoneal route. Moderate acute toxicity by ingestion. An experimental carcinogen, teratogen, and neoplastigen. When heated to decomposition it emits very toxic fumes of NO_x and SO_x.

N-METHYL-p-(m-TOLYLAZO)ANILINE   HR: 3
CAS: 2058-62-0   NIOSH: BY 6350000
mf: $C_{14}H_{15}N_3$ mw: 225.32

SYNS: N-METHYL-3'-METHYL-P-AMINOAZOBEN-ZENE * N-METHYL-3'-METHYL-4-AMINOAZO-BENZENE * 3'-METHYL-4-MONOMETHYLAMI-NOAZOBENZENE

THR: An experimental tumorigen. Mutagenic data. When heated to decomposition it emits toxic fumes of NO_x.

METHYLTRICHLOROSILANE   HR: 3
CAS: 75-79-6   NIOSH: VV 4550000
mf: CH_3Cl_3Si mw: 149.48

SYN: METHYLTRICHLORSILAN (CZECH)

DOT Classification: Label: Flammable Liquid

THR: Poison by inhalation and intraperitoneal routes. Moderately toxic by ingestion. A moderate skin and severe eye irritant. See also chlorosilanes. When heated to decomposition it emits toxic fumes of Cl^-.

5-METHYL-2-TRIFLUOROMETHYLOXAZOLIDINE   HR: 3
CAS: 31185-56-5   NIOSH: RQ 2130000
mf: $C_5H_8F_3NO$ mw: 155.14

THR: An experimental carcinogen. Moderate acute toxicity by intraperitoneal route. See also fluorides. When heated to decomposition it emits very toxic fumes of F^- and NO_x.

4-METHYL TRIMETHYLENE SULFITE   HR: 3
CAS: 4426-51-1   NIOSH: EK 1225000
mf: $C_4H_8O_3S$ mw: 136.18

SYNS: 1,3-BUTANEDIOL, CYCLIC SULFITE
* NSC-60195

THR: Poison by intravenous and intraperitoneal routes. When heated to decomposition it emits toxic fumes of SO_x.

METHYLTRIOCTYLAMMONIUM CHLORIDE   HR: 3
CAS: 5137-55-3   NIOSH: BR 8575000
mf: $C_{25}H_{54}N \cdot Cl$ mw: 404.25

SYNS: TRIOCTYLMETHYLAMMONIUM CHLORIDE
* METHYLTRICAPRYLYLAMMONIUMCHLORIDE
* TRICAPRYLMETHYLAMMONIUM CHLORIDE
* TRICAPRYLYLMETHYLAMMONIUM CHLORIDE

THR: Poison by ingestion and intraperitoneal routes. See also chlorides. When heated to decomposition it emits very toxic fumes of NO_x and Cl^-.

METHYL TRITHION   HR: 3
CAS: 953-17-3   NIOSH: TD 5425000
mf: $C_9H_{12}ClO_2PS_3$ mw: 314.81

SYNS: S-(((P-CHLOROPHENYL)THIO)METHYL) O,O-DIMETHYL PHOSPHORODITHIOATE * DIMETHYL P-CHLOROPHENYLTHIOMETHYL DITHIOPHOSPHATE * O,O-DIMETHYL-S-(P-CHLOROPHENYLTHIOMETHYL)PHOSPHORODITHIOATE * O,O-DIMETHYLTHIOPHOSPHORIC ACID P-CHLOROPHENYL ESTER * ENT 25,599 * METHYLCARBOPHENOTHION

THR: Poison by ingestion, skin contact, and other routes. A cholinesterase inhibitor. See also parathion. When heated to decomposition it emits very toxic fumes of Cl^-, PO_x, and SO_x.

METHYL VINYL ETHER   HR: 3
CAS: 107-25-5   NIOSH: KO 2300000
mf: C_3H_6O mw: 58.09

PROP: Colorless, easily liquefied gas or colorless liquid, bp: 6.0°, d: 0.7500, vap d: 2.0, fp: −121.6°, vap press: 1052 mm @ 20°; flash p: −68.8°F; lel = 2.6%; uel = 39.0%.

SYN: VINYL METHYL ETHER

THR: Mildly toxic by ingestion. See also ethers. Very dangerous fire hazard when exposed to heat, flame, or oxidizers. Explosive details are not listed. Very dangerous; can react vigorously with oxidizing materials. To fight fire, stop flow of gas. Incompatible with acids; halogens.

METHYL VIOLET 6B   HR: 3
CAS: 5974-19-6   NIOSH: BO 6475000
mf: $C_{31}H_{34}N_3 \cdot Cl$ mw: 484.13

SYN: PENTAMETHYLBENZYL-P-ROSANILINE CHLORIDE

THR: A poison via oral route. When heated to decomposition it emits very toxic fumes of NO_x and Cl^-.

METHYSERGIDE DIMALEATE   HR: 3
CAS: 29605-96-7   NIOSH: KE 5400000
mf: $C_{21}H_{27}N_3O_2 \cdot 2C_4H_4O_4$ mw: 585.67

PROP: Decomp above 165°; sol in methanol; less sol in H_2O; insol in absolute ethanol.

SYN: SANSERT

THR: Poison by ingestion and intravenous routes. When heated to decomposition it emits toxic fumes of NO_x.

METIAPINE   HR: 3
CAS: 5800-19-1   NIOSH: HQ 2275000
mf: $C_{19}H_{21}N_3S$ mw: 323.49

SYN: 2-METHYL-11-(4-METHYL-1-PIPERAZINYL)-DIBENZO(B,F)(1,4)THIAZEPINE

THR: Poison by intraperitoneal route. Moderately toxic by ingestion. When heated to decomposition it emits very toxic fumes of NO_x and SO_x.

METOMIDATE   HR: 3
CAS: 5377-20-8   NIOSH: NI 4025000
mf: $C_{13}H_{14}N_2O_2$ mw: 230.29

SYN: METHYL 1-(ALPHA-METHYLBENZYL)IMIDAZOLE-5-CARBOXYLATE

THR: Poison by ingestion. When heated to decomposition it emits toxic fumes of NO_x.

METOPIMAZINE   HR: 3
CAS: 14008-44-7   NIOSH: NS 5075000
mf: $C_{22}H_{27}N_3O_3S_2$ mw: 445.6

PROP: Solid; mp: 170°-171°.

SYNS: 1-(3-(2-(METHYLSULFONYL)PHENOTHIAZIN-10-YL)PROPYL)ISONIPECOTAMIDE
* 2-METHYLSULFONYL-10-(3-(4CARBAMOYLPIPERIDINO)PROPYL)PHENOTHIAZINE

THR: Poison by intravenous, intraperitoneal, and subcutaneous routes. When heated to decomposition it emits very toxic fumes of SO_x and NO_x.

MEVASIN HYDROCHLORIDE HR: 3
CAS: 826-39-1   NIOSH: RB 6900000
mf: $C_{11}H_{21}N \cdot ClH$ mw: 203.79

SYNS: 3-METHYLAMINOISOCAMPHANE HYDRO-CHLORIDE * N-METHYL-DL-ISOBORNYLAMINE HYDROCHLORIDE * N,2,3,3-TETRAMETHYL-2-NORBORNANAMINE HYDROCHLORIDE

THR: Poison by ingestion, intraperitoneal, intravenous, and subcutaneous routes. When heated to decomposition it emits very toxic fumes of HCl and NO_x.

MEVINPHOS
CAS: 7786-34-7   NIOSH: GQ 5250000
mf: $C_7H_{13}O_6P$ mw: 224.17   HR: 3

SYNS: ALPHA-2-CARBOMETHOXY-1-METHYLVI-NYLDIMETHYL PHOSPHATE * 2-CARBOME-THOXY-1-PROPEN-2-YL DIMETHYL PHOSPHATE * O,O-DIMETHYL-O-(2-CARBOMETHOXY-1-METHYLVINYL) PHOSPHATE * DIMETHYL-1-CARBOMETHOXY-1-PROPEN-2-YL PHOSPHATE * DIMETHYL 2-METHOXYCARBONYL-1-METHYL-VINYL PHOSPHATE * DIMETHYL METHOXYCAR-BONYLPROPENYL PHOSPHATE * DIMETHYL (1-METHOXYCARBOXYPROPEN-2-YL)PHOSPHATE * O,O-DIMETHYL O-(1-METHYL-2-CARBOXYVI-NYL) PHOSPHATE * DIMETHYL PHOSPHATE OF-METHYL-3-HYDROXY-CIS-CROTONATE * 3-HY-DROXYCROTONIC ACIDMETHYL ESTER DIMETHYLPHOSPHATE * (2-METHOXYCAR-BONYL-1-METHYL-VINYL)-DIMETHYL-FOSFAAT (DUTCH) * (2-METHOXYCARBONYL-1-METHYL-VINYL)-DIMETHYL-PHOSPHAT (GERMAN) * 2-METHOXYCARBONYL-1-METHYLVINYL DI-METHYLPHOSPHATE * CIS-2-METHOXYCAR-BONYL-1-METHYLVINYL DIMETHYLPHOSPHATE * METHYL-3-(DIMETHOXYPHOSPHINYLOXY)-CROTONATE * (2-METOSSICARBONIL-1-METIL-VINIL)-DIMETIL-FOSFATO (ITALIAN) * CIS-PHOSDRIN * PHOSPHATE DE DIMETHYLE ET DE 2-METHOXYCARBONYL-1 METHYLVINYLE (FRENCH)

OSHA PEL: TWA 400 ug/m^3

THR: Poison by ingestion, inhalation, skin contact, and intraperitoneal routes. Peripheral nervous system effects in humans. When heated to decomposition it emits toxic fumes of PO_x. For further information see Vol. 6, No. 1 of *DPIM Report*.

MICA HR: 3
CAS: 12001-26-2   NIOSH: VV 8760000

PROP: Containing less than 1% crystalline silica.

SYN: MICA SILICATE

OSHA PEL: TWA 20 mppcf

THR: No specific toxicity data. See also silicates and silica.

MICHLER'S BASE HR: 3
CAS: 101-61-1   NIOSH: BY 5250000
mf: $C_{17}H_{22}N_2$ mw: 254.41

SYNS: P,P'-BIS(DIMETHYLAMINO)DIPHENYL-METHANE * 4,4'-BIS(DIMETHYLAMINO)-DIPHENYLMETHANE * BIS(P-DIMETHYLAMINO-PHENYL)METHANE * BIS(P-(N,N-DIMETHYL-AMINO)PHENYL)METHANE * P,P'-BIS(N,N-DIMETHYLAMINOPHENYL)METHANE * P,P-DIMETHYLAMINODIPHENYLMETHANE * 4,4'-METHYLENEBIS(N,N-DIMETHYL)BENZENA-MINE * 4,4'-METHYLENEBIS(N,N-DIMETHYL ANILINE) * NCI-C01990 * 4,4'-TETRA-METHYLDIAMINODIPHENYLMETHANE * P,P-TETRAMETHYLDIAMINODIPHENYLMETHANE

THR: An experimental carcinogen, neoplastigen, and tumorigen. Mutagenic data. Moderately toxic by ingestion. When heated to decomposition it emits toxic fumes of NO_x.

MICHLER'S KETONE HR: 3
CAS: 90-94-8   NIOSH: DJ 0250000
mf: $C_{17}H_{20}N_2O$ mw: 268.39

PROP: Leaves from ethanol; mp: 172°; bp: >360° decomp; insol in water; very sol in benzene; sol in alc; very sltly sol in ether.

SYNS: P,P'-BIS(N,N-DIMETHYLAMINO)BENZO-PHENONE * 4,4'-BIS(DIMETHYLAMINO)BENZO-PHENONE * BIS(P-(N,N-DIMETHYLAMINO)PHE-NYL)KETONE * BIS(4-(DIMETHYLAMINO)-PHENYL)METHANONE * P,P'-MICHLER'S KETONE * NCI-C02006

THR: An experimental carcinogen and neoplastigen. When heated to decomposition it emits toxic fumes of NO_x.

MILTOWN HR: 3
CAS: 57-53-4   NIOSH: TZ 0175000
mf: $C_9H_{18}N_2O_4$ mw: 218.29

SYNS: MEPROBAM * MEPROBAMAT (GERMAN) * MEPROBAMATE * EQUANIL * EQUANIL SUSPENSION * 2,2-DI(CARBAMOYLOXYMETH-

YL)-PENTANE * 2-METHYL-2-N-PROPYL-1,3-
PROPANEDIOL DICARBAMATE

THR: Poison by intravenous, intraperitoneal, and possibly other routes. Moderately toxic by ingestion. Human central nervous system effects by ingestion. An experimental teratogen. Implicated in aplastic anemia. When heated to decomposition it emits toxic fumes of NO_x.

MIMOSA TANNIN   HR: 3
NIOSH: PY 7955000

SYNS: ACACIA MOLLISSIMA TANNIN * TANNIN FROM MIMOSA

THR: An experimental tumorigen. See also tannin. When heated to decomposition it emits smoke and acrid fumes. For further information see Vol. 1, No. 1 of *DPIM Report*.

MINERAL DUSTS   HR: 3
THR: Variable. From the economic and toxicity standpoints, the most important are those containing free silica, which can cause silicosis upon inhalation of sufficient quantity. These include sand, sandstone, quartz and flint. They consist mainly of silica in the form of quartz; diatomaceous earth, which is essentially amorphous silica; and granite, which contains 20-40% quartz. Minerals that contain combined silica in the form of silicates but no free silica are generally less capable of causing silicosis. Asbestos, however, can cause a fibrotic lung condition of its own, known as asbestosis, and lung cancer. (See asbestos.) Mica and talc dust are also considered somewhat hazardous. Nonsiliceous minerals, like limestone, marble, dolomite, etc., which do not contain toxic elements, do not ordinarily present any significant dust hazard, although minerals containing toxic elements, such as cryolite, which contains fluorine, and pyrolusite, which contains manganese, may cause systemic poisoning upon inhalation or ingestion in sufficient quantity. In any event, the minerals are usually less reactive than synthetic compounds of the same elements, and in fact, may be relatively inert by comparison. These are common air contaminants.

MINERAL OIL   HR: 3
CAS: 8012-95-1   NIOSH: PY 8030000

PROP: Colorless, oily liquid, practically tasteless and odorless, insol in water and alcohol, sol in benzene, chloroform and ether. A mixture of liquid hydrocarbons from petroleum. d: 0.83-0.86 (light), 0.875-0.905 (heavy), flash p: 444°F (OC), ulc: 10-20.

SYNS: ALBOLINE * KAYDOL * LIGNITE OIL * NUJOL * PARAFFIN OIL * PETROLATUM, LIQUID * SAXOL * WHITE MINERAL OIL

THR: An experimental tumorigen and suspected carcinogen. Inhalation of vapor or particulates can cause aspiration pneumonia. A laxative. Combustible. To fight fire, use dry chemical, CO_2, foam. For further information see Vol. 1, No. 2 of *DPIM Report*.

MIREX   HR: 3
CAS: 2385-85-5   NIOSH: PC 8225000
mf: $C_{10}Cl_{12}$ mw: 545.50

PROP: Very white odorless crystals; water insol; sol in dioxane and benzene; decomp @ 485°.

SYNS: DODECACHLOROOCTAHYDRO-1,3,4-METHENO-2H-CYCLOBUTA(C,D)PENTALENE * DODECACHLOROPENTACYCLODECANE * DODECACHLOROPENTACYCLO(3,2,2,0(SUP 2, 6),0(SUP 3,9),0(SUP 5,10))DECANE * ENT 25,719 * FERRIAMICIDE * HEXACHLORO-CYCLOPENTADIENEDIMER * 1,2,3,4,5,5-HEXACHLORO-1,3-CYCLOPENTADIENE DIMER * NCI-C06428 * PERCHLOROPENTACYCLO-DECANE * PERCHLOROPENTACYCLO(5.2.1.0-(SUP 2,6).0(SUP 3,9).0(SUP 5,8))DECANE

THR: Poison by ingestion. Moderately toxic by inhalation and skin contact. An experimental teratogen and carcinogen. A persistent insecticide which is toxic to non-target species. It can bioaccumulate. Dangerous. See also chlorides. For further information see Vol. 6, No. 1 of *DPIM Report*.

MITHRAMYCIN   HR: 3
CAS: 18378-89-7   NIOSH: PZ 2800000
mf: $C_{52}H_{76}O_{24}$ mw: 1085.28

PROP: Antibiotic substance isolated from the fermentation broth of three strains of an unidentified *Streptomyces* species.

SYNS: AUREOLIC ACID * MITRAMYCIN * NSC 24559

THR: Poison by intravenous, intraperitoneal, subcutaneous, and other routes. Mutagenic data. Human skin (systemic) effects. An experimental teratogen. When heated to decomposition it emits acrid smoke and fumes.

MOCA   HR: 3
CAS: 101-14-4   NIOSH: CY 1050000
mf: $C_{13}H_{12}Cl_2N_2$ mw: 267.17

SYNS: DI(-4-AMINO-3-CHLOROPHENYL)METHANE
* DI-(4-AMINO-3-CLOROFENIL)METANO (ITAL-
IAN) * 4,4'-DIAMINO-3,3'-DICHLORODIPHENYL-
METHANE * 3,3'-DICHLOR-4,4'-DIAMINODIPHE-
NYLMETHAN (GERMAN) * 3,3'-DICHLORO-4,
4'-DIAMINODIPHENYLMETHANE * 3,3'-DICLO-
RO-4,4'-DIAMINODIFENILMETANO(ITALIAN)
* 4,4'-METHYLENE(BIS)-CHLOROANILINE
* METHYLENE 4,4'-BIS (O-CHLOROANILINE)
* P,P'-METHYLENEBIS(ALPHA-CHLOROANILINE)
* 4,4'-METHYLENEBIS(O-CHLOROANILINE)
* P,P'-METHYLENEBIS(O-CHLOROANILINE)
* 4,4'-METHYLENEBIS(2-CHLOROANILINE)
* 4,4'-METHYLENEBIS-2-CHLOROBENZENAMINE
* 4,4-METILENE-BIS-O-CLOROANILINA (ITALIAN)

THR: An experimental carcinogen. Mutagenic
data. When heated to decomposition it emits
very toxic fumes of Cl^- and NO_x. For further
information see Vol. 5, No. 2 of *DPIM Report*.

MOLYBDENUM   HR: 3
CAS: 7439-98-7   NIOSH: QA 4680000
af: Mo   aw: 95.94

PROP: Cubic, silver-white metallic crystals or
gray-black powder; mp: 2622°; bp: approx
4825°; d: 10.2; vap press: 1 mm @ 3102°.
ACGIH TLV: TWA (soluble compounds) 5 mg/
m³; (insoluble compounds) 10 mg/m³
DFG MAK: (insoluble compounds) 15 mg/m³;
(soluble compounds) 5 mg/m³

THR: Poison by intraperitoneal and intratracheal
routes. See also molybdenum compounds.
Flammable in the form of dust; when exposed
to heat or flame, violent reaction with BrF_3;
ClF_3; F_2; PbO_2. See also iron dust. A mild explosive in the form of dust, when exposed to flame.
See also powdered metals. Incompatible with
oxidants.

MOLYBDENUM COMPOUNDS   HR: 3
THR: Poison by subcutaneous and intraperitoneal routes. Molybdenum and its compounds
are highly toxic based upon animal experiments.
But in spite of their considerable use in industry,
human industrial poisoning by molybdenum inhalation has yet to be reported. It is suggested
that suitable precautions should be taken against
human inhalation of significant amounts of the
more soluble molybdenum compounds. Molybdenum is rapidly excreted by the body. Recent
studies have shown that molybdenum has importance as a trace element in the normal growth
and development of certain forms of plant life.
It is found also in animal tissue, although its
precise function is unknown. It is a common
air contaminant.

MOLYBDENUM PENTA-
CHLORIDE   HR: 3
CAS: 10241-05-1   NIOSH: QA 4690000
mf: Cl_5Mo mw: 273.19

PROP: Green-black solid, dark-red as liquid
or vapor; hygroscopic, reacting with water and
air, sol in dry ether, dry alc and other anhydrous
organic solvents; mp: 194°; bp: 268°; d: 2.9.
DOT Classification: ORM-B, Label: None

THR: Poison. See also HCl and molybdenum
compounds. Dangerous. See also chloride.

MOLYBDENUM TRIOXIDE   HR: 3
CAS: 1313-27-5   NIOSH: QA 4725000
mf: MoO_3 mw: 143.94

PROP: White or yellow to sltly bluish powder
or granules; sol in 1000 parts water (low solubility in H_2O), in conc mineral acids and solutions
of alkali hydroxides, ammonia or potassium bitartrate which solidifies to a yellowish-white
mass and sublimes at higher temp; mp: 795°;
bp: 1155°; d: 4.696 @ 26°/4°

SYNS: MOLYBDIC ANHYDRIDE * MOLYBDIC
TRIOXIDE

OSHA PEL: TWA 5 mg(Mo)/m³

THR: Poison by ingestion, inhalation, subcutaneous, and intraperitoneal routes. An experimental neoplastigen. Human pulmonary effects.
A powerful irritant. See also molybdenum. Violent reaction with ClF_3; Li; Mg; K; Na. Incompatible with interhalogens; metals.

MONOCHLORODIFLUORO-
METHANE   HR: 1
CAS: 75-45-6   NIOSH: PA 6390000
mf: $CHClF_2$ mw: 86.47

PROP: Gas; d: 3.87 air @ 0°; mp: −146°;
bp: −40.8°; autoign temp: 1170°F.

SYNS: CHLORODIFLUOROMETHANE * DIFLUO-
ROCHLOROMETHANE * DIFLUOROMONOCHLORO-
METHANE * FLUOROCARBON-22 * FREON

* GENETRON 22 * PROPELLANT 22 * REFRIGERANT 22

DOT Classification: Label: Nonflammable Gas

THR: Mildly toxic by inhalation. Mutagenic data. Asphyxiant in high concentrations. Combustible. Dangerous disaster hazard; see also chlorides and fluorides.

MONOCHLOROMETHANE   HR: 1
CAS: 74-87-3   NIOSH: PA 6300000
mf: CH_3Cl mw: 50.49

PROP: Colorless gas; bp: $-23.7°$, lel = 10.7%, uel = 17.2%, fp: $-97.7°$, flash p: $<32°F$ (OC), d: 0.918 @ $20°/4°$, autoign temp: $1170°F$, vap d: 1.78.

SYNS: CHLOOR-METHAAN (DUTCH) * CHLOR-METHAN (GERMAN) * CHLOROMETHANE * CHLORURE DE METHYLE (FRENCH) * CLOROMETANO (ITALIAN) * CLORURO DI METILE (ITALIAN) * METHYLCHLORID (GERMAN) * METHYL CHLORIDE * METYLU CHLOREK (POLISH)

OSHA PEL: TWA 100 ppm; CL 200; Pk 300/5M/3H
DOT Classification: Label: Flammable Gas

THR: A slight irritant by inhalation. A weaker narcotic than chloroform. Acute poisoning, characterized by the narcotic effect, is rare in industry. Repeated exposure to low concentrations can cause damage to the central nervous system, liver, kidneys, bone marrow, and cardiovascular system. Hemorrhages into the lungs, intestinal tract and dura have been reported. In exposures to high concentrations death may be immediate, and if the exposure is not fatal recovery is slow and degenerative changes in the central nervous system are not uncommon. In repeated exposures to lower concentrations there is usually fatigue, loss of appetite, muscular weakness, drowsiness and dimness of vision. After-effects are commonly the result of damage to the central nervous system, with visual changes and attacks of depression and other psychic disturbances being reported. Very dangerous fire hazard when exposed to heat, flame, or powerful oxidizers. Moderately explosive when exposed to heat or flame. Violent reactions with Al; Mg; K; Na; NaK. Dangerous; when heated to decomposition it emits highly toxic fumes of chlorides; can react vigorously with oxidizing materials. To fight fire, stop flow of gas; CO_2, dry chemical, or water spray.

MONOCHLOROMONOFLUORO-METHANE   HR: 3
CAS: 593-70-4   NIOSH: PA 6408000
mf: CH_2ClF mw: 68.48

SYN: FREON 31

THR: An experimental teratogen. Moderately toxic by inhalation. When heated to decomposition it emits very toxic fumes of Cl^- and F^-.

MONOCROTALINE   HR: 3
CAS: 315-22-0   NIOSH: UY 8225000
mf: $C_{16}H_{23}NO_6$ mw: 325.35

PROP: Prisms from absolute ethanol; decomp @ $197°-198°$.

SYNS: 14,19-DIHYDRO-12,13-DIHYDROXY(13-ALPHA,14-ALPHA)-20-NORCROTALANAN-11, 15-DIONE * 12-BETA,13-BETA-DIHYDROXY-12-ALPHA,13-ALPHA,14-ALPHA-TRIMETHYLCRO-TAL-1-ENINE * NCI-C56462 * NSC 28693

THR: Poison by ingestion, intravenous, intraperitoneal, subcutaneous, and other routes. Mutagenic data. An experimental carcinogen. When heated to decomposition it emits toxic fumes of NO_x.

MONOETHANOLAMINE   HR: 3
CAS: 141-43-5   NIOSH: KJ 5775000
mf: C_2H_7NO mw: 61.10

PROP: Colorless liquid, ammoniacal odor. Hygroscopic; bp: $170.5°$, fp: $10.5°$, flash p: $200°F$ (OC), d: 1.0180 @ $20°/4°$, vap press: 6 mm @ $60°$, vap d: 2.11. Misc in water and alc; sltly sol in benzene; sol in chloroform.

SYNS: AETHANOLAMIN (GERMAN) * 2-AMINOAETHANOL (GERMAN) * 2-AMINOETANOLO (ITALIAN) * 2-AMINOETHANOL * BETA-AMINOETHYL ALCOHOL * ETANOLAMINA (ITALIAN) * BETA-ETHANOLAMINE * ETHYLOLAMINE * 2-HYDROXYETHYLAMINE * BETA-HYDROXYETHYLAMINE * MONOAETHANOLAMIN (GERMAN) * MONOETHANOLAMINE (DOT) * USAF EK-1597

OSHA PEL: TWA 3 ppm
DOT Classification: Label: Corrosive

THR: Poison by intraperitoneal route. Moderately toxic by several other routes. Moderate skin and severe eye irritant. Flammable when exposed to heat or flame; can react violently with acetic acid; acetic anhydride; acrolein; acrylic acid; acrylonitrile; chlorosulfonic acid; epichlorohydrin; HCl; HF; mesityl oxide; HNO_3; oleum; H_2SO_4; β-propiolactone; vinyl acetate. To fight fire, use foam, alcohol foam, dry chemical. For further information see ethanolamine, Vol. 4, No. 1 of *DPIM Report*.

8-MONOHYDRO MIREX   HR: 3
CAS: 39801-14-4   NIOSH: PC 8475000
mf: $C_{10}HCl_{11}$ mw: 511.06

SYNS: HYDROMIREX * PHOTOMIREX * 1,2,3,4,5,5,5,6,7,9,10,10-UNDECACHLORO-PENTACYCLO(5.3.0.0(SUP 2,6).0(SUP 3,9).0-(SUP 4,8))DECANE

THR: Poison by ingestion. An experimental teratogen. When heated to decomposition it emits toxic fumes of Cl^-.

MONOMETHYLHYDRAZINE
NITRATE   HR: 3
CAS: 29674-96-2   NIOSH: MV 7300000
mf: CH_6N_2 • HNO_3 mw: 109.11

THR: Poison by ingestion, skin contact, intravenous, and intraperitoneal routes. When heated to decomposition it emits toxic fumes of NO_x.

MONONITROSOPIPERAZINE   HR: 3
CAS: 5632-47-3   NIOSH: TM 2450000
mf: $C_4H_9N_3O$ mw: 115.16

SYNS: N-NITROSOPIPERAZINE * 1-NITROSOPIPERAZINE

THR: An experimental carcinogen and neoplastigen. Mutagenic data. Moderate acute toxicity by ingestion. When heated to decomposition it emits toxic fumes of NO_x.

MONOSODIUM GLUTAMATE   HR: 3
CAS: 142-47-2   NIOSH: MA 1575000
mf: $C_5H_9NO_4$ • Na   mw: 170.14

PROP: White or almost white crystals or powder. Slight peptone-like odor. Meal-like taste. Very sol in H_2O; sltly sol in alc.

SYNS: ACCENT * CHINESE SEASONING * GLUTAMMATO MONOSODICO (ITALIAN) * MONOSODIOGLUTAMMATO (ITALIAN) * MONOSODIUM-L-GLUTAMATE * MSG * NATRIUMGLUTAMINAT (GERMAN) * L(+) SODIUM GLUTAMATE

THR: An experimental teratogen. Strong systemic effects in humans by ingestion. Mildly toxic by ingestion and other routes. When heated to decomposition it emits toxic fumes of NO_x and Na_2O.

MONOSODIUM METHYLARSO-
NATE   HR: 3
CAS: 2163-80-6   NIOSH: PA 2625000
mf: CH_4AsO_3 • Na   mw: 161.96

SYNS: METHYLARSENIC ACID, SODIUM SALT * MONOSODIUM ACID METHANEARSONATE * MONOSODIUM ACID METHARSONATE * MONOSODIUM METHANEARSONATE * MONOSODIUM METHANEARSONIC ACID * NCI-C60071 * SODIUM METHANEARSONATE * SODIUM ACID METHANEARSONATE

THR: Poison by unspecified routes. A human carcinogen. A skin and eye irritant. See also arsenic. When heated to decomposition it emits toxic fumes of As and Na_2O.

MORIN   HR: 3
CAS: 480-16-0   NIOSH: LK 8749000
mf: $C_{15}H_{10}O_7$ mw: 302.23

PROP: Anhydrous needles from absolute alc, decomp @ 285°-290°; crystalizes with 1 or 2 mols H_2O; slightly sol in H_2O, ether, acetic acid; very sol in alc.

SYNS: AURANTICA * BOIS D'ARC (FRENCH) * CALICO YELLOW * C.I. NATURAL YELLOW 8 * 2'-HYDROXYPELARGIDENOLON 1522 * OSAGE ORANGE * OSAGE ORANGE CRYSTALS * OSAGE ORANGE EXTRACT * 2',3,4',5,7-PENTAHYDROXYFLAVONE

THR: Mutagenic data. When heated to decomposition it emits acrid smoke and fumes.

(−)-MORPHINE   HR: 3
CAS: 57-27-2   NIOSH: QC 7875000
mf: $C_{17}H_{19}NO_3$ mw: 285.37

PROP: White crystalline alkaloid. Short, orthorhombic, columnar prisms from anisole. Decomp @ 254°; mp: 197°; sublimes @ 190°-200°.

SYNS: 4A,5,7A,8-TETRAHYDRO-12-METHYL-9H-9,9C-IMINOETHANOPHENANTHRO(4,5-bcd)FURAN-3,5-DIOL * MORPHIA * MORPHINA * MORPHINE * MORPHINISM * MORPHINUM * MORPHIUM

THR: Poison by ingestion, intracerebral, intra-peritoneal, subcutaneous, intravenous, and other routes. Morphine is the constituent of opium most responsible for its toxic effects. When taken orally, the effects of morphine poisoning begin to appear in 20-40 minutes; if taken hypodermically, the symptoms appear much earlier and narcotism is more likely to follow the early symptoms. Individual susceptibility varies greatly and children are more susceptible than adults. When heated to decomposition it emits toxic fumes of NO_x.

MORPHINE HYDROCHLORIDE   HR: 3
CAS: 52-26-6   NIOSH: QC 8575000
mf: $C_{17}H_{19}NO_3 \cdot ClH$ mw: 321.83

PROP: Trihydrate: White flakes or crystalline powder; bitter taste, mp: approx 200° (decomp). Dissolves slowly in glycerol; insol in chloroform, ether.

SYNS: MORPHINE CHLORHYDRATE * MOR-PHINE CHLORIDE

THR: Poison by intraperitoneal, intravenous, parenteral, and subcutaneous routes. See also morphine. Abuse leads to habituation or addiction. When heated to decomposition it emits very toxic fumes of NO_x and HCl.

MORPHINE SULFATE   HR: 3
CAS: 64-31-3   NIOSH: QC 8750000
mf: $C_{34}H_{38}N_2O_6 \cdot H_2O_4S$ mw: 668.76

THR: Poison by subcutaneous, intravenous, intraperitoneal, and intramuscular routes. An experimental teratogen. See also morphine and sulfates. When heated to decomposition it emits very toxic fumes of NO_x and SO_x.

MORPHOLINE   HR: 3
CAS: 110-91-8   NIOSH: QD 6475000
DOT: 2054
mf: C_4H_9NO mw: 87.14

PROP: Colorless, hygroscopic oil. Amine odor, bp: 128.9°, fp: −7.5°, flash p: 100°F (OC), autoign temp: 590°F, vap press: 10 mm @ 23°, vap d: 3.00; mp: −4.9°; d: 1.007 @ 20°/4°. Volatile with steam; misc with H_2O, evolving some heat; misc with acetone, benzene, ether, castor oil, methanol, ethanol, ethylene, glycol, linseed oil, turpentine, pine oil. Immiscible with concd NaOH solns.

SYNS: DIETHYLENEIMIDE OXIDE * DIETHYLENE IMIDOXIDE * DIETHYLENE OXIMIDE * DIETH-YLENIMIDEOXIDE * 1-OXA-4-AZACYCLOHEX-ANE * TETRAHYDRO-1,4-ISOXAZINE * TET-RAHYDRO-1,4-OXAZINE * TETRAHYDRO-2H-1, 4-OXAZINE

OSHA PEL: TWA 20 ppm (skin)
ACGIH TLV: TWA 20 ppm
DFG MAK: 20 ppm (70 mg/m^3)
DOT Classification: Flammable and
     Corrosive Liquid

THR: Poison by intraperitoneal route. Moderately toxic by several other routes. A severe skin, mucous membrane, and eye irritant. Can cause kidney damage. An experimental neoplastigen. Dangerous fire hazard when exposed to flame, heat, or oxidizers. Dangerous; when heated to decomposition it emits highly toxic fumes of NO_x; can react with oxidizing materials. To fight fire, use alcohol foam, CO_2, dry chemical. Incompatible with cellulose nitrate; nitromethane. For further information see Vol. 1, No. 8 of *DPIM Report*.

4-MORPHOLINENONYLIC ACID   HR: 3
CAS: 5299-64-9   NIOSH: QE 8050000
mf: $C_{13}H_{25}NO_3$ mw: 243.39

SYN: PELARGONIC MORPHOLIDE

THR: Moderately toxic by ingestion, inhalation, and intravenous routes. Human systemic irritant effects. An eye and intense mucous membrane irritant. When heated to decomposition it emits toxic fumes of NO_x.

MORPHOTHION   HR: 3
CAS: 14-41-2   NIOSH: TE 1400000
mf: $C_8H_{16}NO_4PS_2$ mw: 285.34

PROP: Colorless solid; mp: 65°; sol in acetone, dioxane, acetonitrile.

SYNS: O,O-DIMETHYL-S-((MORFOLINO-CAR-BONYL)-METHYL)-DITHIOFOSFAAT (DUTCH) * O,O-DIMETHYL S-(MORPHOLINOCARBAMOYL-METHYL) DITHIOPHOSPHATE * O,O-DIMETHYL MORPHOLINOCARBONYLMETHYL PHOSPHORODI-THIOATE * O,O-DIMETHYL-S-((MORPHOLINO-CARBONYL)-METHYL)-DITHIOPHOSPHAT (GER-MAN) * O,O-DIMETHYL S-(MORPHOLINOCAR-BONYLMETHYL) PHOSPHORODITHIOATE * DIMETHYL S-(MORPHOLINOCARBONYLMETHYL) PHOSPHOROTHIOLOTHIONATE * O,O-DIMETIL-S-((MORFOLINO-CARBONIL)-METIL)-DITIOFOSFATO (ITALIAN) * DITHIOPHOSPHATE DE O,O-DI-

METHYLEETDES-((MORPHOLINOCARBONYLE)-
METHYLE) (FRENCH) * 4-(MERCAPTOACE-
TYL)MORPHOLINE-O,O-DIMETHYL PHOSPHO-
RODITHIOATE * MORFOTHION (DUTCH)
* PHOSPHORODITHIOIC ACID O,O-DIMETHYL
S-(MORPHOLINOCARBONYLMETHYL) ESTER

THR: Poison by ingestion, skin contact, and other routes. A cholinesterase inhibitor. See also parathion. When heated to decomposition it emits very toxic fumes of PO_x, SO_x, and NO_x.

MUCOMYCIN HR: 3
CAS: 992-21-2 NIOSH: OL 5615000
mf: $C_{29}H_{33}N_4O_{10}$ mw: 597.66

SYNS: N(SUP 2)-(((+)-5-AMINO-5-
CARBOXYPENTYLAMINO)METHYL)TETRACYCLINE
* N-LYSINOMETHYLTETRACYCLINE * TETRA-
CYCLINE-L-METHYLENE LYSINE

THR: Poison by intravenous and other routes. When heated to decomposition it emits toxic fumes of NO_x.

MULTERGAN METHYL SULFATE HR: 3
CAS: 58-34-4 NIOSH: SP 0175000
mf: $C_{18}H_{23}N_2S \cdot CH_3O_4S$ mw: 410.59

SYNS: METHYLPHENAZONIUM METHOSULFATE
* N-(BETA-(10-PHENOTHIAZINYL)PROPYL)TRI-
METHYLAMMONIUM METHYL SULFATE * N,N,
N-ALPHA-TETRAMETHYL-10H-PHENOTHIAZINE-
10-ETHANAMINIUM METHYL SULFATE
* THIAZINAMIUM METHYL SULFATE
* TRIMETHYL (1-METHYL-2-PHENOTHIAZIN-
10-YLETHYL)AMMONIUM METHYL SULFATE
* TRIMETHYL(1-METHYL-2-(10-PHENOTHIAZI-
NYL)ETHYL)AMMONIUM METHYL SULFATE

THR: Poison by ingestion and intraperitoneal routes. When heated to decomposition it emits very toxic fumes of SO_x and NO_x.

MUSCARINE HR: 3
CAS: 300-54-9 NIOSH: QG 3325000
mf: $C_9H_{20}NO_2$ mw: 174.30

SYNS: MUSCARINE (THE ALKALOID) * MUS-
KARIN

THR: Poison by ingestion, intravenous, subcutaneous, and other routes. When heated to decomposition it emits toxic fumes of NO_x.

MYCOTOXIN T-2 HR: 3
CAS: 21259-2-0-1 NIOSH: YD 0100000
mf: $C_{24}H_{34}O_9$ mw: 466.58

PROP: A strain of *F. Tricinctum* isolated from infected corn.

SYNS: 4,15-DIACETOXY-8-(3-METHYLBUTYR-
YLOXY)-12,13-EPOXY-DELTA-9-TRICHOTHECEN-
3-OL * FUSARIOTOXIN T-2' * INSARIOTOXIN
* MYCOTOXIN

THR: Poison by ingestion, intravenous, and intraperitoneal routes. Mutagenic data. An experimental teratogen. A skin and eye irritant. When heated to decomposition it emits acrid smoke and fumes.

MYRISTIC ACID HR: 3
CAS: 544-63-8 NIOSH: QH 4375000
mf: $C_{14}H_{28}O_2$ mw: 228.36

PROP: Colorless leaves; d: 0.853 @ 70/4°; mp: 54.2°; bp: 250.5° @ 100 mm; insol in water; very sol in benzene, ether, alc absolute. Crystals from methanol; mp: 58.5°; bp: 250.5° @ 100 mm; d: 0.8622 @ 54°/4°; methanol, ether, petroleum, benzene, chloroform.

SYNS: N-TETRADECOIC ACID * TETRADECA-
NOIC ACID * 1-TRIDECANECARBOXYLIC ACID

THR: Poison by intravenous route. A human skin irritant. When heated to decomposition it emits acrid smoke.

MYRTAN TANNIN HR: 3
NIOSH: QH 6000000

SYNS: EUCALYPTUS REDUNCA TANNIN
* TANNIN FROM MYRTAN

THR: An experimental carcinogen. For further information see Vol. 1, No. 1 of *DPIM Report*.

N

NABAM
HR: 3
CAS: 142-59-6 NIOSH: FA 6825000
mf: $C_4H_6N_2S_4 \cdot 2Na$ mw: 256.34

PROP: Crystals, sol in water.

SYNS: DI-NATRIUM-AETHYLENBISDITHIOCARBA-
MAT (GERMAN) * DINATRIUM-(N,N'-AETH-
YLEN-BIS(DITHIOCARBAMAT)) (GERMAN)
* DINATRIUM-(N,N'-ETHYLEEN-BIS(DITHIO-
CARBAMAAT)) (DUTCH) * DISODIUM ETH-
YLENEBIS(DITHIOCARBAMATE) * N,N'-ETHYL-
ENE BIS(DITHIOCARBAMATE DE SODIUM)
(FRENCH) * ETHYLENEBIS(DITHIOCARBAMATE),
DISODIUM SALT * ETHYLENEBIS(DITHIOCAR-
BAMIC ACID) DISODIUM SALT * N,N'-ETILEN-
BIS(DITIOCARBAMMATO) DI SODIO (ITALIAN)
* DISODIUM ETHYLENE-1,2-BISDITHIOCARBA-
MATE * NABAME (FRENCH) * PARZATE

THR: Poison by ingestion. Moderately toxic
by intraperitoneal route. See also carbamates.
When heated to decomposition it emits very
toxic fumes of Na_2O, NO_x and SO_x.

NAPHAZOLINE
HR: 3
CAS: 835-31-4 NIOSH: NJ 4200000
mf: $C_{14}H_{14}N_2$ mw: 210.27

SYNS: ALPHA-NAPHTHYLMETHYL IMIDAZOLINE
* 2-(ALPHA-NAPHTHYLMETHYL)-IMIDAZOLINE
* 2-(1-NAPHTHYLMETHYL)-2-IMIDAZOLINE

THR: Poison by subcutaneous and intravenous
routes. When heated to decomposition it emits
toxic fumes of NO_x.

NAPHTHA, COAL TAR
HR: 2
CAS: 8030-30-6 NIOSH: QI 9450000

PROP: Dark straw-colored to colorless liquid.
Sol in benzene, toluene, xylene, etc.; bp: 149°-
216°, flash p: 107°F (CC), d: 0.862-0.892, autoign temp: 531°F.

SYNS: BENZIN * 160 DEGREE BENZOL
* COAL TAR NAPHTHA DISTILLATE * LIGHT
LIGROIN * NAFTA (POLISH) * NAPHTHA
* NAPHTHA, PETROLEUM * PETROLEUM BEN-
ZIN * PETROLEUM NAPHTHA

OSHA PEL: TWA 100 ppm

THR: Moderately toxic by inhalation. Can cause
unconsciousness which may go to coma, stentorious breathing, and bluish tint to the skin. Recovery follows removal from exposure. In mild
form, intoxication resembles drunkenness. On
a chronic basis, no true poisoning; sometimes
headache, lack of appetite, dizziness, sleepless-
ness, indigestion, and nausea. A common air
contaminant. See also mineral oils. Flammable
when exposed to heat or flame; can react with
oxidizing materials. Keep containers tightly
closed. Slight explosion hazard. To fight fire,
use foam, CO_2, dry chemical.

NAPHTHALENE
HR: 3
CAS: 91-20-3 NIOSH: QJ 0525000
DOT: 1334/2304
mf: $C_{10}H_8$ mw: 128.18

PROP: Aromatic odor, white, crystalline, volatile flakes; mp: 80.1°, bp: 217.9°, flash p: 174°F
(OC), d: 1.162, lel = 0.9%, uel = 5.9%,
vap press: 1 mm @ 52.6°, vap d: 4.42; autoign temp: 1053°F (567°C); sol in alc, benzene. Insol in water; very sol in ether, CCl_4,
CS_2 hydronaphthalenes, in fixed and volatile
oils.

SYNS: CAMPHOR TAR * MOTH BALLS
* MOTH FLAKES * NAFTALEN (POLISH)
* NAPHTHALINE * NAPHTHENE * NCI-
C52904 * TAR CAMPHOR * WHITE TAR

OSHA PEL: TWA 10 ppm
ACGIH TLV: TWA 10 ppm; STEL 15 ppm
DFG MAK: 10 ppm (50 mg/m³)
DOT Classification: ORM-A (IMO
    Flammable Solid); Label: None

THR: Poison by ingestion, intravenous, and intraperitoneal routes. An eye and skin irritant.
Can cause nausea, headache, diaphoresis, hematuria, fever, anemia, liver damage, vomiting,
convulsions, and coma. Poisoning may occur
by ingestion of large doses, inhalation, or skin
absorption. Flammable when exposed to heat
or flame; reacts with oxidizing materials. Reacts
violently with CrO_3. Moderate explosion hazard
in the form of dust, when exposed to heat or
flame. To fight fire, use water, CO_2, dry chemical. Incompatible with dinitrogen pentaoxide.
For further information see Vol. 5, No. 4 of
DPIM Report.

1,5-NAPHTHALENEDIAMINE   HR: 3
CAS: 2243-62-1   NIOSH: QJ 3400000
mf: $C_{10}H_{10}N_2$ mw: 158.22

SYNS: 1,5-DIAMINONAPHTHALENE * NCI-
c03021

THR: An experimental neoplastigen and carcinogen. Mutagenic data. See also amines.
When heated to decomposition it emits toxic
fumes of NO_x.

2-NAPHTHALENETHIOL   HR: 3
CAS: 91-60-1   NIOSH: QK 3930000
mf: $C_{10}H_8S$ mw: 160.24

PROP: Crystals from ethanol; disagreeable
odor; mp: 81°; bp: 286°; very sol in ethanol,
ether, petroleum ether, sltly sol in H_2O; sltly
volatile with steam.

SYNS: 2-MERCAPTONAPHTHALENE * NAPH-
THALENE-2-THIOL * BETA-NAPHTHYL MER-
CAPTAN * 2-NAPHTHYL MERCAPTAN * 2-
NAPHTHYL THIOL * THIO-BETA-NAPHTHOL
* BETA-THIONAPHTHOL * USAF CY-4

THR: Poison by ingestion and intraperitoneal
routes. A mosquito larvicide. See also mercaptans. Dangerous; when heated to decomposition
it emits highly toxic fumes of SO_x.

NAPHTHALOXIMIDODIETHYL THIO-
PHOSPHATE   HR: 3
CAS: 2668-92-0   NIOSH: QK 5950000
mf: $C_{16}H_{16}NO_5PS$ mw: 365.36

SYNS: O,O-DIETHYL-O-NAPHTHALIMIDE PHOS-
PHOROTHIOATE * O,O-DIETHYL-O-NAPHTHAL-
OXIMIDO PHOSPHOROTHIOATE * O,O-DIETHYL-
O-NAPHTHALOXIMIDOPHOSPHOROTHIONATE
* O,O-DIETHYL-O-NAPHTHYLAMIDOPHOSPHO-
ROTHIOATE * ENT 24970 * NAPHTHALOXI-
MIDE-O,O-DIETHYL PHOSPHOROTHIOATE
* N-HYDROXYNAPHTHALIMIDE-O,O-DIETHYL
PHOSPHOROTHIOATE * PHOSPHOROTHIOIC
ACID-O,O-DIETHYL ESTER,-O-NAPHTHALIMIDO
DERIVATIVE * PHOSPHOROTHIOIC ACID-O,O-
DIETHYL-O-NAPHTHYLAMIDOESTER

THR: A poison by ingestion. See also esters.
When heated to decomposition it emits very
toxic SO_x, PO_x and NO_x.

NAPHTHENIC ACID, COPPER
SALT   HR: 3
CAS: 1338-02-9   NIOSH: QK 9100000

PROP: A solid; flash p: 100°F, d: 1.055. Contains 8% copper.

SYNS: COPPER NAPHTHENATE * COPPER UVER-
SOL * CUPRINOL

THR: Poison by ingestion and inhalation. See
also copper compounds. Flammable when exposed to heat or flame; can react with oxidizing
materials. To fight fire, use foam, CO_2, dry
chemical. For further information see Vol. 3,
No. 1 of *DPIM Report*.

NAPHTHENIC ACID, LEAD
SALT   HR: 2
CAS: 61790-14-5   NIOSH: QK 9150000
mf: $C_7H_{12}O_2$•7Pb   mw: 1578.52

PROP: Contains 24% lead.

SYNS: CYCLOHEXANECARBOXYLIC ACID, LEAD
SALT * LEAD NAPHTHENATE

THR: Moderately toxic by ingestion and intraperitoneal route. An experimental carcinogen.
See also lead compounds. When heated to decomposition it emits toxic fumes of lead.

NAPHTHO (1,2,3,4-def) CHRYS-
ENE   HR: 3
CAS: 192-65-4   NIOSH: QL 0175000
mf: $C_{24}H_{14}$ mw: 302.38

SYNS: DIBENZO(Q,E)PYRENE * 1,2,4,5-DI-
BENZOPYRENE

THR: An experimental carcinogen and neoplastigen. When heated to decomposition it emits
acrid smoke and fumes.

NAPHTHOL   HR: 3
CAS: 1321-67-1   NIOSH: QL 2350000
mf: $C_{10}H_8O$ mw: 144.18

PROP: Monoclinic; d: 1.217 @ 4°; mp: 122°-
123°; bp: 285°-286°; sol in cold and hot water
and in chloroform; very sol in alc and benzene.

THR: Poison by subcutaneous route. When
heated to decomposition it emits acrid smoke
and fumes.

1-NAPHTHOL   HR: 2
CAS: 90-15-3   NIOSH: QL 2800000
mf: $C_{10}H_8O$ mw: 144.18

PROP: Colorless crystals, disagreeable taste.
Odor of phenol; mp: 96°, bp: 282.5°, d: 1.0954
@ 98.7°/4°, vap press: 1 mm @ 94.0°. Very
sltly sol in water; sol in alc and ether.

SYNS: C.I. 76605 * ALPHA-HYDROXYNAPH-
THALENE * 1-HYDROXYNAPHTHALENE
* ALPHA-NAPHTHOL

THR: Moderately toxic by ingestion, skin contact, and possibly other routes. A severe eye
and skin irritant. Ingestion of large amounts
can cause nephritis, vomiting, diarrhea, circulatory collapse, anemia, convulsions, and death.
Can cause kidney irritation and injury to cornea
and lens of the eye. Combustible.

2-NAPHTHOL   HR: 2
CAS: 135-19-3   NIOSH: QL 2975000
mf: $C_{10}H_8O$ mw: 144.18

PROP: White to yellowish-white crystals, slight
phenolic odor. flash p: 307°F, vap press: 10
mm @ 145.5°, vap d: 4.97 mp: 121°-123°; bp:
285°-286°; d: 1.22. Darkens with age or exposure to light. Sublimes when heated; distills in
vacuo; sltly sol in water; more sol in boiling
water, glycerol, olive oil, solns of alkali hydroxides. Protect from light. Very sol in alc, ether.
Sol in chloroform.

SYNS: BETA-HYDROXYNAPHTHALENE * 2-HY-
DROXYNAPHTHALENE * ISONAPHTHOL
* BETA-MONOXYNAPHTHALENE * BETA-NAF-
TOL (DUTCH) * 2-NAFTOL (DUTCH) * BETA-
NAFTOLO (ITALIAN) * 2-NAFTOLO (ITALIAN)
* BETA-NAPHTHOL * 2-NAPHTOL (FRENCH)
* BETA-NAPHTHYL ALCOHOL * BETA-NAPH-
THYL HYDROXIDE * BETA-NAPHTOL (GERMAN)

THR: Moderately toxic by ingestion and subcutaneous routes. See also 1-naphthol. A skin and
eye irritant. Combustible when exposed to heat
or flame. To fight fire, use CO_2 dry chemical.
Incompatible with antipyrine; camphor; phenol;
ferric salts; menthol; potassium permanganate
and other oxidizing materials; urethane. For further information see Vol. 3, No. 6 of *DPIM
Report*.

1,4-NAPHTHOQUINONE   HR: 3
CAS: 130-15-4   NIOSH: QL 7175000
mf: $C_{10}H_6O_2$ mw: 158.16

PROP: Yellow triclinic; odor of benzoquinone;
mp: 125°-126°; very sltly sol in cold water; very
sol in hot alc; sol in ether, benzene, chloroform,
carbon bisulfide, acetic acid, alkali hydroxide
solns; d: 1.422; Volatile with steam.

SYNS: 1,4-DIHYDRO-1,4-DIKETONAPHTHALENE
* 1,4-NAPHTHALENEDIONE * ALPHA-NAPH-
THOQUINONE * USAF CY-10

THR: Poison by ingestion and intraperitoneal
routes. An experimental carcinogen. Slight disaster hazard; when heated, emits toxic fumes
and smoke. For further information see Vol.
4, No. 2 of *DPIM Report*.

1-NAPHTHYLAMINE   HR: 3
CAS: 134-32-7   NIOSH: QM 1400000
mf: $C_{10}H_9N$ mw: 143.20

PROP: White crystals, reddening on exposure
to air. Mp: 50°, bp: 300.8°, flash p: 315°F, d:
1.131, vap press: 1 mm @ 104.3°, vap d: 4.93.
Unpleasant odor. Sublimes, volatile with steam.
Sol in 590 parts H_2O; very sol in alc, ether.
Keep well closed and away from light.

SYNS: ALFANAFTILAMINA (ITALIAN) * ALFA-
NAFTYLOAMINA (POLISH) * 1-AMINONAFTALEN
(CZECH) * 1-AMINONAPHTHALENE * 1-NAF-
TILAMINA (SPANISH) * ALPHA-NAFTYLAMIN
(CZECH) * 1-NAFTYLAMINE (DUTCH)
* NAPHTHALIDINE * 1-NAPHTHYLAMIN
(GERMAN) * ALPHA-NAPHTHYLAMINE

THR: Poison by subcutaneous route. Moderately
toxic by ingestion. Along with β-naphthylamine
and benzidine, it has been incriminated as a
cause of urinary bladder cancer. See also 2-
naphthylamine. See also aromatic amines. A
suspected human carcinogen and tumorigen.
Mutagenic data. Combustible. To fight fire, use
dry chemical, CO_2, mist, spray. When heated
to decomposition it emits toxic fumes of NO_x.
For further information see Vol. 4, No. 3 of
DPIM Report.

2-NAPHTHYLAMINE   HR: 3
CAS: 91-59-8   NIOSH: QM 2100000
DOT: 1650
mf: $C_{10}H_9N$ mw: 143.20

PROP: White to faint pink, lustrous leaflets;
faint aromatic odor. Mp: 111.5°, d: 1.061 @
98°/4°, vap press: 1 mm @ 108.0°, bp: 306°
also listed as 294°; sol in hot water, alc, ether.

SYNS: 2-AMINONAFTALEN (CZECH) * 2-AMI-
NONAPHTHALENE * BETA-NAFTYLOAMINA
(POLISH) * BETA-NAFTILAMINA (ITALIAN)
* 2-NAFTYLAMINE (DUTCH) * BETA-NAF-
TYLOAMINA (POLISH) * 2-NAPHTHALENAMINE
* BETA-NAFTYLAMIN (CZECH) * BETA-NAPH-
THYLAMIN (GERMAN) * 2-NAPHTHYLAMIN
(GERMAN) * BETA-NAPHTHYLAMINE * 6-
NAPHTHYLAMINE * BETA-NAPHTYLAMIN (GER-

MAN) * 2-NAPHTHYLAMINE MUSTARD * USAF CB-22

DOT Classification: Poison B

THR: Poison by intraperitoneal route. Moderately toxic by ingestion. A very toxic chemical in any of its physical forms, such as flake, lump, dust, liquid, or vapor. It can be absorbed into the body through the lungs, the gastrointestinal tract, or the skin. Long and continued exposure to even small amounts may produce tumors and cancers of the bladder. A human carcinogen by ingestion and subcutaneous routes. Combustible; at elevated temperatures it evolves a vapor which is flammable and explosive. The explosive limits of these vapors has not yet been determined. See also aromatic amines. See also α-naphthylamine. Mutagenic data. Flammable when exposed to heat or flame. When heated to decomposition it emits toxic fumes of NO_x. For further information see Vol. 3, No. 6 of *DPIM Report*.

2-NAPHTHYLHYDROXYLAMINE   HR: 3
CAS: 613-47-8   NIOSH: NC 4200000
mf: $C_{10}H_9NO$ mw: 159.20

SYNS: NHA * N-HYDROXY-2-AMINONA-PHTHALENE * N-HYDROXY-2-NAPHTHYLAMINE

THR: An experimental tumorigen, neoplastigen, and carcinogen. Mutagenic data. When heated to decomposition it emits very toxic NO_x.

2-NAPHTHYL-p-PHENYLENEDI-AMINE   HR: 3
CAS: 93-46-9   NIOSH: ST 2100000
mf: $C_{26}H_{20}N_2$ mw: 360.48

SYNS: AGERITE WHITE * DI-BETA-NAPHTHYL-P-PHENYLDIAMINE * DI-BETA-NAPHTHYL-P-PHE-NYLENEDIAMINE * N,N′-DI-BETA-NAPHTHYL-P-PHENYLENEDIAMINE * SYM-DI-BETA-NA-PHTHYL-P-PHENYLENEDIAMINE

THR: An experimental carcinogen. A skin and eye irritant. When heated to decomposition it emits toxic fumes of NO_x.

NCI-C56122   HR: 3
CAS: 989-38-8   NIOSH: DH 0175000
mf: $C_{28}H_{30}N_2O_3$ • ClH   mw: 479.06

SYNS: C.I. 45160 * C.I. BASIC RED 1, MONOHYDROCHLORIDE * RHODAMINE 6G (BIOLOGICAL STAIN) * RHODAMINE 6G EXTRA BASE * RHODAMINE 6GEX ETHYL ESTER

THR: Poison by intraperitoneal route. An experimental carcinogen. Mutagenic data. When heated to decomposition it emits very toxic fumes of Cl^- and NO_x.

NEFOPAM HYDROCHLORIDE   HR: 3
CAS: 23327-57-3   NIOSH: DM 4200000
mf: $C_{17}H_{19}NO$ • ClH   mw: 289.83

SYNS: FENAZOXINE HYDROCHLORIDE * 5-METHYL-1-PHENYL-3,4,5,6-TETRAHYDRO-1H-2,5-BENZOXAZOCINE HYDROCHLORIDE

THR: Poison by ingestion, intravenous, intraperitoneal, and intramuscular routes. When heated to decomposition it emits very toxic fumes of HCl and NO_x.

NEMBUTAL   HR: 3
CAS: 76-74-4   NIOSH: CQ 5775000
mf: $C_{11}H_{18}N_2O_3$ mw: 226.31

SYNS: 5-ETHYL-5-(1-METHYLBUTYL)BARBITU-RIC ACID * 5-ETHYL-5-(1-METHYLBUTYL)MA-LONYLUREA * PENTABARBITONE * PENTO-BARBITAL * PENTOBARBITURIC ACID

THR: Poison by ingestion, intraperitoneal, intravenous, and subcutaneous routes. When heated to decomposition it emits toxic fumes of NO_x.

NEMBUTAL SODIUM   HR: 3
CAS: 57-33-0   NIOSH: CQ 6125000
mf: $C_{11}H_{17}N_2O_3$ • Na   mw: 248.29

PROP: White crystal powder; sol in water and alc; insol in ether.

SYNS: 2,4,6(1H,3H,5H)-PYRIMIDINETRIONE, 5-ETHYL-5-(1-METHYLBUTYL)-, MONOSODIUM SALT (9CI) * PENTONAL * 5-ETHYL-5-(1-METHYL-BUTYL)BARBITURIC ACID SODIUM SALT * MEBUBARBITAL SODIUM * MEBUMAL NA-TRIUM * MEBUMAL SODIUM * PROPYL-METHYLCARBINYLETHYL BARBITURIC ACID SODIUM SALT * SODIUM 5-ETHYL-5-(1-METHYLBUTYL)BARBITURATE * SODIUM PENTABARBITAL * SODIUM PENTABARBITONE * SODIUM PENTOBARBITAL * SODIUM PENTO-BARBITONE * SODIUM PENTOBARBITURATE

THR: Poison by ingestion, intraperitoneal, subcutaneous, intravenous, intraduodenal, intramuscular, intracerebral, and parenteral routes. Human psychotropic and other central nervous system effects by ingestion. When heated to decomposition it emits toxic fumes of NO_x and Na_2O.

NEOANTERGAN PHOSPHATE   HR: 3
CAS: 63681-05-0   NIOSH: UT 1300000
mf: $C_{17}H_{23}N_3O \cdot H_3O_4P$ mw: 383.43

SYN: N-ALPHA-PYRIDYL-N-P-METHOXYBENZYL-
N′,N′ -DIMETHYLETHYLENEDIAMINE PHOSPHATE

THR: Poison by intravenous, intraperitoneal, and subcutaneous routes. When heated to decomposition it emits very toxic fumes of PO_x and NO_x.

NEOCARZINOSTATIN   HR: 3
CAS: 9014-02-2   NIOSH: QO 6780000

PROP: An acidic single-chain polypeptide with a molecular weight of approximately 10,700 isolated from the culture filtrate of *Streptomyces carzinostaticus Var F41*.

SYN: NEOCARCINOSTATIN

THR: Poison by intraperitoneal route. Mutagenic data.

NEOCHROMIUM   HR: 3
CAS: 64093-79-4   NIOSH: QO 6800000
mf: $CrHO_5S$ mw: 165.07

SYNS: BASIC CHROMIC SULFATE * BASIC CHROMIUM SULFATE * CHROMIUM HYDROXIDE SULFATE * CHROMIUM SULFATE * CHROMIUM SULFATE, BASIC * MONOBASIC CHROMIUM SULFATE

THR: An experimental neoplastigen and carcinogen. See also chromium compounds. When heated to decomposition it emits toxic fumes of SO_x.

NEODYMIUM   HR: 3
CAS: 7440-00-8   NIOSH: QO 8575000
af: Nd   aw: 144.24

PROP: It is a bright, silvery, lustrous, very reactive rare earth metal. Bp: 3127°, d: 7.003; mp: approx 1024°.

THR: Poison by intravenous and intracerebral routes. It may be an anticoagulant lanthanon. Care in handling is advised. See also rare earths. Flammable in the form of dust, when exposed to heat or flame. See also powdered metals. Can react violently with air; halogens; N_2; P. Slight explosion hazard in the form of dust when exposed to flame. See also powdered metals. Incompatible with phosphorus.

NEODYMIUM CHLORIDE   HR: 3
CAS: 10024-93-8   NIOSH: QO 8750000
mf: Cl_3Nd mw: 250.59

PROP: Large purple prisms; sol in H_2O, alc.

THR: Poison by intravenous, intraperitoneal, and subcutaneous route. A skin and eye irritant. See also rare earths. When heated to decomposition it emits very toxic fumes of Cl^-.

NEODYMIUM (III) NITRATE (1:3)   HR: 3
CAS: 10045-95-1   NIOSH: QO 9800000
mf: $N_3O_9 \cdot Nd$ mw: 330.27

PROP: Triclinic crystals.

SYN: NITRIC ACID, NEODYMIUM SALT

THR: Poison by intraperitoneal and intravenous routes. Moderately toxic by ingestion. See also nitrates and neodymium. When heated to decomposition it emits very toxic fumes of NO_x.

NEODYMIUM (III) NITRATE, HEXAHYDRATE (1:3:6)   HR: 3
CAS: 16454-60-7   NIOSH: QP 0175000
mf: $N_3O_9 \cdot Nd \cdot 6H_2O$ mw: 438.39

SYN: NITRIC ACID, NEODYMIUM (3+) SALT, HEXAHYDRATE

THR: Poison by intraperitoneal and intravenous routes. Moderately toxic by ingestion. See also rare earths. When heated to decomposition it emits toxic fumes of NO_x.

NEODYMIUM OXIDE   HR: 2
CAS: 1313-97-9   NIOSH: QP 0350000
mf: Nd_2O_3 mw: 336.48

PROP: Blue powder, hexagonal structure; very stable; sol in dil acids.

THR: Moderately toxic by intraperitoneal route. See also neodymium, rare earths.

NEOMYCIN   HR: 3
CAS: 1404-04-2   NIOSH: QP 3850000

PROP: An antibiotic.

SYNS: MYACYNE * MYCIFRADIN * NEOMCIN * NIVEMYCIN * VONAMYCIN POWDER V

THR: Poison by intraperitoneal and subcutaneous routes. Moderately toxic by ingestion. Systemic effects by ingestion. When heated to decomposition it emits acrid smoke and fumes.

NEOMYCIN SULFATE   HR: 3
CAS: 1405-10-3   NIOSH: QP 4375000

SYNS: NEOBIOTIC * OTOBIOTIC * USAF CB-19

THR: Poison by intraperitoneal, intramuscular, intravenous, and subcutaneous routes. Human skin irritant. Human systemic effects by ingestion. When heated to decomposition it emits very toxic fumes of SO_x.

NEON   HR: 1
CAS: 7440-01-9   NIOSH: QP 4450000
af: Ne   aw: 20.18

PROP: Colorless, inert gaseous element; odorless mp: $-248.67°$, bp: $-245.9°$, d (liquid): 1.204 @ $-245.9°$, d (gas): 0.89994 g/L @ $0°$.

SYN: NEON (DOT)

DOT Classification: Label: Nonflammable Gas

THR: Inert asphyxiant gas. See also argon.

NEOPRENE   HR: 3
CAS: 126-99-8   NIOSH: EI 9625000
mf: C_4H_5Cl mw: 88.54

PROP: Colorless liquid; d: 0.958 @ 20°/20°; bp: 59.4°; flash p. $-4°F$; lel = 4.0%; uel = 20%; vap d: 3.0. Sltly sol in water; misc in alc and ether; brittle point: $-35°$ softens @ approx 80°. An oil-resistant synthetic rubber made by the polymerization of chloroprene.

SYNS: CHLOROBUTADIENE * 2-CHLOROBUTA-1,3-DIENE * 2-CHLOOR-1,3-BUTADIEEN (DUTCH) * 2-CHLOR-1,3-BUTADIEN (GERMAN) * 2-CHLORO-1,3-BUTADIENE * CHLOROPREEN (DUTCH) * CHLOROPREN (GERMAN, POLISH) * CHLOROPRENE * BETA-CHLOROPRENE * 2-CLORO-1,3-BUTADIENE (ITALIAN) * CLOROPRENE (ITALIAN)

OSHA PEL: TWA 25 ppm (skin)

THR: Poison by ingestion, intravenous, and subcutaneous routes. A suspected human carcinogen. Mutagenic data. Human exposure has caused dermatitis, conjunctivitis, corneal necrosis, anemia, temporary loss of hair, nervousness, and irritability. Exposure to the vapor can cause respiratory tract irritation leading to asphyxia. Other effects are central nervous system depression, drop in blood pressure, severe degenerative changes in the liver, kidneys, lungs, and other vital organs. Dangerous fire hazard when exposed to heat or flame. To fight fire, use alcohol foam. Incompatible with F_2, liquid F. When heated to decomposition it emits toxic fumes of Cl^-. For further information see Chloroprene, Vol. 1, No. 4 of *DPIM Report*.

NEOPSICAINE HYDROCHLO-RIDE   HR: 3
                          NIOSH: CL 5603000
mf: $C_{19}H_{25}NO_4 • ClH$ mw: 367.91

SYNS: 3-(BENZOYLOXY)-8-METHYL-8-AZABICYCLO(3.2.1)OCTANE-2-CARBOXYLIC ACID PROPYL ESTER, HYDROCHLORIDE (1R-(2-ENDO,3-EXO)) * PSICAINE-NEU HYDROCHLORIDE

THR: Poison by intraperitoneal and subcutaneous routes. An eye irritant. When heated to decomposition it emits very toxic fumes of NO_x and HCl.

NEOSALVARSAN   HR: 3
CAS: 457-60-3   NIOSH: WS 3680000
mf: $C_{13}H_{13}As_2N_2H_2O_4S • Na$ mw: 466.17

PROP: Yellow powder.

SYNS: ((5-(3-AMINO-4-HYDROXYPHENYL)ARSENO)-2-HYDROXYANILINO)METHANOL SULFOXYLATE SODIUM * ARSPHENAMINE METHYLENESULFOXYLIC ACID SODIUM SALT * 3,3'-DIAMINO-4,4'-DIHYDROXYARSENOBENZENEMETHYLENESULFOXYLATE SODIUM

THR: Poison by intravenous and subcutaneous routes. See also arsenic compounds. When heated to decomposition it emits very toxic fumes of As, NO_x and SO_x.

NEOSTIGMINE MONOMETHYLSULFATE   HR: 3
CAS: 51-60-5   NIOSH: CY 1225000
mf: $C_{12}H_{19}N_2O_2 • CH_3O_4S$ mw: 334.43

SYNS: DIMETHYLCARBAMIC ESTER OF 3-OXY-PHENYLTRIMETHYLAMMONIUM METHYLSULFATE * (M-HYDROXYPHENYL)TRIMETHYLAMMONIUM METHYL SULFATE DIMETHYLCARBAMATE * (3-HYDROXYPHENYL)TRIMETHYLAMMONIUM METHYL SULFATE DIMETHYLCARBAMIC ESTER * 3-(DIMETHYLCARBAMOXY)PHENYL TRIMETHYLAMMONIUM METHYL SULFATE * PROSTIGMINE METHYLSULFATE

THR: Poison by ingestion, intravenous, subcutaneous, intraperitoneal, and intramuscular routes. When heated to decomposition it emits very toxic fumes of SO_x and NO_x.

NEOSYNEPHRINE   HR: 3
CAS: 59-42-7   NIOSH: DO 7175000
mf: $C_9H_{13}NO_2$ mw: 167.23

SYNS: 1-ALPHA-HYDROXY-BETA-METHYLAMINO-3-HYDROXY-1-ETHYLBENZENE * 1-M-HYDROXY-ALPHA-((METHYLAMINO)METHYL)-BENZYL ALCOHOL * (−)-M-HYDROXY-ALPHA-(METHYLAMINOMETHYL)BENZYL ALCOHOL * 1-1-(M-HYDROXYPHENYL)-2-METHYLAMINOETHANOL * 1-(3-HYDROXYPHENYL)-N-METHYLETHANOLAMINE * METASYNEPHRINE * M-METHYLAMINOETHANOLPHENOL * M-OXEDRINE * (−)-M-OXEDRINE * (−)-PHENYLEPHRINE

THR: Poison by ingestion, subcutaneous, intravenous, and intraduodenal routes. When heated to decomposition it emits toxic fumes such as NO_x.

NEPTUNIUM   HR: 3
PROP: Exists in alpha, beta and gamma forms. Np; mp: 640°; bp: 3902°; d: 20.45 @ 20°. The first synthetic transuranium element discovered. It is a silvery, radioactive chemically active metal.

THR: Strong radiotoxicity.

NERIINE   HR: 3
CAS: 546-06-5   NIOSH: GK 7620000
mf: $C_{24}H_{40}N_2$ mw: 356.66

SYNS: 3-BETA-DIMETHYLAMINOCON-5-ENEDIHYDROBROMIDE * ROQUESSINE

THR: A poison by ingestion, intraperitoneal, and intravenous routes. When heated to decomposition it emits toxic fumes of NO_x.

NESSLER REAGENT   HR: 3
CAS: 7783-33-7   NIOSH: OU 9670000
mf: HgI_4K_2 mw: 786.39

SYNS: CHANNING'S SOLUTION * MERCURIC POTASSIUM IODIDE * MERCURY(II) POTASSIUM IODIDE * POTASSIUM MERCURIC IODIDE * POTASSIUM TETRAIODOMERCURATE (II) * SOLUTION POTASSIUM IODOHYDRAGYRATE

DOT Classification: Label: Poison

THR: Poison by ingestion. Moderately toxic by skin contact and intraperitoneal routes. See also mercury and iodine. When heated to decomposition it emits very toxic fumes of Hg and I^-.

NGAI CAMPHOR   HR: 2
CAS: 464-45-9   NIOSH: DT 5095000
mf: $C_{10}H_{18}O$ mw: 154.28

SYNS: 1-2-BORNANOL * (−)-BORNEOL * 1-BORNEOL * 1-BORNYL ALCOHOL * 1,2-CAMPHANOL

THR: Slightly toxic by ingestion. Irritant to skin. When heated to decomposition it emits acrid smoke.

NIACINAMIDE   HR: 3
CAS: 98-92-0   NIOSH: QS 3675000
mf: $C_6H_6N_2O$ mw: 122.14

PROP: Colorless needles, very sol in water, ethanol and glycerol; mp: 129°, d: 1.40.

SYNS: ACID AMIDE * AMINICOTIN * NICOTAMIDE * NICOTINIC ACID AMIDE * NICOTINIC AMIDE * NICOTINSAUREAMID (GERMAN) * NIKOTINSAUREAMID (GERMAN) * NIOCINAMIDE * PYRIDINE-3-CARBOXYLIC ACID AMIDE * 3-PYRIDINECARBOXYLIC ACID AMIDE * VITAMIN B3 * VITAMIN PP

THR: Moderately toxic by ingestion, intravenous,and subcutaneous routes. When heated to decomposition it emits toxic fumes of NO_x.

NICKEL   HR: 3
CAS: 7440-02-0   NIOSH: QR 5950000
af: Ni   aw: 58.71

PROP: A silvery-white, hard, malleable and ductile metal; d: 8.90 @ 25°, vap press: 1 mm @ 1810°. Crystallizes as metallic cubes; mp: 1455°; bp: 2730°. Stable in air @ room temp.

SYNS: C.I. 77775 * NICKEL CATALYST, WET (DOT) * NICHEL (ITALIAN) * NICKEL SPONGE * PULVERIZED NICKEL * RANEY ALLOY * RANEY NICKEL

OSHA PEL: TWA 1 mg/m³ (skin)
ACGIH TLV: TWA (metal) 1 mg/m³; (soluble compounds, as Ni) 0.1 mg/m³
TRK: Nickel and compounds (except nickel carbonyl) as respirable dusts and fumes 0.5 mg/m³; as respirable droplets 0.05 mg/m³
DOT Classification: Label: Flammable Solid

THR: Poison by ingestion, intratracheal, and intravenous routes. An experimental carcinogen, tumorigen, and neoplastigen. Ingestion of soluble salts causes nausea, vomiting, diarrhea. Mutagenic data. Reacts violently with F_2; NH_4NO_3; hydrazine; NH_3; (H_2 + dioxane); performic acid; P; Se; S; (Ti + $KClO_3$). Incompatible with aluminum; aluminum trichloride, ethylene; p-dioxan; hydrogen; methanol; non-metals; oxidants; sulfur compounds. May cause dermatitis in sensitive individuals. For further information see Vol. 3, No. 3 of *DPIM Report*.

NICKEL(II) ACETATE (1:2)   HR: 3
CAS: 373-02-4   NIOSH: QR 6125000
mf: $C_4H_6O_4 \cdot Ni$ mw: 176.81

PROP: Green prisms, mp: decomp, d: 1.798.

SYNS: ACETIC ACID, NICKEL(2+) SALT ∗ NICKELOUS ACETATE

OSHA PEL: TWA 1 mg(Ni)/m^3

THR: Poison by ingestion, intraperitoneal, and subcutaneous routes. An experimental neoplastigen, and tumorigen. See also nickel. When heated to decomposition it emits irritating fumes.

NICKEL (II) CARBONATE (1:1)   HR: 3
CAS: 3333-67-3   NIOSH: QR 6200000
mf: $CNiO_3$ mw: 118.72

PROP: Rhombic, light green crystals; mp: decomp.

SYNS: BASIC NICKEL CARBONATE ∗ CARBONIC ACID, NICKEL SALT (1:1) ∗ C.I. 77779 ∗ NICKELOUS CARBONATE

THR: Poison by subcutaneous route. An experimental carcinogen.

NICKEL CARBONYL   HR: 3
CAS: 13463-39-3   NIOSH: QR 6300000
DOT: 1259
mf: C_4NiO_4 mw: 170.75

PROP: Colorless, volatile liquid or needles; bp: 43°, lel = 2% @ 20°, d: 1.3185 @ 17°, vap press: 400 mm @ 25.8°, flash p: $<-4°$. Oxidizes in air; explodes @ about 60°; mp: -19.3°. Sol in alc, benzene, chloroform, acetone, carbon tetrachloride.

SYNS: NICKEL CARBONYL (DOT) ∗ NICKEL CARBONYLE (FRENCH) ∗ NICHEL TETRACARBONILE (ITALIAN) ∗ NICKEL TETRACARBONYL ∗ NIKKELTETRACARBONYL (DUTCH) ∗ NICKEL TETRACARBONYLE (FRENCH)

OSHA PEL: TWA 7 ug/m^3
ACGIH TLV: TWA 0.05 ppm
TRK: 0.1 ppm
DOT Classification: Label: Flammable Liquid and Poison

THR: Poison by inhalation, intravenous, subcutaneous, and intraperitoneal routes. An experimental carcinogen, tumorigen, and teratogen. Vapors may cause irritation, congestion, and edema of lungs. Toxicity by inhalation is believed to be caused by both the nickel and carbon monoxide liberated in the lungs. Chronic exposure may cause cancer of lungs, nasal sinuses. See also nickel compounds and carbonyls. Sensitization dermatitis is fairly common. A common air contaminant. Dangerous fire hazard when exposed to heat, flame, or oxidizers. Moderate explosion when exposed to heat or flame. Can react violently with air; O_2; Br_2 (n-butane + O_2). Dangerous; when heated or on contact with acid or acid fumes it emits highly toxic carbon monoxide fumes; can react with oxidizing materials. To fight fire, use water, foam, CO_2, dry chemical. For further information see Vol. 5, No. 4 of *DPIM Report*.

NICKEL COMPOUNDS   HR: 3
THR: Nickel and most of its salts are generally considered not to cause acute systemic poisoning. However, ingestion of large doses of nickel compounds (1-3 mg/kg) has been shown to cause intestinal disorders, convulsions, and asphyxia. Many nickel compounds are experimental carcinogens and some are human carcinogens by inhalation. All airborne nickel contaminating dusts are regarded as carcinogenic by inhalation. The most common effect resulting from exposure to nickel compounds is the development of the "nickel itch." This form of dermatitis occurs chiefly in persons doing nickel-plating. There is marked variation in individual susceptibility to the dermatitis. It occurs more frequently under conditions of high temperature and humidity, when the skin is moist, and chiefly affects the hands and arms. Nickel carbonyl is highly irritating to the lungs and also can produce asphyxia by decomposing with the formation of carbon monoxide. These compounds are common air contaminants.

NICKEL CYANIDE, (SOLID)   HR: 3
CAS: 557-19-7   NIOSH: QR 6495000
DOT: 1653
mf: C_2N_2Ni mw: 110.75

PROP: Apple-green plates or powder.

SYN: NICKEL CYANIDE, SOLID (DOT)

DOT Classification: Label: Poison

THR: A poison. See cyanides and nickel compounds. Reacts violently with magnesium. When heated to decomposition it emits very toxic fumes of CN^-.

NICKEL (II) FLUOBORATE   HR: 3
CAS: 14708-14-6   NIOSH: QR 6650000

SYN: TL 1091

THR: Poison by inhalation. Moderately toxic by ingestion. See also fluorides, boron and nickel compounds. When heated to decomposition it emits very toxic fumes of F^-.

NICKEL (II) FLUORIDE (1:2)   HR: 3
CAS: 10028-18-9   NIOSH: QR 6825000
mf: F_2Ni mw: 96.71

PROP: Green crystals, slightly water-sol, decomp by boiling water, d: 4.63. Insol in alc, ether.

SYN: NICKELOUS FLUORIDE

OSHA PEL: TWA 1 mg(Ni)/m^3

THR: Poison by intravenous route. See also fluorides and nickel compounds. Reacts violently with potassium. When heated to decomposition it emits very toxic fumes of F^- and Ni dusts. Caution: Chronic exposure may cause mottling of teeth, changes in bones.

NICKEL (II) FLUOSILICATE (1:1)   HR: 3
CAS: 26043-11-8   NIOSH: QR 7000000
mf: $F_6Si \cdot Ni$ mw: 200.80

SYN: HEXAFLUOROSILICATE (2−), NICKEL

THR: Poison by ingestion. See also silicofluoric acid and nickel compounds. When heated to decomposition it emits toxic fumes of F^- and Ni dust.

NICKEL (II) HYDROXIDE   HR: 3
CAS: 12054-48-7   NIOSH: QR 7040000
mf: H_2NiO_2 mw: 92.73

PROP: Light green crystals or amorphous.

SYN: NICKELOUS HYDROXIDE

THR: Poison by subcutaneous route. An experimental carcinogen. See also nickel compounds.

NICKEL MONOXIDE   HR: 3
CAS: 1313-99-1   NIOSH: QR 8400000
mf: NiO   mw: 74.71

PROP: Cubic green-black crystals, mp: 1900°, d: 7.45. Yellow when hot; insol in H_2O. Sol in acids.

SYNS: NICKEL (II) OXIDE (1:1) * BUNSENITE * C.I. 77777 * GREEN NICKEL OXIDE * NICKELOUS OXIDE * NICKEL PROTOXIDE * NICKEL OXIDE

OSHA PEL: TWA 1 mg(Ni)/m^3

THR: Poison by intratracheal and subcutaneous routes. An experimental carcinogen and tumorigen. See also nickel compounds. Can react violently with I_2; H_2S; (BaO + air).

NICKEL (II) NITRATE (1:2)   HR: 3
CAS: 13138-45-9   NIOSH: QR 7200000
mf: $N_2O_6 \cdot Ni$ mw: 182.73

PROP: Green deliquescent crystals, mp: 56.7°, bp: 136.7°, d: 2.05.

SYN: NITRIC ACID, NICKEL(II) SALT

THR: Poison by intravenous route. See nickel compounds and nitrates. When heated to decomposition it emits very toxic fumes of NO_x.

NICKELOUS CHLORIDE   HR: 3
CAS: 7718-54-9   NIOSH: QR 6475000
mf: Cl_2Ni mw: 129.61

SYN: NICKEL(II) CHLORIDE (1:2)

OSHA PEL: TWA 1 mg(Ni)/m^3

THR: Poison by ingestion, intravenous, intramuscular, and intraperitoneal routes. See also nickel compounds. Mutagenic data. An experimental teratogen. When heated to decomposition it emits very toxic fumes of Cl^-.

NICKEL PEROXIDE   HR: 3
CAS: 1314-06-3   NIOSH: QR 8420000
mf: Ni_2O_3 mw: 165.42

PROP: Gray-black powder, mp: $-O_2$ @ 600°, d: 4.83. Decomp about 600° into NiO and O_2. Insol in H_2O, very sltly sol in cold acid; dissolved by hot HCl evolving Cl_2. Dissolved by hot H_2SO_4 or HNO_3 evolving O_2.

SYNS: DINICKEL TRIOXIDE * NICKELIC OXIDE * NICKEL OXIDE * NICKEL OXIDE PEROXIDE * NICKEL SESQUIOXIDE

THR: Poison by subcutaneous route. See nickel compounds and peroxides.

NICKEL REFINERY DUST   HR: 3
NIOSH: QR 8925000

PROP: *Analysis:* Cupric oxide (3.4%), nickel sulfate (20.0%), nickel sulfide (57.0%), nickel oxide (6.3%), cobalt oxide (1.0%), ferric oxide (1.8%), silicon dioxide (1.2%), misc. (2.0%), water (7.3%).

THR: A human carcinogen. An experimental neoplastigen. Moderately toxic (acute) by intramuscular route. When heated to decomposition it emits toxic fumes of SO_x.

NICKEL SUBSULFIDE   HR: 3
CAS: 12035-72-2   NIOSH: QR 9800000
mf: Ni_3S_2 mw: 240.25

SYNS: HEAZLEWOODITE * NICKEL SULPHIDE * NICKEL TRITADISULPHIDE

THR: Poison by intraperitoneal route. A human and experimental carcinogen, neoplastigen, and tumorigen. Mutagenic data. See also nickel compounds. When heated to decomposition it emits toxic fumes of SO_x.

NICKEL SULFATE   HR: 3
CAS: 7786-81-4   NIOSH: QR 9400000
mf: $O_4S \cdot Ni$ mw: 154.77

PROP: Cubic yellow crystals, mp: $-SO_3$ @ 840°, d: 3.68.

SYNS: NICKEL (II) SULFATE- * NICKEL(2+) SULFATE(1:1) * NCI-C60344 * NICKELOUS SULFATE

OSHA PEL: TWA 1 mg(Ni)/m^3

THR: Poison by intravenous, intraperitoneal, and subcutaneous routes. An experimental tumorigen by implant. See also nickel compounds and sulfates. Mutagenic data. When heated to decomposition it emits very toxic fumes of SO_x.

NICOTERGOLINE   HR: 3
CAS: 27848-84-6   NIOSH: KE 6341000
mf: $C_{24}H_{26}BrN_3O_3$ mw: 484.44

SYNS: 10-METHOXY-1,6-DIMETHYL ERGOLINE-8-BETA-METHANOL-5-BROMO-3-PYRIDINECAR-BOXYLATE (ESTER) * NICERGOLIN (GERMAN)

THR: Poison by intravenous route. When heated to decomposition it emits very toxic fumes such as Br^- and NO_x.

NICOTINE   HR: 3
CAS: 54-11-5   NIOSH: QS 5250000
DOT: 1654
mf: $C_{10}H_{14}N_2$ mw: 162.26

PROP: In its pure state, a colorless and almost odorless oil, sharp burning taste, alkaloid from tobacco; mp: $<-80°$, bp: 247.3°, lel = 0.75%, uel = 4.0%, d: 1.0092 @ 20°, autoign temp: 471°F, vap press: 1 mm @ 61.8°, vap d: 5.61. Partial decomp @ bp. Volatile with steam; misc with H_2O below 60°; very sol in alc, chloroform ether, petroleum ether, kerosene oils.

SYNS: 1-METHYL-2-(3-PYRIDYL)PYRROLIDINE
* L-3-(1-METHYL-2-PYRROLIDYL)PYRIDINE
* (−)-3-(1-METHYL-2-PYRROLIDYL)PYRIDINE
* NICOTINA (ITALIAN) * NICOTINE ALKALOID
* NIKOTIN (GERMAN) * NIKOTYNA (POLISH)
* ORTHO N-4 AND N-5 DUSTS * BETA-PYRI-DYL-ALPHA-N-METHYLPYRROLIDINE

OSHA PEL: TWA 500 ug/m^3 (skin)
ACGIH TLV: TWA 0.5 mg/m^3
DFG MAK: 0.07 ppm; 0.5 mg/m^3
DOT Classification: Label: Poison

THR: Poison by all routes. Causes nausea, vomiting, diarrhea, mental disturbances, and convulsions. Human blood pressure effects. May be absorbed by intact skin. An experimental carcinogen and teratogen. Mutagenic data. Combustible when exposed to heat or flame. Moderately explosive when exposed to heat or flame. Dangerous; when heated to decomposition it emits NO_x, CO, and other highly toxic fumes; can react with oxidizing materials. To fight fire, use alcohol foam, dry chemical, CO_2. For further information see Vol. 1, No. 8 of *DPIM Report*.

NICOTINE MONOSALICYLATE   HR: 3
CAS: 29790-52-1   NIOSH: QS 9600000
DOT: 1657
mf: $C_{10}H_{14}N_2 \cdot C_7H_6O_3$ mw: 300.39

PROP: mp: 118°; very sol in H_2O, alc.

SYN: NICOTINE SALICYLATE (DOT)

DOT Classification: Poison B, Label: Poison

THR: A poison. See also nicotine. When heated to decomposition it emits toxic fumes of NO_x and CO. Symptoms of exposure are extreme nausea, vomiting, evacuation of bowel and bladder, mental confusion, twitching, convulsions. Base is readily absorbed through mucous membranes and intact skin. Institute medical treatment immediately. For further information see Vol. 5, No. 4 of *DPIM Report*.

NICOTINE SULFATE   HR: 3
CAS: 65-30-5   NIOSH: QS 9625000
mf: $C_{20}H_{26}N_4 \cdot O_4S$ mw: 418.56

SYNS: NICOTINE SULFATE (2:1) * 1-1-METHYL-2-(3-PYRIDYL)-PYRROLIDINE SULFATE * (S)-3-(1-METHYL-2-PYRROLIDINYL)PYRIDINE SULFATE (2:1) * 1,3-(1-METHYL-2-PYRROLIDYL)PYRIDINE SULFATE * SULFATE DE NICOTINE (FRENCH)

DOT Classification: Label: Poison

THR: Poison by ingestion and skin contact. See also nicotine, sulfates. When heated to decomposition it emits very toxic fumes of SO_x and organic fumes. For further information see Vol. 5, No. 4 of *DPIM Report*.

NICOTINE TARTRATE (1:2)   HR: 3
CAS: 65-31-6   NIOSH: QT 0350000
mf: $C_{10}H_{14}N_2 \cdot 2C_4H_6O_6$ mw:462.46

SYNS: NICOTINE ACID TARTRATE * NICOTINE BITARTRATE * (−)-NICOTINE HYDROGEN TARTRATE * NICOTINE TARTRATE (DOT) * TARTRATE DE NICOTINE (FRENCH)

DOT Classification: Label: Poison

THR: Poison by ingestion, intravenous, intraperitoneal, and subcutaneous routes. See also nicotine. When heated to decomposition it emits toxic fumes of NO_x and CO.

NICOTINIC ACID   HR: 3
CAS: 59-67-6   NIOSH: QT 0525000
mf: $C_6H_5NO_2$ mw: 123.12

PROP: The anti-pellagra vitamin, colorless needles, odorless, sol in water and alc, insol in most lipid solvents. Mp: 236°, sublimes above mp, d: 1.473. Nonhygroscopic and stable in air. Sublimes without decomp.

SYNS: ANTI-PELLAGRA VITAMIN * 3-CARBOXYPYRIDINE * DAVITAMON PP * NIACIN * NICACID * NICONACID * NICOTINE ACID * NICOTINSAURE (GERMAN) * PELLAGRAMIN * PELLAGRA PREVENTIVE FACTOR * PELONIN * P.P. FACTOR-PELLAGRA PREVENTIVE FACTOR * PYRIDINE-3-CARBONIC ACID * PYRIDINE-BETA-CARBOXYLIC ACID * PYRIDINE-3-CARBOXYLIC ACID

THR: An experimental carcinogen by ingestion. Moderately toxic by subcutaneous route. When heated to decomposition it emits toxic fumes of NO_x.

NIFURADENE   HR: 3
CAS: 555-84-0   NIOSH: NJ 0875000
mf: $C_8H_8N_4O_4$ mw: 224.20

PROP: Lemon-yellow solid from nitromethane; mp: 261.5°-263° (decomp).

SYNS: N-(5-NITRO-2-FURFURYLIDENE)-1-AMINO-2-IMIDAZOLIDONE * N-(5-NITRO-2-FURFURYLIDENEAMINO)-2-IMIDAZOAIDINONE * 1-((5-NITROFURFURYLIDENE)AMINO)-2-IMIDAZOLIDINONE

THR: An experimental carcinogen. Moderate acute toxicity by ingestion. When heated to decomposition it emits toxic fumes of NO_x.

NIFURTHIAZOLE   HR: 3
CAS: 3570-75-0   NIOSH: LQ 9275000
mf: $C_8H_6N_4O_4S$ mw: 254.24

PROP: Bright yellow plates; mp: 215.5°.

SYN: 2-(2-FORMYLHYDRAZINO)-4-(5-NITRO-2-FURYL)THIAZOLE NEFURTHIAZOLE

THR: An experimental carcinogen and neoplastigen. Mutagenic data. When heated to decomposition it emits very toxic fumes of NO_x and SO_x.

NIOBIUM   HR: 1
af: Nb   aw: 92.906

PROP: Steel gray cubic crystals. Occurs throughout nature. Mp: 2468 ± 10°; bp: 4742°; d: 8.57. Sol in aqua regia, fused alkali.

THR: Some niobium is found in all parts of the body. No reports of human intoxication. An eye and severe skin irritant. Can cause kidney damage. Flammable in the form of dust, when exposed to flame or by chemical reaction. See also silicofluoric acid. Incompatible with BrF_3; Cl_2; F_2. Moderately explosive in the form of dust when exposed to flame.

NIOBIUM CHLORIDE   HR: 3
CAS: 10026-12-7   NIOSH: QU 0350000
mf: Cl_5Nb mw: 270.20

PROP: Yellow, very deliquescent, monoclinic crystals. Decomp in moist air evolving HCl; d: 2.75; mp: 204.7°-209.5°; bp: approx 250°; sublimes @ 125°. Sol in HCl, carbon tetrachloride.

SYNS: COLUMBIUM PENTACHLORIDE ∗ NIO-
BIUM PENTACHLORIDE

THR: Poison by intraperitoneal route. Moderately toxic by ingestion. Causes kidney injury.
See HCl and niobium. When heated to decomposition it emits very toxic fumes of HCl.

NITRALAMINE HYDROCHLO-
RIDE   HR: 3
CAS: 1432-75-3   NIOSH: KR 3150000
mf: $C_{10}H_{13}ClN_2O_2S \cdot ClH$ mw: 297.22

SYNS: 2-((O-CHLORO-
ALPHA-(NITROMETHYL)BENZYL)THIO)ETHYLAMINE
HYDROCHLORIDE ∗ ((ALPHA-NITROMETHYL)-O-
CHLOROBENZYLTHIO)ETHYLAMINE HYDROCHLO-
RIDE

THR: Poison by intravenous and intraperitoneal
routes. When heated to decomposition it emits
very toxic fumes of HCl, SO_x, NO_x, and Cl^-.

NITRATES   HR: 3
PROP: Organic nitrates are usually termed nitro
compounds. These compounds are a combination of the nitro ($-NO_2$) group and an organic
radical. However, this term is often used to
denote nitric acid esters of an organic material.
Inorganic nitrates are compounds of metals
which are combined with the mono-valent
$-NO_3$ radical.

THR: Organic nitrates are usually termed nitro
compounds. These compounds are a combination of the nitro (NO_2) group and an organic
radical. However, this term is often used to
denote nitric acid esters of an organic material.
Inorganic nitrates are compounds of metals
which are combined with the mono-valent$-NO_3$
radical. Large amounts taken by mouth may
have serious or even fatal effects. The symptoms
are dizziness, abdominal cramps, vomiting,
bloody diarrhea, weakness, convulsions and collapse. Small, repeated doses may lead to weakness, general depression, headache and mental
impairment. Also there is some implication of
increased cancer incidence among those exposed. Flammable by spontaneous chemical reaction; practically all nitrates are powerful oxidizing agents. Nitrates may explode when
shocked, exposed to heat or flame or by spontaneous chemical reaction (See also explosives,
high). All the inorganic nitrates act as oxygen
carriers; under proper conditions these can give
up their oxygen to other materials, which may

in turn detonate. Ammonium nitrate has all the
properties of the other nitrates, but is also able
to detonate by itself under certain conditions.
It is therefore a high explosive, although very
insensitive to impact and difficult to detonate.
In the pure state, it requires a combination of
an initiator and a high explosive. It is a relatively
safe high explosive which, however, must be
stored in a cool, ventilated place, away from
acute fire hazards and easily oxidized materials.
Ammonium nitrate must not be confined, because if a fire should start, confinement can
cause detonation with extremely violent results.
Also reacts violently with Al; BP; cyanides;
esters; PN_2H; P; NaCN; $SnCl_2$; sodium hypophosphite; thiocyanates. Dangerous disaster
hazard due to fire and explosion hazard. On
decomposition, they emit toxic fumes. They
are powerful oxidizing agents which may cause
violent reaction with reducing materials. Nitrates should be protected carefully in storage.

NITRIC ACID   HR: 3
CAS: 7697-37-2
DOT: 2031   NIOSH: QU 5775000
mf: HNO_3 mw: 63.02

PROP: Transparent colorless or yellowish, fuming, suffocating, caustic and corrosive liquid.
Mp: $-42°$, bp: $86°$, d: 1.50269 @ 25°/4°.

SYNS: ACIDE NITRIQUE (FRENCH) ∗ ACIDO NI-
TRICO (ITALIAN) ∗ AQUA FORTIS ∗ AZOTIC
ACID ∗ AZOTOWY KWAS (POLISH) ∗ HYDRO-
GEN NITRATE ∗ NITRIC ACID (DOT) ∗ SAL-
PETERSAURE (GERMAN) ∗ SALPETERZUURO-
PLOSSINGEN (DUTCH)

OSHA PEL: TWA 2 ppm
ACGIH TLV: TWA 2 ppm; STEL 4 ppm
DFG MAK: 10 ppm (25 mg/m³)
DOT Classification: Label: Oxidizer and
  Corrosive

THR: Poison by ingestion and possibly other
routes. Corrosive to eyes, skin, and mucous
membranes and teeth. Causes upper respiratory
irritation which may seem to clear up only to
return in a few hours and more severely. Depending on environmental factors the vapor will
consist of a mixture of the various oxides of
nitrogen and nitric acid. Flammable by chemical
reaction with reducing agents. It is a powerful
oxidizing agent. Incompatible with acetic acid

[with sodium hexahydroxyplatinate(IV)]; acetic anhydride; acetone [with acetic acid]; acetone [with sulfuric acid]; acetonitrile; acetylene; acrolein; acrylonitrile; alcohols; allyl alcohol; allyl chloride; 2-amino ethanol; 2-aminothiazole [with sulfuric acid]; ammonia; ammonium hydroxide; aniline; anilinium nitrate; anion exchange resins; antimony; SbH_3; aromatic amines; arsenic hydride; arsine-borontri-bromide; benzo[b]thiophene derivatives; bismuth; boron; boron decahydride; B_4H_{10}; boron phosphide; BrF_5; bromine pentafluoride; butanethiol; 2,6-di-*tert*-butyl phenol; calcium hypophosphite; carbon; $C_2H_5PH_2$; cellulose; Cs_2C_2; chlorine trifluoride; 4-chloro-2-nitro-aniline; chlorosulfonic acid; CuN_3; Cu_3N_2; copper(I) nitride; cresol; crotonaldehyde; cumene; cyanides; cyclic ketones; cyclohexanol; cyclohexanone; cyclohexylamine; 1,2-diaminoethane-bistrimethylgold; diborane; dichloromethane; dichromate [with anion exchange resins]; diethyl ether; diisopropylether; 1,1-dimethyl hydrazine; unsdimethyl hydrazine; diphenyldistibene; disodium phenylorthophosphate; divinyl ether; epichlorohydrin; ethanol; *m*-ethylaniline; ethylene diamine; ethylene imine; 5-ethyl-2-methyl-pyridine; 5-ethyl-2-picoline; fluorine; furfuryl alcohol; germanium; glyoxal; hexalithium disilicide; 2,2,4,4,6,6-hexamethyltrithiane; hydrazine; hydrocarbons; hydrogen iodide; HN_3; hydrogen peroxide; hydrogen peroxide [with ketones]; hydrogen peroxide [with mercury (II) oxide]; hydrogen selenide; hydrogen sulfide; H_2Te; indane [with sulfuric acid]; ion exchange resins; FeO; iron (II) oxide; isoprene; ketones [with hydrogen peroxide]; lactic acid [with HF]; lithium; Li_6Si_2; magnesium; Mg_3P_2; magnesium-titanium alloy; manganese; mesityl oxide; mesitylene; metal acetylides; metal hexacyanoferrates (3-) or (4-); metal salicylates; metal thiocyanates; metals; 2-methyl-5-ethyl pyridine; 4-methylcyclohexanone; NdP; nitroaromatics; nitrobenzene; nitrobenzene [with nitric acid and water]; nitromethane; non-metal hydrides; non-metals; *n*-butyraldehyde; oleum; organic matter; organic matter [with perchloric acid]; organic matter [with sulfuric acid]; phenylacetylene [with 1,1-dimethylhydrazine]; phosphine derivatives; phosphorus; PCl_3; PH_3; PH_4I; phosphorus halides; P_4I_3; phthalic acid; phthalic anhydride; phthalic anhydride [with sulfuric acid]; polyalkenes; polydibromosilane; KH_2PO_2; β-propiolactone; propylene oxide; pyridine; pyrotechol; Rb_2C_2; reducants; selenium; selenium iodophosphide; silver [with ethanol]; sodium; sodium hydroxide; NaN_3; sulfamic acid; sulfuric acid; sulfuric acid [with $C_6H_5CH_3$]; sulfuric acid [with glycerides]; sulfuric acid [with terephthalic acid]; sulfur dioxide; sulfur halides; terpenes; thioaldehydes; thiocyanates; thioketones; thiophene; titanium; titanium alloy; titanium-magnesium alloy; toluidine; triazine; tricadmium diphosphide; triethylgallium monoetherate; trimagnesium diphosphide; 2,4,6-trimethyltrioxane; uranium; uranium disulfide; uranium-neodymium alloy; uranium-neodymium-zirconium alloy; vinylacetate; vinylidene chloride; zinc; zirconium-uranium alloys. Dangerous; when heated to decomposition emits highly toxic fumes of NO_x and hydrogen nitrate; will react with water or steam to produce heat and toxic and corrosive fumes. To fight fire, use water. For further information see Vol. 5, No. 3 of *DPIM Report*.

NITRIC ACID (RED FUMING)   HR: 3
NIOSH: QU 5900000

DOT: 2032

PROP: Colorless to yellow to red corrosive liquid, d: > 1.480. Contains from 8 to 17% NO_2.

SYNS: NITRIC ACID, FUMING (DOT) * NITROUS FUMES * RED FUMING NITRIC ACID

DOT Classification: Label: Oxidizer and Poison

THR: Poison by inhalation. Severe irritant to skin, eyes, and mucous membranes. Corrosive. See also nitric acid. Dangerous fire hazard very powerful oxidizing agent. Can react explosively with many reducing agents. Dangerous; when heated to decomposition it emits highly toxic fumes of NO_x; will react with water or steam to produce heat and toxic, corrosive, and flammable vapors.

NITRIC ACID (WHITE FUMING)   HR: 3
NIOSH: QU 6000000

PROP: Contains from 0.1 to 0.4% NO_2.

SYNS: WFNA * WHITE FUMING NITRIC ACID

THR: Poison by inhalation. Severe irritant to skin, eyes, and mucous membranes. Corrosive. See also nitric acid. Dangerous fire hazard very powerful oxidizing agent. Can react explosively with many reducing agents. Dangerous; when heated to decomposition it emits highly toxic fumes of NO_x; will react with water or steam to produce heat and toxic, corrosive, and flammable vapors. See also nitrates.

NITRIC OXIDE   HR: 3
mf: NO   mw: 30.01

PROP: Colorless gas, blue liquid and solid. mp: $-163.6°$; bp: -151.7; d: 1.3402 g/L, liquid; 1.269 @ $-150°$. d: (gas) 1.04.

THR: A poison gas. A severe eye, skin, and mucous membrane irritant. A systemic irritant by inhalation. Exposure may occur whenever nitric acid acts upon organic material, such as wood, sawdust, and refuse; it occurs when nitric acid is heated, and when organic nitro compounds are burned, as for example, celluloid, cellulose nitrate (guncotton), and dynamite. The action of nitric acid upon metals, as in metal etching and pickling, also liberates the fumes. In high temperature welding, as with the oxyacetylene or electric torch, the nitrogen and oxygen of the air unite to form oxides of nitrogen. Exposure may also occur in many manufacturing processes when nitric acid is made or used. See also nitrogen monoxide. Incompatible with Al; B; CS_2; ClO; Cr; F_2; fuels; hydrocarbons; NCl_3; O_3; PH_3; P; Rb_2C_2; Na_2O; uns-dimethyl hydrazine; U; acetic anhydride; NH_3; BaO; BCl_3; $CsHC_2$; $CHCl_3$; 1,2-dichloroethane; dichloroethylene; ethylene; Fe; Mg; Mn; CH_2Cl_2; olefins; PNH_2; K; propylene; Na; S; WC; trichloroethylene; 1,1,1-trichloroethane; uns-tetrachloroethane. Dangerous; when heated to decomposition it emits highly toxic fumes of NO_x; will react with water or steam to produce heat and corrosive fumes; can react vigorously with reducing materials. For further information see Vol. 1, No. 5 of *DPIM Report*.

NITRIDES   HR: 2
PROP: Compounds of $N\equiv$ as the anion, such as Li_3N, Ca_3N_2, etc.

THR: The details of the toxicity of nitrides as a group are unknown. However, many nitrides react with moisture to evolve ammonia. This gas is an irritant to mucous membranes. See ammonia. To the extent that many nitrides evolve flammable ammonia gas upon contact with moisture, nitrides can be fire hazards. See also ammonia. Moderate explosion hazards on decomposition if they emit toxic fumes of ammonia.

NITRILES   HR: 3
PROP: Nitriles are organic compounds containing the $(-CN)$ grouping, e.g., acrylonitrile $(CH_2:CHCN)$.

THR: Nitriles are organic cyanides; acrylonitrile, propionitrile and some others resemble cyanides in toxicity. Other nitriles, such as cyanamides and cyanates, have no cyanide effect. When nitriles are heated to decomposition they emit highly toxic cyanide fumes. See also specific compounds and cyanides. The nitriles may be used as insecticides. Can react violently with $(LiAlH_4 + H_2O)$.

NITRILOTRIACETIC ACID TRISODIUM SALT MONOHYDRATE   HR: 3
CAS: 18662-53-8   NIOSH: AJ 1070000
mf: $C_6H_6NO_6 \cdot 3Na \cdot H_2O$ mw: 275.12

SYNS: N,N-BIS(CARBOXYMETHYL)GLYCINE TRISODIUM SALT MONOHYDRATE * NCI-C01445 * NITRILOACETIC ACID TRISODIUM SALT MONOHYDRATE

THR: An experimental carcinogen. Moderately toxic by intraperitoneal route. When heated to decomposition it emits toxic fumes of NO_x and Na_2O.

NITRITES   HR: 2
THR: Large amounts taken by mouth may produce nausea, vomiting, cyanosis (due to methemoglobin formation), collapse, and coma. Repeated small doses cause a fall in blood pressure, rapid pulse, headache, and visual disturbances. There appears to be some implication of increased cancer incidence associated with repeated ingestion of nitrites. Fire hazards are variable. They are generally powerful oxidizers. On contact with readily oxidized materials, a violent reaction such as a fire or explosion may ensue. Explosion hazards are also variable. Organic nitrites may decompose violently in contact with NH_4; salts; cyanides; KCN. Dangerous; shock may explode them; when heated to decomposition they emit highly toxic fumes of NO_x; can react vigorously with reducing materials.

5-NITROACENAPHTHENE   HR: 3
CAS: 602-87-9   NIOSH: AB 1060000
mf: $C_{12}H_9NO_2$ mw: 199.22

SYNS: 1,2-DIHYDRO-5-NITRO-ACENAPHTHYLENE * NCI-C01967 * 5-NITROACENAPHTHYLENE

∗ 5-NITROACENAPHTHENE ∗ 5-NITRONAPH-
THALENE ETHYLENE

THR: An experimental carcinogen by ingestion.
Mutagenic data. When heated to decomposition
it emits toxic fumes of NO_x.

3′-NITRO-p-ACETOPHENETIDE   HR: 3
CAS: 1777-84-0   NIOSH: AM 4550000
mf: $C_{10}H_{12}N_2O_4$ mw: 224.24

PROP: Yellow needles in water; mp: 103°-104°;
sol in absolute alc; sol in ether and chloroform.

SYNS: 4-ACETAMINO-2-NITROPHENETOLE
∗ N-(4-ETHOXYPHENYL)-3′-NITROACETAMIDE
∗ NCI C01978 ∗ 2-NITRO-4-ACETAMINOFENE-
TOL (CZECH) ∗ 3-NITRO-P-ACETOPHENETIDE
∗ 3′-NITRO-P-ACETOPHENETIDIN

THR: An experimental neoplastigen and carcinogen. A severe eye irritant. See also nitro compounds of aromatic hydrocarbons. When heated
to decomposition it emits toxic fumes of NO_x.

m-NITROANILINE   HR: 3
CAS: 99-09-2   NIOSH: BY 6825000
mf: $C_6H_6N_2O_2$ mw: 138.14

PROP: Yellow rhombic; d: 0.9011 @ 25°/4°;
mp: 114°; bp: 306.4°; sol in water, alc and ether.

SYNS: M-AMINONITROBENZENE ∗ 1-AMINO-
3-NITROBENZENE ∗ C.I. 37030 ∗ M-NITRO-
AMINOBENZENE ∗ M-NITRANILINE ∗ 3-
NITROANILINE ∗ 3-NITROBENZENAMINE
∗ M-NITROPHENYLAMINE

THR: Poison by ingestion and intraperitoneal
routes. Mutagenic data. Absorbed through the
skin and by inhalation of the dust. Acute exposure may cause methemoglobinemia cyanosis.
Chronic exposure may cause liver damage. See
also o-nitroaniline, p-nitroaniline, aniline.
When heated to decomposition it emits toxic
fumes of NO_x. Incompatible with ethylene oxide.

o-NITROANILINE   HR: 2
CAS: 88-74-4   NIOSH: BY 6650000
mf: $C_6H_6N_2O_2$ mw: 138.14

PROP: Orange yellow cryst; mp: 69°-71°; bp:
284.5°; vap press: 1 mm @ 104°; d: 0.9015
@ 25°/4°; slightly sol in cold H_2O; sol in hot
water, alc and ether.

SYNS: 1-AMINO-2-NITROBENZENE ∗ C.I.
37025 ∗ O-NITRANILINE

THR: Moderately toxic by ingestion. See also
m-nitroaniline, o-nitroaniline, nitro compounds
of aromatic hydrocarbons. When heated to decomposition it emits toxic fumes of NO_x. Incompatible with sulfuric acid.

p-NITROANILINE   HR: 3
CAS: 100-01-6   NIOSH: BY 7000000
DOT: 1661
mf: $C_6H_6N_2O_2$ mw: 138.14

PROP: Bright yellow powder; mp: 148.5°; bp:
332°; flash p: 390°F (CC); d: 1.424; vap press:
1 mm @ 142.4°; sol in water, alc, ether, benzene, methanol.

SYNS: P-AMINONITROBENZENE ∗ 1-AMINO-4-
NITROBENZENE ∗ C.I. 37035 ∗ NCI-C60786
∗ P-NITRANILINE ∗ 4-NITRANILINE ∗ P-NI-
TROANILINE (DOT) ∗ 4-NITROANILINE
∗ 4-NITROBENZENAMINE ∗ P-NITROPHENYL-
AMINE

OSHA PEL: TWA 1 ppm (skin)
ACGIH TLV: TWA 3 mg/m³
DFG MAK: 1 ppm (6 mg/m³)
DOT Classification: Label: Poison

THR: Poison by ingestion, intravenous, and intraperitoneal routes. Moderately toxic by other
routes. See also nitro compounds of aromatic
hydrocarbons, m-nitroaniline. Acute symptoms
of exposure are headache, nausea, vomiting,
weakness and stupor, cyanosis and methemoglobinemia. Chronic exposure can cause liver damage. See also aniline. Combustible when exposed to heat or flame. See nitrates for explosion
and disaster hazards. When heated to decomposition it emits toxic NO_x. To fight fire, use water
spray or mist, foam, dry chemical, CO_2. Incompatible with sulfuric acid.

5-NITRO-o-ANISIDINE   HR: 3
CAS: 99-59-2   NIOSH: BZ 7175000
mf: $C_7H_8N_2O_3$ mw: 168.17

PROP: Red needles from alc; d: 1.207 @ 156°;
mp: 118°; sol in hot benzene; sol in alc and
acetic acid; very sltly sol in ligroin.

SYNS: 3-AMINO-4-METHOXYNITROBENZENE
∗ 2-AMINO-4-NITROANISOLE ∗ C.I. 37130
∗ 2-METHOXY-5-NITROANILINE ∗ NCI-C01934
∗ 2-METHOXY-5-NITROBENZENAMINE

THR: An experimental carcinogen and tumorigen by ingestion. Moderate acute toxicity by

ingestion. When heated to decomposition it emits toxic fumes of NO_x.

NITROBENZENE　　　　　　　　HR: 3
CAS: 98-95-3　　　　　NIOSH: DA 6475000
DOT: 1662
mf: $C_6H_5NO_2$　　　mw: 123.12

PROP: Bright yellow crystals or yellow, oily liquid, odor of volatile almond oil; mp: 6°; bp: 210°-211°; ulc: 20-30, lel = 1.8% @ 200°F, flash p: 190°F (CC); d: 1.205 @ 15°/4°, autoign temp: 900°F, vap press: 1 mm @ 44.4°, vap d: 4.25. Volatile with steam; sol in about 500 parts water; very sol in alc, benzene, ether, oils.

SYNS: ESSENCE OF MIRBANE * MIRBANE OIL * NCI-C60082 * NITROBENZEEN (DUTCH) * NITROBENZEN (POLISH) * NITROBENZOL

OSHA PEL: TWA 1 ppm (skin)
ACGIH TLV: TWA 1 ppm
DFG MAK: 1 ppm (5 mg/m^3)
DOT Classification: Label: Poison

THR: Poison by subcutaneous, intravenous, intraperitoneal, and possibly other routes. Moderately toxic by ingestion and skin contact. An eye and skin contaminant. Can cause cyanosis due to formation of methemoglobin. It is absorbed rapidly through the skin. The vapors are hazardous. Flammable when exposed to heat, flame, or oxidizers. Moderate explosion hazard when exposed to heat or flame. Reacts violently with HNO_3; ($AlCl_3$ + C_6H_5OH); (aniline + glycerin); N_2O; $AgClO_4$. See also nitrates for disaster hazard. To fight fire, use water, foam, CO_2, dry chemical. Incompatible with aluminum trichloride; aniline, glycerol; sulfuric acid; oxidants; phosphorous pentachloride; potassium; potassium hydroxide; sulfuric acid.

3-NITROBENZENESULFONIC
ACID　　　　　　　　　　　　HR: 3
CAS: 98-47-5　　　　　NIOSH: DB 7190000
mf: $C_6H_5NO_5S$　　　mw: 203.18

PROP: Hygroscopic leaflets; mp: 70°; bp: decomp; very sol in water; sol in alc and alkali; insol in ether.

SYNS: KYSELINA NITROBENZEN-M-SULFONOVA (CZECH) * M-NITROBENZENESULFONIC ACID

THR: Moderately toxic by skin contact. Skin and severe eye irritant. Decomposes violently at about 200°. When heated to decomposition it emits very toxic fumes of NO_x and SO_x.

4-NITROBENZIMIDAZOLE　　　　HR: 3
CAS: 10597-52-1　　　NIOSH: DD 9790000
mf: $C_7H_5N_3O_2$　　　mw: 163.2

SYN: 4-NITRO-1H-BENZIMIDAZOLE

THR: Mutagenic data. When heated to decomposition it emits toxic fumes of NO_x.

3-NITROBENZOTRIFLUORIDE　　HR: 3
CAS: 98-46-4　　　　　NIOSH: XT 3500000
mf: $C_7H_4F_3NO_2$　　mw: 191.12

PROP: Thin, pale, straw-colored, oily liquid with aromatic odor; mp: −5°; bp: 202.8°; flash p: 217°F (OC); d: 1.437 @ 15.5°/15.5°; vap press: 0.3 mm @ 25°.

SYNS: M-(TRIFLUOROMETHYL)NITROBENZENE * 3-TRIFLUOROMETHYLNITROBENZENE * M-NITROBENZOTRIFLUORIDE * M-NITROTRI-FLUOROTOLUENE * M-NITROTRIFLUORTOLUOL-(GERMAN) * USAF MA-5

THR: Poison by intraperitoneal route. Moderately toxic by ingestion and subcutaneous routes. See also fluorides and nitro compounds of aromatic hydrocarbons. Combustible when exposed to heat or flame. When heated to decomposition it emits very toxic fumes of F^- and NO_x.

o-NITROBIPHENYL　　　　　　HR: 3
CAS: 86-00-0　　　　　NIOSH: DV 5530000
mf: $C_{12}H_9NO_2$　　mw: 199.22

PROP: Light yellow- to reddish-colored liquid or crystalline solid, mp: 35°, bp: 330°, flash p: 354°F, d: 1.189 @ 40°/15.5°, autoign temp: 356°F, vap press: 2 mm @ 140°, vap d: 5.9. Insol in water; sol in methanol, ethanol, tetrahydrofurfuryl alc, acetone, dimethyl-formamide.

SYNS: 2-NITRODIPHENYL * O-NITRODIPHENYL

THR: Moderately toxic by ingestion. An irritant. See aromatic amines. Mutagenic data. Combustible when exposed to heat or flame. See nitrates for disaster hazard. To fight fire, use CO_2, dry chemical.

p-NITROBIPHENYL　　　　　　HR: 3
CAS: 92-93-3　　　　　NIOSH: DV 5600000
mf: $C_{12}H_9NO_2$　　mw: 199.22

PROP: Needles from alc; mp: 113°-114°; bp: 340°; insol in water; slightly sol in cold alcohol; very sol in ether.

SYNS: 4-NITROBIPHENYL * P-NITRODIPHENYL * 4-NITRODIPHENYL * P-PHENYL-NITROBEN-ZENE * 4-PHENYL-NITROBENZENE

THR: An experimental carcinogen and neoplastigen. Moderately toxic by ingestion. Mutagenic data. When heated to decomposition it emits toxic fumes of NO_x.

p-NITRO-o-CHLOROPHENYL DIMETHYL THIONOPHOSPHATE   HR: 3
CAS: 2463-84-5   NIOSH: TE 7875000
mf: $C_8H_9ClNO_5PS$ mw: 297.66

PROP: White solid, insol H_2O but sol in acetone, cyclohexane, xylene, toluene, ethylacetate; mp: 52°.

SYNS: O-(4-CHLOOR-3-NITRO-FENYL)-O,O-DI-METHYLMONOTHIOFOSFAAT (DUTCH) * O-(4-CHLOR-3-NITRO-PHENYL)-O,O-DIMETHYL-MONO-THIOPHOSPHAT (GERMAN) * O-(4-CLORO-3-NI-TRO-FENIL)-O,O-DIMETIL-MONOIIOFOSFATO (ITALIAN) * O-(2-CHLORO-4-NITROPHENYL) O,O-DIMETHYL PHOSPHOROTHIOATE * O,O-DI-METHYL O-2-CHLORO-4-NITROPHENYL PHOSPHO-ROTHIOATE * DIMETHYL-2-CHLORONITROPHE-NYL THIOPHOSPHATE * ENT 17,035 * EXPERIMENTAL INSECTICIDE 4124 * ISO-CHLOORTHION (DUTCH) * ISOMERIC CHLOR-THION * THIOPHOSPHATE DE O,O-DIMETHYLE ET DE O-4-CHLORO-3-NITROPHENYLE (FRENCH)

THR: Poison by ingestion. Moderately toxic by skin contact. See also parathion. When heated to decomposition it emits very toxic fumes of PO_x, SO_x, NO_x, and Cl^-.

NITRO COMPOUNDS OF AROMATIC HYDROCARBONS   HR: 3
THR: The mono-, di-, and trinitrobenzenes are absorbed chiefly through the skin and through inhalation of the dust or vapor when these materials are heated. The dinitrobenzenes are believed to be somewhat more toxic than the mononitrobenzene, and more toxic than aniline. The effect of di- and trinitrobenzene on the body is similar to that of aniline and mononitrobenzene, with reduction of the oxygen-carrying power of the blood and depression of the nervous system being responsible for most of the symptoms following acute exposure. Poisoning with the solid nitro compounds is usually slower and less severe than is the case with the liquid nitro and amino benzenes since absorption is less rapid. Thus, chronic poisoning occurs more frequently than acute, the picture observed in the chronic form being one of anemia, moderate cyanosis, fatigue, slight dizziness, headache, insomnia and loss of weight. Prolonged chronic exposure may result in damage to the liver and kidneys, with production of acute yellow atrophy, toxic hepatitis, and fatty degeneration of the kidneys. The introduction of one or more Cl atoms into the nitrobenzene ring results in the formation of chloronitrobenzene compounds or nitrochlors. The chloro-mono-nitrobenzenes have essentially the same toxic effect as nitrobenzene. The Cl derivatives of dinitrobenzene, on the other hand, while resembling dinitrobenzene in their systemic effects are much more irritant to the skin. They act as direct irritants, and in addition may cause sensitization. For further information see specific nitro compounds. Dangerous; many of these compounds are highly flammable and some are explosive. When heated to decomposition they evolve highly toxic fumes of NO_x.

4-NITRO-M-CRESOL DIMETHYL PHOSPHATE   HR: 3
CAS: 2255-17-6   NIOSH: TC 5260000
mf: $C_9H_{12}NO_6P$ mw: 261.19

SYNS: PHOSPHORIC ACID DIMETHYL-4-NITRO-M-TOLYL ESTER * O,O-DIMETHYL-O-(3-METHYL-4-NITROPHENYL)PHOSPHORATE

THR: Poison by ingestion, intravenous, and intraperitoneal routes. See also esters. When heated to decomposition it emits very toxic fumes of NO_x and PO_x.

NITROETHANE   HR: 3
CAS: 79-24-3   NIOSH: KI 5600000
DOT: 2842
mf: $C_2H_5NO_2$ mw: 75.08

PROP: Oily, colorless liquid. Agreeable odor; mp: −90°; sol in water, acid and alkali; misc in alc, chloroform and ether; bp: 114.0°, fp: −50°, d: 1.052 @ 20°/20°, autoign temp: 778°F, flash p: 106°F; decomp @ 335°-382°; lel = 4.0%; vap press: 15.6 mm @ 20°, vap d: 2.58.
OSHA PEL: TWA 100 ppm
ACGIH TLV: 100 ppm
DFG MAK: 100 ppm (310 mg/m^3)
DOT Classification: Flammable Liquid

THR: Poison by intraperitoneal route. Moderately toxic by ingestion. Causes injury to liver and kidneys. Irritant to eyes and mucous membranes. Dangerous when exposed to heat, flame, or oxidizers. See nitrates for explosion and disaster hazards. To fight fire, use alcohol foam, CO_2, dry chemical; water can blanket fire. Incompatible with $Ca(OH)_2$; hydrocarbons; hydroxides; inorganic bases; KOH; NaOH. Metal oxides. Explodes when heated.

2-NITROFLUORENE   HR: 3
CAS: 607-57-8   NIOSH: LL 8225000
mf: $C_{13}H_9NO_2$ mw: 211.23

PROP: Needles from 50% acetic acid; mp: 157°-158°.

THR: An experimental tumorigen and neoplastigen by ingestion and skin contact. Mutagenic data. When heated to decomposition it emits toxic fumes of NO_x.

5-NITRO-2-FURALDEHYDE
OXIME   HR: 3
CAS: 555-15-7   NIOSH: LT 7525000
mf: $C_5H_4N_2O_4$ mw: 156.11

SYNS: 5-NITRO-2-FURALDOXIME * NITROFUROXIME * USAF EA-5

THR: Poison by intravenous and intraperitoneal routes. Mutagenic data. When heated to decomposition it emits toxic fumes of NO_x.

NITROFURANTOIN   HR: 3
CAS: 67-20-9   NIOSH: MU 2800000
mf: $C_8H_6N_4O_5$ mw: 238.18

SYNS: FURADANTIN * FURADONIN * FURANTOIN * NCI C55196 * NSC 2107 * ORAFURAN * USAF EA-2

THR: Poison by ingestion and intraperitoneal route. Moderately toxic by ingestion. An experimental teratogen. Human central nervous system and systemic effects by ingestion and other routes. Mutagenic data. When heated to decomposition it emits toxic fumes of NO_x.

NITROFURAZONE   HR: 3
CAS: 59-87-0   NIOSH: LT 7700000
mf: $C_6H_6N_4O_4$ mw: 198.16

PROP: Odorless, lemon-yellow crystals. Decomp @ 236°-240°. Bitter aftertaste. Darkens upon prolonged exposure to light. Very sltly sol in water; sltly sol in alc, propylene glycol; sol in alkaline solns. Insol in ether.

SYNS: COCAFURIN * FURACILLIN * FURAN-OFTENO * FURAPLAST * FURASEPTYL * NCI-C56064 * 5-NITROFURALDEHYDE SEMICARBAZIDE * 6-NITROFURALDEHYDE SEMICARBAZIDE * 5-NITRO-2-FURALDEHYDE SEMICARBAZONE * 5-NITROFURAN-2-ALDEHYDE SEMICARBAZONE * 5-NITRO-2-FURANCARBOXALDEHYDE SEMICARBAZONE * 5-NITROFURFURAL SEMICARBAZONE * 2((5-NITRO-2-FURANYL)METHYLENE)HYDRAZINECARBOXAMIDE * (5-NITRO-2-FURFURYLIDENEAMINO)UREA * NSC-2100 * U-6421 * USAF EA-4 * VETERINARY NITROFURAZONE

THR: Poison by ingestion and intraperitoneal routes. An experimental teratogen, neoplastigen, and carcinogen. A human sensitizer. Mutagenic data. When heated to decomposition it emits toxic fumes such as NO_x.

2-(5-NITRO-2-FURYL)-5-AMINO-1,3,4-
THIADIAZOLE   HR: 3
CAS: 712-68-5   NIOSH: XI 3600000
mf: $C_6H_4N_4O_3S$ mw: 212.20

SYNS: FURIDIAZINE 1,3,4-THIADIAZOL-2-AMINE, 5-(5-NITRO-2-FURANYL)- (9CI) * 2-AMINO-5-(5-NITRO-2-FURYL)-1,3,4-THIADIAZOLE * 5-AMINO-2-(5-NITRO-2-FURYL)-1,3,4-THIADIAZOLE * 5-(5-NITRO-2-FURYL)-2-AMINO-1,3,4-THIADIAZOLE * 2-(5-NITRO-2-FURYL)-5-AMINO-1,3,4-THIADIAZOLE

THR: An experimental neoplastigen and carcinogen. See also nitro compounds of aromatic hydrocarbons. When heated to decomposition it emits very toxic fumes of NO_x and SO_x.

N-(4-(5-NITRO-2-FURYL)-2-THIAZOL-
YL)FORMAMIDE   HR: 3
CAS: 24554-26-5   NIOSH: LQ 3150000
mf: $C_8H_5N_3O_4S$ mw: 239.22

SYNS: FANFT * N-(4-(5-NITRO-2-FURYL)-2-THIAZOLYL)FORMAMID (GERMAN)

THR: An experimental tumorigen, neoplastigen, and carcinogen. Mutagenic data. When heated to decomposition it emits very toxic fumes of SO_x and NO_x.

NITROGEN   HR: 1
CAS: 7727-37-9   NIOSH: QW 9700000
DOT: 1066/1977
mf: N_2 mw: 28.02

PROP: Colorless gas, colorless liquid or cubic crystals at low temp., mp: −210.0°, d: 1.2506 g/L @ 0°, d (liquid): 0.808 @ −195.8°. Condenses to a liquid; sltly sol in water; sol in liquid ammonia, alc.

SYNS: NITROGEN (DOT) * NITROGEN GAS

DOT Classification: Label: Nonflammable Gas

THR: Low toxicity. In high concentrations it is a simple asphyxiant. The release of nitrogen from solution in the blood, with formation of small bubbles, is the cause of most of the symptoms and changes found in compressed air illness (caisson disease). It is a narcotic at high concentration and high pressure. Both the narcotic effects and the bends are hazards of compressed air atmospheres such as found in underwater diving. See also argon. Can react violently with lithium; neodymium; titanium under the proper conditions.

NITROGEN (CRYOGENIC LIQUID)
HR: 2
CAS: 7727-37-9   NIOSH: QW 9720000
DOT:

SYN: NITROGEN, CRYOGENIC LIQUID (DOT)

DOT Classification: Label: Nonflammable Gas

THR: Nitrogen (liquid) can explode during use. See also nitrogen.

NITROGEN CHLORIDE
HR: 3
mf: NCl_3 mw: 120.37

PROP: Volatile yellowish oil or rhombic crystals, pungent odor mp: less than −40°; explodes @ 93°; bp: less than 71°; d: 1.653; vap press: 150 mm @ 20°.

THR: An irritant to the eyes, skin, mucous membranes, and a systemic central nervous system irritant. Severe explosion hazard when shocked or exposed to heat or flame or by spontaneous chemical reaction. Certain common materials catalyze its decomposition. It is particularly sensitive when it contains impurities. Intense light can explode it. Reacts violently with NH_3; As; H_2S; NO_2; organic matter; O_3; PH_3; P; KCN; KOH; Se; dibutyl ether; grease; initiators. Dangerous; on decomposition it emits highly toxic fumes of Cl^-; can react vigorously with reducing materials.

NITROGEN DIOXIDE
HR: 3
CAS: 10102-44-0   NIOSH: QW 9800000

DOT: 1067
mf: NO_2 mw: 46.01

PROP: Colorless solid to yellow liquid; mp: −9.3° (yellow liquid), bp: 21° (red-brown gas with decomp), d: 1.491 @ 0°, vap press: 400 mm @ 80°. Liquid below 21.15°; irr odor; sol in concentrated sulfuric acid, nitric acid, corrosive to steel when wet.

SYNS: AZOTE (FRENCH) * AZOTO (ITALIAN) * NITROGEN PEROXIDE * STICKSTOFFDIOXID (GERMAN) * STIKSTOFDIOXYDE (DUTCH) * NO_x

OSHA PEL: TWA 5 ppm
ACGIH TLV: TWA 3 ppm; STEL 5 ppm
DFG MAK: 5 ppm (9 mg/m³)
DOT Classification: Poison A and Oxidizer

THR: Deadly poison to humans by inhalation. Human pulmonary system effects by inhalation. Mutagenic data. See also nitric oxide. When heated to decomposition it emits toxic fumes of NO_x. Violent reaction with cyclohexane; F_2; formaldehyde and alcohols; nitrobenzene; petroleum; toluene.

NITROGEN FLUORIDE OXIDE
HR: 3
CAS: 13847-65-9   NIOSH: QX 0350000
mf: F_3NO mw: 87.01

SYN: TRIFLUOROAMINE OXIDE

THR: Poison irritant by inhalation and intraperitoneal routes. Irritant to skin, eyes, and mucous membranes. See also fluorides. When heated to decomposition it emits very toxic fumes of F^- and NO_x.

NITROGEN MONOXIDE
HR: 3
CAS: 10102-43-9   NIOSH: QX 0525000
DOT: 1660
mf: NO   mw: 30.01

PROP: Colorless gas, blue liquid and solid; mp: −161°, bp: −151.18, d: 1.3402 g/L, liquid: 1.269 @ −150°.

SYNS: BIOXYDE D'AZOTE (FRENCH) * NO_x * OXYDE NITRIQUE (FRENCH) * STICKMONOXYD (GERMAN)

OSHA PEL: TWA 25 ppm
DOT Classification: Poison A, Label: Poison Gas

THR: Poison by inhalation. The vapors are a severe irritant to eyes, skin, and mucous membranes. See also nitric oxide. Can react violently with Al; B; CS_2; ClO; Cr; F_2; fuels; hydrocarbons; NCl_3; O_3; PH_3; P; Rb_2C_2; Na_2O; uns-dimethyl hydrazine; U; acetic anhydride; NH_3; BaO; BCl_3; $CsHC_2$; $CHCl_3$; 1,2-dichloroethane; dichloroethylene; ethylene; Fe; Mg; Mn; CH_2Cl_2; olefins; PNH_2; K; propylene; Na; S; WC; trichloroethylene; 1,1,1-trichloroethane; uns-tetrachloroethane. Dangerous; when heated to decomposition it emits highly toxic fumes of NO_x; will react with water or steam to produce heat and corrosive fumes; can react vigorously with reducing materials.

NITROGEN OXIDE   HR: 3
CAS: 10024-97-2   NIOSH: QX 1350000
mf: N_2O mw: 44.02

PROP: Colorless gas, liquid or cubic crystals; mp: $-90.8°$, bp: $-88.49°$, d: 1.977 g/L, d liq: 1.226 @ $-89°$.

SYNS: DINITROGEN MONOXIDE * FACTITIOUS AIR * HYPONITROUS ACID ANHYDRIDE * LAUGHING GAS * NITROUS OXIDE (DOT) * STICKDIOXYD (GERMAN)

DOT Classification: Label: Nonflammable Gas

THR: An experimental teratogen. Moderately toxic by inhalation. Mutagenic data. Human blood pressure effects. An asphyxiant. Flammable by chemical reaction; it supports combustion. Moderate explosion hazard; it can form an explosive mixture with air. Violent reaction with Al; B; hydrazine; LiH; LiC_6H_5; PH_3; Na; WC. Also self-explodes at high temperatures.

NITROGEN TETROXIDE   HR: 3
CAS: 10544-72-6   NIOSH: QX 1575000
mf: N_2O_4 mw: 92.02

PROP: Nitrogen Tetroxide is a dimer of nitrogen dioxide.

SYN: DINITROGEN TETROXIDE

THR: Poison by inhalation. See also nitrogen monoxide. When heated to decomposition it emits toxic fumes of NO_x.

NITROGEN TRIFLUORIDE   HR: 3
CAS: 7783-54-2   NIOSH: QX 1925000
mf: F_3N mw: 71.01

PROP: Colorless gas, odor of mold; mp: $-208.5°$, bp: $-129°$, d (liquid): 1.537 @ $-129°$; d: (liq @ bp:) 1.885; insol in H_2O.

OSHA PEL: TWA 10 ppm
ACGIH TLV: TWA 10 ppm

THR: Poison by inhalation. Prolonged absorption may cause mottling of teeth, skeletal changes. See also fluorides. Severe explosion hazard by chemical reaction with reducing agents, particularly when under pressure. Dangerous fire hazard; a very powerful oxidizer; otherwise inert at normal temperatures and pressures. Reacts violently when ignited with H_2. When pure (dry) it does not attack glass or mercury at normal temperatures. Can react violently with NH_3; CO; diborane; H_2; H_2S; CH_4; tetrafluorohydrazine. Dangerous; on decomposition it emits highly toxic fumes of fluorides; can react vigorously with reducing materials. Particularly hazardous under pressure. Incompatible with charcoal; hydrogen-containing compounds; tetrafluorohydrazine.

NITROGEN TRIIODIDE
mf: NI_3 mw: 394.7

PROP: Black crystals. Mp: explodes, bp: sublimes in vacuo.

THR: No toxicity data. See also iodides. Severe explosion hazard when shocked, exposed to heat or flame, or by spontaneous chemical reaction. It has no known uses as an explosive because it is far too sensitive in the dry state to store or handle safely. If this material must be worked with, it should be kept wet. A convenient way of keeping it wet is with ether; when it is needed in the dry state, it simply has to be taken out into the open and the ether will evaporate, leaving it perfectly dry. When dry, it will explode when given the slightest touch, vibration, or rise in temperature. Even a puff of air directed into it can cause it to detonate. It is a high explosive and is very violent. It can be destroyed by throwing a quantity of it into large bodies of water, flowing streams, or rivers. Incompatible with O_3; H_2S; Cl_2; Br_2; acids.

NITROGLYCERIN   HR: 3
CAS: 55-63-0   NIOSH: QX 2100000
mf: $C_3H_5N_3O_9$ mw: 227.11

PROP: Colorless to yellow liquid, sweet taste. Mp: 13°, bp: explodes @ 218°, d: 1.599 @ 15°/15°; vap press: 1 mm @ 127°, vap d: 7.84, autoign temp: 518°F. Decomp @ 50°-60°; volatile @ 100°; misc with ether, acetone, glacial acetic acid, ethyl acetate, benzene, nitrobenzene, pyridine, chloroform, ethylene bromide, dichloroethylene; slightly sol in petroleum ether, glycerol.

SYNS: BLASTING GELATIN (DOT) ∗ BLASTING OIL ∗ GLYCEROLTRINTRAAT (DUTCH) ∗ GLYCERINTRINITRATE (CZECH) ∗ GLYCEROL TRINITRATE ∗ GLYCEROL(TRINITRATE DE) (FRENCH) ∗ GLYCEROL, NITRIC ACID TRIESTER ∗ GLYCERYL NITRATE ∗ GLYCERYL TRINITRATE ∗ NITRIC ACID TRIESTER OF GLYCEROL ∗ NITROGLICERINA (ITALIAN) ∗ NITROGLICERYNA (POLISH) ∗ NITROGLYCERINE ∗ NITROGLYCEROL ∗ 1,2,3-PROPANETRIOL, TRINITRATE ∗ 1,2,3-PROPANETRIYL NITRATE ∗ SOUP ∗ TRINITRIN ∗ TRINITROGLYCERIN ∗ TRINITROGLYCEROL

OSHA PEL: TWA 2 mg/m^3 (skin)
ACGIH TLV: TWA 0.05 ppm
DFG MAK: 0.05 ppm (0.5 mg/m^3)
DOT Classification: IMO: Flammable
   Liquid; Label: Explosive A

THR: Poison by ingestion, subcutaneous, intravenous, intramuscular, and possibly other routes. It can cause respiratory difficulties and death due to respiratory paralysis by ingestion. The acute symptoms of nitroglycerin poisoning are headaches, nausea, vomiting, abdominal cramps, convulsions, methemoglobinemia, circulatory collapse and reduced blood pressure, excitement, vertigo, fainting, respiratory rales and cyanosis. Toxic effects may occur by ingestion, inhalation of dust, or absorption through intact skin. Dangerous fire hazard when exposed to heat or flame or by spontaneous chemical reaction. Severe explosion hazard when shocked or exposed to O$_3$, heat or flame. Nitroglycerin is a powerful explosive, very sensitive to mechanical shock. Small quantities of it can readily be detonated by a hammer blow on a hard surface, particularly when it has been absorbed in filter paper. Frozen nitroglycerin is somewhat less sensitive than the liquid. However, a partially thawed out mixture is more sensitive than either one. See also explosives, high and dynamites. Toxic fumes evolve on decomposition. For further information see Vol. 1, No. 4 of *DPIM Report*.

alpha-NITROGUANIDINE   **HR: 3**
CAS: 556-88-7   NIOSH: MF 4600000
mf: CH$_4$N$_4$O$_2$ mw: 104.09

PROP: Yellow solid, high explosive. Usually stable needles, mp: 246°. decomp: 225°-250°; slightly sol in alc, conc acids, cold solns of alkalies; sol in water; very sltly sol in ether.

SYNS: NITROGUANIDINE ∗ BETA-NITROGUANIDINE ∗ PICRITE (THE EXPLOSIVE)

THR: Poison by ingestion. Mutagenic data. Dangerous fire hazard when exposed to heat, flame or by chemical reaction with oxidizers. Severe explosion hazard when shocked or exposed to heat or flame. It is about as powerful as TNT. It is normally mixed with colloided nitrocellulose or ammonium nitrate and paraffin wax. Dangerous; when heated to decomposition it emits highly toxic fumes; can react vigorously with oxidizing materials and the derivatives can be explosive. Incompatible with complex salts of mercury and silver.

3-NITRO-3-HEXENE   **HR: 3**
CAS: 481-22-2   NIOSH: MP 7350000
mf: C$_6$H$_{11}$NO$_2$ mw: 129.18

THR: Poison by ingestion and intraperitoneal routes. Moderately toxic by skin contact. A severe eye and skin irritant. An experimental neoplastigen by inhalation. Dangerous; see nitrates for disaster hazard.

NITROHYDROCHLORIC ACID (DILUTED)   **HR: 3**
                                  NIOSH: QX 4175000

PROP: Yellow fuming corrosive, volatile liquid, suffocating odor. Misc with H$_2$O.

SYN: AQUA REGIA

DOT Classification: Label: Corrosive

THR: Very corrosive and irritating chemical. See also nitric and hydrochloric acids. When heated to decomposition it emits very toxic fumes of NO$_x$ and HCl.

NITROMETHANE   **HR: 3**
CAS: 75-52-5   NIOSH: PA 9800000
DOT: 1261
mf: CH$_3$NO$_2$ mw: 61.05

PROP: An oily liquid; mod strong disagreeable odor; bp: 101°; lel: 7.3%; fp: −29°; flash p: 95°F (CC); d: 1.1322 @ 25°/4°; autoign temp: 785°F; vap press: 27.8 mm @ 20°; vap d: 2.11; sltly sol in water, sol in alc, ether.

SYNS: NITROCARBOL ∗ NITROMETAN (POLISH)

OSHA PEL: TWA 100 ppm
ACGIH TLV: TWA 100 ppm
DFG MAK: 100 ppm (250 mg/m^3)
DOT Classification: Flammable Liquid

THR: Poison by ingestion, inhalation, and intra-peritoneal routes. In humans it may cause an-orexia, nausea, vomiting, diarrhea, and kidney injury and liver damage. Dangerous fire hazard when exposed to heat, oxidizers, or flame. See also 2-nitropropane. When heated to decomposi-tion it emits toxic fumes of NO$_x$. Explosive when shocked or exposed to heat or flame. High temperatures can detonate it. Can react violently with AlCl$_3$ + organic matter; Ca(OH)$_2$; m-methyl aniline; Ca(OCl)$_2$; hexamethylbenzene; hydrocarbons; acids; alkylmetal halides; inor-ganic bases; hydroxides; organic amines; KOH; formaldehyde; nitric acid, metal oxide; 1,2-di-aminomethane; lithium perchlorate.

NITROMIFENE CITRATE   HR: 3
CAS: 5863-35-4   NIOSH: UY 1348500
mf: C$_{27}$H$_{28}$N$_2$O$_4$•C$_6$H$_8$O$_7$ mw: 636.71

SYNS: 1-(2-(P-(ALPHA-(P-METHOXYPHENYL)-BETA-NITROSTYRYL)PHENOXY)ETHYL)PYRROLI-DINE MONOCITRATE * PARKE DAVIS CI-628

THR: Poison by ingestion. When heated to de-composition it emits toxic fumes of NO$_x$.

2-NITRONAPHTHALENE   HR: 3
CAS: 581-89-5   NIOSH: QJ 9760000
mf: C$_{10}$H$_7$NO$_2$ mw: 173.18

PROP: Colorless in ethanol; mp: 79°; bp: 165° @ 15 mm; insol in water; very sol in alc and ether.
TRK: 0.035 ppm

THR: Poison by ingestion. An experimental tu-morigen. Skin and lung irritant. Mutagenic data. For explosion and disaster hazards, see also nitrates. Combustible when exposed to heat or flame.

p-NITROPEROXYBENZOIC ACID HR: 3
CAS: 943-39-5   NIOSH: SD 9500000
mf: C$_7$H$_5$NO$_5$ mw: 183.13

THR: An experimental tumorigen. When heated to decomposition it emits toxic fumes of NO$_x$.

3-NITROPHENOL   HR: 3
CAS: 554-84-7   NIOSH: SM 1925000
mf: C$_6$H$_5$NO$_3$ mw: 139.12

PROP: Monoclinic crystals; mp: 97°; bp: 194° @ 70 mm; d: 1.485 @ 20°/4°. Decomp when distilled @ ordinary pressure. Sol in hot water and dil acids, caustic solns; insol in petroleum ether.

SYN: M-NITROPHENOL

THR: Poison by ingestion, subcutaneous, in-travenous, and possibly other routes. A severe eye and moderate skin irritant. See also nitrates. When heated to decomposition it emits toxic fumes of NO$_x$. For further information see Vol. 6, No. 3 of DPIM Report.

o-NITROPHENOL   HR: 3
CAS: 88-75-5   NIOSH: SM 2100000
mf: C$_6$H$_5$NO$_3$ mw: 139.12

PROP: Light yellow crystals, aromatic odor; mp: 45°; bp: 214.5°; d: 1.495 @ 20°; vap press: 1 mm @ 49.3°; sol in water; very sol in alc, ether, benzene, CS$_2$; volatile with steam.

SYNS: 2-HYDROXYNITROBENZENE * 2-NITRO-PHENOL

THR: Poison by subcutaneous, intravenous, and possibly other routes. Moderately toxic by in-gestion. Can cause liver and kidney damage. It reacts violently with KOH. When heated to de-composition it emits toxic fumes of NO$_x$. See also nitrates. For further information see Vol. 5, No. 3 of DPIM Report.

4-NITROPHENOL   HR: 3
CAS: 100-02-7   NIOSH: SM 2275000
mf: C$_6$H$_5$NO$_3$ mw: 139.12

PROP: Colorless to slightly yellow, odorless crystals, sweet then burning taste; d: 1.270 @ 120°/4°; mp: 113°-114° (sublimes); sltly sol in cold water; very sol in alc, chloroform, ether. Sol in alkali solns, hydroxides and carbonates.

SYNS: 4-HYDROXYNITROBENZENE * NCI-C55992 * 4-NITROFENOL (DUTCH) * P-NI-TROPHENOL * PARANITROFENOL (DUTCH) * PARANITROFENOLO (ITALIAN) * PARANITRO-PHENOL (FRENCH,GERMAN)

THR: Poison by ingestion, subcutaneous, intra-peritoneal, intravenous, intramuscular, and pos-sibly other routes. See also nitrates. When heated to decomposition it emits toxic fumes of NO$_x$. For further information see Vol. 3, No. 3 of DPIM Report.

p-NITROPHENYLDI-n-BUTYLPHOSPHI-NATE   HR: 3

CAS: 1224-64-2   NIOSH: SZ 4600000
mf: $C_{14}H_{22}NO_4P$ mw: 299.34

SYNS: DIBUTYL-PHOSPHINIC ACID, 4-NITROPHE-
NYL ESTER * NIBUFIN * P-NITROPHENYLDI-
BUTYLPHOSPHINATE

THR: Poison by intraperitoneal, intramuscular,
intravenous, and subcutaneous routes. See also
nitrates. When heated to decomposition it emits
very toxic fumes of PO_x and NO_x.

p-NITROPHENYL DIETHYLPHOS-PHATE   HR: 3

CAS: 311-45-5   NIOSH: TC 2275000
mf: $C_{10}H_{14}NO_6P$ mw: 275.22

PROP: Oily liquid, slt odor; bp: 170° @ 1
mm, d: 1.2736 @ 20°/4°; sltly water-sol, freely
sol in organic solvents.

SYNS: O,O'DIETHYL-P-NITROPHENYLPHOSPHAT
(GERMAN) * DIETHYL-P-NITROFENYL ESTER
KYSELINY FOSFORECNE (CZECH) * DIETHYL
P-NITROPHENYL PHOSPHATE * O,O-DIETHYL
O-P-NITROPHENYL PHOSPHATE * DIETHYL
PARAOXON * O,O-DIETHYL PHOSPHORIC ACID
O-P-NITROPHENYL ESTER * O,O-DIETYL-O-P-
NITROFENYLFOSFAT (CZECH) * ENT 16,087
* ETHYL PARAOXON * OXYPARATHION

THR: Poison by ingestion, intraperitoneal, intravenous, subcutaneous, intramuscular and parenteral routes. A cholinesterase inhibitor. See
also parathion, and phosphorous compounds,
NO_x, and PO_x.

p-NITROPHENYL ETHYLBUTYLPHOS-PHONATE   HR: 3

CAS: 3015-74-5   NIOSH: SZ 7015000
mf: $C_{12}H_{18}NO_5P$ mw: 287.28

THR: Poison by intravenous and subcutaneous
routes. See also nitrates. When heated to decomposition it emits very toxic fumes of PO_x and
NO_x.

1-NITROPROPANE   HR: 3

CAS: 108-03-2   NIOSH: TZ 5075000
DOT: 2608
mf: $C_3H_7NO_2$ mw: 89.11

PROP: Sltly sol in water; misc in alc and ether.
Colorless liquid; bp: 132°, fp: −108°, flash p:
93°F (TCC), d: 1.003 @ 20°/20°, autoign temp:

789°F, vap press: 7.5 mm @ 20°, vap d: 3.06,
lel = 2.2%; misc with many organic solvents.

OSHA PEL: TWA 25 ppm
ACGIH TLV: TWA 25 ppm
DFG MAK: 25 ppm (90 mg/m^3)
DOT Classification: Flammable Liquid

THR: Poison by intraperitoneal route and ingestion. Human eye irritant. Systemic irritant effects by inhalation. Flammable when exposed
to heat, open flame, or oxidizers. For disaster
and explosion hazards, see nitrates. Reacts violently with $Ca(OH)_2$; hydrocarbons; hydroxides;
inorganic bases. May explode on heating. To
fight fire, use alcohol foam, CO_2, dry chemical,
water spray. Incompatible with metal oxides.

2-NITROPROPANE   HR: 3

CAS: 79-46-9   NIOSH: TZ 5250000
DOT: 2608
mf: $C_3H_7NO_2$ mw: 89.11

PROP: Sol in water, alcohol and ether. Colorless liquid, bp: 120°, fp: −93°, flash p: 82°F
(TCC), d: 0.992 @ 20°/20°, autoign temp:
802°F, vap press: 10 mm @ 15.8°, vap d: 3.06,
lel = 2.6%; misc organic solvents.

SYNS: DIMETHYLNITROMETHANE * ISONITRO-
PROPANE * NITROISOPROPANE

OSHA PEL: TWA 25 ppm
ACGIH TLV: TWA CL 25 ppm
TRK: 5 ppm
DOT Classification: Flammable Liquid

THR: Poison by inhalation, ingestion, and intraperitoneal route. Mutagenic data. Human gastrointestinal tract effects. An experimental carcinogen by inhalation. Can cause liver and kidney injury, methemoglobinemia, and cyanosis.
See also nitrates. Chronic exposure to 207 ppm
for 6 months can cause hepatocellular carcinoma
by inhalation. Flammable when exposed to heat,
open flame, or oxidizers. May explode on heating; also violent reactions with chlorosulfonic
acid, oleum. For disaster hazard, see nitrates.
To fight fire, use alcohol foam, CO_2, dry chemical, water spray. For further information see
Vol. 4, No. 1 of *DPIM Report*.

4-NITROPYRIDINE-N-OXIDE   HR: 3

CAS: 1124-33-0   NIOSH: UT 6380000
mf: $C_5H_4N_2O_3$ mw: 140.11

THR: Poison by ingestion. Mutagenic data. An experimental carcinogen. When heated to decomposition it emits toxic fumes of NO_x.

8-NITROQUINOLINE   HR: 3
CAS: 507-35-2   NIOSH: VC 1925000
mf: $C_9H_6N_2O_2$ mw: 174.17

PROP: Monoclinic crystals from alcohol; mp: 88°-89°; sol in hot water, alc, ether and benzene.

THR: Poison by intraperitoneal route. Mutagenic data. An experimental carcinogen. When heated to decomposition it emits toxic fumes of NO_x.

4-NITROQUINOLINE-N-OXIDE   HR: 3
CAS:.56-57-5   NIOSH: VC 2100000
mf: $C_9H_6N_2O_3$ mw: 190.17

SYNS: 4-NITROCHINOLIN N-OXID (SWEDISH) * 4-NITROQUINOLINE-1-OXIDE * 4-NQO

THR: An experimental teratogen, carcinogen, neoplastigen, and tumorigen. Mutagenic data. When heated to decomposition it emits toxic fumes of NO_x.

NITROSAMINES   HR: 3
THR: Organic nitrogen compounds suspected of causing cancer of the lung, nasal sinus, brain, esophagus, stomach, liver, bladder, and kidney. They are often produced in food as by products from processing and preparation.

1-NITROSOAZACYCLOTRIDE-CANE   HR: 3
CAS: 40580-89-0   NIOSH: CL 6900000
mf: $C_{12}H_{24}N_2O$ mw: 212.38

SYNS: NITROSODODECAMETHYLENIMINE * N-NITROSODODECAMETHYLENIMINE * N-NITROSODODECAMETHYLENEIMINE

THR: An experimental tumorigen. Mutagenic data. See also nitroso compounds. When heated to decomposition it emits toxic fumes of NO_x.

1-NITROSOAZETIDINE   HR: 3
CAS: 15216-10-1   NIOSH: CM 4375000
mf: $C_3H_6N_2O$ mw: 86.11

SYNS: N-NITROSAZETIDINE * NITROSOAZETIDINE * NITROSOTRIMETHYLENEIMINE * NITROSO-AZETIDIN (GERMAN) * N-NITROSOAZETIDINE * N-NITROSOTRIMETHYLENEIMINE

THR: An experimental tumorigen. Mutagenic

data. Moderately toxic by ingestion. See also nitroso compounds. When heated to decomposition it emits toxic fumes of NO_x.

NITROSO-BAYGON   HR: 3
CAS: 38777-13-8   NIOSH: FC 6480000
mf: $C_{11}H_{14}N_2O_4$ mw: 238.27

SYNS: O-ISOPROPOXYPHENYL METHYLNITROSO-CARBAMATE * O-ISOPROPOXYPHENYL-N-METH-YL-N-NITROSOCARBAMATE * METHYLNITROSO-CARBAMIC ACID-O-ISOPROPOXYPHENYL ESTER * N-METHYL-N-NITROSOCARBAMIC ACID-O-ISO-PROPOXYPHENYL ESTER

THR: An experimental carcinogen. Mutagenic data. When heated to decomposition it emits toxic fumes of NO_x.

N-NITROSOBIS(2-ACETOXYPROPYL)-AMINE   HR: 3
CAS: 60414-81-5   NIOSH: UB 8758000
mf: $C_{10}H_{18}N_2O_5$ mw: 246.30

SYN: BAP

THR: An experimental tumorigen, neoplastigen, and carcinogen. Low toxicity by subcutaneous route. When heated to decomposition it emits toxic fumes of NO_x.

N-NITROSOBIS(2-OXOPROPYL)-AMINE   HR: 3
CAS: 60599-38-4   NIOSH: JL 9678000
mf: $C_6H_{10}N_2O_3$ mw: 158.18

SYN: 2,2'-DIOXOPROPYL-N-PROPYLNITROSAMINE

THR: Poison by ingestion and subcutaneous route. An experimental tumorigen, carcinogen, and neoplastigen. Mutagenic data. See also amines. When heated to decomposition it emits toxic fumes of NO_x.

N-NITROSOCARBARYL   HR: 3
CAS: 7090-25-7   NIOSH: FC 6600000
mf: $C_{12}H_{10}N_2O_3$ mw: 230.24

SYNS: DENAPON, NITROSATED (JAPANESE) * METHYL-NITROSOCARBAMIC ACID-1-NAPHTHYL ESTER * 1-NAPHTHYL N-METHYL-N-NITROSO-CARBAMATE * 1-NAPHTHYL METHYLNITROSO-CARBAMATE

THR: An experimental tumorigen, neoplastigen, and carcinogen. Mutagenic data. When heated to decomposition it emits toxic fumes of NO_x.

NITROSO COMPOUNDS　　　　　HR: 3

THR: Organic nitrogen compounds. Usually highly toxic carcinogens, teratogens, neoplastigens, and mutagens by almost all routes of exposure. When heated to decomposition they emit very toxic NO_x. See also specific compound entries.

N-NITROSODIETHYLAMINE　　　HR: 3

CAS: 55-18-5　　　　　NIOSH: IA 3500000
mf: $C_4H_{10}N_2O$　　mw: 102.16

PROP: Yellow oil; d: 0.9422 @ 20°/4°; bp: 47° @ 5 mm; bp: 176.9°; sol in water, alc and ether.

SYNS: DIAETHYLNITROSAMIN (GERMAN)
* DIETHYLNITROSAMINE * DIETHYLNITROSO-
AMINE * N,N-DIETHYLNITROSAMINE * N-
ETHYL-N-NITROSO-ETHANAMINE * NITROSODI-
ETHYLAMINE * N-NITROSO-DIAETHYLAMINE
(GERMAN) * N-NITROSODIETHYLAMINE

THR: Poison by ingestion, intravenous, intraperitoneal, subcutaneous, and possibly other routes. Mutagenic data. An experimental teratogen, carcinogen, tumorigen, and neoplastigen. An experimental transplacental brain carcinogen. A suspected human carcinogen. When heated to decomposition it emits toxic fumes of NO_x. See also nitroso compounds. For further information see Vol. 5, No. 5 of *DPIM Report*.

1-NITROSO-5,6-DIHYDROURA-
CIL　　　　　　　　　　　　　　　HR: 3

CAS: 16813-36-8　　　NIOSH: MX 9280000
mf: $C_4H_5N_3O_3$　　mw: 143.12

SYNS: NDHU * DIHYDRO-1-NITROSO-2,
4(1H,3H)-PYRIMIDINEDIONE

THR: An experimental tumorigen and carcinogen. Mutagenic data. Moderately toxic by intraperitoneal route. When heated to decomposition it emits toxic fumes of NO_x. See also nitroso compounds.

p-NITROSODIPHENYLAMINE　　HR: 3

CAS: 156-10-5　　　　　NIOSH: JK 0175000
mf: $C_{12}H_{10}N_2O$　　mw: 198.24

PROP: Green plates with bluish luster (from benzene) or steel blue prisms or plates (from ether + H_2O); mp: 144°-145°; sltly sol in water or petroleum ether; very sol in alc, ether, benzene, chloroform.

SYNS: NCI-C02244 * P-NITROSODIFENYLAMIN

(CZECH) * 4-NITROSODIPHENYLAMINE
* P-NITROSO-N-PHENYLANILINE * 4-NITROSO-
N-PHENYLBENZENAMINE

THR: Poison by intravenous route. Moderately toxic by ingestion. An experimental neoplastigen and carcinogen. A severe eye irritant. When heated to decomposition it emits toxic fumes of NO_x. See also nitroso compounds. For further information see Vol. 1, No. 5 of *DPIM Report*.

N-NITROSOEPHEDRINE　　　　HR: 3

CAS: 17608-59-2　　　　NIOSH: DP 0600000
mf: $C_{10}H_{14}N_2O_2$　　mw: 194.26

SYNS: ALPHA-(1-(N-METHYL-N-NITROSO-
AMINO)ETHYL)BENZYL ALCOHOL * 2-(N-
METHYL-N-NITROSOAMINO)-1-PHENYL-1-PRO-
PANOL

THR: Poison by intraperitoneal route. An experimental carcinogen. Mutagenic data. When heated to decomposition it emits toxic fumes of NO_x. See also nitroso compounds.

N-NITROSO-N-ETHYLURETHAN　HR: 3

CAS: 614-95-9　　　　　NIOSH: FB 0350000
mf: $C_5H_{10}N_2O_3$　　mw: 146.17

SYNS: AETHYLNITROSOURETHAN (GERMAN)
* N-ETHYL-N-NITROSOURETHANE * N-ETHYL-
N-NITROSOCARBAMIC ACID ETHYL ESTER
* NITROSOETHYLURETHAN

THR: Poison (acute) by intravenous route. Mutagenic data. An experimental teratogen and tumorigen. See also nitrates and nitroso compounds. When heated to decomposition it emits toxic fumes of NO_x.

N-NITROSO-N-ETHYLVINYL-
AMINE　　　　　　　　　　　　　HR: 3

CAS: 13256-13-8　　　　NIOSH: YZ 0700000
mf: $C_4H_8N_2O$　　mw: 100.14

SYNS: AETHYL-VINYL-NITROSOAMIN (GERMAN)
* N-ETHYL-N-NITROSOVINYLAMINE * ETHYL-
VINYLNITROSAMINE

THR: Poison by ingestion, intravenous, and subcutaneous routes. An experimental carcinogen and tumorigen. When heated to decomposition it emits toxic fumes of NO_x. See also nitroso compounds.

2-NITROSOFLUORENE　　　　　HR: 3

CAS: 2508-20-5　　　　　NIOSH: LL 8400000
mf: $C_{13}H_9NO$　　mw: 195.23

THR: An experimental neoplastigen and carcinogen. Mutagenic data. When heated to decomposition it emits toxic fumes of NO_x. See also nitroso compounds.

N-NITROSOHEXAHYDROAZE-PINE HR: 3
CAS: 932-83-2   NIOSH: CM 3325000
mf: $C_6H_{12}N_2O$ mw: 128.20

SYNS: HEXAHYDRO-1-NITROSO-1H-AZEPINE * N-NITROSOAZACYCLOHEPTANE * NITROSO-HEXAMETHYLENIMINE * N-NITROSOHEXAMETH-YLENEIMINE * N-NITROSOPERHYDROAZEPINE

THR: Poison by ingestion, intraperitoneal, and subcutaneous routes. Mutagenic data. An experimental carcinogen and tumorigen. See also nitroso compounds. When heated to decomposition it emits toxic fumes of NO_x.

N-NITROSOIMIDAZOLIDINE-THIONE HR: 3
CAS: 3715-92-2   NIOSH: NJ 0530000
mf: $C_3H_5ON_3S$ mw: 131.17

SYN: N-NITROSOETHYLENETHIOUREA

THR: Poison by ingestion. An experimental neoplastigen. Mutagenic data. See also nitrates and nitroso compounds. When heated to decomposition it emits very toxic fumes of NO_x and SO_x.

1-NITROSOIMIDAZOLIDINONE HR: 3
CAS: 3844-63-1   NIOSH: NJ 0950000
mf: $C_3H_5N_3O_2$ mw: 115.11

SYNS: N-NITROSO-IMIDAZOLIDON (GERMAN) * N-NITROSOIMIDAZOLIDONE

THR: Poison by ingestion and subcutaneous route. Mutagenic data. An experimental carcinogen. See also nitrates and nitroso compounds. When heated to decomposition it emits toxic fumes of NO_x.

N-NITROSOMETHYLAMINOACETO-NITRILE HR: 3
CAS: 3684-97-7   NIOSH: AM 0950000
mf: $C_3H_5N_3O$ mw: 99.11

SYNS: N-NITROSOMETHYLAMINACETONITRIL (GERMAN) * 2-(N-METHYL-N-NITROSO)AMINO-ACETONITRILE

THR: Poison by ingestion. An experimental tumorigen. See also nitroso compounds. When heated to decomposition it emits toxic fumes of NO_x.

N-NITROSO-N-METHYLCYCLOHEXYL-AMINE HR: 3
CAS: 5432-28-0   NIOSH: GX 1575000
mf: $C_7H_{14}N_2O$ mw: 142.23

SYNS: METHYLCYCLOHEXYLNITROSAMIN (GERMAN) * METHYLCYCLOHEXYLNITROSAMINE * N-METHYL-N-NITROSOCYCLOHEXYLAMINE * N-NITROSOMETHYLCYCLOHEXYLAMINE

THR: Poison by ingestion, intravenous, and intraperitoneal routes. Mutagenic data. An experimental carcinogen. See also nitrates and nitroso compounds. When heated to decomposition it emits toxic fumes of NO_x.

NITROSOMETHYL-n-DODECYL-AMINE HR: 3
CAS: 55090-44-3   NIOSH: JR 6900000
mf: $C_{13}H_{28}N_2O$ mw: 228.43

SYNS: N-METHYL-N-NITROSOLAURYLAMINE * N-NITROSO-N-METHYL-N-DODECYLAMIN (GERMAN)

THR: An experimental carcinogen. Mutagenic data. Moderate acute toxicity by ingestion. When heated to decomposition it emits toxic fumes of NO_x. See also nitroso compounds

1-NITROSO-4-METHYLPIPERA-ZINE HR: 3
CAS: 16339-07-4   NIOSH: TM 1780000
mf: $C_5H_{11}N_3O$ mw: 129.19

SYNS: N'-METHYL-N-NITROSOPIPERAZINE * 1-METHYL-4-NITROSOPIPERAZINE * N-NI-TROSO-N'-METHYLPIPERAZIN (GERMAN) * N-NITROSO-N'-METHYLPIPERAZINE

THR: Poison by ingestion. Mutagenic data. An experimental carcinogen. See also nitrates and nitroso compounds. When heated to decomposition it emits toxic fumes of NO_x.

NITROSOMETHYLPROPYL-AMINE HR: 3
CAS: 924-46-9   NIOSH: UI 3850000
mf: $C_4H_{10}N_2O$ mw: 102.16

SYNS: METHYLPROPYLNITROSOAMINE * METHYL-N-PROPYLNITROSAMINE * NITROSO-METHYL-N-PROPYLAMINE * N-METHYL-N-NITROSO-1-PROPANAMINE

THR: Poison by subcutaneous route. Mutagenic data. An experimental neoplastigen and carcinogen. When heated to decomposition it emits toxic fumes of NO_x.

N-NITROSO-N-METHYLURE-THANE   HR: 3
CAS: 615-53-2   NIOSH: FC 6300000
mf: $C_4H_8N_2O_3$ mw: 132.14

SYNS: ETHYL ESTER OF METHYLNITROSO-CAR-BAMIC ACID * METHYLNITROSOURETHAN (GER-MAN) * N-METHYL-N-NITROSO-URETHANE * NITROSOMETHYLURETHAN (GERMAN) * N-METHYL-N-NITROSOETHYLCARBAMATE

THR: Poison by ingestion, inhalation, intraperi-toneal, subcutaneous, and intravenous routes. Gastrointestinal system effects by ingestion. Mutagenic data. An experimental teratogen, tu-morigen, and carcinogen by ingestion and other routes. See also nitrosamines. Has been impli-cated as a transplacental brain carcinogen. Com-bustible when exposed to heat, sparks, open flame, and powerful oxidizers. When heated to decomposition it emits toxic fumes of NO_x.

4-NITROSOMORPHOLINE   HR: 3
CAS: 59-89-2   NIOSH: QE 7525000
mf: $C_4H_8N_2O_2$ mw: 116.14

SYNS: N-NITROSOMORPHOLIN (GERMAN) * N-NITROSOMORPHOLINE

THR: Poison by ingestion, inhalation, intraperi-toneal, subcutaneous, and intravenous routes. Mutagenic data. An experimental carcinogen, tumorigen, and neoplastigen. A suspected hu-man carcinogen. See also nitroso compounds. When heated to decomposition it emits toxic fumes of NO_x.

N'-NITROSONORNICOTINE   HR: 3
CAS: 16543-55-8   NIOSH: QS 6550000
mf: $C_9H_{11}N_3O$ mw: 177.23

SYNS: 1'-NITROSO-1'-DEMETHYLNICOTINE * 1-NITROSO-2-(3-PYRIDYL)PYRROLIDINE * 3-(1-NITROSO-2-PYRROLIDINYL)PYRIDINE

THR: An experimental tumorigen, neoplastigen, and carcinogen. Mutagenic data. A suspected human carcinogen. When heated to decomposi-tion it emits toxic fumes of NO_x. See also nitroso compounds.

p-NITROSOPHENOL   HR: 3
CAS: 104-9-1-6   NIOSH: SM 4725000
mf: $C_6H_5NO_2$ mw: 123.11

PROP: Pale yellow orthorhombic needles. mp: 144° (decomp). Sltly sol in water; sol in dilute alkalies, sol in alc, ether and acetone.

SYNS: QUINONE MONOXIME * 4-NITROSOPHE-NOL

THR: Poison by parenteral and intraperitoneal routes. Mutagenic data. An irritant and sensi-tizer. Dangerous fire and explosion hazard. When exposed to heat or flame, it burns explo-sively. Contamination by acid or alkali may cause ignition. Can heat spontaneously and cause fire. See also nitrates and nitroso com-pounds.

N-NITROSO-4-PICOLYLETHYL-AMINE   HR: 3
CAS: 13256-23-0   NIOSH: UT 3675000
mf: $C_8H_{11}N_3O$ mw: 165.22

SYNS: AETHYL-4-PICOLYLNITROSAMIN (GER-MAN) * 4-((ETHYLNITROSAMINO)METHYL)-PYRIDINE

THR: Poison by ingestion and intravenous routes. An experimental tumorigen. See also nitrates and nitroso compounds. When heated to decomposition it emits toxic fumes of NO_x.

N-NITROSOPIPERIDINE   HR: 3
CAS: 100-75-4   NIOSH: TN 2100000
mf: $C_5H_{10}N_2O$ mw: 114.17

PROP: Light yellow oil; d: 1.063 @ 18.5°/4°; bp: 217°-218°; sol in water; very sol in acid solns.

SYNS: NITROSOPIPERIDIN (GERMAN) * N-NI-TROSO-PIPERIDIN (GERMAN) * 1-NITROSOPIPERI-DINE * NO-PIP

THR: Acute poison by ingestion, intravenous, and subcutaneous routes. Mutagenic data. An experimental carcinogen, tumorigen, and neo-plastigen. Suspected human carcinogen. See also nitrates and nitroso compounds. When heated to decomposition it emits toxic fumes of NO_x. For further information see Vol. 6, No. 1 of *DPIM Report*.

1-(NITROSOPROPYLAMINO)-2-PROPA-NOL   HR: 3
CAS: 39603-53-7   NIOSH: UI 3150000
mf: $C_6H_{14}N_2O_2$ mw: 146.22

SYNS: BETA-HYDROXYPROPYLPROPYLNITRO-SAMINE * (2-HYDROXYPROPYL)PROPYLNITRO-SOAMINE * N-NITROSO-2-HYDROXY-N-PROPYL-N-PROPYLAMINE

THR: An experimental carcinogen and neoplas-tigen. Mutagenic data. Moderately toxic by subcutaneous route. See also nitrates and nitroso compounds. When heated to decomposition it emits toxic fumes of NO_x.

N-NITROSO-N-PROPYLUREA   HR: 3
CAS: 816-57-9   NIOSH: YT 9740000
mf: $C_4H_9N_3O_2$ mw: 131.16

SYNS: NITROSOPROPYLUREA * N-PROPYLNI-TROSOHARNSTOFF (GERMAN) * N-PROPYLNI-TROSUREA * 1-PROPYL-1-NITROSOUREA

THR: Poison by ingestion. Mutagenic data. An experimental teratogen, tumorigen, and carcinogen. When heated to decomposition it emits toxic fumes of NO_x. See also nitroso compounds.

N-NITROSOPYRROLIDINE   HR: 3
CAS: 930-55-2   NIOSH: UY 1575000
mf: $C_4H_8N_2O$ mw: 100.14

SYN: N-NITROSOPYRROLIDIN (GERMAN)

THR: Poison by ingestion and subcutaneous routes. Mutagenic data. An experimental carcinogen, neoplastigen, and tumorigen. A suspected human carcinogen. See also nitrates and nitroso compounds. When heated to decomposition it emits toxic fumes of NO_x. For further information see Vol. 5, No. 5 of *DPIM Report*.

NITROSTARCH (DRY)
CAS: 9056-38-6   NIOSH: QZ 7825000

PROP: Solid.

SYN: NITROSTARCH, DRY (DOT)

DOT Classification: Label: Explosive A

THR: No toxicity data. See also nitrates. Dangerous fire and explosion hazard when exposed to heat, flame, shock, or oxidizers. It is a powerful high explosive. Nitrostarch is not a definite compound, but a mixture of various nitric acid esters of starch with different degrees of nitration.

NITROSYL CHLORIDE   HR: 3
CAS: 2696-92-6   NIOSH: QZ 7883000
DOT: 1069
mf: ClNO   mw: 65.46

PROP: Yellow gas. Irr odor; mp: $-64.5°$; bp: $-5.8°$; d (liquid): 1.250 @ 30°; vap d: 2.3; vap press: 76 mm @ 50°. Non-explosive; very corrosive. Liquid @ $-5.5°$; solid @ $-61.5°$; sol in fuming H_2SO_4.

SYN: NITROGEN OXYCHLORIDE

DOT Classification: Label: Nonflammable Gas

THR: Poison by inhalation and ingestion. A very toxic irritating and corrosive gas to skin, eyes, and mucous membranes. Inhalation may cause pulmonary edema and hemorrhage. When heated to decomposition it emits very toxic fumes of Cl^- and NO_x.

NITROSYL FLUORIDE   HR: 3
mf: FNO   mw: 49.01

PROP: Colorless gas, often bluish because of impurities; mp: $-132.5°$; bp: $-59.9°$; d: (liquid @ bp) 1.326; (solid) 1.719.

SYN: NITROGEN OXYFLUORIDE

THR: Poison. Highly irritant to skin, eyes, and mucous membranes. Reacts vigorously with glass; corrodes quartz. See fluorides and hydrofluoric acid. Incompatible with haloalkene (unspecified); metals or non-metals; oxygen difluoride; Na; B; P; Si; halogenated olefins. When heated to decomposition it emits highly toxic fumes of F^- and NO_x.

NITROSYL PERCHLORATE
mf: $ClNO_5$ mw: 129.46

THR: No toxicity data. Below 100° it slowly decomposes; @ 115° $-120°$ it speeds up and explodes. A powerful oxidizer. When heated to decomposition it emits very toxic fumes of NO_x and Cl^-. See also perchlorates. Incompatible with acetone and also amines, diethyl ether, metal salts, (cobalt pentammine triazoperchlorate + phenyl isocyanate), organic materials.

NITROSYLSULFURIC ACID   HR: 3
mf: HNO_5S mw: 107.07

PROP: Prisms, decomp @ 73.5°. Sol in sulfuric acid; decomp in water.

SYNS: CHAMBER CRYSTALS * NITROSYL SULFATE

THR: A poison; irritant to skin, eyes, and mucous membranes. Preparative hazard. See also sulfates and nitrates. Violent reaction with dinitroaniline hydrochloride.

NITROTHIAZOLE   HR: 3
CAS: 61-57-4   NIOSH: NJ 1050000
mf: $C_6H_6N_4O_3S$ mw: 214.22

SYNS: 1-(5-NITRO-2-THIAZOLYL)-2-IMIDAZOLI-
DINONE ∗ NITROTHIAMIDAZOLE ∗ 1-(5-NI-
TRO-2-THIAZOLYL)IMIDAZOLIDIN-2-ONE
∗ 1-(5-NITRO-2-THIAZOLYL)-2-IMIDAZOLINONE
∗ 1-(5-NITRO-2-THIAZOLYL)-2-OXOTETRAHY-
DROIMIDAZOL ∗ 1-(5-NITRO-2-THIAZOLYL)-2-
OXOTETRAHYDROIMIDAZOLE

THR: An experimental carcinogen and neoplastigen. Mutagenic data. Moderately toxic by ingestion and intraperitoneal route. When heated
to decomposition it emits very toxic fumes of
SO_x and NO_x.

NITROTRIBROMOMETHANE   HR: 3
CAS: 464-10-8   NIOSH: PB 0100000
mf: CBr_3NO_2 mw: 297.75

PROP: Crystals; bp: 127°; mp: 103°; d: 2.79
@ 18°.

SYNS: BROMOPICRIN ∗ NITROBROMOFORM
∗ TRIBROMONITROMETHANE

THR: Poison by intraperitoneal route. In vapor
form it is highly toxic by ingestion and inhalation
and contact with skin, eyes, and mucous membranes. When heated to decomposition it emits
very toxic fumes of Br^- and NO_x.

NITROUREA
mf: $CH_3N_3O_3$ mw: 105.06

PROP: Crystals.

SYN: M-NITROCARBAMIDE

THR: No toxicity data. Dangerous fire hazard
when exposed to heat or flame. Severe explosion
hazard when shocked or exposed to heat. See
also explosives, high, and nitrates. Dangerous;
When heated to decomposition it emits highly
toxic fumes of NO_x; can react vigorously with
oxidizing materials. It is a high explosive. Incompatible with mercuric and silver salts.

NITROUS ACID   HR: 3
CAS: 7782-77-6   NIOSH: RA 0760000
mf: HNO_2 mw: 47.02

PROP: Pale blue solution.

SYN: NITROSYL HYDROXIDE

THR: Mutagenic data. See also nitric oxide.
Flammable by chemical reaction; a powerful
oxidizer. Reacts violently with PH_3; PCl_3. Dangerous; when heated to decomposition it emits
highly toxic fumes of NO_x. Incompatible with

semicarbazone; silver nitrate; phosphine; phosphorus trichloride.

NITRYL CHLORIDE   HR: 3
mf: $ClNO_2$ mw: 81.46

PROP: Corrosive, toxic, colorless gas. Decomp
> 120°, vap d: 2.81 g/L @ 100°; bp: −14.3°;
mp: −145°; d: (liq): 1.37 @ 0°.

SYN: NITROXYL CHLORIDE

THR: A poison by inhalation and a severe irritant
to skin, eyes, and mucous membranes. See also
chlorine. The gas or liquid may attack organic
matter with explosive violence. Incompatible
with inorganic materials, or organic matter.
Dangerous; when heated to decomposition it
emits highly toxic fumes of chlorides and nitrogen dioxide; will react with water or steam to
produce corrosive fumes. Violent reaction with
NH_3; $SnBr_4$; SnI_4; SO_3.

NITRYL FLUORIDE   HR: 3
mf: FNO_2 mw: 65.01

PROP: Colorless gas, pungent odor. Mp:
−166.0°; bp: −72.4°; d (liq): 1.796 @ bp;. d
(solid): 1.924.

THR: Poison by inhalation. A severe irritant
to skin, eyes, and mucous membranes. See also
fluorine and fluorides. Powerful oxidizing agent.
Dangerous disaster hazard; see NO_x and fluorine. This gas is intensely reactive. It will ignite
spontaneously in I; Se; P; As; B; Sb; Th; Mo;
upon warming, will ignite Pb; Bi; Cr; Mn; Fe;
Ni; W; S; C. Incompatible with metals; non-
metals.

NITRYL HYPOFLUORITE
mf: FNO_3 mw: 81.01

THR: No toxicity data. Probably a poison irritant
to skin, eyes, mucous membranes. See also
fluorides. Dangerous explosion hazard. When
heated to decomposition it emits highly toxic
fumes of F^- and NO_x.

N-NONANE   HR: 3
CAS: 111-84-2   NIOSH: RA 6115000
DOT: 1920
mf: C_9H_{20} mw: 128.29

PROP: Colorless liquid. Mp: −53.7°, bp:
150.7°, lel = 0.8%, uel = 2.9%, flash p: 88°F
(CC), d: 0.718 @ 20°/4°, autoign temp: 374°F,

vap press: 10 mm @ 38.0°, vap d: 4.41. Insol in water; sol in absolute alc; sol in ether.

SYN: SHELLSOL 140

ACGIH TLV: TWA 200 ppm
DOT Classification: Flammable Liquid

THR: Poison by intravenous route. Irritant to respiratory tract. Narcotic in high concentrations. Dangerous fire hazard when exposed to heat or flame; can react with oxidizing materials. Moderately explosive in the form of gas, when exposed to flame. To fight fire, use CO_2, dry chemical.

NONANOIC ACID HR: 3
CAS: 112-05-0 NIOSH: RA 6650000
mf: $C_9H_{18}O_2$ mw: 158.27

PROP: Oily colorless liquid, very sltly sol in water. Bp: 254°, mp: 12°, d: 0.9055 @ 20°/4°.

SYNS: N-NONYLIC ACID * PELARGIC ACID * 1-OCTANECARBOXYLIC ACID * PELARGONIC ACID

THR: Poison by intravenous route. Moderately toxic by ingestion. A severe skin and eye irritant. When heated to decomposition it emits acrid smoke and fumes.

1-NONANOYLAZIRIDINE HR: 3
CAS: 63021-51-2 NIOSH: CM 8255000
mf: $C_{11}H_{21}NO$ mw: 183.33

SYN: NONANOYLETHYLENEIMINE

THR: An experimental neoplastigen and carcinogen. When heated to decomposition it emits toxic fumes of NO_x.

(−)-NOREPHEDRINE HR: 3
CAS: 492-41-1 NIOSH: RC 2275000
mf: $C_9H_{13}NO$ mw: 151.23

SYNS: 2-AMINO-1-PHENYL-1-PROPANOL * USAF CS-6 * MYDRIATIN * PHENYLPRO-PANOLAMINE * PROPADRINE

THR: Poison by intravenous and intraperitoneal routes. When heated to decomposition it emits toxic fumes of NO_x.

l-NOREPINEPHRINE HR: 3
CAS: 51-41-2 NIOSH: DN 5950000
mf: $C_8H_{11}NO_3$ mw: 169.20

PROP: Microcrystals; decomp @ 216.5°-218°

SYNS: LEVONORADRENALINE * 1-2-AMINO-1-(3,4-DIHYDROXYPHENYL)ETHANOL * 1-AL-PHA-(AMINOMETHYL)-3,4-DIHYDROXYBENZYL ALCOHOL * (−)-ALPHA-(AMINOMETHYL)-PROTOCATECHUYL ALCOHOL * 1-1-(3,4-DIHYDROXYPHENYL)-2-AMINOETHANOL * 1-3,4-DIHYDROXYPHENYLETHANOLAMINE * 1-NORADRENALINE * (−)-NOREPINEPHRINE * 1-NOREPINEPHRINE

THR: Poison by ingestion, subcutaneous, and intravenous routes. An experimental teratogen. When heated to decomposition it emits toxic fumes of NO_x.

19-NORETHISTERONE HR: 3
CAS: 68-22-4 NIOSH: RC 8975000
mf: $C_{20}H_{26}O_2$ mw: 298.46

SYNS: 17-ALPHA-ETHINYLESTRA-4-EN-17-BETA-OL-3-ONE * 17-ALPHA-ETHINYL-17-BETA-HYDROXY-DELTA(SUP:4)-ESTREN-3-ONE * 17-ALPHA-ETHINYL-19-NORTESTOSTERONE * 17-ALPHA-ETHYNYL-4-ESTREN-17-OL-3-ONE * 17-ALPHA-ETHYNYL-17-HYDROXY-4-ESTREN-3-ONE * 17-ALPHA-ETHYNYL-17-BETA-HY-DROXY-19-NORANDROST-4-EN-3-ONE * 17-ALPHA-ETHYNYL-17-BETA-HYDROXY-19-NORANDROST-4-EN-3-ONE * 17-ALPHA-ETH-YNYL-19-NORANDROST-4-EN-17-BETA-OL-3-ONE * 17-ALPHA-ETHYNYL-19-NOR-4-ANDROSTEN-17-BETA-OL-3-ONE * 17-ALPHA-ETHYNYL-19-NORTESTOSTERONE * (17-ALPHA)-17-HY-DROXY-19-NORPREGN-4-EN-20-YN-3-ONE * 17-BETA-HYDROXY-19-NORPREGN-4-EN-20-YN-3-ONE * 19-NOR-ETHINYL-4,5-TESTOS-TERONE * 17-HYDROXY-19-NOR-17-ALPHA-PREGN-4-EN-20-YN-3-ONE * 19-NOR-17-AL-PHA-ETHYNYLANDROSTEN-17-BETA-OL-3-ONE * 19-NOR-17-ALPHA-ETHYNYL-17-BETA-HY-DROXY-4-ANDROSTEN-3-ONE * 19-NOR-17-AL-PHA-ETHYNYLTESTOSTERONE * NORLUTIN

THR: A teratogen and glandular toxin in women by ingestion. An experimental teratogen, and carcinogen. When heated to decomposition it emits acrid smoke and fumes.

NORETHISTERONE ENAN-THATE HR: 3
CAS: 3836-23-5 NIOSH: RC 8973000

SYNS: 17-ALPHA-ETHYNYL-17-BETA-HEPTANO-YLOXY-4-ESTREN-3-ONE * 17-ALPHA-ETHYNYL-19-NORTESTOSTERONE ENANTHATE * 17-BETA-HEPTANOYLOXY-19-NOR-17-ALPHA-PREGNEN-20-YNONE

THR: Affects the central nervous system in women by intravenous route. An experimental teratogen, and neoplastigen.

NORGESTREL   HR: 3
CAS: 6533-00-2   NIOSH: JF 8225000
mf: $C_{21}H_{28}O_2$ mw: 312.49

PROP: Crystals from diethyl ether-hexane; mp: 142°-143°.

SYNS: 13-ETHYL-17-ALPHA-ETHYNYLGON-4-EN-17-BETA-OL-3-ONE * 13-ETHYL-17-ALPHA-ETHYNYL-17-BETA-HYDROXY-4-GONEN-3-ONE * (±)- 13-ETHYL-17-ALPHA-ETHYNYL-17-HYDROXYGON-4-EN-3-ONE * 17-ETHYNYL-18-METHYL-19-NORTESTOSTERONE * 17-BETA-HYDROXY-18-METHYL-19-NOR-17-ALPHA-PREGN-4-EN-20-YN-3-ONE * 18-METHYL-17-ALPHA-ETHYNYL-19-NORTESTOSTERONE * D(−)-NORGESTREL * D-NORGESTREL * 19-NORTESTOSTERONE

THR: An experimental neoplastigen and carcinogen. When heated to decomposition it emits acrid smoke and fumes.

19-NORTESTOSTERONE   HR: 3
CAS: 6533-00-2   NIOSH: JF 8260000
mf: $C_{21}H_{28}O_2$ mw: 312.49

SYNS: 13-BETA-ETHYL-17-ALPHA-ETHYNYL-17-BETA-HYDROXYGONE-4-EN-3-ONE * LD NORGESTREL (FRENCH)

THR: An experimental neoplastigen. When heated to decomposition it emits acrid smoke and fumes.

NORTRIPTYLINE   HR: 3
CAS: 72-69-5   NIOSH: HO 9800000
mf: $C_{19}H_{21}N$ mw: 263.41

SYNS: DEMETHYLAMITRIPTYLENE * DESMETHYLAMITRIPTYLINE * 10,11-DIHYDRO-N-METHYL-5H-DIBENZO(A,D)CYCLOHEPTANE-DELTA,GAMMA-PROPYLAMINE * 5-(3-METHYLAMINOPROPYLIDENE)-10,11-DIHYDRO-5H-DIBENZO(A,D)CYCLOHEPTENE * NORAMITRIPTYLINE * 5-(3-(METHYLAMINO)PROPYLIDENE)DIBENZO-(a,e)CYCLOHEPTA(1,5)DIENE

THR: Poison by ingestion, intraperitoneal, and intravenous routes. When heated to decomposition it emits toxic fumes of NO_x.

NSC-129943   HR: 3
CAS: 21416-87-5   NIOSH: TL 6400000
mf: $C_{11}H_{16}N_4O_4$ mw: 268.31

SYNS: (+,−)-1,2-BIS(3,5-DIOXOPIPERAZINE-1-YL)PROPANE * ICRF-159 * NCI-C01627 * 2,6-PIPERAZINEDIONE-4,4'-PROPYLENE DIOXOPIPERAZINE

THR: Human blood and gastrointestinal system effects by ingestion. An experimental carcinogen, teratogen and tumorigen. When heated to decomposition it emits toxic fumes of NO_x.

NSC 37448   HR: 3
CAS: 59-96-1   NIOSH: DP 3500000
mf: $C_{18}H_{22}ClNO$ mw: 303.86

SYNS: 2-(N-BENZYL-2-CHLOROETHYLAMINO)-1--PHENOXYPROPANE * BENZYL(2-CHLOROETHYL)-(1-METHYL-2-PHENOXYETHYL)AMINE * N-(2-CHLOROETHYL)-N-(1-METHYL-2-PHENOXYETHYL)BENZENEMETHANAMINE * N-(2-CHLOROETHYL)-N-(1-METHYL-2-PHENOXYETHYL)BENZYLAMINE * N-PHENOXYISOPROPYL-N-BENZYL-BETA-CHLOROETHYLAMINE

THR: Poison by intravenous route. An experimental carcinogen and neoplastigen. Moderately toxic by ingestion. When heated to decomposition it emits very toxic fumes of Cl^- and NO_x.

NU-1932   HR: 3
CAS: 63039-90-7   NIOSH: TN 7600000
mf: $C_{17}H_{25}NO_2 \cdot ClH$ mw: 311.89

SYNS: 3-BETA-AETHYL-1-METHYL-4-PHENYL-4-ALPHA-PIPERIDYLPROPIONAT HYDROCHLORID (GERMAN) * 3-BETA-AETHYL-1-METHYL-4-PHENYL-4-ALPHA-PROPIONYLOXYPIPERIDIN HYDROCHLORID (GERMAN) * MEPRODINE (GERMAN)

THR: Poison by ingestion, subcutaneous, intravenous, and intraperitoneal routes. When heated to decomposition it emits very toxic fumes of HCl and NO_x.

1-NUCIFERINE   HR: 3
CAS: 475-83-2   NIOSH: CE 0350000
mf: $C_{19}H_{21}NO_2$ mw: 295.41

PROP: Alkaloid obtained from the lotus *Nelumbo Nucifera* and *Nelumbo Lutea*.

SYNS: 1-5, 6-DIMETHOXYAPORPHINE * (R)-1,2-DIMETHOXYAPORPHINE * (−)-NUCIFERINE * 1,2-DIMETHOXY-6a-BETA-APORPHINE

THR: Poison by ingestion. When heated to decomposition it emits toxic fumes of NO_x.

NUISANCE DUSTS AND AEROSOLS

THR: Variable toxicity depending upon composition. Causes local irritation of eyes, nose, throat, and lungs. Some may lead to chronic bronchitis, emphysema, and bronchial asthma. Dermatitis may result from short contact. Asthma, angioneurotic edema, hives, etc., may result from short periods of inhalation. A topic eczema, angioneurotic edema, hives, etc., may also result from prolonged contact. A common air contaminant. Nuisance aerosols do evoke some tissue response in the lung upon inhalation of sufficient amounts. However, this reaction is potentially reversible and leaves no scar tissue.

NUPERCAINE HYDROCHLO-RIDE   HR: 3

CAS: 61-12-1   NIOSH: GD 3325000
mf: $C_{20}H_{29}N_3O_2 \cdot ClH$ mw: 379.98

PROP: Crystals; mp: decomp @ 90°-98°; vap d: 13.1.

SYNS: BENZOLIN HYDROCHLORIDE * BUTOXY-CINCHONINIC ACID DIETHYLETHYLENEDIAMIDE HYDROCHLORIDE * 2-N-BUTOXY-N-(2-DIETHYLAMINOETHYL)CINCHONINAMIDE HYDRO-CHLORIDE * 2-BUTOXY-N-(2-DIETHYLAMI-NOETHYL)CINCHONINAMIDE HYDROCHLORIDE * 2-BUTOXY-N-(2-DIETHYLAMINOETHYL)-CINCHONINIC ACID AMIDE HYDROCHLORIDE * CINCAINE HYDROCHLORIDE * CINCHOCAINE HYDROCHLORIDE * CINCHOCAINIUM CHLORIDE * DIBUCAINE HYDROCHLORIDE * PERCAINE HYDROCHLORIDE

THR: Poison by ingestion, subcutaneous, intravenous, intraspinal, parenteral, and intraperitoneal routes. Severe eye irritant. When heated to decomposition it emits very toxic fumes of HCl and NO_x.

NYLON   HR: 3

PROP: Film used in carcinogenic experiments.

CAS: 63428-83-1   NIOSH: RF 5250000
mf: $(NOC_6H_{11})_x$.

PROP: Crystalline solids; sol in phenol, cresols, xylene, formic acid; insol in alc, esters, ketones, hydrocarbons.

SYNS: CAPROLON * PHRILON * POLYAMID (GERMAN) * SILON

THR: An experimental neoplastigen by implantation. Reacts violently with F_2. When heated to decomposition it emits toxic fumes of NO_x.

NYSTATIN   HR: 3

CAS: 1400-61-9   NIOSH: RF 5950000
mf: $C_{46}H_{83}NO_{18}$ mw: 938.30

PROP: Yellow to light-tan powder, odor suggestive of cereals, sparingly sol in methanol and ethanol, very slightly sol in water, insol in chloroform, ether and benzene. Mp: decomp > 160°.

SYNS: FUNGICIDIN * MYCOSTATIN * MYKO-STATYNA

THR: Poison by intraperitoneal and intravenous routes. Moderately toxic by subcutaneous route. When heated to decomposition it emits toxic fumes of NO_x.

O

OCHRATOXIN A HR: 3
CAS: 303-47-9 NIOSH: AY 4375000
mf: $C_{20}H_{18}ClNO_6$ mw: 403.84

PROP: Crystals, mp: 169°.

SYN: NCI-C56586

THR: Poison by ingestion, intraperitoneal, intravenous, and subcutaneous routes. An experimental teratogen and carcinogen. Mutagenic data. Dangerous; when heated to decomposition it emits toxic fumes of Cl^- and NO_x.

OCTABROMODIPHENYL HR: 3
CAS: 27858-07-7 NIOSH: DV 5700000
mf: $C_{12}H_2Br_8$ mw: 785.42

SYNS: OCTABROMOBIPHENYL
* AR,AR,AR,AR,AR',AR',AR',AR'
OCTABROMO-1,1'-BIPHENYL

THR: Poison by ingestion and skin contact. An experimental teratogen. Irritant to skin, eyes, and mucous membranes. Can cause kidney and liver enlargement and thyroid hyperplasia; it is stored in fatty tissue. See also bromides. When heated to decomposition it emits toxic fumes of Br^-.

OCTACHLORODIBENZODIOXIN HR: 3
CAS: 3268-87-9 NIOSH: HP 3350000
mf: $C_{12}Cl_8O_2$ mw: 459.72

PROP: Colorless crystals, mp: 239°.

SYNS: NCI-C03678 * OCDD * OCTA-
CHLORODIBENZODIOXIN * OCTACHLO-
RODIBENZO(B,E)(1,4)DIOXIN * OCTA-
CHLORODIBENZO-P-DIOXIN * 1,2,3,4,6,7,
8,9-OCTACHLORODIBENZODIOXIN

THR: An experimental carcinogen and teratogen. Irritant to eyes. When heated to decomposition it emits toxic fumes of Cl^-.

1,3,4,5,6,8,8-OCTACHLORO-1,3,3a,4, 7,7a-HEXAHYDRO-4,7-METHANO ISO BENZOFURAN HR: 3
CAS: 297-78-9 NIOSH: PC 1225000
mf: $C_9H_4Cl_8O$ mw: 411.73

SYNS: ENT 25,545 * ENT 25,545-X
* 1,3,4,5,6,7,8,8-OCTACHLORO-2-OXA-
3A,4,7,7A-TETRAHYDRO-4,7-METHANOINDENE

* 1,3,4,5,6,7,10,10-OCTACHLORO-4,7-ENDO-
METHYLENE-4,7,8,9-TETRAHYDROPHTHALAN

THR: Poison by ingestion, skin contact, intraperitoneal, intravenous, and other routes. An experimental carcinogen. When heated to decomposition it emits toxic fumes of Cl^-.

1-OCTADECANOL HR: 3
CAS: 112-92-5 NIOSH: RG 2010000
mf: $C_{18}H_{38}O$ mw: 270.56

PROP: Colorless solid or flakes, mp: 58°, bp: 202° @ 10 mm, d: 0.8124 @ 59°/4°.

SYNS: DECYL OCTYL ALCOHOL * N-OCTADE-
CANOL * N-OCTADECYL ALCOHOL * STEARYL
ALCOHOL * USP XIII STEARYL ALCOHOL

THR: An experimental carcinogen. Mild acute toxicity by ingestion. See alcohols. Flammable when exposed to heat or flame; can react with oxidizing materials. To fight fire, use foam, CO_2, dry chemical.

OCTADECYLAMINE HR: 3
CAS: 124-30-1 NIOSH: RG 4150000
mf: $C_{18}H_{39}N$ mw: 269.58

SYNS: N-OCTADECYLAMINE * OKTADECYLA-
MIN (CZECH) * STEARYLAMINE

THR: Poison by intraperitoneal route. A skin irritant. See also amines. When heated to decomposition it emits toxic fumes of NO_x.

OCTAHYDRO-1-NITROSOAZO-CINE HR: 3
CAS: 20917-49-1 NIOSH: CN 4900000
mf: $C_7H_{14}N_2O$ mw: 142.23

SYNS: N-NITROSOAZACYCLOOCTANE * N-NI-
TROSOHEPTAMETHYLENEIMINE * NITROSO-HEP-
TAMETHYLENIMIN (GERMAN)

THR: Poison by ingestion and subcutaneous routes. Mutagenic data. An experimental carcinogen and tumorigen. When heated to decomposition it emits toxic fumes of NO_x.

OCTAMETHYLPYROPHOSPHOR AMIDE HR: 3
CAS: 152-16-9 NIOSH: UX 5950000
mf: $C_8H_{24}N_4O_3P_2$ mw: 286.30

PROP: Viscous liquid; mp: 20°-21°; bp: 154° @ 2.0 mm; d: 1.09 @ 25°/4°. Miscible with water. Sol in most organic solvents. Almost insol in higher aliphatic hydrocarbons.

SYNS: BIS(DIMETHYLAMINO)PHOSPHONOUS AN-YHYDRIDE * BIS(DIMETHYLAMINO)PHOSPHORIC ANHYDRIDE * BIS-N,N,N′,N′-TETRAMETHYL-PHOSPHORODIAMIDIC ANHYDRIDE * OCTA-METHYL-DIFOSFORZUUR-TETRAMIDE (DUTCH) * OCTAMETHYL-DIPHOSPHORSAEURE-TETRAMID (GERMAN) * OCTAMETHYL PYROPHOSPHORTET-RAMIDE * OCTAMETHYL TETRAMIDO PYRO-PHOSPHATE * OTTOMETIL-PIROFOSFORAMMIDE (ITALIAN) * PYROPHOSPHORIC ACID OCTA-METHYLTETRAAMIDE * PYROPHOSPHORYL-TETRAKISDIMETHYLAMIDE * TETRAKISDIMETH-YLAMINOPHOSPHONOUS ANHYDRIDE

THR: Poison by ingestion, skin contact, intra-peritoneal, and intravenous routes. A cholines-terase inhibitor. Has been found to inhibit periph-eral cholinesterase without pronounced effects on the central nervous system. See parathion. An insecticide. Dangerous; see also phosphates.

OCTANE HR: 1
CAS: 111-65-9 NIOSH: RG 8400000
DOT: 1262
mf: C_8H_{18} mw: 114.26

PROP: Clear liquid. Flammable. Bp: 125.8°, lel = 1.0%, uel = 4.7%; fp: −56.5°, flash p: 56°F, d: 0.7036 @ 20°/4°, autoign temp: 428°F, vap press: 10 mm @ 19.2°, vap d: 3.86; insol in water; sltly sol in alc, ether; misc with ben-zene.

SYNS: OKTAN (POLISH) * OKTANEN (DUTCH) * OTTANI (ITALIAN).

OSHA PEL: TWA 500 ppm
ACGIH TLV: TWA 300 ppm; STEL 375 ppm
DFG MAK: 500 ppm (2350 mg/m³)
DOT Classification: Label: Flammable
  Liquid

THR: May act as a simple asphyxiant. See also argon. A narcotic in high concentration. Human dermal exposure to undiluted octane for five hours resulted in blister formation but no an-esthesia; one hour caused diffuse burning sensa-tion. See also alkanes. Dangerous fire hazard when exposed to heat, flame, or oxidizers. Se-vere explosion hazard when exposed to heat or flame.

1-OCTEN-3-OL HR: 3
CAS: 3391-86-4 NIOSH: RH 3300000
mf: $C_8H_{16}O$ mw: 128.24

SYNS: AMYLVINYLCARBINOL * MATSUTAKE ALCOHOL (JAPANESE)

THR: Poison by ingestion and intravenous routes. Moderately toxic by skin contact. When heated to decomposition it emits acrid smoke and fumes.

OCTIN HYDROCHLORIDE HR: 3
CAS: 6168-86-1 NIOSH: MP 9625000
mf: $C_9H_{19}N \cdot ClH$ mw: 177.75

SYNS: ISOMETHEPTENE HYDROCHLORIDE * 2-METHYLAMINOISOOCTANE HYDROCHLORIDE * METHYLAMINO-METHYLHEPTENE HYDROCHLO-RIDE

THR: Poison by intraperitoneal, intravenous, and subcutaneous routes. When heated to de-composition it emits very toxic fumes of Cl^- and NO_x.

N-OCTYL ALCOHOL HR: 2
CAS: 111-87-5 NIOSH: RH 6550000
mf: $C_8H_{18}O$ mw: 130.26

PROP: Colorless liquid; d: 0.827 @ 20°16/4°; mp: −16.7°; bp: 194.5°; sol water; misc in alc, ether and chloroform. Found in several citrus oils and at least 10 other natural sources.

SYNS: ALCOHOL C-8 * CAPRYL ALCOHOL * CAPRYLIC ALCOHOL * HEPTYL CARBINOL * OCTANOL * N-OCTANOL * 1-OCTANOL * OCTYL ALCOHOL, NORMAL-PRIMARY * PRIMARY OCTYL ALCOHOL

THR: Moderately toxic by ingestion. Skin irri-tant. Flammable when exposed to heat or flame; can react with oxidizing materials. To fight fire, use water foam, fog, alcohol foam, dry chemi-cal, CO_2. For further information see 1-Octanol Vol. 3, No. 2 of *DPIM Report*.

N-OCTYL BICYCLOHEPTENE DICAR-BOXIMIDE HR: 2
CAS: 113-48-4 NIOSH: RB 8575000
mf: $C_{17}H_{25}NO_2$ mw: 275.43

SYNS: BICYCLO(2.2.1)HEPTENE-2-DICAR-BOXYLIC ACID, 2-ETHYLHEXYLIMIDE * ENDOMETHYLENETETRAHYDROPHTHALIC ACID,N-2-ETHYLHEXYL IMIDE * ENT 8,184 * N-(2-ETHYLHEXYL)BICYCLO-(2,2,1)-HEPT-5-

ENE-2,3-DICARBOXIMIDE * N-2-ETHYLHEXYL-
IMIDEENDOMETHYLENETETRAHYDROPHTHALIC
ACID * N-(2-ETHYLHEXYL)-5-NORBORNENE-
2,3-DICARBOXIMIDE * 2-(2-ETHYLHEXYL)-
3A,4,7,7A-TETRAHYDRO-4,7-METHANO-1H-ISO-
INDOLE-1,3(2H)-DIONE * N-OCTYLBICYCLO-
(2.2.1)-5-HEPTENE-2,3-DICARBOXIMIDE

THR: Moderately toxic by ingestion, skin contact, and intraperitoneal routes. Large doses can cause central nervous system stimulation followed by depression. When heated to decomposition it emits toxic fumes of NO_x.

OIL OF CALAMUS   HR: 3
CAS: 8015-79-0   NIOSH: RI 7000000

PROP: Extract of *Acorus Calamus L., Araceae*. Containing: Asarone, Eugenol; esters of acetic and heptylic acids. Volatile oil. Yellow to yellowish-brown liquid (viscid); aromatic odor, bitter taste. D: 0.960-0.9707 @ 20°/20°. Very slightly sol in water, misc with alc. Keep well closed, cool and protected from light.

SYNS: CALAMUS OIL * KALMUS OEL (GERMAN) * OIL OF SWEET FLAG

THR: Poison by intraperitoneal route. An experimental carcinogen. Moderately toxic by ingestion. When heated to decomposition it emits acrid smoke and fumes. For further information see Vol. 1, No. 2 of *DPIM Report*.

OIL GAS   HR: 3
PROP: A gas derived from petroleum. Composition: illuminants 4.2%, carbon monoxide 10.4%, hydrogen 47.6%, methane 27.0%, carbon dioxide 4.6%, nitrogen 5.8%, oxygen 0.4%, lel = 4.8%; uel = 32.5%; autoign temp: 637°F.

THR: Poison. See also carbon monoxide. Dangerous fire hazard when exposed to heat or flame. Moderate explosion hazard when exposed to heat or flame. To fight fire, use CO_2, dry chemical, water spray. Incompatible with oxidizing materials.

OIL MIST (MINERAL)   HR: 3
                         NIOSH: RI 7400000

PROP: Mist of petroleum-base cutting oils or white mineral petroleum oil.

SYN: OIL MIST

OSHA PEL: TWA 5 mg/m^3

ACGIH TLV: TWA 5 mg/m^3; STEL 10 mg/m^3

THR: Poison by inhalation. No toxicity data.

OIL OF ORANGE   HR: 3
CAS: 8008-57-9   NIOSH: RI 8600000

PROP: Yellow to deep orange liquid; characteristic orange taste and odor; d: 0.842-0.846 @ 25°/25°. Slightly sol in water, misc with absolute alc, carbon disulfide. Sol in 2 vols 90% alc, in 1 vol glacial acetic acid. Keep well closed, cool and protected from light. Oil from the peel of citrus sinensis.

SYNS: NEAT OIL OF SWEET ORANGE * OIL OF SWEET ORANGE * ORANGE OIL

THR: An experimental neoplastigen. Skin irritant. For further information see Vol. 1, No. 2 of *DPIM Report*.

OIL RED   HR: 3
CAS: 85-86-9   NIOSH: QK 4250000
mf: $C_{22}H_{16}N_4O$ mw: 352.42

SYNS: BENZENEAZOBENZENEAZO-BETA-NAPHTHOL * CERASINROT * C.I. SOLVENT RED 23 * D & C RED NO. 17 * FETTSCHARLACH * 1-((4-(PHENYLAZO)PHENYL)AZO)-2-NAPHTHALENOL * OIL SCARLET * ORGANOL SCARLET * 1-(P-PHENYLAZOPHENYLAZO)-2-NAPHTHOL * ROUGE CERASINE * SOMALIA RED III * SUDAN III * TETRAZOBENZENE-BETA-NAPHTHOL * TONY RED

THR: Poison by intraperitoneal route. Moderately toxic by subcutaneous and intrapleural routes. An experimental carcinogen. When heated to decomposition it emits toxic fumes of NO_x.

OIL OF SASSAFRAS   HR: 3
CAS: 8006-80-2   NIOSH: RJ 3400000

PROP: Yellow to reddish-yellow liquid; characteristic odor and taste of sassafras. D: 1.065-1.077 @ 25°/25°. Very slightly sol in water, sol in 2 vols 90% alcohol. Keep well closed, cool and protected from light. 80% Safrol.

THR: Poison by ingestion. When heated to decomposition it emits acrid smoke and fumes.

OLEANDRIN   HR: 3
CAS: 465-16-7   NIOSH: RK 0175000
mf: $C_{32}H_{48}O_9$ mw: 576.80

PROP: Crystals from dil methanol; mp: 250°. Almost insol in water; sol in alcohol, chloroform.

SYNS: CORRIGEN * FOLIANDRIN

THR: Poison by subcutaneous and intravenous routes. When heated to decomposition it emits acrid smoke and fumes.

OLEFINS
THR: Unsaturated aliphatic hydrocarbons do not differ greatly from paraffins, particularly insofar as their toxic effect on working personnel is concerned. Ethylene and some of its homologs occur in manufactured and natural gases. Ethylene can be used as an anesthetic, and on inhalation in sufficient quantity it can be an asphyxiant. However, the greatest hazard from its use is the danger of fire and explosion. Prolonged or repeated exposures to high concentrations of various olefins have caused certain toxic effects in animals, such as liver damage and hyperplasia of the bone marrow (due to butene-2), but no corresponding effects have been discoverable in human beings due to industrial exposures. The diolefins, butadiene and isoprene, are more irritating than paraffins or mono-olefins of the same volatility. In general, it may be stated that the olefins are comparatively innocuous materials.

OLEIC ACID   HR: 3
CAS: 112-80-1   NIOSH: RG 2275000
mf: $C_{18}H_{34}O_2$ mw: 282.52

PROP: Colorless, odorless liquid when pure, mp: 6°, bp: 360.0°, flash p: 372°F (CC), d: 0.895 @ 25°/25°, autoign temp: 685°F, vap press: 1 mm @ 176.5°, bp: 286° @ 100 mm. Insol in water; misc in alc and ether.

SYNS: L'ACIDE OLEIQUE (FRENCH) * CIS-DELTA(SUP 9)-OCTADECENOIC ACID * CIS-OCTADEC-9-ENOIC ACID * CIS-9-OCTADECENOIC ACID * 9,10-OCTADECENOIC ACID

THR: Poison by intravenous route. Human skin irritant. An experimental carcinogen. Combustible when exposed to heat or flame. To fight fire, use CO_2, dry chemical. Incompatible with aluminum; perchloric acid.

cis-OLEIC ACID, METHYL ESTER   HR: 3
CAS: 112-62-9   NIOSH: RK 0895000
mf: $C_{19}H_{36}O_2$ mw: 296.55

PROP: Oil; d: 0.874 @ 20°/4°; bp: 168°-170°; insol in water; misc in alc and ether.

SYNS: METHYL-9-OCTADECENOATE * METHYL OLEATE * (Z)-9-OCTADECENOIC ACID,METHYL ESTER

THR: An experimental carcinogen. When heated to decomposition it emits acrid smoke and fumes.

OLIVE OIL   HR: 1
CAS: 8001-25-0   NIOSH: RK 4300000

PROP: Yellow oil, pleasing, delicate flavor; mp: −6°, flash p: 437°F (CC), autoign temp: 650°F, d: 0.909-0.915 @ 25°/25°. Becomes rancid on exposure to air. Slightly sol in alc; misc with ether, chloroform, carbon disulfide. From fruit of *Olea europaea*.

THR: Human skin irritant. Combustible when exposed to heat or flame; can react with oxidizing materials. Some spontaneous heating. To fight fire, use CO_2, dry chemical. When heated to decomposition it emits acrid smoke and fumes.

OMP-2   HR: 3
                            NIOSH: RK 6110000

PROP: Trialkyltin methacrylate polymer.

THR: Poison by ingestion and inhalation. A skin and eye irritant. See also tin compounds. When heated to decomposition it emits acrid smoke and fumes.

OPIUM   HR: 3
PROP: Air dried, milky exudation from incised, unripe capsules of *Papaver somniferum L.* or *P. album Mill.* Morphine is the most important alkaloid and occurs to the extent of 10%-16%.

SYN: GUM OPIUM

THR: Poison by ingestion. May lead to habituation and addiction. A narcotic, sedative, analgesic, hypnotic. Source of morphine, codeine, papaverine, etc. Can cause nausea, vomiting, constipation and respiratory problems. Combustible when exposed to heat or flame.

ORGANOMETALS   HR: 3
PROP: Compounds containing carbon and a metal. Ordinarily metallic carbonates (calcium carbonate, etc.) are excluded and also metallic salts of common organic acids. Examples of

organic metal compounds are Grignard compounds such as methyl magnesium iodide (CH_3MgI) and metallic alkyls such as butyllithium (C_4H_9Li). Also we have many organotin compounds such as monoalkyltins, monoaryltins, dialkyltins, diaryltins, trialkyltins, triaryltins, tetraalkyltins and tetraaryltins.

THR: Poison by ingestion and intravenous route. Irritant to skin, eyes, and mucous membranes. Can damage lung tissue and the liver. See also individual compounds. Triaryltins are most toxic as a group. Next are the dialkyltins and the monoalkyltins. In each major organotin group the ethyltin derivative is the most toxic, followed by the methyltins. This group of compounds is constantly growing in importance but there is relatively little toxicity information on most of them. Alkyl compounds of lead, tin, mercury and aluminum are known to be highly toxic. Less is known about other organometals, but for the most part they are highly reactive chemically and therefore dangerous, if only on direct contact. Until specific toxicity data become available, it is prudent to exercise great caution in handling organometals, particularly the alkyl forms.

ORPHENADRINE CITRATE   HR: 3
CAS: 4724-58-7   NIOSH: KR 6125000
mf: $C_{18}H_{23}NO \cdot C_6H_8O_7$ mw: 461.56

SYNS: N,N-DIMETHYL-2-((O-METHYL-ALPHA-PHENYL-BENZYL)OXY)-ETHYLAMINE CITRATE
* 2-DIMETHYLAMINOETHYL-2-METHYL-BENZHYDRYL ETHER CITRATE

THR: Poison by ingestion, intramuscular, and intravenous routes. When heated to decomposition it emits toxic fumes of NO_x.

ORPHENADRINE HYDROCHLO-
RIDE   HR: 3
CAS: 341-69-5   NIOSH: KR 6300000
mf: $C_{18}H_{23}NO \cdot ClH$ mw: 305.88

PROP: Crystals, mp: 156°-157°. Sol in water, alc, chloroform; sltly sol in acetone, benzene; almost insol in ether.

SYNS: 2-DIMETHYLAMINOETHYL-2METHYL-BENZHYDRYL ETHERHYDROCHLORIDE * N,N-DI-METHYL-2-(O-METHYL-ALPHA-PHENYLBENZ-YLOXY)ETHYLAMINE HYDROCHLORIDE
* MEPHENAMINE HYDROCHLORIDE

THR: Acute poison by ingestion. Poison by intravenous, intraperitoneal, and subcutaneous routes. An experimental teratogen. When heated to decomposition it emits toxic fumes of NO_x, HCl.

OSMIUM and COMPOUNDS   HR: 3
CAS: 7440-04-2   NIOSH: RN 1100000
af: Os   aw: 190.20

PROP: A lustrous, bluish-white, extremely hard and dense, brittle metal; bp: 5027°; d: 22.57; mp: approx 2700°.

SYN: METALLIC OSMIUM

THR: Poison by intravenous route. An irritant to eyes and mucous membranes. The principal effects of exposure are ocular disturbances and an asthmatic condition caused by inhalation. Furthermore, it causes dermatitis and ulceration of the skin upon contact. When osmium is heated, it gives off a pungent, poisonous fume of osmium tetraoxide. One case of osmium poisoning reported in the literature resulted from the inhalation of osmium tetraoxide, which gave rise to a capillary bronchitis and dermatitis. The vapor has a pronounced and nauseating odor which should be taken as a warning of a possibly toxic concentration in the atmosphere, and personnel should immediately remove to an area of fresh air. Osmium compounds, other than the tetraoxide, are probably safe, particularly as ordinarily handled in industry. The metal itself is not highly toxic. Flammable in the form of dust when exposed to heat or flame. Slight explosion hazard in the form of dust when exposed to heat or flame. Os dust reacts violently with ClF_3; OF_2. Incompatible with chlorine trifluoride; fluorine; phosphorus.

OSMIUM TETROXIDE   HR: 3
CAS: 20816-12-0   NIOSH: RN 1140000
DOT: 2471
mf: O_4Os mw: 254.20

PROP: (A) monoclinic, colorless crystals; (B) yellow mass, pungent, chlorine-like odor; mp (A): 39.5°, mp: (B): 41°, bp: 130° (sublimes), d: 4.906 @ 22°, vap press (A): 10 mm @ 26.0°, vap press (B): 10 mm @ 31.3°. Sol in benzene.

SYN: OSMIC ACID

OSHA PEL: TWA 2 ug/m^3
ACGIH TLV: TWA 0.0002 ppm; STEL 0.0006 ppm
DOT Classification: Poison B

THR: Poison by ingestion, inhalation, and intra-peritoneal route. A human eye irritant. Mutagenic data. When heated to decomposition it emits toxic fumes of Os.

OUABAIN   HR: 3
CAS: 630-60-4   NIOSH: RN 3675000
mf: $C_{29}H_{44}O_{12}$ mw: 584.73

PROP: A natural plant product.

SYNS: G-STROPHANTHIN * OUABAGENIN-L-RHAMNOSIDE

THR: Poison by ingestion, intramuscular, intra-peritoneal, intravenous, subcutaneous, and in-traduodenal routes. When heated to decomposition it emits acrid smoke and fumes.

OXALATES   HR: 3
PROP: Salts of oxalic acid.

THR: Poison by ingestion and inhalation. Powerful irritant. See also oxalic acid. Oxalates are corrosive to tissue and produce local irritation. When taken by mouth they have a caustic effect on the mouth, esophagus and stomach. The soluble oxalates are readily absorbed from the gastrointestinal tract and can cause severe damage to the kidneys. Dangerous; when heated to decomposition they emit toxic and irritating fumes.

OXALIC ACID   HR: 3
CAS: 144-62-7   NIOSH: RO 2450000
mf: $C_2H_2O_4$ mw: 90.04

PROP: Colorless rhombic; mp: 101°, 189° (an-hydrous), bp: sol in water; absolute alc and ether; sublimes @ 150°, d: 1.653.

SYNS: ACIDE OXALIQUE (FRENCH) * ACIDO OSSALICO (ITALIAN) * ETHANEDIOIC ACID * ETHANEDIONIC ACID * KYSELINA STAVE-LOVA (CZECH) * NCI-C55209 * OXAALZUUR (DUTCH) * OXALSAEURE (GERMAN)

OSHA PEL: TWA 1 mg/m³
ACGIH TLV: TWA 1 mg/m³; STEL 2 mg/m³

THR: Poison irritant to humans by ingestion. Acute oxalic poisoning results from ingestion of a solution of the acid. There is marked corrosion of the mouth, esophagus and stomach, with symptoms of vomiting, burning and abdominal pain, collapse and sometimes convulsions. Death may follow quickly. The systemic effects are attributed to the removal by the oxalic acid of the calcium in the blood. The renal tubules become obstructed by the insoluble calcium oxa-late, and there is profound kidney disturbance. The inhalation of the dust or vapor may cause chronic symptoms of irritation of the upper respiratory tract, gastrointestinal disturbances, al-buminuria, gradual loss of weight, increasing weakness and nervous system complaints. Oxalic acid has a caustic action on the skin and may cause dermatitis; a case of early gangrene of the fingers resembling that caused by phenol has been described. The chief effects of inhalation of the dusts or vapor are irritation of the eyes and upper respiratory tract, ulceration of the mucous membranes of the nose and throat, epistaxis, headache, irritation and nervousness. More severe cases may show albuminuria, chronic cough, vomiting, pain in the back and gradual emaciation and weakness. The skin lesions are characterized by cracking and fissuring of the skin and the development of slow-healing ulcers. The skin may be bluish in color, and the nails brittle and yellow. Violent reaction with furfuryl alcohol; Ag; $NaClO_3$; NaOCl. Incompatible with silver; sodium chlorite.

OXAMIDE   HR: 3
CAS: 471-46-5   NIOSH: RO 4900000
mf: $C_2H_4N_2O_2$ mw: 88.08

PROP: Triclinic needles; decomp @ 350°; d: 1.667 @ 20°/4°. Sltly sol in hot water, alc.

SYNS: AMID KYSELINY STAVELOVE (CZECH) * ETHANEDIAMIDE * OXALAMIDE * OXALIC ACID DIAMIDE * OXAMIMIDIC ACID

THR: Poison by intraperitoneal route. An eye irritant. When heated to decomposition it emits toxic fumes of NO_x.

1,2-OXATHIOLANE-2,2-DIOXIDE   HR: 3
CAS: 1120-71-4   NIOSH: RP 5425000
mf: $C_3H_6O_3S$ mw: 122.15

SYNS: 3-HYDROXY-1-PROPANESULPHONIC ACID SULTONE * 3-HYDROXY-1-PROPANESULPHONIC ACID SULFONE * 1-PROPANESULFONIC ACID-3-HYDROXY-GAMMA-SULTONE * 1,3-PROPANE SULTONE

THR: Poison by subcutaneous route. An experimental carcinogen. Mutagenic data. Implicated as a brain carcinogen. When heated to decomposition it emits toxic fumes of SO_x. For further information see 1,3-propane sultone, Vol. 4, No. 3 of *DPIM Report*.

OXATIMIDE   **HR: 3**
CAS: 60607-34-3   NIOSH: DE 2276000
mf: $C_{27}H_{30}N_4O$ mw: 426.61

SYNS: OXATOMIDA * 1-(3-(4-(DIPHENYL-
METHYL)-1-PIPERAZINYL)PROPYL)-1,3-DIHYDRO-
2H-BENZ IMIDAZOL-2-ONE

THR: Poison by ingestion and intravenous
routes. When heated to decomposition it emits
toxic fumes of NO_x.

OXETANE   **HR: 3**
CAS: 503-30-0   NIOSH: RQ 6825000
mf: C_3H_6O mw: 58.09

PROP: Oil, agreeable odor, d: 0.8930 @ 25°/
4°, bp: 480 @ 750 mm.

SYNS: 1,3-PROPYLENE OXIDE * TRIMETHYL-
ENOXID (GERMAN) * TRIMETHYLENE OXIDE

THR: An experimental carcinogen. Moderate
acute toxicity by subcutaneous route. May be
narcotic in high concentrations. When heated
to decomposition it emits acrid smoke and
fumes.

2-OXETANONE   **HR: 3**
CAS: 57-57-8   NIOSH: RQ 7350000
mf: $C_3H_4O_2$ mw: 72.07

SYNS: HYDRACRYLIC ACID,BETA,LACTONE
* 3-HYDROXYPROPIONIC ACID LACTONE
* PROPIOLACTONE * BETA-PROPIOLACTONE
* BETA-PROPIONOLACTONE

THR: Poison by intraperitoneal and intravenous
routes. An experimental carcinogen, tumorigen,
and neoplastigen. Mutagenic data. When heated
to decomposition it emits acrid smoke and
fumes. For further information see Vol. 5, No.
6 of *DPIM Report*.

**2-(3-OXO-1-INDANYLIDENE)-1,3-INDAN-
DIONE**   **HR: 3**
CAS: 1707-95-5   NIOSH: NK 6050000
mf: $C_{18}H_{10}O_3$ mw: 274.28

SYN: BINDON

THR: Acute poison by intraperitoneal route. An
experimental teratogen. When heated to decomposition it emits acrid smoke and fumes.

OXOLAMINE CITRATE   **HR: 3**
CAS: 959-14-8   NIOSH: RO 0720000
mf: $C_{14}H_{19}N_3O \cdot C_6H_8O_7$ mw:437.50

PROP: Crystals, sltly sol in water, alc.

SYNS: 5-BETA-DIETHYLAMINOETHYL-3-PHENYL-
1,2,4-OXADIAZOLE CITRATE * 3-PHENYL-5-
(BETA-(DIETHYLAMINO)ETHYL)-1,2,4-OXADIA-
ZOLE CITRATE

THR: An experimental carcinogen and teratogen. When heated to decomposition it emits
toxic fumes of NO_x.

4,4-OXYDIANILINE   **HR: 3**
CAS: 101-80-4   NIOSH: BY 7900000
mf: $C_{12}H_{12}N_2O$ mw: 200.26

PROP: Colorless crystals; mp: 187°; bp: > 300°.

SYNS: P-AMINOPHENYL ETHER * 4-AMINO-
PHENYL ETHER * BIS(4-AMINOPHENYL)ETHER
* BIS(P-AMINOPHENYL)ETHER * 4,4′-DIAMI-
NOBIPHENYLOXIDE * DIAMINODIPHENYL
ETHER * 4,4-DIAMINODIPHENYL ETHER
* P,P′-DIAMINODIPHENYL ETHER * 4,4′-DI-
AMINODIPHENYL OXIDE * 4,4′-DIAMINOPHE-
NYL ETHER * NCI-C50146 * OXYBIS(4-
AMINOBENZENE) * 4,4′-OXYBISANILINE
* P,P′-OXYBIS(ANILINE) * 4,4′-OXYBISBENZ-
ENAMINE * OXYDIANILINE * P,P′-OXYDIANI-
LINE * 4,4′-OXYDIPHENYLAMINE * OXYDI-
P-PHENYLENEDIAMINE

THR: Poison by intraperitoneal route. Moderately toxic by ingestion. Mutagenic data. An
experimental carcinogen, tumorigen. When
heated to decomposition it emits toxic fumes
of NO_x.

OXYDISULFOTON   **HR: 3**
CAS: 2497-07-6   NIOSH: TD 8600000
mf: $C_8H_{19}O_3PS_3$ mw: 290.42

SYNS: BAY 23323 * O,O-DIETHYL-
S-((ETHYLSULFINYL)ETHYL)PHOSPHORODITHIOATE
* O,O-DIETHYL S-(2-(ETHYLSULFINYL)ETHYL)
PHOSPHORODITHIOATE * DISULFOTON DISULIDE
* DISULFOTON SULFOXIDE * DISYSTON SUL-
FOXIDE * ETHYLTHIOMETON SULFOXIDE

THR: Poison by ingestion and skin contact.
When heated to decomposition it emits very
toxic fumes of SO_x and PO_x.

OXYGEN   **HR: 3**
CAS: 7782-44-7   NIOSH: RS 2060000
DOT: 1072/1073
mf: O_2 mw: 32.00

PROP: Colorless, odorless, tasteless gas or liquid or hexagonal crystals. Supports combustion;

d (liquid): 1.14 @ −183.0°, d (solid): 1.426 @ −252.5°, vap d: 1.429 @ 0°. d: (gas) 1.429 g/L @ 0°; mp: −218.4°; bp: −182.96°. One vol gas dissolves in 32 vols water @ 20°; dissolves in 7 vols alc @ 20°. Sol in other organic liquids to a greater extent than water.

DOT Classification: Nonflammable Gas, Label: Oxidizer

THR: Not toxic as gas. In liquid form it can cause severe "burns" and tissue damage on contact with the skin due to extreme cold. Mutagenic data. Though itself nonflammable, it is essential to combustion. Exclusion of O_2 from the neighborhood of a fire is one of the principal methods of extinguishment. Avoid smoking, flames, electric sparks. Liquid O_2 can explode on contact with readily oxidizable material, especially at high temperature, under the proper conditions of temperature, pressure, and reagent concentration it can react violently with Al; $Al(BH_4)_3$; AlH_3; $Be(BH_4)_2$; BAs_2Br_3; B_2H_{10}; BCl_3; CsH; [butane + $Ni(CO)_4$]; Ca; Cs; Ca_3P_2; (chlorotrifluoroethylene + Br_2); $C_{10}H_{14}$; B_2H_6; Ge; ethyl ether; diphenyl ethylene; disilane; ethers; hydrazine; H_2; Li; ($Ni(CO)_4$ + butane); organic matter; (OF_2 + H_2O); PH_3; P; PF_3; P_2O_3; polyurethane; polyvinyl chloride; (O_2 + CO); K_2O_2; Rb; Se; NaH; teflon; tetrafluorohydrazine; tetrasilane; Ti alloy; 1,1,1-trichloroethane; trichloroethylene; trisilane; asphalt; benzene; CO; CCl_4; chlorinated hydrocarbons; cyanogen; fuels; hydrocarbons; LiH; Mg; CH_4; CH_2Cl_2; oil; paraformaldehyde; wood; charcoal. Compressed O_2 is shipped in steel cylinders under high pressure. If these containers are broken due to shock or exposed to high temperature, an explosion and fire may result. Incompatible with acetaldehyde; alcohols; alkali metals; alkaline earth metals; aluminum-titanium alloys; biological matter, ether; carbon disulphide, mercury, anthracene; diboron tetrafluoride; ethers; fibrous fabrics; halocarbons; hydrocarbons; hydrogen; mercury, tetracarbonylnickel; metal hydrides; non-metal hydrides; organic analytical samples; phosphorus trioxide; polymers; rhenium; tetrafluoroethylene; titanium; tricalcium diphosphide. Liquid incompatibilities: acetone; asphalt; carbon; halocarbons; hydrocarbons; liquefied gases; lithium hydride; metals; 1,3,5-trioxane (paraformaldehyde); (wood + charcoal).

OXYGEN DIFLUORIDE   HR: 3
CAS: 7783-41-7   NIOSH: RS 2100000

DOT: 2190
mf: F_2O mw: 54.00

PROP: Colorless gas, yellowish-brown liquid. Reacts slowly with water, d: (liq) 1.90 @ −224°, mp: −223.8°, bp: −144.8°.

SYNS: FLUORINE MONOXIDE * FLUORINE OXIDE

OSHA PEL: TWA 100 ug/m³
DOT Classification: Poison A Gas

THR: Poison irritant to skin, eyes, and mucous membranes. Highly corrosive to tissue. Attacks lungs with delayed appearance of symptoms. See also fluorides. A very powerful oxidizer. Must be kept away from contact with reducing agents. Incompatible with $AlCl_3$; NH_3; As_2O_3; Br_2; CO; Cl_2; (Cl_2 + Cu); CrO_3; H_2; H_2S; I; Ir; CH_4; O_3; (O_2 + H_2O); Pd; P_2O_5; Pt; Rh; Ru; SiO_2. Dangerous; when heated to decomposition it emits highly toxic fumes of fluorine. Incompatible with adsorbents; combustible gases; phosphorus (V) oxide; halogens or metal halides; metals; nitrogen oxide; nitrosyl fluoride; non-metals; water.

OXYPHENONIUM BROMIDE   HR: 3
CAS: 50-10-2   NIOSH: BP 7625000
mf: $C_{21}H_{34}NO_3$ • Br   mw: 428.47

SYNS: ANTRENYL BROMIDE * C 5473 * DIETHYL(2-HYDROXYETHYL)METHYLAMMONIUM BROMIDE-ALPHA-PHENYLCYCLO-HEXANE-GLYCOLATE

THR: Poison by ingestion, intravenous, and subcutaneous routes. When heated to decomposition it emits very toxic fumes of NO_x and Br^-.

beta-OXYPROPYLPROPYLNITROS-AMINE   HR: 3
CAS: 39603-54-8   NIOSH: UI 4550000
mf: $C_6H_{12}N_2O_2$ mw: 144.20

SYNS: N-NITROSO-2-OXO-N-PROPYL-N-PROPYL-AMINE * 1-(NITROSOPROPYLAMINO)-2-PROPANONE * 2-OXI-PROPYL-PROPYLNITROSAMIN (GERMAN) * 2-OXO-PROPYL-PROPYLNITROSA-MINE * (2-OXOPROPYL)PROPYLNITROSOAMINE

THR: An experimental carcinogen and neoplastigen. Mutagenic data. Moderately toxic by subcutaneous route. See also nitrosamines. When heated to decomposition it emits toxic fumes of NO_x.

OZONE   **HR: 3**
CAS: 10028-15-6   NIOSH: RS 8225000
mf: O_3 mw: 48.00

PROP: Colorless gas or dark blue liquid. Unstable; mp: $-193°$, bp: $-111.9°$, d (gas): 2.144 g/L, 1.71 @ $-183°$, d: (liq) 1.614 g/mL @ $-195.4°$.

SYNS: OZON (POLISH) * TRIATOMIC OXYGEN

OSHA PEL: TWA 200 ug/m^3
ACGIH TLV: TWA 0.1 ppm; STEL 0.3 ppm
DFG MAK: 0.1 ppm (0.2 mg/m^3)

THR: Poison by inhalation and very irritating to skin, eyes, upper respiratory system, and mucous membranes. Can be a safe water disinfectant in low concentration. An experimental neoplastigen by inhalation. Concentration of 0.015 ppm of ozone in air produces a barely detectable odor. Concentration of 1 ppm produce a disagreeable sulfur-like odor and may cause headache and irritation of eyes and the upper respiratory tract; symptoms disappear after leaving the exposure. No systemic effects have been reported following industrial exposures. Ozone is a common air contaminant. Affects central nervous system. Mutagenic data. Powerful oxidizing agent. Dangerous chemical reaction with aniline; C_6H_6; Br_2; (diallyl methyl carbinol + acetic acid); diethyl ether; N_2O_5; ethylene; HBr; HI; NO_2; NO; NCl_3; NI_3; nitroglycerin; organic liquids; organic matter; Sb. Severe explosion hazard in liquid form, when shocked, exposed to heat or flame or in concentrated form by chemical reaction with powerful reducing agents. Incompatible with alkenes; aromatic compounds; benzene, rubber; bromine; dicyanogen; diethyl ether; dinitrogen tetraoxide; hydrogen bromide; 4-hydroxy-4-methyl-1,6-heptadiene; nitrogen trichloride; stibine; tetrafluorohydrazine. For further information see Vol. 1, No. 2 of *DPIM Report*.

OZONE MIXED WITH NITROGEN OXIDES (53%:47%)   **HR: 3**
NIOSH: RS 8250000

SYN: NITROGEN OXIDES MIXED WITH OZONE (47%:53%)

THR: Poison by inhalation. Human central nervous system effects.

P

PALLADIUM HR: 2
af: Pd   aw: 106.4

PROP: A steely white, stable metal, can be annealed to be soft and ductile. Mp: 1555°; bp: 3167°; d: 12.02 @ 20°/4. Volatile at high temps.

THR: This metal in the form of palladium chloride has been administered by ingestion in dosage of about 1 grain daily in the treatment of tuberculosis without apparent ill effects. In the laboratory, palladium appears to bind to many cell components; blocks the action of a number of enzymes and interferes with the use of energy by nerves and muscles; induces lung malfunction and produces abnormal fetuses. Lethal intravenous doses cause appetite loss, hemolysis, renal deposition and bone marrow damage. Palladium dust can be a fire and explosion hazard. Combustible in the form of dust, when exposed to heat, or flame. Violent reaction with Al; H_2 with isopropyl alcohol; OF_2; S. Incompatible with arsenic; carbon; ozonides; sodium tetrahydroborate; sulfur.

PALLADIUM (2+) CHLORIDE HR: 3
CAS: 7657-10-1   NIOSH: RT 3500000
mf: Cl_2Pd mw: 177.30

PROP: Dark brown deliquescent crystals, sol in water, alc, acetone and hydrochloric acid. D: 4.0 @ 18°, mp: 678°-680° (decomp).

SYNS: NCI-C60184 * PALLADOUS CHLORIDE

THR: Poison by ingestion, intraperitoneal, intravenous, and intratracheal routes. An experimental carcinogen. A skin irritant. See also palladium. When heated to decomposition it emits highly toxic fumes of Cl^-.

PALMITIC ACID HR: 3
CAS: 57-10-3   NIOSH: RT 4550000
mf: $C_{17}H_{32}O_2$ mw: 256.48

PROP: Colorless plates; d: 0.849 @ 70°/4°; mp: 63°-64°; bp: 271.5° @ 100 mm; insol in water; very sltly sol in petroleum ether; sol in absolute ether and chloroform.

SYNS: HEXADECANOIC ACID * N-HEXADECOIC ACID * 1-PENTADECANECARBOXYLIC ACID

THR: Acute poison by intravenous route. An experimental neoplastigen. Human skin irritant.

When heated to decomposition it emits acrid smoke and fumes.

PAPAIN HR: 3
CAS: 9001-73-4   NIOSH: RU 4950000

PROP: White to gray, slightly hygroscopic powder, sol in water and glycerin, insol in other common organic solvents; the most thermostatic enzyme known; digests protein. Isolated from the latex of the green fruit and leaves of *Carcia Papaya L.*

SYNS: CAROID * PAPAYOTIN * VEGETABLE PEPSIN

THR: Poison by intraperitoneal route. Moderately toxic by ingestion. An allergen. For further information see Vol. 1, No. 7 of *DPIM Report*.

PAPAVERIN CARBOXYLIC ACID, SODIUM SALT HR: 3
NIOSH: NW 7160000
mf: $C_{21}H_{20}NO_6 \cdot Na$ mw: 405.41

SYNS: 6,7-DIMETHOXY-1-VERATRYLISOQUINOLINE-3-CARBOXYLIC ACID SODIUM SALT * 1-(3,4)-DIMETHOXYBENZYL-6,7-DIMETHOXYISOQUINOLINE-3-CARBOXYLIC ACID, SODIUM SALT

THR: Poison by ingestion, intravenous, and subcutaneous routes. Causes blood pressure effects in man. When heated to decomposition it emits toxic fumes of NO_x and Na_2O.

PAPAVERINE HR: 3
CAS: 58-74-2   NIOSH: NW 8450000
mf: $C_{20}H_{21}NO_4$ mw: 339.42

PROP: Colorless, rhombic needles. Mp: 147°, bp: decomp, d: 1.337 @ 20°/4°. Insol in water; sol in hot benzene, glacial acetic acid, acetone, sltly sol in chloroform, carbon tetrachloride, petroleum ether.

SYNS: 1-((3,4-DIMETHOXYPHENYL)METHYL)-6,7-DIMETHOXYISOQUINOLINE * 6,7-DIMETHOXY-1-VERATRYLISOQUINOLINE * PAPANERINE

THR: Poison by ingestion, subcutaneous, intraduodenal, intraperitoneal, and intravenous routes. Considered to be a comparatively mild poison. Its central nervous system action is about midway between morphine and codeine, and even large doses fail to produce the amount of excitement caused by codeine or the soporific action of morphine. See also morphine. Combustible. When heated to decomposition it emits toxic fumes of NO_x.

PARAFFIN   HR: 3
CAS: 8002-74-2   NIOSH: RV 0350000

PROP: Colorless or white, translucent odorless mass; d: approx 0.90; mp: 50°-57°. Insol in water, alc; sol in benzene, chloroform, ether, carbon disulfide, oils; misc with fats.

THR: The effects of the paraffin hydrocarbons vary with the volatility. The gaseous hydrocarbons, such as methane, ethane, etc., have but slight anesthetic effects and are hazardous only when present in sufficient concentration to dilute the oxygen to a point below that which is necessary to sustain life. With the volatile liquid hydrocarbons, or with the next higher fraction, the anesthetic action predominates, and with the higher molecular weights or with the less volatile compounds, the anesthetic increases, but at the same time an irritant action becomes more pronounced. For information concerning toxic and hazardous properties of these materials, see the individual compounds. The semi-refined, fully refined, and the crude paraffins are experimental tumorigens. See also paraffin waxes and petroleum waxes. Paraffins are common air contaminants. Can be a dangerous fire hazard depending on volatility. For further information see Vol. 1, No. 7 of DPIM Report.

PARAFFIN WAX (FUME)   HR: 3
PROP: Colorless or white, somewhat translucent, odorless mass, greasy feel. Mp: 43.3°, bp: < 370°, flash p. 390°F (CC), d: 0.9, autoign temp: 473°.

THR: These materials are experimental carcinogens of the lung, stomach and skin. Combustible when exposed to heat or flame; can react with oxidizing materials. To fight fire, use CO_2, dry chemical.

PARAFORMALDEHYDE   HR: 2
CAS: 30525-89-4
mf: $(CH_2O)_n$

PROP: White crystals, odor of formaldehyde. Sltly sol in cold water, mod sol in hot water yielding formaldehyde; flash p: 158°F; autoign temp: 572°F.

THR: Moderately toxic by ingestion. Mildly toxic by skin contact. See formaldehyde. Flammable when exposed to heat or flame. Moderately dangerous; when heated, it emits formaldehyde gas and oxides of carbon; can react with oxidizing materials. To fight fire, use alcohol foam, CO_2, dry chemical. Incompatible with liquid O_2. For further information see Vol. 3, No. 3 of DPIM Report.

PARALDEHYDE   HR: 2
CAS: 123-63-7   NIOSH: YK 0525000
DOT: 1264
mf: $C_6H_{12}O_3$ mw: 132.18

PROP: Colorless liquid, disagreeable taste, aromatic odor. Mp: 12.6°, lel = 1.3%, bp: 124.4° @ 752 mm, flash p: 62.6°F, d: 0.9943 @ 20°/4°, autoign temp: 460°F, vap d: 4.55. Sol in water; misc with alc, ether, oils, chloroform.

SYNS: ELALDEHYDE * PARACETALDEHYDE * PARALDEHYD (GERMAN) * PARALDEIDE (ITALIAN) * TRIACETALDEHYDE (FRENCH) * 2,4,6-TRIMETHYL-1,3,5TRIOXAAN (DUTCH) * 2,4,6-TRIMETHYL-S-TRIOXANE * 2,4,6-TRIMETHYL-1,3,5TRIOXANE * S-TRIMETHYLTRIOXYMETHYLENE

DOT Classification: Label: Flammable Liquid

THR: Moderately toxic by inhalation, ingestion, and several other routes. A severe eye and moderate skin irritant. Mutagenic data. Low doses produce hypnotic and analgesic effects. Larger doses depress the nervous system with loss of reflexes, coma, and respiratory depression leading to respiratory paralysis and death. Chronic effects include weight loss, muscular weakness, and mental fatigue. However, poisoning is rare. Dangerous fire hazard when exposed to heat, flame, or oxidizers. Slight explosion hazard when exposed to heat or flame. Dangerous; keep away from heat and open flame; emits toxic fumes on heating; can react vigorously with oxidizing materials. To fight fire, use alcohol foam, CO_2, dry chemical. Incompatible with alkalies; hydrocyanic acid; iodides; oxidizers; nitric acid. For further information see Vol. 5, No. 6 of DPIM Report.

PARAQUAT
CAS: 1910-4-2-5   NIOSH: DW 2275000
mf: $C_{12}H_{14}N_2 \cdot 2Cl$ mw: 257.18

PROP: Yellow solid. Sol in water.

SYNS: N,N'-DIMETHYL-4,4'-BIPYRIDINIUM DI-
CHLORIDE * 1,1'-DIMETHYL-4,4'-DIPYRIDY-
LIUM CHLORIDE * PARAQUAT CHLORIDE

OSHA PEL: TWA 500 ug/m^3 (skin)
DFG MAK: 0.1 mg/m^3

THR: Poison by ingestion, inhalation, skin contact, intraperitoneal, intravenous, and subcutaneous routes. An experimental teratogen. Has
a delayed damaging effect on the lung alveoli.
Has caused fatal poisoning in humans with severe injury to lungs. Has been implicated in
aplastic anemia. An eye irritant. Mutagenic
data. Dangerous disaster hazard; see chlorides.
The National Institute of Drug Abuse (NIDA),
USA, has concluded that contamination of marihuana with the herbicide paraquat may pose a
serious threat to marihuana (cannabis) smokers,
and issued a warning that marihuana contaminated with the herbicide paraquat could lead
to permanent lung damage for regular and heavy
users of marihuana, and conceivably other users
as well. The maximum level of contamination
that is permitted for domestic uses is 0.05 ppm;
but paraquat has been found in marihuana samples at levels ranging from 3 ppm to 2,204
ppm, averaging 452 ppm. It tends to concentrate
in lung tissue whether it is ingested or inhaled,
and produces a condition called fibrosis which
reduces the capacity of the lung to absorb oxygen. *Inhalation* of paraquat creates a greater
risk of lung damage than *ingestion* of an identical
amount.

PARASCORBIC ACID   HR: 3
CAS: 10048-32-5   NIOSH: UQ 0525000
mf: C$_6$H$_8$O$_2$ mw: 112.14

PROP: Oily liquid, sweet, aromatic odor. Bp:
104°-105° @ 14 mm, 119°-123° @ 22 mm; d:
1.079 @ 18°/4°. Sol in water, very sol in alc,
ether.

SYNS: (s)-(+)-5,6-DIHYDRO-6-METHYL-2H-PY-
RAN-2-ONE * GAMMA-HEXENOLACTONE
* 2-HEXEN-5,1-OLIDE * 5-HYDROXY-2-HEXE-
NOIC ACID LACTONE * PARASORBIC ACID
* (+)-PARASORBINSAEURE (GERMAN) * SCOR-
BIC OIL

THR: Poison by intraperitoneal and intravenous
routes. An experimental neoplastigen. Heating
causes emission of irritating fumes. When

heated to decomposition it emits acrid smoke
and fumes.

PARATHIAZINE   HR: 3
CAS: 84-08-2   NIOSH: SP 4725000
mf: C$_{18}$H$_{20}$N$_2$S   mw: 296.46

SYNS: PYRATHIAZINE * PYRROLAZOATE
* PYRROLIDINO-AETHYLPHENTHIAZIN (GERMAN)
* 10-(2-(1-PYRROLIDINYL)ETHYL)PHENOTHIA-
ZINE * 10-(2-(1-PYRROLIDYL)ETHYL)PHENO-
THIAZINE

THR: Poison by intraperitoneal route. Moderately toxic by ingestion and subcutaneous
routes. When heated to decomposition it emits
very toxic fumes of NO$_x$ and SO$_x$.

PARATHION   HR: 3
CAS: 56-38-2   NIOSH: TF 4550000
DOT: 2783
mf: C$_{10}$H$_{14}$NO$_5$PS   mw: 291.28

PROP: Pale-yellow liquid. Bp: 375°; mp: 6°.
Very sol in alcs, esters, ethers, ketones, aromatic hydrocarbons; insol in water, petroleum
ether, kerosene.

SYNS: O,O-DIAETHYL-O-(4-NITROPHENYL)-
MONOTHIOPHOSPHAT (GERMAN) * O,O-DI-
ETHYL-O-(4-NITRO-FENIL)-MONOTHIOFOSFAAT
(DUTCH) * O,O-DIETHYL-O-P-NITROFENYLESTER
KYSELINYTHIOFOSFORECNE (CZECH) * O,O-DI-
ETHYL-0-4-NITROPHENYLPHOSPHOROTHIOATE
* DIETHYL-P-NITROPHENYLTHIOPHOSPHATE
* O,O-DIETHY-O-(P-NITROPHENYL) PHOSPHO-
ROTHIOATE * O,O-DIETHYL-O-(4-NITROPHENYL)
PHOSPHOROTHIOATE * DIETHYL 4-NITROPHENYL
PHOSPHOROTHIONATE * O,O-DIETHYL O-P-NI-
TROPHENYL THIOPHOSPHATE * DIETHYL P-NI-
TROPHENYLTHIONOPHOSPHATE * O,O-DI-
ETHYL-O-(P-NITROPHENYL)THIONOPHOSPHATE
* O,O-DIETHYL O-4-NITROPHENYL THIOPHOS-
PHATE * DIETHYLPARATHION * O,O-DIETIL-
O-(4-NITRO-FENIL)-MONOTIOFOSFATO (ITALIAN)
* O,O-DIETHYL-O-P-NITROFENYLTIOFOSFAT
(CZECH) * ENT 15,108 * NCI-C00226
* NITROSTIGMINE * PARATHENE * PARA-
THION, LIQUID (DOT) * PHOSPHOSTIGMINE
* STABILIZED ETHYL PARATHION * THIOPHOS-
PHATE DE O,ODIETHYLE ET DE O-(4-NITROPHE-
NYLE) (FRENCH)

OSHA PEL: TWA 110 ug/m^3 (skin)
ACGIH TLV: TWA 0.1 mg/m^3
DFG MAK: 0.1 mg/m^3
DOT Classification: Poison B, Label: Poison

THR: A deadly poison by all routes. Human central nervous system effects by ingestion. Mutagenic data. An experimental tumorigen and carcinogen. A cholinesterase inhibitor. Parathion, like other organic phosphorus poisons, acts as an irreversible inhibitor of the molecules of the enzyme cholinesterase and thus allows the accumulation of large amounts of acetylcholine. When a critical level of cholinesterase depletion is reached, grave symptoms appear. Whether death is actually caused entirely by cholinesterase depletion or by the disturbance of a number of enzymes is not yet known. Recovery is apparently complete if a poisoned animal or man has time to reform his critical quota of cholinesterase. The organism exposed remains susceptible to relatively low dosages of parathion until the cholinesterase level has regenerated. Small doses at frequent intervals are, therefore, more or less additive. There is not, however, at the present time, any indication that, when recovery from a given exposure is entirely complete, the exposed organism is prejudiced in any way. Combustible when exposed to heat or flame. Decomposed by heat. Violent reaction with endrin. Highly dangerous; i.e., shock can shatter the container, releasing the contents. When heated to decomposition it emits highly toxic fumes of NO_x, PO_x, SO_x. For further information see Vol. 3, No. 3 of *DPIM Report*.

PARATHION MIXTURE, DRY (DOT)   HR: 3
CAS: 56-38-2   NIOSH: TF 4920000
DOT: 2783

SYN: O,O-DIETHYL-O-(P-NITROPHENYL) ESTER (DRY MIXTURE) PHOSPHOROTHIOIC ACID

DOT Classification: Label: Poison

THR: A powerful poison. See also parathion.

PARTERGIN   HR: 3
CAS: 20313-98-8   NIOSH: KE 5425000
mf: $C_{20}H_{25}N_3O_2$ mw: 339.48

SYNS: 9,10-DIDEHYDRO-N-(ALPHA-(HYDROXY-METHYL)PROPYL)-6-METHYL-ERGOLINE-8-BETA-CARBOXAMIDE * METHYLERGOBASINE * METHYLERGOBREVIN * METHYLERGOMETRINE * METHYLERGONOVINE * METHYLERGOMETRIN

THR: Poison by ingestion and intravenous

routes. When heated to decomposition it emits toxic fumes of NO_x.

PEANUT OIL   HR: 2
CAS: 8002-03-7   NIOSH: RX 2830000

PROP: Straw-yellow to greenish-yellow or nearly colorless oil, nutty odor and bland taste. Mp: 2.7°, flash p: 540°F, d: 0.92, autoign temp: 833°F. Misc with ether, petroleum ether, chloroform, carbon disulfide; sol in benzene, carbon tetrachloride, oils; very sltly sol in alc. From seed of *Arachis hypogaea*.

SYNS: ARACHIS OIL * EARTHNUT OIL * GROUNDNUT OIL * KATCHUNG OIL * PECAN SHELL POWDER

THR: An irritant and mild allergen. Mutagenic data. Combustible when exposed to heat or flame; can react with oxidizing materials. Slt spontaneous heating. To fight fire, use CO_2, dry chemical.

PEG-9 NONYL PHENYL ETHER   HR: 2
CAS: 9016-45-9   NIOSH: MD 0905000

SYNS: IGEPAL CO-630 * NEUTRONYX 600 * POLYOXYETHYLENE (9) NONYL PHENYL ETHER * POLYETHYLENE GLYCOL 450 NONYL PHENYL ETHER * TERGITOL TP-9 (NONIONIC)

THR: Moderately toxic by ingestion and skin contact. A severe eye and mild skin irritant in humans. When heated to decomposition it emits acrid smoke and fumes.

PENFLURIDOL   HR: 3
CAS: 26864-56-2   NIOSH: TN 6957000
mf: $C_{28}H_{27}ClF_5NO$ mw: 524.01

PROP: White, microcrystals. Mp: 105°-107°. Sltly sol in water.

SYN: 4-(4-CHLORO-ALPHA,ALPHA,ALPHA-TRI-FLUORO-M-TOLYL)-1-(4,4-BIS(P-FLUORO-PHENYL)BUTYL)-4-PIPERIDINOL

THR: Poison by ingestion and intravenous routes. When heated to decomposition it emits very toxic fumes of HCl, F^- and NO_x.

PENICILLAMINE HYDROCHLORIDE   HR: 3
CAS: 2219-30-9   NIOSH: YV 9450000
mf: $C_5H_{11}NO_2S \cdot ClH$ mw: 185.69

SYNS: D-3-MERCAPTOVALINE HYDROCHLORIDE * USAF A-1705

THR: An experimental teratogen. Moderately toxic by intraperitoneal route. Human skin effects (systemic). When heated to decomposition it emits very toxic fumes of NO_x, SO_x, and HCl.

PENICILLIC ACID   HR: 3
CAS: 90-65-3   NIOSH: MM 2625000
mf: $C_8H_{10}O_4$ mw: 170.18

PROP: Needles from petroleum ether. Mp: 83°-84°. Sltly sol in cold water, hot petroleum ether; very sol in hot water, alc, ether, benzene chloroform; insol in pentane,hexane.

SYNS: 3-METHOXY-5-METHYL-4-OXO-2,5-HEXA-DIENOIC ACID * PENCILLIC ACID

THR: Poison by intravenous, subcutaneous, and intraperitoneal routes. An experimental neoplastigen. Moderately toxic by ingestion. Mutagenic data. When heated to decomposition it emits acrid smoke and fumes.

PENICILLIN   HR: 3
CAS: 1406-05-9   NIOSH: RY 4375000
mf: $(CH_3)_2C_5H_3NSO(COOH)NHCOOR$ (bicyclic).

PROP: A group of isomeric and closely related antibiotic compounds with outstanding bacterial activity. An extract from *Penicillium Notatum*.

SYN: PENIZILLIN (GERMAN)

THR: Poison by intraperitoneal and subcutaneous routes. Moderately toxic by intravenous route. Has been implicated in aplastic anemia. When heated to decomposition it emits very toxic fumes of NO_x and SO_x.

PENTABORANE(9)   HR: 3
CAS: 19624-22-7   NIOSH: RY 8925000
DOT: 1380
mf: B_5H_9 mw: 63.14

PROP: Colorless gas or liquid, bad odor. Mp: −46.6°, d: 0.61 @ 0°, vap d: 2.2; vap press: 66 mm @ 0°, lel = 0.42%, bp: 60°.

SYN: PENTABORANE (DOT)

OSHA PEL: TWA 10 ug/m³
ACGIH TLV: TWA 0.005 ppm; STEL 0.015 ppm
DFG MAK: 0.005 ppm (0.01 mg/m³)
DOT Classification: Label: Poison and Flammable Liquid

THR: Poison by inhalation and intraperitoneal routes. Dangerous fire hazard by chemical reaction; spontaneously flammable in air. Dangerous explosion hazard (details unknown). Dangerous; on decomposition it emits toxic fumes and can react vigorously with oxidizing materials. To fight fire, use special fire-fighting materials; water is not effective; reacts violently with halogenated extinguishing agents. Get instructions from supplier. Incompatible with dimethyl sulfoxide.

PENTABORANE (11)
mf: B_5H_{11} mw: 65.13

PROP: Liquid. Mp: −123°; bp: 63°; unstable.

SYN: DIHYDROPENTABORANE

THR: No toxicity data. See also boron compounds, pentaborane (9). Fire hazard; air-ignition. When heated to decomposition it emits toxic fumes of BO_x. Can react vigorously with oxidizing materials. To fight fire, get instructions from supplier.

PENTACARBONYLIRON   HR: 3
                        NIOSH: NO 4900000
mf: $\overset{\circ}{C}_5FeO_5$ mw: 195.85

PROP: Yellow-dark red viscous liquid. Mp: −25°; bp: 103.0°; flash p: 5°; d: 1.453 @ 25°/4°; vap press: 40 mm @ 30.3°.

SYNS: IRON PENTACARBONYL * IRON CARBONYL

THR: Poison by inhalation, ingestion, skin contact, and intravenous routes. Inhalation causes dizziness, nausea, vomiting, with continued exposure causing unconsciousness, difficult breathing, cyanosis, and circulatory collapse. In fatal cases, death occurs from the fourth to eleventh day with pneumonitis and injury to kidneys, liver, and brain. Dangerous fire hazard; see carbonyls. Pyrophoric in air. When heated to decomposition it emits toxic fumes of CO. To fight fire, use water, foam, CO_2, dry chemical. Incompatible with acetic acid; water; nitrogen oxide; transition metal halides; zinc.

PENTACHLOROACETO-PHENONE   HR: 3
                        NIOSH: AM 9658000
mf: $C_8Cl_5H_3O$ mw: 221.6

SYN: 2′,3′,4′,5′,6′- PENTACHLOROACETOPHE-NONE

THR: Poison by intraperitoneal route. Moderately toxic by ingestion and skin contact. When heated to decomposition it emits toxic fumes of Cl^-.

PENTACHLOROETHANE   HR: 3
CAS: 76-01-7   NIOSH: KI 6300000
DOT: 1669
mf: C_2HCl_5 mw: 202.28

PROP: Colorless liquid, chloroform-like odor; mp: $-29°$, bp: $161°-162°$, d: 1.6728 @ $25°/4°$. Insol in water; misc in alc and ether.

SYNS: ETHANE PENTACHLORIDE * NCI-C53894 * PENTACHLOORETHAAN (DUTCH) * PENTACHLORAETHAN (GERMAN) * PENTACHLORETHANE (FRENCH) * PENTACLOROETANO (ITALIAN) * PENTALIN

DFG MAK: 5 ppm (40 mg/m^3)
DOT Classification: Poison B

THR: Poison by inhalation and intravenous routes. Moderately toxic by ingestion and subcutaneous routes with narcotic effects. An irritant. Flammable when exposed to heat or flame. Moderately explosive by spontaneous chemical reaction. Dehalogenation by reaction with alkalies, metals, etc.; will produce spontaneously explosive chloroacetylenes. Violent reaction with NaK alloy combined with bromoform; K. Dangerous; when heated to decomposition it emits highly toxic fumes of Cl^-. To fight fire, use water, CO_2, dry chemical.

PENTACHLORONAPHTHALENE   HR: 3
CAS: 1321-64-8   NIOSH: QK 0300000
mf: $C_{10}H_3Cl_5$ mw: 300.41

PROP: White solid.

SYN: HALOWAX 1013

OSHA PEL: TWA 500 ppm (skin)
ACGIH TLV: TWA 0.5 mg/m^3
DFG MAK: 0.5 mg/m^3

THR: Poison by ingestion, inhalation, and skin contact. An irritant. Action similar to chlorinated naphthalenes and chlorinated diphenyls. See also polychlorinated biphenyls. Dangerous; when heated to decomposition it emits highly toxic fumes of Cl^-. For further information see Vol. 5, No. 1 of *DPIM Report*.

PENTACHLORONITRO BENZENE   HR: 3
CAS: 82-68-8   NIOSH: DA 6650000
mf: $C_6Cl_5NO_2$ mw: 295.32

PROP: Colorless crystals; mp: $146°$; bp: $328°$; vap press: 0.013 mm @ $25°$.

SYNS: NCI-C00419 * QUINIOZENE

THR: An experimental carcinogen, neoplastigen, and teratogen. Moderately toxic by ingestion. See nitro compounds of aromatic hydrocarbons. Dangerous; when heated to decomposition it emits highly toxic fumes of NO_x and Cl^-. For further information see Vol. 5, No. 5 of *DPIM Report*.

PENTACHLOROPHENOL   HR: 3
CAS: 87-86-5   NIOSH: SM 6300000
DOT: 2020
mf: C_6HCl_5O mw: 266.32

PROP: Dark-colored flakes and sublimed needle-like crystals with a characteristic odor. Mp: $191°$, bp: $310°$ (decomp), d: 1.978, vap press: 40 mm @ $211.2°$. Sol in ether, benzene; very sol in alc; insol in water; sltly sol in cold petroleum ether.

SYNS: NCI-C54933 * NCI-C55378 * NCI-C56655 * PENTACHLOORFENOL (DUTCH) * PENTACHLOROPHENATE * 2,3,4,5,6-PENTACHLOROPHENOL * PENTACLOROFENOLO (ITALIAN) * THOMPSON'S WOOD FIX

OSHA PEL: TWA 500 ug/m^3
ACGIH TLV: TWA 0.5 mg/m^3
DFG MAK: 0.05 ppm (0.5 mg/m^3)
DOT Classification: Label: Poison

THR: Poison by ingestion, inhalation, skin contact, intraperitoneal, and subcutaneous routes. An experimental teratogen. A carcinogen. Human central nervous system effects. Skin irritant. Acute poisoning is marked by weakness and respiratory, blood pressure and urinary output changes. Also causes dermatitis, convulsions and collapse. Chronic exposure can cause liver and kidney injury. See also phenols. Dangerous; when heated to decomposition it emits highly toxic fumes of Cl^-. For further information see Vol. 3, No. 4 of *DPIM Report*.

PENTAERYTHRITOL TETRANITRATE   HR: 2
CAS: 78-11-5   NIOSH: RZ 2620000
DOT: 0411
mf: $C_5H_8N_4O_{12}$ mw: 316.17

PROP: Crystals; mp: $138°-140°$, bp: explodes

@ 205°-215°, d: 1.773 @ 20°/4°. Sol in acetone; insol in water; sltly sol in alc, ether.

SYNS: 2,2-BISDIHYDROXYMETHYL-1,3-PROPAN-EDIOL TETRANITRATE * 2,2-BIS(HYDROXY-METHYL)-1,3-PROPANEDIOL TETRANITRATE * 1,3-DINITRATO-2,2-BIS(NITRATOMETHYL)PRO-PANE * INITIATING EXPLOSIVE PENTAERY-THRITE TETRANITRATE (DOT) * NCI-C55743 * NEOPENTANETETRAYL NITRATE * NITRO-PENTAERYTHRITE * NITROPENTAERYTHRITOL * PENTAERYTHRITE TETRANITRATE * TETRANI-TROPENTAERYTHRITE

DOT Classification: Label: Explosive A

THR: Moderately toxic by ingestion with systemic skin effects in humans. See also nitrates. Effects are similar to nitroglycerin; i.e., headache, weakness, and fall in blood pressure. For fire hazard, see nitrates. Severe explosion hazard when shocked or exposed to heat. It explodes @ 215°. (See explosives, high.) Highly dangerous; on decomposition it emits highly toxic fumes of NO_x; can react vigorously with oxidizing materials.

1,5-PENTAMETHYLENETETRA-ZOLE HR: 3
CAS: 54-95-5   NIOSH: XF 8225000
mf: $C_6H_{10}N_4$ mw: 138.20

PROP: White crystalline powder; mp: 57°-58°; sol in water, alc and ether.

SYNS: ALPHA,BETA-CYCLOPENTAMETHYLENE-TETRAZOLE * PENTYLENETETRAZOL * PENTAMETHYLENETETRAZOL * PENTA-METHYLENE-1,5-TETRAZOLE * 6,7,8,9-TET-RAHYDRO-5-AZEPOTETRAZOLE * 6,7,8,9-TETRAHYDRO-5H-TETRAZOLOAZEPINE * 7,8,9,10-TETRAZABICYCLO(5.3.0)-8,10-DECADIENE * 1,2,3,3A-TETRAZACYCLO-HEPTA-8A,2-CYCLOPENTADIENE

THR: Poison in humans by ingestion and intravenous routes. Human psychotropic effects. Also poison by intraperitoneal, subcutaneous, rectal, and parenteral routes. When heated to decomposition it emits toxic fumes of NO_x.

PENTANE HR: 3
CAS: 109-66-0   NIOSH: RZ 9450000
DOT: 1265
mf: C_5H_{12} mw: 72.17

PROP: Colorless liquid. Bp: 36.1°, flash p:

< −40°F, fp: −129.8°, d: 0.626 @ 20°/4°, autoign temp: 588°F, vap press: 400 mm @ 18.5°, vap d: 2.48, lel = 1.5%, uel = 7.8%. Sol in water; misc in alc, ether, organic solvents.

SYNS: PENTAN (POLISH) * PENTANEN (DUTCH) * PENTANI (ITALIAN)

OSHA PEL: TWA 1000 ppm
ACGIH TLV: TWA 600 ppm; STEL 750 ppm
DFG MAK: 1000 ppm (2950 mg/m³)
DOT Classification: Label: Flammable Liquid

THR: Poison by intravenous route. Narcotic in high concentration. The liquid can cause blisters on contact. Highly dangerous fire hazard when exposed to heat, flame, or oxidizers. Severe explosion hazard when exposed to heat or flame. Highly dangerous; shock can shatter metal containers and release contents. To fight fire, use foam, CO_2, dry chemical.

PENTANEDIAMINE HR: 3
CAS: 462-94-2   NIOSH: SA 0200000
mf: $C_5H_{14}N_2$ mw: 102.21

PROP: Colorless, thick liquid, characteristic odor. Mp: 9°, bp: 178°-180°, d: 0.873 @ 25°/4°. Very sol in water and alc; sltly sol in ether.

SYNS: CADAVERIN * PENTAMETHYLENEDI-AMINE

THR: Poison by intravenous, rectal, and subcutaneous routes. Mutagenic data. Moderately toxic by skin contact. An irritant, sensitizer, and allergen. When heated to decomposition it emits highly toxic fumes of NO_x and CN^-.

1-PENTANETHIOL HR: 2
CAS: 110-66-7   NIOSH: SA 3150000
DOT: 1111
mf: $C_5H_{12}S$ mw: 104.23

PROP: Water white to yellow liquid. D: 0.857 @ 20°, bp: 123.64°, flash p: 65°F, vap press: 13.8 mm @ 25°, vap d: 3.59. Insol in water; misc in alc and ether.

SYNS: AMYL HYDROSULFIDE * N-AMYL MER-CAPTAN * AMYL SULFHYDRATE * AMYL THIOALCOHOL * MERCAPTAN AMYLIQUE (FRENCH) * PENTYL MERCAPTAN

DOT Classification: Label: Flammable Liquid

THR: Moderately toxic by inhalation. See also

mercaptans. A weak sensitizer and allergen. Local contact may cause contact dermatitis. Dangerous fire hazard when exposed to heat or flame. Dangerous; see mercaptans. Reacts with oxidizing materials. To fight fire, use foam, CO_2, dry chemical.

2-PENTANOL   HR: 2
CAS: 6032-29-7   NIOSH: SA 4900000
mf: $C_5H_{12}O$ mw: 88.17

PROP: Colorless liquid. Bp: 119.3°, flash p: 105°F (OC), ulc:40-45, uel = 9.0%, lel = 1.2%, fp: −50°, d: 0.8169 @ 20°/20°, autoign temp: 650°-725°F, vap d: 3.04. Slightly sol in water; misc in alcohol and ether.

SYNS: SEC-AMYL ALCOHOL * METHYL PROPYL CARBINOL * PENTANOL-2 * SEC-PENTYL ALCOHOL

THR: Moderately toxic by ingestion and intraperitoneal routes. A narcotic. A severe eye and moderate skin and mucous membrane irritant. Flammable when exposed to heat or flame; can react with oxidizing materials. To fight fire, use alcohol foam, dry chemical.

3-PENTANOL   HR: 2
CAS: 584-02-1   NIOSH: SA 5075000
mf: $C_5H_{12}O$ mw: 88.17

PROP: Liquid; bp: 115.6°; d: 0.815 @ 25°/4°; flash p: 66°F; lel = 1.2%, uel = 9%; sol alc, ether; sltly sol in water. Acetone-like odor.

SYNS: DIETHYL CARBINOL * PENTANOL-3 * PENTAN-3-OL

THR: Moderately toxic by ingestion, skin contact, and intraperitoneal routes. A severe eye and mild skin irritant. Dangerous fire hazard when exposed to heat, flame, or oxidizing materials. When heated to decomposition it emits acrid smoke and fumes.

2-PENTANONE   HR: 2
CAS: 107-87-9   NIOSH: SA 7875000
DOT: 1249
mf: $C_5H_{10}O$ mw: 86.15

PROP: Sltly sol in water, water-white liquid. D: 0.8, vap d: 3.0, bp: 216°F, flash p: 45°F, autoign temp: 941°F, lel = 1.5%, uel = 8.2%.

SYNS: ETHYL ACETONE * METHYL-PROPYL-CETONE (FRENCH) * METHYLOPROPYLOKETON (POLISH) * METHYL-N-PROPYL KETONE * METHYL PROPYL KETONE (DOT)

OSHA PEL: TWA 200 ppm
DFG MAK: 200 ppm (700 mg/m^3)
DOT Classification: Label: Flammable Liquid

THR: Moderately toxic by ingestion, inhalation, and intraperitoneal routes. Inhalation causes narcosis and irritation of the respiratory passages, eyes, and skin. See also ketones. Dangerous fire hazard when exposed to heat or flame. To fight fire, use alcohol foam. Incompatible with bromine trifluoride.

3-PENTANONE   HR: 2
CAS: 96-22-0   NIOSH: SA 8050000
DOT: 1156
mf: $C_5H_{10}O$ mw: 86.15

PROP: Colorless, mobile liquid, acetone-like odor; mp: −42°, bp: 101°, flash p: 55°F, d: 0.8159 @ 19°/4°, vap d: 2.96, autoign temp: 842°F, lel = 1.6%; sol in water; misc in alc and ether.

SYNS: DEK * DIETHYLCETONE (FRENCH) * DIETHYL KETONE (DOT) * DIMETHYLACETONE * METHACETONE * PROPIONE

DOT Classification: Label: Flammable Liquid

THR: Moderately toxic by ingestion, inhalation, intraperitoneal, and intravenous routes. A skin and eye irritant. See also ketones. Dangerous fire hazard when exposed to heat or flame; can react vigorously with oxidizing materials. To fight fire, use alcohol foam, foam, CO_2, dry chemical. Incompatible with hydrogen peroxide; HNO_3.

PENTHAMIL   HR: 2
CAS: 67-43-6   NIOSH: MB 8205000
mf: $C_{14}H_{23}N_3O_{10}$ mw: 393.40

SYNS: DIETHYLENETRIAMINEPENTAACETIC ACID * 1,1,4,7,7-DIETHYLENETRIAMINEPENTAACETIC ACID * MONAQUEST * PERMA KLEER * (DIETHYLENETRINITRILO)PENTAACETIC ACID

THR: Moderately toxic by intraperitoneal route. When heated to decomposition it emits toxic fumes of NO_x.

PENTHIENATE BROMIDE   HR: 3
CAS: 60-44-6   NIOSH: BP 7525000
mf: $C_{18}H_{30}NO_3S \cdot Br$ mw: 420.46

SYNS: ALPHA-CYCLOPENTYL-2-THIOPHENEGLYCOLATE DIETHYL(2-HYDROXYETHYL)METHYL-

AMMONIUM BROMIDE * 2-DIETHYLAMINO-ETHYL 2-CYCLOPENTYL-2-(2-THIENYL)-HYDROXYACETATE METHOBROMIDE * 2-DIETHYLAMINOETHYL ALPHA-CYCLOPEN-TYL-2-THIOPHENEGLYCOLATE METHOBROMIDE * DIETHYL(2-HYDROXYETHYL)METHYLAMMO-NIUM BROMIDE, ALPHA-CYCLOPENTYL-2-THIO-PHENEGLYCOLATE

THR: Poison by intravenous and subcutaneous routes. Moderately toxic by ingestion. When heated to decomposition it emits very toxic fumes of NO_x, SO_x and Br^-.

PENTOBARBITAL   HR: 3
CAS: 115-58-2   NIOSH: CQ 6748000
mf: $C_{11}H_{18}N_2O_3$ mw: 226.31

SYNS: 5-AETHYL-5-PENTYL-(2')-BARBITURSA-EURE (GERMAN) * 5-ETHYL-5-PENTYLBARBI-TURIC ACID

THR: Poison by intraperitoneal route. When heated to decomposition it emits toxic fumes of NO_x.

PENTOTHAL SODIUM   HR: 3
CAS: 71-73-8   NIOSH: CQ 6475000
mf: $C_{11}H_{17}N_2O_2 \cdot Na$ mw: 264.35

SYNS: MONOSODIUM-5-ETHYL-5-(1-METHYLBU-TYL) THIOBARBITURATE * PENTHIOBARBITAL SODIUM * SODIUM-5-ETHYL-5-(1-METHYLBU-TYL)-2-THIOBARBITURATE * SODIUM PENTO-THAL * SODIUM PENTOTHIOBARBITAL * SODIUM THIOPENTAL * SOLUBLE THIOPEN-TONE * THIOPENTAL SODIUM SALT

THR: Poison by ingestion, intraperitoneal, rectal, subcutaneous, and intravenous routes. Human central nervous system and cardiovascular effects. An experimental teratogen. When heated to decomposition it emits toxic fumes of NO_x and Na_2O.

PENTYL ACETATE   HR: 2
CAS: 628-63-7   NIOSH: AJ 1925000
DOT: 1104
mf: $C_7H_{14}O_2$ mw: 130.21

PROP: Colorless liquid, pear or banana-like odor. Mp: $-78.5°$, bp: 148° @ 737 mm, ulc: 55-60, lel = 1.1%, uel = 7.5%, flash p: 77°F (CC), d: 0.879 @ 20°/20°, autoign temp: 714°F, vap d: 4.5. Very sltly sol in water; misc in alc and ether.

SYNS: ACETATE D'AMYLE (FRENCH) * ACETIC ACID, AMYL ESTER * N-AMYL ACETATE

* AMYL ACETATE (DOT) * AMYL ACETIC ESTER * AMYL ACETIC ETHER * AMYLAZETAT (GERMAN) * BANANA OIL * BIRNENOEL * OCTAN AMYLU (POLISH) * PEAR OIL * PENT-ACETATE * 1-PENTANOL ACETATE * N-PENTYL ACETATE * 1-PENTYL ACETATE * PRIMARY AMYL ACETATE

OSHA PEL: TWA 100 ppm
DOT Classification: Label: Flammable Liquid

THR: Moderately toxic by intraperitoneal route. A human eye irritant. A systemic irritant by inhalation. Mutagenic data. Apparently more toxic than butyl acetate. When inhaled in high concentration, it is a mucous membrane irritant and a narcotic. Chronic toxicity is of a low order. See also esters, amyl alcohol, acetic acid. Dangerous fire hazard when exposed to heat or flame; when heated, it emits acrid fumes; can react with oxidizing materials. Moderately explosive when exposed to flame. To fight fire, use alcohol foam, dry chemical.

PENTYL ALCOHOL   HR: 3
CAS: 71-41-0   NIOSH: SB 9800000
mf: $C_5H_{12}O$ mw: 88.17

PROP: Clear liquid; mp: $-79°$; ulc = 40; bp: 137.8°; flash p: 91°F (CC); d: 0.8168 @ 20°/20°; lel = 1.2%; uel = 10° @ 212°F; autoign temp: 572°F; vap press = 1 mm @ 13.6°, 10 mm @ 44.9°; vap d: 3.04. Sol in water; misc in alc and ether.

SYNS: ALCOOL AMYLIQUE (FRENCH) * AMYL ALCOHOL * N-AMYL ALCOHOL * AMYL AL-COHOL, NORMAL * N-BUTYLCARBINOL * N-AMYLALKOHOL (CZECH) * PENTANOL-1 * N-PENTANOL * PENTAN-1-OL * PENTASOL * PRIMARY AMYL ALCOHOL

THR: Poison by ingestion, intravenous, and intraperitoneal routes. A severe eye and skin irritant. Inhalation can irritate the upper respiratory trat. The amyl alcohols are about four times more toxic than ethanol. However, because of their low volatility and their low solubility in the body fluids, they are absorbed slowly and only to a small extent. Dangerous fire hazard when exposed to heat or flame; can react with oxidizing materials. Moderately explosive when exposed to flame. Hydrogen trisulfide (H_2S_3) can explode it. To fight fire, use alcohol foam, dry chemical. For further information see Amyl alcohol, Vol. 2, No. 3 of *DPIM Report*.

t-PENTYL ALCOHOL   HR: 2
CAS: 75-85-4   NIOSH: SC 0175000
mf: $C_5H_{12}O$ mw: 88.17

PROP: Colorless liquid; mp: $-11.9°$; bp:
101.8°; flash p: 105°F (CC); d: 0.809; autoign
temp: 819°F; vap press: 10 mm @ 17.2°; lel
= 1.2%; uel = 9%; vap d: 3.03; slightly sol
in water; sol in alc and ether.

SYNS: TERT-AMYL ALCOHOL * AMYLENE
HYDRATE * DIMETHYLETHYLCARBINOL
* 2-METHYL BUTANOL-2 * 2-METHYL-2-BU-
TANOL * 3-METHYLBUTAN-3-OL * TERT-PEN-
TANOL

THR: Moderately toxic by ingestion, intraperitoneal, subcutaneous, rectal, and other routes.
Moderate irritation to human mucous membranes. Narcotic in high concentration. Flammable when exposed to heat, flame, or oxidizing
materials. When heated to decomposition it
emits acrid smoke and fumes.

2,n-PENTYLAMINOETHYL-p-AMINO-
BENZOATE   HR: 3
CAS: 2188-67-2   NIOSH: DG 3000000
mf: $C_{14}H_{22}N_2O_2$ mw: 250.38

SYNS: P-AMINOBENZOIC ACID-2-N-AMYLAMINO-
ETHYL ESTER * 2-N-AMYLAMINOETHYL-P-AMI-
NOBENZOATE * AMYLCAINE

THR: Poison by intravenous and subcutaneous
routes. When heated to decomposition it emits
toxic fumes of NO_x.

PENTYL ETHER   HR: 3
CAS: 693-65-2   NIOSH: SC 2900000
mf: $C_{10}H_{22}O$ mw: 158.32

PROP: Liquid, mp: $-69.3°$, bp: 187°, flash
p: 135°F (OC), d: 0.783 @ 20°/4°, vap d: 5.46,
autoign temp: 340°F.

SYNS: 1,1-OXYBIS PENTANE * N-AMYL ETHER
* AMYL ETHER * DIPENTYL ETHER

THR: Poison by intravenous routes. See also
ethers. Flammable when exposed to heat or
flame; reacts with oxidizing materials. To fight
fire, use alcohol foam, dry chemical.

N-PENTYLNITROSOUREA   HR: 3
CAS: 10589-74-9   NIOSH: YT 9720000
mf: $C_6H_{13}N_3O_2$ mw: 159.22

SYNS: 1-AMYL-1-NITROSOUREA * N-AMYLNI-
TROSOUREA

THR: An experimental carcinogen. Mutagenic
data. Moderate acute toxicity by ingestion. See
also nitrates. When heated to decomposition it
emits toxic fumes of NO_x.

PEPPERMINT OIL   HR: 2
CAS: 8006-90-4   NIOSH: SC 6125000

PROP: Colorless to pale yellow liquid; d: 0.896-
0.908 @ 25°/25°.

SYN: PFEFFERMINZ OEL (GERMAN)

THR: Moderately toxic by intraperitoneal route.
An allergen. When heated to decomposition it
emits acrid smoke and fumes.

PEPTICHEMIO   HR: 3
CAS: 9076-25-9   NIOSH: SC 6200000

PROP: Made up of 6 peptides of m-(di-(2-chlo-
roethyl)amino-1-phenylalanine.

SYN: PTC

THR: Poison by intravenous, intraperitoneal,
and subcutaneous routes. Mutagenic data. When
heated to decomposition it emits very toxic
fumes of Cl^- and NO_x.

PERCHLORATES   HR: 1
PROP: Combinations with the monovalent
$^-ClO_4$ radical.

THR: Perchlorates are unstable materials, and
are irritant to the body wherever they come in
contact with it. Avoid skin contact with these
materials. Flammable by chemical reaction;
powerful oxidizers. See also explosives, high.
Moderate explosion hazard when shocked or
exposed to heat or by chemical reaction. Perchlorates, when mixed with carbonaceous material, form explosive mixtures. They are considered a fire and explosive hazard when associated
with carbonaceous materials or finely divided
metals. This is also true of the presence of sulfur,
powdered magnesium and aluminum. See explosives, high. React violently with benzene;
CaH_2; charcoal; olefins; ethanol; SrH_2; S;
H_2SO_4. When heated they emit highly toxic
fumes of Cl^-; they can react with reducing materials. To fight fire, use water or foam.

PERCHLORIC ACID   HR: 3
CAS: 7601-90-3   NIOSH: SC 7500000
mf: $ClHO_4$ mw: 100.46

PROP: Colorless, fuming, unstable liquid; mp:
$-112°$; bp: 19° @ 11 mm; d: 1.768 @ 22°.

SYN: PERCHLORIC ACID (DOT)

DOT Classification: Label: Oxidizer

THR: Poison by inhalation, ingestion,and subcutaneous routes. Severe irritant to the eyes, skin, and mucous membranes. See also perchlorates. A severe explosion hazard, the anhydrous form can explode spontaneously. Reacts violently with acetic acid; acetic acid + acetic anhydride; acetic anhydride; alcohols; aniline + HCHO; antimony compounds; azo pigments; bis-1,2-diaminopropane-cis-dichlorochromium (III) perchlorate; carbon; 1,3-bis(di-n-cyclopentadienyl iron)-2-propen-1-one; bismuth; CH_3OH; CCl_4; cellulose; charcoal; dehydrating agents; dibutyl sulfoxide; dimethyl sulfoxide; ethyl ether; ethylbenzene; F_2; glycolethers; glycols; HNO_3; HCl; H_2; HI; H_2SO_4; hypophosphites; iron sulfate; ketones; PbO + glycerin; 2-methyl cyclohexanone; NI_3; nitrogenous epoxides; nitrosophenol; o-periodic acid; oleic acid; paper; P_2O_5 + $CHCl_3$; phosphine; P_2O_5; P_2Zn_3; pyridine; NaI; sodium phosphinate; steel; SO_3; T1 acetate; trichloroethylene; wood.

PERCHLORIC ACID, AMMONIUM SALT   HR: 2
CAS: 7790-98-9   NIOSH: SC 7520000
DOT: 0402
mf: $ClO_4 \cdot H_4N$ mw: 117.50

PROP: White crystals, mp: decomp, d: 1.95.

SYN: AMMONIUM PERCHLORATE (DOT)

DOT Classification: Label: Oxidizer

THR: Moderately toxic by parenteral route. Flammable when exposed to heat or flame or by spontaneous chemical reaction with reducing materials. A very powerful oxidizer. Ignites violently with combustibles. Severe explosion hazard; decomposes @ 130° and explodes @ 380°. When contaminated by powdered carbon; ferrocene; S; organic matter; powdered metals it becomes impact sensitive. See also perchlorates and explosives, high.

PERCHLORIC ACID, BARIUM SALT · 3H₂O   HR: 3
CAS: 13465-95-7   NIOSH: SC 7550000
DOT: 1447
mf: $Cl_2O_8 \cdot Ba \cdot 3H_2O$ mw: 390.4

PROP: Colorless crystals; mp: decomp @ 400°; d: 2.74.

SYN: BARIUM PERCHLORATE

DOT Classification: Label: Oxidizer

THR: An unstable material. See also perchlorates and barium compounds. When refluxed with an alcohol a highly explosive product is formed. When heated to decomposition it emits toxic fumes of Cl^-.

PERCHLORIC ACID, MAGNESIUM SALT   HR: 2
CAS: 10034-81-8   NIOSH: SC 8925000
DOT: 1475
mf: $Cl_2O_8 \cdot Mg$ mw: 223.21

PROP: White, hygroscopic crystals. Mp: decomp @ 251°, d: 2.60 @ 25°.

SYNS: ANHYDRONE * DEHYDRITE * MAGNESIUM PERCHLORATE * PERCHLORATE DE MAGNESIUM (FRENCH)

DOT Classification: Label: Oxidizer

THR: Moderately toxic by intraperitoneal route. (See also magnesium compounds and perchlorates.) Avoid contact with mineral acids; ammonia; butyl fluorides; P; dimethyl sulfoxide; ethylene oxide; hydrocarbons; organic matter; trimethyl phosphite.

PERCHLOROBUTADIENE   HR: 3
CAS: 87-68-3   NIOSH: EJ 0700000
mf: C_4Cl_6 mw: 260.74

PROP: Autoign temp: 1130°F, vap d: 8.99.

SYNS: HEXACHLOR-1,3-BUTADIEN (CZECH) * HEXACHLORBUTADIENE * 1,1,2,3,4,4-HEXACHLORO-1,3-BUTADIENE

THR: Poison by ingestion, inhalation, intraperitoneal, and possibly other routes. Moderately toxic by skin contact. A skin and eye irritant. An experimental carcinogen and teratogen. Combustible. To fight fire, use dry chemical, CO_2, alcohol foam, water spray, fog, mist. When heated to decomposition it emits very toxic fumes of Cl^-. For further information see hexachlorobutadiene, Vol. 2, No. 5 of *DPIM Report*.

PERCODAN   HR: 3
CAS: 76-42-6   NIOSH: QD 2450000
mf: $C_{18}H_{21}NO_4$ mw: 315.40

SYNS: DIHYDROHYDROXYCODEINONE * DIHYDRO-14-HYDROXYCODEINONE * 14-HYDROXYDIHYDROCODEINONE * OXYCODEINONE

THR: Poison by intravenous and subcutaneous routes. When heated to decomposition it emits toxic fumes of NO_x.

PERILLA KETONE   HR: 3
                                   NIOSH: SA 8935000
mf: $C_{10}H_{14}O_2$ mw: 166.24

PROP: A potent lung toxin from the mint plant, *Perilla frutescens*.

SYNS: 1-(3-FURANYL)-4-METHYL-1-PENTANONE * PURPLE MINT PLANT EXTRACT

THR: Poison by intravenous and intraperitoneal routes. Affects the lungs. A potent pulmonary edemagenic agent (experimental). May also be hazardous to humans. When heated to decomposition it emits acrid smoke and fumes.

o-PERIODIC ACID
mf: HIO_4 mw: 191.91

THR: No toxicity data. See also iodine. Powerful oxidizer. When heated to decomposition it emits toxic fumes of I^-. Incompatible with azopigment; perchloric acid; dimethyl sulfoxide.

PERMANGANATES   HR: 3
PROP: Compounds containing an MnO_4^- radical.

THR: Poison. Many are strong oxidizing agents, hence irritating. See manganese compounds. Flammable by chemical reaction with reducing agents. Moderately explosive when shocked or exposed to heat. Silver permanganate and other metallic permanganates may detonate when exposed to high temp or when they are involved in fires or severely shocked. Store in a cool, ventilated area, away from acute fire hazards and easily oxidized materials. They may be disposed of by dissolving in water. Practically all permanganates are soluble in water. They can react vigorously on contact with reducing materials. Incompatible with acetic acid; acetic anhydride; H_2SO_4 + C_6H_6.

PERNAZINE   HR: 3
CAS: 84-97-9   NIOSH: SP 1140000
mf: $C_{20}H_{25}N_3S$ mw: 339.54

SYNS: N-METHYL-PIPERAZINYL-N'-PROPYL-PHE-NOTHIAZIN (GERMAN) * 10-(3-(4-METHYL-1-PI-PERAZINYL)PROPYL)-10H-PHENOTHIAZINE (9CI) * N-(3-(4-METHYL-1-PIPERAZINYL)PROPYL)-PHENOTHIAZINE

THR: Poison by intravenous and intraperitoneal routes. Moderately toxic by ingestion and subcutaneous routes. When heated to decomposition it emits very toxic fumes of NO_x and SO_x.

PEROXIDES, INORGANIC
THR: Variable toxicity. They may cause injury on contact with skin or mucous membranes. Moderate to dangerous fire hazard by chemical reaction with reducing agents and contaminants; strong oxidizing agents; contact with moisture may produce much heat. See also hydrogen peroxide and sodium peroxide. Moderate explosion hazard; heat, shock, or catalysts can cause violent decomposition. Contact with reducing agents may give rise to explosively violent reactions.

PEROXIDES, ORGANIC   HR: 3
THR: Often highly toxic and irritant to the skin, eyes, and mucous membranes. Dangerous fire hazard by chemical reaction with reducing agents or exposure to heat. They are powerful oxidizers. Severe explosion hazard when shocked, exposed to heat, or by spontaneous chemical reaction. Many peroxides are very unstable. Upon contact with reducing materials, such as organic matter, thiocyanates, an explosion can occur. Handle with great care.

PEROXYACETIC ACID   HR: 3
CAS: 79-21-0   NIOSH: SD 8750000
DOT: 2131
mf: $C_2H_4O_3$ mw: 76.06

PROP: Not over 40% Peracetic Acid and not over 6% Hydrogen Peroxide.

PROP: Colorless liquid, strong odor, water-sol, bp: 105°, explodes @ 110°, flash p: 105°F (OC), d: 1.15 @ 20°. Powerful oxidizer.

SYNS: ACETYL HYDROPEROXIDE * ACIDE PERACETIQUE (FRENCH) * HYDROPEROXIDE, ACETYL * PERACETIC ACID * PERACETIC ACID SOLUTION (DOT)

DOT Classification: Label: Organic Peroxide

THR: Poison by ingestion. An experimental carcinogen by skin contact. Flammable when exposed to heat or flames. See peroxides, organic. A severe eye and skin irritant. Severe explosion hazard when exposed to heat or by spontaneous chemical reaction. A powerful oxidizing agent. Violent reaction with acetic anhydride; olefins;

organic matter. Dangerous; keep away from combustible materials. When heated to decomposition it emits acrid smoke and fumes. To fight fire, use water, foam, CO_2.

PEROXYBENZOIC ACID HR: 3
CAS: 93-59-4 NIOSH: SD 9280000
mf: $C_7H_6O_3$ mw: 138.13

PROP: Leaflets, mp: 42°, bp: explodes @ 80°-100°. Insol in water; sol in alcohol and ether.

SYNS: BENZENECARBOPEROXOIC ACID (9CI) * BENZOYLHYDROGEN PEROXIDE * BENZOYL HYDROPEROXIDE * PERBENZOIC ACID

THR: An experimental carcinogen. Moderately irritating to skin, eyes, and mucous membranes by ingestion and inhalation. See also peroxides, organic. A dangerous fire hazard when exposed to heat, flame, or reducing materials. A powerful oxidizing agent. See peroxides, organic. Severe explosion hazard when exposed to heat or flame. Violent reaction with olefins. Can react vigorously with reducing materials. Avoid evaporation. When heated to decomposition it emits acrid smoke and fumes.

PEROXYDISULFURYL DIFLUORIDE
mf: $F_2O_6S_2$ mw: 198.13

THR: No toxicity data. A powerful irritant and corrosive to skin, eyes, and mucous membranes. A very powerful oxidizer. See also sulfuric acid, fluorides. A dangerous fire hazard. Ignites organic materials immediately on contact. When heated to decomposition it emits very toxic fumes of F^- and SO_x.

PEROXYFORMIC ACID
mf: CH_2O_3 mw: 62.03

THR: No toxicity data. A powerful irritant and an oxidizer. A dangerous fire hazard when exposed to heat, flame, or reducing materials. Unstable; 80% solution explodes; extremely dangerous when moved; also vacuum-distilled material below $-10°C$. When heated to decomposition it emits acrid smoke. Incompatible with metals; non-metals; organic materials.

PEROXYMONOPHOSPHORIC ACID
mf: H_3O_5P mw: 87.0

THR: No toxicity data. Very irritating and corrosive to tissues of skin, eyes, mucous membranes. See also phosphates. Powerful oxidizer.

A dangerous fire hazard when exposed to heat, flame, or reducing materials. When heated to decomposition it emits toxic fumes of PO_x.

PEROXYMONOSULFURIC ACID
mf: H_2O_5S mw: 114.08

THR: No toxicity data. Powerful oxidizer. Strong irritant. When heated to decomposition it emits toxic fumes of SO_x. Incompatible with acetone; alcohols; aromatic compounds; catalysts; fibers.

PEROXYNITRIC ACID
mf: HNO_4 mw: 79.02

THR: A poison. Very irritating and corrosive to tissue. See also nitric acid. When heated to decomposition it emits toxic fumes of NO_x. Decomposes violently. A dangerous fire hazard when exposed to heat, flame, or reducing materials.

PEROXYTRIFLUOROACETIC ACID
mf: $C_2HF_3O_3$ mw: 130.03

THR: A poison. A powerful oxidizer and corrosive and irritating to skin, eyes, mucous membranes. See also fluorides. When heated to decomposition it emits toxic fumes of F^-.

PETASITES JAPONICUS MAXIM HR: 3
NIOSH: SE 5095000

PROP: Dried flower stalk of *Petasites Japonicus Maxim*.

SYNS: COLTS FOOT * FUKI-NO-TOH (JAPANESE)

THR: An experimental carcinogen. When heated to decomposition it emits acrid smoke and fumes.

PETROLEUM HR: 3
NIOSH: SE 7175000
DOT: 1267

PROP: A thick flammable, dark yellow to brown or green-black liquid; d: 0.780-0.970; flash p: 20°-90°F. Insol in water; sol in benzene, chloroform, ether. Consists of a mixture of hydrocarbons from C_2H_6 and up, chiefly of the paraffins, cycloparaffins, or of cyclic aromatic hydrocarbons, with small amounts of benzene hydrocarbons, sulfur, and oxygenated compounds.

SYNS: BASE OIL * COAL LIQUID * COAL OIL * CRUDE OIL * PETROLEUM CRUDE * ROCK OIL * SENECA OIL

DOT Classification: Combustible Liquid, Label: None

THR: An experimental neoplastigen and carcinogen. See also mineral oils. A dangerous fire hazard when exposed to heat, flame, or powerful oxidizers. To fight fire, use foam, CO_2, dry chemical. When heated to decomposition it emits acrid smoke and fumes.

PETROLEUM ASPHALT   HR: 3
                                  NIOSH: SE 7350000

PROP: Steam refined asphalt.

SYNS: ASPHALT, PETROLEUM * PETROLEUM ROOFING TAR * ROAD ASPHALT

THR: An experimental neoplastigen. See also asphalt. When heated to decomposition it emits acrid smoke and fumes.

PETROLEUM SPIRITS   HR: 2
CAS: 8030-30-6   NIOSH: SE 7555000

PROP: Volatile, clear, colorless and non-fluorescent liquid; mp: $< -73°$, bp: 40°-80°, ulc: 95-100, lel = 1.1%, uel = 5.9%, flash P: $<$ 0°F, d: 0.635-0.660, autoign temp: 550°F, vap d: 2.50.

SYNS: BENZINE * BENZOLINE * CANADOL * HERBITOX * LIGROIN * MINERAL SPIRITS * MINERAL THINNER * MINERAL TURPENTINE * PAINTERS' NAPHTHA * REFINED SOLVENT NAPHTHA * SOLVENT NAPHTHA * VARNISH MAKERS' NAPHTHA * VARNISH MAKERS' AND PAINTERS' NAPHTHA * V.M. AND P. NAPHTHA * WHITE SPIRITS

THR: Moderately toxic by inhalation and possibly other routes. A systemic irritant and mucous membrane irritant. Ingestion can cause a burning sensation, vomiting, diarrhea, drowsiness, and, in severe cases, pulmonary edema. Inhalation of concentrated vapors can cause intoxication resembling that from alcohol, headache, nausea, coma, and hemorrhage to various vital organs. Highly dangerous fire hazard when exposed to heat, flame sparks, etc. Moderately explosive when exposed to heat or flame. Highly dangerous; keep away from heat or flame! To fight fire, use foam, CO_2, dry chemical.

PETROLEUM WAXES
THR: An experimental carcinogen of the lung, skin and stomach.

PHENANTHRENE   HR: 3
CAS: 85-01-8   NIOSH: SF 7175000
mf: $C_{14}H_{10}$ mw: 178.24

PROP: Solid or monoclinic crystals; mp: 100°, bp: 339°, d: 1.179 @ 25°, vap press: 1 mm @ 118.3°, vap d: 6.14. Insol in water; sol in CS_2 benzene, hot alcohol; very sol in ether.

SYN: PHENANTHREN (GERMAN)

THR: Poison by intravenous route. Mutagenic data. An experimental neoplastigen and carcinogen. Moderately toxic by ingestion. A human skin photosensitizer. Combustible. To fight fire, use water, foam, CO_2, dry chemical. When heated to decomposition it emits acrid smoke and fumes. For further information see Vol. 6, No. 3 of *DPIM Report*.

PHENARSAZINE CHLORIDE   HR: 3
CAS: 578-94-9   NIOSH: SG 0680000
DOT: 1698
mf: $C_{12}H_9AsClN$ mw: 277.59

PROP: Light yellow to green granules, irr odor. Mp: 195°, bp: 410° (decomp), d: 1.65, vap press: very low @ 20°, vap d: 9.6. Very sltly sol in water; sltly sol in benzene, brass. Corrodes iron, bronze.

SYNS: ADAMSITE * 5-AZA-10-ARSENAAN-THRACENE CHLORIDE * 10-CHLORO-5,10-DIHY-DROARSACRIDINE * 10-CHLORO-5,10-DIHYDRO-PHENARSAZINE * DIPHENYLAMINECHLORO-ARSINE * PHENARSAZINE CHLORIDE

DOT Classification: Label: Irritant

THR: Poison by inhalation and intravenous routes. Human systemic irritant effects. An irritating material. See also arsenic compounds. A vomiting type of poison gas (non-persistent). When heated to decomposition it emits very toxic fumes of As and Cl^-.

PHENAZINE   HR: 3
CAS: 92-82-0   NIOSH: SG 1360000
mf: $C_{12}H_8N_2$ mw: 180.22

PROP: Pale yellow crystals; mp: 171°; bp: $>$ 360° (subl); very sltly sol in water; sol in cold and hot alc; sol in ether.

SYNS: AZOPHENYLENE * DIBENZOPARADIA-ZINE * DIBENZOPYRAZINE

THR: Acute poison by intraperitoneal and intravenous routes. An experimental carcinogen. When heated to decomposition it emits toxic fumes of NO_x.

PHENAZODINE HR: 3
CAS: 94-78-0 NIOSH: US 7700000
mf: $C_{11}H_{11}N_5$ mw: 213.27

SYNS: 2,6-DIAMINO-3-PHENYLAZOPYRIDINE
* 3-(PHENYLAZO)-2,6-PYRIDINEDIAMINE
* PHENAZOPYRIDINE * PYRIPYRIDIUM

THR: An experimental neoplastigen and carcinogen. Moderately toxic by intraperitoneal route. When heated to decomposition it emits toxic fumes of NO_x.

PHENAZOLINE HR: 3
CAS: 91-75-8 NIOSH: NJ 2000000
mf: $C_{17}H_{19}N_3$ mw: 265.39

SYNS: 2-(N-BENZYLANILINOMETHYL)-2-IMIDAZOLINE * 2-PHENYL-BENZYL-AMINO-METHYL-IMIDAZOLIN (GERMAN) * 2-(N-PHENYL-N-BENZYLAMINOMETHYL)IMIDAZOLINE

THR: Poison by ingestion, subcutaneous, and intraperitoneal routes. An eye irritant. When heated to decomposition it emits toxic fumes of NO_x.

PHENAZOPYRIDINIUM CHLORIDE HR: 3
CAS: 135-40-3 NIOSH: US 7875000
mf: $C_{11}H_{11}N_5 \cdot ClH$ mw: 249.73

PROP: Red crystals, sltly bitter taste. Sltly sol in cold water, alc; sol in acetic acid; insol in acetone, benzene, chloroform, ether.

SYNS: 2,6-DIAMINO-3-(PHENYLAZO)PYRIDINE MONOHYDROCHLORIDE * PHENYLAZODIAMINOPYRIDINE HYDROCHLORIDE * PHENYLAZOPYRIDINE HYDROCHLORIDE * 2,6-DIAMINO-3-PHENYLAZOPYRIDINE HYDROCHLORIDE * NCI-c01672 * PHENAZOPYRIDINE HYDROCHLORIDE * BETA-PHENYLAZO-ALPHA,ALPHA'-DIAMINOPYRIDINE HYDROCHLORIDE * 3-PHENYLAZO-2,6-DIAMINOPYRIDINE HYDROCHLORIDE * PHENYLAZO-ALPHA,ALPHA'-DIAMINOPYRIDINE MONOHYDROCHLORIDE * 3-(PHENYLAZO)-2,6-PYRIDINEDIAMINE, HYDROCHLORIDE

THR: An experimental carcinogen and tumorigen, and suspected human carcinogen. Human systemic and red blood cell effects. When heated

to decomposition it emits very toxic fumes of NO_x and HCl.

PHENCAPTON HR: 3
CAS: 2275-15-1 NIOSH: TD 5775000
DOT: 2783
mf: $C_{11}H_{13}Cl_2O_2PS_3$ mw: 375.29

SYNS: O,O-DIAETHYL-S ((2,5-DICHLOR-PHENYL-THIO)-METHYL)-DITHIOPHOSPHAT (GERMAN)
* S-(2,5-DICHLOROPHENYLTHIOMETHYL) O,O-DIETHYL PHOSPHORODITHIOATE * 2,5-DICHLOROPHENYLTHIOMETHYL O,O-DIETHYL PHOSPHORODITHIOATE * S-(2,5-DICHLOROPHENYLTHIOMETHYL) DIETHYL PHOSPHOROTHIOLOTHIONATE * O,O-DIETHYL-S-(2,5-DICHLOROPHENYLTHIOMETHYL) DITHIOPHOSPHATE * O,O-DIETHYL-S-(2,5-DICHLOROPHENYL-THIOMETHYL) DITHIOPHOSPHORAN * O,O-DIETHYL S-(2,5-DICHLOROPHENYLTHIOMETHYL) PHOSPHOROTHIOLOTHIONATE * O,O-DIETHYL S-(2,5-DICHLOROPHENYLTHIOMETHYL) PHOSPHORODITHIOATE * DITHIOPHOSPHATE DE-O,O-DIETHYLE ET DE S(2,5-DICHLOROPHENYL) THIOMETHYLE (FRENCH) * ENT 25,585
* EENKAPTON (DUTCH) * GEIGY G-28029
* PHENCAPTON (DOT) * PRZEDZIORKOFOS (POLISH)

DOT Classification: ORM-A, Label: None

THR: Poison by ingestion and other unspecified routes. Moderately toxic by skin contact. A cholinesterase inhibitor. See also parathion. When heated to decomposition it emits very toxic fumes of Cl^-, PO_x, and SO_x.

PHENETHYL ALCOHOL HR: 3
CAS: 60-12-8 NIOSH: SG 7175000
mf: $C_8H_{10}O$ mw: 122.18

PROP: Colorless liquid, floral odor of roses. Mp: $-27°$, bp: $220°$, flash p: $216°F$, d: 1.0245 @ $15°$, vap d: 4.21. Misc with alc, ether.

SYNS: BENZYL CARBINOL * PHENETHANOL
* BETA-PHENETHYL ALCOHOL * 2-PHENETHYL ALCOHOL * BETA-PHENYLETHANOL * 2-PHENYLETHANOL * BETA-PHENYLETHYL ALCOHOL * 2-PHENYLETHYL ALCOHOL

THR: Poison by ingestion and intraperitoneal routes. Moderately toxic by skin contact. A skin irritant. Being studied for additional oncological information. Causes severe central nervous system injury to experimental animals. A local anesthetic. Combustible when exposed to heat or flame; can react with oxidizing materials. To fight fire, use CO_2, dry chemical. When heated to decomposition it emits acrid smoke and fumes.

alpha-PHENETHYL ALCOHOL   HR: 3
CAS: 98-85-1   NIOSH: DO 9275000
mf: $C_8H_{10}O$ mw: 122.18

PROP: Colorless liquid. Bp: 204°, fp: 21.4°, d: 1.015 @ 20°/20°, vap press: 0.1 mm @ 20°, vap d: 4.21, flash p: 205°F (OC).

SYNS: ALPHA-METHYLBENZYL ALCOHOL * METHYLPHENYLCARBINOL * NCI-C55685 * 1-PHENYLETHANOL * PHENYLMETHYLCARBINOL * STYRALLYL ALCOHOL

THR: Poison by ingestion and moderately toxic by skin contact. A severe eye and moderate skin irritant. Combustible when exposed to heat or flame; can react with oxidizing materials. To fight fire, use alcohol foam, foam, CO_2, dry chemical.

beta-PHENETHYLAMINE   HR: 3
CAS: 64-04-0   NIOSH: SG 8750000
mf: $C_8H_{11}N$ mw: 121.20

PROP: Colorless to slightly yellow liquid. Bp: 194.5°-195°, d: 0.96 @ 15.5°/15.5°, vap d: 4.18. Sol in water; very sol in alc, ether.

SYNS: BETA-AMINOETHYLBENZENE * 1-AMINO-2-PHENYLETHANE * BETA-PHENETHYLAMINE * 1-PHENYL-2-AMINO-ATHAN (GERMAN) * 1-PHENYL-2-AMINOETHANE * BETA-PHENYLAETHYLAMIN (GERMAN) * PHENYLETHYLAMINE * OMEGA-PHENYLETHYLAMINE * 2-PHENYLETHYLAMINE

THR: Poison by ingestion, intraperitoneal, subcutaneous, and intravenous routes. A skin irritant and possible sensitizer. See also amines. When heated to decomposition it emits toxic fumes of NO_x.

(v-PHENETHYLTRIS(OXYETHYLENE))-TRIS(TRIETHYLAMMONIUM-IODIDE)   HR: 3
CAS: 65-29-2   NIOSH: BS 1100000
mf: $C_{30}H_{60}N_3O_3 \cdot 3I$ mw: 891.63

SYNS: TRI(BETA-DIETHYLAMINOETHOXY)-1,2,3-BENZENE TRI IODOETHYLATE * 1,2,3-TRI-(BETA-DIETHYLAMINOETHOXY)BENZENE TRIETHIODIDE * TRIIODOETHYLATE DE GALLAMINE (FRENCH) * TRIIODOETHYLATE OF TRI(DIETHYLAMINOETHYLOXY)-1,2,3-BENZENE * 1,2,3-TRIS(2-DIETHYLAMINOETHOXY)BENZENE TRIETHIODIDE * 1,2,3-TRIS(2-DIETHYLAMINOETHOXY)BENZENE TRIS(ETHYLIODIDE) * 1,2,3-TRIS(2-TRIETHYLAMMONIUM ETHOXY)-BENZENE TRIIODIDE

THR: Poison by ingestion, subcutaneous, intravenous, intraduodenal, intraperitoneal, and intramuscular routes. See also iodides. When heated to decomposition it emits very toxic fumes of NO_x and I^-.

PHENFLUORAMINE HYDROCHLORIDE   HR: 3
CAS: 404-82-0   NIOSH: SH 6825000
mf: $C_{12}H_{16}F_3N \cdot ClH$ mw: 267.75

SYNS: N-ETHYL-ALPHA-METHYL-M-TRIFLUOROMETHYLPHENETHYLAMINE * N-ETHYL-ALPHA-METHYL-M-(TRIFLUOROMETHYL)PHENETHYLAMINE HYDROCHLORIDE * FENFLURAMINE HYDROCHLORIDE * 1-(3-TRIFLUOROMETHYLPHENYL)-2-ETHYLAMINOPROPANE HYDROCHLORIDE

THR: Poison by ingestion, intravenous, and intraperitoneal routes. Central nervous system effects by ingestion. When heated to decomposition it emits very toxic fumes of F^-, NO_x, and HCl.

PHENIPRAZINE   HR: 3
CAS: 55-52-7   NIOSH: MV 7350000
mf: $C_9H_{14}N_2$ mw: 150.25

SYNS: (ALPHA-METHYLPHENETHYL)HYDRAZINE * BETA-PHENYLISOPROPYLHYDRAZINE * 1-PHENYL-2-HYDRAZINOPROPANE * PHENYLISOPROPYLHYDRAZINE

THR: Poison by ingestion, intraperitoneal, subcutaneous, and intravenous routes. When heated to decomposition it emits toxic fumes of NO_x.

PHENMETRAZINE HYDROCHLORIDE   HR: 3
CAS: 1707-14-8   NIOSH: QE 6650000
mf: $C_{11}H_{15}NO \cdot ClH$ mw: 213.73

SYNS: 3-METHYL-2-PHENYLMORPHOLINE HYDROCHLORIDE * USAF GE-1

THR: Poison by ingestion, intravenous, intraperitoneal, and subcutaneous routes. An experimental teratogen by ingestion. Human psychotropic effects by ingestion. When heated to decomposition it emits very toxic fumes of NO_x and HCl.

PHENODIANISYL HYDROCHLORIDE   HR: 3
CAS: 537-05-3   NIOSH: MF 2000000
mf: $C_{23}H_{25}N_3O_3 \cdot ClH$ mw: 427.97

PROP: Crystals, odorless. Mp: 176°. Very sol in alc; insol in water, oils.

SYNS: ALPHA,GAMMA-DI-P-ANISYL-BETA-(ETHOXYPHENYL)GUANIDINE HYDROCHLORIDE * DIANISYL-MONOPHENETHYLGUANIDINE HYDROCHLORIDE * 2-(4-ETHOXYPHENYL)-1,3-BIS(4-METHOXYPHENYL)GUANIDINE HYDROCHLORIDE * GUANICAINE * N,N'-BIS(4-METHOXYPHENYL)-N''-(4-ETHOXYPHENYL)-GUANIDINE HYDROCHLORIDE

THR: Poison by ingestion and subcutaneous routes. Solutions are decomposed by light. When heated to decomposition it emits very toxic fumes of HCl and NO_x.

PHENODODECINIUM BROMIDE   HR: 3
CAS: 538-71-6   NIOSH: BQ 2030000
mf: $C_{22}H_{40}NO \cdot Br$ mw: 414.54

SYNS: DODECYLDIMETHYL(2-PHENOXYETHYL)-AMMONIUM BROMIDE * BETA-PHENOXYETHYLDIMETHYLDODECYLAMMONIUM BROMIDE

THR: Poison by intraperitoneal and intravenous routes. See also bromides. When heated to decomposition it emits very toxic fumes of NO_x and Br^-.

PHENOL   HR: 3
CAS: 108-95-2   NIOSH: SJ 3325000
DOT: 1671/2312/2821
mf: C_6H_6O mw: 94.12

PROP: White, crystalline mass which turns pink or red on standing if not perfectly pure, burning taste, distinctive odor. Mp: 40.6°, bp: 181.9°, flash p: 175°F (CC), d: 1.072, autoign temp: 1319°F, vap press: 1 mm @ 40.1°, vap d: 3.24. Sol in water; misc in alc, ether.

SYNS: ACIDE CARBOLIQUE (FRENCH) * BAKER'S P AND S LIQUID AND OINTMENT * CARBOLIC ACID * CARBOLSAURE (GERMAN) * FENOL (DUTCH, POLISH) * FENOLO (ITALIAN) * HYDROXYBENZENE * MONOHYDROXYBENZENE * NCI-C50124 * OXYBENZENE

* PHENIC ACID * PHENOLE (GERMAN) * PHENYL HYDRATE * PHENYL HYDROXIDE * PHENYLIC ACID * PHENYLIC ALCOHOL

OSHA PEL: TWA 5 ppm (skin)
ACGIH TLV: TWA 5 ppm; BEI: Total phenol in urine at end of shift 250 mg/g creatinine
DFG MAK: 5 ppm (19 mg/m^3)
DOT Classification: Poison B, Label: Poison

THR: Poison by ingestion, subcutaneous, intravenous, parenteral, and intraperitoneal routes. Moderately toxic by skin contact. A severe eye and skin irritant. An experimental carcinogen, neoplastigen, and tumorigen. Mutagenic data. Absorption of phenolic solutions through the skin may be very rapid, and can cause death within 30 minutes to several hours by exposure of as little as 64 square inches of skin. Lesser exposures can cause damage to the kidneys, liver, pancreas, and spleen, and edema of the lungs. Ingestion can cause corrosion of the lips, mouth, throat, esophagus and stomach, and gangrene. Ingestion of 15 g has killed. Chronic exposures can cause death from liver and kidney damage. Dermatitis resulting from contact with phenol or phenol-containing products is fairly common in industry. A common air contaminant. Flammable when exposed to heat, flame, or oxidizers, and reacts violently with ($AlCl_3$ + nitrobenzene); butadiene; peroxy-disulfuric acid; peroxymonosulfuric acid. Dangerous; when heated it emits toxic fumes; can react with oxidizing materials. To fight fire, use alcohol foam, CO_2, dry chemical. For further information see Vol. 3, No. 4 in *DPIM Report*.

PHENOPROPAZINE   HR: 3
CAS: 522-00-9   NIOSH: SO 4900000
mf: $C_{19}H_{24}N_2S$ mw: 312.51

SYNS: 2-DIETHYLAMINO-1-PROPYL-N-DIBENZOPARATHIAZINE * 10-(2-DIETHYLAMINO PROPYL)PHENOTHIAZINE * SKF 2538

THR: Poison by subcutaneous, intraperitoneal, and intravenous routes. Moderately toxic by ingestion. When heated to decomposition it emits very toxic fumes of NO_x and SO_x.

PHENTANYL   HR: 3
CAS: 437-38-7   NIOSH: UE 5550000
mf: $C_{22}H_{28}N_2O$ mw: 336.52

SYNS: FENTANYL * N-PHENETHYL-4-(N-PROPIONYLANILINO)PIPERIDINE * 1-PHENETHYL-4-N-PROPIONYLANILINOPIPERIDINE

THR: Poison by intravenous and intraperitoneal routes. Human pulmonary effects. When heated to decomposition it emits toxic fumes of NO_x. For further information see Fentanyl, Vol. 1, No. 8 of *DPIM Report*.

PHENTANYL CITRATE   HR: 3
CAS: 990-73-8   NIOSH: UE 5600000
mf: $C_{22}H_{28}N_2O \cdot C_6H_8O_7$ mw: 528.66

SYNS: LEPTANAL * SUBLIMAZE * FENTANYL CITRATE * N-(1-PHENETHYL-4-PIPERIDI-NYL)-PROPIONANILIDE DIHYDROGEN CITRATE * N-(1-PHENETHYL-4-PIPERIDYL)PROPIONANILIDE CITRATE * N-(1-PHENETHYL-4-PIPERIDYL)PRO-PIONANILIDE DIHYDROGEN CITRATE * SUBLI-MAZE CITRATE

THR: Poison by ingestion. When heated to decomposition it emits toxic fumes of NO_x.

4'-PHENYLACETANILIDE   HR: 3
CAS: 4075-79-0   NIOSH: AE 6125000
mf: $C_{14}H_{13}NO$ mw: 211.28

SYNS: 4-ACETAMIDOBIPHENYL * 4-ACETYL-AMINOBIPHENYL * 4-BIPHENYLACETAMIDE * N-4-BIPHENYLACETAMIDE * N-(4-BIPHE-NYLYL)ACETAMIDE * P-PHENYLACETANILIDE

THR: An experimental carcinogen and tumorigen by ingestion. When heated to decomposition it emits toxic fumes of NO_x.

PHENYLACETONITRILE   HR: 3
CAS: 140-29-4   NIOSH: AM 1400000
mf: C_8H_7N mw: 117.16

PROP: Oily liquid, aromatic odor. Mp: −23.8°, bp: 233.5°, d: 1.0214 @ 15°/15°, vap press: 1 mm @ 60.0°. insol in water; misc in alc, ether.

SYNS: BENZENEACETONITRILE * BENZYL CYANIDE * BENZYL NITRILE * ALPHA-CYANO-TOLUENE * OMEGA-CYANOTOLUENE * 2-PHE-NYLACETONITRILE * ALPHA-TOLUNITRILE * USAF KF-21

THR: Poison by ingestion, inhalation, subcutaneous, and intraperitoneal routes. See also nitriles. When heated to decomposition it emits very toxic fumes of CN^- and NO_x. Incompatible with sodium hypochlorite.

PHENYLALANINE MUSTARD   HR: 3
CAS: 4213-3-2-5   NIOSH: AY 3950000
mf: $C_{13}H_{18}Cl_2N_2O_2 \cdot ClH$ mw: 341.69

SYNS: NSC-35051 * 4-(BIS(2-CHLOROETH-YL)AMINO)-D-PHENYLALANINE MONOHYDRO-CHLORIDE * 3-(P-(BIS(BETA-CHLOROETHYL)-AMINO)PHENYL)-D-ALANINE HYDROCHLORIDE * D-PHENYLALANINE MUSTARD

THR: An intravenous and intracerebral poison. A powerful irritant. An experimental tumorigen and carcinogen. When heated to decomposition, it emits very toxic fumes of Cl^-, NO_x and HCl.

dl-PHENYLALANINE MUSTARD   HR: 3
CAS: 531-7-6-0   NIOSH: AY 3600000
mf: $C_{13}H_{18}Cl_2N_2O_2$ mw: 305.23

SYNS: 4-(BIS(2-CHLOROETHYL)AMINO)-DL-PHE-NYLALANINE * 3-(P-(BIS(2-CHLOROETHYL)-AMINO)PHENYL)ALANINE * P-DI-(2-CHLORA-ETHYL)-AMINO-DL-PHENYL-ALANIN (GERMAN) * P-DI(2-CHLOROETHYL)AMINO-DL-PHENYL-ALANINE * NCI-CO4944 * NSC-14210 * PHENYLALANIN-LOST (GERMAN) * DL-3-(P-(BIS(2-CHLOROETHYL)AMINO)PHENYL)ALA-NINE * DL-SARCOLYSIN * DL-SARCOLYSINE * SAKOLYSIN (GERMAN)

THR: A poison by ingestion, intraperitoneal, intravenous, intracerebral routes. An experimental tumorigen, carcinogen. Mutagenic data. When heated to decomposition, it emits very toxic fumes of Cl^- and NO_x.

l-PHENYLALANINE MUSTARD   HR: 3
CAS: 148-8-2-3   NIOSH: AY 3675000
mf: $C_{13}H_{18}Cl_2N_2O_2$ mw: 305.23

SYNS: P-N-DI(CHLOROETHYL)AMINOPHENYLALANINE * P-N-BIS(2-CHLOROETHYL)AMINO-L-PHENYLA-LANINE * 3-(P-(P-(BIS(2-CHLOROETHYL)AMI-NO)PHENYL)-L-ALANINE * 4-(BIS(2-CHLORO-ETHYL)AMINO)-L-PHENYLALANINE * P-DI-(2-CHLOROETHYL)AMINO-L-PHENYLALANINE * 3-P-(DI(2-CHLOROETHYL)AMINO)-PHENYL-L-ALANINE * NCI-CO4853 * NSC-8806 * L-3-(P-(BIS(2-CHLOROETHYL)AMINO)PHE-NYL)ALANINE * PHENYLALANINE NITROGEN MUSTARD * L-SARCOLYSIN * P-L-SARCOLYSIN

THR: Poison by intraperitoneal, intracerebral, and subcutaneous routes. An experimental and suspected human carcinogen, neoplastigen, and tumorigen. Mutagenic data. When heated to decomposition, it emits toxic fumes of Cl^- and NO_x. For further information, see Vol. 6, No. 3 of *DPIM Report*.

o-PHENYLALANINE MUSTARD   HR: 3
CAS: 342-9-5-0   NIOSH: AY 3400000
mf: $C_{13}H_{18}Cl_2N_2O_2$ mw: 305.23

SYNS: O-DI-2-CHLOROETHYLAMINO-DL-PHE-
NYLALANINE * NSC-57199 * 3-(O-(BIS(BETA-
CHLOROETHYL)AMINO)PHENYL)-D,L-ALANINE

THR: Poison by intraperitoneal and intravenous
routes. When heated to decomposition, it emits
very toxic fumes of Cl^- and NO_x.

1-PHENYL-4-AMINO-5-CHLORPYRIDAZ-
6-ONE   HR: 3
CAS: 1698-60-8   NIOSH: UR 6125000
mf: $C_{10}H_8ClN_3O$ mw: 221.66

SYNS: 5-AMINO-4-CHLORO-2,3-DIHYDRO-3-
-OXO-2-PHENYLPYRIDAZINE * 5-AMINO-4-
CHLORO-2-PHENYL-3(2H)-PYRIDAZINONE
* BUREX (CZECH) * 1-FENYL-4-AMINO-5-
CHLOR-6-PYRIDAZINON (CZECH) * 1-PHENYL-4-
AMINO-5-CHLOROPYRIDAZIN-6-ONE * 1-PHE-
NYL-4-AMINO-5-CHLORO-6-PYRIDAZONE
* 1-PHENYL-4-AMINO-5-CHLOROPYRIDAZONE-6

THR: Poison by intraperitoneal route. Moderately toxic by ingestion. A severe eye irritant.
When heated to decomposition it emits very
toxic fumes of Cl^- and NO_x.

p-PHENYLANISOLE   HR: 3
CAS: 613-37-6   NIOSH: BZ 8850000
mf: $C_{13}H_{12}O$ mw: 184.25

PROP: Leaves from alc; mp: 90°; sol in hot
alc.

SYNS: P-METHOXYBIPHENYL * 4-METHOXYBI-
PHENYL

THR: An experimental tumorigen. When heated
to decomposition it emits acrid smoke and
fumes.

N-PHENYLANTHRANILIC ACID   HR: 3
CAS: 91-40-7   NIOSH: CB 3730000
mf: $C_{13}H_{11}NO_2$ mw: 213.25

PROP: Needles from alc; mp: 185°-187°; decomp 183°-184°. Very sltly sol in hot water;
sol in hot alc; very sltly sol in ether.

SYNS: O-ANILINOBENZOIC ACID * 2-ANILINO-
BENZOIC ACID * 2-CARBOXYDIPHENYLAMINE
* DIPHENYLAMINE-2-CARBOXYLIC ACID
* 2-(PHENYLAMINO)BENZOIC ACID * PHE-
NYLANTHRANILIC ACID

THR: Poison by intravenous and intraperitoneal
routes. When heated to decomposition it emits
toxic fumes of NO_x.

p-(PHENYLAZO)ANILINE   HR: 3
CAS: 60-09-3   NIOSH: BY 8225000
mf: $C_{12}H_{11}N_3$ mw: 197.26

PROP: Yellow crystals; mp: 128°; bp: 360°;
sltly sol in hot water; sol in hot alc; sol in
ether.

SYNS: AMINOAZOBENZENE * P-AMINOAZO-
BENZENE * 4-AMINOAZOBENZENE * 4-AMI-
NO-1,1'-AZOBENZENE * P-AMINOAZOBENZOL
* 4-AMINOAZOBENZOL * P-AMINODIPHENYLI-
MIDE * ANILINE YELLOW * 4-BENZENEAZO-
ANILINE * C.I. 11000 * PARAPHENOLAZO
ANILINE * 4-(PHENYLAZO)ANILINE
* 4-(PHENYLAZO)BENZENAMINE * P-PHE-
NYLAZOPHENYLAMINE * USAF EK-1375

THR: Poison by intraperitoneal route. Mutagenic data. An experimental carcinogen and
neoplastigen. When heated to decomposition
it emits toxic fumes of NO_x. For further information see p-aminoazobenzene; Vol. I, No. 3 of
DPIM Report.

4-(PHENYLAZO)-o-ANISIDINE   HR: 3
CAS: 3544-23-8   NIOSH: BZ 7350000
mf: $C_{13}H_{13}N_3O$ mw: 227.29

SYN: 3-METHOXY-4-AMINOAZOBENZENE

THR: An experimental tumorigen, carcinogen,
and neoplastigen. Mutagenic data. When heated
to decomposition it emits toxic fumes of NO_x.

N-PHENYLAZO-N-METHYLTAURINE
SODIUM SALT   HR: 3
CAS: 22670-79-7   NIOSH: XY 2785000
mf: $C_9H_{12}N_3O_3S \cdot Na$ mw: 265.29

SYNS: 3-METHYL-1-PHENYL-3-(2-SULFOETH-
YL)TRIAZENE SODIUM SALT * 1-PHENYL-3-
METHYL-3-(2-SULFOAETHYL) NATRIUM SALZ
(GERMAN) * 1-PHENYL-3-METHYL-3-(2-SULFO-
ETHYL)TRIAZENE, SODIUM SALT

THR: Poison by subcutaneous route. An experimental neoplastigen. When heated to decomposition it emits very toxic fumes of Na_2O, NO_x
and SO_x.

1-(PHENYLAZO)-2-NAPHTHOL   HR: 3
CAS: 842-07-9   NIOSH: QL 4900000
mf: $C_{16}H_{12}N_2O$ mw: 248.30

SYNS: BENZENEAZO-BETA-NAPHTHOL * BEN-
ZENE-1-AZO-2-NAPHTHOL * 1-BENZENEAZO-2-
NAPHTHOL * BENZENE-1-AZO-2-NAPHTHOL
* 1-BENZOAZO-2-NAPHTHOL * C.I. 12055
* FAST OIL ORANGE * FAST ORANGE
* NCI-C53929 * ORANGE A L'HUILE
* ORANGE SOLUBLE A L'HUILE * 1-(PHE-
NYLAZO)-2-NAPHTHALENOL * 1-PHENYLAZO-
BETA-NAPHTHOL * SUDAN ORANGE R

THR: An experimental carcinogen and tumorigen. When heated to decomposition it emits
toxic fumes of NO_x.

1-(PHENYLAZO)-2-NAPHTHYLA-
MINE   HR: 3
CAS: 85-84-7   NIOSH: QM 4725000
mf: $C_{16}H_{13}N_3$ mw: 247.32

SYNS: 1-BENZENE-AZO-BETA-NAPHTHYLAMINE
* 1-BENZENEAZO-2-NAPHTHYLAMINE * C.I.
FOOD YELLOW 10 * C.I. SOLVENT YELLOW 5
* FD AND C YELLOW NO. 3 * JAUNE AB
* OIL YELLOW A * 1-(PHENYLAZO)-2-NAPH-
THALENAMINE * YELLOW NO. 2

THR: An experimental carcinogen. Moderately
toxic by ingestion and subcutaneous routes.
When heated to decomposition it emits toxic
fumes of NO_x.

4-PHENYLAZO-m-PHENYLENEDI-
AMINE   HR: 3
CAS: 532-82-1   NIOSH: ST 3380000
mf: $C_{12}H_{12}N_4 \cdot ClH$ mw: 248.74

SYNS: BRILLIANT OIL ORANGE Y BASE
* CHRYSOIDINE * C.I. 11270 * C.I. BASIC
ORANGE 2 * C.I. SOLVENT ORANGE 3
* 2,4-DIAMINOAZOBENZENE HYDROCHLORIDE
* LEATHER ORANGE HR * 4-(PHENYLAZO)-
1,3-BENZENEDIAMINE MONOHYDROCHLORIDE
* 4-(PHENYLAZO)-M-PHENYLENEDIAMINE
MONOHYDROCHLORIDE * PURE CHRYSOIDINE
YBH * PYRACRYL ORANGE Y

THR: An experimental carcinogen. Mutagenic
data. Moderately toxic by ingestion and subcutaneous routes. When heated to decomposition
it emits very toxic fumes of NO_x and HCl.

8-(p-PHENYLBENZYL)ATROPINIUM
BROMIDE   HR: 3
CAS: 511-55-7   NIOSH: CK 2990000
mf: $C_{30}H_{34}NO_3 \cdot Br$ mw: 536.56

PROP: Crystals. Decomp @ 220°-222°.

SYNS: N,4-BIPHENYL-METHYL-DL-TROPEYL-
ALPHA-TROPINIUMBROMIDS (GERMAN)
* P-BIPHENYLMETHYL-(DL-TROPYL-ALPHA-
TROPINIUM)BROMIDE * N-(P-BIPHENYL-
METHYL)-ATROPINIUM BROMIDE * 4-DIPHENYL-
METHYL-DL-TROPYLTROPINIUM BROMIDE
* 4-DIPHENYLMETHYLTROPYLTROPINIUM
BROMIDE * 8-(P-PHENYLBENZYL)ATROPI-
NIUM BROMIDE

THR: Poison by intravenous and intraperitoneal
routes. Moderately toxic by ingestion. See also
bromides. When heated to decomposition it
emits very toxic fumes of NO_x and Br^-.

PHENYL BROMIDE   HR: 3
CAS: 108-86-1   NIOSH: CY 9000000
DOT: 2514
mf: C_6H_5Br mw: 157.02

PROP: Colorless, clear, mobile liquid. Mp:
−30.7, bp: 156.2°, flash p: 124°F, d: 1.497,
vap press: 10 mm @ 40°, vap d: 5.41, autoign
temp: 1051°F.

SYNS: BROMOBENZENE (DOT) * NCI-C55492

DOT Classification: Combustible Liquid,
   Label: None

THR: Moderately toxic by ingestion, inhalation,
and intraperitoneal routes. Eye and mucous
membrane irritant. Mutagenic data. Flammable
when exposed to heat or flame. Dangerous; see
bromides; can react with oxidizing materials.
To fight fire, use water to blanket fire, foam,
CO_2, water spray or mist, dry chemical.

PHENYL CARBITOL   HR: 2
CAS: 104-68-7   NIOSH: KM 0875000
mf: $C_{10}H_{14}O_3$ mw: 182.24

PROP: Liquid. Bp: 207° @ 55 mm, fp: −50°,
d: 1.1158 @ 20°/20°, vap press: <0.01 mm
@ 20°, vap d: 6.28.

SYNS: DIETHYLENE GLYCOL MONOPHENYL
ETHER * DIETHYLENE GLYCOLPHENYL ETHER
* 2-(2-PHENOXYETHOXY)ETHANOL

THR: Moderately toxic by ingestion and skin
contact. See also glycols. Severe skin irritant.
When heated it emits acrid smoke and fumes.
See also ethers.

PHENYL CELLOSOLVE   HR: 2
CAS: 122-99-6   NIOSH: KM 0350000
mf: $C_8H_{10}O_2$ mw: 138.18

PROP: Clear liquid. Mp: 14°, bp: 242°, flash p: 250°F.

SYNS: ETHYLENE GLYCOL MONOPHENYL ETHER * ETHYLENE GLYCOL PHENYL ETHER * 2-FE-NOXYETHANOL (CZECH) * FENYL-CELLOSOLVE (CZECH) * GLYCOL MONOPHENYL ETHER * BETA-HYDROXYETHYL PHENYL ETHER * 1-HYDROXY-2-PHENOXYETHANE * 2-PHE-NOXYETHANOL * PHENOXYETHYL ALCOHOL * PHENYLMONOGLYCOL ETHER

THR: Moderately toxic by ingestion and skin contact. A severe eye and moderate skin irritant. Combustible when exposed to heat or flame. When heated it emits acrid smoke and fumes. To fight fire, use CO_2, dry chemical.

PHENYL CELLOSOLVE ACRY-LATE   HR: 2
CAS: 48146-04-6   NIOSH: KM 0700000
mf: $C_{11}H_{12}O_3$ mw: 192.23

SYN: 2-PHENOXY-ETHANOL ACRYLATE

THR: Moderately toxic by skin contact. Skin irritant. When heated to decomposition it emits acrid smoke and fumes.

PHENYLCYCLOPROMINE SUL-FATE   HR: 3
CAS: 13492-01-8   NIOSH: GZ 2625000
mf: $C_{18}H_{20}N_2 \cdot O_4S$ mw: 360.46

SYNS: TRANS,D,L-2-PHENYLCYCLOPROPYLAMINE SULFATE * TRANCYLPROMINE SULFATE * TRANSAMINE SULFATE * TRANYLCYPROMINE SULFATE

THR: Poison by ingestion, intravenous, and intraperitoneal routes. When heated to decomposition it emits very toxic fumes of SO_x and NO_x.

trans-2-PHENYLCYCLOPROPYL AMINE   HR: 3
CAS: 3721-28-6   NIOSH: GZ 2450000
mf: $C_9H_{11}N$ mw: 133.21

SYNS: 2-PHENYL-1-AMINOCYCLOPROPANE, TRANS- * SKF 385

THR: Poison by ingestion, intraperitoneal, subcutaneous, and other possible routes. When heated to decomposition it emits toxic fumes of NO_x.

1-PHENYL-3,3-DIETHYLTRIA-ZENE   HR: 3
CAS: 13056-98-9   NIOSH: XY 0450000
mf: $C_{10}H_{15}N_3$ mw: 177.28

SYNS: 3,3-DIETHYL-1-PHENYLTRIAZENE * 1-FENYL-3,3-DIETHYLTRIAZEN (CZECH) * 1-PHENYL-3,3-DIAETHYLTRIAZEN (GERMAN)

THR: Poison by subcutaneous route. Moderately toxic by ingestion. Mutagenic data. An experimental carcinogen. When heated to decomposition it emits toxic fumes of NO_x.

o-PHENYL-N,N′-DIMETHYL PHOSPHO-RODIAMIDATE   HR: 3
CAS: 64050-60-8   NIOSH: TD 2800000
mf: $C_8H_{13}N_2O_2P$ mw: 200.20

SYNS: DIAMIDFOS * DOWCO 169

THR: Poison by ingestion and skin contact. When heated to decomposition it emits very toxic fumes of PO_x and NO_x.

m-PHENYLENEDIAMINE   HR: 3
CAS: 108-45-2   NIOSH: SS 7700000
DOT: 1673
mf: $C_6H_8N_2$ mw: 108.16

PROP: White crystals; mp: 63°, bp: 286°, d: 1.139, vap press: 1 mm @ 99.8°. Sol in water, methanol, ethanol, chloroform, acetone; sltly sol in ether, carbon tetrachloride; very slightly sol in benzene, toluene.

SYNS: 3-AMINOANILINE * M-BENZENEDIAMINE * 1,3-BENZENEDIAMINE * C.I. 76025 * M-DIAMINOBENZENE * 1,3-DIAMINOBENZENE * M-FENYLENDIAMIN (CZECH) * METAPHE-NYLENEDIAMINE * 1,3-PHENYLENEDIAMINE

DOT Classification: ORM-A, Label: None

THR: Poison by ingestion, intravenous, subcutaneous, intraperitoneal, and possibly other routes. Mutagenic data. An experimental carcinogen. Slight fire hazard. When heated to decomposition it emits toxic fumes of NO_x.

o-PHENYLENEDIAMINE   HR: 3
CAS: 95-54-5   NIOSH: SS 7875000
mf: $C_6H_8N_2$ mw: 108.16

PROP: Tan crystals; mp: 104°; bp: 257°; sltly sol in water; very sol in alc, chloroform, ether.

SYNS: 2-AMINOANILINE * 1,2-BENZENEDIA-MINE * C.I. OXIDATION BASE 16 * O-DIAMI-NOBENZENE * 1,2-DIAMINOBENZENE

THR: Poison by an unspecified route. Mutagenic data. Moderately toxic by ingestion. When heated to decomposition it emits toxic fumes of NO_x.

p-PHENYLENEDIAMINE   HR: 3
CAS: 106-50-3   NIOSH: SS 8050000
DOT: 1673
mf: $C_6H_8N_2$ mw: 108.16

PROP: White-sltly red crystals; mp: 146°, flash p: 312°F, vap d: 3.72, bp: 267°. Sol in alc, chloroform, ether.

SYNS: P-AMINOANILINE * 4-AMINOANILINE * P-BENZENEDIAMINE * 1,4-BENZENEDIAMINE * C.I. 76060 * C.I. OXIDATION BASE 10 * P-DIAMINOBENZENE * 1,4-DIAMINOBENZENE * FENYLENODWUAMINA (POLISH) * 1,4-PHE-NYLENEDIAMINE * USAF EK-394

OSHA PEL: TWA 100 ug/m^3 (skin)
ACGIH TLV: TWA 0.1 mg/m^3
DFG MAK: 0.1 mg/m^3
DOT Classification: ORM-A, Label: None

THR: Poison by ingestion, subcutaneous, intravenous, and intraperitoneal routes. Moderately toxic by skin contact. Human skin irritant. An experimental carcinogen. Mutagenic data. Implicated in aplastic anemia. Can cause fatal liver damage. The p-form is more toxic and a stronger irritant that the o- and m- isomers. When used as a hair dye it caused vertigo, anemia, gastritis, exfoliative dermatitis, and death. Has caused asthma and other respiratory symptoms in the fur dying industry. Combustible when exposed to heat or flame. Dangerous; when heated, burns and emits highly toxic fumes of nitrogen compounds; can react with oxidizing materials. To fight fire, use water, CO_2, dry chemical.

m-PHENYLENEDIAMINE HYDRO-CHLORIDE   HR: 3
CAS: 541-69-5   NIOSH: SS 9800000
mf: $C_6H_8N_2 \cdot 2ClH$ mw: 181.08

PROP: Colorless needles; very sol in water; sltly sol in alc, ether.

SYNS: M-AMINOANILINE DIHYDROCHLORIDE * 3-AMINOANILINE DIHYDROCHLORIDE * 1,3-BENZENEDIAMINE HYDROCHLORIDE * M-BENZENEDIAMINE DIHYDROCHLORIDE * M-DIAMINOBENZENE DIHYDROCHLORIDE * 1,3-DIAMINOBENZENE DIHYDROCHLORIDE * 1,3-PHENYLENEDIAMINE DIHYDROCHLORIDE * USAF EK-206

THR: Poison by intraperitoneal and other unspecified routes. An experimental carcinogen. When heated to decomposition it emits very toxic fumes of HCl and NO_x.

PHENYLENE THIOCYANATE   HR: 3
CAS: 4044-65-9   NIOSH: NX 9150000
mf: $C_8H_4N_2S_2$ mw: 192.26

PROP: Tasteless, odorless, colorless crystals; mp: 132°.

SYNS: 1,4-DIISOTHIOCYANATOBENZENE * ISOTHIOCYANIC ACID-P-PHENYLENE ESTER * PHENYLENE-1,4-DIISOTHIOCYANATE * 1,4-PHENYLENEDIISOTHIOCYANIC ACID

THR: Poison by ingestion and intraperitoneal routes. Human central nervous system and gastrointestinal effects. When heated to decomposition it emits very toxic fumes of NO_x and SO_x.

2-PHENYLETHYL ACETATE   HR: 3
CAS: 103-45-7   NIOSH: AJ 2220000
mf: $C_{10}H_{12}O_2$ mw: 164.22

PROP: Colorless liquid; mp: 164.2°; bp: 223.6°; fp: $< -20°$; flash p: 230° F; d: 1.032 @ 25°/25°.

SYNS: ACETIC ACID-2-PHENYLETHYL ESTER * BENZYLCARBINYL ACETATE * BETA-PHEN-ETHYL ACETATE * 2-PHENETHYL ACETATE * BETA-PHENYLETHYL ACETATE

THR: An experimental carcinogen. Moderately toxic by ingestion. See also esters. A skin irritant. Combustible. To fight fire, use alcohol foam, CO_2 and dry chemical. When heated to decomposition it emits acrid smoke and fumes.

2-PHENYLETHYLHYDRAZINE   HR: 3
CAS: 51-71-8   NIOSH: MV 8400000
mf: $C_8H_{12}N_2$ mw: 136.22

SYNS: 1-HYDRAZINO-2-PHENYLETHANE * BETA-PHENYLETHYLHYDRAZINE

THR: Poison by ingestion, intraperitoneal, and subcutaneous routes. Human central nervous system effects by ingestion. When heated to decomposition it emits toxic fumes of NO_x.

PHENYLETHYLHYDRAZINE SUL-PHATE   HR: 3
CAS: 156-51-4   NIOSH: MV 8750000
mf: $C_8H_{12}N_2 \cdot H_2O_4S$ mw: 234.30

SYNS: 1-HYDRAZINO-2-PHENYLETHANE HYDROGEN SULPHATE * PHENALZINE HYDROGEN SULPHATE * PHENETHYLHYDRAZINE SULFATE

(1:1) * 2-PHENYLETHYLHYDRAZINE DIHY-
DROGEN SULPHATE * BETA-PHENYLETHYLHY-
DRAZINE HYDROGEN SULPHATE * BETA-PHE-
NYLETHYLHYDRAZINE SULFATE * ALACINE
* PHENELZINE SULFATE * PHENYLAETHYL-
HYDRAZIN * BETA-PHENYLETHYLHYDRAZINE
DIHYDROGEN SULFATE

THR: Poison by ingestion, intraperitoneal, intravenous, and subcutaneous routes. An experimental neoplastigen. Human central nervous
system effects by ingestion. Mutagenic data.
When heated to decomposition it emits very
toxic fumes of SO_x and NO_x.

PHENYLGLYCYDYL ETHER　　HR: 2
CAS: 122-60-1　　　　NIOSH: TZ 3675000
mf: $C_9H_{10}O_2$　　mw: 150.19

SYNS: 1,2-EPOXY-3-PHENOXYPROPANE
* 2,3-EPOXYPROPYLPHENYL ETHER * FENYL-
GLYCIDYLETHER (CZECH) * GLYCIDYL PHENYL
ETHER * PHENOL-GLYCIDAETHER (GERMAN)
* PHENOL GLYCIDYL ETHER * 3-PHENOXY-
1,2-EPOXYPROPANE * PHENOXYPROPENE OXIDE
* PHENOXYPROPYLENE OXIDE * PHENYL-2,3-
EPOXYPROPYL ETHER

OSHA PEL: TWA 10 ppm
ACGIH TLV: TWA 1 ppm
DFG MAK: 1 ppm (6 mg/m^3)

THR: Moderately toxic by ingestion and skin
contact. A severe eye and skin irritant. Mutagenic data. See also ethers. When heated to
decomposition it emits acrid smoke and fumes.

PHENYLHYDRAZINE　　HR: 3
CAS: 100-63-0　　　　NIOSH: MV 8925000
DOT: 2572
mf: $C_6H_8N_2$　　mw: 108.16

PROP: Yellow monoclinic crystals or oil; mp:
19.6°; bp: 243.5° (decomp), flash p: 192°F (CC),
d: 1.0978 @ 20°/4°, vap press: 1 mm @ 71.8°,
vap d: 3.7. Sltly sol in hot water; misc in alc,
chloroform, ether, benzene.

SYNS: FENILIDRAZINA (ITALIAN) * FENYLHY-
DRAZINE (DUTCH) * HYDRAZINOBENEZENE
* PHENYLHYDRAZIN (GERMAN)

OSHA PEL: TWA 5 ppm (skin)
ACGIH TLV: TWA 5 ppm; STEL 10 ppm
DFG MAK: 5 ppm (22 mg/m^3)
DOT Classification: Poison B

THR: Poison by ingestion, subcutaneous, intravenous, and possible other routes. An experimental carcinogen. Mutagenic data. Ingestion
or subcutaneous injection can cause hemolysis
of red blood cells. This effect is used in the
treatment of polycythemia. Other effects are
damage to the spleen, liver, kidneys, and bone
marrow. The most common effect of occupational exposure is the development of dermatitis
which, in sensitized persons, may be quite severe. Systemic effects include anemia and general weakness, gastrointestinal disturbances and
injury to the kidneys. Flammable when exposed
to heat, flame or oxidizers; reacts violently with
PbO_2. Dangerous; when heated to decomposition it emits highly toxic fumes of NO_x; can
react with oxidizing materials. To fight fire,
use alcohol foam. Incompatible with lead (IV)
oxide; perchoryl fluoride.

PHENYLHYDRAZINE HYDRO-
CHLORIDE　　HR: 3
CAS: 59-88-1　　　　NIOSH: MV 9000000
mf: $C_6H_8N_2 \cdot ClH$　　mw: 144.62

PROP: Leaflet crystals. Mp: 245° from alcohol;
very sol in water; sol in alc; insol in ether.

SYNS: PHENYLHYDRAZINE MONOHYDROCHLO-
RIDE * PHENYLHYDRAZIN HYDROCHLORID
(GERMAN) * PHENYLHYDRAZINIUM CHLORIDE

THR: Poison by ingestion and subcutaneous
routes. An experimental neoplastigen and tumorigen. Mutagenic data. When heated to decomposition it emits very toxic fumes of NO_x
and HCl.

beta-PHENYLHYDROXYLAMINE　　HR: 3
CAS: 100-65-2　　　　NIOSH: NC 4900000
mf: C_6H_7NO　　mw: 109.14

PROP: Colorless needles; mp: 81°-82°; sol in
hot and cold water; very sol in alc, ether; very
sltly sol in ligroin.

SYNS: NCI-C60093 * N-PHENYLHYDROXYLA-
MINE

THR: Poison to humans by skin contact. Poison
by ingestion and subcutaneous routes. Preparative hazard. When heated to decomposition it
emits toxic fumes of NO_x.

PHENYL MERCURIC CHLORIDE　　HR: 3
CAS: 100-56-1　　　　NIOSH: OW 1400000
mf: C_6H_5ClHg　　mw: 313.15

PROP: Colorless leaves from benzene; mp: 251°; bp: subl.; insol in water; sltly sol in hot alc; sol in pyridine, ether, benzene.

SYNS: CHLORID FENYLRTUTNATY (CZECH) * (CHLOROMERCURI)BENZENE * FENYLMERCU-RICHLORID (CZECH) * MERCURIPHENYL CHLO-RIDE * PHENYL CHLOROMERCURY * PHENYL-MERCURY CHLORIDE * PHENYLQUECKSILBER-CHLORID (GERMAN)

THR: Poison by ingestion, intraperitoneal, sub-cutaneous, and other unspecified routes. Muta-genic data. See also mercury compounds and chlorides. When heated to decomposition it emits very toxic fumes of Cl⁻ and Hg.

PHENYLMERCURIC DINAPHTHYL-METHANEDISULFONATE   HR: 3
CAS: 12040-56-1   NIOSH: OW 6125000
mf: $C_{33}H_{26}Hg_2O_6S_2$ mw: 983.89

SYNS:
DIPHENYLMERCURIDINAPHTHYLMETHANEDISUL-FONATE * 2-NAPHTHALENESULFONIC ACID, 3,3'-METHYLENEDI PHENYL MERCURY

THR: Poison by ingestion and intraperitoneal routes. See also mercury compounds and sulfo-nates. When heated to decomposition it emits very toxic fumes of SO_x and Hg.

PHENYLMERCURIC DINAPHTHYL-METHANEDISULFONATE   HR: 3
CAS: 14235-86-0   NIOSH: OW 6300000
mf: $C_{33}H_{24}Hg_2O_6S_2$ mw: 981.87

SYNS:
BIS(PHENYLMERCURI)METHYLENEDINAPHTHA-LENESULFONATE * METHYLENEDINAPHTHA-LENESULFONIC ACID BISPHENYLMERCURI SALT * PHENYLMERCURIC 3,3'-METHYLENEBIS(2-NAPHTHALENESULFONATE) * PHENYLMERCURY METHYLENEDINAPHTHALENESULFONATE

THR: Poison by ingestion. A severe eye irritant. See also mercury compounds and sulfonates. When heated to decomposition it emits very toxic fumes of Hg and SO_x.

PHENYL MERCURY UREA   HR: 3
CAS: 2279-64-3   NIOSH: OW 9750000
mf: $C_7H_8HgN_2O$ mw: 336.76

SYNS: LEYTOSAN * PHENYLMERCURIC UREA * PHENYLMERCURIUREA

THR: Poison by unspecified routes. See also mercury compounds. When heated to decom-position it emits very toxic fumes of Hg and NO_x.

1-PHENYL-3-MONOMETHYL-TRIA-ZENE   HR: 3
CAS: 16033-21-9   NIOSH: XY 2800000
mf: $C_7H_9N_3$ mw: 135.19

SYN: PMT

THR: Poison by subcutaneous route. Mutagenic data. An experimental teratogen and neoplasti-gen. When heated to decomposition it emits toxic fumes of NO_x.

PHENYL MORPHOLINE   HR: 3
CAS: 92-53-5   NIOSH: QE 8575000
mf: $C_{10}H_{13}NO$ mw: 163.24

PROP: Crystals from ethanol, ether; d: 1.058 @ 270°; mp: 57°; bp: 259.9°; sol in water, alc, ether.

THR: Poison by skin contact. Moderately toxic by ingestion. An eye irritant. When heated to decomposition it emits toxic fumes of NO_x.

N-PHENYL-2-NAPHTHYLAMINE   HR: 3
CAS: 135-88-6   NIOSH: QM 4550000
mf: $C_{16}H_{13}N$ mw: 219.30

PROP: Rhombic cryst from methanol; mp: 107°-108°; bp: 395.5°; insol in water; sol in hot benzene; very sol in hot alc, ether.

SYNS: 2-ANILINONAPHTHALENE * N-(2-NAPHTHYL)ANILINE * 2-NAPHTHYLPHENYL-AMINE * BETA-NAPHTHYLPHENYLAMINE * NCI-C02915 * 2-PHENYLAMINONAPHTHA-LENE * PHENYL-BETA-NAPHTHYLAMINE * PHENYL-2-NAPHTHYLAMINE * N-PHENYL-BETA-NAPHTHYLAMINE

THR: An experimental neoplastigen and sus-pected carcinogen. Mutagenic data. Moderately toxic by ingestion. When heated to decomposi-tion it emits toxic fumes of NO_x.

PHENYLPHOSPHINE   HR: 3
CAS: 638-21-1   NIOSH: SZ 2100000
mf: C_6H_7P mw: 110.10

PROP: Needles from aq alc; mp: 164°-165°; bp: 305°-308°; insol in water; sol in alkali; very sol in alc and ether.
ACGIH TLV: CL 0.05 ppm

THR: Poison by inhalation. See also phosphine. When heated to decomposition it emits toxic fumes of PO_x.

1-PHENYLPIPERAZINE HR: 3
CAS: 92-54-6 NIOSH: TM 2625000
mf: $C_{10}H_{14}N_2$ mw: 162.26

PROP: Pale yellow oil, insol in water, sol in alc, ether. D: 1.0621 @ 20°/4°, bp: 286.5°, mp: 18.8°, flash p: 285°F.

THR: Poison by ingestion and skin contact. A powerful irritant. See also piperazine. Skin irritant. Combustible when exposed to heat or flame. It supports combustion and decomposes to yield toxic fumes of NO_x. To fight fire, use water, foam, dry chemical.

3-PHENYL-1-PROPANOL CARBA-MATE HR: 3
CAS: 673-31-4 NIOSH: UB 9275000
mf: $C_{10}H_{13}NO_2$ mw: 179.24

SYNS: BENZENEPROPANOL CARBAMATE
* CARBAMIC ACID, 3-PHENYLPROPYL ESTER
* 1-CARBAMOYLOXY-3-PHENYLPROPANE
* GAMMA-PHENYLPROPYLCARBAMAT (GERMAN)
* GAMMA-PHENYLPROPYL CARBAMATE

THR: Poison by intravenous and intraperitoneal routes. Moderately toxic by ingestion. See also esters. When heated to decomposition it emits toxic fumes of NO_x.

1-PHENYL-2-PROPYNYL CARBA-MATE HR: 3
CAS: 3567-38-2 NIOSH: DO 5950000
mf: $C_{10}H_9NO_2$ mw: 175.20

SYNS: ALPHA-ETHYNYLBENZYL CARBAMATE
* PHENYLETHYNLCARBINOL CARBAMATE

THR: Poison by intraperitoneal and other unspecified routes. When heated to decomposition it emits toxic fumes of NO_x.

PROPHENPYRIDAMINE HYDROCHLO-RIDE HR: 3
CAS: 25332-09-6 NIOSH: US 9800000
mf: $C_{16}H_{20}N_2 \cdot ClH$ mw: 276.84

SYN: 1-PHENYL-1-(2-PYRIDYL)-3-DIMETHYLAMI-NOPROPANE HYDROCHLORIDE

THR: Poison by ingestion, intravenous, intraperitoneal, and subcutaneous routes. When heated to decomposition it emits very toxic fumes of NO_x and HCl.

PHENYL SALICYLATE HR: 2
CAS: 118-55-8 NIOSH: VO 6125000
mf: $C_{13}H_{10}O_3$ mw: 214.23

PROP: White, small crystals, pleasant odor and taste. D: 1.250 @ 20°/4°; mp: 41.4°; bp: 172°-173° @ 12 mm; sol in water, ether, benzene; very sol in hot alc.

SYN: SALOL

THR: Moderately toxic by ingestion. See also esters. When heated to decomposition it emits acrid smoke and fumes. Incompatible with bromine water, ferric salts; camphor, monobromated camphor; phenol; chloral hydrate; thymol or urethan.

3-PHENYLSALICYLIC ACID HR: 3
CAS: 304-06-3 NIOSH: VO 5600000
mf: $C_{13}H_{10}O_3$ mw: 214.23

PROP: Rhombic from alc; d: 1.250 @ 20°/4°; mp: 41.4°; bp: 172°-173° @ 12 mm; sol in water, ether, benzene; very sol in hot alc.

SYN: USAF DO-59

THR: Poison by intraperitoneal route. When heated to decomposition it emits acrid smoke and fumes.

1-PHENYLTHIOSEMICARBA-ZIDE HR: 3
CAS: 645-48-7 NIOSH: VT 3850000
mf: $C_7H_9N_3S$ mw: 167.25

PROP: Prisms from alc; mp: 200°-201°; sltly sol in water, ether. Sol in hot alc.

SYNS: USAF EK-5426 * USAF EL-45

THR: Poison by ingestion and intraperitoneal routes. Mutagenic data. When heated to decomposition it emits very toxic fumes of NO_x and SO_x.

1-PHENYL-2-THIOUREA HR: 3
CAS: 103-85-5 NIOSH: YU 1400000
mf: $C_7H_8N_2S$ mw: 152.23

PROP: Needle-like crystals, bitter taste. Mp: 154°, d: 1.3. Sol in water and alc; sol in aq ether. Sol in 400 parts cold water.

SYNS: 1-PHENYLTHIOUREA * NCI-C02017
* PHENYLTHIOCARBAMIDE * N-PHENYL-THIOUREA * USAF EK-1569

THR: Poison by ingestion and intraperitoneal routes. When heated to decomposition or on contact with acid or acid fumes it emits highly toxic fumes of SO_x and NO_x.

2-PHENYLVALERIC ACID-2-(DIETH-YLAMINO)ETHYL ESTER HYDRO-CHLORIDE HR: 3
CAS: 132-45-6 NIOSH: YV 7820000
mf: C$_{17}$H$_{27}$NO$_2$ • ClH mw: 313.91

SYNS: 2-DIETHYLAMINOETHYL-ALPHA-PHENYL-VALERATE HYDROCHLORIDE * 2-DIETHYLAMI-NOETHYL-ALPHA-PROPYLTOLUATE HYDROCHLO-RIDE * ALPHA-PHENYL-VALERATE DU DIETHYLAMINO-ETHANOL CHLORHYDRATE (FRENCH)

THR: Poison by intraperitoneal and intravenous routes. Moderately toxic by ingestion and intra-duodenal routes. See also esters. When heated to decomposition it emits very toxic fumes of NO$_x$ and HCl.

PHENYL XYLYL KETONE HR: 1
CAS: 1322-78-7 NIOSH: DJ 1300000
mf: C$_{15}$H$_{14}$O mw: 210.29

SYN: AR,AR-DIMETHYLBENZOPHENONE

THR: Mildly toxic by ingestion. An eye irritant. See also ketones. When heated to decomposition it emits acrid smoke and fumes.

PHLOROGLUCINOL HR: 2
CAS: 108-73-6 NIOSH: SY 1050000
mf: C$_6$H$_6$O$_3$ mw: 126.12

PROP: White crystals; sweet taste, sltly water sol; mp: 218°. Sol in ether; sublimes with de-comp.

SYNS: BENZENE-S-TRIOL * BENZENE-1,3,5-TRIOL * 1,3,5-BENZENETRIOL * 3,5-DIHY-DROXYPHENOL * 5-OXYRESORCINOL * PHLO-ROGLUCIN * S-TRIHYDROXYBENZENE * SYM-TRIHYDROXYBENZENE * 1,3,5-TRIHY-DROXYBENZENE * 1,3,5-TRIHYDROXYCY-CLOHEXATRIENE

THR: Moderately toxic by subcutaneous and intraperitoneal routes. Sltly toxic by ingestion. Mutagenic data. When heated to decomposition it emits acrid smoke and fumes.

PHORATE HR: 3
CAS: 298-02-2 NIOSH: TD 9450000
mf: C$_7$H$_{17}$O$_2$PS$_3$ mw: 260.39

PROP: Liquid, bp: 118°-120° @ 0.8 mm, d: 1.156. @ 25°/4°; insol in water, misc with car-bon tetrachloride, dioxane, xylene.

SYNS: O,O-DIAETHYL-S-(AETHYLTHIO-METHYL)-DITHIOPHOSPHAT (GERMAN) * O,O-DIETHYL S-

ETHYLMERCAPTOMETHYL DITHIOPHOSPHONATE * O,O-DIETHYL-S-(ETHYLTHIO-METHYL)-DITHIO-FOSFAAT (DUTCH) * O,O-DIETHYL S-ETHYL-THIOMETHYL DITHIOPHOSPHONATE * O,O-DI-ETHYL ETHYLTHIOMETHYL PHOSPHORODITHI-OATE * O,O-DIETHYL S-(ETHYLTHIO)METHYL PHOSPHORODITHIOATE * O,O-DIETHYL S-ETH-YLTHIOMETHYL THIOTHIONOPHOSPHATE * O,O-DIETIL-S-(ETILTIO-METIL)-DITIOFOSFATO (ITALIAN) * DITHIOPHOSPHATE DE O,O-DI-ETHYLE ET D'ETHYLTHIOMETHYLE (FRENCH) * ENT 24,042 * FORAAT (DUTCH)

THR: Poison by ingestion, inhalation, skin con-tact, and intravenous routes. A cholinesterase inhibitor. See parathion for disaster hazard and additional toxicity information.

PHORBOL HR: 3
CAS: 17673-25-5 NIOSH: GZ 0600000
mf: C$_{20}$H$_{27}$O$_6$ mw: 363.47

PROP: Anhyd crystals. Decomp @ 250°-251°. 2 forms: mp: 162°-163° and 233°-234°.

THR: An experimental carcinogen. A skin irri-tant. When heated to decomposition it emits acrid smoke and fumes.

PHORBOL-12,13-DIBENZOATE HR: 3
CAS: 25405-85-0 NIOSH: GZ 0591850
mf: C$_{34}$H$_{36}$O$_8$ mw: 572.70

THR: An experimental tumorigen. See also phorbol. When heated to decomposition it emits acrid smoke and fumes.

PHORBOL-12,13-DIDECANOATE HR: 3
CAS: 24928-17-4 NIOSH: GZ 0653000
mf: C$_{40}$H$_{55}$O$_8$ mw: 663.95

SYN: PDD

THR: An experimental tumorigen. See also phorbol. A skin irritant. When heated to decom-position it emits acrid smoke and fumes.

PHORBOL MYRISTATE ACE-TATE HR: 3
CAS: 16561-29-8 NIOSH: GZ 0630000
mf: C$_{36}$H$_{56}$O$_8$ mw: 616.92

SYNS: 12-O-TETRADECANOYLPHORBOL-13-ACE-TATE * PENTAHYDROXY-TIGLIADIENONE-MONOACETATE(C)MONOMYRISTATE(B) * PHORBOL MONOACETATE MONOMYRISTATE * 12-TETRADECANOYLPHORBOL-13-ACETATE

* 12-O-TETRADEKANOYLPHORBOL-13-ACE-
TAT (GERMAN)

THR: An experimental neoplastigen and tumorigen. Mutagenic data. See also phorbol. A skin
irritant. When heated to decomposition it emits
acrid smoke and fumes.

PHORBOLOL MYRISTATE ACE-
TATE   HR: 3
NIOSH: GZ 0595000
mf: $C_{36}H_{58}O_8$ mw: 618.94

SYNS: PHORBOLOL ACETATE MYRISTATE
* TPA-3-BETA-OL

THR: An experimental carcinogen. A skin irritant. When heated to decomposition it emits
acrid smoke and fumes.

PHORBOL-12-o-TIGLYL-13-BUTY-
RATE   HR: 3
NIOSH: GZ 0665500
mf: $C_{28}H_{40}O_8$ mw: 504.68

SYN: 12-O-TIGLYL-PHORBOL-13-BUTYRATE

THR: An experimental carcinogen. A skin irritant. See also phorbol. When heated to decomposition it emits acrid smoke and fumes.

PHORONE   HR: 2
CAS: 504-20-1   NIOSH: MI 5500000
mf: $C_9H_{14}O$ mw: 138.23

PROP: Solid or greenish liquid. Mp: 28°, flash
p: 185°F (OC), d: 0.879, vap press: 1 mm @
42.0°, vap d: 4.8, bp: 198°-199°. Sol in water,
alc, and ether.

SYNS: DIISOPROPYLIDENE ACETONE * SYM-DI-
ISOPROPYLIDENE ACETONE * 2,6-DIMETHYL-
2,5-HEPTADIEN-4-ONE * PHORON (GERMAN)

THR: Moderately toxic by subcutaneous route.
See also isoacetophorone. When heated to decomposition it emits acrid smoke and fumes.
Flammable when exposed to heat or flame; can
react with oxidizing materials. To fight fire,
use foam, CO_2, dry chemical.

PHOSFOLAN   HR: 3
CAS: 947-02-4   NIOSH: NJ 6475000
mf: $C_7H_{14}NO_3PS_2$ mw: 255.31

SYNS: CI-47031 * CYCLIC ETHYLENE(DIETH-
OXYPHOSPHINOTHIOYL)DITHIOIMIDOCARBONATE
* (DIETHOXYPHOSPHINYL)DITHIOIMIDOCARBONIC
ACID CYCLIC ETHYLENE ESTER * 2-(DIETHOXY-

PHOSPHINYLIMINO)-1,3-DITHIOLANE * P,P-DI-
ETHYL CYCLIC ETHYLENE ESTER OF PHOSPHONO-
DITHIOIMIDOCARBONIC ACID * ENT 25,830
* 1,2-ETHANEDITHIOL, CYCLIC ESTER WITH P,P-
DIETHYL PHOSPHONODITHIOIMIDOCARBONATE

THR: Poison by ingestion and skin contact. See
also esters. When heated to decomposition it
emits very toxic fumes of PO_x, SO_x, and NO_x.

PHOSGENE   HR: 3
CAS: 75-44-5   NIOSH: SY 5600000
DOT: 1076
mf: CCl_2O mw: 98.91

PROP: Colorless, poison gas or volatile liquid,
odor of new mown hay or green corn. Mp:
−118°, bp: 8.3°, d: 1.37 @ 20°, vap press:
1180 mm @ 20°, vap d: 3.4. Very sltly sol in
water; very sol in benzene and acetic acid; decomp sltly in water.

SYNS: CARBONE (OXYCHLORURE DE) (FRENCH)
* CARBON OXYCHLORIDE * CARBONYLCHLO-
RID (GERMAN) * CARBONYL CHLORIDE
* CHLOROFORMYL CHLORIDE * CARBONIO
(OSSICLORURO DI) (ITALIAN) * DIPHOSGENE
* FOSGEEN (DUTCH) * FOSGEN (POLISH)
* FOSGENE (ITALIAN) * KOOLSTOFOXYCHLO-
RIDE (DUTCH) * NCI-C60219 * PHOSGEN
(GERMAN)

OSHA PEL: TWA 0.1 ppm
ACGIH TLV: TWA 0.1 ppm
DFG MAK: 0.1 ppm (0.4 mg/m³)
DOT Classification: Poison A, Label: Poison
Gas

THR: Poison by inhalation. A severe eye irritant.
In the presence of moisture, phosgene decomposes to form hydrochloric acid and carbon
monoxide. This occurs in the bronchioles and
alveoli of the lungs resulting in pulmonary edema,
followed by bronchopneumonia and occasionally lung abscess. There is little irritating effect
upon the respiratory tract, and the warning properties of the gas are therefore very slight. There
may be no immediate warning that dangerous
concentrations are being inhaled. After a latent
period of 2 to 24 hours, the patient complains
of burning in the throat and chest, shortness
of breath and increasing dyspnea. Where the
exposure has been severe, the development of
pulmonary edema may be so rapid that the patient dies within 36 hours after exposure. In
cases where the exposure has been less, pneumo-

nia may develop several days after the occurrence of the accident. In patients who recover, no permanent residual disability is thought to occur. A common air contaminant. Reacts violently with Al; *tert*-butyl azido formate; 2,4-hexadiyn-1,6-diol; isopropyl alcohol; K; Na; hexafluoro isopropylidene; amino lithium; lithium. Highly dangerous; when heated to decomposition, or on contact with water or steam, will react to produce toxic and corrosive fumes. Caution: Arrangements should be made for monitoring its use. For further information see Vol. 3, No. 3 of *DPIM Report*.

PHOSPHATES
THR: No specific toxicity. See individual phosphates. Alkali metal phosphates are strong caustics and therefore powerful irritants. For an example of organic phosphates see parathion.

PHOSPHIDES   HR: 3
PROP: Composition: combination of a cation + elemental phosphorus.

THR: Phosphides are particularly dangerous because they tend to decompose to the very toxic phosphine upon contact with moisture or acids. See also phosphine. Dangerous fire hazard by chemical reaction; they react with water and acids to liberate phosphine. Moderate explosion hazard; see phosphine. Dangerous; when heated they may emit highly toxic fumes of PO_x; they react with water or steam to produce toxic and flammable vapors; on contact with oxidizing materials, they can react vigorously; on contact with acid or acid fumes, they can emit toxic fumes.

PHOSPHINE   HR: 3
CAS: 7803-51-2   NIOSH: SY 7525000
DOT: 2199
mf: H_3P mw: 34.00

PROP: Colorless gas, foul odor; mp: $-132.5°$, bp: $-87.5°$, d: 1.529 g/L @ $0°$, autoign temp: 212°F, lel: 1%. Sltly sol in water.

SYNS: FOSFOROWODOR (POLISH) * HYDROGEN PHOSPHIDE * PHOSPHORUS TRIHYDRIDE * PHOSPHORWASSERSTOFF (GERMAN)

OSHA PEL: TWA 400 ug/m^3
ACGIH TLV: TWA 0.3 ppm; STEL 1 ppm
DFG MAK: 0.1 ppm (0.15 mg/m^3)
DOT Classification: Poison A, Label:
  Flammable Gas and Poison Gas

THR: Poison by inhalation. A very toxic gas whose effects are not completely understood. The chief effects are central nervous system depression and lung irritation. There may be pulmonary edema, dilation of the heart, and hyperemia of the visceral organs. Inhalation can cause coma and convulsions leading to death. However, most cases recover without after-effects. Chronic poisoning, characterized by anemia, bronchitis, gastro-intestinal disturbances and visual, speech and motor disturbances, may result from continued exposure to very low concentrations. Very dangerous fire hazard by spontaneous chemical reaction. Moderately explosive when exposed to flame. Reacts violently with air; BCl_3; Br_2; Cl_2; Cl_2O; $Hg(NO_3)_2$; HNO_3; NO; NCl_3; NO_3; N_2O; HNO_2; O_2; (K + NH_3); $AgNO_3$. Dangerous; when heated to decomposition it emits highly toxic fumes of PO_x; can react vigorously with oxidizing materials. To fight fire, use CO_2, dry chemical or water spray. For further information see Vol. 6, No. 2 of *DPIM Report*.

PHOSPHORAMIDE MUSTARD   HR: 3
CAS: 1566-15-0   NIOSH: TD 2530000
mf: $C_4H_{11}Cl_2N_2O_2P \cdot C_6H_{13} \cdot N$ mw: 320.24

SYN: N,N-BIS(2-CHLOROETHYL)PHOSPHORODIAMIDIC ACID, CYCLOHEXYLAMMONIUM SALT

THR: An experimental teratogen. Mutagenic data. When heated to decomposition it emits very toxic fumes of Cl^-, PO_x, and NO_x.

PHOSPHORIC ACID   HR: 3
CAS: 7664-38-2   NIOSH: TB 6300000
DOT: 1805
mf: H_3O_4P mw: 98.00

PROP: Colorless liquid or rhombic crystals; mp: 42.35°, $-1/2H_2O$ @ 213°, fp: 42.4°, d: 1.864 @ 25°, vap press: 0.0285 mm @ 20°.

SYNS: ACIDE PHOSPHORIQUE (FRENCH) * ACIDO FOSFORICO (ITALIAN) * FOSFORZUUROPLOSSINGEN (DUTCH) * ORTHOPHOSPHORIC ACID * PHOSPHORIC ACID (DOT) * PHOSPHORSAEURELOESUNGEN (GERMAN)

OSHA PEL: TWA 1 mg/m^3
ACGIH TLV: TWA 1 mg/m^3; STEL 3 mg/m^3
DOT Classification: Corrosive Material,
  Label: Corrosive

THR: Poison by an unspecified route in man. Moderately toxic by ingestion and skin contact. A severe eye and skin irritant and a systemic irritant by inhalation. A common air contaminant. Dangerous; when heated to decomposition it emits toxic fumes of PO_x. For further information see Vol. 3, No. 4 of *DPIM Report*.

PHOSPHORODIFLUORIDIC ACID HR: 3
CAS: 13779-41-4 NIOSH: TD 4460000
mf: F_2HO_2P mw: 101.98

SYN: DIFLUOROPHOSPHORIC ACID

THR: Toxic and corrosive material. See also phosphorodifluoridic acid, anhydrous. When heated to decomposition it emits very toxic fumes of PO_x and F^-.

PHOSPHORODIFLUORIDIC ACID (ANHYDROUS) HR: 3
CAS: 13779-41-4 NIOSH: TD 4465000
mf: HPO_2F_2 mw: 102

PROP: Mp: $-75°$; bp: 116°; d: 1.583 @ 25°/4°; vap d: 3.52.

SYN: DIFLUOROPHOSPHORIC ACID, ANHYDROUS (DOT)

DOT Classification: Label: Corrosive

THR: Poison by inhalation, ingestion, and skin contact. Very corrosive and irritating material. See also fluorides and phosphoric acid. When heated to decomposition it emits very toxic fumes of F^- and PO_x.

PHOSPHORODI(ISOPROPYLAMIDIC) FLUORIDE HR: 3
CAS: 371-86-8 NIOSH: TD 3675000
mf: $C_6H_{15}FN_2OP$ mw: 182.21

SYNS: BIS(ISOPROPYLAMIDO) FLUOROPHOSPHATE * BIS(MONOISOPROPYLAMINO)FLUOROPHOSPHATE * BIS(MONOISOPROPYLAMINO)FLUOROPHOSPHINE OXIDE * N,N'-DIISOPROPIL-FOSFORODIAMMIDO-FLUORURO (ITALIAN) * DI(ISOPROPYLAMIDO)-PHOSPHORYLFLUORIDE * N,N'-DIISOPROPYL-DIAMIDO-FOSFORZUUR-FLUORIDE (DUTCH) * N,N'-DIISOPROPYL-DIAMIDO-PHOSPHORSA-EURE-FLUORID (GERMAN) * N,N'-DIISOPROPYL-DIAMIDOPHOSPHORYL FLUORIDE * N,N'-DIISO-PROPYLPHOSPHORODIAMIDIC FLUORIDE * FLUOROBISISOPROPYLAMINO- PHOSPHINE OXIDE * FLUORURE DE N,N'-DIISOPROPYLE PHOSPHORODIAMIDE (FRENCH)

DOT Classification: ORM-A, Label: None

THR: Poison by ingestion, subcutaneous, intraperitoneal, and other unspecified routes. See also parathion, fluorides. When heated to decomposition it emits very toxic fumes of F^-, NO_x, and PO_x.

PHOSPHORODITHIOIC ACID-o,o-DI-METHYL-s-(2-ETHYLTHIO)ETHYL ESTER HR: 3
CAS: 640-15-3 NIOSH: TE 4375000
mf: $C_6H_{15}O_2PS_3$ mw: 246.36

PROP: Colorless liquid; sol in acetone, dioxane, acetonitrile.

SYNS: O,O-DIMETHYL-S-(2-AETHYLTHIO-AETHYL)-DITHIO PHOSPHAT (GERMAN) * O,O-DIMETHYL-S-(CARBONYLMETHYLMORPHOLINO) PHOSPHORODITHIOATE * O,O-DIMETHYL-S-(2-ETHYLMERCAPTOETHYL) DITHIOPHOSPHATE * O,O-DIMETHYL-S-2-ETHYLMERKAPTOETHYLES-TER KYSELINY DITHIOFOSFORECNE (CZECH) * O,O-DIMETHYL-S-(2-ETHYLTHIO-ETHYL)-DI-THIOFOSFAAT (DUTCH) * O,O-DIMETHYL S-(2-(ETHYLTHIO)ETHYL) PHOSPHORODITHIOATE * O,O-DIMETIL-S-(ETILTIO-ETIL)-DITIOFOSFATO (ITALIAN) * DITHIOMETON (FRENCH) * DI-THIOPHOSPHATE DE O,O-DIMETHYLE ET DE S-(2-ETHYLTHIO-ETHYLE) (FRENCH) * 2-ETHYLTHIO-ETHYL O,O-DIMETHYL PHOSPHORODITHIOATE * S-(2-(ETHYLTHIO)ETHYL) O,O-DIMETHYLPHOS-PHORODITHIONATE * S-(2-(ETHYLTHIO)-ETHYL)DIMETHYL PHOSPHOROTHIOLOTHIO-NATE

THR: Poison by ingestion, skin contact, intraperitoneal, intravenous, and other unspecified routes. Mutagenic data. A severe eye and moderate skin irritant. See also esters and parathion. A cholinesterase inhibitor. When heated to decomposition it emits very toxic fumes of PO_x and SO_x.

PHOSPHOROFLUORIDIC ACID
CAS: 13537-32-1 NIOSH: TE 5000000
DOT: 1776
mf: FH_2O_3P mw: 99.99

SYNS: FLUOROPHOSPHORIC ACID, ANHYDROUS * MONOFLUOROPHOSPHORIC ACID, ANHYDROUS

DOT Classification: Corrosive Material, Label: Corrosive

THR: No toxicity data. A corrosive and irritating material to skin, eyes, and mucous membranes.

See also fluorides. When heated to decomposition it emits very toxic fumes of F^- and PO_x.

PHOSPHOROTHIOIC ACID-s-(((1-CYANO-1-METHYL ETHYL)CARBAMOYL)METHYL)-o,o-DIETHYL ESTER   HR: 3
CAS: 3734-95-0   NIOSH: TE 8750000
mf: $C_{10}H_{19}N_2O_4PS$ mw: 294.34

SYNS: ALPHA-CYANOISOPROPYLAMIDE OF THE O,O-DIETHYLTHIOPHOSPHORYL ACETIC ACID * S-(((1-CYANO-1-METHYL-ETHYL)CARBAMOYL)METHYL) O,O-DIETHYL PHOSPHOROTHIOATE * S-N-(1-CYANO-1-METHYLETHYL)CARBAMOYLMETHYL DIETHYL PHOSPHOROTHIOLATE * O,O-DIAETHYL-S-1-METHYL)AETHYL)-CARBAMOYL-METHYL-MONOTHIOPHOSPHAT (GERMAN) * O,O-DIETHYL-S-((2-CYAAN-2-METHYL-ETHYL)-CARBAMOYL)-METHYL-MONOTHIOFOSFAAT (DUTCH) * O,O-DIETHYL-S-N-(A-CYANOISOPROPYL)CARBOMOYLMETHYL PHOSPHOROTHIOATE * O,O-DIETIL-S-((2-CIAN-2-METIL-ETIL)-CARBAMOIL)-METIL-MONOTIOFOSFATO (ITALIAN) PHTHIOPHOSPHATE DE S-N-(1-CYANO-1-METHYLETHYL)CARBAMOYLMETHYLE ET DE O,O-DIETHYLE (FRENCH)

THR: Poison by ingestion and skin contact. See also esters. When heated to decomposition it emits very toxic fumes of PO_x, SO_x, and NO_x.

PHOSPHOROUS ACID TRIS(2-FLUOROETHYLESTER)   HR: 3
CAS: 63980-61-0   NIOSH: KL 1925000
mf: $C_6H_{12}F_3O_3P$ mw: 220.15

SYN: ETHANOL, 2-FLUORO-, PHOSPHITE (3:1)

THR: Poison by inhalation. When heated to decomposition it emits very toxic fumes of F^- and P.

PHOSPHORUS (RED)   HR: 3
CAS: 7723-14-0   NIOSH: TH 3495000
DOT: 1381
af: P   aw: 30.97

PROP: Reddish-brown powder. Bp: 280° (with ignition), mp: 590° @ 43 atm, d: 2.34, autoign temp: 500°F in air, vap d: 4.77.

SYN: PHOSPHORUS, AMORPHOUS, RED (DOT)

ACGIH TLV: TWA 0.1 mg/m³
DFG MAK: 0.1 mg/m³
DOT Classification: Label: Flammable Solid and Poison

THR: Poison to humans. May have white phosphorus as an impurity; see also white phosphorus. Dangerous fire hazard when exposed to heat or by chemical reaction with oxidizers. Moderate explosion hazard by chemical reaction or on contact with organic materials. Dangerous; when heated it emits highly toxic fumes of PO_x; can react with reducing materials. To fight fire, use water. Incompatible with $KClO_3$; $KMnO_4$; peroxides; oxidizing materials. For further information see Vol. 3, No. 4 of *DPIM Report*.

PHOSPHORUS (WHITE)   HR: 3
CAS: 7723-14-0   NIOSH: TH 3500000
DOT: 2447
mf: P_4 mw: 123.88

PROP: Cubic crystals, colorless to yellow, waxlike solid. Mp: 44.1°, bp: 280°, flash p: spontaneously flammable in air, d: 1.82, autoign temp: 86°F, vap press: 1 mm @ 76.6°, vap d: 4.42.

SYNS: COMMON SENSE COCKROACH AND RAT PREPARATIONS * FOSFORO BIANCO (ITALIAN) * GELBER PHOSPHOR (GERMAN) * PHOSPHORE BLANC (FRENCH) * TETRAFOSFOR (DUTCH) * TETRAPHOSPHOR (GERMAN) * WEISS PHOSPHOR (GERMAN) * WHITE PHOSPHORUS * YELLOW PHOSPHORUS

OSHA PEL: TWA 100 ug/m³
DOT Classification: Label: Flammable Solid and Poison

THR: Poison by inhalation, ingestion, skin contact, and subcutaneous routes. Central nervous system and other systemic effects by ingestion. Toxic quantities have an acute effect on the liver, can cause severe eye damage. Inhalation can cause photophobia with myosis, dilation of the pupils, retinal hemorrhage, congestion of the blood vessels, and rarely an optic neuritis. Chronic exposure by inhalation or ingestion can cause anemia, gastrointestinal effects, and brittleness of the long bones leading to spontaneous fractures. The most common symptom, however, of chronic phosphorous poisoning is necrosis of the jaw (phossy-jaw). Dangerous fire hazard when exposed to heat or by chemical reaction with oxidizers. Ignites spontaneously in air. Very reactive. If combustion occurs in a confined space, it will remove the oxygen and cause asphyxiation. Dangerous explosion hazard by chemical reaction with: alkaline hydroxides; NH_4NO_3; SbF_5; $Ba(BrO_3)_2$; Be; BI_3; $Ca(BrO_3)_2$; $Mg(BrO_3)_2$; $K(BrO_3)$; $NaBrO_3$; $Zn(BrO_3)_2$; Br_2;

halogens; BrF_3; BrN_3; (chlorates of Ba, Ca, Mg, K, Na, Zn); (iodates of Ba, Ca, Mg, K, Na, Zn); Ce; Cs; $CsHC_2$; Cs_3N; (charcoal + air); ClO_2; (Cl_2 + heptane); ClO; ClF_3; ClO_3; chlorosulfonic acid; CrO_3; $Cr(OCl)_2$; Cu; NCl; IBr; ICl; IF_5; Fe; La; PbO_2; Li; Li_2C_2; Li_6CS; $Mg(ClO_4)_2$; Mn; HgO; $HgNO_3$; Nd; Ni; nitrates; NBr; NO_2; NBr_3; NCl_3; NOF; FNO_2; O_2; performic acid; Pt; K; KOH; K_3N; $KMnO_4$; K_2O_2; Rb; $RbHC_2$; Se_2Cl_2; $SeOCl_2$; $SeOF_2$; SeF_4; $AgNO_3$; Ag_2O; Na; Na_2C_2; $NaClO_2$; NaOH; Na_2O_2; S; SO_3; H_2SO_4; Th; $VOCl_2$; Zn; peroxyformic acid; chloro sulfuric acid; halogen azides; hexalithium disilicide. Dangerous; when heated to decomposition it emits highly toxic fumes of PO_x; can react vigorously with oxidizing materials. To fight fire, use water. For further information see Vol. 3, No. 4 of *DPIM Report*.

PHOSPHORUS (WHITE IN WATER)   HR: 3
CAS: 7723-14-0   NIOSH: TH 3505000

SYNS: PHOSPHORUS WHITE IN WATER * PHOSPHORUS YELLOW IN WATER

DOT Classification: Label: Flammable Solid and Poison

THR: A poison. Very flammable solid must be kept under water. See also phosphorus (white). When heated to decomposition it emits toxic fumes of PO_x.

PHOSPHORUS CHLORIDE   HR: 3
CAS: 7719-12-2   NIOSH: TH 3675000
DOT: 1809
mf: Cl_3P mw: 137.32

PROP: Clear, colorless, fuming liquid. Mp: $-111.8°$, bp: $76°$, d: 1.574 @ $21°$, vap press: 100 mm @ $21°$, vap d: 4.75. Decomp by H_2O, alc; sol in benzene, chloroform, ether.

SYNS: CHLORIDE OF PHOSPHORUS * FOSFORO-(TRICLORURO DI) (ITALIAN) * FOSFORTRICHLO-RIDE (DUTCH) * PHOSPHORE(TRICHLORURE DE) (FRENCH) * PHOSPHORTRICHLORID (GERMAN) * PHOSPHORUS TRICHLORIDE * TROJCHLOREK FOSFORU (POLISH)

OSHA PEL: TWA 0.5 ppm
ACGIH TLV: TWA 0.2 ppm; STEL 0.5 ppm
DFG MAK: 0.5 ppm (3 mg/m³)
DOT Classification: Label: Corrosive and Poison

THR: Poison by inhalation and poison irritant to skin, ·eyes (@ 2 ppm), and mucous membranes. Moderately toxic by ingestion. Violent fire and explosion reactions with acetic acid; Al; $Cr(OCl)_2$; (diallyl phosphite + allyl alcohol); F_2; dimethyl sulfoxide; hydroxylamine; ICl; PbO_2; HNO_3; HNO_2; organic matter; K; Na; water. Dangerous; when heated to decomposition it emits highly toxic fumes of chlorides and PO_x; will react with water, steam or acids to produce heat and toxic and corrosive fumes; can react with oxidizing materials. To fight fire, use CO_2, dry chemical. For further information see phosphorous trichloride, Vol. 3, No. 4 of *DPIM Report*.

PHOSPHORUS COMPOUNDS, INORGANIC
THR: Variable toxicity. Most inorganic phosphates, except phosphine, have low toxicity but in large doses, they may cause serious disturbances, particularly in calcium metabolism. Metaphosphates may be highly toxic, causing irritation and hemorrhages in the stomach, as well as liver and kidney damage. Common air contaminants.

PHOSPHORUS FLUORIDE   HR: 3
CAS: 7783-55-3   NIOSH: TH 3850000
mf: F_3P mw: 87.97

PROP: Colorless gas; mp: $-152°$, bp: $-102°$, d: 3.907 g/L.

SYN: PHOSPHOROUS TRIFLUORIDE

THR: Poison by inhalation. Severe eye, skin, and mucous membrane irritant. See hydrofluoric acid, fluorides and phosphorus pentafluoride. Violent reaction with diborane; F_2; O_2. Dangerous; when heated to decomposition it emits highly toxic fumes of F^-, PO_x; will react with water or steam to produce toxic and corrosive fumes. Incompatible with dioxygen difluoride; fluorine; ·hexafluoroisopropylidene amino lithium.

PHOSPHORUS HEPTASULFIDE   HR: 3
CAS: 12037-82-0   NIOSH: TH 3870000
DOT: 1339

PROP: Light yellow crystals; light gray powder or fused solid; mp: $310°$; bp: $523°$; d: 2.19 @ $17°$.

DOT Classification: Label: Flammable Solid

THR: Poison by ingestion. See also sulfides and phosphorus. When heated to decomposition it emits very toxic fumes of PO_x and SO_x.

PHOSPHORUS PENTABRO-MIDE   HR: 3
mf: PBr_5 mw: 430.56

PROP: Yellow, crystalline mass. Sol in carbon disulfide. Mp: decomp; bp: decomp @ 106°.

SYN: PHOSPHORIC BROMIDE

THR: A poison. Caustic to body tissues. See bromides. Flammable by chemical reaction. Contact with moisture can cause a violent reaction and evolution of heat. Dangerous; when heated to decomposition it emits highly toxic fumes of Br^-. Incompatible with water or steam to produce heat and toxic and corrosive fumes.

PHOSPHORUS PENTACHLOR-IDE   HR: 3
CAS: 10026-13-8   NIOSH: TB 6125000
DOT: 1806
mf: Cl_5P mw: 208.22

PROP: Yellowish-white, fuming, crystalline mass. Pungent odor. Mp: (under press) 148° decomp, bp: subl @ 160°, d: 4.65 g/L @ 296°, vap press: 1 mm @ 55.5°.

SYNS: FOSFORO(PENTACHLORURO DI) (ITALIAN) * FOSFORPENTACHLORIDE (DUTCH) * PIE-CIOCHLOREK FOSFORU (POLISH) * PHOSPHORE-(PENTACHLORURE DE) (FRENCH) * PHOSPHORIC CHLORIDE * PHOSPHORPENTACHLORID (GER-MAN) * PHOSPHORUS PERCHLORIDE

OSHA PEL: TWA 1 mg/m^3
ACGIH TLV: TWA 0.1 ppm
DFG MAK: 1 mg/m^3
DOT Classification: Label: Corrosive

THR: Poison by inhalation. Moderately toxic by ingestion. A severe eye, skin, and mucous membrane irritant. Corrosive to body tissues. Flammable by chemical reaction. Reacts violently with moisture; ClO_3; F_2; hydroxylamine; MgO; P_2O_3; K; Na. Dangerous; when heated to decomposition it emits highly toxic fumes of Cl^-; will react with water or steam to produce heat and toxic and corrosive fumes. To fight fire, use CO_2, dry chemical. Incompatible with aluminum; chlorine dioxide; chlorine; diphosphorus trioxide; fluorine; hydroxylamine; mag-nesium oxide; 3'-methyl-2-nitrobenzanilide; nitrobenzene; sodium; urea; water.

PHOSPHORUS PENTAFLUOR-IDE   HR: 3
mf: PF_5 mw: 125.98

PROP: Colorless gas, fumes strongly in air. Mp: −93.8°; bp: −84.6°; d: (gas): 5.805 g/L.

THR: Poison. Violently irritating to skin, eyes, and mucous membranes. Inhalation may cause pulmonary edema. See fluorides. Dangerous; when heated to decomposition it emits highly toxic fumes of F^- and phosphorus compounds. Incompatible with water or steam to produce toxic and corrosive fumes.

PHOSPHORUS PENTASULFIDE   HR: 3
CAS: 1314-80-3   NIOSH: TH 4375000
DOT: 1340
mf: P_2S_5 mw: 222.24

PROP: Gray to yellow-green, crystalline, deliquescent mass. Bp: 514°, d: 2.09, autoign temp: 287°F. mp: 286°-290°.

SYNS: PENTASULFURE DE PHOSPHORE (FRENCH) * PHOSPHORIC SULFIDE * PHOSPHORUS PERSULFIDE * SIRNIK FOSFORECNY (CZECH) * THIOPHOSPHORIC ANHYDRIDE

OSHA PEL: TWA 1 mg/m^3
ACGIH TLV: TWA 1 mg/m^3; STEL 3 mg/m^3
DFG MAK: 1 mg/m^3
DOT Classification: Label: Flammable Solid and Dangerous When Wet

THR: Poison by ingestion. A severe eye and mucous membranes irritant. See also hydrogen sulfide. Readily liberates hydrogen sulfide and phosphorus pentoxide on contact with moisture. Dangerous fire hazard in the form of dust when exposed to heat or flame. Evolves heat in contact with moisture. Spontaneous heating in the presence of moisture. Moderate explosion hazard in solid form by spontaneous chemical reaction. Dangerous; when heated to decomposition it emits highly toxic fumes of SO_x and PO_x; will react with water, steam, or acids to produce toxic and flammable vapors; can react vigorously with oxidizing materials. Incompatible with air, alcohols, water. To fight fire use CO_2 snow, dry chemical or sand. For further information see Vol. 3, No. 4 of *DPIM Report*.

PHOSPHORUS PENTOXIDE   HR: 3
CAS: 1314-56-3   NIOSH: TH 3945000

DOT: 1807
mf: O_5P_2 mw: 141.94

PROP: Deliquescent crystals. D: 2.30; mp: 340°; sublimes @ 360°.

SYNS: DIPHOSPHORUS PENTOXIDE * PHOS-PHORIC ANHYDRIDE * PHOSPHORUS (V) OXIDE * PO$_x$

DFG MAK: 1 mg/m^3
DOT Classification: Label: Corrosive

THR: Poison by inhalation. See also phosphorus pentoxide. A corrosive material. Incompatible with formic acid; hydrogen fluoride; inorganic bases; metals; oxidants; water.

PHOSPHORUS SESQUISUL-FIDE   HR: 3
CAS: 1314-85-8   NIOSH: TH 4330000
DOT: 1341
mf: P_4S_3 mw: 220.06

PROP: Yellow cryst mass; mp: 172.5°; bp: 407°; d: 2.03; autoign temp: 212°F.

SYNS: PHOSPHORUS (III) SULFIDE (IV) * SES-QUISULFURE DE PHOSPHORE (FRENCH) * TET-RAPHOSPHORUS TRISULFIDE * TRISULFURATED PHOSPHORUS

DOT Classification: Label: Flammable Solid and Dangerous When Wet

THR: Poison by ingestion. See also sulfides and phosphorus. Moderately flammable by spontaneous ignition. When heated to decomposition it emits very toxic fumes of PO$_x$ and SO$_x$.

PHOSPHORUS TRIAZIDE
mf: N_9P mw: 157.04

THR: No toxicity data; highly explosive. See also azides, phosphides. When heated to decomposition it emits very toxic fumes of NO$_x$, PO$_x$, and PH$_3$.

PHOSPHORUS TRIBROMIDE
mf: Br_3P mw: 270.70

PROP: Mp: −40°; bp: 175.3°; d: 2.852 @ 15°; vap press: 10 mm @ 47.8°.

THR: No toxicity data, but probably highly toxic. See also phosphides, bromides. Corrosive. When heated to decomposition it emits very toxic fumes of Br$^-$ and PO$_x$. Will react with water, steam or acids to produce heat,

toxic and corrosive fumes. Incompatible with water; potassium; sodium; RuO$_4$.

PHOSPHORUS TRIOXIDE   HR: 3
mf: O_3P_2 mw: 110.0

PROP: Transparent, monoclinic crystals, or colorless liquid. D: 2.135 @ 21°/4°; mp: 23.8°; bp: 173.1°. Sol in benzene.

THR: Poisonous; no toxicity data. See also phosphorus pentoxide. When heated to decomposition it emits toxic fumes of PO$_x$. Incompatible with ammonia; disulfur dichloride; halogens; oxygen; phosphorus pentachloride; sulfur; sulfuric acid; water.

PHOSPHORUS TRISULFIDE
CAS: 12165-69-4   NIOSH: TH 4520000
DOT: 1343
mf: P_2S_3 mw: 158.12

PROP: Gray-yellow cryst; mp: 290°; bp: 490°.
DOT Classification: Label: Flammable Solid

THR: No toxicity data. See also sulfides and phosphorus compounds. Can react with oxidizers, water or steam to emit toxic fumes of H$_2$S. When heated to decomposition it emits very toxic fumes of PO$_x$ and SO$_x$.

PHOSPHORYL BROMIDE   HR: 3
CAS: 7789-59-5   NIOSH: TH 4750000
DOT: 1939/2576
mf: Br_3OP mw: 286.70

PROP: Colorless plates; mp: 56°; bp: 190°; d: 2.882.

SYNS: PHOSPHOROUS OXYBROMIDE; PHOSPHO-RYL TRIBROMIDE

DOT Classification: Label: Corrosive

THR: Poison by ingestion, inhalation, and skin contact. Severe irritant to skin, eyes, and mucous membranes. A corrosive material. Reacts with steam, water to produce much heat with toxic fumes. See also bromides and phosphorus compounds. When heated to decomposition it emits very toxic fumes of Br$^-$ and PO$_x$.

PHOSPHORYL CHLORIDE   HR: 3
CAS: 10025-87-3   NIOSH: TH 4897000
DOT: 1810
mf: Cl_3OP mw: 153.32

PROP: Colorless to slightly yellow fuming liquid. Mp: 1.2°, bp: 105.1°, d: 1.685 @ 15.5°, vap press: 40 mm @ 27.3°, vap d: 5.3.

SYNS: PHOSPHORUS OXYCHLORIDE * PHOS-PHORUS OXYTRICHLORIDE

ACGIH TLV: TWA 0.1 ppm; STEL 0.5 ppm
DFG MAK: 0.2 ppm (1 mg/m³)
DOT Classification: Label: Corrosive

THR: Poison by inhalation and ingestion. Severe eye, skin, and mucous membrane irritant. Reacts violently with BI_3; (2,5-dimethyl pyrrole + dimethyl formamide); dimethyl sulfoxide; organic matter; Na; water. Dangerous; when heated to decomposition it emits highly toxic fumes of Cl^- and PO_x; will react with water or steam to produce heat and toxic and corrosive fumes. Incompatible with carbon disulfide; N,N-dimethylformamide; 2,5-dimethylpyrrole; 2,6-dimethylpyridine N-oxide; dimethylsulfoxide; ferrocene-1,l-dicarboxylic acid; water; zinc. For further information see phosphorus oxychloride; Vol. 3, No. 4 of *DPIM Report*.

1(2H)-PHTHALAZINONE HYDRA-ZONE   HR: 3
CAS: 86-54-4   NIOSH: TH 8925000
mf: $C_8H_8N_4$ mw: 160.20

SYNS: CIBA 5968 * HYDRAZINOPHTHALAZINE * 1-HYDRAZINOPHTHALAZINE

THR: Poison by ingestion, intravenous, intraperitoneal, and subcutaneous routes. Mutagenic data. When heated to decomposition it emits toxic fumes of NO_x.

PHTHALIC ACID   HR: 2
CAS: 88-99-3   NIOSH: TH 9625000
mf: $C_8H_6O_4$ mw: 166.14

PROP: Crystals. Mp: >230°; d: 1.59; bp: 155° decomp; sol in water, alc; slightly sol in ether, insol in chloroform.

SYNS: ACIDE PHTALIQUE (FRENCH) * BEN-ZENE-1,2-DICARBOXYLIC ACID * O-BENZENEDI-CARBOXYLIC ACID * 1,2-BENZENEDICARBOX-YLIC ACID * O-DICARBOXYBENZENE

THR: Moderately toxic by intraperitoneal route. Slightly toxic by ingestion. Skin and mucous membrane irritant. Combustible when heated in the form of dust (anhydride) it can explode. Violent reaction with HNO_3. Incompatible with sodium nitrite.

PHTHALIC ANHYDRIDE   HR: 3
CAS: 85-44-9   NIOSH: TI 3150000

DOT: 2214
mf: $C_8H_4O_3$ mw: 148.12

PROP: White crystalline needles. Mp: 131.2°, lel = 1.7%, uel = 10.4%, bp: 295° (sublimes), flash p: 305°F (CC), d: 1.527 @ 4°, autoign temp: 1058°F, vap press: 1 mm @ 96.5°, vap d: 5.10. Very sltly sol in water; sol in alc; sltly sol in ether.

SYNS: ANHYDRIDE PHTALIQUE (FRENCH) * ANIDRIDE FTALICA (ITALIAN) * 1,2-BEN-ZENEDICARBOXYLIC ACID ANHYDRIDE * 1,3-DIOXOPHTHALAN * FTALOWY BEZWODNIK (POLISH) * FTAALZUURANHYDRIDE (DUTCH) * 1,3-ISOBENZOFURANDIONE * NCI-C03601 * 1,3-PHTHALANDIONE * PHTHALIC ACID ANHYDRIDE * PHTHALSAEUREANHYDRID (GERMAN)

OSHA PEL: TWA 2 ppm
ACGIH TLV: TWA 1 ppm;
DFG MAK: 5 mg/m³
DOT Classification: Corrosive

THR: Poison by ingestion. A severe eye and moderate skin irritant. A common air contaminant. Combustible when exposed to heat or flame; can react with oxidizing materials. Moderate explosion hazard in the form of dust, when exposed to flame. Violent reaction with HNO_3. To fight fire, use CO_2, dry chemical. Incompatible with copper oxide; nitric acid; sodium nitrite.

PHTHALIMIDE   HR: 3
CAS: 85-41-6   NIOSH: TI 3920000
mf: $C_8H_5NO_2$ mw: 147.14

PROP: White to light tan powder; mp: 238°; bp: subl; sol in water, alc, alkali; sol in hot ether; insol in benzene.

SYNS: ISOINDOLE-1,3-DIONE * 1,3-ISOINDO-LEDIONE * 1,3-ISOINDOLINEDIONE * O-PHTHALIC IMIDE

THR: An experimental teratogen. When heated to decomposition it emits toxic fumes of NO_x.

PHTHALIMIDOMETHYL-o,o-DIMETHYL PHOSPHORODITHIOATE   HR: 3
CAS: 732-11-6   NIOSH: TE 2275000
mf: $C_{11}H_{12}NO_4PS_2$ mw: 317.33

SYNS: (O,O-DIMETHYL-PHTHALIMIDOMETHYL-DITHIOPHOSPHATE) * O,O-DIMETHYL S-(N-PHTHALIMIDOMETHYL) DITHIOPHOSPHATE * O,O-DIMETHYL S-PHTHALIMIDOMETHYL PHOS-

PHORODITHIOATE * ENT 25,705 * N-(MER-
CAPTOMETHYL)PHTHALIMIDE S-(O,O-DIMETHYL
PHOSPHORODITHIOATE) * PHOSPHORODITHIOIC
ACID, S-((1,3-DIHYDRO-1,3-DIOXO-ISOINDOL-2-
YL)METHYL) O,O-DIMETHYL ESTER * PHTHALI-
MIDO-O,O-DIMETHYL PHOSPHORODITHIOATE

THR: Poison by inhalation, ingestion, and possibly other routes. Moderately toxic by skin contact. An experimental teratogen by ingestion.
Mutagenic data. Human blood system effects
by ingestion. See also esters. When heated to
decomposition it emits very toxic fumes of NO_x,
PO_x, and SO_x.

PHTHALONITRILE   HR: 3
CAS: 91-15-6   NIOSH: TI 8575000
mf: $C_8H_4N_2$ mw: 128.14

SYNS: O-DICYANOBENZENE * 1,2-DICYANO-
BENZENE * PHTHALIC ACID DINITRILE
* PHTHALODINITRILE * O-PHTHALODINITRILE
* USAF ND-09

THR: Poison by ingestion, subcutaneous, and
intraperitoneal routes. An experimental tumorigen. See also nitriles. When heated to decomposition it emits toxic fumes of NO_x.

PHYSOSTIGMINE   HR: 3
CAS: 57-47-6   NIOSH: TJ 2100000
mf: $C_{15}H_{21}N_3O_2$ mw: 275.39

SYNS: METHYL CARBAMIC ACID, ESTER WITH
ESEROLINE * ESEROLEIN, METHYLCARBAMATE
(ESTER)

THR: Poison by ingestion, subcutaneous, intramuscular, intravenous, intraperitoneal, and possibly other routes. Central and peripheral nervous system effects in humans by ingestion.
Normally administered by injection. Poisoning
can occur as a result of a mistake in dosage
or due to hypersensitivity of the patient within
5 to 25 minutes after administration. Death usually results from respiratory paralysis. Combustible when exposed to heat or flame. When heated
to decomposition it emits toxic fumes of NO_x.

PHYSOSTIGMINE SALICYLATE
(1:1)   HR: 3
CAS: 57-64-7   NIOSH: TJ 2450000
mf: $C_{15}H_{21}N_3O_2 \cdot C_7H_6O_3$ mw: 413.52

SYNS: ESERINE SALICYLATE * SALICYLIC
ACID, COMPOUND WITH PHYSOSTIGMINE (1:1)

THR: Poison by ingestion, subcutaneous, intramuscular, intravenous, and intraperitoneal

routes. Human psychotropic effects by ingestion. When heated to decomposition it emits
toxic fumes of NO_x.

PHYTIC ACID   HR: 3
CAS: 83-86-3   NIOSH: NM 7525000
mf: $C_6H_{18}O_{24}P_6$ mw: 660.06

SYNS: FYTIC ACID * HEXAKIS(DIHYDROGEN
PHOSPHATE) MYO-INOSITOL * INOSITHEXA-
PHOSPHORSAURE (GERMAN) * MYO-INOSISTOL
HEXAKISPHOSPHATE * INOSITOL HEXAPHOS-
PHATE * MYO-INOSITOL HEXAPHOSPHATE
* SAURE DES PHYTINS (GERMAN)

THR: Poison by intravenous route. When heated
to decomposition it emits toxic fumes of PO_x.

PICRACONITINE   HR: 3
CAS: 466-24-0   NIOSH: TJ 7849250
mf: $C_{32}H_{45}NO_{10}$ mw: 603.78

SYNS: BENZACONINE * BENZOYLACONINE
* ISACONITINE

THR: Poison by intravenous and intraperitoneal
routes. When heated to decomposition it emits
toxic fumes of NO_x.

PICRIC ACID   HR: 3
CAS: 88-89-1   NIOSH: TJ 7875000
DOT: 0154
mf: $C_6H_3N_3O_7$ mw: 229.12

PROP: Yellow crystals or yellow liquid, very
bitter. Mp: 121.8°, bp: explodes > 300°, flash
p: 302°F, d: 1.763, autoign temp: 572°F, vap
d: 7.90.

SYNS: ACIDE PICRIQUE (FRENCH) * ACIDO PI-
CRICO (ITALIAN) * CARBAZOTIC ACID
* C.I. 10305 * 2-HYDROXY-1,3,5-TRINITRO-
BENZENE * NITROXANTHIC ACID * PHENOL
TRINITRATE * PICRONITRIC ACID * PIKRINE-
ZUUR (DUTCH) * PIKRINSAEURE (GERMAN)
* PIKRYNOWY KWAS (POLISH) * 2,4,6-TRINI-
TROFENOL (DUTCH) * 2,4,6-TRINITROFENOLO
(ITALIAN) * 1,3,5-TRINITROPHENOL * 2,4,6-
TRINITROPHENOL

OSHA PEL: TWA 100 ug/m³ (skin)
ACGIH TLV: TWA 0.1 mg/m³
DFG MAK: 0.1 mg/m³
DOT Classification: Class A Explosive

THR: Poison by ingestion, subcutaneous, and
possibly other routes. Mutagenic data. An irritant and an allergen. Skin contact can cause

local and systemic allergic reactions. A severe explosion hazard when shocked or exposed to heat. It forms salts easily. Many of its salts, known as picrates, are more sensitive explosives than picric acid. It forms unstable salts with concrete; NH_3; bases and metals, particularly copper, lead, and zinc. See also explosives, high. Upon decomposition it emits highly toxic fumes and explodes; can react vigorously with reducing materials.

PICROTOXIN   HR: 3
CAS: 124-87-8   NIOSH: TJ 9100000
mf: $C_{13}H_{18}O_7 \cdot C_{15}H_{16}O_6$ mw: 578.62

PROP: Dried fruit of *Anamerta Cocculus* (L.), containing menispermine, paramenispermine, 1% picrotoxin, pictrotoxic acid, cocculine alkaloid, and 5% fat.

SYNS: COCCULIN * COCCULUS * COCCULUS SOLID (DOT) * COQUES DU LEVANT (FRENCH) * FISH BERRY * INDIAN BERRY * ORIENTAL BERRY * PICROTIN, COMPOUND WITH PICRO-TOXININ (1:1) * PICROTOXINE

DOT Classification: Label: Poison

THR: Poison by ingestion, intraperitoneal, subcutaneous, intravenous, intramuscular, parenteral, intracerebral, and other routes. An alkaloid convulsant poison. When heated to decomposition it emits acrid smoke and fumes.

PILOCARPINE   HR: 3
CAS: 92-13-7   NIOSH: TK 1400000
mf: $C_{11}H_{16}N_2O_2$ mw: 208.29

PROP: Colorless or yellow, hygroscopic, needle-like crystals. Mp: 34°, bp: 260° @ 5 mm.

SYNS: ALMOCARPINE * ALPHA-ETHYL-BETA-(HYDROXYMETHYL)-1-METHYL-IMIDAZOLE-5-BU-TYRIC ACID, GAMMA-LACTONE * PILOCARPOL

THR: Poison by subcutaneous, intravenous, intraperitoneal, and possibly other routes. A very poisonous alkaloid which is used to remove excess fluid accumulations from the body. Its action on the sweat glands makes it a powerful sudorific. It very rarely causes death, but when it does, it is by paralysis of the heart or edema of the lungs. Dangerous; on heating to decomposition it emits toxic fumes of NO_x.

2-PINENE   HR: 3
CAS: 80-56-8   NIOSH: DT 7000000
mf: $C_{10}H_{16}$ mw: 136.26

PROP: Liquid, odor of turpentine. Mp: −55°, bp: 155°, flash p: 91°F, d: 0.8592 @ 20°/4°, vap press: 10 mm @ 37.3°, vap d: 4.7, autoign temp: 491°F. Insol in water; sol in alc, chloroform, ether, glacial acetic acid.

SYNS: ALPHA-PINENE * 2,6,6-TRIMETHYLBICYCLO(3.1.1)-2-HEPT-2-ENE

THR: Poison by inhalation. Moderately toxic by ingestion. An eye, mucous membrane, and severe skin irritant. A dangerous fire hazard when exposed to heat, flame, or oxidizing materials. To fight fire, use foam, CO_2, dry chemical. Incompatible with nitrosyl perchlorate.

PINE OIL   HR: 2
CAS: 8002-09-3   NIOSH: TK 5100000
DOT: 1272

PROP: Pale yellow liquid, penetrating odor. Bp: 200°-220°, flash p: 172°F (CC), d: 0.86, flash p (steam distilled): 138°F. Insol in water; sol in organic solvents.

SYNS: OIL OF PINE * OLEUM ABIETIS

DOT Classification: Combustible Liquid, Label: None

THR: A weak allergen and an irritant to skin and mucous membranes. See also turpentine. Large doses cause central nervous depression. Flammable when exposed to heat or flame; can react with oxidizing materials. Moderate spontaneous heating. To fight fire, use foam, CO_2, dry chemical.

PIPECURIUM BROMIDE   HR: 3
                         NIOSH: TM 3170000
mf: $C_{35}H_{62}N_4O_4 \cdot 2Br$ mw: 762.83

SYN: 2-BETA,16-BETA-(4'-DIMETHYL-1'-PIPERA-ZINO)-3-ALPHA,17-BETA-DIACETOXY-5-ALPHA-ANDROSTANE-2-BR

THR: Poison by ingestion, intramuscular, intravenous, subcutaneous, and intraperitoneal routes. When heated to decomposition it emits very toxic fumes of Br^- and NO_x.

PIPERADROL HYDROCHLO-RIDE   HR: 3
CAS: 71-78-3   NIOSH: TN 0875000
mf: $C_{18}H_{21}NO \cdot ClH$ mw: 303.86

SYNS: ALPHA,ALPHA-DIPHENYL-2-PIPERIDINE-METHANOL HYDROCHLORIDE * ALPHA-(2-PI-PERIDYL)BENZHYDROL HYDROCHLORIDE

* PIPRADOL HYDROCHLORIDE * PIPRADROL
HYDROCHLORIDE

THR: Poison by ingestion, subcutaneous, intravenous, and intraperitoneal routes. When
heated to decomposition it emits very toxic
fumes of HCl and NO$_x$.

PIPERAZINE HR: 2
CAS: 110-85-0 NIOSH: TK 7800000
mf: C$_4$H$_{10}$N$_2$ mw: 86.16

PROP: Colorless rhombic crystals. Mp: 106°,
bp: 146°; flash p: 190°F (OC), d: 1.1, vap d:
3.0. Very sol in water, glycerol; glycols; insol
in ether.

SYNS: 1,4-DIETHYLENEDIAMINE * HEXAHY-
DRO-1,4-DIAZINE * HEXAHYDROPYRAZINE
* PIPERAZIDINE * PIPERAZIN (GERMAN)
* PIPERAZINE, ANHYDROUS * PYRAZINE HEX-
AHYDRIDE

THR: Moderately toxic by ingestion, inhalation,
skin contact, and subcutaneous routes. A skin
and eye irritant. Excessive absorption can cause
urticaria, vomiting, diarrhea, blurred vision, and
weakness. Dangerous; when heated to decomposition it emits highly toxic fumes of NO$_x$. Flammable via heat, flames, oxidizers. To fight fire,
use alcohol foam, mist, dry chemical, water
spray.

PIPERIDINE HR: 3
CAS: 110-89-4 NIOSH: TM 3500000
mf: C$_5$H$_{11}$N mw: 85.17

PROP: Clear, colorless liquid, amine-like odor.
Mp: −7°, bp: 106°, flash p: 37.4°F, d: 0.8622
@ 20°/4°, vap press: 40 mm @ 29.2°, vap d:
3.0. Misc with water; sol in alc, benzene, chloroform.

SYNS: AZACYCLOHEXANE * CYCLOPENTIMINE
* HEXAHYDROPYRIDINE * HEXAZANE
* PENTAMETHYLENEIMINE * PIPERIDIN (GER-
MAN)

THR: Poison by ingestion, skin contact, and
subcutaneous routes. Moderately toxic by inhalation. A skin irritant. Mutagenic data. Dangerous fire hazard when exposed to heat, flame,
or oxidizers. Dangerous; when heated to decomposition it emits highly toxic fumes of NO$_x$;
can react vigorously with oxidizing materials.
To fight fire, use alcohol foam, CO$_2$, dry chemical. Incompatible with 1-perchloryl-piperidine.

PIPERIDINOETHYL-2-HEPTOXYPHE-
NYLCARBAMOATE HYDROCHLO-
RIDE HR: 3
CAS: 55792-21-7 NIOSH: FB 1440000
mf: C$_{21}$H$_{34}$N$_2$O$_3$ • ClH mw: 399.03

SYNS: 2-HEPTYLOXYCARBANILIC ACID-2-(1-PI-
PERIDINYL)ETHYL ESTER HYDROCHLORIDE
* (2-(HEPTYLOXY)PHENYL)CARBAMIC ACID-2-
(1-PIPERIDINYL)ETHYL ESTER HYDROCHLORIDE
* N-(2-(HEPTYLOXYPHENYLCARBAMOYLOXY)-
ETHYL)PIPERIDINIUM CHLORIDE

THR: Poison by intravenous and intraperitoneal
routes. Moderately toxic by subcutaneous route.
When heated to decomposition it emits very
toxic fumes of NO$_x$ and HCl.

2-(1-PIPERIDINO)-2-(2-THENYL)ETH-
YLAMINE MALEATE HR: 3
CAS: 63918-29-6 NIOSH: KR 9770000
mf: C$_{11}$H$_{18}$N$_2$S • C$_4$H$_4$O$_4$ mw: 326.45

SYN: CIBA CO. 2825

THR: Poison by intravenous, intraperitoneal,
and subcutaneous routes. Moderately toxic by
ingestion. When heated to decomposition it
emits very toxic fumes of NO$_x$ and SO$_x$.

PIPEROCAINE HR: 3
CAS: 32248-37-6 NIOSH: TN 3320000
mf: C$_{16}$H$_{23}$NO$_2$ mw: 261.40

SYNS: 3-BENZOXY-1-(2-METHYLPIPERIDINO)-
PROPANE * BENZOYL-GAMMA-(2-METHYLPI-
PERIDINO)PROPANOL * 2-METHYL-1-PI-
PERIDINOPROPANOL BENZOATE
* (2-METHYLPIPERIDINO)PROPYL BENZOATE
* GAMMA-(2-METHYLPIPERIDYL)PROPYL BEN-
ZOATE

THR: Poison by intravenous and intraperitoneal
routes. Moderately toxic by subcutaneous route.
When heated to decomposition it emits toxic
fumes of NO$_x$.

PIPEROCYANOMAZINE HR: 3
CAS: 2622-26-6 NIOSH: SN 7175000
mf: C$_{21}$H$_{23}$N$_3$OS mw: 365.53

SYNS: 2-CYANO-10-(3-(4-HYDROXYPIPERIDINO)-
PROPYL)PHENOTHIAZINE * 2-CYANO-10-(3-
(4-HYDROXY-1-PIPERIDYL)PROPYL)PHENOTHI-
AZINE * CYANO-3 ((HYDROXY-4
PIPERIDYL-1)-3 PROPYL)-10 PHENOTHIAZINE
(FRENCH) * 10-(3-(4-HYDROXYPIPERIDINO)-
PROPYL)PHENOTHIAZINE-2-CARBONITRILE

THR: Poison by ingestion, intraperitoneal, intravenous, and subcutaneous routes. When heated to decomposition it emits very toxic fumes of NO_x and SO_x.

PIPERONAL   HR: 2
CAS: 120-57-0   NIOSH: TO 1575000
mf: $C_8H_6O_3$ mw: 150.14

PROP: Colorless, lustrous crystals. Mp: 37°, bp: 263°, vap press: 1 mm @ 87.0°. Very sol in alc, ether; sol in water.

SYNS: 3,4-DIHYDROXYBENZALDEHYDE METHYLENE KETAL * DIOXYMETHYLENE-PROTOCATECHUIC ALDEHYDE * HELIOTROPIN * 3,4-METHYLENE-DIHYDROXYBENZALDEHYDE * 3,4-METHYLENEDIOXYBENZALDEHYDE * PIPERONALDEHYDE * PIPERONYL ALDEHYDE * PROTOCATECHUIC ALDEHYDE METHYLENE ETHER

THR: Moderately toxic by ingestion. See also aldehydes. Can cause central nervous system depression. A skin irritant. Combustible when exposed to heat or flame; can react with oxidizing materials.

PIPERONYL BUTOXIDE   HR: 3
CAS: 51-03-6   NIOSH: XS 8050000
mf: $C_{19}H_{30}O_5$ mw: 338.49

PROP: Light brown liquid, mild odor. Bp: 180° @ 1 mm, flash p: 340°F, d: 1.04-1.07 @ 20°/20°. Misc with methanol, ethanol, benzene.

SYNS: ALPHA-(2-(2-BUTOXYETHOXY)ETHOXY)-4,5-METHYLENEDIOXY-2-PROPYLTOLUENE * ALPHA-(2-(2-N-BUTOXYETHOXY)-ETHOXY)-4,5-METHYLENEDIOXY-2-PROPYLTOLUENE * 5-((2-(2-BUTOXYETHOXY)ETHOXY)METHYL)-6-PROPYL-1,3-BENZODIOXOLE * BUTYL CARBITOL 6-PROPYLPIPERONYL ETHER * BUTYL-CARBITYL (6-PROPYLPIPERONYL) ETHER * ENT 14,250 * 3,4-METHYLENDIOXY-6-PROPYLBENZYL-N-BUTYL-DIAETHYLENGLYKOLAETHER (GERMAN) * (3,4-METHYLENEDIOXY-6-PROPYLBENZYL) (BUTYL) DIETHYLENE GLICOL ETHER * 3,4-METHYLENEDIOXY-6-PROPYLBENZYL N-BUTYL DIETHYLENEGLYCOL ETHER * NCI-C02813 * 6-(PROPYLPIPERONYL)-BUTYL CARBITYL ETHER * 6-PROPYLPIPERONYL BUTYL DIETHYLENE GLYCOL ETHER * 5-PROPYL-4-(2,5,8-TRIOXA-DODECYL)-1,3-BENZODIOXOL (GERMAN)

THR: Poison by skin contact. Moderately toxic by ingestion and intraperitoneal routes. An ex-

perimental carcinogen. See also ethers. Excessive ingestion can cause gastrointestinal disturbances. Combustible when exposed to heat or flame; can react with oxidizing materials. To fight fire, use foam, CO_2, dry chemical. When heated to decomposition it emits acrid smoke and fumes.

PIPOCTANONE HYDROCHLORIDE   HR: 3
CAS: 18787-40-1   NIOSH: UH 4395000
mf: $C_{22}H_{35}NO \cdot ClH$ mw: 366.04

SYNS: 4'-OCTYL-3-PIPERIDINOPROPIOPHENONE HYDROCHLORIDE * 1-PIPERIDINO-3-(P-OCTYLPHENYL)-3-PROPANONE HYDROCHLORIDE * 1-PIPERIDINO-3-(4'-OCTYLPHENYL)-PROPAN-3-ON-HYDROCHLORID (GERMAN)

THR: Poison by ingestion, intravenous, and intraperitoneal routes. When heated to decomposition it emits very toxic fumes of HCl and NO_x.

PIP-PIP   HR: 3
CAS: 98-77-1   NIOSH: TM 5850000
mf: $C_{11}H_{22}N_2S_2$ mw: 246.47

SYNS: PIPERIDINIUM * PENTAMETHYLENEDITHIOCARBAMATE * "522" RUBBER ACCELERATOR

THR: Poison by ingestion. Human skin irritant. An allergen. When heated to decomposition it emits very toxic fumes of NO_x and SO_x.

PIPTAL   HR: 3
CAS: 125-51-9   NIOSH: TN 5425000
mf: $C_{22}H_{28}NO_3 \cdot Br$ mw: 434.42

SYNS: BENZILIC ACID, ESTER WITH 1-ETHYL-3-HYDROXY-1-METHYLPIPERIDINIUM BROMIDE * 1-ETHYL-3-HYDROXY-1-METHYL-PIPERIDINIUM BROMIDE BENZILATE * N-ETHYL-3-PIPERIDYL-BENZILATE METHOBROMIDE * 1-ETHYL-3-PIPERIDYL BENZILATE METHYLBROMIDE * PIPENZOLATE METHYLBROMIDE

THR: Poison by intravenous route. Moderately toxic by ingestion and subcutaneous routes. See also esters. When heated to decomposition it emits very toxic fumes of Br^- and NO_x.

PIVALIC ACID   HR: 3
CAS: 75-98-9   NIOSH: TO 7700000
mf: $C_5H_{10}O_2$ mw: 102.15

PROP: Crystals; mp: 35.5°; bp: 164°; d: 0.91. Very sol in alc, ether; somewhat sol in water.

SYNS: 2,2-DIMETHYLPROPANOIC ACID
* ALPHA,ALPHA-DIMETHYLPROPIONIC ACID
* 2,2-DIMETHYLPROPIONIC ACID * NEOPENTA-
NOIC ACID * TERT-PENTANOIC ACID * PROP-
ANOIC ACID * TRIMETHYLACETIC ACID

THR: Moderately toxic by ingestion and skin contact. When heated to decomposition it emits acrid smoke and fumes.

2-PIVALOYL-1,3-INDANDIONE HR: 3
CAS: 83-26-1 NIOSH: NK 6300000
DOT: 2472
mf: $C_{14}H_{14}O_3$ mw: 230.28

PROP: Yellow crystals; mp: 108°.

SYNS: 2-(2,2-DIMETHYL-1-OXOPROPYL)-1H-IN-
DENE-1,3(2H)-DIONE * PINDON (DUTCH)
* PIVALDION (ITALIAN) * PIVALDIONE
(FRENCH) * 2-PIVALOYL-INDAAN-1,3-DION
(DUTCH) * 2-PIVALOYL-INDAN-1,3-DION (GER-
MAN) * 2-PIVALOYLINDANE-1,3-DIONE
* 2-PIVALYL-1,3-INDANDIONE * 2-(TRIMETIL-
ACETIL)-INDAN-1,3-DIONE (ITALIAN)

OSHA PEL: TWA 100 ug/m^3
ACGIH TLV: TWA 0.1 mg/m^3
DOT Classification: Poison B

THR: Poison by ingestion and parenteral routes. Causes reduced blood-clotting which leads to hemorrhaging. See also coumadin. When heated to decomposition it emits acrid smoke and fumes.

cis-PLATINOUS DIAMINE DICHLO-
RIDE HR: 3
CAS: 15663-27-1 NIOSH: TP 2450000
mf: $Cl_2H_6N_2Pt$ mw: 300.07

SYNS: CISPLATINO (SPANISH) * CIS-DIAMINE-
DICHLOROPLATINUM * CIS-DICHLORODIAMINE
PLATINUM (II) * NCI-C55776 * NSC-119875
* PEYRONE'S CHLORIDE * CIS-PLATINUM (II)
DIAMINEDICHLORIDE

OSHA PEL: TWA 2 ug(Pt)/m^3

THR: Poison by subcutaneous, intravenous, and intraperitoneal routes. An experimental carcinogen. Mutagenic data. Human skin, blood, gastrointestinal, central nervous system, and other systemic effects by intravenous and intradermal routes. See also platinum compounds. When heated to decomposition it emits very toxic fumes of Cl$^-$ and NO$_x$.

PLATINOUS POTASSIUM
CHLORIDE HR: 3
CAS: 10025-99-7 NIOSH: TP 1850000
mf: $Cl_4Pt \cdot K_2$ mw: 415.09

PROP: Ruby red. Sol in water, mp: decomp @ 250°; d: 3.499 @ 24°.

SYNS: POTASSIUM CHLOROPLATINITE * PO-
TASSIUM PLATINOCHLORIDE * POTASSIUM
TETRACHLOROPLATINATE (II)

THR: Poison to humans by ingestion. Poison by intraperitoneal route. Mutagenic data. Severe human skin irritant by intraduodenal route. See also platinum compounds. When heated to decomposition it emits toxic fumes of Cl$^-$.

PLATINUM HR: 3
CAS: 7440-06-4 NIOSH: TP 2160000
af: Pt aw: 195.09

PROP: Silvery-white, malleable, ductile metal. Stable in air; mp: 1772°; bp: 3827°; d: 21.45 @ 20°.

SYNS: PLATINUM BLACK * PLATINUM SPONGE
* C.I. 77795 * LIQUID BRIGHT PLATINUM
* PLATIN (GERMAN)
ACGIH TLV: TWA (metal) 1 mg/m^3; (soluble salts as Pt) 0.002 mg/m^3
DFG MAK: 0.002 mg/m^3

THR: An experimental tumorigen. See also platinum compounds. Incompatible with acetone, nitrosyl chloride; arsenic; dioxygen difluoride; ethanol; hydrazine; hydrogen; air; hydrogen peroxide; lithium; ozonides; peroxymonosulphuric acid; phosphorus; selenium, tellurium. For further information see Vol. 1, No. 3 of *DPIM Report*.

PLATINUM CHLORIDE HR: 3
CAS: 10025-65-7 NIOSH: TP 2275000
mf: Cl_2Pt mw: 265.99

PROP: Grayish-green powder. D: 5.87. Insol in water, alc, ether, benzene, chloroform.

SYNS: MURIATE OF PLATINUM * PLATINOUS
CHLORIDE

THR: Poison by ingestion and other routes. A skin irritant. Mutagenic data. See also platinum compounds. When heated to decomposition it emits toxic fumes of Cl$^-$.

PLATINUM (IV) CHLORIDE HR: 3
CAS: 13454-96-1 NIOSH: TP 2275500
mf: Cl_4Pt mw: 336.89

SYN: PLATINUM TETRACHLORIDE

THR: Poison by ingestion, intravenous, and intraperitoneal routes. Mutagenic data. A severe skin irritant. See also platinum compounds. When heated to decomposition it emits toxic fumes of Cl^-.

PLATINUM COMPOUNDS

THR: Exposure to complex platinum salts has been shown to cause symptoms of intoxication such as wheezing, coughing, running of the nose, tightness of the chest, shortness of breath and cyanosis; exposure to dust of pure metallic platinum causes no intoxication. Furthermore, many people working with platinum salts are troubled with dermatitis. This seems only to be true of complex platinum salts. It does not include the complex salts of the other precious metals. Platinum amine nitrates and perchlorates either detonate when heated or are impact-sensitive.

PLIOFILM   HR: 3
CAS: 9006-00-2   NIOSH: TP 3710000
mf: $(C_3H_5Cl)_n$

SYNS: PERMASEAL * RUBBER HYDROCHLORIDE * RUBBER HYDROCHLORIDE POLYMER

THR: An experimental carcinogen. When heated to decomposition it emits toxic fumes of Cl^-.

PLUTONIUM
af: Pu   aw: 242

PROP: A silvery radioactive metal, chemically reactive; mp: 641°; bp: 3232°; d: 19.816 @ 20°/4°.

THR: The permissible levels for plutonium for health reasons are the lowest for any of the radioactive elements. This is occasioned by the concentration of plutonium directly on bone surfaces, rather than the more uniform bone distribution shown by other heavy elements. This increases the possibility of damage from equivalent activities of plutonium and has led to the adoption of the extremely low permissible levels given. Radiation Hazard: Artificial isotope ^{238}Pu, $T\frac{1}{2} = 86$ Y, decays to radioactive ^{234}U via alphas of 5.5 MeV. Artificial isotope ^{239}Pu, $T\frac{1}{2} = 24,000$ Y decays to radioactive ^{235}U via alphas of 5.1 MeV. Artificial isotope ^{240}Pu, $T\frac{1}{2} = 6600$ Y decays to radioactive ^{236}Pu (Neptu-

nium Series), $T\frac{1}{2} = 13$ Y decays to radioactive ^{241}Am via betas of 0.02 MeV. Artificial isotope ^{242}Pu, $T\frac{1}{2} = 3.8 \times 10^5$ Y decays to radioactive ^{238}U via alphas of 4.9 MeV. Incompatible with carbon tetrachloride; water; air.

PLUTONIUM COMPOUNDS
THR: The toxicity of plutonium compounds is based first upon the very high radiotoxicity of the plutonium atom and secondly upon whatever atoms or combinations of atoms they might contain. See also plutonium. Very dangerous! Any disaster which could cause quantities of plutonium or plutonium compounds to be scattered about the environment can cause great ecological stress and render areas of the land unfit for public occupancy.

PLUTONIUM NITRATE (SOLUTION)
CAS: 14913-29-2   NIOSH: TP 6670000

SYN: PLUTONIUM NITRATE SOLUTION (DOT)

DOT Classification: Label: Radioactive

THR: See plutonium. When heated to decomposition it emits toxic fumes of NO_x.

PODOPHYLLIN   HR: 3
CAS: 9000-55-9   NIOSH: TP 8925000

PROP: Light yellow powder or small yellow fragile lumps. Bitter, acrid taste.

SYNS: PODOPHYLLUM * PODOPHYLLUM RESIN

THR: Poison by ingestion, subcutaneous, intraperitoneal, and other unspecified routes. An experimental neoplastigen. Combustible. An irritant to skin, eyes, and mucous membranes. When heated to decomposition it emits acrid smoke and fumes. For further information see Vol. 1, No. 3 of *DPIM Report*.

POLONIUM   HR: 3
af: Po   am: 210

PROP: A low melting, radioactive, volatile, naturally occurring metallic element; mp: 254°; bp: 962°; d: 9.4.

SYN: RADIUM F

THR: Severe radiotoxicity. Very dangerous to handle. An experimental carcinogen. See also plutonium. Radiation Hazard: Natural isotope ^{210}Po (radium-F, Uranium Series), $T\frac{1}{2} = 138$ D. Decays to stable ^{206}Pb via alphas of 5.3 MeV.

POLYCAPROLACTAM   HR: 3
CAS: 25038-54-4   NIOSH: TQ 9800000
mf: $(C_6H_{11}NO)_n$

SYNS: POLY(EPSILON-AMINOCAPROIC ACID)
* POLYCAPROAMIDE * POLY(EPSILON-CA-
PROAMIDE) * POLY(IMINOCARBONYLPENTA-
METHYLENE) * POLY(EPSILON-CAPROLACTAM)
* POLY(IMINO(1-OXO-1,6-HEXANEDIYL))
* 6-AMINOHEXANOIC ACID HOMOPOLYMER
* HEXAHYDRO-2H-AZEPIN-2-ONE HOMOPOLYMER
* CAPROLACTAM OLIGOMER * CAPROLACTAM
POLYMER * EPSILON-CAPROLACTAM POLYMER
* EPSILON-CAPROLACTAM POLY-MERE (GERMAN)
* NYLON-6

THR: An experimental neoplastigen and carcinogen. Slightly toxic by inhalation. When
heated to decomposition it emits toxic fumes
of NO_x.

POLYCHLORINATED BI-
PHENYLS   HR: 3
CAS: 1336-36-3   NIOSH: TQ 1350000

PROP: Bp: 340°-375°, flash p: 383°F (COC),
d: 1.44 @ 30°. For toxicity information, see
individual mixtures below. A series of technical
mixtures consisting of many isomers and compounds that vary from mobile oily liquids to
white crystalline solids and hard noncrystalline
resins. Technical products vary in composition,
in the degree of chlorination and possibly according to batch.

SYNS: AROCLOR * CHLOPHEN * CHLORI-
NATED BIPHENYL * CHLORINATED DIPHENYL
* CHLORINATED DIPHENYLENE * CHLOREXTOL
* CHLORO BIPHENYL * CHLORO-1,1-BIPHENYL
* CLOPHEN * KANECHLOR S * NOFLAMOL
* PCBS * PHENOCHLOR * POLYCHLORI-
NATED BIPHENYL * POLYCHLOROBIPHENYL
* PYRALENE * PYRANOL * SANTOTHERM
* THERMINOL FR-1

THR: Poison by ingestion, inhalation, and skin
contact. A suspected human carcinogen. Like
the chlorinated naphthalenes, the chlorinated diphenyls have two distinct actions on the body,
namely, a skin effect and a toxic action on the
liver. This hepato toxic action of the chlorinated
diphenyls appears to be increased if there is
exposure to carbon tetrachloride at the same
time. The higher the chlorine content of the
diphenyl compound, the more toxic is it liable
to be. Oxides of chlorinated diphenyls are more
toxic than the unoxidized materials. In persons
who have suffered systemic intoxication, the
usual signs and symptoms are nausea, vomiting,
loss of weight, jaundice, edema and abdominal
pain. Where the liver damage has been severe
the patient may pass into coma and die. Combustible when exposed to heat or flame. When
heated to decomposition they emit highly toxic
fumes. For further information see PCB's, Vol.
3, No. 4 of *DPIM Report*.

POLYCHLORINATED BIPHENYL
(AROCLOR 1221)   HR: 3
CAS: 11104-28-2   NIOSH: TQ 1352000

SYNS: AROCHLOR 1221 * CHLORODIPHENYL
(21% Cl)

THR: Suspected human carcinogen. Moderately
toxic by ingestion and skin contact. When heated
to decomposition it emits toxic fumes of Cl^-.

POLYCHLORINATED BIPHENYL
(AROCLOR 1232)   HR: 3
CAS: 11141-16-5   NIOSH: TQ 1354000

SYNS: AROCLOR 1232 * CHLORODIPHENYL
(32% Cl)

THR: Suspected human carcinogen. Moderately
toxic by skin contact. When heated to decomposition it emits toxic fumes of Cl^-.

POLYCHLORINATED BIPHENYL
(AROCLOR 1242)   HR: 3
CAS: 53469-21-9   NIOSH: TQ 1356000

SYNS: AROCHLOR 1242 * AROCLOR 1242
* CHLORIERTE BIPHENYLE, CHLORGEHALT 42%
(GERMAN) * CHLORODIPHENYL (42% cl)
* CLORODIFENILI, CLORO 42% (ITALIAN)
* DIPHENYLE CHLORE, 42% DE CHLORE
(FRENCH) * GECHLOREERDEDIFENYL (DUTCH)

OSHA PEL: TWA 1 mg/m^3 (skin)

THR: Poison by subcutaneous route. Suspected
human carcinogen. Human systemic irritant effects by inhalation. Moderately toxic by ingestion. When heated to decomposition it emits
toxic fumes of Cl^-. For further information see
Chlorinated Diphenyls, Vol. 1, No. 3 of *DPIM
Report*.

POLYCHLORINATED BIPHENYL
(AROCLOR 1248)   HR: 3
CAS: 12672-29-6   NIOSH: TQ 1358000

SYNS: AROCLOR 1248 ∗ CHLORODIPHENYL (48% cl)

THR: An experimental teratogen. A suspected human carcinogen. Moderately toxic by skin contact. When heated to decomposition it emits toxic fumes of Cl⁻. For further information see Chlorinated Diphenyls, Vol. 1, No. 3 of *DPIM Report*.

POLYCHLORINATED BIPHENYL (AROCLOR 1254)   HR: 3
CAS: 11097-69-1   NIOSH: TQ 1360000

PROP: Composed of 11% tetra-, 49% penta-, 34% hexa- and 6% heptachlorobiphenyls.

SYNS: AROCHLOR 1254 ∗ AROCLOR 1254 ∗ CHLORIERTE BIPHENYLE, CHLORGEHALT 54% (GERMAN) ∗ CHLORODIPHENYL (54% cl) ∗ CLORODIFENILI, CLORO 54% (ITALIAN) ∗ DIPHENYLE CHLORE, 54% DE CHLORE (FRENCH) ∗ NCI-C02664

OSHA PEL: TWA 500 ug/m³ (skin)

THR: Poison by intravenous route. An experimental neoplastigen and carcinogen. Moderately toxic by ingestion and intraperitoneal route. When heated to decomposition it emits toxic fumes of Cl⁻. For further information see Chlorinated Diphenyls, Vol. 1, No. 3 of *DPIM Report*.

POLYCHLORINATED BIPHENYL (AROCLOR 1260)   HR: 3
CAS: 11096-82-5   NIOSH: TQ 1362000

PROP: Composed of 12% penta-, 38% hexa-, 41% hepta-, 8% octa- and 1% nonachlorobiphenyls.

SYNS: AROCHLOR 1260 ∗ AROCLOR 1260 ∗ CLOPHEN A60 ∗ CHLORODIPHENYL (60% Cl) ∗ PHENOCLOR DP6

THR: A suspected human carcinogen. An experimental carcinogen. Moderately toxic by ingestion and skin contact. When heated to decomposition it emits highly toxic Cl⁻. For further information see Chlorinated Diphenyls, Vol. 1, No. 3 of *DPIM Report*.

POLYCHLORINATED BIPHENYL (AROCLOR 1262)   HR: 3
CAS: 37324-23-5   NIOSH: TQ 1364000

SYNS: AROCLOR 1262 ∗ CHLORODIPHENYL (62% Cl)

THR: A suspected human carcinogen. Moderately toxic by skin contact. When heated to decomposition it emits toxic fumes of Cl⁻. For further information see Chlorinated Diphenyls, Vol. 1, No. 3 of *DPIM Report*.

POLYCHLORINATED BIPHENYL (AROCLOR 1268)   HR: 3
CAS: 11100-14-4   NIOSH: TQ 1366000

SYNS: AROCLOR 1268 ∗ CHLORODIPHENYL (68% Cl)

THR: A suspected human carcinogen. Moderately toxic by skin contact. When heated to decomposition it emits toxic fumes of Cl⁻. For further information see Chlorinated Diphenyls, Vol. 1, No. 3 of *DPIM Report*.

POLYCHLORINATED BIPHENYL (KANECHLOR 300)   HR: 3
CAS: 37353-63-2   NIOSH: TQ 1372000

PROP: Average content: 60% trichlorobiphenyl, 23% tetrachlorobiphenyl, 17% dichlorobiphenyl, 1% pentachlorobiphenyl.

SYN: KANECHLOR 300

THR: A suspected human carcinogen. When heated to decomposition it emits toxic fumes of Cl⁻. For further information see Chlorinated Diphenyls, Vol. 1, No. 3 of *DPIM Report*.

POLYCHLORINATED BIPHENYL (KANECHLOR 400)   HR: 3
CAS: 12737-87-0   NIOSH: TQ 1374000

PROP: Average content: 44% tetrachlorbiphenyl, 33% trichlorobiphenyl, 16% pentachlorobiphenyl, 5% hexachlorobiphenyl, 3% dichlorobiphenyl.

SYN: KANECHLOR 400

THR: A suspected human carcinogen. Human skin systemic effects by ingestion. An experimental neoplastigen. When heated to decomposition it emits toxic fumes of Cl⁻. For further information see Chlorinated Diphenyls, Vol. 1, No. 3 of *DPIM Report*.

POLYCHLORINATED BIPHENYL (KANECHLOR 500)   HR: 3
CAS: 37317-41-2   NIOSH: TQ 1376000

PROP: Average content, 55% pentachlorobiphenyl, 26.5% tetrachlorobiphenyl, 12.8% hexachloro biphenyl and 5% trichlorobiphenyl.

SYNS: KANECHLOR 500 * KC-500

THR: An experimental carcinogen and a suspected human carcinogen. When heated to decomposition it emits toxic fumes of Cl^-. For further information see Chlorinated Diphenyls, Vol. 1, No. 3 of *DPIM Report*.

POLYDIMETHYL SILOXANE HR: 3
CAS: 9016-00-6 NIOSH: TQ 2690000

PROP: A water-insoluble polymer of high viscosity.

SYNS: GUM * LATEX

THR: An experimental neoplastigen. See also polymers, water insoluble.

POLYDIMETHYLSILOXANE RUBBER HR: 3
NIOSH: TQ 2690000

PROP: Vulcanized with 2,4-dichlorbenzoyl peroxide, pure silicon dioxide used as filler and plasticizer.

SYNS: POLYSILICONE * SILASTIC * SILICONE RUBBER

THR: An experimental carcinogen. When heated to decomposition it emits acrid smoke and fumes.

POLYETHYLENE HR: 3
CAS: 9002-88-4 NIOSH: TQ 3325000
mf: $(C_2H_4)_n$

PROP: Odorless. The high molecular wt compounds are tough, white leathery, resinous. D: 0.92 @ 20°/4°; mp: 85°-110°.

SYNS: AGILENE * ALKATHENE * BAKELITE DYNH * DIOTHENE * ETHENE POLYMER * ETHYLENE HOMOPOLYMER * ETHYLENE POLYMERS * HOECHST PA 190 * MICROTHENE * POLYETHYLENE AS * POLYWAX 1000 * TENITE 800

THR: An experimental carcinogen and tumorigen. Reacts violently with F_2. When heated to decomposition it emits acrid smoke and fumes.

POLYETHYLENE GLYCOL HR: 3
CAS: 25322-68-3 NIOSH: TQ 3500000

PROP: Clear liquid or white solid; mw: 285-315; d: 1.110-1.140 @ 20°; mp: 4°-10°; flash p: 471°F. Sol in organic solvents, aromatic hydrocarbons.

SYNS: CARBOWAX * P.E.G. 400 * POLYETHYLENE GLYCOL 400

THR: An experimental carcinogen. Slightly toxic by many routes. A moderate skin and eye irritant. Combustible when exposed to heat or flame. To fight fire, use water, foam, dry chemical. When heated to decomposition it emits acrid smoke and fumes.

POLYETHYLENE GLYCOL DISTEARATE HR: 3
CAS: 9005-08-7 NIOSH: TQ 4375000

PROP: Polyethylene Glycol Distearate, low molecular weight.

SYNS: POLYETHYLENE GLYCOL 300 DISTEARATE * POLYETHYLENE GLYCOL 400 (DI) STEARATE * POLYETHYLENE GLYCOL 600 (DI) STEARATE * POLYGLYCOL DISTEARATE

THR: Poison by intravenous route. See also polyethylene glycol. When heated to decomposition it emits acrid smoke and fumes.

POLYETHYLENE GLYCOL MONOSTEARATE HR: 3
CAS: 9004-99-3 NIOSH: TQ 5400000

SYNS: POLYOXYETHYLENE-8-MONOSTEARATE * POLYOXYETHYLENE(8)STEARATE

THR: An experimental tumorigen. Very slightly toxic by ingestion. When heated to decomposition it emits acrid smoke and fumes.

POLYMERS HR: 3
THR: Water-insoluble: Many are carcinogenic at the site of bodily implantation. Water soluble: A carcinogen of soft tissues around implant, lung, mucosal contact areas, organs and tissues of retention and deposition.

POLYMETHYLENEPOLYPHENYL ISOCYANATE HR: 3
CAS: 9016-87-9 NIOSH: TR 0320000

SYNS: POLY(METHYLENE PHENYLENE ISOCYANATE) * POLYMETHYLENE POLYPHENYLENE ISOCYANATE * POLYMETHYLENEPOLYPHENYLENE ISOCYANATE POLYMER * POLYMETHYLENEPOLYPHENYLENE POLYISOCYANATE

* POLYMETHYLENE POLY(PHENYL ISOCYANATE) * POLYMETHYLENE POLYPHENYL POLYISO- CYANATE * POLYMETHYL POLYPHENYLPOLY- ISOCYANATE * POLY(PHENYLENEMETHYL- ENEISOCYANATE) * POLYPHENYLENE POLYMETHYLENE POLYISOCYANATE * POLYPHE- NYLPOLYMETHYLENE POLYISOCYANATE

THR: Suspected carcinogen. When heated to decomposition it emits toxic fumes of NO_x.

POLYMETHYLMETHACRYLATE   HR: 3
CAS: 9011-14-7   NIOSH: TR 0400000
mf: $(C_5H_8O_2)_n$

SYNS: LUCITE * METHACRYLIC ACID METHYL ESTER POLYMERS * METHYL METHACRYLATE HOMOPOLYMER * PERSPEX * PLEXIGLAS * POLY(METHYL METHACRYLATE) * METHYL METHACRYLATEPOLYMER * METHYL METHAC- RYLATE RESIN * 2-METHYL-2-PROPENOIC ACID- METHYL ESTER HOMOPOLYMER (9CI)

THR: An experimental tumorigen and carcino- gen. When heated to decomposition it emits acrid smoke and fumes.

POLYMYXIN A   HR: 3
CAS: 1404-24-6   NIOSH: TR 0875000

THR: Poison by intraperitoneal, subcutaneous, intravenous, and intracerebral routes. When heated to decomposition it emits acrid smoke and fumes.

POLYPROPYLENE GLYCOL 750   HR: 3
CAS: 25322-69-4   NIOSH: TR 5775000

THR: Poison by ingestion, intraperitoneal, and intravenous route. When heated to decomposi- tion it emits acrid smoke and fumes.

POLYURETHANE FOAM   HR: 3
CAS: 9009-54-5   NIOSH: TR 7875000

SYNS: ETHERON SPONGE * NCI-C56451 * POLYFOAM SPONGE * POLYFOAM PLASTIC SPONGE * POLYURETHANE ESTER FOAM * POLYURETHANE ETHER FOAM * POLYURE- THANE SPONGE

THR: An experimental carcinogen and tumori- gen. When heated to decomposition it emits acrid smoke and fumes of CN^- and NO_x.

POLYURETHANE Y-195   HR: 3
                        NIOSH: TR 8000000

THR: An experimental carcinogen. When heated to decomposition it emits toxic fumes of NO_x.

POLYURETHANE Y-218   HR: 3
                        NIOSH: TR 8010000

THR: An experimental carcinogen. When heated to decomposition it emits very toxic fumes of CN^- and NO_x.

POLYURETHANE Y-221   HR: 3
                        NIOSH: TR 8015000

THR: An experimental carcinogen. When heated to decomposition it emits very toxic fumes of CN^- and NO_x.

POLYURETHANE Y-222   HR: 3
                        NIOSH: TR 8020000

THR: An experimental carcinogen. When heated to decomposition it emits very toxic fumes of CN^- and NO_x.

POLYURETHANE Y-223   HR: 3
                        NIOSH: TR 8025000

THR: An experimental carcinogen. When heated to decomposition it emits very toxic fumes of CN^- and NO_x.

POLYURETHANE Y-224   HR: 3
                        NIOSH: TR 8030000

THR: An experimental carcinogen. When heated to decomposition it emits very toxic fumes of CN^- and NO_x.

POLYURETHANE Y-225   HR: 3
                        NIOSH: TR 8035000

THR: An experimental carcinogen. When heated to decomposition it emits very toxic fumes of CN^- and NO_x.

POLYURETHANE Y-226   HR: 3
                        NIOSH: TR 8040000

THR: An experimental carcinogen. When heated to decomposition it emits very toxic fumes of CN^- and NO_x.

POLYURETHANE Y-227   HR: 3
                        NIOSH: TR 8045000

THR: An experimental carcinogen. When heated to decomposition it emits very toxic fumes of CN^- and NO_x.

POLYURETHANE Y-238   HR: 3
NIOSH: TR 8050000

PROP: Polymer formed from toluene diisocyanate and 1,4-butanediol and cured with 4,4'-methylenebis(o-chloroaniline).

SYNS: CHLORINATED POLYETHER POLYURETHAN * Y-238

THR: An experimental carcinogen. When heated to decomposition it emits very toxic fumes of NO_x, Cl^-, and CN^-.

POLYURETHANE Y-290   HR: 3
NIOSH: TR 8055000

THR: An experimental carcinogen. When heated to decomposition it emits very toxic fumes of CN^- and NO_x.

POLYURETHANE Y-302   HR: 3
NIOSH: TR 8060000

THR: An experimental carcinogen. When heated to decomposition it emits very toxic fumes of CN^- and NO_x.

POLYURETHANE Y-304   HR: 3
NIOSH: TR 8065000

THR: An experimental carcinogen. When heated to decomposition it emits very toxic fumes of CN^- and NO_x.

POLYVINYL ACETATE CHLORIDE
HR: 3
CAS: 34149-92-3   NIOSH: TR 8090000

SYNS: ACETIC ACID, VINYL ESTER, CHLORO-ETHYLENE COPOLYMER * POLYVINYLCHLORIDE ACETATE * VINYL CHLORIDE ACETATE CO-POLYMER * VINYL CHLORIDE VINYL ACETATE COPOLYMER

THR: An experimental carcinogen. When heated to decomposition it emits toxic fumes of Cl^-.

POLYVINYL ALCOHOL   HR: 3
CAS: 9002-89-5   NIOSH: TR 8100000

PROP: Colorless, amorphous powder. mp: decomp over 200°, flash p: 175°F (OC), d: 1.329. Polymer of average molecular weight 120,000.

SYNS: ELVANOL * ETHENOL HOMOPOLYMER (9CI) * GELVATOLS * GOHSENOLS * POLY-(VINYL ALCOHOL) * VINYL ALCOHOL POLYMER

THR: An experimental carcinogen and tumorigen. Not acutely toxic. Moderately toxic by ingestion. Flammable when exposed to heat or flame; can react with oxidizing materials. Slight explosion hazard in the form of dust, when exposed to flame. To fight fire, use alcohol foam, CO_2, dry chemical.

POLYVINYLBROMIDE   HR: 3
CAS: 025951-54-6   NIOSH: KU 8800000
mf: $(C_2H_3Br)_x$

PROP: Commercial PVBR is a 40% aqueous suspension in which PVBR constitutes about 90% of the solids.

SYNS: POLYBROMOETHYLENE * POLY(VINYL-BROMIDE) * PVBR

THR: An experimental carcinogen. When heated to decomposition it emits toxic fumes of Br^-.

POLY(1-VINYL-2-PYRROLIDINONE)
HUEPER'S POLYMER NO. 1   HR: 3
CAS: 9003-39-8   NIOSH: TR 8160000

PROP: Polymer of average molecular weight 20,000.

SYNS: NCI-C60582 * PVP 1

THR: An experimental carcinogen. When heated to decomposition it emits toxic fumes of NO_x.

POLY(1-VINYL-2-PYRROLIDINONE)
HUEPER'S POLYMER NO. 2   HR: 3
CAS: 9003-39-8   NIOSH: TR 8170000

PROP: Polymer of average molecular weight 20,000.

SYNS: NCI-C60582 * PVP 2

THR: An experimental carcinogen and neoplastigen. When heated to decomposition it emits toxic fumes of NO_x.

POLY(1-VINYL-2-PYRROLIDINONE)
HUEPER'S POLYMER NO. 3   HR: 3
CAS: 9003-39-8   NIOSH: TR 8180000

PROP: Polymer of average molecular weight 50,000.

SYNS: NCI-C60582 * PVP 3

THR: An experimental carcinogen. When heated to decomposition it emits toxic fumes of NO_x.

POLY(1-VINYL-2-PYRROLIDINONE) HUEPER'S POLYMER NO. 4   HR: 3
CAS: 9003-39-8   NIOSH: TR 8250000

PROP: Polymer of average molecular weight 300,000.

SYNS: NCI-C60582 * PVP 4

THR: An experimental carcinogen. When heated to decomposition it emits toxic fumes of NO_x.

POLY(1-VINYL-2-PYRROLIDINONE) HUEPER'S POLYMER NO. 5   HR: 3
CAS: 9003-39-8   NIOSH: TR 8300000

PROP: Polymer of average molecular weight 10,000.

SYN: PVP 5

THR: An experimental carcinogen. When heated to decomposition it emits toxic fumes of NO_x.

POLY(1-VINYL-2-PYRROLIDINONE) HUEPER'S POLYMER NO. 6   HR: 3
CAS: 9003-39-8   NIOSH: TR 8350000

PROP: Polymer of average molecular weight 50,000.

SYNS: NCI-C60582 * PVP 6

THR: An experimental carcinogen. When heated to decomposition it emits toxic fumes of NO_x.

POLY(1-VINYL-2-PYRROLIDINONE) HUEPER'S POLYMER NO. 7   HR: 3
CAS: 9003-39-8   NIOSH: TR 8360000

SYNS: NCI-C60582 * PVP 7

THR: An experimental carcinogen and neoplastigen. When heated to decomposition it emits toxic fumes of NO_x.

POLYVINYL SULFATE, POTASSIUM SALT   HR: 3
CAS: 26837-42-3   NIOSH: TR 8400000

SYNS: POTASSIUM SALT OF POLYVINYL SULFATE * PVSK

THR: Poison by intraperitoneal and subcutaneous routes. When heated to decomposition it emits toxic fumes of SO_x and K_2O.

POTASSIUM   HR: 3
CAS: 7440-09-7   NIOSH: TS 6460000
DOT: 2257
af: K   aw: 39.10

PROP: Soft ductile, silvery-white, very reactive metal. Mp: 63.65°, bp: 774°, d: 0.862 @ 20°.

SYN: POTASSIUM, METAL (DOT)

DOT Classification: Label: Flammable Solid and Dangerous When Wet

THR: The toxicity of potassium compounds is almost always that of the anion. Dangerous fire hazard. Metallic potassium reacts violently with moisture to form potassium hydroxide and hydrogen. The reaction evolves much heat, causing the potassium to melt and spatter. It also ignites the hydrogen, which burns, or; if there is any confinement; an explosion can occur. Burning potassium is difficult to extinguish; dry powdered soda ash or graphite or special mixtures of dry chemical are recommended. It can ignite spont in moist air. Violent explosion hazard with the following materials under required conditions of temperature, pressure, state of division: C_2H_2; air; (moist air); $AlBr_3$; metallic halides; ammonium chlorocuprate; NH_4Br; NH_4I; $(NH_4)_2SO_4 + NH_4NO_3$; Sb and As halides; $AsH_3 + NH_3$; Bi_2O_3; boric acid; BBr_3; Br_2; C; CO_2; CS_2; $CO + O_2$; CCl_4; charcoal; chlorinated hydrocarbons; Cl_2; ClO; ClF_3; $CHCl_3$; $CrCl_4$; CrO_3; Cu_2OCl_2; CuO; dichloromethane; ethylene oxide; F_2; graphite; graphite + air; I; graphite + K_2O_2; HI; H_2O_2; IBr; ICl; IF_5; Pb_2OCl_2; PbO_2; $PbSO_4$; maleic anhydride; Hg_2O; CH_3Cl; MoO_3; NO_2; P_2NF; peroxides; $COCl_2$; $PH_3 + NH_3$; P; PCl_5; P_2O_5; PBr_3; PCl_3; potassium chlorocuprate; K oxides; KO_3; K_2O_2; KO_2; Se; $SeOCl_2$; $SiCl_4$; $AgIO_3$; $NaIO_3$; $NH_3 + NaNO_2$; Na_2O_2; $SnI_4 + S$; SnO_2; S; SBr_2; SCl_2; Te; tetrachloroethane; thiophosphoryl fluoride; $VOCl_2$; H_2O. Potassium metal will form the peroxide (K_2O_2) and the superoxide (KO_3 or K_2O_4) at room temperature even when stored under mineral oil. Metal which has oxidized on storage under oil may explode violently when handled or cut. Oxide-coated potassium should be destroyed by burning. Dangerous; a highly reactive alkali metal. See sodium and lithium. In the presence of moist air it can spontaneously catch fire and burn with great intensity. It may even explode. Reacts violently with moisture, acid fumes and oxidizers.

POTASSIUM (LIQUID ALLOY)
CAS: 7440-09-7   NIOSH: TS 6465000
DOT: 1420

SYN: POTASSIUM, METAL LIQUID ALLOY (DOT)

DOT Classification: Label: Flammable Solid and Dangerous When Wet

THR: See potassium. When heated to decomposition in air it emits toxic fumes of K_2O.

POTASSIUM ACID FLUORIDE   HR: 3
CAS: 7789-29-9   NIOSH: TS 6650000
mf: FK • FH   mw: 78.11

PROP: Colorless crystals; mp: decomp.

SYNS: BIFLUORURE DE POTASSIUM (FRENCH) * POTASSIUM BIFLUORIDE * POTASSIUM HYDROGEN FLUORIDE

THR: A poison by all routes. Corrosive to the eyes, skin, and mucous membranes. A very reactive, dangerous material. See also fluorides. When heated to decomposition it emits toxic fumes of F^-.

POTASSIUM ARSENITE   HR: 3
CAS: 10124-50-2   NIOSH: CG 3800000
DOT: 1678
mf: AsH_3O_3 • 7K   mw: 399.65

PROP: White hygroscopic powder. Sol in water.

SYNS: ARSONIC ACID, POTASSIUM SALT (9CI)
* ARSENENOUS ACID, POTASSIUM SALT
* ARSENITE DE POTASSIUM (FRENCH) * FOWLER'S SOLUTION * KALIUMARSENIT (GERMAN)
* NSC 3060 * POTASSIUM METAARSENITE

OSHA PEL: TWA 500 ug(As)/m^3
DOT Classification: Poison B, Label: Poison

THR: Poison by ingestion, skin contact, subcutaneous, and intravenous routes. Mutagenic data. A human carcinogen. Human systemic effects. See also arsenic compounds. When heated to decomposition it emits toxic fumes of As. For further information see Vol. 3, No. 4 of *DPIM Report*.

POTASSIUM BICHROMATE   HR: 3
CAS: 7778-50-9   NIOSH: HX 7680000
DOT: 1479
mf: $Cr_2K_2O_7$ mw: 294.20

PROP: Bright, yellowish-red, transparent crystals, bitter, metallic taste. Mp: 398°, bp: decomp @ 500°, d: 2.69.

SYNS: BICHROMATE OF POTASH * DIPOTASSIUM DICHROMATE * IOPEZITE * KALIUMDICHROMAT (GERMAN) * POTASSIUM DICHROMATE (VI)

DOT Classification: ORM-A, Label: None

THR: Poison by ingestion, intravenous, and subcutaneous routes. A human carcinogen. Mutagenic data. See also chromium. Flammable by chemical reaction. A powerful oxidizer. Reacts violently with (H_2SO_4 + acetone); hydrazine; hydroxylamine.

POTASSIUM BIS(2-HYDROXYETHYL)DITHIOCARBAMATE   HR: 3
CAS: 23746-34-1   NIOSH: EY 9450000
mf: $C_5H_{10}NO_2S_2$ • K   mw: 219.38

SYNS: BIS(2-HYDROXYETHYL)CARBAMODITHIOIC ACID, MONOPOTASSIUM SALT * BIS(2-HYDROXYETHYL)DITHIOCARBAMIC ACID, MONOPOTASSIUM SALT

THR: An experimental carcinogen. When heated to decomposition it emits very toxic fumes of SO_x and NO_x.

POTASSIUM-tert-BUTOXIDE
mf: C_4H_9KO mw: 112.20

THR: No toxicity data. Probably very toxic and irritating to skin, eyes, and mucous membranes. When heated to decomposition it emits toxic fumes of K_2O. Ignites on contact with acetone; butanone; butyl acetate; CH_3COOH; C_2H_5OH; C_3H_7OH; CCl_4; CH_2Cl_2; CH_3OH; $CHCl_3$; carbon tetrachloride; 1-chloro-2,3-epoxypropane; chloroform; dichloromethane-1,2; diethyl sulfate; dimethyl carbonate; epichlorohydrin; ethyl acetate; ethyl methyl ketone; H_2SO_4; isopropanol; 4-methyl-2-butanone; methyl isobutyl ketone; n-butyl acetate; n-propyl formate; propanol; propyl formate. When heated to decomposition it emits very toxic K_2O.

POTASSIUM CARBONATE (2:1)   HR: 2
CAS: 584-08-7   NIOSH: TS 7750000
mf: CO_3 • 2K   mw: 138.21

PROP: White, deliquescent, granular, translucent powder, sol in water, insol in alcohol. (*a*) K_2CO_3, mw: 138; d: 2.428 @ 19°, mp: 891°, bp: decomposes.

SYNS: CARBONIC ACID, DIPOTASSIUM SALT
* KALIUMCARBONAT (GERMAN) * PEARL ASH
* POTASH

THR: Moderately toxic by ingestion. A strong caustic. Incompatible with KCO; ClF$_3$; Mg.

POTASSIUM CHLORATE   HR: 3
CAS: 3811-04-9   NIOSH: FO 0350000
DOT: 1485/2427
mf: ClO$_3$ • K   mw: 122.55

PROP: Transparent colorless crystals or white powder, cooling, saline taste. Mp: 368.4°, bp: decomp @ 400°, d: 2.32.

SYNS: CHLORATE DE POTASSIUM (FRENCH) * CHLORATE OF POTASH * FEKABIT * KALIUMCHLORAAT (DUTCH) * KALIUMCHLORAT (GERMAN) * OXYMURIATE OF POTASH * POTASH CHLORATE (DOT) * POTASSIO (CHLORATO DI) (ITALIAN) * POTASSIUM CHLORATE (DOT) * POTASSIUM (CHLORATE DE) (FRENCH) * POTASSIUM OXYMURIATE

DOT Classification: Label: Oxidizer

THR: Poison by an unspecified route in humans. A gastrointestinal tract and kidney irritant. Can cause hemolysis of red blood cells and methemoglobinemia. Toxic dose to a human approx 5 g. Combination of iodine and perchlorate not recommended. See chlorates. A very reactive material. When heated to decomposition it emits very toxic fumes of Cl$^-$ and K$_2$O. Reacts violently with Al; NH$_3$; NH$_4$Cl; NH$_4$+ salts; (NH$_4$)$_2$SO$_4$; Sb$_2$S$_3$; As; barium hypophosphite; BaS; B; calcium hypophosphite; CaS; C; charcoal; Cr; Cu; Cu$_3$P$_2$; gallic acid; Ge; HI; Mg; (Mg + CuSO$_4$ (anhydrous) + NH$_4$NO$_3$ + H$_2$O); MnO$_2$; Hg$_3$P$_4$; metal sulfides; dibasic organic acids; organic matter; P; Ag$_2$S; NaNH$_2$; S; SO$_2$; H$_2$SO$_4$; Zn; Zr; Ti; thiocyanates; thorium dicarbide; sodium amide; fabrics; KOH; HI; metal phosphides; metal hypophosphites; metals; metal thiocyanates; non-metals; aqua regia; ruthenium; CN$^-$; Ni$_2$O$_3$.

POTASSIUM CHROMATE (VI)   HR: 3
CAS: 7789-00-6   NIOSH: GB 2940000
mf: CrO$_4$ • 2K   mw: 194.20

PROP: Rhombic, yellow crystals; mp: 975°; d: 2.73 @ 18°. Sol in water; insol in alc.

SYNS: BIPOTASSIUM CHROMATE * DIPOTASSIUM CHROMATE * DIPOTASSIUM MONOCHROMATE * NEUTRAL POTASSIUM CHROMATE

OSHA PEL: TWA CL 100 ug(CrO3)/m^3

THR: Poison by subcutaneous, intravenous, and intramuscular routes. A human carcinogen and an experimental tumorigen. Mutagenic data. A powerful oxidizer. See also chromium compounds. For further information see Potassium Chromate, Vol. 1, No. 7 of *DPIM Report*.

POTASSIUM CYANATE   HR: 2
CAS: 590-28-3   NIOSH: GS 6825000
mf: CNO • K   mw: 81.12

PROP: Colorless crystals, mp: 700°-900° (decomp), d: 2.056 @ 20°. Sol in water, very sltly sol in alc.

SYNS: AERO CYANATE * KALIUMCYANAT (GERMAN)

THR: Moderately toxic via oral route. An herbicide. Ingestion can cause irritation of the gastrointestinal tract. It is said to be slowly metabolized in the body to cyanide but does not have high toxicity of cyanides. When heated to decomposition it emits very toxic fumes of CN$^-$ and K$_2$O. For further information see Vol. 1, No. 7 of *DPIM Report*.

POTASSIUM CYANIDE
(SOLUTION)   HR: 3
CAS: 151-50-8   NIOSH: TS 8750000
DOT: 1680
mf: CN • K   mw: 65.12

PROP: Colorless water soln. Slight odor of bitter almonds.

SYNS: CYANIDE OF POTASSIUM (SOLN) * CYANURE DE POTASSIUM (FRENCH) * HYDROCYANIC ACID, POTASSIUM SALT (SOLN) * POTASSIUM CYANIDE SOLUTION (DOT)

OSHA PEL: TWA 5 mg(CN)/m^3 (skin)
DOT Classification: Poison B, Label: Poison

THR: A deadly poison by ingestion, subcutaneous, intravenous, intramuscular, and intraperitoneal routes. Mutagenic data. Reacts with acids or acid fumes to liberate deadly HCN. When heated to decomposition it emits very toxic fumes of CN$^-$ and NO$_x$.

POTASSIUM CYANIDE (SOLID)   HR: 3
CAS: 151-50-8   NIOSH: TS 8760000
mf: CN • K   mw: 65.12

PROP: White, deliquescent crystals. Faint odor of almonds (bitter); mp: 622.5°; d: 1.52 @ 16°; sol in HOH, glycerol; sltly sol in alc; bp: 1625°.

SYN: POTASSIUM CYANIDE, SOLID (DOT)

DOT Classification: Poison B, Label: Poison

THR: A deadly poison by an unspecified route in humans. Ingestion, absorption through injured skin or inhalation may cause poisoning. Strong solutions are corrosive to skin, eyes, and mucous membranes. Reacts with acids and acid fumes to liberate deadly HCN. When heated to decomposition it emits very toxic fumes of CN^- and NO_x. Incompatible with nitrogen trichloride; perchloryl fluoride; sodium nitrite; acids; alkaloids; chloral hydrate; iodine. For further information see Vol. 3, No. 6 of *DPIM Report*.

POTASSIUM DICHLOROISO-CYANURATE HR: 2
CAS: 2244-21-5   NIOSH: XZ 1850000
DOT: 2465
mf: $C_3HCl_2N_3O_3 \cdot K$ mw: 237.07

PROP: White, sltly hygroscopic crystalline powder or granules, chlorine odor. Mp: 250° (decomp).

SYNS: DICHLOROISOCYANURIC ACID POTASSIUM SALT * DICHLORO-S-TRIAZINE-2,4,6(1H,3H,5H)-TRIONE POTASSIUM DERIV * DICHLOR-S-TRIAZIN-2,4,6(1H,3H,5H)TRIONE POTASSIUM * POTASSIUM DICHLORO-S-TRIAZINETRIONE

DOT Classification: Label: Oxidizer

THR: Moderately toxic by ingestion. A severe eye and mild skin irritant. See also cyanurates. Causes emaciation, weakness, lethargy, diarrhea, weight loss. Autopsy indicates gastrointestinal tract irritation, tissue edema, liver and kidney congestion. A powerful oxidizer. When heated to decomposition it emits very toxic fumes of Cl^- and NO_x.

POTASSIUM FLUORIDE HR: 3
CAS: 7789-23-3   NIOSH: TT 0700000
DOT: 1812
mf: FK   mw: 58.10

PROP: White, crystalline, deliquescent powder. Sharp saline taste. Bp: 1500°; d: 2.48; vap press: 1 mm @ 885°; mp: 859.9°; very sol in boiling water.

SYNS: FLUORURE DE POTASSIUM (FRENCH) * POTASSIUM FLUORIDE (DOT) * POTASSIUM FLUORURE (FRENCH)

OSHA PEL: TWA 2500 ug(F)/m^3
DOT Classification: Label: Corrosive

THR: Poison by ingestion, intraperitoneal, and subcutaneous routes. A very reactive and irritating material. See also fluorides and hydrofluoric acid. When heated to decomposition it emits toxic fumes of F^-.

POTASSIUM FLUOROACETATE HR: 3
CAS: 23745-86-0   NIOSH: AH 8800000

PROP: The potassium salt of monofluoroacetic acid was once designated as potassium cymonate.

SYNS: DICHAPETULUM CYMOSUM (HOOK) ENGL * GIFBLAAR

THR: Poison by ingestion, intravenous, intraperitoneal, and subcutaneous routes. See also fluorides. When heated to decomposition it emits toxic fumes of F^-.

POTASSIUM HEXAFLUOROTITA-NATE HR: 3
CAS: 16919-27-0   NIOSH: TT 1575000
mf: $F_6Ti \cdot 2K$ mw: 240.10

SYNS: FLUOTITANATE DE POTASSIUM (FRENCH) * TITANIUM POTASSIUM FLUORIDE

THR: Poison by subcutaneous route. See also fluorides. When heated to decomposition it emits toxic fumes of F^-.

POTASSIUM HYDRIDE
mf: HK   mw: 40.11

PROP: White needles. Mp: decomp; d: 1.43-1.47

THR: See also potassium and hydrides. Dangerous fire hazard by chemical reaction. Moderate explosion hazard when exposed to heat or by chemical reaction. Dangerous; when heated to decomposition it emits highly toxic fumes of KO_x. Will react with water, steam or acids to produce H_2; can react vigorously with oxidizing materials. To fight fire, use CO_2, dry chemical. Incompatible with air; Cl_2; F_2; acetic acid; acrolein; acrylonitrile; ($CaC + Cl_2$); ClO_2; (H_2O_2 + Cl_2); ($CHFl_3 + CH_3OH$); 1,2-dichloroethylene; maleic anhydride; (n-methyl-n-nitrosourea + CH_2Cl_2); nitroethane; NCl_3; nitromethane; nitroparaffins; o-nitrophenol; nitropropane; n-nitrosomethylurea; (nitrosomethylurea + CH_2Cl_2); H_2O; trichloroethylene; tetrahydrofuran; tetrachlorethane.

POTASSIUM HYDROXIDE   HR: 3
CAS: 1310-58-3   NIOSH: TT 2100000
DOT: 1813/1814
mf: HKO   mw: 56.11

PROP: White, deliquescent pieces, lumps or sticks having crystalline fracture. Mp: 360° ± 7°; bp: 1320°; d: 2.044. Violent, exothermic reaction with water.

SYNS: CAUSTIC POTASH * HYDROXYDE DE PO-TASSIUM (FRENCH) * KALIUMHYDROXID (GER-MAN) * KALIUMHYDROXYDE (DUTCH) * LYE * POTASSA * POTASSE CAUSTIQUE (FRENCH) * POTASSIO (IDROSSIDO DI) (ITAL-IAN) * POTASSIUM HYDRATE

DOT Classification: Label: Corrosive

THR: Poison by ingestion. A severe human skin irritant. Very corrosive to the eyes, skin, and mucous membranes. Mutagenic data. See also sodium hydroxide. Ingestion may cause violent pain in throat and epigastrium, hematemesis, collapse. Stricture of esophagus may result if not immediately fatal. Incompatible with acids; ammonium hexachloroplatinate (2-); chlorine dioxide; germanium; hyponitrous acid; maleic anhydride; nitroalkanes; nitrobenzene; nitrogen trichloride; potassium peroxodisulphate; 2, 2,3,3-tetrafluoropropanol; tetrahydrofuran; thorium dicarbide; 2,4,6-trinitrotoluene.

POTASSIUM IODATE   HR: 3
CAS: 7758-05-6   NIOSH: NN 1350000
mf: IO$_3$ • K   mw: 214.00

PROP: Colorless crystals. Mp: 560°, d: 3.89. Insol in alc.

SYN: IODIC ACID, POTASSIUM SALT

THR: Poison by ingestion and intraperitoneal routes. A trace mineral added to animal feeds. Violent reaction with Al; As; C; Cu; metal sul-fides; organic matter; P; S. See also iodates, oxidizable matter. When heated to decomposition it emits very toxic fumes of I$^-$ and K$_2$O.

POTASSIUM IODIDE   HR: 3
CAS: 7681-11-0   NIOSH: TT 2975000
mf: IK   mw: 166.00

PROP: Colorless or white granules. Mp: 723°, bp: 1420°, d: 3.13, vap press: 1 mm @ 745°.

THR: Poison by intravenous route. Moderately toxic by ingestion and intraperitoneal routes. Mutagenic data. See also iodides. A trace min-eral added to animal feeds; a nutrient and/or dietary supplement food additive. When heated to decomposition it emits very toxic fumes of K$_2$O and I$^-$. Incompatible with diazonium salts; diisopropyl peroxydicarbonate; oxidants; BrF$_3$; ClF$_3$; FClO, metallic salts; calomel.

POTASSIUM NITRATE   HR: 3
CAS: 7757-79-1   NIOSH: TT 3700000
DOT: 1486
mf: KNO$_3$ mw: 101.11

PROP: Transparent, colorless or white crystal-line powder or crystals, cooling, pungent, saline taste. Mp: 334°, bp: decomp @ 400°, d: 2.109 @ 16°. Sol in glycerol, water.

SYNS: KALIUMNITRAT (GERMAN) * NITER * NITRE * NITRIC ACID, POTASSIUM SALT * SALTPETER

DOT Classification: Label: Oxidizer

THR: Poison by intravenous route. Moderately toxic by ingestion. See nitrates. A food additive permitted in food for human consumption. In-gestion of large quantities may cause gastroenteri-tis. Chronic exposure can cause anemia, nephri-tis and methemoglobinemia. Can react violently with Sb; Sb$_2$S$_3$; As; AsS$_2$; BaS; B; BP; CaS; F$_2$; charcoal; Cu$_3$P$_2$; Ge; GeS; Na acetate; Na hypophosphite; (Na$_2$O$_2$ + dextrose); (S + As$_2$S$_3$); Ti; TiS$_2$; Zn; trichloroethylene; Zr; met-als; phosphides; reductants; thorium dicarbide. When heated to decomposition it emits very toxic fumes of NO$_x$ and K$_2$O.

POTASSIUM NITRITE (1:1)   HR: 3
CAS: 7758-09-0   NIOSH: TT 3750000
DOT: 1488
mf: NO$_2$ • K   mw: 85.11

PROP: White or slightly yellowish, deliques-cent prisms or sticks. Mp: 387°; bp: decomp, d: 1.915.

SYNS: NITROUS ACID, POTASSIUM SALT * POTASSIUM NITRITE (DOT)

DOT Classification: Label: Oxidizer

THR: Poison by ingestion. An experimental te-ratogen. See nitrites. Flammable; an oxidizing material. Slight explosion hazard when exposed to heat. It will explode @ 1000°F or when mixed with cyanide salts and heated. Also reacts vio-lently with B; (NH$_4$)$_2$SO$_4$; potassium amide. Dangerous. See nitrites for disaster hazard.

POTASSIUM PERCHLORATE   HR: 3
CAS: 7778-74-7   NIOSH: SC 9700000
DOT: 1489
mf: $ClO_4 \cdot K$ mw: 138.55

PROP: Colorless crystals or white powder. Decomp @ 400° and with organic matter. D: 2.52, mp: 610° ± 10°. Insol in alc.

SYNS: PERIODIN * POTASSIUM HYPERCHLORIDE * PERCHLORIC ACID, POTASSIUM SALT (1:1)

DOT Classification: Label: Oxidizer

THR: An experimental teratogen. Powerful oxidizer. Severe irritant to skin, eyes, and mucous membranes. Has been implicated in aplastic anemia. Absorption can cause methemoglobinemia and kidney injury. When heated to decomposition it emits very toxic fumes of K_2O and Cl^-. Incompatible with (Al + Mg); charcoal; F_2, Mg; (Ni + Ti); reducing agents; S. Do not store close to flammable matter. When heated to decomposition it emits toxic fumes of Cl^- and K_2O.

POTASSIUM PERMANGANATE   HR: 3
CAS: 7722-64-7   NIOSH: SD 6475000
DOT: 1490
mf: $MnO_4 \cdot K$ mw: 158.04

PROP: Dark purple crystals with a blue metallic sheen, sweetish astringent taste. Mp: decomp @ <240°, d: 2.703.

SYNS: CHAMELEON MINERAL * C.I. 77755 * CONDY'S CRYSTALS * KALIUMPERMANGANAAT (DUTCH) * KALIUMPERMANGANAT (GERMAN) * PERMANGANATE DE POTASSIUM (FRENCH) * PERMANGANATE OF POTASH * POTASSIO (PERMANGANATO DI) (ITALIAN) * POTASSIUM (PERMANGANATE DE) (FRENCH) * POTASSIUM PERMANGANATE (DOT)

DOT Classification: Label: Oxidizer

THR: Poison by intravenous and subcutaneous routes. Moderately toxic by ingestion. Gastrointestinal effects by ingestion in humans. Mutagenic data. A strong irritant due to its oxidizing properties. See also manganese compounds. Flammable by chemical reaction. A powerful oxidizer which is spontaneously flammable on contact with Al_4C_3; As; dimethyl sulfoxide; ethylene glycol; glycerin; H_2O_2; H_2S_3; HCl; H_2SO_4; (H_2SO_4 + organic matter); (H_2SO_4 + KCl); NH_4ClO_4; NH_3; NH_4; NO_3; NH_2OH; or-

ganic matter; P; polypropylene; S; Sb; Ti; wood. A dangerous explosion hazard; handle with care. Explosions may occur in contact with organic or readily oxidizable materials, either when dry or in solution. Dangerous; keep away from combustible materials. See also permanganates.

POTASSIUM PEROXIDE
CAS: 17014-71-0   NIOSH: TT 4450000
DOT: 1491
mf: KO_2 mw: 71.1

PROP: Yellow, amorphous mass (white crystals). Mp: 490°.

DOT Classification: Label: Oxidizer

THR: See peroxides, inorganic. Dangerous fire hazard by spontaneous chemical reaction. It is a very powerful oxidizer. Fires of this material should be handled like sodium peroxide fires. Moderate explosion hazard by spontaneous chemical reaction. Also violent reactions with air; Sb; As; O_2; K; water. Dangerous; will react with water or steam to produce heat; on contact with reducing material, can react vigorously; on contact with acid or acid fumes, it can emit toxic fumes. Incompatible with carbon; diselenium dichloride; ethanol; hydrocarbons; metals.

POTASSIUM PHOSPHIDE   HR: 3
CAS: 20770-41-6   NIOSH: TT 4890000
mf: K_3P mw: 148.27

PROP: A solid.

SYN: PHOSPHURE DE POTASSIUM (FRENCH)

THR: Poison by inhalation. See also phosphides and phosphine which is released upon contact of phosphide with water, steam. When heated to decomposition it emits very toxic fumes of PO_x and PH_3.

POTASSIUM SELENATE   HR: 3
CAS: 7790-59-2   NIOSH: VS 6600000
mf: $O_4Se \cdot 2K$ mw: 221.16

PROP: Colorless crystals. D: 3.07; sol in water.

SYN: SELENIC ACID, DIPOTASSIUM SALT

THR: Poison by ingestion and intravenous routes. See also selenium compounds. When heated to decomposition it emits toxic fumes of Se.

POTASSIUM SODIUM ALLOY   HR:
CAS: 11135-81-2   NIOSH: TT 5790000
DOT: 1422

SYNS: SODIUM POTASSIUM ALLOY (DOT)
* NAK

DOT Classification: Label: Flammable Solid and Dangerous When Wet

THR: A low melting alloy of Na and K. Its toxicity is due to either Na or K alone. Corrosive to the eyes, skin, and mucous membranes. Upon contact with moisture it reacts violently to evolve H_2; much heat; and a highly caustic residue of NaOH or KOH. Oxidation forms Na_2O and K_2O which are powerful caustics. See sodium and potassium. A dangerous fire hazard; in the presence of O_2; moisture; halogens; oxidizers; acids or acid fumes, etc., it will react violently, giving off much heat, often spattering either red-hot particles or actually flaming particles. A severe explosion hazard; will react explosively under many conditions, such as contact with moisture; CO_2; CCl_4; $CHCl_3$; $CHBr_3$; (ammonium sulfate + NH4 + NO3); CH_2Cl_2; CH_2I_2; HgO; CH_3Cl; oxalyl bromide; oxalyl chloride; pentachloroethane; K oxides; KO_2; Si; Ag halides; $NaHCO_3$; polytetrafluoroethylene; tetrachloroethane; 1,1,1-trichloroethane; trichlorotrifluoroethane. Dangerous; when heated it emits highly toxic fumes of Na_2O, K_2O; will react explosively with water, steam, acid, acid fumes or mists to produce heat, hydrogen and toxic and corrosive fumes; can react vigorously with oxidizing materials. See also sodium and potassium. To fight fire, use G-1 powder, dry sodium chloride or dry soda ash. Never use water or foam. Incompatible with air; carbon dioxide; or carbon disulfide; halocarbons; metal oxides; nitrogen-containing explosives.

POTASSIUM SULFIDE (2:1)   HR: 3
CAS: 1312-73-8   NIOSH: TT 6000000
DOT: 1382
mf: K_2S mw: 110.26

PROP: Red cryst mass; deliquescent in air; mp: 912°; d: 1.805 @ 14°.

SYNS: POTASSIUM MONOSULFIDE * HEPAR SULFUROUS

DOT Classification: Label: Flammable Solid

THR: Poison by ingestion and inhalation. See also sulfides. Emits H_2S in contact with acids, steam. A flammable solid. Unstable. When heated to decomposition it emits very toxic fumes of H_2S and SO_x. May explode on percussion or rapid heating. Incompatible with NO_x.

POTASSIUM THIOCYANATE   HR: 3
CAS: 333-20-0   NIOSH: XL 1925000
mf: CNS•K   mw: 97.18

PROP: Colorless, deliquescent crystals. D: 1.89; mp: about 173°.

SYNS: POTASSIUM RHODANATE * POTASSIUM RHODANIDE * POTASSIUM SULFOCYANATE * POTASSIUM THIOCYANIDE

THR: Poison by ingestion, intramuscular, subcutaneous, and intravenous routes. See also thiocyanates. Large doses can cause skin eruptions, psychoses and collapse. When heated to decomposition it emits very toxic fumes of CN^-, SO_x, and NO_x. Incompatible with perchloryl fluoride.

PRASEODYMIUM   HR: 2
PROP: Pr; At wt: 140.9077. Yellowish metal. Mp: 935°; bp: 3290°; d: (a) 6.772; d: (b) 6.64.

THR: As a lanthanon, it may depress coagulation of the blood. See also lanthanum and Rare Earths. Limited animal experiments suggest moderately toxic. Flammable in the form of dust, when exposed to heat or flame or by chemical reaction. Fine dust ignites readily. See also iron dust. Incompatible with air; halogens.

PRASEODYMIUM CHLORIDE   HR: 3
CAS: 10361-79-2   NIOSH: TU 0175000
mf: Cl_3Pr mw: 247.26

THR: Poison by intraperitoneal and intravenous routes. Moderately toxic by subcutaneous route. See also praseodymium compounds. A skin and eye irritant. When heated to decomposition it emits toxic fumes of Cl^-.

PRASEODYMIUM (III) NITRATE (1:3)   HR: 3
CAS: 10361-80-5   NIOSH: TU 1225000
mf: N_3O_9•Pr   mw: 326.94

SYN: NITRIC ACID, PRASEODYMIUM(3+) SALT

THR: Poison by intraperitoneal and intravenous routes. Moderately toxic by ingestion. See also praseodymium and nitrates. When heated to decomposition it emits toxic fumes of NO_x.

PREDNISONE   HR: 3
CAS: 53-03-2   NIOSH: TU 4154100
mf: $C_{21}H_{26}O_5$ mw: 358.47

PROP: White, odorless, crystalline powder, very sltly sol in water, sltly sol in alcohol, chloroform, methanol and dioxane. Mp: 235° (with some decomp).

SYNS: DELTA(SUP 1)-CORTISONE * 1-DEHY-DROCORTISONE * DELTA CORTISONE * 17,21-DIHYDROXYPREGNA-1,4-DIENE-3,11,20-TRI-ONE * 1,4-PREGNADIENE-17-ALPHA,21-DIOL-3,11,20-TRIONE

THR: An experimental carcinogen. Systemic skin effects in man by an unspecified route. Has been implicated in aplastic anemia. A food additive permitted in food for human consumption.

PREDONIN HR: 3
CAS: 50-24-8   NIOSH: TU 4152000
mf: $C_{21}H_{28}O_5$ mw: 360.49

SYNS: DELTA(SUP 1)-CORTISOL * DELTA(SUP 1)-DEHYDROCORTISOL * DELTA(SUP 1)-DEHY-DROHYDROCORTISONE * 1-DEHYDROHYDRO-CORTISONE * DELTA(SUP 1)-HYDROCORTISONE * PREDNISOLONE * 1,4-PREGNADIENE-3,20-DIONE-11-BETA,17-ALPHA,21-TRIOL * 1,4-PREGNADIENE-11-BETA,17-ALPHA,21-TRIOL-3,20-DIONE * 11-BETA,17,21-TRIHY-DROXYPREGNA-1,4-DIENE-3,20-DIONE * 11-BETA,17-ALPHA,21-TRIHYDROXYPREGNA-1,4-DIENE-3,20-DIONE * 11-BETA,17-ALPHA,21-TRIHYDROXY-1,4-PREGNADIENE-3,20-DIONE

THR: Acute poison by subcutaneous route. An experimental teratogen. When heated to decomposition it emits acrid smoke and fumes.

PRIDINOL HR: 3
CAS: 968-58-1   NIOSH: TN 2975000
mf: $C_{20}H_{25}NO \cdot ClH$ mw: 331.92

PROP: Crystals. Decomp @ 238°; sol in alc.

SYNS: ALPHA,ALPHA-DIPHENYL-1-PIPERIDINE-PROPANOL HYDROCHLORIDE * 1,1-DIPHENYL-3-PIPERIDINO-1-PROPANOL HYDROCHLORIDE * 1,1-DIPHENYL-3-(1-PIPERIDYL)-1-PROPANOL HYDROCHLORIDE * 3-PIPERIDINO-1,1-DIPHE-NYL-1-PROPANOL HYDROCHLORIDE * 3-(N-PI-PERIDYL)-1,1-DIPHENYL-1-PROPANOL HYDRO-CHLORIDE * ALPHA-(2-PIPERIDYLETHYL)BENZHYDROL HYDROCHLORIDE

THR: Poison by intraperitoneal and intravenous routes. When heated to decomposition it emits very toxic fumes of NO_x and HCl.

PROCARBAZINE HR: 3
CAS: 671-1-6-9   NIOSH: XS 4550000
mf: $C_{12}H_{19}N_3O$ mw: 221.34

SYNS: IBENZMETHYZINE * 2-(P-ISOPROPYL CARBAMOYL BENZYL)-1-METHYLHYDRAZINE * N-ISOPROPYL-ALPHA-(2-METHYLHYDRAZI-NO)-P-TOLUAMIDE * MATULANE * 4-((2-METHYLHYDRAZINO)METHYL)-N-ISOPROPYL-BENZAMIDE * 1-METHYL-2-(-ISOPROPYLCAR-BAMOYL)BENZYL)HYDRAZINE * MIH * NATULAN * NSC-77213 * PCB * RO 4-6467

THR: Poison by intravenous route. Mutagenic data. An experimental teratogen, carcinogen, neoplastigen, and tumorigen. Has been implicated as a brain carcinogen. Moderately toxic by intraperitoneal route. When heated to decomposition it emits toxic fumes of NO_x.

PROCHLORPERIZINE MALEATE HR: 3
CAS: 84-02-6   NIOSH: SO 3150000
mf: $C_{20}H_{24}ClN_3S \cdot 2C_4H_4O_4$ mw: 606.14

SYNS: 2-CHLORO-10-(3-(1-METHYL-4-PIPERAZINYL)PROPYL)PHENOTHIAZINE, DIMA-LEATE * 2-CHLORO-10-(3-(4-METHYL-1-PIPERAZINYL)PROPYL)PHENOTHIAZINE DIMA-LEATE * PROCHLOROPROAZINE HYDRO-GEN MALEATE * PROCHLORPERAZINE BIMA-LEATE * PROCHLORPERAZINE DIMALEATE * PROCHLORPERAZINE MALEATE * PROCHLOR-PERAZINE HYDROGEN MALEATE

THR: Poison by ingestion and subcutaneous route. When heated to decomposition it emits very toxic fumes of Cl^-, NO_x, and SO_x.

PROCHLORPROMAZINE HR: 3
CAS: 58-38-8   NIOSH: SO 2700000
mf: $C_{20}H_{24}ClN_3S$ mw: 373.98

SYNS: CHLORO-3 (N-METHYLPIPERAZINYL-3 PROPYL)-10 PHENOTHIAZINE (FRENCH) * 2-CHLORO-10-(3-(1-METHYL-4-PIPERAZINYL)-PRO-PYL)-PHENOTHIAZINE * 2-CHLORO-10-(3-(4-METHYL-1-PIPERAZINYL)PROPYL)PHENOTHIAZINE * 3-CHLORO-10-(3-(1-METHYL-4-PIPERAZINYL)PROPYL)PHENOTHIAZINE * N-(GAMMA-(4'-METHYLPIPERAZINYL-1')PROPYL)-3-CHLOROPHENOTHIAZINE * PRO-CHLOROPERAZINE * PROCHLORPEMAZINE * PROCHLORPERAZINE

THR: Poison by ingestion, intravenous, and intraperitoneal routes. An experimental teratogen. Implicated in aplastic anemia. When heated to decomposition it emits very toxic fumes of SO_x, NO_x, and Cl^-.

PRODUCER GAS   HR: 3
PROP: lel = 20%-30%; uel = 70%-80%.

THR: Poison. See also carbon monoxide, methane, and hydrogen. Dangerous fire hazard when exposed to flame. Explosive by spark or flame when mixed with air in the range of 20.7%-73.7%. Dangerous; can react vigorously with oxidizing materials. To fight fire, use CO_2, dry chemical, water spray.

PROFLAVINE DIHYDROCHLO-
RIDE   HR: 3
CAS: 531-73-7   NIOSH: AR 8700000
mf: $C_{13}H_{11}N_3 \cdot 2ClH$ mw: 282.16

PROP: Dihydrate; orange-red crystalline powder. Very sltly sol in ether, chloroform; sol in water.

SYNS: 3,6-DIAMINOACRIDINE DIHYDROCHLORIDE * 2,8-DIAMINOACRIDINIUM CHLORIDE HYDROCHLORIDE * 3,6-DIAMINOACRIDINIUM CHLORIDE HYDROCHLORIDE

THR: An experimental carcinogen. When heated to decomposition it emits very toxic fumes of NO_x and HCl.

PROFLAVINE MONOHYDROCHLO-
RIDE   HR: 3
CAS: 952-23-8   NIOSH: AR 9064000
mf: $C_{13}H_{11}N_3 \cdot ClH$

SYNS: 3,6-ACRIDINEDIAMINE, MONOHYDRO-CHLORIDE (9CI) * 3,6-DIAMINOACRIDINE MONOHYDROCHLORIDE * 3,6-DIAMINOACRIDINIUM CHLORIDE * 3,6-DIAMINOACRIDINIUM CHLORIDE HYDROCHLORIDE * 2,8-DIAMINO-ACRIDINIUM CHLORIDE MONOHYDROCHLORIDE * PROFLAVINE HYDROCHLORIDE

THR: An experimental carcinogen. When heated to decomposition it emits very toxic fumes of NO_x and HCl.

PROGESTERONE   HR: 3
CAS: 57-83-0   NIOSH: TW 0175000
mf: $C_{31}H_{30}O_2$ mw: 314.51

PROP: A female sex hormone. White crystalline powder, odorless, practically insol in water, sol in alc, acetone and dioxane, sparingly sol in oils. D: 1.166 @ 23°; mp: 127°-131°.

SYNS: CORPUS LUTEUM HORMONE * GLANDUCORPIN * HORMOFLAVEINE * HORMOLUTON * LUTEAL HORMONE * LUTEOHORMONE

* NSC-9704 * 3,20-PREGNENE-4 * PREGNENEDIONE * PREGNENE-3,20-DIONE * PREGN-4-ENE-3,20-DIONE * 4-PREGNENE-3,20-DIONE * DELTA(SUP 4)-PREGNENE-3,20-DIONE * PROGESTEROL * BETA-PROGESTERONE

THR: Poison by intravenous route. Mutagenic data. An experimental teratogen, neoplastigen, tumorigen, and carcinogen. A food additive permitted in the feed and drinking water of animals and/or for the treatment of food-producing animals. Also permitted in food for human consumption.

PROMAZINE   HR: 3
CAS: 58-40-2   NIOSH: SO 7000000
mf: $C_{17}H_{20}N_2S$ mw: 284.45

SYNS: 10H-PHENOTHIAZINE-10-PROPANAMINE, N,N-DIMETHYL * 10-(3-(DIMETHYLAMINO)-PROPYL)PHENOTHIAZINE * PROTACTYL

THR: Poison by ingestion, subcutaneous, intravenous, and intraperitoneal routes. When heated to decomposition it emits very toxic fumes of NO_x and SO_x.

PROMAZINE HYDROCHLORIDE   HR: 3
CAS: 53-60-1   NIOSH: SO 8575000
mf: $C_{17}H_{20}N_2S \cdot ClH$ mw: 320.91

PROP: White to slightly yellow, practically odorless crystalline powder. Decomp @ 181°; sol in water, methanol, ethanol, chloroform; insol in ether, benzene.

SYNS: 10-(GAMMA-DIMETHYLAMINO-N-PROPYL)PHENOTHIAZINE HYDROCHLORIDE * 10-(3-(DIMETHYLAMINO)PROPYL)PHENOTHIAZINE HYDROCHLORIDE * SPARINE HYDROCHLORIDE

THR: Poison by ingestion, subcutaneous, intravenous, intraperitoneal, and intramuscular routes. Human central nervous system effects. A food additive permitted in food for human consumption; also permitted in the feed and drinking water of animals and/or for the treatment of food-producing animals. See chlorides, oxides of nitrogen, sulfur oxide, and hydrochloric acid for disaster hazard.

PROMETHIUM   HR: 3
af: Pm   aw: 147

PROP: Metallic solid. Mp: 1080°; bp: 2460°. d: 7.22. A rare earth.

THR: Poison. Radiotoxic metal. See also Rare Earths.

PROPANE HR: 1

CAS: 74-98-6 NIOSH: TX 2275000
DOT: 1075/1978
mf: C_3H_8 mw: 44.11

PROP: Colorless gas. Bp: $-42.1°$, lel = 2.3%, uel = 9.5%, fp: $-187.1°$, flash p: $-156°F$, d: 0.5852 @ $-44.5°/4°$, autoign temp: $842°F$, vap d: 1.56. Sol in water, alc, ether.

SYNS: DIMETHYLMETHANE * PROPYL HYDRIDE

OSHA PEL: TWA 1000 ppm
DOT Classification: Label: Flammable Gas

THR: Central nervous system effects at high concentrations. A general-purpose food additive. An asphyxiant. Highly dangerous fire hazard when exposed to heat, flame or oxidizers. Severe explosion hazard when exposed to flame or ClO_2. Dangerous; can react vigorously with oxidizing materials. To fight fire, stop flow of gas.

PROPANEDIAL HR: 3

CAS: 542-78-9 NIOSH: TX 6575000
mf: $C_3H_4O_2$ mw: 72.07

SYNS: MALONALDEHYDE * MALONDIALDE-HYDE * MALONIC ALDEHYDE * MALONIC DIALDEHYDE * MALONYLDIALDEHYDE * NCI-C54842

THR: An experimental carcinogen. Mutagenic data. Moderately toxic by ingestion. When heated to decomposition it emits acrid smoke.

1,2-PROPANEDIOL HR: 3

CAS: 57-55-6 NIOSH: TY 2000000
mf: $C_3H_8O_2$ mw: 76.11

PROP: Colorless liquid, practically odorless, hygroscopic. Bp: $188.2°$, flash p: $210°F$ (OC), lel = 2.6%, uel = 12.6%, d: 1.0362 @ $25°/25°$, autoign temp: $700°F$, vap press: 0.08 mm @ $20°$, vap d: 2.62, fp: $-59°$.

SYNS: 1,2-DIHYDROXYPROPANE * METHYL-ETHYLENE GLYCOL * MONOPROPYLENE GLY-COL * PROPANE-1,2-DIOL * PROPYLENE GLY-COL * ALPHA-PROPYLENEGLYCOL * 1,2-PRO-PYLENE GLYCOL * TRIMETHYL GLYCOL

THR: Human central nervous system effects. A skin and eye irritant. Sltly toxic by ingestion,

skin contact, intraperitoneal, subcutaneous, and intramuscular routes. Combustible when exposed to heat or flame; can react with oxidizing materials. Moderate explosion hazard when exposed to flame. To fight fire, use alcohol foam.

PROPARGYL BROMIDE HR: 3

CAS: 106-96-7 NIOSH: UK 4375000
mf: C_3H_3Br mw: 118.97

PROP: An almost colorless liquid, sharp odor. Bp: $88°-90°$, fp: $-61.07°$, flash p: $65°F$ (COC), d: 1.564-1.570, vap d: 6.87.

SYNS: GAMMA-BROMOALLYLENE * 3-BROMO-PROPYNE * 3-BROMO-1-PROPYNE

THR: Deadly poison by ingestion. A dangerous explosion hazard. It can detonate at $220°$ or by impact (especially when mixed with chloropicrin or other chemicals). Dangerous; when heated to decomposition it emits highly toxic fumes of bromides; can react vigorously with oxidizing materials. To fight fire, use water, foam, CO_2, dry chemical.

PROPENE HR: 1

CAS: 115-07-1 NIOSH: UC 6740000
DOT: 1075/1077
mf: C_3H_6 mw: 42.09

PROP: A gas, d (gas): 1.49 (air = 1.0). Mp: $-185°$, bp: $-47.7°$, d (liquid): 0.581 @ $0°$, autoign temp: $860°F$, vap press: 10 atm @ $19.8°$, lel = 2.4%, uel = 10.1%, vap d: 1.5, flash p: $-162°F$.

SYNS: METHYLETHENE * METHYLETHYLENE * NCI-C50077 * 1-PROPENE (9CI) * PRO-PYLENE (DOT)

DOT Classification: Label: Flammable Gas

THR: A simple asphyxiant. No irritant effects from high concentrations in gaseous form. When compressed to liquid form, can cause skin burns from freezing effects on tissue of rapid evaporation. For effects of simple asphyxiants, see argon. Very dangerous fire hazard when exposed to heat flame, or oxidizers. Moderate explosion hazard when exposed to heat or flame. Under unusual conditions, i.e., 955 atmospheres pressure and $327°$, it has been known to explode. Reacts violently with NO_2; N_2O_4; N_2O; $LiNO_3$; SO_2; trifluoro methyl hypofluorite. Dangerous; can react vigorously with oxidizing materials. To fight fire, stop flow of gas.

PROPENE POLYMERS   HR: 3
CAS: 9003-07-0   NIOSH: UD 1842000
mf: $(C_3H_6)n$

PROP: Solid material; mp: about 165°; d: 0.90-0.92. Insol in organic materials.

SYNS: ICI 543 * POLYPROPENE * POLYPRO-PYLENE * ISOTACTIC POLYPROPYLENE * 1-PROPENE HOMOPOLYMER (9CI) * PROPYL-ENE POLYMER

THR: An experimental carcinogen. When heated to decomposition it emits acrid smoke and fumes.

p-PROPENYL ANISOLE   HR: 3
CAS: 104-46-1   NIOSH: BZ 8925000
mf: $C_{10}H_{12}O$ mw: 148.22

PROP: Leaves from alc; d: 0.991 @ 20°/20°; mp: 22.5°; bp: 235.3°; very slightly sol in water; misc in absolute alc; ether.

SYNS: ANETHOLE * ANISE CAMPHOR * ISOESTRAGOLE * P-METHOXY-BETA-METH-YLSTYRENE * 1-METHOXY-4-PROPENYLBEN-ZENE * 4-METHOXYPROPENYLBENZENE * NAULI "GUM" * OIL OF ANISEED * P-PROPENYL ANISOLE * P-1-PROPENYL-ANISOLE * 4-PROPENYLANISOLE * P-PROPE-NYLPHENYL METHYL ETHER

THR: An experimental carcinogen. Moderately toxic by ingestion. When heated to decomposition it emits acrid smoke and fumes.

PROPENYL CHLORIDE   HR: 3
CAS: 590-21-6   NIOSH: UC 7175000
mf: C_3H_5Cl mw: 76.53

PROP: Liquid. Mp: −137.4°; bp: 22.65°; flash p: <21°F; d: 0.9189°; lel = 4.5%; uel = 16%; insol in water.

SYNS: 1-CHLOROPROPENE * 1-CHLORO-1-PRO-PENE

THR: An experimental neoplastigen. Moderately toxic by inhalation and ingestion. Mutagenic data. An eye irritant. Very dangerous fire hazard by heat, flames (sparks) or oxidizers. To fight fire, use alcohol foam, dry chemical, mist spray, fog. When heated to decomposition it emits toxic fumes of Cl^-. For further information see Vol. 6, No. 2 of *DPIM Report*.

((2-PROPENYLOXY)METHYL)OXI-RANE   HR: 3
CAS: 106-92-3   NIOSH: RR 0875000
mf: $C_6H_{10}O_2$ mw: 114.16

PROP: Bp: 153.9°, fp: −100° (forms glass), flash p: 135°F (OC), d: 0.9698 @ 20°/4°, vap press: 21.59 mm @ 60°, vap d: 3.94.

SYNS: ALLIL-GLICIDIL-ETERE (ITALIAN) * ALLYLGLYCIDAETHER (GERMAN) * ALLYL GLYCIDYL ETHER * ALLYL-2,3-EPOXYPROPYL ETHER * 1-ALLYLOXY-2,3-EPOXY-PROPAAN (DUTCH) * 1-ALLYLOXY-2,3-EPOXYPROPAN (GERMAN) * 1-(ALLYLOXY)-2,3-EPOXYPRO-PANE * NCI-C56666 * OXYDE D'ALLYLE ET DE GLYCIDYLE (FRENCH)

OSHA PEL: TWA CL 10 ppm

THR: Poison by ingestion and inhalation. Mutagenic data. Severe skin and eye irritant. Can cause central nervous system depression and pulmonary edema. Flammable when exposed to heat or flame; can react with oxidizing materials. To fight fire, use foam, CO_2, dry chemical.

PROPIONIC ACID   HR: 3
CAS: 79-09-4   NIOSH: UE 5950000
mf: $C_3H_6O_2$ mw: 74.09

PROP: Oily liquid, pungent, disagreeable, rancid odor; d: 0.998 @ 15°/4°; mp: −21.5°; bp: 141.1°; vap press: 10 mm @ 39.7°; vap d: 2.56; autoign temp: 955°F. Misc in water, alc, ether, chloroform.

SYNS: ACIDE PROPIONIQUE (FRENCH) * CAR-BOXYETHANE * ETHANECARBOXYLIC ACID * ETHYLFORMIC ACID * METHYL ACETIC ACID * PROPANOIC ACID * PROPIONIC ACID GRAIN PRESERVER * PSEUDOACETIC ACID

DOT Classification: Label: Corrosive

THR: Poison by skin contact and intraperitoneal routes. A severe eye and skin irritant. When heated it emits acrid smoke and fumes. Flammable when exposed to heat, flame or oxidizers. To fight fire, use alcohol foam.

PROPIONONITRILE   HR: 3
CAS: 107-12-0   NIOSH: UF 9625000
mf: C_3H_5N mw: 55.09

PROP: Colorless liquid, ethereal odor. Bp: 97.1°, d: 0.783 @ 21°/4°, vap d: 1.9, flash p: 36°F, lel = 3.1%; mp: −91.8°. Misc with alc, ether.

SYNS: CYANOETHANE * ETHYL CYANIDE * HYDROCYANIC ETHER * PROPANENITRILE * PROPIONIC NITRILE

THR: Poison by ingestion, skin contact, inhalation, subcutaneous, and intraperitoneal routes. See also nitrites. An eye irritant. Poisonous when heated to decomposition or on contact with acids. Dangerous fire hazard by heat, flame (sparks), oxidizers. To fight fire, use water spray, foam, mist, CO_2, dry chemical. When heated to decomposition it emits toxic fumes of NO_x and CN^-. See also nitriles.

d-PROPOXYPHENE HYDROCHLO-RIDE HR: 3
CAS: 1639-60-7   NIOSH: EL 3000000
mf: $C_{22}H_{29}NO_2 \cdot ClH$ mw: 375.98

PROP: Bitter crystals. Mp: 163°-168.5°; sol in water, alc, chloroform, acetone, insol in benzene, ether.

SYNS: DARVON * DEXTROPROPOXYPHENE HYDROCHLORIDE * D-4-DIMETHYLAMINO-3-METHYL-1,2-DIPHENYL-2-BUTANOL PROPIONATE HYDROCHLORIDE * S-ALPHA-(2-(DIMETHYLAMINO)-1-METHYLETHYL)-ALPHA-PHENYLBENZENEETHANOL PROPIOATE HYDROCHLORIDE * (+)-1,2-DIPHENYL-2-PROPIONOXY-3-METHYL-4-DIMETHYLAMINOBUTANE HYDROCHLORIDE * PROPOXYCHEL * PROPOXYPHENE HYDROCHLORIDE * (+)-PROPOXYPHENE HYDROCHLORIDE

THR: Poison to humans by ingestion. Poison by intraperitoneal, subcutaneous, intravenous, and intramuscular routes. Human central nervous system and pulmonary effects. When heated to decomposition it emits very toxic fumes of HCl and NO_x.

PROPYL ACETATE HR: 2
CAS: 109-60-4   NIOSH: AJ 3675000
DOT: 1276
mf: $C_5H_{10}O_2$ mw: 102.15

PROP: Clear, colorless liquid, pleasant odor. Mp: -92.5°, bp: 101.6°, flash p: 58°F, lel = 2.0%, uel = 8.0%, d: 0.887, autoign temp: 842°F, vap press: 40 mm @ 28.8°, vap d: 3.52. Misc with alc, ether. Water sol.

SYNS: ACETATE DE PROPYLE NORMAL (FRENCH) * 1-ACETOXYPROPANE * OCTAN PROPYLU (POLISH) * N-PROPYL ACETATE * 1-PROPYL ACETATE

OSHA PEL: TWA 200 ppm
ACGIH TLV: TWA 200 ppm; STEL 250 ppm
DFG MAK: 200 ppm (840 mg/m^3)
DOT Classification: Label: Flammable Liquid

THR: A skin irritant and a systemic irritant by inhalation. A narcotic at high concentrations. Isopropyl acetate is slightly less narcotic than normal propyl acetate. It can cause death by inhalation, but no industrial injury has been reported as occurring to workmen exposed to it. Dangerous fire hazard when exposed to heat, flame, or oxidizers. Moderate explosion hazard when exposed to heat or flame. Dangerous, upon exposure to heat or flame; can react vigorously with oxidizing materials. To fight fire, use alcohol foam, CO_2, dry chemical.

N-PROPYLAJMALINE BITARTRATE HR: 3
CAS: 2589-47-1   NIOSH: AX 7750000
mf: $C_{23}H_{32}N_2O_2 \cdot C_4H_6O_6$ mw: 518.67

SYNS: 17R,21-ALPHA-DIHYDROXY-4-PROPYLAJMALANIUM HYDROGEN TARTRATE * N-PROPYLAJMALINE HYDROGEN TARTRATE * N-PROPYLAJMALINIUM BITARTRATE * N(SUP 4)-PROPYLAJMALINIUM HYDROGEN TARTRATE

THR: Poison by ingestion and intravenous routes. Central nervous system effects in humans. When heated to decomposition it emits toxic fumes of NO_x.

PROPYL ALCOHOL HR: 3
CAS: 71-23-8   NIOSH: UH 8225000
DOT: 1274
mf: C_3H_8O mw: 60.11

PROP: Clear, odorless liquid, alcohol-like odor. Mp: -127°, bp: 97.19°, flash p: 59°F (CC), ulc: 55-60, d: 0.8044 @ 20°/4°, lel = 2.1%, uel = 13.5%, autoign temp: 824°F, vap press: 10 mm @ 14.7°, vap d: 2.07. Misc water, alc, ether.

SYNS: ALCOOL PROPILICO (ITALIAN) * ALCOOL PROPYLIQUE (FRENCH) * ETHYL CARBINOL * N-PROPANOL * PROPANOL-1 * 1-PROPANOL * PROPANOLE (GERMAN) * PROPANOLEN (DUTCH) * PROPANOLI (ITALIAN) * N-PROPYL ALCOHOL * 1-PROPYL ALCOHOL * N-PROPYL ALKOHOL (GERMAN) * PROPYLIC ALCOHOL * PROPYLOWY ALKOHOL (POLISH)

OSHA PEL: TWA 200 ppm
ACGIH TLV: TWA 200 ppm; STEL 250 ppm
DOT Classification: Label: Flammable
  Liquid

THR: Poison by ingestion, subcutaneous, and intravenous routes. Moderately toxic by inhalation. A severe eye and moderate skin irritant. An experimental carcinogen. Dangerous fire hazard when exposed to heat, flame, or oxidizers. Moderate explosion hazard when exposed to flame. Reacts violently with potassium-*tert*-butoxide. Dangerous, upon exposure to heat or flame; can react vigorously with oxidizing materials. To fight fire, use alcohol foam, CO_2, dry chemical.

PROPYLAMINE   HR: 2
CAS: 107-10-8   NIOSH: UH 9100000
DOT: 1277
mf: C_3H_9N mw: 59.13

PROP: Colorless, alkaline liquid, strong ammonia odor; d: 0.7191 @ 20°/20°, mp: −83°, bp: 48°-49°, vap press: 248 mm @ 20°, flash p: −35°F, autoign temp: 604°F, lel = 2.0%, uel = 10.4%. Misc water, alc, ether.

SYNS: 1-AMINOPROPANE * MONO-N-PROPYLA-MINE * PROPANAMINE * N-PROPYLAMINE

DOT Classification: Label: Flammable
  Liquid

THR: Moderately toxic by inhalation, ingestion, and skin contact. A severe eye and moderate skin irritant. Possibly a skin sensitizer. See also amines. Very dangerous fire hazard when exposed to heat, flame, or oxidizers. To fight fire, use alcohol foam. When heated to decomposition it emits toxic fumes of NO_x. Incompatible with triethynyl aluminum.

s-PROPYL BUTYLETHYLTHIOCARBA-MATE   HR: 3
CAS: 1114-71-2   NIOSH: EZ 0400000
mf: $C_{10}H_{21}NOS$ mw: 203.38

PROP: Liquid; bp: 142° @ 20 mm.

SYNS: S-PROPYL-N-AETHYL-N-BUTYL-THIOCARBAMAT (GERMAN) * PROPYL-ETHYLBUTYLTHIOCARBAMATE * PROPYL ETHYLBUTYLTHIOLCARBAMATE * N-PROPYL-N-ETHYL-N-(N-BUTYL)THIOCARBAMATE * PROPYLETHYL-N-BUTYLTHIOCARBAMATE * PROPYL N-ETHYL-N-BUTYLTHIOCARBAMATE * S-(N-PROPYL)-N-ETHYL-N-N-BUTYLTHIOCAR-BAMATE * N-PROPYL-N-ETHYL-N-(N-BUTYL)-THIOLCARAMATE * TILLAM (RUSSIAN)

THR: An experimental carcinogen. Moderately toxic by ingestion and inhalation. Causes violent vomiting when accompanied by alcohol ingestion. Dangerous. When heated to decomposition it emits highly toxic fumes of SO_x and NO_x.

PROPYL CARBAMATE   HR: 3
CAS: 627-12-3   NIOSH: FD 0875000
mf: $C_4H_9NO_2$ mw: 103.14

PROP: Crystals. Bp: 196°, mp: 60°, vap press: 1 mm @ 52.4°. Very sol in water, alc, ether.

SYNS: N-PROPYL CARBAMATE * PROPYL URE-THANE

THR: An experimental carcinogen, teratogen, and neoplastigen. Moderately toxic by subcutaneous route. When heated to decomposition it emits toxic fumes of NO_x.

PROPYLENE GLYCOL MONOMETHYL ETHER   HR: 2
CAS: 107-98-2   NIOSH: UB 7700000
mf: $C_4H_{10}O_2$ mw: 90.14

PROP: Colorless liquid. Mp: −96.7°, bp: 120°, flash p: 100°F, d: 0.919 @ 25°/25°.

SYNS: DOWANOL 33B * DOWTHERM 209 * METHOXY ETHER OF PROPYLENE GLYCOL * 1-METHOXY-2-PROPANOL * PROPYLENE GLYCOL METHYL ETHER * ALPHA-PROPYLENE GLYCOL MONOMETHYL ETHER * PROPYLEN-GLYKOL-MONOMETHYLAETHER (GERMAN)
ACGIH TLV: TWA 100 ppm; STEL: 150 ppm
DFG MAK: 100 ppm (375 mg/m³)

THR: Mildly toxic by ingestion and inhalation. No cases of human toxicity known. See also ethylene glycol, glycol, and ethers. A skin and eye irritant. Flammable when exposed to heat or flame; can react with oxidizing materials. To fight fire, use foam, CO_2, dry chemical.

N-PROPYL GALLATE   HR: 3
CAS: 121-79-9   NIOSH: LW 8400000
mf: $C_{10}H_{12}O_5$ mw: 212.22

PROP: Odorless, fine, ivory powder or crystals; mp: 147°-149°.

SYNS: NCI-C50588 * PROPYL GALLATE * N-PROPYL 3,4,5-TRIHYDROXYBENZOATE

* 3,4,5-TRIHYDROXYBENZENE-1-PROPYLCAR-
BOXYLATE

THR: Poison by ingestion and intraperitoneal routes. Used in food as an antioxidant. Combustible when exposed to heat or flame; can react with oxidizing materials. When heated to decomposition it emits toxic fumes and smoke.

N-PROPYL NITRATE   HR: 3
CAS: 627-13-4   NIOSH: UK 0350000
mf: $C_3H_7NO_3$ mw: 105.09

PROP: Liquid. Pale yellow, sickly odor; bp: 110.5°, d: 1.054 @ 20°/4°, flash p: 68°F, autoign temp: 347°F (in air), lel = 2%, uel = 100%. Very sltly sol in water; sol in alc, ether.

SYNS: NITRATE DE PROPYLE NORMAL (FRENCH) * NITRIC ACID, PROPYL ESTER * PROPYL NITRATE

OSHA PEL: TWA 25 ppm
ACGIH TLV: TWA 25 ppm; STEL 40 ppm
DFG MAK: 25 ppm (110 mg/m^3)

THR: Poison by inhalation and intravenous route. See nitrates, esters. Inhalation can cause a hypotension and methemoglobinemia. Dangerous fire hazard when exposed to heat, flame, or oxidizers. See also nitrates. Dangerous explosion hazard; may explode on heating. When heated to decomposition it emits toxic fumes of NO_x.

PROPYLNITROSAMINOMETHYL
ACETATE   HR: 3
CAS: 66017-91-2   NIOSH: AJ 3700000
mf: $C_6H_{12}N_2O_3$ mw: 160.20

SYNS: N-PROPYL-N-(ACETOXYMETHYL)-NITROSAMINE * N-(ACETOXY)METHYL-N-N-PROPYLNITROSAMINE * ACETOXYMETHYLPROPYLNITROSAMINE * N-NITROSO-N-(1-ACETOXYMETHYL)PROPYL AMINE * PROPYL ACETOXYMETHYLNITROSAMINE

THR: An experimental carcinogen. Moderately toxic by ingestion. Mutagenic data. When heated to decomposition it emits toxic fumes of NO_x.

6-PROPYL-2-THIOURACIL   HR: 3
CAS: 51-52-5   NIOSH: YR 1400000
mf: $C_7H_{10}N_2OS$ mw: 170.25

PROP: White, bitter crystalline powder. Mp: 219°-221°; insol in ether, chloroform, benzene;

very sol in aq solns of ammonia; very sltly sol in water.

SYNS: 6-N-PROPYLTHIOURACIL * 2,3-DIHYDRO-6-PROPYL-2-THIOXO-4(1H)-PYRIMIDINONE * 2-MERCAPTO-4-HYDROXY-6-N-PROPYLPYRIMIDINE * 2-MERCAPTO-6-PROPYL-4-PYRIMIDONE * 2-MERCAPTO-6-PROPYLPYRIMID-4-ONE * 6-PROPYL-2-THIO-2,4(1H,3H)PYRIMIDINEDIONE * 4-PROPYL-2-THIOURACIL * 6-N-PROPYL-2-THIOURACIL * 2-THIO-4-OXO-6-PROPYL-1,3-PYRIMIDINE * 2-THIO-6-PROPYL-1,3-PYRIMIDIN-4-ONE * 6-THIO-4-PROPYLURACIL

THR: An experimental carcinogen, teratogen, and neoplastigen. When heated to decomposition it emits very toxic fumes of SO_x and NO_x. For further information see Vol. 6, No. 5 of *DPIM Report*.

N-PROPYLTRICHLOROSILANE   HR: 3
CAS: 141-57-1   NIOSH: VV 5300000
DOT: 1816
mf: $C_3H_7Cl_3Si$ mw: 177.54

PROP: Vap d: 6.15; flash p: 100°F.

SYN: TRICHLOROPROPYLSILANE

DOT Classification: Label: Corrosive

THR: A corrosive and irritating material to skin, eyes, and mucous membranes. See also chlorosilanes. A dangerous fire hazard when exposed to heat or flame. When heated to decomposition it emits toxic fumes of Cl^-. Will react with water or steam to produce toxic and corrosive fumes; can react with oxidizing materials. To fight fire, use foam, CO_2, dry chemical.

PROPYNE   HR: 1
CAS: 74-99-7   NIOSH: UK 4250000
mf: C_3H_4 mw: 40.07

PROP: Gas. Mp: −104°, lel = 1.7%, bp: −23.3°, vap press: 3876 mm @ 20°, d: 1.787 g/L @ 0°, vap d: 1.38.

SYNS: METHYL ACETYLENE * PROPINE

OSHA PEL: TWA 1000 ppm

THR: This compound is a simple anesthetic and in high concentration is an asphyxiant. Dangerous fire hazard when exposed to heat or flame. Moderate explosion hazard when exposed to flame. Can decompose explosively. Moderately dangerous; can react vigorously with oxidizing materials. To fight fire, stop flow of gas.

2-PROPYN-1-OL   HR: 3
CAS: 107-19-7   NIOSH: UK 5075000
mf: C_3H_4O mw: 56.07

PROP: Mod volatile liquid, geranium odor. D:
0.9715 @ 20°/4°; mp: −48° to −52°; bp: 114°-
115°; flash p: 33°C (97°F)(OC); vap press: 11.6
mm @ 20°; vap d: 1.93.

SYNS: PROPARGYL ALCOHOL * 1-PROPYNE-
3-OL

OSHA PEL: TWA 1 ppm (skin)

THR: Poison by ingestion and skin contact.
Moderately toxic by inhalation. A central nervous system depressant. A skin and mucous
membrane irritant. When heated to decomposition it emits acrid smoke and fumes. Dangerous
fire hazard when exposed to heat or flame; can
ignite. To fight fire, use foam, CO_2, dry chemical. Incompatible with alkalis; mercury (II) sulfate; H_2SO_4; oxidizing materials; P_2O_5.

PROSTAGLANDIN F2-ALPHA   HR: 3
CAS: 551-11-1   NIOSH: UK 8020000
mf: $C_{20}H_{34}O_5$ mw: 354.54

SYNS: 7-(3,5-DIHYDROXY-2-(3-HYDROXY-1-OC-
TENYL)CYCLOPENTYL)-5-HEPTENOICACID
* (5z,9-ALPHA,11-ALPHA,13E,15S)-9,11,15-
TRIHYDROXYPROSTA-5,13-DIEN-1-OIC ACID
* U-14583

THR: Poison by subcutaneous, intravenous, and
intramuscular routes. Moderately toxic by ingestion. When heated to decomposition it emits
acrid smoke and fumes.

PROSTIGMINE BROMIDE   HR: 3
CAS: 114-80-7   NIOSH: BR 3150000
mf: $C_{12}H_{19}N_2O_2 \cdot Br$ mw: 303.24
SYNS:
(M-HYDROXYPHENYL)TRIMETHYLAMMONIUM
BROMIDE DIMETHYLCARBAMATE
* 3-HYDROXYPHENYLTRIMETHYLAMMONIUM
BROMIDE DIMETHYLCARBAMIC ESTER * 3-
DIMETHYLCARBAMOXYPHENYLTRIMETHYLAM-
MONIUM BROMIDE * SYNTHOSTIGMINE BRO-
MIDE

THR: Poison by ingestion, subcutaneous, intravenous, and intraperitoneal routes. When
heated to decomposition it emits very toxic
fumes of Br^- and NO_x.

PROTACTINIUM   HR: 3
af: Pa   aw: 231

PROP: A bright, lustrous metal; mp: 1600°;
d: 15.37; vap press: 5×10^{-5} mm @ 1927°.

THR: A highly radiotoxic metallic element. An
alpha emitter. It is a general hazard if absorbed
systemically. The dust and fumes are hazardous
if inhaled. A carcinogen. Radiation Hazard: Nat-
ural isotope ^{231}Pa (Actinium series), $T_{\frac{1}{2}} = 3$
$\times 10^4$ Y., decays to radioactive ^{227}Ac via alphas
of 5.0 MeV. Artificial isotope ^{233}Pa (Neptunium
Series), $T_{\frac{1}{2}} = 27D$, decays to radioactive ^{233}U
via betas of 0.15 (37%), 0.26 (58%), 0.57 (5%)
MeV; emits gammas of 0.02-0.42 MeV. Natural
isotope ^{234}Pa (Uranium Series), $T_{\frac{1}{2}} = 6.7H$,
decays to radioactive 234U via betas of 0.23-
1.36 MeV, emits gammas of 0.04-0.8 MeV.

PROTRIPTYLINE HYDROCHLO-
RIDE   HR: 3
CAS: 1225-55-4   NIOSH: HP 1400000
mf: $C_{19}H_{21}N \cdot ClH$ mw: 299.87

SYNS: N-METHYL-5H-DIBENZO(a,d)-
CYCLOHEPTENE-5-PROPYLAMINE HYDROCHLORIDE
* 5-(3-METHYLAMINOPROPYL)-5H-DIBEN-
ZO(a,d)CYCLOHEPTENE HYDROCHLORIDE

THR: Poison by ingestion, intraperitoneal, intravenous, and subcutaneous routes. When
heated to decomposition it emits very toxic
fumes of NO_x and HCl.

PSEUDOACONITINE   HR: 3
mf: $C_{36}H_{51}NO_{12}$ mw: 689.78

PROP: White crystals or syrupy mass. Mp: 214°
(decomp). Insol in water; sol in alc, ether.

THR: Poison by ingestion, inhalation, and skin
contact. Dangerous; when heated it emits highly
toxic fumes of NO_x.

PSILOCYBIN   HR: 3
CAS: 520-52-5   NIOSH: NM 3150000
mf: $C_{12}H_{17}N_2O_4P$ mw: 284.28

SYNS: 3-2'-DIMETHYLAMINOETHYLINDOL-4-
PHOSPHATE * 3-(2-DIMETHYLAMINOETHYL)IN-
DOL-4-YL DIHYDROGEN PHOSPHATE * PSILOCIN
PHOSPHATE ESTER * O-PHOSPHORYL-4-HY-
DROXY-N,N-DIMETHYLTRYPTAMINE * PSILOTSI-
BIN

THR: Poison by intravenous route. Human central nervous system and psychotropic effects.
When heated to decomposition it emits very
toxic fumes of NO_x and PO_x.

PURAPURIDINE   HR: 3
CAS: 126-17-0   NIOSH: WF 1300000
mf: $C_{27}H_{43}NO_2$ mw: 413.71

SYNS: SOLASOD-5-EN-3-BETA-OL * SOLAN-
CARPIDINE

THR: Poison by intraperitoneal route. Moderately toxic by ingestion. An experimental teratogen. When heated to decomposition it emits
toxic fumes of NO_x.

PURINE-6-THIOL   HR: 3
CAS: 50-44-2   NIOSH: UO 9800000
mf: $C_5H_4N_4S$ mw: 152.19

SYNS: LEUKERAN * 6-MERCAPTOPURINE
* 7-MERCAPTO-1,3,4,6-TETRAZAINDENE
* NCI-C04886 * NSC 755 * 6-PURINETHIOL

THR: Poison by ingestion, intraperitoneal, subcutaneous, parenteral, intravenous, and other
unspecified routes. Mutagenic data. An experimental teratogen and tumorigen. Human toxicity
of an unspecified source. When heated to decomposition it emits very toxic fumes of SO_x
and NO_x.

PYRAZINECARBOXAMIDE   HR: 3
CAS: 98-96-4   NIOSH: UQ 2275000
mf: $C_5H_5N_3O$ mw: 123.13

SYNS: 2-CARBAMYL PYRAZINE * NCI-C01785
* PYRAZINAMIDE * PYRAZINEAMIDE
* PYRAZINE CARBOXYLAMIDE * PYRAZINOIC
ACID AMIDE

THR: A carcinogen and an experimental tumorigen. Mutagenic data. Moderately toxic by intraperitoneal route. When heated to decomposition
it emits toxic fumes of NO_x.

PYRENE   HR: 3
CAS: 129-00-0   NIOSH: UR 2450000
mf: $C_{16}H_{10}$ mw: 202.26

PROP: Colorless solid, solutions have a slight
blue color, insol in water, fairly sol in organic
solvents. (a condensed ring hydrocarbon), mp:
156°, d: 1.271 @ 23°, bp: 404°.

SYNS: BENZO(DEF)PHENANTHRENE * PYREN
(GERMAN)

THR: An experimental carcinogen. Mutagenic
data. A skin irritant. When heated to decomposition it emits acrid smoke and fumes.

PYRETHRIN I   HR: 3
CAS: 8003-34-7   NIOSH: UR 4200000
mf: $C_{21}H_{28}O_3$ mw: 328.4

PROP: Viscous liquid; bp: 170° @ 0.1 mm
(decomp).

SYNS: CINERIN I OR II * JASMOLIN I OR II
* PYRETHRIN I OR II * CHRYSANTHEMUM CIN-
ERAREAEFOLIUM * DALMATION INSECT FLOW-
ERS * INSECT POWDER * PYRETHRUM (INSEC-
TICIDE) * TRIESTE FLOWERS

OSHA PEL: TWA 5 mg/m^3
ACGIH TLV: TWA 5 mg/m^3
DFG MAK: 5 mg/m^3

THR: Poison by ingestion. An allergen. A highly
insecticidal extract with low mammalian toxicity. It is rapidly detoxified in the gastrointestinal
tract, but can cause gastrointestinal, respiratory,
and central nervous system effects. A dose of
15 g has caused the death of a child. Chronic
exposures can cause liver damage. See esters.
Combustible when exposed to heat or flame.
When heated to decomposition it emits acrid
smoke and fumes.

PYRETHRIN II   HR: 2
CAS: 121-29-9   NIOSH: GZ 0700000
mf: $C_{22}H_{28}O_5$ mw: 372.50

PROP: Viscous liquid; bp: 200° @ 0.1 mm
(decomp).

SYNS: ENT 7,543 * PYRETHROLONE CHRY-
SANTHEMUM DICARBOXLIC ACIDMETHYL ESTER
ESTER * PYRETHROLONE ESTER OF CHRYSAN-
THEMUMDICARBOXYLIC ACID MONOMETHYL ES-
TER * PYRETRIN II

THR: Moderately toxic by ingestion and possibly other routes. See also pyrethrin I. An allergen. When heated to decomposition it emits
acrid smoke and fumes.

PYRIDAPHENTHION   HR: 3
CAS: 119-12-0   NIOSH: TF 2275000
mf: $C_{14}H_{17}N_2O_4PS$ mw: 340.36

SYNS: O,O-DIETHYL O-(2,3-DIHYDRO-3-OXO-2-
PHENYL-6-PYRIDAZINYL)PHOSPHOROTHIOATE
* O,O-DIETHYLPHOSPHOROTHIOATE, O-ESTER
WITH 6-HYDROXY-2-PHENYL-3(2H)-PYRIDAZI-
NONE * O-(1,6)-DIHYDRO-6-OXO-1-PHENYLPY-
RIDAZIN-3-LY), O,O-DIETHYL PHOSPHOROTHIOATE
* ENT 23,968

THR: Poison by intraperitoneal route. Moderately toxic by ingestion and skin contact. When
heated to decomposition it emits very toxic
fumes of SO_x, PO_x, and NO_x.

PYRIDINE　　　　　　　　　　　HR: 2
CAS: 110-86-1　　　　NIOSH: UR 8400000
DOT: 1282
mf: C_5H_5N　　　mw: 79.11

PROP: Colorless liquid, sharp, penetrating, empyreumatic odor, burning taste. Flammable; bp: 115.3°, lel = 1.8%, uel = 12.4%, fp: −42°, flash p: 68°F (CC), d: 0.982, autoign temp: 900°F, vap press: 10 mm @ 13.2°, vap d: 2.73. Volatile with steam. Misc with water, alc, ether.

SYNS: AZABENZENE * NCI-C55301 * PYRIDIN (GERMAN) * PIRIDINA (ITALIAN) * PYRIDINE (DOT) * PIRYDYNA (POLISH)

OSHA PEL: TWA 5 ppm
ACGIH TLV: TWA 5 ppm
DFG MAK: 5 ppm (15 mg/m^3)
DOT Classification: Label: Flammable
　Liquid

THR: Moderately toxic by inhalation, ingestion, skin contact, and subcutaneous routes. A severe eye and mild skin irritant. Mutagenic data. Can cause central nervous system depression, gastrointestinal upset, and liver and kidney damage. Dangerous fire hazard when exposed to heat, flame, or oxidizers. Severe explosion hazard in the form of vapor, when exposed to flame or spark. Reacts violently with chlorosulfonic acid; CrO_3; maleic anhydride; HNO_3; oleum; perchromates; β-propiolactone; $AgClO_4$; H_2SO_4; formamide; SO_3; I. Dangerous; when heated to decomposition it emits highly toxic fumes of cyanides; can react vigorously with oxidizing materials. To fight fire, use alcohol foam.

4-PYRIDINEETHANOL　　　　　HR: 3
CAS: 5344-27-4　　　NIOSH: UT 2971000
mf: C_7H_9NO　　mw: 123.17

SYN: 4-ETHANOLPYRIDINE

THR: Poison by intravenous routes. When heated to decomposition it emits toxic fumes of NO$_x$.

PYRIDINIUM-2-ALDOXIME-N-METHYL IODIDE　　　　　　　　　　　　HR: 3
CAS: 94-63-3　　　NIOSH: UU 4375000
mf: $C_7H_9N_2O$ • I　　mw: 264.08

PROP: Water sol crystals; mp: 214°.

SYNS: 2-FORMYL-1-METHYLPYRIDINIUM IODIDE OXIME * 2-FORMYL-N-METHYLPYRIDINIUM OX-

IME IODIDE * 2-HYDROXYIMINOMETHYL-1-METHYLPYRIDINIUM IODIDE * 1-METHYL-2-ALDOXIMINOPYRIDINIUM IODIDE * 1-METHYL-2-HYDROXYIMINOMETHYLPYRIDINIUM IODIDE * N-METHYLPYRIDINE-2-ALDOXIME IODIDE * N-METHYLPYRIDINIUM-2-ALDOXIME IODIDE * PAM (CZECH) * 2-PAM IODIDE * 2-PYRIDINALDOXIM METHOJODID (GERMAN) * PYRIDIN-2-ALDOXIN (CZECH) * 2-PYRIDINE ALDOXIME IODOMETHYLATE * PYRIDINE-2-ALDOXIME METHIODIDE * PYRIDINE-2-ALDOXIME METHYL IODIDE

THR: Poison by subcutaneous, intravenous, intramuscular, and intraperitoneal routes. Moderately toxic by ingestion. Effectively used as an antidote to the cholinesterase inhibitors of the parathion group. Dangerous; when heated to decomposition it·emits highly toxic fumes of NO$_x$ and I$^−$.

PYRIDOXOL　　　　　　　　　　HR: 2
CAS: 65-23-6　　　　NIOSH: UV 1300000
mf: $C_8H_{11}NO_3$　　mw: 169.20

SYNS: 3-HYDROXY-4,5-DIMETHYLOL-ALPHA-PICOLINE * 5-HYDROXY-6-METHYL-3,4-PYRIDINEDIMETHANOL * 2-METHYL-4,5-BIS-(HYDROXYMETHYL)-3-HYDROXYPYRIDINE * 2-METHYL-3-HYDROXY-4,5-BIS(HYDROXYMETHYL)PYRIDINE * 2-METHYL-3-HYDROXY-4,5-DIHYDROXYMETHYL-PYRIDIN (GERMAN) * 2-METHYL-3-HYDROXY-4,5-DI(HYDROXYMETHYL)PYRIDINE * PYRIDOXINE * VITAMIN B6

THR: Moderately toxic by ingestion, subcutaneous, intravenous, and intraperitoneal routes. When heated to decomposition it emits toxic fumes of NO$_x$.

PYRIDOXOL HYDROCHLORIDE
　　　　　　　　　　　　　　　　HR: 3
CAS: 58-56-0　　　NIOSH: UV 1350000
mf: $C_8H_{11}NO_3$ • ClH　　mw: 205.66

PROP: Commercial form of pyridoxine (Vitamin B$_6$), colorless-white platelets, sol in water, alc, acetone, sltly sol in other organic solvents. Mp: 204°-206°.

SYNS: 3-HYDROXY-4,5-DIMETHYLOL-ALPHA-PICOLINE HYDROCHLORIDE * 5-HYDROXY-6-METHYL-3,4-PYRIDINEDICARBINOL HYDROCHLORIDE * 5-HYDROXY-6-METHYL-3,4-PYRIDINEDIMETHANOL HYDROCHLORIDE * 2-METHYL-3-HYDROXY-4,5-BIS(HYDROXY-

METHYL)PYRIDINE HYDROCHLORIDE
* PYRIDOXINE HYDROCHLORIDE * PYRI-
DOXINIUM CHLORIDE * VITAMIN B6-
HYDROCHLORIDE

THR: Poison by intravenous route. Moderately
toxic by ingestion, intramuscular, and subcutaneous routes. When heated to decomposition
it emits very toxic fumes of NO_x and HCl.

1-(PYRIDYL-3)-3,3-DIMETHYL TRI-AZENE   HR: 3
CAS: 19992-69-9   NIOSH: XY 2275000
mf: $C_7H_{10}N_4$ mw: 150.21

SYNS: 1-(PYRIDYL-3)-3,3-DIMETHYL-TRIAZEN
(GERMAN) * 1-(META-PYRIDYL)-3,3-DIMETH-
YL-TRIAZENE

THR: Poison by ingestion and subcutaneous
routes. Mutagenic data. An experimental neoplastigen and teratogen. When heated to decomposition it emits toxic fumes of NO_x.

PYRIMINYL   HR: 3
CAS: 53558-25-1   NIOSH: YT 9690000
mf: $C_{13}H_{12}N_4O_3$ mw: 272.29

SYNS: N-3-PYRIDYLMETHYL-N′-P-NITROPHE-
NYLUREA * N-(4-NITROPHENYL)-N′-(3-PYRIDI-
NYLMETHYL)UREA

THR: Poison by ingestion. Human central nervous system and gastrointestinal tract system
effects. See also nitrates. When heated to decomposition it emits toxic fumes of NO_x.

PYROCATECHOL   HR: 3
CAS: 120-80-9   NIOSH: UX 1050000
mf: $C_6H_6O_2$ mw: 110.12

PROP: Colorless crystals. Mp: 105°, bp: 246°,
flash p: 261°F (CC), d: 1.341 @ 15°, vap press:
10 mm @ 118.3°, vap d: 3.79. Sol in water,
chloroform, benzene; very sol in alc, ether.

SYNS: o-BENZENEDIOL * 1,2-BENZENEDIOL
* CATECHIN * CATECHOL * C.I. 76500
* o-DIHYDROXYBENZENE * 1,2-DIHYDROXY-
BENZENE * o-DIOXYBENZENE * o-DIPHENOL
* o-HYDROQUINONE * o-HYDROXYPHENOL
* 2-HYDROXYPHENOL * NCI-C55856
* OXYPHENIC ACID * o-PHENYLENEDIOL
* PYROCATECHIN * PYROCATECHINIC ACID
* PYROCATECHUIC ACID

THR: Poison by ingestion, subcutaneous, intraperitoneal, intravenous, and parenteral routes.

Mutagenic data. An experimental carcinogen.
Systemic effects similar to phenol. An allergen.
Can cause convulsions and injury to blood. See
also phenol. Can cause dermatitis on skin contact. Combustible when exposed to heat or
flame. Dangerous; when heated it emits highly
toxic fumes; can react with oxidizing materials.
To fight fire, use water, CO_2, dry chemical.

PYROGALLOL   HR: 3
CAS: 87-66-1   NIOSH: UX 2800000
mf: $C_6H_6O_3$ mw: 126.12

PROP: White, lustrous crystals. Bp: 309°, d:
1.453 @ 4°/4°, vap press: 10 mm @ 167.7°;
mp: 131°-133°. Sltly sol in benzene, chloroform.

SYNS: 1,2,3-BENZENETRIOL * C.I. 76515
* PYROGALLIC ACID * 1,2,3-TRIHYDROXY-
BENZEN (CZECH) * 1,2,3-TRIHYDROXYBENZENE

THR: Poison by ingestion, subcutaneous, intravenous, intraperitoneal, and possibly other
routes. A severe eye and skin irritant. An experimental tumorigen. Mutagenic data. Readily absorbed through the skin. Can cause convulsions,
circulatory collapse, hemolysis, methemoglobinemia, kidney injury, liver damage, and
death. Incompatible with alkalies; NH_3; antipyrine; camphor; phenol; iron and lead salts; iodine; lime water; menthol; $KMnO_4$.

PYROSULFURYL CHLORIDE   HR: 3
CAS: 7791-27-7   NIOSH: UX 8310000
DOT: 1817
mf: $Cl_2O_5S_2$ mw: 215.02

PROP: Colorless, mobile, fuming liquid. Mp:
$-37°$; bp: 151°; d: 1.83; (gas): 9.6 g/L. Violent
decomp by water.

SYNS: CHLOROSULFONIC ANHYDRIDE * DISUL-
FUR PENTOXYDICHLORIDE * DISULFURYL CHLO-
RIDE

DOT Classification: Label: Corrosive

THR: A very poisonous material which is also
corrosive to the eyes, skin, and mucous membranes. See chlorosulfuric acid. When heated
to decomposition it emits very toxic fumes of
Cl^- and SO_x.

PYRROLE   HR: 3
CAS: 109-97-7   NIOSH: UX 9275000
mf: C_4H_5N mw: 67.10

PROP: Colorless liquid, darkens on standing,
mild odor. fp: $-24°$, flash p: 102°F (TCC), d:

0.968 @ 20°/4°, vap d: 2.31; bp: 130°-131° @ 761 mm. Sltly sol in water; very sol in alc, benzene, ether. Insol in alkali.

SYNS: 1-AZA-2,4-CYCLOPENTADIENE * AZOLE * DIVINYLENIMINE * IMIDOLE * MONOPYR-ROLE

THR: Poison by subcutaneous, intraperitoneal, and possibly other routes. Flammable when exposed to heat, flame or oxidizers. Dangerous; when heated to decomposition it emits highly toxic fumes of NO$_x$; can react with oxidizing materials. To fight fire, use foam, CO$_2$, dry chemical.

PYRROLIDINE   HR: 3
CAS: 123-75-1   NIOSH: UX 9650000
DOT: 1922
mf: C$_4$H$_9$N   mw: 71.14

PROP: Colorless, mobile liquid, penetrating amine-like odor. fp: −63°, flash p: 37°F (TCC), d: 0.8618 @ 20°/4°, vap press: 128 mm @ 39°, vap d: 2.45; bp: 88.5°-89°. Fumes in air. Misc with water; sol in alc, ether, chloroform.

SYNS: AZACYCLOPENTANE * TETRAHYDRO-PYRROLE * TETRAMETHYLENIMINE

DOT Classification: Label: Flammable Liquid

THR: Poison by ingestion, inhalation, and intravenous routes. Dangerous fire hazard when exposed to heat, flame, or oxidizers. Dangerous; when heated to decomposition it emits highly toxic fumes of NO$_x$; can react vigorously with oxidizing materials. To fight fire, use alcohol foam, CO$_2$, dry chemical.

Q

QM-1143   **HR: 3**
NIOSH: DL 8250000
mf: $C_{15}H_{20}N_4S \cdot ClH$ mw: 324.91

SYN: 4-(4-METHYL-1-PIPERAZINYL)-5,6,7,8-TETRAHYDRO-(1)-BENZOTHIENO(2,3-d) PYRIMIDINE HYDROCHLORIDE

THR: A poison by ingestion, intraperitoneal, and intravenous routes. When heated to decomposition it emits very toxic fumes of HCl, SO_x and NO_x.

QUAALUDE   **HR: 3**
CAS: 72-44-6   NIOSH: VA 3850000
mf: $C_{16}H_{14}N_2O$ mw: 250.32

SYNS: METHAQUALONE * METHAQUALONEINONE * 2-METHYL-3-(2-METHYLPHENYL)-4-QUINAZOLINONE * 2-METHYL-3-(2-METHYLPHENYL)-4(3H)-QUINAZOLINONE * 2-METHYL-3-O-TOLYL-4(3H)-CHINAZOLINON (GERMAN) * 2-METHYL-3-O-TOLYL-4(3H)-CHINAZOLONE * (2-METHYL-3-(O-TOLYL)-3,4-DIHYDRO-4-(QUINAZOLINONE) * 2-METHYL-3-(O-TOLYL)-3,4-DIHYDRO-4-QUINAZOLINONE * 2-METHYL-3-TOLYL-4-OXYBENSDIAZINE * 2-METHYL-3-O-TOLYL-4(3H)-QUINAZOLINONE * 2-METHYL-3-O-TOLYL-4-QUINAZOLONE * 2-METHYL-3-(2-TOLYL)QUINAZOL-4-ONE

THR: Poison by ingestion, intravenous, intraperitoneal, and possibly other routes. An experimental teratogen. Can cause central nervous system effects in man. A controlled drug which is often abused. When heated to decomposition it emits toxic fumes of NO_x.

QUATERNIUM-12   **HR: 3**
CAS: 7173-51-5   NIOSH: BP 6560000
mf: $C_{22}H_{48}N \cdot Cl$ mw: 362.16

SYNS: DIMETHYLDIDECYLAMMONIUM CHLORIDE * DIDECYL DIMETHYL AMMONIUM CHLORIDE

THR: Poison by ingestion and intraperitoneal route. When heated to decomposition it emits very toxic fumes of NO_x and Cl^-.

QUATRESIN   **HR: 3**
CAS: 2748-88-1   NIOSH: TJ 7825000
mf: $C_{20}H_{36}N \cdot Cl$ mw: 326.02

SYNS: MYRISTYL-GAMMA-PICOLINIUM CHLORIDE * WET-TONE B

THR: Poison by ingestion, intraperitoneal, intravenous, and subcutaneous routes. When heated to decomposition it emits very toxic fumes of NO_x and Cl^-.

QUEBRACHO TANNIN   **HR: 3**
NIOSH: UZ 3400000

SYNS: SCHINOPSIS LORENTZII TANNIN * TANNIN FROM QUEBRACHO

THR: An experimental tumorigen. When heated to decomposition it emits acrid smoke and fumes.

QUELAMYCIN   **HR: 3**
CAS: 64719-39-7   NIOSH: UZ 3495000
mf: $C_{27}H_{27}O_{11} \cdot 2Fe(2^+) \cdot Fe(3^+)$ mw: 709.07

SYNS: NSC-267703 * TRIFERRIC ADRIAMYCIN * TRIFERRIC DOXORUBICIN

THR: Poison by intravenous and intraperitoneal routes. Human blood effects. See also iron dust. When heated to decomposition it emits acrid smoke and fumes.

QUERCETIN   **HR: 3**
CAS: 117-39-5   NIOSH: LK 8750000
mf: $C_{15}H_{10}O_7$ mw: 302.25

SYNS: C.I. NATURAL YELLOW 10 * CYAN DELONON 1522 * MELETIN * NCI-C60106 * 3,5,7,3',4'-PENTAHYDROXYFLAVONE * QUERCETINE * QUERCETOL * QUERTINE * SOPHORETIN * 3',4',5,7-TETRAHYDROXYFLAVAN-3-OL * XANTHAURINE

THR: Poison by ingestion, subcutaneous, and intravenous routes. Mutagenic data. When heated to decomposition it emits acrid smoke and fumes.

QUINACRINE ETHYL MUSTARD
                          HR: 3
CAS: 10072-25-0   NIOSH: AR 7525000
mf: $C_{20}H_{22}Cl_3N_3O \cdot 2ClH \cdot H_2O$ mw: 517.74

SYNS: ICR-48B * NSC-34372 * 9-(2-(DI(2-CHLOROETHYL)AMINO)ETHYLAMINO)-6-CHLORO-2-METHOXY-ACRIDINE

THR: An experimental carcinogen. When heated to decomposition it emits very toxic fumes of Cl^-, NO_x and HCl.

QUINACRINE MUSTARD   HR: 3
CAS: 64046-79-3   NIOSH: AR 7570000
mf: $C_{23}H_{28}Cl_3N_3O$ mw: 468.89

SYNS: NSC-3424 * 9-(4-(BIS-BETA-CHLORO-ETHYLAMINO)-1-METHYLBUTYLAMINO)-6-CHLO-RO-2-METHOXYACRIDINE

THR: Poison by intravenous route. Mutagenic data. When heated to decomposition, it emits very toxic fumes of Cl^- and NO_x.

QUINACRINE MUSTARD DIHYDRO-CHLORIDE   HR: 3
CAS: 4213-45-0   NIOSH: AR 7580000
mf: $C_{23}H_{28}Cl_3N_3O \cdot 2ClH$ mw: 541.81

SYNS: 2-METHOXY-6-CHLORO-9-(4-BIS(2-CHLO-ROETHYL)AMINO-1-METHYLBUTYL-AMINO)ACRIDINE DIHYDROCHLORIDE * 9-(4-BIS(2-CHLOROETHYL)AMINO-1-METHYL-BUTYLAMINO)-6-CHLORO-2-METHOXYACRIDINE DIHYDROCHLORIDE

THR: An experimental neoplastigen. Mutagenic data. When heated to decomposition, it emits very toxic fumes of Cl^-, NO_x, and HCl.

QUINIDINE   HR: 3
CAS: 56-54-2   NIOSH: VA 4725000
mf: $C_{20}H_{24}N_2O_2$ mw: 324.46

SYNS: CHINIDIN (GERMAN) * CONQUININE * 6-METHOXY-ALPHA-(5-VINYL-2-QUINUCLI-DINYL)-4-QUINOLINEMETHANOL * NCI-C56246 * QUINICARDINE * (+)-QUINIDINE * BETA-QUININE

THR: Poison by ingestion, subcutaneous, intravenous, intramuscular, and intraperitoneal routes. Implicated in aplastic anemia. When heated to decomposition it emits toxic fumes of NO_x.

QUININE   HR: 3
CAS: 130-95-0   NIOSH: VA 6020000
mf: $C_{20}H_{24}N_2O_2$ mw: 324.46

PROP: Bulky, white amorphous powder or crystals, bitter taste. Mp: 174.9°.

SYNS: CHININ (GERMAN) * 6-METHOXYCIN-CHONINE * ALPHA-(6-METHOXY-4-QUINOYL)-5-VINYL-2-QUINCLIDINEMETHANOL * (−)-QUI-NINE

THR: Poison by ingestion, subcutaneous, intravenous, intramuscular, and possibly other routes. An experimental teratogen by ingestion. Mutagenic data. Can cause temporary loss of vision. Quinine dermatitis is an occupational hazard to barbers particularly and generally to people who work with quinine tonics, medicaments, or cosmetics. An irritant to mucous membranes. Combustible when exposed to heat or flame. When heated to decomposition it emits toxic fumes of NO_x.

QUININE ETHIODIDE   HR: 3
CAS: 73771-81-0   NIOSH: GD 2902000
mf: $C_{22}H_{29}N_2O_2 \cdot I$ mw: 480.43

SYNS: 6-(1-HYDROXY-1-(6-METHOXY-4-QUINO-LINYL)METHYL-1-ETHYL-3-VINYLQUINUCLIDIN-IUM, IODIDE * (8-ALPHA,9R)-1-ETHYL-9-HY-DROXY-6'-METHOXYCINCHONAN-1-IUM IODIDE * 6-(HYDROXY(6-METHOXY-4-QUINOLINYL)-METHYL)-1-ETHYL-3-VINYL-QUINUCLIDINIUM, IODIDE

THR: A poison by intravenous route. When heated to decomposition it emits very toxic fumes of I^- and NO_x.

QUININE HYDROCHLORIDE   HR: 3
CAS: 130-89-2   NIOSH: VA 7700000
mf: $C_{20}H_{24}N_2O_2 \cdot ClH$ mw: 360.92

SYNS: QUININE MONOHYDROCHLORIDE * QUININE CHLORIDE * QUININE MURIATE

THR: Poison by ingestion, subcutaneous, intravenous, intramuscular, intraperitoneal, and possibly other routes. Human central nervous system effects by intravenous route. When heated to decomposition it emits very toxic fumes of NO_x and HCl.

QUININE SULFATE   HR: 3
CAS: 804-63-7   NIOSH: VA 8440000
mf: $C_{20}H_{24}N_2O_2 \cdot O_4S$ mw: 420.52

SYNS: QUININE BISULFATE * QUININE HYDRO-GEN SULFATE

THR: Poison to humans by ingestion. Human central nervous system and blood effects. An experimental teratogen. When heated to decomposition it emits very toxic fumes of SO_x and NO_x.

QUINOLINE   HR: 3
CAS: 91-22-5   NIOSH: VA 9275000
mf: C_9H_7N mw: 129.17

PROP: Refractive, colorless liquid, peculiar odor; mp: $-14.5°$, bp: $237.7°$, d: 1.0900 @ $25°/4°$, autoign temp: $896°F$, vap press: 1 mm @ $59.7°$, vap d: 4.45. Sol in water, CS_2; misc in alc, ether.

SYNS: 1-AZANAPHTHALENE * BENZO(B)PYRI-DINE * CHINOLIN (CZECH) * CHINOLINE * LEUCOLINE * USAF EK-218

THR: Poison by ingestion, subcutaneous, and intraperitoneal routes. A severe eye and mild skin irritant. An experimental neoplastigen and tumorigen by ingestion. Mutagenic data. It can cause retinitis similar to that caused by naphthalene but without causing opacity of the lens. Combustible when exposed to heat. Violent reaction with perchromates. Dangerous; when heated to decomposition it emits toxic fumes of NO_x. Incompatible with dinitrogen tetraoxide; linseed oil; thionyl chloride; maleic anhydride. Unpredictably violent.

8-QUINOLINOL   HR: 3
CAS: 148-24-3   NIOSH: VC 4200000
mf: C_9H_7NO mw: 145.17

PROP: White crystals or powder. Mp: $76°$, bp: $267°$. Very sltly sol in cold water; sltly sol in ether; sol in alc, dilute alkali. Nearly insol in water.

SYNS: HYDROXYBENZOPYRIDINE * 8-HY-DROXYQUINOLINE * NCI-C55298 * OXYBEN-ZOPYRIDINE * OXYCHINOLIN * 8-OXYQUINO-LINE * PHENOPYRIDINE * USAF EK-794

THR: Poison by intraperitoneal route. Moderately toxic by ingestion. A central nervous system stimulant. An experimental carcinogen, neoplastigen, and tumorigen of the uterus, rectum, brain, and bladder by ingestion, implant, and intravenous routes. Mutagenic data. Combustible. Dangerous; when heated to decomposition it emits highly toxic fumes of NO_x.

R

RACEMIC CLOMIPHENE CITRATE
HR: 3

CAS: 43054-45-1   NIOSH: YE 0875000

mf: $C_{26}H_{28}ClNO \cdot C_6H_8O_7$ mw: 598.14

SYNS: 2-(P-(2-CHLORO-1,2-DIPHENYL VINYL)-PHENOXY)TRIETHYLAMINE CITRATE (1:1)
∗ CLOMIPHENE CITRATE ∗ CLOMIPHENE DIHYDROGEN CITRATE ∗ 1-(P-(BETA-DIETHYLAMINO ETHOXY)PHENYL)-1,2-DIPHENYL- ∗ 2-CHLORO-ETHYLENE CITRATE

THR: An experimental teratogen and carcinogen. Moderately toxic by ingestion. When heated to decomposition it emits very toxic fumes of Cl^- and NO_x.

RADIATION

Energy in the form of electromagnetic waves (also called *radiant energy*). It is emitted from matter in the form of photons (quanta), each having an associated electromagnetic wave having frequency (ν) and wavelength (λ). The various forms of radiant energy are characterized by their wavelength, and together they comprise the electromagnetic spectrum, the components of which are as follows (see Table 1): (1) cosmic rays; (2) gamma rays from radioactive disintegration of atomic nuclei; (3) x-rays; (4) ultraviolet rays; (5) visible light rays; (6) infrared; (7) microwave; and (8) radio (Hertzian) and electric rays. All these are identical in every way except wavelength those having the shortest wavelength being the most penetrating. They are not electrically charged and have no mass; their velocity of propagation is the same; all display the properties characteristic of light and have a dual nature (wave-like and corpuscular). Thus, Infrared Radiation is that part of the EM spectrum between visible light and the microwave region, i.e., 7000 Å to 2.2×10^6 Å. All objects at a temperature greater than 0°K emit IR radiation to cooler surfaces and the hotter the emitter the shorter the emitted IR wavelength. When the emitter is hot enough, visible (4000 Å to 7000 Å) and even UV (100 Å to 4000 Å) radiation is also emitted. The main physical effect of exposure to IR is heating. This is also true for biological tissue. In the case of the eye, there is very sensitive tissue available for exposure to IR radiation, particularly in indus-

try, and this IR known as "near IR," (7800 Å to 14000 Å) is blamed for many cataracts of the eyes. It is easy to protect the eyes by wearing goggles. Ultraviolet (UV) radiation is that part of the EM spectrum between 100 Å and 4000 Å. Thus the UV-A band of UV extends from 3150 Å to 4000 Å and is called, "Black light," or "near UV." This band can cause thermal skin burns, skin pigmentation and photoreactions. It does not, in general, cause eye injury. From 2800 Å to 3150 Å is "mid-UV," or erythemal region. This band produces photokeratitis and possibly skin cancer. The UV band from 1000 Å to 2800 Å is the UV-C band. It is known as "far UV" or "short UV." This band of UV is germicidal and viricidal and destroys molds and yeasts as well. There is a sub-region of UV-C from 1700 Å to 2200 Å which produces ozone. The whole UV region can damage human skin, eyes. In eyes it can cause blepharitis, conjunctivitis, keratitis, keratoconjunctivitis. Skin exposure to solar UV can cause erythema, tanning; chronic skin exposure to solar UV leads to tanning, elastosis (dry, leathery, deeply wrinkled skin) and an incidence of non melanoma skin cancer. In a looser sense the term "radiation" also includes energy emitted in the form of particles which possess mass and may or may not be electrically charged, i.e., alpha (positive) and beta (negative); also neutrons. Beams of such particles may be considered as "rays." The charged particles may all be accelerated and high energy imparted to "beams" in particle accelerators such as cyclotrons, betatrons, synchrotrons and linear accelerators.

Type of radiation	Wavelength Å
cosmic	0.0005-0.005
gamma	0.005 -1.4
X	0.1 -100
UV	100 -4000
visible	4000 -7000
infrared	7000 -2,000,000

THR: *Radiation, ionizing*: Extremely short-wavelength, highly energetic penetrating rays of the following types: (a) gamma rays emitted by radioactive elements and radioisotopes (decay of atomic nucleus); (b) x-rays, generated by sudden stoppage of fast-moving electrons; (c) subatomic charged particles (electrons, protons, deuterons) when accelerated in a cyclotron or betatron. The term is restricted to electromagnetic radiation at least as energetic as x-rays, and to charged particles of similar energies. Neutrons also may induce ionization. Such radiation is strong enough to remove electrons from any atoms in its path, leading to the formation of free radicals. These short-lived but highly reactive particles initiate decomposition of many organic compounds. Thus ionizing radiation can cause mutations in DNA and in cell nuclei; adversely affect protein and amino acid mechanisms; impair or destroy body tissue; and attack bone marrow, the source of red blood cells. Exposure to ionizing radiation for even a short period is highly dangerous, and for an extended period may be lethal. The study of the chemical effects of such radiation is called radiation chemistry or (in the case of body reactions) radiation biochemistry.

RADIUM   HR: 3
af: Ra   aw: 226

PROP: A radioactive earth metal. Brilliant white, tarnishes in air. Decomp in water; mp: 700°; bp: 1737°; d: 5.5.

THR: A highly radiotoxic element. 1 g emits 3.7×10^{10} dps. Inhalation, ingestion, or bodily exposure can lead to lung cancer, bone cancer, osteitis, skin damage, and blood dyscrasias. A common air contaminant. Ra replaces calcium in the bone structure and is a source of irradiation to the blood forming organs. The ingestion of luminous dial paint prepared from radium was the cause of death of many of the early dial painters before the hazard was fully understood. The data on these workers has been the source of many of the radiation precautions and the maximum permissible levels for internal emitters which are now accepted. ^{226}Ra is the parent of radon and the precautions described under ^{222}Rn should be followed. ^{228}Ra is a member of the thorium series. It was a common constituent of luminous paints, and while its low beta energy was not a hazard, its daughters in the series may have been a causative agent in the deaths of the radium dial painters following World War I. Its metabolism is the same as any other radium isotope and it is a source of

thoron. The precautions recommended under ^{220}Rn should be followed. Highly dangerous; must be kept heavily shielded and stored away from possible dissemination by explosion, flood, etc. Radiation Hazard: Natural isotope ^{223}Ra (Actinium-X, Actinium Series), $T\frac{1}{2} =$ 11.4 D, decays to radioactive ^{219}Rn by alphas of 5.5-5.7 MeV. Natural isotope ^{224}Ra (Thorium-X, Thorium Series), $T\frac{1}{2} = 3.6$ D, decays to radioactive ^{220}Rn by alphas of 5.7 MeV. Natural isotope ^{226}Ra (Uranium Series), $T\frac{1}{2} =$ 1600 Y, decays to radioactive ^{222}Rn by alphas of 4.8 MeV. Natural isotope ^{228}Ra (Mesothorium = 1, Thorium Series), $T\frac{1}{2} = 6.7$ Y, decays to radioactive ^{228}Ac by betas of 0.05 MeV.

RADON   HR: 3
mf: Rn   mw: 86

PROP: Colorless, odorless, inert gas, very dense. Bp: $-62°$; d (gas @ 1 atm and 0°):9.73 g/L, (liq @ bp): 4.4.

THR: A common air contaminant. Radiation Hazard: Natural isotope ^{220}Rn (Thoron, Thorium Series), $T\frac{1}{2} = 55s$, decays to radioactive ^{216}Po by alphas of 6.3 MeV. Natural isotope ^{222}Rn (Uranium Series), $T\frac{1}{2} = 3.8$ D, decays to radioactive ^{218}Po by alphas of 5.5 MeV. The permissible levels are given for ^{222}Rn in equilibrium with its daughters. The chief hazard from this isotope is inhalation of the gaseous element and its solid daughters, which are collected on the normal dust of the air. This material is deposited in the lungs and has been considered to be a major causative agent in the high incidence of lung cancer found in uranium miners. Radon and its daughters build up to an equilibrium value in about a month from radium compounds, while the build-up from uranium compounds is negligible. Good ventilation of areas where radium is handled or stored is recommended to prevent accumulation of hazardous concentrations of Rn and its daughters.

RED SQUILL   HR: 3
NIOSH: VF 3750000

PROP: Sea onion bulbs contain a potent concentration of Scilliroside, a glycoside bearing a close chemical resemblance to the Scillarens.

SYNS: BONIDE TOPZOL RAT BAITS AND KILLING SYRUP * SCILLIROSIDE GLYCOSIDE * SQUILL

THR: Poison by ingestion. Human gastrointestinal tract effects. When heated to decomposition it emits acrid smoke and fumes.

RESERPINE   HR: 3
CAS: 50-55-5   NIOSH: AG 0350000
mf: $C_{33}H_{40}N_2O_9$ mw: 608.75

PROP: White or pale buff to slightly yellow powder, odorless, insol in water, very sltly sol in alc; sol in chloroform and acetic acid. Mp: 264°-265° (decomp).

SYNS: ENT 50146 * SERPASIL APRESOLINE * RAUSERPIN * METHYLRESERPATE 3,4,5-TRI-METHOXYBENZOIC ACID * METHYL RESERPATE 3,4,5-TRIMETHOXYBENZOIC ACID ESTER * NCI-C50157 * YOHIMBAN-16-CARBOXYLIC ACID DERIVATIVE OF BENZ(G)INDOLO(2,3-A)-QUINOLIZINE * 3,4,5-TRIMETHOXYBENZOYL METHYL RESERPATE * USAF CB-27 * SERPASIL * RAUWOLFEA

THR: Poison by ingestion, intravenous, subcutaneous, and intraperitoneal routes. Mutagenic data. An experimental teratogen and carcinogen. A suspected human carcinogen. In humans, 0.014 mg/kg causes psychotropic effects. A medicine with side effects. Used as a food additive permitted in the feed and drinking water of animals and/or for the treatment of food-producing animals. Also permitted in food for human consumption. A sedative. When heated to decomposition it emits toxic fumes of NO_x. For further information see Vol. 1, No. 4 of *DPIM Report*.

RESORCINOL   HR: 3
CAS: 108-46-3   NIOSH: VG 9625000
DOT: 2876
mf: $C_6H_6O_2$ mw: 110.12

PROP: Very white crystals, become pink on exposure to light when not perfectly pure, unpleasant sweet taste. Mp: 110°, bp: 280.5°, flash p: 261°F (CC), d: 1.285 @ 15°, autoign temp: 1126°F, vap press: 1 mm @ 108.4°, vap d: 3.79. Very sol in alc, ether, glycerol; sltly sol in chloroform. Sol in water.

SYNS: M-BENZENEDIOL * 1,3-BENZENEDIOL * C.I. 76505 * M-DIHYDROXYBENZENE * 1,3-DIHYDROXYBENZENE * M-DIOXYBENZENE * M-HYDROQUINONE * 3-HYDROXYCYCLOHEXADIEN-1-ONE * M-HYDROXYPHENOL * 3-HYDROXYPHENOL * NCI-C05970 * RESORCIN

ACGIH TLV: TWA 10 ppm; STEL 20 ppm
DOT Classification: ORM-E (IMO: Poison B)

THR: Poison by ingestion, intraperitoneal, and subcutaneous routes. Mutagenic data. An experimental carcinogen and tumorigen. It is primarily a skin and eye irritant, however, it can cause systemic poisoning by acting both as a blood and nerve poison. In a suitable solvent, this material can readily be absorbed through human skin and can cause local hyperemia, itching, dermatitis, edema, and corrosion associated with enlargement of regional lymph glands as well as serious systemic disorders such as restlessness, methemoglobinemia, cyanosis, convulsions, tachycardia, dyspnea and death. These same symptoms can be induced by ingestion of the material. For poisoning, treat symptomatically. Get medical advice. Combustible when exposed to heat or flame; can react with oxidizing materials. To fight fire, use water, CO_2, dry chemical. Incompatible with acetanilide, albumin, alkalies, antipyrine, camphor, ferric salts, menthol, spirit nitrous ether, urethan. For further information see Vol. 1, No. 2 of *DPIM Report*.

RESORCINOL DIGLYCIDYL ETHER   HR: 3
CAS: 101-90-6   NIOSH: VH 1050000
mf: $C_{12}H_{14}O_4$ mw: 222.26

SYNS: M-BIS(2,3-EPOXYPROPOXY)BENZENE * 1,3-BIS(2,3-EPOXYPROPOXY)BENZENE * DIGLYCIDYL RESORCINOL ETHER * NCI-C54966 * OXIRANE, 2,2'-(1,3-PHENYLENEBIS-(OXYMETHYLENE))BIS- * 2,2'-(1,3-PHENYLENEBIS(OXYMETHYLENE))BISOXIRANE * RESORCINYL DIGLYCIDYL ETHER

THR: Poison by intraperitoneal route. Moderately toxic by ingestion. An experimental carcinogen. See also ethers.

RETINOL ACETATE   HR: 3
CAS: 127-47-9   NIOSH: VH 6825000
mf: $C_{22}H_{32}O_2$ mw: 328.54

SYNS: RETINYL ACETATE * VITAMIN A ACETATE * TRANS-VITAMIN A ACETATE * VITAMIN A ALCOHOL ACETATE

THR: An experimental teratogen. Moderately toxic by ingestion. When heated to decomposition it emits acrid smoke and fumes.

RETRORSINE HR: 3
CAS: 480-54-6 NIOSH: VH 7525000
mf: $C_{18}H_{25}NO_6$ mw: 351.44

SYNS: 12,18-DIHYDROXY-SENECIONAN-11,16-
DIONE * BETA-LONGILOBINE * CIS-RETRONE-
CIC ACID ESTER OF RETRONECINE

THR: Poison by ingestion, intraperitoneal, intravenous, and other unspecified routes. Mutagenic data. An experimental carcinogen and
neoplastigen. When heated to decomposition
it emits toxic fumes of NO_x.

RETRORSINE-N-OXIDE HR: 3
CAS: 15503-86-3 NIOSH: VH 7700000
mf: $C_{18}H_{25}NO_7$ mw: 367.44

SYNS: ISATIDINE * CIS-RETRONECIC ACID ES-
TER OF RETRONECINE-N-OXIDE

THR: Poison by ingestion and intraperitoneal
routes. Mutagenic data. An experimental carcinogen and neoplastigen. Moderately toxic by
intravenous route. See also esters. When heated
to decomposition it emits toxic fumes of NO_x.

RHENIUM
af: Re aw: 186.2

PROP: Hexagonal close-packed crystals of
metal, black to silver gray. Mp: 3180°; bp: approx 5900°; d: 21.02.

THR: No reported cases of human toxicity. See
also Rare Earths. Radiation Hazard: Natural
(63%) isotope ^{187}Re, $T_{\frac{1}{2}} = 4 \times 10^{10}$ Y, decays
to stable ^{187}Os by betas of less than 0.10 MeV.
Moderate fire hazard in the form of dust, when
exposed to heat or flame. Violent reaction with
F_2 @ 125°. Incompatible with oxygen.

RHENIUM TRICHLORIDE HR: 3
CAS: 13569-63-6 NIOSH: VI 0875000
mf: Cl_3Re mw: 292.55

THR: Poison by intraperitoneal and other unspecified routes. See also rhenium compounds
and chlorides. When heated to decomposition
it emits toxic fumes of Cl^-.

RHODIUM (METAL FUME AND DUSTS)
CAS: 7440-16-6 NIOSH: VI 9355000
af: Rh aw: 102.91

PROP: A silvery white metallic element. Mp:
1966°, bp: 3727°, d: 2.41 @ 20°.
OSHA PEL: TWA 100 ug(Rh)/m^3

ACGIH TLV: TWA metal 1 mg/m^3, insoluble
compounds as Rh 1 mg/m^3, soluble compounds as Rh 0.01 mg/m^3

THR: Toxic by inhalation. Soluble compounds
much more toxic. Flammable when exposed
to heat or flame. Violent reaction with ClF_3,
OF_2.

RHODIUM (III) CHLORIDE (1:3) HR: 3
CAS: 10049-07-7 NIOSH: VI 9275000
mf: Cl_3Rh mw: 209.26

SYNS: RHODIUM CHLORIDE * RHODIUM TRI-
CHLORIDE

OSHA PEL: TWA 1 ug/m^3

THR: Poison by intraperitoneal and intravenous
routes. Mutagenic data. An experimental carcinogen. See also rhodium and chlorides. When
heated to decomposition it emits toxic fumes
of Cl^-. Incompatible with penta carbonyl iron;
zinc.

RIBOFLAVINE HR: 2
CAS: 83-88-5 NIOSH: VJ 1400000
mf: $C_{17}H_{20}N_4O_6$ mw: 376.41

PROP: Orange to yellow crystals, sltly sol in
water and alcohols, insol in lipid solvents. Mp:
282° (decomp).

SYNS: 6,7-DIMETHYL-9-D-RIBITYLISOALLOXA-
ZINE * ISOALLOXAZINE, 7,8-DIMETHYL-10-D-
RIBITYL- * ISOALLOXAZINE, 7,8-DIMETHYL-
10-(D-RIBO-2,3,4,5-TETRAHYDROXYPENTYL)-
* RIBOFLAVIN * VITAMIN B2 * VITAMIN G

THR: Moderately toxic by intraperitoneal route.
When heated to decomposition it emits toxic
fumes of NO_x.

RIFAMYCIN AMP HR: 3
CAS: 13292-46-1 NIOSH: VJ 7000000
mf: $C_{43}H_{58}N_4O_{12}$ mw: 823.05

SYNS: 3-(4-METHYLPIPERAZINYLIMINOMETHYL)-
RIFAMYCIN SV * NSC 113926 * 8-(4-METH-
YLPIPERAZINYLIMINOMETHYL) RIFAMYCIN SV
* 8-(((4-METHYL-1-PIPERAZINYL)IMINO)-
METHYL)RIFAMYCIN SV

THR: Poison by intraperitoneal route. Mutagenic data. An experimental teratogen and neoplastigen. Human skin and eye systemic effects.
When heated to decomposition it emits toxic
fumes of NO_x.

ROTENONE   **HR: 3**
CAS: 83-79-4   NIOSH: DJ 2800000
mf: $C_{23}H_{22}O_6$ mw: 394.45

PROP: Orthorhombic plates. Mp: 165°-166° (dimorphic form mp: 185°-186°). D: 1.27 @ 20°. Almost insol in water; sol in alc, acetone, carbon tetrachloride, chloroform, ether and other organic solvents. Slow decomp on exposure to light and air.

SYNS: CUBE * CUBE EXTRACT * CUBE ROOT * DERRIS * NCI-C55210 * ROTENONA (SPANISH) * TUBATOXIN

OSHA PEL: TWA 5 mg/m^3
ACGIH TLV: TWA 5 mg/m^3
DFG MAK: 5 mg/m^3

THR: It is more toxic when inhaled than when ingested. Poison by ingestion, inhalation, and intraperitoneal routes. An experimental neoplastigen and carcinogen. An insecticide, fish poison. Estimated LD (oral, man) = 200 mg/kg. Acute poisoning causes numbness, nausea, vomiting and tremors. A skin irritant. Chronic exposure injures liver and kidneys. It is toxic to animals and very toxic to fish but leaves no harmful residue on vegetable crops. When heated to decomposition it emits acrid smoke and fumes. For further information see Vol. 1, No. 2 of *DPIM Report*.

RUBIDIUM
CAS: 7440-17-7   NIOSH: VL 8500000
DOT: 1423
af: Rb   aw: 85.47

PROP: Soft, silvery-white metal. Mp: 38.89°, bp: 688°, d (solid): 1.532 @ 20°, d(liquid): 1.475 @ 39°.

SYN: RUBIDIUM METAL

DOT Classification: Label: Flammable Solid

THR: An alkaline metal which much resembles potassium and cesium. Dangerous fire hazard when exposed to heat or flame or by chemical reaction with oxidizers. See also sodium. Ignites spontaneously in O_2. Moderate explosion hazard reacts explosively with air, halogens, mercury, non-metals, vanadium chloride oxide, moisture, acids and oxidizers. See also sodium and NaK. Violent reaction with Cl_2O_2; P; water. Dangerous; when heated it emits toxic fumes of rubidium oxide; will react with water or steam

to produce hydrogen and flam vapors; reacts vigorously with oxidizing materials. Storage and Handling: Keep under benzene, petroleum, or other liquids not containing O_2.

RUBIDIUM CHLORIDE   **HR: 2**
CAS: 7791-11-9   NIOSH: VL 8575000
mf: ClRb   mw: 120.92

PROP: White crystalline powder. Mp: 715°; bp: 1390°; d: 2.76.

THR: Moderately toxic by ingestion and intraperitoneal routes. Reacts violently with BrF_3. See also rubidium and chlorides. When heated to decomposition it emits toxic fumes of Cl^-, RbCl and Rb_2O.

RUBIDIUM DICHROMATE
CAS: 13446-73-6   NIOSH: VL 8600000
mf: $Cr_2O_7Rb_2$ mw: 386.94

PROP: Cryst. D: 3.02-3.13

THR: A poison. A powerful oxidizer. See also dichromates and rubidium. When heated to decomposition it emits acrid smoke and fumes of Rb_2O.

RUBIDIUM FLUORIDE   **HR: 3**
CAS: 13446-74-7   NIOSH: VL 8740000
mf: FRb   mw: 104.47

PROP: Colorless crystals. Mp: 775°, bp: 1410°, d: 3.557, vap press: 1 mm @ 921°.

THR: A poison. Poison as a sol fluoride. See also rubidium and fluorides. When heated to decomposition it emits toxic fumes of F^-.

RUM (DENATURED)
                                    NIOSH: VM 1655000

PROP: Flash p: 77°F (CC).
DOT Classification: Label: Flammable
  Liquid

THR: Variable. See ethanol. Dangerous fire hazard when exposed to heat or flame. Dangerous, upon exposure to heat or flame; can react vigorously with oxidizing materials. To fight fire, use CO_2, dry chemical.

RUSSIAN COMFREY ROOTS   **HR: 3**
                                    NIOSH: VM 2060000

PROP: Fresh roots dried, milled and mixed with diet.

SYNS: COMFREY, RUSSIAN * SYMPHYTUM OF-FICINALE L

THR: An experimental carcinogen. When heated to decomposition it emits acrid smoke and fumes.

RUTHENIUM
af: Ru   aw: 101.07

PROP: Lustrous, hard metal, hexagonal crystallization. D: 12.45 @ 20°/4°; mp: approx 2450°; bp: approx 4150° Stable in air.

THR: See ruthenium compounds. Flammable in the form of dust, when exposed to heat or flame. Violent reaction with ruthenium oxide. When heated to decomposition it emits toxic fumes of RuO_x. Incompatible with aqua regia, potassium chlorate.

RUTHENIUM CHLORIDE   HR: 3
NIOSH: VM 2650000
mf: Cl_3Ru mw: 207.42

PROP: α form: Black lustrous crystals. Insol in alc, water. β form: dark brown, fluffy hexagonal crystals. Sol in alc.

SYN: RUTHENIUM TRICHLORIDE

THR: Poison by intraperitoneal route. See also ruthenium compounds. When heated to decomposition it emits toxic fumes of Cl^-. Incompatible with penta carbonyl iron; zinc.

RUTHENIUM COMPOUNDS
THR: Details unknown. Probably toxic, but such small amounts are used industrially that it does not constitute a hazard. It resembles osmium in that when it is heated in air, it evolves fumes which are injurious to the eyes and lungs. Dangerous; when heated to decomposition it emits toxic fumes of ruthenium oxide.

RYANIA   HR: 3
CAS: 15662-33-6   NIOSH: VM 4025000
mf: $C_{25}H_{35}NO_9$ mw: 493.61

PROP: The powdered stem of *Ryania Speciosa*, of proven insecticidal activity.

SYNS: BONIDE RYATOX * GROUND RYANIA SPECISA(VAHL) STEMWOOD (ALKOLOID RYANO-DINE) * RYANIA POWDER * RYANIA SPECIOSA

THR: Poison to humans by ingestion resulting in weakness, respiratory changes, diarrhea, gastrointestinal disturbances, tremors, convulsions, coma and death. An insecticide. No tolerance limits have been established or proposed by the government for residues on foods. Flammable when exposed to heat or flame. To fight fire, use CO_2, mist, spray, foam. When heated to decomposition it emits toxic fumes of NO_x.

S

SACCHARIN
HR: 3

CAS: 128-44-9   NIOSH: DE 4550000
mf: $C_7H_4NO_3S \cdot Na$ mw: 205.17

SYNS: SAXIN * ARTIFICIAL SWEETENING SUB-
STANZ GENDORF 450 * CRYSTALLOSE
* SACCHARIN SOLUBLE * SACCHAROIDUM NA-
TRICUM * SODIUM-1,2 BENZISOTHIAZOLIN-3-
ONE-1,1-DIOXIDE * SODIUM-O-BENZOSULFIMIDE
* SODIUM-O-BENZOSULPHIMIDE * SODIUM-2-
BENZOSULPHIMIDE * SODIUM SACCHARIDE
* SODIUM SACCHARIN * SODIUM SACCHARI-
NATE * SODIUM SACCHARINE * SOLUBLE
SACCHARIN * SOLUBLE GLUSIDE * O-SULFON-
BENZOIC ACID IMIDE SODIUM SALT * SULPHO-
BENZOIC IMIDE, SODIUM SALT

THR: An experimental neoplastigen and carcinogen. Mutagenic data. When heated to decomposition it emits very toxic fumes of SO_x
and NO_x. An artificial sweetening agent.

SAFROL
HR: 3

CAS: 94-59-7   NIOSH: CY 2800000
mf: $C_{10}H_{10}O_2$ mw: 162.20

PROP: Colorless liquid or crystals. Sassafras
odor. Mp: 11°; bp: 234.5°; d: 1.0960 @ 20°;
vap press: 1 mm @ 63.8°. Insol in water; very
sol in alc; misc with chloroform, ether.

SYNS: 5-ALLYL-1,3-BENZODIOXOLE * ALLYL-
CATECHOL METHYLENE ETHER * ALLYLDIOXY-
BENZENE METHYLENE ETHER * 1-ALLYL-3,
4-METHYLENEDIOXYBENZENE * 4-ALLYL-1,2-
METHYLENEDIOXYBENZENE * M-ALLYLPYRO-
CATECHIN METHYLENE ETHER * 4-ALLYL-
PYROCATECHOL FORMALDEHYDE ACETAL
* ALLYLPYROCATECHOL METHYLENE ETHER
* 1,2-METHYLENEDIOXY-4-ALLYLBENZENE
* 3,4-METHYLENEDIOXY-ALLYBENZENE
* 5-(2-PROPENYL)-1,3-BENZODIOXOLE
* SAFROLE * SHIKIMOLE

THR: An experimental carcinogen, neoplastigen, and tumorigen. Mutagenic data. A powerful irritant by ingestion. Affects liver. Combustible when exposed to heat or flame. When heated
to decomposition it emits acrid smoke, fumes.

SALICYLAMIDE
HR: 3

CAS: 65-45-2   NIOSH: VN 6475000
mf: $C_7H_7NO_2$ mw: 137.15

PROP: White to slightly pink crystals or powder, somewhat bitter taste. Mp: 140°. Sol in
hot water, alc, chloroform, ether.

SYNS: O-HYDROXYBENZAMIDE * RASPBERIN

THR: Poison by intravenous route. Moderately
toxic by ingestion and intraperitoneal route. An
experimental teratogen. See also amides. Can
cause dizziness, drowsiness, nausea, vomiting,
epigastric distress, allergic reactions and blood
dyscrasias in average to large doses. Dangerous;
when heated to decomposition it emits toxic
fumes of NO_x.

SALICYLIC ACID
HR: 3

CAS: 69-72-7   NIOSH: VO 0525000
mf: $C_7H_6O_3$ mw: 138.13

PROP: D: 1.443 @ 20°/4°; mp: 158.3°; bp:
211° @ 20 mm ±; sol in water, alc, ether.

SYNS: O-HYDROXYBENZOIC ACID * 2-HY-
DROXYBENZOIC ACID

THR: An experimental teratogen. Mutagenic
data. Moderately toxic by ingestion, subcutaneous, and intravenous routes. A skin and eye
irritant. When heated to decomposition it emits
acrid smoke. Incompatible with iron salts; spirit
nitrous ether; lead acetate; iodine.

SALITHION
HR: 3

CAS: 3811-49-2   NIOSH: DF 4375000
mf: $C_8H_9O_3PS$ mw: 216.20

SYNS: K-9 * 2-METHOXY-4H-1,2,3-BENZODI-
OXAPHOSPHORINE-2-SULFIDE * SALITHION-SUM-
ITOMO

THR: Poison by ingestion and subcutaneous
routes. An insecticide. See also parathion. When
heated to decomposition it emits very toxic
fumes of SO_x and PO_x.

SALVARSAN
HR: 3

CAS: 139-93-5   NIOSH: SJ 7175000
mf: $C_{12}H_{12}As_2N_2O_2 \cdot 2ClH$ mw: 439.02

SYNS: ARSENPHENOLAMINE HYDROCHLORIDE
* ARSPHENAMINE * 3,3'-DIAMINO-4,4'-DIHY-
DROXYARSENOBENZENE DIHYDROCHLORIDE
* EHRLICH 606 * PHENARSENAMINE

THR: Poison by intravenous route. See also
arsenic. Implicated in aplastic anemia. When

heated to decomposition it emits very toxic fumes of As, NO_x, and HCl.

SAMARIUM
af: Sm   aw: 150.35

PROP: Bright, yellow, lustrous, stable metal; mp: 1072°; bp: 1778°; d(α): 7.536, d(β): 7.40.

THR: No toxicity data. As a lanthanon it may cause impairment of blood clotting. See also lanthanum and rare earths. Flammable in the form of dust when exposed to flame or by spontaneous chemical reaction with oxidizers. See also powdered metals. Ignites at 150° in air, also releases H_2 in contact with water. Can react violently with halogens. Incompatible with 1,1,2-trichlorotrifluoroethane.

SAMARIUM ACETATE   HR: 3
CAS: 10465-27-7   NIOSH: AJ 3890000
mf: $C_6H_9O_6 \cdot Sm$ mw: 327.50

SYN: SAMARIUMACETAT (GERMAN)

THR: Poison by intravenous route. See also samarium and rare earths. When heated to decomposition it emits acrid smoke and fumes.

SAMARIUM (III) CHLORIDE   HR: 3
CAS: 10361-82-7   NIOSH: VP 2625000
mf: Cl_3Sm mw: 256.70

PROP: White-yellowish powder. D: 4.465; mp: 686°.

THR: Poison by intraperitoneal route. Moderately toxic by subcutaneous route. A skin and eye irritant. See also samarium and rare earths. When heated to decomposition it emits toxic fumes of Cl^-.

SAMARIUM (III) NITRATE, HEXAHYDRATE (1:3:6)   HR: 3
CAS: 13759-83-6   NIOSH: VP 3150000
mf: $N_3O_9 \cdot Sm \cdot 6H_2O$ mw: 444.50

PROP: Pale yellow crystals; mp: 78°-79°; d: 2.375.

SYNS: NITRIC ACID, SAMARIUM(3+) SALT, HEXAHYDRATE * SAMARIUM NITRAT (GERMAN)

THR: Poison by intraperitoneal and intravenous routes. Moderately toxic by ingestion and subcutaneous routes. See also samarium and nitrates. When heated to decomposition it emits toxic fumes of NO_x.

SARAN   HR: 3
CAS: 8013-77-2   NIOSH: VQ 2000000
mf: $(C_4H_5Cl_3)_n$

THR: An experimental carcinogen. See also polymers. When heated to decomposition it emits toxic fumes of Cl^-.

SARKOMYCIN   HR: 3
CAS: 11031-48-4   NIOSH: VQ 4200000
mf: $C_7H_8O_3$ mw: 140.15

PROP: Oily liquid. Sol in water, methanol, ethanol, butanol, ethyl acetate; slightly sol in ether. Isolated from *Streptomyces sp.*

SYN: SARCOMYCIN

THR: An experimental teratogen and carcinogen. Moderately toxic by intravenous, subcutaneous, and intraperitoneal routes. Mutagenic data. When heated to decomposition it emits acrid smoke and fumes.

SASSAFRAS   HR: 3
NIOSH: VQ 5946000

PROP: A yellowish, reddish volatile oil, pungent, aromatic odor and taste. Sol alc, ether, chloroform, glacial acetic acid, CS_2; d: 1.065-1.077 @ 25°/25°. Safrole-Free ethanol extract of sassafras albidum root bark.

SYN: SASSAFRAS ALBIDUM

THR: An experimental neoplastigen. When heated to decomposition it emits acrid smoke and fumes.

SCANDIUM
af: Sc   aw: 44.9559

PROP: A naturally occurring element.

THR: See Rare Earths. Should be handled carefully. Flammable in the form of dust, when exposed to heat or flame or by chemical reaction with oxidizers. See also powdered metals, and Rare Earths. Can react violently with halogens; air.

SCANDIUM (3$^+$) CHLORIDE   HR: 3
CAS: 10361-84-9   NIOSH: VQ 8925000
mf: Cl_3Sc mw: 151.31

PROP: White deliquescent solid. Mp: 960°. sol in water; insol in alc.

SYN: SC G3

THR: Poison by intraperitoneal route. Moderately toxic by ingestion. See also scandium. When heated to decomposition it emits toxic fumes of Cl^-.

SCARLET RED   HR: 3
CAS: 85-83-6   NIOSH: QL 5775000
mf: $C_{24}H_{20}N_4O$ mw: 380.48

SYNS: BRASILAZINA OIL RED B * CALCO OIL RED D * C.I. 258 * C.I. SOLVENT RED 24 * 2',3-DIMETHYL-4-(2-HYDROXYNAPHTHYLAZO)AZOBENZENE * FAST OIL RED B * FAT RED B * 1-((2-METHYL-4-((2-METHYLPHENYL)AZO)PHENYL)AZO)-2-NAPHTHALENOL * PHENOPLASTE ORGANOL RED B * RUBRUM SCARLATINUM * 1-((4-(O-TOLYLAZO)-O-TOLYL)AZO)-2-NAPHTHOL) * O-TOLUENEAZO-O-TOLUENEAZO-BETA-NAPHTHOL * O-TOLUENEAZO-O-TOLUENE-BETA-NAPHTHOL * O-TOLYLAZO-O-TOLYLAZO-BETA-NAPHTHOL * O-TOLYLAZO-O-TOLYLAZO-2-NAPHTHOL

THR: An experimental carcinogen. Mutagenic data. When heated to decomposition it emits toxic fumes of NO_x.

SCOPOLAMINE   HR: 3
CAS: 51-34-3   NIOSH: VR 3675000
mf: $C_{17}H_{21}NO_4$ mw: 303.39

PROP: Thick, colorless, syrupy liquid alkaloid. Mp: 55°. Very sol in hot water, alc, ether, chloroform, acetone; slightly sol in benzene, petroleum ether. Decomp on standing.

SYNS: 6-BETA,7-BETA-EPOXY-3-ALPHA-TROPANYL S-(−)-TROPATE * EPOXYTROPINE TROPATE * HYOSCINE * (−)-HYOSCINE * SCOPINE TROPATE * (−)-SCOPOLAMINE * TROPIC ACID, ESTER WITH SCOPINE * TROPIC ACID, 9-METHYL-3-OXA-9-AZATRICYCLO(3.3.1.0(SUP 2, 4))NON-7-YL ESTER

THR: Poisonous alkaloid by intravenous and subcutaneous routes. Human central nervous system effects. See also esters. It can produce profound depression of the central nervous system and occasionally it causes excitement. It can cause hallucinations and the individual affected loses a certain amount of his normal inhibitory control. It is for that reason that it has been called "truth serum." In many cases of poisoning from this material, and even to a certain extent following its medical application, there is retention of the urine caused by paralysis of the bladder, and catheterization is necessary. The fatal dose is variable. Death has occurred from as little as 0.6 mg, while recovery has occurred from doses of 7-15 mg. Dangerous; when heated to decomposition it emits highly toxic fumes of NO_x.

SECONAL   HR: 3
CAS: 76-73-3   NIOSH: CP 9450000
mf: $C_{12}H_{18}N_2O_3$ mw: 238.32

SYNS: 5-ALLYL-5-(1-METHYLBUTYL)BARBITURIC ACID * 5-ALLYL-5-(1-METHYLBUTYL)-MALONYLUREA * QUINALBARBITAL * QUINALBARBITONE * SECOBARBITAL * SECOBARBITONE

THR: Poison by ingestion, intraperitoneal, subcutaneous, and intravenous routes. Human central nervous system effects. See also barbiturates. When heated to decomposition it emits toxic fumes of NO_x.

SELENIUM   HR: 3
CAS: 7782-49-2   NIOSH: VS 7700000
DOT: 2658
af: Se   aw: 78.96

PROP: Steel gray, non-metallic element; mp: 170°-217°; bp: 690°; d: 4.81-4.26; vap press: 1 mm @ 356°.

SYNS: SELENIUM ALLOY * SELENIUM BASE * SELENIUM HOMOPOLYMER * C.I. 77805 * ELEMENTAL SELENIUM * SELEN (POLISH) * SELENIUM DUST

OSHA PEL: TWA 200 ug(Se)/m^3
ACGIH TLV: TWA 0.2 mg/m^3
DFG MAK: 0.1 mg/m^3
DOT Classification: Poison B, Label: Poison

THR: Poison by inhalation, intravenous, and other unknown routes. An experimental carcinogen. See also selenium compounds. When heated to decomposition it emits toxic fumes of Se. Can react violently with barium carbide; bromine pentafluoride; calcium carbide; chlorates; chlorine trifluoride; chromic oxide (CrO_3); fluorine; lithium carbide; lithium silicon (Li_6 Si_2); nickel; nitric acid; sodium; nitrogen trichloride; oxygen; potassium; potassium bromate; rubidium carbide; zinc; silver bromate; strontium carbide; thorium carbide; uranium. For further information see Vol. 1, No. 3 of *DPIM Report*.

SELENIUM COMPOUNDS HR: 3

THR: Poison by inhalation and intravenous routes. An experimental carcinogen. Selenium in small amounts is essential for normal growth of some animals. Deficiency or excess is associated with serious disease in livestock. Longterm exposure may be a cause of amyotrophic lateral sclerosis in humans, just as it may cause "blind staggers" in cattle. Elemental selenium has low acute systemic toxicity, but dust or fumes can cause serious irritation of the respiratory tract. Hydrogen selenide resembles other hydrides in being highly toxic, and selenium oxychloride is a vesicant. Some organoselenium compounds have the high toxicity of other organometals. Inorganic selenium compounds can cause dermatitis. Garlic odor of breath is a common symptom. Pallor, nervousness, depression and digestive disturbances have been reported in cases of chronic exposure. Selenium compounds are common air contaminants.

SELENIUM DIMETHYLDITHIO-CARBAMATE HR: 3

CAS: 144-34-3   NIOSH: VT 0780000
mf: $C_{12}H_{24}N_4S_8 \cdot Se$ mw: 559.84

PROP: Yellow powder, cryst; d: 1.58; M range: 140°-172°.

SYNS: METHYL SELENAC * TETRAKIS-(DIMETHYLCARBAMODITHIOATO-S,S')SELENIUM

THR: An experimental carcinogen. See also selenium compounds and carbamates. When heated to decomposition it emits very toxic fumes of Se, SO_x, and NO_x.

SELENIUM DISULFIDE (2.5%) SHAMPOO HR: 2

NIOSH: VS 9285000

SYN: SELENIUM(IV) DISULFIDE SHAMPOO (2.5%)

THR: An eye irritant. See also selenium compounds and sulfides. When heated to decomposition it emits very toxic fumes of Se and SO_x.

SELENIUM HEXAFLUORIDE HR: 3

CAS: 7783-79-1   NIOSH: VS 9450000
DOT: 2194
mf: F_6Se mw: 192.96

PROP: Colorless gas; mp: −39° (subl @ −40.6°); bp: −34.5°; d: 3.25 @ −25°.

SYN: SELENIUM FLUORIDE

OSHA PEL: TWA 400 ug/m³
ACGIH TLV: TWA 0.05 ppm
DOT Classification: Poison A

THR: Poison by inhalation. See also selenium compounds and fluorides. When heated to decomposition it emits very toxic fumes of F⁻ and Se.

SELENIUM MONOSULFIDE HR: 3

CAS: 7446-34-6   NIOSH: VT 0525000
mf: SSe   mw: 111.02

PROP: Orange-yellow tablets or powder; mp: 111.03°; bp: decomp @ 118°-119°; d: 3.056 @ 0°.

SYNS: SELENIUM SULFIDE * NCI-C50033

OSHA PEL: TWA 200 ug(Se)/m³

THR: Poison by ingestion. An experimental carcinogen. See also selenium compounds and sulfides. When heated to decomposition it emits very toxic fumes of SO_x and Se.

SELSUN HR: 2

NIOSH: VT 2480000

PROP: 2.4% w/v selenium sulfide in aqueous suspension, also contains bentonite, sodium alkyl aryl sulfonate, sodium phosphate, glycerol monoricinoleate, citric acid, captan and perfume.

SYN: NCI-C54546

THR: An eye irritant. See also selenium compounds, bentonite, tri-o-sodium phosphate, citric acid, captan, glycerol monoricinoleate. When heated to decomposition it emits very toxic fumes of Se, SO_x, PO_x, and NO_x.

SEMICARBAZIDE HYDROCHLORIDE HR: 3

CAS: 563-41-7   NIOSH: VT 3500000
mf: $CH_5N_3O \cdot ClH$ mw: 111.55

PROP: Prisms from dilute alcohol. Decomp @ 175°-185°; mp: 176° (decomp). Very sol in water; very slightly sol in hot alc; insol in anhydrous ether.

SYNS: AMIDOUREA HYDROCHLORIDE * AMINOUREA HYDROCHLORIDE * CARBAMYLHYDRA-

ZINE HYDROCHLORIDE * HYDRAZINECARBOX-
AMIDE MONOHYDROCHLORIDE

THR: Poison by ingestion. An experimental carcinogen, teratogen, and neoplastigen. When
heated to decomposition it emits very toxic
fumes of NO_x and HCl. For further information
see Vol. 6, No. 4 of *DPIM Report*.

SENECIPHYLLINE   HR: 3
CAS: 480-81-9   NIOSH: UY 8050000
mf: $C_{18}H_{23}NO_5$ mw: 333.42

PROP: An alkaloid isolated from *S. Stenoce-
phalus*.

SYNS: NCI-C61165 * JACOBINE * NSC
30622 * 6-METHYL-5-METHYLENE-
(1,6)DIOXACYCLODODECINO-(2,3,4-GH)-
PYRROLIZINE-2,7-DIONE

THR: Poison by ingestion, intravenous, intraperitoneal, and other routes. A carcinogen.
When heated to decomposition it emits toxic
fumes of NO_x.

SENKIRKINE   HR: 3
CAS: 2318-18-5   NIOSH: VT 5960000
mf: $C_{19}H_{28}NO_6$ mw: 366.48

SYNS: NSC-89945 * RENARDINE * 2,12-
DIHYDROXY-4-METHYL-11,16-DIOXOSENECIONA-
NIUM

THR: Poison by ingestion and intraperitoneal
routes. An experimental neoplastigen. Mutagenic data. When heated to decomposition it
emits toxic fumes of NO_x.

S-SEVEN   HR: 3
CAS: 3792-59-4   NIOSH: TB 1735000
mf: $C_{14}H_{13}Cl_2O_2PS$ mw: 347.20

SYNS: O-ETHYL-O-2,4-DICHLOROPHENYL THIO-
NOBENZENEPHOSPHONATE * PHENYLPHOSPHO-
NOTHIOIC ACID O-(2,4-DICHLOROPHENYL)
O-ETHYL ESTER

THR: Poison by ingestion. Moderately toxic
by subcutaneous route. See also esters. When
heated to decomposition it emits very toxic
fumes of Cl^-, SO_x, and PO_x.

SHIKIMIC ACID   HR: 3
CAS: 138-59-0   NIOSH: GW 4600000
mf: $C_7H_{10}O_5$ mw: 174.17

PROP: Isolated from Bracken.

SYNS: BRACKEN FERN TOXIC COMPONENT
* SHIKIMATE * 3,4,5-TRIHYDROXY-1-CYCLO-
HEXENE-1-CARBOXYLIC ACID

THR: An experimental tumorigen. Mutagenic
data. Moderately toxic by intraperitoneal route.
When heated to decomposition it emits acrid
smoke and fumes.

SILICA, AMORPHOUS FUMED   HR: 2
CAS: 7631-86-9   NIOSH: VV 7310000
mf: O_2Si mw: 60.09

PROP: A finely powdered microcellular silica
foam with minimum SiO_2 content of 89.5%.

SYNS: SILICA, AMORPHOUS * AMORPHOUS
SILICA DUST * COLLOIDAL SILICA * COLLOI-
DAL SILICON DIOXIDE * ENT 25,550 * FOS-
SIL FLOUR * FUMED SILICA * FUMED SILICON
DIOXIDE * SILICIC ANHYDRIDE

OSHA PEL: TWA 80 mg/m^3/%SiO$_2$

THR: Moderately toxic by ingestion, intravenous, intraperitoneal, and intratracheal routes.
See also silica. Much less toxic than crystalline
forms. Does not cause silicosis.

SILICA, AMORPHOUS FUSED   HR: 3
CAS: 60676-86-0   NIOSH: VV 7320000
mf: O_2Si mw: 60.09

PROP: Made up of spherical submicroscopic
particles under 0.1 micron in size.

SYNS: AMORPHOUS FUSED SILICA * FUSED
QUARTZ * FUSED SILICA * QUARTZ GLASS
* SILICA, VITREOUS * SILICON DIOXIDE
* VITREOUS QUARTZ

OSHA PEL: TWA 80 mg/m^3/%SiO$_2$

THR: Poison by intraperitoneal, intravenous,
and intratracheal routes. See also silica.

SILICA, AMORPHOUS HY-
DRATED   HR: 1
CAS: 763 18- 6-9   NIOSH: VV 7322000
mf: O_2Si mw: 60.09

SYNS: SILICA AEROGEL * SILICA XEROGEL
* SILICA GEL * SILICIC ACID

OSHA PEL: TWA 80 mg/M^3/%SiO$_2$

THR: The pure unaltered form is considered
nontoxic. Some deposits contain small amounts
of crystalline quartz which is therefore fibrogenic. When diatomaceous earth is calcined
(with or without fluxing agents) some silica

is converted to cristobalite and is therefore fibrogenic. Tridymite has never been detected in calcined diatomaceous earth. See also silica.

SILICA (CRYSTALLINE)   HR: 2
mf: SiO$_2$ mw: 60.09

PROP: mp: 1710°; bp: 2230°; d (amorphous): 2.2; d (crystalline): 2.6; vap press: 10 mm @ 1732°.

SYNS: AGATE * AMETHYST * CHALCEDONY * CHERTS * FLINT * GLASS * ONYX * PURE QUARTZ * ROSE QUARTZ * SAND * SILICON DIOXIDE * TRIDYMITE * SILICA FLOUR * CRISTOBALITE

THR: Moderately toxic as an acute irritating dust. From the point of view of numbers of men exposed and cases of disability produced, silica is the chief cause of pulmonary dust disease. The prolonged inhalation of dusts containing free silica may result in the development of a disabling pulmonary fibrosis known as silicosis. The Committee on Pneumoconiosis of the American Public Health Association defines silicosis as "a disease due to the breathing of air containing silica (SiO$_2$), characterized by generalized fibrotic changes and the development of miliary nodules in both lungs, and clinically by shortness of breath, decreased chest expansion, lessened capacity for work, absence of fever, increased susceptibility to tuberculosis (some or all of which symptoms may be present), and characteristic x-ray findings."

Silica occurs in the pure state in nature as highly fibrogenic quartz. It is the main constituent of relatively much less toxic sand, sandstone, tripoli and diatomaceous earth. It is present in crystalline form in high amounts (up to 35%) in granite. Exposure to silica occurs in hard rock mining, in foundries, in manufacture of porcelain and pottery, in the spraying of vitreous enamels, in sandblasting, in granite-cutting and tombstone-making, in the manufacture of silica firebrick and other refractories, in grinding and polishing operations where natural abrasive wheels are used and other occupations.

The duration of exposure which is associated with the development of silicosis varies widely for different occupations. Thus, the average duration of exposure required for the development of silicosis in sand-blasters is 2-10 years, in moulders and granite cutters, about 30 years, and in hard rock miners 10-15 years. There is also much variation in individual susceptibility, certain workers showing radiological evidence of the disease years before their fellow workmen who are similarly exposed. Such susceptible individuals are fortunately rather rare.

The action of crystalline silica on the lungs results in the production of a diffuse, nodular fibrosis in which the parenchyma and the lymphatic system are involved. This fibrosis is, to a certain extent, progressive, and may continue to increase for several years after exposure is terminated. Where the pulmonary reserve is sufficiently reduced, the worker complains of shortness of breath on exertion. This is the first and most common symptom in cases of uncomplicated silicosis. If severe, it may incapacitate the worker for heavy, or even light, physical exertion, and in extreme cases there may be shortness of breath even while at rest. The most common physical sign of silicosis is a limitation of expansion of the chest. There may be a dry cough, sometimes very troublesome. The characteristic radiographic appearance is one of diffuse, discrete nodulation, scattered throughout both lung fields. Where the disease advances, the shortness of breath becomes worse, and the cough more productive and troublesome. There is no fever or other evidence of systemic reaction. Further progress of the disease results in marked fatigue, extreme dyspnea and cyanosis, loss of appetite, pleuritic pain and total incapacity to work. If tuberculosis does not supervene, the condition may eventually cause death either from cardiac failure or from destruction of lung tissue, with resultant anoxemia. In the later stages, the x-ray may show large conglomerate shadows, due to the coalescence of the silicotic nodules, with areas of emphysema between them.

Silica in some forms is used as a food additive permitted in the feed and drinking water of animals and/or for the treatment of food-producing animals. It is also permitted in food for human consumption. It is a common air contaminant. Reacts violently with ClF$_3$; MnF$_3$; OF$_2$.

SILICA, CRYSTALLINE-CRISTOBALITE   HR: 3
CAS: 14464-46-1   NIOSH: VV 7325000
mf: O$_2$Si   mw: 60.09

PROP: White cubic-system crystals formed from QUARTZ at temperatures above 1470°C.

SYNS: CALCINED DIATOMITE * CRISTOBALITE

OSHA PEL: TWA 5 mg/m^3/(%SiO$_2$+2) (Respirable)

THR: Poison by intratracheal route. An experimental carcinogen. Human pulmonary system effects. See also silica, but about twice as toxic in causing silicosis.

SILICA, CRYSTALLINE-QUARTZ HR: 3
CAS: 14808-60-7   NIOSH: VV 7330000
mf: O$_2$Si   mw: 60.09

SYNS: AGATE * AMETHYST * CHALCEDONY * CHERTS * FIBERGLASS * FIBROUS GLASS * FLINT * GLASS * ONYX * PURE QUARTZ * QUARTZ * QUAZO PURO (ITALIAN) * ROSE QUARTZ * SAND * SILICA FLOUR (POWDERED CRYSTALLINE SILICA) * SILICIC ANHYDRIDE

OSHA PEL: TWA 10 mg/m^3/%SiO$_2$+2 (Respirable)

THR: Poison to humans by inhalation. Poison by intratracheal and intravenous routes. An experimental carcinogen, tumorigen, and neoplastigen. Human pulmonary system effects. See also silica. Incompatible with OF$_2$; vinylacetate.

SILICA, CRYSTALLINE-TRIDY-MITE   HR: 3
CAS: 15468-32-3   NIOSH: VV 7335000
mf: O$_2$Si   mw: 60.09

PROP: White or colorless platelets or orthorhombic (cryst) formed from quartz @ temperatures >870°.

SYNS: TRIDIMITE (FRENCH) * TRIDYMITE

OSHA PEL: TWA 5 mg/m^3/(%SiO$_2$+2) (Respirable)

THR: Poison by intratracheal route. An experimental tumorigen. Human pulmonary system effects. See also silica. About twice as toxic as silica in causing silicosis.

SILICA FLOUR   HR: 3
PROP: A finely ground *crystalline* silica sometimes marketed as "Amorphous." It is not amorphous. It has shown a very high incidence of silicosis among "silica flour" workers.

THR: Poison by inhalation. See also silica.

SILICA, PRECIPITATED (AMOR-PHOUS)   HR: 3
CAS: 7699-41-4   NIOSH: VV 8850000
mf: H$_2$O$_3$Si   mw: 78.11

SYNS: KIESELSAURE (GERMAN) * METASILICIC ACID * PRECIPITATED SILICA * SILICA GEL * SILICA HYDROUS GEL

THR: Poison by intravenous route. An eye irritant. See also silica and silicates.

SILICATES   HR: 3
THR: Soluble alkaline silicates act locally like mild alkalies. The dust of certain silicates, such as asbestos (hydrated magnesium silicate) and talc, can produce fibrotic changes in the lungs and are implicated as experimental carcinogens. React violently with Li. See also specific silicates.

SILICIC ACID, BERYLLIUM SALT   HR: 3
CAS: 15191-85-2   NIOSH: VV 8925000
mf: O$_4$Si • 2Be   mw: 110.11

PROP: Colorless crystal. D: 3.0.

SYNS: BERYLLIUM SILICATE * BERYLLIUM SILICIC ACID * BERYLLIUM ORTHOSILICATE * PHENACITE

OSHA PEL: TWA 2 ug/m^3; CL 5; Pk 25/30M/8H

THR: An experimental tumorigen. A suspected human carcinogen. See also beryllium compounds and silicates.

SILICOFLUORIC ACID   HR: 3
CAS: 16961-83-4   NIOSH: VV 8225000
mf: F$_6$Si • 2H   mw: 144.11

PROP: Transparent, colorless, fuming liquid. Bp: decomp.

SYNS: ACIDE FLUOROSILICIQUE (FRENCH) * ACIDE FLUOSILICIQUE (FRENCH) * ACIDO FLUOSILICICO (ITALIAN) * FLUOROSILICIC ACID * FLUOSILICIC ACID * HEXAFLUOROKIESELSAIURE (GERMAN) * HEXAFLUOROKIEZELZUUR (DUTCH) * HEXAFLUOSILICIC ACID * HYDROFLUOSILICIC ACID * HYDROGEN HEXAFLUOROSILICATE * HYDROSILICOFLUORIC ACID * KIEZELFLUORWATERSTOFZUUR (DUTCH) * SAND ACID * HEXAFLUOROSILICATE(2-) DIHYDROGEN

DOT Classification: Label: Corrosive

THR: A poison by subcutaneous route; an irritant to skin, eyes and mucous membranes and by inhalation. See also fluorides. Dangerous; when heated to decomposition it emits highly toxic and corrosive fumes of fluorides; will react with water or steam to produce toxic and corrosive fumes. When heated to decomposition it emits toxic fumes of F^-.

SILICON
CAS: 7440-21-3   NIOSH: WM 0400000
DOT: 1346
mf: Si   mw: 28.09

PROP: Cubic, steel-gray crystals or dark brown powder. Mp: 1420°, bp: 2600°, d: 2.42 or 2.3 @ 20°, vap press: 1 mm @ 1724°. Almost insol in water; sol in molten alkali oxides.
DOT Classification: Flammable Solid

THR: Does not occur free in nature, but is found as silicon dioxide (silica), and as various silicates. See also silica and silicates. Flammable when exposed to flame or by chemical reaction with oxidizers. See also powdered metals. Violent reactions with alkali carbonates; oxidants; (Al + PbO); Ca; Cs_2C_2; Cl_2; CoF_2; F_2; IF_5; MnF_3; Rb_2C_2; FNO; AgF; NaK alloy. Dangerous; when heated, will react with water or steam to produce H_2; can react with oxidizing materials.

SILICON CARBIDE   HR: 1
CAS: 409-21-2   NIOSH: VW 0450000
mf: CSi   mw: 40.10

PROP: Bluish-black, iridescent crystals. Mp: 2600°; bp: subl > 2000°; decomp @ 2210°; d: 3.17.

SYNS: CARBON SILICIDE * CARBORUNDUM * SILICON MONOCARBIDE

THR: A mild irritant by inhalation..

SILICON CHLORIDE   HR: 2
CAS: 10026-04-7   NIOSH: VW 0525000
DOT: 1818
mf: Cl_4Si mw: 169.89

PROP: Colorless, clear, mobile fuming liquid; suffocating odor. D: 1.52 @ 0°/4°; mp: −70°; bp: 59°. Misc with benzene, ether, chloroform, petroleum ether.

SYNS: CHLORID KREMICITY (CZECH) * SILICIO(TETRACLORURO DI) * SILICIUMTETRACHLORID (GERMAN) * SILICIUMTETRACHLORIDE (DUTCH) * SILICIUM(TETRACHLORURE DE) (FRENCH) * SILICON TETRACHLORIDE * TETRACHLOROSILANE * TETRACHLORURE DE SILICIUM (FRENCH)

DOT Classification: Label: Corrosive

THR: Moderately toxic by inhalation. Severe eye and skin irritant. Decomposed by water with much heat into silicic acid and HCl. Dangerous; when heated to decomposition it emits highly toxic fumes of HCl; will react with water or steam to produce heat and toxic and corrosive fumes. Incompatible with dimethyl sulfoxide; K; Na.

SILICONES   HR: 1
SYN: SILOXANES

THR: Generally slight toxicity. Most of the silicones that have been studied should have low toxicity or none at all and little or no irritating effects. May be spontaneously flammable in air. There can be toxicity due to contamination of silicones by components of manufacture.

SILICON FLUORIDE   HR: 3
CAS: 7783-61-1   NIOSH: VW 2327000
DOT: 1859
mf: F_4Si mw: 104.09

PROP: Colorless gas, very pungent odor; mp: −77°; bp: −65° @ 181 mm; d: 4.67.
DOT Classification: Label: Nonflammable Gas

THR: No toxicity data. See also fluorides and hydrofluoric acid. Very irritating to skin, eyes, and mucous membranes. When heated to decomposition it emits toxic fumes of F^-.

SILK   HR: 3
NIOSH: VW 2700000

THR: An experimental carcinogen. In the form of dust it is an allergen and a nuisance dust. Flammable and a moderate explosion hazard. When heated to decomposition it emits acrid smoke and fumes.

SILVER   HR: 3
CAS: 7440-22-4   NIOSH: VW 3500000
af: Ag   aw: 107.87

PROP: Soft, ductile, malleable, lustrous, white metal. Mp: 961.93°, bp: 2212°, d: 10.50 @ 20°.

SYNS: ARGENTUM * C.I. 77820 * SHELL SILVER * SILBER (GERMAN) * SILVER ATOM

OSHA PEL: TWA 10 ug/m^3
ACGIH TLV: TWA (metal) 0.1 mg/m^3, (soluble compounds as Ag) 0.01 mg/m^3
DFG MAK: 0.01 mg/m^3

THR: An experimental tumorigen. Human systemic skin effects. See also silver compounds. Flammable in the form of dust, when exposed to flame or by chemical reaction with C_2H_2; NH_3; bromoazide; ClF_3; ethylene imine; H_2O_2; oxalic acid; H_2SO_4; tartaric acid. See also powdered metals. For further information see Vol. 1, No. 1 of *DPIM Report*.

SILVER, COLLOIDAL   HR: 3
CAS: 7440-22-4   NIOSH: VW 3675000
af: Ag   aw: 107.87

SYNS: ARGENTIUM CREDE * COLLARGOL

OSHA PEL: TWA 10 ug/m^3

THR: Poison by ingestion and intravenous route. See also silver and silver compounds.

SILVER COMPOUNDS

The absorption of silver compounds into the circulation and the subsequent deposition of the reduced silver in various tissues of the body may result in the production of a generalized greyish pigmentation of the skin and mucous membranes; a condition known as argyria. The introduction of fine particles of silver through breaks in the skin produces a local pigmentation at the site of the injury. 1 mg/m^3 of silver dust causes skin effects. The condition develops slowly, usually after 2-25 years of exposure. Pigmentation is noticeable first in conjunctivae, and later in the mucous membranes of the mouth and gums and in the skin. There are no constitutional symptoms or physical disability. Persons exhibiting the condition, and who subsequently died from unrelated disease, showed, on autopsy, a deposition of silver in the blood vessel walls, kidneys, testes, pituitary, choroid plexus, and mucous membranes of the nose, maxillary antra, trachea and bronchi. Once deposited, there is no known method by which the silver can be eliminated; the pigmentation is permanent. These compounds may be irritating to the skin and mucous membranes. See also silver. For further information see Vol. 1, No. 1 of *DPIM Report*.

SILVER CYANIDE   HR: 3
CAS: 506-64-9   NIOSH: VW 3850000
DOT: 1684
mf: CAgN   mw: 133.89

PROP: White, odorless, tasteless powder which darkens upon exposure to light; mp: 320° (decomp); d: 3.95.

SYNS: CYANURE D'ARGENT (FRENCH) * KYANID STRIBRNY (CZECH)

DOT Classification: Poison B, Label: Poison

THR: A deadly poison by ingestion. A skin and eye irritant. See also silver compounds and cyanides. When heated to decomposition it emits very toxic fumes of CN^- and NO_x. Incompatible with phosphorus tricyanide, fluorine.

SILVER (II) FLUORIDE   HR: 3
CAS: 7783-95-1   NIOSH: VW 4200000
mf: AgF$_2$ mw: 145.87

PROP: White when pure; usually a grey-black or brownish solid. D: 4.7; mp: 690°.

SYNS: ARGENT FLUORURE (FRENCH) * ARGENTIC FLUORIDE * SILVER DIFLUORIDE

THR: Poison by subcutaneous route. See also fluorides and silver compounds. Powerful oxidizing agent. When heated to decomposition it emits toxic fumes of F^-.

SILVER (I) NITRATE (1:1)   HR: 3
CAS: 7761-88-8   NIOSH: VW 4725000
DOT: 1493
mf: NO$_3$ • Ag   mw: 169.88

PROP: mp: 212°; bp: 444° (decomp); d: 4.352 @ 19°. Very sol in ammonia water; slightly sol in ether.

SYNS: LUNAR CAUSTIC * NITRATE D'ARGENT (FRENCH) * NITRIC ACID, SILVER(1+) SALT * SILBERNITRAT

OSHA PEL: TWA 10 ug(Ag)/m^3
DOT Classification: Label: Oxidizer

THR: Poison by ingestion, intravenous, subcutaneous, intraperitoneal, and other unknown routes. Mutagenic data. An experimental carcinogen. See also silver compounds and nitrates.

A powerful caustic and irritating to skin and mucous membranes. Swallowing can cause severe gastroenteritis that may be fatal. Incompatible with acetylene; acetylides; alkalies; aluminum; antimony salts; arsenic; arsenites; bromides; carbon; carbonates; chlorides; ClF_3; chlorosulfuric acid; copper; creosote; ethanol; ferrous salts; hypophosphites; iodides; Mg powder with H_2O; morphine salts; NH_3 with KOH to yield black Ag_3N; oils; PH_3; phosphates; phosphonium iodide; phosphorous; plastics; sulfur; tannic acid; tartrates; thiocyanates; vegetable decoctions and extracts; zinc with NH_3 with KOH. When heated to decomposition it emits toxic fumes of NO_x. For further information see Vol. 1, No. 1 of *DPIM Report*.

SLUDGE ACID (DOT)　　　　　HR: 2
NIOSH: AR 2310000
DOT: 1906
DOT Classification: Label: Corrosive

THR: No toxicity data. A corrosive material; irritating to skin, eyes, mucous membranes. See also acids, nitric and sulfuric. When heated to decomposition it emits very toxic fumes of SO_x and NO_x.

SMOG　　　　　　　　　　　HR: 3
PROP: An atmospheric combination of smoke, fog, and industrial gases. Composition: Contents vary, but sulfur dioxide is a common component; others are sulfides, fluorides, chlorides, carbon particles, oxides of nitrogen and various hydrocarbons may be found in smog.

THR: Moderately irritating to eyes and mucous membranes. Chronic effects are presently under study. A common air contaminant. Possibly carcinogenic.

SMOKE CONDENSATE, CIGA-RETTE　　　　　　　　　　　　HR: 3
NIOSH: VX 6615000

SYN: CIGARETTE SMOKE CONDENSATE

THR: An experimental neoplastigen and carcinogen. Mutagenic data.

SODIUM　　　　　　　　　　HR: 3
CAS: 7440-23-5　　　　NIOSH: VY 0686000
DOT: 1428/1429
af: Na　　　aw: 22.99

PROP: Light, soft, ductile, malleable, silver-white metal. Mp: 97.81°, bp: 881.4°, d: 0.9710 @ 20°, autoign temp: > 115° in dry air, vap press: 1.2 mm @ 400°.

SYNS: NATRIUM ＊ SODIUM METAL (DOT)

DOT Classification: Label: Flammable Solid and Dangerous When Wet

THR: Sodium in elemental form is highly reactive, particularly with moisture, with which it reacts violently and therefore attacks living tissue. Also, Na with HOH yields NaOH. See also sodium hydroxide. Metallic sodium reacts exothermally with the moisture of body or tissue surfaces, causing thermal and chemical burns due to the reaction with sodium and the sodium hydroxide formed. Dangerous fire hazard when exposed to heat and moisture. In dry air it reacts very slowly up to 550° or by chemical reaction with moisture; air; $AlBr_3$; $AlCl_3$; AlF_3; NH_4 chlorocuprate; NH_4NO_3; $SbBr_3$; $SbCl_3$; SbI_3; $AsCl_3$; AsI_3; $BiBr_3$; $BiCl_3$; BiI_3; Bi_2O_3; BBr_3; bromoazide; CO_2; CO + NH_3; CCl_4; Cl_2; ClF_3; $CrCl_4$; CrO_3; CoBr; CoCl; $CuCl_2$; CuO; $FeBr_3$; $FeCl_3$; $FeBr_2$; $FeCl_2$; FeI_2; hydrazine hydrate; H_2O_2; H_2S; HCl; HF; F_2; 1,2-dichloroethylene; dichloromethane; Br_2; hydroxylamine; iodine; iodine monochloride; iodine pentafluoride; lead oxide; maleic anhydride; manganous chloride; mercuric bromide; mercuric chloride; mercuric fluoride; mercuric iodide; mercurous chloride; mercurous oxide; methyl chloride; molybdenum trioxide; monoammonium phosphate; nitric acid; nitrogen peroxide; nitrosyl fluoride; nitrous oxide; phosgene; phosphorus; phosphorous pentafluoride; phosphorus pentoxide; phosphorus tribromide; phosphorus trichloride; phosphoryl chloride; potassium oxides; potassium ozonide; potassium superoxide; selenium; silicon tetrachloride; silver bromide; silver chloride; silver fluoride; silver iodide; sodium peroxide; stannic chloride; stannic iodide with sulfur; stannic oxide; stannous chloride; sulfur; sulfur dibromide; sulfur dichloride; sulfur dioxide; sulfuric acid; tellurium; tetrachloroethane; thallous bromide; thiophosphoryl bromide; trichlorethylene; vanadium pentachloride; vanadyl chloride; zinc bromide; or any oxidizing material; decomposes moisture to evolve hydrogen and heat; reacts exothermally with the halogens; acids; halogenated hydrocarbons. Heated sodium is spontaneously flammable in air. Can be safely stored under liquid hydrocarbons. Dangerous explosion hazard when exposed to moisture in any form! Keep dry at all times! Dangerous; when heated in air, emits toxic fumes of sodium oxide; will react with water or steam to produce heat, hydrogen, and flammable vapors; can react vigorously to explosively with oxidizing materials. See hydrogen. To fight fire, use soda ash, dry sodium chloride or graphite, in order of preference. For further information see Vol. 1, No. 8 of *DPIM Report*.

SODIUM (SOLUTION)   HR: 3
CAS: 7440-23-5   NIOSH: VY 0690000

PROP: SODIUM DISPERSIONS. Finely divided metallic sodium suspended in toluene, xylene, naphtha, kerosene, etc.

SYN: SODIUM, METAL DISPERSION IN ORGANIC SOLVENT

DOT Classification: Label: Flammable solid and Dangerous When Wet

THR: Poison. See sodium and individual dispersant. Dangerous fire hazard when exposed to heat or flame or by chemical reaction. These are very reactive forms of sodium, which if carelessly handled may catch fire. To extinguish, see sodium. After sodium has been extinguished, the burning organic vapor can be dealt with by very cautious use of a carbon dioxide extinguisher. Do not use carbon tetrachloride. Moderate explosion hazard by chemical reaction. See also sodium. Dangerous; when heated, it loses the solvent and emits highly toxic fumes of sodium, sodium oxide, etc.; will react with water or steam to produce heat and hydrogen; on contact with oxidizing materials, can react vigorously, and on contact with acid or acid fumes, can emit toxic fumes.

SODIUM AMIDE
CAS: 7782-92-5   NIOSH: VY 2775000
DOT: 1425
mf: H_2NNa mw: 39.02

PROP: White crystalline powder. Mp: 210°, bp: 400°.

SYNS: SODIUM AMIDE (DOT) * SODAMIDE

DOT Classification: Label: Flammable Solid and Dangerous When Wet

THR: See sodium hydroxide and ammonia, both of which are liberated by this material in the presence of moisture. An intense irritant to tissue, skin, and eyes. Flammable by chemical reaction. See also ammonia. Moderate explo-

sion hazard when exposed to moisture or by chemical reaction with CrO_3; $KClO_3$; 1,1-diethoxy-2-chloroethane; air. Can become explosive in storage. Dangerous; when heated to decomposition it emits highly toxic fumes of ammonia and sodium oxide; will react with water or steam to produce heat and toxic and corrosive fumes; can react with oxidizing materials.

SODIUM ARSENITE   HR: 3
CAS: 7784-46-5   NIOSH: CG 3675000
mf: $AsO_2 \cdot Na$ mw: 129.91

PROP: White or grayish white powder. Commercially: 95%-98% pure. Very sol in water, sltly sol in alc.

SYNS: ARSENENOUS ACID, SODIUM SALT * ARSENITE DE SODIUM (FRENCH) * ARSENIOUS ACID, SODIUM SALT * SODIUM METAARSENITE

OSHA PEL: TWA 500 ug(As)/m^3

THR: A deadly poison by ingestion, skin contact, intravenous, and intraperitoneal routes. Mutagenic data. A human carcinogen. See also arsenic compounds. When heated to decomposition it emits toxic fumes of As and Na_2O.

SODIUM ARSENITE (LIQUID)   HR: 3
CAS: 7784-46-5   NIOSH: VY 7705000
DOT Classification: Poison B, Label: Poison

THR: A deadly poison. See also sodium arsenite. When heated to decomposition it emits toxic fumes of As and Na_2O.

SODIUM AZIDE   HR: 3
CAS: 26628-22-8   NIOSH: VY 8050000
DOT: 1687
mf: N_3Na mw: 65.02

PROP: Colorless hexagonal crystals; mp: decomp, d: 1.846. Insol in ether; sol in liquid ammonia.

SYNS: AZOTURE DE SODIUM (FRENCH) * NATRIUMAZID (GERMAN) * NATRIUMMAZIDE (DUTCH) * NCI-C06462 * NSC 3072 * SODIUM, AZOTURE DE (FRENCH) * SODIUM, AZOTURO DI (ITALIAN)

ACGIH TLV: TWA ceiling limit 0.1 ppm
DFG MAK: 0.07 ppm (0.2 mg/m^3)
DOT Classification: Poison B, Label: Poison

THR: Poison by ingestion, skin contact, intraperitoneal, intravenous, subcutaneous, and other unspecified routes. Mutagenic data. Human central nervous system effects by ingestion. See azides and sodium hydroxide. Unstable, explosive. Violent reaction with benzoyl chloride combined with KOH; Br_2; CS_2; $Cr(OCl)_2$; Cu; Pb; HNO_3; $BaCO_3$; H_2SO_4; water; $(CH_3)_2SO_4$; dibromomalononitrile. When heated to decomposition it emits very toxic fumes of NO_x; explosive. For further information see Vol. 2, No. 6 of *DPIM Report*.

SODIUM BISULFITE (1:1)   HR: 3
CAS: 7631-90-5   NIOSH: VZ 2000000
DOT: 2693
mf: $HO_3S \cdot Na$ mw: 104.06

PROP: White, crystalline powder, odor of SO_2, disagreeable taste. D: 1.48; mp: decomp. Very sol in hot or cold water, sltly sol in alc.

SYNS: BISULFITE DE SODIUM (FRENCH) * SODIUM ACID SULFITE * SODIUM BISULPHITE * SODIUM HYDROGEN SULFITE * SULFUROUS ACID, MONOSODIUM SALT

ACGIH TLV: TWA 5 mg/m^3
DOT Classification: ORM-B, Label: None

THR: Poison by intravenous and intraperitoneal routes. Mutagenic data. An allergen. See also sulfurous acid. Concentrated solutions are irritating to skin and mucous membranes. When heated to decomposition it emits toxic fumes of SO_x and Na_2O.

SODIUM BOROHYDRIDE   HR: 3
CAS: 16940-66-2   NIOSH: ED 3325000
mf: $BH_4 \cdot Na$ mw: 37.84

PROP: White to gray-white microcrystalline powder or lumps. Reacts with hot water, sol in liquid ammonia and ''Cellosolve'' ether. Mp: 36°, d: 1.07.

SYNS: BOROHYDRURE DE SODIUM (FRENCH) * SODIUM TETRAHYDROBORATE(1-)

THR: Poison by ingestion. Ignites in air. See boron compounds and hydrides. Flammable when exposed to heat or flame or by chemical reaction with oxidizers. Dangerous; when heated to decomposition it emits toxic fumes; will react with water or steam to produce hydrogen; on contact with acid fumes, it can emit flammable vapors.

SODIUM CHLORATE   HR: 3
CAS: 7775-09-9   NIOSH: FO 0525000
mf: $ClO_3 \cdot Na$ mw: 106.44

PROP: Colorless, odorless crystals, cooling saline taste. Mp: 248°-261°; bp: decomp: d: 2.490 @ 15°.

SYNS: CHLORATE OF SODA (DOT) * DROP LEAF * NATRIUM CHLORAAT (DUTCH) * NATRIUM CHLORAT (GERMAN) * SODIO (CLORATO DI) (ITALIAN) * SODIUM (CHLORATE DE) (FRENCH)

THR: Poison to humans by an unspecified route. Moderately toxic by ingestion and intraperitoneal routes in animals. A skin, mucous membrane, and eye irritant. Mutagenic data. Damages the red blood corpuscles of humans. An herbicide. Can react violently with Al; $NH_4S_2O_3$; Sb_2S_3; As; As_2O_3; C; charcoal; MnO_2; P; KCN; S; H_2SO_4; thiocyanates; Zn. Can also react violently with nitrobenzene; paper; metal sulfides; dibasic organic acids; organic matter; static electricity. When heated to decomposition it emits toxic fumes of Cl^-, Na_2O.

SODIUM CHLORIDE   HR: 2
CAS: 7647-14-5   NIOSH: VZ 4725000
mf: ClNa   mw: 58.44

PROP: Colorless, transparent crystals or white crystalline powder. Mp: 801°, bp: 1413°, d: 2.165, vap press: 1 mm @ 865°.

SYNS: COMMON SALT * HALITE * NATRIUM-CHLORID (GERMAN) * ROCK SALT * SALINE * SALT * SEA SALT * TABLE SALT

THR: Moderately toxic by ingestion, intravenous, intraperitoneal, and subcutaneous routes. A skin and severe eye irritant. Human blood pressure effects. When bulk sodium chloride is heated to high temperature, a vapor is emitted which is irritating, particularly to the eyes. Ingestion of large amounts of sodium chloride can cause irritation of the stomach. Improper use of salt tablets may produce this effect. Violent reaction with BrF_3; Li. When heated to decomposition it emits toxic fumes of Cl^- and Na_2O. For further information see Vol. 1, No. 5 of *DPIM Report*.

SODIUM CHLORITE   HR: 3
CAS: 7758-19-2   NIOSH: VZ 4800000
DOT: 1496
mf: $ClNaO_2$ mw: 90.44

PROP: White crystals or crystalline powder. Bp: decomp @ 180°-200°

DOT Classification: Label: Oxidizer

THR: An experimental teratogen. May act as an irritant due to its oxidizing power. A powerful oxidizing agent; ignited by friction, heat or shock. Dangerous explosion hazard from exposure to percussion; acids; organic matter; oxalic acid; P; S; sodium dithionate. Dangerous; when heated it emits highly toxic fumes of Cl^- and Na_2O and may explode; can react vigorously on contact with reducing materials.

SODIUM COMPOUNDS

THR: Variable toxicity. Sodium ion as such is practically nontoxic. The toxicity of sodium compounds is frequently, though not always, due to the anion involved. The hydroxide is very corrosive, being strongly basic. Even here it is the concentration of hydroxyl ion which is responsible for the caustic action of this material.

SODIUM CYANIDE   HR: 3
CAS: 143-33-9   NIOSH: VZ 7525000
DOT: 1689
mf: CNNa   mw: 49.01

PROP: White, deliquescent crystalline powder. Mp: 563.7°; bp: 1496°; vap press: 1 mm @ 817°.

SYNS: CIANURO DI SODIO (ITALIAN) * CYANIDE OF SODIUM * CYANURE DE SODIUM (FRENCH) * HYDROCYANIC ACID, SODIUM SALT * KYANID SODNY (CZECH)

OSHA PEL: TWA 5 mg(CN)/m³ (skin)
DOT Classification: Poison B, Label: Poison

THR: Poison to humans by ingestion and other unspecified routes. Poison to animals by ingestion, intraperitoneal, subcutaneous, intravenous and parenteral routes. Violent reaction with nitrates and nitrites. See also cyanides. The volatile cyanides resemble hydrocyanic acid physiologically, inhibiting tissue oxidation and causing death through asphyxia. Cyanogen is probably as toxic as hydrocyanic acid; the nitriles are generally considered somewhat less toxic, probably because of their lower volatility. The nonvolatile cyanide salts appear to be relatively nonhazardous systemically, so long as they are not ingested and care is taken to prevent the formation of hydrocyanic acid. Workers, such as electroplaters and picklers, who are daily exposed to cyanide solutions may develop a "cyanide" rash, characterized by itching, and by macular, papular, and vesicular eruptions. Frequently there is secondary infection. Exposure to small amounts of cyanide compounds over long periods of time is reported to cause loss of appetite, headache, weakness, nausea, dizziness and symptoms of irritation of the upper respiratory tract and eyes. See also specific compounds. Flammable by chemical reaction with heat, moisture, acid. Many cyanides evolve hydrocyanic acid rather easily. This is a flammable gas and is highly toxic. Carbon dioxide from the air is sufficiently acidic to liberate hydrocyanic acid from cyanide solutions. See also hydrocyanic acid. Explodes if melted with nitrite or chlorate @ about 450°. Violent reaction with F_2; Mg; nitrates; HNO_3; nitrites. Dangerous; on contact with acid, acid fumes, water or steam, they will produce toxic and flammable vapors of CN^- and Na_2O. For further information see Vol. 3, No. 6 of *DPIM Report*.

SODIUM CYCLAMATE   HR: 3
CAS: 139-0-5-9   NIOSH: GV 7350000
mf: $C_6H_{12}NO_3$ • Na   mw: 169.18

PROP: White, crystalline powder, practically odorless, almost insol in alc, benzene, chloroform, ether; sol in water.

SYNS: CYCLAMATES * CYCLAMATE SODIUM * CYCLOHEXANESULFAMIC ACID, MONOSODIUM SALT * CYCLOHEXYL SULFAMATE SODIUM * SODIUM CYCLOHEXANESULFAMATE * SODIUM CYCLOHEXYL SULFAMIDATE * SODIUM SUCARYL * SUCARYL SODIUM

THR: An experimental carcinogen, neoplastigen, and teratogen. Mutagenic data. Moderately toxic by intravenous and intraperitoneal routes. When heated to decomposition it emits very toxic fumes of Na_2O and NO_x.

SODIUM DIBUTYLDITHIOCAR-
BAMATE   HR: 3
CAS: 136-30-1   NIOSH: EZ 3880000
mf: $C_9H_{18}NS_2$ • Na   mw: 227.39

SYNS: BUTYL NAMATE * DIBUTYLDITHIOCARBAMIC ACID SODIUM SALT * TEPIDONE RUBBER ACCELERATOR * USAF B-35

THR: Poison by intraperitoneal route. When heated to decomposition it emits very toxic fumes of NO_x, SO_x, and Na_2O.

SODIUM DICHLOROACETATE   HR: 2
CAS: 2156-56-1   NIOSH: AG 9275000
mf: $C_2HCl_2O_2 \cdot Na$ mw: 150.92

SYNS: DICHLOROACETATE SODIUM SALT
* DICHLOROACETIC ACID SODIUM SALT
* DICHLOROCTAN SODNY (CZECH)

THR: Mutagenic data. Moderately toxic by intravenous route. When heated to decomposition it emits toxic fumes of Cl^- and Na_2O.

SODIUM DICHLOROCYAN-
URATE   HR: 3
CAS: 2893-78-9   NIOSH: XZ 1900000
DOT: 2465
mf: $C_3HCl_2N_3O_3 \cdot Na$ mw: 220.96

PROP: White crystals; water sol, chlorine odor. Mp: 230°-250°.

SYNS: DICHLOROISOCYANURIC ACID SODIUM SALT * SODIUM DICHLORISOCYANURATE
* SODIUM DICHLOROISOCYANURATE * 1-SODIUM-3,5-DICHLORO-S-TRIAZINE-2,4,6-TRIONE
* SODIUM SALT OF DICHLORO-S-TRIAZINETRIONE

DOT Classification: Label: Oxidizer

THR: Moderately toxic by ingestion. A skin and eye irritant. The main toxic effects were associated with gastrointestinal irritation, including salivation, lachrymation, dyspnoea, weakness, emaciation, lethargy, diarrhea, coma and (following very high dosage) deaths after 1-8 days, showing irritation of stomach and gastrointestinal tract, liver dysfunction and lung congestion. The concentrated material may be a little more toxic, due to greater gastrointestinal irritation. In the dry form, it is not appreciably irritating to dry skin. However, when moist, the concentrated material is irritating to skin, and also may cause severe eye irritation. When heated to decomposition it emits very toxic fumes of Cl^-, NO_x, CN^-, and Na_2O.

SODIUM-2,4-DICHLOROPHENOXYACE-
TATE   HR: 3
CAS: 2702-72-9   NIOSH: AG 8925000
mf: $C_8H_5Cl_2O_3 \cdot Na$ mw: 243.02

SYNS: 2,4-DICHLOROPHENOXYACETIC ACID, SODIUM SALT * SODIUM 2,4-D * SPRITZ-HORMIT

THR: Poison by ingestion, intraperitoneal, subcutaneous, and intravenous routes. Human central nervous system effects by inhalation. When heated to decomposition it emits toxic fumes of Cl^- and Na_2O.

SODIUM DICHROMATE   HR: 3
CAS: 10588-0-1-9   NIOSH: HX 7700000
DOT: 1479
mf: $Cr_2O_7 \cdot 2Na$ mw: 261.98

PROP: Anhydrous; mp: 356.7°; decomp @ about 400°. Very sol in water. D: 2.35 @ 13°.

SYNS: BICHROMATE OF SODA * DISODIUM CHROMIC ACID * CHROMIUM SODIUM OXIDE
* DISODIUM DICHROMATE * NATRIUMBICHRO-MAAT (DUTCH) * SODIO (DICROMATO DI) (ITALIAN) * SODIUM DICHROMATE DE (FRENCH)

OSHA PEL: CL 100 ug/(CrO_3)/m^3
DOT Classification: ORM-A, Label: None

THR: Poison by ingestion, skin contact, intravenous, intraperitoneal, and subcutaneous routes. Mutagenic data. An experimental carcinogen. See also chromium compounds. A caustic and irritant. A powerful oxidizer. When heated to decomposition it emits toxic fumes of Na_2O. Incompatible with hydrazine; acetic anhydride; ethanol; sulfuric acid; hydroxylamine; trinitrotoluene. For further information see Vol. 3, No. 6 of DPIM Report.

SODIUM DIETHYLDITHIOCAR-
BAMATE   HR: 3
CAS: 148-18-5   NIOSH: EZ 6475000
mf: $C_5H_{10}NS_2 \cdot Na$ mw: 171.27

PROP: Crystals; mp: 95°; d: 1.1 @ 20°/20°; vap d: 5.9.

SYNS: DIETHYLCARBAMODITHIOIC ACID, SODIUM SALT * DIETHYLDITHIOCARBAMATE SODIUM
* DIETHYLDITHIOCARBAMIC ACID SODIUM
* DIETHYL SODIUM DITHIOCARBAMATE
* NCI CO2835 * SODIUM-N,N-DIETHYLDITHIO-CARBAMATE * USAF EK-2596

THR: An experimental neoplastigen and carcinogen. Mutagenic data. Moderately toxic by ingestion and subcutaneous route. When heated to decomposition it emits very toxic fumes of NO_x, SO_x, and Na_2O.

SODIUM DISPERSIONS   HR: 3
PROP: Finely divided metallic sodium suspended in toluene, xylene, naphtha, kerosene, etc.

THR: Poison. See also sodium and individual dispersant. Dangerous, when exposed to heat or flame or by chemical reaction. These are very reactive forms of sodium, which if carelessly handled may catch fire. To extinguish, see sodium. After sodium has been extinguished, the burning organic vapor can be dealt with by very cautious use of a carbon dioxide extinguisher. Do not use carbon tetrachloride. Moderate explosion by chemical reaction. Dangerous, when heated, it loses the solvent and emits highly toxic fumes of sodium, sodium oxide, etc; will react with water or steam to produce heat and hydrogen; on contact with oxidizing materials, can react vigorously, and on contact with acid or acid fumes, can emit toxic fumes .

SODIUM FLUORIDE   HR: 3
CAS: 7681-49-4   NIOSH: WB 0350000
DOT: 1690
mf: FNa   mw: 41.99

PROP: Clear, lustrous crystals or white powder or balls. Mp: 993°, bp: 1700°, d: 2 @ 41°, vap press: 1 mm @ 1077°.

SYNS: FLUORID SODNY (CZECH) * FLUORURE DE SODIUM (FRENCH) * NCI-C55221 * ROACH SALT * SODIUM FLUORIDE, SOLID (DOT) * SODIUM FLUORURE (FRENCH)

OSHA PEL: TWA 2500 ug(F)/m^3
DOT Classification: ORM-B, Label: None

THR: Poison to humans by ingestion. Poison to animals by skin contact, ingestion, intraperitoneal, subcutaneous, and intramuscular routes. An eye irritant. An experimental teratogen. Human central nervous system effects. Mutagenic data. See also fluorides. It is very phytotoxic. When heated to decomposition it emits toxic fumes of F$^-$ and Na$_2$O. For further information see Vol. 2, No. 1 of *DPIM Report*.

SODIUM FLUOROACETATE   HR: 3
CAS: 62-7-4-8   NIOSH: AH 9100000
DOT: 2629
mf: C$_2$H$_2$FO$_2$ • Na   mw: 100.03

PROP: Fine white powder. Sol in water.

SYNS: FLUOROACETIC ACID, SODIUM SALT * COMPOUND #1080 * FLUORESSIGAEURE (GERMAN) * FLUORACETATO DI (ITALIAN) * SODIUM MONOFLUOROACETATE * NATRIUM FLUORACETATE (DUTCH)

OSHA PEL: TWA 50 ug/m^3 (skin)

ACGIH TLV: TWA 0.05 mg/m^3 (skin)
DFG MAK: 0.05 mg/m^3
DOT Classification: Poison B

THR: Poison by ingestion, inhalation, skin contact, intraperitoneal, subcutaneous, and implant routes. A very highly toxic water-soluble salt used mainly as a rodenticide. It is rapidly absorbed by the gastrointestinal tract but slowly by skin unless the skin is abraided or cut. It operates by blocking the Krebs cycle by formation of fluorocitric acid, which inhibits aconitase. It has an effect on either or both the cardiovascular and nervous systems in all species and, in some species, the skeletal muscles. Man gives a mixed response with the cardiac feature predominating. By a direct action on the heart, contractile power is lost which leads to declining blood pressure. Ventricular premature contractions and arrhythmias are seen in all species including man. The central nervous system is directly attacked by sodium fluoroacetate. In man, the action on the central nervous system produces epileptiform convulsive seizures followed by severe depression. The dangerous dose for man is 0.5-2 mg/kg. Other species vary considerably in their response to this material with primates and birds being the most resistant and carnivora and rodents being the most susceptible. Most domestic animals show a susceptibility falling between the two extremes indicated above. When heated to decomposition it emits highly toxic fumes of Na$_2$O and F$^-$.

SODIUM HEXACYCLONATE   HR: 3
CAS: 7009-49-6   NIOSH: GU 6650000
mf: C$_9$H$_{15}$O$_3$ • Na   mw: 194.23

SYNS: 1-(HYDROXYMETHYL)CYCLOHEXANEACETIC ACID, SODIUM SALT * 1-(HYDROXYMETHYL)CYCLOHEXANEACETIC ACID SODIUM SALT * SODIUM-1-(HYDROXYMETHYL)CYCLOHEXANEACETATE * SODIUM-BETA,BETA-PENTAMETHYLENE-GAMMA-HYDROXYBUTYRATE

THR: A poison by ingestion, intraperitoneal, and intravenous routes. When heated to decomposition it emits acrid smoke and fumes.

SODIUM HEXAMETAPHOS-
PHATE   HR: 3
CAS: 10124-56-8   NIOSH: OY 3675000
mf: O$_{18}$P$_6$ • 6Na   mw: 611.76

PROP: White powder or flakes; water sol.

SYNS: CALGON * CHEMI-CHARL

THR: Poison by intravenous route. Moderately toxic by intraperitoneal and subcutaneous routes. See also phosphates. When heated to decomposition it emits toxic fumes of PO_x and Na_2O.

SODIUM HYDROSULFIDE   HR: 3
CAS: 16721-80-5   NIOSH: WE 1900000
mf: HNaS   mw: 56.06

SYNS: SODIUM BISULFIDE ∗ SODIUM HYDRO-GEN SULFIDE ∗ SODIUM MERCAPTAN ∗ SO-DIUM MERCAPTIDE ∗ SODIUM SULFHYDRATE

THR: Poison by intraperitoneal route. See also sulfides. Reacts violently with diazonium salts. Readily yields H_2S. When heated to decomposition it emits toxic fumes of SO_x and Na_2O.

SODIUM HYDROXIDE   HR: 3
CAS: 1310-73-2   NIOSH: WB 4900000
DOT: 1823
mf: HNaO   mw: 40.00

PROP: White deliquescent pieces, lumps or sticks. Mp: 318.4°, bp: 1390°, d: 2.120 @ 20°/4°, vap press: 1 mm @ 739°.

SYNS: CAUSTIC SODA ∗ CAUSTIC SODA, BEAD (DOT) ∗ CAUSTIC SODA, DRY (DOT) ∗ CAUS-TIC SODA, FLAKE (DOT) ∗ CAUSTIC SODA, GRANULAR (DOT) ∗ CAUSTIC SODA, SOLID (DOT) ∗ HYDROXYDE DE SODIUM (FRENCH) ∗ LYE ∗ NATRIUMHYDROXID (GERMAN) ∗ NATRIUMHYDROXYDE (DUTCH) ∗ SODA LYE ∗ SODIO(IDROSSIDO DI) (ITALIAN) ∗ SODIUM HYDRATE (DOT) ∗ SODIUM HYDROXIDE, BEAD (DOT) ∗ SODIUM HYDROXIDE, DRY (DOT) ∗ SODIUM HYDROXIDE, FLAKE (DOT) ∗ SO-DIUM HYDROXIDE, GRANULAR (DOT) ∗ SODIUM HYDROXIDE, SOLID (DOT) ∗ SODIUM(HYDROX-YDE DE) (FRENCH) ∗ WHITE CAUSTIC

OSHA PEL: TWA 2 mg/m^3
ACGIH TLV: CL 2 mg/m^3
DFG MAK: 2 mg/m^3
DOT Classification: Label: Corrosive

THR: Poison by ingestion and intraperitoneal routes. A skin and eye irritant. This material, both solid and in solution, has a markedly corro-sive action upon all body tissue causing burns and frequently deep ulceration, with ultimate scarring. Mists, vapors, and dusts of this com-pound cause small burns, and contact with the eyes rapidly causes severe damage to the delicate

tissue. Ingestion causes very serious damage to the mucous membranes or other tissues with which contact is made. It can cause perforation and scarring. Inhalation of the dust or concen-trated mist can cause damage to the upper respi-ratory tract and to lung tissue, depending upon the severity of the exposure. Thus, effects of inhalation may vary from mild irritation of the mucous membranes to a severe pneumonitis. Caution: Under the proper conditions of tem-perature, pressure, and state of division, it can react violently with acetic acid; acetaldehyde; acetic anhydride; acrolein; acrylonitrile; allyl alcohol; allyl chloride; Al; ClF_3; ($CHCl_3$ + CH_3OH); chlorohydrin; chloronitro-toluenes; chlorosulfonic acid; 1,2-dichloroethylene; eth-ylene cyanhydrin; glyoxal; HCl; HF; hydroqui-none; maleic anhydride; HNO_3; nitroethane; ni-tromethane; nitroparaffins; nitropropane; pentol; oleum; P; P_2O_5; β-propiolactone; H_2SO_4; (CH_3OH + tetrachloro-benzene); tetrahydrofu-ran; trichloroethylene; water; 4-chloro-2-meth-ylphenol; cinnamaldehyde; cyanogen azide; diborane; 4-methyl-2-nitrophenol; 3-methyl-2-penten-4-yn-1-ol; 1,2,4,5-tetrachlorobenzene; 1,1,1-trichloroethanol; trichloronitromethane; zinc; zirconium. Dangerous material to handle. For further information see Vol. 4, No. 3 of *DPIM Report*.

SODIUM HYDROXIDE (LIQUID)   HR: 3
CAS: 1310-73-2   NIOSH: WB 4905000
mf: HNaO   mw: 40.00

SYNS: CAUSTIC SODA SOLUTION ∗ LYE SOLU-TION ∗ SODA LYE ∗ SODIUM HYDRATE SOLU-TION ∗ SODIUM HYDROXIDE SOLUTION ∗ WHITE CAUSTIC SOLUTION

DOT Classification: Label: Corrosive

THR: Poison by intraperitoneal route. Moder-ately toxic by ingestion. A corrosive and irritant to skin, eyes, and mucous membranes. Muta-genic data.

SODIUM HYPOCHLORITE   HR: 3
CAS: 7681-52-9   NIOSH: NH 3486300
mf: ClHO•Na   mw: 75.45

PROP: Mp: decomp, bp: decomp.

SYNS: CARREL-DAKIN SOLUTION ∗ CHLOROX ∗ CLOROX ∗ DAKINS SOLUTION

THR: Corrosive and irritant by ingestion and inhalation. The anhydrous salt is highly explosive. Violent reaction with amines; ammonium acetate; $(NH_4)_2CO_3$; aziridine; methanol; phenyl acetonitrile; ammonium nitrate; ammonium oxalate; $(NH_4)_3PO_4$; cellulose; ethylene imine. Mutagenic data. For further information see Vol. 3, No. 6 of *DPIM Report*.

SODIUM IODIDE   HR: 2
CAS: 7681-82-5   NIOSH: WB 6475000
mf: INa   mw: 149.89

PROP: Cubic colorless crystals; mp: 651°, bp: 1300°, d: 3.667, vap press: 1 mm @ 767°.

SYNS: JODID SODNY * NATRIUMJODID (GERMAN) * SODIUM IODINE * SODIUM MONOIODIDE

THR: Moderately toxic by ingestion, intravenous, and intraperitoneal routes. A skin and eye irritant. See also iodides. Reacts violently with BrF_3; $HClO_4$; oxidants. When heated to decomposition it emits toxic fumes of I^- and Na_2O.

SODIUM ISOTHIOCYANATE   HR: 3
CAS: 540-72-7   NIOSH: XL 2275000
mf: CNS•Na   mw: 81.07

PROP: Colorless, deliquescent crystals or white powder; mp: 287°.

SYNS: NATRIUMRHODANID (GERMAN) * SODIUM RHODANATE * SODIUM RHODANIDE * SODIUM SULFOCYANATE * SODIUM SULFOCYANIDE * SODIUM THIOCYANATE * SODIUM THIOCYANIDE * USAF EK-T-434

THR: Poison by ingestion, intravenous, and subcutaneous routes. Moderately toxic by intraperitoneal route. Large doses taken internally cause vomiting, convulsions. Chronic poisoning is manifested by weakness, confusion, diarrhea, and skin rashes. When heated to decomposition it emits very toxic fumes of CN^-, NO_x, SO_x, and Na_2O.

SODIUM LUMINAL   HR: 3
CAS: 57-30-7   NIOSH: CQ 7000000
mf: $C_{12}H_{11}N_2O_3$•Na   mw: 254.24

PROP: White crystals.

SYNS: 2,4,6(1H,3H,5H)-PYRIMIDINETRIONE, 5-ETHYL-5-PHENYL-, MONOSODIUM SALT (9CI)
* SODIUM PHENYLETHYLMALONYLUREA
* 5-ETHYL-5-PHENYLBARBITURIC ACID SODIUM
* 5-ETHYL-5-PHENYLBARBITURIC ACID SODIUM

SALT * 5-ETHYL-5-PHENYL-2,4,6-(1H,3H,5H)PYRIMIDINETRIONE MONOSODIUM SALT * GARDENAL SODIUM * LUMINAL SODIUM * PHENOBARBITAL ELIXIR * PHENOBARBITAL SODIUM * PHENYLETHYLBARBITURIC ACID, SODIUM SALT * SODIUM-5-ETHYL-5-PHENYLBARBITURATE * SODIUM-5-ETHYL-5-PHENYLBARBITURATE * SODIUM-PHENOBARBITAL * SODIUM-PHENOBARBITONE * SODIUM-PHENYLETHYLBARBITURATE * SOLUBLE PHENOBARBITAL * SOLUBLE PHENOBARBITONE

THR: Poison by ingestion, intravenous, intraperitoneal and subcutaneous routes. An experimental teratogen, neoplastigen, and carcinogen. Central nervous system effects. When heated to decomposition it emits toxic fumes of NO_x and Na_2O.

SODIUM MERCAPTOMERIN   HR: 3
CAS: 21259-76-7   NIOSH: OV 8700000
mf: $C_{16}H_{27}HgNO_6S$•2Na   mw: 608.07

SYNS: DIUCARDYN SODIUM * DISODIUM-N-(3-(CARBOXYMETHYLTHIOMERCURI)-2-METHOXYPROPYL)-ALPHA- CAMPHORAMATE * THIOMERIN SODIUM

THR: Poison by subcutaneous, intramuscular, and intravenous routes. See also mercury. When heated to decomposition it emits very toxic fumes of Hg, NO_x, and SO_x.

SODIUM MERSALYL   HR: 3
CAS: 492-18-2   NIOSH: OV 7875000
mf: $C_{13}H_{17}HgNO_6$•Na   mw: 506.89

SYNS: MERCURY-3-(ALPHA-CARBOXY-O-ANISAMIDO)-2-METHOXYPROPYL HYDROXY-, MONOSODIUM SALT * O-((3-HYDROXYMERCURI-2-METHOXYPROPYL)CARBAMOYL)PHENOXYACETIC ACID MONOSODIUM SALT * MERCURAMIDE * SODIUM 0-((3-(HYDROXYMERCURI)-2-METHOXYPROPYL)CARBAMOYL)PHENOXY ACETATE * SODIUM SALICYL-(GAMMA-HYDROXYMERCURI-BETA-METHOXYPROPYL)AMIDE-O-ACETATE

THR: Poison by intravenous, intramuscular, and intraperitoneal routes. See also mercury compounds. When heated to decomposition it emits very toxic fumes of Hg, NO_x, and Na_2O.

SODIUM METAPHOSPHATE   HR: 3
CAS: 50813-16-6   NIOSH: OY 3750000
mf: O_3P•Na   mw: 101.96

PROP: Sodium metaphosphate is known as a number of different molecular species, some of which exhibit various crystalline forms. The vitreous sodium phosphates having a Na_2O/P_2O_3 mole ratio near unity are classified as sodium-meta-phosphates. The term also extends to short-chain vitreous compositions, the compounds of which exhibit the polyphosphate formula $Na_{n+2}P_nO_{3n+1}$ with n as low as 4-5. In such as $(NaPO_3)$, n may be a small integer <3 (cyclic molecules) or a large number (polymers).

SYNS: GRAHAM'S SALT * METAFOS

THR: Poison by intravenous route. Moderately toxic by intraperitoneal route. See also phosphates. Dangerous. See phosphates and Na_2O for disaster hazard.

SODIUM MONOHYDROGEN PHOS-PHATE (2:1:1)   HR: 3
CAS: 7558-79-4   NIOSH: WC 4500000
mf: $HO_4P \cdot 2Na$ mw: 141.96

PROP: Colorless, translucent crystals or white powder, sol in water, very sltly sol in alcohol.

SYNS: DIBASIC SODIUM PHOSPHATE * DISO-DIUM HYDROGEN PHOSPHATE * DISODIUM MONOHYDROGEN PHOSPHATE * DISODIUM ORTHOPHOSPHATE * DISODIUM PHOSPHATE * DISODIUM PHOSPHORIC ACID * EXSICCATED SODIUM PHOSPHATE * NATRIUMPHOSPHAT (GERMAN) * SODA PHOSPHATE * SODIUM HYDROGEN PHOSPHATE * PHOSPHORIC ACID, DISODIUM SALT

THR: Poison by intravenous route. Moderately toxic by intraperitoneal, subcutaneous, and intramuscular routes. A skin and eye irritant. See also phosphates. When heated to decomposition it emits toxic fumes of PO_x and Na_2O.

SODIUM MONOXIDE   HR: 3
CAS: 12401-86-4   NIOSH: WC 4800000
DOT: 1825
mf: Na_2O mw: 61.98

PROP: White-gray deliquescent crystals. Bp: 1275° (sublimes), d. 2.27.

SYNS: CALCINED SODA * DISODIUM MONOX-IDE * DISODIUM OXIDE * SODIUM OXIDE

DOT Classification: Label: Corrosive

THR: Very corrosive and irritating to skin, eyes, and mucous membranes. Can react violently with NO above 100°; and also with water. See also sodium hydroxide. When heated to decom-position it emits Na_2O fumes. Incompatible with P_2O_5; H_2O.

SODIUM (I) NITRATE (1:1)   HR: 3
CAS: 7631-99-4   NIOSH: WC 5600000
DOT: 1498
mf: $NO_3 \cdot Na$ mw: 85.01

PROP: Colorless, transparent, odorless crystals, saline, slightly bitter taste. Mp: 306.8°, bp: decomp @ 380°, d: 2.261.

SYNS: NITRATE DE SODIUM (FRENCH) * NI-TRIC ACID, SODIUM SALT * SODA NITER

DOT Classification: Label: Oxidizer

THR: Poison by ingestion. Mutagenic data. Explodes when heated to over 1000°F, or when mixed with cyanides; (S + charcoal); sodium hypophosphite; BP; Sb. Flammable when mixed with organic matter, it will ignite on friction. See also nitrates. Dangerous disaster hazard; Incompatible with acetic anhydride; aluminum, or aluminum oxide; barium thiocyanate; fibrous material; non-metals; sodium; sodium phosphi-nate; sodium thiosulfate; wood. For further information see Vol. 3, No. 6 of *DPIM Report*.

SODIUM NITRITE   HR: 3
CAS: 7632-00-0   NIOSH: RA 1225000
DOT: 1500
mf: $NO_2 \cdot Na$ mw: 69.00

PROP: Slightly yellowish or white crystals, sticks or powder. Mp: 271°, bp: decomp @ 320°, d: 2.168.

SYNS: DIAZOTIZING SALTS * DUSITAN SODNY (CZECH) * NATRIUM NITRIT (GERMAN) * NCI-C02084 * NITRITE DE SODIUM (FRENCH)

DOT Classification: Label: Oxidizer

THR: Poison by ingestion, subcutaneous, intravenous, and intraperitoneal routes. Mutagenic data. An eye irritant. Human central nervous system effects. An experimental neoplastigen. See also nitrites. Flammable; a strong oxidizing agent. In contact with organic matter, will ignite by friction. See also nitrites. Explodes when heated to over 1000°F or on contact with cyanides; NH_4+ salts; cellulose; Li; (K + NH_3); $Na_2S_2O_3$. Dangerous; see also nitrites. When heated to decomposition it emits toxic fumes of NO_x and Na_2O. Incompatible with amino-guanidine salts; butadiene; phthalic acid;

phthalic anhydride; reducants; sodium amide; sodium disulphite; sodium thiocyanate; urea; wood.

SODIUM NITROFERRICYANIDE   HR: 3
CAS: 14402-89-2   NIOSH: LJ 8750000
mf: $C_5FeN_6O \cdot 2Na$ mw: 261.94

SYNS: SODIUM NITROPRUSSATE * SODIUM NITROPRUSSIDE * SODIUM NITROSYLPENTACYANOFERRATE (III)

THR: Poison to humans by inhalation, intravenous, and other unspecified routes. Human central nervous system effects. An experimental teratogen. When heated to decomposition it emits toxic fumes of NO_x, CN^- and Na_2O.

SODIUM PENTACHLOROPHE-
NATE   HR: 3
CAS: 131-52-2   NIOSH: SM 6650000
mf: $C_6Cl_5O \cdot Na$ mw: 288.30

PROP: Tan powder.

SYNS: PENTACHLOROPHENATE SODIUM
* PENTACHLOROPHENOL, SODIUM SALT
* PENTACHLOROPHENOXY SODIUM * PENTAPHENATE * SODIUM PENTACHLOROPHENOL
* SODIUM PENTACHLOROPHENOLATE
* SODIUM PENTACHLOROPHENOXIDE

DOT Classification: ORM-A, Label: None

THR: Poison by ingestion, skin contact, intravenous, subcutaneous, and intratracheal routes. Mutagenic data. When heated to decomposition it emits toxic fumes of Cl^- and Na_2O.

SODIUM PERMANGANATE   HR: 3
CAS: 10101-50-5   NIOSH: SD 6650000
DOT: 1503
mf: $MnO_4 \cdot Na$ mw: 141.93

PROP: Purple to red-black crystals. Mp: decomp

SYN: PERMANGANATE DE SODIUM (FRENCH)

DOT Classification: Label: Oxidizer

THR: Poison by intravenous route. See also manganese compounds. A powerful oxidizer and fire hazard. Reacts vigorously with combustibles; acetic acid; acetic anhydride.

SODIUM PEROXIDE   HR: 3
CAS: 1313-60-6   NIOSH: WD 3450000
DOT: 1504
mf: Na_2O_2 mw: 77.98

PROP: White powder, turning yellow when heated. Mp: decomp @ 460°, bp: decomp. D: 2.805.

SYNS: DISODIUM DIOXIDE * DISODIUM PEROXIDE * SODIUM DIOXIDE * SODIUM SUPEROXIDE

DOT Classification: Label: Oxidizer

THR: Poison irritant to skin, eyes, and mucous membranes. See also sodium hydroxide and peroxides, inorganic. Dangerous fire hazard by chemical reaction; a powerful oxidizing agent. Reacts violently with water; acids; powdered metals; acetic anhydride; Al; (Al + CO_2); $(NH_4)_2S_2O_8$; aniline; Sb; As; benzene; CaC_2; charcoal; Cu; (KNO_3 + dextrose); ethyl ether; H_2S; glycerin; hexamethylene-tetramine; Mg; non-metal halides; fibrous matter; (Mg + CO_2); MnO_2; organic matter; P; K; Se_2Cl_2; BN; $AlCl_3$; (AgCl + charcoal); Na; SCl; Sn; Zn; peroxyformic acid. Dangerous; will react with water or steam to produce heat and toxic fumes; can react vigorously with reducing materials. To fight fire, use carbon dioxide or dry chemical. Combustible materials ignited by contact with sodium peroxide should be smothered with soda ash, salt or dolomite mixtures. Chemical fire extinguishers should not be used. If the fire cannot be smothered, it should be flooded with large quantities of water from a hose.

SODIUM PERSULFATE   HR: 3
CAS: 7775-27-1   NIOSH: SE 0525000
DOT: 1505
mf: $O_8S_2 \cdot 2Na$ mw: 238.10

PROP: White cryst powder; sol in water, decomp by alc.

SYNS: PERSULFATE DE SODIUM (FRENCH)
* SODIUM PEROXYDISULFATE

DOT Classification: Label: Oxidizer

THR: Poison by intraperitoneal and intravenous routes. An oxidizer; can cause fires. See also sulfates. When heated to decomposition it emits toxic fumes of SO_x and Na_2O.

SODIUM PHENOXIDE   HR: 3
CAS: 139-02-6   NIOSH: SM 8780000
DOT: 2497
mf: $C_6H_5O \cdot Na$ mw: 116.10

PROP: White, deliquescent crystals.

SYNS: PHENOL SODIUM SALT * SODIUM CAR-
BOLATE * SODIUM PHENATE * SODIUM PHE-
NOLATE, SOLID (DOT)

DOT Classification: Label: Corrosive

THR: Poison by subcutaneous route. See also
phenol, sodium hydroxide. When heated to decomposition it emits Na_2O and acrid smoke.
Dangerous when heated.

SODIUM PHOSPHIDE HR: 3
CAS: 12058-85-4   NIOSH: WD 6475000
DOT: 1432
mf: PNa_3 mw: 99.94

PROP: Red cryst; mp: decomp.

SYN: PHOSPHURE DE SODIUM (FRENCH)

DOT Classification: Label: Flammable Solid
and Dangerous When Wet

THR: Poison by inhalation. See also phosphides.
Reacts violently with H_2O to yield phosphine.
When heated to decomposition it emits toxic
fumes of PO_x, PH_3 and Na_2O.

SODIUM SALICYLATE HR: 3
CAS: 54-21-7   NIOSH: VO 5075000
mf: $C_7H_5O_3 \cdot Na$ mw: 160.11

PROP: White, odorless crystals, scales or powder.

SYNS: O-HYDROXYBENZOIC SODIUM SALT
* SALICYLIC ACID, SODIUM SALT * SODIUM-
O-HYDROXYBENZOATE * SODIUM SALICYLIC
ACID

THR: Poison to humans by subcutaneous route.
Moderately toxic to humans by ingestion. Mutagenic data. An experimental teratogen. A powerful irritant; affects the central nervous system.
When heated to decomposition it emits acrid
smoke and Na_2O fumes. Incompatible with ferric salts; lime water; spirit nitrous ether; mineral
acids; iodine; lead acetate; silver nitrate; sodium
phosphate powder.

SODIUM SELENITE HR: 3
CAS: 10102-18-8   NIOSH: VS 7350000
mf: $O_3Se \cdot 2Na$ mw: 172.94

PROP: White cryst.

SYNS: SELENIOUS ACID, DISODIUM SALT
* DISODIUM SELENITE * NATRIUMSELENIT
(GERMAN)

OSHA PEL: TWA 200 ug(Se)/m^3

THR: Poison by ingestion, intraperitoneal, intravenous, subcutaneous, and intramuscular
routes. Mutagenic data. An experimental carcinogen. See also selenium compounds. When
heated to decomposition it emits toxic fumes
of Se and Na_2O.

SODIUM SILICATE HR: 3
CAS: 6834-92-0   NIOSH: VV 9275000
mf: $O_3Si \cdot 2Na$ mw: 122.07

SYNS: DISODIUM METASILICATE * DISODIUM
MONOSILICATE * SODIUM METASILICATE
* SODIUM METASILICATE, ANHYDROUS
* SOLUBLE GLASS * WATER GLASS

THR: Poison by ingestion and intraperitoneal
routes. See also silicates. Caustic material, irritating to skin and mucous membranes. Ingestion
causes gastrointstinal tract upset. Violent reaction with F_2.

SODIUM SULFATE (2:1) HR: 2
CAS: 7767-82-6   NIOSH: WE 1650000
mf: $O_4S \cdot 2Na$ mw: 142.04

PROP: Odorless white cryst or powd; sol in
water, glycerol; insol alc; mp: 888°; d: 2.671;
A substance which migrates to food from packaging material. Violent reaction with Al.

SYNS: DISODIUM SULFATE * NATRIUMSUFAT
(GERMAN) * SALT CAKE * SODIUM SULFATE
ANHYDROUS * SULFURIC ACID, DISODIUM SALT

THR: Moderately toxic by intravenous route.
Slightly toxic by ingestion. See also sulfates.
When heated to decomposition it emits toxic
fumes of SO_x and Na_2O. Violent reaction with
Al.

SODIUM SULFIDE (ANHY-
DROUS) HR: 3
CAS: 1313-82-2   NIOSH: WE 1905000
DOT: 1385
mf: Na_2S mw: 78.04

PROP: Amorphous, yellow-pink or white deliquescent crystals. Mp: 1180°, d: 1.856 @ 14°.

SYNS: SODIUM MONOSULFIDE * SODIUM SUL-
PHIDE

DOT Classification: Label: Flammable Solid

THR: Poison by intravenous route. Reacts violently with C; diazonium salts; n,n-dichlorome-
thylamine; o-nitroaniline diazonium salt; water.
This material is unstable and can explode on

rapid heating or percussion. See also sulfides. When heated to decomposition it emits toxic fumes of SO_x and Na_2O.

SODIUM SULFITE (2:1)　　HR: 3
CAS: 7757-83-7　　NIOSH: WE 2150000
mf: $O_3S \cdot 2Na$　　mw: 126.04

PROP: Hexagonal prisms or white powd; bp: decomp; d: 2.633 @ 15.4°.

SYNS: ANHYDROUS SODIUM SULFITE * DISODIUM SULFITE * EXSICATED SODIUM SULFITE * NATRIUMSULFID (GERMAN) * SULFUROUS ACID, SODIUM SALT (1:2)

THR: Poison by intravenous and subcutaneous routes. Moderately toxic by ingestion. See also sulfites. Mutagenic data. When heated to decomposition it emits very toxic fumes of Na_2O and SO_x.

SODIUM TELLURITE　　HR: 3
CAS: 10102-20-2　　NIOSH: WY 2450000
mf: $O_3Te \cdot 2Na$　　mw: 221.58

SYNS: SODIUM TELLURATE (IV) * TELLUROUS ACID, DISODIUM SALT

OSHA PEL: TWA 100 ug(Te)/m^3

THR: Poison to humans by ingestion and parenteral route. Poison by intravenous and intraperitoneal routes experimentally. Mutagenic data. See also tellurium. When heated to decomposition it emits toxic fumes of Te and Na_2O.

SODIUM TETRAVANADATE　　HR: 3
CAS: 12058-74-1　　NIOSH: WE 5270000
mf: $O_{11}V_4 \cdot 2Na$　　mw: 425.74

THR: Poison to humans by intravenous routes. Poison by subcutaneous, intramuscular, and parenteral routes experimentally. See also vanadium compounds.

SODIUM THIOGLYCOLATE　　HR: 3
CAS: 367-51-1　　NIOSH: AI 7700000
mf: $C_2H_3O_2S \cdot Na$　　mw: 114.10

PROP: Hygroscopic crystals.

SYNS: MERCAPTOACETIC ACID SODIUM SALT * SODIUM MERCAPTOACETATE * THIOGLYCOLATESODIUM * THIOGLYCOLLIC ACID, SODIUM SALT * USAF EK-5199

THR: Poison by intravenous and intraperitoneal routes. Human skin irritant. See also sulfides.

This material yields hydrogen sulfide on decomposition. The literature contains the report of a death attributed to the absorption of toxic decomposition products from the use of this material in a permanent waving solution. Dangerous. Emits toxic SO_x and Na_2O upon decomposition.

SODIUM THIOSULFATE, PENTAHYDRATE　　HR: 2
CAS: 10102-17-7　　NIOSH: WE 6660000
mf: $O_3S_2 \cdot 2Na \cdot 5H_2O$　　mw: 248.20

PROP: Monoclinic, colorless, odorless crystals. Mp: 48° (rapid heating), d: 1.69.

SYNS: ANTICHLOR * HYPO * NSC-45624 * SODIUM HYPOSULFITE * THIOSULFURIC ACID, DISODIUM SALT, PENTAHYDRATE

THR: Moderately toxic by intravenous route. Human blood effects. See thiosulfates. Large doses internally have a cathartic action. Violent reaction with $NaNO_2$. Dangerous; see thiosulfates and Na_2O.

SODIUM TRIMETAPHOSPHATE　　HR: 3
CAS: 7785-84-4　　NIOSH: OY 4025000
mf: $O_9P_3 \cdot 3Na$　　mw: 305.88

SYN: TRIMETAPHOSPHATE SODIUM

THR: Poison by intravenous route. Moderately toxic by intraperitoneal route. See also phosphates. When heated to decomposition it emits toxic fumes of PO_x and Na_2O.

SODIUM TRIPOLYPHOSPHATE　　HR: 3
CAS: 13573-18-7　　NIOSH: YK 4900000
mf: $O_{10}P_3 \cdot 5Na$　　mw: 367.86

PROP: Slightly hygroscopic granules.

SYNS: TRIPHOSPHORIC ACID, SODIUM SALT * NATRIUMTRIPOLYPHOSPHAT (GERMAN) * PENTASODIUM TRIPHOSPHATE * SODIUM TRIPHOSPHATE

THR: Poison by intravenous route. Moderately toxic by ingestion, subcutaneous, and intraperitoneal routes. Skin irritant. Ingestion of large doses of sodium phosphates causes catharsis. Sodium m- and pyrophosphates can cause hemorrhages from the intestine if taken internally in large doses. When heated to decomposition it emits toxic fumes of PO_x and Na_2O. For further information see Vol. 3, No. 1 of *DPIM Report*.

SODIUM TUNGSTATE   HR: 3
CAS: 13472-45-2   NIOSH: YO 7875000
mf: $O_4W \cdot 2Na$ mw: 293.83

PROP: White, rhombic cryst; mp: 698°; d: 4.179.

SYN: TUNGSTIC ACID, DISODIUM SALT

THR: Poison by ingestion and subcutaneous routes. See also tungsten compounds. When heated to decomposition it emits toxic fumes of Na_2O.

SODIUM VANADATE   HR: 3
CAS: 13718-26-8   NIOSH: YW 1050000
mf: $O_3V \cdot Na$ mw: 121.93

SYNS: VANADIC ACID, MONOSODIUM SALT
* SODIUM METAVANADATE

THR: Poison by ingestion, intraperitoneal, subcutaneous, and other unspecified routes. See also vanadium compounds. When heated to decomposition it emits acrid smoke and Na_2O fumes.

SODIUM-o-VANADATE   HR: 3
CAS: 13721-39-6   NIOSH: YW 1120000
mf: $O_4V \cdot 3Na$ mw: 183.91

PROP: Colorless, hexagonal prisms. Mp: 850°-866°.

SYN: VANADIC (II) ACID, TRISODIUM SALT

THR: Poison by ingestion, intravenous, and subcutaneous routes. See also vanadium compounds. When heated to decomposition it emits acrid smoke and Na_2O fumes.

SOMAN   HR: 3
CAS: 96-64-0   NIOSH: TA 8750000
mf: $C_7H_{16}FO_2P$ mw: 182.20

SYNS: METHYLPHOSPHONOFLUORIDIC ACID, 3,3-DIMETHYL-2-BUTYL ESTER * METHYL PINACOLYLOXY PHOSPHORYLFLUORIDE * METHYL PINACOLYL PHOSPHONOFLUORIDATE * PINACOLOXYMETHYLPHOSPHORYL FLUORIDE * PINACOLYL METHYLFLUOROPHOSPHONATE * PINACOLYL METHYLPHOSPHONOFLUORIDATE * PINACOLYLOXY METHYLPHOSPHORYL FLUORIDE * 3,3-DIMETHYL-N-BUT-2-YL METHYLPHOSPHONOFLUORIDATE * 1,2,2-TRIMETHYLPROPYL METHYLPHOSPHONOFLUORIDATE

THR: Poison by inhalation, skin contact, subcutaneous, intravenous, and intraperitoneal routes. An extremely toxic military nerve gas. When heated to decomposition it emits very toxic fumes of F^- and PO_x. For further information see Vol. 1, No. 2 of *DPIM Report*.

SOOT   HR: 3
Soot was defined as a brown-to-black substance incidentally produced during the incomplete and uncontrolled combustion of any carbonaceous material. It is a mixture of colloidal carbon, organic tars and refractory inorganics whose composition depends on combustion conditions. It is not unusual for the tarry component to account for more than 50 wt% of the soot, particularly, when produced by inefficient combustion of coal or wood. The tarry component and, to a lesser extent, trace inorganic impurities, are believed responsible for the known *health hazards attributed to soot,* i.e., chronic contact or long term inhalation can lead to cancers of the skin, scrotum, or lung. Can be distinguished from carbon black on the basis of differences in physical and chemical properties.

THR: An experimental carcinogen and nuisance dust. See also carbon.

"SPEED"   HR: 3
CAS: 826-10-8   NIOSH: SH 5250000
mf: $C_{10}H_{15}N \cdot ClH$ mw: 185.72

PROP: Crystals; bitter taste; mp: 170°-175°; sol in H_2O, alc and chloroform, almost insol in ether.

SYNS: 1-DESOXYEPHEDRINE HYDROCHLORIDE
* N,ALPHA-DIMETHYLPHENETHYLAMINE HYDROCHLORIDE, (−)- * 1-N-METHYL-BETA-PHENYLISOPROPYLAMINE HYDROCHLORIDE * 1-METHAMPHETAMINE HYDROCHLORIDE * "METH"
* ADIPEX * SYNDROX

THR: Poison by ingestion, intravenous, intraperitoneal, and subcutaneous routes. Caution: Excessive use may lead to tolerance, habituation. An experimental teratogen. When heated to decomposition it emits very toxic fumes of HCl and NO_x.

SPIRIT OF GLYCERYL TRINITRATE   HR: 2
NIOSH: WG 9532700
DOT: 1204

PROP: Clear, colorless liquid. Composition: 1.0-1.1% glycerol trinitrate in alcoholic solution. D: 0.814-0.820 @ 25°.

SYNS: SOLUTION GLYCERYL TRINITRATE
* SPIRIT OF GLONOIN * SPIRITS OF NITRO-
GLYCERIN (DOT) * SPIRIT OF TRINITROGLYC-
ERIN

DOT Classification: Label: Flammable
Liquid

THR: Moderately toxic by ingestion and inhalation. See also nitroglycerin. Dangerous, when
exposed to heat or flame. See also ethyl alcohol.
See nitroglycerin and ethyl alcohol for explosion
hazard. If the alcohol evaporates, the residue
is nitroglycerin. Dangerous; when dried out,
shock will explode it; when heated to decomposition it emits highly toxic fumes; on contact
with oxidizing materials the mixture can react
vigorously.

SPRENGEL EXPLOSIVES
PROP: This type of explosive is a mixture of
nitrobenzene and fuming nitric acid. It is a powerful and cheap explosive and would have many
uses except that it is limited by practical disad-
vantages. The components have to be mixed
in glass shortly before the explosive is used.
This requires preparation and equipment not always available at the site of the explosion. This
material can be destroyed by throwing it into
large quantities of water, or possibly by burning
in small quantities at a time. See also explosives,
high.

STARCH DUST   HR: 1
THR: A nuisance dust. An allergen. Flammable
when exposed to flame, can react with oxidizing
materials. Moderately explosive when exposed
to flame.

STEARIC ACID   HR: 3
CAS: 57-11-4   NIOSH: WI 2800000
mf: $C_{18}H_{36}O_2$ mw: 284.54

PROP: White, amorphous solid; mp: 69.3°, bp:
383°, flash p: 385°F (CC), d: 0.847, autoign
temp: 743°F, vap press: 1 mm @ 173.7°, vap
d: 9.80. sol in water, alc, ether, acetone, CCl_4.

SYNS: 1-HEPTADECANECARBOXYLIC ACID
* NEO-FAT 18-61 * OCTADECANOIC ACID
* PEARL STEARIC * STEAREX BEADS
* STEAROPHANIC ACID

THR: Poison by intravenous route. An experimental carcinogen. Human skin irritant. Combustible when exposed to heat or flame. Heats

spontaneously. To fight fire, use CO_2, dry chemical. When heated to decomposition it emits
acrid smoke and fumes.

STEARIC ACID-2,3-EPOXYPROPYL
ESTER   HR: 3
CAS: 7460-84-6   NIOSH: WI 3500000
mf: $C_{21}H_{40}O_3$ mw: 340.61

SYNS: 2,3-EPOXY-1-PROPANOL STEARATE
* 2,3-EPOXYPROPYL ESTER OF STEARIC ACID
* 2,3-EPOXYPROPYL STEARATE * GLYCIDOL
STEARATE * GLYCIDYL OCTADECANOATE
* GLYCIDYL STEARATE * OXIRANYLMETHYL
ESTER OF OCTADECANOIC ACID

THR: An experimental tumorigen. Mutagenic
data. See also esters. When heated to decomposition it emits acrid smoke and fumes.

STERIGMATOCYSTIN   HR: 3
CAS: 10048-13-2   NIOSH: LV 1750000
mf: $C_{18}H_{12}O_6$ mw: 324.30

PROP: A metabolite of *aspergillus versicolor*.

SYN: 3A,12C-DIHYDRO-8-HYDROXY-6-
METHOXY-7H-FURO(3',2':4,5)FURO(2,3-C)-
XANTHEN-7-ONE

THR: Poison by ingestion and intraperitoneal
route. An experimental carcinogen. Mutagenic
data. When heated to decomposition it emits
acrid smoke and fumes. For further information
see Vol. 1, No. 4 of *DPIM Report*.

STIBINE   HR: 3
CAS: 7803-52-3   NIOSH: WJ 0700000
DOT: 2676
mf: H_3Sb mw: 124.78

PROP: Colorless gas, disagreeable odor. Mp:
−88°; bp: −18.4°; d: 2.204 g/mL @ bp: Gas
is slightly sol in water. Very sol in alc, carbon
disulfide, organic solvents.

SYNS: ANTIMONWASSERSTOFFES (GERMAN)
* ANTIMONY HYDRIDE * ANTIMONY TRIHY-
DRIDE * ANTYMONOWODOR (POLISH)
* HYDROGEN ANTIMONIDE

OSHA PEL: TWA 500 ug/m^3
ACGIH TLV: TWA 0.1 ppm
DFG MAK: 0.1 ppm (0.5 mg/m^3)
DOT Classification: Poison A, Poison Gas

THR: Poison by inhalation and intravenous
routes. See also antimony compounds and hy-

drides. Very poisonous. Quickly destroyed @ 200°. Flammable when exposed to flame. Violent reaction with Cl_2; HNO_3; O_3; NH_3. Dangerous; when heated to decomposition it emits toxic fumes of Sb. The decomposition products are also hydrogen and metallic antimony.

STILBENE   HR: 3
CAS: 588-59-0   NIOSH: WJ 4925000
mf: $C_{14}H_{12}$ mw: 180.26

PROP: Colorless or slightly yellow crystals, insol in water, sol in 90 parts cold alcohol and 13 parts boiling alcohol, freely sol in benzene and ether. Mp: 124°-125°, bp: 306°-307°, d: 0.9707. insol in water; sol in absolute alc; sol in ether.

SYNS: STILBEN (GERMAN) * DIPHENYLETHYL-ENE

THR: Poison by intravenous route. Moderately toxic by intraperitoneal route. Violent reaction with O_2. When heated to decomposition it emits acrid smoke and fumes.

STODDARD SOLVENT   HR: 1
CAS: 8052-41-3   NIOSH: WJ 8925000

PROP: Clear, colorless liquid. Composed of 85% nonane and 15% trimethyl benzene; bp: 220°-300°, flash p: 100°-110°F, lel = 1.1%, uel = 6%, autoign temp: 450°F, d: 1.0. Insol in water. misc with abs alc, benzene, ether, chloroform, carbon tetrachloride, carbon disulfide, and some oils (not castor oil). Stoddard solvent to a first approximation contains 85% nonane and 15% trimethylbenzene.

SYNS: VARNOLINE * NAPHTHA SAFETY SOLVENT * WHITE SPIRITS

OSHA PEL: TWA 500 ppm
ACGIH TLV: TWA 100 ppm

THR: Slightly toxic by inhalation. A human eye irritant. See also nonane, trimethyl benzene and mixed isomers. Flammable when exposed to heat or flame. Moderately explosive when exposed to flame. When heated to decomposition it emits acrid fumes and may explode; can react with oxidizing materials. To fight fire, use foam, CO_2, dry chemical.

STRAMONIUM   HR: 3
CAS: 8063-18-1   NIOSH: WK 0900000

PROP: *Datura Stramonium* have 0.25-0.45% alkaloids consisting of Atropine, Hyoscyamine and Scopolamine

SYNS: ANGEL TULIP * DATURA STRAMONIUM * DEVIL'S APPLE * DHUTRA * JAMESTOWN WEED * JIMSON WEED * POMME EPINEUSE (FRENCH) * STECKAPFUL (GERMAN) * STRAMONA (ITALIAN) * THORN APPLE

THR: Poison to humans by ingestion. When heated to decomposition it emits irritating smoke and fumes.

STREPTOMYCIN   HR: 3
CAS: 57-92-1   NIOSH: WK 4375000
mf: $C_{21}H_{39}N_7O_{12}$ mw: 581.67

PROP: An antibiotic, it is a base and readily forms salts with anions.

SYNS: NSC 14083 * STREPTOMYCIN A * STREPTOMYZIN (GERMAN)

THR: Poison by intravenous and subcutaneous routes. Moderately toxic by ingestion and intraperitoneal routes. Mutagenic data. An experimental teratogen. Human pulmonary system effects. Toxic to kidneys and central nervous system. Has been implicated in aplastic anemia. When heated to decomposition it emits toxic fumes of NO_x.

STREPTOZOTICIN   HR: 3
CAS: 18883-66-4   NIOSH: LZ 5775000
mf: $C_8H_{15}N_3O_7$ mw: 265.26

PROP: Plateletes; mp: 115° (decomp).

SYNS: 2-DEOXY-2-(3-METHYL-3-NITROSOUREIDO)-D-GLUCOPYRANOSE * N-D-GLUCOSYL(2)-N'-NITROSOMETHYLHARNSTOFF (GERMAN) * N-D-GLUCOSYL-(2)-N'-NITROSOMETHYLUREA * NCI-C03167 * NSC-85998 * STREPTOZO-CIN

THR: Poison by ingestion, intravenous, parenteral, and intraperitoneal routes. An experimental neoplastigen and tumorigen. Gastrointestinal tract and systemic effects in humans. A suspected human carcinogen. Mutagenic data. When heated to decomposition it emits toxic fumes of NO_x. For further information see Vol. 1, No. 5 of *DPIM Report*.

STRONTIUM   HR: 2
af: Sr   aw: 87.62

PROP: Silvery-white metal. D: 2.6, mp: 757° ± 1°, bp: 1366°; vap press: 10 mm @ 898°.

THR: Moderately toxic by ingestion and inhalation. See also specific compounds. It resembles calcium in its metabolism and behavior. The stable form has low toxicity. Ignites spontaneously in air. Dangerous; in the form of dust; also when exposed to flame. See also powdered metals. Moderately explosive in the form of dust, by spontaneous chemical reaction. Reacts with water to evolve hydrogen. Can be stored under liquid hydrocarbons. Highly dangerous in the form of radioactive isotopes; will react with water or steam to produce heat and hydrogen; on contact with oxidizing materials, can react vigorously.

STRONTIUM CHROMATE (1:1)   HR: 3
CAS: 7789-06-2   NIOSH: GB 3240000
DOT: 9149
mf: $CrO_4 \cdot Sr$ mw: 203.62

PROP: Monoclinic yellow crystals. D: 3.895 @ 15°.

SYNS: c.i. pigment yellow 32 * strontium chromate (vi) * strontium chromate 12170

TRK: 0.1 mg/m^3
DOT Classification: ORM-E

THR: An experimental carcinogen and tumorigen. See also chromium compounds. For further information see Vol. 1, No. 7 of *DPIM Report*.

STRONTIUM COMPOUNDS   HR: 1
THR: The strontium ion has a low order of toxicity. It is chemically and biologically similar to calcium. The oxides and hydroxides are moderately caustic materials. As with other compounds, the toxicity of a given compound may be a function of the anion.

STRONTIUM FLUORIDE   HR: 2
CAS: 7783-48-4   NIOSH: WK 8925000
mf: F_2Sr mw: 125.62

PROP: Cubic odorless, crystals or white powder; mp: 1190°; d: 4.24.

THR: Moderately toxic by intravenous route. Slightly toxic by ingestion and intraperitoneal routes. See also fluorides and strontium compounds. When heated to decomposition it emits toxic fumes of F$^-$.

STRONTIUM SULFIDE   HR: 3
CAS: 1314-96-1   NIOSH: WL 0400000
mf: SSr   mw: 119.68

PROP: Cubic, light gray cryst. D: 3.70 @ 15°.

SYNS: c.i. 77847 * strontium monosulfide * strontium sulphide

THR: Poison by inhalation and ingestion. Readily yields H$_2$S. See also sulfides and strontium compounds. When heated to decomposition it emits toxic fumes of SO$_x$ and H$_2$S. Incompatible with PbO$_2$.

STRYCHNINE   HR: 3
CAS: 57-24-9   NIOSH: WL 2275000
DOT: 1692
mf: $C_{21}H_{22}N_2O_2$ mw: 334.45

PROP: Hard, white, crystalline alkaloid, very bitter taste; mp: 268°-290°, bp: 270° @ 5mm, d: 1.359 @ 18°.

SYNS: mole death * pied piper mouse seed * stricnina (italian) * strychnin (german)

OSHA PEL: TWA 150 ug/m^3
ACGIH TLV: TWA 0.15 mg/m^3
DFG MAK: 0.15 mg/m^3
DOT Classification: Poison B

THR: Very poisonous by ingestion, intravenous, subcutaneous, and other unspecified routes. An allergen. Lethal dose to man: 30-60 mg/kg. If ingested, the time of action depends upon the condition of the stomach, that is, whether empty or full, and the nature of the food present. If taken by subcutaneous injection, the place of administration of the injection will affect the time of action. The first symptoms are a feeling of uneasiness with a heightened reflex of irritability, followed by muscular twitching in some parts of the body. With larger doses, this is followed by a sense of impending suffocation. Convulsive movements begin which have the effect of mechanically causing the patient to cry out or to shriek; then follow the characteristic spasms which set in with violence. These are at first clonic and then tonic. There are successive attacks of spasms. With each successive attack, the symptoms become more violent, eventually resulting in death. A rodenticide. Dangerous when heated it emits highly toxic fumes. For further information see Vol. 2, No. 2 of *DPIM Report*.

STRYCHNINE MONONITRATE   HR: 3
CAS: 66-32-0   NIOSH: WL 2450000
mf: $C_{21}H_{22}N_2O_2 \cdot HNO_3$ mw: 397.47

PROP: Colorless, odorless needles or white, crystalline powder. Insol in ether.

SYN: STRYCHNINE NITRATE

THR: Poison by intravenous and subcutaneous routes. See also strychnine. When heated to decomposition it emits very toxic fumes of NO_x and HNO_3.

STRYCHNINE SALT (solid)   HR: 3
NIOSH: WL 2465000
DOT: 1692

SYN: STRYCHNINE SALT, SOLID (DOT)

DOT Classification: Poison B, Label: Poison

THR: A deadly poison. See also strychnine.

STRYCHNINE SULFATE (2:1)   HR: 3
CAS: 60-41-3   NIOSH: WL 2550000
mf: $C_{21}H_{22}N_2O_2 \cdot 1/2H_2O_4S$ mw: 383.49

SYNS: STRYCHININE SULFATE * STRYCHNIDIN-10-ONE, SULFATE (2:1)

THR: Poison by ingestion, intraperitoneal, intravenous, and subcutaneous routes. See also strychnine and sulfates. When heated to decomposition it emits very toxic fumes of SO_x and NO_x.

STYRENE   HR: 3
CAS: 100-42-5   NIOSH: WL 3675000
DOT: 2055
mf: C_8H_8 mw: 104.16

PROP: Colorless, refractive, oily liquid; mp: −31°, bp: 146°, lel = 1.1%, uel = 6.1%, flash p: 88°F, d: 0.9074 @ 20°/4°, autoign temp: 914°F, vap d: 3.6, fp: −33°, ulc: 40-50. Very sltly sol in water; misc in alc, ether.

SYNS: CINNAMENE * CINNAMENOL * NCI-c02200 * PHENYLETHENE * PHENYLETH-YLENE * STIROLO (ITALIAN) * STYREEN (DUTCH) * STYREN (CZECH) * STYROL (GERMAN) * VINYLBENZEN (CZECH) * VINYLBEN-ZENE * VINYLBENZOL

OSHA PEL: TWA 100 ppm; CL 200; Pk 600/5M/3H

ACGIH TLV: TWA 50 ppm; STEL: 100 ppm; BEI: mandelic acid in urine at end of shift 1 g/L, styrene in mixed-exhaled air prior to shift 40 ppb, styrene in mixed-exhaled air during shift 18 ppm, styrene in blood end of shift 0.55 mg/L, styrene in blood prior to shift 0.02 mg/L

DFG MAK: 100 ppm ($420 \ mg/m^3$)
DOT Classification: Flammable Liquid

THR: Poison by ingestion and intravenous route. Moderately toxic by inhalation and intraperitoneal route. Mutagenic data. A skin and eye irritant. An experimental carcinogen. Human systemic irritant and central nervous system effects. It can cause irritation, violent itching of the eyes @ 200 ppm, lachrymation, and severe human eye injuries. Its toxic effects are usually transient and result in irritation and possible narcosis. Dangerous fire hazard when exposed to flame, heat or oxidants. Reacts violently with chlorosulfonic acid; oleum; H_2SO_4; O_2; alkali metal-graphite. Dangerous on decomposition it emits acrid fumes; can react vigorously with oxidizing materials. To fight fire, use foam, CO_2, dry chemical. For further information see Vol. 6, No. 2 of *DPIM Report*.

SUCCINIC ANHYDRIDE   HR: 3
CAS: 108-30-5   NIOSH: WN 0875000
mf: $C_4H_4O_3$ mw: 100.08

PROP: Colorless needles; mp: 119.6°; d: 1.104; vap press: 1 mm @ 92.0°; very sltly sol in water, petroleum ether; sltly sol in ether. Bp: 261°.

SYNS: BERNSTEINSAURE-ANHYDRID (GERMAN) * BUTANEDIOIC ANHYDRIDE * DIHYDRO-2,5-FURANDIONE * 2,5-DIKETOTETRAHYDROFURAN * NCI-c55696 * SUCCINIC ACID ANHYDRIDE * SUCCINYL OXIDE * TETRAHYDRO-2,5-DIOX-OFURAN

THR: An experimental neoplastigen and suspected carcinogen. An eye irritant. See also anhydrides. When heated to decomposition it emits acrid smoke and fumes.

SUCCINIC PEROXIDE   HR: 3
CAS: 123-23-9   NIOSH: UF 2031000
mf: $C_8H_{10}O_8$ mw: 234.18

PROP: Fine white powder, odorless with tart taste, mod sol in water. Mp: 125° (decomp).

SYNS: BIS(3-CARBOXYPROPIONYL) PEROXIDE * SUCCINIC ACID PEROXIDE (DOT) * 3,3′-(DIOXYDICARBONYL)DIPROPIONIC ACID * SUCCINYL PEROXIDE

DOT Classification: Label: Organic Peroxide

THR: Poison by intraperitoneal route. An irritant. See also peroxides, organic. When heated to decomposition it emits acrid smoke and fumes.

SUCCINONITRILE HR: 3
CAS: 110-61-2   NIOSH: WN 3850000
mf: $C_4H_4N_2$ mw: 80.10

PROP: Colorless, odorless, waxy material. Mp: 58.1°, bp: 267°, flash p: 270°F, d: 1.022 @ 25°, vap press: 2 mm @ 100°, vap d: 2.1. Slightly sol in ether, water, alc; sol in acetone.

SYNS: BUTANEDINITRILE * S-DICYANOETHANE * ETHYLENE CYANIDE * ETHYLENE DICYANIDE * SUCCINIC ACID DINITRILE * SUCCINIC DINITRILE * SUCCINODINITRILE * USAF A-9442

THR: Poison by intraperitoneal and other unspecified routes. See also nitriles. Combustible when exposed to heat or flame. Dangerous when heated to decomposition or on contact with acid or acid fumes, emits highly toxic fumes of NO_x, cyanides; can react with oxidizing materials. To fight fire, use alcohol foam, CO_2, dry chemical.

SUCCINOYLCHOLINE CHLORIDE HR: 3
CAS: 71-2-7-2   NIOSH: GA 2360000
mf: $C_{14}H_{30}N_2O_4 \cdot 2Cl$ mw: 361.36

SYNS: BIS(2-DIMETHYLAMINOETHYL)SUCCINATE BIS(METHOCHLORIDE) * BIS(SUCCINYLDICHLOROCHOLINE) * CHOLINE SUCCINATE DICHLORIDE * DIACETYLCHOLINE DICHLORIDE * 2-DIMETHYLAMINOETHYL SUCCINATE DIMETHOCHLORIDE * SUCCINIC ACID DIESTER WITH CHOLINE CHLORIDE * SUCCINIC ACID BIS(BETA-DIMETHYLAMINOETHYL) ESTER, DIHYDROCHLORIDE * SUCCINIC ACID BIS(BETA-DIMETHYLAMINOETHYL)ESTER DIMETHOCHLORIDE * (2-HYDROXYETHYL)TRIMETHYLAMMONIUM CHLORIDE SUCCINATE * SUCCINYL BISCHOLINE CHLORIDE * SUCCINYLCHOLINE CHLORIDE * SUCCINYLCHOLINE DICHLORIDE * SUCCINYLCHOLINE HYDROCHLORIDE * SUCCINYLDICHOLINE CHLORIDE

THR: A poison by intravenous route. Affects the human cardiovascular system. When heated to decomposition it emits very toxic fumes of NO_x and Cl^-.

SUCROSE HR: 1
CAS: 57-50-1   NIOSH: WN 6500000
mf: $C_{12}H_{22}O_{11}$ mw: 342.34

PROP: White crystals, sweet taste; d: 1.587. @ 25°/4°. Mp: 170°-186° decomp; Sol in water, alc; insol in ether.

SYNS: BEET SUGAR * CANE SUGAR * CONFECTIONER'S SUGAR * ALPHA-D-GLUCOPYRANOSYL BETA-D-FRUCTOFURANOSIDE * (ALPHA-D-GLUCOSIDO)-BETA-D-FRUCTOFURANOSIDE * GRANULATED SUGAR * NCI-C56597 * ROCK CANDY * SACCHAROSE * SACCHARUM * SUGAR

THR: Slightly toxic by ingestion. When heated to decomposition it emits acrid smoke and fumes.

SULFADIMETHYLDIAZINE HR: 3
CAS: 57-68-1   NIOSH: WO 9275000
mf: $C_{12}H_{14}N_4O_2S$ mw: 278.36

SYNS: 2-(P-AMINOBENZENESULFONAMIDO)-4,6-DIMETHYLPYRIMIDINE * 6-(4'-AMINOBENZOL-SULFONAMIDO)-2,4-DIMETHYLPYRIMIDIN (GERMAN) * (P-AMINOBENZOLSULFONYL)-2-AMINO-4,6-DIMETHYLPYRIMIDIN (GERMAN) * N-(4,6-DIMETHYL-2-PYRIMIDYL)SULFANILAMIDE * N(SUP 1)-(4,6-DIMETHYL-2-PYRIMIDYL)SULFANILAMIDE * 4,6-DIMETHYL-2-SULFANILAMIDOPYRIMIDINE * NCI-C56600 * N(SUP 1)-(4,6-DIMETHYL-2-PYRIMIDINYL)SULFANILAMIDE * SULFADIMETHYLPYRIMIDINE * SULFAISODIMIDINE * 2-SULFANILAMIDO-4,6-DIMETHYLPYRIMIDINE * SULPHADIMETHYLPYRIMIDINE

THR: An experimental carcinogen. Moderately toxic by intravenous route. When heated to decomposition it emits very toxic fumes of SO_x and NO_x.

SULFAMETHOXAZOL HR: 3
CAS: 723-46-6   NIOSH: WP 0700000
mf: $C_{10}H_{11}N_3O_3S$ mw: 253.30

SYNS: 3-(PARA-AMINOPHENYLSULPHONAMIDO)-5-METHYLISOXAZOLE * 4-AMINO-N-(5-METHYL-3-ISOXAZOLYL)BENZENESULFONAMIDE * N(SUP 1)-(5-METHYL-3-ISOXAZOLYL)SULPHANILAMIDE * 5-METHYL-3-SULPHANIL-AMIDOISOXAZOLE * N'-(5-METHYL-3-ISOXAZOLE)SULFANILAMIDE * N'-(5-METHYL-3-ISOXAZOLYL)SULFANILAMIDE * 5-METHYL-3-SULFANILAMIDOISOXAZOLE * 5-METHYL-3-SULFANYLAMIDOISOXAZOLE * SULFAMETHYLISOXAZOLE * 3-SULFANILAMIDO-5-METHYLISOXAZOLE * N'-(5-METHYL-ISOXAZOL-3-YL)SULPHANILAMIDE * SULPHA-

METHALAZOLE * SULPHAMETHOXAZOL
* SULPHAMETHOXAZOLE * SULPHA-METH-
OXIZOLE * SULPHAMETHYLISOXAZOLE
* 3-SULPHANILAMIDO-5-METHYLISOXAZOLE

THR: An experimental carcinogen. Moderately toxic by ingestion. When heated to decomposition it emits very toxic fumes of NO_x and SO_x.

SULFAMIC ACID, MONOAMMONIUM SALT   HR: 2
CAS: 7773-06-0   NIOSH: WO 6125000
mf: $H_2NO_3S \cdot H_4N$ mw: 114.14

PROP: Deliquescent crystalline material (white crystalline solid). Bp: 160° (decomp), mp: 131°.

SYNS: AMMONIUM AMIDOSULFONATE * AM-MONIUM AMIDOSULPHATE * AMMONIUM SUL-FAMATE * AMMONIUM SULPHAMATE * AM-MONIUMSALZ DER AMIDOSULFONSAURE (GERMAN) * MONOAMMONIUM SULFAMATE * SULFAMINSAURE (GERMAN)

OSHA PEL: TWA 15 mg/m³

THR: Moderately toxic by ingestion and intraperitoneal routes. See also sulfamates. Slight explosion hazard when exposed to heat or by spontaneous chemical reaction (hydrolysis); in a hot acid solution this material can undergo spontaneous hydrolysis, liberating much heat. Dangerous; see sulfonates. When heated to decomposition it emits very toxic fumes of NH_3, NO_x, and SO_x. For further information see ammonium sulfamate, Vol. 2, No. 3 of *DPIM Report*.

SULFANILAMIDE   HR: 3
CAS: 63-74-1   NIOSH: WO 8400000
mf: $C_6H_8N_2O_2S$ mw: 172.22

PROP: Crystals; mp: 164.5°-166.5°. Sol in glycerol, propylene glycol, HCl; almost insol in chloroform, ether, benzene, petroleum ether.

SYNS: P-AMINOBENZENESULFAMIDE * P-AMI-NOBENZENESULFONAMIDE * 4-AMINOBENZENE-SULFONAMIDE * P-AMINOPHENYLSULFONAMIDE * 4-AMINOPHENYLSULFONAMIDE * P-ANILINE-SULFONAMIDE * ANILINE-P-SULFONIC AMIDE * P-SULFAMIDOANILINE * SULFONAMIDE * SULFONAMIDE P * SULPHANILAMIDE

THR: An experimental carcinogen. Mutagenic data. Moderately toxic by ingestion, intraperitoneal, intravenous, and subcutaneous routes. Implicated in aplastic anemia. When heated to de-

composition it emits very toxic fumes of NO_x and SO_x.

SULFATES
THR: Variable. In general the toxicity qualities of substances containing the sulfate radical is that of the material (cation) with which the sulfate (anion) is combined. See specific compound. Violent reaction with Al; Mg.

SULFIDES
THR: Variable. The alkaline sulfides (potassium, calcium, ammonium and sodium) are similar in action to alkalies. They cause softening and irritation of the skin. If ingested, they are corrosive and irritating through the liberation of hydrogen sulfide and free alkali. Hydrogen sulfide is especially toxic. See also hydrogen sulfide. Sulfides of the heavy metals are generally insoluble and hence have little toxic action except through the liberation of hydrogen sulfide. Sulfides are used as fungicides. Flammable when exposed to flame or by spontaneous chemical reaction. Many sulfides ignite easily in air at room temperature. Others require a higher temperature or the presence of an oxidizer. Upon contact with moisture or acids, hydrogen sulfide is evolved. Many powerful oxidizers on contact with sulfides ignite violently. See also hydrogen sulfide. Many sulfides react violently and explosively on contact with powerful oxidizers. Hydrogen sulfide evolved can form explosive mixtures with air. Dangerous when heated to decomposition they emit highly toxic fumes of SO_x; they react with water, steam or acids to produce toxic and flammable vapors of hydrogen sulfide.

SULFITES   HR: 2
THR: Fairly large doses of sulfites can be tolerated since they are rapidly oxidized to sulfates, although if swallowed they may cause irritation of the stomach by liberating sulfurous acid. Experimentally, large doses of sodium sulfite have been shown to cause retarded growth, nerve irritation, atrophy of bone marrow, depression and paralysis. Dangerous; when heated to decomposition they emit highly toxic fumes of sulfur dioxide; they will react with water, steam or acids to produce a toxic and corrosive material.

SULFONATES
THR: Variable. See specific compounds. Usually irritating. Dangerous; when heated to de-

composition or on contact with acid or acid fumes, they emit highly toxic fumes of SO_x.

4,4′-SULFONYLDIANILINE   HR: 3
CAS: 80-08-0   NIOSH: BY 8925000
mf: $C_{12}H_{12}N_2O_2S$ mw: 248.32

PROP: Crystals, nearly insol in water; sol in acetone, alc.; mp: 176°; vap d: 8.3.

SYNS: BIS(P-AMINOPHENYL)SULPHONE
* BIS(4-AMINOPHENYL)SULPHONE * DIAMINO-4,4′-DIPHENYL SULFONE * DIAMINO-4,4′-DIPHENYL SULPHONE * P,P-DIAMINODIPHENYL SULPHONE * DI(P-AMINOPHENYL)SULPHONE * DI(4-AMINOPHENYL)SULPHONE * DIAPHENYLSULPHON * DIAPHENYLSULPHONE * 1,1′-SULFONYLBIS(4-AMINOBENZENE) * 4,4′-SULFONYLBISANILINE * P,P-SULFONYLBISBENZAMINE * P,P-SULFONYLBISBENZENAMINE * 1,1′-SULPHONYLBIS(4-AMINOBENZENE) * P,P-SULPHONYLBISBENZAMINE * 4,4′-SULPHONYLBISBENZAMINE * P,P-SULPHONYLBISBENZENAMINE * 4,4′-SULPHONYLBISBENZENAMINE * SULPHONYLDIANILINE * P,P-SULPHONYLDIANILINE * BIS(P-AMINOPHENYL) SULFONE * BIS(4-AMINOPHENYL) SULFONE * P,P′-DIAMINODIPHENYL SULFONE * 4,4′-DIAMINODIPHENYL SULFONE * DI(P-AMINOPHENYL) SULFONE * DI(4-AMINOPHENYL)SULFONE * DIAPHENYLSULFONE * NCI-C01718 * NSC 6091D * P,P′-SULFONYLDIANILINE

THR: Poison by ingestion. Moderately toxic by intraperitoneal route. An experimental carcinogen and neoplastigen. Human systemic and red blood cell effects. See also sulfonates. Can cause hepatitis, dermatitis and neuritis. A leprosy treatment and in veterinary medicine. When heated to decomposition it emits very toxic fumes of NO_x and SO_x.

SULFUR   HR: 1
CAS: 7704-34-9   NIOSH: WS 4250000
DOT: 1350/2448
af: S   aw: 32.06

PROP: Rhombic yellow crystals or yellow powder. Mp: 119°, bp: 444.6°, flash p: 405°F (CC), d: 2.07; d(liquid): 1.803; autoign temp: 450°F, vap press: 1 mm @ 183.8°. Insol in water; sltly sol in alc, ether; sol in carbon disulfide, benzene, toluene.

SYNS: BRIMSTONE * COLLOIDAL SULFUR * FLOWERS OF SULPHUR * PRECIPITATED SULFUR * SUBLIMED SULFUR * SULFUR, SOLID

DOT Classification: ORM-C, Label: None

THR: Slightly toxic. A human eye irritant @ 6 ppm. See also nuisance dusts. A fungicide. Chronic inhalation can cause irritation of mucous membranes. Combustible when exposed to heat or flame or by chemical reaction with oxidizers. Explosive in the form of dust when exposed to flame. Can react violently with halogens; carbides; halogenates; halogenites; zinc; uranium; tin; sodium; lithium; nickel; palladium; phosphorus; potassium; indium; calcium; boron; aluminum; (aluminum + niobium pentoxide); ammonia; ammonium nitrate; ammonium perchlorate; BrF_5; BrF_3; (Ca + VO + H_2O); $Ca(OCl)_2$; Ca_3P_2; Cs_3N; charcoal; (Cu + chlorates); ClO_2; ClO; ClF_3; CrO_3; $Cr(OCl)_2$; hydrocarbons; IF_5; IO_5; PbO_2; $Hg(NO_3)_2$; HgO; Hg_2O; NO_2; P_2O_3; (KNO_3 + As_2S_3); K_3N; $KMnO_4$; $AgNO_3$; Ag_2O; NaH; ($NaNO_3$ + charcoal); (Na + SnI_4); SCl_2; Tl_2O_3; F_2. Dangerous; when heated it burns and emits highly toxic fumes of SO_x. Can react with oxidizing materials. To fight fire, use water or special mixtures of dry chemical. For further information see Vol. 3, No. 2 of *DPIM Report*.

SULFUR CHLORIDE   HR: 3
CAS: 10025-67-9   NIOSH: WS 4300000
DOT: 1828
mf: Cl_2S_2 mw: 135.02

PROP: Amber to yellowish-red, oily, fuming liquid, penetrating odor, decomp in water. Mp: −80°, bp: 138.0°, flash p: 245°F (CC), d: 1.6885 @ 15.5°/15.5°, autoign temp: 453°F, vap press: 10 mm @ 27.5°, vap d: 4.66.

SYNS: CHLORIDE OF SULFUR (DOT) * DISULFUR DICHLORIDE * SIARKI CHLOREK (POLISH) * SULFUR MONOCHLORIDE * SULFUR SUBCHLORIDE * THIOSULFUROUS DICHLORIDE

OSHA PEL: TWA 1 ppm
DFG MAK: 1 ppm (6mg/m^3)
DOT Classification: Label: Corrosive

THR: Poison by ingestion and inhalation. A fuming, corrosive liquid which is very irritating to skin, eyes, and mucous membranes. It decomposes on contact with water to form hydrogen chloride, thiosulfuric acid, and sulfur. These decomposition products are highly irritating. Its toxic effects are irritating to the upper respiratory tract, although the results of intoxication are usually transitory in nature. However, if hydrolysis is not complete in the upper respiratory tract, injury to the bronchioles and alveoli can result. A fire hazard when in contact with organic matter; P_2O_3; Na_2O_2; water; $Cr(OCl)_2$. Low flammability by heat or flame. Dangerous; when heated to decomposition it emits highly toxic fumes of chlorides and SO_x; will react with water or steam to produce heat and toxic and corrosive fumes; can react with oxidizing materials. To fight fire, use CO_2, dry chemical. For further information see Vol. 5, No. 6 of *DPIM Report*.

SULFUR DICHLORIDE   HR: 3
CAS: 10545-99-0   NIOSH: WS 4500000
DOT: 1828
mf: Cl_2S mw: 102.96

PROP: Reddish-brown liquid, pungent odor. Mp: $-78°$; bp: $59°$; d: 1.621 @ $15°/15°$; vap d: 3.55.

SYN: CHLORIDE OF SULFUR (DOT)

DOT Classification: Label: Corrosive

THR: Poison irritant and corrosive to skin, eyes, and mucous membranes. Flammable when exposed to heat or flame. Reacts violently with Al; NH_3; K; Na; acetone; dimethyl sulfoxide; water; oxidants; metals; hexafluoro isopropylidene amino lithium. Reactive with H_2O; steam. When heated to decomposition it emits very toxic fumes of SO_x and Cl^-.

SULFUR DIOXIDE   HR: 3
CAS: 7446-09-5   NIOSH: WS 4550000
DOT: 1079
mf: O_2S mw: 64.06

PROP: Colorless gas or liquid, pungent odor, mp: $-75.5°$, bp: $-10.0°$, d (liquid): 1.434 @ $0°$, vap d: 2.264 @ $0°$, vap press: 2538 mm @ $21.1°$.

SYNS: SULFUR OXIDE * SIARKI DWUTLENEK (POLISH) * SULFUROUS ACID ANHYDRIDE * SULFUROUS ANHYDRIDE * SULFUROUS OXIDE

OSHA PEL: TWA 5 ppm
ACGIH TLV: TWA 2 ppm
DFG MAK: 2 ppm (5 mg/m^3)
DOT Classification: Label: Nonflammable Gas (IMO: Poison A)

THR: Poison to humans by inhalation. An eye, skin, and mucous membrane irritant and corro-

sive. Mutagenic data. An experimental carcinogen. Human pulmonary system effects by inhalation. It chiefly affects the upper respiratory tract and the bronchi. It may cause edema of the lungs or glottis, and can produce respiratory paralysis. This material is so irritating that it provides its own warning of toxic concentration. 400-500 ppm is immediately dangerous to life and 50-100 ppm is considered to be the maximum permissible concentration for exposures of 30-60 minutes. Excessive exposures to high enough concentration of this material can be fatal. Its toxicity is comparable to that of hydrogen chloride. However, less than fatal concentration can be borne for fair periods of time with no apparent permanent damage. It is a common air contaminant. It reacts violently with acrolein; Al; $CsHC_2$; Cs_2O; chlorates; ClF_3; Cr; FeO; F_2; Mn; KHC_2; $KClO_3$; Rb_2C_2; Na; Na_2C_2; SnO; lithium acetylene carbide diammino. Dangerous; will react with water or steam to produce toxic and corrosive fumes. Incompatible with halogens, or interhalogens; lithium nitrate; metal acetylides; metal oxides; metals; polymeric tubing; potassium chlorate; sodium hydride. For further information see Vol. 1, No. 3 of *DPIM Report*.

SULFUR FLUORIDE   HR: 1
CAS: 2551-62-4   NIOSH: WS 4900000
DOT: 1080
mf: F_6S mw: 146.06

PROP: Colorless gas; mp: $-51°$ (sublimes @ $-64°$), vap d: 6.602, d(liquid): 1.67 @ $-100°$.

SYNS: HEXAFLUORURE DE SOUFRE (FRENCH) * SULFUR HEXAFLUORIDE (DOT)

OSHA PEL: TWA 1000 ppm
ACGIH TLV: TWA 1,000 ppm
DFG MAK: 1,000 ppm (6,000 mg/m^3)
DOT Classification: Label: Nonflammable Gas

THR: This material is chemically inert in the pure state and is considered to be physiologically inert as well. However, as it is ordinarily obtainable, it can contain variable quantities of the low sulfur fluorides. Some of these are toxic, very reactive chemically and corrosive in nature. These materials can hydrolyze on contact with water to yield hydrogen fluoride, which is highly toxic and very corrosive. In high concentrations and when pure it may act as a simple asphyxiant. Vigorous reaction with disilane. May explode;

when heated to decomposition emits highly toxic fumes of F^- and SO_x. Incompatible with disilane.

SULFURIC ACID   HR: 3
CAS: 7664-93-9   NIOSH: WS 5600000
DOT: 1830/1832
mf: H_2O_4S mw: 98.08

PROP: Colorless, oily liquid, odorless, mp: 10.49°, d: 1.834, vap press 1 mm @ 145.8°. Bp: 290°, decomp @ 340°. Misc with water and alc (liberating great heat).

SYNS: ACIDE SULFURIQUE (FRENCH) * ACIDO SOLFORICO (ITALIAN) * DIPPING ACID * NORDHAUSEN ACID (DOT) * OIL OF VITRIOL * SULPHURIC ACID * SCHWEFELSAEURELOE-SUNGEN (GERMAN) * VITRIOL BROWN OIL * ZWAVELZUUROPLOSSINGEN (DUTCH)

OSHA PEL: TWA 1 mg/m^3
ACGIH TLV: TWA 1 mg/m^3
DFG MAK: 1 mg/m^3
DOT Classification: Label: Corrosive

THR: Poison by inhalation and other unspecified routes. An eye irritant and extremely irritating, corrosive, and toxic to tissue resulting in rapid destruction of tissue, causing severe burns. If much of the skin is involved, it is accompanied by shock, collapse and symptoms similar to those seen in severe burns. See also sulfates. There are systemic effects secondary to tissue damage caused by contact with it. However, repeated contact with dilute solutions can cause a dermatitis, and repeated or prolonged inhalation of a mist of sulfuric acid can cause an inflammation of the upper respiratory tract leading to chronic bronchitis. Sensitivity to sulfuric acid or mists or vapors varies with individuals. Normally 0.125-0.50 ppm may be mildly annoying and 1.5-2.5 ppm can be definitely unpleasant. 10-20 ppm is unbearable. Workers exposed to low concentrations of the vapor gradually lose their sensitivity to its irritating action. Inhalation of concentrated vapor or mists from hot acid or oleum can cause rapid loss of consciousness with serious damage to lung tissue. Severe exposure may cause a chemical pneumonitis; erosion of the teeth due to exposure to strong acid fumes has been recognized in industry. A common air contaminant. This is a very powerful, acidic oxidizer which can ignite or even explode on contact with many materials; i.e., acetic acid; acetone cyanhydrin; (acetone +

HNO_3); (acetone + $K_2Cr_2O_7$); acetonitrile; acrolein; acrylonitrile; (acrylonitrile + H_2O); (alcohols + H_2O_2); allyl alcohol; allyl chloride; NH_4OH; 2-amino ethanol; NH_4triperchromate; aniline; (bromates + metals); BrF_5; n-butyraldehyde; carbides; $CoHC_2$; chlorates; (metals + chlorates); ClF_3; chlorosulfonic acid; Cu_3N; diisobutylene; (dimethyl benzylcarbinol + H_2O_2); epichlorohydrin; ethylene cyanhydrin; ethylene diamine; ethylene glycol; ethylene imine; fulminates; HCl; H_2; IF_7; (indene + HNO_3); Fe; isoprene; Li_6Si_2; Hg_3N_2; mesityl oxide; metals; (HNO_3 + glycerides); p-nitrotoluene; perchlorates; $HClO_4$; (C_6H_6 + permanganates); pentasilver trihydroxydiamino phosphate; (1-phenyl-2-methyl propyl alcohol + H_2O_2); P; $P(OCN)_3$; picrates; potassium-$tert$-butoxide; $KClO_3$; $KMnO_4$; ($KMnO_4$ + KCl); ($KMnO_4$ + H_2O); β-propiolactone; $RbHC_2$; propylene oxide; pyridine; Na; Na_2CO_3; NaOH; steel; styrene monomer; water; vinyl acetate; (HNO_3 + toluene). Dangerous; when heated, emits highly toxic fumes; will react with water or steam to produce heat; can react with oxidizing or reducing materials. Emits toxic fumes of SO_x. For further information see Vol. 5, No. 3 of *DPIM Report*.

SULFURIC ACID, DIMETHYL ESTER   HR: 3
CAS: 77-78-1   NIOSH: WS 8225000
DOT: 1595
mf: $C_2H_6O_4S$ mw: 126.14

PROP: Colorless, odorless liquid; mp: −31.8°, bp: 188°, flash p: 182°F (OC), d: 1.3322 @ 20°/4°, vap d: 4.35, autoign temp: 370°F. poisonous oil.

SYNS: DIMETHYLESTER KYSELINY SIROVE (CZECH) * DIMETHYL MONOSULFATE * DI-METHYLSULFAAT (DUTCH) * DIMETHYLSULFAT (CZECH) * DIMETHYL SULFATE * DIMETIL-SOLFATO (ITALIAN) * DWUMETYLOWY SIAR-CZAN (POLISH) * METHYL SULFATE * SUL-FATE DE METHYLE (FRENCH) * SULFATE DIMETHYLIQUE (FRENCH)

OSHA PEL: TWA 1 ppm (skin)
DOT Classification: Label: Corrosive

THR: Poison to humans by inhalation. Poison by ingestion, intravenous, and subcutaneous routes experimentally. An experimental tumorigen and transplacental brain carcinogen. Mutagenic data. A severe skin, mucous membrane, and eye irritant. There is no odor or initial irritation to give warning of exposure. On brief, mild exposures, conjunctivitis, catarrhal inflammation of the mucous membranes of the nose, throat, larynx and trachea and possibly some reddening of the skin develop after the latent period. With longer, heavier exposures, the cornea shows clouding, the irritation changes to the nasopharynx are more marked and after 6 to 8 hours pulmonary edema may develop. Death may occur in 3 or 4 days. The liver and kidneys are frequently damaged. Spilling of the liquid on the skin can cause ulceration and local necrosis. In patients surviving severe exposures, there may be serious injury of the liver and kidneys, with suppression of urine, jaundice, albuminuria and hematuria appearing. Death, resulting from the kidney or liver damage, may be delayed for several weeks. Flammable when exposed to heat, flame or oxidizers. Dangerous; see sulfates. Can react with oxidizing materials. Violent reaction with NH_4OH; NaN_3. To fight fire, use water, foam, CO_2, dry chemical. For further information see dimethyl sulfate, Vol. 1, No. 5 of *DPIM Report*.

SULFURIC ACID, FUMING   HR: 3
CAS: 8014-95-7   NIOSH: WS 5605000
DOT: 1831
mf: $H_2O_4S \cdot O_3S$ mw: 178.14

PROP: Heavy, fuming yellow liquid. H_2SO_4 + up to 80% SO_3. A solution of sulfuric anhydride (sulfur trioxide) in anhydrous sulfuric acid.

SYNS: DISULPHURIC ACID * DITHIONIC ACID
* FUMING SULFURIC ACID * OLEUM
* PYROSULPHURIC ACID

DOT Classification: Label: Corrosive

THR: Poison by inhalation. Severe irritant to skin, eyes, and mucous membranes by ingestion and inhalation. See also sulfurous acid. Dangerous fire hazard by chemical reaction with reducing agents and carbohydrates. Severe explosion hazard by chemical reaction with acetic acid; acetic anhydride; acetonitrile; acrolein; acrylic acid; acrylonitrile; allyl alcohol; allyl chloride; 2-amino ethanol; NH_4OH; aniline; cresol; *n*-butyraldehyde; cumene; dichloroethyl ether; diethylene glycol monomethyl ether; diisobutylene; epichlorohydrin; ethyl acetate; ethylene cyanohydrin; ethylene diamine; ethylene glycol; ethylene glycol monoethyl ether acetate;

ethylene imine; glyoxal; HCl; HF; isoprene; isopropyl alcohol; mesityl oxide; methyl ethyl ketone; HNO_3; 2-nitropropane; β-propiolactone; propylene oxide; pyridine; NaOH; styrene monomer; vinylidene chloride; sulfolane; vinyl acetate. Dangerous; when heated to decomposition it emits highly toxic fumes of SO_x; will react with water or steam to produce heat and toxic and corrosive fumes; can react vigorously with reducing materials.

SULFURIC ACID, MIST   HR: 3
CAS: 7664-93-9   NIOSH: WS 5608000
mf: H_2O_4S mw: 98.08

PROP: The airborne form of sulfuric acid, is an aerosol of droplets of varying diameter of aq sulfuric acid solution.

THR: Poison by inhalation. Human mouth effects. See also sulfuric acid.

SULFURIC ACID, MONODODECYL ESTER, SODIUM SALT   HR: 3
CAS: 151-21-3   NIOSH: WT 1050000
mf: $C_{12}H_{26}O_4S \cdot Na$ mw: 289.43

PROP: White to cream-colored crystals, flakes or powder. Water sol.

SYNS: DODECYL ALCOHOL, HYDROGEN SULFATE, SODIUM SALT * DODECYL SODIUM SULFATE * DODECYL SULFATE, SODIUM SALT
* DREFT * DUPONAL * LAURYL SODIUM SULFATE * LAURYL SULFATE, SODIUM SALT
* SODIUM DODECYLSULFATE * SODIUM LAURYLSULFATE * MAPROFIX 563 * NCI-C50191

THR: Poison by intraperitoneal route. Moderately toxic by ingestion. Human skin irritant. An eye irritant. A mild allergen. See also esters and sulfates. Mutagenic data. When heated to decomposition it emits toxic fumes of SO_x. For further information see sodium lauryl sulfate, Vol. 2, No. 1 of *DPIM Report*.

SULFUROUS ACID   HR: 3
CAS: 7782-99-2   NIOSH: WT 2775000
DOT: 1833
mf: H_2SO_3 mw: 82.08

PROP: Colorless liquid, suffocating sulfur odor (in solution only). D: approx 1.03.

SYNS: SULFUR DIOXIDE SOLUTION * SCHWEFLIGE SAURE (GERMAN)

DOT Classification: Label: Corrosive

THR: Poison by ingestion and inhalation with irritation to skin, eyes, and mucous membranes. See also SO$_x$. Dangerous; when heated to decomposition it emits highly toxic fumes of SO$_x$.

SULFUROUS ACID, 2-(p-t-BUTYL-PHENOXY)-1-METHYLETHYL-2-CHLOROETHYL ESTER   HR: 3
NIOSH: WT 2975000
mf: C$_{15}$H$_{23}$ClO$_4$S   mw: 334.89

PROP: Liquid. Misc with many organic solvents; insol in water; d: 1.145-1.1620; mp: −31.7°; bp: 175° @ 0.1 mm; vap press: <10 mm @ 25°.

SYNS: ACRYLSAEUREAETHYLESTER (GERMAN) * AETHYLACRYLATE (GERMAN) * 2-(P-T-BUTYLPHENOXY)-1-METHYLETHYL-2-CHLOROETHYL SULFITE * SULFUROUS ACID, 2-(P-TERT-BUTYLPHENOXY)-1-METHYLETHYL-2-CHLOROETHYL ESTER * ACARACIDE * BUTYLPHENOXYISOPROPYL CHLOROETHYL SULFITE * 2-(P-BUTYLPHENOXY)ISOPROPYL 2-CHLOROETHYL SULFITE * 2-(4-T-BUTYLPHENOXY)ISOPROPYL-2-CHLOROETHYL SULFITE * 2-(P-T-BUTYLPHENOXY)ISOPROPYL 2'-CHLOROETHYL SULPHITE * 2-(P-T-BUTYLPHENOXY)-1-METHYLETHYL 2-CHLOROETHYL ESTER OF SULPHUROUS ACID * 2-(P-BUTYLPHENOXY)-1-METHYLETHYL 2-CHLOROETHYL SULFITE * 2-(P-T-BUTYLPHENOXY)-1-METHYLETHYL 2'-CHLOROETHYL SULPHITE * 2-(P-T-BUTYLPHENOXY)-1-METHYLETHYL SULPHITE OF 2-CHLOROETHANOL * BETA-CHLOROETHYL-BETA'-(P-T-BUTYLPHENOXY)-ALPHA'- METHYLETHYL SULFITE * BETA-CHLOROETHYL-BETA-(P-T-BUTYLPHENOXY)-ALPHA-METHYLETHYL SULPHITE * 2-CHLOROETHYL 1-METHYL-2-(P-T-BUTYLPHENOXY)ETHYL SULPHATE * 2-CHLOROETHYL SULPHITE OF 1-(P-T-BUTYLPHENOXY)-2-PROPANOL * ENT 16,519

THR: Poison to humans by ingestion. Poison experimentally by intraperitoneal route. An experimental carcinogen, neoplastigen, and tumorigen. See also esters and sulfurous acid. A pesticide, and chlorinated hydrocarbons. When heated to decomposition it emits toxic fumes of Cl$^-$ and SO$_x$. For further information see Vol. 1, No. 3 of *DPIM Report*.

SULFUR TRIOXIDE   HR: 3
CAS: 7446-11-9   NIOSH: WT 4830000
DOT: 1829
mf: O$_3$S   mw: 80.06

PROP: It exists in 3 forms; the most valuable commercially is the γ form (mp: 16.8°, bp: 44.8°) which has a strong tendency to polymerize to the straight chain β form (mp β: 32.5°) and subsequently to the x-linked α form (mp α: 62°). When the β or α forms are melted they tend to revert to the γ form liquid or ice-like crystals. SO$_3$ (β) asbestos-like crystals; vap press: (β) 433 mm @ 250°, vap press (α): 344 mm, vap d: 2.76.

SYNS: SULFURIC ANHYDRIDE * SULFURIC OXIDE

DOT Classification: Label: Corrosive

THR: Poison by inhalation. Human pulmonary system effects. Severe irritant to skin, eyes, and mucous membranes. A corrosive. See also sulfuric acid. Violent reaction with O$_2$F$_2$; PbO; NClO$_2$; HClO$_4$; P; tetrafluorethylene; acetonitrile; sulphuric acid; dimethyl sulfoxide; dioxan; water; diphenylmercury; formamide; iodine; pyridine; metal oxides. Dangerous; when heated to decomposition it emits highly toxic fumes; reacts with steam to form corrosive, toxic fumes of sulfuric acid. For further information see Vol. 1, No. 5 of *DPIM Report*.

SULFURYL CHLORIDE   HR: 3
CAS: 7791-25-5   NIOSH: WT 4870000
DOT: 1834
mf: Cl$_2$O$_2$S   mw: 134.96

PROP: Colorless liquid, pungent odor; mp: −54.1°, bp: 69.1°, d: 1.6674, vap press: 100 mm @ 17.8°, vap d: 4.65.

SYNS: SULFONYL CHLORIDE * SULFURIC OXYCHLORIDE

DOT Classification: Label: Corrosive

THR: A corrosive material to skin, eyes, and mucous membranes. See sulfuric acid and hydrochloric acid, which are formed upon hydrolysis. Can explode with PbO$_2$. Dangerous; when heated to decomposition it emits highly toxic fumes of chlorides and SO$_x$; will react with water or steam to produce heat and toxic and corrosive fumes. Incompatible with alkalis; diethyl ether; dimethyl sulfoxide; dinitrogen pentaoxide; lead dioxide; phosphorus.

SULFURYL FLUORIDE   HR: 3
CAS: 2699-79-8   NIOSH: WT 5075000
DOT: 2191
mf: F$_2$O$_2$S   mw: 102.06

PROP: Colorless gas; mp: $-137°$; bp: $-55°$; d: 3.72 g/L.

SYNS: FLUORURE DE SULFURYLE (FRENCH) * SULFURIC OXYFLUORIDE

OSHA PEL: TWA 5 ppm
ACGIH TLV: TWA 5 ppm; STEL 10 ppm
DOT Classification: Label: Nonflammable Gas (IMO: Poison A Gas)

THR: Poison by ingestion and inhalation. See also fluorides. Accidental human exposure caused nausea, vomiting, cramps, itching. May be narcotic in high concentration. Can react with water, steam. When heated to decomposition it emits very toxic fumes of F^- and SO_x.

SUNLIGHT   HR: 3
PROP: A source of actinic (i.e. ultra violet) radiation; in the region of 100Å to 3900Å.

THR: Exposure to eyes is dangerous. Over exposure of skin can result in severe skin burns and often skin cancer.

SUPERPREDNOL   HR: 3
CAS: 50-02-2   NIOSH: TU 3980000
mf: $C_{22}H_{29}FO_5$ mw: 392.51

SYNS: DEXAMETHASONE ALCOHOL * 1-DEHY-DRO-16-ALPHA-METHYL-9-ALPHA-FLUOROHYDRO-CORTISONE * DELTA(SUP 1)-9-ALPHA-FLUORO-16-ALPHA-METHYLCORTISOL * 9-ALPHA-FLUORO-16-ALPHA-METHYLPREDNISOLONE * 9-ALPHA-FLUORO-16-ALPHA-METHYL-1,4-PREGNADIENE-11-BETA,17-ALPHA,21-TRIOL-3,20-DIONE * 4-ALPHA-FLUORO-16-ALPHA-METHYL-11-BETA,17,21-TRIHYDROXYPREGNA-1,4-DIENE-3,20-DIONE * 9-FLUORO-11-BETA,17,21-TRIHYDROXY-16-ALPHA-METHYL-PREGNA-1,4-DIENE-3,20-DIONE * 9-ALPHA-FLUORO-11-BETA,17-ALPHA, 21-TRIHYDROXY-16-ALPHA-METHYLPREGNA-1,4-DIENE-3,20-DIONE * 16-ALPHA-METHYL-9-ALPHA-FLUORO-1-DEHYDROCORTISOL * 16-ALPHA-METHYL-9-ALPHA-FLUORO-DELTA(SUP 1)-HYDROCORTISONE * 16-ALPHA-METHYL-9-ALPHA-FLUORO-1,4-PREGNADIENE-11-BETA,17-ALPHA,21-TRIOL-3,20-DIONE * 16-ALPHA-METHYL-9-ALPHA-FLUOROPREDNISOLONE * 16-ALPHA-METHYL-9-ALPHA-FLUORO-11-BETA,17-ALPHA,21-TRIHYDROXYPREGNA-1,4-DIENE-3,20-DIONE

THR: Poison by intraperitoneal and subcutaneous routes. An experimental teratogen. When heated to decomposition it emits toxic fumes of F^-.

SURFACTANTS   HR: 3
THR: Suspected carcinogens of the skin, lungs, alimentary canal, bladder.

SURITAL SODIUM   HR: 3
CAS: 337-47-3   NIOSH: CQ 0350000
mf: $C_{12}H_{17}N_2O_2S \cdot Na$ mw: 276.36

SYNS: 5-ALLYL-5-(1-METHYLBUTYL)-2-THIO-BARBITURATE SODIUM * 5-ALLYL-5-(1-METHYLBUTYL)-2-THIO-BARBITURIC ACID SODIUM SALT * SODIUM-5-ALLYL-5-(1-METHYLBUTYL)-2-THIOBARBITURATE * SURITAL SODIUM SALT * THIOMYLAL SODIUM

THR: Poison by ingestion, subcutaneous, intravenous, and intraperitoneal routes. When heated to decomposition it emits very toxic fumes of NO_x, SO_x, and Na_2O.

SWAT   HR: 3
CAS: 122-10-1   NIOSH: LZ 9450000
mf: $C_9H_{15}O_8P$ mw: 282.21

SYNS: ENT 24,833 * DIMETHYL-1,3-DI(CAR-BOMETHOXY)-1-PROPEN-2-YL PHOSPHATE * DIMETHYL 3-(DIMETHOXYPHOSPHINYLOXY)GLUTACONATE * DIMETHYL 3-HYDROXYGLUTACONATE DIMETHYL PHOSPHATE

THR: Insecticide. Poison by ingestion. When heated to decomposition it emits very toxic fumes of PO_x.

SYDNOPHEN HYDROCHLO-RIDE   HR: 3
CAS: 3441-64-3   NIOSH: WU 7682000
mf: $C_{11}H_{13}N_3O \cdot ClH$ mw: 239.73

SYNS: 3-(BETA-PHENYLISOPROPYL)-SIDNONIMINE HYDROCHLORIDE * 3-(ALPHA-METHYLPHENE-THYL)SYDONE IMINE MONOHYDROCHLORIDE

THR: Poison by intraperitoneal, intravenous, and subcutaneous routes. When heated to decomposition it emits very toxic fumes of NO_x and HCl.

meta-SYNEPHRINE HYDROCHLO-RIDE   HR: 3
CAS: 61-76-7   NIOSH: DO 7525000
mf: $C_9H_{13}NO_2 \cdot ClH$ mw: 203.6

SYNS: M-HYDROXY-ALPHA-((METHYLAMINO)-
METHYL)BENZYLALCOHOL HYDROCHLORIDE, (-)-
∗ 1-1-(M-HYDROXYPHENYL)-2-METHYL-AMINO-
ETHANOLHYDROCHLORIDE ∗ 1-M-HYDROXY-
ALPHA-(METHYLAMINOMETHYL)BENZYL
ALCOHOL HYDROCHLORIDE ∗ M-METHYL-
AMINOETHANOLPHENOL HYDROCHLORIDE
∗ D-(-)-PHENYLEPHRINE HYDROCHLORIDE

THR: A poison by ingestion, intraperitoneal,
subcutaneous, intravenous and intramuscular
routes. Mutagenic data. When heated to decomposition it emits very toxic HCl and NO_x.

SYSTOX SULFONE   HR: 3
CAS: 4891-54-7   NIOSH: TF 2975000
mf: $C_8H_{19}O_5PS_2$ mw: 290.36

SYNS: O,O-DIETHYL-2-ETHYLMERCAPTOETHYL
THIOPHOSPHATE, THIONO ISOMER ∗ O,O-DI-
ETHYL-O-(2-ETHYLSULFONYL-
ETHYL)PHOSPHOROTHIOATE ∗ DIETHYL-2-
ETHYLSULFONYLETHYL THIONOPHOSPHATE
∗ THIONODEMETON SULFONE

THR: Poison by ingestion and possibly other
routes. When heated to decomposition it emits
very toxic fumes of PO_x and SO_x.

T

2,4,5-T
HR: 3

CAS: 93-76-5   NIOSH: AJ 8400000
DOT: 2765
mf: $C_8H_5Cl_3O_3$ mw: 255.48

PROP: Crystals, light tan solid; mp: 151°-153°. The teratogenicity is due in part to 2,3,7,8-TCDD, which is present as a contaminant.

SYNS: ACIDE 2,4,5-TRICHLORO PHENOXYACE-TIQUE (FRENCH) * ACIDO (2,4,5-TRICLORO-FENOSSI)-ACETICO (ITALIAN) * 2,4,5-TRI-CHLOROPHENOXYACETIC ACID * (2,4,5-TRICHLOOR-FENOXY)-AZIJNZUUR (DUTCH) * (2,4,5-TRICHLOR-PHENOXY)-ESSIGSAEURE (GERMAN)

OSHA PEL: TWA 10 mg/m^3
DFG MAK: 10 mg/m^3
DOT Classification: ORM-A

THR: Poison by ingestion. Moderately toxic by unspecified routes. Mutagenic data. An experimental teratogen and neoplastigen. See also chlorophenols. A highly toxic chlorinated phenoxy acid herbicide; rapidly excreted after ingestion. Readily absorbed by inhalation and ingestion, slowly by skin. Signs of intoxication include weakness, lethargy, anorexia, diarrhea, ventricular fibrillation and/or cardiac arrest and death. When heated to decomposition it emits toxic fumes of Cl$^-$. For further information see Vol. 1, No. 4 of *DPIM Report*.

TALC (powder)
HR: 2

CAS: 14807-96-6   NIOSH: WW 2710000
mf: $H_2O_3Si \cdot 3/4Mg$ mw: 96.33

PROP: White to grayish-white, fine, odorless powder. Powdered native hydrous magnesium silicate. Insol in water, cold acids or in alkalies.

SYNS: AGALITE * ALPINE TALC USP, BC 127 * ASBESTINE * CI 77718 * LO MICRON TALC USP, BC 2755 * NCI-C06008 * SNOW-GOOSE * TALCUM

THR: The talc with <1% asbestos is mainly regarded as a nuisance dust. Prolonged or repeated exposure can produce a form of pulmonary fibrosis (talc pneumoconiosis) which may be due to asbestos content. A common air contaminant.

TANNIC ACID
HR: 3

CAS: 1401-55-4   NIOSH: WW 5075000
mf: $C_{76}H_{52}O_{46}$ mw: 1701.28

PROP: Yellowish-white or brown bulky powder or flakes; mp: 200°, flash p: 390°F (OC), autoign temp: 980°F. Very sol in alc, acetone; almost insol in benzene, chloroform, ether, petroleum ether, carbon disulfide.

SYNS: D'ACIDE TANNIQUE (FRENCH) * GALLOTANNIC ACID * GALLOTANNIN * TANNIN

THR: Poison by ingestion, intravenous, and subcutaneous routes. An experimental carcinogen, tumorigen. Moderately toxic by parenteral route. Combustible when exposed to heat or flame. To fight fire, use water. When heated to decomposition it emits acrid smoke and fumes. Incompatible with salts of heavy metals; alkaloids, gelatin; albumin; starch; oxidizing materials; spirit nitrous ether; lime water. For further information see Vol. 2, No. 1 of *DPIM Report*.

TANNIN
HR: 3

CAS: 1401-55-4   NIOSH: WW 5100000
mf: $C_{13}H_{11}O_{11}$ mw: 322.23

PROP: Amorphous powder; mp: decomp @ 200°; sol in water; sltly sol in absolute alc; insol in absolute ether.

THR: An experimental and suspected carcinogen. Slightly toxic by ingestion. See also tannic acid. When heated to decomposition it emits acrid smoke and fumes.

TANNIN-FREE FRACTION OF BRACKEN FERN
HR: 3

NIOSH: EE 1521300

SYN: BRACKEN FERN, TANNIN-FREE

THR: An experimental carcinogen. When heated to decomposition it emits toxic and irritating fumes.

TANTALUM
HR: 3

CAS: 7440-25-7   NIOSH: WW 5505000
af: Ta   aw: 180.95

PROP: Gray, very hard, malleable, ductile metal. Mp: 2996°; bp: 5429°; d: 16.69. Insol in water.

SYN: TANTALUM-181

OSHA PEL: TWA 5 mg/m^3
ACGIH TLV: TWA 5 mg/m^3
DFG MAK: 5 mg/m^3

THR: An experimental carcinogen. See also specific tantalum compounds. Some industrial skin injuries from tantalum have been reported. However, systemic industrial poisoning is apparently not known. The dry powder ignites spontaneously in air. Incompatible with BrF_3; F_2.

TANTALUM CHLORIDE   **HR: 3**
CAS: 7721-01-9   NIOSH: WW 5600000
mf: Cl_5Ta mw: 358.20

SYN: TANTALUM PENTACHLORIDE

THR: Poison by intraperitoneal route. Moderately toxic by ingestion. When heated to decomposition it emits toxic fumes of Cl^-.

TANTALUM FLUORIDE   **HR: 3**
CAS: 7783-71-3   NIOSH: WW 5775000
mf: F_5Ta mw: 275.95

PROP: Deliquescent, refractive prisms. D: 4.74 @ 20°; mp: 96.8°; bp: 229.5°; vap press: 100 mm @ 130°. Sol in water, concentrated nitric acid; very sol in fuming nitric acid; slightly sol in hot carbon disulfide, hot carbon tetrachloride.

SYN: TANTALIUM PENTAFLUORIDE

THR: Poison by intravenous route. See also tantalum and fluoride compounds. When heated to decomposition it emits toxic fumes of F^-.

TCDD   **HR: 3**
CAS: 1746-01-6   NIOSH: HP 3500000
mf: $C_{12}H_4Cl_4O_2$ mw: 321.96

PROP: Colorless needles; mp: 305°.

SYNS: 2,3,7,8-CZTEROCHLORODWUBENZO-P-DWUOKSYNY (POLISH) * DIOXIN (HERBICIDE CONTAMINANT) * NCI-C03714 * TCDBD * 2,3,7,8-TCDD * 2,3,7,8-TETRACHLORODI-BENZO(B,E)(1,4)DIOXAN * 2,3,6,7-TETRA-CHLORODIBENZO-P-DIOXIN (GERMAN) * 2, 3,7,8-TETRACHLORODIBENZO-P-DIOXIN * 2,3,7,8-TETRACHLORODIBENZO-1,4-DIOXIN

THR: A deadly poison by ingestion, skin contact, and intraperitoneal routes. It causes death in rats by hepatic cell necrosis. Death can follow

a lethal dose by weeks. Acute and subacute exposure result in hepatic necrosis, thymic atrophy, hemorrhage, lymphoid depletion, chloracne. An experimental carcinogen, teratogen, tumorigen, and neoplastigen. An eye irritant. For further information see 2,3,7,8-Tetrachlorodibenzo-1,4-dioxin, Vol. 1, No. 2 of *DPIM Report*.

TCTP   **HR: 3**
CAS: 6012-9-7-1   NIOSH: XN 0350000
mf: C_4Cl_4S mw: 221.90

SYNS: 2,3,4,5-CHLOROTHIOPHENE * ENT 25,764 * 2,3,4,5-TETRACHLORO THIOPHENE

THR: Poison by skin contact, inhalation, intravenous, and intraperitoneal routes. Moderately toxic by ingestion. Dangerous; when heated to decomposition it emits very toxic fumes of Cl^- and SO_x.

TEFLON   **HR: 3**
CAS: 9002-84-0   NIOSH: KX 4025000
mf: $(C_2F_4)n$;

PROP: Grayish-white tough plastic. Chemically very inert.

SYNS: EK 1108GY-A * FLUON * FLUORO-FLEX * POLY(ETHYLENE TETRAFLUORIDE) * POLYTETRAFLUOROETHENE * POLYTETRA-FLUOROETHYLENE * TEFLON (VARIOUS) * TETRAFLUOROETHENE HOMOPOLYMER * TETRAFLUOROETHENE POLYMER * TETRA-FLUOROETHYLENE HOMOPOLYMER * TETRA-FLUOROETHYLENE POLYMERS

THR: An experimental carcinogen and tumorigen by implant route. The finished polymerized compound is inert under ordinary conditions. There have been reports of "polymer fume fever" in humans exposed to the unfinished product dust or to pyrolysis products which also are irritants. Smoking should be prohibited in areas where this material is being fabricated or, in general, where there may be dust from it. It appears the chief health-related problem from "Teflon" is exposure to its pyrolysis or decomposition products. Dangerous; when heated to above 750°F it decomposes to yield highly toxic fumes of fluorides.

TELDRIN   **HR: 3**
CAS: 113-92-8   NIOSH: US 6475000
mf: $C_{16}H_{19}ClN_2 \cdot C_4H_4O_4$ mw: 390.90

SYNS: DL-2(-P-CHLORO-ALPHA-2-(DIMETHYL-AMINO)ETHYLBENZYL)PYRIDINE BIMALEATE * CHLORPHENIRAMINE MALEATE * 1-P-CHLOROPHENYL-1-(2-PYRIDYL)-3-DIMETH-YLAMINOPROPANE MALEATE * NCI-C55265

THR: Poison by ingestion, intraperitoneal, intravenous, and subcutaneous routes. Human central nervous system effects. When heated to decomposition it emits very toxic fumes of Cl^- and NO_x.

TELLURIUM HR: 3
CAS: 13494-80-9   NIOSH: WY 2625000
af: Te   aw: 127.60

PROP: Silvery-white, metallic, lustrous element, quite brittle. Mp: 449.5°; bp: 989.8°; d: 6.24 @ 20°; vap press: 1 mm @ 520°. Insol in water, benzene, carbon disulfide.

SYNS: NCI-C60117 * TELLUR (POLISH)
ACGIH TLV: TWA 0.1 mg/m³
DFG MAK: 0.1 mg/m³

THR: Poison by subcutaneous and intratracheal routes. An experimental teratogen. Exposure causes nausea, vomiting, and central nervous system depression; and garlic odor to breath. When heated to decomposition it emits toxic fumes of TeO_x. Incompatible with halogens; interhalogens; metals; hexalithium disilicide.

TELLURIUM COMPOUNDS HR: 2
THR: Elemental tellurium has relatively low toxicity. It is converted in the body to dimethyl telluride which imparts a garlic-like odor to the breath and sweat. Heavy exposures may, in addition, result in headache, drowsiness, metallic taste, loss of appetite and nausea. Various tellurium salts may also produce similar symptoms. Large doses can be fatal, as was the case following accidental administration of sodium tellurite. Dangerous; when heated or on contact with acid or acid fumes; they emit highly toxic fumes.

TELLURIUM (DUST OR FUME)
CAS: 13494-80-9   NIOSH: WY 2705000
OSHA PEL: Air: TWA 100 ug/m³

THR: See also tellurium. When heated to decomposition it emits toxic fumes of Te.

TELLURIUM HEXAFLUORIDE HR: 3
CAS: 7783-80-4   NIOSH: WY 2800000
DOT: 2195
mf: F_6Te mw: 241.60

PROP: Colorless gas, repulsive odor. Mp: −37.6°; bp: −38.9° (subl); d: (solid) 4.006 @ −191°; (liquid) 2.499 @ −10°.
OSHA PEL: TWA 200 ug/m³
ACGIH TLV: 0.02 ppm
DOT Classification: Poison A Gas

THR: Poison by inhalation. Human skin (systemic) effects. See also fluorides and tellurium compounds. When heated to decomposition it emits very toxic fumes of F^- and Te.

TEPA HR: 3
CAS: 545-5-5-1   NIOSH: SZ 1750000
DOT: 2501
mf: $C_6H_{12}N_3OP$ mw: 173.18

PROP: Colorless crystals; mp: 41°; bp: 90° @ 23 mm. Sol in water, alc, ether.

SYNS: APO * 1,1′,1″-PHOSPHINYLIDYNETRIS AZIRIDINE * ENT 24915 * NSC 9717 * PHOSPHORIC ACID TRIETHYLENEIMINE * TEF * TRIS-(1-AZIRIDINYL)PHOSPHINE OXIDE * TRIAETHYLENPHOSPHORSAEUREAMID (GERMAN) * N,N′,N″-TRI-1,2-ETHANEDIYL PHOSPHORIC TRIMIDE * TRIETHYLENE PHOSPHORO-TRIAMIDE

DOT Classification: Label: Corrosive

THR: Poison by ingestion, skin contact, intravenous, intraperitoneal, and other unspecified routes. An experimental carcinogen, neoplastigen, and teratogen. Mutagenic data. Dangerous; when heated to decomposition it emits very toxic fumes of PO_x and NO_x.

TERBIUM HR: 2
af: Tb   aw: 158.9254

PROP: A silvery-gray, soft ductile, malleable metallic element. Easily oxidized in air. Mp: 1356°; bp: 3041°; d: 8.234.

THR: As a lanthanon it may impair blood coagulation. See also lanthanons and rare earths. Fire hazard in the form of dust in air or on contact with halogens. See also powdered metals.

TERBIUM CHLORIDE HR: 3
CAS: 10042-88-3   NIOSH: WY 9100000
mf: Cl_3Tb mw: 265.27

THR: Poison by intraperitoneal route. Moderately toxic by ingestion. A skin and eye irritant. See also terbium, chlorides and rare earths. When heated to decomposition it emits very toxic fumes of Cl^-.

TERBIUM CITRATE   HR: 3
CAS: 13482-49-0   NIOSH: TZ 8600000
mf: $C_6H_8O_7 \cdot Tb$ mw: 351.06

SYN: 2-HYDROXY-1,2,3-PROPANECARBOXYLIC
ACID TERBIUM (3+) SALT (1:1)

THR: Poison by intraperitoneal route. See also
terbium, rare earths. When heated to decomposition it emits acrid smoke and fumes.

TEREPHTHALIC ACID METHYL
ESTER   HR: 3
CAS: 120-61-6   NIOSH: WZ 1225000
mf: $C_{10}H_{10}O_4$ mw: 194.20

SYNS: 1,4-BENZENEDICARBOXYLIC ACID, DI-
METHYL ESTER (9CI) * DIMETHYL-1,4-BEN-
ZENEDICARBOXYLATE * DIMETHYLESTER
KYSELINY ISOFTALOVE (CZECH) * DIMETHYL-
P-PHTHALATE * DIMETHYL TEREPHTHALATE
* METHYL-4-CARBOMETHOXYBENZOATE
* NCI-C50055

THR: An experimental carcinogen. Moderately
toxic by ingestion and intraperitoneal routes.
An eye irritant. See also esters. When heated
to decomposition it emits acrid smoke and
fumes.

TERGITOL 08   HR: 3
CAS: 126-92-1   NIOSH: MP 0700000
mf: $C_8H_{18}O_4S \cdot Na$ mw: 233.31

SYNS: 2-ETHYL-1-HEXANOL HYDROGEN SUL-
FATE, SODIUM SALT * 2-ETHYL-1-HEXANOL
SULFATE SODIUM SALT * 2-ETHYLHEXYL SO-
DIUM SULFATE * NCI-C50204 * SODIUM
ETHASULFATE * SODIUM(2-ETHYLHEXYL)-
ALCOHOL SULFATE

THR: Poison by intraperitoneal route. Moderately toxic by ingestion and subcutaneous
routes. When heated to decomposition it emits
very toxic fumes of SO_x and Na_2O.

TERPENE POLYCHLORINATES   HR: 3
CAS: 8001-50-1   NIOSH: WZ 6400000

PROP: Chlorinated mixed terpenes.

SYNS: DICHLORICIDE MOTHPROOFER * ENT
19,442 * STROBANE

THR: Poison by ingestion and other unspecified
routes. An experimental carcinogen by ingestion. When heated to decomposition it emits
toxic fumes of Cl^-.

TESTOSTERONE   HR: 3
CAS: 58-22-0   NIOSH: XA 3030000
mf: $C_{19}H_{28}O_2$ mw: 288.47

PROP: Crystals. Mp: 155°. Insol in water; sol
in alc, ether.

SYNS: ANDROST-4-EN-17BETA-OL-3-ONE
* DELTA(SUP 4)-ANDROSTEN-17(BETA)-OL-3-
ONE * 17-BETA-HYDROXY-DELTA(SUP
4)ANDROSTEN-3-ONE * 17-BETA-HYDROXY-
ANDROST-4-EN-3-ONE * 17-BETA-HYDROXY-
4-ANDROSTEN-3-ONE * 17-HYDROXY-
(17-BETA)-ANDROST-4-EN-3-ONE
* TRANS-TESTOSTERONE * TESTOSTERONE
HYDRATE * TESTOSTOSTERONE

THR: Poison by intraperitoneal route. Mutagenic data. An experimental carcinogen, neoplastigen, and teratogen. Workers engaged in
manufacture and packaging have shown effects
from this hormone, i.e., enlargement of the
breasts in male workers. When heated to decomposition it emits acrid smoke and fumes. For
further information see Vol. 1, No. 3 of *DPIM
Report*.

TESTOSTERONE PROPIONATE   HR: 3
CAS: 57-85-2   NIOSH: XA 3115000
mf: $C_{22}H_{32}O_3$ mw: 344.54

SYNS: DELTA(SUP 4)-ANDROSTENE-17-BETA-
PROPIONATE-3-ONE * BIO-TESTICULINA
* NSC 9166 * RECTHORMONE TESTOSTERONE
* TESTOSTERON PROPIONATE * TESTOSTER-
ONE-17-PROPIONATE * TESTOSTERONE-17-
BETA-PROPIONATE

THR: An experimental tumorigen and teratogen.
See also esters and testosterone. When heated
to decomposition it emits acrid fumes and
smoke.

TETRABUTYLSTANNANE   HR: 3
CAS: 1461-25-2   NIOSH: WH 8605000
mf: $C_{16}H_{36}Sn$ mw: 347.21

SYNS: TETRA-N-BUTYLCIN (CZECH) * TETRA-
BUTYLTIN

OSHA PEL: TWA 100 ug(Sn)/m^3 (skin)

THR: Poison by intravenous route. Moderately
toxic by skin contact. See also tin compounds.
An eye irritant. When heated to decomposition
it emits acrid smoke and fumes.

TETRACHLOROACETONE HR: 3
NIOSH: UC 3815000
mf: $C_3H_2Cl_4O$ mw: 195.85

PROP: Liquid; bp: 180°-182° sltly decomp; very sol in benzene, alc, ether.

THR: An experimental teratogen. Mutagenic data. When heated to decomposition it emits toxic fumes of Cl^-.

1,1,1,2-TETRACHLOROETHANE HR: 3
CAS: 630-20-6 NIOSH: KI 8450000
mf: $C_2H_2Cl_4$ mw: 167.84

PROP: Liquid; d: 1.588 @ 20°/4°; bp: 129°-130°; sol in water; misc in alc, ether.

SYN: NCI-C52459

THR: A suspected carcinogen. Severe eye irritant and moderate skin irritant. When heated to decomposition it emits very toxic fumes of Cl^-. Incompatible with dinitrogen tetraoxide. For further information see Vol. 4, No. 3 of *DPIM Report*.

1,1,2,2-TETRACHLOROETHYL-ENE HR: 3
CAS: 127-18-4 NIOSH: KX 3850000
DOT: 1897
mf: C_2Cl_4 mw: 165.82

PROP: Colorless liquid, chloroform-like odor. Mp: −23.35°, bp: 121.20°, d: 1.6311 @ 15°/4°, vap press: 15.8 mm @ 22°, vap d: 5.83.

SYNS: CARBON BICHLORIDE * CARBON DI-CHLORIDE * CZTEROCHLOROETYLEN (POLISH) * DOW-PER * ETHYLENE TETRACHLORIDE * NCI-C04580 * PERCHLOORETHYLEEN, PER (DUTCH) * PERCHLORAETHYLEN, PER (GERMAN) * PERCHLORETHYLENE, PER (FRENCH) * PERCHLOROETHYLENE * PERCLENE * PERCLOROETILENE (ITALIAN) * TETRA-CHLOORETHEEN (DUTCH) * TETRACHLO-RAETHEN (GERMAN) * TETRACHLOROETHYLENE (DOT) * TETRACLOROETENE (ITALIAN)

OSHA PEL: TWA 100 ppm; CL 200; Pk 300/5M/3H
DFG MAK: 50 ppm (345 mg/m³); BAT: blood 100 ug/dl
DOT Classification: ORM-A (IMO: Poison B), Label: None

THR: Poison by intravenous route. Moderately toxic by inhalation, ingestion, skin contact, and subcutaneous routes. The liquid can cause injuries to the eyes; however, with proper precautions it can be handled safely. The symptoms of acute intoxication from this material are the result of its effects upon the nervous system. Can cause dermatitis, particularly after repeated or prolonged contact with the skin. Irritates the gastrointestinal tract upon ingestion. An experimental carcinogen. Mutagenic data. It may be handled in the presence or absence of air, water, and light with any of the common construction materials at temperatures up to 140°C. This material is extremely stable and resists hydrolysis. A common air contaminant. Reacts violently with Ba; Be; Li; N_2O_4; metals; NaOH. Dangerous; when heated to decomposition it emits highly toxic fumes of chlorides. For further information see perchloroethylene Vol. 1, No. 2 of *DPIM Report*.

TETRACHLOROISOPHTHALO-NITRILE HR: 3
CAS: 1897-45-6 NIOSH: NT 2600000
mf: $C_8Cl_4N_2$ mw: 265.90

SYNS: CHLORTHALONIL (GERMAN) * NCI-C00102 * M-TETRACHLOROPHTHALONITRILE

THR: An experimental carcinogen. Moderately toxic by intraperitoneal route and slightly toxic by ingestion. When heated to decomposition it emits very toxic fumes of Cl^-, NO_x, and CN^-.

TETRACHLORONAPHTHALENE HR: 3
CAS: 1335-88-2 NIOSH: QK 3700000
mf: $C_{10}H_4Cl_4$ mw: 265.94

PROP: Crystals. Mp: 182°.
OSHA PEL: TWA 2 mg/m³ (skin)

THR: Poison by inhalation and skin contact. See also chlorinated naphthalenes and polychlorinated biphenyls. Dangerous; when heated to decomposition it emits highly toxic fumes of Cl^-. For further information see Vol. 6, No. 5 of *DPIM Report*.

TETRACHLORONITROANISOLE HR: 3
CAS: 2438-88-2 NIOSH: BZ 9625000
mf: $C_7H_3Cl_4NO_3$ mw: 290.91

SYNS: BENZENE, 1,2,4,5-TETRACHLORO-3-ME-THOXY-6-NITRO- (9CI) * ENT 22,335 * NCI-C03032 * 4-NITRO-2,3,5,6-TETRA-

CHLORANISOLE * 2,3,5,6-TETRACHLORO-4-NI-TROANISOLE

THR: Poison by ingestion. When heated to decomposition it emits very toxic fumes of Cl^- and NO_x.

2,4,5,6-TETRACHLOROPHENOL　HR: 3
CAS: 58-90-2　　　　NIOSH: SM 9275000
mf: $C_6H_2Cl_4O$　　mw: 231.88

SYNS: DOWICIDE 6 * 2,3,4,6-TETRACHLORO-PHENOL

THR: Poison by ingestion, skin contact, intraperitoneal, and subcutaneous routes. An experimental carcinogen. When heated to decomposition it emits toxic fumes of Cl^-.

TETRACYCLINE　HR: 3
CAS: 60-54-8　　　　NIOSH: QI 8750000
mf: $C_{22}H_{24}N_2O_8$　　mw: 444.48

PROP: Produced by *Streptomyces Albo-Niger*. Trihydrate, crystals; decomp @ 170°-175°.

SYNS: CYCLOMYCIN * 6-METHYL-1,11-DI-OXY-2-NAPHTHACENECARBOXAMIDE * NEOCYCLINE * OMEGAMYCIN * POLYCYCLINE * SIGMAMYCIN * TETRACYCLINE I

THR: Poison to humans by several routes. Has been implicated in aplastic anemia. An experimental teratogen. Mutagenic data. A systemic toxin in humans. When heated to decomposition it emits toxic fumes of NO_x.

TETRACYCLINE HYDROCHLOR-IDE　HR: 3
CAS: 64-75-5　　　　NIOSH: QI 9100000
mf: $C_{22}H_{24}N_2O_8 \cdot ClH$　　mw: 480.94

PROP: Very sol in water; sol in methanol, ethanol; insol in ether, hydrocarbon solvents.

SYNS: ACHROMYCIN HYDROCHLORIDE * NCI-C55561 * POLYCYCLINE HYDROCHLORIDE * TETRACYCLINE CHLORIDE * U-5965

THR: Poison by intraperitoneal and intravenous routes. When heated to decomposition it emits very toxic fumes of HCl and NO_x.

TETRADECANE　HR: 3
CAS: 64036-86-8　　　NIOSH: XB 8000000
mf: $C_{14}H_{30}$　　mw: 198.44

PROP: Colorless liquid; d: 0.765 @ 20°/4°; mp: 5.5°; bp: 252-255°; lel = 0.5%, flash p: 212°F, vap press: 1 mm @ 76.4°, vap d: 6.83,

autoign temp: 396°F, insol in water; very sol in alc, ether.

SYN: 14H

THR: An experimental carcinogen. Probably irritating and narcotic in high concentrations. Combustible when exposed to heat or flame. Moderate explosion hazard in the form of vapor when exposed to flame. Moderately dangerous; when heated, emits acrid fumes; can react with oxidizing materials. To fight fire, use foam, CO_2, dry chemical.

TETRAETHYL LEAD　HR: 3
CAS: 78-00-2　　　　NIOSH: TP 4550000
DOT: 1649
mf: $C_8H_{20}Pb$　　mw: 323.47

PROP: Colorless, oily liquid, pleasant characteristic odor. Mp: 125°-150°, bp: 198°-202° with decomp, d: 1.659 @ 18°, vap press: 1 mm @ 38.4°, flash p: 200°F. Including flash point for export shipment by water.

SYNS: CZTEROETHLEK OLOWIU (POLISH) * NCI-C54988 * TETRAETHYLPLUMBANE

OSHA PEL: TWA 75 ug(Pb)/m³ (skin)
ACGIH TLV: TWA 0.1 mg/m³
DFG MAK: 0.01 ppm (0.075 mg/m³)
DOT Classification: Poison B, Label: Poison

THR: Poison by ingestion, skin contact, intraperitoneal, intravenous, subcutaneous, parenteral, and other unspecified routes. An experimental carcinogen. See also lead compounds. This material is a solvent for fatty materials. It has some solvent action on rubber as well. The fact that it is a lipoid solvent makes it an industrial hazard because it can cause intoxication not only by inhalation but also by absorption through the skin. Decomposes when exposed to sunlight or allowed to evaporate; forms triethyl lead, which is also a poisonous compound, as one of its decomposition products. May cause lead exposure intoxication by coming in contact with the skin. A common air contaminant. Flammable when exposed to heat, flame, or oxidizers. Dangerous; see lead; can react vigorously with oxidizing materials. To fight fire, use dry chemical, CO_2, mist, foam. For further information see Vol. 5, No. 5 of *DPIM Report*.

TETRAETHYLPYROPHOS-PHATE　HR: 3
CAS: 107-49-3　　　　NIOSH: UX 6825000

DOT: 2783
mf: $C_8H_{20}O_7P_2$ mw: 290.22

PROP: Water white to amber hygroscopic liquid. D: 1.20.

SYNS: BIS-O,O-DIETHYLPHOSPHORIC ANHYDRIDE * ENT 18,771 * PYROPHOSPHATE DE TETRA-ETHYLE (FRENCH) * O,O,O,O-TETRAAETHYL-DIPHOSPHAT, BIS(O,O-DIAETHYLPHOSPHOR-SAEURE-ANHYDRID (GERMAN) * O,O, O,O-TETRAETHYL-DIFOSFAAT (DUTCH) * O,O, O,O-TETRAETIL-PIROFOSFATO (ITALIAN) * TETRAETHYL PYROFOSFAAT (BELGIAN) * TETRAETHYL PYROPHOSPHATE, LIQUID (DOT) * TEPP

OSHA PEL: TWA 50 ug/m^3 (skin)
DOT Classification: Poison B, Label: Poison

THR: Deadly poison to humans by ingestion and intramuscular route. Poison to experimental animals by ingestion, skin contact, intraperitoneal, intramuscular, subcutaneous, ocular, and other routes. Human central nervous system effects. The action is similar to that of parathion. Briefly, the action results in an irreversible inhibition of the cholinesterase molecules and the consequent accumulation of large amounts of acetylcholine. See also parathion. Small doses at frequent intervals are largely additive. When heated to decomposition it emits toxic fumes of PO$_x$. For further information see Vol. 5, No. 4 of *DPIM Report*.

TETRAFLUOROBORATE   HR: 3
CAS: 14874-70-5   NIOSH: ED 2675000
mf: BF_4 mw: 86.81

PROP: Colorless liquid; bp: decomp @ 130°.

SYNS: FLUOBORIC ACID * HYDROFLUOBORIC ACID

DOT Classification: Label: Corrosive

THR: A corrosive material that irritates the skin, eyes, and mucous membranes. See also fluorides. When heated to decomposition it emits toxic fumes of F$^-$ and BO$_x$.

TETRAFLUORO HYDRAZINE   HR: 3
mf: N_2F_4 mw: 104.0

PROP: Colorless gas or liquid; white solid when pure. Mp: −163°; bp: −73°; d (liq): 1.5 @ −100°.

THR: No toxicity data. See also fluorides, hydrofluoric acid and hydrazine. Flammable and

explosive. Highly reactive with reducing agents. Can react explosively with reducing agents at normal temperatures. When ignited with hydrogen it can explode. At high pressures it can explode due to shock or blast. Dangerous; when heated to decomposition it emits highly toxic fumes of F$^-$ and NO$_x$. Heat, shock or blast can detonate it when under pressure. Incompatibles; air; alkenyl nitrates; hydrocarbons; hydrogen; nitrogen trifluoride; organic materials; ozone.

TETRAFLUORO-m-PHENYLENE DI-AMINE DIHYDROCHLORIDE   HR: 3
CAS: 63886-77-1   NIOSH: ST 4025000
mf: $C_6H_4F_4N_2 \cdot 2ClH$ mw: 253.04

THR: Poison by intraperitoneal route. An experimental carcinogen. When heated to decomposition it emits very toxic fumes of F$^-$, NO$_x$, and HCl.

1-trans-delta(sup 8)-TETRAHYDROCAN-NABINOL   HR: 3
CAS: 5957-75-5   NIOSH: HP 8400000
mf: $C_{21}H_{30}O_2$ mw: 314.51

SYNS: (−)-DELTA(SUP 8)-TRANS-TETRAHYDRO-CANNABINOL * DELTA(SUP 6)-THC * DELTA(SUP 8)-THC

THR: Poison by intravenous and intraperitoneal routes. Moderately toxic by ingestion. When heated to decomposition it emits acrid smoke and fumes.

1-trans-delta(sup 9)-TETRAHYDROCAN-NABINOL   HR: 3
CAS: 1972-08-3   NIOSH: HP 8225000
mf: $C_{21}H_{30}O_2$ mw: 314.51

SYNS: (−)-DELTA(SUP 9)-TRANS-TETRAHYDRO-CANNABINOL * DELTA(SUP 9)-THC * 6,6,9-TRIMETHYL-3-PENTYL-7,8,9,10-TETRAHYDRO-6H-DIBENZO(B,D)PYRAN-1-OL

THR: Poison by inhalation, intraperitoneal, and intravenous routes. Moderately toxic by ingestion. An experimental teratogen. When heated to decomposition it emits acrid smoke and fumes.

1,2,3,4-TETRAHYDRODIBENZ(a,j) ANTHRACENE   HR: 3
CAS: 16310-68-2   NIOSH: HN 5775000
mf: $C_{22}H_{18}$ mw: 282.40

THR: An experimental tumorigen. When heated to decomposition it emits acrid smoke and fumes.

1,2,3,4-TETRAHYDRO-3-ALPHA,4-BETA-DIHYDROXY-1-alpha,2, alpha-EPOXY BENZ(a)ANTHRA-CENE (E)   HR: 3
CAS: 64598-80-7   NIOSH: CX 3072000
mf: $C_{18}H_{14}O_3$ mw: 278.32

SYNS: BA 3,4-DIOL-1,2-EPOXIDE 1 * BA 3,4-DIOL-1,2-EPOXIDE-2 * TRANS-3-ALPHA,4-BETA-DIHYDROXY-1-ALPHA,2-ALPHA-EPOXY-1,2,3,4-(+)-(1R,2S,3R,4R)-3,4-DIHYDRO-3,4-DIHYDROXY-1,2-EPOXYBENZ(a)ANTHRACENE

THR: An experimental tumorigen by skin contact. When heated to decomposition it emits acrid smoke and irritating fumes.

TETRAHYDROFURAN   HR: 2
CAS: 109-99-9   NIOSH: LU 5950000
DOT: 2056
mf: C_4H_8O mw: 72.12

PROP: Colorless, mobile liquid, ether-like odor. Bp: 65.4°, flash p: 1.4°F (TCC), lel = 1.8%, uel = 11.8%, fp: −108.5°, d: 0.888 @ 20°/4°, vap press: 114 mm @ 15°, vap d: 2.5, autoign temp: 610°F. Misc with water, alc, ketones, esters, ethers and hydrocarbons.

SYNS: BUTYLENE OXIDE * CYCLOTETRA-METHYLENE OXIDE * DIETHYLENE OXIDE * FURANIDINE * HYDROFURAN * NCI-C60560 * OXACYCLOPENTANE * TETRAHYDROFURAAN (DUTCH) * TETRAHYDROFURAN (DOT) * TETRAIDROFURANO (ITALIAN) * TETRAMETHYLENE OXIDE

OSHA PEL: TWA 200 ppm
ACGIH TLV: 200 ppm; STEL 250 ppm
DFG MAK: 200 ppm (590 mg/m³)
DOT Classification: Label: Flammable Liquid

THR: Moderately toxic by ingestion, inhalation, and intraperitoneal routes. Irritant to eyes and mucous membranes. Narcotic in high concentrations. Reported as causing injury to liver and kidneys. Dangerous fire hazard by heat, flames, oxidizers. In common with other ethers, unstabilized tetrahydrofuran forms thermally explosive peroxides on exposure to air. Stored ether samples must always be tested for peroxide prior to distillation. Peroxides can be removed by treatment with strong ferrous sulfate solution

made slightly acidic with sodium bisulfate. Reacts violently with air on standing; $LiAlH_2$; KOH; $NaAlH_2$; NaOH; lithium tetrahydroaluminate; sodium tetrahydroaluminate. When heated to decomposition it emits irritating fumes; can react with oxidizing materials. To fight fire, use foam, dry chemical, CO_2. For further information see Vol. 5, No. 5 of *DPIM Report*.

TETRAHYDROPHENOBARBITAL HR: 3
CAS: 52-31-3   NIOSH: CQ 2975000
mf: $C_{12}H_{16}N_2O_3$ mw: 236.30

SYNS: CYCLOBARBITAL * CYCLOBARBITOL * CYCLOBARBITONE * CYCLOHEXENYL-ETHYL BARBITURIC ACID * 5-(1-CYCLOHEXENYL)-5-ETHYLBARBITURIC ACID * 5-(1-CYCLOHEXEN-1-YL)-5-ETHYLBARBITURIC ACID * 5-ETHYL-5-CYCLOHEXENYLBARBITURIC ACID

THR: Poison by ingestion, subcutaneous, and intraperitoneal routes. Human psychotropic and pulmonary system effects by ingestion. See also barbiturates. When heated to decomposition it emits toxic fumes of NO_x.

TETRAHYDROPHTHALIC ACID ANHYDRIDE   HR: 2
CAS: 85-43-8   NIOSH: GW 5775000
DOT: 2698
mf: $C_8H_8O_3$ mw: 152.16

PROP: White powder. Mp: 101.9°; bp: 195° @ 50 mm; flash p: 315°F (OC); d: 1.375 @ 25°/20°; vap press: 0.01 mm @ 20°; vap d: 5.25.

SYNS: ANHYDRID KYSELINY TETRAHYDROFTA-LOVE (CZECH) * 1,2,3,6-TETRAHYDRO PHTHALIC ANHYDRIDE * THPA

DOT Classification: Label: Corrosive

THR: Moderately toxic by intraperitoneal route. A skin and eye irritant. Combustible when exposed to heat or flame. Slightly dangerous; when heated it emits acrid fumes; will react with water or steam to produce heat; can react with oxidizing materials. To fight fire, use water, foam, CO_2, dry chemical.

TETRAMETHYLAMMONIUM CHLORIDE   HR: 3
CAS: 75-57-0   NIOSH: BS 7700000
mf: $C_4H_{12}N \cdot Cl$ mw: 109.62

SYN: USAF AN-8

THR: Poison by intraperitoneal, subcutaneous, and other unspecified routes. See also chlorides. When heated to decomposition it emits very toxic fumes of NO_x and Cl^-.

TETRAMETHYLAMMONIUM HYDROX-
IDE   HR: 3
CAS: 75-59-2   NIOSH: PA 0875000
DOT: 1835
mf: $C_4H_{12}N \cdot HO$ mw: 91.18

PROP: Liquid; d: 1.

SYN: HYDROXYDE DE TETRAMETHYLAMMONIUM (FRENCH)

DOT Classification: Label: Corrosive

THR: Poison by subcutaneous route. A powerful caustic. Severe irritant to skin, eyes, and mucous membranes. When heated to decomposition it emits toxic fumes of NO_x.

TETRAMETHYL DIAMINO BENZO-
PHENONE   HR: 3
CAS: 90-94-8   NIOSH: DJ 0250000
mf: $C_{17}H_{20}N_2O$ mw: 268.39

SYNS: 4,4'-BIS(DIMETHYLAMINO)BENZOPHE-NONE * BIS(P-(N,N-DIMETHYL AMINO)PHE-NYL)KETONE * MICHLER'S KETONE * NCI-c02006

THR: An experimental carcinogen and neoplastigen. See also ketones. When heated to decomposition it emits toxic fumes of NO_x.

TETRAMETHYL LEAD   HR: 3
CAS: 75-74-1   NIOSH: TP 4725000
mf: $C_4H_{12}Pb$ mw: 267.35

PROP: Colorless liquid. Mp: $-18°F$, lel = 1.8%, bp: 110°, d: 1.99, vap d: 9.2, flash p: 100°F.

SYN: TETRAMETHYLPLUMBANE

OSHA PEL: TWA 70 ug(Pb)/m^3 (skin)
ACGIH TLV: TWA 0.15 mg/m^3
DFG MAK: 0.01 ppm (0.075 mg/m^3)

THR: Poison by ingestion, intraperitoneal, parenteral, and intravenous routes. An experimental carcinogen. Moderately toxic by skin contact. See also lead compounds. Dangerous fire hazard when exposed to heat, flame or oxidizers. Moderate explosion hazard in the form of vapor when exposed to flame. Dangerous; see lead; can react vigorously with oxidizing materials. Can explode. To fight fire, use water, foam, CO_2, dry chemical.

1,1,3,3-TETRAMETHYLUREA   HR: 3
CAS: 632-22-4   NIOSH: YU 2625000
mf: $C_5H_{12}N_2O$ mw: 116.19

PROP: Liquid, fat odor. Bp: 177°, mp: $-1.2°$, d: 0.969, flash p: 167°F. Very sol in alc, ether; misc with water.

SYN: TMU

THR: An experimental teratogen. Moderately toxic by ingestion and intravenous routes. Flammable by heat, flame, and oxidizers. To fight fire, use foam, mist, spray, dry chemicals. When heated to decomposition it emits toxic fumes of NO_x.

TETRANITROANILINE   HR: 2
CAS: 3698-5-4-2   NIOSH: BY 9275000
mf: $C_6H_3N_5O_8$ mw: 273.14

PROP: Solid. Mp: 170°; bp: Explodes @ 237°.

SYNS: TETRANITRANILINE (FRENCH) * TNA * 2,3,4,6-TETRANITROANILINE

THR: Moderately toxic by subcutaneous route. For fire hazard, see nitrates. Severe explosion hazard when shocked or exposed to heat. Tetranitroaniline is a powerful and sensitive high explosive, similar to tetryl. It deteriorates in the presence of moisture. See also explosives, high and nitrates. Dangerous; shock or heat will explode it; when heated to decomposition it emits toxic fumes of NO_x. Incompatible with reducing materials.

TETRANITROMETHANE   HR: 3
CAS: 509-14-8   NIOSH: PB 4025000
DOT: 1510
mf: CN_4O_8 mw: 196.05

PROP: Colorless or yellow liquid. Mp: 13°, bp: 125.7°, d: 1.650 @ 13°, vap press: 10 mm @ 22.7°. Insol in water; very sol in alc, ether.

SYN: NCI-c55947

OSHA PEL: TWA 1 ppm
ACGIH TLV: TWA 1 ppm
DFG MAK: 1 ppm (8mg/m^3)
DOT Classification: Label: Oxidizer

THR: Poison by ingestion, inhalation, intravenous, and intraperitoneal routes. Irritating to the eyes, skin, mucous membranes, and respiratory passages, and does serious damage to the liver. It occurs as an impurity in crude TNT, and is thought to be mainly responsible for the irritating properties of that material. It can cause pulmonary edema, mild methemoglobinemia and fatty degeneration of the liver and kidneys. Dangerous fire hazard. See nitrates and explosives, high. Severe explosion hazard when shocked or exposed to heat. It can form very powerful explosives when mixed with other nitro high explosives which are somewhat oxygen-deficient, or hydrocarbons. Highly dangerous; shock will explode it; when heated to decomposition it emits highly toxic fumes of NO_x; can react vigorously with oxidizing materials. Incompatible with Al; toluene; cotton; aromatic nitro compounds; hydrocarbons. For further information see Vol. 5, No. 5 of *DPIM Report*.

TETRANITRO NAPHTHALENE
mf: $C_{10}H_4(NO_2)_4$ mw: 308.2

PROP: Mp; 200° (approx); bp: explodes.

PROP: Crystals.

THR: No toxicity data. See also nitrates, organic. Dangerous fire hazard. Severe explosion hazard when shocked or exposed to heat. Tetranitronaphthalene is a much used high explosive equal to but somewhat less sensitive to impact than TNT. See also nitrates and explosives, high. Dangerous; shock or heat will explode it; when heated to decomposition it emits toxic fumes of NO_x. Incompatible with reducing materials.

TETRAPROPYL LEAD   HR: 3
CAS: 3440-75-3   NIOSH: TP 4900000
mf: $C_{12}H_{28}Pb$ mw: 379.59

PROP: Colorless liquid, sol in benzene. D: 1.44, bp: 126° @ 13 mm.

THR: Poison by ingestion and parenteral route. See also lead compounds. When heated to decomposition it emits toxic fumes of Pb.

TETRASODIUM PYROPHOSPHATE,
ANHYDROUS   HR: 3
CAS: 7722-88-5   NIOSH: UX 7350000
mf: $O_7P_2 \cdot 4Na$ mw: 265.90

PROP: White powder. Mp: 988°, d: 2.534.

SYNS: NATRIUMPYROPHOSPHAT * SODIUM PYROPHOSPHATE * TETRANATRIUMPYROPHOSPHAT

(GERMAN) * TETRASODIUM DIPHOSPHATE * TETRASODIUM PYROPHOSPHATE

ACGIH TLV: TWA 5 mg/m^3

THR: Poison by ingestion, intraperitoneal, intravenous, and subcutaneous routes. It is not a cholinesterase inhibitor. When heated to decomposition it emits toxic fumes of PO_x.

TETRAZENE
CAS: 109-27-3   NIOSH: XF 6900000
DOT: 0114
mf: $C_2H_8N_{10}O$ mw: 188.20

PROP: Crystals.

SYNS: 4-AMIDINO-1-(NITROSAMINOAMIDINO)-1-TETRAZENE * GUANYL NITROSAMINO GUANYL TETRAZENE * 1-GUANYL-4-NITROSAMINO GUANYLTETRAZENE

DOT Classification: Class A Explosive, Label: Explosive

THR: See also nitrates, organic. A high explosive. Dangerous fire hazard see also explosives, high. Severe explosion hazard when shocked or exposed to heat. It is a high explosive which evolves much flame. Highly dangerous. When heated to decomposition it emits highly toxic fumes of NO_x and explodes.

TETRON   HR: 3
CAS: 107-49-3   NIOSH: UX 7051000
DOT: 2783
mf: $C_8H_{20}O_7P_2$ mw: 290.22

PROP: Water white to amber; hygroscopic liquid; d: 1.20. Decomp @ 170°-213°. Bp: 138° @ 2.3 mm. Misc with water, acetone, methanol, ethanol, benzene, chloroform.

SYNS: TEP * PYROPHOSPHORIC ACID, TETRAETHYL ESTER (liquid mixture) * VAPTONE * FOSVEX * TETRAETHYL PYROPHOSPHATE MIXTURE, LIQUID (DOT)

DOT Classification: Poison B, Label: Poison

THR: Deadly poison by all routes. A cholinesterase inhibitor type of insecticide. The effects of chronic exposure to small doses is additive. See parathion. Dangerous, when heated it emits highly toxic fumes.

TETRYL   HR: 3
CAS: 479-45-8   NIOSH: BY 6300000
mf: $C_7H_5N_5O_8$ mw: 287.17

PROP: Yellow monoclinic crystals. Mp: 130°, bp: explodes @ 187°, d: 1.57 @ 19°.

SYNS: PICRYLMETHYLNITRAMINE * PICRYLNI-TROMETHYLAMINE * TETRALITE * N-METH-YL-N,2,4,6-TETRANITROANILINE * 2,4,6-TET-RYL * TRINITROPHENYLMETHYLNITRAMINE * 2,4,6-TRINITROPHENYLMETHYLNITRAMINE * 2,4,6-TRINITROPHENYL-N-METHYLNITRAMINE

OSHA PEL: TWA 1500 ug/m^3 (skin)
DOT Classification: Label: Explosive A

THR: Mutagenic data. An irritant, sensitizer, and allergen. The chief effect from exposure is dermatitis. Conjunctivitis is followed by irido-cyclitis, and keratitis can occur. Sensitization produced by exposure may play a part in these symptoms. Gastrointestinal effects and anemia have also been reported. A dangerous fire and disaster hazard. See also nitrates. Severe explo-sion hazard when shocked or exposed to heat or flame. A high explosive more sensitive to shock and friction than TNT. It reacts violently with hydrazine.

THALIDOMIDE   HR: 3
CAS: 50-35-1   NIOSH: TI 4375000
mf: C$_{13}$H$_{10}$N$_2$O$_4$ mw: 258.25

PROP: Needles. Mp: 269°-271°. Sltly sol in water, methanol, ethanol, acetone, ethylacry-late. Very sol in dioxane; sol in ether.

SYNS: 2,6-DIOXO-3-PHTHALIMIDOPIPERIDINE * N-(2,6-DIOXO-3-PIPERIDYL)PHTHALIMIDE * GLUTARIMIDE, 2-PHTHALIMIDO-ALPHA-PHTHALIMIDOGLUTARIMIDE * ALPHA-(N-PHTHALIMIDO)GLUTARIMIDE * 3-PHTHALIMIDO-GLUTARIMIDE * N-PHTHALOYLGLUTAMIMIDE * N-PHTHALYLGLUTAMIC ACID IMIDE * N-PHTHALYL-GLUTAMINSAEURE-IMID (GERMAN) * ALPHA-N-PHTHALYLGLUTARAMIDE

THR: Poison by ingestion. Moderately toxic by skin contact and intraperitoneal routes. Muta-genic data. A human teratogen. When heated to decomposition it emits toxic fumes of NO$_x$. For further information see Vol. 1, No. 2 of *DPIM Report*.

(+)-THALIDOMIDE   HR: 3
CAS: 841-67-8   NIOSH: TI 4900000
mf: C$_{13}$H$_{10}$N$_2$O$_4$ mw: 258.25

SYN: N-(2,6-DIOXO-3-PIPERIDYL)PHTHALIMIDE (+)-

THR: An experimental teratogen. Moderate tox-icity via oral route. When heated to decomposi-tion it emits toxic fumes of NO$_x$.

(+)-THALIDOMIDE   HR: 3
CAS: 2614-06-4   NIOSH: TI 4910000
mf: C$_{13}$H$_{10}$N$_2$O$_4$ mw: 258.25

SYN: N-(2,6-DIOXO-3-PIPERIDYL)PHTHALIMIDE (+)-

THR: Poison by ingestion. An experimental te-ratogen. When heated to decomposition it emits toxic fumes of NO$_x$.

THALLIC OXIDE   HR: 3
CAS: 1314-32-5   NIOSH: XG 2975000
mf: O$_3$Tl$_2$ mw: 456.74

PROP: Hexagonal black crystals, amorphous prisms. Mp: 717° ± 5°, bp: −O$_2$ @ 875°, d(amorphous): 9.65 @ 21°, d(hexagonal): 10.19 @ 22°.

SYNS: THALLIUM PEROXIDE * THALLIUM SES-QUIOXIDE

THR: Poison by ingestion, intraperitoneal, and intravenous routes. See also thallium com-pounds. Evolves O$_2$ @ 875°. When heated to decomposition it emits toxic fumes of Tl.

THALLIUM   HR: 3
CAS: 7440-28-0   NIOSH: XG 3425000
af: Tl   aw: 204.37

PROP: Bluish-white, soft, malleable metal. Mp: 303.5°, bp: 1457°, d: 11.85 @ 20°, vap press: 1 mm @ 825°.

SYN: RAMOR

ACGIH TLV: TWA 0.1 mg/m^3 (skin)
DFG MAK: 0.1 mg/m^3

THR: Poison by unspecified routes. See also thallium compounds. Flammable in the form of dust, when exposed to heat or flame. See also powdered metals. Violent reaction with F$_2$. When heated to decomposition it emits toxic fumes of Tl.

THALLIUM ACETATE   HR: 3
CAS: 563-68-8   NIOSH: AJ 5425000
mf: C$_2$H$_3$O$_2$•Tl   mw: 263.42

PROP: Silk-white crystals. Mp: 110°, d: 3.68. Sol in water, alc.

SYNS: THALLIUM(1+) ACETATE * THAL-LIUM(I) ACETATE * THALLIUM MONOACE-TATE * THALLOUS ACETATE

OSHA PEL: TWA 100 ug(Tl)/m^3

THR: Poison to humans by unspecified routes. Poison experimentally by ingestion, intraperitoneal, and subcutaneous routes. Mutagenic data. See also thallium compounds. When heated to decomposition it emits toxic fumes of Tl.

THALLIUM (I) CHLORIDE   HR: 3
CAS: 7791-12-0   NIOSH: XG 4200000
mf: ClTl   mw: 239.82

PROP: Colorless or white powder. Mp: 430°, bp: 720°, d: 7.00, vap press: 10 mm @ 517°.

SYNS: THALLIUM MONOCHLORIDE * THALLOUS CHLORIDE

OSHA PEL: TWA 100 ug(Tl)/m^3 (skin)

THR: Poison by ingestion and intraperitoneal routes. Mutagenic data. See also thallium compounds and chlorides. When heated to decomposition it emits very toxic fumes of Cl$^-$ and Tl. Incompatible with F$_2$.

THALLIUM COMPOUNDS   HR: 3
THR: Acute poisoning usually follows the ingestion of toxic quantities of a thallium-bearing depilatory, or accidental or suicidal ingestion of rat poison. Acute poisoning results in swelling of the feet and legs, arthralgia, vomiting, insomnia, hyperesthesia and paresthesia of the hands and feet, mental confusion, polyneuritis with severe pains in the legs and loins, partial paralysis of the legs with reaction of degeneration, angina-like pains, nephritis, wasting and weakness, and lymphocytosis and eosinophilia. About the 18th day, complete loss of the hair of the body and head occurs. Fatal poisoning has been known to occur. Industrial poisoning is reported to have caused discoloration of the hair (which later falls out), joint pain, loss of appetite, fatigue, severe pain in the calves of the legs, albuminuria, eosinophilia, lymphocytosis and optic neuritis followed by atrophy. Cases of industrial poisoning are rare, however. Dangerous; when heated, they emit highly toxic fumes of Tl.

THALLIUM (I) NITRATE (1:1)   HR: 3
CAS: 10102-45-1   NIOSH: XG 5950000
mf: NO$_3$•Tl   mw: 266.38

PROP: Cubic crystals. Mp: 206°, bp: 430°, d: 5.55. Decomp @ 450°.

SYNS: NITRIC ACID, THALLIUM(1+) SALT
* THALLOUS NITRATE

THR: Poison by ingestion, intravenous, and subcutaneous routes. Mutagenic data. Human central nervous system effects by ingestion. See also thallium compounds and nitrates. When heated to decomposition it emits very toxic fumes of Tl and NO$_x$.

THALLIUM SULFATE   HR: 3
CAS: 10031-59-1   NIOSH: XG 6600000
mf: O$_4$S•Tl

PROP: Colorless crystals. Mp: 632°, bp: decomp, d: 6.77.

SYN: SULFURIC ACID, THALLIUM SALT

OSHA PEL: TWA 100 ug(Tl)/m^3 (skin)

THR: Poison to humans by ingestion. Poison experimental by ingestion and intravenous routes. Its main hazard is due to its cumulation, especially in liver, brain and skeletal muscle; readily absorbed by gastrointestinal tract and skin. A cellular toxicant like arsenic. Fatal human dose is about 500 mg of Tl. Intake of Tl causes depilation. Many reported fatalities. See thallium compounds and sulfates. When heated to decomposition it emits very toxic fumes of SO$_x$ and Tl.

THALLIUM (I) SULFATE (2:1)   HR: 3
CAS: 7446-18-6   NIOSH: XG 6800000
mf: O$_4$S•2Tl   mw: 504.80

SYNS: THALLIUM SULFATE, SOLID (DOT)
* THALLOUS SULFATE

OSHA PEL: TWA 100 ug(Tl)/m^3 (skin)
DOT Classification: Poison B, Label: Poison

THR: Poison to humans by ingestion. Poison experimentally by ingestion and subcutaneous routes. See also thallium compounds and sulfates. When heated to decomposition it emits very toxic fumes of Tl and SO$_x$. For further information see Vol. 4, No. 1 of *DPIM Report*.

THALLIUM (II) SULFATE (1:1)   HR: 3
CAS: 63906-56-9   NIOSH: XG 6900000
mf: O$_4$S•Tl   mw: 300.43

SYN: SULFURIC ACID, THALLIUM(2+) SALT

THR: Poison by ingestion. See also thallium compounds and sulfates. When heated to decomposition it emits very toxic fumes of Tl and SO$_x$.

THENYLPYRAMINE   HR: 3
CAS: 91-80-5   NIOSH: UT 1400000
mf: C$_{14}$H$_{19}$N$_3$S   mw: 261.42

SYNS: N,N-DIMETHYL-N'-PYRID-2-YL-N'-2-THENYLETHYLENEDIAMINE * NCI-C55550 * N-(ALPHA-PYRIDYL)-N-(ALPHA-THENYL)-N',N'-DIMETHYLETHYLENEDIAMINE * 2-((2-(DIMETHYLAMINO)ETHYL)-2-THENYLAMINO)-PYRIDINE

THR: Poison by ingestion, subcutaneous, intraperitoneal, and intravenous routes. When heated to decomposition it emits very toxic fumes of SO_x and NO_x.

THEOBROMINE   HR: 3
CAS: 83-67-0   NIOSH: XH 2275000
mf: $C_7H_8N_4O_2$ mw: 180.19

PROP: White powder, bitter tasting alkaloid. Mp: 357°, sublimes @ 290°-295°. Mod sol in ammonia; almost insol in benzene, ether, chloroform, carbon tetrachloride.

SYN: 3,7-DIMETHYLXANTHINE

THR: Poison by ingestion. Moderately toxic by subcutaneous route. An experimental teratogen. Human central nervous system effects. When heated to decomposition it emits toxic fumes of NO_x.

THEOPHYLLINE   HR: 3
CAS: 58-55-9   NIOSH: XH 3850000
mf: $C_7H_8N_4O_2$ mw: 180.19

PROP: Monoclinic, odorless needles, bitter taste. Mp: 270°-274°. Sol in hot water, alkali hydroxides, ammonia, dil HCl, HNO_3; sltly sol in ether.

SYNS: AMINOPHYLLINE * 1,3-DIMETHYLXANTHINE * NSC 2066 * PSEUDOTHEOPHYLLINE * THEOPHYLLIN * THEOPHYLLINE, ANHYDROUS

THR: Poison to humans by ingestion, parenteral, intravenous, and rectal routes. Mutagenic data. An experimental teratogen. Human central nervous system effects. When heated to decomposition it emits toxic fumes of NO_x. For further information see Vol. 3, No. 4 of *DPIM Report*.

THERMIC AND OXIDATION PRODUCTS   HR: 3
THR: In the case of vegetable and animal derived oils and fats and also waxes and greases, it is a suspected carcinogen of the lung, alimentary system and bladder.

"THERMIT"
PROP: Composition: Fe_2O_3 + Al.

THR: See aluminum and iron compounds. Dangerous when exposed to heat or flame. The reaction of Fe_2O_3 + Al is typical of a series of oxide-metal reactions. These reactions are very difficult to stop, as they supply their own oxygen. They may attain a temperature of about 2500°. Dangerous; keep away from combustible materials.

THIAMINE DICHLORIDE   HR: 3
CAS: 67-03-8   NIOSH: XI 7350000
mf: $C_{12}H_{17}N_4OS \cdot ClH \cdot Cl$ mw: 337.30

PROP: Small white crystals or crystalline powder, hygroscopic, nut-like odor, sol in water, glycerol; sltly sol in alc, insol in ether, benzene. Mp: 248° (decomp).

SYNS: THIAMIN HYDROCHLORIDE * THIAMINE CHLORIDE HYDROCHLORIDE * THIAMINE HYDROCHLORIDE * THIAMINIUM CHLORIDE HYDROCHLORIDE * USAF CB-20 * VITAMIN B(SUP 1) * VITAMIN B HYDROCHLORIDE

THR: Poison by intravenous and intraperitoneal routes. When heated to decomposition it emits very toxic fumes of HCl, Cl^-, SO_x, and NO_x. The vitamin is destroyed by alkalies and alkaline drugs such as phenobarbital sodium and by oxidizing and reducing agents.

THIAMINE NITRATE   HR: 3
CAS: 532-43-4   NIOSH: XI 7400000
mf: $C_{12}H_{17}N_4OS \cdot NO_3$ mw: 327.40

PROP: White crystals or crystalline powder, non-hygroscopic, sltly sol in water, alcohol and chloroform. Mp: 196°-200° (decomp).

SYNS: THIAMINE NITRATE (SALT) (8CI) * VITAMIN B1 NITRATE * 3-(4-AMINO-2-METHYLPYRIMIDYL-5-METHYL)-4-METHYL-5,BETA-HYDROXYETHYLTHIAZOLIUM NITRATE * THIAMINE MONONITRATE * VITAMIN B1 MONONITRATE

THR: Poison by intravenous and intraperitoneal routes. See also nitrates. A powerful oxidizer. When heated to decomposition it emits very toxic fumes of NO_x, SO_x, and nitrates.

THIOACETAMIDE   HR: 3
CAS: 62-55-5   NIOSH: AC 8925000
mf: C_2H_5NS mw: 75.14

PROP: Colorless leaflets, mercaptan odor. Mp: 113°. Very sol in water; sltly sol in alc, ether.

SYNS: ACETOTHIOAMIDE * ETHANETHIOAMIDE
* THIACETAMIDE * USAF CB-21 * USAF EK-1719

THR: Poison by ingestion and intraperitoneal routes. Mutagenic data. An experimental carcinogen, teratogen, and neoplastigen. Moderately toxic by subcutaneous route. Exposure has caused liver damage. See also sulfides. When heated to decomposition it emits very toxic fumes of NO_x and SO_x. For further information see Vol. 5, No. 5 of *DPIM Report.*

THIOCYANATES   HR: 1
THR: Variable. Thiocyanates are not normally dissociated into cyanide; they have a low acute toxicity. Prolonged absorption may produce various skin eruptions, running nose, and occasionally dizziness, cramps, nausea, vomiting and mild or severe disturbances of the nervous system. Violent reactions have occurred when mixed with chlorates; nitrates; HNO_3; organic peroxides; peroxides; $KClO_3$; $NaClO_3$. Dangerous; when heated to decomposition or on contact with acid or acid fumes they emit highly toxic fumes of cyanides. Metal thiocyanates are oxidized explosively by chlorates, nitrates @ 400° in intimate mixture or spark, flame ignition, HNO_3.

4,4'-THIODIANILINE   HR: 3
CAS: 139-65-1   NIOSH: BY 9625000
mf: $C_{12}H_{12}N_2S$ mw: 216.32

PROP: Needles; mp: 108°.

SYNS: BIS(P-AMINOPHENYL)SULFIDE * BIS(4-AMINOPHENYL) SULFIDE * P,P'-DIAMINODIPHENYL SULFIDE * 4,4'-DIAMINODIPHENYL SULFIDE * DI(P-AMINOPHENYL) SULFIDE * NCI-C01707 * 4,4'-THIOANILINE * 4,4'-THIOBISBENZENAMINE * P,P-THIODIANILINE * THIODI-P-PHENYLENEDIAMINE

THR: Poison by intravenous route. Moderately toxic by ingestion. Mutagenic data. An experimental carcinogen. See also sulfides. When heated to decomposition it emits very toxic fumes of NO_x and SO_x.

beta-THIOGUANINE DEOXYRIBO-SIDE   HR: 3
CAS: 64039-27-6   NIOSH: UP 0175000
mf: $C_{10}H_{13}N_5O_3S \cdot H_2O$ mw: 301.36

SYNS: BETA-DEOXYTHIOGUANOSINE * BETA-2'-DEOXY-6-THIOGUANOSINE MONOHYDRATE
* NCI-C01581 * NSC-71261

THR: Poison by intravenous route. An experimental carcinogen and neoplastigen. When heated to decomposition it emits very toxic fumes of NO_x and SO_x.

THIONYL CHLORIDE   HR: 3
CAS: 7719-09-7   NIOSH: XM 5150000
DOT: 1836
mf: Cl_2OS mw: 118.96

PROP: Colorless to yellow to red liquid. Suffocating odor. Mp: −105°, bp: 78.8° @ 746 mm, d: 1.640 @ 15.5°/15.5°, vap press: 100 mm @ 21.4°. Misc with benzene, chloroform, carbon tetrachloride.

SYN: SULFUROUS OXYCHLORIDE

DOT Classification: Label: Corrosive and Poison

THR: This material has a pungent odor similar to that of sulfur dioxide; it fumes upon exposure to air. In the presence of moisture it decomposes into hydrogen chloride and sulfur dioxide. Both these decomposition products are very toxic and constitute serious toxicity hazards. The material itself is more toxic than sulfur dioxide. It is a corrosive liquid and burns the skin, eyes, and mucous membranes wherever it comes in contact with the body. See also hydrogen chloride and sulfur dioxide. May react violently with dimethyl sulfoxide; 2,4-hexadiyn-1-6-diol; *o*-nitrobenzoyl acetic acid; H_2O; *o*-nitrophenyl-acetic acid. Corrosive. See also hydrochloric acid and sulfur dioxide. Incompatible with ammonia; chloryl perchlorate; dimethyl sulfoxide; hexafluoro isopropylidene amino lithium; linseed oil; quinoline; sodium.

THIONYL FLUORIDE   HR: 3
CAS: 7783-42-8   NIOSH: XM 5425000
mf: F_2OS mw: 86.06

PROP: Colorless gas. Suffocating odor. Bp: −44°, mp: −130°. d (liq): 1.780 @ −100°, (solid): 2.095 @ −183°. Sol in ether, benzene.

SYNS: FLUORURE DE THIONYLE (FRENCH)
* SULFUROUS OXYFLUORIDE * THIONYL DIFLUORIDE

THR: Poison by inhalation. See also fluorides. Severe irritant to skin, eyes, and mucous membranes. Dangerous; when heated, emits highly toxic fumes; will react with water or steam to produce toxic and corrosive fumes of SO_x and F^-.

THIOPHENE **HR: 3**
CAS: 110-02-1   NIOSH: XM 7350000
mf: C_4H_4S mw: 84.14

PROP: Clear, colorless liquid, slt aromatic odor similar to benzene. D: 1.0573 @ 25°/4°; mp: −38.3°; bp: 84.4°; flash p: 21.2°F; vap press: 40 mm @ 12.5°; vap d: 2.9. Insol in water; misc with most organic solvents. May be heated to 850°F without decomposition.

SYNS: DIVINYLENE SULFIDE * THIACYCLO-PENTADIENE * THIAPHENE * THIOFURAN * USAF EK-1860

THR: Poison by intraperitoneal route. Dangerous fire hazard. May explode with HNO_3. Dangerous; when heated to decomposition it emits highly toxic fumes of SO_x. To fight fire, use foam, CO_2, dry chemical. Incompatible with oxidizing materials; HNO_3.

THIOPHOSGENE **HR: 3**
CAS: 463-71-8   NIOSH: XN 2450000
DOT: 2474
mf: CCl_2S mw: 114.97

PROP: Reddish liquid. Bp: 73.5°, d: 1.5085 @ 15°. Decomp in water, alc. Sol in ether.

SYNS: CARBON CHLOROSULFIDE * THIOCAR-BONYL CHLORIDE * THIOFOSGEN (CZECH) * THIOKARBONYLCHLORID (CZECH)

DOT Classification: Poison B, Label: Poison

THR: Poison by inhalation and intravenous routes. Moderately toxic by ingestion. Severe irritant to skin, eyes, and mucous membranes by inhalation. See also phosgene. When heated to decomposition it emits very toxic fumes of Cl^- and SO_x.

THIOPHOSPHAMIDE **HR: 3**
CAS: 52-24-4   NIOSH: SZ 2975000
mf: $C_6H_{12}N_3PS$ mw: 189.24

SYNS: NCI-C01649 * NSC-6396 * 1,1′,1″-PHOSPHINOTHIOYLIDYNETRISAZIRIDINE * THIOTRIETHYLENEPHOSPHORAMIDE * THIO-TEP * TRIAZIRIDINYLPHOSPHINE SULFIDE * N,N′,N″-TRI-1,2-ETHANEDIYLPHOSPHORO-THIOIC TRIAMIDE * N,N′,N″ -TRI-1,2-ETHANE-DIYLTHIOPHOSPHORAMIDE * TRI(ETHYL-ENEIMINO)THIOPHOSPHORAMIDE * N,N′,N″-TRIETHYLENEPHOSPHOROTHIOIC TRIAMIDE * N,N′,N″-TRIETHYLENETHIOPHOSPHAMIDE * N,N′,N″-TRIETHYLENETHIOPHOSPHORAMIDE

* TRIETHYLENETHIOPHOSPHOROTRIAMIDE
* TRIS(1-AZIRIDINYL)PHOSPHINE SULFIDE
* TRIS(ETHYLENIMINO)THIOPHOSPHATE

THR: Poison by ingestion, intraperitoneal, intravenous, and other unspecified routes. Mutagenic data. An experimental carcinogen, neoplastigen, and teratogen. Human central nervous system effects and other toxic effects unspecified as to source. When heated to decomposition it emits very toxic fumes of PO_x, SO_x, and NO_x. For further information see Tris(1-Azridino) phosphine sulfide Vol. 1, No. 2 of *DPIM Report*.

2-THIOPROPANE **HR: 3**
CAS: 75-18-3   NIOSH: PV 5075000
DOT: 1164
mf: C_2H_6S mw: 62.14

PROP: Colorless liquid, disagreeable odor. Mp: −83.2°, lel = 2.2%, uel = 19.7%, flash p: <0°F, bp: 37.5°-38°, d: 0.8458 @ 21°/4°, vap d: 2.14, autoign temp: 403°F. Insol in water; sol in alc, ether.

SYNS: DIMETHYLSULFID (CZECH) * DIMETHYL SULFIDE * METHYL SULPHIDE * METHYL-THIOMETHANE * SULFURE DE METHYLE (FRENCH) * 2-THIAPROPANE

DOT Classification: Label: Flammable Liquid

THR: Poison by inhalation. A skin irritant. See also sulfides. Very dangerous fire hazard when exposed to heat or flame. Moderate explosion hazard. Dangerous; when heated to decomposition it emits highly toxic fumes of SO_x and may explode; can react vigorously with oxidizing materials. To fight fire, use CO_2, dry chemical.

THIOSEMICARBAZIDE **HR: 3**
CAS: 79-19-6   NIOSH: VT 4200000
mf: CH_5N_3S mw: 91.15

PROP: Needles from water; mp: 182°-184°; sol in water, alc.

SYNS: N-AMINOTHIOUREA * USAF EK-1275

THR: Mutagenic data. A poison by ingestion, intraperitoneal, and intravenous routes. When heated to decomposition it emits very toxic fumes of NO_x and SO_x.

THIOSULFATES **HR: 1**
THR: Up to 12 g of sodium thiosulfate can be

taken daily by mouth with no ill effects except catharsis. Most of the thiosulfates are low in acute toxicity. Dangerous; when heated to decomposition they emit highly toxic fumes of SO_x.

2-THIOURACIL   HR: 3
CAS: 141-90-2   NIOSH: YR 1575000
mf: $C_4H_4N_2OS$ mw: 128.16

PROP: Small crystals, bitter taste, practically insol in water, alc, ether, acids; sol in alkalis.

SYNS: 2,3-DIHYDRO-2-THIOXO-4(1H)-PYRIMIDI-NONE * 6-HYDROXY-2-MERCAPTOPYRIMIDINE * 2-MERCAPTO-4-HYDROXYPYRIMIDINE * 2-MERCAPTO-4-PYRIMIDINOL * 2-MER-CAPTO-4-PYRIMIDONE * 2-MERCAPTOPYRIMID-4-ONE * 2-THIO-1,3-PYRIMIDIN-4-ONE * THIOURACIL * 6-THIOURACIL

THR: An experimental neoplastigen and carcinogen. Moderately toxic by ingestion. When heated to decomposition it emits very toxic fumes of NO_x and SO_x.

THORIUM   HR: 3
CAS: 7440-29-1   NIOSH: XO 6400000
DOT: 2975
af: Th   aw: 232.00

PROP: Silvery-white, air stable, soft, ductile metal. D: 11.72; mp: 1842° ± 30°. A radioactive element.

SYN: THORIUM METAL, PYROPHORIC (DOT)

DOT Classification: Label: Radioactive and Flammable Solid

THR: On an acute basis it has caused dermatitis. However, taken internally, as ThO_2, it has proven to be a carcinogen due to its radioactivity. Flammable in the form of dust, when exposed to heat or flame; or by chemical reaction with oxidizers, such as Cl_2; P; S; air; halogens; nitryl fluoride; O_2; peroxyformic acid. A pyrophoric element.

THORIUM CHLORIDE   HR: 3
CAS: 10026-08-1   NIOSH: XO 6475000
mf: Cl_4Th mw: 373.80

PROP: White odorless crystals. D: 4.59, mp: 770°, bp: 921°. Sol in water, alc.

SYN: THORIUM TETRACHLORIDE

THR: Poison by intravenous route. Moderately toxic by intraperitoneal and subcutaneous routes. See also thorium. When heated to decomposition it emits toxic fumes of Cl^-.

THORIUM NITRATE   HR: 3
CAS: 13823-29-5   NIOSH: XO 6825000
DOT: 2976
mf: $N_4O_{12}Th \cdot 4H_2O$ mw: 552.12

PROP: White crystalline mass, sol in water, alc.

SYN: THORIUM (IV) NITRATE TETRAHYDRATE

DOT Classification: Label: Radioactive and Oxidizer

THR: Poison by intraperitoneal, intravenous, and intratracheal routes. See also thorium and nitrates. Dangerous; oxidizing material; when in contact with readily combustible substances, will cause violent combustion or ignition; it emits toxic fumes of NO_x.

THORIUM OXIDE   HR: 3
CAS: 1314-20-1   NIOSH: XO 6950000
mf: O_2Th mw: 264.00

PROP: Heavy, white crystalline powd. D: 9.7; mp: 3390°. Insol in water, alkalies; sol in acids but slowly. Sterile contrast medium made up of 24-26% stabilized colloidal thorium dioxide in 25% aqueous dextrin and 0.15% methyl parasept as preservative.

SYNS: THORIA * THORIUM DIOXIDE * THO-ROTRAST

THR: An experimental carcinogen. See also thorium.

THUJONE   HR: 3
CAS: 546-80-5   NIOSH: XO 9625000
mf: $C_{10}H_{16}O$ mw: 152.26

PROP: A flavor constituent. A major component of Wormwood Oil, (*artemisia absinthium, L*) which is the principal ingredient of Absinthe, a liquer. Occurs as alpha, l, (−) or beta, d, (+) called isothujone.

SYNS: (−)-THUJONE * ALPHA-THUJONE * -(−)-3-THUJANONE, (1S,4R,5R)

THR: Poison by intraperitoneal and subcutaneous routes. Serious physiological consequences from abuse of Absinthe (mainly in France), led to its abolition in 1915. Wormwood is still used in concentrations of less than 10 ppm in flavored wines. Thujon at 30 mg/kg causes convulsions associated with lesions of the cerebral cortex.

Little is known of Thujone metabolism. Both forms occur in Wormwood oil, Oak Moss. The alpha form is major constituent of Cedar Leaf Oil or Oil of Thuja, Sage. The beta form occurs in Tansy, Yarrow. When heated to decomposition it emits acrid smoke and fumes.

THULIUM and THULIUM COMPOUNDS   HR: 3
af: Tm   aw: 168.9342

PROP: A bright, silvery-gray, lustrous, soft, malleable, ductile metallic element. Mp: 1545°; bp: 1727°; d: 9.333.

THR: Poison by intraperitoneal route. As a lanthanide it has probably at least a moderate degree of toxicity. See also Rare Earths. Flammable in the form of dust when exposed to flame. See also powdered metals. Explosive in the form of dust, when exposed to flame or violent reaction with air, halogens. See also powdered metals.

THULIUM CHLORIDE   HR: 3
CAS: 13537-18-3   NIOSH: XP 0525000
mf: Cl_3Tm mw: 275.28

THR: Poison by intraperitoneal route. A skin and eye irritant. See also thulium and thulium compounds, rare earths. When heated to decomposition it emits toxic fumes of Cl^-.

THULIUM (III) NITRATE, HEXAHYDRATE (1:3:6)   HR: 3
CAS: 35725-33-8   NIOSH: XP 1050000
mf: $N_3O_9 \cdot Tm \cdot 6H_2O$ mw: 463.08

SYN: NITRIC ACID, THULIUM(3+) SALT, HEXAHYDRATE

THR: Poison by intraperitoneal route. See also thulium, nitrates, rare earths. When heated to decomposition it emits toxic fumes of NO_x.

THYMOL   HR: 3
CAS: 89-83-8   NIOSH: XP 2275000
mf: $C_{10}H_{14}O$ mw: 150.24

PROP: Colorless, translucent crystals. Pungent, caustic taste. Mp: 51°, bp: 233°, d: 0.972, vap press: 1 mm @ 64°. Sol in water, alkali; very sol in alc, ether, chloroform.

SYNS: 2-ISOPROPYL-5-METHYLPHENOL * M-THYMOL * P-CYMEN-3-OL * 3-P-CYMENOL * 3-HYDROXY-P-CYMENE * 3-HYDROXY-1-METHYL-4-ISOPROPYLBENZENE

* ISOPROPYL CRESOL * 1-METHYL-3-HYDROXY-4-ISOPROPYLBENZENE * 5-METHYL-2-ISOPROPYL-1-PHENOL * THYME CAMPHOR * THYMIC ACID

THR: Poison by ingestion, intravenous, intraperitoneal, and subcutaneous routes. An allergen. When heated to decomposition it emits acrid smoke and fumes. Incompatible with acetanilide; antipyrine; camphor; monobromated camphor; chlorohydrate; menthol; quinine sulfate; salol; urethene; spirit nitrous ether.

TIAMUTIN   HR: 3
CAS: 55297-96-6   NIOSH: AG 9560000
mf: $C_{28}H_{47}NO_4S \cdot C_4H_4O_4$ mw: 609.90

SYNS: SQ 22947 * 14-DEOXY-14-((2-DIETHYLAMINOETHYL)MERCAPTOACETOXY)-MUTILIN HYDROGEN FUMARATE

THR: Poison by intramuscular route. Moderately toxic by ingestion and other routes. When heated to decomposition it emits very toxic fumes of NO_x and SO_x.

TIN   HR: 3
CAS: 7440-31-5   NIOSH: XP 7320000
af: Sn   aw: 118.69

PROP: Cubic, gray, crystalline metallic element. Mp: 231.9°, stabilizes <18°, d: 7.31, vap press: 1 mm @ 1492°; bp: 2507°.

SYNS: SILVER MATT POWDER * TIN (ALPHA) * TIN FLAKE * TIN POWDER * ZINN (GERMAN)

ACGIH TLV: TWA metal 2 mg/m³, oxide and inorganic compounds (except SnH4) as Sn 2 mg/m³, organic compounds 0.1 mg/m³ (skin)
DFG MAK: Inorganic 2 mg/m³, organic 0.1 mg/m³

THR: An experimental tumorigen. See also tin compounds. Combustible in the form of dust when exposed to heat or by spontaneous chemical reaction with Br_2; BrF_3; Cl_2; ClF_3; $Cu(NO_3)$; K_2O_2; S. See also powdered metals. For further information see Vol. 1, No. 3 of *DPIM Report*.

TIN (IV) BROMIDE (1:4)   HR: 3
CAS: 7789-67-5   NIOSH: XP 8300000
mf: Br_4Sn mw: 438.33

PROP: White crystalline mass. Mp: 31°, bp: 202°, d(liquid): 3.340 @ 35°, vap press: 10 mm @ 72.7°.

SYNS: STANNIC BROMIDE ✳ TIN PERBROMIDE ✳ TIN TETRABROMIDE

THR: Poison by intravenous route. See also bromides and tin compounds. Violent reaction with NO_2Cl. When heated to decomposition it emits toxic fumes of Br^-.

TIN (II) CHLORIDE (1:2)   HR: 3
CAS: 7772-99-8   NIOSH: XP 8700000
mf: Cl_2Sn mw: 189.59

PROP: Colorless crystals, sol in less than its own weight of water, very sol in hydrochloric acid (dilute or conc); sol in alc, ethyl acetate, glacial acetic acid, sodium hydroxide solution. D: 2.71, mp: 37°-38°.

SYNS: C.I. 77864 ✳ NCI-C02722 ✳ STANNOUS CHLORIDE ✳ TIN DICHLORIDE ✳ TIN PROTOCHLORIDE

OSHA PEL: TWA 2 mg(Sn)/m^3
DOT Classification: ORM-B, Label: None

THR: Poison by ingestion, intraperitoneal, intravenous, and subcutaneous routes. Violent reactions with BrF_3; CaC_2; ethylene oxide; hydrazine hydrate; nitrates; K; Na; H_2O_2. When heated to decomposition it emits toxic fumes of Cl^-.

TIN (IV) CHLORIDE (1:4)   HR: 3
CAS: 7646-78-8   NIOSH: XP 8750000
DOT: 1827
mf: Cl_4Sn mw: 260.49

PROP: Colorless, fuming caustic liquid or crystals. Mp: −33°, bp: 114.1°, d: 2.232, vap press: 10 mm @ 10°.

SYNS: ETAIN (TETRACHLORURE D') (FRENCH) ✳ LIBAVIUS FUMING SPIRIT ✳ STAGNO (TETRACLORURO DI) (ITALIAN) ✳ STANNIC CHLORIDE, ANHYDROUS ✳ TIN CHLORIDE, FUMING (DOT) ✳ TIN PERCHLORIDE ✳ TIN TETRACHLORIDE, ANHYDROUS ✳ TINTETRACHLORIDE (DUTCH) ✳ ZINNTETRACHLORID (GERMAN)

OSHA PEL: TWA 2 mg(Sn)/m^3
DOT Classification: Label: Corrosive

THR: Poison by intraperitoneal, subcutaneous, and intravenous routes. Corrosive. See also hydrochloric acid. Combustible by chemical reaction. Upon contact with moisture, considerable heat is generated. Violent reaction with K, Na, turpentine, ethylene oxide; alkyl nitrates. Dangerous; hydrochloric acid is liberated on contact with moisture or heat; it emits toxic fumes of Cl^-.

TIN COMPOUNDS   HR: 3-1
THR: Elemental tin is not generally considered toxic. Some inorganic tin salts are irritating or can liberate toxic fumes on decomposition. The latter is particularly true of tin halogens. Alkyl tin compounds may be highly toxic and produce skin rashes. Dust of tin oxides have caused a pneumoconiosis, which is relatively benign. Organic tin compounds are absorbed by the skin, many are highly toxic.

TINDURIN   HR: 3
CAS: 58-14-0   NIOSH: UV 8140000
mf: $C_{12}H_{13}ClN_4$ mw: 248.74

SYNS: 5-(4'-CHLOROPHENYL)-2,4-DIAMINO-6-ETHYLPYRIMIDINE ✳ 5-(4-CHLOROPHENYL)-6-ETHYL-2,4-PYRIMIDINEDIAMINE ✳ 2,4-DIAMINO-5-P-CHLOROPHENYL-6-ETHYLPYRIMIDINE ✳ 2,4-DIAMINO-5-(4-CHLOROPHENYL)-6-ETHYL-PYRIMIDINE ✳ NCI-C01683 ✳ NSC 3061

THR: Poison by ingestion and intraperitoneal routes. Mutagenic data. An experimental and suspected carcinogen, teratogen, and neoplastigen. When heated to decomposition it emits very toxic fumes of Cl^- and NO_x.

TIN (II) IODIDE   HR: 3
CAS: 10294-70-9   NIOSH: XQ 3650000
mf: I_2Sn mw: 372.49

SYN: STANNOUS IODIDE

THR: Poison by intravenous route. See also iodides and tin compounds. When heated to decomposition it emits toxic fumes of I^-.

TIN (IV) IODIDE (1:4)   HR: 3
CAS: 7790-47-8   NIOSH: XQ 3675000
mf: I_4Sn mw: 626.29

PROP: Red cubic crystals. Mp: 144.5°, bp: 364°, d: 4.473 @ 0°.

SYNS: STANNIC IODIDE ✳ TIN TETRAIODIDE

OSHA PEL: TWA 2 mg(Sn)/m^3

THR: Poison by intravenous route. Strong reaction with NO_2Cl, (K + S), (Na + S). See also tin compounds and iodides. When heated to decomposition it emits toxic fumes of I^-.

TIN (IV) PHOSPHIDE   HR: 3
CAS: 25324-56-5   NIOSH: XQ 4050000
DOT: 1433

PROP: Silver-white crystals. SnP, mw: 149.68, d: 6.56.

SYN: STANNIC PHOSPHIDE (DOT)

DOT Classification: Label: Flammable Solid and Dangerous When Wet

THR: See also phosphides and tin compounds. A flammable solid. Reacts with moisture, acid fumes to liberate PH_3 which is highly toxic. When heated to decomposition it emits toxic fumes of PO_x and other P compounds.

TITANIUM (DRY POWDER)   HR: 3
CAS: 7440-32-6   NIOSH: XR 1700000
DOT: 2546/2878
af: Ti   aw: 47.90

PROP: Dark gray amorphous powder or lustrous white metal. D: 4.5 @ 20°, autoign temp: 1200° for massive metal in air, 250° for powder; mp: 1677°, bp: 3277°.

SYNS: C.P. TITANIUM * TITANIUM ALLOY * NCI-C04251 * TITANIUM METAL POWDER, DRY

DOT Classification: Label: Flammable Solid

THR: An experimental tumorigen. See also titanium compounds. Flammable in the form of dust when exposed to heat or flame or by chemical reaction. See also powdered metals. Titanium can burn in an atmosphere of carbon dioxide; nitrogen or air. Also reacts violently with BrF_3; CuO; PbO; (Ni + $KClO_3$); metaloxy salts; halocarbons; halogens; CO_2; metal carbonates; Al; water; AgF; O_2; nitryl fluoride; HNO_3; O_2; $KClO_3$; KNO_3; $KMnO_4$; steam @ 704°; trichloroethylene; trichlorotri-fluoroethane. Ordinary extinguishers are often ineffective against titanium fires. Such fires require the special extinguishers designed for metal fires. See also magnesium. In airtight enclosures, titanium fires can be controlled by the use of argon or helium. When titanium burns in the absence of moisture, it burns slowly but evolves much heat. The application of water to burning titanium can cause an explosion. Finely divided titanium dust and powders, like most metal powders, are potential explosion hazards when exposed to sparks, open flame or high heat sources. For further information see Vol. 1, No. 3 of *DPIM Report*.

TITANIUM CHLORIDE   HR: 3
CAS: 7550-45-0   NIOSH: XR 1925000

DOT: 1838
mf: Cl_4Ti mw: 189.70

PROP: Colorless to light yellow liquid, fumes in moist air. Mp: −30°, bp: 136.4°, d: 1.772 @ 25°/25°, vap press: 10 mm @ 21.3°.

SYNS: TETRACHLORURE DE TITANE (FRENCH) * TITAANTETRACHLORIDE (DUTCH) * TITANIO (TETRACLORURO DI) (ITALIAN) * TITANIUM TETRACHLORIDE * TITANTETRACHLORID (GERMAN)

DOT Classification: Label: Corrosive

THR: Poison by inhalation. See also titanium compounds. Severe irritant to skin, eyes, and mucous membranes by inhalation. Severely corrosive because it liberates heat and hydrochloric acid upon contact with moisture. If spilled on skin, wipe off with dry cloth before applying water. Reacts violently with K, HF. When heated to decomposition it emits toxic fumes of Cl^- and HCl.

TITANIUM COMPOUNDS
THR: This material is considered to be physiologically inert. There are no reported cases in the literature where titanium as such has caused intoxication. The dusts of titanium or titanium compounds such as titanium oxide may be placed in the nuisance category. Titanium tetrachloride, however, is an irritant and corrosive material, because when exposed to moisture, it hydrolyzes to hydrogen chloride. See also hydrochloric acid, and titanium.

TITANIUM OXIDE   HR: 3
CAS: 13463-67-7   NIOSH: XR 2275000
mf: O_2Ti mw: 79.90

PROP: Blue crystals. Mp: 1860° (decomp), d: 4.26.

SYNS: C.I. 77891 * C.I. PIGMENT WHITE 6 * NCI-C04240 * TITANDIOXID (SWEDEN) * TITANIUM DIOXIDE * RUTILE * TRIOXIDE(S) * ATLAS WHITE TITANIUM DIOXIDE

OSHA PEL: TWA 15 mg/m^3

THR: An experimental neoplastigen and carcinogen. Human skin irritant. See also titanium compounds. A common air contaminant and nuisance dust. Violent reaction with Li and other metals. For further information see titanium dioxide, Vol. 3, No. 1 of *DPIM Report*.

TL 1217   **HR: 3**
CAS: 60398-22-3   NIOSH: BR 1840000
mf: $C_{13}H_{21}N_2O_2 \cdot I$ mw: 364.26

SYN: METHYLCARBAMIC ESTER OF
OXYPHENYLMETHYLDIETHYLAMMONIUM IODIDE

THR: Poison by ingestion, intravenous, and subcutaneous routes. See also esters and carbamates. When heated to decomposition it emits
very toxic fumes of NO_x and I^-.

**TOBACCO LEAF, NICOTIANA
GLAUCA**   **HR: 3**
                              NIOSH: XR 7357000

THR: A nicotine-containing dried leaf of the
tobacco plant. The smoke produced by burning
tobacco contains the highly toxic alkaloid, nicotine, tars and phenols, carbon monoxide, cyanides, nitrates, nitrites, carcinogen, co-car-
cinogen and perhaps 100 other chemicals,
α-emitters, etc. Habitual inhalation of tobacco
smoke is considered a leading cause of lung
cancer and circulatory problems, cardiac problems, etc. See also nicotine. An experimental
teratogen. Combustible when exposed to heat
or flame.

TOFRANIL   **HR: 3**
CAS: 50-49-7   NIOSH: HO 1575000
mf: $C_{19}H_{24}N_2$ mw: 280.45

SYNS: 1-(3-DIMETHYLAMINOPROPYL)-4,
5-DIHYDRO-2,3,6,7-DIBENZAZEPINE
* 5-(3-(DIMETHYLAMINO)PROPYL)-10,11-
DIHYDRO-5H-DIBENZ(B,F)AZEPINE * 5-(3-
DIMETHYLAMINOPROPYL)-10,11-DIHYDRO-
5H-DIBENZO(B,F)AZEPINE * 2,2'-(3-
DIMETHYLAMINOPROPYLIMINO)DIBENZYL
* N-(GAMMA-DIMETHYLAMINOPROPYL)IMINO-
DIBENZYL * 2,2'-(3-DIMETHYLAMINOPRO-
PYLAMINO)BIBENZYL * 5,6-DIHYDRO-N-(3-
(DIMETHYLAMINO)PROPYL)-11H-DIBENZ(b,
e)AZEPINE

THR: Poison by ingestion, subcutaneous, intravenous, and intraperitoneal routes. Central
nervous system effects by ingestion. An experimental teratogen by ingestion. When heated to
decomposition it emits toxic fumes of NO_x.

3,3'-TOLIDINE   **HR: 3**
CAS: 119-93-7   NIOSH: DD 1225000
mf: $C_{14}H_{16}N_2$ mw: 212.32

PROP: White to reddish crystals. Mp: 129°-
131°C. Very sltly sol in water; sol in alc, ether,
acetic acid.

SYNS: BIANISIDINE * 4,4'-BI-O-TOLUIDINE
* 4,4'-DIAMINO-3,3'-DIMETHYLBIPHENYL
* 4,4'-DIAMINO-3,3'-DIMETHYLDIPHENYL
* 3,3'-DIMETHYLBENZIDINE * 3,3'-DIMETH-
YL-4,4'-BIPHENYLDIAMINE * 3,3'-DIMETHYL-
BIPHENYL-4,4'-DIAMINE * 3,3'-DIMETHYL-4,
4'-DIPHENYLDIAMINE * 3,3'-DIMETHYLDIPHE-
NYL-4,4'-DIAMINE * 4,4'-DI-O-TOLUIDINE
* 2-TOLIDINA (ITALIAN) * 2-TOLIDIN (GER-
MAN) * O-TOLIDINE * O,O'-TOLIDINE

THR: Poison by ingestion. An experimental carcinogen and tumorigen. Mutagenic data. When
heated to decomposition it emits toxic fumes
of NO_x. For further information see Vol. 5,
No. 3 of *DPIM Report*.

TOLUENE   **HR: 3**
CAS: 108-88-3   NIOSH: XS 5250000
DOT: 1294
mf: C_7H_8 mw: 92.15

PROP: Colorless liquid, benzol-like odor.
Flammable. Mp: −95° to −94.5°, bp: 110.4°,
flash p: 40°F (CC), ulc: 75-80, lel = 1.27%,
uel = 7%, d: 0.866 @ 20°/4°, autoign temp:
896°F, vap press: 36.7 mm @ 30°, vap d: 3.14.
Insol in water; sol in acetone; misc in absolute
alc, ether, chloroform.

SYNS: METHYLBENZENE * METHYLBENZOL
* NCI-C07272 * PHENYLMETHANE * TOL-
UEEN (DUTCH) * TOLUEN (CZECH) * TOLUOL
* TOLUOLO (ITALIAN)

OSHA PEL: TWA 200 ppm; CL 300; Pk 500/
  10M
ACGIH TLV: TWA 100 ppm; STEL 150 ppm;
  BEI: toluene in venous blood end of shift 1
  mg/L
DFG MAK: 100 ppm (375 mg/m³); BAT: blood
  end of shift 340 ug/dl
DOT Classification: Label: Flammable
  Liquid

THR: Poison by intraperitoneal route. Moderately toxic by inhalation and subcutaneous
routes. Mutagenic data. A skin and eye irritant.
Human central nervous system and psychotropic
effects. Toluene is derived from coal tar, and
commercial grades usually contain small
amounts of benzene as an impurity. Inhalation

of 200 ppm of toluene for 8 hours may cause impairment of coordination and reaction time; with higher concentrations (up to 800 ppm) these effects are increased and are observed in a shorter time. In the few cases of acute toluene poisoning reported, the effect has been that of a narcotic, the victim passing through a stage of intoxication into one of coma. Recovery following removal from exposure has been the rule. An occasional report of chronic poisoning describes an anemia and leucopenia, with biopsy showing a bone marrow hypoplasia. These effects, however, are less common in people working with toluene, and they are not as severe. At 200-500 ppm, headache, nausea, eye irritation, loss of appetite, a bad taste, lassitude, impairment of coordination and reaction time are reported, but are not usually accompanied by any laboratory or physical findings of significance. With higher concentrations, the above complaints are increased and in addition, anemia, leucopenia and enlarged liver may be found in rare cases. A common air contaminant. Combustible when exposed to heat, flame or oxidizers. Moderately explosive when exposed to flame or reacted with (H_2SO_4 + HNO_3), N_2O_4, $AgClO_4$, BrF_3, UF_6. Moderately dangerous; when heated it emits irritating fumes; can react vigorously with oxidizing materials. To fight fire, use foam, CO_2, dry chemical. For further information see Vol. 5, No. 5 of *DPIM Report*.

TOLUENE-2,4-DIAMINE   HR: 3
CAS: 95-80-7   NIOSH: XS 9625000
mf: $C_7H_{10}N_2$ mw: 122.19

PROP: Prisms. Mp: 99°, bp: 280°, vap press: 1 mm @ 106.5°.

SYNS: C.I. OXIDATION BASE * M-TOLYLENE-DIAMINE * 3-AMINO-P-TOLUIDINE * 5-AMI-NO-O-TOLUIDINE * C.I. 76035 * 1,3-DI-AMINO-4-METHYLBENZENE * 2,4-DIAMINO-1-METHYLBENZENE * 2,4-DIAMINOTOLUEN (CZECH) * DIAMINOTOLUENE * 2,4-DIAMINO-TOLUENE * 2,4-DIAMINO-1-TOLUENE * 2,4-DIAMINOTOLUOL * 4-METHYL-1,3-BENZENEDI-AMINE * 4-METHYL-M-PHENYLENEDIAMINE * NCI-C02302 * 2,4-TOLAMINE * M-TOLUENEDIAMINE * 2,4-TOLUENEDIAMINE * M-TOLUYLENDIAMIN (CZECH) * M-TOLU-YLENEDIAMINE * 2,4-TOLUYLENEDIAMINE * M-TOLYENEDIAMINE * TOLYLENE-2,4-DI-AMINE * 2,4-TOLYLENEDIAMINE * 4-M-TOLYLENEDIAMINE

THR: Poison by ingestion and subcutaneous route. Mutagenic data. A skin and eye irritant. An experimental carcinogen. This material has a marked toxic action upon the liver and can cause fatty degeneration of that organ. Moderately dangerous; when heated it emits toxic fumes of NO_x. For further information see Vol. 5, No. 5 of *DPIM Report*.

TOLUENE-2,5-DIAMINE   HR: 3
CAS: 95-70-5   NIOSH: XS 9700000
mf: $C_7H_{10}N_2$ mw: 122.19

PROP: Colorless, crystalline tablets. Mp: 64°, bp: 274°.

SYNS: 4-AMINO-2-METHYLANILINE * C.I. 76042 * 2,5-DIAMINOTOLUENE * 2-METHYL-1,4-BENZENEDIAMINE * 2-METHYL-P-PHENYL-ENEDIAMINE * P-TOLUENEDIAMINE * P-TOL-UYLENDIAMINE * TOLUYLENE-2,5-DIAMINE * P,M-TOLYLENEDIAMINE

THR: Poison by ingestion and subcutaneous routes. Mutagenic data. A skin irritant. An experimental carcinogen. Has a toxic action upon the liver and can cause fatty degeneration of that organ. Its total effect upon the body seems to take place three different ways. It is toxic to the central nervous system. It produces jaundice by action on the liver and spleen, and it produces anemia by destruction of the red blood cells. In this action it is quite similar to aniline, although by no means identical with it. Its high boiling point and the fact that the material is solid at room temperature makes it somewhat less hazardous than aniline, particularly at ordinary working temperatures. The literature contains a reference to a permanent injury to an eye due to the use of this material as an eyelash dye. It is considered to be an irritating dye material. Moderately dangerous; when heated it emits toxic fumes of NO_x. For further information see Vol. 5, No. 5 of *DPIM Report*.

TOLUENE DIISOCYANATE   HR: 3
CAS: 584-84-9   NIOSH: CZ 6300000
mf: $C_9H_6N_2O_2$ mw: 174.17

PROP: Liquid at room temp, sharp, pungent odor. Mp: 19.5°-21.5°; d (liq): 1.2244 @ 20°/4°; bp: 251°; flash p: 270°F (OC); vap d: 6.0; lel = 0.9%; uel = 9.5%. Misc with alc (decomp), ether, acetone, carbon tetrachloride, benzene, chlorobenzene, kerosene, olive oil.

SYNS: 2,4-DIISOCYANATO-1-METHYL BENZENE * DI-ISOCYANATE DE TOLUYLENE (FRENCH) * DIISOCYANATOTOLUENE * 2,4-DIISOCYA-NATO-1-METHYL BENZENE (9CL) * DIISOCYA-NAT-TOLUOL (GERMAN) * ISOCYANIC ACID, METHYLPHENYLENE ESTER * NCI-C50533 * TOLUEEN-DIISOCYANAAT (DUTCH) * TOL-UEN-DISOCIANATO (ITALIAN) * TOLUILENOD-WUIZOCYJANIAN (POLISH)

OSHA PEL: TWA CL 0.02 ppm
DOT Classification: Poison B, Label: Poison

THR: Poison by inhalation and intravenous routes. A skin and eye irritant. Mutagenic data. Capable of producing severe dermatitis and bronchial spasm. A common air contaminant. Combustible when exposed to heat or flame. When heated to decomposition it emits very toxic fumes of CN^- and NO_x. To fight fire, use dry chemical, CO_2.

o-TOLUENESULFONAMIDE   HR: 3
CAS: 88-19-7   NIOSH: XT 4900000
mf: $C_7H_9NO_2S$ mw: 171.23

PROP: Tetragonal prisms; mp: 156°; sol in water, alc.

SYNS: TOLUENE-2-SULFONAMIDE * O-METH-YLBENZENESULFONAMIDE * 2-METHYLBEN-ZENESULFONAMIDE * ORTHO-TOLUOL-SULFON-AMID (GERMAN)

THR: An experimental carcinogen. Mutagenic data. An eye irritant. See also sulfonates. When heated to decomposition it emits very toxic fumes of NO_x and SO_x.

alpha-TOLUENETHIOL   HR: 3
CAS: 100-53-8   NIOSH: XT 8650000
mf: C_7H_8S mw: 124.21

PROP: A water white, mobile liquid, strong odor. Bp: 194.8°, flash p: 158°F (CC), d: 1.058 @ 20°, vap d: 4.28.

SYNS: (MERCAPTOMETHYL)BENZENE * ALPHA-MERCAPTOTOLUENE * PHENYLMETHANETHIOL * PHENYLMETHYL MERCAPTAN * ALPHA-TOLUOLTHIOL * ALPHA-TOLYL MERCAPTAN * BENZYL MERCAPTAN * BENZYLTHIOL * THIOBENZYL ALCOHOL * USAF EK-1509

THR: Poison by intraperitoneal route. Moderately toxic by ingestion. An experimental carcinogen. An eye irritant. See also sulfides. Flammable when exposed to heat or flame.

Dangerous; when heated to decomposition and on contact with acid or acid fumes it emits highly toxic fumes; can react vigorously with oxidizing materials. When heated to decomposition it emits toxic fumes of SO_x. To fight fire, use foam, CO_2, dry chemical, water spray, mist, fog. For further information see Benzyl Mercaptan, Vol. 2, No. 2 of *DPIM Report*.

m-TOLUIDINE   HR: 3
CAS: 108-44-1   NIOSH: XU 2800000
mf: C_7H_9N mw: 107.17

PROP: Colorless liquid. Mp: −50.5°, bp: 203.3°, d: 0.989 @ 20°/4°, vap press: 1 mm @ 41°, vap d: 3.90. Sltly sol in water; sol in alc, ether.

SYNS: 3-AMINOPHENYLMETHANE * M-AMINO-TOLUENE * 3-AMINOTOLUENE * M-METHYL-ANILINE * M-METHYLBENZENAMINE * 3-METHYLBENZENAMINE * 3-TOLUIDINE * M-TOLYLAMINE * 3-AMINO-1-METHYL-BENZENE * 3-AMINOTOLUEN (CZECH) * 3-METHYLANILINE * M-TOLUIDIN (CZECH)

THR: A poison by intraperitoneal route; Moderate ingestion toxicity. A skin, eye irritant. See also aniline. Flammable when exposed to heat or flame. Dangerous; when heated it emits highly toxic fumes; can react vigorously on contact with oxidizing materials. To fight fire, use foam, CO_2, dry chemical.

o-TOLUIDINE   HR: 3
CAS: 95-53-4   NIOSH: XU 2975000
mf: C_7H_9N mw: 107.17

PROP: Colorless liquid. Mp: −16.3°, bp: 200°-202°, ulc: 20-25, flash p: 185° (CC), d: 1.004 @ 20°/4°, autoign temp: 900°F, vap press: 1 mm @ 44°, vap d: 3.69. Sltly sol in water, dilute acid; sol in alc, ether.

SYNS: 1-AMINO-2-METHYLBENZENE * 2-AMINO-1-METHYLBENZENE * O-AMINOTOLUENE * 2-AMINOTOLUENE * 1-METHYL-2-AMINO-BENZENE * 2-METHYL-1-AMINOBENZENE * O-METHYLANILINE * 2-METHYLANILINE * O-METHYLBENZENAMINE * 2-METHYLBEN-ZENAMINE * O-TOLUIDIN (CZECH) * 2-TOLU-IDINE * O-TOLUIDYNA (POLISH) * O-TOL-YLAMINE

OSHA PEL: TWA 5 ppm (skin)
DFG MAK: 5 ppm (22 mg/m³)

THR: Poison by ingestion and intraperitoneal route. Moderately toxic by skin contact. Mutagenic data. A skin and eye irritant. An experimental carcinogen and neoplastigen. A suspected human carcinogen. Human mucous membrane effects. Can produce severe systemic disturbances. The main portal of entry into the body is the respiratory tract, particularly in cases of industrial exposure. The symptoms produced are headache, weakness, difficulty in breathing, air hunger, psychic disturbances, and marked irritation of the kidneys and bladder. The literature does not yield any good data for comparing the toxicity of the o-, m- and p-isomers. Their behavior is generally comparable to that of aniline. It has been determined experimentally that a concentration of about 100 ppm is the maximum endurable for 1 hour without serious consequences and that from 6-23 ppm is endurable for several hours without serious disturbances. See also aniline. Flammable when exposed to heat or flame. Reacts violently with HNO_3. Dangerous; when heated, emits highly toxic fumes; can react with oxidizing materials. Emits toxic fumes of NO_x. To fight fire, use foam, CO_2, dry chemical. For further information see Vol. 2, No. 1 of *DPIM Report*.

p-TOLUIDINE   HR: 3
CAS: 106-49-0   NIOSH: XU 3150000
mf: C_7H_9N mw: 107.17

PROP: Colorless leaflets. Mp: 44.5°, bp: 200.4°, flash p: 188°F (CC), d: 1.046 @ 20°/4°, autoign temp: 900°F, vap press: 1 mm @ 42°, vap d: 3.90. Sol in water, dilute acid, CS_2; very sol in alc, ether.

SYNS: c.i. 37107 ∗ c.i. azoic coupling component 107 ∗ p-methylaniline ∗ p-methylbenzenamine ∗ 4-methylbenzenamine ∗ 4-toluidine ∗ p-tolylamine ∗ 4-amino-1-methylbenzene ∗ p-aminotoluene ∗ 4-aminotoluene ∗ 4-aminotoluen (czech) ∗ 4-methylaniline ∗ p-toluidin (czech) ∗ tolylamine

THR: Poison by ingestion, intraperitoneal, and other unspecified routes. Severe skin and eye irritant. Flammable when exposed to heat or flame or oxidizers. Dangerous; when heated, emits highly toxic fumes; can react vigorously on contact with oxidizing materials. Emits toxic fumes of NO_x. To fight fire, use foam, CO_2, dry chemical.

TOLUIDINE BLUE   HR: 3
CAS: 3209-30-1   NIOSH: XT 6650000
mf: $C_{28}H_{22}N_2O_{10}S_2$ • 2Na   mw: 656.62

SYNS: c. i. 63340 ∗ 6,6′-((4,8-dihydroxy-1,5anthraquinonylene)diimino) di-m-toluene sulfonic acid disodium salt

THR: Poison by intravenous route. Mutagenic data. When heated to decomposition it emits very toxic fumes of NO_x and SO_x.

o-TOLUIDINE HYDROCHLORIDE   HR: 3
CAS: 636-21-5   NIOSH: XU 7350000
mf: C_7H_9N • ClH   mw: 143.63

PROP: Monoclinic prisms; mp: 218-220°; bp: 242°; sol in water; sltly sol in alc.

SYNS: 1-amino-2-methylbenzene hydrochloride ∗ 2-amino-1-methylbenzene hydrochloride ∗ 2-aminotoluene hydrochloride ∗ o-aminotoluene hydrochloride ∗ 1-methyl-2-aminobenzene hydrochloride ∗ 2-methyl-1-aminobenzene hydrochloride ∗ o-methylaniline hydrochloride ∗ 2-methylaniline hydrochloride ∗ o-methylbenzenamine hydrochloride ∗ 2-methylbenzenamine hydrochloride ∗ nci-c02335 ∗ 2-toluidine hydrochloride ∗ o-tolylamine hydrochloride

THR: Poison by intraperitoneal route. Moderately toxic by ingestion. An experimental carcinogen and neoplastigen. A suspected human carcinogen. When heated to decomposition it emits very toxic fumes of HCl and NO_x.

p-TOLUIDINE HYDROCHLORIDE   HR: 3
CAS: 540-23-8   NIOSH: XU 7525000
mf: C_7H_9N • ClH   mw: 143.63

PROP: Needles from acetic ether; mp: 243°; bp: 257.5°; sol in water, alc; insol in ether, benzene.

SYNS: 4-aminotoluene hydrochloride ∗ p-toluidinium chloride ∗ 4-methylaniline hydrochloride

THR: Poison by intraperitoneal route. Moderately toxic by ingestion. An experimental carcinogen. When heated to decomposition it emits very toxic fumes of NO_x and HCl.

1-(o-TOLYLAZO)-2-NAPHTHYL-
AMINE   HR: 3
CAS: 131-79-3   NIOSH: QM 5400000
mf: $C_{17}H_{15}N_3$ mw: 261.35

SYNS: C.I. 11390 * C.I. FOOD YELLOW 11
* FD AND C YELLOW NO. 4 * 1-(2-METHYL-
PHENYL)AZO-2-NAPHTHALENAMINE * 1-((2-
METHYLPHENYL)AZO)-2-NAPHTHALENAMINE
* 1-(2-METHYLPHENYL)AZO-2-NAPHTHYLAMINE
* O-TOLUENE-1-AZO-2-NAPHTHYLAMINE

THR: An experimental carcinogen. Moderately
toxic by ingestion, intraperitoneal, and subcutaneous routes. See also aromatic amines. When
heated to decomposition it emits toxic fumes
of NO_x.

TOLYL CHLORIDE   HR: 3
CAS: 100-44-7   NIOSH: XS 8925000
DOT: 1738
mf: C_7H_7Cl mw: 126.59

PROP: Colorless liquid, very refractive, irr,
unpleasant odor. Mp: $-43°$, bp: $179°$, lel: 1.1%,
flash p: 153°F, d: 1.1026 @ 18°/4°, autoign
temp: 1085°F, vap d: 4.36.

SYNS: BENZILE (CLORURO DI) (ITALIAN)
* BENZYL CHLORIDE * BENZYLE (CHLORURE
DE) (FRENCH) * BENZYLCHLORID (GERMAN)
* CHLOROMETHYLBENZENE * CHLOROPHENYL-
METHANE * ALPHA-CHLOROTOLUENE
* OMEGA-CHLOROTOLUENE * ALPHA-CHLOR-
TOLUOL (GERMAN) * NCI-C06360

OSHA PEL: TWA 1 ppm
DOT Classification: Label: Corrosive

THR: Poison by inhalation. Moderately toxic
by ingestion and subcutaneous routes. Mutagenic data. An experimental carcinogen and tumorigen. Human toxic effects unspecified as
to source. Flammable when exposed to heat
or flame. Moderately explosive when exposed
to flame. The decomposition rate can reach explosive violence in presence of metals such as
iron. Dangerous; see also chlorides. Will react
with water or steam to produce toxic and corrosive fumes; can react vigorously with oxidizing
materials. When heated to decomposition it
emits toxic fumes of Cl^-. For further information see Benzyl Chloride, Vol. 2, No. 2 of
DPIM Report.

p-TOLYLUREA   HR: 2
CAS: 622-51-5   NIOSH: YU 4200000
mf: $C_8H_{10}N_2O$ mw: 150.20

PROP: Plates from alcohol; mp: 188°. Very
sltly sol in cold water; sol in hot alcohol.

SYN: NCI-C02153

THR: Moderately toxic by ingestion. When
heated to decomposition it emits toxic fumes
of NO_x.

TOMATINE   HR: 3
CAS: 17406-45-0   NIOSH: XW 1050000
mf: $C_{50}H_{83}NO_{21}$ mw: 1034.34

PROP: Antifungal substance in wilt-resistant
tomato plants. Needles. Mp: 263°-268°. Sol in
ethanol, methanol, dioxane, propylene alc; almost insol in water, ether, petroleum ether.

SYNS: LYCOPERSICIN * ALPHA-TOMATINE
* TOMATIDINE GLYCOSIDE

THR: Poison by intravenous and intraperitoneal
routes. Moderately toxic by ingestion. When
heated to decomposition it emits toxic fumes
of NO_x.

TOXAPHENE   HR: 3
CAS: 8001-35-2   NIOSH: XW 5250000
DOT: 2761
mf: $C_{10}H_{10}Cl_8$ mw: 413.80

PROP: Yellow, waxy solid, pleasant piney
odor. Mp: 65°-90°. Almost insol in water; very
sol in aromatic hydrocarbons.

SYNS: CHLORINATED CAMPHENE * CHLORO-
CAMPHENE * ENT 9,735 * NCI-C00259
* OCTACHLOROCAMPHENE * POLYCHLORCAM-
PHENE * POLYCHLORINATED CAMPHENES
* POLYCHLOROCAMPHENE * TOXAFEEN
(DUTCH) * TOXAPHEN (GERMAN)

OSHA PEL: TWA 500 ug/m³ (skin)
ACGIH TLV: TWA 0.5 mg/m³ (skin)
DOT Classification: ORM-A, Label: None

THR: Poison to humans by ingestion and other
unspecified routes. Poison experimentally by
ingestion, inhalation, and intraperitoneal routes.
Mutagenic data. A skin irritant; absorbed thru
the skin. An experimental carcinogen. Causes
central nervous system stimulation with tremors,
convulsions, death. Liver injury has been reported. Lethal amounts of toxaphene can enter
the body through the mouth, lungs and skin.
Systemic absorption of the insecticide is increased by the presence of digestible oils, and
liquid preparations of the insecticide which pen-

etrate the skin more readily than do dusts and wettable powders. A toxic mixture of organochlorine pesticides stored to some extent in body fat. It resembles chlordane and, to some extent, camphor in its physiological action. It causes diffuse stimulation of the brain and spinal cord resulting in generalized convulsions of a tonic or clonic character. Death usually results from respiratory failure. Detoxification appears to occur in the liver. The lethal ingestion dose for man is estimated to be 2-7 g, a toxicity of about 4 times that of DDT. At least seven human deaths have been reported due to toxaphene, all in children. Two families have been made ill by eating vegetables containing a large residue of toxaphene. When heated to decomposition it emits toxic fumes of Cl⁻. For further information see Vol. 2, No. 2 of *DPIM Report*.

TRENTADIL HYDROCHLORIDE HR: 3
CAS: 3736-86-5   NIOSH: XH 4920000
mf: $C_{20}H_{27}N_5O_3 \cdot ClH$ mw: 421.98

SYN: 8-BENZYL-7-(2-(ETHYL(2-HYDROXYETHYL)AMINO)ETHYL)THEOPHYLLINE, HYDROCHLORIDE

THR: Poison by ingestion, intraperitoneal, and intravenous routes. When heated to decomposition it emits very toxic fumes of NO_x and HCl.

TRIALLYLAMINE HR: 3
CAS: 102-70-5   NIOSH: XX 5950000
mf: $C_9H_{15}N$ mw: 137.25

PROP: Liquid; d: 0.800 @ 20°/4°; mp: < −70°; bp: 150°-151°; flash p: 103°F (TOC).

THR: Poison by intraperitoneal route. Moderately toxic by ingestion and inhalation. Severe skin irritant. Human pulmonary system effects by inhalation. See also amines. Flammable by heat, flame, or oxidizers. To fight fire, use foam, alcohol foam, fog. When heated to decomposition it emits toxic fumes of NO_x.

TRIALLYL CYANAURATE HR: 3
CAS: 101-37-1   NIOSH: XZ 2080000
mf: $C_{12}H_{15}N_3O_3$ mw: 243.24

PROP: Bp: 120° @ 5 mm; fp: 27.3°; flash p: >176°F (TOC); d: 1.1133 @ 30°; vap press: 1 mm @ 100°.

SYNS: 2,4,6-TRIS(ALLYLOXY)TRIAZINE * TRIPROPARGYL CYANURATE * 2,4,6-TRIPROP-2-YNYLOXY-S-TRIAZINE

THR: Poison by intravenous route. See also esters. Flammable when exposed to heat, flame, or oxidizers. Dangerous; when heated to decomposition and on contact with acid or acid fumes it emits highly toxic fumes of CN⁻ and NO_x. To fight fire, spray, foam, dry chemical.

TRIAMINOTRIPHENYLMETHANE HR: 3
CAS: 548-61-8   NIOSH: PB 7350000
mf: $C_{19}H_{19}N_3$ mw: 289.41

PROP: Leaves from water; mp: 148°; sltly sol in cold water; sol in absolute alc and in benzene.

SYNS: LEUCOPARAFUCHSIN * P,P′,P″-TRIAMINOTRIPHENYLMETHANE * 4,4′,4″-TRIAMINOTRIPHENYLMETHANE * TRIS-4-AMINOFENYLMETHAN (CZECH)

THR: An experimental tumorigen. Moderately toxic by ingestion. An eye irritant. See also amines. When heated to decomposition it emits toxic fumes of NO_x.

TRIAZINATE HR: 3
CAS: 41191-04-2   NIOSH: XS 3650000
mf: $C_{19}H_{21}ClN_6O_2 \cdot C_2H_6O_3S$ mw: 511.05

SYNS: BAKER'S ANTIFOLANTE * NSC 139105

THR: Human gastrointestinal, white blood cell, eye, skin, and central nervous system effects by intravenous route. When heated to decomposition it emits very toxic fumes of Cl⁻, NO_x, and SO_x.

TRIBROMOETHANOL HR: 3
CAS: 1329-86-8   NIOSH: KM 3500000
mf: $C_2H_3Br_4O$ mw: 282.78

PROP: Crystals, ethereal odor, aromatic taste; bp: 92° @ 10 mm, mp: 79°-82°. Decomp @ 70°. Sltly water sol; sol in alc, organic solvents.

SYN: AVERTIN

THR: Poison by intravenous route. Moderately toxic by intraperitoneal route. When heated to decomposition it emits toxic fumes of Br⁻.

TRIBUTYLAMINE HR: 3
CAS: 102-82-9   NIOSH: YA 0350000
mf: $C_{12}H_{27}N$ mw: 185.40

PROP: A colorless liquid; mp: −70°; bp: 213°; flash p: 187°F (OC) d: 0.78-0.79; vap d: 6.38; insol in water sol in alc, ether.

SYNS: TRIS-N-BUTYLAMINE * TRI-N-BUTYLAMINE

THR: Poison by inhalation, skin contact, and subcutaneous routes. Moderately toxic by ingestion. See also amines. A central nervous system stimulant, irritant, and sensitizer. Flammable when exposed to heat, flame, or oxidizers. Moderately dangerous; when heated to decomposition it emits toxic fumes of NO_x; can react with oxidizing materials. To fight fire, use foam, CO_2, dry chemical.

TRI-n-BUTYL BORANE　　　HR: 3
CAS: 122-56-5　　　　NIOSH: ED 1850000
mf: $C_{12}H_{27}B$　　mw: 182.20

PROP: Colorless pyroforic liquid; mp: 34°, bp: 170° @ 222 mm, d: 0.747 @ 25°, vap press: 1 mm @ 20°, flash p: −32°F. Insol in water, sol in most organic solvents.

SYNS: TBB　＊　TRIBUTYLBORINE

THR: Poison by intravenous route. Moderately toxic by ingestion. See also boranes. Dangerous fire hazard; can ignite spontaneously.

TRI-n-BUTYL BORATE　　　HR: 2
CAS: 688-74-4　　　　NIOSH: ED 4900000
mf: $C_{12}H_{27}BO_3$　　mw: 230.20

PROP: Colorless mobile liquid, odor like n-butanol; bp: 230°, fp: <−70°, flash p: 200°F (COC), d: 0.847 @ 28°, vap d: 7.95.

SYNS: BUTYL BORATE　＊　N-BUTYL BORATE ＊　TRI-N-BUTOXYBORANE　＊　TRIBUTYL BORATE

THR: Moderately toxic by ingestion and intraperitoneal routes. Eye irritant. Flammable when exposed to heat, open flame, or oxidizers. Moderately dangerous; when heated to decomposition or on contact with acid or acid fumes, can emit toxic fumes; on contact with oxidizing materials, can react vigorously. To fight fire, use foam, CO_2, dry chemical.

TRI-n-BUTYL PHOSPHATE　　　HR: 3
CAS: 126-73-8　　　　NIOSH: TC 7700000
mf: $C_{12}H_{27}O_4P$　　mw: 266.36

PROP: Colorless odorless liquid; bp: 289° (decomp); mp: <−80°; flash p: 295°F (COC); d: 0.982 @ 20°; vap d: 9.20; sol in water; misc in alc, ether.

SYNS: TRIBUTILFOSFATO (ITALIAN)　＊　TRIBUTYLE (PHOSPHATE DE) (FRENCH)　＊　TRIBUTYL-FOSFAAT (DUTCH)　＊　TRIBUTYLPHOSPHAT (GERMAN)　＊　TRIBUTYL PHOSPHATE

OSHA PEL: TWA 5 ppm
ACGIH TLV: TWA 0.2 ppm

THR: Poison by intraperitoneal and intravenous routes. Moderately toxic by ingestion. Human central nervous system effects. A skin, eye, and mucous membrane irritant. Combustible. To fight fire, use CO_2, dry chemical, fog, mist. When heated to decomposition it emits toxic fumes of PO_x.

TRIBUTYL(8-QUINOLINOLATO) TIN　　　HR: 3
NIOSH: WH 8598000
mf: $C_{21}H_{33}NOSn$　　mw: 434.24

SYN: (8-QUINOLINOLATO)TRIBUTYL-TANNANE

THR: A poison by intravenous route. See also tin compounds. When heated to decomposition it emits toxic fumes of NO_x.

TRIBUTYLTIN NEODECANDATE　HR: 3
CAS: 28801-69-6　　　NIOSH: WH 8588000
mf: $C_{22}H_{46}O_2Sn$　　mw: 461.37

SYNS: (NEODECANOYLOXY)TRIBUTYLSTANNANE ＊　(4,4-DIMETHYLOCTANOYLOXY)TRIBUTYL-STANNANE　＊　HYDROXYTRIBUTYLSTANNANE-4,4-DIMETHYLOCTANOATE

OSHA PEL: TWA 100 ug(Sn)/m³ (skin)

THR: A poison by intravenous route. See also tin compounds. When heated to decomposition it emits acrid smoke and fumes.

S,S,S-TRIBUTYL TRITHIOPHOS-PHITE　　　HR: 3
CAS: 150-50-5　　　NIOSH: TG 5600000
mf: $C_{12}H_{27}PS_3$　　mw: 298.54

PROP: Colorless liquid, mild characteristic odor. Bp: 142°-145° @ 4.5 mm, flash p: 295°F (COC), d: 0.987 @ 20°/4°.

SYNS: PHOSPHOROTRITHIOUS ACID, S,S,S-TRIBUTYL ESTER　＊　TRIBUTYL PHOSPHOROTRITHIOITE ＊　S,S,S-TRIBUTYL PHOSPHOROTRITHIOITE

THR: Poison by intraperitoneal and other unspecified routes. A cholinesterase inhibitor, see also parathion. Combustible when exposed to heat or flame. Dangerous; when heated to decomposition it emits highly toxic fumes of PO_x and SO_x; can react vigorously with oxidizing materials.

TRIBUTYRIN　　　　　　　　　**HR: 3**
CAS: 60-01-5　　　　NIOSH: ET 7350000
mf: $C_{15}H_{26}O_6$　　mw: 302.41

PROP: Oily liquid, bitter taste; mp: −75°, d: 1.0356 @ 20°/20°. Bp: 305°-310°. Insol in water, very sol in alc, ether.

SYNS: BUTANOIC ACID, 1,2,3-PROPANETRIYL ESTER * GLYCEROL TRIBUTYRATE * BUTYRYL TRIGLYCERIDE * GLYCERYL TRIBUTYRATE * TRIBUTYROIN

THR: Poison by intravenous route. Moderately toxic by ingestion. An experimental tumorigen. See also esters. Combustible when exposed to heat or flame; can react with oxidizing materials. When heated to decomposition it emits acrid smoke and fumes.

TRICHLORACETIC ACID　　　**HR: 2**
CAS: 76-03-9
DOT: 1839/2564　　　NIOSH: AJ 7875000
mf: $C_2HCl_3O_2$　　mw: 163.38

PROP: Colorless, rhombic deliquescent crystals. Bp: 197.5°; fp: 57.7°; flash p: none. D: 1.6298 @ 61°/4°; vap press: 1 mm @ 51.0°.

SYNS: ACIDE TRICHLORACETIQUE (FRENCH) * ACIDO TRICLOROACETICO (ITALIAN) * TRICHLOORAZIJNZUUR (DUTCH) * TRICHLORESSIGSAEURE (GERMAN) * TRICHLOROACETIC ACID SOLUTION (DOT) * TRICHLOROETHANOIC ACID
ACGIH TLV: TWA 1 ppm
DOT Classification: Corrosive

THR: Moderately toxic by intraperitoneal route. Mutagenic data. A corrosive and irritating material to skin, eyes, and mucous membranes. When heated to decomposition it emits toxic fumes of Cl^-.

1,1,1-TRICHLOROETHANE　　　**HR: 2**
CAS: 71-55-6　　　NIOSH: KJ 2975000
DOT: 2831
mf: $C_2H_3Cl_3$　　mw: 133.40

PROP: Colorless liquid; bp: 74.1°, fp: −32.5°, flash p: none, d: 1.3376 @ 20°/4°, vap press: 100 mm @ 20.0°. Insol in water; sol in acetone, benzene, carbon tetrachloride, methanol, ether.

SYNS: CHLOROETHENE * CHLOROTHANE NU * CHLOROTHENE * METHYL CHLOROFORM * METHYLTRICHLOROMETHANE * NCI-C04626 * 1,1,1-TRICHLOORETHAAN (DUTCH) * 1,1,1-

TRICHLORAETHAN (GERMAN) * TRICHLORO-1,1,1-ETHANE (FRENCH) * ALPHA-TRICHLOROETHANE * 1,1,1-TRICLOROETANO (ITALIAN)

OSHA PEL: TWA 350 ppm
DFG MAK: 200 ppm (2080 mg/m³); BAT: blood 55 ug/dl
DOT Classification: ORM-A (IMO: Poison B)

THR: Moderately toxic by ingestion and intraperitoneal routes. Human psychotropic, gastrointestinal tract, and central nervous system effects. A moderate skin and severe eye irritant. Narcotic in high concentrations. Causes a proarrhythmic activity which sensitizes the heart to epinephrine-induced arrhythmias. This sometimes will cause a cardiac arrest particularly when this material is massively inhaled as in drug abuse for euphoria. Reacts violently with N_2O_4; O_2; O_2 liquid; Na; NaOH; Na-K alloy. Dangerous; see also chlorides. For further information see methyl chloroform, Vol. 2, No. 5 of *DPIM Report*.

1,1,2-TRICHLOROETHANE　　　**HR: 3**
CAS: 79-00-5　　　NIOSH: KJ 3150000
mf: $C_2H_3Cl_3$　　mw: 133.40

PROP: Liquid, pleasant odor; bp: 114°, fp: −35°, d: 1.4416 @ 20°/4°, vap press: 40 mm @ 35.2°.

SYNS: ETHANE TRICHLORIDE * NCI-C04579 * BETA-TRICHLOROETHANE * 1,2,2-TRICHLOROETHANE * TROJCHLOROETAN(1,1,2) (POLISH) * VINYL TRICHLORIDE

OSHA PEL: TWA 10 ppm (skin)
ACGIH TLV: TWA 10 ppm (skin)
DFG MAK: 10 ppm (55 mg/m³)

THR: Poison by intravenous and subcutaneous routes. Moderately toxic by ingestion, inhalation, skin contact, and intraperitoneal routes. Moderate skin and severe eye irritant. Has narcotic properties and acts as a local irritant to the eyes, nose and lungs. It may also be injurious to the liver and kidneys. An experimental carcinogen. Mutagenic data. Dangerous; see chlorides. Incompatible with potassium. For further information see Vol. 5, No. 3 of *DPIM Report*.

TRICHLOROETHENYLSILANE　　　**HR: 2**
CAS: 75-94-5　　　NIOSH: VV 6125000
DOT: 1305
mf: $C_2H_3Cl_3Si$　　mw: 161.49

PROP: Fuming liquid; bp: 90.6°; d: 1.265 @ 25°/25°; flash p: 16°F.

SYNS: TRICHLORO(VINYL)SILANE * TRICHLO-ROVINYL SILICANE * VINYLSILICON TRICHLO-RIDE * VINYL TRICHLOROSILANE

DOT Classification: Label: Flammable Liquid

THR: Moderately toxic by ingestion, inhalation, and skin contact. A skin and eye irritant. See also chlorosilanes. Dangerous fire hazard; reacts violently with water, moist air. When heated to decomposition it emits toxic fumes of Cl⁻. Will react with water or steam to produce toxic and corrosive fumes.

TRICHLOROETHYLENE   HR: 3
CAS: 79-01-6   NIOSH: KX 4550000
DOT: 1710
mf: C_2HCl_3 mw: 131.38

PROP: Mobile liquid; characteristic odor of chloroform. D: 1.4649 @ 20°/4°; bp: 86.7°; flash p: 89.6°F; lel = 12.5%; uel = 90% @ > 30°; mp: −73°; fp: −86.8°; autoign temp: 788°F; vap press: 100 mm @ 32°; vap d: 4.53.

SYNS: ACETYLENE TRICHLORIDE * 1-CHLORO-2,2-DICHLOROETHYLENE * 1,1-DICHLORO-2-CHLOROETHYLENE * DOW-TRI * ETHYLENE TRICHLORIDE * NCI-CO4546 * TRICHLOORE-THEEN (DUTCH) * TRICHLORAETHEN (GERMAN) * TRI-CLENE * TRICLORETENE (ITALIAN) * VESTROL

OSHA PEL: TWA 100 ppm; C1200; Pk 300/5M/2H
ACGIH TLV: TWA 50 ppm; STEL 200 ppm; BEI: trichloroethanol in urine end of shift 320 mg/g creatinine, trichloroethylene in end-exhaled air prior to shift and end of work week 0.5 ppm
DFG MAK: 50 ppm (260 mg/m³); BAT: blood end of work week and end of shift 500 ug/dL
DOT Classification: ORM-A (IMO: Poison B), Label: None

THR: Poison by inhalation, intravenous and subcutaneous routes. Moderately toxic by ingestion. Mutagenic data. An experimental teratogen, carcinogen, and tumorigen. A strong skin and eye irritant. Inhalation of high concentrations causes narcosis and anesthesia. A form of addiction has been observed in exposed workers. Prolonged inhalation of moderate concentrations causes headache and drowsiness. Fatalities following severe, acute exposure have been attributed to ventricular fibrillation resulting in cardiac failure. There is damage to liver and other organs from chronic exposure. A common air contaminant. Combustible when exposed to heat or flame. High concentrations of trichloroethylene vapor in high-temperature air can be made to burn mildly if plied with a strong flame. Though such a condition is difficult to produce, flames or arcs should not be used in closed equipment which contains any solvent residue or vapor. Dangerous. When heated to decomposition it emits toxic fumes of Cl⁻. See chlorides. Can react violently with Al; Ba; N_2O_4; Li; Mg; liquid O_2; O_3; KOH; KNO_3; Na; NaOH; Ti. For further information see Vol. 3, No. 1 of *DPIM Report*.

TRICHLOROFLUOROMETHANE   HR: 3
CAS: 75-69-4   NIOSH: PB 6125000
mf: CCl_3F mw: 137.36

PROP: Colorless liquid; mp: −111°, bp: 24.1°, d: 1.484 @ 17.2°.

SYNS: FLUOROTRICHLOROMETHANE * FLUO-ROTROJCHLOROMETAN (POLISH) * FREON 11 * MONOFLUOROTRICHLOROMETHANE * NCI-c04637 * TRICHLOROMONOFLUOROMETHANE

OSHA PEL: TWA 1000 ppm
ACGIH TLV: CL 1,000 ppm

THR: An experimental carcinogen. Moderately toxic by intraperitoneal route. Human eye (systemic) and peripheral nervous system effects. Reacts violently with Al, Li. High concentrations cause narcosis and anesthesia. Dangerous; when heated to decomposition it emits highly toxic fumes of F⁻ and Cl⁻.

((2,2,2-TRICHLORO-1-HYDROXYETHYL) DIMETHYLPHOSPHONATE)   HR: 3
CAS: 52-68-6   NIOSH: TA 0700000
mf: $C_4H_8Cl_3O_4P$ mw: 257.44

SYNS: CLOROFOS * DIMETHOXY-2,2,2-TRI-CHLORO-1-HYDROXY-ETHYL-PHOSPHINE OXIDE * O,O-DIMETHYL-(1-HYDROXY-2,2,2-TRICHLO-RAETHYL)PHOSPHONSAEURE ESTER (GERMAN) * O,O-DIMETHYL-(1-HYDROXY-2,2,2-TRICHLO-RATHYL)-PHOSPHAT (GERMAN) * O,O-DI-METHYL-(1-HYDROXY-2,2,2-TRICHLORO)ETHYL

PHOSPHATE * DIMETHYL 1-HYDROXY-2,2,2-TRICHLOROETHYL PHOSPHONATE * O,O-DIMETHYL (1-HYDROXY-2,2,2-TRICHLOROETHYL)-PHOSPHONATE * O,O-DIMETHYL-1-OXY-2,2,2-TRICHLOROETHYL PHOSPHONATE * O,O-DIMETHYL-(2,2,2-TRICHLOOR-1-HYDROXY-ETHYL)-FOSFONAAT (DUTCH) * O,O-DIMETHYL-(2,2,2-TRICHLOR-1-HYDROXY-AETHYL)PHOSPHONAT (GERMAN) * DIMETHYLTRICHLOROHYDROXY-ETHYL PHOSPHONATE * DIMETHYL 2,2,2-TRICHLORO-1-HYDROXYETHYLPHOSPHONATE * O,O-DIMETHYL-2,2,2-TRICHLORO-1-HYDROXY-ETHYL PHOSPHONATE * O,O-DIMETIL-(2,2,2-TRICLORO-1-IDROSSI-ETIL)-FOSFONATO (ITALIAN) * 1-HYDROXY-2,2,2-TRICHLOROETHYLPHOSPHONIC ACID DIMETHYL ESTER * NCI-C54831 * 2,2,2-TRICHLORO-1-HYDROXYETHYL-PHOSPHONATE, DIMETHYL ESTER * (2,2,2-TRICHLORO-1-HYDROXYETHYL)PHOSPHONIC ACID DIMETHYL ESTER

THR: Poison by inhalation, ingestion, intraperitoneal, subcutaneous, intravenous, and intramuscular routes. Mutagenic data. An eye irritant. An experimental carcinogen and teratogen. When heated to decomposition it emits very toxic fumes of Cl^- and PO_x.

TRICHLOROMETAFOS   HR: 3
CAS: 299-84-3   NIOSH: TG 0525000
mf: $C_8H_8Cl_3O_3PS$ mw: 321.54

PROP: White powder; mp: 41°, vap press: 8 × 10^{-4} mm.

SYNS: O,O-DIMETHYL O-2,4,5-TRICHLOROPHENYL PHOSPHOROTHIOATE * O,O-DIMETHYL O-(2,4,5-TRICHLOROPHENYL)THIOPHOSPHATE * ENT 23,284 * FENCHLOORFOS (DUTCH) * FENCHLORFOS * FENCHLOROPHOS * RONNEL * THIOPHOSPHATE DE O,O-DIMETHYLE ET DE O-(2,4,5-TRICHLOROPHENYLE) (FRENCH) * O-(2,4,5-TRICHLOOR-FENYL)-O,O-DIMETHYL-MONOTHIOFOSFAAT (DUTCH) * O-(2,4,5-TRICHLOR-PHENYL)-O,O-DIMETHYL-MONOTHIOPHOSPHAT (GERMAN) * O-(2,4,5-TRICLORO-FENIL)-O,O-DIMETIL-MONOTIOFOSFATO (ITALIAN)

OSHA PEL: TWA 10 mg/m³

THR: Poison by ingestion, intraperitoneal, and other unspecified routes. Moderately toxic by skin contact. A cholinesterase inhibitor. See also parathion. An experimental teratogen. When heated to decomposition it emits very toxic fumes of Cl^-, PO_x, and SO_x.

TRICHLORONITROMETHANE   HR: 3
CAS: 76-06-2   NIOSH: PB 6300000
DOT: 1580/1583
mf: CCl_3NO_2 mw: 164.37

PROP: Slightly oily, colorless liquid, d: 1.651 @ 22.8°/4°; mp: −64°; bp: 112.3 @ 766 mm; sol in water, alc, ether; vap press: 40 mm @ 33.80; vap d: 6.69.

SYNS: CHLOORPIKRINE (DUTCH) * CHLOROPICRIN * CHLOROPICRINE (FRENCH) * CHLORPIKRIN (GERMAN) * CLOROPICRINA (ITALIAN) * NCI-C00533 * NITROCHLOROFORM * NITROTRICHLOROMETHANE * TRICHLOORNITROMETHAAN (DUTCH) * TRICHLORNITROMETHAN (GERMAN) * TRICLORO-NITROMETANO (ITALIAN)

OSHA PEL: TWA 0.1 ppm
ACGIH TLV: TWA 0.1 ppm; STEL 0.3 ppm
DOT Classification: Poison B, Label: Poison

THR: Poison by ingestion, inhalation, and intraperitoneal routes. An experimental carcinogen. Human irritant and eye systemic effects. When heated to decomposition it emits very toxic fumes of Cl^- and NO_x. Reacts violently with propargyl bromide. Above a critical volume it can be shock detonated. A powerful irritant that affects all body surfaces. It causes lachrymation, vomiting, bronchitis, and pulmonary edema. Irritating to gastrointestinal and respiratory tracts. An additional toxic effect is its reaction with SH-groups in hemoglobin thus interfering with oxygen transport. Photochemical transformation of chloropicrin into phosgene (carboxy chloride, $COCl_2$) has been reported. A concentration of 1 ppm causes a smarting pain in the eyes and therefore in itself constitutes a good warning of exposure. It causes vomiting, probably due to swallowing saliva in which small amounts of chloropicrin have dissolved. Its primary lethal effect is to produce lung injury and it is a difficult gas to protect oneself against because it is chemically inert and does not react with the usual chemicals used in gas masks. Four ppm is sufficient to render a man unfit for action and 20 ppm, when breathed from 1 to 2 minutes, causes definite bronchial or pulmonary lesions. Industrially it is used as a warning agent in commercial fumigants. It is more toxic than chlorine but less so than phosgene. Dangerous; when heated to decomposition it emits highly toxic fumes of Cl^- and NO_x. Incompatible with aniline;

NaOH; NaOCH$_3$. Can be shocked into detonation. For further information see Chloropicrin, Vol. 2, No. 2 of *DPIM Report*.

2,4,5-TRICHLOROPHENOL   HR: 3
CAS: 95-95-4   NIOSH: SN 1400000
mf: C$_6$H$_3$Cl$_3$O   mw: 197.44   .

PROP: Colorless needles or gray flakes; bp: 252°, fp: 57.0°, d: 1.678 @ 25°/4°, vap press: 1 mm @ 72.0°. Mp: 61°-63°. Insol in water; sol in CCl$_4$, alc, benzene, ether.

SYNS: NCI-C61187 * DOWICIDE 2 * DOWICIDE B

THR: Poison by intraperitoneal, intravenous and other unspecified routes. Moderately toxic by ingestion and subcutaneous routes. An experimental carcinogen and neoplastigen. See also chlorophenols. When heated to decomposition it emits toxic fumes of Cl$^-$ and explodes. For further information see Vol. 5, No. 1 of *DPIM Report*.

2,4,6-TRICHLOROPHENOL   HR: 3
CAS: 88-06-2   NIOSH: SN 1575000
mf: C$_6$H$_3$Cl$_3$O   mw: 197.44

PROP: Colorless needles or yellow solid, strong phenolic odor. Mp: 68°, bp: 244.5°, fp: 62°, d: 1.490 @ 75°/4°, vap press: 1 mm @ 76.5°. Sol in water; very sol in alc, ether.

SYNS: DOWICIDE 2S * NCI-C02904 * 2,4,6-TRICHLORFENOL (CZECH)

THR: Poison by intraperitoneal route. Moderately toxic by ingestion. Mutagenic data. A skin and eye irritant. An experimental carcinogen. See also chlorophenols. When heated to decomposition it emits toxic fumes of Cl$^-$. For further information see Vol. 4, No. 5 of *DPIM Report*.

TRICHLOROPHENYLSILANE   HR: 3
CAS: 98-13-5   NIOSH: VV 6650000
DOT: 1804
mf: C$_6$H$_5$Cl$_3$Si   mw: 211.55

SYNS: PHENYLSILICON TRICHLORIDE * PHENYL TRICHLOROSILANE (DOT) * SILICON PHENYL TRICHLORIDE

DOT Classification: Label: Corrosive

THR: Poison by inhalation and intravenous routes. Moderately toxic by ingestion and skin contact. A skin and eye irritant. See also silanes. When heated to decomposition it emits toxic fumes of Cl$^-$.

TRICHLOROSILANE   HR: 2
CAS: 10025-78-2   NIOSH: VV 5950000
DOT: 1295
mf: Cl$_3$HSi   mw: 135.45

PROP: Colorless, very volatile liquid, decomp in water; mp: −126.5°, bp: 31.8°, flash p: −18.4°F (OC), d: 1.35 @ 0°, vap press: 400 mm @ 14.5°, vap d: 4.7, autoign temp: 219°F. Sol in benzene, carbon disulfide, chloroform, carbon tetrachloride. Generates fumes in air.

SYNS: SILICI-CHLOROFORME (FRENCH) * SILICIUMCHLOROFORM (GERMAN) * SILICO-CHLOROFORM * TRICHLOORSILAAN (DUTCH) * TRICHLOROMONOSILANE * TRICHLORSILAN (GERMAN) * TRICLOROSILANO (ITALIAN)

DOT Classification: Label: Flammable Liquid

THR: Moderately toxic by ingestion and inhalation. See also chlorosilanes. Very dangerous fire hazard when exposed to flame or by chemical reaction. Spontaneously flammable in air. Dangerous; when heated to decomposition it emits highly toxic fumes of chlorides; will react with water or steam to produce heat and toxic and corrosive fumes; on contact with oxidizing materials, can react vigorously. To fight fire, use CO$_2$, dry chemical. When heated to decomposition it emits toxic fumes of Cl$^-$.

TRICHOTHECIN   HR: 3
CAS: 6379-69-7   NIOSH: YD 0175000
mf: C$_{19}$H$_{24}$O$_5$ mw: 332.43

PROP: Needles, mp: 118°. Sltly sol in water, very sol in organic solvents.

THR: Poison by intravenous and subcutaneous routes. When heated to decomposition it emits acrid smoke and fumes.

TRICYCLOQUINAZOLINE   HR: 3
CAS: 195-84-6   NIOSH: YD 2800000
mf: C$_{21}$H$_{12}$N$_4$ mw: 320.37

THR: An experimental carcinogen and neoplastigen. When heated to decomposition it emits toxic fumes of NO$_x$.

TRIETHYLAMINE   HR: 2
CAS: 121-44-8   NIOSH: YE 0175000
DOT: 1296
mf: C$_6$H$_{15}$N   mw: 101.22

PROP: Colorless liquid, ammonia odor. Mp: −114.8°, bp: 89.5°, flash p: 20°F (OC), d:

0.7255 @ 25°/4°, vap d: 3.48, lel = 1.2%, uel = 8.0%. Misc in water, alc, ether.

SYNS: (DIETHYLAMINO)ETHANE * TRIAETH-YLAMIN (GERMAN) * TRIETILAMINA (ITALIAN)

OSHA PEL: TWA 25 ppm
ACGIH TLV: TWA 10 ppm; STEL 15 ppm
DFG MAK: 10 ppm (40 mg/m^3)
DOT Classification: Label: Flammable
  Liquid

THR: Moderately toxic by ingestion, inhalation, and skin contact. A mild skin and severe eye irritant. Can cause kidney and liver damage. Dangerous fire hazard when exposed to heat, flame or oxidizers. Highly dangerous; keep away from heat or open flame; can react with oxidizing materials. When heated to decomposition it emits toxic fumes of NO_x. To fight fire, use CO_2, dry chemical, alcohol foam. Incompatible with N_2O_4.

TRIETHYLBORANE   HR: 3
CAS: 97-94-9   NIOSH: ED 2100000
mf: $C_6H_{15}B$ mw: 98.02

PROP: Colorless liquid; mp: −93°, d: 0.6961 @ 23°.

SYN: TRIETHYLBORINE

THR: Poison by ingestion. Moderately toxic by inhalation. Animal experiments show that the vapor is a poison and leading to pulmonary irritation and convulsions. Dangerous fire hazard by spontaneous chemical reaction with oxidizers. Spontaneously flammable in air. Reacts violently with (triethyl aluminum + air). Highly dangerous; when heated to decomposition or upon contact with air it emits toxic fumes; will react with water or steam to produce toxic and flammable vapors; can react vigorously with oxidizing materials. To fight fire, do NOT use halogenated extinguishing agents.

TRIETHYLENE GLYCOL   HR: 3
CAS: 112-27-6   NIOSH: YE 4550000
mf: $C_6H_{14}O_4$ mw: 150.20

PROP: Odorless, colorless liquid, hygroscopic; fp: −7.3°, flash p: 350°F, d: 1.122 @ 25°/25°, lel = 0.9%, uel = 9.2%, autoign temp: 700°F, vap press: 1 mm @ 114°, vap d: 5.17; bp: 285°. Misc in water, alc, benzene; insol in petroleum ether; very sltly sol in ether.

SYNS: DI-BETA-HYDROXYETHOXYETHANE

* 3,6-DIOXAOCTANE-1,8-DIOL * 2,2′-ETH-YLENEDIOXYDIETHANOL * 2,2′-ETHYL-ENEDIOXYETHANOL * ETHYLENE GLYCOL DIHYDROXYDIETHYL ETHER * GLYCOL BIS-(HYDROXYETHYL) ETHER * TRIGLYCOL

THR: Poison by intravenous route. Mild toxicity by several routes. See also esters, glycols. Combustible when exposed to heat or flame; can react with oxidizing materials. Moderately explosive in the form of vapor when exposed to flame, spark or heat source. To fight fire, use alcohol foam, dry chemical. When heated to decomposition it emits acrid smoke and fumes. For further information see Vol. 4, No. 3 of *DPIM Report*.

TRIETHYLENEPHOSPHOROTRI-AMIDE   HR: 3
CAS: 545-55-1   NIOSH: SZ 1750000
DOT: 2501
mf: $C_6H_{12}N_3OP$ mw: 173.18

PROP: Colorless crystals; mp: 41°; bp: 90° @ 23 mm. Sol in water, alc, ether.

SYNS: ENT 24915 * NSC 9717 * 1,1′,1″-PHOSPHINYLIDYNETRISAZIRIDINE * PHOSPHORIC ACID TRIETHYLENE IMIDE * PHOSPHORIC ACID TRIETHYLENEIMINE (DOT) * TRIAETHYLEN-PHOSPHORSAEUREAMID (GERMAN) * N,N′,N″-TRI-1,2-ETHANEDIYLPHOSPHORIC TRIAMIDE
* N,N′,N″-TRIETHYLENEPHOSPHORAMIDE
* TRIETHYLENEPHOSPHORIC TRIAMIDE
* N,N′,N″-TRIETHYLENEPHOSPHORIC TRIAMIDE
* TRI(1-AZIRIDINYL)PHOSPHINE OXIDE
* TRIS(1-AZIRIDINYL)PHOSPHINE OXIDE
* TRIS(N-ETHYLENE)PHOSPHOROTRIAMIDATE

DOT Classification: Label: Corrosive

THR: Poison by ingestion, skin contact, intravenous, and intraperitoneal routes. Mutagenic data. An experimental carcinogen, teratogen, and neoplastigen. When heated to decomposition it emits very toxic fumes of PO_x and NO_x.

TRIETHYLLEAD FLUOROACE-TATE   HR: 3
CAS: 562-95-8   NIOSH: AH 9670000
mf: $C_2H_2FO_2 \cdot C_6H_{15}Pb$ mw: 371.44

THR: Poison by subcutaneous route. Human pulmonary system effects. See also fluorides and lead compounds. When heated to decomposition it emits very toxic fumes of F^- and Pb.

2,2,2-TRIFLUOROETHYL VINYL ETHER   HR: 3
CAS: 406-90-6   NIOSH: KO 4250000
mf: $C_4H_5F_3O$ mw: 126.09

SYNS: FLOROXENE * FLUROXENE * FLU-
ROXENE

THR: Poison to humans by ingestion. Poison
experimentally by inhalation. Mutagenic data.
When heated to decomposition it emits toxic
fumes of F^-.

2,2,2-TRIFLUORO-N-(9-OXOFLUOREN-2-YL)ACETAMIDE   HR: 3
CAS: 318-22-9   NIOSH: AD 0175000
mf: $C_{15}H_8F_3NO_2$ mw: 291.24

SYN: 2-TRIFLUOROACETYLAMINOFLUOREN-9-ONE

THR: An experimental carcinogen. See also
fluorides. When heated to decomposition it
emits very toxic fumes of F^- and NO_x.

TRIFLUPERIDOL   HR: 3
CAS: 749-13-3   NIOSH: EU 3500000
mf: $C_{22}H_{23}F_4NO_2$ mw: 409.46

SYNS: 4'-FLUORO-4-(4-HYDROXY-4-(ALPHA,
ALPHA,ALPHA-TRIFLUORO-M-TOLYL)PIPERIDINO)-
BUTYROPHENONE * TRIPERIDOL

THR: Poison by ingestion and intraperitoneal
routes. An experimental teratogen. See also
fluorides. When heated to decomposition it
emits very toxic fumes of F^- and NO_x.

TRIFLUPROMAZINE   HR: 3
CAS: 146-54-3   NIOSH: SO 8850000
mf: $C_{18}H_{19}F_3N_2S$ mw: 352.45

SYN: 10-(3-(DIMETHYLAMINO)PROPYL-2-(TRI-
FLUOROMETHYL) PHENOTHIAZINE

THR: Poison by ingestion, intravenous, and intraperitoneal routes. Mutagenic data. See also
fluorides. When heated to decomposition it
emits very toxic fumes of F^-, NO_x and SO_x.

TRIFLURALINE   HR: 3
CAS: 1582-0-9-8   NIOSH: XU 9275000
mf: $C_{13}H_{16}F_3N_3O_4$ mw: 335.32

PROP: Technical product contains 84-88 ppm
diproplynitrosoamine

SYNS: 2,6-DINITRO-N,N-DI-N-PROPYL-AL-
PHA,ALPHA,ALPHA-TRIFLURO-P-TOLUIDINE

* 4-(DI-N-PROPYLAMINO)-3,5-DINITRO-1-TRI-
FLUOROMETHYLBENZENE * N,N-DI-N-PROPYL-
2,6-DINITRO-4-TRIFLUOROMETHYLANILINE
* 2,6-DINITRO-4-TRIFLUORMETHYL-N,N-DIPRO-
PYLANILIN (GERMAN) * N,N-DIPROPYL-4-TRI-
FLUOROMETHYL-2,6-DINITROANILINE * NCI-
C00442 * 2,6-DINITRO-N,N-DIPROPYL-4-(TRI-
FLUOROMETHYL)BENZENAMINE * ALPHA,
ALPHA,ALPHA-TRIFLUORO-2,6-DINITRO-N,N-
DIPROPYL-P-TOLUIDINE

THR: An experimental carcinogen and teratogen. Moderately toxic by ingestion and intraperitoneal routes. Mutagenic data. See also
fluorides and dipropylnitroso amine. When
heated to decomposition it emits very toxic
fumes of F^- and NO_x. For further information see alpha,alpha,alpha-trifluoro-2,6-dini-
tro-N,N-dipropyl-p-toluidine in Vol. 1, No. 2
of *DPIM Report*.

TRIHYDROXYTRIETHYLAMINE   HR: 3
CAS: 102-71-6   NIOSH: KL 9275000
mf: $C_6H_{15}NO_3$ mw: 480.22

PROP: Pale yellow viscous liquid; mp: 21.2°,
bp: 360°, flash p: 355°F (CC), d: 1.1258 @
20°/20°, vap press: 10 mm @ 205°, vap d: 5.14.

SYNS: NITRILO-2,2',2''-TRIETHANOL * 2,
2',2''-NITRILOTRIETHANOL * TRIAETHANOL-
AMIN-NG * TRIETHANOLAMIN * TRIETHANOL-
AMINE * TRI(HYDROXYETHYL)AMINE
* TRIS(2-HYDROXYETHYL)AMINE

THR: An experimental carcinogen. Moderately
toxic by intraperitoneal route. Mildly toxic by
ingestion. Liver and kidney damage has been
demonstrated in animals from chronic exposure.
A skin and eye irritant. Combustible when exposed to heat or flame. Dangerous; when heated
to decomposition it emits toxic fumes of NO_x;
can react vigorously with oxidizing materials.
To fight fire, use alcohol foam, CO_2, dry chemical.

TRIISO BUTYLTIN CHLORIDE   HR: 3
CAS: 7342-38-3   NIOSH: WH 6845000
mf: $C_{12}H_{27}ClSn$ mw: 325.53

PROP: D: 1.1290 @ 34°; mp: 30.2°; bp: 174°
@ 13 mm. Solid.

SYN: CHLORO(TRIISOBUTYL)STANNANE

THR: Poison by intravenous route. See also
tin compounds and chlorides. When heated to

decomposition it emits toxic fumes of Cl⁻ and SnO$_x$.

TRIISOPROPYLTIN ACETATE HR: 3
CAS: 19464-55-2   NIOSH: WH 6300000
mf: $C_{11}H_{24}O_2Sn$ mw: 307.04

SYN: ACETOXYTRIISOPROPYLSTANNANE

THR: Poison by ingestion and intravenous routes. See also tin compounds and esters.

TRIMECAINE HR: 3
CAS: 1027-1-4-1   NIOSH: AE 2100000
mf: $C_{15}H_{24}N_2O \cdot ClH$ mw: 284.87

SYNS: 2-DIETHYLAMINO-2',4',6'-TRIMETHYL-ACETANILIDE HYDROCHLORIDE * N-SYM-TRIMETHYLPHENYLDIETHYLAMINOACETAMIDE HYDROCHLORIDE * DIETHYLAMINOACETYL-2,4,6-TRIMETHYL-ANILINE HYDROCHLORIDE HYDROCHLORIDE

THR: Poison by intraperitoneal and subcutaneous routes. When heated to decomposition it emits very toxic fumes of NO$_x$ and HCl.

TRIMELLIC ACID ANHYDRIDE HR: 2
CAS: 552-30-7   NIOSH: DC 2050000
mf: $C_9H_4O_5$ mw: 192.13

PROP: Crystals, mp: 162°, bp: 240°-245° @ 14 mm, sol in acetone, ethyl acetate, dimethylformamide.

SYNS: 1,2,4-BENZENETRICARBOXYLIC ACID ANHYDRIDE * 1,2,4-BENZENETRICARBOXYLIC ACID, CYCLIC 1,2-ANHYDRIDE * 1,2,4-BENZENETRICARBOXYLIC ANHYDRIDE * 4-CARBOXYPHTHALIC ANHYDRIDE * 1,3-DIHYDRO-1,3-DIOXO-5-ISOBENZOFURANCARBOXYLIC ACID * 1,3-DIOXO-5-PHTHALANCARBOXYLIC ACID * DIPHENYLMETHANE-4,4'-DIISOCYANATE-TRIMELLIC ANHYDRIDE-ETHOMID HT POLYMER * NCI-C56633 * TRIMELLIC ACID-1,2-ANHYDRIDE * TRIMELLITIC ACID CYCLIC-1,2-ANHYDRIDE * TRIMELLITIC ANHYDRIDE
ACGIH TLV: TWA 0.005 ppm
DFG MAK: 0.005 ppm (0.04 mg/m³)

THR: Moderately toxic. Has caused pulmonary edema from inhalation. Irritant to lungs and air passages. May be a powerful allergen. Typical attack consists of breathlessness, wheezing, cough, running nose, immunological sensitization and asthma symptoms.

TRIMETHADIONE HR: 3
CAS: 127-48-0   NIOSH: RQ 2100000
mf: $C_6H_9NO_3$ mw: 143.16

SYNS: 3,3,5-TRIMETHYL-2,4-DIKETOOXAZOLI-DINE * 3,5,5-TRIMETHYL-2,4-OXAZOLIDINE-DIONE

THR: Poison to humans by unspecified routes. Moderately toxic experimentally by ingestion, subcutaneous, intraperitoneal, and intravenous routes. See also ketones. When heated to decomposition it emits toxic fumes of NO$_x$.

3,4,5-TRIMETHOXYAMPHETAMINE HYDROCHLORIDE HR: 3
CAS: 5688-80-2   NIOSH: SI 2285000
mf: $C_{12}H_{19}NO_3 \cdot ClH$ mw: 261.78

SYNS: ALPHA-METHYL-3,4,5-TRIMETHOXYPHEN-ETHYLAMINE HYDROCHLORIDE * 3,4,5-TRIME-THOXY-ALPHA-METHYL-BETA-PHENYLETHYL-AMINE HYDROCHLORIDE * 1-(3,4,5-TRIME-THOXYPHENYL)-2-AMINOPROPANE

THR: Poison by intraperitoneal and intravenous routes. Human central nervous system effects. When heated to decomposition it emits very toxic fumes of NO$_x$ and HCl.

TRIMETHYLAMINE HR: 3
CAS: 75-50-3   NIOSH: PA 0350000
mf: C_3H_9N mw: 59.13

PROP: Colorless gas. Pungent, fishy, ammoniacal· odor, saline taste; bp: 2.87°, lel = 2%, uel = 11.6%, fp: −117.1°, d: 0.662 @ −5°, autoign temp: 374°F, vap d: 2.0, flash p: 20°F (CC). Misc with alc; sol in ether, benzene, toluene, xylene, chloroform.

SYN: TMA

ACGIH TLV: TWA 10 ppm; STEL 15 ppm

THR: Poison by intravenous route. Moderately toxic by inhalation, subcutaneous, and rectal routes. Very dangerous fire hazard when exposed to flame. Moderate explosion hazard when exposed to spark or flame. Moderately dangerous; when heated to decomposition it emits toxic fumes of NO$_x$. Can react with oxidizing materials. To fight fire, stop flow of gas. Incompatible with ethylene oxide. For further information see Vol. 5, No. 6 of *DPIM Report*.

2,4,5-TRIMETHYLANILINE HR: 3
CAS: 137-17-7   NIOSH: BZ 0520000
mf: $C_9H_{13}N$ mw: 135.23

SYNS: PSI-CUMIDINE * NCI-C02299 * PSEU-
DOCUMIDINE * 2,4,5-TRIMETHYLANILIN
(CZECH) * 2,4,5-TRIMETHYLBENZENAMINE

THR: An experimental carcinogen. Moderately
toxic by ingestion. Mutagenic data. When
heated to decomposition it emits toxic fumes
of NO_x.

2,4,6-TRIMETHYLANILINE   HR: 3
CAS: 88-05-1   NIOSH: BZ 0700000
mf: $C_9H_{13}N$ mw: 135.23

SYNS: AMINOMESITYLENE * 1-AMINO-2,4,6-
TRIMETHYLBENZEN (CZECH) * MESIDIN
(CZECH) * MESIDINE * MESITYLAMINE

THR: Poison by unspecified route. Moderately
toxic by ingestion and other unspecified routes.
Mutagenic data. A moderate skin and severe
eye irritant. An experimental neoplastigen and
carcinogen. When heated to decomposition it
emits toxic fumes of NO_x.

2,4,5-TRIMETHYLANILINE HYDRO-
CHLORIDE   HR: 3
CAS: 21436-97-5   NIOSH: BZ 0875000
mf: $C_9H_{13}N \cdot ClH$ mw: 171.69

SYN: PSEUDOCUMIDINE HYDROCHLORIDE

THR: Poison by intraperitoneal route. Moderately toxic by ingestion. An experimental neoplastigen and carcinogen. When heated to decomposition it emits very toxic fumes of NO_x
and HCl.

2,4,6-TRIMETHYLANILINE HYDRO-
CHLORIDE   HR: 3
CAS: 6334-11-8   NIOSH: BZ 1050000
mf: $C_9H_{13}N \cdot ClH$ mw: 171.69

SYN: MESIDINE HYDROCHLORIDE

THR: Poison by intraperitoneal route. Moderately toxic by ingestion. An experimental carcinogen and neoplastigen. When heated to decomposition it emits very toxic fumes of NO_x
and HCl.

7,9,11-TRIMETHYLBENZ(c)ACRI-
DINE   HR: 3
CAS: 51787-42-9   NIOSH: CU 4345000
mf: $C_{20}H_{17}N$ mw: 271.38

SYN: 1,3,10-TRIMETHYL-7,8-BENZACRIDINE
(FRENCH)

THR: An experimental tumorigen. Mutagenic
data. When heated to decomposition it emits
toxic fumes of NO_x.

4,9,10-TRIMETHYL-1,2,-BENZ-
ANTHRACENE   HR: 3
CAS: 20627-33-2   NIOSH: CX 4200000
mf: $C_{21}H_{18}$ mw: 270.39

SYN: 6,7,12-TRIMETHYLBENZ(A)ANTHRACENE

THR: An experimental tumorigen. Mutagenic
data. When heated to decomposition it emits
acrid smoke.

7,8,12-TRIMETHYLBENZ(a)
ANTHRACENE   HR: 3
CAS: 13345-64-7   NIOSH: CX 4550000
mf: $C_{21}H_{18}$ mw: 270.39

SYNS: 7,8,12-TMBA * 5:9:10-TRIMETHYL-1:
2-BENZANTHRACENE

THR: Poison by intravenous route. An experimental tumorigen. Mutagenic data. When
heated to decomposition it emits acrid smoke
and fumes.

1,2,4-TRIMETHYLBENZENE   HR: 2
CAS: 95-63-6   NIOSH: DC 3325000
mf: C_9H_{12} mw: 120.21

PROP: Liquid, insol in water, sol in alc, benzene, ether; mp: 120.19°, d: 0.888 @ 4°/4°,
fp: −61°, bp: 168.89°, flash p: 130°F, autoign
temp: 959°F.

SYNS: PSI-CUMENE * PSEUDOCUMENE
* PSEUDOCUMOL * AS-TRIMETHYLBENZENE

THR: Moderately toxic by ingestion, skin contact, and intraperitoneal routes. Can cause central nervous system depression, anemia, bronchitis. Flammable when exposed to heat, flame,
or oxidizers. To fight fire, use foam, alcohol
foam, mist.

TRIMETHYL BISMUTH   HR: 3
CAS: 593-91-9   NIOSH: EB 3150000
mf: C_3H_9Bi mw: 254.06

PROP: Liquid. Bp: 110°, d: 2.300 @ 18°.

THR: Poison by ingestion, skin contact, intravenous, and subcutaneous routes. Can cause narcosis and central nervous system depression.
Prolonged exposure can cause encephalopathy
similar to that of organic lead compounds. See
also bismuth compounds. Flammable when exposed to heat or flame. Spontaneously flammable in air. Dangerous; when heated to decomposition it emits toxic fumes of bismuth and reacts
with oxidizing materials.

TRIMETHYL BORATE　　　　HR: 2
CAS: 121-43-7　　　　NIOSH: ED 5600000
mf: $C_3H_9BO_3$　　mw: 103.93

PROP: Colorless liquid, decomp in water, misc in alc, ether. Mp: −29°, bp: 68°, flash p: <73°F, d: 0.92 @ 20°, vap d: 3.59.

SYNS: METHYL BORATE * TRIMETHOXY-BORINE

THR: Moderately toxic by ingestion and skin contact. See also esters and boron compounds. Dangerous fire hazard when exposed to heat, flame or oxidizers. Moderately explosive when exposed to flame. Dangerous; will react with water or steam to produce toxic and flammable vapors; can react vigorously with oxidizing materials. Emits smoke and fumes of BO_x. To fight fire, use dry chemical, CO_2, spray, foam.

TRIMETHYL CHLOROSILANE　　HR: 3
CAS: 75-77-4　　　　NIOSH: VV 2710000
DOT: 1298
mf: C_3H_9ClSi　　mw: 108.66

PROP: Colorless liquid; sol in benzene, ether, perchloroethylene, bp: 57°, d: 0.854 @ 25°/25°, flash p: −18°F.
DOT Classification: Label: Flammable
　Liquid

THR: Poison by inhalation. Moderately toxic by intraperitoneal route. An experimental neoplastigen. See also silanes. Dangerous fire hazard when exposed to heat or flame. Violent reaction with water. Dangerous; see chlorides. When heated to decomposition it emits toxic fumes of Cl^-. To fight fire, use foam, alcohol foam and fog.

TRIMETHYLCOLCHICINIC ACID　　HR: 3
CAS: 3482-37-9　　　　NIOSH: GH 0960000
mf: $C_{19}H_{21}NO_5$　　mw: 343.41

SYNS: DEACETYLCHOLCHICEINE * DESACETYL-CHOLCHICEINE

THR: Poison by intraperitoneal and other unspecified routes. When heated to decomposition it emits toxic fumes of NO_x.

1,3-TRIMETHYLENEDINITRILE　　HR: 3
CAS: 544-13-8　　　　NIOSH: YI 3500000
mf: $C_5H_6N_2$　　mw: 94.13

PROP: Colorless liquid, sol in water, insol in ether. D: 0.989 @ 15°/4°, mp: −29°, bp: 286.4°.

SYNS: 1,3-DICYANOPROPANE * GLUTARIC ACID DINITRILE * GLUTARODINITRILE * PENTANEDINITRILE (9CI) * PYROTARTARIC ACID NITRILE

THR: Poison by unspecified routes. See also nitriles. When heated to decomposition it emits very toxic fumes of NO_x and CN^-.

TRIMETHYLENETRINITRAMINE　　HR: 3
CAS: 121-82-4　　　　NIOSH: XY 9450000
DOT: 0072/0118
mf: $C_3H_6N_6O_6$　　mw: 222.15

PROP: White, crystalline powder. Mp: 202°.

SYNS: HEXOGEN (EXPLOSIVE) * CYCLONITE * CYCLOTRIMETHYLENENITRAMINE * CYCLO-TRIMETHYLENETRINITRAMINE * ESAIDRO-1,3,5-TRINITRO-1,3,5-TRIAZINA (ITALIAN) * HEXA-HYDRO-1,3,5-TRINITRO-S-TRIAZINE * HEXAHY-DRO-1,3,5-TRINITRO-1,3,5-TRIAZIN (GERMAN) * HEXAHYDRO-1,3,5-TRINITRO-1,3,5-TRIAZINE * HEXOGEEN (DUTCH) * HEXOLITE * TRI-METHYLEENTRINITRAMINE (DUTCH) * SYM-TRIMETHYLENETRINITRAMINE * TRINITRO-CYCLOTRIMETHYLENE TRIAMINE * 1,3,5-TRINITRO-1,3,5-TRIAZACYCLOHEXANE

OSHA PEL: TWA 1500 ug/m^3 (skin)
DOT Classification: Class A Explosive,
　Label: Explosive A

THR: Poison by ingestion, intraperitoneal, and intravenous routes. Cases of epileptiform convulsions have been reported from exposure. See nitrates for fire and disaster hazards. It is one of the most powerful high explosives in use today. See explosives, high. Has more shattering power than TNT and is often mixed with TNT as a bursting charge for aerial bombs, mines and torpedoes. Because it is easily initiated by mercury fulminate it may be used as a booster.

TRIMETHYL-o-FORMATE　　HR: 2
CAS: 149-73-5　　　　NIOSH: RM 6650000
mf: $C_4H_{10}O_3$　　mw: 106.14

PROP: Colorless liquid, pungent odor, vap d: 3.67, flash p: 59°F.

SYNS: ORTHOFORMIC ACID, TRIMETHYL ESTER * METHYLESTER KYSELINY ORTHOMRAVENCI (CZECH) * METHYL ORTHOFORMATE * OR-THOMRAVENCAN METHYLNATY (CZECH) * TRIMETHOXYMETHANE

THR: Moderately toxic by inhalation and ingestion. A skin and eye irritant. See also esters. Combustible when exposed to heat or flame; can react with oxidizing materials. To fight fire, use CO_2, fog, haze. When heated to decomposition it emits acrid smoke and fumes. Hazardous to prepare.

1,1,3-TRIMETHYL-3-NITROSO-UREA   HR: 3
CAS: 3475-63-6   NIOSH: YU 4725000
mf: $C_4H_9N_3O_2$ mw: 131.16

SYNS: N-TRIMETHYL-N-NITROSOUREA * N-NITROSO-TRIMETHYLHARNSTOFF (GERMAN) * N-NITROSOTRIMETHYLUREA * TRIMETHYLNITROSOHARNSTOFF (GERMAN)

THR: Poison by ingestion and intravenous routes. Mutagenic data. An experimental teratogen and tumorigen. When heated to decomposition it emits toxic fumes of NO_x.

2,2,4-TRIMETHYLPENTANE   HR: 2
CAS: 540-84-1   NIOSH: SA 3320000
DOT: 1262
mf: C_8H_{18} mw: 114.26

PROP: Clear liquid, odor of gasoline; bp: 99.2°, fp: −116°, flash p: 10°F, d: 0.692 @ 20°/4°, autoign temp: 779°F, vap press: 40.6 mm @ 21°, vap d: 3.93, lel = 1.1%, uel = 6.0%.

SYNS: ISOBUTYLTRIMETHYLETHANE * ISOOCTANE

DOT Classification: Label: Flammable Liquid

THR: Moderately toxic by ingestion and inhalation. High concentrations can cause narcosis. Dangerous fire hazard when exposed to heat, flame, oxidizers. Dangerous; keep away from heat and open flame; can react vigorously with reducing materials. When heated to decomposition it emits acrid smoke. To fight fire, use CO_2, dry chemical.

3,4,5-TRIMETHYLPHENYL METHYL-CARBAMATE   HR: 3
CAS: 2686-99-9   NIOSH: FC 8575000
mf: $C_{11}H_{15}NO_2$ mw: 193.27

SYNS: ENT-25,843 * LANDRIN * SD 8530

THR: Poison by ingestion, intraperitoneal, intravenous, and intramuscular routes. See also carbamates. When heated to decomposition it emits toxic fumes of NO_x.

TRIMETHYL PHOSPHATE   HR: 3
CAS: 512-56-1   NIOSH: TC 8225000
mf: $C_3H_9O_4P$ mw: 140.09

PROP: Liquid; d: 1.97 @ 19.5°/0°; bp: 197.2°; sol in alc, water, ether.

SYNS: METHYL PHOSPHATE * NCI-C03781

THR: An experimental neoplastigen, tumorigen, and carcinogen. Moderately toxic by ingestion, skin contact, intraperitoneal and other unspecified routes. Mutagenic data. See also esters. When heated to decomposition it emits toxic fumes of PO_x.

N,N,2′-TRIMETHYL-4-STILBEN-AMINE   HR: 3
CAS: 63019-09-0   NIOSH: WJ 4723000
mf: $C_{17}H_{19}N$ mw: 237.37

SYNS: 4-DIMETHYLAMINO-2′-METHYLSTILBENE * N,N-DIMETHYL-2′-METHYLSTILBENAMINE * 2′-METHYL-4-DIMETHYLAMINOSTILBENE

THR: An experimental tumorigen and neoplastigen. See also amines. When heated to decomposition it emits toxic fumes of NO_x.

1,1,3-TRIMETHYL-2-THIOUREA   HR: 3
CAS: 2489-77-2   NIOSH: YU 4900000
mf: $C_4H_{10}N_2S$ mw: 118.22

PROP: Trimethylthiourea tested contained 15% 1,3-dimethyl-2-thiourea and 5% Zeolex 80.

SYNS: NCI-C02186 * TRIMETHYLTHIOUREA * N,N,N′-TRIMETHYLTHIOUREA

THR: Poison by ingestion. An experimental carcinogen. See also isothiourea. When heated to decomposition it emits very toxic fumes of NO_x and SO_x.

TRIMETHYLTIN ACETATE   HR: 3
CAS: 1118-14-5   NIOSH: WH 6475000
mf: $C_5H_{12}O_2Sn$ mw: 222.86

SYN: ACETOXYTRIMETHYLSTANNANE

THR: Poison by ingestion. See also tin compounds. When heated to decomposition it emits acrid smoke and fumes of SnO_x.

TRIMETHYLTIN SULPHATE   HR: 3
CAS: 63869-87-4   NIOSH: XQ 7225000
mf: $C_3H_{10}O_4SSn$ mw: 260.88

SYN: TRIMETHYLSTANNANE SULPHATE

THR: Poison by ingestion and intraperitoneal routes. See also sulfates and tin compounds. When heated to decomposition it emits toxic fumes of SO_x.

TRIMETON   HR: 3
CAS: 86-21-5   NIOSH: US 9625000
mf: $C_{16}H_{20}N_2$ mw: 240.38

SYNS: P-AMINOSALICYLSAURES SALZ (GERMAN) * 2-(ALPHA-(2-DIMETHYLAMINOETHYL)BENZYL)PYRIDINE * 2-(3-DIMETHYLAMINO-1-PHENYLPROPYL)PYRIDINE * N,N-DIMETHYL-3-PHENYL-3-(2-PYRIDYL)PROPYLAMINE * NCI-C60695 * 1-PHENYL-1-(2-PYRIDYL)-3-DIMETHYLAMINOPROPANE * 3-PHENYL-3-(2-PYRIDYL)-N,N-DIMETHYLPROPYLAMINE

THR: Poison by ingestion, intraperitoneal, and intravenous routes. Human central nervous system effects. When heated to decomposition it emits toxic fumes of NO_x.

TRIMUSTINE   HR: 3
CAS: 817-09-4   NIOSH: YE 2800000
mf: $C_6H_{12}Cl_3N \cdot ClH$ mw: 241.00

SYNS: NSC-30211 * SINALOST * TRI(BETA-CHLOROETHYL)AMINE HYDROCHLORIDE * TRI-(2-CHLOROETHYL)AMINE HYDROCHLORIDE * 2,2',2''-TRICHLOROTRIETHYLAMINE HYDROCHLORIDE * TRIS(2-CHLOROETHYL)AMMONIUM CHLORIDE * TRIMUSTINE HYDROCHLORIDE * TRIS(BETA-CHLOROETHYL)AMINE HYDROCHLORIDE * TRIS(2-CHLOROETHYL)AMINE HYDROCHLORIDE * TRIS(2-CHLOROETHYL)AMINE MONOHYDROCHLORIDE * TRIS-N-LOST

THR: Poison by ingestion, subcutaneous, intravenous, and intraperitoneal route. Mutagenic data. An experimental carcinogen. Human blood and central nervous system effects. When heated to decomposition it emits very toxic fumes of HCl, Cl^-, and NO_x.

1,3,5-TRINITROBENZENE   HR: 3
CAS: 99-35-4   NIOSH: DC 3850000
DOT: 0214
mf: $C_6H_3N_3O_6$ mw: 213.12

PROP: Yellow crystals; mp: 122°, bp: decomp, d: 1.760 @ 20°/4°.

SYNS: TRINITROBENZEEN (DUTCH) * TRINITROBENZENE * TRINITROBENZOL (GERMAN)

DOT Classification: Class A Explosive, Label: Explosive A

THR: Poison by ingestion and intravenous routes. See also nitro compounds of aromatic hydrocarbons and nitrates for fire hazard. Severe explosion hazard when shocked or exposed to heat. Trinitrobenzene is considered a powerful high explosive and has more shattering power than TNT. Although it is less sensitive to impact than TNT, it is not used much because it is difficult to produce. Highly dangerous; when heated to decomposition it emits highly toxic fumes of NO_x and explodes; can react vigorously with reducing materials.

2,4,6-TRINITROBENZOIC ACID (DRY)
CAS: 129-66-8   NIOSH: DI 0920000
DOT: 0215
mf: $C_7H_3N_3O_8$ mw: 257.09

PROP: Orthorhombic crystals; mp: 228.7°. Sol @ 25° (2.05% in water, 26.6% in alc, 14.7% in ether), sol in methanol, sltly sol in benzene.
DOT Classification: Class A Explosive, Label: Explosive A

THR: No toxicity data. See also nitrates, and explosives, high. When heated to decomposition it emits toxic fumes of NO_x. A hazard in preparation.

2,4,6-TRINITRO-meta-CRESOL   HR: 3
CAS: 602-99-3   NIOSH: GP 3675000
mf: $C_7H_5N_3O_7$ mw: 243.15

PROP: Yellow crystals. Mp: 106°, bp: explodes @ 150°.

SYN: CRESYLITE

THR: Poison by intraperitoneal route. See also nitro compounds of aromatic hydrocarbons. See also nitrates for fire hazard. Severe explosion hazard when shocked or exposed to heat. Trinitrocresol is not as powerful a high explosive as TNT or picric acid. When heated to decomposition it emits highly toxic fumes of NO_x and explodes; can react vigorously with oxidizing materials.

TRINITROMETHANE   HR: 3
CAS: 517-25-9   NIOSH: PB 7200000
mf: CHN_3O_6 mw: 151.05

PROP: Mp: 15°, d: 1.469, bp: decomp > 25°. Sol in water.

SYN: NITROFORM

THR: Poison by intraperitoneal route. See also nitrates. Irritant to skin, eyes, and mucous membranes. Inhalation can cause headache and nausea. Causes mild narcosis. Dangerous explosion hazard; explodes when heated rapidly. Concentration of more than 50% dissolve exothermally and can explode; 90% concentration + 10% isopropyl alcohol in polyethylene bottles have exploded. Can explode during distillation. When heated to decomposition it emits toxic fumes of NO_x. Incompatible with divinyl ketone; 2-propanol.

sym-TRINITROTOLUENE   HR: 3
CAS: 118-96-7   NIOSH: XU 0175000
DOT: 0209
mf: $C_7H_5N_3O_6$ mw: 227.15

PROP: Colorless monoclinic crystals. Mp: 80.7°, bp: 240° explodes, flash p: explodes, d: 1.654. Sol in hot water, alc, ether.

SYNS: ALPHA-TNT * SYM-TRINITROTOLUOL * NCI-C56155 * TNT * 2,4,6-TRINITROTO-LUEEN (DUTCH) * TRINITROTOLUENE * TRINI-TROTOLUENE, DRY (DOT) * S-TRINITROTOL-UENE * 2,4,6-TRINITROTOLUENE * S-TRI-NITROTOLUOL * 2,4,6-TRINITROTOLUOL (GERMAN) * TROJNITROTOLUEN (POLISH) * TROTYL * TROJNITROTOLUEN (POLISH)

OSHA PEL: TWA 1500 ug/m³ (skin)
ACGIH TLV: TWA 0.5 mg/m³ (skin)
DFG MAK: 0.15 ppm (1.5 mg/m³)
DOT Classification: Class A Explosive,
   Label: Explosive

THR: Poison by subcutaneous route. Moderately toxic by ingestion. Mutagenic data. Severe eye irritant. See also explosives, high and nitrates. Has been implicated in aplastic anemia. Can cause headache, weakness, anemia, liver injury. May be absorbed through skin. For fire hazard see nitrates. Moderate explosion hazard; will detonate under strong shock. See explosives, high. It detonates at around 240°C but can be distilled safely under reduced pressure. It is a comparatively insensitive explosive. In small quantities it will burn quietly if not confined. However, sudden heating of any quantity will cause it to detonate; the accumulation of heat when large quantities are burning will cause detonation. In other respects it is one of the most stable of all high explosives and there are but few restrictions to its handling. It is

for this reason, from the military standpoint, that TNT is quantitatively the most used. It requires a fall of 130 cm for a 2 kg weight to detonate it. It is one of the most powerful high explosives. It can be detonated by the usual detonators and blasting caps (at least a No. 6). For full efficiency, the use of a high velocity initiator, such as tetryl, is required. TNT is one of those explosives containing an oxygen deficiency. In other words, the addition of products which are oxygen rich can enhance its explosive power. Also mono- and dinitrotoluene may be added for reduction of the temperature of the explosion and to make the explosion flash-less. Various materials are added to TNT to make what is known as permissible explosives. TNT may be regarded as the equivalent of 40% dynamite and can be used under water. It is also used in the manufacture of detonator fuse known as Cordeau Detonant. For the military, TNT finds use in all types of bursting charges, including armor-piercing types, although it is somewhat too sensitive to be ideal for this purpose, and has since been replaced to a great extent by ammonium picrate. It is a relatively expensive explosive and does not compete seriously with dynamite for general commercial use. Highly dangerous; shock will explode it; when heated to decomposition it emits highly toxic fumes of NO_x; can react vigorously with reducing materials. Incompatible with sodium dichromate; sulfuric acid. For further information see Vol. 2, No. 5 of *DPIM Report*.

TRI-n-OCTYL BORATE   HR: 2
CAS: 2467-12-1   NIOSH: ED 5630000
mf: $C_{24}H_{51}BO_3$ mw: 398.56

PROP: Colorless liquid, odor of octyl alcohol, bp: 192°-194° @ 2 mm, flash p: 370°F (COC), d: 0.846 @ 23°, vap d: 13.7.

THR: Moderately toxic by ingestion. See esters and boron compounds. An eye irritant. Combustible when exposed to heat or flame; can react with oxidizing materials. To fight fire, use foam, CO_2, dry chemical. When heated to decomposition it emits acrid smoke and fumes of BO_x.

s-TRIOXANE   HR: 2
CAS: 110-83-3   NIOSH: YK 0400000
mf: $C_3H_6O_3$ mw: 90.09

PROP: Stable, cyclic trimer of formaldehyde, having characteristic ethanol and chloroform-like odors. Crystalline solid. Mp: 64°; bp: 114.5°; sublimes readily; lel = 3.6%; uel = 28.7%; flash p: 113°F (OC); d: 1.17, @ 65°; autoign temp: 777°F; vap press: 13 mm @ 25°; vap d: 3.1. Very sol in water, alc, ketones, ether, acetone, chlorinated and aromatic hydrocarbons, organic solvents; sltly sol in pentane, petroleum ether.

SYNS: TRIOXANE * 1,3,5-TRIOXANE * TRIOXYMETHYLEEN (DUTCH) * TRIOSSIMETHLENE (ITALIAN) * TRIOXYMETHYLEN (GERMAN) * TRIOXYMETHYLENE * POLYOXYMETHYLENE

THR: Moderately toxic by ingestion. Mildly toxic by skin contact. Severe eye and skin irritant. Can evolve formaldehyde when heated strongly or in contact with strong acids. See also formaldehyde. Flammable when exposed to heat, flame or oxidizers. Moderately dangerous; can explode when heated. Incompatible with oxidizing materials; on contact with acid or acid fumes, can emit toxic fumes. To fight fire, use foam, CO_2, dry chemical.

TRIPELENNAMINE   HR: 3
CAS: 91-81-6   NIOSH: US 2800000
mf: $C_{16}H_{21}N_3$ mw: 255.40

PROP: Oily liquid, amine odor; bp: 167°-172° @ 0.1 mm. Misc with water.

SYNS: BETA-DIMETHYLAMINO ETHYL-2-PYRIDYL AMINO TOLUENE * 2-(BENZYL(2-DIMETHYL AMINOETHYL)AMINO)PYRIDINE * N-BENZYL-N′,N′-DIMETHYL-N-2-PYRIDYLETHYLENE DIAMINE * BENZYL-(ALPHA-PYRIDYL)-DIMETHYLAETHYLENDIAMIN (GERMAN) * NCI-C60662 * PYRIBENZAMINE * PYRIBENZAMINE * N,N-DIMETHYL-N′-BENZYL-N′-(ALPHA-PYRIDYL)ETHYLENEDIAMINE

THR: Poison by ingestion, subcutaneous, intravenous, and intraperitoneal routes. Addicts have added it to paregoric to make "blue velvet," which can cause a euphoria by injection. Has been implicated in aplastic anemia. When heated to decomposition it emits toxic fumes of NO_x.

2,3,3-TRIPHENYLACRYLONITRILE   HR: 3
CAS: 6304-33-2   NIOSH: AT 7100000
mf: $C_{21}H_{15}N$ mw: 281.37

SYNS: ALPHA,BETA-DIPHENYLCINNAMONITRILE * TRIPHENYLCYANOETHYLENE * TRIPHENYL-ACRYLONITRILE * ALPHA,BETA,BETA-TRIPHENYLACRYLONITRILE

THR: Poison by ingestion and intravenous routes. An experimental carcinogen. See also nitriles. When heated to decomposition it emits toxic fumes of NO_x.

TRIPHENYLETHYLENE   HR: 3
CAS: 58-72-0   NIOSH: KX 4920000
mf: $C_{20}H_{16}$ mw: 256.36

THR: An experimental tumorigen. Mutagenic data. When heated to decomposition it emits acrid smoke and fumes. For further information see Vol. 1, No. 2 of *DPIM Report*.

TRIPHENYL PHOSPHATE   HR: 3
CAS: 115-86-6   NIOSH: TC 8400000
mf: $C_{18}H_{15}O_4P$ mw: 326.30

PROP: Colorless, odorless, crystalline solid. Mp: 49°-50°, bp: 245° @ 11 mm, flash p: 428°F (CC), d: 1.268 @ 60°, vap press: 1 mm @ 193.5°. Insol in water; sol in alc, benzene, ether, chloroform and acetone.

SYN: CELLUFLEX TPP

OSHA PEL: TWA 3 mg/m^3
ACGIH TLV: TWA 3 mg/m^3

THR: Poison by subcutaneous route. Moderately toxic by ingestion and other unspecified routes. See also phosphates. Absorbed (slowly) particularly by skin. Not a potent cholinesterase inhibitor. See also tritolyl phosphate. Combustible when exposed to heat or flame. Dangerous; see phosphates. When heated to decomposition it emits toxic fumes of PO_x. To fight fire, use CO_2, dry chemical. For further information see Vol. 6, No. 4 of *DPIM Report*.

TRIPHENYLPHOSPHINE   HR: 2
CAS: 603-35-0   NIOSH: SZ 3500000
mf: $C_{18}H_{15}P$ mw: 262.30

PROP: Odorless, crystals; mp: 79°, bp: >360°, d: 1.194, flash p: 356°F (OC), vap d: 9.0. Insol in water; sol in HCl, benzene; sltly sol in alc; very sol in ether.

THR: Moderately toxic by ingestion and inhalation. See also phosphine and phenol. Combustible when exposed to heat or flame. Slight explosion hazard in the form of vapor when exposed to flame. Dangerous; when heated to decomposition it emits highly toxic fumes of phosphine and PO_x; can react vigorously with oxidizing materials. To fight fire, use dry chemical, fog, CO_2.

TRIPHENYL PHOSPHITE   HR: 3
CAS: 101-02-0   NIOSH: TH 1575000
mf: $C_{18}H_{15}O_3P$ mw: 310.30

PROP: Water white to pale yellow solid or oily liquid, clean and pleasant odor, insol in water, d: 1.184 @ 25°/25°, mp: 22°-25°, bp: 155°-160° @ 0.1 mm, flash p: 425°F (OC).

SYNS: o,o-DIMETHYL S-(2-(ETHYLSULFINYL)-ISOPROPYL) THIOPHOSPHATE * TRIFENOXYFOS-FIN (CZECH) * TRIFENYLFOSFIT (CZECH)

THR: Poison by intraperitoneal and subcutaneous routes. Moderately toxic by ingestion. Human skin irritant. An eye irritant. See also phenol and phosphites. Combustible when exposed to heat or flame. To fight fire, use CO_2, mist, dry chemical. When heated to decomposition it emits toxic fumes of PO_x.

TRIPOLI
PROP: Finely granulated white or gray siliceous rock. An amorphous form of SiO_2.

THR: A nuisance dust which may contain silica. See also silica. Effects due to silica content may be enough to produce pulmonary fibrosis.

TRI-n-PROPYLAMINE   HR: 3
CAS: 102-69-2   NIOSH: TX 1575000
mf: $C_9H_{21}N$ mw: 143.31

PROP: Liquid, very sltly sol in water. Mp: −93°, bp: 156°, flash p: 105°F (OC), d: 0.75, vap d: 4.9.

THR: Poison by ingestion and inhalation. Moderately toxic by skin contact. See also amines. Flammable when exposed to heat, flame or oxidizers. Moderately dangerous; when heated to decomposition it emits toxic fumes of NO_x; can react with oxidizing materials. To fight fire, use foam, CO_2, dry chemical.

TRIS   HR: 3
CAS: 126-72-7   NIOSH: UB 0350000
mf: $C_9H_{15}Br_6O_4P$ mw: 697.67

PROP: Crystals, d: 2.24, flash p: > 112°.

SYNS: 2,3-DIBROMO-1-PROPANOL, PHOSPHATE (3:1) * 2,3-DIBROMO-1-PROPANOL PHOSPHATE * (2,3-DIBROMOPROPYL) PHOSPHATE * FIRE-MASTER T23P-LV * NCI-C03270 * TRIS(DI-BROMOPROPYL)PHOSPHATE * TRIS(2,3-DI-BROMOPROPYL) PHOSPHATE * TRIS(2,3-DI-BROMOPROPYL) PHOSPHORIC ACID ESTER * TRIS-2,3-DIBROMPROPYL ESTER KYSELINY FOSFORECNE (CZECH) * USAF DO-41

THR: Poison by intraperitoneal route. Moderately toxic by ingestion. Mutagenic data. An experimental carcinogen, neoplastigen, tumorigen, and teratogen. A suspected human carcinogen. Used to control flammability of cloth (children's sleepwear). Can be absorbed by human skin, or chewed or sucked off of sleepwear by infants. Can cause testicular atrophy and sterility. When heated to decomposition it emits very toxic fumes of Br^- and PO_x.

TRISAZIRIDINYLTRIAZINE   HR: 3
CAS: 51-18-3   NIOSH: XZ 2100000
mf: $C_9H_{12}N_6$ mw: 204.27

PROP: Small crystals, water-sol, decomp @ 139°.

SYNS: ENT 25,296 * NSC 9706 * TRIS-(ETHYLENEIMINO)TRIAZINE * 1,1′,1″-S-TRIAZINE-2,4,6-TRIYLTRISAZIRIDINE * TRI-AZIRIDINYL TRIAZINE * TRIETHANOMELAMINE * 2,4,6-TRI(ETHYLENEIMINO)-1,3,5-TRIAZINE * 2,4,6-TRIETHYLENEIMINO-S-TRIAZINE * TRIETHYLENEMELAMINE * 2,4,6-TRIETHYLENIMINO-S-TRIAZINE * 2,4,6-TRIETHYLENIMINO-1,3,5-TRIAZINE * 2,4,6-TRIS(1-AZIRIDINYL)-S-TRIAZINE * 2,4,6-TRIS(1′-AZIRIDINYL)-1,3,5-TRIAZINE * 2,4,6-TRIS(ETHYLENEIMINO)-S-TRIAZINE * 2,4,6-TRIS(ETHYLENIMINO)-S-TRIAZINE * TRISETHYLENEIMINO-1,3,5-TRIAZINE

THR: Poison by ingestion, intraperitoneal, intramuscular, and intravenous routes. Mutagenic data. An experimental teratogen, carcinogen, and neoplastigen. Can cause gastrointestinal tract disturbances and bone marrow depression. Dangerous; when heated to decomposition it emits highly toxic fumes of NO_x.

TRISETHYLENEIMINOQUINONE   HR: 3
CAS: 68-76-8   NIOSH: DK 7175000
mf: $C_{12}H_{13}N_3O_2$ mw: 231.28

SYNS: BAYER 3231 * 1,1′,1″ -(3,6-DIOXO-1,4-CYCLOHEXADIENE-1,2,4-TRIYL)TRISAZIRIDINE * TRIAZICHON (GERMAN) * 2,3,5-TRI-(1-AZI-RIDINYL)-P-BENZOQUINONE * 2,3,5-TRIETHYL-

ENEIMINO-1,4-BENZOQUINONE * TRISAETHYL-ENIMINOBENZOCHINON (GERMAN) * 2,3,5-TRIS(AZIRIDINO)-1,4-BENZOQUINONE * 2,3,5-TRIS(1-AZIRIDINO)-P-BENZOQUINONE * TRIS(AZIRIDINYL)-P-BENZOQUINONE * 2,3,5-TRIS(AZIRIDINYL)-1,4-BENZOQUINONE * 2,3,5-TRIS(1-AZIRIDINYL)-2,5-CYCLOHEXA-DIENE-1,4-DIONE * TRIS(1-AZIRIDINYL)-P-BENZOQUINONE * 2,3,5-TRIS(ETHYLENIMINO)-P-BENZOQUINONE

THR: Poison by intraperitoneal route. Moderately toxic by intravenous and parenteral routes. An experimental carcinogen. Mutagenic data. When heated to decomposition it emits toxic fumes of NO_x.

TRIS(1-METHYLETHYLENE)PHOS-PHORIC TRIAMIDE   HR: 3
CAS: 57-39-6   NIOSH: SZ 1925000
mf: $C_9H_{18}N_3OP$ mw: 215.27

PROP: Amber-colored liquid, amine odor, miscible with water and all organic solvents. Bp: 118°-125° @ 1mm, d: 1.079 @ 25°/25°.

SYNS: ENT 50,003 * 1,1',1'' -PHOSPHINYL-IDYNETRIS(2-METHYL)AZRIDINE * TRIS(2-METHYL-1-AZIRIDINYL)PHOSPHINE OXIDE * TRIS(2-METHYLAZIRIDIN-1-YL)PHOSPHINE OXIDE * N,N',N''-TRIS(1-METHYLETHYLENE)-PHOSPHORAMIDE

THR: Poison by ingestion, skin contact, intraperitoneal, and subcutaneous routes. Mutagenic data. An experimental teratogen and carcinogen. Animal experiments suggest cholinesterase inhibition, possibly due to metabolic products of this material in body. When heated to decomposition it emits very toxic fumes of NO_x and PO_x.

TRISODIUM EDETATE   HR: 3
CAS: 150-38-9   NIOSH: AH 5250000
mf: $C_{10}H_{13}N_2O_8$ • 3Na   mw: 358.22

SYNS: ETHYLENEDIAMINEACETIC ACID TRISO-DIUM SALT * ETHYLENEDIAMINETETRAACETI-CACID, TRISODIUM SALT * NCI-C03974 * SEQUESTRENE TRISODIUM SALT * TRISO-DIUM ETHYLENEDIAMINETETRAACETATE * TRISODIUM EDTA * TRISODIUM HYDROGEN ETHYLENEDIAMINETETRAACETATE * TRISO-DIUM HYDROGEN (ETHYLENEDINITRILO)TETRA-ACETATE * TRISODIUM VERSENATE

THR: Poison by intraperitoneal route. Moderately toxic by ingestion. When heated to decomposition it emits toxic fumes of NO_x.

TRISODIUM PHOSPHATE   HR: 2
CAS: 76-54-9
mf: Na_3PO_4 • $12H_2O$ mw: 380.21

PROP: Colorless crystals. D: 1.62 @ 20°; mp: 73.3°-76.7°. A strong alkali. Decomp @ 100° $(-12H_2O)$; sol in cold water, very sol in hot water.

SYNS: TRISODIUM-O-PHOSPHATE DODECAHY-DRATE * TRISODIUM-O-PHOSPHATE * TSP

THR: Inhalation of mist or dust can be very irritating to skin, eyes, and mucous membranes @ concentrations of 7-10 mg/m^3 for even short periods (minutes). A TWA (8 hour) of 0.1-1.5 mg/m^3 for respirable dust has been used (AIHA WEEL guide). A strong caustic material. A sequestrant, general-purpose, nutrient, dietary supplement food additive. When heated to decomposition it emits toxic fumes of PO_x.

TRITHION   HR: 3
CAS: 786-19-6   NIOSH: TD 5250000
mf: $C_{11}H_{16}ClO_2PS_3$ mw: 342.87

PROP: Amber liquid, essentially insol in water, misc in common solvents; bp: 82° @ 0.1 mm, d: 1.29 @ 20°.

SYNS: CARBOFENOTHION (DUTCH) * CARBO-FENOTHION * S-((P-CHLOROPHENYLTHIO)-METHYL)-O,O-DIETHYL PHOSPHORODITHIOATE * S-(4-CHLOROPHENYLTHIOMETHYL)DIETHYL PHOSPHOROTHIOLOTHIONATE * O,O-DIAETHYL-S-((4-CHLOR-PHENYL-THIO)-METHYL)DITHIO-PHOSPHAT (GERMAN) * O,O-DIETHYL-S-(4-CHLOOR-FENYL-THIO)-METHYL)-DITHIOFOSFAAT (DUTCH) * O,O-DIETHYL-S-P-CHLORFENYL-THIOMETHYLESTER KYSELINY DITHIOFOSFORECNE (CZECH) * O,O-DIETHYL P-CHLOROPHENYLMER-CAPTOMETHYL DITHIOPHOSPHATE * O,O-DI-ETHYL S-P-CHLORLPHENYLTHIOMETHYL DITHIOPHOSPHATE * O,O-DIETHYL S-(4-CHLOROPHENYLTHIOMETHYL) DITHIOPHOSPHATE * O,O-DIETHYL S-(P-CHLOROPHENYLTHIO-METHYL) PHOSPHORODITHIOATE * O,O-DI-ETHYL DITHIOPHOSPHORIC ACID P-CHLOROPHE-NYLTHIOMETHYL ESTER * O,O-DIETIL-S-((4-CLORO-FENIL-TIO)-METILE)-DITIOFOSFATO (ITALIAN) * DITHIOPHOSPHATE DE O,O-DIETHYLE ET DE (4-CHLORO-PHENYL) THIO-METHYLE (FRENCH) * ENT 23,708

THR: Poison by ingestion, skin contact, intraperitoneal, and subcutaneous routes. See also esters. A cholinesterase inhibitor. When heated to decomposition it emits very toxic fumes of SO_x, PO_x, and Cl^-. For further information see Carbofenothion, Vol. 2, No. 4 of *DPIM Report*.

TRITOLYL PHOSPHATE   HR: 3
CAS: 1330-78-5   NIOSH: TD 0175000
mf: $C_{21}H_{21}O_4P$ mw: 368.39

SYNS: CRESYL PHOSPHATE * FYRQUEL 150 * NCI-C61041 * PHOSPHATE DE TRICRESYLE (FRENCH) * TRICRESILFOSFATI (ITALIAN) * TRICRESYLFOSFATEN (DUTCH) * TRICRESYL PHOSPHATE * TRIKRESYLPHOSPHATE (GERMAN) * TRIS(TOLYLOXY)PHOSPHINE OXIDE

THR: A poison by ingestion. An eye irritant. See also tri-2-tolyl phosphate and phosphates. When heated to decomposition it emits toxic fumes of PO_x. For further information see Vol. 2, No. 3 of *DPIM Report*.

TRI-2-TOLYL PHOSPHATE   HR: 3
CAS: 78-30-8   NIOSH: TD 0350000
mf: $C_{21}H_{21}O_4P$ mw: 368.39

PROP: Colorless liquid; mp: $-25°$ to $-30°$, bp: $410°$ (slt decomp), flash p: $437°F$, d: 1.17, autoign temp: $725°F$, vap d: 12.7. Insol in water; sol in alc and ether.

SYNS: O-CRESYL PHOSPHATE * PHOSPHORIC ACID, TRI-O-CRESYL ESTER * PHOSPHORIC ACID, TRIS(2-METHYLPHENYL) ESTER * O-TOLYL PHOSPHATE * TRI-O-CRESYL PHOSPHATE * O-TRIKESYLPHOSPHATE (GERMAN) * TRIORTHOCRESYL PHOSPHATE * TRI 2-METHYLPHENYL PHOSPHATE * TRI-O-TOLYL PHOSPHATE * TROJKREZYLU FOSFORAN (POLISH)

OSHA PEL: TWA 0.1 mg/m³

THR: Poison by subcutaneous, intramuscular, intravenous, and intraperitoneal routes. Moderately toxic to humans by ingestion. Human central nervous system effects. See also phosphates. Most of the cases of tri-*o*-cresyl phosphate poisoning have followed its ingestion. In 1930, some 15,000 persons were affected in the United States, and of these, 10 died. The responsible material was found to be an alcoholic drink known as Jamaica ginger, or "jake." This beverage had been adulterated with about 2% of tri-*o*-cresyl phosphate. The affected persons developed a polyneuritis, which progressed, in many cases, with degeneration of the peripheral motor nerves, the anterior horn cells and the pyramidal tracts. Sensory changes were absent. Since 1930 there have been several other outbreaks of poisoning following ingestion of the material. Tri-*o*-cresyl phosphate is more toxic than the *m*-form, and much more so than tri-*p*-cresyl phosphate or triphenyl phosphate. Combustible when exposed to heat or flame. Dangerous; when heated to decomposition it emits highly toxic fumes of PO_x; can react with oxidizing materials. To fight fire, use CO_2, dry chemical. For further information see Tri-o-tolyl ester phosphoric acid, Vol. 2, No. 2 of the *DPIM Report*.

TRITON X   HR: 3
CAS: 9002-93-1   NIOSH: MD 0907500
mf: $(C_2H_4O)n$ $C_{14}H_{22}O$

SYNS: HYDROL SW * HYONIC PE-250 * IGEPAL CA-630 * OCTYL PHENOL CONDENSED WITH 12-13 MOLES ETHYLENE OXIDE * P-TERT-OCTYLPHENOXYPOLYETHOXYETHANOL * POLYETHYLENE GLYCOLMONOETHER WITH P-TERT-OCTYLPHENYL * POLYETHYLENE GLYCOL MONO(4-OCTYLPHENYL) ETHER * POLYETHYLENE GLYCOL MONO(4-TERT-OCTYLPHENYL) ETHER * POLYETHYLENE GLYCOLMONO(P-(1,1,3,3-TETRAMETHYLBUTYL)PHENYL) ETHER * POLYETHYLENE GLYCOL OCTYLPHENOL ETHER * POLYETHYLENE GLYCOL 450 OCTYL PHENYL ETHER * POLYETHYLENE GLYCOL P-OCTYLPHENYL ETHER * POLYETHYLENE GLYCOL P-TERT-OCTYLPHENYL ETHER * POLYETHYLENE GLYCOL P-1,1,3,3-TETRAMETHYLBUTYLPHENYL ETHER * POLYOXYETHYLENE MONO(OCTYLPHENYL) ETHER * POLYOXYETHYLENE (9) OCTYLPHENYL ETHER * POLYOXYETHYLENE (13) OCTYLPHENYL ETHER * POLY(OXYETHYLENE)P-TERT-OCTYLPHENYL ETHER

THR: Moderately toxic by ingestion. Human skin irritant. Severe eye irritant. When heated to decomposition it emits acrid smoke and fumes.

TROPHOSPHAMIDE   HR: 3
CAS: 22089-22-1   NIOSH: RP 3450000
mf: $C_9H_{18}Cl_3N_2P_2$ mw: 322.61

SYNS: TETRAHYDRO-N,N,3-TRIS(2-CHLOROETHYL)-2H-1,3,2-OXAPHOSPHORIN-2-AMINE-2-OXIDE * 2-(BIS(2-CHLOROETHYL)AMINO)-

3-(2-CHLOROETHYL)TETRAHYDRO-2H-1,3,2-OXA-PHOSPHORINE-2-OXIDE * NSC 109723 * N,N,N'-TRIS(2-CHLOROETHYL)-N',O-PROPYL-ENE PHOSPHORIC ACID ESTER DIAMIDE * N,N,3-TRIS(2-CHLOROETHYL)TETRAHYDRO-2H-1,3,2-OXAPHOSPHORIN-2-AMINE-2-OXIDE

THR: Poison by intraperitoneal route. Mutagenic data. Human blood and gastrointestinal tract effects. When heated to decomposition it emits very toxic fumes of Cl^-, NO_x, and PO_x.

TROPINE BENZOHYDRYL ETHER METHANESULFONATE   HR: 3
CAS: 132-17-2   NIOSH: YM 3150000
mf: $C_{21}H_{25}NO \cdot CH_4O_3S$ mw: 403.58

SYNS: BENZATROPINE METHANESULFONATE * BENZOTROPINE MESYLATE * BENZOTROPINE METHANESULFONATE * BENZTROPINE MESYLATE * BENZTROPINE METHANESULFONATE * 3-DIPHENYLMETHOXYTROPANE MESYLATE * 3-DIPHENYLMETHOXYTROPANE METHANESULFONATE

THR: Poison by ingestion, intravenous, subcutaneous, and intraperitoneal routes. Human psychotropic effects. See also sulfonates, ethers. When heated to decomposition it emits very toxic fumes of NO_x and SO_x.

TSUMACIDE   HR: 3
CAS: 1129-41-5   NIOSH: FC 8050000
mf: $C_9H_{11}NO_2$ mw: 165.21

SYNS: M-CRESYL ESTER OF N-METHYLCARBAMIC ACID * M-CRESYL METHYLCARBAMATE * 3-METHYLPHENYL-N-METHYLCARBAMATE * M-TOLYL-N-METHYLCARBAMATE * 3-TOLYL-N-METHYLCARBAMATE * METHYLCARBAMIC ACID-M-TOLYL ESTER

THR: Poison by ingestion and skin contact. Moderately toxic by inhalation. See also carbamates. When heated to decomposition it emits toxic fumes of NO_x.

TUBOCURARINE HYDROCHLORIDE   HR: 3
CAS: 57-94-3   NIOSH: YO 4900000
mf: $C_{38}H_{44}N_2O_6 \cdot 2Cl$ mw: 594.74

SYNS: DEXTROTUBOCURARINE CHLORIDE * TUBOCURARINE, CHLORIDE, HYDROCHLORIDE, (+)- (8CI) * (+)-TUBOCURARINE HYDROCHLORIDE * D-7',12'-DIHYDROXY-6,6'-DIMETHOXY-2,2',2'-TRIMETHYLTUBOCURARANIUM CHLORIDE * D-PARACURARINE CHLORIDE * TUBOCURARINE CHLORIDE * (+)-TUBOCURARINE CHLORIDE * D-TUBOCURARINE CHLORIDE * D-TUBOCURARINE DICHLORIDE * D-TUBOCURARINE HYDROCHLORIDE

THR: Poison by ingestion, intravenous, intraperitoneal, and subcutaneous routes. When heated to decomposition it emits very toxic fumes of NO_x and Cl^-. Human toxicity: large doses and overdoses may cause respiratory paralysis and hypotension.

TUNG NUT MEALS
THR: Toxic by ingestion. Contact causes dermatitis. Ingestion causes nausea, vomiting, cramps, diarrhea and tenesmus, thirst, dizziness, lethargy and disorientation. Large doses can cause fever, tachycardia and respiratory effects. See also saponin. Flammable in the form of dust when exposed to heat or flame; process material and cool thoroughly before storage so as not to over dry; can react with oxidizing materials.

TUNGSTEN   HR: 1
CAS: 7440-33-7   NIOSH: YO 7175000
af: W   aw: 183.85

PROP: A steely-gray to white, cuttable, forgeable and spinnable metal. Mp: 3410°, d: 19.3 @ 20°. Bp: 5900°.

SYN: WOLFRAM

ACGIH TLV: TWA insoluble compounds 5 mg/m^3, soluble compounds 1 mg/m^3

THR: Slightly toxic by intraperitoneal route. See also tungsten compounds. Flammable in the form of dust when exposed to flame. See also powdered metals. Incompatible with air, oxidants.

TUNGSTEN COMPOUNDS   HR: 2
THR: Tungsten compounds are considered somewhat more toxic than those of molybdenum. However, industrially, this element does not constitute an important health hazard. Exposure is related chiefly to the dust arising from the crushing and milling of the two chief ores of tungsten, namely, scheelite and wolframite. There is very little published with reference to its toxicity. The feeding of 2, 5, and 10% of diet as tungsten metal over a period of 70 days has been shown to be without marked effect

upon the growth of rats, as measured in terms of gain in weight. Ammonium-*p*-tungstate has been found to be much less toxic to rats upon ingestion than either tungstic oxide or sodium tungstate. Recent studies have failed to indicate any serious toxic effect following the inhalation or ingestion of various tungsten compounds, although heavy exposure to the dust or the ingestion of large amounts of the soluble compounds produces a certain rate of mortality in experimental animals.

TURPENTINE   HR: 3
CAS: 8006-64-2   NIOSH: YO 8400000
DOT: 1299

PROP: Colorless liquid, characteristic odor. Bp: 154°-170°, lel = 0.8%, flash p: 95°F (CC), d: 0.854-0.868 @ 25°/25°, autoign temp: 488°F, vap d: 4.84, ulc: 40-50.

SYNS: OIL OF TURPENTINE * OIL OF TURPEN-TINE, RECTIFIED * SPIRIT OF TURPENTINE * SPIRITS OF TURPENTINE * TEREBENTHINE (FRENCH) * TERPENTIN OEL (GERMAN) * TURPENTINE STEAM DISTILLED

OSHA PEL: TWA 100 ppm
ACGIH TLV: TWA 100 ppm
DFG MAK: 100 ppm (560 mg/m^3)
DOT Classification: Flammable Liquid

THR: Poison by inhalation (aspiration). Moderately toxic by intravenous route. Mildly toxic by ingestion. A human eye irritant. An experimental carcinogen. Human central nervous system, systemic, and irritant effects. Irritant to skin and mucous membranes. Can cause serious irritation of kidneys. A common air contaminant. Flammable when exposed to heat or flame. Avoid impregnation of leakage with combustibles. Keep cool and ventilated. Spontaneous Heating: Yes. Moderate explosion hazard in the form of vapor when exposed to flame; can react violently with $Ca(OCl)_2$; Cl_2; CrO_3; $Cr(OCl)_2$; $SnCl_4$; hexachloromelamine; trichloromelamine. When heated it emits acrid fumes; can react with oxidizing materials. To fight fire, use foam, CO_2, dry chemical. For further information see Vol. 2, No. 2 of *DPIM Report*.

TWEEN 60   HR: 3
CAS: 9005-67-8   NIOSH: WG 2934000

SYN: POLYOXYETHYLENE SORBITAN MONOSTEA-RATE

THR: An experimental tumorigen. Moderately toxic by intravenous route. See also surfactants. When heated to decomposition it emits acrid smoke and fumes. For further information see Sorbitan Monostearate, Vol. 1, No. 2 of *DPIM Report*.

U

UNDECANE
HR: 2
CAS: 1120-21-4   NIOSH: YQ 1525000
mf: $C_{11}H_{24}$ mw: 156.35

PROP: Colorless liquid; insol in water; d: 0.7402 @ 20°/4°; fp: −25.75°; bp: 195.6°; flash p: 149°F (OC); vap d: 5.4.

SYN: HENDECANE

THR: Moderately toxic by intravenous route. See also aliphatic hydrocarbons. Flammable when exposed to heat or flame or oxidizers. To fight fire, use foam, mist, dry chemical. When heated to decomposition it emits acrid smoke and fumes.

2-UNDECANONE
HR: 2
CAS: 112-12-9   NIOSH: YQ 2820000
mf: $C_{11}H_{22}O$ mw: 170.33

PROP: Colorless liquid, insol in water, mp: 12°, bp: 223°, flash p: 192°F (CC), d: 0.829 @ 30°, vap d: 5.9.

SYNS: 2-HENDECANONE * METHYL NONYL KETONE * METHYL-N-NONYL KETONE * MGK DOG AND CAT REPELLENT * NONYL METHYL KETONE

THR: Moderately toxic by ingestion. See also ketones. Flammable when exposed to heat or flame; can react with oxidizing materials. To fight fire, use CO_2, dry chemical.

URANIUM
HR: 3
CAS: 7440-61-1   NIOSH: YR 3490000
DOT: 9175
af: U   aw: 238.00

PROP: A heavy, silvery-white, malleable, ductile, softer-than-steel metal. Mp: 1132°, bp: 3818°, d: 18.95 (ca).

SYN: URANIUM METAL, PYROPHORIC (DOT)

ACGIH TLV: TWA soluble and insoluble compounds 0.2 mg/m^3
DFG MAK: 0.25 mg/m^3
DOT Classification: Radioactive and Flammable Solid

THR: A highly toxic element on an acute basis. The permissible levels for soluble compounds are based on chemical toxicity, while the permis-sible body level for insoluble compounds is based on radiotoxicity. The high chemical toxicity of uranium and its salts is largely shown in kidney damage, and acute necrotic arterial lesions. The rapid passage of soluble uranium compounds through the body tends to allow relatively large amounts to be taken in. The high toxicity effect of insoluble compounds is largely due to lung irradiation by inhaled parti-cles. This material is transferred from the lungs of animals quite slowly. Dangerous fire hazard in the form of a solid or dust when exposed to heat or flame. It can react violently with air; Cl_2; F_2; HNO_3; NO; Se; S; water; NH_3; BrF_3; trichloroethylene; nitryl fluoride. During storage may form a pyroforic surface due to effects of air, moisture.

URANIUM FLUORIDE (fissile)
HR: 3
CAS: 7783-81-5   NIOSH: YR 4720000
DOT: 2977
mf: F_6U mw: 352.00

PROP: Containing more than 0.7% U-235

SYN: URANIUM HEXAFLUORIDE, FISSILE (DOT)

DOT Classification: Label: Radioactive and Corrosive

THR: Radioactive poison; see also uranium. See also fluorides. When heated to decomposition it emits toxic fumes of F^-.

URANIUM FLUORIDE (low specific activity)
HR: 3
PROP: Containing 0.7% or less U-235

CAS: 7783-81-5   NIOSH: YR 4722000
DOT: 2978
mf: F_6U mw: 352.00

SYN: URANIUM HEXAFLUORIDE, LOW SPECIFIC ACTIVITY (DOT)

DOT Classification: Label: Radioactive and Corrosive

THR: Radioactive and chemical toxicity. See also uranium and fluorides. When heated to de-composition it emits toxic fumes of F^-.

URANIUM (III) HYDRIDE
mf: H_3U mw: 241.06

THR: No toxicity data. See also hydrides, uranium. Ignites in air.

URANIUM OXYFLUORIDE   HR: 3
CAS: 13536-84-0   NIOSH: YR 4700000
mf: F_2O_2U mw: 308.00

SYNS: URANIUM FLUORIDE OXIDE   ∗   URANYL FLUORIDE

OSHA PEL: TWA 50 ug(U)/m^3

THR: Poison by intravenous route. Radio-toxic material. See also fluorides. See also uranium. When heated to decomposition it emits toxic fumes of F^-.

URANIUM TETRACHLORIDE   HR: 3
CAS: 10026-10-5   NIOSH: YR 4025000
mf: Cl_4U mw: 379.80

PROP: Cubic, dark green-gray deliquescent crystals; mp: 590°; bp: 791°; d: 4.725 @ 25°/4°.

SYN: URANIUM (IV) CHLORIDE

THR: Poison by intraperitoneal route. See also uranium. When heated to decomposition it emits toxic fumes of Cl^-.

URANYL NITRATE HEXAHY-DRATE   HR: 3
CAS: 13520-83-7   NIOSH: YR 3850000
DOT: 2980
mf: $N_2O_8U \cdot 6H_2O$ mw: 502.14

PROP: Rhombic, yellow cryst; deliquescent; mp: 60.2° decomp @ 100°; d: 2.807 @ 13°.

SYN: URANYL NITRATE HEXAHYDRATE SOLUTION (DOT)

OSHA PEL: TWA 50 ug(U)/m^3
DOT Classification: Label: Radioactive and Corrosive

THR: Poison by ingestion, subcutaneous, intravenous, and intraperitoneal routes. See also uranium. When heated to decomposition it emits toxic fumes of NO_x.

URANYL NITRATE, SOLID (DOT)   HR: 3
CAS: 10102-06-4   NIOSH: YR 3805000
DOT: 2981
mf: N_2O_8U mw: 394.02
DOT Classification: Label: Radioactive and Oxidizer

THR: Poison by ingestion and inhalation. A corrosive and irritating material. See also uranyl nitrate, hexahydrate; uranium. When heated to decomposition it emits toxic fumes of NO_x.

UREA   HR: 3
CAS: 57-13-6   NIOSH: YR 6250000
mf: CH_4N_2O mw: 60.07

PROP: White crystals; mp: 132.7°; bp: decomp; d: (solid) 1.335; sol in water, alc; sltly sol in ether.

SYNS: CARBAMIDE RESIN ∗ CARBAMIMIDIC ACID ∗ CARBONYL DIAMIDE ∗ ISOUREA ∗ PSEUDOUREA ∗ UREAPHIL ∗ CARBAMIDE ∗ CARBONYLDIAMINE ∗ NCI-C02119

THR: An experimental neoplastigen and carcinogen. Mutagenic data. Moderately toxic by ingestion, intravenous, and subcutaneous routes. Human skin irritant. When heated to decomposition it emits toxic fumes of NO_x. Incompatible with $NaNO_2$; P_2Cl_5; nitrosyl perchlorate. For further information see Vol. 2, No. 2 of *DPIM Report*.

URETHANE   HR: 3
CAS: 51-79-6   NIOSH: FA 8400000
mf: $C_3H_7NO_2$ mw: 89.11

PROP: Colorless, odorless crystals. Mp: 49°, bp: 184°, d: 0.9862, vap press: 10 mm @ 77.8°, vap d: 3.07. Very sol in water, alc, ether.

SYNS: AETHYLCARBAMAT (GERMAN) ∗ AETHYLURETHAN (GERMAN) ∗ CARBAMIDSAEURE-AETHYLESTER (GERMAN) ∗ ETHYL CARBAMATE ∗ ETHYL URETHANE ∗ O-ETHYLURETHANE ∗ LEUCOTHANE ∗ NSC 746 ∗ URETHAN

THR: An experimental carcinogen, teratogen, neoplastigen, and tumorigen. Mutagenic data. Moderately toxic by ingestion intraperitoneal, subcutaneous, intramuscular, parenteral, intravenous, and other unspecified routes. Causes depression of bone marrow and occasionally focal degeneration in the brain. Can also produce central nervous system depression, nausea and vomiting. See also carbamates. When heated it emits toxic fumes of NO_x.

UV ABSORBER-2   HR: 3
CAS: 5232-99-5   NIOSH: AT 6200000
mf: $C_{18}H_{15}NO_2$ mw: 277.34

SYNS: 2-CYANO-3,3-DIPHENYLACRYLIC ACID, ETHYL ESTER * ALPHA-CARBETHOXY-BETA,BETA-BISCYCLOPROPYL ACRYLONITRILE * 2-CYANO-3,3-DIPHENYL-2-PROPENOIC ACID, ETHYL ESTER * USAF A -15972 * 3,3-DICY-CLOPROPYL-2-(ETHOXYCARBONYL)ACRYLONI-TRILE

THR: Poison by intraperitoneal route. See also esters and nitriles. When heated to decomposition it emits toxic fumes of NO_x.

V

n-VALERALDEHYDE HR: 2

CAS: 110-62-3   NIOSH: YV 3600000
DOT: 2058
mf: $C_5H_{10}O$ mw: 86.15

PROP: Liquid, flash p = 53.6°F; bp: 102°-103°;
d: 0.8095 @ 20°/4°. Very sltly sol in water;
misc with organic solvents.

SYNS: N-PENTANAL * VALERIC ACID ALDE-
HYDE * VALERYLALDEHYDE * AMYL ALDE-
HYDE * BUTYL FORMAL * PENTANAL
* VALERIC ALDEHYDE * VALERIANIC ALDE-
HYDE
ACGIH TLV: TWA 50 ppm
DOT Classification: Flammable Liquid

THR: Moderately toxic by ingestion. See also
aldehydes. A mild irritant. When heated to decomposition it emits acrid smoke and fumes.
Dangerous fire hazard.

VALETHAMATE BROMIDE HR: 3

CAS: 90-22-2   NIOSH: BP 7600000
mf: $C_{19}H_{32}NO_2 \cdot Br$ mw: 386.43

SYNS: 2-DIETHYLAMINOETHYL 3-METHYL-
2-PHENYLVALERATE METHYLBROMIDE
* 2-DIETHYLAMINOETHYL 2-PHENYL-3-
METHYLVALERATEMETHYL BROMIDE * DI-
ETHYL(2-HYDROXYETHYL)METHYLAMMONIUM
3-METHYL-2-PHENYLVALERATE BROMIDE
* 3-METHYL-2-PHENYLVALERIC ACID 2-DI-
ETHYLAMINOETHYL ESTER METHYL BROMIDE
* 3-METHYL-2-PHENYLVALERIC ACID
* DIETHYL(2-HYDROXYETHYL) METHYLAM-
MONIUMBROMIDE ESTER * PHENYLMETH-
YLVALERIANSAEURE-BETA-DIAETHYLAMINO-
AETHYLESTER-BROMMET HYLAT (GERMAN)
* DIETHYL(2-HYDROXYETHYL)METHYL-AM-
MONIUM BROMIDE 3-METHYL-2-PHENYL-
VALERATE

THR: Poison by ingestion, subcutaneous, and
intravenous routes. See also esters, bromides.
When heated to decomposition it emits very
toxic fumes of NO_x and Br^-.

VANADIUM HR: 3

CAS: 7440-62-2   NIOSH: YW 1355000
af: V   aw: 50.94

PROP: A bright white, soft ductile metal; sltly

radioactive; bp: 3000°; d: 6.11 @ 18.7°; mp:
1917°. Insol in water.

THR: An experimental carcinogen. See also vanadium compounds. Flammable in dust form
from heat or flame, sparks. Violent reaction
with BrF_3; Cl_2; Li; oxidants.

VANADIUM COMPOUNDS

THR: Variable toxicity. Vanadium compounds
act chiefly as an irritant to the conjunctivae and
respiratory tract. Prolonged exposures may lead
to pulmonary involvement. There is still some
controversy as to the effects of industrial exposure on other systems of the body. Responses
are acute, never chronic. The first report of
human vanadium poisoning described rather
widespread systemic effects, consisting of polycythemia, followed by red blood cell destruction
and anemia, loss of appetite, pallor and emaciation, albuminuria and hematuria, gastrointestinal disorders, nervous complaints and cough,
sometimes severe enough to cause hemoptysis.
More recent reports describe symptoms which,
for the most part, are restricted to the conjunctivae and respiratory system, no evidence being
found of disturbances of the gastrointestinal
tract, kidneys, blood or central nervous system.
Though certain workers believe that it is only
the pentoxide which is harmful, other investigators have found that patronite dust (chiefly vanadium sulfide) is quite toxic to animals, causing
acute pulmonary edema. The fumes are highly
toxic. See also specific compounds. These are
common air contaminants.

VANADIUM OXYTRICHLORIDE HR: 3

CAS: 7727-18-6   NIOSH: YW 2975000
DOT: 2443
mf: Cl_3OV mw: 173.29

PROP: Yellow deliquescent liquid; mp: −77°
± 2°; bp: 126.7°; d: 1.811 @ 32°.

SYN: VANADYL TRICHLORIDE

DOT Classification: Label: Corrosive

THR: Poison by ingestion. See also vanadium
compounds and hydrochloric acid. Can react
violently with P. When heated to decomposition
it emits toxic fumes of Cl^-. For further information see Vol. 2, No. 2 of *DPIM Report*.

VANADIUM PENTOXIDE (DUST)   HR: 3
CAS: 1314-62-1   NIOSH: YW 2450000
DOT: 2862
mf: O_4V_2 mw: 181.88

PROP: Yellow to red crystalline powder. Mp: 690°; bp: decomp @ 1750°; d: 3.357 @ 18°.

SYNS: ANHYDRIDE VANADIQUE (FRENCH) * C.I. 77938 * VANADIC ANHYDRIDE * VANADIUM OXIDE * VANADIO, PENTOSSIDO DI (ITALIAN) * VANADIUMPENTOXID (GERMAN) * VANADIUMPENTOXYDE (DUTCH) * VANADIUM, PENTOXYDE DE (FRENCH) * WANADU PIECIOTLENEK (POLISH)

OSHA PEL: CL 500 ug/m³
ACGIH TLV: TWA (respirable dust & fume) 0.05 mg/m³
DFG MAK: (fine dust) 0.05 mg/m³
DOT Classification: ORM-E (IMO: Poison B)

THR: Poison by ingestion, inhalation, intraperitoneal, subcutaneous, intratracheal, and intravenous routes. Mutagenic data. Human pulmonary system and allergenic effects. See also vanadium compounds. A respiratory irritant; causes skin pallor, greenish-black tongue, chest pain, cough, dyspnea, palpitation, lung changes. When ingested, causes gastrointestinal tract disturbances. May also cause a papular skin rash. When heated to decomposition it emits acrid smoke and fumes of VO_x. Incompatible with ClF_3; Li; peroxyformic acid; (Ca + S + H_2O). For further information see Vol. 2, No. 2 of *DPIM Report*.

VANADIUM SESQUIOXIDE   HR: 3
CAS: 1314-34-7   NIOSH: YW 3050000
mf: O_3V_2 mw: 149.88

PROP: Black crystals; mp: 1970°; d: 4.87 @ 18°.

SYNS: VANADIUM TRIOXIDE * VANADIC OXIDE * VANADIUM OXIDE

THR: Poison by ingestion, subcutaneous, and intratracheal routes. See also vanadium compounds. Can self-ignite in air upon heating. When heated to decomposition it emits toxic fumes of VO_x.

VANADIUM TETRACHLORIDE   HR: 3
CAS: 7632-51-1   NIOSH: YW 2625000

DOT: 2444
mf: Cl_4V mw: 192.74

PROP: Reddish brown liquid; mp: −28 ± 2°; bp: 148.5°; d: 1.816 @ 30°.

SYN: VANADIUM CHLORIDE

DOT Classification: Label: Corrosive

THR: Poison by ingestion. See also vanadium compounds and hydrochloric acid. When heated to decomposition it emits toxic fumes of Cl^-.

VANADIUM TRICHLORIDE   HR: 3
CAS: 7718-98-1   NIOSH: YW 2800000
mf: Cl_3V mw: 157.29

PROP: Pink crystals; mp: decomp; d: 3.00 @ 18°.

SYN: VANADIUM (III) CHLORIDE

THR: Poison by ingestion and subcutaneous routes. See also vanadium compounds and hydrochloric acid. When heated to decomposition it emits toxic fumes of Cl^-. Incompatible with methyl magnesium iodide.

VANADYL SULFATE   HR: 3
CAS: 27774-13-6   NIOSH: YW 1925000
mf: O_5SV mw: 163.00

PROP: Blue crystals.

SYN: C.I. 77940

THR: Poison by intravenous, intraperitoneal, and subcutaneous routes. See also sulfates and vanadium compounds. When heated to decomposition it emits toxic fumes of SO_x. For further information see Vol. 2, No. 1 of *DPIM Report*.

VANCIDE   HR: 3
CAS: 12427-38-2   NIOSH: OP 0700000
mf: $C_4H_7N_2S_4 \cdot Mn$ mw: 266.31

PROP: Yellow powder or crystals; water sol.

SYNS: ENT 14,875 * 1,2-ETHANEDIYL-BIS(CARBAMODITHIOATO)(2−)MANGANESE * 1,2-ETHANEDIYLBISCARBAMODITHIOIC ACID, MANGANESE COMPLEX * 1,2-ETHANEDIYLBIS-CARBAMODITHIOIC ACID, MANGANESE(2+) SALT (1:1) * 1,2-ETHANEDIYLBISMANEB, MANGANESE (2+) SALT (1:1) * ETHYLENEBISDITHIO-CARBAMATE MANGANESE * N,N'-ETHYLENE BIS(DITHIOCARBAMATE MANGANEUX) (FRENCH) * ETHYLENEBIS(DITHIOCARBAMATO), MAN-

GANESE * ETHYLENEBIS(DITHIOCARBAMIC ACID), MANGANESE SALT * ETHYLENEBIS-(DITHIOCARBAMIC ACID) MANGANOUS SALT * 1,2-ETHYLENEDIYLBIS(CARBAMODITHIOATO)-MANGANESE * N,N'-ETILEN-BIS(DITIOCARBAM-MATO) DI MANGANESE(ITALIAN) * MANGAAN (II)-(N,N'-ETHYLEEN-BIS(DITHIOCARBAMAAT)) (DUTCH) * MANGAN (II)-(N,N'-AETHYLEN-BIS-(DITHIOCARBAMATE)) (GERMAN) * MANGA-NESE ETHYLENE-1,2-BISDITHIOCARBAMATE * MANGANESE (II) ETHYLENE DI(DITHIOCAR-BAMATE) * MANZATE MANEB FUNGICIDE * MANGANESE (II) ETHYLENEBIS(DITHIOCAR-BAMATE)

THR: An experimental carcinogen, tumorigen, and teratogen. Moderately toxic by ingestion and possibly other routes. Mutagenic data. See also manganese compounds and thiocarbamates. Dangerous, when heated to decomposition it emits highly toxic fumes of NO_x and SO_x.

VANILLIN   HR: 2
CAS: 121-33-5   NIOSH: YW 5775000
mf: $C_8H_8O_3$ mw: 152.16

PROP: White crystalline needles, pleasant odor, sol in 125 parts water, 20 parts glycerol 2 parts 95% alc, chloroform and ether; d: 1.056; bp: 285°; mp: 80°-81°.

SYNS: 4-HYDROXY-3-METHOXYBENZALDEHYDE * 3-METHOXY-4-HYDROXYBENZALDEHYDE * VANILLA * VANILLALDEHYDE * VANILLIC ALDEHYDE

THR: Moderately toxic by ingestion, intraperitoneal, subcutaneous, and intravenous routes. Mutagenic data. Can react violently with Br_2; $HClO_4$; K-tert-butoxide; (tert-chlorobenzene + NaOH); (formic acid + $Tl(NO_3)_3$). See also aldehydes. When heated to decomposition it emits acrid smoke and fumes.

VAPAM   HR: 3
CAS: 137-42-8   NIOSH: FC 2100000
mf: $C_2H_4NS_2 \cdot Na$ mw: 129.18

SYNS: METAM-SODIUM (DUTCH, FRENCH, GER-MAN, ITALIAN) * METHAM SODIUM * N-METHYLDITHIOCARBAMATE DE SODIUM (FRENCH) * METHYLDITHIOCARBAMIC ACID, SODIUM SALT * N-METIL-DITIOCARBAMMATO DI SODIO (ITALIAN) * NATRIUM-N-METHYL-DITHIOCARBAMAAT (DUTCH) * NATRIUM-N-METHYL-DITHIOCARBAMAT (GERMAN) * SO-DIUM METHYLDITHIOCARBAMATE * SODIUM-N-METHYLDITHIOCARBAMATE * N-METHYL-DITHIOCARBAMIC ACID, SODIUM SALT

THR: Poison by ingestion and possibly other routes. Irritant to skin and mucous membranes. Accompanied by alcohol intake, causes violent vomiting and shock. When heated to decomposition it emits very toxic fumes of NO_x.

VASOTONIN   HR: 3
CAS: 51-43-4   NIOSH: DO 2625000
mf: $C_9H_{13}NO_3$ mw: 183.23

SYNS: 1-ADRENALIN * (R)-EPINEPHRINE * ADRENALIN * ADRENALINE * (−)-ADRE-NALINE * 1-ADRENALINE * ANTIASTHMA-TIQUE * 3,4-DIHYDROXY-ALPHA-((METHYLAMI-NO)METHYL)BENZYL ALCOHOL * 1-1-(3,4-DIHYDROXYPHENYL)-2-METHYLAMINOETHANOL * EPINEPHRINE * (−)-EPINEPHRINE * 1-EPI-NEPHRINE * 1-EPINEPHRINE (SYNTHETIC) * SUPRARENIN * VAPONEFRIN * VASOCON-STRICTOR

THR: Poison to humans by subcutaneous route. Poison experimentally by ingestion, skin contact, intraperitoneal, intravenous, and intramuscular routes. Mutagenic data. An experimental teratogen. When heated to decomposition it emits toxic fumes of NO_x.

VERATRIDINE   HR: 3
CAS: 71-62-5   NIOSH: YX 5600000
mf: $C_{36}H_{51}NO_{11}$ mw: 673.88

PROP: Yellow-white powder; mp: 180°. Sol in water; sltly sol in ether.

SYNS: VERATRINE (AMORPHOUS) * 4,9-EPOX-YCEVANE-3,4,12,14,16,17,20-HEPTOL 3-(3,4-DIMETHOXYBENZOATE) * 3-VERATROYLVERA-CEVINE

THR: Poison by intraperitoneal and intravenous routes. See also veratrine. Combustible. When heated to decomposition it emits toxic fumes of NO_x.

VERATRINE   HR: 3
CAS: 8051-02-3   NIOSH: YX 5950000

PROP: From the plant Schoenocaulon officinale. Botanical insecticide. The active ingredients are a group of alkaloids known as veratrin, i.e., cevadine and veratridine. A powder.

SYNS: ASAGRAEA OFFICINALIS * CEVADILLA * CEVADINE * ENT 123 * SABACIDE * SABADILLA * SABANE DUST * VERATRIN (GERMAN)

THR: Poison by ingestion, intraperitoneal, and other unspecified routes. Ingestion causes severe gastrointestinal tract disturbances, burning in the mouth, vomiting, diarrhea and cramps. Also produces headache, dizziness, slow pulse and weakness. Large doses cause death by circulatory and respiratory failure. It is a powerful irritant to skin and mucous membranes and is less toxic than rotenone. Inhalation causes violent sneezing. Dangerous; when heated to decomposition it emits toxic fumes of NO_x.

VERILOID   HR: 3
CAS: 8002-39-9   NIOSH: YX 7875000

SYNS: VERATRUM VIRIDE * VERATRUM VIRIDE ALKALOIDS EXTRACT

THR: Poison by ingestion, intravenous, and intraperitoneal routes. An experimental teratogen.

VINBARBITAL SODIUM   HR: 3
CAS: 125-44-0   NIOSH: CQ 5600000
mf: $C_{11}H_{15}N_2O_3 \cdot Na$ mw: 246.27

SYNS: DELVINAL SODIUM * 5-ETHYL-5-(1-METHYL-1-BUTENYL)BARBITURIC ACID SODIUM SALT * SODIUM-5-ETHYL-5-(1-METHYL-1-BUTENYL) BARBITURATE * SODIUM VINBARBITAL

THR: Poison by ingestion and intraperitoneal routes. When heated to decomposition it emits toxic fumes of Na_2O, and NO_x.

VINBLASTINE SULFATE   HR: 3
CAS: 143-67-9   NIOSH: YY 8400000
mf: $C_{46}H_{58}N_4O_9 \cdot H_2O_4S$ mw: 909.16

SYNS: NSC 49842 * VINCALEUKOBLASTINE SULFATE (1:1), (SALT) * VINCALEUKOBLASTINE SULFATE

THR: Poison by intraperitoneal route. Mutagenic data. An experimental teratogen. Human blood effects. See also sulfates. When heated to decomposition it emits very toxic fumes of NO_x and SO_x.

VINCALEUKOBLASTINE   HR: 3
CAS: 865-21-4   NIOSH: YY 8050000
mf: $C_{46}H_{58}N_4O_9$ mw: 811.08

SYNS: NCI-C04842 * NINCALUICOLFLASTINE * NSC 47842 * VINBLASTINE * VINCALEUCOBLASTINE

THR: Poison by intravenous, subcutaneous, and other unspecified routes. Mutagenic data. An experimental carcinogen and teratogen. Human eye (systemic), central nervous system, and other toxic effects unspecified in source. When heated to decomposition it emits toxic fumes of NO_x.

VINYL BROMIDE   HR: 3
CAS: 593-60-2   NIOSH: KU 8400000
DOT: 1085
mf: C_2H_3Br mw: 106.96

PROP: A gas; mp: $-138°$, bp: $15.6°$, d: 1.51. Insol in water; misc in alc, ether.

SYNS: BROMOETHENE * BROMOETHYLENE * BROMURE DE VINYLE (FRENCH) * VINILE (BROMURO DI) (ITALIAN) * VINYLBROMID (GERMAN)
ACGIH TLV: TWA 5 ppm
DOT Classification: Flammable Gas

THR: An experimental carcinogen, mutagenic, and neoplastigen. Moderately toxic by ingestion. Dangerous fire hazard when exposed to heat or flame. Dangerous; can react vigorously with oxidizing materials. See also bromides. To fight fire, use CO_2, dry chemical or water spray. For further information see Vol. 4, No. 5 of *DPIM Report*.

VINYL BUTYL ETHER   HR: 2
CAS: 111-34-2   NIOSH: KN 5950000
mf: $C_6H_{12}O$ mw: 100.18

PROP: Liquid. Mp: $-112.7°$, bp: $94.1°$, flash p: $15°F$ (OC), d: 0.7803 @ $20°/20°$, vap d: 3.45.

SYN: VINYL-N-BUTYL ETHER

THR: Moderately toxic by skin contact. Mildly toxic by ingestion. Skin and eye irritant. Dangerous fire hazard when exposed to heat or flame. See also ethers for explosion hazard. Highly dangerous, when exposed to heat or flame; can react with oxidizing materials. To fight fire, use foam, CO_2, dry chemical.

VINYL BUTYRATE   HR: 2
CAS: 123-20-6   NIOSH: ET 7000000
mf: $C_6H_{10}O_2$ mw: 114.16

PROP: D: 0.9, vap d: 4.0, bp: 116°, flash p: 68°F (OC), lel = 1.4%, uel = 8.8%.

SYN: BUTYRIC ACID, VINYL ESTER

THR: A skin and eye irritant. Mildly toxic by inhalation and ingestion. See also esters. Danger fire hazard when exposed to heat, flame or oxidizers. To fight fire, use alcohol foam, fog, mist, CO_2.

VINYL CARBAMATE   HR: 3
CAS: 15805-73-9   NIOSH: FD 1995000
mf: $C_3H_5NO_2$ mw: 87.09

SYN: CARBAMIC ACID, VINYL ESTER

THR: Poison by intraperitoneal route. Mutagenic data. An experimental neoplastigen. See also esters. When heated to decomposition it emits toxic fumes of NO_x.

VINYL CHLORIDE   HR: 3
CAS: 75-01-4   NIOSH: KU 9625000
DOT: 1086
mf: C_2H_3Cl mw: 62.50

PROP: Colorless liquid or gas (when inhibited), faintly sweet odor. Mp: −160°; bp: −13.9°, lel = 4%, uel = 22%; flash p: 17.6°F (COC), fp: −159.7°, d(liquid): 0.9195 @ 15°/4°, vap press: 2600 mm @ 25°, vap d: 2.15, autoign temp: 882°F. Sltly sol in water; sol in alc; very sol in ether.

SYNS: CHLOROETHENE * CHLOROETHYLENE * CHLORURE DE VINYLE (FRENCH) * CLORURO DI VINILE (ITALIAN) * ETHYLENE MONOCHLORIDE * MONOCHLOROETHENE * MONOCHLOROETHYLENE (DOT) * VINYLCHLORID (GERMAN) * VINYL CHLORIDE (DOT) * VINYL CHLORIDE MONOMER * VINYL C MONOMER * WINYLU CHLOREK (POLISH)

OSHA PEL: TWA 1 ppm; CL 5 ppm/15M
ACGIH TLV: TWA 5 ppm
TRK: 3 ppm
DOT Classification: Label: Flammable Gas

THR: A human brain carcinogen. An experimental neoplastigen and tumorigen by inhalation. Severe irritant by inhalation to skin, eyes, and mucous membranes. Causes skin burns by rapid evaporation and consequent freezing. In high concentration, it acts as an anesthetic. Chronic exposure has shown liver injury. Circulatory and bone changes in the fingertips reported in workers handling unpolymerized materials. Dangerous fire hazard when exposed to heat, flame or oxidizers. Large fires of this material are practically inextinguishable. Severe explosion hazard in the form of vapor, when exposed to heat or flame. Also, on standing, forms peroxides in air and can then explode. Very dangerous; when heated to decomposition it emits highly toxic fumes of phosgene; can react vigorously with oxidizing materials. Before storing or handling this material, instructions for its use should be obtained from the supplier. To fight fire, stop flow of gas. For further information see Vol. 6, No. 4 of *DPIM Report*.

VINYL CYCLOHEXENE DIOXIDE   HR: 3
CAS: 106-87-6   NIOSH: RN 8640000
mf: $C_8H_{12}O_2$ mw: 140.20

PROP: Colorless liquid; d: 1.098 @ 20°/20°, bp: 227°, flash p: 230°F.

SYNS: 1,2-EPOXY-4-(EPOXYETHYL)CYCLOHEXANE * 1-EPOXYETHYL-3,4-EPOXYCYCLOHEXANE * 3-(EPOXYETHYL)-7-OXABICYCLO(4.1.0)HEPTANE * 3-(1,2-EPOXYETHYL)-7-OXABICYCLO(4.1.0)HEPTANE * 4-(1,2-EPOXYETHYL)-7-OXABICYCLO-(4.1.0)HEPTANE * 4-(EPOXYETHYL)-7-OXABICYCLO(4.1.0)HEPTANE * 1-ETHYLENEOXY-3,4-EPOXYCYCLOHEXANE * NCI-C60139 * 3-OXIRANYL-7-OXABICYCLO(4.1.0)HEPTENE * 4-VINYL-1-CYCLOHEXENE DIEPOXIDE * 4-VINYL-1,2-CYCLOHEXENE DIEPOXIDE * 1-VINYL-3-CÝCLOHEXENE DIOXIDE * 4-VINLYCYCLOHEXENE DIOXIDE * 4-VINYL-1-CYCLOHEXENE DIOXIDE
ACGIH TLV: TWA 10 ppm

THR: An experimental carcinogen and tumorigen. Moderately toxic by ingestion, inhalation, and skin contact. Mutagenic data. Combustible when exposed to heat or flame. To fight fire, use water, foam, dry chemical.

VINYL ETHER   HR: 2
CAS: 109-93-3   NIOSH: YZ 6700000
DOT: 1167
mf: C_4H_6O mw: 70.10

PROP: Colorless liquid, very volatile, flammable. Bp: 29°, ulc: 100, lel = 1.7%, uel = 27%, flash p: <−22°F (CC), d: 0.774 @ 20°/20°, autoign temp: 680°F, vap d: 2.41. Very sltly sol in water; misc in alc, ether.

SYNS: DIVINYL ETHER * VINESTHESIN * VINETHEN * VINETHENE * VINETHER * VINIDYL * DIVINYL ETHER (DOT) * DIVINYL OXIDE

DOT Classification: Label: Flammable Liquid

THR: Mildly toxic by inhalation. See also ethers. Prolonged exposure causes liver injury. Dangerous fire hazard when exposed to heat, flame or oxidizers. Severe explosion hazard in the form of vapor, when exposed to heat or flame. Violent reaction with air, O_2. Highly dangerous; when exposed to heat or flame, can react vigorously with oxidizing materials. To fight fire, use CO_2, dry chemical. For further information see Vol. 1, No. 7 of *DPIM Report*.

VINYL FLUORIDE HR: 3
CAS: 75-02-5 NIOSH: YZ 7351000
DOT: 1860
mf: CH_2:CHF mw: 46

PROP: Colorless gas; mp: $-160.5°$; bp: $-72°$; lel = 2.6%; uel = 21.7%; insol in water; sol in alc, ether.

SYNS: FLUOROETHENE * FLUOROETHYLENE * MONOFLUOROETHYLENE

DOT Classification: Label: Flammable Gas

THR: A poison. See also fluorides. Ignites in presence of heat or sources of ignition. Highly dangerous; see fluorides. To fight fire, stop flow of gas.

VINYLIDENE CHLORIDE HR: 3
CAS: 75-35-4 NIOSH: YZ 8061000
DOT: 1150/1303
mf: CH_2CCl_2 mw: 97.0

PROP: Colorless volatile liquid; bp: $31.6°$; lel = 7.3%; uel = 16.0%; fp: $-122°$; flash p: $0°F$ (OC); d: 1.213 @ 20°/4°; autoign temp: 1058°F.

SYNS: 1,1-DICHLOROETHENE * 1,1-DICHLOROETHYLENE
ACGIH TLV: TWA 5 ppm; STEL 20 ppm
DFG MAK: 2 ppm (8 mg/m³)
DOT Classification: Label: Flammable Liquid

THR: A poison. Mutagenic data. Highly dangerous fire hazard when exposed to heat or flame. Moderately explosive in the form of gas, when exposed to heat or flame. Also can explode spontaneously; reacts violently with chlorosul-

fonic acid; HNO_3; oleum. Highly dangerous; see chlorides; can react vigorously with oxidizing materials. To fight fire, use alcohol foam, CO_2, dry chemical. For further information see Vol. 2, No. 6 of *DPIM Report*.

VINYLIDENE FLUORIDE HR: 3
CAS: 75-38-7 NIOSH: KW 0560000
mf: $C_2H_2F_2$ mw: 64.04

PROP: Colorless gas; bp: $<-70°$, lel = 5.5%, uel = 21.3%.

SYNS: 1,1-DIFLUOROETHYLENE * NCI-C60208

THR: An experimental neoplastigen and mutagen. Mildly toxic by inhalation. Very dangerous fire hazard by heat, flames or oxidizers. Dangerous; when heated to decomposition it emits toxic fumes of F^-. To fight fire, stop flow of gas.

VINYL PROPIONATE HR: 3
CAS: 105-38-4 NIOSH: UF 8575000
mf: $C_5H_8O_2$ mw: 100.13

PROP: Liquid, almost insol in water. D: 0.9173 @ 20°/20°, bp: $95°$, fp: $-81.1°$, flash p: 34°F (OC), vap d: 3.3.

SYN: PROPANOIC ACID, ETHENYL ESTER

THR: Poison by inhalation. A skin and eye irritant. See also esters. A dangerous fire hazard. To fight fire, use alcohol foam, mist, fog.

VITAMIN A PALMITATE HR: 3
CAS: 79-81-2 NIOSH: VH 6860000
mf: $C_{36}H_{60}O_2$ mw: 524.96

SYN: 211-TRANS-RETINOL PALMITATE

THR: An experimental teratogen. Mutagenic data. When heated to decomposition it emits acrid smoke and fumes.

VITAMIN B$_6$ HYDROCHLORIDE HR: 3
CAS: 65-22-5 NIOSH: UV 1225000
mf: $C_8H_9NO_3 \cdot ClH$ mw: 203.64

SYNS: PYRIDOXAL, HYDROCHLORIDE * 3-HYDROXY-5-(HYDROXYMETHYL)-2-METHYLISONICOTINALDEHYDE, HYDROCHLORIDE * 2-METHYL-3-HYDROXY-4-FORMYL-5-HYDROXYMETHYLPYRIDINE HYDROCHLORIDE

THR: Poison by intramuscular, intravenous, and intraperitoneal routes. When heated to decomposition it emits very toxic fumes of NO_x and HCl.

VITAMIN B$_{12}$ COMPLEX   HR: 3
CAS: 68-19-9   NIOSH: GG 3750000
mf: $C_{63}H_{88}CoN_{14}O_{14}P$ mw: 1355.55

PROP: The anti-pernicious anemia vitamin; all vitamin B$_{12}$ compounds contain the cobalt atom in its trivalent state. There are at least three active forms: cyanocobalamin, hydroxycobalamin and nitrocobalamin. Dark red crystals.

SYNS: COBINAMIDE, CYANIDE PHOSPHATE 3'-ESTER WITH 5,6-DIMETHYL-1-ALPHA-D-RIBOFURANOSYLBENZIMIDAZOLE, INNER SALT * B-12 * COBALIN * CYANO-B12 * CYANOCOBALAMIN * 5,6-DIMETHYLBENZIMIDAZOLYLCOBAMIDE CYANIDE * 5,6-DIMETHYL BENZIMIDAZOL * DIMETHYLBENZIMIDAZOYLCOBAMIDE * DISTIVIT (B12 PEPTIDE) * ERYTHROTIN * EXTRINSIC FACTOR * LACTOBACILLUS LACTIS DORNER FACTOR * MEGABION (INDIAN) * VITAMIN B12

THR: Poison by intraperitoneal and subcutaneous routes. See also cobalt. When heated to decomposition it emits very toxic fumes of PO$_x$ and NO$_x$.

VITAMIN K1   HR: 3
CAS: 84-80-0   NIOSH: QJ 5800000
mf: $C_{31}H_{46}O_2$ mw: 450.77

SYNS: 2-METHYL-3-(3,7,11,15-TETRAMETHYL-2-HEXADECENYL)-1,4-NAPHTHALENEDIONE * 2-METHYL-3-PHYTHYL-1,4-NAPHTHOCHINON (GERMAN) * 2-METHYL-3-PHYTYL-1,4-NAPH-THOCHINON (GERMAN) * PHYLLOCHINON (GERMAN) * PHYLLOQUINONE * ALPHA-PHYLLOQUINONE * TRANS-PHYLLOQUINONE * PHYTOMENADIONE * ANTIHEMORRHAGIC VITAMIN

THR: Poison by ingestion. Moderately toxic by subcutaneous route. When heated to decomposition it emits acrid smoke and fumes.

VOLIDAN   HR: 3
CAS: 595-33-5   NIOSH: TU 4075000
mf: $C_{24}H_{32}O_4$ mw: 384.56

SYNS: 17-ALPHA-ACETOXY-6-DEHYDRO-6-METHYLPROGESTERONE * 17-ACETOXY-6-METHYLPREGNA-4,6-DIENE-3,20-DIONE * 17-ALPHA-ACETOXY-6-METHYLPREGNA-4,6-DIENE-3,20-DIONE * 17-ALPHA-ACETOXY-6-METHYL-4,6-PREGNADIENE-3,20-DIONE * 6-DEHYDRO-6-METHYL-17-ALPHA-ACETOXY-PROGESTERONE * 17-HYDROXY-6-METHYL-PREGNA-4,6-DIENE-3,20-DIONE ACETATE * 6-METHYL-6-DEHYDRO-17-ALPHA-ACETOXY-PROGESTERONE * 6-METHYL-6-DEHYDRO-17-ALPHA-ACETYLPROGESTERONE * 6-METHYL-17-ALPHA-HYDROXY-DELTA(SUP 6)-PROGESTERONE ACETATE * 6-METHYL-DELTA(SUP 4,6)-PREGNADIEN-17-ALPHA-OL-3,20-DIONE ACETATE

THR: Poison by intravenous routes. An experimental carcinogen. When heated to decomposition it emits acrid smoke and fumes.

W

WAIT'S GREEN MOUNTAIN ANTIHISTA-MINE HR: 3
CAS: 91-84-9 NIOSH: UT 0875000
mf: $C_{17}H_{23}N_3O$ mw: 285.43

SYNS: N-DIMETHYLAMINO-AETHYL-N-P-METHOXY-BENZYL-ALPHA-AMINO-PYRIDIN-MALEAT (GERMAN) ∗ 2-((2-(DIMETHYLAMI-NO)ETHYL)-(P-METHOXYBENZYL)AMINO)PYRI-DINE ∗ N-P-METHOXYBENZYL-N′,N′-DI-METHYL-N-ALPHA-PYRIDYLETHYLENEDIAMINE ∗ N-(P-METHOXYBENZYL)-N′,N′-DIMETHYL-N-2-PYRIDYLETHYLENEDIAMINE ∗ NCI-C60651

THR: Poison by ingestion, intraperitoneal, subcutaneous and other unspecified routes. Human gastrointestinal tract effects. An eye irritant. When heated to decomposition it emits toxic fumes of NO_x.

WALNUT EXTRACT HR: 3
CAS: 481-39-0 NIOSH: QJ 5775000
mf: $C_{10}H_6O_3$ mw: 174.16

SYNS: C.I. 75500 ∗ 5-HYDROXY-1,4-NAPH-THALENEDIONE ∗ C.I. NATURAL BROWN 7 ∗ 5-HYDROXY-1,4-NAPHTHOQUINONE

THR: An experimental neoplastigen. A poison by ingestion. When heated to decomposition it emits acrid smoke and fumes.

WELDING FUMES HR: 3
ACGIH TLV: TWA 5 mg/m³

THR: When welding is done on a surface coated with cadmium, toxic fumes of cadmium are evolved. When zinc-coated surfaces are welded, toxic quantities of zinc oxide may be liberated. When painted surfaces are welded, lead or other pigment fumes may be liberated. And when fluoride fluxes are used in welding, very toxic fluoride fumes are evolved. When oily surfaces are welded, offensive and toxic fumes can be liberated, and when the welding torch is improperly ignited, carbon monoxide, which is very toxic may be evolved. Also, NO_x is formed. It is therefore considered hazardous to inhale excessive amounts of welding fumes. It is also possible to inhale sufficient quantities of iron oxide from welding to cause siderosis. Metal fume fever is a common reaction. It is characterized by chills, fever, sweating, and leucocytosis coming on several hours after exposure. Recovery is usually complete in 24-48 hours and there are no significant after effects. Safety goggles are required to protect against spatter. Light-filtering goggles are required to shield the eyes against the intense UV light from the welding arc.

WHISKEY HR: 2
PROP: Light yellow-amber liquid. Pleasant to fruity odor. D: 0.923-0.935 @ 15.56°; 47%-53% of ethanol, by volume; flash p: 80.0°F (CC). Made by distillation of fermented malted grains, i.e., corn, rye or barley. After distillation, whiskey is aged in wooden containers for up to several years. The aging extracts from the wood such components as acids, esters; promotes oxidation of components of raw whiskey and some reactions between organic components to form new flavors.

THR: The whiskey or wine equivalent of 1 ounce of pure ethanol per capita per day is often cited as healthful to adults to relieve stress and promote relaxation. However, it is often abused which can lead to habituation with consequent liver damage, malnutrition and a wide variety of other physical and mental problems. See also ethanol. Severe fire hazard. To fight fire, use water, water spray, alcohol foam, CO_2, dry chemical.

WINE HR: 2
PROP: An alcoholic beverage made from the fermented juice of grapes, other fruits or plants. Contains from 7-20% ethanol by volume. Concentrations of alcohol higher than those produced naturally are obtained by fortifying with pure ethanol. The distinctive colors, tastes, bouquets of wines are usually produced by adding coloring matter, sugar, acetic acid, salts and higher fatty acids.

THR: Moderately toxic. See also whiskey. Some of the additives to wines have been known to cause allergic reactions in humans.

X

XANTHACRIDINE HR: 3
CAS: 86-40-8 NIOSH: AR 9625000
mf: $C_{14}H_{14}N_3 \cdot Cl$ mw: 259.76

SYNS: 3,6-DIAMINO-10-METHYLACRIDINIUM CHLORIDE * C.I. 46000 * 2,8-DIAMINO-10-METHYLACRIDINIUM CHLORIDE * ACRIFLAVINE NEUTRAL

THR: Poison to humans by intravenous route. Poison by intraperitoneal, intravenous, and subcutaneous routes. Mutagenic data. When heated to decomposition it emits very toxic fumes of NO_x and Cl^-.

XANTHINE HR: 3
CAS: 69-89-6 NIOSH: ZD 7700000
mf: $C_5H_4N_4O_2$ mw: 152.13

PROP: Scales or plates. Decomp on heating without melting, partial sublimation. Sol in water; less sol in alc; sol in mineral acids; very sol in NH_4OH and NaOH solns.

SYNS: PSEUDOXANTHINE * 2,6-DIOXOPURINE * ISOXANTHINE * PURINE-2,6-DIOL * 9H-PURINE-2,6-DIOL * 2,6(1,3)-PURINEDION * PURINE-2,6-(1H,3H)-DIONE * USAF CB-17 * XANTHIC OXIDE

THR: An experimental neoplastigen. Moderately toxic by intraperitoneal route. When heated to decomposition it emits toxic fumes of NO_x. For further information see Vol. 2, No. 2 of *DPIM Report*.

XANTHINE BROMIDE HR: 3
CAS: 53-46-3 NIOSH: BP 7632500
mf: $C_{21}H_{26}NO_3 \cdot Br$ mw: 420.39

SYNS: XANTHENE-9-CARBOXYLIC ACID,ESTER WITH DIETHYL(2-HYDROXYETHYL) METHYL AMMONIUM BROMIDE * BETA-DIETHYLAMINOETHYL XANTHENE-9-CARBOXYLATE METHOBROMIDE * BETA-DIETHYLAMINOETHYL 9-XANTHENECARBOXYLATEMETHOBROMIDE * DIETHYL(2-HYDROXYETHYL)METHYLAMMONIUM BROMIDE XANTHENE-9-CARBOXYLATE

THR: Poison by intraperitoneal and intravenous routes. Moderately toxic by ingestion and subcutaneous routes. When heated to decomposition it emits very toxic fumes of NO_x and Br^-.

XANTHOTOXIN HR: 3
CAS: 298-81-7 NIOSH: LV 1400000
mf: $C_{12}H_8O_4$ mw: 216.20

SYNS: MELADININ * 8-METHOXY-(FURANO-3'.2':6.7-COUMARIN) * 8-METHOXY-2',3',6,7-FUROCOUMARIN * 8-METHOXY-4',5',6,7-FUROCOUMARIN * 8-METHOXYPSORALEN * NCI-C55903

THR: An experimental carcinogen. Moderately toxic by ingestion, intraperitoneal, and subcutaneous routes. Mutagenic data. When heated to decomposition it emits acrid smoke and fumes.

XENON HR: 1
CAS: 7440-63-3 NIOSH: ZE 1280000
DOT: 2036/2591
af: Xe aw: 131.30

PROP: Colorless, gaseous nearly inert (noble) element; d (gas): 5.8878 g/L; d (liq): 3.52 @ $-109°$; mp: $-112°$; bp: $-107°$.
DOT Classification: Label: Nonflammable Gas

THR: A simple asphyxiant. For a discussion of toxicity effects see argon. A common air contaminant. For further information see Vol. 2, No. 2 of *DPIM Report*.

XYLENE HR: 3
CAS: 1330-20-7 NIOSH: ZE 2100000
DOT: 1307 (NIOSH: ZE 2190000)
mf: C_8H_{10} mw: 106.18

PROP: A clear liquid; bp: 138.5°, flash p: 100°F (TOC), d: 0.864 @ 20°/4°, vap press: 6.72 mm @ 21°. Composition: as nonaromatics 0.07%, toluene 14%, ethyl benzene 19.27%, p-xylene 7.84%, m-xylene 65.01%, o-xylene 7.63%, C9 and aromatics 0.04%.

SYNS: AROMATIC HYDROCARBONS, MIXED * NCI-C55232 * DIMETHYLBENZENE * KSYLEN (POLISH) * XILOLI (ITALIAN) * XYLENEN (DUTCH) * XYLOL * XYLOLE (GERMAN)

OSHA PEL: TWA 100 ppm
ACGIH TLV: TWA (all isomers) 100 ppm; STEL 150 ppm; BEI: methyl hippuric acids in urine end of shift,1.5 g/g creatinine

DFG MAK: (all isomers) 100 ppm (440 mg/ m^3);BAT: blood end of shift,150 ug/dL; urine 2 g/L
DOT Classification: Flammable Liquid

THR: Poison by intraperitoneal route. Moderately toxic by inhalation, ingestion, and subcutaneous routes. A severe human eye irritant. Some temporary corneal effects are noted, as well as some conjunctival irritation by instillation. Irritation can start @ 200 ppm. A moderate skin irritant. Human irritant (systemic) effects. Flammable in the presence of heat or flame; can react with oxidizing materials. To fight fire, use foam, CO_2, dry chemical. When heated to decomposition it emits acrid smoke and fumes. For further information see Vol. 6, No. 4 of *DPIM Report*.

m-XYLENE   **HR: 3**
CAS: 108-38-3   NIOSH: ZE 2275000
mf: C_8H_{10} mw: 106.18

PROP: Colorless liquid; mp: −47.9°; bp: 139°; lel = 1.1%; uel = 7.0%; flash p: 77°F; d: 0.864 @ 20°/4°; vap press: 10 mm @ 28.3°; vap d: 3.66; autoign temp: 986°F. Insol in water; misc with alc, ether and some organic solvents.

SYNS: M-DIMETHYLBENZENE * 1,3-XYLENE * 1,3-DIMETHYLBENZENE * M-XYLOL

THR: Poison by ingestion and inhalation. A common air contaminant. An eye irritant. Severe skin irritant. Dangerous fire hazard when exposed to heat or flame, can react with oxidizing materials. Moderately explosive in the form of vapor when exposed to heat or flame. Dangerous; keep away from open flame. When heated to decomposition it emits acrid smoke. To fight fire, use foam, CO_2, dry chemical. For further information see Vol. 1, No. 7 of *DPIM Report*.

o-XYLENE   **HR: 3**
CAS: 95-47-6   NIOSH: ZE 2450000
mf: C_8H_{10} mw: 106.18

PROP: Colorless liquid; d: 0.880 @ 20°/4°; mp: −25.2°; bp: 144.4°; flash p: 62.6°F. Lel = 1.0%; uel = 6.0%. Insol in water; misc in absolute alc; ether.

SYNS: O-DIMETHYLBENZENE * O-METHYLTOLUENE * 1,2-XYLENE * 1,2-DIMETHYLBENZENE * O-XYLOL

THR: Poison by ingestion and inhalation. An eye irritant. A common air contaminant. Dangerous fire hazard when exposed to heat or flame. Slight explosion hazard in the form of vapor, when exposed to heat or flame. When heated to decomposition it emits acrid smoke and fumes. To fight fire, use foam, CO_2, dry chemical. Incompatible with oxidizing materials. For further information see Vol. 4, No. 5 of *DPIM Report*.

p-XYLENE   **HR: 2**
CAS: 106-42-3   NIOSH: ZE 2625000
mf: C_8H_{10} mw: 106.18

PROP: Clear plates; bp: 138.3°; lel: 1.1%; uel = 7.0%; flash p: 77°F (CC); d: 0.8611 @ 20°/4°; vap press: 10 mm @ 27.3°; vap d: 3.66; autoign temp: 986°F. Mp: 13°-14°. Insol in water; sol in alc, ether, organic solvents.

SYNS: P-DIMETHYLBENZENE * P-METHYLTOLUENE * 1,4-XYLENE * 1,4-DIMETHYLBENZENE * P-XYLOL

THR: Mildly toxic by ingestion and inhalation. An eye irritant. May be narcotic in high concentrations. Chronic toxicity not established; but is less toxic than benzene. Dangerous fire hazard when exposed to heat or flame; can react with oxidizing materials. Moderately explosive in the form of vapor, when exposed to heat or flame. When heated to decomposition it emits acrid smoke and fumes. To fight fire, use foam, CO_2, dry chemical. Incompatible with acetic acid + air; HNO_3; 1,3-dichloro-5,5-dimethyl-2,4-imidazolidindione. For further information see Vol. 4, No. 5 of *DPIM Report*.

3,5-XYLENOL   **HR: 3**
CAS: 108-68-9   NIOSH: ZE 6475000
mf: $C_8H_{10}O$ mw: 122.18

PROP: White crystals; mp: 64°; bp: 219.5°; d: 1.0362; vap press: 1 mm @ 62°; sltly sol in water; sol in alc.

SYN: 2,5-DIMETHYLPHENOL

THR: An experimental carcinogen. Moderately toxic by ingestion. An eye irritant. When heated to decomposition it emits acrid smoke and fumes. For further information see Vol. 4, No. 1 of *DPIM Report*.

XYLIDINE   **HR: 3**
CAS: 1300-73-8   NIOSH: ZE 8575000
DOT: 1711
mf: $C_8H_{11}N$ mw: 121.20

PROP: Usually liquid (except for *o*-4-xylidine). Bp: 213°-226°, flash p: 206° (CC), d: 0.97-0.99, vap d: 4.17. Sltly sol in water; sol in alc.

SYNS: AMINODIMETHYLBENZENE * DIMETHYLANILINE * DIMETHYLPHENYLAMINE * XILIDINE (ITALIAN) * XYLIDINEN (DUTCH)

OSHA PEL: TWA 5 ppm (skin)
ACGIH TLV: TWA 2 ppm (skin)
DFG MAK: (all isomers except 2,4-xylidene) 5 ppm (25 mg/m^3)
DOT Classification: Poison B

THR: Poison by inhalation, skin contact, and intravenous routes. Moderately toxic by ingestion. This material, which so closely resembles aniline in its toxic effects, is actually twice as toxic as aniline. It can cause injury to the blood and the liver. It does not necessarily give any alarm or warning, such as cyanosis, headache, and dizziness which characterizes aniline poisoning. Thus it may be considered a more insidious poison than aniline, and severe and possibly fatal intoxication may come about through skin absorption. Combustible when exposed to heat or flame. Dangerous; when heated to decomposition it emits highly toxic fumes; can react vigorously with oxidizing materials. When heated to decomposition emits toxic fumes of NO$_x$. To fight fire, use foam, CO$_2$, dry chemical.

2,5-XYLIDINE HR: 3
CAS: 95-78-3   NIOSH: ZE 9100000
mf: C$_8$H$_{11}$N   mw: 121.20

PROP: Colorless oil; bp: 214°; d: 0.979 @ 21°/4°; mp: 155°; very sltly sol in water.

SYNS: 2,5-DIMETHYLPHENYLAMINE * 1-AMINO-2,5-DIMETHYLBENZENE * 3-AMINO-1,4-DIMETHYLBENZENE * 2-AMINO-1,4-XYLENE * 2,5-DIMETHYLANILINE * 2,5-DIMETHYLBENZENAMINE * 5-METHYL-O-TOLUIDINE * 6-METHYL-M-TOLUIDINE * P-XYLIDINE

THR: An experimental and suspected carcinogen. Moderately toxic by ingestion. Mutagenic data. When heated to decomposition it emits toxic fumes of NO$_x$.

2,6-XYLIDINE HR: 2
CAS: 87-62-7   NIOSH: ZE 9275000
mf: C$_8$H$_{11}$N   mw: 121.20

PROP: Liquid; d: 0.980 @ 15°; mp: 10-12°; bp: 216°-217°.

SYNS: 2,6-DIMETHYLBENZENAMINE * O-XYLIDINE * 2,6-XYLYLAMINE * 2,6-DIMETHYLANILINE * NCI-c56188

THR: Moderately toxic by ingestion. When heated to decomposition it emits toxic fumes of NO$_x$.

3,5-XYLIDINE HR: 2
CAS: 108-69-0   NIOSH: ZE 9625000
mf: C$_8$H$_{11}$N   mw: 121.20

PROP: An oil; d: 0.972 @ 20°/4°; bp: 221°-222°.

SYNS: 3,5-DIMETHYLBENZENAMINE * 3,5-DIMETHYLPHENYLAMINE * 3,5-XYLYLAMINE * 3,5-DIMETHYLANILINE

THR: Moderately toxic by ingestion. When heated to decomposition it emits toxic fumes of NO$_x$.

2,4-XYLIDINE HYDROCHLORIDE HR: 3
CAS: 21436-96-4   NIOSH: ZF 0175000
mf: C$_8$H$_{11}$N • ClH   mw: 157.66

SYNS: 1-AMINO-2,4-DIMETHYLBENZENE HYDROCHLORIDE * 4-AMINO-1,3-DIMETHYLBENZENE HYDROCHLORIDE * 4-AMINO-3-METHYLTOLUENE HYDROCHLORIDE * 4-AMINO-1,3-XYLENE HYDROCHLORIDE * 2,4-DIMETHYLANILINE HYDROCHLORIDE * 2,4-DIMETHYLBENZENAMINE HYDROCHLORIDE * 4-METHYL-O-TOLUIDINE HYDROCHLORIDE * 2-METHYL-P-TOLUIDINE HYDROCHLORIDE * M-XYLIDINE HYDROCHLORIDE

THR: An experimental neoplastigen and suspected carcinogen. Moderately toxic by ingestion and intraperitoneal routes. When heated to decomposition it emits very toxic fumes of NO$_x$ and HCl.

2,5-XYLIDINE HYDROCHLORIDE HR: 3
CAS: 51786-53-9   NIOSH: ZF 0350000
mf: C$_8$H$_{11}$N • ClH   mw: 157.66

SYNS: 1-AMINO-2,5-DIMETHYLBENZENE HYDROCHLORIDE * 3-AMINO-1,4-DIMETHYLBENZENE HYDROCHLORIDE * 5-AMINO-1,4-DIMETHYLBENZENE HYDROCHLORIDE * 2-AMINO-4-METHYLTOLUENE HYDROCHLORIDE * 2-AMINO-1,4-XYLENE HYDROCHLORIDE

* 2,5-DIMETHYL ANILINE HYDROCHLORIDE
* 2,5-DIMETHYLANILINE HYDROCHLORIDE
* 2,5-DIMETHYLBENZENAMINE HYDRO-
CHLORIDE * 5-METHYL-O-TOLUIDINE HYDRO-
CHLORIDE * 6-METHYL-M-TOLUIDINE
HYDROCHLORIDE * PARA-XYLIDINE HYDRO-
CHLORIDE

THR: An experimental tumorigen and carcinogen. Moderately toxic by intraperitoneal route.
When heated to decomposition it emits very
toxic fumes of NO_x and HCl.

XYLITOL   HR: 3
CAS: 87-99-0   NIOSH: ZF 0800000
mf: $C_5H_{12}O_5$ mw: 152.17

PROP: A syrup, sol in water.

SYN: XYLITE (SUGAR)

THR: An experimental tumorigen. Moderately
toxic by intravenous route. Mildly toxic by ingestion. When heated to decomposition it emits
acrid smoke and fumes.

XYLOCAIN   HR: 3
CAS: 137-58-6   NIOSH: AN 7525000
mf: $C_{14}H_{22}N_2O$ mw: 234.38

SYNS: ALPHA-DIETHYLAMINOACETO-2,6-XYLI-
DIDE * ALPHA-DIETHYLAMINO-2,6-ACETOXYLI-
DIDE * DIETHYLAMINOACET-2,6-XYLIDIDE
* DIETHYLAMINOACETO-2,6-XYLIDIDE
* ALPHA-DIETHYLAMINO-2,6-DIMETHYL-
ACETANILIDE * OMEGA-DIETHYLAMINO-2,6-
DIMETHYLACETANILIDE * LIDOCAINE
* 2-(DIETHYLAMINO)-2',6'-ACETOXYLIDIDE

THR: Poison by intravenous, intraperitoneal,
and subcutaneous routes. Moderately toxic by
ingestion. When heated to decomposition it
emits toxic fumes of NO_x.

XYLOCAINE HYDROCHLORIDE   HR: 3
CAS: 73-78-9   NIOSH: AN 7600000
mf: $C_{14}H_{22}N_2O \cdot ClH$ mw: 270.84

SYNS: OMEGA-DIETHYLAMINO-2,6-DIMETHYL-
ACETANILIDE HYDROCHLORIDE * LIDOCAINE
HYDROCHLORIDE * LIGNOCAINE HYDROCHLO-
RIDE * 2-DIETHYLAMINO-2',6'-ACETOXYLI-
DIDEHYDROCHLORIDE

THR: Poison by ingestion, intraperitoneal, intravenous, subcutaneous, intramuscular, and intratracheal routes. When heated to decomposition it emits very toxic fumes of NO_x and HCl.

1-(2,4-XYLYLAZO)-2-NAPHTHOL   HR: 3
CAS: 3118-97-6   NIOSH: QL 5850000
mf: $C_{18}H_{16}N_2O$ mw: 276.36

SYNS: C.I. 12140 * C.I. SOLVENT ORANGE 7
* 1-((2,4-DIMETHYLPHENYL)AZO)-2-NAPHTHA-
LENOL * EXTRACT D AND C RED NO. 14
* FAST OIL ORANGE II * 1-XYLYLAZO-2-
NAPHTHOL * 1-(O-XYLYLAZO)-2-NAPHTHOL

THR: An experimental carcinogen. Mutagenic
data. When heated to decomposition it emits
toxic fumes of NO_x.

Y

YOHIMBINE HR: 3
CAS: 146-48-5 NIOSH: ZG 1000000
mf: $C_{21}H_{26}N_2O_3$ mw: 354.49

PROP: Colorless needles from water and alc, mp: 234°. Sltly sol in water, ether; sol in alc, chloroform, hot benzene.

SYNS: APHRODINE * APHROSOL * CORYNINE * QUEBRACHINE * YOHIMBIC ACID METHYL ESTER

THR: Poison by ingestion, subcutaneous, intravenous, and intraperitoneal routes. Cases of poisoning have occurred from its use as an aphrodisiac. Upon local application, it produces anesthesia. However, absorption of it can give rise to toxic symptoms, such as salivation, increased respiration, and repeated defecation. With reference to the circulatory system, there may be a fall in blood pressure and sometimes myocardial damage, involving particularly the conduction system of the heart, with a resultant decrease in the efficiency of the heart. Moderately dangerous; when heated to decomposition it emits toxic fumes of NO_x.

YOSHI 864 HR: 3
CAS: 3458-22-8 NIOSH: UB 6620000
mf: $C_8H_{19}NO_6S_2 \cdot ClH$ mw: 325.86

SYNS: 3,3'-IMIDODI-1-PROPANOL, DIMETHANE-SULFONATE (ESTER), HYDROCHLORIDE * NCI-c01547 * NSC 102627

THR: An experimental carcinogen and neoplastigen. Human gastrointestinal tract effects. When heated to decomposition it emits very toxic fumes of NO_x, SO_x, and HCl.

YTTERBIUM HR: 3
CAS: 7440-64-4 NIOSH: ZG 1925000
af: Yb aw: 173.04

PROP: A bright, silvery, lustrous soft, malleable, ductile and fairly stable element. Mp: 824°; bp: 1193°; d: 6.977. A rare earth.

THR: An experimental carcinogen. As a lanthanon it may have an anticoagulant action on blood. See also lanthanum and rare earths. Flammable in the form of dust, when reacted with air, halogens.

YTTERBIUM CHLORIDE HR: 3
CAS: 10361-91-8 NIOSH: ZG 2100000
mf: Cl_3Yb mw: 279.39

PROP: Hexahydrate, deliquescent needles, crystals. D: 2.575; mp: 150°-155°.

SYN: YTTERBIUM TRICHLORIDE

THR: Poison by intraperitoneal route. A skin and eye irritant. See also ytterbium, rare earths and chlorides. When heated to decomposition it emits toxic fumes of Cl^-.

YTTERBIUM (III) NITRATE, HEXAHYDRATE (1:3:6) HR: 3
CAS: 13839-85-5 NIOSH: ZG 2800000
mf: $N_3O_9 \cdot Yb \cdot 6HOH$ mw: 467.19

SYN: NITRIC ACID, YTTERBIUM(3+) SALT, HEXAHYDRATE

THR: Poison by ingestion and intraperitoneal route. See also nitrates and ytterbium and rare earths. When heated to decomposition it emits toxic fumes of NO_x.

YTTRIUM AND COMPOUNDS
CAS: 7440-65-5 NIOSH: ZG 2980000
af: Y aw: 88.91

PROP: Hexagonal, gray-black metallic rare earth element; mp: 1509°; bp: 3200°; d: 4.472.

SYN: YTTRIUM-89

ACGIH TLV: TWA 1 mg/m^3
DFG MAK: 5 mg/m^3

THR: No toxicity data. As a lanthanon, it may have an anticoagulant effect on the blood. See also lanthanum and rare earths. Flammable in the form of dust, when reacted with air, halogens.

YTTRIUM CHLORIDE HR: 3
CAS: 10361-92-9 NIOSH: ZG 3150000
mf: Cl_3Y mw: 195.26

PROP: Hexahydrate, colorless, deliquescent crystals. Sol in water, alc.

SYN: YTTRIUM TRICHLORIDE

OSHA PEL: TWA 1 mg(Yt)/m^3

THR: Poison by intraperitoneal route. See also yttrium and rare earths. When heated to decomposition it emits toxic fumes of Cl^-.

YTTRIUM (III) NITRATE (1:3)   HR: 3
CAS: 10361-93-0   NIOSH: ZG 3675000
mf: $N_3O_9 \cdot Y$ mw: 274.94

PROP: Hexahydrate, deliquescent crystals. Sol in water.

SYN: NITRIC ACID, YTTRIUM(3+) SALT

OSHA PEL: TWA 1 mg(Yt)/m^3

THR: Poison by intraperitoneal route. An experimental carcinogen. See also nitrates, yttrium and rare earths. When heated to decomposition it emits toxic fumes of NO$_x$.

YTTRIUM OXIDE   HR: 2
CAS: 1314-36-9   NIOSH: ZG 3850000
mf: O_3Y_2 mw: 225.82

PROP: White powder; d: 4.84.

SYN: YTTRIA

OSHA PEL: TWA 1 mg(Yt)/m^3

THR: Moderately toxic by intraperitoneal route. See also yttrium and rare earths.

Z

ZINC
HR: 2
CAS: 7440-66-6   NIOSH: ZG 8600000
af: Zn   aw: 65.37

PROP: Bluish-white, lustrous metal; mp: 419.8°; bp: 908°; d: 7.14 @ 25°; vap press: 1 mm @ 487°.

SYNS: BLUE POWDER * C.I. 77945 * C.I. PIGMENT BLACK 16 * GRANULAR ZINC * ZINC DUST * ZINC POWDER

THR: Human skin irritant and pulmonary system effects. See also zinc compounds. Pure zinc powder, dust, fume is relatively non-toxic to humans by inhalation. The difficulty arises from oxidation of zinc fumes prior to inhalation or presence of impurities such as Cd, Sb, As, Pb. Flammable in the form of dust when exposed to heat or flame. Explosion in the form of dust when reacted with acids. Incompatible with NH_4NO_3; BaO_2; $Ba(NO_3)_2$; Cd; CS_2; chlorates; Cl_2, ClF_3; CrO_3; (ethyl acetoacetate + tribromoneopentyl alcohol); F_2; hydrazine mononitrate; hydroxylamine; $Pb(N_3)_2$; $(Mg + Ba(NO_3)_2 + BaO_2)$; $MnCl_2$; HNO_3; performic acid; $KClO_3$; KNO_3; K_2O_2; Se; $NaClO_3$; Na_2O_2; S; Te; H_2O; $(NH_4)_2S$; As_2O_3; CS_2; $CaCl_2$; NaOH; chlorinated rubber; catalytic metals; halocarbons; o-nitroanisole; nitrobenzene; non-metals; oxidants; paint primer base; pentacarbonyliron; transition metal halides; seleninyl bromide. To fight fire, use special mixtures of dry chemical. For further information see Vol. 1, No. 7 of *DPIM Report*.

ZINC ARSENATE
HR: 3
CAS: 1303-39-5   NIOSH: ZG 9180000
mf: $As_4O_{15} \cdot 5Zn$ mw: 866.53

PROP: White, odorless powder.

SYNS: ARSENIC ACID, ZINC SALT * ZINC ARSENATE, BASIC

DOT Classification: Label: Poison

THR: A poison. See also arsenic and zinc compounds. When heated to decomposition it emits toxic fumes of As and ZnO_x.

ZINC-m-ARSENITE
HR: 3
CAS: 10326-24-6   NIOSH: CG 3890000
DOT: 1712
mf: $AsHO_2 \cdot 1/2Zn$ mw: 140.61

PROP: A white powder.

SYNS: ARSENIOUS ACID, ZINC SALT * ZINC ARSENITE, SOLID (DOT) * ZINC METHARSENITE

DOT Classification: Poison B, Label: Poison

THR: Poison. See also arsenic compounds; zinc compounds. When heated to decomposition it emits toxic fumes of As and ZnO.

ZINC CHLORIDE
HR: 3
CAS: 7646-85-7   NIOSH: ZH 1400000
DOT: 1840/2331
mf: Cl_2Zn mw: 136.27

PROP: Odorless, cubic white, deliquescent crystals; mp: 290°; bp: 732°; d: 2.91 @ 25°; vap press: 1 mm @ 428°.

SYNS: BUTTER OF ZINC * CHLORURE DE ZINC (FRENCH) * ZINC DICHLORIDE * ZINCO (CLORURO DI) (ITALIAN) * ZINKCHLORID (GERMAN) * ZINKCHLORIDE (DUTCH)

OSHA PEL: TWA 1 mg(fume)/m^3
ACGIH TLV: TWA 1 mg/m^3; STEL 2 mg/m^3
DOT Classification: Label: Corrosive

THR: Poison by ingestion, intravenous, and intraperitoneal routes. Mutagenic data. An experimental teratogen and carcinogen. Human pulmonary system effects. See also zinc compounds and chlorides. Exposure to $ZnCl_2$ fumes or dusts can cause dermatitis, boils, conjunctivitis, gastrointestinal tract upsets. The fumes are highly toxic. A poison by inhalation too. Incompatible with potassium. When heated to decomposition it emits toxic fumes of Cl^- and ZnO. For further information see Vol. 5, No. 3 of *DPIM Report*.

ZINC COMPOUNDS
HR: 3
THR: Variable toxicity, but generally of low toxicity. However, zinc salts, such as chromates and arsenates, are experimental carcinogens. Zinc is not inherently a toxic element. However, when heated, it evolves a fume of zinc oxide which, when inhaled fresh, can cause a disease known as "brass founders" "ague," or "brass chills," sweet taste, throat dryness, cough, weakness, generalized aching, fever, nausea, vomiting. It is possible for people to become immune to it, but this immunity can be broken

by cessation of exposure of only a few days. Zinc oxide dust which is not freshly formed is virtually innocuous. There is no cumulative effect from the inhalation of zinc fumes. Fatalities however, have resulted from lung damage caused by the inhalation of zinc chloride fumes. Soluble salts of zinc have a harsh metallic taste; small doses can cause nausea and vomiting, while larger doses cause violent vomiting and purging. So far as can be determined, the continued administration of zinc salts in small doses has no effect in man except those of disordered digestion and constipation. Exposure to zinc chloride fumes can cause damage to the mucous membranes of the nasopharnyx and respiratory tract and give rise to a pale gray cyanosis. Workers in zinc refining have been reported as suffering from a variety of non-specific intestinal, respiratory and nervous symptoms. Ulceration of the nasal septum and eczematous dermatosis are also reported. It has been stated that zinc oxide or zinc stearate dust can block the ducts of the sebaceous glands and give rise to a papular, pustular eczema in men engaged in packing these compounds into barrels. Sensitivity to zinc oxide in man is extremely rare. Zinc chloride and zinc sulfate, because of caustic action, can cause ulceration of the fingers, hands and forearms of those who use them as a flux in soldering or other industrial use. This condition has even been observed in men who handle railway ties which have been impregnated with this material. Common air contaminants.

ZINC CYANIDE HR: 3
CAS: 557-21-1   NIOSH: ZH 1575000
mf: C_2N_2Zn mw: 117.41

PROP: Rhombic colorless crystals; mp: decomp @ 800°. Insol in water; sol in solns of alkali cyanides; decomp by dil mineral acid.

SYNS: ZINC DICYANIDE * CYANURE DE ZINC (FRENCH)

DOT Classification: Poison B, Label: Poison

THR: A deadly poison by intraperitoneal route. See also cyanide and zinc compounds. Can react violently with Mg. When heated to decomposition it emits toxic fumes of CN^-, Zn and NO_x. For further information see Vol. 4, No. 2 of *DPIM Report.*

ZINC FLUORIDE HR: 3
CAS: 7783-49-5   NIOSH: ZH 3500000
mf: F_2Zn mw: 103.37

PROP: Tetragonal needles or white crystalline mass. D: 5.00 @ 25°; mp: 872°; bp: 1500°; vap press: 1mm @ 970°. Sltly sol in aq HF, sol in HCl, HNO_3, NH_4OH.

SYN: ZINC FLUORURE (FRENCH)

THR: Poison by subcutaneous route. See also fluorides and zinc compounds. Can react violently with potassium. When heated to decomposition it emits toxic fumes of F^- and ZnO. For further information see Vol. 3, No. 6 of *DPIM Report.*

ZINC FLUOSILICATE HR: 3
CAS: 16871-71-9   NIOSH: ZH 3675000
mf: $F_6Si \cdot Zn$ mw: 207.46

SYNS: FLUOSILICATE DE ZINC * ZINC HEXA-FLUOROSILICATE

THR: Poison by ingestion and subcutaneous routes. See also zinc and silicofluoric acid. When heated to decomposition it emits toxic fumes of F^- and ZnO. For further information see Vol. 3, No. 6 of *DPIM Report.*

ZINC NITRATE
CAS: 7779-88-6   NIOSH: ZH 4772000
mf: $N_2O_6 \cdot Zn$ mw: 189.39

PROP: A: needles; B: tetragonal colorless crystals; A: trihydrate; B: hexahydrate; d: (B) = 2.065 @ 14°; mp: (A) 42.5°; mp: (B): =36.4°; bp: (B): $-6H_2O$ @ 105°-131°. Very sol in alc; sol in water.

SYNS: NITRATE DE ZINC (FRENCH) * NITRIC ACID, ZINC SALT

DOT Classification: Label: Oxidizer

THR: No toxicity data. See also nitrates and zinc compounds; can react violently with C; Cu; metal sulfides; organic matter; P; S. When heated to decomposition it emits toxic fumes of NO_x, ZnO. For further information see Vol. 5, No. 3 of *DPIM Report.*

ZINC OXIDE HR: 3
CAS: 1314-13-2   NIOSH: ZH 4810000
mf: OZn   mw: 81.37

PROP: Odorless, white or yellowish powder. Mp: >1800°; d: 5.47. Almost insol in water; sol in dil acetic or mineral acids, ammonia.

SYNS: CHINESE WHITE * C.I. 77947 * CYNKU TLENEK (POLISH) * FLOWERS OF ZINC * ZINC OXIDE FUME * ZINC WHITE

OSHA PEL: TWA 5 mg(fume)/m^3
ACGIH TLV: TWA (fume) 5 mg/m^3; STEL 10
mg/m^3
DFG MAK: 5 mg/m^3

THR: Human pulmonary system effects. A skin
and eye irritant. See also zinc compounds. Has
exploded when mixed with chlorinated rubber.
Violent reaction with Mg, linseed oil. Human
Toxicity: Freshly formed fumes, as from welding, may cause metal fume fever with chills,
fever, tightness in chest, cough and leukocytes.

ZINC PANTOTHENATE   HR: 3
NIOSH: AY 5100000
mf: C$_{18}$H$_{32}$N$_2$O$_{10}$•Zn   mw: 501.89

SYNS: PANTOTHENIC ACID, ZINC SALT
∗ (R)-N-(2,4-DIHYDROXY-3,3-DIMETHYLBUTY-
RYL)-BETA-ALANINE ZINC SALT (2:1)

THR: Poison by intraperitoneal route. Moderately toxic by ingestion. When heated to decomposition it emits toxic fumes of NO$_x$ and Zn.

ZINC PEROXIDE
CAS: 1314-22-3   NIOSH: ZH 4865000
mf: O$_2$Zn   mw: 97.37

PROP: Odorless, yellow-white powder. D:
1.571 (theoretical). Decomp >150°. Sol in dil
acids.

SYN: ZINC SUPEROXIDE

DOT Classification: Label: Oxidizer

THR: No toxicity data. Systemic toxicity is similar to zinc oxide. See also peroxides and zinc
compounds. Flammable when exposed to heat
or by chemical reaction with reducing materials.
Finely divided powder is slightly soluble in water, decomposes rapidly at 150°. It is not dangerous unless mixed with highly combustible materials. Dangerous explosion hazard when
exposed to heat. Can react violently with Al;
Zn. Explodes at 212°. For peroxide prepared
from ZnSO$_4$, NH$_3$, H$_2$O$_2$; or 4ZnO•
3HOOH•HOH product is very dangerous;
explodes when heated; will react with water
or steam to produce heat; on contact with reducing material, can react vigorously.

ZINC PHOSPHIDE   HR: 3
CAS: 1314-84-7   NIOSH: ZH 4900000
mf: P$_2$Zn$_3$ mw: 258.05

PROP: Cubic dark gray crystals or powder;
mp: 420°; bp: 1100°; d: 4.55 @ 13°. Insol in
water, alc; sol in benzene, carbon disulfide.

SYNS: PHOSPHURE DE ZINC (FRENCH) ∗ PHOS-
VIN ∗ ZINCO(FOSFURO DI) (ITALIAN) ∗ ZINK-
FOSFIDE (DUTCH) ∗ ZINKPHOSPHID (GERMAN)

DOT Classification: Label: Flammable Solid
and Dangerous When Wet

THR: Poison to humans by ingestion. See also
phosphides and zinc compounds. This material
is stable while kept dry. In moist air, it decomposes slowly. Reacts violently with acids or
acid fumes to emit the highly toxic and flammable phosphine. When heated to decomposition
it emits toxic fumes of PO$_x$ and ZnO. Incompatible with HCl; H$_2$SO$_4$ (violently with concentrated H$_2$SO$_4$; HNO$_3$; oxidizing materials). For
further information see Vol. 5, No. 5 of *DPIM
Report*.

ZINC STEARATE   HR: 3
CAS: 557-05-1   NIOSH: ZH 5200000
mf: Zn(C$_{18}$H$_{35}$O$_2$)$_2$ mw: 632.30

PROP: White powder. Mp: 130°, flash p: 530°F
(OC), autoign temp: 790°F. Insol in water, alc,
ether; sol in benzene. Decomp by dil acids.

SYNS: DIBASIC ZINC STEARATE ∗ OCTADECA-
NOIC ACID, ZINC SALT ∗ STEARIC ACID, ZINC
SALT ∗ ZINC DISTERATE ∗ ZINC OCTADECA-
NOATE

THR: Poison by intratracheal route. See also
zinc compounds. Inhalation of zinc stearate has
been reported as causing pulmonary fibrosis.
A nuisance dust. Combustible when exposed
to heat or flame. To fight fire, use water, foam,
CO$_2$, dry chemical. When heated to decomposition it emits acrid smoke and fumes of ZnO.

ZINC SULFATE   HR: 3
CAS: 7733-02-0   NIOSH: ZH 5260000
mf: O$_4$S•Zn   mw: 161.43

PROP: Rhombic colorless crystals; mp: decomp
@ 740°; d: 3.74 @ 15°. Sol in water; almost
insol in alc.

SYNS: SULFATE DE ZINC (FRENCH) ∗ SUL-
FURIC ACID, ZINC SALT (1:1) ∗ WHITE COP-
PERAS ∗ WHITE VITRIOL ∗ ZINC VITRIOL

THR: Poison by intraperitoneal, subcutaneous,
and intravenous routes. Moderately toxic by ingestion. An eye irritant. Mutagenic data. An ex-

perimental teratogen and tumorigen. Human gastrointestinal tract, systemic, and blood pressure effects. See also sulfates and zinc compounds. When heated to decomposition it emits toxic fumes of SO_x and ZnO. For further information see Vol. 5, No. 5 of *DPIM Report*.

ZIRCONIUM
CAS: 7440-67-7   NIOSH: ZH 7074000
DOT: 1308/1358/2008/2009/2858
af: Zr   aw: 91.22

PROP: A grayish-white lustrous metal, very sltly radioactive. Mp: 1852°, bp: 3577°, d: 6.506 @ 20°.

SYN: ZIRCONIUM METAL

DFG MAK: 5 mg/m^3
DOT Classification: Flammable Solid

THR: See also zirconium compounds. Dangerous fire hazard in the form of dust, when exposed to heat or flame or by chemical reaction with oxidizers. Dangerous explosion hazard in the form of dust, by chemical reaction with air; alkali hydroxides; alkali metal chromates; dichromates; molybdates; salts; sulfates; tungstates; borax; CCl_4; CuO; Pb; PbO; P; $KClO_3$; KNO_3; nitrylfluoride. Explosive range: 0.16 g/L in air. To fight fire, use special mixtures, dry chemical, salt or dry sand.

ZIRCONIUM (DRY)
CAS: 7440-67-7   NIOSH: ZH 7070000
af: Zr;aw: 91.22

PROP: Chemically produced, finer than 20 mesh particle size or mechanically produced, finer than 270 mesh particle size. See also zirconium.

SYN: ZIRCONIUM METAL, DRY (DOT)

DOT Classification: Label: Flammable Solid

THR: No toxicity data. See also zirconium compounds. Dangerous fire hazard in the form of dust, when exposed to heat or flame or by chemical reaction with oxidizers. Dangerous explosion hazard in the form of dust, by chemical reaction with air; alkali hydroxides; alkali metal chromates; dichromates; molybdates; salts; sulfates; tungstates; borax; CCl_4; CuO; Pb; PbO; P; $KClO_3$; KNO_3. Explosive range: 0.16 g/L in air. To fight fire, use special mixtures, dry chemical, salt or dry sand.

ZIRCONIUM CHLORIDE   HR: 2
CAS: 10026-11-6   NIOSH: ZH 7175000
mf: Cl_4Zr mw: 233.02

PROP: White lustrous crystals; mp: sublimes @ 300°; bp: 331°; d: 2.80; vap press: 1 mm @ 190°.

SYN: ZIRCONIUM TETRACHLORIDE

OSHA PEL: TWA 5 mg(Zr)/m^3
DOT Classification: Label: Poison

THR: Moderately toxic by ingestion. See also zirconium compounds and hydrochloric acid. When heated to decomposition it emits toxic fumes of Cl^-. Self-ignites in air. For further information see Zirconium tetrachloride, Vol. 3, No. 4 in *DPIM Report*.

ZIRCONIUM COMPOUNDS   HR: 2
THR: Zirconium is not an important industrial poison. Deaths in rabbits have been caused by intravenous injection of 150 mg/kg of body weight. Most zirconium compounds in common use are insoluble and considered inert. Pulmonary granuloma in zirconium workers has been reported and sodium zirconium lactate has been held responsible for skin granulomas. Avoid inhalation of Zr-containing aerosols, which can cause lung granulomas. Zirconium-containing drugs or cosmetic products are being controlled by the FDA.

ZIRCONIUM FLUORIDE   HR: 3
CAS: 7783-64-4   NIOSH: ZH 7875000
mf: F_4Zr mw: 167.22

PROP: Refractive crystals, water-sol. D: 4.6 @ 16°, sublimes @ 600°. Very sol in HF.

SYN: ZIRCONIUM TETRAFLUORIDE

OSHA PEL: TWA 2500 ug(F)/m^3

THR: Poison by intravenous route. See also zirconium compounds and fluorides. When heated to decomposition it emits toxic fumes of F^-.

ZIRCONIUM HYDRIDE
CAS: 7704-99-6   NIOSH: ZH 8015000
mf: H_2Zr mw: 93.24

PROP: Metallic dark gray to black powder; d: 5.6; autoign temp: 270° (in air).
DOT Classification: Label: Flammable Solid and Dangerous When Wet

THR: No toxicity data. See also hydrides and zirconium compounds. Incandesces when heated in air. Flammable when wet. Very dangerous to handle; can explode.

ZIRCONIUM NITRATE   HR: 3
CAS: 13746-89-9   NIOSH: ZH 8750000
mf: $N_4O_{12} \cdot Zr$ mw: 339.26

PROP: White crystals.

SYN: DUSICNAN ZIRKONICITY (CZECH)

OSHA PEL: TWA 5 mg(Zr)/m^3

THR: Poison by inhalation. Moderately toxic by ingestion. See also nitrates and zirconium compounds. When heated to decomposition it emits toxic fumes of NO_x. For further information see Vol. 3, No. 6 of *DPIM Report.*

ZIRCONIUM OXYCHLORIDE   HR: 3
CAS: 7699-43-6   NIOSH: ZH 7700000
mf: Cl_2OZr mw: 178.12

PROP: Crystals. D: 1.91. Very sol in water, alc.

SYNS: BASIC ZIRCONIUM CHLORIDE * CHLOROZIRCONYL * NCI-C60811 * ZIRCONYL CHLORIDE

OSHA PEL: TWA 5 mg(Zr)/m^3

THR: Poison by intraperitoneal route. Moderately toxic by ingestion and subcutaneous route. An experimental neoplastigen. See also zirconium compounds, and chlorides. When heated to decomposition it emits toxic fumes of Cl^-.

ZIRCONIUM (IV) SULFATE (1:2)   HR: 3
CAS: 14644-61-2   NIOSH: ZH 9100000
mf: $O_8S_2 \cdot Zr$ mw: 283.34

PROP: Tetrahydrate, crystalline solid.

SYNS: DISULFATOZIRCONIC ACID * SULFURIC ACID, ZIRCONIUM(4+) SALT (2:1) * ZIRCONYL SULFATE

OSHA PEL: TWA 5 mg(Zr)/m^3

THR: Poison by intraperitoneal route. Moderately toxic by ingestion and subcutaneous routes. See also sulfates and zirconium compounds. When heated to decomposition it emits toxic fumes of SO_x. For further information see zirconium sulfate, Vol. 3, No. 6 in *DPIM Report.*

ZYGOSPORIN A   HR: 3
CAS: 22144-77-0   NIOSH: GZ 4850000
mf: $C_{30}H_{37}NO_6$ mw: 507.68

SYNS: 3-BENZYL-3,3-ALPHA,4,5,6,6-ALPHA,9,10,12,15-DECAHYDRO-6,12,15-TRIHYDROXY-4,10,12-TRIMETHYL-5-METHYLENE-1H-CYCLOUNDEC(D)ISOINDOLE-1,11(2H)-DIONE, 15-ACETATE * CYTOCHALASIN D

THR: Poison by ingestion, subcutaneous, and intraperitoneal routes. An experimental teratogen. When heated to decomposition it emits toxic fumes of NO_x.

Appendix I

ALPHABETICAL SYNONYM CROSS-REFERENCE

ABBOTT-28440 see 2,6-DICHLORO-N-CYCLOPROPYL-N-ETHYL ISONI-COTINAMIDE

ABBOTT-30360 see 4'-FLUORO-4-(8-FLUORO-2,3,4,5-TETRAHYDRO-1H-PYRIDO(4,3-b)INDOL-2-YL)BUTYROPHENONE HYDROCHLORIDE

ABRINS see ABRIN

ABS see ALKYLBENZENESULFONATE

ABSOLUTE ETHANOL see ETHANOL

ABS (PYROLYSIS PRODUCTS) see ACRYLONITRILE, POLYMER WITH 1,3-BUTADIENE, AND STYRENE, COMBUSTION PRODUCTS

ACACIA see ARABIC GUM

ACACIA DEALBATA GUM see ARABIC GUM

ACACIA GUM see ARABIC GUM

ACACIA MOLLISSIMA TANNIN see MIMOSA TANNIN

ACACIA SENEGAL see ARABIC GUM

ACACIA SYRUP see ARABIC GUM

ACANTHOPHIS ANTARCTICUS (AUSTRALIA) VENOM see A. ANTRACTI-CUS (AUSTRALIA) VENOM

ACARACIDE see SULFUROUS ACID, 2-(p-t-BUTYLPHENOXY)-1-METH-YLETHYL-2-CHLOROETHYL ESTER

ACCENT see MONOSODIUM GLUTAMATE

4,10-ACE-1,2-BENZANTHRACENE see 1,2-DIHYDROBENZ(e)ACEAN-THRYLENE

ACEDE CRESYLIQUE (FRENCH) see CRESOL

13H-ACENAPHTHO(1,8-AB)PHENANTHRENE see 13H-DIBENZ(bc,j)-ACEANTHRYLENE

ACEPHAT (GERMAN) see O,S-DIMETHYLACETYLPHOSPHOROAMIDO-THIOATE

ACEPROMAZINA see ACETOPROMAZINE

ACEPROMAZINE see ACETOPROMAZINE

ACEPROMIZINA see ACETOPROMAZINE

ACETAAL (DUTCH) see ACETAL

ACETALDEHYD (GERMAN) see ACETALDEHYDE

ACETALDEHYDE AMINE SALT see ALDEHYDE AMMONIA

ACETALDEHYDE AMMONIA see ALDEHYDE AMMONIA

ACETAL DIETHYLIQUE (FRENCH) see ACETAL

ACETALDOXIME see ACETALDEHYDE OXIME

ACETALE (ITALIAN) see ACETAL

ACETAMIDOBENZENE see ACETANILIDE

4-ACETAMIDOBIPHENYL see 4'-PHENYLACETANILIDE

1-ACETAMIDO-4-ETHOXYBENZENE see p-ACETOPHENETIDIDE

2-ACETAMIDOFLUORENE see N-FLUOREN-2-YL ACETAMIDE

1-(N-ACETAMIDOFLUOROMETHYL)-NAPHTHALENE see N-METHYL-N-(1-NAPHTHYL)FLUOROACETAMIDE

3-ACETAMIDO-4-HYDROXYBENZENEARSONIC ACID see N-ACETYL-4-HYDROXY-m-ARSANILIC ACID

3-ACETAMIDO-4-HYDROXYBENZENEARSONIC ACID see ACETPHENAR-SINE

3-ACETAMIDO-4-HYDROXY-PHENYLARSONIC ACID see ACETPHENAR-SINE

3-ACETAMIDO-4-HYDROXY-PHENYLARSONIC ACID see N-ACETYL-4-HYDROXY-m-ARSANILIC ACID

7-ACETAMIDO-10-HYDROXY-1,2,3-TRIMETHOXY-6,7-DIHYDROBEN-ZO(A)HEPTALEN-9(5H)-ONE see N-ACETYL TRIMETHYLCOLCHICINIC ACID

P-ACETAMIDOPHENACYL CHLORIDE see 4'-CHLOROACETYL ACETANI-LIDE

4-ACETAMIDOPHENOL see 4'-HYDROXYACETANILIDE

P-ACETAMIDOPHENOL see 4'-HYDROXYACETANILIDE

2-ACETAMIDO-5-SULFONAMIDO-1,3,4-THIADIAZOLE see 5-ACET-AMIDE-1,3,4-THIADIAZOLE-2-SULFONAMIDE

ACETAMIDOTHIADIAZOLESULFONAMIDE see 5-ACETAMIDE-1,3,4-THIADIAZOLE-2-SULFONAMIDE

P-ACETAMIDOTOLUENE see p-ACETOTOLUIDIDE

M-ACETAMINOANILINE see 3'-AMINOACETANILIDE

2-ACETAMINO-4-(5-NITRO-2-FURYL)THIAZOLE see 2-ACETAMIDO-4-(5-NITRO-2-FURYL)THIAZOLE

4-ACETAMINO-2-NITROPHENETOLE see 3'-NITRO-p-ACETOPHENETIDE

ACETAMINOPHEN see 4'-HYDROXYACETANILIDE

2-ACETAMINOPHENANTHRENE see 2-ACETAMIDOPHENANTHRENE

P-ACETAMINOPHENOL see 4'-HYDROXYACETANILIDE

ACETAMOX see 5-ACETAMIDE-1,3,4-THIADIAZOLE-2-SULFONAMIDE

ACETANIL see ACETANILIDE

ACETARSOL see ACETPHENARSINE

ACETARSONE see ACETPHENARSINE

ACETATE D'AMYLE (FRENCH) see PENTYL ACETATE

ACETATE DE BUTYLE (FRENCH) see N-BUTYL ACETATE

ACETATE DE BUTYLE SECONDAIRE (FRENCH) see sec-BUTYL ACE-TATE

ACETATE DE L'ETHER MONOMETHYLIQUE DE L'ETHYLENE-GLYCOL (FRENCH) see ETHYLENE GLYCOL MONOMETHYL ETHER ACE-TATE

ACETATE DE METHYLE (FRENCH) see ACETIC ACID METHYL ESTER

ACETATE DE METHYLE GLYCOL (FRENCH) see ETHYLENE GLYCOL MONOMETHYL ETHER ACETATE

ACETATE DE PLOMB (FRENCH) see LEAD ACETATE

ACETATE DE PROPYLE NORMAL (FRENCH) see PROPYL ACETATE

ACETATE D'ETHYLGLYCOL (FRENCH) see ''CELLOSOLVE'' ACETATE

ACETATE DE TRIPHENYL-ETAIN (FRENCH) see ACETOXYTRIPHENYL-STANNANE

ACETATE D'ISOBUTYLE (FRENCH) see ISOBUTYL ACETATE

ACETATE D'ISOPROPYLE (FRENCH) see ACETIC ACID ISOPROPYL ESTER

ACETATE PHENYLMERCURIQUE (FRENCH) see ACETOXYPHENYLMER-CURY

ACETATO(2-AMINO-5-NITROPHENYL)MERCURY see 2-(ACETOXYMER-CURI)-4-NITROANILINE

(ACETATO)(P-AMINOPHENYL)MERCURY see p-(ACETOXYMERCURI)-ANILINE

ACETATO DI CELLOSOLVE (ITALIAN) see ''CELLOSOLVE'' ACETATE

ACETATO DI METIL CELLOSOLVE (ITALIAN) see ETHYLENE GLYCOL MONOMETHYL ETHER ACETATE

ACETATO DI STAGNO TRIFENILE (ITALIAN) see ACETOXYTRIPHENYL-STANNANE

ACETATO(2-METHOXYETHYL)MERCURY see METHOXYETHYL MER-CURIC ACETATE

(ACETATO-O)(TRIMETAARSENITO)DICOPPER see CUPRIC ACETOAR-SENITE

(ACETATO)PHENYLMERCURY see ACETOXYPHENYLMERCURY

ACETATOTRIPHENYLSTANNANE see ACETOXYTRIPHENYLSTANNANE

ACETAZINE see ACETOPROMAZINE

ACETAZOLAMID see 5-ACETAMIDE-1,3,4-THIADIAZOLE-2-SULFON-AMIDE

ACETAZOLAMIDE see 5-ACETAMIDE-1,3,4-THIADIAZOLE-2-SULFON-AMIDE

ACETAZOLAMIDE SODIUM SALT see ACETAZOLAMIDE SODIUM

ACETAZOLEAMIDE see 5-ACETAMIDE-1,3,4-THIADIAZOLE-2-SULFON-AMIDE

ACETDIMETHYLAMIDE see N,N-DIMETHYLACETAMIDE

ACETENE see ETHYLENE

ACETETHYLANILIDE see N-ETHYLACETANILIDE

ACETHYDRAZIDE see ACETYL HYDRAZIDE

ACETHYLPROMAZIN see ACETOPROMAZINE

ACETIC ACID (N-ACETYL-N-(4-BIPHENYL)AMINO) ESTER see N-ACE-TOXY-4-ACETAMIDOBIPHENYL

ACETIC ACID ALLYL ESTER see ALLYL ACETATE

ACETIC ACID AMIDE see ACETAMIDE

ACETIC ACID, AMMONIUM SALT see AMMONIUM ACETATE

ACETIC ACID, AMYL ESTER see AMYL ACETATE (MIXED ISOMERS)

ACETIC ACID, AMYL ESTER see PENTYL ACETATE

ACETIC ACID, ANHYDRIDE see ACETIC ANHYDRIDE

ACETIC ACID, ANHYDRIDE WITH NITRIC ACID (1:1) see ACETYL NITRATE

ACETIC ACID ANILIDE see ACETANILIDE

ACETIC ACID (AQUEOUS SOLUTION) (DOT) see ACETIC ACID

ACETIC ACID, BARIUM SALT see BARIUM ACETATE

ACETIC ACID, BENZ(A)ANTHRACENE-7,12-DIMETHANOL DIESTER see BENZ(a)ANTHRACENE-7,12-DIMETHANOLDIACETATE

ACETIC ACID, BENZ(A)ANTHRACENE-7-METHANOL ESTER see BENZ(a)ANTHRACENE-7-METHANOL ACETATE

ACETIC ACID BENZYL ESTER see BENZYL ACETATE

ACETIC ACID, 1,3-BUTADIENYL ESTER see 1-ACETOXY-1,3-BUTADIENE

ACETIC ACID-2-BUTOXY ESTER see sec-BUTYL ACETATE

ACETIC ACID CHLORIDE see ACETYL CHLORIDE

ACETIC ACID, COBALT(2+) SALT see COBALT DIACETATE

ACETIC ACID, CUPRIC SALT see COPPER ACETATE

ACETIC ACID, DIMETHYLAMIDE see N,N-DIMETHYLACETAMIDE

ACETIC ACID, 2,6-DIMETHYL-M-DIOXAN-4-YL ESTER see ACETO-METHOXANE

ACETIC ACID-1,1-DIMETHYLETHYL ESTER see ACETIC ACID-tert-BU-TYL ESTER

ACETIC ACID, (2,4-DINITRO-6-S-BUTYLPHENYL) ESTER see O-ACE-TYL-2-sec-BUTYL-4,6-DINITROPHENOL

ACETIC ACID, (4,6-DINITRO-2-S-BUTYLPHENYL) ESTER see O-ACE-TYL-2-sec-BUTYL-4,6-DINITROPHENOL

ACETIC ACID, ESTER WITH N-4-BIPHENYLYLACETOHYDROXAMIC ACID see N-ACETOXY-4-ACETAMIDOBIPHENYL

ACETIC ACID ESTER WITH N-(FLUOREN-3-YL)ACETOHYDROXAMIC ACID see N-ACETOXYFLUORENYLACETAMIDE

ACETIC ACID ETHENYL ESTER see ACETIC ACID VINYL ESTER

ACETIC ACID, GLACIAL (DOT) see ACETIC ACID

ACETIC ACID, ISOBUTYL ESTER see ISOBUTYL ACETATE

ACETIC ACID LEAD (2+) SALT see LEAD ACETATE

ACETIC ACID, LEAD(+2) SALT TRIHYDRATE see LEAD ACETATE (II), TRIHYDRATE

ACETIC ACID, MAGNESIUM SALT see MAGNESIUM ACETATE

ACETIC ACID, MANGANESE(II) SALT (2:1) see MANGANESE ACETATE

ACETIC ACID, MERCURY(2+) SALT see MERCURIC ACETATE

ACETIC ACID, 1-METHYLETHYL ESTER (9CI) see ACETIC ACID ISO-PROPYL ESTER

ACETIC ACID, 2-METHYL-6-METHYLENE-7-OCTEN-2-YL ESTER see ACETIC ACID MYRCENYL ESTER

ACETIC ACID, 4-METHYLPHENYL ESTER see 4-METHYLPHENYL ACETATE

ACETIC ACID-2-METHYLPROPYL ESTER see ISOBUTYL ACETATE

ACETIC ACID-1-METHYLPROPYL ESTER (9CI) see sec-BUTYL ACETATE

ACETIC ACID, NICKEL(2+) SALT see NICKEL(II) ACETATE (1:2)

ACETIC ACID-2-PHENYLETHYL ESTER see 2-PHENYLETHYL ACETATE

ACETIC ACID, PHENYLMERCURY DERIV. see ACETOXYPHENYLMER-CURY

ACETIC ACID PHENYLMETHYL ESTER see BENZYL ACETATE

ACETIC ACID-2-PROPENYL ESTER see ALLYL ACETATE

ACETIC ACID, 1-(PROPYLNITROSAMINO)PROPYL ESTER see 1-ACET-OXY-N-NITROSODIPROPYLAMINE

ACETIC ACID, VINYL ESTER, CHLOROETHYLENE COPOLYMER see POLYVINYL ACETATE CHLORIDE

ACETIC ALDEHYDE see ACETALDEHYDE

ACETIC CHLORIDE see ACETYL CHLORIDE

ACETIC ETHER see ETHYL ACETATE

ACETIC OXIDE see ACETIC ANHYDRIDE

ACETIDIN see ETHYL ACETATE

ACETIMIDIC ACID see ACETAMIDE

ACETOACETAMIDOBENZENE see ACETOACETANILIDE

O-ACETOACETANISIDE see ACETOACETYL-O-ANISIDINE

ACETOACET-O-ANISIDIN (CZECH) see ACETOACETYL-O-ANISIDINE

ACETOACET-P-CHLOROANILIDE see p-CHLORO ACETO ACETANILIDE

ACETOACETIC ACID ANILIDE see ACETOACETANILIDE

ACETOACETIC ACID-O-ANISIDIDE see ACETOACETYL-O-ANISIDINE

ACETOACETIC ANILIDE see ACETOACETANILIDE

ACETOACETIC ESTER see ETHYL ACETYL ACETATE

ACETOACETIC METHYL ESTER see METHYL ACETYLACETATE

ACETOACETONE see ACETYL ACETONE

2-ACETOACETYLAMINOANISOLE see ACETOACETYL-O-ANISIDINE

((ACETOACETYL)AMINO)BENZENE see ACETOACETANILIDE

2-ACETOACETYLAMINOTOLUENE see ACETOACET-o-TOLUIDIDE

ACETOACETYLANILINE see ACETOACETANILIDE

ACETOACETYL-O-ANISIDE see ACETOACETYL-O-ANISIDINE

ACETOACETYL-O-ANISINE see ACETOACETYL-O-ANISIDINE

ACETOACETYL-4-CHLOROANILIDE see p-CHLORO ACETO ACETANI-LIDE

ACETOACETYL-2-METHYLANILIDE see ACETOACET-o-TOLUIDIDE

ACETOANILIDE see ACETANILIDE

ACETOARSENITE DE CUIVRE (FRENCH) see CUPRIC ACETOARSENITE

ACETOHYDRAZIDE see ACETYL HYDRAZIDE

ACETOMETHOXAN see ACETOMETHOXANE

ACETOMORFINE see HEROIN

ACETOMORPHINE see HEROIN

2'-ACETONAPHTHONE, 3'-ETHYL-5',6',7',8'-TETRAHYDRO-5',5',8'-TETRAMETHYL- see ACETYL ETHYL TETRAMETHYL TETRALIN

ACETONCIANHIDRINEI (ROUMANIAN) see 2-METHYLLACTONITRILE

ACETONCIANIDRINA (ITALIAN) see 2-METHYLLACTONITRILE

ACETONCYAANHYDRINE (DUTCH) see 2-METHYLLACTONITRILE

ACETONCYANHYDRIN (GERMAN) see 2-METHYLLACTONITRILE

ACETONECYANHYDRINE (FRENCH) see 2-METHYLLACTONITRILE

ACETONE CYANOHYDRIN see 2-METHYLLACTONITRILE

ACETONE CYANOHYDRIN (DOT) see 2-METHYLLACTONITRILE

ACETONE, CYCLIC (HYDROXYMETHYL)ETHYLENE ACETAL see DIOX-OLAN

ACETON (GERMAN, DUTCH, POLISH) see ACETONE

ACETONIC ACID see LACTIC ACID

ACETONITRIL (GERMAN, DUTCH) see ACETONITRILE

ACETONKYANHYDRIN (CZECH) see 2-METHYLLACTONITRILE

3-(alpha-ACETONYLBENZYL)-4-HYDROXYCOUMARIN see COUMADIN

3-(alpha-ACETONYLBENZYL)-4-HYDROXY-COUMARIN SODIUM SALT see COUMADIN SODIUM

ACETONYL CHLORIDE see CHLOROACETONE

ACETOPHEN see ACETOL (2)

ACETO-PARA-PHENALIDE see p-ACETOPHENETIDIDE

P-ACETOPHENETIDE see p-ACETOPHENETIDIDE

PARA-ACETOPHENETIDIDE see p-ACETOPHENETIDIDE

ACETO-PARA-PHENETIDIDE see p-ACETOPHENETIDIDE

ACETOPHENETIDIN see p-ACETOPHENETIDIDE

ACETOPHENETIDINE see p-ACETOPHENETIDIDE

ACETO-4-PHENETIDINE see p-ACETOPHENETIDIDE

P-ACETOPHENETIDINE see p-ACETOPHENETIDIDE

ACETOPHENETIN see p-ACETOPHENETIDIDE

ACETOPHOS see O,O-DIETHYL S-(CARBETHOXY)METHYL PHOSPHORO-THIOLATE

ACETOSALIC ACID see ACETOL (2)

ACETOTHIOAMIDE see THIOACETAMIDE

4-ACETOTOLUIDE see p-ACETOTOLUIDIDE

P-ACETOTOLUIDE see p-ACETOTOLUIDIDE

ACETOXON see O,O-DIETHYL S-(CARBETHOXY)METHYL PHOSPHORO-THIOLATE

N-ACETOXY-2-ACETAMIDOFLUORENE see N-ACETOXY-N-ACETYL-2-
AMINOFLUORENE

N-ACETOXY-2-ACETYLAMINOFLUORENE see N-ACETOXY-N-ACETYL-
2-AMINOFLUORENE

ALPHA-ACETOXYACRYLONITRILE see 2-ACETOXYACRYLONITRILE

2-ACETOXYBENZOIC ACID see ACETOL (2)

O-ACETOXYBENZOIC ACID see ACETOL (2)

N-ACETOXY-4-BIPHENYLACETAMIDE see N-ACETOXY-4-ACETAMIDO-
BIPHENYL

17-ALPHA-ACETOXY-6-CHLORO-6-DEHYDROPROGESTERONE see CAP

17-ALPHA-ACETOXY-6-CHLORO-6,7-DEHYDROPROGESTERONE see
CAP

17-ALPHA-ACETOXY-6-CHLOROPREGNA-4,6-DIENE-3,20-DIONE see
CAP

17-ALPHA-ACETOXY-6-CHLORO-4,6-PREGNADIENE-3,20-DIONE see
CAP

17-ALPHA-ACETOXY-6-DEHYDRO-6-METHYLPROGESTERONE see
VOLIDAN

2-ACETOXY-3-DIETHYLCARBAMYL-9,10-DIMETHOXY-1,2,3,4,6,7-
HEXAHYDRO -11B-BENZO(A)QUINOLIZINE see BENZOGUANAMINE

ACETOXYDIETHYLPHENYLSTANNANE see DIETHYL PHENYLTIN ACE-
TATE

1-ACETOXY-1,4-DIHYDRO-4-HYDROXYAMINO QUINOLINE ACETATE
(ESTER) see 1-ACETOXY-1,4-DIHYDRO-4-(HYDROXYAMINO)QUI-
NOLINE ACETATE (ESTER)

6-ACETOXY-2,4-DIMETHYL-M-DIOXANE see ACETOMETHOXANE

ALPHA-ACETOXY DIMETHYLNITROSAMINE see ACETIC ACID METHYL-
NITROSAMINOMETHYL ESTER

3-(2-(5-ACETOXY-3,5-DIMETHYL-2-OXOCYCLOHEXYL)-2-HYDROXY-
ETHYL)GLUTARIMIDE see ACETOXYCYCLOHEXIMIDE

ACETOXYETHANE see ETHYL ACETATE

1-ACETOXYETHYLENE see ACETIC ACID VINYL ESTER

N-ACETOXY-2-FLUORENYLACETAMIDE see N-ACETOXY-N-ACETYL-2-
AMINOFLUORENE

N-ACETOXY-3-FLUORENYLACETAMIDE see N-ACETOXYFLUORENYL-
ACETAMIDE

2-ACETOXYISOSUCCINODINITRILE see METHYLACETOXYMALONONI-
TRILE

(ACETOXYMERCURI)BENZENE see ACETOXYPHENYLMERCURY

10-ACETOXYMETHYL-1,2-BENZANTHRACENE see BENZ(a)ANTHRA-
CENE-7-METHANOL ACETATE

ACETOXYMETHYLBUTYLNITROSAMINE see BUTYL NITROS AMINO
METHYL ACETATE

N-(ACETOXY)METHYL-N-N-BUTYLNITROSAMINE see BUTYL NITROS
AMINO METHYL ACETATE

ACETOXYMETHYLETHYLNITROSAMINE see (ETHYL NITROSAMINO)-
METHYL ACETATE

ACETOXYMETHYL-METHYL-NITROSAMIN (GERMAN) see ACETIC ACID
METHYLNITROSAMINOMETHYL ESTER

ACETOXYMETHYL METHYLNITROSAMINE see ACETIC ACID METHYL-
NITROSAMINOMETHYL ESTER

N-ALPHA-ACETOXYMETHYL-N-METHYLNITROSAMINE see ACETIC
ACID METHYLNITROSAMINOMETHYL ESTER

N-ACETOXYMETHYL-N-NITROSOETHYLAMINE see (ETHYL NITRO-
SAMINO)METHYL ACETATE

17-ACETOXY-6-METHYLPREGNA-4,6-DIENE-3,20-DIONE see VOLIDAN

17-ALPHA-ACETOXY-6-METHYLPREGNA-4,6-DIENE-3,20-DIONE see
VOLIDAN

17-ALPHA-ACETOXY-6-METHYL-4,6-PREGNADIENE-3,20-DIONE see
VOLIDAN

17-ALPHA-ACETOXY-6-ALPHA-METHYLPREGN-4-ENE-3,20-DIONE see
MEDROXYPROGESTERONE ACETATE

17-ACETOXY-6-ALPHA-METHYLPROGESTERONE see MEDROXYPRO-
GESTERONE ACETATE

ACETOXYMETHYLPROPYLNITROSAMINE see PROPYLNITROSAMINO-
METHYL ACETATE

N-(ACETOXY)METHYL-N-N-PROPYLNITROSAMINE see PROPYLNITRO-
SAMINOMETHYL ACETATE

1-ACETOXY-N-NITROSODIMETHYLAMINE see ACETIC ACID METHYL-
NITROSAMINOMETHYL ESTER

17-BETA-ACETOXY-19-NOR-17-ALPHA-PREGN-4-EN-20-YN-3-ONE see
17-ACETOXY-19-NOR-17-alpha-PREGN-4-EN-20-YN-3-ONE

1-ACETOXYPROPANE see PROPYL ACETATE

2-ACETOXYPROPANE see ACETIC ACID ISOPROPYL ESTER

3-ACETOXYPROPENE see ALLYL ACETATE

N-(ALPHA-ACETOXY)PROPYL-N-N-PROPYLNITROSAMINE see 1-ACET-
OXY-N-NITROSODIPROPYLAMINE

3-ACETOXYQUINUCLIDINE GLAUCOSTAT see ACECLIDINE

1'-ACETOXYSAFROLE see 5-(1-ACETYLOXY-2-PROPENYL)-1,3-BEN-
ZODIOXOLE

4-ACETOXYTOLUENE see 4-METHYLPHENYL ACETATE

P-ACETOXYTOLUENE see 4-METHYLPHENYL ACETATE

ALPHA-ACETOXYTOLUENE see BENZYL ACETATE

ACETOXYTRIISOPROPYLSTANNANE see TRIISOPROPYLTIN ACETATE

ACETOXYTRIMETHYLSTANNANE see TRIMETHYLTIN ACETATE

ACETOXY-TRIPHENYL-STANNAN (GERMAN) see ACETOXYTRIPHENYL-
STANNANE

ACETOXY-TRIPHENYLSTANNANE see ACETOXYTRIPHENYLSTANNANE

ACETOZALAMIDE see 5-ACETAMIDE-1,3,4-THIADIAZOLE-2-SULFON-
AMIDE

ACET-P-PHENALIDE see p-ACETOPHENETIDIDE

ACETPHENARSINE see N-ACETYL-4-HYDROXY-m-ARSANILIC ACID

ACETPHENARSINE see ACETPHENARSINE

ACETPHENETIDIN see p-ACETOPHENETIDIDE

P-ACETPHENETIDIN see p-ACETOPHENETIDIDE

ACET-P-PHENETIDIN see p-ACETOPHENETIDIDE

ACETYLACETANILIDE see ACETOACETANILIDE

ALPHA-ACETYLACETANILIDE see ACETOACETANILIDE

N-(ACETYLACETYL)ANILINE see ACETOACETANILIDE

3-ACETYLAMINOANILINE see 3'-AMINOACETANILIDE

M-(ACETYLAMINO)ANILINE see 3'-AMINOACETANILIDE

ACETYLAMINOBENZENE see ACETANILIDE

4-ACETYLAMINOBIPHENYL see 4'-PHENYLACETANILIDE

4-ACETYLAMINOFLUOREN (GERMAN) see 4-ACETYLAMINOFLUORENE

N-ACETYL-2-AMINOFLUORENE see N-FLUOREN-2-YL ACETAMIDE

2-ACETYLAMINO-9-FLUORENONE see 2-ACETYLAMINOFLUORENONE

(3-(ACETYLAMINO)-4-HYDROXYPHENYL)- (9CI) see ACETPHENARSINE

3-ACETYLAMINO-4-HYDROXYPHENYLARSONIC ACID see ACETPHE-
NARSINE

3-ACETYLAMINO-4-HYDROXYPHENYLARSONIC ACID see N-ACETYL-
4-HYDROXY-m-ARSANILIC ACID

2-ACETYLAMINO-4-(5-NITRO-2- FURYL)THIAZOLE see 2-ACETAMIDO-
4-(5-NITRO-2-FURYL)THIAZOLE

P-(ACETYLAMINO)PHENACYL CHLORIDE see 4'-CHLOROACETYL
ACETANILIDE

2-ACETYLAMINOPHENANTHRENE see 2-ACETAMIDOPHENATHRENE

P-ACETYLAMINOPHENOL see 4'-HYDROXYACETANILIDE

N-ACETYL-P-AMINOPHENOL see 4'-HYDROXYACETANILIDE

P-ACETYLAMINOPHENYL DERIVATIVE OF NITROGEN MUSTARD see
4'-(BIS(2-CHLOROETHYL)AMINO)ACETANILIDE

4-(ACETYLAMINO)TOLUENE see p-ACETOTOLUIDIDE

ACETYL ANHYDRIDE see ACETIC ANHYDRIDE

ACETYLANILINE see ACETANILIDE

3-ACETYLANILINE see 3'-AMINOACETOPHENONE

4-ACETYLANILINE see p-AMINO ACETOPHENONE

M-ACETYLANILINE see 3'-AMINOACETOPHENONE

N-ACETYLANILINE see ACETANILIDE

ACETYLBENZENE see ACETOPHENONE

ACETYL BENZOYL ACONINE see ACONITINE

ACETYLBROMODIETHYLACETYLCARBAMIDE see 1-ACETYL-3-(2-
BROMO-2-ETHYLBUTYRYL)UREA

N-ACETYL-N-BROMODIETHYLACETYLCARBAMIDE see 1-ACETYL-3-(2-
BROMO-2-ETHYLBUTYRYL)UREA

N-ACETYL-N-BROMODIETHYLACETYLUREA see 1-ACETYL-3-(2-
BROMO-2-ETHYLBUTYRYL)UREA

N-ACETYL-N'-ALPHA-BROMO-ALPHA-ETHYLBUTYRYLCARBAMIDE see
1-ACETYL-3-(2-BROMO-2-ETHYLBUTYRYL)UREA

1-ACETYL-3- (ALPHA-BROMO-ALPHA-ETHYLBUTYRYL)UREA see 1-
ACETYL-3-(2-BROMO-2-ETHYLBUTYRYL)UREA

ACETYLCHOLINE HYDROCHLORIDE see 2-ACETOXYETHYLTRIMETHYL-
AMMONIUM CHLORIDE

ACETYLCHOLINIUM CHLORIDE see 2-ACETOXYETHYLTRIMETHYLAM-
MONIUM CHLORIDE

ACETYLCYSTEINE see N-ACETYL-L-CYSTEINE

N-ACETYLCYSTEINE see N-ACETYL-L-CYSTEINE

N-ACETYL-N-CYSTEINE see N-ACETYL-L-CYSTEINE

1-ACETYL-9,10-DIDEHYDRO-N,N-DIETHYL-6-METHYLERGOLINE-8-
BETA-CARBOXAMIDE BITARTRATE see 1-ACETYLLYSERGIC ACID
DIETHYLAMIDE BITARTRATE

1-ACETYL-9,10-DIDEHYDRO-N-ETHYL-6-METHYLERGOLINE-8-BETA-
CARBOXAMIDE see d-1-ACETYL LYSERGIC ACID MONOETHYL-
AMIDE

ALPHA-ACETYLDIGITOXIN see ACETYL-DIGITOXIN-alpha

ALPHA-ACETYLDIGOXIN see ACETYL-DIGOXIN-alpha

BETA-ACETYLDIGOXIN see ACETYL-DIGITOXIN-beta

BETA-ACETYLDIGOXIN see ACETYL-DIGOXIN-beta

EPSILON-ACETYLDIGOXIN (GERMAN) see epsilon-DIGOXIN ACETATE

3-ACETYL-10-(3-DIMETHYLAMINOPROPYL)PHENOTHIAZINE see ACE-
TOPROMAZINE

ACETYLEN see ACETYLENE

ACETYLENE BLACK see CARBON BLACK

ACETYLENEDICARBOXYLIC ACID DIAMIDE see ACETYLENEDICAR-
BOXAMIDE

ACETYLENE TRICHLORIDE see TRICHLOROETHYLENE

ACETYL ETHER see ACETIC ANHYDRIDE

ACETYLETHYLENEIMINE see 1-ACETYLAZIRIDINE

7-ACETYL-6-ETHYL-1,1,4,4-TETRAMETHYL-1,2,3,4-TETRAHYDRO-
NAPHTHALENE see 3'-ETHYL-5',6',7',8'-TETRAHYDRO-5',5',8',
8'-TETRAMETHYL-2'-ACETONAPHTHONE

ACETYLETHYLTETRAMETHYLTETRALIN see 3'-ETHYL-5',6',7',8'-
TETRAHYDRO-5',5',8',8'-TETRAMETHYL-2'-ACETONAPHTHONE

ACETYLETHYL TETRAMETHYL-TETRALIN see ACETYL ETHYL TETRA-
METHYL TETRALIN

N-ACETYL-M-FENYLENEDIAMIN (CZECH) see 3'-AMINOACETANILIDE

N-ACETYLHYDRAZINE see ACETYL HYDRAZIDE

ACETYL HYDROPEROXIDE see PEROXYACETIC ACID

N-ACETYL-4-HYDROXY-M-ARSANILIC ACID see ACETPHENARSINE

N-ACETYL-4-HYDROXY-M-ARSANILIC ACID, CALCIUM SALT see CAL-
CIUM ACETARSONE

1-ACETYLLYSERGIC ACID ETHYLAMIDE see d-1-ACETYL LYSERGIC
ACID MONOETHYLAMIDE

ACETYL METHYL CARBINOL see ACETOIN

ACETYL-METHYL-NITROSO-HARNSTOFF (GERMAN) see ACETYL-
METHYLNITROSOUREA

N'-ACETYL-METHYLNITROSOUREA see ACETYLMETHYLNITROSOUREA

3-ACETYL-6-METHYLPYRANDIONE-2,4 see METHYLACETOPYRONONE

3-ACETYL-6-METHYL-2,4-PYRANDIONE see METHYLACETOPYRONONE

3-ACETYL-6-METHYL-2H-PYRAN-2,4(3H)-DIONE see METHYLACETO-
PYRONONE

ACETYL OXIDE see ACETIC ANHYDRIDE

2-(ACETYLOXY)BENZOIC ACID, MIXED WITH 3,7-DIHYDRO-1,3,7-
TRIMETHYL-1H-PURINE-2,6-DIONE AND N-(4-ETHOXYPHENYL)-
ACETAMIDE see ASPIRIN, PHENACETIN AND CAFFEINE

17-(ACETYLOXY)-6-CHLOROPREGNA-4,6-DIENE-3,20-DIONE see CAP

4-ACETYLOXY-12,13-EPOXY-3,7,15-TRIHYDROXY-(3-ALPHA,4-BETA,
7-BETA)-TRICHOTHEC-9-EN-8-ONE see 4-(ACETYLOXY)-12,13-
EPOXY-3,7,15-TRIHYDROXY-TRICHOTHEC-9-EN-8-ONE-(3-alpha,4-
beta,7-beta)

4-ACETYLOXY-12,13-EPOXY-3,7,15-TRIHYDROXY-(3-alpha,4-beta,7-
beta)-TRICHOTHEC-9-EN-8-ONE see FUSARENONE X

(6-ALPHA)-17-(ACETYLOXY)-6-METHYLPREG-4-ENE-3,20-DIONE see
MEDROXYPROGESTERONE ACETATE

(17-ALPHA)-17-(ACETYLOXY)-19-NORPREGN-4-EN-20-YN-3-ONE see
17-ACETOXY-19-NOR-17-alpha-PREGN-4-EN-20-YN-3-ONE

17-ACETYLOXY(17-ALPHA)-19-NORPREGN-4-ESTREN-17-BETA-OL-
ACETATE-3-ONE see 17-ACETOXY-19-NOR-17-alpha-PREGN-4-EN-
20-YN-3-ONE

ACETYLPHENETIDIN see p-ACETOPHENETIDIDE

N-ACETYL-P-PHENETIDINE see p-ACETOPHENETIDIDE

N-ACETYL-M-PHENYLENEDIAMINE see 3'-AMINOACETANILIDE

ACETYLPHOSPHORAMIDOTHIOIC ACID-O,S-DIMETHYL ESTER see O,S-
DIMETHYLACETYLPHOSPHOROAMIDOTHIOATE

1-ACETYL-2-PICOLINOYLHYDRAZINE see 1-ACETYL-2-PICOLINOLHY-
DRAZINE

ACETYLPROMAZINE see ACETOPROMAZINE

ACETYLPROMAZINE MALEATE (1:1) see ACEPROMAZINE MALEATE

ACETYLSALICYLIC ACID see ACETOL (2)

ACETYLSALICYLSAURE (GERMAN) see ACETOL (2)

6-ACETYL-1,1,4,4-TETRAMETHYL-7-ETHYL-1,2,3,4,-TETRALIN see
ACETYL ETHYL TETRAMETHYL TETRALIN

6-ACETYL-1,1,4,4-TETRAMETHYL-7-ETHYL-1,2,3,4-TETRALIN see
3'-ETHYL-5',6',7',8'-TETRAHYDRO-5',5',8',8'-TETRAMETHYL-2'-
ACETONAPHTHONE

7-ACETYL-1,1,4,4-TETRAMETHYL-1,2,3,4-TETRAHYDRONAPHTHA-
LENE see ACETYL ETHYL TETRAMETHYL TETRALIN

7-ALPHA-ACETYLTHIO-3-OXO-17-ALPHA-PREGN-4-ENE-21,17-BETA-
CARBOLACTONE see ALDACTAZIDE

7-ALPHA-ACETYLTHIO-3-OXO-17-BETA-PREGN-4-ENE-21,17-BETA-
CARBOLACTONE see ALDACTAZIDE

1-ACETYL-2-THIOUREA see ACETYL THIOUREA

N-ACETYL-P-TOLUIDIDE see p-ACETOTOLUIIDIDE

ACETYL-P-TOLUIDINE see p-ACETOTOLUIIDIDE

ACHROMYCIN HYDROCHLORIDE see TETRACYCLINE HYDROCHLORIDE

ACID see N,N-DIETHYLLYSERGAMIDE

ACIDAL GREEN G see ACID GREEN 3

ACID AMARANTH see FOOD RED 2

ACID AMIDE see NIACINAMIDE

ACID AMMONIUM CARBONATE see AMMONIUM BICARBONATE (1:1)

ACID BLUE A see ACID BLUE 92

ACID BRILLIANT SCARLET 3R see FOOD RED 7

ACID BUTYL PHOSPHATE see ACID BUTYL PHOSPHATE

ACID BUTYL PHOSPHATE (DOT) see BUTYL ESTER PHOSPHORIC ACID

ACID COPPER ARSENITE see COPPER ORTHOARSENITE

ACIDE ACETIQUE (FRENCH) see ACETIC ACID

ACIDE ACETYLSALICYLIQUE (FRENCH) see ACETOL (2)

ACIDE ARSENIEUX (FRENCH) see ARSENIC TRIOXIDE

ACIDE ARSENIQUE LIQUIDE (FRENCH) see o-ARSENIC ACID

ACIDE BENZOIQUE (FRENCH) see BENZOIC ACID

ACIDE BROMHYDRIQUE (FRENCH) see HYDROBROMIC ACID

ACIDE CACODYLIQUE (FRENCH) see HYDROXYDIMETHYLARSINE
OXIDE

ACIDE CARBOLIQUE (FRENCH) see PHENOL

ACIDE CHLORHYDRIQUE (FRENCH) see HYDROCHLORIC ACID

ACIDE CHROMIQUE (FRENCH) see CHROMIC ACID

ACIDE CYANACETIQUE (FRENCH) see CYANOACETIC ACID

ACIDE CYANHYDRIQUE (FRENCH) see HYDROCYANIC ACID

ACIDE 2,4-DICHLORO PHENOXYACETIQUE (FRENCH) see DICHLORO-
PHENOXYACETIC ACID

ACIDE DIMETHYLARSINIQUE (FRENCH) see HYDROXYDIMETHYLAR-
SINE OXIDE

ACIDE ETHYLENEDIAMINETETRACETIQUE (FRENCH) see ETHYLENE
DIAMINE TETRAACETIC ACID

ACIDE FLUORHYDRIQUE (FRENCH) see HYDROFLUORIC ACID

ACIDE FLUOROSILICIQUE (FRENCH) see SILICOFLUORIC ACID

ACIDE FLUOSILICIQUE (FRENCH) see SILICOFLUORIC ACID

ACIDE FORMIQUE (FRENCH) see FORMIC ACID

ACIDE MONOCHLORACETIQUE (FRENCH) see CHLOROACETIC ACID

ACIDE-MONOFLUORACETIQUE (FRENCH) see FLUORO ETHANOIC ACID

ACIDE NITRIQUE (FRENCH) see NITRIC ACID

ACIDE OXALIQUE (FRENCH) see OXALIC ACID

ACIDE PERACETIQUE (FRENCH) see PEROXYACETIC ACID

ACIDE PHOSPHORIQUE (FRENCH) see PHOSPHORIC ACID

ACIDE PHTALIQUE (FRENCH) see PHTHALIC ACID

ACIDE PICRIQUE (FRENCH) see PICRIC ACID

ACIDE PROPIONIQUE (FRENCH) see PROPIONIC ACID

ACIDE SULFHYDRIQUE (FRENCH) see HYDROGEN SULFIDE

ACIDE SULFURIQUE (FRENCH) see SULFURIC ACID

ACIDE THIOGLYCOLIQUE (FRENCH) see 2-MERCAPTOACETIC ACID

ACIDE TRICHLORACETIQUE (FRENCH) see TRICHLORACETIC ACID

ACIDE 2,4,5-TRICHLORO PHENOXYACETIQUE (FRENCH) see 2,4,5-T

ACID FAST ORANGE EGG see HISPACID FAST ORANGE 2G

ACID GREEN see ACID GREEN 3

ACID LEAD ARSENATE see LEAD ACID ARSENATE

ACID LEAD ORTHOARSENATE see LEAD ACID ARSENATE

ACID LEATHER BLUE R see ACID BLUE 92

ACID LIGHT ORANGE G see HISPACID FAST ORANGE 2G

ACIDO ACETICO (ITALIAN) see ACETIC ACID

ACIDO BROMIDRICO (ITALIAN) see HYDROBROMIC ACID

ACIDO CIANIDRICO (ITALIAN) see HYDROCYANIC ACID

ACIDO CLORIDRICO (ITALIAN) see HYDROCHLORIC ACID

ACIDO (2,4-DICLORO-FENOSSI)-ACETICO (ITALIAN) see DICHLORO-
    PHENOXYACETIC ACID

ACIDO FLUORIDRICO (ITALIAN) see HYDROFLUORIC ACID

ACIDO FLUOSILICICO (ITALIAN) see SILICOFLUORIC ACID

ACIDO FORMICO (ITALIAN) see FORMIC ACID

ACIDO FOSFORICO (ITALIAN) see PHOSPHORIC ACID

ACIDOMONOCLOROACETICO (ITALIAN) see CHLOROACETIC ACID

ACIDO MONOFLUOROACETIO (ITALIAN) see FLUORO ETHANOIC ACID

ACIDO NITRICO (ITALIAN) see NITRIC ACID

ACIDO OSSALICO (ITALIAN) see OXALIC ACID

ACIDO PICRICO (ITALIAN) see PICRIC ACID

ACID ORANGE 10 see HISPACID FAST ORANGE 2G

ACIDO SOLFORICO (ITALIAN) see SULFURIC ACID

ACIDO TRICLOROACETICO (ITALIAN) see TRICHLORACETIC ACID

ACIDO (2,4,5-TRICLORO-FENOSSI)-ACETICO (ITALIAN) see 2,4,5-T

ACID WOOL BLUE RL see ACID BLUE 92

ACONITANE see ACONITINE (CRYSTALLINE)

ACONITIN CRISTALLISAT (GERMAN) see ACONITINE (CRYSTALLINE)

ACONITUM see ACONITE

ACP-M-728 see 3-AMINO-2,5-DICHLOROBENZOIC ACID

ACRALDEHYDE see ACROLEIN

ACRAMINE RED see 2,6-DIAMINOACRIDINE

9-ACRIDINAMINE see 9-AMINOACRIDINE

2,6-ACRIDINEDIAMINE see 2,6-DIAMINOACRIDINE

3,6-ACRIDINEDIAMINE see 3,6-DIAMINOACRIDINIUM

3,6-ACRIDINEDIAMINE, MONOHYDROCHLORIDE (9CI) see PROFLA-
    VINE MONOHYDROCHLORIDE

3,6-ACRIDINEDIAMINE SULFATE (1:1) see 3,6-DIAMINOACRIDINE
    SULPHATE (1:1)

3,6-ACRIDINEDIAMINE SULFATE (2:1) see 3,6-DIAMINOACRIDINE
    SULPHATE (1:1)

ACRIDINE ORANGE see 3,6-BIS(DIMETHYL AMINO)ACRIDINE

ACRIDINE YELLOW BASE see 3,6-DIAMINO-2,7-DIMETHYLACRIDINE

4'-(9-ACRIDINYLAMINO)METHYLSULFONYL-M-ANISIDINE see 4'-(9-
    ACRIDINYLAMINO)METHANESULPHON-m-ANISIDIDE

ACRIFLAVINE MIXTURE WITH PROFLAVINE see ACRIFLAVINIUM
    CHLORIDE

ACRIFLAVINE NEUTRAL see XANTHACRIDINE

ACROLEIC ACID see ACRYLIC ACID

ACROLEINA (ITALIAN) see ACROLEIN

ACROLEIN CYANOHYDRIN see 2-HYDROXY-3-BUTENENITRILE

ACROLEINE (DUTCH, FRENCH) see ACROLEIN

ACRYLALDEHYD (GERMAN) see ACROLEIN

ACRYLALDEHYDE see ACROLEIN

ACRYLATE DE METHYLE (FRENCH) see METHYL ACRYLATE

ACRYLATE D'ETHYLE (FRENCH) see ETHYL ACRYLATE

ACRYLIC ACID-N-BUTYL ESTER see BUTYL-2-PROPENOATE

ACRYLIC ACID-2-CYANOETHYL ESTER see ACRYLIC ACID ESTER
    WITH HYDRACRYLONITRILE

ACRYLIC ACID, 2,3-EPOXYPROPYL ESTER see 2,3-EPOXYPROPYL
    ACRYLATE

ACRYLIC ACID, ETHYLENE ESTER see ETHYLENE ACRYLATE

ACRYLIC ACID, ETHYLENE GLYCOL DIESTER see ETHYLENE ACRY-
    LATE

ACRYLIC ACID ETHYL ESTER see ETHYL ACRYLATE

ACRYLIC ACID, GLACIAL see ACRYLIC ACID

ACRYLIC ACID, 2-METHOXYETHYL ESTER see METHYL CELLOSOLVE
    ACRYLATE

ACRYLIC ACID METHYL ESTER see METHYL ACRYLATE

ACRYLIC ALDEHYDE see ACROLEIN

ACRYLIC AMIDE see ACRYLAMIDE

ACRYLNITRIL (GERMAN, DUTCH) see ACRYLONITRILE

ACRYLONITRILE-BUTADIENE-STYRENE COPOLYMER see ACRYLONI-
    TRILE, POLYMER WITH 1,3-BUTADIENE AND STYRENE

ACRYLONITRILE-1,3-BUTADIENE-STYRENE COPOLYMER see ACRYLO-
    NITRILE, POLYMER WITH 1,3-BUTADIENE AND STYRENE

ACRYLONITRILE-BUTADIENE-STYRENE POLYMER see ACRYLONI-
    TRILE, POLYMER WITH 1,3-BUTADIENE AND STYRENE

ACRYLONITRILE-1,3-BUTADIENE-STYRENE POLYMER see ACRYLO-
    NITRILE, POLYMER WITH 1,3-BUTADIENE AND STYRENE

ACRYLONITRILE-BUTADIENE-STYRENE (PYROLYSIS PRODUCTS) see
    ACRYLONITRILE, POLYMER WITH 1,3-BUTADIENE, AND STYRENE,
    COMBUSTION PRODUCTS

ACRYLONITRILE-BUTADIENE-STYRENE RESIN see ACRYLONITRILE,
    POLYMER WITH 1,3-BUTADIENE AND STYRENE

ACRYLONITRILE-BUTADIENE-STYRENE TERPOLYMER see ACRYLONI-
    TRILE, POLYMER WITH 1,3-BUTADIENE AND STYRENE

ACRYLONITRILE MONOMER see ACRYLONITRILE

ACRYLONITRILE-STYRENE-BUTADIENE RESIN see ACRYLONITRILE,
    POLYMER WITH 1,3-BUTADIENE AND STYRENE

ACRYLONITRILE-STYRENE COPOLYMER see ACRYLONITRILE POLYMER
    WITH STYRENE

ACRYLONITRILE-STYRENEPOLYMER see ACRYLONITRILE POLYMER
    WITH STYRENE

ACRYLONITRILE-STYRENE RESIN see ACRYLONITRILE POLYMER WITH
    STYRENE

ACRYLSAEUREAETHYLESTER (GERMAN) see ETHYL ACRYLATE

ACRYLSAEUREAETHYLESTER (GERMAN) see SULFUROUS ACID, 2-(p-
    t-BUTYLPHENOXY)-1-METHYLETHYL-2-CHLOROETHYL ESTER

ACRYLSAEUREMETHYLESTER (GERMAN) see METHYL ACRYLATE

ACTI-AID see CYCLOHEXIMIDE

ACTIDIONE see CYCLOHEXIMIDE

ACTIDONE see CYCLOHEXIMIDE

ACTINOLITE see ASBESTOS

ACTINOMYCIN 1048A see ACTINOMYCIN S

ACTINOMYCIN 2104L see ACTINOMYCIN L

ACTINOMYCIN I see ACTINOMYCIN D

ACTIOQUINONE LIGHT YELLOW see ACETAMINE YELLOW CG

ACTIVATED ALUMINUM OXIDE see ALUMINUM OXIDE (2:3)

ACTIVE ACETYL ACETATE see ETHYL ACETYL ACETATE

ACYLANID see ACETYL-DIGITOXIN-alpha

ADAMANTANAMINE HYDROCHLORIDE see 1-ADAMANTANAMINE HY-
    DROCHLORIDE

ADAMANTINE HYDROCHLORIDE see 1-ADAMANTANAMINE HYDRO-
    CHLORIDE

ADAMANTYLAMINE HYDROCHLORIDE see 1-ADAMANTANAMINE HY-
    DROCHLORIDE

1-ADAMANTYLAMINE HYDROCHLORIDE see 1-ADAMANTANAMINE
    HYDROCHLORIDE

ADAMSITE see PHENARSAZINE CHLORIDE

ADANON see METHADONE

ADANON HYDROCHLORIDE see METHADONE HYDROCHLORIDE

ADC MALACHITE GREEN CRYSTALS see AIZEN MALACHITE GREEN

ADENINE ARABINOSIDE see 9-beta-d-ARABINO FURANOSYL ADENINE

ADENINIMINE see ADENINE

ADIPEX see ''SPEED''

ADIPIC ACID DIAMIDE see ADIPAMIDE

ADIPIC ACID DINITRILE see ADIPONITRILE

ADIPIC ACID NITRILE see ADIPONITRILE

ADIPIC DIAMIDE see ADIPAMIDE

ADIPIC KETONE see CYCLOPENTANONE

ADIPINIC ACID see ADIPIC ACID

ADIPODINITRILE see ADIPONITRILE

ADRENALIN see VASOTONIN

1-ADRENALIN see VASOTONIN

ADRENALINE see VASOTONIN

(−)-ADRENALINE see VASOTONIN

1-ADRENALINE see VASOTONIN

L-(+)-ADRENALINE see d-ADRENALINE

(-)-ADRENALINE HYDROCHLORIDE see l-ADRENALINE CHLORIDE

l-ADRENALINE HYDROCHLORIDE see l-ADRENALINE CHLORIDE

ADRENALIN HYDROCHLORIDE see l-ADRENALINE CHLORIDE

ADRIAMYCIN-HCl see ADRIAMYCIN

AERO CYANATE see POTASSIUM CYANATE

AERO LIQUID HCN see HYDROCYANIC ACID

AESCIN SODIUM SALT see ESCIN, SODIUM SALT

AESCULETIN DIMETHYL ETHER see 6,7-DIMETHYLESCULETIN

AESCUSAN SODIUM SALT see ESCIN, SODIUM SALT

AET DICHLORIDE see 2-AMINOETHYL-2-THIOPSEUDOUREADICHLO-
RIDE

AETHALDIAMIN (GERMAN) see 1,2-ETHANEDIAMINE

AETHANETHIOL (GERMAN) see ETHANETHIOL

AETHANOL (GERMAN) see ETHANOL

AETHANOLAMIN (GERMAN) see MONOETHANOLAMINE

AETHER see ETHYL ETHER

AETHOBROMID DES ALPHA,ALPHA-DICYCLOPENTYLESSIGSAEURE-
BETA'-DIAETHYLAMINO AETHYLESTER (GERMAN) see (2-(DICY-
CLOPENTYLACETOXY)ETHYL)TRIETHYLAMMONIUM BROMIDE

2-AETHOXY-AETHYLACETAT (GERMAN) see ''CELLOSOLVE'' ACE-
TATE

p-AETHOXYPHENYLHARNSTOFF (GERMAN) see 4-ETHOXY PHENYL
UREA

AETHYLACETAT (GERMAN) see ETHYL ACETATE

AETHYL ACETOXYMETHYLNITROSAMIN (GERMAN) see (ETHYL NI-
TROSAMINO)METHYL ACETATE

AETHYLACRYLAT (GERMAN) see ETHYL ACRYLATE

AETHYLACRYLATE (GERMAN) see SULFUROUS ACID, 2-(p-t-BUTYL-
PHENOXY)-1-METHYLETHYL-2-CHLOROETHYL ESTER

AETHYL-AETHANOL-NITROSOAMIN (GERMAN) see ETHYL-2-HYDROXY
ETHYL NITROSAMINE

AETHYLALKOHOL (GERMAN) see ETHANOL

AETHYLAMINE (GERMAN) see ETHANAMINE

AETHYLANILIN (GERMAN) see N-ETHYL ANILINE

AETHYLBENZOL (GERMAN) see ETHYL BENZENE

AETHYLBUTYLKETON (GERMAN) see 3-HEPTANONE

AETHYL-N-BUTYL-NITROSOAMIN (GERMAN) see ETHYL-n-BUTYL
NITROS AMINE

AETHYLCARBAMAT (GERMAN) see URETHANE

AETHYLCHLORID (GERMAN) see ETHYL CHLORIDE

AETHYL-CHLORVYNOL see 1-CHLORO-3-ETHYL-1-PENTEN-4-YN-3-OL

O-AETHYL-S-(2-DIMETHYLAMINOAETHYL)-METHYLPHOSPHONOTHIO-
ATE (GERMAN) see O-ETHYL-S-(2-DIMETHYL AMINO ETHYL)-
METHYL PHOSPHONO THIOATE

S-AETHYL-N,N-DIPROPYLTHIOLCARBAMAT (GERMAN) see S-ETHYL-
N,N-DI-n-PROPYLTHIOCARBAMATE

AETHYLEN-BIS-THIURAMMONOSULFID (GERMAN) see ISOTHIOCYANIC
ACID, ETHYLENE ESTER

AETHYLENBROMID (GERMAN) see 1,2-ETHYLENE DIBROMIDE

AETHYLENCHLORID (GERMAN) see 1,2-DICHLORETHANE

AETHYLENCHLORID (GERMAN) see ETHYLENE DICHLORIDE

AETHYLENECHLORHYDRIN (GERMAN) see 2-CHLOROETHYL ALCO-
HOL

AETHYLENEDIAMIN (GERMAN) see 1,2-ETHANEDIAMINE

AETHYLENGLYKOLMETHYLAETHERACETAT (GERMAN) see ETHYLENE
GLYCOL MONOMETHYL ETHER ACETATE

AETHYLENGLYKOL-MONOMETHYLAETHER (GERMAN) see ETHYLENE
GLYCOL METHYL ETHER

AETHYLENIMIN (GERMAN) see ETHYLENIMINE

AETHYLENOXID (GERMAN) see ETHYLENE OXIDE

AETHYLENSULFID (GERMAN) see ETHYLENE SULFIDE

AETHYLFORMIAT (GERMAN) see ETHYL FORMATE

AETHYLHARNSTOFF UND NATRIUMNITRIT (GERMAN) see ETHYLUREA
and SODIUM NITRITE (2:1)

AETHYLHARNSTOFF UND NITRIT (GERMAN) see ETHYLUREA and SO-
DIUM NITRITE (2:1)

AETHYLIDENCHLORID (GERMAN) see ETHYLIDENE DICHLORIDE

AETHYLIS CHLORIDUM see ETHYL CHLORIDE

2-(O-AETHYL-N-ISOPROPYLAMIDOTHIOPHOSPHORYLOXY)-BENZOSA-
EURE-ISOPROPYLESTER(GERMAN) see ISOFENPHOS

AETHYL-ISOPROPYL-NITROSOAMIN (GERMAN) see ETHYL ISOPROPYL
NITROSOAMINE

AETHYLMERCAPTAN (GERMAN) see ETHANETHIOL

AETHYLMETHYLKETON (GERMAN) see 2-BUTANONE

O-AETHYL-O-(3-METHYL-4-METHYLTHIOPHENYL)-ISOPROPYLAMIDO-
PHOSPHOR SAEURE ESTER (GERMAN) see FENAMIPHOS

3-BETA-AETHYL-1-METHYL-4-PHENYL-4-ALPHA-PIPERIDYLPROPIO-
NAT HYDROCHLORID (GERMAN) see NU-1932

3-BETA-AETHYL-1-METHYL-4-PHENYL-4-ALPHA-PROPIONYLOXYPI-
PERIDIN HYDROCHLORID (GERMAN) see NU-1932

N-AETHYL-N'-NITRO-N-NITROSOGUANIDIN (GERMAN) see N-ETHYL-
N-NITROSO-N'-NITROGUANIDINE

O-AETHYL-O-(4-NITRO-PHENYL)-PHENYL-MONOTHIOPHOSPHONAT
(GERMAN) see ETHOXY-4-NITRO PHENOXY PHENYL PHOSPHINE
SULFIDE

AETHYLNITROSO-HARNSTOFF (GERMAN) see 1-ETHYL-1-NITROSO-
UREA

AETHYLNITROSOURETHAN (GERMAN) see N-NITROSO-N-ETHYLURE-
THAN

5-AETHYL-5-PENTYL-(2')-BARBITURSAEURE (GERMAN) see PENTO-
BARBITAL

O-AETHYL-S-PHENYL-AETHYL-DITHIOPHOSPHONAT (GERMAN) see
DYPHONATE

AETHYL-4-PICOLYLNITROSAMIN (GERMAN) see N-NITROSO-4-PICO-
LYLETHYLAMINE

N-AETHYLPIPERIDIN (GERMAN) see 1-ETHYLPIPERIDINE

AETHYLRHODANID (GERMAN) see ETHYL THIOCYANATE

AETHYLSENFOEL (GERMAN) see ISOTHIOCYANIC ACID, ETHYLENE-
ESTER

O-AETHYL-S,S-DIPHENYL-DITHIOPHOSPHAT (GERMAN) see O-ETHYL-
S,S-DIPHENYL DITHIO PHOSPHATE

S-2-AETHYLSULFINYL-1-METHYLAETHYL-O,O DIMETHYL-MONOTHIO-
PHOSPHAT (GERMAN) see METASYSTOX-S

AETHYLURETHAN (GERMAN) see URETHANE

AETHYL-VINYL-NITROSOAMIN (GERMAN) see N-NITROSO-N-ETHYL-
VINYLAMINE

AETT see ACETYL ETHYL TETRAMETHYL TETRALIN

AFBI see AFLATOXIN B1

A.F. GREEN NO. 2 see FD AND C GREEN NO. 2

AFL 1082 see FLUOROACETANILIDE

AFLATOXICOL see AFLATOXIN RO

AFLATOXIN B see AFLATOXIN B1

A.F. RED NO. 1 see FD AND C RED NO. 1

AGALITE see TALC (powder)

AGAR-AGAR see AGAR

AGAR AGAR FLAKE see AGAR

AGAR-AGAR GUM see AGAR

AGATE see SILICA (CRYSTALLINE)

AGATE see SILICA, CRYSTALLINE-QUARTZ

AGERITE see 1,4-BIS(PHENYL AMINO)BENZENE

AGERITE ALBA see AGERITE

AGERITE WHITE see 2-NAPHTHYL-p-PHENYLENEDIAMINE

AGGLUTININ see ABRIN

AGILENE see POLYETHYLENE

AGILENE see 2,6-DIMETHYL-1,1-DIETHYLPIPERIDINIUM BROMIDE

AIREDALE BLUE 2BD see C.I. DIRECT BLUE 6, TETRASODIUM SALT

AIREDALE BLUE FFD see C.I. DIRECT BLUE 1, TETRASODIUM SALT

AIREDALE BLUE RL see ACID BLUE 92

AIZEN CRYSTAL VIOLET EXTRA PURE see ANILINE VIOLET

AIZEN DIRECT BLUE 2BH see C.I. DIRECT BLUE 6, TETRASODIUM
SALT

AIZEN FOOD GREEN NO. 3 see FD AND C GREEN NO. 3

AIZEN URANINE see FLUORESCEIN SODIUM

AKROLEIN (CZECH) see ACROLEIN

AKROLEINA (POLISH) see ACROLEIN

AKRYLAMID (CZECH) see ACRYLAMIDE

AKRYLONITRYL (POLISH) see ACRYLONITRILE

AL-1021 see 1-(3-(4-FLUORO BENZOYL)PROPYL)-4-PIPERIDYL
    N-ISOPROPYL CARBAMATE
ALABASTER see CALCIUM (II) SULFATE DIHYDRATE (1:1:2)
ALACINE see PHENYLETHYLHYDRAZINE SULPHATE
ALANINE MUSTARD see N,N-BIS(beta-CHLOROETHYL)-D,L-ALANINE
    HYDROCHLORIDE
ALAUN (GERMAN) see ALUMINUM
ALBAGEL PREMIUM USP 4444 see BENTONITE
ALBOLINE see MINERAL OIL
ALBONE see HYDROGEN PEROXIDE, 90%
ALBSAPOGENIN see GYPSOGENIN
ALCOHOL see ETHANOL
ALCOHOL, ANHYDROUS see ETHANOL
ALCOHOL C-8 see N-OCTYL ALCOHOL
ALCOHOL C-10 see DECYL ALCOHOL
ALCOHOL C-12 see DODECYL ALCOHOL
ALCOHOL DEHYDRATED see ETHANOL
ALCOOL ALLILCO (ITALIAN) see ALLYL ALCOHOL
ALCOOL ALLYLIQUE (FRENCH) see ALLYL ALCOHOL
ALCOOL AMILICO (ITALIAN) see ISOAMYL ALCOHOL
ALCOOL AMYLIQUE (FRENCH) see AMYL ALCOHOL
ALCOOL AMYLIQUE (FRENCH) see PENTYL ALCOHOL
ALCOOL BUTYLIQUE (FRENCH) see N-BUTYL ALCOHOL
ALCOOL BUTYLIQUE TERTIAIRE (FRENCH) see tert-BUTYL ALCOHOL
ALCOOL BUTYLIQUE SECONDAIRE (FRENCH) see sec-BUTYL ALCO-
    HOL
ALCOOL ETHYLIQUE (FRENCH) see ETHANOL
ALCOOL ETILICO (ITALIAN) see ETHANOL
ALCOOL ISOAMYLIQUE (FRENCH) see ISOAMYL ALCOHOL
ALCOOL ISOBUTYLIQUE (FRENCH) see ISOBUTYL ALCOHOL
ALCOOL ISOPROPILICO (ITALIAN) see ISOPROPYL ALCOHOL
ALCOOL ISOPROPYLIQUE (FRENCH) see ISOPROPYL ALCOHOL
ALCOOL METHYL AMYLIQUE (FRENCH) see ISOBUTYLMETHYLCAR-
    BINOL
ALCOOL METHYLIQUE (FRENCH) see METHANOL
ALCOOL METILICO (ITALIAN) see METHANOL
ALCOOL PROPILICO (ITALIAN) see PROPYL ALCOHOL
ALCOOL PROPYLIQUE (FRENCH) see PROPYL ALCOHOL
ALDACTIDE see ALDACTAZIDE
ALDACTONE see ALDACTAZIDE
ALDEHYDE ACETIQUE (FRENCH) see ACETALDEHYDE
ALDEHYDE ACRYLIQUE (FRENCH) see ACROLEIN
ALDEHYDE BUTYRIQUE (FRENCH) see N-BUTYRALDEHYDE
ALDEHYDE C-6 see 1-HEXANAL
ALDEHYDE C10 see 1-DECANAL
ALDEHYDE COLLIDINE see 5-ETHYL-alpha-PICOLINE
ALDEHYDE CROTONIQUE (FRENCH) see trans-2-BUTENAL
ALDEHYDE FORMIQUE (FRENCH) see FORMALDEHYDE
ALDEIDE ACETICA (ITALIAN) see ACETALDEHYDE
ALDEIDE ACRILICA (ITALIAN) see ACROLEIN
ALDEIDE BUTIRRICA (ITALIAN) see N-BUTYRALDEHYDE
ALDEIDE FORMICA (ITALIAN) see FORMALDEHYDE
ALDICARBE (FRENCH) see CARBANOLATE
ALDOL see ACETALDOL
ALDOXIME see ACETALDEHYDE OXIME
ALETAMINE HYDROCHLORIDE see alpha-ALLYL PHENETHYLAMINE-
    HYDROCHLORIDE
ALFANAFTILAMINA (ITALIAN) see 1-NAPHTHYLAMINE
ALFA-NAFTYLOAMINA (POLISH) see 1-NAPHTHYLAMINE
ALGAROBA see LOCUST BEAN GUM
ALGIDON see dl-METHADONE HYDROCHLORIDE
ALGOLYSIN see dl-METHADONE HYDROCHLORIDE
ALKALOID II see AJMALICINE
ALKALOIDS see ALKALOID POISONS, ALSO THEIR SALTS, LIQUID,
    NOS, OR ALKALOID POISONS, AND THEIR SALTS, SOLID, NOS
ALKALOID V see CYCLOPAMINE
ALKATHENE see POLYETHYLENE
ALKOHOL (GERMAN) see ETHANOL
ALKYL(C9-15)TOLYL METHYLTRIMETHYL AMMONIUM CHLORIDE see
    DODECYL-p-TOLYL TRIMETHYL AMMONIUM CHLORIDE

ALKYL(C$_8$H$_{17}$ to C$_{18}$H$_{37}$) DIMETHYL-3,4-DICHLOROBENZYL AMMONIUM
    CHLORIDE see ALKYL(C$_8$C$_{18}$)DIMETHYL-3,4-DICHLOROBEN-
    ZYLAMMONIUM CHLORIDE
ALKYLDIMETHYL(PHENYLMETHYL)QUATERNARY AMMONIUM CHLO-
    RIDES see ALKYL DIMETHYLBENZYL AMMONIUM CHLORIDE
(+)-ALLELRETHONYL (+)-CIS,TRANS-CHRYSANTHEMATE see AL-
    LETHRIN
D-TRANS ALLETHRIN see BIOALLETHRIN
D-ALLETHROLONE CHRYSANTHEMUMATE see trans-(+)-ALLETHRIN
(+)-ALLETHRONYL (+)-TRANS-CHRYSANTHEMUMATE see trans-
    (+)-ALLETHRIN
ALLILE (CLORURO DI) (ITALIAN) see ALLYL CHLORIDE
ALLIL-GLICIDIL-ETERE (ITALIAN) see ((2-PROPENYLOXY)METHYL)-
    OXIRANE
ALLILOWY ALKOHOL (POLISH) see ALLYL ALCOHOL
ALLIONAL see ALLONAL
ALLOBARBITONE see ALLOBARBITAL
ALLOXAN MONOHYDRATE see MESOXALYLUREA MONOHYDRATE
2-ALLOXYETHANOL (CZECH) see 2-ALLYLOXYETHANOL
ALLUMINIO(CLORURO DI) (ITALIAN) see ALUMINUM CHLORIDE
ALLUMINIO DIISOBUTIL-MONOCLORURO (ITALIAN) see CHLORO DI-
    ISOBUTYL ALUMINUM
ALLYL ALDEHYDE see ACROLEIN
ALLYLALKOHOL (GERMAN) see ALLYL ALCOHOL
ALLYLBARBITAL see ALLYLISOBUTYLBARBITURATE
ALLYLBARBITONE see ALLYLISOBUTYLBARBITURATE
ALLYLBARBITURAL see ALLOBARBITAL
5-ALLYL-1,3-BENZODIOXOLE see SAFROL
5-ALLYL-5-SEC-BUTYL-2-THIOBARBITURIC ACID see ALLYL-sec-BU-
    TYL THIOBARBITURIC ACID
ALLYLCARBAMIDE see ALLYLUREA
ALLYLCATECHOL METHYLENE ETHER see SAFROL
ALLYLCHLORID (GERMAN) see ALLYL CHLORIDE
ALLYL CHLOROCARBONATE (DOT) see ALLYL CHLOROCARBONATE
ALLYL CHLOROFORMATE see ALLYL CHLOROCARBONATE
ALLYL CINERIN see ALLETHRIN
ALLYL CYANIDE see 3-BUTENE NITRILE
N-ALLYL-7,8-DEHYDRO-4,5-EPOXY-3,6-DIHYDROXYMORPHINAN see
    ALLORPHINE
17-ALPHA-ALLYL-3-DEOXY-19-NORTESTOSTERONE see ALLYLESTRE-
    NOL
N-ALLYL-N-DESMETHYLMORPHINE see ALLORPHINE
1-ALLYL-3,4-DIMETHOXYBENZENE see 4-ALLYL-1,2-DIMETHOXY-
    BENZENE
ALLYLDIOXYBENZENE METHYLENE ETHER see SAFROL
ALLYLE (CHLORURE D') (FRENCH) see ALLYL CHLORIDE
ALLYL-2,3-EPOXYPROPYL ETHER see ((2-PROPENYLOXY)METHYL)-
    OXIRANE
17-ALPHA-ALLYL-4-ESTREN-17-BETA-OL see ALLYLESTRENOL
17-ALPHA-ALLYLESTR-4-EN-17-BETA-OL see ALLYLESTRENOL
ALLYLETHER see DIALLYL ETHER
ALLYLGLYCIDAETHER (GERMAN) see ((2-PROPENYLOXY)METHYL)-
    OXIRANE
ALLYL GLYCIDYL ETHER see ((2-PROPENYLOXY)METHYL)OXIRANE
4-ALLYLGUAIACOL see EUGENOL
ALLYL HOMOLOG OF CINERIN I see ALLETHRIN
17-ALPHA-ALLYL-17-BETA-HYDROXY-DELTA(SUP 4)-ESTREN see AL-
    LYLESTRENOL
17A-ALLYL-17B-HYDROXY-4-ESTRENE see ALLYLESTRENOL
17-ALPHA-ALLYL-17-BETA-HYDROXY-4-ESTRENE see ALLYLESTRE-
    NOL
4-ALLYL-1-HYDROXY-2-METHOXYBENZENE see EUGENOL
ALLYLIC ALCOHOL see ALLYL ALCOHOL
ALLYLIDENE DIACETATE see ACROLEIN DIACETATE
ALLYLISOBUTYLBARBITAL see ALLYLISOBUTYLBARBITURATE
5-ALLYL-5-ISOBUTYLBARBITURIC ACID see ALLYLISOBUTYLBAR-
    BITURATE
5-ALLYL-5-ISOPROPYLBARBITURATE see ALLONAL
ALLYLISOPROPYLBARBITURIC ACID see ALLONAL
5-ALLYL-5-ISOPROPYLBARBITURIC ACID see ALLONAL
ALLYLISOPROPYLMALONYLUREA see ALLONAL

ALLYL ISOSULFOCYANATE see ALLYL ISOTHIOCYANATE

3-ALLYL-4-KETO-2-METHYLCYCLOPENTENYL CHRYSANTHEMUM-
MONOCARBOXYLATE see ALLETHRIN

4-ALLYL-1-METHOXYBENZENE see p-ALLYLANISOLE

4-ALLYL-2-METHOXYPHENOL see EUGENOL

5-ALLYL-5-(1-METHYLBUTYL)BARBITURIC ACID see SECONAL

5-ALLYL-5-(1-METHYLBUTYL)MALONYLUREA see SECONAL

5-ALLYL-5-(1-METHYLBUTYL)-2-THIOBARBITURATE SODIUM see
SURITAL SODIUM

5-ALLYL-5-(1-METHYLBUTYL)-2-THIO-BARBITURIC ACID SODIUM
SALT see SURITAL SODIUM

1-ALLYL-3,4-METHYLENEDIOXYBENZENE see SAFROL

4-ALLYL-1,2-METHYLENEDIOXYBENZENE see SAFROL

5-ALLYL-1-METHYL-5-(1-METHYL-2-PENTYNYL)BARBITURIC ACID
SODIUM SALT see METHOHEXITAL SODIUM

3-ALLYL-2-METHYL-4-OXO-2-CYCLOPENTEN-1-YL CHRYSANTHE-
MATE see ALLETHRIN

DL-3-ALLYL-2-METHYL-4-OXOCYCLOPENT-2-ENYL-DL-CIS TRANS-
CHRYSANTHEMATE see ALLETHRIN

5-ALLYL-5-(1-METHYLPROPYL) BARBITURIC ACID see 5-ALLYL-5-
sec-BUTYLBARBITURIC ACID

5-ALLYL-5-(2'-METHYL-N-PROPYL) BARBITURIC ACID see ALLYLISO-
BUTYLBARBITURATE

ALLYL MONOSULFIDE see ALLYL SULFIDE

ALLYL MUSTARD OIL see ALLYL ISOTHIOCYANATE

N-ALLYLNORMORPHINE see ALLORPHINE

17-ALPHA-ALLYL-4-OESTRENE-17-BETA-OL see ALLYLESTRENOL

1-ALLYLOXY-2,3-EPOXY-PROPAAN (DUTCH) see ((2-PROPENYLOXY)-
METHYL)OXIRANE

1-ALLYLOXY-2,3-EPOXYPROPAN (GERMAN) see ((2-PROPENYLOXY)-
METHYL)OXIRANE

1-(ALLYLOXY)-2,3-EPOXYPROPANE see ((2-PROPENYLOXY)METHYL)-
OXIRANE

2-ALLYL PHENOL see o-ALLYL PHENOL

ALLYLPROPYMAL see ALLONAL

M-ALLYLPYROCATECHIN METHYLENE ETHER see SAFROL

4-ALLYLPYROCATECHOL FORMALDEHYDE ACETAL see SAFROL

ALLYLPYROCATECHOL METHYLENE ETHER see SAFROL

ALLYLRHODANID (GERMAN) see ALLYL THIOCYANATE

ALLYLSENFOEL (GERMAN) see ALLYL ISOTHIOCYANATE

ALLYL SULFOCYANIDE see ALLYL THIOCYANATE

ALLYLTHEOBROMINE see 1-ALLYLTHEOBROMINE

ALLYLTHIOCARBAMIDE see 1-ALLYL-2-THIOUREA

1-ALLYLTHIOUREA see 1-ALLYL-2-THIOUREA

N-ALLYLTHIOUREA see 1-ALLYL-2-THIOUREA

4-ALLYLVERATROLE see 4-ALLYL-1,2-DIMETHOXYBENZENE

(ALL-Z)-5,8,11,14-EICOSATETRAENOIC ACID see ARACHIDONIC ACID

ALMOCARPINE see PILOCARPINE

ALMOND ARTIFICIAL ESSENTIAL OIL see BENZALDEHYDE

ALMOND OIL EXPRESSED see ALMOND OIL

ALMOND OIL SWEET see ALMOND OIL

ALOCHLOR see 2-CHLORO-2',6'-DIETHYL-N-(METHOXY METHYL)-
ACETANILIDE

ALPINE TALC USP, BC 127 see TALC (powder)

ALUM see ALUMINUM SULFATE (2:3)

ALUMIGEL see ALUMINUM HYDROXIDE

ALUMINA see ALUMINUM OXIDE (2:3)

ALPHA-ALUMINA see ALUMINUM OXIDE (2:3)

BETA-ALUMINA see ALUMINUM OXIDE (2:3)

GAMMA-ALUMINA see ALUMINUM OXIDE (2:3)

ALUMINA FIBRE see ALUMINUM

ALUMINA HYDRATE see ALUMINUM HYDROXIDE

ALUMINA HYDRATED see ALUMINUM HYDROXIDE

ALUMINA TRIHYDRATE see ALUMINUM HYDROXIDE

ALPHA-ALUMINA TRIHYDRATE see ALUMINUM HYDROXIDE

ALUMINIC ACID see ALUMINUM HYDROXIDE

ALUMINIUM ALLOY, Al,Be see BERYLLIUM ALUMINUM ALLOY

ALUMINIUMCHLORID (GERMAN) see ALUMINUM CHLORIDE

ALUMINIUM FLAKE see ALUMINUM

ALUMINIUM FLUORURE (FRENCH) see ALUMINUM FLUORIDE

ALUMINIUM FOSFIDE (DUTCH) see ALUMINUM PHOSPHIDE

ALUMINIUM HYDROXIDE see ALUMINUM HYDROXIDE

ALUMINIUM OXIDE see ALUMINUM OXIDE (2:3)

ALUMINIUM PHOSPHIDE see ALUMINUM PHOSPHIDE

ALUMINUM BERYLLIUM ALLOY see BERYLLIUM ALUMINUM ALLOY

ALUMINUM BROMIDE (ANHYDROUS) see ALUMINUM BROMIDE

ALUMINUM CHLORHYDRATE see ALUMINUM CHLORIDE HYDROXIDE

ALUMINUM CHLORHYDROL see ALUMINUM CHLORIDE HYDROXIDE

ALUMINUM CHLORHYDROXIDE see ALUMINUM CHLORIDE HYDROX-
IDE

ALUMINUM CHLORIDE (1:3) see ALUMINUM CHLORIDE

ALUMINUM(III) CHLORIDE,HEXAHYDRATE see ALUMINUM CHLORIDE
HEXAHYDRATE

ALUMINUM CHLOROHYDROXIDE see ALUMINUM CHLORIDE HYDROX-
IDE

ALUMINUM DEHYDRATED see ALUMINUM

ALUMINUM HYDRATE see ALUMINUM HYDROXIDE

ALUMINUM(III) HYDROXIDE see ALUMINUM HYDROXIDE

ALUMINUM HYDROXIDE CHLORIDE see ALUMINUM CHLORIDE HY-
DROXIDE

ALUMINUM HYDROXIDE GEL see ALUMINUM HYDROXIDE

ALUMINUM HYDROXYCHLORIDE see ALUMINUM CHLORIDE HYDROX-
IDE

ALUMINUM, METALLIC, POWDER (DOT) see ALUMINUM

ALUMINUM MONOPHOSPHIDE see ALUMINUM PHOSPHIDE

ALUMINUM NITRATE (DOT) see ALUMINUM (III) NITRATE (1:3)

ALUMINUM OXIDE see ALUMINUM OXIDE (2:3)

ALUMINUM OXIDE-3H2O see ALUMINUM HYDROXIDE

ALPHA-ALUMINUM OXIDE see ALUMINUM OXIDE (2:3)

BETA-ALUMINUM OXIDE see ALUMINUM OXIDE (2:3)

GAMMA-ALUMINUM OXIDE see ALUMINUM OXIDE (2:3)

ALUMINUM OXIDE HYDRATE see ALUMINUM HYDROXIDE

ALUMINUM OXIDE TRIHYDRATE see ALUMINUM HYDROXIDE

ALUMINUM PHOSPHIDE (DOT) see ALUMINUM PHOSPHIDE

ALUMINUM POWDER see ALUMINUM

ALUMINUM SESQUIOXIDE see ALUMINUM OXIDE (2:3)

ALUMINUM SULFATE (2:3) see ALUMINUM SULFATE (2:3)

ALUMINUM TETRAHYDROBORATE see ALUMINUM BOROHYDRIDE

ALUMINUM TRIBROMIDE see ALUMINUM BROMIDE

ALUMINUM TRICHLORIDE see ALUMINUM CHLORIDE

ALUMINUM TRICHLORIDEHEXAHYDRATE see ALUMINUM CHLORIDE
HEXAHYDRATE

ALUMINUM TRIFLUORIDE see ALUMINUM FLUORIDE

ALUMINUM TRIHYDRAT see ALUMINUM HYDROXIDE

ALUMINUM TRIHYDRIDE see ALUMINUM HYDRIDE

ALPHA-ALUMINUM TRIHYDRIDE see ALUMINUM HYDRIDE

ALUMINUM TRIHYDROXIDE see ALUMINUM HYDROXIDE

ALUMINUM TRINITRATE see ALUMINUM (III) NITRATE (1:3)

ALUMINUM TRISULFATE see ALUMINUM SULFATE (2:3)

ALYPINE see ALYPIN

AMANIL SKY BLUE 6B see C.I. DIRECT BLUE 1, TETRASODIUM SALT

AMANTADINE HYDROCHLORIDE see 1-ADAMANTANAMINE HYDRO-
CHLORIDE

AMARANTH see FOOD RED 2

AMARANTHE USP (BIOLOGICAL STAIN) see FOOD RED 2

AMARSAN see ACETPHENARSINE

AMBER see AMBERGRIS TINCTURE

AMBRA see AMBERGRIS TINCTURE

AMEBICIDE see 1-EMETINE DIHYDROCHLORIDE

AMEISENSAEURE (GERMAN) see FORMIC ACID

AMERICAN CYANAMID-43073 see N-METHYL-O,O-DIMETHYLTHIOLO-
PHOSPHORYL-5-THIA-3-METHYL-2-VALERAMIDE

AMERICAN PENICILLIN see BENZYL PENICILLINIC ACID SODIUM SALT

AMETHOPTERIN see METHOTREXATE

AMETHYST see SILICA (CRYSTALLINE)

AMETHYST see SILICA, CRYSTALLINE-QUARTZ

AMFETAMINA see DL-AMPHETAMINE SULFATE

AMFETAMINE see DL-AMPHETAMINE SULFATE

AMIANTHUS see ASBESTOS

4-AMIDINO-1-(NITROSAMINOAMIDINO)-1-TETRAZENE see TETRAZENE

AMID KYSELINY STAVELOVE (CZECH) see OXAMIDE

O-AMIDOAZOTOLUOL (GERMAN) see 2-AMINO-5-AZOTOLUENE

ORTHO-AMIDOBENZOIC ACID see ANTHRANILIC ACID

AMIDOCYANOGEN see CYANAMIDE

AMIDONE see METHADONE

AMIDONE HYDROCHLORIDE see METHADONE HYDROCHLORIDE

AMIDOUREA HYDROCHLORIDE see SEMICARBAZIDE HYDROCHLORIDE

AMINICOTIN see NIACINAMIDE

2-AMINO-4-ACETAMINIFENETOL (CZECH) see 3-AMINO-4-ETHOXY-
ACETANILIDE

3-AMINOACETANILID (CZECH) see 3'-AMINOACETANILIDE

M-AMINOACETANILIDE see 3'-AMINOACETANILIDE

AMINOACETONITRILE BISULFATE see AMINOACETONITRILE SULFATE .

M-AMINOACETOPHENONE see 3'-AMINOACETOPHENONE

P-AMINO ACETOPHENONE see p-AMINO ACETOPHENONE

4'-AMINOACETOPHENONE see p-AMINO ACETOPHENONE

BETA-AMINOACETOPHENONE see 3'-AMINOACETOPHENONE

OMEGA-AMINOACETOPHENONE see 2-AMINOACETOPHENONE

M-AMINOACETYLBENZENE see 3'-AMINOACETOPHENONE

P-AMINOACETYLBENZENE see p-AMINO ACETOPHENONE

5-AMINOACRIDINE see 9-AMINOACRIDINE

AMINOADAMANTANE HYDROCHLORIDE see 1-ADAMANTANAMINE
HYDROCHLORIDE

1-AMINOADAMANTENE HYDROCHLORIDE see 1-ADAMANTANAMINE
HYDROCHLORIDE

2-AMINOAETHANOL (GERMAN) see MONOETHANOLAMINE

BETA-AMINOAETHYL-ISOTHIURONIUM DIHYDROBROMID(GERMAN)
see 2-beta-AMINOETHYLISOTHIOUREA

BETA-AMINOAETHYL-MORPHOLIN (GERMAN) see N-AMINOETHYL-
MORPHOLINE

2-AMINOANILINE see o-PHENYLENEDIAMINE

3-AMINOANILINE see m-PHENYLENEDIAMINE

4-AMINOANILINE see p-PHENYLENEDIAMINE

P-AMINOANILINE see p-PHENYLENEDIAMINE

3-AMINOANILINE DIHYDROCHLORIDE see m-PHENYLENEDIAMINE
HYDROCHLORIDE

M-AMINOANILINE DIHYDROCHLORIDE see m-PHENYLENEDIAMINE
HYDROCHLORIDE

4-AMINOANISOLE see p-ANISIDINE

P-AMINOANISOLE see p-ANISIDINE

2-AMINOANTHRACENE see 2-ANTHRACENAMINE

BETA-AMINOANTHRACENE see 2-ANTHRACENAMINE

1-AMINO-9,10-ANTHRACENEDIONE see 1-AMINOANTHRAQUINONE

2-AMINO-9,10-ANTHRACENEDIONE see 2-AMINOANTHRAQUINONE

1-AMINOANTHRACHINON (CZECH) see 1-AMINOANTHRAQUINONE

1-AMINO-9,10-ANTHRAQUINONE see 1-AMINOANTHRAQUINONE

ALPHA-AMINOANTHRAQUINONE see 1-AMINOANTHRAQUINONE

BETA-AMINOANTHRAQUINONE see 2-AMINOANTHRAQUINONE

4-AMINO-1-ARABINOFURANOSYL-2-OXO-1,2-DIHYDROPYRIMIDINE
see ARABINOCYTIDINE

2-AMINO-4-ARSENOSOPHENOL HYDROCHLORIDE see ARSPHENOXIDE

AMINOAZOBENZENE see p-(PHENYLAZO)ANILINE

4-AMINOAZOBENZENE see p-(PHENYLAZO)ANILINE

P-AMINOAZOBENZENE see p-(PHENYLAZO)ANILINE

4-AMINO-1,1'-AZOBENZENE see p-(PHENYLAZO)ANILINE

4-AMINOAZOBENZOL see p-(PHENYLAZO)ANILINE

P-AMINOAZOBENZOL see p-(PHENYLAZO)ANILINE

O-AMINOAZOTOLUENE see 2-AMINO-5-AZOTOLUENE

4'-AMINO-2:3'-AZOTOLUENE see 2-AMINO-5-AZOTOLUENE

4'-AMINO-2,3'-AZOTOLUENE see 2-AMINO-5-AZOTOLUENE

AMINOAZOTOLUENE (INDICATOR) see 2-AMINO-5-AZOTOLUENE

O-AMINOAZOTOLUENO (SPANISH) see 2-AMINO-5-AZOTOLUENE

O-AMINOAZOTOLUOL see 2-AMINO-5-AZOTOLUENE

AMINOBENZ see 3-AMINO-4-HYDROXYBENZOIC ACID METHYL ESTER

AMINOBENZENE see ANILINE

4-AMINOBENZENEARSONIC ACID see ARSANILIC ACID

P-AMINOBENZENEARSONIC ACID see ARSANILIC ACID

P-AMINOBENZENESULFAMIDE see SULFANILAMIDE

4-AMINOBENZENESULFONAMIDE see SULFANILAMIDE

P-AMINOBENZENESULFONAMIDE see SULFANILAMIDE

2-(P-AMINOBENZENESULFONAMIDO)-4,6-DIMETHYLPYRIMIDINE see
SULFADIMETHYLDIAZINE

N-(4-AMINOBENZENESULFONYL)-N'-BUTYLUREA see 1-BUTYL-3-SUL-
FANILYL UREA

2-AMINOBENZOIC ACID see ANTHRANILIC ACID

4-AMINOBENZOIC ACID see p-AMINOBENZOIC ACID

O-AMINOBENZOIC ACID see ANTHRANILIC ACID

GAMMA-AMINOBENZOIC ACID see p-AMINOBENZOIC ACID

ORTHO-AMINOBENZOIC ACID see ANTHRANILIC ACID

P-AMINOBENZOIC ACID-2-N-AMYLAMINOETHYL ESTER see
2,n-PENTYLAMINOETHYL-p-AMINOBENZOATE

P-AMINOBENZOIC ACID-3-(DIBUTYLAMINO)PROPYL ESTER HYDRO-
CHLORIDE see AMINOBENZOYLDIBUTYLAMINOPROPANOL
HYDROCHLORIDE

4-AMINOBENZOIC ACID DIETHYLAMINOETHYL ESTER see p-AMINO-
BENZOIC ACID-2-DIETHYLAMINOETHYL ESTER

P-AMINOBENZOIC ACID-2-DIETHYLAMINOETHYL ESTER HYDROCHLO-
RIDE see p-AMINOBENZOYLDIETHYLAMINOETHANOL HYDROCHLO-
RIDE

P-AMINOBENZOIC ACID ETHYL ESTER see ETHYL-4-AMINOBENZOATE

2-AMINOBENZOIC ACID METHYL ESTER see ANTHRANILIC ACID,
METHYL ESTER

O-AMINOBENZOIC ACID METHYL ESTER see ANTHRANILIC ACID,
METHYL ESTER

2-AMINOBENZOIC ACID-3-PHENYL-2-PROPENYL ESTER see ANTHRA-
NILIC ACID, CINNAMYL ESTER

P-AMINOBENZOIC DIETHYLAMINOETHYLAMIDE see p-AMINO-n-(2-
DIETHYLAMINOETHYL)BENZAMIDE

6-(4'-AMINOBENZOL-SULFONAMIDO)-2,4-DIMETHYLPYRIMIDIN (GER-
MAN) see SULFADIMETHYLDIAZINE

(P-AMINOBENZOLSULFONYL)-2-AMINO-4,6-DIMETHYLPYRIMIDIN
(GERMAN) see SULFADIMETHYLDIAZINE

2-AMINOBENZONITRILE see ANTHRANILONITRILE

O-AMINOBENZONITRILE see ANTHRANILONITRILE

3-AMINOBENZOTRIFLUORIDE see m-AMINOBENZAL FLUORIDE

M-AMINOBENZOTRIFLUORIDE see m-AMINOBENZAL FLUORIDE

3-(P-AMINOBENZOXY)-1-DI-N-BUTYLAMINOPROPANE see BUTACAINE

3-(P-AMINOBENZOXY)-1-DI-N-BUTYLAMINOPROPANE SULFATE see
BUTACAINE SULFATE

P-AMINOBENZOYLDIBUTYLAMINOPROPANOL see BUTACAINE

P-AMINOBENZOYLDIBUTYLAMINOPROPANOL SULFATE see BUTA-
CAINE SULFATE

P-AMINOBENZOYLDIETHYLAMINOETHANOL see p-AMINOBENZOIC
ACID-2-DIETHYLAMINOETHYL ESTER

O-AMINOBENZOYL DI(ISOPROPYLAMINO)ETHANOL HYDROCHLORIDE
see ISOCAINE

4-(4-AMINOBENZYL)ANILINE see 4,4'-METHYLENEDIANILINE

2-AMINOBIPHENYL see 2-BIPHENYLAMINE

4-AMINOBIPHENYL see 4-BIPHENYLAMINE

O-AMINOBIPHENYL see 2-BIPHENYLAMINE

p-AMINOBIPHENYL see 4-BIPHENYLAMINE

5-AMINO-1-BIS(DIMETHYLAMIDO)PHOSPHORYL-3-PHENYL-1,2,4-TRI-
AZOLE see 5-AMINO-1-BIS(DIMETHYLAMIDE)PHOSPHORYL-3-
PHENYL-1,2,4-TRIAZOLE

5-AMINO-1-(BIS(DIMETHYLAMINO)PHOSPHINYL)-3-PHENYL-1,2,4-
TRIAZOLE see 5-AMINO-1-BIS(DIMETHYLAMIDE)PHOSPHORYL-3-
PHENYL-1,2,4-TRIAZOLE

1-AMINO-BUTAAN (DUTCH) see N-BUTYLAMINE

1-AMINOBUTAN (GERMAN) see N-BUTYLAMINE

1-AMINOBUTANE see N-BUTYLAMINE

2-AMINOBUTANE see sec-BUTYLAMINE

2-AMINO-1-BUTANOL see 2-AMINOBUTAN-1-OL

OMEGA-AMINOCAPROIC ACID see 6-AMINOCAPROIC ACID

EPSILON-AMINOCAPROIC ACID see 6-AMINOCAPROIC ACID

AMINOCAPROIC LACTAM see HEXAHYDRO-2H-AZEPIN-2-ONE

1-AMINO-2-CARBOXYBENZENE see ANTHRANILIC ACID

1-AMINO-4-CARBOXYBENZENE see p-AMINOBENZOIC ACID

4-AMINO-5-CHLORO-N-(2-(DIETHYLAMINO) ETHYL)-2-METHOXYBEN-
ZAMIDE see 4-AMINO-5-CHLORO-N-(2-(DIETHYLAMINO)ETHYL)-
n-ANISAMIDE

5-AMINO-4-CHLORO-2,3-DIHYDRO- 3-OXO-2-PHENYLPYRIDAZINE see 1-PHENYL-4-AMINO-5-CHLORPYRIDAZ-6-ONE

1-AMINO-3-CHLORO-6-METHYLBENZENE see 2-AMINO-4-CHLORO-TOLUENE

5-AMINO-4-CHLORO-2-PHENYL-3(2H)-PYRIDAZINONE see 1-PHENYL-4-AMINO-5-CHLORPYRIDAZ-6-ONE

3-AMINO-p-CRESOL METHYL ETHER see 5-METHYL-o-ANISIDINE

AMINOCYCLOHEXANE see CYCLOHEXYLAMINE

1-AMINO-1-CYCLOPENTANECARBOXYLIC ACID see 1-AMINOCYCLO-PENTANE-1-CARBOXYLIC ACID

1-AMINODECANE see 1-DECANEAMINE

4-AMINO-4-DEOXY-N(SUP 10)-METHYLPTEROYLGLUTAMATE see METHOTREXATE

4-AMINO-4-DEOXY-N(SUP 10)-METHYLPTEROYLGLUTAMIC ACID see METHOTREXATE

4-AMINO-4-DEOXYPTEROYLGLUTAMATE see AMINOPTERIDINE

3-AMINODIBENZOFURAN see 2-DIBENZOFURANAMINE

2-AMINO-4-DICHLOROARSINOPHENOL HYDROCHLORIDE see DICHLO-ROPHENARSINE HYDROCHLORIDE

4-AMINODIFENIL (SPANISH) see 4-BIPHENYLAMINE

6-AMINO-1,2-DIHYDRO-1-HYDROXY-2-IMINO-4-PIPERIDINOPYRIMI-DINE see 2,4-DIAMINO-6-PIPERIDINOPYRIMIDINE-3-OXIDE

1-2-AMINO-1-(3,4-DIHYDROXYPHENYL)ETHANOL see L-NOREPINEPH-RINE

2-AMINO-1-(2,5-DIMETHOXYPHENYL)-1-PROPANOL HYDROCHLORIDE see METHOXAMINE HYDROCHLORIDE

4-AMINO-2',3-DIMETHYLAZOBENZENE see 2-AMINO-5-AZOTOLUENE

4'-AMINO-2,3'-DIMETHYLAZOBENZENE see 2-AMINO-5-AZOTOLUENE

AMINODIMETHYLBENZENE see XYLIDINE

1-AMINO-2,5-DIMETHYLBENZENE see 2,5-XYLIDINE

3-AMINO-1,4-DIMETHYLBENZENE see 2,5-XYLIDINE

1-AMINO-2,4-DIMETHYLBENZENE HYDROCHLORIDE see 2,4-XYLI-DINE HYDROCHLORIDE

1-AMINO-2,5-DIMETHYLBENZENE HYDROCHLORIDE see 2,5-XYLI-DINE HYDROCHLORIDE

3-AMINO-1,4-DIMETHYLBENZENE HYDROCHLORIDE see 2,5-XYLI-DINE HYDROCHLORIDE

4-AMINO-1,3-DIMETHYLBENZENE HYDROCHLORIDE see 2,4-XYLI-DINE HYDROCHLORIDE

5-AMINO-1,4-DIMETHYLBENZENE HYDROCHLORIDE see 2,5-XYLI-DINE HYDROCHLORIDE

2-AMINODIPHENYL see 2-BIPHENYLAMINE

4-AMINODIPHENYL see 4-BIPHENYLAMINE

O-AMINODIPHENYL see 2-BIPHENYLAMINE

P-AMINODIPHENYL see 4-BIPHENYLAMINE

2-AMINODIPHENYLENE OXIDE see 2-DIBENZOFURANAMINE

P-AMINODIPHENYLIMIDE see p-(PHENYLAZO)ANILINE

2-AMINOETANOLO (ITALIAN) see MONOETHANOLAMINE

AMINOETHANE see ETHANAMINE

1-AMINO ETHANOL see ALDEHYDE AMMONIA

2-AMINOETHANOL see MONOETHANOLAMINE

ALPHA-AMINO ETHANOL see ALDEHYDE AMMONIA

2-(2-AMINOETHOXY)ETHANOL see 2-AMINOETHOXYETHANOL

BETA-AMINOETHYL ALCOHOL see MONOETHANOLAMINE

O-AMINOETHYLBENZENE see 2-ETHYL ANILINE

1-AMINO-4-ETHYLBENZENE see 4-ETHYL ANILINE

BETA-AMINOETHYLBENZENE see beta-PHENETHYLAMINE

3-(2-AMINOETHYL)-1-BENZYL-5-METHOXY-2-METHYLINDOLE HY-DROCHLORIDE see 1-BENZYL-2,5-DIMETHYL SEROTONIN HYDRO-CHLORIDE

3-AMINO-N-ETHYLCARBAZOLE see 3-AMINO-9-ETHYLCARBAZOLE

ALPHA-(1-AMINOETHYL)-2,5-DIMETHOXYBENZYL ALCOHOL HYDRO-CHLORIDE see METHOXAMINE HYDROCHLORIDE

AMINOETHYLENE see ETHYLENIMINE

AMINOETHYLETHANDIAMINE see DIETHYLENETRIAMINE

1-ALPHA-(1-AMINOETHYL)-M-HYDROXYBENZYL ALCOHOL see HY-DROXYNOREPHEDRINE

S-2-AMINOETHYLISOTHIOURONIUM BROMIDE HYDROBROMIDE see 2-beta-AMINOETHYLISOTHIOUREA

2-(BETA-AMINOETHYL)ISOTHIOURONIUM BROMIDE HYDROBROMIDE see 2-beta-AMINOETHYLISOTHIOUREA

2-AMINOETHYLISOTHIOURONIUM DIBROMIDE see 2-beta-AMINO-ETHYLISOTHIOUREA

2-AMINOETHYLISOTHIURONIUM BROMIDE HYDROBROMIDE see 2-beta-AMINOETHYLISOTHIOUREA

S-AMINOETHYLISOTHIURONIUM BROMIDE HYDROBROMIDE see 2-beta-AMINOETHYLISOTHIOUREA

S-(2-AMINOETHYL)ISOTHIURONIUM BROMIDE HYDROBROMIDE see 2-beta-AMINOETHYLISOTHIOUREA

BETA-AMINOETHYLISOTHIURONIUM BROMIDE HYDROBROMIDE see 2-beta-AMINOETHYLISOTHIOUREA

BETA-AMINOETHYLISOTHIURONIUM BROMIDE HYDROBROMIDE see 2-beta-AMINOETHYLISOTHIOUREA

S-(BETA-AMINOETHYL)ISOTHIURONIUM BROMIDE HYDROBROMIDE see 2-beta-AMINOETHYLISOTHIOUREA

2-AMINOETHYLISOTHIURONIUM DIHYDROBROMIDE see 2-beta-AMINOETHYLISOTHIOUREA

2-AMINOETHYL MERCAPTAN see 2-AMINOETHANETHIOL

3-(2-AMINOETHYL)-6-METHOXYINDOLE HYDROCHLORIDE see 6-METHOXYTRYPTAMINE

1-(2-AMINOETHYL)PIPERAZINE see N-AMINOETHYLPIPERAZINE

N-(2-AMINOETHYL)PIPERAZINE see N-AMINOETHYLPIPERAZINE

N-(BETA-AMINOETHYL)PIPERAZINE see N-AMINOETHYLPIPERAZINE

S-(2-AMINOETHYL)PSEUDOTHIOUREA DIHYDROBROMIDE see 2-beta-AMINOETHYLISOTHIOUREA

4-(2-AMINOETHYL)PYROCATECHOL see DOPAMINE

2-AMINOETHYLTHIOPSEUDOUREA DIHYDROBROMIDE see 2-beta-AMINOETHYLISOTHIOUREA

2-(2-AMINOETHYL)-2-THIOPSEUDOUREA HYDROBROMIDE see 2-beta-AMINOETHYLISOTHIOUREA

5-AMINO-3-FENIL-1-BIS(-DIMETILAMINO)-FOSFORIL-1,2,4-TRIAZOLO (ITALIAN) see 5-AMINO-1-BIS(DIMETHYLAMIDE)PHOSPHORYL-3-PHENYL-1,2,4-TRIAZOLE

M-AMINOFENOL (CZECH) see m-AMINOPHENOL

P-AMINOFENOL (CZECH) see 4-AMINOPHENOL

5-AMINO-3-FENYL-1-BIS(DIMETHYL-AMINO)-FOSFORYL-1,2,4-TRIAZOOL (DUTCH) see 5-AMINO-1-BIS(DIMETHYLAMIDE)-PHOSPHORYL-3-PHENYL-1,2,4-TRIAZOLE

AMINOFLUOREN (GERMAN) see FLUOREN-2-AMINE

2-AMINOFLUORENE see FLUOREN-2-AMINE

AMINOFORMAMIDINE see GUANIDINE

L-2-AMINOGLUTARIC ACID see L-GLUTAMIC ACID

alpha-AMINOGLUTARIC ACID see L-GLUTAMIC ACID

AMINOGLYCOL see 2-AMINO-2-METHYL-1,3-PROPANEDIOL

AMINOHEXAHYDROBENZENE see CYCLOHEXYLAMINE

1-AMINOHEXANE see HEXYLAMINE

OMEGA-AMINOHEXANOIC ACID see 6-AMINOCAPROIC ACID

EPSILON-AMINOHEXANOIC ACID see 6-AMINOCAPROIC ACID

6-AMINOHEXANOIC ACID CYCLIC LACTAM see HEXAHYDRO-2H-AZE-PIN-2-ONE

6-AMINOHEXANOIC ACID HOMOPOLYMER see POLYCAPROLACTAM

2-AMINO-1-HYDROXYBENZENE see 2-AMINOPHENOL

3-AMINO-1-HYDROXYBENZENE see m-AMINOPHENOL

4-AMINO-1-HYDROXYBENZENE see 4-AMINOPHENOL

4-AMINO-2-HYDROXYBENZOIC ACID see 4-AMINOSALICYLIC ACID

5-AMINO-2-HYDROXYBENZOIC ACID see 5-AMINOSALICYLIC ACID

4-AMINO-2-HYDROXYBENZOIC ACID, 2-(DIETHYLAMINO)ETHYL ES-TER, HYDROCHLORIDE (9CI) see p-AMINOSALICYLIC ACID, 2-(DI-ETHYLAMINO)ETHYL ESTERHYDROCHLORIDE

2-AMINO-3-HYDROXY-1,7-DIHYDRO-8-METHYL-6H-PURIN-6-ONE see 3-HYDROXY-9-METHYLGUANINE

3-AMINO-4-HYDROXY-PHENARSINE HYDROCHLORIDE see ARSPHE-NOXIDE

((5-(3-AMINO-4-HYDROXYPHENYL)ARSENO)-2-HYDROXYANILINO)-METHANOL SULFOXYLATE SODIUM see NEOSALVARSAN

3-AMINO-4-HYDROXYPHENYLARSINE OXIDE HYDROCHLORIDE see ARSPHENOXIDE

3-AMINO-4-HYDROXYPHENYL ARSINOXIDE HYDROCHLORIDE see ARSPHENOXIDE

3-AMINO-4-HYDROXYPHENYL DICHLORARSINE HYDROCHLORIDE see DICHLOROPHENARSINE HYDROCHLORIDE

2-AMINO-PROPANO (ITALIAN) see ISOPROPYLAMINE

1-AMINO-2-PROPANOL see 1-AMINOPROPAN-2-OL

3-AMINOPROPENE see ALLYLAMINE

BETA-AMINOPROPIONITRILE see 3-AMINOPROPIONITRILE

AMINOPROPYLAMINO)ACRIDINE DIHYDROCHLORIDE see 4-METH-OXY-9-(3-(ETHYL-2-CHLOROETHYL)

BETA-AMINOPROPYLBENZENE see AMPHETAMINE

3-AMINOPROPYLENE see ALLYLAMINE

3-(GAMMA-AMINOPROPYL)-INDOLE HYDROCHLORIDE see 3-(gamma-AMINOPROPYL)-INDOLEHYDROCHLORIDE

N-AMINOPROPYLMORPHOLINE (DOT) see 4-AMINOPROPYLMORPHO-LINE

alpha-(1-AMINOPROPYL)PROTOCATECHUYL ALCOHOL HYDROCHLO-RIDE see ETHYL NORADRENALINE HYDROCHLORIDE

AMINOPTERIN see AMINOPTERIDINE

4-AMINOPTEROYLGLUTAMIC ACID see AMINOPTERIDINE

6-AMINOPURINE see ADENINE

6-AMINO-1H-PURINE see ADENINE

6-AMINO-3H-PURINE see ADENINE

6-AMINO-9H-PURINE see ADENINE

2-AMINOPYRIDINE see o-AMINOPYRIDINE

AMINO-2-PYRIDINE see o-AMINOPYRIDINE

AMINO-4-PYRIDINE see 4-AMINOPYRIDINE

ALPHA-AMINOPYRIDINE see o-AMINOPYRIDINE

AMINOPYRINE see DIMETHYLAMINOANTIPYRINE

AMINOREXFUMARATE see 2-AMINO-5-PHENYL-OXAZOLINE FORMATE

4-AMINO-1-BETA-D-RIBOFURANOSYL-D-TRIAZIN-2(1H)-ONE see AZACYTIDINE

AMINOSALICYLIC ACID see 4-AMINOSALICYLIC ACID

M-AMINOSALICYLIC ACID see 5-AMINOSALICYLIC ACID

P-AMINOSALICYLIC ACID see 4-AMINOSALICYLIC ACID

P-AMINOSALICYLSAEUREDIAETHYLAMINOAETHYLESTER-CHLORHY-DRAT (GERMAN) see p-AMINOSALICYLIC ACID, 2-(DIETHYL-AMINO)ETHYL ESTERHYDROCHLORIDE

P-AMINOSALICYLSAURES SALZ (GERMAN) see TRIMETON

1-AMINO-4-SULFONAPHTHALENE see 4-AMINO-1-NAPHTHALENESUL-FONIC ACID

1-AMINO-6-SULFONAPHTHALENE see 5-AMINO-2-NAPHTHALENESUL-FONIC ACID

5-AMINO-2,2,4,4-TETRAKIS(TRIFLUOROMETHYL)IMIDAZOLIDINE see 4-AMINO-2,2,5,5-TETRAKIS(TRIFLUOROMETHYL)-3-IMIDAZOLINE

AMINOTHIAZOLE see 2-AMINOTHIAZOLE

2-AMINOTHIOPHENOL see 2-AMINOBENZENETHIOL

O-AMINOTHIOPHENOL see 2-AMINOBENZENETHIOL

N-AMINOTHIOUREA see THIOSEMICARBAZIDE

3-AMINOTOLUEN (CZECH) see m-TOLUIDINE

4-AMINOTOLUEN (CZECH) see p-TOLUIDINE

2-AMINOTOLUENE see o-TOLUIDINE

3-AMINOTOLUENE see m-TOLUIDINE

4-AMINOTOLUENE see p-TOLUIDINE

M-AMINOTOLUENE see m-TOLUIDINE

O-AMINOTOLUENE see o-TOLUIDINE

P-AMINOTOLUENE see p-TOLUIDINE

2-AMINOTOLUENE HYDROCHLORIDE see o-TOLUIDINE HYDROCHLO-RIDE

4-AMINOTOLUENE HYDROCHLORIDE see p-TOLUIDINE HYDROCHLO-RIDE

O-AMINOTOLUENE HYDROCHLORIDE see o-TOLUIDINE HYDROCHLO-RIDE

3-AMINO-P-TOLUIDINE see TOLUENE-2,4-DIAMINE

5-AMINO-O-TOLUIDINE see TOLUENE-2,4-DIAMINE

3-AMINO-S-TRIAZOLE see 3-AMINOTRIAZOLE

2-AMINO-1,3,4-TRIAZOLE see 3-AMINOTRIAZOLE

3-AMINO-1,2,4-TRIAZOLE see 3-AMINOTRIAZOLE

3-AMINO-1H-1,2,4-TRIAZOLE see 3-AMINOTRIAZOLE

4-AMINO-3,5,6-TRICHLORO-2PICOLINIC ACID see 4-AMINO-3,5,6-TRICHLOROPICOLINIC ACID

4-AMINO-3,5,6-TRICHLORPICOLINSAEURE (GERMAN) see 4-AMINO-3,5,6-TRICHLOROPICOLINIC ACID

1-AMINO-2,4,6-TRIMETHYLBENZEN (CZECH) see 2,4,6-TRIMETHYL-ANILINE

11-AMINOUNDECANOIC ACID see AMINOUNDECANOIC ACID

AMINOURACIL MUSTARD see 5-(BIS(2-CHLOROETHYL)AMINO)URACIL

AMINOUREA see HYDRAZINECARBOXAMIDE

AMINOUREA HYDROCHLORIDE see SEMICARBAZIDE HYDROCHLORIDE

2-AMINO-1,4-XYLENE see 2,5-XYLIDINE

2-AMINO-1,4-XYLENE HYDROCHLORIDE see 2,5-XYLIDINE HYDRO-CHLORIDE

4-AMINO-1,3-XYLENE HYDROCHLORIDE see 2,4-XYLIDINE HYDRO-CHLORIDE

1,2,-AMINOZOPHENYLENE see 1H-BENZOTRIAZOLE

AMITRIPTYLIN (GERMAN) see ELAVIL

AMITRIPTYLINE see ELAVIL

AMITRIPTYLINE CHLORIDE see ELAVIL HYDROCHLORIDE

AMITROLE see 3-AMINOTRIAZOLE

AMITRYPTYLINE HYDROCHLORIDE see ELAVIL HYDROCHLORIDE

AMMATE see AMMONIUM SULFAMATE

AMMONIA ANHYDROUS see AMMONIA

AMMONIA AQUEOUS see AMMONIUM HYDROXIDE

AMMONIAC (FRENCH) see AMMONIA

AMMONIACA (ITALIAN) see AMMONIA

AMMONIA GAS see AMMONIA

AMMONIAK (GERMAN) see AMMONIA

AMMONIA SOLUTION (DOT) see AMMONIUM HYDROXIDE

AMMONIO (DICROMATO DI) (ITALIAN) see AMMONIUM BICHROMATE

AMMONIOFORMALDEHYDE see HEXAMETHYLENETETRAMINE

AMMONIUM ACID ARSENATE see DIAMMONIUM HYDROGEN ARSE-NATE

AMMONIUM AMIDOSULFONATE see SULFAMIC ACID, MONOAM-MONIUM SALT

AMMONIUM AMIDOSULPHATE see SULFAMIC ACID, MONOAMMONIUM SALT

AMMONIUM AMINOFORMATE see AMMONIUM CARBAMATE

AMMONIUM ARSENATE, SOLID (DOT) see DIAMMONIUM HYDROGEN ARSENATE

AMMONIUM AURINTRICARBOXYLATE see ALUMINON

AMMONIUMBICHROMAAT (DUTCH) see AMMONIUM BICHROMATE

AMMONIUM BIFLUORIDE see AMMONIUM HYDROGEN FLUORIDE

AMMONIUM BISULFIDE see AMMONIUM SULFIDE

AMMONIUM BOROFLUORIDE see AMMONIUM FLUOBORATE

AMMONIUMCARBONAT (GERMAN) see AMMONIUM CARBONATE

AMMONIUM CARBONATE see AMMONIUM BICARBONATE (1:1)

AMMONIUMCHLORID (GERMAN) see AMMONIUM CHLORIDE

AMMONIUMDICHROMAAT (DUTCH) see AMMONIUM BICHROMATE

AMMONIUMDICHROMAT (GERMAN) see AMMONIUM BICHROMATE

AMMONIUM DICHROMATE see AMMONIUM BICHROMATE

AMMONIUM DICHROMATE (VI) see AMMONIUM BICHROMATE

AMMONIUM FLUOROBORATE see AMMONIUM FLUOBORATE

AMMONIUM FLUORURE (FRENCH) see AMMONIUM FLUORIDE

AMMONIUM FLUOSILICATE see CRYPTOHALITE

AMMONIUM FLUOSILICATE see AMMONIUM SILICO FLUORIDE

AMMONIUM HEXACHLOROPALLADATE see AMMONIUM CHLOROPAL-LADATE (IV)

AMMONIUM HEXACHLOROPLATINATE(IV) see AMMONIUM CHLORO-PLATINATE

AMMONIUM HEXAFLUOROSILICATE see AMMONIUM SILICO FLUORIDE

AMMONIUM HEXAFLUOROSILICATE see CRYPTOHALITE

AMMONIUM HYDROGEN CARBONATE see AMMONIUM BICARBONATE (1:1)

AMMONIUM HYDROGEN FLUORIDE, SOLID see AMMONIUM HYDRO-GEN FLUORIDE

AMMONIUM HYDROGEN FLUORIDE SOLUTION (DOT) see AMMONIUM HYDROGEN FLUORIDE (SOLUTION)

AMMONIUM HYDROGEN SULFIDE see AMMONIUM SULFIDE

AMMONIUM HYDROSULFIDE see AMMONIUM SULFIDE

AMMONIUM HYDROSULFIDE SOLUTION (DOT) see AMMONIUM HY-DROSULFIDE (SOLUTION)

AMMONIUM MERCAPTAN see AMMONIUM SULFIDE

AMMONIUM METAVANADATE see AMMONIUM VANADATE

AMMONIUM MURIATE see AMMONIUM CHLORIDE

AMMONIUM NITRATE see AMMONIUM(I) NITRATE(1:1)

AMMONIUM NITRATE (DOT) see AMMONIUM(I) NITRATE(1:1)

AMMONIUM ORTHOPHOSPHITE see AMMONIUM PHOSPHITE

AMMONIUM PARAMOLYBDATE see AMMONIUM MOLYBDATE

AMMONIUM PERCHLORATE (DOT) see PERCHLORIC ACID, AMMONIUM
    SALT

AMMONIUM PEROXYDISULFATE see AMMONIUM PERSULFATE

AMMONIUM PHOSPHATE see AMMONIUM PHOSPHATE DIBASIC

AMMONIUM PICRATE, WET (DOT) see AMMONIUM PICRATE (WET)

AMMONIUM PICRONITRATE see AMMONIUM PICRATE

AMMONIUM PLATINIC CHLORIDE see AMMONIUM CHLOROPLATINATE

AMMONIUM RHODANIDE see AMMONIUM THIOCYANATE

AMMONIUMSALZ DER AMIDOSULFONSAURE (GERMAN) see SULFAMIC
    ACID, MONOAMMONIUM SALT

AMMONIUM SILICOFLUORIDE see CRYPTOHALITE

AMMONIUM SULFAMATE see SULFAMIC ACID, MONOAMMONIUM
    SALT

AMMONIUM SULFHYDRATE see AMMONIUM SULFIDE

AMMONIUM SULFOCYANATE see AMMONIUM THIOCYANATE

AMMONIUM SULFOCYANIDE see AMMONIUM THIOCYANATE

AMMONIUM SULPHAMATE see SULFAMIC ACID, MONOAMMONIUM
    SALT

AMMONIUM SULPHATE see AMMONIUM SULFATE (2:1)

AMMONIUM TETRAFLUOROBORATE see AMMONIUM FLUOBORATE

AMMONIUM TETRAFLUOROBORATE(1-) see AMMONIUM FLUOBORATE

AMMONIUM THIOGLYCOLATE see AMMONIUM MERCAPTOACETATE

AMMONIUM THIOGLYCOLLATE see AMMONIUM MERCAPTOACETATE

AMOBARBITAL see AMITAL

AMOEBAL see ACETPHENARSINE

AMONIAK (POLISH) see AMMONIA

AMORPHOUS FUSED SILICA see SILICA, AMORPHOUS FUSED

AMORPHOUS SILICA DUST see SILICA, AMORPHOUS FUMED

AMOSITE see ASBESTOS

AMOSITE ASBESTOS see AMOSITE (SEE ALSO ASBESTOS)

AMOTRIL see ATROMID S

AMPHEDRINE see d-BENZEDRINE SULFATE

AMPHEREX see d-BENZEDRINE SULFATE

(+)-AMPHETAMINE see D-AMPHETAMINE

DL-AMPHETAMINE see BENZEDRINE

AMPHETAMINE, 2,5-DIMETHOXY-4-METHYL-, HYDROCHLORIDE see
    2,5-DIMETHOXY-4-METHYLAMPHETAMINE HYDROCHLORIDE

DL-AMPHETAMINE PHOSPHATE see AMPHETANE PHOSPHATE

(-)-AMPHETAMINE SULFATE see l-BENZEDRINE SULFATE

(+MPHETAMINE SULFATE see d-BENZEDRINE SULFATE

(+-)-AMPHETAMINE SULFATE see DL-AMPHETAMINE SULFATE

D-AMPHETAMINE SULFATE see d-BENZEDRINE SULFATE

L-AMPHETAMINE SULFATE see l-BENZEDRINE SULFATE

AMPHIBOLE see ASBESTOS

AMPHOMORONAL see AMPHOTERICIN B

AMPHOTERICINE B see AMPHOTERICIN B

AMYGDALIC ACID see MANDELIC ACID

AMYGDALINIC ACID see MANDELIC ACID

AMYGDALONITRILE see MANDELIC ACID NITRILE

N-AMYL ACETATE see PENTYL ACETATE

AMYL ACETATE (DOT) see PENTYL ACETATE

AMYL ACETIC ESTER see PENTYL ACETATE

AMYL ACETIC ETHER see PENTYL ACETATE

AMYL ALCOHOL see PENTYL ALCOHOL

N-AMYL ALCOHOL see PENTYL ALCOHOL

N-AMYL ALCOHOL see AMYL ALCOHOL

SEC-AMYL ALCOHOL see 2-PENTANOL

TERT-AMYL ALCOHOL see t-PENTYL ALCOHOL

TERT-AMYL ALCOHOL see tert-N-AMYL ALCOHOL, REFINED

AMYL ALCOHOL, NORMAL see AMYL ALCOHOL

AMYL ALCOHOL, NORMAL see PENTYL ALCOHOL

AMYL ALDEHYDE see n-VALERALDEHYDE

N-AMYLALKOHOL (CZECH) see PENTYL ALCOHOL

N-AMYLALKOHOL (CZECH) see AMYL ALCOHOL

2-N-AMYLAMINOETHYL-P-AMINOBENZOATE see 2,n-PENTYLAMINO-
    ETHYL-p-AMINOBENZOATE

AMYLAZETAT (GERMAN) see PENTYL ACETATE

AMYLBARBITONE see AMITAL

AMYLCAINE see 2,n-PENTYLAMINOETHYL-p-AMINOBENZOATE

AMYLCARBINOL see N-HEXYL ALCOHOL

ALPHA-AMYL CINNAMIC ALDEHYDE see alpha-AMYL CINNAMALDE-
    HYDE

AMYLEINE see AMYLOCAINE

AMYLENE HYDRATE see tert-N-AMYL ALCOHOL, REFINED

AMYLENE HYDRATE see t-PENTYL ALCOHOL

AMYL ETHER see PENTYL ETHER

N-AMYL ETHER see PENTYL ETHER

AMYL FORMATE (DOT) see N-AMYL FORMATE

AMYL HYDROSULFIDE see 1-PENTANETHIOL

N-AMYL MERCAPTAN see 1-PENTANETHIOL

AMYL METHYL CARBINOL see 2-HEPTANOL

AMYL-METHYL-CETONE (FRENCH) see 2-HEPTANONE

N-AMYL METHYL KETONE see 2-HEPTANONE

AMYL NITRITE (DOT) see N-AMYL NITRITE

N-AMYLNITROSOUREA see N-PENTYLNITROSOUREA

1-AMYL-1-NITROSOUREA see N-PENTYLNITROSOUREA

AMYLOBARBITAL see AMITAL

AMYLOBARBITONE see AMITAL

AMYLOWY ALKOHOL (POLISH) see ISOAMYL ALCOHOL

O-AMYL PHENOL see 4-n-AMYLPHENOL

O-SEC AMYL PHENOL see 2-sec-AMYLPHENOL

P-SEC AMYL PHENOL see 4-sec-AMYLPHENOL

P-TERT-AMYLPHENOL see 4-tert-AMYLPHENOL

ALPHA-AMYL-BETA-PHENYLACROLEIN see alpha-AMYL CINNAMAL-
    DEHYDE

AMYL SULFHYDRATE see 1-PENTANETHIOL

AMYL THIOALCOHOL see 1-PENTANETHIOL

AMYLVINYLCARBINOL see 1-OCTEN-3-OL

AMYTAL see AMITAL

ANAESTHETIC ETHER see ETHYL ETHER

ANATRAN see ACETOPROMAZINE

ANATROPIN MIXED WITH MESTRANOL (10:1) see ANAGESTONE
    ACETATE MIXED WITH MESTRANOL (10:1)

ANDROMETH see 17-METHYL TESTOSTERONE

ANDROSAN see 17-METHYL TESTOSTERONE

ANDROSAN (TABLETS) see 17-METHYL TESTOSTERONE

ANDROSTEN see 17-METHYL TESTOSTERONE

DELTA(SUP 4)-ANDROSTENE-17-BETA-PROPIONATE-3-ONE see TES-
    TOSTERONE PROPIONATE

ANDROST-4-EN-17BETA-OL-3-ONE see TESTOSTERONE

DELTA(SUP 4)-ANDROSTEN-17(BETA)-OL-3-ONE see TESTOSTERONE

ANDROST-4-EN-3-ONE, 17-HYDROXY-17-METHYL-, (17-BETA)-(9CI)
    see 17-METHYL TESTOSTERONE

ANERGAN see ACETOPROMAZINE

ANERTAN see 17-METHYL TESTOSTERONE

ANERTAN (TABLETS) see 17-METHYL TESTOSTERONE

ANESTHESIN see ETHYL-4-AMINOBENZOATE

ANESTHESOL see p-AMINOBENZOYLDIETHYLAMINOETHANOL HYDRO-
    CHLORIDE

ANESTHETIC ETHER see ETHYL ETHER

ANETHOLE see p-PROPENYL ANISOLE

ANFETAMINA see DL-AMPHETAMINE SULFATE

ANG 66 see ANGUIDIN

ANGEL TULIP see STRAMONIUM

ANGLISLITE see LEAD (II) SULFATE (1:1)

ANG.-STERANTHREN (GERMAN) see ANG-STERANTHRENE

ANHYDRIDE ACETIQUE (FRENCH) see ACETIC ANHYDRIDE

ANHYDRIDE ARSENIEUX (FRENCH) see ARSENIC TRIOXIDE

ANHYDRIDE ARSENIQUE (FRENCH) see ARSENIC PENTOXIDE

ANHYDRIDE CARBONIQUE (FRENCH) see CARBON DIOXIDE

ANHYDRIDE CHROMIQUE (FRENCH) see CHROMIUM (VI) OXIDE (1:3)

ANHYDRIDE PHTALIQUE (FRENCH) see PHTHALIC ANHYDRIDE

ANHYDRIDE VANADIQUE (FRENCH) see VANADIUM PENTOXIDE
    (DUST)

ANHYDRID KYSELINY TETRAHYDROFTALOVE (CZECH) see TETRA-
HYDROPHTHALIC ACID ANHYDRIDE
ANHYDRO-4,4'-BIS(DIETHYLAMINO)TRIPHENYLMETHANOL-2',4''-
DISULPHONIC ACID,MONOSODIUM SALT see ACID BLUE 1
3,6-ANHYDRO-D-GALACTAN see CARRAGEEN
ANHYDROGITALIN see GITOXIN
ANHYDROGLUCOCHLORAL see alpha-d-GLUCOCHLORALOSE
ANHYDROL see ETHANOL
ANHYDRONE see PERCHLORIC ACID, MAGNESIUM SALT
ANHYDRO-O-SULFAMINE BENZOIC ACID see 1,2-BENZISOTHIAZOL-
3(2H)-ONE-1,1-DIOXIDE
ANHYDROUS CHLOROBUTANOL see ACETONE CHLOROFORM
ANHYDROUS DEXTROSE see D-GLUCOSE
ANHYDROUS HYDRIODIC ACID see HYDRIODIC ACID
ANHYDROUS HYDROBROMIC ACID see HYDROBROMIC ACID
ANHYDROUS IRON OXIDE see IRON (III) OXIDE
ANHYDROUS SODIUM ARSANILATE see ARSANILIC ACID, MONO-
SODIUM SALT
ANHYDROUS SODIUM SULFITE see SODIUM SULFITE (2:1)
ANIDRIDE ACETICA (ITALIAN) see ACETIC ANHYDRIDE
ANIDRIDE CROMICA (ITALIAN) see CHROMIUM (VI) OXIDE (1:3)
ANIDRIDE FTALICA (ITALIAN) see PHTHALIC ANHYDRIDE
ANILIN (CZECH) see ANILINE
ANILINA (ITALIAN, POLISH) see ANILINE
P-ANILINEARSONIC ACID see ARSANILIC ACID
ANILINE CHLORIDE see ANILINE HYDROCHLORIDE
ANILINE GREEN see BENZALDEHYDE GREEN
ANILINE GREEN see AIZEN MALACHITE GREEN
ANILINE HYDROCHLORIDE see BENZENAMINE HYDROCHLORIDE
ANILINE OIL see ANILINE
"ANILINE SALT" see BENZENAMINE HYDROCHLORIDE
P-ANILINESULFONAMIDE see SULFANILAMIDE
ANILINE-P-SULFONIC AMIDE see SULFANILAMIDE
ANILINE YELLOW see p-(PHENYLAZO)ANILINE
ANILINOBENZENE see DIPHENYLAMINE
2-ANILINOBENZOIC ACID see N-PHENYLANTHRANILIC ACID
O-ANILINOBENZOIC ACID see N-PHENYLANTHRANILIC ACID
ANILINOETHANE see N-ETHYL ANILINE
ANILINOMETHANE see METHYLANILINE
2-ANILINONAPHTHALENE see N-PHENYL-2-NAPHTHYLAMINE
4-ANILINOPHENOL see p-ANILINOPHENOL
ANIMAL OIL see BONE OIL
ANISE CAMPHOR see p-PROPENYL ANISOLE
ANISEED OIL see ANISE OIL
ANISIC ACID HYDRAZIDE see p-ANISIC ACID, HYDRAZIDE
ANISIC ALDEHYDE see p-ANISALDEHYDE
ANISIC HYDRAZIDE see p-ANISIC ACID, HYDRAZIDE
4-ANISIDINE see p-ANISIDINE
ANIS OEL (GERMAN) see ANISE OIL
ANISOPIROL see HALVISOL
ANISOYLHYDRAZINE see p-ANISIC ACID, HYDRAZIDE
P-ANISOYLHYDRAZINE see p-ANISIC ACID, HYDRAZIDE
2-(P-ANISYL)ACETIC ACID see p-METHOXYPHENYLACETIC ACID
P-ANISYLAMINE see p-ANISIDINE
P-ANISYL CHLORIDE see ANISOYL CHLORIDE
ANISYL FORMATE see p-METHOXYPHENYLACETIC ACID
(6)ANNULENE see BENZENE
ANTABUSE see BIS(DIETHYL THIO CARBAMOYL) DISULFIDE
ANTHANTHREN (GERMAN) see ANTHANTHRENE
ANTHIOLIMINE see LITHIUM ANTIMONY THIOMALATE
ANTHIOMALINE see LITHIUM ANTIMONY THIOMALATE
ANTHOPHYLLITE see ASBESTOS
ANTHOPHYLLITE see ANTHOPHYLITE (SEE ALSO ASBESTOS)
ANTHRACEN (GERMAN) see ANTHRACENE
9,10-ANTHRACENEDIONE see ANTHRAQUINONE
ANTHRACIN see ANTHRACENE
2-ANTHRACYLAMINE see 2-ANTHRACENAMINE
ANTHRADIONE see ANTHRAQUINONE
ANTHRALIN see 1,8,9-ANTHRACENETRIOL
2-ANTHRAMINE see 2-ANTHRACENAMINE

ANTHRANTHRENE see ANTHANTHRENE
9,10-ANTHRAQUINONE see ANTHRAQUINONE
ALPHA-ANTHRAQUINONYLAMINE see 1-AMINOANTHRAQUINONE
1,8,9-ANTHRATRIOL see 1,8,9-ANTHRACENETRIOL
2-ANTHRYLAMINE see 2-ANTHRACENAMINE
ANTIASTHMATIQUE see VASOTONIN
ANTIBIOTIC U 18496 see AZACYTIDINE
ANTI-BP-7,8-DIHYDRODIOL-9,10-OXIDE see anti-BENZO(a)PYRENE-
7,8-DIHYDRODIOL-9,10-OXIDE
ANTICHLOR see SODIUM THIOSULFATE, PENTAHYDRATE
ANTI-CHROMOTRICHIA FACTOR see p-AMINOBENZOIC ACID
ANTIFEBRIN see ACETANILIDE
ANTIHEMORRHAGIC VITAMIN see VITAMIN K1
ANTI-INFLAMMATORY HORMONE see CORTISOL
ANTIMOINE FLUORURE (FRENCH) see ANTIMONY (III) FLUORIDE
(1:3)
ANTIMONIAL SAFFRON see ANTIMONY PENTASULFIDE
ANTIMONIC "ACID" see ANTIMONY PENTOXIDE
ANTIMONIC CHLORIDE see ANTIMONY (V) CHLORIDE
ANTIMONIC CHLORIDE see ANTIMONY (V) CHLORIDE (SOLUTION)
ANTIMONIC OXIDE see ANTIMONY PENTOXIDE
ANTIMONIC SULFIDE see ANTIMONY PENTASULFIDE
ANTIMONIO (PENTACLORURO DI) (ITALIAN) see ANTIMONY (V)
CHLORIDE
ANTIMONIO (TRICLORURO DI) (ITALIAN) see ANTIMONY (III) CHLO-
RIDE
ANTIMONIOUS OXIDE see ANTIMONY OXIDE
ANTIMONOUS CHLORIDE see ANTIMONY (III) CHLORIDE
ANTIMONOUS FLUORIDE see ANTIMONY (III) FLUORIDE (1:3)
ANTIMONOUS SULFIDE see ANTIMONY TRISULFIDE
ANTIMONPENTACHLORID (GERMAN) see ANTIMONY (V) CHLORIDE
ANTIMONTRICHLORID (GERMAN) see ANTIMONY (III) CHLORIDE
ANTIMONWASSERSTOFFES (GERMAN) see STIBINE
ANTIMONY BLACK see ANTIMONY
ANTIMONY BROMIDE see ANTIMONY TRIBROMIDE
ANTIMONY CHLORIDE see ANTIMONY (III) CHLORIDE
ANTIMONY EMETINE IODIDE see EMETINE ANTIMONY IODIDE
ANTIMONY FLUORIDE see ANTIMONY (V) PENTAFLUORIDE
ANTIMONY GLANCE see ANTIMONY TRISULFIDE
ANTIMONY HYDRIDE see STIBINE
ANTIMONYL ANILINE TARTRATE see ANILINE ANTIMONYL TARTRATE
ANTIMONYL POTASSIUM TARTRATE see ANTIMONY POTASSIUM TAR-
TRATE
ANTIMONY ORANGE see ANTIMONY TRISULFIDE
ANTIMONY PENTACHLORIDE see ANTIMONY (V) CHLORIDE
ANTIMONY PENTACHLORIDE (DOT) see ANTIMONY (V) CHLORIDE
ANTIMONY PENTACHLORIDE SOLUTION (DOT) see ANTIMONY (V)
CHLORIDE (SOLUTION)
ANTIMONY PENTAOXIDE see ANTIMONY PENTOXIDE
ANTIMONY PERCHLORIDE see ANTIMONY (V) CHLORIDE (SOLUTION)
ANTIMONY PERCHLORIDE see ANTIMONY (V) CHLORIDE
ANTIMONY PEROXIDE see ANTIMONY OXIDE
ANTIMONY RED see ANTIMONY PENTASULFIDE
ANTIMONY REGULUS see ANTIMONY
ANTIMONY SESQUIOXIDE see ANTIMONY OXIDE
ANTIMONY SODIUM OXIDE l-(+)-TARTRATE see ANTIMONY SODIUM
TARTRATE
ANTIMONY SULFIDE see ANTIMONY TRISULFIDE
ANTIMONY SULFIDE see ANTIMONY PENTASULFIDE
ANTIMONY TELLURIDE see ANTIMONY TRITELLURIDE
ANTIMONY TRICHLORIDE, SOLID (DOT) see ANTIMONY (III) CHLO-
RIDE
ANTIMONY TRIFLUORIDE see ANTIMONY (III) FLUORIDE (1:3)
ANTIMONY TRIHYDRIDE see STIBINE
ANTIMONY TRIOXIDE see ANTIMONY OXIDE
ANTIMONY, TRIPHENYLDICHLORIDE see DICHLOROTRIPHENYLANTI-
MONY
ANTIMONY (V) FLUORIDE see ANTIMONY (V) PENTAFLUORIDE
ANTIMONY WHITE see ANTIMONY OXIDE
ANTIMOONPENTACHLORIDE (DUTCH) see ANTIMONY (V) CHLORIDE

ANTIMOONTRICHLRIDE (DUTCH) see ANTIMONY (III) CHLORIDE

ANTI-PELLAGRA VITAMIN see NICOTINIC ACID

ANTI-PICA see FLUANISONE HYDROCHLORIDE

(ANTIPYRINYLMETHYLAMINO)METHANESULFONIC ACID SODIUM
    SALT see AMINOPYRINE SODIUM SULFONATE

ANTOXYLIC ACID see ARSANILIC ACID

ANTRENYL BROMIDE see OXYPHENONIUM BROMIDE

ANTYMON (POLISH) see ANTIMONY

ANTYMONOWODOR (POLISH) see STIBINE

AOM see AZOXYMETHANE

APGA see AMINOPTERIDINE

APHRODINE see YOHIMBINE

APHROSOL see YOHIMBINE

APIGENOL see CHAMOMILE

BETA-APN see beta-AMINOPROPIONITRILE FUMARATE

APO see TEPA

APOMORFIN see APORMORPHINE

APOMORPHINE see APORMORPHINE

6A-BETA-APORPHINE-10,11-DIOL see APORMORPHINE

APROBARBITAL SODIUM see BUTALBITAL SODIUM

APROBARBITONE see ALLONAL

APROBARBITONE SODIUM see BUTALBITAL SODIUM

AQUA AMMONIA see AMMONIUM HYDROXIDE

AQUACHLORAL see CHLORAL HYDRATE

AQUA FORTIS see NITRIC ACID

AQUAMYCIN see ACETYLENEDICARBOXAMIDE

AQUA REGIA see HYDROCHLORIC ACID, MIXED WITH NITRIC ACID
    (3:1)

AQUA REGIA see NITROHYDROCHLORIC ACID (DILUTED)

AQUATAG see BENZOTHIAZIDE

1-BETA-D-ARABINOFURANOSYL-4-AMINO-2(1H)PYRIMIDINONE see
    ARABINOCYTIDINE

1-ARABINOFURANOSYLCYTOSINE see ARABINOCYTIDINE

1-BETA-ARABINOFURANOSYLCYTOSINE see ARABINOCYTIDINE

1-(BETA-D-ARABINOFURANOSYL)CYTOSINE see ARABINOCYTIDINE

ARABINOSYLADENINE see 9-beta-d-ARABINO FURANOSYL ADENINE

9-ARABINOSYLADENINE see 9-beta-d-ARABINO FURANOSYL ADENINE

BETA-D-ARABINOSYLADENINE see 9-beta-d-ARABINO FURANOSYL
    ADENINE

BETA-D-ARABINOSYLCYTOSINE see ARABINOCYTIDINE

ARACHIC ACID see EICOSANOIC ACID

ARACHIDIC ACID see EICOSANOIC ACID

ARACHIS OIL see PEANUT OIL

ARAKONIUM CHLORIDE see ALKYL DIMETHYL-3,4-DICHLOROBEN-
    ZENE AMMONIUM CHLORIDE

ARANCIO CROMO (ITALIAN) see LEAD CHROMATE, BASIC

ARCHIDONATE see ARACHIDONIC ACID

ARCTUVIN see HYDROQUINONE

ARECA CATECHU see BETEL NUT

ARECAIDINE METHYL ESTER see ARECOLINE

ARECA NUT see BETEL NUT

ARECOLINE BASE see ARECOLINE

ARGENT FLUORURE (FRENCH) see SILVER (II) FLUORIDE

ARGENTIC FLUORIDE see SILVER (II) FLUORIDE

ARGENTIUM CREDE see SILVER, COLLOIDAL

ARGENTUM see SILVER

ARGON-40 see ARGON

ARISTOLOCHIC ACID see ARISTOLOCHINE

ARMEEN 16D see 1-HEXADECANAMINE

AROCHLOR 1221 see POLYCHLORINATED BIPHENYL (AROCLOR 1221)

AROCHLOR 1242 see POLYCHLORINATED BIPHENYL (AROCLOR 1242)

AROCHLOR 1254 see POLYCHLORINATED BIPHENYL (AROCLOR 1254)

AROCHLOR 1260 see POLYCHLORINATED BIPHENYL (AROCLOR 1260)

AROCLOR see POLYCHLORINATED BIPHENYLS

AROCLOR 1232 see POLYCHLORINATED BIPHENYL (AROCLOR 1232)

AROCLOR 1242 see POLYCHLORINATED BIPHENYL (AROCLOR 1242)

AROCLOR 1248 see POLYCHLORINATED BIPHENYL (AROCLOR 1248)

AROCLOR 1254 see POLYCHLORINATED BIPHENYL (AROCLOR 1254)

AROCLOR 1260 see POLYCHLORINATED BIPHENYL (AROCLOR 1260)

AROCLOR 1262 see POLYCHLORINATED BIPHENYL (AROCLOR 1262)

AROCLOR 1268 see POLYCHLORINATED BIPHENYL (AROCLOR 1268)

AROMATIC CASTOR OIL see CASTOR OIL

AROMATIC HYDROCARBONS, MIXED see XYLENE

AROMATIC SOLVENT see BENZIN

4-ARSANILIC ACID see ARSANILIC ACID

P-ARSANILIC ACID see ARSANILIC ACID

M-ARSANILIC ACID, N-ACETYL-4-HYDROXY- see ACETPHENARSINE

ARSANILIC ACID SODIUM SALT see ARSANILIC ACID, MONOSODIUM
    SALT

ARSEN (GERMAN, POLISH) see ARSENIC

ARSENATE OF IRON, FERRIC see IRON (III) ARSENATE (1:1)

ARSENATE OF IRON, FERROUS see IRON (II) ARSENATE (3:2)

ARSENENOUS ACID, POTASSIUM SALT see POTASSIUM ARSENITE

ARSENENOUS ACID, SODIUM SALT see SODIUM ARSENITE

ARSENIATE DE CALCIUM (FRENCH) see ARSENIC ACID, CALCIUM
    SALT (2:3)

ARSENIATE DE MAGNESIUM (FRENCH) see ARSENIC ACID, MAGNE-
    SIUM SALT

ARSENIATE DE PLOMB (FRENCH) see ARSENIC ACID, LEAD SALT

ARSENIC-75 see ARSENIC

ARSENIC ACID see ARSENIC PENTOXIDE

ARSENIC ACID ANHYDRIDE see ARSENIC PENTOXIDE

ARSENIC ACID, SOLID (DOT) see O-ARSENIC ACID, HEMIHYDRATE

ARSENIC ACID, ZINC SALT see ZINC ARSENATE

ARSENICAL DIP, LIQUID (DOT) see ARSENICAL DIP

ARSENICAL FLUE DUST see ARSENICAL DUST

ARSENICALS see ARSENIC COMPOUNDS

ARSENICALS see ARSENIC

ARSENIC ANHYDRIDE see ARSENIC PENTOXIDE

ARSENIC BLACK see ARSENIC

ARSENIC BLANC (FRENCH) see ARSENIC TRIOXIDE

ARSENIC BUTTER see ARSENIC CHLORIDE

ARSENIC(III) CHLORIDE see ARSENIC CHLORIDE

ARSENIC DICHLOROETHANE see DICHLOROETHYLARSINE

ARSENIC FLUORIDE see ARSENIC TRIFLUORIDE

ARSENIC HYDRIDE see ARSINE

ARSENIC (V) OXIDE see ARSENIC PENTOXIDE

ARSENIC OXIDE see ARSENIC PENTOXIDE

ARSENIC OXIDE see ARSENIC TRIOXIDE

ARSENIC (III) OXIDE see ARSENIC TRIOXIDE

ARSENIC SESQUIOXIDE see ARSENIC TRIOXIDE

ARSENIC SESQUISULFIDE see ARSENIC SULFIDE

ARSENIC SULFIDE see ARSENIC BISULFIDE

ARSENIC SULFIDE YELLOW see ARSENIC SULFIDE

ARSENIC TERSULPHIDE see ARSENIC SULFIDE

ARSENIC TRIBROMIDE see ARSENIC(III) BROMIDE

ARSENIC TRIHYDRIDE see ARSINE

ARSENIC TRIIODIDE see ARSENIC IODIDE

ARSENIC TRISULFIDE see ARSENIC SULFIDE

ARSENIC YELLOW see ARSENIC SULFIDE

ARSENIGEN SAURE (GERMAN) see ARSENIC TRIOXIDE

ARSENIOUS ACID see ARSENIC TRIOXIDE

ARSENIOUS ACID, DISODIUM SALT see DISODIUM ARSENITE

ARSENIOUS ACID, SODIUM SALT see SODIUM ARSENITE

ARSENIOUS ACID, SODIUM SALT POLYMERS see ARSENIOUS ACID SO-
    DIUM SALT

ARSENIOUS ACID, ZINC SALT see ZINC-m-ARSENITE

ARSENIOUS AND MERCURIC IODIDE SOLUTION (DOT) see ARSENIC
    TRIIODIDE MIXED WITH MERCURIC IODIDE

ARSENIOUS CHLORIDE see ARSENIC CHLORIDE

ARSENIOUS OXIDE see ARSENIC TRIOXIDE

ARSENIOUS SULPHIDE see ARSENIC SULFIDE

ARSENIOUS TRIOXIDE see ARSENIC TRIOXIDE

ARSENITE DE POTASSIUM (FRENCH) see POTASSIUM ARSENITE

ARSENITE DE SODIUM (FRENCH) see SODIUM ARSENITE

ARSENIURETTED HYDROGEN see ARSINE

ARSENOMARCASITE see ARSENOPYRITE

ARSENOUS ACID see ARSENIC TRIOXIDE

ARSENOUS ACID ANHYDRIDE see ARSENIC TRIOXIDE

ARSENOUS ANHYDRIDE see ARSENIC TRIOXIDE

ARSENOUS BROMIDE see ARSENIC(III) BROMIDE

ARSENOUS CHLORIDE see ARSENIC CHLORIDE

ARSENOUS FLUORIDE see ARSENIC TRIFLUORIDE

ARSENOUS HYDRIDE see ARSINE

ARSENOUS IODIDE see ARSENIC IODIDE

ARSENOUS OXIDE see ARSENIC TRIOXIDE

ARSENOUS OXIDE ANHYDRIDE see ARSENIC TRIOXIDE

ARSENOUS SULFIDE see ARSENIC SULFIDE

ARSENOUS TRIBROMIDE see ARSENIC(III) BROMIDE

ARSENOUS TRICHLORIDE (9CI) see ARSENIC CHLORIDE

ARSENOUS TRIIODIDE (9CI) see ARSENIC IODIDE

ARSENOWODOR (POLISH) see ARSINE

ARSENPHENOLAMINE HYDROCHLORIDE see SALVARSAN

ARSENWASSERSTOFF (GERMAN) see ARSINE

ARSINOTRIS PIPERIDINIUM TRICHLORIDE see ARSINE-TRI-1-PIPERIDI-
NIUM CHLORIDE

ARSONIC ACID see ACETPHENARSINE

ARSONIC ACID, POTASSIUM SALT (9CI) see POTASSIUM ARSENITE

ARSONIC ACID, SODIUM SALT (9CI) see ARSENIOUS ACID SODIUM
SALT

ARSONINE see ACETPHENARSINE

4-ARSONOPHENYLGLYCINAMIDE see N-(CARBAMOYL METHYL)-
ARSANILIC ACID

P-ARSONOPHENYLUREA see N-CARBAMOYL ARSANILIC ACID

ARSPHEN see ACETPHENARSINE

ARSPHENAMINE see SALVARSAN

ARSPHENAMINE METHYLENESULFOXYLIC ACID SODIUM SALT see
NEOSALVARSAN

ARTANE HYDROCHLORIDE see BENZHEXOL HYDROCHLORIDE

ARTIC see CHLOROMETHANE

ARTIFICIAL ALMOND OIL see BENZALDEHYDE

ARTIFICIAL ANT OIL see 2-FURALDEHYDE

ARTIFICIAL BARITE see BARIUM SULFATE

ARTIFICIAL GUM see DEXTRINS

ARTIFICIAL HEAVY SPAR see BARIUM SULFATE

ARTIFICIAL SWEETENING SUBSTANZ GENDORF 450 see SACCHARIN

ASAGRAEA OFFICINALIS see VERATRINE

ASBESTINE see TALC (powder)

ASBESTOSE (GERMAN) see ASBESTOS

ASBESTOS FIBER see ASBESTOS

ASCARISIN see ASCARIDOLE

ASCARITE see ASBESTOS

ASCLEPIN see 3′-O-ACETYLCALOTROPIN

ASCORBIC ACID see L-ASCORBIC ACID

L(+)-ASCORBIC ACID see L-ASCORBIC ACID

ASCORBUTINA see L-ASCORBIC ACID

AS-DIMETHYL SULPHATE see METHYL MESYLATE

ASEPTICHROME see MERCUROCHROME

AS-METHYLPHENYLETHYLENE see alpha-METHYLSTYRENE

ASPARAGINASE see L-ASPARAGINASE

1-ASPARAGINASE X see L-ASPARAGINASE

L-ASPARAGINASI (ITALIAN) see L-ASPARAGINASE

L-ASPARAGINE AMIDOHYDROLASE see L-ASPARAGINASE

ASPHALT, PETROLEUM see PETROLEUM ASPHALT

ASPHALTUM see ASPHALT

ASPIRIN see ACETOL (2)

ASPRON see HEROIN

ASTRAFER see IRON-DEXTRIN COMPLEX

AS-TRIMETHYLBENZENE see 1,2,4-TRIMETHYLBENZENE

ASYMMETRIC DIMETHYLHYDRAZINE see 1,1-DIMETHYLHYDRAZINE

ASYMMETRIN see N-FORMYL-N-HYDROXYGLYCINE

ATACTIC POLY(VINYL CHLORIDE) see BAKELITE

ATARAX see 1-(p-CHLORO-alpha-PHENYLBENZYL)-4-(2-((2-HY-
DROXYETHOXY)ETHYL)PIPERAZINE

ATA-sb see ANTIMONY AMMONIA TRIACETIC ACID

ATERAX see 1-(p-CHLORO-alpha-PHENYLBENZYL)-4-(2-((2-HY-
DROXYETHOXY)ETHYL)PIPERAZINE

ATHYLEN (GERMAN) see ETHYLENE

ATHYLENGLYKOL (GERMAN) see ETHYLENE GLYCOL

ATLAS WHITE TITANIUM DIOXIDE see TITANIUM OXIDE

ATOXICOCAINE see p-AMINOBENZOYLDIETHYLAMINOETHANOL HY-
DROCHLORIDE

ATOXYL see ARSANILIC ACID, MONOSODIUM SALT

ATOXYLIC ACID see ARSANILIC ACID

ATRAVET see ACETOPROMAZINE

ATROPAMINE see APOATROPINE

ATROPIN (GERMAN) see ATROPINE

(−)-ATROPINE see (−)-HYOSCYAMINE

ATROPIN SIRAN (CZECH) see ATROPINE SULFATE (2:1)

ATROPINSULFAT (GERMAN) see ATROPINE SULFATE (2:1)

ATROPYLTROPEINE see APOATROPINE

ATSETOZIN see ACETOPROMAZINE

AURANTICA see MORIN

AUREOLIC ACID see MITHRAMYCIN

AUREOMYCIN see CHLORTETRACYCLINE

AURICIDINE see GOLD SODIUM THIOSULFATE DIHYDRATE

AURINTRICARBOXYLIC ACID AMMONIUM SALT see ALUMINON

AUROLIN see GOLD SODIUM THIOSULFATE DIHYDRATE

AUROMYOSE see 1-AUROTHIO-D-GLUCOPYRANOSE

AUROPEX see GOLD SODIUM THIOSULFATE DIHYDRATE

AUROPIN see GOLD SODIUM THIOSULFATE DIHYDRATE

AUROSAN see GOLD SODIUM THIOSULFATE DIHYDRATE

AUROTAN see 1-AUROTHIO-D-GLUCOPYRANOSE

AUROTHIOGLUCOSE see 1-AUROTHIO-D-GLUCOPYRANOSE

AUROTHION see GOLD SODIUM THIOSULFATE DIHYDRATE

AURUMINE see 1-AUROTHIO-D-GLUCOPYRANOSE

AUSTRALIAN DEATH ADDER SNAKE VENOM see A. ANTRACTICUS
(AUSTRALIA) VENOM

AUSTRALIAN GUM see ARABIC GUM

AUSTRIAN CINNABAR see LEAD CHROMATE, BASIC

AUTHRON see 1-AUROTHIO-D-GLUCOPYRANOSE

AVERSAN see BIS(DIETHYL THIO CARBAMOYL) DISULFIDE

AVERTIN see TRIBROMOETHANOL

AY-57,062 see ACETOPROMAZINE

9-AZAANTHRACENE see ACRIDINE

10-AZAANTHRACENE see ACRIDINE

5-AZA-10-ARSENAANTHRACENE CHLORIDE see PHENARSAZINE CHLO-
RIDE

12-AZABENZ(A)ANTHRACENE see BENZ(c)ACRIDINE

AZABENZENE see PYRIDINE

AZACITIDINE see AZACYTIDINE

2-AZACYCLOHEPTANONE see HEXAHYDRO-2H-AZEPIN-2-ONE

AZACYCLOHEXANE see PIPERIDINE

1-AZA-2,4-CYCLOPENTADIENE see PYRROLE

AZACYCLOPENTANE see PYRROLIDINE

AZACYCLOPROPANE see ETHYLENIMINE

5-AZACYTIDINE see AZACYTIDINE

5′-AZACYTIDINE see AZACYTIDINE

7-AZADIBENZ(A,J)ANTHRACENE see DIBENZ(a,j)ACRIDINE

14-AZADIBENZ(A,J)ANTHRACENE see DIBENZ(c,h)ACRIDINE

9-AZAFLUORENE see CARBAZOLE

1-AZAINDENE see INDOLE

1-AZANAPHTHALENE see QUINOLINE

2-AZANAPHTHALENE see ISOQUINOLINE

3-AZAPENTANE-1,5-DIAMINE see DIETHYLENETRIAMINE

AZBOLEN ASBESTOS see ANTHOPHYLITE (SEE ALSO ASBESTOS)

AZEPERONE see 4′-FLUORO-4-(4-(2-PYRIDYL)-1-PIPERAZINYL)BU-
TYROPHENONE

AZEPROMAZINE see ACETOPROMAZINE

AZIJNZUUR (DUTCH) see ACETIC ACID

AZIJNZUURANHYDRIDE (DUTCH) see ACETIC ANHYDRIDE

AZIMETHYLENE see DIAZOMETHANE

AZIMIDOBENZENE see 1H-BENZOTRIAZOLE

AZIMINOBENZENE see 1H-BENZOTRIAZOLE

AZINFOS-ETHYL (DUTCH) see ETHYL GUTHION

AZINPHOS-AETHYL (GERMAN) see ETHYL GUTHION

AZINPHOS ETHYL see ETHYL GUTHION

AZINPHOS-ETILE (ITALIAN) see ETHYL GUTHION

AZIRIDIN (GERMAN) see ETHYLENIMINE

AZIRIDINE see ETHYLENIMINE

AZIRIDINE-1,3,5,2,4,6-TRIAZATRIPHOSPHORINE DERIVATIVE see APHOLATE

2-(1-AZIRIDINYL)ETHANOL see 1-AZIRIDINE ETHANOL

1-(1-AZIRIDINYL)-N-(P-METHOXYPHENYL)FORMAMIDE see N-(p-METHOXYPHENYL)-1-AZIRIDINECARBOXAMIDE

1-AZIRIDINYLPHOSPHONITRILE TRIMER see APHOLATE

AZIRIDYL BENZOQUINONE see BENZOQUINONE AZIRIDINE

AZOAETHAN (GERMAN) see AZO ETHANE

AZOBENZEEN (DUTCH) see AZOBENZENE

AZOBENZENE OXIDE see AZOXYBENZENE

AZOBENZIDE see AZOBENZENE

AZOBENZOL see AZOBENZENE

AZOBISBENZENE see AZOBENZENE

1,1'-AZOBISCARBAMIDE see AZODICARBAMIDE

AZOBISCARBONAMIDE see AZODICARBAMIDE

AZOBISCARBOXAMIDE see AZODICARBAMIDE

1,1'-AZOBIS(FORMAMIDE) see AZODICARBAMIDE

AZODIBENZENE see AZOBENZENE

AZODIBENZENEAZOFUME see AZOBENZENE

AZODICARBOAMIDE see AZODICARBAMIDE

AZODICARBONAMIDE see AZODICARBAMIDE

AZODICARBOXAMIDE see AZODICARBAMIDE

AZODICARBOXYLIC ACID DIAMIDE see AZODICARBAMIDE

AZOIMIDE see HYDRAZOIC ACID

AZOLE see PYRROLE

AZOPHENYLENE see PHENAZINE

AZO RED R see FOOD RED 2

AZO RUBY S see FOOD RED 2

AZOSSIBENZENE (ITALIAN) see AZOXYBENZENE

AZOTE (FRENCH) see NITROGEN DIOXIDE

AZOTHIOPRINE see AZATHIOPRINE

AZOTIC ACID see NITRIC ACID

AZOTO (ITALIAN) see NITROGEN DIOXIDE

2:3'-AZOTOLUENE see 3,6'-DIMETHYLAZOBENZENE

AZOTOWY KWAS (POLISH) see NITRIC ACID

AZOTURE DE SODIUM (FRENCH) see SODIUM AZIDE

AZOXYAETHAN (GERMAN) see AZOXYETHANE

AZOXYBENZEEN (DUTCH) see AZOXYBENZENE

AZOXYBENZIDE see AZOXYBENZENE

AZOXYBENZOL (GERMAN) see AZOXYBENZENE

AZOXYDIBENZENE see AZOXYBENZENE

Am-NINOPTERIN see AMINOPTERIDINE

B-12 see VITAMIN B_{12} COMPLEX

B-45 see 4-AMYL-N-BENZOHYDRYLPYRIDINIUM BROMIDE

BA see BENZ(a)ANTHRACENE

BACILLOMYCIN (8CI,9CI) see BACILLUS SUBTILIS BPN

BACILLOMYCIN R see BACILLUS SUBTILIS BPN

BACILLOPEPTIDASE A see BACILLUS SUBTILIS CARLSBERG

BACILLOPEPTIDASE B see BACILLUS SUBTILIS CARLSBERG

BACTERIAL VITAMIN H1 see p-AMINOBENZOIC ACID

BA-1,2-DIHYDRODIOL see BENZ(a)ANTHRACENE-1,2-DIHYDRODIOL

BA-5,6-DIHYDRODIOL see BENZ(a)ANTHRACENE-5,6-DIHYDRODIOL

BA-8,9-DIHYDRODIOL see trans-BENZ(a)ANTHRACENE-8,9-DIHYDRO-DIOL

BA-10,11-DIHYDRODIOL see BENZ(a)ANTHRACENE-10,11-DIHYDRO-DIOL

BA 3,4-DIOL-1,2-EPOXIDE 1 see 1,2,3,4-TETRAHYDRO-3-ALPHA,4-BETA-DIHYDROXY-1-alpha,2,alpha-EPOXY BENZ(a)ANTHRA-CENE (E)

BA 3,4-DIOL-1,2-EPOXIDE-2 see 1,2,3,4-TETRAHYDRO-3-ALPHA,4-BETA-DIHYDROXY-1-alpha,2,alpha-EPOXY BENZ(a)ANTHRA-CENE (E)

BAKELITE see ACRYLONITRILE, POLYMER WITH 1,3-BUTADIENE AND STYRENE

BAKELITE DYNH see POLYETHYLENE

BAKER'S ANTIFOLANTE see TRIAZINATE

BAKER'S P AND S LIQUID AND OINTMENT see PHENOL

BANANA OIL see PENTYL ACETATE

BANANA OIL see ISOPENTYL ALCOHOL ACETATE

BAP see N-NITROSOBIS(2-ACETOXYPROPYL)AMINE

BAPN FUMARATE see beta-AMINOPROPIONITRILE FUMARATE

BARBIPHENYL see 5-ETHYL-5-PHENYLBARBITURIC ACID

BARBITAL see BARBITURATES

BARBITAL SODIUM see BARBITURATES

BARBITONE see BARBITURATES

BARBITONE SODIUM see BARBITAL SODIUM

BARIO (PEROSSIDO DI) (ITALIAN) see BARIUM PEROXIDE

BARIUM BICHROMATE see BARIUM DICHROMATE

BARIUM BINOXIDE see BARIUM PEROXIDE

BARIUM BIS(TETRAFLUOROBORATE) see BARIUM FLUOBORATE

BARIUM CHROMATE (1:1) see BARIUM CHROMATE (VI)

BARIUM CHROMATE OXIDE see BARIUM CHROMATE (VI)

BARIUM DIACETATE see BARIUM ACETATE

BARIUM DICHLORIDE see BARIUM CHLORIDE

BARIUM DICYANIDE see BARIUM CYANIDE

BARIUM DINITRATE see BARIUM(II) NITRATE (1:2)

BARIUM DIOXIDE see BARIUM PEROXIDE

BARIUM FLUOROSILICATE see BARIUM SILICOFLUORIDE

BARIUM FLUOSILICATE see BARIUM SILICOFLUORIDE

BARIUM HEXAFLUOROSILICATE see BARIUM SILICOFLUORIDE

BARIUM HEXAFLUOROSILICATE(2-) see BARIUM SILICOFLUORIDE

BARIUM MONOXIDE see BARIUM OXIDE

BARIUM PERCHLORATE see PERCHLORIC ACID, BARIUM SALT • $3H_2O$

BARIUMPEROXID (GERMAN) see BARIUM PEROXIDE

BARIUMPEROXYDE (DUTCH) see BARIUM PEROXIDE

BARIUM PROTOXIDE see BARIUM OXIDE

BARIUM SILICON FLUORIDE see BARIUM SILICOFLUORIDE

BARIUM SUPEROXIDE see BARIUM PEROXIDE

BARIUM TETRAFLUOROBORATE. see BARIUM FLUOBORATE

BARIUM ZIRCONATE see BARIUM ZIRCONIUM(IV) OXIDE

BARIUM ZIRCONIUM TRIOXIDE see BARIUM ZIRCONIUM(IV) OXIDE

BAROS CAMPHOR see BORNEOL

BARYTA see BARIUM OXIDE

BARYTA YELLOW see BARIUM CHROMATE (VI)

BARYTES see BARIUM SULFATE

BARYUM FLUORURE (FRENCH) see BARIUM FLUORIDE

BAS-3050 see 2-METHYL-N-PHENYLBENZAMIDE

BASE OIL see PETROLEUM

BASF URSOL 3GA see 2-AMINOPHENOL

BASF URSOL P BASE see 4-AMINOPHENOL

BASIC ALUMINUM CHLORATE see ALUMINUM CHLORIDE HYDROXIDE

BASIC CHROMIC SULFATE see NEOCHROMIUM

BASIC CHROMIUM SULFATE see NEOCHROMIUM

BASIC COPPER CARBONATE see COPPER (II) CARBONATE HYDROXIDE (2:1:2)

BASIC LEAD ACETATE see LEAD ACETATE, BASIC

BASIC LEAD CHROMATE see LEAD CHROMATE, BASIC

BASIC NICKEL CARBONATE see NICKEL (II) CARBONATE (1:1)

BASIC ZIRCONIUM CHLORIDE see ZIRCONIUM OXYCHLORIDE

BASOFORTINA see METHYLERGONOVINE MALEATE

BAUXITE RESIDUE see IRON (III) OXIDE

BAY 23323 see OXYDISULFOTON

BAY 37342 see O,O-DIMETHYL O-4-(METHYLTHIO)-3,5-XYLYL PHOS-PHOROTHIOATE

BAY 68138 see FENAMIPHOS

BAY 105807 see METHYLCARBAMIC ACID-O-CUMENYL ESTER

BAYER 3231 see TRISETHYLENEIMINOQUINONE

BAYER 25820 see N-HYDROXYNAPHTHALIMIDE, DIETHYL PHOS-PHATE

BAYER 39007 see BAYGON

BAY LEAF OIL see BAY OIL

BAYRITES see BARIUM SULFATE

BEET SUGAR see SUCROSE

BELT see CHLORDANE

BENACTIZINE HYDROCHLORIDE see BENZILIC ACID-beta-DIETHYL-AMINOETHYLESTER HYDROCHLORIDE

BENADRYL see BENZHYDRYL

BENGAL GELATIN see AGAR

BENGAL ISINGLASS see AGAR

BENNIE see DL-AMPHETAMINE SULFATE

BENODAINE HYDROCHLORIDE see BENZODIOXANE HYDROCHLORIDE

BENZ(E)ACEANTHRENE see 1,2-DIHYDROBENZ(e)ACEANTHRYLENE

BENZ(J)ACEANTHRYLENE see CHOLANTHRENE

BENZ(K)ACEPHENANTHRENE see ACENAPHTHANTHRACENE

3,4-BENZ(E)ACEPHENANTHRYLENE see BENZ(e)ACEPHENANTHRYLENE

BENZACONINE see PICRACONITINE

3,4-BENZACRIDINE see BENZ(c)ACRIDINE

7,8-BENZACRIDINE (FRENCH) see BENZ(c)ACRIDINE

3,4-BENZACRIDINE-9-ALDEHYDE see BENZ(c)ACRIDINE-7-CARBOXALDEHYDE

BENZ(A)ACRIDINE, 8,10-DIMETHYL- see 5,7-DIMETHYL-1,2-BENZACRIDINE

BENZAHEX see BENZENE HEXACHLORIDE (MIXED ISOMERS)

BENZAL ALCOHOL see BENZYL ALCOHOL

BENZALDEHYDE CYANOHYDRIN see MANDELIC ACID NITRILE

BENZALDEHYDE GREEN see AIZEN MALACHITE GREEN

BENZALDEHYDKYANHYDRIN (CZECH) see MANDELIC ACID NITRILE

BENZALKONIUM CHLORIDE see ALKYL DIMETHYLBENZYL AMMONIUM CHLORIDE

BENZAMIDE, 4-AMINO-N-(2-(DIETHYLAMINO)ETHYL)- (9CI) see p-AMINO-n-(2-DIETHYLAMINOETHYL)BENZAMIDE

BENZAMIDE, N-(1-METHYLETHYL)-4-((2-METHYLHYDRAZINO)METHYL)- (9CI) see 1-METHYL-2-(p-(ISOPROPYLCARBAMOYL)BENZYL)HYDRAZINE

1-BENZAMIDO-1-PHENYL-3-PIPERIDINOPROPANE HYDROCHLORIDE see DIGAMMACAINE

N-(3-BENZAMIDO-3-PHENYL)PROPYL PIPERIDINE HYDROCHLORIDE see DIGAMMACAINE

BENZAMPHETAMINE see DL-AMPHETAMINE SULFATE

BENZANTHRACENE see BENZ(a)ANTHRACENE

1,2-BENZANTHRACENE see BENZ(a)ANTHRACENE

1,2-BENZ(A)ANTHRACENE see BENZ(a)ANTHRACENE

BENZ(A)ANTHRACENE, 7-CYANO-12-METHYL- see 9-METHYL-10-CYANO-1,2-BENZANTHRACENE

BENZ(A)ANTHRACENE, 7,12-DIETHYL see 9,10-DIETHYL-1,2-BENZANTHRACENE

1,2-BENZANTHRAZEN (GERMAN) see BENZ(a)ANTHRACENE

BENZANTHRENE see BENZ(a)ANTHRACENE

1,2-BENZANTHRENE see BENZ(a)ANTHRACENE

BENZATROPINE METHANESULFONATE see TROPINE BENZOHYDRYL ETHER METHANESULFONATE

2-BENZAZINE see ISOQUINOLINE

1-BENZAZOLE see INDOLE

15,16-BENZDEHYDROCHOLANTHRENE see DIBENZ(a,j)ACEANTHRYLENE

(+-)-BENZEDRINE see BENZEDRINE

DL-BENZEDRINE see BENZEDRINE

BENZEDRYNA see DL-AMPHETAMINE SULFATE

BENZEEN (DUTCH) see BENZENE

BENZEN (POLISH) see BENZENE

BENZENAMINE see ANILINE

BENZENEACETONITRILE see PHENYLACETONITRILE

4-BENZENEAZOANILINE see p-(PHENYLAZO)ANILINE

BENZENEAZOBENZENE see AZOBENZENE

BENZENEAZOBENZENEAZO-BETA-NAPHTHOL see OIL RED

BENZENE-1-AZO-2-NAPHTHOL see 1-(PHENYLAZO)-2-NAPHTHOL

1-BENZENEAZO-2-NAPHTHOL see 1-(PHENYLAZO)-2-NAPHTHOL

BENZENE-1-AZO-2-NAPHTHOL see 1-(PHENYLAZO)-2-NAPHTHOL

BENZENEAZO-BETA-NAPHTHOL see 1-(PHENYLAZO)-2-NAPHTHOL

1-BENZENEAZO-2-NAPHTHYLAMINE see 1-(PHENYLAZO)-2-NAPHTHYLAMINE

1-BENZENE-AZO-BETA-NAPHTHYLAMINE see 1-(PHENYLAZO)-2-NAPHTHYLAMINE

P-BENZENEAZOPHENOL see 4-HYDROXYAZOBENZENE

BENZENECARBINOL see BENZYL ALCOHOL

BENZENECARBONAL see BENZALDEHYDE

BENZENECARBONYL CHLORIDE see BENZOYL CHLORIDE

BENZENECARBOPEROXOIC ACID (9CI) see PEROXYBENZOIC ACID

BENZENECARBOXYLIC ACID see BENZOIC ACID

M-BENZENEDIAMINE see m-PHENYLENEDIAMINE

P-BENZENEDIAMINE see p-PHENYLENEDIAMINE

1,2-BENZENEDIAMINE see o-PHENYLENEDIAMINE

1,3-BENZENEDIAMINE see m-PHENYLENEDIAMINE

1,4-BENZENEDIAMINE see p-PHENYLENEDIAMINE

M-BENZENEDIAMINE DIHYDROCHLORIDE see m-PHENYLENEDIAMINE HYDROCHLORIDE

1,4-BENZENEDIAMINE, 2METHYL-, SULFATE (1:1) (9CI) see 2,5-DIAMINOTOLUENE SULFATE

1,3-BENZENEDIAMINE HYDROCHLORIDE see m-PHENYLENEDIAMINE HYDROCHLORIDE

1,4-BENZENEDIAMINE, 2-METHYL-, DIHYDROCHLORIDE (9CI) see 2,5-DIAMINOTOLUENE DIHYDROCHLORIDE

O-BENZENEDICARBOXYLIC ACID see PHTHALIC ACID

BENZENE-1,2-DICARBOXYLIC ACID see PHTHALIC ACID

1,2-BENZENEDICARBOXYLIC ACID see PHTHALIC ACID

1,2-BENZENEDICARBOXYLIC ACID ANHYDRIDE see PHTHALIC ANHYDRIDE

O-BENZENEDICARBOXYLIC ACID, DIBUTYL ESTER see DI-n-BUTYL PHTHALATE

BENZENE-O-DICARBOXYLIC ACID DI-N-BUTYL ESTER see DI-n-BUTYL PHTHALATE

1,2-BENZENEDICARBOXYLIC ACID, DIETHYL ESTER see DIETHYL-O-PHTHALATE

1,2-BENZENEDICARBOXYLIC ACID, DIMETHYL ESTER see DIMETHYL PHTHALATE

1,4-BENZENE DICARBOXYLIC ACID, DIMETHYL ESTER (9CI) see DIMETHYL TEREPHTHALATE

1,4-BENZENEDICARBOXYLIC ACID, DIMETHYL ESTER (9CI) see TEREPHTHALIC ACID METHYL ESTER

O-BENZENEDICARBOXYLIC ACID, DIOCTYL ESTER see n-DIOCTYL PHTHALATE

M-BENZENEDIOL see RESORCINOL

O-BENZENEDIOL see PYROCATECHOL

P-BENZENEDIOL see HYDROQUINONE

1,2-BENZENEDIOL see PYROCATECHOL

1,3-BENZENEDIOL see RESORCINOL

1,4-BENZENEDIOL see HYDROQUINONE

1,2-BENZENEDIOL, 4-(1-HYDROXY-2-(METHYLAMINO)ETHYL)-, HYDROCHLORIDE, (R)- (9CI) see 1-ADRENALINE CHLORIDE

1,2-BENZENEDIOL, 3-METHYL- see 2,3-DIHYDROXYTOLUENE

ALPHA-BENZENEHEXACHLORIDE see BENZENE HEXACHLORIDE-alpha-isomer

BETA-BENZENEHEXACHLORIDE see trans-alpha-BENZENEHEXACHLORIDE

DELTA-BENZENEHEXACHLORIDE see delta-BHC

GAMMA-BENZENE HEXACHLORIDE see BENZENE HEXACHLORIDE-gamma isomer

BENZENE-1-ISOTHIOCYANATE see ISOTHIOCYANIC ACID, PHENYL ESTER

BENZENEMETHANOL see BENZYL ALCOHOL

BENZENEPROPANAL see HYDROCINNAMALDEHYDE

BENZENEPROPANOL CARBAMATE see 3-PHENYL-1-PROPANOL CARBAMATE

BENZENESULFONATE DE 4-CHLOROPHENYLE (FRENCH) see 4-CHLOROPHENYL BENZENESULFONATE

BENZENE SULFONECHLORIDE see BENZENESULFONYL CHLORIDE

BENZENESULFONIC ACID, ALKYL DERIV see ALKYLBENZENESULFONATE

BENZENESULFONIC (ACID) CHLORIDE see BENZENESULFONYL CHLORIDE

BENZENESULFONIC ACID, 4-CHLOROPHENYL ESTER see 4-CHLOROPHENYL BENZENESULFONATE

BENZENESULPHONIC ACID, 2,4-DICHLOROPHENYL ESTER see 2,4-DICHLOROPHENOL BENZENESULFONATE

BENZENE, 1,2,4,5-TETRACHLORO-3-METHOXY-6-NITRO- (9CI) see TETRACHLORONITROANISOLE

1,2,4-BENZENETRICARBOXYLIC ACID ANHYDRIDE see TRIMELLIC ACID ANHYDRIDE

1,2,4-BENZENETRICARBOXYLIC ACID, CYCLIC 1,2-ANHYDRIDE see TRIMELLIC ACID ANHYDRIDE

1,2,4-BENZENETRICARBOXYLIC ANHYDRIDE see TRIMELLIC ACID ANHYDRIDE

BENZENE-S-TRIOL see PHLOROGLUCINOL

1,2,3-BENZENETRIOL see PYROGALLOL

1,3,5-BENZENETRIOL see PHLOROGLUCINOL

BENZENE-1,3,5-TRIOL see PHLOROGLUCINOL

BENZENOSULFOCHLOREK (POLISH) see BENZENESULFONYL CHLORIDE

BENZENOSULPHOCHLORIDE see BENZENESULFONYL CHLORIDE

BENZENYL CHLORIDE see BENZYL TRICHLORIDE

BENZENYL FLUORIDE see BENZOTRIFLUORIDE

BENZENYL TRICHLORIDE see BENZYL TRICHLORIDE

2,3-BENZFLUORANTHENE see BENZ(e)ACEPHENANTHRYLENE

3,4-BENZFLUORANTHENE see BENZ(e)ACEPHENANTHRYLENE

10,11-BENZFLUORANTHENE see BENZO(j)FLUORANTHENE

BENZHEXOL CHLORIDE see BENZHEXOL HYDROCHLORIDE

BENZHYDRAMINE see BENZHYDRYL

BENZHYDRAMINE HYDROCHLORIDE see BENADRYL HYDROCHLORIDE

BENZHYDRIL see BENZHYDRYL

O-BENZHYDRYLDIMETHYLAMINOETHANOL see BENZHYDRYL

O-BENZHYDRYLDIMETHYLAMINOETHANOL-8-CHLOROTHEOPHYLLINATE see DRAMAMINE

2-(BENZHYDRYLOXY)-N,N-DIMETHYLETHYLAMINE see BENZHYDRYL

2-(BENZHYDRYLOXY)-N,N-DIMETHYLETHYLAMINE COMPOUND WITH 8-CHLOROTHEOPHYLLINE see DRAMAMINE

2-(BENZHYDRYLOXY)-N,N-DIMETHYLETHYLAMINEHYDROCHLORIDE see BENADRYL HYDROCHLORIDE

4-(BENZHYDRYLOXY)-1-METHYLPIPERIDINE see LYSSIPOLL

BENZIDIN (CZECH) see BENZIDINE

BENZIDINA (ITALIAN) see BENZIDINE

3,3'-BENZIDINEDICARBOXYLIC ACID see 5,5'-BIANTHRANILIC ACID

BENZIES see DL-AMPHETAMINE SULFATE

BENZILATE DU DIETHYLAMINO-ETHANOL CHLORHYDRATE (FRENCH) see BENZILIC ACID-beta-DIETHYLAMINOETHYLESTER HYDROCHLORIDE

BENZILE (CLORURO DI) (ITALIAN) see TOLYL CHLORIDE

BENZILIC ACID, ESTER WITH 1-ETHYL-3-HYDROXY-1-METHYLPIPERIDINIUM BROMIDE see PIPTAL

BENZILIC ACID-1-METHYL-3-PIPERIDYL ESTER see N-METHYL-3-PIPERIDYL BENZILATE

O-BENZIMIDAZOLE see BENZIMIDAZOLE

1H-BENZIMIDAZOLE (9CI) see BENZIMIDAZOLE

BENZIMIDAZOLE MUSTARD see BENZIMIDAZOLE METHYLENE MUSTARD

BENZIN see NAPHTHA, COAL TAR

BENZINDAMINE HYDROCHLORIDE see BENZIDAMINE HYDROCHLORIDE

BENZINE see PETROLEUM SPIRITS

BENZINOFORM see CARBON TETRACHLORIDE

3-BENZISOTHIAZOLINONE-1,1-DIOXIDE see 1,2-BENZISOTHIAZOL-3(2H)-ONE-1,1-DIOXIDE

1,2-BENZISOTHIAZOL-3(2H)-ONE, 1,1-DIOXIDE, CALCIUM SALT see CALCIUM-O-BENZOSULFIMIDE

BENZISOTRIAZOLE see 1H-BENZOTRIAZOLE

3,4-BENZOACRIDINE see BENZ(c)ACRIDINE

BENZOANTHRACENE see BENZ(a)ANTHRACENE

BENZO(A)ANTHRACENE see BENZ(a)ANTHRACENE

1,2-BENZOANTHRACENE see BENZ(a)ANTHRACENE

BENZOATE D'OESTRADIOL (FRENCH) see ESTRADIOL-3-BENZOATE

BENZOATE D'OESTRONE (FRENCH) see ESTRONE BENZOATE

1-BENZOAZO-2-NAPHTHOL see 1-(PHENYLAZO)-2-NAPHTHOL

BENZO(E)(1)BENZOTHIOPYRANO(4,3-b)INDOLE,6,13-DIHYDRO- see 6,13-DIHYDROBENZO(e)(1)BENZOTHIOPYRANO(4,3-b)INDOLE

BENZOCAINE see ETHYL-4-AMINOBENZOATE

BENZO-CHINON (GERMAN) see p-BENZOQUINONE

2,3-BENZOCHRYSENE see BENZO(b)CHRYSENE

BENZO(D,E,F)CHRYSENE see BENZO(a)PYRENE

1,3-BENZODIAZOLE see BENZIMIDAZOLE

1,2-BENZODIHYDROPYRONE see HYDROCOUMARIN

1-(1,4-BENZODIOXAN-2-YLMETHYL)PIPERIDINEHYDROCHLORIDE see BENZODIOXANE HYDROCHLORIDE

2,4,3-BENZODIOXATHIEPIN-3OXIDE see BENZOEPIN

BENZO(B)FLUORANTHENE see BENZ(e)ACEPHENANTHRYLENE

2,3-BENZOFLUORANTHENE see BENZ(e)ACEPHENANTHRYLENE

7,8-BENZOFLUORANTHENE see BENZO(j)FLUORANTHENE

BENZO(JK)FLUORENE see FLUORANTHENE

BENZOHYDROQUINONE see HYDROQUINONE

2-(BENZOHYDRYLOXY)-N,N-DIMETHYLETHYLAMINE see BENZHYDRYL

BENZOIC ACID-P-(CHLOROMERCURI) see p-CHLORO-MERCURIC BENZOIC ACID

BENZOIC ACID ESTRADIOL see ESTRADIOL-3-BENZOATE

BENZOIC ACID NITRILE see BENZONITRILE

BENZOIC ACID, 3,4,5-TRIMETHOXY-, 2-(4-(3-(2-CHLOROPHENOTHIAZIN-10-YL)PROPYL)-1-PIPERAZINYL)ETHYL ESTER, DIFUMARATE see METHOPHENAZINE DIFUMARATE

BENZOIC ALDEHYDE see BENZALDEHYDE

BENZOIC ETHER see ETHYL BENZOATE

BENZOIMIDAZOLE see BENZIMIDAZOLE

BENZOL see BENZENE

BENZOLENE see BENZENE

BENZOLINE see PETROLEUM SPIRITS

BENZOLIN HYDROCHLORIDE see NUPERCAINE HYDROCHLORIDE

BENZOLO (ITALIAN) see BENZENE

BENZOPENICILLIN see BENZYL-6-AMINOPENICILLINIC ACID

BENZO(A)PHENANTHRENE see BENZ(a)ANTHRACENE

BENZO(B)PHENANTHRENE see BENZ(a)ANTHRACENE

2,3-BENZOPHENANTHRENE see BENZ(a)ANTHRACENE

3,4-BENZOPHENANTHRENE see BENZO(c)PHENANTHRENE

BENZO(DEF)PHENANTHRENE see PYRENE

3,4-BENZOPIRENE (ITALIAN) see BENZO(a)PYRENE

2H-1-BENZOPYRAN-3-CARBOXYLIC ACID, 8-(3-(HYDROXYMERCURI)-2-METHOXYPROPYL)-2-OXO-, SODIUM SALT, COMPOUND WITH THEOPHYLLINE (1:1) see MERCUMATILIN SODIUM

2H-1-BENZOPYRAN-2-ONE see COUMARIN

1,2-BENZOPYRENE see BENZO(e)PYRENE

3,4-BENZOPYRENE see BENZO(a)PYRENE

4,5-BENZOPYRENE see BENZO(e)PYRENE

6,7-BENZOPYRENE see BENZO(a)PYRENE

BENZO(A)PYRENE-7,8-DIHYDRODIOL-9,10-EXPOXIDE (ANTI) see anti-BENZO(a)PYRENE-7,8-DIHYDRODIOL-9,10-OXIDE

BENZO(A)PYRENE-4,5-EPOXIDE see BENZO(a)PYRENE-4,5-OXIDE

BENZO(A)PYRENE-7,8-EPOXIDE see BENZO(a)PYRENE-7,8-OXIDE

BENZO(B)PYRIDINE see QUINOLINE

BENZO(C)PYRIDINE see ISOQUINOLINE

1,2-BENZOPYRONE see COUMARIN

BENZOPYRROLE see INDOLE

2,3-BENZOPYRROLE see INDOLE

BENZOQUINAMIDE see BENZOGUANAMINE

1,4-BENZOQUINE see p-BENZOQUINONE

BENZOQUINOL see HYDROQUINONE

BENZO(B)QUINOLINE see ACRIDINE

2,3-BENZOQUINOLINE see ACRIDINE

1,4-BENZOQUINONE see p-BENZOQUINONE

P-BENZOQUINONE, 2,5-DICHLORO-3,6-DIHYDROXY-, MERCURY SALT see (2,5-DICHLORO-3,6-DIHYDROXY-p-BENZOQUINOLATO)MERCURY

P-BENZOQUINONE OXIME BENZOYLHYDRAZONE see 1,4-BENZOQUINONE-N'-BENZOYLHYDRAZONE OXIME

O-BENZOSULFIMIDE see 1,2-BENZISOTHIAZOL-3(2H)-ONE-1,1-DIOXIDE

BENZOSULFONAZOLE see BENZOTHIAZOLE

BENZO-2-SULPHIMIDE see 1,2-BENZISOTHIAZOL-3(2H)-ONE-1,1-DIOXIDE

BENZO-SULPHINIDE see 1,2-BENZISOTHIAZOL-3(2H)-ONE-1,1-DIOXIDE

3,4-BENZOTETRACENE see BENZO(b)CHRYSENE

3,4-BENZOTETRAPHENE see BENZO(b)CHRYSENE

2-BENZOTHIAZOLETHIOL, ZINC SALT (2:1) see BIS(2-BENZOTHIA-ZOLYTHIO)ZINC

2-BENZOTHIAZOLYL DISULFIDE see BENZOTHIAZOLE DISULFIDE

2-BENZOTHIAZOLYLSULFENYL MORPHOLINE see 2-BENZOTHIAZOLYL-N-MORPHOLINOSULFIDE

4-(2-BENZOTHIAZOLYLTHIO)MORPHOLINE see 2-BENZOTHIAZOLYL-N-MORPHOLINOSULFIDE

BENZOTRIAZINE DERIVATIVE OF AN ETHYL DITHIOPHOSPHATE see ETHYL GUTHION

BENZOTRIAZINEDITHIOPHOSPHORIC ACID DIMETHOXY ESTER see AZINPHOS METHYL

1,2,3-BENZOTRIAZOLE see 1H-BENZOTRIAZOLE

BENZOTRICHLORIDE see BENZYL TRICHLORIDE

BENZOTROPINE MESYLATE see TROPINE BENZOHYDRYL ETHER METHANESULFONATE

BENZOTROPINE METHANESULFONATE see TROPINE BENZOHYDRYL ETHER METHANESULFONATE

3-BENZOXY-1-(2-METHYLPIPERIDINO)PROPANE see PIPEROCAINE

3-BENZOXY-1-(2-METHYLPIPERIDINO)PROPANE HYDROCHLORIDE see ISOCAINE

DL-3-BENZOXY-1-(2-METHYLPIPERIDINO)PROPANE HYDROCHLORIDE see ISOCAINE

BENZOYLACONINE see PICRACONITINE

BENZOYL ALCOHOL see BENZYL ALCOHOL

BENZOYLBENZENE see BENZOPHENONE

BENZOYL CYANIDE-O-(DIETHOXYPHOSPHINOTHIOYL)OXIME see BAY-THION

BENZOYL HYDRAZIDE see BENZHYDRAZIDE

BENZOYLHYDROGEN PEROXIDE see PEROXYBENZOIC ACID

BENZOYL HYDROPEROXIDE see PEROXYBENZOIC ACID

BENZOYL METHIDE see ACETOPHENONE

BENZOYLMETHYLECGONINE see COCAINE

BENZOYL-gamma-(2-METHYLPIPERIDINE)-PROPANOL HYDROCHLO-RIDE see ISOCAINE

BENZOYL-GAMMA-(2-METHYLPIPERIDINO)PROPANOL see PIPERO-CAINE

3-(BENZOYLOXY)ESTRA-1,3,5(10)-TRIEN-17-ONE see ESTRONE BEN-ZOATE

3-(BENZOYLOXY)-8-METHYL-8-AZABICYCLO(3.2.1)OCTANE-2-CAR-BOXYLIC ACID PROPYL ESTER, HYDROCHLORIDE (1R-(2-ENDO,3-EXO)) see NEOPSICAINE HYDROCHLORIDE

BENZOYLPEROXID (GERMAN) see BENZOYL PEROXIDE

BENZOYLPEROXYDE (DUTCH) see BENZOYL PEROXIDE

O-BENZOYL SULFIMIDE see 1,2-BENZISOTHIAZOL-3(2H)-ONE-1,1-DIOXIDE

BENZOYL SUPEROXIDE see BENZOYL PEROXIDE

alpha-BENZOYLTRIETHYLAMINE HYDROCHLORIDE see 2-DIETHYLAMINOPROPIOPHENONEHYDROCHLORIDE

2,3-BENZPHENANTHRENE see BENZ(a)ANTHRACENE

3,4-BENZPHENANTHRENE see BENZO(c)PHENANTHRENE

3,4-BENZPYREN (GERMAN) see BENZO(a)PYRENE

BENZ(A)PYRENE see BENZO(a)PYRENE

1,2-BENZPYRENE see BENZO(e)PYRENE

3,4-BENZ(A)PYRENE see BENZO(a)PYRENE

3,4-BENZPYRENE-5-ALDEHYDE see BENZO(a)PYRENE-6-CARBOXYAL-DEHYDE

BENZTROPINE MESYLATE see TROPINE BENZOHYDRYL ETHER METH-ANESULFONATE

BENZTROPINE METHANESULFONATE see TROPINE BENZOHYDRYL ETHER METHANESULFONATE

BENZYDAMINE HYDROCHLORIDE see BENZIDAMINE HYDROCHLORIDE

BENZYDYNA (POLISH) see BENZIDINE

BENZYHYDRYLCYANIDE see DIPHENYLACETONITRILE

BENZYLACETALDEHYDE see HYDROCINNAMALDEHYDE

BENZYL ALCOHOL BENZOIC ESTER see BENZOIC ACID, BENZYL ESTER

BENZYL ALCOHOL, CINNAMIC ESTER see BENZYL CINNAMATE

6-BENZYLAMINOPURINE see 2-AMINO-6-(1'-METHYL-4'-NITRO-5'-IMIDAZOLYL)MERCAPTOPURINE

2-(N-BENZYLANILINOMETHYL)-2-IMIDAZOLINE see PHENAZOLINE

BENZYL BENZENECARBOXYLATE see BENZOIC ACID, BENZYL ESTER

BENZYL BENZOATE see BENZOIC ACID, BENZYL ESTER

BENZYLBIS(BETA-CHLOROETHYL)AMINE see N,N-BIS(2-CHLORO-ETHYL)BENZYLAMINE

BENZYL CARBINOL see PHENETHYL ALCOHOL

BENZYLCARBINYL ACETATE see 2-PHENYLETHYL ACETATE

BENZYLCARBONYL CHLORIDE see BENZYL CHLOROFORMATE

BENZYLCHLORID (GERMAN) see TOLYL CHLORIDE

BENZYL CHLORIDE see TOLYL CHLORIDE

BENZYL CHLOROCARBONATE see BENZYL CHLOROFORMATE

2-(N-BENZYL-2-CHLOROETHYLAMINO)-1-PHENOXYPROPANE see NSC 37448

2-(N-BENZYL-2-CHLOROETHYLAMINO)-1-PHENOXYPROPANE HYDRO-CHLORIDE see DIBENZYLINE HYDROCHLORIDE

BENZYL(2-CHLOROETHYL)(1METHYL-2-PHENOXYETHYL)AMINE HY-DROCHLORIDE see DIBENZYLINE HYDROCHLORIDE

BENZYL(2-CHLOROETHYL)-(1-METHYL-2-PHENOXYETHYL)AMINE see NSC 37448

BENZYL CYANIDE see PHENYLACETONITRILE

3-BENZYL-3,3-ALPHA,4,5,6,6-ALPHA,9,10,12,15-DECAHYDRO-6,12,15-TRIHYDROXY-4,10,12-TRIMETHYL-5-METHYLENE-1H-CY-CLOUNDEC(D)ISOINDOLE-1,11(2H)-DIONE, 15-ACETATE see ZY-GOSPORIN A

BENZYLDIMETHYLAMINE METHIODIDE see BENZYL TRIMETHYL AM-MONIUM IODIDE

2-(BENZYL(2-DIMETHYL AMINOETHYL)AMINO)PYRIDINE see TRIPELENNAMINE

2-(N-BENZYL-N-(2-DIMETHYLAMINOETHYL)AMINO)PYRIDINE see N,N-DIMETHYL-N'-BENZYL-N'-(2-PYRIDYL)ETHYLENEDIAMINE

1-BENZYL-3-(3-(DIMETHYLAMINO)PROPOXY)-1H-INDAZOLE HYDRO-CHLORIDE see BENZIDAMINE HYDROCHLORIDE

1-BENZYL-3-GAMMA-DIMETHYLAMINOPROPOXY-1H-INDAZOLE HY-DROCHLORIDE see BENZIDAMINE HYDROCHLORIDE

BENZYLDIMETHYL(2-(2-(p-(1,1,3,3TETRAMETHYLBUTYL)-PHENOXY)ETHOXY)ETHYL)AMMONIUM CHLORIDE see BENZE-THONIUM CHLORIDE

N-BENZYL-N',N'-DIMETHYL-N-(2-PYRIDYL)ETHYLENEDIAMINE see N,N-DIMETHYL-N'-BENZYL-N'-(2-PYRIDYL)ETHYLENEDIAMINE

N-BENZYL-N',N'-DIMETHYL-N-2-PYRIDYLETHYLENE DIAMINE see TRIPE-LENNAMINE

BENZYLDIMETHYLSTEARYLAMMONIUM CHLORIDE see DIMETHYLOCTADECYLBENZYLAMMONIUM CHLORIDE

BENZYLDIMETHYL-P-(1,1,3,3-TETRAMETHYLBUTYL)PHENOXYETH-OXY-ETHYLAMMONIUM CHLORIDE see BENZETHONIUM CHLORIDE

BENZYLE (CHLORURE DE) (FRENCH) see TOLYL CHLORIDE

BENZYL ETHANOATE see BENZYL ACETATE

5-BENZYL-5-ETHYLBARBITURICACID see BENZYLBARBITAL

8-BENZYL-7-(2-(ETHYL(2-HYDROXYETHYL)AMINO)ETHYL)THEO-PHYLLINE, TRENTADIL HYDROCHLORIDE see TRENTADIL HYDROCHLORIDE

(5-BENZYL-3-FURYL) METHYL-2,2-DIMETHYL-3-(2-METHYLPRO-PENYL)-CYCLOPROPANECARBOXYLATE see BENZOFUROLINE

5-BENZYL-3-FURYL METHYL(+-)-cis,transCHRYSANTHEMATE see BENZOFUROLINE

BENZYL HYDROQUINONE see AGERITE

BENZYLIDENEPHENYLACETONITRILE see 2,3-DIPHENYLACRYLONI-TRILE

BENZYLIDYNE CHLORIDE see BENZYL TRICHLORIDE

BENZYLIDYNE FLUORIDE see BENZOTRIFLUORIDE

BENZYLIMIDAZOLINE HYDROCHLORIDE see BENZAZOLINE HYDRO-CHLORIDE

2-BENZYL-2-IMIDAZOLINE MONOHYDROCHLORIDE see BENZAZOLINE HYDROCHLORIDE

2-BENZYLISOTHIOURONIUM CHLORIDE see BENZYL ISOTHIOUREA HYDROCHLORIDE

BENZYL MERCAPTAN see alpha-TOLUENETHIOL

BENZYL METHANOATE see BENZYL FORMATE

1-BENZYL-2-METHYL-3-(2-AMINOETHYL)-5-METHOXYINDOLE HY-DROCHLORIDE see 1-BENZYL-2,5-DIMETHYL SEROTONIN HY-DROCHLORIDE

1-BENZYL-2-METHYLHYDRAZINE see 1-METHYL-2-BENZYLHYDRA-
ZINE

1-BENZYL-1-(5-METHYL-3-ISOXAZOIYLCARBONYL)HYDRAZINE see
ISOCARBOXAZID

1-BENZYL-2-(5-METHYL-3-ISOXAZOIYL-CARBONYL)HYDRAZINE see
ISOCARBOXAZID

N'-BENZYL N-METHYL-5-ISOXAZOLECARBOXYLHYDRAZIDE-3 see ISO-
CARBOXAZID

1-BENZYL-2-METHYL-5-METHOXYTRYPTAMINE HYDROCHLORIDE see
1-BENZYL-2,5-DIMETHYL SEROTONIN HYDROCHLORIDE

BENZYL MUSTARD OIL see BENZYL THIOCYANATE

BENZYL MUSTARD OIL see BENZYL ISOTHIOCYANATE

BENZYL NITRILE see PHENYLACETONITRILE

BENZYL NOR-MECHLORETHAMINE see N,N-BIS(2-CHLOROETHYL)BEN-
ZYLAMINE

BENZYL OXIDE (CZECH) see BENZYL ETHER

BENZYLOXYCARBONYL CHLORIDE see BENZYL CHLOROFORMATE

P-BENZYLOXYPHENOL see AGERITE

BENZYLPENICILLIN see BENZYL-6-AMINOPENICILLINIC ACID

BENZYLPENICILLIN G see BENZYL-6-AMINOPENICILLINIC ACID

BENZYLPENICILLINIC ACID see BENZYL-6-AMINOPENICILLINIC ACID

BENZYL GAMMA-PHENYLACRYLATE see BENZYL CINNAMATE

BENZYL PHENYLFORMATE see BENZOIC ACID, BENZYL ESTER

BENZYL-ALPHA-PYRIDYL-DIMETHYL-AETHYLENDIAMIN (GERMAN)
see N,N-DIMETHYL-N'-BENZYL-N'-(2-PYRIDYL)ETHYLENEDIAMINE

BENZYL-(ALPHA-PYRIDYL)-DIMETHYLAETHYLENDIAMIN (GERMAN)
see TRIPELENNAMINE

N-BENZYL-N-(2-PYRIDYL)-N',N'-DIMETHYL ETHYLENEDIAMINE
see N,N-DIMETHYL-N'-BENZYL-N'-(2-PYRIDYL)ETHYLENEDIAMINE

BENZYLSENFOEL (GERMAN) see BENZYL ISOTHIOCYANATE

BENZYLSTEARYLDIMETHYLAMMONIUM CHLORIDE see DIMETHYLOC-
TADECYLBENZYLAMMONIUM CHLORIDE

BENZYLTHIOL see alpha-TOLUENETHIOL

3-((BENZYLTHIO)METHYL)-6-CHLORO-1,2,4-BENZOTHIADIAZINE-7-
SULFONAMIDE-1,1-DIOXIDE see BENZOTHIAZIDE

3-BENZYLTHIOMETHYL-6-CHLORO-2H-1,2,4-BENZOTHIADIAZINE-7-
SULFONAMIDE-1,1-DIOXIDE see BENZOTHIAZIDE

3-BENZYLTHIOMETHYL-6-CHLORO-7-SULFAMOYL-1,2,4-BENZOTHIA-
DIAZINE-1,1-DIOXIDE see BENZOTHIAZIDE

3-BENZYLTHIOMETHYL-6-CHLORO-7-SULFAMYL-1,2,4-BENZOTHIA-
DIAZINE-1,1-DIOXIDE see BENZOTHIAZIDE

3-BENZYLTHIOMETHYL-6-CHLORO-7-SULFAMYL-2H-1,2,4-BENZO-
THIADIAZINE-1,1-DIOXIDE see BENZOTHIAZIDE

BENZYL THIOPSEUDOUREA HYDROCHLORIDE see BENZYL ISOTHIO-
UREA HYDROCHLORIDE

S-BENZYLTHIURONIUM CHLORIDE see BENZYL ISOTHIOUREA HYDRO-
CHLORIDE

3,4-BENZYPYRENE see BENZO(a)PYRENE

BERBERIN see BERBERINE

BERNICE see COCAINE

BERNIES see COCAINE

BERNSTEINSAURE-ANHYDRID (GERMAN) see SUCCINIC ANHYDRIDE

BERTHOLITE see CHLORINE

BERYLLIA see BERYLLIUM OXIDE

BERYLLIUM-9 see BERYLLIUM

BERYLLIUM ACETATE, BASIC see BERYLLIUM OXYACETATE

BERYLLIUM ACETATE, NORMAL see BERYLLIUM ACETATE

BERYLLIUM-ALLUMINIUM ALLOY see BERYLLIUM ALUMINUM ALLOY

BERYLLIUM ALUMINIUM SILICATE see BERYL

BERYLLIUM ALUMINOSILICATE see BERYL

BERYLLIUM CARBONATE, BASIC see BERYLLIUM CARBONATE

BERYLLIUM-COPPER ALLOY see COPPER ALLOY, Cu,Be

BERYLLIUM DICHLORIDE see BERYLLIUM CHLORIDE

BERYLLIUM DIFLUORIDE see BERYLLIUM FLUORIDE

BERYLLIUM DIHYDROXIDE see BERYLLIUM HYDROXIDE

BERYLLIUM HYDRATE see BERYLLIUM HYDROXIDE

BERYLLIUM LACTATE see LACTIC ACID, BERYLLIUM SALT

BERYLLIUM MONOXIDE see BERYLLIUM OXIDE

BERYLLIUM ORTHOSILICATE see SILICIC ACID, BERYLLIUM SALT

BERYLLIUM OXIDE ACETATE see BERYLLIUM OXYACETATE

BERYLLIUMOXIDE CARBONATE see BERYLLIUM CARBONATE

BERYLLIUM PHOSPHATE see BERYLLIUM HYDROGEN PHOSPHATE
(1:1)

BERYLLIUM SILICATE see SILICIC ACID, BERYLLIUM SALT

BERYLLIUM SILICATE HYDRATE see BERTRANDITE

BERYLLIUM SILICIC ACID see SILICIC ACID, BERYLLIUM SALT

BERYL ORE see BERYL

BETAFEN see DL-AMPHETAMINE SULFATE

BETULA OIL see METHYL SALICYLATE

BHANG see CANNABIS

BHC see BENZENE HEXACHLORIDE

ALPHA-BHC see BENZENE HEXACHLORIDE-alpha-isomer

BETA-BHC see trans-alpha-BENZENEHEXACHLORIDE

DELTA BHC see delta-BENZENE HEXACHLORIDE

GAMMA-BHC see BENZENE HEXACHLORIDE-gamma isomer

BHIMSAIM CAMPHOR see BORNEOL

BIACETYL see 2,3-BUTANEDIONE

O,P'-BIANILINE see 2,4'-BIPHENYLDIAMINE

BIANISIDINE see 3,3'-TOLIDINE

BIBENZENE see BIPHENYL

BICARBURET OF HYDROGEN see BENZENE

BICHLORIDE OF MERCURY see CORROSIVE SUBLIMATE

1,2-BICHLOROETHANE see ETHYLENE DICHLORIDE

1,2-BICHLOROETHANE see 1,2-DICHLORETHANE

BICHLORURE DE MERCURE (FRENCH) see CORROSIVE SUBLIMATE

BICHLORURE DE PROPYLENE (FRENCH) see 1,2-DICHLOROPROPANE

BICHLORURE D'ETHYLENE (FRENCH) see 1,2-DICHLORETHANE

BICHLORURE D'ETHYLENE (FRENCH) see ETHYLENE DICHLORIDE

BICHROMATE D'AMMONIUM (FRENCH) see AMMONIUM BICHROMATE

BICHROMATE OF POTASH see POTASSIUM BICHROMATE

BICHROMATE OF SODA see SODIUM DICHROMATE

BICYCLO(4.4.0)DECANE see DECAHYDRONAPHTHALENE

BICYCLO(2.2.1)HEPTAN-2-ONE, 1,7,7-TRIMETHYL-, (1R)- see CAM-
PHOR, (1R,4R)-(+)-

BICYCLO(2.2.1)HEPTENE-2-DICARBOXYLIC ACID, 2-ETHYLHEXYLI-
MIDE see N-OCTYL BICYCLOHEPTENE DICAREOXIMIDE

1-BICYCLOHEPTENYL-1-PHENYL-3-PIPERIDINO-PROPANOL-1 see BI-
PERIDEN

ALPHA-(BICYCLO(2.2.1)HEPT-5-EN-2-YL)-ALPHA-PHENYL-1-PIPER-
IDINO PROPANOL see BIPERIDEN

BIETHYLENE see 1,3-BUTADIENE

BIETHYLXANTHOGENTRISULFIDE see BISETHYL XANTHOGEN DISUL-
FIDE

BIFLUORIDEN (DUTCH) see FLUORINE

BIFLUORURE DE POTASSIUM (FRENCH) see POTASSIUM ACID FLUO-
RIDE

BIFORMAL see GLYOXAL

BIG DIPPER see DIPHENYLAMINE

BIGITALIN see GITOXIN

BIGUMAL see 1-(p-CHLOROPHENYL)-5-ISOPROPYLBIGUANIDE

BIHEXYL see DODECANE

BIMETHYL see ETHANE

BINDON see 2-(3-OXO-1-INDANYLIDENE)-1,3-INDANDIONE

BINDON ATHYLATHER see BINDON ETHYL ETHER

BINITROBENZENE see m-DINITROBENZENE

BIOALETRINA (PORTUGUESE) see BIOALLETHRIN

BIO-TESTICULINA see TESTOSTERONE PROPIONATE

2,2'-BIOXIRANE see 1,1'-BI(ETHYLENE OXIDE)

(r*,s*)-2,2'-BIOXIRANE see meso-1,2,3,4-DIEPOXYBUTANE

BIOXYDE D'AZOTE (FRENCH) see NITROGEN MONOXIDE

BIOXYDE DE PLOMB (FRENCH) see LEAD DIOXIDE

BIOXYDE DE PLOMB (FRENCH) see LEAD(IV)OXIDE BROWN

1,1'-BIPHENYL see BIPHENYL

4-BIPHENYLACETAMIDE see 4'-PHENYLACETANILIDE

N-4-BIPHENYLACETAMIDE see 4'-PHENYLACETANILIDE

4-BIPHENYLACETHYDROXAMIC ACID see N-ACETYL-4-BIPHENYLHY-
DROXYLAMINE

BIPHENYLAMINE see 4-BIPHENYLAMINE

O-BIPHENYLAMINE see 2-BIPHENYLAMINE

P-BIPHENYLAMINE see 4-BIPHENYLAMINE

(1,1'-BIPHENYL)-4-AMINE see 4-BIPHENYLAMINE

4,4'-BIPHENYLDIAMINE see BENZIDINE
(1,1'-BIPHENYL)-2,4'-DIAMINE see 2,4'-BIPHENYLDIAMINE
N,N'-(1,1'-BIPHENYL)-4,4'-DIYLBIS-ACETAMIDE 4',4'''-BIACET-
   ANILIDE see 4',4'''-BIACETANILIDE
2,2'-(1,1'-BIPHENYL-4,4' -DIYLBIS(20HYDROXY-4,4-DIMETHYL-
   MORPHOLINIUM DIBROMIDE see HEMICHOLINIUM-3-DIBROMIDE
N-(P-BIPHENYLMETHYL)-ATROPINIUM BROMIDE see 8-(p-PHENYL-
   BENZYL)ATROPINIUM BROMIDE
N,4-BIPHENYL-METHYL-DL-TROPEYL-ALPHA-TROPINIUMBROMIDS
   (GERMAN) see 8-(p-PHENYLBENZYL)ATROPINIUM BROMIDE
P-BIPHENYLMETHYL-(DL-TROPYL-ALPHA-TROPINIUM)BROMIDE see
   8-(p-PHENYLBENZYL)ATROPINIUM BROMIDE
O-BIPHENYLOL see 2-BIPHENYLOL
(1,1'-BIPHENYL)-2-OL see 2-BIPHENYLOL
BIPHENYL OXIDE see DIPHENYL ETHER
N-(4-BIPHENYLYL)ACETAMIDE see 4'-PHENYLACETANILIDE
N-(4-BIPHENYLYL)ACETOHYDROXAMIC ACETATE see N-ACETOXY-4-
   ACETAMIDOBIPHENYL
N-4-BIPHENYLYLACETOHYDROXAMIC ACID see N-ACETYL-4-BIPHE-
   NYLHYDROXYLAMINE
N,N'-4,4'-BIPHENYLYLENEBISACETAMIDE see 4',4'''-BIACETANILIDE
N-4-BIPHENYLYLHYDROXYLAMINE see 4-BIPHENYLHYDROXYLAMINE
BIPOTASSIUM CHROMATE see POTASSIUM CHROMATE (VI)
2,2'-BIPYRIDIN see 2,2'-BIPYRIDINE
ALPHA,ALPHA'-BIPYRIDINE see 2,2'-BIPYRIDINE
2,2'-BIPYRIDYL see 2,2'-BIPYRIDINE
ALPHA,ALPHA'-BIPYRIDYL see 2,2'-BIPYRIDINE
BIRNENOEL see PENTYL ACETATE
BIRTHWORT see ARISTOLOCHINE
BIS(ACETATO)TETRAHYDROXYTRILEAD see LEAD ACETATE, BASIC
BIS(ACETATO)TRIHYDROXYTRILEAD see LEAD ACETATE (II), TRIHY-
   DRATE
BIS(ACETO)DIHYDROXYTRILEAD see LEAD ACETATE, BASIC
BIS(ACETOXY)CADMIUM see CADMIUM (II) ACETATE
9,10-BISACETOXYMETHYL-1,2-BENZANTHRACENE see BENZ(a)AN-
   THRACENE-7,12-DIMETHANOLDIACETATE
BIS(ACETYLOXY)DIBUTYLSTANNANE see DIACETOXY DIBUTYL STAN-
   NANE
BIS(ACETYLOXY)MERCURY see MERCURIC ACETATE
S-(1,2-BIS(AETHOXY-CARBONYL)-AETHYL)-O,O-DIMETHYL-DITHIO-
   PHASPHAT(GERMAN) see CARBETHOXY MALATHION
BIS(2-AMINOETHYL)AMINE see DIETHYLENETRIAMINE
BIS(BETA-AMINOETHYL)AMINE see DIETHYLENETRIAMINE
BIS(BETA-AMINOETHYL)DISULFIDE see beta-MERCAPTOETHYLAMINE
   DISULFIDE
BIS-P-AMINOFENYLMETHAN (CZECH) see 4,4'-METHYLENEDIANILINE
4,4'-BIS(7-(1-AMINO-8-HYDROXY-2,4-DISULFO)NAPHTHYLAZO)-3,3'-
   BITOLYL, TETRASODIUM SALT see 4,4'-BIS(1-AMINO-8-HYDROXY-
   2,4-DISULFO-7-NAPHTHYLAZO)-3,3'-BITOLYL, TETRASODIUM SALT
4,4'-BIS(1-AMINO-8-HYDROXY-2,4-DISULPHO-7-NAPHTHYLAZO)-3,3'-
   BITOLYL, TETRASODIUM SALT see 4,4'-BIS(1-AMINO-8-HYDROXY-
   2,4-DISULFO-7-NAPHTHYLAZO)-3,3'-BITOLYL, TETRASODIUM SALT
BIS-4-AMINO-3-METHYLFENYLMETHAN (CZECH) see 4,4'-METHYLENE
   DI-O-TOLUIDINE
BIS(2-AMINOPHENYL)DISULFIDE see 2,2'-DITHIOBISANILINE
BIS(4-AMINOPHENYL)ETHER see 4,4-OXYDIANILINE
BIS(P-AMINOPHENYL)ETHER see 4,4-OXYDIANILINE
BIS(4-AMINOPHENYL)METHANE see 4,4'-METHYLENEDIANILINE
BIS(P-AMINOPHENYL)METHANE see 4,4'-METHYLENEDIANILINE
BIS(4-AMINOPHENYL) SULFIDE see 4,4'-THIODIANILINE
BIS(P-AMINOPHENYL)SULFIDE see 4,4'-THIODIANILINE
BIS(4-AMINOPHENYL) SULFONE see 4,4'-SULFONYLDIANILINE
BIS(P-AMINOPHENYL)SULFONE see 4,4'-SULFONYLDIANILINE
BIS(4-AMINOPHENYL)SULPHONE see 4,4'-SULFONYLDIANILINE
BIS(P-AMINOPHENYL)SULPHONE see 4,4'-SULFONYLDIANILINE
BIS-(3-AMINOPROPYL)AMINE see AMINOBIS(PROPYLAMINE)
BIS(3-AMINOPROPYL)METHYLAMINE see BIS(gamma-AMINOPROPYL)-
   METHYLAMINE
N,N-BIS(3-AMINOPROPYL)METHYLAMINE see BIS(gamma-AMINOPRO-
   PYL)METHYLAMINE

BIS(OMEGA-AMINOPROPYL)METHYLAMINE see BIS(gamma-AMINO-
   PROPYL)METHYLAMINE
N,N-BIS(GAMMA-AMINOPROPYL)METHYLAMINE see BIS(gamma-
   AMINOPROPYL)METHYLAMINE
2,2-BIS(P-ANISYL)-1,1,1-TRICHLOROETHANE see DIMETHOXY-DDT
2,5-BIS(1-AZIRIDINYL)-3,6-BIS(2-METHOXYETHOXY)-P-BENZOQUI-
   NONE see BENZOQUINONE AZIRIDINE
2,5-BIS(1-AZIRIDINYL)-3,6-BIS(2-METHOXYETHOXY)-2,5-CYCLO-
   HEXADIENE-1,4-DIONE see BENZOQUINONE AZIRIDINE
2,5-BIS(1-AZIRIDINYL)-3-(2-HYDROXY-1-METHOXYETHYL)-6-
   METHYL-P-BENZOQUINONE CARBAMATE (ESTER) see 2,5-
   BIS(1-AZIRIDINYL)-3-(2-CARBAMOYLOXY-1-METHOXYETHYL)-
   6-METHYL-1,4-BENZOQUINONE
BIS(1-AZIRIDINYL)(2-METHYL-3-THIAZOLIDINYL)PHOSPHINE OXIDE
   see IMIPHOS
(BIS(1-AZIRIDINYL)PHOSPHINYL)CARBAMIC ACID, ETHYL ESTER see
   ETHYL(DI-(1-AZIRIDINYL)PHOSPHINYL)CARBAMATE
BIS(BENZOTHIAZOLYL)DISULFIDE see BENZOTHIAZOLE DISULFIDE
BIS(2-BENZOTHIAZYL) DISULFIDE see BENZOTHIAZOLE DISULFIDE
BIS((4-(BIS(2-CHLOROETHYL)AMINO)BENZENE)ACETATE)ESTRA-1,3,
   5(10)-TRIENE-3,17-DIOL(17-beta) see ESTRADIOL MUSTARD
BIS((4-(BIS(2-CHLOROETHYL)AMINO)BENZENE)ACETATE)OESTRA-1,
   3,5(10)-TRIENE-3,17-DIOL(17-beta) see ESTRADIOL MUSTARD
BIS((P-(BIS(2-CHLOROETHYL)AMINO)PHENYL)ACETATE)ESTRADIOL
   see ESTRADIOL MUSTARD
BIS((P-(BIS(2-CHLOROETHYL)AMINO)PHENYL)ACETATE)ESTRA-1,3,
   5(10)-TRIENE-3,17-beta-DIOL see ESTRADIOL MUSTARD
BIS((P-(BIS(2-CHLOROETHYL)AMINO)PHENYL)ACETATE)OESTRADIOL
   see ESTRADIOL MUSTARD
BIS((P-BIS(2-CHLOROETHYL)AMINOPHENYL)ACETATE)OESTRA-1,3,
   5(10)-TRIENE-3,17-beta-DIOL see ESTRADIOL MUSTARD
BIS(CARBONATO(2-))DIHYDROXYTRIBERYLLIUM see BERYLLIUM
   CARBONATE
N,N-BIS(CARBOXYMETHYL)GLYCINE TRISODIUM SALT MONOHY-
   DRATE see NITRILOTRIACETIC ACID TRISODIUM SALT MONOHY-
   DRATE
N,N-BIS(CARBOXYMETHYL)GLYSINE see AMINOTRIACETIC ACID
BIS(CARBOXYMETHYLMERCAPTO)(P-UREIDOPHENYL)ARSINE see
   4-CARBAMIDOPHENYL BIS(CARBOXY METHYL THIO)ARSENITE
BIS(3-CARBOXYPROPIONYL) PEROXIDE see SUCCINIC PEROXIDE
N,N-BIS-(BETA-CHLORAETHYL)-AMIN (GERMAN) see BIS-beta-CHLO-
   ROETHYLAMINE
BIS(5-CHLOR-2-HYDROXYPHENYL)-METHAN (GERMAN) see 2,2'-
   METHYLENEBIS(4-CHLOROPHENOL)
BIS(BETA-CHLOROETHYL)-AMINE HYDROCHLORIDE see BIS(2-CHLO-
   ROETHYL)AMINE HYDROCHLORIDE
4-(BIS(2-CHLOROETHYL)AMINO)BENZENEBUTANOIC ACID see CHLO-
   RAMBUCIL
1,6-BIS(CHLOROETHYLAMINO)-1,6-BIS-DEOXY-D-MANNITOL
   see MANNOMUSTINE
2-(BIS(2-CHLOROETHYL)AMINO)-3-(2-CHLOROETHYL)TETRAHYDRO-
   2H-1,3,2-OXAPHOSPHORINE-2-OXIDE see TROPHOSPHAMIDE
1,6-BIS(CHLOROETHYLAMINO)-1,6-DESOXY-D-MANNITOLDIHYDRO-
   CHLORIDE see MANNOMUSTINE DIHYDROCHLORIDE
1,6-BIS(CHLOROETHYLAMINO)-1,6-DIDEOXY-D-MANNITE see MAN-
   NOMUSTINE
1,6-BIS-(CHLOROETHYLAMINO)-1,6-DIDEOXY-D-MANNITEDIHYDRO-
   CHLORIDE see MANNOMUSTINE DIHYDROCHLORIDE
1,6-BIS((2-CHLOROETHYL)AMINO)-1,6-DIDEOXY-D-MANNITOL see
   MANNOMUSTINE
1,6-BIS((BETA-CHLOROETHYL)AMINO)-1,6-DIDEOXY-D-MANNITOL see
   MANNOMUSTINE
9-(4-BIS(2-CHLOROETHYL)AMINO-1-METHYLBUTYLAMINO)-6-CHLO-
   RO-2-METHOXYACRIDINE DIHYDROCHLORIDE see QUINACRINE
   MUSTARD DIHYDROCHLORIDE
9-(4-(BIS-BETA-CHLOROETHYLAMINO)-1-METHYLBUTYLAMINO)-6-
   CHLORO-2-METHOXYACRIDINE see QUINACRINE MUSTARD
4-((4-(BIS(2-CHLOROETHYL)AMINO)-1-METHYLBUTYL)AMINO-7-
   CHLOROQUINOLINE, DIHYDROCHLORIDE see CHLOROQUINE MUS-
   TARD

2-(BIS(2-CHLOROETHYL)AMINOMETHYL)-5,5-DIMETHYLBENZIMIDA-
ZOLE HYDROCHLORIDE see BENZIMIDAZOLE METHYLENE MUSTARD

2-BIS(2-CHLOROETHYL)AMINONAPHTHALENE see N,N-BIS(2-CHLORO-
ETHYL)-2-NAPHTHYLAMINE

(P-(BIS(2-CHLOROETHYL)AMINO)PHENYL)ACETIC ACID CHOLESTEROL
ESTER see CHOLESTERYL-p-BIS(2-CHLOROETHYL)AMINO PHENYL-
ACETATE

3-(P-(BIS(2-CHLOROETHYL)AMINO)PHENYL)ALANINE see DL-PHENYL-
ALANINE MUSTARD

4-(BIS(2-CHLOROETHYL)AMINO)-L-PHENYLALANINE see L-PHENYL-
ALANINE MUSTARD

4-(BIS(2-CHLOROETHYL)AMINO)-DL-PHENYLALANINE see DL-PHE-
NYLALANINE MUSTARD

L-3-(P-(BIS(2-CHLOROETHYL)AMINO)PHENYL)ALANINE see L-PHE-
NYLALANINE MUSTARD

P-N-BIS(2-CHLOROETHYL)AMINO-L-PHENYLALANINE see L-PHENYL-
ALANINE MUSTARD

3-(P-(P-(BIS(2-CHLOROETHYL)AMINO)PHENYL)-L-ALANINE see
L-PHENYLALANINE MUSTARD

DL-3-(P-(BIS(2-CHLOROETHYL)AMINO)PHENYL)ALANINE see
DL-PHENYLALANINE MUSTARD

3-(P-(BIS(BETA-CHLOROETHYL)AMINO)PHENYL)-D-ALANINE HYDRO-
CHLORIDE see PHENYLALANINE MUSTARD

4-(BIS(2-CHLOROETHYL)AMINO)-D-PHENYLALANINE MONOHYDRO-
CHLORIDE see PHENYLALANINE MUSTARD

4-(BIS(2-CHLOROETHYL)AMINO)DL-PHENYLALANINE MONOHYDRO-
CHLORIDE see 3-(p-(BIS(beta-CHLOROETHYL)AMINO)-
PHENYL)-D,L-ALANINE HYDROCHLORIDE

4-(P-(BIS(2-CHLOROETHYL)AMINO)PHENYL)BUTYRIC ACID see
CHLORAMBUCIL

4(P-BIS(BETA-CHLOROETHYL)AMINOPHENYL)BUTYRIC ACID see
CHLORAMBUCIL

GAMMA-(P-BIS(2-CHLOROETHYL)AMINOPHENYL)BUTYRIC ACID
see CHLORAMBUCIL

3-(O-(BIS(BETA-CHLOROETHYL)AMINO)PHENYL)-D,L-ALANINE see
O-PHENYLALANINE MUSTARD

5-(BIS(2-CHLOROETHYL)AMINO)-2,4(1H,3H)PYRIMIDINEDIONE see
5-(BIS(2-CHLOROETHYL)AMINO)URACIL

2-(BIS(2-CHLOROETHYL)AMINO)TETRAHYDROOXAZAPHOSPHORINE
CYCLOHEXYLAMINE SALT see CYTOXAL ALCOHOL

5-N,N-BIS(2-CHLOROETHYL)AMINOURACIL see 5-(BIS(2-CHLORO-
ETHYL)AMINO)URACIL

N,N-BIS(2-CHLOROETHYL)BENZENEMETHANAMINE see N,N-BIS(2-
CHLOROETHYL)BENZYLAMINE

2,3-(N,N)sup 1)-BIS(2-CHLOROETHYL)DIAMIDO-1,3,2-OXAZAPHOS-
PHORIDINOXY see ISOPHOSPHAMIDE

BIS(BETA-CHLOROETHYL) ETHER see BIS(2-CHLOROETHYL) ETHER

BIS(2-CHLOROETHYL)FORMAL see BIS(beta-CHLOROETHYL)FORMAL

N,N-BIS(beta-CHLOROETHYL)GLYCINE HYDROCHLORIDE see GLYCINE
NITROGEN MUSTARD

N,N-BIS(2-CHLOROETHYL)-N'-(3-HYDROXYPROPYL)PHOSPHORO-
DIAMIDATE, CYCLOHEXYLAMMONIUM SALT see CYTOXAL ALCO-
HOL

N,N-BIS(2-CHLOROETHYL)ISOPROPYLAMINE HYDROCHLORIDE see
ISOPROPYL-S-HYDROCHLORIDE

BIS(2-CHLOROETHYL)METHYLAMINE see BIS(beta-CHLOROETHYL)-
METHYLAMINE

N,N-BIS(2-CHLOROETHYL)METHYLAMINE see BIS(beta-CHLORO-
ETHYL)METHYLAMINE

BIS(2-CHLOROETHYL)-BETA-NAPHTHYLAMINE see N,N-BIS(2-CHLO-
ROETHYL)-2-NAPHTHYLAMINE

BIS(2-CHLOROETHYL)NITROSOUREA see N,N'-BIS(2-CHLOROETHYL)-
N-NITROSOUREA

1,3-BIS-(2-CHLOROETHYL)-1-NITROSOUREA see N,N'-BIS(2-CHLORO-
ETHYL)-N-NITROSOUREA

1,3-BIS(BETA-CHLOROETHYL)-1-NITROSOUREA see N,N'-BIS(2-CHLO-
ROETHYL)-N-NITROSOUREA

N,N-BIS(2-CHLOROETHYL)PHOSPHORODIAMIDIC ACID, CYCLOHEXYL-
AMMONIUM SALT see PHOSPHORAMIDE MUSTARD

BIS(BETA-CHLOROETHYL)SULFIDE see BIS(2-CHLOROETHYL)SULFIDE

BIS(BETA-CHLOROETHYL)SULFONE see BIS(2-CHLOROETHYL)SULFONE

BIS(2-CHLOROETHYL)SULPHIDE see BIS(2-CHLOROETHYL)SULFIDE

5-(3,3-BIS(2-CHLOROETHYL)-1-TRIAZENO)IMIDAZOLE-4-CARBOX-
AMIDE see IMIDAZOLE MUSTARD

BIS(5-CHLORO-2-HYDROXYPHENYL)METHANE see 2,2'-
METHYLENEBIS(4-CHLOROPHENOL)

BIS-1,4-(CHLOROMETHOXY)-P-XYLENE see 1,4-BIS(CHLOROMETH-
OXYMETHYL)BENZENE

1,2-BIS(CHLOROMETHYL)BENZENE see alpha,alpha'-DICHLORO-O-XY-
LENE

BIS(2-CHLORO-1-METHYLETHYL) ETHER see BIS(2-CHLOROISOPRO-
PYL) ETHER

1,1-BIS(P-CHLOROPHENYL)-2,2-DICHLOROETHANE see 1,1-BIS(4-
CHLOROPHENYL)-2,2-DICHLOROETHANE

2,2-BIS(4-CHLOROPHENYL)-1,1-DICHLOROETHANE see 1,1-BIS(4-
CHLOROPHENYL)-2,2-DICHLOROETHANE

2,2-BIS(P-CHLOROPHENYL)-1,1-DICHLOROETHANE see 1,1-BIS(4-
CHLOROPHENYL)-2,2-DICHLOROETHANE

alpha,alpha-BIS(P-CHLOROPHENYL)-beta,beta,beta-TRICHLORETHANE
see DDT

2,2-BIS(p-CHLOROPHENYL)-1,1,1-TRICHLOROETHANE see DDT

1,1-BIS(4-CHLOROPHENYL)-2,2,2-TRICHLOROETHANOL see 1,1-BIS-
(p-CHLOROPHENYL)-2,2,2-TRICHLOROETHANOL

BIS-(2-CYANOETHYL)AMINE see BIS(beta-CYANOETHYL)AMINE

N,N-BIS(2-CYANOETHYL)AMINE see BIS(beta-CYANOETHYL)AMINE

CIS-BIS(CYCLOBUTYLAMMINE)DICHLOROPLATINUM(II) see cis-DICY-
CLOBUTYLAMMINEDICHLOROPLATINUM (II)

BISCYCLOPENTADIENE see BICYCLOPENTADIENE

BIS CYCLOPENTADIENYL IRON see FERROCENE

BIS(DECANOYLOXY)DI-N-BUTYLTIN see BIS(DECANOYLOXY)DI-n-
BUTYLSTANNANE

TRANS-1,4-BIS(2-DICHLOROBENZYLAMINOETHYL)CYCLOHEXANE DI-
CHLORHYDRATE (FRENCH) see trans-N,N'-BIS(2-CHLOROBENZYL)-
1,4-CYCLOHEXANEBIS(METHYLAMINE) DIHYDROCHLORIDE

BIS(DIETHYLAMINO)THIOXOMETHYL)DISULPHIDE see BIS(DIETHYL
THIO CARBAMOYL) DISULFIDE

BIS-O,O-DIETHYLPHOSPHORIC ANHYDRIDE see TETRAETHYLPYRO-
PHOSPHATE

BIS-O,O-DIETHYLPHOSPHOROTHIONIC ANHYDRIDE see ETHYL THIO
PYROPHOSPHATE

BIS(N,N-DIETHYLTHIOCARBAMOYL) DISULFIDE see BIS(DIETHYL THIO
CARBAMOYL) DISULFIDE

BIS(DIETHYLTHIOCARBAMOYL)DISULPHIDE see BIS(DIETHYL THIO
CARBAMOYL) DISULFIDE

BIS(N,N-DIETHYLTHIOCARBAMOYL)DISULPHIDE see BIS(DIETHYL
THIO CARBAMOYL) DISULFIDE

2,2-BISDIHYDROXYMETHYL-1,3-PROPANEDIOL TETRANITRATE see
PENTAERYTHRITOL TETRANITRATE

BIS(DIMETHYLAMIDO)-PHOSPHORYL FLUORIDE see BIS(DIMETHYL
AMIDO)FLUORO PHOSPHATE

2,8-BISDIMETHYLAMINOACRIDINE see 3,6-BIS(DIMETHYL AMINO)-
ACRIDINE

BIS(DIMETHYLAMINO)-3-AMINO-5-PHENYLTRIAZOLYL PHOSPHINE
OXIDE see 5-AMINO-1-BIS(DIMETHYLAMIDE)PHOSPHORYL-3-
PHENYL-1,2,4-TRIAZOLE

4,4'-BIS(DIMETHYLAMINO)BENZOPHENONE see MICHLER'S KETONE

4,4'-BIS(DIMETHYLAMINO)BENZOPHENONE see TETRAMETHYL DI-
AMINO BENZOPHENONE

P,P'-BIS(N,N-DIMETHYLAMINO)BENZOPHENONE see MICHLER'S KE-
TONE

BIS((DIMETHYLAMINO)CARBONOTHIOYL) DISULPHIDE see BIS(DI-
METHYL THIOCARBAMYL)DISULFIDE

4,4'-BIS(DIMETHYLAMINO)DIPHENYLMETHANE see MICHLER'S BASE

P,P'-BIS(DIMETHYLAMINO)DIPHENYLMETHANE see MICHLER'S BASE

BIS(2-DIMETHYLAMINOETHYL)SUCCINATE BIS(METHOCHLORIDE) see
SUCCINOYLCHOLINE CHLORIDE

BIS(DIMETHYLAMINO)FLUOROPHOSPHATE see BIS(DIMETHYL AMI-
DO)FLUORO PHOSPHATE

BISDIMETHYLAMINOFLUOROPHOSPHINE OXIDE see BIS(DIMETHYL
AMIDO)FLUORO PHOSPHATE

3,7-BIS(DIMETHYLAMINO) -4-NITRO-PHENOTHIAZIN-5-IUM, CHLORIDE see METHYLENE GREEN

3,7-BIS(DIMETHYLAMINO)PHENOTHIAZIN-5-IUM CHLORIDE see 3,7-BIS(DIMETHYL AMINO)PHENAZA THIONIUM CHLORIDE

BIS(P-(N,N-DIMETHYLAMINO)PHENYL)KETONE see MICHLER'S KETONE

BIS(P-(N,N-DIMETHYL AMINO)PHENYL)KETONE see TETRAMETHYL DIAMINO BENZOPHENONE

BIS(P-DIMETHYLAMINOPHENYL)METHANE see MICHLER'S BASE

BIS(P-(N,N-DIMETHYLAMINO)PHENYL)METHANE see MICHLER'S BASE

P,P'-BIS(N,N-DIMETHYLAMINOPHENYL)METHANE see MICHLER'S BASE

BIS(4-(DIMETHYLAMINO)PHENYL)METHANONE see MICHLER'S KETONE

BIS(DIMETHYLAMINO)PHOSPHONOUS ANYHYDRIDE see OCTA-METHYLPYROPHOSPHORAMIDE

BIS(DIMETHYLAMINO)PHOSPHORIC ANHYDRIDE see OCTAMETHYLPY-ROPHOSPHORAMIDE

BIS(DIMETHYLCARBAMODITHIOATO-S,S')LEAD see LEAD DIMETHYL-DITHIOCARBAMATE

BIS(DIMETHYLCARBAMODITHIOATO-S,S')ZINC see BIS(DIMETHYL DI-THIO CARBAMATO)ZINC

BIS-DIMETHYLDITHIOCARBAMATE DE ZINC (FRENCH) see BIS(DI-METHYL DITHIO CARBAMATO)ZINC

BIS(DIMETHYLDITHIOCARBAMIATO)LEAD see LEAD DIMETHYLDITHIO-CARBAMATE

1,1'-BIS(2-(3,5-DIMETHYL-4-MORPHOLINYL)-2-OXOETHYL)-4,4'-BIPYRIDINIUM DICHLORIDE see 1,1'-BIS(3,5-DIMETHYL MORPHO-LINO CARBONYL METHYL)-4,4'-BIPYRIDYNIUM DICHLORIDE

BIS(DIMETHYLTHIOCARBAMOYL) DISULFIDE see BIS(DIMETHYL THIO-CARBAMYL)DISULFIDE

BIS(DIMETHYLTHIOCARBAMYL) MONOSULFIDE see BIS(DIMETHYL THIO CARBAMOYL)SULFIDE

BIS(N,N-DIMETIL-DITIOCARBAMMATO) DI ZINCO (ITALIAN) see BIS-(DIMETHYL DITHIO CARBAMATO)ZINC

(+,−)-1,2-BIS(3,5-DIOXOPIPERAZINE-1-YL)PROPANE see NSC-129943

BIS(DITHIOPHOSPHATE DE O,O-DIETHYLE) DE S,S'-METHYLENE (FRENCH) see ETHYL METHYLENE PHOSPHORO DITHIOATE

BIS(2,3-EPOXYCYCLOPENTYL) ETHER see BIS(2,3-EPOXY CYCLOPEN-TYL) ETHER

M-BIS(2,3-EPOXYPROPOXY)BENZENE see RESORCINOL DIGLYCIDYL ETHER

1,3-BIS(2,3-EPOXYPROPOXY)BENZENE see RESORCINOL DIGLYCIDYL ETHER

2,2-BIS(4-(2,3-EPOXYPROPYLOXY)PHENYL)PROPANE see BISPHENOL A DIGLYCIDYL ETHER

S-(1,2-BIS(ETHOXY-CARBONYL)-O,O-DIMETHYL-DITHIOFOSFAAT (DUTCH) see CARBETHOXY MALATHION

S-(1,2-BIS(ETHOXYCARBONYL)ETHYL) O,O-DIMETHYL PHOSPHORODI-THIOATE see CARBETHOXY MALATHION

S-1,2-BIS(ETHOXYCARBONYL)ETHYL-O,O-DIMETHYL THIOPHOSPHATE see CARBETHOXY MALATHION

BIS(ETHYLENIMIDO)PHOSPHORYLURETHAN see ETHYL(DI-(1-AZIRIDI-NYL)PHOSPHINYL)CARBAMATE

BIS(2-ETHYLHEXYL)-1,2-BENZENEDICARBOXYLATE see BIS(2-ETHYL-HEXYL)PHTHALATE

BIS(ETHYLHEXYL) ESTER OF SODIUM SULFOSUCCINIC ACID see DIO-CTYL SODIUM SULFOSUCCINATE

BIS(2-ETHYLHEXYL) FUMARATE see DIOCTYL FUMARATE

BIS(2-ETHYLHEXYL)HYDROGEN PHOSPHATE see BIS(2-ETHYLHEXYL)-PHOSPHATE

BIS(2-ETHYLHEXYL)ORTHOPHOSPHORIC ACID see BIS(2-ETHYL-HEXYL)PHOSPHATE

BIS(2-ETHYLHEXYL)PHOSPHORIC ACID see BIS(2-ETHYLHEXYL)PHOS-PHATE

BIS(2-ETHYLHEXYL)SODIUM SULFOSUCCINATE see DIOCTYL SO-DIUM SULFOSUCCINATE

BIS(2-ETHYLHEXYL)-S-SODIUM SULFOSUCCINATE see DIOCTYL SO-DIUM SULFOSUCCINATE

1,4-BIS(2-ETHYLHEXYL) SODIUM SULFOSUCCINATE see DIOCTYL SO-DIUM SULFOSUCCINATE

1,1-BIS(P-ETHYLPHENYL)-2,2-DICHLOROETHANE see DIETHYL DIPHE-NYL DICHLOROETHANE

2,2-BIS(P-ETHYLPHENYL)-1,1-DICHLOROETHANE see DIETHYL DIPHE-NYL DICHLOROETHANE

BIS(N-ETHYL-N-PHENYL)UREA see 1,3-DIETHYL-1,3-DIPHENYLUREA

BIS(ETHYLXANTHIC)DISULFIDE see BISETHYL XANTHOGEN DISULFIDE

BIS(ETHYLXANTHOGEN) TRISULFIDE see DI(ETHYLXANTHOGEN)TRI-SULFIDE

S-(1,2-BIS(ETOSSI-CARBONIL)-ETIL)-O,O-DIMETIL-DITIOFOSFATO (ITALIAN) see CARBETHOXY MALATHION

1,6-BIS(9 FLUORENYLDIMETHYL-AMMONIUM)HEXANE BROMIDE see HEXAFLURONIUM BROMIDE

2-(BIS(FURFURYLIDENAMINO))METHYLFURAN see FURFURAMIDE

BIS(4-GLYCIDYLOXYPHENYL)DIMETHYAMETHANE see BISPHENOL A DIGLYCIDYL ETHER

2,2-BIS(p-GLYCIDYLOXYPHENYL)PROPANE see BISPHENOL A DIGLY-CIDYL ETHER

BIS(HYDROXYAETHYL)-AETHER-DINITRAT (GERMAN) see DIETHYLENE GLYCOL DINITRATE

BIS(4-HYDROXY-5-TERT-BUTYL-2-METHYLPHENYL) SULFIDE see BIS-(3-tert-BUTYL-4-HYDROXY-6-METHYLPHENYL) SULFIDE

BIS-2-HYDROXY-5-CHLORFENYLMETHAN (CZECH) see 2,2'-METHYL-ENEBIS(4-CHLOROPHENOL)

BIS(2-HYDROXY-5-CHLOROPHENYL)METHANE see 2,2'-METHYLENE-BIS(4-CHLOROPHENOL)

BIS-3,3'-(4-HYDROXYCOUMARINYL)ACETIC ACID ETHYL ESTER see BIS(4-HYDROXY-3-COUMARIN) ACETIC ACID ETHYL ESTER

BIS-(4-HYDROXY-3-COUMARINYL)ETHYL ACETATE see BIS(4-HY-DROXY-3-COUMARIN) ACETIC ACID ETHYL ESTER

BIS(4-HYDROXYCOUMARIN-3-YL)METHANE see BISHYDROXY COU-MARIN

BIS(2-HYDROXYETHYL)CARBAMODITHIOIC ACID, MONOPOTASSIUM SALT see POTASSIUM BIS(2-HYDROXYETHYL)DITHIOCARBAMATE

BIS(2-HYDROXYETHYL)DITHIOCARBAMIC ACID, MONOPOTASSIUM SALT see POTASSIUM BIS(2-HYDROXYETHYL)DITHIOCARBAMATE

BIS(2-HYDROXYETHYL) ETHER see DIETHYLENE GLYCOL

BIS(2-HYDROXYETHYL)LAURAMIDE see N,N-BIS(2-HYDROXY ETHYL)-DODECAN AMIDE

N,N-BIS(HYDROXYETHYL)LAURAMIDE see N,N-BIS(2-HYDROXY ETHYL)DODECAN AMIDE

N,N-BIS(2-HYDROXYETHYL)LAURAMIDE see N,N-BIS(2-HYDROXY ETHYL)DODECAN AMIDE

N,N-BIS(BETA-HYDROXYETHYL)LAURAMIDE see N,N-BIS(2-HYDROXY ETHYL)DODECAN AMIDE

2,2-BIS-4'-HYDROXYFENYLPROPAN (CZECH) see BISPHENOL A

BIS(4-HYDROXYIMINOMETHYLPYRIDINIUM-1-METHYL)ETHER DICHLO-RIDE see 1,3-BIS(4-ALDOX IMINOPYRIDINIUM)DIMETHYL ETHER BI-CHLORIDE

3-BIS(HYDROXYMETHYL)AMINO-6-(5-NITRO-2-FURYLETHENYL)-1,2,4-TRIAZINE see BIS(HYDROXY METHYL)FURATRIZINE

2,2-BIS(HYDROXYMETHYL)-1,3-PROPANEDIOL TETRANITRATE see PENTAERYTHRITOL TETRANITRATE

BIS(4-HYDROXY-2-OXO-2H-1-BENZOPYRAN-3-YL)ACETIC ACID ETHYL ESTER see BIS(4-HYDROXY-3-COUMARIN) ACETIC ACID ETHYL ESTER

BIS(4-HYDROXYPHENYL) DIMETHYLMETHANE see BISPHENOL A

BIS(4-HYDROXYPHENYL)DIMETHYLMETHANE DIGLYCIDYL ETHER see BISPHENOL A DIGLYCIDYL ETHER

3,4-BIS(P-HYDROXYPHENYL)-2,4-HEXADIENE see DEHYDROSTILBES-TROL

3,4-BIS(P-HYDROXYPHENYL)HEXANE see DIHYDROSTILBESTROL

MESO-3,4-BIS(P-HYDROXYPHENYL)-N-HEXANE see DIHYDROSTILBES-TROL

3,4-BIS(P-HYDROXYPHENYL)-3-HEXENE see DIETHYLSTILBESTEROL

BIS(4-HYDROXYPHENYL)PROPANE see BISPHENOL A

2,2-BIS(4-HYDROXYPHENYL)PROPANE see BISPHENOL A

2,2-BIS(p-HYDROXYPHENYL)PROPANE see BISPHENOL A

2,2-BIS(4-HYDROXYPHENYL)PROPANE, DIGLYCIDYL ETHER see BIS-PHENOL A DIGLYCIDYL ETHER

2,2-BIS(p-HYDROXYPHENYL)PROPANE, DIGLYCIDYL ETHER see BIS-PHENOL A DIGLYCIDYL ETHER

N-BIS(2-HYDROXYPROPYL)NITROSAMINE see DI(2-HYDROXY-n-PRO-PYL)AMINE

2,2'-BISHYDROXYPROPYLNITROSAMINE see DI(2-HYDROXY-n-PRO-PYL)AMINE

BIS(2-HYDROXY-3,5,6-TRICHLOROPHENYL)METHANE see HEXA-CHLOROPHENE

BIS(ISOBUTYL)ALUMINUM CHLORIDE see CHLORO DIISOBUTYL ALUMINUM

BIS(4-ISOCYANATOPHENYL)METHANE see DIPHENYL METHANE DI-ISOCYANATE

BIS(P-ISOCYANATOPHENYL)METHANE see DIPHENYL METHANE DI-ISOCYANATE

BIS(1,4-ISOCYANATOPHENYL)METHANE see DIPHENYL METHANE DI-ISOCYANATE

BIS(ISOPROPYLAMIDO) FLUOROPHOSPHATE see PHOSPHORODI(ISO-PROPYLAMIDIC) FLUORIDE

BIS(LAUROYLOXY)DI(N-BUTYL)STANNANE see DIBUTYLBIS(LAUROYL-OXY)STANNANE

BIS(L-HISTIDINATO)COBALT see BIS(L-HISTIDINE)COBALT

BISMARSEN see BISMUTH ARSPHENAMINE SULFONATE

BISMATE see BISMUTH DIMETHYL DITHIOCARBAMATE

BIS(MERCAPTOBENZOTHIAZOLATO)ZINC see BIS(2-BENZOTHIAZOLY-THIO)ZINC

1,4-BIS(METHANESULFONOXY)BUTANE see 1,4-BUTANEDIOL DI-METHYL SULFONATE

(1,4-BIS(METHANESULFONYLOXY)BUTANE) see 1,4-BUTANEDIOL DI-METHYL SULFONATE

2,5-BISMETHOXYETHOXY-3,6-BISETHYLENEIMINO-1,4-BENZOQUI-NONE see BENZOQUINONE AZIRIDINE

3,6-BIS (BETA-METHOXYETHOXY)-2,5-BIS(ETHYLENEIMINO)-P-BEN-ZOQUINONE see BENZOQUINONE AZIRIDINE

3,6-BIS(BETA-METHOXYETHOXY)-2,5-BIS(ETHYLENIMINO)-P-BENZO-QUINONE see BENZOQUINONE AZIRIDINE

3,4-BIS(P-METHOXYPHENYL)-3-HEXENE see 3,4-DIANISYL-3-HEXENE

N,N'-BIS(4-METHOXYPHENYL)-N''-(4-ETHOXYPHENYL)GUANIDINE HYDROCHLORIDE see PHENODIANISYL HYDROCHLORIDE

1,1-BIS(P-METHOXYPHENYL)-2,2,2-TRICHLOROETHANE see DIMETH-OXY-DDT

2,2-BIS(P-METHOXYPHENYL)-1,1,1-TRICHLOROETHANE see DIMETH-OXY-DDT

N,N-BIS(N-METHYL-N-PHENYL-TERT-BUTYLACETAMIDO)-BETA-HY-DROXYETHYLAMINE see 2-DI(N-METHYL-N-PHENYL-tert-BUTYL CARBAMOYLMETHYL)AMINOETHANOL

N-BISMETHYLPTEROYLGLUTAMIC ACID see METHOTREXATE

1,6-BIS-0-METHYLSULFONYL-d-MANNITOL see BIS(METHANE SULFO-NYL)-D-MANNITOL

BIS(MONOISOPROPYLAMINO)FLUOROPHOSPHATE see PHOSPHORODI-(ISOPROPYLAMIDIC) FLUORIDE

BIS(MONOISOPROPYLAMINO)FLUOROPHOSPHINE OXIDE see PHOSPHO-RODI(ISOPROPYLAMIDIC) FLUORIDE

BISMORPHOLINO DISULFIDE see N,N'-BISMORPHOLINE DISULFIDE

BIS(MORPHOLINO-)METHAN (GERMAN) see 4,4'-METHYLENEDIMOR-PHOLINE

BISMORPHOLINO METHANE see 4,4'-METHYLENEDIMORPHOLINE

BISMUTH-209 see BISMUTH

BISMUTH EMETINE IODIDE see EMETINE BISMUTH IODIDE

BIS(2,4-PENTANEDIONATO)COPPER see BIS(ACETYL ACETONE)COPPER

BIS(2,4-PENTANEDIONATO)TITANIUM OXIDE see BIS(ACETYLACETO-NATO) TITANIUM OXIDE

BIS(PHENYLMERCURI)METHYLENEDINAPHTHALENESULFONATE see PHENYLMERCURIC DINAPHTHYLMETHANEDISULFONATE

BIS(8-QUINOLINATO)COPPER see BIS(8-OXYQUINOLINE)COPPER

BIS(8-QUINOLINOLATO)COPPER see BIS(8-OXYQUINOLINE)COPPER

BIS(S-DIETHOXYPHOSPHINOTHIOYL)MERCAPTO)METHANE see ETHYL METHYLENE PHOSPHORO DITHIOATE

BIS(SUCCINYLDICHLOROCHOLINE) see SUCCINOYLCHOLINE CHLORIDE

BIS-N,N,N',N'-TETRAMETHYLPHOSPHORODIAMIDIC ANHYDRIDE see OCTAMETHYLPYROPHOSPHORAMIDE

BIS-1,4-p-TOLYLAMINOANTHRCHINON (CZECH) see 1,4-BIS(p-TOLY-AMINO)ANTHRAQUINONE

BIS-(TRI-N-BUTYLCIN)OXID (CZECH) see BIS(TRIBUTYL TIN)OXIDE

BIS(TRI-N-BUTYLZINN)-OXYD (GERMAN) see BIS(TRIBUTYL TIN)OXIDE

BIS-2,3,5-TRICHLOR-6-HYDROXYFENYLMETHAN (CZECH) see HEXA-CHLOROPHENE

BIS(3,5,6-TRICHLORO-2-HYDROXYPHENYL)METHANE see HEXACHLO-ROPHENE

BIS-TRICHLOROMETHYL-TRISULFID (CZECH) see BIS(TRICHLORO METHYL)TRISULFIDE

BIS(TRIFLUOROETHYL)ETHER see HEXAFLUORODIETHYL ETHER

BIS(2,2,2-TRIFLUOROETHYL)ETHER see HEXAFLUORODIETHYL ETHER

2,2'-BIS(1,6,7-TRIHYDROXY-3-METHYL-5-ISOPROPYL-8-ALDEHYDO-NAPHTHALENE see GOSSYPOL

ALPHA,OMEGA-BIS(TRIMETHYL AMMONIUM)HEXANE DIBROMIDE see HEXAMETHONIUM BROMIDE

BISULFITE DE SODIUM (FRENCH) see SODIUM BISULFITE (1:1)

4,4'-BI-O-TOLUIDINE see 3,3'-TOLIDINE

BITTER SALTS see MAGNESIUM SULFATE HEPTAHYDRATE

BITUMEN see ASPHALT

DELTA(1,1')-BIUREA see AZODICARBAMIDE

BIVINYL see 1,3-BUTADIENE

BLACK BLASTING POWDER see EXPLOSIVES, LOW

BLACK OXIDE OF IRON see IRON (III) OXIDE

BLACK PEARLS see CARBON

BLADAN BASE see HEXAETHYL TETRAPHOSPHATE

BLANC FIXE see BARIUM SULFATE

BLANCOSOLV see ACETYLENE TRICHLORIDE

BLASTING GELATIN (DOT) see NITROGLYCERIN

BLASTING OIL see NITROGLYCERIN

"A" BLASTING POWDER see EXPLOSIVES, LOW

"B" BLASTING POWDER see EXPLOSIVES, LOW

BLAUSAEURE (GERMAN) see HYDROCYANIC ACID

BLAUWZUUR (DUTCH) see HYDROCYANIC ACID

BLEACHING POWDER (DOT) see HYPOCHLOROUS ACID, CALCIUM SALT

BLEIACETAT (GERMAN) see LEAD ACETATE

BLEIPHOSPHAT (GERMAN) see LEAD (II) PHOSPHATE (3:2)

BLEISULFAT (GERMAN) see LEAD (II) SULFATE (1:1)

BLENDED RED OXIDES OF IRON see IRON (III) OXIDE

BLISTERING BEETLES see CANTHARIDES

BLISTERING FLIES see CANTHARIDES

BLOOD STONE see HEMATITE

BLOOD SUGAR see D-GLUCOSE

BLUE ASBESTOS see CROCIDOLITE (see ASBESTOS)

BLUE COPPERRAS see COPPER (II) SULFATE PENTAHYDRATE (1:1:5)

BLUE CROSS see CHLORO DIPHENYL ARSINE

BLUE DYE NUMBER 1 FOOD ADDITIVE see FD AND C BLUE 1

BLUE OIL see ANILINE

BLUE POWDER see ZINC

BLUESTONE see COPPER (II) SULFATE PENTAHYDRATE (1:1:5)

BLUE VITRIOL see COPPER (II) SULFATE PENTAHYDRATE (1:1:5)

7-BMBA see 7-BROMO METHYL BENZ(a)ANTHRACENE

BMC see n-BUTYLMERCURIC CHLORIDE

BOIS D'ARC (FRENCH) see MORIN

BOIS D'INDE see BAY OIL

BONE OIL (DOT) see BONE OIL

BONIDE RYATOX see RYANIA

BONIDE TOPZOL RAT BAITS AND KILLING SYRUP see RED SQUILL

BORACIC ACID see BORIC ACID

BORANE, COMPOUND WITH DIMETHYLAMINE (1:1) see DIMETHYL-AMINE BORANE

BORAZOLE see BORAZINE

BORDEAUX see FOOD RED 2

BORIC ANHYDRIDE see BORON OXIDE

1-2-BORNANOL see NGAI CAMPHOR

2-BORNANONE see CAMPHOR

(+)-2-BORNANONE see CAMPHOR, (1R,4R)-(+)-

D-2-BORNANONE see CAMPHOR, (1R,4R)-(+)-
BORNEO CAMPHOR see BORNEOL
(−)-BORNEOL see NGAI CAMPHOR
1-BORNEOL see NGAI CAMPHOR
TRANS-BORNEOL see BORNEOL
BORNYL ALCOHOL see BORNEOL
1-BORNYL ALCOHOL see NGAI CAMPHOR
BOROETHANE see BORON HYDRIDE
BOROHYDRURE DE SODIUM (FRENCH) see SODIUM BOROHYDRIDE
BORON CHLORIDE see BORON TRICHLORIDE
BORON HYDRIDE see DECABORANE(14)
BORON SESQUIOXIDE see BORON OXIDE
BORON TRIBROMIDE (DOT) see BORON BROMIDE
BORON TRIFLUORIDE see BORON FLUORIDE
BORON TRIOXIDE see BORON OXIDE
BORSAURE (GERMAN) see BORIC ACID
BOY see HEROIN
BP 400 see 9-(1-METHYL-4-PIPERIDYLIDENE)THIOXANTHENE
BP-7,8-DIHYDRODIOL-9,10-EPOXIDE(ANTI) see anti-BENZO(a)PY-
  RENE-7,8-DIHYDRODIOL-9,10-OXIDE
(−)-BP 7-ALPHA,8-BETA-DIOL-9-BETA,10-BETA-EPOXIDE 2 see sym-
  DIOLEPOXIDE
BP-4,5-EPOXIDE see BENZO(a)PYRENE-4,5-OXIDE
BP-9,10-OXIDE see BENZO(a)PYRENE-9,10-OXIDE
BP-11,12-OXIDE see BENZO(a)PYRENE-11,12-OXIDE
BRACKEN FERN, TANNIN-FREE see TANNIN-FREE FRACTION OF
  BRACKEN FERN
BRACKEN FERN TOXIC COMPONENT see SHIKIMIC ACID
BRADYKININ (SYNTHETIC) see BRADYKININ
BRAIN SUGAR see GALACTOSE
BRASILAZINA OIL RED B see SCARLET RED
BRENOL see 1-AUROTHIO-D-GLUCOPYRANOSE
BRESTAN see ACETOXYTRIPHENYLSTANNANE
BRETYLIUM-P-TOLUENESULFONATE see (o-BROMO BENZYL)ETHYL
  DIMETHYL AMMONIUM-p-TOLUENE SULFONATE
BRETYLIUM TOSYLATE see (o-BROMO BENZYL)ETHYL DIMETHYL
  AMMONIUM-p-TOLUENE SULFONATE
BREVITAL SODIUM see METHOHEXITAL SODIUM
BRILLIANT BLUE FCD NO. 1 see FD AND C BLUE 1
BRILLIANT OIL ORANGE Y BASE see 4-PHENYLAZO-m-PHENYLENEDI-
  AMINE
BRILLIANT PONCEAU 3R see FOOD RED 7
BRILLIANT SCARLET see FOOD RED 7
BRIMSTONE see SULFUR
BRITISH ANTILEWISITE see BAL
BROCIDE see ETHYLENE DICHLORIDE
BROM (GERMAN) see BROMINE
BROMACETYLENE see BROMOACETYLENE
BROMALLYLENE see ALLYL BROMIDE
BROMBENZYL CYANIDE see BROMOBENZYLNITRILE
BROMCARBAMIDE see 2-BROMO-3-METHYL BUTYRYL UREA
BROME (FRENCH) see BROMINE
BROMELIN see BROMELAIN
BROMINE CYANIDE see CYANOGEN BROMIDE
ALPHA-BROMISOVALERYLUREA see 2-BROMO-3-METHYL BUTYRYL
  UREA
BROM LSD see 2-BROMO-D-LYSERGIC ACID DIETHYLAMIDE
BROMLYSERGAMIDE see 2-BROMO-D-LYSERGIC ACID DIETHYLAMIDE
2-BROM-D-LYSERGIC ACID DIETHYLAMINE see 2-BROMO-D-LYSERGIC
  ACID DIETHYLAMIDE
BROM-METHAN (GERMAN) see BROMO METHANE
BROMO (ITALIAN) see BROMINE
BROMOACETIC ACID ETHYLENE ESTER see 1,2-BIS(BROMOACET-
  OXY)ETHANE
BROMOACETIC ACID, ETHYL ESTER see ETHYL BROMACETATE
BROMOACETONE see BROMO-2-PROPANONE
5-(2-BROMOALLYL)-5-SEC-BUTYLBARBITURIC ACID see BUTALLY-
  LONAL
GAMMA-BROMOALLYLENE see PROPARGYL BROMIDE

5-(2'-BROMOALLYL)-5-(1'-METHYL-N-PROPYL)BARBITURIC ACID see
  BUTALLYLONAL
BROMOAZIDE see BROMINE AZIDE
BROMOBENZENE (DOT) see PHENYL BROMIDE
BETA-(P-BROMOBENZHYDRYLOXY)ETHYLDIMETHYLAMINE HYDRO-
  CHLORIDE see BROMO PHENYL HYDRAMINE HYDROCHLORIDE
2-(4-BROMOBENZOHYDRYLOXY)-ETHYL-DIMETHYLAMINE HYDRO-
  CHLORIDE see BROMO PHENYL HYDRAMINE HYDROCHLORIDE
ALPHA-BROMOBENZYL CYANIDE see BROMOBENZYLNITRILE
ALPHA-BROMOBENZYLNITRILE see BROMOBENZYLNITRILE
BROMOCARBAMIDE see 2-BROMO-3-METHYL BUTYRYL UREA
3-BROMO-N-(2-CHLOROMERCURICYCLOHEXYL)PROPIONAMIDE see
  trans-CHLORO(2-(3-BROMO PROPION AMIDO)CYCLO HEXYL)MER-
  CURY
2-BROMO-2-CHLORO-1,1,1-TRIFLUOROETHANE see HALOTHANE
BROMOCYANOGEN see CYANOGEN BROMIDE
5-BROMODEOXYURIDINE see 5-BROMO-2'-DEOXY URIDINE
5-BROMO-2-DEOXYURIDINE see 5-BROMO-2'-DEOXY URIDINE
5-BROMODESOXYURIDINE see 5-BROMO-2'-DEOXY URIDINE
O-(4-BROMO-2,5-DICHLOROPHENYL)-O,O-DIETHYL PHOSPHOROTHIO-
  ATE see ETHYL BROMOPHOS
O-(4-BROMO-2,5 DICHLOROPHENYL)-O,O-DIETHYLPHOSPHOROTHIO-
  NATE see ETHYL BROMOPHOS
O-(4-BROMO-2,5-DICHLOROPHENYL)-O-METHYL PHENYLPHOSPHONO-
  THIOATE see LEPTOPHOS
2-BROMO-9,10-DIDEHYDRO-N,N-DIETHYL-6-METHYLERGOLINE-8-
  BETA-CARBOXAMIDE see 2-BROMO-D-LYSERGIC ACID DIETHYL-
  AMIDE
BROMODIETHYLACETYLCARBAMIDE see 2-BROMO-2-ETHYL BUTYRYL
  UREA
BROMODIETHYLACETYLUREA see 2-BROMO-2-ETHYL BUTYRYL UREA
ALPHA-BROMO-BETA-DIMETHYLPROPANOYLUREA see 2-BROMO-3-
  METHYL BUTYRYL UREA
2-(1-(4-BROMODIPHENYL)ETHOXY)-N,N-DIMETHYLETHYLAMINE HY-
  DROCHLORIDE see BROMADRYL
2-BROMOERGOCRYPTINE see BROMOCRIPTINE
2-BROMO-ALPHA-ERGOKRYPTIN see BROMOCRIPTINE
BROMOETHANIOC ACID see alpha-BROMOACETIC ACID
BROMOETHENE see VINYL BROMIDE
(ALPHA-BROMO-ALPHA-ETHYLBUTYRYL)CARBAMIDE see 2-BROMO-2-
  ETHYL BUTYRYL UREA
1-BROMO-ETHYL-BUTYRYL-UREA see 2-BROMO-2-ETHYL BUTYRYL
  UREA
(ALPHA-BROMO-ALPHA-ETHYLBUTYRYL)UREA see 2-BROMO-2-ETHYL
  BUTYRYL UREA
BROMOETHYLENE see VINYL BROMIDE
BROMOETHYNE see BROMOACETYLENE
BROMOFLUORESCEIC ACID see BROMO EOSINE
BROMOFLUOROFORM see BROMO TRIFLUORO METHANE
BROMOFORME (FRENCH) see BROMOFORM
BROMOFORMIO (ITALIAN) see BROMOFORM
BROMOFUME see 1,2-ETHYLENE DIBROMIDE
2-BROMO-12'-HYDROXY-2'-(1-METHYLETHYL)-5'-ALPHA-(2-METHYL-
  PROPYL)ERGOTAMIN-3',6',18-TRIONE see BROMOCRIPTINE
2-BROMOISOBUTANE see tert-BUTYL BROMIDE
ALPHA-BROMOISOVALERIC ACID UREIDE see 2-BROMO-3-METHYL
  BUTYRYL UREA
ALPHA-BROMOISOVALEROYLUREA see 2-BROMO-3-METHYL BU-
  TYRYL UREA
(ALPHA-BROMOISOVALERYL)UREA see 2-BROMO-3-METHYL BUTYRYL
  UREA
BROMOMETANO (ITALIAN) see BROMO METHANE
(BROMOMETHYL)BENZENE see BENZYL BROMIDE
P-BROMO-ALPHA-METHYLBENZHYDRYL-2-DIMETHYLAMINOETHYL
  ETHER HYDROCHLORIDE see BROMADRYL
2-((P-BROMO-ALPHA-METHYL-ALPHA-PHENYLBENZYL)OXY)-N,N-
  DIMETHYLETHYLAMINE HYDROCHLORIDE see BROMADRYL
2-BROMO-2-METHYLPROPANE see tert-BUTYL BROMIDE
2-BROMO-2-NITROPROPAN-1,3-DIOL see 2-BROMO-2-NITRO-1,3-PRO-
  PANEDIOL

P-BROMOPHENOL see 4-BROMOPHENOL

ALPHA-BROMOPHENYLACETONITRILE see BROMOBENZYLNITRILE

BROMOPHENYLMETHANE see BENZYL BROMIDE

1-(P-BROMOPHENYL)-1-PHENYL-1-(2-DIMETHYLAMINOETHOXY)-
ETHANE HYDROCHLORIDE see BROMADRYL

2-(1-(4-BROMOPHENYL)-1-PHENYLETHOXY)-N,N-DIMETHYLETHAN-
AMINE HYDROCHLORIDE see BROMADRYL

(2-(1-P-BROMOPHENYL-1-PHENYLETHOXY)ETHYL)DIMETHYLETHYL-
AMINE HYDROCHLORIDE see BROMADRYL

BROMOPICRIN see NITROTRIBROMOMETHANE

3-BROMOPROPENE see ALLYL BROMIDE

BETA-BROMOPROPIONIC ACID see 3-BROMOPROPIONIC ACID

3-BROMOPROPYLENE see ALLYL BROMIDE

3-BROMOPROPYNE see PROPARGYL BROMIDE

3-BROMO-1-PROPYNE see PROPARGYL BROMIDE

BROMO SELTZER see p-ACETOPHENETIDIDE

N-BROMOSUCCIMIDE see N-BROMO SUCCINIMIDE

ALPHA-BROMOTOLUENE see BENZYL BROMIDE

OMEGA-BROMOTOLUENE see BENZYL BROMIDE

ALPHA-BROMO-ALPHA-TOLUNITRILE see BROMOBENZYLNITRILE

5-BROMOURACIL DEOXYRIBOSIDE see 5-BROMO-2'-DEOXY URIDINE

5-BROMOURACIL-2-DEOXYRIBOSIDE see 5-BROMO-2'-DEOXY URIDINE

BROMOVALEROCARBAMIDE see 2-BROMO-3-METHYL BUTYRYL UREA

BROMOWODOR (POLISH) see HYDROBROMIC ACID

BROMOXYNIL see 3,5-DIBROMO-4-HYDROXYBENZONITRILE

BROMOXYNIL OCTANOATE see 2,6-DIBROMO-4-CYANOPHENYL OC-
TANOATE

BROMURE DE CYANOGEN (FRENCH) see CYANOGEN BROMIDE

BROMURE DE METHYLE (FRENCH) see BROMO METHANE

BROMURE D'ETHYLE see BROMOETHANE

BROMURE DE VINYLE (FRENCH) see VINYL BROMIDE

BROMURO DI ETILE (ITALIAN) see 1,2-ETHYLENE DIBROMIDE

BROMURO DI METILE (ITALIAN) see BROMO METHANE

BROMWASSERSTOFF (GERMAN) see HYDROBROMIC ACID

BRONZE POWDER see COPPER

BROOM (DUTCH) see BROMINE

BROOMMETHAAN (DUTCH) see BROMO METHANE

BROOMWATERSTOF (DUTCH) see HYDROBROMIC ACID

BROWN ACETATE see CALCIUM ACETATE

BROWN COPPER OXIDE see COPPER (I) OXIDE

BROWN HEMATITE see LIMONITE

BROWN IRON ORE see LIMONITE

BROWN IRONSTONE CLAY see LIMONITE

BRUCIN (GERMAN) see BRUCINE

BRUCINA (ITALIAN) see BRUCINE

BRUCINE ALKALOID see BRUCINE

BRUCINE IODOMETHYLATE see BRUCINE METHIODIDE

BRUCINE IODOMETHYLE (FRENCH) see BRUCINE METHIODIDE

SECBUBARBITAL SODIUM see BUTISOL SODIUM

BUCACID FAST ORANGE G see HISPACID FAST ORANGE 2G

BUFONAMIN see BUFORMIN HYDROCHLORIDE

BUFOTALIN see BUFOTALINE

BUFOTENIN see 3-(2-DIMETHYLAMINOETHYL)-5-INDOLOL

BU-MDI see 2-n-BUTYL-3-DIMETHYLAMINO-5,6-METHYLENE DIOXY
INDENE HYDROCHLORIDE

BUNSENITE see NICKEL MONOXIDE

DL-BUPIVACAINE see 1-BUTYL-2',6'-PIPECOLOXYLIDIDE

BURESE see COCAINE

BUREX (CZECH) see 1-PHENYL-4-AMINO-5-CHLORPYRIDAŻ-6-ONE

BURNISH GOLD see GOLD

BURNT LIME see CALCIUM OXIDE

BURNT SIENNA see IRON (III) OXIDE

BURNT UMBER see IRON (III) OXIDE

BUTABARBITAL see BUTISOL

SECBUTABARBITAL see BUTISOL

BUTABARBITAL SODIUM see BUTISOL SODIUM

BUTADIEEN (DUTCH) see 1,3-BUTADIENE

BUTA-1,3-DIEEN (DUTCH) see 1,3-BUTADIENE

BUTADIEN (POLISH) see 1,3-BUTADIENE

BUTA-1,3-DIEN (GERMAN) see 1,3-BUTADIENE

BUTA-1,3-DIENE see 1,3-BUTADIENE

ALPHA-GAMMA-BUTADIENE see 1,3-BUTADIENE

BUTADIENE-ACRYLONITRILE-STYRENE COPOLYMER see ACRYLONI-
TRILE, POLYMER WITH 1,3-BUTADIENE AND STYRENE

BUTADIENE-ACRYLONITRILE-STYRENE TERPOLYMER see ACRYLONI-
TRILE, POLYMER WITH 1,3-BUTADIENE AND STYRENE

BUTADIENE DIMER see CYCLOHEXENYLETHYLENE

BUTADIENE DIOXIDE see 1,1'-BI(ETHYLENE OXIDE)

DL-BUTADIENE DIOXIDE see d,l-DIEPOXYBUTANE

BUTADIENE MONOXIDE see 3,4-EPOXY-1-BUTENE

BUTADIENE-STYRENE-ACRYLONITRILE COPOLYMER see ACRYLONI-
TRILE, POLYMER WITH 1,3-BUTADIENE AND STYRENE

BUTALBARBITAL see ALLYLISOBUTYLBARBITURATE

BUTALGIN see dl-METHADONE HYDROCHLORIDE

N-BUTANAL (CZECH) see N-BUTYRALDEHYDE

BUTANAL OXIME see m-BUTYRALDEHYDE OXIME

1-BUTANAMINE see N-BUTYLAMINE

1,4-BUTANEDICARBOXAMIDE see ADIPAMIDE

1,4-BUTANEDICARBOXYLIC ACID see ADIPIC ACID

BUTANEDINITRILE see SUCCINONITRILE

BUTANEDIOIC ACID MONO(2,2-DIMETHYLHYDRAZIDE) see DIMETHYL-
AMINOSUCCINAMIC ACID

BUTANEDIOIC ANHYDRIDE see SUCCINIC ANHYDRIDE

BUTANE-1,3-DIOL see 1,3-BUTANEDIOL

BUTANE-1,4-DIOL see 1,4-BUTANEDIOL

1,3-BUTANEDIOL, CYCLIC SULFITE see 4-METHYL TRIMETHYLENE
SULFITE

1,4-BUTANEDIOL DIMETHANESULPHONATE see 1,4-BUTANEDIOL
DIMETHYL SULFONATE

BUTANEN (DUTCH) see N-BUTANE

BUTANENITRILE see BUTYRONITRILE

1,4-BUTANESULTONE see BUTANESULTONE

DELTA-BUTANE SULTONE see BUTANESULTONE

5H,6H-6,5A,13A,14-(1,2,3,4)BUTANETETRAYCYCLOOCTA(1,2-B:
5,6-B')DINAPHTHALENE see LUTEOSKYRIN

BUTANETHIOIC ACID, 4-FLUORO-3-HYDROXY-, METHYL ESTER see
METHYL-gamma-FLUORO-beta-HYDROXYTHIOLBUTYRATE

TERT-BUTANETHIOL see 2-METHYL-2-PROPANETHIOL

BUTANI (ITALIAN) see N-BUTANE

BUTANOIC ACID see N-BUTYRIC ACID

BUTANOIC ACID, 1,2,3-PROPANETRIYL ESTER see TRIBUTYRIN

1-BUTANOL see N-BUTYL ALCOHOL

BUTAN-1-OL see N-BUTYL ALCOHOL

2-BUTANOL see sec-BUTYL ALCOHOL

BUTAN-2-OL see sec-BUTYL ALCOHOL

N-BUTANOL see N-BUTYL ALCOHOL

T-BUTANOL see tert-BUTYL ALCOHOL

SEC-BUTANOL see sec-BUTYL ALCOHOL

BUTANOL (FRENCH) see N-BUTYL ALCOHOL

BUTANOL-2-AMINE see 2-AMINOBUTAN-1-OL

2-BUTANOL, 1-(DIMETHYLAMINO)-2-((DIMETHYLAMINO)METHYL)-,
BENZOATE,(ESTER) see ALYPIN

2-BUTANOL, 1-(DIMETHYLAMINO)-2-METHYL-, BENZOATE (ESTER)
see AMYLOCAINE

BUTANOLEN (DUTCH) see N-BUTYL ALCOHOL

BUTANOLO (ITALIAN) see N-BUTYL ALCOHOL

2-BUTANOL-3-ONE see ACETOIN

BUTANONE-2 (FRENCH) see 2-BUTANONE

2-BUTANONE, SEMICARBAZONE see METHYL ETHYL KETONE SEMI-
CARBAZONE

2-BUTENAL, (E)- see trans-2-BUTENAL

BUTENEDIOIC ACID, (Z)- see cis-BUTENEDIOIC ACID

2-BUTENEDIOIC ACID, DIBUTYL ESTER see DIBUTYL MALEATE

CIS-BUTENEDIOIC ANHYDRIDE see MALEIC ANHYDRIDE

1-BUTENE-4-NITRILE see 3-BUTENE NITRILE

3-BUTENE-2-ONE see 3-BUTEN-2-ONE

1,2-BUTENE OXIDE see 1-BUTENE OXIDE

2-BUTENOIC ACID see CROTONIC ACID

alpha-BUTENOIC ACID see CROTONIC ACID

2-BUTENOIC ACID, 3-((DIMETHOXYPHOSPHINYL) OXY)-, 1-PHENYL-ETHYL ESTER, (E) - (9CI) see CROTOXYPHOS

2-BUTENOL see 2-BUTEN-1-OL

BETA-BUTENONITRILE see 3-BUTENE NITRILE

2-BUTENYL ALCOHOL see 2-BUTEN-1-OL

BUTETHAMINE HYDROCHLORIDE see IBYLCAINE HYDROCHLORIDE

N-BUTILAMINA (ITALIAN) see N-BUTYLAMINE

BUTILE (ACETATI DI) (ITALIAN) see N-BUTYL ACETATE

BUTIL METACRILATO (ITALIAN) see 2-METHYL BUTYLACRYLATE

BUTIWAS-SIMPLE see 4-BUTYL-1,2-DIPHENYL-3,5-DIOXO PYRAZOLI-DINE

BUTOBARBITONE see BUTOBARBITAL

BUTOBARBITURAL see BUTOBARBITAL

BUTOKSYETYLOWY ALKOHOL (POLISH) see BUTYL CELLOSOLVE

2-BUTOSSI-ETANOLO (ITALIAN) see BUTYL CELLOSOLVE

2-BUTOXY-AETHANOL (GERMAN) see BUTYL CELLOSOLVE

2-BUTOXY-3-AMINOBENZOIC ACID BETA-DIETHYLAMINOETHYL ESTER HYDROCHLORIDE see 3-AMINO-2-BUTOXYBENZOIC ACID-2-DIETHYLAMINOETHYL ESTER HYDROCHLORIDE

1-BUTOXYBUTANE see n-BUTYL ETHER

BUTOXYCINCHONINIC ACID DIETHYLETHYLENEDIAMIDE HYDRO-CHLORIDE see NUPERCAINE HYDROCHLORIDE

2-BUTOXY-N-(2-(DIETHYLAMINO)ETHYL)CINCHONINAMIDE see DIBUCAINE

2-BUTOXY-N-(beta-DIETHYLAMINOETHYL)CINCHONINAMIDE see DIBUCAINE

2-BUTOXY-N-(2-DIETHYLAMINOETHYL)CINCHONINAMIDE HYDRO-CHLORIDE see NUPERCAINE HYDROCHLORIDE

2-N-BUTOXY-N-(2-DIETHYLAMINOETHYL)CINCHONINAMIDE HYDRO-CHLORIDE see NUPERCAINE HYDROCHLORIDE

2-BUTOXY-N-(2-DIETHYLAMINOETHYL)CINCHONINIC ACID AMIDE HY-DROCHLORIDE see NUPERCAINE HYDROCHLORIDE

2-BUTOXY-1-ETHANOL see BUTYL CELLOSOLVE

2-BUTOXYETHANOL ACETATE see 2-BUTOXYETHYL ACETATE

2-(2-BUTOXYETHOXY)ETHANOL ACETATE see BUTYL CARBITOL ACE-TATE

ALPHA-(2-(2-BUTOXYETHOXY)ETHOXY)-4,5-METHYLENEDIOXY-2-PROPYLTOLUENE see PIPERONYL BUTOXIDE

ALPHA-(2-(2-N-BUTOXYETHOXY)-ETHOXY)-4,5-METHYLENEDIOXY-2-PROPYLTOLUENE see PIPERONYL BUTOXIDE

5-((2-(2-BUTOXYETHOXY)ETHOXY)METHYL)-6-PROPYL-1,3-BENZO-DIOXOLE see PIPERONYL BUTOXIDE

2-(2-BUTOXYETHOXY)ETHYL ACETATE see BUTYL CARBITOL ACE-TATE

2-(2-(BUTOXY)ETHOXY)ETHYL THIOCYANIC ACID ESTER see 2-(2-BU-TOXY ETHOXY)ETHYL THIOCYANATE

1-(2-BUTOXYETHOXY)-2-PROPANOL see 1-BUTOXY ETHOXY-2-PRO-PANOL

BUTOXYPHENYL see BUTYL PHENYL ETHER

4-BUTOXYPHENYLACETOHYDROXAMIC ACID see p-BUTOXY PHENYL ACETOHYDROXAMIC ACID

4-N-BUTOXY-BETA-(1-PIPERIDYL)PROPIOPHENONE HYDROCHLORIDE see 4'-BUTOXY-3-PIPERIDINO PROPIOPHENONE HYDROCHLORIDE

2-BUTOXYQUINOLINE-4-CARBOXYLIC ACID DIETHYLAMINOETHYL-AMIDE see DIBUCAINE

BUTOXYRHODANODIETHYL ETHER see 2-(2-BUTOXY ETHOXY)ETHYL THIOCYANATE

2-BUTOXY-2'-THIOCYANODIETHYL ETHER see 2-(2-BUTOXY ETH-OXY)ETHYL THIOCYANATE

BETA-BUTOXY-BETA'-THIOCYANODIETHYL ETHER see 2-(2-BUTOXY ETHOXY)ETHYL THIOCYANATE

1-BUTOXY-2-(2-THIOCYANOETHOXY)ETHANE see 2-(2-BUTOXY ETH-OXY)ETHYL THIOCYANATE

BUTTERCUP YELLOW see CHROMIUM(6+)ZINC OXIDE HYDRATE (1:2:6:1)

BUTTERCUP YELLOW see BASIC ZINC CHROMATE

BUTTER OF ANTIMONY see ANTIMONY (V) CHLORIDE

BUTTER OF ANTIMONY see ANTIMONY (III) CHLORIDE

BUTTER OF ZINC see ZINC CHLORIDE

BUTTERSAEURE (GERMAN) see N-BUTYRIC ACID

BUTTER YELLOW see 2-AMINO-5-AZOTOLUENE

BUTYLACETAT (GERMAN) see N-BUTYL ACETATE

1-BUTYL ACETATE see N-BUTYL ACETATE

2-BUTYL ACETATE see sec-BUTYL ACETATE

T-BUTYL ACETATE see ACETIC ACID-tert-BUTYL ESTER

BUTYLACETATEN (DUTCH) see N-BUTYL ACETATE

BUTYL ACETOXYMETHYLNITROSAMINE see BUTYL NITROS AMINO METHYL ACETATE

N-BUTYL-N-(ACETOXYMETHYL)NITROSAMINE see BUTYL NITROS AMINO METHYL ACETATE

N-BUTYL ACID PHOSPHATE see ACID BUTYL PHOSPHATE

N-BUTYL ACID PHOSPHATE (DOT) see BUTYL ESTER PHOSPHORIC ACID

N-BUTYL ACRYLATE see BUTYL-2-PROPENOATE

2-BUTYL ALCOHOL see sec-BUTYL ALCOHOL

SEC-BUTYL ALCOHOL ACETATE see sec-BUTYL ACETATE

N-BUTYL ALDEHYDE see N-BUTYRALDEHYDE

SEC-BUTYL ALLYL BARBITURIC ACID see 5-ALLYL-5-sec-BUTYLBAR-BITURIC ACID

N-BUTYLAMIN (GERMAN) see N-BUTYLAMINE

P-(BUTYLAMINO)BENZOIC ACID, 2-(DIMETHYLAMINO)ETHYL ESTER see p-(BUTYL AMINO)BENZOIC ACID-2(DIMETHYL AMINO)ETHYL ESTER

P-BUTYLAMINOBENZOYL-2-DIMETHYLAMINOETHANOL see p-(BUTYL AMINO)BENZOIC ACID-2(DIMETHYL AMINO)ETHYL ESTER

1-(BUTYLAMINO)CYCLOHEXYLPHOSPHONIC ACID DIBUTYL ESTER see AMINOPHON

2-(TERT-BUTYLAMINO)ETHYL METHACRYLATE see tert-BUTYL AMINO ETHYL METHACRYLATE

2-(TERT-BUTYLAMINO)-1-(4-HYDROXY-3-HYDROXYMETHYLPHE-NYL)ETHANOL see alpha'-((tert-BUTYL AMINO)METHYL)-4-HYDROXY-m-XYLENE-alpha, alpha'-DIOL

BUTYLAMYLNITROSAMIN (GERMAN) see N-BUTYL-N-NITROSO AMYL AMINE

BUTYLATED HYDROXYTOLUENE see BHT (FOOD GRADE)

2-BUTYL-3-BENZOFURANYL P-((2-DIETHYLAMINO)ETHOXY)-M,M-DIIODOPHENYL KETONE see AMINODARONE

BUTYL BENZYL PHTHALATE see BENZYL BUTYL PHTHALATE

1-BUTYLBIGUANIDE HYDROCHLORIDE see N-BUTYLBIGUANIDE HY-DROCHLORIDE

BUTYL BORATE see TRI-n-BUTYL BORATE

N-BUTYL BORATE see TRI-n-BUTYL BORATE

SEC-BUTYL-BROM-ALLYL BARBITURIC ACID SODIUM SALT see BU-TALLYLONAL SODIUM

SEC-BUTYL BROMIDE see 2-BROMOBUTANE

5-SEC-BUTYL-5-(BETA-BROMOALLYL)BARBITURIC ACID see BUTAL-LYLONAL

BUTYL-BUTANOL(4)-NITROSAMIN (GERMAN) see BUTANOL-4-BUTYL NITROSAMINE

BUTYL-BUTANOL-NITROSAMINE see BUTANOL-4-BUTYL NITROSA-MINE

N-BUTYL-N-BUTYRATE see n-BUTYL-n-BUTANOATE

1-(BUTYLCARBAMOYL)-2-BENZIMIDAZOLECARBAMIC ACID, METHYL ESTER see METHYL-1-(BUTYLCARBAMOYL)-2-BENZIMIDAZO-LYLCARBAMATE

1-(N-BUTYLCARBAMOYL)-2-(METHOXY-CARBOXAMIDO)-BENZIMIDA-ZOL (GERMAN) see METHYL-1-(BUTYLCARBAMOYL)-2-BENZIMI-DAZOLYLCARBAMATE

N'-(BUTYLCARBAMOYL)SULFANILAMIDE see 1-BUTYL-3-SULFANILYL UREA

N-BUTYLCARBINOL see AMYL ALCOHOL

N-BUTYLCARBINOL see PENTYL ALCOHOL

SEC-BUTYLCARBINOL see 2-METHYL BUTANOL-1

BUTYL CARBITOL 6-PROPYLPIPERONYL ETHER see PIPERONYL BUTOXIDE

BUTYL CARBITOL RHODANATE see 2-(2-BUTOXY ETHOXY)ETHYL THIOCYANATE

BUTYL CARBITOL THIOCYANATE see 2-(2-BUTOXY ETHOXY)ETHYL THIOCYANATE

BUTYL-CARBITYL (6-PROPYLPIPERONYL) ETHER see PIPERONYL BUTOXIDE

4-6-BUTYLCATECHOL see 4-tert-BUTYLPYROCATECHOL

BUTYL CELLOSOLVE ACETATE see 2-BUTOXYETHYL ACETATE

O-(4-TERT BUTYL-2-CHLOOR-FENYL)-O-METHYL-FOSFORZUUR-N-METHYL-AMIDE (DUTCH) see 4-tert-BUTYL-2-CHLORO PHENYL METHYL METHYL PHOSPHORAMIDATE

sec-BUTYL CHLORIDE see 2-CHLOROBUTANE

4-TERT. BUTYL 2-CHLOROPHENYL METHYLPHOSPHORAMIDATE DE METHYLE (FRENCH) see 4-tert-BUTYL-2-CHLORO PHENYL METHYL METHYL PHOSPHORAMIDATE

O-(4-TERT-BUTYL-2-CHLOR-PHENYL)-O-METHYL-PHOSPHORSAEURE-N-METHYL AMID (GERMAN) see 4-tert-BUTYL-2-CHLORO PHENYL METHYL METHYL PHOSPHORAMIDATE

2-t-BUTYL-p-CRESOL see 2-tert-BUTYL-p-CRESOL

BUTYL 2,4-D see BUTYL DICHLORO PHENOXY ACETATE

BUTYL (2,4-DICHLOROPHENOXY)ACETATE see BUTYL DICHLORO PHENOXY ACETATE

N-BUTYLDIETHANOLAMINE see N-BUTYL-N,N-BIS(HYDROXY ETHYL)-AMINE

1-BUTYLDIGUANIDE HYDROCHLORIDE see N-BUTYLBIGUANIDE HYDROCHLORIDE

BUTYL-3,4-DIHYDRO-2,2-DIMETHYL-4-OXO-2H-PYRAN-6-CARBOXY-LATE see n-BUTYL MESITYL OXIDE OXALATE

2-BUTYL-3-(3,5-DIIODO-4-(2-DIETHYLAMINOETHOXY)BENZOYL)BENZOFURAN see AMINO-DARONE

2-N-BUTYL-3′,5′-DIIODO-4′-N-DIETHYLAMINOETHOXY-3-BENZOYL-BENZOFURAN see AMINODARONE

3-BUTYL-1-(2-(DIMETHYLAMINO)ETHOXY)ISOQUINOLINE HYDRO-CHLORIDE see DIMETHISOQUIN HYDROCHLORIDE

6-BUTYL-5-DIMETHYLAMINO-5H-INDENO(5,6-D)-1,3-DIOXOLE HYDROCHLORIDE see 2-n-BUTYL-3-DIMETHYLAMINO-5,6-METHYLENE DIOXY INDENE HYDROCHLORIDE

2-SEC-BUTYL-4,6-DINITROPHENOL, AMMONIUM SALT see BUTOPHEN

2-SEC-BUTYL-4,6-DINITROPHENYLACETATE see O-ACETYL-2-sec-BU-TYL-4,6-DINITROPHENOL

6-SEC-BUTYL-2,4-DINITROPHENYLACETATE see O-ACETYL-2-sec-BU-TYL-4,6-DINITROPHENOL

2-SEC-BUTYL-4,6-DINITROPHENYL-3,3-DIMETHYLACRYLATE see BINAPACRYL

2-SEC-BUTYL-4,6-DINITROPHENYL-3-METHYL-2-BUTENOATE see BINAPACRYL

2-SEC-BUTYL-4,6-DINITROPHENYL-3-METHYLCROTONATE see BINAPACRYL

4-BUTYL-1,2-DIPHENYL-3,5-PYRAZOLIDINEDIONE see 4-BUTYL-1,2-DIPHENYL-3,5-DIOXO PYRAZOLIDINE

4-BUTYL-1,2-DIPHENYLPYRAZOLIDINE-3,5-DIONE see 4-BUTYL-1,2-DIPHENYL-3,5-DIOXO PYRAZOLIDINE

4-BUTYL-1,2-DIPHENYL-3,5-PYRAZOLIDINEDIONE SODIUM SALT see BUTAZOLIDINE SODIUM

BUTYLE (ACETATE DE) (FRENCH) see N-BUTYL ACETATE

ALPHA BUTYLENE see 1-BUTENE

1,2-BUTYLENE GLYCOL see 1,2-BUTANEDIOL

1,3-BUTYLENE GLYCOL see 1,3-BUTANEDIOL

1,4-BUTYLENE GLYCOL see 1,4-BUTANEDIOL

2,3-BUTYLENE GLYCOL see 2,3-BUTANEDIOL

BETA-BUTYLENE GLYCOL see 1,3-BUTANEDIOL

BUTYLENE HYDRATE see sec-BUTYL ALCOHOL

BUTYLENE OXIDE see TETRAHYDROFURAN

1,2-BUTYLENE OXIDE see 1-BUTENE OXIDE

1,4-BUTYLENE SULFONE see BUTANESULTONE

2,4,5-T-N-BUTYL ESTER see BUTYL-2,4,5-TRICHLORO PHENOXY ACETATE

N-BUTYLESTER KYSELINY 2,4,5-TRICHLORFENOXYOCTOVE (CZECH) see BUTYL-2,4,5-TRICHLORO PHENOXY ACETATE

BUTYL ESTER OF DICHLOROPHENOXYACETIC ACID see BUTYL DI-CHLORO PHENOXY ACETATE

N-BUTYL ESTER OF 3,4-DIHYDRO-2,2-DIMETHYL-4-OXO-2H-PYRAN-6-CARBOXYLIC ACID see n-BUTYL MESITYL OXIDE OXALATE

BUTYL ETHANOATE see N-BUTYL ACETATE

5-BUTYL-5-ETHYLBARBITURIC ACID see BUTOBARBITAL

5-SEC-BUTYL-5-ETHYLBARBITURIC ACID see BUTISOL

5-SEC-BUTYL-5-ETHYLBARBITURIC ACID SODIUM SALT see BUTISOL SODIUM

O-BUTYL ETHYLENE GLYCOL see BUTYL CELLOSOLVE

N-BUTYL ETHYL KETONE see 3-HEPTANONE

5-SEC-BUTYL-5-ETHYLMALONYL UREA see BUTISOL

2-sec-BUTYLFENOL (CZECH) see o-sec-BUTYLPHENOL

p-tert-BUTYLFENOL (CZECH) see 4-tert-BUTYLPHENOL

BUTYL FORMAL see n-VALERALDEHYDE

BUTYL FORMATE (DOT) see N-BUTYL FORMATE

N-BUTYL GLYCIDYL ETHER see BUTYL GLYCIDYL ETHER

BUTYLGLYCOL (FRENCH, GERMAN) see BUTYL CELLOSOLVE

BUTYL HYDROXIDE see N-BUTYL ALCOHOL

T-BUTYL HYDROXIDE see tert-BUTYL ALCOHOL

BUTYLHYDROXYANISOLE see BUTYLATED HYDROXY ANISOLE

TERT-BUTYLHYDROXYANISOLE see BUTYLATED HYDROXY ANISOLE

TERT-BUTYL-4-HYDROXYANISOLE see BUTYLATED HYDROXY ANISOLE

N-BUTYL-(4-HYDROXYBUTYL)NITROSAMINE see BUTANOL-4-BUTYL NITROSAMINE

N-BUTYL-N-(4-HYDROXYBUTYL)NITROSAMINE see BUTANOL-4-BUTYL NITROSAMINE

BUTYLHYDROXYOXOSTANNANE see BUTYL STANNOIC ACID

4-BUTYL-2-(4-HYDROXYPHENYL)-1-PHENYL-3,5-DIOXOPYRAZOLIDINE see p-HYDROXYPHENYLBUTAZONE

4-BUTYL-1-(4-HYDROXYPHENYL)-2-PHENYL-3,5-PYRAZOLIDINEDIONE see p-HYDROXYPHENYLBUTAZONE

4-BUTYL-1-(P-HYDROXYPHENYL)-2-PHENYL-3,5-PYRAZOLIDINEDIONE see p-HYDROXYPHENYLBUTAZONE

4-BUTYL-2-(P-HYDROXYPHENYL)-1-PHENYL-3,5-PYRAZOLIDINEDIONE see p-HYDROXYPHENYLBUTAZONE

BUTYLHYDROXYTOLUENE see BHT (FOOD GRADE)

N-BUTYL-IMIDODICARBONIMIDIC DIAMIDE MONOHYDROCHLORIDE (9CI) see N-BUTYLBIGUANIDE HYDROCHLORIDE

N-BUTYL-2,2′-IMINODIETHANOL see N-BUTYL-N,N-BIS(HYDROXY ETHYL)AMINE

2-tert-BUTYL-p-KRESOL (CZECH) see 2-tert-BUTYL-p-CRESOL

BUTYL MERCAPTAN see n-BUTANETHIOL

TERT-BUTYL MERCAPTAN see 2-METHYL-2-PROPANETHIOL

N-BUTYLMESITYLOXID OXALATE see n-BUTYL MESITYL OXIDE OXALATE

BUTYLMETHACRYLAAT (DUTCH) see 2-METHYL BUTYLACRYLATE

BUTYL-2-METHACRYLATE see 2-METHYL BUTYLACRYLATE

N-BUTYL METHACRYLATE see 2-METHYL BUTYLACRYLATE

N-BUTYL METHYL KETONE see 2-HEXANONE

2-T-BUTYL-4-METHYLPHENOL see 2-tert-BUTYL-p-CRESOL

1-BUTYL-3-(P-METHYLPHENYLSULFONYL)UREA see 1-BUTYL-3-(p-TOLYL SULFONYL)UREA

BUTYL-2-METHYL-2-PROPENOATE see 2-METHYL BUTYLACRYLATE

N-BUTYL-2-METHYL-2-PROPYL-1,3-PROPANEDIOL DICARBAMATE see 2-METHYL-2-PROPYLTRIMETHYLENE BUTYLCARBAMATE CARBA-MATE

N-N-BUTYL-2-METHYL-2-PROPYL-1,3-PROPANEDIOL DICARBAMATE see 2-METHYL-2-PROPYLTRIMETHYLENE BUTYLCARBAMATE CARBAMATE

BUTYL NAMATE see SODIUM DIBUTYLDITHIOCARBAMATE

BUTYL NITRITE see n-BUTYL NITRITE

4-(BUTYLNITROSAMINO)-1-BUTANOL see BUTANOL-4-BUTYL NITRO-SAMINE

4-(N-BUTYLNITROSAMINO)-1-BUTANOL see BUTANOL-4-BUTYL NITROSAMINE

4-(BUTYLNITROSOAMINO)BUTANOIC ACID see N-BUTYL-(3-CARBOXY PROPYL)NITROSAMINE

BUTYLNITROSOHARNSTOFF (GERMAN) see n-BUTYL NITROSO UREA

N-BUTYL-N-NITROSOPENTYLAMINE see N-BUTYL-N-NITROSO AMYL AMINE

1-BUTYL-1-NITROSOUREA see n-BUTYL NITROSO UREA

N-BUTYL-N-NITROSOUREA see n-BUTYL NITROSO UREA

N-BUTYL-N-NITROSOURETHAN see N-BUTYL-N-NITROSO ETHYL CARBAMATE

2-BUTYLOCTYL ALCOHOL see 2-BUTYL-1-OCTANOL

BUTYLOWY ALKOHOL (POLISH) see N-BUTYL ALCOHOL

alpha-BUTYLOXYCINCHONINIC ACID DIETHYLETHYLENEDIAMIDE see DIBUCAINE

N-BUTYL-N-PENTYLINITROSAMINE see N-BUTYL-N-NITROSO AMYL AMINE

TERT-BUTYLPERBENZOAN (CZECH) see tert-BUTYL PERBENZOATE

T-BUTYL PEROXYACETATE see tert-BUTYL PERACETATE

T-BUTYL PEROXY BENZOATE see tert-BUTYL PERBENZOATE

P-TERT-BUTYLPHENOL see 4-tert-BUTYLPHENOL

BUTYLPHENOXYISOPROPYL CHLOROETHYL SULFITE see SULFUROUS ACID, 2-(p-t-BUTYLPHENOXY)-1-METHYLETHYL-2-CHLOROETHYL ESTER

2-(P-BUTYLPHENOXY)ISOPROPYL 2-CHLOROETHYL SULFITE see SULFUROUS ACID, 2-(p-t-BUTYLPHENOXY)-1-METHYLETHYL-2-CHLOROETHYL ESTER

2-(4-T-BUTYLPHENOXY)ISOPROPYL-2-CHLOROETHYL SULFITE see SULFUROUS ACID, 2-(p-t-BUTYLPHENOXY)-1-METHYLETHYL-2-CHLOROETHYL ESTER

2-(P-T-BUTYLPHENOXY)ISOPROPYL 2'-CHLOROETHYL SULPHITE see SULFUROUS ACID, 2-(p-t-BUTYLPHENOXY)-1-METHYLETHYL-2-CHLOROETHYL ESTER

2-(P-T-BUTYLPHENOXY)-1-METHYLETHYL 2-CHLOROETHYL ESTER OF SULPHUROUS ACID see SULFUROUS ACID, 2-(p-t-BUTYLPHENOXY)-1-METHYLETHYL-2-CHLOROETHYL ESTER

2-(P-BUTYLPHENOXY)-1-METHYLETHYL 2-CHLOROETHYL SULFITE see SULFUROUS ACID, 2-(p-t-BUTYLPHENOXY)-1-METHYLETHYL-2-CHLOROETHYL ESTER

2-(P-T-BUTYLPHENOXY)-1-METHYLETHYL-2-CHLOROETHYL SULFITE see SULFUROUS ACID, 2-(p-t-BUTYLPHENOXY)-1-METHYLETHYL-2-CHLOROETHYL ESTER

2-(P-T-BUTYLPHENOXY)-1-METHYLETHYL 2'-CHLOROETHYL SULPHITE see SULFUROUS ACID, 2-(p-t-BUTYLPHENOXY)-1-METHYLETHYL-2-CHLOROETHYL ESTER

2-(P-T-BUTYLPHENOXY)-1-METHYLETHYL SULPHITE OF 2-CHLOROETHANOL see SULFUROUS ACID, 2-(p-t-BUTYLPHENOXY)-1-METHYLETHYL-2-CHLOROETHYL ESTER

O-SEC-BUTYLPHENYL METHYLCARBAMATE see 2-(1-METHYLPROPYL)PHENYL METHYLCARBAMATE

2-SEC-BUTYLPHENYL-N-METHYLCARBAMATE see 2-(1-METHYLPROPYL)PHENYL METHYLCARBAMATE

BUTYL PHOSPHORIC ACID see ACID BUTYL PHOSPHATE

BUTYL PHOSPHORIC ACID (DOT) see BUTYL ESTER PHOSPHORIC ACID

BUTYL PHTHALATE BUTYL GLYCOLATE see BUTYL CARBO BUTOXY METHYL PHTHALATE

BUTYL PHTHALYL BUTYL GLYCOLATE see BUTYL CARBO BUTOXY METHYL PHTHALATE

1-BUTYL-2',6'-PIPECOLOXYLIDIDE, HYDROCHLORIDE (±) see BUPICAINE HYDROCHLORIDE (±)

BUTYL PROPIONATE see BUTYL PROPANOATE

N-BUTYL PROPIONATE see BUTYL PROPANOATE

5-BUTYL-2-PYRIDINECARBOXYLIC ACID see 5-BUTYL PICOLINIC ACID

N-BUTYL RHODANINE see n-BUTYL THIOCYANATE

N-BUTYL-N'-P-TOLUENESULFONYLUREA see 1-BUTYL-3-(p-TOLYL SULFONYL)UREA

1-BUTYL-3-TOSYLUREA see 1-BUTYL-3-(p-TOLYL SULFONYL)UREA

N-N-BUTYL-N'-TOSYLUREA see 1-BUTYL-3-(p-TOLYL SULFONYL)UREA

4-N-BUTYL-4H-1,2,4-TRIAZOLE see BUTRIZOL

N-BUTYL (2,4,5-TRICHLOROPHENOXY)ACETATE see BUTYL-2,4,5-TRICHLORO PHENOXY ACETATE

BUTYN see BUTACAINE

1-BUTYNE see ETHYL ACETYLENE

2-BUTYNE see CROTONYLENE (DOT)

2-BUTYNEDIAMIDE see ACETYLENEDICARBOXAMIDE

1,4-BUTYNEDIOL see 2-BUTYNE-1,4-DIOL

BUTYN SULFATE see BUTACAINE SULFATE

2-BUTYNYL-4-CHLORO-M-CHLOROCARBANILATE see m-CHLORO CARBANILIC ACID-4-CHLORO-2-BUTYNYL ESTER

BUTYRAL see N-BUTYRALDEHYDE

BUTYRALDEHYD (GERMAN) see N-BUTYRALDEHYDE

BUTYRALDEHYDE (CZECH) see N-BUTYRALDEHYDE

N-BUTYRALDOXIME see m-BUTYRALDEHYDE OXIME

BUTYRHODANID (GERMAN) see n-BUTYL THIOCYANATE

BUTYRIC ACID, GAMMA-FLUORO-BETA-HYDROXY-, METHYL ESTER see METHYL-gamma-FLUORO-beta-HYDROXYBUTYRATE

BUTYRIC ACID LACTONE see 4-BUTANOLIDE

BUTYRIC ACID, VINYL ESTER see VINYL BUTYRATE

BUTYRIC ALDEHYDE see N-BUTYRALDEHYDE

ALPHA-BUTYROLACTONE see 4-BUTANOLIDE

GAMMA-BUTYROLACTONE see 4-BUTANOLIDE

BUTYRONE see 4-HEPTANONE

BUTYRYLETHYLENEIMINE see 1-n-BUTYRYLAZIRIDINE

BUTYRYLETHYLENIMINE see 1-n-BUTYRYLAZIRIDINE

BUTYRYL LACTONE see 4-BUTANOLIDE

BUTYRYL TRIGLYCERIDE see TRIBUTYRIN

C 3067 see 2-CHLORO-6-METHYLCARBANILIC ACID-2-(PYRROLIDINYL)ETHYL ESTER HYDROCHLORIDE

C 3235 see N-(2-(DIETHYLAMINO)ETHYL)-2,4,6-TRIMETHYLBENZAMIDE, HYDROCHLORIDE

C 5420 see 5'-CHLORO-2-(METHYL(2-(PYRROLIDINYL)ETHYL)-AMINO)-o-ACETOTOLUIDIDE DIHYDROCHLORIDE

C 5473 see OXYPHENONIUM BROMIDE

C ACID see 3-AMINO-1,5-NAPHTHALENEDISULFONIC ACID

CACODYL HYDRIDE see DIMETHYLARSINE

CACODYLIC ACID see HYDROXYDIMETHYLARSINE OXIDE

CADAVERIN see PENTANEDIAMINE

CADDY see CADMIUM CHLORIDE

CADMIUM DIACETATE see CADMIUM (II) ACETATE

CADMIUM DICHLORIDE see CADMIUM CHLORIDE

CADMIUM DIETHYL DITHIOCARBAMATE see BIS(DIETHYL DITHIO CARBAMATO)CADMIUM

CADMIUM FLUORURE (FRENCH) see CADMIUM FLUORIDE

CADMIUM FUME see CADMIUM OXIDE FUME

CADMIUM SULFATE OCTAHYDRATE see CADMIUM SULFATE (1:1) HYDRATE (3:8)

CADMIUM SULPHATE see CADMIUM SULFATE (1:1) HYDRATE (3:8)

CADMIUM SULPHIDE see CADMIUM SULFIDE

CA-DTPA see CALCIUM TRISODIUM DIETHYLENE TRIAMINE PENTAACETATE

CAFFEIN see CAFFEINE

CAFFEINE BROMIDE see CAFFEINE HYDROBROMIDE

CAKE ALUM see ALUMINUM SULFATE (2:3)

CALAMUS OIL see OIL OF CALAMUS

CALCIA see CALCIUM OXIDE

CALCIC LIVER OF SULFUR see CALCIUM SULFIDE

CALCIFEROL see ERGOCALCIFEROL

CALCINED BARYTA see BARIUM OXIDE

CALCINED BRUCITE see MAGNESIUM OXIDE

CALCINED DIATOMITE see SILICA, CRYSTALLINE-CRISTOBALITE

CALCINED MAGNESIA see MAGNESIUM OXIDE

CALCINED MAGNESITE see MAGNESIUM OXIDE

CALCINED SODA see SODIUM MONOXIDE

CALCITRIOL see DIHYDROXYVITAMIN D3

CALCIUMARSENAT see ARSENIC ACID, CALCIUM SALT (2:3)

CALCIUM ARSENATE see ARSENIC ACID, CALCIUM SALT (2:3)

CALCIUM ARSENITE, SOLID (DOT) see CALCIUM ARSENITE

CALCIUM-O-BENZOSULPHIMIDE see CALCIUM-O-BENZOSULFIMIDE

CALCIUM BISULFITE SOLUTION (DOT) see CALCIUM BISULFITE (solution)

CALCIUM CARBIMIDE see CALCIUM CYANAMIDE

CALCIUM CHLORATE (DOT) see CALCIUM CHLORATE

CALCIUM CHLOROHYDROCHLORITE see HYPOCHLOROUS ACID, CALCIUM SALT

CALCIUM CHROMATE (CaCRO4) see CALCIUM CHROMATE

CALCIUM CHROME YELLOW see CALCIUM CHROMATE (VI) DIHYDRATE

CALCIUM CHROME YELLOW see CALCIUM CHROMATE

CALCIUM CHROMIUM OXIDE (CaCrO4) see CALCIUM CHROMATE

CALCIUM CYANAMID see CALCIUM CYANAMIDE

CALCIUM CYANIDE MIXTURE, SOLID (DOT) see CALCIUM CYANIDE (mixture)

CALCIUM CYANIDE, SOLID (DOT) see CALCIUM CYANIDE

CALCIUM CYCLOHEXANESULFAMATE see CALCIUM CYCLOHEXYL-SULPHAMATE

CALCIUM CYCLOHEXYLSULFAMATE see CALCIUM CYCLOHEXYL-SULPHAMATE

CALCIUM DIACETATE see CALCIUM ACETATE

CALCIUM D(+)-N-(ALPHA,GAMMA-DIHYDROXY-BETA,BETA-DIMETHYLBUTYRYL)-BETA-ALANINATE see CALCIUM-d-PANTOTHENATE

CALCIUM FLUOSILICATE see CALCIUM SILICOFLUORIDE

CALCIUM HEXAFLUOROSILICATE see CALCIUM SILICOFLUORIDE

CALCIUM HYDRATE see CALCIUM HYDROXIDE

CALCIUM HYDROGEN SULFITE SOLUTION (DOT) see CALCIUM BISULFITE (solution)

CALCIUM HYPOCHLORIDE see HYPOCHLOROUS ACID, CALCIUM SALT

CALCIUM HYPOCHLORITE see HYPOCHLOROUS ACID, CALCIUM SALT

CALCIUM, METAL (DOT) see CALCIUM

CALCIUM, METAL, CRYSTALLINE (DOT) see CALCIUM

CALCIUM MONOCHROMATE see CALCIUM CHROMATE

CALCIUM NITRATE (DOT) see CALCIUM (II) NITRATE (1:2)

CALCIUM ORTHOARSENATE see ARSENIC ACID, CALCIUM SALT (2:3)

CALCIUM OXYCHLORIDE see HYPOCHLOROUS ACID, CALCIUM SALT

CALCIUM PANTOTHENATE see CALCIUM-d-PANTOTHENATE

D-CALCIUM PANTOTHENATE see CALCIUM-d-PANTOTHENATE

CALCIUMRHODANID (GERMAN) see CALCIUM THIOCYANATE

CALCIUM SACCHARIN see CALCIUM-o-BENZOSULFIMIDE

CALCIUM SACCHARINA see CALCIUM-o-BENZOSULFIMIDE

CALCIUM SACCHARINATE see CALCIUM-o-BENZOSULFIMIDE

CALCIUM SALT OF DIETHYLENETRIAMINEPENTAACETIC ACID see CALCIUM TRISODIUM DIETHYLENE TRIAMINE PENTAACETATE

CALCIUM TRISODIUM SALT OF DIETHYLENE TRIAMINE PENTA ACETIC ACID see CALCIUM TRISODIUM DIETHYLENE TRIAMINE PENTA-ACETATE

CALCOCID FAST BLUE SR see ACID BLUE 92

CALCOCID FAST LIGHT ORANGE 2G see HISPACID FAST ORANGE 2G

CALCOCID URANINE B4315 see FLUORESCEIN SODIUM

CALCO OIL RED D see SCARLET RED

CALCYANIDE see CALCIUM CYANIDE

CALGON see SODIUM HEXAMETAPHOSPHATE

CALICO YELLOW see MORIN

CALMIXENE see 9-(1-METHYL-4-PIPERIDYLIDENE)THIOXANTHENE

CALOMEL see MERCUROUS CHLORIDE

CALOMELANO (ITALIAN) see MERCUROUS CHLORIDE

CALX see CALCIUM OXIDE

2-CAMPHANOL see BORNEOL

1-2-CAMPHANOL see NGAI CAMPHOR

2-CAMPHANONE see CAMPHOR

D-2-CAMPHANONE see CAMPHOR, (1R,4R)-(+)-

CAMPHOGEN see p-CYMENE

CAMPHOR, (+)- see CAMPHOR, (1R,4R)-(+)-

(+)-CAMPHOR see CAMPHOR, (1R,4R)-(+)-

D-CAMPHOR see CAMPHOR, (1R,4R)-(+)-

D-(+)-CAMPHOR see CAMPHOR, (1R,4R)-(+)-

CAMPHOR-NATURAL see CAMPHOR

CAMPHOR OIL, RECTIFIED see CAMPHOR OIL

CAMPHOR OIL YELLOW see CAMPHOR OIL

CAMPHOR TAR see NAPHTHALENE

CAMPHOR USP see CAMPHOR, (1R,4R)-(+)-

CANADOL see PETROLEUM SPIRITS

CANARY CHROME YELLOW 40-2250 see LEAD CHROMATE

CANDEPTIN see CANDICIDIN

CANE SUGAR see SUCROSE

CANTHARIDES CAMPHOR see CANTHARIDINE

CANTHARIDIN see CANTHARIDES CAMPHOR

CANTHARIDIN see CANTHARIDINE

CANTHARONE see CANTHARIDES CAMPHOR

CANTHARONE see CANTHARIDINE

CAPRALDEHYDE see 1-DECANAL

N-CAPRIC ACID see DECANOIC ACID

CAPRIC ALCOHOL see DECYL ALCOHOL

CAPRINIC ALCOHOL see DECYL ALCOHOL

CAPROALDEHYDE see 1-HEXANAL

CAPROIC ALDEHYDE see 1-HEXANAL

CAPROLACTAM see HEXAHYDRO-2H-AZEPIN-2-ONE

6-CAPROLACTAM see HEXAHYDRO-2H-AZEPIN-2-ONE

OMEGA-CAPROLACTAM see HEXAHYDRO-2H-AZEPIN-2-ONE

EPSILON-CAPROLACTAM see HEXAHYDRO-2H-AZEPIN-2-ONE

CAPROLACTAM OLIGOMER see POLYCAPROLACTAM

CAPROLACTAM POLYMER see POLYCAPROLACTAM

EPSILON-CAPROLACTAM POLYMER see POLYCAPROLACTAM

EPSILON-CAPROLACTAM POLY-MERE (GERMAN) see POLYCAPRO-LACTAM

CAPROLATTAME (FRENCH) see HEXAHYDRO-2H-AZEPIN-2-ONE

CAPROLON see NYLON

CAPRONALDEHYDE see 1-HEXANAL

CAPRONITRILE see HEXANENITRILE

CAPROYL ALCOHOL see N-HEXYL ALCOHOL

N-CAPROYLALDEHYDE see 1-HEXANAL

1-CAPROYLAZIRIDINE see 1-HEXANOYLAZIRIDINE

CAPROYLETHYLENEIMINE see 1-HEXANOYLAZIRIDINE

CAPRYL ALCOHOL see N-OCTYL ALCOHOL

CAPRYLDINITROPHENYL CROTONATE see ARATHANE

2-CAPRYL-4,6-DINITROPHENYL CROTONATE see ARATHANE

CAPRYLIC ALCOHOL see N-OCTYL ALCOHOL

CAPSAICINE see CAPSAICIN

CAPUT MORTUUM see IRON (III) OXIDE

CARBACHOLINE CHLORIDE see CARBACHOL CHLORIDE

CARBAETHOXYDIGOXIN (GERMAN) see ETHOXY CARBONYL DIGOXIN

CARBAMALDEHYDE see FORMAMIDE

CARBAMAMIDINE see GUANIDINE

CARBAMIC ACID, 3-DIMETHYLAMINOPHENYL ESTER, METHIODIDE see (m-HYDROXYPHENYL)TRIMETHYLAMMONIUM IODIDE, METHYLCARBAMATE

CARBAMIC ACID, N,N-DIMETHYL-, 8-QUINOLINYL ESTER, METHO-SULFATE see DIMETHYLCARBAMIC ESTER of 8OXYMETHYLQUINOLINIUM METHYLSULFATE

CARBAMIC ACID HYDRAZIDE see HYDRAZINECARBOXAMIDE

CARBAMIC ACID, N-METHYL-3-DIMETHYLAMINOPHENYL ESTER METHIODIDE see (m-HYDROXYPHENYL)TRIMETHYLAM-MONIUM IODIDE, METHYLCARBAMATE

CARBAMIC ACID-1-METHYLETHYL ESTER see ISOPROPYL CARBA-MATE

CARBAMIC ACID, 3-PHENYLPROPYL ESTER see 3-PHENYL-1-PROPANOL CARBAMATE

CARBAMIC ACID, (3-TRIMETHYLAMMONIOPHENYL) ESTER, IODIDE see (m-HYDROXYPHENYL)TRIMETHYLAMMONIUM IODIDE, METHYLCARBAMATE

CARBAMIC ACID, VINYL ESTER see VINYL CARBAMATE

CARBAMIDE see UREA

CARBAMIDE RESIN see UREA

CARBAMIDINE see GUANIDINE

P-CARBAMIDOBENZENEARSONIC ACID see N-CARBAMOYL ARSANILIC ACID

CARBAMIDSAEURE-AETHYLESTER (GERMAN) see URETHANE

CARBAMIMIDIC ACID see UREA

CARBAMINOPHENYL-P-ARSONIC ACID see N-CARBAMOYL ARSANILIC ACID

P-CARBAMINO PHENYL ARSONIC ACID see N-CARBAMOYL ARSANILIC ACID

CARBAMINOYLCHOLINE CHLORIDE see CARBACHOL CHLORIDE

CARBAMOHYDROXAMIC ACID see HYDROXYUREA

CARBAMOHYDROXIMIC ACID see HYDROXYUREA

CARBAMOHYDROXYAMIC ACID see HYDROXYUREA

CARBAMONITRILE see CYANAMIDE

CARBAMOYLCHOLINE CHLORIDE see CARBACHOL CHLORIDE
GAMMA-CARBAMOYL CHOLINE CHLORIDE see CARBACHOL CHLORIDE
5-CARBAMOYL-5H-DIBENZ(B,F)AZEPINE see 5H-DIBENZ(b,f)AZEPINE-5-CARBOXAMIDE
5-CARBAMOYLDIBENZO(B,F)AZEPINE see 5H-DIBENZ(b,f)AZEPINE-5-CARBOXAMIDE
5-CARBAMOYL-5H-DIBENZO(B,F)AZEPINE see 5H-DIBENZ(b,f)AZEPINE-5-CARBOXAMIDE
CARBAMOYL OXIME see HYDROXYUREA
3-CARBAMOYLOXY-3-METHYL-4-PENTYNE see METHYLPENTYNOL CARBAMATE
1-CARBAMOYLOXY-3-PHENYLPROPANE see 3-PHENYL-1-PROPANOL CARBAMATE
1-CARBAMYL-2-PHENYLHYDRAZINE see 1-CARBAMYL-2-PHENYL HYDRAZINE
4-CARBAMYLAMINOPHENYLARSONIC ACID see N-CARBAMOYL ARSANILIC ACID
N-CARBAMYL ARSANILIC ACID see N-CARBAMOYL ARSANILIC ACID
CARBAMYLCHOLINE CHLORIDE see CARBACHOL CHLORIDE
5-CARBAMYLDIBENZO(B,F)AZEPINE see 5H-DIBENZ(b,f)AZEPINE-5-CARBOXAMIDE
5-CARBAMYL-5H-DIBENZO(B,F)AZEPINE see 5H-DIBENZ(b,f)AZEPINE-5-CARBOXAMIDE
CARBAMYLHYDRAZINE see HYDRAZINECARBOXAMIDE
CARBAMYLHYDRAZINE HYDROCHLORIDE see SEMICARBAZIDE HYDROCHLORIDE
CARBAMYL HYDROXAMATE see HYDROXYUREA
2-CARBAMYL PYRAZINE see PYRAZINECARBOXAMIDE
CARBATOX-60 see CARBARYL
CARBAZAMIDE see HYDRAZINECARBOXAMIDE
CARBAZOTIC ACID see PICRIC ACID
CARBETHOXYACETIC ESTER see ETHYL MALONATE
ALPHA-CARBETHOXY-BETA,BETA-BISCYCLOPROPYL ACRYLONITRILE see UV ABSORBER-2
N-CARBETHOXYETHYLENIMINE see AZIRIDINE CARBOXYLIC ACID ETHYL ESTER
CARBIDE 6-12 see ETHYL HEXYLENE GLYCOL
CARBIMIDE see CYANAMIDE
CARBINOL see METHANOL
CARBITOL see CARBITOL CELLOSOLVE
CARBITOL SOLVENT see CARBITOL CELLOSOLVE
CARBOBENZYLOXY CHLORIDE see BENZYL CHLOROFORMATE
2-CARBO-N-BUTOXY-6,6-DIMETHYL-5,6-DIHYDRO-1,4-PYRONE see n-BUTYL MESITYL OXIDE OXALATE
CARBOFENOTHION see TRITHION
CARBOFENOTHION (DUTCH) see TRITHION
CARBOLIC ACID see PHENOL
CARBOLSAURE (GERMAN) see PHENOL
CARBOMETHENE see KETENE
2-CARBOMETHOXYANILINE see ANTHRANILIC ACID, METHYL ESTER
O-CARBOMETHOXYANILINE see ANTHRANILIC ACID, METHYL ESTER
2-beta-CARBOMETHOXY-3-beta-BENZOXYTROPANE see COCAINE
ALPHA-2-CARBOMETHOXY-1-METHYLVINYLDIMETHYL PHOSPHATE see MEVINPHOS
2-CARBOMETHOXY-1-PROPEN-2-YL DIMETHYL PHOSPHATE see MEVINPHOS
CARBONA see CARBON TETRACHLORIDE
(CARBONATO)DIHYDROXYDICOPPER see COPPER (II) CARBONATE HYDROXIDE (2:1:2)
CARBON BICHLORIDE see 1,1,2,2-TETRACHLOROETHYLENE
CARBON BISULFIDE see CARBON DISULFIDE
CARBON BROMIDE see CARBON TETRABROMIDE
CARBON CHLORIDE see CARBON TETRACHLORIDE
CARBON CHLOROSULFIDE see THIOPHOSGENE
CARBON DICHLORIDE see 1,1,2,2-TETRACHLOROETHYLENE
CARBON DIOXIDE-NITROUS OXIDE MIXTURE (DOT) see CARBON DIOXIDE MIXED WITH NITROUS OXIDE
CARBON DIOXIDE-OXYGEN MIXTURE (DOT) see CARBON DIOXIDE MIXED WITH OXYGEN
CARBONE (ITALIAN) see CARBON

CARBONE (OXYCHLORURE DE) (FRENCH) see PHOSGENE
CARBONE (OXYDE DE) (FRENCH) see CARBON MONOXIDE
CARBONE (SUFURE DE) (FRENCH) see CARBON DISULFIDE
CARBON FERROCHROMIUM see FERROCHROME (EXOTHERMIC)
CARBON HEXACHLORIDE see 1,1,1,2,2,2-HEXACHLOROETHANE
CARBONIC ACID, AMMONIUM SALT see AMMONIUM CARBONATE
CARBONIC ACID, BARIUM SALT (1:1) see BARIUM CARBONATE (1:1)
CARBONIC ACID BERYLLIUM SALT (1:1) see BERYLLIUM CARBONATE (1:1)
CARBONIC ACID, DIAMMONIUM SALT see AMMONIUM CARBONATE
CARBONIC ACID, DILITHIUM SALT see LITHIUM CARBONATE (2:1)
CARBONIC ACID, DIPOTASSIUM SALT see POTASSIUM CARBONATE (2:1)
CARBONIC ACID GAS see CARBON DIOXIDE
CARBONIC ACID, LEAD(2+) SALT (1:1) see LEAD CARBONATE
CARBONIC ACID LITHIUM SALT see LITHIUM CARBONATE (2:1)
CARBONIC ACID, MONOAMMONIUM SALT see AMMONIUM BICARBONATE (1:1)
CARBONIC ACID, NICKEL SALT (1:1) see NICKEL (II) CARBONATE (1:1)
CARBONIC ANHYDRASE INHIBITOR NO. 6063 see 5-ACETAMIDE-1,3,4-THIADIAZOLE-2-SULFONAMIDE
CARBONIC ANHYDRIDE see CARBON DIOXIDE
CARBONIC OXIDE see CARBON MONOXIDE
CARBONIO (OSSICLORURO DI) (ITALIAN) see PHOSGENE
CARBONIO (OSSIDO DI) (ITALIAN) see CARBON MONOXIDE
CARBONIO (SOLFURO DI) (ITALIAN) see CARBON DISULFIDE
CARBON OIL see BENZENE
CARBON OXIDE SULFIDE see CARBONYL SULFIDE
CARBON OXYCHLORIDE see PHOSGENE
CARBON OXYFLUORIDE see CARBONYL FLUORIDE
CARBON OXYSULFIDE see CARBONYL SULFIDE
CARBON SILICIDE see SILICON CARBIDE
CARBON SULFIDE see CARBON DISULFIDE
CARBON TET see CARBON TETRACHLORIDE
CARBONYLCHLORID (GERMAN) see PHOSGENE
CARBONYL CHLORIDE see PHOSGENE
CARBONYL DIAMIDE see UREA
CARBONYLDIAMINE see UREA
CARBONYL DIFLUORIDE see CARBONYL FLUORIDE
CARBORUNDUM see SILICON CARBIDE
CARBOWAX see POLYETHYLENE GLYCOL
CARBOXYANILINE see ANTHRANILIC ACID
2-CARBOXYANILINE see ANTHRANILIC ACID
4-CARBOXYANILINE see p-AMINOBENZOIC ACID
O-CARBOXYANILINE see ANTHRANILIC ACID
2-CARBOXYDIPHENYLAMINE see N-PHENYLANTHRANILIC ACID
CARBOXYETHANE see PROPIONIC ACID
3-CARBOXY-1-ETHYL-7-METHYL-1,8-NAPHTHIDIN-4-ONE see 1-ETHYL-1,4-DIHYDRO-7-METHYL-4-OXO-1,8-NAPHTHYRIDINE-3-CARBOXYLIC ACID
2'-CARBOXY-2-HYDROXY-4-METHOXYBENZOPHENONE(O-(2-HYDROXY-P-ANISOYL)BENZOIC ACID) see O-(2-HYDROXY-4-METHOXYBENZOYL)BENZOIC ACID
3,3'-(CARBOXYMETHYLENE)BIS(4-HYDROXYCOUMARIN) ETHYL ESTER see BIS(4-HYDROXY-3-COUMARIN) ACETIC ACID ETHYL ESTER
(CARBOXYMETHYLTHIO)ACETIC ACID see MERCAPTODIACETIC ACID
O-CARBOXYPHENYL ACETATE see ACETOL (2)
P-CARBOXYPHENYLAMINE see p-AMINOBENZOIC ACID
9-O-CARBOXYPHENYL-6-DIETHYLAMINO-3-ETHYLIMINO-3-ISOXANTHENE, 3-ETHOCHLORIDE see (9-(O-CARBOXYPHENYL)-6-(DIETHYLAMINO)-3H-XANTHEN-3-YLIDENE) DIETHYLAMMONIUM CHLORIDE
9-(O-CARBOXYPHENYL)-6-HYDROXY-3-ISOXANTHENONE see FLUORESCEIN
((O-CARBOXYPHENYL)THIO)ETHYLMERCURY SODIUM SALT see MERTHIOLATE SODIUM
4-CARBOXYPHTHALIC ANHYDRIDE see TRIMELLIC ACID ANHYDRIDE
3-CARBOXYPYRIDINE see NICOTINIC ACID

CARBYNE see m-CHLORO CARBANILIC ACID-4-CHLORO-2-BUTYNYL ESTER

CARDIORYTHMINE see AJMALINE

CARMINE BLUE (BIOLOGICAL STAIN) see DISODIUM INDIGO-5,5-DISULFONATE

CARMINOMYCIN see KARMINOMYCIN

CAROB BEAN GUM see LOCUST BEAN GUM

CAROB FLOUR see LOCUST BEAN GUM

CAROID see PAPAIN

CARRAGEENAN GUM see CARRAGEEN

CARRAGHEANIN see CARRAGEEN

CARRAGHEEN see CARRAGEEN

CARREL-DAKIN SOLUTION see SODIUM HYPOCHLORITE

CARTWHEELS see DL-AMPHETAMINE SULFATE

BETA-CARYOPHYLLENE see CARYOPHYLLENE

CARYOPHYLLIC ACID see EUGENOL

CASSEL BROWN see MANGANESE (III) OXIDE

CASSEL GREEN see MANGANESE (II) OXIDE

CASTANEA SATIVA MILL TANNIN see CHESTNUT TANNIN

CASTOR BEANS (DOT) see CASTOR BEAN

CASTOR OIL AROMATIC see CASTOR OIL

CASTOR POMACE (DOT) see CASTOR BEAN

CATECHIN see PYROCATECHOL

CATECHOL see PYROCATECHOL

CAUSTIC POTASH see POTASSIUM HYDROXIDE

CAUSTIC SODA see SODIUM HYDROXIDE

CAUSTIC SODA, BEAD (DOT) see SODIUM HYDROXIDE

CAUSTIC SODA, DRY (DOT) see SODIUM HYDROXIDE

CAUSTIC SODA, FLAKE (DOT) see SODIUM HYDROXIDE

CAUSTIC SODA, GRANULAR (DOT) see SODIUM HYDROXIDE

CAUSTIC SODA, SOLID (DOT) see SODIUM HYDROXIDE

CAUSTIC SODA SOLUTION see SODIUM HYDROXIDE (LIQUID)

1522 CB see ACETOPROMAZINE

CBN see m-CHLORO CARBANILIC ACID-4-CHLORO-2-BUTYNYL ESTER

"C" CARRIE see COCAINE

CCNU see 1-(2-CHLOROETHYL)-3-CYCLOHEXYL-1-NITROSOUREA

CECIL see COCAINE

CEDRO OIL see LEMON OIL

CEEPRYN see CETYL PYRIDINIUM CHLORIDE MONOHYDRATE

CEEPRYN CHLORIDE see CEPACOL CHLORIDE

CELLOCIDIN see ACETYLENEDICARBOXAMIDE

CELLOIDIN see CELLULOSE TETRANITRATE

CELLOSOLVE see CELLOSOLVE SOLVENT

CELLOSOLVE ACRYLATE see ACRYLIC ACID-2-ETHOXYETHYL ESTER

CELLUFLEX TPP see TRIPHENYL PHOSPHATE

CELLULOSE NITRATE see CELLULOSE TETRANITRATE

CEMENT BLACK see MANGANESE DIOXIDE

CEMENT, LEATHER (DOT) see CEMENT (LEATHER)

CEPHALEXIN see 7-(D-alpha-AMINOPHENYLACETAMIDO)DESACET-OXYCEPHALOSPORANIC ACID

CEPOREXINE see 7-(D-alpha-AMINOPHENYLACETAMIDO)DESACET-OXYCEPHALOSPORANIC ACID

CERAMIC FIBRE see ALUMINUM(III) SILICATE (2:1)

CERASINROT see OIL RED

CEREBROSE see GALACTOSE

CERESAN see CHLORO ETHYL MERCURY

CERESAN see ACETOXYPHENYLMERCURY

CERESAN UNIVERSAL NAZBEIZE see 2-METHOXYETHYLMERCURY CHLORIDE

CERIUM DIOXIDE see CERIC OXIDE

CERIUM FLUORURE (FRENCH) see CERIUM FLUORIDE

CERIUM NITRATE, HEXAHYDRATE see CERIUM (III) NITRATE, HEXA-HYDRATE (1:3:6)

CERIUM TRIACETATE see CERIUM ACETATE

CERIUM TRICHLORIDE see CERIUM CHLORIDE

CERIUM TRIFLUORIDE see CERIUM FLUORIDE

CEROUS ACETATE see CERIUM ACETATE

CEROUS CHLORIDE see CERIUM CHLORIDE

CEROUS NITRATE HEXAHYDRATE see CERIUM (III) NITRATE, HEXA-HYDRATE (1:3:6)

CERTIQUAL FLUORESCEINE see FLUORESCEIN SODIUM

CESIUM HYDRATE see CESIUM HYDROXIDE

CETYLAMINE see 1-HEXADECANAMINE

CETYL GAMMA-AMINOBUTYRATE see gamma-AMINOBUTYRIC ACID CETYL ESTER

CETYL GABA see gamma-AMINOBUTYRIC ACID CETYL ESTER

CETYLPYRIDINIUM CHLORIDE see CEPACOL CHLORIDE

CETYLTRIMETHYLAMMONIUM BROMIDE see HEXADECYL TRIMETHYL AMMONIUM BROMIDE

N-CETYLTRIMETHYLAMMONIUM BROMIDE see HEXADECYL TRI-METHYL AMMONIUM BROMIDE

CEVADILLA see VERATRINE

CEVADINE see VERATRINE

CEVITAMIC ACID see L-ASCORBIC ACID

CEVITAMIN see L-ASCORBIC ACID

CEYLON ISINGLASS see AGAR

CGA 15324 see O-(4-BROMO-2-CHLOROPHENYL)-O-ETHYL-S-PROPYL PHOSPHOROTHIOATE

CHALCEDONY see SILICA (CRYSTALLINE)

CHALCEDONY see SILICA, CRYSTALLINE-QUARTZ

CHALK see CALCIUM (II) CARBONATE (1:1)

CHAMBER CRYSTALS see NITROSYLSULFURIC ACID

CHAMELEON MINERAL see POTASSIUM PERMANGANATE

CHANNING'S SOLUTION see NESSLER REAGENT

CHARAS see CANNABIS

CHARCOAL, ACTIVATED see ACTIVATED CARBON

CHARCOAL BLACK see CARBON

CHARCOAL BRIQUETTES (DOT) see CHARCOAL (BRIQUETTES)

CHAVICOL METHYL ETHER see p-ALLYLANISOLE

CHEMAGRO B-1843 see trans-1,2-BIS(n-PROPYL SULFONYL)-ETHYLENE

CHEMI-CHARL see SODIUM HEXAMETAPHOSPHATE

CHEMOTHERAPY CENTER NO. 914 see 4-CARBAMIDOPHENYL BIS(CARBOXY METHYL THIO)ARSENITE

CHERTS see SILICA (CRYSTALLINE)

CHERTS see SILICA, CRYSTALLINE-QUARTZ

CHINA GREEN (BIOLOGICAL STAIN) see AIZEN MALACHITE GREEN

CHINESE ISINGLASS see AGAR

CHINESE SEASONING see MONOSODIUM GLUTAMATE

CHINESE WHITE see ZINC OXIDE

CHINIDIN (GERMAN) see QUINIDINE

CHININ (GERMAN) see QUININE

CHINOLIN (CZECH) see QUINOLINE

CHINOLINE see QUINOLINE

CHINON (DUTCH, GERMAN) see p-BENZOQUINONE

CHINONOXIMEBENZOYLHYDRAZONE see 1,4-BENZOQUINONE-N'-BEN-ZOYLHYDRAZONE OXIME

CHLOOR (DUTCH) see CHLORINE

2-CHLORBENZALDEHYDE (DUTCH) see o-CHLOROBENZALDEHYDE

CHLOORBENZEEN (DUTCH) see BENZENE CHLORIDE

(4-CHLOOR-BENZYL)-(4-CHLOOR-FENYL)-SULFIDE (DUTCH) see p-CHLOROBENZYL-p-CHLOROPHENYL SULFIDE

2-CHLOOR-1,3-BUTADIEEN (DUTCH) see NEOPRENE

(4-CHLOOR-BUT-2-YN-YL)-N-(3-CHLOOR-FENYL)-CARBAMAAT (DUTCH) see m-CHLORO CARBANILIC ACID-4-CHLORO-2-BUTYNYL ESTER

CHLOORDAAN (DUTCH) see CHLORDANE

O-2-CHLOOR-1-(2,4-DICHLOOR-FENYL)-VINYL-O,O-DIETHYLFOSFAAT (DUTCH) see CHLORFENVINFOS

(2-CHLOOR-3-DIETHYLAMINO-1-METHYL-3-OXO-PROP-1-EN-YL)-DIMETHYL-FOSFAAT (DUTCH) see DIMECRON

2-CHLOOR-4-DIMETHYLAMINO-6-METHYL-PYRIMIDINE (DUTCH) see CASTRIX

1-CHLOOR-2,4-DINITROBENZEEN (DUTCH) see 1-CHLORO-2,4-DINITROBENZENE

1-CHLOOR-2,3-EPOXY-PROPAAN (DUTCH) see 1-CHLORO-2,3-EPOXY PROPANE

CHLOORETHAAN (DUTCH) see ETHYL CHLORIDE

2-CHLOORETHANOL (DUTCH) see 2-CHLOROETHYL ALCOHOL

CHLOORFACINON (DUTCH) see CHLOROPHACINONE

CHLOORFENSON (DUTCH) see 4-CHLOROPHENYL 4-CHLOROBENZENE-SULFONATE

(4-CHLOOR-FENYL)-BENZEEN-SULFONAAT (DUTCH) see 4-CHLORO-PHENYL BENZENESULFONATE

(4-CHLOOR-FENYL)-4-CHLOOR-BENZEEN-SULFONAAT (DUTCH) see 4-CHLOROPHENYL 4-CHLOROBENZENESULFONATE

3-(4-CHLOOR-FENYL)-1,1-DIMETHYLUREUM (DUTCH) see 3-(p-CHLOROPHENYL)-1,1-DIMETHYLUREA

2(2-(4-CHLOOR-FENYL-2-FENYL)-ACETYL)-INDAAN-1,3-DION (DUTCH) see·CHLOROPHACINONE

N-(3-CHLOOR-FENYL)-ISOPROPYL CARBAMAAT (DUTCH) see N-3-CHLOROPHENYL ISOPROPYL CARBAMATE

CHLOOR-METHAAN (DUTCH) see MONOCHLOROMETHANE

1-CHLOOR-4-NITROBENZEEN (DUTCH) see 1-CHLORO-4-NITROBEN-ZENE

O-(3-CHLOOR-4-NITRO-FENYL)-O,O-DIMETHYL-MONOTHIOFOSFAAT (DUTCH) see METHYLCHLOROTHION

O-(4-CHLOOR-3-NITRO-FENYL)-O,O-DIMETHYLMONOTHIOFOSFAAT (DUTCH) see p-NITRO-O-CHLOROPHENYL DIMETHYL THIONOPHOSP-HATE

CHLOORPIKRINE (DUTCH) see TRICHLORONITROMETHANE

CHLOORTHION (DUTCH) see METHYLCHLOROTHION

CHLOORWATERSTOF (DUTCH) see HYDROCHLORIC ACID

CHLOPHEN see POLYCHLORINATED BIPHENYLS

CHLOR (GERMAN) see CHLORINE

CHLORACETAMID (GERMAN) see 2-CHLORO ACETAMIDE

CHLORACETIC ACID see CHLOROACETIC ACID

CHLORACETONE (FRENCH) see ACETONYL CHLORIDE

CHLORACETONE (FRENCH) see CHLOROACETONE

2-CHLORAETHANOL (GERMAN) see 2-CHLOROETHYL ALCOHOL

CHLORALLYLENE see ALLYL CHLORIDE

CHLORAMEISENSAEURE METHYLESTER (GERMAN) see METHYL CHLOROCARBONATE

CHLORAMINE see CHLORAMINE-T

CHLORAMINE see BIS(2-CHLOROETHYL)METHYLAMINE HYDRO-CHLORIDE

CHLORAMINE see BIS(beta-CHLOROETHYL)METHYLAMINE

CHLORAMINE BLACK C see APOMINE BLACK GX

CHLORAMINE (THE NITROGEN MUSTARD) see BIS(beta-CHLORO-ETHYL)METHYLAMINE

CHLORAMINOPHENE see CHLORAMBUCIL

CHLORAMP (RUSSIAN) see 4-AMINO-3,5,6-TRICHLOROPICOLINIC ACID

D-CHLORAMPHENICOL see CHLORAMPHENICOL

1-CHLORANTHRACHINON (CZECH) see 1-CHLORO ANTHRA QUINONE

CHLORATE DE CALCIUM (FRENCH) see CALCIUM CHLORATE

CHLORATE DE POTASSIUM (FRENCH) see POTASSIUM CHLORATE

CHLORATE OF POTASH see POTASSIUM CHLORATE

CHLORATE OF SODA (DOT) see SODIUM CHLORATE

CHLORAZOL BLUE 3B see C.I. DIRECT BLUE 14, TETRASODIUM SALT

CHLORAZOL VIOLET N see C.I. DIRECT VIOLET 1, DISODIUM SALT

CHLORBENSID (GERMAN) see p-CHLOROBENZYL-p-CHLOROPHENYL SULFIDE

2-CHLORBENZALDEHYD (GERMAN) see o-CHLOROBENZALDEHYDE

CHLORBENZENE see BENZENE CHLORIDE

CHLORBENZIDE see p-CHLOROBENZYL-p-CHLOROPHENYL SULFIDE

CHLORBENZOL see BENZENE CHLORIDE

o-CHLORBENZONITRIL (CZECH) see o-CHLOROBENZONITRILE

N-P-CHLORBENZOYL-5-METHOXY-2-METHYLINDOLE-3-ACETIC ACID see INDOMETHACIN

(4-CHLOR-BENZYL)-(4-CHLOR-PHENYL)-SULFID (GERMAN) see p-CHLOROBENZYL-p-CHLOROPHENYL SULFIDE

CHLORBICYCLENE (FRENCH) see DOLANTIN HYDROCHLORIDE

2-CHLOR-1,3-BUTADIEN (GERMAN) see NEOPRENE

CHLORBUTANOL see ACETONE CHLOROFORM

(4-CHLOR-BUT-2-IN-YL)-N-(3-CHLOR-PHENYL)-CARBAMAT (GERMAN) see m-CHLORO CARBANILIC ACID-4-CHLORO-2-BUTYNYL ESTER

CHLORBUTOL see ACETONE CHLOROFORM

p-CHLOR-m-CRESOL see 4-CHLORO-m-CRESOL

CHLORCYAN see CYANOGEN CHLORIDE

CHLORCYCLINE see CHLOROCYCLINE

CHLORCYCLIZINE see CHLOROCYCLINE

CHLORCYCLIZINIUM CHLORIDE see CHLORCYCLIZINE HYDROCHLO-RIDE

GAMMA-CHLORDAN see CHLORDANE

CHLORDECONE see KEPONE

7-CHLOR-4-(4-(DIAETHYLAMINO)-1-METHYLBUTYLAMINO)-CHINO-LINDIPHOSPHAT (GERMAN) see CHLOROQUINE DIPHOSPHATE

(2-CHLOR-3-DIAETHYLAMINO-1-METHYL-3-OXO-PROP-1-EN-YL)-DIMETHYL-PHOSPHAT (GERMAN) see DIMECRON

CHLORDIAZEPOXIDE MONOHYDROCHLORIDE see LIBRIUM HYDRO-CHLORIDE

O-2-CHLOR-1-(2,4-DICHLOR-PHENYL)-VINYL-O,O-DIAETHYLPHOS-PHAT (GERMAN) see CHLORFENVINFOS

CHLORDIMEFORM see CHLOROPHENAMIDINE

CHLORDIMETHYLETHER (CZECH) see CHLOROMETHYL METHYL ETHER

CHLORE (FRENCH) see CHLORINE

CHLOREFENIZON (FRENCH) see 4-CHLOROPHENYL 4-CHLOROBEN-ZENESULFONATE

1-CHLOR-2,3-EPOXY-PROPAN (GERMAN) see 1-CHLORO-2,3-EPOXY PROPANE

CHLORETHAMINACIL see 5-(BIS(2-CHLOROETHYL)AMINO)URACIL

CHLOREXTOL see POLYCHLORINATED BIPHENYLS

CHLORFACINON (GERMAN) see CHLOROPHACINONE

P-CHLORFENOL (CZECH) see 4-CHLOROPHENOL

CHLORFENSON see 4-CHLOROPHENYL 4-CHLOROBENZENESULFONATE

CHLORFENVINPHOS see CHLORFENVINFOS

P-CHLORFENYLISOKYANAT (CZECH) see p-CHLOROPHENYL ISO-CYANATE

CHLOR-N-(2-FURYLMETHYL)-5-SULFAMYLANTHRANILSAEURE (GERMAN) see 4-CHLORO-N-FURFURYL-5-SULFAMOYL ANTHRA-NILIC ACID

CHLORHYDRATE D'ANILINE (FRENCH) see BENZENAMINE HYDRO-CHLORIDE

CHLORIC ACID, BARIUM SALT see BARIUM CHLORATE

CHLORIC ACID, BARIUM SALT (WET) see BARIUM CHLORATE, WET (DOT)

CHLORID AMONNY (CZECH) see AMMONIUM CHLORIDE

CHLORID ANILINU (CZECH) see BENZENAMINE HYDROCHLORIDE

CHLORID ANTIMONITY (CZECH) see ANTIMONY (III) CHLORIDE

CHLORID-N-BUTYLCINICITY (CZECH) see BUTYL TRICHLORO STAN-NANE

CHLORID DI-N-BUTYLCINICITY (CZECH) see DIBUTYLDICHLOROSTAN-NANE

CHLORIDEAZEPOXIDE HYDROCHLORIDE see LIBRIUM HYDROCHLO-RIDE

CHLORIDE OF LIME (DOT) see HYPOCHLOROUS ACID, CALCIUM SALT

CHLORIDE OF PHOSPHORUS see PHOSPHORUS CHLORIDE

CHLORIDE OF SULFUR (DOT) see SULFUR DICHLORIDE

CHLORIDE OF SULFUR (DOT) see SULFUR CHLORIDE

CHLORID FENYLRTUTNATY (CZECH) see PHENYL MERCURIC CHLO-RIDE

CHLORIDIAZEPOXIDE see LIBRIUM

CHLORID KREMICITY (CZECH) see SILICON CHLORIDE

CHLORID KYSELINY CHLORMETHANSULFONOVE (CZECH) see CHLOROMETHANE SULFONYL CHLORIDE

CHLORID MEDNY (CZECH) see COPPER (I) CHLORIDE

CHLORID RTUTNATY (CZECH) see CORROSIVE SUBLIMATE

CHLORID TRI-N-BUTYLCINICITY (CZECH) see CHLORO TRIBUTYL STANNANE

CHLORIERTE BIPHENYLE, CHLORGEHALT 42% (GERMAN) see POLY-CHLORINATED BIPHENYL (AROCLOR 1242)

CHLORIERTE BIPHENYLE, CHLORGEHALT 54% (GERMAN) see POLY-CHLORINATED BIPHENYL (AROCLOR 1254)

CHLORINATED BIPHENYL see POLYCHLORINATED BIPHENYLS

CHLORINATED CAMPHENE see TOXAPHENE

CHLORINATED DIPHENYL see POLYCHLORINATED BIPHENYLS

CHLORINATED DIPHENYLENE see POLYCHLORINATED BIPHENYLS

CHLORINATED HC, ALIPHATIC see CHLORINATED HYDROCARBONS, ALIPHATIC

CHLORINATED HC AROMATIC see CHLORINATED HYDROCARBONS, AROMATIC

CHLORINATED HYDROCHLORIC ETHER see ETHYLIDENE DICHLORIDE

CHLORINATED LIME (DOT) see HYPOCHLOROUS ACID, CALCIUM SALT

CHLORINATED POLYETHER POLYURETHAN see POLYURETHANE Y-238

CHLORINE CYANIDE see CYANOGEN CHLORIDE

CHLORINE DIOXIDE see CHLORINE OXIDE

CHLORINE DIOXIDE HYDRATE, FROZEN (DOT) see CHLORIC ACID

CHLORINE PEROXIDE see CHLORINE OXIDE

CHLORINE TETROXYFLUORIDE see FLUORINE PERCHLORATE

CHLORINE(IV) OXIDE see CHLORINE OXIDE

CHLORMADINONE ACETATE MIXED WITH MESTRANOL see C-QUENS

CHLOR-METHAN (GERMAN) see MONOCHLOROMETHANE

CHLORMETHANSULFOCHLORID (CZECH) see CHLOROMETHANE SUL-FONYL CHLORIDE

3-CHLOR-2-METHYL-PROP-1-EN (GERMAN) see 3-CHLORO-2-METHYL-PROPENE

1-CHLOR-4-NITROBENZOL (GERMAN) see 1-CHLORO-4-NITROBEN-ZENE

O-(3-CHLOR-4-NITRO-PHENYL)-O,O-DIMETHYL-MONOTHIOPHOSPHAT (GERMAN) see METHYLCHLOROTHION

O-(4-CHLOR-3-NITRO-PHENYL)-O,O-DIMETHYL-MONOTHIOPHOSPHAT (GERMAN) see p-NITRO-o-CHLOROPHENYL DIMETHYL THIONO-PHOSPHATE

CHLOROACETALDEHYDE MONOMER see 2-CHLOROACETALDEHYDE

CHLOROACETAMIDE see 2-CHLORO ACETAMIDE

ALPHA-CHLOROACETAMIDE see 2-CHLORO ACETAMIDE

ALPHA-CHLOROACETIC ACID see CHLOROACETIC ACID

CHLOROACETIC ACID CHLORIDE see CHLOROACETYL CHLORIDE

CHLOROACETIC ACID, ETHYL ESTER see ETHYL CHLORACETATE

CHLOROACETIC CHLORIDE see CHLOROACETYL CHLORIDE

4'-CHLOROACETOACETANILIDE see p-CHLORO ACETO ACETANILIDE

CHLOROACETONE see ACETONYL CHLORIDE

2-CHLOROACETONITRILE see CHLORACETONITRILE

ALPHA-CHLOROACETONITRILE see CHLORACETONITRILE

1-CHLOROACETOPHENONE see 2-CHLOROACETOPHENONE

ALPHA-CHLOROACETOPHENONE see 2-CHLOROACETOPHENONE

OMEGA-CHLOROACETOPHENONE see 2-CHLOROACETOPHENONE

6-CHLORO-17-ALPHA-ACETOXY-4,6-PREGNADIENE-3,20-DIONE see CAP

6-CHLORO-DELTA(SUP 6)-17-ACETOXYPROGESTERONE see ÇAP

6-CHLORO-DELTA(SUP 6)-(17-ALPHA)ACETOXYPROGESTERONE see CAP

DELTA(SUP 6)-6-CHLORO-17-ALPHA-ACETOXYPROGESTERONE see CAP

CHLOROAETHAN (GERMAN) see ETHYL CHLORIDE

gamma-CHLOROALLYL CHLORIDE see alpha-CHLOROALLYL CHLO-RIDE

2-CHLOROALLYL-N,N-DIETHYLDITHIOCARBAMATE see 2-CHLO-RALLYL DIETHYLDITHIOCARBAMATE

4-CHLORO-2-AMINOTOLUENE see 2-AMINO-4-CHLOROTOLUENE

1-CHLORO-9,10-ANTHRAQUINONE see 1-CHLORO ANTHRA QUINONE

alpha-CHLOROANTHRAQUINONE see 1-CHLORO ANTHRA QUINONE

CHLOR(O)AZIDE see CHLORINE AZIDE

2-CHLOROBENZALDEHYDE see o-CHLOROBENZALDEHYDE

2-CHLOROBENZALMALONONITRILE see (o-CHLORO BENZAL)MALONO NITRILE

CHLOROBENZEN (POLISH) see BENZENE CHLORIDE

CHLOROBENZENE see BENZENE CHLORIDE

O-CHLOROBENZENECARBOXALDEHYDE see o-CHLOROBENZALDEHYDE

4-CHLOROBENZENESULFONATE DE 4-CHLOROPHENYLE (FRENCH) see 4-CHLOROPHENYL 4-CHLOROBENZENESULFONATE

P-CHLOROBENZENESULFONIC ACID-P-CHLOROPHENYL ESTER see 4-CHLOROPHENYL 4-CHLOROBENZENESULFONATE

1-(P-CHLOROBENZENESULFONYL)-3-PROPYLUREA see 1-(p-CHLORO-PHENYLSULFONYL)-3-PROPYLUREA

1-(P-CHLOROBENZHYDRYL)-4-(2-(2-HYDROXYETHOXY)ETHYL)DIETHYLENEDIAMINE see 1-(p-CHLORO-alpha-PHENYLBENZYL)-4-(2-((2-HYDROXYETHOXY)-ETHYL)PIPERAZINE

N-(4-CHLOROBENZHYDRYL)-N'-(HYDROXYETHOXYETHYL)PIPERAZINE see 1-(p-CHLORO-alpha-PHENYLBENZYL)-4-(2-((2-HYDROXY-ETHOXY)ETHYL)PIPERAZINE

1-(P-CHLOROBENZHYDRYL)-4-(2-(2-HYDROXYETHOXY)ETHYL)PIPER-AZINE see 1-(p-CHLORO-alpha-PHENYLBENZYL)-4-(2-((2-HYDROXYETHOXY)ETHYL)PIPERAZINE

1-(P-CHLOROBENZHYDRYL)-4-(M-METHYLBENZYL)DIETHYLENEDI-AMINE see HISTAMETHIZINE

1-P-CHLOROBENZHYDRYL-4-M-METHYLBENZYLPIPERAZINE see HISTAMETHIZINE

1-(4-CHLOROBENZHYDRYL)-4-METHYLPIPERAZINE see CHLOROCY-CLINE

1-(4-CHLOROBENZHYDRYL)-4-METHYLPIPERAZINE DIHYDROCHLO-RIDE see CHLORCYCLIZINE DIHYDROCHLORIDE

1-(P-CHLOROBENZHYDRYL)-4-METHYLPIPERAZINE HYDROCHLORIDE see CHLORCYCLIZINE HYDROCHLORIDE

1-(P-CHLOROBENZOYL)-5-METHOXY-2-METHYLINDOLE-3-ACETIC ACID see INDOMETHACIN

1-(P-CHLOROBENZOYL)-2-METHYL-5-METHOXYINDOLE-3-ACETIC ACID see INDOMETHACIN

1-(P-CHLOROBENZOYL)-2-METHYL-5-METHOXY-3-INDOLE-ACETIC ACID see INDOMETHACIN

alpha-(1-(P-CHLOROBENZOYL)-2-METHYL-5-METHOXY-3-INDOLYL)-ACETIC ACID see INDOMETHACIN

P-CHLOROBENZOYL PEROXIDE see BIS(p-CHLOROBENZOYL) PEROXIDE

4-CHLOROBENZYL-4-CHLOROPHENYL SULPHIDE see p-CHLOROBEN-ZYL-p-CHLOROPHENYL SULFIDE

P-CHLOROBENZYL-P-CHLOROPHENYL SULPHIDE see p-CHLOROBEN-ZYL-p-CHLOROPHENYL SULFIDE

O-CHLOROBENZYLIDENE MALONITRILE see (o-CHLORO BENZAL)-MALONO NITRILE

2-CHLOROBENZYLIDENE MALONONITRILE see (o-CHLORO BENZAL)-MALONO NITRILE

CHLORO BIPHENYL see POLYCHLORINATED BIPHENYLS

CHLORO-1,1-BIPHENYL see POLYCHLORINATED BIPHENYLS

2-CHLORO-4,6-BIS(ETHYLAMINO)-S-TRIAZINE see 2,4-BIS(ETHYLAM-INO)-6-CHLORO-S-TRIAZINE

1-CHLORO, 3,5-BISETHYLAMINO-2,4,6-TRIAZINE see 2,4-BIS(ETHYL-AMINO)-6-CHLORO-S-TRIAZINE

2-CHLORO-4,6-BIS(ETHYLAMINO)-1,3,5-TRIAZINE see 2,4-BIS(ETHYL-AMINO)-6-CHLORO-S-TRIAZINE

CHLOROBROMOMETHANE see BROMOCHLOROMETHANE

CHLORO(2-(3-BROMOPROPIONAMIDO)CYCLOHEXYL), MERCURY (E)-see trans-CHLORO(2-(3-BROMO PROPION AMIDO)CYCLO HEXYL)-MERCURY

CHLOROBUTADIENE see NEOPRENE

1-CHLORO-1,3-BUTADIENE see 1-CHLOROBUTADIENE

2-CHLORO-1,3-BUTADIENE see NEOPRENE

2-CHLOROBUTA-1,3-DIENE see NEOPRENE

2-CHLORO-1,3-BUTADIENE HOMOPOLYMER (9CI) see 2-CHLORO-1,3-BUTADIENE POLYMER

CHLOROBUTADIENE POLYMER see 2-CHLORO-1,3-BUTADIENE POLY-MER

1-CHLOROBUTANE see N-BÚTYL CHLORIDE

CHLOROBUTANOL see ACETONE CHLOROFORM

CHLOROBUTINE see CHLORAMBUCIL

4-CHLORO-2-BUTYNYL-M-CHLOROCARBANILATE see m-CHLORO CARBANILIC ACID-4-CHLORO-2-BUTYNYL ESTER

4-CHLOROBUT-2-YNYL-3-CHLOROPHENYLCARBAMATE see m-CHLORO CARBANILIC ACID-4-CHLORO-2-BUTYNYL ESTER

4-CHLORO-2-BUTYNYL-N-(3-CHLOROPHENYL)CARBAMATE see m-CHLORO CARBANILIC ACID-4-CHLORO-2-BUTYNYL ESTER

CHLOROCAMPHENE see TOXAPHENE

M-CHLOROCARBANILIC ACID, 4-CHLORO-2-BUTYNYL ESTER see m-CHLORO CARBANILIC ACID-4-CHLORO-2-BUTYNYL ESTER

CHLOROCARBONATE DE METHYLE (FRENCH) see METHYL CHLORO-CARBONATE

CHLOROCARBONATE D'ETHYLE (FRENCH) see ETHYL CHLORO FOR-MATE

3-CHLOROCHLORDENE see HEPTACHLOR

CHLOROETHENE see 1,1,1-TRICHLOROETHANE

CHLOROETHENE HOMOPOLYMER see BAKELITE

1,1',1''-(1-CHLORO-1-ETHENYL-2-YLIDENE)-TRIS(4-METHOXYBEN-ZENE) see CHLOROTRIANISENE

CHLOROETHYL ACRYLATE see ACRYLIC ACID-beta-CHLOROETHYL ESTER

2-CHLOROETHYL ACRYLATE see ACRYLIC ACID-beta-CHLOROETHYL ESTER

BETA-CHLOROETHYL ACRYLATE see ACRYLIC ACID-beta-CHLORO-ETHYL ESTER

beta-CHLOROETHYL ALCOHOL see 2-CHLOROETHYL ALCOHOL

2-(4-CHLORO-6-ETHYLAMINO-s-TRIAZINE-2-YLAMINO)-2-METHYL-PROPIONITRILE see BLADEX

2-(4-CHLORO-6-ETHYLAMINO-1,3,5-TRIAZINE-2-YLAMINO)-2-METHYLPROPIONITRILE see BLADEX

2-((4-CHLORO-6-(ETHYLAMINO)-s-TRIAZIN-2-YL)AMINO)-2-METHYL-PROPIONITRILE see BLADEX

CHLOROETHYLBENZENE see ETHYL CHLORO BENZENE

BETA-CHLOROETHYL BETA-(BIS(BETA-HYDROXYETHYL)SULFONIUM) ETHYL SULFIDE CHLORIDE see BIS(2-HYDROXYETHYL)-2-(2-CHLORO ETHYL THIO)ETHYL SULFONIUM, CHLORIDE

BETA-CHLOROETHYL-BETA'-(P-T-BUTYLPHENOXY)-ALPHA'-METHYLETHYL SULFITE see SULFUROUS ACID, 2-(p-t-BUTYLPHENOXY)-1-METHYLETHYL-2-CHLOROETHYL ESTER

BETA-CHLOROETHYL-BETA-(P-T-BUTYLPHENOXY)-ALPHA-METHYL-ETHYL SULPHITE see SULFUROUS ACID, 2-(p-t-BUTYLPHENOXY)-1-METHYLETHYL-2-CHLOROETHYL ESTER

3-(2-CHLOROETHYL)-2-((2-CHLOROETHYL)AMINO) TETRAHYDRO-1,3,2-OXAZAPHOSPHORINE-2-OXIDE see ISOPHOSPHAMIDE

N-(2-CHLOROETHYL)DIBENZYLAMINE HYDROCHLORIDE see DIBEN-AMINE HYDROCHLORIDE

(2-CHLOROETHYL)DIETHYLAMINE see N-(2-CHLORO ETHYL)DIETHYL AMINE

(2-CHLOROETHYL)DIMETHYLAMINE see N-(2-CHLORO ETHYL)DI-METHYL AMINE

CHLOROETHYLENE see VINYL CHLORIDE

2-CHLOROETHYLENE CITRATE see RACEMIC CLOMIPHENE CITRATE

N-(2-CHLOROETHYL)-N-(1METHYL-2-PHENOXYETHYL)BENZENEMETHANAMINE HYDROCHLORIDE see DIBENZYLINE HYDROCHLORIDE

N-(2-CHLOROETHYL)-N-(1METHYL-2-PHENOXYETHYL)BENZYLAMINE, HYDROCHLORIDE see DIBENZYLINE HYDROCHLORIDE

2-CHLOROETHYL ETHYL SULFIDE see CHLOROETHYL ETHYL SUL-FIDE

2-CHLOROETHYL ETHYL THIOETHER see CHLOROETHYL ETHYL SULFIDE

beta-CHLOROETHYLMETHANESULFONATE see 2-CHLORO ETHYL METHANE SULFONATE

2-CHLOROETHYL-1-METHYL-2-(P-T-BUTYLPHENOXY)ETHYL SUL-PHATE see SULFUROUS ACID, 2-(p-t-BUTYLPHENOXY)-1-METHYLETHYL-2-CHLOROETHYL ESTER

N-(2-CHLOROETHYL)-N-(1-METHYL-2-PHENOXYETHYL)BENZENE-METHANAMINE see NSC 37448

N-(2-CHLOROETHYL)-N-(1-METHYL-2-PHENOXYETHYL)BENZYLAMINE see NSC 37448

2-((((2-CHLOROETHYL)NITROSOAMINO)CARBONYL)AMINO)-2-DEOXY-D-GLUCOSE see CHLOROZOTOCIN

2-(3-(2-CHLOROETHYL)-3-NITROSOUREIDO)-2-DEOXY-D-GLUCOSO-PYRANOSE see CHLOROZOTOCIN

N-(beta-CHLOROETHYL)-N-NITROSOURETHAN see 2-CHLORO ETHYL-N-NITROSO URETHANE

CHLOROETHYLOWY ALKOHOL (POLISH) see 2-CHLOROETHYL ALCO-HOL

2-CHLOROETHYL SULPHITE OF 1-(P-T-BUTYLPHENOXY)-2-PROPANOL see SULFUROUS ACID, 2-(p-t-BUTYLPHENOXY)-1-METHYLETHYL-2-CHLOROETHYL ESTER

1-CHLORO-2-(ETHYLTHIO)ETHANE see CHLOROETHYL ETHYL SULFIDE

N-(2-CHLOROETHYL)-N'-(TRANS-4-METHYLCYCLOHEXYL)-N-NITRO SOUREA see 1-(2-CHLOROETHYL)-3-(4-METHYL-CYCLOHEXYL)-1-NITROSOUREA

CHLOROETHYNE see ACETYLENE CHLORIDE

CHLOROFENVINPHOS see CHLORFENVINFOS

P-CHLOROFENYLESTER KYSELINY BENZENSULFONOVE (CZECH) see 4-CHLOROPHENYL BENZENESULFONATE

CHLOROFORME (FRENCH) see CHLOROFORM

CHLOROFORMIC ACID BENZYL ESTER see BENZYL CHLOROFORMATE

CHLOROFORMIC ACID DIMETHYLAMIDE see (DIMETHYLAMINO)CAR-BONYL CHLORIDE

CHLOROFORMIC ACID ISOPROPYL ESTER see ISOPROPYL CHLOROCAR-BONATE

CHLOROFORMYL CHLORIDE see PHOSGENE

4-CHLORO-N-(2-FURYLMETHYL)-5-SULFAMOYLANTHRANILIC ACID see 4-CHLORO-N-FURFURYL-5-SULFAMOYL ANTHRANILIC ACID

CHLOROGUANIDE see 1-(p-CHLOROPHENYL)-5-ISOPROPYLBIGUANIDE

CHLOROHYDRIC ACID see HYDROCHLORIC ACID

ALPHA-CHLOROHYDRIN see CHLORHYDRIN

2-CHLORO-10-3-(1-(2-HYDROXYETHYL)-4-PIPERAZINYL)PROPYL PHE-NOTHIAZINE see 4-(3-(2-CHLOROPHENOTHIAZIN-10-YL)PROPYL)-1-PIPERAZINEETHANOL

5-CHLORO-8-HYDROXY-7-IODOQUINOLINE see 5-CHLORO-7-IODO-8-QUINOLINOL

3-CHLORO-7-HYDROXY-4-METHYL-COUMARIN-O,O-DIETHYL PHOS-PHOROTHIOATE see COUMAPHOS

3-CHLORO-7-HYDROXY-4-METHYL-COUMARIN-O-ESTER WITH-O,O-DIETHYL PHOSPHOROTHIOATE see COUMAPHOS

5-CHLORO-2-((2-HYDROXY-1-NAPHTHALENYL)AZO)-4-METHYLBEN-ZENESULFONICACID, BARIUM SALT (2:1) see 5-CHLORO-2-((2-HY-DROXY-1-NAPHTHYL)AZO)-p-TOLUENESULFONIC ACID, BARIUM SALT

5-CHLORO-2-((2-HYDROXY-1-NAPHTHALENYL)AZO)-4-METHYLBEN-ZENESULPHONICACID, BARIUM SALT see 5-CHLORO-2-((2-HY-DROXY-1-NAPHTHYL)AZO)-p-TOLUENESULFONIC ACID, BARIUM SALT

5-CHLORO-2-((2-HYDROXY-1-NAPHTHYL)AZO)-P-TOLUENESULPHONIC ACID, BARIUM SALT see 5-CHLORO-2-((2-HYDROXY-1-NAPH-THYL)AZO)-p-TOLUENESULFONIC ACID, BARIUM SALT

7-CHLORO-3-HYDROXY-5-PHENYL-1,3-DIHYDRO-2H-1,4-BENZODI-AZEPIN-2-ONE see 7-CHLORO-1,3-DIHYDRO-3-HYDROXY-5-PHENYL-2H-1,4-BENZODIAZEPINE-2-ONE

(3-CHLORO-4-HYDROXYPHENYL)HYDROXYMERCURY see 2-CHLORO-4-(HYDROXY MERCURI)PHENOL

6-CHLORO-17-ALPHA-HYDROXYPREGNA-4,6-DIENE-3,20-DIONE ACE-TATE see CAP

6-CHLORO-17-ALPHA-HYDROXY-DELTA(SUP 6)-PROGESTERONE ACETATE see CAP

2-CHLORO-HYDROXYTOLUENE see 4-CHLORO-m-CRESOL

6-CHLORO-3-HYDROXYTOLUENE see 4-CHLORO-m-CRESOL

5-CHLORO-7-IODO-8-HYDROXYQUINOLINE see 5-CHLORO-7-IODO-8-QUINOLINOL

2-CHLOROISOBUTANE see tert-BUTYL CHLORIDE

gamma-CHLOROISOBUTYLENE see 3-CHLORO-2-METHYLPROPENE

alpha-CHLORO-N-ISOPROPYLACETANILIDE see 2-CHLORO-N-ISOPRO-PYLACETANILIDE

2-CHLORO-N-ISOPROPYL-N-PHENYLACETAMIDE see 2-CHLORO-N-ISO-PROPYLACETANILIDE

S-(6-CHLORO-3-(MERCAPTOMETHYL)-2-BENZOXAZOLINONE)- O,O-DIETHYL PHOSPHORODITHIOATE see S-((3-BENZOXAZOLINYL-6-CHLORO-2-OXO)METHYL) O,O-DIETHYLPHOSPHORODITHIOATE

(CHLOROMERCURI)BENZENE see PHENYL MERCURIC CHLORIDE

CHLOROMETHANE see MONOCHLOROMETHANE

CHLOROMETHOXY ETHANE see CHLOROMETHYL ETHYL ETHER

CHLORO(2-METHOXYETHYL)MERCURY see 2-METHOXYETHYLMER-CURY CHLORIDE

3-CHLORO-7-METHOXY-9-(1-METHYL-4-DIETHYLAMINOBUTYL-AMINO)ACRIDINE see ATABRINE

5-CHLORO-2-METHOXYPROCAINAMIDE see 4-AMINO-5-CHLORO-N-(2-(DIETHYLAMINO)ETHYL)-n-ANISAMIDE

7-CHLORO-2-METHYLAMINO-5-PHENYL -3H-1,4-BENZODIAZEPINE 4-OXIDE see LIBRIUM

3-CHLORO-6-METHYLANILINE see 2-AMINO-4-CHLOROTOLUENE

CHLOROMETHYLBENZENE see TOLYL CHLORIDE

3-CHLORO-4-METHYL-7-COUMARINYL DIETHYL PHOSPHOROTHIOATE see COUMAPHOS

O-3-CHLORO-4-METHYL-7-COUMARINYL-O,O-DIETHYL PHOSPHORO-THIOATE see COUMAPHOS

CHLOROMETHYL CYANIDE see CHLORACETONITRILE

2-CHLORO-4-METHYL-6-DIMETHYLAMINOPYRIMIDINE see CASTRIX

6-CHLORO-1,2-alpha-METHYLENE-6-DEHYDRO-17-alpha-HYDROXY-PROGESTERONE ACETATE see CYPROSTERONE ACETATE

6-CHLORO-1,2-alpha-METHYLENE-17-alpha-HYDROXY-delta(SUP 6)-PROGESTERONE ACETATE see CYPROSTERONE ACETATE

6-CHLORO-delta(SUP 6)-1,2-alpha-METHYLENE-17-alpha-HYDROXY-PROGESTERONE ACETATE see CYPROSTERONE ACETATE

D-2-CHLORO-6-METHYLERGOLINE-8-beta-ACETONITRILE METHANE-SULFONIC ACID SALT see LERGOTRILE MESYLATE

(CHLOROMETHYL)ETHYLENE OXIDE see 1-CHLORO-2,3-EPOXY PROPANE

(2-CHLORO-1-METHYLETHYL) ETHER see BIS(2-CHLOROISO-PROPYL) ETHER

3-CHLORO-4-METHYL-7-HYDROXYCOUMARIN DIETHYL THIOPHOS-PHORIC ACID ESTER see COUMAPHOS

CHLOROMETHYLMERCURY see MERCURYMETHYLCHLORIDE

2-(CHLOROMETHYL)OXIRANE see 1-CHLORO-2,3-EPOXY PROPANE

4-CHLORO-3-METHYLPHENOL see 4-CHLORO-m-CRESOL

2-(2-CHLORO-4-METHYLPHENYL)AMINO-1,3-DIAZACYCLOPENT-2-ENE see 2-(2-CHLORO-p-TOLUIDINO)-2-IMIDAZOLIDINE

N'-(4-CHLORO-2-METHYLPHENYL)-N,N-DIMETHYLMETHANIMIDAMIDE see CHLOROPHENAMIDINE

CHLOROMETHYL PHENYL KETONE see 2-CHLOROACETOPHENONE

2-CHLORO-11-(4-METHYL-1PIPERAZINYL)-DIBENZO(B,F)(1,4)OXOAZE-PINE see DIBENZACEPIN

2-CHLORO-11-(4-METHYL-1-PIPERAZINYL)DIBENZO(B,F)(1,4)-THIAZEPINE see 2-CHLORO-11-(4-METHYLPIPERAZINO)DI-BENZO(b,f)(1,4)THIAZEPINE

2-CHLORO-10-(3-(1-METHYL-4-PIPERAZINYL)-PROPYL)-PHENO-THIAZINE see PROCHLORPROMAZINE

2-CHLORO-10-(3-(4-METHYL-1-PIPERAZINYL)PROPYL)PHENOTHIAZINE see PROCHLORPROMAZINE

3-CHLORO-10-(3-(1-METHYL-4-PIPERAZINYL)PROPYL)PHENOTHIAZINE see PROCHLORPROMAZINE

CHLORO-3 (N-METHYLPIPERAZINYL-3 PROPYL)-10 PHENOTHIAZINE (FRENCH) see PROCHLORPROMAZINE

2-CHLORO-10-(3-(1-METHYL-4-PIPERAZINYL)PROPYL)PHENOTHI-AZINE, DIMALEATE see PROCHLORPERIZINE MALEATE

2-CHLORO-10-(3-(4-METHYL-1-PIPERAZINYL)PROPYL)PHENOTHIAZINE DIMALEATE see PROCHLORPERIZINE MALEATE

2-CHLORO-2-METHYLPROPANE see tert-BUTYL CHLORIDE

3-CHLORO-2-METHYL-1-PROPENE see 3-CHLORO-2-METHYLPRO-PENE

5'-CHLORO-2-(METHYL(2-(PYRROLIDINYL)ETHYL)AMINO)-O-ACETO-TOLUIDIDE DIHYDROCHLORIDE see 5'-CHLORO-2-(METHYL(2-(PYRROLIDINYL)ETHYL)AMINO)-O-ACETOTOLUIDIDE DIHY-DROCHLORIDE

3-CHLORO-4-METHYLUMBELLIFERONE-O-ESTER WITH-O,O-DIETHYL PHOSPHOROTHIOATE see COUMAPHOS

CHLOROMYCETIN see CHLORAMPHENICOL

CHLORONITRIN see CHLORAMPHENICOL

CHLORO-M-NITROBENZENE see 1-CHLORO-3-NITROBENZENE

M-CHLORONITROBENZENE see 1-CHLORO-3-NITROBENZENE

O-CHLORONITROBENZENE see CHLORO-O-NITROBENZENE

P-CHLORONITROBENZENE see 1-CHLORO-4-NITROBENZENE

1-CHLORO-2-NITROBENZENE see CHLORO-O-NITROBENZENE

2-CHLORO-1-NITROBENZENE see CHLORO-O-NITROBENZENE

4-CHLORO-1-NITROBENZENE see 1-CHLORO-4-NITROBENZENE

2-((O-CHLORO-ALPHA-(NITROMETHYL)BENZYL)THIO)ETHYLAMINE HYDROCHLORIDE see NITRALAMINE HYDROCHLORIDE

2-CHLORO-4-NITROPHENYLAMIDE-6-CHLOROSALICYLIC ACID see 2',5-DICHLORO-4'-NITROSALICYLANILIDE

N-(2-CHLORO-4-NITROPHENYL)-5-CHLOROSALICYLAMIDE see 2',5-DICHLORO-4'-NITROSALICYLANILIDE

O-(2-CHLORO-4-NITROPHENYL) O,O-DIMETHYL PHOSPHOROTHIOATE see p-NITRO-O-CHLOROPHENYL DIMETHYL THIONOPHOSPHATE

O-(3-CHLORO-4-NITROPHENYL) O,O-DIMETHYL PHOSPHOROTHIOATE see METHYLCHLOROTHION

CHLORONITROPROPAN (POLISH) see CHLORONITROPROPANE

1-CHLORO-2-NITROPROPANE see CHLORONITROPROPANE

3-(6-CHLORO-2-OXOBENZOXAZOLIN-3-YL)METHYL-O,O-DIETHYL PHOSPHOROTHIOLOTHIONATE see S-((3-BENZOXAZOLINYL-6-CHLORO-2-OXO)METHYL) O,O-DIETHYLPHOSPHORODITHIOATE

CHLOROPENTAHYDROXYDIALUMINUM see ALUMINUM CHLORIDE HY-DROXIDE

M-CHLOROPHENOL see 3-CHLOROPHENOL

O-CHLOROPHENOL see 2-CHLOROPHENOL

P-CHLOROPHENOL see 4-CHLOROPHENOL

CHLOROPHENOTHANE see DDT

4-(3-(2-CHLOROPHENOTHIAZIN-10-YL)PROPYL-1-PIPERAZINEETHA-NOL, 3,4,5-TRIMETHOXYBENZOATE DIFUMARATE see DIFUMARATE

(4-CHLOROPHENOXY)ACETIC ACID see p-CHLORO PHENOXY ACETIC ACID

ALPHA-P-CHLOROPHENOXYISOBUTYRYL ETHYL ESTER see ATROMID S

2-(4-CHLOROPHENOXY)-2-METHYLPROPANOIC ACID ETHYL ESTER see ATROMID S

2-(P-CHLOROPHENOXY)-2-METHYLPROPIONIC ACID ETHYL ESTER see ATROMID S

2-(ALPHA-P-CHLOROPHENYLACETYL)INDANE-1,3-DIONE see CHLO-ROPHACINONE

P-CHLOROPHENYL BENZENESULFONATE see 4-CHLOROPHENYL BEN-ZENESULFONATE

4-CHLOROPHENYL BENZENESULPHONATE see 4-CHLOROPHENYL BEN-ZENESULFONATE

1-(P-CHLORO-alpha-PHENYLBENZYL)-4-(M-METHYLBENZYL)PIPER-AZINE see HISTAMETHIZINE

1-(P-CHLORO-ALPHA-PHENYLBENZYL)-4-(M-METHYLBENZYL)PIPER-AZINE HYDROCHLORIDE see MECLIZINE HYDROCHLORIDE

1-(p-CHLORO-alpha-PHENYLBENZYL)-4-METHYLPIPERAZINE see CHLOROCYCLINE

2-(2-(4-(P-CHLORO-ALPHA-PHENYLBENZYL)-1-PIPERAZINYL)ETH-OXY)ETHANOL see 1-(p-CHLORO-alpha-PHENYLBENZYL)-4-(2-((2-HYDROXYETHOXY)ETHYL)PIPERAZINE

N-(3-CHLORO PHENYL) CARBAMATE DE 4-CHLORO 2-BUTYNYLE (FRENCH) see m-CHLORO CARBANILIC ACID-4-CHLORO-2-BUTYNYL ESTER

N-(3-CHLORO PHENYL) CARBAMATE D'ISOPROPYLE (FRENCH) see N-3-CHLOROPHENYL ISOPROPYL CARBAMATE

N-(3-CHLOROPHENYL)CARBAMIC ACID,-ISOPROPYL ESTER see N-3-CHLOROPHENYL ISOPROPYL CARBAMATE

(3-CHLOROPHENYL)CARBAMIC ACID,-1-METHYLETHYL ESTER see N-3-CHLOROPHENYL ISOPROPYL CARBAMATE

P-CHLOROPHENYL-P-CHLOROBENZENESULFONATE see 4-CHLORO-PHENYL 4-CHLOROBENZENESULFONATE

4-CHLOROPHENYL-4-CHLOROBENZENESULPHONATE see 4-CHLORO-PHENYL 4-CHLOROBENZENESULFONATE

P-CHLOROPHENYL-P-CHLOROBENZENESULPHONATE see 4-CHLORO-PHENYL 4-CHLOROBENZENESULFONATE

4-CHLOROPHENYL-4'-CHLOROBENZYL SULFIDE see p-CHLOROBEN-ZYL-p-CHLOROPHENYL SULFIDE

2-(O-CHLOROPHENYL)-2-(P-CHLOROPHENYL)-1,1-DICHLOROETHANE see CHLODITHANE

5-(4'-CHLOROPHENYL)-2,4-DIAMINO-6-ETHYLPYRIMIDINE see TIN-DURIN

beta-(p-CHLOROPHENYL)-alpha,alpha-DIMETHYLETHYLAMINE see CHLORPHENTERMINE

S-(P-CHLOROPHENYL)-O,O-DIMETHYL PHOSPHOTHIOATE see FUJITHION

1-(P-CHLOROPHENYL)-3,3-DIMETHYLUREA see 3-(p-CHLORO-PHENYL)-1,1-DIMETHYLUREA

3-(4-CHLOROPHENYL)-1,1-DIMETHYLUREA see 3-(p-CHLOROPHE-NYL)-1,1-DIMETHYLUREA

N-(P-CHLOROPHENYL)-N',N'-DIMETHYLUREA see 3-(p-CHLORO-PHENYL)-1,1-DIMETHYLUREA

1-(4-CHLORO PHENYL)-3,3-DIMETHYLUREE (FRENCH) see 3-(p-CHLO-ROPHENYL)-1,1-DIMETHYLUREA

5-(4-CHLOROPHENYL)-6-ETHYL-2,4-PYRIMIDINEDIAMINE see TIN-DURIN

4-(2-CHLOROPHENYLHYDRAZONE)-3-METHYL-5-ISOXAZOLONE see 3-METHYL-4,5-ISOXAZOLEDIONE,-4-((O-CHLOROPHENYL) HYDRA-ZONE)

N'-(4-CHLOROPHENYL)-N-ISOBUTINYL-N-METHYLUREA see 3-(p-CHLOROPHENYL)-1-METHYL-1-(1-METHYL-2-PROPYNYL)UREA

CHLOROPHENYLMETHANE see TOLYL CHLORIDE

1-(P-CHLOROPHENYL)-2-METHYL-2-AMINOPROPANE HYDROCHLORIDE see AVICOL

2-CHLOROPHENYL-N-METHYLCARBAMATE see O-CHLOROPHENYL METHYLCARBAMATE

2-((P-CHLOROPHENYL)PHENYLACETYL)-1,3-INDANDIONE see CHLO-ROPHACINONE

2(2-(4-CHLOROPHENYL)-2-PHENYLACETYL)INDAN-1,3-DIONE see CHLOROPHACINONE

1-P-CHLOROPHENYL-1-(2-PYRIDYL)-3-DIMETHYLAMINOPROPANE MA-LEATE see TELDRIN

S-((P-CHLOROPHENYLTHIO)METHYL)-O,O-DIETHYL PHOSPHORODI-THIOATE see TRITHION

S-((P-CHLOROPHENYLTHIO)METHYL) O,O-DIETHYL PHOSPHORODI-THIOATE see DANIFOS

S-(4-CHLOROPHENYLTHIOMETHYL)DIETHYL PHOSPHOROTHIOLOTHI-ONATE see TRITHION

S-(((P-CHLOROPHENYL)THIO)METHYL) O,O-DIMETHYL PHOSPHORO-DITHIOATE see METHYL TRITHION

4-CHLOROPHENYL, 2,4,5-TRICHLOROPHENYL SULFONE see p-CHLO-ROPHENYL-2,4,5-TRICHLOROPHENYL SULFONE

CHLOROPHOSPHORIC ACID, DIETHYL ESTER see DIETHYL CHLORO-PHOSPHATE

CHLOROPICRIN see TRICHLORONITROMETHANE

CHLOROPICRINE (FRENCH) see TRICHLORONITROMETHANE

CHLOROPLATINIC (IV) ACID see CHLOROPLATINIC ACID

CHLOROPREEN (DUTCH) see NEOPRENE

6-CHLORO-DELTA(SUP 4,6)-PREGNADIENE-17-ALPHA-OL-3,20-DIONE-17-ACETATE see CAP

6-CHLORO-PREGNA-4,6-DIEN-17-ALPHA-OL-3,20-DIONE ACETATE see CAP

CHLOROPREN (GERMAN, POLISH) see NEOPRENE

CHLOROPRENE see NEOPRENE

3-CHLOROPRENE see ALLYL CHLORIDE

BETA-CHLOROPRENE see NEOPRENE

1-CHLOROPROPANE-2,3-DIOL see CHLORHYDRIN

3-CHLOROPROPANE-1,2-DIOL see CHLORHYDRIN

3-CHLORO-1,2-PROPANEDIOL see CHLORHYDRIN

CHLOROPROPANONE see CHLOROACETONE

CHLOROPROPANONE see ACETONYL CHLORIDE

1-CHLORO-2-PROPANONE see ACETONYL CHLORIDE

1-CHLORO-2-PROPANONE see CHLOROACETONE

3-CHLOROPROPANONITRILE see 3-CHLOROPROPIONITRILE

1-CHLOROPROPENE see PROPENYL CHLORIDE

3-CHLOROPROPENE see ALLYL CHLORIDE

1-CHLORO-1-PROPENE see PROPENYL CHLORIDE

1-CHLORO-2-PROPENE see ALLYL CHLORIDE

1-CHLORO PROPENE-2 see ALLYL CHLORIDE

2-CHLOROPROPENE (DOT) see 2-CHLORO-1-PROPENE

beta-CHLOROPROPIONITRILE see 3-CHLOROPROPIONITRILE

2-CHLOROPROPYL ALCOHOL see 2-CHLORO-1-PROPANOL

3-CHLOROPROPYLENE see ALLYL CHLORIDE

3-CHLOROPROPYLENE GYLCOL see CHLORHYDRIN

3-CHLORO-1,2-PROPYLENE OXIDE see 1-CHLORO-2,3-EPOXY PRO-PANE

gamma-CHLOROPROPYLENE OXIDE see 1-CHLORO-2,3-EPOXY PRO-PANE

2'-CHLORO-4-STILBENYL-N,N-DIMETHYLAMINE see 2'-CHLORO-N,N-DIMETHYL-4-STILBENAMINE

6-CHLORO-7-SULFAMOYL-3,4-DIHYDRO-2H-1,2,4-BENZOTHIADI-AZINE-1,1-DIOXIDE see 6-CHLORO-3,4-DIHYDRO-2H-1,2,4-BENZO-THIADIAZINE-7-SULFONAMIDE- 1,1-DIOXIDE

CHLOROSULFONIC ACID see CHLOROSULFURIC ACID

CHLOROSULFONIC ANHYDRIDE see PYROSULFURYL CHLORIDE

1-CHLOROSULFONYL-5-DIMETHYLAMINONAPHTHALENE see 5-(DI-METHYLAMINO)-1-NAPHTHALENESULFONYL CHLORIDE

1-(4-CHLORO-O-SULFO-5-TOLYLAZO)-2-NAPHTHOL,BARIUM SALT see 5-CHLORO-2-((2-HYDROXY-1-NAPHTHYL)AZO)-p-TOLUENESUL-FONIC ACID, BARIUM SALT

7-CHLOROTETRACYCLINE see CHLORTETRACYCLINE

CHLOROTHANE NU see 1,1,1-TRICHLOROETHANE

CHLOROTHENE see 1,1,1-TRICHLOROETHANE

2-((5-CHLORO-2-THENYL)(2-DIMETHYLAMINOETHYL)AMINO)PYRI-DINE see CHLOROMETHAPYRILENE

CHLOROTHENYLPYRAMINE see CHLOROMETHAPYRILENE

CHLOROTHIAMIDE see 2,6-DICHLOROTHIOBENZAMIDE

2,3,4,5-CHLOROTHIOPHENE see TCTP

ALPHA-CHLOROTOLUENE see TOLYL CHLORIDE

OMEGA-CHLOROTOLUENE see TOLYL CHLORIDE

5-CHLORO-O-TOLUIDINE see 2-AMINO-4-CHLOROTOLUENE

N'-(4-CHLORO-O-TOLYL)-N,N-DIMETHYLFORMAMIDINE see CHLORO-PHENAMIDINE

2-CHLOROTRIETHYLAMINE see N-(2-CHLORO ETHYL)DIETHYL AMINE

beta-CHLOROTRIETHYLAMINE see N-(2-CHLORO ETHYL)DIETHYL AMINE

4-(4-CHLORO-ALPHA,ALPHA,ALPHA-TRIFLUORO-M-TOLYL)-1-(4,4-BIS(P-FLUOROPHENYL)BUTYL)-4-PIPERIDINOL see PENFLURIDOL

CHLORO(TRIISOBUTYL)STANNANE see TRIISO BUTYLTIN CHLORIDE

CHLOROTRIMETHYLTIN see CHLOROTRIMETHYLSTANNANE

CHLOROTRIPHENYLTIN see CHLOROTRIPHENYLSTANNANE

CHLOROTRIS(P-METHOXYPHENYL)ETHYLENE see CHLOROTRIANISENE

beta-CHLOROVINYLBICHLOROARSINE see CHLOROVINYLARSINE DI-CHLORIDE

2-CHLOROVINYLDICHLOROARSINE see CHLOROVINYLARSINE DICHLO-RIDE

beta-CHLOROVINYL ETHYLETHYNYL CARBINOL see 1-CHLORO-3-ETHYL-1-PENTEN-4-YN-3-OL

3-(beta-CHLOROVINYL)-1-PENTYN-3-OL see 1-CHLORO-3-ETHYL-1-PENTEN-4-YN-3-OL

CHLOROWODOR (POLISH) see HYDROCHLORIC ACID

CHLOROX see SODIUM HYPOCHLORITE

CHLORO XYLENOL see 4-CHLORO-3,5-XYLENOL

p-CHLORO-m-XYLENOL see 4-CHLORO-3,5-XYLENOL

CHLOROZIRCONYL see ZIRCONIUM OXYCHLORIDE

CHLORPHACINON (ITALIAN) see CHLOROPHACINONE

CHLORPHENIRAMINE MALEATE see TELDRIN

CHLORPHENTERAMINE see CHLORPHENTERMINE

CHLORPHENVINFOS see CHLORFENVINFOS

CHLORPHENVINPHOS see CHLORFENVINFOS

(4-CHLOR-PHENYL)-BENZOLSULFONAT (GERMAN) see 4-CHLORO-PHENYL BENZENESULFONATE

(4-CHLOR-PHENYL)-4-CHLOR-BENZOL-SULFONATE (GERMAN) see 4-CHLOROPHENYL 4-CHLOROBENZENESULFONATE

3-(4-CHLOR-PHENYL)-1,1-DIMETHYL-HARNSTOFF (GERMAN) see 3-(p-CHLOROPHENYL)-1,1-DIMETHYLUREA

N-(3-CHLOR-PHENYL)-ISOPROPYL-CARBAMAT (GERMAN) see N-3-CHLOROPHENYL ISOPROPYL CARBAMATE

3-(4-CHLORPHENYL)-1-METHOXY-1-METHYLHARNSTOFF (GERMAN) see 3-(4-CHLOROPHENYL)-1-METHOXY-1-METHYLUREA

N-(4-CHLORPHENYL)-N'-METHYL-N'-ISOBUTINYLHARNSTOFF (GERMAN) see 3-(p-CHLOROPHENYL)-1-METHYL-1-(1-METHYL-2-PROPYNYL)UREA

((4-CHLORPHENYL)-1-PHENYL)-ACETYL-1,3-INDANDION (GERMAN) see CHLOROPHACINONE

2(2-(4-CHLOR-PHENYL-2-PHENYL)ACETYL)INDAN-1,3-DION (GER-MAN) see CHLOROPHACINONE

CHLORPIKRIN (GERMAN) see TRICHLORONITROMETHANE

CHLORPROMAZIN see CHLOROPROMAZINE

3-CHLORPROPEN (GERMAN) see ALLYL CHLORIDE

CHLORPROPHAME (FRENCH) see N-3-CHLOROPHENYL ISOPROPYL
CARBAMATE
CHLORPROPHENPYRIDAMINE MALEATE see CHLORPHENIRAMINE MA-
LEATE
CHLORTHALONIL (GERMAN) see TETRACHLOROISOPHTHALONITRILE
CHLORTHION METHYL see METHYLCHLOROTHION
CHLORTION (CZECH) see METHYLCHLOROTHION
ALPHA-CHLORTOLUOL (GERMAN) see TOLYL CHLORIDE
N'-(4-CHLOR-O-TOLYL)-N,N-DIMETHYLFORMAMIDIN (GERMAN) see
CHLOROPHENAMIDINE
CHLORTRIFLUORAETHYLEN (GERMAN) see CHLOROTRIFLUORO-
ETHYLENE
CHLORURE ANTIMONIEUX (FRENCH) see ANTIMONY (III) CHLORIDE
CHLORURE ARSENIEUX (FRENCH) see ARSENIC CHLORIDE
CHLORURE D'ALUMINIUM (FRENCH) see ALUMINUM CHLORIDE
CHLORURE D'ARSENIC (FRENCH) see ARSENIC CHLORIDE
CHLORURE DE BENZENYLE (FRENCH) see BENZYL TRICHLORIDE
CHLORURE DE BORE (FRENCH) see BORON TRICHLORIDE
CHLORURE DE BUTYLE (FRENCH) see N-BUTYL CHLORIDE
CHLORURE DE CHLORACETYLE (FRENCH) see CHLOROACETYL
CHLORIDE
CHLORURE DE CHROMYLE (FRENCH) see CHROMIUM OXYCHLORIDE
CHLORURE DE CYANOGENE (FRENCH) see CYANOGEN CHLORIDE
CHLORURE DE FUMARYLE (FRENCH) see FUMARYL CHLORIDE
CHLORURE DE METHALLYLE (FRENCH) see 3-CHLORO-2-METHYL-
PROPENE
CHLORURE DE METHYLE (FRENCH) see MONOCHLOROMETHANE
CHLORURE DE METHYLENE (FRENCH) see METHANE DICHLORIDE
CHLORURE D'ETHYLE (FRENCH) see ETHYL CHLORIDE
CHLORURE D'ETHYLENE (FRENCH) see 1,2-DICHLORETHANE
CHLORURE D'ETHYLIDENE (FRENCH) see ETHYLIDENE DICHLORIDE
CHLORURE DE VINYLE (FRENCH) see VINYL CHLORIDE
CHLORURE DE VINYLIDENE (FRENCH) see 1,1-DICHLOROETHYLENE
CHLORURE DE ZINC (FRENCH) see ZINC CHLORIDE
CHLORURE E'ETHYLENE (FRENCH) see ETHYLENE DICHLORIDE
CHLORURE FERRIQUE (FRENCH) see FERRIC CHLORIDE
CHLORURE MERCUREUX (FRENCH) see MERCUROUS CHLORIDE
CHLORWASSERSTOFF (GERMAN) see HYDROCHLORIC ACID
CHLORYL ANESTHETIC see ETHYL CHLORIDE
2-CHOLANTHRENONE, 3-METHYL- see 3-METHYLCHOLANTHRENE-2-
ONE
CHOLEIC ACID see DEOXYCHOLATIC ACID
3-BETA-CHOLESTANOL see DIHYDROCHOLESTEROL
CHOLEST-5-EN-3-beta-OL see CHOLESTEROL
5-CHOLESTEN-3-beta-OL see CHOLESTEROL
5:6-CHOLESTEN-3-beta-OL see CHOLESTEROL
CHOLESTENONE see CHOLEST-5-EN-3-ONE
5-CHOLESTEN-3-ONE see CHOLEST-5-EN-3-ONE
DELTA(SUP 5)-CHOLESTENONE see CHOLEST-5-EN-3-ONE
CHOLESTERIN see CHOLESTEROL
CHOLESTEROL, (P-(BIS(2-CHLOROETHYL)AMINO)PHENYL) ACETATE
see CHOLESTERYL-P-BIS(2-CHLOROETHYL)AMINO PHENYLACETATE
CHOLESTEROL-alpha-EPOXIDE see EPOXYCHOLESTEROL
CHOLESTEROL 5-alpha,6-alpha-EPOXIDE see EPOXYCHOLESTEROL
CHOLESTEROL alpha-OXIDE see EPOXYCHOLESTEROL
CHOLINE CARBAMATE CHLORIDE see CARBACHOL CHLORIDE
CHOLINE CHLORINE CARBAMATE see CARBACHOL CHLORIDE
CHOLINE, IODIDE, SUCCINATE (2:1) see BIS(beta-DIMETHYL AMINO
ETHYL)SUCCINATE BIS(METHYL IODIDE)
CHOLINE SUCCINATE DICHLORIDE see SUCCINOYLCHOLINE CHLORIDE
CHOLINE SUCCINATE (ESTER) see CHOLINE SUCCINATE (2:1) (ESTER)
CHOLLY see COCAINE
CHOLSAEURE (GERMAN) see CHOLIC ACID (HYDRATE)
CHONDRUS EXTRACT see CARRAGEEN
CHRLICH 594 see ACETPHENARSINE
CHROMATE DE PLOMB (FRENCH) see LEAD CHROMATE
CHROMATE OF SODA see DISODIUM CHROMATE
CHROME see CHROMIUM
CHROME FERROALLOY see FERROCHROME (EXOTHERMIC)
CHROME GREEN see LEAD CHROMATE

CHROME LEATHER BLUE 3B see C.I. DIRECT BLUE 14, TETRASODIUM
SALT
CHROME LEATHER SKY BLUE see C.I. DIRECT BLUE 1, TETRASODIUM
SALT
CHROME ORE see CHROMITE (MINERAL)
CHROME (TRIOXYDE DE) (FRENCH) see CHROMIUM (VI) OXIDE (1:3)
CHROMIC ACID see CHROMIUM (VI) OXIDE (1:3)
CHROMIC ACID, BIS(TRIPHENYLSILYL) ESTER see BIS(TRIPHENYL
SILYL)CHROMATE
CHROMIC ACID, CALCIUM SALT (1:1) see CALCIUM CHROMATE
CHROMIC ACID, CALCIUM SALT (1:1), DIHYDRATE see CALCIUM
CHROMATE (VI) DIHYDRATE
CHROMIC ACID, CHROMIUM (3+)SALT (3:2) see CHROMIC CHRO-
MATE
CHROMIC ACID, DI-t-BUTYL ESTER see tert-BUTYL CHROMATE
CHROMIC ACID, LEAD AND MOLYBDENUM SALT see LEAD-MOLYB-
DENUM CHROMATE
CHROMIC ACID, LEAD(2+) SALT (1:1) see LEAD CHROMATE
CHROMIC ACID LEAD SALT WITH LEAD MOLYBDATE see LEAD-
MOLYBDENUM CHROMATE
CHROMIC ACID, SOLID see CHROMIUM (VI) OXIDE (1:3)
CHROMIC ACID, SOLID (DOT) see CHROMIUM (VI) OXIDE (1:3)
CHROMIC ACID SOLUTION (DOT) see CHROMIC ACID (SOLUTION)
CHROMIC ACID, ZINC SALT see BASIC ZINC CHROMATE
CHROMIC ANHYDRIDE see CHROMIC ACID (MIXTURE)
CHROMIC ANHYDRIDE see CHROMIUM (VI) OXIDE (1:3)
CHROMIC ANHYDRIDE (DOT) see CHROMIUM (VI) OXIDE (1:3)
CHROMIC CHLORIDE see CHROMIUM (III) CHLORIDE
CHROMIC (VI) ACID see CHROMIUM (VI) OXIDE (1:3)
CHROMIC (VI) ACID see CHROMIC ACID
CHROMIC OXYCHLORIDE see CHROMIUM OXYCHLORIDE
CHROMIC TRIOXIDE see CHROMIUM (VI) OXIDE (1:3)
CHROMIC TRIOXIDE (DOT) see CHROMIUM (VI) OXIDE (1:3)
CHROMITE ORE see CHROMITE (MINERAL)
CHROMIUM(III) ACETATE see CHROMIC ACETATE
CHROMIUM ALLOY see FERROCHROME (EXOTHERMIC)
CHROMIUM ALLOY, BASE, CT,C,FE,N,SI (FERROCHROMIUM) see
FERROCHROME (EXOTHERMIC)
CHROMIUM CHLORIDE see CHROMIUM (III) CHLORIDE
CHROMIUM CHLORIDE, ANHYDROUS see CHROMIUM (III) CHLORIDE
CHROMIUM CHLORIDE OXIDE see CHROMIUM OXYCHLORIDE
CHROMIUM CHROMATE see CHROMIC CHROMATE
CHROMIUM DICHLORIDE DIOXIDE see CHROMIUM OXYCHLORIDE
CHROMIUM DIOXIDE DICHLORIDE see CHROMIUM OXYCHLORIDE
CHROMIUM DISODIUM OXIDE see DISODIUM CHROMATE
CHROMIUM HYDROXIDE SULFATE see NEOCHROMIUM
CHROMIUM (VI) DIOXYCHLORIDE see CHROMIUM OXYCHLORIDE
CHROMIUM LEAD OXIDE see LEAD CHROMATE, BASIC
CHROMIUM OXIDE see CHROMIUM (VI) OXIDE (1:3)
CHROMIUM SODIUM OXIDE see SODIUM DICHROMATE
CHROMIUM SULFATE see NEOCHROMIUM
CHROMIUM SULFATE, BASIC see NEOCHROMIUM
CHROMIUM TRIACETATE see CHROMIC ACETATE
CHROMIUM TRICHLORIDE see CHROMIUM (III) CHLORIDE
CHROMIUM (6+) TRIOXID see CHROMIUM (VI) OXIDE (1:3)
CHROMIUM TRIOXIDE see CHROMIC ACID (MIXTURE)
CHROMIUM TRIOXIDE see CHROMIUM (VI) OXIDE (1:3)
CHROMIUM (VI) OXIDE see CHROMIUM (VI) OXIDE (1:3)
CHROMOTRICHIA FACTOR see p-AMINOBENZOIC ACID
CHROMO (TRIOSSIDO DI) (ITALIAN) see CHROMIUM (VI) OXIDE (1:3)
CHROMOXYLCHLORIDE (DUTCH) see CHROMIUM OXYCHLORIDE
CHROMSAEUREANHYDRID (GERMAN) see CHROMIUM (VI) OXIDE
(1:3)
CHROMTRIOXID (GERMAN) see CHROMIUM (VI) OXIDE
(1:3)
CHROMYLCHLORID (GERMAN) see CHROMIUM OXYCHLORIDE
CHROMYL CHLORIDE see CHROMIUM OXYCHLORIDE
CHROOMTRIOXYDE (DUTCH) see CHROMIUM (VI) OXIDE (1:3)
CHROOMZUURANHYDRIDE (DUTCH) see CHROMIUM (VI) OXIDE (1:3)
CHRYSANTHEMUM CINERAREAEFOLIUM see PYRETHRIN I

(+)-TRANS-CHRYSANTHEMUMIC ACID ESTER OF (+−)-ALLETHRO-LONE see BIOALLETHRIN

ALPHA-CHRYSIDINE see BENZ(c)ACRIDINE

CHRYSOIDINE see 4-PHENYLAZO-m-PHENYLENEDIAMINE

CHRYSOTILE see ASBESTOS

C.I. 27 see HISPACID FAST ORANGE 2G

C.I. 258 see SCARLET RED

CI395 see ANGEL DUST

C.I. 456 see 4-CHLORO-N-METHYL-3-(METHYLSULFAMOYL)-BENZAMIDE

CI 581 see 2-(o-CHLOROPHENYL)-2-(METHYLAMINO)CYCLO-HEXANONE HYDROCHLORIDE

C.I. 10305 see PICRIC ACID

C.I. 10355 see DIPHENYLAMINE

C.I. 11000 see p-(PHENYLAZO)ANILINE

C.I. 11020 see N,N-DIMETHYL-p-PHENYLAZOANILINE

C.I. 11160 see 2-AMINO-5-AZOTOLUENE

C.I. 11270 see 4-PHENYLAZO-m-PHENYLENEDIAMINE

C.I. 11390 see 1-(o-TOLYLAZO)-2-NAPHTHYLAMINE

C.I. 12055 see 1-(PHENYLAZO)-2-NAPHTHOL

C.I. 12140 see 1-(2,4-XYLYLAZO)-2-NAPHTHOL

C.I. 13390 see ACID BLUE 92

C.I. 14700 see FOOD RED 4

C.I. 15985 see FOOD YELLOW 3

C.I. 16045 see FOOD RED 4

C.I. 16155 see FD AND C RED NO. 1

C.I. 16185 see FOOD RED 2

C.I. 16255 see FOOD RED 7

C.I. 22570 see C.I. DIRECT VIOLET 1, DISODIUM SALT

C.I. 22610 see C.I. DIRECT BLUE 6, TETRASODIUM SALT

C.I. 23850 see C.I. DIRECT BLUE 14, TETRASODIUM SALT

C.I. 23860 see 4,4′-BIS(1-AMINO-8-HYDROXY-2,4-DISULFO-7-NAPHTHYLAZO)-3,3′-BITOLYL, TETRASODIUM SALT

C.I. 24410 see C.I. DIRECT BLUE 1, TETRASODIUM SALT

C.I. 26050 see N-ETHYL-1-((p-(PHENYLAZO)PHENYL)AZO)-2-NAPHTHYLAMINE

C.I. 30145 see C.I. DIRECT BROWN

C.I. 30235 see APOMINE BLACK GX

C.I. 37025 see o-NITROANILINE

C.I. 37030 see m-NITROANILINE

C.I. 37035 see p-NITROANILINE

C.I. 37107 see p-TOLUIDINE

C.I. 37115 see o-ANISIDINE HYDROCHLORIDE

C.I. 37130 see 5-NITRO-o-ANISIDINE

C.I. 37275 see 1-AMINOANTHRAQUINONE

C.I. 42000 see AIZEN MALACHITE GREEN

C.I. 42040 see BENZALDEHYDE GREEN

C.I. 42053 see FD AND C GREEN NO. 3

C.I. 42090 see FD AND C BLUE 1

C.I. 42090 see FD AND C BLUE NO. 1

C.I. 42095 see FD AND C GREEN NO. 2

C.I. 44090 see ACID BRILLIANT GREEN BS

C.I. 45160 see RHODAMINE 6G EXTRA BASE

C.I. 45380 see BROMO EOSINE

C.I. 46000 see XANTHACRIDINE

C.I. 46005 see 3,6-BIS(DIMETHYL AMINO)ACRIDINE

CI-47031 see PHOSFOLAN

C.I. 59825 see JADE GREEN BASE

C.I. 60700 see 1-AMINO-2-METHYLANTHRAQUINONE

C.I. 60710 see 1-AMINO-4-HYDROXYANTHRAQUINONE

C.I. 61200 see BRILLIANT BLUE R

C.I. 61565 see 1,4-BIS(p-TOLYAMINO)ANTHRAQUINONE

C. I. 63340 see TOLUIDINE BLUE

C.I. 73015 see DISODIUM INDIGO-5,5-DISULFONATE

C.I. 75500 see WALNUT EXTRACT

C.I. 76000 see ANILINE

C.I. 76020 see 2-AMINO-4-NITROANILINE

C.I. 76025 see m-PHENYLENEDIAMINE

C.I. 76035 see TOLUENE-2,4-DIAMINE

C.I. 76042 see TOLUENE-2,5-DIAMINE

C.I. 76043 see 2,5-DIAMINOTOLUENE SULFATE

C.I. 76060 see p-PHENYLENEDIAMINE

C.I. 76070 see 4-AMINO-2-NITROANILINE

C.I. 76500 see PYROCATECHOL

C.I. 76505 see RESORCINOL

C.I. 76515 see PYROGALLOL

C.I. 76520 see 2-AMINOPHENOL

C.I. 76545 see m-AMINOPHENOL

C.I. 76605 see 1-NAPHTHOL

C.I. 77000 see ALUMINUM

C.I. 77002 see ALUMINUM HYDROXIDE

C.I. 77050 see ANTIMONY

C.I. 77056 see ANTIMONY (III) CHLORIDE

C.I. 77060 see ANTIMONY TRISULFIDE

C.I. 77061 see ANTIMONY PENTASULFIDE

C.I. 77086 see ARSENIC SULFIDE

C.I. 77099 see BARIUM CARBONATE (1:1)

C.I. 77103 see BARIUM CHROMATE (VI)

C.I. 77120 see BARIUM SULFATE

C.I. 77180 see CADMIUM

C.I. 77185 see CADMIUM (II) ACETATE

C.I. 77199 see CADMIUM SULFIDE

C.I. 77223 see CALCIUM CHROMATE (VI) DIHYDRATE

C.I. 77231 see CALCIUM (II) SULFATE DIHYDRATE (1:1:2)

C.I. 77266 see CARBON

C.I. 77295 see CHROMIUM (III) CHLORIDE

C.I. 77400 see COPPER

C.I. 77491 see IRON (III) OXIDE

C.I. 77575 see LEAD

C.I. 77577 see LEAD MONOXIDE

C.I. 77578 see LEAD OXIDE RED

C.I. 77580 see LEAD OXIDE

C.I. 77580 see LEAD(IV)OXIDE BROWN

C.I. 77600 see LEAD CHROMATE

C.I. 77601 see LEAD CHROMATE, BASIC

C.I. 77610 see LEAD (II) CYANIDE

C.I. 77622 see LEAD (II) PHOSPHATE (3:2)

C.I. 77630 see LEAD (II) SULFATE (1:1)

CI 77718 see TALC (powder)

C.I. 77726 see MANGANESE (II) OXIDE

C.I. 77727 see MANGANESE (III) OXIDE

C.I. 77728 see MANGANESE DIOXIDE

C.I. 77755 see POTASSIUM PERMANGANATE

C.I. 77760 see MERCURIC OXIDE

C.I. 77764 see MERCUROUS CHLORIDE

C.I. 77775 see NICKEL

C.I. 77777 see NICKEL MONOXIDE

C.I. 77779 see NICKEL (II) CARBONATE (1:1)

C.I. 77795 see PLATINUM

C.I. 77805 see SELENIUM

C.I. 77820 see SILVER

C.I. 77847 see STRONTIUM SULFIDE

C.I. 77864 see TIN (II) CHLORIDE (1:2)

C.I. 77891 see TITANIUM OXIDE

C.I. 77938 see VANADIUM PENTOXIDE (DUST)

C.I. 77940 see VANADYL SULFATE

C.I. 77945 see ZINC

C.I. 77947 see ZINC OXIDE

C.I. 11160B see 2-AMINO-5-AZOTOLUENE

C.I. 52 015 (CZECH) see 3,7-BIS(DIMETHYL AMINO)PHENAZA THIONIUM CHLORIDE

C.I. ACID BLUE 9, DIAMMONIUM SALT see FD AND C BLUE NO. 1

C.I. ACID BLUE 9, DISODIUM SALT see FD AND C BLUE 1

C.I. ACID BLUE 1, SODIUM SALT see ACID BLUE 1

C.I. ACID GREEN 5 see FD AND C GREEN NO. 2

C.I. ACID GREEN 5, DISODIUM SALT see FD AND C GREEN NO. 2

C.I. ACID GREEN 3, MONOSODIUM SALT see ACID GREEN 3

C.I. ACID GREEN 50, MONOSODIUM SALT see ACID BRILLIANT GREEN BS

C.I. ACID RED 2 see 2-CARBOXY-4′-(DIMETHYL AMINO)AZOBENZENE

C.I. ACID RED 26 see D&C RED NO. 5

C.I. ACID RED 14, DISODIUM SALT see C.I. FOOD RED 3

CIANURO DI SODIO (ITALIAN) see SODIUM CYANIDE

CIANURO DI VINILE (ITALIAN) see ACRYLONITRILE

C.I. AZOIC COUPLING COMPONENT 107 see p-TOLUIDINE

CIBA 5968 see 1(2H)-PHTHALAZINONE HYDRAZONE

CIBA C-768 see DIMETHYL PHOSPHATE ESTER WITH 2-CHLORO-N-METHYL-3-HYDROXYCROTONAMIDE

CIBA CO. 2825 see 2-(1-PIPERIDINO)-2-(2-THENYL)ETHYLAMINE MALEATE

C.I. BASIC GREEN 4 see AIZEN MALACHITE GREEN

C.I. BASIC GREEN 1, SULFATE (1:1) see BENZALDEHYDE GREEN

C.I. BASIC ORANGE 2 see 4-PHENYLAZO-m-PHENYLENEDIAMINE

C.I. BASIC RED 1, MONOHYDROCHLORIDE see RHODAMINE 6G EXTRA BASE

C.I. BASIC RED 9, MONOHYDROCHLORIDE see BASIC PARAFUCHSINE

CICLOESANO (ITALIAN) see CYCLOHEXANE

CICLOESANONE (ITALIAN) see CYCLOHEXANONE

6-CICLOESIL-2,4-DINITR-FENOLO (ITALIAN) see 2-CYCLOHEXYL-4,6-DINITROPHENOL

C.I. DIRECT BLUE 1 see C.I. DIRECT BLUE 1, TETRASODIUM SALT

C.I. DIRECT BLUE 14 see C.I. DIRECT BLUE 14, TETRASODIUM SALT

C.I. DIRECT BROWN 78, DIAMMONIUM SALT see FD AND C BLUE NO. 1

C.I. DISPERSE BLACK 6 DIHYDROCHLORIDE see 3,3'-DIMETHOXY-BENZIDINE DIHYDROCHLORIDE

C.I. DISPERSE YELLOW 3 see ACETAMINE YELLOW CG

CIDOXEPIN HYDROCHLORIDE see ADAPIN

C.I. FOOD BLUE 2 see FD AND C BLUE NO. 1

C.I. FOOD BLUE 2 see FD AND C BLUE 1

C.I. FOOD GREEN 1 see ACID GREEN 3

C.I. FOOD GREEN 2 see FD AND C GREEN NO. 2

C.I. FOOD GREEN 3 see FD AND C GREEN NO. 3

C.I. FOOD GREEN 4 see ACID BRILLIANT GREEN BS

C.I. FOOD RED 5 see D&C RED NO. 5

C.I. FOOD RED 15 see (9-(o-CARBOXYPHENYL)-6-(DIETHYLAMINO)-3H-XANTHEN-3-YLIDENE) DIETHYLAMMONIUM CHLORIDE

C.I. FOOD RED 1, DISODIUM SALT see FOOD RED 4

C.I. FOOD RED 6, DISODIUM SALT see FD AND C RED NO. 1

C.I. FOOD YELLOW 10 see 1-(PHENYLAZO)-2-NAPHTHYLAMINE

C.I. FOOD YELLOW 11 see 1-(o-TOLYLAZO)-2-NAPHTHYLAMINE

C.I. FOOD YELLOW 3, DISODIUM SALT see FOOD YELLOW 3

C.I. 45350 (FREE ACID) see FLUORESCEIN

CIGARETTE SMOKE CONDENSATE see SMOKE CONDENSATE, CIGARETTE

C.I. NATURAL BROWN 7 see WALNUT EXTRACT

C.I. NATURAL YELLOW 1 see CHAMOMILE

C.I. NATURAL YELLOW 8 see MORIN

C.I. NATURAL YELLOW 10 see QUERCETIN

CINCAINE HYDROCHLORIDE see NUPERCAINE HYDROCHLORIDE

CINCHOCAINE HYDROCHLORIDE see NUPERCAINE HYDROCHLORIDE

CINCHOCAINIUM CHLORIDE see NUPERCAINE HYDROCHLORIDE

1,8-CINEOLE see CAJEPUTOL

CINERIN I OR II see PYRETHRIN I

CINNAMEIN see BENZYL CINNAMATE

CINNAMENE see STYRENE

CINNAMENOL see STYRENE

CINNAMIC ALDEHYDE see CINNAMMALDEHYDE

CINNAMYL ALCOHOL ANTHRANILATE see ANTHRANILIC ACID, CINNAMYL ESTER

CINNAMYL ALDEHYDE see CINNAMMALDEHYDE

CINNAMYL-2-AMINOBENZOATE see ANTHRANILIC ACID, CINNAMYL ESTER

CINNAMYL-O-AMINOBENZOATE see ANTHRANILIC ACID, CINNAMYL ESTER

CINNAMYL ANTHRANILATE see ANTHRANILIC ACID, CINNAMYL ESTER

CINNAMYLEPHEDRINE HYDROCHLORIDE, DEXTRO see d-CINNAMYLEPHEDRINE HYDROCHLORIDE

C.I. NO. 46005:1 see 3,6-BIS(DIMETHYL AMINO)ACRIDINE

CIODRIN VINYL PHOSPHATE see CROTOXYPHOS

C.I. OXIDATION BASE see TOLUENE-2,4-DIAMINE

C.I. OXIDATION BASE 7 see m-AMINOPHENOL

C.I. OXIDATION BASE 10 see p-PHENYLENEDIAMINE

C.I. OXIDATION BASE 16 see o-PHENYLENEDIAMINE

C.I. OXIDATION BASE 17 see 2-AMINOPHENOL

C.I. OXIDATION BASE 6A see 4-AMINOPHENOL

C.I. PIGMENT BLACK 14 see MANGANESE DIOXIDE

C.I. PIGMENT BLACK 16 see ZINC

C.I. PIGMENT BROWN 8 see MANGANESE DIOXIDE

CI PIGMENT RED 104 see LEAD-MOLYBDENUM CHROMATE

C.I. PIGMENT WHITE 6 see TITANIUM OXIDE

C.I. PIGMENT WHITE 11 see ANTIMONY OXIDE

C.I. PIGMENT YELLOW 32 see STRONTIUM CHROMATE (1:1)

C.I. PIGMENT YELLOW 33 see CALCIUM CHROMATE (VI) DIHYDRATE

C.I. PIGMENT YELLOW 33 see CALCIUM CHROMATE

C.I. PIGMENT YELLOW 48 see LEAD (II) CYANIDE

CIPROMID see CYPROMID

CIRCOSOLV see ACETYLENE TRICHLORIDE

CIS,CIS-9,12-OCTADECADIENOIC ACID see LINOLEIC ACID

EXO-1,2-CIS-DIMETHYL-3,6-EPOXYHEXAHYDROPHTHALIC ANHYDRIDE see CANTHARIDINE

CIS-9,12-OCTADECADIENOIC ACID see LINOLEIC ACID

C.I. SOLVENT GREEN 3 see 1,4-BIS(p-TOLYAMINO)ANTHRAQUINONE

C.I. SOLVENT ORANGE 3 see 4-PHENYLAZO-m-PHENYLENEDIAMINE

C.I. SOLVENT ORANGE 7 see 1-(2,4-XYLYLAZO)-2-NAPHTHOL

C.I. SOLVENT RED 23 see OIL RED

C.I. SOLVENT RED 24 see SCARLET RED

C.I. SOLVENT RED 41 see MAGENTA BASE

C.I. SOLVENT YELLOW 2 see N,N-DIMETHYL-p-PHENYLAZOANILINE

C.I. SOLVENT YELLOW 3 see 2-AMINO-5-AZOTOLUENE

C.I. SOLVENT YELLOW 5 see 1-(PHENYLAZO)-2-NAPHTHYLAMINE

C.I. SOLVENT YELLOW 7 see 4-HYDROXYAZOBENZENE

CISPLATINO (SPANISH) see cis-PLATINOUS DIAMINE DICHLORIDE

CITRIC ACID, DYSPROSIUM(3+) salt (1:1) see DYSPROSIUM CITRATE

CITRIC ACID, GALLIUM SALT (1:1) see GALLIUM CITRATE and COMPOUNDS

CLEVE'S ACID-1,6 see 5-AMINO-2-NAPHTHALENESULFONIC ACID

CLEVE'S BETA-ACID see 5-AMINO-2-NAPHTHALENESULFONIC ACID

CHLORAMEISENSAEUREAETHYLESTER (GERMAN) see ETHYL CHLORO FORMATE

CLIFT see MEPROSCILLARIN

CLIRADON HYDROCHLORIDE see KETOBEMIDONE HYDROCHLORIDE

CLOBBER see CYPROMID

CLOFLUPEROL HYDROCHLORIDE see 4-(4-(4-CHLORO-alpha,alpha,-alpha-TRIFLUORO-m-TOLYL)-4-HYDROXYPIPERIDINO)BUTYRO-PHENONE-4'-FLUORO-,HYDROCHLORIDE

CLOMIPHENE CITRATE see RACEMIC CLOMIPHENE CITRATE

CLOMIPHENE DIHYDROGEN CITRATE see RACEMIC CLOMIPHENE CITRATE

CLONAZEPAM see CLOAZEPAM

CLOPHEN see POLYCHLORINATED BIPHENYLS

CLOPHEN A60 see POLYCHLORINATED BIPHENYL (AROCLOR 1260)

CLORDAN (ITALIAN) see CHLORDANE

CLOREPIN see 7-CHLORO-1-METHYL-5-PHENYL-1H-1,5-BENZODIAZE-PINE-2,4(3H,5H)-DIONE

CLORO (ITALIAN) see CHLORINE

2-CLOROBENZALDEIDE (ITALIAN) see o-CHLOROBENZALDEHYDE

CLOROBENZENE (ITALIAN) see BENZENE CHLORIDE

(4-CLORO-BENZIL)-(4-CLORO-FENIL)-SOLFURO (ITALIAN) see p-CHLOROBENZYL-p-CHLOROPHENYL SULFIDE

2-CLORO-1,3-BUTADIENE (ITALIAN) see NEOPRENE

(4-CLORO-BUT-2-IN-IL)-N-(3-CLORO-FENIL)-CARBAMMATO (ITALIAN) see m-CHLORO CARBANILIC ACID-4-CHLORO-2-BUTYNYL ESTER

O-2-CLORO-1-(2,4-DICLORO-FENIL)-VINYL-O,O-DIETILFOSFATO (ITALIAN) see CHLORFENVINFOS

(2-CLORO-3-DIETILAMINO-1-METIL-3-OXO-PROP-1-EN-IL)-DIMETIL-FOSFATO (ITALIAN) see DIMECRON

CLORODIFENILI, CLORO 42% (ITALIAN) see POLYCHLORINATED BIPHENYL (AROCLOR 1242)

CLORODIFENILI, CLORO 54% (ITALIAN) see POLYCHLORINATED BIPHENYL (AROCLOR 1254)

2-CLORO-4-DIMETILAMINO-6-METIL-PIRIMIDINA (ITALIAN) see CASTRIX

1-CLORO-2,4-DINITROBENZENE (ITALIAN) see 1-CHLORO-2,4-DINITROBENZENE

1-CLORO-2,3-EPOSSIPROPANO (ITALIAN) see 1-CHLORO-2,3-EPOXY PROPANE

CLOROETANO (ITALIAN) see ETHYL CHLORIDE

2-CLOROETHANOLO (ITALIAN) see 2-CHLOROETHYL ALCOHOL

(4-CLORO-FENIL)-BENZOL-SOLFONATO (ITALIAN) see 4-CHLORO-PHENYL BENZENESULFONATE

(4-CLORO-FENIL)-4-CLORO-VENZOL-SOLFONATO (ITALIAN) see 4-CHLOROPHENYL 4-CHLOROBENZENESULFONATE

3-(4-CLORO-FENIL)-1,1-DIMETIL-UREA (ITALIAN) see 3-(p-CHLORO-PHENYL)-1,1-DIMETHYLUREA

2(2-(4-CLORO-FENIL-2FENIL)-ACETIL)INDAN-1,3-DIONE (ITALIAN) see CHLOROPHACINONE

N-(3-CLORO-FENIL)-ISOPROPIL-CARBAMMATO (ITALIAN) see N-3-CHLOROPHENYL ISOPROPYL CARBAMATE

CLOROFORMIO (ITALIAN) see CHLOROFORM

CLOROFOS see ((2,2,2-TRICHLORO-1-HYDROXYETHYL) DIMETHYL-PHOSPHONATE)

CLOROMETANO (ITALIAN) see MONOCHLOROMETHANE

3-CLORO-2-METIL-PROP-1-ENE (ITALIAN) see 3-CHLORO-2-METHYL-PROPENE

1-CLORO-4-NITROBENZENE (ITALIAN) see 1-CHLORO-4-NITROBEN-ZENE

O-(4-CLORO-3-NITRO-FENIL)-O,O-DIMETIL-MONOIIOFOSFATO (ITALIAN) see p-NITRO-o-CHLOROPHENYL DIMETHYL THIONOPHOSPHATE

O-(3-CLORO-4-NITRO-FENIL)-O,O-DIMETIL-MONOTIOFOSFATO (ITALIAN) see METHYLCHLOROTHION

CLOROPICRINA (ITALIAN) see TRICHLORONITROMETHANE

CLOROPRENE (ITALIAN) see NEOPRENE

CLOROX see SODIUM HYPOCHLORITE

CLORURO DI ETHENE (ITALIAN) see ETHYLENE DICHLORIDE

CLORURO DI ETHENE (ITALIAN) see 1,2-DICHLORETHANE

CLORURO DI ETILE (ITALIAN) see ETHYL CHLORIDE

CLORURO DI ETILIDENE (ITALIAN) see ETHYLIDENE DICHLORIDE

CLORURO DI METALLILE (ITALIAN) see 3-CHLORO-2-METHYLPRO-PENE

CLORURO DI METILE (ITALIAN) see MONOCHLOROMETHANE

CLORURO DI VINILE (ITALIAN) see VINYL CHLORIDE

CLORURO MERCUROSO (ITALIAN) see MERCUROUS CHLORIDE

CME see CANNABIS

COAL DUST see ANTHRACITE PARTICLES

COAL FACINGS see COAL, GROUND BITUMINOUS (DOT)

COAL GAS see HYDROCARBON GAS (NONLIQUEFIED)

COAL LIQUID see PETROLEUM

COAL NAPHTHA see BENZENE

COAL OIL see PETROLEUM

COAL OIL see KEROSENE

COAL TAR NAPHTHA DISTILLATE see NAPHTHA, COAL TAR

COBALIN see VITAMIN B$_{12}$ COMPLEX

COBALT ACETATE see COBALT DIACETATE

COBALT(2+) ACETATE see COBALT DIACETATE

COBALTATE(2−), ((ETHYLENEDINITRILO)TETRAACETATO(2−))-, COBALT(2+) SALT see DICOBALT EDETATE

COBALT(II) ACETATE see COBALT DIACETATE

COBALT CHLORIDE see COBALTOUS CHLORIDE

COBALT-CHROMIUM ALLOY see COBALT ALLOY, CO,CR

COBALT DICHLORIDE see COBALTOUS CHLORIDE

COBALT DINITRATE see COBALTOUS NITRATE

COBALT (II) NITRATE (1:2) see COBALTOUS NITRATE

COBALT MURIATE see COBALTOUS CHLORIDE

COBALT NITROSOPENTACYANOFERRATE(3) see COBALT NITROPRUS-SIDE

COBALT OCTACARBONYL see COBALT CARBONYL

COBALTOUS DIACETATE see COBALT DIACETATE

COBALT TETRACARBONYL see COBALT CARBONYL

COBINAMIDE, CYANIDE PHOSPHATE 3′-ESTER WITH 5,6-DIMETHYL-1-ALPHA-D-RIBOFURANOSYLBENZIMIDAZOLE, INNER SALT see VITAMIN B$_{12}$ COMPLEX

COCAFURIN see NITROFURAZONE

COCAIN-CHLORHYDRAT (GERMAN) see COCAINE CHLORIDE

(-)-COCAINE see COCAINE

1-COCAINE see COCAINE

L-COCAINE see COCAINE

beta-COCAINE see COCAINE

COCAINE HYDROCHLORIDE see COCAINE CHLORIDE

(-)-COCAINE HYDROCHLORIDE see COCAINE CHLORIDE

L-COCAINE HYDROCHLORIDE see COCAINE CHLORIDE

COCAINE MURIATE see COCAINE CHLORIDE

COCCULIN see PICROTOXIN

COCCULUS see PICROTOXIN

COCCULUS SOLID (DOT) see PICROTOXIN

COCHINEAL RED A SPECIALLY PURE see FOOD RED 7

COCONUT OIL AMIDE OF DIETHANOLAMINE see N,N-BIS(2-HYDROXY ETHYL)DODECAN AMIDE

COFFEINE see CAFFEINE

COKE see COCAINE

COLACID BLUE A see ACID BLUE 92

COLCHINE, N-DEACETYL-NMETHYL see N-METHYL-N-DESACETYL-COLCHICINE

COLLARGOL see SILVER, COLLOIDAL

COLLODION COTTON see CELLULOSE TETRANITRATE

COLLOIDAL ARSENIC see ARSENIC

COLLOIDAL FERRIC OXIDE see IRON (III) OXIDE

COLLOIDAL GOLD see GOLD

COLLOIDAL MANGANESE see MANGANESE

COLLOIDAL MERCURY see MERCURY

COLLOIDAL SILICA see SILICA, AMORPHOUS FUMED

COLLOIDAL SILICON DIOXIDE see SILICA, AMORPHOUS FUMED

COLLOIDAL SULFUR see SULFUR

COLLOXYLIN see CELLULOSE TETRANITRATE

COLOGNE EARTH see MANGANESE (III) OXIDE

COLOGNE SPIRIT see ETHANOL

COLOGNE SPIRITS (DOT) see ETHANOL

COLOGNE UMBER see MANGANESE (III) OXIDE

COLOGNE YELLOW see LEAD CHROMATE

COLONIAL SPIRIT see METHANOL

COLTS FOOT see PETASITES JAPONICUS MAXIM

COLUMBIAN CARBON see CARBON

COLUMBIAN SPIRIT see METHANOL

COLUMBIAN SPIRITS (DOT) see METHANOL

COLUMBIUM PENTACHLORIDE see NIOBIUM CHLORIDE

COMBUSTION IMPROVER -2 see MANGANESE TRICARBONYL METHYL-CYCLOPENTADIENYL

COMFREY, RUSSIAN see RUSSIAN COMFREY ROOTS

COMMON SALT see SODIUM CHLORIDE

COMMON SENSE COCKROACH AND RAT PREPARATIONS see PHOS-PHORUS (WHITE)

COMPOUND-666 see BENZENE HEXACHLORIDE

COMPOUND 1081 see FLUOROACETAMIDE

COMPOUND-4018 see COPPER DIMETHYLDITHIOCARBAMATE

COMPOUND 4049 see CARBETHOXY MALATHION

COMPOUND #1080 see SODIUM FLUOROACETATE

COMPOUND G-11 see HEXACHLOROPHENE

COMPOUND 6-12 INSECT REPELLENT see ETHYL HEXYLENE GLYCOL

CONDY'S CRYSTALS see POTASSIUM PERMANGANATE

CONFECTIONER'S SUGAR see SUCROSE

CONQUININE see QUINIDINE

COOMASSIE BLUE see ACID BLUE 92

COPPER(2+) ACETATE see COPPER ACETATE

COPPER ACETO ARSENITE see CUPRIC ACETOARSENITE

COPPER ARSENATE (BASIC) see COPPER ARSENATE HYDROXIDE

COPPER ARSENITE, SOLID (DOT) see COPPER ORTHOARSENITE

COPPERAS see IRON (II) SULFATE (1:1)

COPPER BIS(ACETYLACETONATE) see BIS(ACETYL ACETONE)COPPER

COPPER BIS(ACETYLACETONE) see BIS(ACETYL ACETONE)COPPER

COPPER BIS(2,4-PENTANEDIONATE) see BIS(ACETYL ACETONE)COPPER

COPPER BRONZE see COPPER

COPPER CARBONATE HYDROXIDE see COPPER (II) CARBONATE HYDROXIDE (2:1:2)

COPPER(II) ACETATE see COPPER ACETATE

COPPER(II) ACETYLACETONATE see BIS(ACETYL ACETONE)COPPER

COPPER CHLORIDE (DOT) see COPPER (II) CHLORIDE (1:2)

COPPER CYANIDE (DOT) see COPPER (II) CYANIDE

COPPER DIACETATE see COPPER ACETATE

COPPER(2+) DIACETATE see COPPER ACETATE

COPPER DIACETYLACETONATE see BIS(ACETYL ACETONE)COPPER

COPPER-8-HYDROXYQUINOLATE see BIS(8-OXYQUINOLINE)COPPER

COPPER-8-HYDROXYQUINOLINATE see BIS(8-OXYQUINOLINE)COPPER

COPPER-8-HYDROXYQUINOLINE see BIS(8-OXYQUINOLINE)COPPER

COPPER NAPHTHENATE see NAPHTHENIC ACID, COPPER SALT

COPPER (2+) OXINATE see BIS(8-OXYQUINOLINE)COPPER

COPPER OXYQUINOLATE see BIS(8-OXYQUINOLINE)COPPER

COPPER OXYQUINOLINE see BIS(8-OXYQUINOLINE)COPPER

COPPER QUINOLATE see BIS(8-OXYQUINOLINE)COPPER

COPPER-8-QUINOLATE see BIS(8-OXYQUINOLINE)COPPER

COPPER-8-QUINOLINOL see BIS(8-OXYQUINOLINE)COPPER

COPPER QUINOLINOLATE see BIS(8-OXYQUINOLINE)COPPER

COPPER-8-QUINOLINOLATE see BIS(8-OXYQUINOLINE)COPPER

COPPER SULFIDE see COPPER (I) SULFIDE

COPPER UVERSOL see NAPHTHENIC ACID, COPPER SALT

COQUES DU LEVANT (FRENCH) see PICROTOXIN

CORICIDIN see p-ACETOPHENETIDIDE

CORINE see COCAINE

CORNOCENTIN see D-LYSERGIC ACID-L,2-PROPANOLAMIDE

CORNOTONE see N,N-DIETHYLNICOTINAMIDE

CORN SUGAR see D-GLUCOSE

COROTONIN see N,N-DIETHYLNICOTINAMIDE

CORPUS LUTEUM HORMONE see PROGESTERONE

CORRIGEN see OLEANDRIN

CORROSIVE MERCURY CHLORIDE see CORROSIVE SUBLIMATE

DELTA(SUP 1)-CORTISOL see PREDONIN

CORTISOL ALCOHOL see CORTISOL

CORTISOL SUCCINATE, SODIUM SALT see HYDROCORTISONE SODIUM SUCCINATE

DELTA CORTISONE see PREDNISONE

DELTA(SUP 1)-CORTISONE see PREDNISONE

CORTISPRAY see CORTISOL

CORYDININE see FUMARINE

CORYNINE see YOHIMBINE

COSMETIC BLUE LAKE see FD AND C BLUE 1

COSMETIC GREEN BLUE R25396 see ACID BLUE 1

COUMAFENE see COUMADIN

COUMAFURYL see 3-(alpha-ACETONYLFURFURYL)-4-HYDROXY-COUMARIN

COUMAPHOS-O-ANALOG see 3-CHLORO-4-METHYL-7-COUMARINYL DIETHYL PHOSPHATE

cis-o-COUMARINIC ACID LACTONE see COUMARIN

COUMARINIC ANHYDRIDE see COUMARIN

C.P. TITANIUM see TITANIUM (DRY POWDER)

CRAB'S EYES see ABRIN

CRAG FUNGICIDE 974 see DIMETHYLFORMOCARBOTHIALDINE

CRAG SEVIN see CARBARYL

M-CRESIDINE see 2-METHYL-p-ANISIDINE

2-CRESOL see o-CRESOL

3-CRESOL see m-CRESOL

4-CRESOL see p-CRESOL

para-CRESOL see p-CRESOL

P-CRESOL ACETATE see 4-METHYLPHENYL ACETATE

O-CRESOL GLYCERYL ETHER see GLYCERYL-O-TOLYL ETHER

CRESOLI (ITALIAN) see CRESOL

CRESOSSIDIOLO see GLYCERYL-O-TOLYL ETHER

CRESOSSIPROPANDIOLO see GLYCERYL-O-TOLYL ETHER

CRESOXYDIOL see GLYCERYL-O-TOLYL ETHER

CRESOXYPROPANEDIOL see GLYCERYL-O-TOLYL ETHER

P-CRESYL ACETATE see 4-METHYLPHENYL ACETATE

M-CRESYL ESTER OF N-METHYLCARBAMIC ACID see TSUMACIDE

O-CRESYL ALPHA-GLYCERYL ETHER see GLYCERYL-O-TOLYL ETHER

CRESYLIC ACID see CRESOL

M-CRESYLIC ACID see m-CRESOL

O-CRESYLIC ACID see o-CRESOL

P-CRESYLIC ACID see p-CRESOL

CRESYLITE see 2,4,6-TRINITRO-meta-CRESOL

M-CRESYL METHYLCARBAMATE see TSUMACIDE

CRESYL PHOSPHATE see TRITOLYL PHOSPHATE

O-CRESYL PHOSPHATE see TRI-2-TOLYL PHOSPHATE

CRIMIDIN (GERMAN) see CASTRIX

CRIMIDINA (ITALIAN) see CASTRIX

CRIMIDINE see CASTRIX

CRIMSON ANTIMONY see ANTIMONY TRISULFIDE

CRISPATINE see FULVINE

CRISTOBALITE see SILICA (CRYSTALLINE)

CRISTOBALITE see SILICA, CRYSTALLINE-CRISTOBALITE

CROCIDOLITE see ASBESTOS

CROCIODOLITE see CROCIDOLITE (see ASBESTOS)

CROCOITE see LEAD CHROMATE

CROMILE, CLORURO DI (ITALIAN) see CHROMIUM OXYCHLORIDE

CROMO, OSSICLORURO DI (ITALIAN) see CHROMIUM OXYCHLORIDE

CROTENALDEHYDE see trans-2-BUTENAL

CROTONALDEHYDE see trans-2-BUTENAL

CROTONATE DE 2,4-DINITRO 6-(1-METHYL-HEPTYL)-PHENYLE (FRENCH) see ARATHANE

alpha-CROTONIC ACID see CROTONIC ACID

CROTONIC ACID, 4-FLUORO-, METHYL ESTER see METHYL-gamma-FLUOROCROTONATE

CROTONIC ACID, 3-HYDROXY-, alpha-METHYLBENZYL ESTER, DIMETHYL PHOSPHATE, (E) - see CROTOXYPHOS

CROTONIC ALDEHYDE see trans-2-BUTENAL

CROTONOEL (GERMAN) see CROTON OIL

CROTON RESIN see CROTON OIL

CROTON TIGLIUM L. OIL see CROTON OIL

CROTONYL ALCOHOL see 2-BUTEN-1-OL

CROTYL ALCOHOL see 2-BUTEN-1-OL

CRUDE ARSENIC see ARSENIC TRIOXIDE

CRUDE OIL see PETROLEUM

CRYSTALLIZED VERDIGRIS see COPPER ACETATE

CRYSTALLOSE see SACCHARIN

CRYSTALS OF VENUS see COPPER ACETATE

CT 4436 see 5-METHOXY-3-(2-PYRROLIDINOETHYL)INDOLE

CUBE see ROTENONE

CUBE EXTRACT see ROTENONE

CUBE ROOT see ROTENONE

CUBES see N,N-DIETHYLLYSERGAMIDE

CULLEN EARTH see MANGANESE (III) OXIDE

CUMAFOS (DUTCH) see COUMAPHOS

CUMATE see COPPER DIMETHYLDITHIOCARBAMATE

CUMEEN (DUTCH) see CUMENE

CUMEENHYDROPEROXYDE (DUTCH) see ISOPROPYLBENZENE HYDROPEROXIDE

CUMENE HYDROPEROXIDE see ISOPROPYLBENZENE HYDROPEROXIDE

CUMENE HYDROPEROXIDE (DOT) see ISOPROPYLBENZENE HYDROPEROXIDE

CUMENYL HYDROPEROXIDE see ISOPROPYLBENZENE HYDROPEROXIDE

m-CUMENYL METHYLCARBAMATE see m-CUMENOL METHYLCARBAMATE

CUMERTILIN SODIUM see MERCUMATILIN SODIUM

p-CUMIC ALDEHYDE see CUMALDEHYDE

CUMINALDEHYDE see CUMALDEHYDE

CUMINIC ALDEHYDE see CUMALDEHYDE

CUMINYL ALDEHYDE see CUMALDEHYDE

CUMOL see CUMENE

CUMOLHYDROPEROXID (GERMAN) see ISOPROPYLBENZENE HYDRO-
PEROXIDE

CUMYL HYDROPEROXIDE see ISOPROPYLBENZENE HYDROPEROXIDE

alpha-CUMYL HYDROPEROXIDE see ISOPROPYLBENZENE HYDRO-
PEROXIDE

alpha-CUMYL HYDROPEROXIDE see ISOPROPYLBENZENE HYDRO-
PEROXIDE

CUPFERRON see AMMONIUM-N-NITROSOPHENYLHYDROXYLAMINE

CUPRASULFIDE see COPPER (I) SULFIDE

CUPRAVIT BLUE see COPPER HYDROXIDE

CUPRIC ACETATE see COPPER ACETATE

CUPRIC ACETYLACETONATE see BIS(ACETYL ACETONE)COPPER

CUPRIC ARSENITE see COPPER ORTHOARSENITE

CUPRIC CARBONATE see COPPER (II) CARBONATE HYDROXIDE
(2:1:2)

CUPRIC CHLORIDE see COPPER (II) CHLORIDE (1:2)

CUPRIC CYANIDE (DOT) see COPPER (II) CYANIDE

CUPRIC DIACETATE see COPPER ACETATE

CUPRIC GREEN see COPPER ORTHOARSENITE

CUPRIC HYDROXIDE see COPPER HYDROXIDE

CUPRIC-8-HYDROXYQUINOLATE see BIS(8-OXYQUINOLINE)COPPER

CUPRICIN see COPPER CYANIDE

CUPRIC-8-QUINOLINOLATE see BIS(8-OXYQUINOLINE)COPPER

CUPRIC SULFATE see COPPER (II) SULFATE (1:1)

CUPRIC SULFATE PENTAHYDRATE see COPPER (II) SULFATE PENTA-
HYDRATE (1:1:5)

CUPRINOL see NAPHTHENIC ACID, COPPER SALT

CUPROUS ARSENATE, BASIC see COPPER ARSENATE HYDROXIDE

CUPROUS CHLORIDE see COPPER (I) CHLORIDE

CUPROUS CYANIDE see COPPER CYANIDE

CUPROUS OXIDE see COPPER (I) OXIDE

CUPROUS SULFIDE see COPPER (I) SULFIDE

CYAANWATERSTOF (DUTCH) see HYDROCYANIC ACID

CYAMEPROMAZINE MALEATE see CYANOTRIMEPRAZINE MALEATE

CYANACETATE ETHYLE (GERMAN) see ETHYL CYANOACETATE

CYANACETHYDRAZIDE see CYANACETIC ACID HYDRAZIDE

CYANACETOHYDRAZIDE see CYANACETIC ACID HYDRAZIDE

CYANACETYLHYDRAZIDE see CYANACETIC ACID HYDRAZIDE

CYANAMIDE see CALCIUM CYANAMIDE

CYANAMIDE CALCIQUE (FRENCH) see CALCIUM CYANAMIDE

CYANAMIDE, CALCIUM SALT (1:1) see CALCIUM CYANAMIDE

CYANAZINE see BLADEX

CYANESSIGSAEURE (GERMAN) see CYANOACETIC ACID

CYANHYDRINE D'ACETONE (FRENCH) see 2-METHYLLACTONITRILE

CYANIDELONON 1522 see QUERCETIN

CYANIDE OF POTASSIUM (SOLN) see POTASSIUM CYANIDE (soln)

CYANIDE OF SODIUM see SODIUM CYANIDE

CYANINE ACID BLUE R see ACID BLUE 92

CYANITE see ALUMINUM(III) SILICATE (2:1)

CYANOACETHYDRAZIDE see CYANACETIC ACID HYDRAZIDE

CYANOACETIC ACID ETHYL ESTER see ETHYL CYANOACETATE

CYANOACETIC ACID HYDRAZIDE see CYANACETIC ACID HYDRAZIDE

CYANOACETIC ESTER see ETHYL CYANOACETATE

CYANOACETOHYDRAZIDE see CYANACETIC ACID HYDRAZIDE

ALPHA-CYANOACETOHYDRAZIDE see CYANACETIC ACID HYDRAZIDE

CYANOACETONITRILE see MALONONITRILE

(CYANOACETYL)HYDRAZIDE see CYANACETIC ACID HYDRAZIDE

CYANOACETYLHYDRAZIDE see CYANACETIC ACID HYDRAZIDE

2-CYANOANILINE see ANTHRANILONITRILE

O-CYANOANILINE see ANTHRANILONITRILE

CYANO-B12 see VITAMIN B_{12} COMPLEX

CYANOBENZENE see BENZONITRILE

4-CYANOBENZONITRILE see p-BENZENEDINITRILE

CYANOCOBALAMIN see VITAMIN B_{12} COMPLEX

4-CYANO-2,6-DIIODOPHENOL see 4-HYDROXY-3,5-DIIODOBENZONI-
TRILE

4-CYANO-2,6-DIJODPHENOL (GERMAN) see 4-HYDROXY-3,5-DIIODO-
BENZONITRILE

4-CYANO-2,6-DIJODPHENOL CAPRYSAEUREESTER (GERMAN) see
IOXYNIL OCTANOATE

CYANO-3 (DIMETHYLAMINO-3 METHYL-2 PROPYL)-10PHENOTHIAZINE
MALEATE see CYANOTRIMEPRAZINE MALEATE

2-CYANO-3,3-DIPHENYLACRYLIC ACID, ETHYL ESTER see UV ABSOR-
BER-2

2-CYANO-3,3-DIPHENYL-2-PROPENOIC ACID, ETHYL ESTER see UV
ABSORBER-2

1-(3-CYANO-3,3-DIPHENYLPROPYL)-4-PHENYLISONIPECOTIC ACID
ETHYL ESTER HYDROCHLORIDE see LOMOTIL

CYANOETHANE see PROPIONONITRILE

2-CYANOETHANOL see HYDRACRYLONITRILE

CYANOETHYDRAZIDE see CYANACETIC ACID HYDRAZIDE

CYANOETHYL ACRYLATE see ACRYLIC ACID ESTER WITH HYDRA-
CRYLONITRILE

2-CYANOETHYL ACRYLATE see ACRYLIC ACID ESTER WITH HYDRA-
CRYLONITRILE

2-CYANOETHYL ALCOHOL see HYDRACRYLONITRILE

BETA-CYANOETHYLAMINE see 3-AMINOPROPIONITRILE

CYANOETHYLENE see ACRYLONITRILE

2-CYANOETHYL-2'-FLUOROETHYLETHER see 2-CYANO-2'-FLUORO-
DIETHYL ETHER

2-CYANOETHYL PROPENOATE see ACRYLIC ACID ESTER WITH HY-
DRACRYLONITRILE

CYANOGAS see CALCIUM CYANIDE

CYANOGENAMIDE see CYANAMIDE

CYANOGENE (FRENCH) see CYANOGEN

CYANOGEN GAS (DOT) see CYANOGEN

CYANOGEN NITRIDE see CYANAMIDE

2-CYANO-10-(3-(4-HYDROXYPIPERIDINO)PROPYL)PHENOTHIAZINE see
PIPEROCYANOMAZINE

2-CYANO-10-(3-(4-HYDROXY-1-PIPERIDYL)PROPYL)PHENOTHIAZINE
see PIPEROCYANOMAZINE

CYANO-3 ((HYDROXY-4 PIPERIDYL-1)-3 PROPYL)-10 PHENOTHIAZINE
(FRENCH) see PIPEROCYANOMAZINE

ALPHA-CYANOISOPROPYLAMIDE OF THE O,O-DIETHYLTHIOPHOS-
PHORYL ACETIC ACID see PHOSPHOROTHIOIC ACID-S-(((1-CYANO-
1-METHYL-ETHYL)CARBAMOYL)METHYL)-O,O-DIETHYL ESTER

CYANOMETHANE see ACETONITRILE

CYANOMETHANOL see HYDROXYACETONITRILE

S-N-(1-CYANO-1-METHYLETHYL)CARBAMOYLMETHYL DIETHYL PHOS-
PHOROTHIOLATE see PHOSPHOROTHIOIC ACID-S-(((1-CYANO-
1-METHYL-ETHYL)CARBAMOYL)METHYL)-O,O-DIETHYL ESTER

S-(((1-CYANO-1-METHYL-ETHYL)CARBAMOYL)METHYL) O,O-DIETHYL
PHOSPHOROTHIOATE see PHOSPHOROTHIOIC ACID-S-(((1-CYANO-1-
METHYL-ETHYL)CARBAMOYL)METHYL)-O,O-DIETHYL ESTER

3-(CYANOMETHYL)INDOLE see 3-INDOLYLACETONITRILE

CYANO(METHYLMERCURI)GUANIDINE see METHYLMERCURIC DICY-
ANDIAMIDE

CYANOPHENYLMETHYL-beta-D-GLUCOPYRANOSIDURONIC ACID see
LAETRILE

2-CYANOPROPENE-1 see METHYLACRYLONITRILE

ALPHA-CYANOSTILBENE see 2,3-DIPHENYLACRYLONITRILE

ALPHA-CYANOTOLUENE see PHENYLACETONITRILE

OMEGA-CYANOTOLUENE see PHENYLACETONITRILE

ALPHA-CYANOVINYL ACETATE see 2-ACETOXYACRYLONITRILE

CYANURAMIDE see MELAMINE

CYANURE (FRENCH) see CYANIDES

CYANURE D'ARGENT (FRENCH) see SILVER CYANIDE

CYANURE DE CALCIUM (FRENCH) see CALCIUM CYANIDE

CYANURE DE CUIVRE (FRENCH) see COPPER (II) CYANIDE

CYANURE DE MERCURE (FRENCH) see MERCURY (II) CYANIDE

CYANURE DE METHYL (FRENCH) see ACETONITRILE

CYANURE DE PLOMB (FRENCH) see LEAD (II) CYANIDE

CYANURE DE POTASSIUM (FRENCH) see POTASSIUM CYANIDE (soln)

CYANURE DE SODIUM (FRENCH) see SODIUM CYANIDE

CYANURE DE VINYLE (FRENCH) see ACRYLONITRILE

CYANURE DE ZINC (FRENCH) see ZINC CYANIDE

CYANUROTRIAMIDE see MELAMINE

CYANUROTRIAMINE see MELAMINE

CYANWASSERSTOFF (GERMAN) see HYDROCYANIC ACID

CYCAS CIRCINALIS HUSK see CYCAD HUSK

CYCAS REVOLUTA GLUCOSIDE see CYCASIN
CYCLAMATE CALCIUM see CALCIUM CYCLOHEXYLSULPHAMATE
CYCLAMATES see SODIUM CYCLAMATE
CYCLAMATE SODIUM see SODIUM CYCLAMATE
CYCLIC ETHYLENE (DIETHOXYPHOSPHINOTHIOYL)DITHIOIMIDOCAR-BONATE see DITHIOLANE IMINOPHOSPHATE
CYCLIC ETHYLENE(DIETHOXYPHOSPHINOTHIOYL)DITHIOIMIDOCAR-BONATE see PHOSFOLAN
CYCLIC ETHYLENE ESTER OF (DIETHOXYPHOSPHINOTHIOYL)DITHIO-IMIDOCARBONIC ACID see DITHIOLANE IMINOPHOSPHATE
CYCLIC PROPYLENE (DIETHOXYPHOSPHINYL)DITHIOIMIDOCARBONATE see MEPHOSFOLAN
CYCLOBARBITAL see TETRAHYDROPHENOBARBITAL
CYCLOBARBITOL see TETRAHYDROPHENOBARBITAL
CYCLOBARBITONE see TETRAHYDROPHENOBARBITAL
CYCLOBUTYLENE see CYCLOBUTENE
5-(1-CYCLOHEPTEN-1-YL)-5-ETHYLBARBITURIC ACID see CYCLO-HEPTENYL ETHYLBARBITURIC ACID
CYCLOHEPTENYLETHYLMALONYLUREA see CYCLOHEPTENYL ETHYL-BARBITURIC ACID
CYCLOHEXAAN (DUTCH) see CYCLOHEXANE
1,4-CYCLOHEXADIENEDIONE see p-BENZOQUINONE
2,5-CYCLOHEXADIENE-1,4-DIONE see p-BENZOQUINONE
1,4-CYCLOHEXADIENE DIOXIDE see p-BENZOQUINONE
CYCLOHEXAN (GERMAN) see CYCLOHEXANE
CYCLOHEXANAMINE see CYCLOHEXYLAMINE
CYCLOHEXANECARBAMIC ACID, 1,1-DIPHENYL-2-PROPYNYL ESTER see 1,1-DIPHENYL-2-PROPYNYL-N-CYCLOHEXYLCARBAMATE
CYCLOHEXANECARBOXYLIC ACID, LEAD SALT see NAPHTHENIC ACID, LEAD SALT
CYCLOHEXANESULFAMIC ACID, CALCIUM SALT see CALCIUM CY-CLOHEXYLSULPHAMATE
CYCLOHEXANESULFAMIC ACID, MONOSODIUM SALT see SODIUM CYCLAMATE
CYCLOHEXANOL ACETATE see CYCLOHEXYL ACETATE
CYCLOHEXANOLAZETAT (GERMAN) see CYCLOHEXYL ACETATE
CYCLOHEXANON (DUTCH) see CYCLOHEXANONE
CYCLOHEXANONE ISO-OXIME see HEXAHYDRO-2H-AZEPIN-2-ONE
CYCLOHEXANYL ACETATE see CYCLOHEXYL ACETATE
CYCLOHEXATRIENE see BENZENE
CYCLOHEXENE EPOXIDE see CYCLOHEXENE OXIDE
CYCLOHEXENE-1-OXIDE see CYCLOHEXENE OXIDE
1,2-CYCLOHEXENE OXIDE see CYCLOHEXENE OXIDE
CYCLOHEXENONE see 2-CYCLOHEXEN-1-ONE
5-(1-CYCLOHEXEN-1-YL)-1,5-DIMETHYLBARBITURIC ACID see EVIPAL
CYCLOHEXENYL-ETHYL BARBITURIC ACID see TETRAHYDROPHENO-BARBITAL
5-(1-CYCLOHEXENYL)-5-ETHYLBARBITURIC ACID see TETRAHYDRO-PHENOBARBITAL
5-(1-CYCLOHEXEN-1-YL)-5-ETHYLBARBITURIC ACID see TETRAHY-DROPHENOBARBITAL
5-(1-CYCLOHEXENYL-1)-1-METHYL-5-METHYLBARBITURIC ACID see EVIPAL
5-(DELTA-1,2-CYCLOHEXENYL)-5-METHYL-N-METHYL-BARBITURSA-EURE (GERMAN) see EVIPAL
N-CYCLOHEXYLCYCLOHEXANAMINE see N,N-DICYCLOHEXYLAMINE
2-CYCLOHEXYL-4,6-DINITROFENOL (DUTCH) see 2-CYCLOHEXYL-4,6-DINITROPHENOL
6-CYCLOHEXYL-2,4-DINITROPHENOL see 2-CYCLOHEXYL-4,6-DINI-TROPHENOL
TRANS-N,N'-(1,4-CYCLOHEXYLENEDIMETHYLENE)BIS(2-CHLORO-BENZYLAMINE) DIHYDROCHLORIDE see trans-N,N'-BIS(2-CHLOROBENZYL)-1,4-CYCLOHEXANEBIS(METHYL-AMINE) DIHYDROCHLORIDE
CYCLOHEXYLENE OXIDE see CYCLOHEXENE OXIDE
ALPHA-CYCLOHEXYL-ALPHA-PHENYL-1-PIPERIDINEPROPANOL HY-DROCHLORIDE see BENZHEXOL HYDROCHLORIDE
1-CYCLOHEXYL-1-PHENYL-3-(1-PYRROLIDINYL)-1-PROPANOL see 1-CYCLOHEXYL-1-PHENYL-3-PYRROLIDINO-1-PROPANOL

CYCLOHEXYL SULFAMATE SODIUM see SODIUM CYCLAMATE
CYCLOLEUCINE see 1-AMINOCYCLOPENTANE-1-CARBOXYLIC ACID
CYCLOMYCIN see TETRACYCLINE
CYCLON see HYDROCYANIC ACID
CYCLONAL see EVIPAL
CYCLONE B see HYDROCYANIC ACID
CYCLONITE see TRIMETHYLENETRINITRAMINE
CYCLOPENTADIENE see 1,3-CYCLOPENTADIENE
1,3-CYCLOPENTADIENE, DIMER see BICYCLOPENTADIENE
ALPHA,BETA-CYCLOPENTAMETHYLENETETRAZOLE see 1,5-PENTA-METHYLENETETRAZOLE
CYCLOPENTANE (DOT) see CYCLOPENTANE
2-CYCLOPENTEN-1-ONE, 2,5-DIHYDROXY-3-DIMETHYLAMINO-5-METHYL- see DMA
CYCLOPENTIMINE see PIPERIDINE
ALPHA-CYCLOPENTYL-2-THIOPHENEGLYCOLATE DIETHYL(2-HY-DROXYETHYL)METHYLAMMONIUM BROMIDE see PENTHIENATE BROMIDE
CYCLOPROPANECARBOXYLIC ACID, 2,2-DIMETHYL-3-(2-METHYL-PROPENYL)-, ESTER WITH 2-ALLYL-4-HYDROXY-3-METHYL-2-CYCLOPENTEN-ONE, (+)-(z)- see (+)-cis-ALLETHRIN
2-CYCLOPROPYLMETHYL-5,9-DIMETHYL-2'-HYDROXY-6,7-BENEO-MORPHAN see CYCLAZOCINE
3-(CYCLOPROPYLMETHYL)1-1,2,3,4,5,6-HEXAHYDRO-6,11-DI-METHYL-2,6-METHANO-3-BENZAZOCIN-8-OL see CYCLAZOCINE
2-CYCLOPROPYLMETHYL-2'-HYDROXY-5,9-DIMETHYL-6,7-BENZO-MORPHAN see CYCLAZOCINE
3-CYCLOPROPYLMETHYL-6(eq),11(ax)-DIMETHYL-2,6-METHANO-3-BENZAZOCIN-8-OL see CYCLAZOCINE
CYCLOTETRAMETHYLENE OXIDE see TETRAHYDROFURAN
CYCLOTRIMETHYLENENITRAMINE see TRIMETHYLENETRINITRAMINE
CYCLOTRIMETHYLENETRINITRAMINE see TRIMETHYLENETRINITRA-MINE
CYJANOWODOR (POLISH) see HYDROCYANIC ACID
CYKAZINE see CYCASIN
CYKLOHEKSAN (POLISH) see CYCLOHEXANE
CYKLOHEKSANON (POLISH) see CYCLOHEXANONE
2-P-CYMENOL see CARVACROL
3-P-CYMENOL see THYMOL
P-CYMEN-3-OL see THYMOL
CYMOL see p-CYMENE
CYMONIC ACID see FLUORO ETHANOIC ACID
CYNKU TLENEK (POLISH) see ZINC OXIDE
CYSTAMINE see beta-MERCAPTOETHYLAMINE DISULFIDE
CYSTEAMIDE see 2-AMINOETHANETHIOL
CYSTEAMINE see 2-AMINOETHANETHIOL
CYSTEIN see L-CYSTEINE
CYSTEINAMINE DISULFIDE see beta-MERCAPTOETHYLAMINE DISUL-FIDE
CYSTEINE see L-CYSTEINE
L-(+)-CYSTEINE see L-CYSTEINE
CYSTINAMIN (GERMAN) see beta-MERCAPTOETHYLAMINE DISULFIDE
CYSTINEAMINE see beta-MERCAPTOETHYLAMINE DISULFIDE
CYTHION see CARBETHOXY MALATHION
CYTOCHALASIN D see ZYGOSPORIN A
CYTOSINE-BETA-ARABINOSIDE see ARABINOCYTIDINE
CYTOSINE BETA-D-ARABINOSIDE see ARABINOCYTIDINE
CYTOXYL ALCOHOL CYCLOHEXYLAMMONIUM SALT see CYTOXAL ALCOHOL
CZTEROCHLOREK WEGLA (POLISH) see CARBON TETRACHLORIDE
2,3,7,8-CZTEROCHLORODWUBENZO-P-DWUOKSYNY (POLISH) see TCDD
1,1,2,2-CZTEROCHLOROETAN (POLISH) see ACETYLENE TETRACHLO-RIDE
CZTEROCHLOROETYLEN (POLISH) see 1,1,2,2-TETRACHLORO-ETHYLENE
CZTEROETHLEK OLOWIU (POLISH) see TETRAETHYL LEAD
D₂ see DEUTERIUM
D & C RED NO. 17 see OIL RED
D&C RED NO. 22 see BROMO EOSINE

2-DEOXY-2-(3-METHYL-3- NITROSOUREIDO)-D-GLUCOPYRANOSE see STREPTOZOTICIN

DEOXYNOREPHEDRINE see DL-AMPHETAMINE SULFATE

2-DEOXYPHENOBARBITAL see 2-DESOXY PHENOBARBITAL

1-beta-D-2'-DEOXYRIBOFURANOSYL-5-FLUROURACIL see 2'-DEOXY-5-FLUORO URIDINE

1-beta-D-2'-DEOXYRIBOFURANOSYL-5-IODOURACIL see 2'-DEOXY-5-IODO URIDINE

4-DEOXYTETRONIC ACID see 4-BUTANOLIDE

BETA-DEOXYTHIOGUANOSINE see beta-THIOGUANINE DEOXYRIBO-SIDE

BETA-2'-DEOXY-6-THIOGUANOSINE MONOHYDRATE see beta-THIO-GUANINE DEOXYRIBOSIDE

DEPO-PROVERA see MEDROXYPROGESTERONE ACETATE

DEPRIDOL see dl-METHADONE HYDROCHLORIDE

DERIVATIVES OF BARBITURIC ACID see BARBITURATES

DERRIS see ROTENONE

DERRIS RESINS see DERRIS

DERRIS ROOT see DERRIS

DESACETYLCHOLCHICEINE see TRIMETHYLCOLCHICINIC ACID

N-DESACETYL-N-METHYLCOLCHICINE see N-METHYL-N-DESACETYL-COLCHICINE

DESMETHYLAMITRIPTYLINE see NORTRIPTYLINE

DESMETHYLIMIPRAMINE HYDROCHLORIDE see DEMETHYLIMIPRA-MINE HYDROCHLORIDE

DESOXYCHOLIC ACID see DEOXYCHOLATIC ACID

DESOXYCHOLSAEURE (GERMAN) see DEOXYCHOLATIC ACID

DESOXYEPHEDRINE see DESOXYN

1-DESOXYEPHEDRINE HYDROCHLORIDE see ''SPEED''

D-DESOXYEPHEDRINE HYDROCHLORIDE see d-METHAPHETAMINE HY-DROCHLORIDE

DESOXYNOREPHEDRINE see DL-AMPHETAMINE SULFATE

DESOXYNOREPHEDRINE see AMPHETAMINE

(+-)-DESOXYNOREPHEDRINE see BENZEDRINE

DESOXYPHENOBARBITONE see 2-DESOXY PHENOBARBITAL

DES (SYNTHETIC ESTROGEN) see DIETHYLSTILBESTEROL

DEVEGAN see ACETPHENARSINE

DEVIL'S APPLE see STRAMONIUM

DEXAMETHASONE ALCOHOL see SUPERPREDNOL

DEXAMPHETAMINE see D-AMPHETAMINE

DEXAMPHETAMINE SULFATE see d-BENZEDRINE SULFATE

DEXAMYL see d-BENZEDRINE SULFATE

DEXEDRINA see d-BENZEDRINE SULFATE

DEXEDRINE see D-AMPHETAMINE

DEXEDRINE SULFATE see d-BENZEDRINE SULFATE

DEXIES see d-BENZEDRINE SULFATE

DEXTRAN see DEXTRINS

DEXTRAN IRON COMPLEX see IRON-DEXTRAN COMPLEX

DEXTRIFERRON see IRON-DEXTRIN COMPLEX

DEXTRIFERRON INJECTION see IRON-DEXTRIN COMPLEX

DEXTROAMPHETAMINE SULFATE see d-BENZEDRINE SULFATE

DEXTRO CALCIUM PANTOTHENATE see CALCIUM-d-PANTOTHENATE

DEXTRO-ALPHA-METHYLPHENETHYLAMINE SULFATE see d-BENZE-DRINE SULFATE

DEXTROMYCETIN see CHLORAMPHENICOL

DEXTRO-1-PHENYL-2-AMINOPROPANE SULFATE see d-BENZEDRINE SULFATE

DEXTRO-BETA-PHENYLISOPROPYLAMINE SULFATE see d-BENZEDRINE SULFATE

DEXTROPROPOXYPHENE see CARVON

DEXTROPROPOXYPHENE HYDROCHLORIDE see d-PROPOXYPHENE HY-DROCHLORIDE

DEXTROSE see D-GLUCOSE

DEXTROSOL see D-GLUCOSE

DEXTROTUBOCURARINE CHLORIDE see TUBOCURARINE HYDROCHLO-RIDE

DGE see DIGLYCIDYL ETHER

(D-GLUCOPYRANOSYLTHIO)GOLD see 1-AUROTHIO-D-GLUCOPYRA-NOSE

DHUTRA see STRAMONIUM

DIABASIC MALACHITE GREEN see AIZEN MALACHITE GREEN

DIABRIN see BUFORMIN HYDROCHLORIDE

DIACEPHIN see HEROIN

2,7-DIACETAMIDOFLUORENE see 2,7-BIS(ACETAMIDO)FLUORENE

DIACETIC ETHER see ETHYL ACETYL ACETATE

2,3-DIACETIN see 1,3-DIACETIN

DIACETONALCOHOL (DUTCH) see 2-METHYL-2-PENTANOL-4-ONE

DIACETONALCOOL (ITALIAN) see 2-METHYL-2-PENTANOL-4-ONE

DIACETONALKOHOL (GERMAN) see 2-METHYL-2-PENTANOL-4-ONE

DIACETONE ALCOHOL (DOT) see 2-METHYL-2-PENTANOL-4-ONE

DIACETONE-ALCOOL (FRENCH) see 2-METHYL-2-PENTANOL-4-ONE

DIACETOTOLUIDE see N-ACETYL-N-(2-METHYL-4-((2-METHYL-PHENYL)AZO)PHENYL)ACETAMIDE

DIACETOXYBUTYLTIN see DIACETOXY DIBUTYL STANNANE

DIACETOXYDIBUTYLTIN see DIACETOXY DIBUTYL STANNANE

DIACETOXYDIMETHYLPUMBANE see DIMETHYL LEAD DIACETATE

3-BETA, 17-BETA-DIACETOXY-17-ALPHA-ETHYNYL-4-OESTRENE see ETHYNODIOL ACETATE

4-beta,15-DIACETOXY-3-alpha-HYDROXY-12,13-EPOXYTRICHOTHEC-9-ENE see DIACETOXYSCIRPENOL

4-BETA,15-DIACETOXY-3-ALPHA-HYDROXY-12,13-EPOXYTRICHO-THEC-9-ENE see ANGUIDIN

DIACETOXYMERCURY see MERCURIC ACETATE

4,15-DIACETOXY-8-(3-METHYLBUTYRYLOXY)-12,13-EPOXY-DELTA-9-TRICHOTHECEN-3-OL see MYCOTOXIN T-2

4,15-DIACETOXY-8-(3-METHYLBUTYRYLOXY)-12,13-EPOXY-delta-9-TRICHOTHECEN-3-OL see FUSARIOTOXIN T 2

3-BETA,17-BETA-DIACETOXY-19-NOR-17-ALPHA-PREGN-4-EN-20-YNE see ETHYNODIOL ACETATE

DIACETOXYPROPENE see ACROLEIN DIACETATE

3,3-DIACETOXYPROPENE see ACROLEIN DIACETATE

1,1-DIACETOXYPROPENE-2 see ACROLEIN DIACETATE

DIACETOXYSCIRPENOL see ANGUIDIN

DIACETYL see 2,3-BUTANEDIONE

DIACETYLAMINOAZOTOLUENE see N-ACETYL-N-(2-METHYL-4-((2-METHYLPHENYL)AZO)PHENYL)ACETAMIDE

4,4'-DIACETYLAMINOBIPHENYL see 4',4'''-BIACETANILIDE

2-DIACETYLAMINOFLUORENE see 2-DIACETAMIDO FLUORENE

2,7-DIACETYLAMINOFLUORENE see 2,7-BIS(ACETAMIDO)FLUORENE

N-DIACETYL-2-AMINOFLUORENE see 2-DIACETAMIDO FLUORENE

N,N-DIACETYL-2-AMINOFLUORENE see 2-DIACETAMIDO FLUORENE

4,4'-DIACETYLBENZIDINE see 4',4'''-BIACETANILIDE

N,N'-DIACETYL BENZIDINE see 4',4'''-BIACETANILIDE

DIACETYLCHOLINE see CHOLINE SUCCINATE (2:1) (ESTER)

DIACETYLCHOLINE DICHLORIDE see SUCCINOYLCHOLINE CHLORIDE

DIACETYLCHOLINE DIIODIDE see BIS(beta-DIMETHYL AMINO ETHYL)-SUCCINATE BIS(METHYL IODIDE)

N,N-DIACETYL-2-FLUORENAMINE see 2-DIACETAMIDO FLUORENE

DIACETYL GLYCERINE see 1,3-DIACETIN

DIACETYLGLYCEROL see 1,3-DIACETIN

DIACETYLMANGANESE see MANGANESE ACETATE

DIACETYLMETHANE see ACETYL ACETONE

DIACETYLMORFIN see HEROIN

DIACETYLMORPHINE see HEROIN

DIACETYL PEROXIDE see ACETYL PEROXIDE

DIACETYL PEROXIDE (SOLUTION) see ACETYL PEROXIDE SOLUTION

N,N-DIACETYL-O-TOLYLAZO-O-TOLUIDINE see N-ACETYL-N-(2-METHYL-4-((2-METHYLPHENYL)AZO)PHENYL)ACETAMIDE

DIETHANOLAMIN (GERMAN) see BIS(2-HYDROXY ETHYL)AMINE

1,1-DIAETHOXY-AETHAN (GERMAN) see ACETAL

DIAETHYLACETAL (GERMAN) see ACETAL

DIAETHYLAETHER (GERMAN) see ETHYL ETHER

O,O-DIAETHYL-S-(2-AETHYLTHIO-AETHYL)-DITHIOPHOSPHAT (GER-MAN) see DITHIOSYSTOX

O,O-DIAETHYL-S(2-AETHYLTHIO-AETHYL)-MONOTHIOPHOSPHAT (GERMAN) see DEMETON-S

O,O-DIAETHYL-S-(AETHYLTHIO-METHYL)-DITHIOPHOSPHAT (GER-MAN) see PHORATE

DIAETHYLALLYLACETAMIDE (GERMAN) see 2,2-DIETHYL-4-PENTEN-AMIDE

2,4-DIAMINO-5-P-CHLOROPHENYL-6- ETHYLPYRIMIDINE see TIN-
DURIN

DI(-4-AMINO-3-CHLOROPHENYL)METHANE see MOCA

DI-(4-AMINO-3-CLOROFENIL)METANO (ITALIAN) see MOCA

4,4'-DIAMINO-3,3'-DICHLOROBIPHENYL see DICHLOROBENZIDINE
BASE

4,4'-DIAMINO-3,3'-DICHLORODIPHENYLMETHANE see MOCA

2,2'-DIAMINODIETHYLAMINE see DIETHYLENETRIAMINE

BETA,BETA'-DIAMINODIETHYL DISULFIDE see beta-MERCAPTOETHYL-
AMINE DISULFIDE

P,P'-DIAMINODIFENYLMETHAN (CZECH) see 4,4'-METHYLENEDIANI-
LINE

1,5-DIAMINO-4,8-DIHYDROXYANTHRAQUINONE see 1,5-DIAMINOAN-
THRARUFIN

4,8-DIAMINO-1,5-DIHYDROXYANTHRAQUINONE see 1,5-DIAMINOAN-
THRARUFIN

3,3'-DIAMINO-4,4'-DIHYDROXYARSENOBENZENE DIHYDROCHLORIDE
see SALVARSAN

3,3'-DIAMINO-4,4'-DIHYDROXYARSENOBENZENEMETHYLENESULF-
OXYLATE SODIUM see NEOSALVARSAN

2,4-DIAMINO-6-DIMETHOXYPHOSPHINOTHIONYLTHIOMETHYL-S-
TRIAZINE see AZIDITHION

2,8-DIAMINO-3,7-DIMETHYLACRIDINE see 3,6-DIAMINO-2,7-
DIMETHYLACRIDINE

4,4'-DIAMINO-3,3'-DIMETHYLBIPHENYL see 3,3'-TOLIDINE

4,4'-DIAMINO-3,3'-DIMETHYLDIPHENYL see 3,3'-TOLIDINE

P-DIAMINODIPHENYL see BENZIDINE

2,4'-DIAMINODIPHENYL see 2,4'-BIPHENYLDIAMINE

4,4'-DIAMINODIPHENYL see BENZIDINE

O,O'-DIAMINO DIPHENYL DISULFIDE see 2,2'-DITHIOBISANILINE

DIAMINODIPHENYL ETHER see 4,4-OXYDIANILINE

4,4-DIAMINODIPHENYL ETHER see 4,4-OXYDIANILINE

P,P'-DIAMINODIPHENYL ETHER see 4,4-OXYDIANILINE

4,4'-DIAMINODIPHENYLMETHAN (GERMAN) see 4,4'-METHYLENEDI-
ANILINE

DIAMINODIPHENYLMETHANE see 4,4'-METHYLENEDIANILINE

4,4'-DIAMINODIPHENYLMETHANE see 4,4'-METHYLENEDIANILINE

P,P'-DIAMINODIPHENYLMETHANE see 4,4'-METHYLENEDIANILINE

4,4'-DIAMINODIPHENYL OXIDE see 4,4-OXYDIANILINE

4,4'-DIAMINODIPHENYL SULFIDE see 4,4'-THIODIANILINE

P,P'-DIAMINODIPHENYL SULFIDE see 4,4'-THIODIANILINE

4,4'-DIAMINODIPHENYL SULFONE see 4,4'-SULFONYLDIANILINE

DIAMINO-4,4'-DIPHENYL SULFONE see 4,4'-SULFONYLDIANILINE

P,P'-DIAMINODIPHENYL SULFONE see 4,4'-SULFONYLDIANILINE

P,P-DIAMINODIPHENYL SULPHONE see 4,4'-SULFONYLDIANILINE

DIAMINO-4,4'-DIPHENYL SULPHONE see 4,4'-SULFONYLDIANILINE

3,3-DIAMINODIPROPYLAMINE see AMINOBIS(PROPYLAMINE)

3,3'-DIAMINODIPROPYLAMINE see AMINOBIS(PROPYLAMINE)

1,2-DIAMINO-ETHAAN (DUTCH) see 1,2-ETHANEDIAMINE

1,2-DIAMINOETHANE see 1,2-ETHANEDIAMINE

1,2-DIAMINO-ETHANO (ITALIAN) see 1,2-ETHANEDIAMINE

4,6-DIAMINO-2-HYDROXY-1,3-CYCLOHEXANE 3,6'-DIAMINO-3,6'-
DIDEOXYDI-alpha-D-GLUCOSIDE see KANAMYCIN A

2,4-DIAMINO-1-METHOXYBENZENE see 2,4-DIAMINOANISOLE SUL-
PHATE

2,8-DIAMINO-10-METHYLACRIDINIUM CHLORIDE see XANTHACRIDINE

3,6-DIAMINO-10-METHYLACRIDINIUM CHLORIDE see XANTHACRIDINE

3,6-DIAMINO-10-METHYLACRIDINIUM CHLORIDE MIXTURE WITH 3,6-
ACRIDINEDIAMINE see ACRIFLAVINIUM CHLORIDE

2,8-DIAMINO-10-METHYLACRIDINIUM CHLORIDE MIXTURE WITH 2,8-
DIAMINOACRIDINE see ACRIFLAVINIUM CHLORIDE

1,3-DIAMINO-4-METHYLBENZENE see TOLUENE-2,4-DIAMINE

2,4-DIAMINO-1-METHYLBENZENE see TOLUENE-2,4-DIAMINE

3,7'-DIAMINO-N-METHYLDIPROPYLAMINE see BIS(gamma-AMINOPRO-
PYL)METHYLAMINE

DIAMINON see METHADONE

1,5-DIAMINONAPHTHALENE see 1,5-NAPHTHALENEDIAMINE

DIAMINON HYDROCHLORIDE see METHADONE HYDROCHLORIDE

1,2-DIAMINO-4-NITROBENZENE see 2-AMINO-4-NITROANILINE

1,4-DIAMINO-2-NITROBENZENE see 4-AMINO-2-NITROANILINE

2,6-DIAMINO-3-PHENYLAZOPYRIDINE see PHENAZODINE

2,6-DIAMINO-3-PHENYLAZOPYRIDINE HYDROCHLORIDE see PHE-
NAZOPYRIDINIUM CHLORIDE

2,6-DIAMINO-3-(PHENYLAZO)PYRIDINE MONOHYDROCHLORIDE see
PHENAZOPYRIDINIUM CHLORIDE

4,4'-DIAMINOPHENYL ETHER see 4,4-OXYDIANILINE

2,7-DIAMINO-9-PHENYL-10-ETHYLPHENANTHRIDINIUM BROMIDE see
2,7-DIAMINO-10-ETHYL-9-PHENYLPHENANTHRIDINIUM BROMIDE

DI-(4-AMINOPHENYL)METHANE see 4,4'-METHYLENEDIANILINE

2,7-DIAMINO-9-PHENYLPHENANTHRIDINE ETHOBROMIDE see 2,7-
DIAMINO-10-ETHYL-9-PHENYLPHENANTHRIDINIUM BROMIDE

DI(P-AMINOPHENYL) SULFIDE see 4,4'-THIODIANILINE

DI(4-AMINOPHENYL)SULFONE see 4,4'-SULFONYLDIANILINE

DI(P-AMINOPHENYL) SULFONE see 4,4'-SULFONYLDIANILINE

DI(4-AMINOPHENYL)SULPHONE see 4,4'-SULFONYLDIANILINE

DI(P-AMINOPHENYL)SULPHONE see 4,4'-SULFONYLDIANILINE

2,4-DIAMINOTOLUEN (CZECH) see TOLUENE-2,4-DIAMINE

DIAMINOTOLUENE see TOLUENE-2,4-DIAMINE

2,4-DIAMINOTOLUENE see TOLUENE-2,4-DIAMINE

2,5-DIAMINOTOLUENE see TOLUENE-2,5-DIAMINE

2,4-DIAMINO-1-TOLUENE see TOLUENE-2,4-DIAMINE

P-DIAMINOTOLUENE SULFATE see 2,5-DIAMINOTOLUENE SULFATE

2,5-DIAMINOTOLUENE SULPHATE see 2,5-DIAMINOTOLUENE SUL-
FATE

2,4-DIAMINOTOLUOL see TOLUENE-2,4-DIAMINE

S-(4,6-DIAMINO-1,3,5-TRIAZIN-2-YL)-METHYL)-O,O-DIMETHYL-
DITHIOFOSFAAT (DUTCH) see AZIDITHION

S-(4,6-DIAMINO-1,3,5-TRIAZIN-2-YL)-METHYL)-O,O-DIMETHYL-
DITHIOPHOSPHAT (GERMAN) see AZIDITHION

S-((4,6-DIAMINO-S-TRIAZIN-2-YL)METHYL)-O,O-DIMETHYL PHOS-
PHORODITHIOATE see AZIDITHION

4,6-DIAMINO-1,3,5-TRIAZIN-2-YLMETHYL-O,O-DIMETHYL PHOS-
PHORODITHIOATE see AZIDITHION

S-(4,6-DIAMINO-1,3,5-TRIAZIN-2-YLMETHYL)-O,O-DIMETHYL PHOS-
PHORODITHIOATE see AZIDITHION

S-(4,6-DIAMINO-1,3,5-TRIAZIN-2-YLMETHYL) DIMETHYL PHOS-
PHOROTHIOLOTHIONATE see AZIDITHION

TRANS-DIAMMINEDICHLOROPLATINUM (II) see trans-DICHLORO
DIAMMINE PLATINUM (II)

DIAMMONIUM ARSENATE see DIAMMONIUM HYDROGEN ARSENATE

DIAMMONIUM CARBONATE see AMMONIUM CARBONATE

DIAMMONIUM HEXACHLOROPALLADATE see AMMONIUM CHLORO-
PALLADATE (IV)

DIAMMONIUM HEXACHLOROPLATINATE (2-) see AMMONIUM CHLO-
ROPLATINATE

DIAMMONIUM HEXAFLUOROSILICATE see AMMONIUM SILICO FLUO-
RIDE

DIAMMONIUM HEXAFLUOROSILICATE see CRYPTOHALITE

DIAMMONIUM HYDROGEN PHOSPHATE see AMMONIUM PHOSPHATE
DIBASIC

DIAMMONIUM MOLYBDATE see AMMONIUM MOLYBDATE

DIAMMONIUM MONOHYDROGEN ARSENATE see DIAMMONIUM HY-
DROGEN ARSENATE

DIAMMONIUM SULFATE see AMMONIUM SULFATE (2:1)

DIAMORFINA see HEROIN

DIAMORPHINE see HEROIN

DIAMPHETAMINE SULFATE see BENZEDRINE SULFATE

DI-N-AMYLAMINE see DIAMYL AMINE

DIAMYLNITROSAMIN (GERMAN) see DI-n-AMYLNITROSAMINE

DIANABOL see 17-METHYL TESTOSTERONE

DIANALINEMETHANE see 4,4'-METHYLENEDIANILINE

1,2:3,4-DIANHYDROERYTHRITOL see meso-1,2,3,4-DIEPOXYBUTANE

1,2:5,6-DIANHYDROGALACTITOL see DIANHYDROGALACTITOL

1,2:3,4-DIANHYDRO-DL-THREITOL see d,l-DIEPOXYBUTANE

P,P'-DIANILINE see BENZIDINE

O-DIANISIDIN (CZECH, GERMAN) see o-DIANISIDINE

O-DIANISIDINA (ITALIAN) see o-DIANISIDINE

O,O'-DIANISIDINE see o-DIANISIDINE

ALPHA,GAMMA-DI-P-ANISYL-BETA-(ETHOXYPHENYL)GUANIDINE HY-
DROCHLORIDE see PHENODIANISYL HYDROCHLORIDE

DIANISYL-MONOPHENETHYLGUANIDINE HYDROCHLORIDE see PHENO-
DIANISYL HYDROCHLORIDE

2,2-DI-P-ANISYL-1,1,1-TRICHLOROETHANE see DIMETHOXY-DDT

DIANTIMONY PENTOXIDE see ANTIMONY PENTOXIDE

DIANTIMONY TRIOXIDE see ANTIMONY OXIDE

DIAPAMIDE see 4-CHLORO-N-METHYL-3-(METHYLSULFAMOYL)BEN-
ZAMIDE

DIAPHEN (NEUROPLEGIC) see DIAMINOPHEN

DIAPHENYLSULFONE see 4,4'-SULFONYLDIANILINE

DIAPHENYLSULPHON see 4,4'-SULFONYLDIANILINE

DIAPHENYLSULPHONE see 4,4'-SULFONYLDIANILINE

DIAPHORM see HEROIN

CIS-DIAQUODIAMMINEPLATINUM(II) DINITRATE see DIAQUODIAM-
MINEPLATINUM DINITRATE

DIARSENIC PENTOXIDE see ARSENIC PENTOXIDE

DIARSENIC TRIOXIDE see ARSENIC TRIOXIDE

DIARSENIC TRISULFIDE see ARSENIC SULFIDE

DIASETIELMORFIEN see HEROIN

DIASETILMORFIN see HEROIN

DIASETYLMORFIIMI see HEROIN

DIASONE HYDROCHLORIDE see METHADONE HYDROCHLORIDE

DIATOMITE see DIATOMACEOUS EARTH

1,3-DIAZA-2,4-CYCLOPENTADIENE see IMIDAZOLE

1,3-DIAZAINDENE see BENZIMIDAZOLE

2,3-DIAZAINDOLE see 1H-BENZOTRIAZOLE

DIAZENEDICARBOXAMIDE see AZODICARBAMIDE

DIAZETYLMORPHINE see HEROIN

DIAZINON see DIAZIDE

DIAZIRINE see DIAZOMETHANE

DIAZOACETIC ACID, ETHYL ESTER see DIAZOACETIC ESTER

N-DIAZOACETILGLICINA-AMIDE (ITALIAN) see N-(CARBAMOYL-
METHYL)-2-DIAZOACETAMIDE

N-DIAZOACETILGLICINA-IDRAZIDE (ITALIAN) see N-(DIAZOACETYL)-
GLYCINE HYDRAZINE

DIAZOACETYLGLYCINAMIDE see N-(CARBAMOYLMETHYL)-2-DIAZO-
ACETAMIDE

N-(DIAZOACETYL)GLYCINAMIDE see N-(CARBAMOYLMETHYL)-2-DIA-
ZOACETAMIDE

N-DIAZOACETYLGLYCINE AMIDE see N-(CARBAMOYLMETHYL)-2-DIA-
ZOACETAMIDE

DIAZOACETYLGLYCINE HYDRAZIDE see N-(DIAZOACETYL)GLYCINE
HYDRAZINE

N-DIAZOACETYL GLYCYLHYDRAZIDE see N-(DIAZOACETYL)GLYCINE
HYDRAZINE

DIAZOAMINOBENZEN (CZECH) see 1,3-DIPHENYLTRIAZENE

DIAZOAMINOBENZENE see 1,3-DIPHENYLTRIAZENE

P-DIAZOAMINOBENZENE see 1,3-DIPHENYLTRIAZENE

DIAZOAMINOBENZOL (GERMAN) see 1,3-DIPHENYLTRIAZENE

DIAZOBENZENE see AZOBENZENE

DIAZOESSIGSAEURE-AETHYLESTER (GERMAN) see DIAZOACETIC
ESTER

DIAZOIMIDE see HYDRAZOIC ACID

1,3-DIAZOLE see IMIDAZOLE

DIAZOTIZING SALTS see SODIUM NITRITE

DIBASIC AMMONIUM ARSENATE see DIAMMONIUM HYDROGEN AR-
SENATE

DIBASIC AMMONIUM PHOSPHATE see AMMONIUM PHOSPHATE
DIBASIC

DIBASIC LEAD ARSENATE see LEAD ACID ARSENATE

DIBASIC LEAD CARBONATE see LEAD CARBONATE

DIBASIC SODIUM PHOSPHATE see SODIUM MONOHYDROGEN PHOS-
PHATE (2:1:1)

DIBASIC ZINC STEARATE see ZINC STEARATE

1,2,5,6-DIBENZACRIDINE see DIBENZ(a,h)ACRIDINE

1,2,7,8-DIBENZACRIDINE see DIBENZ(a,j)ACRIDINE

3,4,5,6-DIBENZACRIDINE see DIBENZ(a,j)ACRIDINE

3,4:5,6-DIBENZACRIDINE see DIBENZ(c,h)ACRIDINE

1,2,7,8-DIBENZACRIDINE (FRENCH) see DIBENZ(c,h)ACRIDINE

3,4:5,6-DIBENZACRIDINE METHOSULFATE see DIBENZ(a,j)ACRIDINE
METHOSULFATE

1,2,5,6-DIBENZANTHRACEEN (DUTCH) see DIBENZ(a,h)ANTHRACENE

DIBENZ(A,C)ANTHRACENE see BENZO(b)TRIPHENYLENE

1,2:3,4-DIBENZANTHRACENE see BENZO(b)TRIPHENYLENE

1,2:5,6-DIBENZANTHRACENE see DIBENZ(a,h)ANTHRACENE

1,2:7,8-DIBENZANTHRACENE see DIBENZ(a,j)ANTHRACENE

4-(3-(5H-DIBENZ(B,F)AZEPIN-5-YL)PROPYL)-1-PIPERAZINEETHANOL
DIHYDROCHLORIDE see INSIDON DIHYDROCHLORIDE

3,4,5,6-DIBENZCARBAZOLE see 7H-DIBENZO(c,g)CARBAZOLE

DIBENZEPIN HYDROCHLORIDE see DIBENZEPINE HYDROCHLORIDE

DIBENZO(A,H)ANTHRACENE see DIBENZ(a,h)ANTHRACENE

DIBENZOAZEPINE see DIBENZACEPIN

DIBENZO-(DEF,MNO)CHRYSENE see ANTHANTHRENE

DIBENZODIOXIN see DIBENZO-p-DIOXIN

DIBENZO(1,4)DIOXIN see DIBENZO-p-DIOXIN

DIBENZO(B,E)(1,4)DIOXIN see DIBENZO-p-DIOXIN

1,2,5,6-DIBENZOFLUORENE see 13H-DIBENZO(a,g)FLUORENE

DIBENZO(A,J,K)FLUORENE see BENZO(j)FLUORANTHENE

DIBENZOPARADIAZINE see PHENAZINE

1,2,3,4-DIBENZOPHENANTHRENE see BENZO(g)CHRYSENE

DIBENZO-2,3,7,8-PHENANTHRENE see BENZO(b)CHRYSENE

DIBENZOPYRAZINE see PHENAZINE

DIBENZO(A,D)PYRENE see DIBENZO(def,p)CHRYSENE

DIBENZO(A,H)PYRENE see DIBENZO(b,def)CHRYSENE

DIBENZO(A,I)PYRENE see BENZO(rst)PENTAPHENE

DIBENZO(A,L)PYRENE see DIBENZO(def,p)CHRYSENE

DIBENZO(Q,E)PYRENE see NAPHTHO (1,2,3,4-def) CHRYSENE

1,2:3,4-DIBENZOPYRENE see DIBENZO(def,p)CHRYSENE

1,2,4,5-DIBENZOPYRENE see NAPHTHO (1,2,3,4-def) CHRYSENE

1,2,6,7-DIBENZOPYRENE see DIBENZO(b,def)CHRYSENE

1,2,7,8-DIBENZOPYRENE see BENZO(rst)PENTAPHENE

3,4,8,9-DIBENZOPYRENE see DIBENZO(b,def)CHRYSENE

DIBENZO(CD,MK)PYRENE see ANTHANTHRENE

1,2,9,10-DIBENZOPYRENE see DIBENZO(def,p)CHRYSENE

3,4:9,10-DIBENZOPYRENE see BENZO(rst)PENTAPHENE

DIBENZO(B,E)PYRIDINE see ACRIDINE

DIBENZO(B,D)PYRROLE see CARBAZOLE

DIBENZOTHIAZEPINE see 2-CHLORO-11-(4-METHYLPIPERAZINO)DI-
BENZO(b,f)(1,4)THIAZEPINE

DI-2-BENZOTHIAZOLYLDISULFIDE see BENZOTHIAZOLE DISULFIDE

2,2'-DIBENZOTHIAZYLDISULFIDE see BENZOTHIAZOLE DISULFIDE

DIBENZ(B,E)OXEPIN-delta(SUP 11(6H)),gamma-PROPYLAMINE see
DESMETHYLDOXEPIN

DIBENZOYLPEROXID (GERMAN) see BENZOYL PEROXIDE

DIBENZOYLPEROXYDE (DUTCH) see BENZOYL PEROXIDE

DIBENZOYLTHIAZYL DISULFIDE see BENZOTHIAZOLE DISULFIDE

1,2,3,4-DIBENZPHENANTHRENE see BENZO(g)CHRYSENE

1,2,5,6-DIBENZPHENANTHRENE see BENZO(c)CHRYSENE

1,2,3,4-DIBENZPYRENE see DIBENZO(def,p)CHRYSENE

3,4,8,9-DIBENZPYRENE see DIBENZO(b,def)CHRYSENE

DIBENZTHIAZYL DISULFIDE see BENZOTHIAZOLE DISULFIDE

DIBENZYL see BIBENZYL

N,N-DIBENZYLAMINOETHYL CHLORIDE HYDROCHLORIDE see DIBEN-
AMINE HYDROCHLORIDE

DIBENZYLCHLORETHYLAMINE HYDROCHLORIDE see DIBENAMINE
HYDROCHLORIDE

N,N-DIBENZYL-2-CHLOROETHYLAMINE HYDROCHLORIDE see DIBEN-
AMINE HYDROCHLORIDE

N,N-DIBENZYL-beta-CHLOROETHYLAMINE HYDROCHLORIDE see
DIBENAMINE HYDROCHLORIDE

DIBENZYLETHER (CZECH) see BENZYL ETHER

DIBORANE see BORON HYDRIDE

DIBORANE(6) see BORON HYDRIDE

DIBORON HEXAHYDRIDE see BORON HYDRIDE

1,2-DIBROMAETHAN (GERMAN) see 1,2-ETHYLENE DIBROMIDE

DIBROMCHLORPROPAN (GERMAN) see DIBROMOCHLOROPROPANE

1,2-DIBROM-3-CHLOR-PROPAN (GERMAN) see DIBROMOCHLOROPRO-
PANE

O-(1,2-DIBROM-2,2-DICHLORAETHYL)-O,O- DIMETHYL-PHOSPHAT
(GERMAN) see DIMETHYL-1,2-DIBROMO-2,2-DICHLORO-
ETHYL PHOSPHATE

alpha,alpha'-DIBROMOBIACETYL see 2,2'-DIBROMOBIACETYL

DIBROMOCHLOROMETHANE see CHLORO DIBROMO METHANE

1,2-DIBROMO-3-CHLOROPROPANE see DIBROMOCHLOROPROPANE

1,2-DIBROMO-3-CLORO-PROPANO (ITALIAN) see DIBROMOCHLORO-PROPANE

DIBROMODIBUTYLTIN see DIBROMODIBUTYLSTANNANE

1,2-DIBROMO-2,2-DICHLOROETHYL DIMETHYL PHOSPHATE see DIMETHYL-1,2-DIBROMO-2,2-DICHLOROETHYL PHOSPHATE

O-(1,2-DIBROMO-2,2-DICLORO-ETIL)-O,O-DIMETIL-FOSTATO (ITALIAN) see DIMETHYL-1,2-DIBROMO-2,2-DICHLOROETHYL PHOSPHATE

1,6-DIBROMO-1,6-DIDEOXYDULCITOL see DIBROMDULCITOL

1,6-DIBROMO-1,6-DIDEOXYGALACTITOL see DIBROMDULCITOL

1,6-DIBROMO-1,6-DIDEOXY-D-MANNITOL see 1,6-DIBROMOMANNITOL

1,6-DIBROMO-1,6-D-DIDESOXYMANNITOL see 1,6-DIBROMOMANNITOL

DIBROMODIMETHYL STANNANE see DIMETHYLTIN DIBROMIDE

1,6-DIBROMODULCITOL see DIBROMDULCITOL

1,2-DIBROMOETANO (ITALIAN) see 1,2-ETHYLENE DIBROMIDE

1,2-DIBROMOETHANE see 1,2-ETHYLENE DIBROMIDE

SYM-DIBROMOETHANE see 1,2-ETHYLENE DIBROMIDE

ALPHA,BETA-DIBROMOETHANE see 1,2-ETHYLENE DIBROMIDE

2,7-DIBROMO-4-HYDROXYMERCURIFLUORESCEINE DISODIUM SALT see MERCUROCHROME

3,5-DIBROMO-4-OCTANOYLOXY-BENZONITRILE see 2,6-DIBROMO-4-CYANOPHENYL OCTANOATE

2,3-DIBROMO-1-PROPANOL PHOSPHATE see TRIS

2,3-DIBROMO-1-PROPANOL, PHOSPHATE (3:1) see TRIS

(2,3-DIBROMOPROPYL) PHOSPHATE see TRIS

DIBROMURE D'ETHYLENE (FRENCH) see 1,2-ETHYLENE DIBROMIDE

1,2-DIBROOM-3-CHLOORPROPAAN (DUTCH) see DIBROMOCHLOROPROPANE

O-(1,2-DIBROOM-2,2-DICHLOOR-ETHYL)-O,O-DIMETHYL-FOSFAAT (DUTCH) see DIMETHYL-1,2-DIBROMO-2,2-DICHLOROETHYL PHOSPHATE

1,2-DIBROOMETHAAN (DUTCH) see 1,2-ETHYLENE DIBROMIDE

DIBUCAINE HYDROCHLORIDE see NUPERCAINE HYDROCHLORIDE

DI-N-BUTYLAMINE see N-DIBUTYLAMINE

2-N-DIBUTYLAMINOETHANOL see DIBUTYLAMINOETHANOL

BETA-N-DIBUTYLAMINOETHYL ALCOHOL see DIBUTYLAMINOETHANOL

3-(DIBUTYLAMINO)-1-PROPANOL-P-AMINOBENZOATE see BUTACAINE

3-DIBUTYLAMINOPROPYL-P-AMINOBENZOATE see BUTACAINE

DIBUTYLAMINOPROPYL-P-AMINOBENZOATE SULFATE see BUTACAINE SULFATE

3'-DIBUTYLAMINOPROPYL-4-AMINOBENZOATE SULFATE see BUTACAINE SULFATE

DIBUTYL-1,2-BENZENEDICARBOXYLATE see DI-n-BUTYL PHTHALATE

DIBUTYLBIS(LAUROYLOXY)TIN see DIBUTYLBIS(LAUROYLOXY)STANNANE

O,O-DIBUTYL-1-BUTYLAMINO-CYCLOHEXYLPHOSPHONATE see AMINOPHON

DIBUTYL BUTYLPHOSPHONATE see DIBUTYL BUTANEPHOSPHONATE

(2-DIBUTYLCARBAMYLOXYETHYL)-DIMETHYLETHYLAMMONIUM SULFATE see DI-n-BUTYL-CARBAMYLCHOLINE SULPHATE

DIBUTYL-O-(O-CARBOXYBENZOYL) GLYCOLATE see BUTYL CARBO BUTOXY METHYL PHTHALATE

DIBUTYL-O-CARBOXYBENZOYLOXYACETATE see BUTYL CARBO BUTOXY METHYL PHTHALATE

2,6-DI-TERT-BUTYL-P-CRESOL see BHT (FOOD GRADE)

N,N'-DIBUTYL-N,N'-DICARBOXYMORPHOLIDE-ETHYLENEDIAMINE see DIMORPHOLAMINE

DIBUTYLDICHLOROTIN see DIBUTYLDICHLOROSTANNANE

DIBUTYLDITHIOCARBAMIC ACID, NICKEL SALT see BIS(DIBUTYL DITHIO CARBAMATO)NICKEL

DIBUTYLDITHIOCARBAMIC ACID SODIUM SALT see SODIUM DIBUTYLDITHIOCARBAMATE

DIBUTYLDITHIOCARBAMIC ACID ZINC SALT see BIS(DIBUTYL DITHIO CARBAMATO)ZINC

N,N-DIBUTYLETHANOLAMINE see DIBUTYLAMINOETHANOL

DI-N-BUTYL ETHER see n-BUTYL ETHER

2,6-DI-SEC-BUTYLFENOL (CZECH) see 2,6-DI-sec-BUTYLPHENOL

DI-sec-BUTYLFLUOROPHOSPHATE see DI-sec-BUTYL FLUOROPHOSPHONATE

N,N-DIBUTYL-N-(2-HYDROXYETHYL)AMINE see DIBUTYLAMINOETHANOL

2,6-DI-TERT-BUTYL-1-HYDROXY-4-METHYLBENZENE see BHT (FOOD GRADE)

3,5-DI-TERT-BUTYL-4-HYDROXYTOLUENE see BHT (FOOD GRADE)

2,6-DI-TERC-BUTYL-P-KRESOL (CZECH) see BHT (FOOD GRADE)

2,6-DI-TERT-BUTYL-4-METHYLPHENOL see BHT (FOOD GRADE)

2,6-DI-TERT-BUTYL-P-METHYLPHENOL see BHT (FOOD GRADE)

DI-N-BUTYLNITROSAMIN (GERMAN) see N-BUTYL-N-NITROSO-1-BUTAMINE

DI-N-BUTYLNITROSAMINE see N-BUTYL-N-NITROSO-1-BUTAMINE

N,N-DI-N-BUTYLNITROSAMINE see N-BUTYL-N-NITROSO-1-BUTAMINE

N,N-DIBUTYLNITROSOAMINE see N-BUTYL-N-NITROSO-1-BUTAMINE

DIBUTYL OXIDE see n-BUTYL ETHER

DIBUTYLOXOTIN see DIBUTYLOXOSTANNANE

DI-tert-BUTYLPEROXID (GERMAN) see tert-BUTYL PEROXIDE

DI-t-BUTYL PEROXIDE see tert-BUTYL PEROXIDE

DI-tert-BUTYL PEROXYDE (DUTCH) see tert-BUTYL PEROXIDE

DIBUTYL-PHOSPHINIC ACID, 4-NITROPHENYL ESTER see p-NITROPHENYLDI-n-BUTYLPHOSPHINATE

DIBUTYL PHTHALATE see DI-n-BUTYL PHTHALATE

DIBUTYLSTANNANE OXIDE see DIBUTYLOXOSTANNANE

DI-N-BUTYLSULFAT (GERMAN) see DIBUTYL ESTER SULFURIC ACID

DIBUTYL SULFATE see DIBUTYL ESTER SULFURIC ACID

DIBUTYL TIN DIACETATE see DIACETOXY DIBUTYL STANNANE

DIBUTYL TIN DIBROMIDE see DIBROMODIBUTYLSTANNANE

DIBUTYLTIN DICHLORIDE see DIBUTYLDICHLOROSTANNANE

DI-N-BUTYLTIN DICHLORIDE see DIBUTYLDICHLOROSTANNANE

DI-N-BUTYLTIN DI-2-ETHYLHEXANOATE see BIS(2-ETHYL HEXANOYL OXY)DIBUTYL STANNANE

DIBUTYLTIN DILAURATE see DIBUTYLBIS(LAUROYLOXY)STANNANE

DI-N-BUTYLTIN DI(MONOBUTYL)MALEATE see BIS(BUTOXYMALEOYLOXY)DIBUTYLSTANNANE

DIBUTYLTIN OXIDE see DIBUTYLOXOSTANNANE

DI-N-BUTYLTIN OXIDE see DIBUTYLOXOSTANNANE

DI-N-BUTYL-ZINN-DICHLORID (GERMAN) see DIBUTYLDICHLOROSTANNANE

DIBUTYL-ZINN-DILAURAT (GERMAN) see DIBUTYLBIS(LAUROYLOXY)STANNANE

DI-N-BUTYL-ZINN-DI(MONOBUTYL)MALEINAT (GERMAN) see BIS(BUTOXYMALEOYLOXY)DIBUTYLSTANNANE

DI-N-BUTYL-ZINN-OXYD (GERMAN) see DIBUTYLOXOSTANNANE

DICACODYL SULFIDE see CACODYL SULFIDE

2,2-DI(CARBAMOYLOXYMETHYL)-PENTANE see MILTOWN

DICARBETHOXYMETHANE see ETHYL MALONATE

DICARBOETHOXYETHYL-O,O-DIMETHYL PHOSPHORODITHIOATE see CARBETHOXY MALATHION

2,3-DICARBONITRILO-1,4-DIATHIAANTHRACHINON (GERMAN) see 5,10-DIHYDRO-5,10-DIOXONAPHTHO(2,3-b)-p-DITHIIN-2,3-DICARBONITRILE

O-DICARBOXYBENZENE see PHTHALIC ACID

3,3'-DICARBOXYBENZIDINE see 5,5'-BIANTHRANILIC ACID

((1,2-DICARBOXYETHYL)THIO)GOLD DISODIUM SALT see GOLD SODIUM THIOMALATE

DICESIUM CARBONATE see CESIUM CARBONATE

DICHAPETULUM CYMOSUM (HOOK) ENGL see POTASSIUM FLUOROACETATE

DICHLOFENTHION see DICHLOROFENTHION

1,4-DICHLOORBENZEEN (DUTCH) see p-DICHLOROBENZENE

1,1-DICHLOOR-2,2-BIS(4-CHLOOR FENYL)-ETHAAN (DUTCH) see 1,1-BIS(4-CHLOROPHENYL)-2,2-DICHLOROETHANE

1,1-DICHLOORETHAAN (DUTCH) see ETHYLIDENE DICHLORIDE

1,2-DICHLOORETHAAN (DUTCH) see 1,2-DICHLORETHANE

1,2-DICHLOORETHAAN (DUTCH) see ETHYLENE DICHLORIDE

2,2′-DICHLOORETHYLETHER (DUTCH) see BIS(2-CHLOROETHYL) ETHER

DICHLOORFEEN (DUTCH) see 2,2′-METHYLENEBIS(4-CHLOROPHENOL)

(2,4-DICHLOOR-FENOXY)-AZIJNZUUR (DUTCH) see DICHLOROPHE-NOXYACETIC ACID

(3,4-DICHLOOR-FENYL-AZO)-THIOUREUM (DUTCH) see 3,4-DICHLO-ROBENZENEDIAZOTHIOUREA

3-(3,4-DICHLOOR-FENYL)-1,1-DIMETHYLUREUM (DUTCH) see 3-(3,4-DICHLOROPHENYL)-1,1-DIMETHYLUREA

3-(3,4-DICHLOOR-FENYL)-1-METHOXY-1-METHYLUREUM(DUTCH) see 3-(3,4-DICHLOROPHENYL)-1-METHOXYMETHYLUREA

1,1-DICHLOOR-1-NITROEITHAAN (DUTCH) see 1,1-DICHLORO-1-NI-TROETHANE

(2,2-DICHLOOR-VINYL)-DIMETHYL-FOSFAAT (DUTCH) see DIMETHYL DICHLOROVINYL PHOSPHATE

DICHLOORVO (DUTCH) see DIMETHYL DICHLOROVINYL PHOSPHATE

1,1-DICHLORAETHAN (GERMAN) see ETHYLIDENE DICHLORIDE

1,2-DICHLOR-AETHAN (GERMAN) see ETHYLENE DICHLORIDE

1,2-DICHLOR-AETHAN (GERMAN) see 1,2-DICHLORETHANE

1,2-DICHLOR-AETHEN (GERMAN) see ACETYLENE DICHLORIDE

P-DI-(2-CHLORAETHYL)-AMINO-DL-PHENYL-ALANIN (GERMAN) see DL-PHENYLALANINE MUSTARD

S-(2,3-DICHLOR-ALLYL)-N,N-DIISOPROPYL-MONOTHIOCARBAMAAT (DUTCH) see DIALLATE

DICHLORAN see ALKYL DIMETHYL-3,4-DICHLOROBENZENE AMMO-NIUM CHLORIDE

3,3′-DICHLORBENZIDIN (CZECH) see DICHLOROBENZIDINE BASE

4,4′-DICHLORBENZILSAEUREAETHYLESTER (GERMAN) see ETHYL-4,4′-DICHLOROBENZILATE

O-DICHLOR BENZOL see O-DICHLOROBENZENE

P-DICHLORBENZOL (GERMAN) see P-DICHLOROBENZENE

1,1-DICHLOR-2,2-BIS(4-CHLOR-PHENYL)-AETHAN (GERMAN) see 1,1-BIS(4-CHLOROPHENYL)-2,2-DICHLOROETHANE

2,2′-DICHLOR-DIAETHYLAETHER (GERMAN) see BIS(2-CHLORO-ETHYL) ETHER

3,3′-DICHLOR-4,4′-DIAMINO-DIPHENYLAETHER (GERMAN) see BIS(4-AMINO-3-CHLOROPHENYL) ETHER

3,3′-DICHLOR-4,4′-DIAMINODIPHENYLMETHAN (GERMAN) see MOCA

DICHLORDIMETHYLAETHER (GERMAN) see BIS(CHLOROMETHYL) ETHER

DICHLOREMULSION see ETHYLENE DICHLORIDE

DICHLOREN (GERMAN) see BIS(beta-CHLOROETHYL)METHYLAMINE

1,2-DICHLORETHANE see 1,2-DICHLORETHANE

1,2-DICHLORETHANE see ETHYLENE DICHLORIDE

DICHLORETHANOIC ACID see DICHLORACETIC ACID

2,2′-DICHLORETHYL ETHER see BIS(2-CHLOROETHYL) ETHER

2-(alpha,beta-DICHLORETHYL)-PYRIDINE HYDROCHLORIDE see CLEP

BETA,BETA-DICHLOR-ETHYL-SULPHIDE see BIS(2-CHLOROETHYL)-SULFIDE

2,6-DICHLORFENOL (CZECH) see 2,6-DICHLOROPHENOL

DICHLORFOS (POLISH) see DIMETHYL DICHLOROVINYL PHOSPHATE

DICHLORICIDE MOTHPROOFER see TERPENE POLYCHLORINATES

DICHLORID KYSELINY FUMAROVE (CZECH) see FUMARYL CHLORIDE

2,3-DICHLOR-1,4-NAPHTHOCHINON (GERMAN) see 2,3-DICHLORO-1,4-NAPHTHOQUINONE

1,1-DICHLOR-1-NITROAETHAN (GERMAN) see 1,1-DICHLORO-1-NITROETHANE

2,4-DICHLOR-6-NITROFENOL (CZECH) see 2,4-DICHLORO-6-NITRO-PHENOL

2′,5-DICHLOR-4′-NITRO-SALIZYLSAEUREANILID (GERMAN) see 2′,5-DICHLORO-4′-NITROSALICYLANILIDE

DICHLOROACETATE SODIUM SALT see SODIUM DICHLOROACETATE

2,2-DICHLOROACETIC ACID see DICHLORACETIC ACID

DICHLOROACETIC ACID METHYL ESTER see METHYL DICHLORO ACE-TATE

DICHLOROACETIC ACID SODIUM SALT see SODIUM DICHLOROACE-TATE

S-(2,3-DICHLORO-ALLIL)-N,N-DIISOPROPIL-MONOTIOCARBAMMATO (ITALIAN) see DIALLATE

5-2,3-DICHLOROALLYLDIISOPROPYLTHIOCARBAMATE see DIALLATE

2,3-DICHLOROALLYL N,N-DIISOPROPYLTHIOLCARBAMATE see DIAL-LATE

3′,5′-DICHLOROAMETHOPTERIN see 3′5′-DICHLOROMETHOTREXATE

2,5-DICHLORO-3-AMINOBENZOIC ACID see 3-AMINO-2,5-DICHLORO-BENZOIC ACID

3′,5′-DICHLORO-4-AMINO-4-DEOXY-N(SUP 10)-METHYLPTEROGLU-TAMIC ACID see 3′5′-DICHLOROMETHOTREXATE

O-DICHLOROBENZENE see O-DICHLOROBENZENE

1,2-DICHLOROBENZENE see O-DICHLOROBENZENE

1,4-DICHLOROBENZENE see P-DICHLOROBENZENE

2,6-DICHLOROBENZENECARBOTHIOAMIDE see 2,6-DICHLOROTHIO-BENZAMIDE

3,4-DICHLOROBENZENE DIAZOTHIOCARBAMID see 3,4-DICHLORO-BENZENEDIAZOTHIOUREA

3,3′-DICHLOROBENZIDENE see DICHLOROBENZIDINE BASE

3,3′-DICHLOROBENZIDINA (SPANISH) see DICHLOROBENZIDINE BASE

O,O′-DICHLOROBENZIDINE see DICHLOROBENZIDINE BASE

4,4′-DICHLOROBENZILIC ACID ETHYL ESTER see ETHYL-4,4′-DICHLOROBENZILATE

P-DICHLOROBENZOL see P-DICHLOROBENZENE

P,P′-DICHLOROBENZOYL PEROXIDE see BIS(P-CHLOROBENZOYL) PEROXIDE

3,3′-DICHLOROBIPHENYL-4,4′-DIAMINE see DICHLOROBENZIDINE BASE

3,3′-DICHLORO-4,4′-BIPHENYLDIAMINE see DICHLOROBENZIDINE BASE

1,1-DICHLORO-2,2-BIS(P-CHLOROPHENYL)ETHANE see 1,1-BIS(4-CHLOROPHENYL)-2,2-DICHLOROETHANE

1,1-DICHLORO-2,2-BIS(4-CHLOROPHENYL)-ETHANE (FRENCH) see 1,1-BIS(4-CHLOROPHENYL)-2,2-DICHLOROETHANE

1,1-DICHLORO-2,2-BIS(P-CHLOROPHENYL)ETHYLENE see 2,2-BIS(P-CHLOROPHENYL)-1,1-DICHLOROETHYLENE

CIS-DICHLOROBIS(CYCLOBUTYLAMMINE)PLATINUM(II) see CIS-DICYCLOBUTYLAMMINEDICHLOROPLATINUM (II)

CIS-DICHLOROBIS(CYCLOPENTYLAMMINE)PLATINUM(II) see CIS-BIS(CYCLOPENTYLAMMINE)PLATINUM(II)

1,1-DICHLORO-2,2-BIS(2,4′-DICHLOROPHENYL)ETHANE see CHLODI-THANE

DICHLOROBIS(ETA(SUP 5)-2,4-CYCLOPENTADIEN-1-YLTITANIUM see DICHLOROTITANOCENE

1,1-DICHLORO-2,2-BIS(P-ETHYLPHENYL)ETHANE see DIETHYL DIPHE-NYL DICHLOROETHANE

2,2-DICHLORO-1,1-BIS(P-ETHYLPHENYL)ETHANE see DIETHYL DIPHE-NYL DICHLOROETHANE

1,1-DICHLORO-2,2-BIS(PARACHLOROPHENYL)ETHANE see 1,1-BIS(4-CHLOROPHENYL)-2,2-DICHLOROETHANE

DICHLOROBIS(eta-CYCLOPENTADIENYL)HAFNIUM see HAFNOCENE DI-CHLORIDE

DICHLOROBROMOMETHANE see BROMO DICHLORO METHANE

O-(2,5-DICHLORO-4-BROMOPHENYL)-O-METHYL PHENYLTHIOPHOS-PHONATE see LEPTOPHOS

1,4-DICHLORO-2-BUTENE see 2-BUTYLENE DICHLORIDE

1,4-DICHLOROBUTENE-2 (TRANS) see 2-BUTYLENE DICHLORIDE

1,4-DICHLOROBUTYNE see 1,4-DICHLORO-2-BUTYNE

1,1-DICHLORO-2-CHLOROETHYLENE see ACETYLENE TRICHLORIDE

1,1-DICHLORO-2-CHLOROETHYLENE see TRICHLOROETHYLENE

1,1-DICHLORO-2-(O-CHLOROPHENYL)-2- (P-CHLOROPHENYL)ETHANE see CHLODITHANE

DICHLORO(2-CHLOROVINYL)ARSINE see CHLOROVINYLARSINE DICHLORIDE

DICHLOROCTAN SODNY (CZECH) see SODIUM DICHLOROACETATE

3,4′-DICHLOROCYCLOPROPANECARBOXANILIDE see CYPROMID

CIS-DICHLORODIAMINE PLATINUM (II) see CIS-PLATINOUS DIAMINE DICHLORIDE

3,3′-DICHLORO-4,4′-DIAMINOBIPHENYL see DICHLOROBENZIDINE BASE

3,3′-DICHLORO-4,4′-DIAMINODIPHENYL ETHER see BIS(4-AMINO-3-CHLOROPHENYL) ETHER

3,3′-DICHLORO-4,4′-DIAMINODIPHENYL METHANE see MOCA

N-(3,5-DICHLORO-4-((2,4-DIAMINO-6- PTERIDINYLMETHYL) METHYLAMINO)BENZOYL)GLUTAMIC ACID see 3′5′-DICHLORO-METHOTREXATE

2,7-DICHLORODIBENZODIOXIN see DCDD

2,7-DICHLORODIBENZO-p-DIOXIN see DCDD

DICHLORODIBUTYLSTANNANE see DIBUTYLDICHLOROSTANNANE

DICHLORODIBUTYLTIN see DIBUTYLDICHLOROSTANNANE

1,1-DICHLORO-2,2-DICHLOROETHANE see ACETYLENE TETRACHLO-RIDE

1,1-DICHLORO-2,2-DI(4-CHLOROPHENYL)ETHANE see 1,1-BIS(4-CHLOROPHENYL)-2,2-DICHLOROETHANE

DICHLORODICYCLOPENTADIENYLHAFNIUM see HAFNOCENE DICHLO-RIDE

DICHLORODICYCLOPENTADIENYLTITANIUM see DICHLOROTITANO-CENE

2,2′-DICHLORO DIETHYLAMINE HYDROCHLORIDE see BIS(2-CHLOROETHYL)AMINE HYDROCHLORIDE

BETA,BETA-DICHLORODIETHYL ETHER see BIS(2-CHLOROETHYL) ETHER

BETA,BETA′-DICHLORODIETHYL-N-METHYLAMINE see BIS(beta-CHLOROETHYL)METHYLAMINE

BETA,BETA′-DICHLORODIETHYL-N-METHYLAMINE HYDROCHLORIDE see BIS(2-CHLOROETHYL)METHYLAMINE HYDROCHLORIDE

2,2′-DICHLORODIETHYL SULFIDE see BIS(2-CHLOROETHYL)SULFIDE

DICHLORODIETHYLTIN see DICHLORODIETHYLSTANNANE

DICHLORODIFLUOROMETHANE (DOT) see DICHLORODIFLUORO-METHANE

(2,5-DICHLORO-3,6-DIHYDROXY-P-BENZOQUINONE), MERCURY SALT see (2,5-DICHLORO-3,6-DIHYDROXY-p-BENZOQUINOLATO)MER-CURY

5,5′-DICHLORO-2,2′-DIHYDROXYDIPHENYLMETHANE see 2,2′-METHYLENEBIS(4-CHLOROPHENOL)

DICHLORODIISOPROPYL ETHER see BIS(2-CHLOROISOPROPYL) ETHER

DICHLORODIISOPROPYLSTANNANE see DIISOPROPYLTIN DICHLORIDE

SYM-DICHLORO-DIMETHYL ETHER see BIS(CHLOROMETHYL) ETHER

DICHLORODIMETHYLTIN see DICHLORODIMETHYLSTANNANE

4,4′-DICHLORO-6,6′-DINITRO-O,O′-BIPHENOL see 5,5′-DICHLORO-2,2′-DIHYDROXY-3,3′-DINITROBIPHENYL

5,5′-DICHLORO-3,3′-DINITRO(1,1′-BIPHENYL)-2,2′-DIOL see 5,5′-DI-CHLORO-2,2′-DIHYDROXY-3,3′-DINITROBIPHENYL

DICHLORODIOXOCHROMIUM see CHROMIUM OXYCHLORIDE

DICHLORODIPENTYLSTANNANE see DIPENTYLTIN DICHLORIDE

O,P′-DICHLORODIPHENYLDICHLOROETHANE see CHLODITHANE

P,P′-DICHLORODIPHENYLDICHLOROETHANE see 1,1-BIS(4-CHLORO-PHENYL)-2,2-DICHLOROETHANE

P,P′-DICHLORODIPHENYL DICHLOROETHYLENE see 2,2-BIS(p-CHLO-ROPHENYL)-1,1-DICHLOROETHYLENE

P,P′-DICHLORODIPHENYL SULFIDE see p-CHLOROBENZYL-p-CHLORO-PHENYL SULFIDE

4,4′-DICHLORODIPHENYLTRICHLOROETHANE see DDT

p,p′-DICHLORODIPHENYLTRICHLOROETHANE see DDT

DICHLORODI-PI-CYCLOPENTADIENYLTITANIUM see DICHLOROTITANO-CENE

DICHLORODIPROPYLTIN see DICHLORODIPROPYLSTANNANE

DICHLORODI-pi-CYCLOPENTADIENYLHAFNIUM see HAFNOCENE DICHLORIDE

1,1-DICHLOROETHANE see ETHYLIDENE DICHLORIDE

1,2-DICHLOROETHANE see ETHYLENE DICHLORIDE

1,2-DICHLOROETHANE see 1,2-DICHLORETHANE

sym-DICHLOROETHANE see ETHYLENE DICHLORIDE

SYM.-DICHLOROETHANE see 1,2-DICHLORETHANE

DICHLORO-1,2-ETHANE (FRENCH) see 1,2-DICHLORETHANE

ALPHA,BETA-DICHLOROETHANE see 1,2-DICHLORETHANE

DICHLOROETHANOIC ACID see DICHLORACETIC ACID

1,1-DICHLOROETHENE see 1,1-DICHLOROETHYLENE

1,1-DICHLOROETHENE see VINYLIDENE CHLORIDE

1,1-DICHLOROETHENE POLYMER WITH CHLOROETHENE see CHLORO ETHYLENE-1,1-DICHLORO ETHYLENE POLYMER

2,2-DICHLOROETHENYL DIMETHYL PHOSPHATE see DIMETHYL DICHLOROVINYL PHOSPHATE

1,1′-DICHLOROETHENYLIDENE)BIS(4-CHLOROBENZENE) see 2,2-BIS(p-CHLOROPHENYL)-1,1-DICHLOROETHYLENE

2,2-DICHLOROETHENYL PHOSPHORIC ACID DIMETHYL ESTER see DIMETHYL DICHLOROVINYL PHOSPHATE

9-(2-(DI(2-CHLOROETHYL)AMINO)ETHYLAMINO)-6-CHLORO-2-METH-OXY-ACRIDINE see QUINACRINE ETHYL MUSTARD

2-(DI-2-CHLOROETHYL)AMINOMETHYL-5,6-DIMETHYLBENZIMIDA-ZOLE see BENZIMIDAZOLE METHYLENE MUSTARD

P-N-DI(CHLOROETHYL)AMINOPHENYLALANINE see L-PHENYLALANINE MUSTARD

P-DI-(2-CHLOROETHYL)AMINO-L-PHENYLALANINE see L-PHENYLALA-NINE MUSTARD

3-P-(DI(2-CHLOROETHYL)AMINO)-PHENYL-L-ALANINE see L-PHE-NYLALANINE MUSTARD

O-DI-2-CHLOROETHYLAMINO-DL-PHENYLALANINE see O-PHENYLALA-NINE MUSTARD

P-DI(2-CHLOROETHYL)AMINO-DL-PHENYLALANINE see DL-PHENYL-ALANINE MUSTARD

P-(N,N-DI-2-CHLOROETHYL)AMINOPHENYL BUTYRIC ACID see CHLO-RAMBUCIL

GAMMA-(P-DI(2-CHLOROETHYL)AMINOPHENYL)BUTYRIC ACID see CHLORAMBUCIL

P-N,N-DI-(BETA-CHLOROETHYL)AMINOPHENYL BUTYRIC ACID see CHLORAMBUCIL

N,N-DI-2-CHLOROETHYL-GAMMA-P-AMINOPHENYLBUTYRIC ACID see CHLORAMBUCIL

5-(DI-2-CHLOROETHYL)AMINOURACIL see 5-(BIS(2-CHLOROETHYL)-AMINO)URACIL

5-(DI-(beta-CHLOROETHYL)AMINO)URACIL see 5-(BIS(2-CHLORO-ETHYL)AMINO)URACIL

O,O-DI(2-CHLOROETHYL)-7-(3-CHLORO-4-METHYLCOUMARINYL)-PHOSPHATE see 3-CHLORO-4-METHYL-7-COUMARINYL DIETHYL PHOSPHATE

DICHLOROETHYLENE see ETHYLENE DICHLORIDE

1,1-DICHLOROETHYLENE see VINYLIDENE CHLORIDE

1,2-DICHLOROETHYLENE see ACETYLENE DICHLORIDE

SYM-DICHLOROETHYLENE see ACETYLENE DICHLORIDE

TRANS-1,2-DICHLOROETHYLENE see trans-ACETYLENE DICHLORIDE

DICHLORO-1,2-ETHYLENE (FRENCH) see ACETYLENE DICHLORIDE

1,1-DICHLOROETHYLENE-MONOCHLOROETHYLENE POLYMER see CHLORO ETHYLENE-1,1-DICHLORO ETHYLENE POLYMER

1,1-DICHLOROETHYLENEPOLYMER WITH CHLOROETHYLENE see CHLORO ETHYLENE-1,1-DICHLORO ETHYLENE POLYMER

S-(1,2-DICHLOROETHYLENEYL)-L-CYSTEINE see S-DICHLOROVINYL-L-CYSTEINE

DICHLOROETHYL ETHER see BIS(2-CHLOROETHYL) ETHER

2,2′-DICHLOROETHYL ETHER see BIS(2-CHLOROETHYL) ETHER

SYM-DICHLOROETHYL ETHER see BIS(2-CHLOROETHYL) ETHER

DI(BETA-CHLOROETHYL)ETHER see BIS(2-CHLOROETHYL) ETHER

BETA,BETA′-DICHLOROETHYL ETHER see BIS(2-CHLOROETHYL) ETHER

DICHLOROETHYL FORMAL see BIS(beta-CHLOROETHYL)FORMAL

DI-2-CHLOROETHYL FORMAL see BIS(beta-CHLOROETHYL)FORMAL

DI(2-CHLOROETHYL)METHYLAMINE see BIS(beta-CHLOROETHYL)-METHYLAMINE

DI(2-CHLOROETHYL)METHYLAMINE HYDROCHLORIDE see BIS(2-CHLOROETHYL)METHYLAMINE HYDROCHLORIDE

2-N,N-DI(2-CHLOROETHYL)NAPHTHYLAMINE see N,N-BIS(2-CHLORO-ETHYL)-2-NAPHTHYLAMINE

DICHLOROETHYL-BETA-NAPHTHYLAMINE see N,N-BIS(2-CHLORO-ETHYL)-2-NAPHTHYLAMINE

DI(2-CHLOROETHYL)-BETA-NAPHTHYLAMINE see N,N-BIS(2-CHLORO-ETHYL)-2-NAPHTHYLAMINE

N,N-DI(2-CHLOROETHYL)-BETA-NAPHTHYLAMINE see N,N-BIS(2-CHLOROETHYL)-2-NAPHTHYLAMINE

DICHLOROETHYL OXIDE see BIS(2-CHLOROETHYL) ETHER

2-(1,2-DICHLOROETHYL)PYRIDINE HYDROCHLORIDE see CLEP

DI-2-CHLOROETHYL SULFIDE see BIS(2-CHLOROETHYL)SULFIDE

BETA,BETA′-DICHLOROETHYL SULFIDE see BIS(2-CHLOROETHYL)-SULFIDE

CIS-1,3-DICHLOROPROPYLENE see cis-1,3-DICHLOROPROPENE

alpha,gamma-DICHLOROPROPYLENE see alpha-CHLOROALLYL CHLO-
RIDE

SYM-DICHLOROTETRAFLUOROACETONE see DICHLOROTETRAFLUORO-
ACETONE

1,3-DICHLORO-1,1,3,3-TETRAFLUORO-2-PROPANONE see DICHLORO-
TETRAFLUOROACETONE

3,4-DICHLOROTETRAHYDROTHIOPHENE-1,1-DIOXIDE see 3,4-DICHLO-
ROSULFOLANE

DICHLOROTHIOLANE DIOXIDE see 3,4-DICHLOROSULFOLANE

DICHLORO-S-TRIAZINE-2,4,6(1H,3H,5H)-TRIONE POTASSIUM DERIV
see POTASSIUM DICHLOROISOCYANURATE

4,4′-DICHLORO-ALPHA-(TRICHLOROMETHYL)BENZHYDROL see 1,1-
BIS(p-CHLOROPHENYL)-2,2,2-TRICHLOROETHANOL

5,6-DICHLORO-2-TRIFLUOROMETHYLBENZIMIDAZOLE-1-CARBOXYL-
ATE see 5,6-DICHLORO-1-PHENOXYCARBONYL-2-TRIFLUORO-
METHYLBENZIMIDAZOLE

DICHLOROTRIPHENYLSTIBINE see DICHLOROTRIPHENYLANTIMONY

(2,2-DICHLORO-VINIL)DIMETILFOSFATO (ITALIAN) see DIMETHYL DI-
CHLOROVINYL PHOSPHATE

2,2-DICHLOROVINYL DIMETHYL PHOSPHATE see DIMETHYL DICHLO-
ROVINYL PHOSPHATE

2,2-DICHLOROVINYL DIMETHYL PHOSPHORIC ACID ESTER see
DIMETHYL DICHLOROVINYL PHOSPHATE

L-3-((1,2-DICHLOROVINYL)THIO)ALANINE see S-DICHLOROVINYL-L-
CYSTEINE

DICHLOROVOS see DIMETHYL DICHLOROVINYL PHOSPHATE

2,4-DICHLORPHENOXYACETIC ACID see DICHLOROPHENOXYACETIC
ACID

(2,4-DICHLOR-PHENOXY)-ESSIGSAEURE (GERMAN) see DICHLORO-
PHENOXYACETIC ACID

(3,4-DICHLOR-PHENYL-AZO)-THIOHARNSTOFF (GERMAN) see 3,4-DI-
CHLOROBENZENEDIAZOTHIOUREA

3-(3,4-DICHLOR-PHENYL)-1,1-DIMETHYL-HARNSTOFF (GERMAN) see
3-(3,4-DICHLOROPHENYL)-1,1-DIMETHYLUREA

3-(3,4-DICHLOR-PHENYL)-1-METHOXY-1-METHYL-HARNSTOFF (GER-
MAN) see 3-(3,4-DICHLOROPHENYL)-1-METHOXYMETHYLUREA

DICHLOR-S-TRIAZIN-2,4,6(1H,3H,5H)TRIONE POTASSIUM see POTAS-
SIUM DICHLOROISOCYANURATE

(2,2-DICHLOR-VINYL)-DIMETHYL-PHOSPHAT (GERMAN) see DI-
METHYL DICHLOROVINYL PHOSPHATE

O-(2,2-DICHLORVINYL)-O,O-DIMETHYLPHOSPHAT (GERMAN) see
DIMETHYL DICHLOROVINYL PHOSPHATE

DICK (GERMAN) see DICHLOROETHYLARSINE

1,4-DICLOROBENZENE (ITALIAN) see p-DICHLOROBENZENE

1,1-DICLORO-2,2-BIS(4-CLORO-FENIL)-ETANO (ITALIAN) see 1,1-
BIS(4-CHLOROPHENYL)-2,2-DICHLOROETHANE

3,3′-DICLORO-4,4′-DIAMINODIFENILMETANO(ITALIAN) see MOCA

1,1-DICLOROETANO (ITALIAN) see ETHYLIDENE DICHLORIDE

1,2-DICLOROETANO (ITALIAN) see 1,2-DICHLORETHANE

1,2-DICLOROETANO (ITALIAN) see ETHYLENE DICHLORIDE

2,2′-DICLOROETILETERE (ITALIAN) see BIS(2-CHLOROETHYL) ETHER

(3,4-DICLORO-FENIL-AZO)-TIOUREA (ITALIAN) see 3,4-DICHLORO-
BENZENEDIAZOTHIOUREA

3-(3,4-DICLORO-FENYL)-1,1-DIMETIL-UREA (ITALIAN) see 3-(3,4-DI-
CHLOROPHENYL)-1,1-DIMETHYLUREA

1,1-DICLORO-1-NITROETANO (ITALIAN) see 1,1-DICHLORO-1-NITRO-
ETHANE

DICOBALT CARBONYL see COBALT CARBONYL

DICOBALT EDTA see DICOBALT EDETATE

DICOPPER MONOSULFIDE see COPPER (I) SULFIDE

DICOPPER SULFIDE see COPPER (I) SULFIDE

DICUMENYLCHROMIUM see DICUMENE CHROMIUM

M-DICYANOBENZENE see 1,3-BENZENEDICARBONITRILE

O-DICYANOBENZENE see PHTHALONITRILE

P-DICYANOBENZENE see p-BENZENEDINITRILE

1,2-DICYANOBENZENE see PHTHALONITRILE

1,3-DICYANOBENZENE see 1,3-BENZENEDICARBONITRILE

1,4-DICYANOBENZENE see p-BENZENEDINITRILE

1,4-DICYANOBUTANE see ADIPONITRILE

beta,beta-DICYANO-O-CHLOROSTYRENE see (O-CHLORO BENZAL)-
MALONO NITRILE

beta,beta′-DICYANODIETHYL SULFIDE see DI(2-CYANOETHYL)-
SULFIDE

2,3-DICYANO-1,4-DITHIA-ANTHRAQUINONE see 5,10-DIHYDRO-5,10-
DIOXONAPHTHO(2,3-b)-p-DITHIIN-2,3-DICARBONITRILE

S-DICYANOETHANE see SUCCINONITRILE

DI-(2-CYANOETHYL)AMINE see BIS(beta-CYANOETHYL)AMINE

DICYANOGEN see CYANOGEN

DICYANOMETHANE see MALONONITRILE

1,3-DICYANOPROPANE see 1,3-TRIMETHYLENEDINITRILE

DICYCLOHEXYLAMINONITRITE see DICYCLOHEXYLAMINE NITRITE

DICYCLOHEXYLAMMONIUM NITRITE see DICYCLOHEXYLAMINE
NITRITE

DICYCLOPENTADIENE see BICYCLOPENTADIENE

DICYCLOPENTADIENYLDICHLOROTITANIUM see DICHLOROTITANO-
CENE

DICYCLOPENTADIENYLHAFNIUM DICHLORIDE see HAFNOCENE-
DICHLORIDE

DI-2,4-CYCLOPENTADIEN-1-YL IRON see FERROCENE

DICYCLOPENTADIENYLTITANIUMDICHLORIDE see DICHLOROTITANO-
CENE

2-((DICYCLOPENTYLACETYL)OXY)-N,N,N-TRIETHYL-ETHANAMINIUM
BROMIDE see (2-(DICYCLOPENTYLACETOXY)ETHYL)TRIETHYLAM-
MONIUM BROMIDE

CIS-DICYCLOPENTYLAMMINEDICHLOROPLATINUM (II) see cis-
BIS(CYCLOPENTYLAMMINE)PLATINUM(II)

3,3-DICYCLOPROPYL-2-(ETHOXYCARBONYL)ACRYLONITRILE see UV
ABSORBER-2

DICYKLOHEXYLAMIN (CZECH) see N,N-DICYCLOHEXYLAMINE

DICYKLOHEXYLAMIN NITRIT (CZECH) see DICYCLOHEXYLAMINE
NITRITE

DIDECYL DIMETHYL AMMONIUM CHLORIDE see QUATERNIUM-12

9,10-DIDEHYDRO-N,N-DIETHYL-2-BROMO-6-METHYLERGOLINE-8-
BETA-CARBOXAMIDE see 2-BROMO-D-LYSERGIC ACID DIETHYL-
AMIDE

9,10-DIDEHYDRO-N,N-DIETHYL-6-METHYL-ERGOLINE-8-beta-CAR-
BOXAMIDE see N,N-DIETHYLLYSERGAMIDE

7,8-DIDEHYDRO-4,5-alpha-EPOXY-3-ETHOXY-17-METHYLMOR-
PHINAN-6-alpha-OL HYDROCHLORIDE DIHYDRATE see
ETHYL MORPHINE HYDROCHLORIDE DIHYDRATE

9,10-DIDEHYDRO-N-ETHYL-1,6-DIMETHYLERGOLINE-8-BETA-
CARBOXAMIDE see 1-METHYLLYSERGIC ACID ETHYLAMIDE

9,10-DIDEHYDRO-N-ETHYL-6-METHYLERGOLINE-8-beta-CARBOX-
AMIDE see LYSERGIC ACID ETHYLAMIDE

9,10-DIDEHYDRO-N-(alpha-(HYDROXYMETHYL)ETHYL)-6METHYL-
ERGOLINE-8-beta-CARBOXAMIDE see D-LYSERGIC ACID-
L,2-PROPANOLAMIDE

9,10-DIDEHYDRO-N-(ALPHA-(HYDROXYMETHYL)PROPYL)-6-METHYL-
ERGOLINE-8-BETA-CARBOXAMIDE see PARTERGIN

9,10-DIDEHYDRO-N,N,6-TRIMETHYLERGOLINE-8-beta-CARBOXAMIDE
see D-LYSERGIC ACID DIMETHYLAMIDE

4,4′-DIDEMETHYL-4,4′-DI-2-PROPENYLTOXIFERINE I DICHLORIDE see
N,N′-DIALLYLNORTOXIFERINIUM DICHLORIDE

1,6-DIDEOXY-1,6-DI(2-CHLOROETHYLAMINO)-D-MANNITOLDIHYDRO-
CHLORIDE see MANNOMUSTINE DIHYDROCHLORIDE

3′,4′-DIDEOXYKANAMYCIN B see DIBEKACIN

4,4′-DI(DIETHYLAMINO)-4′,6′-DISULPHOTRIPHENYLMETHANOL
ANHYDRIDE, SODIUM SALT see ACID BLUE 1

3,6-DI(DIMETHYLAMINO)ACRIDINE see 3,6-BIS(DIMETHYL AMINO)-
ACRIDINE

DIELDRINE (FRENCH) see DIELDRIN

DIENESTROL see DEHYDROSTILBESTROL

L-DIEPOXYBUTANE see L-BUTADIENE DIEPOXIDE

2,4-DIEPOXYBUTANE see 1,1′-BI(ETHYLENE OXIDE)

1,2:3,4-DIEPOXYBUTANE see 1,1′-BI(ETHYLENE OXIDE)

(±)-1,2:3,4-DIEPOXYBUTANE see d,l-DIEPOXYBUTANE

(R*,S*)-DIEPOXYBUTANE see meso-1,2,3,4-DIEPOXYBUTANE

(2S,3S)-DIEPOXYBUTANE see L-BUTADIENE DIEPOXIDE

MESO-DIEPOXYBUTANE see meso-1,2,3,4-DIEPOXYBUTANE

3-(2-(DIETHYLAMINO)ETHYL)-5-ETHYL -5-PHENYLBARBITURIC ACID HYDROCHLORIDE see HEXAMID

DIETHYLAMINOETHYL-3-HYDROXY-4-AMINOBENZOATE see METAHYDROXYPROCAINE

N-(DIETHYLAMINOETHYL)-2-METHOXY-4-AMINO-5-CHLOROBENZAMIDE see 4-AMINO-5-CHLORO-N-(2-(DIETHYLAMINO)ETHYL)-n-ANISAMIDE

N-(2-(DIETHYLAMINO)ETHYL)-2-(P-METHOXYPHENOXY)ACETAMIDE see MEPHEXAMIDE

2-DIETHYLAMINOETHYL 3-METHYL-2-PHENYLVALERATE METHYLBROMIDE see VALETHAMATE BROMIDE

2-DIETHYLAMINOETHYL 2-PHENYL-3-METHYLVALERATEMETHYL BROMIDE see VALETHAMATE BROMIDE

5-BETA-DIETHYLAMINOETHYL-3-PHENYL-1,2,4-OXADIAZOLE CITRATE see OXOLAMINE CITRATE

2-DIETHYLAMINOETHYL-ALPHA-PHENYLVALERATE HYDROCHLORIDE see 2-PHENYLVALERIC ACID-2-(DIETHYLAMINO)ETHYL ESTER HYDROCHLORIDE

(2-DIETHYLAMINO)ETHYLPHOSPHOROTHIOIC ACID-O,O-DIETHYL ESTER see O,O-DIETHYL-S-(2-DIETHYLAMINOETHYL) THIOPHOSPHATE

2-DIETHYLAMINOETHYL-ALPHA-PROPYLTOLUATE HYDROCHLORIDE see 2-PHENYLVALERIC ACID-2-(DIETHYLAMINO)ETHYL ESTER HYDROCHLORIDE

BETA-DIETHYLAMINOETHYL XANTHENE-9-CARBOXYLATE METHOBROMIDE see XANTHINE BROMIDE

BETA-DIETHYLAMINOETHYL 9-XANTHENECARBOXYLATEMETHOBROMIDE see XANTHINE BROMIDE

2-(DIETHYLAMINO)-4-METHYL-1-PENTANOL-P-AMINOBENZOATE (ESTER) see LEUCINOCAINE

DIETHYL-M-AMINO-PHENOLPHTHALEIN HYDROCHLORIDE see (9-(o-CARBOXYPHENYL)-6-(DIETHYLAMINO)-3H-XANTHEN-3-YLIDENE) DIETHYLAMMONIUM CHLORIDE

10-DIETHYLAMINOPROPIONYL-3-TRIFLUOROMETHYL PHENOTHIAZINE HYDROCHLORIDE see FLUORACIZINE

2-DIETHYLAMINO-1-PROPYL-N-DIBENZOPARATHIAZINE see PHENOPROPAZINE

10-(2-DIETHYLAMINO PROPYL)PHENOTHIAZINE see PHENOPROPAZINE

DIETHYLAMINO STILBENE see N,N-DIETHYL-4-STILBENAMINE

2-DIETHYLAMINO-2',4',6'-TRIMETHYLACETANILIDE HYDROCHLORIDE see TRIMECAINE

DIETHYLBARBITONE see BARBITAL

5,5-DIETHYLBARBITURIC ACID see BARBITAL

O,O-DIETHYL-O-(4-BROOM-2,5-DICHLOOR-FENYL)-MONOTHIOFOSFAAT (DUTCH) see ETHYL BROMOPHOS

DIETHYLCARBAMAZANE CITRATE see DIETHYLCARBAMAZINE ACID CITRATE

DIETHYLCARBAMAZINE CITRATE see DIETHYLCARBAMAZINE ACID CITRATE

DIETHYLCARBAMAZINE HYDROGEN CITRATE see DIETHYLCARBAMAZINE ACID CITRATE

DIETHYLCARBAMODITHIOIC ACID, SODIUM SALT see SODIUM DIETHYLDITHIOCARBAMATE

1-DIETHYLCARBAMOYL-4-METHYLPIPERAZINE DIHYDROGEN CITRATE see DIETHYLCARBAMAZINE ACID CITRATE

N,N-DIETHYLCARBANILIDE see 1,3-DIETHYL-1,3-DIPHENYLUREA

DIETHYL CARBINOL see 3-PENTANOL

O,O-DIETHYL-S-CARBOETHOXYMETHYL PHOSPHOROTHIOATE see O,O-DIETHYL S-(CARBETHOXY)METHYL PHOSPHOROTHIOLATE

O,O-DIETHYL-S-CARBOETHOXYMETHYL THIOPHOSPHATE see O,O-DIETHYL S-(CARBETHOXY)METHYL PHOSPHOROTHIOLATE

DIETHYLCETONE (FRENCH) see 3-PENTANONE

O,O-DIETHYL-S-(4-CHLOOR-FENYL-THIO)-METHYL)-DITHIOFOSFAAT (DUTCH) see TRITHION

O,O-DIETHYL-O-(3-CHLOOR-4-METHYL-CUMARIN-7-YL)MONOTHIOFOSFAAT (DUTCH) see COUMAPHOS

O,O-DIETHYL-S-(((6-CHLOOR-2-OXO-BENZOXAZOLIN-3-YL)-METHYL)-DITHIO FOSFAAT (DUTCH) see S-((3-BENZOXAZOLINYL-6-CHLORO-2-OXO)METHYL) O,O-DIETHYLPHOSPHORODITHIOATE

O,O-DIETHYL-S-P-CHLORFENYLTHIOMETHYLESTER KYSELINY DITHIOFOSFORECNE (CZECH) see TRITHION

O,O-DIETHYL S-P-CHLORLPHENYLTHIOMETHYL DITHIOPHOSPHATE see TRITHION

O,O-DIETHYL- S-(6-CHLOROBENZOXAZOLINYL-3-METHYL)DITHIOPHOSPHATE see S-((3-BENZOXAZOLINYL-6-CHLORO-2-OXO)-METHYL) O,O-DIETHYLPHOSPHORODITHIOATE

O,O-DIETHYL O-(2-CHLORO-1-(2',4'-DICHLOROPHENYL)VINYL) PHOSPHATE see CHLORFENVINFOS

DIETHYL(2-CHLOROETHYL)AMINE see N-(2-CHLORO ETHYL)DIETHYL AMINE

DIETHYL 3-CHLORO-4-METHYL-7-COUMARINYL PHOSPHATE see 3-CHLORO-4-METHYL-7-COUMARINYL DIETHYL PHOSPHATE

O,O-DIETHYL-O-(3-CHLORO-4-METHYLCOUMARIN-7-YL) PHOSPHATE see 3-CHLORO-4-METHYL-7-COUMARINYL DIETHYL PHOSPHATE

O,O-DIETHYL O-(3-CHLORO-4-METHYL-7-COUMARINYL)PHOSPHOROTHIOATE see COUMAPHOS

O,O-DIETHYL O-(3-CHLORO-4-METHYLCOUMARINYL-7) THIOPHOSPHATE see COUMAPHOS

O,O-DIETHYL O-(3-CHLORO-4-METHYL-2-OXO-2H-BENZOPYRAN-7-YL)PHOSPHOROTHIOATE see COUMAPHOS

O,O-DIETHYL 3-CHLORO-4-METHYL-7-UMBELLIFERONE THIOPHOSPHATE see COUMAPHOS

O,O-DIETHYL O-(3-CHLORO-4-METHYLUMBELLIFERYL)PHOSPHOROTHIOATE see COUMAPHOS

DIETHYL 3-CHLORO-4-METHYLUMBELLIFERYL THIONOPHOSPHATE see COUMAPHOS

O,O-DIETHYL- S-((6-CHLORO-2-OXOBENZOXAZOLIN-3-YL)METHYL) PHOSPHORODITHIOATE see S-((3-BENZOXAZOLINYL-6-CHLORO-2-OXO)METHYL) O,O-DIETHYLPHOSPHORODITHIOATE

O,O-DIETHYL-S-(6-CHLORO-2-OXO-BENZOXAZOLIN-3-YL)METHYL-PHOSPHORO THIOLOTHIONATE see S-((3-BENZOXAZOLINYL-6-CHLORO-2-OXO)METHYL) O,O-DIETHYLPHOSPHORODITHIOATE

O,O-DIETHYL P-CHLOROPHENYLMERCAPTOMETHYL DITHIOPHOSPHATE see TRITHION

O,O-DIETHYL S-(4-CHLOROPHENYLTHIOMETHYL) DITHIOPHOSPHATE see TRITHION

O,O-DIETHYL S-(P-CHLOROPHENYLTHIOMETHYL) PHOSPHORODITHIOATE see TRITHION

O,O-DIETHYL-S-p-CHLOROPHENYL THIOMETHYLPHOSPHOROTHIOATE see DANIFOS

O,O-DIETHYL-S-((2-CYAAN-2-METHYL-ETHYL)-CARBAMOYL)-METHYL-MONOTHIOFOSFAAT (DUTCH) see PHOSPHOROTHIOIC ACID-S-(((1-CYANO-1-METHYL-ETHYL)CARBAMOYL)METHYL)-O,O-DIETHYL ESTER

DIETHYLCYANOPHOSPHATE see DIETHYL PHOSPHOROCYANIDATE

DIETHYL CYANOPHOSPHONATE see DIETHYL PHOSPHOROCYANIDATE

P,P-DIETHYL CYCLIC ETHYLENE ESTER OF PHOSPHONODITHIOIMIDOCARBONIC ACID see PHOSFOLAN

P,P-DIETHYL CYCLIC PROPYLENE ESTER OF PHOSPHONODITHIOIMIDOCARBONIC ACID see MEPHOSFOLAN

N-DIETHYL CYSTEAMINE see N,N-DIETHYL CYSTEAMINE

DIETHYLDIAZENE-1-OXIDE see AZOXYETHANE

O,O-DIETHYL-O-(2,4-DICHLOOR-FENYL)-MONOTHIOFOSFAAT (DUTCH) see DICHLOROFENTHION

O,O-DIETHYL O-2,5-DICHLORO-4-BROMOPHENYL-PHOSPHOROTHIOATE see ETHYL BROMOPHOS

O,O-DIETHYL O-(2,5-DICHLORO-4-BROMOPHENYL) THIOPHOSPHATE see ETHYL BROMOPHOS

DIETHYL-1-(2,4-DICHLOROPHENYL)-2-CHLOROVINYL PHOSPHATE see CHLORFENVINFOS

O,O-DIETHYL O-(2,4-DICHLOROPHENYL) PHOSPHOROTHIOATE see DICHLOROFENTHION

DIETHYL 2,4-DICHLOROPHENYL PHOSPHOROTHIONATE see DICHLOROFENTHION

O,O-DIETHYL S-(2,5-DICHLOROPHENYLTHIOMETHYL) DITHIOPHOSPHATE see PHENCAPTON

O,O-DIETHYL-S-(2,5-DICHLOROPHENYLTHIOMETHYL) DITHIOPHOSPHORAN see PHENCAPTON

O,O-DIETHYL S-(2,5-DICHLOROPHENYLTHIOMETHYL) PHOSPHORODITHIOATE see PHENCAPTON

O,O-DIETHYL S-(2,5-DICHLOROPHENYLTHIOMETHYL) PHOSPHOROTHIOLOTHIONATE see PHENCAPTON

O,O-DIETHYL O-2,4-DICHLOROPHENYL THIOPHOSPHATE see DICHLOROFENTHION

DIETHYLDICHLOROSILANE (DOT) see DICHLORODIETHYLSILANE

DIETHYLDICHLOROSTANNANE see DICHLORODIETHYLSTANNANE

DIETHYL-S-2-DIETHYLAMINOETHYL PHOSPHOROTHIOATE see O,O-DIETHYL-S-(2-DIETHYLAMINOETHYL) THIOPHOSPHATE

O,O-DIETHYL-S-2-DIETHYLAMINOETHYL PHOSPHOROTHIOATE see O,O-DIETHYL-S-(2-DIETHYLAMINOETHYL) THIOPHOSPHATE

O,O-DIETHYL-S-DIETHYLAMINOETHYL PHOSPHOROTHIOLATE see O,O-DIETHYL-S-(2-DIETHYLAMINOETHYL) THIOPHOSPHATE

O,O-DIETHYL-S-2-DIETHYLAMINOETHYL PHOSPHOROTHIOLATE see O,O-DIETHYL-S-(2-DIETHYLAMINOETHYL) THIOPHOSPHATE

O,O-DIETHYL-S-(BETA-DIETHYLAMINO)ETHYL PHOSPHOROTHIOLATE see O,O-DIETHYL-S-(2-DIETHYLAMINOETHYL) THIOPHOSPHATE

5,5-DIETHYLDIHYDRO-2H-1,3-OXAZINE-2,4(3H)-DIONE see 5,5-DIETHYL-1,3-OXAZIN-2,4-DIONE

O,O-DIETHYL O-(2,3-DIHYDRO-3-OXO-2-PHENYL-6-PYRIDAZINYL)-PHOSPHOROTHIOATE see PYRIDAPHENTHION

alpha,alpha'-DIETHYL-4,4'-DIMETHOXYSTILBENE see 3,4-DIANISYL-3-HEXENE

TRANS-ALPHA,ALPHA'-DIETHYL-4,4'-DIMETHOXYSTILBENE see 3,4-DIANISYL-3-HEXENE

SYM-DIETHYLDIPHENYLUREA see 1,3-DIETHYL-1,3-DIPHENYLUREA

N,N'-DIETHYL-N,N'-DIPHENYLUREA see 1,3-DIETHYL-1,3-DIPHENYLUREA

DIETHYLDITHIO BIS(THIONOFORMATE) see BISETHYL XANTHOGEN DISULFIDE

DIETHYLDITHIOCARBAMATE SODIUM see SODIUM DIETHYLDITHIOCARBAMATE

DIETHYLDITHIOCARBAMIC ACID-2-CHLOROALLYL ESTER see 2-CHLORALLYL DIETHYLDITHIOCARBAMATE

DIETHYLDITHIOCARBAMIC ACID SODIUM see SODIUM DIETHYLDITHIOCARBAMATE

DIETHYLDITHIO CARBAMIC ACID TELLURIUM SALT see ETHYL TELLURAC

DIETHYLDITHIOCARBAMIC ACID ZINC SALT see BIS(DIETHYL DITHIO CARBAMATO)ZINC

DIETHYLDITHIOCARBAMIC ANHYDRIDE OF O,O-DIISOPROPYL THIONOPHOSPHORIC ACID see DIETHYLDITHIOCARBAMIC ANHYDROSULFIDE

DIETHYL-N-1,3-DITHIOLANYL-2-IMINO PHOSPHATE see DITHIOLANE IMINOPHOSPHATE

O,O-DIETHYL DITHIOPHOSPHORIC ACID P-CHLOROPHENYLTHIOMETHYL ESTER see TRITHION

O,O-DIETHYLDITHIOPHOSPHORYLACETIC ACID, N-MONOISOPROPYL-AMIDE see ISOPROPYL DIETHYL DITHIOPHOSPHORYL ACETAMIDE

3-DIETHYLDITHIOPHOSPHORYLMETHYL-6-CHLOROBENZOXAZOLONE-2 see S-((3-BENZOXAZOLINYL-6-CHLORO-2-OXO)METHYL) O,O-DIETHYLPHOSPHORODITHIOATE

1,4-DIETHYLENEDIAMINE see PIPERAZINE

DIETHYLENE DIOXIDE see p-DIOXANE

1,4-DIETHYLENE DIOXIDE see p-DIOXANE

DIETHYLENE ETHER see p-DIOXANE

DIETHYLENE GLYCOL, MONOESTER WITH STEARIC ACID see 2-(2-HYDROXYETHOXY)ETHYL ESTER STEARIC ACID

DIETHYLENE GLYCOL MONOETHYL ETHER see CARBITOL CELLOSOLVE

DIETHYLENE GLYCOL MONOETHYL ETHER ACETATE see CARBITOL ACETATE

DIETHYLENE GLYCOL MONOPHENYL ETHER see PHENYL CARBITOL

DIETHYLENE GLYCOL MONOSTEARATE see 2-(2-HYDROXYETHOXY)-ETHYL ESTER STEARIC ACID

DIETHYLENE GLYCOLPHENYL ETHER see PHENYL CARBITOL

DIETHYLENE GLYCOL STEARATE see 2-(2-HYDROXYETHOXY)ETHYL ESTER STEARIC ACID

DIETHYLENEIMIDE OXIDE see MORPHOLINE

DIETHYLENE IMIDOXIDE see MORPHOLINE

DIETHYLENE OXIDE see TETRAHYDROFURAN

DI(ETHYLENE OXIDE) see p-DIOXANE

DIETHYLENE OXIMIDE see MORPHOLINE

DIETHYLENETRIAMINEPENTAACETIC ACID see PENTHAMIL

1,1,4,7,7-DIETHYLENETRIAMINEPENTAACETIC ACID see PENTHAMIL

(DIETHYLENETRINITRILO)PENTAACETIC ACID see PENTHAMIL

DIETHYLENGLYKOLDINITRATE (CZECH) see DIETHYLENE GLYCOL DINITRATE

DIETHYLENIMIDE OXIDE see MORPHOLINE

DIETHYL ETHANEDIOATE see DIETHYL OXALATE

4,4'-(1,2-DIETHYL-1,2-ETHENEDIYL)BIS-PHENOL see DIETHYLSTILBESTEROL

TRANS-4,4'-(1,2-DIETHYL-1,2-ETHENEDIYL)BISPHENOL see DIETHYLSTILBESTEROL

TRANS-4,4'-(1,2-DIETHYL-1,2-ETHENEDIYL)BISPHENOL DIPROPIONATE see DIETHYLSTILBESTROL DIPROPIONATE

DIETHYL ETHER see ETHYL ETHER

DIETHYL S-(2-ETHIOETHYL)THIOPHOSPHATE see DEMETON-S

O,O-DIETHYL S-(N-ETHOXYCARBONYL-N-METHYLCARBAMOYL-METHYL) PHOSPHOROTHIOLOTHIONATE see O,O-DIETHYL-S-(N-ETHOXYCARBONYL-N-METHYLCARBAMOYLMETHYL) PHOSPHORO DITHIOATE

O,O-DIETHYL-S-(2-ETHTHIOETHYL) PHOSPHORODITHIOATE see DITHIOSYSTOX

O,O-DIETHYL S-(2-ETHTHIOETHYL)PHOSPHOROTHIOATE see DEMETON-S

O,O-DIETHYL-S-(2-ETHTHIOETHYL) THIOTHIONOPHOSPHATE see DITHIOSYSTOX

O,O-DIETHYL-S-(2-ETHTHIONYLETHYL) PHOSPHOROTHIOATE see ISO SYSTOX SULFOXIDE

DIETHYL-S-(2-ETHTHIONYLETHYL) THIOPHOSPHATE see ISO SYSTOX SULFOXIDE

4,4'-(1,2-DIETHYLETHYLENE)DIPHENOL see DIHYDROSTILBESTROL

O,O-DIETHYL S-ETHYL-2-ETHYLMERCAPTOPHOSPHOROTHIOLATE see DEMETON-S

O,O-DIETHYL-S-ETHYL-2-ETHYLMERCAPTO PHOSPHOROTHIOLATE SULFOXIDE see ISO SYSTOX SULFOXIDE

O,O-DIETHYL-S-(2-ETHYLMERCAPTOETHYL) DITHIOPHOSPHATE see DITHIOSYSTOX

O,O-DIETHYL 2-ETHYLMERCAPTOETHYL THIOPHOSPHATE see DEMETON-O + DEMETON-S

O,O-DIETHYL-2-ETHYLMERCAPTOETHYL THIOPHOSPHATE, THIONO ISOMER see SYSTOX SULFONE

O,O-DIETHYL S-ETHYLMERCAPTOMETHYL DITHIOPHOSPHONATE see PHORATE

O,O-DIETHYL-S-((ETHYLSULFINYL)ETHYL)PHOSPHORODITHIOATE see OXYDISULFOTON

O,O-DIETHYL S-(2-(ETHYLSULFINYL)ETHYL) PHOSPHORODITHIOATE see OXYDISULFOTON

O,O-DIETHYL-O-(2-ETHYLSULFONYLETHYL)PHOSPHOROTHIOATE see SYSTOX SULFONE

DIETHYL-2-ETHYLSULFONYLETHYL THIONOPHOSPHATE see SYSTOX SULFONE

O,O-DIETHYL-S-(2-ETHYLTHIO-ETHYL) -DITHIOFOSFAAT (DUTCH) see DITHIOSYSTOX

O,O-DIETHYL-S-(2-ETHYLTHIO-ETHYL)-MONOTHIOFOSFAAT (DUTCH) see DEMETON-S

O,O-DIETHYL-2-ETHYLTHIOETHYL PHOSPHORODITHIOATE see DITHIOSYSTOX

O,O-DIETHYL-S-2-(ETHYLTHIO)ETHYL PHOSPHORODITHIOATE see DITHIOSYSTOX

O,O-DIETHYL S-2-(ETHYLTHIO)ETHYL PHOSPHOROTHIOATE see DEMETON-S

O,O-DIETHYL S-(2-(ETHYLTHIO)ETHYL) PHOSPHOROTHIOLATE see DEMETON-S

O,O-DIETHYL-S-(ETHYLTHIO-METHYL) -DITHIOFOSFAAT (DUTCH) see PHORATE

O,O-DIETHYL-O-P-NITROFENYLTIOFOSFAT (CZECH) see PARATHION

O,O-DIETHYL-O-(P-NITROPHENYL) ESTER (DRY MIXTURE) PHOSPHO-ROTHIOIC ACID see PARATHION MIXTURE, DRY (DOT)

DIETHYL P-NITROPHENYL PHOSPHATE see p-NITROPHENYL DIETHYL-PHOSPHATE

O,O-DIETHYL O-P-NITROPHENYL PHOSPHATE see p-NITROPHENYL DIETHYLPHOSPHATE

O,O-DIETHYL-0-4-NITROPHENYLPHOSPHOROTHIOATE see PARATHION

O,O-DIETHYL-O-(4-NITROPHENYL) PHOSPHOROTHIOATE see PARA-THION

O,O-DIETHYL O-(P-NITROPHENYL) PHOSPHOROTHIOATE see PARA-THION

DIETHYL 4-NITROPHENYL PHOSPHOROTHIONATE see PARATHION

DIETHYL P-NITROPHENYLTHIONOPHOSPHATE see PARATHION

O,O-DIETHYL-O-(P-NITROPHENYL)THIONOPHOSPHATE see PARATHION

DIETHYL-P-NITROPHENYLTHIOPHOSPHATE see PARATHION

O,O-DIETHYL 0-4-NITROPHENYL THIOPHOSPHATE see PARATHION

O,O-DIETHYL O-P-NITROPHENYL THIOPHOSPHATE see PARATHION

DIETHYLNITROSAMINE see N-NITROSO-DIETHYLAMINE

N,N-DIETHYLNITROSAMINE see N-NITROSO-DIETHYLAMINE

DIETHYLNITROSOAMINE see N-NITROSO-DIETHYLAMINE

O,O-DIETHYL O(AND S)-2-(ETHYLTHIO)ETHYL PHOSPHOROTHIOATE MIXTURE see DEMETON-O + DEMETON-S

O,O-DIETHYL O,2-PYRAZINYL PHOSPHOROTHIOATE see ETHYL PY-RAZINYL PHOSPHOROTHIOATE

5,5-DIETHYL-1,3-OXAZINE-2,4-DIONE see 5,5-DIETHYL-1,3-OXAZIN-2,4-DIONE

DIETHYL OXIDE see ETHYL ETHER

O,O-DIETHYL-S-(4-OXO-3H-1,2,3-BENZOTRIAZINE-3-YL)-METHYL-DI-THIOPHOSPHATE see ETHYL GUTHION

O,O-DIETHYL S-(4-OXOBENZOTRIAZINO-3-METHYL)PHOSPHORODI-THIOATE see ETHYL GUTHION

O,O-DIETHYL-S-((4-OXO-3H-1,2,3-BENZOTRIAZIN-3-YL)-METHYL)-DI-THIO FOSFAAT (DUTCH) see ETHYL GUTHION

DIETHYL PARAOXON see p-NITROPHENYL DIETHYLPHOSPHATE

DIETHYLPARATHION see PARATHION

3,3'-DIETHYLPENTAMETHINETHIACYANINE IODIDE see DITHIAZANINE IODIDE

DI(P-ETHYLPHENYL)DICHLOROETHANE see DIETHYL DIPHENYL DI-CHLOROETHANE

3,3-DIETHYL-1-PHENYLTRIAZENE see 1-PHENYL-3,3-DIETHYLTRIA-ZENE

O,O-DIETHYL PHOSPHORIC ACID O-P-NITROPHENYL ESTER see p-NITROPHENYL DIETHYLPHOSPHATE

O,O-DIETHYL PHOSPHORODITHIOATE S-ESTER WITH 3-(MERCAPTO-METHYL)-1,2,3-BENZOTRIAZIN-4(3H)-ONE see ETHYL GUTHION

O,O-DIETHYLPHOSPHOROTHIOATE, O-ESTER WITH 6-HYDROXY-2-PHENYL-3(2H)-PYRIDAZINONE see PYRIDAPHENTHION

DIETHYL PROPANEDIOATE see ETHYL MALONATE

DIETHYLPROPIONE HYDROCHLORIDE see 2-DIETHYLAMINOPROPIO-PHENONEHYDROCHLORIDE

DIETHYLPROPION HYDROCHLORIDE see 2-DIETHYLAMINOPROPIOPHE-NONEHYDROCHLORIDE

DIETHYL O-2-PYRAZINYL PHOSPHOROTHIONATE see ETHYL PYRA-ZINYL PHOSPHOROTHIOATE

O,O-DIETHYL O-2-PYRAZINYL PHOSPHOTHIONATE see ETHYL PYRA-ZINYL PHOSPHOROTHIOATE

O,O-DIETHYL O-PYRAZINYL THIOPHOSPHATE see ETHYL PYRAZINYL PHOSPHOROTHIOATE

O,O-DIETHYL-O-(2-QUINOXALINYL) PHOSPHOROTHIOATE see O,O-DIETHYL-O-2-QUINOXALYL THIOPHOSPHATE

O,O-DIETHYL O-QUINOXALIN-2-YL PHOSPHOROTHIOATE see O,O-DIETHYL-O-2-QUINOXALYL THIOPHOSPHATE

O,O-DIETHYL-O-(2-QUINOXALYL) PHOSPHOROTHIOATE see O,O-DIETHYL-O-2-QUINOXALYL THIOPHOSPHATE

DIETHYL SODIUM DITHIOCARBAMATE see SODIUM DIETHYLDITHIO-CARBAMATE

DIETHYLSTANNYL DICHLORIDE see DICHLORODIETHYLSTANNANE

ALPHA,ALPHA'-DIETHYLSTILBENEDIOL see DIETHYLSTILBESTEROL

ALPHA,ALPHA'-DIETHYL-4,4'-STILBENEDIOL see DIETHYLSTILBES-TEROL

TRANS-ALPHA,ALPHA'-DIETHYL-4,4'-STILBENEDIOL see DIETHYL-STILBESTEROL

alpha,alpha'-DIETHYL-4,4'-STILBENEDIOL DIPALMITATE see DIETHYLSTILBESTROL DIPALMITATE

ALPHA,ALPHA'-DIETHYL-4,4'-STILBENEDIOL, DIPROPIONATE see DIETHYLSTILBESTROL DIPROPIONATE

ALPHA,ALPHA'-DIETHYL-4,4'-STILBENEDIOL TRANS-DIPROPIONATE see DIETHYLSTILBESTROL DIPROPIONATE

TRANS-ALPHA,ALPHA'-DIETHYL-4,4'-STILBENEDIOL DIPROPIONATE see DIETHYLSTILBESTROL DIPROPIONATE

ALPHA,ALPHA'-DIETHYL-4,4'-STILBENEDIOL DIPROPIONYL ESTER see DIETHYLSTILBESTROL DIPROPIONATE

alpha,alpha'-DIETHYL-4,4'-STILBENEDIOL DISODIUM SALT see DES DISODIUM SALT

DIETHYLSTILBENE DIPROPIONATE see DIETHYLSTILBESTROL DIPRO-PIONATE

TRANS-DIETHYLSTILBESTEROL see DIETHYLSTILBESTEROL

DIETHYLSTILBESTEROL DIPROPIONATE see DIETHYLSTILBESTROL DIPROPIONATE

DIETHYLSTILBESTROL see DIETHYLSTILBESTEROL

TRANS-DIETHYLSTILBESTROL see DIETHYLSTILBESTEROL

DIETHYLSTILBESTROL DIMETHYL ETHER see 3,4-DIANISYL-3-HEXENE

DIETHYLSTILBESTROL PROPIONATE see DIETHYLSTILBESTROL DIPRO-PIONATE

DIETHYLSTILBOESTEROL see DIETHYLSTILBESTEROL

TRANS-DIETHYLSTILBOESTEROL see DIETHYLSTILBESTEROL

DIETHYL SULFATE see ETHYL SULFATE

DIETHYL SULFATE see DIETHYL ESTER SULFURIC ACID

DIETHYLSULFID (CZECH) see ETHYL SULFIDE

DIETHYL SULFIDE see ETHYL SULFIDE

5,5-DIETHYLTETRAHYDRO-2H-1,3-OXAZINE-2,4(3H)-DIONE see 5,5-DIETHYL-1,3-OXAZIN-2,4-DIONE

O,O-DIETHYL O-(7,8,9,10-TETRAHYDRO-6-OXOBENZO(C)CHROMAN-3-YL)PHOSPHOROTHIOATE see DITHION

O,O-DIETHYL O-(7,8,9,10-TETRAHYDRO-6-OXO-6H-DIBEN-ZO(B,D)PYRAN-3-YL)PHOSPHOROTHIOATE see DITHION

O,O-DIETHYL O-(3,4-TETRAMETHYLENECOUMARINYL-7) THIOPHOS-PHATE see DITHION

3,3'-DIETHYLTHIADICARBOCYANINE IODIDE see DITHIAZANINE IO-DIDE

N,N'-DIETHYLTHIOCARBAMIDE see 1,3-DIETHYLTHIOUREA

N,N-DIETHYLTHIOCARBAMYL-O,O-DIISOPROPYLDITHIOPHOSPHATE see DIETHYLDITHIOCARBAMIC ANHYDROSULFIDE

DIETHYLTHIOETHER see ETHYL SULFIDE

DIETHYL THIOPHOSPHORIC ACIDESTER OF 3-CHLORO-4-METHYL-7-HYDROXYCOUMARIN see COUMAPHOS

1,3-DIETHYL-2-THIOUREA see 1,3-DIETHYLTHIOUREA

N,N'-DIETHYLTHIOUREA see 1,3-DIETHYLTHIOUREA

DIETHYLTIN DICAPRYLATE see DIETHYLBIS(OCTANOYLOXY)STAN-NANE

DIETHYLTIN DICHLORIDE see DICHLORODIETHYLSTANNANE

DIETHYLTIN DIIODIDE see DIETHYLDIIODOSTANNANE

DIETHYLTIN DIOCTANOATE see DIETHYLBIS(OCTANOYLOXY)STAN-NANE

DIETHYLTOLUAMIDE see DIETHYL-m-TOLUAMIDE

DIETHYL-M-TOLUAMIDE see DIETHYL-m-TOLUAMIDE

N,N-DIETHYL-M-TOLUAMIDE see DIETHYL-m-TOLUAMIDE

O,O-DIETHYL-O-3,5,6-TRICHLORO-2-PYRIDYL PHOSPHOROTHIOATE see DOWCO 179

DIETHYLXANTHOGEN DISULFIDE see BISETHYL XANTHOGEN DISUL-FIDE

DIETILAMINA (ITALIAN) see DIETHYLAMINE

O,O-DIETIL-O-(4-BROMO-2,5 DICLORO-FENIL)-MONOTIOFOSFATO (ITALIAN) see ETHYL BROMOPHOS

O,O-DIETIL-S-((2-CIAN-2-METIL-ETIL) -CARBAMOIL)-METIL-MONOTIO-FOSFATO (ITALIAN) see PHOSPHOROTHIOIC ACID-S-(((1-CYANO-1-METHYL-ETHYL)CARBAMOYL)METHYL)-O,O-DIETHYL ESTER

O,O-DIETIL-S-((4-CLORO-FENIL-TIO) -METILE)-DITIOFOSFATO
(ITALIAN) see TRITHION

O,O-DIETIL-O-(3-CLORO-4-METIL-CUMARIN-7-IL-MONOTIOFOSFATO)
(ITALIAN) see COUMAPHOS

O,O-DIETIL-S-((6-CLORO-2-OXO-BENZOSSAZOLIN-3-IL)-METIL)-DITIO-
FOSFATO (ITALIAN) see S-((3-BENZOXAZOLINYL-6-CHLORO-2-
OXO)METHYL) O,O-DIETHYLPHOSPHORODITHIOATE

O,O-DIETIL-O-(2,4-DICLORO-FENIL)-MONOTIOFOSFATO (ITALIAN) see
DICHLOROFENTHION

DIETILESTILBESTROL (SPANISH) see DIETHYLSTILBESTEROL

O,O-DIETIL-S-(2-ETILTIO-ETIL)-DITIOFOSFATO (ITALIAN) see DITHIO-
SYSTOX

O,O-DIETIL-S-(2-ETILTIO-ETIL)-MONOTIOFOSFATO (ITALIAN) see
DEMETON-S

O,O-DIETIL-S-(ETILTIO-METIL)-DITIOFOSFATO (ITALIAN) see PHORATE

O,O-DIETIL-S-(N-ETOSSI-CARBONIL-N-METIL-CARBAMOIL-METIL)-
DITIOFOSFATO (ITALIAN) see O,O-DIETHYL-S-(N-ETHOXYCAR-
BONYL-N-METHYLCARBAMOYLMETHYL) PHOSPHORO DITHIOATE

O,O-DIETIL-O-(2-ISOPROPIL-4METIL-PIRIMIDIN-6-IL)-MONOTIOFOS-
FATO(ITALIAN) see DIAZIDE

O,O-DIETIL-O-(4-METILCUMARIN-7-IL)-MONOTIOFOSFATO (ITALIAN)
see DIETHYL (4-METHYLUMBELLIFERYL) THIONOPHOSPHATE

O,O-DIETIL-O-(3-METIL-1H-PIRAZOL-5-IL)-FOSFATO (ITALIAN) see
METHYLPYRAZOLYL DIETHYLPHOSPHATE

O,O-DIETIL-O-(4-NITRO-FENIL)-MONOTIOFOSFATO (ITALIAN) see
PARATHION

O,O-DIETIL-S-((4-OXO-3H-1,2,3-BENZOTRIAZIN-3-IL)-METIL)-DITIO-
FOSFATO (ITALIAN) see ETHYL GUTHION

1,1-DIETOSSIETANO (ITALIAN) see ACETAL

O,O-DIETYL-S-2-ETYLMERKAPTOETYLTIOFOSFAT (CZECH) see DEME-
TON-S

O,O-DIETYL-O-4-METHYLKUMARINYL(7)TIOFOSFAT (CZECH) see
DIETHYL (4-METHYLUMBELLIFERYL) THIONOPHOSPHATE

O,O-DIETYL-O-P-NITROFENYLFOSFAT (CZECH) see P-NITROPHENYL
DIETHYLPHOSPHATE

DIFENILHIDANTOINA (SPANISH) see DIHYDANTOIN

DIFENIL-METAN-DIISOCIANATO (ITALIAN) see DIPHENYL METHANE
DIISOCYANATE

N,N′-DIFENYL-p-FENYLENDIAMIN (CZECH) see 1,4-BIS(PHENYL
AMINO)BENZENE

2-(DIFENYL-HYDROXYACETOXY)ETHYL-DIETHYLAMMONIUMCHLORID
(CZECH) see BENZILIC ACID-beta-DIETHYLAMINOETHYLESTER
HYDROCHLORIDE

DIFENYLMETHAAN-DISSOCYANAAT (DUTCH) see DIPHENYL METHANE
DIISOCYANATE

DIFLAVINE (ACRIDINE) see 2,6-DIAMINOACRIDINE

DIFLUOROCHLOROMETHANE see MONOCHLORODIFLUOROMETHANE

DIFLUORODICHLOROMETHANE see DICHLORODIFLUOROMETHANE

1,1-DIFLUOROETHYLENE see VINYLIDENE FLUORIDE

2′,4′-DIFLUORO-4-HYDROXY-3-BIPHENYLCARBOXYLIC ACID see 5--
(2,4-DIFLUOROPHENYL)SALICYLIC ACID

2′,4′-DIFLUORO-4-HYDROXY-(1,1′-BIPHENYL)-3-CARBOXYLIC ACID
see 5-(2,4-DIFLUOROPHENYL)SALICYLIC ACID

2′,4′-DIFLUORO-4-HYDROXY-(1′,1-DIPHENYL)-3-CARBOXYLIC ACID
see 5-(2,4-DIFLUOROPHENYL)SALICYLIC ACID

DIFLUOROMONOCHLOROMETHANE see MONOCHLORODIFLUORO-
METHANE

DIFLUOROPHOSPHORIC ACID see PHOSPHORODIFLUORIDIC ACID

DIFLUOROPHOSPHORIC ACID, ANHYDROUS (DOT) see PHOSPHORODI-
FLUORIDIC ACID (ANHYDROUS)

3,3-DIFLUORO-2-(TRIFLUOROMETHYL)ACRYLIC ACID, METHYL ESTER
see METHYL PERFLUOROMETHACRYLATE

DIFORMAL see GLYOXAL

DIGENEA SIMPLEX MUCILAGE see AGAR

DIGILANID C see LANATOSIDE C

DIGITALIN see DIGITOXIN

DIGITALINE CRISTALLISEE see DIGITOXIN

DIGITALINE NATIVELLE see DIGITOXIN

DIGITALINUM VERUM see DIGITOXIN

DIGITIN see DIGITONIN

DIGITOPHYLLIN see DIGITOXIN

DIGITOXIGENIN TRIDIGITOXOSIDE see DIGITOXIN

DIGLYCIDYL BISPHENOL A ETHER see BISPHENOL A DIGLYCIDYL
ETHER

DIGLYCIDYL ETHER OF 2,2-BIS(4-HYDROXYPHENYL)PROPANE see
BISPHENOL A DIGLYCIDYL ETHER

DIGLYCIDYL ETHER OF 2,2-BIS(p-HYDROXYPHENYL)PROPANE see
BISPHENOL A DIGLYCIDYL ETHER

DIGLYCIDYL ETHER OF BISPHENOL A see BISPHENOL A DIGLYCIDYL
ETHER

DIGLYCIDYL ETHER OF 4,4′-ISOPROPYLIDENEDIPHENOL see BIS-
PHENOL A DIGLYCIDYL ETHER

DIGLYCIDYL RESORCINOL ETHER see RESORCINOL DIGLYCIDYL
ETHER

DIGLYCOL see DIETHYLENE GLYCOL

DIGLYCOLAMINE see 2-AMINOETHOXYETHANOL

DIGLYCOLDINITRAAT (DUTCH) see DIETHYLENE GLYCOL DINITRATE

DIGLYCOL MONOBUTYL ETHER ACETATE see BUTYL CARBITOL
ACETATE

DIGLYCOL MONOETHYL ETHER ACETATE see CARBITOL ACETATE

DIGLYCOL MONOSTEARATE see 2-(2-HYDROXYETHOXY)ETHYL ESTER
STEARIC ACID

DIGLYCOL STEARATE see 2-(2-HYDROXYETHOXY)ETHYL ESTER
STEARIC ACID

DIGLYKOLDINITRAT (GERMAN) see DIETHYLENE GLYCOL DINITRATE

DIGORID A see ACETYL-DIGOXIN-alpha

DIGOXINE see DIGOXIN

DIHEXYL see DODECANE

DI-N-HEXYLAMINE see DIHEXYLAMINE

DIHEXYLTIN DICHLORIDE see DICHLORODIHEXYLSTANNANE

DIHIDROBENZPERIDOL see DROPERIDOL

DIHIDROCLORURO DE BENZIDINA (SPANISH) see BENZIDINE HYDRO-
CHLORIDE

DIHYDRIN see DIHYDROCODEINE

DIHYDROAFLATOXIN B1 see AFLATOXIN B2

DIHYDRO-1H-AZIRINE see ETHYLENIMINE

4,5-DIHYDROBENZ(K)ACEPHENANTHRYLENE see ACENAPHTHAN-
THRACENE

1,2-DIHYDROBENZ(A)ANTHRACENE see 1,2-DIHYDROBENZO(a)AN-
THRACENE

3,4-DIHYDROBENZ(A)ANTHRACENE see 3,4-DIHYDROBENZO(a)AN-
THRACENE

6-BETA,7-ALPHA-DIHYDROBENZO(10,11)CHRYSENO(1,2-B)OXIRENE
see BENZO(a)PYRENE-7,8-OXIDE

9,10-DIHYDROBENZO(A)PYRENE-9,10-DIOL see 9,10-DIHYDRO-9,10-
DIHYDROXYBENZO(a)PYRENE

trans-4,5-DIHYDROBENZO(A)PYRENE-4,5-DIOL see trans-4,5-DIHY-
DRO-4,5-DIHYDROXYBENZO(a)PYRENE

3,4-DIHYDRO-6-CHLORO-7-SULFAMYL-1,2,4-BENZOTHIADIAZINE-1,1-
DIOXIDE see 6-CHLORO-3,4-DIHYDRO-2H-1,2,4-BENZOTHIADIA-
ZINE-7-SULFONAMIDE- 1,1-DIOXIDE

DIHYDROCHLOROTHIAZIDE see 6-CHLORO-3,4-DIHYDRO-2H-1,2,4-
BENZOTHIADIAZINE-7-SULFONAMIDE- 1,1-DIOXIDE

3,4-DIHYDROCHLOROTHIAZIDE see 6-CHLORO-3,4-DIHYDRO-2H-
1,2,4-BENZOTHIADIAZINE-7-SULFONAMIDE- 1,1-DIOXIDE

6,12B-DIHYDROCHOLANTHRENE see meso-DIHYDROCHOLANTHRENE

DIHYDROCINNAMALDEHYDE see HYDROCINNAMALDEHYDE

7,8-DIHYDROCODEINE see DIHYDROCODEINE

DIHYDROCODEINE ACID TARTRATE see DIHYDROCODEINE BITAR-
TRATE

DIHYDRO-CODEINE TARTRATE (1:1) see DIHYDROCODEINE BITAR-
TRATE

DIHYDROCOUMARIN see HYDROCOUMARIN

3,4-DIHYDROCOUMARIN see HYDROCOUMARIN

9,10-DIHYDRO-8A,10,-DIAZONIAPHENANTHRENE DIBROMIDE see DI-
QUAT DIBROMIDE

9,10-DIHYDRO-8A,10A-DIAZONIAPHENANTHRENE(1,1′-ETHYLENE-
2,2′-BIPYRIDYLIUM)DIBROMIDE see DIQUAT DIBROMIDE

7,14-DIHYDRODIBENZ(A,B)ANTHRACENE see 9,10-DIHYDRO-1,2,5,6-
DIBENZANTHRACENE

10,11-DIHYDRO-5H-DIBENZO(A,D)CYCLOHEPTEN-5-ONE OXIME see DIBENZOSUBERONE OXIME

DIHYDRODIETHYLSTILBESTROL see DIHYDROSTILBESTROL

TRANS-3,4-DIHYDRO-3,4-DIHYDROXYBENZO(A)ANTHRACENE see BENZ(a)ANTHRACENE-3,4-DIHYDRODIOL

trans-3,4-DIHYDRO-3,4-DIHYDROXY-7,12-DIMETHYLBENZ(a)AN-THRACENE see trans-3,4-DIHYDRO-3,4-DIHYDROXY-DMBA

trans-8,9-DIHYDRO-8,9-DIHYDROXY-7,12-DIMETHYLBENZ(a)AN-THRACENE see trans-8,9-DIHYDRO-8,9-DIHYDROXY-DMBA

TRANS-3,4-DIHYDRO-3,4-DIHYDROXY-7-METHYLBENZ(a)ANTHRA-CENE see 7-MBA-3,4-DIHYDRODIOL

14,19-DIHYDRO-12,13-DIHYDROXY(13-ALPHA,14-ALPHA)-20-NOR-CROTALANAN-11,15-DIONE see MONOCROTALINE

1,4-DIHYDRO-1,4-DIKETONAPHTHALENE see 1,4-NAPHTHOQUINONE

5,6-DIHYDRO-9,10-DIMETHOXYBENZO(G)-1,3-BENZODIOXOLO-(5,6-A)QUINOLIZINIUM SULFATE TRIHYDRATE see BERBERINE SULFATE TRIHYDRATE

5,6-DIHYDRO-N-(3-(DIMETHYLAMINO)PROPYL)-11H-DIBENZ(b,e)-AZEPINE see TOFRANIL

2,3-DIHYDRO-2,2-DIMETHYL-7-BENZOFURANYL METHYLCARBAMATE see FURADAN

2,3-DIHYDRO-2,2-DIMETHYLBENZOFURANYL-7-N-METHYLCARBA-MATE see FURADAN

10,11-DIHYDRO-N,N-DIMETHYL-5H-DIBENZO(A,D)HEPTALENE-DELTA-(SUP 5),GAMMA-PROPYLAMINE see ELAVIL

10,11-DIHYDRO-N,N-DIMETHYL-5,10-METHANO-5H-DIBENZO(A,D)-CYCLOHEPTENE-12-METHANAMINE HCl see anti-8-(N,N-DIMETHYL-AMINOMETHYL)DIBENZOBICYCLO(3.2.1)OCTADIENE HYDROCHLO-RIDE

3,4-DIHYDRO-2,2-DIMETHYL-4-OXO-2H-PYRAN-6-CARBOXYLIC ACID-N-BUTYL ESTER see n-BUTYL MESITYL OXIDE OXALATE

(±)-9,10,DIHYDRO-N,N-DIMETHYL2-(TRIFLUORMETHYL)-9ENTHRA-CENPROPANAMIN (GERMAN) see (±)-9,10-DIHYDRO-N,N-DI-METHYL-2-TRIFLUOROMETHYL)-9-ANTHRACENE PROPANAMINE

1,3-DIHYDRO-1,3-DIOXO-5-ISOBENZOFURANCARBOXYLIC ACID see TRIMELLIC ACID ANHYDRIDE

5,10-DIHYDRO-5,10-DIOXONAPHTHO(2,3-b)-p-DITHIIN-2,3-DICARBO-NITRILE see 1,4-DITHIAANTHRAQUINONE-2,3-DINITRILE

5,6-DIHYDRO-DIPYRIDO(1,2A;2,1C)PYRAZINIUM DIBROMIDE see DI-QUAT DIBROMIDE

7,8-DIHYDRO-4,5-alpha-EPOXY-17-METHYLMORPHINAN-3,6-alpha-DIOL DIACETATE see HEROIN

DIHYDROESTRIN BENZOATE see ESTRADIOL-3-BENZOATE

1,4-DIHYDRO-1-ETHYL-7-METHYL-4-OXO-1,8-NAPHTHYRIDINE-3-CARBOXYLIC ACID see 1-ETHYL-1,4-DIHYDRO-7-METHYL-4-OXO-1,8-NAPHTHYRIDINE-3-CARBOXYLIC ACID

DIHYDROFOLLICULAR HORMONE see ESTRADIOL (β)

DIHYDROFOLLICULIN see ESTRADIOL (β)

DIHYDROFOLLICULIN BENZOATE see ESTRADIOL-3-BENZOATE

DIHYDRO-2,5-FURANDIONE see SUCCINIC ANHYDRIDE

DIHYDRO-2(3H)-FURANONE see 4-BUTANOLIDE

DIHYDROGEN HEXACHLOROPLATINATE see CHLOROPLATINIC ACID

(DIHYDROGEN MERCAPTOSUCCINATO)GOLD DISODIUM SALT see GOLD SODIUM THIOMALATE

DIHYDROHYDROXYCODEINONE see PERCODAN

DIHYDRO-14-HYDROXYCODEINONE see PERCODAN

3A,12C-DIHYDRO-8-HYDROXY-6-METHOXY-7H-FURO(3',2':4,5)FURO(2,3-C)XANTHEN-7-ONE see STERIGMATOCYSTIN

3,7-DIHYDRO-1-(3-HYDROXYPROPYL)-3,7-DIMETHYL-1H-PURINE-2,6-DIONE see 1-(3-HYDROXYPROPYL)THEOBROMINE

4,5-DIHYDROIMIDAZOLE-2(3H)-THIONE see 2-IMIDAZOLIDINETHIONE

1,6-DIHYDRO-6-IMINOPURINE see ADENINE

3,6-DIHYDRO-6-IMINOPURINE see ADENINE

DIHYDROISOCODEINE TARTRATE see DIHYDROISOCODEINE ACID TAR-TRATE

DIHYDRO ISOCODEINE TARTRATE (1:1) see DIHYDROISOCODEINE ACID TARTRATE

1,2-DIHYDRO-2-KETOBENZISOSULPHONAZOLE see 1,2-BENZISOTHIA-ZOL-3(2H)-ONE-1,1-DIOXIDE

DIHYDROMENFORMON see ESTRADIOL (β)

S-(2,3-DIHYDRO-5-METHOXY-2-OXO-1,3,4-THIADIAZOL-3-METHYL) DIMETHYL PHOSPHOROTHIOLOTHIONATE see O,O-DIMETHYL-S-(5-METHOXY-1,3,4-THIADIAZOLINYL-3-METHYL)DITHIO PHOS-PHATE

3,12-DIHYDRO-6-METHOXY-3,3,12-TRIMETHYL-7HPYRANO(2,3-C)-ACRIDIN-7-ONE see ACRONYCINE

10,11-DIHYDRO-5-(3-METHYLAMINOPROPYL)-5H-DIBENZ(b,f)-AZEPINE see DIMETHYLIMIPRAMINE

1,2-DIHYDRO-3-METHYL-BENZ(J)ACEANTHRYLENE see 3-METHYL-CHOLANTHRENE

1,2-DIHYDRO-3-METHYLBENZ(J)ACEANTHRYLENE COMPOUND WITH 2,4,6-TRINITROPHENOL (1:1) see 20-METHYLCHOLANTHRENE PIC-RATE

4,5-DIHYDRO-2-((2-METHYLBENZO(B)THIEN-3-YL)METHYL)-1H-IMIDAZOLE HYDROCHLORIDE see 2-((2-METHYLBENZO(b)-THIEN-3-YL)METHYL)-2-IMIDAZOLINE HYDROCHLORIDE

7,8-DIHYDRO-5-METHYL-7,8-CHRYSENEDIOL see 7,8-DIHYDRO-7,8-DIHYDROXY-5-METHYLCHRYSENE

15,16-DIHYDRO-11-METHYL-17H-CYCLOPENTA(A)PHENANTHREN-17-ONE see 11-METHYL-15,16-DIHYDRO-17-OXOCYCLOPENTA(a)PHE-NANTHRENE

10,11-DIHYDRO-N-METHYL-5H-DIBENZO(A,D)CYCLOHEPTANE-DELTA,GAMMA-PROPYLAMINE see NORTRIPTYLINE

6,11-DIHYDRO-11-(1-METHYL-4-PIPERIDYLIDENE)5H-BENZO(5,6)-CYCLOHEPTA(1,2-b)PYRIDINE DIMALEATE see AZATADINE MALEATE

(S)-(+)-5,6-DIHYDRO-6-METHYL-2H-PYRAN-2-ONE see PARASCORBIC ACID

2,3-DIHYDRO-6-METHYL-2-THIOXO-4(1H)-PYRIMIDINONE see 6-METHYLTHIOURACIL

DIHYDROMORPHINONE HYDROCHLORIDE see DILAUDID

1,2-DIHYDRO-5-NITRO-ACENAPHTHYLENE see 5-NITROACENAPH-THENE

1,3-DIHYDRO-7-NITRO-5-(2-CHLOROPHENYL)-2H-1,4-BENZODIAZE-PIN-2-ONE see CLOAZEPAM

1,3-DIHYDRO-7-NITRO-5-PHENYL-2H-1,4-BENZODIAZEPIN-2-ONE see BENZALIN

DIHYDRO-1-NITROSO-2,4(1H,3H)-PYRIMIDINEDIONE see 1-NITROSO-5,6-DIHYDROURACIL

DIHYDROOXIRENE see ETHYLENE OXIDE

2,3-DIHYDRO-3-OXOBENZISOSULFONAZOLE see 1,2-BENZISOTHIA-ZOL-3(2H)-ONE-1,1-DIOXIDE

3,4-DIHYDRO-4-OXO-3-BENZOTRIAZINYLMETHYL O,O-DIETHYL PHOS-PHORODITHIOATE see ETHYL GUTHION

S-(3,4-DIHYDRO-4-OXO-1,2,3-BENZOTRIAZIN-3-YLMETHYL) O,O-DIETHYL PHOSPHORODITHIOATE see ETHYL GUTHION

S-(3,4-DIHYDRO-4-OXO-1,2,3-BENZOTRIAZIN-3-YLMETHYL)-O,O-DIMETHYL PHOSPHORODITHIOATE see AZINPHOS METHYL

S-(3,4-DIHYDRO-4-OXO-BENZO(ALPHA)(1,2,3)TRIAZIN-3-YLMETHYL)-O,O-DIMETHYL PHOSPHORODITHIOATE see AZINPHOS METHYL

O-(1,6)-DIHYDRO-6-OXO-1-PHENYLPYRIDAZIN-3-LY), O,O-DIETHYL PHOSPHOROTHIOATE see PYRIDAPHENTHION

DIHYDROOXYCODEINONE HYDROCHLORIDE see DIHYDRONE HYDRO-CHLORIDE

DIHYDROPENTABORANE see PENTABORANE (11)

5-(2-(3,6-DIHYDRO-4-PHENYL-1(2H)-PYRIDYL)ETHYL)-3-METHYL-2-OXAZOL IDINONE see AHR-1680

2,3-DIHYDROPHORBOL ACETATE MYRISTATE see 2,3-DIHYDROPHOR-BOL MYRISTATE ACETATE

2,3-DIHYDRO-6-PROPYL-2-THIOXO-4(1H)-PYRIMIDINONE see 6-PRO-PYL-2-THIOURACIL

1,2-DIHYDROPYRIDAZINE-3,6-DIONE see MALEIC HYDRAZIDE

1,2-DIHYDRO-3,6-PYRIDAZINEDIONE see MALEIC HYDRAZIDE

6,7-DIHYDROPYRIDO(1,2-A;2',1'-C)PYRAZINEDIUM DIBROMIDE see DIQUAT DIBROMIDE

6,7-DIHYDRO-1,2,3,10-TETRAMETHOXY-7- (METHYLAMINO)-BENZO-

(ALPHA)HEPTALEN-9(5H)-ONE see N-METHYL-N-DESACETYLCOL-CHICINE

DIHYDROTHEELIN see ESTRADIOL (β)

2,3-DIHYDROTHIIRENE see ETHYLENE SULFIDE

2,3-DIHYDRO-2-THIOXO-4(1H)-PYRIMIDINONE see 2-THIOURACIL

3,7-DIHYDRO-1,3,7-TRIMETHYL-1H-PURINE-2,6-DIONE, MONOHY-DROBROMIDE see CAFFEINE HYDROBROMIDE

1,4-DIHYDROXYANTHRACHINON (CZECH) see 1,4-DIHYDROXY-ANTHRAQUINONE

DIHYDROXY-ANTHRANOL see 1,8,9-ANTHRACENETRIOL

1,8-DIHYDROXYANTHRANOL see 1,8,9-ANTHRACENETRIOL

1,8-DIHYDROXY-9-ANTHRANOL see 1,8,9-ANTHRACENETRIOL

1,4-DIHYDROXY-9,10-ANTHRAQUINONE see 1,4-DIHYDROXYAN-THRAQUINONE

1,8-DIHYDROXY-9-ANTHRONE see 1,8,9-ANTHRACENETRIOL

3,4-DIHYDROXYBENZALDEHYDE METHYLENE KETAL see PIPERONAL

1,4-DIHYDROXY-BENZEEN (DUTCH) see HYDROQUINONE

1,4-DIHYDROXYBENZEN (CZECH) see HYDROQUINONE

DIHYDROXYBENZENE see HYDROQUINONE

M-DIHYDROXYBENZENE see RESORCINOL

O-DIHYDROXYBENZENE see PYROCATECHOL

P-DIHYDROXYBENZENE see HYDROQUINONE

1,2-DIHYDROXYBENZENE see PYROCATECHOL

1,3-DIHYDROXYBENZENE see RESORCINOL

1,4-DIHYDROXYBENZENE see HYDROQUINONE

1,4-DIHYDROXY-BENZOL (GERMAN) see HYDROQUINONE

2,6-DIHYDROXY-5-BIS(2-CHLOROETHYL)AMINOPYRAMIDINE see 5-(BIS(2-CHLOROETHYL)AMINO)URACIL

1,3-DIHYDROXYBUTANE see 1,3-BUTANEDIOL

1,4-DIHYDROXYBUTANE see 1,4-BUTANEDIOL

2,3-DIHYDROXYBUTANE see 2,3-BUTANEDIOL

3,12-DIHYDROXYCHOLANIC ACID see DEOXYCHOLATIC ACID

3-alpha,12-alpha-DIHYDROXYCHOLANIC ACID see DEOXYCHOLATIC ACID

3-alpha,12-alpha-DIHYDROXY-5-beta-CHOLAN-24-OIC ACID see DE-OXYCHOLATIC ACID

3-alpha,12-alpha-DIHYDROXY-5-beta-CHOLANOIC ACID see DEOXY-CHOLATIC ACID

3-alpha,12-alpha-DIHYDROXYCHOLANSAEURE (GERMAN) see DE-OXYCHOLATIC ACID

1a,25-DIHYDROXYCHOLECALCIFEROL see DIHYDROXYVITAMIN D3

DIHYDROXYCODEINONE HYDROCHLORIDE see DIHYDRONE HYDRO-CHLORIDE

DI-(4-HYDROXY-3-COUMARINYL)METHANE see BISHYDROXY COU-MARIN

1,5-DIHYDROXY-4,8-DIAMINOANTHRACHINON (CZECH) see 1,5-DIAMINOANTHRARUFIN

1,5-DIHYDROXY-4,8-DIAMINOANTHRAQUINONE see 1,5-DIAMINOAN-THRARUFIN

2,2'-DIHYDROXY-5,5'-DICHLORODIPHENYLMETHANE see 2,2'-METHYLENEBIS(4-CHLOROPHENOL)

2,2'-DIHYDROXYDIETHYLAMINE see BIS(2-HYDROXY ETHYL)AMINE

4,4'-DIHYDROXY-ALPHA,BETA-DIETHYLDIPHENYLETHANE see DIHY-DROSTILBESTROL

BETA,BETA'-DIHYDROXYDIETHYL ETHER see DIETHYLENE GLYCOL

4,4'-DIHYDROXYDIETHYLSTILBENE see DIETHYLSTILBESTEROL

4,4'-DIHYDROXY-ALPHA,BETA-DIETHYLSTILBENE see DIETHYLSTIL-BESTEROL

DIHYDROXYDIETHYLSTILBENE DIPROPIONATE see DIETHYLSTILBES-TROL DIPROPIONATE

4,4'-DIHYDROXY-ALPHA,BETA-DIETHYLSTILBENE DIPROPIONATE see DIETHYLSTILBESTROL DIPROPIONATE

4,4'-DIHYDROXY-alpha,beta-DIETHYLSTILBENE PALMITATE see DIETHYLSTILBESTROL DIPALMITATE

TRANS-1,2-DIHYDROXY-1,2-DIHYDROBENZ(A)ANTHRACENE see BENZ(a)ANTHRACENE-1,2-DIHYDRODIOL

TRANS-3,4-DIHYDROXY-3,4-DIHYDROBENZ(A)ANTHRACENE see BENZ(a)ANTHRACENE-3,4-DIHYDRODIOL

TRANS-5,6-DIHYDROXY-5,6-DIHYDROBENZ(A)ANTHRACENE see BENZ(a)ANTHRACENE-5,6-DIHYDRODIOL

TRANS-8,9-DIHYDROXY-8,9-DIHYDROBENZ(A)ANTHRACENE see trans-BENZ(a)ANTHRACENE-8,9-DIHYDRODIOL

TRANS-10,11-DIHYDROXY-10,11-DIHYDROBENZ(A)ANTHRACENE see BENZ(a)ANTHRACENE-10,11-DIHYDRODIOL

(−)-trans-7,8-DIHYDROXY-7,8-DIHYDROBENZO(a)PYRENE see BP-7,8-DIHYDRODIOL

(+,−)-trans-7,8-DIHYDROXY-7,8-DIHYDROBENZO(a)PYRENE see BP-7,8-DIHYDRODIOL

(+,−)-trans-7,8-DIHYDROXY-7,8-DIHYDROBENZO(a)PYRENE see BP-7,8-DIHYDRODIOL

D-7',12'-DIHYDROXY-6,6'-DIMETHOXY-2,2',2'-TRIMETHYLTUBOCU-RARANIUM CHLORIDE see TUBOCURARINE HYDROCHLORIDE

3,4-DIHYDROXY-ALPHA-(DIMETHYLAMINOMETHYL) BENZYL ALCO-HOL see N-METHYLEPINEPHRINE

N-(2,4-DIHYDROXY-3,3-DIMETHYLBUTYRYL)-BETA-ALANINE CAL-CIUM see CALCIUM-d-PANTOTHENATE

4,4'-DIHYDROXYDIPHENYLDIMETHYLMETHANE see BISPHENOL A

P,P'-DIHYDROXYDIPHENYLDIMETHYLMETHANE see BISPHENOL A

4,4'-DIHYDROXYDIPHENYLDIMETHYLMETHANE DIGLYCIDYL ETHER see BISPHENOL A DIGLYCIDYL ETHER

P,P'-DIHYDROXYDIPHENYLDIMETHYLMETHANE DIGLYCIDYL ETHER see BISPHENOL A DIGLYCIDYL ETHER

4,4'-DIHYDROXY-GAMMA,DELTA-DIPHENYLHEXANE see DIHYDRO-STILBESTROL

4,4'-DIHYDROXYDIPHENYLPROPANE see BISPHENOL A

P,P'-DIHYDROXYDIPHENYLPROPANE see BISPHENOL A

2,2-(4,4'-DIHYDROXYDIPHENYL)PROPANE see BISPHENOL A

4,4'-DIHYDROXY-2,2-DIPHENYLPROPANE see BISPHENOL A

4,4'-DIHYDROXYDIPHENYL-2,2-PROPANE see BISPHENOL A

2,2'-DIHYDROXY-DI-N-PROPYLNITROSOAMINE see DI(2-HYDROXY-n-PROPYL)AMINE

TRANS-3-ALPHA,4-BETA-DIHYDROXY-1-ALPHA,2-ALPHA-EPOXY-1,2, 3,4-(+)-(1R,2S,3R,4R)-3,4-DIHYDRO-3,4-DIHYDROXY-1,2-EPOXY-BENZ(a)ANTHRACENE see 1,2,3,4-TETRAHYDRO-3-ALPHA,4-BETA-DIHYDROXY-1-alpha,2,alpha-EPOXY BENZ(a)ANTHRACENE (E)

3,17-beta-DIHYDROXY-1,3,5(10)-ESTRATRIENE see ESTRADIOL (β)

3,17-beta-DIHYDROXYESTRA-1,3,5(10)-TRIENE see ESTRADIOL (β)

1,2-DIHYDROXYETHANE see ETHYLENE GLYCOL

DI-BETA-HYDROXYETHOXYETHANE see TRIETHYLENE GLYCOL

DI(2-HYDROXYETHYL)AMINE see BIS(2-HYDROXY ETHYL)AMINE

DI(HYDROXYETHYL) ETHER DINITRATE see DIETHYLENE GLYCOL DINITRATE

3,17-BETA-DIHYDROXY-17-ALPHA-ETHYNYL-1,3,5(10)-ESTRATRIENE see ETHINYL ESTRADIOL

3,17-BETA-DIHYDROXY-17-ALPHA-ETHYNYL-1,3,5(10)-OESTRA-TRIENE see ETHINYL ESTRADIOL

3',6'-DIHYDROXYFLUORAN see FLUORESCEIN

DIHYDROXYFLUORANE see FLUORESCEIN

2,2'-DIHYDROXY-3,3',5,5',6,6'-HEXACHLORODIPHENYLMETHANE see HEXACHLOROPHENE

2,2'-DIHYDROXY-3,5,6,3',5',6'-HEXACHLORODIPHENYLMETHANE see HEXACHLOROPHENE

7-(3,5-DIHYDROXY-2-(3-HYDROXY-1-OCTENYL)CYCLOPENTYL)-5-HEPTENOICACID see PROSTAGLANDIN F2-ALPHA

2,3-DIHYDROXY-2-((4-HYDROXYPHENYL)METHYL)BUTANEDIOIC ACID see (p-HYDROXYBENZYL)TARTARIC ACID

3,4-DIHYDROXY-alpha-((ISOPROPYLAMINO)METHYL)BENZYLALCOHOL see ISOPROPYLADRENALINE

3,4-DIHYDROXY-alpha-((ISOPROPYLAMINO)METHYL)BENZYL ALCO-HOL see ISOPROPYLADRENALINE

3,4-DIHYDROXY-ALPHA-((METHYLAMINO)METHYL)BENZYL ALCOHOL see VASOTONIN

3-DI(HYDROXYMETHYL)AMINO-6-(5-NITRO-2-FURYLETHENYL)-1,2,4-TRIAZINE see BIS(HYDROXY METHYL)FURATRIZINE

1,3-DIHYDROXY-5-METHYLBENZENE see 5-METHYLRESORCINOL

2,12-DIHYDROXY-4-METHYL-11,16-DIOXOSENECIONANIUM see SEN-KIRKINE

DI-4-HYDROXY-3,3'-METHYLENEDICOUMARIN see BISHYDROXY COU-MARIN

1,2-DIHYDROXY-3-(2-METHYLPHENOXY)PROPANE see GLYCERYL-O-
    TOLYL ETHER
ALPHA,BETA-DIHYDROXY-GAMMA-(2-METHYLPHENOXY)PROPANE see
    GLYCERYL-O-TOLYL ETHER
3,4-DIHYDROXYNOREPHEDRINE HYDROCHLORIDE see AMINOPROPA-
    NOL PYROCATECHOLHYDROCHLORIDE
6,6'-((4,8-DIHYDROXY-1,5ANTHRAQUINONYLENE)DIIMINO) DI-m-
    TOLUENE SULFONIC ACID DISODIUM SALT see TOLUIDINE BLUE
3,17-beta-DIHYDROXYOESTRA-1,3,5-TRIENE see ESTRADIOL (β)
3,17-beta-DIHYDROXY-1,3,5(10)-OESTRATRIENE see ESTRADIOL (β)
DIHYDROXYOESTRIN see ESTRADIOL (β)
(5Z,11-ALPHA,13E,15S)-11,15-DIHYDROXY-9-OXOPROSTA-5,13-
    DIEN-1-OIC ACID see DINOPROSTONE
3,5-DIHYDROXYPHENOL see PHLOROGLUCINOL
L-DIHYDROXYPHENYLALANINE see L-DOPA
DIHYDROXY-L-PHENYLALANINE see L-DOPA
(−)-3,4-DIHYDROXYPHENYLALANINE see L-DOPA
3,4-DIHYDROXYPHENYLALANINE see L-DOPA
3,4-DIHYDROXY-L-PHENYLALANINE see L-DOPA
3,4-DIHYDROXYPHENYL-L-ALANINE see L-DOPA
L-3,4-DIHYDROXYPHENYLALANINE see L-DOPA
3-(3,4-DIHYDROXYPHENYL)-L-ALANINE see L-DOPA
L-3-(3,4-DIHYDROXYPHENYL)ALANINE see L-DOPA HYDROCHLORIDE
L-ALPHA-DIHYDROXYPHENYLALANINE see L-DOPA
BETA-(3,4-DIHYDROXYPHENYL)-L-ALANINE see L-DOPA
L-3,4-DIHYDROXYPHENYL-ALPHA-ALANINE see L-DOPA
L-BETA-(3,4-DIHYDROXYPHENYL)ALANINE see L-DOPA
1-(3,4-DIHYDROXYPHENYL)-2-AMINO-1-BUTANOL HYDROCHLORIDE
    see ETHYL NORADRENALINE HYDROCHLORIDE
1-1-(3,4-DIHYDROXYPHENYL)-2-AMINOETHANOL see L-NOREPINEPH-
    RINE
3,4-DIHYDROXYPHENYLAMINOPROPANOL HYDROCHLORIDE see
    AMINOPROPANOL PYROCATECHOLHYDROCHLORIDE
ALPHA-(3,4-DIHYDROXYPHENYL)-BETA-DIMETHYLAMINOETHANOL
    see N-METHYLEPINEPHRINE
1-3,4-DIHYDROXYPHENYLETHANOLAMINE see L-NOREPINEPHRINE
3,4-DIHYDROXYPHENYLETHYLMETHYLAMINE HYDROCHLORIDE see
    EPHININE HYDROCHLORIDE
MESO-3,4-DI(P-HYDROXYPHENYL)-N-HEXANE see DIHYDROSTILBES-
    TROL
GAMMA,DELTA-DI(P-HYDROXYPHENYL)-HEXANE see DIHYDROSTIL-
    BESTROL
3,4'(4,4'-DIHYDROXYPHENYL)HEX-3-ENE see DIETHYLSTILBESTEROL
1-(3,4-DIHYDROXYPHENYL)-1-HYDROXY-2-AMINOBUTANE HYDRO-
    CHLORIDE see ETHYL NORADRENALINE HYDROCHLORIDE
ALPHA-(3,4-DIHYDROXYPHENYL)-ALPHA-HYDROXY-BETA-DIMETHYL-
    AMINOETHANE see N-METHYLEPINEPHRINE
7-(3-(2-(3,5-DIHYDROXYPHENYL-2-HYDROXY-ETHYLAMINO)PROPYL)-
    THEOPHYLLINE HYDROCHLORIDE see BRONCHOSPASMIN
DL-alpha-3,4-DIHYDROXYPHENYL-beta-ISOPROPYLAMINOETHANOL
    SULFATE see (±)-ISOPROTERENOL SULFATE
L-(−)-3-(3,4-DIHYDROXYPHENYL)-2-METHYLALANINE see ALDOMET
L(−)-BETA-(3,4-DIHYDROXYPHENYL)-ALPHA-METHYLALANINE see
    ALDOMET
3,4-DIHYDROXYPHENYL-1-METHYLAMINO-2-ETHANE HYDROCHLO-
    RIDE see EPHININE HYDROCHLORIDE
1-1-(3,4-DIHYDROXYPHENYL)-2-METHYLAMINOETHANOL see VASO-
    TONIN
1-1-(3,4-DIHYDROXYPHENYL)-2-METHYLAMINO-1-ETHANOL HYDRO-
    CHLORIDE see 1-ADRENALINE CHLORIDE
2,2-DI(4-HYDROXYPHENYL)PROPANE see BISPHENOL A
BETA-DI-p-HYDROXYPHENYLPROPANE see BISPHENOL A
3,4-DIHYDROXYPHENYLPROPANOLAMINE HYDROCHLORIDE see
    AMINOPROPANOL PYROCATECHOLHYDROCHLORIDE
17,21-DIHYDROXYPREGNA-1,4-DIENE-3,11,20-TRIONE see PREDNI-
    SONE
1,2-DIHYDROXYPROPANE see 1,2-PROPANEDIOL
17R,21-ALPHA-DIHYDROXY-4-PROPYLAJMALANIUM HYDROGEN TAR-
    TRATE see N-PROPYLAJMALINE BITARTRATE
2,3-DIHYDROXYPROPYL CHLORIDE see CHLORHYDRIN

2,4-DIHYDROXY-2H-PYRAN-delta-3(6H),alpha-ACETIC ACID-3,4-LAC-
    TONE see CLAVACIN
(2,4-DIHYDROXY-2H-PYRAN-3(6H)-YLIDENE)ACETIC ACID-3,4-LAC-
    TONE see CLAVACIN
4,8-DIHYDROXYQUINALDINIC ACID see 4,8-DIHYDROXYQUINALDIC
    ACID
4,8-DIHYDROXYQUINOLINE-2-CARBOXYLIC ACID see 4,8-DIHY-
    DROXYQUINALDIC ACID
8,8'-DIHYDROXY-RUGULOSIN see LUTEOSKYRIN
12,18-DIHYDROXY-SENECIONAN-11,16-DIONE see RETRORSINE
3,5-DIHYDROXYTOLUENE see 5-METHYLRESORCINOL
12-BETA,13-BETA-DIHYDROXY-12-ALPHA,13-ALPHA,14-ALPHA-
    TRIMETHYLCROTAL-1-ENINE see MONOCROTALINE
1-ALPHA-DIHYDROXYVITAMIN D3 see 1-HYDROXYCHOLECALCIFEROL
1,4-DIIDROBENZENE (ITALIAN) see HYDROQUINONE
1,3-DIIMINOISOINDOLIN (CZECH) see 1,3-DIIMINOISOINDOLINE
1,3-DIIMINOISOINDOLINE see 1,3-DIIMINOISOINDOLINE
3,5-DIIODO-4-HYDROXYBENZONITRILE see 4-HYDROXY-3,5-DIIODO-
    BENZONITRILE
3,5-DIIODO-4-HYDROXYBENZONITRILE OCTANOATE see IOXYNIL
    OCTANOATE
DIIODOHYDROXYQUINOLINE see DIIODOHYDROXYQUIN
5,7-DIIODO-8-HYDROXYQUINOLINE see DIIODOHYDROXYQUIN
3,5-DIIODO-4-OCTANOYLOXYBENZONITRILE see IOXYNIL OCTANO-
    ATE
5,7-DIIODO-OXINE see DIIODOHYDROXYQUIN
5,7-DIIODO-8-QUINOLINOL see DIIODOHYDROXYQUIN
O,O-DIISOPROPYL DITHIOPHOSPHORIC ACID ESTER OF-N,N-S-
    DIETHYLTHIOCARBAMOYL-O,O-DIISOPROPYL PHOSPHORO-
    THIOATE see DIETHYLDITHIOCARBAMIC ANHYDROSULFIDE
DIIRON TRISULFATE see FERRIC SULFATE
DIISOBUTILCHETONE (ITALIAN) see DIISOBUTYL KETONE
DIISOBUTYLALUMINUM CHLORIDE see CHLORO DIISOBUTYL ALU-
    MINUM
DIISOBUTYLALUMINUM MONOCHLORIDE see CHLORO DIISOBUTYL
    ALUMINUM
DI-ISOBUTYLCETONE (FRENCH) see DIISOBUTYL KETONE
DIISOBUTYLCHLOROALUMINUM see CHLORO DIISOBUTYL ALUMINUM
DIISOBUTYLKETON (DUTCH, GERMAN) see DIISOBUTYL KETONE
DIISOBUTYL KETONE (DOT) see DIISOBUTYL KETONE
DIISOBUTYLPHENOXYETHOXYETHYLDIMETHYL BENZYL AMMONIUM
    CHLORIDE see BENZETHONIUM CHLORIDE
DIISOBUTYLTIN OXIDE see DIISOBUTYLOXOSTANNANE
4-4'-DIISOCYANATE OF DIPHENYLMETHANE (FRENCH) see DIPHENYL
    METHANE DIISOCYANATE
DI-ISOCYANATE DE TOLUYLENE (FRENCH) see DI-ISO-CYANATOL-
    UENE
DI-ISOCYANATE DE TOLUYLENE (FRENCH) see TOLUENE DIISOCYA-
    NATE
4,4'-DIISOCYANATO-3,3'-DIMETHOXY-1,1'-BIPHENYL see
    DIANISIDINE DIISOCYANATE
4,4'-DIISOCYANATODIPHENYLMETHANE see DIPHENYL METHANE
    DIISOCYANATE
2,4-DIISOCYANATO-1-METHYL BENZENE see TOLUENE DIISOCYA-
    NATE
2,4-DIISOCYANATO-1-METHYL BENZENE (9CL) see TOLUENE DIISO-
    CYANATE
DIISOCYANATOTOLUENE see TOLUENE DIISOCYANATE
2,4-DIISOCYANATOTOLUENE see DI-ISO-CYANATOLUENE
DIISOCYANAT-TOLUOL (GERMAN) see DI-ISO-CYANATOLUENE
DIISOCYANAT-TOLUOL (GERMAN) see TOLUENE DIISOCYANATE
2,3-DIISONITROSOBUTANE see DIACETYL DIOXIME
DIISOOCTYL ((DIOCTYLSTANNYLENE)DITHIO)DIACETATE see BIS(ISO-
    OCTYL OXYCARBONYL METHYL THIO)DIOCTYL STANNANE
DIISOPHENOL see 2,6-DIIODO-4-NITROPHENOL
DIISOPROPANOLNITROSAMINE see DI(2-HYDROXY-n-PROPYL)AMINE
N,N'-DIISOPROPIL-FOSFORODIAMMIDO-FLUORURO (ITALIAN) see
    PHOSPHORODI((ISOPROPYLAMIDIC) FLUORIDE
DIISOPROPOXYPHOSPHORYL FLUORIDE see ISOPROPYL PHOSPHORO-
    FLUORIDATE
S-DIISOPROPYLACETONE see DIISOBUTYL KETONE

DI(ISOPROPYLAMIDO)PHOSPHORYLFLUORIDE see PHOSPHORODI(ISO-PROPYLAMIDIC) FLUORIDE

N,N'-DIISOPROPYL-DIAMIDO-FOSFORZUUR-FLUORIDE (DUTCH) see PHOSPHORODI(ISOPROPYLAMIDIC) FLUORIDE

N,N'-DIISOPROPYL-DIAMIDO-PHOSPHORSAEURE-FLUORID (GERMAN) see PHOSPHORODI(ISOPROPYLAMIDIC) FLUORIDE

N,N'-DIISOPROPYLDIAMIDOPHOSPHORYL FLUORIDE see PHOSPHO-RODI(ISOPROPYLAMIDIC) FLUORIDE

O,O-DIISOPROPYL-S-DIETHYLDITHIOCARBAMOYLPHOSPHORODITHIO-ATE see DIETHYLDITHIOCARBAMIC ANHYDROSULFIDE

DIISOPROPYL ESTER OF DITHIOCARBAMYL PHOSPHOROTHIOIC ACID see DIETHYLDITHIOCARBAMIC ANHYDROSULFIDE

DIISOPROPYL FLUOROPHOSPHATE see ISOPROPYL PHOSPHOROFLUORI-DATE

O,O-DIISOPROPYL FLUOROPHOSPHATE see ISOPROPYL PHOSPHORO-FLUORIDATE

DIISOPROPYL FLUOROPHOSPHONATE see ISOPROPYL PHOSPHORO-FLUORIDATE

DIISOPROPYLFLUOROPHOSPHORICACID ESTER see ISOPROPYL PHOS-PHOROFLUORIDATE

DIISOPROPYLIDENE ACETONE see PHORONE

SYM-DIISOPROPYLIDENE ACETONE see PHORONE

DIISOPROPYL-P-NITROPHENYL PHOSPHATE see DIISOPROPYL PARA-OXON

DIISOPROPYL OXIDE see DIISOPROPYL ETHER

DIISOPROPYL PHOSPHOFLUORIDATE see ISOPROPYL PHOSPHORO-FLUORIDATE

N,N'-DIISOPROPYLPHOSPHORODIAMIDIC FLUORIDE see PHOSPHO-RODI(ISOPROPYLAMIDIC) FLUORIDE

DIISOPROPYL PHOSPHOROFLUORIDATE see ISOPROPYL PHOSPHORO-FLUORIDATE

O,O'-DIISOPROPYL PHOSPHORYL FLUORIDE see ISOPROPYL PHOS-PHOROFLUORIDATE

DI-ISOPROPYLSULFAT (GERMAN) see DIISOPROPYL ESTER SULFURIC ACID

DI-ISOPROPYLSULFATE see DIISOPROPYL ESTER SULFURIC ACID

DI-ISOPROPYLTHIOLOCARBAMATE DE S-(2,3-DICHLOROALLYLE) (FRENCH) see DIALLATE

DIISOPROPYLTIN OXIDE see DIISOPROPYLOXOSTANNANE

O,O-DIISOPROPYL-O,p-NITROPHENYL PHOSPHATE see DIISOPROPYL PARAOXON

1,4-DIISOTHIOCYANATOBENZENE see PHENYLENE THIOCYANATE

3,5-DIJOD-4-HYDROXY-BENZONITRIL (GERMAN) see 4-HYDROXY-3,5-DIIODOBENZONITRILE

3,5-DIJOD-4-HYDROXY-BENZONITRIL CAPRYSAEUREESTER (GERMAN) see IOXYNIL OCTANOATE

2,3-DIKETOBUTANE see 2,3-BUTANEDIONE

1,3-DIKETOHYDRINDENE see 1,3-INDANDIONE

DIKETONE ALCOHOL see 2-METHYL-2-PENTANOL-4-ONE

2,5-DIKETOTETRAHYDROFURAN see SUCCINIC ANHYDRIDE

DILANTIN see DIHYDANTOIN

DILANTIN SODIUM see DILANTIN

DILAUDID HYDROCHLORIDE see DIHYDROMORPHINE HYDROCHLORIDE

DILAUROYL PEROXIDE see LAUROYL PEROXIDE

DILITHIUM CARBONATE see LITHIUM CARBONATE (2:1)

1,3-DIMALEIMIDOBENZENE see 1,3-BISMALEIMIDO BENZENE

DIMANGANESE TRIOXIDE see MANGANESE (III) OXIDE

DIMEMORFAN PHOSPHATE see d-3-METHYL-N-METHYLMORPHINAN PHOSPHATE

DIMENFORMON BENZOATE see ESTRADIOL-3-BENZOATE

DIMENFORMON DIPROPIONATE see ESTRADIOL DIPROPIONATE

DIMENFORMON PROLONGATUM see ESTRADIOL (β)

DIMERCAPROL PROPANOL see BAL

DIMERCAPTOL see BAL

2,3-DIMERCAPTOL-1-PROPANOL see BAL

4,5-DI(MERCAPTOMETHYL)-2-METHYL-3-PYRIDINOL DITHIOACETATE HYDROBROMIDE see 3,4-DI(ACETYLTHIOMETHYL)-5-HYDROXY-6-METHYLPYRIDINE HYDROBROMIDE

2,3-DIMERCAPTOPROPANOL see BAL

2,3-DIMERCAPTOPROPAN-1-OL see BAL

4,5-DIMERCAPTOPYRIDOXINDI-THIOACETAT HYDROBROMID

(GERMAN) see 3,4-DI(ACETYLTHIOMETHYL)-5-HYDROXY-6-METHYLPYRIDINE HYDROBROMIDE

DIMER CYKLOPENTADIENU (CZECH) see BICYCLOPENTADIENE

1,6-DIMESYL-d-MANNITOL see BIS(METHANE SULFONYL)-D-MANNI-TOL

DIMETACRINE BITARTRATE see 9,9-DIMETHYL-10-DIMETHYLAMINOPRO-PYLACRIDAN HYDROGEN TARTRATE

DIMETACRIN HYDROGEN TARTRATE see 9,9-DIMETHYL-10-DIMETHYL-AMINOPROPYLACRIDAN HYDROGEN TARTRATE

DIMETHACRINE TARTRATE see 9,9-DIMETHYL-10-DIMETHYLAMINOPRO-PYLACRIDAN HYDROGEN TARTRATE

DIMETHACHLON see N-(3,5-DICHLOROPHENYL)SUCCINIMIDE

2,5-DIMETHANESULFOMYLOXYHEXANE see DIMETHYLMYLERAN

1,6-DIMETHANESULFONATE-d-MANNITOL see BIS(METHANE SUL-FONYL)-D-MANNITOL

1,4-DIMETHANESULFONOXYBUTANE see 1,4-BUTANEDIOL DIMETHYL SULFONATE

1,4-DIMETHANESULFONOXY-1,4-DIMETHYLBUTANE see DIMETHYL-MYLERAN

1,6-DIMETHANE-SULFONOXY-d-MANNITOL see BIS(METHANE SULFO-NYL)-D-MANNITOL

1,4-DI(METHANESULFONYLOXY)BUTANE see 1,4-BUTANEDIOL DIMETHYL SULFONATE

1,6-DIMETHANESULPHONOXY-1,6-DIDEOXY-d-MANNITOL see BIS(METHANE SULFONYL)-D-MANNITOL

1,4-DIMETHANESULPHONYLOXYBUTANE see 1,4-BUTANEDIOL DIMETHYL SULFONATE

DIMETHOAAT (DUTCH) see O,O-DIMETHYL METHYLCARBAMOYLMETHYL PHOSPHORODITHIOATE

DIMETHOATE see O,O-DIMETHYL METHYLCARBAMOYLMETHYL PHOS-PHORODITHIOATE

DIMETHOXANE see ACETOMETHOXANE

1,2-DIMETHOXY-4-ALLYLBENZENE see 4-ALLYL-1,2-DIMETHOXY-BENZENE

2,5-DIMETHOXY-4'-AMINOSTILBENE(TRANS) see 2',5'-DIMETHOXY-STILBENAMINE

2,5-DIMETHOXYAMPHETAMINE HYDROCHLORIDE see 1-(2,5-DIMETH-OXYPHENYL)-2-AMINOPROPANE

3,4-DIMETHOXYAMPHETAMINE HYDROCHLORIDE see 1-(3,4-DIMETH-OXYPHENYL)-2-AMINOPROPANE

1-5, 6-DIMETHOXYAPORPHINE see 1-NUCIFERINE

(R)-1,2-DIMETHOXYAPORPHINE see 1-NUCIFERINE

AR,AR-DIMETHYLBENZOPHENONE see PHENYL XYLYL KETONE

1,2-DIMETHOXY-6a-BETA-APORPHINE see 1-NUCIFERINE

3,3'-DIMETHOXYBENZIDIN (CZECH) see O-DIANISIDINE

3,3'-DIMETHOXYBENZIDINE see O-DIANISIDINE

3,3'-DIMETHOXYBENZIDINE-4,4'-DIISOCYANATE see DIANISIDINE DIISOCYANATE

1-(3,4)-DIMETHOXYBENZYL-6,7-DIMETHOXYISOQUINOLINE-3-CAR-BOXYLIC ACID, SODIUM SALT see PAPAVERIN CARBOXYLIC ACID, SODIUM SALT

3,3'-DIMETHOXY-4,4'-BIPHENYLENE DIISOCYANATE see DIANISIDINE DIISOCYANATE

6,7-DIMETHOXYCOUMARIN see 6,7-DIMETHYLESCULETIN

4,4'-DIMETHOXY-ALPHA,BETA-DIETHYLSTILBENE see 3,4-DIANISYL-3-HEXENE

2,5-DIMETHOXY-ALPHA,4-DIMETHYLPHENETHYLAMINE HYDROCHLO-RIDE see 2,5-DIMETHOXY-4-METHYLAMPHETAMINE HYDROCHLO-RIDE

P,P'-DIMETHOXYDIPHENYLTRICHLOROETHANE see DIMETHOXY-DDT

9,10-DIMETHOXY-2,3-(METHYLENEDIOXY)-7,8,13,13A-TETRAHY-DROBERBERINIUM see BERBERINE

5,8-DIMETHOXY-2-METHYL-6,7-FURANOCHROMONE see AMICARDINE

5,8-DIMETHOXY-2-METHYL-4',5'-FURANO-6,7-CHROMONE see AMI-CARDINE

5,8-DIMETHOXY-2-METHYL-4',5'-FURO-6,7-CHROMONE see AMICAR-DINE

4,9-DIMETHOXY-7-METHYL-5H-FURO(3,2-G)(1)BENZOPYRAN-5-ONE see AMICARDINE

4,9-DIMETHOXY-7-METHYL-5-OXO-1,8-DIOXABENZ-(F)INDENE see AMICARDINE

4,9-DIMETHOXY-7-METHYL-5-OXOFURO(3,2-G)(1)BENZOPYRAN see AMICARDINE

4,9-DIMETHOXY-7-METHYL-5-OXOFURO(3,2-G)-1,2-CHROMENE see AMICARDINE

2,5-DIMETHOXY-alpha-METHYL-beta-PHENYLETHYLAMINEHYDRO-CHLORIDE see 1-(2,5-DIMETHOXYPHENYL)-2-AMINOPROPANE

3,4-DIMETHOXY-ALPHAMETHYL-BETA-PHENYLETHYLAMINEHYDRO-CHLORIDE see 1-(3,4-DIMETHOXYPHENYL)-2-AMINOPROPANE

3,4-DIMETHOXYPHENETHYLAMINE see 3,4-DIMETHOXYDOPAMINE

3,4-DIMETHOXY-BETA-PHENETHYLAMINE see 3,4-DIMETHOXYDOPA-MINE

4-(2,5-DIMETHOXYPHENETHYL)ANILINE see 2',5'-DIMETHOXYSTIL-BENAMINE

5-((3,4-DIMETHOXYPHENETHYL)METHYLAMINO)-2-(3,4-DIMETHOXY-PHENYL)-2-ISOPROPYLVALERONITRILE see ISOPTIN

DIMETHOXYPHENYLETHYLAMINE see 3,4-DIMETHOXYDOPAMINE

3,4-DIMETHOXYPHENYLETHYLAMINE see 3,4-DIMETHOXYDOPAMINE

2-(3,4-DIMETHOXYPHENYL)ETHYLAMINE see 3,4-DIMETHOXYDOPA-MINE

3,4-DIMETHOXY-BETA-PHENYLETHYLAMINE see 3,4-DIMETHOXYDOP-AMINE

BETA-(3,4-DIMETHOXYPHENYL)ETHYLAMINE see 3,4-DIMETHOXY-DOPAMINE

3,4-DIMETHOXYPHENYLETHYLAMINE (BASE) see 3,4-DIMETHOXY-DOPAMINE

3,4-DIMETHOXY-BETA-PHENYLETHYLAMINE HYDROCHLORIDE see 3,4-DIMETHOXYPHENETHYLAMINE HYDROCHLORIDE

1-((3,4-DIMETHOXYPHENYL)METHYL)-6,7-DIMETHOXYISOQUINOLINE see PAPAVERINE

1-(3,4-DIMETHOXYPHENYL)-2-PROPENE see 4-ALLYL-1,2-DIMETH-OXYBENZENE

DI(P-METHOXYPHENYL)-TRICHLOROMETHYL METHANE see DIMETH-OXY-DDT

3-(DIMETHOXYPHOSPHINYLOXY)-N,N-DIMETHYL-CIS-CROTONAMIDE see 3-HYDROXYDIMETHYL CROTONAMIDE DIMETHYL PHOSPHATE

3-(DIMETHOXYPHOSPHINYLOXY)-N,N DIMETHYLISOCROTONAMIDE see 3-HYDROXYDIMETHYL CROTONAMIDE DIMETHYL PHOSPHATE

4-(2,5-DIMETHOXY)STILBENAMINE see 2',5'-DIMETHOXYSTILBEN-AMINE

DIMETHOXY STRYCHNINE see BRUCINE

DIMETHOXY-2,2,2-TRICHLORO-1-N-BUTYRYLOXY-ETHYLPHOSPHINE OXIDE see BUTONATE

DIMETHOXY-2,2,2-TRICHLORO-1-HYDROXY-ETHYL-PHOSPHINE OXIDE see ((2,2,2-TRICHLORO-1-HYDROXYETHYL) DIMETHYLPHOSPHO-NATE)

6,7-DIMETHOXY-1-VERATRYLISOQUINOLINE see PAPAVERINE

6,7-DIMETHOXY-1-VERATRYLISOQUINOLINE-3-CARBOXYLIC ACID SO-DIUM SALT see PAPAVERIN CARBOXYLIC ACID, SODIUM SALT

DIMETHOXYVIOLANTHRONE see JADE GREEN BASE

16,17-DIMETHOXYVIOLANTHRONE see JADE GREEN BASE

DIMETHYL see ETHANE

DIMETHYLACETIC ACID see ISOBUTYRIC ACID

DIMETHYLACETONE see 3-PENTANONE

DIMETHYLACETONE AMIDE see N,N-DIMETHYLACETAMIDE

O,O-DIMETHYL S-(2-ACETYLAMINOETHYL) PHOSPHORODITHIOATE see O,O-DIMETHYL-S-(2-(ACETYLAMINO)ETHYL) DITHIOPHOSPHATE

3,3-DIMETHYL-ACRYLATE DE 2,4-DINITRO-6-(1-METHYLPROPYLE) PHENYLE (FRENCH) see BINAPACRYL

O,O-DIMETHYL-S-(2-AETHYLTHIO-AETHYL)-DITHIO PHOSPHAT (GER-MAN) see PHOSPHORODITHIOIC ACID-O,O-DIMETHYL-S-(2-ETHYLTHIO)ETHYL ESTER

O,O-DIMETHYL-O-(2-AETHYLTHIO-AETHYL MONOTHIOPHOSPHAT (GERMAN) see DEMETON-O-METHYL

O,O-DIMETHYL-S-(2-AETHYLTHIO-AETHYL)-MONOTHIOPHOSPHAT (GERMAN) see DEMETON-S-METHYL

2-(3,3-DIMETHYLALLYL)-2',2'-HYDROXY-5,9-DIMETHYL-6,7-BENZO-MORPHAN see 2-(3,3-DIMETHYLALLYL)CYCLAZOCINE

DIMETHYLAMIDE ACETATE see N,N-DIMETHYLACETAMIDE

DIMETHYLAMIDOETHOXYPHOSPHORYL CYANIDE see ETHYL DIMETHYL AMIDO CYANO PHOSPHATE

DIMETHYLAMINE AQUEOUS SOLUTION (DOT) see DIMETHYLAMINE

DIMETHYLAMINE BENZHYDRYL ESTER HYDROCHLORIDE see BENA-DRYL HYDROCHLORIDE

N',N'-DIMETHYL-4'-AMINO-N-ACETYL-N-MONOMETHYL-4-AMINO-AZOBENZENE see 4-((p-(DIMETHYLAMINO)PHENYL)AZO)-N-METHYLACETANILIDE

BETA-DIMETHYLAMINO-AETHYL-BENZHYDRYL-AETHER (GERMAN) see BENZHYDRYL

N-DIMETHYLAMINO-AETHYL-N-P-METHOXY-BENZYL-ALPHA-AMINO-PYRIDIN-MALEAT (GERMAN) see WAIT'S GREEN MOUNTAIN ANTIHISTAMINE

5-(DIMETHYLAMINOAETHYL-OXYIMINO)-5H-DIBENZO(A,D)CYCLO-HEPTA-1,4-DIENHYDROCHLORID (GERMAN) see 5-DIMETHYL-AMINOETHYLOXYIMINO-5H-DIBENZO(a,d)CYCLOHEPTA-1,4-DIENE HYDROCHLORIDE

4-(DIMETHYLAMINO)ANTIPYRINE see DIMETHYLAMINOANTIPYRINE

P-DIMETHYLAMINOAZOBENZEN (CZECH) see N,N-DIMETHYL-p-PHENYLAZOANILINE

DIMETHYLAMINOAZOBENZENE see N,N-DIMETHYL-p-PHENYLAZO-ANILINE

4-DIMETHYLAMINOAZOBENZENE see N,N-DIMETHYL-p-PHENYLAZO-ANILINE

P-DIMETHYLAMINOAZOBENZENE see N,N-DIMETHYL-p-PHENYLAZO-ANILINE

4-(N,N-DIMETHYLAMINO)AZOBENZENE see N,N-DIMETHYL-p-PHE-NYLAZOANILINE

N,N-DIMETHYL-4-AMINOAZOBENZENE see N,N-DIMETHYL-p-PHE-NYLAZOANILINE

N,N-DIMETHYL-P-AMINOAZOBENZENE see N,N-DIMETHYL-p-PHENYL-AZOANILINE

2',3-DIMETHYL-4-AMINOAZOBENZENE see 2-AMINO-5-AZOTOLUENE

4-DIMETHYLAMINOAZOBENZENE AMINE-N-OXIDE see N,N-DIMETHYL-p-PHENYLAZOANILINE-N-OXIDE

4'-DIMETHYLAMINOAZOBENZENE-2-CARBOXYLIC ACID see 2-CAR-BOXY-4'-(DIMETHYL AMINO)AZOBENZENE

N,N-DIMETHYLAMINOAZOBENZENE-N-OXIDE see N,N-DIMETHYL-p-PHENYLAZOANILINE-N-OXIDE

DIMETHYLAMINOAZOBENZOL see N,N-DIMETHYL-p-PHENYLAZOANI-LINE

4-DIMETHYLAMINOAZOBENZOL see N,N-DIMETHYL-p-PHENYLAZO-ANILINE

P-DIMETHYLAMINO-AZOBENZOL (GERMAN) see N,N-DIMETHYL-p-PHENYLAZOANILINE

DIMETHYLAMINOAZOPHENE see DIMETHYLAMINOANTIPYRINE

5-(P-DIMETHYLAMINOBENZAL)RHODANINE see p-DIMETHYLAMINO-BENZALRHODANINE

P-(DIMETHYLAMINO)BENZAL-5-RHODANINE see p-DIMETHYLAMINO-BENZALRHODANINE

(DIMETHYLAMINO)BENZENE see N,N-DIMETHYLANILINE

P-DIMETHYLAMINOBENZENEAZO-1-NAPHTHALENE see N,N-DIMETHYL-p-(1-NAPHTHYLAZO)ANILINE

P-DIMETHYLAMINOBENZENE-1-AZO-1-NAPHTHALENE see N,N-DIMETHYL-p-(1-NAPHTHYLAZO)ANILINE

P-DIMETHYLAMINOBENZENE DIAZO SODIUM SULFONATE see METHYL ORANGE

P-DIMETHYLAMINOBENZENEDIAZOSODIUM SULPHONATE see METHYL ORANGE

P-(DIMETHYLAMINO)BENZENEDIAZOSULFONATE see METHYL ORANGE

P-(DIMETHYLAMINO)BENZENEDIAZOSULPHONATE see METHYL ORANGE

P-DIMETHYLAMINOBENZOLDIAZOSULFONAT (NATRIUMSALZ) (GERMAN) see METHYL ORANGE

5-(P-DIMETHYLAMINOBENZOYLIDENE)RHODANINE see p-DIMETHYL-AMINOBENZALRHODANINE

(4-DIMETHYLAMINOBENZYLIDENE)INDENE see 1-(4-DIMETHYL-AMINOBENZAL)INDENE

P-DIMETHYLAMINOBENZYLIDENE RHODAMINE see p-DIMETHYL-AMINOBENZALRHODAMINE

N-DIMETHYL AMINO-BETA-CARBAMYL PROPIONIC ACID see DIMETHYLAMINOSUCCINAMIC ACID

3-BETA-DIMETHYLAMINOCON-5 -ENEDIHYDROBROMIDE see NERIINE

4-DIMETHYLAMINO-3-CRESYL METHYLCARBAMATE see 4-DIMETHYLAMINE-m-CRESYLMETHYLCARBAMATE

4-(DIMETHYLAMINO)-3,5-DIMETHYLPHENOL METHYLCARBAMATE (ESTER) see 4-(DIMETHYLAMINE)-3,5-XYLYL N-METHYL-CARBAMATE

4-(DIMETHYLAMINO)-3,5-DIMETHYLPHENYL N-METHYLCARBAMATE see 4-(DIMETHYLAMINE)-3,5-XYLYL N-METHYLCARBAMATE

4-DIMETHYLAMINO-2,3-DIMETHYL-1-PHENYL-3-PYRAZOLIN-5-ONE see DIMETHYLAMINOANTIPYRINE

4-DIMETHYLAMINO-2,3-DIMETHYL-1-PHENYL-5-PYRAZOLONE see DIMETHYLAMINOANTIPYRINE

3-DIMETHYLAMINO-1,2-DIMETHYLPROPYL P-AMINOBENZOATE HYDROCHLORIDE see p-AMINOBENZOYLDIMETHYLAMINO-1,2-DIMETHYLPROPANOL HYDROCHLORIDE

3,2'-DIMETHYL-4-AMINODIPHENYL see (m,o'-BITOLYL)-4-AMINE

6-DIMETHYLAMINO-4,4-DIPHENYL-3-HEPTANONE HYDROCHLORIDE see METHADONE HYDROCHLORIDE

1-6-DIMETHYLAMINO-4,4-DIPHENYL-3-HEPTANONE HYDROCHLORIDE see L-METHADONE HYDROCHLORIDE

P,P-DIMETHYLAMINODIPHENYLMETHANE see MICHLER'S BASE

ALPHA-(+)-4-DIMETHYLAMINO-1,2-DIPHENYL-3-METHYL-2-BUTANOL PROPIONATE ESTER see CARVON

BETA-DIMETHYLAMINOETHANOL DIPHENYLMETHYL ETHER see BENZHYDRYL

1-(BETA-DIMETHYLAMINOETHOXY)-3-N-BUTYLISOQUINOLINE MONOHYDROCHLORIDE see DIMETHISOQUIN HYDROCHLORIDE

ALPHA-(2-DIMETHYLAMINOETHOXY)DIPHENYLMETHANE see BENZHYDRYL

BETA-DIMETHYLAMINOETHYLBENZHYDRYLETHER see BENZHYDRYL

BETA-DIMETHYLAMINOETHYL BENZHYDRYL ETHER HYDROCHLORIDE see BENADRYL HYDROCHLORIDE

DIMETHYLAMINOETHYL BENZILATE, HYDROCHLORIDE see BENZACINE HYDROCHLORIDE

2-(DIMETHYLAMINO)ETHYL BENZILATE HYDROCHLORIDE see BENZACINE HYDROCHLORIDE

BETA-DIMETHYLAMINOETHYL BENZILATE HYDROCHLORIDE see BENZACINE HYDROCHLORIDE

DIMETHYLAMINOETHYL BENZYLATE HYDROCHLORIDE see BENZACINE HYDROCHLORIDE

2-(ALPHA-(2-DIMETHYLAMINOETHYL)BENZYL)PYRIDINE see TRIMETON

BETA-DIMETHYLAMINOETHYL-P-BROMO-ALPHA-METHYLBENZHYDRYL ETHER HYDROCHLORIDE see BROMADRYL

DIMETHYLAMINOETHYL CHLORIDE see N-(2-CHLORO ETHYL)-DIMETHYL AMINE

2-DIMETHYLAMINOETHYLCHLORIDE see N-(2-CHLORO ETHYL)-DIMETHYL AMINE

beta-(DIMETHYLAMINO)ETHYL CHLORIDE see N-(2-CHLORO ETHYL)-DIMETHYL AMINE

10-(2-(DIMETHYLAMINO)ETHYL)-5,10-DIHYDRO-5-METHYL-11H-DIBENZO(B,E)(1,4)DIAZEPIN-11-ONE see DIBENZEPIN

DIMETHYLAMINOETHYLDIPHENYLHYDROXYACETATEHYDROCHLORIDE see BENZACINE HYDROCHLORIDE

2-DIMETHYLAMINOETHYL-2METHYLBENZHYDRYL ETHERHYDROCHLORIDE see ORPHENADRINE HYDROCHLORIDE

3-(BETA-DIMETHYLAMINOETHYL)-5-HYDROXYINDOLE see 3-(2-DIMETHYLAMINOETHYL)-5-INDOLOL

3-(2-(DIMETHYLAMINO)ETHYL)INDOL-4-OL see 4-HYDROXY-N,N-DIMETHYLTRYPTAMINE

3-2'-DIMETHYLAMINOETHYLINDOL-4-PHOSPHATE see PSILOCYBIN

3-(2-DIMETHYLAMINOETHYL)INDOL-4-YL DIHYDROGEN PHOSPHATE see PSILOCYBIN

2-DIMETHYLAMINOETHYL METHACRYLATE see DIMETHYLAMINOETHYL METHACRYLATE

2-(DIMETHYLAMINO)ETHYL METHACRYLATE see DIMETHYLAMINOETHYL METHACRYLATE

N,N-DIMETHYLAMINOETHYL METHACRYLATE see DIMETHYLAMINOETHYL METHACRYLATE

BETA-DIMETHYLAMINOETHYL METHACRYLATE see DIMETHYLAMINOETHYL METHACRYLATE

N-DIMETHYLAMINOETHYL-N-P-METHOXY-alpha-AMINOPYRIDINE MALEATE see DIAMINIDE MALEATE

2-((2-(DIMETHYLAMINO)ETHYL)-(P-METHOXYBENZYL)AMINO)PYRIDINE see WAIT'S GREEN MOUNTAIN ANTIHISTAMINE

2-((2-(DIMETHYLAMINO)ETHYL)(P-METHOXYBENZYL)AMINO)PYRIDINE BIMALEATE see DIAMINIDE MALEATE

2-((2-(DIMETHYLAMINO)ETHYL)(P-METHOXYBENZYL)AMINO)PYRIDINE MALEATE see DIAMINIDE MALEATE

2-DIMETHYLAMINOETHYL-2-METHYL-BENZHYDRYL ETHER CITRATE see ORPHENADRINE CITRATE

DIMETHYLAMINOETHYL-PBUTYL-AMINOBENZOATE see p-(BUTYL AMINO)BENZOIC ACID-2(DIMETHYL AMINO)ETHYL ESTER

2-DIMETHYLAMINOETHYL-PBUTYLAMINOBENZOATE see p-(BUTYL AMINO)BENZOIC ACID-2(DIMETHYL AMINO)ETHYL ESTER

BETA-DIMETHYLAMINO ETHYL-2-PYRIDYL AMINO TOLUENE see TRIPELENNAMINE

2-DIMETHYLAMINOETHYL SUCCINATE DIMETHOCHLORIDE see SUCCINOYLCHOLINE CHLORIDE

2-((2-(DIMETHYLAMINO)ETHYL)-2-THENYLAMINO)PYRIDINE see THENYLPYRAMINE

2-DIMETHYLAMINO-FLUOREN (GERMAN) see 2-DIMETHYLAMINO-FLUORENE

N,N-DIMETHYL-2-AMINOFLUORENE see 2-DIMETHYLAMINOFLUORENE

DIMETHYLAMINO HEXOSE REDUCTIONE see DMA

DIMETHYLAMINO-ISOPROPYL-PHENTHIAZIN (GERMAN) see 10-(2-(DIMETHYLAMINO)PROPYL)PHENOTHIAZINE

4-(N,N-DIMETHYLAMINO)-3'-METHYLAZOBENZENE see N,N-DIMETHYL-p-(m-TOLYLAZO)ANILINE

D-4-DIMETHYLAMINO-3-METHYL-1,2-DIPHENYL-2-BUTANOL PROPIONATE HYDROCHLORIDE see d-PROPOXYPHENE HYDROCHLORIDE

6-DIMETHYLAMINO-5-METHYL-4,4-DIPHENYL-3-HEXANONE see ISOMETHADONE

M-(((DIMETHYLAMINO)METHYLENE)AMINO)PHENYLMETHYL CARBAMATE,HYDROCHLORIDE see N,N-DIMETHYL-N'-(((METHYLAMINO)CARBONYL)OXY)PHENYLMETHANIMIDAMIDE MONOHYDROCHLORIDE

3-DIMETHYLAMINOMETHYLENEIMINOPHENYL-N-METHYLCARBAMATE, HYDROCHLORIDE see N,N-DIMETHYL-N'-(((METHYLAMINO)CARBONYL)OXY)PHENYLMETHANIMIDAMIDE MONOHYDROCHLORIDE

(2-DIMETHYLAMINO-2-METHYL)ETHYL-N-DIBENZOPARATHIAZINE see 10-(2-(DIMETHYLAMINO)PROPYL)PHENOTHIAZINE

10-(2-(DIMETHYLAMINO)-2-METHYLETHYL)PHENOTHIAZINE see 10-(2-(DIMETHYLAMINO)PROPYL)PHENOTHIAZINE

N-(2'-DIMETHYLAMINO-2'-METHYL)ETHYLPHENOTHIAZINE see 10-(2-(DIMETHYLAMINO)PROPYL)PHENOTHIAZINE

S-ALPHA-(2-(DIMETHYLAMINO)-1-METHYLETHYL)-ALPHA-PHENYL-BENZENEETHANOL PROPIOATE HYDROCHLORIDE see d-PROPOXYPHENE HYDROCHLORIDE

N-DIMETHYLAMINO-2-METHYLETHYL THIODIPHENYLAMINE see 10-(2-(DIMETHYLAMINO)PROPYL)PHENOTHIAZINE

8-(DIMETHYLAMINOMETHYL)-7-METHOXY-3-METHYLFLAVONE see DIMEFLINE

3-(DIMETHYLAMINO)-1-METHYL-3-OXO-1-PROPENYL DIMETHYL PHOSPHATE see 3-HYDROXYDIMETHYL CROTONAMIDE DIMETHYL PHOSPHATE

(4-DIMETHYLAMINO-3-METHYL-PHENYL)N-METHYL-CARBAMAAT (DUTCH) see 4-DIMETHYLAMINE-m-CRESYLMETHYLCARBAMATE

(4-DIMETHYLAMINO-3-METHYL-PHENYL)N-METHYL-CARBAMAT (GERMAN) see 4-DIMETHYLAMINE-m-CRESYLMETHYLCARBAMATE

(4-DIMETHYLAMINO-3-METHYL-PHENYL)N-METHYL-CARBAMATE see 4-DIMETHYLAMINE-m-CRESYLMETHYLCARBAMATE

10-(3-(DIMETHYLAMINO)-2-METHYLPROPYL)-PHENOTHIAZINE-2-CARBONITRILE MALEATE see CYANOTRIMEPRAZINE MALEATE

ALPHA-(DIMETHYLAMINOMETHYL)PROTOCATECHUYL ALCOHOL see N-METHYLEPINEPHRINE

4-DIMETHYLAMINO-2'-METHYLSTILBENE see N,N,2'-TRIMETHYL-4-STILBENAMINE

2-DIMETHYLAMINO-1-(METHYLTHIO)GLYOXAL-O-METHYLCARBA-

MOYLMONOXIME see N′,N′-DIMETHYL-N-((METHYLCAR-
BAMOYL)OXY)-1-METHYLTHIOOXAMIMIDIC ACID
1-DIMETHYLAMINONAPHTHALENE see N,N-DIMETHYL-1-NAPHTHYL-
AMINE
DIMETHYLAMINONAPHTHALENESULFONYL CHLORIDE see 5-(DIMETH-
YLAMINO)-1-NAPHTHALENESULFONYL CHLORIDE
1-DIMETHYLAMINONAPHTHALENE-5-SULFONYL CHLORIDE see
5-(DIMETHYLAMINO)-1-NAPHTHALENESULFONYL CHLORIDE
1-(DIMETHYLAMINO)-5-NAPHTHALENESULFONYLCHLORIDE see
5-(DIMETHYLAMINO)-1-NAPHTHALENESULFONYL CHLORIDE
5-DIMETHYLAMINONAPHTHYL-5-SULFONYL CHLORIDE see
5-(DIMETHYLAMINO)-1-NAPHTHALENESULFONYL CHLORIDE
4-(DIMETHYLAMINO)NITROSOBENZENE see N,N-DIMETHYL-p-NITRO-
SOANILINE
P-(DIMETHYLAMINO)NITROSOBENZENE see N,N-DIMETHYL-p-NITRO-
SOANILINE
DIMETHYLAMINOPHENAZON (GERMAN) see DIMETHYLAMINOANTIPY-
RINE
DIMETHYLAMINOPHENAZONE see DIMETHYLAMINOANTIPYRINE
4-DIMETHYLAMINOPHENAZONE see DIMETHYLAMINOANTIPYRINE
4-DIMETHYLAMINOPHENYLAZOBENZENE see N,N-DIMETHYL-p-
PHENYLAZOANILINE
2-((4-DIMETHYLAMINO)PHENYLAZO)BENZOIC ACID see 2-CARBOXY-
4′-(DIMETHYL AMINO)AZOBENZENE
4-((4-(DIMETHYLAMINO)PHENYL)AZO)-2-PICOLINE 1-OXIDE see N,N-
DIMETHYL-4-(2-METHYL-4-PYRIDYLAZO)ANILINE-N-OXIDE
4-DIMETHYLAMINO-1-PHENYL-2,3-DIMETHYLPYRAZOLONE see
DIMETHYLAMINOANTIPYRINE
2-(3-DIMETHYLAMINO-1-PHENYLPROPYL)PYRIDINE see TRIMETON
BETA-DIMETHYLAMINOPROPIONITRILE see 3-(DIMETHYLAMINO)PRO-
PIONITRILE
3-(DIMETHYLAMINO)PROPYLAMINE see 1-AMINO-3-DIMETHYLAMINO-
PROPANE
N,N-DIMETHYL-N-(3-AMINOPROPYL)AMINE see 1-AMINO-3-DIMETH-
YLAMINOPROPANE
2,2′-(3-DIMETHYLAMINOPROPYLAMINO)BIBENZYL see TOFRANIL
10-(3-DIMETHYLAMINOPROPYL)-2-CHLOROPHENOTHIAZINE MONOHY-
DROCHLORIDE see CHLOROPROMAZINE HYDROCHLORIDE
1-(3-DIMETHYLAMINOPROPYL)-4,5-DIHYDRO-2,3,6,7-DIBENZAZEPINE
see TOFRANIL
5-(3-DIMETHYLAMINOPROPYL)-10,11-DIHYDRO-5H-DIBENZ(B,F)-
AZEPINEHYDROCHLORIDE see N-(gamma-DIMETHYLAMINOPRO-
PYL)IMINODIBENZYL HYDROCHLORIDE
5-(3-DIMETHYLAMINOPROPYL)-10,11-DIHYDRO-5H-DIBENZ(B,F)-
AZEPINE-5-OXIDE see IMIPRAMINE-N-OXIDE
5-(3-DIMETHYLAMINOPROPYL)-10,11-DIHYDRO-5H-DIBENZ(B,F)-
AZEPINE see TOFRANIL
5-(3-DIMETHYLAMINOPROPYL)-10,11-DIHYDRO-5H-DIBENZO(B,F)-
AZEPINE see TOFRANIL
10-(3-(DIMETHYLAMINO)PROPYL)-9,9-DIMETHYLACRIDAN TARTRATE
(1:1) see 9,9-DIMETHYL-10-DIMETHYLAMINOPROPYLACRIDAN HY-
DROGEN TARTRATE
5-(3-(DIMETHYLAMINO)PROPYL)-2-HYDROXY-10,11-DIHYDRO-5H-DI-
BENZ(b,f)AZEPINE see 2-HYDROXYINIPRAMINE
3-(3-DIMETHYLAMINOPROPYLIDENE)-1:2-4: 5-DIBENZOCYCLOHEP-
TA-1:4-DIENE see ELAVIL HYDROCHLORIDE
5-(3-DIMETHYLAMINOPROPYLIDENE)DIBENZO(A,D)(1,4)CYCLOHEP-
TADIENE HYDROCHLORIDE see ELAVIL HYDROCHLORIDE
11-DIMETHYLAMINO PROPYLIDENE 6H-DIBENZ(B,E)OXEPIN see ADA-
PIN
5-(3-DIMETHYLAMINOPROPYLIDENE)-10,11-DIHYDRO-5H-DIBEN-
ZO(A,D)CYCLOHEPTENE see ELAVIL
5-(GAMMA-DIMETHYLAMINOPROPYLIDENE)-10,11-DIHYDRO-5H-
DIBENZO(A,D)CYCLOHEPTENE see ELAVIL
11-(3-DIMETHYLAMINOPROPYLIDENE)-6,11-DIHYDRODIBENZ(B,E)-
OXIPIN see DEXEPIN
2,2′-(3-DIMETHYLAMINOPROPYLIMINO)DIBENZYL see TOFRANIL
N-(GAMMA-DIMETHYLAMINOPROPYL)IMINODIBENZYL see TOFRANIL
N-(3-DIMETHYLAMINOPROPYL)IMINODIBENZYL HYDROCHLORIDE see
N-(gamma-DIMETHYLAMINOPROPYL)IMINODIBENZYL HYDROCHLO-
RIDE

10-(3-DIMETHYLAMINOPROPYL)-2-METHOXYPHENOTHIAZINE see
2-METHOXYPROMAZINE
10-(3-(DIMETHYLAMINO)PROPYL)-2-METHOXY)PHENOTHIAZINE, MA-
LEATE see METHOXYPROMAZINE MALEATE
10-(3-DIMETHYLAMINOPROPYL)PHENOTHIAZINE see PROMAZINE
10-(3-DIMETHYLAMINOPROPYL)PHENOTHIAZINE-3-ETHYLONE see
ACETOPROMAZINE
10-(3-(DIMETHYLAMINO)PROPYL)PHENOTHIAZINE HYDROCHLORIDE
see PROMAZINE HYDROCHLORIDE
10-(GAMMA-DIMETHYLAMINO-N-PROPYL)PHENOTHIAZINE HYDRO-
CHLORIDE see PROMAZINE HYDROCHLORIDE
(DIMETHYLAMINO-2-PROPYL-10-PHENOTHIAZINE HYDROCHLORIDE
(FRENCH) see 10-(2-(DIMETHYLAMINO)PROPYL)PHENOTHIAZINE
1-(10-(3-(DIMETHYLAMINO)PROPYL)-10H-PHENOTHIAZIN-2-YL)ETHA-
NONE see ACETOPROMAZINE
10-(3-DIMETHYLAMINOPROPYL)PHENOTHIAZIN-3-YLMETHYL KETONE
see ACETOPROMAZINE
10-(3-(DIMETHYLAMINO)PROPYL-2-(TRIFLUOROMETHYL) PHENOTHI-
AZINE see TRIFLUPROMAZINE
4-DIMETHYLAMINOSTILBEN (GERMAN) see N,N-DIMETHYL-4-STIL-
BENAMINE
TRANS-4-DIMETHYLAMINOSTILBENE see (E)-N,N-DIMETHYL-4-STIL-
BENAMINE
TRANS-P-(DIMETHYLAMINO)STILBENE see (E)-N,N-DIMETHYL-4-STIL-
BENAMINE
N-(DIMETHYLAMINO)SUCCINAMIC ACID see DIMETHYLAMINOSUCCI-
NAMIC ACID
N-DIMETHYLAMINO-SUCCINAMIDSAEURE (GERMAN) see DIMETHYL-
AMINOSUCCINAMIC ACID
O-(4-((DIMETHYLAMINO)SULFONYL)PHENYL) O,O-DIMETHYL PHOS-
PHOROTHIOATE see O,O-DIMETHYL-O-(p-(N,N-DIMETHYLSULFA-
MOYL) PHENYL)PHOSPHOROTHIOATE
4-(DIMETHYLAMINO)-M-TOLYL METHYLCARBAMATE see 4-
DIMETHYLAMINE-m-CRESYLMETHYLCARBAMATE
4-(DIMETHYLAMINO)-3,5-XYLENOL, METHYLCARBAMATE (ESTER)
see 4-(DIMETHYLAMINE)-3,5-XYLYL N-METHYLCARBAMATE
4-DIMETHYLAMINO-3,5-XYLYL METHYLCARBAMATE see 4-(DIMETH-
YLAMINE)-3,5-XYLYL N-METHYLCARBAMATE
4-DIMETHYLAMINO-3,5-XYLYL N-METHYLCARBAMATE see 4-(DI-
METHYLAMINE)-3,5-XYLYL N-METHYLCARBAMATE
4-(N,N-DIMETHYLAMINO)-3,5-XYLYL N-METHYLCARBAMATE see
4-(DIMETHYLAMINE)-3,5-XYLYL N-METHYLCARBAMATE
DIMETHYLANILINE see XYLIDINE
2,5-DIMETHYLANILINE see 2,5-XYLIDINE
2,6-DIMETHYLANILINE see 2,6-XYLIDINE
3,5-DIMETHYLANILINE see 3,5-XYLIDINE
2,4-DIMETHYLANILINE HYDROCHLORIDE see 2,4-XYLIDINE HYDRO-
CHLORIDE
2,5-DIMETHYLANILINE HYDROCHLORIDE see 2,5-XYLIDINE HYDRO-
CHLORIDE
2,5-DIMETHYL ANILINE HYDROCHLORIDE see 2,5-XYLIDINE HYDRO-
CHLORIDE
DIMETHYLARSENIC ACID see HYDROXYDIMETHYLARSINE OXIDE
DIMETHYLARSINIC ACID see HYDROXYDIMETHYLARSINE OXIDE
N,N-DIMETHYL-P-AZOANILINE see N,N-DIMETHYL-p-PHENYLAZO-
ANILINE
2,3′-DIMETHYLAZOBENZENE see 3,6′-DIMETHYLAZOBENZENE
7,10-DIMETHYLBENZ(C)ACRIDINE see 6,9-DIMETHYL-1,2-BENZACRI-
DINE
2,10-DIMETHYL-7,8-BENZACRIDINE (FRENCH) see 6,9-DIMETHYL-
1,2-BENZACRIDINE
3,10-DIMETHYL-7,8-BENZACRIDINE (FRENCH) see 7,9-DIMETHYL-
BENZ(c)ACRIDINE
7,12-DIMETHYLBENZANTHRACENE see DIMETHYLBENZANTHRACENE
6,8-DIMETHYL-1,2-BENZANTHRACENE see 6,8-DIMETHYLBENZ(a)AN-
THRACENE
7,12-DIMETHYLBENZ(A)ANTHRACENE see DIMETHYLBENZANTHRA-
CENE
9,10-DIMETHYLBENZ(A)ANTHRACENE see 6,7-DIMETHYL-1,2-BENZ-
ANTHRACENE

9,10-DIMETHYLBENZ(A)ANTHRACENE see DIMETHYLBENZANTHRA-
CENE

9,10-DIMETHYL-1,2-BENZANTHRACENE see DIMETHYLBENZANTHRA-
CENE

9,10-DIMETHYL-1,2-BENZANTHRAZEN (GERMAN) see DIMETHYL-
BENZANTHRACENE

O,O-DIMETHYL-S-(BENZAZIMINOMETHYL) DITHIOPHOSPHATE see
AZINPHOS METHYL

2,5-DIMETHYLBENZENAMINE see 2,5-XYLIDINE

2,6-DIMETHYLBENZENAMINE see 2,6-XYLIDINE

3,5-DIMETHYLBENZENAMINE see 3,5-XYLIDINE

2,4-DIMETHYLBENZENAMINE HYDROCHLORIDE see 2,4-XYLIDINE HY-
DROCHLORIDE

2,5-DIMETHYLBENZENAMINE HYDROCHLORIDE see 2,5-XYLIDINE HY-
DROCHLORIDE

DIMETHYLBENZENE see XYLENE

M-DIMETHYLBENZENE see m-XYLENE

O-DIMETHYLBENZENE see o-XYLENE

P-DIMETHYLBENZENE see p-XYLENE

1,2-DIMETHYLBENZENE see o-XYLENE

1,3-DIMETHYLBENZENE see m-XYLENE

1,4-DIMETHYLBENZENE see p-XYLENE

N,N-DIMETHYLBENZENEAMINE see N,N-DIMETHYLANILINE

DIMETHYL-1,2-BENZENEDICARBOXYLATE see DIMETHYL PHTHALATE

DIMETHYL-1,4-BENZENE DICARBOXYLATE see DIMETHYL TEREPH-
THALATE

DIMETHYL-1,4-BENZENEDICARBOXYLATE see TEREPHTHALIC ACID
METHYL ESTER

DIMETHYL BENZENEORTHODICARBOXYLATE see DIMETHYL PHTHA-
LATE

3,3'-DIMETHYLBENZIDINE see 3,3'-TOLIDINE

5,6-DIMETHYL BENZIMIDAZOL see VITAMIN B$_{12}$ COMPLEX

5,6-DIMETHYLBENZIMIDAZOLYLCOBAMIDE CYANIDE see VITAMIN B$_{12}$
COMPLEX

DIMETHYLBENZIMIDAZOYLCOBAMIDE see VITAMIN B$_{12}$ COMPLEX

7,12-DIMETHYLBENZO(A)ANTHRACENE see DIMETHYLBENZANTHRA-
CENE

AR,AR-DIMETHYLBENZOPHENONE see PHENYL XYLYL KETONE

O,O-DIMETHYL-S-(1,2,3-BENZOTRIAZINYL-4-KETO)METHYL PHOS-
PHORODITHIOATE see AZINPHOS METHYL

1,4-DIMETHYL-2,3-BENZPHENANTHRENE see DIMETHYLBENZAN-
THRACENE

DIMETHYLBENZYLAMMONIUM CHLORIDE see DIMETHYLBENZYL-
AMINE HYDROCHLORIDE

alpha,alpha-DIMETHYLBENZYL HYDROPEROXIDE see ISOPROPYLBEN-
ZENE HYDROPEROXIDE

DIMETHYLBENZYLOCTADECYLAMMONIUM CHLORIDE see
DIMETHYLOCTADECYLBENZYLAMMONIUM CHLORIDE

N,N-DIMETHYL-N'-BENZYL-N'-(ALPHA-PYRIDYL)ETHYLENEDIAMINE
see TRIPELENNAMINE

N,N-DIMETHYLBIGUANIDE see 1,1-DIMETHYLBIGUANIDE

3,2'-DIMETHYL-4-BIPHENYLAMINE see (m,o'-BITOLYL)-4-AMINE

3,3'-DIMETHYLBIPHENYL-4,4'-DIAMINE see 3,3'-TOLIDINE

3,3'-DIMETHYL-4,4'-BIPHENYLDIAMINE see 3,3'-TOLIDINE

N,N'-DIMETHYL-4,4'-BIPYRIDINIUM DICHLORIDE see PARAQUAT

O,O-DIMETHYL-S-(1,2-BIS(ETHOXYCARBONYL)ETHYL)DITHIOPHOS-
PHATE see CARBETHOXY MALATHION

DIMETHYL BIS(p-HYDROXYPHENYL)METHANE see BISPHENOL A

1,3-DIMETHYL BUTANOL see AMYL METHYL ALCOHOL

5-(1,3-DIMETHYL-2-BUTENYL)-5ETHYL-2,4,6(1H,3H,5H)PYRIMI-
DINETRIONE see 5-(1,3-DIMETHYL-2-BUTENYL)-5-ETHYLBAR-
BITURIC ACID

3,3-DIMETHYL-N-BUT-2-YL METHYLPHOSPHONOFLUORIDATE see SO-
MAN

O,O-DIMETHYL-(1-BUTYRYLOXY-2,2,2-TRICHLOROETHYL) PHOSPHO-
NATE see BUTONATE

DIMETHYLCARBAMATE DE 5,5-DIMETHYL DIHYDRORESORCINOL
(FRENCH) see 5,5-DIMETHYLDIHYDRORESORCINOL DIMETHYL-
CARBAMATE

DIMETHYLCARBAMATE-D'L-ISOPROPYL-3-METHYL-5-PYRAZOLYLE

(FRENCH) see DIMETHYL-5-(1-ISOPROPYL-3-METHYL-PYRAZOLYL)-
CARBAMATE

DIMETHYLCARBAMIC ACID CHLORIDE see (DIMETHYLAMINO)CAR-
BONYL CHLORIDE

DIMETHYLCARBAMIC CHLORIDE see (DIMETHYLAMINO)CARBONYL
CHLORIDE

DIMETHYLCARBAMIC ESTER OF 3-OXYPHENYLTRIMETHYLAMMONIUM
METHYLSULFATE see NEOSTIGMINE MONOMETHYLSULFATE

DIMETHYLCARBAMIDOYL CHLORIDE see (DIMETHYLAMINO)CAR-
BONYL CHLORIDE

DIMETHYLCARBAMODITHIOIC ACID, IRON(3+) SALT see FERBAM

DIMETHYLCARBAMODITHIOIC ACID, ZINC SALT see BIS(DIMETHYL
DITHIO CARBAMATO)ZINC

3-DIMETHYLCARBAMOXYPHENYLTRIMETHYLAMMONIUM BROMIDE
see PROSTIGMINE BROMIDE

3-(DIMETHYLCARBAMOXY)PHENYL TRIMETHYLAMMONIUM METHYL
SULFATE see NEOSTIGMINE MONOMETHYLSULFATE

DIMETHYLCARBAMOYL CHLORIDE see (DIMETHYLAMINO)CARBONYL
CHLORIDE

N,N-DIMETHYLCARBAMOYL CHLORIDE see (DIMETHYLAMINO)CAR-
BONYL CHLORIDE

CIS-2-DIMETHYLCARBAMOYL-1-METHYLVINYL DIMETHYLPHOSPHATE
see 3-HYDROXYDIMETHYL CROTONAMIDE DIMETHYL PHOSPHATE

DIMETHYLCARBAMYL CHLORIDE see (DIMETHYLAMINO)CARBONYL
CHLORIDE

N,N-DIMETHYLCARBAMYL CHLORIDE see (DIMETHYLAMINO)CAR-
BONYL CHLORIDE

DIMETHYLCARBINOL see ISOPROPYL ALCOHOL

2,2-DIMETHYL-6-CARBOBUTOXY-2,3-DIHYDRO-4-PYRONE see
n-BUTYL MESITYL OXIDE OXALATE

ALPHA,ALPHA-DIMETHYL-ALPHA'-CARBOBUTOXY-DIHYDRO-GAMMA-
PYRONE see n-BUTYL MESITYL OXIDE OXALATE

O,O-DIMETHYL S-(1-CARBOETHOXYBENZYL) DITHIOPHOSPHATE see
(O,O-DIMETHYLDITHIOPHOSPHORYLPHENYL)ACETIC ACID ETHYL
ESTER

O,O-DIMETHYL-O-(2-CARBOMETHOXY-1-METHYLVINYL) PHOSPHATE
see MEVINPHOS

DIMETHYL-1-CARBOMETHOXY-1-PROPEN-2-YL PHOSPHATE see
MEVINPHOS

O,O-DIMETHYL-S-(CARBONYLMETHYLMORPHOLINO) PHOSPHORODI-
THIOATE see PHOSPHORODITHIOIC ACID-O,O-DIMETHYL-S-
(2-ETHYLTHIO)ETHYL ESTER

O,O-DIMETHYL-O-3-CHLOR-4-NITROFENYLTIOFOSFAT (CZECH) see
METHYLCHLOROTHION

O,O-DIMETHYL-O-(3-CHLOR-4-NITROPHENYL)-MONOTHIOPHOSPHAT
(GERMAN) see METHYLCHLOROTHION

DIMETHYL-2-CHLORO-2-DIETHYLCARBAMOYL-1-METHYLVINYL PHOS-
PHATE see DIMECRON

O,O-DIMETHYL-O-(2-CHLORO-2(N,N-DIETHYLCARBAMOYL)-1-
METHYLVINYL)PHOSPHATE see DIMECRON

DIMETHYLCHLOROETHER see CHLOROMETHYL METHYL ETHER

DIMETHYL(2-CHLOROETHYL)AMINE see N-(2-CHLORO ETHYL)-
DIMETHYL AMINE

O,O-DIMETHYL O-2-CHLORO-4-NITROPHENYL PHOSPHOROTHIOATE
see p-NITRO-o-CHLOROPHENYL DIMETHYL THIONOPHOSPHATE

O,O-DIMETHYL-O-(3-CHLORO-4-NITROPHENYL) PHOSPHOROTHIOATE
see METHYLCHLOROTHION

DIMETHYL-3-CHLORO-4-NITROPHENYL THIONOPHOSPHATE see
METHYLCHLOROTHION

DIMETHYL-2-CHLORONITROPHENYL THIOPHOSPHATE see p-NITRO-o-
CHLOROPHENYL DIMETHYL THIONOPHOSPHATE

O,O-DIMETHYL-O-(3-CHLORO-4-NITROPHENYL) THIOPHOSPHATE see
METHYLCHLOROTHION

ALPHA,ALPHA-DIMETHYL-P-CHLOROPHENETHYLAMINE HYDROCHLO-
RIDE see AVICOL

N,N-DIMETHYL-p-((p-CHLOROPHENYL)AZO)ANILINE see P-CHLORO-
DIMETHYLAMINOAZOBENZENE

O,O-DIMETHYL-S-p-CHLOROPHENYL PHOSPHOROTHIOATE see FUJI-
THION

DIMETHYL P-CHLOROPHENYLTHIOMETHYL DITHIOPHOSPHATE see METHYL TRITHION

O,O-DIMETHYL-S-(P-CHLOROPHENYLTHIOMETHYL)PHOSPHORODITHIO-ATE see METHYL TRITHION

1,1-DIMETHYL-3-(P-CHLOROPHENYL)UREA see 3-(p-CHLOROPHE-NYL)-1,1-DIMETHYLUREA

N,N-DIMETHYL-N'-(4-CHLOROPHENYL)UREA see 3-(p-CHLORO-PHENYL)-1,1-DIMETHYLUREA

5,7-DIMETHYLCHRYSENE see 1,11-DIMETHYLCHRYSENE

DIMETHYLCYANOARSINE see CYANODIMETHYLARSINE

1,5-DIMETHYL-5-(1-CYCLOHEXENYL)BARBITURIC ACID see EVIPAL

DIMETHYLCYSTEINE see d,3-MERCAPTOVALINE

3,3'-DIMETHYL-4,4'-DIAMINODIPHENYLMETHANE see 4,4'-METH-YLENE DI-O-TOLUIDINE

N,N-DIMETHYL-1,3-DIAMINOPROPANE see 1-AMINO-3-DIMETHYL-AMINOPROPANE

O,O-DIMETHYL-S-(4,6-DIAMINO-1,3,5-TRIAZINYL-2-METHYL) DITHI-OPHOSPHATE see AZIDITHION

O,O-DIMETHYL- S-(4,6-DIAMINO-S-TRIAZIN-2-YLMETHYL)PHOSPHO-RODITHIOATE see AZIDITHION

O,O-DIMETHYL-S-(4,6-DIAMINO-1,3,5-TRIAZIN-2-YL)METHYL PHOS-PHORODITHIOATE see AZIDITHION

O,O-DIMETHYL-S-(4,6-DIAMINO-1,3,5-TRIAZIN-2-YL)METHYL PHOS-PHOROTHIOLOTHIONATE see AZIDITHION

1,5-DIMETHYL-1,5-D-DIAZAUNDECAMETHYLENE POLYMETHOBRO-MIDE see HEXADIMETHRINE BROMIDE

O,O-DIMETHYL-O-(1,2-DIBROMO-2,2-DICHLOROETHYL)PHOSPHATE see DIMETHYL-1,2-DIBROMO-2,2-DICHLOROETHYL PHOSPHATE

O,O-DIMETHYL-S-(1,2-DICARBETHOXYETHYL) DITHIOPHOSPHATE see CARBETHOXY MALATHION

O,O-DIMETHYL-S-(1,2-DICARBETHOXYETHYL)PHOSPHORODITHIOATE see CARBETHOXY MALATHION

O,O-DIMETHYL-S-(1,2-DICARBETHOXYETHYL) THIOTHIONOPHOS-PHATE see CARBETHOXY MALATHION

DIMETHYL-1,3-DI(CARBOMETHOXY)-1-PROPEN-2-YL PHOSPHATE see SWAT

O,O-DIMETHYL O-2,2-DICHLORO-1,2-DIBROMOETHYL PHOSPHATE see DIMETHYL-1,2-DIBROMO-2,2-DICHLOROETHYL PHOSPHATE

DIMETHYL 2,2-DICHLOROETHENYL PHOSPHATE see DIMETHYL DICHLOROVINYL PHOSPHATE

DIMETHYL-1,1'-DICHLOROETHER see BIS(CHLOROMETHYL) ETHER

1,1-DIMETHYL-3-(3,4-DICHLOROPHENYL)UREA see 3-(3,4-DICHLO-ROPHENYL)-1,1-DIMETHYLUREA

DIMETHYLDICHLOROSILANE (DOT) see DICHLORODIMETHYLSILANE

DIMETHYLDICHLOROSTANNANE see DICHLORODIMETHYLSTANNANE

DIMETHYLDICHLOROTIN see DICHLORODIMETHYLSTANNANE

DIMETHYL 2,2-DICHLOROVINYL PHOSPHATE see DIMETHYL DICHLO-ROVINYL PHOSPHATE

O,O-DIMETHYL DICHLOROVINYL PHOSPHATE see DIMETHYL DICHLO-ROVINYL PHOSPHATE

O,O-DIMETHYL O-2,2-DICHLOROVINYL PHOSPHATE see DIMETHYL DI-CHLOROVINYL PHOSPHATE

DIMETHYL-DICHLORSILAN (CZECH) see DICHLORODIMETHYLSILANE

O,O-DIMETHYL-O-(2,2-DICHLOR-VINYL)-PHOSPHAT (GERMAN) see DI-METHYL DICHLOROVINYL PHOSPHATE

DIMETHYLDIDECYLAMMONIUM CHLORIDE see QUATERNIUM-12

O,O-DIMETHYL-S-1,2-DI(ETHOXYCARBAMYL)ETHYL PHOSPHORODI-THIOATE see CARBETHOXY MALATHION

DIMETHYL DIETHYLAMIDO-1-CHLOROCROTONYL (2) PHOSPHATE see DIMECRON

N,N-DIMETHYLDIGUANIDE see 1,1-DIMETHYLBIGUANIDE

2,2-DIMETHYL-2,2-DIHYDROBENZOFURANYL-7-N-METHYLCARBA-MATE see FURADAN

11,17-DIMETHYL-16,17-DIHYDRO-15H-CYCLOPENTA(A)PHENAN-THRENE see 16,17-DIHYDRO-11,17-DIMETHYLCYCLOPENTA(a)-PHENANTHRENE

5,5-DIMETHYL-4,5-DIHYDRO-3RESORCYL-DIMETHYL-CARBAMAT (GERMAN) see 5,5-DIMETHYLDIHYDRORESORCINOL DIMETHYLCARBAMATE

O,O-DIMETHYL-S-(3,4-DIHYDRO-4-KETO-1,2,3-BENZOTRIAZINYL-3-METHYL) DITHIOPHOSPHATE see AZINPHOS METHYL

O,O-DIMETHYL-S-1,2-DIKARBETOXYLETHYLDITIOFOSFAT (CZECH) see CARBETHOXY MALATHION

DIMETHYL DIKETONE see 2,3-BUTANEDIONE

DIMETHYL 3-(DIMETHOXYPHOSPHINYLOXY)GLUTACONATE see SWAT

1,5-DIMETHYL-4-DIMETHYLAMINO-2-PHENYL-3-PYRAZOLONE see DIMETHYLAMINOANTIPYRINE

2,3-DIMETHYL-4-DIMETHYLAMINO-1-PHENYL-5-PYRAZOLONE see DIMETHYLAMINOANTIPYRINE

9,9-DIMETHYL-10-(3-DIMETHYLAMINO)PROPYLACRIDINE TARTRATE see 9,9-DIMETHYL-10-DIMETHYLAMINOPROPYLACRIDAN HYDROGEN TARTRATE

3',4'-DIMETHYL-4-DIMETHYLAMINOZOBENZENE see N,N-DIMETHYL-p-(3,4-XYLYLAZO)ANILINE

O,O-DIMETHYL O-(N,N-DIMETHYLCARBAMOYL-1METHYLVINYL) PHOSPHATE see 3-HYDROXYDIMETHYL CROTONAMIDE DIMETHYL PHOSPHATE

O,O-DIMETHYL-O-(2-DIMETHYL-CARBAMOYL-1-METHYL-VINYL)PHOS-PHAT (GERMAN) see 3-HYDROXYDIMETHYL CROTONAMIDE DIMETHYL PHOSPHATE

O,O-DIMETHYL O-(3,5-DIMETHYL-4-METHYLTHIOPHENYL) PHOSPHO-ROTHIOATE see O,O-DIMETHYL O-4-(METHYLTHIO)-3,5-XYLYL PHOSPHOROTHIOATE

O,O-DIMETHYL-O-(1,4-DIMETHYL-3-OXO-4-AZA-PENT-1-ENYL)FOS FAAT (DUTCH) see 3-HYDROXYDIMETHYL CROTONAMIDE DIMETHYL PHOSPHATE

O,O-DIMETHYL-O-(1,4-DIMETHYL-3-OXO-4-AZA-PENT-1-ENYL)PHOS-PHATE see 3-HYDROXYDIMETHYL CROTONAMIDE DIMETHYL PHOSPHATE

N,N-DIMETHYL-4-(4'-(2',6'-DIMETHYLPYRIDYL-1'-OXIDE)AZO)ANI-LINE see 4-((4-(DIMETHYLAMINO)PHENYL)AZO)-2,6-LUTIDINE-1-OXIDE

N,N'-DIMETHYL-N,N'-DINITROSO-1,3-PROPANEDIAMINE see DINITROSODIMETHYLPROPANEDIAMINE

2,6-DIMETHYL-m-DIOXAN-4-OL ACETATE see ACETOMETHOXANE

2,6-DIMETHYL-m-DIOXAN-4-YL ACETATE see ACETOMETHOXANE

2,2-DIMETHYL-1,3-DIOXOLANE-4-METHANOL see DIOXOLAN

3,3'-DIMETHYL-4,4'-DIPHENYLDIAMINE see 3,3'-TOLIDINE

3,3'-DIMETHYLDIPHENYL-4,4'-DIAMINE see 3,3'-TOLIDINE

N,N-DIMETHYL-1,2-DIPHENYLETHYLAMINE HYDROCHLORIDE, (R) (−)- see 1,2-DIPHENYL-1-(DIMETHYLAMINO)ETHANE

1,1'-DIMETHYL-4,4'-DIPYRIDYLIUM CHLORIDE see PARAQUAT

DIMETHYLDITHIOCARBAMATE ZINC SALT see BIS(DIMETHYL DITHIO CARBAMATO)ZINC

DIMETHYLDITHIOCARBAMIC ACID COPPER SALT see COPPER DI-METHYLDITHIOCARBAMATE

DIMETHYLDITHIOCARBAMIC ACID, IRON SALT see FERBAM

DIMETHYLDITHIOCARBAMIC ACID, IRON(3+) SALT see FERBAM

O,O-DIMETHYLDITHIOPHOSPHATE DIETHYLMERCAPTOSUCCINATE see CARBETHOXY MALATHION

DIMETHYLDITHIOPHOSPHORIC-ACID N-METHYLBENZAZIMIDE ESTER see AZINPHOS METHYL

O,O-DIMETHYLDITHIOPHOSPHORYLACETIC ACID, N-MONOMETHYL-AMIDE SALT see O,O-DIMETHYL METHYLCARBAMOYLMETHYL PHOS-PHORODITHIOATE

2,5-DIMETHYL-DNPZ see 2,5-DIMETHYLDINITROSOPIPERAZINE

2,6-DIMETHYL-DNPZ see 2,6-DIMETHYLDINITROSOPIPERAZINE

7,8-DIMETHYLENEBENZ(A)ANTHRACENE see CHOLANTHRENE

3:4-DIMETHYLENE-1:2-BENZANTHRACENE see ACENAPHTHANTHRA-CENE

DIMETHYLENEDIAMINE see 1,2-ETHANEDIAMINE

DIMETHYLENE GLYCOL see 2,3-BUTANEDIOL

DIMETHYLENEIMINE see ETHYLENIMINE

DIMETHYLENE OXIDE see ETHYLENE OXIDE

DIMETHYLENIMINE see ETHYLENIMINE

3,7-DIMETHYL-(E)-2,6-OCTADIEN-1-OL see GERANIOL

DIMETHYLESTER KYSELINY ISOFTALOVE (CZECH) see TEREPHTHALIC ACID METHYL ESTER

DIMETHYLESTER KYSELINY SIROVE (CZECH) see SULFURIC ACID, DIMETHYL ESTER

1,1-DIMETHYLETHANOL see tert-BUTYL ALCOHOL

DIMETHYL ETHER see METHYL ETHER

O,O-DIMETHYL S-ALPHAETHOXY-CARBONYLBENZYL PHOSPHORODITHIOATE see (O,O-DIMETHYLDITHIOPHOSPHORYLPHENYL)-ACETIC ACID ETHYL ESTER

O,O-DIMETHYL S-(2-ETHSULFONYLETHYL)PHOSPHOROTHIOATE see DEMETON-S-METHYL SULPHONE

DIMETHYL S-(2-ETHSULFONYLETHYL)THIOPHOSPHATE see DEMETON-S-METHYL SULPHONE

O,O-DIMETHYL S-(2-ETHTHIOETHYL)PHOSPHOROTHIOATE see DEMETON-S-METHYL

DIMETHYL-S-(2-ETHTHIOETHYL)THIOPHOSPHATE see DEMETON-S-METHYL

O,O-DIMETHYL S-(2-ETHTHIONYLETHYL) PHOSPHOROTHIOATE see DEMETON-O-METHYL SULFOXIDE

DIMETHYL-S-(2-ETHTHIONYLETHYL) THIOPHOSPHATE see DEMETON-O-METHYL SULFOXIDE

1,1-DIMETHYLETHYLAMINE see tert-BUTYLAMINE

N,N-DIMETHYL-4′-ETHYL-4-AMINOAZOBENZENE see p-((p-ETHYLPHENYL)AZO)-N,N-DIMETHYLANILINE

O,O-DIMETHYL S-(N-ETHYLCARBAMOYLMETHYL) DITHIOPHOSPHATE see DIMETHOATE-ETHYL

O,O-DIMETHYL S-(N-ETHYLCARBAMOYLMETHYL) PHOSPHORODITHIOATE see DIMETHOATE-ETHYL

DIMETHYL ETHYL CARBINOL see tert-N-AMYL ALCOHOL, REFINED

DIMETHYLETHYLCARBINOL see t-PENTYL ALCOHOL

DIMETHYLETHYLENE see cis-2-BUTENE

1,1-DIMETHYL ETHYL HYDRO PEROXIDE. see 6-BUTYL HYDROPEROXIDE

DIMETHYL-ETHYL-BETA-HYDROXYETHYL-AMMONIUM-SULFATE-DI-N-BUTYL-CARBAMATE see DI-n-BUTYL-CARBAMYLCHOLINE SULPHATE

DIMETHYLETHYL-BETA-HYDROXYETHYLAMMONIUM SULFATE DIBUTYLURETHAN see DI-n-BUTYL-CARBAMYLCHOLINE SULPHATE

O,O-DIMETHYL-S-(2-ETHYLMERCAPTOETHYL) DITHIOPHOSPHATE see PHOSPHORODITHIOIC ACID-O,O-DIMETHYL-S-(2-ETHYLTHIO)ETHYL ESTER

O,O-DIMETHYL-O-ETHYLMERCAPTOETHYL THIOPHOSPHATE see DEMETON-O-METHYL

O,O-DIMETHYL S-ETHYLMERCAPTOETHYL THIOPHOSPHATE see DEMETON-S-METHYL

O,O-DIMETHYL-S-2-ETHYLMERKAPTOETHYLESTER KYSELINY DITHIOFOSFORECNE (CZECH) see PHOSPHORODITHIOIC ACID-O,O-DIMETHYL-S-(2-ETHYLTHIO)ETHYL ESTER

N,N-DIMETHYL-p-((4-ETHYLPHENYL)AZO)ANILINE see p-((p-ETHYLPHENYL)AZO)-N,N-DIMETHYLANILINE

O,O-DIMETHYL-S-(2-(ETHYLSULFINYL)ETHYL)-O,O-DIMETHYL ESTER see DEMETON-O-METHYL SULFOXIDE

O,O-DIMETHYL-S-(2-ETHYLSULFINYL-ETHYL)-MONOTHIOFOSFAAT (DUTCH) see DEMETON-O-METHYL SULFOXIDE

O,O-DIMETHYL S-(2-ETHYLSULFINYL)ETHYL THIOPHOSPHATE see DEMETON-O-METHYL SULFOXIDE

O,O-DIMETHYL-S-((ETHYLSULFINYL)ISOPROPYL) PHOSPHOROTHIOATE- see O,O-DIMETHYL-S-ISOPROPYL-2-SULFINYLETHYLPHOSPHOROTHIOATE

O,O-DIMETHYL-S-(ETHYLSULFINYL-(2-ISOPROPYL)) PHOSPHOROTHIOATE see METASYSTOX-S

O,O-DIMETHYL S-(2-(ETHYLSULFINYL)ISOPROPYL) THIOPHOSPHATE see TRIPHENYL PHOSPHITE

O,O-DIMETHYL S-ETHYL-2-SULFONYLETHYL PHOSPHOROTHIOLATE see DEMETON-S-METHYL SULPHONE

O,O-DIMETHYL S-ETHYLSULPHINYLETHYL PHOSPHOROTHIOLATE see DEMETON-O-METHYL SULFOXIDE

O,O-DIMETHYL S-ETHYLSULPHONYLETHYL PHOSPHOROTHIOLATE see DEMETON-S-METHYL SULPHONE

O,O-DIMETHYL-S-(2-ETHYLTHIO-ETHYL)-DITHIOFOSFAAT (DUTCH)

see PHOSPHORODITHIOIC ACID-O,O-DIMETHYL-S-(2-ETHYLTHIO)-ETHYL ESTER

O,O-DIMETHYL-O-(2-ETHYL-THIO-ETHYL)-MONOTHIOFOSFAAT (DUTCH) see DEMETON-O-METHYL

O,O-DIMETHYL-S-(2-ETHYLTHIO-ETHYL)-MONOTHIOFOSFAAT (DUTCH) see DEMETON-S-METHYL

O,O-DIMETHYL S-(2-(ETHYLTHIO)ETHYL) PHOSPHORODITHIOATE see PHOSPHORODITHIOIC ACID-O,O-DIMETHYL-S-(2-ETHYLTHIO)ETHYL ESTER

O,O-DIMETHYL-O-2-(ETHYLTHIO)ETHYL PHOSPHOROTHIOATE see DEMETON-O-METHYL

O,O-DIMETHYL S-(2-(ETHYLTHIO)ETHYL)PHOSPHOROTHIOATE see DEMETON-S-METHYL

2,10-DIMETHYL-3-FLUORO-5,6-BENZACRIDINE see 10-FLUORO-9,12-DIMETHYLBENZ(a)ACRIDINE

DIMETHYLFORMALDEHYDE see ACETONE

DIMETHYLFORMAMID (GERMAN) see DIMETHYL FORMAMIDE

N,N-DIMETHYL FORMAMIDE see DIMETHYL FORMAMIDE

O,O-DIMETHYL S-(N-FORMYL-N-METHYLCARBAMOYLMETHYL) PHOSPHORODITHIOATE see O,O-DIMETHYL DITHIOPHOSPHORYLACETIC ACID-N-METHYL-N-FORMYLAMIDE

DIMETHYLGLYOXAL see 2,3-BUTANEDIONE

DIMETHYLGLYOXIME see DIACETYL DIOXIME

N,N′-DIMETHYLHARNSTOFF (GERMAN) see 1,3-DIMETHYLUREA

2,6-DIMETHYL-2,5-HEPTADIEN-4-ONE see PHORONE

2,6-DIMETHYL-4-HEPTANOL see DIISOBUTYL CARBINOL

2,6-DIMETHYL HEPTANOL-4 see DIISOBUTYL CARBINOL

2,6-DIMETHYL-HEPTAN-4-ON (DUTCH, GERMAN) see DIISOBUTYL KETONE

2,6-DIMETHYL-4-HEPTANONE see DIISOBUTYL KETONE

2,6-DIMETHYLHEPTAN-4-ONE see DIISOBUTYL KETONE

1,5-DIMETHYLHEXYLAMINE see 2-ISOOCTYL AMINE

ALPHA, EPSILON-DIMETHYLHEXYLAMINE see 2-ISOOCTYL AMINE

1,2-DIMETHYLHYDRAZIN (GERMAN) see 1,2-DIMETHYL HYDRAZINE

DIMETHYLHYDRAZINE see 1,1-DIMETHYLHYDRAZINE

1,1-DIMETHYL HYDRAZINE see 1,1-DIMETHYLHYDRAZINE

N,N-DIMETHYLHYDRAZINE see 1,1-DIMETHYLHYDRAZINE

SYM-DIMETHYLHYDRAZINE see 1,2-DIMETHYL HYDRAZINE

UNSYM-DIMETHYLHYDRAZINE see 1,1-DIMETHYLHYDRAZINE

1,1-DIMETHYLHYDRAZINE (GERMAN) see 1,1-DIMETHYLHYDRAZINE

N,N′-DIMETHYLHYDRAZINE DIHYDROCHLORIDE see 1,2-DIMETHYL-HYDRAZINE DIHYDROCHLORIDE

SYM-DIMETHYLHYDRAZINE DIHYDROCHLORIDE see 1,2-DIMETHYL-HYDRAZINE DIHYDROCHLORIDE

SYM-DIMETHYLHYDRAZINE HYDROCHLORIDE (DMH) see 1,2-DIMETHYLHYDRAZINE HYDROCHLORIDE

DIMETHYLHYDRAZINE UNSYMMETRICAL (DOT) see 1,1-DIMETHYLHYDRAZINE

2-(2,2-DIMETHYLHYDRAZINO)-4-(5-NITRO-2-FURYL)THIAZOLE see DMNT

DIMETHYL 3-HYDROXYGLUTACONATE DIMETHYL PHOSPHATE see SWAT

2,2-DIMETHYL-5-HYDROXYMETHYL-1,3-DIOXOLANE see DIOXOLAN

2′,3-DIMETHYL-4-(2-HYDROXYNAPHTHYLAZO)AZOBENZENE see SCARLET RED

1,1-DIMETHYL-3-HYDROXYPYRROLIDINIUM BROMIDE-alpha-CYCLOPENTYLMANDELATE see GLYCOPYRRONIUM BROMIDE

O,O-DIMETHYL-(1-HYDROXY-2,2,2-TRICHLORAETHYL)PHOSPHONSAERE ESTER (GERMAN) see ((2,2,2-TRICHLORO-1-HYDROXYETHYL) DIMETHYLPHOSPHONATE)

O,O-DIMETHYL-(1-HYDROXY-2,2,2-TRICHLORATHYL)-PHOSPHAT (GERMAN) see ((2,2,2-TRICHLORO-1-HYDROXYETHYL) DIMETHYLPHOSPHONATE)

O,O-DIMETHYL-(1-HYDROXY-2,2,2-TRICHLORO)ETHYL PHOSPHATE see ((2,2,2-TRICHLORO-1-HYDROXYETHYL) DIMETHYLPHOSPHONATE)

DIMETHYL 1-HYDROXY-2,2,2-TRICHLOROETHYL PHOSPHONATE see ((2,2,2-TRICHLORO-1-HYDROXYETHYL) DIMETHYLPHOSPHONATE)

O,O-DIMETHYL (1-HYDROXY-2,2,2-TRICHLOROETHYL)PHOSPHONATE

see ((2,2,2-TRICHLORO-1-HYDROXYETHYL) DIMETHYLPHOSPHO-
NATE)

N,N-DIMETHYL-5-HYDROXYTRYPTAMINE see 3-(2-DIMETHYLAMINO-
ETHYL)-5-INDOLOL

DIMETHYLIMIPRAMINE see 10,11-DIHYDRO-5-(3-METHYLAMINOPRO-
PYL)-5H-DIBENZ(b,f)AZEPINE

N,N-DIMETHYL-ALPHA-INDOLYLIDENE-P-TOLUIDINE see 1-(4-
DIMETHYLAMINOBENZAL)INDENE

O,O-DIMETHYL-S-2-(ISOPROPYLTHIO)ETHYLPHOSPHORODITHIOATE see
HOSDON GRANULE

N,N′-DIMETHYL-4-(5′-ISOQUINOLINYLAZO)ANILINE see
5-((p-(DIMETHYLAMINO)PHENYL)AZO)ISOQUINOLINE

DIMETHYLKETAL see ACETONE

DIMETHYLKETOL see ACETOIN

DIMETHYL KETONE see ACETONE

DIMETHYLMESCALINE see 3,4-DIMETHOXYDOPAMINE

DIMETHYLMETHANE see PROPANE

DIMETHYL 2-METHOXYCARBONYL-1-METHYLVINYL PHOSPHATE see
MEVINPHOS

DIMETHYL METHOXYCARBONYLPROPENYL PHOSPHATE see MEVIN-
PHOS

DIMETHYL (1-METHOXYCARBOXYPROPEN-2-YL)PHOSPHATE see
MEVINPHOS

O,O-DIMETHYL-S-(2-METHOXYETHYLCARBAMOYLMETHYL)DITHIO-
PHOSPHATE see AMIDITHION

O,O-DIMETHYL-S-(2-METHOXYETHYLCARBAMOYL METHYL)PHOS-
PHORODITHIOATE see AMIDITHION

O,O-DIMETHYL S-(5-METHOXY-4-OXO-4H-PYRAN-2-YL)PHOSPHORO-
THIOATE see ENDOTHION

O,O-DIMETHYL S-(5-METHOXYPYRONYL-2-METHYL) THIOPHOSPHATE
see ENDOTHION

(O,O-DIMETHYL)-S-(-2-METHOXY-delta(SUP 2)-1,3,4-THIADIAZO-
LIN-5-ON-4-YLMETHYL)DITHIOPHOSPHATE see O,O-DIMETHYL-S-
(5-METHOXY-1,3,4-THIADIAZOLINYL-3-METHYL)DITHIO
PHOSPHATE

O,O-DIMETHYL-S-(2-METHOXY-1,3,4-THIADIAZOL-5-(4H)-ONYL-(4)-
METHYL)-DITHIOPHOSPHAT (GERMAN) see O,O-DIMETHYL-S-(5-
METHOXY-1,3,4-THIADIAZOLINYL-3-METHYL)DITHIO PHOS-
PHATE

O,O-DIMETHYL-S-(2-METHOXY-1,3,4-THIADIAZOL-5(4H)-ONYL-
(4)-METHYL) PHOSPHORODITHIOATE see O,O-DIMETHYL-S-
(5-METHOXY-1,3,4-THIADIAZOLINYL-3-METHYL)DITHIO
PHOSPHATE

O,O-DIMETHYL-S-((2-METHOXY-1,3,4 (4H)-THIODIAZOL-5-ON-4-YL)-
METHYL)-DITHIOFOSFAAT (DUTCH) see O,O-DIMETHYL-S-(5-
METHOXY-1,3,4-THIADIAZOLINYL-3-METHYL)DITHIO PHOS-
PHATE

O,O-DIMETHYL S-(2-(METHYLAMINO)-2-OXOETHYL) PHOSPHORODI-
THIOATE see O,O-DIMETHYL METHYLCARBAMOYLMETHYL PHOS-
PHORODITHIOATE

O,O-DIMETHYL S-2-(1-N-METHYLCARBAMOYLETHYLMERCAPTO)-
ETHYL THIOPHOSPHATE see N-METHYL-O,O-DIMETHYLTHIOLO-
PHOSPHORYL-5-THIA-3-METHYL-2-VALERAMIDE

DIMETHYL S-(2-(1-METHYLCARBAMOYLETHYLTHIO ETHYL) PHOS-
PHOROTHIOLATE see N-METHYL-O,O-DIMETHYLTHIOLOPHOS-
PHORYL-5-THIA-3-METHYL-2-VALERAMIDE

O,O-DIMETHYL S-(2-(1-METHYLCARBAMOYLETHYLTHIO)ETHYL)
PHOSPHOROTHIOATE see N-METHYL-O,O-DIMETHYLTHIOLO-
PHOSPHORYL-5-THIA-3-METHYL-2-VALERAMIDE

O,O-DIMETHYL S-(N-METHYLCARBAMOYLMETHYL) DITHIOPHOSPHATE
see O,O-DIMETHYL METHYLCARBAMOYLMETHYL PHOSPHORODI-
THIOATE

O,O-DIMETHYL-S-((N-METHYL-CARBAMOYL)-METHYL)-MONOTHIO-
FOSFAAT (DUTCH) see DIMETHOATE OXYGEN ANALOG

O,O-DIMETHYL-S-(N-METHYL-CARBAMOYL)-METHYL-MONOTHIO-
PHOSPHAT (GERMAN) see DIMETHOATE OXYGEN ANALOG

O,O-DIMETHYL-S-((METHYLCARBAMOYL)METHYL)PHOSPHOROTHIO-
ATE see DIMETHOATE OXYGEN ANALOG

O,O-DIMETHYL-S-(N-METHYLCARBAMOYLMETHYL)PHOSPHOROTHIO-
ATE see DIMETHOATE OXYGEN ANALOG

DIMETHYL-S-(N-METHYL-CARBAMOYL-METHYL)PHOSPHOROTHIOLATE
see DIMETHOATE OXYGEN ANALOG

O,O-DIMETHYL S-(N-METHYLCARBAMOYLMETHYL) PHOSPHOROTHIO-
LATE see DIMETHOATE OXYGEN ANALOG

O,O-DIMETHYL S-(N-METHYLCARBAMOYLMETHYL) THIOPHOSPHATE
see DIMETHOATE OXYGEN ANALOG

O,O-DIMETHYL-O-(2-N-METHYLCARBAMOYL-1-METHYL-VINYL)-FOS-
FAAT (DUTCH) see 3-(DIMETHOXYPHOSPHINYLOXY)N-METHYL-cis-
CROTONAMIDE

O,O-DIMETHYL-O-(2-N-METHYLCARBAMOYL-1-METHYL)-VINYL-
PHOSPHAT (GERMAN) see 3-(DIMETHOXYPHOSPHINYLOXY)-
N-METHYL-cis-CROTONAMIDE

O,O-DIMETHYL-O-(2-N-METHYLCARBAMOYL-1-METHYL-VINYL)
PHOSPHATE see 3-(DIMETHOXYPHOSPHINYLOXY)N-
METHYL-cis-CROTONAMIDE

O,O-DIMETHYL S-(N-METHYLCARBAMYLMETHYL) THIOTHIONOPHOS-
PHATE see O,O-DIMETHYL METHYLCARBAMOYLMETHYL PHOS-
PHORODITHIOATE

O,O-DIMETHYL O-(1-METHYL-2-CARBOXYVINYL) PHOSPHATE see
MEVINPHOS

N,N-DIMETHYL-N′-(2-METHYL-4-CHLOROPHENYL)-FORMAMIDINE see
CHLOROPHENAMIDINE

O,O-DIMETHYL-S-(3-METHYL-2,4-DIOXO-3-AZA-BUTYL)-DITHIOFOS-
FAAT (DUTCH) see O,O-DIMETHYL DITHIOPHOSPHORYLACETIC
ACID-N-METHYL-N-FORMYLAMIDE

O,O-DIMETHYL-S-(3-METHYL-2,4-DIOXO-3-AZA-BUTYL)-DITHIOPHOS-
PHAT (GERMAN) see O,O-DIMETHYL DITHIOPHOSPHORYLACETIC
ACID-N-METHYL-N-FORMYLAMIDE

DIMETHYLMETHYLENE-p,p′-DIPHENOL see BISPHENOL A

O,O-DIMETHYL-S-(N-METHYL-N-FORMYL-CARBAMOYLMETHYL)-
DITHIOPHOSPHAT (GERMAN) see O,O-DIMETHYL DITHIOPHOS-
PHORYLACETIC ACID-N-METHYL-N-FORMYLAMIDE

O,O-DIMETHYL S-(N-METHYL-N-FORMYLCARBAMOYLMETHYL)PHOS-
PHORODITHIOATE see O,O-DIMETHYL DITHIOPHOSPHORYLACETIC
ACID-N-METHYL-N-FORMYLAMIDE

O,O-DIMETHYL O-4-(METHYLMERCAPTO)-3-METHYLPHENYL PHOS-
PHOROTHIOATE see O,O-DIMETHYLPHOSPHOROTHIOIC ACID-O-
(4-METHYLTHIO)-m-TOLYLESTER

O,O-DIMETHYL-O-4-(METHYLMERCAPTO)-3-METHYLPHENYL THIO-
PHOSPHATE see O,O-DIMETHYLPHOSPHOROTHIOIC ACID-O-
(4-METHYLTHIO)-m-TOLYLESTER

O,O-DIMETHYL O-(4-METHYLMERCAPTOPHENYL) PHOSPHATE see
4-METHYLTHIOPHENYLDIMETHYL PHOSPHATE

DIMETHYL-1-METHYL-2-(METHYLCARBAMOYL)VINYLPHOSPHATE,
CIS see 3-(DIMETHOXYPHOSPHINYLOXY)N-METHYL-cis-CROTON-
AMIDE

O,O-DIMETHYL O-(3-METHYL-4-METHYLMERCAPTOPHENYL)PHOS-
PHOROTHIOATE see O,O-DIMETHYLPHOSPHOROTHIOIC ACID-O-
(4-METHYLTHIO)-m-TOLYLESTER

O,O-DIMETHYL-O-(3-METHYL-4-METHYLTHIO-FENYL)-MONOTHIOFOS-
FAAT (DUTCH) see O,O-DIMETHYLPHOSPHOROTHIOIC ACID-O-(4-
METHYLTHIO)-m-TOLYLESTER

O,O-DIMETHYL-O-(3-METHYL-4-METHYLTHIOPHENYL)-MONOTHIO-
PHOSPHAT (GERMAN) see O,O-DIMETHYLPHOSPHOROTHIOIC
ACID-O-(4-METHYLTHIO)-m-TOLYLESTER

O,O-DIMETHYL-O-3-METHYL-4-METHYLTHIOPHENYL PHOSPHORO-
THIOATE see O,O-DIMETHYLPHOSPHOROTHIOIC ACID-O-(4-
METHYLTHIO)-m-TOLYLESTER

O,O-DIMETHYL-O-(3-METHYL-4-NITROFENYL)-MONOTHIOFOSFAAT
(DUTCH) see O,O-DIMETHYL-O-(3-METHYL) PHOSPHOROTHIO-
ATE

O,O-DIMETHYL-O-(3-METHYL-4-NITRO-PHENYL)- MONOTHIOPHOS-
PHAT (GERMAN) see O,O-DIMETHYL-O-(3-METHYL) PHOSPHORO-
THIOATE

O,O-DIMETHYL-O-(3-METHYL-4-NITROPHENYL) PHOSPHORATE see
4-NITRO-M-CRESOL DIMETHYL PHOSPHATE

O,O-DIMETHYL-O-(3-METHYL-4-NITROPHENYL) PHOSPHOROTHIOATE
see O,O-DIMETHYL-O-(3-METHYL) PHOSPHOROTHIOATE

DIMETHYL-3-METHYL-4-NITRO PHENYLPHOSPHOROTHIONATE see
O,O-DIMETHYL-O-(3-METHYL) PHOSPHOROTHIOATE

O,O-DIMETHYL-O-(3-METHYL-4- NITROPHENYL) THIOPHOSPHATE see O,O-DIMETHYL-O-(3-METHYL) PHOSPHOROTHIOATE

N,N-DIMETHYL-2-((O-METHYL-ALPHA-PHENYL-BENZYL)OXY)-ETHYL-AMINE CITRATE see ORPHENADRINE CITRATE

N,N-DIMETHYL-2-(O-METHYL-ALPHA-PHENYLBENZYLOXY)ETHYL-AMINE HYDROCHLORIDE see ORPHENADRINE HYDROCHLORIDE

3,3-DIMETHYL-1-(O-METHYLPHENYL)TRIAZENE see 1-(O-METHYL-PHENYL)-3,3-DIMETHYL-TRIAZENE

DIMETHYL-3-(2-METHYL-1-PROPENYL)CYCLOPROPANECARBOXYLATE see BENZOFUROLINE

N,N-DIMETHYL-4-(4'-(2'-METHYLPYRIDYL-1'-OXIDE)AZO)ANILINE see N,N-DIMETHYL-4-(2-METHYL-4-PYRIDYLAZO)ANILINE-N-OXIDE

N,N-DIMETHYL-4-(4'-(2'METHYLPYRIDYL-1OXIDE)AZO)-O-TOLUIDINE see 4-((4-(DIMETHYLAMINO)-O-TOLYL)AZO)-2PICOLINE 1-OXIDE

N,N-DIMETHYL-4-(5'-(7'-METHYLQUINOLYL)AZO)ANILINE see 5-((p-(DIMETHYLAMINO)PHENYL)AZO)-7-METHYLQUINOLINE

N,N-DIMETHYL-2'-METHYLSTILBENAMINE see N,N,2'-TRIMETHYL-4-STILBENAMINE

O,O-DIMETHYL-O-(4-METHYLTHIO-3-METHYLPHENYL) PHOSPHORO-THIOATE see O,O-DIMETHYLPHOSPHOROTHIOIC ACID-O-(4-METHYLTHIO)-m-TOLYLESTER

3,5-DIMETHYL-4-(METHYLTHIO)PHENOLMETHYLCARBAMATE see 4-METHYLMERCAPTO-3,5-XYLYL METHYLCARBAMATE

3,5-DIMETHYL-4-METHYLTHIOPHENYL N-METHYLCARBAMATE see 4-METHYLMERCAPTO-3,5-XYLYL METHYLCARBAMATE

DIMETHYL P-(METHYLTHIO)PHENYL PHOSPHATE see 4-METHYLTHIO-PHENYLDIMETHYL PHOSPHATE

O,O-DIMETHYL-O-(4-(METHYLTHIO)-m-TOLYL) PHOSPHOROTHIOATE see O,O-DIMETHYLPHOSPHOROTHIOIC ACID-O-(4-METHYLTHIO)-m-TOLYLESTER

O,O-DIMETHYL O-((4-METHYLTHIO)-m-TOLYL) PHOSPHOROTHIOATE SULFONE see O,O-DIMETHYL-O-(4-(METHYLSULFONYL)-m-TOLYL) PHOSPHOROTHIOATE

O,O-DIMETHYL-S-(N-MONOMETHYL)-CARBAMYL METHYLDITHIOPHOS-PHATE see O,O-DIMETHYL METHYLCARBAMOYLMETHYL PHOS-PHORODITHIOATE

DIMETHYL MONOSULFATE see SULFURIC ACID, DIMETHYL ESTER

O,O-DIMETHYL-S-((MORFOLINO-CARBONYL)-METHYL)-DITHIOFOS-FAAT (DUTCH) see MORPHOTHION

3,17-DIMETHYL-9-ALPHA,13-ALPHA,14-ALPHA-MORPHINAN PHOS-PHATE see d-3-METHYL-N-METHYLMORPHINAN PHOSPHATE

O,O-DIMETHYL S-(MORPHOLINOCARBAMOYLMETHYL) DITHIOPHOS-PHATE see MORPHOTHION

O,O-DIMETHYL-S-((MORPHOLINO-CARBONYL)-METHYL)-DITHIOPHOS-PHAT (GERMAN) see MORPHOTHION

O,O-DIMETHYL MORPHOLINOCARBONYLMETHYL PHOSPHORODITHIO-ATE see MORPHOTHION

O,O-DIMETHYL S-(MORPHOLINOCARBONYLMETHYL) PHOSPHORODI-THIOATE see MORPHOTHION

DIMETHYL S-(MORPHOLINOCARBONYLMETHYL) PHOSPHOROTHIOLO-THIONATE see MORPHOTHION

DIMETHYL-ALPHA-NAPHTHYLAMINE see N,N-DIMETHYL-1-NAPH-THYLAMINE

ALPHA-DIMETHYLNAPHTHYLAMINE see N,N-DIMETHYL-1-NAPH-THYLAMINE

N,N-DIMETHYL-ALPHA-NAPHTHYLAMINE see N,N-DIMETHYL-1-NAPH-THYLAMINE

DIMETHYLNITRAMIN (GERMAN) see DIMETHYLNITRAMINE

DIMETHYLNITROAMINE see DIMETHYLNITRAMINE

O,O-DIMETHYL-P-NITRO-M-CHLOROPHENYL THIOPHOSPHATE see METHYLCHLOROTHION

O,O-DIMETHYL-O-4-NITRO-3-CHLOROPHENYL THIOPHOSPHATE see METHYLCHLOROTHION

O,O-DIMETHYL-O-P-NITROFENYLESTER KYSELINY THIOFOSFORECNE (CZECH) see METHYL PARATHION

O,O-DIMETHYL-S-P-NITROFENYL ESTER KYSELINY THIOFOSFORECNE (CZECH) see DIMETHYL PARANITROPHENYL THIONOPHOSPHATE

O,O-DIMETHYL-O-(4-NITROFENYL)-MONOTHIOFOSFAAT (DUTCH) see METHYL PARATHION

DIMETHYLNITROMETHANE see 2-NITROPROPANE

O,O-DIMETHYL-O-(4-NITRO-3-METHYLPHENYL)THIOPHOSPHATE see O,O-DIMETHYL-O-(3-METHYL) PHOSPHOROTHIOATE

O,O-DIMETHYL-O-(4-NITRO-PHENYL)-MONOTHIOPHOSPHAT (GERMAN) see METHYL PARATHION

DIMETHYL P-NITROPHENYL MONOTHIOPHOSPHATE see METHYL PARATHION

O,O-DIMETHYL-O-(4-NITROPHENYL) PHOSPHOROTHIOATE see METHYL PARATHION

O,O-DIMETHYL O-(P-NITROPHENYL) PHOSPHOROTHIOATE see METHYL PARATHION

O,O-DIMETHYL S-(P-NITROPHENYL) PHOSPHOROTHIOATE see DIMETHYL PARANITROPHENYL THIONOPHOSPHATE

O,O-DIMETHYL-O-(P-NITROPHENYL)-THIONOPHOSPHAT (GERMAN) see METHYL PARATHION

O,O-DIMETHYL O-(P-NITROPHENYL) THIONOPHOSPHATE see METHYL PARATHION

DIMETHYL-P-NITROPHENYL THIONPHOSPHATE see METHYL PARA-THION

DIMETHYL P-NITROPHENYL THIOPHOSPHATE see METHYL PARA-THION

O,O-DIMETHYL O-P-NITROPHENYL THIOPHOSPHATE see METHYL PARATHION

O,O-DIMETHYL S-(4-NITROPHENYL)THIOPHOSPHATE see DIMETHYL PARANITROPHENYL THIONOPHOSPHATE

DIMETHYLNITROSAMIN (GERMAN) see DIMETHYLNITROSAMINE

N,N-DIMETHYLNITROSAMINE see DIMETHYLNITROSAMINE

DIMETHYLNITROSOAMINE see DIMETHYLNITROSAMINE

DIMETHYLNITROSOHARNSTOFF (GERMAN) see 1,3-DIMETHYLNITRO-SOUREA

DIMETHYL(P-NITROSOPHENYL)AMINE see N,N-DIMETHYL-p-NITROSO-ANILINE

1,3-DIMETHYL-N-NITROSOUREA see 1,3-DIMETHYLNITROSOUREA

N,N'-DIMETHYLNITROSOUREA see 1,3-DIMETHYLNITROSOUREA

O,O-DIMETHYL-O-4-NITRO-M-TOLYL PHOSPHOROTHIOATE see O,O-DI-METHYL-O-(3-METHYL) PHOSPHOROTHIOATE

2,6-DIMETHYL-TRANS-2,6-OCTADIEN-8-OL see GERANIOL

3,7-DIMETHYL-TRANS-2,6-OCTADIEN-1-OL see GERANIOL

(4,4-DIMETHYLOCTANOYLOXY)TRIBUTYLSTANNANE see TRIBUTYL-TIN NEODECANDATE

2,6-DIMETHYL-2-OCTEN-8-OL see CITRONELLOL

3,7-DIMETHYL-6-OCTEN-1-OL see CITRONELLOL

2,3-DIMETHYL-7-OXABICYCLO(2.2.1)HEPTANE-2,3-DICARBOXYLIC ANHYDRIDE see CANTHARIDINE

2,3-DIMETHYL-7-OXABICYCLO(2.2.1)HEPTANE-2,3-DICARBOXYLIC ANHYDRIDE see CANTHARIDES CAMPHOR

3,3-DIMETHYL-2-OXETANONE see 3,3-DIMETHYL-2-OXETHANONE

O,O-DIMETHYL-S-(2-OXO-3-AZABUTYL)-MONOTHIOPHOSPHATE see DIMETHOATE OXYGEN ANALOG

O,O-DIMETHYL-S-(4-OXOBENZOTRIAZINO-3-METHYL)PHOSPHORODI-THIOATE see AZINPHOS METHYL

O,O-DIMETHYL-S-(4-OXO-1,2,3-BENZOTRIAZINO(3)-METHYL) THIO-THIONOPHOSPHATE see AZINPHOS METHYL

O,O-DIMETHYL-S-((4-OXO-3H-1,2,3-BENZOTRIAZIN-3-YL)-METHYL)-DITHIOFOSFAAT (DUTCH) see AZINPHOS METHYL

O,O-DIMETHYL-S-((4-OXO-3H-1,2,3-BENZOTRIAZIN-3-YL)-METHYL)-DITHIOPHOSPHAT (GERMAN) see AZINPHOS METHYL

O,O-DIMETHYL-S-4-OXO-1,2,3-BENZOTRIAZIN-3(4H)-YLMETHYL PHOSPHORODITHIOATE see AZINPHOS METHYL

O,O-DIMETHYL-S-(4-OXO-3H-1,2,3-BENZOTRIZIANE-3-METHYL)PHOS-PHORODITHIOATE see AZINPHOS METHYL

(5,5-DIMETHYL-3-OXO-CYCLOHEX-1-EN-YL)-N,N-DIMETHYL-CARBA-MAAT (DUTCH) see 5,5-DIMETHYLDIHYDRORESORCINOL DIMETHYLCARBAMATE

(5,5-DIMETHYL-3-OXO-CYCLOHEX-1-EN-YL)-N,N-DIMETHYL-CARBA-MAT (GERMAN) see 5,5-DIMETHYLDIHYDRORESORCINOL DIMETHYLCARBAMATE

5,5-DIMETHYL-3-OXOCYCLOHEX-1-ENYL DIMETHYLCARBAMATE see 5,5-DIMETHYLDIHYDRORESORCINOL DIMETHYLCARBAMATE

5,5-DIMETHYL-3-OXO-1-CYCLOHEXEN-1-YL DIMETHYLCARBAMATE see 5,5-DIMETHYLDIHYDRORESORCINOL DIMETHYLCARBAMATE

3-(2-(3,5-DIMETHYL-2-OXOCYCLOHEXYL)-2-HYDROXYETHYL)GLU-TARIMIDE see CYCLOHEXIMIDE

2-(2,2-DIMETHYL-1-OXOPROPYL)-1H-INDENE-1,3(2H)-DIONE see 2-PIVALOYL-1,3-INDANDIONE

O,O-DIMETHYL-S-(3-OXO-3-THIA-PENTYL)-MONOTHIOPHOSPHAT (GERMAN) see DEMETON-O-METHYL SULFOXIDE

2,2-DIMETHYL-4-OXYMETHYL-1,3-DIOXOLANE see DIOXOLAN

DIMETHYLOXYQUINAZINE see ANTIPYRINE

O,O-DIMETHYL-1-OXY-2,2,2-TRICHLOROETHYL PHOSPHONATE see ((2,2,2-TRICHLORO-1-HYDROXYETHYL) DIMETHYLPHOSPHONATE)

DIMETHYL PARATHION see METHYL PARATHION

3,5-DIMETHYLPERHYDRO-1,3,5-THIADIAZIN-2-THION (CZECH, GERMAN) see DIMETHYLFORMOCARBOTHIALDINE

N,N-DIMETHYL-P((M-ETHYLPHENYL)AZO)ANILINE see p-((m-ETHYL-PHENYL)AZO)-N,N-DIMETHYLANILINE

N,alpha-DIMETHYLPHENETHYLAMINE HYDROCHLORIDE see dl-DESOXY EPHEDRINE HYDROCHLORIDE

N,ALPHA-DIMETHYLPHENETHYLAMINE HYDROCHLORIDE, (−)- see "SPEED"

(+)-N,ALPHA-DIMETHYLPHENETHYLAMINE HYDROCHLORIDE see d-METHAPHETAMINE HYDROCHLORIDE

D-N,ALPHA-DIMETHYLPHENETHYLAMINE HYDROCHLORIDE see d-METHAPHETAMINE HYDROCHLORIDE

2,5-DIMETHYLPHENOL see 3,5-XYLENOL

O,O-DIMETHYL-S-(PHENYLACETIC ACID ETHYL ESTER) PHOSPHORO-DITHIOATE see (O,O-DIMETHYLDITHIOPHOSPHORYLPHENYL)-ACETIC ACID ETHYL ESTER

DIMETHYLPHENYLAMINE see XYLIDINE

2,5-DIMETHYLPHENYLAMINE see 2,5-XYLIDINE

3,5-DIMETHYLPHENYLAMINE see 3,5-XYLIDINE

N,N-DIMETHYLPHENYLAMINE see N,N-DIMETHYLANILINE

N,N-DIMETHYL-4-(PHENYLAZO)BENZAMINE see N,N-DIMETHYL-p-PHENYLAZOANILINE

N,N-DIMETHYL-4-(PHENYLAZO)BENZENAMINE see N,N-DIMETHYL-p-PHENYLAZOANILINE

4-((2,4-DIMETHYLPHENYL)AZO)-3-HYDROXY-2,7-NAPHTHALENEDI-SULFONIC ACID, DISODIUM SALT see D&C RED NO. 5

1-((2,4-DIMETHYLPHENYL)AZO)-2-NAPHTHALENOL see 1-(2,4-XYLY-LAZO)-2-NAPHTHOL

O,O-DIMETHYL-S-(PHENYL)(CARBOETHOXY)METHYL PHOSPHORODI-THIOATE see (O,O-DIMETHYLDITHIOPHOSPHORYLPHENYL)-ACETIC ACID ETHYL ESTER

(DIMETHYL S-(PHENYLETHOXYCARBONYLMETHYL)PHOSPHOROTHIO-LOTHIONATE) see (O,O-DIMETHYLDITHIOPHOSPHORYLPHENYL)-ACETIC ACID ETHYL ESTER

alpha,alpha-DIMETHYL-beta-PHENYLETHYLAMINE see alpha,alpha-DIMETHYLPHENETHYLAMINE

BETA-(2,4-DIMETHYLPHENYL)ETHYLHYDRAZINE DIHYDROGEN SUL-PHATE see o,p-DIMETHYL-beta-PHENYLETHYLHYDRAZINE DIHY-DROGENSULFATE

3,4-DIMETHYL-2-PHENYLMORPHOLINE BITARTRATE see DIETROL

1,1-DIMETHYL-4-PHENYLPIPERAZINIUM IODIDE see 1,1-DIMETHYL-4-PHENYLPIPERAZINE IODIDE

2,3-DIMETHYL-1-PHENYL-3-PYRAZOLIN-5-ONE see ANTIPYRINE

2,3-DIMETHYL-1-PHENYL-5-PYRAZOLONE see ANTIPYRINE

N,N-DIMETHYL-3-PHENYL-3-(2-PYRIDYL)PROPYLAMINE see TRIME-TON

N,N-DIMETHYL-2-(alpha-PHENYL-o-TOLOXY)-ETHYLAMINE HYDRO-CHLORIDE see BRISTAMIN HYDROCHLORIDE

DIMETHYL PHOSPHATE ESTER OF 3-HYDROXY-N-METHYL-CIS-CRO-TONAMIDE see 3-(DIMETHOXYPHOSPHINYLOXY)N-METHYL-cis-CROTONAMIDE

DIMETHYLPHOSPHATE ESTER WITH (E)-3-HYDROXY-N,N-DIMETHYL-CROTONAMIDE see 3-HYDROXYDIMETHYL CROTONAMIDE DIMETHYL PHOSPHATE

DIMETHYL PHOSPHATE ESTER WITH 3-HYDROXY-N,N-DIMETHYL-CIS-CROTONAMIDE see 3-HYDROXYDIMETHYL CROTONAMIDE DIMETHYL PHOSPHATE

DIMETHYL PHOSPHATE OF 2-CHLORO-N,N-DIETHYL-3-HYDROXY-CROTONAMIDE see DIMECRON

DIMETHYL PHOSPHATE OF 3-HYDROXY-N,N-DIMETHYL-CIS-CROTON-AMIDE see 3-HYDROXYDIMETHYL CROTONAMIDE DIMETHYL PHOSPHATE

DIMETHYL PHOSPHATE OF 3-HYDROXY-N-METHYL-CIS-CROTONAMINE see 3-(DIMETHOXYPHOSPHINYLOXY)N-METHYL-cis-CROTONAMIDE

DIMETHYL PHOSPHATE OF alpha-METHYLBENZYL-3-HYDROXY-cis-CROTONATE see CROTOXYPHOS

DIMETHYL PHOSPHATE OFMETHYL-3-HYDROXY-CIS-CROTONATE see MEVINPHOS

DIMETHYLPHOSPHORAMIDOCYANIDIC ACID, ETHYL ESTER see ETHYL DIMETHYL AMIDO CYANO PHOSPHATE

O,O-DIMETHYL PHOSPHORODITHIOATE N-FORMYL-2-MERCAPTO-N-METHYLACETAMIDE S-ESTER see O,O-DIMETHYL DITHIOPHOS-PHORYLACETIC ACID-N-METHYL-N-FORMYLAMIDE

N-((O,O-DIMETHYLPHOSPHORODITHIOYL)ETHYL)ACETAMIDE see O,O-DIMETHYL-S-(2-(ACETYLAMINO)ETHYL) DITHIOPHOSPHATE

O,O-DIMETHYL PHOSPHOROTHIOATE-O,O-DIESTER WITH 4,4'-THIODI-PHENOL see ABATE

DIMETHYL-P-PHTHALATE see TEREPHTHALIC ACID METHYL ESTER

(O,O-DIMETHYL-PHTHALIMIDIOMETHYL-DITHIOPHOSPHATE) see PHTHALIMIDOMETHYL-O,O-DIMETHYL PHOSPHORODITHIOATE

O,O-DIMETHYL S-(N-PHTHALIMIDOMETHYL) DITHIOPHOSPHATE see PHTHALIMIDOMETHYL-O,O-DIMETHYL PHOSPHORODITHIOATE

O,O-DIMETHYL S-PHTHALIMIDOMETHYL PHOSPHORODITHIOATE see PHTHALIMIDOMETHYL-O,O-DIMETHYL PHOSPHORODITHIOATE

2-BETA,16-BETA-(4'-DIMETHYL-1'-PIPERAZINO)-3-ALPHA,17-BETA-DIACETOXY-5-ALPHA-ANDROSTANE 2BR see PIPECURIUM BROMIDE

N,N-DIMETHYL-P(5'-QUINOLYLAZO)ANILINE see 5-((p-DIMETHYL-AMINO)PHENYL)AZO)QUINOLINE

N,N-DIMETHYL-1,3-PROPANEDIAMINE see 1-AMINO-3-DIMETHYL-AMINOPROPANE

2,2-DIMETHYLPROPANOIC ACID see PIVALIC ACID

DIMETHYL PROPIOLACTONE see 3,3-DIMETHYL-2-OXETHANONE

3,3-DIMETHYL-BETA-PROPIOLACTONE see 3,3-DIMETHYL-2-OXETHA-NONE

2,2-DIMETHYLPROPIONIC ACID see PIVALIC ACID

ALPHA,ALPHA-DIMETHYLPROPIONIC ACID see PIVALIC ACID

2,2-DIMETHYLPROPIONYL CHLORIDE see 2,2-DIMETHYLPROPANOYL CHLORIDE

N,N-DIMETHYL-1,3-PROPYLENEDIAMINE see 1-AMINO-3-DIMETHYL-AMINOPROPANE

5-(3-DIMETHYLPROPYLIDENE)DIBENZO(A,D)(1,4)CYCLOHEPTADIENE see ELAVIL

P-(1,1-DIMETHYLPROPYL)PHENOL see 4-tert-AMYLPHENOL

P-(ALPHA,ALPHA-DIMETHYLPROPYL)PHENOL see 4-tert-AMYLPHENOL

N-(1,1-DIMETHYLPROPYNYL)-3,5-DICHLOROBENZAMIDE see KERB

3,4-DIMETHYLPYRIDINE see 3,4-LUTIDINE

2,6-DIMETHYLPYRIDINE-1-OXIDE-4-AZO-P-DIMETHYL-ANILINE see 4-((4-(DIMETHYLAMINO)PHENYL)AZO)-2,6-LUTIDINE-1-OXIDE

N,N-DIMETHYL-N'-(2-PYRIDYL)-N'-(5-CHLORO-2-THENYL)ETHYL-ENEDIAMINE see CHLOROMETHAPYRILENE

N,N-DIMETHYL-N'-PYRID-2-YL-N'-2-THENYLETHYLENEDIAMINE see THENYLPYRAMINE

N-(4,6-DIMETHYL-2-PYRIMIDYL)SULFANILAMIDE see SULFADI-METHYLDIAZINE

1,3-DIMETHYL PYROGALLATE see 2,6-DIMETHOXYPHENOL

N,N-DIMETHYL-4-(5'-QUINOLYLAZO)ANILINE see 5-((p-DIMETHYL-AMINO)PHENYL)AZO)QUINOLINE

N,N-DIMETHYL-4-(6'-QUINOLYLAZO)ANILINE see 6-((p-(DIMETHYL-AMINO)PHENYL)AZO)QUINOLINE

6,7-DIMETHYL-9-D-RIBITYLISOALLOXAZINE see RIBOFLAVINE

N,N-DIMETHYLSEROTONIN see 3-(2-DIMETHYLAMINOETHYL)-5-INDOLOL

TRANS-N,N-DIMETHYL-4-STILBENAMINE see (E)-N,N-DIMETHYL-4-STILBENAMINE

N,N-DIMETHYL-P-STYRYLANILINE see N,N-DIMETHYL-4-STILBEN-AMINE

DIMETHYLSULFAAT (DUTCH) see SULFURIC ACID, DIMETHYL ESTER

4,6-DIMETHYL-2-SULFANILAMIDOPYRIMIDINE see SULFADIMETHYL-
DIAZINE

DIMETHYLSULFAT (CZECH) see SULFURIC ACID, DIMETHYL ESTER

DIMETHYL SULFATE see SULFURIC ACID, DIMETHYL ESTER

DIMETHYLSULFID (CZECH) see 2-THIOPROPANE

DIMETHYL SULFIDE see 2-THIOPROPANE

DIMETHYLSULFIDE-ALPHA,ALPHA′-DICARBOXYLIC ACID see MERCAP-
TODIACETIC ACID

1,4-DIMETHYLSULFONOXYBUTANE see 1,4-BUTANEDIOL DIMETHYL
SULFONATE

3-((2,4-DIMETHYL-5-SULFOPHENYL)AZO)-4-HYDROXY-1-NAPH-
THALENESULFONICACID, DISODIUM SALT see FOOD RED 4

DIMETHYL SULPHOXIDE see DIMETHYL SULFOXIDE

DIMETHYL TEREPHTHALATE see TEREPHTHALIC ACID METHYL ESTER

3,5-DIMETHYLTETRAHYDRO-1,3,5-THIADIAZINE-2-THIONE see
DIMETHYLFORMOCARBOTHIALDINE

3,5-DIMETHYLTETRAHYDRO-1,3,5-2H-THIADIAZINE-2-THIONE see DI-
METHYLFORMOCARBOTHIALDINE

3,5-DIMETHYL-1,3,5-2H-TETRAHYDROTHIADIAZINE-2-THIONE see DI-
METHYLFORMOCARBOTHIALDINE

3,5-DIMETHYLTETRAHYDRO-2H-1,3,5-THIADIAZINE-2-THIONE see DI-
METHYLFORMOCARBOTHIALDINE

3,5-DIMETHYL-1,2,3,5-TETRAHYDRO-1,3,5-THIADIAZINETHIONE-2
see DIMETHYLFORMOCARBOTHIALDINE

O,O-DIMETHYL-S-(3-THIA-PENTYL)-MONOTHIOPHOSPHAT (GERMAN)
see DEMETON-S-METHYL

2,2′-DIMETHYLTHIOCARBANILIDE see DI-O-TOLYLTHIOUREA

3,5-DIMETHYL-2-THIONOTETRAHYDRO-1,3,5-THIADIAZINE see
DIMETHYLFORMOCARBOTHIALDINE

O,O-DIMETHYLTHIOPHOSPHORIC ACID P-CHLOROPHENYL ESTER see
METHYL TRITHION

DIMETHYLTIN DICHLORIDE see DICHLORODIMETHYLSTANNANE

DIMETHYLTIN DIFLUORIDE see DIFLUORODIMETHYLSTANNANE

DIMETHYLTIN FLUORIDE see DIFLUORODIMETHYLSTANNANE

DIMETHYLTIN OXIDE see DIMETHYLOXOSTANNANE

3,3-DIMETHYL-1-(M-TOLYL)TRIAZENE see 3,3-DIMETHYL-1-(m-
METHYLPHENYL)TRIAZENE

3,3-DIMETHYL-1-(O-TOLYL)TRIAZENE see 1-(O-METHYLPHENYL)-3,3-
DIMETHYL-TRIAZENE

4-(3,3-DIMETHYL-1-TRIAZENO)IMIDAZOLE-5-CARBOXAMIDE see DA-
CARBAZINE

5-(3,3-DIMETHYL-1-TRIAZENO)IMIDAZOLE-4-CARBOXAMIDE see DA-
CARBAZINE

4-(5)-(3,3-DIMETHYL-1-TRIAZENO)IMIDAZOLE-5(4)-CARBOXAMIDE
see DACARBAZINE

3-(3′,3′-DIMETHYLTRIAZENO)-PYRIDINE-N-OXIDE see (3,3-
DIMETHYL-1-(m-PYRIDYL-N-OXIDE))TRIAZENE

3-(3′,3′-DIMETHYLTRIAZENO)-PYRIDIN-N-OXID (GERMAN) see
(3,3-DIMETHYL-1-(m-PYRIDYL-N-OXIDE))TRIAZENE

O,O-DIMETHYL-(2,2,2-TRICHLOOR-1-HYDROXY-ETHYL)-FOSFONAAT
(DUTCH) see ((2,2,2-TRICHLORO-1-HYDROXYETHYL) DIMETHYL-
PHOSPHONATE)

O,O-DIMETHYL-(2,2,2-TRICHLOR-1-HYDROXY-AETHYL)PHOSPHONAT
(GERMAN) see ((2,2,2-TRICHLORO-1-HYDROXYETHYL) DIMETHYL-
PHOSPHONATE)

O,O-DIMETHYL 2,2,2-TRICHLORO-1-(N-BUTYRYLOXY)ETHYLPHOS-
PHONATE see BUTONATE

DIMETHYLTRICHLOROHYDROXYETHYL PHOSPHONATE see ((2,2,2-
TRICHLORO-1-HYDROXYETHYL) DIMETHYLPHOSPHONATE)

DIMETHYL 2,2,2-TRICHLORO-1-HYDROXYETHYLPHOSPHONATE see
((2,2,2-TRICHLORO-1-HYDROXYETHYL) DIMETHYLPHOSPHONATE)

O,O-DIMETHYL-2,2,2-TRICHLORO-1-HYDROXYETHYL PHOSPHONATE
see ((2,2,2-TRICHLORO-1-HYDROXYETHYL) DIMETHYLPHOSPHO-
NATE)

O,O-DIMETHYL O-2,4,5-TRICHLOROPHENYL PHOSPHOROTHIOATE see
TRICHLOROMETAFOS

O,O-DIMETHYL O-(2,4,5-TRICHLOROPHENYL)THIOPHOSPHATE see
TRICHLOROMETAFOS

N,N-DIMETHYLTRYPTAMINE see 3-(2-(DIMETHYLAMINO)ETHYL)IN-
DOLE

DIMETHYL TUBOCURARINE see O,O′-DIMETHYLTUBOCURARINE

D,O-DIMETHYLTUBOCURARINE see O,O′-DIMETHYLTUBOCURARINE

N,N′-DIMETHYLUREA see 1,3-DIMETHYLUREA

SYM-DIMETHYLUREA see 1,3-DIMETHYLUREA

1,3-DIMETHYLXANTHINE see THEOPHYLLINE

3,7-DIMETHYLXANTHINE see THEOBROMINE

10,11-DIMETHYSTRYCHNINE see BRUCINE

5-(DIMETILAMINOETILOXIMINO-5H-DIBENZO(A,D)CICLOEPTA-1,4-
DIENE) CLORIDRATO(ITALIAN) see 5-DIMETHYLAMINOETHYL-
OXYIMINO-5H-DIBENZO(a,d)CYCLOHEPTA-1,4-DIENE HYDRO-
CHLORIDE

(4-DIMETILAMINO-3-METIL-FENIL)-N-METIL-CARBAMMATO (ITALIAN)
see 4-DIMETHYLAMINE-m-CRESYLMETHYLCARBAMATE

O,O-DIMETIL-O-(1,4-DIMETIL-3-OXO-4-AZA-PENT-1-ENIL)-FOSFATO
(ITALIAN) see 3-HYDROXYDIMETHYL CROTONAMIDE DIMETHYL
PHOSPHATE

2,6-DIMETIL-EPTAN-4-ONE (ITALIAN) see DIISOBUTYL KETONE

O,O-DIMETIL-S-(2-ETILITIO-ETIL)-MONOTIOFOSFATO (ITALIAN) see
DEMETON-S-METHYL

O,O-DIMETIL-S-(2-ETIL-SOLFINIL-ETIL)-MONOTIOFOSFATO (ITALIAN)
see DEMETON-O-METHYL SULFOXIDE

O,O-DIMETIL-S-(ETILTIO-ETIL)-DITIOFOSFATO (ITALIAN) see PHOS-
PHORODITHIOIC ACID-O,O-DIMETHYL-S-(2-ETHYLTHIO)ETHYL
ESTER

O,O-DIMETIL-O-(2-ETILTIO-ETIL)-MONOTIOFOSFATO (ITALIAN) see
DEMETON-O-METHYL

DIMETILFORMAMIDE (ITALIAN) see DIMETHYL FORMAMIDE

O,O-DIMETIL-S-(N-FORMIL-N-METIL-CARBAMOIL-METIL)-DITIOFOS-
FATO (ITALIAN) see O,O-DIMETHYL DITHIOPHOSPHORYLACETIC
ACID-N-METHYL-N-FORMYLAMIDE

O,O-DIMETIL-S-(N-METIL-CARBAMOIL-METIL)-DITIOFOSFATO
(ITALIAN) see O,O-DIMETHYL METHYLCARBAMOYLMETHYL PHOS-
PHORODITHIOATE

O,O-DIMETIL-S-(N-METIL-CARBAMOIL)-METIL-MONOTIOFOSFATO
(ITALIAN) see DIMETHOATE OXYGEN ANALOG

O,O-DIMETIL-O-(2-N-METILCARBAMOIL-1-METIL-VINIL)-FOSFATO
(ITALIAN) see 3-(DIMETHOXYPHOSPHINYLOXY)N-METHYL-cis-
CROTONAMIDE

O,O-DIMETIL-O-(3-METIL-4-METILTIO-FENIL)-MONOTIOFOSFATO
(ITALIAN) see O,O-DIMETHYLPHOSPHOROTHIOIC ACID-O-(4-
METHYLTHIO)-m-TOLYLESTER

O,O-DIMETIL-O-(3-METIL-4-NITRO-FENIL)-MONOTIOFOSFATO
(ITALIAN) see O,O-DIMETHYL-O-(3-METHYL) PHOSPHO-
ROTHIOATE

O,O-DIMETIL-S-((2-METOSSI-1,3,4-(4H)-TIADIZAOL-5-ON-4-IL)-
METIL)-DITIFOSFATO (ITALIAN) see O,O-DIMETHYL-S-(5-METH-
OXY-1,3,4-THIADIAZOLINYL-3-METHYL)DITHIO PHOSPHATE

O,O-DIMETIL-S-((MORFOLINO-CARBONIL)-METIL)-DITIOFOSFATO
(ITALIAN) see MORPHOTHION

O,O-DIMETIL-O-(4-NITRO-FENIL)-MONOTIOFOSFATO (ITALIAN) see
METHYL PARATHION

O,O-DIMETIL-S-((4-OXO-3H-1,2,3-BENZOTRIAZIN-3-IL)-METIL)-DITIO-
FOSFATO (ITALIAN) see AZINPHOS METHYL

(5,5-DIMETIL-3-OXO-CICLOES-1-EN-IL)-N,N-DIMETIL-CARBAMMATO
(ITALIAN) see 5,5-DIMETHYLDIHYDRORESORCINOL DIMETHYLCAR-
BAMATE

3,5-DIMETIL-PERIDRO-1,3,5-TIHADIAZIN-2-TIONE (ITALIAN) see
DIMETHYLFORMOCARBOTHIALDINE

DIMETILSOLFATO (ITALIAN) see SULFURIC ACID, DIMETHYL ESTER

O,O-DIMETIL-(2,2,2-TRICLORO-1-IDROSSI-ETIL)-FOSFONATO
(ITALIAN) see ((2,2,2-TRICLORO-1-HYDROXYETHYL)
DIMETHYLPHOSPHONATE)

3,3′-DIMETOSSIBENZODINA (ITALIAN) see O-DIANISIDINE

DIMETRIDAZOLE see 1,2-DIMETHYL-5-NITROIMIDAZOLE

DIMORPHOLINE DISULFIDE see N,N′-BISMORPHOLINE DISULFIDE

DIMORPHOLINO DISULFIDE see N,N′-BISMORPHOLINE DISULFIDE

3,4,5,6-DINAPHTHACARBAZOLE see 7H-DIBENZO(c,g)CARBAZOLE

1,2,5,6-DINAPHTHACRIDINE see DIBENZ(a,h)ACRIDINE

3,4,6,7-DINAPHTHACRIDINE see DIBENZ(a,j)ACRIDINE

DI-BETA-NAPHTHYL-P-PHENYLDIAMINE see 2-NAPHTHYL-p-PHENYL-
ENEDIAMINE

DI-BETA-NAPHTHYL-P-PHENYLENEDIAMINE see 2-NAPHTHYL-p-PHE-
NYLENEDIAMINE

N,N'-DI-BETA-NAPHTHYL-P-PHENYLENEDIAMINE see 2-NAPHTHYL-p-
PHENYLENEDIAMINE

SYM-DI-BETA-NAPHTHYL-P-PHENYLENEDIAMINE see 2-NAPHTHYL-p-
PHENYLENEDIAMINE

DI-NATRIUM-AETHYLENBISDITHIOCARBAMAT (GERMAN) see NABAM

DINATRIUM-(N,N'-AETHYLEN-BIS(DITHIOCARBAMAT)) (GERMAN) see
NABAM

DINATRIUM-(3,6-EPOXY-CYCLOHEXAAN-1,2-DICARBOXYLAAT)
(DUTCH) see DISODIUM-3,6-ENDOXOHEXAHYDROPHTHALATE

DINATRIUM-(3,6-EPOXY-CYCLOHEXAN-1,2-DICARBOXYLAT)
(GERMAN) see DISODIUM-3,6-ENDOXOHEXAHYDROPHTHALATE

DINATRIUM-(N,N'-ETHYLEEN-BIS(DITHIOCARBAMAAT)) (DUTCH) see
NABAM

DINATRIUMPYROPHOSPHAT (GERMAN) see DISODIUM PYROPHOS-
PHATE

DINICKEL TRIOXIDE see NICKEL PEROXIDE

2,4-DINITRANILINE see 2,4-DINITROANILINE

DINITRATE DE DIETHYLENE-GLYCOL (FRENCH) see DIETHYLENE
GLYCOL DINITRATE

1,3-DINITRATO-2,2-BIS(NITRATOMETHYL)PROPANE see PENTAERY-
THRITOL TETRANITRATE

2,3-DINITRILO-1,4-DITHIA-ANTHRAQUINONE see 5,10-DIHYDRO-
5,10-DIOXONAPHTHO(2,3-b)-p-DITHIIN-2,3-DICARBONITRILE

2,3-DINITRILO-1,4-DITHIOANTHRACHINON (GERMAN) see 5,10-DIHY-
DRO-5,10-DIOXONAPHTHO(2,3-b)-p-DITHIIN-2,3-DICARBONITRILE

2,4-DINITROANILIN (GERMAN) see 2,4-DINITROANILINE

2,4-DINITROANILINA (ITALIAN) see 2,4-DINITROANILINE

2,4-DINITROANISOLE see 2,4-DINITROANISOL

ALPHA-DINITROANISOLE see 2,4-DINITROANISOL

1,3-DINITROBENZENE see m-DINITROBENZENE

2,4-DINITROBENZENE see m-DINITROBENZENE

O-DINITROBENZOL see o-DINITROBENZENE

1,3-DINITROBENZOL see m-DINITROBENZENE

4,4'-DINITROBIFENYL (CZECH) see 4,4'-DINITROBIPHENYL

4,6-DINITRO-2-SEC.BUTYLFENOL (CZECH) see 2-sec-BUTYL-4,6-DINI-
TRO PHENOL

4,6-DINITRO-2-SEC.BUTYLFENOLATE AMMONY (CZECH) see BUTO-
PHEN

2,4-DINITRO-6-S-BUTYLFENYLESTER KYSELINY OCTOVE (CZECH) see
O-ACETYL-2-sec-BUTYL-4,6-DINITROPHENOL

DINITROBUTYLPHENOL see 2-sec-BUTYL-4,6-DINITRO PHENOL

2,4-DINITRO-6-SEC-BUTYLPHENOL see 2-sec-BUTYL-4,6-DINITRO
PHENOL

4,6-DINITRO-2-SEC-BUTYLPHENOL see 2-sec-BUTYL-4,6-DINITRO
PHENOL

4,6-DINITRO-O-SEC-BUTYLPHENOL see 2-sec-BUTYL-4,6-DINITRO
PHENOL

4,6-DINITRO-2-SEC-BUTYLPHENOL AMMONIUM SALT see BUTOPHEN

4,6-DINITRO-O-SEC-BUTYLPHENOL AMMONIUM SALT see BUTOPHEN

4,6-DINITRO-2-S-BUTYLPHENYL ACETATE see O-ACETYL-2-sec-
BUTYL-4,6-DINITROPHENOL

2,4-DINITRO-6-SEC-BUTYLPHENYL-2-METHYLCROTONATE see BINAP
ACRYL

4,6-DINITRO-2-CAPRYLPHENYL CROTONATE see ARATHANE

4,6-DINITRO-2-(2-CAPRYL)PHENYL CROTONATE see ARATHANE

2,4-DINITROCHLOROBENZENE see 1-CHLORO-2,4-DINITROBENZENE

1,3-DINITRO-4-CHLOROBENZENE see 1-CHLORO-2,4-DINITROBEN-
ZENE

2,4-DINITRO-1-CHLOROBENZENE see 1-CHLORO-2,4-DINITROBEN-
ZENE

DINITRO-P-CRESOL see 2,6-DINITRO-p-CRESOL

2,4-DINITRO-O-CRESOL see 4,6-DINITRO-O-CRESOL

4,6-DINITRO-O-CRESOLO (ITALIAN) see 4,6-DINITRO-O-CRESOL

4,6-DINITRO-O-CRESOL SODIUM SALT see 3,5-DINITRO-O-CRESOL SO-
DIUM SALT

DINITROCYCLOHEXYLPHENOL see 2-CYCLOHEXYL-4,6-DINITROPHE-
NOL

DINITRO-O-CYCLOHEXYLPHENOL see 2-CYCLOHEXYL-4,6-DINITRO-
PHENOL

2,4-DINITRO-6-CYCLOHEXYLPHENOL see 2-CYCLOHEXYL-4,6-DINI-
TROPHENOL

4,6-DINITRO-O-CYCLOHEXYLPHENOL see 2-CYCLOHEXYL-4,6-DINI-
TROPHENOL

DINITROCYCLOHEXYLPHENOL (DOT) see 2-CYCLOHEXYL-4,6-DINI-
TROPHENOL

DINITRODIGLICOL (ITALIAN) see DIETHYLENE GLYCOL DINITRATE

DINITRODIGLYKOL (CZECH) see DIETHYLENE GLYCOL DINITRATE

2,6-DINITRO-N,N-DIPROPYL-4-(TRIFLUOROMETHYL)BENZENAMINE
see TRIFLURALINE

2,6-DINITRO-N,N-DI-N-PROPYL-ALPHA,ALPHA,ALPHA-TRIFLURO-P-
TOLUIDINE see TRIFLURALINE

2,4-DINITROFENOL (DUTCH) see 2,4-DINITROPHENOL

DINITROFENOLO (ITALIAN) see 2,4-DINITROPHENOL

DINITROGEN MONOXIDE see NITROGEN OXIDE

DINITROGEN TETROXIDE see NITROGEN TETROXIDE

DINITROGLICOL (ITALIAN) see ETHYLENE GLYCOL DINITRATE

DINITROGLYCOL see ETHYLENE GLYCOL DINITRATE

3,5-DINITRO-2-HYDROXYTOLUENE see 4,6-DINITRO-o-CRESOL

4,6-DINITROKRESOL (DUTCH) see 4,6-DINITRO-o-CRESOL

4,6-DINITRO-O-KRESOL (CZECH) see 4,6-DINITRO-o-CRESOL

4,6-DINITRO-O-KRESYLESTER KYSELINY OCTOVE (CZECH) see
ACETIC ACID-4,6-DINITRO-o-CRESYL ESTER

2,6-DINITRO-3-METHOXY-4-tert-BUTYLTOLUENE see 6-tert-BUTYL-3-
METHYL-2,4-DINITRO ANISOLE

DINITROMETHYL CYCLOHEXYLTRIENOL see 4,6-DINITRO-o-CRESOL

DINITRO(1-METHYLHEPTYL)PHENYL CROTONATE see ARATHANE

2,4-DINITRO-6-(1-METHYLHEPTYL)PHENYL CROTONATE see ARA-
THANE

2,4-DINITRO-6-METHYLPHENOL see 4,6-DINITRO-o-CRESOL

2,4-DINITRO-6-METHYLPHENOL SODIUM SALT see 3,5-DINITRO-o-
CRESOL SODIUM SALT

4,6-DINITRO-2-(1-METHYL-N-PROPYL)PHENOL see 2-sec-BUTYL-4,6-
DINITRO PHENOL

2,4-DINITRO-6-(1-METHYL-PROPYL)PHENOL (FRENCH) see 2-sec-BU-
TYL-4,6-DINITRO PHENOL

2-4 DINITRO-alpha-NAPHTOL (FRENCH) see 2,4-DINITRO-1-NAPHTHOL

2,4-DINITRO-6-(2-OCTYL)PHENYL CROTONATE see ARATHANE

ALPHA-DINITROPHENOL see 2,4-DINITROPHENOL

beta-DINITROPHENOL see 2,6-DINITROPHENOL

gamma-DINITROPHENOL see 2,5-DINITROPHENOL

4,6-DINITROPHENYL-2-SEC-BUTYL-3-METHYL-2-BUTENONATE see
BINAPACRYL

2,4-DINITROPHENYL ETHER OF MORPHINE see 2,4-DINITROPHENYL-
MORPHINEHYDROCHLORIDE

2,4-DINITROPHENYLMETHYL ETHER see 2,4-DINITROANISOL

N,N'-DINITROSO-N,N'-DIETHYLETHYLENEDIAMINE see N,N'-DIETHYL-
N,N'-DINITROSOETHYLENEDIAMINE

N,N'-DINITROSO-N,N'-DIMETHYL-1,3-PROPANEDIAMINE see
DINITROSODIMETHYLPROPANEDIAMINE

N,N-DINITROSOPENTAMETHYLENETETRAMINE see
DINITROSOPENTAMETHYLENETETRAMINE

3,4-DI-N-NITROSOPENTAMETHYLENETETRAMINE see
DINITROSOPENTAMETHYLENETETRAMINE

3,7-DI-N-NITROSOPENTAMETHYLENETETRAMINE see
DINITROSOPENTAMETHYLENETETRAMINE

N,N'-DINITROSOPENTAMETHYLENETETRAMINE see
DINITROSOPENTAMETHYLENETETRAMINE

N(SUP 1),N(SUP 3)-DINITROSOPENTAMETHYLENETETRAMINE see
DINITROSOPENTAMETHYLENETETRAMINE

DINITROSOPIPERAZIN (GERMAN) see DINITROSOPIPERAZINE

1,4-DINITROSOPIPERAZINE see DINITROSOPIPERAZINE

N,N'-DINITROSOPIPERAZINE see DINITROSOPIPERAZINE

3,7-DINITROSO-1,3,5,7-TETRAAZABICYCLO-(3,3,1)-NONANE see
DINITROSOPENTAMETHYLENETETRAMINE

2,4-DINITROTOLUOL see 2,4-DINITROTOLUENE

2,6-DINITRO-4-TRIFLUORMETHYL-N,N-DIPROPYLANILIN (GERMAN)
see TRIFLURALINE

DINKUM OIL see EUCALYPTUS OIL

DINOSEBE (FRENCH) see 2-sec-BUTYL-4,6-DINITRO PHENOL

DIOCTYL-O-BENZENEDICARBOXYLATE see n-DIOCTYL PHTHALATE

2,2-DIOCTYL-1,3-DIOXA-2-STANNA-7-THIADECAN-4,10-DIONE see DIOCTYLTIN-3,3′-THIODIPROPIONATE

DIOCTYL ESTER OF SODIUM SULFOSUCCINATE see DIOCTYL SODIUM SULFOSUCCINATE

DIOCTYL PHTHALATE see n-DIOCTYL PHTHALATE

DI-SEC-OCTYL PHTHALATE see BIS(2-ETHYLHEXYL)PHTHALATE

DIOCTYL SEBACATE see BIS(2-ETHYLHEXYL)SEBACATE

DI-N-OCTYLTIN BIS(BUTYL MALEATE) see BIS(BUTOXYMALEOYLOXY)DIOCTYLSTANNANE

DIOCTYLTIN BIS(ISOOCTYL MERCAPTOACETATE) see BIS(ISOOCTYL OXYCARBONYL METHYL THIO)DIOCTYL STANNANE

DIOCTYLTIN BIS(ISOOCTYL THIOGLYCOLATE) see BIS(ISOOCTYL OXYCARBONYL METHYL THIO)DIOCTYL STANNANE

DIOCTYLTIN DICHLORIDE see DICHLORODIOCTYLSTANNANE

DI-N-OCTYLTINDICHLORIDE see DICHLORODIOCTYLSTANNANE

DI-N-OCTYLTIN DIISOOCTYL THIOGLYCOLATE see BIS(ISOOCTYL OXYCARBONYL METHYL THIO)DIOCTYL STANNANE

DI-N-OCTYLTIN DIMONOBUTYLMALEATE see BIS(BUTOXYMALEOYLOXY)DIOCTYLSTANNANE

DI-N-OCTYLTIN DI(1,2-PROPYLENEGLYCOLMALEATE) see DIOCTYL(1,2-PROPYLENEDIOXYBIS(MALEOYLDIOXY))STANNANE

DIOCTYLTIN-S,S′-BIS(ISOOCTYL MERCAPTOACETATE) see BIS(ISOOCTYL OXYCARBONYL METHYL THIO)DIOCTYL STANNANE

DI-N-OCTYL-ZINN DICHLORID (GERMAN) see DICHLORODIOCTYLSTANNANE

DI-N-OCTYL-ZINN-DI-ISOOCTYLTHIOGLYKOLAT (GERMAN) see BIS-(ISOOCTYL OXYCARBONYL METHYL THIO)DIOCTYL STANNANE

DI-N-OCTYLZINN-DIMONOBUTYLMALEINAT (GERMAN) see BIS(BUTOXYMALEOYLOXY)DIOCTYLSTANNANE

DI-N-OCTYL-ZINN-DI-(1,2-PROPYLENGLYKOLMALEINAT)(GERMAN) see DIOCTYL(1,2-PROPYLENEDIOXYBIS(MALEOYLDIOXY))STANNANE

DIODOHYDROXYQUIN see DIIODOHYDROXYQUIN

DIOGYNETS see ESTRADIOL (β)

DIOKSAN (POLISH) see p-DIOXANE

DIONIN see ETHYL MORPHINE HYDROCHLORIDE DIHYDRATE

DIORTHOTOLYLGUANIDINE see DI-O-TOLYLGUANIDINE

DIOSSANO-1,4 (ITALIAN) see p-DIOXANE

1,4-DIOSSIBENZENE (ITALIAN) see p-BENZOQUINONE

2,4-DIOSSI-5-DIAZOPIRIMIDINA (ITALIAN) see DIAZOURACIL

DIOTHENE see POLYETHYLENE

DIOXAAN-1,4 (DUTCH) see p-DIOXANE

1,4-DIOXACYCLOHEXANE see p-DIOXANE

P-DIOXAN (CZECH) see p-DIOXANE

DIOXAN-1,4 (GERMAN) see p-DIOXANE

1,4-DIOXANE see p-DIOXANE

DIOXANNE (FRENCH) see p-DIOXANE

3,6-DIOXAOCTANE-1,8-DIOL see TRIETHYLENE GLYCOL

DIOXIME-1,4-CYCLOHEXADIENEDIONE see DIOXIME-p-BENZOQUINONE

DIOXIME-2,5-CYCLOHEXADIENE-1,4-DIONE see DIOXIME-p-BENZOQUINONE

DIOXIN (BACTERICIDE) (OBS.) see ACETOMETHOXANE

DIOXIN (HERBICIDE CONTAMINANT) see TCDD

DIOXITOL see CARBITOL CELLOSOLVE

9,10-DIOXOANTHRACENE see ANTHRAQUINONE

P-DIOXOBENZENE see HYDROQUINONE

1,1′,1″-(3,6-DIOXO-1,4-CYCLOHEXADIENE-1,2,4-TRIYL)TRISAZIRIDINE see TRISETHYLENEIMINOQUINONE

DIOXODICHLOROCHROMIUM see CHROMIUM OXYCHLORIDE

5,5-DIOXO-10-(2-(DIMETHYLAMINO)PROPYL)PHENOTHIAZINE HYDROCHLORIDE see DIOXOPROMETHAZINE HYDROCHLORIDE

3,5-DIOXO-1,2-DIPHENYL-4-N-BUTYLPYRAZOLIDENE see 4-BUTYL-1,2-DIPHENYL-3,5-DIOXO PYRAZOLIDINE

3,5-DIOXO-1,2-DIPHENYL-4-N-BUTYL-PYRAZOLIDIN see 4-BUTYL-1,2-DIPHENYL-3,5-DIOXO PYRAZOLIDINE

3,5-DIOXO-1,2-DIPHENYL-4-M-BUTYL-PYRAZOLIDINE see 4-BUTYL-1,2-DIPHENYL-3,5-DIOXO PYRAZOLIDINE

3,5-DIOXO-1,2-DIPHENYL-4-N-BUTYLPYRAZOLIDIN SODIUM see BUTAZOLIDINE SODIUM

DIOXOLANE (DOT) see DIOXOLAN

2-(1,3-DIOXOLANE-2-YL)PHENYL N-METHYLCARBAMATE see O-(1,3-DIOXOLAN-2-YL)PHENYL METHYLCARBAMATE

((1,3-DIOXOLAN-4-YL)METHYL)TRIMETHYLAMMONIUM IODIDE see FORMAL-gamma-TRIMETHYL AMMONIUM PROPANEDIOL

2-(1,3-DIOXOLAN-2-YL)PHENYL-N-METHYLCARBAMAT see O-(1,3-DIOXOLAN-2-YL)PHENYL METHYLCARBAMATE

2,6-DIOXO-4-METHYL-4-ETHYLPIPERIDINE see 3-METHYL-3-ETHYL-GLUTARIMIDE

3,5-DIOXO-1-PHENYL-2-(P-HYDROXYPHENYL)-4-N-BUTYLPYRAZOLIDENE see p-HYDROXYPHENYLBUTAZONE

1,3-DIOXOPHTHALAN see PHTHALIC ANHYDRIDE

1,3-DIOXO-5-PHTHALANCARBOXYLIC ACID see TRIMELLIC ACID ANHYDRIDE

2,6-DIOXO-3-PHTHALIMIDOPIPERIDINE see THALIDOMIDE

N-(2,6-DIOXO-3-PIPERIDYL)PHTHALIMIDE see THALIDOMIDE

N-(2,6-DIOXO-3-PIPERIDYL)PHTHALIMIDE (+)- see (+)-THALIDOMIDE

N-(2,6-DIOXO-3-PIPERIDYL)PHTHALIMIDE (+)- see (+)-THALIDOMIDE

2,2′-DIOXOPROPYL-N-PROPYLNITROSAMINE see N-NITROSOBIS(2-OXOPROPYL)AMINE

2,6-DIOXOPURINE see XANTHINE

DIOXYANTHRANOL see 1,8,9-ANTHRACENETRIOL

1,4-DIOXYANTHRAQUINONE (RUSSIAN) see 1,4-DIHYDROXYANTHRAQUINONE

M-DIOXYBENZENE see RESORCINOL

O-DIOXYBENZENE see PYROCATECHOL

1,4-DIOXYBENZENE see p-BENZOQUINONE

3,3′-DIOXYBENZIDINE see 3,3′-DIHYDROXYBENZIDINE

1,4-DIOXY-BENZOL (GERMAN) see p-BENZOQUINONE

DIOXYDE DE BARYUM (FRENCH) see BARIUM PEROXIDE

DIOXYDEMETON-S-METHYL see DEMETON-S-METHYL SULPHONE

3,3′-(DIOXYDICARBONYL)DIPROPIONIC ACID see SUCCINIC PEROXIDE

DIOXYETHYLENE ETHER see p-DIOXANE

DIOXYMETHYLENE-PROTOCATECHUIC ALDEHYDE see PIPERONAL

DI(P-OXYPHENYL)-2,4-HEXADIENE see DEHYDROSTILBESTROL

DIPARALENE HYDROCHLORIDE see CHLORCYCLIZINE HYDROCHLORIDE

DI-PARALENE-2-HYDROCHLORIDE see CHLORCYCLIZINE DIHYDROCHLORIDE

DI(PENTANOYLOXY)DIBUTYLSTANNANE see DIBUTYL DIPENTANOYLOXY STANNANE

DIPENTENE DIOXIDE see LIMONENE DIOXIDE

DIPENTYLAMINE see DIAMYL AMINE

DIPENTYL ETHER see PENTYL ETHER

DIPENTYLNITROSAMINE see DI-n-AMYLNITROSAMINE

DI-N-PENTYLNITROSAMINE see DI-n-AMYLNITROSAMINE

O-DIPHENOL see PYROCATECHOL

DIPHENOXYLATE HYDROCHLORIDE see LOMOTIL

DIPHENYL see BIPHENYL

2-DIPHENYLACETYL-1,3-DIKETOHYDRINDENE see DIPHENADIONE

2-(DIPHENYLACETYL)INDAN-1,3-DIONE see DIPHENADIONE

2-DIPHENYLACETYL-1,3-INDANDIONE see DIPHENADIONE

2-(DIPHENYLACETYL)-1H-INDENE-1,3(2H)-DIONE see DIPHENADIONE

2,3-DIPHENYLACRYLONITRILE see 2,3-DIPHENYLACRYLONITRILE

ALPHA,BETA-DIPHENYLACRYLONITRILE see 2,3-DIPHENYLACRYLONITRILE

DIPHENYLAMINE-2-CARBOXYLIC ACID see N-PHENYLANTHRANILIC ACID

DIPHENYLAMINECHLOROARSINE see PHENARSAZINE CHLORIDE

DIPHENYL BLUE 2B see C.I. DIRECT BLUE 6, TETRASODIUM SALT

DIPHENYLBUTAZONE see 4-BUTYL-1,2-DIPHENYL-3,5-DIOXO PYRAZOLIDINE

1,2-DIPHENYL-4-BUTYL-3,5-DIOXOPYRAZOLIDINE see 4-BUTYL-1,2-DIPHENYL-3,5-DIOXO PYRAZOLIDINE

1,2-DIPHENYL-4-BUTYL-3,5-PYRAZOLIDINEDIONE see 4-BUTYL-1,2-DIPHENYL-3,5-DIOXO PYRAZOLIDINE

DIPHENYLCHLOORARSINE (DUTCH) see CHLORO DIPHENYL ARSINE

DIPHENYLCHLOROARSINE see CHLORO DIPHENYL ARSINE

ALPHA,BETA-DIPHENYLCINNAMONITRILE see 2,3,3-TRIPHENYL-ACRYLONITRILE

DIPHENYL-ALPHA-CYANOMETHANE see DIPHENYLACETONITRILE

2,4′-DIPHENYLDIAMINE see 2,4′-BIPHENYLDIAMINE

DIPHENYLDIAZENE see AZOBENZENE

1,2-DIPHENYLDIAZENE see AZOBENZENE

DIPHENYLDIIMIDE see AZOBENZENE

2,2-DIPHENYL-N,N-DIMETHYLACETAMIDE see N,N-DIMETHYL-2,2-DIPHENYLACETAMIDE

4,4-DIPHENYL-6-DIMETHYLAMINO-3-HEPTANONE HYDROCHLORIDE (±)- see dl-METHADONE HYDROCHLORIDE

1,1-DIPHENYL-1-(DIMETHYLAMINOISOPROPYL)BUTANONE-2 see ISO-METHADONE

1,1-DIPHENYL-1-(2-DIMETHYLAMINOPROPYL)-2-BUTANONE HYDRO-CHLORIDE, (±)- see dl-METHADONE HYDROCHLORIDE

1,1-DIPHENYL-1-(BETA-DIMETHYLAMINOPROPYL)BUTANONE-2 HY-DROCHLORIDE see METHADONE HYDROCHLORIDE

ALPHA,ALPHA-DIPHENYL-GAMMA-DIMETHYLAMINOVALERAMIDE see 4-(DIMETHYLAMINO)-2,2-DIPHENYLVALERAMIDE

1,2-DIPHENYL-3,5-DIOXO-4-BUTYLPYRAZOLIDINE see 4-BUTYL-1,2-DIPHENYL-3,5-DIOXO PYRAZOLIDINE

DIPHENYLDIOXOBUTYLPYRAZOLIDINE-BUTAZOLIDINE-SODIUM see BUTAZOLIDINE SODIUM

1,2-DIPHENYL-2,3-DIOXO-4-N-BUTYLPYRAZOLINE see 4-BUTYL-1,2-DIPHENYL-3,5-DIOXO PYRAZOLIDINE

DIPHENYLE CHLORE, 42% DE CHLORE (FRENCH) see POLYCHLORI-NATED BIPHENYL (AROCLOR 1242)

DIPHENYLE CHLORE, 54% DE CHLORE (FRENCH) see POLYCHLORI-NATED BIPHENYL (AROCLOR 1254)

DIPHENYLENE DIOXIDE see DIBENZO-p-DIOXIN

DIPHENYLENEIMINE see CARBAZOLE

DIPHENYLENIMIDE see CARBAZOLE

DIPHENYLENIMINE see CARBAZOLE

1,2-DIPHENYLETHANE see BIBENZYL

DIPHENYLETHYLENE see STILBENE

DIPHENYL FAST BROWN BRL see C.I. DIRECT BROWN

DIPHENYLGLYCOLLIC ACID-2-(DIETHYLAMINO)ETHYL ESTER HYDRO-CHLORIDE see BENZILIC ACID-beta-DIETHYLAMINOETHYLESTER HYDROCHLORIDE

1,3-DIPHENYLGUANIDINE see DIPHENYLGUANIDINE

N,N′-DIPHENYLGUANIDINE see DIPHENYLGUANIDINE

DIPHENYLHYDANTOIN see DIHYDANTOIN

5,5-DIPHENYLHYDANTOIN see DIHYDANTOIN

DIPHENYLHYDANTOIN SODIUM see DILANTIN

5,5-DIPHENYLHYDANTOIN SODIUM see DILANTIN

DIPHENYLHYDRAMINE see BENZHYDRYL

DIPHENYLHYDRAMINE HYDROCHLORIDE see BENADRYL HYDROCHLO-RIDE

1,2-DIPHENYLHYDRAZINE see HYDRAZOBENZENE

2,2-DIPHENYL-3-HYDROXYPROPIONIC ACID LACTONE see 3,3-DIPHE-NYL-2-OXETANONE

5,5-DIPHENYLIMIDAZOLIDIN-2,4-DIONE see DIHYDANTOIN

DIPHENYLINE see 2,4′-BIPHENYLDIAMINE

DIPHENYL KETONE see BENZOPHENONE

DIPHENYLMERCURIDINAPHTHYLMETHANEDISULFONATE see PHENYL-MERCURIC DINAPHTHYLMETHANEDISULFONATE

DIPHENYLMETHAN-4,4′-DIISOCYANAT (GERMAN) see DIPHENYL METHANE DIISOCYANATE

4,4′-DIPHENYLMETHANEDIAMINE see 4,4′-METHYLENEDIANILINE

DIPHENYLMETHANE 4,4′-DIISOCYANATE see DIPHENYL METHANE DIISOCYANATE

4,4′-DIPHENYLMETHANE DIISOCYANATE see DIPHENYL METHANE DIISOCYANATE

P,P′-DIPHENYLMETHANE DIISOCYANATE see DIPHENYL METHANE DIISOCYANATE

DIPHENYLMETHANE-4,4′-DIISOCYANATE-TRIMELLIC ANHYDRIDE-ETHOMID HT POLYMER see TRIMELLIC ACID ANHYDRIDE

DIPHENYLMETHANONE see BENZOPHENONE

2-(DIPHENYLMETHOXY)-N,N-DIMETHYLETHYLAMINE see BENZHY-DRYL

2-DIPHENYLMETHOXY-N,N-DIMETHYLETHYLAMINE HYDROCHLORIDE see BENADRYL HYDROCHLORIDE

4-(DIPHENYLMETHOXY)-1-METHYLPIPERIDINE see LYSSIPOLL

3-DIPHENYLMETHOXYTROPANE MESYLATE see TROPINE BENZOHY-DRYL ETHER METHANESULFONATE

3-DIPHENYLMETHOXYTROPANE METHANESULFONATE see TROPINE BENZOHYDRYL ETHER METHANESULFONATE

DIPHENYL METHYL BROMIDE, SOLID (DOT) see BROMO DIPHENYL METHANE

DIPHENYLMETHYLCYANIDE see DIPHENYLACETONITRILE

1-(3-(4-(DIPHENYLMETHYL)-1-PIPERAZINYL)PROPYL)-1,3-DIHYDRO-2H-BENZ IMIDAZOL-2-ONE see OXATIMIDE

4-DIPHENYLMETHYLTROPYLTROPINIUM BROMIDE see 8-(p-PHENYL-BENZYL)ATROPINIUM BROMIDE

4-DIPHENYLMETHYL-DL-TROPYLTROPINIUM BROMIDE see 8-(p-PHE-NYLBENZYL)ATROPINIUM BROMIDE

DIPHENYL MIXED WITH DIPHENYL OXIDE see BIPHENYL, mixed with BIPHENYL OXIDE (3:7)

DIPHENYLNITROSAMIN (GERMAN) see DIPHENYLNITROSAMINE

2,2-DI(4-PHENYLOL)PROPANE see BISPHENOL A

DIPHENYL-p-PHENYLENEDIAMINE see 1,4-BIS(PHENYL AMINO)BEN-ZENE

N,N′-DIPHENYL-p-PHENYLENEDIAMINE see 1,4-BIS(PHENYL AMINO)-BENZENE

1,2-DIPHENYL-4-(2′-PHENYLSULFINETHYL)-3,5-PYRAZOLIDINEDIONE see DIPHENYLPYRAZONE

ALPHA,ALPHA-DIPHENYL-ALPHA-2-PIPERIDINEMETHANOL HYDROCHLORIDE see PIPERADROL HYDROCHLORIDE

ALPHA,ALPHA-DIPHENYL-1-PIPERIDINEPROPANOL HYDROCHLORIDE see PRIDINOL

1,1-DIPHENYL-3-PIPERIDINO-1-PROPANOL HYDROCHLORIDE see PRIDINOL

1,1-DIPHENYL-3-(1-PIPERIDYL)-1-PROPANOL HYDROCHLORIDE see PRIDINOL

alpha,alpha-DIPHENYL-beta-PROPIOLACTONE see 3,3-DIPHENYL-2-OXETANONE

(+)-1,2-DIPHENYL-2-PROPIONOXY-3-METHYL-4-DIMETHYLAMINO-BUTANE HYDROCHLORIDE see d-PROPOXYPHENE HYDROCHLORIDE

DIPHENYLPYRILENE see LYSSIPOLL

S-DIPHENYLTHIOCARBAMIDE see DIPHENYLTHIOUREA

1,3-DIPHENYLTHIOUREA see DIPHENYLTHIOUREA

1,3-DIPHENYL-2-THIOUREA see DIPHENYLTHIOUREA

N,N′-DIPHENYLTHIOUREA see DIPHENYLTHIOUREA

SYM-DIPHENYLTHIOUREA see DIPHENYLTHIOUREA

DIPHENYLTRICHLOROETHANE see DDT

DIPHOSGENE see PHOSGENE

DIPHOSPHAMIDE see ISOPHOSPHAMIDE

DIPHOSPHORIC ACID, DISODIUM SALT see DISODIUM PYROPHOSPHATE

DIPHOSPHORUS PENTOXIDE see PHOSPHORUS PENTOXIDE

DIPIPERONE see FLUOROBUTYROPHENONE

DIPOTASSIUM CHROMATE see POTASSIUM CHROMATE (VI)

DIPOTASSIUM DICHROMATE see POTASSIUM BICHROMATE

DIPOTASSIUM MONOCHROMATE see POTASSIUM CHROMATE (VI)

DIPPEL'S OIL see BONE OIL

DIPPING ACID see SULFURIC ACID

DI-2-PROPENYLAMINE see DIALLYLAMINE

DIPROPIONATO DE ESTILBENE (SPANISH) see DIETHYLSTILBESTROL DIPROPIONATE

P,P′-DIPROPIONOXY-TRANS-ALPHA,BETA -DIETHYLSTILBENE see DI-ETHYLSTILBESTROL DIPROPIONATE

4-(DI-N-PROPYLAMINO)-3,5-DINITRO-1 -TRIFLUOROMETHYLBENZENE see TRIFLURALINE

N,N-DI-N-PROPYL-2,6-DINITRO-4-TRIFLUOROMETHYLANILINE see TRIFLURALINE

DIPROPYLENETRIAMINE see AMINOBIS(PROPYLAMINE)

DIPROPYL KETONE see 4-HEPTANONE

DIPROPYL METHANE see N-HEPTANE

P-(DIPROPYLSULFAMOYL)BENZOIC ACID, SODIUM SALT see p-(DIPROPYLSULFAMOYL)BENZOIC ACID SODIUM SALT

P-(DI-N-PROPYLSULFAMYL)BENZOIC ACID, SODIUM SALT see p-(DIPROPYLSULFAMOYL)BENZOIC ACID SODIUM SALT

N,N-DIPROPYLTHIOCARBAMIC ACID S-ETHYL ESTER see S-ETHYL-N,N-DI-n-PROPYLTHIOCARBAMATE

DIPROPYLTIN CHLORIDE see DICHLORODIPROPYLSTANNANE

DIPROPYLTIN DICHLORIDE see DICHLORODIPROPYLSTANNANE

DI-N-PROPYLTIN DICHLORIDE see DICHLORODIPROPYLSTANNANE

DIPROPYLTIN OXIDE see DIPROPYLOXOSTANNANE

N,N-DIPROPYL-4-TRIFLUOROMETHYL-2,6-DINITROANILINE see TRIFLURALINE

2,2'-DIPYRIDYL see 2,2'-BIPYRIDINE

ALPHA,ALPHA'-DIPYRIDYL see 2,2'-BIPYRIDINE

DIPYRIDYL PHOSPHATE see DIPYRIDYL HYDROGEN PHOSPHATE

CIS-DIPYRROLIDINEDICHLOROPLATINUM (II) see cis-DICHLOROBIS-(PYRROLIDINE)PLATINUM (II)

DIRECT BLUE 6 see C.I. DIRECT BLUE 6, TETRASODIUM SALT

DISODIUM ARSENATE see ARSENIC ACID, DISODIUM SALT

DISODIUM ARSENATE, HEPTAHYDRATE see ARSENIC ACID, DISODIUM SALT, HEPTAHYDRATE

DISODIUM ARSENIC ACID see ARSENIC ACID, DISODIUM SALT

DISODIUM ARSENIC ACID see ARSENIC ACID, DISODIUM SALT

DISODIUM AUROTHIOMALATE see GOLD SODIUM THIOMALATE

DISODIUM-N-(3-(CARBOXYMETHYLTHIOMERCURI)-2-METHOXYPROPYL)-alpha- CAMPHORAMATE see SODIUM MERCAPTOMERIN

DISODIUM CHROMIC ACID see SODIUM DICHROMATE

DISODIUM DIACID ETHYLENEDIAMINETETRAACETATE see EDATHAMIL DISODIUM

DISODIUM 2,7-DIBROM-4-HYDROXY-MERCURI-FLUORESCEIN see MERCUROCHROME

DISODIUM 2',7'-DIBROMO-4'-(HYDROXYMERCURY)FLUORESCEIN see MERCUROCHROME

DISODIUM DICHROMATE see SODIUM DICHROMATE

DISODIUM DIHYDROGEN ETHYLENEDIAMINETETRAACETATE see EDATHAMIL DISODIUM

DISODIUM DIHYDROGEN(ETHYLENEDINITRILO)TETRAACETATE see EDATHAMIL DISODIUM

DISODIUM DIHYDROGEN PYROPHOSPHATE see DISODIUM PYROPHOSPHATE

DISODIUM (2,4-DIMETHYLPHENYLAZO)-2-HYDROXYNAPHTHALENE-3,6-DISULFONATE see D&C RED NO. 5

DISODIUM (2,4-DIMETHYLPHENYLAZO)-2-HYDROXYNAPHTHALENE-3,6-DISULPHONATE see D&C RED NO. 5

DISODIUM DIOXIDE see SODIUM PEROXIDE

DISODIUM DIPHOSPHATE see DISODIUM PYROPHOSPHATE

DISODIUM EDETATE see EDATHAMIL DISODIUM

DISODIUM EDTA see EDATHAMIL DISODIUM

DISODIUM EOSIN see BROMO EOSINE

DISODIUM 3,6-EPOXYCYCLOHEXANE-1,2-DICARBOXYLATE see DISODIUM-3,6-ENDOXOHEXAHYDROPHTHALATE

DISODIUM ETHYLENEBIS(DITHIOCARBAMATE) see NABAM

DISODIUM ETHYLENE-1,2-BISDITHIOCARBAMATE see NABAM

DISODIUM ETHYLENEDIAMINETETRAACETATE see EDATHAMIL DISODIUM

DISODIUM ETHYLENEDIAMINETETRAACETIC ACID see EDATHAMIL DISODIUM

DISODIUM (ETHYLENEDINITRILO)TETRAACETATE see EDATHAMIL DISODIUM

DISODIUM (ETHYLENEDINITRILO)TETRAACETIC ACID see EDATHAMIL DISODIUM

DISODIUM HEXAFLUOROSILICATE (2-) see DISODIUM HEXAFLUOROSILICATE

DISODIUM HYDROGEN ARSENATE see ARSENIC ACID, DISODIUM SALT

DISODIUM HYDROGEN ARSENATE see ARSENIC ACID, DISODIUM SALT

DISODIUM HYDROGEN ORTHOARSENATE see ARSENIC ACID, DISODIUM SALT

DISODIUM HYDROGEN ORTHOARSENATE see ARSENIC ACID, DISODIUM SALT

DISODIUM HYDROGEN PHOSPHATE see SODIUM MONOHYDROGEN PHOSPHATE (2:1:1)

DISODIUM-6-HYDROXY-3-OXO-9-XANTHENE-O-BENZOATE see FLUORESCEIN SODIUM

DISODIUM-3-HYDROXY-4-((2,4,5-TRIMETHYLPHENYL)AZO)-2,7-NAPHTHALENEDISULFONIC ACID see FD AND C RED NO. 1

DISODIUM-3-HYDROXY-4-((2,4,5-TRIMETHYLPHENYL)AZO)-2,7-NAPHTHALENEDISULFONATE see FD AND C RED NO. 1

DISODIUM-3-HYDROXY-4-((2,4,5-TRIMETHYLPHENYL)AZO)-2,7-NAPHTHALENEDISULPHONATE see FD AND C RED NO. 1

DISODIUM-3-HYDROXY-4-((2,4,5-TRIMETHYLPHENYL)AZO)-2,7-NAPHTHALENEDISULPHONIC ACID see FD AND C RED NO. 1

DISODIUM METASILICATE see SODIUM SILICATE

DISODIUM METHANEARSONATE see DISODIUM METHANEARSENATE

DISODIUM METHYLARSENATE see DISODIUM METHANEARSENATE

DISODIUM METHYLARSONATE see DISODIUM METHANEARSENATE

DISODIUM MONOHYDROGEN ARSENATE see ARSENIC ACID, DISODIUM SALT

DISODIUM MONOHYDROGEN ARSENATE see ARSENIC ACID, DISODIUM SALT

DISODIUM MONOHYDROGEN PHOSPHATE see SODIUM MONOHYDROGEN PHOSPHATE (2:1:1)

DISODIUM MONOMETHYLARSONATE see DISODIUM METHANEARSENATE

DISODIUM MONOSILICATE see SODIUM SILICATE

DISODIUM MONOXIDE see SODIUM MONOXIDE

DISODIUM ORTHOPHOSPHATE see SODIUM MONOHYDROGEN PHOSPHATE (2:1:1)

DISODIUM 7-OXABICYCLO(2.2.1)HEPTANE-2,3-DICARBOXYLATE see DISODIUM-3,6-ENDOXOHEXAHYDROPHTHALATE

DISODIUM OXIDE see SODIUM MONOXIDE

DISODIUM PEROXIDE see SODIUM PEROXIDE

DISODIUM PHOSPHATE see SODIUM MONOHYDROGEN PHOSPHATE (2:1:1)

DISODIUM PHOSPHORIC ACID see SODIUM MONOHYDROGEN PHOSPHATE (2:1:1)

DISODIUM SALT OF 1-INDIGOTIN-S,S'-DISULPHONIC ACID see DISODIUM INDIGO-5,5-DISULFONATE

DISODIUM SALT OF 7-OXABICYCLO(2.2.1)HEPTANE-2,3-DICARBOXYLIC ACID see DISODIUM-3,6-ENDOXOHEXAHYDROPHTHALATE

DISODIUM SALT OF 2-(4-SULPHO-1-NAPHTHYLAZO)-1-NAPHTHOL-4-SULPHONIC ACID see C.I. FOOD RED 3

DISODIUM SALT OF 1-P-SULPHOPHENYLAZO-2-NAPHTHOL-6-SULPHONIC ACID see FOOD YELLOW 3

DISODIUM SALT OF ENDOTHALL see DISODIUM-3,6-ENDOXOHEXAHYDROPHTHALATE

DISODIUM SALT OF 1-(2,4-XYLYLAZO)-2-NAPHTHOL-3,6-DISULFONIC ACID see D&C RED NO. 5

DISODIUM SALT OF 1-(2,4-XYLYLAZO)-2-NAPHTHOL-3,6-DISULPHONIC ACID see D&C RED NO. 5

DISODIUM SELENITE see SODIUM SELENITE

DISODIUM SEQUESTRENE see EDATHAMIL DISODIUM

DISODIUM SULFATE see SODIUM SULFATE (2:1)

DISODIUM SULFITE see SODIUM SULFITE (2:1)

DISODIUM 2-(4-SULFO-1-NAPHTHYLAZO)-1-NAPHTHOL-4-SULFONATE see C.I. FOOD RED 3

DISODIUM 2-(4-SULPHO-1-NAPHTHYLAZO)-1-NAPHTHOL-4-SULPHONATE see C.I. FOOD RED 3

DISODIUM TETRACEMATE see EDATHAMIL DISODIUM

DISODIUM VERSENATE see EDATHAMIL DISODIUM

DISODIUM VERSENE see EDATHAMIL DISODIUM

DISOLFURO DI TETRAMETILTIOURAME (ITALIAN) see BIS(DIMETHYL THIOCARBAMYL)DISULFIDE

2,4-D ISOOCTYL ESTER see ISOOCTYL-2,4-DICHLOROPHENOXYACETATE

2,4-D, ISOPROPYL ESTER see ISOPROPYL-2,4-D ESTER

DISPARICIDA see ACETPHENARSINE

DISTHENE see ALUMINUM(III) SILICATE (2:1)

DISTILLATE see FUEL OIL

DISTILLED MUSTARD see BIS(2-CHLOROETHYL)SULFIDE

DISTIVIT (B12 PEPTIDE) see VITAMIN B$_{12}$ COMPLEX

DISULFATOZIRCONIC ACID see ZIRCONIUM (IV) SULFATE (1:2)

DISULFIRAM see BIS(DIETHYL THIO CARBAMOYL) DISULFIDE

4,8-DISULFO-2-NAPHTHALAMINE see 3-AMINO-1,5-NAPHTHALENEDI-
SULFONIC ACID

DISULFOTON DISULIDE see OXYDISULFOTON

DISULFOTON SULFOXIDE see OXYDISULFOTON

DISULFUR DICHLORIDE see SULFUR CHLORIDE

DISULFURE DE TETRAMETHYLTHIOURAME (FRENCH) see BIS(DI-
METHYL THIOCARBAMYL)DISULFIDE

DISULFUR PENTOXYDICHLORIDE see PYROSULFURYL CHLORIDE

DISULFURYL CHLORIDE see PYROSULFURYL CHLORIDE

DISULPHURIC ACID see SULFURIC ACID, FUMING

DISYSTON SULFOXIDE see OXYDISULFOTON

DITHANE A-4 see p-DINITROBENZENE

1,4-DITHIAANTHRAQUINONE-2,3-DINITRILE see 5,10-DIHYDRO)-5,10-
DIOXONAPHTHO(2,3-b)-p-DITHIIN-2,3-DICARBONITRILE

3-(DI-2-THIENYLMETHYLENE)-1-METHYL-PIPERIDINE see BITIODIN

O,O-DITHIO-BIS-ANILINE see 2,2'-DITHIOBISANILINE

2,2'-DITHIOBIS(BENZOTHIAZOLE) see BENZOTHIAZOLE DISULFIDE

1,1'-DITHIOBIS(N,N-DIETHYLTHIOFORMAMIDE) see BIS(DIETHYL THIO
CARBAMOYL) DISULFIDE

ALPHA,ALPHA'-DITHIOBIS(DIMETHYLTHIO)FORMAMIDE see BIS(DI-
METHYL THIOCARBAMYL)DISULFIDE

2,2'-DITHIOBIS(ETHYLAMINE) see beta-MERCAPTOETHYLAMINE DI-
SULFIDE

DITHIOBISMORPHOLINE see N,N'-BISMORPHOLINE DISULFIDE

4,4'-DITHIOBIS(MORPHOLINE) see N,N'-BISMORPHOLINE DISULFIDE

DITHIOBIS(THIOFORMIC ACID)-O,O-DIETHYL ESTER see BISETHYL
XANTHOGEN DISULFIDE

DITHIOCARBONIC ANHYDRIDE see CARBON DISULFIDE

DITHIODEMETON see DITHIOSYSTOX

N,N'-(DITHIODICARBONOTHIOYL)BIS(N-METHYLMETHANAMINE) see
BIS(DIMETHYL THIOCARBAMYL)DISULFIDE

N,N-DITHIODIMORPHOLINE see N,N'-BISMORPHOLINE DISULFIDE

4,4'-DITHIODIMORPHOLINE see N,N'-BISMORPHOLINE DISULFIDE

DITHIODIPHOSPHORIC ACID, TETRAETHYL ESTER see ETHYL THIO
PYROPHOSPHATE

1,2-DITHIOGLYCEROL see BAL

DITHIOMETON (FRENCH) see PHOSPHORODITHIOIC ACID-O,O-
DIMETHYL-S-(2-ETHYLTHIO)ETHYL ESTER

4,4'-DITHIOMORPHOLINE see N,N'-BISMORPHOLINE DISULFIDE

DITHIONE see DITHION

DITHIONIC ACID see SULFURIC ACID, FUMING

DITHIOPHOSPHATE DE O,O-DIETHYLE ET DE (4-CHLORO-PHENYL)
THIOMETHYLE (FRENCH) see TRITHION

DITHIOPHOSPHATE DE O,O-DIETHYLE ET DE S-(2-ETHYLTHIO-
ETHYLE) (FRENCH) see DITHIOSYSTOX

DITHIOPHOSPHATE DE O,O-DIETHYLE ET DE S-N-METHYL N-CARBO-
ETHOXY CARBAMOYLMETHYLE (FRENCH) see O,O-DIETHYL-S-(N-
ETHOXYCARBONYL-N-METHYLCARBAMOYLMETHYL) PHOSPHORO
DITHIOATE

DITHIOPHOSPHATE DE-O,O-DIETHYLE ET DE S(2,5-DICHLOROPHENYL)
THIOMETHYLE (FRENCH) see PHENCAPTON

DITHIOPHOSPHATE DE O,O-DIETHYLE ET D'ETHYLTHIOMETHYLE
(FRENCH) see PHORATE

DITHIOPHOSPHATE DE O,O-DIMETHYLE ET DE S-(4,6-DIAMINO-1,3,5-
TRIAZINE-2-YL)-METHYLE) (FRENCH) see AZIDITHION

DITHIOPHOSPHATE DE O,O-DIMETHYLE ET DE S-(1,2-DICARBOETH-
OXYETHYLE) (FRENCH) see CARBETHOXY MALATHION

DITHIOPHOSPHATE DE O,O-DIMETHYLE ET DE S-(2-ETHYLTHIO-
ETHYLE) (FRENCH) see PHOSPHORODITHIOIC ACID-O,O-
DIMETHYL-S-(2-ETHYLTHIO)ETHYL ESTER

DITHIOPHOSPHATE DE O,O-DIMETHYLE ET DE S-((MORPHOLINOCAR-
BONYLE)-METHYLE) (FRENCH) see MORPHOTHION

DI(THIOPHOSPHORIC) ACID, TETRAETHYL ESTER see ETHYL THIO
PYROPHOSPHATE

DITHIOPHOSPHORSAEURE-O-AETHYL-S,S-DIPHENYLESTER (GERMAN)
see O-ETHYL-S,S-DIPHENYL DITHIO PHOSPHATE

2,3-DITHIOPROPANOL see BAL

DITHIOPYROPHOSPHATE DE TETRAETHYLE (FRENCH) see ETHYL THIO
PYROPHOSPHATE

DITHIOXAMIDE see DITHIOOXAMIDE

DITHIZON see DIPHENYLTHIOCARBAZONE

DITHIZONE see DIPHENYLTHIOCARBAZONE

4,4'-DI-O-TOLUIDINE see 3,3'-TOLIDINE

DI-O-TOLUYLTHIOUREA see DI-O-TOLYLTHIOUREA

DIUCARDYN SODIUM see DISODIUM-N-(3-(CARBOXYMETHYLTHIOMER-
CURI)-2-METHOXYPROPYL)-alpha-CAMPHORAMATE

DIUCARDYN SODIUM see SODIUM MERCAPTOMERIN

1,1-DIUREIDISOBUTANE see ISOBUTYLIDENEDIUREA

DIUREIDOISOBUTANE see ISOBUTYLIDENEDIUREA

DIVINYLENE OXIDE see FURAN

DIVINYLENE SULFIDE see THIOPHENE

DIVINYLENIMINE see PYRROLE

DIVINYL ETHER see VINYL ETHER

DIVINYL ETHER (DOT) see VINYL ETHER

DIVYNYL OXIDE see VINYL ETHER

DKB SULFATE see 3',4'-DIDEOXYKANAMYCIN B SULFATE

(DL)-alpha-EPICHLOROHYDRIN see 1-CHLORO-2,3-EPOXY PROPANE

DMA see DIMETHYLAMINE

DMP see DIMETHYL PHTHALATE

DMSO see DIMETHYL SULFOXIDE

DNOK (CZECH) see 4,6-DINITRO-o-CRESOL

2,5-DNP see 2,5-DINITROPHENOL

DNPC see 2,6-DINITRO-p-CRESOL

DOA see ANGEL DUST

DODECACHLOROOCTAHYDRO-1,3,4-METHENO-2H-CYCLOBUTA(C,D)-
PENTALENE see MIREX

DODECACHLOROPENTACYCLODECANE see MIREX

DODECACHLOROPENTACYCLO(3,2,2,0(SUP 2,6),0(SUP 3,9),0(SUP
5,10))DECANE see MIREX

DODECAHYDRODIPHENYLAMINE see N,N-DICYCLOHEXYLAMINE

DODECAHYDROPHENYLAMINE NITRITE see DICYCLOHEXYLAMINE NI-
TRITE

N-DODECAN (GERMAN) see DODECANE

DODECANOIC ACID see LAURIC ACID

1-DODECANOL see DODECYL ALCOHOL

N-DODECANOL see DODECYL ALCOHOL

DODECANOYL PEROXIDE see LAUROYL PEROXIDE

N-DODECYL ALCOHOL see DODECYL ALCOHOL

DODECYL ALCOHOL, HYDROGEN SULFATE, SODIUM SALT see SUL-
FURIC ACID, MONODODECYL ESTER, SODIUM SALT

DODECYLBENZENSULFONAN SODNY (CZECH) see DODECYL BENZENE
SODIUM SULFONATE

DODECYLDIMETHYL(2-PHENOXYETHYL)AMMONIUM BROMIDE see
PHENODODECINIUM BROMIDE

DODECYLGUANIDINE ACETATE see N-DODECYLGUANIDINE ACETATE

2-DODECYLISOQUINOLINIUM BROMIDE see LAURYLISOQUINOLINIUM
BROMIDE

TERT-DODECYLMERCAPTAN see t-DODECANETHIOL

DODECYL SODIUM SULFATE see SULFURIC ACID, MONODODECYL
ESTER, SODIUM SALT

DODECYL SULFATE, SODIUM SALT see SULFURIC ACID, MONODODE-
CYL ESTER, SODIUM SALT

TERT-DODECYLTHIOL see t-DODECANETHIOL

DODINE ACETATE see N-DODECYLGUANIDINE ACETATE

DODINE, MIXTURE WITH GLYODIN see N-DODECYLGUANIDINE ACE-
TATE

DOLOPHINE see METHADONE

D-DOLOPHINE HYDROCHLORIDE see d-METHADONE HYDROCHLORIDE

DOLOPHIN HYDROCHLORIDE see dl-METHADONE HYDROCHLORIDE

DOM see 2,5-DIMETHOXY-4-METHYLAMPHETAMINE

DOMOLITE see CALCIUM (II) CARBONATE (1:1)

DONOVAN'S SOLUTION see ARSENIC IODIDE MIXED WITH MERCURIC
IODIDE

DOOJE see HEROIN

DOP see BIS(2-ETHYLHEXYL)PHTHALATE

DOVENIX see 4-HYDROXY-3-IODO-5-NITROBENZONITRILE

DOW 1329 see O-(2,4-DICHLOROPHENYL)-O-METHYLISOPROPYLPHOS-
PHORAMIDOTHIOATE

DOWANOL see CARBITOL CELLOSOLVE

DOWANOL 33B see PROPYLENE GLYCOL MONOMETHYL ETHER

DOWANOL EB see BUTYL CELLOSOLVE

DOWANOL EE see CELLOSOLVE SOLVENT

DOWANOL EM see ETHYLENE GLYCOL METHYL ETHER

DOWCO-163 see 2-CHLORO-6-(TRICHLOROMETHYL)PYRIDINE

DOWCO 169 see O-PHENYL-N,N'-DIMETHYL PHOSPHORODIAMIDATE

DOWFUME 40 see 1,2-ETHYLENE DIBROMIDE

DOWFUME N see DICHLOROPROPANE-DICHLOROPROPENE MIXTURE

DOWICIDE see 2-BIPHENYLOL, SODIUM SALT

DOWICIDE 2 see 2,4,5-TRICHLOROPHENOL

DOWICIDE 6 see 2,4,5,6-TETRACHLOROPHENOL

DOWICIDE 2S see 2,4,6-TRICHLOROPHENOL

DOWICIDE B see 2,4,5-TRICHLOROPHENOL

DOW-PER see 1,1,2,2-TETRACHLOROETHYLENE

DOWTHERM 209 see PROPYLENE GLYCOL MONOMETHYL ETHER

DOWTHERM A see BIPHENYL, mixed with BIPHENYL OXIDE (3:7)

DOW-TRI see TRICHLOROETHYLENE

DPG see DIPHENYLGUANIDINE

DPMA see 2,3-DIHYDROPHORBOL MYRISTATE ACETATE

DREFT see SULFURIC ACID, MONODODECYL ESTER, SODIUM SALT

DROMORAN HYDROBROMIDE see METHORPHINAN HYDROBROMIDE

DROP LEAF see SODIUM CHLORATE

DRY ICE see CARBON DIOXIDE

DRYOBALANOPS CAMPHOR see BORNEOL

DTIC CITRATE see 5-(3,3-DIMETHYL-1-TRIAZENO)IMIDAZOLE-4-CAR-
BOXAMIDE CITRATE

DUAZOMYCIN B see AZOTOMYCIN

DULCINE see 4-ETHOXY PHENYL UREA

DULCITOLDIEPOXIDE see DIANHYDROGALACTITOL

DUODECANE see DODECANE

DUODECYL ALCOHOL see DODECYL ALCOHOL

DUODECYLIC ACID see LAURIC ACID

DUPONAL see SULFURIC ACID, MONODODECYL ESTER, SODIUM SALT

DURAX see N-CYCLOHEXYL-2-BENZOTHIAZOLESULFENAMIDE

DUSICNAN BARNATY (CZECH) see BARIUM(II) NITRATE (1:2)

DUSICNAN KADEMNATY (CZECH) see CADMIUM (II) NITRATE TETRA-
HYDRATE (1:2:4)

DUSICNAN ZIRKONICITY (CZECH) see ZIRCONIUM NITRATE

DUSITAN DICYKLOHEXYLAMINU (CZECH) see DICYCLOHEXYLAMINE
NITRITE

DUSITAN SODNY (CZECH) see SODIUM NITRITE

DUTCH LIQUID see ETHYLENE DICHLORIDE

DUTCH OIL see ETHYLENE DICHLORIDE

DWUBROMOETAN (POLISH) see 1,2-ETHYLENE DIBROMIDE

DWUCHLOROCZTEROFLUOROETAN (POLISH) see DICHLOROTETRA-
FLUOROETHANE

DWUCHLORODWUETYLOWY ETER (POLISH) see BIS(2-CHLOROETHYL)
ETHER

DWUCHLORODWUFLUOROMETAN (POLISH) see DICHLORODIFLUORO-
METHANE

DWUCHLOROPROPAN (POLISH) see 1,2-DICHLOROPROPANE

DWUETYLOAMINA (POLISH) see DIETHYLAMINE

DWUETYLOWY ETER (POLISH) see ETHYL ETHER

DWUMETHYLOFORMAMID (POLISH) see DIMETHYL FORMAMIDE

DWUMETYLOANILINA (POLISH) see N,N-DIMETHYLANILINE

DWUMETYLOWY SIARCZAN (POLISH) see SULFURIC ACID, DIMETHYL
ESTER

DWUNITROBENZEN (POLISH) see m-DINITROBENZENE

DWUNITRO-O-KREZOL (POLISH) see 4,6-DINITRO-O-CRESOL

DYCLONINE HYDROCLORIDE see 4'-BUTOXY-3-PIPERIDINO PROPIO-
PHENONE HYDROCHLORIDE

DYE FD & C RED NO. 4 see FOOD RED 4

DYE FD & C YELLOW NO. 6 see FOOD YELLOW 3

DYE FDC RED 2 see FOOD RED 2

DYNAMITE (DOT) see DYNAMITE

DYNARSAN see ACETPHENARSINE

EA 3547 see DIBENZ(b,f)(1,4)OXAZEPINE

E-73 ACETATE see ACETOXYCYCLOHEXIMIDE

EARTHNUT OIL see PEANUT OIL

EASTMAN INHIBITOR DHPB see 2,4-DIHYDROXYBENZOPHENONE

ECGONINE, METHYL ESTER, BENZOATE (ESTER) see COCAINE

ECLORION see HEROIN

ECTYLCARBAMIDE see cis-(2-ETHYLCROTONYL) UREA

ECTYLUREA see cis-(2-ETHYLCROTONYL) UREA

EDC see ETHYLENE DICHLORIDE

EDETATE DISODIUM see EDATHAMIL DISODIUM

EDETIC ACID see ETHYLENE DIAMINE TETRAACETIC ACID

EDETIC ACID TETRASODIUM SALT see EDATHANIL TETRASODIUM

(E)-1,1'-(1,2-DIETHYL-1,2-ETHENE-DIYL)BIS(4-METHOXYBENZENE)
see 3,4-DIANISYL-3-HEXENE

EDIFENPHOS see O-ETHYL-S,S-DIPHENYL DITHIO PHOSPHATE

(−)-(E)-7-ALPHA,8-BETA-DIHYDROXY-9-BETA,10-BETA-EPOXY-7,
8,9,10-TETRAHYDROBENZO(A)PYRENE see sym-DIOLEPOXIDE

(E)-DIMETHYL 1-METHYL-3-(METHYLAMINO)-3-OXO-1-PROPENYL
PHOSPHATE see 3-(DIMETHOXYPHOSPHINYLOXY)N-METHYL-cis-
CROTONAMIDE

EDTA ACID see ETHYLENE DIAMINE TETRAACETIC ACID

EDTA (CHELATING AGENT) see ETHYLENE DIAMINE TETRAACETIC
ACID

EDTA, DISODIUM SALT see EDATHAMIL DISODIUM

EDTA TETRASODIUM SALT see EDATHANIL TETRASODIUM

EENKAPTON (DUTCH) see PHENCAPTON

EHRLICH 594 see ACETPHENARSINE

EHRLICH 606 see SALVARSAN

EISENDEXTRAN (GERMAN) see IRON-DEXTRAN COMPLEX

EISENDIMETHYLDITHIOCARBAMAT (GERMAN) see FERBAM

EISEN(III)-TRIS(N,N-DIMETHYLDITHIOCARBAMAT) (GERMAN) see FER-
BAM

E-KAPROLAKTAM (CZECH) see HEXAHYDRO-2H-AZEPIN-2-ONE

EKTASOLVE EIB see ISOBUTYL CELLOSOLVE

EKTYLCARBAMID see cis-(2-ETHYLCROTONYL) UREA

EK 1108GY-A see TEFLON

ELALDEHYDE see PARALDEHYDE

ELANIL see ELAVIL

ELAYL see ETHYLENE

ELEMENTAL SELENIUM see SELENIUM

ELEPHANT TRANQUILIZER see ANGEL DUST

ELOCRON see O-(1,3-DIOXOLAN-2-YL)PHENYL METHYLCARBAMATE

ELVANOL see POLYVINYL ALCOHOL

ELYSION see ANGEL DUST

EMERALD GREEN see CUPRIC ACETOARSENITE

EMERY see CORUNDUM

(−)-EMETINE see EMETINE

(−)-EMETINE DIHYDROCHLORIDE see 1-EMETINE DIHYDROCHLORIDE

EMETIQUE (FRENCH) see ANTIMONY POTASSIUM TARTRATE

EMPIRIN see ACETOL (2)

EMPIRIN COMPOUND see p-ACETOPHENETIDIDE

ENALLYNYMAL SODIUM see METHOHEXITAL SODIUM

ENAMEL WHITE see BARIUM SULFATE

1,5-ENDOMETHYLENE-3,7-DINITROSO-1,3,5,7-TETRAAZACYCLOOC-
TANE see DINITROSOPENTAMETHYLENETETRAMINE

3,6-ENDOXOHEXAHYDROPHTHALIC ACID DISODIUM SALT see DISO-
DIUM-3,6-ENDOXOHEXAHYDROPHTHALATE

ENDRATE DISODIUM see EDATHAMIL DISODIUM

ENDRATE TETRASODIUM see EDATHANIL TETRASODIUM

ENDRINE (FRENCH) see ENDRIN

ENOVID see ENAVID

ENT 6 see 2-(2-BUTOXY ETHOXY)ETHYL THIOCYANATE

ENT 9 see n-BUTYL MESITYL OXIDE OXALATE

17-ENT see 17-ACETOXY-19-NOR-17-alpha-PREGN-4-EN-20-YN-
3-ONE

ENT 54 see ACRYLONITRILE

ENT 92 see ISOBORNYL THIOCYANATO ACETATE

ENT 123 see VERATRINE
ENT 157 see 2-CYCLOHEXYL-4,6-DINITROPHENOL
ENT 262 see DIMETHYL PHTHALATE
ENT 988 see BIS(DIMETHYL DITHIO CARBAMATO)ZINC
ENT 1,122 see 2-sec-BUTYL-4,6-DINITRO PHENOL
ENT 1,506 see DDT
ENT 1,656 see ETHYLENE DICHLORIDE
ENT 1,656 see 1,2-DICHLORETHANE
ENT 1,716 see DIMETHOXY-DDT
ENT 3,776 see 2,3-DICHLORO-1,4-NAPHTHOQUINONE
ENT 4,225 see 1,1-BIS(4-CHLOROPHENYL)-2,2-DICHLOROETHANE
ENT 4,504 see BIS(2-CHLOROETHYL) ETHER
ENT 4,705 see CARBON TETRACHLORIDE
ENT 7,543 see PYRETHRIN II
ENT 7,796 see BENZENE HEXACHLORIDE-gamma isomer
ENT 8,184 see N-OCTYL BICYCLOHEPTENE DICARBOXIMIDE
ENT 8,420 see dichloropropane-DICHLOROPROPENE MIXTURE
ENT 8,601 see BENZENE HEXACHLORIDE
ENT 9,232 see BENZENE HEXACHLORIDE-alpha-isomer
ENT 9,234 see delta-BHC
ENT 9,236 see delta-BENZENE HEXACHLORIDE
ENT 9,735 see TOXAPHENE
ENT 9,932 see CHLORDANE
ENT 14,250 see PIPERONYL BUTOXIDE
ENT 14,611 see AZOBENZENE
ENT 14,874 see ETHYLENE BIS(DITHIO CARBAMATO)ZINC
ENT 14,875 see VANCIDE
ENT 15,108 see PARATHION
ENT 15,152 see HEPTACHLOR
ENT 15,349 see 1,2-ETHYLENE DIBROMIDE
ENT 15,406 see 1,2-DICHLOROPROPANE
ENT 15,949 see ALDRIN
ENT 16,087 see p-NITROPHENYL DIETHYLPHOSPHATE
ENT 16,225 see DIELDRIN
ENT 16,273 see ETHYL THIO PYROPHOSPHATE
ENT 16275 see BIOALLETHRIN
ENT 16,358 see 4-CHLOROPHENYL 4-CHLOROBENZENESULFONATE
ENT 16,391 see KEPONE
ENT 16,436 see N-DODECYLGUANIDINE ACETATE
ENT 16,519 see SULFUROUS ACID, 2-(p-t-BUTYLPHENOXY)-
    1-METHYLETHYL-2-CHLOROETHYL ESTER
ENT 17,034 see CARBETHOXY MALATHION
ENT 17,035 see p-NITRO-o-CHLOROPHENYL DIMETHYL THIONOPHOS-
    PHATE
ENT 17,251 see ENDRIN
ENT 17,292 see METHYL PARATHION
ENT 17,295 see DEMETON-O + DEMETON-S
ENT 17470 see DICHLOROFENTHION
ENT 17,510 see ALLETHRIN
ENT 17,798 see ETHOXY-4-NITRO PHENOXY PHENYL PHOSPHINE
    SULFIDE
ENT 18,060 see N-3-CHLOROPHENYL ISOPROPYL CARBAMATE
ENT 18,771 see TETRAETHYLPYROPHOSPHATE
ENT 18,861 see METHYLCHLOROTHION
ENT 18,862 see DEMETON-O-METHYL
ENT 18,870 see MALEIC HYDRAZIDE
ENT 19,060 see DIMETHYL-5-(1-ISOPROPYL-3-METHYL-PYRAZOLYL)-
    CARBAMATE
ENT 19,109 see BIS(DIMETHYL AMIDO)FLUORO PHOSPHATE
ENT 19,244 see ISODRIN
ENT 19,442 see TERPENE POLYCHLORINATES
ENT 19,507 see DIAZIDE
ENT 20218 see DIETHYL-m-TOLUAMIDE
ENT 20,696 see p-CHLOROBENZYL-p-CHLOROPHENYL SULFIDE
ENT 20738 see DIMETHYL DICHLOROVINYL PHOSPHATE
ENT 20,852 see BUTONATE
ENT 22,014 see ETHYL GUTHION
ENT 22,335 see TETRACHLORONITROANISOLE
ENT 22542 see DIETHYL-m-TOLUAMIDE
ENT 22,865 see O,O-DIETHYL-S-2-ISOPROPYLMERCAPTOMETHYLDI-

THIOPHOSPHATE
ENT 23,233 see AZINPHOS METHYL
ENT 23,284 see TRICHLOROMETAFOS
ENT 23,437 see DITHIOSYSTOX
ENT 23,648 see 1,1-BIS(p-CHLOROPHENYL)-2,2,2-TRICHLOROETHA-
    NOL
ENT 23,708 see TRITHION
ENT 23,737 see p-CHLOROPHENYL-2,4,5-TRICHLOROPHENYL SUL-
    FONE
ENT 23,968 see PYRIDAPHENTHION
ENT 23,969 see CARBARYL
ENT 23,979 see BENZOEPIN
ENT 24,042 see PHORATE
ENT 24,105 see ETHYL METHYLENE PHOSPHORO DITHIOATE
ENT 24,482 see 3-HYDROXYDIMETHYL CROTONAMIDE DIMETHYL
    PHOSPHATE
ENT 24,650 see O,O-DIMETHYL METHYLCARBAMOYLMETHYL
    PHOSPHORODITHIOATE
ENT 24,652 see ISOPROPYL DIETHYL DITHIOPHOSPHORYL ACET-
    AMIDE
ENT 24,653 see ENDOTHION
ENT 24,717 see CROTOXYPHOS
ENT 24,723 see METHYLPYRAZOLYL DIETHYLPHOSPHATE
ENT 24,725 see DIETHYLDITHIOCARBAMIC ANHYDROSULFIDE
ENT 24727 see ARATHANE
ENT 24,738 see 5,5-DIMETHYLDIHYDRORESORCINOL DIMETHYLCAR-
    BAMATE
ENT 24,833 see SWAT
ENT 24915 see TRIETHYLENEPHOSPHOROTRIAMIDE
ENT 24915 see TEPA
ENT 24,945 see O,O-DIETHYL O-P-(METHYLSULFINYL)PHENYL THIO-
    PHOSPHATE
ENT 24,964 see DEMETON-O-METHYL SULFOXIDE
ENT 24969 see CHLORFENVINFOS
ENT 24970 see NAPHTHALOXIMIDODIETHYL THIOPHOSPHATE
ENT 24,979 see BIS(TRIBUTYL TIN)OXIDE
ENT-24986 see DITHION
ENT 24988 see DIMETHYL-1,2-DIBROMO-2,2-DICHLOROETHYL PHOS-
    PHATE
ENT 25208 see ACETOXYTRIPHENYLSTANNANE
ENT 25,296 see TRISAZIRIDINYLTRIAZINE
ENT 25445 see 3-AMINOTRIAZOLE
ENT 25,500 see m-CUMENOL METHYLCARBAMATE
ENT-25,506 see DIMETHOATE-ETHYL
ENT 25515 see DIMECRON
ENT 25,540 see O,O-DIMETHYLPHOSPHOROTHIOIC ACID-O-(4-
    METHYLTHIO)-m-TOLYLESTER
ENT 25,543 see m-CUMENOL METHYLCARBAMATE
ENT 25,545 see 1,3,4,5,6,8,8-OCTACHLORO-1,3,3a,4,7,7a-HEXA-
    HYDRO-4,7-METHANO ISO BENZOFURAN
ENT 25,550 see SILICA, AMORPHOUS FUMED
ENT 25,567 see N-HYDROXYNAPHTHALIMIDE, DIETHYL PHOSPHATE
ENT 25,580 see ETHYL PYRAZINYL PHOSPHOROTHIOATE
ENT 25,584 see EPOXYHEPTACHLOR
ENT 25,585 see PHENCAPTON
ENT 25,599 see METHYL TRITHION
ENT 25,644 see O,O-DIMETHYL-O-(p-(N,N-DIMETHYLSULFAMOYL)
    PHENYL)PHOSPHOROTHIOATE
ENT 25647 see O-(2,4-DICHLOROPHENYL)-O-METHYLISOPROPYLPHOS-
    PHORAMIDOTHIOATE
ENT 25670 see METHYLCARBAMIC ACID-O-CUMENYL ESTER
ENT 25,671 see BAYGON
ENT 25,674 see METASYSTOX-S
ENT 25684 see O,O-DIMETHYL O-4-(METHYLTHIO)-3,5-XYLYL PHOS-
    PHOROTHIOATE
ENT 25,705 see PHTHALIMIDOMETHYL-O,O-DIMETHYL PHOSPHORO-
    DITHIOATE
ENT 25,712 see ETHYL TRICHLORO PHENYL ETHYL PHOSPHONO-
    THIOATE
ENT 25,715 see O,O-DIMETHYL-O-(3-METHYL) PHOSPHOROTHIOATE

ENT 25,719 see MIREX

ENT 25,726 see 4-METHYLMERCAPTO-3,5-XYLYL METHYLCARBA-
MATE

ENT 25734 see 4-METHYLTHIOPHENYLDIMETHYL PHOSPHATE

ENT 25,737 see S-(4,6-DIMETHYL-2-PYRIMIDINYL)-O,O-DIETHYL
PHOSPHORODITHIOATE

ENT 25,760 see AZIDITHION

ENT 25,764 see TCTP

ENT 25766 see 4-(DIMETHYLAMINE)-3,5-XYLYL N-METHYLCARBA-
MATE

ENT 25,776 see DIMETHOATE OXYGEN ANALOG

ENT 25,784 see 4-DIMETHYLAMINE-m-CRESYLMETHYLCARBAMATE

ENT 25,796 see DYPHONATE

ENT 25,809 see DITHIOLANE IMINOPHOSPHATE

ENT 25,830 see PHOSFOLAN

ENT-25,843 see 3,4,5-TRIMETHYLPHENYL METHYLCARBAMATE

ENT-25,991 see MEPHOSFOLAN

ENT-26079 see AMINOPTERIDINE

ENT 26,316 see APHOLATE

ENT 26,538 see CAPTAN

ENT 26,613 see N-METHYL-O,O-DIMETHYLTHIOLOPHOSPHORYL-
5-THIA-3-METHYL-2-VALERAMIDE

ENT 27,093 see CARBANOLATE

ENT 27,102 see O,O-DIETHYL-O-(2-CHLORO-1,2,5-DICHLOROPHE-
NYLVINYL)PHOSPHODITHIOATE

ENT 27,128 see 3-(1-METHYLPROPYL)-6-CHLOROPHENYL METHYL-
CARBAMATE

ENT 27,129 see 3-(DIMETHOXYPHOSPHINYLOXY)N-METHYL-cis-CRO-
TONAMIDE

ENT 27, 160 see AMIDITHION

ENT 27,163 see S-((3-BENZOXAZOLINYL-6-CHLORO-2-OXO)METHYL)
O,O-DIETHYLPHOSPHORODITHIOATE

ENT 27,164 see FURADAN

ENT 27, 165 see ABATE

ENT 27,180 see METHYLPHOSPHODITHIOIC ACID-S-(((p-CHLOROPHE-
NYL)THIO)METHYL)-O-METHYL ESTER

ENT 27193 see O,O-DIMETHYL-S-(5-METHOXY-1,3,4-THIADIAZOLI-
NYL-3-METHYL)DITHIO PHOSPHATE

ENT 27,223 see 5-AMINO-1-BIS(DIMETHYLAMIDE)PHOSPHORYL-3-
PHENYL-1,2,4-TRIAZOLE

ENT 27,257 see O,O-DIMETHYL DITHIOPHOSPHORYLACETIC ACID-N-
METHYL-N-FORMYLAMIDE

ENT 27,258 see ETHYL BROMOPHOS

ENT 27,267 see BIS(DIETHYL THIO)CHLORO METHYL PHOSPHONATE

ENT 27311 see DOWCO 179

ENT 27,318 see O-ETHYL-S,S-DIPROPYL PHOSPHORO DITHIOATE

ENT 27,324 see 2,3-DIHYDRO-2-METHYLBENZOPYRANYL-7,N-
METHYLCARBAMATE

ENT 27,346 see O,O-DIMETHYL-S-(2-(ACETYLAMINO)ETHYL) DITHIO-
PHOSPHATE

ENT 27,357 see DIMETHYL PHOSPHATE ESTER WITH 2-CHLORO-N-
METHYL-3-HYDROXYCROTONAMIDE

ENT 27,389 see O-(1,3-DIOXOLAN-2-YL)PHENYL METHYLCARBA-
MATE

ENT 27394 see O,O-DIETHYL-O-2-QUINOXALYL THIOPHOSPHATE

ENT 27,396 see O,S-DIMETHYL PHOSPHORAMIDOTHIOATE

ENT 27,407 see 8-(METHYLQUINOLYL)-N-METHYL CARBAMATE

ENT 27438 see 5,6-DICHLORO-1-PHENOXYCARBONYL-2-TRIFLUORO-
METHYLBENZIMIDAZOLE

ENT 27474 see BENZOFUROLINE

ENT-27566 see N,N-DIMETHYL-
N'-(((METHYLAMINO)CARBONYL)OXY)PHENYLMETHANIMIDAMIDE
MONOHYDROCHLORIDE

ENT 27572 see FENAMIPHOS

ENT 27625 see 3-(DIMETHOXYPHOSPHINYLOXY)-N-METHYL-N-METH-
OXY-cis-CROTONAMIDE

ENT 27822 see O,S-DIMETHYLACETYLPHOSPHOROAMIDOTHIOATE

ENT 28009 see HYDROXY TRIPHENYL STANNANE

ENT 50,003 see TRIS(1-METHYLETHYLENE)PHOSPHORIC TRIAMIDE

ENT 50146 see RESERPINE

ENT 50,434 see ANTIMONY POTASSIUM TARTRATE

ENT 50439 see 5-(BIS(2-CHLOROETHYL)AMINO)URACIL

ENT-50825 see N-METHYLMITOMYCIN C

ENT 50852 see HEXAMETHYLMELAMINE

ENT 50882 see HEXAMETHYL PHOSPHORAMIDE

ENT-61241 see ACETYL HYDRAZIDE

ENT 27386GC see (O,O-DIMETHYLDITHIOPHOSPHORYLPHENYL)ACETIC
ACID ETHYL ESTER

ENT 24,980-X see O,O-DIETHYL-S-(2-DIETHYLAMINOETHYL) THIO-
PHOSPHATE

ENT 25,545-X see 1,3,4,5,6,8,8-OCTACHLORO-1,3,3a,4,7,7a-HEXA-
HYDRO-4,7-METHANO ISO BENZOFURAN

ENT 25,552-X see CHLORDANE

ENT 25,555-X see O,O-DIETHYL-S-(3,4-DICHLOROPHENYLTHIO)-
METHYL PHOSPHOROTHIOATE

ENT 25,602-X see 4-tert-BUTYL-2-CHLORO PHENYL METHYL METHYL
PHOSPHORAMIDATE

ENU see 1-ETHYL-1-NITROSOUREA

EOSINE SODIUM SALT see BROMO EOSINE

EOSIN GELBLICH (GERMAN) see BROMO EOSINE

EPHEDRIN see EPHEDRINE

1-EPHEDRINE see EPHEDRINE

L(-)-EPHEDRINE see EPHEDRINE

EPIBENZALIN see BENZALIN

EPIBROMOHYDRIN see 3-BROMO-1,2-EPOXYPROPANE

EPICHLOORHYDRINE (DUTCH) see 1-CHLORO-2,3-EPOXY PROPANE

EPICHLORHYDRIN (GERMAN) see 1-CHLORO-2,3-EPOXY PROPANE

EPICHLORHYDRINE (FRENCH) see 1-CHLORO-2,3-EPOXY PROPANE

EPICHLOROHYDRIN see 1-CHLORO-2,3-EPOXY PROPANE

EPICHLOROHYDRIN (DOT) see 1-CHLORO-2,3-EPOXY PROPANE

alpha-EPICHLOROHYDRIN see 1-CHLORO-2,3-EPOXY PROPANE

EPICHLOROHYDRYNA (POLISH) see 1-CHLORO-2,3-EPOXY PROPANE

EPICLORIDRINA (ITALIAN) see 1-CHLORO-2,3-EPOXY PROPANE

3,17-EPIDIHYDROXYESTRATRIENE see ESTRADIOL (β)

3,17-EPIDIHYDROXYOESTRATRIENE see ESTRADIOL (β)

EPIHYDRIN ALCOHOL see 2,3-EPOXY-1-PROPANOL

EPIHYDRINALDEHYDE see GLYCIDALDEHYDE

EPIHYDRINE ALDEHYDE see GLYCIDALDEHYDE

EPILEO PETIT MAL see 2-ETHYL-2-METHYLSUCCINIMIDE

(−)-EPINEPHRINE see VASOTONIN

EPINEPHRINE see VASOTONIN

1-EPINEPHRINE see VASOTONIN

D-EPINEPHRINE see d-ADRENALINE

(R)-EPINEPHRINE see VASOTONIN

1-EPINEPHRINE CHLORIDE see 1-ADRENALINE CHLORIDE

(-)-EPINEPHRINE HYDROCHLORIDE see 1-ADRENALINE CHLORIDE

1-EPINEPHRINE HYDROCHLORIDE see 1-ADRENALINE CHLORIDE

EPINEPHRINE RACEMIC see dl-EPINEPHRINE

1-EPINEPHRINE (SYNTHETIC) see VASOTONIN

EPINOVAL see 2,2-DIETHYL-4-PENTENAMIDE

EPN see ETHOXY-4-NITRO PHENOXY PHENYL PHOSPHINE SULFIDE

(3,6-EPOSSI-CICLOESAN-1,2-DICARBOSSILATO) DISODICO (ITALIAN)
see DISODIUM-3,6-ENDOXOHEXAHYDROPHTHALATE

EPOXIDE 269 see LIMONENE DIOXIDE

1,2-EPOXYAETHAN (GERMAN) see ETHYLENE OXIDE

1,2-EPOXYBUTANE see 1-BUTENE OXIDE

1,2-EPOXYBUTENE-3 see 3,4-EPOXY-1-BUTENE

4,9-EPOXYCEVANE-3,4,12,14,16,17,20-HEPTOL 3-(3,4-DIMETHOXY-
BENZOATE) see VERATRIDINE

1,2-EPOXY-3-CHLOROPROPANE see 1-CHLORO-2,3-EPOXY PROPANE

5-alpha,6-alpha-EPOXYCHOLESTANOL see EPOXYCHOLESTEROL

1,2-EPOXYCYCLOHEXANE see CYCLOHEXENE OXIDE

3,6-EPOXY-CYCLOHEXANE 1,2-CARBOXYLATE DISODIQUE (FRENCH)
see DISODIUM-3,6-ENDOXOHEXAHYDROPHTHALATE

12,13-EPOXY-4-BETA,15-DIAZETOXY-3-ALPHA-HYDROXYTRICHO-
THEC-9-ENE see ANGUIDIN

15,20-EPOXY-15,30-DIHYDRO-12-HYDROXY SENECIONAN-11,16-
DIONE see JACOBINE

5,6-EPOXY-5,6-DIHYDRO-7-METHYLBENZ(A) ANTHRACENE see 7-
METHYLBENZ(a)ANTHRACENE-5,6-OXIDE

1,2-EPOXYDODECANE see DODECENE EPOXIDE

1,2-EPOXY-4-(EPOXYETHYL)CYCLOHEXANE see VINYL CYCLOHEX-
ENE DIOXIDE

1,2-EPOXY-7,8-EPOXYOCTANE see 1,2,7,8-DIEPOXYOCTANE

1,2-EPOXYETHANE see ETHYLENE OXIDE

EPOXYETHYLBENZENE (8CI) see 1,2-EPOXYETHYLBENZENE

1-EPOXYETHYL-3,4-EPOXYCYCLOHEXANE see VINYL CYCLOHEXENE
DIOXIDE

3-(EPOXYETHYL)-7-OXABICYCLO(4.1.0)HEPTANE see VINYL CYCLO-
HEXENE DIOXIDE

4-(EPOXYETHYL)-7-OXABICYCLO(4.1.0)HEPTANE see VINYL CYCLO-
HEXENE DIOXIDE

3-(1,2-EPOXYETHYL)-7-OXABICYCLO(4.1.0)HEPTANE see VINYL CY-
CLOHEXENE DIOXIDE

4-(1,2-EPOXYETHYL)-7-OXABICYCLO(4.1.0)HEPTANE see VINYL CY-
CLOHEXENE DIOXIDE

4,5-EPOXY-14-HYDROXY-3-METHOXY-17-METHYL-MORPHINAN-6-
ONE, HYDROCHLORIDE, (5-ALPHA)- (9CI) see DIHYDRONE HYDRO-
CHLORIDE

1,8-EPOXY-P-MENTHANE see CAJEPUTOL

4-(1,2-EPOXY-1-METHYLETHYL)-1-METHYL-7-OXABICYCLO(4.1.0)-
HEPTANE see LIMONENE DIOXIDE

cis-9,10-EPOXYOCTADECANOATE see cis-9,10-EPOXYOCTADECANOIC
ACID

EPOXYOLEIC ACID see cis-9,10-EPOXYOCTADECANOIC ACID

1,2-EPOXY-3-PHENOXYPROPANE see PHENYLGLYCIDYL ETHER

2,3-EPOXY-1-PROPANAL see GLYCIDALDEHYDE

1,2-EPOXYPROPANE see EPOXYPROPANE

2,3-EPOXY-1-PROPANOL OLEATE see 2,3-EPOXYPROPYL OLEATE

2,3-EPOXY-1-PROPANOL STEARATE see STEARIC ACID-2,3-EPOXY-
PROPYL ESTER

2,3-EPOXYPROPIONALDEHYDE see GLYCIDALDEHYDE

2,3-EPOXYPROPYL BUTYL ETHER see BUTYL GLYCIDYL ETHER

2,3-EPOXYPROPYL CHLORIDE see 1-CHLORO-2,3-EPOXY PROPANE

2,3-EPOXYPROPYL ESTER OF OLEIC ACID see 2,3-EPOXYPROPYL
OLEATE

2,3-EPOXYPROPYL ESTER OF STEARIC ACID see STEARIC ACID-2,3-
EPOXYPROPYL ESTER

2,3-EPOXYPROPYLPHENYL ETHER see PHENYLGLYCIDYL ETHER

2,3-EPOXYPROPYL STEARATE see STEARIC ACID-2,3-EPOXYPROPYL
ESTER

9,10-EPOXYSTEARIC ACID see cis-9,10-EPOXYOCTADECANOIC ACID

cis-9,10-EPOXYSTEARIC ACID see cis-9,10-EPOXYOCTADECANOIC
ACID

EPOXYSTYRENE see 1,2-EPOXYETHYLBENZENE

ALPHA,BETA-EPOXYSTYRENE see 1,2-EPOXYETHYLBENZENE

12,13-EPOXY-TRICHOTHEC-9-ENE-3-alpha,4-beta,15-TRIOL-4,15-DI-
ACETATE see DIACETOXYSCIRPENOL

6-BETA,7-BETA-EPOXY-3-ALPHA-TROPANYL S-(−)-TROPATE see
SCOPOLAMINE

EPOXYTROPINE TROPATE see SCOPOLAMINE

EPSICAPRON see 6-AMINOCAPROIC ACID

EPSOM SALTS see MAGNESIUM SULFATE HEPTAHYDRATE

EPSYLON KAPROLAKTAM (POLISH) see HEXAHYDRO-2H-AZEPIN-
2-ONE

EPTACLORO (ITALIAN) see HEPTACHLOR

EPTANI (ITALIAN) see N-HEPTANE

EPTAN-3-ONE (ITALIAN) see 3-HEPTANONE

EQUANIL see MILTOWN

EQUANIL SUSPENSION see MILTOWN

EQUILENINA (SPANISH) see EQUILENIN

EQUILENINE see EQUILENIN

ERAMIN see HISTAMINE

ERBIUM TRICHLORIDE see ERBIUM CHLORIDE

ERGAMINE see HISTAMINE

ERGOBASINE see D-LYSERGIC ACID-L,2-PROPANOLAMIDE

ERGOMETRINE see D-LYSERGIC ACID-L,2-PROPANOLAMIDE

ERGOMETRINE ACID MALEATE see ERGOT

ERGOMETRINE MALEATE see ERGOT

ERGONOVINE see D-LYSERGIC ACID-L,2-PROPANOLAMIDE

ERGONOVINE, MALEATE (1:1) (SALT) see ERGOT

ERGOSTEROL ACTIVATED see ERGOCALCIFEROL

ERGOSTEROL, IRRADIATED see ERGOCALCIFEROL

ERGOTAMINE BITARTRATE see ERGOTAMINE TARTRATE

ERGOTIDINE see HISTAMINE

ERGOTRATE see D-LYSERGIC ACID-L,2-PROPANOLAMIDE

ERGOTRATE see ERGOT

ERGOTRATE MALEATE see ERGOT

ERITROXILINA see COCAINE

ERMETRINE see D-LYSERGIC ACID-L,2-PROPANOLAMIDE

EROINA see HEROIN

ERYTHRITOL ANHYDRIDE see meso-1,2,3,4-DIEPOXYBUTANE

ERYTHRITOL ANHYDRIDE see 1,1′-BI(ETHYLENE OXIDE)

ERYTHROCIN see ERYTHROMYCIN

ERYTHROGRAN see ERYTHROMYCIN

ERYTHROMYCIN A see ERYTHROMYCIN

ERYTHROTIN see VITAMIN B_{12} COMPLEX

ERYTROXYLIN see COCAINE

ESAIDRO-1,3,5-TRINITRO-1,3,5-TRIAZINA (ITALIAN) see TRIMETHYL-
ENETRINITRAMINE

ESANI (ITALIAN) see N-HEXANE

ESCULETIN DIMETHYL ETHER see 6,7-DIMETHYLESCULETIN

ESERINE SALICYLATE see PHYSOSTIGMINE SALICYLATE (1:1)

ESEROLEIN, METHYLCARBAMATE (ESTER) see PHYSOSTIGMINE

ESSENCE OF MIRBANE see NITROBENZENE

ESSENCE OF NIOBE see ETHYL BENZOATE

ESSIGESTER (GERMAN) see ETHYL ACETATE

ESSIGSAEURE (GERMAN) see ACETIC ACID

ESSIGSAEUREANHYDRID (GERMAN) see ACETIC ANHYDRIDE

ESTERONE see ESTRONE

D-ESTRADIOL see ESTRADIOL (β)

cis-ESTRADIOL see ESTRADIOL (β)

alpha-ESTRADIOL see ESTRADIOL (β)

beta-ESTRADIOL see ESTRADIOL (β)

ESTRADIOL-17-BETA see ESTRADIOL (β)

17-beta-ESTRADIOL see ESTRADIOL (β)

3,17-beta-ESTRADIOL see ESTRADIOL (β)

D-3,17-beta-ESTRADIOL see ESTRADIOL (β)

beta-ESTRADIOL BENZOATE see ESTRADIOL-3-BENZOATE

beta-ESTRADIOL-3-BENZOATE see ESTRADIOL-3-BENZOATE

17-beta-ESTRADIOL BENZOATE see ESTRADIOL-3-BENZOATE

ESTRADIOL-17-beta-BENZOATE see ESTRADIOL-3-BENZOATE

17-beta-ESTRADIOL-3-BENZOATE see ESTRADIOL-3-BENZOATE

ESTRADIOL-17-beta-3-BENZOATE see ESTRADIOL-3-BENZOATE

ESTRADIOL-3,17-DIPROPIONATE see ESTRADIOL DIPROPIONATE

beta-ESTRADIOL DIPROPIONATE see ESTRADIOL DIPROPIONATE

17-beta-ESTRADIOL DIPROPIONATE see ESTRADIOL DIPROPIONATE

3,17-beta-ESTRADIOL DIPROPIONATE see ESTRADIOL DIPROPIONATE

beta-ESTRADIOL-3,17-DIPROPIONATE see ESTRADIOL DIPROPIONATE

ESTRADIOL MONOBENZOATE see ESTRADIOL-3-BENZOATE

17-beta-ESTRADIOL MONOBENZOATE see ESTRADIOL-3-BENZOATE

ESTRADIOL PHOSPHATE POLYMER see ESTRADIOL POLYESTER WITH
PHOSPHORIC ACID

ESTRAL-3,5(10)-TRIENE-3,17-DIOL (17-beta)-DIPROPIONATE see ES-
TRADIOL DIPROPIONATE

1,3,5,7-ESTRATETRAEN-3-OL-17-ONE see EQUILIN

1,3,5-ESTRATRIENE-3,17-beta-DIOL see ESTRADIOL (β)

ESTRA-1,3,5(10)-TRIENE-3,17-beta-DIOL see ESTRADIOL (β)

17-beta-ESTRA-1,3,5(10)-TRIENE-3,17-DIOL see ESTRADIOL (β)

1,3,5(10)-ESTRATRIENE-3,17-beta-DIOL 3-BENZOATE see ESTRA-
DIOL-3-BENZOATE

ESTRA-1,3,5(10)-TRIENE-3,17-beta-DIOL, 3-BENZOATE see ESTRA-
DIOL-3-BENZOATE

ESTRA-1,3,5(10)-TRIENE-3,17-DIOL (17-beta)-3-BENZOATE see ES-
TRADIOL-3-BENZOATE

1,3,5(10)-ESTRATRIENE-3,17-beta-DIOL DIPROPIONATE see ESTRA-
DIOL DIPROPIONATE

ESTRA-1,3,5(10)-TRIENE-3,17 DIOL (17-beta)-, POLYMER WITH
PHOSPHORIC ACID see ESTRADIOL POLYESTER WITH PHOSPHORIC
ACID

(17-beta)-ESTRA-1,3,5(10)-TRIENE- 3,17-DIOL POLYMER WITH PHOS-
PHORIC ACID see ESTRADIOL POLYESTER WITH PHOSPHORIC ACID
ESTRA-1,3,5(10)-TRIENE-3,16-alpha,17-beta-TRIOL see ESTRIOL
1,3,5-ESTRATRIENE-3-beta,16-alpha,17-beta-TRIOL see ESTRIOL
(16-alpha,17-beta)-ESTRA-1,3,5(10)-TRIENE-3,16,17-TRIOL see ES-
TRIOL
1,3,5-ESTRATRIEN-3-OL-17-ONE see ESTRONE
1,3,5(10)-ESTRATRIEN-3-OL-17-ONE see ESTRONE
delta-1,3,5-ESTRATRIEN-3-beta-OL-17-ONE see ESTRONE
ESTRATRIOL see ESTRIOL
16-alpha,17-beta-ESTRIOL see ESTRIOL
3,16-alpha,17-beta-ESTRIOL see ESTRIOL
ESTROGEN see ETHINYL ESTRADIOL
ESTRONA (SPANISH) see ESTRONE
ETAIN (TETRACHLORURE D') (FRENCH) see TIN (IV) CHLORIDE (1:4)
ETANOLAMINA (ITALIAN) see MONOETHANOLAMINE
ETANOLO (ITALIAN) see ETHANOL
ETANTIOLO (ITALIAN) see ETHANETHIOL
ETERE ETILICO (ITALIAN) see ETHYL ETHER
ETHAANTHIOL (DUTCH) see ETHANETHIOL
ETHANAL see ACETALDEHYDE
ETHANAL OXIME see ACETALDEHYDE OXIME
ETHANAMIDE see ACETAMIDE
ETHANECARBOXYLIC ACID see PROPIONIC ACID
ETHANEDIAL see GLYOXAL
ETHANEDIAMIDE see OXAMIDE
ETHANEDINITRILE see CYANOGEN
ETHANEDIOIC ACID see OXALIC ACID
1,2-ETHANEDIOL see ETHYLENE GLYCOL
ETHANEDIOL DINITRATE see ETHYLENE GLYCOL DINITRATE
1,2-ETHANEDIONE see GLYOXAL
ETHANEDIONIC ACID see OXALIC ACID
ETHANEDITHIOAMIDE see DITHIOOXAMIDE
1,2-ETHANEDITHIOL, CYCLIC ESTER WITH P,P-DIETHYL PHOS-
PHONODITHIOIMIDOCARBONATE see PHOSFOLAN
1,2-ETHANEDIYLBIS(CARBAMODITHIOATO)(2−)-MANGANESE see
VANCIDE
1,2-ETHANEDIYLBIS(CARBAMODITHIOATO) (2-)-S,S′-ZINC see ETHYL-
ENE BIS(DITHIO CARBAMATO)ZINC
1,2-ETHANEDIYLBISCARBAMODITHIOIC ACID, MANGANESE COMPLEX
see VANCIDE
1,2-ETHANEDIYLBISCARBAMODITHIOIC ACID, MANGANESE(2+) SALT
(1:1) see VANCIDE
1,2-ETHANEDIYLBISCARBAMODITHIOIC ACID, ZINC COMPLEX see
ETHYLENE BIS(DITHIO CARBAMATO)ZINC
1,2-ETHANEDIYLBISCARBAMOTHIOIC ACID, ZINC SALT see ETHYLENE
BIS(DITHIO CARBAMATO)ZINC
1,2-ETHANEDIYLBISMANEB, MANGANESE (2+) SALT (1:1) see VAN-
CIDE
ETHANE HEXACHLORIDE see 1,1,1,2,2,2-HEXACHLOROETHANE
ETHANEHYDRAZONIC ACID see ACETYL HYDRAZIDE
ETHANENITRILE see ACETONITRILE
ETHANE PENTACHLORIDE see PENTACHLOROETHANE
ETHANETHIOAMIDE see THIOACETAMIDE
ETHANETHIOL, 2-(METHYLTHIO)-, O,O-DIMETHYL PHOSPHOROTHIO-
ATE see METHYL DEMETON METHYL
ETHANETHIOL, 2-(METHYLTHIO)-, S-ESTER WITH O,O-DIMETHYL
PHOSPHOROTHIOATE see METHYL DEMETON METHYL
ETHANE TRICHLORIDE see 1,1,2-TRICHLOROETHANE
ETHANOIC ACID see ACETIC ACID
ETHANOIC ANHYDRATE see ACETIC ANHYDRIDE
BETA-ETHANOLAMINE see MONOETHANOLAMINE
ETHANOL, 2-FLUORO-, PHOSPHITE (3:1) see PHOSPHOROUS ACID
TRIS(2-FLUOROETHYLESTER)
ETHANOL, 2,2′-IMINODI-, COMPOUND WITH 1,2-DIHYDRO-3,6-PYRI-
DAZINEDIONE (1:1) see DIETHANOLAMMONIUM MALEIC HYDRA-
ZIDE
ETHANOL 200 PROOF see ETHANOL
4-ETHANOLPYRIDINE see 4-PYRIDINEETHANOL
1-ETHANOL-2-THIOL see 2-MERCAPTOETHANOL

ETHANONE, 1-(3-ETHYL-5,6,7,8-TETRAHYDRO-5,5,8,8-TETRA-
METHYL-2-NAPHTHALENYL)(9CI) see ACETYL ETHYL TETRA-
METHYL TETRALIN
ETHANOYL CHLORIDE see ACETYL CHLORIDE
ETHENE see ETHYLENE
ETHENE POLYMER see POLYETHYLENE
ETHENOL HOMOPOLYMER (9CI) see POLYVINYL ALCOHOL
ETHENONE see KETENE
ETHENYL ACETATE see ACETIC ACID VINYL ESTER
4-ETHENYL-1-CYCLOHEXENE see CYCLOHEXENYLETHYLENE
ETHER BUTYLIQUE (FRENCH) see n-BUTYL ETHER
ETHER DICHLORE (FRENCH) see BIS(2-CHLOROETHYL) ETHER
ETHER ETHYLBUTYLIQUE (FRENCH) see ETHYL BUTYL ETHER
ETHER ETHYLIQUE (FRENCH) see ETHYL ETHER
ETHER HYDROCHLORIC see ETHYL CHLORIDE
ETHER ISOPROPYLIQUE (FRENCH) see DIISOPROPYL ETHER
ETHER METHYLIQUE MONOCHLORE (FRENCH) see CHLOROMETHYL
METHYL ETHER
ETHER MONOETHYLIQUE DE L'ETHYLENE-GLYCOL (FRENCH) see
CELLOSOLVE SOLVENT
ETHER MONOMETHYLIQUE DE L'ETHYLENE-GLYCOL (FRENCH) see
ETHYLENE GLYCOL METHYL ETHER
ETHER MURIATIC see ETHYL CHLORIDE
ETHERON SPONGE see POLYURETHANE FOAM
ETHINE see ACETYLENE
17-ALPHA-ETHINYL-3,17-DIHYDROXY-DELTA(SUP 1,3,5)-ESTRA-
TRIENE see ETHINYL ESTRADIOL
17-ALPHA-ETHINYL-3,17-DIHYDROXY-DELTA(SUP. 1,3,5)OESTRA-
TRIENE see ETHINYL ESTRADIOL
17-alpha-ETHINYL-delta(SUPER)-5,10-19-NORTESTOSTERONE see 17-
alpha-ETHINYL-5,10-ESTRENOLONE
17-ETHINYLESTRADIOL see ETHINYL ESTRADIOL
17-ETHINYL-3,17-ESTRADIOL see ETHINYL ESTRADIOL
17-ALPHA-ETHINYLESTRADIOL see ETHINYL ESTRADIOL
17-ALPHA-ETHINYL-17-BETA-ESTRADIOL see ETHINYL ESTRADIOL
ETHINYLESTRADIOL-3-METHYL ETHER see 3-METHYLETHYNYLESTRA-
DIOL
17-ALPHA-ETHINYL ESTRADIOL 3-METHYL ETHER see 3-METHYL-
ETHYNYLESTRADIOL
17-alpha-ETHINYL-ESTRA(5,10)ENEOLONE see 17-alpha-ETHINYL-
5,10-ESTRENOLONE
17-ALPHA-ETHINYLESTRA-4-EN-17-BETA-OL-3-ONE see 19-NORETHI-
STERONE
17-ALPHA-ETHINYLESTRA-1,3,5(10)-TRIENE-3,17-BETA-DIOL see
ETHINYL ESTRADIOL
ETHINYLESTRIOL see ETHINYL ESTRADIOL
17-ALPHA-ETHINYL-17-BETA-HYDROXY-DELTA(SUP:4)-ESTREN-3-
ONE see 19-NORETHISTERONE
17-ALPHA-ETHINYL-19-NORTESTOSTERONE see 19-NORETHISTERONE
17-ALPHA-ETHINYL-19-NORTESTOSTERONE ACETATE see 17-ACET-
OXY-19-NOR-17-alpha-PREGN-4-EN-20-YN-3-ONE
17-ALPHA-ETHINYL-19-NORTESTOSTERONE-17-BETA-ACETATE see
17-ACETOXY-19-NOR-17-alpha-PREGN-4-EN-20-YN-3-ONE
17-ALPHA-ETHINYL-19-NORTESTOSTERONE ENANTHATE see NORETHI-
STERONE ENANTHATE
ETHINYLOESTRADIOL see ETHINYL ESTRADIOL
17-ETHINYL-3,17-OESTRADIOL see ETHINYL ESTRADIOL
ETHINYLOESTRADIOL-3-METHYL ETHER see 3-METHYLETHYNYLES-
TRADIOL
17-ALPHA-ETHINYL OESTRADIOL-3-METHYL ETHER see 3-METHYL-
ETHYNYLESTRADIOL
ETHINYL-OESTRANOL see ETHINYL ESTRADIOL
17-ALPHA-ETHINYLOESTRA-1,3,5(10)-TRIENE-3,17-BETA-DIOL see
ETHINYL ESTRADIOL
17-ALPHA-ETHINYL-DELTA(SUP 1,3,5(10))OESTRATRIENE-3,17-BETA-
DIOL see ETHINYL ESTRADIOL
ETHINYLOESTRIOL see ETHINYL ESTRADIOL
ETHINYL TRICHLORIDE see ACETYLENE TRICHLORIDE
(+ −)-ETHIONINE see dl-ETHIONINE
ETHOHEXADIOL see ETHYL HEXYLENE GLYCOL

ETHOPROP see O-ETHYL-S,S-DIPROPYL PHOSPHORO DITHIOATE

ETHOPROPHOS see O-ETHYL-S,S-DIPROPYL PHOSPHORO DITHIOATE

ETHOSUCCIMIDE see 2-ETHYL-2-METHYLSUCCINIMIDE

ETHOSUCCIMIDE see 2-ETHYL-2-METHYLSUCCINIMIDE

ETHOSUXIMIDE see 2-ETHYL-2-METHYLSUCCINIMIDE

PARA-ETHOXY-ACETANILID see p-ACETOPHENETIDIDE

4-ETHOXYACETANILIDE see p-ACETOPHENETIDIDE

P-ETHOXYACETANILIDE see p-ACETOPHENETIDIDE

4′-ETHOXYACETANILIDE see p-ACETOPHENETIDIDE

ETHOXY ACETATE see "CELLOSOLVE" ACETATE

N-(ETHOXYCARBONYL)AZIRIDINE see AZIRIDINE CARBOXYLIC ACID ETHYL ESTER

S-alpha-ETHOXYCARBONYLBENZYL O,O-DIMETHYL PHOSPHORODITHIOATE see (O,O-DIMETHYLDITHIOPHOSPHORYLPHENYL)-ACETIC ACID ETHYL ESTER

S-alpha-ETHOXYCARBONYLBENZYL DIMETHYL PHOSPHOROTHIOLOTHIONATE see (O,O-DIMETHYLDITHIOPHOSPHORYLPHENYL)-ACETIC ACID ETHYL ESTER

ETHOXYCARBONYLDIAZO-METHANE see DIAZOACETIC ESTER

ETHOXYCARBONYLETHYLENE see ETHYL ACRYLATE

N-ETHOXYCARBONYLETHYLENEIMINE see AZIRIDINE CARBOXYLIC ACID ETHYL ESTER

ETHOXYCARBONYL-1-ETHYLENIMINE see AZIRIDINE CARBOXYLIC ACID ETHYL ESTER

4-ETHOXYCARBONYL-1-(2-HYDROXY-3-PHENOXYPROPYL) 4-PHENYLPIPERIDINE HYDROCHLORIDE see ALIDINE DIHYDROCHLORIDE

ETHOXYCARBONYLMETHYL BROMIDE see ETHYL BROMACETATE

S-((ETHOXYCARBONYL)METHYLCARBAMOYL)METHYL-O,O-DIETHYL PHOSPHORODITHIOATE see O,O-DIETHYL-S-(N-ETHOXYCARBONYL-N-METHYLCARBAMOYLMETHYL) PHOSPHORO DITHIOATE

N-ETHOXYCARBONYL-N-METHYLCARBAMOYLMETHYL-O,O-DIETHYL PHOSPHORODITHIOATE see O,O-DIETHYL-S-(N-ETHOXYCARBONYL-N-METHYLCARBAMOYLMETHYL) PHOSPHORO DITHIOATE

2-ETHOXY-5-CHLOROBENZYL-BIS(BETA-CHLOROETHYL) AMINE HYDROCHLORIDE see N,N-BIS(2-CHLOROETHYL)-3-CHLORO-6-ETHOXYBENZYLAMINE HYDROCHLORIDE

ETHOXY CHLOROMETHANE see CHLOROMETHYL ETHYL ETHER

ETHOXY DIGLYCOL see CARBITOL CELLOSOLVE

2-ETHOXY-3,4-DIHYDRO-1,2-PYRAN see 2-ETHOXY-3,4-DIHYDRO-2H-PYRAN

2-ETHOXY-2,3-DIHYDRO-gamma-PYRAN see 2-ETHOXY-3,4-DIHYDRO-2H-PYRAN

2-ETHOXY DIHYDROPYRAN, IN PREGNANCY DIAGNOSIS see 2-ETHOXY-3,4-DIHYDRO-2H-PYRAN

ETHOXYETHANE see ETHYL ETHER

2-ETHOXYETHANOL see CELLOSOLVE SOLVENT

2-ETHOXYETHANOL, ESTER WITH ACETIC ACID see "CELLOSOLVE" ACETATE

2-(2-ETHOXYETHOXY)ETHANOL see CARBITOL CELLOSOLVE

2-(2-(2-ETHOXYETHOXY)ETHOXY)ETHANOL see ETHOXYTRIGLYCOL

2-ETHOXY-ETHYLACETAAT (DUTCH) see "CELLOSOLVE" ACETATE

ETHOXYETHYL ACETATE see "CELLOSOLVE" ACETATE

BETA-ETHOXYETHYL ACETATE see "CELLOSOLVE" ACETATE

ETHOXYETHYL ACRYLATE see ACRYLIC ACID-2-ETHOXYETHYL ESTER

2-ETHOXYETHYL ACRYLATE see ACRYLIC ACID-2-ETHOXYETHYL ESTER

2-ETHOXYETHYLE, ACETATE DE (FRENCH) see "CELLOSOLVE" ACETATE

2-ETHOXYETHYL-2-PROPENOATE see ACRYLIC ACID-2-ETHOXYETHYL ESTER

ETHOXYFORMIC ANHYDRIDE see DIETHYL CARBONATE

3-ETHOXY-4-HYDROXYBENZALDEHYDE see ETHYL VANILLIN

2-(3-ETHOXY-1-INDANYLIDENE)-1,3-DINDANDIONE see BINDON ETHYL ETHER

ETHOXY METHYL CHLORIDE see CHLOROMETHYL ETHYL ETHER

N-(4-ETHOXYPHENYL)ACETAMIDE see p-ACETOPHENETIDIDE

N-PARA-ETHOXYPHENYLACETAMIDE see p-ACETOPHENETIDIDE

2-(4-ETHOXYPHENYL)-1,3-BIS(4-METHOXYPHENYL)GUANIDINE HYDROCHLORIDE see PHENODIANISYL HYDROCHLORIDE

N-(4-ETHOXYPHENYL)-3′-NITROACETAMIDE see 3′-NITRO-p-ACETOPHENETIDE

N-(4-ETHOXYPHENYL)UREA see 4-ETHOXY PHENYL UREA

ETHOXYTRIETHYLENE GLYCOL see ETHOXYTRIGLYCOL

ETHYLACETAAT (DUTCH) see ETHYL ACETATE

ETHYLACETANILIDE see N-ETHYLACETANILIDE

ETHYLACETIC ACID see N-BUTYRIC ACID

ETHYL ACETIC ESTER see ETHYL ACETATE

ETHYL ACETOACETATE see ETHYL ACETYL ACETATE

ETHYL ACETONE see 2-PENTANONE

ETHYL ACETOXYMETHYLNITROSAMINE see (ETHYL NITROSAMINO)-METHYL ACETATE

N-ETHYL-N-(ACETOXYMETHYL)NITROSAMINE see (ETHYL NITROSAMINO)METHYL ACETATE

ETHYL ACETYLACETONATE see ETHYL ACETYL ACETATE

ETHYLACRYLAAT (DUTCH) see ETHYL ACRYLATE

ETHYL ACRYLATE, INHIBITED see ACRYLIC ACID ETHYL ESTER (INHIBITED)

ETHYLAKRYLAT (CZECH) see ETHYL ACRYLATE

ETHYL ALCOHOL see ETHANOL

ETHYLALCOHOL (DUTCH) see ETHANOL

ETHYL ALCOHOL ANHYDROUS see ETHANOL

ETHYL ALDEHYDE see ACETALDEHYDE

ETHYLAMINE see ETHANAMINE

ETHYLAMINE, 1-ALLYL-2-PHENYL-, HYDROCHLORIDE see alpha-ALLYL PHENETHYLAMINEHYDROCHLORIDE

ETHYLAMINE-2-(DIPHENYLMETHOXY)-N,N-DIMETHYL-, COMPOUND WITH 8-CHLOROTHEOPHYLLINE (1:1) see DRAMAMINE

N-ETHYLAMINOBENZENE see N-ETHYL ANILINE

ETHYL-p-AMINOBENZOATE see ETHYL-4-AMINOBENZOATE

ETHYL-1-(P-AMINOPHENETHYL)-4-PHENYLISONIPECOTATE see N-beta-(p-AMINOPHENYL)ETHYLNORMEPERIDINE

ETHYL- 1-(P-AMINOPHENETHYL)-4-PHENYLISONIPECOTATEDIHYDROCHLORIDE see ALIDINE DIHYDROCHLORIDE

ETHYL 1-(4-AMINOPHENYLETHYL)-4-PHENYLISONIPECOTATE DIHYDROCHLORIDE see ALIDINE DIHYDROCHLORIDE

ETHYL 1-(4-AMINOPHENYLETHYL)-4-PHENYLISONIPECOTATE DIHYDROCHLORIDE see ALIDINE DIHYDROCHLORIDE

ETHYL AMYL KETONE see AMYL ETHYL KETONE

ETHYLANILINE see N-ETHYL ANILINE

O-ETHYLANILINE see 2-ETHYL ANILINE

P-ETHYLANILINE see 4-ETHYL ANILINE

ETHYL ANISATE see p-ANISIC ACID, ETHYL ESTER

ETHYL-P-ANISATE see p-ANISIC ACID, ETHYL ESTER

ETHYL AZIRIDINECARBOXYLATE see AZIRIDINE CARBOXYLIC ACID ETHYL ESTER

ETHYL-1-AZIRIDINECARBOXYLATE see AZIRIDINE CARBOXYLIC ACID ETHYL ESTER

ETHYL AZIRIDINOCARBOXYLATE see AZIRIDINE CARBOXYLIC ACID ETHYL ESTER

ETHYL-1-AZIRIDINYLCARBOXYLATE see AZIRIDINE CARBOXYLIC ACID ETHYL ESTER

ETHYL AZIRIDINYLFORMATE see AZIRIDINE CARBOXYLIC ACID ETHYL ESTER

7-ETHYLBENZ(C)ACRIDINE see BENZENECARBOXALDEHYDE

9-ETHYL-3,4-BENZACRIDINE see BENZENECARBOXALDEHYDE

ETHYLBENZEEN (DUTCH) see ETHYL BENZENE

2-ETHYLBENZENAMINE see 2-ETHYL ANILINE

N-ETHYLBENZENAMINO see N-ETHYL ANILINE

ETHYLBENZOL see ETHYL BENZENE

ETHYLBENZYLBARBITURIC ACID see BENZYLBARBITAL

ETHYL (BIS(1-AZIRIDINYL)PHOSPHINYL)CARBAMATE see ETHYL(DI-(1-AZIRIDINYL)PHOSPHINYL)CARBAMATE

ETHYLBIS(2-CHLOROETHYL)AMINE see BIS(2-CHLOROETHYL)ETHYLAMINE

ETHYLBIS(BETA-CHLOROETHYL)AMINE see BIS(2-CHLOROETHYL)-ETHYLAMINE

ETHYL BIS(4-HYDROXYCOUMARINYL)ACETATE see BIS(4-HYDROXY-3-COUMARIN) ACETIC ACID ETHYL ESTER

ETHYL BIS(4-HYDROXY-3-COUMARINYL) ACETATE see BIS(4-HY-
DROXY-3-COUMARIN) ACETIC ACID ETHYL ESTER
ETHYLBORATE (DOT) see BORIC ACID, ETHYL ESTER
ETHYL BROMIDE see BROMOETHANE
ETHYL BROMOACETATE see ETHYL BROMACETATE
ETHYL-alpha-BROMOACETATE see ETHYL BROMACETATE
2-ETHYL-1-BUTANAMINE see 2-ETHYL BUTYLAMINE
2-ETHYL BUTANOIC ACID see DIETHYLACETIC ACID
2-ETHYLBUTANOL-1 see 2-ETHYLBUTANOL
2-ETHYL-1-BUTANOL see 2-ETHYLBUTANOL
ETHYL BUTYLACETATE (DOT) see ETHYL CAPROATE
2-ETHYLBUTYL ALCOHOL see 2-ETHYLBUTANOL
5-ETHYL-5-N-BUTYLBARBITURIC ACID see BUTOBARBITAL
ETHYLBUTYLCETONE (FRENCH) see 3-HEPTANONE
ETHYLBUTYLKETON (DUTCH) see 3-HEPTANONE
2-ETHYLBUTYRIC ACID see DIETHYLACETIC ACID
ALPHA-ETHYLBUTYRIC ACID see DIETHYLACETIC ACID
ETHYL CADMATE see BIS(DIETHYL DITHIO CARBAMATO)CADMIUM
ALPHA-ETHYLCAPROALDEHYDE see BUTYL ETHYL ACETALDEHYDE
ALPHA-ETHYLCAPROIC ACID see BUTYL ETHYL ACETIC ACID
ETHYL CARBAMATE see URETHANE
S-(N-ETHYLCARBAMOYLMETHYL) DIMETHYL PHOSPHORODITHIOATE
see DIMETHOATE-ETHYL
ETHYL CARBANILATE see CARBANILIC ACID ETHYL ESTER
ETHYL CARBINOL see PROPYL ALCOHOL
ETHYL CARBITOL see CARBITOL CELLOSOLVE
ETHYL CARBONATE see DIETHYL CARBONATE
ETHYL CELLOSOLVE see CELLOSOLVE SOLVENT
ETHYL CELLOSOLVE ACETAAT (DUTCH) see ''CELLOSOLVE'' ACE-
TATE
ETHYLCHLOORFORMIAAT (DUTCH) see ETHYL CHLORO FORMATE
ETHYL CHLOROACETATE see ETHYL CHLORACETATE
ETHYL-ALPHA-CHLOROACETATE see ETHYL CHLORACETATE
ETHYL CHLOROCARBONATE see ETHYL CHLORO FORMATE
ETHYL CHLOROETHANOATE see ETHYL CHLORACETATE
9-(3-ETHYL-2-CHLOROETHYL)AMINO PROPYLAMINO)-4-METHOXY
ACRIDINE DIHYDROCHLORIDE see 4-METHOXY-9-(3-(ETHYL-2-
CHLOROETHYL)
ETHYL N-(beta-CHLOROETHYL)-N-NITROSOCARBAMATE see 2-
CHLORO ETHYL-N-NITROSO URETHANE
ETHYL-2-CHLOROETHYL SULFIDE see CHLOROETHYL ETHYL SULFIDE
ETHYL CHLOROPHENOXYISOBUTYRATE see ATROMID S
ETHYL-2-(P-CHLOROPHENOXY)ISOBUTYRATE see ATROMID S
ETHYL-PARA-CHLOROPHENOXYISOBUTYRATE see ATROMID S
ETHYL-ALPHA-(4-CHLOROPHENOXY)ISOBUTYRATE see ATROMID S
ETHYL-ALPHA-P-CHLOROPHENOXYISOBUTYRATE see ATROMID S
ETHYL 2-(4-CHLOROPHENOXY)-2-METHYLPROPIONATE see
ATROMID S
ETHYL 2-(P-CHLOROPHENOXY)-2-METHYLPROPIONATE see
ATROMID S
ETHYL-ALPHA-(4-CHLOROPHENOXY)-ALPHA-METHYLPROPIONATE see
ATROMID S
ETHYL-ALPHA-(P-CHLOROPHENOXY)-ALPHA-METHYLPROPIONATE see
ATROMID S
ETHYL CHLOROTHIOFORMATE (DOT) see CHLOROTHIOFORMIC ACID
ETHYL ESTER
ETHYL-beta-CHLOROVINYLETHYNYL CARBINOL see 1-CHLORO-3-
ETHYL-1-PENTEN-4-YN-3-OL
ETHYLCINNAMATE see ETHYL-trans-CINNAMATE
ETHYL CLOFIBRATE see ATROMID S
2-ETHYLCROTONYLUREA see cis-(2-ETHYLCROTONYL) UREA
ETHYL CYANIDE see PROPIONONITRILE
ETHYL CYANOETHANOATE see ETHYL CYANOACETATE
5-ETHYL-5-(1'-CYCLOHEPTENYL)-BARBITURIC ACID see CYCLOHEP-
TENYL ETHYLBARBITURIC ACID
5-ETHYL-5-CYCLOHEXENYLBARBITURIC ACID see TETRAHYDROPHE-
NOBARBITAL
N-ETHYL-CYCLOHEXYLAMINE see N-ETHYL(CYCLO HEXYL)AMINE
ETHYL DIAZOACETATE see DIAZOACETIC ESTER
ETHYLDICHLOROARSINE see DICHLOROETHYLARSINE

ETHYL-p,p'-DICHLOROBENZILATE see ETHYL-4,4'-DICHLOROBEN-
ZILATE
ETHYL-4,4'-DICHLORODIPHENYL GLYCOLLATE see ETHYL-4,4'-
DICHLOROBENZILATE
N-ETHYLDICHLOROMALEINIMIDE see 2,3-DICHLORO-N-ETHYLMA-
LEINIMIDE
ETHYL-4,4'-DICHLOROPHENYL GLYCOLLATE see ETHYL-4,4'-
DICHLOROBENZILATE
O-ETHYL-O-2,4-DICHLOROPHENYL THIONOBENZENEPHOSPHONATE
see S-SEVEN
ETHYL DICHLOROSILANE (DOT) see DICHLOROETHYLSILANE
ETHYLDICOUMAROL ACETATE see BIS(4-HYDROXY-3-COUMARIN)
ACETIC ACID ETHYL ESTER
ETHYL DIETHYLENE GLYCOL see CARBITOL CELLOSOLVE
5-ETHYLDIHYDRO-5-PHENYL-4,6(1H,5H)-PYRIMIDINEDIONE see
2-DESOXY PHENOBARBITAL
ETHYL 4,4'-DIHYDROXYDICOUMARINYL-3,3'-ACETATE see BIS(4-HY-
DROXY-3-COUMARIN) ACETIC ACID ETHYL ESTER
ETHYL alpha-((DIMETHOXYPHOSPHENOTHIOYL)THIO)BENZENEACE-
TATE see (O,O-DIMETHYLDITHIOPHOSPHORYLPHENYL)ACETIC
ACID ETHYL ESTER
3'-ETHYL-4-DIMETHYLAMINOAZOBENZENE see p-((m-ETHYLPHE-
NYL)AZO)-N,N-DIMETHYLANILINE
4'-ETHYL-4-DIMETHYLAMINOAZOBENZENE see p-((p-ETHYLPHE-
NYL)AZO)-N,N-DIMETHYLANILINE
N-ETHYL-N-1-DIMETHYL-3,3-DI-2-THIENYLALLYLAMINE HYDRO-
CHLORIDE see ETHYLMETHIAMBUTENE HYDROCHLORIDE
N-ETHYL-N-1-DIMETHYL-3,3-DI-2-THIENYL-2-PROPENAMINE HYDRO-
CHLORIDE see ETHYLMETHIAMBUTENE HYDROCHLORIDE
ETHYL-N,N-DIMETHYLPHOSPHORAMIDOCYANIDATE see ETHYL
DIMETHYL AMIDO CYANO PHOSPHATE
ETHYL-O,O-DIMETHYL PHOSPHORODITHIOYLPHENYL ACETATE see
(O,O-DIMETHYLDITHIOPHOSPHORYLPHENYL)ACETIC ACID ETHYL
ESTER
ETHYLDIOL ACRILATE (RUSSIAN) see ETHYLENE ACRYLATE
S-ETHYL-N,N-DIPROPYLTHIOCARBAMATE see S-ETHYL-N,N-DI-n-PRO-
PYLTHIOCARBAMATE
ETHYL DI-N-PROPYLTHIOLCARBAMATE see S-ETHYL-N,N-DI-n-PRO-
PYLTHIOCARBAMATE
ETHYL-N,N-DI-N-PROPYLTHIOLCARBAMATE see S-ETHYL-N,N-DI-n-
PROPYLTHIOCARBAMATE
ETHYLE (ACETATE D') (FRENCH) see ETHYL ACETATE
ETHYLE, CHLOROFORMIAT D' (FRENCH) see ETHYL CHLORO FOR-
MATE
ETHYLEEN-CHLOORHYDRINE (DUTCH) see 2-CHLOROETHYL ALCOHOL
ETHYLEENDIAMINE (DUTCH) see 1,2-ETHANEDIAMINE
ETHYLEENDICHLORIDE (DUTCH) see 1,2-DICHLORETHANE
ETHYLEENDICHLORIDE (DUTCH) see ETHYLENE DICHLORIDE
ETHYLEENIMINE (DUTCH) see ETHYLENIMINE
ETHYLEENOXIDE (DUTCH) see ETHYLENE OXIDE
ETHYLE (FORMIATE D') (FRENCH) see ETHYL FORMATE
ETHYLENE ACETATE see ETHYLENE DIACETATE
ETHYLENE ALCOHOL see ETHYLENE GLYCOL
ETHYLENE ALDEHYDE see ACROLEIN
1,1'-ETHYLENE-2,2'-BIPYRIDYLIUM DIBROMIDE see DIQUAT DIBRO-
MIDE
ETHYLENE BIS(BROMOACETATE) see 1,2-BIS(BROMOACETOXY)-
ETHANE
N,N'-ETHYLENEBIS(N-BUTYL-4-MORPHOLINECARBOXAMIDE) see
DIMORPHOLAMINE
N,N'-ETHYLENE BIS(DITHIOCARBAMATE DE SODIUM) (FRENCH) see
NABAM
ETHYLENEBIS(DITHIOCARBAMATE), DISODIUM SALT see NABAM
ETHYLENEBISDITHIOCARBAMATE MANGANESE see VANCIDE
N,N'-ETHYLENE BIS(DITHIOCARBAMATE MANGANEUX) (FRENCH) see
VANCIDE
ETHYLENEBIS(DITHIOCARBAMATO), MANGANESE see VANCIDE
ETHYLENEBIS(DITHIOCARBAMIC ACID) DISODIUM SALT see NABAM
ETHYLENEBIS(DITHIOCARBAMIC ACID), MANGANESE SALT see VAN-
CIDE

ETHYLENEBIS(DITHIOCARBAMIC ACID) MANGANOUS SALT see VAN-CIDE

ETHYLENEBIS(DITHIOCARBAMIC ACID), ZINC SALT see ETHYLENE BIS(DITHIO CARBAMATO)ZINC

N,N'-ETHYLENEBIS(3-FLUOROSALICYLIDENEIMINATO) COBALT (II) see FLUOMINE DUST

N,N'-ETHYLENE BIS(3-FLUORO SALICYLIDENEIMINATO)COBALT (II) see FLUOMINE DUST

ETHYLENEBIS(IMINODIACETIC ACID) TETRASODIUM SALT see EDATHANIL TETRASODIUM

ETHYLENE BIS ISOTHIOCYANATE see ISOTHIOCYANIC ACID, ETHYLENE ESTER

2,2'-(ETHYLENEBIS(NITROSOIMINO))BISBUTANOL see D-N,N'-(1-HYDROXYMETHYLPROPYL)ETHYLENEDINITROSAMINE

2,2'-ETHYLENE-BIS-(2-THIOPSEUDOUREA), DIHYDROBROMIDE see ETHYLENE DIISO THIOURONIUM DIBROMIDE

ETHYLENE-BIS-THIURAMMONO-SULFIDE see ISOTHIOCYANIC ACID, ETHYLENE ESTER

ETHYLENE BROMIDE GLYCOL DIBROMIDE NCI-C00522 see 1,2-ETHYLENE DIBROMIDE

ETHYLENE BROMOACETATE see 1,2-BIS(BROMOACETOXY)ETHANE

ETHYLENEBROMOHYDRIN see 2-BROMO ETHANOL

ETHYLENECARBOXAMIDE see ACRYLAMIDE

ETHYLENECARBOXYLIC ACID see ACRYLIC ACID

ETHYLENE CHLORIDE see ETHYLENE DICHLORIDE

ETHYLENE CHLOROHYDRIN see 2-CHLOROETHYL ALCOHOL

ETHYLENE CYANIDE see SUCCINONITRILE

ETHYLENE CYANOHYDRIN see HYDRACRYLONITRILE

ETHYLENE DIACRYLATE see ETHYLENE ACRYLATE

ETHYLENEDIAMINE see 1,2-ETHANEDIAMINE

1,2-ETHYLENEDIAMINE see 1,2-ETHANEDIAMINE

ETHYLENE-DIAMINE (FRENCH) see 1,2-ETHANEDIAMINE

ETHYLENEDIAMINEACETIC ACID TRISODIUM SALT see TRISODIUM EDETATE

N,N'-ETHYLENE DIAMINE DIACETIC ACID TETRASODIUM SALT see EDATHANIL TETRASODIUM

ETHYLENEDIAMINETETRAACETATE see ETHYLENE DIAMINE TETRAACETIC ACID

ETHYLENEDIAMINETETRAACETATE DISODIUM SALT see EDATHAMIL DISODIUM

ETHYLENEDIAMINE-N,N,N',N'-TETRAACETIC ACID see ETHYLENE DIAMINE TETRAACETIC ACID

ETHYLENE DIAMINE TETRAACETIC ACID DISODIUM SALT see EDATHAMIL DISODIUM

ETHYLENEDIAMINETETRAACETICACID, TRISODIUM SALT see TRISODIUM EDETATE

CIS-1,2-ETHYLENEDICARBOXYLIC ACID see cis-BUTENEDIOIC ACID

ETHYLENE DICHLORIDE see 1,2-DICHLORETHANE

1,2-ETHYLENE DICHLORIDE see ETHYLENE DICHLORIDE

ETHYLENE DICYANIDE see SUCCINONITRILE

ETHYLENE DIGLYCOL see DIETHYLENE GLYCOL

ETHYLENE DINITRATE see ETHYLENE GLYCOL DINITRATE

ETHYLENEDINITRILOTETRAACETIC ACID see ETHYLENE DIAMINE TETRAACETIC ACID

(ETHYLENEDINITRILO)TETRA-ACETIC ACID COPPER(II) COMPLEX see COPPER EDTA COMPLEX

2,2'-ETHYLENEDIOXYDIETHANOL see TRIETHYLENE GLYCOL

2,2'-ETHYLENEDIOXYETHANOL see TRIETHYLENE GLYCOL

1,1'-ETHYLENE-2,2'-DIPYRIDINIUM DICHLORIDE see DIQUAT DICHLORIDE

ETHYLENE DIPYRIDYLIUM DIBROMIDE see DIQUAT DIBROMIDE

1,1-ETHYLENE 2,2-DIPYRIDYLIUM DIBROMIDE see DIQUAT DIBROMIDE

1,1'-ETHYLENE-2,2'-DIPYRIDYLIUM DIBROMIDE see DIQUAT DIBROMIDE

1,2-ETHYLENEDIYLBIS(CARBAMODITHIOATO)MANGANESE see VANCIDE

ETHYLENE EPISULFIDE see ETHYLENE SULFIDE

ETHYLENE FORMATE see ETHYLENE GLYCOL DIFORMATE

ETHYLENE GLYCOL ACETATE see ETHYLENE DIACETATE

ETHYLENE GLYCOL, BIS(BROMOACETATE) see 1,2-BIS(BROMOACETOXY)ETHANE

ETHYLENE GLYCOL BIS(CHLOROMETHYL)ETHER see BIS-1,2-(CHLOROMETHOXY)ETHANE

ETHYLENE GLYCOL-N-BUTYL ETHER see BUTYL CELLOSOLVE

ETHYLENE GLYCOL, CYCLIC SULFATE see ETHYLENE SULFATE

ETHYLENE GLYCOL DIACRYLATE see ETHYLENE ACRYLATE

ETHYLENE GLYCOL DIHYDROXYDIETHYL ETHER see TRIETHYLENE GLYCOL

ETHYLENE GLYCOL ETHYL ETHER ACETATE see "CELLOSOLVE" ACETATE

ETHYLENE GLYCOL ISOPROPYL ETHER see ISOPROPYL GLYCOL

ETHYLENE GLYCOL MONOBUTYL ETHER ACETATE see 2-BUTOXYETHYL ACETATE

ETHYLENE GLYCOL MONOETHYL ETHER (DOT) see CELLOSOLVE SOLVENT

ETHYLENE GLYCOL MONOETHYL ETHER ACETATE see "CELLOSOLVE" ACETATE

ETHYLENE GLYCOL MONOETHYL ETHER ACETATE (DOT) see "CELLOSOLVE" ACETATE

ETHYLENE GLYCOL MONOETHYL ETHER ACRYLATE see ACRYLIC-ACID-2-ETHOXYETHYL ESTER

ETHYLENE GLYCOL MONOETHYL ETHER PROPENOATE see ACRYLIC ACID-2-ETHOXYETHYL ESTER

ETHYLENE GLYCOL MONOISOBUTYL ETHER see ISOBUTYL CELLOSOLVE

ETHYLENE GLYCOL MONOMETHYL ETHER ACETATE(DOT) see ETHYLENE GLYCOL MONOMETHYL ETHER ACETATE

ETHYLENE GLYCOL MONOMETHYL ETHER ACRYLATE see METHYLCELLOSOLVE ACRYLATE

ETHYLENE GLYCOL MONOPHENYL ETHER see PHENYL CELLOSOLVE

ETHYLENE GLYCOL, MONOSTEARATE see ETHYLENE GLYCOL STEARATE

ETHYLENE GLYCOL PHENYL ETHER see PHENYL CELLOSOLVE

ETHYLENE HEXACHLORIDE see 1,1,1,2,2,2-HEXACHLOROETHANE

ETHYLENE HOMOPOLYMER see POLYETHYLENE

ETHYLENEIMINE see ETHYLENIMINE

ETHYLENE MONOCHLORIDE see VINYL CHLORIDE

ETHYLENE OXIDE (DOT) see ETHYLENE OXIDE

ETHYLENE (OXYDE D') (FRENCH) see ETHYLENE OXIDE

1-ETHYLENEOXY-3,4-EPOXYCYCLOHEXANE see VINYL CYCLOHEXENE DIOXIDE

ETHYLENE POLYMERS see POLYETHYLENE

ETHYLENE SULFITE see CYCLIC ETHYLENE SULFITE

1,2-ETHYLENE SULFITE see CYCLIC ETHYLENE SULFITE

ETHYLENE TETRACHLORIDE see 1,1,2,2-TETRACHLOROETHYLENE

ETHYLENE THIOUREA see 2-IMIDAZOLIDINETHIONE

1,3-ETHYLENE-2-THIOUREA see 2-IMIDAZOLIDINETHIONE

N,N'-ETHYLENETHIOUREA see 2-IMIDAZOLIDINETHIONE

ETHYLENE TRICHLORIDE see ACETYLENE TRICHLORIDE

ETHYLENE TRICHLORIDE see TRICHLOROETHYLENE

ETHYLENGLYKOLDINITRAT (CZECH) see ETHYLENE GLYCOL DINITRATE

ETHYLESTER KYSELINY ORTHOMRAVENCI (CZECH) see ETHYL ORTHOFORMATE

ETHYL ESTER OF N-ACETYL-DL-SARCOLYSYL-L-PHENYLALANINE see N-(N-ACETYL-3-(p-(BIS(2-CHLOROETHYL)AMINO)PHENYL)ALANYL-3PHENYLALANINE ETHYL ESTER

ETHYL ESTER OF N-ACETYL-DL-SARCOSYLYL-DL-VALINE see ASALIN

ETHYL ESTER OF O,O-DIMETHYLDITHIOPHOSPHORYL alpha-PHENYL ACETATE ACID see (O,O-DIMETHYLDITHIOPHOSPHORYLPHENYL)-ACETIC ACID ETHYL ESTER

ETHYL ESTER OF METHANESULFONIC ACID see ETHYL METHANESULFONATE

ETHYL ESTER OF METHYLNITROSO-CARBAMIC ACID see N-NITROSO-N-METHYLURETHANE

ETHYL ESTER OF METHYLSULFONIC ACID see ETHYL METHANESULFONATE

ETHYL ETHANOATE see ETHYL ACETATE

3-ETHYL-2-(5-(3-ETHYL-2-BENZO THIAZOLINYLIDENE)-1,3-PENTADIE-NYL)BENZOTHIAZOLIUM IODIDE see DITHIAZANINE IODIDE

5-ETHYL-5-(1-ETHYLPROPYL)2,4,6(1H,3H,5H)-PYRIMIDINETRIONE see 5-ETHYL-5-(1-ETHYLPROPYL)BARBITURIC ACID

O,O-ETHYL-S-2(ETHYL THIO)ETHYL PHOSPHORO DITHIOATE see DI-THIOSYSTOX

13-ETHYL-17-ALPHA-ETHYNYLGON-4-EN-17-BETA-OL-3-ONE see NORGESTREL

13-BETA-ETHYL-17-ALPHA-ETHYNYL-17-BETA-HYDROXYGONE-4-EN-3-ONE see 19-NORTESTOSTERONE

(±)- 13-ETHYL-17-ALPHA-ETHYNYL-17-HYDROXYGON-4-EN-3-ONE see NORGESTREL

13-ETHYL-17-ALPHA-ETHYNYL-17-BETA-HYDROXY-4-GONEN-3-ONE see NORGESTREL

ETHYL-OMEGA-FLUORODECANOATE see ETHYL-10-FLUORODECANO-ATE

ETHYL-9-FLUORONONANECARBOXYLATE see ETHYL-10-FLUORO-DECANOATE

ETHYL-OMEGA-FLUOROOCTANOATE see ETHYL-8-FLUORO OCTANO-ATE

ETHYLFORMIAAT (DUTCH) see ETHYL FORMATE

ETHYLFORMIC ACID see PROPIONIC ACID

ETHYL FORMIC ESTER see ETHYL FORMATE

ETHYLGLYKOLACETAT (GERMAN) see "CELLOSOLVE" ACETATE

5-ETHYLHEXAHYDRO-4,6-DIOXO-5-PHENYLPYRIMIDINE see 2-DES-OXY PHENOBARBITAL

5-ETHYLHEXAHYDRO-5-PHENYLPYRIMIDINE-4,6-DIONE see 2-DES-OXY PHENOBARBITAL

2-ETHYLHEXALDEHYDE see BUTYL ETHYL ACETALDEHYDE

ETHYLHEXALDEHYDE (DOT) see BUTYL ETHYL ACETALDEHYDE

2-ETHYLHEXANAL see BUTYL ETHYL ACETALDEHYDE

ETHYL HEXANEDIOL see ETHYL HEXYLENE GLYCOL

2-ETHYL-1,3-HEXANEDIOL see ETHYL HEXYLENE GLYCOL

ETHYL HEXANOATE see ETHYL CAPROATE

2-ETHYLHEXANOIC ACID see BUTYL ETHYL ACETIC ACID

2-ETHYL-1-HEXANOL HYDROGEN PHOSPHATE see BIS(2-ETHYL-HEXYL)PHOSPHATE

2-ETHYL-1-HEXANOL HYDROGEN SULFATE, SODIUM SALT see TER-GITOL 08

2-ETHYL-1-HEXANOL SULFATE SODIUM SALT see TERGITOL 08

2-ETHYL HEXENE-1 see 2-ETHYL-1-HEXENE

2-ETHYLHEXOIC ACID see BUTYL ETHYL ACETIC ACID

2-ETHYLHEXYL ACRYLATE see ACRYLIC ACID-2-ETHYLHEXYL ESTER

N-(2-ETHYLHEXYL)BICYCLO-(2,2,1)-HEPT-5-ENE-2,3-DICARBOXIMIDE see N-OCTYL BICYCLOHEPTENE DICARBOXIMIDE

2-ETHYLHEXYL FUMARATE see DIOCTYL FUMARATE

N-2-ETHYLHEXYLIMIDEENDOMETHYLENETETRAHYDROPHTHALIC ACID see N-OCTYL BICYCLOHEPTENE DICARBOXIMIDE

N-(2-ETHYLHEXYL)-5-NORBORNENE-2,3-DICARBOXIMIDE see N-OCTYL BICYCLOHEPTENE DICARBOXIMIDE

2-ETHYLHEXYL PHTHALATE see BIS(2-ETHYLHEXYL)PHTHALATE

2-ETHYLHEXYL-2-PROPENOATE see ACRYLIC ACID-2-ETHYLHEXYL ESTER

2-ETHYLHEXYL SEBACATE see BIS(2-ETHYLHEXYL)SEBACATE

2-ETHYLHEXYL SODIUM SULFATE see TERGITOL 08

2-(2-ETHYLHEXYL)-3A,4,7,7A-TETRAHYDRO-4,7-METHANO-1H-ISO-INDOLE-1,3(2H)-DIONE see N-OCTYL BICYCLOHEPTENE DICAR-BOXIMIDE

ETHYL HYDRATE see ETHANOL

ETHYL HYDRIDE see ETHANE

ETHYL HYDROSULFIDE see ETHANETHIOL

ETHYL HYDROXIDE see ETHANOL

ETHYL-2-HYDROXY-2,2-BIS(4-CHLOROPHENYL)ACETATE see ETHYL-4,4'-DICHLOROBENZILATE

N-ETHYL-N-HYDROXYETHYLNITROSAMINE see ETHYL-2-HYDROXY ETHYL NITROSAMINE

(8-ALPHA,9R)-1-ETHYL-9-HYDROXY-6'-METHOXYCINCHONAN-1-IUM IODIDE see QUININE ETHIODIDE

ALPHA-ETHYL-BETA-(HYDROXYMETHYL)-1-METHYL-IMIDAZOLE-5-BUTYRIC ACID, GAMMA-LACTONE see PILOCARPINE

1-ETHYL-3-HYDROXY-1-METHYL-PIPERIDINIUM BROMIDE BENZILATE see PIPTAL

5-ETHYL-2'-HYDROXY-2(N)-(3-METHYL-2-BUTENYL)-9-METHYL-6,7-BENZOMORPHAN see 2-(3,3-DIMETHYLALLYL)-5-ETHYL-2'-HY-DROXY-9-METHYL-6,7-BENZOMORPHAN

ETHYLIC ACID see ACETIC ACID

ETHYLIDENE CHLORIDE see ETHYLIDENE DICHLORIDE

ETHYLIDENE DIETHYL ETHER see ACETAL

ETHYLIDENE GYROMITRIN see ACETALDEHYDE-N-METHYL-N-FOR-MYLHYDRAZONE

ETHYLIDENEHYDROXYLAMINE see ACETALDEHYDE OXIME

ETHYLIDENELACTIC ACID see LACTIC ACID

ETHYLIMINE see ETHYLENIMINE

5-ETHYL-5-ISOAMYLBARBITURIC ACID see AMITAL

5-ETHYL-5-ISOAMYLMALONYL UREA see AMITAL

ETHYL ISOBUTANOATE see ETHYL ISOBUTYRATE

2-ETHYLISONICOTINIC ACID THIOAMIDE see 2-ETHYL THIO ISONI-COTINAMIDE

alpha-ETHYLISONICOTINIC ACID THIOAMIDE see 2-ETHYL THIO ISO-NICOTINAMIDE

2-ETHYLISONICOTINIC THIOAMIDE see 2-ETHYL THIO ISONICOTIN-AMIDE

alpha-ETHYLISONICOTINOYLTHIOAMIDE see 2-ETHYL THIO ISONICO-TINAMIDE

ETHYLISOPENTYLBARBITURIC ACID see AMITAL

5-ETHYL-5-ISOPENTYLBARBITURIC ACID see AMITAL

5-ETHYL-5-ISOPENTYLBARBITURIC ACID SODIUM SALT see AMYTAL SODIUM

O-ETHYL-O-(2-ISOPROPOXY-CARBONYL)-PHENYL ISOPROPYLPHOS-PHORAMIDOTHIOATE see ISOFENPHOS

5-ETHYL-5-ISOPROPYLBARBITURIC ACID see ETHYL ISOPROPYL BAR-BITURIC ACID

ETHYLISOTHIAMIDE see 2-ETHYL THIO ISONICOTINAMIDE

2-ETHYLISOTHIONICOTINAMIDE see 2-ETHYL THIO ISONICOTINAMIDE

ALPHA-ETHYLISOTHIONICOTINAMIDE see 2-ETHYL THIO ISONICOTIN-AMIDE

N-ETHYLMALEIMIDE see MALEIC ACID-N-ETHYLIMIDE

ETHYLMERCAPTAAN (DUTCH) see ETHANETHIOL

ETHYL MERCAPTAN see ETHANETHIOL

ETHYL MERCAPTAN (DOT) see ETHANETHIOL

beta-ETHYLMERCAPTOETHYL DIMETHYL THIONOPHOSPHATE see DEMETON-O-METHYL

ETHYL MERCAPTOPHENYLACETATE O,O-DIMETHYL PHOSPHOROCI-THIOATE see (O,O-DIMETHYLDITHIOPHOSPHORYLPHENYL)-ACETIC ACID ETHYL ESTER

ETHYLMERCURIC CHLORIDE see CHLORO ETHYL MERCURY

ETHYLMERCURIC PHOSPHATE see BIS(ETHYL MERCURI)PHOSPHATE

N-(ETHYLMERCURI)-1,4,5,6,7,7-HEXACHLOROBICYCLO(2.2.1)HEPT-5-ENE- 2,3-DICARBOXIMIDE see ETHYL MERCURI CHLORENDIMIDE

N-ETHYLMERCURI-3,4,5,6,7,7-HEXACHLORO-3,6-ENDOMETHYLENE-1,2,3,6- TETRAHYDROPHTHALIMIDE see ETHYL MERCURI CHLO-RENDIMIDE

N-ETHYLMERCURI-N-PHENYL-P-TOLUENESULFONAMIDE see ETHYL-MERCURY-p-TOLUENE SULFONAMIDE

N-ETHYLMERCURI-1,2,3,6-TETRAHYDRO-3,6-ENDOMETHANO- 3,4,5, 6,7,7-HEXACHLOROPHTHALIMIDE see ETHYL MERCURI CHLOREN-DIMIDE

O-(ETHYLMERCURITHIO)BENZOIC ACID SODIUM SALT see MERTHIO-LATE SODIUM

ETHYLMERCURITHIOSALICYLIC ACID SODIUM SALT see MERTHIOLATE SODIUM

N-(ETHYLMERCURI)-P-TOLUENESULFONANILIDE see ETHYLMERCURY-p-TOLUENE SULFONAMIDE

N-(ETHYLMERCURI)-P-TOLUENESULPHONANILIDE see ETHYLMER-CURY-p-TOLUENE SULFONAMIDE

ETHYLMERCURY CHLORIDE see CHLORO ETHYL MERCURY

ETHYLMERCURY PHOSPHATE see BIS(ETHYL MERCURI)PHOSPHATE

ETHYLMERCURY P-TOLUENESULFANILIDE see ETHYLMERCURY-p-TOLUENE SULFONAMIDE

ETHYLMERCURY-P-TOLUENESULFONANILIDE see ETHYLMERCURY-p-
TOLUENE SULFONAMIDE

ETHYLMERKAPTAN (CZECH) see ETHANETHIOL

BETA-ETHYLMERKAPTOETHYLCHLORID (CZECH) see CHLOROETHYL
ETHYL SULFIDE

ETHYL METHANOATE see ETHYL FORMATE

ETHYL METHANSULFONATE see ETHYL METHANESULFONATE

ETHYL-4-METHOXYBENZOATE see p-ANISIC ACID, ETHYL ESTER

ETHYL-P-METHOXYBENZOATE see p-ANISIC ACID, ETHYL ESTER

ETHYL-2-METHYLACRYLATE see ETHYL METHACRYLATE

ETHYL-ALPHA-METHYL ACRYLATE see ETHYL METHACRYLATE

4-ETHYLMETHYLAMINOAZOBENZENE see N-ETHYL-N-METHYL-p-
(PHENYLAZO)ANILINE

P-ETHYLMETHYLAMINOAZOBENZENE see N-ETHYL-N-METHYL-p-
(PHENYLAZO)ANILINE

4'-ETHYL-N-METHYL-4-AMINOAZOBENZENE see p-(4-ETHYLPHENYL-
AZO)-N-METHYLANILINE

3-ETHYLMETHYLAMINO-1,1-DI(2'-THIENYL)BUT-1-ENE HYDROCHLO-
RIDE see ETHYLMETHIAMBUTENE HYDROCHLORIDE

ETHYL-P-METHYL BENZENESULFONATE see ETHYL TOSYLATE

5-ETHYL-5-(1-METHYL-1-BUTENYL)BARBITURIC ACID see 5-ETHYL-
5-(1-METHYL-1-BUTENYL)BARBITURATE

5-ETHYL-5-(1-METHYL-1-BUTENYL)BARBITURIC ACID SODIUM SALT
see VINBARBITAL SODIUM

5-ETHYL-5-(1-METHYLBUTYL)BARBITURIC ACID see NEMBUTAL

5-ETHYL-5-(3-METHYLBUTYL)BARBITURIC ACID see AMITAL

5-ETHYL-5-(3-METHYLBUTYL)BARBITURIC ACID SODIUM DERIVATIVE
see AMYTAL SODIUM

5-ETHYL-5-(1-METHYLBUTYL)BARBITURIC ACID SODIUM SALT see
NEMBUTAL SODIUM

5-ETHYL-5-(1-METHYLBUTYL)MALONYLUREA see NEMBUTAL

ETHYLMETHYL CARBINOL see sec-BUTYL ALCOHOL

ETHYL METHYL CETONE (FRENCH) see 2-BUTANONE

1-ETHYL-7-METHYL-1,4-DIHYDRO-1,8-NAPHTHYRIDIN-4-ONE-3-CAR-
BOXYLICACID see 1-ETHYL-1,4-DIHYDRO-7-METHYL-4-OXO-1,8-
NAPHTHYRIDINE-3-CARBOXYLIC ACID

4-ETHYL-4-METHYL-2,6-DIOXOPIPERIDINE see 3-METHYL-3-ETHYL-
GLUTARIMIDE

3-ETHYL-3-METHYLGLUTARIMIDE see 3-METHYL-3-ETHYLGLUTARI-
MIDE

BETA-ETHYL-BETA-METHYLGLUTARIMIDE see 3-METHYL-3-ETHYL-
GLUTARIMIDE

ETHYLMETHYLKETON (DUTCH) see 2-BUTANONE

ETHYL METHYL KETONE see 2-BUTANONE

1-ETHYL-2-METHYL-7-METHOXY-1,2,3,4-TETRAHYDROPHENAN-
THRYL-2-CARBOXYLIC ACID see BIS DEHYDRO ISYNOLIC
ACID METHYL ESTER

ETHYL 3-METHYL-4-(METHYLTHIO)PHENYL(1-METHYLETHYL)PHOS-
PHORAMIDATE see FENAMIPHOS

ETHYLMETHYLNITROSAMINE see N,N-METHYLETHYLNITROSAMINE

1-ETHYL-7-METHYL-4-OXO-1,4-DIHYDRO-1,8-NAPHTHYRIDINE-3-
CARBOXYLIC ACID see 1-ETHYL-1,4-DIHYDRO-7-METHYL-4-
OXO-1,8-NAPHTHYRIDINE-3-CARBOXYLIC ACID

N-ETHYLMETHYLPHENYLBARBITURIC ACID see 5-ETHYL-N-METHYL-
5-PHENYL BARBITURIC ACID

5-ETHYL-3-METHYL-5-PHENYLHYDANTOIN see 3-METHYL-5-ETHYL-
5-PHENYLHYDANTOIN

5-ETHYL-3-METHYL-5-PHENYL-2,4(3H,5H)-IMIDAZOLEDIONE see
3-METHYL-5-ETHYL-5-PHENYLHYDANTOIN

5-ETHYL-3-METHYL-5-PHENYLIMIDAZOLIDIN-2,4-DIONE see
3-METHYL-5-ETHYL-5-PHENYLHYDANTOIN

ETHYL-1-METHYL-4-PHENYLISONIPECOTATE see DEMAROL

ETHYL-1-METHYL-4-PHENYLISONIPECOTATE HYDROCHLORIDE see
DOLANTIN HYDROCHLORIDE

ETHYL-1-METHYL-4-PHENYLPIPERIDINE-4-CARBOXYLATE see
DEMAROL

ETHYL-1-METHYL-4-PHENYLPIPERIDINE-4-CARBOXYLATE HYDRO-
CHLORIDE see DOLANTIN HYDROCHLORIDE

ETHYL-1-METHYL-4-PHENYLPIPERIDYL-4-CARBOXYLATE HYDRO-
CHLORIDE see DOLANTIN HYDROCHLORIDE

4-ETHYL-4-METHYL-2,6-PIPERIDINEDIONE see 3-METHYL-3-ETHYL-
GLUTARIMIDE

ETHYL-2-METHYLPROPANOATE see ETHYL ISOBUTYRATE

ETHYL-2-METHYL-2-PROPENOATE see ETHYL METHACRYLATE

ETHYL-2-METHYLPROPIONATE see ETHYL ISOBUTYRATE

5-ETHYL-5-(1-METHYLPROPYL)BARBITURATE see BUTISOL

5-ETHYL-5-(1-METHYLPROPYL)BARBITURIC ACID see BUTISOL

5-ETHYL-5-(1-METHYLPROPYL)BARBITURIC ACID SODIUM SALT see
BUTISOL SODIUM

1-ETHYL-1-METHYL-2-PROPYNYL CARBAMATE see METHYLPENTY-
NOL CARBAMATE

3-ETHYL-6-METHYLPYRIDINE see 5-ETHYL-alpha-PICOLINE

5-ETHYL-2-METHYLPYRIDINE see 5-ETHYL-alpha-PICOLINE

3-ETHYL-3-METHYLPYRROLIDINE-2,5-DIONE see 2-ETHYL-2-METHYL-
SUCCINIMIDE

alpha-ETHYL-alpha-METHYLSUCCINIMIDE see 2-ETHYL-2-METHYL-
SUCCINIMIDE

ETHYLMETHYLTHIAMBUTENE HYDROCHLORIDE see ETHYLMETHIAM-
BUTENE HYDROCHLORIDE

ETHYL-4-(METHYLTHIO)-m-TOLYL ISOPROPYL PHOSPHOR AMIDATE
see FENAMIPHOS

N-ETHYL-ALPHA-METHYL-M-TRIFLUOROMETHYLPHENETHYLAMINE
see PHENFLUORAMINE HYDROCHLORIDE

N-ETHYL-ALPHA-METHYL-M-(TRIFLUOROMETHYL)PHENETHYLAMINE
HYDROCHLORIDE see PHENFLUORAMINE HYDROCHLORIDE

ETHYL MONOBROMOACETATE see ETHYL BROMACETATE

ETHYL MONOCHLORACETATE see ETHYL CHLORACETATE

ETHYL MONOCHLOROACETATE see ETHYL CHLORACETATE

ETHYL MONOSULFIDE see ETHYL SULFIDE

ETHYLMORPHINE HYDROCHLORIDE see ETHYL MORPHINE HYDRO-
CHLORIDE DIHYDRATE

N-ETHYL MORPHOLINE see 4-ETHYLMORPHOLINE

ETHYL NITRILE see ACETONITRILE

ETHYL NITRITE (DOT) see ETHYL NITRITE

O-ETHYL-O-((4-NITRO-FENYL)-FENYL)-MONOTHIOFOSFONAAT(DUTCH)
see ETHOXY-4-NITRO PHENOXY PHENYL PHOSPHINE SULFIDE

N-ETHYL-N'-NITRO-N-NITROSOGUANIDINE see N-ETHYL-N-NITROSO-
N'-NITROGUANIDINE

ETHYL P-NITROPHENYL BENZENETHIONOPHOSPHONATE see ETHOXY-
4-NITRO PHENOXY PHENYL PHOSPHINE SULFIDE

O-ETHYL O-(4-NITROPHENYL)BENZENETHIONOPHOSPHONATE see
ETHOXY-4-NITRO PHENOXY PHENYL PHOSPHINE SULFIDE

ETHYL P-NITROPHENYL BENZENETHIOPHOSPHATE see ETHOXY-4-
NITRO PHENOXY PHENYL PHOSPHINE SULFIDE

ETHYL P-NITROPHENYL BENZENETHIOPHOSPHONATE see ETHOXY-
4-NITRO PHENOXY PHENYL PHOSPHINE SULFIDE

ETHYL P-NITROPHENYL PHENYLPHOSPHONOTHIOATE see ETHOXY-
4-NITRO PHENOXY PHENYL PHOSPHINE SULFIDE

O-ETHYL O-(4-NITROPHENYL) PHENYLPHOSPHONOTHIOATE see ETH-
OXY-4-NITRO PHENOXY PHENYL PHOSPHINE SULFIDE

O-ETHYL O-P-NITROPHENYL PHENYLPHOSPHONOTHIOLATE see ETH-
OXY-4-NITRO PHENOXY PHENYL PHOSPHINE SULFIDE

O-ETHYL O-P-NITROPHENYL PHENYLPHOSPHOROTHIOATE see ETH-
OXY-4-NITRO PHENOXY PHENYL PHOSPHINE SULFIDE

ETHYL P-NITROPHENYL THIONOBENZENEPHOSPHATE see ETHOXY-
4-NITRO PHENOXY PHENYL PHOSPHINE SULFIDE

ETHYL P-NITROPHENYL THIONOBENZENEPHOSPHONATE see ETHOXY-
4-NITRO PHENOXY PHENYL PHOSPHINE SULFIDE

2-(ETHYLNITROSAMINO)ETHANOL see ETHYL-2-HYDROXY ETHYL NI-
TROSAMINE

4-((ETHYLNITROSAMINO)METHYL)PYRIDINE see N-NITROSO-4-PICO-
LYLETHYLAMINE

ETHYL NITROSO BIURET see N-ETHYL-N-NITROSOBIURET

N-ETHYL-N-NITROSOBUTYLAMINE see ETHYL-n-BUTYL NITROS
AMINE

N-ETHYL-N-NITROSOCARBAMIC ACID ETHYL ESTER see N-NITROSO-
N-ETHYLURETHAN

N-ETHYL-N-NITROSO-ETHANAMINE see N-NITROSO-DIETHYLAMINE

ETHYLNITROSOUREA see 1-ETHYL-1-NITROSOUREA

N-ETHYL-N-NITROSO-UREA see 1-ETHYL-1-NITROSOUREA

17-ALPHA-ETHYNYL-17-BETA-ACETOXY-19-NORANDROST-4-EN-3-
ONE see 17-ACETOXY-19-NOR-17-alpha-PREGN-4-EN-20-YN-3-ONE

ALPHA-ETHYNYLBENZYL CARBAMATE see 1-PHENYL-2-PROPYNYL
CARBAMATE

2-ETHYNYL-2-BUTYL CARBAMATE see METHYLPENTYNOL CARBA-
MATE

17-ALPHA-ETHYNYL-3,17-DIHYDROXY-4-ESTRENE DIACETATE see
ETHYNODIOL ACETATE

17-ETHYNYL-3,17-DIHYDROXY-1,3,5-OESTRATRIENE see ETHINYL
ESTRADIOL

ETHYNYLESTRADIOL see ETHINYL ESTRADIOL

17-ALPHA-ETHYNYLESTRADIOL see ETHINYL ESTRADIOL

17-ALPHA-ETHYNYLESTRADIOL-17-BETA see ETHINYL ESTRADIOL

17-ETHYNYLESTRADIOL-3METHYL ETHER see 3-METHYLETHYNYLES-
TRADIOL

ETHYNYLESTRADIOL-3-METHYL ETHER see 3-METHYLETHYNYLES-
TRADIOL

17-ALPHA-ETHYNYLESTRADIOL-3-METHYL ETHER see 3-METHYL-
ETHYNYLESTRADIOL

17-ALPHA-ETHYNYLESTRA-1,3,5(10)-TRIENE-3,17-BETA-DIOL see
ETHINYL ESTRADIOL

17-ALPHA-ETHYNYL-1,3,5(10)-ESTRATRIENE-3,17-BETA-DIOL see
ETHINYL ESTRADIOL

17-ALPHA-ETHYNYLESTR-4-ENE-3-BETA,17-BETA-DIOL ACETATE see
ETHYNODIOL ACETATE

17-ALPHA-ETHYNYL-4-ESTRENE-3-BETA,17-DIOL DIACETATE see
ETHYNODIOL ACETATE

17-ALPHA-ETHYNYL-4-ESTRENE-3-BETA,17-BETA-DIOL DIACETATE
see ETHYNODIOL ACETATE

17-ALPHA-ETHYNYL-4-ESTREN-17-OL-3-ONE see 19-NORETHISTER-
ONE

17-alpha-ETHYNYL-5(10)-ESTREN-17-OL-3-ONE see 17-alpha-ETHI-
NYL-5,10-ESTRENOLONE

17-ALPHA-ETHYNYLESTR-5(10)-EN-17-BETA-OL-3-ONE see 17-alpha-
ETHINYL-5,10-ESTRENOLONE

17-ALPHA-ETHYNYL-ESTR-5(10)-EN-3-ON-17-BETA-OL see 17-alpha-
ETHINYL-5,10-ESTRENOLONE

17-ALPHA-ETHYNYL-17-BETA-HEPTANOYLOXY-4-ESTREN-3-ONE see
NORETHISTERONE ENANTHATE

17-ALPHA-ETHYNYL-17-BETA-HYDROXYESTER-5(10)-EN-3-ONE see
17-alpha-ETHINYL-5,10-ESTRENOLONE

17-ALPHA-ETHYNYL-17-HYDROXY-4-ESTREN-3-ONE see 19-NOR-
ETHISTERONE

17-ALPHA-ETHYNYL-17-HYDROXY-5(10)-ESTREN-3-ONE see 17-
alpha-ETHYNYL-5,10-ESTRENOLONE

17-ALPHA-ETHYNYL-17-BETA-HYDROXY-DELTA(SUP 5(10))-ESTREN-
3-ONE see 17-alpha-ETHINYL-5,10-ESTRENOLONE

17-ALPHA-ETHYNYL-17-BETA-HYDROXY-DELTA(SUP. 5(10))-ESTREN-
3-ONE see 17-alpha-ETHINYL-5,10-ESTRENOLONE

17-ALPHA-ETHYNYL-17-HYDROXYESTR-4-EN-3-ONE ACETATE see 17-
ACETOXY-19-NOR-17-alpha-PREGN-4-EN-20-YN-3-ONE

(+)-17- alpha-ETHYNYL-17-BETA-HYDROXY-3-METHOXY-1,3,5(10)-
ESTRATRIENE see 3-METHYLETHYNYLESTRADIOL

(+)-17- alpha-ETHYNYL-17-BETA-HYDROXY-3-METHOXY-1,3,5(10)-
OESTRATRIENE see 3-METHYLETHYNYLESTRADIOL

17-ALPHA-ETHYNYL-17-BETA-HYDROXY-19-NORANDROST-4-EN-3-
ONE see 19-NORETHISTERONE

17-ALPHA-ETHYNYL-17-BETA-HYDROXY-19-NORANDROST-4-EN-3-
ONE see 19-NORETHISTERONE

17-ALPHA-ETHYNYL-17-BETA-HYDROXY-3-OXO-DELTA(SUP5(10))-ES-
TRENE see 17-alpha-ETHINYL-5,10-ESTRENOLONE

17-ETHYNYL-3-METHOXY1,3,5(10)-ESTRATRIEN-17-BETA-OL see
3-METHYLETHYNYLESTRADIOL

17-ALPHA-ETHYNYL-3-METHOXY-1,3,5(10)-ESTRATRIEN-17-BETA-OL
see 3-METHYLETHYNYLESTRADIOL

17-ALPHA-ETHYNYL-3-METHOXY-17-BETA-HYDROXY-DELTA-
1,3,5(10)-ESTRATRIENE see 3-METHYLETHYNYLESTRADIOL

17-ALPHA-ETHYNYL-3-METHOXY-17-BETA-HYDROXY-DELTA-
1,3,5(10)-OESTRATRIENE see 3-METHYLETHYNYLESTRADIOL

17-ETHYNYL-3-METHOXY-1,3,5(10)-OESTRATIEN-17-BETA-OL see
3-METHYLETHYNYLESTRADIOL

17-ETHYNYL-18-METHYL-19-NORTESTOSTERONE see NORGESTREL

17-ALPHA-ETHYNYL-19-NORANDROST-4-ENE-3-BETA,17-BETA-DIOL
DIACETATE see ETHYNODIOL ACETATE

17-ALPHA-ETHYNYL-19-NOR-4-ANDROSTEN-17-BETA-OL-3-ONE see
19-NORETHISTERONE

17-ALPHA-ETHYNYL-19-NORANDROST-4-EN-17-BETA-OL-3-ONE see
19-NORETHISTERONE

17-ALPHA-ETHYNYL-19-NOR-5(10)-ANDROSTEN-17-BETA-OL-3-ONE
see 17-alpha-ETHINYL-5,10-ESTRENOLONE

17-ALPHA-ETHYNYL-19-NORTESTOSTERONE see 19-NORETHISTERONE

17-ALPHA-ETHYNYL-19-NORTESTOSTERONE ACETATE see 17-ACET-
OXY-19-NOR-17-alpha-PREGN-4-EN-20-YN-3-ONE

ETHYNYLOESTRADIOL see ETHINYL ESTRADIOL

17-ETHYNYLOESTRADIOL see ETHINYL ESTRADIOL

17-ALPHA-ETHYNYLOESTRADIOL see ETHINYL ESTRADIOL

17-ALPHA-ETHYNYL-17-BETA-OESTRADIOL see ETHINYL ESTRADIOL

17-ALPHA-ETHYNYLOESTRADIOL-17-BETA see ETHINYL ESTRADIOL

17-ETHYNYLOESTRADIOL 3METHYL ETHER see 3-METHYLETHYNYL-
ESTRADIOL

17-ALPHA-ETHYNYLOESTRADIOL METHYL ETHER see 3-METHYLETH-
YNYLESTRADIOL

17-ALPHA-ETHYNYLOESTRADIOL 3-METHYL ETHER see 3-METHYL-
ETHYNYLESTRADIOL

17-ETHYNYLOESTRA-1,3,5(10)-TRIENE-3,17-BETA-DIOL see ETHINYL
ESTRADIOL

17-ALPHA-ETHYNYLOESTRA-1,3,5(10)-TRIENE-3,17-BETA-DIOL see
ETHINYL ESTRADIOL

17-ALPHA-ETHYNYL-1,3,5(10)-OESTRATRIENE-3,17-BETA-DIOL see
ETHINYL ESTRADIOL

17-ALPHA-ETHYNYL-1,3,5-OESTRATRIENE-3,17-BETA-DIOL see ETHI-
NYL ESTRADIOL

ETHYONOMIDE see 2-ETHYL THIO ISONICOTINAMIDE

ETIL ACRILATO (ITALIAN) see ETHYL ACRYLATE

ETILACRILATULUI (ROUMANIAN) see ETHYL ACRYLATE

ETILAMINA (ITALIAN) see ETHANAMINE

ETILBENZENE (ITALIAN) see ETHYL BENZENE

ETILBUTILCHETONE (ITALIAN) see 3-HEPTANONE

ETIL CLOROCARBONATO (ITALIAN) see ETHYL CHLORO FORMATE

ETIL CLOROFORMIATO (ITALIAN) see ETHYL CHLORO FORMATE

ETILE (ACETATO DI) (ITALIAN) see ETHYL ACETATE

ETILE (FORMIATO DI) (ITALIAN) see ETHYL FORMATE

N,N'-ETILEN-BIS(DITIOCARBAMMATO) DI MANGANESE(ITALIAN) see
VANCIDE

N,N'-ETILEN-BIS(DITIOCARBAMMATO) DI SODIO (ITALIAN) see NA-
BAM

ETILENE (OSSIDO DI) (ITALIAN) see ETHYLENE OXIDE

ETILENIMINA (ITALIAN) see ETHYLENIMINE

ETILMERCAPTANO (ITALIAN) see ETHANETHIOL

O-ETIL-O-((4-NTIRO-FENIL)-FENIL)-MONOTIOFOSFONATO (ITALIAN)
see ETHOXY-4-NITRO PHENOXY PHENYL PHOSPHINE SULFIDE

S-2-ETIL-SULFINIL-1-METIL-ETIL-O,O-DIMETIL-MONOTIOFOSFATO
(ITALIAN) see METASYSTOX-S

ETIZOLAM see 6-(O-CHLOROPHENYL)-8-ETHYL-1-METHYL-4H-S-
TRIAZOLO(3,4-c)THIENO(2,3-e)(1,4)-DIAZEPINE

ETOKSYETYLOWY ALKOHOL (POLISH) see CELLOSOLVE SOLVENT

2-ETOSSIETIL-ACETATO (ITALIAN) see "CELLOSOLVE" ACETATE

ETRYPTAMINE ACETATE see 3-(2-AMINOBUTYL)INDOLE ACETATE

ETYLENU TLENEK (POLISH) see ETHYLENE OXIDE

ETYLOAMINA (POLISH) see ETHANAMINE

ETYLOBENZEN (POLISH) see ETHYL BENZENE

ETYLOWY ALKOHOL (POLISH) see ETHANOL

ETYLU BROMEK (POLISH) see BROMOETHANE

ETYLU CHLOREK (POLISH) see ETHYL CHLORIDE

EUCALYPTOL see CAJEPUTOL

EUCALYPTUS REDUNCA TANNIN see MYRTAN TANNIN

EUGENIC ACID see EUGENOL

1,3,4-EUGENOL METHYL ETHER see 4-ALLYL-1,2-DIMETHOXYBEN-
ZENE

EUGENYL METHYL ETHER see 4-ALLYL-1,2-DIMETHOXYBENZENE
EUKALYPTUS OEL (GERMAN) see EUCALYPTUS OIL
EVIPAN see EVIPAL
EXHAUST GAS see CARBON MONOXIDE
EXOTHERMIC FERROCHROME (DOT) see FERROCHROME (EXOTHER-MIC)
EXP 338 see 4-AMINO-2,2,5,5-TETRAKIS(TRIFLUOROMETHYL)-3-IMIDAZOLINE
EXPANSIN see CLAVACIN
EXPERIMENTAL INSECTICIDE 269 see ENDRIN
EXPERIMENTAL INSECTICIDE 4049 see CARBETHOXY MALATHION
EXPERIMENTAL INSECTICIDE 4124 see p-NITRO-o-CHLOROPHENYL DIMETHYL THIONOPHOSPHATE
EXPERIMENTAL INSECTICIDE 7744 see CARBARYL
EXPLOSIVE D see AMMONIUM PICRATE (WET)
EXPLOSIVE HIGH see LEAD NITRORESORCINATE
EXSICATED SODIUM SULFITE see SODIUM SULFITE (2:1)
EXSICCATED FERROUS SULFATE see IRON (II) SULFATE (1:1)
EXSICCATED SODIUM PHOSPHATE see SODIUM MONOHYDROGEN PHOSPHATE (2:1:1)
EXT. D AND C RED NO. 15 see FD AND C RED NO. 1
EXTRACT D AND C RED NO. 10 see C.I. FOOD RED 3
EXTRACT D AND C RED NO. 14 see 1-(2,4-XYLYLAZO)-2-NAPHTHOL
EXTRINSIC FACTOR see VITAMIN B$_{12}$ COMPLEX
F-33 see 8-(4-p-FLUORO PHENYL-4-OXOBUTYL)-2-METHYL-2,8-DIAZASPIRO(4.5)DECANE-1,3-DIONE
F 190 see ACETPHENARSINE
190 F see ACETPHENARSINE
FACTITIOUS AIR see NITROGEN OXIDE
FANFT see N-(4-(5-NITRO-2-FURYL)-2-THIAZOLYL)FORMAMIDE
FANNOFORM see FORMALDEHYDE
FAST ACID GREEN N see FD AND C GREEN NO. 2
FASTBALLS see d-BENZEDRINE SULFATE
FAST GREEN FCF see FD AND C GREEN NO. 3
FAST OIL ORANGE see 1-(PHENYLAZO)-2-NAPHTHOL
FAST OIL ORANGE II see 1-(2,4-XYLYLAZO)-2-NAPHTHOL
FAST OIL RED B see SCARLET RED
FAST ORANGE see 1-(PHENYLAZO)-2-NAPHTHOL
FAST RED see FOOD RED 2
FAST WOOL BLUE R see ACID BLUE 92
FAT RED 7B see N-ETHYL-1-((p-(PHENYLAZO)PHENYL)AZO)-2-NAPH-THYLAMINE
FAT RED B see SCARLET RED
FD & C NO. 6 see FOOD YELLOW 3
FD & C YELLOW 6 see FOOD YELLOW 3
FD & C YELLOW NO. 6 Aluminum LAKE see FOOD YELLOW 3
FD AND C BLUE NO. 2 see DISODIUM INDIGO-5,5-DISULFONATE
FD AND C BLUE NO. 1-ALUMINUM LAKE see FD AND C BLUE NO. 1
FD AND C GREEN NO. 1 see ACID GREEN 3
FD AND C RED NO. 2 see FOOD RED 2
FD AND C RED NO. 19 see (9-(o-CARBOXYPHENYL)-6-(DIETHYLAM-INO)-3H-XANTHEN-3-YLIDENE) DIETHYLAMMONIUM CHLORIDE
FD AND C RED NO. 2-Aluminum LAKE see FOOD RED 2
FD AND C YELLOW NO. 3 see 1-(PHENYLAZO)-2-NAPHTHYLAMINE
FD AND C YELLOW NO. 4 see 1-(o-TOLYLAZO)-2-NAPHTHYLAMINE
F D + C BLUE 1 see FD AND C BLUE NO. 1
FDC GREEN 1 see ACID GREEN 3
FE 158/C see IRIMONT
FEKABIT see POTASSIUM CHLORATE
FENADONE see dl-METHADONE HYDROCHLORIDE
FENAZO BLUE SR see ACID BLUE 92
FENAZOXINE HYDROCHLORIDE see NEFOPAM HYDROCHLORIDE
FENCHLOORFOS (DUTCH) see TRICHLOROMETAFOS
FENCHLORFOS see TRICHLOROMETAFOS
FENCHLOROPHOS see TRICHLOROMETAFOS
FENESTERIN see CHOLESTERYL-p-BIS(2-CHLOROETHYL)AMINO PHENYLACETATE
FENFLURAMINE see N-ETHYL-alpha-METHYL-m-(TRIFLUOROMETHYL)-PHENETHYLAMINE
FENFLURAMINE HYDROCHLORIDE see PHENFLUORAMINE HYDRO-CHLORIDE

FENILIDRAZINA (ITALIAN) see PHENYLHYDRAZINE
2-FENILPROPANO (ITALIAN) see CUMENE
FENIZON (FRENCH) see 4-CHLOROPHENYL BENZENESULFONATE
FENOL (DUTCH, POLISH) see PHENOL
FENOLO (ITALIAN) see PHENOL
2-FENOXYETHANOL (CZECH) see PHENYL CELLOSOLVE
FENTANYL see PHENTANYL
FENTANYL CITRATE see PHENTANYL CITRATE
FENTIN ACETAAT (DUTCH) see ACETOXYTRIPHENYLSTANNANE
FENTIN ACETAT (GERMAN) see ACETOXYTRIPHENYLSTANNANE
FENTIN ACETATE see ACETOXYTRIPHENYLSTANNANE
FENTIN CHLORIDE see CHLOROTRIPHENYLSTANNANE
FENTINE ACETATE (FRENCH) see ACETOXYTRIPHENYLSTANNANE
FENTIN HYDROXIDE see HYDROXY TRIPHENYL STANNANE
1-FENYL-4-AMINO-5-CHLOR-6-PYRIDAZINON (CZECH) see 1-PHENYL-4-AMINO-5-CHLORPYRIDAZ-6-ONE
FENYL-CELLOSOLVE (CZECH) see PHENYL CELLOSOLVE
1-FENYL-3,3-DIETHYLTRIAZEN (CZECH) see 1-PHENYL-3,3-DIETHYL-TRIAZENE
1-FENYL-3,3-DIMETHYLTRIAZIN see 3,3-DIMETHYL-1-PHENYLTRI-AZENE
M-FENYLENDIAMIN (CZECH) see m-PHENYLENEDIAMINE
FENYLENODWUAMINA (POLISH) see p-PHENYLENEDIAMINE
FENYL-GLYCIDYLETHER (CZECH) see PHENYLGLYCYDYL ETHER
FENYLHYDRAZINE (DUTCH) see PHENYLHYDRAZINE
FENYLMERCURIACETAT (CZECH) see ACETOXYPHENYLMERCURY
FENYLMERCURICHLORID (CZECH) see PHENYL MERCURIC CHLORIDE
2-FENYL-PROPAAN (DUTCH) see CUMENE
FENZEN (CZECH) see BENZENE
FEOJECTIN see IRON OXIDE, SACCHARATED
FERGON PREPARATIONS see FERROUS GLUCONATE
FERLUCON see FERROUS GLUCONATE
FERMATE FERBAM FUNGICIDE see FERBAM
FERMENTATION ALCOHOL see ETHANOL
FERMENTATION AMYL ALCOHOL see ISOAMYL ALCOHOL
FERMENTATION BUTYL ALCOHOL see ISOBUTYL ALCOHOL
FERMOCIDE see FERBAM
FER PENTACARBONYLE (FRENCH) see IRON CARBONYL
FERRADOW see FERBAM
FERRIAMICIDE see MIREX
FERRIC ARSENATE see IRON (III) ARSENATE (1:1)
FERRIC ARSENITE see IRON (III)-o-ARSENITE PENTAHYDRATE
FERRIC ARSENITE, BASIC see IRON (III)-o-ARSENITE PENTAHYDRATE
FERRIC ARSENITE, SOLID (DOT) see IRON (III)-o-ARSENITE PENTAHY-DRATE
FERRIC DIMETHYLDITHIOCARBAMATE see FERBAM
FERRIC OXIDE see IRON (III) OXIDE
FERRIC OXIDE, SACCHARATED see IRON OXIDE, SACCHARATED
FERRIC SACCHARATE—IRON OXIDE (MIX.) see IRON OXIDE, SACCHARATED
FERRIGEN see IRON-DEXTRIN COMPLEX
FERRIVENIN see IRON OXIDE, SACCHARATED
FERROANTHOPHYLLITE see ANTHOPHYLITE (SEE ALSO ASBESTOS)
FERROCHROMIUM see FERROCHROME (EXOTHERMIC)
FERRONICUM see FERROUS GLUCONATE
FERROSULFAT (GERMAN) see IRON (II) SULFATE (1:1)
FERROUS ARSENATE see IRON (II) ARSENATE (3:2)
FERROUS LACTATE see LACTIC ACID, IRON(2+) SALT (2:1)
FERROUS SULFATE see IRON (II) SULFATE (1:1)
FETTSCHARLACH see OIL RED
FHIOFURAN see THIOPHENE
FIBERGLASS see SILICA, CRYSTALLINE-QUARTZ
FIBROUS GLASS see SILICA, CRYSTALLINE-QUARTZ
FINTIN ACETATO (ITALIAN) see ACETOXYTRIPHENYLSTANNANE
FINTINE HYDROXYDE (FRENCH) see HYDROXY TRIPHENYL STANNANE
FINTIN HYDROXID (GERMAN) see HYDROXY TRIPHENYL STANNANE
FINTIN HYDROXYDE (DUTCH) see HYDROXY TRIPHENYL STANNANE
FINTIN IDROSSIDO (ITALIAN) see HYDROXY TRIPHENYL STANNANE
FIRE DAMP see METHANE
FIREMASTER T23P-LV see TRIS
FISH BERRY see PICROTOXIN

FLAVOMYCELIN see LUTEOSKYRIN

FLAVUROL see MERCUROCHROME

FLINT see SILICA (CRYSTALLINE)

FLINT see SILICA, CRYSTALLINE-QUARTZ

FLORALTONE see GIBBERELLIC ACID

FLORES MARTIS see FERRIC CHLORIDE

FLOROXENE see 2,2,2-TRIFLUOROETHYL VINYL ETHER

FLOWERS OF ANTIMONY see ANTIMONY OXIDE

FLOWERS OF SULPHUR see SULFUR

FLOWERS OF ZINC see ZINC OXIDE

FLUE DUST, ARSENIC-CONTG. see ARSENICAL DUST

FLUE GAS see CARBON MONOXIDE

FLUOBORIC ACID see TETRAFLUOROBORATE

FLUON see TEFLON

FLUOPHOSGENE see CARBONYL FLUORIDE

FLUOPHOSPHORIC ACID DI(DIMETHYLAMIDE) see BIS(DIMETHYL AMIDO)FLUORO PHOSPHATE

FLUOPHOSPHORIC ACID, DIISOPROPYL ESTER see ISOPROPYL PHOSPHOROFLUORIDATE

FLUOPHOSPHORIC ACID, DIMETHYL ESTER see DIMETHYLFLUORO-PHOSPHATE

FLUOR (DUTCH, FRENCH, GERMAN, POLISH) see FLUORINE

FLUORACETATO DI (ITALIAN) see SODIUM FLUOROACETATE

5-FLUORACIL (GERMAN) see FLUOROURACIL

3,6-FLUORANDIOL see FLUORESCEIN

FLUORANE 114 see DICHLOROTETRAFLUOROETHANE

4-FLUORANILIN (CZECH) see 4-FLUOROANILINE

2-FLUORENAMINE see FLUOREN-2-AMINE

2-FLUORENEAMINE see FLUOREN-2-AMINE

N-4-FLUORENYLACETAMIDE see 4-ACETYLAMINOFLUORENE

N-FLUOREN-4-YLACETAMIDE see 4-ACETYLAMINOFLUORENE

FLUORENYL-2-ACETHYDROXAMIC ACID see N-HYDROXY-N-ACETYL-2-AMINOFLUORENE

N-(FLUOREN-2-YL)ACETOHYDROXAMIC ACETAMIDE see N-ACETOXY-N-ACETYL-2-AMINOFLUORENE

N-(FLUOREN-3-YL)ACETOHYDROXAMIC ACETATE see N-ACETOXY-FLUORENYLACETAMIDE

N-FLUOREN-2-YLACETOHYDROXAMIC ACID see N-HYDROXY-N-ACETYL-2-AMINOFLUORENE

N-2-FLUORENYLACETOHYDROXAMIC ACID see N-HYDROXY-N-ACETYL-2-AMINOFLUORENE

2,7-FLUORENYLBISACETAMIDE see 2,7-BIS(ACETAMIDO)FLUORENE

N,N'-FLUOREN-2,7-YLBISACETAMIDE see 2,7-BIS(ACETAMIDO)FLUO-RENE

2-FLUORENYLDIACETAMIDE see 2-DIACETAMIDO FLUORENE

N-FLUOREN-2-YLDIACETAMIDE see 2-DIACETAMIDO FLUORENE

2-FLUORENYLDIMETHYLAMINE see 2-DIMETHYLAMINOFLUORENE

N,N'-FLUOREN-2,7-YLENEBISACETAMIDE see 2,7-BIS(ACETAMIDO)-FLUORENE

N,N'-2,7-FLUORENYLENEBISACETAMIDE see 2,7-BIS(ACETAMIDO)-FLUORENE

N,N'-(FLUOREN-2,7-YLENE)BIS(ACETYLAMINE) see 2,7-BIS(ACET-AMIDO)FLUORENE

N,N'-2,7-FLUORENYLENEDIACETAMIDE see 2,7-BIS(ACETAMIDO)-FLUORENE

N-(2-FLUORENYL)-2,2,2-TRIFLUOROACETAMIDE see N-FLUOREN-2-YL-2,2,2-TRIFLUORO ACETAMIDE

FLUORESCEIN SODIUM B.P see FLUORESCEIN SODIUM

FLUORESSIGAEURE (GERMAN) see SODIUM FLUOROACETATE

FLUORIC ACID see HYDROFLUORIC ACID (SOLUTION)

FLUORID HLINITY (CZECH) see ALUMINUM FLUORIDE

FLUORID SODNY (CZECH) see SODIUM FLUORIDE

FLUORINE (DOT) see FLUORINE

FLUORINE MONOXIDE see OXYGEN DIFLUORIDE

FLUORINE OXIDE see OXYGEN DIFLUORIDE

FLUORO (ITALIAN) see FLUORINE

2-FLUOROACETAMIDE see FLUOROACETAMIDE

FLUOROACETIC ACID see FLUORO ETHANOIC ACID

2-FLUOROACETIC ACID see FLUORO ETHANOIC ACID

FLUOROACETIC ACID AMIDE see FLUOROACETAMIDE

FLUOROACETIC ACID METHYL ESTER see METHYL FLUOROACETATE

FLUOROACETIC ACID, SODIUM SALT see SODIUM FLUOROACETATE

7-FLUORO-2-ACETYLAMINOFLUORENE see 7-FLUORO-2-ACETAMIDO-FLUORENE

P-FLUOROACETYLAMINOPHENYL DERIVATIVE OF NITROGEN MUSTARD see 4'-(BIS(2-CHLOROETHYL)AMINO)-2-FLUORO ACETANILIDE

4'-FLUORO-4-AMINODIPHENYL see 4-AMINO-4'-FLUORODIPHENYL

4-FLUOROBENZ(A)ANTHRACENE see 4-FLUORO BENZ ANTHRACENE

4'-FLUORO-1,2-BENZANTHRACENE see 4-FLUORO BENZ ANTHRACENE

1-(4'-FLUOROBENZOIL)-3-PIRROLIDINOPROPANO MALEATO (ITALIAN) see 1-(4'-FLUORO BENZOYL)-3-PYRROLIDINYL PROPANE MALEATE

1'-(3-(P-FLUOROBENZOYL)PROPYL)(1,4'-BIPIPERIDINE 1-4'-CARBOX-AMIDE see FLUOROBUTYROPHENONE

1-(3-P-FLUOROBENZOYLPROPYL)-4-P-CHLOROPHENYL-4-HYDROXY-PIPERIDINE see gamma-(4-(p-CHLORPHENYL)-4-HYDROXPIPER-IDINO)-p-FLUORBUTYRO-PHENONE

1-(1-(3-(P-FLUOROBENZOYL)PROPYL)-1,2,3,6-TETRAHYDRO-4-PYRI-DYL)-2 -BENZIMIDAZOLINONE see DROPERIDOL

4'-FLUORO-4-BIPHENYLAMINE see 4-AMINO-4'-FLUORODIPHENYL

FLUOROBISISOPROPYLAMINO- PHOSPHINE OXIDE see PHOSPHORODI-(ISOPROPYLAMIDIC) FLUORIDE

FLUOROBORATE see CADMIUM FLUOBORATE

FLUOROCARBON-12 see DICHLORODIFLUOROMETHANE

FLUOROCARBON-22 see MONOCHLORODIFLUOROMETHANE

FLUOROCARBON 113 see FREON 113

FLUOROCHROME see MERCUROCHROME

2-FLUORO-2'-CYANODIETHYL ETHER see 2-CYANO-2'-FLUORODI-ETHYL ETHER

5-FLUORODEOXYCYTIDINE see 5-FLUORO-2'-DEOXYCYTIDINE

FLUORODEOXYURIDINE see 2'-DEOXY-5-FLUORO URIDINE

5-FLUORODEOXYURIDINE see 2'-DEOXY-5-FLUORO URIDINE

5-FLUORO-2-DEOXYURIDINE see 2'-DEOXY-5-FLUORO URIDINE

5-FLUORO-2'-DEOXYURIDINE see 2'-DEOXY-5-FLUORO URIDINE

FLUORODIISOPROPYL PHOSPHATE see ISOPROPYL PHOSPHOROFLUORI-DATE

3'-FLUORO-4-DIMETHYLAMINOAZOBENZENE see N,N-DIMETHYL-p-(3-FLUOROPHENYLAZO)ANILINE

4'-FLUORO-4-DIMETHYLAMINOAZOBENZENE see N,N-DIMETHYL-p-((p-FLUOROPHENYL)AZO)ANILINE

4'-FLUORO-P-DIMETHYLAMINOAZOBENZENE see N,N-DIMETHYL-p-((p-FLUOROPHENYL)AZO)ANILINE

4'-FLUORO-N,N-DIMETHYL-4-AMINOAZOBENZENE see N,N-DIMETHYL-p-((p-FLUOROPHENYL)AZO)ANILINE

3-FLUORO-2,10-DIMETHYL-5,6-BENZACRIDINE see 10-FLUORO-9,12-DIMETHYLBENZ(a)ACRIDINE

5-FLUORO-7,12-DIMETHYLBENZ(A)ANTHRACENE see 7,12-DIMETHYL-5-FLUOROBENZ(a)ANTHRACENE

11-FLUORO-7,12-DIMETHYLBENZ(A)ANTHRACENE see 7,12-DIMETHYL-11-FLUOROBENZ(a)ANTHRACENE

4'-FLUORO-N,N-DIMETHYL-P-PHENYLAZOANILINE see N,N-DIMETHYL-p-((p-FLUOROPHENYL)AZO)ANILINE

1,2,4-FLUORODINITROBENZENE see 2,4-DINITRO-1-FLUOROBENZENE

1-FLUORO-2,4-DINITROBENZENE see 2,4-DINITRO-1-FLUOROBEN-ZENE

beta-FLUOROETHANOL see 2-FLUOROETHANOL

FLUOROETHENE see VINYL FLUORIDE

2-FLUOROETHYL (4-BIPHENYLYL)ACETATE see 2-FLUORETHYL (1,1'-BIPHENYL)-4-ACETATE

FLUOROETHYLENE see VINYL FLUORIDE

2-FLUOROETHYL FLUOROACETATE see beta-FLUOROETHYL FLUORO-ACETATE

beta-FLUOROETHYL-gamma-FLUOROBUTYRATE see 2-FLUORO ETHYL gamma-FLUORO BUTYRATE

BETA-FLUOROETHYLIC ESTER OF XENYLACETIC ACID see 2-FLUOR-ETHYL (1,1'-BIPHENYL)-4-ACETATE

1-FLUORO-2-FAA see 1-FLUORO-2-ACETYL AMINO FLUORENE

FLUOROFLEX see TEFLON

N-(7-FLUOROFLUORENE-2-YL)ACETAMIDE see 7-FLUORO-2-ACET-AMIDO-FLUORENE

N-(1-FLUOROFLUOREN-2-YL)ACETAMIDE see 1-FLUORO-2-ACETYL
AMINO FLUORENE

FLUOROFORM see CARBON TRIFLUORIDE

4'-FLUORO-4-(4-HYDROXY-4-(4'-CHLOROPHENYL)PIPERIDINO)BUTY-
ROPHENONE see gamma-(4-(p-CHLORPHENYL)-4-HYDROXPIPER-
IDINO)-p-FLUORBUTYRO-PHENONE

9-ALPHA-FLUORO-16-ALPHA-HYDROXYPREDNISOLONE see ARISTO-
CORT

9-ALPHA-FLUORO-16-HYDROXYPREDNISOLONE ACETONIDE see
ARISTOCORT ACETONIDE

4'-FLUORO-4-(4-HYDROXY-4-P-TOLYLPIPERIDINO)BUTYROPHENONE,
HYDROCHLORIDE see METHYLPERIDOL HYDROCHLORIDE

4'-FLUORO-4-(4-HYDROXY-4-(ALPHA,ALPHA,ALPHA-TRIFLUORO-
M-TOLYL)PIPERIDINO)BUTYROPHENONE see TRIFLUPERIDOL

9-ALPHA-FLUORO-16-ALPHA-17-ALPHA-ISOPROPYLEDENE DIOXY
PREDNISOLONE see ARISTOCORT ACETONIDE

4'-FLUORO-4-(4-(O-METHOXYPHENYL)-1-PIPERAZINYL)BUTYROPHE-
NONE,HYDROCHLORIDE see FLUANISONE HYDROCHLORIDE

2-FLUORO-N-METHYL-N-1NAPHTHYLACETAMIDE see N-METHYL-
N-(1-NAPHTHYL)FLUOROACETAMIDE

3-FLUORO-10-METHYL-7,8-BENZACRIDINE (FRENCH) see 7-METHYL-
9-FLUOROBENZ(c)ACRIDINE

10-FLUORO-7-METHYLBENZ(A)ANTHRACENE see 7-FLUORO-10-
METHYL-1,2-BENZANTHRACENE

3'-FLUORO-10-METHYL-1,2-BENZANTHRACENE see 3-FLUORO-
7-METHYLBENZ(a)ANTHRACENE

DELTA(SUP 1)-9-ALPHA-FLUORO-16-ALPHA-METHYLCORTISOL see
SUPERPREDNOL

FLUOROMETHYL CYANIDE see FLUOROACETONITRILE

9-ALPHA-FLUORO-16-ALPHA-METHYLPREDNISOLONE see SUPERPRED-
NOL

9-ALPHA-FLUORO-16-BETA-METHYLPREDNISOLONE see BETAMETHA-
SONE

9-ALPHA-FLUORO-16-ALPHA-METHYL-1,4-PREGNADIENE-11-BETA,-
17-ALPHA,21-TRIOL-3,20-DIONE see SUPERPREDNOL

9-ALPHA-FLUORO-16-BETA-METHYL-1,4-PREGNADIENE-11-BETA,17-
ALPHA,21-TRIOL-3,20-DIONE see BETAMETHASONE

4-ALPHA-FLUORO-16-ALPHA-METHYL-11-BETA,17,21-TRIHYDROXY-
PREGNA-1,4-DIENE-3,20-DIONE see SUPERPREDNOL

8-FLUOROOCTANOIC ACID, ETHYL ESTER see ETHYL-8-FLUORO
OCTANOATE

5-FLUOROPENTYLAMINE see 5-FLUORO AMYLAMINE

5-FLUOROPENTYL THIOCYANATE see 5-FLUOROAMYL THIOCYANATE

P-((P-FLUOROPHENYL)AZO)-N,N-DIMETHYLANILINE see N,N-
DIMETHYL-p-((p-FLUOROPHENYL)AZO)ANILINE

4-((4-FLUOROPHENYL)AZO)-N,N-DIMETHYLBENZENAMINE see N,N-
DIMETHYL-p-((p-FLUOROPHENYL)AZO)ANILINE

(+ -)-alpha-(P-FLUOROPHENYL)-4-(O-METHOXYPHENYL)-1-PIPER-
AZINEBUTANOL see HALVISOL

DL-1-(4-FLUOROPHENYL)-4-(1-(4-(2-METHOXY-PHENYL))-PIPER-
AZINYL)BUTANOL see HALVISOL

1-(1-(4-(P-FLUOROPHENYL-4-OXOBUTYL)-1,2,3,6-TETRAHYDRO-
4-PYRIDYL)-2-BENZIMIDAZOLINONE see DROPERIDOL

FLUOROPHOSGENE see CARBONYL FLUORIDE

FLUOROPHOSPHORIC ACID, ANHYDROUS see PHOSPHOROFLUORIDIC
ACID

4'-FLUORO-4-(4-N-PIPERIDINO-4-CARBAMIDOPIPERIDINO)BUTYRO-
PHENONE see FLUOROBUTYROPHENONE

P-FLUORO-GAMMA-(4-PIPERIDINO-4-CARBAMOYLPIPERIDINO)BUTYRO-
PHENONE see FLUOROBUTYROPHENONE

3-FLUOROPROPENE see ALLYL FLUORIDE

4'-FLUORO-4-(4-(2-PYRIDYL)-1-PIPERAZINYL)BUTYROPHENONE see
4'-FLUORO-4-(4-(2-PYRIDYL)-1-PIPERAZINYL)BUTYROPHENONE

4'-FLUORO-4-(1-PYRROLIDINYL)BUTYROPHENONE MALEATE see
1-(4'-FLUORO BENZOYL)-3-PYRROLIDINYL PROPANE MALEATE

FLUOROSILICIC ACID see SILICOFLUORIC ACID

FLUOROSUFONIC ACID see FLUOROSULFURIC ACID

9-ALPHA-FLUORO-11-BETA,16-ALPHA,17,21-TETRAHYDROXY-1,4-
PREGNADIENE-3,20-DIONE see ARISTOCORT

9-ALPHA-FLUORO-11-BETA,16-ALPHA,17,21-TETRAHYDROXYPREG-
NA-1,4-DIENE-3,20-DIONE see ARISTOCORT

9-ALPHA-FLUORO-11-BETA,16-ALPHA,17-ALPHA,21-TETRAHY-
DROXYPREGNA-1,4-DIENE-3,20-DIONE FLUOXYPREDNI-
SOLONE see ARISTOCORT

FLUOROTRICHLOROMETHANE see TRICHLOROFLUOROMETHANE

4-FLUORO-4'-TRIFLUOROMETHYLBENZOPHENONE GUANYLHYDRA-
ZONEHYDROCHLORIDE see FTBG

9-FLUORO-11-BETA,17,21-TRIHYDROXY-16-ALPHA-METHYLPREGNA-
1,4-DIENE-3,20-DIONE see SUPERPREDNOL

9-FLUORO-11-BETA,17,21-TRIHYDROXY-16-BETA-METHYLPREGNA-
1,4-DIENE-3,20-DIONE see BETAMETHASONE

9-ALPHA-FLUORO-11-BETA,17-ALPHA,21-TRIHYDROXY-16-BETA-
METHYLPREGNA-1,4-DIENE-3,20-DIONE see BETAMETHASONE

9-ALPHA-FLUORO-11-BETA,17,21-TRIHYDROXY-16-BETA-METHYL-
PREGNA-1,4-DIENE- 3,20-DIONE see BETAMETHASONE

9-ALPHA-FLUORO-11-BETA,17-ALPHA, 21-TRIHYDROXY-16-ALPHA-
METHYLPREGNA-1,4-DIENE-3,20-DIONE see SUPERPREDNOL

FLUOROTROJCHLOROMETAN (POLISH) see TRICHLOROFLUORO-
METHANE

3-FLUOROTYROSINE see 3-FLUOROTYROSIN

M-FLUOROTYROSINE see 3-FLUOROTYROSIN

5-FLUOROURACIL see FLUOROURACIL

5-FLUOROURACIL DEOXYRIBOSIDE see 2'-DEOXY-5-FLUORO URIDINE

5-FLUOROURACIL-2'-DEOXYRIBOSIDE see 2'-DEOXY-5-FLUORO
URIDINE

FLUOROWODOR (POLISH) see HYDROFLUORIC ACID

FLUOROXENE see 2,2,2-TRIFLUOROETHYL VINYL ETHER

5-FLUORPROPYRIMIDINE-2,4-DIONE see FLUOROURACIL

FLUORSPAR see CALCIUM FLUORIDE

FLUORTHYRIN see 3-FLUOROTYROSIN

3-FLUORTYROSIN (GERMAN) see 3-FLUOROTYROSIN

FLUORURE DE BORE (FRENCH) see BORON FLUORIDE

FLUORURE DE N,N'-DIISOPROPYLE PHOSPHORODIAMIDE (FRENCH) see
PHOSPHORODI(ISOPROPYLAMIDIC) FLUORIDE

FLUORURE DE POTASSIUM (FRENCH) see POTASSIUM FLUORIDE
(SOLID)

FLUORURE DE SODIUM (FRENCH) see SODIUM FLUORIDE

FLUORURE DE SULFURYLE (FRENCH) see SULFURYL FLUORIDE

FLUORURE DE N,N,N',N'-TETRAMETHYLE PHOSPHORO-DIAMIDE
(FRENCH) see BIS(DIMETHYL AMIDO)FLUORO PHOSPHATE

FLUORURE DE THIONYLE (FRENCH) see THIONYL FLUORIDE

FLUORURES ACIDE (FRENCH) see FLUORINE

FLUORURI ACIDI (ITALIAN) see FLUORINE

FLUORWASSERSTOFF (GERMAN) see HYDROFLUORIC ACID

FLUORWATERSTOF (DUTCH) see HYDROFLUORIC ACID

FLUOSILICATE DE AMMONIUM (FRENCH) see CRYPTOHALITE

FLUOSILICATE DE AMMONIUM (FRENCH) see AMMONIUM SILICO
FLUORIDE

FLUOSILICATE DE MAGNESIUM (FRENCH) see MAGNESIUM HEXA-
FLUOROSILICATE

FLUOSILICATE DE SODIUM see DISODIUM HEXAFLUOROSILICATE

FLUOSILICATE DE ZINC see ZINC FLUOSILICATE

FLUOSILICIC ACID see SILICOFLUORIC ACID

FLUOSULFONIC ACID see FLUOROSULFURIC ACID

FLUOTHANE see HALOTHANE

FLUOTITANATE DE POTASSIUM (FRENCH) see POTASSIUM HEXA-
FLUOROTITANATE

FLURAZEPAM HYDROCHLORIDE see DALMANE

FLUROXENE see 2,2,2-TRIFLUOROETHYL VINYL ETHER

FOLACIN see FOLIC ACID

FOLCYSTEINE see FOLIC ACID

FOLIANDRIN see OLEANDRIN

FOLLICULAR HORMONE see ESTRONE

FOLLICULAR HORMONE HYDRATE see ESTRIOL

ENDOFOLLICULINA see ESTRONE

FOOD BLUE 1 see FD AND C BLUE NO. 1

FOOD BLUE 1 see FD AND C BLUE 1

FOOD BLUE 2 see FD AND C BLUE 1

FOOD BLUE DYE NO. 1 see FD AND C BLUE 1

FOOD RED 5 see C.I. FOOD RED 3

FOOD RED NO. 102 see FOOD RED 7

FOOD YELLOW 6 see FOOD YELLOW 3

FORAAT (DUTCH) see PHORATE

FORMAL see DIMETHOXYMETHANE

FORMALDEHYD (CZECH, POLISH) see FORMALDEHYDE

FORMALDEHYDE, AS FORMALIN SOLUTION (DOT) see FORMALDE-
HYDE

FORMALDEHYDE BIS(BETA-CHLOROETHYL) ACETAL see BIS(beta-
CHLOROETHYL)FORMAL

FORMALDEHYDE CYANOHYDRIN see HYDROXYACETONITRILE

FORMALDEHYDE DIMETHYLACETAL see DIMETHOXYMETHANE

FORMALIN see FORMALDEHYDE

FORMALINA (ITALIAN) see FORMALDEHYDE

FORMALINE (GERMAN) see FORMALDEHYDE

FORMALIN-LOESUNGEN (GERMAN) see FORMALDEHYDE

FORMALITH see FORMALDEHYDE

FORMAMINE see HEXAMETHYLENETETRAMINE

FORMIATE DE METHYLE (FRENCH) see METHYL FORMATE

FORMIC ACID, ALLYL ESTER see ALLYL FORMATE

FORMIC ALDEHYDE see FORMALDEHYDE

FORMIC ETHER see ETHYL FORMATE

FORMOL see FORMALDEHYDE

FORMOSA CAMPHOR see CAMPHOR

FORMOSA CAMPHOR OIL see CAMPHOR OIL

FORMOSE OIL OF CAMPHOR see CAMPHOR OIL

7-FORMYLBENZ(C)ACRIDINE see BENZ(c)ACRIDINE-7-CARBOXALDE-
HYDE

7-FORMYLBENZO(C)ACRIDINE see BENZ(c)ACRIDINE-7-CARBOXALDE-
HYDE

6-FORMYLBENZO(A)PYRENE see BENZO(a)PYRENE-6-CARBOXYALDE-
HYDE

N-FORMYLDIMETHYLAMINE see DIMETHYL FORMAMIDE

2-(2-FORMYLHYDRAZINO)-4-(5-NITRO-2-FURYL)THIAZOLE NEFUR-
THIAZOLE see NIFURTHIAZOLE

N-FORMYL HYDROXYAMINOACETIC ACID see N-FORMYL-N-HY-
DROXYGLYCINE

FORMYLIC ACID see FORMIC ACID

S-(2-(FORMYLMETHYLAMINO)-2-OXOETHYL) O,O-DIMETHYLPHOS-
PHORODITHIOATE see O,O-DIMETHYL DITHIOPHOSPHORYLACETIC
ACID-N-METHYL-N-FORMYLAMIDE

N-FORMYL-N-METHYLCARBAMOYLMETHYL O,O-DIMETHYL PHOS-
PHORODITHIOATE see O,O-DIMETHYL DITHIOPHOSPHORYLACETIC
ACID-N-METHYL-N-FORMYLAMIDE

S-(N-FORMYL-N-METHYLCARBAMOYLMETHYL) O,O-DIMETHYL PHOS-
PHORODITHIOATE see O,O-DIMETHYL DITHIOPHOSPHORYLACETIC
ACID-N-METHYL-N-FORMYLAMIDE

S-(N-FORMYL-N-METHYLCARBAMOYLMETHYL) DIMETHYL PHOS-
PHOROTHIOLOTHIONATE see O,O-DIMETHYL DITHIOPHOSPHORYL-
ACETIC ACID-N-METHYL-N-FORMYLAMIDE

2-FORMYL-1-METHYLPYRIDINIUM IODIDE OXIME see PYRIDINIUM-
2-ALDOXIME-N-METHYLIODIDE

2-FORMYL-N-METHYLPYRIDINIUM OXIME CHLORIDE see 2-FORMYL-
1-METHYLPYRIDINIUM CHLORIDE OXIME

2-FORMYL-N-METHYLPYRIDINIUM OXIME IODIDE see PYRIDINIUM-
2-ALDOXIME-N-METHYLIODIDE

N-FORMYL-L-P-SARCOLYSIN see L-3-(p-(BIS(2-CHLOROETHYL)-
AMINO)PHENYL)-N-FORMYLALANINE

FORMYL TRICHLORIDE see CHLOROFORM

8-FORMYL-1,6,7-TRIHYDROXY-5-ISOPROPYL-3-METHYL-2,2′- BIS-
NAPHTHALENE see GOSSYPOL

FOSFAMIDON (DUTCH) see DIMECRON

FOSFAMIDONE (ITALIAN) see DIMECRON

FOSFORO BIANCO (ITALIAN) see PHOSPHORUS (WHITE)

FOSFORO(PENTACHLORURO DI) (ITALIAN) see PHOSPHORUS PENTA-
CHLORIDE

FOSFORO(TRICLORURO DI) (ITALIAN) see PHOSPHORUS CHLORIDE

FOSFOROWODOR (POLISH) see PHOSPHINE

FOSFORPENTACHLORIDE (DUTCH) see PHOSPHORUS PENTACHLO-
RIDE

FOSFORTRICHLORIDE (DUTCH) see PHOSPHORUS CHLORIDE

FOSFORZUUROPLOSSINGEN (DUTCH) see PHOSPHORIC ACID

FOSFURI DI ALLUMINIO (ITALIAN) see ALUMINUM PHOSPHIDE

FOSFURI DI MAGNESIO (ITALIAN) see MAGNESIUM PHOSPHIDE

FOSGEEN (DUTCH) see PHOSGENE

FOSGEN (POLISH) see PHOSGENE

FOSGENE (ITALIAN) see PHOSGENE

FOSSIL FLOUR see SILICA, AMORPHOUS FUMED

FOSVEL see LEPTOPHOS

FOSVEX see TETRON

FOUMARIN see 3-(alpha-ACETONYLFURFURYL)-4-HYDROXYCOU-
MARIN

FOURAMINE EG see m-AMINOPHENOL

FOURAMINE P see 4-AMINOPHENOL

FOURNEAU 190 see ACETPHENARSINE

FOWLER'S SOLUTION see POTASSIUM ARSENITE

FOXGLOVE see DIGITALIS

FR-33 see 8-(4-p-FLUORO PHENYL-4-OXOBUTYL)-2-METHYL-2,8-
DIAZASPIRO(4.5)DECANE-1,3-DIONE

FREE BENZYLPENICILLIN see BENZYL-6-AMINOPENICILLINIC ACID

FREE HISTAMINE see HISTAMINE

FREEMANS WHITE LEAD see LEAD (II) SULFATE (1:1)

FRENCH GREEN see CUPRIC ACETOARSENITE

FREON see MONOCHLORODIFLUOROMETHANE

FREON 10 see CARBON TETRACHLORIDE

FREON 11 see TRICHLOROFLUOROMETHANE

FREON 13 see CHLORO TRIFLUORO METHANE

FREON 14 see CARBON TETRAFLUORIDE

FREON 20 see CHLOROFORM

FREON 30 see METHANE DICHLORIDE

FREON 31 see MONOCHLOROMONOFLUOROMETHANE

FREON 114 see DICHLOROTETRAFLUOROETHANE

FREON 115 see CHLORO PENTAFLUORO ETHANE

FREON 150 see 1,2-DICHLORETHANE

FREON C-318 see CYCLOOCTAFLUOROBUTANE

FREON F-12 see DICHLORODIFLUOROMETHANE

FREON F-23 see CARBON TRIFLUORIDE

FRIARS COWL see ACONITE

FRUIT RED A GEIGY see FOOD RED 2

FRUIT SUGAR see FRUCTOSE

FRUTABS see FRUCTOSE

FTAALZUURANHYDRIDE (DUTCH) see PHTHALIC ANHYDRIDE

FTALOWY BEZWODNIK (POLISH) see PHTHALIC ANHYDRIDE

FTOROTAN (RUSSIAN) see HALOTHANE

FUCHSINE BASE see MAGENTA BASE

FUKI-NO-TOH (JAPANESE) see PETASITES JAPONICUS MAXIM

FULMINATE OF MERCURY, DRY see MERCURY FULMINATE (DRY)

FULMINATE OF MERCURY, WET see MERCURY FULMINATE (WET)

FUMARYLCHLORID (CZECH) see FUMARYL CHLORIDE

FUMARYL CHLORIDE (DOT) see FUMARYL CHLORIDE

FUMED SILICA see SILICA, AMORPHOUS FUMED

FUMED SILICON DIOXIDE see SILICA, AMORPHOUS FUMED

FUMING LIQUID ARSENIC see ARSENIC CHLORIDE

FUMING SULFURIC ACID see SULFURIC ACID, FUMING

FUNGICIDE 1991 see METHYL-1-(BUTYLCARBAMOYL)-2-BENZIMIDA-
ZOLYLCARBAMATE

FUNGICIDIN see NYSTATIN

FUNGILIN see AMPHOTERICIN B

FUNGISONE see AMPHOTERICIN B

FUNGOCIN see BACILLUS SUBTILIS BPN

5-FUR see 5-FLUOROURIDINE

FURACILLIN see NITROFURAZONE

FURADANTIN see NITROFURANTOIN

FURADONIN see NITROFURANTOIN

FURAL see 2-FURALDEHYDE

2-FURANALDEHYDE see 2-FURALDEHYDE

2-FURANCARBINOL see FURFURYL ALCOHOL

2-FURANCARBONAL see 2-FURALDEHYDE

2-FURANCARBOXALDEHYDE see 2-FURALDEHYDE

2,5-FURANDIONE see MALEIC ANHYDRIDE

FURANIDINE see TETRAHYDROFURAN

2-FURANMETHANOL see FURFURYL ALCOHOL

FURAN-OFTENO see NITROFURAZONE

FURANTOIN see NITROFURANTOIN

1-(3-FURANYL)-4-METHYL-1-PENTANONE see PERILLA KETONE

FURAPLAST see NITROFURAZONE

FURASEPTYL see NITROFURAZONE

FURAZOSIN see 1-(4-AMINO-6,7-DIMETHOXY-2-QUINAZOLINYL-4-(2-
FURANYLCARBONYL) PIPERAZINE

FURFURAL see 2-FURALDEHYDE

FURFURAL (DOT) see 2-FURALDEHYDE

FURFURAL ALCOHOL see FURFURYL ALCOHOL

FURFURALDEHYDE see 2-FURALDEHYDE

FURFURALE (ITALIAN) see 2-FURALDEHYDE

FURFURAN see FURAN

2-FURFURYLALKOHOL (CZECH) see FURFURYL ALCOHOL

FURIDIAZINE see 5-(5-NITRO-2-FURYL)-2-AMINO-1,3,4-THIADIAZOLE

2-FURIL-METANALE (ITALIAN) see 2-FURALDEHYDE

FURNACE BLACK see CARBON BLACK

FUROSEMIDE see 4-CHLORO-N-FURFURYL-5-SULFAMOYL ANTHRA-
NILIC ACID

2-(4-(2-FUROYL)PIPERAZIN-1-YL)-4-AMINO-6,7-DIMETHOXYQUINA-
ZOLINE see 1-(4-AMINO-6,7-DIMETHOXY-2-QUINAZOLINYL-
4-(2-FURANYLCARBONYL) PIPERAZINE

3-(1-FURYL-3-ACETYLETHYL)-4-HYDROXYCOUMARIN see 3-(alpha-
ACETONYLFURFURYL)-4-HYDROXYCOUMARIN

FURYL ALCOHOL see FURFURYL ALCOHOL

FURYLAMIDE see 2-(2-FURYL)-3-(5-NITRO-2-FURYL)ACRYLAMIDE

2-FURYLCARBINOL see FURFURYL ALCOHOL

ALPHA-FURYLCARBINOL see FURFURYL ALCOHOL

FURYLFURAMIDE see 2-(2-FURYL)-3-(5-NITRO-2-FURYL)ACRYLAMIDE

2-FURYL-METHANAL see 2-FURALDEHYDE

FUSARIC ACID see 5-BUTYL PICOLINIC ACID

FUSARIOTOXIN T-2′ see MYCOTOXIN T-2

FUSED BORIC ACID see BORON OXIDE

FUSED QUARTZ see SILICA, AMORPHOUS FUSED

FUSED SILICA see SILICA, AMORPHOUS FUSED

FUSELOEL (GERMAN) see FUSEL OIL

FYDE see FORMALDEHYDE

FYRQUEL 150 see TRITOLYL PHOSPHATE

FYTIC ACID see PHYTIC ACID

G-11 see HEXACHLOROPHENE

GADOLINIUM TRICHLORIDE see GADOLINIUM CHLORIDE

4-(beta-D-GALACTOSIDO)-D-GLUCOSE see LACTOSE

GALENA see LEAD SULFIDE

GALLOCHROME see MERCUROCHROME

GALLOTANNIC ACID see TANNIC ACID

GALLOTANNIN see TANNIC ACID

GAMASOL 90 see DIMETHYL SULFOXIDE

GAMMAHEXANE see BENZENE HEXACHLORIDE-gamma isomer

GAMMEXANE see BENZENE HEXACHLORIDE

GANJA see CANNABIS

GARAMYCIN see GENTAMICIN

GARDENAL SODIUM see SODIUM LUMINAL

GAS OIL see FUEL OIL

GAULTHERIA OIL, ARTIFICIAL see METHYL SALICYLATE

GBL see 4-HEPTANONE

GC 7787 see HEXAFLUORO ACETONE TRIHYDRATE

GC-9160 see KELEVAN

GECHLOREERDEDIFENYL (DUTCH) see POLYCHLORINATED BIPHENYL
(AROCLOR 1242)

GEIGY G-28029 see PHENCAPTON

GELATINE DYNAMITE see DYNAMITE

GELATIN-EPINEPHRINE see 1-ADRENALINE CHLORIDE

GELBER PHOSPHOR (GERMAN) see PHOSPHORUS (WHITE)

GELBIN see CALCIUM CHROMATE

GELBIN YELLOW ULTRAMARINE see CALCIUM CHROMATE (VI) DIHY-
DRATE

G-ELEVEN see HEXACHLOROPHENE

GELSEMIN see GELSEMINE

GELVATOLS see POLYVINYL ALCOHOL

GENETRON 12 see DICHLORODIFLUOROMETHANE

GENETRON 22 see MONOCHLORODIFLUOROMETHANE

GENETRON 1113 see CHLOROTRIFLUOROETHYLENE

GENTAMYCIN see GENTAMICIN

GENTAMYCIN-CREME (GERMAN) see GENTAMICIN

GENTIAN VIOLET see ANILINE VIOLET

GERANIOL ALCOHOL see GERANIOL

GERANIUM CRYSTALS see DIPHENYL ETHER

GERANYL ALCOHOL see GERANIOL

GERMANIUM HYDRIDE see GERMANE

GERMANIUM TETRABROMIDE see GERMANIUM BROMIDE

GERMANIUM TETRACHLORIDE see GERMANIUM CHLORIDE

GERMANIUM TETRAHYDRIDE see GERMANE

G.I. 77640 see LEAD SULFIDE

GIALLO CROMO (ITALIAN) see LEAD CHROMATE

GIBBERELLIN see GIBBERELLIC ACID

GIEGY GS-13798 see 8-(METHYLQUINOLYL)-N-METHYL CARBAMATE

GIFBLAAR see POTASSIUM FLUOROACETATE

GIFBLAAR POISON see FLUORO ETHANOIC ACID

GINARSOL see ACETPHENARSINE

GIRL see COCAINE

GLACIAL ACETIC ACID see ACETIC ACID

GLANDUCORPIN see PROGESTERONE

GLASS see SILICA, CRYSTALLINE-QUARTZ

GLASS see SILICA (CRYSTALLINE)

GLICOL MONOCLORIDRINA (ITALIAN) see 2-CHLOROETHYL ALCOHOL

GLUCINUM see BERYLLIUM

GLUCO-FERRUM see FERROUS GLUCONATE

GLUCOLIN see D-GLUCOSE

GLUCONATE DE CALCIUM (FRENCH) see CALCIUM GLUCONATE

4-(ALPHA-D-GLUCOPYRANOSIDO)-ALPHA-GLUCOPYRANOSE see MAL-
TOSE

ALPHA-D-GLUCOPYRANOSYL BETA-D-FRUCTOFURANOSIDE see SU-
CROSE

GLUCOSE see D-GLUCOSE

D-GLUCOSE, ANHYDROUS see D-GLUCOSE

GLUCOSE LIQUID see D-GLUCOSE

(ALPHA-D-GLUCOSIDO)-BETA-D-FRUCTOFURANOSIDE see SUCROSE

4-(ALPHA-D-GLUCOSIDO)-D-GLUCOSE see MALTOSE

N-D-GLUCOSYL(2)-N′-NITROSOMETHYLHARNSTOFF (GERMAN) see
STREPTOZOTICIN

N-D-GLUCOSYL-(2)-N′-NITROSOMETHYLUREA see STREPTOZOTICIN

beta-D-GLUCOSYLOXYAZOXYMETHANE see CYCASIN

(1-D-GLUCOSYLTHIO)GOLD see 1-AUROTHIO-D-GLUCOPYRANOSE

GLUTACID see L-GLUTAMIC ACID

alpha-GLUTAMIC ACID see L-GLUTAMIC ACID

L-GLUTAMINIC ACID see L-GLUTAMIC ACID

GLUTAMINOL see L-GLUTAMIC ACID

GLUTAMMATO MONOSODICO (ITALIAN) see MONOSODIUM GLUTA-
MATE

GLUTARIC ACID DINITRILE see 1,3-TRIMETHYLENEDINITRILE

GLUTARIMIDE, 2-PHTHALIMIDO-ALPHA-PHTHALIMIDOGLUTARIMIDE
see THALIDOMIDE

GLUTARODINITRILE see 1,3-TRIMETHYLENEDINITRILE

GLUTATON see L-GLUTAMIC ACID

GLYCERIN, ANHYDROUS see GLYCERINE

GLYCERIN DIACETATE see 1,3-DIACETIN

GLYCERIN-ALPHA-MONOCHLORHYDRIN see CHLORHYDRIN

GLYCERINTRINITRATE (CZECH) see NITROGLYCERIN

GLYCERITOL see GLYCERINE

GLYCEROLACETONE see DIOXOLAN

GLYCEROL CHLOROHYDRIN see CHLORHYDRIN

GLYCEROL-ALPHA-CHLOROHYDRIN see CHLORHYDRIN

GLYCEROL DIACETATE see 1,3-DIACETIN

GLYCEROL alpha,beta-DICHLOROHYDRIN see 2,3-DICHLOROPROPA-
NOL

GLYCEROL DIMETHYLKETAL see DIOXOLAN

GLYCEROL EPICHLORHYDRIN see 1-CHLORO-2,3-EPOXY PROPANE

GLYCEROL-ALPHA-MONOCHLOROHYDRIN see CHLORHYDRIN

GLYCEROL, NITRIC ACID TRIESTER see NITROGLYCERIN

GLYCEROL TRIBUTYRATE see TRIBUTYRIN

GLYCEROL TRICHLOROHYDRIN see ALLYL TRICHLORIDE

GLYCEROL TRINITRATE see NITROGLYCERIN

GLYCEROL(TRINITRATE DE) (FRENCH) see NITROGLYCERIN

GLYCEROLTRINTRAAT (DUTCH) see NITROGLYCERIN

GLYCERYL-ALPHA-CHLOROHYDRIN see CHLORHYDRIN

GLYCERYL-1,3-DIACETATE see 1,3-DIACETIN

GLYCERYL NITRATE see NITROGLYCERIN

GLYCERYL TRIBUTYRATE see TRIBUTYRIN

GLYCERYL TRICHLOROHYDRIN see ALLYL TRICHLORIDE

GLYCERYL TRINITRATE see NITROGLYCERIN

GLYCIDAL see GLYCIDALDEHYDE

GLYCIDOL see 2,3-EPOXY-1-PROPANOL

GLYCIDOL STEARATE see STEARIC ACID-2,3-EPOXYPROPYL ESTER

GLYCIDYL ACRYLATE see 2,3-EPOXYPROPYL ACRYLATE

GLYCIDYL ALCOHOL see 2,3-EPOXY-1-PROPANOL

GLYCIDYL BUTYL ETHER see BUTYL GLYCIDYL ETHER

GLYCIDYL OCTADECANOATE see STEARIC ACID-2,3-EPOXYPROPYL
   ESTER

GLYCIDYL OCTADECENOATE see 2,3-EPOXYPROPYL OLEATE

GLYCIDYL OLEATE see 2,3-EPOXYPROPYL OLEATE

GLYCIDYL PHENYL ETHER see PHENYLGLYCYDYL ETHER

GLYCIDYL PROPENATE see 2,3-EPOXYPROPYL ACRYLATE

GLYCIDYL STEARATE see STEARIC ACID-2,3-EPOXYPROPYL ESTER

GLYCINE, N,N'-1,2-ETHANEDIYLBIS(N-(CARBOXYMETHYL)- (9CI) see
   ETHYLENE DIAMINE TETRAACETIC ACID

GLYCINE MUSTARD see GLYCINE NITROGEN MUSTARD

GLYCOL see ETHYLENE GLYCOL

GLYCOL ALCOHOL see ETHYLENE GLYCOL

GLYCOL BIS(HYDROXYETHYL) ETHER see TRIETHYLENE GLYCOL

GLYCOL BROMOHYDRIN see 2-BROMO ETHANOL

GLYCOL CHLOROHYDRIN see 2-CHLOROETHYL ALCOHOL

GLYCOL CYANOHYDRIN see HYDRACRYLONITRILE

GLYCOL DIACETATE see ETHYLENE DIACETATE

GLYCOL DICHLORIDE see ETHYLENE DICHLORIDE

GLYCOL DICHLORIDE see 1,2-DICHLORETHANE

GLYCOL DIFORMATE see ETHYLENE GLYCOL DIFORMATE

GLYCOLDINITRAAT (DUTCH) see ETHYLENE GLYCOL DINITRATE

GLYCOL DINITRATE see ETHYLENE GLYCOL DINITRATE

GLYCOL (DINITRATE DE) (FRENCH) see ETHYLENE GLYCOL DINI-
   TRATE

GLYCOL ETHYLENE ETHER see p-DIOXANE

GLYCOL ETHYL ETHER see DIETHYLENE GLYCOL

GLYCOLIC NITRILE see HYDROXYACETONITRILE

GLYCOLMETHYL ETHER see ETHYLENE GLYCOL METHYL ETHER

GLYCOL MONOBUTYL ETHER see BUTYL CELLOSOLVE

GLYCOL MONOBUTYL ETHERACETATE see 2-BUTOXYETHYL ACETATE

GLYCOLMONOCHLOORHYDRINE (DUTCH) see 2-CHLOROETHYL
   ALCOHOL

GLYCOL MONOCHLOROHYDRIN see 2-CHLOROETHYL ALCOHOL

GLYCOL MONOETHYL ETHER see CELLOSOLVE SOLVENT

GLYCOL MONOETHYL ETHER ACETATE see "CELLOSOLVE" ACETATE

GLYCOL MONOMETHYL ETHER see ETHYLENE GLYCOL METHYL
   ETHER

GLYCOL MONOMETHYL ETHER ACETATE see ETHYLENE GLYCOL
   MONOMETHYL ETHER ACETATE

GLYCOL MONOMETHYL ETHER ACRYLATE see METHYL CELLOSOLVE
   ACRYLATE

GLYCOL MONOPHENYL ETHER see PHENYL CELLOSOLVE

GLYCOL MONOSTEARATE see ETHYLENE GLYCOL STEARATE

GLYCOLONITRILE (8CI) see HYDROXYACETONITRILE

GLYCOL STEARATE see ETHYLENE GLYCOL STEARATE

GLYCOL SULFATE see ETHYLENE SULFATE

GLYCOL SULFITE see CYCLIC ETHYLENE SULFITE

GLYCONITRILE see HYDROXYACETONITRILE

GLYCYL ALCOHOL see GLYCERINE

GLYKOLDINITRAT (GERMAN) see ETHYLENE GLYCOL DINITRATE

GLYOXALIN see IMIDAZOLE

GLYOXYLALDEHYDE see GLYOXAL

GLYSANOL B see 1-AUROTHIO-D-GLUCOPYRANOSE

GM SULFATE see GENTAMYCIN SULFATE

GOHSENOLS see POLYVINYL ALCOHOL

GOLD BRONZE see COPPER

GOLD DUST see COCAINE

GOLDEN ANTIMONY SULFIDE see ANTIMONY PENTASULFIDE

GOLDEN YELLOW see 2,4-DINITRO-1-NAPHTHOL

GOLD FLAKE see GOLD

GOLD LEAF see GOLD

GOLD POWDER see GOLD

GOLD THIOGLUCOSE see 1-AUROTHIO-D-GLUCOPYRANOSE

GOTAMINE TARTRATE see ERGOTAMINE TARTRATE

GOYL see ACETPHENARSINE

GP 33679 see 2-HYDROXYINIPRAMINE

GP 38383 see IMPIRAMINE-N-OXIDE

GRAHAM'S SALT see SODIUM METAPHOSPHATE

GRAIN ALCOHOL see ETHANOL

GRANULAR ZINC see ZINC

GRANULATED SUGAR see SUCROSE

GRAPE SUGAR see D-GLUCOSE

GRAY ACETATE see CALCIUM ACETATE

GRAY AMBER see AMBERGRIS TINCTURE

GREEN NICKEL OXIDE see NICKEL MONOXIDE

GREEN OIL see ANTHRACENE

GREEN VITRIOL see IRON (II) SULFATE (1:1)

GREY ARSENIC see ARSENIC

GRISCOFULVIN see GRISOFULVIN

GRISEOFULVIN see GRISOFULVIN

(+)-GRISEOFULVIN see GRISOFULVIN

GRISEOFULVIN-FORTE see GRISOFULVIN

GRISEOFULVINUM see GRISOFULVIN

GROUNDNUT OIL see PEANUT OIL

GROUND RYANIA SPECISA(VAHL) STEMWOOD (ALKOLOID RYANO-
   DINE) see RYANIA

GS-13,798 see 8-(METHYLQUINOLYL)-N-METHYL CARBAMATE

G-STROPHANTHIN see OUABAIN

GTG see 1-AUROTHIO-D-GLUCOPYRANOSE

GUAICOL see GUAIACOL

GUANAZOLE see 3,5-DIAMINO-S-TRIAZOLE

GUANERAN see 2-AMINO-6-(1'-METHYL-4'-NITRO-5'-IMIDAZOLYL)
   MERCAPTOPURINE

GUANICAINE see PHENODIANISYL HYDROCHLORIDE

GUANIDINE, CYANO-, METHYLMERCURY DERIV. see METHYLMER-
   CURIC DICYANDIAMIDE

N-(2-GUANIDINO ETHYL)HEPTAMETHYLENIMINE SULFATE see
   GUANETHIDINE MONOSULFATE

GUANIOL see GERANIOL

GUANYL NITROSAMINO GUANYL TETRAZENE see TETRAZENE

1-GUANYL-4-NITROSAMINOGUANYLTETRAZENE see TETRAZENE

GUINEA GREEN B see ACID GREEN 3

GUM see POLYDIMETHYL SILOXANE

GUM ARABIC see ACACIA GUM

GUM ARABIC see ARABIC GUM

GUM CAMPHOR see CAMPHOR

GUM CHON 2 see CARRAGEEN

GUM CHROND see CARRAGEEN

GUM OPIUM see OPIUM

GUM OVALINE see ARABIC GUM

GUM SENEGAL see ARABIC GUM

GUNCOTTON see CELLULOSE TETRANITRATE

GUNPOWDER see EXPLOSIVES, LOW

GYNERGEN see ERGOTAMINE TARTRATE

GYNOCHROME see MERCUROCHROME

GYNOPLIX see ACETPHENARSINE

GYPSOPHILASAPOGENIN see GYPSOGENIN

GYPSOPHILASAPONIN see GYPSOGENIN

GYPSUM see CALCIUM (II) SULFATE DIHYDRATE (1:1:2)

GYPSUM STONE see CALCIUM (II) SULFATE DIHYDRATE (1:1:2)

"H" see HEROIN

14H see TETRADECANE

HADACIDINE see N-FORMYL-N-HYDROXYGLYCINE

HADACIN see N-FORMYL-N-HYDROXYGLYCINE

HAEMATITE see HEMATITE

HAFNIUM CHLORIDE OXIDE see HAFNIUM OXYCHLORIDE

HAFNIUM DICYCLOPENTADIENE DICHLORIDE see HAFNOCENE
DICHLORIDE

HAFNIUM METAL, DRY (DOT) see HAFNIUM

HAIRY see HEROIN

HALF-CYSTEINE see L-CYSTEINE

HALF-MUSTARD GAS see CHLOROETHYL ETHYL SULFIDE

HALF-MYLERAN see ETHYL METHANESULFONATE

HALITE see SODIUM CHLORIDE

HALLUCINOGEN see N,N-DIMETHYLACETAMIDE

HALOANISONE COMPOSITUM see FLUANISONE HYDROCHLORIDE

HALON see DICHLORODIFLUOROMETHANE

HALON 14 see CARBON TETRAFLUORIDE

HALON 1011 see BROMOCHLOROMETHANE

HALON 1301 see BROMO TRIFLUORO METHANE

HALOWAX 1013 see PENTACHLORONAPHTHALENE

HALOWAX 1014 see HEXACHLORONAPHTHALENE

HAPPY DUST see COCAINE

HARRY see HEROIN

HASACH see CANNABIS

HASHISH see CANNABIS

HASTELLOY C see COBALT ALLOY, CO,CR

HAYNES STELLITE 21 see COBALT ALLOY, CO,CR

9,10-H2 B(E)P see 9,10-DIHYDROBENZO(e)PYRENE

BETA-HCH see trans-alpha-BENZENEHEXACHLORIDE

DELTA HCH see delta-BENZENE HEXACHLORIDE

HC RED NO. 3 see 2-((4-AMINO-2-NITROPHENYL)AMINO)ETHANOL

HCl SALZ DES P-AMINO-SALICYLSAEURE-DIAETHYLAMINOAETHYL-
ESTER (GERMAN) see p-AMINOSALICYLIC ACID, 2-(DIETHYL-
AMINO)ETHYL ESTERHYDROCHLORIDE

HCl SALZ DES P-N-N-BUTYLAMINOSALICYLSAEUREDIAETHYLAMINO-
AETHYLESTER (GERMAN) see p-BUTYLAMINO SALICYLIC ACID,
2-(DIETHYLAMINO)ETHYL ESTER HYDROCHLORIDE

HEARTS see DL-AMPHETAMINE SULFATE

HEARTS see d-BENZEDRINE SULFATE

HEAVENLY BLUE see N,N-DIETHYLLYSERGAMIDE

HEAZLEWOODITE see NICKEL SUBSULFIDE

HED-HEPARIN see HEPARIN

HEKSAN (POLISH) see N-HEXANE

HELIOTRIDINE ESTER WITH LASIOCARPUM AND ANGELIC ACID see
LASIOCARPINE

HELIOTRON see HELIOTRINE

HELIOTROPIN see PIPERONAL

HEMICHOLINIUM BROMIDE see HEMICHOLINIUM-3-DIBROMIDE

HEMICHOLINIUM-3-BROMIDE see HEMICHOLINIUM-3-DIBROMIDE

HEMICHOLINIUM DIBROMIDE see HEMICHOLINIUM-3-DIBROMIDE

HENDECANE see UNDECANE

2-HENDECANONE see 2-UNDECANONE

HENU see 1-(2-HYDROXYETHYL)-1-NITROSOUREA

HEPAR CALCIS see CALCIUM SULFIDE

HEPARINATE see HEPARIN

HEPAR SULFUROUS see POTASSIUM SULFIDE (2:1)

HEPTABARBITAL see CYCLOHEPTENYL ETHYLBARBITURIC ACID

HEPTABARBITONE see CYCLOHEPTENYL ETHYLBARBITURIC ACID

HEPTACHLOOR (DUTCH) see HEPTACHLOR

1,4,5,6,7,8,8-HEPTACHLOOR-3A,4,7,7A-TETRAHYDRO-4,7-ENDO-
METHANO-INDEEN (DUTCH) see HEPTACHLOR

HEPTACHLOR see HEPTACHLOR

HEPTACHLORE (FRENCH) see HEPTACHLOR

HEPTACHLOR EPOXIDE see EPOXYHEPTACHLOR

3,4,5,6,7,8,8-HEPTACHLORODICYCLOPENTADIENE see HEPTACHLOR

3,4,5,6,7,8,8A-HEPTACHLORODICYCLOPENTADIENE see HEPTACHLOR

1,4,5,6,7,8,8-HEPTACHLORO-2,3-EPOXY-2,3,3A,4,7,7A-HEXAHY-
DRO-4,7-METHANOINDENE see EPOXYHEPTACHLOR

1,4,5,6,7,8,8-HEPTACHLORO-2,3-EPOXY-3A,4,7,7A-TETRAHYDRO-
4,7-METHANOINDAN see EPOXYHEPTACHLOR

2,3,4,5,6,7,7-HEPTACHLORO-1A,1B,5,5A,6,6A-HEXAHYDRO-2,5-
METHANO-2H-INDENO(1,2-B)OXIRENE see EPOXYHEPTACHLOR

1,4,5,6,7,8,8-HEPTACHLORO-3A,4,7,7A-TETRAHYDRO-4,7-ENDO-
METHANOINDENE see HEPTACHLOR

1,4,5,6,7,10,10-HEPTACHLORO-4,7,8,9-TETRAHYDRO-4,7-ENDO-
METHYLENEINDENE see HEPTACHLOR

1,4,5,6,7,8,8A-HEPTACHLORO-3A,4,7,7A-TETRAHYDRO-4,7-METHA-
NOINDANE see HEPTACHLOR

1(3A),4,5,6,7,8,8-HEPTACHLORO-3A(1),4,7,7A-TETRAHYDRO-4,7-
METHANOINDENE see HEPTACHLOR

1,4,5,6,7,8,8-HEPTACHLORO-3A,4,7,7A-TETRAHYDRO-4,7-METHA-
NOINDENE see HEPTACHLOR

3A,4,5,6,7,8,8-HEPTACHLORO-3A,4,7,7A-TETRAHYDRO-4,7-METHA-
NOINDENE see HEPTACHLOR

1,4,5,6,7,8,8-HEPTACHLORO-3A,4,7,7A-TETRAHYDRO-4,7-METHA-
NOL-1H-INDENE see HEPTACHLOR

1,4,5,6,7,10,10-HEPTACHLORO-4,7,8,9-TETRAHYDRO-4,7-METHY-
LENEINDENE see HEPTACHLOR

1,4,5,6,7,8,8-HEPTACHLORO-3A,4,7,7A-TETRAHYDRO-4,7-METHY-
LENE INDENE see HEPTACHLOR

1-HEPTADECANECARBOXYLIC ACID see STEARIC ACID

2-(8-HEPTADECENYL)-2-IMIDAZOLINE-1-ETHANOL see AMINE 220

HEPTADONE see METHADONE

HEPTAN (POLISH) see N-HEPTANE

HEPTANEN (DUTCH) see N-HEPTANE

HEPTANOL-2 see 2-HEPTANOL

HEPTANON see METHADONE

HEPTAN-3-ON (DUTCH, GERMAN) see 3-HEPTANONE

HEPTAN-3-ONE see 3-HEPTANONE

HEPTAN-4-ONE see 4-HEPTANONE

17-BETA-HEPTANOYLOXY-19-NOR-17-ALPHA-PREGNEN-20-YNONE
see NORETHISTERONE ENANTHATE

HEPTYL CARBINOL see N-OCTYL ALCOHOL

HEPTYL HYDRIDE see N-HEPTANE

HEPTYL MERCAPTAN see 1-HEPTANETHIOL

N-HEPTYLMERCAPTAN see 1-HEPTANETHIOL

2-HEPTYLOXYCARBANILIC ACID-2-(1-PIPERIDINYL)ETHYL ESTER HY-
DROCHLORIDE see PIPERIDINOETHYL-2-HEPTOXYPHENYLCARBA-
MOATE HYDROCHLORIDE

(2-(HEPTYLOXY)PHENYL)CARBAMIC ACID-2-(1-PIPERIDINYL)ETHYL
ESTER HYDROCHLORIDE see PIPERIDINOETHYL-2-HEPTOXYPHE-
NYLCARBAMOATE HYDROCHLORIDE

N-(2-(HEPTYLOXYPHENYLCARBAMOYLOXY)ETHYL)PIPERIDINIUM
CHLORIDE see PIPERIDINOETHYL-2-HEPTOXYPHENYLCARBA-
MOATE HYDROCHLORIDE

HERBITOX see PETROLEUM SPIRITS

HEROIEN see HEROIN

HEROIIN see HEROIN

HEROLAN see HEROIN

HETEROAUXIN see 1H-INDOLE-3-ACETIC ACID

HEXA(1-AZIRIDINYL)TRIPHOSPHOTRIAZINE see APHOLATE

HEXABARBITAL see EVIPAL

2,4,5,2',4',5'-HEXABROMOBIPHENYL see FIREMASTER FFI

HEXABROMOBIPHENYL (TECHNICAL GRADE) see FIREMASTER BP-6

HEXABUTYLDISTANNOXANE see BIS(TRIBUTYL TIN)OXIDE

HEXABUTYLDITIN see BIS(TRIBUTYL TIN)OXIDE

HEXACHLOR-AETHAN (GERMAN) see 1,1,1,2,2,2-HEXACHLOROETHANE

HEXACHLORAN see BENZENE HEXACHLORIDE-gamma isomer

GAMMA-HEXACHLORAN see BENZENE HEXACHLORIDE-gamma isomer

GAMMA-HEXACHLORANE see BENZENE HEXACHLORIDE-gamma isomer

HEXACHLORBENZOL (GERMAN) see HEXACHLOROBENZENE

HEXACHLOR-1,3-BUTADIEN (CZECH) see PERCHLOROBUTADIENE

HEXACHLORBUTADIENE see PERCHLOROBUTADIENE

2,3,4,4,5,6-HEXACHLORCYKLOHEXA-2,5-DIEN-1-ON (CZECH) see
HEXACHLORO-2,5-CYCLOHEXADIEN-1-ONE

HEXACHLORFENOL (CZECH) see HEXACHLORO-2,5-CYCLOHEXADIEN-
1-ONE

BETA-HEXACHLOROBENZENE see trans-alpha-BENZENEHEXACHLO-
RIDE

GAMMA-HEXACHLOROBENZENE see BENZENE HEXACHLORIDE-
gamma isomer

1,2,3,4,7,7-HEXACHLOROBICYCLO(2.2.1)HEPTEN-5,6-BIOXYMETHY-
LENESULFITE see BENZOEPIN
ALPHA,BETA-1,2,3,4,7,7-HEXACHLOROBICYCLO(2.2.1)-2-HEPTENE-
5,6-BISOXYMETHYLENE SULFITE see BENZOEPIN
1,1,2,3,4,4-HEXACHLORO-1,3-BUTADIENE see PERCHLOROBUTA-
DIENE
HEXACHLOROCYCLOHEXANE see BENZENE HEXACHLORIDE
ALPHA-HEXACHLOROCYCLOHEXANE see BENZENE HEXACHLORIDE-
alpha-isomer
BETA-HEXACHLOROCYCLOHEXANE see trans-alpha-BENZENEHEXA-
CHLORIDE
DELTA-HEXACHLOROCYCLOHEXANE see delta-BHC
GAMMA-HEXACHLOROCYCLOHEXANE see BENZENE HEXACHLORIDE-
gamma isomer
1,2,3,4,5,6-HEXACHLOROCYCLOHEXANE see BENZENE HEXACHLO-
RIDE
ALPHA-1,2,3,4,5,6-HEXACHLOROCYCLOHEXANE see BENZENE HEXA-
CHLORIDE-alpha-isomer
BETA-1,2,3,4,5,6-HEXACHLOROCYCLOHEXANE see trans-alpha-BEN-
ZENEHEXACHLORIDE
DELTA-1,2,3,4,5,6-HEXACHLOROCYCLOHEXANE see delta-BHC
1-ALPHA,2-ALPHA,3-ALPHA,4-BETA,5-ALPHA,6-BETA-HEXACHLORO-
CYCLOHEXANE see delta-BHC
1-ALPHA,2-ALPHA,3-BETA,4-ALPHA,5-ALPHA,6-BETA-HEXACHLORO-
CYCLOHEXANE see BENZENE HEXACHLORIDE-gamma isomer
1-ALPHA,2-ALPHA,3-BETA,4-BETA,5-ALPHA,6-BETA-HEXACHLORO-
CYCLOHEXANE see BENZENE HEXACHLORIDE-alpha-isomer
1-ALPHA,2-BETA,3-ALPHA,4-BETA,5-ALPHA,6-BETA-HEXACHLORO-
CYCLOHEXANE see trans-alpha-BENZENEHEXACHLORIDE
1,2,3,4,5,6-HEXACHLORO CYCLOHEXANE-DELTA ISOMER see delta-
BENZENE HEXACHLORIDE
1,2,3,4,5,6-HEXACHLOROCYCLOHEXANE, GAMMA-ISOMER see BEN-
ZENE HEXACHLORIDE-gamma isomer
1,2,3,4,5,6-HEXACHLOROCYCLOHEXANE (MIXTURE OF ISOMERS) see
BENZENE HEXACHLORIDE (MIXED ISOMERS)
HEXACHLOROCYCLOPENTADIENEDIMER see MIREX
1,2,3,4,5,5-HEXACHLORO-1,3-CYCLOPENTADIENE DIMER see MIREX
HEXACHLORODIBENZO-p-DIOXIN see HCDD
2,2',3,3',5,5'-HEXACHLORO-6,6'-DIHYDROXYDIPHENYLMETHANE
see HEXACHLOROPHENE
HEXACHLORO EPOXY OCTAHYDRO-ENDO,ENDO-DIMETHANO NAPH-
THALENE see ENDRIN
HEXACHLOROEPOXYOCTAHYDRO-ENDO,EXO-DIMETHANONAPH-
THALENE see DIELDRIN
HEXACHLOROETHANE (DOT) see 1,1,1,2,2,2-HEXACHLOROETHANE
HEXACHLOROETHYLENE see 1,1,1,2,2,2-HEXACHLOROETHANE
HEXACHLOROFEN (CZECH) see HEXACHLOROPHENE
1,2,3,4,10,10-HEXACHLORO-1,4,4a,5,8,8a-HEXAHYDRO-1,4,5,8-DI-
METHANONAPHTHALENE see ALDRIN
1,2,3,4,10,10-HEXACHLORO-1,4,4a,5,8,8a-HEXAHYDRO-1,4,5,8-
ENDO,ENDO-DIMETHANONAPHTHALENE see ISODRIN
HEXACHLOROHEXAHYDRO-ENDO-EXO-DIMETHANONAPHTHALENE see
ALDRIN
1,2,3,4,10,10-HEXACHLORO-1,4,4a,5,8,8a-HEXAHYDRO-1,4-ENDO-
EXO-5,8-DIMETHANONAPHTHALENE see ALDRIN
1,2,3,4,10,10-HEXACHLORO-1,4,4a,5,8,8a-HEXAHYDRO-EXO-1,4,
ENDO-5,8-DIMETHANONAPHTHALENE see ALDRIN
HEXACHLOROHEXAHYDROMETHANO see BENZOEPIN
6,7,8,9,10,10-HEXACHLORO-1,5,5a,6,9,9a-HEXAHYDRO-6,9-METH-
ANO-2,4,3-BENZODIOXATHIEPIN-3-OXIDE see BENZOEPIN
1,4,5,6,7,7-HEXACHLORO-5-NORBORNENE-2,3-DICARBOXYLIC ACID
see CHLORENDIC ACID
1,4,5,6,7,7-HEXACHLORO-5-NORBORNENE-2,3-DIMETHANOL CYCLIC
SULFITE see BENZOEPIN
HEXACHLOROPHANE see HEXACHLOROPHENE
HEXACHLOROPHEN see HEXACHLOROPHENE
HEXACHLOROPLATINIC (IV) ACID see CHLOROPLATINIC ACID
HEXACHLOROPROPYLENE see HEXACHLOROPROPENE
HEXADECANOIC ACID see PALMITIC ACID
HEXADECENE EPOXIDE see 1,2-EPOXYHEXADECANE

N-HEXADECOIC ACID see PALMITIC ACID
N-HEXADECYLAMINE see 1-HEXADECANAMINE
1-HEXADECYLPYRIDINIUM CHLORIDE see CEPACOL CHLORIDE
N-HEXADECYLPYRIDINIUM CHLORIDE see CEPACOL CHLORIDE
(1-HEXADECYL)TRIMETHYLAMMONIUM BROMIDE see HEXADECYL
TRIMETHYL AMMONIUM BROMIDE
N-HEXADECYLTRIMETHYLAMMONIUM BROMIDE see HEXADECYL
TRIMETHYL AMMONIUM BROMIDE
N-HEXADECYL-N,N,N-TRIMETHYLAMMONIUM BROMIDE see HEXA-
DECYL TRIMETHYL AMMONIUM BROMIDE
HEXAETHYL TETRAPHOSPHATE, LIQUID (DOT) see HEXAETHYL TET-
RAPHOSPHATE
HEXAFLUORENIUM DIBROMIDE see HEXAFLURONIUM BROMIDE
HEXAFLUOROACETONE SESQUIHYDRATE see HEXAFLUORO-2-PROPA-
NONE SESQUIHYDRATE
HEXAFLUOROISOBUTYRIC ACID METHYL ESTER see METHYL HEXA-
FLUOROISOBUTYRATE
HEXAFLUOROKIESELSAIURE (GERMAN) see SILICOFLUORIC ACID
HEXAFLUOROKIEZELZUUR (DUTCH) see SILICOFLUORIC ACID
1,1,1,3,3,3-HEXAFLUORO-2-PROPANOL see HEXAFLUOROISOPROPA-
NOL
HEXAFLUOROSILICATE(2-) DIHYDROGEN see SILICOFLUORIC ACID
HEXAFLUOROSILICATE (2-1) LEAD (II) SALT DIHYDRATE see
LEAD(II) FLUOROSILICATE
HEXAFLUOROSILICATE (2−), NICKEL see NICKEL (II) FLUOSILICATE
(1:1)
HEXAFLUORO VANADATE (3-) TRIAMMONIUM SALT see AMMONIUM
HEXAFLUOROVANADATE
HEXAFLUORURE DE SOUFRE (FRENCH) see SULFUR FLUORIDE
HEXAFLUOSILICIC ACID see SILICOFLUORIC ACID
HEXAFORM see HEXAMETHYLENETETRAMINE
HEXAHYDROANILINE see CYCLOHEXYLAMINE
HEXAHYDRO-2-AZEPINONE see HEXAHYDRO-2H-AZEPIN-2-ONE
HEXAHYDRO-2H-AZEPIN-2-ONE HOMOPOLYMER see POLYCAPROLAC-
TAM
(2-(HEXAHYDRO-1(2H)-AZOCINYL)ETHYL) GUANIDINE HYDROGEN
SULFATE see GUANETHIDINE MONOSULFATE
HEXAHYDROBENZENAMINE see CYCLOHEXYLAMINE
HEXAHYDROBENZENE see CYCLOHEXANE
HEXAHYDRO-P-CYMENE HYDROPEROXIDE see 1-ISOPROPYL-4-
METHYLCYCLOHEXANE HYDROPEROXIDE
HEXAHYDRO-1,4-DIAZINE see PIPERAZINE
HEXAHYDRO-3A,7A-DIMETHYL-4,7-EPOXYISOBENZOFURAN-1,3-
DIONE see CANTHARIDINE
HEXAHYDRO-3A,7A-DIMETHYL-4,7-EPOXYISOBENZOFURAN-1,3-
DIONE see CANTHARIDES CAMPHOR
2,2,4,4,6,6-HEXAHYDRO-2,2,4,4,6,6-HEXAKIS(1-AZIRIDINYL)-1,
3,5,2,4,6-TRIAZATRIPHOSPHORINE see APHOLATE
1,2,3,4,5,6-HEXAHYDRO-3-(3-METHYL-2-BUTENYL)-6,11-DIMETHYL-
8-HYDROXY-2,6-METHANO-3-BENZAZOCINE see 2-(3,3-DIMETHY-
LALLYL)CYCLAZOCINE
1,3,4,9,10,10A-HEXAHYDRO-11-METHYL-2H-10,4A-IMINOETHANO-
PHENANTHREN-6-OL, DL-MIXTURE see N-METHYL-3-HYDROXY-
MORPHINAN
HEXAHYDRO-1-NITROSO-1H-AZEPINE see N-NITROSOHEXAHYDRO-
AZEPINE
HEXAHYDROPHENOL see CYCLOHEXANOL
HEXAHYDROPYRAZINE see PIPERAZINE
HEXAHYDROPYRIDINE see PIPERIDINE
HEXAHYDROTOLUENE see CYCLOHEXYLMETHANE
HEXAHYDRO-1,3,5-TRINITROSO-s-TRIAZINE see HEXAHYDRO-1,3,5-
s-TRIAZINE
HEXAHYDRO-1,3,5-TRINITROSO-1,3,5-TRIAZINE see HEXAHYDRO-
1,3,5-s-TRIAZINE
HEXAHYDRO-1,3,5-TRINITRO-1,3,5-TRIAZIN (GERMAN) see TRI-
METHYLENETRINITRAMINE
HEXAHYDRO-1,3,5-TRINITRO-s-TRIAZINE see TRIMETHYLENETRINI-
TRAMINE
HEXAHYDRO-1,3,5-TRINITRO-1,3,5-TRIAZINE see TRIMETHYLENE-
TRINITRAMINE

2,2,4,4,6,6-HEXAKIS(1-AZIRIDINYL) CYCLOTRIPHOSPHAZA-1,3,5-TRIENE see APHOLATE

2,2,4,4,6,6-HEXAKIS(1-AZIRIDINYL)-2,2,4,4,6,6-HEXAHYDRO-1,3,5,2,4,6-TRIAZATRIPHOSPHORINE see APHOLATE

HEXAKIS(AZIRIDINYL)PHOSPHOTRIAZINE see APHOLATE

HEXAKIS(DIHYDROGEN PHOSPHATE) MYO-INOSITOL see PHYTIC ACID

HEXAKIS(HYDROXYMETHYL)MELAMINE see HEXA(HYDROXY-METHYL)MELAMINE

HEXAKIS(MU-ACETATO)-MU(SUP 4)-OXOTETRABERYLLIUM see BERYLLIUM OXYACETATE

HEXAKIS(MU-ACETATO-O:O'))-MU(SUP 4)-OXOTETRABERYLLIUM see BERYLLIUM OXYACETATE

HEXALDEHYDE (DOT) see 1-HEXANAL

HEXAMETHIONIUM BROMIDE see HEXAMETHONIUM BROMIDE

HEXAMETHONIUM DIBROMIDE see HEXAMETHONIUM BROMIDE

HEXAMETHONIUM IODIDE see HEXAMETHONIUM DIIODIDE

HEXAMETHYLDISTANNANE see HEXAMETHYLDITIN

HEXAMETHYLENAMINE see HEXAMETHYLENETETRAMINE

HEXAMETHYLENE see CYCLOHEXANE

HEXAMETHYLENEBIS(DIMETHYL-9-FLUORENYLAMMONIUM BROMIDE) see HEXAFLURONIUM BROMIDE

HEXAMETHYLENEBIS(FLUOREN-9-YLDIMETHYLAMMONIUM BROMIDE) see HEXAFLURONIUM BROMIDE

HEXAMETHYLENE BIS(9-FLUORENYLDIMETHYLAMMONIUM)DIBRO-MIDE see HEXAFLURONIUM BROMIDE

HEXAMETHYLENEBIS(TRIMETHYLAMMONIUM) BROMIDE see HEXA-METHONIUM BROMIDE

HEXAMETHYLENEBIS(TRIMETHYLAMMONIUM IODIDE) see HEXA-METHONIUM DIIODIDE

1,6-HEXAMETHYLENE DIISOCYANATE see 1,6-DIISOCYANATOHEX-ANE

HEXAMETHYLENE-1,6-DIISOCYANATE see 1,6-DIISOCYANATOHEX-ANE

2-(beta-HEXAMETHYLENIMINOAETHYL)CYCLOHEXANON-2-CARBON-SAUREBENZYLESTER-HYDROCHLORIDE (GERMAN) see 1-(2-HEXA-METHYLENEIMINOETHYL)-2-OXOCYCLOHEXANECARBOXYLIC ACID BENZYL ESTER HYDROCHLORIDE

HEXAMETHYLENTETRAMIN (GERMAN) see HEXAMETHYLENETETRA-MINE

2,3,3,4,4,5-HEXAMETHYL-2-HEXANETHIOL see t-DODECANETHIOL

HEXAMETHYLOLMELAMIN (CZECH) see HEXA(HYDROXYMETHYL)-MELAMINE

HEXAMETHYLOLMELAMINE see HEXA(HYDROXYMETHYL)MELAMINE

HEXAMETHYLPHOSPHORIC ACID TRIAMIDE see HEXAMETHYL PHOS-PHORAMIDE

HEXAMETHYLPHOSPHORIC TRIAMIDE see HEXAMETHYL PHOSPHOR-AMIDE

N,N,N,N,N,N-HEXAMETHYLPHOSPHORIC TRIAMIDE see HEXAMETHYL PHOSPHORAMIDE

HEXAMETHYLPHOSPHOROTRIAMIDE see HEXAMETHYL PHOSPHOR-AMIDE

HEXAMETHYLPHOSPHOTRIAMIDE see HEXAMETHYL PHOSPHOR-AMIDE

HEXAMETHYL P-ROSANILINE HYDROCHLORIDE see ANILINE VIOLET

HEXAMETHYL VIOLET see ANILINE VIOLET

HEXAMIDINE (THE ANTISPASMODIC) see 2-DESOXY PHENOBARBITAL

HEXAMINE see HEXAMETHYLENETETRAMINE

1-HEXANAMINE see HEXYLAMINE

HEXANAPHTHENE see CYCLOHEXANE

HEXANE (DOT) see N-HEXANE

HEXANEDIAMIDE (9CI) see ADIPAMIDE

HEXANEDINITRILE see ADIPONITRILE

1,6-HEXANEDIOIC ACID see ADIPIC ACID

HEXANEDIOIC ACID DINITRILE see ADIPONITRILE

1,6-HEXANEDIOL DIISOCYANATE see 1,6-DIISOCYANATOHEXANE

3,3'-(1,6-HEXANEDIYLBIS-((METHYLIMINO)CARBONYL)OXY)BIS(1-METHYLPYRIDINIUMDIBROMIDE) see DISTIGMINE BROMIDE

HEXANEN (DUTCH) see N-HEXANE

HEXANITRODIPHENYLSULFIDE see BIS(TRINITROPHENYL)SULFIDE

HEXANITROL see MANNITOL HEXANITRATE

N-HEXANOL see N-HEXYL ALCOHOL

HEXANONE ISOXIME see HEXAHYDRO-2H-AZEPIN-2-ONE

HEXANONISOXIM (GERMAN) see HEXAHYDRO-2H-AZEPIN-2-ONE

HEXANOYLETHYLENEIMINE see 1-HEXANOYLAZIRIDINE

17-alpha-HEXANOYLOXYPREGN-4-ENE-3,20-DIONE see HYDROXY-PROGESTERONE CAPROATE

1,1,1,3,3,3-HEXAPHENYLDISTANNTHIANE see BIS(TRIPHENYL TIN)-SULFIDE

HEXAPLAS M/1B see DIISOBUTYL PHTHALATE

HEXAZANE see PIPERIDINE

HEX-2-ENAL see 2-HEXENAL

HEX-2-EN-1-AL see 2-HEXENAL

HEXENAL (BARBITURATE) see EVIPAL

GAMMA-HEXENOLACTONE see PARASCORBIC ACID

2-HEXEN-5,1-OLIDE see PARASCORBIC ACID

HEXILMETHYLENAMINE see HEXAMETHYLENETETRAMINE

HEXMETHYLPHOSPHORAMIDE see HEXAMETHYL PHOSPHORAMIDE

HEXOBARBITAL see EVIPAL

HEXOBARBITONE see EVIPAL

HEXOGEEN (DUTCH) see TRIMETHYLENETRINITRAMINE

HEXOGEN (EXPLOSIVE) see TRIMETHYLENETRINITRAMINE

1,6-HEXOLACTAM see HEXAHYDRO-2H-AZEPIN-2-ONE

HEXOLITE see TRIMETHYLENETRINITRAMINE

HEXOPHENE see HEXACHLOROPHENE

sec-HEXYL ALCOHOL see 2-ETHYLBUTANOL

N-HEXYLAMINE see HEXYLAMINE

4-HEXYL-1,3-DIHYDROXYBENZENE see HEXYLRESORCINOL

HEXYLENIC ALDEHYDE see 2-HEXENAL

HEXYL MERCAPTAN see 1-HEXANETHIOL

4-HEXYLRESORCINOL see HEXYLRESORCINOL

P-HEXYLRESORCINOL see HEXYLRESORCINOL

4-N-HEXYLRESORCINOL see HEXYLRESORCINOL

3-HEXYL-7,8,9,10-TETRAHYDRO-6,6,9-TRIMETHYL-6H-DIBEN-ZO(B,D)PYRAN-1-OL see 3-HOMOTETRA HYDRO CANNIBINOL

HFCB see HEXAFLUORODICHLOROBUTENE

HIDACID FLUORESCEIN see FLUORESCEIN

HIDACID URANINE see FLUORESCEIN SODIUM

HIDRO-COLISONA see CORTISOL

HI-FLASH NAPHTHAETHYLEN see BENZIN

HISTAMETHINE see HISTAMETHIZINE

HISTAMETIZINE see HISTAMETHIZINE

HISTAMETIZYNE see HISTAMETHIZINE

HNT see 2-HYDRAZINO-4-(5-NITRO-2-FURYL)THIAZOLE

HOCH see FORMALDEHYDE

HOECHST PA 190 see POLYETHYLENE

HOG see ANGEL DUST

HOME HEATING OIL NO. 2 see FUEL OIL

HOMOANISIC ACID see p-METHOXYPHENYLACETIC ACID

HOMOCHLORCYCLIZINE DIHYDROCHLORIDE see 1-(p-CHLORO-alpha-PHENYLBENZYL)HEXAHYDRO-4-METHYL-1H-1,4-DIAZEPINEDIHY-DROCHLORIDE

HOMOCHLOROCYCLIZINE DIHYDROCHLORIDE see 1-(p-CHLORO-alpha-PHENYLBENZYL)HEXAHYDRO-4-METHYL-1H-1,4-DIAZEPINEDIHYDROCHLORIDE

HOMOTRYPTAMINE HYDROCHLORIDE see 3-(gamma-AMINOPROPYL)-INDOLEHYDROCHLORIDE

HOMOVERATRYLAMINE see 3,4-DIMETHOXYDOPAMINE

HOPCIDE see O-CHLOROPHENYL METHYLCARBAMATE

HORMEX ROOTING POWDER see 1H-INDOLE-3-BUTANOIC ACID

HORMODIN see 1H-INDOLE-3-BUTANOIC ACID

HORMOFLAVEINE see PROGESTERONE

HORMOLUTON see PROGESTERONE

HORSE see HEROIN

BETA-HPN see HYDRACRYLONITRILE

HTH see HYPOCHLOROUS ACID, CALCIUM SALT

1-alpha-H,5-alpha-H-TROPANE-2-beta-CARBOXYLIC ACID, 3-beta-HYDROXY-,METHYL ESTER, BENZOATE see COCAINE

1-alpha-H,5-alpha-H-TROPANE-2-beta-CARBOXYLIC ACID, 3-beta-
 HYDROXY-,METHYL ESTER, BENZOATE (ESTER), HYDROCHLORIDE
 see COCAINE CHLORIDE
HUILE D'ANILINE (FRENCH) see ANILINE
HUILE DE CAMPHRE (FRENCH) see CAMPHOR
HUILE DE FUSEL (FRENCH) see FUSEL OIL
HYAMINE 1622 see BENZETHONIUM CHLORIDE
HYAMINE 2389 see DODECYL-p-TOLYL TRIMETHYL AMMONIUM
 CHLORIDE
HYAMINE 3500 see ALKYL DIMETHYLBENZYL AMMONIUM CHLORIDE
HYCANTHONE MESYLATE see HYCANTHONE METHANESULFONATE
HYCANTHONE MONOMETHANESULPHONATE see HYCANTHONE METH-
 ANESULFONATE
HY-CHLOR see HYPOCHLOROUS ACID, CALCIUM SALT
HYDANTOIN see DILANTIN
HYDRACRYLIC ACID,BETA,LACTONE see 2-OXETANONE
HYDRACRYLONITRILE ACRYLATE see ACRYLIC ACID ESTER WITH
 HYDRACRYLONITRILE
HYDRALAZINE CHLORIDE see HYDRALAZINE HYDROCHLORIDE
HYDRALAZINE MONOHYDROCHLORIDE see HYDRALAZINE HYDRO-
 CHLORIDE
HYDRARGYRUM BIJODATUM (GERMAN) see MERCURY (II) IODIDE
HYDRATED LIME see CALCIUM HYDROXIDE
HYDRAZIDE see ISONICOTINIC ACID HYDRAZIDE
HYDRAZINE, ANHYDROUS (DOT) see HYDRAZINE
HYDRAZINE BASE see HYDRAZINE
HYDRAZINECARBOXAMIDE MONOHYDROCHLORIDE see SEMICARBA-
 ZIDE HYDROCHLORIDE
HYDRAZINE SULPHATE see HYDRAZINE SULFATE (1:1)
HYDRAZINOBENEZENE see PHENYLHYDRAZINE
1-HYDRAZINO-2-PHENYLETHANE see 2-PHENYLETHYLHYDRAZINE
1-HYDRAZINO-2-PHENYLETHANE HYDROGEN SULPHATE see PHENYL-
 ETHYLHYDRAZINE SULPHATE
HYDRAZINOPHTHALAZINE see 1(2H)-PHTHALAZINONE HYDRAZONE
1-HYDRAZINOPHTHALAZINE see 1(2H)-PHTHALAZINONE HYDRAZONE
1-HYDRAZINOPHTHALAZINE HYDROCHLORIDE see HYDRALAZINE HY-
 DROCHLORIDE
1-HYDRAZINOPHTHLAZINE MONOHYDROCHLORIDE see HYDRALAZINE
 HYDROCHLORIDE
HYDRAZOBENZEN (CZECH) see HYDRAZOBENZENE
HYDRAZOETHANE see 1,2-DIETHYLHYDRAZINE
HYDRAZOMETHANE see METHYLHYDRAZINE
HYDRIODIC ACID (DOT) see HYDRIODIC ACID
HYDROBROMIC ACID MONOAMMONIATE see AMMONIUM BROMIDE
HYDROCHINON (CZECH, POLISH) see HYDROQUINONE
HYDROCHLORIC ACID (DOT) see HYDROCHLORIC ACID
HYDROCHLORIC ACID, ANHYDROUS (DOT) see HYDROGEN CHLORIDE
HYDROCHLORIDE see HYDROCHLORIC ACID
HYDROCINNAMIC ALDEHYDE see HYDROCINNAMALDEHYDE
11-BETA-HYDROCORTISONE see CORTISOL
DELTA(SUP 1)-HYDROCORTISONE see PREDONIN
HYDROCORTISONE FREE ALCOHOL see CORTISOL
HYDROCORTISYL see CORTISOL
HYDROCORTONE see CORTISOL
HYDROCYANIC ACID, LIQUEFIED (DOT) see HYDROCYANIC ACID
HYDROCYANIC ACID, POTASSIUM SALT (SOLN) see POTASSIUM CY-
 ANIDE (soln)
HYDROCYANIC ACID, SODIUM SALT see SODIUM CYANIDE
HYDROCYANIC ACID, SOLUTION (DOT) see HYDROCYANIC ACID (SO-
 LUTION)
HYDROCYANIC ETHER see PROPIONONITRILE
HYDRO-DIURIL see 6-CHLORO-3,4-DIHYDRO-2H-1,2,4-BENZOTHIADI-
 AZINE-7-SULFONAMIDE- 1,1-DIOXIDE
HYDROFLUOBORIC ACID see TETRAFLUOROBORATE
HYDROFLUORIC ACID, ANHYDROUS (DOT) see HYDROFLUORIC
 ACID
HYDROFLUORIC ACID SOLUTION see HYDROFLUORIC ACID (SOLU-
 TION)
HYDROFLUOSILICIC ACID see SILICOFLUORIC ACID

HYDROFURAMIDE see FURFURAMIDE
HYDROFURAN see TETRAHYDROFURAN
HYDROGEN ANTIMONIDE see STIBINE
HYDROGEN ARSENIDE see ARSINE
HYDROGEN AZIDE see HYDRAZOIC ACID
HYDROGEN BROMIDE see HYDROBROMIC ACID
HYDROGEN CARBOXYLIC ACID see FORMIC ACID
HYDROGEN CYANAMIDE see CYANAMIDE
HYDROGEN CYANIDE see HYDROCYANIC ACID
HYDROGENE SULFURE (FRENCH) see HYDROGEN SULFIDE
HYDROGEN FLUORIDE (DOT) see HYDROFLUORIC ACID
HYDROGEN HEXACHLOROPLATINATE(4+) see CHLOROPLATINIC
 ACID
HYDROGEN HEXAFLUOROPHOSPHATE see HEXAFLUOROPHOSPHORIC
 ACID
HYDROGEN HEXAFLUOROSILICATE see SILICOFLUORIC ACID
HYDROGEN IODIDE see HYDRIODIC ACID
HYDROGEN IODIDE SOLUTION (DOT) see HYDRIODIC ACID
HYDROGEN NITRATE see NITRIC ACID
HYDROGEN PEROXIDE SOLUTION (DOT) see HYDROGEN PEROXIDE,
 90%
HYDROGEN PHOSPHIDE see PHOSPHINE
HYDROGEN SULFIDE (DOT) see HYDROGEN SULFIDE
HYDROL SW see TRITON X
HYDROMIREX see 8-MONOHYDRO MIREX
HYDROMORPHONE see DIHYDROMORPHINONE
HYDROMORPHONE HYDROCHLORIDE see DILAUDID
HYDRONITRIC ACID see HYDRAZOIC ACID
HYDROPEROXIDE, ACETYL see PEROXYACETIC ACID
HYDROPEROXYDE DE BUTYLE TERTIAIRE (FRENCH) see 6-BUTYL HY-
 DROPEROXIDE
HYDROPEROXYDE DE CUMENE (FRENCH) see ISOPROPYLBENZENE HY-
 DROPEROXIDE
HYDROPEROXYDE DE CUMYLE (FRENCH) see ISOPROPYLBENZENE
 HYDROPEROXIDE
HYDROQUINOL see HYDROQUINONE
M-HYDROQUINONE see RESORCINOL
O-HYDROQUINONE see PYROCATECHOL
P-HYDROQUINONE see HYDROQUINONE
ALPHA-HYDROQUINONE see HYDROQUINONE
HYDROQUINONE BENZYL ETHER see AGERITE
HYDROQUINONE MONOBENZYL ETHER see AGERITE
HYDROQUINONE MUSTARD see 2,5-BIS(BIS-(2-CHLOROETHYL)-
 AMINOMETHYL)HYDROQUINONE
HYDRORUBEANIC ACID see DITHIOOXAMIDE
HYDROSARPAN see AJMALICINE
HYDROSCINE HYDROBROMIDE see HYOSCINE HYDROBROMIDE
HYDROSILICOFLUORIC ACID see SILICOFLUORIC ACID
N-HYDROXY-4-ACETAMIDOBIPHENYL see N-ACETYL-4-BIPHENYLHY-
 DROXYLAMINE
N-4-(N-HYDROXYACETAMIDO)BIPHENYL see N-ACETYL-4-BIPHENYL-
 HYDROXYLAMINE
N-HYDROXY-4-ACETAMIDODIPHENYL see N-ACETYL-4-BIPHENYLHY-
 DROXYLAMINE
2-(N-HYDROXYACETAMIDO)FLUORENE see N-HYDROXY-N-ACETYL-2-
 AMINOFLUORENE
N-HYDROXY-2-ACETAMIDOFLUORENE see N-HYDROXY-N-ACETYL-2-
 AMINOFLUORENE
4-HYDROXYACETANILIDE see 4'-HYDROXYACETANILIDE
P-HYDROXYACETANILIDE see 4'-HYDROXYACETANILIDE
2-HYDROXYACETONITRILE see HYDROXYACETONITRILE
N-HYDROXY-4-ACETYLAMINOBIPHENYL see N-ACETYL-4-BIPHENYL-
 HYDROXYLAMINE
N-HYDROXY-2-ACETYLAMINOFLUORENE see N-HYDROXY-N-ACETYL-
 2-AMINOFLUORENE
3-(2-HYDROXY-1-ADAMANTYL)-N-METHYLPROPLYAMINE HYDRO-
 CHLORIDE see 1-(3-METHYLAMINOPROPYL)-2-ADAMANTANOL
 HYDROCHLORIDE
4-HYDROXYAFLATOXIN Bl see AFLATOXIN Ml

1-HYDROXY-4-AMINOANTHRAQUINONE see 1-AMINO-4-HYDROXYAN-
THRAQUINONE

2-HYDROXY-4-AMINOBENZOIC ACID see 4-AMINOSALICYLIC ACID

5-HYDROXY-3-(BETA-AMINOETHYL)INDOLE see 3-(2-AMINOETHYL)-
INDOL-5-OL

3-HYDROXY-5-AMINOMETHYLISOXAZOLE-AGARIN see 5-AMINO-
METHYL-3-ISOXYZOLE

N-HYDROXY-2-AMINONAPHTHALENE see 2-NAPHTHYLHYDROXYL-
AMINE

17-BETA-HYDROXY-4-ANDROSTEN-3-ONE see TESTOSTERONE

17-BETA-HYDROXYANDROST-4-EN-3-ONE see TESTOSTERONE

17-HYDROXY-(17-BETA)-ANDROST-4-EN-3-ONE see TESTOSTERONE

17-BETA-HYDROXY-DELTA(SUP 4)ANDROSTEN-3-ONE see TESTOS-
TERONE

3-HYDROXYANILINE see m-AMINOPHENOL

4-HYDROXYANILINE see 4-AMINOPHENOL

P-HYDROXYANILINE see 4-AMINOPHENOL

2-HYDROXYANISOLE see GUAIACOL

O-HYDROXYANISOLE see GUAIACOL

3-HYDROXYANTHRANILIC ACID see 2-AMINO-3-HYDROXYBENZOIC
ACID

3-HYDROXY-ANTHRANILSAEURE (GERMAN) see 2-AMINO-3-HY-
DROXYBENZOIC ACID

4-HYDROXY-1-ANTHRAQUINONYLAMINE see 1-AMINO-4-HYDROXY-
ANTHRAQUINONE

P-HYDROXYAZOBENZENE see 4-HYDROXYAZOBENZENE

4-HYDROXY-3,4'-AZODI-1-NAPHTHALENESULFONIC ACID, DISODIUM
SALT see C.I. FOOD RED 3

4-HYDROXY-3,4'-AZODI-1-NAPHTHALENESULPHONIC ACID, DISO-
DIUM SALT see C.I. FOOD RED 3

2-HYDROXY-1,1'-AZONAPHTHALENE-3,6,4'-TRISULFONIC ACID TRI-
SODIUM SALT see FOOD RED 2

O-HYDROXYBENZAMIDE see SALICYLAMIDE

HYDROXYBENZENE see PHENOL

3-HYDROXY BENZISOTHIAZOL-S,S-DIOXIDE see 1,2-BENZISOTHIAZOL-
3(2H)-ONE-1,1-DIOXIDE

2-HYDROXYBENZOIC ACID see SALICYLIC ACID

O-HYDROXYBENZOIC ACID see SALICYLIC ACID

O-HYDROXYBENZOIC ACID, METHYL ESTER see METHYL SALICYLATE

O-HYDROXYBENZOIC SODIUM SALT see SODIUM SALICYLATE

2-HYDROXYBENZO(A)PYRENE see BENZO(a)PYREN-2-OL

3-HYDROXYBENZO(A)PYRENE see BENZO(a)PYREN-3-OL

6-HYDROXYBENZO(A)PYRENE see BENZO(a)PYREN-6-OL

7-HYDROXYBENZO(A)PYRENE see BENZO(a)PYREN-7-OL

HYDROXYBENZOPYRIDINE see 8-QUINOLINOL

8-HYDROXY-3,4-BENZPYRENE see BENZO(a)PYREN-3-OL

2-HYDROXYBIFENYL (CZECH) see 2-BIPHENYLOL

2-HYDROXYBIPHENYL see 2-BIPHENYLOL

4-HYDROXYBIPHENYL see 4-BIPHENYLOL

O-HYDROXYBIPHENYL see 2-BIPHENYLOL

P-HYDROXYBIPHENYL see 4-BIPHENYLOL

3-HYDROXYBUTANAL see ACETALDOL

1-HYDROXYBUTANE see N-BUTYL ALCOHOL

2-HYDROXYBUTANE see sec-BUTYL ALCOHOL

4-HYDROXYBUTANOIC ACID LACTONE see 4-BUTANOLIDE

3-HYDROXYBUTANOIC ACID, BETA-LACTONE see beta-BUTYROLAC-
TONE

3-HYDROXY-2-BUTANONE see ACETOIN

1-HYDROXY-4-TERT-BUTYLBENZENE see 4-tert-BUTYLPHENOL

4-HYDROXYBUTYLBUTYLNITROSAMINE see BUTANOL-4-BUTYL
NITROSAMINE

3-HYDROXYBUTYRALDEHYDE see ACETALDOL

BETA-HYDROXYBUTYRALDEHYDE see ACETALDOL

GAMMA-HYDROXYBUTYRIC ACID CYCLIC ESTER see 4-BUTANOLIDE

HYDROXYBUTYRIC ACID LACTONE see beta-BUTYROLACTONE

4-HYDROXYBUTYRIC ACID, GAMMA-LACTONE see 4-BUTANOLIDE

GAMMA-HYDROXYBUTYROLACTONE see 4-BUTANOLIDE

2-HYDROXYCAMPHANE see BORNEOL

3-HYDROXY-4-CARBOXYANILINE see 4-AMINOSALICYLIC ACID

2-HYDROXY-5-CHLORO-N-(2-CHLORO-4-NITROPHENYL)BENZAMIDE
see 2',5-DICHLORO-4'-NITROSALICYLANILIDE

4-(4-HYDROXY-4'-CHLORO-4-PHENYLPIPERIDINO)-4'-FLUOROBUTY
ROPHENONE see gamma-(4-(p-CHLORPHENYL)-4-HYDROXPI-
PERIDINO)-p-FLUORBUTYRO-PHENONE

HYDROXYCHOLECALCIFEROL see 1-HYDROXYCHOLECALCIFEROL

1-ALPHA-HYDROXYCHOLECALCIFEROL see 1-HYDROXYCHOLECALCI-
FEROL

3-BETA-HYDROXYCHOLESTANE see DIHYDROCHOLESTEROL

3-beta-HYDROXYCHOLEST-5-ENE see CHOLESTEROL

O-HYDROXYCINNAMIC ACID LACTONE see COUMARIN

17-HYDROXYCORTICOSTERONE see CORTISOL

11-beta-HYDROXYCORTISONE see CORTISOL

3-HYDROXYCROTONIC ACIDMETHYL ESTER DIMETHYLPHOSPHATE
see MEVINPHOS

3-HYDROXYCYCLOHEXADIEN-1-ONE see RESORCINOL

2-HYDROXY-P-CYMENE see CARVACROL

3-HYDROXY-P-CYMENE see THYMOL

14-HYDROXYDAUNOMYCIN see ADRIAMYCIN

14'-HYDROXYDAUNOMYCIN see ADRIAMYCIN

14-HYDROXYDAUNORUBICINE see ADRIAMYCIN

HYDROXYDE DE POTASSIUM (FRENCH) see POTASSIUM HYDROXIDE

HYDROXYDE DE SODIUM (FRENCH) see SODIUM HYDROXIDE

HYDROXYDE DE TETRAMETHYLAMMONIUM (FRENCH) see TETRA-
METHYLAMMONIUM HYDROXIDE

HYDROXYDE DE TRIPHENYL-ETAIN (FRENCH) see HYDROXY TRI-
PHENYL STANNANE

3-ALPHA-HYDROXY-4-BETA,15-DIACETOXY-8-ALPHA(3-METHYLBU-
TYRYLOXY)-12,13-EPOXY-TRICHLOTHEC-9-ENE see 8-(3-
METHYLBUTYRYLOXY) DIACETOXYSCIRPENOL

3-HYDROXY-4,15-DIACETOXY-8-(3-METHYLBUTYRYLOXY)-12,13-
EPOXY-delta (SUP 9)-TRICHOTHECENE see FUSARIOTOXIN T 2

4-HYDROXY-3,5-DIBROMOBENZONITRILE see 3,5-DIBROMO-4-HY-
DROXYBENZONITRILE

4-HYDROXY-3,5-DI-TERT-BUTYLTOLUENE see BHT (FOOD GRADE)

P-HYDROXYDIFENYLAMIN (CZECH) see p-ANILINOPHENOL

PARA-HYDROXYDIFENYLAMIN (CZECH) see p-ANILINOPHENOL

2-(HYDROXY)-5-(2,4-DIFLUOROPHENYL)BENZOIC ACID see 5-(2,4-
DIFLUOROPHENYL)SALICYLIC ACID

14-HYDROXYDIHYDROCODEINONE see PERCODAN

8-HYDROXY-5,7-DIIODOQUINOLINE see DIIODOHYDROXYQUIN

3-HYDROXY-N,N-DIMETHYL-CIS-CROTONAMIDE DIMETHYL PHOS-
PHATE see 3-HYDROXYDIMETHYL CROTONAMIDE DIMETHYL
PHOSPHATE

2'-HYDROXY-5,9-DIMETHYL-2-(3,3-DIMETHYLALLYL)-6,7-BENZO-
MORPHAN see 2-(3,3-DIMETHYLALLYL)CYCLAZOCINE

DL-2'-HYDROXY-5,9-DIMETHYL-2-(3,3-DIMETHYLALLYL)-6,7-BEN-
ZOMORPHAN see 2-(3,3-DIMETHYLALLYL)CYCLAZOCINE

3-HYDROXY-4,5-DIMETHYOL-ALPHA-PICOLINE see PYRIDOXOL

3-HYDROXY-4,5-DIMETHYOL-ALPHA-PICOLINE HYDROCHLORIDE see
PYRIDOXOL HYDROCHLORIDE

3-HYDROXY-1,1-DIMETHYLPYRROLIDINIUM BROMIDE-alpha-CYCLO-
PENTYLMANDELATE see GLYCOPYRRONIUM BROMIDE

5-HYDROXY-N,N-DIMETHYLTRYPTAMINE see 3-(2-DIMETHYLAMINO-
ETHYL)-5-INDOLOL

1-HYDROXY-2,4-DINITROBENZENE see 2,4-DINITROPHENOL

BETA-(2-HYDROXY-3,5-DINITROPHENYL)BUTANE ACETATE see
O-ACETYL-2-sec-BUTYL-4,6-DINITROPHENOL

2-HYDROXYDIPHENYL see 2-BIPHENYLOL

4-HYDROXYDIPHENYL see 4-BIPHENYLOL

O-HYDROXYDIPHENYL see 2-BIPHENYLOL

P-HYDROXYDIPHENYL see 4-BIPHENYLOL

4-HYDROXYDIPHENYLAMINE see p-ANILINOPHENOL

P-HYDROXYDIPHENYLAMINE see p-ANILINOPHENOL

4,4'-HYDROXY-gamma,delta-DIPHENYL-beta,delta-HEXADIENE see
DEHYDROSTILBESTROL

ALPHA-HYDROXYDIPHENYLMETHANE-BETA-DIMETHYLAMINOETHYL
ETHER HYDROCHLORIDE see BENADRYL HYDROCHLORIDE

2-HYDROXYDIPHENYL SODIUM see 2-BIPHENYLOL, SODIUM SALT

3-HYDROXY-1,2-EPOXYPROPANE see 2,3-EPOXY-1-PROPANOL

1-M-HYDROXY-ALPHA-((METHYL AMINO)METHYL)-BENZYL ALCOHOL see NEOSYNEPHRINE

M-HYDROXY-ALPHA-((METHYLAMINO)METHYL)BENZYLALCOHOL HYDROCHLORIDE, (-)- see meta-SYNEPHRINE HYDROCHLORIDE

1-M-HYDROXY-ALPHA-(METHYLAMINOMETHYL)BENZYL ALCOHOL HYDROCHLORIDE see meta-SYNEPHRINE HYDROCHLORIDE

1-HYDROXY-2-METHYLAMINO-1-PHENYLPROPANE see EPHEDRINE

2-HYDROXY-1-(3-METHYLAMINOPROPYL)ADAMANTANE HYDROCHLORIDE see 1-(3-METHYLAMINOPROPYL)-2-ADAMANTANOL HYDROCHLORIDE

17-BETA-HYDROXY-17-METHYLANDROST-4-EN-3-ONE see 17-METHYL TESTOSTERONE

10-HYDROXYMETHYL-1,2-BENZANTHRACENE see BENZ(a)ANTHRACENE-7-METHANOL

1-HYDROXY-2-METHYLBENZENE see o-CRESOL

1-HYDROXY-3-METHYLBENZENE see m-CRESOL

1-HYDROXY-4-METHYLBENZENE see p-CRESOL

6-HYDROXYMETHYLBENZO(A)PYRENE see BENZO(a)PYRENE-6-METHANOL

3-HYDROXY-N-METHYL-CIS-CROTONAMIDE DIMETHYL PHOSPHATE see 3-(DIMETHOXYPHOSPHINYLOXY)N-METHYL-cis-CROTONAMIDE

1-(HYDROXYMETHYL)CYCLOHEXANEACETIC ACID SODIUM SALT see SODIUM HEXACYCLONATE

1-(HYDROXYMETHYL)CYCLOHEXANEACETIC ACID, SODIUM SALT see SODIUM HEXACYCLONATE

4-HYDROXY-3-METHYL-2-CYCLOPENTEN-1-ONE, CIS- MIXED WITH TRANS-2,2-DIMETHYL-3-(2-METHYL-PROPENYL)CYCLOPROPANECARBOXYLIC ACID ESTER WITH 2-ALLYL-4-HYDROXY-3-METHYL-2-CYCLOPENTEN-1-ONE (1:4) see ALLETHRIN RACEMIC MIXTURE

4-HYDROXYMETHYL-2,2-DIMETHYL-1,3-DIOXOLANE see DIOXOLAN

(HYDROXYMETHYL)ETHYLENE ACETATE see 1,3-DIACETIN

N-(1-(HYDROXYMETHYL)ETHYL)-D-LYSERGOMIDE see D-LYSERGIC ACID-L,2-PROPANOLAMIDE

N-(alpha-(HYDROXYMETHYL)ETHYL)-D-LYSERGOMIDE see D-LYSERGIC ACID-L,2-PROPANOLAMIDE

2-HYDROXYMETHYLFURAN see FURFURYL ALCOHOL

3-HYDROXYMETHYL-N-HEPTAN-4-OL see ETHYL HEXYLENE GLYCOL

HYDROXYMETHYLINITRILE see HYDROXYACETONITRILE

3-HYDROXY-1-METHYL-4-ISOPROPYLBENZENE see THYMOL

2-(HYDROXYMETHYL)-2-(METHYLPENTYL) BUTYLCARBAMATE CARBAMATE see 2-METHYL-2-PROPYLTRIMETHYLENE BUTYLCARBAMATE CARBAMATE

(−)-3-HYDROXY-N-METHYLMORPHINAN see LEVORPHANOL

(+-)-3-HYDROXY-N-METHYLMORPHINAN see N-METHYL-3-HYDROXYMORPHINAN

DL-3-HYDROXY-N-METHYLMORPHINAN see N-METHYL-3-HYDROXYMORPHINAN

DL-3-HYDROXY-N-METHYLMORPHINAN HYDROBROMIDE see METHORPHINAN HYDROBROMIDE

17-BETA-HYDROXY-18-METHYL-19-NOR-17-ALPHA-PREGN-4-EN-20-YN-3-ONE see NORGESTREL

4-HYDROXY-4-METHYL-PENTAN-2-ON (GERMAN, DUTCH) see 2-METHYL-2-PENTANOL-4-ONE

4-HYDROXY-4-METHYL PENTAN-2-ONE see 2-METHYL-2-PENTANOL-4-ONE

4-HYDROXY-4-METHYL-2-PENTANONE see 2-METHYL-2-PENTANOL-4-ONE

N-(4-((2-HYDROXY-5-METHYLPHENYL)AZO)PHENYL)ACETAMIDE see ACETAMINE YELLOW CG

P-HYDROXY-ALPHA-(1-((1-METHYL-3-PHENYLPROPYL)AMINO)ETHYL) BENZYL ALCOHOL HYDROCHLORIDE see DILATOL HYDROCHLORIDE

17-HYDROXY-6-METHYLPREGNA-4,6-DIENE-3,20-DIONE ACETATE see VOLIDAN

17-HYDROXY-6-ALPHA-METHYLPREGN-4-ENE-3,20-DIONE ACETATE see MEDROXYPROGESTERONE ACETATE

17-ALPHA-HYDROXY-6-ALPHA-METHYLPREGN-4-ENE-3,20-DIONE ACETATE see MEDROXYPROGESTERONE ACETATE

17-ALPHA-HYDROXY-6-ALPHA-METHYLPROGESTERONE ACETATE see MEDROXYPROGESTERONE ACETATE

1-HYDROXYMETHYLPROPANE see ISOBUTYL ALCOHOL

2-HYDROXY-2-METHYLPROPIONITRILE see 2-METHYLLACTONITRILE

5-HYDROXY-6-METHYL-3,4-PYRIDINEDICARBINOL HYDROCHLORIDE see PYRIDOXOL HYDROCHLORIDE

5-HYDROXY-6-METHYL-3,4-PYRIDINEDIMETHANOL see PYRIDOXOL

5-HYDROXY-6-METHYL-3,4-PYRIDINEDIMETHANOL HYDROCHLORIDE see PYRIDOXOL HYDROCHLORIDE

3-HYDROXY-1-METHYLPYRIDINIUM BROMIDE HEXAMETHYLENEBIS-(METHYLCARBAMATE) see DISTIGMINE BROMIDE

1-HYDROXYNAPHTHALENE see 1-NAPHTHOL

2-HYDROXYNAPHTHALENE see 2-NAPHTHOL

ALPHA-HYDROXYNAPHTHALENE see 1-NAPHTHOL

BETA-HYDROXYNAPHTHALENE see 2-NAPHTHOL

5-HYDROXY-1,4-NAPHTHALENEDIONE see WALNUT EXTRACT

N-HYDROXYNAPHTHALIMIDE-O,O-DIETHYL PHOSPHOROTHIOATE see NAPHTHALOXIMIDODIETHYL THIOPHOSPHATE

5-HYDROXY-1,4-NAPHTHOQUINONE see WALNUT EXTRACT

N-HYDROXY-2-NAPHTHYLAMINE see 2-NAPHTHYLHYDROXYLAMINE

1-HYDROXY-2-NAPHTHYLAMINE HYDROCHLORIDE see 2-AMINO-1-NAPHTHOL HYDROCHLORIDE

2-HYDROXY-1-NAPHTHYLAMINE HYDROCHLORIDE see 1-AMINO-2-NAPHTHOL HYDROCHLORIDE

4-HYDROXY-3-NITROANILINE see 4-AMINO-2-NITROPHENOL

2-HYDROXYNITROBENZENE see o-NITROPHENOL

4-HYDROXYNITROBENZENE see 4-NITROPHENOL

N-HYDROXY-N-NITROSO-BENZENAMINE, AMMONIUM SALT see AMMONIUM-N-NITROSOPHENYLHYDROXYLAMINE

M-HYDROXY NOREPHEDRINE see HYDROXYNOREPHEDRINE

17-BETA-HYDROXY-19-NORPREGN-4-EN-20-YN-3-ONE see 19-NORETHISTERONE

17-HYDROXY(17-ALPHA)-19-NORPREGN-5(10)-EN-20-YN-3-ONE see 17-alpha-ETHINYL-5,10-ESTRENOLONE

17-HYDROXY-19-NOR-17-ALPHA-PREGN-4-EN-20-YN-3-ONE see 19-NORETHISTERONE

17-HYDROXY-19-NOR-17-alpha-PREGN-5(10)-EN-20-YN-3-ONE see 17-alpha-ETHINYL-5,10-ESTRENOLONE

(17-ALPHA)-17-HYDROXY-19-NORPREGN-4-EN-20-YN-3-ONE see 19-NORETHISTERONE

(17-ALPHA)-17-HYDROXY-19-NORPREGN-5(10)-EN-20-YN-3-ONE see 17-alpha-ETHINYL-5,10-ESTRENOLONE

17-HYDROXY-19-NOR-17-ALPHA-PREGN-4-EN-20-YN-3-ONE ACETATE see 17-ACETOXY-19-NOR-17-alpha-PREGN-4-EN-20-YN-3-ONE

17-BETA-HYDROXY-19-NOR-17-ALPHA-PREGN-4-EN-20-YN-3-ONE ACETATE see 17-ACETOXY-19-NOR-17-alpha-PREGN-4-EN-20-YN-3--ONE

3-HYDROXY-OESTRA-1,3,5(10)-TRIEN-17-ONE see ESTRONE

3-HYDROXY-1,3,5(10)-OESTRATRIEN-17-ONE see ESTRONE

GAMMA-HYDROXY-BETA-OXOBUTANE see ACETOIN

4-HYDROXY-3-(3-OXO-1-FENYL-BUTYL) CUMARINE (DUTCH) see COUMADIN

3-beta-HYDROXY-23-OXO-OLEAN-12-EN-28-OIC ACID see GYPSOGENIN

4-HYDROXY-3-(3-OXO-1-PHENYL-BUTYL)-CUMARIN (GERMAN) see COUMADIN

2′-HYDROXYPELARGIDENOLON 1522 see MORIN

2-(6-(beta-HYDROXYPHENETHYL)-1-METHYL-2-PIPERIDYL)ACETOPHENONEHYDROCHLORIDE see LOBELINE HYDROCHLORIDE

2-HYDROXYPHENOL see PYROCATECHOL

3-HYDROXYPHENOL see RESORCINOL

M-HYDROXYPHENOL see RESORCINOL

O-HYDROXYPHENOL see PYROCATECHOL

P-HYDROXYPHENOL see HYDROQUINONE

1-HYDROXY-2-PHENOXYETHANE see PHENYL CELLOSOLVE

P-(4-HYDROXYPHENYL)ACETAMIDE see 4′-HYDROXYACETANILIDE

ALPHA-HYDROXYPHENYLACETIC ACID see MANDELIC ACID

1-(M-HYDROXYPHENYL)-2-AMINOETHANOL see alpha-(AMINO-METHYL)-m-HYDROXYBENZYL ALCOHOL

3-HYDROXY-4-(2,4-XYLYLAZO)-2,7 -NAPHTHALENEDISULFONIC ACID, DISODIUM SALT see D&C RED NO. 5

HYDRURE DE LITHIUM (FRENCH) see LITHIUM HYDRIDE

HYMECROMONE SODIUM see 7-HYDROXY-4-METHYLCOUMARIN SODIUM

HYMORPHAN see DIHYDROMORPHINONE

HYOCINE F HYDROBROMIDE see HYOSCINE HYDROBROMIDE

HYONIC PE-250 see TRITON X

(−)-HYOSCINE see SCOPOLAMINE

HYOSCINE see SCOPOLAMINE

HYOSCINE BROMIDE see HYOSCINE HYDROBROMIDE

(−)-HYOSCINE HYDROBROMIDE see HYOSCINE HYDROBROMIDE

l-HYOSCINE HYDROBROMIDE see HYOSCINE HYDROBROMIDE

HYOSCYAMINE see (−)-HYOSCYAMINE

l-HYOSCYAMINE see (−)-HYOSCYAMINE

DL-HYOSCYAMINE see ATROPINE

HYOSCYINE HYDROBROMIDE see HYOSCINE HYDROBROMIDE

HYPNONE see ACETOPHENONE

HYPO see SODIUM THIOSULFATE, PENTAHYDRATE

HYPOGLYCINE see 2-METHYLENECYCLOPROPANYLALANINE

HYPOGLYCINE A see 2-METHYLENECYCLOPROPANYLALANINE

HYPONITROUS ACID ANHYDRIDE see NITROGEN OXIDE

HYPOTHIAZIDE see 6-CHLORO-3,4-DIHYDRO-2H-1,2,4-BENZOTHIA-DIAZINE-7-SULFONAMIDE- 1,1-DIOXIDE

TERTIARY BUTYL ISOPROPYL BENZENE HYDROPEROXIDE (DOT) see tert-BUTYL ISOPROPYL BENZENE HYDROPEROXIDE

IBENZMETHYZINE see PROCARBAZINE

IBIOZEDRINE see DL-AMPHETAMINE SULFATE

I-BUTYL BROMIDE see 1-BROMO-2-METHYL PROPANE

ICG see CARDIO-GREEN

ICI 543 see PROPENE POLYMERS

ICI 28257 see ATROMID S

I.C.I. 33,828 see 1-METHYL-6-(1-METHYLALLYL)-2,5-DITHIOBIUREA

ICI 38174 see alpha-((ISOPROPYLAMINO)METHYL)NAPHTHALENE-METHANOL, HYDROCHLORIDE

ICI 45520 see INDERAL

ICR-25A see CHLOROQUINE MUSTARD

ICR 377 see 4-METHOXY-9-(3-(ETHYL-2-CHLOROETHYL)

ICR 451 see 7-CHLOROMETHYL BENZ(a)ANTHRACENE

ICR-48B see QUINACRINE ETHYL MUSTARD

ICR 498 see 7-BROMO METHYL BENZ(a)ANTHRACENE

ICR 502 see 7-BROMO METHYL-12-METHYL BENZ(a)ANTHRACENE

ICRF-159 see NSC-129943

IDRAZINA IDRATA (ITALIAN) see HYDRAZINE HYDRATE

IDRAZINA SOLFATO (ITALIAN) see HYDRAZINE SULFATE (1:1)

IDROCHINONE (ITALIAN) see HYDROQUINONE

IDROGENO SOLFORATO (ITALIAN) see HYDROGEN SULFIDE

IDROPEROSSIDO DI CUMENE (ITALIAN) see ISOPROPYLBENZENE HYDROPEROXIDE

IDROPEROSSIDO DI CUMOLO (ITALIAN) see ISOPROPYLBENZENE HYDROPEROXIDE

IDROSSIDO DI STAGNO TRIFENILE (ITALIAN) see HYDROXY TRIPHENYL STANNANE

4-IDROSSI-4-METIL-PENTAN-2-ONE (ITALIAN) see 2-METHYL-2-PENTANOL-4-ONE

D-N,N'-(1-IDROSSIMETIL PROPIL)-ETILENDINITROSAMINA (ITALIAN) see D-N,N'-(1-HYDROXYMETHYLPROPYL)ETHYLENEDINITROSAMINE

4-IDROSSI-3-(3-OXO-)-FENIL-BUTIL)-CUMARINE (ITALIAN) see COUMADIN

IDRYL see FLUORANTHENE

IEROIN see HEROIN

IGEPAL CA-630 see TRITON X

IGEPAL CO-630 see PEG-9 NONYL PHENYL ETHER

IMFERON see IRON-DEXTRAN COMPLEX

IMIDAZOLE-4-ETHYLAMINE see HISTAMINE

4-IMIDAZOLEETHYLAMINE see HISTAMINE

5-IMIDAZOLEETHYLAMINE see HISTAMINE

2-(4-IMIDAZOLYL)ETHYLAMINE see HISTAMINE

beta-IMIDAZOLYL-4-ETHYLAMINE see HISTAMINE

3,3'-IMIDODI-1-PROPANOL, DIMETHANESULFONATE (ESTER), HYDROCHLORIDE see YOSHI 864

IMIDOLE see PYRROLE

IMINAZOLE see IMIDAZOLE

IMINOBIS(PROPYLAMINE) see AMINOBIS(PROPYLAMINE)

3,3'-IMINOBIS(PROPYLAMINE) see AMINOBIS(PROPYLAMINE)

4,4'-((4-IMINO-2,5-CYCLOHEXADIEN-1-YLIDENE)METHYLENE) DIANILINE MONOHYDROCHLORIDE see BASIC PARAFUCHSINE

2,2'-IMINODIETHANOL see BIS(2-HYDROXY ETHYL)AMINE

3,3'-IMINODIPROPIONITRILE see BIS(beta-CYANOETHYL)AMINE

BETA,BETA'-IMINODIPROPIONITRILE see BIS(beta-CYANOETHYL)-AMINE

BETA,BETA-IMINODIPROPIONITRILE see BIS(beta-CYANOETHYL)-AMINE

IMINO-BETA,BETA'-DIPROPIONITRILE see BIS(beta-CYANOETHYL)-AMINE

2H-10, 4A-IMINOETHANOPHENANTHREN-6-OL, 1,3,4,9,10,10A-HEXAHYDRO-11METHYL-, DL see N-METHYL-3-HYDROXY-MORPHINAN

IMINOUREA see GUANIDINE

IMPERIAL GREEN see CUPRIC ACETOARSENITE

1H-INDENE-1,3(2H)-DIONE see 1,3-INDANDIONE

4-(1H-INDEN-1-YLIDENEMETHYL)-N,N-DIMETHYLBENZENAMINE see 1-(4-DIMETHYLAMINOBENZAL)INDENE

INDIAN BERRY see PICROTOXIN

INDIAN CANNABIS see CANNABIS

INDIAN GUM see ARABIC GUM

INDIAN HEMP see CANNABIS

INDIAN LICORICE SEED see ABRIN

INDIGO BLUE 2B see C.I. DIRECT BLUE 6, TETRASODIUM SALT

INDIGO CARMINE see DISODIUM INDIGO-5,5-DISULFONATE

INDIGO CARMINE (BIOLOGICAL STAIN) see DISODIUM INDIGO-5,5-DISULFONATE

INDIGO CARMINE DISODIUM SALT see DISODIUM INDIGO-5,5-DISULFONATE

INDIGO EXTRACT see DISODIUM INDIGO-5,5-DISULFONATE

5,5'-INDIGOTINDISULFONIC ACID see DISODIUM INDIGO-5,5-DISULFONATE

INDIGOTINE DISODIUM SALT see DISODIUM INDIGO-5,5-DISULFONATE

INDISULFAT (GERMAN) see INDIUM SULFATE

INDIUM CHLORIDE see INDIUM TRICHLORIDE

INDOL (GERMAN) see INDOLE

3-INDOLACETONITRILE see 3-INDOLYLACETONITRILE

BETA-INDOLAETHYLAMIN-CHLORHYDRAT (GERMAN) see 3-(2-AMINO-ETHYL)INDOLE HYDROCHLORIDE

3-INDOLEACETIC ACID see 1H-INDOLE-3-ACETIC ACID

BETA-INDOLEACETIC ACID see 1H-INDOLE-3-ACETIC ACID

BETA-INDOLE-3-ACETIC ACID see 1H-INDOLE-3-ACETIC ACID

INDOLEACETONITRILE see 3-INDOLYLACETONITRILE

INDOLE-3-ACETONITRILE see 3-INDOLYLACETONITRILE

1H-INDOLE-3-ACETONITRILE see 3-INDOLYLACETONITRILE

INDOLE-3-(2-AMINOBUTYL) ACETATE see 3-(2-AMINOBUTYL)INDOLE ACETATE

3-INDOLEBUTYRIC ACID see 1H-INDOLE-3-BUTANOIC ACID

BETA-INDOLEBUTYRIC ACID see 1H-INDOLE-3-BUTANOIC ACID

GAMMA-(INDOLE-3)-BUTYRIC ACID see 1H-INDOLE-3-BUTANOIC ACID

INDOLE-3-ETHYLAMINE HYDROCHLORIDE see 3-(2-AMINOETHYL)INDOLE HYDROCHLORIDE

BETA-INDOLE-ETHYLAMINE HYDROCHLORIDE see 3-(2-AMINO-ETHYL)INDOLE HYDROCHLORIDE

INDOLE-3-PROPYLAMINE HYDROCHLORIDE see 3-(gamma-AMINOPROPYL)-INDOLEHYDROCHLORIDE

INDOLYL-3-ACETIC ACID see 1H-INDOLE-3-ACETIC ACID

3-INDOLYLACETIC ACID see 1H-INDOLE-3-ACETIC ACID

BETA-INDOLYLACETIC ACID see 1H-INDOLE-3-ACETIC ACID

ALPHA-INDOL-3-YL-ACETIC ACID see 1H-INDOLE-3-ACETIC ACID

INDOLYLACETONITRILE see 3-INDOLYLACETONITRILE

3-INDOLYLACRYLIC ACID see INDOLE-3-ACRYLIC ACID

INDOLYL-3-BUTYRIC ACID see 1H-INDOLE-3-BUTANOIC ACID

4-(INDOL-3-YL)BUTYRIC ACID see 1H-INDOLE-3-BUTANOIC ACID

3-INDOLYL-GAMMA-BUTYRIC ACID see 1H-INDOLE-3-BUTANOIC ACID

GAMMA-(3-INDOLYL)BUTYRIC ACID see 1H-INDOLE-3-BUTANOIC ACID

GAMMA-(INDOL-3-YL)BUTYRIC ACID see 1H-INDOLE-3-BUTANOIC ACID

BETA-3-INDOLYLETHYLAMINE HYDROCHLORIDE see 3-(2-AMINO-ETHYL)INDOLE HYDROCHLORIDE

3-(1-H-INDOL-3-YL)-2-PROPENOIC ACID see INDOLE-3-ACRYLIC ACID

GAMMA-3-INDOLYLPROPYLAMINE HYDROCHLORIDE see 3-(gamma-AMINOPROPYL)-INDOLEHYDROCHLORIDE

INDOMETHAZINE see INDOMETHACIN

INDONAPHTHENE see INDENE

INDOPAN see 3-(2-AMINOPROPYL)INDOLE

INFUSORIAL EARTH see DIATOMACEOUS EARTH

INHIBINE see HYDROGEN PEROXIDE, 90%

INITIATING EXPLOSIVE FULMINATE OF MERCURY (DOT) see MER-CURY FULMINATE (WET)

INITIATING EXPLOSIVE IMINOBISPROPYLAMINE (DOT) see AMINOBIS-(PROPYLAMINE)

INITIATING EXPLOSIVE LEAD AZIDE (DOT) see LEAD AZIDE (II)

INITIATING EXPLOSIVE LEAD MONONITRORESORCINATE (DOT) see LEAD NITRORESORCINATE

INITIATING EXPLOSIVE LEAD STYPHNATE (DOT) see LEAD TRINITRO-RESORCINATE

INITIATING EXPLOSIVE PENTAERYTHRITE TETRANITRATE (DOT) see PENTAERYTHRITOL TETRANITRATE

INK ORANGE JSN see HISPACID FAST ORANGE 2G

INOSITHEXAPHOSPHORSAURE (GERMAN) see PHYTIC ACID

INOSITOL HEXAPHOSPHATE see PHYTIC ACID

INSARIOTOXIN see MYCOTOXIN T-2

INSECT POWDER see PYRETHRIN I

INSOLUBLE SACCHARINE see 1,2-BENZISOTHIAZOL-3(2H)-ONE-1,1-DIOXIDE

INSULAMIN see BUFORMIN HYDROCHLORIDE

INSULIN NOVO LENTE see LENTE INSULIN

INSULIN ZINC SUSPENSION see LENTE INSULIN

INTENSE BLUE see DISODIUM INDIGO-5,5-DISULFONATE

INTEXSAN LQ75 see LAURYLISOQUINOLINIUM BROMIDE

INTOCOSTRINE see CURARE

INTRASPERSE YELLOW GBA EXTRA see ACETAMINE YELLOW CG

IODE (FRENCH) see IODINE

IODIC ACID, POTASSIUM SALT see POTASSIUM IODATE

IODINE CHLORIDE see IODINE MONOCHLORIDE

IODINE CRYSTALS see IODINE

IODINE CYANIDE see CYANOGEN IODIDE

IODINE PENTACHLORIDE (DOT) see IODINE (V) CHLORIDE

IODINE SUBLIMED see IODINE

IODIO (ITALIAN) see IODINE

2-IODOACETAMIDE see IODOACETAMIDE

alpha-IODOACETAMIDE see IODOACETAMIDE

IODOACETATE see IODOACETIC ACID

1-IODOBUTANE see n-BUTYL IODIDE

IODOCHLORHYDROXYQUIN see 5-CHLORO-7-IODO-8-QUINOLINOL

IODOCHLORHYDROXYQUINOL see 5-CHLORO-7-IODO-8-QUINOLINOL

IODOCHLORHYDROXYQUINOLINE see 5-CHLORO-7-IODO-8-QUINOLI-NOL

7-IODO-5-CHLORO-8-HYDROXYQUINOLINE see 5-CHLORO-7-IODO-8-QUINOLINOL

5-IODODEOXYURIDINE see 2′-DEOXY-5-IODO URIDINE

5-IODO-2′-DEOXYURIDINE see 2′-DEOXY-5-IODO URIDINE

IODOMETANO (ITALIAN) see IODOMETHANE

IODOTRIMETHYLSTANNANE see IODOTRIMETHYLTIN

5-IODOURACIL DEOXYRIBOSIDE see 2′-DEOXY-5-IODO URIDINE

IODURE DE MERCURE (FRENCH) see MERCURY (I) IODIDE

IODURE DE METHYLE (FRENCH) see IODOMETHANE

IOPEZITE see POTASSIUM BICHROMATE

IPECACUANHA see IPECAC SYRUP

IPNO see IMPIRAMINE-N-OXIDE

IPRATROPIUMBROMID (GERMAN) see IPRATROPIUM BROMIDE

IPRONIAZID see ISONICOTINIC ACID-2-ISOPROPYLHYDRAZIDE

IPROVERATRIL see ISOPTIN

IRC 453 see 7-CHLOROMETHYL-12-METHYL BENZ(a)ANTHRACENE

IRISH MOSS GELOSE see CARRAGEEN

IROINI see HEROIN

IRON ARSENATE see IRON (II) ARSENATE (3:2)

IRON BIS(CYCLOPENTADIENE) see FERROCENE

IRON CARBOHYDRATE COMPLEX see IRON-DEXTRIN COMPLEX

IRON CARBONYL see PENTACARBONYLIRON

IRON (III) CHLORIDE see FERRIC CHLORIDE

IRON CHROMITE see CHROMITE (MINERAL)

IRON DEXTRAN INJECTION see IRON-DEXTRAN COMPLEX

IRON DEXTRIN INJECTION see IRON-DEXTRIN COMPLEX

IRON DICHLORIDE see FERROUS CHLORIDE

IRON DICYCLOPENTADIENYL see FERROCENE

IRON DIMETHYLDITHIOCARBAMATE see FERBAM

IRON FLUORIDE see FERRIC FLUORIDE

IRON GLUCONATE see FERROUS GLUCONATE

IRON(II) CHLORIDE (1:2) see FERROUS CHLORIDE

IRON(2+) LACTATE see LACTIC ACID, IRON(2+) SALT (2:1)

IRON ORE see HEMATITE

IRONORM INJECTION see IRON-DEXTRAN COMPLEX

IRON OXIDE RED see IRON (III) OXIDE

IRON OXIDE SACCHARATED see IRON OXIDE, SACCHARATED

IRON PENTACARBONYL see PENTACARBONYLIRON

IRON PERSULFATE see FERRIC SULFATE

IRON PROTOCHLORIDE see FERROUS CHLORIDE

IRON PROTOSULFATE see IRON (II) SULFATE (1:1)

IRON PYRITES see IRON DISULFIDE

IRON SACCHARATE see IRON OXIDE, SACCHARATED

IRON SESQUIOXIDE see IRON (III) OXIDE

IRON SESQUIOXIDE HYDRATED see LIMONITE

IRON SESQUISULFATE see FERRIC SULFATE

IRON SORBITEX see IRON SORBITOL CITRATE

IRON SUGAR see IRON OXIDE, SACCHARATED

IRON SULFATE (2:3) see FERRIC SULFATE

IRON SULFIDE see IRON DISULFIDE

IRON TERSULFATE see FERRIC SULFATE

IRON TRICHLORIDE see FERRIC CHLORIDE

IRON TRIFLUORIDE see FERRIC FLUORIDE

IRON(II) SULFATE (1:1), HEPTAHYDRATE see FERROUS SULFATE

IRON VITRIOL see IRON (II) SULFATE (1:1)

IROX (GADOR) see FERROUS GLUCONATE

IRRADIATED ERGOSTA-5,7,22-TRIEN-3-BETA-OL see ERGOCAL-CIFEROL

ISACONITINE see PICRACONITINE

ISATIDINE see RETRORSINE-N-OXIDE

ISLANDITOXIN see CYCLOCHLOROTINE

ISOADRENALINE HYDROCHLORIDE see AMINOPROPANOL PYROCATE-CHOLHYDROCHLORIDE

ISOALLOXAZINE, 7,8-DIMETHYL-10-(D-RIBO-2,3,4,5-TETRAHY-DROXYPENTYL)- see RIBOFLAVINE

ISOALLOXAZINE, 7,8-DIMETHYL-10-D-RIBITYL- see RIBOFLAVINE

ISOAMIDONE II see ISOMETHADONE

ISOAMYL ACETATE see ISOPENTYL ALCOHOL ACETATE

ISOAMYL ALKOHOL (CZECH) see ISOAMYL ALCOHOL

ISO-AMYLALKOHOL (GERMAN) see ISOAMYL ALCOHOL

BETA-ISO-AMYLENE see α,η-AMYLENE

ISOAMYL ETHANOATE see ISOPENTYL ALCOHOL ACETATE

ISOAMYLETHYLBARBITURIC ACID see AMITAL

5-ISOAMYL-5-ETHYLBARBITURIC ACID see AMITAL

5-ISOAMYL-5-ETHYLBARBITURIC ACID, SODIUM DERIV see AMYTAL SODIUM

ISOAMYLHYDRIDE see ETHYL DIMETHYL METHANE

ISOAMYLOL see ISOAMYL ALCOHOL

ISOANETHOLE see p-ALLYLANISOLE

1,3-ISOBENZOFURANDIONE see PHTHALIC ANHYDRIDE

ISOBORNEOL THIOCYANATOACETATE see ISOBORNYL THIOCYANATO ACETATE

ISOBORNYL THIOCYANOACETATE see ISOBORNYL THIOCYANATO
  ACETATE
ISOBUTANDIOL-2-AMINE see 2-AMINO-2-METHYL-1,3-PROPANEDIOL
ISOBUTANE see 2-METHYLPROPANE
ISOBUTANOL see ISOBUTYL ALCOHOL
ISOBUTENAL see METHYLACRYLALDEHYDE
ISOBUTENYL CHLORIDE see 3-CHLORO-2-METHYLPROPENE
ISOBUTENYL METHYL KETONE see MESITYL OXIDE
ISOBUTYLALDEHYDE see ISOBUTYRALDEHYDE
ISOBUTYLALKOHOL (CZECH) see ISOBUTYL ALCOHOL
ISOBUTYLALLYLBARTURIC ACID see ALLYLISOBUTYLBARBITURATE
2-(ISOBUTYLAMINO)ETHYL-p-AMINOBENZOATE HYDROCHLORIDE see
  IBYLCAINE HYDROCHLORIDE
ISOBUTYL BROMIDE see 1-BROMO-2-METHYL PROPANE
ISOBUTYLCARBINOL see ISOAMYL ALCOHOL
ISOBUTYLDIUREA see ISOBUTYLIDENEDIUREA
ISOBUTYLENE see ISOBUTENE
1,1'-ISOBUTYLIDENEBISUREA see ISOBUTYLIDENEDIUREA
ISOBUTYL KETONE see DIISOBUTYL KETONE
ISOBUTYL-alpha-METHACRYLATE see ISOBUTYL METHACRYLATE
ISOBUTYL-METHYLKETON (CZECH) see HEXONE
ISOBUTYL METHYL KETONE see HEXONE
ISOBUTYLMETHYLMETHANOL see ISOBUTYLMETHYLCARBINOL
ISOBUTYLTRIMETHYLETHANE see 2,2,4-TRIMETHYLPENTANE
ISOBUTYRALDEHYD (CZECH) see ISOBUTYRALDEHYDE
ISOBUTYRIC ACID, ETHYL ESTER see ETHYL ISOBUTYRATE
ISOBUTYRIC ACID, ISOBUTYL ESTER see ISOBUTYL ISOBUTYRATE
ISOBUTYRIC ALDEHYDE see ISOBUTYRALDEHYDE
ISOCARBONAZID see ISOCARBOXAZID
ISOCARBOSSAZIDE see ISOCARBOXAZID
ISOCARBOXAZIDE see ISOCARBOXAZID
ISOCHLOORTHION (DUTCH) see p-NITRO-o-CHLOROPHENYL
  DIMETHYL THIONOPHOSPHATE
ISOCYANATE DE METHYLE (FRENCH) see METHYL ISOCYANATE
ISOCYANATOETHANE see ETHYL ISOCYANATE
ISO-CYANATOMETHANE see METHYL ISOCYANATE
ISOCYANIC ACID, BUTYL ESTER see n-BUTYL ISOCYANATE
ISOCYANIC ACID-P-CHLOROPHENYL ESTER see p-CHLOROPHENYL
  ISOCYANATE
ISOCYANIC ACID, ETHYL ESTER see ETHYL ISOCYANATE
ISOCYANIC ACID, METHYL ESTER see METHYL ISOCYANATE
ISOCYANIC ACID, METHYLPHENYLENE ESTER see TOLUENE DIISO-
  CYANATE
ISODEMETON see DEMETON-S
BETA-ISODURYLAMIDE, N-(2-(DIETHYLAMINO)ETHYL)-,HYDROCHLO-
  RIDE see N-(2-(DIETHYLAMINO)ETHYL)-2,4,6-TRIMETHYL-
  BENZAMIDE, HYDROCHLORIDE
ISOESTRAGOLE see p-PROPENYL ANISOLE
ISOFEDROL see 1-EPHEDRINE SULFATE
ISOFORONE (ITALIAN) see ISOACETOPHORONE
ISOFTALODINITRIL (CZECH) see 1,3-BENZENEDICARBONITRILE
ISOHEXYL ALCOHOL see AMYL METHYL ALCOHOL
ISOHOL see ISOPROPYL ALCOHOL
ISOINDOLE-1,3-DIONE see PHTHALIMIDE
1,3-ISOINDOLEDIONE see PHTHALIMIDE
1,3-ISOINDOLINEDIONE see PHTHALIMIDE
ISOLANE (FRENCH) see DIMETHYL-5-(1-ISOPROPYL-3-METHYL-PYRA-
  ZOLYL)-CARBAMATE
ISOLANID see LANATOSIDE C
ISOMEBUMAL see 5-ETHYL-5-(1-ETHYLPROPYL)BARBITURIC ACID
ISOMEPROBAMATE see ISOPROPYL MEPROBAMATE
ISOMERIC CHLORTHION see p-NITRO-o-CHLOROPHENYL DIMETHYL
  THIONOPHOSPHATE
ISOMETASYSTOX see DEMETON-S-METHYL
ISOMETASYSTOX SULFONE see DEMETON-S-METHYL SULPHONE
ISOMETHEPTENE see ISONYL
ISOMETHEPTENE HYDROCHLORIDE see OCTIN HYDROCHLORIDE
ISOMETHYLSYSTOX see DEMETON-S-METHYL
ISOMETHYLSYSTOX SULFONE see DEMETON-S-METHYL SULPHONE
ISOMETHYLSYSTOX SULFOXIDE see DEMETON-O-METHYL SULFOXIDE

ISOMYST see ISOPROPYL MYRISTATE
ISONAPHTHOL see 2-NAPHTHOL
ISONICOTINHYDRAZID see ISONICOTINIC ACID HYDRAZIDE
ISONICOTINIC ACID, ETHYL ESTER see ETHYL ISONICOTINATE
ISONICOTINOYL HYDRAZID see ISONICOTINIC ACID HYDRAZIDE
1-ISONICOTINOYL-2-ISOPROPYLHYDRAZINE see ISONICOTINIC ACID-
  2-ISOPROPYLHYDRAZIDE
ISONICOTINYL HYDRAZIDE see ISONICOTINIC ACID HYDRAZIDE
1-ISONICOTINYL-2-ISOPROPYLHYDRAZINE see ISONICOTINIC ACID-
  2-ISOPROPYLHYDRAZIDE
ISONITROPROPANE see 2-NITROPROPANE
ISOOCTANE see 2,2,4-TRIMETHYLPENTANE
ISOOCTANOL see ISOOCTYL ALCOHOL
ISOOCTYL ALCOHOL (2,4-DICHLOROPHENOXY)ACETATE see ISO-
  OCTYL-2,4-DICHLOROPHENOXYACETATE
ISOPENTANE see ETHYL DIMETHYL METHANE
ISOPENTANOIC ACID see ISOVALERIC ACID
ISOPENTANOL see ISOAMYL ALCOHOL
ISOPENTYL ACETATE see ISOPENTYL ALCOHOL ACETATE
ISOPENTYL ALCOHOL see ISOAMYL ALCOHOL
ISOPHORONE see ISOACETOPHORONE
ISOPHTHALODINITRILE see 1,3-BENZENEDICARBONITRILE
ISOPHTHALONITRILE see 1,3-BENZENEDICARBONITRILE
(±)-ISOPRENALINE SULFATE see (±)-ISOPROTERENOL SULFATE
DL-ISOPRENALINE SULFATE see (±)-ISOPROTERENOL SULFATE
ISOPROPANOL see ISOPROPYL ALCOHOL
ISOPROPANOLAMINE see 1-AMINOPROPAN-2-OL
ISOPROPENE CYANIDE see METHYLACRYLONITRILE
ISOPROPENIL-BENZOLO (ITALIAN) see alpha-METHYLSTYRENE
ISOPROPENYL-BENZEEN (DUTCH) see alpha-METHYLSTYRENE
ISOPROPENYLBENZENE see alpha-METHYLSTYRENE
ISOPROPENYL-BENZOL (GERMAN) see alpha-METHYLSTYRENE
ISOPROPENYLNITRILE see METHYLACRYLONITRILE
ISOPROPILAMINA (ITALIAN) see ISOPROPYLAMINE
ISOPROPILBENZENE (ITALIAN) see CUMENE
ISOPROPILE (ACETATO DI) (ITALIAN) see ACETIC ACID ISOPROPYL
  ESTER
ISOPROPIL-N-FENIL-CARBAMMATO (ITALIAN) see CARBANILIC ACID
  ISOPROPYL ESTER
(1-ISOPROPIL-3-METIL-1H-PIRAZOL-5-IL)-N,N-DIMETIL-CARBAMMATO
  (ITALIAN) see DIMETHYL-5-(1-ISOPROPYL-3-METHYL-PYRAZOLYL)-
  CARBAMATE
2-ISOPROPOXYETHANOL see ISOPROPYL GLYCOL
O-ISOPROPOXYPHENYL METHYLCARBAMATE see BAYGON
O-ISOPROPOXYPHENYL METHYLNITROSOCARBAMATE see NITROSO-
  BAYGON
O-ISOPROPOXYPHENYL-N-METHYL-N-NITROSOCARBAMATE see
  NITROSO-BAYGON
2-ISOPROPOXYPHENYL-NMETHYLCARBAMATE see BAYGON
O-ISOPROPOXYPHENYL-NMETHYLCARBAMATE see BAYGON
2-ISOPROPOXYPROPANE see DIISOPROPYL ETHER
ISOPROPYLACETAAT (DUTCH) see ACETIC ACID ISOPROPYL ESTER
ISOPROPYLACETAT (GERMAN) see ACETIC ACID ISOPROPYL ESTER
ISOPROPYL ACETATE (DOT) see ACETIC ACID ISOPROPYL ESTER
ISOPROPYLACETIC ACID see ISOVALERIC ACID
ISOPROPYLACETONE see HEXONE
ISO-PROPYLALKOHOL (GERMAN) see ISOPROPYL ALCOHOL
ISOPROPYLALLYLBARBITURIC ACID see ALLONAL
4-(2-ISOPROPYLAMINE-1-HYDROXYETHYL)METHANESULFOANILIDE
  HYDROCHLORIDE see beta-CARDONE
2-ISOPROPYLAMINO-1-(2NAPHTHYL)ETHANOL HYDROCHLORIDE see
  alpha-((ISOPROPYLAMINO)METHYL)NAPHTHALENEMETHANOL, HY-
  DROCHLORIDE
1-ISOPROPYLAMINO-3-(1NAPHTHYLOXY)-2-PROPANOL see INDERAL
ISOPROPYLAMINO-O-ETHYL-(4-METHYLMERCAPTO-3-METHYLPHE-
  NYL)PHOSPHATE see FENAMIPHOS
ISOPROPYLAMINOHYDROXYETHYLMETHANESULFONALIDE HYDRO-
  CHLORIDE see beta-CARDONE
ISOPROPYLAMINOMETHYL-3,4-DIHYDROXYPHENYL CARBINOL see
  ISOPROPYLADRENALINE

ISOUREA see UREA

ISOVALERONE see DIISOBUTYL KETONE

ISOXANTHINE see XANTHINE

IZOACRIDINA see 9-AMINOACRIDINE

IZOFORON (POLISH) see ISOACETOPHORONE

IZOPROPYLOWY ETER (POLISH) see DIISOPROPYL ETHER

JACOBINE see SENECIPHYLLINE

JAMESTOWN WEED see STRAMONIUM

JAPACONITINA B (ITALIAN) see JAPACONITINE B

JAPAN AGAR see AGAR

JAPAN CAMPHOR see CAMPHOR

JAPANESE CAMPHOR see CAMPHOR, (1R,4R)-(+)-

JAPANESE CAMPHOR OIL see CAMPHOR OIL

JAPANESE, OIL OF CAMPHOR see CAMPHOR OIL

JAPAN ISINGLASS see AGAR

JASMINALDEHYDE see alpha-AMYL CINNAMALDEHYDE

JASMOLIN I OR II see PYRETHRIN I

JAUNE AB see 1-(PHENYLAZO)-2-NAPHTHYLAMINE

JAVA ORANGE 2G see HISPACID FAST ORANGE 2G

JAVA SCARLET 3R see FOOD RED 7

JB 329 see DITRAN

JB 336 see N-METHYL-3-PIPERIDYL BENZILATE

JEWELER'S ROUGE see IRON (III) OXIDE

JIMSON WEED see STRAMONIUM

JOD (GERMAN, POLISH) see IODINE

JODCYAN see CYANOGEN IODIDE

JODID SODNY see SODIUM IODIDE

JOD-METHAN (GERMAN) see IODOMETHANE

JOOD (DUTCH) see IODINE

JOODMETHAAN (DUTCH) see IODOMETHANE

JOY POWDER see HEROIN

JUDEAN PITCH see ASPHALT

JULIN'S CARBON CHLORIDE see HEXACHLOROBENZENE

JUMBLE BEAD see ABRIN

K-9 see SALITHION

KADMIUM (GERMAN) see CADMIUM

KADMIUMCHLORID (GERMAN) see CADMIUM CHLORIDE

KADMU TLENEK (POLISH) see CADMIUM OXIDE

KALIUMARSENIT (GERMAN) see POTASSIUM ARSENITE

KALIUMCARBONAT (GERMAN) see POTASSIUM CARBONATE (2:1)

KALIUMCHLORAAT (DUTCH) see POTASSIUM CHLORATE

KALIUMCHLORAT (GERMAN) see POTASSIUM CHLORATE

KALIUMCYANAT (GERMAN) see POTASSIUM CYANATE

KALIUMDICHROMAT (GERMAN) see POTASSIUM BICHROMATE

KALIUMHYDROXID (GERMAN) see POTASSIUM HYDROXIDE

KALIUMHYDROXYDE (DUTCH) see POTASSIUM HYDROXIDE

KALIUMNITRAT (GERMAN) see POTASSIUM NITRATE

KALIUMPERMANGANAAT (DUTCH) see POTASSIUM PERMANGANATE

KALIUMPERMANGANAT (GERMAN) see POTASSIUM PERMANGANATE

KALLIDIN see BRADYKININ

KALMUS OEL (GERMAN) see OIL OF CALAMUS

KALOMEL (GERMAN) see MERCUROUS CHLORIDE

KALZIUMARSENIAT (GERMAN) see ARSENIC ACID, CALCIUM SALT (2:3)

KALZIUMZYKLAMATE (GERMAN) see CALCIUM CYCLOHEXYLSULPHA-MATE

KAMPFER (GERMAN) see CAMPHOR

KAMPSTOFF "LOST" see BIS(2-CHLOROETHYL)SULFIDE

KANECHLOR 300 see POLYCHLORINATED BIPHENYL (KANECHLOR 300)

KANECHLOR 400 see POLYCHLORINATED BIPHENYL (KANECHLOR 400)

KANECHLOR 500 see POLYCHLORINATED BIPHENYL (KANECHLOR 500)

KANECHLOR S see POLYCHLORINATED BIPHENYLS

KANNASYN see KANAMYCIN SULFATE

KAN-TO-KA (JAPANESE) see COLTSFOOT

KANTREX SULFATE see KANAMYCIN SULFATE

KAPRYLAN DI-N-BUTYLCINICITY (CZECH) see BIS(OCTANOYLOXY)DI-n-BUTYL STANNANE

KARBARYL (POLISH) see CARBARYL

KARSAN see FORMALDEHYDE

KASSIA OEL (GERMAN) see CASSIA OIL

KATCHUNG OIL see PEANUT OIL

KAYDOL see MINERAL OIL

KC-500 see POLYCHLORINATED BIPHENYL (KANECHLOR 500)

KELENE see ETHYL CHLORIDE

KELINCOR see AMICARDINE

KELTHANETHANOL see 1,1-BIS(p-CHLOROPHENYL)-2,2,2-TRICHLO-ROETHANOL

KEROSINE see KEROSENE

KESSCOMIR see ISOPROPYL MYRISTATE

KETALGIN see METHADONE

KETALGIN HYDROCHLORIDE see dl-METHADONE HYDROCHLORIDE

KETAMINE HYDROCHLORIDE see 2-(o-CHLOROPHENYL)-2-(METHYL-AMINO)CYCLOHEXANONE HYDROCHLORIDE

3-(3-KETO-7-ALPHA-ACETYLTHIO-17-BETA-HYDROXY-4-ANDROSTEN-17-ALPHA-YL)PROPIONIC ACID LACTONE see ALDACTAZIDE

BETA-KETOBUTYRANILIDE see ACETOACETANILIDE

KETOCYCLOHEPTANE see CYCLOHEPTANONE

KETOCYCLOPENTANE see CYCLOPENTANONE

3-KETO-1,5-DIMETHYL-4-DIMETHYLAMINO-2-PHENYL-2,3-DIHYDRO-PYRAZOLE see DIMETHYLAMINOANTIPYRINE

KETO-ETHYLENE see KETENE

3-KETO-L-GULOFURANOLACTONE see L-ASCORBIC ACID

KETOHEPTAMETHYLENE see CYCLOHEPTANONE

KETOHEXAMETHYLENE see CYCLOHEXANONE

2-KETOHEXAMETHYLENIMINE see HEXAHYDRO-2H-AZEPIN-2-ONE

KETOHYDROXYESTRIN BENZOATE see ESTRONE BENZOATE

KETOLE see INDOLE

15-KETO-20-METHYLCHOLANTHRENE see 20-METHYLCHOLANTHREN-15-ONE

KETONE, ISOBUTYL METHYL see HEXONE

KETONE PROPANE see ACETONE

KETOPENTAMETHYLENE see CYCLOPENTANONE

BETA-KETOPROPANE see ACETONE

L-3-KETOTHREOHEXURONIC ACID LACTONE see L-ASCORBIC ACID

2-KETO-1,7,7-TRIMETHYLNORCAMPHANE see CAMPHOR

KHAROPHEN see ACETPHENARSINE

KIESELGUHR see DIATOMACEOUS EARTH

KIESELSAURE (GERMAN) see SILICA, PRECIPITATED (AMORPHOUS)

KIEZELFLUORWATERSTOFZUUR (DUTCH) see SILICOFLUORIC ACID

KING'S GREEN see CUPRIC ACETOARSENITE

KING'S YELLOW see ARSENIC SULFIDE

KING'S YELLOW see LEAD CHROMATE

KIPCA, OIL SOLUBLE see 2-METHYL-1,4-NAPHTHOQUINONE

KM (THE ANTIBIOTIC) see KANAMYCIN A

KOBALT CHLORID (GERMAN) see COBALTOUS CHLORIDE

KOBALT-EDTA (GERMAN) see DICOBALT EDETATE

KOBALT HISTIDIN (GERMAN) see BIS(L-HISTIDINE)COBALT

KOCIDE see COPPER HYDROXIDE

KOHLENDISULFID (SCHWEFELKOHLENSTOFF) (GERMAN) see CARBON DISULFIDE

KOHLENMONOXID (GERMAN) see CARBON MONOXIDE

KOHLENSAURE (GERMAN) see CARBON DIOXIDE

KOKAIN see COCAINE

KOKAN see COCAINE

KOKAYEEN see COCAINE

KOOLMONOXYDE (DUTCH) see CARBON MONOXIDE

KOOLSTOFDISULFIDE (ZWAVELKOOLSTOF) (DUTCH) see CARBON DISULFIDE

KOOLSTOFOXYCHLORIDE (DUTCH) see PHOSGENE

KOPROSTERIN (GERMAN) see DIHYDROCHOLESTEROL

KORAX see 1-CHLORO-1-NITROPROPANE

KORUND see CORUNDUM

KRAMERIA IXINA see CADIA DEL PERRO

M-KRESOL see m-CRESOL

P-KRESOL see p-CRESOL

O-KRESOL (GERMAN) see o-CRESOL

KRESOLE (GERMAN) see CRESOL

KRESOLEN (DUTCH) see CRESOL

KREZOL (POLISH) see CRESOL

KROKYDOLITH (GERMAN) see CROCIDOLITE (see ASBESTOS)

KSYLEN (POLISH) see XYLENE

KUBARSOL see ACETPHENARSINE

KUEMMEL OIL (GERMAN) see CARAWAY OIL

KUPFERRON (CZECH) see AMMONIUM-N-NITROSOPHENYLHYDROXYL-AMINE

KUPFERSULFAT-PENTAHYDRAT (GERMAN) see COPPER (II) SULFATE PENTAHYDRATE (1:1:5)

KWIK (DUTCH) see MERCURY

KYANID SODNY (CZECH) see SODIUM CYANIDE

KYANID STRIBRNY (CZECH) see SILVER CYANIDE

KYANITE see ALUMINUM(III) SILICATE (2:1)

KYSELINA ADIPOVA (CZECH) see ADIPIC ACID

KYSELINA P-AMINOSALICYLOVA (CZECH) see 4-AMINOSALICYLIC ACID

KYSELINA BENZOOVA (CZECH) see BENZOIC ACID

KYSELINA CLEVE (CZECH) see 5-AMINO-2-NAPHTHALENESULFONIC ACID

KYSELINA 3,6-ENDOMETHYLEN-3,4,5,6,7,7-HEXACHLOR-DELTA(SUP 4)-TETRAHYDROFTALOVA (CZECH) see CHLORENDIC ACID

KYSELINA HET (CZECH) see CHLORENDIC ACID

KYSELINA MLECNA (CZECH) see LACTIC ACID

KYSELINA-2-NAFTYLAMIN-4,8-DISULFONOVA (CZECH) see 3-AMINO-1,5-NAPHTHALENEDISULFONIC ACID

KYSELINA-1-NAFTYLAMIN-6-SULFONOVA (CZECH) see 5-AMINO-2-NAPHTHALENESULFONIC ACID

KYSELINA NITROBENZEN-M-SULFONOVA (CZECH) see 3-NITROBEN-ZENESULFONIC ACID

KYSELINA STAVELOVA (CZECH) see OXALIC ACID

KYSLICNIK DI-N-BUTYLCINICITY (CZECH) see DIBUTYLOXOSTAN-NANE

KYSLICNIK DIISOBUTYLCINICITY (CZECH) see DIISOBUTYLOXOSTAN-NANE

KYSLICNIK DIISOPROPYLCINICITY (CZECH) see DIISOPROPYLOXO-STANNANE

KYSLICNIK DI-N-PROPYLCINICITY (CZECH) see DIPROPYLOXOSTAN-NANE

KYSLICNIK TRI-N-BUTYLCINICITY (CZECH) see BIS(TRIBUTYL TIN)-OXIDE

L'ACIDE OLEIQUE (FRENCH) see OLEIC ACID

DL-LACTIC ACID see LACTIC ACID

LACTOBACILLUS LACTIS DORNER FACTOR see VITAMIN B_{12} COM-PLEX

LACTOBIOSE see LACTOSE

D-LACTOSE see LACTOSE

LAE-32 see LYSERGIC ACID ETHYLAMIDE

LAEVORAL see FRUCTOSE

LAEVOSAN see FRUCTOSE

LAMP BLACK see CARBON BLACK

LAND PLASTER see CALCIUM (II) SULFATE DIHYDRATE (1:1:2)

LANDRIN see 3,4,5-TRIMETHYLPHENYL METHYLCARBAMATE

LANOXIN see DIGOXIN

LANTHANACETAT (GERMAN) see LANTHANUM ACETATE

LANTHANUM TRIACETATE see LANTHANUM ACETATE

LASEX see 4-CHLORO-N-FURFURYL-5-SULFAMOYL ANTHRANILIC ACID

LASIX see 4-CHLORO-N-FURFURYL-5-SULFAMOYL ANTHRANILIC ACID

LASSO see 2-CHLORO-2',6'-DIETHYL-N-(METHOXY METHYL)-ACETANILIDE

LATEX see POLYDIMETHYL SILOXANE

LAUDICON see DIHYDROMORPHINONE

LAUDRAN DI-N-BUTYLCINICITY (CZECH) see DIBUTYLBIS(LAU-ROYLOXY)STANNANE

LAUGHING GAS see NITROGEN OXIDE

LAUREL CAMPHOR see CAMPHOR

LAUREL LEAF OIL see BAY OIL

LAURIC ACID, DIBUTYLSTANNYLENE SALT see DIBUTYLBIS(LAU-ROYLOXY)STANNANE

LAURIC ACID DIETHANOLAMIDE see N,N-BIS(2-HYDROXY ETHYL)-DODECAN AMIDE

LAURIC ALCOHOL see DODECYL ALCOHOL

LAURINIC ALCOHOL see DODECYL ALCOHOL

LAUROSTEARIC ACID see LAURIC ACID

LAUROYL DIETHANOLAMIDE see N,N-BIS(2-HYDROXY ETHYL)DO-DECAN AMIDE

LAURYL 24 see DODECYL ALCOHOL

LAURYL ALCOHOL see DODECYL ALCOHOL

N-LAURYL ALCOHOL, PRIMARY see DODECYL ALCOHOL

LAURYLAMINE see DODECYLAMINE

LAURYL DIETHANOLAMIDE see N,N-BIS(2-HYDROXY ETHYL)DO-DECAN AMIDE

LAURYLGUANIDINE ACETATE see N-DODECYLGUANIDINE ACETATE

LAURYL SODIUM SULFATE see SULFURIC ACID, MONODODECYL ESTER, SODIUM SALT

LAURYL SULFATE, SODIUM SALT see SULFURIC ACID, MONODO-DECYL ESTER, SODIUM SALT

LAYOR CARANG see AGAR

LD NORGESTREL (FRENCH) see 19-NORTESTOSTERONE

LEAD (2+) ACETATE see LEAD ACETATE

LEAD ACETATE TRIHYDRATE see LEAD ACETATE (II), TRIHYDRATE

LEAD ARSENATE see ARSENIC ACID, LEAD SALT

LEAD ARSENATE, SOLID (DOT) see LEAD ACID ARSENATE

LEAD ARSENATE (STANDARD) see LEAD ACID ARSENATE

LEAD ARSENITE, SOLID (DOT) see LEAD(II) ARSENITE

LEAD BOTTOMS see LEAD (II) SULFATE (1:1)

LEAD BROWN see LEAD DIOXIDE

LEAD BROWN see LEAD(IV)OXIDE BROWN

LEAD(2+) CARBONATE see LEAD CARBONATE

LEAD(II) ACETATE see LEAD ACETATE

LEAD (2+) CHLORIDE see LEAD CHLORIDE

LEAD CHROMATE see LEAD CHROMATE

LEAD CHROMATE(VI) see LEAD CHROMATE

LEAD CHROMATE OXIDE see LEAD CHROMATE, BASIC

LEAD CHROMATE, RED see LEAD CHROMATE, BASIC

LEAD CHROMATE, SULPHATE AND MOLYBDATE see LEAD-MOLYBDE-NUM CHROMATE

LEAD CYANIDE (DOT) see LEAD (II) CYANIDE

LEAD DIACETATE see LEAD ACETATE

LEAD DIACETATE TRIHYDRATE see LEAD ACETATE (II), TRIHY-DRATE

LEAD DIBASIC ACETATE see LEAD ACETATE

LEAD DICHLORIDE see LEAD CHLORIDE

LEAD DIFLUORIDE see LEAD (II) FLUORIDE

LEAD DINITRATE see LEAD (II) NITRATE (1:2)

LEAD DIOXIDE see LEAD(IV)OXIDE BROWN

LEAD FLAKE see LEAD

LEAD (II) CHLORIDE see LEAD CHLORIDE

LEAD (II) NITRATE see LEAD (II) NITRATE (1:2)

LEAD MONONITRORESORCINATE see LEAD NITRORESORCINATE

LEAD MONOSUBACETATE see LEAD ACETATE, BASIC

LEAD NAPHTHENATE see NAPHTHENIC ACID, LEAD SALT

LEAD NITRATE see LEAD (II) NITRATE (1:2)

LEAD (2+) NITRATE see LEAD (II) NITRATE (1:2)

LEAD ORTHOPHOSPHATE see LEAD (II) PHOSPHATE (3:2)

LEAD ORTHOPLUMBATE see LEAD OXIDE RED

LEAD OXIDE see LEAD MONOXIDE

LEAD OXIDE BROWN see LEAD DIOXIDE

LEAD PEROXIDE see LEAD DIOXIDE

LEAD PEROXIDE see LEAD(IV)OXIDE BROWN

LEAD PEROXIDE (DOT) see LEAD DIOXIDE

LEAD PHOSPHATE see LEAD (II) PHOSPHATE (3:2)

LEAD (2+) PHOSPHATE see LEAD (II) PHOSPHATE (3:2)

LEAD PHOSPHATE (3:2) see LEAD (II) PHOSPHATE (3:2)

LEAD PROTOXIDE see LEAD MONOXIDE

LEAD S2 see LEAD

LEAD STYPHNATE see LEAD TRINITRORESORCINATE
LEAD SUBACETATE see LEAD ACETATE, BASIC
LEAD SUPEROXIDE see LEAD DIOXIDE
LEAD SUPEROXIDE see LEAD(IV)OXIDE BROWN
LEAD TETRAOXIDE see LEAD OXIDE RED
LEAD(II) OXIDE see LEAD MONOXIDE
LEAF ALDEHYDE see 2-HEXENAL
LEATHER ORANGE HR see 4-PHENYLAZO-m-PHENYLENEDIAMINE
LE CAPTANE (FRENCH) see CAPTAN
LEDATE see LEAD DIMETHYLDITHOCARBAMATE
LE DINITROCRESOL-4,6 (FRENCH) see 4,6-DINITRO-O-CRESOL
LEINOLEIC ACID see LINOLEIC ACID
LEIPZIG YELLOW see LEAD CHROMATE
LEMONENE see BIPHENYL
LEMONOL see GERANIOL
LEMON YELLOW see LEAD CHROMATE
LEMON YELLOW see BARIUM CHROMATE (VI)
LENAMYCIN see ACETYLENEDICARBOXAMIDE
LENTE ILETIN see LENTE INSULIN
LENTINE (FRENCH) see CARBACHOL CHLORIDE
LEPTANAL see PHENTANYL CITRATE
L'ETHYLENE THIOUREE (FRENCH) see 2-IMIDAZOLIDINETHIONE
LEUCINE see L-LEUCINE
EPSILON-LEUCINE see 6-AMINOCAPROIC ACID
LEUCO-1,5-DIAMINO-4,8-DIHYDROXYANTHRAQUINONE see 1,5-DI-
    AMINOANTHRARUFIN
LEUCOGEN see L-ASPARAGINASE
LEUCOHARMINE see HARMINE
LEUCOLINE see ISOQUINOLINE
LEUCOLINE see QUINOLINE
LEUCOPARAFUCHSIN see TRIAMINOTRIPHENYLMETHANE
LEUCOTHANE see URETHANE
LEUKAEMOMYCIN C see DAUNOMYCIN
LEUKERAN see PURINE-6-THIOL
LEVADONE see L-METHADONE HYDROCHLORIDE
LEVEDRINE see l-BENZEDRINE SULFATE
LEVO-DROMORAN see LEVORPHANOL
LEVONORADRENALINE see L-NOREPINEPHRINE
LEVOPHACETOPERAN HYDROCHLORIDE see LIDEPRAN HYDROCHLO-
    RIDE
LEVORPHAN see LEVORPHANOL
LEVOTHYL see L-METHADONE HYDROCHLORIDE
LEVUGEN see FRUCTOSE
LEVULOSE see FRUCTOSE
LEWISITE (ARSENIC COMPOUND) see CHLOROVINYLARSINE DICHLO-
    RIDE
LEWISITE I OXIDE see DICHLORO(2-CHLOROVINYL)ARSINE OXIDE
LEWISITE II see BIS(2-CHLOROVINYL)CHLORO ARSINE
LEYTOSAN see PHENYL MERCURY UREA
LIATRIX OLEORESIN see DEERTONGUE INCOLORE
LIBAVIUS FUMING SPIRIT see TIN (IV) CHLORIDE (1:4)
LIBRAX see LIBRIUM
LIBRININ see LIBRIUM
LIDOCAINE see XYLOCAIN
LIDOCAINE HYDROCHLORIDE see XYLOCAINE HYDROCHLORIDE
LIGHT CAMPHOR OIL see CAMPHOR OIL
LIGHT LIGROIN see NAPHTHA, COAL TAR
LIGHT OIL OF CAMPHOR see CAMPHOR OIL
LIGHT SPAR see CALCIUM (II) SULFATE DIHYDRATE (1:1:2)
LIGNASAN FUNGICIDE see BIS(ETHYL MERCURI)PHOSPHATE
LIGNITE OIL see MINERAL OIL
LIGNOCAINE HYDROCHLORIDE see XYLOCAINE HYDROCHLORIDE
LIGROIN see PETROLEUM SPIRITS
LIMARSOL MALAGRIDE see ACETPHENARSINE
LIME see CALCIUM OXIDE
LIME ACETATE see CALCIUM ACETATE
LIME, BURNED see CALCIUM OXIDE
LIME CHLORIDE see HYPOCHLOROUS ACID, CALCIUM SALT
LIMED ROSIN see CALCIUM RESINATE
LIMED ROSIN, FUSED see CALCIUM RESINATE (FUSED)

LIME-NITROGEN see CALCIUM CYANAMIDE
LIME PYROLIGNITE see CALCIUM ACETATE
LIMESTONE see CALCIUM (II) CARBONATE (1:1)
LIME, UNSLAKED (DOT) see CALCIUM OXIDE
LIME WATER see CALCIUM HYDROXIDE
D-(+)-LIMONENE see d-LIMONENE
(+)-R-LIMONENE see d-LIMONENE
LIMONENE OXIDE see CAJEPUTOL
LINAMPHETA see DL-AMPHETAMINE SULFATE
LINDANE see BENZENE HEXACHLORIDE-gamma isomer
ALPHA-LINDANE see BENZENE HEXACHLORIDE-alpha-isomer
BETA-LINDANE see trans-alpha-BENZENEHEXACHLORIDE
DELTA-LINDANE see delta-BHC
DELTA LINDANE see delta-BENZENE HEXACHLORIDE
9,12-LINOLEIC ACID see LINOLEIC ACID
LIQUAEMIN SODIUM see HEPARIN
LIQUID BRIGHT PLATINUM see PLATINUM
LIQUID CAMPHOR see CAMPHOR OIL
LIQUID CARBONIC GAS see LIQUEFIED CARBON DIOXIDE
LIQUID ETHYENE see ETHYLENE
LISERGAN see ACETOPROMAZINE
LITHAMIDE see LITHIUM AMIDE
LITHARGE see LEAD MONOXIDE
LITHIUM AMIDE, POWDER see LITHIUM AMIDE
LITHIUM ANTIMONIOTHIOMALATE see LITHIUM ANTIMONY THIOMA-
    LATE
LITHIUM FLUORURE (FRENCH) see LITHIUM FLUORIDE
LITHIUM METAL see LITHIUM
LO MICRON TALC USP, BC 2755 see TALC (powder)
LON 41 see o,p-DIMETHYL-beta-PHENYLETHYLHYDRAZINE DIHYDRO-
    GENSULFATE
LONDON PURPLE, SOLID (DOT) see LONDON PURPLE
BETA-LONGILOBINE see RETRORSINE
N-LOST see BIS(2-CHLOROETHYL)METHYLAMINE HYDROCHLORIDE
S-LOST see BIS(2-CHLOROETHYL)SULFIDE
LOSUNGSMITTEL APV see CARBITOL CELLOSOLVE
LSD see N,N-DIETHYLLYSERGAMIDE
D-LSD see N,N-DIETHYLLYSERGAMIDE
LSD-25 see N,N-DIETHYLLYSERGAMIDE
LSD-25-PYRROLIDATE see LYSERGIC ACID PYROLIDATE
LUCEL see CHLORQUINOX
LUCIDOL see BENZOYL PEROXIDE
LUCITE see POLYMETHYLMETHACRYLATE
LUMINAL SODIUM see SODIUM LUMINAL
LUNAR CAUSTIC see SILVER (I) NITRATE (1:1)
LUTEAL HORMONE see PROGESTERONE
LUTEOHORMONE see PROGESTERONE
LUTESTRAL (FRENCH) see C-QUENS
LYCOPERSICIN see TOMATINE
LYE see SODIUM HYDROXIDE
LYE see POTASSIUM HYDROXIDE
LYE SOLUTION see SODIUM HYDROXIDE (LIQUID)
LYSERGAMID see N,N-DIETHYLLYSERGAMIDE
LYSERGAMIDE, N,N-DIETHYL- see N,N-DIETHYLLYSERGAMIDE
LYSERGAURE DIETHYLAMID see N,N-DIETHYLLYSERGAMIDE
D-LYSERGIC ACID DIETHYLAMIDE see N,N-DIETHYLLYSERGAMIDE
LYSERGIC ACID DIETHYLAMIDE-25 see N,N-DIETHYLLYSERGAMIDE
LYSERGIC ACID DIETHYLAMIDE, 1-ISOMER see D-LYSERGIC ACID
    DIETHYLAMIDE
D-LYSERGIC ACID-1-HYDROXYMETHYLETHYLAMIDE see D-LYSERGIC
    ACID-L,2-PROPANOLAMIDE
D-LYSERGIC ACID-L,2-PROPANOLAMIDE see D-LYSERGIC ACID-
    L,2-PROPANOLAMIDE
D-LYSERGIC ACID MONOETHYLAMIDE see LYSERGIC ACID ETHYL-
    AMIDE
LYSERGIC ACID PROPANOLAMIDE see D-LYSERGIC ACID-L,2-PRO-
    PANOLAMIDE
D-LYSERGIC ACID PYRROLIDIDE see LYSERGIC ACID PYROLIDATE
LYSERGIDE see N,N-DIETHYLLYSERGAMIDE

LYSERGSAUEREDIAETHYLAMID see N,N-DIETHYLLYSERGAMIDE

N-LYSINOMETHYLTETRACYCLINE see MUCOMYCIN

1-N-2-MA (RUSSIAN) see 2-METHYL-1-NITROANTHRAQUINONE

MACE (LACRYMATOR) see 2-CHLOROACETOPHENONE

MACQUER'S SALT see ARSENIC ACID, MONOPOTASSIUM SALT

MAGBOND see BENTONITE

PARA-MAGENTA see BASIC PARAFUCHSINE

MAGIC METHYL see METHYL FLUOROSULFATE

MAGNESIA see MAGNESIUM OXIDE

MAGNESIAAREVEDSONITE see AMPHIBOLE (SEE ALSO ASBESTOS)

MAGNESIAARPHVEDSONITE see AMPHIBOLE (SEE ALSO ASBESTOS)

MAGNESIA USTA see MAGNESIUM OXIDE

MAGNESIO (ITALIAN) see MAGNESIUM

MAGNESIUM ALUMINUM PHOSPHIDE (DOT) see ALUMINUM MAGNE-
SIUM PHOSPHIDE

MAGNESIUM ARSENATE see ARSENIC ACID, MAGNESIUM SALT

MAGNESIUM ARSENATE PHOSPHOR see ARSENIC ACID, MAGNESIUM
SALT

MAGNESIUM BORINGS see MAGNESIUM SHAVINGS

MAGNESIUM CLIPPINGS see MAGNESIUM SHAVINGS

MAGNESIUM DIACETATE see MAGNESIUM ACETATE

MAGNESIUM FLUOSILICATE see MAGNESIUM HEXAFLUOROSILICATE

MAGNESIUMFOSFIDE (DUTCH) see MAGNESIUM PHOSPHIDE

MAGNESIUM GOLD PURPLE see GOLD

MAGNESIUM METAL (DOT) see MAGNESIUM

MAGNESIUM PELLETS see MAGNESIUM

MAGNESIUM PERCHLORATE see PERCHLORIC ACID, MAGNESIUM
SALT

MAGNESIUM PEROXIDE, SOLID (DOT) see MAGNESIUM PEROXIDE

MAGNESIUM POWDERED see MAGNESIUM

MAGNESIUM RIBBONS see MAGNESIUM

MAGNESIUM SCALPINGS see MAGNESIUM SHAVINGS

MAGNESIUM SCRAP (DOT) see MAGNESIUM SHAVINGS

MAGNESIUM SHEET see MAGNESIUM SHAVINGS

MAGNESIUM SILICOFLUORIDE see MAGNESIUM HEXAFLUOROSILICATE

MAGNESIUM TURNINGS see MAGNESIUM SHAVINGS

MAGNESIUM TURNINGS see MAGNESIUM

MAGNEZU TLENEK (POLISH) see MAGNESIUM OXIDE

MALACHITE see COPPER (II) CARBONATE HYDROXIDE (2:1:2)

MALATHION see CARBETHOXY MALATHION

MALATHION (DOT) see CARBETHOXY MALATHION

MALATHION LV CONCENTRATE see CARBETHOXY MALATHION

MALAYAN CAMPHOR see BORNEOL

MALEATE ACIDE DE L'ACETYL-3-DIMETHYLAMINO-3-PROPYL-10-
PHENOTHIAZINE (FRENCH) see ACEPROMAZINE MALEATE

MALEIC ACID see cis-BUTENEDIOIC ACID

MALEIC ACID ANHYDRIDE see MALEIC ANHYDRIDE

MALEIC ACID HYDRAZIDE see MALEIC HYDRAZIDE

MALEIC ACID, METHYL ERGONOVINE see METHYLERGONOVINE MA-
LEATE

MALEIC HYDRAZIDE DIETHANOLAMINE SALT see DIETHANOLAMMO-
NIUM MALEIC HYDRAZIDE

MALEINIC ACID see cis-BUTENEDIOIC ACID

MALEINIMIDE see MALEIMIDE

MALENIC ACID see cis-BUTENEDIOIC ACID

N,N-MALEOYLHYDRAZINE see MALEIC HYDRAZIDE

MALIVAN see DIMEFLINE

MALONALDEHYDE see PROPANEDIAL

MALONDIALDEHYDE see PROPANEDIAL

MALONIC ACID ETHYL ESTER NITRILE see ETHYL CYANOACETATE

MALONIC ALDEHYDE see PROPANEDIAL

MALONIC DIALDEHYDE see PROPANEDIAL

MALONIC DINITRILE see MALONONITRILE

MALONIC ESTER see ETHYL MALONATE

MALONIC MONONITRILE see CYANOACETIC ACID

MALONITRILE HYDRAZIDE see CYANACETIC ACID HYDRAZIDE

MALONONITRILE HYDRAZIDE see CYANACETIC ACID HYDRAZIDE

MALONYLDIALDEHYDE see PROPANEDIAL

MALTOBIOSE see MALTOSE

D-MALTOSE see MALTOSE

MALT SUGAR see MALTOSE

ALPHA-MALT SUGAR see MALTOSE

MANCHESTER YELLOW see 2,4-DINITRO-1-NAPHTHOL

1-MANDELONITRILE-beta-GLUCURONIC ACID see LAETRILE

MANGAAN (II)-(N,N'-ETHYLEEN-BIS(DITHIOCARBAMAAT)) (DUTCH) see
VANCIDE

MANGAN (POLISH) see MANGANESE

MANGAN (II)-(N,N'-AETHYLEN-BIS(DITHIOCARBAMATE)) (GERMAN)
see VANCIDE

MANGANESE(2+) ACETATE see MANGANESE ACETATE

MANGANESE BLACK see MANGANESE DIOXIDE

MANGANESE(II) ACETATE see MANGANESE ACETATE

MANGANESE CYCLOPENTADIENYLTRICARBONYL see CYCLO PENTA-
DIENYLMANGANESE TRICARBONYL

MANGANESE DIACETATE see MANGANESE ACETATE

MANGANESE DICHLORIDE see MANGANESE (II) CHLORIDE (1:2)

MANGANESE ETHYLENE-1,2-BISDITHIOCARBAMATE see VANCIDE

MANGANESE MANGANATE see MANGANESE (III) OXIDE

MANGANESE MONOXIDE see MANGANESE (II) OXIDE

MANGANESE TETROXIDE see MANGANESE OXIDE

MANGANESE (II) ETHYLENEBIS(DITHIOCARBAMATE) see VANCIDE

MANGANESE (II) ETHYLENE DI(DITHIOCARBAMATE) see VANCIDE

MANGANESE TRIOXIDE see MANGANESE (III) OXIDE

MANGANESE ZINC BERYLLIUM SILICATE see BERYLLIUM MANGANESE
ZINC SILICATE

MANGANIC OXIDE see MANGANESE (III) OXIDE

MANGANOMANGANIC OXIDE see MANGANESE OXIDE

MANGANOUS ACETATE see MANGANESE ACETATE

MANGANOUS CHLORIDE see MANGANESE (II) CHLORIDE (1:2)

MANGANOUS OXIDE see MANGANESE (II) OXIDE

MANGANOUS SULFATE see MANGANESE (II) SULFATE (1:1)

MANNIT-LOST (GERMAN) see MANNOMUSTINE

MANNIT-MUSTARD (GERMAN) see MANNOMUSTINE

D-MANNITOL BUSULFAN see BIS(METHANE SULFONYL)-D-MANNITOL

D-MANNITOL HEXANITRATE see MANNITOL HEXANITRATE

MANNITOL MUSTARD DIHYDROCHLORIDE see MANNOMUSTINE DIHY-
DROCHLORIDE

MANNITOL MYLERAN see BIS(METHANE SULFONYL)-D-MANNITOL

MANNITOL NITROGEN MUSTARD see MANNOMUSTINE

MANUFACTURED IRON OXIDES see IRON (III) OXIDE

MANZATE MANEB FUNGICIDE see VANCIDE

MAPLE AMARANTH see FOOD RED 2

MAPLE INDIGO CARMINE see DISODIUM INDIGO-5,5-DISULFONATE

MAPROFIX 563 see SULFURIC ACID, MONODODECYL ESTER, SODIUM
SALT

MARBLE see CALCIUM (II) CARBONATE (1:1)

MARCOPHANE see IMIPHOS

MAREVAN (SODIUM SALT) see COUMADIN SODIUM

MARIHUANA see CANNABIS

MARITUS YELLOW see 2,4-DINITRO-1-NAPHTHOL

MARKOFANE see IMIPHOS

MARSH GAS see METHANE

MATRICARIA CAMPHOR see CAMPHOR

MATSUTAKE ALCOHOL (JAPANESE) see 1-OCTEN-3-OL

MATULANE see PROCARBAZINE

MAYT see MAYTANSINE

7-MBA see 10-METHYL-1,2-BENZANTHRACENE

MCN-485 see 2-AMINO-5-CHLOROBENZOXAZOLE

MCNEIL 481 see 5-(1,3-DIMETHYL-2-BUTENYL)-5-ETHYLBARBITURIC
ACID

MDS see 2,2'-DITHIOBIS(PYRIDINE-1-OXIDE)MAGNESIUM SULFATE
TRIHYDRATE

MEADOW GREEN see CUPRIC ACETOARSENITE

MEBUBARBITAL SODIUM see NEMBUTAL SODIUM

MEBUMAL NATRIUM see NEMBUTAL SODIUM

MEBUMAL SODIUM see NEMBUTAL SODIUM

MECHLORETHAMINE OXIDE see 2-CHLORO-N-(2-CHLOROETHYL)-
N-METHYL ETHANAMINE-N-OXIDE

MECHLORETHAMINE OXIDE HYDROCHLORIDE see 2-CHLORO-N-(2-

CHLORO ETHYL)-N-METHYL ETHANAMINE-N-OXIDE HYDROCHLO-
RIDE

MECODIN see dl-METHADONE HYDROCHLORIDE

MECODIN see METHADONE

MEGABION (INDIAN) see VITAMIN B$_{12}$ COMPLEX

MEK see 2-BUTANONE

MELADININ see XANTHOTOXIN

MELANILINE see DIPHENYLGUANIDINE

MELETIN see QUERCETIN

MELILOTIN see HYDROCOUMARIN

MELLERIL see 10-(2-(1-METHYL-2-PIPERIDYL)ETHYL)-2-(METHYL-
THIO)PHENOTHIAZINE

MELLARIL HYDROCHLORIDE see 10-(2-(1-METHYL-2-PIPERIDYL)
ETHYL)-2-METHYLTHIOPHENOTHIAZINE HYDROCHLORIDE

MELPHALAN (RUSSIAN) see 3-(p-(BIS(beta-CHLOROETHYL)AMINO)-
PHENYL)-D,L-ALANINE HYDROCHLORIDE

MENOCIL see 2-AMINO-5-PHENYL-OXAZOLINE FORMATE

P-MENTHANE HYDROPEROXIDE see p-MENTHANE-8-HYDROPEROXIDE

P-MENTHAN-3-OL see MENTHOL

DL-3-P-MENTHANOL see dl-MENTHOL

3-P-MENTHOL see dl-MENTHOL

U.S.P. MENTHOL see l-MENTHOL

MENTHOL RACEMIC see dl-MENTHOL

MEPERIDIDE see 4′-FLUORO-4-(n-(4-PYRROLIDINAMIDO-4-m-TOLYPI-
PERIDINO)-BUTYROPHENONE

MEPHENAMINE HYDROCHLORIDE see ORPHENADRINE HYDROCHLO-
RIDE

MEPHOBARBITAL see 5-ETHYL-N-METHYL-5-PHENYL BARBITURIC
ACID

MEPHOBARBITONE see 5-ETHYL-N-METHYL-5-PHENYL BARBITURIC
ACID

MEPROBAM see MILTOWN

MEPROBAMAT (GERMAN) see MILTOWN

MEPROBAMATE see MILTOWN

MEPRODINE (GERMAN) see NU-1932

MERBAPHEN see (4-(CARBOXY METHOXY)-3-CHLOROPHENYL)(5,5-
DIETHYL-2,4,6(1H,3H,5H)-PYRIMIDINETRIONATO-O(sup 2)MER-
CURY, MONOSODIUM SALT

MERBROMIN see MERCUROCHROME

MERCAPTAMINE see 2-AMINOETHANETHIOL

MERCAPTAN AMYLIQUE (FRENCH) see 1-PENTANETHIOL

MERCAPTAN METHYLIQUE (FRENCH) see METHANETHIOL

MERCAPTAZOLE see 2-MERCAPTO-1-METHYLIMIDAZOLE

2-MERCAPTOACETANILIDE see alpha-MERCAPTOACETANILIDE

MERCAPTOACETIC ACID see 2-MERCAPTOACETIC ACID

ALPHA-MERCAPTOACETIC ACID see 2-MERCAPTOACETIC ACID

MERCAPTOACETIC ACID, DIESTER WITH DITHIO-P-UREIDOBENZENE-
ARSONOUS ACID see 4-CARBAMIDOPHENYL BIS(CARBOXY
METHYL THIO)ARSENITE

MERCAPTOACETIC ACID DIESTER WITH DITHIO-P-UREIDOBENZENE-
ARSONOUS ACID see 4-CARBAMIDOPHENYL BIS(CARBOXY
METHYL THIO)ARSENITE

MERCAPTOACETIC ACID SODIUM SALT see SODIUM THIOGLYCOLATE

4-(MERCAPTOACETYL)MORPHOLINE O,O-DIMETHYL PHOSPHORODI-
THIOATE see MORPHOTHION

BETA-MERCAPTOALANINE see L-CYSTEINE

O-MERCAPTOANILINE see 2-AMINOBENZENETHIOL

2-MERCAPTOBENZIMIDAZOLE see 2-BENZIMIDAZOLETHIOL

MERCAPTOBENZOIMIDAZOLE see 2-BENZIMIDAZOLETHIOL

2-MERCAPTOBENZOIMIDAZOLE see 2-BENZIMIDAZOLETHIOL

2-MERCAPTOBENZOTHIAZOLE see 2-BENZOTHIAZOLETHIOL

2-MERCAPTOBENZOTHIAZOLEDISULFIDE see BENZOTHIAZOLE DISUL-
FIDE

2-MERCAPTOBENZOTHIAZOLE ZINC SALT see BIS(2-BENZOTHIAZO-
LYTHIO)ZINC

2-MERCAPTOBENZOTHIAZYLDISULFIDE see BENZOTHIAZOLE DISUL-
FIDE

(MERCAPTOBUTANEDIOATO(1-))GOLD DISODIUM SALT see GOLD SO-
DIUM THIOMALATE

BETA-MERCAPTOETHANOL see 2-MERCAPTOETHANOL

(2-MERCAPTOETHYL)AMINE see 2-AMINOETHANETHIOL

2-MERCAPTOETHYLAMINE (OXIDIZED) see beta-MERCAPTOETHYL-
AMINE DISULFIDE

2-MERCAPTO-4-HYDROXY-6METHYLPYRIMIDINE see 6-METHYLTHIO-
URACIL

2-MERCAPTO-4-HYDROXY-6-N-PROPYLPYRIMIDINE see 6-PROPYL-2-
THIOURACIL

2-MERCAPTO-4-HYDROXYPYRIMIDINE see 2-THIOURACIL

2-MERCAPTOIMIDAZOLINE see 2-IMIDAZOLINETHIOL

(MERCAPTOMETHYL)BENZENE see alpha-TOLUENETHIOL

3-(MERCAPTOMETHYL)-1,2,3-BENZOTRIAZIN-4(3H)-ONE-O,O-
DIMETHYL PHOSPHORODITHIOATE-S-ESTER see AZINPHOS METHYL

N-(MERCAPTOMETHYL)PHTHALIMIDE S-(O,O-DIMETHYL PHOS-
PHORODITHIOATE) see PHTHALIMIDOMETHYL-O,O-DIMETHYL PHOS-
PHORODITHIOATE

2-MERCAPTO-6-METHYL-4-PYRIMIDONE see 6-METHYLTHIOURACIL

2-MERCAPTO-6-METHYLPYRIMID-4-ONE see 6-METHYLTHIOURACIL

2-MERCAPTONAPHTHALENE see 2-NAPHTHALENETHIOL

MERCAPTOPHOS see DEMETON-O + DEMETON-S

2-MERCAPTO-6-PROPYLPYRIMID-4-ONE see 6-PROPYL-2-THIOURACIL

2-MERCAPTO-6-PROPYL-4-PYRIMIDONE see 6-PROPYL-2-THIOURACIL

6-MERCAPTOPURINE see PURINE-6-THIOL

6-MERCAPTOPURINE RIBOSIDE see MERCAPTOPURINE RIBONUCLEOSIDE

2-MERCAPTO-4-PYRIMIDINOL see 2-THIOURACIL

2-MERCAPTOPYRIMID-4-ONE see 2-THIOURACIL

2-MERCAPTO-4-PYRIMIDONE see 2-THIOURACIL

MERCAPTOSUCCINIC ACID ANTIMONATE (III) HEXALITHIUM SALT see
LITHIUM ANTIMONY THIOMALATE

MERCAPTOSUCCINIC ACID-S-ANTIMONY DERIVATIVE LITHIUM SALT see
LITHIUM ANTIMONY THIOMALATE

7-MERCAPTO-1,3,4,6-TETRAZAINDENE see PURINE-6-THIOL

ALPHA-MERCAPTOTOLUENE see alpha-TOLUENETHIOL

D-3-MERCAPTOVALINE HYDROCHLORIDE see PENICILLAMINE HYDRO-
CHLORIDE

MERCAPTURIC ACID see N-ACETYL-L-CYSTEINE

MERCURAMIDE see SODIUM MERSALYL

MERCURAN see METHOXYETHYL MERCURIC ACETATE

MERCURANINE see MERCUROCHROME

MERCURE (FRENCH) see MERCURY

MERCURIACETATE see MERCURIC ACETATE

MERCURIC AMMONIUM CHLORIDE, SOLID see MERCURY AMIDE CHLO-
RIDE

MERCURIC ARSENATE see MERCURY(II)-orthoARSENATE

MERCURIC BENZOATE see MERCURY (II) BENZOATE

MERCURIC BENZOATE, SOLID (DOT) see MERCURY (II) BENZOATE

MERCURIC BICHLORIDE see CORROSIVE SUBLIMATE

MERCURIC BROMIDE see MERCURY (II) BROMIDE (1:2)

MERCURIC BROMIDE, SOLID (DOT) see MERCURY (II) BROMIDE (1:2)

MERCURIC CHLORIDE, AMMONIATED see MERCURY AMIDE CHLORIDE

MERCURIC CYANIDE, SOLID (DOT) see MERCURY (II) CYANIDE

MERCURIC DIACETATE see MERCURIC ACETATE

MERCURIC IODIDE see MERCURY (II) IODIDE

MERCURIC IODIDE, RED see MERCURY (II) IODIDE

MERCURIC IODIDE, SOLID (DOT) see MERCURY (II) IODIDE

MERCURIC NITRATE see MERCURY (II) NITRATE (1:2)

MERCURIC OXIDE, RED see MERCURIC OXIDE

MERCURIC OXIDE, SOLID (DOT) see MERCURIC OXIDE

MERCURIC OXIDE, YELLOW see MERCURIC OXIDE

MERCURIC POTASSIUM IODIDE see NESSLER REAGENT

MERCURIC SULFATE, SOLID (DOT) see MERCURY(II) SULFATE (1:1)

MERCURIC SULFOCYANIDE see MERCURIC SULFOCYANATE

MERCURIC THIOCYANATE see MERCURIC SULFOCYANATE

MERCURIO (ITALIAN) see MERCURY

MERCURIPHENYL ACETATE see ACETOXYPHENYLMERCURY

MERCURIPHENYL CHLORIDE see PHENYL MERCURIC CHLORIDE

MERCUROCHLORIDE (DUTCH) see MERCUROUS CHLORIDE

MERCUROCHROME-220 SOLUBLE see MERCUROCHROME

MERCUROCOL see MERCUROCHROME

MERCUROME see MERCUROCHROME

MERCUROPHAGE see MERCUROCHROME

MERCUROTHIOLATE see MERTHIOLATE SODIUM

METHEGRIN see METHYLERGONOVINE MALEATE

METHENAMINE see HEXAMETHYLENETETRAMINE

N,N'-METHENYL-O-PHENYLENEDIAMINE see BENZIMIDAZOLE

METHENYL TRIBROMIDE see BROMOFORM

METHENYL TRICHLORIDE see CHLOROFORM

METHIAMAZOLE see 2-MERCAPTO-1-METHYLIMIDAZOLE

METHOHEXITONE SODIUM see METHOHEXITAL SODIUM

METHOLENE 2218 see METHYL STEARATE

METHOPROMAZINE MALEATE see METHOXYPROMAZINE MALEATE

METHOPTERIN see METHOTREXATE

METHORPHINAN see N-METHYL-3-HYDROXYMORPHINAN

METHOTEXTRATE see METHOTREXATE

2-METHOXYACETOACETANILIDE see ACETOACETYL-O-ANISIDINE

O-METHOXYACETOACETANILIDE see ACETOACETYL-O-ANISIDINE

2'-METHOXYACETOACETANILIDE see ACETOACETYL-O-ANISIDINE

2-METHOXY-AETHANOL (GERMAN) see ETHYLENE GLYCOL METHYL ETHER

2-METHOXYAETHYLACETAT (GERMAN) see ETHYLENE GLYCOL MONOMETHYL ETHER ACETATE

1-METHOXY-AETHYL-AETHYLNITROSAMIN (GERMAN) see 1-METHOXY ETHYL ETHYLNITROSAMINE

1-METHOXY-AETHYL-METHYLNITROSAMIN (GERMAN) see 1-METHOXY ETHYL METHYLNITROSAMINE

METHOXYAETHYLQUECKSILBERCHLORID (GERMAN) see 2-METHOXYETHYLMERCURY CHLORIDE

P-METHOXYALLYLBENZENE see p-ALLYLANISOLE

3-METHOXY-4-AMINOAZOBENZENE see 4-(PHENYLAZO)-O-ANISIDINE

4-METHOXYAMPHETAMINE HYDROCHLORIDE see dl,4-METHOXYAMPHETAMINE HYDROCHLORIDE

4-METHOXYANILINE see p-ANISIDINE

P-METHOXYANILINE see p-ANISIDINE

2-METHOXYANILINE HYDROCHLORIDE see O-ANISIDINE HYDROCHLORIDE

4-METHOXYBENZALDEHYDE see p-ANISALDEHYDE

P-METHOXYBENZALDEHYDE see p-ANISALDEHYDE

4-METHOXYBENZENAMINE see p-ANISIDINE

METHOXYBENZENE see ANISOLE

4-METHOXYBENZENEACETIC ACID see p-METHOXYPHENYLACETIC ACID

4-METHOXYBENZENEAMINE see p-ANISIDINE

4-METHOXY-1,3-BENZENEDIAMINE SULFATE see 2,4-DIAMINOANISOLE SULPHATE

4-METHOXY-1,3-BENZENEDIAMINE SULPHATE see 2,4-DIAMINOANISOLE SULPHATE

2-METHOXY-4H-1,2,3-BENZODIOXAPHOSPHORINE-2-SULFIDE see SALITHION

O-METHOXYBENZOHYDRAZIDE see O-ANISIC ACID, HYDRAZIDE

3-METHOXYBENZOIC ACID see m-ANISIC ACID

M-METHOXYBENZOIC ACID see m-ANISIC ACID

2-METHOXYBENZOIC ACID HYDRAZIDE see O-ANISIC ACID, HYDRAZIDE

4-METHOXYBENZOIC ACID HYDRAZIDE see p-ANISIC ACID, HYDRAZIDE

O-METHOXYBENZOIC ACID HYDRAZIDE see O-ANISIC ACID, HYDRAZIDE

P-METHOXYBENZOIC ACID HYDRAZIDE see p-ANISIC ACID, HYDRAZIDE

P-METHOXYBENZOIC HYDRAZIDE see p-ANISIC ACID, HYDRAZIDE

METHOXYBENZOYL CHLORIDE see ANISOYL CHLORIDE

2-METHOXYBENZOYL HYDRAZIDE see O-ANISIC ACID, HYDRAZIDE

4-METHOXYBENZOYL HYDRAZIDE see p-ANISIC ACID, HYDRAZIDE

O-METHOXYBENZOYLHYDRAZIDE see O-ANISIC ACID, HYDRAZIDE

2-METHOXYBENZOYLHYDRAZINE see O-ANISIC ACID, HYDRAZIDE

4-METHOXYBENZOYLHYDRAZINE see p-ANISIC ACID, HYDRAZIDE

(P-METHOXYBENZOYL)HYDRAZINE see p-ANISIC ACID, HYDRAZIDE

N-(P-METHOXYBENZYL)-N',N'-DIMETHYL-N-2-PYRIDYLETHYLENEDIAMINE see WAIT'S GREEN MOUNTAIN ANTIHISTAMINE

N-P-METHOXYBENZYL-N',N'-DIMETHYL-N-ALPHA-PYRIDYLETHYLENEDIAMINE see WAIT'S GREEN MOUNTAIN ANTIHISTAMINE

N-P-METHOXYBENZYL-N'-N'-DIMETHYL-N-alpha-PYRIDYLETHYLENEDIAMINE MALEATE see DIAMINIDE MALEATE

P-METHOXYBENZYL FORMATE see p-METHOXYPHENYLACETIC ACID

4-METHOXYBIPHENYL see p-PHENYLANISOLE

P-METHOXYBIPHENYL see p-PHENYLANISOLE

2-(METHOXYCARBONYL)ANILINE see ANTHRANILIC ACID, METHYL ESTER

METHOXYCARBONYLETHYLENE see METHYL ACRYLATE

(2-METHOXYCARBONYL-1-METHYL-VINYL)-DIMETHYL-FOSFAAT (DUTCH) see MEVINPHOS

(2-METHOXYCARBONYL-1-METHYL-VINYL)-DIMETHYL-PHOSPHAT (GERMAN) see MEVINPHOS

2-METHOXYCARBONYL-1-METHYLVINYL DIMETHYLPHOSPHATE see MEVINPHOS

CIS-2-METHOXYCARBONYL-1-METHYLVINYL DIMETHYLPHOSPHATE see MEVINPHOS

P,P'-METHOXYCHLOR see DIMETHOXY-DDT

2-METHOXY-6-CHLORO-9-(4-BIS(2-CHLOROETHYL)AMINO-1-METHYL-BUTYLAMINO)ACRIDINE DIHYDROCHLORIDE see QUINACRINE MUSTARD DIHYDROCHLORIDE

2-METHOXY-6-CHLORO-9-DIETHYLAMINOPENTYLAMINOACRIDINE see ATABRINE

2-METHOXY-5-CHLOROPROCAINAMIDE see 4-AMINO-5-CHLORO-N-(2-(DIETHYLAMINO)ETHYL)-n-ANISAMIDE

6-METHOXYCINCHONINE see QUININE

3-METHOXY-DBA see 5-METHOXYDIBENZ(a,h)ANTHRACENE

9-METHOXY-DBA see 7-METHOXYDIBENZ(a,h)ANTHRACENE

3-METHOXY-1,2:5,6-DIBENZANTHRACENE see 5-METHOXYDIBENZ(a,h)ANTHRACENE

9-METHOXY-1,2,5,6-DIBENZANTHRACENE see 7-METHOXYDIBENZ(a,h)ANTHRACENE

5-METHOXY-2-(DIMETHOXYPHOSPHINYLTHIOMETHYL)PYRONE-4 see ENDOTHION

2-METHOXY-10-(3'-DIMETHYLAMINOPROPYL)PHENOTHIAZINE see 2-METHOXYPROMAZINE

10-METHOXY-1,6-DIMETHYL ERGOLINE-8-BETA-METHANOL-5-BROMO-3-PYRIDINECARBOXYLATE (ESTER) see NICOTERGOLINE

1-METHOXY-2,4-DINITROBENZENE see 2,4-DINITROANISOL

2-METHOXYETHANOL see ETHYLENE GLYCOL METHYL ETHER

2-METHOXYETHANOL see ETHYLENE GLYCOL MONOMETHYL ETHER

2-METHOXYETHANOL, ACRYLATE see METHYL CELLOSOLVE ACRYLATE

METHOXY ETHER OF PROPYLENE GLYCOL see PROPYLENE GLYCOL MONOMETHYL ETHER

3-METHOXY-17-ALPHA-ETHINYLESTRADIOL see 3-METHYLETHYNYLESTRADIOL

3-METHOXY-17-ALPHA-ETHINYLOESTRADIOL see 3-METHYLETHYNYLESTRADIOL

2-METHOXY-ETHYL ACETAAT (DUTCH) see ETHYLENE GLYCOL MONOMETHYL ETHER ACETATE

2-METHOXYETHYL ACETATE see ETHYLENE GLYCOL MONOMETHYL ETHER ACETATE

S-(N-2-METHOXYETHYLCARBAMOYLMETHYL)DIMETHYL PHOPHOROTHIOLOTHIONATE see AMIDITHION

2-METHOXYETHYLE, ACETATE DE (FRENCH) see ETHYLENE GLYCOL MONOMETHYL ETHER ACETATE

METHOXYETHYL MERCURIC CHLORIDE see 2-METHOXYETHYLMERCURY CHLORIDE

2-METHOXYETHYLMERCURIC CHLORIDE see 2-METHOXYETHYLMERCURY CHLORIDE

(BETA-METHOXYETHYL)MERCURIC CHLORIDE see 2-METHOXYETHYLMERCURY CHLORIDE

METHOXYETHYLMERCURY CHLORIDE see 2-METHOXYETHYLMERCURY CHLORIDE

BETA-METHOXYETHYLMERCURY CHLORIDE see 2-METHOXYETHYLMERCURY CHLORIDE

3-METHOXYETHYNYLESTRADIOL see 3-METHYLETHYNYLESTRADIOL

3-METHOXY-17-ALPHA-ETHYNYLESTRADIOL see 3-METHYLETHYNYLESTRADIOL

3-METHOXY-17-ALPHA-ETHYNYL-1,3,5(10)-ESTRATRIEN-17-BETA-OL see 3-METHYLETHYNYLESTRADIOL

3-METHOXYETHYNYLOESTRADIOL see 3-METHYLETHYNYLESTRADIOL

3-METHOXY-17-ALPHA-ETHYNYL-1,3,5(10)-OESTRATRIEN-17-BETA-OL see 3-METHYLETHYNYLESTRADIOL

1-P-METHOXYFENYL-3,3-DIMETHYLTRIAZEN (CZECH) see 3,3-DIMETHYL-1,p-METHOXYPHENYLTRIAZENE

1-METHOXYFLUOREN-2-AMINE HYDROCHLORIDE see 1-METHOXY-2-FLUORENAMINE HYDROCHLORIDE

7-METHOXY-N-2-FLUORENYLACETAMIDE see 7-METHOXY-2-FAA

N-(7-METHOXYFLUOREN-2-YL)ACETAMIDE see 7-METHOXY-2-FAA

N-(7-METHOXY-2-FLUORENYL)ACETAMIDE see 7-METHOXY-2-FAA

8-METHOXY-(FURANO-3'.2':6.7-COUMARIN) see XANTHOTOXIN

8-METHOXY-2',3',6,7-FUROCOUMARIN see XANTHOTOXIN

8-METHOXY-4',5',6,7-FUROCOUMARIN see XANTHOTOXIN

6-METHOXYHARMAN see HARMINE

3-METHOXY-4-HYDROXYBENZALDEHYDE see VANILLIN

3-METHOXY-4-HYDROXYBENZOIC ACID DIETHYLAMIDE see N,N-DIETHYLVANILLAMIDE

METHOXYHYDROXYETHANE see ETHYLENE GLYCOL METHYL ETHER

METHOXYHYDROXYMERCURIPROPYLSUCCINYLUREA see METHOXYOXIMERCURIPROPYLSUCCINYL UREA

2-METHOXY-10-(3-(4-HYDROXYPIPERIDINO)-2-METHYLPROPYL)PHENOTHIAZINE see LEPTRYL

3-METHOXY-10-(3-(4-HYDROXYPIPERIDYL)-2-METHYLPROPYL)PHENOTHIAZINE see LEPTRYL

METHOXYMETHYL-AETHYLNITROSAMINE (GERMAN) see METHOXYMETHYL ETHYL NITROSAMINE

3-METHOXYMETHYLAMINOAZOBENZENE see N-METHYL-4-(PHENYLAZO)-o-ANISIDINE

2-METHOXY-5-METHYLANILINE see 5-METHYL-o-ANISIDINE

4-METHOXY-2-METHYLANILINE see 2-METHYL-p-ANISIDINE

2-METHOXY-5-METHYLBENZENAMINE see 5-METHYL-o-ANISIDINE

1-METHOXY-1-METHYL-3-(3,4-DICHLOROPHENYL)UREA see 3-(3,4-DICHLOROPHENYL)-1-METHOXYMETHYLUREA

4-METHOXYMETHYL-5-HYDROXY-6-METHYL-3-PYRIDINEMETHANOL HYDROCHLORIDE see 4-METHOXYMETHYLPYRIDOXINE HYDROCHLORIDE

2-METHOXY-10-(2-METHYL-3-(4-HYDROXYPIPERIDINO)PROPYL)PHENOTHIAZINE see LEPTRYL

METHOXYMETHYL-METHYLNITROSAMIN (GERMAN) see METHOXYMETHYL METHYLNITROSAMINE

3-METHOXY-5-METHYL-4-OXO-2,5-HEXADIENOIC ACID see PENICILLIC ACID

N-(7-METHOXY-3-METHYL-4-OXO-2-PHENYL-4H-CHROMEN-8-YL)-METHYL-N,N-DIMETHYLAMINE see DIMEFLINE

4-METHOXYMETHYLPYRIDOXOL HYDROCHLORIDE see 4-METHOXYMETHYLPYRIDOXINE HYDROCHLORIDE

7-METHOXY-1-METHYL-9H-PYRIDO(3,4-b)INDOLE see HARMINE

P-METHOXY-BETA-METHYLSTYRENE see p-PROPENYL ANISOLE

3-METHOXY-4-MONOMETHYLAMINOAZOBENZENE see N-METHYL-4-(PHENYLAZO)-o-ANISIDINE

2-METHOXY-5-NITROANILINE see 5-NITRO-o-ANISIDINE

2-METHOXY-5-NITROBENZENAMINE see 5-NITRO-o-ANISIDINE

8-METHOXY-6-NITROPHENANTHOL (3,4-d) 1,3-DIOXOLE-5-CARBOXYLIC ACID see ARISTOLOCHINE

3-METHOXY-17-ALPHA-19-NORPREGNA-K,3,5(10)-TRIEN-20-YN-17-OL see 3-METHYLETHYNYLESTRADIOL

3-METHOXY-19-NOR-17-ALPHAPREGNA-1,3,5(10)-TRIEN-10-YN-17-OL see 3-METHYLETHYNYLESTRADIOL

(17-ALPHA)-3-METHOXY-19-NORPREGN-1,3,5(10)-TRIEN-20-YN-17-OL see 3-METHYLETHYNYLESTRADIOL

S-5-METHOXY-4-OXOPYRAN-2-YLMETHYL DIMETHYL PHOSPHOROTHIOATE see ENDOTHION

S-((5-METHOXY-2-OXO-1,3,4-THIADIAZOL-3(2H)-YL)METHYL) O,O-DIMETHYL PHOSPHORODITHIOATE see O,O-DIMETHYL-S-(5-METHOXY-1,3,4-THIADIAZOLINYL-3-METHYL)DITHIO PHOSPHATE

2-METHOXYPHENOL see GUAIACOL

O-METHOXYPHENOL see GUAIACOL

1-(3-(2-METHOXYPHENOTHIAZIN-10-YL)-2-METHYLPROPYL)-4-PIPERIDINOL see LEPTRYL

2-(P-METHOXYPHENOXY)-N-(2-(DIETHYLAMINO)ETHYL)ACETAMIDE see MEPHEXAMIDE

4-METHOXYPHENYLACETIC ACID see p-METHOXYPHENYLACETIC ACID

P-METHOXYPHENYLAMINE see p-ANISIDINE

P-METHOXYPHENYL-N-CARBAMOYLAZIRIDINE see N-(p-METHOXYPHENYL)-1-AZIRIDINECARBOXAMIDE

1-(P-METHOXYPHENYL)-3,3-DIMETHYL-TRIAZENE see 3,3-DIMETHYL-1,p-METHOXYPHENYLTRIAZENE

4-METHOXY-M-PHENYLENEDIAMINE SULFATE see 2,4-DIAMINOANISOLE SULFATE

4-METHOXY-M-PHENYLENEDIAMINE SULPHATE see 2,4-DIAMINOANISOLE SULPHATE

1-(2-(P-(ALPHA-(P-METHOXYPHENYL)-BETA-NITROSTYRYL)PHENOXY)ETHYL)PYRROLIDINE MONOCITRATE see NITROMIFENE CITRATE

1-METHOXY-2-PROPANOL see PROPYLENE GLYCOL MONOMETHYL ETHER

4-METHOXYPROPENYLBENZENE see p-PROPENYL ANISOLE

1-METHOXY-4-PROPENYLBENZENE see p-PROPENYL ANISOLE

2-METHOXY-4-PROPENYLPHENOL see GUAIACOL

2-METHOXY-4-PROP-2-ENYLPHENOL see EUGENOL

2-METHOXY-4-(2-PROPENYL)PHENOL see EUGENOL

8-METHOXYPSORALEN see XANTHOTOXIN

S-((5-METHOXY-4H-PYRON-2-YL)-METHYL)-O,O-DIMETHYL-MONOTHIOFOSFAAT (DUTCH) see ENDOTHION

S-((5-METHOXY-4H-PYRON-2-YL)-METHYL)-O,O-DIMETHYL-MONOTHIOPHOSPHAT(GERMAN) see ENDOTHION

S-(5-METHOXY-4-PYRON-2-YLMETHYL) DIMETHYL PHOSPHOROTHIOATE see ENDOTHION

METHOXY-5-PYRROLIDINO-2'-ETHYL-3-INDOLE see 5-METHOXY-3-(2-PYRROLIDINOETHYL)INDOLE

ALPHA-(6-METHOXY-4-QUINOYL)-5-VINYL-2-QUINCLIDINEMETHANOL see QUININE

4-METHOXY-M-TOLUIDINE see 5-METHYL-o-ANISIDINE

6-METHOXY-ALPHA-(5-VINYL-2-QUINUCLIDINYL)-4-QUINOLINEMETHANOL see QUINIDINE

METHVTIOLO (ITALIAN) see METHANETHIOL

12-METHYLBENZ(A)ANTHRACENE-7-METHANOL see 7-HYDROXYMETHYL-12-METHYLBENZ(a)ANTHRACENE

METHYCAINE HYDROCHLORIDE see ISOCAINE

METHYL 'CELLOSOLVE' see ETHYLENE GLYCOL MONOMETHYL ETHER

METHYLACETAAT (DUTCH) see ACETIC ACID METHYL ESTER

N-METHYLACETAMIDE see METHYLACETAMIDE

4-METHYLACETANILIDE see p-ACETOTOLUIDIDE

P-METHYLACETANILIDE see p-ACETOTOLUIDIDE

4'-METHYLACETANILIDE see p-ACETOTOLUIDIDE

METHYLACETAT (GERMAN) see ACETIC ACID METHYL ESTER

METHYL ACETATE (DOT) see ACETIC ACID METHYL ESTER

METHYL ACETIC ACID see PROPIONIC ACID

2'-METHYLACETOACETANILIDE see ACETOACET-o-TOLUIDIDE

METHYLACETOACETATE see METHYL ACETYLACETATE

METHYL ACETONE see 2-BUTANONE

6-ALPHA-METHYL-17-ALPHA-ACETOXYPREGN-4-ENE-3,20-DIONE see MEDROXYPROGESTERONE ACETATE

6-ALPHA-METHYL-17-ALPHA-ACETOXYPROGESTERONE see MEDROXYPROGESTERONE ACETATE

METHYL ACETYLACETONATE see METHYL ACETYLACETATE

METHYLACETYL CHOLINE see O-ACETYL-beta-METHYLCHOLINE CHLORIDE

BETA-METHYLACETYLCHOLINE CHLORIDE see O-ACETYL-beta-METHYLCHOLINE CHLORIDE

METHYL ACETYLENE see PROPYNE

2-METHYLACROLEIN see METHYLACRYLALDEHYDE

ALPHA-METHYLACROLEIN see METHYLACRYLALDEHYDE

BETA-METHYL ACROLEIN see trans-2-BUTENAL

METHYLACRYLAAT (DUTCH) see METHYL ACRYLATE

2-METHYLACRYLAMIDE see METHACRYLIC ACID AMIDE

METHYL-ACRYLAT (GERMAN) see METHYL ACRYLATE

3-METHYLACRYLIC ACID see CROTONIC ACID

ALPHA-METHYLACRYLIC ACID see METHACRYLIC ACID

beta-METHYLACRYLIC ACID see CROTONIC ACID

ALPHA-METHYL ACRYLIC AMIDE see METHACRYLIC ACID AMIDE

ALPHA-METHYLACRYLONITRILE see METHYLACRYLONITRILE

N-METHYLADRENALINE see N-METHYLEPINEPHRINE

METHYLAETHYLNITROSAMIN (GERMAN) see N,N-METHYLETHYLNI-TROSAMINE

METHYLAL (DOT) see DIMETHOXYMETHANE

METHYL ALCOHOL see METHANOL

METHYL ALCOHOL (DOT) see METHANOL

METHYL ALDEHYDE see FORMALDEHYDE

1-METHYL-2-ALDOXIMINOPYRIDINIUM CHLORIDE see 2-FORMYL-1-METHYLPYRIDINIUM CHLORIDE OXIME

1-METHYL-2-ALDOXIMINOPYRIDINIUM IODIDE see PYRIDINIUM-2-AL-DOXIME-N-METHYLIODIDE

METHYLALKOHOL (GERMAN) see METHANOL

2-METHYL-ALLYLCHLORID (GERMAN) see 3-CHLORO-2-METHYLPRO-PENE

2-METHYLALLYL CHLORIDE see 3-CHLORO-2-METHYLPROPENE

BETA-METHYLALLYL CHLORIDE see 3-CHLORO-2-METHYLPROPENE

METHYLALLYLNITROSAMIN (GERMAN) see N-METHYL-N-NITROSOAL-LYLAMINE

METHYLALLYLNITROSAMINE see N-METHYL-N-NITROSOALLYLAMINE

1-ALPHA-METHYLALLYLTHIOCARBAMOYL-2-METHYLTHIOCARBA-MOYLHYDRAZINE see 1-METHYL-6-(1-METHYLALLYL)-2,5-DITHIOBIUREA

N-((1-METHYLALLYL)THIOCARBAMOYL)-N′-(METHYLTHIOCARBA-MOYL)HYDRAZINE see 1-METHYL-6-(1-METHYLALLYL)-2,5-DITHIOBIUREA

METHYLAMINEN (DUTCH) see METHYLAMINE

METHYLAMINOANTIPYRINE SODIUM METHANESULFONATE see AMINOPYRINE SODIUM SULFONATE

4-(METHYLAMINO)AZOBENZENE see N-METHYL-p-(PHENYLAZO)-ANILINE

N-METHYL-4-AMINOAZOBENZENE see N-METHYL-p-(PHENYLAZO)-ANILINE

N-METHYL-P-AMINOAZOBENZENE see N-METHYL-p-(PHENYLAZO)-ANILINE

(METHYLAMINO)BENZENE see METHYLANILINE

N-METHYLAMINOBENZENE see METHYLANILINE

1-METHYL-2-AMINOBENZENE see o-TOLUIDINE

2-METHYL-1-AMINOBENZENE see o-TOLUIDINE

1-METHYL-2-AMINOBENZENE HYDROCHLORIDE see o-TOLUIDINE HY-DROCHLORIDE

2-METHYL-1-AMINOBENZENE HYDROCHLORIDE see o-TOLUIDINE HY-DROCHLORIDE

METHYL-2-AMINOBENZOATE see ANTHRANILIC ACID, METHYL ESTER

METHYL-O-AMINOBENZOATE see ANTHRANILIC ACID, METHYL ESTER

4-METHYLAMINO-1,5-DIMETHYL-2-PHENYL-3-PYRAZOLONE SODIUM METHANESULFONATE see AMINOPYRINE SODIUM SULFONATE

BETA-(METHYLAMINO)ETHANOL see 2-METHYLAMINOETHANOL

1-METHYLAMINOETHANOLCATHECHOL HYDROCHLORIDE see 1-ADRENALINE CHLORIDE

M-METHYLAMINOETHANOLPHENOL see NEOSYNEPHRINE

M-METHYLAMINOETHANOLPHENOL HYDROCHLORIDE see meta-SYNEPHRINE HYDROCHLORIDE

(−)-alpha-(1-METHYLAMINOETHYL)BENZYL ALCOHOL see EPHE-DRINE

1-alpha-(1-METHYLAMINOETHYL)BENZYL ALCOHOL see EPHEDRINE

dl-alpha-(1-(METHYLAMINO)ETHYL) BENZYL ALCOHOL HYDROCHLO-RIDE see dl-EPHEDRINE HYDROCHLORIDE

1-ALPHA-(1-(METHYLAMINO)ETHYL)BENZYL ALCOHOL SULFATE see 1-EPHEDRINE SULFATE

ALPHA-(1-METHYLAMINOETHYL)-P-HYDROXYBENZYL ALCOHOL see p-HYDROXYEPHEDRINE

4-(2-METHYLAMINOETHYL)PYROCATECHOL HYDROCHLORIDE see EPHININE HYDROCHLORIDE

2-METHYL-6-AMINOHEPTANE see 2-ISOOCTYL AMINE

3-METHYLAMINOISOCAMPHANE HYDROCHLORIDE see MEVASIN HY-DROCHLORIDE

2-METHYLAMINOISOOCTANE HYDROCHLORIDE see OCTIN HYDRO-CHLORIDE

6-METHYLAMINO-2-METHYLHEPTENE see ISONYL

METHYLAMINO-METHYLHEPTENE HYDROCHLORIDE see OCTIN HY-DROCHLORIDE

2-METHYL-4-AMINO-1-NAPHTHOL see 4-AMINO-2-METHYL-1-NAPH-THOL

METHYLAMINOPHENYLDIMETHYLPYRAZOLONE METHANESULFONATE SODIUM see AMINOPYRINE SODIUM SULFONATE

(+)-2-METHYLAMINO-2-PHENYLPROPANE HYDROCHLORIDE see d-METHAPHETAMINE HYDROCHLORIDE

1-2-METHYLAMINO-1-PHENYLPROPANOL see EPHEDRINE

5-(3-METHYLAMINOPROPYL)-5H-DIBENZO(a,d)CYCLOHEPTENE HY-DROCHLORIDE see PROTRIPTYLINE HYDROCHLORIDE

5-(3-(METHYLAMINO)PROPYLIDENE)DIBENZO(a,e)CYCLOHEPTA-(1,5)DIENE see NORTRIPTYLINE

5-(3-METHYLAMINOPROPYLIDENE)-10,11-DIHYDRO-5H-DIBEN-ZO(A,D)CYCLOHEPTENE see NORTRIPTYLINE

METHYLAMINOPROPYLIMINODIBENZYL see 10,11-DIHYDRO-5-(3-METHYLAMINOPROPYL)-5H-DIBENZ(b,f)AZEPINE

N-(GAMMA-METHYLAMINOPROPYL)IMINODIBENZYL HYDROCHLORIDE see DEMETHYLIMIPRAMINE HYDROCHLORIDE

METHYLAMINOPTERIN see METHOTREXATE

4-METHYL-2-AMINOPYRIDINE see 2-AMINO-4-METHYLPYRIDINE

METHYL-4-AMINO-2-PYRIDINE see 2-AMINO-4-METHYLPYRIDINE

N-METHYLAMPHETAMINE see DESOXYN

METHYLAMPHETAMINE HYDROCHLORIDE see dl-DESOXY EPHEDRINE HYDROCHLORIDE

(+)-N-METHYLAMPHETAMINE HYDROCHLORIDE see d-METHAPHET-AMINE HYDROCHLORIDE

METHYLAMYL ALCOHOL see AMYL METHYL ALCOHOL

METHYL AMYL ALCOHOL see ISOBUTYLMETHYLCARBINOL

METHYL AMYL CARBINOL see 2-HEPTANOL

METHYL-AMYL-CETONE (FRENCH) see 2-HEPTANONE

METHYL-N-AMYL KETONE see 2-HEPTANONE

METHYL AMYL KETONE (DOT) see 2-HEPTANONE

METHYLAMYLNITROSAMIN (GERMAN) see N-AMYL-N-METHYLNITRO-SAMINE

METHYLAMYLNITROSAMINE see N-AMYL-N-METHYLNITROSAMINE

METHYL-N-AMYLNITROSAMINE see N-AMYL-N-METHYLNITROSAMINE

2-METHYLANILINE see o-TOLUIDINE

3-METHYLANILINE see m-TOLUIDINE

4-METHYLANILINE see p-TOLUIDINE

M-METHYLANILINE see m-TOLUIDINE

N-METHYLANILINE see METHYLANILINE

O-METHYLANILINE see o-TOLUIDINE

P-METHYLANILINE see p-TOLUIDINE

2-METHYLANILINE HYDROCHLORIDE see o-TOLUIDINE HYDROCHLO-RIDE

4-METHYLANILINE HYDROCHLORIDE see p-TOLUIDINE HYDROCHLO-RIDE

O-METHYLANILINE HYDROCHLORIDE see o-TOLUIDINE HYDROCHLO-RIDE

METHYL ANTHRANILATE see ANTHRANILIC ACID, METHYL ESTER

2-METHYL-1-ANTHRAQUINONYLAMINE see 1-AMINO-2-METHYLAN-THRAQUINONE

METHYLARSENIC ACID, SODIUM SALT see MONOSODIUM METHYLAR-SONATE

METHYLARSINE DICHLORIDE see DICHLOROMETHYLARSINE

METHYLARSINE SULFIDE see METHYLARSENIC SULFIDE

METHYLARSINIC SULFIDE see METHYLARSENIC SULFIDE

2-METHYLAZACYCLOPROPANE see 2-METHYLAZIRIDINE

METHYLAZOXYMETHANOL see 1-HYDROXYMETHYL-2-METHYLDIA-MIDE-2-OXIDE

METHYLAZOXYMETHANOL ACETATE see METHYL AZOXYMETHYL ACETATE

METHYLAZOXYMETHANOL GLUCOSIDE see CYCASIN
METHYLAZOXYMETHANOL-beta-d-GLUCOSIDE see CYCASIN
3-METHYLBENZ(J)ACEANTHRYLENE see 1,2-DEHYDRO-3-METHYL-CHOLANTHRENE
10-METHYL-1,2-BENZANTHRACEN (GERMAN) see 10-METHYL-1,2-BENZANTHRACENE
7-METHYLBENZ(a)ANTHRACENE see 10-METHYL-1,2-BENZANTHRACENE
3-METHYL-1,2-BENZANTHRACENE see 5-METHYLBENZ(a)ANTHRACENE
4-METHYL-1,2-BENZANTHRACENE see 6-METHYLBENZ(a)ANTHRACENE
5-METHYL-1,2-BENZANTHRACENE see 8-METHYLBENZ(a)ANTHRACENE
6-METHYL-1,2-BENZANTHRACENE see 9-METHYLBENZ(a)ANTHRACENE
7-METHYL-1,2-BENZANTHRACENE see 10-METHYLBENZ(a)ANTHRACENE
9-METHYL-1,2-BENZANTHRACENE see 12-METHYLBENZ(a)ANTHRACENE
4'-METHYL-1:2-BENZANTHRACENE see 4-METHYLBENZ(a)ANTHRACENE
7-METHYLBENZ(A)ANTHRACENE-12-METHANOL see 12-HYDROXYMETHYL-7-METHYLBENZ(a)ANTHRACENE
12-METHYLBENZ(A)ANTHRACENE-7-METHANOL ACETATE (ESTER) see 7-ACETOXYMETHYL-12-METHYLBENZ(a)ANTHRACENE
2-METHYLBENZENAMINE see o-TOLUIDINE
3-METHYLBENZENAMINE see m-TOLUIDINE
4-METHYLBENZENAMINE see p-TOLUIDINE
M-METHYLBENZENAMINE see m-TOLUIDINE
N-METHYLBENZENAMINE see METHYLANILINE
O-METHYLBENZENAMINE see o-TOLUIDINE
P-METHYLBENZENAMINE see p-TOLUIDINE
2-METHYLBENZENAMINE HYDROCHLORIDE see o-TOLUIDINE HYDROCHLORIDE
O-METHYLBENZENAMINE HYDROCHLORIDE see o-TOLUIDINE HYDROCHLORIDE
METHYLBENZENE see TOLUENE
METHYL BENZENECARBOXYLATE see METHYL BENZOATE
2-METHYL-1,4-BENZENEDIAMINE see TOLUENE-2,5-DIAMINE
4-METHYL-1,3-BENZENEDIAMINE see TOLUENE-2,4-DIAMINE
2-METHYL-1,4-BENZENEDIAMINE SULFATE see 2,5-DIAMINOTOLUENE SULFATE
2-METHYLBENZENESULFONAMIDE see o-TOLUENESULFONAMIDE
O-METHYLBENZENESULFONAMIDE see o-TOLUENESULFONAMIDE
2-METHYLBENZIMIDAZOLE see METHYL-2-BENZIMIDAZOLE
4-METHYLBENZOIC ACID METHYL ESTER see 4-METHYLPHENYL ACETATE
METHYLBENZOL see TOLUENE
4-METHYLBENZO(C)PHENANTHRENE see 8-METHYL-3:4-BENZPHENANTHRENE
6-METHYLBENZO(A)PYRENE see 5-METHYL-3,4-BENZPYRENE
7-METHYLBENZO(A)PYRENE see 4'-METHYLBENZO(a)PYRENE
5-METHYL-3,4-BENZOPYRENE see 5-METHYL-3,4-BENZPYRENE
METHYL-P-BENZOQUINONE see 2-METHYL-p-BENZOQUINONE
METHYL-1,4-BENZOQUINONE see 2-METHYL-p-BENZOQUINONE
2-METHYLBENZOQUINONE-1,4 see 2-METHYL-p-BENZOQUINONE
2-METHYL-3,4-BENZPHENANTHRENE see 5-METHYLBENZO(c)PHENANTHRENE
4'-METHYL-3:4-BENZPYRENE see 4'-METHYLBENZO(a)PYRENE
ALPHA-METHYLBENZYL ALCOHOL see alpha-PHENETHYL ALCOHOL
1-METHYLBENZYL-3-(DIMETHOXYPHOSPHINYLOXO) ISOCROTONATE see CROTOXYPHOS
alpha-METHYL BENZYL-3-(DIMETHOXY-PHOSPHINYLOXY)-cis-CROTONATE see CROTOXYPHOS
alpha-METHYLBENZYL-3-HYDROXY-CROTONATE DIMETHYL PHOSPHATE see CROTOXYPHOS
METHYL-BENZYL-NITROSOAMIN (GERMAN) see N-METHYL-N-BENZYLNITROSAMINE

METHYLBIS(3-AMINOPROPYL)AMINE see BIS(gamma-AMINOPROPYL)-METHYLAMINE
N-METHYL-BIS-CHLORAETHYLAMIN (GERMAN) see BIS(beta-CHLOROETHYL)METHYLAMINE
gamma(1-METHYL-5-BIS(beta-CHLORAETHYL)AMINOBENZIMIDAZO-LYL-)BUTTERSAEUREHYDRO-CHLORID(GERMAN) see CYTOSTASAN
N-METHYL-BIS-BETA-CHLORETHYLAMINE HYDROCHLORIDE see BIS(2-CHLOROETHYL)METHYLAMINE HYDROCHLORIDE
gamma-(1-METHYL-5-BIS(beta-CHLOROAETHYL) AMINOBENZIMIDA-ZOYL) BUTTERSAUERHYDROCHLORID(GERMAN) see CYTOSTASAN
METHYLBIS(BETA-CHLOROETHYL)AMINE see BIS(beta-CHLOROETHYL)METHYLAMINE
N-METHYL-BIS(BETA-CHLOROETHYL)AMINE see BIS(beta-CHLOROETHYL)METHYLAMINE
METHYLBIS(2-CHLOROETHYL)AMINE HYDROCHLORIDE see BIS(2-CHLOROETHYL)METHYLAMINE HYDROCHLORIDE
METHYLBIS(BETA-CHLOROETHYL)AMINE-N-OXIDE see 2-CHLORO-N-(2-CHLOROETHYL)-N-METHYL ETHANAMINE-N-OXIDE
N-METHYLBIS(2-CHLOROETHYL)AMINE-N-OXIDE HYDROCHLORIDE see 2-CHLORO-N-(2-CHLORO ETHYL)-N-METHYL ETHANAMINE-N-OXIDE HYDROCHLORIDE
METHYLBIS(beta-CHLOROETHYL)AMINE-N-OXIDE HYDROCHLORIDE see 2-CHLORO-N-(2-CHLORO ETHYL)-N-METHYL ETHANAMINE-N-OXIDE HYDROCHLORIDE
7-METHYLBISDEHYDRODOISYNOLIC ACID see BIS DEHYDRO ISYNOLIC ACID METHYL ESTER
2-METHYL-4,5-BIS(HYDROXYMETHYL)-3-HYDROXYPYRIDINE see PYRIDOXOL
beta-METHYLBIVINYL see ISOPRENE
METHYL BORATE see TRIMETHYL BORATE
METHYLBROMID (GERMAN) see BROMO METHANE
METHYL BROMIDE see BROMO METHANE
2-METHYLBUTADIENE see ISOPRENE
2-METHYL-1,3-BUTADIENE see ISOPRENE
2-METHYLBUTANE see ETHYL DIMETHYL METHANE
3-METHYLBUTANOIC ACID see ISOVALERIC ACID
2-METHYLBUTANOL see 2-METHYL BUTANOL-1
3-METHYL BUTANOL see ISOAMYL ALCOHOL
2-METHYL BUTANOL-2 see tert-N-AMYL ALCOHOL, REFINED
2-METHYL-2-BUTANOL see tert-N-AMYL ALCOHOL, REFINED
2-METHYL BUTANOL-2 see t-PENTYL ALCOHOL
2-METHYL-2-BUTANOL see t-PENTYL ALCOHOL
2-METHYL-4-BUTANOL see ISOAMYL ALCOHOL
3-METHYLBUTAN-1-OL see ISOAMYL ALCOHOL
3-METHYLBUTAN-3-OL see t-PENTYL ALCOHOL
3-METHYLBUTAN-3-OL see tert-N-AMYL ALCOHOL, REFINED
3-METHYL-1-BUTANOL (CZECH) see ISOAMYL ALCOHOL
3-METHYL-2-BUTANONE see METHYL ISOPROPYL KETONE
2-METHYL-2-BUTENE see α,η-AMYLENE
3-METHYL-3-BUTEN-2-ON (GERMAN) see ISOPROPENYL METHYL KETONE
2-METHYL-1-BUTEN-3-ONE see ISOPROPENYL METHYL KETONE
3-(3-METHYL-2-BUTENYL)-1,2,3,4,5,6-HEXAHYDRO-6,11-DIMETHYL-2,6-METHANO-3-BENZAZOCIN-8-OL see 2-(3,3-DIMETHYLALLYL)-CYCLAZOCINE
3-(3-METHYL-2-BUTENYL)-1,2,3,4,5,6-HEXAHYDRO-8-HYDROXY-6,11-DIMETHYL-2,6-METHANO-3-BENZAZOCINE see 2-(3,3-DIMETHYLALLYL)CYCLAZOCINE
3'-METHYLBUTTERGELB (GERMAN) see N,N-DIMETHYL-p-(m-TOLYL-AZO)ANILINE
3-METHYLBUTYL ACETATE see ISOPENTYL ALCOHOL ACETATE
3-METHYL-1-BUTYL ACETATE see ISOPENTYL ALCOHOL ACETATE
2-METHYL-BUTYLACRYLAAT (DUTCH) see 2-METHYL BUTYLACRYLATE
2-METHYL-BUTYLACRYLAT (GERMAN) see 2-METHYL BUTYLACRYLATE
p-METHYL-tert-BUTYLBENZENE see p-tert-BUTYLTOLUENE
1-METHYL-4-tert-BUTYLBENZENE see p-tert-BUTYLTOLUENE
3-METHYLBUTYL ETHANOATE see ISOPENTYL ALCOHOL ACETATE

METHYL N-BUTYL KETONE see 2-HEXANONE

METHYL-BUTYL-NITROSAMIN (GERMAN) see METHYLBUTYLNITROSA-
MINE

METHYL-N-BUTYLNITROSAMINE see METHYLBUTYLNITROSAMINE

BETA-METHYLBUTYRIC ACID see ISOVALERIC ACID

N-METHYLCARBAMATE DE 4-DIMETHYLAMINO-3-METHYL PHENYLE
(FRENCH) see 4-DIMETHYLAMINE-m-CRESYLMETHYLCARBA-
MATE

N-METHYLCARBAMATE DE 1-NAPHTYLE (FRENCH) see CARBARYL

METHYLCARBAMATE-1-NAPHTHALENOL see CARBARYL

METHYLCARBAMATE-1-NAPHTHOL see CARBARYL

METHYLCARBAMIC ACID-3-SEC-BUTYL-6-CHLOROPHENYL ESTER see
3-(1-METHYLPROPYL)-6-CHLOROPHENYL METHYLCARBAMATE

METHYLCARBAMIC ACID, 4-(DIMETHYLAMINO)-3,5-XYLYL ESTER
see 4-(DIMETHYLAMINE)-3,5-XYLYL N-METHYLCARBAMATE

METHYL-CARBAMIC ACID, ESTER WITH ESEROLINE see PHYSOSTIG-
MINE

METHYLCARBAMIC ACID-m-(1-METHYL)BUTYL)PHENYL ESTER MIXED
WITH CARBAMIC ACID, METHYL-m-(1-ETHYLPROPYL)PHENYL
ESTER (3:1) see BUX-TEN

METHYL CARBAMIC ACID-4-(METHYLTHIO)-3,5-XYLYL ESTER see 4-
METHYLMERCAPTO-3,5-XYLYL METHYLCARBAMATE

METHYLCARBAMIC ACID-1-NAPHTHYL ESTER see CARBARYL

METHYLCARBAMIC ACID-1-NAPHTHYL ESTER see CARBARYL

METHYLCARBAMIC ACID-M-TOLYL ESTER see TSUMACIDE

METHYLCARBAMIC ACID, 5-(TRIMETHYLAMMONIO)-O-TOLYL ESTER,
CHLORIDE see (3-HYDROXY-p-TOLYL)TRIMETHYLAMMONIUM
CHLORIDE,METHYLCARBAMATE

METHYLCARBAMIC ESTER OF OXYPHENYLMETHYLDIETHYLAMMO-
NIUM IODIDE see TL 1217

S-METHYLCARBAMOYLMETHYL O,O-DIMETHYL PHOSPHORODITHIO-
ATE see O,O-DIMETHYL METHYLCARBAMOYLMETHYL PHOSPHO-
RODITHIOATE

1-METHYL-4-CARBETHOXY-4-PHENYLPIPERIDINE HYDROCHLORIDE
see DOLANTIN HYDROCHLORIDE

METHYLCARBINOL see ETHANOL

METHYL-4-CARBOMETHOXY BENZOATE see DIMETHYL TEREPHTHA-
LATE

METHYL-4-CARBOMETHOXYBENZOATE see TEREPHTHALIC ACID
METHYL ESTER

METHYLCARBONYL FLUORIDE see ACETYL FLUORIDE

METHYLCARBOPHENOTHION see METHYL TRITHION

17-beta-(1-METHYL-3-CARBOXYPROPYL)-ETIOCHOLANE-3-alpha,12-
alpha-DIOL see DEOXYCHOLATIC ACID

3-METHYLCATECHOL see 2,3-DIHYDROXYTOLUENE

METHYL CELLOSOLVE (DOT) see ETHYLENE GLYCOL METHYL ETHER

METHYL CELLOSOLVE ACETATE see ETHYLENE GLYCOL MONO-
METHYL ETHER ACETATE

METHYL CELLOSOLYE ACETAAT (DUTCH) see ETHYLENE GLYCOL
MONOMETHYL ETHER ACETATE

METHYL CHAVICOL see p-ALLYLANISOLE

METHYLCHLOORFORMIAT (DUTCH) see METHYL CHLOROCARBONATE

METHYL-2-CHLORAETHYLNITROSAMIN (GERMAN) see 2-CHLORO-N-
METHYL-N-NITROSOETHYLAMINE

METHYLCHLORID (GERMAN) see MONOCHLOROMETHANE

METHYL CHLORIDE see MONOCHLOROMETHANE

METHYL CHLORIDE see CHLOROMETHANE

N-METHYL-N'-(4-CHLOROBENZHYDRYL)PIPERAZINE DIHYDROCHLO-
RIDE see CHLORCYCLIZINE DIHYDROCHLORIDE

METHYL CHLOROFORM see 1,1,1-TRICHLOROETHANE

METHYL CHLOROFORMATE see METHYL CHLOROCARBONATE

METHYL CHLOROFORMATE (DOT) see METHYL CHLOROCARBONATE

METHYL CHLOROMETHYL ETHER, ANHYDROUS (DOT) see CHLORO-
METHYL METHYL ETHER

3-METHYL-4-CHLOROPHENOL see 4-CHLORO-m-CRESOL

O-METHYL-O-2-CHLORO-4-TERT-BUTYLPHENYL-N-METHYLAMIDO-
PHOSPHATE see 4-tert-BUTYL-2-CHLORO PHENYL METHYL
METHYL PHOSPHORAMIDATE

METHYLCHOLANTHRENE see 3-METHYLCHOLANTHRENE

20-METHYLCHOLANTHRENE see 3-METHYLCHOLANTHRENE

3-METHYLCHOLANTHRENE COMPOUND WITH PICRIC ACID (1:1) see
20-METHYLCHOLANTHRENE PICRATE

3-METHYLCHOLANTHREN-2-OL see 2-HYDROXY-3-METHYLCHOLAN-
THRENE

3-METHYLCHOLANTHREN-1-ONE see 20-METHYLCHOLANTHREN-15-
ONE

3-METHYLCHOLANTHREN-2-ONE see 3-METHYLCHOLANTHRENE-2-
ONE

3-METHYLCHOLANTHRYLENE see 1,2-DEHYDRO-3-METHYLCHOLAN-
THRENE

20-METHYLCHOLANTHRYLENE see 1,2-DEHYDRO-3-METHYLCHOLAN-
THRENE

METHYLCOLCHAMINONE see METHYLAMINOCOLCHICIDE

METHYL CYANIDE see ACETONITRILE

METHYLCYCLOHEXANE see CYCLOHEXYLMETHANE

2-METHYL-CYCLOHEXANON (GERMAN, DUTCH) see 2-METHYLCY-
CLOHEXANONE

O-METHYLCYCLOHEXANONE see 2-METHYLCYCLOHEXANONE

N-METHYL-5-CYCLOHEXENYL-5-METHYLBARBITURIC ACID see
EVIPAL

METHYLCYCLOHEXYLNITROSAMIN (GERMAN) see N-NITROSO-N-
METHYLCYCLOHEXYLAMINE

METHYLCYCLOHEXYLNITROSAMINE see N-NITROSO-N-METHYLCY-
CLOHEXYLAMINE

METHYL CYCLOPENTADIENYL MANGANESE TRICARBONYL see MAN-
GANESE TRICARBONYL METHYLCYCLOPENTADIENYL

2-METHYL CYCLOPENTADIENYL MANGANESE TRICARBONYL see
MANGANESE TRICARBONYL METHYLCYCLOPENTADIENYL

6-METHYL-6-DEHYDRO-17-ALPHA-ACETOXYPROGESTERONE see
VOLIDAN

6-METHYL-6-DEHYDRO-17-ALPHA-ACETYLPROGESTERONE see VOLI-
DAN

METHYL DEMETON see METASYSTOX

O-METHYLDEMETON see DEMETON-O-METHYL

METHYL DEMETON THIOESTER see DEMETON-S-METHYL

6-METHYL DIBENZ(A,H)ANTHRACENE see 4-METHYL-1,2,5,6-DIBEN-
ZANTHRACENE

3'-METHYL-1:2:5:6-DIBENZANTHRACENE see 3-METHYLDIBENZ-
(a,h)ANTHRACENE

N-METHYL-5H-DIBENZO(A,D)CYCLOHEPTENE-5-PROPYLAMINE HY-
DROCHLORIDE see PROTRIPTYLINE HYDROCHLORIDE

METHYLDI-TERT-BUTYLPHENOL see BHT (FOOD GRADE)

4-METHYL-2,6-DI-TERT-BUTYLPHENOL see BHT (FOOD GRADE)

METHYLDICHLOROARSINE see DICHLOROMETHYLARSINE

N-METHYL-2,2'-DICHLORODIETHYLAMINE see BIS(beta-CHLORO-
ETHYL)METHYLAMINE

N-METHYL-2,2'-DICHLORODIETHYLAMINE HYDROCHLORIDE see
BIS(2-CHLOROETHYL)METHYLAMINE HYDROCHLORIDE

N-METHYL-2,2'-DICHLORODIETHYLAMINE-N-OXIDE HYDROCHLORIDE
see 2-CHLORO-N-(2-CHLORO ETHYL)-N-METHYL ETHANAMINE-
N-OXIDE HYDROCHLORIDE

METHYL DICHLORO-ENTHANOATE see METHYL DICHLORO ACETATE

METHYLDI(2-CHLOROETHYL)AMINE see BIS(beta-CHLOROETHYL)-
METHYLAMINE

METHYLDI(2-CHLOROETHYL)AMINE HYDROCHLORIDE see BIS(2-
CHLOROETHYL)METHYLAMINE HYDROCHLORIDE

N-METHYL-DI-2-CHLOROETHYLAMINE HYDROCHLORIDE see BIS(2-
CHLOROETHYL)METHYLAMINE HYDROCHLORIDE

METHYLDI(BETA-CHLOROETHYL)AMINE HYDROCHLORIDE see BIS(2-
CHLOROETHYL)METHYLAMINE HYDROCHLORIDE

N-METHYL-DI-2-CHLOROETHYLAMINE-N-OXIDE see 2-CHLORO-N-(2-
CHLOROETHYL)-N-METHYL ETHANAMINE-N-OXIDE

METHYLDI(2-CHLOROETHYL)AMINE-N-OXIDE HYDROCHLORIDE see
2-CHLORO-N-(2-CHLORO ETHYL)-N-METHYL ETHANAMINE-N-OXIDE
HYDROCHLORIDE

N-METHYLDICHLOROMALEINIMIDE see DICHLORO-N-METHYLMALEI-
MIDE

METHYL DICHLOROSILANE (DOT) see DICHLOROMETHYLSILANE

METHYL-DICHLORSILAN (CZECH) see DICHLOROMETHYLSILANE

3-METHYL-N,N-DIETHYLBENZAMIDE see DIETHYL-m-TOLUAMIDE

11-METHYL-15,16-DIHYDRO-17H-CYCLOPENTA(A)PHENANTHREN-17-
ONE see 11-METHYL-15,16-DIHYDRO-17-OXOCYCLOPENTA(a)PHE-
NANTHRENE
ALPHA-METHYL-L-3,4-DIHYDROXYPHENYLALANINE see ALDOMET
L-ALPHA-METHYL-3,4-DIHYDROXYPHENYLALANINE see ALDOMET
ALPHA-METHYL-BETA-(3,4-DIHYDROXYPHENYL)-L-ALANINE see AL-
DOMET
L-(−)-ALPHA-METHYL-BETA-(3,4-DIHYDROXYPHENYL)ALANINE see
ALDOMET
METHYL-(beta-(3,4-DIHYDROXY PHENYL ETHYL) AMINE HYDRO-
CHLORIDE see EPHININE HYDROCHLORIDE
METHYL-3-(DIMETHOXYPHOSPHINYLOXY)CROTONATE see MEVIN-
PHOS
5-METHYL-10-BETA-DIMETHYLAMINOAETHYL-10,11-DIHYDRO-11-
OXO-5-DIBENZO(B,E)(1,4)DIAZEPIN see DIBENZEPINE HYDRO-
CHLORIDE
3′-METHYL-4-DIMETHYLAMINOAZOBENZEN (CZECH) see N,N-
DIMETHYL-p-(m-TOLYLAZO)ANILINE
3′-METHYL-4-DIMETHYLAMINOAZOBENZENE see N,N-DIMETHYL-p-
(m-TOLYLAZO)ANILINE
3′-METHYL-N,N-DIMETHYL-4-AMINOAZOBENZENE see N,N-
DIMETHYL-p-(m-TOLYLAZO)ANILINE
3′-METHYLDIMETHYLAMINOAZOBENZOL (GERMAN) see N,N-
DIMETHYL-p-(m-TOLYLAZO)ANILINE
3-METHYL-4-DIMETHYLAMINO-2,2-DIPHENYLBUTYRAMIDE see
4-(DIMETHYLAMINO)-2,2-DIPHENYLVALERAMIDE
METHYL-2-(DIMETHYLAMINO)-N-(((METHYLAMINO)CARBONYL)-
OXY)-2-OXOETHANIMIDOTHIOATE see N′,N′-DIMETHYL-N-
((METHYLCARBAMOYL)OXY)-1-METHYLTHIOOXAMIMIDIC ACID
7′-METHYL-5′-(P-DIMETHYLAMINOPHENYLAZO)QUINOLINE see
5-((p(DIMETHYLAMINO)PHENYL)AZO)-7-METHYLQUINOLINE
2′-METHYL-4-DIMETHYLAMINOSTILBENE see N,N,2′-TRIMETHYL-
4-STILBENAMINE
METHYL-4-DIMETHYLAMINO-3,5-XYLYL CARBAMATE see
4-(DIMETHYLAMINE)-3,5-XYLYL N-METHYLCARBAMATE
METHYL-4-DIMETHYLAMINO-3,5-XYLYL ESTER OF CARBAMIC ACID
see 4-(DIMETHYLAMINE)-3,5-XYLYL N-METHYLCARBAMATE
S-METHYL-1-(DIMETHYLCARBAMOYL)-N-((METHYLCARBAMOYL)-
OXY)THIOFORMIMIDATE see N′,N′-DIMETHYL-N-((METHYL-
CARBAMOYL)OXY)-1-METHYLTHIOOXAMIMIDIC ACID
METHYL-N′,N′-DIMETHYL-N-((METHYLCARBAMOYL)OXY)-1-THIO-
OXAMIMIDATE see N′,N′-DIMETHYL-N-((METHYLCARBAMOYL)-
OXY)-1-METHYLTHIOOXAMIMIDIC ACID
2-METHYL-3,5-DINITROBENZAMIDE see 3,5-DINITRO-O-TOLUAMIDE
1-METHYL-2,4-DINITROBENZENE see 2,4-DINITROTOLUENE
2-METHYL-4,6-DINITROPHENOL see 4,6-DINITRO-O-CRESOL
6-METHYL-1,11-DIOXY-2-NAPHTHACENECARBOXAMIDE see TETRA-
CYCLINE
4-METHYL-2,6-DI-TERC. BUTYLFENOL (CZECH) see BHT (FOOD
GRADE)
N-METHYLDITHIOCARBAMATE DE SODIUM (FRENCH) see VAPAM
METHYLDITHIOCARBAMIC ACID, SODIUM SALT see VAPAM
N-METHYLDITHIOCARBAMIC ACID, SODIUM SALT see VAPAM
METHYL DODECYL BENZYL AMMONIUM CHLORIDE see DODECYL-p-
TOLYL TRIMETHYL AMMONIUM CHLORIDE
METHYLDOPA see ALDOMET
ALPHA-METHYL-L-DOPA see ALDOMET
L-ALPHA-METHYLDOPA see ALDOMET
N-METHYLDOPAMINE HYDROCHLORIDE see EPHININE HYDROCHLO-
RIDE
METHYLE (ACETATE DE) (FRENCH) see ACETIC ACID METHYL ESTER
O,O-METHYLEEN-BIS(4-CHLOORFENOL) (DUTCH) see 2,2′-METHY-
LENEBIS(4-CHLOROPHENOL)
3,3′-METHYLEEN-BIS(4-HYDROXY-CUMARINE) (DUTCH) see BISHY-
DROXY COUMARIN
METHYLEEN-S,S′-BIS(O,O-DIETHYL-DITHIOFOSFAAT) (DUTCH) see
ETHYL METHYLENE PHOSPHORO DITHIOATE
3,3′-METHYLEN-BIS(4-HYDROXY-CUMARIN) (GERMAN) see BISHY-
DROXY COUMARIN

3,4-METHYLENDIOXY-6-PROPYLBENZYL-N-BUTYL-DIAETHYLENGLY-
KOLAETHER (GERMAN) see PIPERONYL BUTOXIDE
METHYLENDIRHODANID (CZECH) (GERMAN) see METHYLENE DITHIO-
CYANATE
METHYLENE BICHLORIDE see METHANE DICHLORIDE
METHYLENEBIS(ANILINE) see 4,4′-METHYLENEDIANILINE
4,4′-METHYLENEBISANILINE see 4,4′-METHYLENEDIANILINE
4,4′-METHYLENE(BIS)-CHLOROANILINE see MOCA
4,4′-METHYLENEBIS(2-CHLOROANILINE) see MOCA
4,4′-METHYLENEBIS(O-CHLOROANILINE) see MOCA
METHYLENE 4,4′-BIS (O-CHLOROANILINE) see MOCA
P,P′-METHYLENEBIS(O-CHLOROANILINE) see MOCA
P,P′-METHYLENEBIS(ALPHA-CHLOROANILINE) see MOCA
4,4′-METHYLENEBIS-2-CHLOROBENZENAMINE see MOCA
4,4′-METHYLENEBIS(N,N-DIMETHYL ANILINE) see MICHLER'S BASE
4,4′-METHYLENEBIS(N,N-DIMETHYL)BENZENAMINE see MICHLER'S
BASE
N,N-METHYLENE-BIS(ETHYL CARBAMATE) see METHYLENE DIURE-
THAN
3,3′-METHYLENEBIS(4-HYDROXY-1,2-BENZOPYRONE) see BISHY-
DROXY COUMARIN
3,3′-METHYLENEBIS(4-HYDROXYCOUMARIN) see BISHYDROXY COU-
MARIN
3,3′-METHYLENE-BIS(4-HYDROXYCOUMARINE) (FRENCH) see BISHY-
DROXY COUMARIN
METHYLENEBIS(4-ISOCYANATOBENZENE) see DIPHENYL METHANE
DIISOCYANATE
1,1-METHYLENEBIS(4-ISOCYANATOBENZENE) see DIPHENYL METH-
ANE DIISOCYANATE
METHYLENE BIS(METHANESULFONATE) see METHYLENE DIMETHANE-
SULFONATE
4,4′-METHYLENEBIS(2-METHYLANILINE) see 4,4′-METHYLENE DI-O-
TOLUIDINE
4,4′-METHYLENEBIS(2-METHYLBENZENAMINE) see 4,4′-METHYLENE
DI-O-TOLUIDINE
METHYLENEBIS(P-PHENYLENE ISOCYANATE) see DIPHENYL METHANE
DIISOCYANATE
METHYLENEBIS(4-PHENYL ISOCYANATE) see DIPHENYL METHANE
DIISOCYANATE
METHYLENEBIS(P-PHENYL ISOCYANATE) see DIPHENYL METHANE
DIISOCYANATE
4,4′-METHYLENEBIS(PHENYL ISOCYANATE) see DIPHENYL METHANE
DIISOCYANATE
P,P′-METHYLENEBIS(PHENYL ISOCYANATE) see DIPHENYL METHANE
DIISOCYANATE
2,2′-METHYLENEBIS(3,4,6-TRICHLOROPHENOL) see HEXACHLORO-
PHENE
METHYLENE BLUE USP XII (MEDICINAL) see 3,7-BIS(DIMETHYL AMI-
NO)PHENAZA THIONIUM CHLORIDE
METHYLENE BROMIDE see DIBROMOMETHANE
(4-(2-METHYLENEBUTYRYL)-2,3-DICHLOROPHENOXY)ACETIC ACID
see ETHACRYNIC ACID
METHYLENEBUTYRYL PHENOXYACETIC ACID see ETHACRYNIC ACID
METHYLENE CHLORIDE (DOT) see METHANE DICHLORIDE
METHYLENE CHLOROBROMIDE see BROMOCHLOROMETHANE
1,2-alpha-METHYLENE-6-CHLORO-delta(SUP 6)-17-alpha-HYDROXY-
PROGESTERONE ACETATE see CYPROSTERONE ACETATE
1,2-alpha-METHYLENE-6-CHLORO-PREGNA-4,6-DIENE-3,20-DIONE
17-alpha-ACETATE see CYPROSTERONE ACETATE
1,2-alpha-METHYLENE-6-CHLORO-delta-(SUP 4,6)-PREGNADIENE-17-
alpha-OL-3,20-DIONE ACETATE see CYPROSTERONE ACETATE
1,2-alpha-METHYLENE-6-CHLORO-(SUP 4,6)-PREGNADIENE-17-al-
pha-OL-3,20-DIONE 17-alpha-ACETATE see CYPROSTERONE ACE-
TATE
METHYLENEDIANILINE see 4,4′-METHYLENEDIANILINE
P,P′-METHYLENEDIANILINE see 4,4′-METHYLENEDIANILINE
1′,9-METHYLENE-1,2:5,6-DIBENZANTHRACENE see 13H-DI-
BENZ(bc,j)ACEANTHRYLENE
METHYLENE DIBROMIDE see DIBROMOMETHANE
METHYLENE DICHLORIDE see METHANE DICHLORIDE

3,4-METHYLENE-DIHYDROXYBENZALDEHYDE see PIPERONAL

METHYLENE DIMETHYL ETHER see DIMETHOXYMETHANE

METHYLENEDINAPHTHALENESULFONIC ACID BISPHENYLMERCURI
SALT see PHENYLMERCURIC DINAPHTHYLMETHANEDISULFONATE

1,5-METHYLENE-3,7-DINITROSO-1,3,5,7-TETRAAZACYCLOOCTANE
see DINITROSOPENTAMETHYLENETETRAMINE

3,4-METHYLENEDIOXY-ALLYBENZENE see SAFROL

1,2-METHYLENEDIOXY-4-ALLYLBENZENE see SAFROL

3,4-METHYLENEDIOXYBENZALDEHYDE see PIPERONAL

1,2-METHYLENEDIOXY-4-(1-HYDROXYALLYL)BENZENE see 1,3-BEN-
ZODIOXOLE-5-(2-PROPEN-1-OL)

3,4-METHYLENEDIOXY-ALPHA-METHYL-BETA-PHENYLETHYLAMINE
HYDROCHLORIDE see 1-(3,4-METHYLENEDIOXYPHENYL)-2-AMINO-
PROPANE

1,2-(METHYLENEDIOXY)-4-(2-(OCTYLSULFINYL)PROPYL)BENZENE
see ISOSAFROLE-n-OCTYLSULFOXIDE

1,2-METHYLENEDIOXY-4-PROPENYLBENZENE see ISOSAFROLE

3,4-METHYLENEDIOXY-1-PROPENYL BENZENE see ISOSAFROLE

1,2-(METHYLENEDIOXY)-4-PROPYLBENZENE see DIHYDROSAFROLE

(3,4-METHYLENEDIOXY-6-PROPYLBENZYL) (BUTYL) DIETHYLENE
GLICOL ETHER see PIPERONYL BUTOXIDE

3,4-METHYLENEDIOXY-6-PROPYLBENZYL N-BUTYL DIETHYLENEGLY-
COL ETHER see PIPERONYL BUTOXIDE

4,4'-METHYLENEDIPHENYL DIISOCYANATE see DIPHENYL METHANE
DIISOCYANATE

METHYLENEDI-P-PHENYLENE DIISOCYANATE see DIPHENYL METHANE
DIISOCYANATE

4,4'-METHYLENEDIPHENYLENE ISOCYANATE see DIPHENYL METHANE
DIISOCYANATE

3-METHYLENE-7-METHYL-1-OCTEN-7-YL ACETATE see ACETIC ACID
MYRCENYL ESTER

METHYLENE OXIDE see FORMALDEHYDE

METHYLENE-S,S'-BIS(O,O-DIAETHYL-DITHIOPHOSPHAT) (GERMAN) see
ETHYL METHYLENE PHOSPHORO DITHIOATE

ENDOMETHYLENETETRAHYDROPHTHALIC ACID,N-2-ETHYLHEXYL
IMIDE see N-OCTYL BICYCLOHEPTENE DICARBOXIMIDE

8-METHYLENE-4,11,11-(TRIMETHYL)BICYCLO(7.2.0)UNDEC-4-ENE
see CARYOPHYLLENE

METHYLENO-BIS-FENYLOIZOCYJANIAN (POLISH) see 1,6-DIISOCY
ANATOHEXANE

METHYLEPHEDRIN (GERMAN) see METHYLEPHEDRINE

METHYLERGOBASINE see PARTERGIN

METHYLERGOBREVIN see PARTERGIN

METHYLERGOMETRIN see PARTERGIN

METHYLERGOMETRINE see PARTERGIN

METHYLERGONOVINE see PARTERGIN

METHYLESTER KISELINY OCTOVE (CZECH) see ACETIC ACID METHYL
ESTER

METHYLESTER KYSELINY ORTHOMRAVENCI (CZECH) see TRIMETHYL-
O-FORMATE

METHYLESTER KYSELINY P-TOLUENSULFONOVE (CZECH) see
METHYL-p-METHYLBENZENESULFONATE

METHYL ESTER OF P-HYDROXYBENZOIC ACID see p-HYDROXYBEN-
ZOIC ACID METHYL ESTER

METHYL ESTER STEARIC ACID see METHYL STEARATE

1,1'-(METHYLETHANEDILIDENEDINITRILO)BIGUANIDINE DIHYDRO-
CHLORIDE DIHYDRATE see METHYL GAG

N-METHYL ETHANOANTHRACENE-9(10H)-METHYLAMINE see BENZ-
OCTAMINE

METHYLETHANOLAMINE see 2-METHYLAMINOETHANOL

N-METHYLETHANOLAMINE see 2-METHYLAMINOETHANOL

METHYLETHENE see PROPENE

2-(1-METHYLETHOXY)PHENOL METHYLCARBAMATE see BAYGON

4-(METHYLETHYL)AMINOAZOBENZENE see N-ETHYL-N-METHYL-p-
(PHENYLAZO)ANILINE

N-METHYL-N-ETHYL-P-AMINOAZOBENZENE see N-ETHYL-N-METHYL-
p-(PHENYLAZO)ANILINE

N-METHYL-4'-ETHYL-P-AMINOAZOBENZENE see p-(4-ETHYLPHENYL-
AZO)-N-METHYLANILINE

ALPHA-(((1-METHYLETHYL)AMINO)METHYL)-2-NAPHTHALENEMETH-

ANOL, HYDROCHLORIDE see alpha-((ISOPROPYLAMINO)METHYL)-
NAPHTHALENEMETHANOL, HYDROCHLORIDE

METHYLETHYLBROMOMETHANE see 2-BROMOBUTANE

METHYLETHYLCARBINOL see sec-BUTYL ALCOHOL

4-METHYL-4-ETHYL-2,6-DIOXOPIPERIDINE see 3-METHYL-3-ETHYL-
GLUTARIMIDE

METHYLETHYLENE see PROPENE

METHYLETHYLENE GLYCOL see 1,2-PROPANEDIOL

METHYL ETHYLENE OXIDE see EPOXYPROPANE

METHYLETHYLENIMINE see 2-METHYLAZIRIDINE

2-METHYLETHYLENIMINE see 2-METHYLAZIRIDINE

1-METHYLETHYL-2-((ETHOXY((1-METHYLETHYL)AMINO)PHOS-
PHINOTHIOYL)OXY)BENZOATE see ISOFENPHOS

1-(METHYLETHYL)-ETHYL 3-METHYL-4-(METHYLTHIO)PHENYL PHOS-
PHORAMIDE see FENAMIPHOS

BETA-METHYL-BETA-ETHYLGLUTARIMIDE see 3-METHYL-3-ETHYL-
GLUTARIMIDE

METHYL ETHYL KETONE see 2-BUTANONE

METHYL ETHYL KETONE (DOT) see 2-BUTANONE

METHYLETHYLKETONHYDROPEROXIDE see METHYL ETHYL KETONE
PEROXIDE

METHYLETHYLMETHANE see N-BUTANE

N-(1-METHYLETHYL)-N'-(1-NITRO-9-ACRIDINYL)-1,3-PROPANEDI-
AMINE DIHYDROCHLORIDE see 9-((3-(ISOPROPYLAMINO)PRO-
PYL)AMINO)-1-NITROACRIDINE DIHYDROCHLORIDE

1-METHYL-5-ETHYL-5-PHENYLBARBITURIC ACID see 5-ETHYL-N-
METHYL-5-PHENYL BARBITURIC ACID

2-METHYL-5-ETHYLPYRIDINE see 5-ETHYL-alpha-PICOLINE

6-METHYL-3-ETHYLPYRIDINE see 5-ETHYL-alpha-PICOLINE

METHYL ETHYL PYRIDINE (DOT) see 5-ETHYL-alpha-PICOLINE

gamma-METHYL-gamma-ETHYL-SUCCINIMIDE see 2-ETHYL-2-
METHYLSUCCINIMIDE

18-METHYL-17-ALPHA-ETHYNYL-19-NORTESTOSTERONE see NORGES-
TREL

3-METHYLETHYNYLOESTRADIOL see 3-METHYLETHYNYLESTRADIOL

METHYL EUGENOL see 4-ALLYL-1,2-DIMETHOXYBENZENE

METHYL-GAMMA-FLUOROBUTYRATE see METHYL-4-FLUOROBUTY-
RATE

METHYL-OMEGA-FLUOROBUTYRATE see METHYL-4-FLUOROBUTY-
RATE

16-ALPHA-METHYL-9-ALPHA-FLUORO-1-DEHYDROCORTISOL see
SUPERPREDNOL

16-ALPHA-METHYL-9-ALPHA-FLUORO-DELTA(SUP 1)-HYDROCORTI-
SONE see SUPERPREDNOL

16-ALPHA-METHYL-9-ALPHA-FLUOROPREDNISOLONE see SUPERPRED-
NOL

16-ALPHA-METHYL-9-ALPHA-FLUORO-1,4-PREGNADIENE-11-BETA,-
17-ALPHA,21-TRIOL-3,20-DIONE see SUPERPREDNOL

METHYL FLUOROSULFONATE see METHYL FLUOROSULFATE

16-ALPHA-METHYL-9-ALPHA-FLUORO-11-BETA,17-ALPHA,21-TRIHY-
DROXYPREGNA-1,4-DIENE-3,20-DIONE see SUPERPREDNOL

METHYL FORMATE (DOT) see METHYL FORMATE

METHYLFORMIAAT (DUTCH) see METHYL FORMATE

METHYLFORMIAT (GERMAN) see METHYL FORMATE

N-METHYL-N-FORMLYHYDRAZINE see N-FORMYL-N-METHYLHYDRA-
ZINE

N-METHYL-N-FORMYL HYDRAZONE OF ACETALDEHYDE see ACETAL-
DEHYDE-N-METHYL-N-FORMYLHYDRAZONE

METHYLFURAN see 2-METHYLFURAN

METHYL GLYCOL see ETHYLENE GLYCOL METHYL ETHER

METHYL GLYCOL ACETATE see ETHYLENE GLYCOL MONO-
METHYL ETHER ACETATE

METHYL GLYCOL MONOACETATE see ETHYLENE GLYCOL MONO-
METHYL ETHER ACETATE

METHYLGLYKOL (GERMAN) see ETHYLENE GLYCOL METHYL ETHER

METHYLGLYKOLACETAT(GERMAN) see ETHYLENE GLYCOL MONO-
METHYL ETHER ACETATE

METHYLGUANIDIN (GERMAN) see METHYLGUANIDINE

METHYL GUTHION see AZINPHOS METHYL

3-METHYL-5-HEPTANONE see AMYL ETHYL KETONE

5-METHYL-3-HEPTANONE see AMYL ETHYL KETONE

6-METHYL-2-HEPTYLAMINE see 2-ISOOCTYL AMINE

(6-(1-METHYL-HEPTYL)-2,4-DINITRO-FENYL)-CROTONAAT (DUTCH) see ARATHANE

(6-(1-METHYL-HEPTYL)-2,3-DINITRO-PHENYL)-CROTONAT (GERMAN) see ARATHANE

2-(1-METHYLHEPTYL)-4,6-DINITROPHENYL CROTONATE see ARATHANE

METHYLHEPTYLNITROSAMIN (GERMAN) see HEPTYLMETHYLINITROSAMINE

1,-METHYLHYDRAZINE see METHYLHYDRAZINE

METHYLHYDRAZINE (DOT) see METHYLHYDRAZINE

4-((2-METHYLHYDRAZINO)METHYL)-N-ISOPROPYLBENZAMIDE see 1-METHYL-2-(p-(ISOPROPYLCARBAMOYL)BENZYL)HYDRAZINE

4-((2-METHYLHYDRAZINO)METHYL)-N-ISOPROPYLBENZAMIDE see PROCARBAZINE

METHYL HYDRIDE see METHANE

DELTA(SUP 1)-6-ALPHA-METHYLHYDROCORTISONE see METHYL-PREDNISOLONE

METHYL HYDROXIDE see METHANOL

1-METHYL-4-HYDROXYBENZENE see p-CRESOL

METHYL-O-HYDROXYBENZOATE see METHYL SALICYLATE

METHYL P-HYDROXYBENZOATE see p-HYDROXYBENZOIC ACID METHYL ESTER

2-METHYL-3-HYDROXY-4,5-BIS(HYDROXYMETHYL)PYRIDINE see PYRIDOXOL

2-METHYL-3-HYDROXY-4,5-BIS(HYDROXYMETHYL)PYRIDINE HYDROCHLORIDE see PYRIDOXOL HYDROCHLORIDE

4-METHYL-7-HYDROXYCOUMARIN DIETHOXYTHIOPHOSPHATE see DIETHYL (4-METHYLUMBELLIFERYL) THIONOPHOSPHATE

2-METHYL-3-HYDROXY-4,5-DIHYDROXYMETHYL-PYRIDIN (GERMAN) see PYRIDOXOL

2-METHYL-3-HYDROXY-4,5-DI(HYDROXYMETHYL)PYRIDINE see PYRIDOXOL

METHYL(BETA-HYDROXYETHYL)AMINE see 2-METHYLAMINOETHANOL

2-METHYL-1-(2-HYDROXYETHYL)-5-NITROIMIDAZOLE see 2-METHYL-5-NITROIMIDAZOLE-1-ETHANOL

2-METHYL-3-(2-HYDROXYETHYL)-4-NITROIMIDAZOLE see 2-METHYL-5-NITROIMIDAZOLE-1-ETHANOL

2-METHYL-3-HYDROXY-4-FORMYL-5-HYDROXYMETHYLPYRIDINE HYDROCHLORIDE see VITAMIN B$_6$ HYDROCHLORIDE

1-METHYL-2-HYDROXYIMINOMETHYLPYRIDINIUM IODIDE see PYRIDINIUM-2-ALDOXIME-N-METHYLIODIDE

1-METHYL-3-HYDROXY-4-ISOPROPYLBENZENE see THYMOL

1-METHYL-4-(m-HYDROXYPHENYL)PIPERIDINE-4-ETHYLKETONE HYDROCHLORIDE see KETOBEMIDONE HYDROCHLORIDE

6-ALPHA-METHYL-17-ALPHA-HYDROXYPROGESTERONE ACETATE see MEDROXYPROGESTERONE ACETATE

6-METHYL-17-ALPHA-HYDROXY-DELTA(SUP 6)-PROGESTERONE ACETATE see VOLIDAN

1-METHYLIMIDAZOLE-2-THIOL see 2-MERCAPTO-1-METHYLIMIDAZOLE

2-METHYL-3-(DELTA(SUP 2)-IMIDAZOLINYLMETHYL)BENZO(B)THIOPHENE HYDROCHLORIDE see 2-((2-METHYLBENZO(b)THIEN-3-YL)-METHYL)-2-IMIDAZOLINE HYDROCHLORIDE

ALPHA-METHYL-BETA-INDOLAETHYLAMINE (GERMAN) see 3-(2-AMINOPROPYL)INDOLE

3-METHYLINDOLE see beta-METHYLINDOLE

ALPHA-METHYL-BETA-INDOLEETHYLAMINE see 3-(2-AMINOPROPYL)-INDOLE

METHYL IODIDE see IODOMETHANE

N-METHYL-DL-ISOBORNYLAMINE HYDROCHLORIDE see MEVASIN HYDROCHLORIDE

METHYL ISOBUTENYL KETONE see MESITYL OXIDE

METHYLISOBUTYL CARBINOL see ISOBUTYLMETHYLCARBINOL

METHYL ISOBUTYL CARBINOL see AMYL METHYL ALCOHOL

METHYL-ISOBUTYL-CETONE (FRENCH) see HEXONE

METHYLISOBUTYLKETON (DUTCH, GERMAN) see HEXONE

METHYL ISOBUTYL KETONE see HEXONE

METHYLISOCYANAAT (DUTCH) see METHYL ISOCYANATE

METHYL ISOCYANAT (GERMAN) see METHYL ISOCYANATE

METHYLISOOCTENYLAMINE see ISONYL

METHYL ISOPROPENYL KETONE see ISOPROPENYL METHYL KETONE

N-METHYL-2-ISOPROPOXYPHENYLCARBAMATE see BAYGON

P-METHYLISOPROPYL BENZENE see p-CYMENE

1-METHYL-4-ISOPROPYLBENZENE see p-CYMENE

1-METHYL-2-(-ISOPROPYLCARBAMOYL)BENZYL)HYDRAZINE see PROCARBAZINE

2-METHYL-5-ISOPROPYLPHENOL see CARVACROL

5-METHYL-2-ISOPROPYL-1-PHENOL see THYMOL

n-METHYL-3-ISOPROPYLPHENYL CARBAMATE see m-CUMENOL METHYLCARBAMATE

n-METHYL-m-ISOPROPYLPHENYL CARBAMATE see m-CUMENOL METHYLCARBAMATE

5-METHYL-2-ISOPROPYL-3-PYRAZOLYL DIMETHYLCARBAMATE see DIMETHYL-5-(1-ISOPROPYL-3-METHYL-PYRAZOLYL)-CARBAMATE

METHYL ISOSYSTOX see DEMETON-S-METHYL

METHYLISOTHIOCYANAAT (DUTCH) see ISOTHIOCYANATOMETHANE

METHYL-ISOTHIOCYANAT (GERMAN) see ISOTHIOCYANATOMETHANE

METHYL ISOTHIOCYANATE see ISOTHIOCYANATOMETHANE

5-METHYL-3-ISOXAZOLECARBOXYLIC ACID-2-BENZYLHYDRAZIDE see ISOCARBOXAZID

3-METHYL-4,5-ISOXAZOLEDIONE 4-((2-CHLOROPHENYL)HYDRAZONE) see 3-METHYL-4,5-ISOXAZOLEDIONE,-4-((o-CHLOROPHENYL) HYDRAZONE)

N'-(5-METHYL-3-ISOXAZOLE)SULFANILAMIDE see SULFAMETHOXAZOL

N'-(5-METHYL-3-ISOXAZOLYL)SULFANILAMIDE see SULFAMETHOXAZOL

N'-(5-METHYLISOXAZOL-3-YL)SULPHANILAMIDE see SULFAMETHOXAZOL

N(SUP 1)-(5-METHYL-3-ISOXAZOLYL)SULPHANILAMIDE see SULFAMETHOXAZOL

METHYLJODID (GERMAN) see IODOMETHANE

METHYLJODIDE (DUTCH) see IODOMETHANE

METHYL KETONE see ACETONE

N-METHYL-LOST see BIS(beta-CHLOROETHYL)METHYLAMINE

D-1-METHYL LYSERGIC ACID MONOETHYLAMIDE see 1-METHYLLYSERGIC ACID ETHYLAMIDE

METHYLMERCAPTAAN (DUTCH) see METHANETHIOL

METHYL MERCAPTAN (DOT) see METHANETHIOL

4-METHYLMERCAPTO-3,5-DIMETHYLPHENYL N-METHYLCARBAMATE see 4-METHYLMERCAPTO-3,5-XYLYL METHYLCARBAMATE

METHYL-MERCAPTOFOS TEOLOVY see DEMETON-S-METHYL

1-METHYL-2-MERCAPTOIMIDAZOLE see 2-MERCAPTO-1-METHYLIMIDAZOLE

2-METHYLMERCAPTO-10-(2-N-METHYL-2-PIPERIDYL)ETHYL)PHENOTHIAZINE see 10-(2-(1-METHYL-2-PIPERIDYL)ETHYL)-2-(METHYLTHIO)PHENOTHIAZINE

2-METHYLMERCAPTO-10-(2-(N-METHYL-2-PIPERIDYL)ETHYLPHENOTHIAZINE HYDROCHLORIDE see 10-(2-(1-METHYL-2-PIPERIDYL) ETHYL)-2-METHYLTHIOPHENOTHIAZINE HYDROCHLORIDE

METHYL-MERCAPTOPHOS see METASYSTOX

METHYLMERCAPTOPHOS see DEMETON-O-METHYL

6-METHYLMERCAPTOPURINE RIBOSIDE see METHYLTHIOINOSINE

METHYLMERCURIC CHLORIDE see MERCURYMETHYLCHLORIDE

METHYLMERCURIC CYANOGUANIDINE see METHYLMERCURIC DICYANDIAMIDE

N-(METHYLMERCURI)-1,4,5,6,7,7-HEXACHLOROBICYCLO(2.2.1)-HEPT-5-ENE-2,3-DICARBOXIMIDE see METHYLMERCURICHLORENDIMIDE

8-(METHYLMERCURIOXY)QUINOLINE see METHYLMERCURY QUINOLINOLATE

N-METHYLMERCURI-1,2,3,6-TETRAHYDRO-3,6-ENDOMETHANO-3,4,5,6,7,7-HEXACHLOROPHTHALIMIDE see METHYLMERCURICHLORENDIMIDE

N-METHYLMERCURI -1,2,3,6-TETRAHYDRO-3,6-METHANO-3,4,5,6,7,7-HEXACHLOROPHTHALIMIDE see METHYLMERCURICHLORENDIMIDE

METHYLMERCURY CHLORIDE see MERCURYMETHYLCHLORIDE

METHYLMERCURY DICYANDIAMIDE see METHYLMERCURIC DICYAN-
DIAMIDE

METHYLMERCURY BETA-HYDROXYQUINOLATE see METHYLMERCURY
QUINOLINOLATE

METHYLMERCURY 8-HYDROXYQUINOLINATE see METHYLMERCURY
QUINOLINOLATE

METHYLMERCURY OXINATE see METHYLMERCURY QUINOLINOLATE

METHYLMERCURY OXYQUINOLINATE see METHYLMERCURY QUINO-
LINOLATE

METHYLMETHACRYLAAT (DUTCH) see METHYL METHACRYLATE

METHYL-METHACRYLAT (GERMAN) see METHYL METHACRYLATE

METHYL METHACRYLATE HOMOPOLYMER see POLYMETHYLMETH-
ACRYLATE

METHYL METHACRYLATE MONOMER, UNINHIBITED (DOT) see
METHYL METHACRYLATE

METHYL METHACRYLATEPOLYMER see POLYMETHYLMETHACRYLATE

METHYL METHACRYLATE RESIN see POLYMETHYLMETHACRYLATE

METHYLMETHANE see ETHANE

METHYL METHANESULFONATE see METHYL MESYLATE

METHYL METHANOATE see METHYL FORMATE

METHYLMETHANSULFONAT (GERMAN) see METHYL MESYLATE

METHYL METHANSULFONATE see METHYL MESYLATE

METHYL METHANSULPHONATE see METHYL MESYLATE

1-METHYL-7-METHOXY-BETA-CARBOLINE see HARMINE

9-METHYL-10-METHOXYMETHYL-1,2-BENZANTHRACENE see
7-METHOXYMETHYL-12-METHYLBENZ(a)ANTHRACENE

METHYL-ALPHA-METHYLACRYLATE see METHYL METHACRYLATE

1-METHYL-6-(1-METHYLALLYL)DITHIOBIUREA see 1-METHYL-6-(1-
METHYLALLYL)-2,5-DITHIOBIUREA

N-METHYL-3'-METHYL-4-AMINOAZOBENZENE see N-METHYL-P-(m-
TOLYLAZO)ANILINE

N-METHYL-3'-METHYL-P-AMINOAZOBENZENE see N-METHYL-P-(m-
TOLYLAZO)ANILINE

METHYL N-((METHYLAMINO)CARBONYL)OXY)ETHANIMIDO)THIOATE
see S-METHYL N-(METHYLCARBAMOYLOXY)THIOACETIMIDATE

2-METHYL-6-METHYLAMINO-2HEPTENE see ISONYL

METHYL-4-METHYLBENZENESULFONATE see METHYL-P-METHYLBEN-
ZENESULFONATE

METHYL 1-(ALPHA-METHYLBENZYL)IMIDAZOLE-5-CARBOXYLATE see
METOMIDATE

METHYL-N-((METHYLCARBAMOYL)OXY)THIOACETIMIDATE see
S-METHYL N-(METHYLCARBAMOYLOXY)THIOACETIMIDATE

CIS-1-METHYL-2-METHYL CARBAMOYL VINYL PHOSPHATE see
3-(DIMETHOXYPHOSPHINYLOXY)N-METHYL-cis-CROTONAMIDE

6-METHYL-5-METHYLENE(1,6)DIOXACYCLODODECINO-(2,3,4-GH)-
PYRROLIZINE-2,7-DIONE see SENECIPHYLLINE

ALPHA-METHYL-3,4-METHYLENEDIOXYPHENETHYLAMINE HYDRO-
CHLORIDE see 1-(3,4-METHYLENEDIOXYPHENYL)-2-AMINOPRO-
PANE

1-METHYL-2-(3,4-METHYLENEDIOXYPHENYL)ETHYL OCTYLSULFOX-
IDE see ISOSAFROLE-n-OCTYLSULFOXIDE

2-METHYL-6-METHYLENE-7-OCTEN-2-OL ACETATE see ACETIC ACID
MYRCENYL ESTER

2-METHYL-6-METHYLENE-7-OCTEN-2-YL ACETATE see ACETIC ACID
MYRCENYL ESTER

1-METHYL-4-(1-METHYLETHENYL)-CYCLOHEXENE, (R)- see d-LIMO-
NENE

2-METHYL-4-((2-METHYLPHENYL)AZO)BENZENAMINE see 2-AMINO-
5-AZOTOLUENE

1-((2-METHYL-4-((2-METHYLPHENYL)AZO)PHENYL)AZO)-2-NAPH-
THALENOL see SCARLET RED

2-METHYL-3-(2-METHYLPHENYL)-4-QUINAZOLINONE see QUAALUDE

2-METHYL-3-(2-METHYLPHENYL)-4(3H)-QUINAZOLINONE see QUA-
ALUDE

2-METHYL-11-(4-METHYL-1-PIPERAZINYL)-DIBENZO(B,F)(1,4)THIAZE-
PINE see METIAPINE

METHYL-2-METHYL-2-PROPENOATE see METHYL METHACRYLATE

2-METHYL-2-(METHYLTHIO)PROPANAL, O-((METHYLAMINO)CAR-
BONYL)OXIME see CARBANOLATE

2-METHYL-2-(METHYLTHIO)PROPIONALDEHYDE O-(METHYLCARBA-
MOYL)OXIME see CARBANOLATE

2-METHYL-2-METHYLTHIO-PROPIONALDEHYD-O-(N-METHYL-CARBA-
MOYL)-OXIM(GERMAN) see CARBANOLATE

3'-METHYL-4-MONOMETHYLAMINOAZOBENZENE see N-METHYL-P-
(m-TOLYLAZO)ANILINE

METHYLMORPHINE see CODEINE

METHYL MUSTARD OIL see ISOTHIOCYANATOMETHANE

N-METHYL-1-NAFTYL-CARBAMAAT (DUTCH) see CARBARYL

2-METHYL-1,4-NAPHTHOCHINON (GERMAN) see 2-METHYL-1,4-
NAPHTHOQUINONE

3-METHYL-1,4-NAPHTHOQUINONE see 2-METHYL-1,4-NAPHTHOQUI-
NONE

N-METHYL-1-NAPHTHYL-CARBAMAT (GERMAN) see CARBARYL

N-METHYL-1-NAPHTHYL CARBAMATE see CARBARYL

N-METHYL-ALPHA-NAPHTHYLCARBAMATE see CARBARYL

N-METHYL-N-(1-NAPHTHYL)MONOFLUOROACETAMIDE see
N-METHYL-N-(1-NAPHTHYL)FLUOROACETAMIDE

N-METHYL-ALPHA-NAPHTHYLURETHAN see CARBARYL

2-METHYL-1-NITRO-9,10-ANTHRACENEDIONE see 2-METHYL-1-NI-
TROANTHRAQUINONE

2-METHYLNITROBENZENE see O-METHYLNITROBENZENE

3-METHYLNITROBENZENE see m-METHYLNITROBENZENE

4-METHYLNITROBENZENE see p-METHYL NITROBENZENE

P-METHYLNITROBENZENE see p-METHYL NITROBENZENE

2-METHYL-5-NITROIMIDAZOLE-1-ETHANOL see 2-METHYL-5-NITRO-
IMIDAZOLE-1-ETHANOL

METHYLNITROIMIDAZOLYLMERCAPTOPURINE see AZATHIOPRINE

6-(METHYL-P-NITRO-5-IMIDAZOLYL)-THIOPURINE see AZATHIOPRINE

6-(1-METHYL-4-NITROIMIDAZOL-5-YLTHIO)PURINE see AZATHIO-
PRINE

6-((1-METHYL-4-NITROIMIDAZOL-5-YL)THIO)PURINE see AZATHIO-
PRINE

6-(1-METHYL-P-NITRO-5-IMIDAZOLYL)-THIOPURINE see AZATHIO-
PRINE

1-METHYL-3-NITRO-1-NITROSOGUANIDINE see N-METHYL-N'-
NITRO-N-NITROSOGUANIDINE

2-METHYL-4-NITROQUINOLINE-1-OXIDE see 2-METHYL-4-NITROQUI-
NOLINE-1-OXIDE

METHYLNITROSOACETAMID (GERMAN) see METHYLNITROSOACET-
AMIDE

METHYLNITROSO-ACETAMIDE see METHYLNITROSOACETAMIDE

N-METHYL-N-NITROSOACETAMIDE see METHYLNITROSOACETAMIDE

1-METHYL-1-NITROSOACETYLUREA see ACETYLMETHYLNITROSO-
UREA

N-METHYL-N-NITROSO-N'-ACETYLUREA see ACETYLMETHYLNITRO-
SOUREA

2-(N-METHYL-N-NITROSO)AMINOACETONITRILE see N-NITROSO-
METHYLAMINOACETONITRILE

ALPHA-(1-(N-METHYL-N-NITROSOAMINO)ETHYL)BENZYL ALCOHOL
see N-NITROSOEPHEDRINE

2-(N-METHYL-N-NITROSOAMINO)-1-PHENYL-1-PROPANOL see
N-NITROSOEPHEDRINE

4-(N-METHYL-N-NITROSOAMINO)-4-(3-PYRIDYL)-1-BUTANONE see
4-(N-METHYL-N-NITROSAMINO)-1-(3-PYRIDYL)-1-BUTANONE

N-METHYL-N-NITROSOANILINE see N-METHYL-N-NITROSO-
ANILINE

N-METHYL-N-NITROSOBENZENAMINE see N-METHYL-N-NITROSO-
ANILINE

N-METHYL-N-NITROSOBENZYLAMINE see N-METHYL-N-BENZYLNI-
TROSAMINE

1-METHYL-1-NITROSOBIURET see N-METHYL-N-NITROSOBIURET

N-METHYL-N-NITROSOBUTYLAMINE see METHYLBUTYLNITROSAMINE

METHYLNITROSOCARBAMIC ACID-O-ISOPROPOXYPHENYL ESTER see
NITROSO-BAYGON

N-METHYL-N-NITROSOCARBAMIC ACID-O-ISOPROPOXYPHENYL ESTER
see NITROSO-BAYGON

METHYL-NITROSOCARBAMIC ACID-1-NAPHTHYL ESTER see N-NITRO-
SOCARBARYL

N-METHYL-N-NITROSO-N'-CARBAMOYLUREA see N-METHYL-N-NITROSOBIURET

N-METHYL-N-NITROSOCYCLOHEXYLAMINE see N-NITROSO-N-METHYLCYCLOHEXYLAMINE

1-METHYL-N-NITROSODIETHYLAMINE see ETHYL ISOPROPYL NITROSOAMINE

N-METHYL-N-NITROSO-ETHAMINE see N,N-METHYLETHYLNITROSAMINE

N-METHYL-N-NITROSOETHYLAMINE see N,N-METHYLETHYLNITROSAMINE

N-METHYL-N-NITROSOETHYLCARBAMATE see N-NITROSO-N-METHYL-URETHANE

METHYLNITROSO-HARNSTOFF (GERMAN) see N-METHYL-N-NITRO-SOUREA

N-METHYL-N-NITROSO-HARNSTOFF (GERMAN) see N-METHYL-N-NITROSOUREA

N-METHYL-N-NITROSOHEPTYLAMINE see HEPTYLMETHYLINITROSAMINE

N-METHYL-N-NITROSOLAURYLAMINE see NITROSOMETHYL-n-DODECYLAMINE

N-METHYL-N-NITROSOMETHANAMINE see DIMETHYLNITROSAMINE

N-METHYL-N-NITROSONITROGUANIDIN (GERMAN) see N-METHYL-N'-NITRO-N-NITROSOGUANIDINE

N-METHYL-N-NITROSO-N'-NITROGUANIDINE see N-METHYL-N'-NITRO-N-NITROSOGUANIDINE

N-METHYL-N-NITROSO-4-OXO-4-(3-PYRIDYL)BUTYL AMINE see 4-(N-METHYL-N-NITROSAMINO)-1-(3-PYRIDYL)-1-BUTANONE

N-METHYL-N-NITROSOPENTYLAMINE see N-AMYL-N-METHYLNITRO-SAMINE

N-METHYL-N-NITROSOPHENETHYLAMINE see METHYL PHENYLETHYL NITROSAMINE

1-METHYL-4-NITROSOPIPERAZINE see 1-NITROSO-4-METHYLPIPER-AZINE

N'-METHYL-N-NITROSOPIPERAZINE see 1-NITROSO-4-METHYLPIPER-AZINE

N-METHYL-N-NITROSO-1-PROPANAMINE see NITROSOMETHYLPRO-PYLAMINE

METHYLNITROSO-PROPIONAMIDE see N-METHYL-N-NITROSOPRO-PIONAMIDE

METHYL-NITROSOPROPIONSAEUREAMID (GERMAN) see N-METHYL-N-NITROSOPROPIONAMIDE

METHYLNITROSOPROPIONYLAMIDE see N-METHYL-N-NITROSOPRO-PIONAMIDE

N-METHYL-N-NITROSO-4-(2-(4-QUINOLINYL)ETHENYL)BENZENAMINE see 4-(4-N-METHYL-N-NITROSAMINOSTYRYL)QUINOLINE

METHYLNITROSOUREA see N-METHYL-N-NITROSOUREA

METHYLNITROSOUREE (FRENCH) see N-METHYL-N-NITROSOUREA

METHYLNITROSOURETHAN (GERMAN) see N-NITROSO-N-METHYLURE-THANE

N-METHYL-N-NITROSO-URETHANE see N-NITROSO-N-METHYLURE-THANE

METHYL NONYL KETONE see 2-UNDECANONE

METHYL-N-NONYL KETONE see 2-UNDECANONE

ALPHA-METHYLNORADRENALINE HYDROCHLORIDE see AMINOPRO-PANOL PYROCATECHOLHYDROCHLORIDE

METHYL OCTADECANOATE see METHYL STEARATE

METHYL-9-OCTADECENOATE see cis-OLEIC ACID, METHYL ESTER

METHYLOCTENYLAMINE see ISONYL

METHYL OLEATE see cis-OLEIC ACID, METHYL ESTER

3-METHYLOLPENTANE see 2-ETHYLBUTANOL

METHYLOLPROPANE see N-BUTYL ALCOHOL

METHYL-4-OMBELLIFERONE SODEE (FRENCH) see 7-HYDROXY-4-METHYLCOUMARIN SODIUM

(METHYL-ONN-AZOXY)METHANOL, ACETATE (ESTER) see METHYL AZOXYMETHYL ACETATE

METHYLOPROPYLOKETON (POLISH) see 2-PENTANONE

METHYL ORTHOFORMATE see TRIMETHYL-O-FORMATE

4-METHYL-2-OXETANONE see beta-BUTYROLACTONE

METHYL OXIRANE see EPOXYPROPANE

METHYL-3-OXOBUTYRATE see METHYL ACETYLACETATE

METHYL P-OXYBENZOATE see p-HYDROXYBENZOIC ACID METHYL ESTER

1-(4-METHYLOXYPHENYL)-3,3-DIMETHYLTRIAZINE see 3,3-DIMETHYL-1,p-METHOXYPHENYLTRIAZENE

METHYLPARABEN see p-HYDROXYBENZOIC ACID METHYL ESTER

METHYLPARAFYNOL CARBAMATE see METHYLPENTYNOL CARBA-MATE

METHYL PARAHYDROXYBENZOATE see p-HYDROXYBENZOIC ACID METHYL ESTER

2-METHYLPENTANOL-1 see AMYL METHYL ALCOHOL

2-METHYL-4-PENTANOL see ISOBUTYLMETHYLCARBINOL

4-METHYL-2-PENTANOL see ISOBUTYLMETHYLCARBINOL

4-METHYLPENTANOL-2 see ISOBUTYLMETHYLCARBINOL

4-METHYL-2-PENTANON (CZECH) see HEXONE

4-METHYL-PENTAN-2-ON (DUTCH, GERMAN) see HEXONE

2-METHYL-4-PENTANONE see HEXONE

4-METHYL-2-PENTANONE see HEXONE

4-METHYL-3-PENTENE-2-ONE see MESITYL OXIDE

4-METHYL-3-PENTEN-2-ON (DUTCH, GERMAN) see MESITYL OXIDE

2-METHYL-2-PENTEN-4-ONE see MESITYL OXIDE

4-METHYL-3-PENTEN-2-ONE see MESITYL OXIDE

3-METHYL-PENTIN-(1)-OL-(3) (GERMAN) see METHYLPENTYNOL CARBAMATE

METHYL-N-PENTYLNITROSAMINE see N-AMYL-N-METHYLNITROSA-MINE

3-METHYL-1-PENTYN-3-OL CARBAMATE see METHYLPENTYNOL CAR-BAMATE

METHYLPERIDIDE see 4'-FLUORO-4-(n-(4-PYRROLIDINAMIDO-4-m-TOLYPIPERIDINO)-BUTYROPHENONE

METHYLPERIDOL see METHYLPERIDOL HYDROCHLORIDE

METHYLPERONE HYDROCHLORIDE see 4'-FLUORO-4-(4-METHYL PIPERIDINO)BUTYRO PHENONE HYDROCHLORIDE

METHYLPHENAZONIUM METHOSULFATE see MULTERGAN METHYL SULFATE

ALPHA-METHYLPHENETHYLAMINE see AMPHETAMINE

(+-)-ALPHA-METHYLPHENETHYLAMINE see BENZEDRINE

ALPHA-METHYLPHENETHYLAMINE PHOSPHATE, DL-MIXTURE see AM-PHETANE PHOSPHATE

ALPHA-METHYLPHENETHYLAMINE SULFATE, (+-)- see DL-AMPHETA-MINE SULFATE

D-ALPHA-METHYLPHENETHYLAMINE SULFATE see d-BENZEDRINE SULFATE

DL-ALPHA-METHYLPHENETHYLAMINE SULFATE see BENZEDRINE SUL-FATE

(ALPHA-METHYLPHENETHYL)HYDRAZINE see PHENIPRAZINE

3-(ALPHA-METHYLPHENETHYL)SYDONE IMINE MONOHYDROCHLORIDE see SYDNOPHEN HYDROCHLORIDE

METHYLPHENOBARBITAL see 5-ETHYL-N-METHYL-5-PHENYL BAR-BITURIC ACID

1-METHYLPHENOBARBITAL see 5-ETHYL-N-METHYL-5-PHENYL BAR-BITURIC ACID

N-METHYLPHENOBARBITAL see 5-ETHYL-N-METHYL-5-PHENYL BAR-BITURIC ACID

N-METHYLPHENOBARBITAL see 5-ETHYL-N-METHYL-5-PHENYL BAR-BITURIC ACID

2-METHYLPHENOL see o-CRESOL

3-METHYLPHENOL see m-CRESOL

4-METHYLPHENOL see p-CRESOL

M-METHYLPHENOL see m-CRESOL

O-METHYLPHENOL see o-CRESOL

P-METHYLPHENOL see p-CRESOL

N-METHYLPHENOLBARBITOL see 5-ETHYL-N-METHYL-5-PHENYL BAR-BITURIC ACID

2-(2-(4-(2-METHYL-3-PHENOTHIAZIN-10-YLPROPYL)-1-PIPERAZINYL)-ETHOXY)ETHANOL DIHYDROCHLORIDE see DIXYRAZINE DIHYDRO-CHLORIDE

3-(2-METHYLPHENOXY)-1,2-PROPANEDIOL see GLYCERYL-O-TOLYL ETHER

P-METHYLPHENYL ACETATE see 4-METHYLPHENYL ACETATE

METHYL(2-PHENYLAETHYL)NITROSAMIN (GERMAN) see METHYL
PHENYLETHYL NITROSAMINE

METHYLPHENYLAMINE see METHYLANILINE

N-METHYLPHENYLAMINE see METHYLANILINE

1-((2-METHYLPHENYL)AZO)-2-NAPHTHALENAMINE see 1-(O-TOLYL-
AZO)-2-NAPHTHYLAMINE

1-(2-METHYLPHENYL)AZO-2-NAPHTHALENAMINE see 1-(O-TOLYL-
AZO)-2-NAPHTHYLAMINE

1-(2-METHYLPHENYL)AZO-2-NAPHTHYLAMINE see 1-(O-TOLYLAZO)-
2-NAPHTHYLAMINE

N-METHYL-N-(P-(PHENYLAZO)PHENYL)HYDROXYLAMINE see N-HY-
DROXY-N-METHYL-4-AMINOAZOBENZENE

METHYLPHENYLBARBITURIC ACID see 5-ETHYL-N-METHYL-5-PHENYL
BARBITURIC ACID

5-METHYL-7-PHENYL-1:2-BENZACRIDINE see 7-METHYL-9-PHENYL-
BENZ(C)ACRIDINE

METHYLPHENYLCARBAMIC ESTER OF 3-OXYPHENYLTRIMETHYLAM-
MONIUM METHYLSULFATE see (m-HYDROXYPHENYL)TRI-
METHYLAMMONIUM METHYLSULFATE METHYLPHENYLCARBA-
MATE

N-METHYL-4-PHENYL-4-CARBETHOXYPIPERIDINE see DEMAROL

N-METHYL-4-PHENYL-4-CARBETHOXYPIPERIDINE HYDROCHLORIDE
see DOLANTIN HYDROCHLORIDE

METHYLPHENYLCARBINOL see alpha-PHENETHYL ALCOHOL

1-METHYL-4-PHENYL-4-CARBOETHOXYPIPERIDINE HYDROCHLORIDE
see DOLANTIN HYDROCHLORIDE

1-(O-METHYLPHENYL)-3,3-DIMETHYL-TRIAZEN (GERMAN) see 1-(O-
METHYLPHENYL)-3,3-DIMETHYL-TRIAZENE

1-(2-METHYLPHENYL)-3,3-DIMETHYLTRIAZENE see 1-(O-METHYL-
PHENYL)-3,3-DIMETHYL-TRIAZENE

1-(3-METHYLPHENYL)-3,3-DIMETHYLTRIAZENE see 3,3-DIMETHYL-
1-(m-METHYLPHENYL)TRIAZENE

1-(M-METHYLPHENYL)-3,3-DIMETHYL-TRIAZENE see 3,3-DIMETHYL-
1-(m-METHYLPHENYL)TRIAZENE

2-METHYL-P-PHENYLENEDIAMINE see TOLUENE-2,5-DIAMINE

4-METHYL-M-PHENYLENEDIAMINE see TOLUENE-2,4-DIAMINE

2-METHYL-p-PHENYLENEDIAMINE SULPHATE see 2,5-DIAMINOTOL-
UENE SULFATE

4-METHYL-PHENYLENE DIISOCYANATE see DI-ISO-CYANATOLU-
ENE

4-METHYL-PHENYLENE ISOCYANATE see DI-ISO-CYANATOLUENE

METHYL PHENYL ETHER see ANISOLE

N-METHYL-5-PHENYL-5-ETHYLBARBITAL see 5-ETHYL-N-METHYL-
5PHENYL BARBITURIC ACID

1-METHYL-5-PHENYL-5-ETHYLBARBITURIC ACID see 5-ETHYL-N-
METHYL-5-PHENYL BARBITURIC ACID

3-METHYL-5,5-PHENYLETHYLHYDANTOIN see 3-METHYL-5-ETHYL-
5-PHENYLHYDANTOIN

1-METHYL-4-PHENYLISONIPECOTIC ACID, ETHYL ESTER see DEMA-
ROL

1-METHYL-4-PHENYLISONIPECOTIC ACID ETHYL ESTER HYDROCHLO-
RIDE see DOLANTIN HYDROCHLORIDE

N-METHYL-beta-PHENYLISOPROPYLAMIN (GERMAN) see DESOXYN

N-METHYL-beta-PHENYLISOPROPYLAMINE see DESOXYN

1-N-METHYL-BETA-PHENYLISOPROPYLAMINE HYDROCHLORIDE see
"SPEED"

D-N-METHYL-BETA-PHENYLISOPROPYLAMINE HYDROCHLORIDE see
d-METHAPHETAMINE HYDROCHLORIDE

N-METHYL-beta-PHENYLISOPROPYLAMINHYDROCHLORID (GERMAN)
see dl-DESOXY EPHEDRINE HYDROCHLORIDE

METHYL PHENYL KETONE see ACETOPHENONE

3-METHYLPHENYL-N-METHYLCARBAMATE see TSUMACIDE

3-METHYL-2-PHENYLMORPHOLINE HYDROCHLORIDE see PHEN-
METRAZINE HYDROCHLORIDE

METHYLPHENYLNITROSAMINE see N-METHYL-N-NITROSOANILINE

1-METHYL-4-PHENYL-PIPERIDIN-4-CARBON-SAEURE-AETHYLESTER-
HYDROCHLORID (GERMAN) see DEMAROL

1-METHYL-4-PHENYLPIPERIDINE-4-CARBOXYLIC ACID ETHYL ESTER
see DEMAROL

METHYL ALPHA-PHENYL-ALPHA-(2-PIPERIDYL)ACETATE see METHYL
PHENIDYL ACETATE

2-METHYL-2-PHENYLPROPANE see tert-BUTYLBENZENE

METHYL-5 PHENYL-1-(PYRROLIDINYL-1)-2 IMIDAZOLE (FRENCH) see
5-METHYL-1-PHENYL-2-(PYRROLIDINYL)IMIDAZOLE

3-METHYL-1-PHENYL-3-(2-SULFOETHYL)TRIAZENE SODIUM SALT see
N-PHENYLAZO-N-METHYLTAURINE SODIUM SALT

5-METHYL-1-PHENYL-3,4,5,6-TETRAHYDRO-1H-2,5-BENZOXAZOCINE
HYDROCHLORIDE see NEFOPAM HYDROCHLORIDE

3-METHYL-2-PHENYLVALERIC ACID see VALETHAMATE BROMIDE

3-METHYL-2-PHENYLVALERIC ACID 2-DIETHYLAMINOETHYL ESTER
METHYL BROMIDE see VALETHAMATE BROMIDE

METHYL PHOSPHATE see TRIMETHYL PHOSPHATE

METHYLPHOSPHONOFLUORIDIC ACID, 3,3-DIMETHYL-2-BUTYL ESTER
see SOMAN

METHYLPHOSPHONOFLUORIDIC ACID ISOPROPYL ESTER see ISOPRO-
PYL METHANEFLUOROPHOSPHONATE

METHYL PHTHALATE see DIMETHYL PHTHALATE

2-METHYL-3-PHYTHYL-1,4-NAPHTHOCHINON (GERMAN) see VITAMIN
K1

2-METHYL-3-PHYTYL-1,4-NAPHTHOCHINON (GERMAN) see VITAMIN
K1

METHYL PINACOLYLOXY PHOSPHORYLFLUORIDE see SOMAN

METHYL PINACOLYL PHOSPHONOFLUORIDATE see SOMAN

3-(4-METHYLPIPERAZINYLIMINOMETHYL)-RIFAMYCIN SV see RIFA-
MYCIN AMP

8-(4-METHYLPIPERAZINYLIMINOMETHYL) RIFAMYCIN SV see RIFA-
MYCIN AMP

8-(((4-METHYL-1-PIPERAZINYL)IMINO)METHYL)RIFAMYCIN SV see
RIFAMYCIN AMP

N-(GAMMA-(4'-METHYLPIPERAZINYL-1')PROPYL)-3-CHLOROPHENO-
THIAZINE see PROCHLORPROMAZINE

N-METHYL-PIPERAZINYL-N'-PROPYL-PHENOTHIAZIN (GERMAN) see
PERNAZINE

N-(3-(4-METHYL-1-PIPERAZINYL)PROPYL)PHENOTHIAZINE see PER-
NAZINE

10-(3-(4-METHYL-1-PIPERAZINYL)PROPYL)-10H-PHENOTHIAZINE
(9CI) see PERNAZINE

4-(4-METHYL-1-PIPERAZINYL)-5,6,7,8-TETRAHYDRO-(1)-
BENZOTHIENO(2,3-d) PYRIMIDINE HYDROCHLORIDE see
QM-1143

2-METHYL-1-PIPERIDINOPROPANOL, BENZOATE see PIPEROCAINE

(2-METHYLPIPERIDINO)PROPYL BENZOATE see PIPEROCAINE

3-(2-METHYLPIPERIDINO)PROPYL BENZOATE HYDROCHLORIDE see
ISOCAINE

DL-(2-METHYLPIPERIDINO)PROPYL BENZOATE HYDROCHLORIDE see
ISOCAINE

gamma-(2-METHYLPIPERIDINO)PROPYL BENZOATE HYDROCHLORIDE
see ISOCAINE

N-METHYLPIPERIDYL-(4)-BENZHYDRYLAETHER SALZSAUREN SALZE
(GERMAN) see LYSSIPOLL

1-METHYL-3-PIPERIDYLIDENEDI(2-THIENYL)METHANE see BITIODIN

9-(1-METHYL-PIPERIDYL-(2)-METHYL)-CARBAZOL (GERMAN) see
9(METHYL-2-PIPERIDYL)METHYLCARBAZOLE

GAMMA-(2-METHYLPIPERIDYL)PROPYL BENZOATE see PIPEROCAINE

(+−)-gamma-(2-METHYLPIPERIDYL)PROPYL BENZOATE HYDRO-
CHLORIDE see ISOCAINE

6-ALPHA-METHYLPREDNISOLONE see METHYLPREDNISOLONE

16-BETA-METHYL-1,4-PREGNADIENE-9-ALPHA-FLUORO-11-BETA,17-
ALPHA,21-TRIOL- 3,20-DIONE see BETAMETHASONE

6-METHYL-DELTA(SUP 4,6)-PREGNADIEN-17-ALPHA-OL-3,20-DIONE
ACETATE see VOLIDAN

6-ALPHA-METHYL-4-PREGNENE-3,20-DION-17-ALPHA-OL ACETATE
see MEDROXYPROGESTERONE ACETATE

2-METHYLPROPANAL see ISOBUTYRALDEHYDE

2-METHYL-1-PROPANAL see ISOBUTYRALDEHYDE

2-METHYLPROPANOIC ACID see ISOBUTYRIC ACID

2-METHYL PROPANOL see ISOBUTYL ALCOHOL

2-METHYL-1-PROPANOL see ISOBUTYL ALCOHOL

2-METHYLPROPAN-1-OL see ISOBUTYL ALCOHOL

N-METHYL-N-PROPARGYL-3-(2,4-DICHLOROPHENOXY)PROPYLAMINE HYDROCHLORIDE see CLORGYLINE HYDROCHLORIDE

2-METHYLPROPENAL (CZECH) see METHYLACRYLALDEHYDE

2-METHYLPROPENAMIDE see METHACRYLIC ACID AMIDE

METHYL PROPENATE see METHYL ACRYLATE

METHYL PROPENE see ISOBUTENE

2-METHYLPROPENENITRILE see METHYLACRYLONITRILE

METHYL PROPENOATE see METHYL ACRYLATE

METHYL-2-PROPENOATE see METHYL ACRYLATE

2-METHYLPROPENOIC ACID see METHACRYLIC ACID

2-METHYL-2-PROPENOIC ACIDMETHYL ESTER HOMOPOLYMER (9CI) see POLYMETHYLMETHACRYLATE

2-METHYLPROPIONALDEHYDE see ISOBUTYRALDEHYDE

2-METHYLPROPIONIC ACID see ISOBUTYRIC ACID

alpha-METHYLPROPIONIC ACID see ISOBUTYRIC ACID

2-METHYLPROPIONITRILE see ISOBUTYRONITRILE

2-METHYLPROPYL ACETATE see ISOBUTYL ACETATE

2-METHYL-1-PROPYL ACETATE see ISOBUTYL ACETATE

2-METHYLPROPYL ALCOHOL see ISOBUTYL ALCOHOL

METHYL PROPYL CARBINOL see 2-PENTANOL

METHYL-PROPYL-CETONE (FRENCH) see 2-PENTANONE

6-(1-METHYL-PROPYL)-2,4-DINITROFENOL (DUTCH) see 2-sec-BUTYL-4,6-DINITRO PHENOL

(6-(1-METHYL-PROPYL)-2,4-DINITRO-FENYL)-3,3-DIMETHYL-ACRYLAAT (DUTCH) see BINAPACRYL

2-(1-METHYLPROPYL)-4,6-DINITROPHENOL see 2-sec-BUTYL-4,6-DINITRO PHENOL

2-(1-METHYL-N-PROPYL) 4,6-DINITROPHENOL, AMMONIUM SALT see BUTOPHEN

2-(1-METHYLPROPYL)-4,6-DINITROPHENYL ACETATE see O-ACETYL-2-sec-BUTYL-4,6-DINITROPHENOL

2-(1-METHYLPROPYL)-4,6-DINITROPHENYL BETA,BETA-DIMETHACRYLATE see BINAPACRYL

(6-(1-METHYL-PROPYL)-2,4-DINITRO-PHENYL)-3,3-DIMETHYL-ACRYLAT (GERMAN) see BINAPACRYL

BETA-METHYLPROPYL ETHANOATE see ISOBUTYL ACETATE

2-METHYL-2-PROPYLETHANOL see AMYL METHYL ALCOHOL

2-METHYLPROPYL ISOBUTYRATE see ISOBUTYL ISOBUTYRATE

METHYL-N-PROPYL KETONE see 2-PENTANONE

METHYL PROPYL KETONE (DOT) see 2-PENTANONE

2-METHYLPROPYL METHACRYLATE see ISOBUTYL METHACRYLATE

METHYL-N-PROPYLNITROSAMINE see NITROSOMETHYLPROPYLAMINE

METHYLPROPYLNITROSOAMINE see NITROSOMETHYLPROPYLAMINE

2-METHYL-2-PROPYL-1,3-PROPANEDIOL BUTYLCARBAMATE CARBAMATE see 2-METHYL-2-PROPYLTRIMETHYLENE BUTYLCARBAMATE CARBAMATE

2-METHYL-2-PROPYL-1,3-PROPANEDIOL CARBAMATE ISOPROPYLCARBAMATE see ISOPROPYL MEPROBAMATE

2-METHYL-2-N-PROPYL-1,3-PROPANEDIOL DICARBAMATE see MILTOWN

3-METHYLPYRAZOLYL-5-DIETHYLPHOSPHATE see METHYLPYRAZOLYL DIETHYLPHOSPHATE

ALPHA-METHYLPYRIDINE see 2-METHYL PYRIDINE

N-METHYLPYRIDINE-2-ALDOXIME IODIDE see PYRIDINIUM-2-ALDOXIME-N-METHYLIODIDE

2-METHYLPYRIDINE-1-OXIDE-4-AZO-p-DIMETHYLANILINE see N,N-DIMETHYL-4-(2-METHYL-4-PYRIDYLAZO)ANILINE-N-OXIDE

1-METHYL-2-PYRIDINIUM ALDOXIME CHLORIDE see 2-FORMYL-1-METHYLPYRIDINIUM CHLORIDE OXIME

N-METHYLPYRIDINIUM-2-ALDOXIME IODIDE see PYRIDINIUM-2-ALDOXIME-N-METHYLIODIDE

N-METHYLPYRIDINIUM CHLORIDE-2-ALDOXIME see 2-FORMYL-1-METHYLPYRIDINIUM CHLORIDE OXIME

1-METHYL-2-(3-PYRIDYL)PYRROLIDINE see NICOTINE

1-1-METHYL-2-(3-PYRIDYL)-PYRROLIDINE SULFATE see NICOTINE SULFATE

3-METHYLPYROCATECHOL see 2,3-DIHYDROXYTOLUENE

10-((1-METHYL-3-PYRROLIDINYL)METHYL)-PHENOTHIAZINE see METHDILAZINE

1-METHYL-3-PYRROLIDINYL alpha-PHENYLCYCLOPENTANEGLYCOLATE METHOBROMIDE see GLYCOPYRRONIUM BROMIDE

(S)-3-(1-METHYL-2-PYRROLIDINYL)PYRIDINE SULFATE (2:1) see-NICOTINE SULFATE

beta-1-METHYL-3-PYRROLIDYL-alpha-CYCLOPENTYLMANDELATE METHOBROMIDE see GLYCOPYRRONIUM BROMIDE

(−)-3-(1-METHYL-2-PYRROLIDYL)PYRIDINE see NICOTINE

L-3-(1-METHYL-2-PYRROLIDYL)PYRIDINE see NICOTINE

1-3-(1-METHYL-2-PYRROLIDYL)PYRIDINE SULFATE see NICOTINE SULFATE

METHYLQUINAZOLONE HYDROCHLORIDE see METHAQUALONE HYDROCHLORIDE

2-METHYL-1,4-QUINONE see 2-METHYL-p-BENZOQUINONE

METHYLRESERPATE 3,4,5-TRIMETHOXYBENZOIC ACID see RESERPINE

METHYL RESERPATE 3,4,5-TRIMETHOXYBENZOIC ACID ESTER see RESERPINE

METHYLRHODANID (GERMAN) see METHYL THIOCYANATE

METHYLROSANILINE CHLORIDE see ANILINE VIOLET

METHYL SELENAC see SELENIUM DIMETHYLDITHIOCARBAMATE

METHYLSENFOEL (GERMAN) see ISOTHIOCYANATOMETHANE

METHYL SILICATE see METHYL-o-SILICATE

ALPHA-METHYLSTYREEN (DUTCH) see alpha-METHYLSTYRENE

ALPHA-METHYL-STYROL (GERMAN) see alpha-METHYLSTYRENE

5-METHYL-3-SULFANILAMIDOISOXAZOLE see SULFAMETHOXAZOL

5-METHYL-3-SULFANYLAMIDOISOXAZOLE see SULFAMETHOXAZOL

METHYL SULFATE see SULFURIC ACID, DIMETHYL ESTER

METHYL SULFOCYANATE see METHYL THIOCYANATE

METHYLSULFONIC ACID, ETHYL ESTER see ETHYL METHANESULFONATE

2-METHYLSULFONYL-10-(3-(4CARBAMOYLPIPERIDINO)PROPYL)-PHENOTHIAZINE see METOPIMAZINE

1-(3-(2-(METHYLSULFONYL)PHENOTHIAZIN-10-YL)PROPYL)ISONIPECOTAMIDE see METOPIMAZINE

METHYL SULFOXIDE see DIMETHYL SULFOXIDE

5-METHYL-3-SULPHANIL-AMIDOISOXAZOLE see SULFAMETHOXAZOL

METHYL SULPHIDE see 2-THIOPROPANE

METHYL SYSTOX see METASYSTOX

METHYLSYSTOX see DEMETON-O-METHYL

METHYLTESTOSTERONE see 17-METHYL TESTOSTERONE

17-ALPHA-METHYLTESTOSTERONE see 17-METHYL TESTOSTERONE

METHYL-1,2,5,6-TETRAHYDRO-1-METHYLNICOTINATE see ARECOLINE

2-METHYL-2-(4-(1,2,3,4-TETRAHYDRO-1-NAPHTHALENYL)PHENOXY)PROPANOIC ACID see MELIPAN

2-METHYL-2-(4-(1,2,3,4-TETRAHYDRO-1-NAPHTHYL)PHENOXY)PROPANOIC ACID see MELIPAN

2-METHYL-2-(P-(1,2,3,4-TETRAHYDRO-1-NAPHTHYL)PHENOXY)PROPIONIC ACID see MELIPAN

ALPHA-METHYL-ALPHA-(P-1,2,3,4-TETRAHYDRONAPHTH-1-YLPHENOXY)PROPIONIC ACID see MELIPAN

N-METHYL-DELTA-TETRAHYDRONICOTINIC ACID METHYL ESTER see ARECOLINE

N-METHYL-1,2,3,6-TETRAHYDROPHTHALIMIDE see N-METHYL-4-CYCLOHEXENE-1,2-DICARBOXIMIDE

N-METHYLTETRAHYDROPYRIDINE-BETA-CARBOXYLIC ACID METHYL ESTER see ARECOLINE

2-METHYL-3-(3,7,11,15-TETRAMETHYL-2-HEXADECENYL)-1,4-NAPHTHALENEDIONE see VITAMIN K1

N-METHYL-N,2,4,6-TETRANITROANILINE see TETRYL

METHYLTHEOBROMIDE see CAFFEINE

N-METHYL-3-THIA-2-METHYL-VALERAMID DER O,O-DIMETHYLTHIOLOPHOSPHORSAEURE (GERMAN) see N-METHYL-O,O-DIMETHYL-THIOLOPHOSPHORYL-5-THIA-3-METHYL-2-VALERAMIDE

4-METHYLTHIO-3,5-DIMETHYLPHENYL METHYLCARBAMATE see 4-METHYLMERCAPTO-3,5-XYLYL METHYLCARBAMATE

METHYLTHIOMETHANE see 2-THIOPROPANE

METHYLTHIONINE CHLORIDE see 3,7-BIS(DIMETHYL AMINO)PHENAZA THIONIUM CHLORIDE

METHYLTHIONIUM CHLORIDE see 3,7-BIS(DIMETHYL AMINO)PHE-
NAZA THIONIUM CHLORIDE

2-METHYLTHIO-PROPIONALDEHYD-O-(METHYLCARBAMOYL)-OXIM
(GERMAN) see S-METHYL N-(METHYLCARBAMOYLOXY)THIO-
ACETIMIDATE

6-METHYL-2-THIO-2,4-(1H3H)PYRIMIDINEDIONE see 6-METHYLTHI-
OURACIL

METHYLTHIOURACIL see 6-METHYLTHIOURACIL

4-METHYL-2-THIOURACIL see 6-METHYLTHIOURACIL

6-METHYL-2-THIOURACIL see 6-METHYLTHIOURACIL

METHYLTHIOXOARSINE see METHYLARSENIC SULFIDE

4-(METHYLTHIO)-3,5-XYLYL ISOMETHYLCARBAMATE see 4-METHYL-
MERCAPTO-3,5-XYLYL METHYLCARBAMATE

METHYL THIURAMDISULFIDE see BIS(DIMETHYL THIOCARBAMYL)-
DISULFIDE

O-METHYLTOLUENE see O-XYLENE

P-METHYLTOLUENE see p-XYLENE

METHYL TOLUENE-4-SULFONATE see METHYL-p-METHYLBENZENE-
SULFONATE

METHYL-P-TOLUENESULFONATE see METHYL-p-METHYLBENZENE-
SULFONATE

5-METHYL-O-TOLUIDINE see 2,5-XYLIDINE

6-METHYL-M-TOLUIDINE see 2,5-XYLIDINE

2-METHYL-P-TOLUIDINE HYDROCHLORIDE see 2,4-XYLIDINE HYDRO-
CHLORIDE

4-METHYL-O-TOLUIDINE HYDROCHLORIDE see 2,4-XYLIDINE HYDRO-
CHLORIDE

5-METHYL-O-TOLUIDINE HYDROCHLORIDE see 2,5-XYLIDINE HYDRO-
CHLORIDE

6-METHYL-M-TOLUIDINE HYDROCHLORIDE see 2,5-XYLIDINE HYDRO-
CHLORIDE

2-METHYL-3-O-TOLYL-4(3H)-CHINAZOLINON (GERMAN) see QUA-
ALUDE

2-METHYL-3-O-TOLYL-4(3H)-CHINAZOLONE see QUAALUDE

2-METHYL-3-TOLYLCHINAZOLON-4 HYDROCHLORIDE (GERMAN) see
METHAQUALONE HYDROCHLORIDE

(2-METHYL-3-(O-TOLYL)-3,4-DIHYDRO-4-(QUINAZOLINONE) see
QUAALUDE

2-METHYL-3-(O-TOLYL)-3,4-DIHYDRO-4-QUINAZOLINONE see QUA-
ALUDE

2-METHYL-3-TOLYL-4-OXYBENSDIAZINE see QUAALUDE

2-METHYL-3-O-TOLYL-4(3H)-QUINAZOLINONE see QUAALUDE

2-METHYL-3-O-TOLYL-4(3H)-QUINAZOLINONE HYDROCHLORIDE see
METHAQUALONE HYDROCHLORIDE

2-METHYL-3-(2-TOLYL)QUINAZOL-4-ONE see QUAALUDE

2-METHYL-3-O-TOLYL-4-QUINAZOLONE see QUAALUDE

2-METHYL-3-(O-TOLYL)-4-QUINAZOLONE HYDROCHLORIDE see
METHAQUALONE HYDROCHLORIDE

METHYL TOSYLATE see METHYL-p-METHYLBENZENESULFONATE

METHYL-P-TOSYLATE see METHYL-p-METHYLBENZENESULFONATE

METHYLTRICAPRYLYLAMMONIUMCHLORIDE see METHYLTRIOCTYL-
AMMONIUM CHLORIDE

METHYL TRICHLORIDE see CHLOROFORM

METHYLTRICHLOROMETHANE see 1,1,1-TRICHLOROETHANE

METHYL-TRICHLORSILAN (CZECH) see METHYLTRICHLOROSILANE

METHYL TRIFLUORIDE see CARBON TRIFLUORIDE

METHYL-3,3,3-TRIFLUORO-2-(TRIFLUOROMETHYL)PROPIONATE see-
METHYL HEXAFLUOROISOBUTYRATE

ALPHA-METHYL-3,4,5-TRIMETHOXYPHENETHYLAMINE HYDROCHLO-
RIDE see 3,4,5-TRIMETHOXYAMPHETAMINE HYDROCHLORIDE

ALPHA-METHYLTRYPTAMINE see 3-(2-AMINOPROPYL)INDOLE

4-METHYLUMBELLIFERONE-O,O-DIETHYL THIOPHOSPHATE see
DIETHYL (4-METHYLUMBELLIFERYL) THIONOPHOSPHATE

METHYL-4-UMBELLIFERONE SODIUM see 7-HYDROXY-4-METHYL-
COUMARIN SODIUM

4-METHYLURACIL see 6-METHYLTHIOURACIL

METHYLURETHAN see METHYL CARBAMATE

N-METHYL URETHAN see ETHYL-N-METHYL CARBAMATE

METHYLURETHANE see METHYL CARBAMATE

TRANS-8-METHYL-N-VANILLYL-6-NONEAMIDE see CAPSAICIN

METHYL-VINYL-CETONE (FRENCH) see 3-BUTEN-2-ONE

METHYLVINYLKETON (GERMAN) see 3-BUTEN-2-ONE

METHYLVINYL KETONE see 3-BUTEN-2-ONE

METILACRILATO (ITALIAN) see METHYL ACRYLATE

METILAMIL ALCOHOL (ITALIAN) see ISOBUTYLMETHYLCARBINOL

METILAMINE (ITALIAN) see METHYLAMINE

2-METIL-6-AMINO-EPTANO (ITALIAN) see 2-ISOOCTYL AMINE

3-METIL-BUTANOLO (ITALIAN) see ISOAMYL ALCOHOL

METIL CELLOSOLVE (ITALIAN) see ETHYLENE GLYCOL METHYL
ETHER

2-METILCICLOESANONE (ITALIAN) see 2-METHYLCYCLOHEXANONE

METILCLOROFORMIATO (ITALIAN) see METHYL CHLOROCARBONATE

N-METIL-DITIOCARBAMMATO DI SODIO (ITALIAN) see VAPAM

METILE (ACETATO DI) (ITALIAN) see ACETIC ACID METHYL ESTER

O,O-METILEN-BIS(4-CLOROFENOLO) (ITALIAN) see 2,2'-METHYLENE-
BIS(4-CHLOROPHENOL)

3,3'-METILEN-BIS(4-IDROSSI-CUMARINA) (ITALIAN) see BISHY-
DROXY COUMARIN

4,4-METILENE-BIS-O-CLOROANILINA (ITALIAN) see MOCA

METILEN-S,S'-BIS(O,O-DIETIL-DITIOFOSFATO) (ITALIAN) see ETHYL
METHYLENE PHOSPHORO DITHIOATE

(6-(1-METIL-EPITL)-2,4-DINITRO-FENIL)-CROTONATO (ITALIAN) see
ARATHANE

METILESTER DEL ACIDO BISDEHIDROISYNOLICO (SPANISH) see BIS
DEHYDRO ISYNOLIC ACID METHYL ESTER

METILETILCHETONE (ITALIAN) see 2-BUTANONE

METIL (FORMIATO DI) (ITALIAN) see METHYL FORMATE

METILISOBUTILCHETONE (ITALIAN) see HEXONE

METIL ISOCIANATO (ITALIAN) see METHYL ISOCYANATE

METILMERCAPTANO (ITALIAN) see METHANETHIOL

METIL METACRILATO (ITALIAN) see METHYL METHACRYLATE

N-METIL-1-NAFTIL-CARBAMMATO (ITALIAN) see CARBARYL

4-METILPENTAN-2-OLO (ITALIAN) see ISOBUTYLMETHYLCARBINOL

4-METILPENTAN-2-ONE (ITALIAN) see HEXONE

4-METIL-3-PENTEN-2-ONE (ITALIAN) see MESITYL OXIDE

(6-(1-METIL-PROPIL)-2,4-DINITRO-FENIL)-3,3-DIMETIL-ACRILATO
(ITALIAN) see BINAPACRYL

6-(1-METIL-PROPIL)-2,4-DINITRO-FENOLO (ITALIAN) see 2-sec-
BUTYL-4,6-DINITRO PHENOL

ALPHA-METIL-STIROLO (ITALIAN) see alpha-METHYLSTYRENE

2-METIL-2-TIOMETIL-PROPIONALDEID-O-(N-METIL-CARBAMOIL)-
OSSIMA (ITALIAN) see CARBANOLATE

METILTRIAZOTION see AZINPHOS METHYL

METOKSYCHLOR (POLISH) see DIMETHOXY-DDT

METOKSYETYLOWY ALKOHOL (POLISH) see ETHYLENE GLYCOL
METHYL ETHER

(2-METOSSICARBONIL-1-METIL-VINIL)-DIMETIL-FOSFATO (ITALIAN)
see MEVINPHOS

2-METOSSIETANOLO (ITALIAN) see ETHYLENE GLYCOL METHYL
ETHER

2-METOSSIETILACETATO(ITALIAN) see ETHYLENE GLYCOL MONO-
METHYL ETHER ACETATE

S-((5-METOSSI-4H-PIRON-2-IL)-METIL)-O,O-DIMETIL-MONOTIOFOS-
FATO (ITALIAN) see ENDOTHION

METYLAL (POLISH) see DIMETHOXYMETHANE

METYLENU CHLOREK (POLISH) see METHANE DICHLORIDE

METYLESTER KYSELINY SALICYLOVE (CZECH) see METHYL SALICY-
LATE

METYLOAMINA (POLISH) see METHYLAMINE

METYLOCYKLOHEKSAN (POLISH) see CYCLOHEXYLMETHANE

METYLOCYKLOHEKSANON (POLISH) see METHYLCYCLOHEXANONE

METYLOETYLOKETON (POLISH) see 2-BUTANONE

METYLOIZOBUTYLOKETON (POLISH) see HEXONE

METYLOWY ALKOHOL (POLISH) see METHANOL

METYLPARATION (CZECH) see METHYL PARATHION

METYLU BROMEK (POLISH) see BROMO METHANE

METYLU CHLOREK (POLISH) see MONOCHLOROMETHANE

METYLU JODEK (POLISH) see IODOMETHANE

MEXEPHENAMIDE see MEPHEXAMIDE

MEXYL see ACETPHENARSINE

MEZCALINE see MESCALINE

MG 18037 see N-(4-tert-BUTYL CYCLOHEXYL)-3,3-DIPHENYL PRO-PYLAMINE HYDROCHLORIDE

MGK DOG AND CAT REPELLENT see 2-UNDECANONE

MIADONE see dl-METHADONE HYDROCHLORIDE

MICA SILICATE see MICA

MICHLER'S KETONE see TETRAMETHYL DIAMINO BENZOPHENONE

P,P'-MICHLER'S KETONE see MICHLER'S KETONE

MICROSETILE YELLOW GR see ACETAMINE YELLOW CG

MICROTHENE see POLYETHYLENE

MIEDZ (POLISH) see COPPER FUME

MIERENZUUR (DUTCH) see FORMIC ACID

MIH see PROCARBAZINE

MILCHSAURE (GERMAN) see LACTIC ACID

MILD MERCURY CHLORIDE see MERCUROUS CHLORIDE

MILD SILVER PROTEIN see ARGYROL

MIL-DU-RID see 2-BIPHENYLOL, SODIUM SALT

MILK ACID see LACTIC ACID

MILK SUGAR see LACTOSE

MINERAL GREEN see CUPRIC ACETOARSENITE

MINERAL NAPHTHA see BENZENE

MINERAL ORANGE see LEAD OXIDE RED

MINERAL PITCH see ASPHALT

MINERAL RED see LEAD OXIDE RED

MINERAL SPIRITS see PETROLEUM SPIRITS

MINERAL THINNER see PETROLEUM SPIRITS

MINERAL TURPENTINE see PETROLEUM SPIRITS

MINERAL WHITE see CALCIUM (II) SULFATE DIHYDRATE (1:1:2)

MINIUM see LEAD OXIDE RED

MIPK see METHYL ISOPROPYL KETONE

MIRBANE OIL see NITROBENZENE

MISPICKEL see ARSENOPYRITE

MITRAMYCIN see MITHRAMYCIN

MITSUI DIRECT BLACK GX see APOMINE BLACK GX

MLA-74 see 1-METHYLLYSERGIC ACID ETHYLAMIDE

M'-METHYL-P-DIMETHYLAMINOAZOBENZENE see N,N-DIMETHYL-p-(m-TOLYLAZO)ANILINE

MMH see METHYLHYDRAZINE

MMS see METHYL METHANE SULFONATE

MNC see METHYLNITROSOCYANAMIDE

MNFA see N-METHYL-N-(1-NAPHTHYL)FLUOROACETAMIDE

MNU AND CYCLOPHOSPHAMIDE (2:1) see CYCLOPHOSPHAMIDE AND MNU (1:2)

MOLASSES ALCOHOL see ETHANOL

MOLE DEATH see STRYCHNINE

MOLTEN ADIPIC ACID see ADIPIC ACID

MOLYBDENUM-LEAD CHROMATE see LEAD-MOLYBDENUM CHRO-MATE

MOLYBDENUM ORANGE see LEAD-MOLYBDENUM CHROMATE

MOLYBDIC ACID DIAMMONIUM SALT see AMMONIUM MOLYBDATE

MOLYBDIC ANHYDRIDE see MOLYBDENUM TRIOXIDE

MOLYBDIC TRIOXIDE see MOLYBDENUM TRIOXIDE

MONAQUEST see PENTHAMIL

MONARGAN see ACETPHENARSINE

MONKSHOOD see ACONITE

MONOACETYLHYDRAZINE see ACETYL HYDRAZIDE

MONOAETHANOLAMIN (GERMAN) see MONOETHANOLAMINE

MONOALLYLAMINE see ALLYLAMINE

MONOALLYLUREA see ALLYLUREA

MONOAMMONIUM CARBONATE see AMMONIUM BICARBONATE (1:1)

MONOAMMONIUM SULFAMATE see SULFAMIC ACID, MONOAMMO-NIUM SALT

MONOAMMONIUM SULFIDE see AMMONIUM SULFIDE

MONOBASIC CHROMIUM SULFATE see NEOCHROMIUM

MONOBASIC LEAD ACETATE see LEAD ACETATE, BASIC

MONOBASIC DL-ALPHA-METHYLPHENETHYLAMINE PHOSPHATE see AMPHETANE PHOSPHATE

MONOBASIC RACEMIC AMPHETAMINE PHOSPHATE see AMPHETANE PHOSPHATE

MONOBENZYL ETHER HYDROQUINONE see AGERITE

MONOBENZYL HYDROQUINONE see AGERITE

MONOBROMOACETIC ACID see alpha-BROMOACETIC ACID

MONOBROMOETHANE see BROMOETHANE

MONOBROMOISOVALERYLUREA see 2-BROMO-3-METHYL BUTYRYL UREA

MONOBROMOMETHANE see BROMO METHANE

MONOBROMOTRIFLUORMETHANE (DOT) see BROMO TRIFLUORO METHANE

MONO-N-BUTYLAMINE see N-BUTYLAMINE

MONOCAINE HYDROCHLORIDE see IBYLCAINE HYDROCHLORIDE

MONOCALCIUM ARSENITE see CALCIUM ARSENITE

MONOCHLOORAZIJNZUUR (DUTCH) see CHLOROACETIC ACID

MONOCHLOORBENZEEN (DUTCH) see BENZENE CHLORIDE

MONOCHLORACETIC ACID see CHLOROACETIC ACID

MONOCHLORACETONE see CHLOROACETONE

MONOCHLORBENZENE see BENZENE CHLORIDE

MONOCHLORBENZOL (GERMAN) see BENZENE CHLORIDE

MONOCHLORESSIGSAEURE (GERMAN) see CHLOROACETIC ACID

MONOCHLORETHANE see ETHYL CHLORIDE

MONOCHLORHYDRINE DU GLYCOL (FRENCH) see 2-CHLOROETHYL ALCOHOL

MONOCHLOROACETALDEHYDE see 2-CHLOROACETALDEHYDE

MONOCHLOROACETIC ACID see CHLOROACETIC ACID

MONOCHLOROACETONE see CHLOROACETONE

MONOCHLOROACETONE see ACETONYL CHLORIDE

MONOCHLOROACETONITRILE see CHLORACETONITRILE

MONOCHLOROACETYL CHLORIDE see CHLOROACETYL CHLORIDE

alpha-MONOCHLOROANTHRAQUINONE see 1-CHLORO ANTHRA QUI-NONE

MONOCHLOROBENZENE see BENZENE CHLORIDE

MONOCHLOROETHANOIC ACID see CHLOROACETIC ACID

2-MONOCHLOROETHANOL see 2-CHLOROETHYL ALCOHOL

MONOCHLOROETHENE see VINYL CHLORIDE

MONOCHLOROETHYLENE (DOT) see VINYL CHLORIDE

MONOCHLOROHYDRIN see CHLORHYDRIN

ALPHA-MONOCHLOROHYDRIN see CHLORHYDRIN

MONOCHLOROMETHANE see CHLOROMETHANE

MONOCHLOROMETHYL CYANIDE see CHLORACETONITRILE

MONO-CHLORO-MONO-BROMO-METHANE see BROMOCHLORO-METHANE

MONOCHLOROPENTAFLUOROETHANE see CHLORO PENTAFLUORO ETHANE

MONOCHLOROTRIFLUOROMETHANE (DOT) see CHLORO TRIFLUORO METHANE

MONOCHROMIUM OXIDE) see CHROMIUM (VI) OXIDE (1:3)

MONOCHROMIUM TRIOXIDE see CHROMIUM (VI) OXIDE (1:3)

"MONOCITE" METHACRYLATE MONOMER see METHYL METHACRY-LATE

MONOCLOROBENZENE (ITALIAN) see BENZENE CHLORIDE

MONOCROTOPHOS see "AZODRIN"

MONOCYANOACETIC ACID see CYANOACETIC ACID

MONODEMETHYLIMIPRAMINE see 10,11-DIHYDRO-5-(3-METHYL-AMINOPROPYL)-5H-DIBENZ(b,f)AZEPINE

MONOETHANOLAMINE (DOT) see MONOETHANOLAMINE

MONOETHANOLETHYLENEDIAMINE see N-AMINOETHYL ETHANOL-AMINE

N-MONOETHYLAMIDE OF O,O-DIMETHYLDITHIOPHOSPHORYLACETIC ACID see DIMETHOATE-ETHYL

MONOETHYLAMINE see ETHANAMINE

2-N-MONOETHYLAMINOETHANOL see 2-ETHYL AMINO ETHANOL

MONOETHYLENE GLYCOL see ETHYLENE GLYCOL

MONOETHYLETHANOLAMINE see 2-METHYLAMINOETHANOL

MONOETHYL ETHER OF DIETHYLENE GLYCOL see CARBITOL CELLO-SOLVE

MONOFLUORAZIJNZUUR (DUTCH) see FLUORO ETHANOIC ACID

MONOFLUORESSIGSAURE (GERMAN) see FLUORO ETHANOIC ACID

MONOFLUORETHANOL see FLUOROETHANOL

MONOFLUOROACETAMIDE see FLUOROACETAMIDE

MONOFLUOROACETATE see FLUORO ETHANOIC ACID

MONOFLUOROACETIC ACID see FLUORO ETHANOIC ACID

MONOFLUOROETHANOL see FLUOROETHANOL

MONOFLUOROETHYLENE see VINYL FLUORIDE

MONOFLUOROPHOSPHORIC ACID, ANHYDROUS see PHOSPHOROFLUORIDIC ACID

MONOFLUOROTRICHLOROMETHANE see TRICHLOROFLUOROMETHANE

MONOGERMANE see GERMANE

MONO-N-HEXYLAMINE see HEXYLAMINE

MONOHYDROXYBENZENE see PHENOL

MONOHYDROXYMETHANE see METHANOL

MONOIODOACETAMIDE see IODOACETAMIDE

MONOIODOACETATE see IODOACETIC ACID

MONOIODOACETIC ACID see IODOACETIC ACID

MONOIODURO DI METILE (ITALIAN) see IODOMETHANE

MONOISOBUTYLAMINE see ISOBUTYLAMINE

MONO-ISO-PROPANOLAMINE see 1-AMINOPROPAN-2-OL

N-MONOISOPROPYLAMIDE OF O,O-DIETHYLDITHIOPHOSPHORYL-ACETIC ACID see ISOPROPYL DIETHYL DITHIOPHOSPHORYL ACETAMIDE

MONOISOPROPYLAMINE see ISOPROPYLAMINE

MONOLINURON see 3-(4-CHLOROPHENYL)-1-METHOXY-1-METHYL-UREA

MONOMETHYLACETAMIDE see METHYLACETAMIDE

N-MONOMETHYLAMIDE OF O,O-DIMETHYLDITHIOPHOSPHORYLACETIC ACID see O,O-DIMETHYL METHYLCARBAMOYLMETHYL PHOSPHO-RODITHIOATE

MONOMETHYLAMINE see METHYLAMINE

MONOMETHYLAMINE, ANHYDROUS (DOT) see ETHANAMINE (ANHYDROUS)

MONOMETHYLAMINE, AQUEOUS SOLUTION (DOT) see ETHANAMINE, (AQUEOUS SOLUTION)

MONOMETHYL-AMINOAETHANOL (GERMAN) see 2-METHYLAMINO-ETHANOL

4-MONOMETHYLAMINOAZOBENZENE see N-METHYL-p-(PHENYLAZO)-ANILINE

P-MONOMETHYLAMINOAZOBENZENE see N-METHYL-p-(PHENYLAZO)-ANILINE

MONOMETHYLAMINOETHANOL see 2-METHYLAMINOETHANOL

N-MONOMETHYLAMINOETHANOL see 2-METHYLAMINOETHANOL

MONOMETHYLANILINE see METHYLANILINE

N-MONOMETHYLANILINE see METHYLANILINE

MONOMETHYLFORMAMIDE see N-METHYLFORMAMIDE

MONOMETHYL GUANIDIN (GERMAN) see METHYLGUANIDINE

MONOMETHYLGUANIDINE see METHYLGUANIDINE

MONOMETHYLHYDRAZINE see METHYLHYDRAZINE

MONOMETHYL MERCURY CHLORIDE see MERCURYMETHYLCHLORIDE

MONOPOTASSIUM ARSENATE see ARSENIC ACID, MONOPOTASSIUM SALT

MONOPOTASSIUM ARSENATE see ARSENIC ACID, MONOPOTASSIUM SALT

MONOPOTASSIUM DIHYDROGEN ARSENATE see ARSENIC ACID, MONOPOTASSIUM SALT

MONOPOTASSIUM SALT OF ACETYLENEDICARBOXYLIC ACID see ACE-TYLENEDICARBOXYLIC ACID MONOPOTASSIUM SALT

MONO-N-PROPYLAMINE see PROPYLAMINE

MONOPROPYLENE GLYCOL see 1,2-PROPANEDIOL

MONOPYRROLE see PYRROLE

MONOSODIOGLUTAMMATO (ITALIAN) see MONOSODIUM GLUTAMATE

MONOSODIUM ACID METHANEARSONATE see MONOSODIUM METHYL-ARSONATE

MONOSODIUM ACID METHARSONATE see MONOSODIUM METHYLAR-SONATE

MONOSODIUM-5-ETHYL-5-(1-METHYLBUTYL) THIOBARBITURATE see PENTOTHAL SODIUM

MONOSODIUM-L-GLUTAMATE see MONOSODIUM GLUTAMATE

MONOSODIUM METHANEARSONATE see MONOSODIUM METHYLARSO-NATE

MONOSODIUM METHANEARSONIC ACID see MONOSODIUM METHYL-ARSONATE

MONOTRICHLOR-AETHYLIDEN-alpha-GLUCOSE (GERMAN) see alpha-d-GLUCOCHLORALOSE

BETA-MONOXYNAPHTHALENE see 2-NAPHTHOL

MONTMORILLONITE see BENTONITE

MONURON see 3-(p-CHLOROPHENYL)-1,1-DIMETHYLUREA

MORFOTHION (DUTCH) see MORPHOTHION

MORPHACETIN see HEROIN

MORPHANQUAT DICHLORIDE see 1,1'-BIS(3,5-DIMETHYL MORPHO-LINO CARBONYL METHYL)-4,4'-BIPYRIDYNIUM DICHLORIDE

MORPHIA see (−)-MORPHINE

MORPHINA see (−)-MORPHINE

MORPHINE see (−)-MORPHINE

MORPHINE CHLORHYDRATE see MORPHINE HYDROCHLORIDE

MORPHINE CHLORIDE see MORPHINE HYDROCHLORIDE

MORPHINE DIACETATE see HEROIN

MORPHINE-3-METHYL ETHER see CODEINE

MORPHINE MONOMETHYL ETHER see CODEINE

MORPHINISM see (−)-MORPHINE

MORPHINUM see (−)-MORPHINE

MORPHIUM see (−)-MORPHINE

MORPHOLINE DISULFIDE see N,N'-BISMORPHOLINE DISULFIDE

4-MORPHOLINEETHANAMINE see N-AMINOETHYLMORPHOLINE

MORPHOLINODISULFIDE see N,N'-BISMORPHOLINE DISULFIDE

2-(MORPHOLINOTHIO)BENZOTHIAZOLE see 2-BENZOTHIAZOLYL-N-MORPHOLINOSULFIDE

MORPHOLINYLMERCAPTOBENZOTHIAZOLE see 2-BENZOTHIAZOLYL-N-MORPHOLINOSULFIDE

2-(4-MORPHOLINYLTHIO)BENZOTHIAZOLE see 2-BENZOTHIAZOLYL-N-MORPHOLINOSULFIDE

MOSS GREEN see CUPRIC ACETOARSENITE

MOTH BALLS see NAPHTHALENE

MOTH FLAKES see NAPHTHALENE

MOTOR BENZOL see BENZENE

MOUNTAIN TOBACCO see ARNICA

MOUSEBANE see ACONITE

3-MPA see 3-METHOXYPROPYLAMINE

MRAVENCAN DI-N-BUTYLCINICITY (CZECH) see DIBUTYL(DIFORMYL-OXY)STANNANE

MRAVENCAN VAPENATY (CZECH) see CALCIUM FORMATE

MROWCZAN ETYLU (POLISH) see ETHYL FORMATE

MSG see MONOSODIUM GLUTAMATE

MURIATE OF PLATINUM see PLATINUM CHLORIDE

MURIATIC ACID (DOT) see HYDROCHLORIC ACID

MURIATIC ETHER see ETHYL CHLORIDE

MURVESCO see 4-CHLOROPHENYL BENZENESULFONATE

MUSCARINE (THE ALKALOID) see MUSCARINE

MUSCIMOL see 5-AMINOMETHYL-3-ISOXYZOLE

MUSCULARON see IRON-POLYSACCHARIDE COMPLEX

MUSK 36A see ACETYL ETHYL TETRAMETHYL TETRALIN

MUSK AMBRETTE see 6-tert-BUTYL-3-METHYL-2,4-DINITRO ANISOLE

MUSKARIN see MUSCARINE

MUSTARD GAS see BIS(2-CHLOROETHYL)SULFIDE

MUSTARD GAS SULFONE see BIS(2-CHLOROETHYL)SULFONE

MUSTARD HD see BIS(2-CHLOROETHYL)SULFIDE

MUSTARD OIL see ALLYL ISOTHIOCYANATE

MUSTARD SULFONE see BIS(2-CHLOROETHYL)SULFONE

MUSTARD VAPOR see BIS(2-CHLOROETHYL)SULFIDE

MUSTINE see BIS(beta-CHLOROETHYL)METHYLAMINE

MUTAMYCIN(MITOMYCIN FOR INJECTION) see AMETYCIN

MUTHMANN'S LIQUID see ACETYLENE TETRABROMIDE

MYACYNE see NEOMYCIN

MYCIFRADIN see NEOMYCIN

MYCOSTATIN see NYSTATIN

MYCOTOXIN see MYCOTOXIN T-2

T-2 MYCOTOXIN see FUSARIOTOXIN T 2

MYDRIATIN see (−)-NOREPHEDRINE

MYKOSTATYNA see NYSTATIN

MYLERAN see 1,4-BUTANEDIOL DIMETHYL SULFONATE

MYLON (CZECH) see DIMETHYLFORMOCARBOTHIALDINE

MYO-INOSISTOL HEXAKISPHOSPHATE see PHYTIC ACID

MYO-INOSITOL HEXAPHOSPHATE see PHYTIC ACID

MYRCENYL ACETATE see ACETIC ACID MYRCENYL ESTER

MYRCIA OIL see BAY OIL

MYRICIA OIL see BAY OIL

MYRISTYL-GAMMA-PICOLINIUM CHLORIDE see QUATRESIN

MYSORITE see AMOSITE (SEE ALSO ASBESTOS)

MYTOMYCIN see AMETYCIN

N 4548 see METHYLPHOSPHODITHIOIC ACID-S-(((p-CHLOROPHENYL)-THIO)METHYL)-O-METHYL ESTER

NA-AESCINAT see ESCIN, SODIUM SALT

NABAME (FRENCH) see NABAM

NACCANOL NR see DODECYL BENZENE SODIUM SULFONATE

NACCONATE 300 see DIPHENYL METHANE DIISOCYANATE

NACELAN FAST YELLOW CG see ACETAMINE YELLOW CG

NAFTA (POLISH) see NAPHTHA, COAL TAR

NAFTALEN (POLISH) see NAPHTHALENE

1-NAFTILAMINA (SPANISH) see 1-NAPHTHYLAMINE

BETA-NAFTILAMINA (ITALIAN) see 2-NAPHTHYLAMINE

2-NAFTOL (DUTCH) see 2-NAPHTHOL

BETA-NAFTOL (DUTCH) see 2-NAPHTHOL

2-NAFTOLO (ITALIAN) see 2-NAPHTHOL

BETA-NAFTOLO (ITALIAN) see 2-NAPHTHOL

ALPHA-NAFTYLAMIN (CZECH) see 1-NAPHTHYLAMINE

BETA-NAFTYLAMIN (CZECH) see 2-NAPHTHYLAMINE

1-NAFTYLAMINE (DUTCH) see 1-NAPHTHYLAMINE

2-NAFTYLAMINE (DUTCH) see 2-NAPHTHYLAMINE

ALPHA-NAFTYL-N-METHYLKARBAMAT (CZECH) see CARBARYL

BETA-NAFTYLOAMINA (POLISH) see 2-NAPHTHYLAMINE

BETA-NAFTYLOAMINA (POLISH) see 2-NAPHTHYLAMINE

NAK see POTASSIUM SODIUM ALLOY

N(SUP 2)-(((+)-5-AMINO-5-CARBOXYPENTYLAMINO)METHYL)TET-RACYCLINE see MUCOMYCIN

NAPHTAMINE BLUE 2B see C.I. DIRECT BLUE 6, TETRASODIUM SALT

NAPHTHA see BENZIN

NAPHTHA see NAPHTHA, COAL TAR

NAPHTHA, COAL TAR see COAL TAR NAPHTHA (DOT)

ALPHA-NAPHTHACRIDINE see BENZ(c)ACRIDINE

2-NAPHTHALENAMINE see 2-NAPHTHYLAMINE

1,4-NAPHTHALENEDIONE see 1,4-NAPHTHOQUINONE

2-NAPHTHALENESULFONIC ACID, 3,3'-METHYLENEDI-PHENYL-MER-CURY see PHENYLMERCURIC DINAPHTHYLMETHANEDISULFONATE

NAPHTHALENE-2-THIOL see 2-NAPHTHALENETHIOL

NAPHTHALIDINE see 1-NAPHTHYLAMINE

NAPHTHALINE see NAPHTHALENE

NAPHTHALOXIMIDE-O,O-DIETHYL PHOSPHOROTHIOATE see NAPHTHA-LOXIMIDODIETHYL THIOPHOSPHATE

NAPHTHANTHRACENE see BENZ(a)ANTHRACENE

NAPHTHA, PETROLEUM see NAPHTHA, COAL TAR

NAPHTHA SAFETY SOLVENT see STODDARD SOLVENT

NAPHTHA SOLVENT see COAL TAR NAPHTHA (DOT)

NAPHTHENE see NAPHTHALENE

1,4-NAPHTHIONIC ACID see 4-AMINO-1-NAPHTHALENESULFONIC ACID

ALPHA-NAPHTHOL see 1-NAPHTHOL

BETA-NAPHTHOL see 2-NAPHTHOL

1-NAPHTHOL-N-METHYLCARBAMATE see CARBARYL

NAPHTHOL YELLOW see 2,4-DINITRO-1-NAPHTHOL

ALPHA-NAPHTHOQUINONE see 1,4-NAPHTHOQUINONE

BETA-NAPHTHYL ALCOHOL see 2-NAPHTHOL

1-NAPHTHYLAMIN (GERMAN) see 1-NAPHTHYLAMINE

2-NAPHTHYLAMIN (GERMAN) see 2-NAPHTHYLAMINE

BETA-NAPHTHYLAMIN (GERMAN) see 2-NAPHTHYLAMINE

6-NAPHTHYLAMINE see 2-NAPHTHYLAMINE

ALPHA-NAPHTHYLAMINE see 1-NAPHTHYLAMINE

BETA-NAPHTHYLAMINE see 2-NAPHTHYLAMINE

2-NAPHTHYLAMINE-4,8-DISULFONIC ACID see 3-AMINO-1,5-NAPH-THALENEDISULFONIC ACID

BETA-NAPHTHYLAMINEDISULFONIC ACID see 3-AMINO-1,5-NAPH-THALENEDISULFONIC ACID

BETA-NAPHTHYLAMINE-4,8-DISULFONIC ACID see 3-AMINO-1,5-NAPHTHALENEDISULFONIC ACID

NAPHTHYLAMINE MUSTARD see N,N-BIS(2-CHLOROETHYL)-2-NAPH-THYLAMINE

2-NAPHTHYLAMINE MUSTARD see 2-NAPHTHYLAMINE

1-NAPHTHYLAMINE-4-SULFONIC ACID see 4-AMINO-1-NAPHTHALENE-SULFONIC ACID

1-NAPHTHYLAMINE-6-SULFONIC ACID see 5-AMINO-2-NAPHTHALENE-SULFONIC ACID

5-NAPHTHYLAMINE-2-SULFONIC ACID see 5-AMINO-2-NAPHTHALENE-SULFONIC ACID

ALPHA-NAPHTHYLAMINE-P-SULFONIC ACID see 4-AMINO-1-NAPH-THALENESULFONIC ACID

N-(2-NAPHTHYL)ANILINE see N-PHENYL-2-NAPHTHYLAMINE

2-NAPHTHYLBIS(2-CHLOROETHYL)AMINE see N,N-BIS(2-CHLORO-ETHYL)-2-NAPHTHYLAMINE

BETA-NAPHTHYL-BIS-(BETA-CHLOROETHYL)AMINE see N,N-BIS(2-CHLOROETHYL)-2-NAPHTHYLAMINE

BETA-NAPHTHYL-DI-(2-CHLOROETHYL)AMINE see N,N-BIS(2-CHLORO-ETHYL)-2-NAPHTHYLAMINE

1,2-(1,8-NAPHTHYLENE)BENZENE see FLUORANTHENE

NAPHTHYLENE YELLOW see 2,4-DINITRO-1-NAPHTHOL

BETA-NAPHTHYL HYDROXIDE see 2-NAPHTHOL

NAPHTHYLISOPROTERENOL HYDROCHLORIDE see alpha-((ISOPROPYL-AMINO)METHYL)NAPHTHALENEMETHANOL, HYDROCHLORIDE

2-NAPHTHYL MERCAPTAN see 2-NAPHTHALENETHIOL

BETA-NAPHTHYL MERCAPTAN see 2-NAPHTHALENETHIOL

1-NAPHTHYL METHYLCARBAMATE see CARBARYL

1-NAPHTHYL-N-METHYLCARBAMATE see CARBARYL

ALPHA-NAPHTHYL, N-METHYLCARBAMATE see CARBARYL

2-(1-NAPHTHYLMETHYL)-2-IMIDAZOLINE see NAPHAZOLINE

ALPHA-NAPHTHYLMETHYL IMIDAZOLINE see NAPHAZOLINE

2-(ALPHA-NAPHTHYLMETHYL)-IMIDAZOLINE see NAPHAZOLINE

1-NAPHTHYL METHYLNITROSOCARBAMATE see N-NITROSOCARBARYL

1-NAPHTHYL N-METHYL-N-NITROSOCARBAMATE see N-NITROSOCAR-BARYL

2-NAPHTHYLPHENYLAMINE see N-PHENYL-2-NAPHTHYLAMINE

BETA-NAPHTHYLPHENYLAMINE see N-PHENYL-2-NAPHTHYLAMINE

2-NAPHTHYL THIOL see 2-NAPHTHALENETHIOL

2-NAPHTOL (FRENCH) see 2-NAPHTHOL

BETA-NAPHTOL (GERMAN) see 2-NAPHTHOL

BETA-NAPHTYLAMIN (GERMAN) see 2-NAPHTHYLAMINE

NATIVE CALCIUM SULFATE see CALCIUM (II) SULFATE DIHYDRATE (1:1:2)

NATRASCORB INJECTABLE see L-ASCORBIC ACID

NATRIUM see SODIUM

NATRIUMANTIMONYLTARTRAT (GERMAN) see ANTIMONY SODIUM TARTRATE

NATRIUMARSENIT (GERMAN) see ARSENIOUS ACID SODIUM SALT

NATRIUMAZID (GERMAN) see SODIUM AZIDE

NATRIUMBARBITALS (GERMAN) see BARBITAL SODIUM

NATRIUMBICHROMAAT (DUTCH) see SODIUM DICHROMATE

NATRIUM CHLORAAT (DUTCH) see SODIUM CHLORATE

NATRIUM CHLORAT (GERMAN) see SODIUM CHLORATE

NATRIUMCHLORID (GERMAN) see SODIUM CHLORIDE

NATRIUM-2,4-DICHLORPHENOXYATHYLSULFAT (GERMAN) see CRAG HERBICIDE

NATRIUM FLUORACETATE (DUTCH) see SODIUM FLUOROACETATE

NATRIUMGLUTAMINAT (GERMAN) see MONOSODIUM GLUTAMATE

NATRIUMHYDROXID (GERMAN) see SODIUM HYDROXIDE

NATRIUMHYDROXYDE (DUTCH) see SODIUM HYDROXIDE

NATRIUMJODID (GERMAN) see SODIUM IODIDE

NATRIUMMAZIDE (DUTCH) see SODIUM AZIDE

NATRIUM-N-METHYL-DITHIOCARBAMAAT (DUTCH) see VAPAM

NATRIUM-N-METHYL-DITHIOCARBAMAT (GERMAN) see VAPAM

NATRIUMMOLYBDAT (GERMAN) see DISODIUM MOLYBDATE

NATRIUM NITRIT (GERMAN) see SODIUM NITRITE

NATRIUMPHOSPHAT (GERMAN) see SODIUM MONOHYDROGEN PHOS-PHATE (2:1:1)

NATRIUMPYROPHOSPHAT see TETRASODIUM PYROPHOSPHATE, AN-HYDROUS

NATRIUMRHODANID (GERMAN) see SODIUM ISOTHIOCYANATE

NATRIUMSELENIAT (GERMAN) see DISODIUM SELENATE

NATRIUMSELENIT (GERMAN) see SODIUM SELENITE

NATRIUMSUFAT (GERMAN) see SODIUM SULFATE (2:1)
NATRIUMSULFID (GERMAN) see SODIUM SULFITE (2:1)
NATRIUMTRIPOLYPHOSPHAT (GERMAN) see SODIUM TRIPOLYPHOS-
   PHATE
NATULAN see PROCARBAZINE
NATURAL CALCIUM CARBONATE see CALCIUM (II) CARBONATE (1:1)
NATURAL IRON OXIDES see IRON (III) OXIDE
NATURAL LEAD SULFIDE see LEAD SULFIDE
NATURAL RED OXIDE see IRON (III) OXIDE
NATURAL WINTERGREEN OIL see METHYL SALICYLATE
NAULI "GUM" see p-PROPENYL ANISOLE
NC5 see HEXANENITRILE
NC-262 see O,O-DIMETHYL METHYLCARBAMOYLMETHYL PHOSPHO-
   RODITHIOATE
N(SUP 4)-(7-CHLORO-4-QUINOLINYL)-N(SUP 1),N(SUP 1)-DIETHYL-
   1,4-PENTANEDIAMINE see CHLOROQUINE
NCI see CYANOGEN IODIDE
NCI-C00044 see ALDRIN
NCI-C00055 see 3-AMINO-2,5-DICHLOROBENZOIC ACID
NCI-C00066 see AZINPHOS METHYL
NCI-C00099 see CHLORDANE
NCI-C00102 see TETRACHLOROISOPHTHALONITRILE
NCI-C00113 see DIMETHYL DICHLOROVINYL PHOSPHATE
NCI-C00124 see DIELDRIN
NCI-C00157 see ENDRIN
NCI-C00191 see KEPONE
NCI-C00215 see CARBETHOXY MALATHION
NCI-C00226 see PARATHION
NCI-C00237 see 4-AMINO-3,5,6-TRICHLOROPICOLINIC ACID
NCI-C00259 see TOXAPHENE
NCI-C00260 see HYDROXY TRIPHENYL STANNANE
NCI-C00408 see ETHYL-4,4'-DICHLOROBENZILATE
NCI-C00419 see PENTACHLORONITRO BENZENE
NCI-C00420 see 2,4-DICHLORO-4'-NITRODIPHENYL ETHER
NCI-C00431 see 2',5-DICHLORO-4'-NITROSALICYLANILIDE, 2-
   AMINOETHANOL SALT
NCI-C00442 see TRIFLURALINE
NCI-C00453 see 2-CHLORALLYL DIETHYLDITHIOCARBAMATE
NCI-C00464 see DDT
NCI-C00475 see 1,1-BIS(4-CHLOROPHENYL)-2,2-DICHLOROETHANE
NCI-C00486 see 1,1-BIS(p-CHLOROPHENYL)-2,2,2-TRICHLOROETHA-
   NOL
NCI-C00497 see DIMETHOXY-DDT
NCI-C00500 see DIBROMOCHLOROPROPANE
NCI-C00511 see 1,2-DICHLORETHANE
NCI-C00511 see ETHYLENE DICHLORIDE
NCI-C00533 see TRICHLORONITROMETHANE
NCI-C00544 see 4-(DIMETHYLAMINE)-3,5-XYLYL N-METHYLCARBA-
   MATE
NCI-C00555 see 2,2-BIS(p-CHLOROPHENYL)-1,1-DICHLOROETHYLENE
NCI-C00566 see BENZOEPIN
NCI-C00588 see DIMECRON
NCI-C00920 see ETHYLENE GLYCOL
NCI-C01445 see NITRILOTRIACETIC ACID TRISODIUM SALT MONOHY-
   DRATE
NCI-C01478 see LASIOCARPINE
NCI-C01514 see ADRIAMYCIN
NCI-C01536 see ACRONYCINE
NCI-C01547 see YOSHI 864
NCI-C01558 see CHOLESTERYL-p-BIS(2-CHLOROETHYL)AMINO
   PHENYLACETATE
NCI-C01569 see AZACYTIDINE
NCI-C01570 see ESTRADIOL MUSTARD
NCI-C01581 see beta-THIOGUANINE DEOXYRIBOSIDE
NCI-C01592 see 1,4-BUTANEDIOL DIMETHYL SULFONATE
NCI-C01616 see IMIDAZOLE MUSTARD
NCI-C01627 see NSC-129943
NCI-C01638 see ISOPHOSPHAMIDE
NCI-C01649 see THIOPHOSPHAMIDE
NCI-C01661 see DIBENZYLINE HYDROCHLORIDE

NCI-C01672 see PHENAZOPYRIDINIUM CHLORIDE
NCI-C01683 see TINDURIN
NCI-C01694 see 2-ETHYL THIO ISONICOTINAMIDE
NCI-C01707 see 4,4'-THIODIANILINE
NCI-C01718 see 4,4'-SULFONYLDIANILINE
NCI-C01730 see ANTHRANILIC ACID
NCI-C01785 see PYRAZINECARBOXAMIDE
NCI-C01821 see N,N-DIMETHYL-p-NITROSOANILINE
NCI-C01832 see 2,5-DIAMINOTOLUENE SULFATE
NCI-C01854 see HYDRAZOBENZENE
NCI-C01865 see 2,4-DINITROTOLUENE
NCI-C01876 see 2-AMINOANTHRAQUINONE
NCI-C01887 see 3-AMINO-4-ETHOXYACETANILIDE
NCI-C01901 see 1-AMINO-2-METHYLANTHRAQUINONE
NCI-C01923 see 2-METHYL-1-NITROANTHRAQUINONE
NCI-C01934 see 5-NITRO-o-ANISIDINE
NCI-C01967 see 5-NITROACENAPHTHENE
NCI-C01978 see 3'-NITRO-p-ACETOPHENETIDE
NCI-C01989 see 2,4-DIAMINOANISOLE SULPHATE
NCI-C01990 see MICHLER'S BASE
NCI-C02006 see MICHLER'S KETONE
NCI-C02006 see TETRAMETHYL DIAMINO BENZOPHENONE
NCI-C02017 see 1-PHENYL-2-THIOUREA
NCI-C02028 see DIACETOXY DIBUTYL STANNANE
NCI-C02073 see 4-ETHOXY PHENYL UREA
NCI-C02084 see SODIUM NITRITE
NCI-C02095 see ADIPAMIDE
NCI-C02108 see ACETAMIDE
NCI-C02119 see UREA
NCI-C02142 see HEXANAMIDE
NCI-C02153 see p-TOLYLUREA
NCI-C02175 see DIANISIDINE DIISOCYANATE
NCI-C02186 see 1,1,3-TRIMETHYL-2-THIOUREA
NCI-C02200 see STYRENE
NCI-C02222 see 4-AMINO-2-NITROANILINE
NCI-C02244 see p-NITROSODIPHENYLAMINE
NCI-C02299 see 2,4,5-TRIMETHYLANILINE
NCI-C02302 see TOLUENE-2,4-DIAMINE
NCI-C02335 see o-TOLUIDINE HYDROCHLORIDE
NCI-C02551 see CADMIUM OXIDE
NCI-C02653 see HEXACHLOROPHENE
NCI-C02664 see POLYCHLORINATED BIPHENYL (AROCLOR 1254)
NCI-C02697 see ASPIRIN, PHENACETIN AND CAFFEINE
NCI-C02711 see CADMIUM SULFIDE
NCI-C02722 see TIN (II) CHLORIDE (1:2)
NCI-C02733 see CAFFEINE
NCI-C02766 see AMINOTRIACETIC ACID
NCI-C02799 see FORMALDEHYDE
NCI-C02813 see PIPERONYL BUTOXIDE
NCI-C02824 see ISOSAFROLE-n-OCTYLSULFOXIDE
NCI-C02846 see 3-(p-CHLOROPHENYL)-1,1-DIMETHYLUREA
NCI-C02857 see ETHYL TELLURAC
NCI-C02868 see DIETHYL DIPHENYL DICHLOROETHANE
NCI-C02880 see DIPHENYLNITROSAMINE
NCI-C02891 see LEAD DIMETHYLDITHIOCARBAMATE
NCI-C02904 see 2,4,6-TRICHLOROPHENOL
NCI-C02915 see N-PHENYL-2-NAPHTHYLAMINE
NCI-C02926 see AZOBENZENE
NCI-C02937 see CALCIUM CYANAMIDE
NCI-C02959 see BIS(DIETHYL THIO CARBAMOYL) DISULFIDE
NCI-C02971 see METHYL PARATHION
NCI-C02982 see 5-METHYL-o-ANISIDINE
NCI-C02993 see 2-METHYL-p-ANISIDINE
NCI-C03010 see METHYL ORANGE
NCI-C03021 see 1,5-NAPHTHALENEDIAMINE
NCI-C03032 see TETRACHLORONITROANISOLE
NCI-C03043 see 3-AMINO-9-ETHYLCARBAZOLEHYDROCHLORIDE
NCI-C03065 see 2-AMINO-5-NITROTHIAZOLE
NCI-C03134 see ETHANOL
NCI-C03167 see STREPTOZOTICIN

NCI-C54706 see FLUORESCEIN SODIUM
NCI-C54739 see BASIC PARAFUCHSINE
NCI-C54808 see L-ASCORBIC ACID
NCI-C54820 see 3-CHLORO-2-METHYLPROPENE
NCI-C54831 see ((2,2,2-TRICHLORO-1-HYDROXYETHYL) DIMETHYL-PHOSPHONATE)
NCI-C54842 see PROPANEDIAL
NCI-C54853 see CELLOSOLVE SOLVENT
NCI-C54886 see BENZENE CHLORIDE
NCI-C54922 see 2-((4-AMINO-2-NITROPHENYL)AMINO)ETHANOL
NCI-C54933 see PENTACHLOROPHENOL
NCI-C54944 see o-DICHLOROBENZENE
NCI-C54955 see p-DICHLOROBENZENE
NCI-C54966 see RESORCINOL DIGLYCIDYL ETHER
NCI-C54977 see 1,2-EPOXYETHYLBENZENE
NCI-C54988 see TETRAETHYL LEAD
NCI-C54999 see CYCLOHEXENYLETHYLENE
NCI-C55005 see CYCLOHEXANONE
NCI-C55072 see CHLORENDIC ACID
NCI-C55107 see 2-CHLOROACETOPHENONE
NCI-C55118 see (o-CHLORO BENZAL)MALONO NITRILE
NCI-C55130 see BROMOFORM
NCI-C55141 see 1,2-DICHLOROPROPANE
NCI-C55152 see ANTIMONY OXIDE
NCI-C55163 see CASTOR OIL
NCI-C55174 see BIS(2-HYDROXY ETHYL)AMINE
NCI-C55185 see GILSONITE
NCI C55196 see NITROFURANTOIN
NCI-C55209 see OXALIC ACID
NCI-C55210 see ROTENONE
NCI-C55221 see SODIUM FLUORIDE
NCI-C55232 see XYLENE
NCI-C55243 see BROMO DICHLORO METHANE
NCI-C55254 see CHLORO DIBROMO METHANE
NCI-C55265 see TELDRIN
NCI-C55276 see BENZENE
NCI-C55298 see 8-QUINOLINOL
NCI-C55301 see PYRIDINE
NCI-C55323 see N,N-BIS(2-HYDROXY ETHYL)DODECAN AMIDE
NCI-C55345 see 2,4-DICHLOROPHENOL
NCI-C55367 see tert-BUTYL ALCOHOL
NCI-C55378 see PENTACHLOROPHENOL
NCI-C55447 see METHYL ETHYL KETONE PEROXIDE
NCI-C55481 see BROMOETHANE
NCI-C55492 see PHENYL BROMIDE
NCI-C55516 see 2,2-BIS(2-BROMOETHYL)-1,3-PROPANEDIOL
NCI-C55527 see 1-BUTENE OXIDE
NCI-C55538 see 1,2-EPOXYHEXADECANE
NCI-C55549 see 2,3-EPOXY-1-PROPANOL
NCI-C55550 see THENYLPYRAMINE
NCI-C55561 see TETRACYCLINE HYDROCHLORIDE
NCI-C55572 see d-LIMONENE
NCI-C55594 see METHYL CARBAMATE
NCI-C55618 see ISOACETOPHORONE
NCI-C55652 see 1-EPHEDRINE SULFATE
NCI-C55663 see 1-ADRENALINE CHLORIDE
NCI-C55685 see alpha-PHENETHYL ALCOHOL
NCI-C55696 see SUCCINIC ANHYDRIDE
NCI-C55709 see CHLORAMPHENICOL
NCI-C55710 see DL-AMPHETAMINE SULFATE
NCI-C55721 see ALDOMET
NCI-C55743 see PENTAERYTHRITOL TETRANITRATE
NCI-C55765 see DIHYDANTOIN
NCI-C55776 see cis-PLATINOUS DIAMINE DICHLORIDE
NCI-C55787 see HEXYLRESORCINOL
NCI-C55801 see 4'-HYDROXYACETANILIDE
NCI-C55823 see GIBBERELLIC ACID
NCI-C55834 see HYDROQUINONE
NCI-C55845 see p-BENZOQUINONE
NCI-C55856 see PYROCATECHOL

NCI-C55878 see 4-BUTANOLIDE
NCI-C55890 see HYDROCOUMARIN
NCI-C55903 see XANTHOTOXIN
NCI-C55925 see 6-CHLORO-3,4-DIHYDRO-2H-1,2,4-BENZOTHIADIA-ZINE-7-SULFONAMIDE- 1,1-DIOXIDE
NCI-C55936 see 4-CHLORO-N-FURFURYL-5-SULFAMOYL ANTHRANI-LIC ACID
NCI-C55947 see TETRANITROMETHANE
NCI-C55969 see ANILINE VIOLET
NCI-C55981 see AZODICARBAMIDE
NCI-C55992 see 4-NITROPHENOL
NCI-C56031 see ACETYLENE DICHLORIDE
NCI-C56064 see NITROFURAZONE
NCI-C56075 see BENADRYL HYDROCHLORIDE
NCI-C56122 see RHODAMINE 6G EXTRA BASE
NCI-C56133 see BENZALDEHYDE
NCI-C56144 see INDOMETHACIN
NCI-C56155 see sym-TRINITROTOLUENE
NCI-C56177 see 2-FURALDEHYDE
NCI-C56188 see 2,6-XYLIDINE
NCI-C56199 see 1-ETHYL-1,4-DIHYDRO-7-METHYL-4-OXO-1,8-NAPHTHYRIDINE-3-CARBOXYLIC ACID
NCI-C56202 see FURAN
NCI-C56213 see ACETOMETHOXANE
NCI-C56224 see FURFURYL ALCOHOL
NCI-C56235 see ISOPROPYL MEPROBAMATE
NCI-C56246 see QUINIDINE
NCI-C56279 see trans-2-BUTENAL
NCI-C56280 see METHYL PHENIDYL ACETATE
NCI-C56291 see N-BUTYRALDEHYDE
NCI-C56326 see ACETALDEHYDE
NCI-C56360 see 2-DESOXY PHENOBARBITAL
NCI-C56382 see BIS(2-CHLOROETHYL)METHYLAMINE HYDROCHLO-RIDE
NCI-C56393 see ETHYL BENZENE
NCI-C56417 see BORIC ACID
NCI-C56428 see N,N-DIMETHYLANILINE
NCI-C56439 see ISOPROPYL GLYCIDYL ETHER
NCI-C56440 see HEXAFLUOROACETONE
NCI-C56451 see POLYURETHANE FOAM
NCI-C56462 see MONOCROTALINE
NCI-C56473 see 5-HYDROXYTETRACYCLINE
NCI-C56519 see 2-BENZOTHIAZOLETHIOL
NCI-C56531 see 4-BUTYL-1,2-DIPHENYL-3,5-DIOXO PYRAZOLIDINE
NCI-C56553 see n-BUTYL NITRITE
NCI-C56564 see CAPSAICIN
NCI-C56575 see CAJEPUTOL
NCI-C56586 see (-)-N-((5-CHLORO-8-HYDROXY-3-METHYL-1-OXO-7-ISOCHROMANYL)CARBONYL)-3-PHENYLALANINE
NCI-C56586 see OCHRATOXIN A
NCI-C56597 see SUCROSE
NCI-C56600 see SULFADIMETHYLDIAZINE
NCI-C56633 see TRIMELLIC ACID ANHYDRIDE
NCI-C56655 see PENTACHLOROPHENOL
NCI-C56666 see ((2-PROPENYLOXY)METHYL)OXIRANE
NCI-C60048 see DIETHYL-O-PHTHALATE
NCI-C60071 see MONOSODIUM METHYLARSONATE
NCI-C60082 see NITROBENZENE
NCI-C60093 see beta-PHENYLHYDROXYLAMINE
NCI-C60106 see QUERCETIN
NCI-C60106 see ISOPROPYL FORMATE
NCI-C60117 see TELLURIUM
NCI-C60128 see 2-CHLORO ETHANOL PHOSPHATE
NCI-C60139 see VINYL CYCLOHEXENE DIOXIDE
NCI-C60173 see CORROSIVE SUBLIMATE
NCI-C60184 see PALLADIUM(2+) CHLORIDE
NCI-C60208 see VINYLIDENE FLUORIDE
NCI-C60219 see PHOSGENE
NCI-C60220 see ALLYL TRICHLORIDE
NCI-C60231 see CHLOROACETIC ACID

NCI-C60297 see COUMARIN
NCI-C60344 see NICKEL SULFATE
NCI-C60399 see MERCURY
NCI-C60402 see 1,2-ETHANEDIAMINE
NCI-C60413 see ETHYL-4,4'-DICHLOROBENZILATE
NCI-C60537 see p-METHYL NITROBENZENE
NCI-C60559 see CHLOROMETHAPYRILENE
NCI-C60560 see TETRAHYDROFURAN
NCI-C60571 see N-HEXANE
NCI-C60582 see POLY(1-VINYL-2-PYRROLIDINONE) HUEPER'S POLY-
MER NO. 1
NCI-C60582 see POLY(1-VINYL-2-PYRROLIDINONE) HUEPER'S POLY-
MER NO. 2
NCI-C60582 see POLY(1-VINYL-2-PYRROLIDINONE) HUEPER'S POLY-
MER NO. 3
NCI-C60582 see POLY(1-VINYL-2-PYRROLIDINONE) HUEPER'S POLY-
MER NO. 7
NCI-C60582 see POLY(1-VINYL-2-PYRROLIDINONE) HUEPER'S POLY-
MER NO. 6
NCI-C60582 see POLY(1-VINYL-2-PYRROLIDINONE) HUEPER'S POLY-
MER NO. 4
NCI-C60639 see DRAMAMINE
NCI-C60651 see WAIT'S GREEN MOUNTAIN ANTIHISTAMINE
NCI-C60662 see TRIPELENNAMINE
NCI-C60673 see 10-(2-(DIMETHYLAMINO)PROPYL)PHENOTHIAZINE
NCI-C60695 see TRIMETON
NCI-C60720 see METHDILAZINE
NCI-C60753 see 2,4-DINITROANILINE
NCI-C60786 see p-NITROANILINE
NCI-C60797 see BAKELITE
NCI-C60811 see ZIRCONIUM OXYCHLORIDE
NCI-C60822 see ACETONITRILE
NCI-C60866 see n-BUTANETHIOL
NCI-C60899 see DIETHYL CARBONATE
NCI-C60913 see DIMETHYL FORMAMIDE
NCI-C60924 see DIPHENYLGUANIDINE
NCI-C60968 see ISOBUTYRALDEHYDE
NCI-C60980 see 2-BENZIMIDAZOLETHIOL
NCI-C61041 see TRITOLYL PHOSPHATE
NCI-C61052 see ISOBUTYL NITRITE
NCI-C61074 see BARIUM CHLORIDE
NCI-C61109 see C.I. DIRECT BLUE 1, TETRASODIUM SALT
NCI-C61165 see SENECIPHYLLINE
NCI-C61176 see ARSANILIC ACID, MONOSODIUM SALT
NCI-C61187 see 2,4,5-TRICHLOROPHENOL
NCI-C60077 see CAPTAN
NCI-CO1752 see 1-(p-CHLOROPHENYLSULFONYL)-3-PROPYLUREA
NCI-CO1763 see 1-BUTYL-3-(p-TOLYL SULFONYL)UREA
NCI-CO2051 see 2-AMINO-4-CHLOROTOLUENE
NCI-CO2131 see N-BUTYLUREA
NCI-CO2686 see CHLOROFORM
NCI-CO2835 see SODIUM DIETHYLDITHIOCARBAMATE
NCI-CO3736 see ANILINE
NCI-CO3736 see BENZENAMINE HYDROCHLORIDE
NCI-CO4035 see 2-CHLORO-N,N-DIALLYLACETAMIDE
NCI-CO4546 see TRICHLOROETHYLENE
NCI-CO4740 see 1-(2-CHLOROETHYL)-3-CYCLOHEXYL-1-NITROSO-
UREA
NCI-CO4820 see 5-(BIS(2-CHLOROETHYL)AMINO)URACIL
NCI-CO4853 see L-PHENYLALANINE MUSTARD
NCI-CO4944 see DL-PHENYLALANINE MUSTARD
NCI-CO5210 see CHLOROPROMAZINE HYDROCHLORIDE
NCR CE EE DOV7 see 2,6-DIBROMO-4-CYANOPHENYL OCTANOATE
NCS-14210 see 3-(p-(BIS(beta-CHLOROETHYL)AMINO)PHENYL)-D,L-
ALANINE HYDROCHLORIDE
NCl-C00180 see HEPTACHLOR
NDHU see 1-NITROSO-5,6-DIHYDROURACIL
N,N,DIMETHYL-1,2-DIPHENYLACETAMIDE see N,N-DIMETHYL-2,2-DI-
PHENYLACETAMIDE
N,N'DIMETHYLHYDRAZINE see 1,2-DIMETHYL HYDRAZINE

N,N'DIMETHYL-3-METHYL-4(4'-(2'-METHYLPYRIDYL-
1'OXIDE)AZO)ANILINE see 4-((4-(DIMETHYLAMINO)-O-TOLYL)-
AZO)-2PICOLINE 1-OXIDE
N(sup 1)-(4,6-DIMETHYL-2-PYRIMIDINYL)SULFANILAMIDE see SUL-
FADIMETHYLDIAZINE
N(SUP 1)-(4,6-DIMETHYL-2-PYRIMIDYL)SULFANILAMIDE see SUL-
FADIMETHYLDIAZINE
NEAT OIL OF SWEET ORANGE see OIL OF ORANGE
NEEDLE ANTIMONY see ANTIMONY TRISULFIDE
NEFOPAM see FENAZOXINE
NEMAGON see DIBROMOCHLOROPROPANE
NEMATOCIDE see DIBROMOCHLOROPROPANE
NEMATOCIDE see ETHYL PYRAZINYL PHOSPHOROTHIOATE
NEMBUTAL CALCIUM see CALCIUM PENTOBARBITAL
NEOBIOTIC see NEOMYCIN SULFATE
NEOCARCINOSTATIN see NEOCARZINOSTATIN
NEOCYCLINE see TETRACYCLINE
NEOCYCLOHEXIMIDE see CYCLOHEXIMIDE
(NEODECANOYLOXY)TRIBUTYLSTANNANE see TRIBUTYLTIN NEODE-
CANDATE
NEO-ERGOTIN see ERGOTAMINE TARTRATE
NEO-FAT 10 see DECANOIC ACID
NEO-FAT 12 see LAURIC ACID
NEO-FAT 18-61 see STEARIC ACID
NEOFEMERGEN see D-LYSERGIC ACID-L,2-PROPANOLAMIDE
NEO-FERRUM see IRON OXIDE, SACCHARATED
NEOHEXANE (DOT) see 2,2-DIMETHYLBUTANE
NEOLOID see CASTOR OIL
NEOMCIN see NEOMYCIN
NEON (DOT) see NEON
NEO-OESTRANOL 1 see DIETHYLSTILBESTEROL
NEOPANTANOYL CHLORIDE see 2,2-DIMETHYLPROPANOYL CHLORIDE
NEOPENTANETETRAYL NITRATE see PENTAERYTHRITOL TETRANI-
TRATE
NEOPENTANOIC ACID see PIVALIC ACID
NEOPRENE see 2-CHLORO-1,3-BUTADIENE POLYMER
NEOTHESIN HYDROCHLORIDE see ISOCAINE
NEUROCAINE see COCAINE
NEUTRAL AMMONIUM FLUORIDE see AMMONIUM FLUORIDE
NEUTRAL POTASSIUM CHROMATE see POTASSIUM CHROMATE (VI)
NEUTRAL SODIUM CHROMATE see DISODIUM CHROMATE
NEUTRAL VERDIGRIS see COPPER ACETATE
NEUTRONYX 600 see PEG-9 NONYL PHENYL ETHER
NEW COCCINE EXTRA PURE A see FOOD RED 7
NHA see 2-NAPHTHYLHYDROXYLAMINE
NH-LOST see BIS-beta-CHLOROETHYLAMINE
NIACIN see NICOTINIC ACID
NIAGARA 1240 see ETHYL METHYLENE PHOSPHORO DITHIOATE
NIAGARA BLUE 2B see C.I. DIRECT BLUE 6, TETRASODIUM SALT
NIAX FLAME RETARDANT 3 CF see 2-CHLORO ETHANOL PHOSPHATE
NIBUFIN see p-NITROPHENYLDI-n-BUTYLPHOSPHINATE
NICACID see NICOTINIC ACID
NICERGOLIN (GERMAN) see NICOTERGOLINE
NICHEL (ITALIAN) see NICKEL
NICHEL TETRACARBONILE (ITALIAN) see NICKEL CARBONYL
NICKEL, BIS(TRIPHENYLPHOSPHINE)-, DITHIOCYANATE see BIS(TRI-
PHENYL PHOSPHINE)NICKEL DITHIO CYANATE
NICKEL CARBONYL (DOT) see NICKEL CARBONYL
NICKEL CARBONYLE (FRENCH) see NICKEL CARBONYL
NICKEL CATALYST, WET (DOT) see NICKEL
NICKEL CYANIDE, SOLID (DOT) see NICKEL CYANIDE, (SOLID)
NICKEL DIBUTYLDITHIOCARBAMATE see BIS(DIBUTYL DITHIO CAR-
BAMATO)NICKEL
NICKEL(II) CHLORIDE (1:2) see NICKELOUS CHLORIDE
NICKELIC OXIDE see NICKEL PEROXIDE
NICKELOUS ACETATE see NICKEL(II) ACETATE (1:2)
NICKELOUS CARBONATE see NICKEL (II) CARBONATE (1:1)
NICKELOUS FLUORIDE see NICKEL (II) FLUORIDE (1:2)
NICKELOUS HYDROXIDE see NICKEL (II) HYDROXIDE
NICKELOUS OXIDE see NICKEL MONOXIDE

P-NITROPHENYLDIMETHYLTHIONOPHOSPHATE see METHYL PARA-THION

1-(P-NITROPHENYL-3,3-DIMETHYL-TRIAZEN (GERMAN) see 3,3-DIMETHYL-1-(p-NITROPHENYL)TRIAZENE

1-(4-NITROPHENYL)-3,3-DIMETHYLTRIAZENE see 3,3-DIMETHYL-1-(p-NITROPHENYL)TRIAZENE

1-(P-NITROPHENYL)-3,3-DIMETHYL-TRIAZENE see 3,3-DIMETHYL-1-(p-NITROPHENYL)TRIAZENE

NITRO-P-PHENYLENEDIAMINE see 4-AMINO-2-NITROANILINE

4-NITRO-O-PHENYLENE-DIAMINE see 2-AMINO-4-NITROANILINE

O-NITRO-P-PHENYLENEDIAMINE see 4-AMINO-2-NITROANILINE

P-NITRO-O-PHENYLENEDIAMINE see 2-AMINO-4-NITROANILINE

2-NITRO-1,4-PHENYLENEDIAMINE see 4-AMINO-2-NITROANILINE

4-NITRO-1,2-PHENYLENEDIAMINE see 2-AMINO-4-NITROANILINE

N-(4-NITROPHENYL)-N'-(3-PYRIDINYLMETHYL)UREA see PYRIMINYL

4-NITROQUINALDINE-N-OXIDE see 2-METHYL-4-NITROQUINOLINE-1-OXIDE

4-NITROQUINOLINE-1-OXIDE see 4-NITROQUINOLINE-N-OXIDE

N-NITROSAZETIDINE see 1-NITROSOAZETIDINE

N-NITROSO-N-(1-ACETOXYMETHYL)BUTYLAMINE see BUTYL NITROS AMINO METHYL ACETATE

N-NITROSO-N-(1-ACETOXYMETHYL)PROPYL AMINE see PROPYLNI-TROSAMINOMETHYL ACETATE

N-NITROSOAETHYLAETHANOLAMIN (GERMAN) see ETHYL-2-HY-DROXY ETHYL NITROSAMINE

N-NITROSOALLYLMETHYLAMINE see N-METHYL-N-NITROSOALLYL-AMINE

4-(NITROSOAMINO-N-METHYL)-1-(3-PYRIDYL)-1-BUTANONE see 4-(N-METHYL-N-NITROSAMINO)-1-(3-PYRIDYL)-1-BUTANONE

N-NITROSOAZACYCLOHEPTANE see N-NITROSOHEXAHYDROAZEPINE

N-NITROSOAZACYCLOOCTANE see OCTAHYDRO-1-NITROSOAZOCINE

NITROSO-AZETIDIN (GERMAN) see 1-NITROSOAZETIDINE

NITROSOAZETIDINE see 1-NITROSOAZETIDINE

N-NITROSOAZETIDINE see 1-NITROSOAZETIDINE

N-NITROSOBENZYLMETHYLAMINE see N-METHYL-N-BENZYLNITROSA-MINE

NITROSOBIS(2-CHLOROPROPYL)AMINE see 2,2'-DICHLORO-N-NITRO-SODIPROPYLAMINE

N-NITROSOBIS(2-HYDROXYPROPYL)AMINE see DI(2-HYDROXY-n-PRO-PYL)AMINE

N-NITROSO-N-BUTYL-N-(3-CARBOXYPROPYL)AMINE see N-BUTYL-(3-CARBOXY PROPYL)NITROSAMINE

N-NITROSO-N-BUTYLETHYLAMINE see ETHYL-n-BUTYL NITROS AMINE

N-NITROSO-n-BUTYL-(4-HYDROXYBUTYL)AMINE see BUTANOL-4-BUTYL NITROSAMINE

N-NITROSO-N-BUTYLMETHYLAMINE see METHYLBUTYLNITROSAMINE

N-NITROSO-N-BUTYLPENTYLAMINE see N-BUTYL-N-NITROSO AMYL AMINE

N-NITROSO-N-BUTYL-N-PENTYLAMINE see N-BUTYL-N-NITROSO AMYL AMINE

N-NITROSOBUTYLUREA see n-BUTYL NITROSO UREA

1'-NITROSO-1'-DEMETHYLNICOTINE see N'-NITROSONORNICOTINE

N-NITROSO-DIAETHYLAMINE (GERMAN) see N-NITROSO-DIETHYL-AMINE

N-NITROSO-3,4-DIBROMOPIPERIDINE see 3,4-DIBROMO NITROSO PIPERIDINE

N-NITROSO-DI-N-BUTYLAMINE see N-BUTYL-N-NITROSO-1-BUTAMINE

N-NITROSO-3,4-DICHLOROPIPERIDINE see 3,4-DICHLORONITROSOPI-PERIDINE

NITROSODIETHYLAMINE see N-NITROSO-DIETHYLAMINE

N-NITROSODIETHYLAMINE see N-NITROSO-DIETHYLAMINE

NITROSO-1,1-DIETHYL-3-METHYLUREA see 1,1-DIETHYL-3-METHYL-3-NITROSOUREA

N-NITROSODIFENYLAMIN (CZECH) see DIPHENYLNITROSAMINE

P-NITROSODIFENYLAMIN (CZECH) see p-NITROSODIPHENYLAMINE

N-NITROSODIMETHYLAMINE see DIMETHYLNITROSAMINE

4-NITROSODIMETHYLANILINE see N,N-DIMETHYL-p-NITROSOANILINE

P-NITROSO-N,N-DIMETHYLANILINE see N,N-DIMETHYL-p-NITROSOANI-LINE

N-NITROSO-2,2'-DIMETHYLDI-N-PROPYLAMINE see 2,2'-DIMETHYLDI-PROPYLINITROSOAMINE

NITROSO-2,6-DIMETHYLMORPHOLINE see 2,6-DIMETHYLNITROSO-MORPHOLINE

N-NITROSODI-N-PENTYLAMINE see DI-n-AMYLNITROSAMINE

4-NITROSODIPHENYLAMINE see p-NITROSODIPHENYLAMINE

N-NITROSODIPHENYLAMINE see DIPHENYLNITROSAMINE

N-NITROSODI-N-PROPYLAMINE see DI-n-PROPYLNITROSAMINE

N-NITROSODODECAMETHYLENEIMINE see 1-NITROSOAZACYCLOTRI-DECANE

NITROSODODECAMETHYLENIMINE see 1-NITROSOAZACYCLOTRIDE-CANE

N-NITROSODODECAMETHYLENIMINE see 1-NITROSOAZACYCLOTRIDE-CANE

N-NITROSO-N-ETHYL BIURET see N-ETHYL-N-NITROSOBIURET

N-NITROSOETHYL-N-BUTYLAMINE see ETHYL-n-BUTYL NITROS AMINE

N-NITROSOETHYLENETHIOUREA see N-NITROSOIMIDAZOLIDINETHIONE

N-NITROSOETHYLETHANOLAMINE see ETHYL-2-HYDROXY ETHYL NI-TROSAMINE

N-NITROSO-N-ETHYL-N-(2-HYDROXYETHYL)AMINE see ETHYL-2-HY-DROXY ETHYL NITROSAMINE

N-NITROSOETHYLISOPROPYLAMINE see ETHYL ISOPROPYL NITROSOA-MINE

N-NITROSOETHYLMETHYLAMINE see N,N-METHYLETHYLNITROSA-MINE

NITROSOETHYLUREA see 1-ETHYL-1-NITROSOUREA

NITROSOETHYLURETHAN see N-NITROSO-N-ETHYLURETHAN

N-NITROSOFENYLHYDROXYLAMIN AMONNY (CZECH) see AMMONIUM-N-NITROSOPHENYLHYDROXYLAMINE

N-NITROSOHEPTAMETHYLENEIMINE see OCTAHYDRO-1-NITROSOAZO-CINE

NITROSO-HEPTAMETHYLENIMIN (GERMAN) see OCTAHYDRO-1-NITRO-SOAZOCINE

N-NITROSOHEXAMETHYLENEIMINE see N-NITROSOHEXAHYDROAZE-PINE

NITROSOHEXAMETHYLENIMINE see N-NITROSOHEXAHYDROAZEPINE

N-NITROSO(2-HYDROXYPROPYL)(2-OXOPROPYL)AMINE see 1-((2-HY-DROXYPROPYL)-NITROSO)AMINO)ACETONE

N-NITROSO-2-HYDROXY-N-PROPYL-N-PROPYLAMINE see 1-(NITROSO-PROPYLAMINO)-2-PROPANOL

N-NITROSO-IMIDAZOLIDON (GERMAN) see 1-NITROSOIMIDAZOLIDIN-ONE

N-NITROSOIMIDAZOLIDONE see 1-NITROSOIMIDAZOLIDINONE

N-NITROSO-N-METHOXYMETHYLMETHYLAMINE see METHOXYMETHYL METHYLNITROSAMINE

N-NITROSO-N-METHYLACETAMIDE see METHYLNITROSOACETAMIDE

N-NITROSO-N-METHYL-N-ACETOXYMETHYLAMINE see ACETIC ACID METHYLNITROSAMINOMETHYL ESTER

NITROSOMETHYLALLYLAMINE see N-METHYL-N-NITROSOALLYLAMINE

N-NITROSOMETHYLALLYLAMINE see N-METHYL-N-NITROSOALLYL-AMINE

N-NITROSOMETHYLAMINACETONITRIL (GERMAN) see N-NITROSO-METHYLAMINOACETONITRILE

4-(N-NITROSO-N-METHYLAMINO)-1-(3-PYRIDYL)-1-BUTANONE see 4-(N-METHYL-N-NITROSAMINO)-1-(3-PYRIDYL)-1-BUTANONE

NITROSOMETHYLANILINE see N-METHYL-N-NITROSOANILINE

N-NITROSO-N-METHYLANILINE see N-METHYL-N-NITROSOANILINE

N-NITROSOMETHYLBENZYLAMINE see N-METHYL-N-BENZYLNITROSA-MINE

N-NITROSO-N-METHYLBIURET see N-METHYL-N-NITROSOBIURET

N-NITROSOMETHYL-N-BUTYLAMINE see METHYLBUTYLNITROSAMINE

N-NITROSO-N-METHYLCARBAMIDE see N-METHYL-N-NITROSOUREA

N-NITROSOMETHYL-2-CHLOROETHYLAMINE see 2-CHLORO-N-METHYL-N-NITROSOETHYLAMINE

N-NITROSOMETHYLCYCLOHEXYLAMINE see N-NITROSO-N-METHYL-CYCLOHEXYLAMINE

NITROSOMETHYLDIAETHYLHARNSTOFF see 1,1-DIETHYL-3-METHYL-3-NITROSOUREA

NITROSOMETHYLDIETHYLUREA see 1,1-DIETHYL-3-METHYL-3-NITRO-SOUREA

1-NITROSO-1-METHYL-3,3-DIETHYLUREA see 1,1-DIETHYL-3-METHYL-3-NITROSOUREA

N-NITROSO-N-METHYL-N-DODECYLAMIN (GERMAN) see NITROSO-METHYL-n-DODECYLAMINE

N-NITROSOMETHYLETHYLAMINE see N,N-METHYLETHYLNITROSAMINE

N-NITROSO-N-METHYLHEPTYLAMINE see HEPTYLMETHYLINITROSAMINE

NITROSOMETHYL-N-PENTYLAMINE see N-AMYL-N-METHYLNITROSAMINE

N-NITROSOMETHYLPHENYLAMINE see N-METHYL-N-NITROSOANILINE

N-NITROSO-N-METHYL-2-PHENYLETHYLAMINE see METHYL PHENYL-ETHYL NITROSAMINE

NITROSOMETHYLPHENYLUREA see 1-METHYL-1-NITROSO-3-PHENYL-UREA

N-NITROSO-N'-METHYLPIPERAZIN (GERMAN) see 1-NITROSO-4-METHYLPIPERAZINE

N-NITROSO-N'-METHYLPIPERAZINE see 1-NITROSO-4-METHYLPIPERAZINE

NITROSOMETHYL-N-PROPYLAMINE see NITROSOMETHYLPROPYLAMINE

NITROSOMETHYLUREA see N-METHYL-N-NITROSOUREA

N-NITROSO-N-METHYLUREA see N-METHYL-N-NITROSOUREA

NITROSOMETHYLURETHAN (GERMAN) see N-NITROSO-N-METHYLURE-THANE

N-NITROSOMORPHOLIN (GERMAN) see 4-NITROSOMORPHOLINE

N-NITROSOMORPHOLINE see 4-NITROSOMORPHOLINE

N-NITROSO-N-(2-OXOBUTYL)BUTYLAMINE see N-BUTYL-N-(2-OXOBU-TYL)NITROSAMINE

N-NITROSO-2-OXO-N-PROPYL-N-PROPYLAMINE see beta-OXYPRO-PYLPROPYLNITROSAMINE

N-NITROSOPERHYDROAZEPINE see N-NITROSOHEXAHYDROAZEPINE

4-NITROSOPHENOL see p-NITROSOPHENOL

P-NITROSO-N-PHENYLANILINE see p-NITROSODIPHENYLAMINE

4-NITROSO-N-PHENYLBENZENAMINE see p-NITROSODIPHENYLAMINE

N-NITROSOPHENYLHYDROXYLAMIN AMMONIUM SALZ (GERMAN) see AMMONIUM-N-NITROSOPHENYLHYDROXYLAMINE

N-NITROSOPHENYLHYDROXYLAMINE AMMONIUM SALT see AMMO-NIUM-N-NITROSOPHENYLHYDROXYLAMINE

1-NITROSOPIPERAZINE see MONONITROSOPIPERAZINE

N-NITROSOPIPERAZINE see MONONITROSOPIPERAZINE

NITROSOPIPERIDIN (GERMAN) see N-NITROSOPIPERIDINE

N-NITROSO-PIPERIDIN (GERMAN) see N-NITROSOPIPERIDINE

1-NITROSOPIPERIDINE see N-NITROSOPIPERIDINE

1-(NITROSOPROPYLAMINO)-2-PROPANONE see beta-OXYPROPYLPRO-PYLNITROSAMINE

N-NITROSO-N-PROPYL-1-PROPANAMINE see DI-n-PROPYLNITROSA-MINE

NITROSOPROPYLUREA see N-NITROSO-N-PROPYL-UREA

1-NITROSO-2-(3-PYRIDYL)PYRROLIDINE see N'-NITROSONORNICOTINE

N-NITROSOPYRROLIDIN (GERMAN) see N-NITROSOPYRROLIDINE

3-(1-NITROSO-2-PYRROLIDINYL)PYRIDINE see N'-NITROSONORNICO-TINE

n-NITROSO-2,2,4-TRIMETHYL-1,2-DIHYDROQUINOLINE,POLYMER see CURETARD

NITROSOTRIMETHYLENEIMINE see 1-NITROSOAZETIDINE

N-NITROSOTRIMETHYLENEIMINE see 1-NITROSOAZETIDINE

N-NITROSO-TRIMETHYLHARNSTOFF (GERMAN) see 1,1,3-TRIMETHYL-3-NITROSOUREA

1-NITROSO-2,2,4-TRIMETHYL-1(2H)-QUINOLINE, POLYMER see CURE-TARD

N-NITROSOTRIMETHYLUREA see 1,1,3-TRIMETHYL-3-NITROSOUREA

NITROSTARCH, DRY (DOT) see NITROSTARCH (DRY)

NITROSTIGMINE see PARATHION

NITROSYL ETHOXIDE see ETHYL NITRITE

NITROSYL HYDROXIDE see NITROUS ACID

NITROSYL SULFATE see NITROSYLSULFURIC ACID

4-NITRO-2,3,5,6-TETRACHLORANISOLE see TETRACHLORONITROANI-SOLE

NITROTHIAMIDAZOLE see NITROTHIAZOLE

1-(5-NITRO-2-THIAZOLYL)-2-IMIDAZOLIDINONE see NITROTHIAZOLE

1-(5-NITRO-2-THIAZOLYL)IMIDAZOLIDIN-2-ONE see NITROTHIAZOLE

1-(5-NITRO-2-THIAZOLYL)-2-IMIDAZOLINONE see NITROTHIAZOLE

1-(5-NITRO-2-THIAZOLYL)-2-OXOTETRAHYDROIMIDAZOL see NITRO-THIAZOLE

1-(5-NITRO-2-THIAZOLYL)-2-OXOTETRAHYDROIMIDAZOLE see NITRO-THIAZOLE

2-NITROTOLUENE see o-METHYLNITROBENZENE

3-NITROTOLUENE see m-METHYLNITROBENZENE

4-NITROTOLUENE see p-METHYL NITROBENZENE

M-NITROTOLUENE see m-METHYLNITROBENZENE

O-NITROTOLUENE see o-METHYLNITROBENZENE

P-NITROTOLUENE see p-METHYL NITROBENZENE

3-NITROTOLUOL see m-METHYLNITROBENZENE

NITROTRICHLOROMETHANE see TRICHLORONITROMETHANE

M-NITROTRIFLUOROTOLUENE see 3-NITROBENZOTRIFLUORIDE

M-NITROTRIFLUORTOLUOL(GERMAN) see 3-NITROBENZOTRIFLUORIDE

NITROUS ACID, N-BUTYL ESTER see n-BUTYL NITRITE

NITROUS ACID, sec-BUTYL ESTER see sec-BUTYL NITRITE

NITROUS ACID ETHYL ESTER see ETHYL NITRITE

NITROUS ACID, ISOBUTYL ESTER see ISOBUTYL NITRITE

NITROUS ACID, METHYL ESTER see METHYL NITRITE

NITROUS ACID, 1-METHYL PROPYL ESTER see sec-BUTYL NITRITE

NITROUS ACID, PENTYL ESTER see N-AMYL NITRITE

NITROUS ACID, POTASSIUM SALT see POTASSIUM NITRITE (1:1)

NITROUS ETHER see ETHYL NITRITE

NITROUS ETHER (DOT) see ETHYL NITRITE

NITROUS ETHYL ETHER see ETHYL NITRITE

NITROUS FUMES see NITRIC ACID (RED FUMING)

NITROUS OXIDE (DOT) see NITROGEN OXIDE

NITROXANTHIC ACID see PICRIC ACID

NITROXYL CHLORIDE see NITRYL CHLORIDE

NITROXYNIL see 4-HYDROXY-3-IODO-5-NITROBENZONITRILE

NIVALENOL-4-O-ACETATE see FUSARENONE X

NIVEMYCIN see NEOMYCIN

NOFLAMOL see POLYCHLORINATED BIPHENYLS

1-NONANECARBOXYLIC ACID see DECANOIC ACID

NONANOYLETHYLENEIMINE see 1-NONANOYLAZIRIDINE

NONYLCARBINOL see DECYL ALCOHOL

N-NONYLIC ACID see NONANOIC ACID

NONYL METHYL KETONE see 2-UNDECANONE

NO-PIP see N-NITROSOPIPERIDINE

1-NORADRENALINE see L-NOREPINEPHRINE

NORAMITRIPTYLINE see NORTRIPTYLINE

NORCAIN see ETHYL-4-AMINOBENZOATE

NORDHAUSEN ACID (DOT) see SULFURIC ACID

NOREPHEDRANE see DL-AMPHETAMINE SULFATE

(−)-NOREPINEPHRINE see L-NOREPINEPHRINE

1-NOREPINEPHRINE see L-NOREPINEPHRINE

NORETHINDRONE-17-ACETATE see 17-ACETOXY-19-NOR-17-alpha-PREGN-4-EN-20-YN-3-ONE

19-NOR-ETHINYL-4,5-TESTOSTERONE see 19-NORETHISTERONE

19-NOR-ETHINYL-5,10-TESTOSTERONE see 17-alpha-ETHYNYL-5,10-ESTRENOLONE

19-NORETHISTERONE ACETATE see 17-ACETOXY-19-NOR-17-alpha-PREGN-4-EN-20-YN-3-ONE

NORETHYNODREL MIXED WITH MESTRANOL see ENAVID

19-NOR-17-ALPHA-ETHYNYLANDROSTEN-17-BETA-OL-3-ONE see 19-NORETHISTERONE

19-NOR-17-ALPHA-ETHYNYL-17-BETA-HYDROXY-4-ANDROSTEN-3-ONE see 19-NORETHISTERONE

19-NOR-17-ALPHA-ETHYNYLTESTOSTERONE see 19-NORETHISTERONE

19-NORETHYNYLTESTOSTERONE ACETATE see 17-ACETOXY-19-NOR-17-alpha-PREGN-4-EN-20-YN-3-ONE

NORETHYSTERONE ACETATE see 17-ACETOXY-19-NOR-17-alpha-PREGN-4-EN-20-YN-3-ONE

D-NORGESTREL see NORGESTREL

D(−)-NORGESTREL see NORGESTREL

NORHOMOEPINEPHRINE HYDROCHLORIDE see AMINOPROPANOL PYROCATECHOL-HYDROCHLORIDE

EPSILON-NORLEUCINE see 6-AMINOCAPROIC ACID

NOR-LOST HYDROCHLORID (GERMAN) see BIS(2-CHLOROETHYL)-
AMINE HYDROCHLORIDE
NORLUTIN see 19-NORETHISTERONE
NORMAL LEAD ACETATE see LEAD ACETATE
NORMAL LEAD ORTHOPHOSPHATE see LEAD (II) PHOSPHATE (3:2)
NOR-NITROGEN MUSTARD see BIS-beta-CHLOROETHYLAMINE
NORNITROGEN MUSTARD HYDROCHLORIDE see BIS(2-CHLO-
ROETHYL)AMINE HYDROCHLORIDE
(17-ALPHA)-19-NORPREGNA-1,3,5(10)-TRIEN-20-YNE-3,17,DIOL see
ETHINYL ESTRADIOL
19-NOR-17-alpha-PREGNA-1,3,5(10)-TRIEN-2-YNE-3,17-DIOL see
ETHINYL ESTRADIOL
19-NOR-17-alpha-PREGN-4-EN-20-YNE-3-beta,17-DIOL DIACETATE
see ETHYNODIOL ACETATE
(3-BETA,17-ALPHA)-19-NORPREGN-4-EN-20-YNE-3,17-DIOL DIACE-
TATE see ETHYNODIOL ACETATE
19-NORTESTOSTERONE see NORGESTREL
NOTENQUIL see ACETOPROMAZINE
NOTENSIL see ACETOPROMAZINE
NOTESIL see ACETOPROMAZINE
NOVASUROL see (4-(CARBOXY METHOXY)-3-CHLOROPHENYL)(5,5-
DIETHYL-2,4,6(1H,3H,5H)-PYRIMIDINETRIONATO-O(sup 2)MER-
CURY, MONOSODIUM SALT
NOVOCAINAMIDE see p-AMINO-n-(2-DIETHYLAMINOETHYL)BENZA-
MIDE
NOVOCAIN-CHLORHYDRAT (GERMAN) see p-AMINOBENZOYLDI-
ETHYLAMINOETHANOL HYDROCHLORIDE
NOVOCAINE see p-AMINOBENZOIC ACID-2-DIETHYLAMINOETHYL ES-
TER
NOVOCAINE HYDROCHLORIDE see p-AMINOBENZOYLDIETHYLAMINO-
ETHANOL HYDROCHLORIDE
NOVOCAIN HYDROCHLORID (GERMAN) see p-AMINOBENZOYLDI-
ETHYLAMINOETHANOL HYDROCHLORIDE
NO$_x$ see NITROGEN MONOXIDE
NO$_x$ see NITROGEN DIOXIDE
N(SUP 4)-PROPYLAJMALINIUM HYDROGEN TARTRATE see N-PROPYL-
AJMALINE BITARTRATE
4-NQO see 4-NITROQUINOLINE-N-OXIDE
NSC-185 see CYCLOHEXIMIDE
NSC 339 see DINITROSOPIPERAZINE
NSC 739 see AMINOPTERIDINE
NSC-740 see METHOTREXATE
NSC 746 see URETHANE
NSC-750 see 1,4-BUTANEDIOL DIMETHYL SULFONATE
NSC 751 see dl-ETHIONINE
NSC 755 see PURINE-6-THIOL
NSC 759 see BENZIMIDAZOLE
NSC 762 see BIS(beta-CHLOROETHYL)METHYLAMINE
NSC 1026 see 1-AMINOCYCLOPENTANE-1-CARBOXYLIC ACID
NSC 1532 see 2,4-DINITROPHENOL
NSC 1895 see 3,5-DIAMINO-S-TRIAZOLE
NSC 2066 see THEOPHYLLINE
NSC-2100 see NITROFURAZONE
NSC 2107 see NITROFURANTOIN
NSC 3051 see N-METHYLFORMAMIDE
NSC 3053 see ACTINOMYCIN D
NSC-3058 see 1H-BENZOTRIAZOLE
NSC 3060 see POTASSIUM ARSENITE
NSC 3061 see TINDURIN
NSC 3069 see CHLORAMPHENICOL
NSC-3070 see DIETHYLSTILBESTEROL
NSC 3072 see SODIUM AZIDE
NSC 3073 see FOLIC ACID
NSC-3088 see CHLORAMBUCIL
NSC 3096 see N-METHYL-N-DESACETYLCOLCHICINE
NSC 3138 see N,N-DIMETHYLACETAMIDE
NSC-3424 see QUINACRINE MUSTARD
NSC 4730 see 2-ETHYLAMINO-1,3,4-THIADIAZOLE
NSC 4911 see MERCAPTOPURINE RIBONUCLEOSIDE

NSC 5356 see DIMETHYL FORMAMIDE
NSC-6396 see THIOPHOSPHAMIDE
NSC-8806 see L-3-(p-(BIS(2-CHLOROETHYL)AMINO)PHENYL)ALANINE
MONOHYDROCHLORIDE
NSC-8806 see L-PHENYLALANINE MUSTARD
NSC 8819 see ACROLEIN
NSC 9166 see TESTOSTERONE PROPIONATE
NSC 9659 see ISONICOTINIC ACID HYDRAZIDE
NSC-9698 see MANNOMUSTINE DIHYDROCHLORIDE
NSC-9704 see PROGESTERONE
NSC 9706 see TRISAZIRIDINYLTRIAZINE
NSC 9717 see TRIETHYLENEPHOSPHOROTRIAMIDE
NSC 9717 see TEPA
NSC-9895 see ESTRADIOL (β)
NSC-10107 see 2-CHLORO-N-(2-CHLORO ETHYL)-N-METHYL ETHANA-
MINE-N-OXIDE HYDROCHLORIDE
NSC 10107 see 2-CHLORO-N-(2-CHLOROETHYL)-N-METHYL ETHANA-
MINE-N-OXIDE
NSC-10108 see CHLOROTRIANISENE
NSC 10483 see CORTISOL
NSC 10815 see AMITAL
NSC-10873 see BIS-beta-CHLOROETHYLAMINE
NSC-12169 see ESTRIOL
NSC 13875 see HEXAMETHYLMELAMINE
NSC 14083 see STREPTOMYCIN
NSC-14210 see DL-PHENYLALANINE MUSTARD
NSC 15193 see 2-DEOXYGLUCOSE
NSC-16498 see o-(4-BIS(beta-CHLOROETHYL)AMINO-o-TOLYLAZO)-
BENZOIC ACID
NSC-17118 see CHLOROQUINE MUSTARD
NSC-17262 see BENZOQUINONE AZIRIDINE
NSC 17661 see GLYCINE NITROGEN MUSTARD
NSC 17663 see N,N-BIS(beta-CHLOROETHYL)-D,L-ALANINE HYDRO-
CHLORIDE
NSC 18016 see 2-CHLORO ETHYL METHANE SULFONATE
NSC 18321 see 2,5-BIS(BIS-(2-CHLOROETHYL)AMINOMETHYL)HY-
DROQUINONE
NSC-19893 see FLUOROURACIL
NSC 21206 see 6-AMINONICOTINAMIDE
NSC-23890 see DIMETHYLMYLERAN
NSC-23892 see BENZIMIDAZOLE METHYLENE MUSTARD
NSC 23909 see N-METHYL-N-NITROSOUREA
NSC 24559 see MITHRAMYCIN
NSC 26805 see ETHYL METHANESULFONATE
NSC-26812 see APHOLATE
NSC 26980 see AMETYCIN
NSC-27640 see 2'-DEOXY-5-FLUORO URIDINE
NSC 28693 see MONOCROTALINE
NSC-29630 see 3'5'-DICHLOROMETHOTREXATE
NSC-30211 see TRIMUSTINE
NSC 30622 see SENECIPHYLLINE
NSC 32065 see HYDROXYUREA
NSC-32606 see L-BUTADIENE DIEPOXIDE
NSC 32946 see METHYL GAG
NSC-33669 see 1-EMETINE DIHYDROCHLORIDE
NSC-33669 see EMETINE
NSC-34372 see QUINACRINE ETHYL MUSTARD
NSC-34462 see 5-(BIS(2-CHLOROETHYL)AMINO)URACIL
NSC 34533 see GRISOFULVIN
NSC-35051 see PHENYLALANINE MUSTARD
NSC 37095 see ETHYL(DI-(1-AZIRIDINYL)PHOSPHINYL)CARBAMATE
NSC-37538 see BIS(METHANE SULFONYL)-D-MANNITOL
NSC 38721 see CHLODITHANE
NSC 39661 see 2'-DEOXY-5-IODO URIDINE
NSC 40774 see METHYLTHIOINOSINE
NSC-45388 see DACARBAZINE
NSC 45403 see 1-ETHYL-1-NITROSOUREA
NSC-45624 see SODIUM THIOSULFATE, PENTAHYDRATE
NSC 47842 see VINCALEUKOBLASTINE

NSC 49842 see VINBLASTINE SULFATE
NSC-50256 see METHYL MESYLATE
NSC-50256 see METHYL METHANE SULFONATE
NSC-50413 see ARISTOLOCHINE
NSC-52695 see CYTOXAL ALCOHOL
NSC-56410 see N-METHYLMITOMYCIN C
NSC-56654 see AZOTOMYCIN
NSC-57199 see O-PHENYLALANINE MUSTARD
NSC-58404 see N-(DIAZOACETYL)GLYCINE HYDRAZINE
NSC-60195 see 4-METHYL TRIMETHYLENE SULFITE
NSC 6091D see 4,4'-SULFONYLDIANILINE
NSC-62209 see N,N-BIS(2-CHLOROETHYL)-2-NAPHTHYLAMINE
NSC 62579 see N,N'-DIETHYL-N,N'-DINITROSOETHYLENEDI-
    AMINE
NSC 62580 see DINITROSODIMETHYLPROPANEDIAMINE
NSC 63878 see ARABINOCYTIDINE
NSC-67574 see LEUROCRISTINE
NSC-68626 see 1-ACETYL-2-PICOLINOLHYDRAZINE
NSC-71261 see beta-THIOGUANINE DEOXYRIBOSIDE
NSC 73599 see DINITROSOPENTAMETHYLENETETRAMINE
NSC-77213 see 1-METHYL-2-(p-(ISOPROPYLCARBAMOYL)BENZYL)-
    HYDRAZINE
NSC-77213 see PROCARBAZINE
NSC-79037 see 1-(2-CHLOROETHYL)-3-CYCLOHEXYL-1-NITRO-
    SOUREA
NSC-80087 see 1-(4-DIMETHYLAMINOBENZAL)INDENE
NSC-82151 see DAUNOMYCIN
NSC-82196 see IMIDAZOLE MUSTARD
NSC 83653 see 1-ADAMANTANAMINE HYDROCHLORIDE
NSC-85998 see STREPTOZOTICIN
NSC 89936 see JACOBINE
NSC-89945 see SENKIRKINE
NSC-94100 see 1,6-DIBROMOMANNITOL
NSC-95441 see 1-(2-CHLOROETHYL)-3-(4-METHYL-CYCLOHEXYL)-1-
    NITROSOUREA
NSC-101984 see 4-(4-N-METHYL-N-NITROSAMINOSTYRYL)QUINO-
    LINE
NSC 102627 see YOSHI 864
NSC-102816 see AZACYTIDINE
NSC 104469 see CHOLESTERYL-p-BIS(2-CHLOROETHYL)AMINO
    PHENYLACETATE
NSC-104800 see DIBROMDULCITOL
NSC-109229 see L-ASPARAGINASE
NSC 109723 see TROPHOSPHAMIDE
NSC-109724 see ISOPHOSPHAMIDE
NSC 112259 see ESTRADIOL MUSTARD
NSC 113926 see RIFAMYCIN AMP
NSC-119875 see cis-PLATINOUS DIAMINE DICHLORIDE
NSC-122819 see ETP
NSC-123127 see ADRIAMYCIN
NSC 132313 see DIANHYDROGALACTITOL
NSC 139105 see TRIAZINATE
NSC 141540 see EPE
NSC 141549 see 4'-(9-ACRIDINYLAMINO)METHANESULPHON-m-ANI-
    SIDIDE
NSC-150014 see HYDRAZINE SULFATE (1:1)
NSC-153858 see MAYTANSINE
NSC 178248 see CHLOROZOTOCIN
NSC 190945 see O,O-DIMETHYL-S-(2-(ACETYLAMINO)ETHYL) DITHI-
    OPHOSPHATE
NSC 190955 see DIMETHYL PHOSPHATE ESTER WITH 2-CHLORO-N-
    METHYL-3-HYDROXYCROTONAMIDE
NSC 249992 see 4'-(9-ACRIDINYLAMINO)METHANESULPHON-m-ANI-
    SIDIDE
NSC-256439 see 4-DEMETHOXYDAUNOMYCIN
NSC-267703 see QUELAMYCIN
NSC 403169 see ACRONYCINE
NSC-409962 see N,N'-BIS(2-CHLOROETHYL)-N-NITROSOUREA
NSC 521778 see N-FORMYL-N-HYDROXYGLYCINE

NSC-762 HYDROCHLORIDE see BIS(2-CHLOROETHYL)METHYLAMINE
    HYDROCHLORIDE
N,N,3-TRIS(2-CHLOROETHYL)TETRAHYDRO-2H-1,3,2-OXAPHOSPHO-
    RIN-2-AMINE-2-OXIDE see TROPHOSPHAMIDE
NU 2017 see 3-ACETOXYPHENYLTRIMETHYLAMMONIUM IODIDE
(−)-NUCIFERINE see 1-NUCIFERINE
NUJOL see MINERAL OIL
NUMOQUIN HYDROCHLORIDE see ETHYL HYDROCUPREINE HYDRO-
    CHLORIDE
NUPERCAINE see DIBUCAINE
NYLON-6 see POLYCAPROLACTAM
OBELINE PICRATE see AMMONIUM PICRATE
OBESEDRIN see d-BENZEDRINE SULFATE
OBIDOXIME CHLORIDE see 1,3-BIS(4-ALDOX IMINOPYRIDINIUM)DI-
    METHYL ETHER BICHLORIDE
OCDD see OCTACHLORODIBENZODIOXIN
OCHRATOXIN A see (-)-N-((5-CHLORO-8-HYDROXY-3-METHYL-1-
    OXO-7-ISOCHROMANYL)CARBONYL)-3-PHENYLALANINE
OCHRE see IRON (III) OXIDE
OCTABROMOBIPHENYL see OCTABROMODIPHENYL
AR,AR,AR,AR,AR',AR',AR',AR' OCTABROMO-1,1'-BIPHENYL see
    OCTABROMODIPHENYL
1,2,4,5,6,7,8,8-OCTACHLOOR-3A,4,7,7A-TETRAHYDRO-4,7-ENDO-
    METHANO-INDAAN (DUTCH) see CHLORDANE
OCTACHLOR see CHLORDANE
OCTACHLOROCAMPHENE see TOXAPHENE
OCTACHLORODIBENZODIOXIN see OCTACHLORODIBENZODIOXIN
OCTACHLORODIBENZO-P-DIOXIN see OCTACHLORODIBENZODIOXIN
OCTACHLORODIBENZO(B,E)(1,4)DIOXIN see OCTACHLORODIBENZODI-
    OXIN
1,2,3,4,6,7,8,9-OCTACHLORODIBENZODIOXIN see OCTACHLORODI-
    BENZODIOXIN
OCTACHLORODIHYDRODICYCLOPENTADIENE see CHLORDANE
1,3,4,5,6,7,10,10-OCTACHLORO-4,7-ENDO-METHYLENE-4,7,8,9-
    TETRAHYDROPHTHALAN see 1,3,4,5,6,8,8-OCTACHLORO-1,3
    ,3a,4,7,7a-HEXAHYDRO-4,7-METHANO ISO BENZOFURAN
1,2,4,5,6,7,8,8-OCTACHLORO-2,3,3A,4,7,7A-HEXAHYDRO-4,7-
    METHANOINDENE see CHLORDANE
1,2,4,5,6,7,8,8-OCTACHLORO-2,3,3A,4,7,7A-HEXAHYDRO-4,7-
    METHANO-1H-INDENE see CHLORDANE
1,2,4,5,6,7,8,8-OCTACHLORO-3A,4,7,7A-HEXAHYDRO-4,7-METHY-
    LENE INDANE see CHLORDANE
OCTACHLORO-4,7-METHANOHYDROINDANE see CHLORDANE
OCTACHLORO-4,7-METHANOTETRAHYDROINDANE see CHLORDANE
1,2,4,5,6,7,8,8-OCTACHLORO-4,7-METHANO-3A,4,7,7A-TETRAHY-
    DROINDANE see CHLORDANE
1,3,4,5,6,7,8,8-OCTACHLORO-2-OXA-3A,4,7,7A-TETRAHYDRO-4,7-
    METHANOINDENE see 1,3,4,5,6,8,8-OCTACHLORO-1,3,3a,4,7,7a-
    HEXAHYDRO-4,7-METHANO ISO BENZOFURAN
1,2,4,5,6,7,8,8-OCTACHLORO-3A,4,7,7A-TETRAHYDRO-4,7-METHA-
    NOINDANE see CHLORDANE
1,2,4,5,6,7,8,8-OCTACHLORO-3A,4,7,7A-TETRAHYDRO-4,7-METHA-
    NOINDANE see CHLORDANE
1,2,4,5,6,7,10,10-OCTACHLORO-4,7,8,9-TETRAHYDRO-4,7-METHYL-
    ENEINDANE see CHLORDANE
1,2,4,5,6,7,8,8-OCTACHLOR-3A,4,7,7A-TETRAHYDRO-4,7-ENDO-
    METHANO-INDAN (GERMAN) see CHLORDANE
9,12-OCTADECADIENOIC ACID see LINOLEIC ACID
OCTADECANOIC ACID see STEARIC ACID
OCTADECANOIC ACID, METHYL ESTER see METHYL STEARATE
OCTADECANOIC ACID, ZINC SALT see ZINC STEARATE
N-OCTADECANOL see 1-OCTADECANOL
9,10-OCTADECENOIC ACID see OLEIC ACID
CIS-9-OCTADECENOIC ACID see OLEIC ACID
CIS-OCTADEC-9-ENOIC ACID see OLEIC ACID
CIS-DELTA(SUP 9)-OCTADECENOIC ACID see OLEIC ACID
(Z)-9-OCTADECENOIC ACID, BUTYL ESTER see BUTYL OLEATE
(Z)-9-OCTADECENOIC ACID, METHYL ESTER see cis-OLEIC ACID,
    METHYL ESTER

N-OCTADECYL ALCOHOL see 1-OCTADECANOL

N-OCTADECYLAMINE see OCTADECYLAMINE

N-OCTADECYL-N-BENZYL-N,N-DIMETHYLAMMONIUMCHLORIDE see DIMETHYLOCTADECYLBENZYLAMMONIUM CHLORIDE

OCTADECYLDIMETHYLBENZYLAMMONIUM CHLORIDE see DIMETHYLOCTADECYLBENZYLAMMONIUM CHLORIDE

OCTAFLUOROCYCLOBUTANE see CYCLOOCTAFLUOROBUTANE

2-(OCTAHYDRO-1-AZOCINYL)ETHYL GUANIDINE SULPHATE see GUANETHIDINE MONOSULFATE

OCTAMETHYL-DIFOSFORZUUR-TETRAMIDE (DUTCH) see OCTAMETH-YLPYROPHOSPHORAMIDE

OCTAMETHYL-DIPHOSPHORSAEURE-TETRAMID (GERMAN) see OCTA-METHYLPYROPHOSPHORAMIDE

OCTAMETHYL PYROPHOSPHORTETRAMIDE see OCTAMETHYLPYRO-PHOSPHORAMIDE

OCTAMETHYL TETRAMIDO PYROPHOSPHATE see OCTAMETHYLPYRO-PHOSPHORAMIDE

OCTAN AMYLU (POLISH) see PENTYL ACETATE

OCTAN BARNATY (CZECH) see BARIUM ACETATE

OCTAN N-BUTYLU (POLISH) see N-BUTYL ACETATE

1-OCTANECARBOXYLIC ACID see NONANOIC ACID

OCTAN ETOKSYETYLU (POLISH) see "CELLOSOLVE" ACETATE

OCTAN ETYLU (POLISH) see ETHYL ACETATE

OCTAN FENYLRTUTNATY (CZECH) see ACETOXYPHENYLMERCURY

OCTAN MANGANATY (CZECH) see MANGANESE ACETATE

OCTAN MEDNATY (CZECH) see COPPER ACETATE

OCTAN METYLU (POLISH) see ACETIC ACID METHYL ESTER

OCTANOIC ACID, CADMIUM SALT (2:1) see CADMIUM CAPRYLATE

OCTANOL see N-OCTYL ALCOHOL

1-OCTANOL see N-OCTYL ALCOHOL

N-OCTANOL see N-OCTYL ALCOHOL

OCTAN PROPYLU (POLISH) see PROPYL ACETATE

OCTAN WINYLU (POLISH) see ACETIC ACID VINYL ESTER

OCTOGEN see CYCLOTETRAMETHYLENETETRANITRAMINE

OCTOIL see BIS(2-ETHYLHEXYL)PHTHALATE

OCTOWY ALDEHYD (POLISH) see ACETALDEHYDE

OCTOWY BEZWODNIK (POLISH) see ACETIC ANHYDRIDE

OCTOWY KWAS (POLISH) see ACETIC ACID

OCTYL ACRYLATE see ACRYLIC ACID-2-ETHYLHEXYL ESTER

OCTYL ALCOHOL, NORMAL-PRIMARY see N-OCTYL ALCOHOL

N-OCTYLBICYCLO-(2.2.1)-5-HEPTENE-2,3-DICARBOXIMIDE see N-OCTYL BICYCLOHEPTENE DICARBOXIMIDE

OCTYLENE GLYCOL see ETHYL HEXYLENE GLYCOL

N-OCTYLISOSAFROLE SULFOXIDE see ISOSAFROLE-n-OCTYLSULFOX-IDE

cis-3-OCTYL-OXIRANEOCTANOIC ACID see cis-9,10-EPOXYOCTADE-CANOIC ACID

OCTYL PHENOL CONDENSED WITH 12-13 MOLES ETHYLENE OXIDE see TRITON X

P-TERT-OCTYLPHENOXYETHOXYETHYLDIMETHYLBENZYLAMMONIUM CHLORIDE see BENZETHONIUM CHLORIDE

P-TERT-OCTYLPHENOXYPOLYETHOXYETHANOL see TRITON X

OCTYL PHTHALATE see n-DIOCTYL PHTHALATE

4'-OCTYL-3-PIPERIDINOPROPIOPHENONE HYDROCHLORIDE see PIPOCTANONE HYDROCHLORIDE

OCTYL SEBACATE see BIS(2-ETHYLHEXYL)SEBACATE

OESTRADIOL see ESTRADIOL (β)

D-OESTRADIOL see ESTRADIOL (β)

cis-OESTRADIOL see ESTRADIOL (β)

alpha-OESTRADIOL see ESTRADIOL (β)

beta-OESTRADIOL see ESTRADIOL (β)

OESTRADIOL-17-beta see ESTRADIOL (β)

3,17-beta-OESTRADIOL see ESTRADIOL (β)

D-3,17-beta-OESTRADIOL see ESTRADIOL (β)

OESTRADIOL-3-BENZOATE see ESTRADIOL-3-BENZOATE

beta-OESTRADIOL BENZOATE see ESTRADIOL-3-BENZOATE

beta-OESTRADIOL 3-BENZOATE see ESTRADIOL-3-BENZOATE

17-beta-OESTRADIOL 3-BENZOATE see ESTRADIOL-3-BENZOATE

OESTRADIOL DIPROPIONATE see ESTRADIOL DIPROPIONATE

OESTRADIOL-3,17-DIPROPIONATE see ESTRADIOL DIPROPIONATE

beta-OESTRADIOL DIPROPIONATE see ESTRADIOL DIPROPIONATE

17-beta-OESTRADIOL DIPROPIONATE see ESTRADIOL DIPROPIONATE

3,17-beta-OESTRADIOL DIPROPIONATE see ESTRADIOL DIPROPIONATE

OESTRADIOL MONOBENZOATE see ESTRADIOL-3-BENZOATE

OESTRADIOL MUSTARD see ESTRADIOL MUSTARD

OESTRADIOL PHOSPHATE POLYMER see ESTRADIOL POLYESTER WITH PHOSPHORIC ACID

OESTRADIOL POLYESTER WITH PHOSPHORIC ACID see ESTRADIOL POLYESTER WITH PHOSPHORIC ACID

OESTRA-1,3,5(10)-TRIENE-3,17-beta-DIOL see ESTRADIOL (β)

17-beta-OESTRA-1,3,5(10)-TRIENE-3,17-DIOL see ESTRADIOL (β)

1,3,5(10)-OESTRATRIENE-3,17-beta-DIOL 3-BENZOATE see ESTRA-DIOL-3-BENZOATE

OESTRA-1,3,5(10)-TRIENE-3,16-alpha,17-beta-TRIOL see ESTRIOL

1,3,5-OESTRATRIENE-3-beta-3,16-alpha,17-beta-TRIOL see ESTRIOL

(16-alpha,17-beta)-OESTRA-1,3,5(10)-TRIENE-3,16,17-TRIOL see ES-TRIOL

1,3,5-OESTRATRIEN-3-OL-17-ONE see ESTRONE

1,3,5(10)-OESTRATRIEN-3-OL-17-ONE see ESTRONE

delta-1,3,5-OESTRATRIEN-3-beta-OL-17-ONE see ESTRONE

16-alpha,17-beta-OESTRIOL see ESTRIOL

3,16-alpha,17-beta-OESTRIOL see ESTRIOL

OESTRONBENZOAT (GERMAN) see ESTRONE BENZOATE

17-beta-OH-ESTRADIOL see ESTRADIOL (β)

7-OHM-MBA see 7-HYDROXYMETHYL-12-METHYLBENZ(a)ANTHRA-CENE

7-OHM-12-MBA see 7-HYDROXYMETHYL-12-METHYLBENZ(a)AN-THRACENE

17-beta-OH-OESTRADIOL see ESTRADIOL (β)

OIL CAMPHOR SASSAFRASSY see CAMPHOR OIL

OIL GARLIC see ALLYL SULFIDE

OIL MIST see OIL MIST (MINERAL)

OIL OF ABIES ALBA see ABIES ALBA OIL

OIL OF ANISE see ANISE OIL

OIL OF ANISEED see p-PROPENYL ANISOLE

OIL OF BAY see BAY OIL

OIL OF CAMPHOR RECTIFIED see CAMPHOR OIL

OIL OF CAMPHOR WHITE see CAMPHOR OIL

OIL OF CARAWAY see CARAWAY OIL

OIL OF CASSIA see CASSIA OIL

OIL OF EUCALYPTUS see EUCALYPTUS OIL

OIL OF FUR see ABIES ALBA OIL

OIL OF HARTSHORN see BONE OIL

OIL OF JUNIPER BERRY see JUNIPER BERRY OIL

OIL OF LEMON see LEMON OIL

OIL OF MYRCIA see BAY OIL

OIL OF NIOBE see METHYL BENZOATE

OIL OF PALMA CHRISTI see CASTOR OIL

OIL OF PINE see PINE OIL

OIL OF SILVER FIR see ABIES ALBA OIL

OIL OF SILVER PINE see ABIES ALBA OIL

OIL OF SWEET FLAG see OIL OF CALAMUS

OIL OF SWEET ORANGE see OIL OF ORANGE

OIL OF TURPENTINE see TURPENTINE

OIL OF TURPENTINE, RECTIFIED see TURPENTINE

OIL OF VITRIOL see SULFURIC ACID

OIL SCARLET see OIL RED

OIL THUJA see CEDAR LEAF OIL

OIL VIOLET see N-ETHYL-1-((p-(PHENYLAZO)PHENYL)AZO)-2-NAPH-THYLAMINE

OIL YELLOW A see 1-(PHENYLAZO)-2-NAPHTHYLAMINE

OKTADECYLAMIN (CZECH) see OCTADECYLAMINE

OKTAN (POLISH) see OCTANE

OKTANEN (DUTCH) see OCTANE

OLDHAMITE see CALCIUM SULFIDE

OLEFIANT GAS see ETHYLENE

OLEIC ACID GLYCIDYL ESTER see 2,3-EPOXYPROPYL OLEATE

OLEOCUIVRE see COPPER (I) OXIDE

OLEO NORDOX see COPPER (I) OXIDE

OLEOPHOSPHOTHION see CARBETHOXY MALATHION

OLEOVITAMIN D see ERGOCALCIFEROL

OLEOVITAMIN D3 see CHOLECALCIFEROL

OLEUM see SULFURIC ACID, FUMING

OLEUM ABIETIS see PINE OIL

OLEUM SINAPIS VOLATILE see ALLYL ISOTHIOCYANATE

OLEUM TIGLII see CROTON OIL

OLIO DI CROTON (ITALIAN) see CROTON OIL

OLOW (POLISH) see LEAD

OMEGAMYCIN see TETRACYCLINE

ONCOSTATIN K see ACTINOMYCIN D

ONION OIL see ALLYL PROPYL DISULFIDE

ONYX see SILICA, CRYSTALLINE-QUARTZ

ONYX see SILICA (CRYSTALLINE)

OPIPRAMOL DIHYDROCHLORIDE see INSIDON DIHYDROCHLORIDE

OPLOSSINGEN (DUTCH) see FORMALDEHYDE

OPTOQUINHYDROCHLORIDE see ETHYL HYDROCUPREINE HYDRO-
CHLORIDE

ORAFURAN see NITROFURANTOIN

ORALCID see ACETPHENARSINE

ORANGE A L'HUILE see 1-(PHENYLAZO)-2-NAPHTHOL

ORANGE G DYE see HISPACID FAST ORANGE 2G

ORANGE LEAD see LEAD OXIDE RED

ORANGE OIL see OIL OF ORANGE

ORANGES see d-BENZEDRINE SULFATE

ORANGE SOLUBLE A L'HUILE see 1-(PHENYLAZO)-2-NAPHTHOL

ORARSAN see ACETPHENARSINE

ORCINOL see 5-METHYLRESORCINOL

ORDINARY AZOXYBENZENE see AZOXYBENZENE

ORGANOL SCARLET see OIL RED

ORIENTAL BERRY see PICROTOXIN

ORINASE see 1-BUTYL-3-(p-TOLYL SULFONYL)UREA

ORONOL see 1-AUROTHIO-D-GLUCOPYRANOSE

ORPIMENT see ARSENIC SULFIDE

ORTHO 5353 see BUX-TEN

ORTHO 5865 see DIFOLATAN

ORTHO N-4 AND N-5 DUSTS see NICOTINE

ORTHOARSENIC ACID see ARSENIC ACID (SOLUTION)

ORTHOARSENIC ACID see O-ARSENIC ACID

ORTHOARSENIC ACID HEMIHYDRATE see O-ARSENIC ACID, HEMIHY-
DRATE

ORTHOBORIC ACID see BORIC ACID

ORTHOCAINE see 3-AMINO-4-HYDROXYBENZOIC ACID METHYL ESTER

ORTHOCRESOL see o-CRESOL

ORTHODERM see 3-AMINO-4-HYDROXYBENZOIC ACID METHYL ESTER

ORTHOFORM see 3-AMINO-4-HYDROXYBENZOIC ACID METHYL ESTER

ORTHOFORMIC ACID, TRIETHYL ESTER see ETHYL ORTHOFORMATE

ORTHOFORMIC ACID, TRIMETHYL ESTER see TRIMETHYL-O-FORMATE

ORTHOHYDROXYDIPHENYL see 2-BIPHENYLOL

ORTHOMRAVENCAN ETHYLNATY (CZECH) see ETHYL ORTHOFOR-
MATE

ORTHOMRAVENCAN METHYLNATY (CZECH) see TRIMETHYL-O-FOR-
MATE

ORTHOPHENYLPHENOL see 2-BIPHENYLOL

ORTHOPHOSPHORIC ACID see PHOSPHORIC ACID

ORTHOXENOL see 2-BIPHENYLOL

OSAGE ORANGE see MORIN

OSAGE ORANGE CRYSTALS see MORIN

OSAGE ORANGE EXTRACT see MORIN

OSARSAL see ACETPHENARSINE

OSARSOL see ACETPHENARSINE

OSARSOLE see ACETPHENARSINE

O,S-DIETHYL-O-(4-NITROPHENYL)PHOSPHOROTHIOATE see O,S-
DIETHYL-O-(4-NITROPHENYL)THIOPHOSPHATE

O,S-DIETHYL-O-(P-NITROPHENYL) PHOSPHOROTHIOATE see O,S-DI-
ETHYL-O-(4-NITROPHENYL)THIOPHOSPHATE

O,S-DIMETHYL ESTER AMIDE OF AMIDOTHIOATE see O,S-DIMETHYL
PHOSPHORAMIDOTHIOATE

OSMIC ACID see OSMIUM TETROXIDE

OSSIDO DI MESITILE (ITALIAN) see MESITYL OXIDE

OSVARSAN see ACETPHENARSINE

OSVARSON see ACETPHENARSINE

OTBE (FRENCH) see BIS(TRIBUTYL TIN)OXIDE

OTOBIOTIC see NEOMYCIN SULFATE

OTTANI (ITALIAN). see OCTANE

1,2,4,5,6,7,8,8-OTTOCHLORO-3A,4,7,7A-TETRAIDRO-4,7-ENDO-
METANO-INDANO (ITALIAN) see CHLORDANE

OTTOMETIL-PIROFOSFORAMMIDE (ITALIAN) see OCTAMETHYLPYRO-
PHOSPHORAMIDE

OUABAGENIN-L-RHAMNOSIDE see OUABAIN

OURARI see CURARE

OVAHORMON BENZOATE see ESTRADIOL-3-BENZOATE

OVOCYCLIN BENZOATE see ESTRADIOL-3-BENZOATE

OXAALZUUR (DUTCH) see OXALIC ACID

1-OXA-4-AZACYCLOHEXANE see MORPHOLINE

7-OXABICYCLO(4.1.0)HEPTANE see CYCLOHEXENE OXIDE

OXACYCLOPENTADIENE see FURAN

OXACYCLOPENTANE see TETRAHYDROFURAN

OXACYCLOPROPANE see ETHYLENE OXIDE

OXAL see GLYOXAL

OXALALDEHYDE see GLYOXAL

OXALAMIDE see OXAMIDE

OXALIC ACID DIAMIDE see OXAMIDE

OXALONITRILE see CYANOGEN

OXALSAEURE (GERMAN) see OXALIC ACID

OXAMIMIDIC ACID see OXAMIDE

OXAMMONIUM see HYDROXYLAMINE

OXANTHRENE see DIBENZO-p-DIOXIN

3-OXAPENTANE-1,5-DIOL see DIETHYLENE GLYCOL

3-OXA-1,5-PENTANEDIOL see DIETHYLENE GLYCOL

OXAPROPANIUM IODIDE see FORMAL-gamma-TRIMETHYL AMMO-
NIUM PROPANEDIOL

OXATOMIDA see OXATIMIDE

M-OXEDRINE see NEOSYNEPHRINE

(−)-M-OXEDRINE see NEOSYNEPHRINE

OXIDATION BASE 25 see 4-AMINO-2-NITROPHENOL

ALPHA,BETA-OXIDOETHANE see ETHYLENE OXIDE

1,8-OXIDO-P-MENTHANE see CAJEPUTOL

2-OXI-PROPYL-PROPYLNITROSAMIN (GERMAN) see beta-OXYPROPYL-
PROPYLNITROSAMINE

OXIRAAN (DUTCH) see ETHYLENE OXIDE

OXIRANE see ETHYLENE OXIDE

OXIRANE-CARBOXALDEHYDE see GLYCIDALDEHYDE

OXIRANE, 2,2'-(1,3-PHENYLENEBIS(OXYMETHYLENE))BIS- see
RESORCINOL DIGLYCIDYL ETHER

OXIRANYLMETHYL ESTER OF OCTADECANOIC ACID see STEARIC
ACID-2,3-EPOXYPROPYL ESTER

OXIRANYLMETHYL ESTER OF 9-OCTADECENOIC ACID see 2,3-EPOXY-
PROPYL OLEATE

3-OXIRANYL-7-OXABICYCLO(4.1.0)HEPTENE see VINYL CYCLO-
HEXENE DIOXIDE

OXITETRACYCLIN see 5-HYDROXYTETRACYCLINE

3'-(3-OXO-7-ALPHA-ACETYLTHIO-17-BETA-HYDROXYANDROST-4-EN-
17-BETA-YL)PROPIONIC ACID LACTONE see ALDACTAZIDE

2-OXO-1,2-BENZOPYRAN see COUMARIN

N-(6-OXO-6H-DIBENZO(B,D)PYRAN-1-YL)ACETAMIDE see N-6-(3,4-
BENZOCOUMARINYL)ACETAMIDE

ALPHA-OXODIPHENYLMETHANE see BENZOPHENONE

9-OXO-2-FLUORENYLACETAMIDE see 2-ACETYLAMINOFLUORENONE

N-(9-OXO-2-FLUORENYL)ACETAMIDE see 2-ACETYLAMINOFLUOREN-
ONE

3-OXO-L-GULOFURANOLACTONE see L-ASCORBIC ACID

2-OXOHEXAMETHYLENIMINE see HEXAHYDRO-2H-AZEPIN-2-ONE

17-((1-OXOHEXYL)OXY)PREGN-4-ENE-3,20-DIONE see HYDROXYPRO-
GESTERONE CAPROATE

OXOLE see FURAN

OXOMETHANE see FORMALDEHYDE

OXOPHENARSINE HYDROCHLORIDE see ARSPHENOXIDE

2-OXO-PROPYL-PROPYLNITROSAMINE see
  beta-OXYPROPYLPROPYLNITROSAMINE
(2-OXOPROPYL)PROPYLNITROSOAMINE see
  beta-OXYPROPYLPROPYLNITROSAMINE
3-OXYANTHRANILIC ACID see 2-AMINO-3-HYDROXYBENZOIC ACID
OXYBENZENE see PHENOL
OXYBENZOPYRIDINE see 8-QUINOLINOL
OXYBIS(4-AMINOBENZENE) see 4,4-OXYDIANILINE
4,4'-OXYBISANILINE see 4,4-OXYDIANILINE
P,P'-OXYBIS(ANILINE) see 4,4-OXYDIANILINE
4,4'-OXYBISBENZENAMINE see 4,4-OXYDIANILINE
4,4'-OXYBIS(2-CHLOROANILINE) see BIS(4-AMINO-3-CHLORO-
  PHENYL) ETHER
4,4'-OXYBIS(2-CHLORO-BENZENAMINE) see BIS(4-AMINO-3-CHLORO-
  PHENYL) ETHER
1,1'-OXYBIS(2-CHLORO)ETHANE see BIS(2-CHLOROETHYL) ETHER
OXYBIS(CHLOROMETHANE) see BIS(CHLOROMETHYL) ETHER
1,1'-OXYBISETHANE see ETHYL ETHER
2,2'-OXYBISETHANOL see DIETHYLENE GLYCOL
2,2'-OXYBIS-6-OXABICYCLO-(3.1.0)HEXANE see BIS(2,3-EPOXY CY-
  CLOPENTYL) ETHER
1,1-OXYBIS PENTANE see PENTYL ETHER
OXYBIS(TRIBUTYLTIN) see BIS(TRIBUTYL TIN)OXIDE
OXYBUTYRIC ALDEHYDE see ACETALDOL
OXYCHINOLIN see 8-QUINOLINOL
OXYCHLORURE CHROMIQUE (FRENCH) see CHROMIUM OXYCHLORIDE
OXYCODEINONE see PERCODAN
OXYDE D'ALLYLE ET DE GLYCIDYLE (FRENCH) see ((2-PROPENYL-
  OXY)METHYL)OXIRANE
OXYDE DE BARYUM (FRENCH) see BARIUM OXIDE
OXYDE DE CALCIUM (FRENCH) see CALCIUM OXIDE
OXYDE DE CARBONE (FRENCH) see CARBON MONOXIDE
OXYDE DE CHLORETHYLE (FRENCH) see BIS(2-CHLOROETHYL) ETHER
OXYDE DE MERCURE (FRENCH) see MERCURIC OXIDE
OXYDE DE MESITYLE (FRENCH) see MESITYL OXIDE
OXYDE DE PROPYLENE (FRENCH) see EPOXYPROPANE
OXYDE D'ETHYLE (FRENCH) see ETHYL ETHER
OXYDE DE TRIBUTYLETAIN see BIS(TRIBUTYL TIN)OXIDE
OXYDEMETON-METILE (ITALIAN) see DEMETON-O-METHYL SULFOX-
  IDE
OXYDE NITRIQUE (FRENCH) see NITROGEN MONOXIDE
OXYDIANILINE see 4,4-OXYDIANILINE
P,P'-OXYDIANILINE see 4,4-OXYDIANILINE
2,2'-OXYDIETHANOL see DIETHYLENE GLYCOL
N-(OXYDIETHYLENE)BENZOTHIAZOLE-2-SULFENAMIDE see 2-BENZO-
  THIAZOLYL-N-MORPHOLINOSULFIDE
OXYDIMETHYLQUINAZINE see ANTIPYRINE
P-OXYDIPHENYLAMINE see p-ANILINOPHENOL
4,4'-OXYDIPHENYLAMINE see 4,4-OXYDIANILINE
OXYDI-P-PHENYLENEDIAMINE see 4,4-OXYDIANILINE
N-OXYD-LOST see 2-CHLORO-N-(2-CHLORO ETHYL)-N-METHYL ETH-
  ANAMINE-N-OXIDE HYDROCHLORIDE
N-OXYD-LOST (GERMAN) see 2-CHLORO-N-(2-CHLOROETHYL)-N-
  METHYL ETHANAMINE-N-OXIDE
N-OXYD-MUSTARD (GERMAN) see 2-CHLORO-N-(2-CHLOROETHYL)-N-
  METHYL ETHANAMINE-N-OXIDE
OXYMETHYLENE see FORMALDEHYDE
OXYMURIATE OF POTASH see POTASSIUM CHLORATE
OXYPARATHION see p-NITROPHENYL DIETHYLPHOSPHATE
OXYPHENBUTAZONE see p-HYDROXYPHENYLBUTAZONE
OXYPHENIC ACID see PYROCATECHOL
OXYPHENYLBUTAZONE see p-HYDROXYPHENYLBUTAZONE
P-OXYPROPIOPHENONE see ETHYL-p-HYDROXY PHENYL KETONE
gamma-OXYPROPYLTHEOBROMIN (GERMAN) see 1-(3-HYDROXY-
  PROPYL)THEOBROMINE
gamma-(gamma-OXYPROPYL)-THEOBROMIN (GERMAN) see 1-(3-HY-
  DROXYPROPYL)THEOBROMINE
8-OXYQUINOLINE see 8-QUINOLINOL
OXYQUINOLINOLEATE DE CUIVRE (FRENCH) see BIS(8-OXYQUINO-
  LINE)COPPER

5-OXYRESORCINOL see PHLOROGLUCINOL
OXYTETRACYCLINE AMPHOTERIC see 5-HYDROXYTETRACYCLINE
M-OXYTOLUENE see m-CRESOL
O-OXYTOLUENE see o-CRESOL
P-OXYTOLUENE see p-CRESOL
OZON (POLISH) see OZONE
P 284 see 4-(DIPHENYLMETHOXY)-1-METHYLPIPERIDINE CHLORO-
  THEOPHYLLINE
PABA see p-AMINOBENZOIC ACID
PAINTERS' NAPHTHA see PETROLEUM SPIRITS
PALLADOUS CHLORIDE see PALLADIUM(2+) CHLORIDE
PALLICID see ACETPHENARSINE
PALMITYLAMINE see 1-HEXADECANAMINE
PAM (CZECH) see PYRIDINIUM-2-ALDOXIME-N-METHYLIODIDE
2-PAM IODIDE see PYRIDINIUM-2-ALDOXIME-N-METHYLIODIDE
PANTHER CREEK BENTONITE see BENTONITE
PANTHERINE see 5-AMINOMETHYL-3-ISOXYZOLE
PANTOTHENATE CALCIUM see CALCIUM-d-PANTOTHENATE
(+)-PANTOTHENIC ACIDCALCIUM SALT see CALCIUM-d-PANTOTHE-
  NATE
PANTOTHENIC ACID, ZINC SALT see ZINC PANTOTHENATE
PANTOZOL 1 see CROTOXYPHOS
PANWARFIN see COUMADIN SODIUM
PAPANERINE see PAPAVERINE
PAPAYOTIN see PAPAIN
PAPP see p-AMINOPROPIOPHENONE
PARAAMINODIPHENYL see 4-BIPHENYLAMINE
PARACETALDEHYDE see PARALDEHYDE
PARACETOPHENETIDIN see p-ACETOPHENETIDIDE
PARACHLOROPHENOL see 4-CHLOROPHENOL
PARACRESYL ACETATE see 4-METHYLPHENYL ACETATE
D-PARACURARINE CHLORIDE see TUBOCURARINE HYDROCHLORIDE
PARADICHLOROBENZOL see p-DICHLOROBENZENE
PARAFFIN OIL see MINERAL OIL
PARAFUCHSIN (GERMAN) see BASIC PARAFUCHSINE
PARAFUCHSINE see BASIC PARAFUCHSINE
PARAHEXYL see 3-HOMOTETRA HYDRO CANNIBINOL
PARALDEHYD (GERMAN) see PARALDEHYDE
PARALDEIDE (ITALIAN) see PARALDEHYDE
PARAMANDELIC ACID see MANDELIC ACID
PARAMENTHANE HYDROPEROXIDE see 1-ISOPROPYL-4-METHYLCY-
  CLOHEXANE HYDROPEROXIDE
PARAMETHYL PHENOL see p-CRESOL
PARAMORFAN see DIHYDROMORPHINE HYDROCHLORIDE
PARANAPHTHALENE see ANTHRACENE
PARANITROFENOL (DUTCH) see 4-NITROPHENOL
PARANITROFENOLO (ITALIAN) see 4-NITROPHENOL
PARANITROPHENOL (FRENCH,GERMAN) see 4-NITROPHENOL
PARANITROSODIMETHYLANILIDE see N,N-DIMETHYL-p-NITROSOANI-
  LINE
PARAPHENOLAZO ANILINE see p-(PHENYLAZO)ANILINE
PARAQUAT CHLORIDE see PARAQUAT
PARAROSANILINE HYDROCHLORIDE see BASIC PARAFUCHSINE
PARASORBIC ACID see PARASCORBIC ACID
(+)-PARASORBINSAEURE (GERMAN) see PARASCORBIC ACID
PARATHENE see PARATHION
PARATHESIN see ETHYL-4-AMINOBENZOATE
PARATHION, LIQUID (DOT) see PARATHION
PARATHION METHYL see METHYL PARATHION
PARATHION-METILE (ITALIAN) see METHYL PARATHION
PARAXENOL see 4-BIPHENYLOL
PARBENDAZOLE see 5-BUTYL-2-BENZIMID AZOLECARBAMIC ACID
  METHYL ESTER
PARENTRACIN see BACITRACIN
PARIS GREEN see CUPRIC ACETOARSENITE
PARIS RED see LEAD OXIDE RED
PARIS YELLOW see LEAD CHROMATE
PARKE DAVIS CI-628 see NITROMIFENE CITRATE
PAROXYL see ACETPHENARSINE

PAROXYPROPIONE see ETHYL-p-HYDROXY PHENYL KETONE

PARROT GREEN see CUPRIC ACETOARSENITE

PARZATE see NABAM

SYMPATEDRINE see BENZEDRINE

PATULIN see CLAVACIN

PBB see FIREMASTER BP-6

PCB see PROCARBAZINE

PCBS see POLYCHLORINATED BIPHENYLS

PCP see ANGEL DUST

PDD see PHORBOL-12,13-DIDECANOATE

P'-DIETHYL P-DIMETHYL THIOPYROPHOSPHATE see p-DIETHYL p'-
DIMETHYLTHIOPYROPHOSPHATE

PEACE PILL see ANGEL DUST

PEACHES see DL-AMPHETAMINE SULFATE

PEARL ASH see POTASSIUM CARBONATE (2:1)

PEARLPUSS see CARRAGEEN

PEARL STEARIC see STEARIC ACID

PEARLY GATES see N,N-DIETHYLLYSERGAMIDE

PEAR OIL see ISOPENTYL ALCOHOL ACETATE

PEAR OIL see PENTYL ACETATE

PECAN SHELL POWDER see PEANUT OIL

PEDIAMYCIN see ERYTHROMYCIN

P.E.G. 400 see POLYETHYLENE GLYCOL

PELARGIC ACID see NONANOIC ACID

PELARGONIC ACID see NONANOIC ACID

PELARGONIC MORPHOLIDE see 4-MORPHOLINENONYLIC ACID

PELLAGRAMIN see NICOTINIC ACID

PELLAGRA PREVENTIVE FACTOR see NICOTINIC ACID

PELLUGEL see CARRAGEEN

PELONIN see NICOTINIC ACID

(S)-PENICILLAMIN see d,3-MERCAPTOVALINE

PENCILLIC ACID see PENICILLIC ACID

PENICILLAMINE see d,3-MERCAPTOVALINE

D-PENICILLAMINE see d,3-MERCAPTOVALINE

PENICILLIN G see BENZYL-6-AMINOPENICILLINIC ACID

PENIZILLIN (GERMAN) see PENICILLIN

PENTAACETYLGITOXIN see GITOXIN PENTA ACETATE

PENTABARBITONE see NEMBUTAL

PENTABORANE (DOT) see PENTABORANE(9)

PENTACARBONYLIRON see IRON CARBONYL

PENT-ACETATE see PENTYL ACETATE

PENTACHLOORETHAAN (DUTCH) see PENTACHLOROETHANE

PENTACHLOORFENOL (DUTCH) see PENTACHLOROPHENOL

PENTACHLORAETHAN (GERMAN) see PENTACHLOROETHANE

PENTACHLORETHANE (FRENCH) see PENTACHLOROETHANE

2',3',4',5',6'- PENTACHLOROACETOPHENONE see PENTACHLORO-
ACETOPHENONE

PENTACHLOROANTIMONY see ANTIMONY (V) CHLORIDE

PENTACHLOROPHENATE see PENTACHLOROPHENOL

PENTACHLOROPHENATE SODIUM see SODIUM PENTACHLOROPHENATE

2,3,4,5,6-PENTACHLOROPHENOL see PENTACHLOROPHENOL

PENTACHLOROPHENOL, SODIUM SALT see SODIUM PENTACHLORO-
PHENATE

PENTACHLOROPHENOXY SODIUM see SODIUM PENTACHLOROPHENATE

PENTACHLOROPHENYL CHLORIDE see HEXACHLOROBENZENE

PENTACHLORURE D'ANTIMOINE (FRENCH) see ANTIMONY (V) CHLO-
RIDE

PENTACLOROETANO (ITALIAN) see PENTACHLOROETHANE

PENTACLOROFENOLO (ITALIAN) see PENTACHLOROPHENOL

1-PENTADECANECARBOXYLIC ACID see PALMITIC ACID

PENTAERYTHRITE TETRANITRATE see PENTAERYTHRITOL TETRANI-
TRATE

PENTAERYTHRITOL DICHLOROHYDRIN see 2-AMINO-2-METHYL-1,3-
PROPANEDIOL

PENTAFLUOROANTIMONY see ANTIMONY (V) PENTAFLUORIDE

2',3,4',5,7-PENTAHYDROXYFLAVONE see MORIN

3,5,7,3',4'-PENTAHYDROXYFLAVONE see QUERCETIN

PENTAHYDROXY-TIGLIADIENONE-MONOACETATE(C)MONOMYRI-
STATE(B) see PHORBOL MYRISTATE ACETATE

PENTALIN see PENTACHLOROETHANE

PENTAMETHYLBENZYL-P-ROSANILINE CHLORIDE see METHYL VIOLET
6B

N,N,N',N'-3-PENTAMETHYL-N,N'-DIAETHYL-3-AZA-PENTAN-1,5-DI-
AMMONIUM-DIBROMID (GERMAN) see ((METHYLIMINO)DIETH-
YLENE)BIS)DIMETHYLETHYLAMMONIUM DIBROMIDE

N,N,N',N'-PENTAMETHYL-N,N'-DIETHYL-3-AZA-PENTANE-1,5-DIAM-
MONIUM DIBROMIDE see ((METHYLIMINO)DIETHYLENE)BIS)DI-
METHYLETHYLAMMONIUM DIBROMIDE

PENTAMETHYLENE see CYCLOPENTANE

PENTAMETHYLENEDIAMINE see PENTANEDIAMINE

P,P'-(PENTAMETHYLENEDIOXY)DIBENZAMIDINE BIS(BETA-HY-
DROXYETHANESULFONATE) see DIAMIDINE

PENTAMETHYLENEDITHIOCARBAMATE see PIP-PIP

PENTAMETHYLENEIMINE see PIPERIDINE

PENTAMETHYLENETETRAZOL see 1,5-PENTAMETHYLENETETRAZOLE

PENTAMETHYLENE-1,5-TETRAZOLE see 1,5-PENTAMETHYLENE-
TETRAZOLE

PENTAMIDINE DIISETHIONATE see DIAMIDINE

PENTAMIDINE ISETHIONATE see DIAMIDINE

PENTAN (POLISH) see PENTANE

PENTANAL see n-VALERALDEHYDE

N-PENTANAL see n-VALERALDEHYDE

3-PENTANECARBOXYLIC ACID see DIETHYLACETIC ACID

PENTANEDINITRILE (9CI) see 1,3-TRIMETHYLENEDINITRILE

PENTANEDIONE see ACETYL ACETONE

PENTANEDIONE-2,4 see ACETYL ACETONE

N-PENTANE + 2 ISOMERS, N-HEXANE + 4 ISOMERS, N-HEPTANE +
8 ISOMERS, N-OCTANE + 17 ISOMERS see ALKANES

PENTANEN (DUTCH) see PENTANE

PENTANI (ITALIAN) see PENTANE

TERT-PENTANOIC ACID see PIVALIC ACID

PENTANOL-1 see PENTYL ALCOHOL

PENTAN-1-OL see PENTYL ALCOHOL

PENTANOL-1 see AMYL ALCOHOL

PENTAN-1-OL see AMYL ALCOHOL

PENTANOL-2 see 2-PENTANOL

PENTAN-3-OL see 3-PENTANOL

PENTANOL-3 see 3-PENTANOL

N-PENTANOL see PENTYL ALCOHOL

N-PENTANOL see AMYL ALCOHOL

TERT-PENTANOL see tert-N-AMYL ALCOHOL, REFINED

TERT-PENTANOL see t-PENTYL ALCOHOL

1-PENTANOL ACETATE see PENTYL ACETATE

PENTAPHENATE see SODIUM PENTACHLOROPHENATE

PENTASODIUM TRIPHOSPHATE see SODIUM TRIPOLYPHOSPHATE

PENTASOL see AMYL ALCOHOL

PENTASOL see PENTYL ALCOHOL

PENTASULFURE DE PHOSPHORE (FRENCH) see PHOSPHORUS PENTA-
SULFIDE

PENTHIOBARBITAL SODIUM see PENTOTHAL SODIUM

PENTOBARBITAL see NEMBUTAL

PENTOBARBITAL CALCIUM see CALCIUM PENTOBARBITAL

PENTOBARBITURIC ACID see NEMBUTAL

PENTONAL see NEMBUTAL SODIUM

1-PENTYL ACETATE see PENTYL ACETATE

N-PENTYL ACETATE see PENTYL ACETATE

PENTYL ALCOHOL see AMYL ALCOHOL

SEC-PENTYL ALCOHOL see 2-PENTANOL

8-PENTYLBENZ(a)ANTHRACENE see 5-n-AMYL-1:2-BENZANTHRACENE

PENTYLCARBINOL see N-HEXYL ALCOHOL

3-PENTYLCARBINOL see 2-ETHYLBUTANOL

SEC-PENTYLCARBINOL see 2-ETHYLBUTANOL

ALPHA-PENTYLCINNAMALDEHYDE see alpha-AMYL CINNAMALDE-
HYDE

PENTYLENETETRAZOL see 1,5-PENTAMETHYLENETETRAZOLE

PENTYL FORMATE see N-AMYL FORMATE

M-PENTYL FORMATE see N-AMYL FORMATE

PENTYL MERCAPTAN see 1-PENTANETHIOL

PENTYL NITRITE see N-AMYL NITRITE

PENTYL PENTYLAMINE see DIAMYL AMINE

P-T-PENTYLPHENOL see 4-tert-AMYLPHENOL

PERACETIC ACID see PEROXYACETIC ACID

PERACETIC ACID SOLUTION (DOT) see PEROXYACETIC ACID

PERAZIL DIHYDROCHLORIDE see CHLORCYCLIZINE DIHYDROCHLO-
RIDE

PERBENZOATE DE BUTYLE TERTIAIRE (FRENCH) see tert-BUTYL
PERBENZOATE

PERBENZOIC ACID see PEROXYBENZOIC ACID

PERCAINE HYDROCHLORIDE see NUPERCAINE HYDROCHLORIDE

PERCHLOORETHYLEEN, PER (DUTCH) see 1,1,2,2-TETRACHLORO-
ETHYLENE

PERCHLORAETHYLEN, PER (GERMAN) see 1,1,2,2-TETRACHLORO-
ETHYLENE

PERCHLORATE DE MAGNESIUM (FRENCH) see PERCHLORIC ACID,
MAGNESIUM SALT

PERCHLORETHYLENE, PER (FRENCH) see 1,1,2,2-TETRACHLORO-
ETHYLENE

PERCHLORIC ACID (DOT) see PERCHLORIC ACID

PERCHLORIC ACID, POTASSIUM SALT (1:1) see POTASSIUM PERCHLO-
RATE

PERCHLORIDE OF MERCURY see CORROSIVE SUBLIMATE

PERCHLOROBENZENE see HEXACHLOROBENZENE

PERCHLOROETHANE see 1,1,1,2,2,2-HEXACHLOROETHANE

PERCHLOROETHYLENE see 1,1,2,2-TETRACHLOROETHYLENE

PERCHLOROMETHANE see CARBON TETRACHLORIDE

PERCHLOROPENTACYCLODECANE see MIREX

PERCHLOROPENTACYCLO(5.2.1.0(SUP 2,6).0(SUP 3,9).0(SUP
5,8))DECANE see MIREX

PERCHLORURE D'ANTIMOINE (FRENCH) see ANTIMONY (V) CHLORIDE

PERCHLORURE DE FER (FRENCH) see FERRIC CHLORIDE

PERCLENE see 1,1,2,2-TETRACHLOROETHYLENE

PERCLOROETILENE (ITALIAN) see 1,1,2,2-TETRACHLOROETHYLENE

PERCODAN see p-ACETOPHENETIDIDE

PERFLUOROCYCLOBUTANE see CYCLOOCTAFLUOROBUTANE

2-PERHYDROAZEPINONE see HEXAHYDRO-2H-AZEPIN-2-ONE

PERHYDROL see HYDROGEN PEROXIDE, 90%

PERHYDRONAPHTHALENE see DECAHYDRONAPHTHALENE

PERIODIN see POTASSIUM PERCHLORATE

PERMA KLEER see PENTHAMIL

PERMANGANATE DE POTASSIUM (FRENCH) see POTASSIUM PERMAN-
GANATE

PERMANGANATE DE SODIUM (FRENCH) see SODIUM PERMANGANATE

PERMANGANATE OF POTASH see POTASSIUM PERMANGANATE

PERMASEAL see PLIOFILM

PERMETRINA (PORTUGUESE) see AMBUSH

PEROSSIDO DI BENZOILE(ITALIAN) see BENZOYL PEROXIDE

PEROSSIDO DI BUTILE TERZIARIO (ITALIAN) see tert-BUTYL PEROXIDE

PEROSSIDO DI IDROGENO (ITALIAN) see HYDROGEN PEROXIDE, 90%

PEROXAN see HYDROGEN PEROXIDE, 90%

1,4-PEROXIDO-p-MENTHENE-2 see ASCARIDOLE

PEROXYDE DE BENZOYLE (FRENCH) see BENZOYL PEROXIDE

PEROXYDE DE BUTYLE TERTIAIRE (FRENCH) see tert-BUTYL PEROX-
IDE

PEROXYDE DE LAUROYLE (FRENCH) see LAUROYL PEROXIDE

PEROXYDE DE PLOMB (FRENCH) see LEAD DIOXIDE

PEROXYDE D'HYDROGENE (FRENCH) see HYDROGEN PEROXIDE, 90%

PERSPEX see POLYMETHYLMETHACRYLATE

PERSULFATE D'AMMONIUM (FRENCH) see AMMONIUM PERSULFATE

PERSULFATE DE SODIUM (FRENCH) see SODIUM PERSULFATE

PETROHOL see ISOPROPYL ALCOHOL

PETROL see GASOLINE (from 50-100 octane)

PETROLATUM, LIQUID see MINERAL OIL

PETROLEUM BENZIN see NAPHTHA, COAL TAR

PETROLEUM BENZIN see BENZIN

PETROLEUM CRUDE see PETROLEUM

PETROLEUM DISTILLATES (NAPHTHA) see BENZIN

PETROLEUM ETHER see BENZIN

PETROLEUM NAPHTHA see NAPHTHA, COAL TAR

PETROLEUM PITCH see ASPHALT

PETROLEUM ROOFING TAR see PETROLEUM ASPHALT

PEYRONE'S CHLORIDE see cis-PLATINOUS DIAMINE DICHLORIDE

PFEFFERMINZ OEL (GERMAN) see PEPPERMINT OIL

PHARMEDRINE see DL-AMPHETAMINE SULFATE

PARA-PHENACETIN see p-ACETOPHENETIDIDE

PHENACITE see SILICIC ACID, BERYLLIUM SALT

PHENACYLAMINE see 2-AMINOACETOPHENONE

PHENACYL CHLORIDE see 2-CHLOROACETOPHENONE

PHENACYLCHLORIDE see p-CHLOROACETOPHENONE

PHENADONE see METHADONE

PHENADONE HYDROCHLORIDE see dl-METHADONE HYDROCHLORIDE

PHENALZINE HYDROGEN SULPHATE see PHENYLETHYLHYDRAZINE
SULPHATE

PHENAMINE see DL-AMPHETAMINE SULFATE

PHENANTHREN (GERMAN) see PHENANTHRENE

2-PHENANTHRYLACETAMIDE see 2-ACETAMIDOPHENATHRENE

N-(2-PHENANTHRYL)ACETAMIDE see 2-ACETAMIDOPHENATHRENE

N-2-PHENANTHRYLACETAMIDE see 2-ACETAMIDOPHENATHRENE

PHENARSAZINE CHLORIDE see PHENARSAZINE CHLORIDE

PHENARSENAMINE see SALVARSAN

PHENAZONE (PHARMACEUTICAL) see ANTIPYRINE

PHENAZOPYRIDINE see PHENAZODINE

PHENAZOPYRIDINE HYDROCHLORIDE see PHENAZOPYRIDINIUM CHLO-
RIDE

PHENCAPTON (DOT) see PHENCAPTON

PHENCYCLIDINE HYDROCHLORIDE see ANGEL DUST

PHENDIMETRAZINE BITARTRATE see DIETROL

PHENDIMETRAZINE HYDROCHLORIDE see 3,4-DIMETHYL-2-PHENYL-
MORPHOLINEHYDROCHLORIDE

PHENEDRINE see d-BENZEDRINE SULFATE

PHENEDRINE see DL-AMPHETAMINE SULFATE

PHENEENE GERMICIDAL SOLUTION AND TINCTURE see ALKYL DI-
METHYLBENZYL AMMONIUM CHLORIDE

PHENELZINE SULFATE see PHENYLETHYLHYDRAZINE SULPHATE

PHENETHANOL see PHENETHYL ALCOHOL

2-PHENETHYL ACETATE see 2-PHENYLETHYL ACETATE

BETA-PHENETHYL ACETATE see 2-PHENYLETHYL ACETATE

2-PHENETHYL ALCOHOL see PHENETHYL ALCOHOL

BETA-PHENETHYL ALCOHOL see PHENETHYL ALCOHOL

BETA-PHENETHYLAMINE see beta-PHENETHYLAMINE

PHENETHYLAMINE, ALPHAMETHYL-, SULFATE (2:1) see BENZEDRINE
SULFATE

PHENETHYLCARBAMID (GERMAN) see 4-ETHOXY PHENYL UREA

PHENETHYLHYDRAZINE SULFATE (1:1) see PHENYLETHYLHYDRAZINE
SULPHATE

N-(1-PHENETHYL-4-PIPERIDI-NYL)PROPIONANILIDE DIHYDROGEN CI-
TRATE see PHENTANYL CITRATE

N-(1-PHENETHYL-4-PIPERIDYL)PROPIONANILIDE CITRATE see PHEN-
TANYL CITRATE

N-(1-PHENETHYL-4-PIPERIDYL)PROPIONANILIDE DIHYDROGEN CI-
TRATE see PHENTANYL CITRATE

1-PHENETHYL-4-N-PROPIONYLANILINOPIPERIDINE see PHENTANYL

N-PHENETHYL-4-(N-PROPIONYLANILINO)PIPERIDINE see PHENTANYL

p-PHENETOLCARBAMID (GERMAN) see 4-ETHOXY PHENYL UREA

p-PHENETOLCARBAMIDE see 4-ETHOXY PHENYL UREA

p-PHENETYLUREA see 4-ETHOXY PHENYL UREA

PHENIC ACID see PHENOL

(+-)-PHENISOPROPYLAMINE SULFATE see DL-AMPHETAMINE SUL-
FATE

PHENMETHYL-TRIMETHYLAMMONIUM IODIDE see BENZYL TRI-
METHYL AMMONIUM IODIDE

PHENOBARBITAL ELIXIR see SODIUM LUMINAL

PHENOBARBITAL SODIUM see SODIUM LUMINAL

PHENOCHLOR see POLYCHLORINATED BIPHENYLS

PHENOCLOR DP6 see POLYCHLORINATED BIPHENYL (AROCLOR 1260)

PHENODIOXIN see DIBENZO-p-DIOXIN

PHENOLCARBINOL see BENZYL ALCOHOL

PHENOLE (GERMAN) see PHENOL

PHENOL-GLYCIDAETHER (GERMAN) see PHENYLGLYCYDYL ETHER

PHENOL GLYCIDYL ETHER see PHENYLGLYCYDYL ETHER

PHENOL SODIUM SALT see SODIUM PHENOXIDE

PHENOL TRINITRATE see PICRIC ACID

PHENOPLASTE ORGANOL RED B see SCARLET RED

PHENOPROMIN see d-BENZEDRINE SULFATE

PHENOPYRIDINE see 8-QUINOLINOL

PHENOTHIAZINE HYDROCHLORIDE see CHLOROPROMAZINE HYDRO-CHLORIDE

10H-PHENOTHIAZINE-10-PROPANAMINE, N,N-DIMETHYL see PROMAZINE

N-(BETA-(10-PHENOTHIAZINYL)PROPYL)TRIMETHYLAMMONIUM METHYL SULFATE see MULTERGAN METHYL SULFATE

PHENOXAZOLE see 2-IMINO-5-PHENYL-4-OXAZOLIDINONE

PHENOXYBENZENE see DIPHENYL ETHER

3-PHENOXYBENZYL (+-)-3-(2,2-DICHLOROVINYL)-2,2-DIMETHYL-CYCLOPROPANECARBOXYLATE see AMBUSH

3-PHENOXY-1,2-EPOXYPROPANE see PHENYLGLYCYDYL ETHER

2-PHENOXYETHANOL see PHENYL CELLOSOLVE

2-PHENOXY-ETHANOL, ACRYLATE see PHENYL CELLOSOLVE ACRYLATE

2-(2-PHENOXYETHOXY)ETHANOL see PHENYL CARBITOL

PHENOXYETHYL ALCOHOL see PHENYL CELLOSOLVE

BETA-PHENOXYETHYLDIMETHYLDODECYLAMMONIUM BROMIDE see PHENODODECINIUM BROMIDE

N-PHENOXYISOPROPYL-N-BENZYL-BETA-CHLOROETHYLAMINE see NSC 37448

N-2-PHENOXYISOPROPYL-N-BENZYL-CHLOROETHYLAMINE HYDRO-CHLORIDE see DIBENZYLINE HYDROCHLORIDE

N-PHENOXYISOPROPYL-N-BENZYL-beta-CHLOROETHYLAMINE HYDRO-CHLORIDE see DIBENZYLINE HYDROCHLORIDE

(3-PHENOXYPHENYL)METHYL 3-(2,2-DICHLORETHENYL)-2,2-DIMETHYLCYCLOPROPANECARBOXYLATE see AMBUSH

PHENOXYPROPENE OXIDE see PHENYLGLYCYDYL ETHER

PHENOXYPROPYLENE OXIDE see PHENYLGLYCYDYL ETHER

PHENTIN ACETATE see ACETOXYTRIPHENYLSTANNANE

PHENTINOACETATE see ACETOXYTRIPHENYLSTANNANE

PHENTYRIN see 0-(4-(BIS(2-CHLOROETHYL)AMINO)PHENYL-DL-TYROSINE

N-PHENYLACETAMIDE see ACETANILIDE

PHENYLACETAMIDOPENICILLANIC ACID see BENZYL-6-AMINOPENICILLINIC ACID

P-PHENYLACETANILIDE see 4'-PHENYLACETANILIDE

N-PHENYLACETOACETAMIDE see ACETOACETANILIDE

2-PHENYLACETONITRILE see PHENYLACETONITRILE

3-(1'-PHENYL-2'-ACETYLETHYL)-4-HYDROXYCOUMARIN see COUMADIN

3-(alpha-PHENYL-beta-ACETYLETHYL)-4-HYDROXYCOUMARIN see COUMADIN

(PHENYL-1 ACETYL-2 ETHYL)-3-HYDROXY-4 COUMARINE (FRENCH) see COUMADIN

BETA-PHENYLAETHYLAMIN (GERMAN) see beta-PHENETHYLAMINE

5,5-PHENYL-AETHYL-3-(BETA-DIAETHYLAMINO-AETHYL)-2,4,6-TRI-OXO-HEXAHYDROPYRIMIDIN-HCl (GERMAN) see HEXAMID

PHENYLAETHYL-HYDRAZIN see PHENYLETHYLHYDRAZINE SULPHATE

D-PHENYLALANINE MUSTARD see PHENYLALANINE MUSTARD

L-PHENYLALANINE MUSTARD HYDROCHLORIDE see L-3-(p-(BIS(2-CHLOROETHYL)AMINO)PHENYL)ALANINE MONOHYDROCHLORIDE

PHENYLALANINE NITROGEN MUSTARD see L-PHENYLALANINE MUSTARD

PHENYLALANIN-LOST (GERMAN) see DL-PHENYLALANINE MUSTARD

PHENYLAMINE see ANILINE

1-PHENYL-2-AMINO-ATHAN (GERMAN) see beta-PHENETHYLAMINE

2-(PHENYLAMINO)BENZOIC ACID see N-PHENYLANTHRANILIC ACID

1-PHENYL-4-AMINO-5-CHLOROPYRIDAZIN-6-ONE see 1-PHENYL-4-AMINO-5-CHLORPYRIDAZ-6-ONE

1-PHENYL-4-AMINO-5-CHLOROPYRIDAZONE-6 see 1-PHENYL-4-AMINO-5-CHLORPYRIDAZ-6-ONE

1-PHENYL-4-AMINO-5-CHLORO-6-PYRIDAZONE see 1-PHENYL-4-AMINO-5-CHLORPYRIDAZ-6-ONE

2-PHENYL-1-AMINOCYCLOPROPANE, TRANS- see trans-2-PHENYLCYCLOPROPYLAMINE

4-PHENYLAMINODIPHENYLAMINE see 1,4-BIS(PHENYL AMINO)BENZENE

P-PHENYLAMINODIPHENYLAMINE see 1,4-BIS(PHENYL AMINO)BENZENE

1-PHENYL-2-AMINOETHANE see beta-PHENETHYLAMINE

2-(PHENYLAMINO)ETHANOL see 2-ANILINOETHANOL

2-PHENYLAMINONAPHTHALENE see N-PHENYL-2-NAPHTHYLAMINE

N-PHENYL-P-AMINOPHENOL see p-ANILINOPHENOL

1-PHENYL-2-AMINO-PROPAN (GERMAN) see AMPHETAMINE

D-1-PHENYL-2-AMINOPROPAN (GERMAN) see D-AMPHETAMINE

1-PHENYL-2-AMINOPROPANE see AMPHETAMINE

D-1-PHENYL-2-AMINOPROPANE see D-AMPHETAMINE

DL-1-PHENYL-2-AMINOPROPANE see BENZEDRINE

1-PHENYL-2-AMINOPROPANE SULFATE see BENZEDRINE SULFATE

D-1-PHENYL-2-AMINOPROPANE SULFATE see d-BENZEDRINE SULFATE

L-1-PHENYL-2-AMINOPROPANE SULFATE see l-BENZEDRINE SULFATE

N-PHENYLANILINE see DIPHENYLAMINE

O-PHENYLANILINE see 2-BIPHENYLAMINE

P-PHENYLANILINE see 4-BIPHENYLAMINE

PHENYLANTHRANILIC ACID see N-PHENYLANTHRANILIC ACID

PHENYL ARSENIC ACID see BENZENEARSONIC ACID

PHENYLARSINEDICHLORIDE see DICHLOROPHENYLARSINE

PHENYLARSONIC ACID see BENZENEARSONIC ACID

4-(PHENYLAZO)ANILINE see p-(PHENYLAZO)ANILINE

4-(PHENYLAZO)BENZENAMINE see p-(PHENYLAZO)ANILINE

4-(PHENYLAZO)-1,3-BENZENEDIAMINE MONOHYDROCHLORIDE see 4-PHENYLAZO-m-PHENYLENEDIAMINE

PHENYLAZODIAMINOPYRIDINE HYDROCHLORIDE see PHENAZOPYRIDINIUM CHLORIDE

3-PHENYLAZO-2,6-DIAMINOPYRIDINE HYDROCHLORIDE see PHENAZOPYRIDINIUM CHLORIDE

BETA-PHENYLAZO-ALPHA,ALPHA'-DIAMINOPYRIDINE HYDROCHLORIDE see PHENAZOPYRIDINIUM CHLORIDE

PHENYLAZO-ALPHA,ALPHA'-DIAMINOPYRIDINE MONOHYDROCHLORIDE see PHENAZOPYRIDINIUM CHLORIDE

1-(PHENYLAZO)-2-NAPHTHALENAMINE see 1-(PHENYLAZO)-2-NAPHTHYLAMINE

1-(PHENYLAZO)-2-NAPHTHALENOL see 1-(PHENYLAZO)-2-NAPHTHOL

1-PHENYLAZO-BETA-NAPHTHOL see 1-(PHENYLAZO)-2-NAPHTHOL

1-PHENYLAZO-2-NAPHTHOL-6,8-DISULFONIC ACID, DISODIUM SALT see HISPACID FAST ORANGE 2G

4-PHENYLAZOPHENOL see 4-HYDROXYAZOBENZENE

P-PHENYLAZOPHENOL see 4-HYDROXYAZOBENZENE

P-PHENYLAZOPHENYLAMINE see p-(PHENYLAZO)ANILINE

1-(4-PHENYLAZO-PHENYLAZO)-2-ETHYLAMINONAPHTHALENE see N-ETHYL-1-((p-(PHENYLAZO)PHENYL)AZO)-2-NAPHTHYLAMINE

(PHENYLAZO-4-PHENYLAZO)-1-ETHYLAMINO-2-NAPHTHALENE see N-ETHYL-1-((p-(PHENYLAZO)PHENYL)AZO)-2-NAPHTHYLAMINE

1-((4-(PHENYLAZO)PHENYL)AZO)-2-NAPHTHALENOL see OIL RED

1-(P-PHENYLAZOPHENYLAZO)-2-NAPHTHOL see OIL RED

4-(PHENYLAZO)-m-PHENYLENEDIAMINE MONOHYDROCHLORIDE see 4-PHENYLAZO-m-PHENYLENEDIAMINE

3-(PHENYLAZO)-2,6-PYRIDINEDIAMINE see PHENAZODINE

3-(PHENYLAZO)-2,6-PYRIDINEDIAMINE, HYDROCHLORIDE see PHENAZOPYRIDINIUM CHLORIDE

PHENYLAZOPYRIDINE HYDROCHLORIDE see PHENAZOPYRIDINIUM CHLORIDE

PHENYLBENZENE see BIPHENYL

2-PHENYL-BENZYL-AMINO-METHYLIMIDAZOLIN (GERMAN) see PHENAZOLINE

2-(N-PHENYL-N-BENZYLAMINOMETHYL) IMIDAZOLINE see PHENAZOLINE

8-(P-PHENYLBENZYL)ATROPINIUM BROMIDE see 8-(p-PHENYLBENZYL)ATROPINIUM BROMIDE

ALPHA-PHENYLBENZYLCYANIDE see DIPHENYLACETONITRILE

1-PHENYLBUTANE see n-BUTYLBENZENE

2-PHENYLBUTANE see sec-BUTYLBENZENE

PHENYLBUTYRIC ACID NITROGEN MUSTARD see CHLORAMBUCIL

N-PHENYLCARBAMATE D'ISO PROPYLE (FRENCH) see CARBANILIC ACID ISOPROPYL ESTER

PHENYLCARBINOL see BENZYL ALCOHOL

PHENYL CARBONATE see DIPHENYL CARBONATE

PHENYL CARBOXYLIC ACID see BENZOIC ACID

PHENYL CHLORIDE see BENZENE CHLORIDE

PHENYL CHLOROFORM see BENZYL TRICHLORIDE

PHENYL CHLOROMERCURY see PHENYL MERCURIC CHLORIDE

PHENYLCHLOROMETHYLKETONE see p-CHLOROACETOPHENONE

PHENYL CHLOROMETHYL KETONE see 2-CHLOROACETOPHENONE

PHENYLCHLOROMETHYLKETONE see 2-CHLOROACETOPHENONE

2-(2-PHENYL-2-(4-CHLOROPHENYL)ACETYL)-1,3-INDANDIONE see CHLOROPHACINONE

ALPHA-PHENYLCINNAMONITRILE see 2,3-DIPHENYLACRYLONITRILE

PHENYL CYANIDE see BENZONITRILE

1-(1-PHENYLCYCLOHEXYL)PIPERIDINE HYDROCHLORIDE see ANGEL DUST

1-PHENYL-1-CYCLOHEXYL-3-PIPERIDYL-1-PROPANOL HYDROCHLORIDE see BENZHEXOL HYDROCHLORIDE

TRANS,D,L-2-PHENYLCYCLOPROPYLAMINE SULFATE see PHENYLCYCLOPROMINE SULFATE

1-PHENYL-3,3-DIAETHYLTRIAZEN (GERMAN) see 1-PHENYL-3,3-DIETHYLTRIAZENE

2-PHENYLDIAZENECARBOXAMIDE see 1-CARBAMYL-2-PHENYL HYDRAZINE

PHENYLDIBROMOARSINE see DIBROMOPHENYLARSINE

PHENYL DICHLOROARSINE see DICHLOROPHENYLARSINE

PHENYL-5,6-DICHLORO-2-TRIFLUOROMETHYL-BENZIMIDAZOLE-1-CARBOXYLATE see 5,6-DICHLORO-1-PHENOXYCARBONYL-2-TRIFLUOROMETHYLBENZIMIDAZOLE

3-PHENYL-5-(BETA-(DIETHYLAMINO)ETHYL)-1,2,4-OXADIAZOLE CITRATE see OXOLAMINE CITRATE

1-PHENYL-2-DIETHYLAMINOPROPANONE-1 HYDROCHLORIDE see 2-DIETHYLAMINOPROPIOPHENONEHYDROCHLORIDE

1-PHENYL-2-DIETHYLAMINO-1-PROPANONE HYDROCHLORIDE see 2-DIETHYLAMINOPROPIOPHENONEHYDROCHLORIDE

PHENYLDIFLUOROARSINE see DIFLUOROPHENYLARSINE

1-PHENYL-2-DIMETHYLAMINOPROPANOL see METHYLEPHEDRINE

1-PHENYL-2,3-DIMETHYL-4-DIMETHYLAMINOPYRAZOL-5-ONE see DIMETHYLAMINOANTIPYRINE

1-PHENYL-2,3-DIMETHYL-4-DIMETHYLAMINOPYRAZOLONE-5 see DIMETHYLAMINOANTIPYRINE

1-PHENYL-2,3-DIMETHYL-4-ISOPROPYL-3-PYRAZOLIN-5-ONE see 4-ISOPROPYLANTIPYRINE

1-PHENYL-2,3-DIMETHYL-4-ISOPROPYLPYRAZOL-5-ONE see 4-ISOPROPYLANTIPYRINE

D-2-PHENYL-3,4-DIMETHYLMORPHOLINE HYDROCHLORIDE see 3,4-DIMETHYL-2-PHENYLMORPHOLINEHYDROCHLORIDE

1-PHENYL-2,3-DIMETHYLPYRAZOLE-5-ONE see ANTIPYRINE

1-PHENYL-2,3-DIMETHYL-5-PYRAZOLONE see ANTIPYRINE

1-PHENYL-2,3-DIMETHYL-5-PYRAZOLONE-4-METHYLAMINOMETHANESULFONATESODIUM see AMINOPYRINE SODIUM SULFONATE

1-PHENYL-2,3-DIMETHYLPYRAZOLONE-(5)-4-METHYLAMINOMETHANESULFONICACID SODIUM see AMINOPYRINE SODIUM SULFONATE

PHENYL DIMETHYL PYRAZOLON METHYL AMINOMETHANE SODIUM SULFONATE see AMINOPYRINE SODIUM SULFONATE

1-PHENYL-3,3-DIMETHYLTRIAZENE see 3,3-DIMETHYL-1-PHENYLTRIAZENE

N,N'-(M-PHENYLENE)BISMALEIMIDE see 1,3-BISMALEIMIDO BENZENE

2,2'-(1,3-PHENYLENEBIS(OXYMETHYLENE))BISOXIRANE see RESORCINOL DIGLYCIDYL ETHER

1,3-PHENYLENEDIAMINE see m-PHENYLENEDIAMINE

1,4-PHENYLENEDIAMINE see p-PHENYLENEDIAMINE

1,3-PHENYLENEDIAMINE DIHYDROCHLORIDE see m-PHENYLENEDIAMINE HYDROCHLORIDE

PHENYLENE-1,4-DIISOTHIOCYANATE see PHENYLENE THIOCYANATE

1,4-PHENYLENEDIISOTHIOCYANIC ACID see PHENYLENE THIOCYANATE

N,N'-(M-PHENYLENEDIMALEIMIDE) see 1,3-BISMALEIMIDO BENZENE

O-PHENYLENEDIOL see PYROCATECHOL

O-PHENYLENETHIOUREA see 2-BENZIMIDAZOLETHIOL

(−)-PHENYLEPHRINE see NEOSYNEPHRINE

D-(-)-PHENYLEPHRINE HYDROCHLORIDE see meta-SYNEPHRINE HYDROCHLORIDE

1-PHENYL-1,2-EPOXYETHANE see 1,2-EPOXYETHYLBENZENE

PHENYL-2,3-EPOXYPROPYL ETHER see PHENYLGLYCIDYL ETHER

PHENYLETHANE see ETHYL BENZENE

1-PHENYLETHANOL see alpha-PHENETHYL ALCOHOL

2-PHENYLETHANOL see PHENETHYL ALCOHOL

BETA-PHENYLETHANOL see PHENETHYL ALCOHOL

PHENYL ETHANOLAMINE see 2-ANILINOETHANOL

N-PHENYLETHANOLAMINE see 2-ANILINOETHANOL

1-PHENYLETHANONE see ACETOPHENONE

PHENYLETHENE see STYRENE

PHENYL ETHER see DIPHENYL ETHER

PHENYL ETHER-DIPHENYL MIXTURE see BIPHENYL, mixed with BIPHENYL OXIDE (3:7)

cis-2-(1-PHENYLETHOXY) CARBONYL-1-METHYLVINYL DIMETHYLPHOSPHATE see CROTOXYPHOS

BETA-PHENYLETHYL ACETATE see 2-PHENYLETHYL ACETATE

2-PHENYLETHYL ALCOHOL see PHENETHYL ALCOHOL

BETA-PHENYLETHYL ALCOHOL see PHENETHYL ALCOHOL

PHENYLETHYLAMINE see beta-PHENETHYLAMINE

2-PHENYLETHYLAMINE see beta-PHENETHYLAMINE

OMEGA-PHENYLETHYLAMINE see beta-PHENETHYLAMINE

PHENYLETHYLBARBITURATE see 5-ETHYL-5-PHENYLBARBITURIC ACID

PHENYL-ETHYL-BARBITURIC ACID see 5-ETHYL-5-PHENYLBARBITURIC ACID

5-PHENYL-5-ETHYLBARBITURIC ACID see 5-ETHYL-5-PHENYLBARBITURIC ACID

PHENYLETHYLBARBITURIC ACID, SODIUM SALT see SODIUM LUMINAL

PHENYLETHYL CARBAMATE see CARBANILIC ACID ETHYL ESTER

3-PHENYL-3-ETHYL-2,6-DIKETOPIPERIDINE see DORIDEN

3-PHENYL-3-ETHYL-2,6-DIOXOPIPERIDINE see DORIDEN

PHENYLETHYLENE see STYRENE

PHENYLETHYLENE OXIDE see 1,2-EPOXYETHYLBENZENE

2-PHENYL-2-ETHYLGLUTARIC ACID IMIDE see DORIDEN

alpha-PHENYL-alpha-ETHYLGLUTARIMIDE see DORIDEN

5-PHENYL-5-ETHYL-HEXAHYDROPYRIMIDINE-4,6-DIONE see 2-DESOXY PHENOBARBITAL

BETA-PHENYLETHYLHYDRAZINE see 2-PHENYLETHYLHYDRAZINE

BETA-PHENYLETHYLHYDRAZINE DIHYDROGEN SULFATE see PHENYLETHYLHYDRAZINE SULPHATE

2-PHENYLETHYLHYDRAZINE DIHYDROGEN SULPHATE see PHENYLETHYLHYDRAZINE SULPHATE

BETA-PHENYLETHYLHYDRAZINE HYDROGEN SULPHATE see PHENYLETHYLHYDRAZINE SULPHATE

BETA-PHENYLETHYLHYDRAZINE SULFATE see PHENYLETHYLHYDRAZINE SULPHATE

PHENYLETHYLMALONYLUREA see 5-ETHYL-5-PHENYLBARBITURIC ACID

5-PHENYL-5-ETHYL-3-METHYLBARBITURIC ACID see 5-ETHYL-N-METHYL-5-PHENYL BARBITURIC ACID

PHENYLETHYLMETHYLHYDANTOIN see 3-METHYL-5-ETHYL-5-PHENYLHYDANTOIN

PHENYLETHYNLCARBINOL CARBAMATE see 1-PHENYL-2-PROPYNYL CARBAMATE

PHENYLFLUOROFORM see BENZOTRIFLUORIDE

PHENYLFORMIC ACID see BENZOIC ACID

PHENYLGLYCOLIC ACID see MANDELIC ACID

PHENYLGLYOXYLONITRILE OXIME-O,O-DIETHYL PHOSPHOROTHIOATE see BAYTHION

2-PHENYLHYDRACRYLIC ACID-3-ALPHA-TROPANYL ESTER see ATROPINE

PHENYL HYDRATE see PHENOL

2-PHENYLHYDRAZIDE, CARBAMIC ACID see 1-CARBAMYL-2-PHENYL HYDRAZINE

PHENYLHYDRAZIN (GERMAN) see PHENYLHYDRAZINE

1-PHENYLHYDRAZINE CARBOXAMIDE see 1-CARBAMYL-2-PHENYL HYDRAZINE

2-PHENYLHYDRAZINECARBOXAMIDE see 1-CARBAMYL-2-PHENYL HYDRAZINE

PHENYLHYDRAZINE MONOHYDROCHLORIDE see PHENYLHYDRAZINE HYDROCHLORIDE

PHENYLHYDRAZIN HYDROCHLORID (GERMAN) see PHENYLHYDRAZINE HYDROCHLORIDE

PHENYLHYDRAZINIUM CHLORIDE see PHENYLHYDRAZINE HYDROCHLORIDE

1-PHENYL-2-HYDRAZINOPROPANE see PHENIPRAZINE

PHENYL HYDRIDE see BENZENE

PHENYL HYDROXIDE see PHENOL

PHENYLHYDROXYACETIC ACID see MANDELIC ACID

N-PHENYLHYDROXYLAMINE see beta-PHENYLHYDROXYLAMINE

1-PHENYL-2-(P-HYDROXYPHENYL)-3,5-DIOXO-4-BUTYLPYRAZOLIDINE see p-HYDROXYPHENYLBUTAZONE

PHENYLIC ACID see PHENOL

PHENYLIC ALCOHOL see PHENOL

5-PHENYL-2-IMINO-4-OXAZOLIDINONE see 2-IMINO-5-PHENYL-4-OXAZOLIDINONE

5-PHENYL-2-IMINO-4-OXOOXAZOLIDINE see 2-IMINO-5-PHENYL-4-OXAZOLIDINONE

PHENYL ISOHYDANTOIN see 2-IMINO-5-PHENYL-4-OXAZOLIDINONE

BETA-PHENYLISOPROPYLAMIN (GERMAN) see AMPHETAMINE

(PHENYLISOPROPYL)AMINE see AMPHETAMINE

BETA-PHENYLISOPROPYLAMINE see AMPHETAMINE

BETA-PHENYL ISOPROPYLAMINE SULFATE see DL-AMPHETAMINE SULFATE

D-BETA-PHENYLISOPROPYLAMINE SULFATE see d-BENZEDRINE SULFATE

N-PHENYL ISOPROPYL CARBAMATE see CARBANILIC ACID ISOPROPYL ESTER

PHENYLISOPROPYLHYDRAZINE see PHENIPRAZINE

BETA-PHENYLISOPROPYLHYDRAZINE see PHENIPRAZINE

3-(BETA-PHENYLISOPROPYL)-SIDNONIMINE HYDROCHLORIDE see SYDNOPHEN HYDROCHLORIDE

PHENYL ISOTHIOCYANATE see ISOTHIOCYANIC ACID, PHENYL ESTER

PHENYL KETONE see BENZOPHENONE

PHENYL MERCAPTAN see BENZENETHIOL

PHENYLMERCURIACETATE see ACETOXYPHENYLMERCURY

PHENYL MERCURIC ACETATE see ACETOXYPHENYLMERCURY

PHENYLMERCURIC 3,3'-METHYLENEBIS(2-NAPHTHALENESULFONATE) see PHENYLMERCURIC DINAPHTHYLMETHANEDISULFONATE

PHENYLMERCURIC NITRATE see MERCURIPHENYL NITRATE

PHENYLMERCURIC UREA see PHENYL MERCURY UREA

PHENYLMERCURIUREA see PHENYL MERCURY UREA

PHENYLMERCURY ACETATE see ACETOXYPHENYLMERCURY

PHENYLMERCURY CHLORIDE see PHENYL MERCURIC CHLORIDE

PHENYLMERCURY METHYLENEDINAPHTHALENESULFONATE see PHENYLMERCURIC DINAPHTHYLMETHANEDISULFONATE

PHENYLMERCURY NITRATE see MERCURIPHENYL NITRATE

PHENYLMETHANAL see BENZENECARBOXALDEHYDE

PHENYLMETHANE see TOLUENE

PHENYLMETHANETHIOL see alpha-TOLUENETHIOL

PHENYLMETHANOL see BENZYL ALCOHOL

PHENYLMETHYL ALCOHOL see BENZYL ALCOHOL

N-PHENYLMETHYLAMINE see METHYLANILINE

1-PHENYL-2-METHYLAMINE-PROPANOL-1-SULFATE see 1-EPHEDRINE SULFATE

1-PHENYL-2-METHYLAMINOPROPAN (GERMAN) see DESOXYN

1-PHENYL-2-METHYLAMINOPROPANE see DESOXYN

alpha-PHENYL-beta-METHYLAMINOPROPANE see DESOXYN

1-PHENYL-2-METHYLAMINOPROPANOL see EPHEDRINE

1-PHENYL-2-METHYLAMINOPROPANOL-1 see dl-EPHEDRINE HYDROCHLORIDE

3 PHENYL, 10-METHYL,7:8 BENZACRIDINE (FRENCH) see 7-METHYL-9-PHENYLBENZ(C)ACRIDINE

PHENYLMETHYLCARBINOL see alpha-PHENETHYL ALCOHOL

1-PHENYL-5-METHYL-8-CHLORO-1,2,4,5-TETRAHYDRO-2,4-DIOXO-

3H-1,5-BENZODIAZEPINE see 7-CHLORO-1-METHYL-5-PHENYL-1H-1,5-BENZODIAZEPINE-2,4(3H,5H)-DIONE

PHENYL METHYL ETHER see ANISOLE

1-PHENYL-3-METHYL-3-(2-HYDROXYAETHYL)-TRIAZEN (GERMAN) see 3-(2-HYDROXYETHYL)-3-METHYL-1-PHENYLTRIAZENE

1-PHENYL-3-METHYL-3-(2-HYDROXYETHYL)TRIAZENE see 3-(2-HYDROXYETHYL)-3-METHYL-1-PHENYLTRIAZENE

PHENYL METHYL KETONE see ACETOPHENONE

PHENYLMETHYL MERCAPTAN see alpha-TOLUENETHIOL

PHENYLMETHYLNITROSAMINE see N-METHYL-N-NITROSOANILINE

PHENYLMETHYLNITROSAMINE see N-METHYL-N-NITROSOANILINE

3-PHENYL-1-METHYL-1-NITROSOHARNSTOFF (GERMAN) see 1-METHYL-1-NITROSO-3-PHENYLUREA

(PHENYLMETHYL) PENICILLINIC ACID see BENZYL-6-AMINOPENICILLINIC ACID

1-PHENYL-3-METHYL-3-(2-SULFOAETHYL) NATRIUM SALZ (GERMAN) see N-PHENYLAZO-N-METHYLTAURINE SODIUM SALT

1-PHENYL-3-METHYL-3-(2-SULFOETHYL)TRIAZENE, SODIUM SALT see N-PHENYLAZO-N-METHYLTAURINE SODIUM SALT

PHENYLMETHYLVALERIANSAEURE-BETA-DIAETHYLAMINOAETHYLESTER-BROMMET HYLAT (GERMAN) see VALETHAMATE BROMIDE

PHENYLMONOGLYCOL ETHER see PHENYL CELLOSOLVE

PHENYL MUSTARD OIL see ISOTHIOCYANIC ACID, PHENYL ESTER

PHENYL-2-NAPHTHYLAMINE see N-PHENYL-2-NAPHTHYLAMINE

PHENYL-BETA-NAPHTHYLAMINE see N-PHENYL-2-NAPHTHYLAMINE

N-PHENYL-BETA-NAPHTHYLAMINE see N-PHENYL-2-NAPHTHYLAMINE

4-PHENYL-NITROBENZENE see p-NITROBIPHENYL

P-PHENYL-NITROBENZENE see p-NITROBIPHENYL

2-PHENYLOXIRANE see 1,2-EPOXYETHYLBENZENE

BETA-PHENYL-GAMMA-OXYPROPIONSAEURE-TROPYL-ESTER (GERMAN) see ATROPINE

PHENYL PERCHLORYL see HEXACHLOROBENZENE

2-PHENYLPHENOL see 2-BIPHENYLOL

4-PHENYLPHENOL see 4-BIPHENYLOL

O-PHENYLPHENOL see 2-BIPHENYLOL

P-PHENYLPHENOL see 4-BIPHENYLOL

ALPHA-PHENYLPHENYLACETONITRILE see DIPHENYLACETONITRILE

PHENYLPHOSPHONOTHIOIC ACID O-(2,4-DICHLOROPHENYL) O-ETHYL ESTER see S-SEVEN

ALPHA-PHENYL-2-PIPERIDINEACETIC ACID METHYL ESTER see METHYL PHENIDYL ACETATE

alpha-PHENYL-2-PIPERIDINEMETHANOL ACETATE HYDROCHLORIDE see LIDEPRAN HYDROCHLORIDE

1-PHENYL-1-(2-PIPERIDYL)-1-ACETOXYMETHANE HYDROCHLORIDE see LIDEPRAN HYDROCHLORIDE

PHENYL-(2-PIPERIDYL)METHYL ACETATE HYDROCHLORIDE see LIDEPRAN HYDROCHLORIDE

ALPHA-PHENYL-2-PIPERODINEACETIC ACID METHYL ESTER see METHYL PHENIDYL ACETATE

3-PHENYL-1-PROPANAL see HYDROCINNAMALDEHYDE

2-PHENYLPROPANE see CUMENE

PHENYLPROPANOLAMINE see (−)-NOREPHEDRINE

3-PHENYL PROPENAL see CINNAMMALDEHYDE

2-PHENYLPROPENE see alpha-METHYLSTYRENE

BETA-PHENYLPROPENE see alpha-METHYLSTYRENE

3-PHENYL-2-PROPENYLANTHRANILATE see ANTHRANILIC ACID, CINNAMYL ESTER

3-PHENYL-2-PROPEN-1-YL ANTHRANILATE see ANTHRANILIC ACID, CINNAMYL ESTER

3-PHENYLPROPIONALDEHYDE see HYDROCINNAMALDEHYDE

3-PHENYLPROPYL ALDEHYDE see HYDROCINNAMALDEHYDE

GAMMA-PHENYLPROPYLCARBAMAT (GERMAN) see 3-PHENYL-1-PROPANOL CARBAMATE

GAMMA-PHENYLPROPYL CARBAMATE see 3-PHENYL-1-PROPANOL CARBAMATE

2-PHENYLPROPYLENE see alpha-METHYLSTYRENE

BETA-PHENYLPROPYLENE see alpha-METHYLSTYRENE

PHENYLPSEUDOHYDANTOIN see 2-IMINO-5-PHENYL-4-OXAZOLIDINONE

1-PHENYL-1-(2-PYRIDYL)-3-DIMETHYL AMINOPROPANE see TRIMETON

1-PHENYL-1-(2-PYRIDYL)-3-DIMETHYL AMINOPROPANE HYDROCHLO-
RIDE see PROPHENPYRIDAMINE HYDROCHLORIDE

3-PHENYL-3-(2-PYRIDYL)-N,N-DIMETHYLPROPYLAMINE see TRIMETON

1-PHENYL-3-(P-2-PYRIDYLSULFAMOYLANILINO)-1,3-PROPANEDISUL-
FONIC ACID DISODIUM SALT see DISODIUM-2-(p-(gamma-PHENYL-
PROPYLAMINO)BENZENESULFONAMIDO) PYRIDINE

PHENYLQUECKSILBERACETAT (GERMAN) see ACETOXYPHENYLMER-
CURY

PHENYLQUECKSILBERCHLORID (GERMAN) see PHENYL MERCURIC
CHLORIDE

1-PHENYLSEMICARBAZIDE see 1-CARBAMYL-2-PHENYL HYDRAZINE

PHENYLSENFOEL (GERMAN) see ISOTHIOCYANIC ACID, PHENYL
ESTER

PHENYLSILICON TRICHLORIDE see TRICHLOROPHENYLSILANE

PHENYLSULFONIC ACID see BENZENESULFONIC ACID

4-(PHENYLSULFOXYETHYL)-1,2-DIPHENYL-3,5-PYRAZOLIDINEDIONE
see DIPHENYLPYRAZONE

6-PHENYL-2,3,5,6-TETRAHYDROIMIDAZO(2,1-B)THIAZOLE see LEVA-
MISOLE

PHENYLTHIOCARBAMIDE see 1-PHENYL-2-THIOUREA

PHENYLTHIOPHOSPHONATE DE O-ETHYLE ET O-4-NITROPHENYLE
(FRENCH) see ETHOXY-4-NITRO PHENOXY PHENYL PHOSPHINE SUL-
FIDE

1-PHENYLTHIOUREA see 1-PHENYL-2-THIOUREA

N-PHENYLTHIOUREA see 1-PHENYL-2-THIOUREA

PHENYLTOLOXAMINE HYDROCHLORIDE see BRISTAMIN HYDROCHLO-
RIDE

(N-PHENYL-P-TOLUENESULFONAMIDO)ETHYLMERCURY see ETHYL-
MERCURY-p-TOLUENE SULFONAMIDE

PHENYL-P-TOLYL KETONE see 4-METHYL-p-BENZOPHENONE

PHENYLTRICHLOROMETHANE see BENZYL TRICHLORIDE

PHENYL TRICHLOROSILANE (DOT) see TRICHLOROPHENYLSILANE

PHENYL UREA-P-DI(CARBOXYMETHYL) THIOARSENITE see 4-CARBA-
MIDOPHENYL BIS(CARBOXY METHYL THIO)ARSENITE

N-PHENYLURETHANE see CARBANILIC ACID ETHYL ESTER

ALPHA-PHENYL-VALERATE DU DIETHYLAMINO-ETHANOL CHLORHY-
DRATE (FRENCH) see 2-PHENYLVALERIC ACID-2-(DIETHYLAMINO)-
ETHYL ESTER HYDROCHLORIDE

2-PHENYL-tert-BUTYLAMINE see alpha,alpha-DIMETHYLPHENETHYL-
AMINE

PHENYTOIN SODIUM see DILANTIN

PHISOHEX see HEXACHLOROPHENE

PHLOROGLUCIN see PHLOROGLUCINOL

PHORBOL MONOACETATE MONOMYRISTATE see PHORBOL MYRI-
STATE ACETATE

PHORBOLOL ACETATE MYRISTATE see PHORBOLOL MYRISTATE ACE-
TATE

PHORBYOL see CASTOR OIL

PHORON (GERMAN) see PHORONE

CIS-PHOSDRIN see MEVINPHOS

PHOSGEN (GERMAN) see PHOSGENE

PHOSPHAMID see O,O-DIMETHYL METHYLCARBAMOYLMETHYL PHOS-
PHORODITHIOATE

PHOSPHAMIDON see DIMECRON

PHOSPHATE DE O,O-DIETHYLE ET DE O-2-CHLORO-1-(2,4-DICHLORO-
PHENYL) VINYLE (FRENCH) see CHLORFENVINFOS

PHOSPHATE DE DIETHYLE ET DE 3-METHYL-5-PYRAZOLYLE (FRENCH)
see METHYLPYRAZOLYL DIETHYLPHOSPHATE

PHOSPHATE DE O,O-DIMETHLE ET DE O-(1,2-DIBROMO-2,2-DICHLOR-
ETHYLE) (FRENCH) see DIMETHYL-1,2-DIBROMO-2,2-DICHLORO-
ETHYL PHOSPHATE

PHOSPHATE DE DIMETHYLE ET DE (2-CHLORO-2-DIETHYLCARBA-
MOYL-1-METHYL-VINYLE) (FRENCH) see DIMECRON

PHOSPHATE DE DIMETHYLE ET DE 2,2-DICHLOROVINYLE (FRENCH)
see DIMETHYL DICHLOROVINYL PHOSPHATE

PHOSPHATE DE DIMETHYLE ET DE 2-DIMETHYLCARBAMOYL 1-
METHYL VINYLE (FRENCH) see 3-HYDROXYDIMETHYL CROTON-
AMIDE DIMETHYL PHOSPHATE

PHOSPHATE DE DIMETHYLE ET DE 2-METHOXYCARBONYL-1
METHYLVINYLE (FRENCH) see MEVINPHOS

PHOSPHATE DE DIMETHYLE ET DE 2-METHYLCARBAMOYL 1-METHYL

VINYLE (FRENCH) see 3-(DIMETHOXYPHOSPHINYLOXY)N-METHYL-
cis-CROTONAMIDE

PHOSPHATE DE TRICRESYLE (FRENCH) see TRITOLYL PHOSPHATE

PHOSPHINE, NICKEL BIS(TRIPHENYL-, DITHIOCYANATE see BIS(TRI-
PHENYL PHOSPHINE)NICKEL DITHIO CYANATE

1,1',1''-PHOSPHINOTHIOYLIDYNETRISAZIRIDINE see THIOPHOSPHA-
MIDE

1,1',1''-PHOSPHINYLIDYNETRISAZIRIDINE see TRIETHYLENEPHOS-
PHOROTRIAMIDE

1,1',1''-PHOSPHINYLIDYNETRIS AZIRIDINE see TEPA

1,1',1'' -PHOSPHINYLIDYNETRIS(2-METHYL)AZIRIDINE see TRIS
(1-METHYLETHYLENE)PHOSPHORIC TRIAMIDE

PHOSPHORE BLANC (FRENCH) see PHOSPHORUS (WHITE)

PHOSPHORE(PENTACHLORURE DE) (FRENCH) see PHOSPHORUS PEN-
TACHLORIDE

PHOSPHORE(TRICHLORURE DE) (FRENCH) see PHOSPHORUS CHLORIDE

PHOSPHORIC ACID (DOT) see PHOSPHORIC ACID

PHOSPHORIC ACID, BERYLLIUM SALT (1:1) see BERYLLIUM HYDRO-
GEN PHOSPHATE (1:1)

PHOSPHORIC ACID, DIETHYL-(3-METHYL-5-PYRAZOLYL) ESTER see
METHYLPYRAZOLYL DIETHYLPHOSPHATE

PHOSPHORIC ACID, DIMETHYL ESTER, ESTER WITH CIS-3-HYDROXY-
N-METHYLCROTONAMIDE see 3-(DIMETHOXYPHOSPHINYLOXY)N-
METHYL-cis-CROTONAMIDE

PHOSPHORIC ACID DIMETHYL-P-(METHYLTHIO)PHENYL ESTER see 4-
METHYLTHIOPHENYLDIMETHYL PHOSPHATE

PHOSPHORIC ACID DIMETHYL-4-NITRO-M-TOLYL ESTER see 4-NITRO-
M-CRESOL DIMETHYL PHOSPHATE

PHOSPHORIC ACID, DISODIUM SALT see SODIUM MONOHYDROGEN
PHOSPHATE (2:1:1)

PHOSPHORIC ACID, LEAD (2+) SALT (2:3) see LEAD (II) PHOSPHATE
(3:2)

PHOSPHORIC ACID, TRI-O-CRESYL ESTER see TRI-2-TOLYL PHOS-
PHATE

PHOSPHORIC ACID TRIETHYLENE IMIDE see TRIETHYLENEPHOSPHO-
ROTRIAMIDE

PHOSPHORIC ACID TRIETHYLENEIMINE see TEPA

PHOSPHORIC ACID TRIETHYLENEIMINE (DOT) see TRIETHYLENEPHOS-
PHOROTRIAMIDE

PHOSPHORIC ACID, TRIS(2-METHYLPHENYL) ESTER see TRI-2-TOLYL
PHOSPHATE

PHOSPHORIC ANHYDRIDE see PHOSPHORUS PENTOXIDE

PHOSPHORIC BROMIDE see PHOSPHORUS PENTABROMIDE

PHOSPHORIC CHLORIDE see PHOSPHORUS PENTACHLORIDE

PHOSPHORIC SULFIDE see PHOSPHORUS PENTASULFIDE

PHOSPHORIC TRIS(DIMETHYLAMIDE) see HEXAMETHYL PHOSPHORA-
MIDE

PHOSPHORODIAMIDIC ACID, N,N-BIS(2-CHLOROETHYL)-N'-3-HY-
DROXYLPROPYLCYCLOHEXYLAMINE SALT see CYTOXAL ALCOHOL

PHOSPHORODITHIOIC ACID, O,O-DIETHYL ESTER, S,S-DIESTER WITH
METHANEDITHIOL see ETHYL METHYLENE PHOSPHORO DITHIOATE

PHOSPHORODITHIOIC ACID, S-((1,3-DIHYDRO-1,3-DIOXO-ISOINDOL-2-
YL)METHYL) O,O-DIMETHYL ESTER see PHTHALIMIDOMETHYL-
O,O-DIMETHYL PHOSPHORODITHIOATE

PHOSPHORODITHIOIC ACID, O,O-DIMETHYL ESTER, S-ESTER WITH DI-
ETHYL MERCAPTOSUCCINATE see CARBETHOXY MALATHION

PHOSPHORODITHIOIC ACID, O,O-DIMETHYL ESTER, S-ESTER WITH N-
(2-MERCAPTOETHYL)ACETAMIDE see O,O-DIMETHYL-S-(2
-(ACETYLAMINO)ETHYL) DITHIOPHOSPHATE

PHOSPHORODITHIOIC ACID, O,O-DIMETHYL-S-(2-(METHYLAMINO)-2-
OXOETHYL) ESTER (9CI) see O,O-DIMETHYL METHYLCARBAMOYL-
METHYL PHOSPHORODITHIOATE

PHOSPHORODITHIOIC ACID, O,O-DIMETHYL S-(MORPHOLINOCARBO-
NYLMETHYL) ESTER see MORPHOTHION

PHOSPHOROFLUORIDIC ACID,DIISOPROPYL ESTER see ISOPROPYL
PHOSPHOROFLUORIDATE

PHOSPHOROTHIOIC ACID,-O,O-DIETHYL ESTER, S-ESTER WITH ETHYL
MERCAPTOACETATE see O,O-DIETHYL S-(CARBETHOXY)METHYL
PHOSPHOROTHIOLATE

PHOSPHOROTHIOIC ACID-O,O-DIETHYL ESTER,-O-NAPHTHALIMIDO
DERIVATIVE see NAPHTHALOXIMIDODIETHYL THIOPHOSPHATE

PHOSPHOROTHIOIC ACID, O,O-DIETHYL O-(2-(ETHYLTHIO)ETHYL) ES-TER, MIXED WITH O,O-DIETHYL S-(2-(ETHYLTHIO)ETHYL) ESTER (7:3) see DEMETON-O + DEMETON-S

PHOSPHOROTHIOIC ACID-O,O-DIETHYL-O-NAPHTHYLAMIDOESTER see NAPHTHALOXIMIDODIETHYL THIOPHOSPHATE

PHOSPHOROTHIOIC ACID, O,O-DIETHYL O-2-PYRAZINYL ESTER see ETHYL PYRAZINYL PHOSPHOROTHIOATE

PHOSPHOROTHIOIC ACID, O,O-DIMETHYL S-ETHYLMETHYLSULFONIO-ETHYL ESTER, INNER SALT see O,O-DIMETHYL-S-ETHYLSUL-PHONIOETHYLMETHYL PHOSPHOROTHIOLATE

PHOSPHOROTHIOIC ACID, O,O-DIMETHYL S-(ETHYLSULFINYL-(2-ISO-PROPYL)) ESTER see METASYSTOX-S

PHOSPHOROTHIOIC ACID, O,O-DIMETHYL S-(2-(METHYLAMINO)-2-OXOETHYL) ESTER see DIMETHOATE OXYGEN ANALOG

PHOSPHOROTRITHIOUS ACID, S,S,S-TRIBUTYL ESTER see S,S,S-TRI-BUTYL TRITHIOPHOSPHITE

PHOSPHOROUS ACID, BERYLLIUM SALT see BERYLLIUM HYDROGEN PHOSPHATE (1:1)

PHOSPHOROUS OXYBROMIDE see PHOSPHORYL BROMIDE

PHOSPHOROUS TRIFLUORIDE see PHOSPHORUS FLUORIDE

PHOSPHORPENTACHLORID (GERMAN) see PHOSPHORUS PENTACHLO-RIDE

PHOSPHORSAEURELOESUNGEN (GERMAN) see PHOSPHORIC ACID

PHOSPHORTRICHLORID (GERMAN) see PHOSPHORUS CHLORIDE

PHOSPHORUS, AMORPHOUS, RED (DOT) see PHOSPHORUS (RED)

PHOSPHORUS(V) OXIDE see PHOSPHORUS PENTOXIDE

PHOSPHORUS OXYCHLORIDE see PHOSPHORYL CHLORIDE

PHOSPHOROUS OXYTRICHLORIDE see PHOSPHORYL CHLORIDE

PHOSPHORUS PERCHLORIDE see PHOSPHORUS PENTACHLORIDE

PHOSPHORUS PERSULFIDE see PHOSPHORUS PENTASULFIDE

PHOSPHORUS (III) SULFIDE (IV) see PHOSPHORUS SESQUISULFIDE

PHOSPHORUS TRICHLORIDE see PHOSPHORUS CHLORIDE

PHOSPHORUS TRIHYDRIDE see PHOSPHINE

PHOSPHORUS WHITE IN WATER see PHOSPHORUS (WHITE IN WATER)

PHOSPHORUS YELLOW IN WATER see PHOSPHORUS (WHITE IN WA-TER)

PHOSPHORWASSERSTOFF (GERMAN) see PHOSPHINE

PHOSPHORYL HEXAMETHYLTRIAMIDE see HEXAMETHYL PHOSPHOR-AMIDE

O-PHOSPHORYL-4-HYDROXY-N,N-DIMETHYLTRYPTAMINE see PSILO-CYBIN

PHOSPHORYL TRIBROMIDE see PHOSPHORYL BROMIDE

PHOSPHOSTIGMINE see PARATHION

PHOSPHOTHION see CARBETHOXY MALATHION

PHOSPHURE DE MAGNESIUM (FRENCH) see MAGNESIUM PHOSPHIDE

PHOSPHURE DE POTASSIUM (FRENCH) see POTASSIUM PHOSPHIDE

PHOSPHURE DE SODIUM (FRENCH) see SODIUM PHOSPHIDE

PHOSPHURE DE ZINC (FRENCH) see ZINC PHOSPHIDE

PHOSPHURES D'ALUMIUM (FRENCH) see ALUMINUM PHOSPHIDE

PHOSVEL see LEPTOPHOS

PHOSVIN see ZINC PHOSPHIDE

PHOTOMIREX see 8-MONOHYDRO MIREX

PHRILON see NYLON

1,3-PHTHALANDIONE see PHTHALIC ANHYDRIDE

1(2H)-PHTHALAZINONE HYDRAZONE HYDROCHLORIDE see HYDRALA-ZINE HYDROCHLORIDE

PHTHALIC ACID ANHYDRIDE see PHTHALIC ANHYDRIDE

PHTHALIC ACID DINITRILE see PHTHALONITRILE

PHTHALIC ACID METHYL ESTER see DIMETHYL PHTHALATE

O-PHTHALIC IMIDE see PHTHALIMIDE

PHTHALIMIDIMIDE see 1,3-DIIMINOISOINDOLINE

PHTHALIMIDO-O,O-DIMETHYL PHOSPHORODITHIOATE see PHTHALIMI-DOMETHYL-O,O-DIMETHYL PHOSPHORODITHIOATE

3-PHTHALIMIDOGLUTARIMIDE see THALIDOMIDE

ALPHA-(N-PHTHALIMIDO)GLUTARIMIDE see THALIDOMIDE

PHTHALODINITRILE see PHTHALONITRILE

M-PHTHALODINITRILE see 1,3-BENZENEDICARBONITRILE

O-PHTHALODINITRILE see PHTHALONITRILE

P-PHTHALODINITRILE see p-BENZENEDINITRILE

N-PHTHALOYLGLUTAMIMIDE see THALIDOMIDE

PHTHALSAEUREANHYDRID (GERMAN) see PHTHALIC ANHYDRIDE

PHTHALSAEUREDIMETHYLESTER (GERMAN) see DIMETHYL PHTHAL-ATE

N-PHTHALYLGLUTAMIC ACID IMIDE see THALIDOMIDE

N-PHTHALYL-GLUTAMINSAEURE-IMID (GERMAN) see THALIDOMIDE

ALPHA-N-PHTHALYLGLUTARAMIDE see THALIDOMIDE

PHYLLOCHINON (GERMAN) see VITAMIN K1

PHYLLOQUINONE see VITAMIN K1

ALPHA-PHYLLOQUINONE see VITAMIN K1

TRANS-PHYLLOQUINONE see VITAMIN K1

PHYTOMENADIONE see VITAMIN K1

SYMPHYTUM OFFICINALE L see RUSSIAN COMFREY ROOTS

PICLORAM see 4-AMINO-3,5,6-TRICHLOROPICOLINIC ACID

2-PICOLINE see 2-METHYL PYRIDINE

4-PICOLINE see 4-METHYLPYRIDINE

ALPHA-PICOLINE see 2-METHYL PYRIDINE

GAMMA-PICOLINE see 4-METHYLPYRIDINE

4-PICOLYLAMINE see 2-AMINO-4-METHYLPYRIDINE

PICRATE OF AMMONIA (DOT) see AMMONIUM PICRATE

PICRATOL see AMMONIUM PICRATE (WET)

PICRIC ACID, AMMONIUM SALT see AMMONIUM PICRATE (WET)

PICRIC ACID, AMMONIUM SALT see AMMONIUM PICRATE

PICRITE (THE EXPLOSIVE) see alpha-NITROGUANIDINE

PICRONITRIC ACID see PICRIC ACID

PICROTIN, COMPOUND WITH PICROTOXININ (1:1) see PICROTOXIN

PICROTOXINE see PICROTOXIN

PICRYLMETHYLNITRAMINE see TETRYL

PICRYLNITROMETHYLAMINE see TETRYL

PICRYL SULFIDE see BIS(TRINITROPHENYL)SULFIDE

PIECIOCHLOREK FOSFORU (POLISH) see PHOSPHORUS PENTACHLO-RIDE

PIED PIPER MOUSE SEED see STRYCHNINE

PIGMENT YELLOW 33 see CALCIUM CHROMATE (VI) DIHYDRATE

PIG-WRACK see CARRAGEEN

PIKRINEZUUR (DUTCH) see PICRIC ACID

PIKRINSAEURE (GERMAN) see PICRIC ACID

PIKRYNOWY KWAS (POLISH) see PICRIC ACID

PILOCARPOL see PILOCARPINE

PIMELIC KETONE see CYCLOHEXANONE

PINACOLOXYMETHYLPHOSPHORYL FLUORIDE see SOMAN

PINACOLYL METHYLFLUOROPHOSPHONATE see SOMAN

PINACOLYL METHYLPHOSPHONOFLUORIDATE see SOMAN

PINACOLYLOXY METHYLPHOSPHORYL FLUORIDE see SOMAN

PINANG see BETEL NUT

PINDON (DUTCH) see 2-PIVALOYL-1,3-INDANDIONE

ALPHA-PINENE see 2-PINENE

PIPENZOLATE METHYLBROMIDE see PIPTAL

PIPERAZIDINE see PIPERAZINE

PIPERAZIN (GERMAN) see PIPERAZINE

PIPERAZINE, ANHYDROUS see PIPERAZINE

2,6-PIPERAZINEDIONE-4,4'-PROPYLENE DIOXOPIPERAZINE see NSC-129943

1,4-PIPERAZINEDIYLBIS(BIS(1-AZIRIDINYL)PHOSPHINE OXIDE see 1,4-BIS(N,N'-DIETHYLENE PHOSPHAMIDE)PIPERAZINE

PIPERIDIN (GERMAN) see PIPERIDINE

PIPERIDINE, 1-(2-(4-BUTOXYBENZOYL)ETHYL), HYDROCHLORIDE see 4'-BUTOXY-3-PIPERIDINO PROPIOPHENONE HYDROCHLORIDE

PIPERIDINIUM see PIP-PIP

3-PIPERIDINO-1,1-DIPHENYL-1-PROPANOL HYDROCHLORIDE see PRI-DINOL

2-PIPERIDINOMETHYL-1,4-BENZODIOXAN HYDROCHLORIDE see BEN-ZODIOXANE HYDROCHLORIDE

1-PIPERIDINO-3-(P-OCTYLPHENYL)-3-PROPANONE HYDROCHLORIDE see PIPOCTANONE HYDROCHLORIDE

1-PIPERIDINO-3-(4'-OCTYLPHENYL)-PROPAN-3-ON-HYDROCHLORID (GERMAN) see PIPOCTANONE HYDROCHLORIDE

3-PIPERIDINO-1-PHENYL-1-BICYCLOHEPTENYL-1-PROPANOL see BI-PERIDEN

3-PIPERIDINO-1-PHENYL-1-BICYCLOHEPTENYL -1-PROPANOL see BI-PERIDEN

3-PIPERIDINO-1-PHENYL-1-BICYCLO (2.2.1)HEPTEN-(5)-YL-PROPANOL-(1)(GERMAN) see BIPERIDEN

4-(3-PIPERIDINOPROPIONAMIDO) SALICYCLIC ACID METHYL ESTER, METHIODIDE see METHYL-4-(3-PIPERIDINOPROPIONYLAMINO)SALICYLATE, METHIODIDE

ALPHA-(2-PIPERIDYL)BENZHYDROL see alpha,alpha-DIPHENYL-2-PIPERIDINEMETHANOL

ALPHA-(2-PIPERIDYL)BENZHYDROL HYDROCHLORIDE see PIPERADROL HYDROCHLORIDE

3-(1-PIPERIDYL)-1-CYCLOHEXYL-1-PHENYL-1-PROPANOL HYDROCHLORIDE see BENZHEXOL HYDROCHLORIDE

3-(N-PIPERIDYL)-1,1-DIPHENYL-1-PROPANOL HYDROCHLORIDE see PRIDINOL

ALPHA-(2-PIPERIDYLETHYL) BENZHYDROL HYDROCHLORIDE see PRIDINOL

2-(1-PIPERIDYLMETHYL)-1,4-BENZODIOXAN HYDROCHLORIDE see BENZODIOXANE HYDROCHLORIDE

PIPEROCAINE HYDROCHLORIDE see ISOCAINE

PIPERONALDEHYDE see PIPERONAL

PIPERONYL see FLUOROBUTYROPHENONE

PIPERONYL ALDEHYDE see PIPERONAL

PIPERONYL SULFOXIDE see ISOSAFROLE-n-OCTYLSULFOXIDE

PIPEROXANE HYDROCHLORIDE see BENZODIOXANE HYDROCHLORIDE

ALPHA-PIPRADOL see alpha,alpha-DIPHENYL-2-PIPERIDINEMETHANOL

PIPRADOL HYDROCHLORIDE see PIPERADROL HYDROCHLORIDE

PIPRADROL HYDROCHLORIDE see PIPERADROL HYDROCHLORIDE

PIRAZOXON (ITALIAN) see METHYLPYRAZOLYL DIETHYLPHOSPHATE

PIRIA'S ACID see 4-AMINO-1-NAPHTHALENESULFONIC ACID

PIRIDINA (ITALIAN) see PYRIDINE

PIRYDYNA (POLISH) see PYRIDINE

PISCIDEIN see (p-HYDROXYBENZYL)TARTARIC ACID

PISCIDIC ACID see (p-HYDROXYBENZYL)TARTARIC ACID

PIVALDION (ITALIAN) see 2-PIVALOYL-1,3-INDANDIONE

PIVALDIONE (FRENCH) see 2-PIVALOYL-1,3-INDANDIONE

PIVALIC ACID CHLORIDE see 2,2-DIMETHYLPROPANOYL CHLORIDE

PIVALIC ACID LACTONE see 3,3-DIMETHYL-2-OXETHANONE

PIVALOLYL CHLORIDE see 2,2-DIMETHYLPROPANOYL CHLORIDE

PIVALOYL CHLORIDE (8CI) see 2,2-DIMETHYLPROPANOYL CHLORIDE

2-PIVALOYL-INDAAN-1,3-DION (DUTCH) see 2-PIVALOYL-1,3-INDANDIONE

2-PIVALOYL-INDAN-1,3-DION (GERMAN) see 2-PIVALOYL-1,3-INDANDIONE

2-PIVALOYLINDANE-1,3-DIONE see 2-PIVALOYL-1,3-INDANDIONE

PIVALYL CHLORIDE see 2,2-DIMETHYLPROPANOYL CHLORIDE

2-PIVALYL-1,3-INDANDIONE see 2-PIVALOYL-1,3-INDANDIONE

PLANOCHROME see MERCUROCHROME

PLANT PROTEASE CONCENTRATE see BROMELAIN

PLATIN (GERMAN) see PLATINUM

PLATINIC AMMONIUM CHLORIDE see AMMONIUM CHLOROPLATINATE

PLATINIC CHLORIDE see CHLOROPLATINIC ACID

PLATINOUS CHLORIDE see PLATINUM CHLORIDE

PLATINUM BLACK see PLATINUM

CIS-PLATINUM(II) DIAMINEDICHLORIDE see cis-PLATINOUS DIAMINE DICHLORIDE

PLATINUM SPONGE see PLATINUM

PLATINUM TETRACHLORIDE see PLATINUM (IV) CHLORIDE

PLEGECYL see ACETOPROMAZINE

PLEGICIN see ACETOPROMAZINE

PLEXIGLAS see POLYMETHYLMETHACRYLATE

PLIVAPHEN see ACETOPROMAZINE

PLOMB FLUORURE (FRENCH) see LEAD (II) FLUORIDE

PLUMBOPLUMBIC OXIDE see LEAD OXIDE RED

PLUMBOUS ACETATE see LEAD ACETATE

PLUMBOUS CHLORIDE see LEAD CHLORIDE

PLUMBOUS CHROMATE see LEAD CHROMATE

PLUMBOUS FLUORIDE see LEAD (II) FLUORIDE

PLUMBOUS OXIDE see LEAD MONOXIDE

PLUMBOUS PHOSPHATE see LEAD (II) PHOSPHATE (3:2)

PLUMBOUS SULFIDE see LEAD SULFIDE

PLUTONIUM NITRATE SOLUTION (DOT) see PLUTONIUM NITRATE (SOLUTION)

PMT see 1-PHENYL-3-MONOMETHYL-TRIAZENE

P,M-TOLYLENEDIAMINE see TOLUENE-2,5-DIAMINE

PODOPHYLLUM see PODOPHYLLIN

PODOPHYLLUM RESIN see PODOPHYLLIN

POLAMIDON see L-METHADONE HYDROCHLORIDE

POLAMIDONE see METHADONE

POLYACRYLONITRILE see ACRYLONITRILE, POLYMER WITH 1,3-BUTADIENE, AND STYRENE, COMBUSTION PRODUCTS

POLYACRYLONITRILE WITH 1,3-BUTADIENE AND ETHENYLBENZENE see ACRYLONITRILE, POLYMER WITH 1,3-BUTADIENE AND STYRENE

POLYAMID (GERMAN) see NYLON

POLY(EPSILON-AMINOCAPROIC ACID) see POLYCAPROLACTAM

POLYBREME see HEXADIMETHRINE BROMIDE

POLYBROMINATED BIPHENYL see FIREMASTER FFI

POLYBROMINATED BIPHENYLS see FIREMASTER BP-6

POLYBROMOETHYLENE see POLYVINYLBROMIDE

POLYCAPROAMIDE see POLYCAPROLACTAM

POLY(EPSILON-CAPROAMIDE) see POLYCAPROLACTAM

POLY(EPSILON-CAPROLACTAM) see POLYCAPROLACTAM

POLYCHLORCAMPHENE see TOXAPHENE

POLYCHLORINATED BIPHENYL see POLYCHLORINATED BIPHENYLS

POLYCHLORINATED CAMPHENES see TOXAPHENE

POLYCHLOROBIPHENYL see POLYCHLORINATED BIPHENYLS

POLY(2-CHLOROBUTADIENE) see 2-CHLORO-1,3-BUTADIENE POLYMER

POLY(2-CHLORO-1,3-BUTADIENE) see 2-CHLORO-1,3-BUTADIENE POLYMER

POLYCHLOROCAMPHENE see TOXAPHENE

POLY(CHLOROETHYLENE) see BAKELITE

POLYCHLOROPRENE see 2-CHLORO-1,3-BUTADIENE POLYMER

POLYCRON see O-(4-BROMO-2-CHLOROPHENYL)-O-ETHYL-S-PROPYL PHOSPHOROTHIOATE

POLYCYCLIC MUSK see ACETYL ETHYL TETRAMETHYL TETRALIN

POLYCYCLIC MUSK see 3'-ETHYL-5',6',7',8'-TETRAHYDRO-5',5',8',8'-TETRAMETHYL-2'-ACETONAPHTHONE

POLYCYCLINE see TETRACYCLINE

POLYCYCLINE HYDROCHLORIDE see TETRACYCLINE HYDROCHLORIDE

POLY(ESTRADIOL PHOSPHATE) see ESTRADIOL POLYESTER WITH PHOSPHORIC ACID

POLYETHYLENE AS see POLYETHYLENE

POLYETHYLENE GLYCOL 400 see POLYETHYLENE GLYCOL

POLYETHYLENE GLYCOL 300 DISTEARATE see POLYETHYLENE GLYCOL DISTEARATE

POLYETHYLENE GLYCOL 400 (DI) STEARATE see POLYETHYLENE GLYCOL DISTEARATE

POLYETHYLENE GLYCOL 600 (DI) STEARATE see POLYETHYLENE GLYCOL DISTEARATE

POLYETHYLENE GLYCOLMONOETHER WITH P-TERT-OCTYLPHENYL see TRITON X

POLYETHYLENE GLYCOL MONO(4-OCTYLPHENYL) ETHER see TRITON X

POLYETHYLENE GLYCOL MONO(4-TERT-OCTYLPHENYL) ETHER see TRITON X

POLYETHYLENE GLYCOLMONO(P-(1,1,3,3-TETRAMETHYLBUTYL)-PHENYL) ETHER see TRITON X

POLYETHYLENE GLYCOL 450 NONYL PHENYL ETHER see PEG-9 NONYL PHENYL ETHER

POLYETHYLENE GLYCOL OCTYLPHENOL ETHER see TRITON X

POLYETHYLENE GLYCOL P-OCTYLPHENYL ETHER see TRITON X

POLYETHYLENE GLYCOL 450 OCTYL PHENYL ETHER see TRITON X

POLYETHYLENE GLYCOL P-TERT-OCTYLPHENYL ETHER see TRITON X

POLYETHYLENE GLYCOL P-1,1,3,3-TETRAMETHYLBUTYLPHENYL ETHER see TRITON X

POLY(ETHYLENE TETRAFLUORIDE) see TEFLON

POLYFOAM PLASTIC SPONGE see POLYURETHANE FOAM

POLYFOAM SPONGE see POLYURETHANE FOAM

POLYGLYCOL DISTEARATE see POLYETHYLENE GLYCOL DISTEARATE

POLY(IMINOCARBONYLPENTAMETHYLENE) see POLYCAPROLACTAM

POLY(IMINO(1-OXO-1,6-HEXANEDIYL)) see POLYCAPROLACTAM

POLYMERS OF EPICHLOROHYDRIN AND 2,2-BIS(4-HYDROXY PHENYL)PIPERAZINE see EPOXY RESINS, UNCURED

POLY(METHYLENE PHENYLENE ISOCYANATE) see POLYMETHYLENE-POLYPHENYL ISOCYANATE

POLYMETHYLENE POLYPHENYLENE ISOCYANATE see POLYMETHY-LENEPOLYPHENYL ISOCYANATE

POLYMETHYLENEPOLYPHENYLENE ISOCYANATE POLYMER see POLY-METHYLENEPOLYPHENYL ISOCYANATE

POLYMETHYLENEPOLYPHENYLENE POLYISOCYANATE see POLYMETH-YLENEPOLYPHENYL ISOCYANATE

POLYMETHYLENE POLY(PHENYL ISOCYANATE) see POLYMETHYLENE-POLYPHENYL ISOCYANATE

POLYMETHYLENE POLYPHENYL POLYISOCYANATE see POLYMETHY-LENEPOLYPHENYL ISOCYANATE

POLY(METHYL METHACRYLATE) see POLYMETHYLMETHACRYLATE

POLYMETHYL POLYPHENYLPOLYISOCYANATE see POLYMETHYLENE-POLYPHENYL ISOCYANATE

POLYOXYETHYLENE MONO(OCTYLPHENYL) ETHER see TRITON X

POLYOXYETHYLENE-8-MONOSTEARATE see POLYETHYLENE GLYCOL MONOSTEARATE

POLYOXYETHYLENE (9) NONYL PHENYL ETHER see PEG-9 NONYL PHENYL ETHER

POLYOXYETHYLENE (9) OCTYLPHENYL ETHER see TRITON X

POLYOXYETHYLENE (13) OCTYLPHENYL ETHER see TRITON X

POLY(OXYETHYLENE)P-TERT-OCTYLPHENYL ETHER see TRITON X

POLYOXYETHYLENE SORBITAN MONOSTEARATE see TWEEN 60

POLYOXYETHYLENE(8)STEARATE see POLYETHYLENE GLYCOL MONOSTEARATE

POLYOXYMETHYLENE see s-TRIOXANE

POLY(PHENYLENEMETHYLENEISOCYANATE) see POLYMETHYLENE POLYPHENYL ISOCYANATE

POLYPHENYLENE POLYMETHYLENE POLYISOCYANATE see POLY-METHYLENEPOLYPHENYL ISOCYANATE

POLYPHENYLPOLYMETHYLENE POLYISOCYANATE see POLYMETHY-LENEPOLYPHENYL ISOCYANATE

POLYPROPENE see PROPENE POLYMERS

POLYPROPYLENE see PROPENE POLYMERS

POLYSILICONE see POLYDIMETHYLSILOXANE RUBBER

POLY-SOLV see CARBITOL CELLOSOLVE

POLY-SOLV EE see CELLOSOLVE SOLVENT

POLY-SOLV EE ACETATE see "CELLOSOLVE" ACETATE

POLYSTYRENE-ACRYLONITRILE see ACRYLONITRILE POLYMER WITH STYRENE

POLYTETRAFLUOROETHENE see TEFLON

POLYTETRAFLUOROETHYLENE see TEFLON

POLYURETHANE ESTER FOAM see POLYURETHANE FOAM

POLYURETHANE ETHER FOAM see POLYURETHANE FOAM

POLYURETHANE SPONGE see POLYURETHANE FOAM

POLY(VINYL ALCOHOL) see POLYVINYL ALCOHOL

POLY(VINYLBROMIDE) see POLYVINYLBROMIDE

POLYVINYLCHLORID (GERMAN) see BAKELITE

POLY(VINYL CHLORIDE) see BAKELITE

POLYVINYL CHLORIDE see BAKELITE

POLYVINYLCHLORIDE ACETATE see POLYVINYL ACETATE CHLORIDE

POLYWAX 1000 see POLYETHYLENE

POMME EPINEUSE (FRENCH) see STRAMONIUM

PONTACYL FAST BLUE R see ACID BLUE 92

PONTACYL GREEN BL see ACID GREEN 3

PORPHYROMYCIN see N-METHYLMITOMYCIN C

PORTLAND STONE see CALCIUM (II) CARBONATE (1:1)

POTASH see POTASSIUM CARBONATE (2:1)

POTASH CHLORATE (DOT) see POTASSIUM CHLORATE

POTASSA see POTASSIUM HYDROXIDE

POTASSE CAUSTIQUE (FRENCH) see POTASSIUM HYDROXIDE

POTASSIO (CHLORATO DI) (ITALIAN) see POTASSIUM CHLORATE

POTASSIO (IDROSSIDO DI) (ITALIAN) see POTASSIUM HYDROXIDE

POTASSIO (PERMANGANATO DI) (ITALIAN) see POTASSIUM PERMAN-GANATE

POTASSIUM ACID ARSENATE see ARSENIC ACID, MONOPOTASSIUM SALT

POTASSIUM ANTIMONYL-D,L-TARTRATE see dl-ANTIMONY POTAS-SIUM TARTRATE

POTASSIUM ANTIMONYL TARTRATE see ANTIMONY POTASSIUM TAR-TRATE

POTASSIUM ANTIMONYL-D-TARTRATE see ANTIMONY POTASSIUM TARTRATE

POTASSIUM ANTIMONYL-MESO-TARTRATE see meso-ANTIMONY PO-TASSIUM TARTRATE

POTASSIUM ANTIMONY TARTRATE see ANTIMONY POTASSIUM TAR-TRATE

POTASSIUM ARSENATE see ARSENIC ACID, MONOPOTASSIUM SALT

POTASSIUM BIFLUORIDE see POTASSIUM ACID FLUORIDE

POTASSIUM BROMATE (DOT) see BROMIC ACID, POTASSIUM SALT

POTASSIUM CHLORATE (DOT) see POTASSIUM CHLORATE

POTASSIUM (CHLORATE DE) (FRENCH) see POTASSIUM CHLORATE

POTASSIUM CHLOROPLATINITE see PLATINOUS POTASSIUM CHLORIDE

POTASSIUM CYANIDE, SOLID (DOT) see POTASSIUM CYANIDE (SOLID)

POTASSIUM CYANIDE SOLUTION (DOT) see POTASSIUM CYANIDE (soln)

POTASSIUM DICHLORO-S-TRIAZINETRIONE see POTASSIUM DI-CHLOROISOCYANURATE

POTASSIUM DICHROMATE (VI) see POTASSIUM BICHROMATE

POTASSIUM DIHYDROGEN ARSENATE see ARSENIC ACID, MONOPO-TASSIUM SALT

POTASSIUM FLUORIDE (DOT) see POTASSIUM FLUORIDE (SOLID)

POTASSIUM FLUORURE (FRENCH) see POTASSIUM FLUORIDE (SOLID)

POTASSIUM HYDRATE see POTASSIUM HYDROXIDE

POTASSIUM HYDROGEN ARSENATE see ARSENIC ACID, MONOPOTAS-SIUM SALT

POTASSIUM HYDROGEN FLUORIDE see POTASSIUM ACID FLUORIDE

POTASSIUM HYPERCHLORIDE see POTASSIUM PERCHLORATE

POTASSIUM MERCURIC IODIDE see NESSLER REAGENT

POTASSIUM METAARSENITE see POTASSIUM ARSENITE

POTASSIUM, METAL (DOT) see POTASSIUM

POTASSIUM, METAL LIQUID ALLOY (DOT) see POTASSIUM (LIQUID ALLOY)

POTASSIUM MONOSULFIDE see POTASSIUM SULFIDE (2:1)

POTASSIUM NITRITE (DOT) see POTASSIUM NITRITE (1:1)

POTASSIUM OXYMURIATE see POTASSIUM CHLORATE

POTASSIUM PERMANGANATE (DOT) see POTASSIUM PERMANGANATE

POTASSIUM (PERMANGANATE DE) (FRENCH) see POTASSIUM PER-MANGANATE

POTASSIUM PEROXYDISULFATE see DIPOTASSIUM PERSULFATE

POTASSIUM PEROXYDISULPHATE see DIPOTASSIUM PERSULFATE

POTASSIUM PERSULFATE see DIPOTASSIUM PERSULFATE

POTASSIUM PLATINOCHLORIDE see PLATINOUS POTASSIUM CHLO-RIDE

POTASSIUM RHODANATE see POTASSIUM THIOCYANATE

POTASSIUM RHODANIDE see POTASSIUM THIOCYANATE

POTASSIUM SALT OF POLYVINYL SULFATE see POLYVINYL SULFATE, POTASSIUM SALT

POTASSIUM SULFOCYANATE see POTASSIUM THIOCYANATE

POTASSIUM TETRACHLOROPLATINATE(II) see PLATINOUS POTASSIUM CHLORIDE

POTASSIUM TETRAIODOMERCURATE (II) see NESSLER REAGENT

POTASSIUM THIOCYANIDE see POTASSIUM THIOCYANATE

POTATO ALCOHOL see ETHANOL

POUNCE see AMBUSH

POX see PHOSPHORUS PENTOXIDE

P.P. FACTOR-PELLAGRA PREVENTIVE FACTOR see NICOTINIC ACID

PRAYER BEAD see ABRIN

PRECIPITATED BARIUM SULPHATE see BARIUM SULFATE

PRECIPITATED CALCIUM SULFATE see CALCIUM (II) SULFATE DIHY-DRATE (1:1:2)

PRECIPITATED SILICA see SILICA, PRECIPITATED (AMORPHOUS)

PRECIPITATED SULFUR see SULFUR

PREDNISOLONE see PREDONIN

1,4-PREGNADIENE-17-ALPHA,21- DIOL-3,11,20-TRIONE see PREDNI-
SONE

1,4-PREGNADIENE-3,20-DIONE-11-BETA,17-ALPHA,21-TRIOL see
PREDONIN

1,4-PREGNADIENE-11-BETA,17-ALPHA,21-TRIOL-3,20-DIONE see
PREDONIN

3,20-PREGNENE-4 see PROGESTERONE

17-ALPHA-PREGN-4-ENE-21-CARBOXYLIC ACID, 1-HYDROXY-7-
ALPHA-MERCAPTO-3-OXO-ALPHA-LACTONE see ALDACTAZIDE

PREGNENEDIONE see PROGESTERONE

PREGNENE-3,20-DIONE see PROGESTERONE

4-PREGNENE-3,20-DIONE see PROGESTERONE

PREGN-4-ENE-3,20-DIONE see PROGESTERONE

DELTA(SUP 4)-PREGNENE-3,20-DIONE see PROGESTERONE

4-PREGNENE-11-beta,17-alpha,21-TRIOL 3,20-DIONE see CORTISOL

PRESSOMIN HYDROCHLORIDE see METHOXAMINE HYDROCHLORIDE

PREZA see HEROIN

PRIMACAINE HYDROCHLORIDE see 3-AMINO-2-BUTOXYBENZOIC
ACID-2-DIETHYLAMINOETHYL ESTER HYDROCHLORIDE

PRIMARY AMYL ACETATE see PENTYL ACETATE

PRIMARY AMYL ALCOHOL see PENTYL ALCOHOL

PRIMARY AMYL ALCOHOL see AMYL ALCOHOL

PRIMARY DECYL ALCOHOL see DECYL ALCOHOL

PRIMARY OCTYL ALCOHOL see N-OCTYL ALCOHOL

PRIMROSE YELLOW see BASIC ZINC CHROMATE

PROBARBITAL see ETHYL ISOPROPYL BARBITURIC ACID

PROBARBITONE see ETHYL ISOPROPYL BARBITURIC ACID

PROBENECID SODIUM SALT see p-(DIPROPYLSULFAMOYL)BENZOIC
ACID SODIUM SALT

PROCAINAMIDE see p-AMINO-n-(2-DIETHYLAMINOETHYL)BENZAMIDE

PROCAINE see p-AMINOBENZOIC ACID-2-DIETHYLAMINOETHYL ESTER

PROCAINE HYDROCHLORIDE see p-AMINOBENZOYLDIETHYLAMINO-
ETHANOL HYDROCHLORIDE

PROCARBAZIN (GERMAN) see 1-METHYL-2-(p-(ISOPROPYLCARBA-
MOYL)BENZYL)HYDRAZINE

PROCHLOROPERAZINE see PROCHLORPROMAZINE

PROCHLOROPROAZINE HYDROGEN MALEATE see PROCHLORPERIZINE
MALEATE

PROCHLORPEMAZINE see PROCHLORPROMAZINE

PROCHLORPERAZINE see PROCHLORPROMAZINE

PROCHLORPERAZINE BIMALEATE see PROCHLORPERIZINE MALEATE

PROCHLORPERAZINE DIMALEATE see PROCHLORPERIZINE MALEATE

PROCHLORPERAZINE HYDROGEN MALEATE see PROCHLORPERIZINE
MALEATE

PROCHLORPERAZINE MALEATE see PROCHLORPERIZINE MALEATE

PROCYCLIDINE see 1-CYCLOHEXYL-1-PHENYL-3-PYRROLIDINO-1-
PROPANOL

PRODILIDINE see 1,2-DIMETHYL-3-PHENYL-3-PYRROLIDYLPROPIO-
NATE

PROFAMINA see BENZEDRINE

PROFERRIN see IRON OXIDE, SACCHARATED

PROFLAVINE see 3,6-DIAMINOACRIDINIUM

PROFLAVINE HEMISULFATE see 3,6-DIAMINOACRIDINE SULPHATE
(1:1)

PROFLAVINE HYDROCHLORIDE see PROFLAVINE MONOHYDROCHLO-
RIDE

PROGESTEROL see PROGESTERONE

BETA-PROGESTERONE see PROGESTERONE

PROGESTERONE CAPROATE see HYDROXYPROGESTERONE CAPROATE

PROHEPTADIEN MONOHYDROCHLORIDE see ELAVIL HYDROCHLORIDE

PROMAMIDE see KERB

PRONAMIDE see KERB

PROPADRINE see (−)-NOREPHEDRINE

PROPANALOL see INDERAL

PROPANAMINE see PROPYLAMINE

1-PROPANECARBOXYLIC ACID see N-BUTYRIC ACID

PROPANEDINITRILE, ((2-CHLOROPHENYL)METHYLENE) see
(O-CHLORO BENZAL)MALONO NITRILE

PROPANEDIOIC ACID, DIETHYL ESTER see ETHYL MALONATE

PROPANE-1,2-DIOL see 1,2-PROPANEDIOL

1,2-PROPANEDIOL-1-ACRYLATE see ACRYLIC ACID-2-HYDROXY-
PROPYL ESTER

PROPANENITRILE see PROPIONONITRILE

1-PROPANESULFONIC ACID-3-HYDROXY-GAMMA-SULTONE see 1,2-
OXATHIOLANE-2,2-DIOXIDE

1,3-PROPANE SULTONE see 1,2-OXATHIOLANE-2,2-DIOXIDE

1,2,3-PROPANETRIOL see GLYCERINE

1,2,3-PROPANETRIOL, DIACETATE see 1,3-DIACETIN

1,2,3-PROPANETRIOL, TRINITRATE see NITROGLYCERIN

1,2,3-PROPANETRIYL NITRATE see NITROGLYCERIN

PROPANOIC ACID see PROPIONIC ACID

PROPANOIC ACID see PIVALIC ACID

PROPANOIC ACID, BUTYLESTER (9CI) see BUTYL PROPANOATE

PROPANOIC ACID, ETHENYL ESTER see VINYL PROPIONATE

PROPANOIC ACID, METHYL ESTER see METHYL PROPIONATE

PROPANOL-1 see PROPYL ALCOHOL

1-PROPANOL see PROPYL ALCOHOL

2-PROPANOL see ISOPROPYL ALCOHOL

PROPAN-2-OL see ISOPROPYL ALCOHOL

N-PROPANOL see PROPYL ALCOHOL

1-PROPANOL (GERMAN) see ISOPROPYL ALCOHOL

PROPANOLE (GERMAN) see PROPYL ALCOHOL

PROPANOLEN (DUTCH) see PROPYL ALCOHOL

PROPANOLI (ITALIAN) see PROPYL ALCOHOL

PROPANONE see ACETONE

2-PROPANONE see ACETONE

PROPARGYL ALCOHOL see 2-PROPYN-1-OL

PROPELLANT 22 see MONOCHLORODIFLUOROMETHANE

PROPELLANT C318 see CYCLOOCTAFLUOROBUTANE

2-PROPENAL see ACROLEIN

PROP-2-EN-1-AL see ACROLEIN

PROPENAL (CZECH) see ACROLEIN

PROPENAMIDE see ACRYLAMIDE

2-PROPENAMIDE see ACRYLAMIDE

2-PROPENAMINE see ALLYLAMINE

2-PROPEN-1-AMINE see ALLYLAMINE

1-PROPENE (9CI) see PROPENE

PROPENE ACID see ACRYLIC ACID

1-PROPENE HOMOPOLYMER (9CI) see PROPENE POLYMERS

PROPENENITRILE see ACRYLONITRILE

2-PROPENENITRILE see ACRYLONITRILE

2-PROPENENITRILE HOMOPOLYMER (9CI) see ACRYLONITRILE, POLY-
MER WITH 1,3-BUTADIENE, AND STYRENE, COMBUSTION PROD-
UCTS

2-PROPENENITRILE POLYMER WITH 1,3-BUTADIENE AND ETHENYL-
BENZENE see ACRYLONITRILE, POLYMER WITH 1,3-BUTADIENE
AND STYRENE

2-PROPENENITRILE POLYMER WITH ETHENYLBENZENE see ACRYLONI-
TRILE POLYMER WITH STYRENE

PROPENE OXIDE see EPOXYPROPANE

2-PROPENOIC ACID see ACRYLIC ACID

2-PROPENOIC ACID-2-CHLOROETHYL ESTER see ACRYLIC ACID-beta-
CHLOROETHYL ESTER

2-PROPENOIC ACID, 2-CYANOETHYL ESTER see ACRYLIC ACID ESTER
WITH HYDRACRYLONITRILE

2-PROPENOIC ACID-1,2-ETHANEDIYL ESTER see ETHYLENE ACRYLATE

2-PROPENOIC ACID-2-ETHOXYETHYL ESTER see ACRYLIC ACID-2-
ETHOXYETHYL ESTER

2-PROPENOIC ACID, 2-ETHYLBUTYL ESTER see 2-ETHYL BUTYL
ACRYLATE

2-PROPENOIC ACID-2-ETHYLHEXYL ESTER see ACRYLIC ACID-2-
ETHYLHEXYL ESTER

2-PROPENOIC ACID-2-HYDROXYPROPYL ESTER see ACRYLIC ACID-2-
HYDROXYPROPYL ESTER

PROPENOIC ACID METHYL ESTER see METHYL ACRYLATE

2-PROPENOIC ACID METHYL ESTER see METHYL ACRYLATE

2-PROPENOIC ACID, OXIRANYLMETHYL ESTER see 2,3-EPOXYPROPYL
ACRYLATE

PROPENOL see ALLYL ALCOHOL

PROPEN-1-OL-3 see ALLYL ALCOHOL

1-PROPEN-3-OL see ALLYL ALCOHOL

2-PROPEN-1-OL see ALLYL ALCOHOL

2-PROPEN-1-ONE see ACROLEIN

PROPENYL ALCOHOL see ALLYL ALCOHOL

2-PROPENYL ALCOHOL see ALLYL ALCOHOL

4-PROPENYLANISOLE see p-PROPENYL ANISOLE

P-PROPENYL ANISOLE see p-PROPENYL ANISOLE

P-1-PROPENYLANISOLE see p-PROPENYL ANISOLE

5-(1-PROPENYL)-1,3-BENZODIOXOLE see ISOSAFROLE

5-(2-PROPENYL)-1,3-BENZODIOXOLE see SAFROL

4-PROPENYLCATECHOL METHYLENE ETHER see ISOSAFROLE

PROPENYL ETHER see DIALLYL ETHER

PROPENYLGUAIACOL see GUAIACOL

4-PROPENYLGUAIACOL see GUAIACOL

2-PROPENYL ISOTHIOCYANATE see ALLYL ISOTHIOCYANATE

4-PROPENYL-1,2-METHYLENEDIOXYBENZENE see ISOSAFROLE

(2-PROPENYLOXY)BENZENE see ALLYL PHENYL ETHER

P-PROPENYLPHENYL METHYL ETHER see p-PROPENYL ANISOLE

PROPINE see PROPYNE

PROPIOCINE see ERYTHROMYCIN

PROPIOLACTONE see 2-OXETANONE

BETA-PROPIOLACTONE see 2-OXETANONE

PROPIONATE DE METHYLE (FRENCH) see METHYL PROPIONATE

PROPIONATE D'ETHYLE (FRENCH) see ETHYL PROPIONATE

PROPIONE see 3-PENTANONE

PROPIONIC ACID GRAIN PRESERVER see PROPIONIC ACID

PROPIONIC ETHER see ETHYL PROPIONATE

PROPIONIC NITRILE see PROPIONONITRILE

BETA-PROPIONOLACTONE see 2-OXETANONE

P-PROPIONYLPHENOL see ETHYL-p-HYDROXY PHENYL KETONE

PROPOXYCHEL see d-PROPOXYPHENE HYDROCHLORIDE

PROPOXYPHENE, (+)- see CARVON

PROPOXYPHENE HYDROCHLORIDE see d-PROPOXYPHENE HYDROCHLO-
RIDE

(+)-PROPOXYPHENE HYDROCHLORIDE see d-PROPOXYPHENE HYDRO-
CHLORIDE

1-PROPYL ACETATE see PROPYL ACETATE

2-PROPYL ACETATE see ACETIC ACID ISOPROPYL ESTER

N-PROPYL ACETATE see PROPYL ACETATE

PROPYL ACETOXYMETHYLNITROSAMINE see PROPYLNITROSAMINO-
METHYL ACETATE

N-PROPYL-N-(ACETOXYMETHYL)NITROSAMINE see PROPYLNITROSA-
MINOMETHYL ACETATE

S-PROPYL-N-AETHYL-N-BUTYL-THIOCARBAMAT (GERMAN) see
S-PROPYL BUTYLETHYLTHIOCARBAMATE

N-PROPYLAJMALINE HYDROGEN TARTRATE see N-PROPYLAJMALINE
BITARTRATE

N-PROPYLAJMALINIUM BITARTRATE see N-PROPYLAJMALINE BITAR-
TRATE

1-PROPYL ALCOHOL see PROPYL ALCOHOL

N-PROPYL ALCOHOL see PROPYL ALCOHOL

SEC-PROPYL ALCOHOL see ISOPROPYL ALCOHOL

1-PROPYLALKOHOL (GERMAN) see ISOPROPYL ALCOHOL

N-PROPYL ALKOHOL (GERMAN) see PROPYL ALCOHOL

N-PROPYLAMINE see PROPYLAMINE

5-PROPYL-1,3-BENZODIOXOLE see DIHYDROSAFROLE

PROPYL BROMIDE see 1-BROMOPROPANE

N-PROPYL CARBAMATE see PROPYL CARBAMATE

PROPYLCARBINOL see N-BUTYL ALCOHOL

N-PROPYLCARBINYL CHLORIDE see N-BUTYL CHLORIDE

N-PROPYL CHLORIDE see 1-CHLOROPROPANE

1-PROPYL-3-(P-CHLOROBENZENESULFONYL)UREA see 1-(p-CHLORO-
PHENYLSULFONYL)-3-PROPYLUREA

N-PROPYL-N'-(P-CHLOROBENZENESULFONYL)UREA see 1-(p-CHLORO-
PHENYLSULFONYL)-3-PROPYLUREA

N-PROPYL-N'-P-CHLORPHENYLSULFONYLCARBAMIDE see 1-(p-
CHLOROPHENYLSULFONYL)-3-PROPYLUREA

PROPYL CYANIDE see BUTYRONITRILE

2-PROPYL-3-DIMETHYLAMINO-5,6-METHYLENEDIOXYINDENE HYDRO-

CHLORIDE see 5-DIMETHYLAMINO-6-PROPYL-5H-INDENO-
(5,6-d)-1,3-DIOXOLE HYDROCHLORIDE

2-N-PROPYL-3-DIMETHYLAMINO-5,6-METHYLENEDIOXYINDENE HY-
DROCHLORIDE see 5-DIMETHYLAMINO-6-PROPYL-5H-INDENO-
(5,6-d)-1,3-DIOXOLE HYDROCHLORIDE

PROPYLENE (DOT) see PROPENE

PROPYLENE ALDEHYDE see trans-2-BUTENAL

PROPYLENECHLOROHYDRIN see 2-CHLORO-1-PROPANOL

PROPYLENE DICHLORIDE see 1,2-DICHLOROPROPANE

alpha,beta-PROPYLENE DICHLORIDE see 1,2-DICHLOROPROPANE

PROPYLENE GLYCOL see 1,2-PROPANEDIOL

1,2-PROPYLENE GLYCOL see 1,2-PROPANEDIOL

ALPHA-PROPYLENEGLYCOL see 1,2-PROPANEDIOL

PROPYLENE GLYCOL METHYL ETHER see PROPYLENE GLYCOL MONO-
METHYL ETHER

PROPYLENE GLYCOL MONOACRYLATE see ACRYLIC ACID-2-HY-
DROXYPROPYL ESTER

ALPHA-PROPYLENE GLYCOL MONOMETHYL ETHER see PROPYLENE
GLYCOL MONOMETHYL ETHER

1,2-PROPYLENEIMINE see 2-METHYLAZIRIDINE

1,2-PROPYLENE OXIDE see EPOXYPROPANE

1,3-PROPYLENE OXIDE see OXETANE

PROPYLENE POLYMER see PROPENE POLYMERS

PROPYLENGLYKOL-MONOMETHYLAETHER (GERMAN) see PROPYLENE
GLYCOL MONOMETHYL ETHER

PROPYL-ETHYLBUTYLTHIOCARBAMATE see S-PROPYL BUTYLETHYL-
THIOCARBAMATE

PROPYLETHYL-N-BUTYLTHIOCARBAMATE see S-PROPYL BUTYLETHYL-
THIOCARBAMATE

PROPYL N-ETHYL-N-BUTYLTHIOCARBAMATE see S-PROPYL BUTYL-
ETHYLTHIOCARBAMATE

N-PROPYL-N-ETHYL-N-(N-BUTYL)THIOCARBAMATE see S-PROPYL BU-
TYLETHYLTHIOCARBAMATE

S-(N-PROPYL)-N-ETHYL-N-N-BUTYLTHIOCARBAMATE see S-PROPYL
BUTYLETHYLTHIOCARBAMATE

N-PROPYL-N-ETHYL-N-(N-BUTYL)THIOLCARAMATE see S-PROPYL BU-
TYLETHYLTHIOCARBAMATE

PROPYL ETHYLBUTYLTHIOLCARBAMATE see S-PROPYL BUTYLETHYL-
THIOCARBAMATE

PROPYLFORMIC ACID see N-BUTYRIC ACID

PROPYL GALLATE see N-PROPYL GALLATE

PROPYL HYDRIDE see PROPANE

PROPYLIC ALCOHOL see PROPYL ALCOHOL

PROPYL KETONE see 4-HEPTANONE

PROPYLMETHANOL see N-BUTYL ALCOHOL

PROPYLMETHYLCARBINYLETHYL BARBITURIC ACID SODIUM SALT see
NEMBUTAL SODIUM

4-PROPYL-1,2-METHYLENEDIOXYBENZENE see DIHYDROSAFROLE

PROPYL NITRATE see N-PROPYL NITRATE

1-(PROPYLNITROSAMINO)PROPYL ACETATE see 1-ACETOXY-N-NITRO-
SODIPROPYLAMINE

N-PROPYLNITROSOHARNSTOFF (GERMAN) see N-NITROSO-N-PROPYL-
UREA

1-PROPYL-1-NITROSOUREA see N-NITROSO-N-PROPYL-UREA

N-PROPYLNITROSUREA see N-NITROSO-N-PROPYL-UREA

PROPYLOWY ALKOHOL (POLISH) see PROPYL ALCOHOL

6-(PROPYLPIPERONYL)-BUTYL CARBITYL ETHER see PIPERONYL BUT-
OXIDE

6-PROPYLPIPERONYL BUTYL DIETHYLENE GLYCOL ETHER see PI-
PERONYL BUTOXIDE

6-PROPYL-2-THIO-2,4(1H,3H)PYRIMIDINEDIONE see 6-PROPYL-2-
THIOURACIL

4-PROPYL-2-THIOURACIL see 6-PROPYL-2-THIOURACIL

6-N-PROPYLTHIOURACIL see 6-PROPYL-2-THIOURACIL

6-N-PROPYL-2-THIOURACIL see 6-PROPYL-2-THIOURACIL

N-PROPYL 3,4,5-TRIHYDROXYBENZOATE see N-PROPYL GALLATE

5-PROPYL-4-(2,5,8-TRIOXA-DODECYL)-1,3-BENZODIOXOL (GERMAN)
see PIPERONYL BUTOXIDE

PROPYL URETHANE see PROPYL CARBAMATE

1-PROPYNE-3-OL see 2-PROPYN-1-OL

PROPYZAMIDE see KERB

PROSTAGLANDIN E2 see DINOPROSTONE

PROSTIGMINE METHYLSULFATE see NEOSTIGMINE MONOMETHYLSUL-
FATE

PROTACTYL see PROMAZINE

PROTHROMBIN see COUMADIN SODIUM

PROTOCATECHUIC ALDEHYDE ETHYL ETHER see ETHYL VANILLIN

PROTOCATECHUIC ALDEHYDE METHYLENE ETHER see PIPERONAL

PROTOCHLORURE D'IODE (FRENCH) see IODINE MONOCHLORIDE

PROTOPINE see FUMARINE

PROVITAMIN D see CHOLESTEROL

PRUSSIC ACID see HYDROCYANIC ACID

PRUSSIC ACID SOLUTION see HYDROCYANIC ACID (SOLUTION)

PRZEDZIORKOFOS (POLISH) see PHENCAPTON

PSEUDOACETIC ACID see PROPIONIC ACID

PSEUDOBUTYLBENZENE see tert-BUTYLBENZENE

PSEUDO-BUTYLENE see cis-2-BUTENE

PSEUDOCUMENE see 1,2,4-TRIMETHYLBENZENE

PSEUDOCUMIDINE see 2,4,5-TRIMETHYLANILINE

PSEUDOCUMIDINE HYDROCHLORIDE see 2,4,5-TRIMETHYLANILINE
HYDROCHLORIDE

PSEUDOCUMOL see 1,2,4-TRIMETHYLBENZENE

PSEUDODIGITOXIN see GITOXIN

PSEUDOHEXYL ALCOHOL see 2-ETHYLBUTANOL

PSEUDOTHEOPHYLLINE see THEOPHYLLINE

PSEUDOTHIOUREA see ISOTHIOUREA

PSEUDOUREA see UREA

PSEUDOUREA, 2,2'-ETHYLENEBIS(2-THIO-, DIHYDROBROMIDE see
ETHYLENE DIISO THIOURONIUM DIBROMIDE

PSEUDOXANTHINE see XANTHINE

PSICAINE-NEU HYDROCHLORIDE see NEOPSICAINE HYDROCHLORIDE

PSI-CUMENE see 1,2,4-TRIMETHYLBENZENE

PSI-CUMIDINE see 2,4,5-TRIMETHYLANILINE

PSILOCINE see 4-HYDROXY-N,N-DIMETHYLTRYPTAMINE

PSILOCIN PHOSPHATE ESTER see PSILOCYBIN

PSILOTSIBIN see PSILOCYBIN

PSILOTSIN see 4-HYDROXY-N,N-DIMETHYLTRYPTAMINE

PSYCHEDRINE see BENZEDRINE

PSYCHEDRINUM see DL-AMPHETAMINE SULFATE

PSYCHEDRYNA see DL-AMPHETAMINE SULFATE

PTC see PEPTICHEMIO

PTEGLU see FOLIC ACID

PTERIDIUM AQUILINUM see BRACKEN FERN, DRIED

PTERIDIUM AQUILINUM TANNIN see BRACKEN FERN TANNIN

PTERIS AQUALINA see BRACKEN FERN, DRIED

PTEROYL-L-GLUTAMIC ACID see FOLIC ACID

PULVERIZED NICKEL see NICKEL

PURATRONIC CHROMIUM CHLORIDE see CHROMIUM (III) CHLORIDE

PURATRONIC CHROMIUM TRIOXIDE see CHROMIUM (VI) OXIDE (1:3)

PURE CHRYSOIDINE YBH see 4-PHENYLAZO-m-PHENYLENEDIAMINE

PURE QUARTZ see SILICA (CRYSTALLINE)

PURE QUARTZ see SILICA, CRYSTALLINE-QUARTZ

PURE ZINC CHROME see BASIC ZINC CHROMATE

PURIFIED CHARCOAL see CARBON

1H-PURIN-6-AMINE see ADENINE

PURINE-2,6-DIOL see XANTHINE

9H-PURINE-2,6-DIOL see XANTHINE

2,6(1,3)-PURINEDION see XANTHINE

PURINE-2,6-(1H,3H)-DIONE see XANTHINE

6-PURINETHIOL see PURINE-6-THIOL

PURPLE MINT PLANT EXTRACT see PERILLA KETONE

PVBR see POLYVINYLBROMIDE

PVP 1 see POLY(1-VINYL-2-PYRROLIDINONE) HUEPER'S POLYMER
NO. 1

PVP 2 see POLY(1-VINYL-2-PYRROLIDINONE) HUEPER'S POLYMER
NO. 2

PVP 3 see POLY(1-VINYL-2-PYRROLIDINONE) HUEPER'S POLYMER
NO. 3

PVP 4 see POLY(1-VINYL-2-PYRROLIDINONE) HUEPER'S POLYMER
NO. 4

PVP 5 see POLY(1-VINYL-2-PYRROLIDINONE) HUEPER'S POLYMER
NO. 5

PVP 6 see POLY(1-VINYL-2-PYRROLIDINONE) HUEPER'S POLYMER
NO. 6

PVP 7 see POLY(1-VINYL-2-PYRROLIDINONE) HUEPER'S POLYMER
NO. 7

PVSK see POLYVINYL SULFATE, POTASSIUM SALT

PYRACRYL ORANGE Y see 4-PHENYLAZO-m-PHENYLENEDIAMINE

PYRAHEXYL see 3-HOMOTETRA HYDRO CANNIBINOL

PYRALENE see POLYCHLORINATED BIPHENYLS

PYRALIN see CELLULOSE TETRANITRATE

PYRAMIDONE see DIMETHYLAMINOANTIPYRINE

PYRANISAMINE MALEATE see DIAMINIDE MALEATE

PYRANOL see POLYCHLORINATED BIPHENYLS

PYRATHIAZINE see PARATHIAZINE

PYRAZINAMIDE see PYRAZINECARBOXAMIDE

PYRAZINEAMIDE see PYRAZINECARBOXAMIDE

PYRAZINE CARBOXYLAMIDE see PYRAZINECARBOXAMIDE

PYRAZINE HEXAHYDRIDE see PIPERAZINE

PYRAZINOIC ACID AMIDE see PYRAZINECARBOXAMIDE

PYREN (GERMAN) see PYRENE

PYRETHRIN I OR II see PYRETHRIN I

PYRETHROLONE CHRYSANTHEMUM DICARBOXLIC ACIDMETHYL
ESTER ESTER see PYRETHRIN II

PYRETHROLONE ESTER OF CHRYSANTHEMUMDICARBOXYLIC ACID
MONOMETHYL ESTER see PYRETHRIN II

PYRETHRUM (INSECTICIDE) see PYRETHRIN I

PYRETRIN II see PYRETHRIN II

PYRIBENZAMINE see TRIPELENNAMINE

PYRIBENZAMINE see TRIPELENNAMINE

PYRIDIN (GERMAN) see PYRIDINE

2-PYRIDINALDOXIM METHOJODID (GERMAN) see PYRIDINIUM-2-
ALDOXIME-N-METHYLIODIDE

PYRIDIN-2-ALDOXIN (CZECH) see PYRIDINIUM-2-ALDOXIME-N-
METHYLIODIDE

4-PYRIDINAMINE see 4-AMINOPYRIDINE

PYRIDINE (DOT) see PYRIDINE

2-PYRIDINE-ALDOXIME CHLORIDE see 2-FORMYL-1-METHYLPYRIDI-
NIUM CHLORIDE OXIME

2-PYRIDINE ALDOXIME IODOMETHYLATE see PYRIDINIUM-2-ALDOXI-
ME-N-METHYLIODIDE

PYRIDINE-2-ALDOXIME METHIODIDE see PYRIDINIUM-2-ALDOXIME-N-
METHYLIODIDE

PYRIDINE-2-ALDOXIME METHOCHLORIDE see 2-FORMYL-1-METHYL-
PYRIDINIUM CHLORIDE OXIME

2-PYRIDINE ALDOXIME METHYL CHLORIDE see 2-FORMYL-1-METHYL-
PYRIDINIUM CHLORIDE OXIME

PYRIDINE-2-ALDOXIME METHYL IODIDE see PYRIDINIUM-2-ALDOXI-
ME-N-METHYLIODIDE

PYRIDINE-3-CARBONIC ACID see NICOTINIC ACID

PYRIDINE-3-CARBOXYDIETHYLAMIDE see N,N-DIETHYLNICOTINAMIDE

PYRIDINE-3-CARBOXYLIC ACID see NICOTINIC ACID

PYRIDINE-BETA-CARBOXYLIC ACID see NICOTINIC ACID

3-PYRIDINECARBOXYLIC ACID AMIDE see NIACINAMIDE

PYRIDINE-3-CARBOXYLIC ACID AMIDE see NIACINAMIDE

PYRIDINE-3-CARBOXYLIC ACID DIETHYLAMIDE see N,N-DIETHYLNI-
COTINAMIDE

4-PYRIDINECARBOXYLIC ACID, ETHYL ESTER see ETHYL ISONICOTI-
NATE

4-PYRIDINECARBOXYLIC ACID, HYDRAZIDE see ISONICOTINIC ACID
HYDRAZIDE

PYRIDINIUM ALDOXIME METHOCHLORIDE see 2-FORMYL-1-METHYL-
PYRIDINIUM CHLORIDE OXIME

PYRIDOXAL, HYDROCHLORIDE see VITAMIN B6 HYDROCHLORIDE

PYRIDOXINE see PYRIDOXOL

PYRIDOXINE HYDROCHLORIDE see PYRIDOXOL HYDROCHLORIDE

PYRIDOXINIUM CHLORIDE see PYRIDOXOL HYDROCHLORIDE

PYRIDROL see alpha,alpha-DIPHENYL-2-PIPERIDINEMETHANOL

1-(PYRIDYL-3-)-3,3-DIAETHYL-TRIAZEN (GERMAN) see 3,3-DIETHYL-
1-(m-PYRIDYL)TRIAZENE

M-PYRIDYL-DIETHYL-TRIAZENE see 3,3-DIETHYL-1-(m-PYRIDYL)-TRIAZENE

1-PYRIDYL-3,3-DIETHYLTRIAZENE see 3,3-DIETHYL-1-(m-PYRIDYL)-TRIAZENE

1-(3-PYRIDYL)-3,3-DIETHYLTRIAZENE see 3,3-DIETHYL-1-(m-PYRIDYL)TRIAZENE

1-(PYRIDYL-3)-3,3-DIETHYLTRIAZENE see 3,3-DIETHYL-1-(m-PYRIDYL)TRIAZENE

1-(PYRIDYL-3)-3,3-DIMETHYL-TRIAZEN (GERMAN) see 1-(PYRIDYL-3)-3,3-DIMETHYL TRIAZENE

N-ALPHA-PYRIDYL-N-P-METHOXYBENZYL-N',N'-DIMETHYLETHYL-ENEDIAMINE PHOSPHATE see NEOANTERGAN PHOSPHATE

N-3-PYRIDYLMETHYL-N'-P-NITROPHENYLUREA see PYRIMINYL

BETA-PYRIDYL-ALPHA-N-METHYLPYRROLIDINE see NICOTINE

1-(PYRIDYL-3-N-OXID)-3,3-DIMETHYL-TRIAZEN (GERMAN) see (3,3-DIMETHYL-1-(m-PYRIDYL-N-OXIDE))TRIAZINE

1-(PYRIDYL-3-N-OXIDE)-3,3-DIMETHYLTRIAZENE see (3,3-DIMETHYL-1-(m-PYRIDYL-N-OXIDE))TRIAZENE

N-(ALPHA-PYRIDYL)-N-(ALPHA-THENYL)-N',N'-DIMETHYLETHYLENE-DIAMINE see THENYLPYRAMINE

2,4,5,6(1H,3H)-PYRIMIDINETETRONE see ALLOXAN

2,4,5,6(1H,3H)-PYRIMIDINETETRONE HYDRATE see MESOXALYLUREA MONOHYDRATE

2,4,6(1H,3H,5H)-PYRIMIDINETRIONE, 5-(2-BROMO-2-PROPENYL)-5-(1-METHYLPROPYL)-(9CI) see BUTALLYLONAL

2,4,6(1H,3H,5H)-PYRIMIDINETRIONE, 5-BUTYL-5-ETHYL- (9CI) see BUTOBARBITAL

2,4,6(1H,3H,5H)-PYRIMIDINETRIONE, 5-(1-CYCLOHEPTEN-1-YL)-5-ETHYL-(9CI) see CYCLOHEPTENYL ETHYLBARBITURIC ACID

2,4,6(1H,3H,5H)-PYRIMIDINETRIONE, 5,5-DI-2-PROPENYL- (9CI) see ALLOBARBITAL

2,4,6(1H,3H,5H)-PYRIMIDINETRIONE, 5-ETHYL-5-(1METHYLBUTYL)-, CALCIUM SALT see CALCIUM PENTOBARBITAL

2,4,6(1H,3H,5H)-PYRIMIDINETRIONE, 5-ETHYL-5-(1-METHYL-1-BU-TENYL)- (9CI) see 5-ETHYL-5-(1-METHYL-1-BUTENYL)BARBITU-RATE

2,4,6(1H,3H,5H)-PYRIMIDINETRIONE, 5-ETHYL-5-(1-METHYL-BUTYL)-, MONOSODIUM SALT (9CI) see NEMBUTAL SODIUM

2,4,5(1H,3H,5H)-PYRIMIDINETRIONE, 5-ETHYL-5-(1-METHYLETHYL)-(9CI) see ETHYL ISOPROPYL BARBITURIC ACID

2,4,6(1H,3H,5H)-PYRIMIDINETRIONE, 5-ETHYL-5-(1-METHYLPRO-PYL)-(9CI) see BUTISOL

2,4,6(1H,3H,5H)-PYRIMIDINETRIONE, 5-ETHYL-5-(1-METHYLPRO-PYL)-, MONOSODIUM SALT see BUTISOL SODIUM

2,4,6(1H,3H,5H)-PYRIMIDINETRIONE, 5-ETHYL-5-(PHENYLMETHYL)-(9CI) see BENZYLBARBITAL

2,4,6(1H,3H,5H)-PYRIMIDINETRIONE, 5-ETHYL-5-PHENYL-, MONOSO-DIUM SALT (9CI) see SODIUM LUMINAL

2,4,6(1H,3H,5H)-PYRIMIDINETRIONE, 5-(1-METHYLPROPYL)-5-(2-PROPENYL)- (9CI) see 5-ALLYL-5-sec-BUTYLBARBITURIC ACID

2,4,6(1H,3H,5H)-PYRIMIDINETRIONE, 5-(2-METHYLPROPYL)-5-(2-PROPENYL)- (9CI) see ALLYLISOBUTYLBARBITURATE

2(1H)-PYRIMIDINONE, 4-AMINO-1-BETA-D-ARABINOFURANOSYL-(9CI) see ARABINOCYTIDINE

2,4,5,6-PYRIMIDINTETRON (CZECH) see ALLOXAN

PYRIPYRIDIUM see PHENAZODINE

PYROACETIC ACID see ACETONE

PYROACETIC ETHER see ACETONE

PYROBENZOLE see BENZENE

PYROCATECHIN see PYROCATECHOL

PYROCATECHINIC ACID see PYROCATECHOL

PYROCATECHUIC ACID see PYROCATECHOL

PYROGALLIC ACID see PYROGALLOL

PYROGALLOL DIMETHYLETHER see 2,6-DIMETHOXYPHENOL

PYROGALLOL-1,3-DIMETHYL ETHER see 2,6-DIMETHOXYPHENOL

PYROGUAIAC ACID see GUAIACOL

PYROLUSITE BROWN see MANGANESE DIOXIDE

PYROMUCIC ALDEHYDE see 2-FURALDEHYDE

PYROPENTYLENE see 1,3-CYCLOPENTADIENE

PYROPHOSPHATE DE TETRAETHYLE (FRENCH) see TETRAETHYLPYRO-PHOSPHATE

PYROPHOSPHORIC ACID OCTAMETHYLTETRAAMIDE see OCTAMETHYL-PYROPHOSPHORAMIDE

PYROPHOSPHORIC ACID, TETRAETHYL ESTER (liquid mixture) see TETRON

PYROPHOSPHORODITHIOIC ACID, TETRAETHYL ESTER see ETHYL THIO PYROPHOSPHATE

PYROPHOSPHORYLTETRAKISDIMETHYLAMIDE see OCTAMETHYLPYRO-PHOSPHORAMIDE

PYROSULPHURIC ACID see SULFURIC ACID, FUMING

PYROTARTARIC ACID NITRILE see 1,3-TRIMETHYLENEDINITRILE

PYROXYLIC SPIRIT see METHANOL

PYROXYLIN see CELLULOSE TETRANITRATE

PYROXYLIN PLASTICS (DOT) see CELLULOSE TETRANITRATE

PYROXYLIN PLASTIC SCRAP (DOT) see CELLULOSE TETRANITRATE

PYRRO(B)MONAZOLE see IMIDAZOLE

PYRROLAZOATE see PARATHIAZINE

PYRROLE-2,5-DIONE see MALEIMIDE

PYRROLIDINO-AETHYLPHENTHIAZIN (GERMAN) see PARATHIAZINE

2-(PYRROLIDINYL)ETHYL 2-CHLORO-6-METHYLCARBANILATE HY-DROCHLORIDE see 2-CHLORO-6-METHYLCARBANILIC ACID-2-(PYRROLIDINYL)ETHYL ESTER HYDROCHLORIDE

10-(2-(1-PYRROLIDINYL)ETHYL)PHENOTHIAZINE see PARATHIAZINE

10-(2-(1-PYRROLIDYL)ETHYL)PHENOTHIAZINE see PARATHIAZINE

3-PYRROLINE-2,5-DIONE see MALEIMIDE

1H-PYRROLIZINE-7-METHANOL 2,3-DIHYDRO-1-HYDROXY- (R)- see DEHYDRORETRONECINE

PYRROLYLENE see 1,3-BUTADIENE

PY-TETRAHYDROSERPENTINE see AJMALICINE

QUANTROVANIL see ETHYL VANILLIN

QUARTZ see SILICA, CRYSTALLINE-QUARTZ

QUARTZ GLASS see SILICA, AMORPHOUS FUSED

QUATERNARY AMMONIUM COMPOUNDS, ALKYLBENZYLDIMETHYL, CHLORIDES see ALKYL DIMETHYLBENZYL AMMONIUM CHLO-RIDE

QUAZO PURO (ITALIAN) see SILICA, CRYSTALLINE-QUARTZ

QUEBRACHINE see YOHIMBINE

QUECKSILBER (GERMAN) see MERCURY

QUECKSILBER CHLORID (GERMAN) see CORROSIVE SUBLIMATE

QUECKSILBER(I)-CHLORID (GERMAN) see MERCUROUS CHLORIDE

QUERCETINE see QUERCETIN

QUERCETOL see QUERCETIN

QUERCUS FALCATA PAGODAEFOLIA see CHERRY BARK OAK

QUERTINE see QUERCETIN

QUICKLIME see CALCIUM OXIDE

QUICK SILVER see MERCURY

QUINACRINE see ATABRINE

QUINALBARBITAL see SECONAL

QUINALBARBITONE see SECONAL

QUINICARDINE see QUINIDINE

(+)-QUINIDINE see QUINIDINE

(−)-QUININE see QUININE

BETA-QUININE see QUININE

QUININE BISULFATE see QUININE SULFATE

QUININE CHLORIDE see QUININE HYDROCHLORIDE

QUININE HYDROGEN SULFATE see QUININE SULFATE

QUININE MONOHYDROCHLORIDE see QUININE HYDROCHLORIDE

QUININE MURIATE see QUININE HYDROCHLORIDE

QUINIOZENE see PENTACHLORONITRO BENZENE

QUINIZARIN see 1,4-DIHYDROXYANTHRAQUINONE

BETA-QUINOL see HYDROQUINONE

QUINOLINE-6-AZO-P-DIMETHYLANILINE see 6-((p-(DIMETHYLAMINO)-PHENYL)AZO)QUINOLINE

QUINOLINE, 1-NITROSO-2,2,4-TRIMETHYL-1,2,-DIHYDRO-,(POLY-MER) see CURETARD

QUINOLINIUM, 8-HYDROXY-1-METHYL-, METHYLSULFATE, DI-METHYLCARBAMATE see DIMETHYLCARBAMIC ESTER of 8OXYMETHYLQUINOLINIUM METHYLSULFATE

8-(QUINOLINOLATO)METHYL MERCURY see METHYLMERCURY QUINO-LINOLATE

(8-QUINOLINOLATO)TRIBUTYL-TANNANE see TRIBUTYL(8-QUINO-LINOLATO)TIN

8-QUINOLINOL, MERCURY COMPLEX see METHYLMERCURY QUINO-LINOLATE

QUINONE see p-BENZOQUINONE

P-QUINONE see p-BENZOQUINONE

P-QUINONE DIOXIME see DIOXIME-p-BENZOQUINONE

QUINONE MONOXIME see p-NITROSOPHENOL

QUINONE OXIME BENZOYLHYDRAZONE see 1,4-BENZOQUINONE-N'-BENZOYLHYDRAZONE OXIME

3-QUINUCLIDINOL ACETATE see ACECLIDINE

QUOTANE HYDROCHLORIDE see DIMETHISOQUIN HYDROCHLORIDE

R 7158 see 8-(4-p-FLUORO PHENYL-4-OXOBUTYL)-2-METHYL-2,8-DIAZASPIRO(4.5)DECANE-1,3-DIONE

R 9298 see 4-(4-(4-CHLORO-alpha,alpha,alpha-TRIFLUORO-m-TO-LYL)-4-HYDROXYPIPERIDINO)BUTYROPHENONE-4'-FLUORO-,HY-DROCHLORIDE

RACEMIC LACTIC ACID see LACTIC ACID

RACEMIC MANDELIC ACID see MANDELIC ACID

RACEMORPHAN HYDROBROMIDE see METHORPHINAN HYDROBRO-MIDE

RACEPHEDRINE HYDROCHLORIDE see dl-EPHEDRINE HYDROCHLO-RIDE

RACEPHEN see DL-AMPHETAMINE SULFATE

RADIUM F see POLONIUM

RADOSAN see METHOXYETHYL MERCURIC ACETATE

RAMOR see THALLIUM

RANEY ALLOY see NICKEL

RANEY NICKEL see NICKEL

RASPBERIN see SALICYLAMIDE

RASPBERRY RED FOR JELLIES see FOOD RED 2

RAUBASINE see AJMALICINE

RAUGALLINE see AJMALINE

RAUMALINA see AJMALICINE

RAUSERPIN see RESERPINE

RAUWOLFEA see RESERPINE

RAUWOLFIN see AJMALINE

RAUWOLFINE see AJMALINE

REALGAR. see ARSENIC BISULFIDE

REANIMIL see DIMEFLINE

RECTHORMONE TESTOSTERONE see TESTOSTERONE PROPIONATE

1306 RED see FOOD RED 4

12101 RED see FOOD RED 4

RED DYE NO. 2 see FOOD RED 2

RED FUMING NITRIC ACID see NITRIC ACID (RED FUMING)

RED IRON ORE see HEMATITE

RED IRON OXIDE see IRON (III) OXIDE

RED LEAD see LEAD OXIDE RED

RED LEAD OXIDE see LEAD OXIDE RED

RED MERCURIC IODIDE see MERCURY (II) IODIDE

RED NO. 1 see FOOD RED 4

RED NO. 4 see FOOD RED 4

RED OXIDE OF MERCURY see MERCURIC OXIDE

RED PRECIPITATE see MERCURIC OXIDE

REDUCED-D-PENICILLAMINE see d,3-MERCAPTOVALINE

REFINED SOLVENT NAPHTHA see PETROLEUM SPIRITS

REFOSPOREN see CEFAZEDONE

REFRIGERANT 22 see MONOCHLORODIFLUOROMETHANE

REICHSTEIN'S SUBSTANCE M see CORTISOL

REMAZOL BRILLIANT BLUE R see BRILLIANT BLUE R

REMEFLIN see DIMEFLINE

RENARDINE see SENKIRKINE

REPARIL SODIUM SALT see ESCIN, SODIUM SALT

REPELLENT 612 see ETHYL HEXYLENE GLYCOL

REPROTEROL-HCl see BRONCHOSPASMIN

RESERPILIN-24-OIC ACID, 2-(DIMETHYLAMINO) ETHYL ESTER see 2-(DIMETHYLAMINO) RESERPILINATE

RESMETRINA (PORTUGUESE) see BENZOFUROLINE

RESORCIN see RESORCINOL

RESORCINOLPHTHALEIN see FLUORESCEIN

RESORCINOL PHTHALEIN SODIUM see FLUORESCEIN SODIUM

RESORCINYL DIGLYCIDYL ETHER see RESORCINOL DIGLYCIDYL ETHER

211-TRANS-RETINOL, PALMITATE see VITAMIN A PALMITATE

RETINYL ACETATE see RETINOL ACETATE

CIS-RETRONECIC ACID ESTER OF RETRONECINE see RETRORSINE

CIS-RETRONECIC ACID ESTER OF RETRONECINE-N-OXIDE see RETROR-SINE-N-OXIDE

RHODAMINE 6G (BIOLOGICAL STAIN) see RHODAMINE 6G EXTRA BASE

RHODAMINE 6GEX ETHYL ESTER see RHODAMINE 6G EXTRA BASE

RHODIUM CHLORIDE see RHODIUM(III) CHLORIDE (1:3)

RHODIUM TRICHLORIDE see RHODIUM(III) CHLORIDE (1:3)

RHYTMATON see AJMALINE

RIBOFLAVIN see RIBOFLAVINE

RIBOFURANOSIDE, 9H-PURINE-6-THIOL-9 see MERCAPTOPURINE RIBO-NUCLEOSIDE

RICINUS OIL see CASTOR OIL

RICIRUS OIL see CASTOR OIL

RO 2-3599 see 6,7-DIHYDRO-6-(2-HYDROXYETHYL)-5H-DI-BENZ(c,e)AZEPINE

RO 4-6467 see PROCARBAZINE

ROACH SALT see SODIUM FLUORIDE

ROAD ASPHALT see PETROLEUM ASPHALT

ROAD ASPHALT (DOT) see ASPHALT (CUT BACK)

ROAD ASPHALT (DOT) see ASPHALT

ROAD TAR (DOT) see ASPHALT

ROAD TAR, LIQUID (DOT) see ASPHALT (CUT BACK)

ROCK CANDY see SUCROSE

ROCK OIL see PETROLEUM

ROCK SALT see SODIUM CHLORIDE

RODANIN S-62 (CZECH) see 2-IMIDAZOLINETHIOL

RODINOL see CITRONELLOL

ROMAN VITRIOL see COPPER (II) SULFATE (1:1)

ROMAN VITRIOL see COPPER (II) SULFATE PENTAHYDRATE (1:1:5)

ROMOSOL see 1-AUROTHIO-D-GLUCOPYRANOSE

RONNEL see TRICHLOROMETAFOS

TERTROPHENE GREEN M see AIZEN MALACHITE GREEN

ROQUESSINE see NERIINE

ROSANILINE BASE see MAGENTA BASE

ROSE QUARTZ see SILICA (CRYSTALLINE)

ROSE QUARTZ see SILICA, CRYSTALLINE-QUARTZ

ROSES see DL-AMPHETAMINE SULFATE

ROTENONA (SPANISH) see ROTENONE

ROUGE CERASINE see OIL RED

ROYAL BLUE see N,N-DIETHYLLYSERGAMIDE

RTEC (POLISH) see MERCURY

"522" RUBBER ACCELERATOR see PIP-PIP

RUBBER HYDROCHLORIDE see PLIOFILM

RUBBER HYDROCHLORIDE POLYMER see PLIOFILM

RUBEANIC ACID see DITHIOOXAMIDE

RUBENS BROWN see MANGANESE (III) OXIDE

RUBIDIUM METAL see RUBIDIUM

RUBRUM SCARLATINUM see SCARLET RED

RUTGERS 612 see ETHYL HEXYLENE GLYCOL

RUTHENIUM TRICHLORIDE see RUTHENIUM CHLORIDE

RUTILE see TITANIUM OXIDE

RYANIA POWDER see RYANIA

RYANIA SPECIOSA see RYANIA

SABACIDE see VERATRINE

SABADILLA see VERATRINE

SABANE DUST see VERATRINE

SACCAHARIMIDE see 1,2-BENZISOTHIAZOL-3(2H)-ONE-1,1-DIOXIDE

SACCHARATED IRON see IRON OXIDE, SACCHARATED

SACCHARINA see 1,2-BENZISOTHIAZOL-3(2H)-ONE-1,1-DIOXIDE

SACCHARIN ACID see 1,2-BENZISOTHIAZOL-3(2H)-ONE-1,1-DIOXIDE

SACCHARINE see 1,2-BENZISOTHIAZOL-3(2H)-ONE-1,1-DIOXIDE

SACCHARINOL see 1,2-BENZISOTHIAZOL-3(2H)-ONE-1,1-DIOXIDE

SACCHARINOSE see 1,2-BENZISOTHIAZOL-3(2H)-ONE-1,1-DIOXIDE

SACCHARIN SOLUBLE see SACCHARIN

SACCHAROIDUM NATRICUM see SACCHARIN

SACCHAROL see 1,2-BENZISOTHIAZOL-3(2H)-ONE-1,1-DIOXIDE

SACCHAROSE see SUCROSE

SACCHARUM see SUCROSE

SACCHARUM LACTIN see LACTOSE

SACHSISCHBLAU see DISODIUM INDIGO-5,5-DISULFONATE

SAEURE FLUORIDE (GERMAN) see FLUORINE

SAFFRON YELLOW see 2,4-DINITRO-1-NAPHTHOL

SAFROLE see SAFROL

SAKOLYSIN (GERMAN) see DL-PHENYLALANINE MUSTARD

SAL AMMONIA see AMMONIUM CHLORIDE

SAL AMMONIAC see AMMONIUM CHLORIDE

SAL DE MERCK see COCAINE CHLORIDE

SALICYLIC ACID, ACETATE see ACETOL (2)

SALICYLIC ACID, COMPOUND WITH PHYSOSTIGMINE (1:1) see PHY-
SOSTIGMINE SALICYLATE (1:1)

SALICYLIC ACID, FLUOROACETATE see O-(FLUOROACETYL)SALICYLIC
ACID

SALICYLIC ACID, SODIUM SALT see SODIUM SALICYLATE

SALINE see SODIUM CHLORIDE

SALITHION-SUMITOMO see SALITHION

SALOL see PHENYL SALICYLATE

SALPETERSAURE (GERMAN) see NITRIC ACID

SALPETERZUUROPLOSSINGEN (DUTCH) see NITRIC ACID

SALT see SODIUM CHLORIDE

SALT CAKE see SODIUM SULFATE (2:1)

SALT OF SATURN see LEAD ACETATE

SALTPETER see POTASSIUM NITRATE

SALZBURG VITRIOL see COPPER (II) SULFATE PENTAHYDRATE
(1:1:5)

SAMARIUMACETAT (GERMAN) see SAMARIUM ACETATE

SAMARIUM NITRAT (GERMAN) see SAMARIUM (III) NITRATE, HEXA-
HYDRATE (1:3:6)

SAND see SILICA, CRYSTALLINE-QUARTZ

SAND see SILICA (CRYSTALLINE)

SAND ACID see SILICOFLUORIC ACID

SANSERT see METHYSERGIDE DIMALEATE

SANTAVY'S SUBSTANCE F see N-METHYL-N-DESACETYLCOLCHICINE

SANTOTHERM see POLYCHLORINATED BIPHENYLS

L-SARCOLYSIN see L-PHENYLALANINE MUSTARD

DL-SARCOLYSIN see DL-PHENYLALANINE MUSTARD

P-L-SARCOLYSIN see L-PHENYLALANINE MUSTARD

DL-SARCOLYSINE see DL-PHENYLALANINE MUSTARD

DL-SARCOLYSINE HYDROCHLORIDE see 3-(p-(BIS(beta-CHLORO-
ETHYL)AMINO)PHENYL)-D,L-ALANINE HYDROCHLORIDE

SARCOLYSIN HYDROCHLORIDE see 3-(p-(BIS(beta-CHLOROETHYL)-
AMINO)PHENYL)-D,L-ALANINE HYDROCHLORIDE

SARCOMYCIN see SARKOMYCIN

SARIN see ISOPROPYL METHANEFLUOROPHOSPHONATE

SARIN II see ISOPROPYL METHANEFLUOROPHOSPHONATE

SARKOKLORIN see 3-(p-(BIS(beta-CHLOROETHYL)AMINO)PHENYL)-D,
L-ALANINE HYDROCHLORIDE

SASSAFRAS ALBIDUM see SASSAFRAS

SATURN BROWN LBR see C.I. DIRECT BROWN

SAURE DES PHYTINS (GERMAN) see PHYTIC ACID

SAXIN see SACCHARIN

SAXOL see MINERAL OIL

SBa 0108E see BARIUM CHLORIDE

SC-1950 see 2,6-DIMETHYL-1,1-DIETHYLPIPERIDINIUM BROMIDE

SCATOLE see beta-METHYLINDOLE

SC G3 see SCANDIUM (3+) CHLORIDE

SCH 6673 see ACETOPHENAZINE

SCH 10649 see AZATADINE MALEATE

SCHEELES GREEN see COPPER ORTHOARSENITE

SCHEELE'S MINERAL see COPPER ORTHOARSENITE

SCHINOPSIS LORENTZII TANNIN see QUEBRACHO TANNIN

SCHWEFELKOHLENSTOFF (GERMAN) see CARBON DISULFIDE

SCHWEFEL-LOST see BIS(2-CHLOROETHYL)SULFIDE

SCHWEFELSAEURELOESUNGEN (GERMAN) see SULFURIC ACID

SCHWEFELWASSERSTOFF (GERMAN) see HYDROGEN SULFIDE

SCHWEFLIGE SAURE (GERMAN) see SULFUROUS ACID

SCHWEINFURTERGRUN see CUPRIC ACETOARSENITE

SCHWEINFURT GREEN see CUPRIC ACETOARSENITE

SCILLAGLYKOSID A (GERMAN) see GLUCO PROSCILLARIDIN A

SCILLIROSIDE GLYCOSIDE see RED SQUILL

3-beta-SCILLOBIOSIDO-14-beta-HYDROXY-delta-4,20,22-BUFATRI-
ENOLID(GERMAN) see GLUCO PROSCILLARIDIN A

SCOPARONE see 6,7-DIMETHYLESCULETIN

SCOPINE TROPATE see SCOPOLAMINE

(−)-SCOPOLAMINE see SCOPOLAMINE

SCOPOLAMINE BROMIDE see HYOSCINE HYDROBROMIDE

(−)-SCOPOLAMINE BROMIDE see HYOSCINE HYDROBROMIDE

(−)-SCOPOLAMINE HYDROBROMIDE see HYOSCINE HYDROBROMIDE

SCORBIC OIL see PARASCORBIC ACID

SCOT see HEROIN

SD 4294 see CROTOXYPHOS

SD 8530 see 3,4,5-TRIMETHYLPHENYL METHYLCARBAMATE

SEA ANEMONE TOXIN II see ATX II

SEA COAL see COAL, GROUND BITUMINOUS (DOT)

SEA SALT see SODIUM CHLORIDE

SEAWATER MAGNESIA see MAGNESIUM OXIDE

SECACORNIN see D-LYSERGIC ACID-L,2-PROPANOLAMIDE

SECOBARBITAL see SECONAL

SECOBARBITONE see SECONAL

9,10-SECOCHOLESTA-5,7,10(19)-TRIENE-1-ALPHA,3-BETA-DIOL see
1-HYDROXYCHOLECALCIFEROL

9,10-SECOCHOLESTA-5,7,10(19)-TRIEN-3-BETA-OL see CHOLECAL-
CIFEROL

9,10,SECOERGOSTA-5,7,10(19),22-TETRAEN 3-BETA-OL see ERGO-
CALCIFEROL

SECONDARY AMMONIUM ARSENATE see DIAMMONIUM HYDROGEN
ARSENATE

SECONDARY AMMONIUM PHOSPHATE see AMMONIUM PHOSPHATE
DIBASIC

16,17-SECO-13-ALPHA-ESTRA-1,3,5,6,7,9-PENTAEN-17-OIC ACID,
METHYL ESTER see BIS DEHYDRO ISYNOLIC ACID METHYL ESTER

SEDEVAL see BARBITAL

S. EGRELTRI ATUNUN (TURKISH) see BRACKEN FERN, DRIED

SELEN (POLISH) see SELENIUM

SELENIC ACID, DIPOTASSIUM SALT see POTASSIUM SELENATE

SELENIC ACID, DISODIUM SALT see DISODIUM SELENATE

SELENIOUS ACID, DISODIUM SALT see SODIUM SELENITE

SELENIUM ALLOY see SELENIUM

SELENIUM BASE see SELENIUM

SELENIUM CYSTINE see 3,3'-DISELENODIALANINE

SELENIUM DUST see SELENIUM

SELENIUM FLUORIDE see SELENIUM HEXAFLUORIDE

SELENIUM HOMOPOLYMER see SELENIUM

SELENIUM HYDRIDE see HYDROGEN SELENIDE

SELENIUM(IV) DISULFIDE SHAMPOO (2.5%) see SELENIUM DISULFIDE
(2.5%) SHAMPOO

SELENIUM SULFIDE see SELENIUM MONOSULFIDE

SELENOCYSTINE see 3,3'-DISELENODIALANINE

SELF ROCK MOSS see CARRAGEEN

SEMESAN see 2-CHLORO-4-(HYDROXY MERCURI)PHENOL

SEMICARBAZIDE see HYDRAZINECARBOXAMIDE

SEMUSTINE see 1-(2-CHLOROETHYL)-3-(4-METHYL-CYCLOHEXYL)-1-
NITROSOUREA

SENECA OIL see PETROLEUM

SENECIONINE see AUREINE

SENEGAL GUM see ARABIC GUM

SENF OEL (GERMAN) see ALLYL ISOTHIOCYANATE

SEQUESTRENE TRISODIUM SALT see TRISODIUM EDETATE

STAUFFER N-4548 see METHYLPHOSPHODITHIOIC ACID-S-(((p-
CHLOROPHENYL)THIO)METHYL)-O-METHYL ESTER

SERNYL see ANGEL DUST

SERNYLAN see ANGEL DUST

SERNYL HYDROCHLORIDE see ANGEL DUST

SEROTONIN see 3-(2-AMINOETHYL)INDOL-5-OL

SERPASIL see RESERPINE

SERPASIL APRESOLINE see RESERPINE

STAUFFER R-3413 see S-(4,6-DIMETHYL-2-PYRIMIDINYL)-O,O-DIETHYL PHOSPHORODITHIOATE

N-SERVE NITROGEN STABILIZER see 2-CHLORO-6-(TRICHLOROMETHYL)PYRIDINE

SESQUISULFURE DE PHOSPHORE (FRENCH) see PHOSPHORUS SESQUISULFIDE

SEVIN see CARBARYL

SHEEP DIP see ARSENICAL DIP

SHELL GOLD see GOLD

SHELL SD 4294 see CROTOXYPHOS

SHELL SILVER see SILVER

SHELLSOL 140 see N-NONANE

SHIKIMATE see SHIKIMIC ACID

SHIKIMOLE see SAFROL

SIARKI CHLOREK (POLISH) see SULFUR CHLORIDE

SIARKI DWUTLENEK (POLISH) see SULFUR DIOXIDE

SIARKOWODOR (POLISH) see HYDROGEN SULFIDE

SIENNA see IRON (III) OXIDE

SIGMAMYCIN see TETRACYCLINE

SILASTIC see POLYDIMETHYLSILOXANE RUBBER

SILBER (GERMAN) see SILVER

SILBERNITRAT see SILVER (I) NITRATE (1:1)

SILICA AEROGEL see SILICA, AMORPHOUS HYDRATED

SILICA, AMORPHOUS see SILICA, AMORPHOUS FUMED

SILICA FLOUR see SILICA (CRYSTALLINE)

SILICA FLOUR (POWDERED CRYSTALLINE SILICA) see SILICA, CRYSTALLINE-QUARTZ

SILICA GEL see SILICA, PRECIPITATED (AMORPHOUS)

SILICA GEL see SILICA, AMORPHOUS HYDRATED

SILICA HYDROUS GEL see SILICA, PRECIPITATED (AMORPHOUS)

SILICA, VITREOUS see SILICA, AMORPHOUS FUSED

SILICA XEROGEL see SILICA, AMORPHOUS HYDRATED

SILICIC ACID see SILICA, AMORPHOUS HYDRATED

SILICIC ACID ALUMINUM SALT see ALUMINUM(III) SILICATE (2:1)

SILICIC ANHYDRIDE see SILICA, AMORPHOUS FUMED

SILICIC ANHYDRIDE see SILICA, CRYSTALLINE-QUARTZ

SILICI-CHLOROFORME (FRENCH) see TRICHLOROSILANE

SILICIO(TETRACLORURO DI) see SILICON CHLORIDE

SILICIUMCHLOROFORM (GERMAN) see TRICHLOROSILANE

SILICIUMTETRACHLORID (GERMAN) see SILICON CHLORIDE

SILICIUMTETRACHLORIDE (DUTCH) see SILICON CHLORIDE

SILICIUM(TETRACHLORURE DE) (FRENCH) see SILICON CHLORIDE

SILICOCHLOROFORM see TRICHLOROSILANE

SILICOETHANE see DISILANE

SILICON DIOXIDE see SILICA (CRYSTALLINE)

SILICON DIOXIDE see SILICA, AMORPHOUS FUSED

SILICONE RUBBER see POLYDIMETHYLSILOXANE RUBBER

SILICON MONOCARBIDE see SILICON CARBIDE

SILICON PHENYL TRICHLORIDE see TRICHLOROPHENYLSILANE

SILICON SODIUM FLUORIDE see DISODIUM HEXAFLUOROSILICATE

SILICON TETRACHLORIDE see SILICON CHLORIDE

SILICON TETRACHLORIDE see SILICON CHLORIDE

SILON see NYLON

SILOXANES see SILICONES

SILVAN (CZECH) see 2-METHYLFURAN

SILVER ATOM see SILVER

SILVER DIFLUORIDE see SILVER (II) FLUORIDE

SILVER FIR NEEDLE OIL see ABIES ALBA OIL

SILVER FIR OIL see ABIES ALBA OIL

SILVER MATT POWDER see TIN

SILVER PINE OIL see ABIES ALBA OIL

SINALOST see TRIMUSTINE

SINUTAB see p-ACETOPHENETIDIDE

SIRAN HYDRAZINU (CZECH) see HYDRAZINE SULFATE (1:1)

SIRNIK AMONNY (CZECH) see AMMONIUM SULFIDE

SIRNIK FOSFORECNY (CZECH) see PHOSPHORUS PENTASULFIDE

SIRUP see D-GLUCOSE

SK&F 36914 see CHLORO(TRIETHYLPHOSPHINE)GOLD

SKATOLE see beta-METHYLINDOLE

OMEGA-SKATOLE CARBOXYLIC ACID see 1H-INDOLE-3-ACETIC ACID

SKF 385 see trans-2-PHENYLCYCLOPROPYLAMINE

SKF 2538 see PHENOPROPAZINE

SKF-2601 see CHLOROPROMAZINE

SKF 25971 see (±)-9,10-DIHYDRO-N,N-DIMETHYL-2-TRIFLUOROMETHYL)-9-ANTHRACENE PROPANAMINE

SKF-525-A see 2-DIETHYLAMINOETHYLPROPYLDIPHENYLACETATE

SKLERO-TABLINEN see ATROMID S

SLAKED LIME see CALCIUM HYDROXIDE

SNAPPING HAZEL see HAMAMELIS

SNEEZING GAS see CHLORO DIPHENYL ARSINE

SNOWGOOSE see TALC (powder)

SOAP YELLOW F see FLUORESCEIN

SODA LYE see SODIUM HYDROXIDE (LIQUID)

SODA LYE see SODIUM HYDROXIDE

SODAMIDE see SODIUM AMIDE

SODA NITER see SODIUM (I) NITRATE (1:1)

SODA PHOSPHATE see SODIUM MONOHYDROGEN PHOSPHATE (2:1:1)

SODIO (CLORATO DI) (ITALIAN) see SODIUM CHLORATE

SODIO (DICROMATO DI) (ITALIAN) see SODIUM DICHROMATE

SODIO(IDROSSIDO DI) (ITALIAN) see SODIUM HYDROXIDE

SODIUM ACETAZOLAMIDE see ACETAZOLAMIDE SODIUM

SODIUM ACID ARSENATE see ARSENIC ACID, DISODIUM SALT

SODIUM ACID ARSENATE, HEPTAHYDRATE see ARSENIC ACID, DISODIUM SALT, HEPTAHYDRATE

SODIUM ACID METHANEARSONATE see MONOSODIUM METHYLARSONATE

SODIUM ACID PYROPHOSPHATE see DISODIUM PYROPHOSPHATE

SODIUM ACID SULFITE see SODIUM BISULFITE (1:1)

SODIUM-5-ALLYL-5-ISOPROPYLBARBITURATE see BUTALBITAL SODIUM

SODIUM-5-ALLYL-5-(1-METHYLBUTYL)-2-THIOBARBITURATE see SURITAL SODIUM

SODIUM-DL-5-ALLYL-1-METHYL-5-(1-METHYL-2-PENTYNYL)BARBITURATE see METHOHEXITAL SODIUM

SODIUM AMAZOLENE see ACID BLUE 92

SODIUM AMIDE (DOT) see SODIUM AMIDE

SODIUM AMINARSONATE see ARSANILIC ACID, MONOSODIUM SALT

SODIUM-P-AMINOBENZENEARSONATE see ARSANILIC ACID, MONOSODIUM SALT

SODIUM AMINOPHENOL ARSONATE see ARSANILIC ACID, MONOSODIUM SALT

SODIUM-P-AMINOPHENYLARSONATE see ARSANILIC ACID, MONOSODIUM SALT

SODIUM AMYLOBARBITONE see AMYTAL SODIUM

SODIUM ANAZOLENE see ACID BLUE 92

SODIUM ANILARSONATE see ARSANILIC ACID, MONOSODIUM SALT

SODIUM-ANILINE ARSONATE see ARSANILIC ACID, MONOSODIUM SALT

SODIUM ANTIMONY GLUCONATE see ANTIMONY (III) SODIUM GLUCONATE

SODIUM ANTIMONY (III) GLUCONATE see ANTIMONY (III) SODIUM GLUCONATE

SODIUM ANTIMONYL DIMETHYLCYSTEINE TARTRATE see ANTIMONY SODIUM DIMETHYL CYSTEINO TARTRATE

SODIUM ANTIMONYL TARTRATE see ANTIMONY SODIUM TARTRATE

SODIUM ANTIMONY (V) TARTRATE see ANTIMONY (V) SODIUM TARTRATE

SODIUM ANTIMONY TARTRATE see ANTIMONY SODIUM TARTRATE

SODIUM ANTIMONY (V) GLUCONATE see ANTIMONY (V) SODIUM GLUCONATE

SODIUM ARSANILATE see ARSANILIC ACID, MONOSODIUM SALT

SODIUM-P-ARSANILATE see ARSANILIC ACID, MONOSODIUM SALT

SODIUM ARSENATE see ARSENIC ACID, DISODIUM SALT

SODIUM ARSENATE (DOT) see ARSENIC ACID, SODIUM SALT

SODIUM ARSENATE DIBASIC, ANHYDROUS see ARSENIC ACID, DISODIUM SALT

SODIUM ARSENATE, DIBASIC, HEPTAHYDRATE see ARSENIC ACID, DISODIUM SALT, HEPTAHYDRATE

SODIUM ARSENATE HEPTAHYDRATE see ARSENIC ACID, DISODIUM SALT, HEPTAHYDRATE

SODIUM ARSONILATE see ARSANILIC ACID, MONOSODIUM SALT

SODIUM AUROTHIOSULPHATE DIHYDRATE see GOLD SODIUM THIO-SULFATE DIHYDRATE

SODIUM, AZOTURE DE (FRENCH) see SODIUM AZIDE

SODIUM, AZOTURO DI (ITALIAN) see SODIUM AZIDE

SODIUM-1,2 BENZISOTHIAZOLIN-3-ONE-1,1-DIOXIDE see SACCHARIN

SODIUM-O-BENZOSULFIMIDE see SACCHARIN

SODIUM-2-BENZOSULPHIMIDE see SACCHARIN

SODIUM-O-BENZOSULPHIMIDE see SACCHARIN

SODIUM BENZYLPENICILLINATE see BENZYL PENICILLINIC ACID SO-DIUM SALT

SODIUM BENZYLPENICILLIN G see BENZYL PENICILLINIC ACID SO-DIUM SALT

SODIUM BIS(2-ETHYLHEXYL) SULFOSUCCINATE see DIOCTYL SODIUM SULFOSUCCINATE

SODIUM BISMUTH THIOGLYCOLATE see BISMUTH SODIUM THIOGLY-COLLATE

SODIUM BISMUTH THIOGLYCOLLATE see BISMUTH SODIUM THIOGLY-COLLATE

SODIUM BISULFIDE see SODIUM HYDROSULFIDE

SODIUM BISULPHITE see SODIUM BISULFITE (1:1)

SODIUM-5-(2-BROMOALLYL)-5-SEC-BUTYLBARBITURATE see BUTAL-LYLONAL SODIUM

SODIUM BUTABARBITAL see BUTISOL SODIUM

SODIUM-5-SEC-BUTYL-5-ETHYLBARBITURATE see BUTISOL SODIUM

SODIUM CARBOLATE see SODIUM PHENOXIDE

SODIUM (CHLORATE DE) (FRENCH) see SODIUM CHLORATE

SODIUM CHROMATE see DISODIUM CHROMATE

SODIUM COUMADIN see COUMADIN SODIUM

SODIUM CUMENEAZO-BETA-NAPHTHOL DISULFONATE see FD AND C RED NO. 1

SODIUM CUMENEAZO-BETA-NAPHTHOL DISULPHONATE see FD AND C RED NO. 1

SODIUM CYCLOHEXANESULFAMATE see SODIUM CYCLAMATE

SODIUM CYCLOHEXYL SULFAMIDATE see SODIUM CYCLAMATE

SODIUM 2,4-D see SODIUM-2,4-DICHLOROPHENOXYACETATE

SODIUM DICHLORISOCYANURATE see SODIUM DICHLOROCYANURATE

SODIUM DICHLOROISOCYANURATE see SODIUM DICHLOROCYANU-RATE

SODIUM-2-(2,4-DICHLOROPHENOXY)ETHYL SULFATE see CRAG HER-BICIDE

SODIUM-2,4-DICHLOROPHENOXYETHYL SULPHATE see CRAG HERBI-CIDE

SODIUM-2,4-DICHLOROPHENYL CELLOSOLVE SULFATE see CRAG HER-BICIDE

1-SODIUM-3,5-DICHLORO-S-TRIAZINE-2,4,6-TRIONE see SODIUM DI-CHLOROCYANURATE

SODIUM DICHROMATE DE (FRENCH) see SODIUM DICHROMATE

SODIUM DIETHYLBARBITURATE see BARBITAL SODIUM

SODIUM-5,5-DIETHYLBARBITURATE see BARBITAL SODIUM

SODIUM-N,N-DIETHYLDITHIOCARBAMATE see SODIUM DIETHYLDI-THIOCARBAMATE

SODIUM DI-(2-ETHYLHEXYL) SULFOSUCCINATE see DIOCTYL SODIUM SULFOSUCCINATE

SODIUM 4-(DIMETHYLAMINO)BENZENEDIAZOSULFONATE see METHYL ORANGE

SODIUM P-(DIMETHYLAMINO)BENZENEDIAZOSULFONATE see METHYL ORANGE

SODIUM 4-(DIMETHYLAMINO)BENZENEDIAZOSULPHONATE see METHYL ORANGE

SODIUM P-(DIMETHYLAMINO)BENZENEDIAZOSULPHONATE see METHYL ORANGE

SODIUM (4-(DIMETHYLAMINO)PHENYL)DIAZENESULFONATE see METHYL ORANGE

SODIUM-4,6-DINITRO-O-CRESOXIDE see 3,5-DINITRO-O-CRESOL SO-DIUM SALT

SODIUM DIOCTYL SULFOSUCCINATE see DIOCTYL SODIUM SULFO-SUCCINATE

SODIUM DIOXIDE see SODIUM PEROXIDE

SODIUM DIPHENYL-4,4'-BIS-AZO-2''-8''-AMINO-1''-NAPHTHOL-3'',6''-DISULPHONATE see C.I. DIRECT BLUE 6, TETRASODIUM SALT

SODIUM DIPHENYLHYDANTOIN see DILANTIN

SODIUM DIPHENYL HYDANTOINATE see DILANTIN

SODIUM-5,5-DIPHENYLHYDANTOINATE see DILANTIN

SODIUM-5,5-DIPHENYL-2,4-IMIDAZOLIDINEDIONE see DILANTIN

SODIUM DITOLYLDIAZOBIS-8-AMINO-1-NAPHTHOL-3,6-DISULFONATE see C.I. DIRECT BLUE 14, TETRASODIUM SALT

SODIUM DODECYLBENZENESULFONATE see DODECYL BENZENE SO-DIUM SULFONATE

SODIUM DODECYLSULFATE see SULFURIC ACID, MONODODECYL ES-TER, SODIUM SALT

SODIUM EOSINATE see BROMO EOSINE

SODIUM ETHASULFATE see TERGITOL 08

SODIUM ETHYLBARBITAL see BARBITAL SODIUM

SODIUM-5-ETHYL-5-SEC-BUTYLBARBITURATE see BUTISOL SODIUM

SODIUM ETHYLENEDIAMINETETRAACETATE see EDATHANIL TETRA-SODIUM

SODIUM ETHYLENEDIAMINETETRAACETIC ACID see EDATHANIL TET-RASODIUM

SODIUM(2-ETHYLHEXYL)ALCOHOL SULFATE see TERGITOL 08

SODIUM ETHYLISOAMYLBARBITURATE see AMYTAL SODIUM

SODIUM ETHYLMERCURIC THIOSALICYLATE see MERTHIOLATE SO-DIUM

SODIUM-O-(ETHYLMERCURITHIO)BENZOATE see MERTHIOLATE SO-DIUM

SODIUM ETHYLMERCURITHIOSALICYLATE see MERTHIOLATE SODIUM

SODIUM-5-ETHYL-5-(1-METHYL-1-BUTENYL) BARBITURATE see VIN-BARBITAL SODIUM

SODIUM 5-ETHYL-5-(1-METHYLBUTYL)BARBITURATE see NEMBUTAL SODIUM

SODIUM-5-ETHYL-5-(1-METHYLBUTYL)-2-THIOBARBITURATE see PEN-TOTHAL SODIUM

SODIUM-5-ETHYL-5-(1-METHYLPROPYL)BARBITURATE see BUTISOL SODIUM

SODIUM-5-ETHYL-5-PHENYLBARBITURATE see SODIUM LUMINAL

SODIUM-5-ETHYL-5-PHENYLBARBITURATE see SODIUM LUMINAL

SODIUM FLUORESCEIN see FLUORESCEIN SODIUM

SODIUM FLUORIDE, SOLID (DOT) see SODIUM FLUORIDE

SODIUM FLUOROSILICATE see DISODIUM HEXAFLUOROSILICATE

SODIUM FLUORURE (FRENCH) see SODIUM FLUORIDE

SODIUM FLUOSILICATE see DISODIUM HEXAFLUOROSILICATE

SODIUM FUMARATE see DISODIUM FUMARATE

L(+) SODIUM GLUTAMATE see MONOSODIUM GLUTAMATE

SODIUM HEPARIN see HEPARIN

SODIUM HEXAFLUOROSILICATE see DISODIUM HEXAFLUOROSILICATE

SODIUM HYDRATE (DOT) see SODIUM HYDROXIDE

SODIUM HYDRATE SOLUTION see SODIUM HYDROXIDE (LIQUID)

SODIUM HYDROGEN PHOSPHATE see SODIUM MONOHYDROGEN PHOS-PHATE (2:1:1)

SODIUM HYDROGEN SULFIDE see SODIUM HYDROSULFIDE

SODIUM HYDROGEN SULFITE see SODIUM BISULFITE (1:1)

SODIUM HYDROXIDE, BEAD (DOT) see SODIUM HYDROXIDE

SODIUM HYDROXIDE, DRY (DOT) see SODIUM HYDROXIDE

SODIUM HYDROXIDE, FLAKE (DOT) see SODIUM HYDROXIDE

SODIUM HYDROXIDE, GRANULAR (DOT) see SODIUM HYDROXIDE

SODIUM HYDROXIDE, SOLID (DOT) see SODIUM HYDROXIDE

SODIUM HYDROXIDE SOLUTION see SODIUM HYDROXIDE (LIQUID)

SODIUM-O-HYDROXYBENZOATE see SODIUM SALICYLATE

SODIUM(HYDROXYDE DE) (FRENCH) see SODIUM HYDROXIDE

SODIUM-2-HYDROXYDIPHENYL see 2-BIPHENYLOL, SODIUM SALT

SODIUM 0-((3-(HYDROXYMERCURI) -2-METHOXYPROPYL)CARBA-MOYL)PHENOXY ACETATE see SODIUM MERSALYL

SODIUM-1-(HYDROXYMETHYL) CYCLOHEXANEACETATE see SODIUM HEXACYCLONATE

SODIUM HYPOSULFITE see SODIUM THIOSULFATE, PENTAHYDRATE

SODIUM 5,5'-INDIGOTIDISULFONATE see DISODIUM INDIGO-5,5-DISULFONATE

SODIUM IODINE see SODIUM IODIDE

SODIUM ISOAMYLETHYL BARBITURATE see AMYTAL SODIUM

SODIUM LAURYLBENZENESULFONATE see DODECYL BENZENE SODIUM SULFONATE

SODIUM LAURYLSULFATE see SULFURIC ACID, MONODODECYL ESTER, SODIUM SALT

SODIUM MALONYLUREA see BARBITAL SODIUM

SODIUM MERCAPTAN see SODIUM HYDROSULFIDE

SODIUM MERCAPTIDE see SODIUM HYDROSULFIDE

SODIUM MERCAPTOACETATE see SODIUM THIOGLYCOLATE

SODIUM MERCAPTOMERIN see DISODIUM-N-(3-(CARBOXYMETHYL THIOMERCURI)-2-METHOXYPROPYL)-alpha-CAMPHORAMATE

SODIUM MERTHIOLATE see MERTHIOLATE SODIUM

SODIUM METAARSENITE see SODIUM ARSENITE

SODIUM METAL (DOT) see SODIUM

SODIUM, METAL DISPERSION IN ORGANIC SOLVENT see SODIUM, (SOLUTION)

SODIUM METASILICATE see SODIUM SILICATE

SODIUM METASILICATE, ANHYDROUS see SODIUM SILICATE

SODIUM METAVANADATE see SODIUM VANADATE

SODIUM METHANEARSONATE see MONOSODIUM METHYLARSONATE

SODIUM METHANEARSONATE see DISODIUM METHANEARSENATE

4-SODIUM METHANESULFONATE METHYLAMINE-ANTIPYRINE see AMINOPYRINE SODIUM SULFONATE

SODIUM METHARSONATE see DISODIUM METHANEARSENATE

SODIUM A-DL-1-METHYL-5-ALLYL-5-(1-METHYL-2-PENTYNYL)BARBITURATE see METHOHEXITAL SODIUM

SODIUM METHYLAMINOANTIPYRINE METHANESULFONATE see AMINOPYRINE SODIUM SULFONATE

SODIUM-4-METHYLAMINO-1,5-DIMETHYL-2-PHENYL-3-PYRAZOLONE 4-METHANESULFONATE see AMINOPYRINE SODIUM SULFONATE

SODIUM METHYLARSONATE see DISODIUM METHANEARSENATE

SODIUM METHYLDITHIOCARBAMATE see VAPAM

SODIUM-N-METHYLDITHIOCARBAMATE see VAPAM

SODIUM MOLYBDATE(VI) see DISODIUM MOLYBDATE

SODIUM MONOFLUOROACETATE see SODIUM FLUOROACETATE

SODIUM MONOIODIDE see SODIUM IODIDE

SODIUM MONOSULFIDE see SODIUM SULFIDE (ANHYDROUS)

SODIUM NITROPRUSSATE see SODIUM NITROFERRICYANIDE

SODIUM NITROPRUSSIDE see SODIUM NITROFERRICYANIDE

SODIUM NITROSYLPENTACYANOFERRATE (III) see SODIUM NITROFERRICYANIDE

SODIUM NORAMIDOPYRINE METHANESULFONATE see AMINOPYRINE SODIUM SULFONATE

SODIUM ORTHOARSENATE see ARSENIC ACID, SODIUM SALT

SODIUM ORTHOARSENITE see ARSENIOUS ACID SODIUM SALT

SODIUM OXIDE see SODIUM MONOXIDE

SODIUM PENICILLIN G see BENZYL PENICILLINIC ACID SODIUM SALT

SODIUM PENICILLIN II see BENZYL PENICILLINIC ACID SODIUM SALT

SODIUM PENTABARBITAL see NEMBUTAL SODIUM

SODIUM PENTABARBITONE see NEMBUTAL SODIUM

SODIUM PENTACHLOROPHENOL see SODIUM PENTACHLOROPHENATE

SODIUM PENTACHLOROPHENOLATE see SODIUM PENTACHLOROPHENATE

SODIUM PENTACHLOROPHENOXIDE see SODIUM PENTACHLOROPHENATE

SODIUM-BETA,BETA-PENTAMETHYLENE-GAMMA-HYDROXYBUTYRATE see SODIUM HEXACYCLONATE

SODIUM PENTOBARBITAL see NEMBUTAL SODIUM

SODIUM PENTOBARBITONE see NEMBUTAL SODIUM

SODIUM PENTOBARBITURATE see NEMBUTAL SODIUM

SODIUM PENTOTHAL see PENTOTHAL SODIUM

SODIUM PENTOTHIOBARBITAL see PENTOTHAL SODIUM

SODIUM PEROXYDISULFATE see SODIUM PERSULFATE

SODIUM PHENATE see SODIUM PHENOXIDE

SODIUM-PHENOBARBITAL see SODIUM LUMINAL

SODIUM-PHENOBARBITONE see SODIUM LUMINAL

SODIUM PHENOLATE, SOLID (DOT) see SODIUM PHENOXIDE

SODIUM PHENYLBUTAZONE see BUTAZOLIDINE SODIUM

SODIUM-1-PHENYL-2,3-DIMETHYL-4-METHYLAMINOPYRAZOLON-N-METHANESULFONATE see AMINOPYRINE SODIUM SULFONATE

SODIUM-1-PHENYL-2,3-DIMETHYL-5-PYRAZOLONE-4-METHYLAMINO METHANESULFONATE see AMINOPYRINE SODIUM SULFONATE

SODIUM PHENYLDIMETHYLPYRAZOLON-METHYLAMINO-METHANE SULFONATE see AMINOPYRINE SODIUM SULFONATE

SODIUM-PHENYLETHYLBARBITURATE see SODIUM LUMINAL

SODIUM PHENYLETHYLMALONYLUREA see SODIUM LUMINAL

SODIUM-N-PHENYLGLYCINAMIDE-P-ARSONATE see N-(CARBAMOYL METHYL)ARSANILIC ACID

SODIUM-2-PHENYLPHENATE see 2-BIPHENYLOL, SODIUM SALT

SODIUM-O-PHENYLPHENATE see 2-BIPHENYLOL, SODIUM SALT

SODIUM-O-PHENYLPHENOLATE see 2-BIPHENYLOL, SODIUM SALT

SODIUM-O-PHENYLPHENOXIDE see 2-BIPHENYLOL, SODIUM SALT

SODIUM POTASSIUM ALLOY (DOT) see POTASSIUM SODIUM ALLOY

SODIUM PYROPHOSPHATE see TETRASODIUM PYROPHOSPHATE, ANHYDROUS

SODIUM PYROPHOSPHATE see DISODIUM PYROPHOSPHATE

SODIUM RHODANATE see SODIUM ISOTHIOCYANATE

SODIUM RHODANIDE see SODIUM ISOTHIOCYANATE

SODIUM SACCHARIDE see SACCHARIN

SODIUM SACCHARIN see SACCHARIN

SODIUM SACCHARINATE see SACCHARIN

SODIUM SACCHARINE see SACCHARIN

SODIUM SALICYL-(GAMMA-HYDROXYMERCURI-BETA-METHOXYPROPYL)AMIDE-O-ACETATE see SODIUM MERSALYL

SODIUM SALICYLIC ACID see SODIUM SALICYLATE

SODIUM SALT OF DICHLORO-S-TRIAZINETRIONE see SODIUM DICHLOROCYANURATE

SODIUM SALT OF ETHYLENEDIAMINETETRAACETIC ACID TETRASODIUM EDETATE see EDATHANIL TETRASODIUM

SODIUM SALT OF HYDROXY-O-CARBOXY-PHENYL-FLUORONE see FLUORESCEIN SODIUM

SODIUM SELENATE see DISODIUM SELENATE

SODIUM SILICOFLUORIDE see DISODIUM HEXAFLUOROSILICATE

SODIUM SUCARYL see SODIUM CYCLAMATE

SODIUM SULFATE ANHYDROUS see SODIUM SULFATE (2:1)

SODIUM SULFHYDRATE see SODIUM HYDROSULFIDE

SODIUM SULFOCYANATE see SODIUM ISOTHIOCYANATE

SODIUM SULFOCYANIDE see SODIUM ISOTHIOCYANATE

SODIUM SULFODI-(2-ETHYLHEXYL) SULFOSUCCINATE see DIOCTYL SODIUM SULFOSUCCINATE

SODIUM SULPHIDE see SODIUM SULFIDE (ANHYDROUS)

SODIUM SUPEROXIDE see SODIUM PEROXIDE

SODIUM TELLURATE (IV) see SODIUM TELLURITE

SODIUM TETRAHYDROBORATE(1-) see SODIUM BOROHYDRIDE

SODIUM THIOCYANATE see SODIUM ISOTHIOCYANATE

SODIUM THIOCYANIDE see SODIUM ISOTHIOCYANATE

SODIUM THIOPENTAL see PENTOTHAL SODIUM

SODIUM-P-TOLUENE SULFON CHLORAMIDE see CHLORAMINE-T

SODIUM TRIPHOSPHATE see SODIUM TRIPOLYPHOSPHATE

SODIUM VERONAL see BARBITAL SODIUM

SODIUM VINBARBITAL see VINBARBITAL SODIUM

SODIUM WARFARIN see COUMADIN SODIUM

SOHNHOFEN STONE see CALCIUM (II) CARBONATE (1:1)

SOLANCARPIDINE see PURAPURIDINE

SOLASOD-5-EN-3-BETA-OL see PURAPURIDINE

SOLGANAL see 1-AUROTHIO-D-GLUCOPYRANOSE

SOLGANAL B see 1-AUROTHIO-D-GLUCOPYRANOSE

SOLID CROTONIC ACID see CROTONIC ACID

SOL SODOWA KWASU LAURYLOBENZENOSULFONOWEGO(POLISH) see DODECYL BENZENE SODIUM SULFONATE

SOLUBLE BARBITAL see BARBITAL SODIUM

SOLUBLE FLUORESCEIN see FLUORESCEIN SODIUM

SOLUBLE GLASS see SODIUM SILICATE

SOLUBLE GLUSIDE see SACCHARIN

SOLUBLE GUN COTTON see CELLULOSE TETRANITRATE

SOLUBLE INDIGO see DISODIUM INDIGO-5,5-DISULFONATE

SOLUBLE PHENOBARBITAL see SODIUM LUMINAL

SOLUBLE PHENOBARBITONE see SODIUM LUMINAL

SOLUBLE SACCHARIN see SACCHARIN

SOLUBLE THIOPENTONE see PENTOTHAL SODIUM

SOLUTION GLYCERYL TRINITRATE see SPIRIT OF GLYCERYL TRINI-TRATE

SOLUTION POTASSIUM IODOHYDRAGYRATE see NESSLER REAGENT

SOLVENT ETHER see ETHYL ETHER

SOLVENT NAPHTHA see PETROLEUM SPIRITS

SOMALIA RED III see OIL RED

SOPHORETIN see QUERCETIN

SOPRINTIN see ACETOPROMAZINE

SOPRONTIN see ACETOPROMAZINE

SOPROTIN see ACETOPROMAZINE

SORBO-CALCIAN see CALCIUM ACETATE

SOUP see NITROGLYCERIN

SOUTHERN BENTONITE see BENTONITE

SPANISH FLY see CANTHARIDES

SPARINE HYDROCHLORIDE see PROMAZINE HYDROCHLORIDE

SPECULAR IRON see IRON (III) OXIDE

SPEED see dl-DESOXY EPHEDRINE HYDROCHLORIDE

SPHERES OR ''TEARS'' see ACACIA GUM

SPHEROIDINE see FUGU POISON

SPIRIT OF GLONOIN see SPIRIT OF GLYCERYL TRINITRATE

SPIRIT OF HARTSHORN see AMMONIA

SPIRIT OF TRINITROGLYCERIN see SPIRIT OF GLYCERYL TRINITRATE

SPIRIT OF TURPENTINE see TURPENTINE

SPIRITS OF NITROGLYCERIN (DOT) see SPIRIT OF GLYCERYL TRINI-TRATE

SPIRITS OF SALT (DOT) see HYDROCHLORIC ACID

SPIRITS OF TURPENTINE see TURPENTINE

SPIRITS OF WINE see ETHANOL

SPIROCID see ACETPHENARSINE

SPIRO(17H-CYCLOPENTA(A)PHENAUTHRENE-17,2'-(3'H)-FURAN) see ALDACTAZIDE

SPIROFULVIN see GRISOFULVIN

SPIROZID see ACETPHENARSINE

SPOTTED ALDER see HAMAMELIS

SPRITZ-HORMIT see SODIUM-2,4-DICHLOROPHENOXYACETATE

SPURRED RYE see ERGOT

SQ 22947 see TIAMUTIN

SQUILL see RED SQUILL

S,S'-METHYLENE O,O,O',O'-TETRAETHYL PHOSPHORODITHIOATE see ETHYL METHYLENE PHOSPHORO DITHIOATE

S,S,S-TRIBUTYL PHOSPHOROTRITHIOATE see BUTYL PHOSPHORO TRI-THIOATE

S,S,S-TRIBUTYL PHOSPHOROTRITHIOITE see S,S,S-TRIBUTYL TRITHIO-PHOSPHITE

S,S,S-TRIBUTYL TRITHIOPHOSPHATE see BUTYL PHOSPHORO TRITHIO-ATE

STABILIZED ETHYL PARATHION see PARATHION

STAGNO (TETRACLORURO DI) (ITALIAN) see TIN (IV) CHLORIDE (1:4)

STANNANE, TRIETHYLHYDROXY-, SULFATE (2:1) (8CI) see BIS(TRI-ETHYL TIN) SULFATE

STANNIC BROMIDE see TIN (IV) BROMIDE (1:4)

STANNIC CHLORIDE, ANHYDROUS see TIN (IV) CHLORIDE (1:4)

STANNIC IODIDE see TIN (IV) IODIDE (1:4)

STANNIC PHOSPHIDE (DOT) see TIN (IV) PHOSPHIDE

STANNORAM see DECAFENTIN

STANNOUS CHLORIDE see TIN (II) CHLORIDE (1:2)

STANNOUS IODIDE see TIN (II) IODIDE

STAR ANISE OIL see ANISE OIL

STARCH GUM see DEXTRINS

STAR DUST see COCAINE

STARSOL NO. 1 see ARABIC GUM

STEAREX BEADS see STEARIC ACID

STEARIC ACID, 2-HYDROXYETHYL ESTER see ETHYLENE GLYCOL STEARATE

STEARIC ACID, MONOESTER WITH ETHYLENE GLYCOL see ETHYLENE GLYCOL STEARATE

STEARIC ACID, ZINC SALT see ZINC STEARATE

STEAROPHANIC ACID see STEARIC ACID

STEARYL ALCOHOL see 1-OCTADECANOL

USP XIII STEARYL ALCOHOL see 1-OCTADECANOL

STEARYLAMINE see OCTADECYLAMINE

STEARYLDIMETHYLBENZYLAMMONIUM CHLORIDE see DIMETHYLOCTADECYLBENZYLAMMONIUM CHLORIDE

STECKAPFUL (GERMAN) see STRAMONIUM

STEINBUHL YELLOW see BARIUM CHROMATE (VI)

STEINBUHL YELLOW see CALCIUM CHROMATE (VI) DIHYDRATE

STIBIC ANHYDRIDE see ANTIMONY PENTOXIDE

STIBIUM see ANTIMONY

STICKDIOXYD (GERMAN) see NITROGEN OXIDE

STICKMONOXYD (GERMAN) see NITROGEN MONOXIDE

STICKSTOFFDIOXID (GERMAN) see NITROGEN DIOXIDE

STIKSTOFDIOXYDE (DUTCH) see NITROGEN DIOXIDE

STILBEN (GERMAN) see STILBENE

TRANS-4-N-STILBENAMINE see trans-4-AMINOSTILBENE

TRANS-4-STILBENE see trans-4-AMINOSTILBENE

ALPHA-STILBENECARBONITRILE see 2,3-DIPHENYLACRYLONITRILE

4-STILBENYL-N,N-DIETHYLAMINE see N,N-DIETHYL-4-STILBENAMINE

STILBENYL-N,N-DIMETHYLAMINE see N,N-DIMETHYL-4-STILBENA-MINE

STILBESTROL see DIETHYLSTILBESTEROL

STILBESTROL DIMETHYL ETHER see 3,4-DIANISYL-3-HEXENE

STILBESTROL DIPROPIONATE see DIETHYLSTILBESTROL DIPROPIO-NATE

STIMULAN see DL-AMPHETAMINE SULFATE

STINK DAMP see HYDROGEN SULFIDE

STIROLO (ITALIAN) see STYRENE

ST. JOHN'S BREAD see LOCUST BEAN GUM

STOVAINE see AMYLOCAINE

STOVARSAL see ACETPHENARSINE

STOVARSOL see N-ACETYL-4-HYDROXY-m-ARSANILIC ACID

STOVARSOL see ACETPHENARSINE

STOVARSOLAN see ACETPHENARSINE

STP see 2,5-DIMETHOXY-4-METHYLAMPHETAMINE

STRAMONA (ITALIAN) see STRAMONIUM

STRAWBERRY RED A GEIGY see FOOD RED 7

STREPTOMYCES PEUCETIUS see DAUNOMYCIN

STREPTOMYCIN A see STREPTOMYCIN

STREPTOMYZIN (GERMAN) see STREPTOMYCIN

STREPTOZOCIN see STREPTOZOTICIN

STRICNINA (ITALIAN) see STRYCHNINE

STRIPED ALDER see HAMAMELIS

STROBANE see TERPENE POLYCHLORINATES

STRONTIUM CHROMATE (VI) see STRONTIUM CHROMATE (1:1)

STRONTIUM CHROMATE 12170 see STRONTIUM CHROMATE (1:1)

STRONTIUM MONOSULFIDE see STRONTIUM SULFIDE

STRONTIUM SULPHIDE see STRONTIUM SULFIDE

STRYCHNINE SULFATE see STRYCHNINE SULFATE (2:1)

STRYCHNIDIN-10-ONE, SULFATE (2:1) see STRYCHNINE SULFATE (2:1)

STRYCHNIN (GERMAN) see STRYCHNINE

STRYCHNINE NITRATE see STRYCHNINE MONONITRATE

STRYCHNINE SALT, SOLID (DOT) see STRYCHNINE SALT (solid)

STYRALLYL ALCOHOL see alpha-PHENETHYL ALCOHOL

STYREEN (DUTCH) see STYRENE

STYREN (CZECH) see STYRENE

STYREN-ACRYLONITRILEPOLYMER see ACRYLONITRILE POLYMER WITH STYRENE

STYRENE-ACRYLONITRILE-BUTADIENE COPOLYMER see ACRYLONI-TRILE, POLYMER WITH 1,3-BUTADIENE AND STYRENE

STYRENE-ACRYLONITRILE-BUTADIENE POLYMER see ACRYLONITRILE, POLYMER WITH 1,3-BUTADIENE AND STYRENE

STYRENE-ACRYLONITRILE-BUTADIENE RESIN see ACRYLONITRILE, POLYMER WITH 1,3-BUTADIENE AND STYRENE

STYRENE-ACRYLONITRILE-BUTADIENE TERPOLYMER see ACRYLONI-TRILE, POLYMER WITH 1,3-BUTADIENE AND STYRENE

STYRENE-ACRYLONITRILE COPOLYMER see ACRYLONITRILE POLYMER WITH STYRENE

STYRENE-BUTADIENE-ACRYLONITRILE COPOLYMER see ACRYLONI-TRILE, POLYMER WITH 1,3-BUTADIENE AND STYRENE

STYRENE EPOXIDE see 1,2-EPOXYETHYLBENZENE
STYRENE OXIDE see 1,2-EPOXYETHYLBENZENE
STYROL (GERMAN) see STYRENE
STYRYL 430 see 2-(p-AMINOSTYRYL)-6-(p-ACETYLAMINOBENZOYL-AMINO)QUINOLINE METHOACETATE
STYRYL OXIDE see 1,2-EPOXYETHYLBENZENE
SUBCHLORIDE OF MERCURY see MERCUROUS CHLORIDE
SUBERANE see CYCLOHEPTANE
SUBERONE see CYCLOHEPTANONE
SUBLIMAT (CZECH) see CORROSIVE SUBLIMATE
SUBLIMAZE see PHENTANYL CITRATE
SUBLIMAZE CITRATE see PHENTANYL CITRATE
SUBLIMED SULFUR see SULFUR
SUBSTANCE II see AJMALICINE
SUBTILISIN (9CI) see BACILLUS SUBTILIS CARLSBERG
SUBTILISIN BPN see BACILLUS SUBTILIS BPN
SUBTILOPEPTIDASE BPN' see BACILLUS SUBTILIS CARLSBERG
SUCARYL CALCIUM see CALCIUM CYCLOHEXYLSULPHAMATE
SUCARYL SODIUM see SODIUM CYCLAMATE
SUCCINBROMIMIDE see N-BROMO SUCCINIMIDE
SUCCINIBROMIMIDE see N-BROMO SUCCINIMIDE
SUCCINIC ACID ANHYDRIDE see SUCCINIC ANHYDRIDE
SUCCINIC ACID BIS(BETA-DIMETHYLAMINOETHYL) ESTER BISMETH-IODIDE see BIS(beta-DIMETHYL AMINO ETHYL)SUCCINATE BIS-(METHYL IODIDE)
SUCCINIC ACID BIS(BETA-DIMETHYLAMINOETHYL) ESTER, DIHYDRO-CHLORIDE see SUCCINOYLCHOLINE CHLORIDE
SUCCINIC ACID BIS(BETA-DIMETHYLAMINOETHYL)ESTER DIMETHO-CHLORIDE see SUCCINOYLCHOLINE CHLORIDE
SUCCINIC ACID DIESTER WITH CHOLINE see CHOLINE SUCCINATE (2:1) (ESTER)
SUCCINIC ACID DIESTER WITH CHOLINE CHLORIDE see SUCCINOYL-CHOLINE CHLORIDE
SUCCINIC ACID, DIESTER WITH CHOLINE IODIDE see BIS(beta-DI-METHYL AMINO ETHYL)SUCCINATE BIS(METHYL IODIDE)
SUCCINIC ACID-2,2-DIMETHYLHYDRAZIDE see DIMETHYLAMINOSUC-CINAMIC ACID
SUCCINIC ACID DINITRILE see SUCCINONITRILE
SUCCINIC ACID PEROXIDE (DOT) see SUCCINIC PEROXIDE
SUCCINIC-1,1-DIMETHYL HYDRAZIDE see DIMETHYLAMINOSUCCINA-MIC ACID
SUCCINIC DINITRILE see SUCCINONITRILE
SUCCINODINITRILE see SUCCINONITRILE
SUCCINOYLCHOLINE see CHOLINE SUCCINATE (2:1) (ESTER)
SUCCINYLBISCHOLINE see CHOLINE SUCCINATE (2:1) (ESTER)
SUCCINYL BISCHOLINE CHLORIDE see SUCCINOYLCHOLINE CHLORIDE
SUCCINYLCHOLINE CHLORIDE see SUCCINOYLCHOLINE CHLORIDE
SUCCINYLCHOLINE DICHLORIDE see SUCCINOYLCHOLINE CHLORIDE
SUCCINYLCHOLINE HYDROCHLORIDE see SUCCINOYLCHOLINE CHLO-RIDE
SUCCINYLDICHOLINE see CHOLINE SUCCINATE (2:1) (ESTER)
SUCCINYLDICHOLINE CHLORIDE see SUCCINOYLCHOLINE CHLORIDE
SUCCINYLDICHOLINE IODIDE see BIS(beta-DIMETHYL AMINO ETHYL)-SUCCINATE BIS(METHYL IODIDE)
O-O-SUCCINYLDICHOLINE IODIDE see BIS(beta-DIMETHYL AMINO ETHYL)SUCCINATE BIS(METHYL IODIDE)
SUCCINYL OXIDE see SUCCINIC ANHYDRIDE
SUCCINYL PEROXIDE see SUCCINIC PEROXIDE
SUCRE EDULCOR see 1,2-BENZISOTHIAZOL-3(2H)-ONE-1,1-DIOXIDE
SUCRETS see HEXYLRESORCINOL
SUDAN III see OIL RED
SUDAN ORANGE R see 1-(PHENYLAZO)-2-NAPHTHOL
SUGAR see SUCROSE
SUGAR OF LEAD see LEAD ACETATE
SULFACID BRILLIANT GREEN 1B see ACID GREEN 3
SULFADIMETHYLPYRIMIDINE see SULFADIMETHYLDIAZINE
SULFAISODIMIDINE see SULFADIMETHYLDIAZINE
SULFAMETHYLISOXAZOLE see SULFAMETHOXAZOL
P-SULFAMIDOANILINE see SULFANILAMIDE
SULFAMINSAURE (GERMAN) see SULFAMIC ACID, MONOAMMONIUM SALT

2-SULFANILAMIDO-4,6-DIMETHYLPYRIMIDINE see SULFADIMETHYL-DIAZINE
3-SULFANILAMIDO-5-METHYLISOXAZOLE see SULFAMETHOXAZOL
N(SUP 1)-SULFANILYL-N(SUP 2)-BUTYLCARBAMIDE see 1-BUTYL-3-SULFANILYL UREA
N(SUP 1)-SULFANILYL-N(SUP 2)-BUTYLUREA see 1-BUTYL-3-SUL-FANILYL UREA
SULFAPOLU (POLISH) see DODECYL BENZENE SODIUM SULFONATE
SULFARSPHENAMINE BISMUTH see BISMUTH ARSPHENAMINE SULFO-NATE
SULFATE D'ATROPINE (FRENCH) see ATROPINE SULFATE (2:1)
SULFATE DE CUIVRE (FRENCH) see COPPER (II) SULFATE (1:1)
SULFATE DE METHYLE (FRENCH) see SULFURIC ACID, DIMETHYL ES-TER
SULFATE DE NICOTINE (FRENCH) see NICOTINE SULFATE
SULFATE DE PLOMB (FRENCH) see LEAD (II) SULFATE (1:1)
SULFATE DE ZINC (FRENCH) see ZINC SULFATE
SULFATE DIMETHYLIQUE (FRENCH) see SULFURIC ACID, DIMETHYL ESTER
SULFATE MERCURIQUE (FRENCH) see MERCURY(II) SULFATE (1:1)
SULFENAMIDE TS see N-CYCLOHEXYL-2-BENZOTHIAZOLESULFENA-MIDE
SULFINPYRAZINE see DIPHENYLPYRAZONE
O-SULFOBENZIMIDE see 1,2-BENZISOTHIAZOL-3(2H)-ONE-1,1-DIOX-IDE
O-SULFOBENZOIC ACID IMIDE see 1,2-BENZISOTHIAZOL-3(2H)-ONE-1,1-DIOXIDE
SULFOCARBANILIDE see DIPHENYLTHIOUREA
SULFODOR (CZECH) see ETHYL SULFIDE
SULFONAMIDE see SULFANILAMIDE
SULFONAMIDE P see SULFANILAMIDE
1-(4-SULFO-1-NAPHTHYLAZO)-2-NAPHTHOL-3,6-DISULFONIC ACID TRISODIUM SALT see FOOD RED 2
2-(4-SULFO-1-NAPHTHYLAZO)-1-NAPHTHOL-4-SULFONIC ACID, DISO-DIUM SALT see C.I. FOOD RED 3
O-SULFONBENZOIC ACID IMIDE SODIUM SALT see SACCHARIN
SULFONIC ACID, MONOCHLORIDE see CHLOROSULFURIC ACID
SULFONIMIDE see DIFOLATAN
1,1'-SULFONYLBIS(4-AMINOBENZENE) see 4,4'-SULFONYLDIANILINE
4,4'-SULFONYLBISANILINE see 4,4'-SULFONYLDIANILINE
P,P-SULFONYLBISBENZAMINE see 4,4'-SULFONYLDIANILINE
P,P-SULFONYLBISBENZENAMINE see 4,4'-SULFONYLDIANILINE
SULFONYL CHLORIDE see SULFURYL CHLORIDE
P,P'-SULFONYLDIANILINE see 4,4'-SULFONYLDIANILINE
N-(SULFONYL-P-METHYLBENZENE)-N'-N-BUTYLUREA see 1-BUTYL-3-(p-TOLYL SULFONYL)UREA
1-P-SULFOPHENYLAZO-2-HYDROXYNAPHTHALENE-6-SULFONATE, DI-SODIUM SALT see FOOD YELLOW 3
1-P-SULFOPHENYLAZO-2-NAPHTHOL-6-SULFONIC ACID,DISODIUM SALT see FOOD YELLOW 3
2-(6-SULFO-2,4-XYLYLAZO)-1-NAPHTHOL-4-SULFONIC ACID,DISO-DIUM SALT see FOOD RED 4
SULFOXYPHENYLPYRAZOLIDINE see DIPHENYLPYRAZONE
SULFUR DIOXIDE SOLUTION see SULFUROUS ACID
SULFURE DE 4-CHLOROBENZYLE ET DE 4-CHLOROPHENYLE (FRENCH) see p-CHLOROBENZYL-p-CHLOROPHENYL SULFIDE
SULFURE DE METHYLE (FRENCH) see 2-THIOPROPANE
SULFURETED HYDROGEN see HYDROGEN SULFIDE
SULFUR FLUORIDE OXIDE see BISPENTA FLUORO SULFUR OXIDE
SULFUR HEXAFLUORIDE (DOT) see SULFUR FLUORIDE
SULFUR HYDRIDE see HYDROGEN SULFIDE
SULFURIC ACID, ALUMINUM SALT (3:2) see ALUMINUM SULFATE (2:3)
SULFURIC ACID, BARIUM SALT (1:1) see BARIUM SULFATE
SULFURIC ACID, BERYLLIUM SALT (1:1) see BERYLLIUM SULFATE (1:1)
SULFURIC ACID, CALCIUM(2+) SALT, DIHYDRATE see CALCIUM (II) SULFATE DIHYDRATE (1:1:2)
SULFURIC ACID, COPPER(2+) SALT (1:1) see COPPER (II) SULFATE (1:1)

SULFURIC ACID, COPPER(2+) SALT, PENTAHYDRATE see COPPER (II) SULFATE PENTAHYDRATE (1:1:5)

SULFURIC ACID, CYCLIC ETHYLENE ESTER see ETHYLENE SULFATE

SULFURIC ACID, DIAMMONIUM SALT see AMMONIUM SULFATE (2:1)

SULFURIC ACID DIETHYL ESTER see ETHYL SULFATE

SULFURIC ACID, DISODIUM SALT see SODIUM SULFATE (2:1)

SULFURIC ACID, INDIUM SALT see INDIUM SULFATE

SULFURIC ACID, IRON (3*) SALT (3:2) see FERRIC SULFATE

SULFURIC ACID, IRON(2+) SALT (1:1) see IRON (II) SULFATE (1:1)

SULFURIC ACID, LEAD (2+) SALT (1:1) see LEAD (II) SULFATE (1:1)

SULFURIC ACID, MANGANESE(2+) SALT see MANGANESE (II) SULFATE (1:1)

SULFURIC ACID, MERCURY(2+) SALT (1:1) see MERCURY(II) SULFATE (1:1)

SULFURIC ACID, THALLIUM SALT see THALLIUM SULFATE

SULFURIC ACID, THALLIUM(2+) SALT see THALLIUM (II) SULFATE (1:1)

SULFURIC ACID, ZINC SALT (1:1) see ZINC SULFATE

SULFURIC ACID, ZIRCONIUM(4+) SALT (2:1) see ZIRCONIUM (IV) SULFATE (1:2)

SULFURIC ANHYDRIDE see SULFUR TRIOXIDE

SULFURIC CHLOROHYDRIN see CHLOROSULFURIC ACID

SULFURIC OXIDE see SULFUR TRIOXIDE

SULFURIC OXYCHLORIDE see SULFURYL CHLORIDE

SULFURIC OXYFLUORIDE see SULFURYL FLUORIDE

SULFUR MONOCHLORIDE see SULFUR CHLORIDE

SULFUR MUSTARD see BIS(2-CHLOROETHYL)SULFIDE

SULFUR MUSTARD GAS see BIS(2-CHLOROETHYL)SULFIDE

SULFUROUS ACID ANHYDRIDE see SULFUR DIOXIDE

SULFUROUS ACID, 2-(P-TERT-BUTYLPHENOXY)-1-METHYLETHYL-2-CHLOROETHYL ESTER see SULFUROUS ACID, 2-(p-t-BUTYLPHEN-OXY)-1-METHYLETHYL-2-CHLOROETHYL ESTER

SULFUROUS ACID, MONOSODIUM SALT see SODIUM BISULFITE (1:1)

SULFUROUS ACID, SODIUM SALT (1:2) see SODIUM SULFITE (2:1)

SULFUROUS ANHYDRIDE see SULFUR DIOXIDE

SULFUROUS OXIDE see SULFUR DIOXIDE

SULFUROUS OXYCHLORIDE see THIONYL CHLORIDE

SULFUROUS OXYFLUORIDE see THIONYL FLUORIDE

SULFUR OXIDE see SULFUR DIOXIDE

SULFUR, SOLID see SULFUR

SULFUR SUBCHLORIDE see SULFUR CHLORIDE

SULPHADIMETHYLPYRIMIDINE see SULFADIMETHYLDIAZINE

SULPHAMETHALAZOLE see SULFAMETHOXAZOL

SULPHAMETHOXAZOL see SULFAMETHOXAZOL

SULPHAMETHOXAZOLE see SULFAMETHOXAZOL

SULPHA-METHOXIZOLE see SULFAMETHOXAZOL

SULPHAMETHYLISOXAZOLE see SULFAMETHOXAZOL

SULPHANILAMIDE see SULFANILAMIDE

3-SULPHANILAMIDO-5-METHYLISOXAZOLE see SULFAMETHOXAZOL

SULPHOBENZOIC IMIDE CALCIUM SALT see CALCIUM-O-BENZOSULFI-MIDE

SULPHOBENZOIC IMIDE, SODIUM SALT see SACCHARIN

SULPHOCARBONIC ANHYDRIDE see CARBON DISULFIDE

1-(4-SULPHO-1-NAPHTHYLAZO)-2-NAPHTHOL-3,6-DISULPHONIC ACID, TRISODIUM SALT see FOOD RED 2

1-(4-SULPHO-1-NAPHTHYLAZO)-2-NAPHTHOL-6,8-DISULPHONIC ACID, TRISODIUM SALT see FOOD RED 7

1,1'-SULPHONYLBIS(4-AMINOBENZENE) see 4,4'-SULFONYLDIANILINE

P,P-SULPHONYLBISBENZAMINE see 4,4'-SULFONYLDIANILINE

4,4'-SULPHONYLBISBENZAMINE see 4,4'-SULFONYLDIANILINE

P,P-SULPHONYLBISBENZENAMINE see 4,4'-SULFONYLDIANILINE

4,4'-SULPHONYLBISBENZENAMINE see 4,4'-SULFONYLDIANILINE

SULPHONYLDIANILINE see 4,4'-SULFONYLDIANILINE

P,P-SULPHONYLDIANILINE see 4,4'-SULFONYLDIANILINE

SULPHURIC ACID see SULFURIC ACID

SULPHURIC ACID, CADMIUM SALT (1:1) see CADMIUM SULFATE (1:1)

SULPHUR MUSTARD GAS see BIS(2-CHLOROETHYL)SULFIDE

SUMATRA CAMPHOR see BORNEOL

SUN YELLOW EXTRA PURE A see FOOD YELLOW 3

SUPRARENIN see VASOTONIN

SURITAL SODIUM SALT see SURITAL SODIUM

SV-1522 see ACETOPROMAZINE

SVC see ACETPHENARSINE

SWEDISH GREEN see COPPER ORTHOARSENITE

SWEET BIRCH OIL see METHYL SALICYLATE

(SYM)-DIPHENYLHYDRAZINE see HYDRAZOBENZENE

N-SYM-TRIMETHYLPHENYLDIETHYLAMINOACETAMIDE HYDROCHLO-RIDE see TRIMECAINE

SYNDROX see "SPEED"

SYNHEXYL see 3-HOMOTETRA HYDRO CANNIBINOL

SYNS: see ETHYL PYRAZINYL PHOSPHOROTHIOATE

SYNSTIGMINE see 3,4-DICHLOROBENZOIC ACID

SYNTHETIC BRADYKININ see BRADYKININ

SYNTHETIC EUGENOL see EUGENOL

SYNTHETIC GLYCERIN see GLYCERINE

SYNTHETIC IRON OXIDE see IRON (III) OXIDE

SYNTHETIC MUSTARD OIL see ALLYL ISOTHIOCYANATE

SYNTHETIC PYRETHRINS see ALLETHRIN

SYNTHETIC WINTERGREEN OIL see METHYL SALICYLATE

SYNTHOSTIGMINE BROMIDE see PROSTIGMINE BROMIDE

SYNTOMETRINE see D-LYSERGIC ACID-L,2-PROPANOLAMIDE

SYNTOSTIGMIN see 3,4-DICHLOROBENZOIC ACID

SYSTOX see DEMETON-O + DEMETON-S

SZESCIOMETYLENODWUIZOCYJANIAN (POLISH) see 1,6-DIISOCYANA-TOHEXANE

T-1036 see DIETHYL FLUOROPHOSPHATE

TABLE SALT see SODIUM CHLORIDE

TABUN see ETHYL DIMETHYL AMIDO CYANO PHOSPHATE

TACITIN see BENZOCTAMINE

TAJMALIN see AJMALINE

TAKYCOR see AJMALINE

TALCUM see TALC (powder)

TALLOW BENZYL DIMETHYLAMMONIUM CHLORIDE see DIMETHYLOCTADECYLBENZYLAMMONIUM CHLORIDE

TAMARON see O,S-DIMETHYL PHOSPHORAMIDOTHIOATE

TANGANTANGAN OIL see CASTOR OIL

TANNIN see TANNIC ACID

TANNIN FROM BRACKEN FERN see BRACKEN FERN TANNIN

TANNIN FROM CHERRY BARK OAK see CHERRY BARK OAK

TANNIN FROM CHESTNUT see CHESTNUT TANNIN

TANNIN FROM MIMOSA see MIMOSA TANNIN

TANNIN FROM MYRTAN see MYRTAN TANNIN

TANNIN FROM QUEBRACHO see QUEBRACHO TANNIN

TANTALIUM PENTAFLUORIDE see TANTALUM FLUORIDE

TANTALUM-181 see TANTALUM

TANTALUM PENTACHLORIDE see TANTALUM CHLORIDE

TAPIOCA see DEXTRINS

TAR CAMPHOR see NAPHTHALENE

TARICHATOXIN see FUGU POISON

TARRAGON see P-ALLYLANISOLE

TARTAR EMETIC see ANTIMONY POTASSIUM TARTRATE

DL-TARTARIC ACID, ANTIMONY POTASSIUM SALT see dl-ANTIMONY POTASSIUM TARTRATE

TARTARIZED ANTIMONY see ANTIMONY POTASSIUM TARTRATE

TARTRATE ANTIMONIO-POTASSIQUE (FRENCH) see ANTIMONY POTAS-SIUM TARTRATE

TARTRATED ANTIMONY see ANTIMONY POTASSIUM TARTRATE

TARTRATE DE NICOTINE (FRENCH) see NICOTINE TARTRATE (1:2)

TAURE(O)DON see GOLD SODIUM THIOMALATE

TAZEPAM see 7-CHLORO-1,3-DIHYDRO-3-HYDROXY-5-PHENYL-2H-1,4-BENZODIAZEPINE-2-ONE

TBB see TRI-n-BUTYL BORANE

TBBA see p-tert-BUTYL BENZOIC ACID

2,4,5-T N-BUTYL ESTER MIXED WITH 2,4-D N-BUTYL ESTER see AGENT ORANGE

TCDBD see TCDD

2,3,7,8-TCDD see TCDD

TEABERRY OIL see METHYL SALICYLATE

TECHNICAL BHC see BENZENE HEXACHLORIDE (MIXED ISOMERS)

90 TECHNICAL GLYCERINE see GLYCERINE

TECHNICAL HCH see BENZENE HEXACHLORIDE (MIXED ISOMERS)

TEF see TEPA

TEFLON (VARIOUS) see TEFLON

TELEPATHINE see HARMINE

TELLUR (POLISH) see TELLURIUM

TELLURIUM DIETHYLDITHIOCARBAMATE see ETHYL TELLURAC

TELLUROUS ACID, DISODIUM SALT see SODIUM TELLURITE

TEMOPHOS see ABATE

TEMPLIN OIL see ABIES ALBA OIL

TENAMENE 2 see N,N'-DI-S-BUTYL-p-PHENYLENEDIAMINE

TENITE 800 see POLYETHYLENE

TENTONE MALEATE see METHOXYPROMAZINE MALEATE

TEP see TETRON

TEPIDONE RUBBER ACCELERATOR see SODIUM DIBUTYLDITHIOCARBAMATE

TEPP see TETRAETHYLPYROPHOSPHATE

TERAMETHYL THIURAM DISULFIDE see BIS(DIMETHYL THIOCARBAMYL)DISULFIDE

TERC. BUTYLHYDROPEROXID (CZECH) see 6-BUTYL HYDROPEROXIDE

TERC.DODECYLMERKAPTAN (CZECH) see t-DODECANETHIOL

TEREBENTHINE (FRENCH) see TURPENTINE

TEREFTALODINITRIL (CZECH) see p-BENZENEDINITRILE

TEREPHTHALIC ACID METHYL ESTER see DIMETHYL TEREPHTHALATE

TEREPHTHALONITRILE see p-BENZENEDINITRILE

TERGITOL TP-9 (NONIONIC) see PEG-9 NONYL PHENYL ETHER

TERPENTIN OEL (GERMAN) see TURPENTINE

TERPINYL THIOCYANOACETATE see ISOBORNYL THIOCYANATO ACETATE

TERRA ALBA see CALCIUM (II) SULFATE DIHYDRATE (1:1:2)

TERRAMYCIN see 5-HYDROXYTETRACYLINE

O-(4-TERZ.-BUTIL-2-CLORO-FENIL)-O-METIL-FOSFORAMMIDE (ITALIAN) see 4-tert-BUTYL-2-CHLORO PHENYL METHYL METHYL PHOSPHORAMIDATE

TESTHORMONE see 17-METHYL TESTOSTERONE

TRANS-TESTOSTERONE see TESTOSTERONE

TESTOSTERONE HYDRATE see TESTOSTERONE

TESTOSTERONE-17-PROPIONATE see TESTOSTERONE PROPIONATE

TESTOSTERONE-17-BETA-PROPIONATE see TESTOSTERONE PROPIONATE

TESTOSTERON PROPIONATE see TESTOSTERONE PROPIONATE

TESTOSTOSTERONE see TESTOSTERONE

O,O,O',O'-TETRAAETHYL-BIS(DITHIOPHOSPHAT) (GERMAN) see ETHYL METHYLENE PHOSPHORO DITHIOATE

O,O,O,O-TETRAAETHYL-DIPHOSPHAT, BIS(O,O-DIAETHYLPHOSPHORSAEURE-ANHYDRID (GERMAN) see TETRAETHYLPYROPHOSPHATE

1,3,5,7-TETRAAZAADAMANTANE see HEXAMETHYLENETETRAMINE

1,1,2,2-TETRABROMAETHAN (GERMAN) see ACETYLENE TETRABROMIDE

TETRABROMOACETYLENE see ACETYLENE TETRABROMIDE

2,4,5,7-TETRABROMO-9-O-CARBOXYPHENYL-6-HYDROXY-3-ISOXANTHONE, DISODIUM SALT see BROMO EOSINE

1,1,2,2-TETRABROMOETANO (ITALIAN) see ACETYLENE TETRABROMIDE

S-TETRABROMOETHANE see ACETYLENE TETRABROMIDE

1,1,2,2-TETRABROMOETHANE see ACETYLENE TETRABROMIDE

2',4',5',7'-TETRABROMOFLUORESCEIN DISODIUM SALT see BROMO EOSINE

TETRABROMOFLUORESCEIN SOLUBLE see BROMO EOSINE

2-(2,4,5,7-TETRABROMO-6-HYDROXY-3-OXO-3H-XANTHENE-9-YL)-BENZOIC ACID, DISODIUM SALT see BROMO EOSINE

TETRABROMOMETHANE see CARBON TETRABROMIDE

1,1,2,2-TETRABROOMETHAAN (DUTCH) see ACETYLENE TETRABROMIDE

TETRA-N-BUTYLCIN (CZECH) see TETRABUTYLSTANNANE

TETRABUTYLTIN see TETRABUTYLSTANNANE

TETRABUTYLTITANATE (CZECH) see BUTYL TITANATE

TETRACAINE see p-(BUTYL AMINO)BENZOIC ACID-2-(DIMETHYL AMINO)ETHYL ESTER

2,4,4',5-TETRACHLOOR-DIFENYL-SULFON (DUTCH) see p-CHLOROPHENYL-2,4,5-TRICHLOROPHENYL SULFONE

1,1,2,2-TETRACHLOORETHAAN (DUTCH) see ACETYLENE TETRACHLORIDE

TETRACHLOORETHEEN (DUTCH) see 1,1,2,2-TETRACHLOROETHYLENE

TETRACHLOORKOOLSTOF (DUTCH) see CARBON TETRACHLORIDE

TETRACHLOORMETAAN (DUTCH) see CARBON TETRACHLORIDE

1,1,2,2-TETRACHLORAETHAN (GERMAN) see ACETYLENE TETRACHLORIDE

TETRACHLORAETHEN (GERMAN) see 1,1,2,2-TETRACHLOROETHYLENE

N-(1,1,2,2-TETRACHLORAETHYLTHIO)-CYCLOHEX-4-EN-1,4-DIACARBOXIMID(GERMAN) see DIFOLATAN

2,4,4',5-TETRACHLOR-DIPHENYL-SULFON (GERMAN) see p-CHLOROPHENYL-2,4,5-TRICHLOROPHENYL SULFONE

1,1,2,2-TETRACHLORETHANE (FRENCH) see ACETYLENE TETRACHLORIDE

TETRACHLORKOHLENSTOFF, TETRA (GERMAN) see CARBON TETRACHLORIDE

TETRACHLORMETHAN (GERMAN) see CARBON TETRACHLORIDE

TETRACHLOROCARBON see CARBON TETRACHLORIDE

2,3,7,8-TETRACHLORODIBENZO(B,E)(1,4)DIOXAN see TCDD

2,3,7,8-TETRACHLORODIBENZO-P-DIOXIN see TCDD

2,3,7,8-TETRACHLORODIBENZO-1,4-DIOXIN see TCDD

2,3,6,7-TETRACHLORODIBENZO-P-DIOXIN (GERMAN) see TCDD

TETRACHLORODIPHENYLETHANE see 1,1-BIS(4-CHLOROPHENYL)-2,2-DICHLOROETHANE

2,4,4',5-TETRACHLORODIPHENYL SULFONE see p-CHLOROPHENYL-2,4,5-TRICHLOROPHENYL SULFONE

2,4,5,4'-TETRACHLORODIPHENYLSULPHONE see p-CHLOROPHENYL-2,4,5-TRICHLOROPHENYL SULFONE

SYM-TETRACHLOROETHANE see ACETYLENE TETRACHLORIDE

1,1,2,2-TETRACHLOROETHANE see ACETYLENE TETRACHLORIDE

TETRACHLOROETHYLENE (DOT) see 1,1,2,2-TETRACHLOROETHYLENE

N-1,1,2,2-TETRACHLOROETHYLMERCAPTO-4-CYCLOHEXENE-1,2-CARBOXIMIDE see DIFOLATAN

N((1,1,2,2-TETRACHLOROETHYL)SULFENYL)-CIS-4-CYCLOHEXENE-1,2-DICARBOXIMIDE see DIFOLATAN

N-(1,1,2,2-TETRACHLOROETHYLTHIO)-4-CYCLOHEXENE-1,2-DICARBOXIMIDE see DIFOLATAN

TETRACHLOROMETHANE see CARBON TETRACHLORIDE

2,3,5,6-TETRACHLORO-4-NITROANISOLE see TETRACHLORONITROANISOLE

2,3,4,6-TETRACHLOROPHENOL see 2,4,5,6-TETRACHLOROPHENOL

M-TETRACHLOROPHTHALONITRILE see TETRACHLOROISOPHTHALONITRILE

5,6,7,8-TETRACHLOROQUINOXALINE see CHLORQUINOX

TETRACHLOROSILANE see SILICON CHLORIDE

TETRACHLOROSILANE see SILICON CHLORIDE

2,3,4,5-TETRACHLORO THIOPHENE see TCTP

TETRACHLORURE D'ACETYLENE (FRENCH) see ACETYLENE TETRACHLORIDE

TETRACHLORURE DE CARBONE (FRENCH) see CARBON TETRACHLORIDE

TETRACHLORURE DE SILICIUM (FRENCH) see SILICON CHLORIDE

TETRACHLORURE DE TITANE (FRENCH) see TITANIUM CHLORIDE

TERTRACID FAST BLUE SR see ACID BLUE 92

2,4,4',5-TETRACLORO-DIFENIL-SOLFONE (ITALIAN) see p-CHLOROPHENYL-2,4,5-TRICHLOROPHENYL SULFONE

1,1,2,2-TETRACLOROETANO (ITALIAN) see ACETYLENE TETRACHLORIDE

TETRACLOROETENE (ITALIAN) see 1,1,2,2-TETRACHLOROETHYLENE

TETRACLOROMETANO (ITALIAN) see CARBON TETRACHLORIDE

TETRACLORURO DI CARBONIO (ITALIAN) see CARBON TETRACHLORIDE

TETRACYANOETHYLENE see ETHENE TETRA CARBO NITRILE

TETRACYCLINE CHLORIDE see TETRACYCLINE HYDROCHLORIDE

TETRACYCLINE I see TETRACYCLINE

TETRACYCLINE-L-METHYLENE LYSINE see MUCOMYCIN

TETRADECANOIC ACID see MYRISTIC ACID

TETRADECANOIC ACID, ISOPROPYL see ISOPROPYL MYRISTATE

12-TETRADECANOYLPHORBOL-13-ACETATE see PHORBOL MYRISTATE ACETATE

12-O-TETRADECANOYLPHORBOL-13-ACETATE see PHORBOL MYRISTATE ACETATE

N-TETRADECOIC ACID see MYRISTIC ACID

7,8,13,13A-TETRADEHYDRO-9,10-DIMETHOXY-2,3-(METHYLENE-DIOXY)BERBINIUM SULFATE TRIHYDRATE see BERBERINE SULFATE TRIHYDRATE

TETRADEHYDRODOISYNOLIC ACID METHYL ETHER see BIS DEHYDRO ISYNOLIC ACID METHYL ESTER

12-O-TETRADEKANOYLPHORBOL-13-ACETAT (GERMAN) see PHORBOL MYRISTATE ACETATE

TETRAETHYLDIAMINO-O-CARBOXY-PHENYL-XANTHENYL CHLORIDE see (9-(O-CARBOXYPHENYL)-6-(DIETHYLAMINO)-3H-XANTHEN-3-YLIDENE) DIETHYLAMMONIUM CHLORIDE

O,O,O,O-TETRAETHYL-DIFOSFAAT (DUTCH) see TETRAETHYLPYRO-PHOSPHATE

O,O,O,O-TETRAETHYL-DITHIO-DIFOSFAAT (DUTCH) see ETHYL THIO PYROPHOSPHATE

TETRAETHYL DITHIONOPYROPHOSPHATE see ETHYL THIO PYROPHOSPHATE

O,O,O,O-TETRAETIL-DITIO-PIROFOSFATO (ITALIAN) see ETHYL THIO PYROPHOSPHATE

TETRAETHYL DITHIOPYROPHOSPHATE see ETHYL THIO PYROPHOSPHATE

O,O,O,O-TETRAETHYL DITHIOPYROPHOSPHATE see ETHYL THIO PYROPHOSPHATE

TETRAETHYL DITHIO PYROPHOSPHATE, LIQUID (DOT) see ETHYL THIO PYROPHOSPHATE

O,O,O,O-TETRAETHYL S,S'-METHYLENEBIS(DITHIOPHOSPHATE) see ETHYL METHYLENE PHOSPHORO DITHIOATE

O,O,O',O'-TETRAETHYL S,S'-METHYLENEBISPHOSPHORDITHIOATE see ETHYL METHYLENE PHOSPHORO DITHIOATE

O,O,O',O'-TETRAETHYL-S,S'-METHYLENEBISPHOSPHORODITHIOATE see ETHYL METHYLENE PHOSPHORO DITHIOATE

O,O,O',O'-TETRAETHYL S,S'-METHYLENE DI(PHOSPHORODITHIOATE) see ETHYL METHYLENE PHOSPHORO DITHIOATE

TETRAETHYLPLUMBANE see TETRAETHYL LEAD

TETRAETHYL PYROFOSFAAT (BELGIAN) see TETRAETHYLPYROPHOSPHATE

TETRAETHYL PYROPHOSPHATE, LIQUID (DOT) see TETRAETHYLPYROPHOSPHATE

TETRAETHYL PYROPHOSPHATE MIXTURE, LIQUID (DOT) see TETRON

TETRAETHYLRHODAMINE see (9-(O-CARBOXYPHENYL)-6-(DIETHYLAMINO)-3H-XANTHEN-3-YLIDENE) DIETHYLAMMONIUM CHLORIDE

TETRAETHYL S,S'-METHYLENE BIS(PHOSPHOROTHIOLOTHIONATE) see ETHYL METHYLENE PHOSPHORO DITHIOATE

TETRAETHYLTHIOPEROXYDICARBONIC DIAMIDE see BIS(DIETHYL THIO CARBAMOYL) DISULFIDE

TETRAETHYLTHIURAM DISULPHIDE see BIS(DIETHYL THIO CARBAMOYL) DISULFIDE

N,N,N',N'-TETRAETHYLTHIURAM DISULPHIDE see BIS(DIETHYL THIO CARBAMOYL) DISULFIDE

O,O,O,O-TETRAETIL-PIROFOSFATO (ITALIAN) see TETRAETHYLPYROPHOSPHATE

TETRAFLUOROETHENE HOMOPOLYMER see TEFLON

TETRAFLUOROETHENE POLYMER see TEFLON

TETRAFLUOROETHYLENE HOMOPOLYMER see TEFLON

TETRAFLUOROETHYLENE POLYMERS see TEFLON

TETRAFLUOROMETHANE see CARBON TETRAFLUORIDE

TETRAFOSFOR (DUTCH) see PHOSPHORUS (WHITE)

6,7,8,9-TETRAHYDRO-5-AZEPOTETRAZOLE see 1,5-PENTAMETHYLENETETRAZOLE

1,2,3,4-TETRAHYDROBENZENE see CYCLOHEXENE

(−)-DELTA(SUP 8)-TRANS-TETRAHYDROCANNABINOL see 1-trans-delta(sup 8)-TETRAHYDROCANNABINOL

(−)-DELTA(SUP 9)-TRANS-TETRAHYDROCANNABINOL see 1-trans-delta(sup 9)-TETRAHYDROCANNABINOL

5,5a,6,7-TETRAHYDRO-4H-DIBENZ(f,g,j)ACEANTHRYLENE see ANGSTERANTHRENE

TETRAHYDRO-2H-3,5-DIMETHYL-1,3,5-THIADIAZINE-2-THIONE see DIMETHYLFORMOCARBOTHIALDINE

TETRAHYDRO-3,5-DIMETHYL-2H-1,3,5-THIADIAZINE-2-THIONE see DIMETHYLFORMOCARBOTHIALDINE

TETRAHYDRO-2,4-DIMETHYLTHIOPHENE 1,1-DIOXIDE see 2,4-DIMETHYL SULFOLANE

TETRAHYDRO-P-DIOXIN see p-DIOXANE

TETRAHYDRO-1,4-DIOXIN see p-DIOXANE

TETRAHYDRO-2,5-DIOXOFURAN see SUCCINIC ANHYDRIDE

1,2,3,6-TETRAHYDRO-3,6-DIOXOPYRIDAZINE see MALEIC HYDRAZIDE

TETRAHYDROFURAAN (DUTCH) see TETRAHYDROFURAN

TETRAHYDROFURAN (DOT) see TETRAHYDROFURAN

TETRAHYDRO-2-FURANONE see 4-BUTANOLIDE

TETRAHYDRO-3-IODOTHIOPHENE-1,1-DIOXIDE see 3-IODOTETRAHYDROTHIOPHENE-1,1-DIOXIDE

TETRAHYDRO-1,4-ISOXAZINE see MORPHOLINE

3A,4,7,7A-TETRAHYDRO-4,7-METHANOINDENE see BICYCLOPENTADIENE

4A,5,7A,8-TETRAHYDRO-12-METHYL-9H-9,9C-IMINOETHANOPHENANTHRO(4,5-bcd)FURAN-3,5-DIOL see (−)-MORPHINE

1,2,5,6-TETRAHYDRO-1-METHYLNICOTINIC ACID, METHYL ESTER see ARECOLINE

3,4,5,6,7-TETRAHYDRO-5-METHYL-1-PHENYL 1H-2,5-BENZOXAZOCINE see FENAZOXINE

TETRAHYDRO-N,N,3-TRIS(2-CHLOROETHYL)-2H-1,3,2-OXAPHOSPHORIN-2-AMINE-2-OXIDE see TROPHOSPHAMIDE

TETRAHYDRO-1,4-OXAZINE see MORPHOLINE

TETRAHYDRO-2H-1,4-OXAZINE see MORPHOLINE

1,2,3,6-TETRAHYDRO PHTHALIC ANHYDRIDE see TETRAHYDROPHTHALIC ACID ANHYDRIDE

TETRAHYDROPYRROLE see PYRROLIDINE

TETRAHYDROSERPENTINE see AJMALICINE

1,2,3,4-TETRAHYDROSTYRENE see CYCLOHEXENYLETHYLENE

6,7,8,9-TETRAHYDRO-5H-TETRAZOLOAZEPINE see 1,5-PENTAMETHYLENETETRAZOLE

3',4',5,7-TETRAHYDROXYFLAVAN-3-OL see QUERCETIN

11-BETA,16-ALPHA,17-ALPHA,21-TETRAHYDROXY-9-ALPHA-FLUORO-1,4-PREGNADIENE-3,20-DIONE TRIAMCINOLONE see ARISTOCORT

TETRAIDROFURANO (ITALIAN) see TETRAHYDROFURAN

TETRAKIS(DIETHYLCARBAMODITHIOATO-S,S')TELLURIUM see ETHYL TELLURAC

TETRAKIS(DIETHYLDITHIOCARBAMATO)TELLURIUM see ETHYL TELLURAC

TETRAKISDIMETHYLAMINOPHOSPHONOUS ANHYDRIDE see OCTAMETHYLPYROPHOSPHORAMIDE

TETRAKIS(DIMETHYLCARBAMODITHIOATO-S,S')SELENIUM see SELENIUM DIMETHYLDITHIOCARBAMATE

TETRALITE see TETRYL

TETRALLOBARBITAL see ALLYLISOBUTYLBARBITURATE

TETRAMETHOXY SILANE see METHYL-O-SILICATE

N,N,N'-TETRAMETHYL-3,6-ACRIDINEDIAMINE see 3,6-BIS(DIMETHYL AMINO)ACRIDINE

N,N,N',N'-TETRAMETHYL-DIAMIDO-FOSFORZUUR-FLUORIDE (DUTCH) see BIS(DIMETHYL AMIDO)FLUORO PHOSPHATE

TETRAMETHYLDIAMIDOPHOSPHORIC FLUORIDE see BIS(DIMETHYL AMIDO)FLUORO PHOSPHATE

N,N,N',N'-TETRAMETHYL-DIAMIDO-PHOSPHORSAEURE-FLUORID (GERMAN) see BIS(DIMETHYL AMIDO)FLUORO PHOSPHATE

P,P-TETRAMETHYLDIAMINODIPHENYLMETHANE see MICHLER'S BASE

4,4'-TETRAMETHYLDIAMINODIPHENYLMETHANE see MICHLER'S BASE

TETRAMETHYL DIAPARA-AMIDO-TRIPHENYL CARBINOL see AIZEN MALACHITE GREEN

TETRAMETHYL DIARSYL see CACODYL

TETRAMETHYLDIURANE SULPHITE see BIS(DIMETHYL THIOCARBAMYL)DISULFIDE

TETRAMETHYLENE see CYCLOBUTANE

TETRAMETHYLENE BIS(METHANESULFONATE) see 1,4-BUTANEDIOL DIMETHYL SULFONATE

TETRAMETHYLENE CYANIDE see ADIPONITRILE

TETRAMETHYLENE DIMETHANE SULFONATE see 1,4-BUTANEDIOL DIMETHYL SULFONATE

1,4-TETRAMETHYLENE GLYCOL see 1,4-BUTANEDIOL

TETRAMETHYLENE OXIDE see TETRAHYDROFURAN

TETRAMETHYLENEOXIRANE see CYCLOHEXENE OXIDE

TETRAMETHYLENETHIURAM DISULPHIDE see BIS(DIMETHYL THIO-CARBAMYL)DISULFIDE

TETRAMETHYLENIMINE see PYRROLIDINE

N,2,3,3-TETRAMETHYL-2-NORBORNANAMINE HYDROCHLORIDE see MEVASIN HYDROCHLORIDE

P-1',1',4',4'-TETRAMETHYLOKTYLBENZENSULFONAN SODNY (CZECH) see DODECYL BENZENE SODIUM SULFONATE

N,N,N-ALPHA-TETRAMETHYL-10H-PHENOTHIAZINE-10-ETHANAMI-NIUM METHYL SULFATE see MULTERGAN METHYL SULFATE

TETRAMETHYLPHOSPHORODIAMIDIC FLUORIDE see BIS(DIMETHYL AMIDO)FLUORO PHOSPHATE

N,N,N-TETRAMETHYLPHOSPHORODIAMIDIC FLUORIDE see BIS(DI-METHYL AMIDO)FLUORO PHOSPHATE

TETRAMETHYLPLUMBANE see TETRAMETHYL LEAD

TETRAMETHYLTHIOCARBAMOYLDISULPHIDE see BIS(DIMETHYL THIO-CARBAMYL)DISULFIDE

O,O,O',O'-TETRAMETHYL-O,O'-THIODI-P-PHENYLENE PHOSPHORO-THIOATE see ABATE

TETRAMETHYLTHIONINE CHLORIDE see 3,7-BIS(DIMETHYL AMINO)-PHENAZA THIONIUM CHLORIDE

TETRAMETHYLTHIOPEROXYDICARBONIC DIAMIDE see BIS(DIMETHYL THIOCARBAMYL)DISULFIDE

TETRAMETHYLTHIORAMDISULFIDE (DUTCH) see BIS(DIMETHYL THIO-CARBAMYL)DISULFIDE

TETRAMETHYL-THIRAM DISULFID (GERMAN) see BIS(DIMETHYL THIO-CARBAMYL)DISULFIDE

TETRAMETHYLTHIURAM BISULFIDE see BIS(DIMETHYL THIOCARBA-MYL)DISULFIDE

TETRAMETHYLTHIURAM DISULFIDE see BIS(DIMETHYL THIOCARBA-MYL)DISULFIDE

N,N,N',N'-TETRAMETHYLTHIURAM DISULFIDE see BIS(DIMETHYL THIOCARBAMYL)DISULFIDE

TETRAMETHYLTHIURAMMONIUM SULFIDE see BIS(DIMETHYL THIO CARBAMOYL)SULFIDE

TETRAMETHYLTHIURAM MONOSULFIDE see BIS(DIMETHYL THIO CAR-BAMOYL)SULFIDE

TETRAMETHYLTHIURAMONOSULFIDE see BIS(DIMETHYL THIO CARBA-MOYL)SULFIDE

TETRAMETHYLTHIURAM SULFIDE see BIS(DIMETHYL THIO CARBA-MOYL)SULFIDE

TETRAMETHYL THIURANE DISULFIDE see BIS(DIMETHYL THIOCARBA-MYL)DISULFIDE

TETRAMETHYLTHIURUM DISULFIDE see BIS(DIMETHYL THIOCARBA-MYL)DISULFIDE

N,N',O,O-TETRAMETHYL-(+)-TUBOCURINE see O,O'-DIMETHYLTUBO-CURARINE

N,N,N',N'-TETRAMETIL-FOSFORODIAMMIDO-FLUORURO (ITALIAN) see BIS(DIMETHYL AMIDO)FLUORO PHOSPHATE

TETRANATRIUMPYROPHOSPHAT (GERMAN) see TETRASODIUM PYRO-PHOSPHATE, ANHYDROUS

TETRANITRANILINE (FRENCH) see TETRANITROANILINE

2,3,4,6-TETRANITROANILINE see TETRANITROANILINE

TETRANITROPENTAERYTHRITE see PENTAERYTHRITOL TETRANITRATE

TETRA OLIVE N2G see ANTHRACENE

2,4,5,6-TETRAOXOHEXAHYDROPYRIMIDINE see ALLOXAN

2,4,5,6-TETRAOXOHEXAHYDROPYRIMIDINE HYDRATE see MESOXA-LYLUREA MONOHYDRATE

TETRAPHENE see BENZ(a)ANTHRACENE

TETRAPHOSPHATE HEXAETHYLIQUE (FRENCH) see HEXAETHYL TETRAPHOSPHATE

TETRAPHOSPHOR (GERMAN) see PHOSPHORUS (WHITE)

TETRAPHOSPHORUS TRISULFIDE see PHOSPHORUS SESQUISULFIDE

TETRASAN see ALKYL DIMETHYL-3,4-DICHLOROBENZENE AMMONIUM CHLORIDE

TETRASODIUM DIPHOSPHATE see TETRASODIUM PYROPHOSPHATE, ANHYDROUS

TETRASODIUM EDTA see EDATHANIL TETRASODIUM

TETRASODIUM ETHYLENEDIAMINETETRAACETATE see EDATHANIL TETRASODIUM

TETRASODIUM ETHYLENEDIAMINETETRACETATE see EDATHANIL TETRASODIUM

TETRASODIUM (ETHYLENEDINITRILO)TETRAACETATE see EDATHANIL TETRASODIUM

TETRASODIUM PYROPHOSPHATE see TETRASODIUM PYROPHOSPHATE, ANHYDROUS

TETRASODIUM SALT EDTA see EDATHANIL TETRASODIUM

TETRASODIUM SALT OF EDTA see EDATHANIL TETRASODIUM

TETRASODIUM SALT OF ETHYLENEDIAMINETETRACETICACID see EDA-THANIL TETRASODIUM

7,8,9,10-TETRAZABICYCLO(5.3.0)-8,10-DECADIENE see 1,5-PENTA-METHYLENETETRAZOLE

1,2,3,3A-TETRAZACYCLOHEPTA-8A,2-CYCLOPENTADIENE see 1,5-PENTAMETHYLENETETRAZOLE

TETRAZOBENZENE-BETA-NAPHTHOL see OIL RED

TETRODONTOXIN see FUGU POISON

TETRODOTOXIN see FUGU POISON

TETRODOXIN see FUGU POISON

TETROLE see FURAN

TETROSAN see ALKYL(C_8C_{18})DIMETHYL-3,4-DICHLOROBENZYLAMMO-NIUM CHLORIDE

2,4,6-TETRYL see TETRYL

THALLIUM(1+) ACETATE see THALLIUM ACETATE

THALLIUM(I) ACETATE see THALLIUM ACETATE

THALLIUM MONOACETATE see THALLIUM ACETATE

THALLIUM MONOCHLORIDE see THALLIUM (I) CHLORIDE

THALLIUM PEROXIDE see THALLIC OXIDE

THALLIUM SESQUIOXIDE see THALLIC OXIDE

THALLIUM SULFATE, SOLID (DOT) see THALLIUM (I) SULFATE (2:1)

THALLOUS ACETATE see THALLIUM ACETATE

THALLOUS CHLORIDE see THALLIUM (I) CHLORIDE

THALLOUS NITRATE see THALLIUM (I) NITRATE (1:1)

THALLOUS SULFATE see THALLIUM (I) SULFATE (2:1)

DELTA(SUP 6)-THC see 1-trans-delta(sup 8)-TETRAHYDROCANNABI-NOL

DELTA(SUP 8)-THC see 1-trans-delta(sup 8)-TETRAHYDROCANNABI-NOL

DELTA(SUP 9)-THC see 1-trans-delta(sup 9)-TETRAHYDROCANNABI-NOL

THEINE see CAFFEINE

THEOPHYLLIN see THEOPHYLLINE

THEOPHYLLINE, ANHYDROUS see THEOPHYLLINE

THERAMINE see HISTAMINE

THERMINOL FR-1 see POLYCHLORINATED BIPHENYLS

1-THIA-3-AZAINDENE see BENZOTHIAZOLE

THIACETAMIDE see THIOACETAMIDE

THIACYCLOPENTADIENE see THIOPHENE

THIACYCLOPROPANE see ETHYLENE SULFIDE

1,3,4-THIADIAZOL-2-AMINE, 5-(5-NITRO-2-FURANYL)- (9CI) see 5-(5-NITRO-2-FURYL)-2-AMINO-1,3,4-THIADIAZOLE

4-THIAHEPTANEDIOIC ACID see BIS(2-CARBOXYETHYL) SULFIDE

THIAMAZOLE see 2-MERCAPTO-1-METHYLIMIDAZOLE

THIAMINE CHLORIDE HYDROCHLORIDE see THIAMINE DICHLORIDE

THIAMINE HYDROCHLORIDE see THIAMINE DICHLORIDE

THIAMINE MONONITRATE see THIAMINE NITRATE

THIAMINE NITRATE (SALT) (8CI) see THIAMINE NITRATE

THIAMIN HYDROCHLORIDE see THIAMINE DICHLORIDE

THIAMINIUM CHLORIDE HYDROCHLORIDE see THIAMINE DICHLORIDE

THIAMIZIDE see 4-CHLORO-N-METHYL-3-(METHYLSULFAMOYL)BEN-ZAMIDE

THIAPHENE see THIOPHENE

2-THIAPROPANE see 2-THIOPROPANE

THIAZINAMIUM METHYL SULFATE see MULTERGAN METHYL SUL-FATE

2-THIAZYLAMINE see 2-AMINOTHIAZOLE

THIOALLYL ETHER see ALLYL SULFIDE

4,4'-THIOANILINE see 4,4'-THIODIANILINE

THIOBENZYL ALCOHOL see alpha-TOLUENETHIOL

4,4'-THIOBISBENZENAMINE see 4,4'-THIODIANILINE

4,4′-THIOBIS(6-TERT-BUTYL-M-CRESOL) see BIS(3-tert-BUTYL-4-HY-DROXY-6-METHYLPHENYL) SULFIDE

4,4′-THIOBIS(2-TERT-BUTYL-5-METHYLPHENOL) see BIS(3-tert-BU-TYL-4-HYDROXY-6-METHYLPHENYL) SULFIDE

4,4′-THIOBIS(6-TERT-BUTYL-3-METHYLPHENOL) see BIS(3-tert-BU-TYL-4-HYDROXY-6-METHYLPHENYL) SULFIDE

1,1′-THIOBIS(2-CHLOROETHANE) see BIS(2-CHLOROETHYL)SULFIDE

4,4′-THIOBIS(3-METHYL-6-TERT-BUTYLPHENOL) see BIS(3-tert-BU-TYL-4-HYDROXY-6-METHYLPHENYL) SULFIDE

1,1′-THIOBIS(2-METHYL-4-HYDROXY-5-TERT-BUTYLBENZENE) see BIS(3-tert-BUTYL-4-HYDROXY-6-METHYLPHENYL) SULFIDE

THIOBISMOL see BISMUTH SODIUM THIOGLYCOLLATE

THIOCARBAMIDE see ISOTHIOUREA

THIOCARBANIL see ISOTHIOCYANIC ACID, PHENYL ESTER

THIOCARBANILIDE see DIPHENYLTHIOUREA

THIOCARBARSONE see 4-CARBAMIDOPHENYL BIS(CARBOXY METHYL THIO)ARSENITE

THIOCARBONYL CHLORIDE see THIOPHOSGENE

THIOCYANATOACETIC ACID ISOBORNYL ESTER see ISOBORNYL THIO-CYANATO ACETATE

THIOCYANIC ACID, ALLYL ESTER see ALLYL THIOCYANATE

THIOCYANIC ACID, AMYL ESTER see n-AMYL THIOCYANATE

THIOCYANIC ACID, CALCIUM SALT (2:1) see CALCIUM THIOCYA-NATE

1-THIOCYANOBUTANE see n-BUTYL THIOCYANATE

P,P-THIODIANILINE see 4,4′-THIODIANILINE

THIODIGLYCOLIC ACID see MERCAPTODIACETIC ACID

2,2′-THIODIGLYCOLIC ACID see MERCAPTODIACETIC ACID

BETA,BETA′-THIODIGLYCOLIC ACID see MERCAPTODIACETIC ACID

THIODIGLYCOLLIC ACID see MERCAPTODIACETIC ACID

THIODIHYDRACRYLIC ACID see BIS(2-CARBOXYETHYL) SULFIDE

2-THIO-3,5-DIMETHYLTETRAHYDRO-1,3,5-THIADIAZINE see DI-METHYLFORMOCARBOTHIALDINE

O,O′-(THIODI-4,1-PHENYLENE)BIS(O,O-DIMETHYL PHOSPHOROTHIO-ATE) see ABATE

THIODI-P-PHENYLENEDIAMINE see 4,4′-THIODIANILINE

THIODIPROPIONIC ACID see BIS(2-CARBOXYETHYL) SULFIDE

3,3′-THIODIPROPIONIC ACID see BIS(2-CARBOXYETHYL) SULFIDE

BETA,BETA′-THIODIPROPIONIC ACID see BIS(2-CARBOXYETHYL) SUL-FIDE

BETA,BETA′-THIODIPROPIONITRILE see DI(2-CYANOETHYL)SULFIDE

THIOETHANOL see ETHANETHIOL

2-THIOETHANOL see 2-MERCAPTOETHANOL

THIOETHANOLAMINE see 2-AMINOETHANETHIOL

THIOETHYL ALCOHOL see ETHANETHIOL

THIOETHYL ETHER see ETHYL SULFIDE

THIOFOSGEN (CZECH) see THIOPHOSGENE

(1-THIO-D-GLUCOPYRANOSATO)GOLD see 1-AUROTHIO-D-GLUCO-PYRANOSE

THIOGLUCOSE D'OR (FRENCH) see 1-AUROTHIO-D-GLUCOPYRANOSE

THIOGLYCOL see 2-MERCAPTOETHANOL

THIOGLYCOLANILIDE see alpha-MERCAPTOACETANILIDE

THIOGLYCOLATESODIUM see SODIUM THIOGLYCOLATE

2-THIOGLYCOLIC ACID see 2-MERCAPTOACETIC ACID

THIOGLYCOLIC ACID (DOT) see 2-MERCAPTOACETIC ACID

THIOGLYCOLIC ACID ANILIDE see alpha-MERCAPTOACETANILIDE

THIOGLYCOLLIC ACID, AMMONIUM SALT see AMMONIUM MERP-CAPTOACETATE

THIOGLYCOLLIC ACID, SODIUM SALT see SODIUM THIOGLYCOLATE

2-THIO-4-HYDRAZINOURACIL see 4-HYDRAZINO-2-THIOURACIL

THIOKARBONYLCHLORID (CZECH) see THIOPHOSGENE

2-THIOL-DIHYDROGLYOXALINE see 2-IMIDAZOLIDINETHIONE

THIOMERIN SODIUM see SODIUM MERCAPTOMERIN

THIOMERIN SODIUM see DISODIUM-N-(3-(CARBOXYMETHYLTHIOMER-CURI)-2-METHOXYPROPYL)-alpha-CAMPHORAMATE

2-THIO-6-METHYL-1,3-PYRIMIDIN-4-ONE see 6-METHYLTHIOURACIL

6-THIO-4-METHYLURACIL see 6-METHYLTHIOURACIL

THIOMYLAL SODIUM see SURITAL SODIUM

BETA-THIONAPHTHOL see 2-NAPHTHALENETHIOL

THIO-BETA-NAPHTHOL see 2-NAPHTHALENETHIOL

THIONODEMETON SULFONE see SYSTOX SULFONE

THIONYL DIFLUORIDE see THIONYL FLUORIDE

2-THIO-4-OXO-6-METHYL-1,3-PYRIMIDINE see 6-METHYLTHIOURA-CIL

2-THIO-4-OXO-6-PROPYL-1,3-PYRIMIDINE see 6-PROPYL-2-THIOURA-CIL

THIOPENTAL SODIUM SALT see PENTOTHAL SODIUM

THIOPHENOL see BENZENETHIOL

THIOPHOSPHATE DE S-N-(1-CYANO-1-METHYLETHYL)CARBAMOYL-METHYLE ET DE O,O-DIETHYLE (FRENCH) see PHOSPHOROTHIOIC ACID-S-(((1-CYANO-1-METHYL-ETHYL)CARBAMOYL)METHYL)-O,O-DIETHYL ESTER

THIOPHOSPHATE DE O,O-DIETHYLE ET DE O-(2,5-DICHLORO-4-BROMO) PHENYLE (FRENCH) see ETHYL BROMOPHOS

THIOPHOSPHATE DE O,O-DIETHYLE ET DE O-2-ETHYLTHIO-ETHYLE (FRENCH) see DEMETON-O-METHYL

THIOPHOSPHATE DE O,O-DIETHYLE ET DE S-(2-ETHYLTHIO-ETHYLE) (FRENCH) see DEMETON-S

THIOPHOSPHATE DE O,O-DIETHYLE ET DE O-2-ISOPROPYL-4-METHYL-6-PYRIMIDYLE (FRENCH) see DIAZIDE

THIOPHOSPHATE DE O,O-DIETHYLE ET DE O-(4-METHYL-7-COUMARI-NYLE) (FRENCH) see DIETHYL (4-METHYLUMBELLIFERYL) THIONO-PHOSPHATE

THIOPHOSPHATE DE O,O-DIMETHYLE ET DE O-3-CHLORO-4-NITRO-PHENYLE (FRENCH) see METHYLCHLOROTHION

THIOPHOSPHATE DE O,O-DIMETHYLE ET DE O-4-CHLORO-3-NITRO-PHENYLE (FRENCH) see p-NITRO-o-CHLOROPHENYL DIMETHYL THIONOPHOSPHATE

THIOPHOSPHATE DE O-DIMETHYLE ET DE S-2-ETHYLSULFINYL-ETHYLE (FRENCH) see DEMETON-O-METHYL SULFOXIDE

THIOPHOSPHATE DE O,O-DIMETHYLE ET DE S-2-ETHYLTHIOETHYLE (FRENCH) see DEMETON-S-METHYL

THIOPHOSPHATE DE O,O-DIMETHYLE ET DE S-2-(ISOPROPYLSUL-FINYL)-ETHYLE (FRENCH) see METASYSTOX-S

THIOPHOSPHATE DE O,O-DIMETHYLE ET DE S-((5-METHOXY-4-PYRO-NYL)-METHYLE)(FRENCH) see ENDOTHION

THIOPHOSPHATE DE O,O-DIMETHYLE ET DE O-(3-METHYL-4-METHYL-THIOPHENYLE) (FRENCH) see O,O-DIMETHYLPHOSPHOROTHIOIC ACID-O-(4-METHYLTHIO)-m-TOLYLESTER

THIOPHOSPHATE DE O,O-DIMETHYLE ET DE O-(3-METHYL-4-NITRO-PHENYLE) (FRENCH) see O,O-DIMETHYL-O-(3-METHYL) PHOSPHO-ROTHIOATE

THIOPHOSPHATE DE O,O-DIMETHYLE ET DE O-(4-NITROPHENYLE) (FRENCH) see METHYL PARATHION

THIOPHOSPHATE DE O,O-DIMETHYLE ET DE O-(2,4,5-TRICHLORO-PHENYLE) (FRENCH) see TRICHLOROMETAFOS

THIOPHOSPHATE DE O,ODIETHYLE ET DE O-(4-NITROPHENYLE) (FRENCH) see PARATHION

THIOPHOSPHORIC ANHYDRIDE see PHOSPHORUS PENTASULFIDE

THIOPHOSPHORSAEURE-O,S-DIMETHYLESTERAMID (GERMAN) see O,S-DIMETHYL PHOSPHORAMIDOTHIOATE

2-THIO-6-PROPYL-1,3-PYRIMIDIN-4-ONE see 6-PROPYL-2-THIOURACIL

6-THIO-4-PROPYLURACIL see 6-PROPYL-2-THIOURACIL

beta-THIOPSEUDOUREA see ISOTHIOUREA

2-THIO-1,3-PYRIMIDIN-4-ONE see 2-THIOURACIL

THIORIDAZINE see 10-(2-(1-METHYL-2-PIPERIDYL)ETHYL)-2-(METHYL-THIO)PHENOTHIAZINE

THIORIDAZINE HYDROCHLORIDE see 10-(2-(1-METHYL-2-PIPERIDYL) ETHYL)-2-METHYLTHIOPHENOTHIAZINE HYDROCHLORIDE

THIOSERINE see L-CYSTEINE

THIOSULFURIC ACID, DISODIUM SALT, PENTAHYDRATE see SODIUM THIOSULFATE, PENTAHYDRATE

THIOSULFUROUS DICHLORIDE see SULFUR CHLORIDE

THIO-TEP see THIOPHOSPHAMIDE

2-THIO-1-(THIOCARBAMOYL)-UREA see DITHIOBIURET

THIOTRIETHYLENEPHOSPHORAMIDE see THIOPHOSPHAMIDE

THIOURACIL see 2-THIOURACIL

6-THIOURACIL see 2-THIOURACIL

2-THIOUREA see ISOTHIOUREA

THIOVANIC ACID see 2-MERCAPTOACETIC ACID

9H-THIOXANTHEN-9-ONE, 1-((2-(DIETHYL AMINO)ETHYL)AMINO)-4-
METHYL- (9CI) see LUCANTHONE

THIRAM (DOT) see BIS(DIMETHYL THIOCARBAMYL)DISULFIDE

THOMPSON'S WOOD FIX see PENTACHLOROPHENOL

THORAZINE see CHLOROPROMAZINE

THORAZINE HYDROCHLORIDE see CHLOROPROMAZINE HYDROCHLO-
RIDE

THORIA see THORIUM OXIDE

THORIUM DIOXIDE see THORIUM OXIDE

THORIUM (IV) NITRATE TETRAHYDRATE see THORIUM NITRATE

THORIUM METAL, PYROPHORIC (DOT) see THORIUM

THORIUM TETRACHLORIDE see THORIUM CHLORIDE

THORN APPLE see STRAMONIUM

THOROTRAST see THORIUM OXIDE

THPA see TETRAHYDROPHTHALIC ACID ANHYDRIDE

D-THREO-CHLORAMPHENICOL see CHLORAMPHENICOL

D-(-)-THREO-CHLORAMPHENICOL see CHLORAMPHENICOL

D-(-)-THREO-2-DICHLOROACETAMIDO-1-P-NITROPHENYL-1,3-PRO-
PANEDIOL see CHLORAMPHENICOL

D-THREO-N-DICHLOROACETYL-1-P-NITROPHENYL-2-AMINO-1,3-PRO-
PANEDIOL see CHLORAMPHENICOL

D-(-)-THREO-2,2-DICHLORO-N-(BETA-HYDROXY-ALPHA-(HYDROXY-
METHYL))-P-NITROPHENETHYLACETAMIDE see CHLORAMPHENICOL

D-THREO-N-(1,1'-DIHYDROXY-1-P-NITROPHENYLISOPROPYL)-
DICHLOROACETAMIDE see CHLORAMPHENICOL

D-THREO-METHOXY-3-(1-OCTENYL-ONN-AZOXY)-2-BUTANOL see
ELAIOMYCIN

D-(-)-THREO-1-P-NITROPHENYL-2-DICHLORACETAMIDO-1,3-PROPANE-
DIOL see CHLORAMPHENICOL

D-THREO-1-(P-NITROPHENYL)-2-(DICHLOROACETYLAMINO)-1,3-PRO-
PANEDIOL see CHLORAMPHENICOL

THROMBOLIQUINE see HEPARIN

-(-)-3-THUJANONE, (1S,4R,5R) see THUJONE

(-)-THUJONE see THUJONE

THUJONE see ABSINTHIUM

ALPHA-THUJONE see THUJONE

THYLOQUINONE see 2-METHYL-1,4-NAPHTHOQUINONE

THYME CAMPHOR see THYMOL

THYMIC ACID see THYMOL

M-THYMOL see THYMOL

O-THYMOL see CARVACROL

12-O-TIGLYL-PHORBOL-13-BUTYRATE see PHORBOL-12-O-TIGLYL-13-
BUTYRATE

TILLAM (RUSSIAN) see S-PROPYL BUTYLETHYLTHIOCARBAMATE

TIN (ALPHA) see TIN

TIN CHLORIDE, FUMING (DOT) see TIN (IV) CHLORIDE (1:4)

TINDAL see ACETOPHENAZINE

TIN DIBUTYL DILAURATE see DIBUTYLBIS(LAUROYLOXY)STANNANE

TIN DICHLORIDE see TIN (II) CHLORIDE (1:2)

TIN FLAKE see TIN

TIN PERBROMIDE see TIN (IV) BROMIDE (1:4)

TIN PERCHLORIDE see TIN (IV) CHLORIDE (1:4)

TIN POWDER see TIN

TIN PROTOCHLORIDE see TIN (II) CHLORIDE (1:2)

TIN TETRABROMIDE see TIN (IV) BROMIDE (1:4)

TINTETRACHLORIDE (DUTCH) see TIN (IV) CHLORIDE (1:4)

TIN TETRACHLORIDE, ANHYDROUS see TIN (IV) CHLORIDE (1:4)

TIN TETRAIODIDE see TIN (IV) IODIDE (1:4)

TIN TRIPHENYL ACETATE see ACETOXYTRIPHENYLSTANNANE

TIRFLUORURE DE CHLORE (FRENCH) see CHLORINE TRIFLUORIDE

TITAANTETRACHLORIDE (DUTCH) see TITANIUM CHLORIDE

TITANDIOXID (SWEDEN) see TITANIUM OXIDE

TITANIO (TETRACLORURO DI) (ITALIAN) see TITANIUM CHLORIDE

TITANIUM ACETONYL ACETONATE see BIS(ACETYLACETONATO) TITA-
NIUM OXIDE

TITANIUM ALLOY see TITANIUM (DRY POWDER)

TITANIUM, DICHLOROBIS(ETA(SUP 5)-2,4-CYCLOPENTADIEN-1-YL-
(9CI) see DICHLOROTITANOCENE

TITANIUM DIOXIDE see TITANIUM OXIDE

TITANIUM METAL POWDER, DRY see TITANIUM (DRY POWDER)

TITANIUM OXIDE BIS(ACETYLACETONATE) see BIS(ACETYLACETO-
NATO) TITANIUM OXIDE

TITANIUM, OXOBIS(2,4-PENTANEDIONATO-O,O') see BIS(ACETYLA-
CETONATO) TITANIUM OXIDE

TITANIUM POTASSIUM FLUORIDE see POTASSIUM HEXAFLUOROTITA-
NATE

TITANIUM TETRACHLORIDE see TITANIUM CHLORIDE

TITANTETRACHLORID (GERMAN) see TITANIUM CHLORIDE

TITANYL BIS(ACETYLACETONATE) see BIS(ACETYLACETONATO)
TITANIUM OXIDE

TIURAM (POLISH) see BIS(DIMETHYL THIOCARBAMYL)DISULFIDE

TL 80 see 1,4-DIBROMO-2-BUTENE

TL 345 see DIETHYL FLUOROPHOSPHATE

TL 797 see DIVINYL SULFONE

TL 1070 see CADMIUM FLUOSILICATE

TL 1091 see NICKEL (II) FLUOBORATE

TL 1183 see METHYL-gamma-FLUOROCROTONATE

TL 1266 see DI-SEC-BUTYL FLUOROPHOSPHONATE

TL 1312 see FLUOROACETANILIDE

TL 1333 see METHYL-gamma-FLUORO-beta-HYDROXYBUTYRATE

TL 1428 see METHYL-beta-ACETOXYETHYL-beta-CHLOROETHYLAMINE

TMA see TRIMETHYLAMINE

7,8,12-TMBA see 7,8,12-TRIMETHYLBENZ(a)ANTHRACENE

TMU see 1,1,3,3-TETRAMETHYLUREA

TNA see TETRANITROANILINE

TNT see sym-TRINITROTOLUENE

ALPHA-TNT see sym-TRINITROTOLUENE

TOBACCO WOOD see HAMAMELIS

2,4-TOLAMINE see TOLUENE-2,4-DIAMINE

TOLBUTAMIDE see 1-BUTYL-3-(p-TOLYL SULFONYL)UREA

2-TOLIDIN (GERMAN) see 3,3'-TOLIDINE

2-TOLIDINA (ITALIAN) see 3,3'-TOLIDINE

O-TOLIDINE see 3,3'-TOLIDINE

O,O'-TOLIDINE see 3,3'-TOLIDINE

3-O-TOLOXY-1,2-PROPANEDIOL see GLYCERYL-O-TOLYL ETHER

TOLUEEN (DUTCH) see TOLUENE

TOLUEEN-DIISOCYANAAT (DUTCH) see TOLUENE DIISOCYANATE

TOLUEEN-DIISOCYANAAT (DUTCH) see DI-ISO-CYANATOLUENE

TOLUEN (CZECH) see TOLUENE

TOLUEN-DISOCIANATO (ITALIAN) see TOLUENE DIISOCYANATE

TOLUEN-DISOCIANATO (ITALIAN) see DI-ISO-CYANATOLUENE

O-TOLUENE-1-AZO-2-NAPHTHYLAMINE see 1-(O-TOLYLAZO)-2-
NAPHTHYLAMINE

O-TOLUENEAZO-O-TOLUENEAZO-BETA-NAPHTHOL see SCARLET RED

O-TOLUENEAZO-O-TOLUENE-BETA-NAPHTHOL see SCARLET RED

O-TOLUENEAZO-O-TOLUIDINE see 2-AMINO-5-AZOTOLUENE

M-TOLUENEDIAMINE see TOLUENE-2,4-DIAMINE

P-TOLUENEDIAMINE see TOLUENE-2,5-DIAMINE

2,4-TOLUENEDIAMINE see TOLUENE-2,4-DIAMINE

P-TOLUENEDIAMINE DIHYDROCHLORIDE see 2,5-DIAMINOTOLUENE
DIHYDROCHLORIDE

2,4-TOLUENEDIAMINE DIHYDROCHLORIDE see 2,4-DIAMINOTOLUENE
DIHYDROCHLORIDE

P-TOLUENEDIAMINE SULFATE see 2,5-DIAMINOTOLUENE SULFATE

2,5-TOLUENEDIAMINE SULFATE see 2,5-DIAMINOTOLUENE SULFATE

TOLUENE-2,5-DIAMINE, SULFATE (1:1) (8CI) see 2,5-DIAMINOTO-
LUENE SULFATE

P-TOLUENEDIAMINE SULPHATE see 2,5-DIAMINOTOLUENE SUL-
FATE

TOLUENE-2,5-DIAMINE SULPHATE see 2,5-DIAMINOTOLUENE SUL-
FATE

TOLUENE DIISOCYANATE see DI-ISO-CYANATOLUENE

TOLUENE-2,4-DIISOCYANATE see DI-ISO-CYANATOLUENE

2,3-TOLUENEDIOL see 2,3-DIHYDROXYTOLUENE

TOLUENE HEXAHYDRIDE see CYCLOHEXYLMETHANE

TOLUENE-2-SULFONAMIDE see O-TOLUENESULFONAMIDE

1-P-TOLUENESULFONYL-3-BUTYLUREA see 1-BUTYL-3-(p-TOLYL
SULFONYL)UREA

TOLUENE TRICHLORIDE see BENZYL TRICHLORIDE

ALPHA-TOLUENOL see BENZYL ALCOHOL

M-TOLUIC ACID DIETHYLAMIDE see DIETHYL-m-TOLUAMIDE

M-TOLUIDIN (CZECH) see m-TOLUIDINE

O-TOLUIDIN (CZECH) see o-TOLUIDINE

P-TOLUIDIN (CZECH) see p-TOLUIDINE

2-TOLUIDINE see o-TOLUIDINE

3-TOLUIDINE see m-TOLUIDINE

4-TOLUIDINE see p-TOLUIDINE

P-TOLUIDINE, N,N-DIMETHYL-ALPHA-INDOLYLIDENE see 1-(4-DI-
METHYLAMINOBENZAL)INDENE

2-TOLUIDINE HYDROCHLORIDE see o-TOLUIDINE HYDROCHLORIDE

P-TOLUIDINIUM CHLORIDE see p-TOLUIDINE HYDROCHLORIDE

O-TOLUIDYNA (POLISH) see o-TOLUIDINE

TOLUILENODWUIZOCYJANIAN (POLISH) see TOLUENE DIISOCYANATE

TOLUILENODWUIZOCYJANIAN (POLISH) see DI-ISO-CYANATOLUENE

ALPHA-TOLUNITRILE see PHENYLACETONITRILE

TOLUOL see TOLUENE

O-TOLUOL-AZO-O-TOLUIDIN (GERMAN) see 2-AMINO-5-AZOTOLUENE

TOLUOLO (ITALIAN) see TOLUENE

ORTHO-TOLUOL-SULFONAMID (GERMAN) see o-TOLUENESULFONA-
MIDE

P-TOLUOLSULFONSAEUREAETHYL ESTER (GERMAN) see ETHYL TOSY-
LATE

P-TOLUOLSULFONSAEURE METHYL ESTER (GERMAN) see METHYL-p-
METHYLBENZENESULFONATE

ALPHA-TOLUOLTHIOL see alpha-TOLUENETHIOL

P-TOLUQUINONE see 2-METHYL-p-BENZOQUINONE

1,4-TOLUQUINONE see 2-METHYL-p-BENZOQUINONE

M-TOLUYLENDIAMIN (CZECH) see TOLUENE-2,4-DIAMINE

P-TOLUYLENDIAMINE see TOLUENE-2,5-DIAMINE

M-TOLUYLENEDIAMINE see TOLUENE-2,4-DIAMINE

2,4-TOLUYLENEDIAMINE see TOLUENE-2,4-DIAMINE

TOLUYLENE-2,5-DIAMINE see TOLUENE-2,5-DIAMINE

P-TOLUYLENEDIAMINE SULPHATE see 2,5-DIAMINOTOLUENE SULFATE

TOLUYLENE-2,5-DIAMINE SULPHATE see 2,5-DIAMINOTOLUENE SUL-
FATE

TOLUYLENE-2,4-DIISOCYANATE see DI-ISO-CYANATOLUENE

M-TOLYENEDIAMINE see TOLUENE-2,4-DIAMINE

P-TOLYL ACETATE see 4-METHYLPHENYL ACETATE

TOLYLAMINE see p-TOLUIDINE

M-TOLYLAMINE see m-TOLUIDINE

O-TOLYLAMINE see o-TOLUIDINE

P-TOLYLAMINE see p-TOLUIDINE

O-TOLYLAMINE HYDROCHLORIDE see o-TOLUIDINE HYDROCHLORIDE

5-(O-TOLYLAZO)-2-AMINOTOLUENE see 2-AMINO-5-AZOTOLUENE

4-O-TOLYLAZO-O-DIACETOTOLUIDE see N-ACETYL-N-(2-METHYL-4-
((2-METHYLPHENYL)AZO)PHENYL)ACETAMIDE

4'-(O-TOLYLAZO)-O-DIACETOTOLUIDIDE see N-ACETYL-N-(2-METH-
YL-4-((2-METHYLPHENYL)AZO)PHENYL)ACETAMIDE

4-(O-TOLYLAZO)-O-TOLUIDINE see 2-AMINO-5-AZOTOLUENE

O-TOLYLAZO-O-TOLYAZO-2-NAPHTHOL see SCARLET RED

1-((4-(O-TOLYLAZO)-O-TOLYL)AZO)-2-NAPHTHOL) see SCARLET RED

O-TOLYLAZO-O-TOLYAZO-BETA-NAPHTHOL see SCARLET RED

M-TOLYLENEDIAMINE see TOLUENE-2,4-DIAMINE

TOLYLENE-2,4-DIAMINE see TOLUENE-2,4-DIAMINE

2,4-TOLYLENEDIAMINE see TOLUENE-2,4-DIAMINE

4-M-TOLYLENEDIAMINE see TOLUENE-2,4-DIAMINE

P-TOLYLENEDIAMINE SULPHATE see 2,5-DIAMINOTOLUENE SULFATE

2,4-TOLYLENEDIISOCYANATE see DI-ISO-CYANATOLUENE

TOLYLENE-2,4-DIISOCYANATE see DI-ISO-CYANATOLUENE

P-TOLYL ETHANOATE see 4-METHYLPHENYL ACETATE

1-ORTHO-TOLYLGLYCEROL ETHER see GLYCERYL-O-TOLYL ETHER

ALPHA-(O-TOLYL)GLYCERYL ETHER see GLYCERYL-O-TOLYL ETHER

ALPHA-TOLYL MERCAPTAN see alpha-TOLUENETHIOL

3-TOLYL-N-METHYLCARBAMATE see TSUMACIDE

M-TOLYL-N-METHYLCARBAMATE see TSUMACIDE

3-(O-TOLYLOXY)PROPANE-1,2-DIOL see GLYCERYL-O-TOLYL ETHER

O-TOLYL PHOSPHATE see TRI-2-TOLYL PHOSPHATE

N-(P-TOLYLSULFONYL)-N'-BUTYLCARBAMIDE see 1-BUTYL-3-(p-
TOLYL SULFONYL)UREA

3-(P-TOLYL-4-SULFONYL)-1-BUTYLUREA see 1-BUTYL-3-(p-TOLYL
SULFONYL)UREA

TOMATIDINE GLYCOSIDE see TOMATINE

ALPHA-TOMATINE see TOMATINE

TOMATOTONE see p-CHLORO PHENOXY ACETIC ACID

TONKA BEAN CAMPHOR see COUMARIN

TONY RED see OIL RED

TOPITRACIN see BACITRACIN

TOTOCAINE HYDROCHLORIDE see p-AMINOBENZOYLDIMETHYLAMINO-
1,2-DIMETHYLPROPANOL HYDROCHLORIDE

TOXAFEEN (DUTCH) see TOXAPHENE

TOXALBUMIN see ABRIN

TOXAPHEN (GERMAN) see TOXAPHENE

TOXILIC ACID see cis-BUTENEDIOIC ACID

TOXILIC ANHYDRIDE see MALEIC ANHYDRIDE

TPA-3-BETA-OL see PHORBOLOL MYRISTATE ACETATE

TRANCYLPROMINE SULFATE see PHENYLCYCLOPROMINE SULFATE

TRANK see ANGEL DUST

TRANSAMINE SULFATE see PHENYLCYCLOPROMINE SULFATE

TRANSVAALIN see GLUCO PROSCILLARIDIN A

TRANYLCYPROMINE SULFATE see PHENYLCYCLOPROMINE SULFATE

TRAZODONE HYDROCHLORIDE see 2-(3-(4-(3-CHLOROPHENYL)-1-PI-
PERAZINYL)PROPYL)-1,2,4-TRIZOLO(4,3-a)PYRIDIN-3(2H)-ONE HY-
DROCHLORIDE

TREMOLITE see ASBESTOS

TRIACETALDEHYDE (FRENCH) see PARALDEHYDE

TRIAETHANOLAMIN-NG see TRIHYDROXYTRIETHYLAMINE

TRIAETHYLAMIN (GERMAN) see TRIETHYLAMINE

TRIAETHYLENPHOSPHORSAEUREAMID (GERMAN) see TEPA

TRIAETHYLENPHOSPHORSAEUREAMID (GERMAN) see TRIETHYLENE-
PHOSPHOROTRIAMIDE

TRIAETHYLZINNACETAT (GERMAN) see ACETOXYTRIETHYLSTANNANE

TRIAMCINOLONE-16,17-ACETONIDE see ARISTOCORT ACETONIDE

TRIAMIFOS (GERMAN, DUTCH, ITALIAN) see 5-AMINO-1-BIS(DI-
METHYLAMIDE)PHOSPHORYL-3-PHENYL-1,2,4-TRIAZOLE

2,4,6-TRIAMINO-s-TRIAZINE see MELAMINE

4,4',4''-TRIAMINOTRIPHENYLMETHANE see TRIAMINOTRIPHENYL-
METHANE

P,P',P''-TRIAMINOTRIPHENYLMETHANE see TRIAMINOTRIPHENYL-
METHANE

4,4'4''-TRIAMINOTRIPHENYLMETHAN-HYDROCHLORID (GERMAN) see
BASIC PARAFUCHSINE

TRI-P-ANISYLCHLOROETHYLENE see CHLOROTRIANISENE

TRIATOMIC OXYGEN see OZONE

1,2,3-TRIAZAINDENE see 1H-BENZOTRIAZOLE

TRIAZICHON (GERMAN) see TRISETHYLENEIMINOQUINONE

(S-TRIAZINE-2,4,6-TRIYLTRINITRILO)HEXAMETHANOL see HEXA(HY-
DROXYMETHYL)MELAMINE

1,1',1''-s-TRIAZINE-2,4,6-TRIYLTRISAZIRIDINE see TRISAZIRIDINYL-
TRIAZINE

2,3,5-TRI-(1-AZIRIDINYL)-P-BENZOQUINONE see TRISETHYLENEI-
MINOQUINONE

TRI-(1-AZIRIDINYL)PHOSPHINE OXIDE see TRIETHYLENEPHOSPHORO-
TRIAMIDE

TRIAZIRIDINYLPHOSPHINE SULFIDE see THIOPHOSPHAMIDE

TRIAZIRIDINYL TRIAZINE see TRISAZIRIDINYLTRIAZINE

TRIAZOIC ACID see HYDRAZOIC ACID

1H-1,2,4-TRIAZOL-3-AMINE see 3-AMINOTRIAZOLE

TRIAZOTION (RUSSIAN) see ETHYL GUTHION

TRIBROMMETHAAN (DUTCH) see BROMOFORM

TRIBROMMETHAN (GERMAN) see BROMOFORM

TRIBROMOALUMINUM see ALUMINUM BROMIDE

TRIBROMOARSINE see ARSENIC(III) BROMIDE

2,2,2-TRIBROMOETHANOL see AVERTIN

2,2,2-TRIBROMOETHYL ALCOHOL see AVERTIN

TRIBROMOMETAN (ITALIAN) see BROMOFORM

TRIBROMOMETHANE see BROMOFORM

TRIBROMONITROMETHANE see NITROTRIBROMOMETHANE

TRIBROMO STIBINE see ANTIMONY TRIBROMIDE

TRIBUTILFOSFATO (ITALIAN) see TRI-N-BUTYL PHOSPHATE

TRI-N-BUTOXYBORANE see TRI-n-BUTYL BORATE

TRI(2-BUTOXYETHYL) PHOSPHATE see 2-BUTOXYETHANOL PHOS-
PHATE

TRI-N-BUTYLAMINE see TRIBUTYLAMINE

TRIBUTYL BORATE see TRI-n-BUTYL BORATE

TRIBUTYLBORINE see TRI-n-BUTYL BORANE

TRIBUTYL CELLOSOLVE PHOSPHATE see 2-BUTOXYETHANOL PHOSPHATE

TRIBUTYLE (PHOSPHATE DE) (FRENCH) see TRI-N-BUTYL PHOSPHATE

TRIBUTYLFOSFAAT (DUTCH) see TRI-N-BUTYL PHOSPHATE

TRIBUTYLPHOSPHAT (GERMAN) see TRI-N-BUTYL PHOSPHATE

TRIBUTYL PHOSPHATE see TRI-N-BUTYL PHOSPHATE

TRIBUTYL PHOSPHOROTRITHIOITE see S,S,S-TRIBUTYL TRITHIOPHOSPHITE

TRIBUTYLTIN BENZOATE see BENZOYLOXYTRIBUTYLSTANNANE

TRIBUTYLTIN CYANATE see CYANATOTRIBUTYLSTANNANE

TRIBUTYLTIN OXIDE see BIS(TRIBUTYL TIN)OXIDE

TRI-N-BUTYL-ZINN BENZOATE (GERMAN) see BENZOYLOXYTRIBUTYL-STANNANE

TRIBUTYROIN see TRIBUTYRIN

TRICALCIUMARSENAT (GERMAN) see ARSENIC ACID, CALCIUM SALT (2:3)

TRICALCIUM ARSENATE see ARSENIC ACID, CALCIUM SALT (2:3)

TRICAPRYLMETHYLAMMONIUM CHLORIDE see METHYLTRIOCTYLAMMONIUM CHLORIDE

TRICAPRYLYLMETHYLAMMONIUM CHLORIDE see METHYLTRIOCTYL-AMMONIUM CHLORIDE

TRICHLOORAZIJNZUUR (DUTCH) see TRICHLORACETIC ACID

1,1,1-TRICHLOOR-2,2-BIS(4-CHLOOR FENYL)-ETHAAN (DUTCH) see DDT

2,2,2-TRICHLOOR-1,1-BIS(4-CHLOOR FENYL)-ETHANOL (DUTCH) see 1,1-BIS(p-CHLOROPHENYL)-2,2,2-TRICHLOROETHANOL

1,1,1-TRICHLOORETHAAN (DUTCH) see 1,1,1-TRICHLOROETHANE

TRICHLOORETHEEN (DUTCH) see TRICHLOROETHYLENE

TRICHLOORETHEEN (DUTCH) see ACETYLENE TRICHLORIDE

(2,4,5-TRICHLOOR-FENOXY)-AZIJNZUUR (DUTCH) see 2,4,5-T

O-(2,4,5-TRICHLOOR-FENYL)-O,O-DIMETHYL-MONOTHIOFOSFAAT (DUTCH) see TRICHLOROMETAFOS

TRICHLOORMETHAAN (DUTCH) see CHLOROFORM

TRICHLOORMETHYLBENZEEN (DUTCH) see BENZYL TRICHLORIDE

TRICHLOORNITROMETHAAN (DUTCH) see TRICHLORONITROMETHANE

TRICHLOORSILAAN (DUTCH) see TRICHLOROSILANE

TRICHLORACETALDEHYD-HYDRAT (GERMAN) see CHLORAL HYDRATE

1,1,1-TRICHLORAETHAN (GERMAN) see 1,1,1-TRICHLOROETHANE

TRICHLORAETHEN (GERMAN) see TRICHLOROETHYLENE

TRICHLORAETHEN (GERMAN) see ACETYLENE TRICHLORIDE

1,1,1-TRICHLOR-2,2-BIS(4-CHLOR-PHENYL)-AETHAN (GERMAN) see DDT

2,2,2-TRICHLOR-1,1-BIS(4-CHLOR-PHENYL)-AETHANOL (GERMAN) see 1,1-BIS(p-CHLOROPHENYL)-2,2,2-TRICHLOROETHANOL

TRICHLORESSIGSAEURE (GERMAN) see TRICHLORACETIC ACID

TRICHLORETHENE (FRENCH) see ACETYLENE TRICHLORIDE

TRICHLORETHYLENE see ACETYLENE TRICHLORIDE

2,4,6-TRICHLORFENOL (CZECH) see 2,4,6-TRICHLOROPHENOL

TRICHLORFENSON see 4-CHLOROPHENYL 4-CHLOROBENZENESULFO-NATE

TRICHLORMETHAN (CZECH) see CHLOROFORM

TRICHLORMETHYLBENZOL (GERMAN) see BENZYL TRICHLORIDE

TRICHLORNITROMETHAN (GERMAN) see TRICHLORONITROMETHANE

TRICHLOROACETALDEHYDEMONOHYDRATE see CHLORAL HYDRATE

TRICHLOROACETIC ACID SOLUTION (DOT) see TRICHLORACETIC ACID

TRICHLOROALUMINUM see ALUMINUM CHLORIDE

3,5,6-TRICHLORO-4-AMINOPICOLINIC ACID see 4-AMINO-3,5,6-TRI-CHLOROPICOLINIC ACID

TRICHLOROARSINE see ARSENIC CHLORIDE

1,1,1-TRICHLORO-2,2-BIS(P-ANISYL)ETHANE see DIMETHOXY-DDT

2,2,2-TRICHLORO-1,1-BIS(4-CHLOROPHENYL)-ETHANOL (FRENCH) see 1,1-BIS(p-CHLOROPHENYL)-2,2,2-TRICHLOROETHANOL

2,2,2-TRICHLORO-1,1-BIS(4-CLORO-FENIL)-ETANOLO (ITALIAN) see 1,1-BIS(p-CHLOROPHENYL)-2,2,2-TRICHLOROETHANOL

1,1,1-TRICHLORO-2,2-BIS(P-METHOXYPHENOL)ETHANOL see DI-METHOXY-DDT

1,1,1-TRICHLORO-2,2-BIS(P-METHOXYPHENYL)ETHANE see DI-METHOXY-DDT

T-TRICHLOROBUTYL ALCOHOL see ACETONE CHLOROFORM

TRICHLORO-T-BUTYL ALCOHOL see ACETONE CHLOROFORM

TRICHLOROCHROMIUM see CHROMIUM (III) CHLORIDE

1,1,1-TRICHLORO-2,2-DI(4-CHLOROPHENYL)-ETHANE see DDT

2,2,2-TRICHLORO-1,1-DI-(4-CHLOROPHENYL)ETHANOL see 1,1-BIS-(p-CHLOROPHENYL)-2,2,2-TRICHLOROETHANOL

1,1,1-TRICHLORO-2,2-DI(4-METHOXYPHENYL)ETHANE see DIMETH-OXY-DDT

1,2,3-TRICHLORO-4,6-DINITROBENZENE see 4,6-DINITRO-1,2,3-TRI-CHLOROBENZENE

TRICHLORODODECYLSILANE see DODECYLTRICHLOROSILANE

1,2,2-TRICHLOROETHANE see 1,1,2-TRICHLOROETHANE

ALPHA-TRICHLOROETHANE see 1,1,1-TRICHLOROETHANE

BETA-TRICHLOROETHANE see 1,1,2-TRICHLOROETHANE

TRICHLORO-1,1,1-ETHANE (FRENCH) see 1,1,1-TRICHLOROETHANE

2,2,2-TRICHLORO-1,1-ETHANEDIOL see CHLORAL HYDRATE

TRICHLOROETHANOIC ACID see TRICHLORACETIC ACID

TRICHLOROETHENE see ACETYLENE TRICHLORIDE

TRI-(2-CHLOROETHYL)AMINE HYDROCHLORIDE see TRIMUSTINE

TRI(BETA-CHLOROETHYL)AMINE HYDROCHLORIDE see TRIMUSTINE

TRICHLOROETHYLENE see ACETYLENE TRICHLORIDE

1,1,2-TRICHLOROETHYLENE see ACETYLENE TRICHLORIDE

1,2,2-TRICHLOROETHYLENE see ACETYLENE TRICHLORIDE

1,1'-(2,2,2-TRICHLOROETHYLIDENE)BIS(4-METHOXYBENZENE) see DI-METHOXY-DDT

1,2-O-(2,2,2-TRICHLOROETHYLIDENE)-alpha-d-GLUCOFURANOSE see alpha-d-GLUCOCHLORALOSE

TRI(2-CHLOROETHYL)PHOSPHATE see 2-CHLORO ETHANOL PHOS-PHATE

TRICHLOROETHYLSILANE see ETHYL TRICHLOROSILANE

TRICHLOROETHYLSILICANE see ETHYL TRICHLOROSILANE

TRICHLOROFORM see CHLOROFORM

2,2,2-TRICHLORO-1-HYDROXYETHYL-PHOSPHONATE, DIMETHYL ES-TER see ((2,2,2-TRICHLORO-1-HYDROXYETHYL) DIMETHYLPHOSPHO-NATE)

(2,2,2-TRICHLORO-1-HYDROXYETHYL)PHOSPHONIC ACID DIMETHYL ESTER see ((2,2,2-TRICHLORO-1-HYDROXYETHYL) DIMETHYL-PHOSPHONATE)

1,1,1-TRICHLOROISOPROPYL ALCOHOL see ISOPRAL

TRICHLOROMETHANE see CHLOROFORM

TRICHLOROMETHYLBENZENE see BENZYL TRICHLORIDE

1-(TRICHLOROMETHYL)BENZENE see BENZYL TRICHLORIDE

N-TRICHLOROMETHYLMERCAPTO-4-CYCLOHEXENE-1,2-DICARBOXI-MIDE see CAPTAN

N-(TRICHLOROMETHYLMERCAPTO)-DELTA(SUP 4)-TETRAHYDROPH-THALIMIDE see CAPTAN

1,1,1-TRICHLORO-2-METHYL-2-PROPANOL see ACETONE CHLORO-FORM

N-TRICHLOROMETHYLTHIO-CIS-DELTA(SUP 4)-CYCLOHEXENE-1,2-DI-CARBOXIMIDE see CAPTAN

N-TRICHLOROMETHYLTHIOCYCLOHEX-4-ENE-1,2-DICARBOXIMIDE see CAPTAN

N-((TRICHLOROMETHYL)THIO)-4-CYCLOHEXENE-1,2-DICARBOXIMIDE see CAPTAN

N-(TRICHLOROMETHYLTHIO)-4-CYCLOHEXENE-1,2-DICARBOXIMIDE see CAPTAN

N-TRICHLOROMETHYLTHIOTETRAHYDROPHTHALIMIDE see CAPTAN

N-((TRICHLOROMETHYL)THIO)TETRAHYDROPHTHALIMIDE see CAPTAN

N-TRICHLOROMETHYLTHIO-3A,4,7,7A-TETRAHYDROPHTHALIMIDE see CAPTAN

TRICHLOROMONOFLUOROMETHANE see TRICHLOROFLUOROMETHANE

TRICHLOROMONOSILANE see TRICHLOROSILANE

TRICHLOROPHENE see HEXACHLOROPHENE

2,4,5-TRICHLORO-PHENOL-O-ESTER WITH O-ETHYL ETHYLPHOSPHO-NOTHIOATE see ETHYL TRICHLORO PHENYL ETHYL PHOSPHONO-THIOATE

2,4,5-TRICHLOROPHENOXYACETIC ACID see 2,4,5-T

TRIFLUOROMETHYL CHLORIDE see CHLORO TRIFLUORO METHANE

3-(TRIFLUOROMETHYL)-N-ETHYL-alpha-METHYL PHENETHYL AMINE see N-ETHYL-alpha-METHYL-m-(TRIFLUOROMETHYL)PHENETHYL-AMINE

2-TRIFLUOROMETHYL-9-(3-(4- (beta-HYDROXYETHYL-1-PIPERAZI-NYL)PROPYLIDENE)THIOXANTHENE see cis (Z)-FLURENTHIXOL

3-TRIFLUOROMETHYLNITROBENZENE see 3-NITROBENZOTRIFLUORIDE

M-(TRIFLUOROMETHYL)NITROBENZENE see 3-NITROBENZOTRIFLUO-RIDE

3-(M-TRIFLUOROMETHYLPHENYL)-1,1-DIMETHYLUREA see 1,1-DI-METHYL-3-(alpha,alpha,alpha-TRIFLUORO-m-TOLYL) UREA

N-(3-TRIFLUOROMETHYLPHENYL)-N'-N'-DIMETHYLUREA see 1,1-DI-METHYL-3-(alpha,alpha,alpha-TRIFLUORO-m-TOLYL) UREA

N-(M-TRIFLUOROMETHYLPHENYL)-N',N'-DIMETHYLUREA see 1,1-DI-METHYL-3-(alpha,alpha,alpha-TRIFLUORO-m-TOLYL) UREA

1-(3-TRIFLUOROMETHYLPHENYL)-2-ETHYLAMINOPROPANE HYDRO-CHLORIDE see PHENFLUORAMINE HYDROCHLORIDE

4-(3-(2-(TRIFLUOROMETHYL)-9H-THIOXANTHEN-9-YLIDENE)PROPYL)-1-PIPERAZINEETHANOL see cis (Z)-FLURENTHIXOL

TRIFLUOROMONOBROMOMETHANE see BROMO TRIFLUORO METHANE

TRIFLUOROMONOCHLOROCARBON see CHLORO TRIFLUORO METHANE

TRIFLUOROMONOCHLOROETHYLENE see CHLOROTRIFLUOROETHYL-ENE

OMEGA-TRIFLUOROTOLUENE see BENZOTRIFLUORIDE

ALPHA,ALPHA,ALPHA-TRIFLUOROTOLUENE see BENZOTRIFLUORIDE

1,1,2-TRIFLUORO-1,2,2-TRICHLOROETHANE see FREON 113

TRIFLUOROVINYL CHLORIDE see CHLOROTRIFLUOROETHYLENE

TRIGLYCINE see AMINOTRIACETIC ACID

TRIGLYCOL see TRIETHYLENE GLYCOL

TRIGLYCOLLAMIC ACID see AMINOTRIACETIC ACID

TRIHEXYLPHENIDYL HYDROCHLORIDE see BENZHEXOL HYDROCHLO-RIDE

TRIHEXYLTIN ACETATE see ACETOXYTRIHEXYLSTANNANE

TRI-N-HEXYLZINNACETAT (GERMAN) see ACETOXYTRIHEXYLSTAN-NANE

3,7,15-TRIHYDROXY-4-ACETOXY-8-OXO-12,13-EPOXY-DELTA(SUP 9)-TRICHOTHECENE see 4-(ACETYLOXY)-12,13-EPOXY-3,7,15-TRIHYDROXY-TRICHOTHEC-9-EN-8-ONE-(3-alpha,4-beta,7-beta)

3,7,15-TRIHYDROXY-4-ACETOXY-8-OXO-12,13-EPOXY-DELTA(SUP 9)-TRICHOTHECENE see FUSARENONE X

1,8,9-TRIHYDROXYANTHRACENE see 1,8,9-ANTHRACENETRIOL

1,2,3-TRIHYDROXYBENZEN (CZECH) see PYROGALLOL

S-TRIHYDROXYBENZENE see PHLOROGLUCINOL

1,2,3-TRIHYDROXYBENZENE see PYROGALLOL

1,2,4-TRIHYDROXYBENZENE see 1,2,4-BENZENETRIOL

1,3,5-TRIHYDROXYBENZENE see PHLOROGLUCINOL

SYM-TRIHYDROXYBENZENE see PHLOROGLUCINOL

3,4,5-TRIHYDROXYBENZENE-1-PROPYLCARBOXYLATE see N-PROPYL GALLATE

3,4,5-TRIHYDROXYBENZOIC ACID see GALLIC ACID

3-BETA-14,16-BETA-TRIHYDROXY-5-BETA-BUFA-20,22-DIENOLIDE, 16-ACETATE see BUFOTALINE

3BETA,12BETA,14BETA-TRIHYDROXY-5-BETA-CARD-20(22) ENOLIDE-3-(4'')-O-METHYL-TRIDIGITOXOSIDE) see beta-METHYLDIGOXIN

3,7,12-TRIHYDROXYCHOLAN-24-OIC ACID (3-alpha,5-beta,7-alpha, 12-alpha see CHOLIC ACID (HYDRATE)

3-alpha,7-alpha,12-alpha-TRIHYDROXY-5-beta-CHOLAN-24-OIC ACID see CHOLIC ACID (HYDRATE)

1,3,5-TRIHYDROXYCYCLOHEXATRIENE see PHLOROGLUCINOL

3,4,5-TRIHYDROXY-1-CYCLOHEXENE-1-CARBOXYLIC ACID see SHIK-IMIC ACID

3,16-alpha,17-beta-TRIHYDROXY-delta-1,3,5-ESTRATRIENE see ES-TRIOL

3,16-alpha,17-beta-TRIHYDROXY ESTRA-1,3,5(10)-TRIENE see ES-TRIOL

TRI(HYDROXYETHYL)AMINE see TRIHYDROXYTRIETHYLAMINE

4',5,7-TRIHYDROXYFLAVAONE see CHAMOMILE

8,12,18-TRIHYDROXY-4-METHYL-11,16-DIOXOSENECIONANIUM see HYDROXYSENKIRKINE

2,4A,7-TRIHYDROXY-1-METHYL-8-METHYLENEGIBB-3-ENE-1,10-CAR-BOXYLICACID 1-4-LACTONE see GIBBERELLIC ACID

11-BETA,17,21-TRIHYDROXY-6-ALPHA-METHYLPREGNA-1,4-DIENE-3,20-DIONE see METHYLPREDNISOLONE

11-BETA,17-ALPHA,21-TRIHYDROXY-6-ALPHA-METHYL-1,4-PREGNA-DIENE-3,20-DIONE see METHYLPREDNISOLONE

3,16-alpha,17-beta-TRIHYDROXYOESTRA-1,3,5(10)-TRIENE see ES-TRIOL

3,16-alpha,17-beta-TRIHYDROXY-delta-1,3,5-OESTRATRIENE see ES-TRIOL

11-BETA,17,21-TRIHYDROXYPREGNA-1,4-DIENE-3,20-DIONE see PRE-DONIN

11-BETA,17-ALPHA,21-TRIHYDROXY-1,4-PREGNADIENE-3,20-DIONE see PREDONIN

11-BETA,17-ALPHA,21-TRIHYDROXYPREGNA-1,4-DIENE-3,20-DIONE see PREDONIN

11-beta,17,21-TRIHYDROXYPREGN-4-ENE-3,20-DIONE see CORTISOL

11-beta,17-alpha-21-TRIHYDROXY-4-PREGNENE-3,20-DIONE see COR-TISOL

1,2,3-TRIHYDROXYPROPANE see GLYCERINE

(5Z,9-ALPHA,11-ALPHA,13E,15S)-9,11,15-TRIHYDROXYPROSTA-5,13-DIEN-1-OIC ACID see PROSTAGLANDIN F2-ALPHA

3,7,15-TRIHYDROXYSCIRP-4-ACETOXY-9-EN-8-ONE see 4-(ACETY-LOXY)-12,13-EPOXY-3,7,15-TRIHYDROXY-TRICHOTHEC-9-EN-8-ONE-(3-alpha,4-beta,7-beta)

3,7,15-TRIHYDROXYSCIRP-4-ACETOXY-9-EN-8-ONE see FUSARE-NONE X

TRIIODOARSINE see ARSENIC IODIDE

TRIIODOETHYLATE DE GALLAMINE (FRENCH) see (V-PHENETHYLTRIS-(OXYETHYLENE))TRIS(TRIETHYLAMMONIUM IODIDE)

TRIIODOETHYLATE OF TRI(DIETHYLAMINOETHYLOXY)-1,2,3-BENZENE see (V-PHENETHYLTRIS(OXYETHYLENE))TRIS(TRIETHYLAMMONIUM IODIDE)

TRIIODOMETHANE see IODOFORM

TRIISOPROPYL BORATE see ISOPROPYL BORATE

O-TRIKESYLPHOSPHATE (GERMAN) see TRI-2-TOLYL PHOSPHATE

TRIKRESYLPHOSPHATE (GERMAN) see TRITOLYL PHOSPHATE

TRILEAD PHOSPHATE see LEAD (II) PHOSPHATE (3:2)

TRILEAD TETROXIDE see LEAD OXIDE RED

TRIMANGANESE TETRAOXIDE see MANGANESE OXIDE

TRIMANGANESE TETROXIDE see MANGANESE OXIDE

TRIMELLIC ACID-1,2-ANHYDRIDE see TRIMELLIC ACID ANHYDRIDE

TRIMELLITIC ACID CYCLIC-1,2-ANHYDRIDE see TRIMELLIC ACID ANHYDRIDE

TRIMELLITIC ANHYDRIDE see TRIMELLIC ACID ANHYDRIDE

TRIMETAPHOSPHATE SODIUM see SODIUM TRIMETAPHOSPHATE

3,4,5-TRIMETHOXYAMPHETAMINE see alpha-METHYLMESCALINE

3,4,5-TRIMETHOXYBENZOYL METHYL RESERPATE see RESERPINE

TRIMETHOXYBORINE see TRIMETHYL BORATE

3,4,5-TRIMETHOXY-BETAPHENYLETHYLAMINE HYDROCHLORIDE see MESCALINE HYDROCHLORIDE

TRIMETHOXYMETHANE see TRIMETHYL-O-FORMATE

3,4,5-TRIMETHOXY-ALPHA-METHYL-BETA-PHENYLETHYLAMINE HYDROCHLORIDE see 3,4,5-TRIMETHOXYAMPHETAMINE HYDRO-CHLORIDE

3,4,5-TRIMETHOXYPHENETHYLAMINE see MESCALINE

3,4,5-TRIMETHOXYPHENETHYLAMINE HYDROCHLORIDE see MESCA-LINE HYDROCHLORIDE

1-(3,4,5-TRIMETHOXYPHENYL)-2-AMINOPROPANE see 3,4,5-TRI-METHOXYAMPHETAMINE HYDROCHLORIDE

TRIMETHOXYPHENYL-BETA-AMINOPROPANE see alpha-METHYLMES-CALINE

3,4,5-TRIMETHOXYPHENYL-BETA-AMINOPROPANE see alpha-ME-THYLMESCALINE

TRIMETHYLACETIC ACID see PIVALIC ACID

TRIMETHYL-BETA-ACETOXYPROPYLAMMONIUM CHLORIDE see O-ACETYL-beta-METHYLCHOLINE CHLORIDE

TRIMETHYL ACETYL CHLORIDE (DOT) see 2,2-DIMETHYLPROPANOYL CHLORIDE

TRIMETHYLAMINOMETHANE see tert-BUTYLAMINE

2,4,5-TRIMETHYLANILIN (CZECH) see 2,4,5-TRIMETHYLANILINE

6,7,12-TRIMETHYLBENZ(A)ANTHRACENE see 4,9,10-TRIMETHYL-1,2,-BENZANTHRACENE

1,3,10-TRIMETHYL-7,8-BENZACRIDINE (FRENCH) see 7,9,11-TRIMETHYLBENZ(C)ACRIDINE

5:9:10-TRIMETHYL-1:2-BENZANTHRACENE see 7,8,12-TRIMETHYL-BENZ(a)ANTHRACENE

2,4,5-TRIMETHYLBENZENAMINE see 2,4,5-TRIMETHYLANILINE

1,3,5-TRIMETHYLBENZENE see MESITYLENE

SYM-TRIMETHYLBENZENE see MESITYLENE

TRIMETHYLBENZOL see MESITYLENE

1,7,7-TRIMETHYLBICYCLO(2.2.1)-2-HEPTANONE see CAMPHOR

2,6,6-TRIMETHYLBICYCLO(3.1.1)-2-HEPT-2-ENE see 2-PINENE

TRIMETHYLBROMOMETHANE see tert-BUTYL BROMIDE

TRIMETHYLCARBINOL see tert-BUTYL ALCOHOL

TRIMETHYLCETYLAMMONIUM BROMIDE see HEXADECYL TRIMETHYL AMMONIUM BROMIDE

TRIMETHYLCHLOROMETHANE see tert-BUTYL CHLORIDE

TRIMETHYLCHLOROSTANNANE see CHLOROTRIMETHYLSTANNANE

TRIMETHYLCHLOROTIN see CHLOROTRIMETHYLSTANNANE

1,1,3-TRIMETHYL-3-CYCLOHEXENE-5-ONE see ISOACETOPHORONE

3,5,5-TRIMETHYL-2-CYCLOHEXENE-1-ONE see ISOACETOPHORONE

3,5,5-TRIMETHYL-2-CYCLOHEXEN-1-ON (GERMAN, DUTCH) see ISO-ACETOPHORONE

3,3,5-TRIMETHYL-2,4-DIKETOOXAZOLIDINE see TRIMETHADIONE

1,3,7-TRIMETHYL-2,6-DIOXOPURINE see CAFFEINE

TRIMETHYLEENTRINITRAMINE (DUTCH) see TRIMETHYLENETRINITRA-MINE

TRIMETHYLENE see CYCLOPROPANE

TRIMETHYLENE DICHLORIDE see 1,3-DICHLOROPROPANE

TRIMETHYLENE OXIDE see OXETANE

SYM-TRIMETHYLENETRINITRAMINE see TRIMETHYLENETRINITRAMINE

TRIMETHYLENOXID (GERMAN) see OXETANE

TRIMETHYLETHYLENE see α,η-AMYLENE

TRIMETHYL GLYCOL see 1,2-PROPANEDIOL

TRIMETHYLHEXADECYLAMMONIUM BROMIDE see HEXADECYL TRIMETHYL AMMONIUM BROMIDE

N-1,5-TRIMETHYL-4-HEXENYLAMINE see ISONYL

TRIMETHYL (1-METHYL-2-PHENOTHIAZIN-10-YLETHYL)AMMONIUM METHYL SULFATE see MULTERGAN METHYL SULFATE

TRIMETHYL(1-METHYL-2-(10-PHENOTHIAZINYL)ETHYL)AMMONIUM METHYL SULFATE see MULTERGAN METHYL SULFATE

TRIMETHYLNITROSOHARNSTOFF (GERMAN) see 1,1,3-TRIMETHYL-3-NITROSOUREA

N-TRIMETHYL-N-NITROSOUREA see 1,1,3-TRIMETHYL-3-NITROSO-UREA

3,5,5-TRIMETHYL-2,4-OXAZOLIDINEDIONE see TRIMETHADIONE

6,6,9-TRIMETHYL-3-PENTYL-7,8,9,10-TETRAHYDRO-6H-DIBENZO(B,D)PYRAN-1-OL see 1-trans-delta(sup 9)-TETRAHYDROCAN-NABINOL

TRIMETHYLPHENYLMETHANE see tert-BUTYLBENZENE

TRI 2-METHYLPHENYL PHOSPHATE see TRI-2-TOLYL PHOSPHATE

1,2,2-TRIMETHYLPROPYL METHYLPHOSPHONOFLUORIDATE see SOMAN

2,4,6-TRIMETHYL-1,3,5TRIOXAAN (DUTCH) see PARALDEHYDE

2,4,6-TRIMETHYL-1,3,5TRIOXANE see PARALDEHYDE

TRIMETHYLSTANNANE SULPHATE see TRIMETHYLTIN SULPHATE

TRIMETHYLSTANNYL CHLORIDE see CHLOROTRIMETHYLSTANNANE

TRIMETHYLSTANNYL IODINE see IODOTRIMETHYLTIN

TRIMETHYL STIBINE see ANTIMONY TRIMETHYL

TRIMETHYLTHIOUREA see 1,1,3-TRIMETHYL-2-THIOUREA

N,N,N'-TRIMETHYLTHIOUREA see 1,1,3-TRIMETHYL-2-THIOUREA

TRIMETHYLTIN CHLORIDE see CHLOROTRIMETHYLSTANNANE

TRIMETHYLTIN IODIDE see IODOTRIMETHYLTIN

2,4,6-TRIMETHYL-S-TRIOXANE see PARALDEHYDE

S-TRIMETHYLTRIOXYMETHYLENE see PARALDEHYDE

1,3,7-TRIMETHYLXANTHINE see CAFFEINE

2-(TRIMETIL-ACETIL)-INDAN-1,3-DIONE (ITALIAN) see 2-PIVALOYL-1,3-INDANDIONE

3,5,5-TRIMETIL-2-CICLOESEN-1-ONE (ITALIAN) see ISOACETOPHO-RONE

TRIMUSTINE HYDROCHLORIDE see TRIMUSTINE

TRINITRIN see NITROGLYCERIN

TRINITROBENZEEN (DUTCH) see 1,3,5-TRINITROBENZENE

TRINITROBENZENE see 1,3,5-TRINITROBENZENE

TRINITROBENZOL (GERMAN) see 1,3,5-TRINITROBENZENE

TRINITROCYCLOTRIMETHYLENE TRIAMINE see TRIMETHYLENETRINI-TRAMINE

2,4,6-TRINITROFENOL (DUTCH) see PICRIC ACID

2,4,6-TRINITROFENOLO (ITALIAN) see PICRIC ACID

TRINITROGLYCERIN see NITROGLYCERIN

TRINITROGLYCEROL see NITROGLYCERIN

1,3,5-TRINITROPHENOL see PICRIC ACID

2,4,6-TRINITROPHENOL see PICRIC ACID

2,4,6-TRINITROPHENOL COMPOUND WITH 1,2-DIHYDRO-3-METHYL-BENZ(J)ACEANTHRYLENE see 20-METHYLCHOLANTHRENE PICRATE

TRINITROPHENYLMETHYLNITRAMINE see TETRYL

2,4,6-TRINITROPHENYLMETHYLNITRAMINE see TETRYL

2,4,6-TRINITROPHENYL-N-METHYLNITRAMINE see TETRYL

1,3,5-TRINITROSO-1,3,5-TRIAZACYCLOHEXANE see HEXAHYDRO-1,3,5-s-TRIAZINE

TRINITROSOTRIMETHYLENETRIAMINE see HEXAHYDRO-1,3,5-s-TRIA-ZINE

TRINITROSOTRIMETHYLENTRIAMIN (GERMAN) see HEXAHYDRO-1,3,5-s-TRIAZINE

2,4,6-TRINITROTOLUEEN (DUTCH) see sym-TRINITROTOLUENE

TRINITROTOLUENE see sym-TRINITROTOLUENE

S-TRINITROTOLUENE see sym-TRINITROTOLUENE

2,4,6-TRINITROTOLUENE see sym-TRINITROTOLUENE

TRINITROTOLUENE, DRY (DOT) see sym-TRINITROTOLUENE

S-TRINITROTOLUOL see sym-TRINITROTOLUENE

SYM-TRINITROTOLUOL see sym-TRINITROTOLUENE

2,4,6-TRINITROTOLUOL (GERMAN) see sym-TRINITROTOLUENE

1,3,5-TRINITRO-1,3,5-TRIAZACYCLOHEXANE see TRIMETHYLENE-TRINITRAMINE

TRIOCTYLMETHYLAMMONIUM CHLORIDE see METHYLTRIOCTYLAM-MONIUM CHLORIDE

TRIORTHOCRESYL PHOSPHATE see TRI-2-TOLYL PHOSPHATE

TRIOSSIMETHLENE (ITALIAN) see s-TRIOXANE

TRIOXANE see s-TRIOXANE

1,3,5-TRIOXANE see s-TRIOXANE

TRIOXIDE(S) see TITANIUM OXIDE

TRIOXYMETHYLEEN (DUTCH) see s-TRIOXANE

TRIOXYMETHYLEN (GERMAN) see s-TRIOXANE

TRIOXYMETHYLENE see s-TRIOXANE

TRIPERIDOL see TRIFLUPERIDOL

TRIPHENYLACETO STANNANE see ACETOXYTRIPHENYLSTANNANE

TRIPHENYLACRYLONITRILE see 2,3,3-TRIPHENYLACRYLONITRILE

ALPHA,BETA,BETA-TRIPHENYLACRYLONITRILE see 2,3,3-TRIPHENYL-ACRYLONITRILE

TRIPHENYLANTIMONY DICHLORIDE see DICHLOROTRIPHENYLANTI-MONY

TRIPHENYLCHLOROSTANNANE see CHLOROTRIPHENYLSTANNANE

TRIPHENYLCHLOROTIN see CHLOROTRIPHENYLSTANNANE

TRIPHENYLCYANOETHYLENE see 2,3,3-TRIPHENYLACRYLONITRILE

TRIPHENYLTIN ACETATE see ACETOXYTRIPHENYLSTANNANE

TRIPHENYLTIN CHLORIDE see CHLOROTRIPHENYLSTANNANE

TRIPHENYLTIN HYDROXIDE see HYDROXY TRIPHENYL STANNANE

TRIPHENYLTIN OXIDE see HYDROXY TRIPHENYL STANNANE

TRIPHENYL-ZINNACETAT (GERMAN) see ACETOXYTRIPHENYLSTAN-NANE

TRIPHENYL-ZINNHYDROXID (GERMAN) see HYDROXY TRIPHENYL STANNANE

TRIPHOSPHORIC ACID, SODIUM SALT see SODIUM TRIPOLYPHOSPHATE

TRIPROPARGYL CYANURATE see TRIALLYL CYANAURATE

TRIPROPYL ALUMINUM see ALUMINUM TRIPROPYL

TRI-N-PROPYLTIN CHLORIDE see CHLOROTRIPROPYLSTANNANE

2,4,6-TRIPROP-2-YNYLOXY-S-TRIAZINE see TRIALLYL CYANAURATE

TRISAETHYLENIMINOBENZOCHINON (GERMAN) see TRISETHYLENEI-MINOQUINONE

2,4,6-TRIS(ALLYLOXY)TRIAZINE see TRIALLYL CYANAURATE

TRIS-4-AMINOFENYLMETHAN (CZECH) see TRIAMINOTRIPHENYL-
METHANE

2,3,5-TRIS(AZIRIDINO)-1,4-BENZOQUINONE see TRISETHYLENEIMINO-
QUINONE

2,3,5-TRIS(1-AZIRIDINO)-P-BENZOQUINONE see TRISETHYLENEIMINO-
QUINONE

TRIS(AZIRIDINYL)-P-BENZOQUINONE see TRISETHYLENEIMINOQUI-
NONE

TRIS(1-AZIRIDINYL)-P-BENZOQUINONE see TRISETHYLENEIMINO-
QUINONE

2,3,5-TRIS(AZIRIDINYL)-1,4-BENZOQUINONE see TRISETHYLENE-
IMINOQUINONE

2,3,5-TRIS(1-AZIRIDINYL)-2,5-CYCLOHEXADIENE-1,4-DIONE see
TRISETHYLENEIMINOQUINONE

TRIS(1-AZIRIDINYL)PHOSPHINE OXIDE see TRIETHYLENEPHOSPHORO-
TRIAMIDE

TRIS-(1-AZIRIDINYL)PHOSPHINE OXIDE see TEPA

TRIS(1-AZIRIDINYL)PHOSPHINE SULFIDE see THIOPHOSPHAMIDE

2,4,6-TRIS(1-AZIRIDINYL)-S-TRIAZINE see TRISAZIRIDINYLTRIAZINE

2,4,6-TRIS(1'-AZIRIDINYL)-1,3,5-TRIAZINE see TRISAZIRIDINYL-
TRIAZINE

TRIS(2-BUTOXYETHYL) PHOSPHATE see 2-BUTOXYETHANOL PHOS-
PHATE

TRIS-N-BUTYLAMINE see TRIBUTYLAMINE

TRIS(2-CHLOROETHYL)AMINE HYDROCHLORIDE see TRIMUSTINE

TRIS(BETA-CHLOROETHYL)AMINE HYDROCHLORIDE see TRIMUSTINE

TRIS(2-CHLOROETHYL)AMINE MONOHYDROCHLORIDE see TRIMUS-
TINE

TRIS(2-CHLOROETHYL)AMMONIUM CHLORIDE see TRIMUSTINE

N,N,N'-TRIS(2-CHLOROETHYL)-N',O-PROPYLENE PHOSPHORIC ACID
ESTER DIAMIDE see TROPHOSPHAMIDE

TRIS(2-CHLOROETHYL) PHOSPHATE see 2-CHLORO ETHANOL PHOS-
PHATE

TRIS(BETA-CHLOROETHYL) PHOSPHATE see 2-CHLORO ETHANOL
PHOSPHATE

TRIS-1,2,3-(CHLOROMETHOXY)PROPANE see GLYCEROL (TRI-
(CHLOROMETHYL))ETHER

TRIS(DIBROMOPROPYL)PHOSPHATE see TRIS

TRIS(2,3-DIBROMOPROPYL) PHOSPHATE see TRIS

TRIS(2,3-DIBROMOPROPYL) PHOSPHORIC ACID ESTER see TRIS

TRIS-2,3-DIBROMPROPYL ESTER KYSELINY FOSFORECNE (CZECH) see
TRIS

1,2,3-TRIS(2-DIETHYLAMINOETHOXY)BENZENE TRIETHIODIDE see
(V-PHENETHYLTRIS(OXYETHYLENE))TRIS(TRIETHYLAMMONIUM IO-
DIDE)

1,2,3-TRIS(2-DIETHYLAMINOETHOXY)BENZENE TRIS(ETHYLIODIDE)
see (V-PHENETHYLTRIS(OXYETHYLENE))TRIS(TRIETHYLAMMONIUM
IODIDE)

2,4,6-TRIS(DI(HYDROXYMETHYL)AMINO)-1,3,5-TRIAZINE see HEXA-
(HYDROXYMETHYL)MELAMINE

TRIS(DIMETHYLAMINO)PHOSPHINE OXIDE see HEXAMETHYL PHOS-
PHORAMIDE

TRIS(DIMETHYLAMINO)PHOSPHORUS OXIDE see HEXAMETHYL PHOS-
PHORAMIDE

2,4,6-TRIS(DIMETHYLAMINO)-S-TRIAZINE see HEXAMETHYLMEL-
AMINE

TRIS(DIMETHYLCARBAMODITHIOATO-S,S')IRON see FERBAM

TRIS(DIMETHYLDITHIOCARBAMATO)BISMUTH see BISMUTH DIMETHYL
DITHIOCARBAMATE

TRIS)DIMETHYLDITHIOCARBAMATO)IRON see FERBAM

TRIS(N,N-DIMETHYLDITHIOCARBAMATO) IRON(111) see FERBAM

TRIS(ETHYLENEIMINO)TRIAZINE see TRISAZIRIDINYLTRIAZINE

TRISETHYLENEIMINO-1,3,5-TRIAZINE see TRISAZIRIDINYLTRIAZINE

2,4,6-TRIS(ETHYLENEIMINO)-S-TRIAZINE see TRISAZIRIDINYLTRIA-
ZINE

TRIS(N-ETHYLENE)PHOSPHOROTRIAMIDATE see TRIETHYLENEPHOS-
PHOROTRIAMIDE

2,3,5-TRIS(ETHYLENIMINO)-P-BENZOQUINONE see TRISETHYLENE-
IMINOQUINONE

TRIS(ETHYLENIMINO)THIOPHOSPHATE see THIOPHOSPHAMIDE

2,4,6-TRIS(ETHYLENIMINO)-S-TRIAZINE see TRISAZIRIDINYLTRIAZINE

TRIS(2-HYDROXYETHYL)AMINE see TRIHYDROXYTRIETHYLAMINE

TRIS-N-LOST see TRIMUSTINE

TRIS(P-METHOXYPHENYL)CHLOROETHYLENE see CHLOROTRIANISENE

TRIS(2-METHYLAZIRIDIN-1-YL)PHOSPHINE OXIDE see TRIS(1-METHYL-
ETHYLENE)PHOSPHORIC TRIAMIDE

TRIS(2-METHYL-1-AZIRIDINYL)PHOSPHINE OXIDE see TRIS(1-
METHYLETHYLENE)PHOSPHORIC TRIAMIDE

N,N',N''-TRIS(1-METHYLETHYLENE)PHOSPHORAMIDE see TRIS(1-
METHYLETHYLENE)PHOSPHORIC TRIAMIDE

TRISODIUM-4'-ANILINO-8-HYDROXY-1,1'-AZONAPHTHALENE-3,6,5'-
TRISULFONATE see ACID BLUE 92

TRISODIUM ARSENATE, HEPTAHYDRATE see ARSENIC(V) ACID, TRI-
SODIUM SALT,HEPTAHYDRATE (1:3:7)

TRISODIUM EDTA see TRISODIUM EDETATE

TRISODIUM ETHYLENEDIAMINETETRAACETATE see TRISODIUM EDE-
TATE

TRISODIUM HYDROGEN ETHYLENEDIAMINETETRAACETATE see TRISO-
DIUM EDETATE

TRISODIUM HYDROGEN (ETHYLENEDINITRILO)TETRAACETATE see
TRISODIUM EDETATE

TRISODIUM-O-PHOSPHATE see TRISODIUM PHOSPHATE

TRISODIUM-O-PHOSPHATE DODECAHYDRATE see TRISODIUM PHOS-
PHATE

TRISODIUM SALT OF 1-(4-SULFO-1-NAPHTHYLAZO)-2-NAPHTHOL-3,6-
DISULFONIC ACID see FOOD RED 2

TRISODIUM SALT OF 1-(4-SULPHO-1-NAPHTHYLAZO)-2-NAPHTHOL-
3,6-DISULPHONICACID see FOOD RED 2

TRISODIUM VERSENATE see TRISODIUM EDETATE

TRIS(TOLYLOXY)PHOSPHINE OXIDE see TRITOLYL PHOSPHATE

1,2,3-TRIS(2-TRIETHYLAMMONIUM ETHOXY)BENZENE TRIIODIDE see
(V-PHENETHYLTRIS(OXYETHYLENE))TRIS(TRIETHYLAMMONIUM IO-
DIDE)

TRISULFURATED PHOSPHORUS see PHOSPHORUS SESQUISULFIDE

TRITHIOBIS(TRICHLOROMETHANE) see BIS(TRICHLORO METHYL)TRI-
SULFIDE

TRI-O-TOLYL PHOSPHATE see TRI-2-TOLYL PHOSPHATE

TRITON K-60 see ALKYL DIMETHYLBENZYL AMMONIUM CHLORIDE

TRITTICO see 2-(3-(4-(3-CHLOROPHENYL)-1-PIPERAZINYL)PROPYL)-
1,2,4-TRIZOLO(4,3-a)PYRIDIN-3(2H)-ONE HYDROCHLORIDE

TRIVALENT SODIUM ANTIMONYL GLUCONATE see ANTIMONY (III) SO-
DIUM GLUCONATE

TRIVINYLTIN CHLORIDE see CHLORO(TRIVINYL)STANNANE

TRODAX see 4-HYDROXY-3-IODO-5-NITROBENZONITRILE

TROJCHLOREK FOSFORU (POLISH) see PHOSPHORUS CHLORIDE

TROJCHLOROETAN(1,1,2) (POLISH) see 1,1,2-TRICHLOROETHANE

TROJKREZYLU FOSFORAN (POLISH) see TRI-2-TOLYL PHOSPHATE

TROJNITROTOLUEN (POLISH) see sym-TRINITROTOLUENE

TROJNITROTOLUEN (POLISH) see sym-TRINITROTOLUENE

1-ALPHA-H,5-ALPHA-H-TROPAN-3-ALPHA-OL, ATROPATE (ESTER)
see APOATROPINE

1-ALPHA-H,5-ALPHA-H-TROPAN-3-ALPHA-OL (+-)-TROPATE (ESTER)
see ATROPINE

1-ALPHA-H,5-ALPHA-H-TROPAN-3-ALPHA-OL (+-)-TROPATE (ESTER),
SULFATE (2:1) SALT see ATROPINE SULFATE (2:1)

3-TROPANYLBENZOATE-2-CARBOXYLIC ACID METHYL ESTER see CO-
CAINE

DL-TROPANYL-2-HYDROXY-1-HENYLPROPIONATE SULFATE see ATRO-
PINE SULFATE (2:1)

DL-TROPANYL-2-HYDROXY-1-PHENYLPROPIONATE see ATROPINE

TROPIC ACID, ESTER WITH SCOPINE see SCOPOLAMINE

TROPIC ACID, (−)-, ESTER WITH TROPINE see (−)-HYOSCYAMINE

TROPIC ACID, ESTER WITH TROPINE see ATROPINE

TROPIC ACID, 9-METHYL-3-OXA-9-AZATRICYCLO(3.3.1.0(SUP 2,
4))NON-7-YL ESTER see SCOPOLAMINE

TROPIC ACID-3-ALPHA-TROPANYL ESTER see ATROPINE

TROPINE TROPATE see ATROPINE

(+,-)-TROPYL TROPATE see ATROPINE

DL-TROPYLTROPATE see ATROPINE

TROTYL see sym-TRINITROTOLUENE

TRP-P-2 see 2-AMINO-1-METHYL-5H-PYRIDO(4,3-b)INDOLE

TRP-P-1 (ACETATE) see 3-AMINO-1,4-DIMETHYL-5H-PYRIDO(4,3-b)-INDOLE ACETATE

TRYPARSAMIDE see N-(CARBAMOYL METHYL)ARSANILIC ACID

TRYPTAMINE HYDROCHLORIDE see 3-(2-AMINOETHYL)INDOLE HYDROCHLORIDE

TRYPTIZOL HYDROCHLORIDE see ELAVIL HYDROCHLORIDE

TSIRAM (RUSSIAN) see BIS(DIMETHYL DITHIO CARBAMATO)ZINC

TSP see TRISODIUM PHOSPHATE

TUBATOXIN see ROTENONE

TUBOCURARINE CHLORIDE see TUBOCURARINE HYDROCHLORIDE

(+)-TUBOCURARINE CHLORIDE see TUBOCURARINE HYDROCHLORIDE

D-TUBOCURARINE CHLORIDE see TUBOCURARINE HYDROCHLORIDE

TUBOCURARINE, CHLORIDE, HYDROCHLORIDE, (+)- (8CI) see TUBOCURARINE HYDROCHLORIDE

D-TUBOCURARINE DICHLORIDE see TUBOCURARINE HYDROCHLORIDE

(+)-TUBOCURARINE HYDROCHLORIDE see TUBOCURARINE HYDROCHLORIDE

D-TUBOCURARINE HYDROCHLORIDE see TUBOCURARINE HYDROCHLORIDE

TULUYLENDIISOCYANAT (GERMAN) see DI-ISO-CYANATOLUENE

TUNGSTIC ACID, DISODIUM SALT see SODIUM TUNGSTATE

TURPENTINE STEAM DISTILLED see TURPENTINE

TUSSILAGO FARFARA L see COLTSFOOT

TUTANE see sec-BUTYLAMINE

TUTOCAINE HYDROCHLORIDE see p-AMINOBENZOYLDIMETHYLAMINO-1,2-DIMETHYLPROPANOL HYDROCHLORIDE

U-4527 see CYCLOHEXIMIDE

U-5965 see TETRACYCLINE HYDROCHLORIDE

U-6421 see NITROFURAZONE

U-14583 see PROSTAGLANDIN F2-ALPHA

U-22394A see 1,2,3,4,5,6-HEXAHYDRO-6-METHYLAZEPINO(4,5-b)-INDOLE HYDROCHLORIDE

UCB 6249 see 2-(2-(DIETHYLAMINO)ETHYL)-2-PHENYL-4-PENTENOIC ACID ETHYL ESTER

UCON FLUOROCARBON 113 see FREON 113

UDMH see 1,1-DIMETHYLHYDRAZINE

UF 1 see 2,4-DIHYDROXYBENZOPHENONE

UINTAITE see GILSONITE

ULTRAWET K see DODECYL BENZENE SODIUM SULFONATE

UMBELLATE SULFATE TRIHYDRATE see BERBERINE SULFATE TRIHYDRATE

UN 2354(DOT) see CHLOROMETHYL ETHYL ETHER

1,2,3,4,5,5,6,7,9,10,10-UNDECACHLOROPENTACYCLO(5.3.0.0(SUP 2,6).0(SUP 3,9).0(SUP 4,8))DECANE see 8-MONOHYDRO MIREX

1-UNDECANECARBOXYLIC ACID see LAURIC ACID

UNS-DIMETHYLHYDRAZINE see 1,1-DIMETHYLHYDRAZINE

URACIL MUSTARD see 5-(BIS(2-CHLOROETHYL)AMINO)URACIL

URAMUSTINE see 5-(BIS(2-CHLOROETHYL)AMINO)URACIL

URANIUM FLUORIDE OXIDE see URANIUM OXYFLUORIDE

URANIUM HEXAFLUORIDE, FISSILE (DOT) see URANIUM FLUORIDE (fissile)

URANIUM HEXAFLUORIDE, LOW SPECIFIC ACTIVITY (DOT) see URANIUM FLUORIDE (low specific activity)

URANIUM (IV) CHLORIDE see URANIUM TETRACHLORIDE

URANIUM METAL, PYROPHORIC (DOT) see URANIUM

URANYL FLUORIDE see URANIUM OXYFLUORIDE

URANYL NITRATE HEXAHYDRATE SOLUTION (DOT) see URANYL NITRATE HEXAHYDRATE

URARI see CURARE

UREAPHIL see UREA

P-UREIDOBENZENEARSONIC ACID see N-CARBAMOYL ARSANILIC ACID

4-UREIDO-1-PHENYLARSONIC ACID see N-CARBAMOYL ARSANILIC ACID

(P-UREIDOPHENYLARSYLENEDITHIO)DIACETIC ACID see 4-CARBAMIDOPHENYL BIS(CARBOXY METHYL THIO)ARSENITE

URETHAN see URETHANE

URETHYLANE see METHYL CARBAMATE

UROMYCINE see GENTAMICIN

UROTROPINE see HEXAMETHYLENETETRAMINE

URSOFERRAN see IRON-DEXTRAN COMPLEX

URSOL P BASE see 4-AMINOPHENOL

USACERT FD & C RED NO. 4 see FOOD RED 4

USACERT FD & C YELLOW NO. 6 see FOOD YELLOW 3

USACERT RED NO. 1 see FD AND C RED NO. 1

USAF D-1 see IODOACETAMIDE

USAF D-3 see p-CHLORO-MERCURIC BENZOIC ACID

USAF D-5 see n-BUTYL THIOUREA

USAF D-9 see CARBANILIC ACID ISOPROPYL ESTER

USAF M-4 see ISOTHIOCYANIC ACID, PHENYL ESTER

USAF M-5 see 4-AMINO-1-NAPHTHALENESULFONIC ACID

USAF M-7 see o-CHLOROBENZALDEHYDE

USAF P-2 see BIS(DIMETHYL DITHIO CARBAMATO)ZINC

USAF P-5 see BIS(DIMETHYL THIOCARBAMYL)DISULFIDE

USAF P-7 see 3-(3,4-DICHLOROPHENYL)-1,1-DIMETHYLUREA

USAF P-8 see 3-(p-CHLOROPHENYL)-1,1-DIMETHYLUREA

USAF P-220 see p-BENZOQUINONE

USAF A-233. see p-AMINOBENZOPHENONE

USAF A-1705 see PENICILLAMINE HYDROCHLORIDE

USAF A-4600 see MALONONITRILE

USAF A-6598 see DI-O-TOLYLGUANIDINE

USAF A-8564 see BIS(beta-CYANOETHYL)AMINE

USAF A-8565 see HYDROXYACETONITRILE

USAF A-8798 see 3-CHLOROPROPIONITRILE

USAF A-9442 see SUCCINONITRILE

USAF A-9789 see 2,3-DIPHENYLACRYLONITRILE

USAF A -15972 see UV ABSORBER-2

USAF AB-315 see 2,2'-DITHIOBISANILINE

USAF AM-5 see ACETALDEHYDE OXIME

USAF AM-6 see m-BUTYRALDEHYDE OXIME

USAF AN-8 see TETRAMETHYLAMMONIUM CHLORIDE

USAF B-7 see 1-ACETYL-2-THIOHYDANTOIN

USAF B-15 see BIS(3-tert-BUTYL-4-HYDROXY-6-METHYLPHENYL)SULFIDE

USAF B-17 see N,N'-BISMORPHOLINE DISULFIDE

USAF B-30 see BIS(DIMETHYL THIOCARBAMYL)DISULFIDE

USAF B-32 see BIS(DIMETHYL THIO CARBAMOYL)SULFIDE

USAF B-35 see SODIUM DIBUTYLDITHIOCARBAMATE

USAF B-43 see DITHIOOXAMIDE

USAF B-100 see N,N-DIETHYLSELENOUREA

USAF B-121 see MALEIC ACID-N-ETHYLIMIDE

USAF BE-0405 see 1-ACETYL-2-THIOHYDANTOIN

USAF CB-2 see ISONICOTINIC ACID HYDRAZIDE

USAF CB-7 see BACITRACIN

USAF CB-13 see FOLIC ACID

USAF CB-17 see XANTHINE

USAF CB-18 see ADENINE

USAF CB-19 see NEOMYCIN SULFATE

USAF CB-20 see THIAMINE DICHLORIDE

USAF CB-21 see THIOACETAMIDE

USAF CB-22 see 2-NAPHTHYLAMINE

USAF CB-27 see RESERPINE

USAF CB-29 see 3-INDOLYLACETONITRILE

USAF CB-35 see 2-MERCAPTOACETIC ACID

USAF CB-36 see MERCAPTODIACETIC ACID

USAF CS-6 see (−)-NOREPHEDRINE

USAF CY-2 see CALCIUM CYANAMIDE

USAF CY-4 see 2-NAPHTHALENETHIOL

USAF CY-7 see 2-BENZOTHIAZOLYL-N-MORPHOLINOSULFIDE

USAF CY-10 see 1,4-NAPHTHOQUINONE

USAF DO-1 see p-CHLOROACETOPHENONE

USAF DO-12 see HYDROCOUMARIN

USAF DO-21 see 2-ETHYL-1-HEXENE

USAF DO-23 see ALLYL PHENYL ETHER

USAF DO-29 see 2-CHLORO ACETAMIDE

USAF DO-36 see DINITROSOPIPERAZINE

USAF DO-41 see TRIS

USAF DO-45 see ACETAL

USAF DO-46 see N-AMINOETHYLPIPERAZINE

USAF DO-47 see 2-ALLYLOXYETHANOL

USAF DO-50 see 2-METHYLAMINOETHANOL

USAF DO-51 see 3-BROMO PROPIONITRILE

USAF DO-54 see 4-METHYL-p-BENZOPHENONE

USAF DO-59 see 3-PHENYLSALICYLIC ACID
USAF DO-65 see HEXACHLORO-2,5-CYCLOHEXADIEN-1-ONE
USAF DO-68 see 3,5-DICHLOROSALICYLIC ACID
USAF E-2 see MERCAPTODIACETIC ACID
USAF EA-2 see NITROFURANTOIN
USAF EA-4 see NITROFURAZONE
USAF EA-5 see 5-NITRO-2-FURALDEHYDE OXIME
USAF EK-3 see ACETANILIDE
USAF EK-206 see m-PHENYLENEDIAMINE HYDROCHLORIDE
USAF EK-218 see QUINOLINE
USAF EK-245 see DIPHENYLTHIOUREA
USAF EK-338 see N,N-DIMETHYL-p-PHENYLAZOANILINE
USAF EK-356 see HYDROQUINONE
USAF EK-394 see p-PHENYLENEDIAMINE
USAF EK-442 see BENZENAMINE HYDROCHLORIDE
USAF EK-488 see ACETONITRILE
USAF EK-496 see ACETOPHENONE
USAF EK-497 see ISOTHIOUREA
USAF EK-600 see CARBAZOLE
USAF EK-704 see AZOBENZENE
USAF EK-794 see 8-QUINOLINOL
USAF EK-1047 see 1,3-DIETHYL-1,3-DIPHENYLUREA
USAF EK-1239 see ACETOACETANILIDE
USAF EK-1270 see DIPHENYLGUANIDINE
USAF EK-1275 see THIOSEMICARBAZIDE
USAF EK-1375 see p-(PHENYLAZO)ANILINE
USAF EK-1509 see alpha-TOLUENETHIOL
USAF EK-1569 see 1-PHENYL-2-THIOUREA
USAF EK-1597 see MONOETHANOLAMINE
USAF EK-1651 see DI-o-TOLYLTHIOUREA
USAF EK-1719 see THIOACETAMIDE
USAF EK-1803 see 1,3-DIETHYLTHIOUREA
USAF EK-1860 see THIOPHENE
USAF EK-1995 see CYANAMIDE
USAF EK-2089 see BIS(DIMETHYL THIOCARBAMYL)DISULFIDE
USAF EK-2122 see 1-HEPTANETHIOL
USAF EK-2124 see BENZYL ISOTHIOUREA HYDROCHLORIDE
USAF EK-2219 see 2-BIPHENYLOL
USAF EK-2596 see SODIUM DIETHYLDITHIOCARBAMATE
USAF EK-3092 see DIPHENYLTHIOCARBAZONE
USAF EK-3302 see ETHYL-p-HYDROXY PHENYL KETONE
USAF EK-4037 see 2-AMINOBENZIMIDAZOLE
USAF EK-4196 see 2-MERCAPTOETHANOL
USAF EK-4376 see 2-AMINOBENZENETHIOL
USAF EK-4394 see DITHIOOXAMIDE
USAF EK-4628 see 1-HEXANETHIOL
USAF EK-4733 see IMIDAZOLE
USAF EK-4812 see BENZOTHIAZOLE
USAF EK-4890 see ACETYL THIOUREA
USAF EK-5185 see 2-METHYL-1,4-NAPHTHOQUINONE
USAF EK-5199 see SODIUM THIOGLYCOLATE
USAF EK-5296 see p-CHLOROPROPIOPHENONE
USAF EK-5426 see 1-PHENYLTHIOSEMICARBAZIDE
USAF EK-6454 see 6-METHYLTHIOURACIL
USAF EK-6540 see 2-BENZIMIDAZOLETHIOL
USAF EK-6561 see 2-AMINO-5-NITROTHIAZOLE
USAF EK-6583 see alpha-MERCAPTOACETANILIDE
USAF EK-P-433 see AMMONIUM THIOCYANATE
USAF EK-T-434 see SODIUM ISOTHIOCYANATE
USAF EK-P-5501 see 2-AMINOTHIAZOLE
USAF EK-P-6255 see BIS(DIMETHYL THIO CARBAMOYL)SULFIDE
USAF EK-T-6645 see N,N'-BISMORPHOLINE DISULFIDE
USAF EL-30 see 2-MERCAPTO-1-METHYLIMIDAZOLE
USAF EL-36 see d-METHAPHETAMINE HYDROCHLORIDE
USAF EL-45 see 1-PHENYLTHIOSEMICARBAZIDE
USAF EL-62 see 2-IMIDAZOLINETHIOL
USAF EL-78 see DIMETHYLBENZYLAMINE HYDROCHLORIDE
USAF EL-101 see BUTYL CARBAMATE
USAF EL-108 see METHYLPENTYNOL CARBAMATE
USAF FO-1 see BUTYL CARBAMATE
USAF GE-1 see PHENMETRAZINE HYDROCHLORIDE

USAF GE-13 see DIPHENYLPYRAZONE
USAF GE-14 see p-HYDROXYPHENYLBUTAZONE
USAF GY-2 see 1,4-BIS(PHENYL AMINO)BENZENE
USAF GY-3 see 2-BENZOTHIAZOLETHIOL
USAF GY-5 see BIS(DIBUTYL DITHIO CARBAMATO)ZINC
USAF GY-7 see BIS(2-BENZOTHIAZOLYTHIO)ZINC
USAF HA-5 see DI(2-CYANOETHYL)SULFIDE
USAF KE-8 see 2-(2-HYDROXYETHOXY)ETHYL ESTER STEARIC ACID
USAF KE-11 see ETHYLENE GLYCOL STEARATE
USAF KF-5 see CHLORACETONITRILE
USAF KF-11 see (o-CHLORO BENZAL)MALONO NITRILE
USAF KF-13 see DIPHENYLACETONITRILE
USAF KF-17 see CYANOACETIC ACID
USAF KF-18 see CYANACETIC ACID HYDRAZIDE
USAF KF-21 see PHENYLACETONITRILE
USAF KF-25 see ETHYL CYANOACETATE
USAF LO-3 see p-MERCAPTO SULFADIAZINE
USAF MA-4 see m-AMINOBENZAL FLUORIDE
USAF MA-5 see 3-NITROBENZOTRIFLUORIDE
USAF MA-12 see 2-AMINO-5-CHLOROBENZOXAZOLE
USAF MA-16 see BENZOTRIFLUORIDE
USAF ME-1 see BAL
USAF MK-6 see DITHIOOXAMIDE
USAF MO-2 see AMMONIUM MERCAPTOACETATE
USAF ND-09 see PHTHALONITRILE
USAF PD-20 see p-DIMETHYLAMINOBENZALRHODANINE
USAF RH-1 see METHACRYLIC ACID AMIDE
USAF RH-3 see DIMETHYLAMINOETHYL METHACRYLATE
USAF RH-7 see HYDRACRYLONITRILE
USAF RH-8 see 2-METHYLLACTONITRILE
USAF SC-2 see GRISOFULVIN
USAF ST-40 see METHYLACRYLONITRILE
USAF SZ-1 see 2-BROMO-D-LYSERGIC ACID DIETHYLAMIDE
USAF SZ-2 see D-LYSERGIC ACID DIETHYLAMIDE
USAF SZ-3 see 10-(2-(1-METHYL-2-PIPERIDYL)ETHYL)-2-METHYLTHIO-
     PHENOTHIAZINE HYDROCHLORIDE
USAF SZ-B see 10-(2-(1-METHYL-2-PIPERIDYL)ETHYL)-2-METHYLTHIO-
     PHENOTHIAZINE HYDROCHLORIDE
USAF TH-9 see DITHIODIGLYCOL
USAF UCTL-8 see METHYLERGONOVINE MALEATE
USAF UCTL-1856 see p-AMINOPROPIOPHENONE
USAF XF-21 see 2-BENZIMIDAZOLETHIOL
USAF XR-10 see DIAMIDINE
USAF XR-22 see 3-AMINOTRIAZOLE
USAF XR-29 see 2-BENZOTHIAZOLETHIOL
USAF XR-31 see 2-beta-AMINOETHYLISOTHIOUREA
USAF XR-41 see 3-(p-CHLOROPHENYL)-1,1-DIMETHYLUREA
USAF XR-42 see 3-(3,4-DICHLOROPHENYL)-1,1-DIMETHYLUREA
VAGISEPT see ACETPHENARSINE
VAGOFLOR see ACETPHENARSINE
VAGOSTIGMIN see 3,4-DICHLOROBENZOIC ACID
VALERAN DI-N-BUTYLCINICITY (CZECH) see DIBUTYL DIPENTAN-
     OYLOXY STANNANE
VALERIANIC ALDEHYDE see n-VALERALDEHYDE
VALERIC ACID ALDEHYDE see n-VALERALDEHYDE
VALERIC ALDEHYDE see n-VALERALDEHYDE
DELTA-VALEROSULTONE see BUTANESULTONE
VALERYLALDEHYDE see n-VALERALDEHYDE
VANADIC ACID, AMMONIUM SALT see AMMONIUM VANADATE
VANADIC ACID, MONOSODIUM SALT see SODIUM VANADATE
VANADIC ANHYDRIDE see VANADIUM PENTOXIDE (DUST)
VANADIC (II) ACID, TRISODIUM SALT see SODIUM-o-VANADATE
VANADIC OXIDE see VANADIUM SESQUIOXIDE
VANADIO, PENTOSSIDO DI (ITALIAN) see VANADIUM PENTOXIDE
     (DUST)
VANADIUM CHLORIDE see VANADIUM TETRACHLORIDE
VANADIUM (III) CHLORIDE see VANADIUM TRICHLORIDE
VANADIUM OXIDE see VANADIUM SESQUIOXIDE
VANADIUM OXIDE see VANADIUM PENTOXIDE (DUST)
VANADIUMPENTOXID (GERMAN) see VANADIUM PENTOXIDE (DUST)
VANADIUMPENTOXYDE (DUTCH) see VANADIUM PENTOXIDE (DUST)

VANADIUM, PENTOXYDE DE (FRENCH) see VANADIUM PENTOXIDE (DUST)

VANADIUM TRIOXIDE see VANADIUM SESQUIOXIDE

VANADYL TRICHLORIDE see VANADIUM OXYTRICHLORIDE

VANCIDE FE95 see FERBAM

VANCIDE PA see trans-1,2-BIS(n-PROPYL SULFONYL)ETHYLENE

VANILLA see VANILLIN

VANILLAL see ETHYL VANILLIN

VANILLALDEHYDE see VANILLIN

VANILLA PLANT see DEERTONGUE INCOLORE

VANILLIC ACID DIETHYLAMIDE see N,N-DIETHYLVANILLAMIDE

VANILLIC ACID-N,N-DIETHYLAMIDE see N,N-DIETHYLVANILLAMIDE

VANILLIC ALDEHYDE see VANILLIN

VANILLINSAEURE-DIAETHYLAMID (GERMAN) see N,N-DIETHYLVANIL-LAMIDE

VANOBID see CANDICIDIN

VAPONA see DIMETHYL DICHLOROVINYL PHOSPHATE

VAPONEFRIN see VASOTONIN

VAPONITE see DIMETHYL DICHLOROVINYL PHOSPHATE

VAPTONE see TETRON

VARNISH MAKERS' AND PAINTERS' NAPHTHA see PETROLEUM SPIRITS

VARNISH MAKERS' NAPHTHA see PETROLEUM SPIRITS

VARNOLINE see STODDARD SOLVENT

VASOCONSTRICTOR see VASOTONIN

VASOXYL HYDROCHLORIDE see METHOXAMINE HYDROCHLORIDE

VEGADEX see 2-CHLORALLYL DIETHYLDITHIOCARBAMATE

VEGETABLE GUM see DEXTRINS

VEGETABLE PEPSIN see PAPAIN

VELSICOL VCS 506 see LEPTOPHOS

VENOM, AUSTRALIAN ELAPIDAE SNAKE, ACANTHOPHIS ANTARCTI-CUS see A. ANTRACTICUS (AUSTRALIA) VENOM

VERATRIN (GERMAN) see VERATRINE

VERATRINE (AMORPHOUS) see VERATRIDINE

VERATROLE METHYL ETHER see 4-ALLYL-1,2-DIMETHOXYBENZENE

3-VERATROYLVERACEVINE see VERATRIDINE

VERATRUM VIRIDE see VERILOID

VERATRUM VIRIDE ALKALOIDS EXTRACT see VERILOID

VERONAL see BARBITAL

VERONAL SODIUM see BARBITAL SODIUM

VERSALIDE see ACETYL ETHYL TETRAMETHYL TETRALIN

VERSENE NTA ACID see AMINOTRIACETIC ACID

VESPARAZ-WIRKSTOFF see 1-(p-CHLORO-alpha-PHENYLBENZYL)-4-(2-((2-HYDROXYETHOXY)ETHYL)PIPERAZINE

VESTROL see TRICHLOROETHYLENE

VETERINARY NITROFURAZONE see NITROFURAZONE

VETRANQUIL see ACETOPROMAZINE

VICTORIA ORANGE see 2,6-DINITRO-p-CRESOL

VICTORIA SCARLET RED see FOOD RED 7

VICTORIA YELLOW see 2,6-DINITRO-p-CRESOL

VIENNA GREEN see CUPRIC ACETOARSENITE

VINBLASTINE see VINCALEUKOBLASTINE

VINCAIN see AJMALICINE

VINCALEUCOBLASTINE see VINCALEUKOBLASTINE

VINCALEUKOBLASTINE SULFATE see VINBLASTINE SULFATE

VINCALEUKOBLASTINE SULFATE (1:1) (SALT) see VINBLASTINE SUL-FATE

VINCEINE see AJMALICINE

VINCIDE PB see 4,6-DINITRO-1,2,3-TRICHLOROBENZENE

VINCRISTINE see LEUROCRISTINE

VINCRISTINE SULFATE ONCORIN see LEUROCRISTINE SULFATE (1:1)

VINEGAR ACID see ACETIC ACID

VINEGAR NAPHTHA see ETHYL ACETATE

VINEGAR SALTS see CALCIUM ACETATE

VINESTHESIN see VINYL ETHER

VINETHEN see VINYL ETHER

VINETHENE see VINYL ETHER

VINETHER see VINYL ETHER

VINIDYL see VINYL ETHER

VINILE (ACETATO DI) (ITALIAN) see ACETIC ACID VINYL ESTER

VINILE (BROMURO DI) (ITALIAN) see VINYL BROMIDE

4-VINLYCYCLOHEXENE DIOXIDE see VINYL CYCLOHEXENE DIOXIDE

VINYLACETAAT (DUTCH) see ACETIC ACID VINYL ESTER

VINYLACETAT (GERMAN) see ACETIC ACID VINYL ESTER

VINYL ACETATE (DOT) see ACETIC ACID VINYL ESTER

VINYLACETONITRILE see 3-BUTENE NITRILE

VINYL ACETYLENE see BUTEN-3-YNE

VINYL ALCOHOL POLYMER see POLYVINYL ALCOHOL

VINYL A MONOMER see ACETIC ACID VINYL ESTER

VINYLBENZEN (CZECH) see STYRENE

VINYLBENZENE see STYRENE

VINYLBENZOL see STYRENE

VINYLBROMID (GERMAN) see VINYL BROMIDE

VINYL-N-BUTYL ETHER see VINYL BUTYL ETHER

VINYLBUTYROLACTAM see 1-ETHENYL-2-PYRROLIDINONE

VINYLCARBINOL see ALLYL ALCOHOL

VINYLCHLORID (GERMAN) see VINYL CHLORIDE

VINYL CHLORIDE (DOT) see VINYL CHLORIDE

VINYL CHLORIDE ACETATE COPOLYMER see POLYVINYL ACETATE CHLORIDE

VINYL CHLORIDE COPOLYMER WITH VINYLIDENE CHLORIDE see CHLORO ETHYLENE-1,1-DICHLORO ETHYLENE POLYMER

VINYL CHLORIDE-1,1-DICHLOROETHYLENE COPOLYMER see CHLORO ETHYLENE-1,1-DICHLORO ETHYLENE POLYMER

VINYL CHLORIDE HOMOPOLYMER see BAKELITE

VINYL CHLORIDE MONOMER see VINYL CHLORIDE

VINYL CHLORIDE POLYMER see BAKELITE

VINYL CHLORIDE VINYL ACETATE COPOLYMER see POLYVINYL ACE-TATE CHLORIDE

VINYL 2-CHLOROETHYL ETHER see 2-CHLOROETHYL VINYL ETHER

VINYL-beta-CHLOROETHYL ETHER see 2-CHLOROETHYL VINYL ETHER

VINYL C MONOMER see VINYL CHLORIDE

VINYL CYANIDE see ACRYLONITRILE

1-VINYLCYCLOHEX-3-ENE see CYCLOHEXENYLETHYLENE

1-VINYLCYCLOHEXENE-3 see CYCLOHEXENYLETHYLENE

4-VINYL-1-CYCLOHEXENE see CYCLOHEXENYLETHYLENE

4-VINYLCYCLOHEXENE-1 see CYCLOHEXENYLETHYLENE

4-VINYL-1-CYCLOHEXENE DIEPOXIDE see VINYL CYCLOHEXENE DIOXIDE

4-VINYL-1,2-CYCLOHEXENE DIEPOXIDE see VINYL CYCLOHEXENE DIOXIDE

1-VINYL-3-CYCLOHEXENE DIOXIDE see VINYL CYCLOHEXENE DIOX-IDE

4-VINYL-1-CYCLOHEXENE DIOXIDE see VINYL CYCLOHEXENE DIOX-IDE

VINYLE (ACETATE DE) (FRENCH) see ACETIC ACID VINYL ESTER

VINYLETHYLENE see 1,3-BUTADIENE

VINYLFORMIC ACID see ACRYLIC ACID

VINYLIDENE CHLORIDE-VINYL CHLORIDE POLYMER see CHLORO ETHYLENE-1,1-DICHLORO ETHYLENE POLYMER

VINYLIDENE DICHLORIDE see 1,1-DICHLOROETHYLENE

VINYL METHYL ETHER see METHYL VINYL ETHER

VINYL METHYL KETONE see 3-BUTEN-2-ONE

ALPHA-VINYLPIPERONYL ALCOHOL see 1,3-BENZODIOXOLE-5-(2-PROPEN-1-OL)

1-VINYLPYRENE see 1-ETHENYL PYRENE

3-VINYLPYRENE see 1-ETHENYL PYRENE

4-VINYLPYRENE see 4-ETHENYL PYRENE

N-VINYLPYRROLIDINONE see 1-ETHENYL-2-PYRROLIDINONE

1-VINYL-2-PYRROLIDINONE see 1-ETHENYL-2-PYRROLIDINONE

N-VINYL-2-PYRROLIDINONE see 1-ETHENYL-2-PYRROLIDINONE

VINYLPYRROLIDONE see 1-ETHENYL-2-PYRROLIDINONE

N-VINYLPYRROLIDONE see 1-ETHENYL-2-PYRROLIDINONE

1-VINYL-2-PYRROLIDONE see 1-ETHENYL-2-PYRROLIDINONE

N-VINYL-2-PYRROLIDONE see 1-ETHENYL-2-PYRROLIDINONE

VINYLSILICON TRICHLORIDE see TRICHLOROETHENYLSILANE

VINYL SULFONE see DIVINYL SULFONE

VINYL TRICHLORIDE see 1,1,2-TRICHLOROETHANE

VINYL TRICHLOROSILANE see TRICHLOROETHENYLSILANE

VITACIN see L-ASCORBIC ACID

VITALLIUM see COBALT ALLOY, CO,Cr
VITAMIN A ACETATE see RETINOL ACETATE
TRANS-VITAMIN A ACETATE see RETINOL ACETATE
VITAMIN A ALCOHOL ACETATE see RETINOL ACETATE
VITAMIN B2 see RIBOFLAVINE
VITAMIN B3 see NIACINAMIDE
VITAMIN B4 see ADENINE
VITAMIN B-5 see CALCIUM-d-PANTOTHENATE
VITAMIN B6 see PYRIDOXOL
VITAMIN B12 see VITAMIN B$_{12}$ COMPLEX
VITAMIN B(SUP 1) see THIAMINE DICHLORIDE
VITAMIN BC see FOLIC ACID
VITAMIN B HYDROCHLORIDE see THIAMINE DICHLORIDE
VITAMIN B6-HYDROCHLORIDE see PYRIDOXOL HYDROCHLORIDE
VITAMIN B1 MONONITRATE see THIAMINE NITRATE
VITAMIN B1 NITRATE see THIAMINE NITRATE
VITAMIN C see L-ASCORBIC ACID
VITAMIN D2 see ERGOCALCIFEROL
VITAMIN D3 see CHOLECALCIFEROL
VITAMIN G see RIBOFLAVINE
VITAMIN H see p-AMINOBENZOIC ACID
VITAMIN K3 see 2-METHYL-1,4-NAPHTHOQUINONE
VITAMIN K5 see 4-AMINO-2-METHYL-1-NAPHTHOL
VITAMIN L see ANTHRANILIC ACID
VITAMIN M see FOLIC ACID
VITAMIN PP see NIACINAMIDE
VITAMISIN see L-ASCORBIC ACID
VITASCORBOL see L-ASCORBIC ACID
VITREOUS QUARTZ see SILICA, AMORPHOUS FUSED
VITRIOL BROWN OIL see SULFURIC ACID
VITRIOL RED see IRON (III) OXIDE
V.M. AND P. NAPHTHA see PETROLEUM SPIRITS
VOLATILE OIL OF MUSTARD see ALLYL ISOTHIOCYANATE
VOLFAZOL see CROTOXYPHOS
VONAMYCIN POWDER V see NEOMYCIN
VULKAZIT see DIPHENYLGUANIDINE
W-2429 see 2,3-DIHYDRO-9H-ISOXAZOLO(3,2-b)QUINAZOLIN-9-ONE
WACHOLDERBEER OEL (GERMAN) see JUNIPER BERRY OIL
WALNUT STAIN see MANGANESE (III) OXIDE
WANADU PIECIOTLENEK (POLISH) see VANADIUM PENTOXIDE (DUST)
WAPNIOWY TLENEK (POLISH) see CALCIUM OXIDE
WARFARIN see COUMADIN
WARFARINE (FRENCH) see COUMADIN
WARFARIN SODIUM see COUMADIN SODIUM
WASSERSTOFFPEROXID (GERMAN) see HYDROGEN PEROXIDE, 90%
WATER GLASS see SODIUM SILICATE
WATERSTOFPEROXYDE (DUTCH) see HYDROGEN PEROXIDE, 90%
WATTLE GUM see ARABIC GUM
WEATHERBEE MUSTARD see 2,5-BIS(BIS-(2-CHLOROETHYL)AMINO-
  METHYL)HYDROQUINONE
WEDDING BELLS see N,N-DIETHYLLYSERGAMIDE
WEEDONE 128 see ISOPROPYL-2,4-D ESTER
WEGLA DWUSIARCZEK (POLISH) see CARBON DISULFIDE
WEGLA TLENEK (POLISH) see CARBON MONOXIDE
WEISS PHOSPHOR (GERMAN) see PHOSPHORUS (WHITE)
WET-TONE B see QUATRESIN
WFNA see NITRIC ACID, (WHITE FUMING)
WHITE ACID (DOT) see AMMONIUM DIFLUORIDE MIXED WITH HYDRO-
  CHLORIC ACID
WHITE ARSENIC see ARSENIC TRIOXIDE
WHITE CAUSTIC see SODIUM HYDROXIDE
WHITE CAUSTIC SOLUTION see SODIUM HYDROXIDE (LIQUID)
WHITE COPPERAS see ZINC SULFATE
WHITE FUMING NITRIC ACID see NITRIC ACID, (WHITE FUMING)
WHITE LEAD see LEAD CARBONATE
WHITE MERCURY PRECIPITATED see MERCURY AMIDE CHLORIDE
WHITE MINERAL OIL see MINERAL OIL
WHITE OIL OF CAMPHOR see CAMPHOR OIL
WHITE PHOSPHORUS see PHOSPHORUS (WHITE)
WHITE PRECIPITATE see MERCURY AMIDE CHLORIDE

WHITE SPIRITS see PETROLEUM SPIRITS
WHITE SPIRITS see STODDARD SOLVENT
WHITE STUFF see HEROIN
WHITE TAR see NAPHTHALENE
WHITE VITRIOL see ZINC SULFATE
WIJS' CHLORIDE see IODINE MONOCHLORIDE
WIN 244 see CHLOROQUINE
WINTER BLOOM see HAMAMELIS
WINTERGREEN OIL, SYNTHETIC see METHYL SALICYLATE
WINYLU CHLOREK (POLISH) see VINYL CHLORIDE
WITCH HAZEL see HAMAMELIS
WOLFRAM see TUNGSTEN
WOLFSBANE see ARNICA
WOLFSBANE see ACONITE
WOOD ALCOHOL see METHANOL
WOOD ETHER see METHYL ETHER
WOOD NAPHTHA see METHANOL
WOOD SPIRIT see METHANOL
WOOL FAST BLUE R see ACID BLUE 92
WOOL GREEN S (BIOLOGICAL STAIN) see ACID BRILLIANT GREEN BS
WOORALI see CURARE
WOORARI see CURARE
WORM GUARD see 5-BUTYL-2-BENZIMID AZOLECARBAMIC ACID
  METHYL ESTER
WORMWOOD see ABSINTHIUM
WOURARA see CURARE
WR 09792 see FTBG
WSQ 1 see METHANESULFONIC ACID
WY-1172 see ACETOPROMAZINE
WY-14,643 see (4-CHLORO-6-(2,3-XYLIDINO)-2-PYRIMIDINYLTHIO)-
  ACETIC ACID
XANTHAURINE see QUERCETIN
XANTHENE-9-CARBOXYLIC ACID,ESTER WITH DIETHYL(2-HYDROXY-
  ETHYL) METHYL AMMONIUM BROMIDE see XANTHINE BROMIDE
XANTHIC OXIDE see XANTHINE
XANTHINE-3-N-OXIDE see 3-HYDROXYXANTHINE
XANTHINE-7-N-OXIDE see 7-HYDROXYXANTHINE
XANTHURENIC ACID see 4,8-DIHYDROXYQUINALDIC ACID
O-XENOL see 2-BIPHENYLOL
XENYLAMIN (CZECH) see 4-BIPHENYLAMINE
XENYLAMINE see 4-BIPHENYLAMINE
XILIDINE (ITALIAN) see XYLIDINE
XILOLI (ITALIAN) see XYLENE
XITIX see L-ASCORBIC ACID
1,2-XYLENE see o-XYLENE
1,3-XYLENE see m-XYLENE
1,4-XYLENE see p-XYLENE
XYLENEN (DUTCH) see XYLENE
3,5-XYLENOL METHYLCARBAMATE see 3,5-DIMETHYLPHENYL-N-
  METHYLCARBAMATE
3,5-XYLENYL-N-METHYLCARBAMATE see 3,5-DIMETHYLPHENYL-N-
  METHYLCARBAMATE
O-XYLIDINE see 2,6-XYLIDINE
P-XYLIDINE see 2,5-XYLIDINE
M-XYLIDINE HYDROCHLORIDE see 2,4-XYLIDINE HYDROCHLORIDE
PARA-XYLIDINE HYDROCHLORIDE see 2,5-XYLIDINE HYDROCHLORIDE
XYLIDINEN (DUTCH) see XYLIDINE
XYLITE (SUGAR) see XYLITOL
L-XYLOASCORBIC ACID see L-ASCORBIC ACID
XYLOIDIN see CELLULOSE TETRANITRATE
XYLOL see XYLENE
M-XYLOL see m-XYLENE
O-XYLOL see o-XYLENE
P-XYLOL see p-XYLENE
XYLOLE (GERMAN) see XYLENE
2,6-XYLYLAMINE see 2,6-XYLIDINE
3,5-XYLYLAMINE see 3,5-XYLIDINE
1-XYLYLAZO-2-NAPHTHOL see 1-(2,4-XYLYLAZO)-2-NAPHTHOL
1-(O-XYLYLAZO)-2-NAPHTHOL see 1-(2,4-XYLYLAZO)-2-NAPHTHOL

1-XYLYLAZO-2-NAPHTHOL-3,6-DISULFONIC ACID, DISODIUM SALT see D&C RED NO. 5

1-(2,4-XYLYLAZO)-2-NAPHTHOL-3,6-DISULPHONIC ACID, DISODIUM SALT see D&C RED NO. 5

O-XYLENE DICHLORIDE see alpha,alpha'-DICHLORO-O-XYLENE

3,5-XYLYL-N-METHYLCARBAMATE see 3,5-DIMETHYLPHENYL-N-METHYLCARBAMATE

Y-238 see POLYURETHANE Y-238

YAGEINE see HARMINE

YAJEINE see HARMINE

YELLOW CROSS LIQUID see BIS(2-CHLOROETHYL)SULFIDE

YELLOW CUPROCIDE see COPPER (I) OXIDE

YELLOW FERRIC OXIDE see IRON (III) OXIDE

YELLOW LEAD OCHER see LEAD MONOXIDE

YELLOW MERCURIC OXIDE see MERCURIC OXIDE

YELLOW MERCURY IODIDE see MERCURY (I) IODIDE

YELLOW NO. 2 see 1-(PHENYLAZO)-2-NAPHTHYLAMINE

YELLOW ORANGE SPECIALLY PURE 85 see FOOD YELLOW 3

YELLOW OXIDE OF IRON see IRON (III) OXIDE

YELLOW OXIDE OF MERCURY see MERCURIC OXIDE

YELLOW PHOSPHORUS see PHOSPHORUS (WHITE)

YELLOW PRECIPITATE see MERCURIC OXIDE

YELLOW SY FOR FOOD see FOOD YELLOW 3

YOHIMBAN-16-CARBOXYLIC ACID DERIVATIVE OF BENZ(G)INDOLO-(2,3-A)QUINOLIZINE see RESERPINE

YOHIMBIC ACID METHYL ESTER see YOHIMBINE

DELTA-YOHIMBINE see AJMALICINE

YPERITE see BIS(2-CHLOROETHYL)SULFIDE

YPERITE SULFONE see BIS(2-CHLOROETHYL)SULFONE

YTTERBIUM TRICHLORIDE see YTTERBIUM CHLORIDE

YTTRIA see YTTRIUM OXIDE

YTTRIUM-89 see YTTRIUM AND COMPOUNDS

YTTRIUM TRICHLORIDE see YTTRIUM CHLORIDE

ZACLONDISCOIDS see HYDROCYANIC ACID

ZAHLREICHE BEZEICHNUNGEN (GERMAN) see 4,6-DINITRO-o-CRESOL

ZELAZA TLENKI (POLISH) see IRON OXIDE FUME

ZELEN OSTANTHRENOVA BRILANTNI FFB (CZECH) see JADE GREEN BASE

ZEPHIRAN CHLORIDE see ALKYL DIMETHYLBENZYL AMMONIUM CHLORIDE

ZINC ARSENATE, BASIC see ZINC ARSENATE

ZINC ARSENITE, SOLID (DOT) see ZINC-m-ARSENITE

ZINC-2-BENZOTHIAZOLETHIOLATE see BIS(2-BENZOTHIAZOLYTHIO)ZINC

ZINC BENZOTHIAZOLYL MERCAPTIDE see BIS(2-BENZOTHIAZOLY-THIO)ZINC

ZINC BENZOTHIAZOL-2-YLTHIOLATE see BIS(2-BENZOTHIAZOLY-THIO)ZINC

ZINC BENZOTHIAZYL-2-MERCAPTIDE see BIS(2-BENZOTHIAZOLY-THIO)ZINC

ZINC BERYLLIUM SILICATE see BERYLLIUM ZINC SILICATE

ZINC-BIBUTYLDITHIOCARBAMATE see BIS(DIBUTYL DITHIO CARBA-MATO)ZINC

ZINC BIS(DIMETHYLDITHIOCARBAMATE) see BIS(DIMETHYL DITHIO-CARBAMATO)ZINC

ZINC BIS(DIMETHYLDITHIOCARBAMOYL)DISULPHIDE see BIS(DI-METHYL DITHIO CARBAMATO)ZINC

ZINC BIS(DIMETHYLTHIOCARBAMOYL)DISULFIDE see BIS(DIMETHYL DITHIO CARBAMATO)ZINC

ZINC CHROMATE see BASIC ZINC CHROMATE

ZINC CHROMATE HYDROXIDE see CHROMIUM(6+)ZINC OXIDE HY-DRATE (1:2:6:1)

ZINC CHROMATE (VI) HYDROXIDE see CHROMIUM(6+)ZINC OXIDE HYDRATE (1:2:6:1)

ZINC-DIBUTYLDITHIOCARBAMATE see BIS(DIBUTYL DITHIO CARBA-MATO)ZINC

ZINC-N,N-DIBUTYLDITHIOCARBAMATE see BIS(DIBUTYL DITHIO CAR-BAMATO)ZINC

ZINC DICHLORIDE see ZINC CHLORIDE

ZINC DICYANIDE see ZINC CYANIDE

ZINC DIETHYLDITHIOCARBAMATE see BIS(DIETHYL DITHIO CARBA-MATO)ZINC

ZINC-N,N-DIETHYLDITHIOCARBAMATE see BIS(DIETHYL DITHIO CAR-BAMATO)ZINC

ZINC DIMETHYLDITHIOCARBAMATE see BIS(DIMETHYL DITHIO CAR-BAMATO)ZINC

ZINC N,N-DIMETHYLDITHIOCARBAMATE see BIS(DIMETHYL DITHIO CARBAMATO)ZINC

ZINC DISTERATE see ZINC STEARATE

ZINC DUST see ZINC

ZINC ETHIDE see DIETHYLZINC

ZINC ETHYL (DOT) see DIETHYLZINC

ZINC ETHYLENE-1,2-BISDITHIOCARBAMATE see ETHYLENE BIS-(DITHIO CARBAMATO)ZINC

ZINC FLUORURE (FRENCH) see ZINC FLUORIDE

ZINC HEXAFLUOROSILICATE see ZINC FLUOSILICATE

ZINC HYDROXYCHROMATE see CHROMIUM(6+)ZINC OXIDE HYDRATE (1::::2:6:1)

ZINC MANGANESE BERYLLIUM SILICATE see BERYLLIUM MANGANESE ZINC SILICATE

ZINC MERCAPTOBENZOTHIAZOLATE see BIS(2-BENZOTHIAZOLYTHIO)-ZINC

ZINC-2-MERCAPTOBENZOTHIAZOLE see BIS(2-BENZOTHIAZOLYTHIO)-ZINC

ZINC MERCAPTOBENZOTHIAZOLE SALT see BIS(2-BENZOTHIAZOLY-THIO)ZINC

ZINC METHARSENITE see ZINC-m-ARSENITE

ZINCO (CLORURO DI) (ITALIAN) see ZINC CHLORIDE

ZINC OCTADECANOATE see ZINC STEARATE

ZINCO(FOSFURO DI) (ITALIAN) see ZINC PHOSPHIDE

ZINC OXIDE FUME see ZINC OXIDE

ZINC POWDER see ZINC

ZINC SUPEROXIDE see ZINC PEROXIDE

ZINC TETRAOXYCHROMATE 76a see BASIC ZINC CHROMATE

ZINC VITRIOL see ZINC SULFATE

ZINC WHITE see ZINC OXIDE

ZINC YELLOW see CHROMIUM(6+)ZINC OXIDE HYDRATE (1:2:6:1)

ZINC YELLOW see BASIC ZINC CHROMATE

ZINEB see ETHYLENE BIS(DITHIO CARBAMATO)ZINC

ZINK-(N,N'-AETHYLEN-BIS(DITHIOCARBAMAT)) (GERMAN) see ETHYL-ENE BIS(DITHIO CARBAMATO)ZINC

ZINK-BIS(N,N-DIMETHYL-DITHIOCARBAMAAT) (DUTCH) see BIS(DI-METHYL DITHIO CARBAMATO)ZINC

ZINK-BIS(N,N-DIMETHYL-DITHIOCARBAMAT) (GERMAN) see BIS(DI-METHYL DITHIO CARBAMATO)ZINC

ZINKCHLORID (GERMAN) see ZINC CHLORIDE

ZINKCHLORIDE (DUTCH) see ZINC CHLORIDE

ZINKFOSFIDE (DUTCH) see ZINC PHOSPHIDE

ZINKPHOSPHID (GERMAN) see ZINC PHOSPHIDE

ZINN (GERMAN) see TIN

ZINNTETRACHLORID (GERMAN) see TIN (IV) CHLORIDE (1:4)

ZIRCONIUM METAL see ZIRCONIUM

ZIRCONIUM METAL, DRY (DOT) see ZIRCONIUM (DRY)

ZIRCONIUM TETRACHLORIDE see ZIRCONIUM CHLORIDE

ZIRCONIUM TETRAFLUORIDE see ZIRCONIUM FLUORIDE

ZIRCONYL CHLORIDE see ZIRCONIUM OXYCHLORIDE

ZIRCONYL SULFATE see ZIRCONIUM (IV) SULFATE (1:2)

ZITRONEN OEL (GERMAN) see LEMON OIL

ZOALENE see 3,5-DINITRO-o-TOLUAMIDE

ZWAVELWATERSTOF (DUTCH) see HYDROGEN SULFIDE

ZWAVELZUUROPLOSSINGEN (DUTCH) see SULFURIC ACID

Appendix II
CAS NUMBER CROSS-REFERENCE

50-00-0 see FORMALDEHYDE
50-02-2 see SUPERPREDNOL
50-06-6 see 5-ETHYL-5-PHENYLBARBITURIC ACID
50-07-7 see AMETYCIN
50-10-2 see OXYPHENONIUM BROMIDE
50-12-4 see 3-METHYL-5-ETHYL-5-PHENYLHYDANTOIN
50-13-5 see DOLANTIN HYDROCHLORIDE
50-14-6 see ERGOCALCIFEROL
50-21-5 see LACTIC ACID
50-23-7 see CORTISOL
50-24-8 see PREDONIN
50-27-1 see ESTRIOL
50-28-2 see ESTRADIOL (β)
50-29-3 see DDT
50-32-8 see BENZO(a)PYRENE
50-33-9 see 4-BUTYL-1,2-DIPHENYL-3,5-DIOXO PYRAZOLIDINE
50-35-1 see THALIDOMIDE
50-36-2 see COCAINE
50-37-3 see N,N-DIETHYLLYSERGAMIDE
50-44-2 see PURINE-6-THIOL
50-47-5 see 10,11-DIHYDRO-5-(3-METHYLAMINOPROPYL)-5H-DI-BENZ(b,f)AZEPINE
50-48-6 see ELAVIL
50-49-7 see TOFRANIL
50-50-0 see ESTRADIOL-3-BENZOATE
50-52-2 see 10-(2-(1-METHYL-2-PIPERIDYL)ETHYL)-2-(METHYLTHIO) PHENOTHIAZINE
50-53-3 see CHLOROPROMAZINE
50-55-5 see RESERPINE
50-63-5 see CHLOROQUINE DIPHOSPHATE
50-65-7 see 2′,5-DICHLORO-4′-NITROSALICYLANILIDE
50-67-9 see 3-(2-AMINOETHYL)INDOL-5-OL
50-71-5 see ALLOXAN
50-76-0 see ACTINOMYCIN D
50-78-2 see ACETOL (2)
50-81-7 see L-ASCORBIC ACID
50-91-9 see 2′-DEOXY-5-FLUORO URIDINE
50-98-6 see EPHEDRINE HYDROCHLORIDE
50-99-7 see D-GLUCOSE
51-02-5 see alpha-((ISOPROPYLAMINO)METHYL)NAPHTHALENE-METHANOL, HYDROCHLORIDE
51-03-6 see PIPERONYL BUTOXIDE
51-05-8 see p-AMINOBENZOYLDIETHYLAMINOETHANOL HYDROCHLO-RIDE
51-06-9 see p-AMINO-n-(2-DIETHYLAMINOETHYL)BENZAMIDE
51-15-0 see 2-FORMYL-1-METHYLPYRIDINIUM CHLORIDE OXIME
51-17-2 see BENZIMIDAZOLE
51-18-3 see TRISAZIRIDINYLTRIAZINE
51-21-8 see FLUOROURACIL
51-28-5 see 2,4-DINITROPHENOL
51-34-3 see SCOPOLAMINE
51-41-2 see L-NOREPINEPHRINE
51-43-4 see VASOTONIN
51-44-5 see 3,4-DICHLOROBENZOIC ACID
51-45-6 see HISTAMINE
51-52-5 see 6-PROPYL-2-THIOURACIL
51-55-8 see ATROPINE
51-57-0 see d-METHAPHETAMINE HYDROCHLORIDE
51-60-5 see NEOSTIGMINE MONOMETHYLSULFATE
51-61-6 see DOPAMINE
51-62-7 see l-BENZEDRINE SULFATE

51-63-8 see d-BENZEDRINE SULFATE
51-64-9 see D-AMPHETAMINE
51-71-8 see 2-PHENYLETHYLHYDRAZINE
51-75-2 see BIS(beta-CHLOROETHYL)METHYLAMINE
51-79-6 see URETHANE
51-83-2 see CARBACHOL CHLORIDE
51-85-4 see beta-MERCAPTOETHYLAMINE DISULFIDE
51-98-9 see 17-ACETOXY-19-NOR-17-alpha-PREGN-4-EN-20-YN-3-ONE
52-01-7 see ALDACTAZIDE
52-24-4 see THIOPHOSPHAMIDE
52-26-6 see MORPHINE HYDROCHLORIDE
52-28-8 see CODEINE PHOSPHATE
52-31-3 see TETRAHYDROPHENOBARBITAL
52-43-7 see ALLOBARBITAL
52-46-0 see APHOLATE
52-49-3 see BENZHEXOL HYDROCHLORIDE
52-51-7 see 2-BROMO-2-NITRO-1,3-PROPANEDIOL
52-52-8 see 1-AMINOCYCLOPENTANE-1-CARBOXYLIC ACID
52-53-9 see ISOPTIN
52-67-5 see d,3-MERCAPTOVALINE
52-68-6 see ((2,2,2-TRICHLORO-1-HYDROXYETHYL) DIMETHYLPHOS-PHONATE)
52-85-7 see O,O-DIMETHYL-O-(p-(N,N-DIMETHYLSULFAMOYL) PHENYL)PHOSPHOROTHIOATE
52-86-8 see gamma-(4-(p-CHLORPHENYL)-4-HYDROXPIPERIDINO)-p-FLUORBUTYRO-PHENONE
52-90-4 see L-CYSTEINE
53-03-2 see PREDNISONE
53-16-7 see ESTRONE
53-19-0 see CHLODITHANE
53-21-4 see COCAINE CHLORIDE
53-46-3 see XANTHINE BROMIDE
53-60-1 see PROMAZINE HYDROCHLORIDE
53-69-0 see 5,7-DIMETHYL-1,2-BENZACRIDINE
53-70-3 see DIBENZ(a,h)ANTHRACENE
53-86-1 see INDOMETHACIN
53-95-2 see N-HYDROXY-N-ACETYL-2-AMINOFLUORENE
53-96-3 see N-FLUOREN-2-YL ACETAMIDE
54-04-6 see MESCALINE
54-05-7 see CHLOROQUINE
54-11-5 see NICOTINE
54-21-7 see SODIUM SALICYLATE
54-31-9 see 4-CHLORO-N-FURFURYL-5-SULFAMOYL ANTHRANILIC ACID
54-42-2 see 2′-DEOXY-5-IODO URIDINE
54-49-9 see HYDROXYNOREPHEDRINE
54-62-6 see AMINOPTERIDINE
54-64-8 see MERTHIOLATE SODIUM
54-77-3 see 1,1-DIMETHYL-4-PHENYLPIPERAZINE IODIDE
54-85-3 see ISONICOTINIC ACID HYDRAZIDE
54-92-2 see ISONICOTINIC ACID-2-ISOPROPYLHYDRAZIDE
54-95-5 see 1,5-PENTAMETHYLENETETRAZOLE
55-18-5 see N-NITROSO-DIETHYLAMINE
55-31-2 see 1-ADRENALINE CHLORIDE
55-37-8 see O,O-DIMETHYL O-4-(METHYLTHIO)-3,5-XYLYL PHOSPHO-ROTHIOATE
55-38-9 see O,O-DIMETHYLPHOSPHOROTHIOIC ACID-O-(4-METHYL-THIO)-m-TOLYLESTER
55-43-6 see DIBENAMINE HYDROCHLORIDE
55-48-1 see ATROPINE SULFATE (2:1)

79-21-0 see PEROXYACETIC ACID
79-22-1 see METHYL CHLOROCARBONATE
79-24-3 see NITROETHANE
79-27-6 see ACETYLENE TETRABROMIDE
79-31-2 see ISOBUTYRIC ACID
79-34-5 see ACETYLENE TETRACHLORIDE
79-38-9 see CHLOROTRIFLUOROETHYLENE
79-39-0 see METHACRYLIC ACID AMIDE
79-40-3 see DITHIOOXAMIDE
79-41-4 see METHACRYLIC ACID
79-43-6 see DICHLORACETIC ACID
79-44-7 see (DIMETHYLAMINO)CARBONYL CHLORIDE
79-46-9 see 2-NITROPROPANE
79-57-2 see 5-HYDROXYTETRACYLINE
79-81-2 see VITAMIN A PALMITATE
79-92-5 see CAMPHENE
80-05-7 see BISPHENOL A
80-08-0 see 4,4′-SULFONYLDIANILINE
80-15-9 see ISOPROPYLBENZENE HYDROPEROXIDE
80-33-1 see 4-CHLOROPHENYL 4-CHLOROBENZENESULFONATE
80-38-6 see 4-CHLOROPHENYL BENZENESULFONATE
80-40-0 see ETHYL TOSYLATE
80-46-6 see 4-tert-AMYLPHENOL
80-47-7 see p-MENTHANE-8-HYDROPEROXIDE
80-48-8 see METHYL-p-METHYLBENZENESULFONATE
80-56-8 see 2-PINENE
80-62-6 see METHYL METHACRYLATE
81-07-2 see 1,2-BENZISOTHIAZOL-3(2H)-ONE-1,1-DIOXIDE
81-25-4 see CHOLIC ACID (HYDRATE)
81-64-1 see 1,4-DIHYDROXYANTHRAQUINONE
81-81-2 see COUMADIN
81-88-9 see (9-(o-CARBOXYPHENYL)-6-(DIETHYLAMINO)-3H-
   XANTHEN-3-YLIDENE) DIETHYLAMMONIUM CHLORIDE
82-02-0 see AMICARDINE
82-28-0 see 1-AMINO-2-METHYLANTHRAQUINONE
82-44-0 see 1-CHLORO ANTHRA QUINONE
82-45-1 see 1-AMINOANTHRAQUINONE
82-66-6 see DIPHENADIONE
82-68-8 see PENTACHLORONITRO BENZENE
82-93-9 see CHLOROCYCLINE
83-26-1 see 2-PIVALOYL-1,3-INDANDIONE
83-34-1 see beta-METHYLINDOLE
83-43-2 see METHYLPREDNISOLONE
83-44-3 see DEOXYCHOLATIC ACID
83-63-6 see N-ACETYL-N-(2-METHYL-4-((2-METHYLPHENYL)AZO)-
   PHENYL)ACETAMIDE
83-66-9 see 6-tert-BUTYL-3-METHYL-2,4-DINITRO ANISOLE
83-67-0 see THEOBROMINE
83-73-8 see DIIODOHYDROXYQUIN
83-79-4 see ROTENONE
83-86-3 see PHYTIC ACID
83-88-5 see RIBOFLAVINE
83-89-6 see ATABRINE
84-02-6 see PROCHLORPERIZINE MALEATE
84-08-2 see PARATHIAZINE
84-16-2 see DIHYDROSTILBESTROL
84-17-3 see DEHYDROSTILBESTROL
84-65-1 see ANTHRAQUINONE
84-66-2 see DIETHYL-o-PHTHALATE
84-69-5 see DIISOBUTYL PHTHALATE
84-74-2 see DI-n-BUTYL PHTHALATE
84-80-0 see VITAMIN K1
84-86-6 see 4-AMINO-1-NAPHTHALENESULFONIC ACID
84-97-9 see PERNAZINE
85-00-7 see DIQUAT DIBROMIDE
85-01-8 see PHENANTHRENE
85-41-6 see PHTHALIMIDE
85-43-8 see TETRAHYDROPHTHALIC ACID ANHYDRIDE
85-44-9 see PHTHALIC ANHYDRIDE
85-68-7 see BENZYL BUTYL PHTHALATE

85-70-1 see BUTYL CARBO BUTOXY METHYL PHTHALATE
85-79-0 see DIBUCAINE
85-83-6 see SCARLET RED
85-84-7 see 1-(PHENYLAZO)-2-NAPHTHYLAMINE
85-86-9 see OIL RED
85-98-3 see 1,3-DIETHYL-1,3-DIPHENYLUREA
86-00-0 see o-NITROBIPHENYL
86-21-5 see TRIMETON
86-29-3 see DIPHENYLACETONITRILE
86-30-6 see DIPHENYLNITROSAMINE
86-40-8 see XANTHACRIDINE
86-50-0 see AZINPHOS METHYL
86-54-4 see 1(2H)-PHTHALAZINONE HYDRAZONE
86-56-6 see N,N-DIMETHYL-1-NAPHTHYLAMINE
86-74-8 see CARBAZOLE
86-85-1 see METHYLMERCURY QUINOLINOLATE
87-29-6 see ANTHRANILIC ACID, CINNAMYL ESTER
87-44-5 see CARYOPHYLLENE
87-51-4 see 1H-INDOLE-3-ACETIC ACID
87-62-7 see 2,6-XYLIDINE
87-65-0 see 2,6-DICHLOROPHENOL
87-66-1 see PYROGALLOL
87-68-3 see PERCHLOROBUTADIENE
87-85-4 see HEXAMETHYLBENZENE
87-86-5 see PENTACHLOROPHENOL
87-99-0 see XYLITOL
88-04-0 see 4-CHLORO-3,5-XYLENOL
88-05-1 see 2,4,6-TRIMETHYLANILINE
88-06-2 see 2,4,6-TRICHLOROPHENOL
88-09-5 see DIETHYLACETIC ACID
88-12-0 see 1-ETHENYL-2-PYRROLIDINONE
88-19-7 see o-TOLUENESULFONAMIDE
88-29-9 see ACETYL ETHYL TETRAMETHYL TETRALIN
88-29-9 see 3′-ETHYL-5′,6′,7′,8′-TETRAHYDRO-5′,5′,8′,8′-TETRA-
   METHYL-2′-ACETONAPHTHONE
88-72-2 see o-METHYLNITROBENZENE
88-73-3 see CHLORO-o-NITROBENZENE
88-74-4 see o-NITROANILINE
88-75-5 see o-NITROPHENOL
88-85-7 see 2-sec-BUTYL-4,6-DINITRO PHENOL
88-89-1 see PICRIC ACID
88-99-3 see PHTHALIC ACID
89-57-6 see 5-AMINOSALICYLIC ACID
89-72-5 see o-sec-BUTYLPHENOL
89-78-1 see MENTHOL
89-83-8 see THYMOL
89-98-5 see o-CHLOROBENZALDEHYDE
90-05-1 see GUAIACOL
90-15-3 see 1-NAPHTHOL
90-22-2 see VALETHAMATE BROMIDE
90-41-5 see 2-BIPHENYLAMINE
90-43-7 see 2-BIPHENYLOL
90-45-9 see 9-AMINOACRIDINE
90-64-2 see MANDELIC ACID
90-65-3 see PENICILLIC ACID
90-94-8 see MICHLER'S KETONE
90-94-8 see TETRAMETHYL DIAMINO BENZOPHENONE
91-10-1 see 2,6-DIMETHOXYPHENOL
91-15-6 see PHTHALONITRILE
91-17-8 see DECAHYDRONAPHTHALENE
91-20-3 see NAPHTHALENE
91-22-5 see QUINOLINE
91-33-8 see BENZOTHIAZIDE
91-40-7 see N-PHENYLANTHRANILIC ACID
91-59-8 see 2-NAPHTHYLAMINE
91-60-1 see 2-NAPHTHALENETHIOL
91-64-5 see COUMARIN
91-75-8 see PHENAZOLINE
91-80-5 see THENYLPYRAMINE
91-81-6 see TRIPELENNAMINE

111-94-4 see BIS(beta-CYANOETHYL)AMINE
111-97-7 see DI(2-CYANOETHYL)SULFIDE
112-05-0 see NONANOIC ACID
112-07-2 see 2-BUTOXYETHYL ACETATE
112-12-9 see 2-UNDECANONE
112-15-2 see CARBITOL ACETATE
112-27-6 see TRIETHYLENE GLYCOL
112-30-1 see DECYL ALCOHOL
112-31-2 see 1-DECANAL
112-40-3 see DODECANE
112-50-5 see ETHOXYTRIGLYCOL
112-53-8 see DODECYL ALCOHOL
112-56-1 see 2-(2-BUTOXY ETHOXY)ETHYL THIOCYANATE
112-61-8 see METHYL STEARATE
112-62-9 see cis-OLEIC ACID, METHYL ESTER
112-80-1 see OLEIC ACID
112-92-5 see 1-OCTADECANOL
113-00-8 see GUANIDINE
113-15-5 see ERGOTAMINE
113-18-8 see 1-CHLORO-3-ETHYL-1-PENTEN-4-YN-3-OL
113-38-2 see ESTRADIOL DIPROPIONATE
113-45-1 see METHYL PHENIDYL ACETATE
113-48-4 see N-OCTYL BICYCLOHEPTENE DICARBOXIMIDE
113-52-0 see N-(gamma-DIMETHYLAMINOPROPYL)IMINODIBENZYL
  HYDROCHLORIDE
113-92-8 see TELDRIN
114-07-8 see ERYTHROMYCIN
114-26-1 see BAYGON
114-45-4 see (±)-ISOPROTERENOL SULFATE
114-49-8 see HYOSCINE HYDROBROMIDE
114-80-7 see PROSTIGMINE BROMIDE
114-90-9 see 1,3-BIS(4-ALDOX IMINOPYRIDINIUM)DIMETHYL ETHER
  BICHLORIDE
115-07-1 see PROPENE
115-09-3 see MERCURYMETHYLCHLORIDE
115-10-6 see METHYL ETHER
115-11-7 see ISOBUTENE
115-21-9 see ETHYL TRICHLOROSILANE
115-25-3 see CYCLOOCTAFLUOROBUTANE
115-26-4 see BIS(DIMETHYL AMIDO)FLUORO PHOSPHATE
115-28-6 see CHLORENDIC ACID
115-29-7 see BENZOEPIN
115-31-1 see ISOBORNYL THIOCYANATO ACETATE
115-32-2 see 1,1-BIS(p-CHLOROPHENYL)-2,2,2-TRICHLOROETHANOL
115-38-8 see 5-ETHYL-N-METHYL-5-PHENYL BARBITURIC ACID
115-44-6 see 5-ALLYL-5-sec-BUTYLBARBITURIC ACID
115-58-2 see PENTOBARBITAL
115-69-5 see 2-AMINO-2-METHYL-1,3-PROPANEDIOL
115-86-6 see TRIPHENYL PHOSPHATE
115-90-2 see O,O-DIETHYL O-P-(METHYLSULFINYL)PHENYL THIO-
  PHOSPHATE
115-96-8 see 2-CHLORO ETHANOL PHOSPHATE
116-01-8 see DIMETHOATE-ETHYL
116-06-3 see CARBANOLATE
116-09-6 see ACETOL (1)
116-29-0 see p-CHLOROPHENYL-2,4,5-TRICHLOROPHENYL SULFONE
116-54-1 see METHYL DICHLORO ACETATE
116-85-8 see 1-AMINO-4-HYDROXYANTHRAQUINONE
117-39-5 see QUERCETIN
117-51-1 see 3-HOMOTETRA HYDRO CANNIBINOL
117-52-2 see 3-(alpha-ACETONYLFURFURYL)-4-HYDROXYCOUMARIN
117-79-3 see 2-AMINOANTHRAQUINONE
117-80-6 see 2,3-DICHLORO-1,4-NAPHTHOQUINONE
117-81-7 see BIS(2-ETHYLHEXYL)PHTHALATE
117-82-8 see DIMETHOXY ETHYL PHTHALATE
117-84-0 see n-DIOCTYL PHTHALATE
118-46-7 see 8-AMINO-2-NAPHTHOL
118-55-8 see PHENYL SALICYLATE
118-68-3 see 3-(2-AMINOBUTYL)INDOLE ACETATE
118-74-1 see HEXACHLOROBENZENE

118-92-3 see ANTHRANILIC ACID
118-96-7 see sym-TRINITROTOLUENE
119-12-0 see PYRIDAPHENTHION
119-27-7 see 2,4-DINITROANISOL
119-34-6 see 4-AMINO-2-NITROPHENOL
119-36-8 see METHYL SALICYLATE
119-38-0 see DIMETHYL-5-(1-ISOPROPYL-3-METHYL-PYRAZOLYL)-
  CARBAMATE
119-48-2 see DIMORPHOLAMINE
119-61-9 see BENZOPHENONE
119-65-3 see ISOQUINOLINE
119-79-9 see 5-AMINO-2-NAPHTHALENESULFONIC ACID
119-84-6 see HYDROCOUMARIN
119-90-4 see o-DIANISIDINE
119-93-7 see 3,3'-TOLIDINE
120-02-5 see 4-CARBAMIDOPHENYL BIS(CARBOXY METHYL THIO)-
  ARSENITE
120-08-1 see 6,7-DIMETHYLESCULETIN
120-12-7 see ANTHRACENE
120-20-7 see 3,4-DIMETHOXYDOPAMINE
120-40-1 see N,N-BIS(2-HYDROXY ETHYL)DODECAN AMIDE
120-51-4 see BENZOIC ACID, BENZYL ESTER
120-57-0 see PIPERONAL
120-58-1 see ISOSAFROLE
120-61-6 see DIMETHYL TEREPHTHALATE
120-61-6 see TEREPHTHALIC ACID METHYL ESTER
120-62-7 see ISOSAFROLE-n-OCTYLSULFOXIDE
120-71-8 see 5-METHYL-o-ANISIDINE
120-72-9 see INDOLE
120-78-5 see BENZOTHIAZOLE DISULFIDE
120-80-9 see PYROCATECHOL
120-83-2 see 2,4-DICHLOROPHENOL
120-92-3 see CYCLOPENTANONE
121-14-2 see 2,4-DINITROTOLUENE
121-29-9 see PYRETHRIN II
121-32-4 see ETHYL VANILLIN
121-33-5 see VANILLIN
121-43-7 see TRIMETHYL BORATE
121-44-8 see TRIETHYLAMINE
121-54-0 see BENZETHONIUM CHLORIDE
121-59-5 see N-CARBAMOYL ARSANILIC ACID
121-66-4 see 2-AMINO-5-NITROTHIAZOLE
121-69-7 see N,N-DIMETHYLANILINE
121-73-3 see 1-CHLORO-3-NITROBENZENE
121-75-5 see CARBETHOXY MALATHION
121-79-9 see N-PROPYL GALLATE
121-82-4 see TRIMETHYLENETRINITRAMINE
122-03-2 see CUMALDEHYDE
122-09-8 see alpha,alpha-DIMETHYLPHENETHYLAMINE
122-10-1 see SWAT
122-14-5 see O,O-DIMETHYL-O-(3-METHYL) PHOSPHOROTHIOATE
122-15-6 see 5,5-DIMETHYLDIHYDRORESORCINOL DIMETHYLCARBA-
  MATE
122-19-0 see DIMETHYLOCTADECYLBENZYLAMMONIUM CHLORIDE
122-34-9 see 2,4-BIS(ETHYLAMINO)-6-CHLORO-s-TRIAZINE
122-37-2 see p-ANILINOPHENOL
122-39-4 see DIPHENYLAMINE
122-40-7 see alpha-AMYL CINNAMALDEHYDE
122-42-9 see CARBANILIC ACID ISOPROPYL ESTER
122-51-0 see ETHYL ORTHOFORMATE
122-60-1 see PHENYLGLYCYDYL ETHER
122-62-3 see BIS(2-ETHYLHEXYL)SEBACATE
122-66-7 see HYDRAZOBENZENE
122-88-3 see p-CHLORO PHENOXY ACETIC ACID
122-98-5 see 2-ANILINOETHANOL
122-99-6 see PHENYL CELLOSOLVE
123-00-2 see 4-AMINOPROPYLMORPHOLINE
123-03-5 see CEPACOL CHLORIDE
123-05-7 see BUTYL ETHYL ACETALDEHYDE
123-11-5 see p-ANISALDEHYDE

123-19-3 see 4-HEPTANONE
123-20-6 see VINYL BUTYRATE
123-23-9 see SUCCINIC PEROXIDE
123-30-8 see 4-AMINOPHENOL
123-31-9 see HYDROQUINONE
123-33-1 see MALEIC HYDRAZIDE
123-39-7 see N-METHYLFORMAMIDE
123-42-2 see 2-METHYL-2-PENTANOL-4-ONE
123-51-3 see ISOAMYL ALCOHOL
123-51-3 see FUSEL OIL
123-54-6 see ACETYL ACETONE
123-63-7 see PARALDEHYDE
123-66-0 see ETHYL CAPROATE
123-72-8 see N-BUTYRALDEHYDE
123-73-9 see trans-2-BUTENAL
123-75-1 see PYRROLIDINE
123-77-3 see AZODICARBAMIDE
123-86-4 see N-BUTYL ACETATE
123-88-6 see 2-METHOXYETHYLMERCURY CHLORIDE
123-91-1 see p-DIOXANE
123-92-2 see ISOPENTYL ALCOHOL ACETATE
123-93-3 see MERCAPTODIACETIC ACID
124-02-7 see DIALLYLAMINE
124-04-9 see ADIPIC ACID
124-16-3 see 1-BUTOXY ETHOXY-2-PROPANOL
124-17-4 see BUTYL CARBITOL ACETATE
124-18-5 see DECANE
124-22-1 see DODECYLAMINE
124-30-1 see OCTADECYLAMINE
124-38-9 see CARBON DIOXIDE
124-38-9 see CARBON DIOXIDE (liquefied)
124-40-3 see DIMETHYLAMINE
124-40-3 see DIMETHYLAMINE (ANHYDROUS)
124-48-1 see CHLORO DIBROMO METHANE
124-87-8 see PICROTOXIN
124-90-3 see DIHYDRONE HYDROCHLORIDE
124-94-7 see ARISTOCORT
124-99-2 see GLUCO PROSCILLARIDIN A
125-04-2 see HYDROCORTISONE SODIUM SUCCINATE
125-28-0 see DIHYDROCODEINE
125-33-7 see 2-DESOXY PHENOBARBITAL
125-40-6 see BUTISOL
125-42-8 see 5-ETHYL-5-(1-METHYL-1-BUTENYL)BARBITURATE
125-44-0 see VINBARBITAL SODIUM
125-51-9 see PIPTAL
125-56-4 see dl-METHADONE HYDROCHLORIDE
125-88-2 see BUTALBITAL SODIUM
126-07-8 see GRISOFULVIN
126-17-0 see PURAPURIDINE
126-22-7 see BUTONATE
126-27-2 see 2-DI(N-METHYL-N-PHENYL-tert-BUTYL CARBAMOYL-METHYL)AMINOETHANOL
126-72-7 see TRIS
126-75-0 see DEMETON-S
126-85-2 see 2-CHLORO-N-(2-CHLOROETHYL)-N-METHYL ETHAN-AMINE-N-OXIDE
126-92-1 see TERGITOL 08
126-98-7 see METHYLACRYLONITRILE
126-99-8 see NEOPRENE
127-07-1 see HYDROXYUREA
127-18-4 see 1,1,2,2-TETRACHLOROETHYLENE
127-19-5 see N,N-DIMETHYLACETAMIDE
127-21-9 see DICHLOROTETRAFLUOROACETONE
127-33-3 see METHYLCHLORTETRACYCLINE
127-47-9 see RETINOL ACETATE
127-48-0 see TRIMETHADIONE
127-85-5 see ARSANILIC ACID, MONOSODIUM SALT
128-08-5 see N-BROMO SUCCINIMIDE
128-37-0 see BHT (FOOD GRADE)
128-44-9 see SACCHARIN

128-53-0 see MALEIC ACID-N-ETHYLIMIDE
128-58-5 see JADE GREEN BASE
128-80-3 see 1,4-BIS(p-TOLYAMINO)ANTHRAQUINONE
129-00-0 see PYRENE
129-06-6 see COUMADIN SODIUM
129-15-7 see 2-METHYL-1-NITROANTHRAQUINONE
129-16-8 see MERCUROCHROME
129-17-9 see ACID BLUE 1
129-18-0 see BUTAZOLIDINE SODIUM
129-20-4 see p-HYDROXYPHENYLBUTAZONE
129-51-1 see ERGOT
129-66-8 see 2,4,6-TRINITROBENZOIC ACID (DRY)
129-67-9 see DISODIUM-3,6-ENDOXOHEXAHYDROPHTHALATE
129-71-5 see CHLORCYCLIZINE DIHYDROCHLORIDE
130-01-8 see AUREINE
130-15-4 see 1,4-NAPHTHOQUINONE
130-24-5 see 4-AMINO-2-METHYL-1-NAPHTHOL
130-26-7 see 5-CHLORO-7-IODO-8-QUINOLINOL
130-61-0 see 10-(2-(1-METHYL-2-PIPERIDYL)ETHYL)-2-METHYLTHIO-PHENOTHIAZINE HYDROCHLORIDE
130-80-3 see DIETHYLSTILBESTROL DIPROPIONATE
130-86-9 see FUMARINE
130-89-2 see QUININE HYDROCHLORIDE
130-95-0 see QUININE
131-11-3 see DIMETHYL PHTHALATE
131-27-1 see 3-AMINO-1,5-NAPHTHALENEDISULFONIC ACID
131-52-2 see SODIUM PENTACHLOROPHENATE
131-56-6 see 2,4-DIHYDROXYBENZOPHENONE
131-74-8 see AMMONIUM PICRATE
131-74-8 see AMMONIUM PICRATE (WET)
131-79-3 see 1-(o-TOLYLAZO)-2-NAPHTHYLAMINE
131-89-5 see 2-CYCLOHEXYL-4,6-DINITROPHENOL
132-17-2 see TROPINE BENZOHYDRYL ETHER METHANESULFONATE
132-27-4 see 2-BIPHENYLOL, SODIUM SALT
132-32-1 see 3-AMINO-9-ETHYLCARBAZOLE
132-45-6 see 2-PHENYLVALERIC ACID-2-(DIETHYLAMINO)ETHYL ESTER HYDROCHLORIDE
132-69-4 see BENZIDAMINE HYDROCHLORIDE
133-06-2 see CAPTAN
133-32-4 see 1H-INDOLE-3-BUTANOIC ACID
133-90-4 see 3-AMINO-2,5-DICHLOROBENZOIC ACID
134-20-3 see ANTHRANILIC ACID, METHYL ESTER
134-29-2 see o-ANISIDINE HYDROCHLORIDE
134-32-7 see 1-NAPHTHYLAMINE
134-62-3 see DIETHYL-m-TOLUAMIDE
134-71-4 see dl-EPHEDRINE HYDROCHLORIDE
134-72-5 see 1-EPHEDRINE SULFATE
134-80-5 see 2-DIETHYLAMINOPROPIOPHENONEHYDROCHLORIDE
134-84-9 see 4-METHYL-p-BENZOPHENONE
135-19-3 see 2-NAPHTHOL
135-20-6 see AMMONIUM-N-NITROSOPHENYLHYDROXYLAMINE
135-40-3 see PHENAZOPYRIDINIUM CHLORIDE
135-87-5 see BENZODIOXANE HYDROCHLORIDE
135-88-6 see N-PHENYL-2-NAPHTHYLAMINE
135-98-8 see sec-BUTYLBENZENE
136-23-2 see BIS(DIBUTYL DITHIO CARBAMATO)ZINC
136-30-1 see SODIUM DIBUTYLDITHIOCARBAMATE
136-35-6 see 1,3-DIPHENYLTRIAZENE
136-60-7 see BUTYL BENZOATE
136-77-6 see HEXYLRESORCINOL
136-78-7 see CRAG HERBICIDE
136-92-5 see DIETHYLDITHIOCARBAMIC ACID SELENIUM (II) SALT
137-07-5 see 2-AMINOBENZENETHIOL
137-08-6 see CALCIUM-d-PANTOTHENATE
137-17-7 see 2,4,5-TRIMETHYLANILINE
137-26-8 see BIS(DIMETHYL THIOCARBAMYL)DISULFIDE
137-29-1 see COPPER DIMETHYLDITHIOCARBAMATE
137-30-4 see BIS(DIMETHYL DITHIO CARBAMATO)ZINC
137-32-6 see 2-METHYL BUTANOL-1
137-42-8 see VAPAM
137-58-6 see XYLOCAIN

371-29-9 see 2-FLUORO ETHYL gamma-FLUORO BUTYRATE
371-40-4 see 4-FLUOROANILINE
371-62-0 see 2-FLUOROETHANOL
371-86-8 see PHOSPHORODI(ISOPROPYLAMIDIC) FLUORIDE
372-09-8 see CYANOACETIC ACID
373-02-4 see NICKEL(II) ACETATE (1:2)
375-22-4 see HEPTAFLUOROBUTYRIC ACID
378-44-9 see BETAMETHASONE
379-79-3 see ERGOTAMINE TARTRATE
386-38-9 see 10-FLUORO-9,12-DIMETHYLBENZ(a)ACRIDINE
388-72-7 see 4-FLUORO BENZ ANTHRACENE
389-08-2 see 1-ETHYL-1,4-DIHYDRO-7-METHYL-4-OXO-1,8-
   NAPHTHYRIDINE-3-CARBOXYLIC ACID
404-82-0 see PHENFLUORAMINE HYDROCHLORIDE
404-86-4 see CAPSAICIN
406-20-2 see METHYL-4-FLUOROBUTYRATE
406-90-6 see 2,2,2-TRIFLUOROETHYL VINYL ETHER
409-21-2 see SILICON CARBIDE
420-04-2 see CYANAMIDE
420-12-2 see ETHYLENE SULFIDE
421-20-5 see METHYL FLUOROSULFATE
427-51-0 see CYPROSTERONE ACETATE
431-03-8 see 2,3-BUTANEDIONE
432-60-0 see ALLYLESTRENOL
437-38-7 see PHENTANYL
438-41-5 see LIBRIUM HYDROCHLORIDE
442-51-3 see HARMINE
443-30-1 see 1-(4-DIMETHYLAMINOBENZAL)INDENE
443-48-1 see 2-METHYL-5-NITROIMIDAZOLE-1-ETHANOL
446-86-6 see AZATHIOPRINE
453-18-9 see METHYL FLUOROACETATE
457-60-3 see NEOSALVARSAN
458-24-2 see N-ETHYL-alpha-METHYL-m-(TRIFLUOROMETHYL)
   PHENETHYLAMINE
459-99-4 see beta-FLUOROETHYL FLUOROACETATE
460-07-1 see 1-ACETYLAZIRIDINE
460-19-5 see CYANOGEN
461-78-9 see CHLORPHENTERMINE
462-94-2 see PENTANEDIAMINE
463-04-7 see N-AMYL NITRITE
463-51-4 see KETENE
463-58-1 see CARBONYL SULFIDE
463-71-8 see THIOPHOSGENE
464-10-8 see NITROTRIBROMOMETHANE
464-45-9 see NGAI CAMPHOR
464-49-3 see CAMPHOR, (1R,4R)-(+)-
465-16-7 see OLEANDRIN
465-73-6 see ISODRIN
466-24-0 see PICRACONITINE
466-40-0 see ISOMETHADONE
466-99-9 see DIHYDROMORPHINONE
467-60-7 see alpha,alpha-DIPHENYL-2-PIPERIDINEMETHANOL
469-59-0 see JERVINE
469-62-5 see CARVON
469-65-8 see (p-HYDROXYBENZYL)TARTARIC ACID
470-82-6 see CAJEPUTOL
470-90-6 see CHLORFENVINFOS
471-03-4 see BIS(2-CHLOROETHYL)SULFONE
471-29-4 see METHYLGUANIDINE
471-46-5 see OXAMIDE
471-95-4 see BUFOTALINE
474-86-2 see EQUILIN
475-83-2 see 1-NUCIFERINE
476-32-4 see CHELIDONINE
477-27-0 see N-ACETYL TRIMETHYLCOLCHICINIC ACID
477-30-5 see N-METHYL-N-DESACETYLCOLCHICINE
478-84-2 see 2-BROMO-D-LYSERGIC ACID DIETHYLAMIDE
478-99-9 see LYSERGIC ACID ETHYLAMIDE
479-23-2 see CHOLANTHRENE
479-45-8 see TETRYL

479-50-5 see LUCANTHONE
479-92-5 see 4-ISOPROPYLANTIPYRINE
480-16-0 see MORIN
480-22-8 see 1,8,9-ANTHRACENETRIOL
480-54-6 see RETRORSINE
480-81-9 see SENECIPHYLLINE
481-22-2 see 3-NITRO-3-HEXENE
481-39-0 see WALNUT EXTRACT
482-41-7 see 7-METHYL-9-FLUOROBENZ(c)ACRIDINE
483-04-5 see AJMALICINE
483-18-1 see EMETINE
485-31-4 see BINAPACRYL
487-10-5 see 1,1'-AZONAPHTHALENE
487-53-6 see METAHYDROXYPROCAINE
487-93-4 see 3-(2-DIMETHYLAMINOETHYL)-5-INDOLOL
488-17-5 see 2,3-DIHYDROXYTOLUENE
488-41-5 see 1,6-DIBROMOMANNITOL
492-17-1 see 2,4'-BIPHENYLDIAMINE
492-18-2 see SODIUM MERSALYL
492-41-1 see (−)-NOREPHEDRINE
493-52-7 see 2-CARBOXY-4'-(DIMETHYL AMINO)AZOBENZENE
494-03-1 see N,N-BIS(2-CHLOROETHYL)-2-NAPHTHYLAMINE
494-38-2 see 3,6-BIS(DIMETHYL AMINO)ACRIDINE
494-47-3 see FURFURAMIDE
495-48-7 see AZOXYBENZENE
495-73-8 see 1,4-BENZOQUINONE-N'-BENZOYLHYDRAZONE OXIME
496-67-3 see 2-BROMO-3-METHYL BUTYRYL UREA
499-75-2 see CARVACROL
500-28-7 see METHYLCHLOROTHION
500-55-0 see APOATROPINE
500-92-5 see 1-(p-CHLOROPHENYL)-5-ISOPROPYLBIGUANIDE
501-53-1 see BENZYL CHLOROFORMATE
502-39-6 see METHYLMERCURIC DICYANDIAMIDE
502-42-1 see CYCLOHEPTANONE
502-55-6 see BISETHYL XANTHOGEN DISULFIDE
503-01-5 see ISONYL
503-17-3 see CROTONYLENE (DOT)
503-20-8 see FLUOROACETONITRILE
503-30-0 see OXETANE
503-74-2 see ISOVALERIC ACID
504-15-4 see 5-METHYLRESORCINOL
504-20-1 see PHORONE
504-24-5 see 4-AMINOPYRIDINE
504-29-0 see o-AMINOPYRIDINE
505-57-7 see 2-HEXENAL
505-60-2 see BIS(2-CHLOROETHYL)SULFIDE
505-75-9 see CICUTOXIN
506-30-9 see EICOSANOIC ACID
506-32-1 see ARACHIDONIC ACID
506-64-9 see SILVER CYANIDE
506-68-3 see CYANOGEN BROMIDE
506-77-4 see CYANOGEN CHLORIDE
506-78-5 see CYANOGEN IODIDE
506-87-6 see AMMONIUM CARBONATE
506-96-7 see ACETYL BROMIDE
507-02-8 see ACETYL IODIDE
507-20-0 see tert-BUTYL CHLORIDE
507-35-2 see 8-NITROQUINOLINE
507-70-0 see BORNEOL
509-14-8 see TETRANITROMETHANE
509-15-9 see GELSEMINE
509-86-4 see CYCLOHEPTENYL ETHYLBARBITURIC ACID
510-15-6 see ETHYL-4,4'-DICHLOROBENZILATE
511-12-6 see DIHYDROERGOTAMINE
511-55-7 see 8-(p-PHENYLBENZYL)ATROPINIUM BROMIDE
512-24-3 see CERIUM CITRATE
512-48-1 see 2,2-DIETHYL-4-PENTENAMIDE
512-56-1 see TRIMETHYL PHOSPHATE
512-85-6 see ASCARIDOLE
513-35-9 see α,η-AMYLENE

513-77-9 see BARIUM CARBONATE (1:1)
513-85-9 see 2,3-BUTANEDIOL
513-86-0 see ACETOIN
514-65-8 see BIPERIDEN
514-73-8 see DITHIAZANINE IODIDE
517-09-9 see EQUILENIN
517-16-8 see ETHYLMERCURY-p-TOLUENE SULFONAMIDE
517-25-9 see TRINITROMETHANE
517-85-1 see ANG-STERANTHRENE
518-47-8 see FLUORESCEIN SODIUM
520-36-5 see CHAMOMILE
520-45-6 see METHYLACETOPYRONONE
520-52-5 see PSILOCYBIN
520-53-6 see 4-HYDROXY-N,N-DIMETHYLTRYPTAMINE
522-00-9 see PHENOPROPAZINE
522-23-6 see METHOPHENAZINE DIFUMARATE
523-87-5 see DRAMAMINE
525-02-0 see 1-BENZYL-2,5-DIMETHYL SEROTONIN HYDROCHLORIDE
525-66-6 see INDERAL
528-29-0 see o-DINITROBENZENE
528-74-5 see 3'5'-DICHLOROMETHOTREXATE
529-65-7 see N-ETHYLACETANILIDE
531-18-0 see HEXA(HYDROXYMETHYL)MELAMINE
531-73-7 see PROFLAVINE DIHYDROCHLORIDE
531-76-0 see DL-PHENYLALANINE MUSTARD
531-82-8 see 2-ACETAMIDO-4-(5-NITRO-2-FURYL)THIAZOLE
531-85-1 see BENZIDINE HYDROCHLORIDE
532-27-4 see 2-CHLOROACETOPHENONE
532-28-5 see MANDELIC ACID NITRILE
532-34-3 see n-BUTYL MESITYL OXIDE OXALATE
532-43-4 see THIAMINE NITRATE
532-49-0 see DI-n-BUTYL-CARBAMYLCHOLINE SULPHATE
532-62-7 see p-AMINOBENZOYLDIMETHYLAMINO-1,2-DIMETHYL-
    PROPANOL HYDROCHLORIDE
532-82-1 see 4-PHENYLAZO-m-PHENYLENEDIAMINE
533-28-8 see ISOCAINE
533-73-3 see 1,2,4-BENZENETRIOL
533-74-4 see DIMETHYLFORMOCARBOTHIALDINE
534-17-8 see CESIUM CARBONATE
534-22-5 see 2-METHYLFURAN
534-52-1 see 4,6-DINITRO-o-CRESOL
535-89-7 see CASTRIX
536-17-4 see p-DIMETHYLAMINOBENZALRHODANINE
536-21-0 see alpha-(AMINOMETHYL)-m-HYDROXYBENZYL ALCOHOL
536-25-4 see 3-AMINO-4-HYDROXYBENZOIC ACID METHYL ESTER
536-29-8 see DICHLOROPHENARSINE HYDROCHLORIDE
536-33-4 see 2-ETHYL THIO ISONICOTINAMIDE
536-43-6 see 4'-BUTOXY-3-PIPERIDINO PROPIOPHENONE HYDRO-
    CHLORIDE
536-69-6 see 5-BUTYL PICOLINIC ACID
537-00-8 see CERIUM ACETATE
537-05-3 see PHENODIANISYL HYDROCHLORIDE
538-03-4 see ARSPHENOXIDE
538-04-5 see 2-CHLORO-4-(HYDROXY MERCURI)PHENOL
538-07-8 see BIS(2-CHLOROETHYL)ETHYLAMINE
538-28-3 see BENZYL ISOTHIOUREA HYDROCHLORIDE
538-71-6 see PHENODODECINIUM BROMIDE
540-23-8 see p-TOLUIDINE HYDROCHLORIDE
540-51-2 see 2-BROMO ETHANOL
540-54-5 see 1-CHLOROPROPANE
540-59-0 see ACETYLENE DICHLORIDE
540-63-6 see 1,2-ETHANEDITHIOL
540-72-7 see SODIUM ISOTHIOCYANATE
540-73-8 see 1,2-DIMETHYLHYDRAZINE HYDROCHLORIDE
540-73-8 see 1,2-DIMETHYL HYDRAZINE
540-84-1 see 2,2,4-TRIMETHYLPENTANE
540-88-5 see ACETIC ACID-tert-BUTYL ESTER
541-19-5 see BIS(beta-DIMETHYL AMINO ETHYL)SUCCINATE BIS-
    (METHYL IODIDE)
541-25-3 see CHLOROVINYLARSINE DICHLORIDE

541-41-3 see ETHYL CHLORO FORMATE
541-53-7 see DITHIOBIURET
541-59-3 see MALEIMIDE
541-66-2 see FORMAL-gamma-TRIMETHYL AMMONIUM PROPANEDIOL
541-69-5 see m-PHENYLENEDIAMINE HYDROCHLORIDE
541-85-5 see AMYL ETHYL KETONE
542-56-3 see ISOBUTYL NITRITE
542-62-1 see BARIUM CYANIDE
542-69-8 see n-BUTYL IODIDE
542-75-6 see alpha-CHLOROALLYL CHLORIDE
542-76-7 see 3-CHLOROPROPIONITRILE
542-78-9 see PROPANEDIAL
542-88-1 see BIS(CHLOROMETHYL) ETHER
542-90-5 see ETHYL THIOCYANATE
542-92-7 see 1,3-CYCLOPENTADIENE
543-21-5 see ACETYLENEDICARBOXAMIDE
543-49-7 see 2-HEPTANOL
543-63-5 see n-BUTYLMERCURIC CHLORIDE
543-80-6 see BARIUM ACETATE
543-81-7 see BERYLLIUM ACETATE
543-82-8 see 2-ISOOCTYL AMINE
543-90-8 see CADMIUM (II) ACETATE
544-13-8 see 1,3-TRIMETHYLENEDINITRILE
544-16-1 see n-BUTYL NITRITE
544-17-2 see CALCIUM FORMATE
544-63-8 see MYRISTIC ACID
545-55-1 see TEPA
545-55-1 see TRIETHYLENEPHOSPHOROTRIAMIDE
546-06-5 see NERIINE
546-80-5 see THUJONE
548-00-5 see BIS(4-HYDROXY-3-COUMARIN) ACETIC ACID ETHYL
    ESTER
548-43-6 see ELYMOCLAVINE
548-61-8 see TRIAMINOTRIPHENYLMETHANE
548-62-9 see ANILINE VIOLET
548-73-2 see DROPERIDOL
548-93-6 see 2-AMINO-3-HYDROXYBENZOIC ACID
549-18-8 see ELAVIL HYDROCHLORIDE
551-11-1 see PROSTAGLANDIN F2-ALPHA
551-74-6 see MANNOMUSTINE DIHYDROCHLORIDE
551-92-8 see 1,2-DIMETHYL-5-NITROIMIDAZOLE
552-30-7 see TRIMELLIC ACID ANHYDRIDE
553-30-0 see 3,6-DIAMINOACRIDINE SULPHATE (1:1)
553-68-4 see IBYLCAINE HYDROCHLORIDE
553-97-9 see 2-METHYL-p-BENZOQUINONE
554-12-1 see METHYL PROPIONATE
554-13-2 see LITHIUM CARBONATE (2:1)
554-84-7 see 3-NITROPHENOL
554-99-4 see N-METHYLEPINEPHRINE
555-15-7 see 5-NITRO-2-FURALDEHYDE OXIME
555-30-6 see ALDOMET
555-84-0 see NIFURADENE
556-52-5 see 2,3-EPOXY-1-PROPANOL
556-61-6 see ISOTHIOCYANATOMETHANE
556-64-9 see METHYL THIOCYANATE
556-88-7 see alpha-NITROGUANIDINE
557-05-1 see ZINC STEARATE
557-11-9 see ALLYLUREA
557-19-7 see NICKEL CYANIDE, (SOLID)
557-20-0 see DIETHYLZINC
557-21-1 see ZINC CYANIDE
557-40-4 see DIALLYL ETHER
557-98-2 see 2-CHLORO-1-PROPENE
557-99-3 see ACETYL FLUORIDE
558-13-4 see CARBON TETRABROMIDE
561-27-3 see HEROIN
562-95-8 see TRIETHYLLEAD FLUOROACETATE
563-12-2 see ETHYL METHYLENE PHOSPHORO DITHIOATE
563-41-7 see SEMICARBAZIDE HYDROCHLORIDE
563-47-3 see 3-CHLORO-2-METHYLPROPENE

563-68-8 see THALLIUM ACETATE
563-80-4 see METHYL ISOPROPYL KETONE
564-00-1 see meso-1,2,3,4-DIEPOXYBUTANE
568-70-7 see 12-HYDROXYMETHYL-7-METHYLBENZ(a)ANTHRACENE
568-75-2 see 7-HYDROXYMETHYL-12-METHYLBENZ(a)ANTHRACENE
569-57-3 see CHLOROTRIANISENE
569-58-4 see ALUMINON
569-61-9 see BASIC PARAFUCHSINE
569-64-2 see AIZEN MALACHITE GREEN
569-65-3 see HISTAMETHIZINE
572-48-5 see DITHION
573-56-8 see 2,6-DINITROPHENOL
576-68-1 see MANNOMUSTINE
577-11-7 see DIOCTYL SODIUM SULFOSUCCINATE
577-71-9 see 3,4-DINITROPHENOL
578-54-1 see 2-ETHYL ANILINE
578-94-9 see PHENARSAZINE CHLORIDE
581-89-5 see 2-NITRONAPHTHALENE
583-15-3 see MERCURY (II) BENZOATE
583-39-1 see 2-BENZIMIDAZOLETHIOL
583-58-4 see 3,4-LUTIDINE
583-60-8 see 2-METHYLCYCLOHEXANONE
584-02-1 see 3-PENTANOL
584-03-2 see 1,2-BUTANEDIOL
584-08-7 see POTASSIUM CARBONATE (2:1)
584-26-9 see 1-ACETYL-2-THIOHYDANTOIN
584-79-2 see BIOALLETHRIN
584-84-0 see TOLUENE DIISOCYANATE
584-84-9 see DI-ISO-CYANATOLUENE
586-11-8 see 3,5-DINITROPHENOL
586-38-9 see m-ANISIC ACID
587-85-9 see DIPHENYLMERCURY
588-59-0 see STILBENE
589-16-2 see 4-ETHYL ANILINE
590-01-2 see BUTYL PROPANOATE
590-21-6 see PROPENYL CHLORIDE
590-28-3 see POTASSIUM CYANATE
590-88-5 see 1,3-BUTANE DIAMINE
590-92-1 see 3-BROMOPROPIONIC ACID
590-96-5 see 1-HYDROXYMETHYL-2-METHYLDIAMIDE-2-OXIDE
591-08-2 see ACETYL THIOUREA
591-09-3 see ACETYL NITRATE
591-27-5 see m-AMINOPHENOL
591-60-6 see BUTYL ACETOACETATE
591-78-6 see 2-HEXANONE
591-87-7 see ALLYL ACETATE
592-01-8 see CALCIUM CYANIDE
592-01-8 see CALCIUM CYANIDE (mixture)
592-04-1 see MERCURY (II) CYANIDE
592-05-2 see LEAD (II) CYANIDE
592-31-4 see N-BUTYLUREA
592-35-8 see BUTYL CARBAMATE
592-62-1 see METHYL AZOXYMETHYL ACETATE
592-79-0 see 5-FLUORO AMYLAMINE
592-84-7 see N-BUTYL FORMATE
592-85-8 see MERCURIC SULFOCYANATE
592-88-1 see ALLYL SULFIDE
593-60-2 see VINYL BROMIDE
593-70-4 see MONOCHLOROMONOFLUOROMETHANE
593-82-8 see 1,1-DIMETHYLHYDRAZINE HYDROCHLORIDE
593-89-5 see DICHLOROMETHYLARSINE
593-91-9 see TRIMETHYL BISMUTH
594-31-0 see DICHLOROTRIPHENYLANTIMONY
594-71-8 see 2-CHLORO-2-NITROPROPANE
594-72-9 see 1,1-DICHLORO-1-NITROETHANE
595-33-5 see VOLIDAN
596-51-0 see GLYCOPYRRONIUM BROMIDE
597-88-6 see O,S-DIETHYL-O-(4-NITROPHENYL)THIOPHOSPHATE
598-14-1 see DICHLOROETHYLARSINE
598-31-2 see BROMO-2-PROPANONE

598-55-0 see METHYL CARBAMATE
598-58-3 see METHYL NITRATE
598-63-0 see LEAD CARBONATE
598-78-7 see alpha-CHLOROPROPIONIC ACID
599-52-0 see HEXACHLORO-2,5-CYCLOHEXADIEN-1-ONE
600-25-9 see 1-CHLORO-1-NITROPROPANE
601-54-7 see CHOLEST-5-EN-3-ONE
602-87-9 see 5-NITROACENAPHTHENE
602-99-3 see 2,4,6-TRINITRO-meta-CRESOL
603-35-0 see TRIPHENYLPHOSPHINE
604-75-1 see 7-CHLORO-1,3-DIHYDRO-3-HYDROXY-5-PHENYL-2H-1,4-BENZODIAZEPINE-2-ONE
605-65-2 see 5-(DIMETHYLAMINO)-1-NAPHTHALENESULFONYL CHLORIDE
605-69-6 see 2,4-DINITRO-1-NAPHTHOL
606-23-5 see 1,3-INDANDIONE
607-57-8 see 2-NITROFLUORENE
607-59-0 see N,N-DIMETHYL-p-(1-NAPHTHYLAZO)ANILINE
608-73-1 see BENZENE HEXACHLORIDE
608-73-1 see BENZENE HEXACHLORIDE (MIXED ISOMERS)
609-89-2 see 2,4-DICHLORO-6-NITROPHENOL
609-93-8 see 2,6-DINITRO-p-CRESOL
612-12-4 see alpha,alpha'-DICHLORO-o-XYLENE
613-13-8 see 2-ANTHRACENAMINE
613-35-4 see 4',4'''-BIACETANILIDE
613-37-6 see p-PHENYLANISOLE
613-47-8 see 2-NAPHTHYLHYDROXYLAMINE
613-89-8 see 2-AMINOACETOPHENONE
613-94-5 see BENZHYDRAZIDE
614-00-6 see N-METHYL-N-NITROSOANILINE
614-45-9 see tert-BUTYL PERBENZOATE
614-95-9 see N-NITROSO-N-ETHYLURETHAN
615-15-6 see METHYL-2-BENZIMIDAZOLE
615-45-2 see 2,5-DIAMINOTOLUENE DIHYDROCHLORIDE
615-50-9 see 2,5-DIAMINOTOLUENE SULFATE
615-53-2 see N-NITROSO-N-METHYLURETHANE
615-67-8 see CHLORO HYDRO QUINONE
616-23-9 see 2,3-DICHLOROPROPANOL
616-91-1 see N-ACETYL-L-CYSTEINE
617-79-8 see 2-ETHYL BUTYLAMINE
618-25-7 see N-(CARBAMOYL METHYL)ARSANILIC ACID
621-64-7 see DI-n-PROPYLNITROSAMINE
621-90-9 see N-METHYL-p-(PHENYLAZO)ANILINE
622-45-7 see CYCLOHEXYL ACETATE
622-51-5 see p-TOLYLUREA
622-78-6 see BENZYL ISOTHIOCYANATE
623-26-7 see p-BENZENEDINITRILE
623-73-4 see DIAZOACETIC ESTER
624-46-4 see METHYL ETHYL KETONE SEMICARBAZONE
624-83-9 see METHYL ISOCYANATE
624-91-9 see METHYL NITRITE
625-17-2 see DI-sec-BUTYL FLUOROPHOSPHONATE
625-22-9 see DIBUTYL ESTER SULFURIC ACID
625-55-8 see ISOPROPYL FORMATE
625-58-1 see ETHYL NITRATE
626-17-5 see 1,3-BENZENEDICARBONITRILE
627-12-3 see PROPYL CARBAMATE
627-13-4 see N-PROPYL NITRATE
627-44-1 see DIETHYL MERCURY
627-63-4 see FUMARYL CHLORIDE
627-72-5 see S-DICHLOROVINYL-L-CYSTEINE
628-02-4 see HEXANAMIDE
628-63-7 see PENTYL ACETATE
628-73-9 see HEXANENITRILE
628-81-9 see ETHYL BUTYL ETHER
628-83-1 see n-BUTYL THIOCYANATE
628-85-3 see DIPROPYL MERCURY
628-86-4 see MERCURY FULMINATE (DRY)
628-86-4 see MERCURY FULMINATE (WET)
628-94-4 see ADIPAMIDE

924-16-3 see N-BUTYL-N-NITROSO-1-BUTAMINE
924-43-6 see sec-BUTYL NITRITE
924-46-9 see NITROSOMETHYLPROPYLAMINE
926-93-2 see 1-METHYL-6-(1-METHYLALLYL)-2,5-DITHIOBIUREA
927-07-1 see tert-BUTYL PEROXY PIVALATE
928-04-1 see ACETYLENEDICARBOXYLIC ACID MONOPOTASSIUM
     SALT
928-65-4 see HEXYLTRICHLOROSILANE
929-06-6 see 2-AMINOETHOXYETHANOL
930-22-3 see 3,4-EPOXY-1-BUTENE
930-55-2 see N-NITROSOPYRROLIDINE
930-69-7 see 2-CYCLOHEXEN-1-ONE
932-83-2 see N-NITROSOHEXAHYDROAZEPINE
934-32-7 see 2-AMINOBENZIMIDAZOLE
937-40-6 see N-METHYL-N-BENZYLNITROSAMINE
943-39-5 see p-NITROPEROXYBENZOIC ACID
944-22-9 see DYPHONATE
947-02-4 see PHOSFOLAN
950-10-7 see MEPHOSFOLAN
950-37-8 see o,o-DIMETHYL-S-(5-METHOXY-1,3,4-THIADIAZOLINYL-
     3-METHYL)DITHIO PHOSPHATE
952-23-8 see PROFLAVINE MONOHYDROCHLORIDE
953-17-3 see METHYL TRITHION
955-50-1 see o-DICHLOROBENZENE
956-90-1 see ANGEL DUST
957-51-7 see N,N-DIMETHYL-2,2-DIPHENYLACETAMIDE
959-14-8 see OXOLAMINE CITRATE
959-24-0 see beta-CARDONE
963-07-5 see ALYPIN
963-89-3 see 7,9-DIMETHYLBENZ(c)ACRIDINE
968-58-1 see PRIDINOL
989-38-8 see RHODAMINE 6G EXTRA BASE
990-73-8 see PHENTANYL CITRATE
992-21-2 see MUCOMYCIN
996-08-7 see DIBROMODIBUTYLSTANNANE
997-95-5 see 2,2'-DIMETHYLDIPROPYLINITROSOAMINE
999-21-3 see DIALLYL MALEATE
999-61-1 see ACRYLIC ACID-2-HYDROXYPROPYL ESTER
1001-55-4 see CYANOMETHYL ACETATE
1002-16-0 see AMYL NITRATE
1003-78-7 see 2,4-DIMETHYL SULFOLANE
1024-57-3 see EPOXYHEPTACHLOR
1027-14-1 see TRIMECAINE
1031-47-6 see 5-AMINO-1-BIS(DIMETHYLAMIDE)PHOSPHORYL-3-
     PHENYL-1,2,4-TRIAZOLE
1066-30-4 see CHROMIC ACETATE
1066-33-7 see AMMONIUM BICARBONATE (1:1)
1066-45-1 see CHLOROTRIMETHYLSTANNANE
1067-33-0 see DIACETOXY DIBUTYL STANNANE
1068-57-1 see ACETYL HYDRAZIDE
1071-39-2 see DIISOPROPYLMERCURY
1072-52-2 see 1-AZIRIDINE ETHANOL
1072-53-3 see ETHYLENE SULFATE
1074-98-2 see 3-METHYL-4-NITROPYRIDINE-1-OXIDE
1078-79-1 see IMIPHOS
1082-88-8 see alpha-METHYLMESCALINE
1095-90-5 see METHADONE HYDROCHLORIDE
1102-47-2 see CHLORPHENIRAMINE MALEATE
1111-39-3 see ACETYL-DIGITOXIN-alpha
1111-78-0 see AMMONIUM CARBAMATE
1113-02-6 see DIMETHOATE OXYGEN ANALOG
1113-14-0 see trans-1,2-BIS(n-PROPYL SULFONYL)ETHYLENE
1114-71-2 see S-PROPYL BUTYLETHYLTHIOCARBAMATE
1118-14-5 see TRIMETHYLTIN ACETATE
1118-39-4 see ACETIC ACID MYRCENYL ESTER
1118-42-9 see DIPENTYLTIN DICHLORIDE
1118-46-3 see BUTYL TRICHLORO STANNANE
1120-21-4 see UNDECANE
1120-71-4 see 1,2-OXATHIOLANE-2,2-DIOXIDE
1123-61-1 see DICHLORO-N-METHYLMALEIMIDE

1124-33-0 see 4-NITROPYRIDINE-N-OXIDE
1125-27-5 see DICHLOROETHYLPHENYLSILANE
1126-79-0 see BUTYL PHENYL ETHER
1129-41-5 see TSUMACIDE
1137-41-3 see p-AMINOBENZOPHENONE
1141-88-4 see 2,2'-DITHIOBISANILINE
1142-70-7 see BUTALLYLONAL
1145-73-9 see N,N-DIMETHYL-4-STILBENAMINE
1155-38-0 see 7-METHYLBENZ(a)ANTHRACENE-5,6-OXIDE
1162-65-8 see AFLATOXIN B1
1165-39-5 see AFLATOXIN G1
1165-48-6 see DIMEFLINE
1169-26-2 see 1-(2-HEXAMETHYLENEIMINOETHYL)-2-OXOCYCLO-
     HEXANECARBOXYLIC ACID BENZYL ESTER HYDROCHLORIDE
1172-18-5 see DALMANE
1184-57-2 see METHYLMERCURY HYDROXIDE
1187-00-4 see BIS(METHANE SULFONYL)-D-MANNITOL
1189-85-1 see tert-BUTYL CHROMATE
1190-53-0 see N-BUTYLBIGUANIDE HYDROCHLORIDE
1193-54-0 see DICHLOROMALEIMIDE
1198-27-2 see 1-AMINO-2-NAPHTHOL HYDROCHLORIDE
1199-85-5 see p-CHLORO-N-METHYLAMPHETAMINE
1204-06-4 see INDOLE-3-ACRYLIC ACID
1215-16-3 see 4'-(BIS(2-CHLOROETHYL)AMINO)ACETANILIDE
1224-64-2 see p-NITROPHENYLDI-n-BUTYLPHOSPHINATE
1225-55-4 see PROTRIPTYLINE HYDROCHLORIDE
1227-61-8 see MEPHEXAMIDE
1229-29-4 see ADAPIN
1239-45-8 see 2,7-DIAMINO-10-ETHYL-9-PHENYLPHENANTHRIDINIUM
     BROMIDE
1250-95-9 see EPOXYCHOLESTEROL
1264-51-3 see ACETYL-DIGITOXIN-beta
1271-19-8 see DICHLOROTITANOCENE
1300-14-7 see ANILINE ANTIMONYL TARTRATE
1300-73-8 see XYLIDINE
1302-52-9 see BERYL
1302-74-5 see CORUNDUM
1302-76-7 see ALUMINUM(III) SILICATE (2:1)
1302-78-9 see BENTONITE
1303-18-0 see ARSENOPYRITE
1303-28-2 see ARSENIC PENTOXIDE
1303-33-9 see ARSENIC SULFIDE
1303-39-5 see ZINC ARSENATE
1303-86-2 see BORON OXIDE
1304-28-5 see BARIUM OXIDE
1304-29-6 see BARIUM PEROXIDE
1304-56-9 see BERYLLIUM OXIDE
1305-62-0 see CALCIUM HYDROXIDE
1305-78-8 see CALCIUM OXIDE
1305-79-9 see CALCIUM PEROXIDE
1305-99-3 see CALCIUM PHOSPHIDE
1306-19-0 see CADMIUM OXIDE
1306-19-0 see CADMIUM OXIDE FUME
1306-23-6 see CADMIUM SULFIDE
1306-38-3 see CERIC OXIDE
1308-31-2 see CHROMITE (MINERAL)
1309-32-6 see CRYPTOHALITE
1309-32-6 see AMMONIUM SILICO FLUORIDE
1309-37-1 see IRON (III) OXIDE
1309-48-4 see MAGNESIUM OXIDE
1309-60-0 see LEAD DIOXIDE
1309-60-0 see LEAD(IV)OXIDE BROWN
1310-53-8 see GERMANIC OXIDE (CRYSTALLINE)
1310-58-3 see POTASSIUM HYDROXIDE
1310-73-2 see SODIUM HYDROXIDE (LIQUID)
1310-73-2 see SODIUM HYDROXIDE
1312-73-8 see POTASSIUM SULFIDE (2:1)
1313-13-9 see MANGANESE DIOXIDE
1313-27-5 see MOLYBDENUM TRIOXIDE
1313-60-6 see SODIUM PEROXIDE

1313-82-2 see SODIUM SULFIDE (ANHYDROUS)
1313-97-9 see NEODYMIUM OXIDE
1313-99-1 see NICKEL MONOXIDE
1314-06-3 see NICKEL PEROXIDE
1314-13-2 see ZINC OXIDE
1314-20-1 see THORIUM OXIDE
1314-22-3 see ZINC PEROXIDE
1314-32-5 see THALLIC OXIDE
1314-34-7 see VANADIUM SESQUIOXIDE
1314-36-9 see YTTRIUM OXIDE
1314-41-6 see LEAD OXIDE RED
1314-56-3 see PHOSPHORUS PENTOXIDE
1314-60-9 see ANTIMONY PENTOXIDE
1314-62-1 see VANADIUM PENTOXIDE (DUST)
1314-80-3 see PHOSPHORUS PENTASULFIDE
1314-84-7 see ZINC PHOSPHIDE
1314-85-8 see PHOSPHORUS SESQUISULFIDE
1314-87-0 see LEAD SULFIDE
1314-96-1 see STRONTIUM SULFIDE
1315-04-4 see ANTIMONY PENTASULFIDE
1317-34-6 see MANGANESE (III) OXIDE
1317-35-7 see MANGANESE OXIDE
1317-36-8 see LEAD MONOXIDE
1317-39-1 see COPPER (I) OXIDE
1317-60-8 see HEMATITE
1317-65-3 see CALCIUM (II) CARBONATE (1:1)
1319-77-3 see CRESOL
1320-37-2 see DICHLOROTETRAFLUOROETHANE
1321-64-8 see PENTACHLORONAPHTHALENE
1321-67-1 see NAPHTHOL
1322-78-7 see PHENYL XYLYL KETONE
1327-33-9 see ANTIMONY OXIDE
1327-53-3 see ARSENIC TRIOXIDE
1328-90-1 see DIHYDRO-beta-ERYTHROIDINE HYDROCHLORIDE
1330-20-7 see XYLENE
1330-78-5 see TRITOLYL PHOSPHATE
1331-11-9 see 3-ETHOXY PROPIONIC ACID
1331-22-2 see METHYLCYCLOHEXANONE
1331-31-3 see ETHYL CHLORO BENZENE
1332-21-4 see ASBESTOS
1332-94-1 see LAETRILE
1333-74-0 see HYDROGEN
1333-82-0 see CHROMIUM (VI) OXIDE (1:3)
1335-32-6 see LEAD ACETATE, BASIC
1335-87-1 see HEXACHLORONAPHTHALENE
1335-88-2 see TETRACHLORONAPHTHALENE
1336-21-6 see AMMONIUM HYDROXIDE
1336-36-3 see POLYCHLORINATED BIPHENYLS
1338-02-9 see NAPHTHENIC ACID, COPPER SALT
1338-16-5 see IRON SORBITOL CITRATE
1338-23-4 see METHYL ETHYL KETONE PEROXIDE
1341-49-7 see AMMONIUM HYDROGEN FLUORIDE (SOLUTION)
1341-49-7 see AMMONIUM HYDROGEN FLUORIDE
1344-28-1 see ALUMINUM OXIDE (2:3)
1344-43-0 see MANGANESE (II) OXIDE
1344-67-8 see COPPER (II) CHLORIDE (1:2)
1345-04-6 see ANTIMONY TRISULFIDE
1395-21-7 see BACILLUS SUBTILIS BPN
1397-89-3 see AMPHOTERICIN B
1398-06-7 see ARISTOLOCHINE
1399-80-0 see DODECYL-p-TOLYL TRIMETHYL AMMONIUM CHLORIDE
1400-05-1 see JAPACONITINE B
1400-61-9 see NYSTATIN
1401-55-4 see TANNIC ACID
1401-55-4 see TANNIN
1402-68-2 see AFLATOXIN
1403-17-4 see LEVORIN
1403-17-4 see CANDICIDIN
1403-66-3 see GENTAMICIN
1404-04-2 see NEOMYCIN

1404-24-6 see POLYMYXIN A
1405-10-3 see NEOMYCIN SULFATE
1405-41-0 see GENTAMYCIN SULFATE
1405-87-4 see BACITRACIN
1405-97-6 see GRAMICIDIN
1406-05-9 see PENICILLIN
1406-65-1 see CHLOROPHYLL
1420-04-8 see 2',5-DICHLORO-4'-NITROSALICYLANILIDE, 2-AMINO-ETHANOL SALT
1420-53-7 see CODEINE SULFATE
1421-28-9 see DIHYDROMORPHINE HYDROCHLORIDE
1422-07-7 see CODEINE HYDROCHLORIDE
1424-27-7 see ACETAZOLAMIDE SODIUM
1432-75-3 see NITRALAMINE HYDROCHLORIDE
1435-55-8 see HYDROQUINIDINE
1455-77-2 see 3,5-DIAMINO-s-TRIAZOLE
1456-28-6 see 2,6-DIMETHYLNITROSOMORPHOLINE
1461-22-9 see CHLORO TRIBUTYL STANNANE
1461-25-2 see TETRABUTYLSTANNANE
1464-43-3 see 3,3'-DISELENODIALANINE
1464-53-5 see 1,1'-BI(ETHYLENE OXIDE)
1465-26-5 see 3-(p-(BIS(beta-CHLOROETHYL)AMINO)PHENYL)-D,L-ALANINE HYDROCHLORIDE
1467-79-4 see DIMETHYLCYANAMIDE
1491-41-4 see N-HYDROXYNAPHTHALIMIDE, DIETHYL PHOSPHATE
1492-93-9 see 4'-(BIS(2-CHLOROETHYL)AMINO)-2-FLUORO ACETANILIDE
1498-51-7 see ETHYL PHOSPHORODICHLORIDATE
1515-76-0 see 1-ACETOXY-1,3-BUTADIENE
1516-32-1 see n-BUTYL THIOUREA
1528-74-1 see 4,4'-DINITROBIPHENYL
1532-19-0 see CLEP
1544-46-3 see FLUOROACETALDEHYDE
1563-66-2 see FURADAN
1563-67-3 see 2,3-DIHYDRO-2-METHYLBENZOPYRANYL-7,N-METHYL-CARBAMATE
1566-15-0 see PHOSPHORAMIDE MUSTARD
1570-45-2 see ETHYL ISONICOTINATE
1582-09-8 see TRIFLURALINE
1596-52-7 see 4,6-DINITROQUINOLINE-1-OXIDE
1596-84-5 see DIMETHYLAMINOSUCCINAMIC ACID
1600-27-7 see MERCURIC ACETATE
1615-80-1 see 1,2-DIETHYLHYDRAZINE
1622-61-3 see CLOAZEPAM
1622-79-3 see 4'-FLUORO-4-(4-METHYL PIPERIDINO)BUTYRO PHENONE HYDROCHLORIDE
1624-02-8 see BIS(TRIPHENYL SILYL)CHROMATE
1632-16-2 see 2-ETHYL-1-HEXENE
1633-83-6 see BUTANESULTONE
1639-09-4 see 1-HEPTANETHIOL
1639-60-7 see d-PROPOXYPHENE HYDROCHLORIDE
1642-54-2 see DIETHYLCARBAMAZINE ACID CITRATE
1649-18-9 see 4'-FLUORO-4-(4-(2-PYRIDYL)-1-PIPERAZINYL)BUTYRO-PHENONE
1653-64-1 see 3,4-METHYLENEDIOXY-beta-PHENYLETHYLAMINE HYDROCHLORIDE
1668-19-5 see DEXEPIN
1675-54-3 see BISPHENOL A DIGLYCIDYL ETHER
1689-82-3 see 4-HYDROXYAZOBENZENE
1689-83-4 see 4-HYDROXY-3,5-DIIODOBENZONITRILE
1689-84-5 see 3,5-DIBROMO-4-HYDROXYBENZONITRILE
1689-89-0 see 4-HYDROXY-3-IODO-5-NITROBENZONITRILE
1689-99-2 see 2,6-DIBROMO-4-CYANOPHENYL OCTANOATE
1698-60-8 see 1-PHENYL-4-AMINO-5-CHLORPYRIDAZ-6-ONE
1705-85-7 see 6-METHYLCHRYSENE
1707-14-8 see PHENMETRAZINE HYDROCHLORIDE
1707-95-5 see 2-(3-OXO-1-INDANYLIDENE)-1,3-INDANDIONE
1709-50-8 see N,N-DIETHYLBENZENESULFONAMIDE
1712-64-7 see ISOPROPYL NITRATE
1719-53-5 see DICHLORODIETHYLSILANE

1738-25-6 see 3-(DIMETHYLAMINO)PROPIONITRILE

1745-81-9 see o-ALLYL PHENOL

1746-01-6 see TCDD

1746-13-0 see ALLYL PHENYL ETHER

1746-77-6 see ISOPROPYL CARBAMATE

1746-81-2 see 3-(4-CHLOROPHENYL)-1-METHOXY-1-METHYLUREA

1757-18-2 see O,O-DIETHYL-O-(2-CHLORO-1,2,5-DICHLOROPHENYLVINYL)PHOSPHOROTHIOATE

1762-95-4 see AMMONIUM THIOCYANATE

1777-84-0 see 3'-NITRO-p-ACETOPHENETIDE

1779-25-5 see CHLORO DIISOBUTYL ALUMINUM

1785-74-6 see DIBENZOSUBERONE OXIME

1789-58-8 see DICHLOROETHYLSILANE

1808-12-4 see BROMO PHENYL HYDRAMINE HYDROCHLORIDE

1824-81-3 see 2-AMINO-6-METHYLPYRIDINE

1836-75-5 see 2,4-DICHLORO-4'-NITRODIPHENYL ETHER

1838-59-1 see ALLYL FORMATE

1851-77-0 see DI(ETHYLXANTHOGEN)TRISULFIDE

1867-66-9 see 2-(o-CHLOROPHENYL)-2-(METHYLAMINO)CYCLOHEXA-NONE HYDROCHLORIDE

1875-92-9 see DIMETHYLBENZYLAMINE HYDROCHLORIDE

1881-76-1 see 7-FLUORO-10-METHYL-1,2-BENZANTHRACENE

1885-29-6 see ANTHRANILONITRILE

1888-71-7 see HEXACHLOROPROPENE

1888-89-7 see 1,2:5,6-DIEPOXYHEXANE

1892-29-1 see DITHIODIGLYCOL

1893-33-0 see FLUOROBUTYROPHENONE

1897-45-6 see TETRACHLOROISOPHTHALONITRILE

1907-13-7 see ACETOXYTRIETHYLSTANNANE

1910-42-5 see PARAQUAT

1918-02-1 see 4-AMINO-3,5,6-TRICHLOROPICOLINIC ACID

1918-13-4 see 2,6-DICHLOROTHIOBENZAMIDE

1918-16-7 see 2-CHLORO-N-ISOPROPYLACETANILIDE

1929-82-4 see 2-CHLORO-6-(TRICHLOROMETHYL)PYRIDINE

1936-15-8 see HISPACID FAST ORANGE 2G

1937-37-7 see APOMINE BLACK GX

1943-16-4 see CHLOROTRINITROMETHANE

1951-25-3 see AMINODARONE

1955-45-9 see 3,3-DIMETHYL-2-OXETHANONE

1972-08-3 see 1-trans-delta(sup 9)-TETRAHYDROCANNABINOL

1977-10-2 see DIBENZACEPIN

1982-36-1 see 1-(p-CHLORO-alpha-PHENYLBENZYL)HEXAHYDRO-4-METHYL-1H-1,4-DIAZEPINEDIHYDROCHLORIDE

1982-37-2 see METHDILAZINE

2001-81-2 see (2-(DICYCLOPENTYLACETOXY)ETHYL)TRIETHYLAM-MONIUM BROMIDE

2016-57-1 see 1-DECANEAMINE

2023-61-2 see 7,12-DIMETHYL-11-FLUOROBENZ(a)ANTHRACENE

2032-59-9 see 4-DIMETHYLAMINE-m-CRESYLMETHYLCARBAMATE

2032-65-7 see 4-METHYLMERCAPTO-3,5-XYLYL METHYLCARBAMATE

2038-03-1 see N-AMINOETHYLMORPHOLINE

2050-92-2 see DIAMYL AMINE

2058-52-8 see 2-CHLORO-11-(4-METHYLPIPERAZINO)DIBENZO(b,f)(1,4)THIAZEPINE

2058-62-0 see N-METHYL-p-(m-TOLYLAZO)ANILINE

2058-66-4 see N-ETHYL-N-METHYL-p-(PHENYLAZO)ANILINE

2068-78-2 see LEUROCRISTINE SULFATE (1:1)

2079-89-2 see beta-AMINOPROPIONITRILE FUMARATE

2086-83-1 see BERBERINE

2092-16-2 see CALCIUM THIOCYANATE

2095-58-1 see ALLYL-sec-BUTYL THIOBARBITURIC ACID

2104-09-8 see 2-AMINO-4-(p-NITROPHENYL)THIAZOLE

2104-64-5 see ETHOXY-4-NITRO PHENOXY PHENYL PHOSPHINE SULFIDE

2114-11-6 see ALLYL CARBAMATE

2130-56-5 see 5,5'-BIANTHRANILIC ACID

2152-34-3 see 2-IMINO-5-PHENYL-4-OXAZOLIDINONE

2156-56-1 see SODIUM DICHLOROACETATE

2163-80-6 see MONOSODIUM METHYLARSONATE

2164-17-2 see 1,1-DIMETHYL-3-(alpha,alpha,alpha-TRIFLUORO-m-TOLYL) UREA

2179-59-1 see ALLYL PROPYL DISULFIDE

2180-92-9 see 1-BUTYL-2',6'-PIPECOLOXYLIDIDE

2188-67-2 see 2,n-PENTYLAMINOETHYL-p-AMINOBENZOATE

2191-10-8 see CADMIUM CAPRYLATE

2192-21-4 see DIALICOR

2206-89-5 see ACRYLIC ACID-beta-CHLOROETHYL ESTER

2207-85-4 see IMPIRAMINE-N-OXIDE

2216-51-5 see 1-MENTHOL

2219-30-9 see PENICILLAMINE HYDROCHLORIDE

2238-07-5 see DIGLYCIDYL ETHER

2243-62-1 see 1,5-NAPHTHALENEDIAMINE

2244-21-5 see POTASSIUM DICHLOROISOCYANURATE

2255-17-6 see 4-NITRO-M-CRESOL DIMETHYL PHOSPHATE

2266-22-0 see 4'-FLUORO-4-(n-(4-PYRROLIDINAMIDO-4-m-TOLY-PIPERIDINO)-BUTYROPHENONE

2270-40-8 see DIACETOXYSCIRPENOL

2270-40-8 see ANGUIDIN

2273-43-0 see BUTYL STANNOIC ACID

2273-45-2 see DIMETHYLOXOSTANNANE

2274-11-5 see ETHYLENE ACRYLATE

2275-15-1 see PHENCAPTON

2275-18-5 see ISOPROPYL DIETHYL DITHIOPHOSPHORYL ACETAMIDE

2275-23-2 see N-METHYL-O,O-DIMETHYLTHIOLOPHOSPHORYL-5-THIA-3-METHYL-2-VALERAMIDE

2279-64-3 see PHENYL MERCURY UREA

2279-76-7 see CHLOROTRIPROPYLSTANNANE

2303-16-4 see DIALLATE

2307-55-3 see 2,4-D AMMONIUM SALT

2310-17-0 see S-((3-BENZOXAZOLINYL-6-CHLORO-2-OXO)METHYL)O,O-DIETHYLPHOSPHORODITHIOATE

2312-76-7 see 3,5-DINITRO-o-CRESOL SODIUM SALT

2318-18-5 see SENKIRKINE

2319-96-2 see 5-METHYLBENZ(a)ANTHRACENE

2321-07-5 see FLUORESCEIN

2353-45-9 see FD AND C GREEN NO. 3

2367-25-1 see METHYL-gamma-FLUOROCROTONATE

2373-98-0 see 3,3'-DIHYDROXYBENZIDINE

2381-15-9 see 10-METHYLBENZ(a)ANTHRACENE

2381-16-0 see 9-METHYLBENZ(a)ANTHRACENE

2381-31-9 see 8-METHYLBENZ(a)ANTHRACENE

2381-39-7 see 5-METHYL-3,4-BENZPYRENE

2381-40-0 see 6,9-DIMETHYL-1,2-BENZACRIDINE

2385-85-5 see MIREX

2385-87-7 see LYSERGIC ACID PYROLIDATE

2386-90-5 see BIS(2,3-EPOXY CYCLOPENTYL) ETHER

2393-53-5 see ESTRONE BENZOATE

2401-85-6 see 2,4-DINITRO-1-CHLORO-NAPHTHALENE

2409-55-4 see 2-tert-BUTYL-p-CRESOL

2417-90-5 see 3-BROMO PROPIONITRILE

2422-79-9 see 12-METHYLBENZ(a)ANTHRACENE

2425-06-1 see DIFOLATAN

2425-25-4 see O,O-DIETHYL S-(CARBETHOXY)METHYL PHOSPHO-ROTHIOLATE

2425-66-3 see CHLORONITROPROPANE

2425-74-3 see tert-BUTYL FORMAMIDE

2426-07-5 see 1,2,7,8-DIEPOXYOCTANE

2426-08-6 see BUTYL GLYCIDYL ETHER

2432-99-7 see AMINOUNDECANOIC ACID

2435-76-9 see DIAZOURACIL

2438-72-4 see p-BUTOXY PHENYL ACETOHYDROXAMIC ACID

2438-88-2 see TETRACHLORONITROANISOLE

2439-10-3 see N-DODECYLGUANIDINE ACETATE

2440-45-1 see BIS(ETHYL MERCURI)PHOSPHATE

2443-39-2 see cis-9,10-EPOXYOCTADECANOIC ACID

2463-84-5 see p-NITRO-o-CHLOROPHENYL DIMETHYL THIONOPHOS-PHATE

2467-12-1 see TRI-n-OCTYL BORATE

2472-17-5 see ATROPINE SULFATE (1:1)

61925-70-0 see N-(4-tert-BUTYL CYCLOHEXYL)-3,3-DIPHENYL
    PROPYLAMINE HYDROCHLORIDE
61947-30-6 see DIISOBUTYLOXOSTANNANE
62019-57-8 see 1-DIETHYLACETYLAZIRIDINE
62207-76-5 see FLUOMINE DUST
63018-62-2 see BENZ(a)ANTHRACENE-7,12-DIMETHANOLDIACETATE
63018-99-5 see 5-n-AMYL-1:2-BENZANTHRACENE
63019-08-9 see DIETHYLSTILBESTROL DIPALMITATE
63019-09-0 see N,N,2'-TRIMETHYL-4-STILBENAMINE
63019-72-7 see 5-METHOXYDIBENZ(a,h)ANTHRACENE
63020-25-7 see 9-METHYL-10-CYANO-1,2-BENZANTHRACENE
63020-47-3 see 5-ISOPROPYL-1:2-BENZANTHRACENE
63020-91-7 see 2'-CHLORO-N,N-DIMETHYL-4-STILBENAMINE
63021-32-9 see BENZENECARBOXALDEHYDE
63021-51-2 see 1-NONANOYLAZIRIDINE
63039-90-7 see NU-1932
63040-64-2 see 5-((p-(DIMETHYLAMINO)PHENYL)AZO)ISOQUINOLINE
63041-72-5 see 7-METHOXYDIBENZ(a,h)ANTHRACENE
63041-77-0 see 4'-METHYLBENZO(a)PYRENE
63041-80-5 see 20-METHYLCHOLANTHRENE PICRATE
63041-84-9 see 3-METHYLDIBENZ(a,h)ANTHRACENE
63041-85-0 see 4-METHYL-1,2,5,6-DIBENZANTHRACENE
63428-83-1 see NYLON
63449-39-8 see CHLOROWAX 500C
63680-76-2 see DIMETHYLCARBAMIC ESTER of 8OXYMETHYLQUINO-
    LINIUM METHYLSULFATE
63681-05-0 see NEOANTERGAN PHOSPHATE
63710-43-0 see 9-((3-(ISOPROPYLAMINO)PROPYL)AMINO)-1-NITRO-
    ACRIDINE DIHYDROCHLORIDE
63732-23-0 see METHYL-gamma-FLUORO-beta-HYDROXYTHIOLBUTY-
    RATE
63732-56-9 see 2,4-DINITROPHENYLMORPHINEHYDROCHLORIDE
63765-78-6 see 2-FLUOROETHYL-5-FLUOROHEXOATE
63815-42-9 see 4-ETHOXY-beta-(1-PIPERIDYL)PROPIO PHENONE HY-
    DROCHLORIDE
63833-90-9 see 2-(N-ETHYL CARBAMOYL HYDROXY METHYL)FURAN
63833-98-7 see CYANOTRIMEPRAZINE MALEATE
63868-62-2 see DIETROL
63869-15-8 see DIMERCUROUS METHANE ARSONATE
63869-87-4 see TRIMETHYLTIN SULPHATE
63885-23-4 see N-ETHYL-N-NITROSO-N'-NITROGUANIDINE
63886-77-1 see TETRAFLUORO-m-PHENYLENE DIAMINE DIHYDRO-
    CHLORIDE
63904-99-4 see METHYL-gamma-FLUORO-beta-HYDROXYBUTYRATE
63905-05-5 see METHYL-beta-CHLOROETHYL-beta-HYDROXYETHYL-
    AMINE HYDROCHLORIDE
63905-54-4 see ALIDINE DIHYDROCHLORIDE
63905-98-6 see 4-AMYL-N-BENZOHYDRYLPYRIDINIUM BROMIDE
63906-56-9 see THALLIUM (II) SULFATE (1:1)
63906-88-7 see DIGAMMACAINE
63916-90-5 see p-ISOBUTYOXYBENZOIC ACID-3-(2'-METHYL-
    PIPERIDINO)PROPYL ESTER
63917-01-1 see LETHANE (SPECIAL)
63917-04-4 see LUTETIUM CITRATE
63917-06-6 see DI-2-CHLOROETHYL MALEATE
63917-71-5 see METHYLAMINOCOLCHICIDE
63917-76-0 see p-AMINOBENZOIC ACID, 3-(beta-DIETHYLAMINO)-
    ETHOXY)PROPYLESTER
63918-29-6 see 2-(1-PIPERIDINO)-2-(2-THENYL)ETHYLAMINE
    MALEATE
63918-74-1 see 6,7-DIHYDRO-6-(2-HYDROXYETHYL)-5H-DIBENZ-
    (c,e)AZEPINE
63918-83-2 see DIBENZ(a,j)ACRIDINE METHOSULFATE
63918-98-9 see 3-ETHOXY PROPION ALDEHYDE
63919-01-7 see FLUOROETHANOL
63919-21-1 see COBALT NITROPRUSSIDE
63937-14-4 see MERCURY (I) GLUCONATE
63938-16-9 see CINNAMIC ACID, NICKEL(II) SALT
63938-24-9 see 1-ACETYLLYSERGIC ACID DIETHYLAMIDE
    BITARTRATE

63976-04-5 see AFLATOXIN B1-2,3-DICHLORIDE
63980-20-1 see DIETHYL TRIAZENE
63980-61-0 see PHOSPHOROUS ACID TRIS(2-FLUOROETHYL-
    ESTER)
63981-09-9 see 4-HYDRAZINO-2-THIOURACIL
63989-69-5 see IRON (III)-O-ARSENITE PENTAHYDRATE
63989-82-2 see 3,5-DINITRO-p-CRESOL
63990-88-5 see BERYLLIUM OXYFLUORIDE
64036-46-0 see DIETHYL PHENYLTIN ACETATE
64036-86-8 see TETRADECANE
64036-91-5 see BIS(2-HYDROXYETHYL)-2-(2-CHLORO ETHYL THIO)
    ETHYL SULFONIUM, CHLORIDE
64037-50-9 see ETHYLMETHIAMBUTENE HYDROCHLORIDE
64038-56-8 see 5-(3,3-DIMETHYL-1-TRIAZENO)IMIDAZOLE-4-CARBOX-
    AMIDE CITRATE
64039-27-6 see beta-THIOGUANINE DEOXYRIBOSIDE
64043-53-4 see d-CINNAMYLEPHEDRINE HYDROCHLORIDE
64046-00-0 see AMMONIUM POTASSIUM SELENIDE MIXED WITH AM-
    MONIUM POTASSIUM SULFIDE
64046-79-3 see QUINACRINE MUSTARD
64046-96-4 see CALCIUM ACETARSONE
64047-35-4 see BUTYL DICHLORO PHENOXY ACETATE
64048-13-1 see p-DIETHYL p'-DIMETHYLTHIOPYROPHOSPHATE
64049-83-8 see ((METHYLIMINO)DIETHYLENE)BIS)DIMETHYLETHYLAM-
    MONIUM DIBROMIDE
64050-03-9 see (3-HYDROXY-p-TOLYL)TRIMETHYLAMMONIUM CHLO-
    RIDE,METHYLCARBAMATE
64050-60-8 see O-PHENYL-N,N'-DIMETHYL PHOSPHORODIAMIDATE
64050-79-9 see (m-HYDROXYPHENYL)TRIMETHYLAMMONIUM METHYL-
    SULFATE METHYLPHENYLCARBAMATE
64059-26-3 see LACTIC ACID, BERYLLIUM SALT
64070-10-6 see meso-ANTIMONY POTASSIUM TARTRATE
64070-12-8 see dl-ANTIMONY POTASSIUM TARTRATE
64070-13-9 see 3',4'-DIDEOXYKANAMYCIN B SULFATE
64070-83-3 see ARSENIC(V) ACID, TRISODIUM SALT,HEPTAHYDRATE
    (1:3:7)
64090-82-0 see m-ANISIDINE ANTIMONYL TARTRATE
64093-79-4 see NEOCHROMIUM
64365-11-3 see CHARCOAL (ACTIVATED)
64521-14-8 see 7-MBA-3,4-DIHYDRODIOL
64521-15-9 see trans-8,9-DIHYDRO-8,9-DIHYDROXY-7-METHYL-
    BENZ(a)ANTHRACENE
64598-80-7 see 1,2,3,4-TETRAHYDRO-3-ALPHA,4-BETA-DIHYDROXY-
    1-alpha,2,alpha-EPOXY BENZ(a)ANTHRACENE (E)
64719-39-7 see QUELAMYCIN
64990-84-1 see LOBELINE HYDROCHLORIDE
65986-80-3 see (ETHYL NITROSAMINO)METHYL ACETATE
65996-79-4 see COAL TAR NAPHTHA (DOT)
66017-91-2 see PROPYLNITROSAMINOMETHYL ACETATE
66104-24-3 see BERYLLIUM CARBONATE
67293-88-3 see DEUTERIOMORPHINE
67360-94-5 see ARSINE-TRI-1-PIPERIDINIUM CHLORIDE
67479-03-2 see p-MERCAPTO SULFADIAZINE
67774-32-7 see FIREMASTER FFI
69226-06-8 see 2,2-DIETHYL-3-THIOMORPHOLINONE
69226-45-5 see DIOCTYL(1,2-PROPYLENEDIOXYBIS(MALEOYLDIOXY))
    STANNANE
69352-97-2 see BERBERINE SULFATE TRIHYDRATE
69382-20-3 see BINDON ETHYL ETHER
72017-60-8 see ANTIMONY AMMONIA TRIACETIC ACID
73622-67-0 see 3,4-DI(ACETYLTHIOMETHYL)-5-HYDROXY-6-METHYL-
    PYRIDINE HYDROBROMIDE
73771-81-0 see QUININE ETHIODIDE
73790-27-9 see METHYL-4-(3-PIPERIDINOPROPIONYLAMINO)SALI-
    CYLATE, METHIODIDE
73840-42-3 see 1-METHYL-4-(PHENYLTHIO)PYRIDINIUM IODIDE
73926-87-1 see trans-CHLORO(2-(3-BROMO PROPION AMIDO)CYCLO
    HEXYL)MERCURY
73987-52-7 see ETHOXY CARBONYL DIGOXIN